PRINCIPLES OF ELECTRONICS

PRINCIPLES OF ELECTRONICS

(For B.E./ B.Tech & other Engg. Examinations)

V.K. MEHTA

ROHIT MEHTA

S.CHAND PUBLISHING

S Chand And Company Limited

(ISO 9001 Certified Company)

S Chand And Company Limited

(ISO 9001 Certified Company)

Head Office: D-92, Sector–2, Noida – 201301, U.P. (India), Ph. 91-120-4682700

S. CHAND PUBLISHING

Registered Office: A-27, 2nd Floor, Mohan Co-operative Industrial Estate, New Delhi – 110 044, Phone: 011-49731800

www.**schandpublishing.com;** e-mail: **info@schandpublishing.com**

Marketing Offices:

Chennai : Ph: 23632120; chennai@schandpublishing.com
Guwahati : Ph: 2738811, 2735640; guwahati@schandpublishing.com
Hyderabad : Ph: 40186018; hyderabad@schandpublishing.com
Jalandhar : Ph: 4645630; jalandhar@schandpublishing.com
Kolkata : Ph: 23357458, 23353914; kolkata@schandpublishing.com
Lucknow : Ph: 4003633; lucknow@schandpublishing.com
Mumbai : Ph: 25000297; mumbai@schandpublishing.com
Patna : Ph: 2260011; patna@schandpublishing.com

First Edition 1980

Subsequent Editions and Reprints 1981, 83, (Twice), 84, 85, (Twice), 86 (Twice), 87 (Twice), 88 (Thrice), 89 (Twice), 90 (Twice), 91 (Twice), 92, 93 (Thrice), 94, 95 (Twice), 97, 98 (Twice), 99, 2000, 2001 (Twice), 2002, 2003, 2004 (Twice), 2005, 2006, 2007, 2008, 2009 (Thrice), 2010 (Thrice), 2012, 2013 (Twice), 2014, 2015, 2016, 2017, 2018, 2019 (Twice)

Low Priced Students' Paperback Edition (LPSPE) 2019

Twelfth Revised Edition 2020; Reprints 2021, 2022 (Twice), 2023, 2024 (Twice)

Reprint 2025

ISBN: 978-93-5283-836-3 **Product Code:** H3POE53ETRN10ENAL20L

PRINTED IN INDIA

By Vikas Publishing House Private Limited, Plot 20/4, Site-IV, Industrial Area Sahibabad, Ghaziabad – 201 010 and Published by S Chand And Company Limited, A-27, 2nd Floor, Mohan Co-operative Industrial Estate, New Delhi – 110 044.

PREFACE TO THE TWELFTH EDITION

The general response to the eleventh edition of the book was very encouraging and the authors feel that their work has been amply rewarded and wish to express their deep sense of gratitude, in general to the large number of readers who have used it, and in particular to those of them who have sent helpful suggestions from time to time for the improvement of the book. The continuous feedback from the readers has helped the authors to make the book more useful.

The new edition has the following notable features :

(*i*) New topics, solved examples alongwith diagrams to increase the technical grasp of the student.

(*ii*) Strengthening of existing subject matter.

(*iii*) Short answer questions which will be very helpful for students in viva-voce, competitive exams etc.

It is hoped that these features will make the book more design and practically oriented and also meet the growing demands of the industry.

Authors wish to thank their colleagues and friends whose valuable suggestions regarding the scope and content-sequence of the book have helped make the text more lucid. Authors are also indebted to M/S S Chand And Company Limited for their resolute support in every way.

Errors might have crept in despite utmost care to avoid them. Authors shall be grateful if these are pointed out alongwith other suggestions for the improvement of the book.

V K MEHTA

ROHIT MEHTA

CONTENTS

18. Silicon Controlled Rectifiers 720–748

19. Power Electronics 749–778

20. Electronic Instruments 779–812

CHAPTER 1

Introduction

Learning Objectives

- Develop an understanding of importance of electronics.
- Understand the atomic structure.
- Develop a strong understanding of constant voltage source and constant current source.
- Recognise the importance of maximum power transfer theorem in electronics.
- Develop an understanding of Thevenin's theorem and Norton's theorem.
- Understand chassis and ground.

General

In this fast developing society, *electronics* has come to stay as the most important branch of engineering. Electronic devices are being used in almost all the industries for quality control and automation and they are fast replacing the present vast army of workers engaged in processing and assembling in the factories. Great strides taken in the industrial applications of electronics during the recent years have demonstrated that this versatile tool can be of great importance in increasing production, efficiency and control.

The rapid growth of electronic technology offers a formidable challenge to the beginner, who may be almost paralysed by the mass of details. However, the mastery of fundamentals can simplify the learning process to a great extent. The purpose of this chapter is to present the elementary knowledge in order to enable the readers to follow the subsequent chapters.

1.1 Electronics

The branch of engineering which deals with current conduction through a vacuum or gas or semiconductor is known as ***electronics.**

Electronics essentially deals with electronic devices and their utilisation. An *electronic device* is that in which current flows through a vacuum or gas or semiconductor. Such devices have valuable properties which enable them to function and behave as the friend of man today.

Importance. Electronics has gained much importance due to its numerous applications in industry. The electronic devices are capable of performing the following functions :

* The word *electronics* derives its name from electron present in all materials.

(*i*) **Rectification.** The conversion of a.c. into d.c. is called *rectification*. Electronic devices can convert a.c. power into d.c. power (See Fig. 1.1) with very high efficiency. This d.c. supply can be used for charging storage batteries, field supply of d.c. generators, electroplating *etc.*

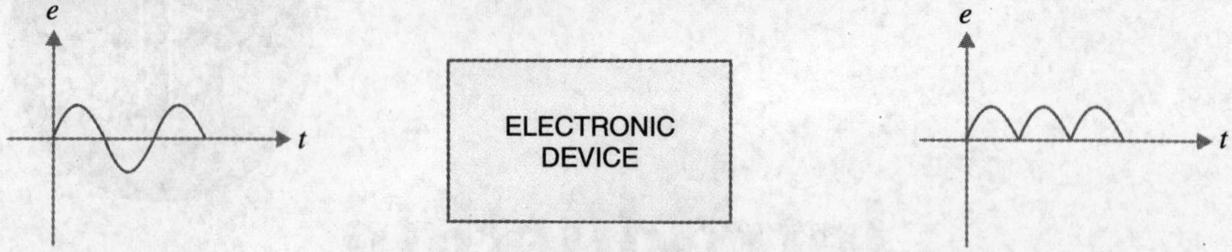

Fig. 1.1

(*ii*) **Amplification.** The process of raising the strength of a weak signal is known as *amplification*. Electronic devices can accomplish the job of amplification and thus act as amplifiers (See Fig. 1.2). The amplifiers are used in a wide variety of ways. For example, an amplifier is used in a radio-set where the weak signal is amplified so that it can be heard loudly. Similarly, amplifiers are used in public address system, television *etc.*

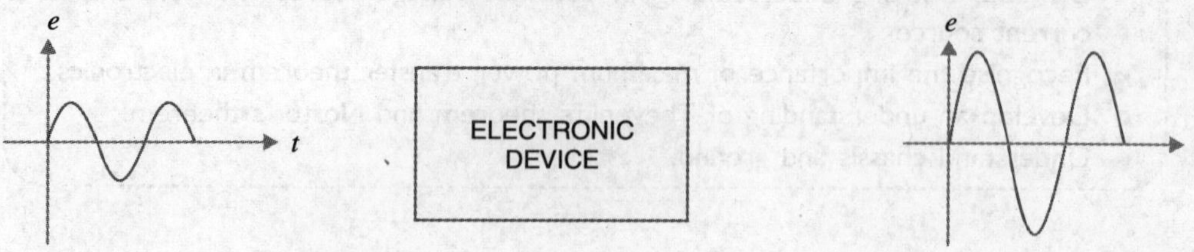

Fig. 1.2

(*iii*) **Control.** Electronic devices find wide applications in automatic control. For example, speed of a motor, voltage across a refrigerator *etc.* can be automatically controlled with the help of such devices.

(*iv*) **Generation.** Electronic devices can convert d.c. power into a.c. power of any frequency (See Fig. 1.3). When performing this function, they are known as *oscillators*. The oscillators are used in a wide variety of ways. For example, electronic high frequency heating is used for annealing and hardening.

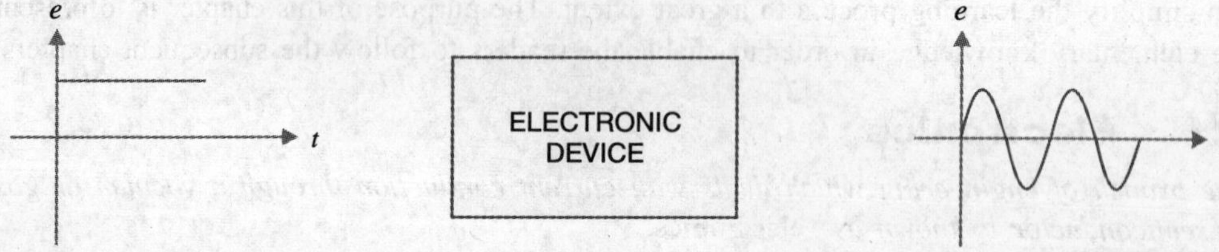

Fig. 1.3

(*v*) **Conversion of light into electricity.** Electronic devices can convert light into electricity. This conversion of light into electricity is known as *photo-electricity*. Photo-electric devices are used in burglar alarms, sound recording on motion pictures *etc.*

(*vi*) **Conversion of electricity into light.** Electronic devices can convert electricity into light. This valuable property is utilised in television and radar.

1.2 Atomic Structure

According to the modern theory, matter is electrical in nature. All the materials are composed of very small particles called *atoms*. The atoms are the building bricks of all matter. An atom consists of a central *nucleus* of positive charge around which small negatively charged particles, called *electrons* revolve in different paths or orbits.

(1) **Nucleus.** It is the central part of an atom and *contains *protons* and *neutrons*. A proton is a positively charged particle, while the neutron has the same mass as the proton, but has no charge. Therefore, the nucleus of an atom is positively charged. The sum of protons and neutrons constitutes the entire weight of an atom and is called atomic weight. It is because the particles in the extra nucleus (*i.e.* electrons) have negligible weight as compared to protons or neutrons.

∴ Atomic weight = No. of protons + No. of neutrons

(2) **Extra nucleus.** It is the outer part of an atom and contains *electrons* only. An electron is a negatively charged particle having negligible mass. The charge on an electron is equal but opposite to that on a proton. Also, the number of electrons is equal to the number of protons in an atom under ordinary conditions. Therefore, an atom is neutral as a whole. The number of electrons or protons in an atom is called *atomic number i.e.*

Atomic number = No. of protons or electrons in an atom

The electrons in an atom revolve around the nucleus in different orbits or paths. The number and arrangement of electrons in any orbit is determined by the following rules :

(*i*) The number of electrons in any orbit is given by $2n^2$ where n is the number of the orbit. For example,

First orbit contains $2 \times 1^2 = 2$ electrons

Second orbit contains $2 \times 2^2 = 8$ electrons

Third orbit contains $2 \times 3^2 = 18$ electrons and so on.

(*ii*) The last orbit cannot have more than 8 electrons.

(*iii*) The last but one orbit cannot have more than 18 electrons.

1.3 Structure of Elements

We have seen that all atoms are made up of protons, neutrons and electrons. The difference among various types of elements is due to the different number and arrangement of these particles within their atoms. For example, the structure** of copper atom is different from that of carbon atom and hence the two elements have different properties.

The atomic structure can be easily built up if we know the atomic weight and atomic number of the element. Thus taking the case of copper atom,

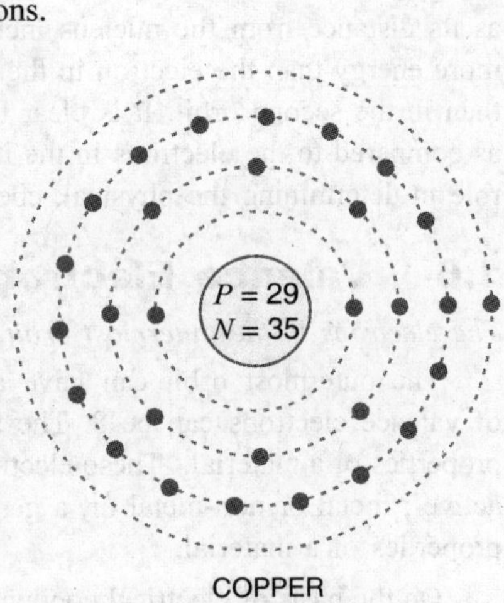

COPPER

Fig. 1.4

* Although the nucleus of an atom is of complex structure, yet for the purpose of understanding electronics, this simplified picture of the nucleus is adequate.

** The number and arrangement of protons, neutrons and electrons.

Atomic weight = 64

Atomic number = 29

∴ No. of protons = No. of electrons = 29

and No. of neutrons = 64 − 29 = 35

Fig. 1.4 shows the structure of copper atom. It has 29 electrons which are arranged in different orbits as follows. The first orbit will have 2 electrons, the second 8 electrons, the third 18 electrons and the fourth orbit will have 1 electron. The atomic structure of all known elements can be shown in this way and the reader is advised to try for a few commonly used elements.

1.4 The Electron

Since electronics deals with tiny particles called electrons, these small particles require detailed study. As discussed before, an electron is a negatively charged particle having negligible mass. Some of the important properties of an electron are :

(*i*) Charge on an electron, e = 1.602×10^{-19} coulomb

(*ii*) Mass of an electron, m = 9.0×10^{-31} kg

(*iii*) Radius of an electron, r = 1.9×10^{-15} metre

The ratio e/m of an electron is 1.77×10^{11} coulombs/kg. This means that mass of an electron is very small as compared to its charge. It is due to this property of an electron that it is very mobile and is greatly influenced by electric or magnetic fields.

1.5 Energy of an Electron

An electron moving around the nucleus possesses two types of energies *viz.* kinetic energy due to its motion and potential energy due to the charge on the nucleus. The total energy of the electron is the sum of these two energies. The energy of an electron increases as its distance from the nucleus increases. Thus, an electron in the second orbit possesses more energy than the electron in the first orbit ; electron in the third orbit has higher energy than in the second orbit. It is clear that electrons in the last orbit possess very high energy as compared to the electrons in the inner orbits. These last orbit electrons play an important role in determining the physical, chemical and electrical properties of a material.

1.6 Valence Electrons

The electrons in the outermost orbit of an atom are known as **valence electrons.**

The outermost orbit can have a maximum of 8 electrons *i.e.* the maximum number of valence electrons can be 8. The valence electrons determine the physical and chemical properties of a material. These electrons determine whether or not the material is chemically active ; metal or non-metal or, a gas or solid. These electrons also determine the electrical properties of a material.

On the basis of electrical conductivity, materials are generally classified into *conductors, insulators* and *semi-conductors.* As a rough rule, one can determine the electrical behaviour of a material from the number of valence electrons as under :

(*i*) When the number of valence electrons of an atom is less than 4 (*i.e.* half of the maximum eight electrons), the material is usually *a metal and a conductor*. Examples are sodium, magnesium and aluminium which have 1, 2 and 3 valence electrons respectively (See Fig. 1.5).

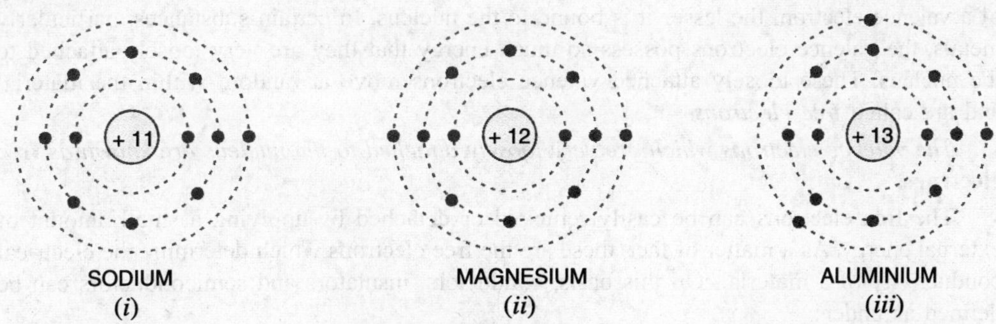

Fig. 1.5

(*ii*) When the number of valence electrons of an atom is more than 4, the material is usually *a non-metal and an insulator*. Examples are nitrogen, sulphur and neon which have 5, 6 and 8 valence electrons respectively (See Fig. 1.6).

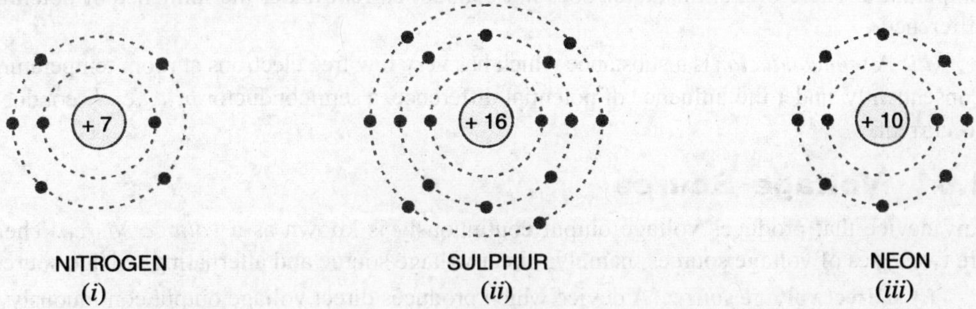

Fig. 1.6

(*iii*) When the number of valence electrons of an atom is 4 (*i.e.* exactly one-half of the maximum 8 electrons), the material has both metal and non-metal properties and is usually a *semi-conductor.* Examples are carbon, silicon and germanium (See Fig. 1.7).

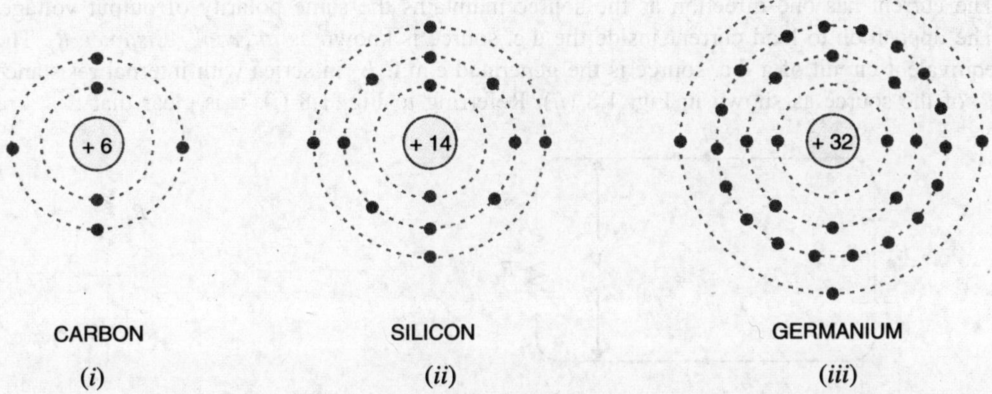

Fig. 1.7

1.7 Free Electrons

The valence electrons of different materials possess different energies. The greater the energy of a valence electron, the lesser it is bound to the nucleus. In certain substances, particularly metals, the valence electrons possess so much energy that they are very loosely attached to the nucleus. These loosely attached valence electrons move at random within the material and are called *free electrons.*

The valence electrons which are very loosely attached to the nucleus are known as **free electrons.**

The free electrons can be easily removed or detached by applying a small amount of external energy. As a matter of fact, these are the free electrons which determine the electrical conductivity of a material. On this basis, conductors, insulators and semiconductors can be defined as under:

(*i*) A *conductor* is a substance which has a large number of free electrons. When potential difference is applied across a conductor, the free electrons move towards the positive terminal of supply, constituting electric current.

(*ii*) An *insulator* is a substance which has practically no free electrons at ordinary temperature. Therefore, an insulator does not conduct current under the influence of potential difference.

(*iii*) A *semiconductor* is a substance which has very few free electrons at room temperature. Consequently, under the influence of potential difference, a semiconductor *practically* conducts no current.

1.8 Voltage Source

Any device that produces voltage output continuously is known as a *voltage source*. There are two types of voltage sources, namely; direct voltage source and alternating voltage source.

(*i*) **Direct voltage source.** A device which produces direct voltage output continuously is called *a direct voltage source*. Common examples are cells and d.c. generators. An important characteristic of a direct voltage source is that it maintains the same polarity of the output voltage *i.e.* positive and negative terminals remain the same. When load resistance R_L is connected across such a source, *current flows from positive terminal to negative terminal *via* the load [See Fig. 1.8 (*i*)]. This is called *direct current* because it has just one direction. The current has one direction as the source maintains the same polarity of output voltage. The opposition to load current inside the d.c. source is known as *internal resistance* R_i. The equivalent circuit of a d.c. source is the generated e.m.f. E_g in series with internal resistance R_i of the source as shown in Fig. 1.8 (*ii*). Referring to Fig. 1.8 (*i*), it is clear that :

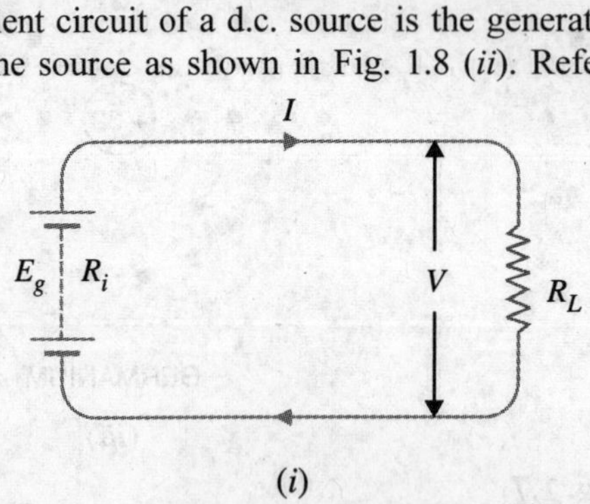

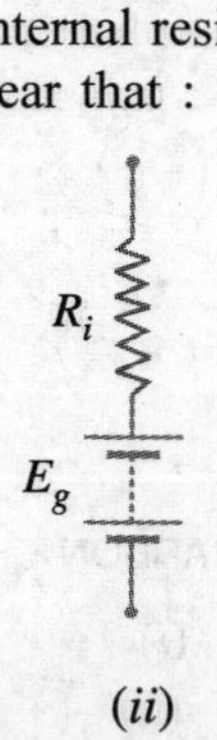

Fig. 1.8

* This is the conventional current. However, the flow of electrons will be in the opposite direction.

$$\text{Load current, } I = \frac{E_g}{R_L + R_i}$$

$$\text{Terminal voltage, } V = (E_g - IR_i) \quad \text{or} \quad IR_L$$

(ii) Alternating voltage source. A device which produces alternating voltage output continuously is known as *alternating voltage source e.g.* a.c. generator. An important characteristic of alternating voltage source is that it periodically reverses the polarity of the output voltage. When load impedance Z_L is connected across such a source, current flows through the circuit that periodically reverses in direction. This is called *alternating current.*

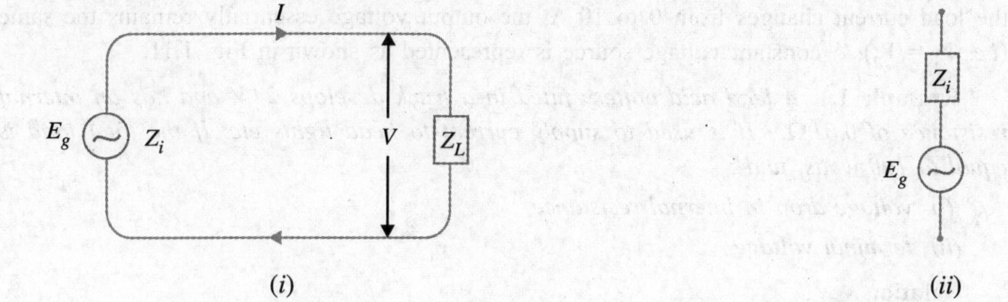

Fig. 1.9

The opposition to load current inside the a.c. source is called its *internal impedance* Z_i. The equivalent circuit of an a.c. source is the generated e.m.f. E_g (r.m.s.) in series with internal impedance Z_i of the source as shown in Fig. 1.9 (*ii*). Referring to Fig. 1.9 (*i*), it is clear that :

$$\text{Load current, } I \text{ (r.m.s.)} = \frac{E_g}{Z_L + Z_i}$$

$$\text{Terminal voltage, } V = (E_g - IZ_i)^* \quad \text{or} \quad IZ_L$$

1.9 Constant Voltage Source

*A voltage source which has very low internal **impedance as compared with external load impedance is known as a* **constant voltage source.**

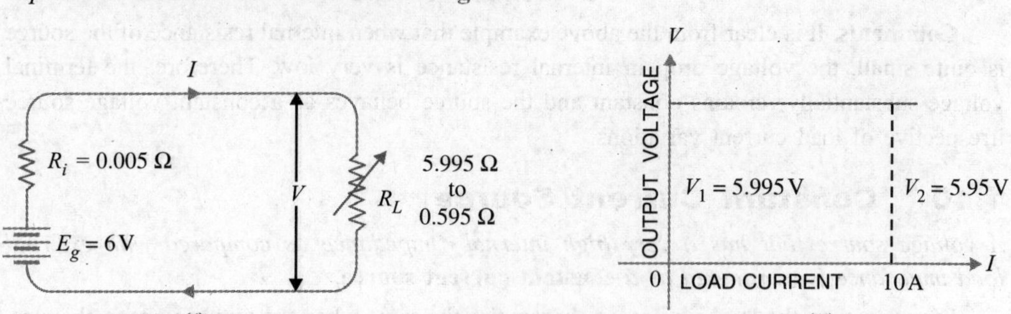

Fig. 1.10

* Phasor difference since a.c. quantities are phasors.

** Resistance in case of a d.c. source.

In such a case, the output voltage nearly remains the same when load current changes. Fig. 1.10 (*i*) illustrates a constant voltage source. It is a d.c. source of 6 V with internal resistance R_i = 0.005 Ω. If the load current varies over a wide range of 1 to 10 A, for any of these values, the internal drop across R_i (= 0.005 Ω) is less than 0.05 volt. Therefore, the voltage output of the source is between 5.995 to 5.95 volts. This can be considered constant voltage compared with the wide variations in load current.

Fig. 1.11

Fig. 1.10 (*ii*) shows the graph for a constant voltage source. It may be seen that the output voltage remains constant inspite of the changes in load current. Thus as the load current changes from 0 to 10 A, the output voltage essentially remains the same (*i.e.* $V_1 = V_2$). A constant voltage source is represented as shown in Fig. 1.11.

Example 1.1. *A lead acid battery fitted in a truck develops 24 V and has an internal resistance of 0.01 Ω. It is used to supply current to head lights etc. If the total load is equal to 100 watts, find :*

(i) voltage drop in internal resistance

(ii) terminal voltage

Solution.

Generated voltage, E_g = 24 V

Internal resistance, R_i = 0.01 Ω

Power supplied, P = 100 watts

(*i*) Let I be the load current.

Now, $P = E_g \times I$ $\qquad$ ($\because$ For an ideal source, $V \simeq E_g$)

$$\therefore \quad I = \frac{P}{E_g} = \frac{100}{24} = 4.17 \text{ A}$$

$\therefore$ Voltage drop in R_i = IR_i = 4.17 × 0.01 = **0.0417 V**

(*ii*) Terminal voltage, $V = E_g - IR_i$

= 24 − 0.0417 = **23.96 V**

Comments. It is clear from the above example that when internal resistance of the source is quite small, the voltage drop in internal resistance is very low. Therefore, the terminal voltage substantially remains constant and the source behaves as a constant voltage source irrespective of load current variations.

1.10 Constant Current Source

*A voltage source that has a very high internal *impedance as compared with external load impedance is considered as a* **constant current source.**

In such a case, the load current nearly remains the same when the output voltage changes. Fig. 1.12 (*i*) illustrates a constant current source. It is a d.c. source of 1000 V with internal resistance R_i = 900 kΩ. Here, load R_L varies over 3 : 1 range from 50 kΩ to 150 kΩ. Over this variation of load R_L, the circuit current I is essentially constant at 1.05 to 0.95 mA or

* Resistance in case of a d.c. source.

approximately 1 mA. It may be noted that output voltage V varies approximately in the same 3 : 1 range as R_L, although load current essentially remains *constant at 1mA. The beautiful example of a constant current source is found in vacuum tube circuits where the tube acts as a generator having internal resistance as high as 1 MΩ.

Fig. 1.12 (*ii*) shows the graph of a constant current source. It is clear that current remains constant even when the output voltage changes substantially. The following points may be noted regarding the constant current source :

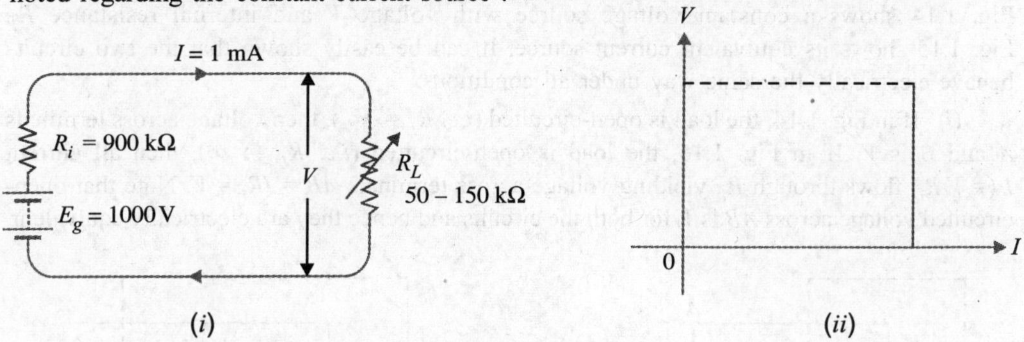

Fig. 1.12

(*i*) Due to high internal resistance of the source, the load current remains essentially constant as the load R_L is varied.

(*ii*) The output voltage varies approximately in the same range as R_L, although current remains constant.

(*iii*) The output voltage V is much less than the generated voltage E_g because of high IR_i drop.

Fig. 1.13 shows the symbol of a constant current source.

Fig. 1.13

Example 1.2. *A d.c. source generating 500 V has an internal resistance of 1000 Ω. Find the load current if load resistance is (i) 10 Ω (ii) 50 Ω and (iii) 100 Ω.*

Solution.

Generated voltage, E_g = 500 V

Internal resistance, R_i = 1000 Ω

(*i*) When R_L = 10 Ω

$$\text{Load current, } I = \frac{E_g}{R_L + R_i} = \frac{500}{10 + 1000} = \mathbf{0.495\ A}$$

(*ii*) When R_L = 50 Ω

$$\text{Load current, } I = \frac{500}{50 + 1000} = \mathbf{0.476\ A}$$

(*iii*) When R_L = 100 Ω

* Now, $I = \dfrac{E_g}{R_L + R_i}$. Since $R_i >> R_L$, $I = \dfrac{E_g}{R_i}$.
As both E_g and R_i are constants, I is constant.

$$\text{Load current, } I = \frac{500}{100 + 1000} = 0.454 \text{ A}$$

It is clear from the above example that load current is essentially constant since $R_i >> R_L$.

1.11 Conversion of Voltage Source into Current Source

Fig. 1.14 shows a constant voltage source with voltage V and internal resistance R_i. Fig. 1.15 shows its equivalent current source. It can be easily shown that the two circuits behave electrically the same way under all conditions.

(*i*) If in Fig. 1.14, the load is open-circuited (*i.e.* $R_L \to \infty$), then voltage across terminals A and B is V. If in Fig. 1.15, the load is open-circuited (*i.e.* $R_L \to \infty$), then all current I (= V/R_i) flows through R_i, yielding voltage across terminals $AB = IR_i = V$. Note that open-circuited voltage across AB is V for both the circuits and hence they are electrically equivalent.

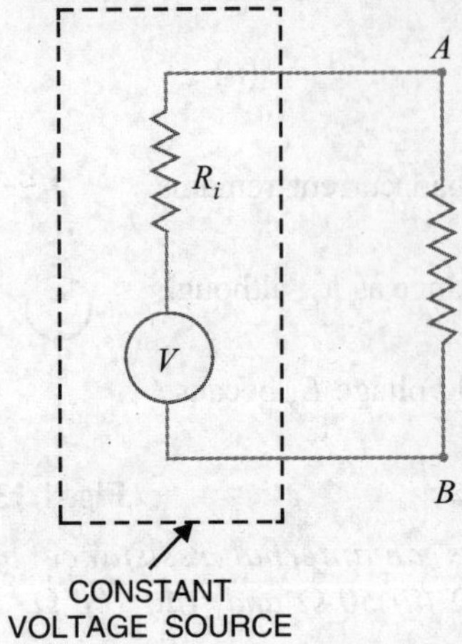

Fig. 1.14

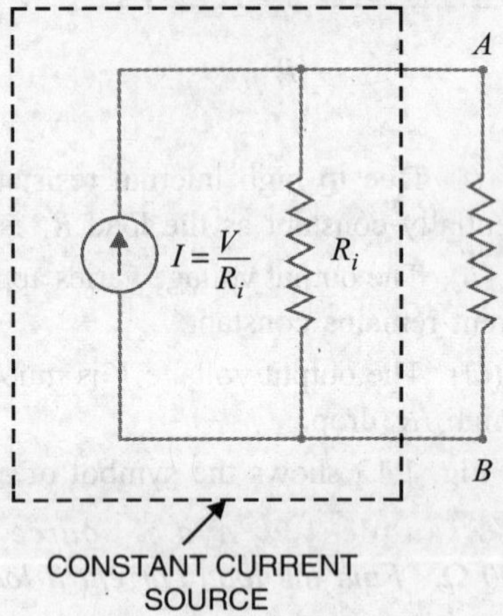

Fig. 1.15

(*ii*) If in Fig. 1.14, the load is short-circuited (*i.e.* $R_L = 0$), the short-circuit current is given by ;

$$I_{short} = \frac{V}{R_i}$$

If in Fig. 1.15, the load is short-circuited (*i.e.* $R_L = 0$), the current I (= V/R_i) bypasses R_i in favour of short-circuit. It is clear that current (= V/R_i) is the same for the two circuits and hence they are electrically equivalent.

Thus to convert a constant voltage source into a constant current source, the following procedure may be adopted :

(*a*) Place a short-circuit across the two terminals in question (terminals AB in the present case) and find the short-circuit current. Let it be I. Then I is the current supplied by the equivalent current source.

(*b*) Measure the resistance at the terminals with load removed and sources of e.m.f.s replaced by their internal resistances if any. Let this resistance be R.

(c) Then equivalent current source can be represented by a single current source of magnitude I in parallel with resistance R.

Note. To convert a current source of magnitude I in parallel with resistance R into voltage source,

$$\text{Voltage of voltage source, } V = IR$$

$$\text{Resistance of voltage source, } R = R$$

Thus voltage source will be represented as voltage V in series with resistance R.

Example 1.3. *Convert the constant voltage source shown in Fig. 1.16 into constant current source.*

Solution. The solution involves the following steps :

(i) Place a short across AB in Fig. 1.16 and find the short-circuit current I.

$$\text{Clearly, } I = 10/10 = 1 \text{ A}$$

Therefore, the equivalent current source has a magnitude of 1 A.

(ii) Measure the resistance at terminals AB with load *removed and 10 V source replaced by its internal resistance. The 10 V source has negligible resistance so that resistance at terminals AB is $R = 10\ \Omega$.

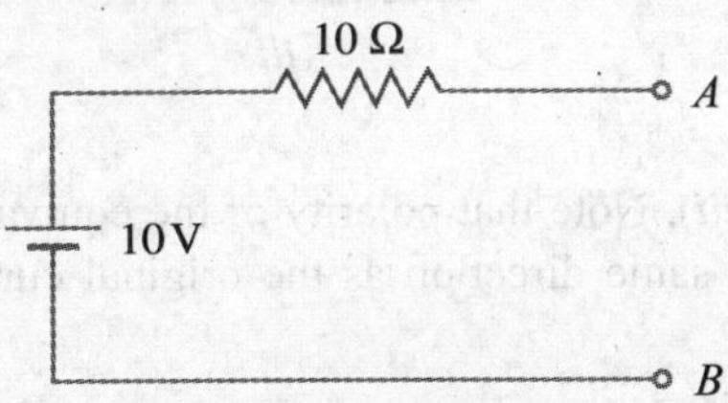

Fig. 1.16

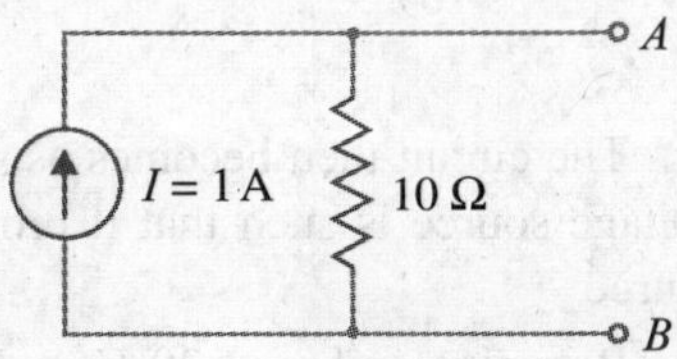

Fig. 1.17

(iii) The equivalent current source is a source of 1 A in parallel with a resistance of 10 Ω as shown in Fig. 1.17.

Example 1.4. *Convert the constant current source in Fig. 1.18 into equivalent voltage source.*

Solution. The solution involves the following steps :

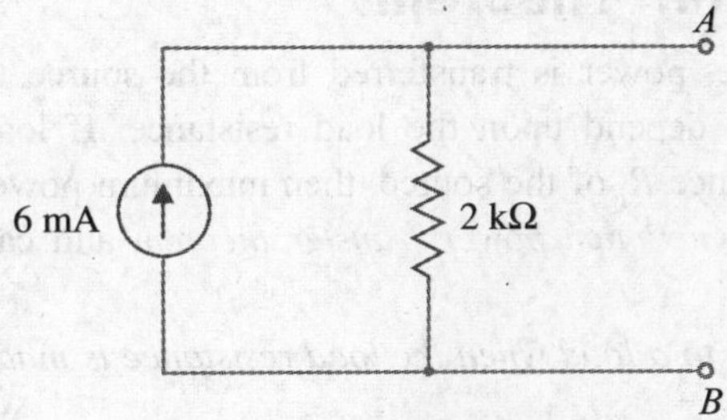

Fig. 1.18

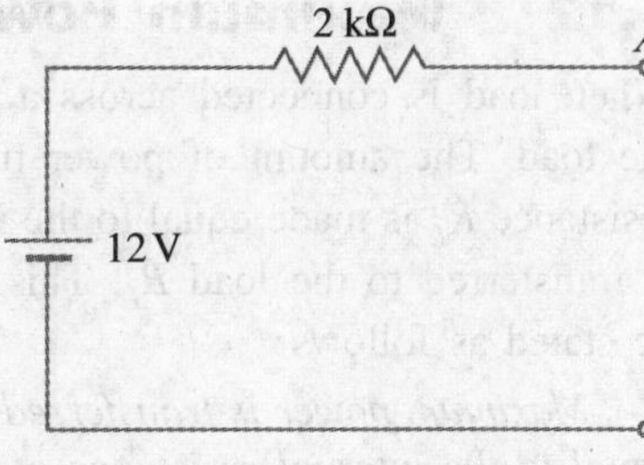

Fig. 1.19

(i) To get the voltage of the voltage source, multiply the current of the current source by its internal resistance *i.e.*

---

* Fortunately, no load is connected across AB. Had there been load across AB, it would have been removed.

Voltage of voltage source = IR = 6 mA × 2 kΩ = 12 V

(*ii*) The internal resistance of voltage source is 2 kΩ.

The equivalent voltage source is a source of 12 V in series with a resistance of 2 kΩ as shown in Fig. 1.19.

Note. The voltage source should be placed with +ve terminal in the direction of current flow.

Example 1.5. *Find the current in 6 kΩ resistor in Fig. 1.20 (i) by converting the current source to a voltage source.*

Solution. Since we want to find the current in 6 kΩ resistor, we use 3 kΩ resistor to convert the current source to an equivalent voltage source. Referring to Fig. 1.20 (*ii*), the voltage of equivalent voltage source is

$$E = 15 \text{ mA} \times 3 \text{ k}\Omega = 45 \text{ V}$$

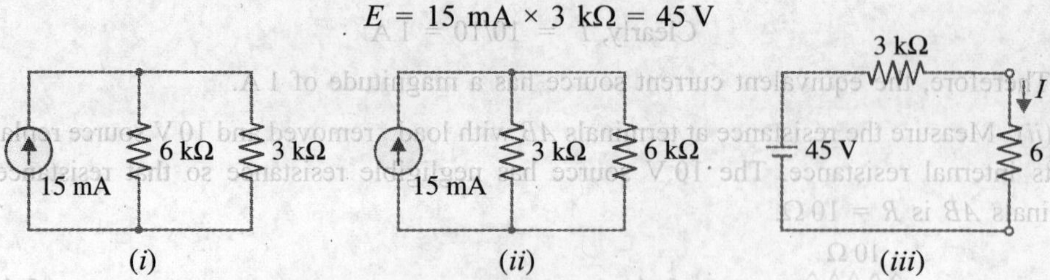

Fig. 1.20

The circuit then becomes as shown in Fig. 1.20 (*iii*). Note that polarity of the equivalent voltage source is such that it produces current in the same direction as the original current source.

Referring to Fig. 1.20 (*iii*), the current in 6 kΩ is

$$I = \frac{45 \text{ V}}{(3+6) \text{ k}\Omega} = 5 \text{ mA}$$

In the series circuit shown in Fig. 1.20 (*iii*), it would appear that current in 3 kΩ resistor is also 5 mA. However, 3 kΩ resistor was involved in source conversion, so we *cannot* conclude that there is 5 mA in the 3 kΩ resistor of the original circuit [See Fig. 1.20 (*i*)]. Verify that the current in the 3 kΩ resistor in that circuit is, in fact, 10 mA.

1.12 Maximum Power Transfer Theorem

When load is connected across a voltage source, power is transferred from the source to the load. The amount of power transferred will depend upon the load resistance. If load resistance R_L is made equal to the internal resistance R_i of the source, then maximum power is transferred to the load R_L. This is known as *maximum power transfer theorem* and can be stated as follows:

Maximum power is transferred from a source to a load when the load resistance is made equal to the internal resistance of the source.

This applies to d.c. as well as a.c. power.*

Illustration. Let us illustrate maximum power transfer theorem with a numerical example. Consider a voltage source E of 100 V (r.m.s.) and internal resistance R_i of 10 Ω delivering

* As power is concerned with resistance only, therefore, this is true for both a.c. and d.c. power.

power to a variable load R_L as shown in Fig. 1.21. The circuit current I is

$$I = \frac{E}{R_i + R_L}$$

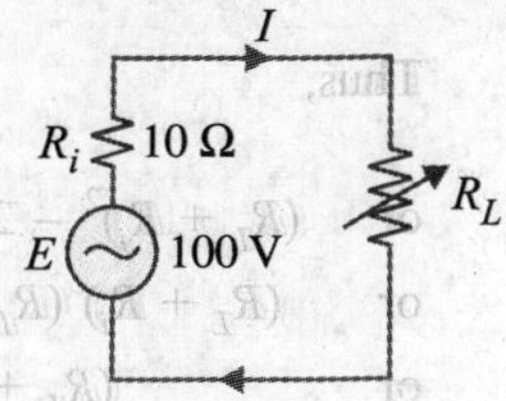

Fig. 1.21

The power P delivered to the load is

$$P = I^2 R_L$$

The calculations below show how load power P varies with R_L.

R_L	Circuit current (I)	Power in R_L ($= I^2 R_L$)
0 Ω	10 A	0 W
5 Ω	6.66 A	222 W
10 Ω	**5 A**	**250 W ..... P_{max}**
15 Ω	4.0 A	240 W
20 Ω	3.3 A	222 W
1000 Ω	0.09 A	8.1 W
$\to \infty$	$\to 0$	$\to 0$

Note that maximum power (*i.e.* 250 W in this case) is delivered to the load when $R_L = R_i$ (= 10 Ω). If R_L is either smaller or larger than R_i, the power delivered to the load becomes smaller.

1.13 Proof of Maximum Power Transfer Theorem

To prove this theorem mathematically, consider a voltage source of generated voltage E and internal resistance R_i and delivering power to a load resistance R_L [See Fig. 1.22 (*i*)]. The current I flowing through the circuit is given by ;

$$I = \frac{E}{R_L + R_i}$$

Power delivered to the load, $P = I^2 R_L = \left(\frac{E}{R_L + R_i}\right)^2 R_L$ ...(*i*)

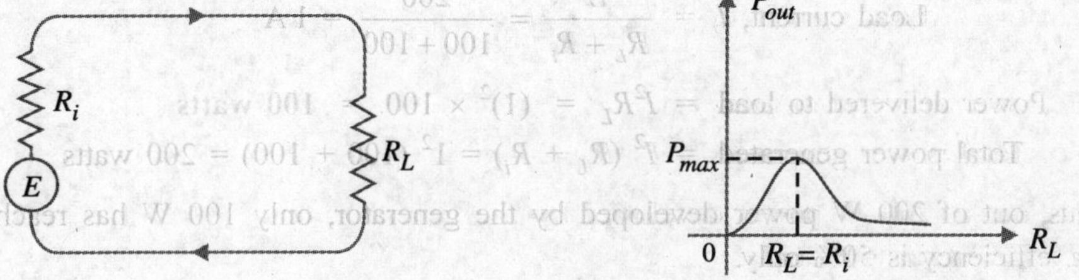

Fig. 1.22

For a given source, generated voltage E and internal resistance R_i are constant. Therefore, power delivered to the load depends upon R_L. In order to find the value of R_L for which the value of P is maximum, it is necessary to differentiate eq. (*i*) *w.r.t.* R_L and set the result equal to zero.

Thus, $$\frac{dP}{dR_L} = E^2 \left[\frac{(R_L + R_i)^2 - 2\,R_L\,(R_L + R_i)}{(R_L + R_i)^4} \right] = 0$$

or $$(R_L + R_i)^2 - 2\,R_L\,(R_L + R_i) = 0$$

or $$(R_L + R_i)\,(R_L + R_i - 2\,R_L) = 0$$

or $$(R_L + R_i)\,(R_i - R_L) = 0$$

Since $(R_L + R_i)$ cannot be zero,

$$\therefore \quad R_i - R_L = 0$$

or $$R_L = R_i$$

i.e. *Load resistance* = *Internal resistance*

Thus, for maximum power transfer, load resistance R_L must be equal to the internal resistance R_i of the source.

Under such conditions, the load is said to be *matched* to the source. Fig. 1.22 (*ii*) shows a graph of power delivered to R_L as a function of R_L. It may be mentioned that efficiency of maximum power transfer is *50% as one-half of the total generated power is dissipated in the internal resistance R_i of the source.

Applications. Electric power systems never operate for maximum power transfer because of low efficiency and high voltage drops between generated voltage and load. However, in the electronic circuits, maximum power transfer is usually desirable. For instance, in a public address system, it is desirable to have load (*i.e.* speaker) "matched" to the amplifier so that there is maximum transference of power from the amplifier to the speaker. In such situations, efficiency is **sacrificed at the cost of high power transfer.

Example 1.6. *A generator develops 200 V and has an internal resistance of 100 Ω. Find the power delivered to a load of (i) 100 Ω (ii) 300 Ω. Comment on the result.*

Solution.

Generated voltage, $E = 200\text{ V}$

Internal resistance, $R_i = 100\,\Omega$

(*i*) When load $R_L = 100\,\Omega$

$$\text{Load current, } I = \frac{E}{R_L + R_i} = \frac{200}{100 + 100} = 1\text{ A}$$

$$\therefore \quad \text{Power delivered to load} = I^2 R_L = (1)^2 \times 100 = \mathbf{100\ watts}$$

$$\text{Total power generated} = I^2\,(R_L + R_i) = 1^2\,(100 + 100) = 200\text{ watts}$$

Thus, out of 200 W power developed by the generator, only 100 W has reached the load *i.e.* efficiency is 50% only.

(*ii*) When load $R_L = 300\,\Omega$

---

* $$\text{Efficiency} = \frac{\text{Output power}}{\text{Input power}} = \frac{I^2 R_L}{I^2\,(R_L + R_i)} = R_L/2\,R_L = 1/2 = 50\% \qquad (\because R_L = R_i)$$

** Electronic devices develop small power. Therefore, if too much efficiency is sought, a large number of such devices will have to be connected in series to get the desired output. This will distort the output as well as increase the cost and size of equipment.

$$\text{Load current, } I = \frac{E}{R_L + R_i} = \frac{200}{300 + 100} = 0.5 \text{ A}$$

$$\text{Power delivered to load} = I^2 R_L = (0.5)^2 \times 300 = \mathbf{75 \text{ watts}}$$

$$\text{Total power generated} = I^2 (R_L + R_i) = (0.5)^2 (300 + 100) = 100 \text{ watts}$$

Thus, out of 100 watts of power produced by the generator, 75 watts is transferred to the load *i.e.* efficiency is 75%.

Comments. Although in case of $R_L = R_i$, a large power (100 W) is transferred to the load, but there is a big wastage of power in the generator. On the other hand, when R_L is ***not*** equal to R_i, the power transfer is less (75 W) but smaller part is wasted in the generator *i.e.* efficiency is high. Thus, it depends upon a particular situation as to what the load should be. If we want to transfer maximum power (*e.g.* in amplifiers) irrespective of efficiency, we should make $R_L = R_i$. However, if efficiency is more important (*e.g.* in power systems), then internal resistance of the source should be considerably smaller than the load resistance.

Example 1.7. *Two identical cells connected in series deliver a maximum power of 1 W to a load resistance of 4 Ω as shown in Fig. 1.23. What is the internal resistance and e.m.f. of each cell?*

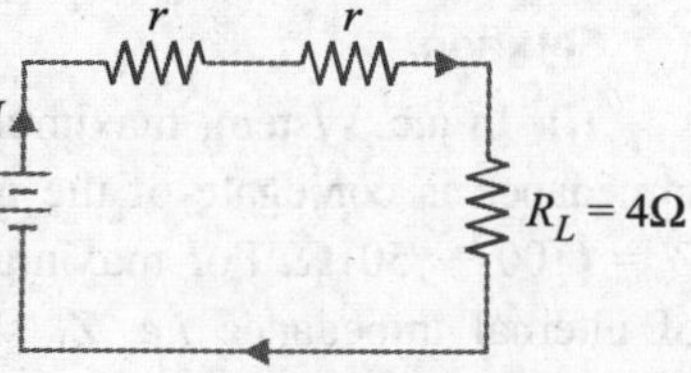

Fig. 1.23

Solution. Let E and r be the e.m.f and internal resistance of each cell respectively. The total internal resistance of the battery is $2r$. For maximum power transfer,

$$2r = R_L \quad \therefore \quad r = R_L/2 = 4/2 = 2\,\Omega$$

Maximum power P_{max} delivered to load R_L (= 4 Ω) is

$$P_{max} = \frac{{}^*(2E)^2}{4R_L} \quad \text{or} \quad 1 \text{ W} = \frac{4E^2}{4 \times 4} \quad \therefore \quad E = \sqrt{4} = 2 \text{ V}$$

Example 1.8. *An audio amplifier produces an alternating output of 12 V before the connection to a load. The amplifier has an equivalent resistance of 15 Ω at the output. What resistance the load need to have to produce maximum power? Also calculate the power output under this condition.*

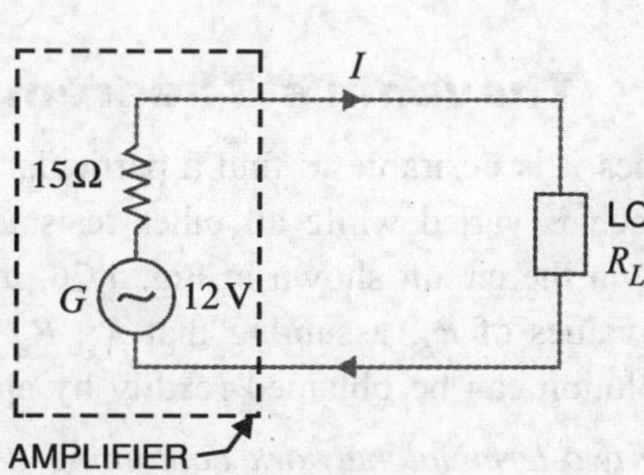

Fig. 1.24

Solution. In order to produce maximum power, the load (*e.g.* a speaker) should have a resistance of 15 Ω to match the amplifier. The equivalent circuit is shown in Fig. 1.24.

$$\therefore \quad \text{Load required, } R_L = \mathbf{15\,\Omega}$$

---

* $I = \dfrac{2E}{2r + R_L}$. For maximum power transfer, $2r = R_L$.

$$\therefore P_{max.} = I^2 R_L = \left(\frac{2E}{2R_L}\right)^2 \times R_L = \frac{(2E)^2}{4R_L}$$

$$\text{Circuit current, } I = \frac{V}{R_T} = \frac{12}{15+15} = 0.4 \text{ A}$$

$$\text{Power delivered to load, } P = I^2 R_L = (0.4)^2 \times 15 = \mathbf{2.4 \text{ W}}$$

Example 1.9. *For the a.c. generator shown in Fig. 1.25 (i), find (i) the value of load so that maximum power is transferred to the load (ii) the value of maximum power.*

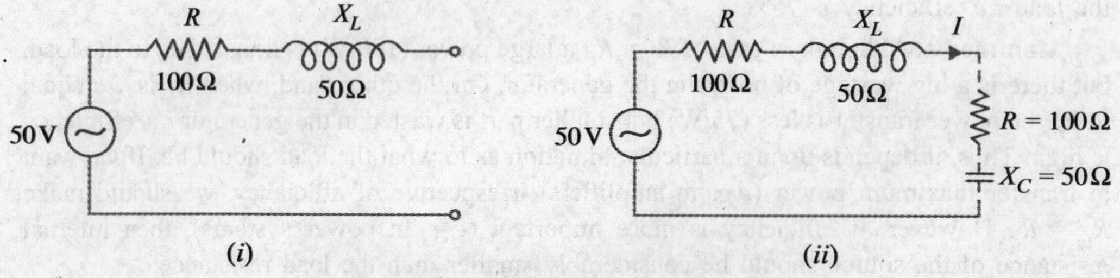

Fig. 1.25

Solution.

(*i*) In a.c. system, maximum power is delivered to the load impedance (Z_L) when load impedance is conjugate of the internal impedance (Z_i) of the source. Now in the problem, $Z_i = (100 + j50)\,\Omega$. For maximum power transfer, the load impedance should be conjugate of internal impedance *i.e.* Z_L should be $(100 - j50)\ \Omega$. This is shown as dotted line in Fig. 1.25 (*ii*).

$$\therefore \qquad Z_L = \mathbf{(100 - j50)\ \Omega}$$

(*ii*) Total impedance, $Z_T = Z_i + Z_L = (100 + j50) + (100 - j50) = 200\ \Omega$*

$$\text{Circuit current, } I = \frac{V}{Z_T} = \frac{50}{200} = 0.25 \text{ A}$$

$$\text{Maximum power transferred to the load} = I^2 R_L = (0.25)^2 \times 100 = \mathbf{6.25 \text{ W}}$$

1.14 Thevenin's Theorem

Sometimes it is desirable to find a particular branch current in a circuit as the resistance of that branch is varied while all other resistances and voltage sources remain constant. For instance, in the circuit shown in Fig. 1.26, it may be desired to find the current through R_L for five values of R_L, assuming that R_1, R_2, R_3 and E remain constant. In such situations, the **solution can be obtained readily by applying *Thevenin's theorem* stated below :

Any two-terminal network containing a number of e.m.f. sources and resistances can be replaced by an equivalent series circuit having a voltage source E_{Th} in series with a resistance R_{Th} where,

E_{Th} = Open-circuited voltage between the two terminals.

R_{Th} = The resistance between two terminals of the circuit obtained by looking

* Note that by making internal impedance and load impedance conjugate, the reactive terms cancel. The circuit then consists of internal and external resistances only. This is quite logical because power is only consumed in resistances as reactances (X_L or X_C) consume no power.

** Solution can also be obtained by applying Kirchhoff's laws but it requires a lot of labour.

"in" at the terminals with load removed and voltage sources replaced by their internal resistances, if any.

To understand the use of this theorem, consider the two-terminal circuit shown in Fig. 1.26. The circuit enclosed in the dotted box can be replaced by one voltage E_{Th} in series with resistance R_{Th} as shown in Fig. 1.27. The behaviour at the terminals AB and $A'B'$ is the same for the two circuits, independent of the values of R_L connected across the terminals.

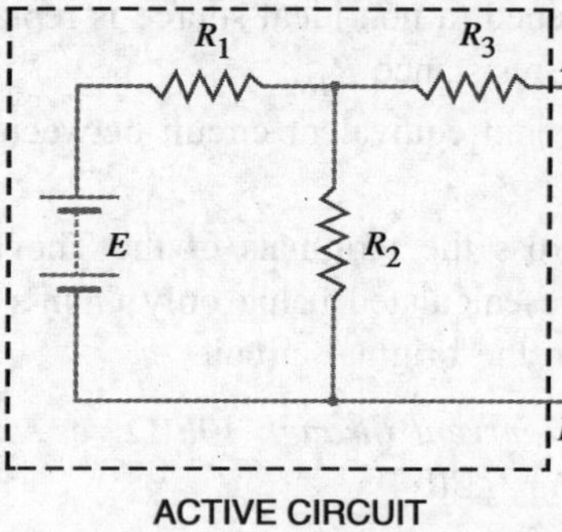

Fig. 1.26

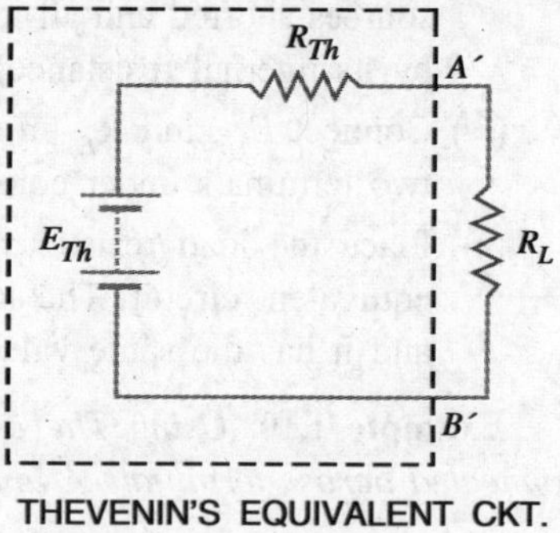

Fig. 1.27

***(i)* Finding E_{Th}.** This is the voltage between terminals A and B of the circuit when load R_L is removed. Fig. 1.28 shows the circuit with load removed. The voltage drop across R_2 is the desired voltage E_{Th}.

$$\text{Current through } R_2 = \frac{E}{R_1 + R_2}$$

$$\therefore \quad \text{Voltage across } R_2,\ E_{Th} = \left(\frac{E}{R_1 + R_2}\right) R_2$$

Thus, voltage E_{Th} is determined.

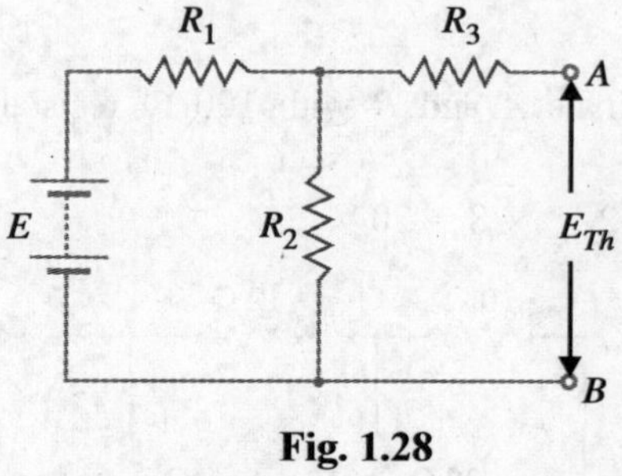

Fig. 1.28

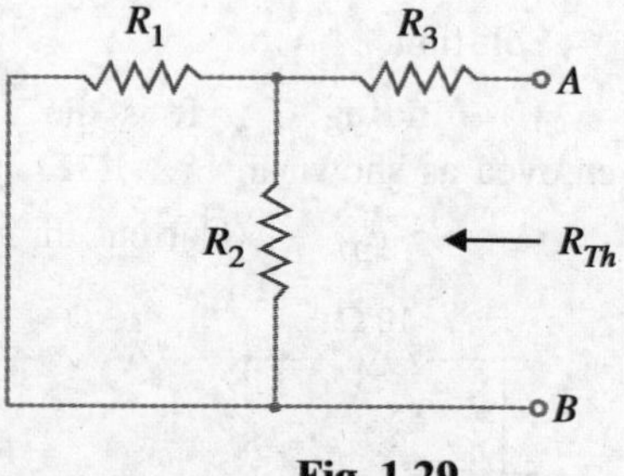

Fig. 1.29

***(ii)* Finding R_{Th}.** This is the resistance between terminals A and B with load removed and e.m.f. reduced to zero (See Fig. 1.29).

$\therefore$ Resistance between terminals A and B in Fig. 1.29 is

$$R_{Th} = \text{Parallel combination of } R_1 \text{ and } R_2 \text{ in series with } R_3$$

$$= \frac{R_1 R_2}{R_1 + R_2} + R_3$$

Thus, the value of R_{Th} is determined. Once the values of E_{Th} and R_{Th} are determined, then the current through the load resistance R_L can be found out easily (Refer to Fig. 1.27).

1.15 Procedure for Finding Thevenin Equivalent Circuit

(i) Open the two terminals (*i.e.* remove any load) between which you want to find Thevenin equivalent circuit.

(ii) Find the open-circuit voltage between the two open terminals. It is called Thevenin voltage E_{Th}.

(iii) Determine the resistance between the two open terminals with all ideal voltage sources shorted and all ideal current sources opened (a non-ideal source is replaced by its internal resistance). It is called Thevenin resistance R_{Th}.

(iv) Connect E_{Th} and R_{Th} in series to produce Thevenin equivalent circuit between the two terminals under consideration.

(v) Place the load resistor removed in step (i) across the terminals of the Thevenin equivalent circuit. The load current can now be calculated using only Ohm's law and it has the same value as the load current in the original circuit.

Example 1.10. *Using Thevenin's theorem, find the current through 100 Ω resistance connected across terminals A and B in the circuit of Fig. 1.30.*

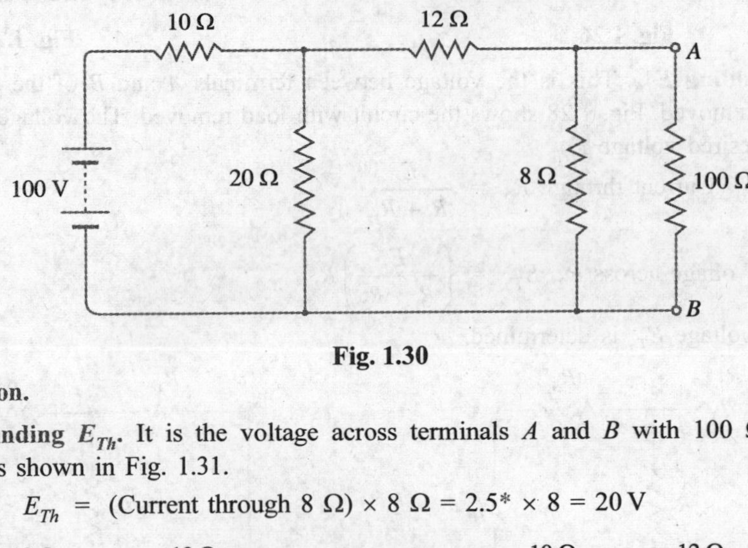

Fig. 1.30

Solution.

(i) **Finding E_{Th}.** It is the voltage across terminals *A* and *B* with 100 Ω resistance removed as shown in Fig. 1.31.

$$E_{Th} = \text{(Current through 8 Ω)} \times 8\ \Omega = 2.5^* \times 8 = 20\ \text{V}$$

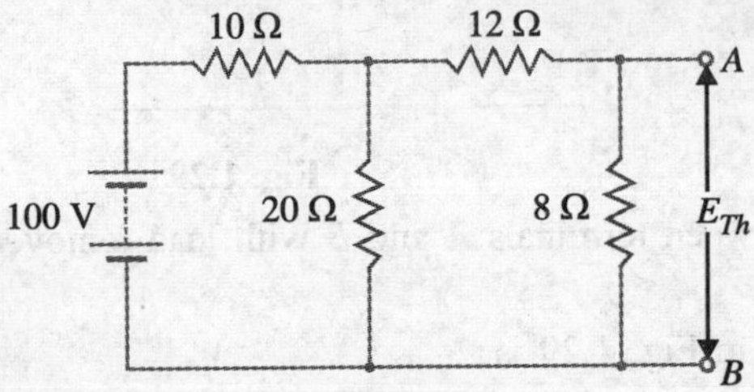

Fig. 1.31

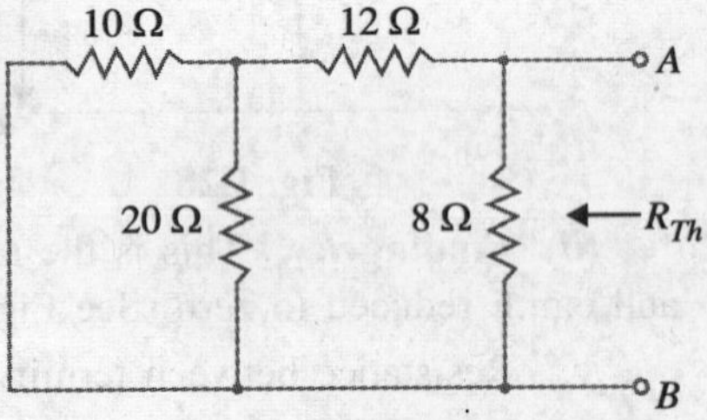

Fig. 1.32

(ii) **Finding R_{Th}.** It is the resistance between terminals *A* and *B* with 100 Ω removed and voltage source short- circuited as shown in Fig. 1.32.

* By solving this series-parallel circuit.

$$R_{Th} = \text{Resistance looking in at terminals } A \text{ and } B \text{ in Fig. 1.32}$$

$$= \frac{\left[\frac{10\times 20}{10+20}+12\right]8}{\left[\frac{10\times 20}{10+20}+12\right]+8}$$

$$= 5.6\ \Omega$$

Therefore, Thevenin's equivalent circuit will be as shown in Fig. 1.33. Now, current through 100 Ω resistance connected across terminals A and B can be found by applying Ohm's law.

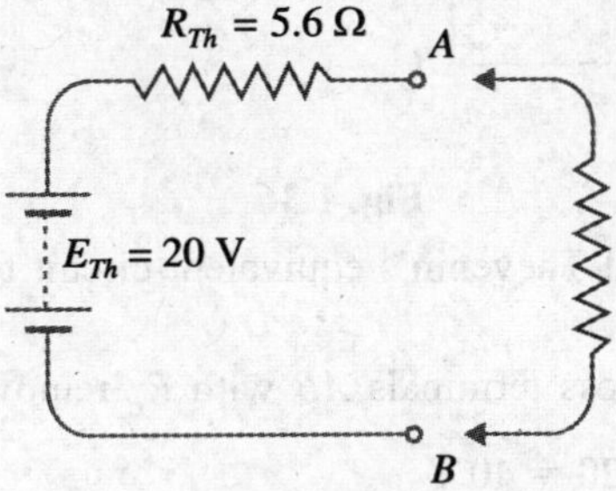

Fig. 1.33

$$\text{Current through 100 }\Omega\text{ resistor} = \frac{E_{Th}}{R_{Th}+R_L} = \frac{20}{5.6+100} = \mathbf{0.19\ A}$$

Example 1.11. *Find the Thevenin's equivalent circuit for Fig. 1.34.*

Solution. The Thevenin's voltage E_{Th} is the voltage across terminals A and B. This voltage is equal to the voltage across R_3. It is because terminals A and B are open-circuited and there is no current flowing through R_2 and hence no voltage drop across it.

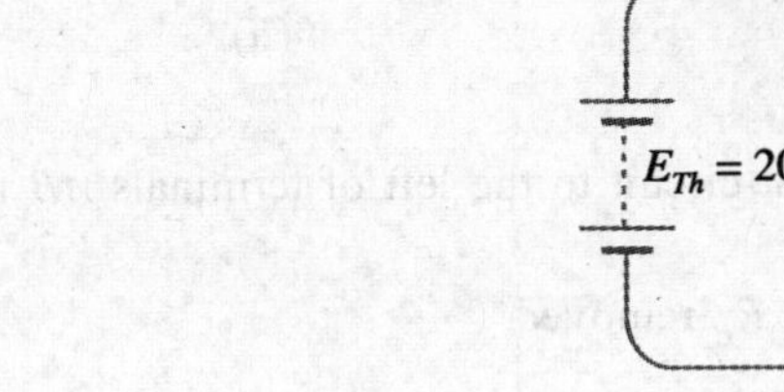

Fig. 1.34

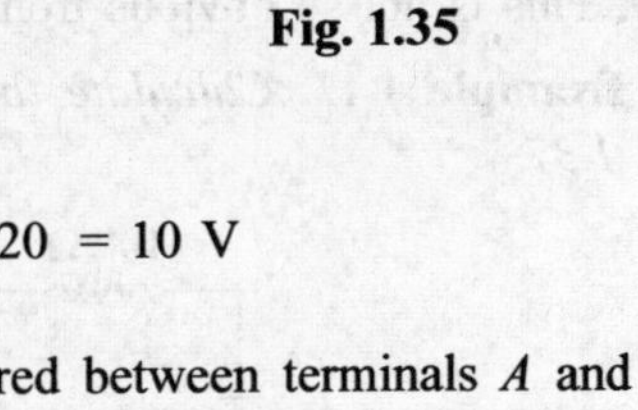

Fig. 1.35

$$\therefore \quad E_{Th} = \text{Voltage across } R_3$$

$$= \frac{R_3}{R_1+R_3}\times V = \frac{1}{1+1}\times 20 = 10\text{ V}$$

The Thevenin's resistance R_{Th} is the resistance measured between terminals A and B with no load (*i.e.* open at terminals A and B) and voltage source replaced by a short circuit.

$$\therefore \quad R_{Th} = R_2 + \frac{R_1 R_3}{R_1+R_3} = 1+\frac{1\times 1}{1+1} = 1.5\text{ k}\Omega$$

Therefore, Thevenin's equivalent circuit will be as shown in Fig. 1.35.

Example 1.12. *Calculate the value of load resistance R_L to which maximum power may be transferred from the circuit shown in Fig. 1.36 (i). Also find the maximum power.*

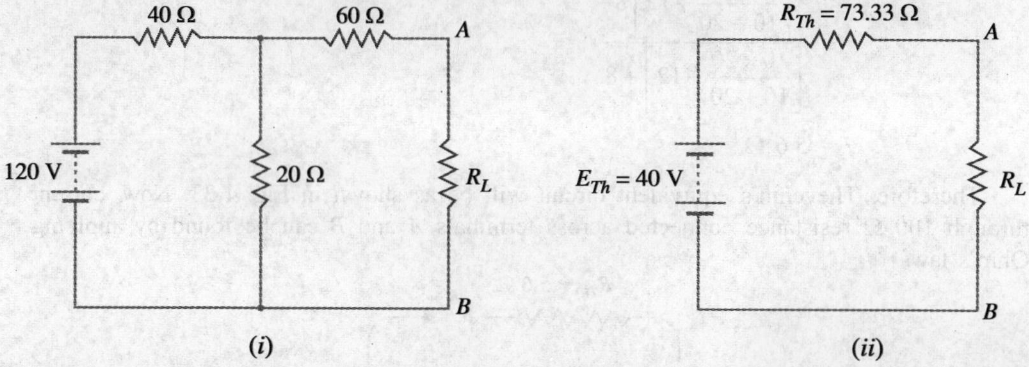

Fig. 1.36

Solution. We shall first find Thevenin's equivalent circuit to the left of terminals *AB* in Fig. 1.36 (*i*).

E_{Th} = Voltage across terminals *AB* with R_L removed

$$= \frac{120}{40+20} \times 20 = 40 \text{ V}$$

R_{Th} = Resistance between terminals *A* and *B* with R_L removed and 120 V source replaced by a short

$$= 60\ \Omega + (40\ \Omega \parallel 20\ \Omega) = 60 + (40 \times 20)/60 = 73.33\ \Omega$$

The Thevenin's equivalent circuit to the left of terminals *AB* in Fig. 1.36 (*i*) is E_{Th} (= 40 V) in series with R_{Th} (= 73.33 Ω). When R_L is connected between terminals *A* and *B*, the circuit becomes as shown in Fig. 1.36 (*ii*). It is clear that maximum power will be transferred when

$$R_L = R_{Th} = 73.33\ \Omega$$

$$\text{Maximum power to load} = \frac{E_{Th}^2}{4\,R_L} = \frac{(40)^2}{4 \times 73.33} = \mathbf{5.45\ W}$$

Comments. This shows another advantage of Thevenin's equivalent circuit of a network. Once Thevenin's equivalent resistance R_{Th} is calculated, it shows at a glance the condition for maximum power transfer. Yet Thevenin's equivalent circuit conveys another information. Thus referring to Fig. 1.36 (*ii*), the maximum voltage that can appear across terminals *A* and *B* is 40 V. This is not so obvious from the original circuit shown in Fig. 1.36 (*i*).

Example 1.13. *Calculate the current in the 50 Ω resistor in the network shown in Fig. 1.37.*

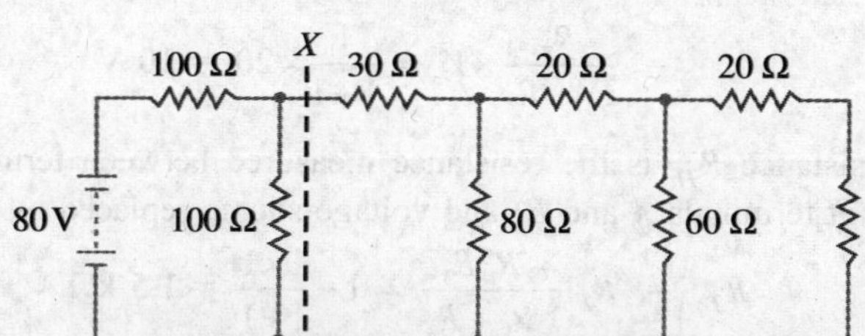

Fig. 1.37

Solution. We shall simplify the circuit shown in Fig. 1.37 by the repeated use of Thevenin's theorem. We first find Thevenin's equivalent circuit to the left of *XX.

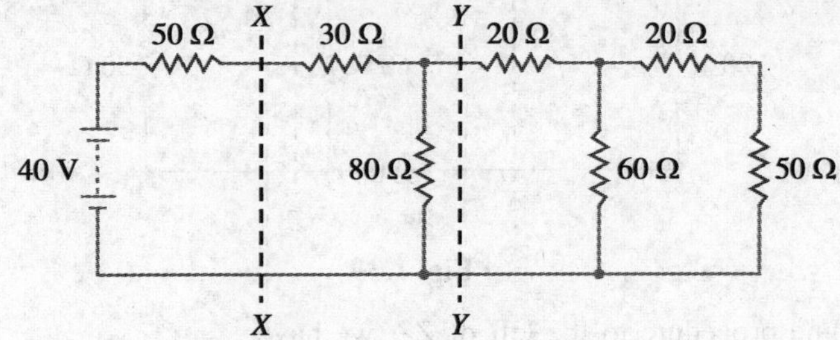

Fig. 1.38

$$E'_{Th} = \frac{40}{50+30+80} \times 80 = 20 \text{ V}$$

$$R'_{Th} = (50 + 30) \parallel 80 = \frac{80 \times 80}{80+80} = 40 \ \Omega$$

Therefore, we can replace the circuit to the left of *XX* in Fig. 1.37 by its Thevenin's equivalent circuit *viz.* E_{Th} (= 40 V) in series with R_{Th} (= 50 Ω). The original circuit of Fig. 1.37 then reduces to the one shown in Fig. 1.38.

We shall now find Thevenin's equivalent circuit to left of *YY* in Fig. 1.38.

$$E_{Th} = \frac{80}{100+100} \times 100 = 40\text{V}$$

$$R_{Th} = 100 \parallel 100 = \frac{100 \times 100}{100+100} = 50 \ \Omega$$

We can again replace the circuit to the left of *YY* in Fig. 1.38 by its Thevenin's equivalent circuit. Therefore, the original circuit reduces to that shown in Fig. 1.39.

---

*

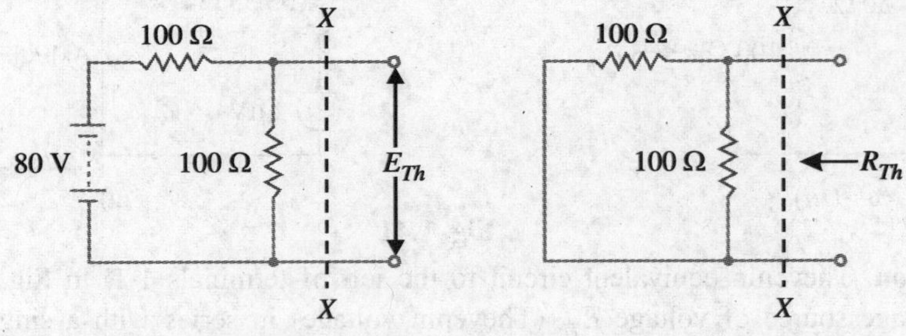

E_{Th} = Current in 100 Ω × 100 Ω = $\frac{80}{100+100} \times 100$ = 40 V [See Fig. (*a*)]

R_{Th} = Resistance looking into the open terminals in Fig. (*b*) = 100 ∥ 100 = $\frac{100 \times 100}{100+100}$ = 50 Ω

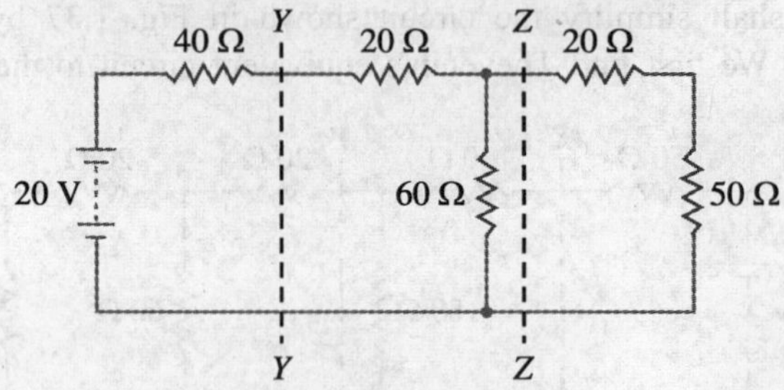

Fig. 1.39

Using the same procedure to the left of *ZZ*, we have,

$$E''_{Th} = \frac{20}{40+20+60} \times 60 = 10 \text{ V}$$

$$R''_{Th} = (40 + 20) \parallel 60 = \frac{60 \times 60}{60+60} = 30 \ \Omega$$

The original circuit then reduces to that shown in Fig. 1.40.

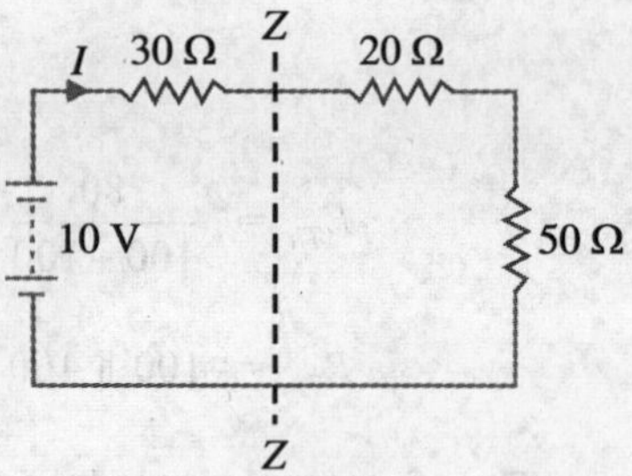

Fig. 1.40

By Ohm's law, current I in 50 Ω resistor is

$$I = \frac{10}{30+20+50} = \mathbf{0.1 \text{ A}}$$

Example 1.14. *Fig. 1.41(i) shows the input circuit of an amplifier having 100 Ω input resistance. Replace the circuit to the left of terminals 1-1′ by Thevenin equivalent circuit.*

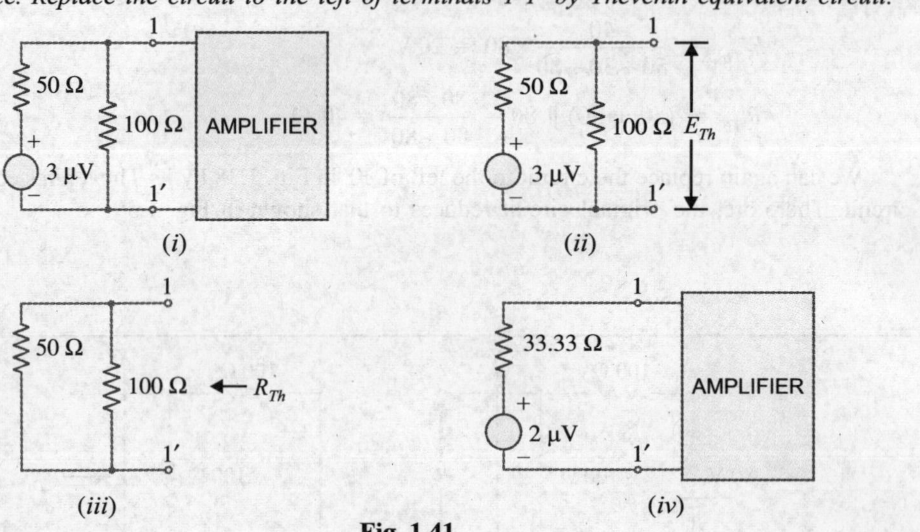

Fig. 1.41

Solution. Thevenin equivalent circuit to the left of terminals 1-1′ in Fig. 1.41 (*i*) is a single voltage source of voltage E_{Th} (Thevenin voltage) in series with a single resistor of resistance R_{Th} (Thevenin resistance). To find E_{Th}, open-circuit terminals 1-1′ as shown in Fig. 1.41 (*ii*). The voltage at the open-circuited terminals in Fig. 1.41 (*ii*) is E_{Th}.

$$E_{Th} = \text{Voltage across } 100 \ \Omega \text{ in Fig. 1.41 } (ii)$$

$$\therefore \quad E_{Th} = \frac{3 \ \mu\text{V}}{(50+100) \ \Omega} \times 100 \ \Omega = 2 \ \mu\text{V}$$

To find R_{Th}, open-circuit terminals 1-1′ in Fig. 1.41 (*i*) and replace the 3 μV source by a short as shown in Fig. 1.41 (*iii*). Then resistance at teminals 1-1′ is R_{Th} *i.e.*

$$R_{Th} = \text{Resistance at terminals 1-1' in Fig. 1.41 } (iii)$$

$$\therefore \quad R_{Th} = 50\ \Omega \parallel 100\ \Omega = \frac{50 \times 100}{50 + 100} = 33.33\ \Omega$$

Therefore, Thevenin equivalent circuit to the left of terminals 1-1′ in Fig. 1.41 (*i*) is a voltage source of 2 μV in series with a resistance of 33.33 Ω as shown in Fig. 1.41 (*iv*). The reader may note that Thevenin's Theorem provides a powerful tool for simplifying a complex circuit for the purpose of circuit analysis.

1.16 Norton's Theorem

Fig. 1.42 (*i*) shows a network enclosed in a box with two terminals *A* and *B* brought out. The network in the box may contain any number of resistors and e.m.f. sources connected in any manner. But according to Norton, the entire circuit behind terminals *A* and *B* can be replaced by a current source of output I_N in parallel with a single resistance R_N as shown in Fig. 1.42 (*ii*). The value of I_N is determined as mentioned in Norton's theorem. The resistance R_N is the same as Thevenin's resistance R_{Th}. Once Norton's equivalent circuit is determined [See Fig. 1.42 (*ii*)], then current through any load R_L connected across terminals *AB* can be readily obtained.

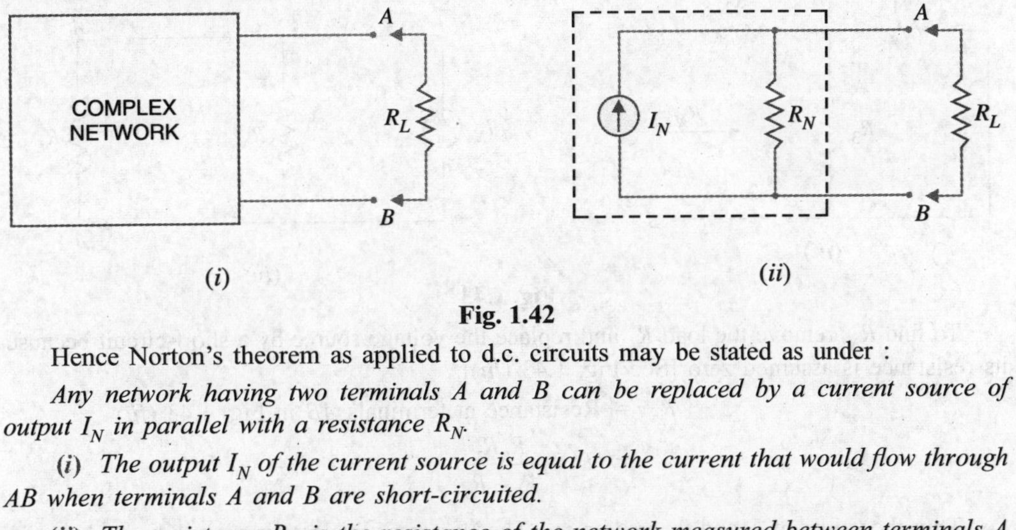

Fig. 1.42

Hence Norton's theorem as applied to d.c. circuits may be stated as under :

Any network having two terminals A and B can be replaced by a current source of output I_N in parallel with a resistance R_N.

(*i*) *The output I_N of the current source is equal to the current that would flow through AB when terminals A and B are short-circuited.*

(*ii*) *The resistance R_N is the resistance of the network measured between terminals A and B with load (R_L) removed and sources of e.m.f. replaced by their internal resistances, if any.*

Norton's theorem is *converse* of Thevenin's theorem in that Norton equivalent circuit uses a current generator instead of voltage generator and resistance R_N (which is the same as R_{Th}) in parallel with the generator instead of being in series with it.

Illustration. Fig. 1.43 illustrates the application of Norton's theorem. As far as circuit behind terminals *AB* is concerned [See Fig. 1.43 (*i*)], it can be replaced by a current source of output I_N in parallel with a resistance R_N as shown in Fig. 1.43 (*iv*). The output I_N of the current generator is equal to the current that would flow through AB when terminals A and B are short-circuited as shown in Fig. 1.43 (*ii*). The load R' on the source when terminals AB are short-circuited is given by ;

$$R' = R_1 + \frac{R_2 R_3}{R_2 + R_3} = \frac{R_1 R_2 + R_1 R_3 + R_2 R_3}{R_2 + R_3}$$

$$\text{Source current, } I' = \frac{V}{R'} = \frac{V(R_2 + R_3)}{R_1 R_2 + R_1 R_3 + R_2 R_3}$$

$$\text{Short-circuit current, } I_N = \text{Current in } R_2 \text{ in Fig. 1.43 } (ii)$$

$$= I' \times \frac{R_3}{R_2 + R_3} = \frac{V R_3}{R_1 R_2 + R_1 R_3 + R_2 R_3}$$

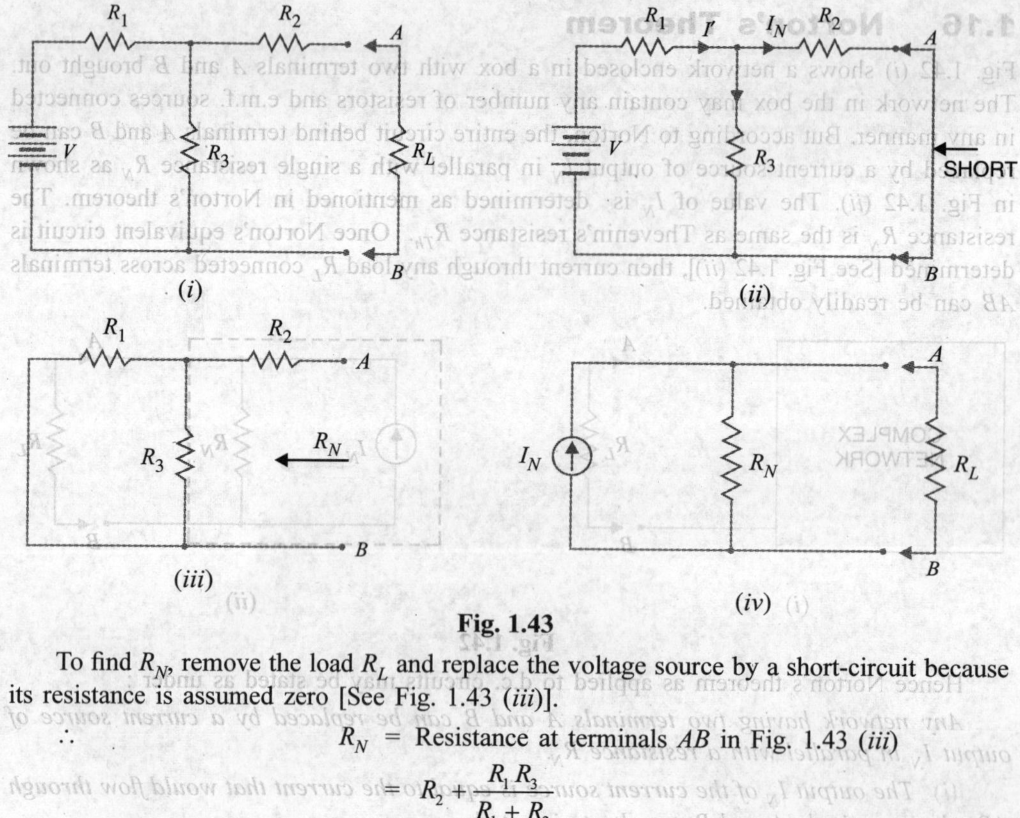

Fig. 1.43

To find R_N, remove the load R_L and replace the voltage source by a short-circuit because its resistance is assumed zero [See Fig. 1.43 (*iii*)].

$$\therefore \quad R_N = \text{Resistance at terminals } AB \text{ in Fig. 1.43 } (iii)$$

$$= R_2 + \frac{R_1 R_3}{R_1 + R_3}$$

Thus the values of I_N and R_N are known. The Norton equivalent circuit will be as shown in Fig. 1.43 (*iv*).

1.17 Procedure for Finding Norton Equivalent Circuit

(*i*) Open the two terminals (*i.e.* remove any load) between which we want to find Norton equivalent circuit.

(*ii*) Put a short-circuit across the terminals under consideration. Find the short-circuit current flowing in the short-circuit. It is called Norton current I_N.

(*iii*) Determine the resistance between the two open terminals with all ideal voltage sources shorted and all ideal current sources opened (a non-ideal source is replaced by its internal resistance). It is called Norton's resistance R_N. It is easy to see that $R_N = R_{Th}$.

(*iv*) Connect I_N and R_N in parallel to produce Norton equivalent circuit between the two terminals under consideration.

(*v*) Place the load resistor removed in step (*i*) across the terminals of the Norton equivalent circuit. The load current can now be calculated by using current-divider rule. This load current will be the same as the load current in the original circuit.

Example 1.15. *Using Norton's theorem, find the current in 8 Ω resistor in the network shown in Fig. 1.44 (i).*

Solution. We shall reduce the network to the left of *AB* in Fig. 1.44 (*i*) to Norton's equivalent circuit. For this purpose, we are required to find I_N and R_N.

(*i*) With load (*i.e.* 8 Ω) removed and terminals *AB* short-circuited [See Fig. 1.44 (*ii*)], the current that flows through *AB* is equal to I_N. Referring to Fig. 1.44 (*ii*),

$$\text{Load on the source} = 4\ \Omega + (5\ \Omega \parallel 6\ \Omega)$$

$$= 4 + \frac{5 \times 6}{5+6} = 6.727\ \Omega$$

$$\text{Source current, } I' = 40/6.727 = 5.94\ \text{A}$$

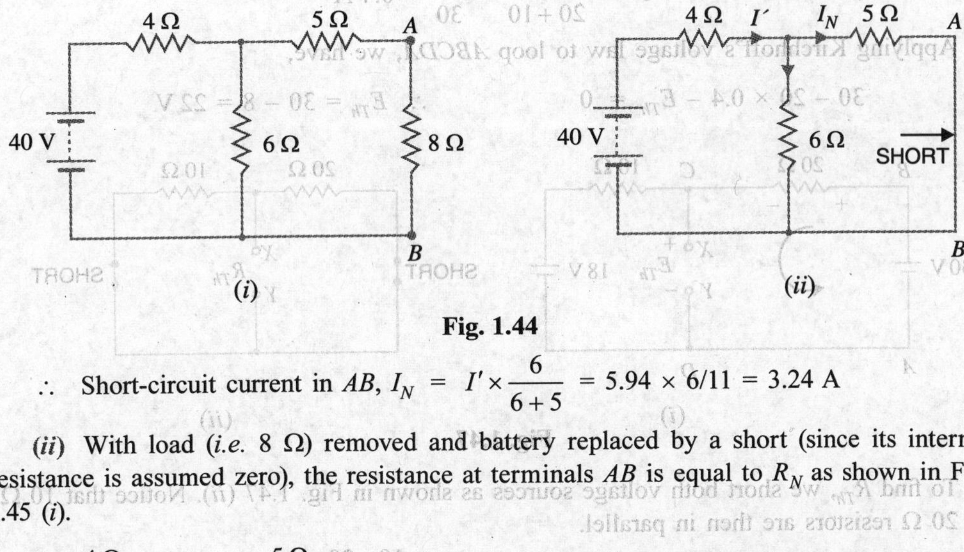

Fig. 1.44

$$\therefore \quad \text{Short-circuit current in } AB,\ I_N = I' \times \frac{6}{6+5} = 5.94 \times 6/11 = 3.24\ \text{A}$$

(*ii*) With load (*i.e.* 8 Ω) removed and battery replaced by a short (since its internal resistance is assumed zero), the resistance at terminals *AB* is equal to R_N as shown in Fig. 1.45 (*i*).

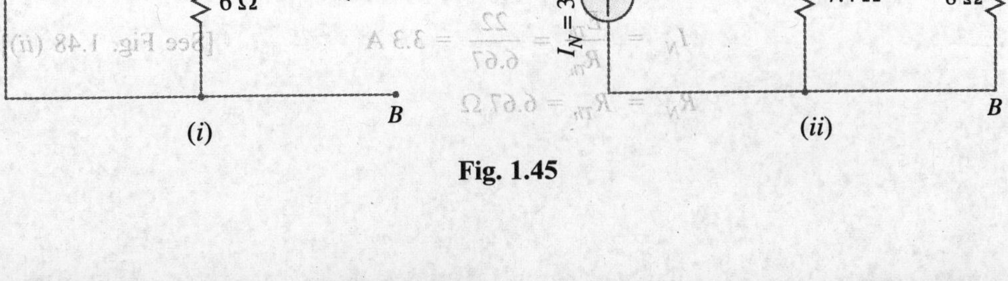

Fig. 1.45

$$R_N = 5\ \Omega + (4\ \Omega \parallel 6\ \Omega) = 5 + \frac{4\times 6}{4+6} = 7.4\ \Omega$$

The Norton's equivalent circuit behind terminals *AB* is I_N (= 3.24 A) in parallel with R_N (= 7.4 Ω). When load (*i.e.* 8 Ω) is connected across terminals *AB*, the circuit becomes as shown in Fig. 1.45 (*ii*). The current source is supplying current to two resistors 7.4 Ω and 8 Ω in parallel.

$$\therefore \quad \text{Current in 8 }\Omega\text{ resistor} = 3.24 \times \frac{7.4}{8+7.4} = \mathbf{1.55\ A}$$

Example 1.16. *Find the Norton equivalent circuit at terminals X – Y in Fig. 1.46.*

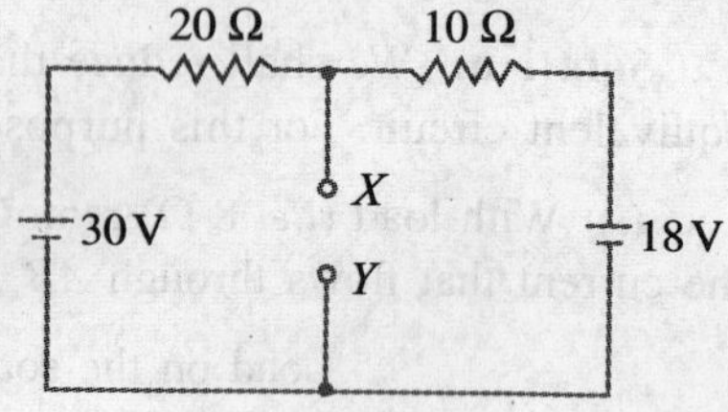

Fig. 1.46

Solution. We shall first find the Thevenin equivalent circuit and then convert it to an equivalent current source. This will then be Norton equivalent circuit.

Finding Thevenin Equivalent circuit. To find E_{Th}, refer to Fig. 1.47 (*i*). Since 30 V and 18 V sources are in opposition, the circuit current *I* is given by ;

$$I = \frac{30-18}{20+10} = \frac{12}{30} = 0.4\ \text{A}$$

Applying Kirchhoff's voltage law to loop *ABCDA*, we have,

$$30 - 20 \times 0.4 - E_{Th} = 0 \qquad \therefore \quad E_{Th} = 30 - 8 = 22\ \text{V}$$

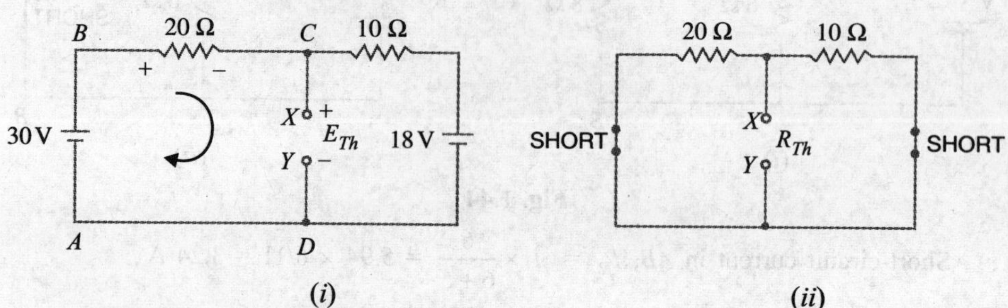

Fig. 1.47

To find R_{Th}, we short both voltage sources as shown in Fig. 1.47 (*ii*). Notice that 10 Ω and 20 Ω resistors are then in parallel.

$$\therefore \quad R_{Th} = 10\ \Omega \parallel 20\ \Omega = \frac{10\times 20}{10+20} = 6.67\ \Omega$$

Therefore, Thevenin equivalent circuit will be as shown in Fig. 1.48 (*i*). Now, it is quite easy to convert it into equivalent current source.

$$I_N = \frac{E_{Th}}{R_{Th}} = \frac{22}{6.67} = 3.3\ \text{A} \qquad \text{[See Fig. 1.48 (}ii\text{)]}$$

$$R_N = R_{Th} = 6.67\ \Omega$$

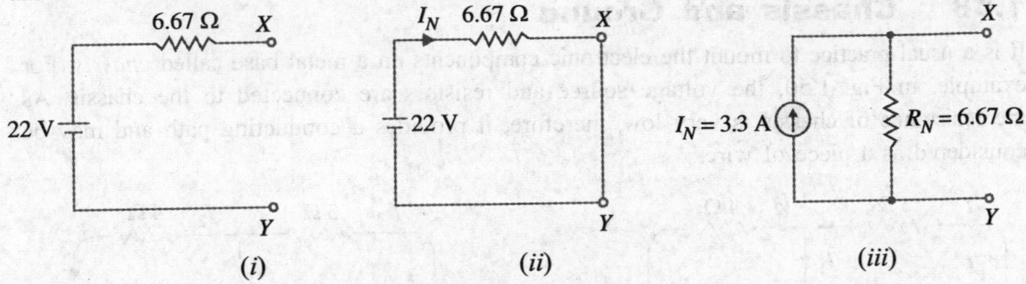

Fig. 1.48

Fig. 1.48 (*iii*) shows Norton equivalent circuit. Observe that the Norton equivalent resistance has the same value as the Thevenin equivalent resistance. Therefore, R_N is found exactly the same way.

Example 1.17. *Show that when Thevenin's equivalent circuit of a network is converted into Norton's equivalent circuit, $I_N = E_{Th}/R_{Th}$ and $R_N = R_{Th}$. Here E_{Th} and R_{Th} are Thevenin voltage and Thevenin resistance respectively.*

Solution. Fig. 1.49 (*i*) shows a network enclosed in a box with two terminals *A* and *B* brought out. Thevenin's equivalent circuit of this network will be as shown in Fig. 1.49 (*ii*). To find Norton's equivalent circuit, we are to find I_N and R_N. Referring to Fig. 1.49 (*ii*),

$$I_N = \text{Current flowing through short-circuited } AB \text{ in Fig. 1.49 (ii)}$$
$$= E_{Th}/R_{Th}$$

$$R_N = \text{Resistance at terminals } AB \text{ in Fig. 1.49 (ii)}$$
$$= R_{Th}$$

Fig. 1.49 (*iii*) shows Norton's equivalent circuit. Hence we arrive at the following two important conclusions :

(*i*) To convert Thevenin's equivalent circuit into Norton's equivalent circuit,

$$I_N = E_{Th}/R_{Th} \quad ; \quad R_N = R_{Th}$$

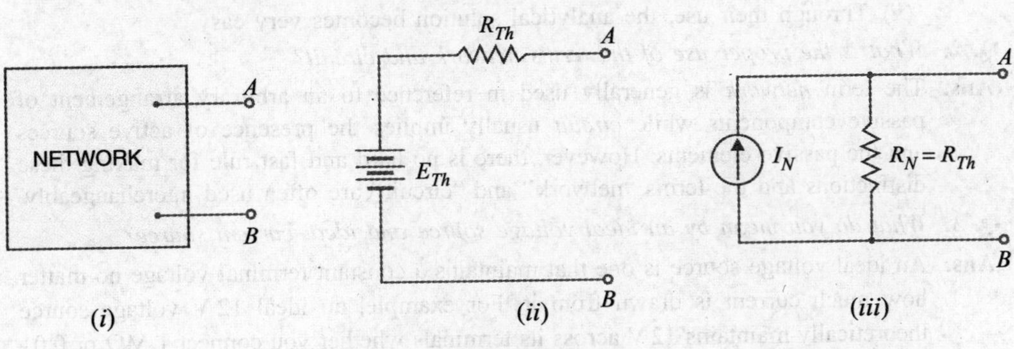

Fig. 1.49

(*ii*) To convert Norton's equivalent circuit into Thevenin's equivalent circuit,

$$E_{Th} = I_N R_N \quad ; \quad R_{Th} = R_N$$

1.18 Chassis and Ground

It is a usual practice to mount the electronic components on a metal base called *chassis*. For example, in Fig. 1.50, the voltage source and resistors are connected to the chassis. As the resistance of chassis is very low, therefore, it provides a conducting path and may be considered as a piece of wire.

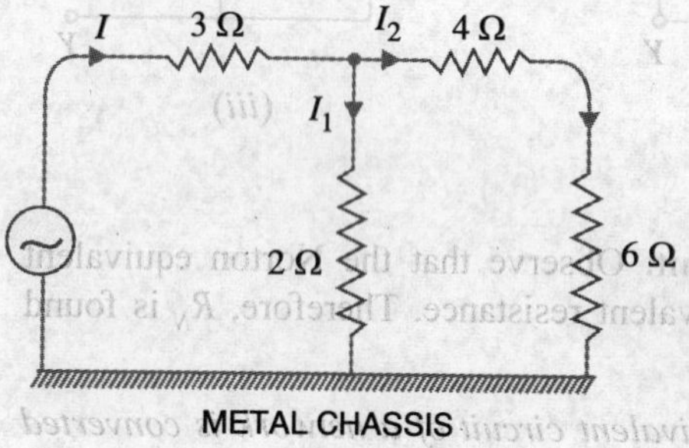

Fig. 1.50

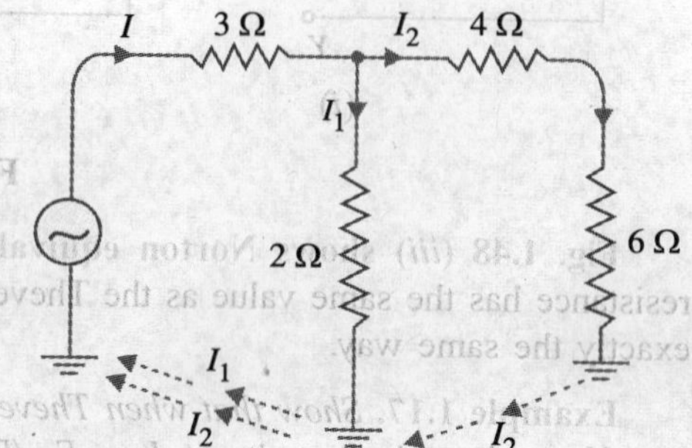

Fig. 1.51

It is customary to refer to the chassis as *ground*. Fig. 1.51 shows the symbol for chassis. It may be seen that all points connected to chassis are shown as grounded and represent the same potential. The adoption of this scheme (*i.e.* showing points of same potential as grounded) often simplifies the electronic circuits. In our further discussion, we shall frequently use this scheme.

Short Answer Questions

Q. 1. *Why do we use network theorems and techniques to solve electrical circuits?*

Ans. Network theorems and techniques are often used to solve electrical circuits for the following reasons :

(*i*) Through their use, a complicated network can be reduced to a simple circuit.

(*ii*) Through their use, the analytical solution becomes very easy.

Q. 2. *What is the proper use of the terms network and circuit?*

Ans. The term *network* is generally used in reference to an arbitrary arrangement of passive components while *circuit* usually implies the presence of active sources and the passive elements. However, there is no hard and fast rule for making these distinctions and the terms "network" and "circuit" are often used interchangeably.

Q. 3. *What do you mean by an ideal voltage source and ideal current source?*

Ans. An ideal voltage source is one that maintains a constant terminal voltage no matter how much current is drawn from it. For example, an ideal 12 V voltage source theoretically maintains 12 V across its terminals whether you connect 1 MΩ or 0.01 Ω resistor across its terminals.

An ideal current source is one which will supply the same amount of current to any resistance connected to its terminals. For example, a 2 A ideal current source will always supply 2 A to any value of resistance connected across its terminals.

Q. 4. *You are given two black boxes A and B as shown in Fig. 1.52 and told that each contains a source of some type. Can you determine what is in each box?*

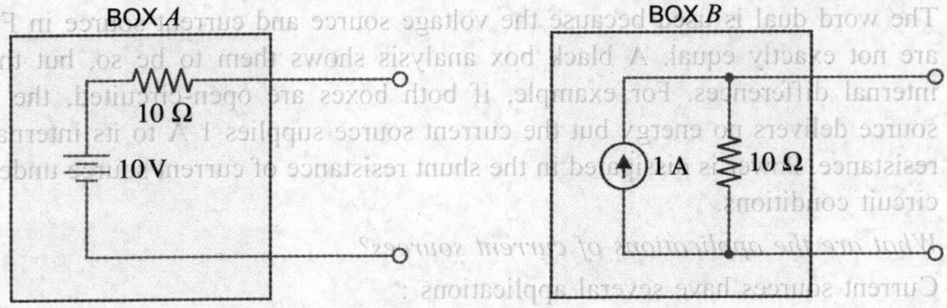

Fig. 1.52

Ans. (*i*) Since there is no way to see inside the boxes, a load is connected to each to determine what it contains. Suppose we connect a load of 10 Ω to each box as shown in Fig. 1.53. Voltmeter and ammeter are also connected to obtain data. In case of box *A*, the 10 Ω load resistor acts in series with 10 Ω internal resistance so that total resistance is 20 Ω.

$$\text{Ammeter reading} = \frac{10}{20} = 0.5 \text{ A}$$

$$\text{Voltmeter reading} = 10 \times 0.5 = 5 \text{ V}$$

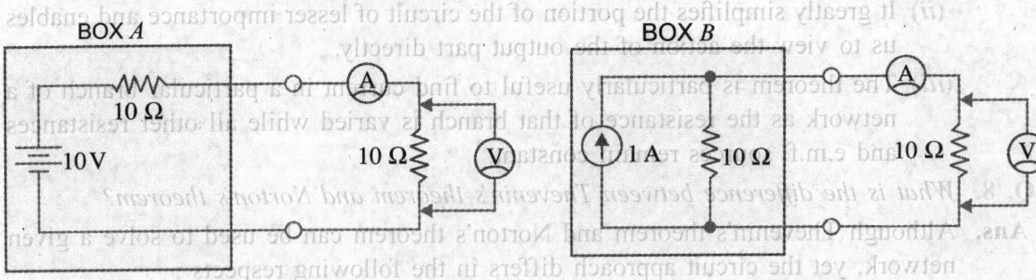

Fig. 1.53

In case of box *B*, two 10 Ω resistors are in parallel and the current of 1 A is divided equally between these resistors.

$$\text{Ammeter reading} = 0.5 \text{ A}$$

$$\text{Voltmeter reading} = 0.5 \times 10 = 5 \text{ V}$$

The meter readings are the same for two boxes so no difference is found as yet.

(*ii*) If the external 10 Ω loads are removed, both voltmeters indicate 10 V and both ammeters read zero.

(*iii*) If the external 10 Ω loads are shorted, both ammeters read 1 A and both voltmeters read 0.

Therefore, the boxes *appear to be identical*. The voltage source and current source of Fig. 1.52 are considered *duals*.

Q. 5. *The voltage source and current source of Fig. 1.52 are called duals. Why is word dual used ?*

Ans. The word dual is used because the voltage source and current source in Fig. 1.52 are not exactly equal. A black box analysis shows them to be so, but there are internal differences. For example, if both boxes are open-circuited, the voltage source delivers no energy but the current source supplies 1 A to its internal shunt resistance. Power is dissipated in the shunt resistance of current source under open-circuit conditions.

Q. 6. *What are the applications of current sources?*

Ans. Current sources have several applications :

(*i*) They are used to represent certain devices that exhibit a constant-current behaviour. Transistors are examples of such devices.

(*ii*) Current sources can also be used to represent equipment such as current-regulated power supplies. These supplies provide a constant current over some limited range of load resistances.

(*iii*) They can be used in certain network analysis procedures.

Q. 7. *What are the advantages of Thevenin's theorem?*

Ans. Thevenin's theorem offers the following advantages :

(*i*) It reduces a complex circuit to a simple circuit *viz.* a single source of e.m.f. E_{Th} in series with a single resistance R_{Th}.

(*ii*) It greatly simplifies the portion of the circuit of lesser importance and enables us to view the action of the output part directly.

(*iii*) The theorem is particularly useful to find current in a particular branch of a network as the resistance of that branch is varied while all other resistances and e.m.f. sources remain constant.

Q. 8. *What is the difference between Thevenin's theorem and Norton's theorem?*

Ans. Although Thevenin's theorem and Norton's theorem can be used to solve a given network, yet the circuit approach differs in the following respects :

(*i*) Norton's theorem is *converse* of Thevenin's theorem in the respect that Norton equivalent circuit uses a current generator instead of voltage generator and the resistance R_N (which is the same as R_{Th}) in parallel with the generator instead of being in series with it.

(*ii*) Thevenin's theorem is a voltage form of an equivalent circuit whereas Norton's theorem is a current form of an equivalent circuit.

(*iii*) The mathematical relation between the two circuit forms is :

$$R_N = R_{Th} \quad ; \quad I_N = E_{Th}/R_{Th}$$

Q. 9. *How does circuit behave under maximum power transfer conditions?*

Ans. Under the conditions of maximum power transfer, the circuit exhibits the following characteristics :

(*i*) The efficiency is 50% only as one-half of the total power generated is dissipated in the internal resistance of the source.

(*ii*) The load voltage is one-half of the open-circuited voltage at the load terminals.

Q. 10. *What are two applications of maximum power transfer theorem?*

Ans. Two important applications of maximum power transfer theorem are given below:

(*i*) In electronic circuits, maximum power transfer is usually desirable. For instance, in a public address system, the circuit is adjusted for maximum power transfer by making load resistance (*i.e.* speaker) equal to source (*i.e.* amplifier) resistance.

(*ii*) Another example of maximum power transfer is found in the starting of a car engine. The power delivered to the starter motor of the car will depend upon the effective resistance of the motor and internal resistance of the battery. If the two resistances are equal (as is the case when the battery is fully charged), maximum power will be transferred to the motor to turn on the engine. This is particularly desirable in winter when every watt that can be extracted from the battery is needed by the starter motor to turn on the cold engine. If the battery is weak, its internal resistance is high and the car does not start.

Q. 11. *Why is power system never operated under maximum power conditions?*

Ans. Under the conditions of maximum power transfer, the efficiency is low (50%) and there is greater voltage drop in the lines. In a power system, the goal is higher efficiency rather than maximum power. For these reasons, maximum power transfer is not desired in a power system.

Q. 12. *Why are electronic circuits generally operated under maximum power transfer conditions?*

Ans. In electronic circuits, it is often desirable to transfer maximum power *e.g.*

(*i*) The signal power available at the receiving antenna is very small. It is very important to recover the maximum possible amount of signal power from the receiving antenna.

(*ii*) In the public address system, it is desired that maximum power is transferred from the amplifier to the speaker (*i.e.* load) in order to operate the speaker.

To meet such situations in electronic circuits, we adjust the circuit for maximum power transfer. The technique is to make the load resistance (*e.g.,* speaker) equal to the source (*e.g.,* amplifier) resistance. The circuit is then said to be *matched.*

Multiple-Choice Questions

1. The outermost orbit of an atom can have a maximum of .... electrons.
(*i*) 8 (*ii*) 6
(*iii*) 4 (*iv*) 3

2. When the outermost orbit of an atom has less than 4 electrons, the material is generally a ....
(*i*) non-metal (*ii*) metal
(*iii*) semiconductor (*iv*) none of above

3. The valence electrons have ....
(*i*) very small energy
(*ii*) least energy
(*iii*) maximum energy
(*iv*) none of the above

4. A large number of free electrons exist in ....
(*i*) semiconductors (*ii*) metals
(*iii*) insulators (*iv*) non-metals

5. An ideal voltage source has .... internal resistance.
(*i*) small (*ii*) large
(*iii*) infinite (*iv*) zero

6. An ideal current source has .... internal resistance.
(*i*) infinite (*ii*) zero
(*iii*) small
(*iv*) none of the above

7. Maximum power is transferred if load resistance is equal to .... of the source.
(*i*) half the internal resistance

(ii) internal resistance
(iii) twice the internal resistance
(iv) none of the above

8. Efficiency at maximum power transfer is ....
(i) 75% (ii) 25%
(iii) 90% (iv) 50%

9. When the outermost orbit of an atom has exactly 4 valence electrons, the material is generally ....
(i) a metal (ii) a non-metal
(iii) a semiconductor
(iv) an insulator

10. Thevenin's theorem replaces a complicated circuit facing a load by an ....
(i) ideal voltage source and parallel resistor
(ii) ideal current source and parallel resistor
(iii) ideal current source and series resistor
(iv) ideal voltage source and series resistor

11. The output voltage of an ideal voltage source is ....
(i) zero (ii) constant
(iii) dependent on load resistance
(iv) dependent on internal resistance

12. The current output of an ideal current source is ....
(i) zero (ii) constant
(iii) dependent on load resistance
(iv) dependent on internal resistance

13. Norton's theorem replaces a complicated circuit facing a load by an ....
(i) ideal voltage source and parallel resistor
(ii) ideal current source and parallel resistor
(iii) ideal voltage source and series resistor
(iv) ideal current source and series resistor

14. The practical example of ideal voltage source is ....
(i) lead-acid cell (ii) dry cell
(iii) Daniel cell
(iv) none of the above

15. The speed of electrons in vacuum is .... than in a conductor.
(i) less (ii) much more
(iii) much less
(iv) none of the above

16. Maximum power will be transferred from a source of 10 Ω resistance to a load of ....
(i) 5 Ω (ii) 20 Ω
(iii) 10 Ω (iv) 40 Ω

17. When the outermost orbit of an atom has more than 4 electrons, the material is generally a ....
(i) metal (ii) non-metal
(iii) semiconductor
(iv) none of the above

18. An ideal source consists of 5 V in series with 10kΩ resistance. The current magnitude of equivalent current source is ....
(i) 2 mA (ii) 3.5 mA
(iii) 0.5 mA
(iv) none of the above

19. To get Thevenin voltage, you have to ....
(i) short the load resistor
(ii) open the load resistor
(iii) short the voltage source
(iv) open the voltage source

20. To get the Norton current, you have to ....
(i) short the load resistor
(ii) open the load resistor
(iii) short the voltage source
(iv) open the voltage source

21. The open-circuited voltage at the terminals of load R_L in a network is 30 V. Under the conditions of maximum power transfer, the load voltage will be ....
(i) 30 V (ii) 10 V
(iii) 5 V (iv) 15 V

22. Under the conditions of maximum power transfer, a voltage source is delivering a power of 30 W to the load. The power produced by the source is ....

(*i*) 45 W (*ii*) 60 W
(*iii*) 30 W (*iv*) 90 W

23. The maximum power transfer theorem is used in ....
(*i*) electronic circuits
(*ii*) power system
(*iii*) home lighting circuits
(*iv*) none of the above

24. The Norton resistance of a network is 20 Ω and the shorted-load current is 2 A. If the network is loaded by a resistance equal to 20 Ω, the current through the load will be ....
(*i*) 2 A (*ii*) 0.5 A
(*iii*) 4 A (*iv*) 1 A

25. The Norton current is sometimes called the ....
(*i*) shorted-load current
(*ii*) open-load current
(*iii*) Thevenin current
(*iv*) Thevenin voltage

Answers to Multiple Choice Questions

1. (*i*)	2. (*ii*)	3. (*iii*)	4. (*ii*)	5. (*iv*)
6. (*i*)	7. (*ii*)	8. (*iv*)	9. (*iii*)	10. (*iv*)
11. (*ii*)	12. (*ii*)	13. (*ii*)	14. (*i*)	15. (*ii*)
16. (*iii*)	17. (*ii*)	18. (*iii*)	19. (*ii*)	20. (*i*)
21. (*iv*)	22. (*ii*)	23. (*i*)	24. (*iv*)	25. (*i*)

Chapter Review Topics

1. What is electronics? Mention some important applications of electronics.
2. Describe briefly the structure of atom.
3. Explain how valence electrons determine the electrical properties of a material.
4. Explain constant voltage and current sources. What is their utility?
5. Derive the condition for transfer of maximum power from a source to a load.
6. State and explain Thevenin's theorem.
7. Write short notes on the following:
(*i*) Atomic structure (*ii*) Valence electrons (*iii*) Free electrons

Problems

1. A dry battery developing 12 V has an internal resistance of 10 Ω. Find the output current if load is (*i*) 100 Ω (*ii*) 10 Ω (*iii*) 2 Ω and (*iv*) 1 Ω. **[(*i*) 0.1 A (*ii*) 0.6 A (*iii*) 1 A (*iv*) 1.1 A]**
2. Convert the current source in Fig. 1.54 into the equivalent voltage source.
[36 V in series with 900 Ω]

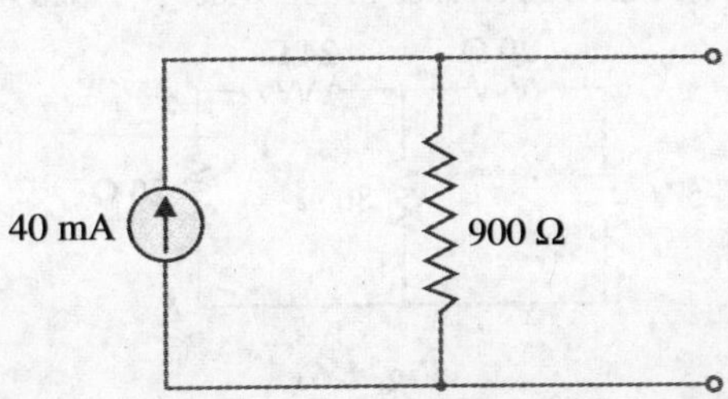

Fig. 1.54

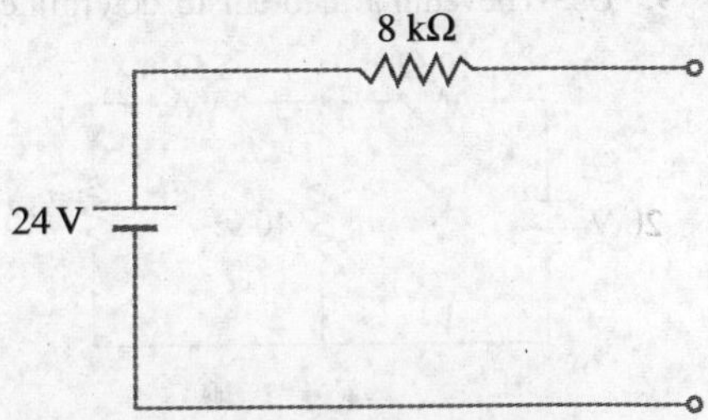

Fig. 1.55

3. Convert the voltage source in Fig. 1.55 into equivalent current source.

[3 mA in parallel with 8 kΩ]

4. Using Norton's Theorem, find the current in branch *AB* containing 6 Ω resistor of the network shown in Fig. 1.56. **[0.466 A]**

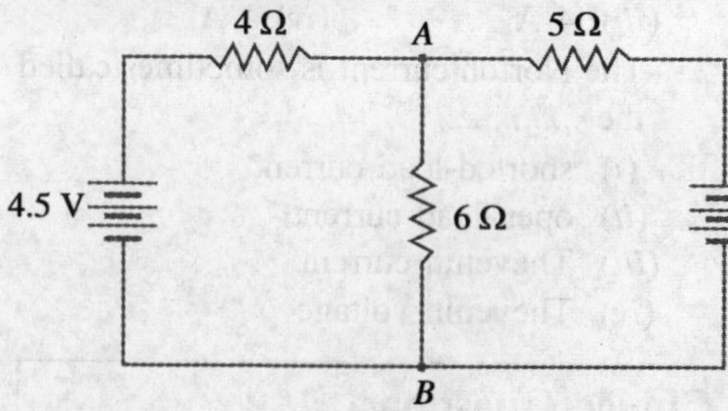

Fig. 1.56

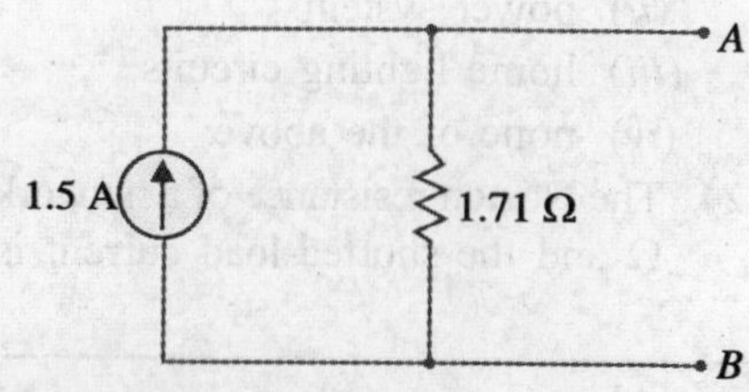

Fig. 1.57

5. Fig. 1.57 shows Norton's equivalent circuit of a network behind terminals *A* and *B*. Convert it into Thevenin's equivalent circuit. **[2.56 V in series with 1.71 Ω]**

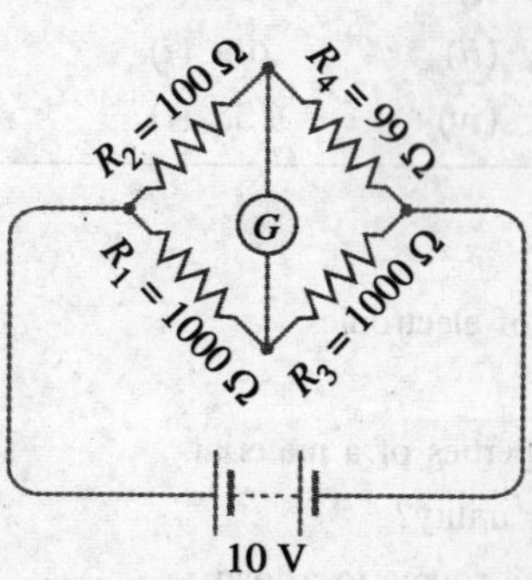

Fig. 1.58

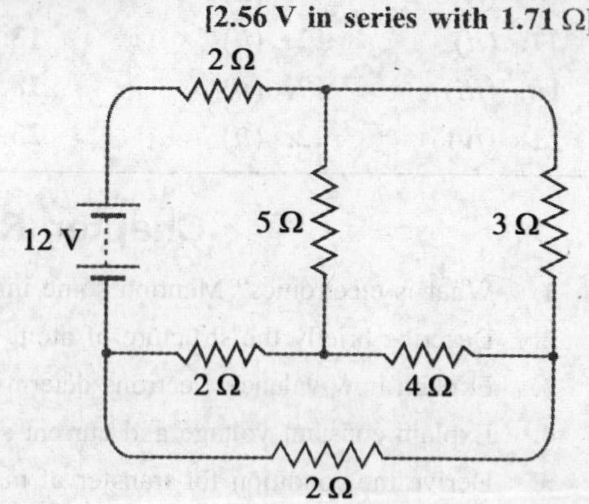

Fig. 1.59

6. A power amplifier has an internal resistance of 5 Ω and develops open-circuited voltage of 12 V. Find the efficiency and power transferred to a load of (*i*) 20 Ω (*ii*) 5 Ω.

[(*i*) 80%, 4.6 W (*ii*) 50%, 7.2 W]

7. Using Thevenin's theorem, find the current through the galvanometer in the Wheatstone bridge shown in Fig. 1.58. **[38.6 μA]**

8. Using Thevenin's theorem, find the current through 4 Ω resistor in the circuit of Fig. 1.59. **[0.305 A]**

9. Use Thevenin's theorem to determine current *I* in the 7 Ω resistor in Fig. 1.60. **[0.8 A]**

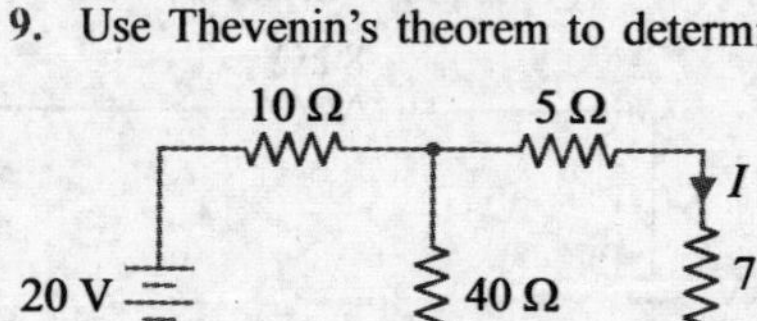

Fig. 1.60

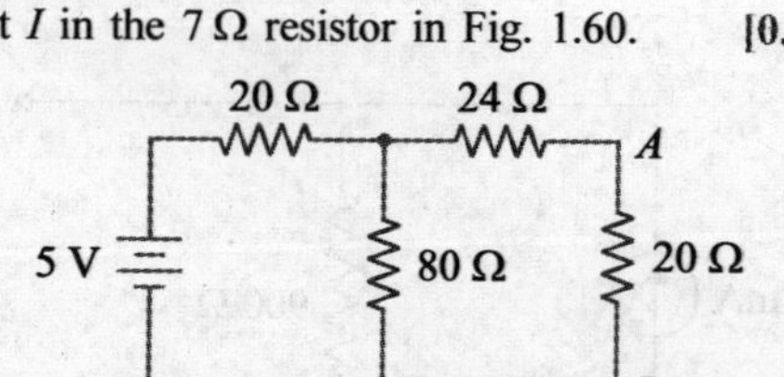

Fig. 1.61

10. Use Norton's theorem to obtain the potential difference between points *A* and *B* in Fig. 1.61. **[1.3 V]**

CHAPTER 2

Atomic Structure

Learning Objectives

- Develop an understanding of Bohr's atomic model.
- Understand energy levels and energy bands.
- Investigate important energy bands in solids.
- Develop a strong understanding of classification of solids on the basis of energy bands.
- Recognise that silicon is superior to germanium in all respects.

Introduction

The study of atomic structure is of considerable importance for electronics engineering. Unfortunately, the size of an atom is so small that it is virtually impossible to see it even with the most powerful microscope. Therefore, we have to employ indirect method for the study of its structure. The method consists of studying the properties of atom experimentally. After this, a *guess* is made regarding the possible structure of atom, which should satisfy the properties studied experimentally.

Various scientists have given different theories regarding the structure of atom. However, for the purpose of understanding electronics, the study of Bohr's atomic model is adequate. Although numerous refinements on Bohr's atomic model have since been made, we still believe in the laws that Bohr applied to the atomic world. In this chapter, we shall deal with Bohr's atomic model in order to understand the problems facing the electronic world.

2.1 Bohr's Atomic Model

In 1913, Neils Bohr, Danish Physicist gave clear explanation of atomic structure. According to Bohr :

(*i*) An atom consists of a positively charged nucleus around which negatively charged electrons revolve in different *circular orbits*.

(*ii*) The electrons can revolve around the nucleus only in certain permitted orbits *i.e.* orbits of certain radii are allowed.

(*iii*) The electrons in each permitted orbit have a certain fixed amount of energy. The larger the orbit (*i.e.* larger radius), the greater is the energy of electrons.

(*iv*) If an electron is given additional energy (*e.g.* heat, light *etc.*), it is lifted to the higher orbit. The atom is said to be in a state of *excitation*. This state does not last long, because the electron soon falls back to the original lower orbit. As it falls, it gives back the acquired energy in the form of heat, light or other radiations.

Fig. 2.1 shows the structure of *silicon atom*. It has 14 electrons. Two electrons revolve in the first orbit, 8 in the second orbit and 4 in the third orbit. The first, second, third orbits *etc.* are also known as *K*, *L*, *M* orbits respectively.

These electrons can revolve only in permitted orbits (*i.e.* orbits of *radii r_1, r_2 and r_3) and not in any arbitrary orbit. Thus, all radii between r_1 and r_2 or between r_2 and r_3 are forbidden. Each orbit has fixed amount of energy associated with it. If an electron in the first orbit is to be lifted to the second orbit, just the **right amount of energy should be supplied to it. When this electron jumps from the second orbit to first, it will give back the acquired energy in the form of electromagnetic radiations.

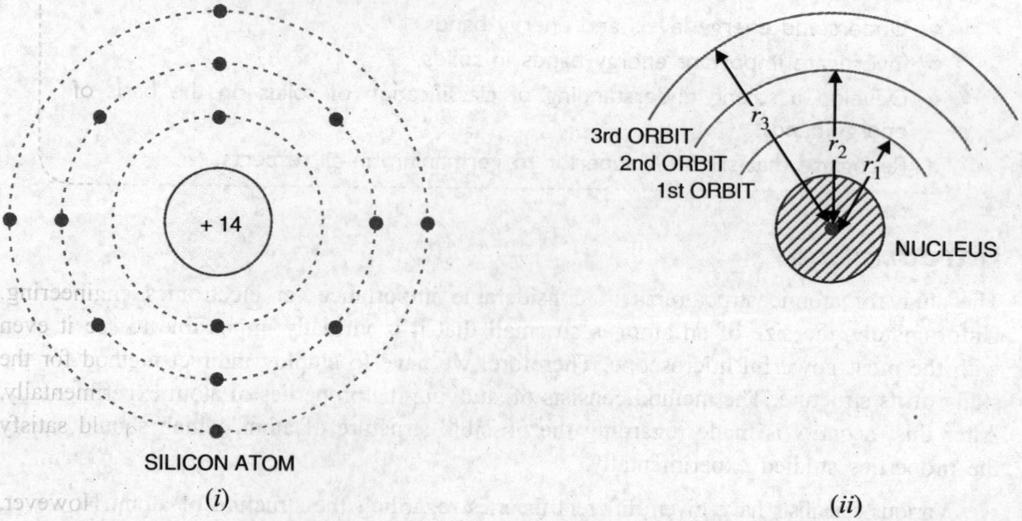

Fig. 2.1

2.2 Energy Levels

It has already been discussed that each orbit has fixed amount of energy associated with it. The electrons moving in a particular orbit possess the energy of that orbit. The larger the orbit, the greater is its energy. It becomes clear that outer orbit electrons possess more energy than the inner orbit electrons.

A convenient way of representing the energy of different orbits is shown in Fig. 2.2 (*ii*). This is known as energy level diagram. The first orbit represents the *first energy level*, the

* The values of radii are determined from quantum considerations.

** So that its total energy is equal to that of second orbit.

second orbit indicates the *second energy level* and so on. The larger the orbit of an electron, the greater is its energy and higher is the energy level.

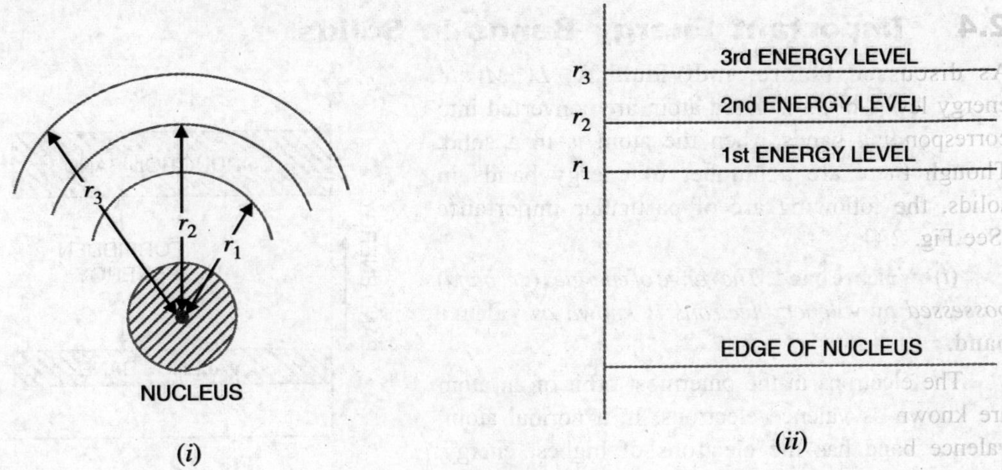

Fig. 2.2

2.3 Energy Bands

In case of a single isolated atom, the electrons in any orbit possess definite energy. However, an atom in a solid is greatly influenced by the closely-packed neighbouring atoms. The result is that the electron in any orbit of such an atom can have a range of energies rather than a single energy. This is known as *energy band.*

The range of energies possessed by an electron in a solid is known as **energy band.**

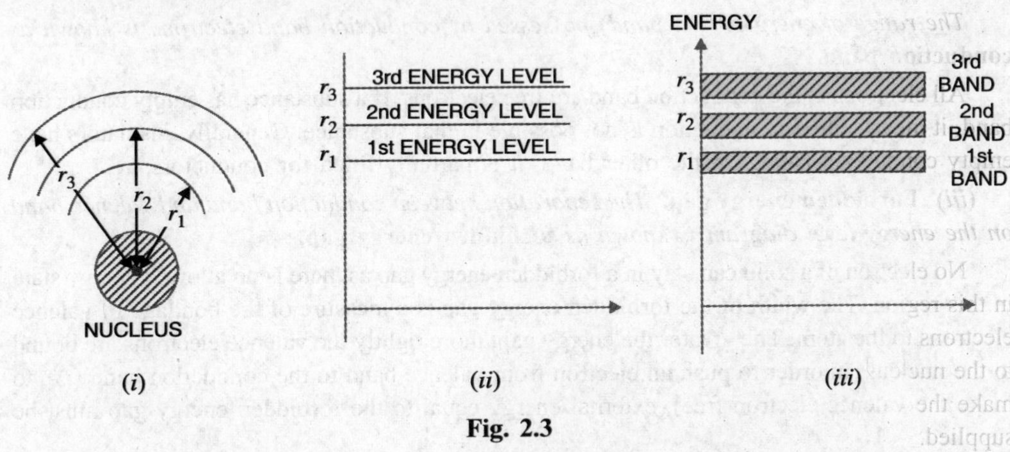

Fig. 2.3

The concept of energy band can be easily understood by referring to Fig. 2.3. Fig. 2.3 (*ii*) shows the energy levels of a single isolated atom of silicon. Each orbit of an atom has a single energy. Therefore, an electron can have only single energy corresponding to the orbit in which it exists. However, when the atom is in a solid, the electron in any orbit can have a range of energies. For instance, electrons in the first orbit have slightly different energies because no two electrons in this orbit see exactly the same charge environment. Since there are millions of first orbit electrons, the slightly different energy levels form a band, called

1st energy band [See Fig. 2.3 (*iii*)]. The electrons in the first orbit can have any energy range in this band. Similarly, second orbit electrons form second energy band and so on.

2.4 Important Energy Bands in Solids

As discussed before, individual *K, L, M etc.* energy levels of an isolated atom are converted into corresponding bands when the atom is in a solid. Though there are a number of energy bands in solids, the following are of particular importance [See Fig. 2.4]:

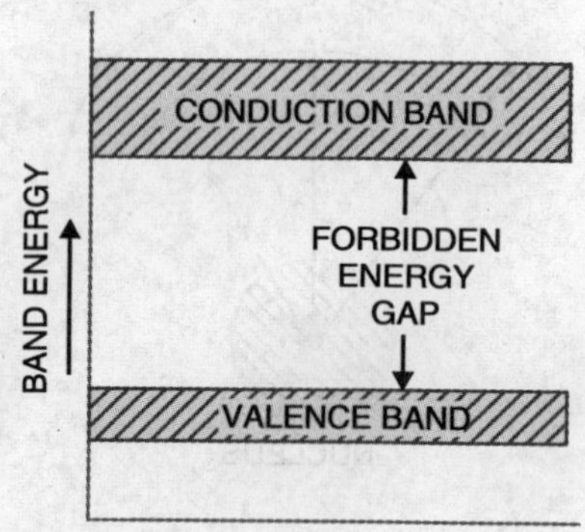

Fig. 2.4

(*i*) **Valence band.** *The range of energies (i.e. band) possessed by valence electrons is known as* **valence band.**

The electrons in the outermost orbit of an atom are known as valence electrons. In a normal atom, valence band has the electrons of highest energy. This band may be completely or partially filled. For instance, in case of inert gases, the valence band is full whereas for other materials, it is only partially filled. The partially filled band can accommodate more electrons.

(*ii*) **Conduction band.** In certain materials (*e.g.* metals), the valence electrons are loosely attached to the nucleus. Even at ordinary temperature, some of the valence electrons may get detached to become free electrons. In fact, it are these free electrons which are responsible for the conduction of current in a conductor. For this reason, they are called *conduction electrons.*

The range of energies (i.e. band) possessed by conduction band electrons is known as **conduction band.**

All electrons in the conduction band are free electrons. If a substance has empty conduction band, it means current conduction is not possible in that substance. Generally, insulators have empty conduction band. On the other hand, it is partially filled for conductors.

(*iii*) **Forbidden energy gap.** *The separation between conduction band and valence band on the energy level diagram is known as* **forbidden energy gap.**

No electron of a solid can stay in a forbidden energy gap as there is no allowed energy state in this region. The width of the forbidden energy gap is a measure of the bondage of valence electrons to the atom. The greater the energy gap, more tightly the valence electrons are bound to the nucleus. In order to push an electron from valence band to the conduction band (*i.e.* to make the valence electron free), external energy equal to the forbidden energy gap must be supplied.

2.5 Classification of Solids and Energy Bands

We know that some solids are good conductors of electricity while others are insulators. There is also an intermediate class of semiconductors. The difference in the behaviour of solids as regards their electrical conductivity can be beautifully explained in terms of energy bands. The electrons in the lower energy bands are tightly bound to the nucleus and play no part in the conduction process. However, the valence and conduction bands are of particular importance in ascertaining the electrical behaviour of various solids.

(*i*) **Insulators.** Insulators (*e.g.* wood, glass *etc.*) are those substances which do not allow the passage of electric current through them. In terms of energy band, the valence band is full while the conduction band is empty. Further, the energy gap between valence and conduction bands is very large ($\simeq$ 15 eV) as shown in Fig. 2.5. Therefore, a very high electric field is required to push the valence electrons to the conduction band. For this reason, the electrical conductivity of such materials is extremely small and may be regarded as nil under ordinary conditions.

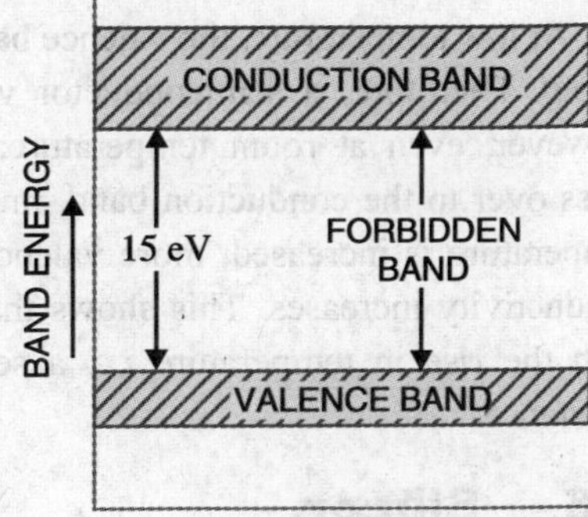

Fig. 2.5

At room temperature, the valence electrons of the insulators do not have enough energy to cross over to the conduction band. However, when the temperature is raised, some of the valence electrons may acquire enough energy to cross over to the conduction band. Hence, the resistance of an insulator decreases with the increase in temperature *i.e.* an insulator has negative temperature co-efficient of resistance.

(*ii*) **Conductors.** Conductors (*e.g.* copper, aluminium) are those substances which easily allow the passage of electric current through them. It is because there are a large number of free electrons available in a conductor. In terms of energy band, the valence and conduction bands overlap each other as shown in Fig. 2.6. Due to this overlapping, a slight potential difference across a conductor causes the free electrons to constitute electric current. Thus, the electrical behaviour of conductors can be satisfactorily explained by the band energy theory of materials.

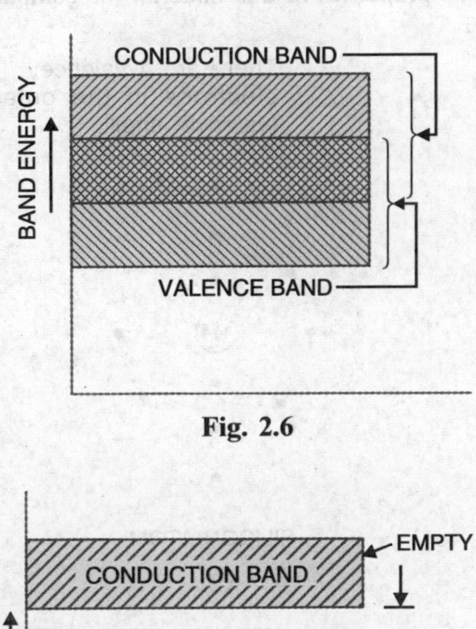

Fig. 2.6

(*iii*) **Semicondutors.** Semiconductors (*e.g.* germanium, silicon *etc.*) are those substances whose electrical conductivity lies inbetween conductors and insulators. In terms of energy band, the valence band is almost filled and conduction band is almost empty. Further, the energy gap between valence and conduction bands is very small as shown in Fig. 2.7. Therefore, comparatively smaller electric field (smaller than insulators but much greater than conductors) is required to push the electrons from the valence band to the conduction band. In short, a semiconductor has :

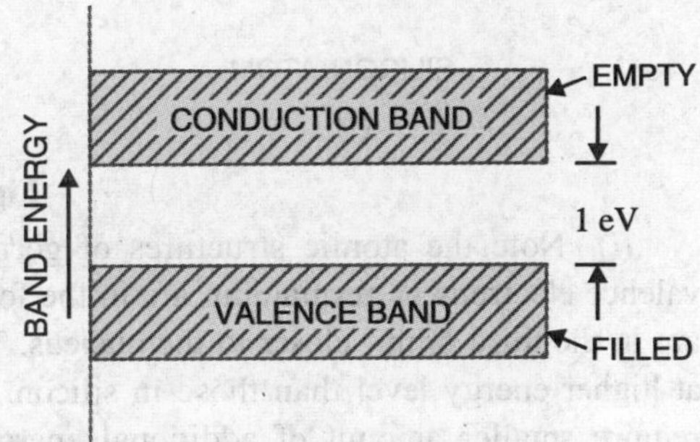

Fig. 2.7

(*a*) almost full valence band

(*b*) almost empty conduction band

(*c*) small energy gap ($\simeq$ 1 eV) between valence and conduction bands.

At low temperature, the valence band is completely full and conduction band is completely empty. Therefore, a semiconductor virtually behaves as an insulator at low temperatures. However, even at room temperature, some electrons (about one electron for 10^{10} atoms) cross over to the conduction band, imparting little conductivity to the semiconductor. As the temperature is increased, more valence electrons cross over to the conduction band and the conductivity increases. This shows that electrical conductivity of a semiconductor increases with the rise in temperature *i.e.* a semiconductor has negative temperature co-efficient of resistance.

2.6 Silicon

During the infancy of electronic industry, both germanium and silicon were used in the manufacture of semiconductor devices. As the electronic field advanced, it was realised that silicon was superior to germanium in many respects. Since silicon is the most widely used material in the manufacture of semiconductor devices, we shall continue to discuss the properties of this material (as compared to germanium) as and when we get the chance.

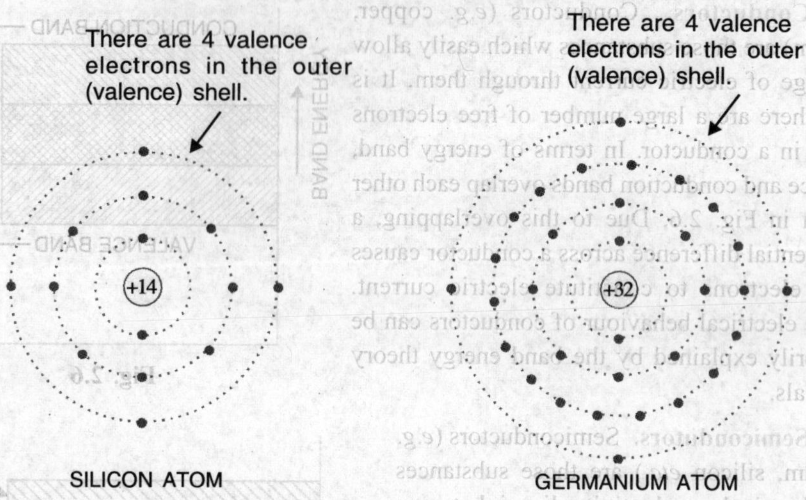

Fig. 2.8

(*i*) Note the atomic structures of germanium and silicon in Fig. 2.8 carefully. The valence electrons in germanium are in the fourth orbit while the valence electrons in silicon are in the third orbit ; closer to the nucleus. Therefore, the germanium valence electrons are at higher energy level than those in silicon. This means that germanium valence electrons require smaller amount of additional energy to escape from the atom and become free electron. What is the effect of this property? This property makes germanium more unstable at high temperatures. This is the basic reason why silicon is widely used as semiconductor material.

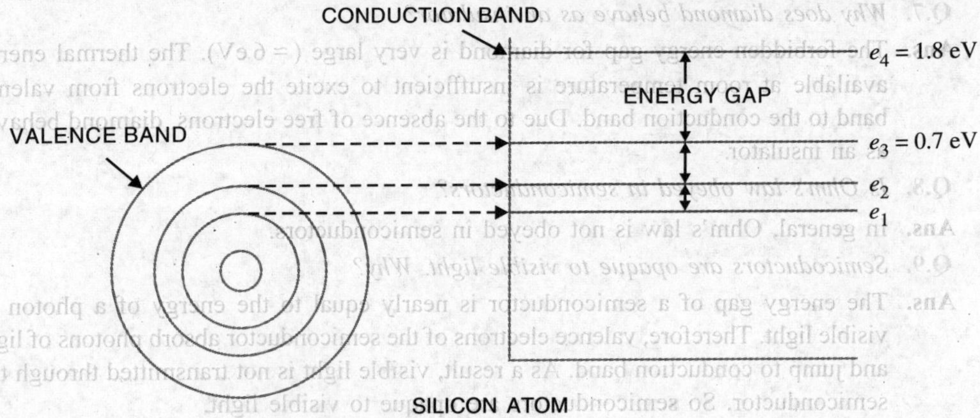

Fig. 2.9

(ii) Fig. 2.9 shows the energy levels/bands of silicon atom. The atomic number of silicon is 14 so that its 14 electrons are distributed in 3 orbits. Each energy level/band is associated with certain amount of energy and is separated from the adjacent bands by energy gap. *No electron can exist in the energy gap.* For an electron to jump from one orbit to the next higher orbit, external energy (*e.g.* heat) equal to the energy difference of the orbits must be supplied. For example, the valence band is shown to have an energy level of 0.7 eV. The conduction band is shown to have an energy level of 1.8 eV. Thus for an electron to jump from the valence band to the conduction band, an energy = 1.8 – 0.7 = 1.1 eV must be supplied. As you will see, the energy band description of substances is very important in understanding many fields of science and engineering including electronics.

Short Answer Questions

Q.1. *How much is the energy gap in (i) a conductor (ii) a semiconductor (iii) an insulator?*

Ans. (*i*) Zero (*ii*) Less than 3 eV (*iii*) More than 3 eV.

Q.2. *What is the energy gap for (i) germanium (ii) silicon?*

Ans. (*i*) 0.7 eV (*ii*) 1.1 eV.

Q.3. *The energy gap for silicon is 1.1 eV. What does it mean?*

Ans. It means that if energy of 1.1 eV is given to its electron in the valence band, the electron will jump to the conduction band.

Q.4. *Why is a semiconductor virtually an insulator at room temperature?*

Ans. Because almost all the valence electrons are engaged in the formation of covalent bonds and there are practically very few free electrons.

Q.5. *Why are germanium and silicon preferred to other semiconductors in electronic devices?*

Ans. It is because energy required to break their covalent bonds is small ; being 0.7 eV for germanium and 1.1 eV for silicon.

Q.6. *How does forbidden energy gap of a semiconductor vary with increase in temperature?*

Ans. It decreases with the increase in temperature. When temperature of a semiconductor is increased, a greater number of electrons in the valence band move to the conduction band. We may consider that the width of forbidden energy gap has decreased.

Q.7. *Why does diamond behave as an insulator?*

Ans. The forbidden energy gap for diamond is very large ($\simeq$ 6 eV). The thermal energy available at room temperature is insufficient to excite the electrons from valence band to the conduction band. Due to the absence of free electrons, diamond behaves as an insulator.

Q.8. *Is Ohm's law obeyed in semiconductors?*

Ans. In general, Ohm's law is not obeyed in semiconductors.

Q.9. *Semicoductors are opaque to visible light. Why?*

Ans. The energy gap of a semiconductor is nearly equal to the energy of a photon of visible light. Therefore, valence electrons of the semiconductor absorb photons of light and jump to conduction band. As a result, visible light is not transmitted through the semiconductor. So semiconductors are opaque to visible light.

Q.10. *At room temperature, name the charge carriers in (i) a conductor (ii) a semiconductor (iii) an insulator.*

Ans. (*i*) Free electrons (*ii*) Free electrons and holes (*iii*) None.

Multiple-Choice Questions

1. The electrons in the third orbit of an atom have .... energy than the electrons in the second orbit.
   (*i*) more (*ii*) less
   (*iii*) the same
   (*iv*) none of the above
2. When an electron jumps from higher orbit to a lower orbit, it .... energy.
   (*i*) absorbs (*ii*) emits
   (*iii*) sometimes emits, sometimes absorbs
   (*iv*) none of the above
3. Which of the following is quantised according to Bohr's theory of atom?
   (*i*) Linear momentum of electron
   (*ii*) Linear velocity of electron
   (*iii*) Angular momentum of electron
   (*iv*) Angular velocity of electron
4. A semiconductor has .... band.
   (*i*) almost empty valence
   (*ii*) almost empty conduction
   (*iii*) almost full conduction
   (*iv*) none of the above
5. The electrons in the conduction band are known as ....
   (*i*) bound electrons
   (*ii*) valence electrons
   (*iii*) free electrons
   (*iv*) none of the above
6. In insulators, the energy gap between valence and conduction bands is ....
   (*i*) very large (*ii*) zero
   (*iii*) very small
   (*iv*) none of the above
7. In a conductor, the energy gap between valence and conduction bands is ....
   (*i*) large (*ii*) very large
   (*iii*) very small
   (*iv*) none of the above
8. According to Bohr's theory of atom, an electron can move in an orbit of ...
   (*i*) any radius
   (*ii*) certain radius
   (*iii*) some range of radii
   (*iv*) none of the above
9. In a semiconductor, the energy gap between valence and conduction bands is about ....
   (*i*) 15 eV (*ii*) 100 eV
   (*iii*) 50 eV (*iv*) 1 eV
10. The energy gap between valence and conduction bands in insulators is about ....
    (*i*) 15 eV (*ii*) 1.5 eV
    (*iii*) zero (*iv*) 0.5 eV

Answers to Multiple-Choice Questions

1. (*i*)	**2.** (*ii*)	**3.** (*iii*)	**4.** (*ii*)	**5.** (*iii*)
6. (*i*)	**7.** (*iii*)	**8.** (*ii*)	**9.** (*iv*)	**10.** (*i*)

Chapter Review Topics

1. Explain the salient features of Bohr's atomic model.
2. Explain the concept of energy bands in solids.
3. Describe the valence band, conduction band and forbidden energy gap with the help of energy level diagram.
4. Give the energy band description of conductors, semiconductors and insulators.

CHAPTER 3

Semiconductor Physics

Learning Objectives

- Understand the properties of semiconductors.
- Commonly used semiconductors in electronics.
- Recognise the effect of temperature on semiconductors.
- Understand the concept of hole current in semiconductors.
- Develop a strong understanding of *p*-type semiconductor and *n*-type semiconductor.
- Role of majority and minority charge carriers in *n*-type and *p*-type semiconductors.
- Develop an understanding of *pn* junction.
- Develop a strong understanding of forward bias and reverse bias operating mode of *pn* junction (*i.e.* semiconductor diode).
- Volt-Ampere (*V-I*) characteristics of *pn* junction.
- Understand the reverse breakdown voltage and knee voltage of *pn* junction.

Introduction

Certain substances like germanium, silicon, carbon *etc.* are neither good conductors like copper nor insulators like glass. In other words, the resistivity of these materials lies inbetween conductors and insulators. Such substances are classified as *semiconductors*. Semiconductors have some useful properties and are being extensively used in electronic circuits. For instance, *transistor*—a semiconductor device is fast replacing bulky vacuum tubes in almost all applications. Transistors are only one of the family of semiconductor devices ; many other semiconductor devices are becoming increasingly popular. In this chapter, we shall focus our attention on the different aspects of semiconductors.

3.1 Semiconductor

It is not easy to define a semiconductor if we want to take into account all its physical characteristics. However, generally, a semiconductor is defined on the basis of electrical conductivity as under :

***A semiconductor** is a substance which has resistivity (10^{-4} to 0.5 Ωm) inbetween conductors and insulators e.g. germanium, silicon, selenium, carbon etc.*

The reader may wonder, when a semiconductor is neither a good conductor nor an insulator, then why not to classify it as a *resistance material* ? The answer shall be readily available if we study the following table :

S.No.	Substance	Nature	Resistivity
1	Copper	good conductor	1.7×10^{-8} Ω m
2	Germanium	semiconductor	0.6 Ω m
3	Glass	insulator	9×10^{11} Ω m
4	Nichrome	resistance material	10^{-4} Ω m

Comparing the resistivities of above materials, it is apparent that the resistivity of germanium (semiconductor) is quite high as compared to copper (conductor) but it is quite low when compared with glass (insulator). This shows that resistivity of a semiconductor lies inbetween conductors and insulators. However, it will be wrong to consider the semiconductor as a resistance material. For example, nichrome, which is one of the highest resistance material, has resistivity much lower than germanium. This shows that electrically germanium cannot be regarded as a conductor or insulator or a resistance material. This gave such substances like germanium the name of semiconductors.

It is interesting to note that it is not the resistivity alone that decides whether a substance is semiconductor or not. For example, it is just possible to prepare an alloy whose resistivity falls within the range of semiconductors but the alloy cannot be regarded as a semiconductor. In fact, semiconductors have a number of peculiar properties which distinguish them from conductors, insulators and resistance materials.

Properties of Semiconductors

(*i*) A semiconductor is formed by covalent bonds.

(*ii*) A semiconductor has crystalline structure.

(*iii*) The resistivity of a semiconductor is less than that of an insulator but more than that of a conductor.

(*iv*) Semiconductors have *negative temperature co-efficient of resistance i.e.* the resistance of a semiconductor decreases with the increase in temperature and *vice-versa*. For example, germanium is actually an insulator at low temperatures but it becomes a good conductor at high temperatures.

(*v*) When a suitable metallic impurity (*e.g.* arsenic, gallium *etc.*) is added to a semiconductor, its current conducting properties change appreciably. This property is most important and is discussed later in detail.

3.2 Bonds in Semiconductors

The atoms of every element are held together by the bonding action of valence electrons. This bonding is due to the fact that it is the tendency of each atom to complete its last orbit by acquiring 8 electrons in it. However, in most of the substances, the last orbit is incomplete *i.e.* the last orbit does not have 8 electrons. This makes the atom active to enter into bargain with other atoms to acquire 8 electrons in the last orbit. To do so, the atom may lose, gain or share valence electrons with other atoms. In semiconductors, bonds are formed by sharing of valence electrons. Such bonds are called *co-valent bonds*. In the formation of a co-valent bond, each atom contributes equal number of valence electrons and the contributed electrons are shared by the atoms engaged in the formation of the bond.

Fig. 3.1 shows the co-valent bonds among germanium atoms. A germanium atom has *4 valence electrons. It is the tendency of each germanium atom to have 8 electrons in the last orbit. To do so, each germanium atom positions itself between four other germanium atoms as shown in Fig. 3.1 (*i*). Each neighbouring atom shares one valence electron with the central atom. In this business of sharing, the central atom completes its last orbit by having 8 electrons revolving around the nucleus. In this way, the central atom sets up co-valent bonds. Fig. 3.1 (*ii*) shows the bonding diagram.

The following points may be noted regarding the co-valent bonds :

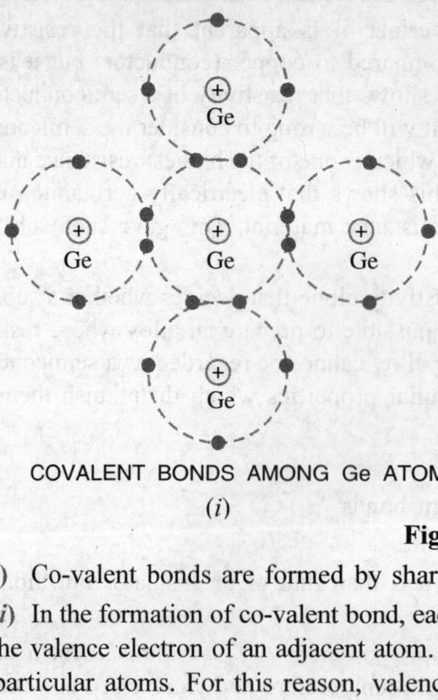

COVALENT BONDS AMONG Ge ATOMS

(*i*)

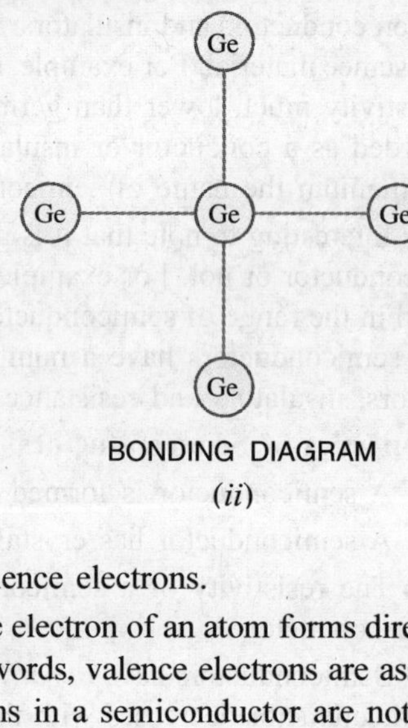

BONDING DIAGRAM

(*ii*)

Fig. 3.1

(*i*) Co-valent bonds are formed by sharing of valence electrons.

(*ii*) In the formation of co-valent bond, each valence electron of an atom forms direct bond with the valence electron of an adjacent atom. In other words, valence electrons are associated with particular atoms. For this reason, valence electrons in a semiconductor are not free.

3.3 Crystals

A substance in which the atoms or molecules are arranged in an orderly pattern is known as a *crystal*. All semi-conductors have crystalline structure. For example, referring to Fig. 3.1, it is clear that each atom is surrounded by neighbouring atoms in a repetitive manner. Therefore, a piece of germanium is generally called germanium crystal.

3.4 Commonly Used Semiconductors

There are many semiconductors available, but very few of them have a practical application in electronics. The two most frequently used materials are *germanium* (Ge) and *silicon* (Si). It is because the energy required to break their co-valent bonds (*i.e.* energy required to release

* A germanium atom has 32 electrons. First orbit has 2 electrons, second 8 electrons, third 18 electrons and the fourth orbit has 4 electrons.

an electron from their valence bands) is very small ; being about 0.7 eV for germanium and about 1.1 eV for silicon. Therefore, we shall discuss these two semiconductors in detail.

(*i*) **Germanium.** Germanium has become the model substance among the semiconductors; the main reason being that it can be purified relatively well and crystallised easily. Germanium is an earth element and was discovered in 1886. It is recovered from the ash of certain coals or from the flue dust of zinc smelters. Generally, recovered germanium is in the form of germanium dioxide powder which is then reduced to pure germanium.

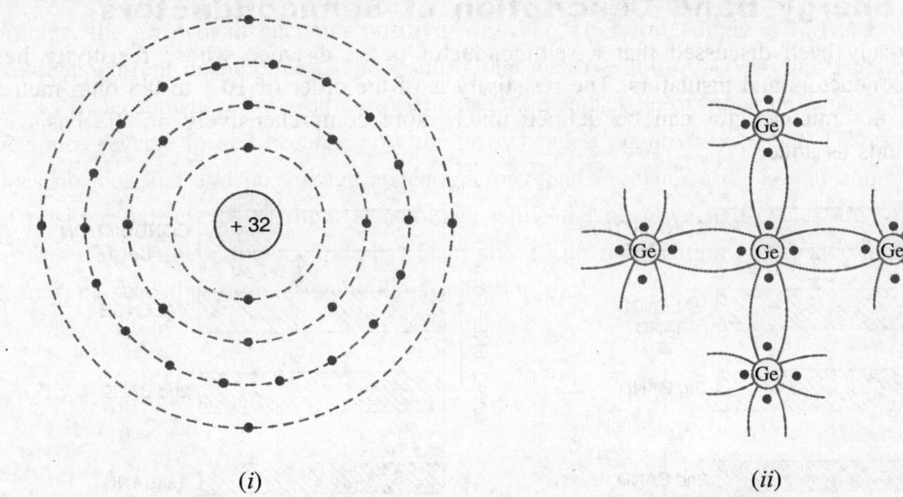

Fig. 3.2

The atomic number of germanium is 32. Therefore, it has 32 protons and 32 electrons. Two electrons are in the first orbit, eight electrons in the second, eighteen electrons in the third and four electrons in the outer or valence orbit [See Fig. 3.2 (*i*)]. It is clear that germanium atom has four valence electrons *i.e.* it is a tetravalent element. Fig. 3.2 (*ii*) shows how the various germanium atoms are held through co-valent bonds. As the atoms are arranged in an orderly pattern, therefore, germanium has crystalline structure.

(*ii*) **Silicon.** Silicon is an element in most of the common rocks. Actually, sand is silicon dioxide. The silicon compounds are chemically reduced to silicon which is 100% pure for use as a semiconductor.

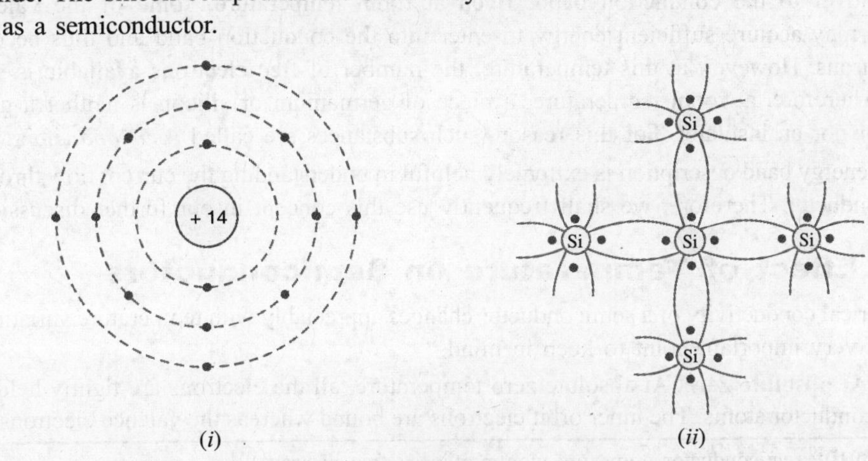

Fig. 3.3

The atomic number of silicon is 14. Therefore, it has 14 protons and 14 electrons. Two electrons are in the first orbit, eight electrons in the second orbit and four electrons in the third orbit [See Fig. 3.3 (*i*)]. It is clear that silicon atom has four valence electrons *i.e.* it is a tetravalent element. Fig. 3.3 (*ii*) shows how various silicon atoms are held through co-valent bonds. Like germanium, silicon atoms are also arranged in an orderly manner. Therefore, silicon has crystalline structure.

3.5 Energy Band Description of Semiconductors

It has already been discussed that a semiconductor is a substance whose resistivity lies between conductors and insulators. The resistivity is of the order of 10^{-4} to 0.5 ohm metre. However, a semiconductor can be defined much more comprehensively on the basis of energy bands as under :

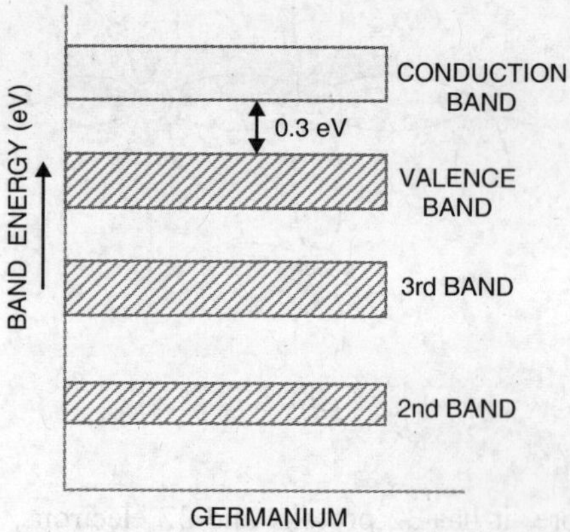

Fig. 3.4

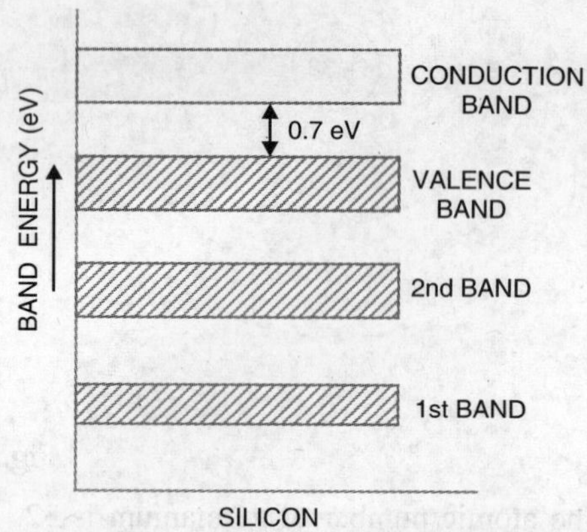

Fig. 3.5

***A semiconductor** is a substance which has almost filled valence band and nearly empty conduction band with a very small energy gap ($\simeq$ 1 eV) separating the two.*

Figs. 3.4 and 3.5 show the energy band diagrams of germanium and silicon respectively. It may be seen that forbidden energy gap is very small ; being 0.7 eV for silicon and 0.3 eV for germanium. Therefore, relatively small energy is needed by their valence electrons to cross over to the conduction band. Even at room temperature, some of the valence electrons may acquire sufficient energy to enter into the conduction band and thus become free electrons. However, at this temperature, the number of free electrons available is very *small. Therefore, at room temperature, a piece of germanium or silicon is neither a good conductor nor an insulator. For this reason, such substances are called *semiconductors.*

The energy band description is extremely helpful in understanding the current flow through a semiconductor. Therefore, we shall frequently use this concept in our further discussion.

3.6 Effect of Temperature on Semiconductors

The electrical conductivity of a semiconductor changes appreciably with temperature variations. This is a very important point to keep in mind.

(*i*) **At absolute zero.** At absolute zero temperature, all the electrons are tightly held by the semiconductor atoms. The inner orbit electrons are bound whereas the valence electrons are

* Out of 10^{10} semiconductor atoms, one atom provides a free electron.

engaged in co-valent bonding. At this temperature, the co-valent bonds are very strong and there are no free electrons. Therefore, the semiconductor crystal behaves as a perfect insulator [See Fig. 3.6 (*i*)].

In terms of energy band description, the valence band is filled and there is a large energy gap between valence band and conduction band. Therefore, no valence electron can reach the conduction band to become free electron. It is due to the non-availability of free electrons that a semiconductor behaves as an insulator.

(*ii*) Above absolute zero. When the temperature is raised, some of the covalent bonds in the semiconductor break due to the thermal energy supplied. The breaking of bonds sets those electrons *free* which are engaged in the formation of these bonds. The result is that a few free electrons exist in the semiconductor. These free electrons can constitute a tiny electric current if potential difference is applied across the semiconductor crystal [See Fig. 3.7 (*i*)]. *This shows that the resistance of a semiconductor decreases with the rise in temperature i.e.* it has negative temperature coefficient of resistance. It may be added that at room temperature, current through a semiconductor is too small to be of any practical value.

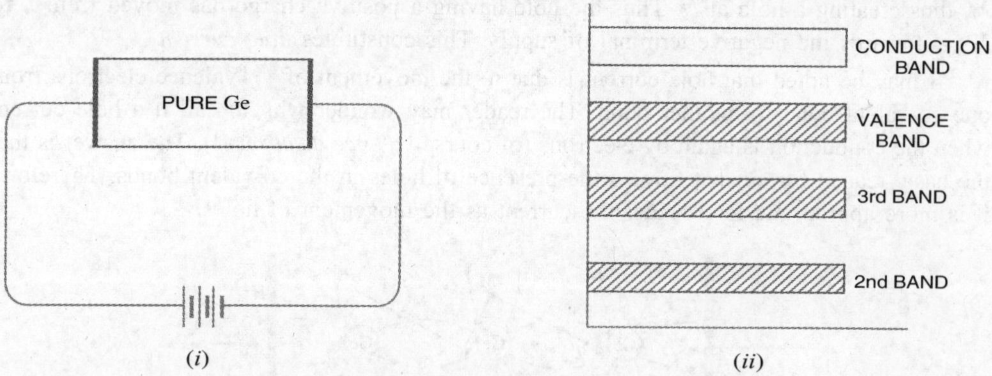

Fig. 3.6

Fig. 3.7 (*ii*) shows the energy band diagram. As the temperature is raised, some of the valence electrons acquire sufficient energy to enter into the conduction band and thus become free electrons. Under the influence of electric field, these free electrons will constitute electric current. It may be noted that each time a valence electron enters into the conduction band, a *hole* is created in the valence band. As we shall see in the next article, holes also contribute to current. In fact, hole current is the most significant concept in semiconductors.

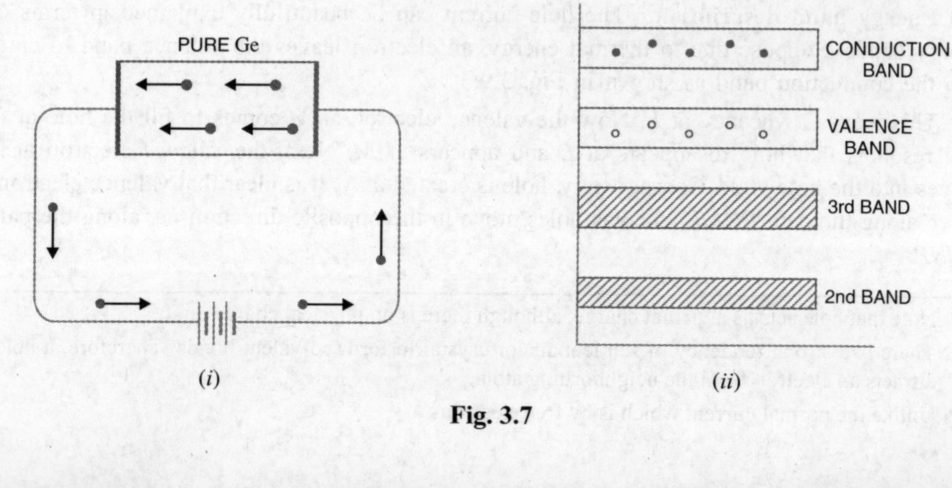

Fig. 3.7

3.7 Hole Current

At room temperature, some of the co-valent bonds in pure semiconductor break, setting up free electrons. Under the influence of electric field, these free electrons constitute electric current. At the same time, another current – the hole current – also flows in the semiconductor. When a covalent bond is broken due to thermal energy, the removal of one electron leaves a vacancy *i.e.* a missing electron in the covalent bond. This missing electron is called a *hole which acts as a positive charge. For one electron set free, one hole is created. Therefore, thermal energy creates *hole-electron pairs;* there being as many holes as the free electrons. The current conduction by holes can be explained as follows.

The hole shows a missing electron. Suppose the valence electron at *L* (See Fig. 3.8) has become free electron due to thermal energy. This creates a hole in the co-valent bond at *L*. The hole is a strong centre of attraction **for the electron. A valence electron (say at *M*) from nearby co-valent bond comes to fill in the hole at *L*. This results in the creation of hole at *M*. Another valence electron (say at *N*) in turn may leave its bond to fill the hole at *M*, thus creating a hole at *N*. Thus the hole having a positive charge has moved from *L* to *N i.e.* towards the negative terminal of supply. This constitutes *hole current.*

It may be noted that hole current is due to the movement of ***valence electrons from one co-valent bond to another bond. The reader may wonder why to call it a hole current when the conduction is again by electrons (of course *valence electrons* !). The answer is that the basic reason for current flow is the presence of holes in the co-valent bonds. Therefore, it is more appropriate to consider the current as the movement of holes.

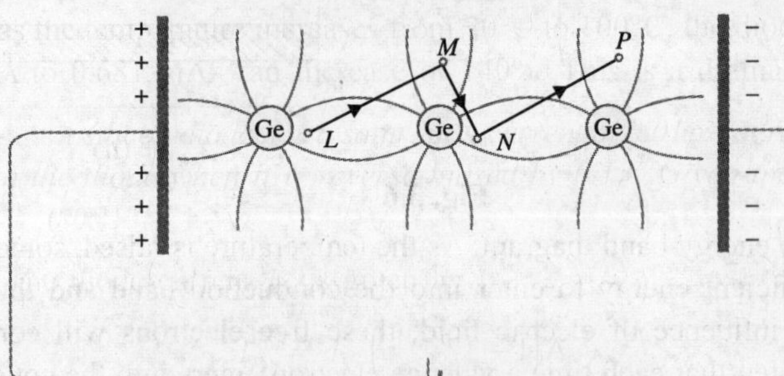

Fig. 3.8

Energy band description. The hole current can be beautifully explained in terms of energy bands. Suppose due to thermal energy, an electron leaves the valence band to enter into the conduction band as shown in Fig. 3.9.

This leaves a vacancy at *L*. Now the valence electron at *M* comes to fill the hole at *L*. The result is that hole disappears at *L* and appears at *M*. Next, the valence electron at *N* moves into the hole at *M*. Consequently, hole is created at *N*. It is clear that valence electrons move along the path *PNML* whereas holes move in the opposite direction *i.e.* along the path *LMNP*.

---

* Note that hole acts as a virtual charge, although there is no physical charge on it.

** There is a strong tendency of semiconductor crystal to form co-valent bonds. Therefore, a hole attracts an electron from the neighbouring atom.

*** Unlike the normal current which is by free electrons.

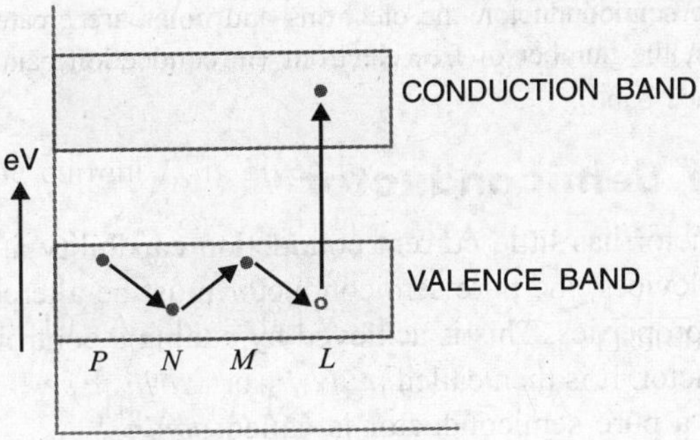

Fig. 3.9

3.8 Intrinsic Semiconductor

A semiconductor in an extremely pure form is known as an **intrinsic semiconductor.**

In an intrinsic semiconductor, even at room temperature, hole-electron pairs are created. When electric field is applied across an intrinsic semiconductor, the current conduction takes place by two processes, namely ; by *free electrons* and *holes* as shown in Fig. 3.10. The free electrons are produced due to the breaking up of some covalent bonds by thermal energy. At the same time, holes are created in the covalent bonds. Under the influence of electric field, conduction through the semiconductor is by both free electrons and holes. Therefore, the total current inside the semiconductor is the sum of currents due to free electrons and holes *i.e.*

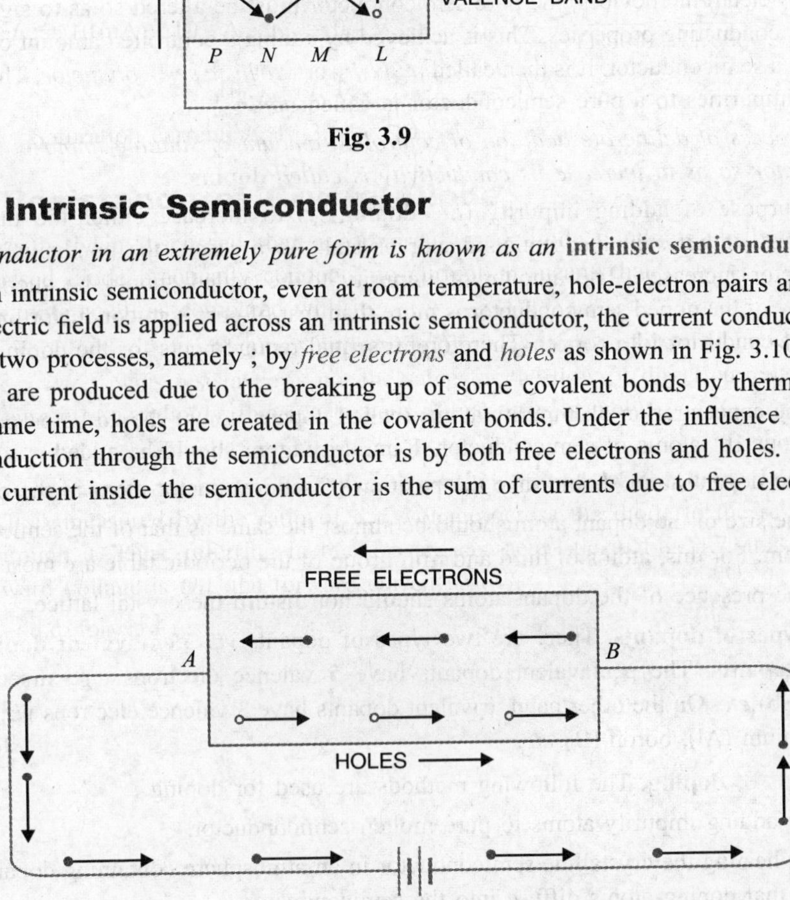

Fig. 3.10

Current inside semiconductor, $I = I_e + I_h$

where I_e = Current due to free electrons

I_h = Current due to holes

It may be noted that current in the external wires is fully electronic *i.e.* by electrons. What about the holes? Referring to Fig. 3.10, holes being positively charged move towards the negative terminal of supply. As the holes reach the negative terminal *B*, electrons enter the semiconductor crystal near the terminal and combine with holes, thus cancelling them. At the same time, the loosely held electrons near the positive terminal *A* are attracted away from their atoms into the positive terminal. This creates new holes near the positive terminal which again drift towards the negative terminal.

Note. In an intrinsic semiconductor, the electrons and holes are created in pairs. Therefore, in an intrinsic semiconductor, the number of free electrons (in conduction band) is exactly equal to the number of holes (in valence band).

3.9 Doping a Semiconductor

The intrinsic semiconductor has little current conduction capability at room temperature. To be useful in electronic devices, the pure semiconductor must be altered so as to significantly increase its conducting properties. This is achieved by adding a controlled amount of suitable impurity to a semiconductor. It is then called *impurity* or *extrinsic semiconductor*. The process of adding impurities to a pure semiconductor is called *doping*.

The process of deliberate addition of controlled amount of suitable impurity to a pure semiconductor so as to increase its conductivity is called **doping**.

The purpose of adding impurity (*i.e.* doping) is to increase either the number of free electrons or holes in the semiconductor crystal. As a result, the conductivity of the semiconductor increases. The amount of impurity to be added should be such that the conductivity of the doped semiconductor is more than that of pure semiconductor but far less than that of conductor like copper. Therefore, essential requirements for the doping process are :

(*i*) The concentration of impurity atoms (called dopant atoms) should be proper *i.e.* for about 10^8 atoms of semiconductor, 1 impurity atom should be added.

(*ii*) The dopant atom should take the position of semiconductor atom in the crystal.

(*iii*) The size of the dopant atom should be almost the same as that of the semiconductor atom. For this, atoms of third and fifth group of the periodic table are most suitable.

(*iv*) The presence of the dopant atoms should not disturb the crystal lattice.

Two types of dopants. There are two types of dopants *viz.* **pentavalent dopants** and **trivalent dopants**. The pentavalent dopants have 5 valence electrons *e.g.*, arsenic (As), antimony (Sb) *etc.* On the other hand, trivalent dopants have 3 valence electrons *viz.*, indium (In), aluminium (Al), boron (B) *etc.*

Methods of doping. The following methods are used for doping :

(*a*) By adding impurity atoms to pure molten semiconductor.

(*b*) By heating the crystalline semiconductor in an atmosphere containing dopant atoms so that doping atoms diffuse into the semiconductor.

(*c*) By bombarding the semiconductor by the doping atoms. In the process, the doping atoms are implanted into the semiconductor.

3.10 Extrinsic Semiconductor

When a controlled amount of suitable impurity is added to a pure semiconductor so as to increase the number of charge carriers (free electrons or holes) in it, the resulting semiconductor is called ***extrinsic semiconductor** or **doped semiconductor**.

The purpose of adding impurity is to increase either the number of free electrons or holes in the semiconductor crystal. As we shall see, if a pentavalent impurity (having 5 valence electrons) is added to the semiconductor, a large number of free electrons are produced in the

* Extrinsic means impure.

semiconductor. On the other hand, addition of trivalent impurity (having 3 valence electrons) creates a large number of holes in the semiconductor crystal. Depending upon the type of impurity added, extrinsic semiconductors are classified into :

(*i*) *n*-type semiconductor (*ii*) *p*-type semiconductor

3.11 n-type Semiconductor

When a small amount of pentavalent impurity is added to a pure semiconductor, it is known as n-**type semiconductor.**

The addition of pentavalent impurity provides a large number of free electrons in the semiconductor crystal. Typical examples of pentavalent impurities are *arsenic* (At. No. 33) and *antimony* (At. No. 51). Such impurities which produce *n*-type semiconductor are known as *donor impurities* because they donate or provide free electrons to the semiconductor crystal.

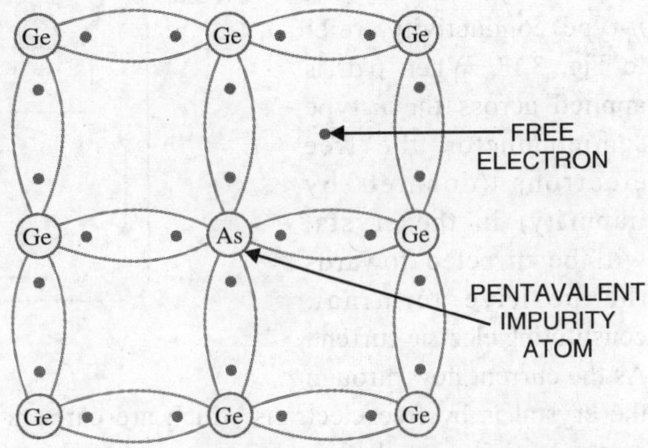

Fig. 3.11

To explain the formation of *n*-type semiconductor, consider a pure germanium crystal. We know that germanium atom has four valence electrons. When a small amount of pentavalent impurity like arsenic is added to germanium crystal, a large number of free electrons become available in the crystal. The reason is simple. Arsenic is pentavalent *i.e.* its atom has five valence electrons. An arsenic atom fits in the germanium crystal in such a way that its four valence electrons form covalent bonds with four germanium atoms. The *fifth* valence electron of arsenic atom finds no place in co-valent bonds and is thus free as shown in Fig. 3.11. Therefore, for each arsenic atom added, one free electron will be available in the germanium crystal. Though each arsenic atom provides one free electron, yet an extremely small amount of arsenic impurity provides enough atoms to supply millions of free electrons.

Fig. 3.12 shows the energy band description of *n*-type semi-conductor. The addition of pentavalent impurity has produced a number of conduction band electrons *i.e.* free electrons. The four valence electrons of pentavalent atom form covalent bonds with four neighbouring germanium atoms. The fifth left over valence electron of the pentavalent atom cannot be accommodated in the valence band and travels to the conduction band. The following points may be noted carefully :

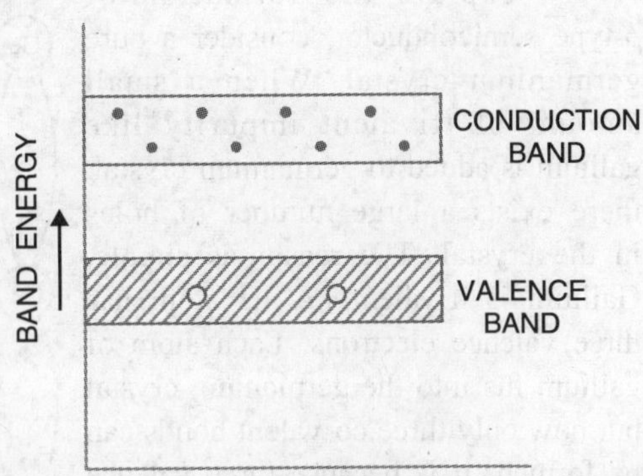

Fig. 3.12

(*i*) Many new free electrons are produced by the addition of pentavalent impurity.

(*ii*) Thermal energy of room temperature still generates a few hole-electron pairs. However, the number of free electrons provided by the pentavalent impurity far exceeds the number of holes. It is due to this predominance of electrons over holes that it is called *n*-type semiconductor (*n* stands for negative).

***n*-type conductivity.** The current conduction in an *n*-type semiconductor is *predominantly* by free electrons *i.e.* negative charges and is called *n-type* or *electron type conductivity*. To understand *n*-type conductivity, refer to Fig. 3.13. When p.d. is applied across the *n*-type semiconductor, the free electrons (donated by impurity) in the crystal will be directed towards the positive terminal, constituting electric current. As the current flow through the crystal is by free electrons which are carriers of negative charge, therefore, this type of conductivity is called negative or *n*-type conductivity. It may be noted that conduction is just as in ordinary metals like copper.

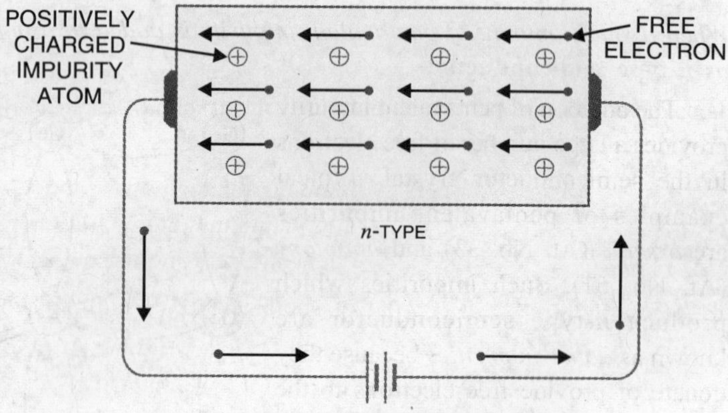

Fig. 3.13

3.12 p-type Semiconductor

When a small amount of trivalent impurity is added to a pure semiconductor, it is called ***p*-type semiconductor.**

The addition of trivalent impurity provides a large number of holes in the semiconductor. Typical examples of trivalent impurities are *gallium* (At. No. 31) and *indium* (At. No. 49). Such impurities which produce *p*-type semiconductor are known as *acceptor impurities* because the holes created can accept the electrons.

To explain the formation of *p*-type semiconductor, consider a pure germanium crystal. When a small amount of trivalent impurity like gallium is added to germanium crystal, there exists a large number of holes in the crystal. The reason is simple. Gallium is trivalent *i.e.* its atom has three valence electrons. Each atom of gallium fits into the germanium crystal but now only three co-valent bonds can be formed. It is because three valence electrons of gallium atom can form only three single co-valent bonds with three germanium atoms as shown in

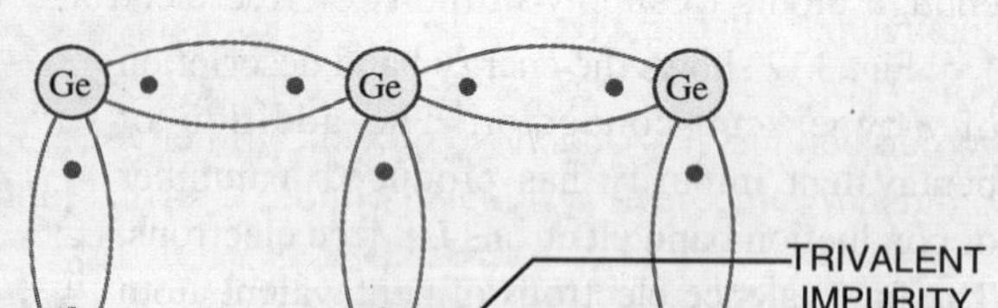

Fig. 3.14

Fig. 3.14. In the fourth co-valent bond, only germanium atom contributes one valence electron while gallium has no valence electron to contribute as all its three valence electrons are already engaged in the co-valent bonds with neighbouring germanium atoms. In other words, fourth bond is incomplete ; being short of one electron. This missing electron is called a *hole*. Therefore, for each gallium atom added, one hole is created. A small amount of gallium provides millions of holes.

Fig. 3.15 shows the energy band description of the *p*-type semiconductor. The addition of trivalent impurity has produced a large number of holes. However, there are a few conduction band electrons due to thermal energy associated with room temperature. But the holes far outnumber the conduction band electrons. It is due to the predominance of holes over free electrons that it is called *p*-type semiconductor (*p* stands for positive).

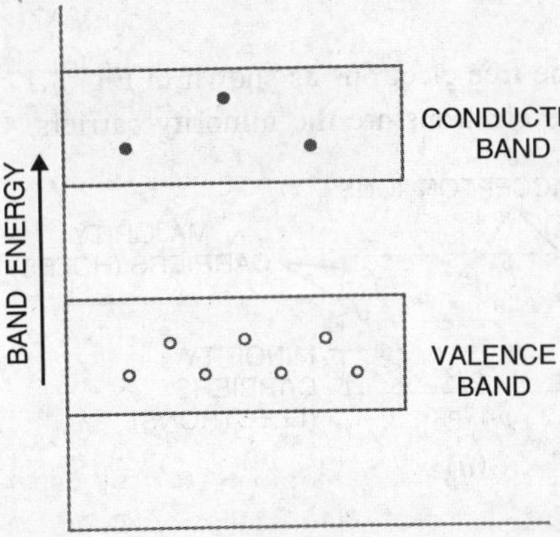

Fig. 3.15

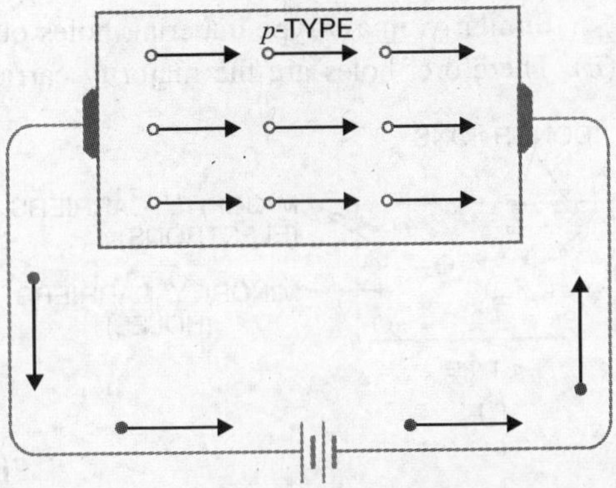

Fig. 3.16

***p*-type conductivity.** The current conduction in *p*-type semiconductor is predominantly by holes *i.e.* positive charges and is called *p-type* or *hole-type conductivity*. To understand *p*-type conductivity, refer to Fig. 3.16. When p.d. is applied to the *p*-type semiconductor, the holes (donated by the impurity) are shifted from one co-valent bond to another. As the holes are positively charged, therefore, they are directed towards the negative terminal, constituting what is known as hole current. It may be noted that in *p*-type conductivity, the valence electrons move from one co-valent bond to another unlike the *n*-type where current conduction is by free electrons.

3.13 Charge on n-type and p-type Semiconductors

As discussed before, in *n*-type semiconductor, current conduction is due to excess of electrons whereas in a *p*-type semiconductor, conduction is by holes. The reader may think that *n*-type material has a net negative charge and *p*-type a net positive charge. But this conclusion is wrong. It is true that *n*-type semiconductor has excess of electrons but these extra electrons were supplied by the atoms of donor impurity and each atom of donor impurity is electrically neutral. When the impurity atom is added, the term "excess electrons" refers to an excess with regard to the number of electrons needed to fill the co-valent bonds in the semiconductor crystal. The extra electrons are free electrons and increase the conductivity of the semiconductor. The situation with regard to *p*-type semiconductor is also similar. *It follows, therefore, that n-type as well as p-type semiconductor is electrically neutral.*

3.14 Majority and Minority Carriers

It has already been discussed that due to the effect of impurity, *n*-type material has a large number of free electrons whereas *p*-type material has a large number of holes. However, it may be recalled that even at room temperature, some of the co-valent bonds break, thus releasing an equal number of free electrons and holes. An *n*-type material has its share of electron-hole pairs (released due to breaking of bonds at room temperature) but in addition has a much larger quantity of free electrons due to the effect of impurity. These impurity-caused free electrons are not associated with holes. Consequently, an *n*-type material has a large number of free electrons and a small number of holes as shown in Fig. 3.17 (*i*). The free electrons in this case are considered *majority carriers* — since the majority portion of current in *n*-type material is by the flow of free electrons — and the holes are the *minority carriers*.

Similarly, in a *p*-type material, holes outnumber the free electrons as shown in Fig. 3.17 (*ii*). Therefore, holes are the majority carriers and free electrons are the minority carriers.

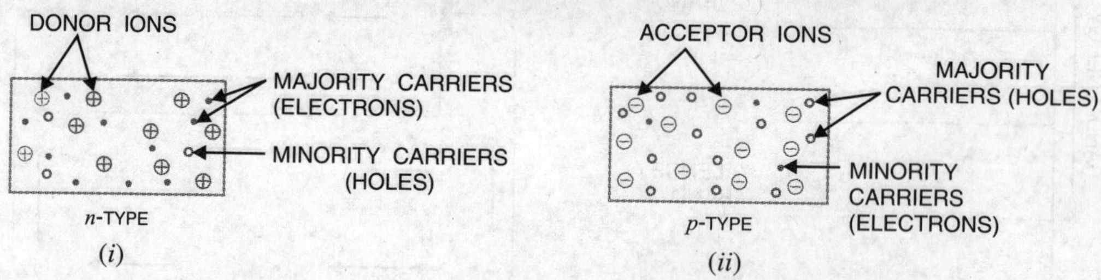

Fig. 3.17

3.15 Comparison of Intrinsic and Extrinsic Semiconductors

S.No.	Intrinsic Semiconductor	Extrinsic Semiconductor
1.	It is a pure semiconductor with no impurity. For this reason, it is called intrinsic (*i.e.* pure) semiconductor.	It is an impure semiconductor *i.e.* a controlled pentavalent or trivalent impurity added to a pure (intrinsic) semiconductor.
2.	The number of free electrons in the conduction band is equal to the number of holes in the valence band.	In an *n*-type semiconductor, free electrons far exceed the holes. In a *p*-type semiconductor, it is the reverse.
3.	It's electrical conductivity is low.	It's electrical conductivity is high.
4.	It's electrical conductivity depends on temperature alone.	It's electrical conductivity depends on tempe-rature and the amount of doping.
5.	It is of no practical use.	It is used in electronic devices.

3.16 Comparison of n-Type and p-Type Semiconductors

S.No.	*n*-type Semiconductor	*p*-type Semiconductor
1.	It is produced by adding controlled amount of pentavalent impurity to a pure (intrinsic) semiconductor.	It is produced by adding controlled amount of trivalent impurity to a pure (intrinsic) semiconductor.
2.	The number of free electrons far exceeds the number of holes.	The number of holes far exceeds the number of free electrons.
3.	The current conduction is predominantly by free electrons.	The current conduction is predominantly by holes.
4.	The donor energy level is just below the bottom of the conduction band.	The acceptor energy level is just above the top of valence band.

3.17 Electrical Conductivity of Semiconductors

In a semiconductor (intrinsic or extrinsic semiconductor), the current flow is due to the movement of free electrons and holes in the opposite directions. Therefore, total current is equal to the sum of two currents. Consider a semiconductor of length l, area of X-section A and connected to a d.c. source of V volts as shown in Fig. 3.18. The magnitude of applied electric field (E) is

$$E = \frac{V}{l}$$

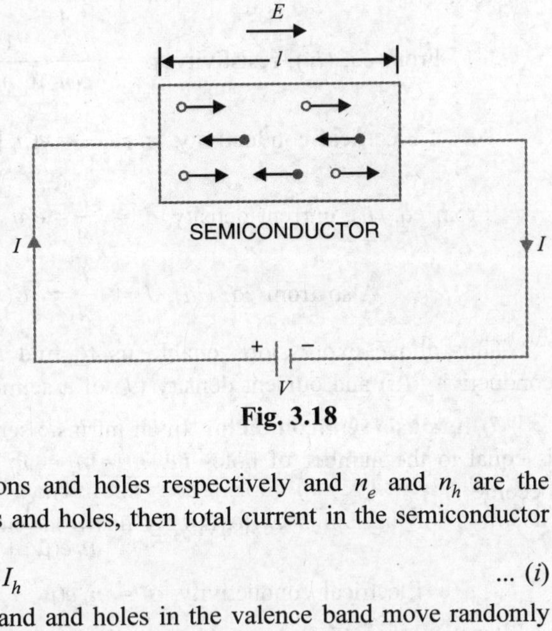

Fig. 3.18

Due to the electric field, both free electrons and holes in the semiconductor move in opposite directions. If v_e and v_h are the drift velocities of free electrons and holes respectively and n_e and n_h are the corresponding *densities of free electrons and holes, then total current in the semiconductor is given by ;

$$I = I_e + I_h \qquad \text{... (i)}$$

As the electrons in the conduction band and holes in the valence band move randomly like free electrons in metals,

$\therefore \qquad I_e = en_e Av_e \text{ and } I_h = en_h Av_h$

Here, $\qquad e = $ Charge on electron or hole $= 1.6 \times 10^{-19}$ C

$\therefore$ Eq. (*i*) becomes : $\qquad I = en_e Av_e + en_h Av_h = eA(n_e v_e + n_h v_h)$

or $\qquad I = eA(n_e v_e + n_h v_h) \qquad \text{... (ii)}$

Now, mobility of a charge carrier (electron or hole) is defined as the velocity imparted to it by unit electric field.

---

* n_e = Number of free electrons per unit volume
n_h = Number of holes per unit volume

∴ *Mobility of electron, $\mu_e = \dfrac{v_e}{E}$ or $v_e = \mu_e E$

Mobility of hole, $\mu_h = \dfrac{v_h}{E}$ or $v_h = \mu_h E$

∴ Eq. (*ii*) becomes :

$$I = eAE(n_e \mu_e + n_h \mu_h) \quad \text{... (iii)}$$

$$= eA\frac{V}{l}(n_e \mu_e + n_h \mu_h) \quad \left[\because E = \frac{V}{l}\right]$$

or

$$\frac{V}{I} = \frac{l}{A} \times \frac{1}{e(n_e \mu_e + n_h \mu_h)}$$

or

$$R = \frac{l}{A} \times \frac{1}{e(n_e \mu_e + n_h \mu_h)} \quad \text{... (iv)}$$

Eq. (iv) gives the expression for the resistance (R) of the semiconductor.

Now, $R = \rho \dfrac{l}{A}$

∴ From eq. (*iv*), resistivity,

$$\rho = \frac{1}{e(n_e \mu_e + n_h \mu_h)} \quad \text{... (v)}$$

Electrical conductivity,

$$\sigma = \frac{1}{\rho} = e(n_e \mu_e + n_h \mu_h) \quad \text{... (vi)}$$

From eq. (*ii*), current density, $J = \dfrac{I}{A} = e(n_e v_e + n_h v_h)$

Also from eq. (*iii*), $J = \dfrac{I}{A} = eE(n_e \mu_e + n_h \mu_h)$

Thus above expressions enable us to find resistance (*R*), resistivity (ρ), electrical conductivity (σ) and current density (*J*) of a semiconductor.

(*i*) **Intrinsic semiconductor.** In an intrinsic semiconductor, the number of free electrons is equal to the number of holes *i.e.* $n_e = n_h = n_i$ (say). Therefore, eqs. (*v*) and (*vi*) above become :

$$\text{Resistivity, } \rho = \frac{1}{n_i e(\mu_e + \mu_h)}$$

$$\text{Electrical conductivity, } \sigma = n_i e(\mu_e + \mu_h)$$

As the value of μ_h is quite **small, the resistivity of intrinsic semiconductor is very high.

(*ii*) **Extrinsic semiconductor.** An extrinsic semiconductor can be *p*-type or *n*-type depending upon the type of impurity added.

(*a*) A practical ***p*-type semiconductor** is so made that $n_h >> n_e$. Now n_h is approximately equal to N_a (= number density of acceptor atoms). Therefore, expressions (*v*) and (*vi*) above become :

---

* The SI unit of mobility is m^2/volt-sec.

** The hole movement is slow because of the involvement of valence elctrons. The hole mobility μ_h is practically half of electron mobility μ_e.

$$\text{Resistivity, } \rho = \frac{1}{en_h \mu_h} = \frac{1}{eN_a \mu_h}$$

$$\text{Electrical conductivity, } \sigma = en_h \mu_h = eN_a \mu_h$$

(*b*) A practical **n-type semiconductor** is so made that $n_e >> n_h$. Now n_e is approximately equal to N_d (= number density of donor atoms). Therefore,

$$\text{Resistivity, } \rho = \frac{1}{en_e \mu_e} = \frac{1}{eN_d \mu_e}$$

$$\text{Electrical conductivity, } \sigma = en_e \mu_e = eN_d \mu_e$$

3.18 Effect of Temperature on Conductivity of Semiconductors

The electrical conductivity σ of a semiconductor is given by ;

$$\sigma = e(n_e \mu_e + n_h \mu_h)$$

As the temperature increases, the values of μ_e (electron mobility) and μ_h (hole mobility) decrease due to increase in their collision frequency. But with the increase in temperature, there is a large increase in charge carrier concentrations (n_e and n_h) due to the increase in breaking of covalent bonds. In fact, the increase in carrier concentrations is so large that the decrease in the values of μ_e and μ_h has no significant effect. Hence, the conductivity of a semiconductor increases with the increase in temperature and *vice-versa*.

Example 3.1. *An intrinsic germanium has a resistivity of 0.47 Ω m at room temperature. Find the intrinsic carrier concentration if the mobilities of electrons and holes are 0.39 m²/volt-sec and 0.19 m²/volt-sec respectively.*

Solution. Let n_i be the intrinsic carrier concentration.

$$\text{Electrical conductivity, } \sigma = \frac{1}{\rho} = \frac{1}{0.47}$$

Now, $\sigma = n_i e(\mu_e + \mu_h)$ ... See section 3.17

or $$n_i = \frac{\sigma}{e(\mu_e + \mu_h)}$$

Here, $\sigma = 1/0.47$ S/m, $\mu_e = 0.39$ m²/volt.sec, $\mu_h = 0.19$ m²/volt.sec, $e = 1.6 \times 10^{-19}$ C

$$\therefore \quad n_i = \frac{1}{0.47 \times 1.6 \times 10^{-19} (0.39 + 0.19)} = \mathbf{2.3 \times 10^{19}/m^3}$$

Example 3.2. *The resistivity of n-type germanium is 0.01 Ω m at room temperature. Find the donor concentration if the mobility of electrons is 0.39 m²/volt-sec.*

Solution. Let N_d be the donor concentration.

$$\text{Conductivity, } \sigma = \frac{1}{\rho} = \frac{1}{0.01} = 100 \text{ S/m}$$

Now, $\sigma = eN_d \mu_e$

$$\text{or Donor concentration, } N_d = \frac{\sigma}{e\mu_e} = \frac{100}{1.6 \times 10^{-19} \times 0.39} = \mathbf{1.6 \times 10^{21}/m^3}$$

Example 3.3. *Mobilities of electrons and holes in a sample of intrinsic germanium at room temperature are 0.36 m²/volt-sec and 0.17 m²/volt-sec respectively. If the electron and hole densities are each equal to 2.5 × 10¹⁹/m³, calculate the conductivity.*

Solution. The conductivity of an intrinsic semiconductor is given by ;

$$\sigma = n_i e (\mu_e + \mu_h)$$

Here, $n_i = 2.5 \times 10^{19}/m^3$; $e = 1.6 \times 10^{-19}C$; $\mu_e = 0.36$ m²/volt-sec ; $\mu_h = 0.17$ m²/volt-sec.

∴ Conductivity, $\sigma = 2.5 \times 10^{19} \times 1.6 \times 10^{-19} (0.36 + 0.17) =$ **2.12 S/m**

Example 3.4. *A semiconductor is known to have an electron concentration of 8×10^{13} per cm³ and a hole concentration of 5×10^{12} per cm³.*

(*i*) *Is the semiconductor n-type or p-type?*

(*ii*) *What is the resistivity of the sample if the electron mobility is 23000 cm²/Vs and hole mobility is 100 cm²/Vs?*

Solution. (*i*) Since electron density $n_e(= 8 \times 10^{13}$ per cm³) is greater than the hole density $n_h(= 5 \times 10^{12}$ per cm³), the semiconductor is ***n*-type.**

(*ii*) The conductivity of the sample is given by ;

$$\sigma = e (n_e \mu_e + n_h \mu_h)$$

Here, $n_e = 8 \times 10^{13}$ per cm³ ; $n_h = 5 \times 10^{12}$ per cm³ ;

$\mu_e = 23000$ cm²/Vs ; $\mu_h = 100$ cm²/Vs

$$\therefore \quad \sigma = 1.6 \times 10^{-19} \times (8 \times 10^{13} \times 23000 + 5 \times 10^{12} \times 100) \text{Scm}^{-1}$$

$$= 1.6 \times 184.05 \times 10^{-3} \text{ Scm}^{-1}$$

∴ Resistivity of the sample is given by ;

$$\rho = \frac{1}{\sigma} = \frac{1}{1.6 \times 184.05 \times 10^{-3}} = \mathbf{3.396\ \Omega\ cm}$$

Example 3.5. *Pure Si at 300 K has equal electron (n_e) and hole (n_h) concentration of 1.5×10^{16} m⁻³. Doping by indium increases n_h to 4.5×10^{22} m⁻³. Calculate n_e in the doped semiconductor.*

Solution. $$n_e n_h = n_i^2 \qquad \therefore \quad n_e = \frac{(n_i)^2}{n_h}$$

Here, $n_i = 1.5 \times 10^{16}$ m⁻³ ; $n_h = 4.5 \times 10^{22}$ m⁻³

$$\therefore \quad n_e = \frac{(1.5 \times 10^{16})^2}{4.5 \times 10^{22}} = \mathbf{5 \times 10^9\ m^{-3}}$$

Example 3.6. *The energy of a photon of sodium light (λ = 589 nm) equals the band gap of a semiconductor material. Find the minimum energy E required to create a hole-electron pair.*

Solution. Wavelength of sodium light, λ = 589 nm = 5890 Å

$$\text{Energy of photon, } E = \frac{12400}{\lambda} \text{eV} \quad \text{... } \lambda \text{ in Å}$$

$$= \frac{12400}{5890} \text{eV} = \mathbf{2.1\ eV}$$

Therefore, the band gap is 2.1 eV. That is, the minimum energy E required to push an electron from the valence band into the conduction band is 2.1 eV. Hence the minimum energy required to create a hole-electron pair is 2.1 eV.

Example 3.7. *In a pure semiconductor, the number of conduction electrons is 6×10^{19} per cubic metre. How many holes are there in a sample of size 1 cm × 1 cm × 1 mm?*

Solution. In a pure semiconductor, hole density is equal to the electron density. Therefore, in the given semiconductor, hole density = $6 \times 10^{19}/m^3$. Now the volume of the given sample = $10^{-2} \times 10^{-2} \times 10^{-3}$ m^3 = 10^{-7}m^3. Therefore, number of holes in the given sample

$$= 6 \times 10^{19} \times 10^{-7} = \mathbf{6 \times 10^{12}}$$

Example 3.8. *Determine the number density of donor atoms which have to be added to an intrinsic germanium semiconductor to produce an n-type semiconductor of conductivity $5\Omega^{-1}$ cm^{-1}, given that mobility of conduction electrons in n-type Ge is 3900 cm^2/Vs. Neglect the contribution of holes to conductivity.*

Solution. The conductivity of a semiconductor is given by ;

$$\sigma = e\,(n_e\,\mu_e + n_h\,\mu_h)$$

Neglecting the contribution of holes to conductivity, we have,

$$\sigma = e n_e\,\mu_e = e N_d\,\mu_e \qquad [\because N_d \simeq n_e]$$

∴ Number density of donor atoms or electron density is

$$N_d = \frac{\sigma}{e\mu_e}$$

Here, $\sigma = 5\Omega^{-1}$ cm^{-1} ; μ_e = 3900 cm^2/Vs ; $e = 1.6 \times 10^{-19}$ C

$$\therefore \quad N_d = \frac{5}{3900 \times 1.6 \times 10^{-19}} = \mathbf{8.01 \times 10^{15}\ cm^{-3}}$$

Example 3.9. *Suppose a pure Si crystal has 5×10^{28} atoms m^{-3}. It is doped by 1 ppm concentration of pentavalent As. Calculate the number of electrons and holes. Given that $n_i = 1.5 \times 10^{16}$ m^{-3}.*

Solution. 1 ppm = 1 part per million = $\frac{1}{10^6}$.

∴ Number of pentavalent atoms doped in Si crystal

$$= \frac{5 \times 10^{28}}{10^6} = 5 \times 10^{22}\ m^{-3}$$

As one pentavalent impurity atom donates 1 free electron to the crystal,

∴ Number of free electrons in the crystal is

$$n_e = \mathbf{5 \times 10^{22}\ m^{-3}}$$

$$\therefore \quad \text{Number of holes, } n_h = \frac{n_i^2}{n_e} = \frac{(1.5 \times 10^{16})^2}{5 \times 10^{22}} = \mathbf{4.5 \times 10^9\ m^{-3}}$$

3.19 pn Junction

When a p-type semiconductor is suitably joined to n-type semiconductor, the contact surface is called ***pn junction.***

Most semiconductor devices contain one or more *pn* junctions. The *pn* junction is of great importance because it is in effect, the *control element* for semiconductor devices. A thorough knowledge of the formation and properties of *pn* junction can enable the reader to understand the semiconductor devices.

Formation of *pn* junction. In actual practice, the characteristic properties of *pn* junction will not be apparent if a *p*-type block is just brought in contact with *n*-type block. In fact, *pn* junction is fabricated by special techniques. One common method of making *pn* junction is called *alloying*. In this method, a small block of indium (trivalent impurity) is placed on an *n*-type germanium slab as shown in Fig. 3.19 (*i*). The system is then heated to a temperature of about 500°C. The indium and some of the germanium melt to form a small puddle of molten germanium-indium mixture as shown in Fig. 3.19 (*ii*). The temperature is then lowered and puddle begins to solidify. Under proper conditions, the atoms of indium impurity will be suitably adjusted in the germanium slab to form a single crystal. The addition of indium overcomes the excess of electrons in the *n*-type germanium to such an extent that it creates a *p*-type region.

As the process goes on, the remaining molten mixture becomes increasingly rich in indium. When all germanium has been redeposited, the remaining material appears as indium button which is frozen on to the outer surface of the crystallised portion as shown in Fig. 3.19 (*iii*). This button serves as a suitable base for soldering on leads.

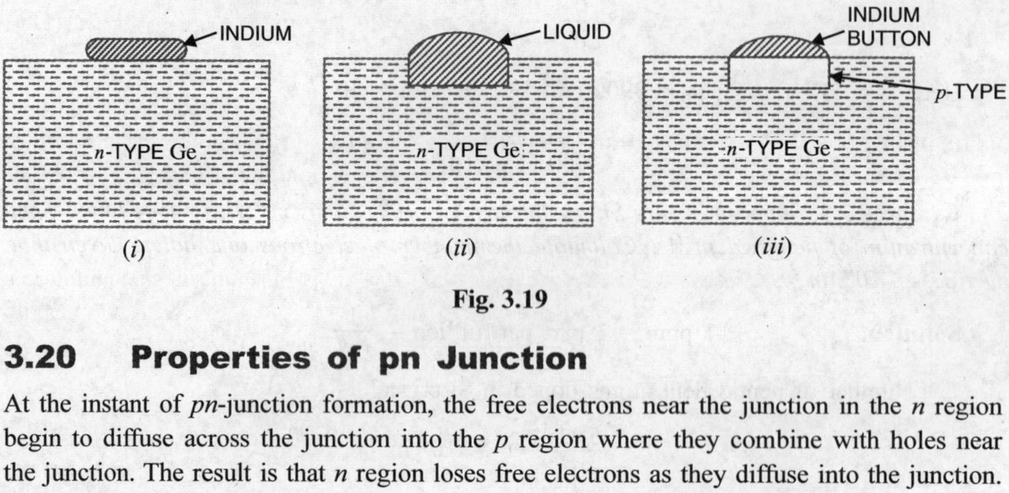

Fig. 3.19

3.20 Properties of pn Junction

At the instant of *pn*-junction formation, the free electrons near the junction in the *n* region begin to diffuse across the junction into the *p* region where they combine with holes near the junction. The result is that *n* region loses free electrons as they diffuse into the junction. This creates a layer of positive charges (pentavalent ions) near the junction. As the electrons move across the junction, the *p* region loses holes as the electrons and holes combine. The result is that there is a layer of negative charges (trivalent ions) near the junction. These two layers of positive and negative charges form the *depletion region* (or *depletion layer*). The term depletion is due to the fact that near the junction, the region is depleted (*i.e.* emptied) of *charge carries* (free electrons and holes) due to diffusion across the junction. It may be noted that depletion layer is formed very quickly and is very thin compared to the *n* region and the *p* region. For clarity, the width of the depletion layer is shown exaggerated.

Once *pn* junction is formed and depletion layer created, the diffusion of free electrons stops. In other words, the depletion region acts as a barrier to the further movement of free electrons across the junction. The positive and negative charges set up an electric field. This is shown by a black arrow in Fig. 3.20 (*ii*). The electric field is a barrier to the free electrons in the *n*-region. There exists a potential difference across the depletion layer and is called **barrier potential** (V_0). The barrier potential of a *pn* junction depends upon several factors including the type of

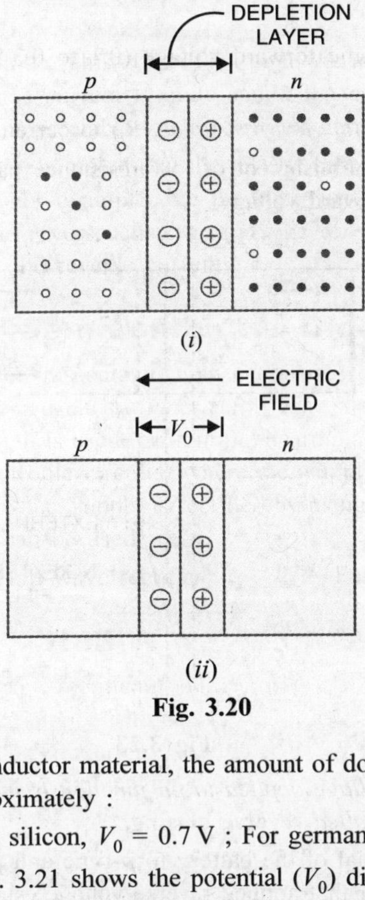

Fig. 3.20

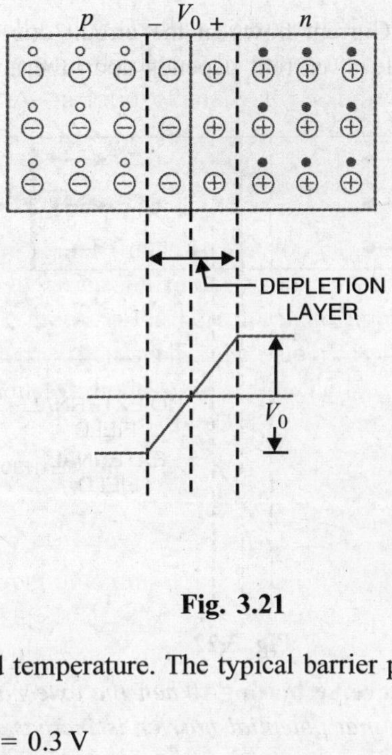

Fig. 3.21

semiconductor material, the amount of doping and temperature. The typical barrier potential is approximately :

For silicon, V_0 = 0.7 V ; For germanium, V_0 = 0.3 V

Fig. 3.21 shows the potential (V_0) distribution curve.

3.21 Applying D.C. Voltage Across pn Junction or Biasing a pn Junction

In electronics, the term bias refers to the use of d.c. voltage to establish certain operating conditions for an electronic device. In relation to a *pn* junction, there are following two bias conditions :

1. **Forward biasing** 2. **Reverse biasing**

1. Forward biasing. *When external d.c. voltage applied to pn junction is in such a direction that it cancels the potential barrier, thus permitting current flow, it is called* **forward biasing.**

To apply forward bias, connect positive terminal of the battery to *p*-type and negative terminal to *n*-type as shown in Fig. 3.22. The applied forward potential establishes an electric field which acts against the field due to potential barrier. Therefore, the resultant field is weakened and the barrier height is reduced at the junction as shown in Fig. 3.22. As potential barrier voltage is very small (0.1 to 0.3 V), therefore, a small forward voltage is sufficient to completely eliminate the barrier. Once the potential barrier is eliminated by the forward voltage, junction resistance becomes almost zero and a low resistance path is established for the entire circuit. Therefore, current flows in the circuit. This is called *forward current*. With forward bias to *pn* junction, the following points are worth noting :

(*i*) The potential barrier is reduced and at some forward voltage (0.1 to 0.3 V), it is eliminated altogether.

(*ii*) The junction offers low resistance (called *forward resistance,* R_f) to current flow.

(*iii*) Current flows in the circuit due to the establishment of low resistance path. The magnitude of current depends upon the applied forward voltage.

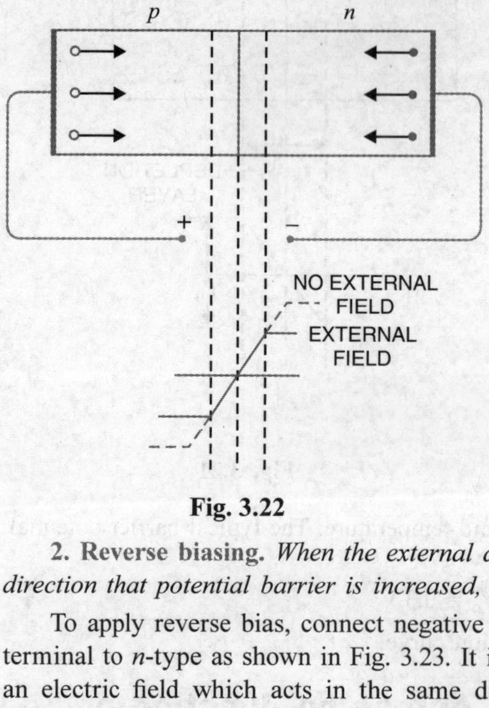

Fig. 3.22

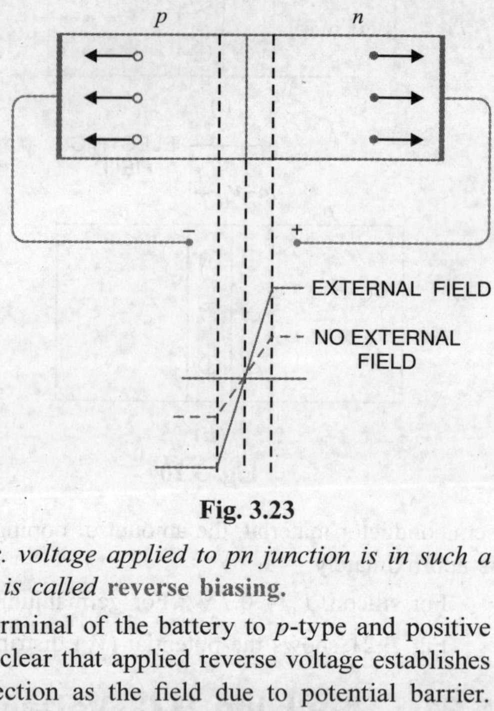

Fig. 3.23

2. Reverse biasing. *When the external d.c. voltage applied to pn junction is in such a direction that potential barrier is increased, it is called* **reverse biasing.**

To apply reverse bias, connect negative terminal of the battery to *p*-type and positive terminal to *n*-type as shown in Fig. 3.23. It is clear that applied reverse voltage establishes an electric field which acts in the same direction as the field due to potential barrier. Therefore, the resultant field at the junction is strengthened and the barrier height is increased as shown in Fig. 3.23. The increased potential barrier prevents the flow of charge carriers across the junction. Thus, a high resistance path is established for the entire circuit and hence the current does not flow. With reverse bias to *pn* junction, the following points are worth noting :

(*i*) The potential barrier is increased.

(*ii*) The junction offers very high resistance (called *reverse resistance,* R_r) to current flow.

(*iii*) No current flows in the circuit due to the establishment of high resistance path.

Conclusion. From the above discussion, it follows that with reverse bias to *pn* junction, a high resistance path is established and hence no current flow occurs. On the other hand, with forward bias to *pn* junction, a low resistance path is set up and hence current flows in the circuit.

3.22 Current Flow in a Forward Biased pn Junction

We shall now see how current flows across *pn* junction when it is forward biased. Fig. 3.24 shows a forward biased *pn* junction. Under the influence of forward voltage, the free electrons in *n*-type move *towards the junction, leaving behind positively charged atoms. However, more electrons arrive from the negative battery terminal and enter the *n*-region to take up their places. As the free electrons reach the junction, they become **valence electrons. As valence electrons, they move through the holes in the *p*-region. The valence electrons move towards left in the *p*-region which is equivalent to the holes moving to right. When the valence electrons reach the left end of the crystal, they flow into the positive terminal of the battery.

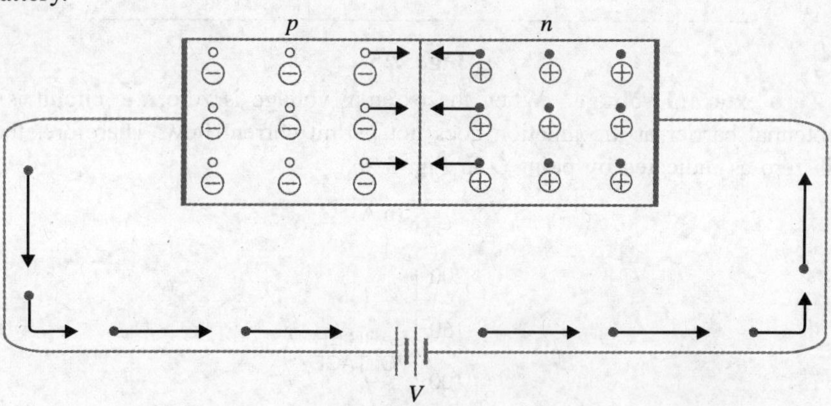

Fig. 3.24

The mechanism of current flow in a forward biased *pn* junction can be summed up as under :

(*i*) The free electrons from the negative terminal continue to pour into the *n*-region while the free electrons in the *n*-region move towards the junction.

(*ii*) The electrons travel through the *n*-region as free-electrons *i.e.* current in *n*-region is by free electrons.

(*iii*) When these electrons reach the junction, they combine with holes and become valence electrons.

(*iv*) The electrons travel through *p*-region as valence electrons *i.e.* current in the *p*-region is by holes.

(*v*) When these valence electrons reach the left end of crystal, they flow into the positive terminal of the battery.

From the above discussion, it is concluded that in *n*-type region, current is carried by free electrons whereas in *p*-type region, it is carried by holes. However, in the external connecting wires, the current is carried by free electrons.

---

* Note that negative terminal of battery is connected to *n*-type. It repels the free electrons in *n*-type towards the junction.

** A hole is in the co-valent bond. When a free electron combines with a hole, it becomes a valence electron.

3.23 Volt-Ampere Characteristics of pn Junction

Volt-ampere or *V-I* characteristic of a *pn* junction (also called a *crystal or semiconductor diode*) is the curve between voltage across the junction and the circuit current. Usually, voltage is taken along *x*-axis and current along *y*-axis. Fig. 3.25 shows the *circuit arrangement for determining the *V-I* characteristics of a *pn* junction. The characteristics can be studied under three heads, namely ; *zero external voltage, forward bias* and *reverse bias.*

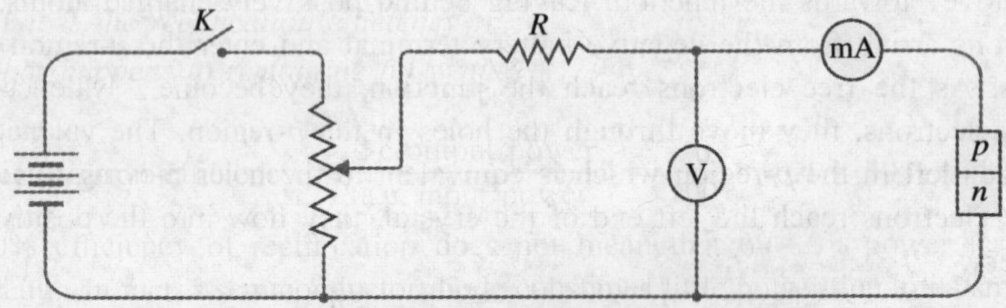

Fig. 3.25

***(i)* Zero external voltage.** When the external voltage is zero, *i.e.* circuit is open at *K*, the potential barrier at the junction does not permit current flow. Therefore, the circuit current is zero as indicated by point *O* in Fig. 3.26.

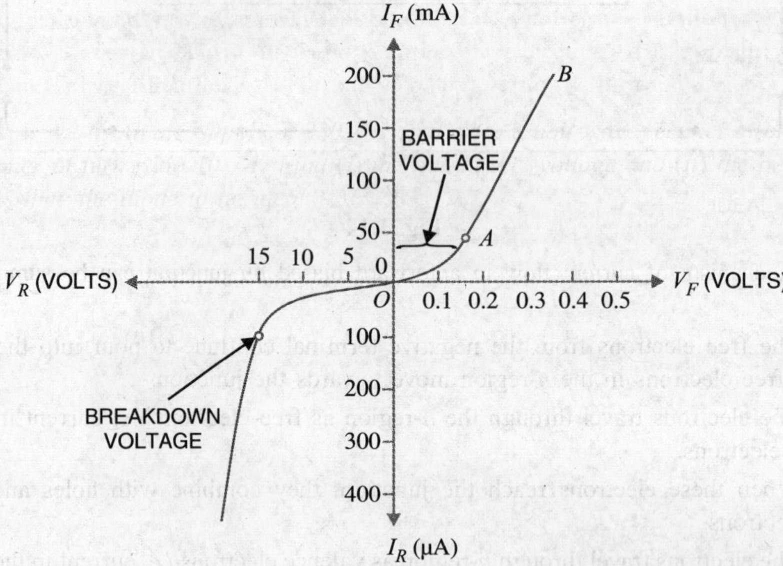

Fig. 3.26

***(ii)* Forward bias.** With forward bias to the *pn* junction *i.e. p*-type connected to positive terminal and *n*-type connected to negative terminal, the potential barrier is reduced. At some forward voltage (0.7 V for Si and 0.3 V for Ge), the potential barrier is altogether eliminated and current starts flowing in the circuit. From now onwards, the current increases with the increase in forward voltage. Thus, a rising curve *OB* is obtained with forward bias as shown in Fig. 3.26. From the forward characteristic, it is seen that at first (*region OA*), the current

* *R* is the current limiting resistance. It prevents the forward current from exceeding the permitted value.

increases very slowly and the curve is non-linear. It is because the external applied voltage is used up in overcoming the potential barrier. However, once the external voltage exceeds the potential barrier voltage, the *pn* junction behaves like an ordinary conductor. Therefore, the current rises very sharply with increase in external voltage (*region AB on the curve*). The curve is almost linear.

(*iii*) **Reverse bias.** With reverse bias to the *pn* junction *i.e.* *p*-type connected to negative terminal and *n*-type connected to positive terminal, potential barrier at the junction is increased. Therefore, the junction resistance becomes very high and practically no current flows through the circuit. However, in practice, a very small current (of the order of μA) flows in the circuit with reverse bias as shown in the reverse characteristic. This is called *reverse *saturation current* (I_s) and is due to the minority carriers. It may be recalled that there are a few free electrons in *p*-type material and a few holes in *n*-type material. These undesirable free electrons in *p*-type and holes in *n*-type are called *minority carriers*. As shown in Fig. 3.27, to these minority carriers, the applied reverse bias appears as forward bias. Therefore, a **small current flows in the reverse direction.

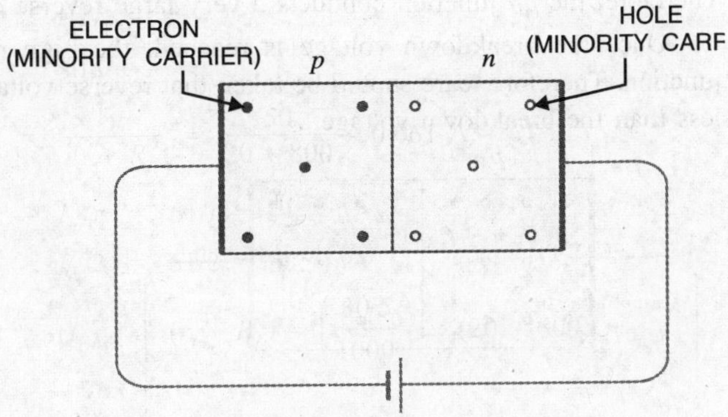

Fig. 3.27

If reverse voltage is increased continuously, the kinetic energy of electrons (minority carriers) may become high enough to knock out electrons from the semiconductor atoms. At this stage *breakdown* of the junction occurs, characterised by a sudden rise of reverse current and a sudden fall of the resistance of barrier region. This may destroy the junction permanently.

Note. The forward current through a *pn* junction is due to the *majority carriers* produced by the impurity. However, reverse current is due to the *minority carriers* produced due to breaking of some co-valent bonds at room temperature.

3.24 Important Terms

Two important terms often used with *pn* junction (*i.e.* crystal diode) are *breakdown voltage* and *knee voltage*. We shall now explain these two terms in detail.

(*i*) **Breakdown voltage.** *It is the minimum reverse voltage at which pn junction breaks down with sudden rise in reverse current.*

Under normal reverse voltage, a very little reverse current flows through a *pn* junction. However, if the reverse voltage attains a high value, the junction may break down with sudden rise in reverse current. For understanding this point, refer to Fig. 3.28. Even at room temperature,

* The term saturation comes from the fact that it reaches its maximum level quickly and does not significantly change with the increase in reverse voltage.

** Reverse current increases with reverse voltage but can generally be regarded as negligible over the working range of voltages.

some hole-electron pairs (minority carriers) are produced in the depletion layer as shown in Fig. 3.28 (*i*). With reverse bias, the electrons move towards the positive terminal of supply. At large reverse voltage, these electrons acquire high enough velocities to dislodge valence electrons from semiconductor atoms as shown in Fig. 3.28 (*ii*). The newly liberated electrons in turn free other valence electrons. In this way, we get an *avalanche* of free electrons. Therefore, the *pn* junction conducts a very large reverse current.

Once the breakdown voltage is reached, the high reverse current may damage the junction. Therefore, care should be taken that reverse voltage across a *pn* junction is always less than the breakdown voltage.

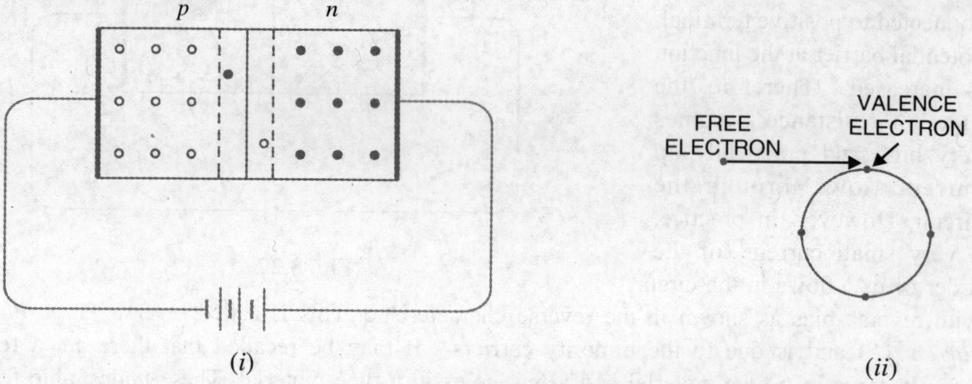

Fig. 3.28

(*ii*) Knee voltage. *It is the forward voltage at which the current through the junction starts to increase rapidly.*

When a diode is forward biased, it conducts current very slowly until we overcome the potential barrier. For silicon *pn* junction, potential barrier is 0.7 V whereas it is 0.3 V for germanium junction. It is clear from Fig. 3.29 that knee voltage for silicon diode is 0.7 V and 0.3 V for germanium diode.

Once the applied forward voltage exceeds the knee voltage, the current starts increasing rapidly. It may be added here that in order to get useful current through a *pn* junction, the applied voltage must be more than the knee voltage.

Note. The potential barrier voltage is also known as *turn-on voltage.* This is obtained by taking the straight line portion of the forward characteristic and extending it back to the horizontal axis.

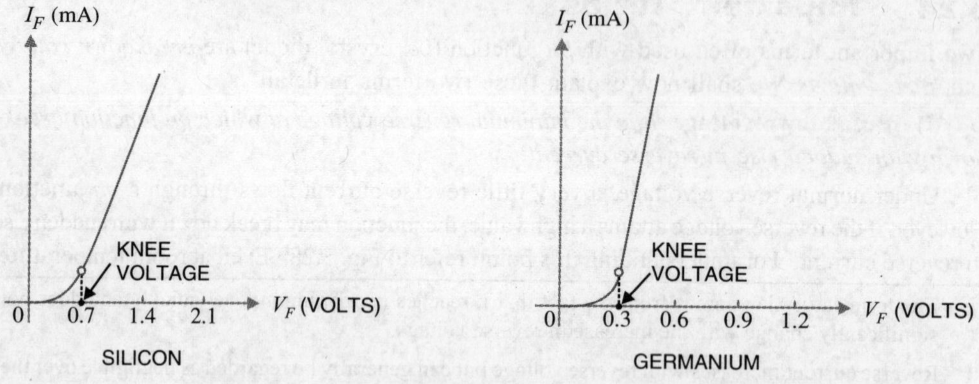

Fig. 3.29

3.25 Limitations in the Operating Conditions of pn Junction

Every *pn* junction has limiting values of *maximum forward current, peak inverse voltage* and *maximum power rating*. The *pn* junction will give satisfactory performance if it is operated within these limiting values. However, if these values are exceeded, the *pn* junction may be destroyed due to excessive heat.

(*i*) **Maximum forward current.** It is the highest instantaneous forward current that a *pn* junction can conduct without damage to the junction. Manufacturer's data sheet usually specifies this rating. If the forward current in a *pn* junction is more than this rating, the junction will be destroyed due to overheating.

(*ii*) **Peak inverse voltage (PIV).** It is the maximum reverse voltage that can be applied to the *pn* junction without damage to the junction. If the reverse voltage across *pn* junction exceeds its *PIV*, the junction may be destroyed due to excessive heat. The peak inverse voltage is of particular importance in rectifier service. A *pn* junction *i.e.* a crystal diode is used as a rectifier to change alternating current into direct current. In such applications, care should be taken that reverse voltage across the diode during negative half-cycle of a.c. does not exceed the *PIV* of diode.

(*iii*) **Maximum power rating.** It is the maximum power that can be dissipated at the junction without damaging it. The power dissipated at the junction is equal to the product of junction current and the voltage across the junction. This is a very important consideration and is invariably specified by the manufacturer in the data sheet.

Short Answer Questions

Q. 1. *Why is a semiconductor virtually an insulator at room temperature?*

Ans. Because almost all the valence electrons are engaged in the formation of bonds and there are practically very few free electrons. At room temperature, some of the covalent bonds of a semiconductor break ; creating some free electrons and an equal number of holes. The current conduction due to these electron-hole pairs is so small (1 μA or so) that we can say that for all practical purposes a pure semiconductor behaves as an insulator.

Q. 2. *Why is silicon preferred to germanium in the manufacture of semiconductor devices?*

Ans. There are many good reasons for preferring silicon to germanium in the manufacture of semiconductor devices :

(*i*) Silicon has small leakage current as compared to germanium. In general, the leakage current in silicon device is 10 to 100 times smaller than with germanium. Typical values at 25°C are :

Silicon 0.01 μA to 1 μA

Germanium 2 μA to 15 μA

(*ii*) Silicon can be used for applications in which the temperature may rise to about 200°C whereas germanium has much lower maximum rating (100°C).

(*iii*) The silicon *pn* junction has, in general, higher *PIV* (peak inverse voltage) and current ratings than a germanium *pn* junction. *PIV* ratings for a silicon *pn* junction can be in the neighbourhood of 1000 V whereas the *PIV* rating for a germanium *pn* junction is close to 400 V.

Due to these reasons, most of the solid state devices are made of silicon.

Q. 3. *How does the addition of trivalent impurity to a pure semiconductor affect the electron-hole pairs at room temperature?*

Ans. At room temperature, electron-hole pairs are created in a pure semiconductor due to thermal energy ; there being as many holes as the number of free electrons. However, with the addition of trivalent impurity, holes far outnumber the free electrons. It is because the trivalent impurity produces many holes whereas the number of free electrons remains at the intrinsic level.

Q. 4. *What happens to depletion layer under forward bias? Why is it called depletion layer?*

Ans. The width of the depletion layer is decreased because the electric field due to external voltage opposes the electric field due to potential barrier. It is called depletion layer because in this region, the charge carriers (*i.e.* free electrons and holes) are depleted *i.e.* emptied.

Q. 5. *Can there be any current flow across a reverse biased pn junction?*

Ans. Yes. This is called reverse or leakage current and is due to the minority carriers. To these minority carriers, the reverse bias acts as a forward bias.

Q. 6. *Which is more sensitive to a change in temperature, forward or reverse current?*

Ans. The reverse current. The reverse current is due to minority carriers whose number is strongly temperature dependent. For germanium devices, reverse (or leakage) current doubles with each 8 to 10°C rise in temperature while in case of silicon it approximately doubles with each 12°C rise in temperature.

The forward current is due to majority carriers and its value is quite large. Therefore, the addition of small leakage current due to temperature rise does not change the value of forward current too much.

Q. 7. *For the same degree of doping, why does n-type semiconductor has more conductivity than p-type semiconductor?*

Ans. The charge carriers in *n*-type material are free electrons whereas these are holes in a *p*-type material. As free electrons have more mobility than holes, therefore, *n*-type semiconductor has higher conductivity than *p*-type semiconductor. As a matter of fact, the resistivity of an *n*-type semiconductor is approximately half that of a *p*-type material for the same degree of doping.

Q. 8. *The hole current is due to the movement of valence electrons from one covalent bond to another. Then why do we call it hole current?*

Ans. The basic reason for current flow is the presence of holes in the covalent bonds. Therefore, it is more appropriate to consider the current as the movement of holes.

Q. 9. *Why is controlled impurity added to a pure semiconductor to make it extrinsic semiconductor?*

Ans. A pure semiconductor virtually behaves as an insulator at room temperature. The controlled impurity added to a pure semiconductor (1 impurity atom for 10^8 atoms of pure semiconductor) introduces either an excess of free electrons or holes in the material to make it *n*-type or *p*-type semiconductor. The idea of adding controlled impurity is that extrinsic material should have conductivity about 20 times greater than that of pure semiconductor. Note that conductivity of copper is thousands of times greater than that of extrinsic semiconductor.

Q. 10. *How is the energy level diagram modified by the addition of impurities?*

Ans. When a pentavalent impurity (*i.e.* donor impurity) is added to a pure semiconductor to make it *n*-type material, an extra energy level is created just below the conduction band as shown in Fig. 3.30 (*i*). The energy required to lift an electron from the donor level to the conduction band is very small. For example, if arsenic is added to Ge, the donor level is 0.0127 eV below the conduction band. Even room temperature is sufficient to provide this energy to push the electrons from the donor level to the conduction band. Consequently, a large number of free electrons exist in an *n*-type material.

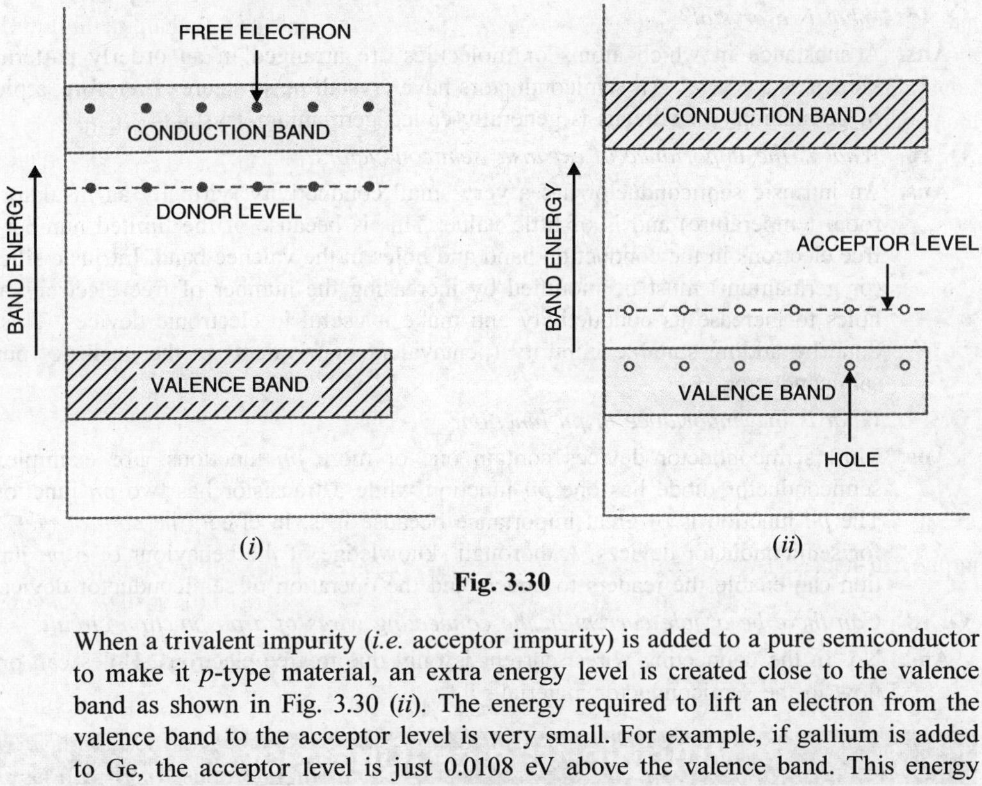

Fig. 3.30

When a trivalent impurity (*i.e.* acceptor impurity) is added to a pure semiconductor to make it *p*-type material, an extra energy level is created close to the valence band as shown in Fig. 3.30 (*ii*). The energy required to lift an electron from the valence band to the acceptor level is very small. For example, if gallium is added to Ge, the acceptor level is just 0.0108 eV above the valence band. This energy is so small that even room temperature can push the electrons from the valence band to the acceptor level. Consequently, a large number of holes are created in a *p*-type material. Note that lifting of electrons from the valence band will create holes in the co-valent bonds.

Q. 11. *When impurity is added to a semiconductor, does the crystal structure change?*

Ans. Since 1 atom of impurity is added for 10^8 atoms of pure semiconductor, the basic crystal structure of extrinsic semiconductor is not very much different from that of the intrinsic crystal.

Q. 12. *What are the possible causes that may destroy a pn junction?*

Ans. A *pn* junction is distroyed due to excessive heat caused by ;

(*i*) Excessive forward bias

(*ii*) Excessive reverse bias

(*iii*) External heat

Q. 13. *What are the factors upon which the potential barrier of a pn junction depends?*

Ans. The factors upon which potential barrier depends are (*i*) type of semiconductor material (*ii*) amount of doping and (*iii*) temperature.

Q. 14. *How do you explain the positive charge on a hole?*

Ans. A hole represents a missing valence electron in a covalent bond. Therefore, a hole can be thought as a positive charge because the absence of an electron leaves a net positive charge on the atom.

Q. 15. *What is a crystal?*

Ans. A substance in which atoms or molecules are arranged in an orderly pattern is known as a *crystal*. All semiconductors have crystalline structure. Therefore, a piece of germanium (or silicon) is generally called germanium crystal.

Q. 16. *What is the importance of extrinsic semiconductors?*

Ans. An intrinsic semiconductor has a very small conductivity (virtually an insulator at room temperature) and is of little value. This is because of the limited number of free electrons in the conduction band and holes in the valence band. Intrinsic silicon (or germanium) must be modified by increasing the number of free electrons and holes to increase its conductivity and make it useful in electronic devices. This is done by adding suitable impurity (pentavalent or trivalent) to the intrinsic (pure) semiconductor.

Q. 17. *What is the importance of pn junction?*

Ans. Most semiconductor devices contain one or more *pn* junctions. For example, a semiconductor diode has one *pn* junction while a transistor has two *pn* junctions. The *pn* junction is of great importance because it is, in effect, the *control element* for semiconductor devices. A thorough knowledge of the behaviour of *a pn* junction can enable the readers to understand the operation of semiconductor devices.

Q. 18. *Can there be a hole current in the connecting wires of a pn junction circuit?*

Ans. No. In the connecting wires, current is only due to free electrons. Holes can only flow in the semiconductor material.

Multiple-Choice Questions

1. A semiconductor is formed by .... bonds.
   (*i*) covalent (*ii*) electrovalent
   (*iii*) co-ordinate
   (*iv*) none of the above

2. A semiconductor has .... temperature coefficient of resistance.
   (*i*) positive (*ii*) zero
   (*iii*) negative
   (*iv*) none of the above

3. The most commonly used semiconductor is ....
   (*i*) germanium (*ii*) silicon
   (*iii*) carbon (*iv*) sulphur

4. A semiconductor has generally .... valence electrons.
   (*i*) 2 (*ii*) 3
   (*iii*) 6 (*iv*) 4

5. The resistivity of pure germanium under standard conditions is about ....

(*i*) $6 \times 10^4\ \Omega$ cm (*ii*) $60\ \Omega$ cm
(*iii*) $3 \times 10^6\ \Omega$ cm (*iv*) $6 \times 10^{-4}\ \Omega$ cm

6. The resistivity of pure silicon is about ....
(*i*) $100\ \Omega$ cm (*ii*) $6000\ \Omega$ cm
(*iii*) $3 \times 10^5\ \Omega$ cm
(*iv*) $1.6 \times 10^{-8}\ \Omega$ cm

7. When a pure semiconductor is heated, its resistance ....
(*i*) goes up (*ii*) goes down
(*iii*) remains the same (*iv*) cannot say

8. The strength of a semiconductor crystal comes from ....
(*i*) forces between nuclei
(*ii*) forces between protons
(*iii*) electron-pair bonds
(*iv*) none of the above

9. When a pentavalent impurity is added to a pure semiconductor, it becomes ....
(*i*) an insulator
(*ii*) an intrinsic semiconductor
(*iii*) *p*-type semiconductor
(*iv*) *n*-type semiconductor

10. Addition of pentavalent impurity to a semiconductor creates many ....
(*i*) free electrons (*ii*) holes
(*iii*) valence electrons
(*iv*) bound electrons

11. A pentavalent impurity has .... valence electrons.
(*i*) 3 (*ii*) 5
(*iii*) 4 (*iv*) 6

12. An *n*-type semiconductor is ....
(*i*) positively charged
(*ii*) negatively charged
(*iii*) electrically neutral
(*iv*) none of the above

13. A trivalent impurity has .... valence electrons.
(*i*) 4 (*ii*) 5
(*iii*) 6 (*iv*) 3

14. Addition of trivalent impurity to a semiconductor creates many ....
(*i*) holes (*ii*) free electrons
(*iii*) valence electrons
(*iv*) bound electrons

15. A hole in a semiconductor is defined as ....
(*i*) a free electron
(*ii*) the incomplete part of an electron pair bond
(*iii*) a free proton
(*iv*) a free neutron

16. The impurity level in an extrinsic semiconductor is about .... of pure semiconductor.
(*i*) 10 atoms for 10^8 atoms
(*ii*) 1 atom for 10^8 atoms
(*iii*) 1 atom for 10^4 atoms
(*iv*) 1 atom for 100 atoms

17. As the doping to a pure semiconductor increases, the bulk resistance of the semiconductor ....
(*i*) remains the same
(*ii*) increases
(*iii*) decreases
(*iv*) none of the above

18. A hole and electron in close proximity would tend to ....
(*i*) repel each other
(*ii*) attract each other
(*iii*) have no effect on each other
(*iv*) none of the above

19. In a semiconductor, current conduction is due ....
(*i*) only to holes
(*ii*) only to free electrons
(*iii*) to holes and free electrons
(*iv*) none of the above

20. The random motion of holes and free electrons due to thermal agitation is called ....
(*i*) diffusion (*ii*) pressure
(*iii*) ionisation
(*iv*) none of the above

21. A forward biased *pn* junction has a resistance of the ....

(*i*) order of Ω (*ii*) order of kΩ
(*iii*) order of MΩ
(*iv*) none of the above

22. The battery connections required to forward bias a *pn* junction are ....
(*i*) +ve terminal to *p* and –ve terminal to *n*
(*ii*) –ve terminal to *p* and +ve terminal to *n*
(*iii*) –ve terminal to *p* and –ve terminal to *n*
(*iv*) none of the above

23. The barrier voltage at a *pn* junction for germanium is about ....
(*i*) 3.5 V (*ii*) 3 V
(*iii*) zero (*iv*) 0.3 V

24. In the depletion region of a *pn* junction, there is a shortage of ....
(*i*) acceptor ions
(*ii*) holes and electrons
(*iii*) donor ions
(*iv*) none of the above

25. A reverse biased *pn* junction has ....
(*i*) very narrow depletion layer
(*ii*) almost no current
(*iii*) very low resistance
(*iv*) large current flow

26. A *pn* junction acts as a ....
(*i*) controlled switch
(*ii*) bidirectional switch
(*iii*) unidirectional switch
(*iv*) none of the above

27. A reverse biased *pn* junction has resistance of the ....
(*i*) order of Ω (*ii*) order of kΩ
(*iii*) order of MΩ
(*iv*) none of the above

28. The leakage current across a *pn* junction is due to ....
(*i*) minority carriers
(*ii*) majority carriers
(*iii*) junction capacitance
(*iv*) none of the above

29. When the temperature of an extrinsic semiconductor is increased, the pronounced effect is on ....
(*i*) junction capacitance
(*ii*) minority carriers
(*iii*) majority carriers
(*iv*) none of the above

30. With forward bias to a *pn* junction, the width of depletion layer ....
(*i*) decreases
(*ii*) increases
(*iii*) remains the same
(*iv*) none of the above

31. The leakage current in a *pn* junction is of the order of ....
(*i*) A (*ii*) mA
(*iii*) kA (*iv*) μA

32. In an intrinsic semiconductor, the number of free electrons ....
(*i*) equals the number of holes
(*ii*) is greater than the number of holes
(*iii*) is less than the number of holes
(*iv*) none of the above

33. At room temperature, an intrinsic semiconductor has ....
(*i*) many holes only
(*ii*) a few free electrons and holes
(*iii*) many free electrons only
(*iv*) no holes or free electrons

34. At absolute zero temperature, an intrinsic semiconductor has ....
(*i*) a few free electrons
(*ii*) many holes
(*iii*) many free electrons
(*iv*) no holes or free electrons

35. At room temperature, an intrinsic silicon crystal acts approximately as ....
(*i*) a battery
(*ii*) a conductor
(*iii*) an insulator
(*iv*) a piece of copper wire

Answers to Multiple-Choice Questions

1. (*i*)	2. (*iii*)	3. (*ii*)	4. (*iv*)	5. (*ii*)
6. (*ii*)	7. (*ii*)	8. (*iii*)	9. (*iv*)	10. (*i*)
11. (*ii*)	12. (*iii*)	13. (*iv*)	14. (*i*)	15. (*ii*)
16. (*ii*)	17. (*iii*)	18. (*ii*)	19. (*iii*)	20. (*i*)
21. (*i*)	22. (*i*)	23. (*iv*)	24. (*ii*)	25. (*ii*)
26. (*iii*)	27. (*iii*)	28. (*i*)	29. (*ii*)	30. (*i*)
31. (*iv*)	32. (*i*)	33. (*ii*)	34. (*iv*)	35. (*iii*)

Chapter Review Topics

1. What do you understand by a semiconductor? Discuss some important properties of semiconductors.
2. Which are the most commonly used semiconductors and why?
3. Give the energy band description of semiconductors.
4. Discuss the effect of temperature on semiconductors.
5. Give the mechanism of hole current flow in a semiconductor.
6. What do you understand by intrinsic and extrinsic semiconductors?
7. What is a *pn* junction? Explain the formation of potential barrier in a *pn* junction.
8. Discuss the behaviour of a *pn* junction under forward and reverse biasing.
9. Draw and explain the *V-I* characteristics of a *pn* junction.
10. Write short notes on the following :
    (*i*) Breakdown voltage
    (*ii*) Knee voltage
    (*iii*) Limitations in the operating conditions of *pn* junction.

Problems

1. A semiconductor has equal electron and hole concentration of 2×10^8 m^{-3}. On doping with a certain impurity, the hole concentration increases to 4×10^{10} m^{-3}.
   (*i*) What type of semiconductor is obtained on doping?
   (*ii*) Calculate the new electron concentration of the semiconductor.
   (*iii*) How does the energy gap vary with doping? **[(*i*) *p*-type (*ii*) 10^6 m^{-3} (*iii*) decreases]**
2. The number density of electrons and holes in intrinsic silicon at a given temperature is 4.94×10^{10} cm^{-3}. Calculate the resistivity and conductivity of the silicon. Given electron mobility = 1000 cm^{2}V^{-1}s^{-1} and hole mobility = 100 cm^{2}V^{-1}s^{-1}. **[9×10^4 Ω cm; 11.1×10^{-6} Sm^{-1}]**
3. Determine the number density of donor atoms which have to be added to an intrinsic germanium semiconductor to produce an *n*-type semiconductor of conductivity 5Ω^{-1} cm^{-1}. Given that mobility of electrons in *n*-type Ge is 3900 cm^{2}V^{-1}s^{-1}. Neglect the contribution of holes to conductivity. **[8.01×10^{15} cm^{-3}]**
4. A semiconductor has the electron concentration of 8×10^{13} cm^{-3} and hole concentration of 4×10^{12} cm^{-3}. Is the semiconductor *p*-type or *n*-type? Also calculate the resistivity of this semiconductor. Given, electron mobility = 24,000 cm^{2}V^{-1}s^{-1} and hole mobility = 200 cm^2 V^{-1}s^{-1}. **[*n*-type; 3.254×10^{-2} Ωm]**

5. A semiconductor has the electron concentration 4×10^{12} cm^{-3} and hole concentration 7×10^{13} cm^{-3}. Is the semiconductor *n*-type or *p*-type? Also calculate the conductivity of this semiconductor. Given, electron mobility = 22,0000 cm^2 V^{-1} s^{-1} and hole mobility = 150 $cm^2V^{-1}s^{-1}$. **[*p*-type; 1.576 Sm^{-1}]**

6. Determine the number density of donor atoms which have to be added to an intrinsic germanium to produce an *n*-type semiconductor of conductivity 0.06 Sm^{-1}. Given the mobility of electron = 0.39 m^2 V^{-1} s^{-1}. Neglect the contribution of the holes to the conductivity. **[96 × 10^{16} m^{-3}]**

CHAPTER 4

Semiconductor Diode

Learning Objectives

- Develop an understanding of the operation of semiconductor diode (*i.e. pn* junction).
- Use diode current equation to determine the current through a diode.
- Differentiate between d.c. resistance and a.c. resistance of a semiconductor diode.
- Develop the equivalent circuit of a semiconductor diode.
- Understand the use of semiconductor diode / diodes for half-wave rectification and full-wave rectification.
- Understand the importance of peak inverse voltage (*PIV*) of a diode in rectifier service.
- Develop expressions of rectification η for full-wave and half-wave rectification.
- Understand ripple factor in rectifier service.
- Use of filter circuits in rectifier service.
- Zener diode as voltage stabiliser.

Introduction

It has already been discussed in the previous chapter that a *pn* junction conducts current easily when forward biased and practically no current flows when it is reverse biased. This unilateral conduction characteristic of *pn* junction (*i.e.* semiconductor diode) is similar to that of a vacuum diode. Therefore, like a vacuum diode, a semiconductor diode can also accomplish the job of *rectification i.e.* change alternating current to direct current. However, semiconductor diodes have become more *popular as they are smaller in size, cheaper and robust and usually operate with greater efficiency. In this chapter, we shall focus our attention on the circuit performance and applications of semiconductor diodes.

4.1 Semiconductor Diode

A pn junction is known as a **semi-conductor diode** *or* ****crystal diode.**

We have already discussed that a *pn* junction (or a semiconductor diode) conducts current

* On the other hand, vacuum diodes can withstand high reverse voltages and can operate at fairly high temperatures.

** So called because *pn* junction is grown out of a crystal.

easily when forward biased and practically no current flows when it is reverse biased. This unilateral current conduction property of a *pn* junction (or a semiconductor diode) permits it to be used for rectification *i.e.* change alternating current to direct current. Semiconductor diodes have become very popular as rectifiers because they are smaller in size, cheaper and robust and usually operate with greater efficiency. A semiconductor or a crystal diode is usually represented by the schematic symbol shown in Fig. 4.1. The anode of diode is of *p*-type material and is represented by an *arrowhead*. The cathode of diode is of *n*-type material and is represented by a thick vertical line called *bar*.

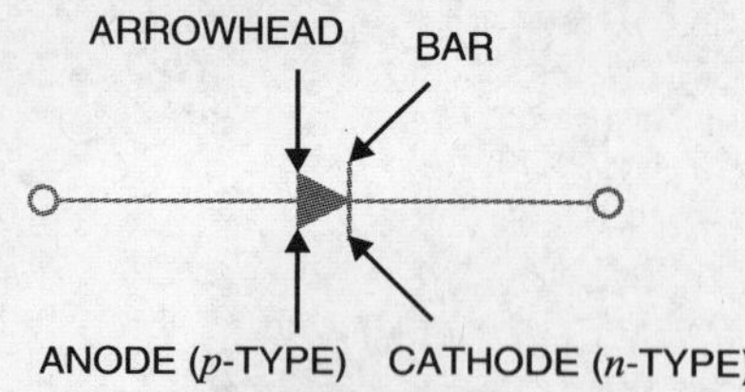

SYMBOL OF SEMICONDUCTOR DIODE

Fig. 4.1

A crystal diode has two terminals. When it is connected in a circuit, one thing to decide is whether the diode is forward or reverse biased. There is an easy rule to ascertain it. If the external circuit is trying to push the conventional current in the direction of arrowhead, the diode is forward biased. On the other hand, if the conventional current is trying to flow opposite to arrowhead, the diode is reverse biased. Putting in simple words :

(*i*) If *arrowhead* of diode symbol is *positive w.r.t. bar* of the symbol, the diode is forward biased [See Fig. 4.2 (*i*)].

(*ii*) If the *arrowhead* of diode symbol is *negative w.r.t. bar*, the diode is reverse biased [See Fig. 4.2 (*ii*)].

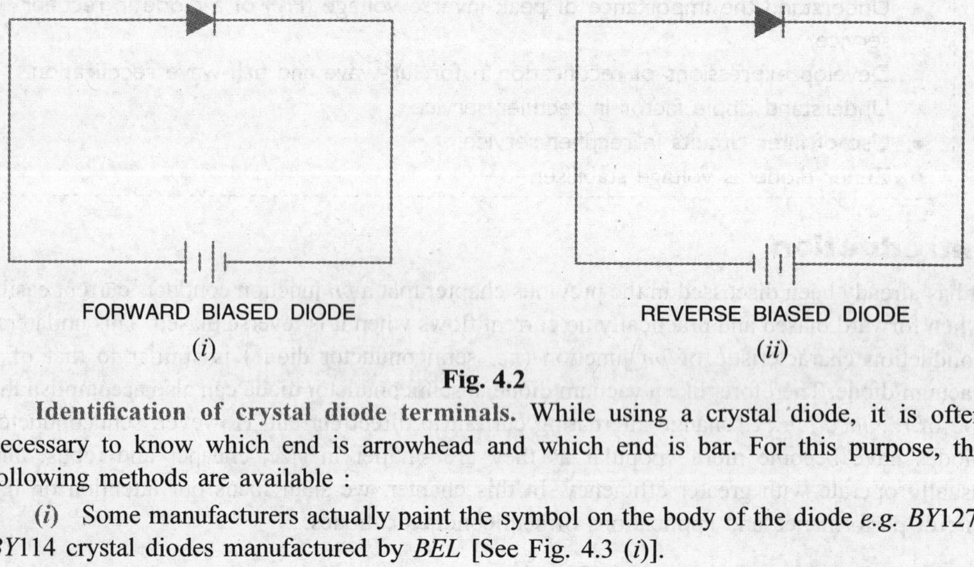

Fig. 4.2

Identification of crystal diode terminals. While using a crystal diode, it is often necessary to know which end is arrowhead and which end is bar. For this purpose, the following methods are available :

(*i*) Some manufacturers actually paint the symbol on the body of the diode *e.g.* *BY*127, *BY*114 crystal diodes manufactured by *BEL* [See Fig. 4.3 (*i*)].

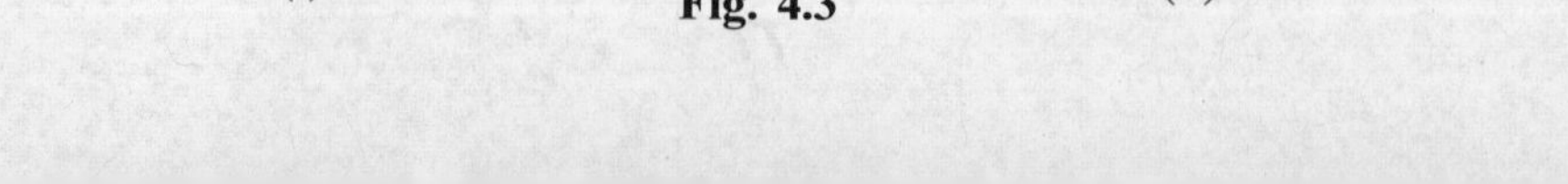

Fig. 4.3

(*ii*) Sometimes, red and blue marks are used on the body of the crystal diode. Red mark denotes arrowhead whereas blue mark indicates bar *e.g. OA*80 crystal diode [See Fig. 4.3 (*ii*)].

4.2 Semiconductor Diode as a Rectifier

Fig. 4.4 illustrates the rectifying action of a semiconductor diode. The a.c. input voltage to be rectified, the diode and load R_L are connected in series. The d.c. output is obtained across the load as explained in the following discussion. During the positive half-cycle of a.c. input voltage, the arrowhead becomes positive *w.r.t.* bar. Therefore, diode is forward biased and conducts current in the circuit. The result is that positive half-cycle of input voltage appears across R_L as shown. However, during the negative half-cycle of input a.c. voltage, the diode becomes reverse biased because now the arrowhead is negative *w.r.t.* bar. Therefore, diode does not conduct and no voltage appears across load R_L. The result is that output consists of positive half-cycles of input a.c. voltage while the negative half-cycles are suppressed. In this way, semiconductor diode has been able to do rectification *i.e.* change a.c. into d.c. It may be seen that output across R_L is pulsating d.c.

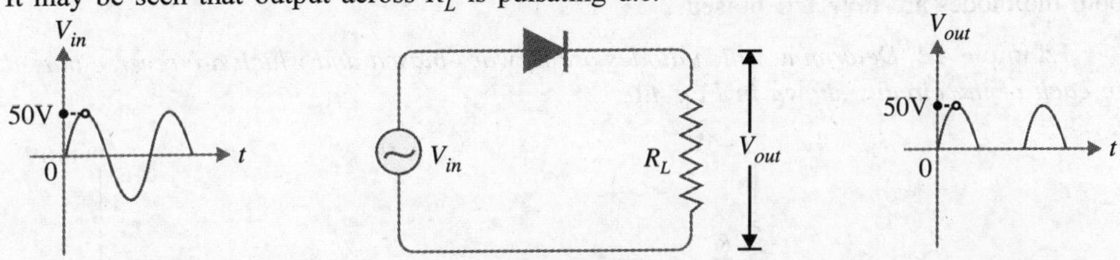

Fig. 4.4

It is interesting to see that behaviour of diode is like a *switch*. When the diode is forward biased, it behaves like a closed switch and connects the a.c. supply to the load R_L. However, when the diode is reverse biased, it behaves like an open switch and disconnects the a.c. supply from the load R_L. This switching action of diode permits only the positive half-cycles of input a.c. voltage to appear across R_L.

Example 4.1. *In each diode circuit of Fig. 4.5, find whether the diodes are forward or reverse biased.*

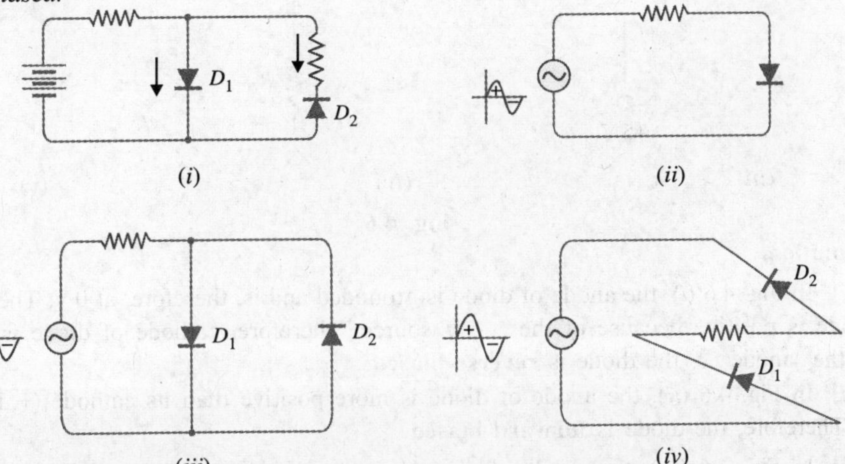

Fig. 4.5

Solution.

(*i*) Refer to Fig. 4.5 (*i*). The conventional current coming out of battery flows in the branch circuits. In diode D_1, the conventional current flows in the direction of arrowhead and hence this diode is **forward biased**. However, in diode D_2, the conventional current flows opposite to arrowhead and hence this diode is **reverse biased**.

(*ii*) Refer to Fig. 4.5 (*ii*). During the positive half-cycle of input a.c. voltage, the conventional current flows in the direction of arrowhead and hence diode is **forward biased**. However, during the negative half-cycle of input a.c. voltage, the diode is **reverse biased**.

(*iii*) Refer to Fig. 4.5 (*iii*). During the positive half-cycle of input a.c. voltage, conventional current flows in the direction of arrowhead in D_1 but it flows opposite to arrowhead in D_2. Therefore, during positive half-cycle, diode D_1 is **forward biased** and diode D_2 is **reverse biased**. However, during the negative half-cycle of input a.c. voltage, diode D_2 is **forward biased** and D_1 is **reverse biased**.

(*iv*) Refer to Fig. 4.5 (*iv*). During the positive half-cycle of input a.c. voltage, both the diodes are **reverse biased**. However, during the negative half-cycle of input a.c. voltage, both the diodes are **forward biased**.

Example 4.2. *Determine which diodes are forward biased and which are reverse biased in each of the circuits shown in Fig. 4.6.*

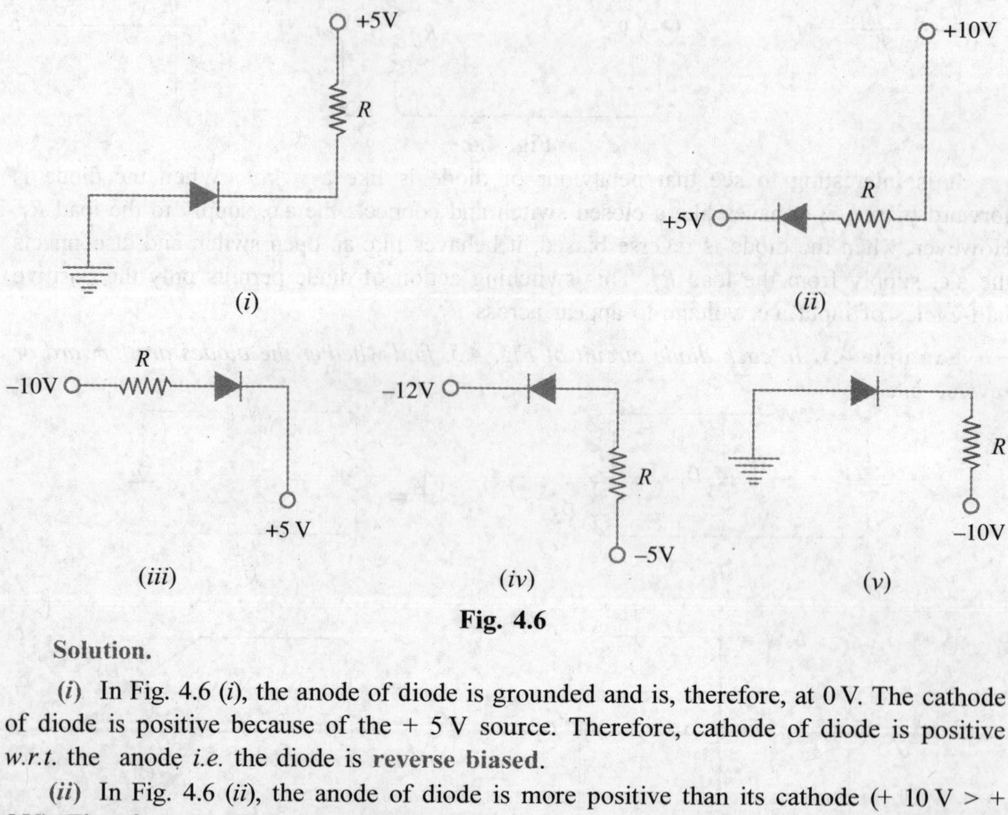

Fig. 4.6

Solution.

(*i*) In Fig. 4.6 (*i*), the anode of diode is grounded and is, therefore, at 0 V. The cathode of diode is positive because of the + 5 V source. Therefore, cathode of diode is positive *w.r.t.* the anode *i.e.* the diode is **reverse biased**.

(*ii*) In Fig. 4.6 (*ii*), the anode of diode is more positive than its cathode (+ 10 V > + 5 V). Therefore, the diode is **forward biased**.

(*iii*) In Fig. 4.6 (*iii*), the anode of diode is more negative than the cathode (– 10 V <

+ 5 V), so the diode is **reverse biased**. Note that no current flows through the circuit. The total reverse voltage across the diode = + 5 V – (– 10 V) = 15 V.

(*iv*) In Fig. 4.6 (*iv*), the cathode of diode is more negative than the anode (– 12 V < – 5 V). Therefore, the diode is **forward biased**.

(*v*) In Fig. 4.6. (*v*), the anode of diode is grounded and is, therefore, at 0 V. The cathode of diode is more negative than the anode (– 10 V < 0 V), so the diode is **forward biased**.

4.3 Diode Current Equation

It can be proved mathematically that current through a semiconductor diode (forward biased or reverse biased) is given by the following formula :

$$\text{Current through diode, } {}^{*}I_D = I_S\left(e^{V_D/\eta V_T}-1\right)\text{A} \qquad \text{... }(i)$$

where, I_D = Forward or reverse diode current in A

I_S = Saturation current for the diode in A

V_D = Voltage in volts applied across the diode. It is positive for forward bias and negative for reverse bias.

** η = Constant. If not given, it is assumed to be 1.

$$V_T = \text{Thermal voltage} = \frac{kT}{q} = \frac{1.38\times10^{-23}\times T}{1.6\times10^{-19}} = \frac{T}{11,600}\text{ volts}$$

where, k = Boltzmann's constant = 1.38×10^{-23} J/K

T = Temperature in kelvin (K)

q = Charge on electron = 1.6×10^{-19} C

Eq. (*i*) is called *universal diode current equation* and can be used to explain the change in the value of diode current I_D due to (*i*) change in both positive and negative voltages applied to the diode (*ii*) change in temperature.

Temperature Effects. The diode current I_D (forward current or reverse current) depends upon temperature because both V_T(= T/11,600 volts) and saturation current I_S are temperature dependent. In fact, the value of I_S is more sensitive to temperature change than is V_T. Therefore, when temperature changes, the change in I_S has more pronounced effect on the diode current I_D. *It has been found experimentally that saturation current I_S doubles for every 10°C increase in temperature.* From this fact, the change in I_S over a given temperature range can be calculated easily. The temperature change (ΔT) is divided by 10 to give the number of 10°C changes (n). If the starting saturation current is represented as $I_S(old)$, then the value of new saturation current is

$$I_S\,(new) = I_S(old) \times 2^n$$

Thus, if temperature increases from 20°C to 80°C, then ΔT = 80°C – 20°C = 60°C and the value of $n = \Delta T/10°C$ = 60°C/10°C = 6.

---

* The derivation of this formula is quite involved.

** η is called *emission coefficient*. Its value varies between 1 and 2, depending upon the level of diode current. It is generally taken as 1 if not given.

Example 4.3. *A silicon diode has a saturation current of 0.1 pA at 20°C. Find current through the diode when it is forward biased by 0.55V. What will be the diode current when the temperature increases to 100°C? Assume emission coefficient η = 1.*

Solution. The forward current I_D through the diode is given by :

$$I_D = I_S(e^{V_D/\eta V_T} - 1)\text{A}$$

Here, $I_S = 0.1 \text{ pA} = 10^{-13} \text{ A}$; $V_D = 0.55 \text{ V}$; $T = 273 + 20 = 293 \text{ K}$; $\eta = 1$;

$$V_T = \frac{T}{11,600} = \frac{293}{11,600} = 0.02526 \text{ V}$$

$$\therefore \quad \text{Diode current, } I_D = 10^{-13}\left(e^{0.55/0.02526} - 1\right)\text{A} = 0.283 \times 10^{-3} \text{ A} = \mathbf{0.283 \text{ mA}}$$

At t = 100°C.

$$V_T = \frac{T}{11,600} = \frac{273+100}{11,600} \text{ volt} = 0.03216 \text{ volt}$$

Temperature increase, $\Delta T = 100°\text{C} - 20°\text{C} = 80°\text{C}$

Number of 10°C changes, $n = \Delta T/10°\text{C} = 80°\text{C}/10°\text{C} = 8$

$$\therefore \quad I_S(new) = I_S(old) \times 2^n = 10^{-13} \times 2^8 \text{ A} = 256 \times 10^{-13} \text{ A}$$

$$\therefore \quad \text{Diode current, } I_D = 256 \times 10^{-13}\,(e^{0.55/0.03216} - 1) \text{ A} = 0.681 \times 10^{-3}\text{A}$$

$$= \mathbf{0.681 \text{ mA}}$$

Note that as the temperature increases from 20°C to 100°C, the diode current I_D increases from 0.283 mA to 0.681 mA – an increase of 140%. This is a dramatic change.

Example 4.4. *A silicon diode has a saturation current of 0.01pA at room temperature. Find current through the diode when it is reverse biased by 0.1 V. Given that emission coefficient η = 1 and V_T = 0.05 V.*

Solution. The diode current I_D is given by ;

$$I_D = I_S\left(e^{V_D/\eta V_T} - 1\right)\text{A}$$

Here, $I_S = 0.01 \text{ pA} = 10^{-14} \text{ A}$; $V_D = -0.1 \text{ V}$; $\eta = 1$; $V_T = 0.05 \text{ V}$

$$\therefore \quad I_D = 10^{-14}\,(e^{-0.1/0.05} - 1)\text{A} = \mathbf{-0.8647 \times 10^{-14} \text{ A}}$$

We find that the value of diode current I_D is negative. It means that I_D flows in a direction opposite to that of the forward current. We call it reverse current.

Note. It is important to note that when the value of V_D is *small and negative* (*i.e.* the diode is reverse biased by a small voltage), the magnitude of the term $e^{V_D/\eta V_T}$ is very small. For instance, if in the above example, $V_D = -0.5$ V, then, $e^{V_D/\eta V_T} = 4.5 \times 10^{-5}$. This value is negligibly small compared to 1 and may be considered as zero.

$$\therefore \quad I_D = I_S(0 - 1) = -I_S$$

Note that this is the maximum (saturation) value of diode current in the reverse direction. For this reason, I_S is called *saturation current* or more precisely *reverse saturation current.*

Example 4.5. *When a forward voltage of 0.4 V is applied to a silicon diode, the forward current is 10 mA. Find the saturation current if the temperature is 293 K. Assume that emission coefficient η = 2.*

Solution. The diode current I_D is given by ;

$$I_D = I_S(e^{V_D/\eta V_T} - 1)\ \text{A}$$

Here, $I_D = 10\ \text{mA} = 10^{-2}\ \text{A}$; $V_D = 0.4\ \text{V}$; $\eta = 2$; $V_T = \frac{T}{11,600} = \frac{293}{11,600} = 0.02526\ \text{V}$

$$\therefore \quad 10^{-2} = I_S\,(e^{0.4/2\,\times\,0.02526} - 1)\ \text{A}$$

or Saturation current, $I_S = 3.64 \times 10^{-6}\ \text{A} = \mathbf{3.64\ \mu A}$

4.4 Resistance of Crystal Diode

It has already been discussed that a forward biased diode conducts easily whereas a reverse biased diode practically conducts no current. It means that *forward resistance* of a diode is quite small as compared with its *reverse resistance*.

1. Forward resistance. The resistance offered by the diode to forward bias is known as *forward resistance*. This resistance is not the same for the flow of direct current as for the changing current. Accordingly; this resistance is of two types, namely; *d.c. forward resistance* and *a.c. forward resistance*.

(i) d.c. forward resistance. It is the opposition offered by the diode to the direct current. It is measured by the ratio of d.c. voltage across the diode to the resulting d.c. current through it. Thus, referring to the forward characteristic in Fig. 4.7, it is clear that when forward voltage is *OA*, the forward current is *OB*.

$$\therefore \quad \text{d.c. forward resistance, } R_f = \frac{OA}{OB}$$

(ii) a.c. forward resistance. It is the opposition offered by the diode to the changing forward current. It is measured by the ratio of change in voltage across diode to the resulting change in current through it *i.e.*

$$\text{a.c. forward resistance, } r_f = \frac{\text{Change in voltage across diode}}{\text{Corresponding change in current through diode}}$$

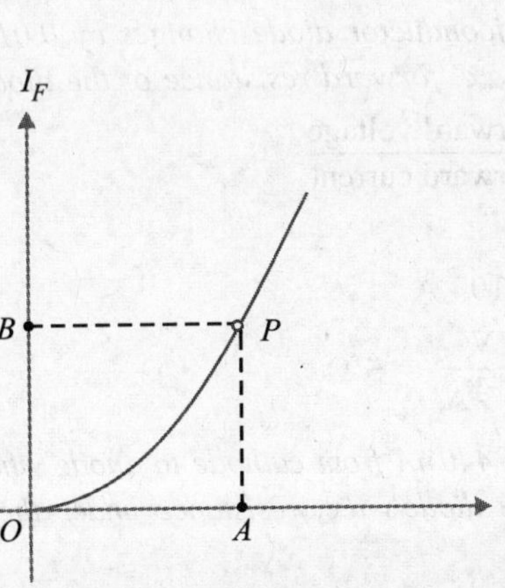

Fig. 4.7

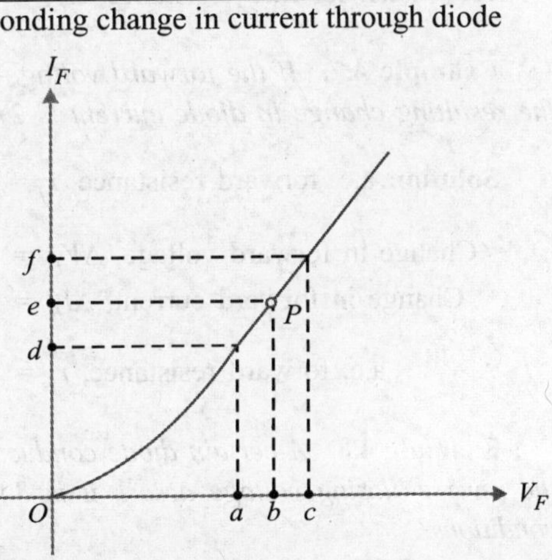

Fig. 4.8

The a.c. forward resistance is more significant as the diodes are generally used with alternating voltages. The a.c. forward resistance can be determined from the forward characteristic as shown in Fig. 4.8. If P is the operating point at any instant, then forward voltage is ob and forward current is oe. To find the a.c. forward resistance, vary the forward voltage on both sides of the operating point equally as shown in Fig. 4.8 where $ab = bc$. It is clear from this figure that :

For forward voltage oa, circuit current is od.

For forward voltage oc, circuit current is of.

$$\therefore \text{ a.c. forward resistance, } r_f = \frac{\text{Change in forward voltage}}{\text{Change in forward current}} = \frac{oc - oa}{of - od} = \frac{ac}{df}$$

It may be mentioned here that forward resistance of a crystal diode is very small, ranging from 1 to 25 Ω.

2. Reverse resistance. The resistance offered by the diode to the reverse bias is known as *reverse resistance*. It can be d.c. reverse resistance or a.c. reverse resistance depending upon whether the reverse bias is direct or changing voltage. Ideally, the reverse resistance of a diode is infinite. However, in practice, the reverse resistance is not infinite because for any value of reverse bias, there does exist a small leakage current. It may be emphasised here that reverse resistance is very large compared to the forward resistance. In germanium diodes, the ratio of reverse to forward resistance is 40000 : 1 while for silicon this ratio is 1000000 : 1.

Example 4.6. *The forward d.c. voltage across a semiconductor diode is 1 V and the resulting direct current is 9.5 mA. What is the d.c. forward resistance of the diode?*

Solution. d.c. forward resistance is

$$R_f = \frac{\text{Forward d.c. voltage}}{\text{Forward direct current}}$$

Forward d.c. voltage, $V_F = 1\text{ V}$

Forward direct current, $I_F = 9.5\text{ mA} = 9.5 \times 10^{-3}\text{ A}$

$$\therefore \quad \text{d.c. forward resistance, } R_f = \frac{V_F}{I_F} = \frac{1\text{ V}}{9.5\times10^{-3}\text{ A}} = \mathbf{105.3\ \Omega}$$

Example 4.7. *If the forward voltage across a semiconductor diode changes by 0.01 V, the resulting change in diode current is 2 mA. Find the a.c. forward resistance of the diode.*

Solution. a.c. forward resistance, $r_f = \dfrac{\text{Change in forward voltage}}{\text{Change in forward current}}$

Change in forward voltage, $\Delta V_F = 0.01\text{ V}$

Change in forward current, $\Delta I_F = 2\text{ mA} = 2 \times 10^{-3}\text{ A}$

$$\therefore \quad \text{a.c. forward resistance, } r_f = \frac{\Delta V_F}{\Delta I_F} = \frac{0.01\text{ V}}{2\times10^{-3}\text{ A}} = 5\ \Omega$$

Example 4.8. *A certain diode conducts a current of 440 nA from cathode to anode when the reverse-biasing voltage across it is 8 V. What is the diode's d.c. resistance under these conditions?*

Solution. We are to find d.c. reverse resistance of the diode.

Reverse d.c. voltage, $V_R = 8\text{ V}$

Reverse direct current, $I_R = 440 \text{ nA} = 440 \times 10^{-9}$ A

$$\therefore \quad \text{d.c. reverse resistance, } R_r = \frac{V_R}{I_R} = \frac{8\text{V}}{440 \times 10^{-9}\text{ A}} = 18.18 \times 10^6 \ \Omega$$

$$= \mathbf{18.18 \ M\Omega}$$

Note that reverse resistance (d.c. as well as a.c.) of a reverse biased diode is extremely large. For this reason, a reverse biased diode practically conducts no current and may be considered as an 'open'.

Example 4.9. *In the circuit shown in Fig. 4.9, the current I is 51.63 mA. What is the voltage drop across the diode? What is the d.c. resistance of the diode?*

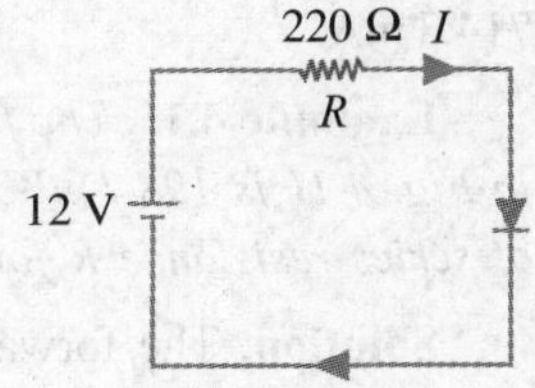

Fig. 4.9

Solution. The diode is forward biased and the circuit current $I = 51.63$ mA (given).

Voltage across 220 Ω resistor $= IR = 51.63 \text{ mA} \times 220 \ \Omega = 11.36$ V

$\therefore$ Voltage drop across diode, $V_D = 12 - 11.36 = \mathbf{0.64 \ V}$

$$\text{d.c. resistance (forward), } R_f = \frac{V_D}{I} = \frac{0.64\text{ V}}{51.63\text{ mA}} = \mathbf{12.39 \ \Omega}$$

Note that forward resistance (d.c. as well as a.c.) of a diode is very small as compared to its reverse resistance.

4.5 Forward Power Rating of a Diode

Like a resistor, a semiconductor diode has a forward power rating. This rating tells us how much power the forward biased diode can safely dissipate without damaging itself.

The maximum power that a forward biased diode can dissipate as heat without damaging itself is called forward power rating of the diode.

Forward power rating, $P_D = V_F I_F (max)$

where, V_F = Forward voltage drop across diode

$I_F (max)$ = Max. permissible forward current

For a silicon diode, $V_F = 0.7$ V and for a germanium diode, $V_F = 0.3$ V.

Let us illustrate forward power rating of a semiconductor diode with a numerical example. The circuit shown in Fig. 4.10 uses a silicon diode. We are to find the forward power rating of the diode for safe operation. The forward current I_F in the diode is

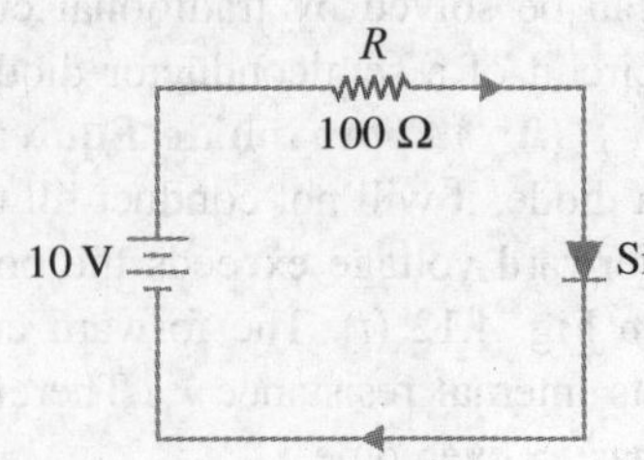

Fig. 4.10

$$I_F = \frac{V - V_F}{R} = \frac{(10 - 0.7)\text{ V}}{100\Omega} = 93 \text{ mA}$$

Power dissipation P_D of the diode is

$$P_D = V_F I_F = 0.7 \text{ V} \times 93 \text{ mA} \simeq 65 \text{ mW}$$

Therefore, the diode used in the circuit in Fig. 4.10 must have forward power rating greater than 65 mW. If the forward power rating of the diode is less than 65 mW, the diode may be burnt due to excessive heat.

Note. Suppose the forward diode current (I_F) becomes such that heat produced becomes 70 mW. The diode is able to dissipate 65 mW as heat and the remaining 5 mW will start building up the temperature of the diode. In a matter of seconds, the diode will be burnt due to excessive heat.

Example 4.10. *A silicon crystal diode has a forward power rating of 500 mW. What is the maximum allowable value of forward current for the device?*

Solution. The forward power rating P_D of the diode is

$$P_D = V_F I_F\,(max)$$

Here, $$P_D = 500\text{ mW}\ ;\ V_F = 0.7\text{ V}\ ;\ I_F\,(max) = ?$$

$$\therefore \quad I_F\,(max) = \frac{P_D}{V_F} = \frac{500\text{ mW}}{0.7\text{ V}} = \mathbf{714.3\ mA}$$

Therefore, for safe operation of the diode, the forward diode current should not exceed 714.3 mA.

Example 4.11. *The forward power rating of silicon diode in Fig. 4.11 is 198.7 mW. What should be the minimum value of series resistance R for the safe operation of the device?*

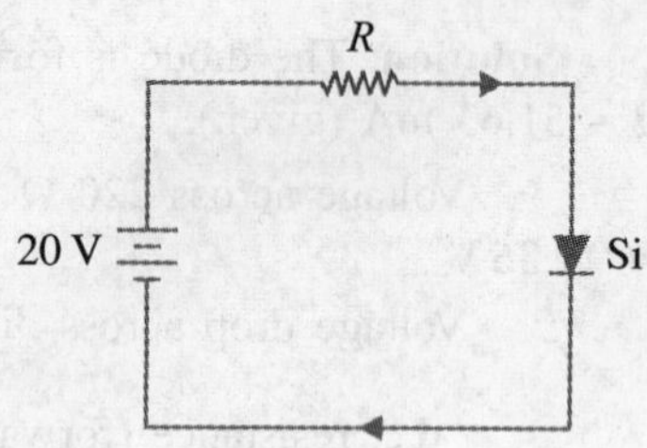

Fig. 4.11

Solution. The forward power rating P_D of the diode is

$$P_D = V_F\, I_F\,(max)$$

$$\therefore \quad I_F(max) = \frac{P_D}{V_F} = \frac{198.7\text{ mW}}{0.7\text{ V}} = 0.284\text{ A}$$

Now, $$I_F(max) = \frac{12400}{5890}\text{eV}$$

$$\therefore \quad R = \frac{V - V_F}{I_F\,(max)} = \frac{20\text{ V} - 0.7\text{ V}}{0.284\text{ A}} = \mathbf{68\ \Omega}$$

Note that the purpose of series resistance R is to limit the forward current of diode to a safe value.

4.6 Equivalent Circuit of Semiconductor Diode

It is generally profitable to replace a device or system by its equivalent circuit. An equivalent circuit of a device (*e.g.* semiconductor diode, transistor *etc.*) is a combination of electric elements, which when connected in a circuit, acts exactly as does the device when connected in the same circuit. Once the device is replaced by its equivalent circuit, the resulting network can be solved by traditional circuit analysis techniques. We shall now find the equivalent circuit of a semiconductor diode.

(*i*) ***Approximate Equivalent circuit.** When the forward voltage V_F is applied across a diode, it will not conduct till the potential barrier V_0 at the junction is overcome. When the forward voltage exceeds the potential barrier voltage, the diode starts conducting as shown in Fig. 4.12 (*i*). The forward current I_f flowing through the diode causes a voltage drop in its internal resistance r_f. Therefore, the forward voltage V_F applied across the *actual* diode has to overcome :

(*a*) potential barrier V_0

(*b*) internal drop $I_f r_f$

$$\therefore \quad V_F = V_0 + I_f\, r_f$$

For a silicon diode, V_0 = 0.7 V whereas for a germanium diode, V_0 = 0.3 V.

Therefore, approximate equivalent circuit for a semiconductor diode is a switch in series

* We assume here that *V*/*I* characteristic of semiconductor diode is linear.

with a battery V_0 and internal resistance r_f as shown in Fig. 4.12 (*ii*). This approximate equivalent circuit of a diode is very helpful in studying the performance of the diode in a circuit.

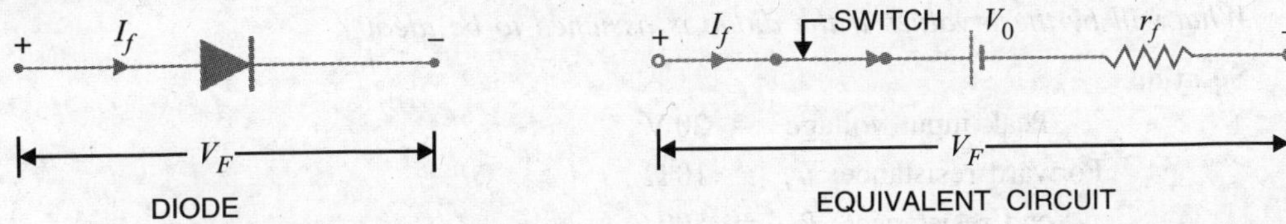

(*i*) (*ii*)

Fig. 4.12

(*ii*) Simplified Equivalent circuit. For most applications, the internal resistance r_f of the semiconductor diode can be ignored in comparison to other elements in the equivalent circuit. The equivalent circuit then reduces to the one shown in Fig. 4.13 (*ii*). This simplified equivalent circuit of the semiconductor diode is frequently used in diode-circuit analysis.

(*i*) (*ii*)

Fig. 4.13

(*iii*) Ideal diode model. An ideal diode is one which behaves as a perfect conductor when forward biased and as a perfect insulator when reverse biased. Obviously, in such a hypothetical situation, forward resistance $r_f = 0$ and potential barrier V_0 is considered negligible. It may be mentioned here that although ideal diode is never found in practice, yet diode circuit analysis is made on this basis. *Therefore, while discussing diode circuits, the diode will be assumed ideal unless and until stated otherwise.*

4.7 Semiconductor Diode Equivalent Circuits

It is desirable to sum up the various models of semiconductor diode equivalent circuit in the tabular form given below :

S.No.	Type	Model	Characteristic
1.	Approximate model	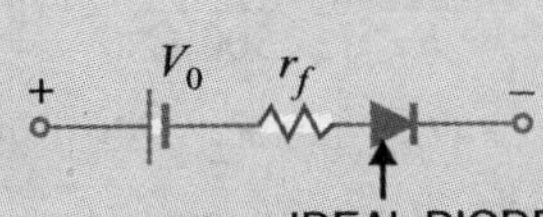	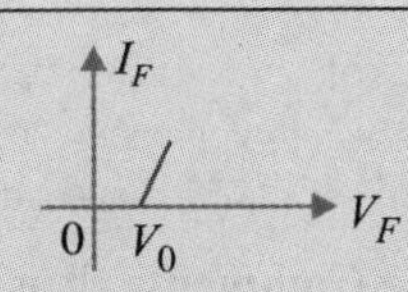
2.	Simplified model	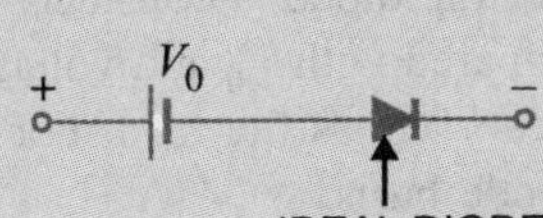	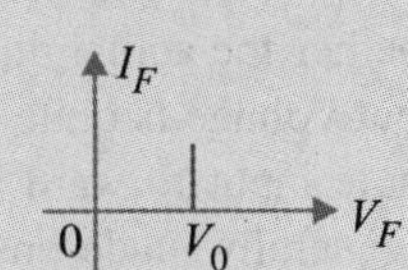
3.	Ideal model		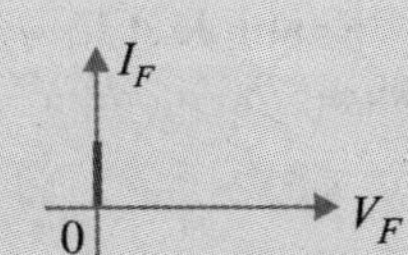

Example 4.12. *An a.c. voltage of peak value 20 V is connected in series with a silicon diode and load resistance of 500 Ω. If the forward resistance of diode is 10 Ω, find :*

(i) peak current through diode *(ii) peak output voltage*

What will be these values if the diode is assumed to be ideal ?

Solution.

$$\text{Peak input voltage} = 20\text{ V}$$
$$\text{Forward resistance, } r_f = 10\ \Omega$$
$$\text{Load resistance, } R_L = 500\ \Omega$$
$$\text{Potential barrier voltage, } V_0 = 0.7\text{ V}$$

The diode will conduct during the positive half-cycles of a.c. input voltage only. The equivalent circuit is shown in Fig. 4.14 (*ii*).

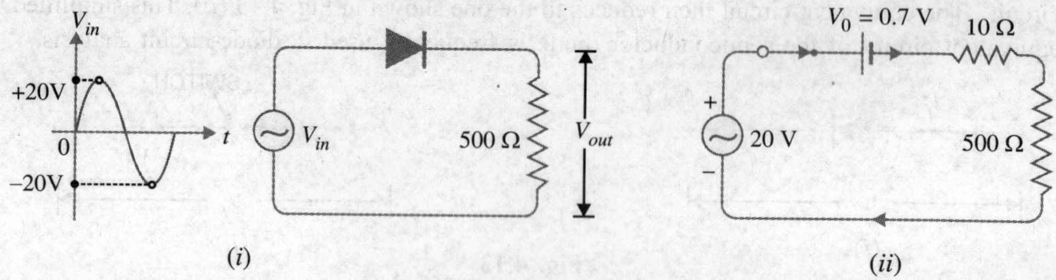

Fig. 4.14

(*i*) The peak current through the diode will occur at the instant when the input voltage reaches positive peak *i.e.* $V_{in} = V_F = 20$ V.

$$\therefore \quad V_F = V_0 + (I_f)_{peak}\,[r_f + R_L] \qquad \text{...(i)}$$

$$\text{or} \quad (I_f)_{peak} = \frac{V_F - V_0}{r_f + R_L} = \frac{20 - 0.7}{10 + 500} = \frac{19.3}{510}\text{ A} = \mathbf{37.8\ mA}$$

(*ii*) Peak output voltage = $(I_f)_{peak} \times R_L$ = 37.8 mA × 500 Ω = **18.9 V**

Ideal diode. For an ideal diode, put $V_0 = 0$ and $r_f = 0$ in equation (*i*).

$$\therefore \quad V_F = (I_f)_{peak} \times R_L$$

$$\text{or} \quad (I_f)_{peak} = \frac{V_F}{R_L} = \frac{20\text{ V}}{500\ \Omega} = \mathbf{40\ mA}$$

Peak output voltage = $(I_f)_{peak} \times R_L$ = 40 mA × 500 Ω = **20 V**

Comments. It is clear from the above example that output voltage is *nearly* the same whether the actual diode is used or the diode is considered ideal. This is due to the fact that input voltage is quite large as compared with V_0 and voltage drop in r_f. Therefore, nearly the whole input forward voltage appears across the load. For this reason, diode circuit analysis is generally made on the ideal diode basis.

Example 4.13. *Find the current through the diode in the circuit shown in Fig. 4.15 (i). Assume the diode to be ideal.*

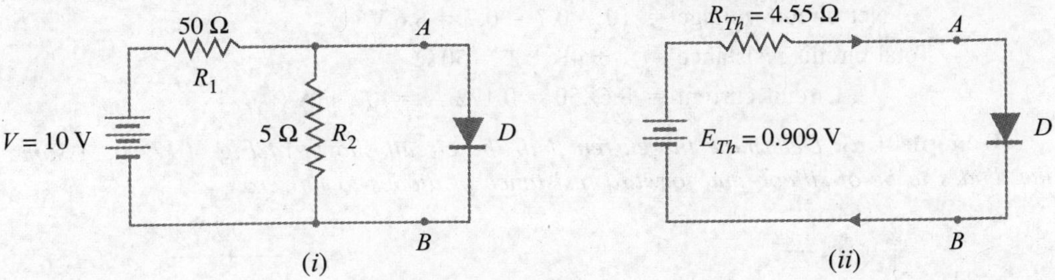

Fig. 4.15

Solution. We shall use Thevenin's theorem to find current in the diode. Referring to Fig. 4.15(*i*),

$$E_{Th} = \text{Thevenin's voltage}$$
$$= \text{Open-circuited voltage across } AB \text{ with diode removed}$$
$$= \frac{R_2}{R_1 + R_2} \times V = \frac{5}{50+5} \times 10 = 0.909 \text{ V}$$

$$R_{Th} = \text{Thevenin's resistance}$$
= Resistance at terminals *AB* with diode removed and battery replaced by a short-circuit

$$= \frac{R_1 R_2}{R_1 + R_2} = \frac{50 \times 5}{50 + 5} = 4.55 \ \Omega$$

Fig. 4.15 (*ii*) shows Thevenin's equivalent circuit. Since the diode is ideal, it has zero resistance.

$$\therefore \quad \text{Current through diode} = \frac{E_{Th}}{R_{Th}} = \frac{0.909}{4.55} = 0.2 \text{ A} = \mathbf{200 \ mA}$$

Example 4.14. *Calculate the current through 48 Ω resistor in the circuit shown in Fig. 4.16 (i). Assume the diodes to be of silicon and forward resistance of each diode is 1 Ω.*

Solution. Diodes D_1 and D_3 are forward biased while diodes D_2 and D_4 are reverse biased. We can, therefore, consider the branches containing diodes D_2 and D_4 as "open". Replacing diodes D_1 and D_3 by their equivalent circuits and making the branches containing diodes D_2 and D_4 open, we get the circuit shown in Fig. 4.16 (*ii*). Note that for a silicon diode, the barrier voltage is 0.7 V.

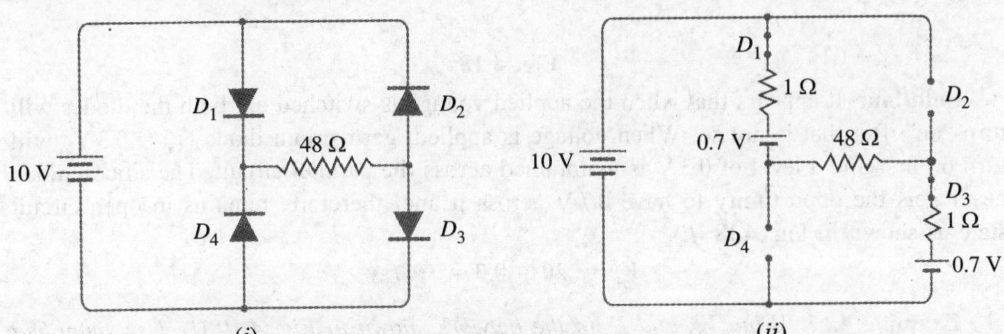

Fig. 4.16

Net circuit voltage = 10 – 0.7 – 0.7 = 8.6 V
Total circuit resistance = 1 + 48 + 1 = 50 Ω
∴ Circuit current = 8.6/50 = 0.172 A = **172 mA**

Example 4.15. *Determine the current I in the circuit shown in Fig. 4.17 (i). Assume the diodes to be of silicon and forward resistance of diodes to be zero.*

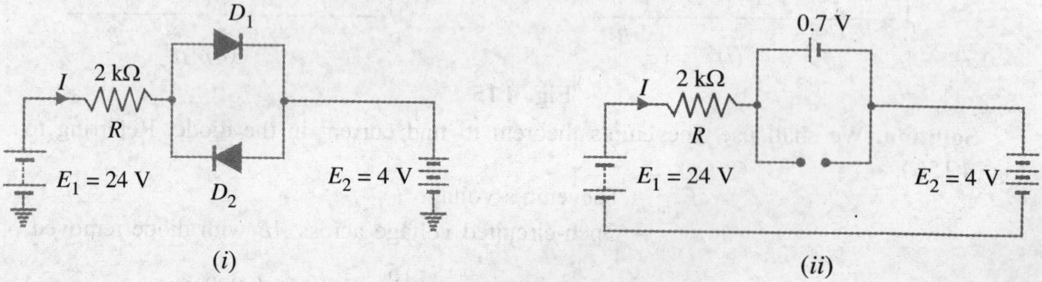

Fig. 4.17

Solution. The conditions of the problem suggest that diode D_1 is forward biased and diode D_2 is reverse biased. We can, therefore, consider the branch containing diode D_2 as open as shown in Fig. 4.17 (*ii*). Further, diode D_1 can be replaced by its simplified equivalent circuit.

$$\therefore \quad I = \frac{E_1 - E_2 - V_0}{R} = \frac{24 - 4 - 0.7}{2\text{ k}\Omega} = \frac{19.3\text{ V}}{2\text{ k}\Omega} = \mathbf{9.65\text{ mA}}$$

Example 4.16. *Find the voltage V_A in the circuit shown in Fig. 4.18 (i). Use simplified model.*

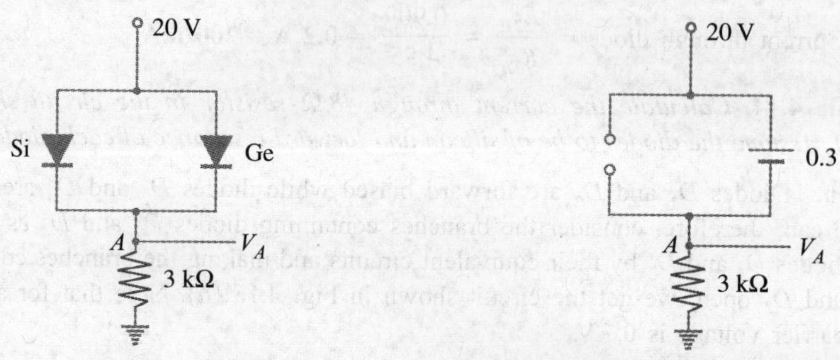

Fig. 4.18

Solution. It appears that when the applied voltage is switched on, both the diodes will turn "on". But that is not so. When voltage is applied, germanium diode (V_0 = 0.3 V) will turn on first and a level of 0.3 V is maintained across the parallel circuit. The silicon diode never gets the opportunity to have 0.7 V across it and, therefore, remains in open-circuit state as shown in Fig. 4.18 (*ii*).

$$\therefore \quad V_A = 20 - 0.3 = \mathbf{19.7\text{ V}}$$

Example 4.17. *Find V_Q and I_D in the network shown in Fig. 4.19 (i). Use simplified model.*

Solution. Replace the diodes by their simplified models. The resulting circuit will be as shown in Fig. 4.19 (*ii*). By symmetry, current in each branch is I_D so that current in branch *CD* is $2I_D$. Applying Kirchhoff's voltage law to the closed circuit *ABCDA*, we have,

$$-0.7 - I_D \times 2 - 2\,I_D \times 2 + 10 = 0 \qquad (I_D \text{ in mA})$$

or $$6\,I_D = 9.3$$

$$\therefore \quad I_D = \frac{9.3}{6} = \mathbf{1.55\ mA}$$

Also, $$V_Q = (2I_D) \times 2\ \text{k}\Omega = (2 \times 1.55\ \text{mA}) \times 2\ \text{k}\Omega = \mathbf{6.2\ V}$$

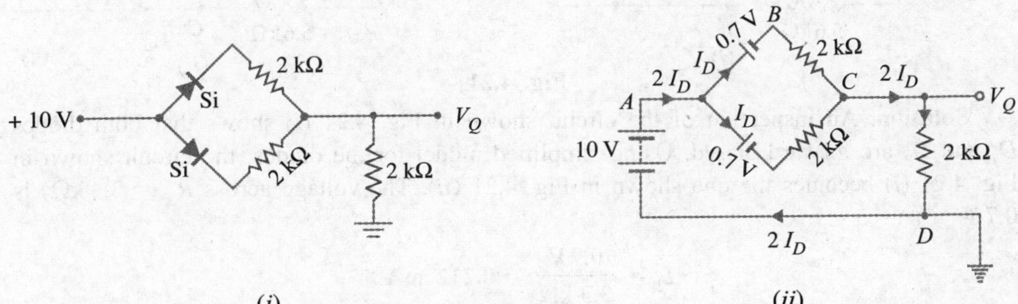

Fig. 4.19

Example 4.18. *Determine current through each diode in the circuit shown in Fig. 4.20 (i) Use simplified model. Assume diodes to be similar.*

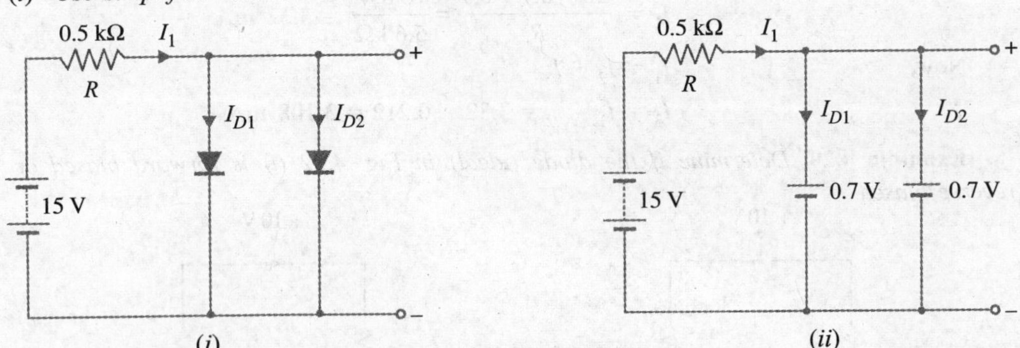

Fig. 4.20

Solution. The applied voltage forward biases each diode so that they conduct current in the same direction. Fig. 4.20 (*ii*) shows the equivalent circuit using simplified model. Referring to Fig. 4.20 (*ii*),

$$I_1 = \frac{\text{Voltage across } R}{R} = \frac{15 - 0.7}{0.5\ \text{k}\Omega} = 28.6\ \text{mA}$$

Since the diodes are similar, $$I_{D1} = I_{D2} = \frac{I_1}{2} = \frac{28.6}{2} = \mathbf{14.3\ mA}$$

Comments. Note the use of placing the diodes in parallel. If the current rating of each diode is 20 mA and a single diode is used in this circuit, a current of 28.6 mA would flow through the diode, thus damaging the device. By placing them in parallel, the current is limited to a safe value of 14.3 mA for the same terminal voltage.

Example 4.19. *Determine the currents I_1, I_2 and I_3 for the network shown in Fig. 4.21(i). Use simplified model for the diodes.*

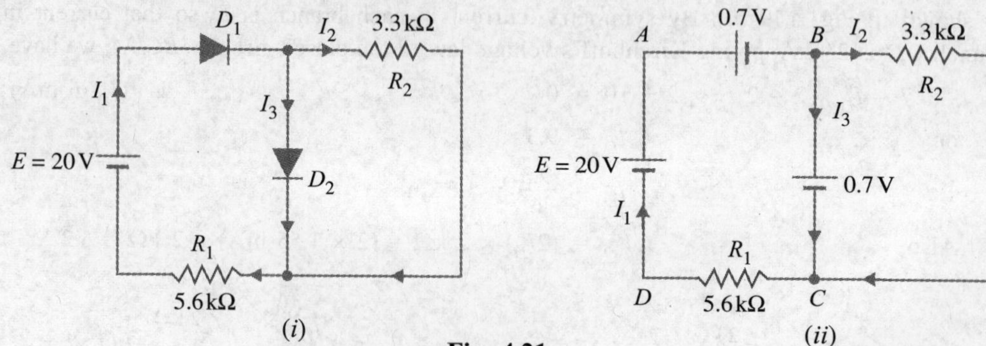

Fig. 4.21

Solution. An inspection of the circuit shown in Fig. 4.21 (*i*) shows that both diodes D_1 and D_2 are forward biased. Using simplified model for the diodes, the circuit shown in Fig. 4.21 (*i*) becomes the one shown in Fig. 4.21 (*ii*). The voltage across R_2 (= 3.3 kΩ) is 0.7 V.

$$\therefore \qquad I_2 = \frac{0.7\text{ V}}{3.3\text{ k}\Omega} = \mathbf{0.212\ mA}$$

Applying Kirchhoff's voltage law to loop *ABCDA* in Fig. 4.21 (*ii*), we have,

$$-0.7 - 0.7 - I_1 R_1 + 20 = 0$$

$$\therefore \qquad I_1 = \frac{20 - 0.7 - 0.7}{R_1} = \frac{18.6\text{ V}}{5.6\text{ k}\Omega} = \mathbf{3.32\ mA}$$

Now, $\qquad I_1 = I_2 + I_3$

$$\therefore \qquad I_3 = I_1 - I_2 = 3.32 - 0.212 = \mathbf{3.108\ mA}$$

Example 4.20. *Determine if the diode (ideal) in Fig. 4.22 (i) is forward biased or reverse biased.*

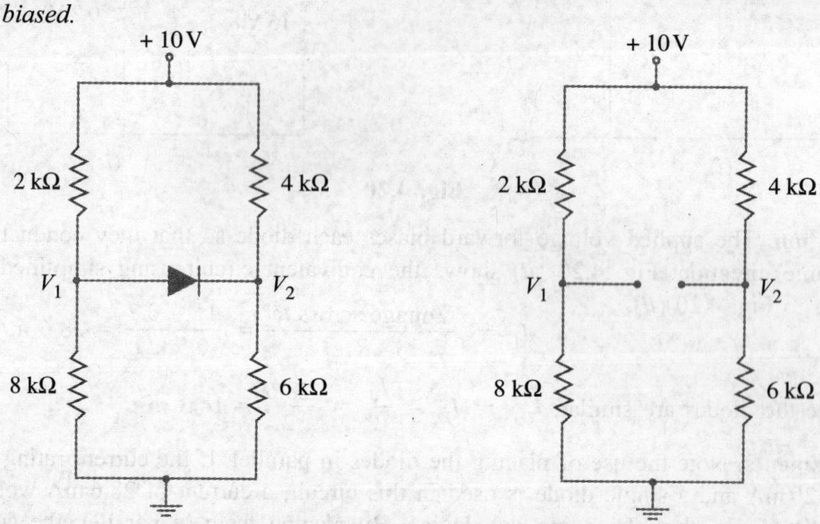

Fig. 4.22

Solution. Let us assume that diode in Fig. 4.22 (*i*) is *OFF i.e.* it is reverse biased. The circuit then becomes as shown in Fig. 4.22 (*ii*). Referring to Fig. 4.22 (*ii*), we have,

$$V_1 = \frac{10\text{ V}}{2\text{ k}\Omega + 8\text{ k}\Omega} \times 8\text{ k}\Omega = 8\text{ V}$$

$$V_2 = \frac{10\text{ V}}{4\text{ k}\Omega + 6\text{ k}\Omega} \times 6\text{ k}\Omega = 6\text{ V}$$

$\therefore$ Voltage across diode $= V_1 - V_2 = 8 - 6 = 2\text{ V}$

Now $V_1 - V_2 = 2\text{ V}$ is enough voltage to make the diode **forward biased.** Therefore, our initial assumption was wrong.

Example 4.21. *Determine the state of diode for the circuit shown in Fig. 4.23 (i) and find I_D and V_D. Assume simplified model for the diode.*

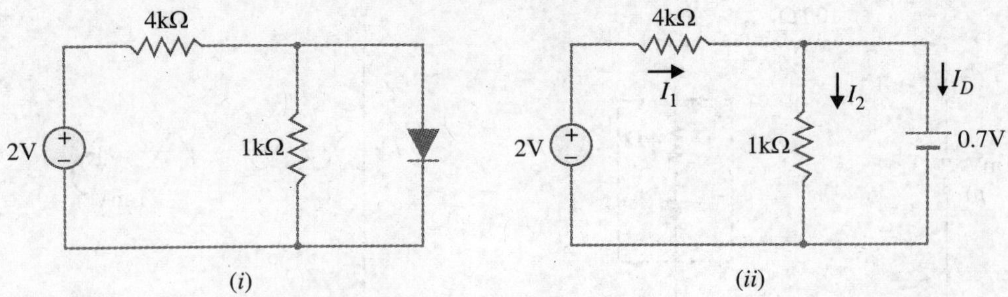

Fig. 4.23

Solution. Let us assume that the diode is *ON*. Therefore, we can replace the diode with a 0.7V battery as shown in Fig. 4.23 (*ii*). Referring to Fig. 4.23 (*ii*), we have,

$$I_1 = \frac{(2-0.7)\text{ V}}{4\text{ k}\Omega} = \frac{1.3\text{ V}}{4\text{ k}\Omega} = 0.325\text{ mA}$$

$$I_2 = \frac{0.7\text{ V}}{1\text{ k}\Omega} = 0.7\text{ mA}$$

Now, $I_D = I_1 - I_2 = 0.325 - 0.7 = -0.375\text{ mA}$

Since the diode current is negative, the diode must be **OFF** and the true value of diode current is $I_D = $ **0 mA**. Our initial assumption was wrong. In order to analyse the circuit properly, we should replace the diode in Fig. 4.23 (*i*) with an open-circuit as shown in Fig. 4.24. The voltage V_D across the diode is

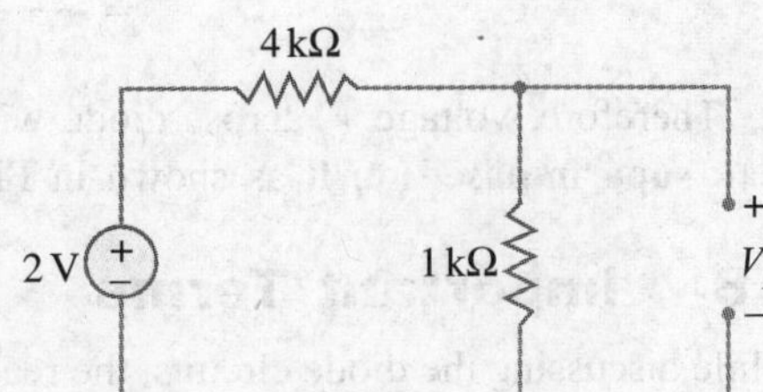

Fig. 4.24

$$V_D = \frac{2\text{ V}}{1\text{ k}\Omega + 4\text{ k}\Omega} \times 1\text{ k}\Omega = \mathbf{0.4\text{ V}}$$

We know that 0.7 V is required to turn *ON* the diode. Since V_D is only 0.4 V, the answer confirms that the diode is *OFF*.

Example 4.22. *Determine the voltage across the silicon diode for the circuit shown in Fig. 4.25 (i). Assume that forward d.c. resistance of the diode is zero and forward a.c. resistance is 4Ω.*

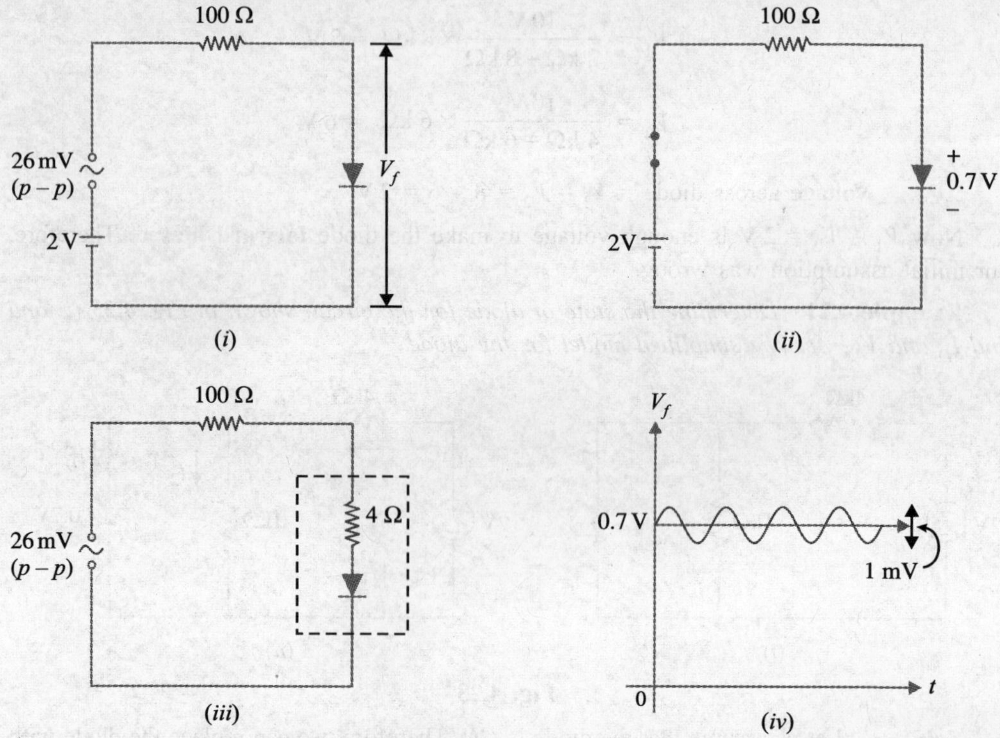

Fig. 4.25

Solution. The 2 V d.c. level ensures that the diode is always forward biased. In order to find the voltage across diode due to d.c. voltage (2 V), consider the d.c. equivalent circuit shown in Fig. 4.25 (*ii*). Since it is a silicon diode, the forward voltage drop $V_F = 0.7$ V and the remaining 1.3 V drops across 100 Ω resistor.

In order to find voltage across diode due to a.c. voltage [26 mV (*p* – *p*)], consider the a.c. equivalent circuit shown in Fig. 4.25 (*iii*). Therefore, forward a.c. voltage across the diode

$$= \frac{26 \text{ mV}(p-p)}{(100+4)\Omega} \times 4\Omega = 1 \text{ mV } (p-p)$$

Therefore, voltage V_f across diode will be **0.7 V with an a.c. voltage of 1 mV peak to peak superimposed on it** as shown in Fig. 4.25 (*iv*).

4.8 Important Terms

While discussing the diode circuits, the reader will generally come across the following terms:

(*i*) **Forward current.** It is the current flowing through a forward biased diode. Every diode has a maximum value of forward current which it can safely carry. If this value is exceeded, the diode may be destroyed due to excessive heat. For this reason, the manufacturers' data sheet specifies the maximum forward current that a diode can handle safely.

(*ii*) **Peak inverse voltage.** *It is the maximum reverse voltage that a diode can withstand without destroying the junction.*

If the reverse voltage across a diode exceeds this value, the reverse current increases sharply and breaks down the junction due to excessive heat. Peak inverse voltage is extremely important when diode is used as a rectifier. In rectifier service, it has to be ensured that reverse voltage across the diode does not exceed its *PIV* during the negative half-cycle of input a.c. voltage. As a matter of fact, *PIV* consideration is generally the deciding factor in diode rectifier circuits. The peak inverse voltage may be between 10 V and 10 kV depending upon the type of diode.

(*iii*) Reverse current or leakage current. It is the current that flows through a reverse biased diode. This current is due to the minority carriers. Under normal operating voltages, the reverse current is quite small. Its value is extremely small (< 1μA) for silicon diodes but it is appreciable (≃ 100 μA) for germanium diodes.

It may be noted that the reverse current is usually very small as compared with forward current. For example, the forward current for a typical diode might range upto 100 mA while the reverse current might be only a few μA—a ratio of many thousands between forward and reverse currents.

4.9 Semiconductor Diode Rectifiers

For reasons associated with economics of generation and transmission, the electric power available is usually an a.c. supply. The supply voltage varies sinusoidally and has a frequency of 50 Hz. It is used for lighting, heating and electric motors. But there are many applications (*e.g.* electronic circuits) where d.c. supply is needed. When such a d.c. supply is required, the mains a.c. supply is rectified by using semiconductor diodes. The following two rectifier circuits can be used :

(*i*) Half-wave rectifier (*ii*) Full-wave rectifier

4.10 Half-Wave Rectifier

In half-wave rectification, the rectifier conducts current only during the positive half-cycles of input a.c. supply. The negative half-cycles of a.c. supply are suppressed *i.e.* during negative half-cycles, no current is conducted and hence no voltage appears across the load. Therefore, current always flows in one direction (*i.e.* d.c.) through the load though after every half-cycle.

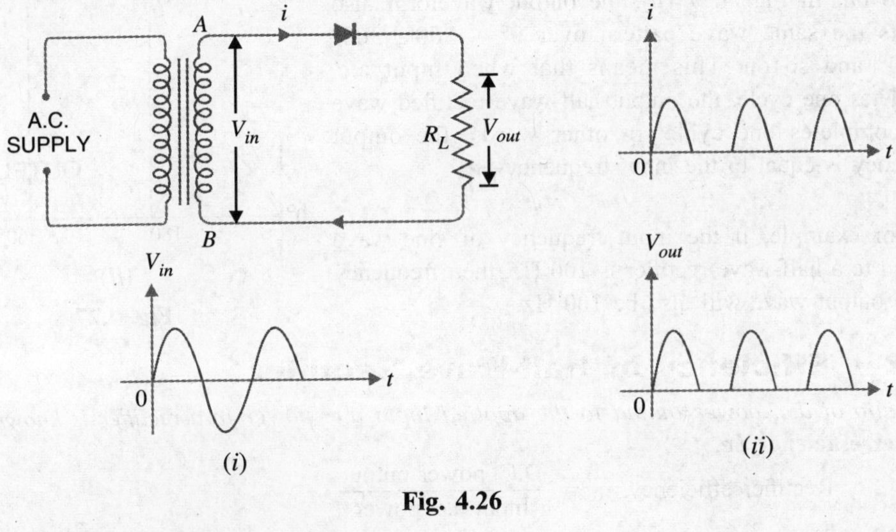

Fig. 4.26

Circuit details. Fig. 4.26 shows the circuit where a single semiconductor diode acts as a half-wave rectifier. The a.c. supply to be rectified is applied in series with the diode and load resistance R_L. Generally, a.c. supply is given through a transformer. The use of transformer permits two advantages. Firstly, it allows us to step-up or step-down the a.c. input voltage as the situation demands. Secondly, the transformer isolates the rectifier circuit from power line and thus reduces the risk of electric shock.

Operation. The a.c. voltage across the secondary winding *AB* changes polarities after every half-cycle. During the positive half-cycle of input a.c. voltage, end *A* becomes positive *w.r.t.* end *B*. This makes the diode forward biased and hence it conducts current. During the negative half-cycle, end *A* is negative *w.r.t.* end *B*. Under this condition, the diode is reverse biased and it conducts no current. Therefore, current flows through the diode during positive half-cycles of input a.c. voltage only ; it is blocked during the negative half-cycles [See Fig. 4.26 (*ii*)]. In this way, current flows through load R_L always in the same direction. Hence d.c. output is obtained across R_L. It may be noted that output across the load is pulsating d.c. These pulsations in the output are further smoothened with the help of *filter circuits* discussed later.

Disadvantages. The main disadvantages of a half-wave rectifier are :

(*i*) The pulsating current in the load contains alternating component whose basic frequency is equal to the supply frequency. Therefore, an elaborate filtering is required to produce steady direct current.

(*ii*) The a.c. supply delivers power only half the time. Therefore, the output is low.

4.11 Output Frequency of Half-Wave Rectifier

The output frequency of a half-wave rectifier is equal to the input frequency (50 Hz). Recall how a complete cycle is defined. A waveform has a complete cycle when it repeats the same wave pattern over a given time. Thus in Fig. 4.27 (*i*), the a.c. input voltage repeats the same wave pattern over 0° – 360°, 360° – 720° and so on. In Fig. 4.27 (*ii*), the output waveform also repeats the same wave pattern over 0° – 360°, 360° – 720° and so on. This means that when input a.c. completes one cycle, the output half-wave rectified wave also completes one cycle. In other words, the output frequency is equal to the input frequency *i.e.*

$$f_{out} = f_{in}$$

For example, if the input frequency of sine wave applied to a half-wave rectifier is 100 Hz, then frequency of the output wave will also be 100 Hz.

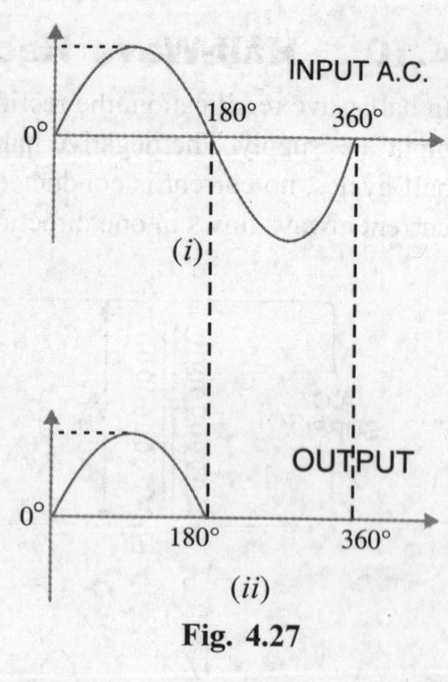

Fig. 4.27

4.12 Efficiency of Half-Wave Rectifier

The ratio of d.c. power output to the applied input a.c. power in a rectifier is known as **rectifier efficiency** *i.e.*

$$\text{Rectifier efficiency, } \eta = \frac{\text{D.C. power output}}{\text{Input a.c. power}}$$

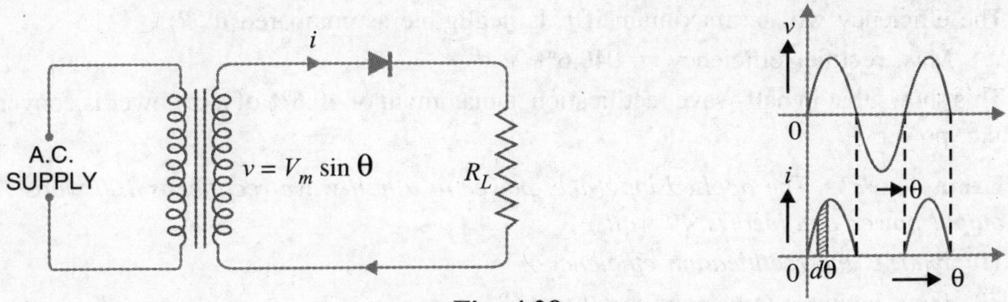

Fig. 4.28

Consider a half-wave rectifier shown in Fig. 4.28. Let $v = V_m \sin\theta$ be the alternating voltage that appears across the secondary winding. Let r_f and R_L be the diode resistance and load resistance respectively. The diode conducts during positive half-cycles of a.c. supply while no current conduction takes place during negative half-cycles.

d.c. power. The output current is pulsating direct current. Therefore, in order to find d.c. power, average current has to be found out.

$$^{*}I_{av} = I_{dc} = \frac{1}{2\pi}\int_0^{\pi} i\, d\theta = \frac{1}{2\pi}\int_0^{\pi} \frac{V_m \sin\theta}{r_f + R_L}\, d\theta$$

$$= \frac{V_m}{2\pi(r_f + R_L)}\int_0^{\pi} \sin\theta\, d\theta = \frac{V_m}{2\pi(r_f + R_L)}[-\cos\theta]_0^{\pi}$$

$$= \frac{V_m}{2\pi(r_f + R_L)} \times 2 = \frac{V_m}{(r_f + R_L)} \times \frac{1}{\pi}$$

$$\overset{**}{=} \frac{I_m}{\pi} \qquad \left[\because I_m = \frac{V_m}{(r_f + R_L)}\right]$$

$$\therefore \quad \text{d.c. power, } P_{dc} = I_{dc}^2 \times R_L = \left(\frac{I_m}{\pi}\right)^2 \times R_L \qquad \text{...}(i)$$

a.c. power input. The a.c. power input is given by ;

$$P_{ac} = I_{rms}^2\,(r_f + R_L)$$

For a half-wave rectified wave, $I_{rms} = I_m/2$

$$\therefore \quad P_{ac} = \left(\frac{I_m}{2}\right)^2 \times (r_f + R_L) \qquad \text{...}(ii)$$

$$\therefore \quad \text{Rectifier efficiency} = \frac{\text{d.c. output power}}{\text{a.c. input power}} = \frac{(I_m/\pi)^2 \times R_L}{(I_m/2)^2\,(r_f + R_L)}$$

$$= \frac{0.406\, R_L}{r_f + R_L} = \frac{0.406}{1 + \dfrac{r_f}{R_L}}$$

---

* Average value $= \dfrac{\text{Area under the curve over a cycle}}{\text{Base}} = \dfrac{\int_0^{\pi} i\, d\theta}{2\pi}$

** It may be remembered that the *area of one-half cycle of a sinusoidal wave* is twice the peak value. Thus in this case, peak value is I_m and, therefore, area of one-half cycle is $2\,I_m$.

$$\therefore \quad I_{av} = I_{dc} = \frac{2I_m}{2\pi} = \frac{I_m}{\pi}$$

The efficiency will be maximum if r_f is negligible as compared to R_L.

$\therefore$ Max. rectifier efficiency $=$ 40.6%

This shows that in half-wave rectification, a maximum of 40.6% of a.c. power is converted into d.c. power.

Example 4.23. *The applied input a.c. power to a half-wave rectifier is 100 watts. The d.c. output power obtained is 40 watts.*

(*i*) *What is the rectification efficiency?*

(*ii*) *What happens to remaining 60 watts?*

Solution.

(*i*) Rectification efficiency $= \dfrac{\text{d.c. output power}}{\text{a.c. input power}} = \dfrac{40}{100} = 0.4 = \mathbf{40\%}$

(*ii*) 40% efficiency of rectification does not mean that 60% of power is lost in the rectifier circuit. In fact, a semiconductor diode consumes little power due to its small internal resistance. The 100 W a.c. power is contained as 50 watts in positive half-cycles and 50 watts in negative half-cycles. The 50 watts in the negative half-cycles are not supplied at all. Only 50 watts in the positive half-cycles are converted into 40 watts.

$$\therefore \quad \text{Power efficiency} = \frac{40}{50} \times 100 = 80\%$$

Although 100 watts of a.c. power was supplied, the half-wave rectifier accepted only 50 watts and converted it into 40 watts d.c. power. Therefore, it is appropriate to say that efficiency of rectification is 40% and *not* 80% which is power efficiency.

Example 4.24. *An a.c. supply of 230 V is applied to a half-wave rectifier circuit through a transformer of turn ratio 10 : 1. Find (i) the output d.c. voltage and (ii) the peak inverse voltage. Assume the diode to be ideal.*

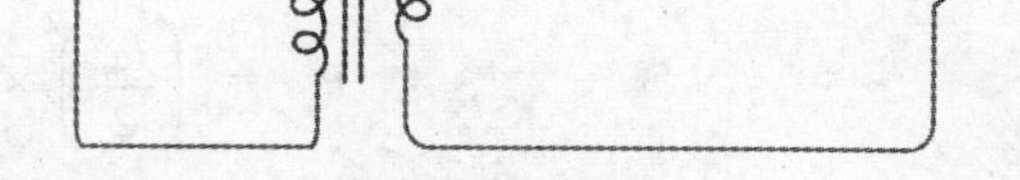

Fig. 4.29

Solution.

Primary to secondary turns is

$$\frac{N_1}{N_2} = 10$$

R.M.S. primary voltage = 230 V

$\therefore$ Max. primary voltage is

$$V_{pm} = (\sqrt{2}) \times \text{r.m.s. primary voltage}$$

$$= (\sqrt{2}) \times 230 = 325.3 \text{ V}$$

Max. secondary voltage is

$$V_{sm} = V_{pm} \times \frac{N_2}{N_1} = 325.3 \times \frac{1}{10} = 32.53 \text{ V}$$

(*i*) $$I_{dc} = \frac{I_m}{\pi}$$

$$\therefore \quad V_{dc} = \frac{I_m}{\pi} \times R_L = \frac{V_{sm}}{\pi} = \frac{32.53}{\pi} = \mathbf{10.36\ V}$$

(*ii*) During the negative half-cycle of a.c. supply, the diode is reverse biased and hence conducts no current. Therefore, the maximum secondary voltage appears across the diode.

$\therefore$ Peak inverse voltage = **32.53 V**

Example 4.25. *A crystal diode having internal resistance $r_f = 20\,\Omega$ is used for half-wave rectification. If the applied voltage $v = 50 \sin \omega t$ and load resistance $R_L = 800\,\Omega$, find :*

(i) I_m, I_{dc}, I_{rms} *(ii) a.c. power input and d.c. power output*

(iii) d.c. output voltage *(iv) efficiency of rectification.*

Solution.

$$v = 50 \sin \omega t$$

$$\therefore \quad \text{Maximum voltage, } V_m = 50\text{ V}$$

(i)

$$I_m = \frac{V_m}{r_f + R_L} = \frac{50}{20 + 800} = 0.061\text{ A} = \mathbf{61\ mA}$$

$$I_{dc} = I_m/\pi = 61/\pi = \mathbf{19.4\ mA}$$

$$I_{rms} = I_m/2 = 61/2 = \mathbf{30.5\ mA}$$

(ii)

$$\text{a.c. power input} = (I_{rms})^2 \times (r_f + R_L) = \left(\frac{30.5}{1000}\right)^2 \times (20 + 800)$$

$$= \mathbf{0.763\ watt}$$

$$\text{d.c. power output} = I^2_{dc} \times R_L = \left(\frac{19.4}{1000}\right)^2 \times 800 = \mathbf{0.301\ watt}$$

(iii)

$$\text{d.c. output voltage} = I_{dc}\, R_L = 19.4\text{ mA} \times 800\ \Omega = \mathbf{15.52\ volts}$$

(iv)

$$\text{Efficiency of rectification} = \frac{0.301}{0.763} \times 100 = \mathbf{39.5\%}$$

Example 4.26. *A half-wave rectifier is used to supply 50 V d.c. to a resistive load of 800 Ω. The diode has a resistance of 25 Ω. Calculate a.c. voltage required.*

Solution.

Output d.c. voltage, $V_{dc} = 50\text{ V}$

Diode resistance, $r_f = 25\,\Omega$

Load resistance, $R_L = 800\,\Omega$

Let V_m be the maximum value of a.c. voltage required.

$$\therefore \quad V_{dc} = I_{dc} \times R_L$$

$$= \frac{I_m}{\pi} \times R_L = \frac{V_m}{\pi(r_f + R_L)} \times R_L \qquad \left[\because I_m = \frac{V_m}{r_f + R_L}\right]$$

or

$$50 = \frac{V_m}{\pi(25 + 800)} \times 800$$

$$\therefore \quad V_m = \frac{\pi \times 825 \times 50}{800} = \mathbf{162\ V}$$

Hence a.c. voltage of maximum value 162 V is required.

Example 4.27. *Fig. 4.30 shows a circuit for charging a 12 V battery. If v_s is a sinusoid with 24 V peak value, find (i) the fraction of each cycle during which the diode conducts (ii) the peak value of the diode current and (iii) the maximum revese-bias voltage that appears across the diode.*

Solution. *(i)* The diode conducts when v_s exceeds 12 V.

Now, 24 sin θ = 12 or sin θ = 1/2 $\therefore$ θ = 30°

$\therefore$ Conduction angle = 180° − 2 × 30° = 120°

Therefore, during the positive half-cycle of v_s, the diode conducts for 120° (*i.e.* from 30° to 150°). However, during the negative half-cycle of v_s, the diode is reverse biased and does not conduct. This means that for 1 cycle (= 360°) of v_s, the diode conducts for 120° **or one-third of 1 cycle of v_s.**

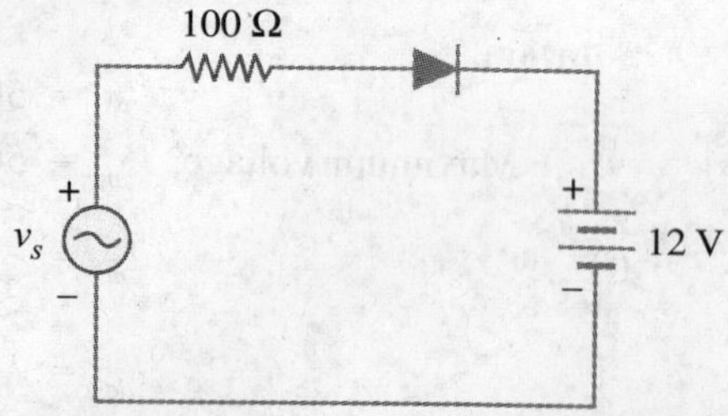

Fig. 4.30

(ii) Peak diode current $= \dfrac{24\text{ V} - 12\text{ V}}{100\,\Omega}$

= **0.12 A**

(iii) During the negative half-cycle of v_s, the maximum reverse voltage across the diode occurs when v_s is at its negative peak. Therefore, maximum reverse voltage across the diode = 24 V + 12 V = **36 V.**

4.13 Full-Wave Rectifier

In full-wave rectification, current flows through the load in the same direction for both half-cycles of input a.c. voltage. This can be achieved with two diodes working alternately. For the positive half-cycle of input voltage, one diode supplies current to the load and for the negative half-cycle, the other diode does so ; current being always in the same direction through the load. Therefore, a full-wave rectifier utilises both half-cycles of input a.c. voltage to produce the d.c. output. The following two circuits are commonly used for full-wave rectification :

(i) Centre-tap full-wave rectifier *(ii)* Full-wave bridge rectifier

4.14 Centre-Tap Full-Wave Rectifier

The circuit employs two diodes D_1 and D_2 as shown in Fig. 4.31. A centre tapped secondary winding *AB* is used with two diodes connected so that each uses one half-cycle of input a.c. voltage. In other words, diode D_1 utilises the a.c. voltage appearing across the upper half (*OA*) of secondary winding for rectification while diode D_2 uses the lower half winding *OB*.

Operation. During the positive half-cycle of secondary voltage, the end *A* of the secondary winding becomes positive and end *B* negative. This makes the diode D_1 forward biased and diode D_2 reverse biased. Therefore, diode D_1 conducts while diode D_2 does not. The conventional current flow is through diode D_1, load resistor R_L and the upper half of secondary winding as shown by the dotted arrows. During the negative half-cycle, end *A* of the secondary winding becomes negative and end *B* positive. Therefore, diode D_2 conducts while diode D_1 does not. The conventional current flow is through diode D_2, load R_L and lower half winding as shown by solid arrows. Referring to Fig. 4.31, it may be seen that current in the load R_L is *in the same direction* for both half-cycles of input a.c. voltage. Therefore, d.c. is obtained across the load R_L. Also, the polarities of the d.c. output across the load should be noted.

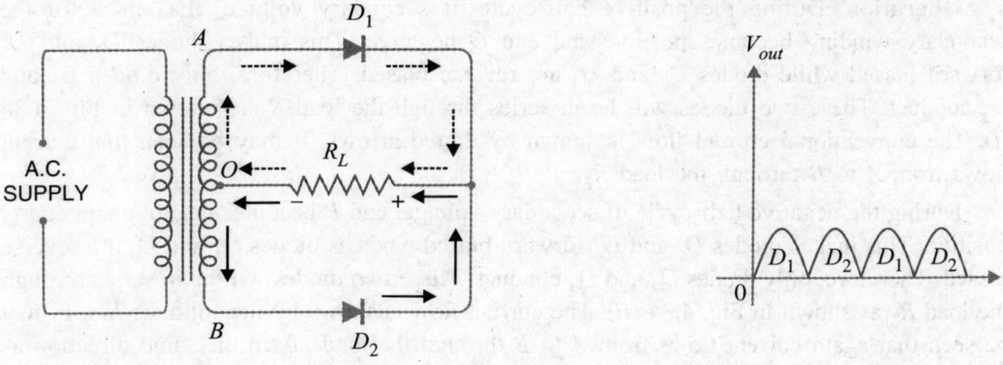

Fig. 4.31

Peak inverse voltage. Suppose V_m is the maximum voltage across the half secondary winding. Fig. 4.32 shows the circuit at the instant secondary voltage reaches its maximum value in the positive direction. At this instant, diode D_1 is conducting while diode D_2 is non-conducting. Therefore, whole of the secondary voltage appears across the non-conducting diode. Consequently, the peak inverse voltage is twice the maximum voltage across the half-secondary winding *i.e.*

$$PIV = 2V_m$$

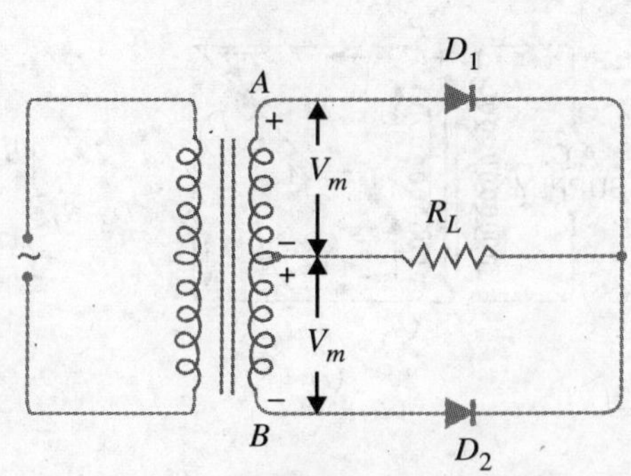

Fig. 4.32

Disadvantages

(*i*) It is difficult to locate the centre tap on the secondary winding.

(*ii*) The d.c. output is small as each diode utilises only one-half of the transformer secondary voltage.

(*iii*) The diodes used must have high peak inverse voltage.

4.15 Full-Wave Bridge Rectifier

The need for a centre tapped power transformer is eliminated in the bridge rectifier. It contains four diodes D_1, D_2, D_3 and D_4 connected to form bridge as shown in Fig. 4.33. The a.c. supply to be rectified is applied to the diagonally opposite ends of the bridge through the transformer. Between other two ends of the bridge, the load resistance R_L is connected.

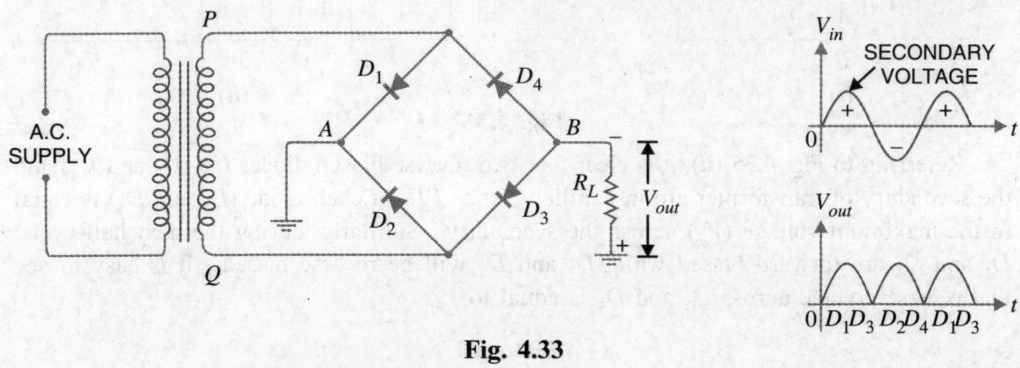

Fig. 4.33

Operation. During the positive half-cycle of secondary voltage, the end *P* of the secondary winding becomes positive and end *Q* negative. This makes diodes D_1 and D_3 forward biased while diodes D_2 and D_4 are reverse biased. Therefore, only diodes D_1 and D_3 conduct. These two diodes will be in series through the load R_L as shown in Fig. 4.34 (*i*). The conventional current flow is shown by dotted arrows. It may be seen that current flows from *A* to *B* through the load R_L.

During the negative half-cycle of secondary voltage, end *P* becomes negative and end *Q* positive. This makes diodes D_2 and D_4 forward biased whereas diodes D_1 and D_3 are reverse biased. Therefore, only diodes D_2 and D_4 conduct. These two diodes will be in series through the load R_L as shown in Fig. 4.34 (*ii*). The current flow is shown by the solid arrows. It may be seen that again current flows from *A* to *B* through the load *i.e.* in the same direction as for the positive half-cycle. Therefore, d.c. output is obtained across load R_L.

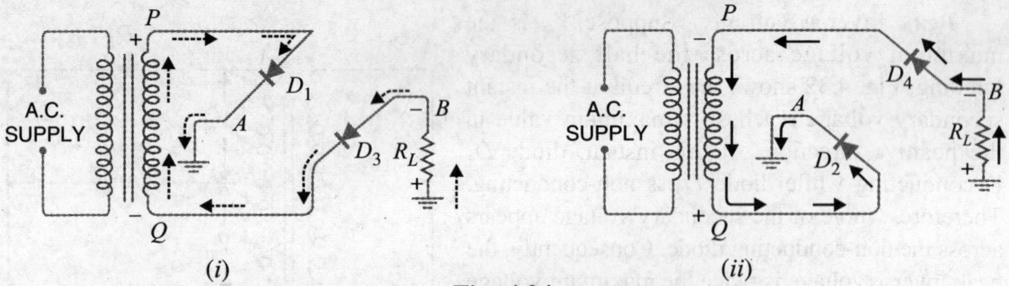

Fig. 4.34

Peak inverse voltage. The peak inverse voltage (*PIV*) of each diode is equal to the maximum secondary voltage of transformer. Suppose during positive half-cycle of input a.c., end *P* of secondary is positive and end *Q* negative. Under such conditions, diodes D_1 and D_3 are forward biased while diodes D_2 and D_4 are reverse biased. Since the diodes are considered ideal, diodes D_1 and D_3 can be replaced by wires as shown in Fig. 4.35 (*i*). This circuit is the same as shown in Fig. 4.35 (*ii*).

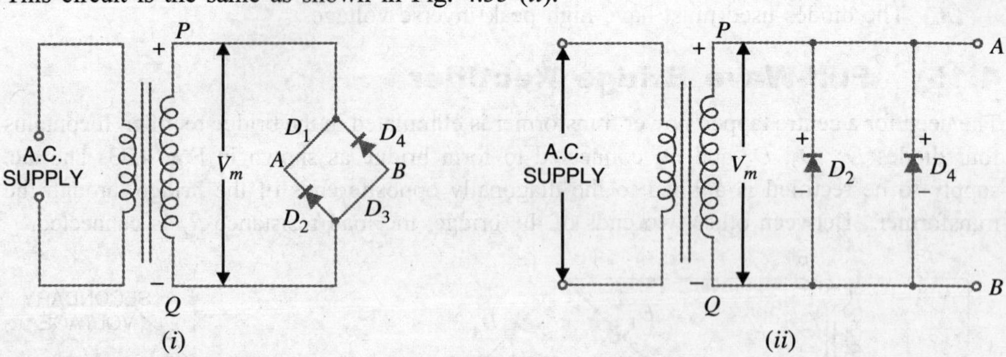

Fig. 4.35

Referring to Fig. 4.35 (*ii*), it is clear that two reverse biased diodes (*i.e.* D_2 and D_4) and the secondary of transformer are in parallel. Hence *PIV* of each diode (D_2 and D_4) is equal to the maximum voltage (V_m) across the secondary. Similarly, during the next half-cycle, D_2 and D_4 are forward biased while D_1 and D_3 will be reverse biased. It is easy to see that reverse voltage across D_1 and D_3 is equal to V_m.

Advantages

(*i*) The need for centre-tapped transformer is eliminated.

(*ii*) The output is twice that of the centre-tap circuit for the same secondary voltage.

(*iii*) The *PIV* is one-half that of the centre-tap circuit (for same d.c. output).

Disadvantages

(*i*) It requires four diodes.

(*ii*) As during each half-cycle of a.c. input two diodes that conduct are in series, therefore, voltage drop in the internal resistance of the rectifying unit will be twice as great as in the centre-tap circuit. This is objectionable when secondary voltage is small.

4.16 Output Frequency of Full-Wave Rectifier

The output frequency of a full-wave rectifier is double the input frequency. Remember that a wave has a complete cycle when it repeats the same pattern. In Fig. 4.36 (*i*), the input a.c. completes one cycle from 0° – 360°. However, the full-wave rectified wave completes 2 cycles in this period [See Fig. 4.36 (*ii*)]. Therefore, output frequency is twice the input frequency *i.e.*

$$f_{out} = 2 f_{in}$$

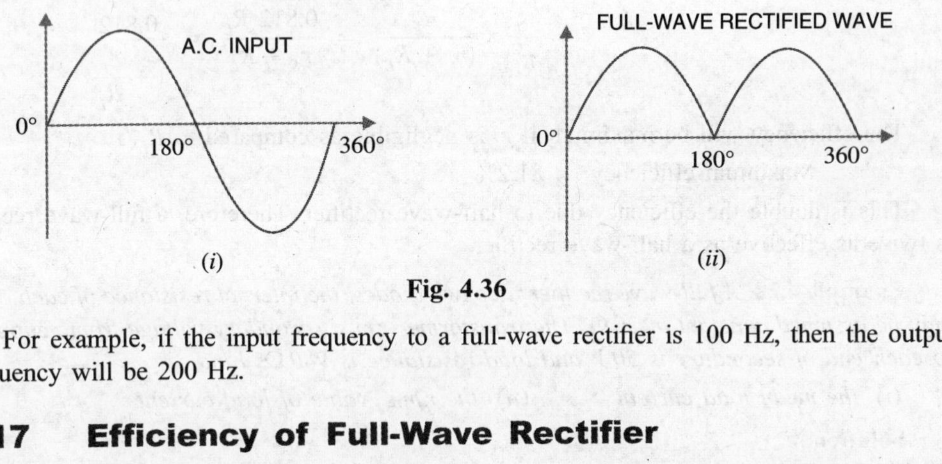

Fig. 4.36

For example, if the input frequency to a full-wave rectifier is 100 Hz, then the output frequency will be 200 Hz.

4.17 Efficiency of Full-Wave Rectifier

Fig. 4.37 shows the process of full-wave rectification. Let $v = V_m \sin\theta$ be the a.c. voltage to be rectified. Let r_f and R_L be the diode resistance and load resistance respectively. Obviously, the rectifier will conduct current through the load in the same direction for both half-cycles of input a.c. voltage. The instantaneous current i is given by ;

$$i = \frac{v}{r_f + R_L} = \frac{V_m \sin\theta}{r_f + R_L}$$

d.c. output power. The output current is pulsating direct current. Therefore, in order to find the d.c. power, average current has to be found

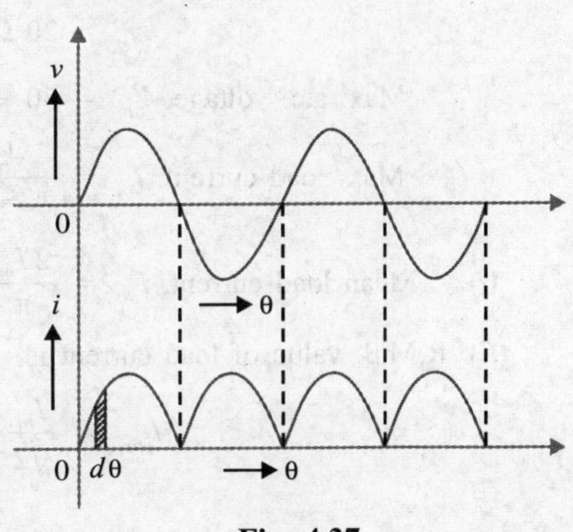

Fig. 4.37

out. From the elementary knowledge of electrical engineering,

$$I_{dc} = \frac{2\,I_m}{\pi}$$

$$\therefore \quad \text{d.c. power output, } P_{dc} = I_{dc}^2 \times R_L = \left(\frac{2\,I_m}{\pi}\right)^2 \times R_L \qquad \text{...}(i)$$

a.c. input power. The a.c. input power is given by ;

$$P_{ac} = I_{rms}^2\,(r_f + R_L)$$

For a full-wave rectified wave, we have,

$$I_{rms} = I_m/\sqrt{2}$$

$$\therefore \quad P_{ac} = \left(\frac{I_m}{\sqrt{2}}\right)^2 (r_f + R_L) \qquad \text{...}(ii)$$

∴ Full-wave rectification efficiency is

$$\eta = \frac{P_{dc}}{P_{ac}} = \frac{(2\,I_m/\pi)^2 R_L}{\left(\frac{I_m}{\sqrt{2}}\right)^2 (r_f + R_L)}$$

$$= \frac{8}{\pi^2} \times \frac{R_L}{(r_f + R_L)} = \frac{0.812\,R_L}{r_f + R_L} = \frac{0.812}{1 + \frac{r_f}{R_L}}$$

The efficiency will be maximum if r_f is negligible as compared to R_L.

∴ Maximum efficiency = 81.2%

This is double the efficiency due to half-wave rectifier. Therefore, a full-wave rectifier is twice as effective as a half-wave rectifier.

Example 4.28. *A full-wave rectifier uses two diodes, the internal resistance of each diode may be assumed constant at 20 Ω. The transformer r.m.s. secondary voltage from centre-tap to each end of secondary is 50 V and load resistance is 980 Ω. Find :*

(*i*) *the mean load current* (*ii*) *the r.m.s. value of load current*

Solution.

$$r_f = 20\ \Omega\ ;\quad R_L = 980\ \Omega$$

$$\text{Max. a.c. voltage, } V_m = 50 \times \sqrt{2} = 70.7\ \text{V}$$

$$\text{Max. load current, } I_m = \frac{V_m}{r_f + R_L} = \frac{70.7\ \text{V}}{(20 + 980)\ \Omega} = 70.7\ \text{mA}$$

(*i*) $$\text{Mean load current, } I_{dc} = \frac{2\,I_m}{\pi} = \frac{2 \times 70.7}{\pi} = \mathbf{45\ mA}$$

(*ii*) R.M.S. value of load current is

$$I_{rms} = \frac{I_m}{\sqrt{2}} = \frac{70.7}{\sqrt{2}} = \mathbf{50\ mA}$$

Example 4.29. *In the centre-tap circuit shown in Fig. 4.38, the diodes are assumed to be ideal i.e. having zero internal resistance. Find :*

(*i*) *d.c. output voltage* (*ii*) *peak inverse voltage* (*iii*) *rectification efficiency.*

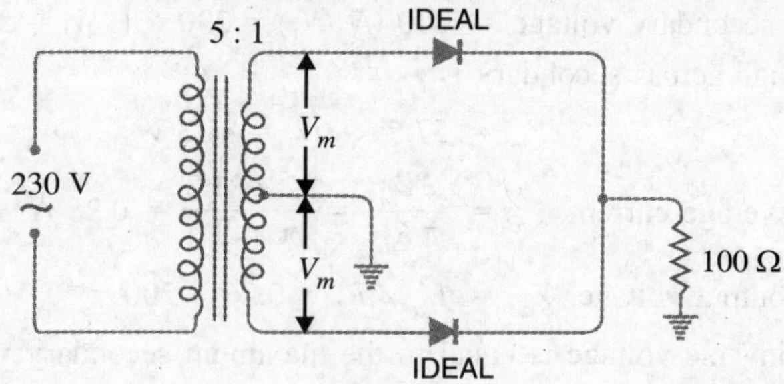

Fig. 4.38

Solution.

Primary to secondary turns, $N_1/N_2 = 5$

R.M.S. primary voltage = 230 V

∴ R.M.S. secondary voltage = 230 × (1/5) = 46 V

Maximum voltage across secondary = $46 \times \sqrt{2}$ = 65 V

Maximum voltage across half secondary winding is

$$V_m = 65/2 = 32.5 \text{ V}$$

(*i*) Average current, $I_{dc} = \dfrac{2V_m}{\pi R_L} = \dfrac{2 \times 32.5}{\pi \times 100} = 0.207 \text{ A}$

∴ d.c. output voltage, $V_{dc} = I_{dc} \times R_L = 0.207 \times 100 =$ **20.7 V**

(*ii*) The peak inverse voltage is equal to the maximum secondary voltage, *i.e.*

$$PIV = \textbf{65 V}$$

(*iii*) Rectification efficiency $= \dfrac{0.812}{1 + \dfrac{r_f}{R_L}}$

Since $r_f = 0$,

∴ Rectification efficiency = **81.2 %**

Example 4.30. *In the bridge type circuit shown in Fig. 4.39, the diodes are assumed to be ideal. Find :*

(*i*) *d.c. output voltage* (*ii*) *peak inverse voltage* (*iii*) *output frequency.*

Assume primary to secondary turns to be 4.

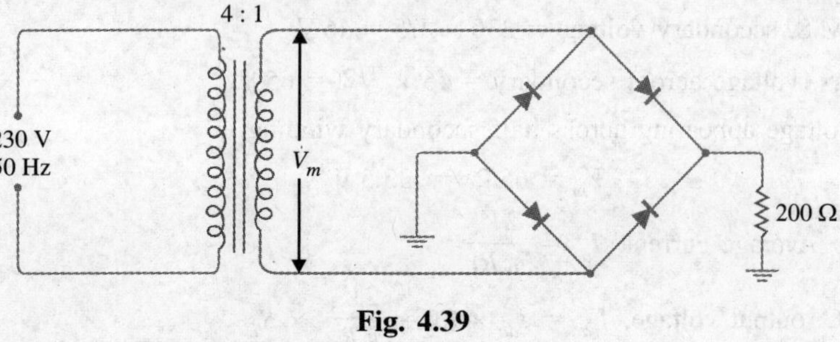

Fig. 4.39

Solution.

Primary/secondary turns, $N_1/N_2 = 4$

R.M.S. primary voltage = 230 V

$\therefore$ R.M.S. secondary voltage = $230 (N_2/N_1) = 230 \times (1/4) = 57.5$ V

Maximum voltage across secondary is

$$V_m = 57.5 \times \sqrt{2} = 81.3 \text{ V}$$

(*i*) Average current, $I_{dc} = \dfrac{2V_m}{\pi R_L} = \dfrac{2 \times 81.3}{\pi \times 200} = 0.26$ A

$\therefore$ d.c. output voltage, $V_{dc} = I_{dc} \times R_L = 0.26 \times 200 =$ **52 V**

(*ii*) The peak inverse voltage is equal to the maximum secondary voltage *i.e.*

$$PIV = \textbf{81.3 V}$$

(*iii*) In full-wave rectification, there are two output pulses for each complete cycle of the input a.c. voltage. Therefore, the output frequency is twice that of the a.c. supply frequency *i.e.*

$$f_{out} = 2 \times f_{in} = 2 \times 50 = \textbf{100 Hz}$$

Example 4.31. *Fig. 4.40 (i) and Fig. 4.40 (ii) show the centre-tap and bridge type circuits having the same load resistance and transformer turn ratio. The primary of each is connected to 230V, 50 Hz supply.*

(*i*) *Find the d.c. voltage in each case.*

(*ii*) *PIV for each case for the same d.c. output. Assume the diodes to be ideal.*

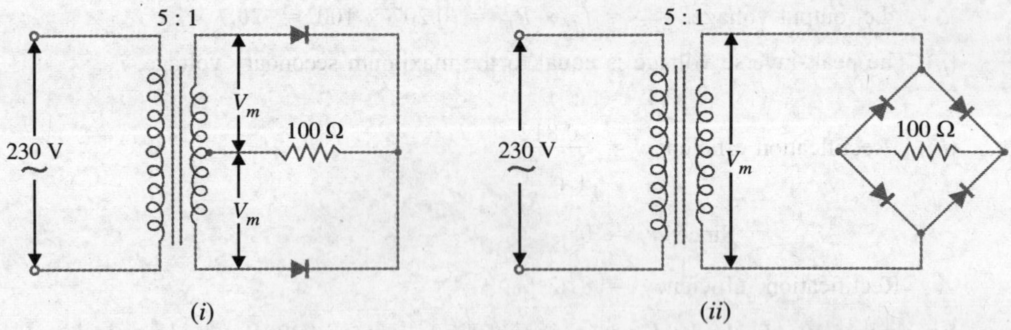

Fig. 4.40

Solution.

(*i*) **D.C. output voltage**

Centre-tap circuit

R.M.S. secondary voltage = 230 × 1/5 = 46 V

Max. voltage across secondary = $46 \times \sqrt{2} = 65$ V

Max. voltage appearing across half secondary winding is

$$V_m = 65/2 = 32.5 \text{ V}$$

Average current, $I_{dc} = \dfrac{2V_m}{\pi R_L}$

D.C. output voltage, $V_{dc} = I_{dc} \times R_L = \dfrac{2V_m}{\pi R_L} \times R_L$

$$= \frac{2V_m}{\pi} = \frac{2 \times 32.5}{\pi} = \mathbf{20.7\ V}$$

Bridge Circuit

Max. voltage across secondary, V_m = 65 V

$$\text{D.C. output voltage, } V_{dc} = I_{dc}\, R_L = \frac{2V_m}{\pi R_L} \times R_L = \frac{2\,V_m}{\pi} = \frac{2 \times 65}{\pi} = \mathbf{41.4\ V}$$

This shows that for the same secondary voltage, the d.c. output voltage of bridge circuit is twice that of the centre-tap circuit.

(*ii*) *PIV* for same d.c. output voltage

The d.c. output voltage of the two circuits will be the same if V_m (*i.e.* max. voltage utilised by each circuit for conversion into d.c.) is the same. For this to happen, the turn ratio of the transformers should be as shown in Fig. 4.41.

Centre-tap circuit

$$\text{R.M.S. secondary voltage} = 230 \times 1/5 = 46\text{ V}$$

$$\text{Max. voltage across secondary} = 46 \times \sqrt{2} = 65\text{ V}$$

Max. voltage across half secondary winding is

$$V_m = 65/2 = 32.5\text{ V}$$

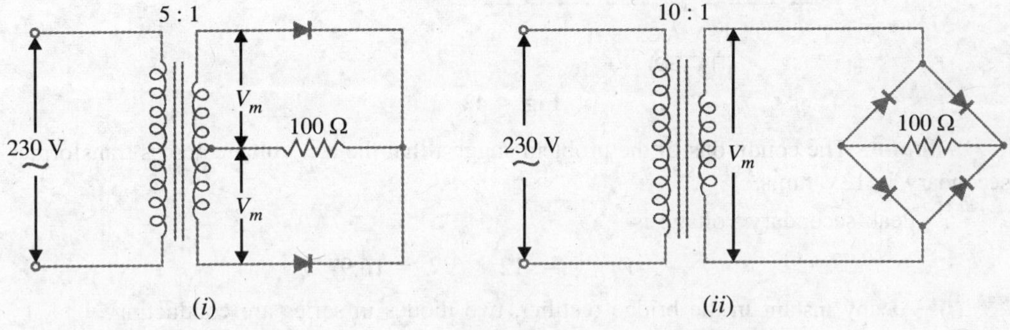

Fig. 4.41

$$\therefore \quad PIV = 2V_m = 2 \times 32.5 = \mathbf{65\ V}$$

Bridge type circuit

$$\text{R.M.S. secondary voltage} = 230 \times 1/10 = 23\text{ V}$$

$$\text{Max. voltage across secondary, } V_m = 23 \times \sqrt{2} = 32.5\text{ V}$$

$$\therefore \quad PIV = V_m = \mathbf{32.5\ V}$$

This shows that for the same d.c. output voltage, *PIV* of bridge circuit is half that of centre-tap circuit. This is a distinct advantage of bridge circuit.

Example 4.32. *The four diodes used in a bridge rectifier circuit have forward resistances which may be considered constant at 1 Ω and infinite reverse resistance. The alternating supply voltage is 240 V r.m.s. and load resistance is 480 Ω. Calculate (i) mean load current and (ii) power dissipated in each diode.*

Solution.

$$\text{Max. a.c. voltage, } V_m = 240 \times \sqrt{2}\text{ V}$$

(*i*) At any instant in the bridge rectifier, two diodes in series are conducting. Therefore, total circuit resistance = $2r_f + R_L$.

$$\text{Max. load current, } I_m = \frac{V_m}{2r_f + R_L} = \frac{240 \times \sqrt{2}}{2 \times 1 + 480} = 0.7\text{ A}$$

$$\therefore \quad \text{Mean load current, } I_{dc} = \frac{2I_m}{\pi} = \frac{2 \times 0.7}{\pi} = \mathbf{0.45\ A}$$

(*ii*) Since each diode conducts only half a cycle, diode r.m.s. current is :

$$I_{rms} = I_m/2 = 0.7/2 = 0.35\text{ A}$$

$$\text{Power dissipated in each diode} = I^2_{rms} \times r_f = (0.35)^2 \times 1 = \mathbf{0.123\ W}$$

Example 4.33. *The bridge rectifier shown in Fig. 4.42 uses silicon diodes. Find (i) d.c. output voltage (ii) d.c. output current. Use simplified model for the diodes.*

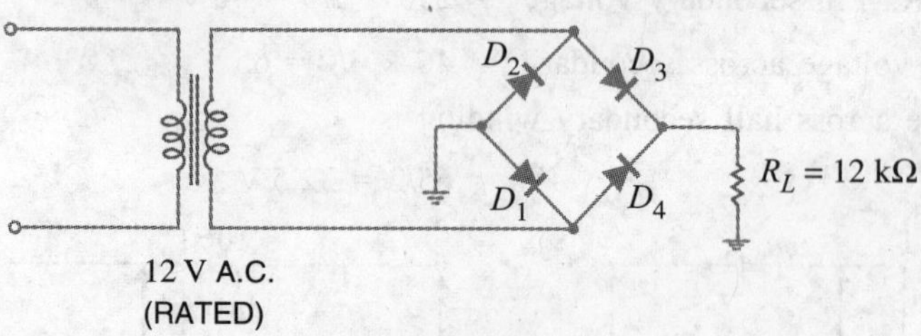

Fig. 4.42

Solution. The conditions of the problem suggest that the a.c. voltage across transformer secondary is 12V r.m.s.

∴ Peak secondary voltage is

$$V_{s\,(pk)} = 12 \times \sqrt{2} = 16.97\text{ V}$$

(*i*) At any instant in the bridge rectifier, two diodes in series are conducting.

∴ Peak output voltage is

$$V_{out\,(pk)} = 16.97 - 2\,(0.7) = 15.57\text{ V}$$

∴ Average (or d.c.) output voltage is

$$V_{av} = V_{dc} = \frac{2\,V_{out\,(pk)}}{\pi} = \frac{2 \times 15.57}{\pi} = \mathbf{9.91\ V}$$

(*ii*) Average (or d.c.) output current is

$$I_{av} = \frac{V_{av}}{R_L} = \frac{9.91\text{V}}{12\text{ k}\Omega} = \mathbf{825.8\ \mu A}$$

Example 4.34. *A full-wave bridge rectifier with 120 V (r.m.s.) sinusoidal input has a load resistor of 1 kΩ. If silicon diodes are used, calculate (i) PIV rating of each diode (ii) maximum current through each diode during conduction (iii) d.c. voltage available at load (iv) power rating of each diode.*

Solution. Input a.c. voltage = 120 V (r.m.s.) ; Load resistance, R_L = 1 kΩ = 1000 Ω

The maximum input a.c. voltage V_m is

$$V_m = \sqrt{2} \times 120 = 169.7\text{ V}$$

(*i*) During each half-cycle of input a.c. voltage, the two non-conducting diodes (*i.e.* reverse biased diodes) are in parallel across the a.c. supply. The voltage drop across each silicon diode is 0.7 V.

∴ *PIV* rating of each diode = 169.7 – 0.7 = **169 V**

(*ii*) During each half-cycle of input a.c. voltage, two silicon diodes are in series with the load.

∴ Maximum current through each diode during conduction is

$$I_m = \frac{V_m - 2\times0.7}{R_L} = \frac{169.7 - 2\times0.7}{1000} = 0.1683\text{ A} = \mathbf{168.3\text{ mA}}$$

(*iii*) Average current, $I_{dc} = \frac{2}{\pi} I_m = \frac{2}{\pi} \times 168.3 = 107.21\text{ mA}$

D.C. voltage available at load is

$$V_{dc} = I_{dc} R_L = (107.21 \times 10^{-3}) \times 1000 = \mathbf{107.21\text{ V}}$$

(*iv*) Power rating P_D of each diode is given by ;

$$P_D = V_F I_{dc} = 0.7\text{ V} \times 107.21\text{ mA} = \mathbf{75\text{ mW}}$$

Example 4.35. *Fig. 4.43 shows the bridge rectifier with two diodes removed. Determine the (i) output waveform (ii) d.c. output voltage (iii) PIV required for each diode. Assume the diodes to be ideal.*

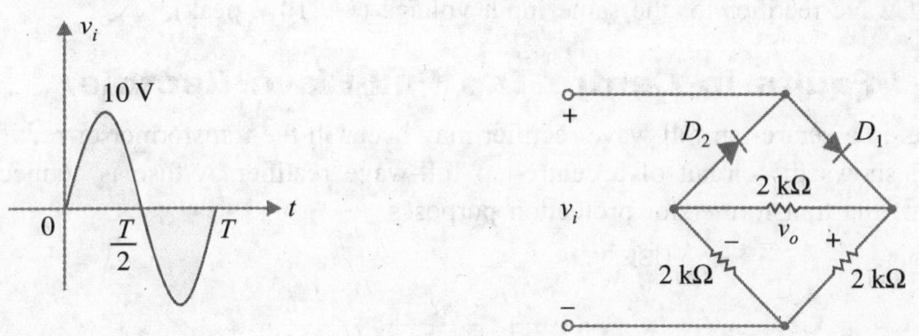

Fig. 4.43

Solution.

(*i*) During the positive half-cycle of input a.c. voltage, diode D_1 is forward biased while diode D_2 is reverse biased and the circuit becomes as shown in Fig. 4.44 (*i*). The circuit shown in Fig. 4.44 (*i*) further reduces to the one shown in Fig. 4.44 (*ii*). It is clear from Fig. 4.44 (*ii*) that the maximum d.c. output voltage V_m is one-half of maximum input a.c. voltage *i.e.*

Max. d.c. output voltage, $V_m = \frac{1}{2} \times 10 = 5\text{ V}$

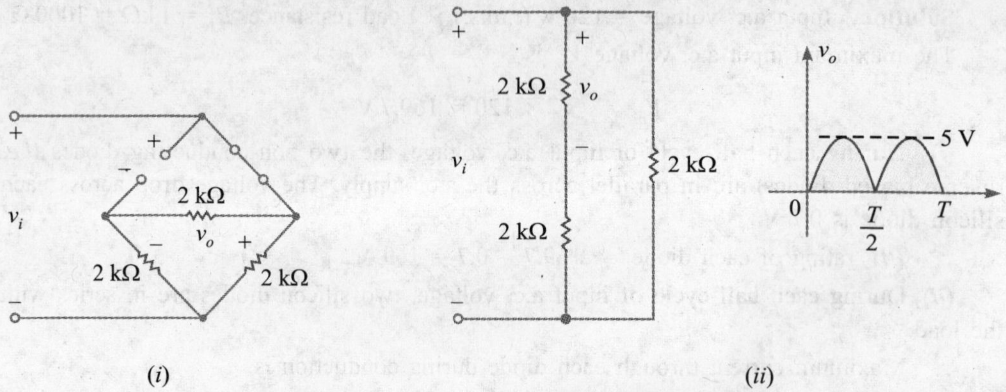

Fig. 4.44

During the negative half-cycle of input a.c. voltage, the roles of the diodes are interchanged *i.e.* now diode D_1 is reverse biased and diode D_2 is forward biased. Again, the maximum d.c. output voltage = (1/2) × 10 V = 5 V. **As a result, full-wave rectification occurs as shown in Fig. 4.44 (*ii*).**

(*ii*) d.c. output voltage, $V_{dc} = \frac{2}{\pi} \times V_m = \frac{2}{\pi} \times 5 =$ **3.18 V**

(*iii*) *PIV* of each diode = **5 V**

Comments. Note the effect of removing two diodes from the bridge rectifier circuit. Now, the d.c. output voltage available is the same as available from a half-wave rectifier with the same input voltage (*i.e.* 10 V peak). However, *PIV* required is half of that required for a half-wave rectifier for the same input voltage (*i.e.* 10 V peak).

4.18 Faults in Centre-Tap Full-Wave Rectifier

The faults in a centre-tap full-wave rectifier may occur in the transformer or rectifier diodes. Fig. 4.45 shows the circuit of a centre-tap full-wave rectifier. A fuse is connected in the primary of the transformer for protection purposes.

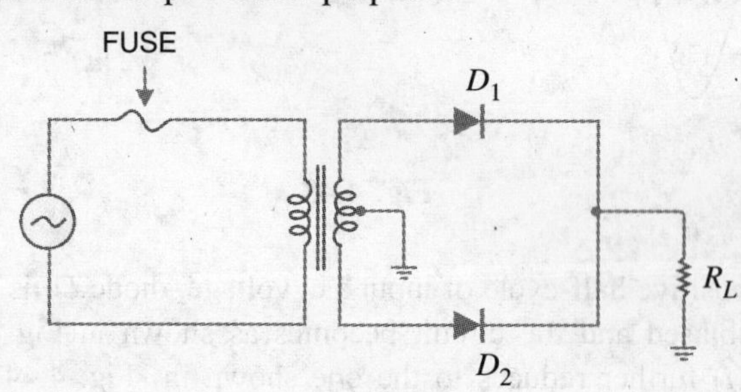

Fig. 4.45

We can divide the rectifier faults into two classes *viz.*

1. Faults in transformer
2. Faults in rectifier diodes

1. Faults in Transformer. The transformer in a rectifier circuit can develop the following faults :

(*i*) A shorted primary or secondary winding.

(*ii*) An open primary or secondary winding.

(*iii*) A short between the primary or secondary winding and the transformer frame.

(*i*) In most cases, a **shorted primary** or **shorted secondary** will cause the fuse in the primary to blow. If the fuse does not blow, the d.c. output from the rectifier will be extremely low and the transformer itself will be very hot.

(*ii*) When the **primary** or **secondary winding of the transformer opens**, the output from the rectifier will drop to zero. In this case, the primary fuse will not blow. If you believe that either transformer winding is open, a simple resistance check will verify your doubt. If either winding reads a very high resistance, the winding is open.

(*iii*) If **either winding shorts to the transformer casing**, the primary fuse will blow. This fault can be checked by measuring the resistances from the winding leads to the transformer casing. A low resistance measurement indicates that a winding-to-case short exists.

2. **Faults in Rectifier Diodes.** If a fault occurs in a rectifier diode, the circuit conditions will indicate the type of fault.

(*i*) If **one diode in the centre-tap full-wave rectifier is shorted**, the primary fuse will blow. The reason is simple. Suppose diode D_2 in Fig. 4.45 is shorted. Then diode D_2 will behave as a wire. When diode D_1 is forward biased, the transformer secondary will be shorted through D_1. This will cause excessive current to flow in the secondary (and hence in the primary), causing the primary fuse to blow.

(*ii*) If **one diode in the centre-tap full-wave rectifier opens**, the output from the rectifier will resemble the output from a half-wave rectifier. The remedy is to replace the diode.

Bridge Rectifier Faults. The transformer faults and their remedies for bridge rectifier circuits are the same as for centre-tap full-wave rectifier. Again symptoms for shorted and open diodes in the bridge rectifier are the same as those for the centre-tap circuit. In the case of bridge circuit, you simply have more diodes that need to be tested.

4.19 Nature of Rectifier Output

It has already been discussed that the output of a rectifier is pulsating d.c. as shown in Fig. 4.46. In fact, if such a waveform is carefully analysed, it will be found that it contains a d.c. component and an a.c. component. The a.c. component is responsible for the *pulsations in the wave. The reader may wonder how a pulsating d.c. voltage can have an a.c. component when the voltage never becomes negative. The answer is that any wave which varies in a regular manner has an a.c. component.

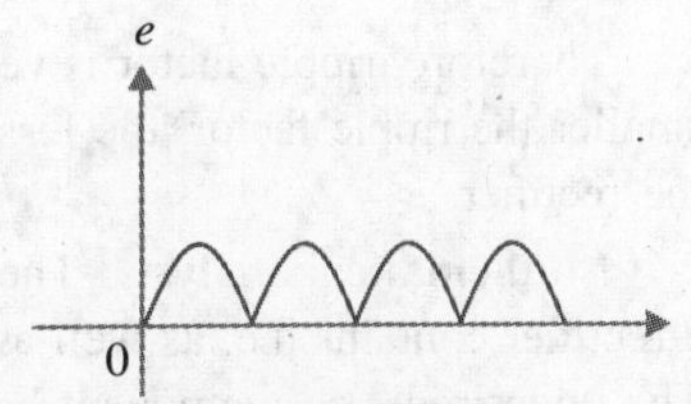

Fig. 4.46

The fact that a pulsating d.c. contains both d.c. and a.c. components can be beautifully illustrated by referring to Fig. 4.47. Fig. 4.47 (*i*) shows a pure d.c. component, whereas Fig. 4.47 (*ii*) shows the **a.c. component. If these two waves are added together, the resulting wave will be as shown in Fig. 4.47 (*iii*). It is clear that the wave shown in Fig. 4.47 (*iii*)

* Means changing output voltage.

** Although the a.c. component is not a sine-wave, yet it is alternating one.

never becomes negative, although it contains both a.c. and d.c. components. The striking resemblance between the rectifier output wave shown in Fig. 4.46 and the wave shown in Fig. 4.47 (*iii*) may be noted.

It follows, therefore, that a pulsating output of a rectifier contains a d.c. component and an a.c. component.

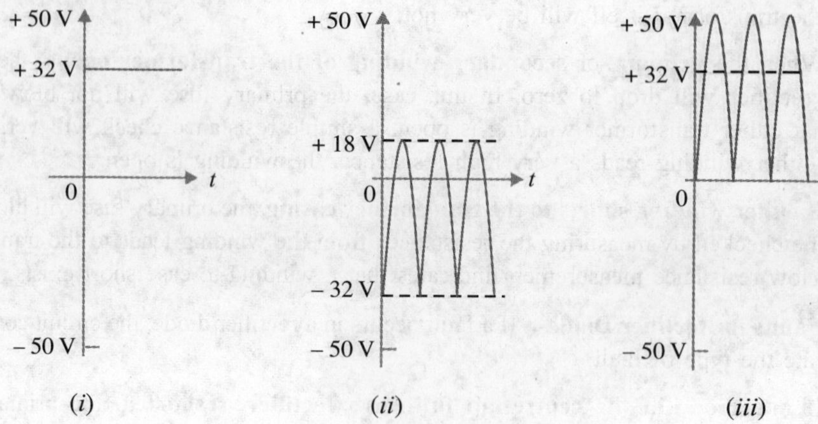

Fig. 4.47

4.20 Ripple Factor

The output of a rectifier consists of a d.c. component and an a.c. component (also known as *ripple*). The a.c. component is undesirable and accounts for the pulsations in the rectifier output. The effectiveness of a rectifier depends upon the magnitude of a.c. component in the output ; the smaller this component, the more effective is the rectifier.

The ratio of r.m.s. value of a.c. component to the d.c. component in the rectifier output is known as ***ripple factor*** *i.e.*

$$\text{Ripple factor} = \frac{\text{r.m.s. value of a.c component}}{\text{value of d.c. component}} = \frac{I_{ac}}{I_{dc}}$$

Therefore, ripple factor is very important in deciding the effectiveness of a rectifier. The smaller the ripple factor, the lesser the effective a.c. component and hence more effective is the rectifier.

Mathematical analysis. The output current of a rectifier contains d.c. as well as a.c. component. The undesired a.c. component has a frequency of 100 Hz (*i.e.* double the supply frequency 50 Hz) and is called the *ripple* (See Fig. 4.48). It is a fluctuation superimposed on the d.c. component.

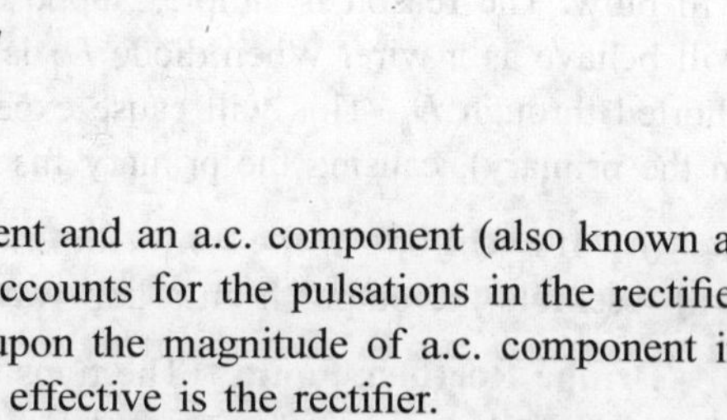

Fig. 4.48

By definition, the effective (*i.e.* r.m.s.) value of total load current is given by ;

$$I_{rms} = \sqrt{I_{dc}^2 + I_{ac}^2}$$

or $$I_{ac} = \sqrt{I_{rms}^2 - I_{dc}^2}$$

Dividing throughout by I_{dc}, we get,

$$\frac{I_{ac}}{I_{dc}} = \frac{1}{I_{dc}}\sqrt{I_{rms}^2 - I_{dc}^2}$$

But I_{ac}/I_{dc} is the ripple factor.

$$\therefore \quad \text{Ripple factor} = \frac{1}{I_{dc}}\sqrt{I_{rms}^2 - I_{dc}^2} = \sqrt{\left(\frac{I_{rms}}{I_{dc}}\right)^2 - 1}$$

(*i*) **For half-wave rectification.** In half-wave rectification,

$$I_{rms} = I_m/2 \; ; \; I_{dc} = I_m/\pi$$

$$\therefore \quad \text{Ripple factor} = \sqrt{\left(\frac{I_m/2}{I_m/\pi}\right)^2 - 1} = 1.21$$

It is clear that a.c. component exceeds the d.c. component in the output of a half-wave rectifier. This results in greater pulsations in the output. Therefore, half-wave rectifier is ineffective for conversion of a.c. into d.c.

(*ii*) **For full-wave rectification.** In full-wave rectification,

$$I_{rms} = \frac{I_m}{\sqrt{2}} \; ; \; I_{dc} = \frac{2\,I_m}{\pi}$$

$$\therefore \quad \text{Ripple factor} = \sqrt{\left(\frac{I_m/\sqrt{2}}{2\,I_m/\pi}\right)^2 - 1} = 0.48$$

i.e. $\dfrac{\text{effective a.c. component}}{\text{d.c. component}} = 0.48$

This shows that in the output of a full-wave rectifier, the d.c. component is more than the a.c. component. Consequently, the pulsations in the output will be less than in half-wave rectifier. For this reason, full-wave rectification is invariably used for conversion of a.c. into d.c.

Example 4.36. *Using d.c. and a.c. voltmeters we measure filter output voltage and obtain readings of 25 V d.c. and 1.5 V r.m.s. Calculate the ripple factor of the filter output voltage.*

Solution. We measure the output voltage of a filter circuit using a d.c. voltmeter and an a.c. voltmeter. The d.c. meter will read only the average or d.c. level of the output voltage. The a.c. voltmeter will read only the r.m.s. value of a.c. component of the output voltage. The smaller the a.c. variation (*i.e.* ripple) *w.r.t.* the d.c. level, the better the filter circuit's operation.

$$\therefore \quad \text{Ripple factor} = \frac{V_{ac}\,(\text{r.m.s.})}{V_{dc}} \times 100 = \frac{1.5\text{ V}}{25\text{ V}} \times 100 = \mathbf{6\%}$$

Example 4.37. *A power supply A delivers 10 V d.c. with a ripple of 0.5 V r.m.s. while the power supply B delivers 25 V d.c. with a ripple of 1 mV r.m.s. Which is better power supply?*

Solution. The lower the ripple factor of a power supply, the better it is.

For power supply A

$$\text{Ripple factor} = \frac{V_{ac(r.m.s.)}}{V_{dc}} \times 100 = \frac{0.5}{10} \times 100 = 5\%$$

For power supply B

$$\text{Ripple factor} = \frac{V_{ac(r.m.s.)}}{V_{dc}} \times 100 = \frac{0.001}{25} \times 100 = 0.004\%$$

Clearly, **power supply *B* is better.**

4.21 Comparison of Rectifiers

Rectifier type : Half-wave   Full-wave Centre-tap   Bridge Rectifier

Schematic diagram :

Typical output waveform :

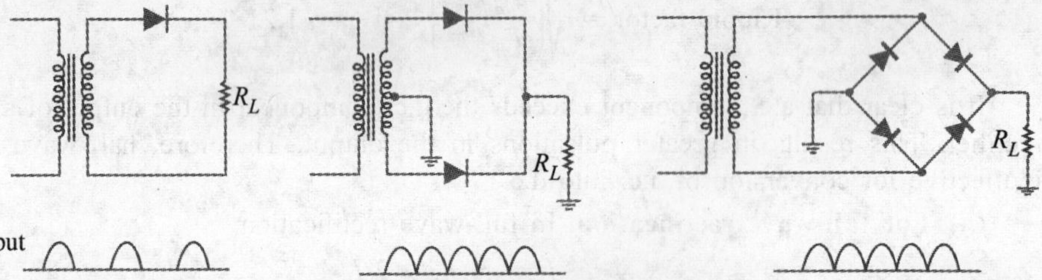

S. No.	Particulars	Half-wave	Centre-tap	Bridge type
1.	No. of diodes	1	2	4
2.	Transformer necessary	no	yes	no
3.	Max. efficiency	40.6%	81.2%	81.2%
4.	Ripple factor	1.21	0.48	0.48
5.	Output frequency	f_{in}	$2\,f_{in}$	$2\,f_{in}$
6.	Peak inverse voltage	V_m	$2\,V_m$	V_m

A comparison among the three rectifier circuits must be made very judiciously. Although bridge circuit has some disadvantages, it is the best circuit from the viewpoint of overall performance. When cost of the transformer is the main consideration in a rectifier assembly, we invariably use the bridge circuit. This is particularly true for large rectifiers which have a low-voltage and a high-current rating.

4.22 Filter Circuits

Generally, a rectifier is required to produce pure d.c. supply for using at various places in the electronic circuits. However, the output of a rectifier has pulsating *character *i.e.* it contains a.c. and d.c. components. The a.c. component is undesirable and must be kept away from the load. To do so, a *filter circuit* is used which removes (or *filters out*) the a.c. component and allows only the d.c. component to reach the load.

A* filter circuit *is a device which removes the a.c. component of rectifier output but allows the d.c. component to reach the load.

Obviously, a filter circuit should be installed between the rectifier and the load as shown in

* If such a d.c. is applied in an electronic circuit, it will produce a *hum*.

Fig. 4.49. A filter circuit is generally a combination of inductors (*L*) and capacitors (*C*). The filtering action of *L* and *C* depends upon the basic electrical principles. A capacitor passes a.c. readily but does not *pass d.c. at all. On the other hand, an inductor †opposes a.c. but allows d.c. to pass through it. It then becomes clear that suitable network of *L* and *C* can effectively remove the a.c. component, allowing the d.c. component to reach the load.

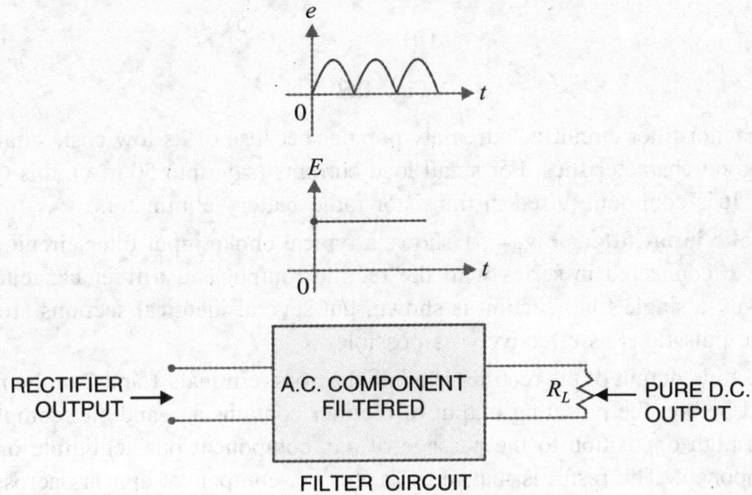

Fig. 4.49

4.23 Types of Filter Circuits

The most commonly used filter circuits are *capacitor filter, choke input filter* and *capacitor input filter or π-filter.* We shall discuss these filters in turn.

(*i*) **Capacitor filter.** Fig. 4.50 (*ii*) shows a typical capacitor filter circuit. It consists of a capacitor *C* placed across the rectifier output in parallel with load R_L. The pulsating direct voltage of the rectifier is applied across the capacitor. As the rectifier voltage increases, it charges the capacitor and also supplies current to the load. At the end of quarter cycle [Point *A* in Fig. 4.50 (*iii*)], the capacitor is charged to the peak value V_m of the rectifier voltage. Now, the rectifier voltage starts to decrease. As this occurs, the capacitor discharges through the load and voltage across it (*i.e.* across parallel combination of *R-C*) decreases as shown by the line *AB* in Fig. 4.50 (*iii*). The voltage across load will decrease only slightly because immediately the next voltage peak comes and recharges the capacitor. This process is repeated again and again and the output voltage waveform becomes *ABCDEFG*. It may be seen that very little ripple is left in the output. Moreover, output voltage is higher as it remains substantially near the peak value of rectifier output voltage.

* A capacitor offers infinite reactance to d.c. For d.c., $f = 0$.

$$\therefore \quad X_C = \frac{1}{2\pi fC} = \frac{1}{2\pi \times 0 \times C} = \infty$$

Hence, a capacitor does not allow d.c. to pass through it.

† We know $X_L = 2\pi fL$. For d.c., $f = 0$ and, therefore, $X_L = 0$. Hence inductor passes d.c. quite readily. For a.c., it offers opposition and drops a part of it.

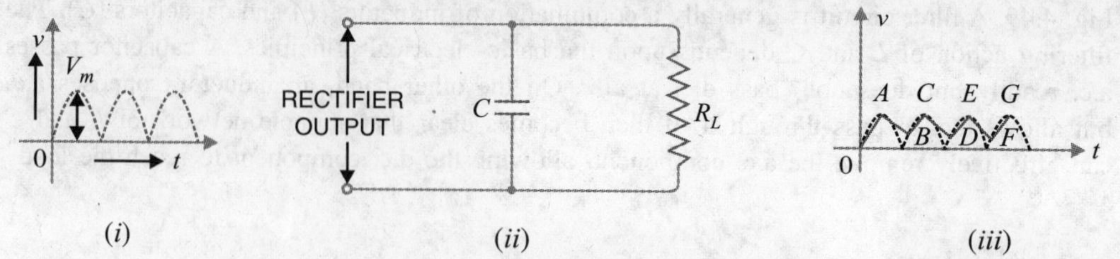

Fig. 4.50

The capacitor filter circuit is extremely popular because of its low cost, small size, little weight and good characteristics. For small load currents (say upto 50 mA), this type of filter is preferred. It is commonly used in transistor radio battery eliminators.

***(ii)* Choke input filter.** Fig. 4.51 shows a typical choke input filter circuit. It consists of a *choke *L* connected in series with the rectifier output and a filter capacitor *C* across the load. Only a single filter section is shown, but several identical sections are often used to reduce the pulsations as effectively as possible.

The pulsating output of the rectifier is applied across terminals 1 and 2 of the filter circuit. As discussed before, the pulsating output of rectifier contains a.c. and d.c. components. The choke offers high opposition to the passage of a.c. component but negligible opposition to the d.c. component. The result is that most of the a.c. component appears across the choke while whole of d.c. component passes through the choke on its way to load. This results in the reduced pulsations at terminal 3.

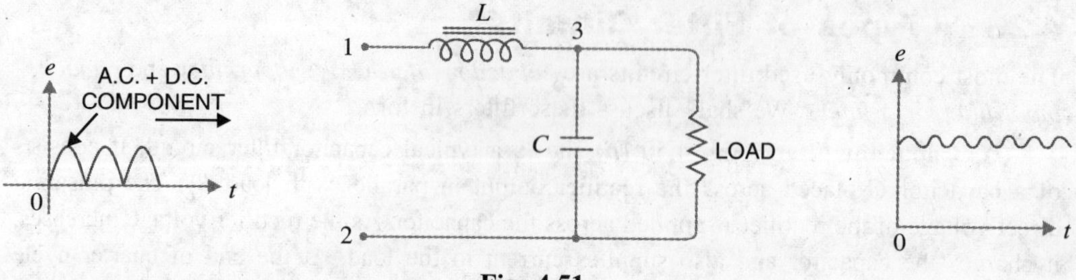

Fig. 4.51

At terminal 3, the rectifier output contains d.c. component and the remaining part of a.c. component which has managed to pass through the choke. Now, the low reactance of filter capacitor bypasses the a.c. component but prevents the d.c. component to flow through it. Therefore, only d.c. component reaches the load. In this way, the filter circuit has filtered out the a.c. component from the rectifier output, allowing d.c. component to reach the load.

***(iii)* Capacitor input filter or π-filter.** Fig. 4.52 shows a typical capacitor input filter or **π-filter. It consists of a filter capacitor C_1 connected across the rectifier output, a choke *L* in series and another filter capacitor C_2 connected across the load. Only one filter section is shown but several identical sections are often used to improve the smoothing action.

The pulsating output from the rectifier is applied across the input terminals (*i.e.* terminals

* The shorthand name of inductor coil is choke.

** The shape of the circuit diagram of this filter circuit appears like Greek letter π (pi) and hence the name π-filter.

1 and 2) of the filter. The filtering action of the three components *viz.* C_1, L and C_2 of this filter is described below :

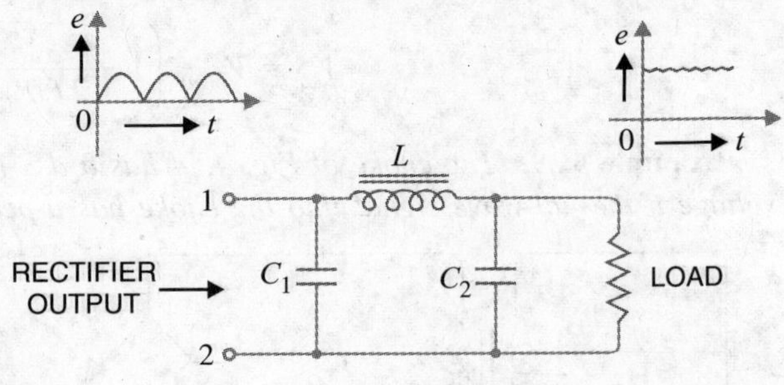

Fig. 4.52

(*a*) The *filter capacitor* C_1 offers low reactance to a.c. component of rectifier output while it offers infinite reactance to the d.c. component. Therefore, capacitor C_1 bypasses an appreciable amount of a.c. component while the d.c. component continues its journey to the choke L.

(*b*) The *choke L* offers high reactance to the a.c. component but it offers almost zero reactance to the d.c. component. Therefore, it allows the d.c. component to flow through it, while the *unbypassed a.c. component is blocked.

(*c*) The *filter capacitor* C_2 bypasses the a.c. component which the choke has failed to block. Therefore, only d.c. component appears across the load and that is what we desire.

Example 4.38. *For the circuit shown in Fig. 4.53, find the output d.c. voltage.*

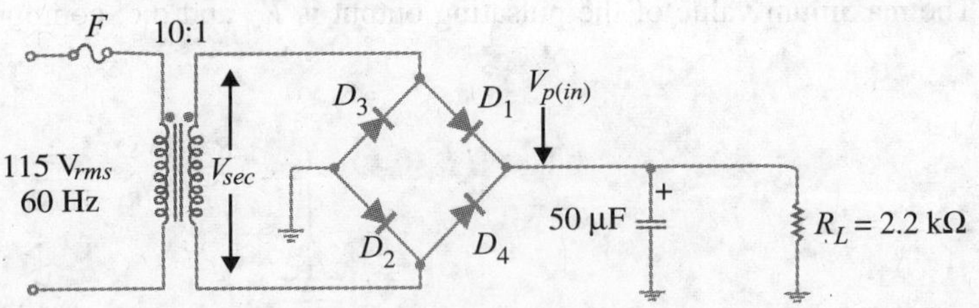

Fig. 4.53

Solution. It can be proved that output d.c. voltage is given by ;

$$V_{dc} = V_{p(in)} \left(1 - \frac{1}{2f\,R_L C}\right)$$

Here, $V_{p(in)}$ = Peak rectified full-wave voltage applied to the filter

f = Output frequency

Peak primary voltage, $V_{p(prim)} = \sqrt{2} \times 115 = 163 \text{ V}$

Peak secondary voltage, $V_{p(sec)} = \left(\frac{1}{10}\right) \times 163 = 16.3 \text{ V}$

Peak full-wave rectified voltage at the filter input is

$$V_{p\,(in)} = V_{p\,(sec)} - 2 \times 0.7 = 16.3 - 1.4 = 14.9 \text{ V}$$

For full-wave rectification, $f = 2f_{in} = 2 \times 60 = 120$ Hz

Now, $$\frac{1}{2f\,R_L C} = \frac{1}{2 \times 120 \times (2.2 \times 10^3) \times (50 \times 10^{-6})} = 0.038$$

* That part of a.c. component which could not be bypassed by capacitor C_1.

$$\therefore \quad V_{dc} = V_{p(in)}\left(1 - \frac{1}{2f\,R_L C}\right) = 14.9\,(1 - 0.038) = \mathbf{14.3\ V}$$

Example 4.39. *The choke of Fig. 4.54 has a d.c. resistance of 25 Ω. What is the d.c. voltage if the full-wave signal into the choke has a peak value of 25.7 V ?*

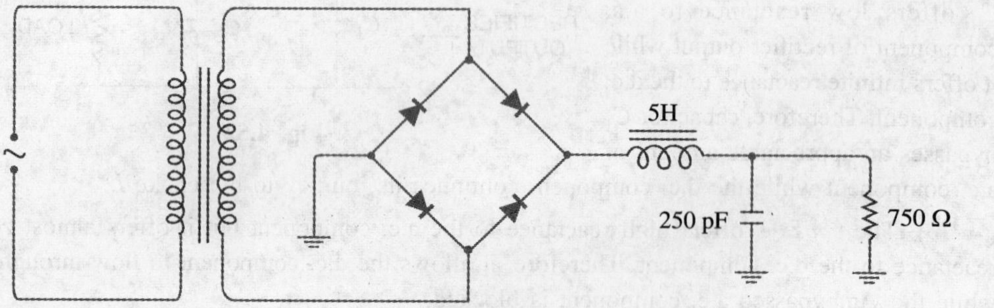

Fig. 4.54

Solution. The output of a full-wave rectifier has a d.c. component and an a.c. component. Due to the presence of a.c. component, the rectifier output has a pulsating character as shown in Fig. 4.55. The maximum value of the pulsating output is V_m and d.c. component is $V'_{dc} = 2V_m/\pi$.

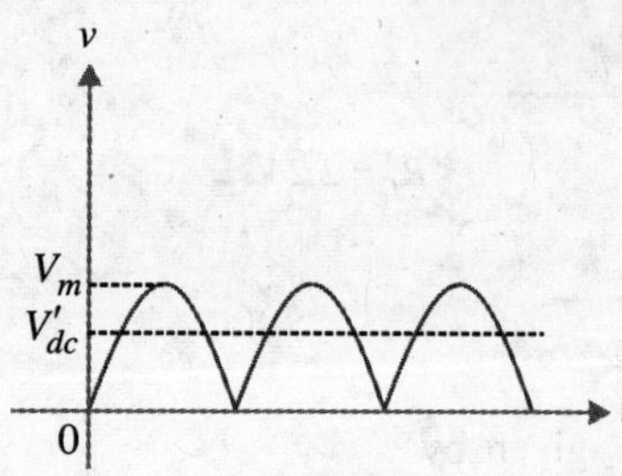

Fig. 4.55

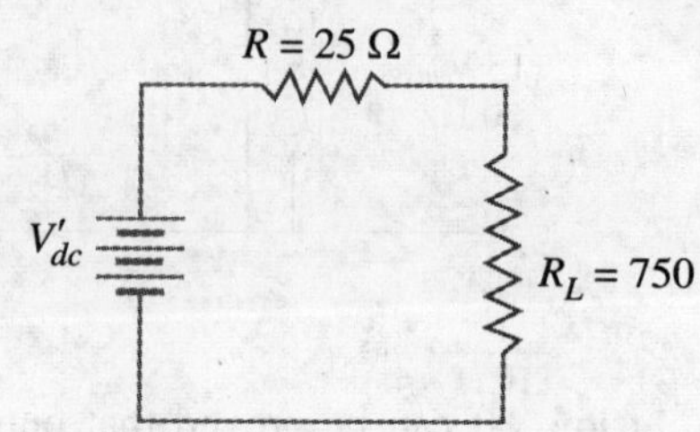

Fig. 4.56

For d.c. component V'_{dc}, the choke resistance is in series with the load as shown in Fig. 4.56.

$$\therefore \quad \text{Voltage across load, } V_{dc} = \frac{V'_{dc}}{R + R_L} \times R_L$$

$$\text{In our example, } V'_{dc} = \frac{2V_m}{\pi} = \frac{2 \times 25.7}{\pi} = 16.4\text{ V}$$

$$\therefore \quad \text{Voltage across load, } V_{dc} = \frac{V'_{dc}}{R + R_L} \times R_L = \frac{16.4}{25 + 750} \times 750 = \mathbf{15.9\ V}$$

The voltage across the load is 15.9 V d.c. *plus* a small ripple.

4.24 Voltage Multipliers

With a diode, we can build a rectifier to produce a d.c. voltage that is nearly equal to the peak value of input a.c. voltage. We can also use diodes and capacitors to build a circuit that will provide a *d.c. output* that is multiple of the *peak input a.c. voltage.* Such a circuit is called a voltage multiplier. For example, a *voltage doubler* will provide a d.c. output that is *twice* the peak input a.c. voltage, a *voltage tripler* will provide a d.c. output that is *three times* the peak input a.c. voltage and so on.

While voltage multipliers provide d.c. output that is much greater than the peak input a.c. voltage, there is no power amplification and law of conservation of energy holds good. When a voltage multiplier *increases* the peak input voltage by a factor *n*, the peak input current is *decreased* by approximately the same factor. Thus the actual power output from a voltage multiplier will *never* be greater than the input power. In fact, there are losses in the circuit (*e.g.* in diodes, capacitors *etc.*) so that the output power will actually be *less than* the input power.

4.25 Half-Wave Voltage Doubler

A half-wave voltage doubler consists of two diodes and two capacitors connected in a manner as shown in Fig. 4.57. It will be shown that if the peak input a.c. voltage is $V_{S(pk)}$, the d.c. output voltage will be $2V_{S(pk)}$ provided the diodes are ideal (this assumption is fairly reasonable). The basic idea in a voltage multiplier is to charge each capacitor to the peak input a.c. voltage and to arrange the capacitors so that their stored voltages will add.

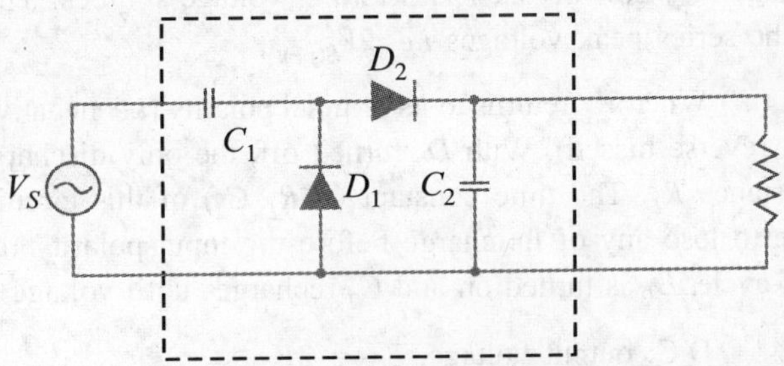

HALF-WAVE VOLTAGE DOUBLER

Fig. 4.57

Circuit action. We now discuss the working of a half-wave voltage doubler.

(*i*) During the negative half-cycle of a.c. input voltage [See Fig. 4.58 (*i*)], diode D_1 is forward biased and diode D_2 is reverse biased [See Fig. 4.58 (*i*)]. Therefore, diode D_1 can be represented by a *short* and diode D_2 as an *open*. The equivalent circuit then becomes as shown in Fig. 4.58 (*ii*).

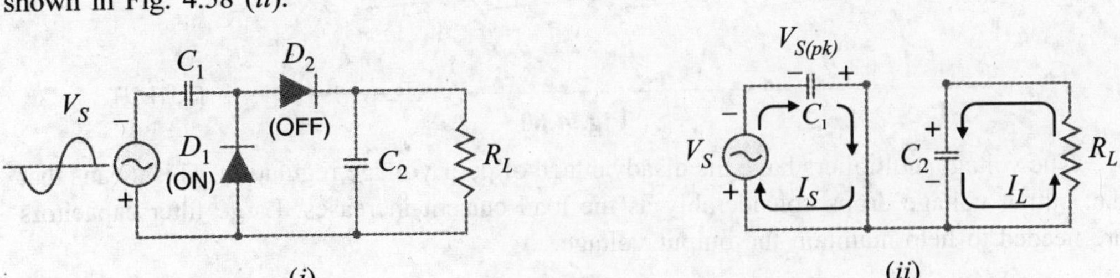

Fig. 4.58

As you can see [See Fig. 4.58 (*ii*)], C_1 will charge until voltage across it becomes equal to peak value of source voltage [$V_{S(pk)}$]. At the same time, C_2 will be in the process of discharging through the load R_L (The source of this charge on C_2 will be explained in a moment). *Note that in all figures electron flow is shown.*

(*ii*) When the polarity of the input a.c. voltage reverses (*i.e.* during positive half-cycle), the circuit conditions become as shown in Fig. 4.59 (*i*). Now D_1 is reverse biased and D_2 is forward biased and the equivalent circuit becomes as shown in Fig. 4.59 (*ii*).

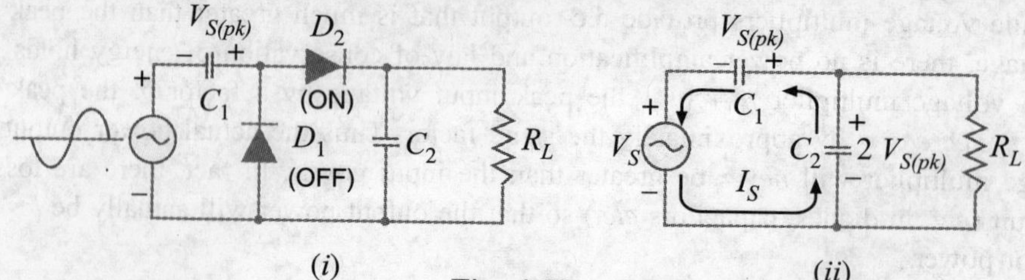

Fig. 4.59

Referring to Fig. 4.59 (*ii*), it is easy to see that C_1 (charged to $V_{S\,(pk)}$) and the source voltage (V_S) now act as *series-aiding* voltage sources. Thus C_2 will be charged to the sum of the series peak voltages *i.e.* $2V_{S(pk)}$.

(*iii*) When V_S returns to its original polarity (*i.e.* negative half-cycle), D_2 is again turned off (*i.e.* reverse biased). With D_2 turned off, the only discharge path for C_2 is through the load resistance R_L. The time constant (= $R_L\ C_2$) of this circuit is so adjusted that C_2 has little time to lose any of its charge before the input polarity reverses again. During the positive half-cycle, D_2 is turned on and C_2 recharges until voltage across it is again equal to $2V_{S\,(pk)}$.

$\therefore$ D.C. output voltage, $V_{dc} = 2V_{S\,(pk)}$

Since C_2 barely discharges between input cycles, the output waveform of the half-wave voltage doubler closely resembles that of a filtered half-wave rectifier. Fig. 4.60 shows the input and output waveforms for a half-wave voltage doubler.

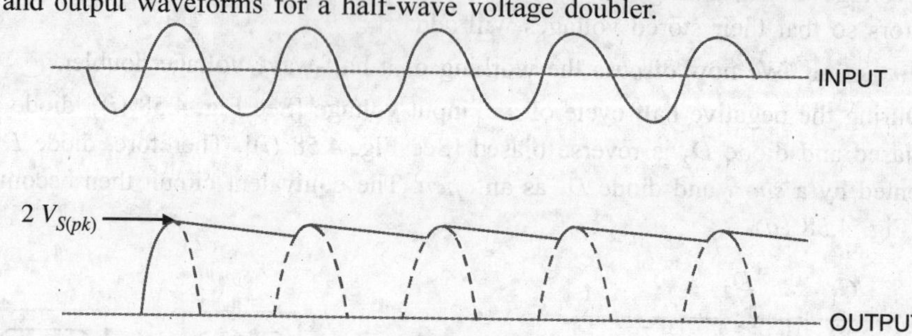

Fig. 4.60

The voltage multipliers have the disadvantage of poor voltage regulation. This means that d.c. output voltage drops considerably as the load current increases. Large filter capacitors are needed to help maintain the output voltage.

4.26 Voltage Stabilisation

A rectifier with an appropriate filter serves as a good source of d.c. output. However, the major disadvantage of such a power supply is that the output voltage changes with the variations in the input voltage or load. Thus, if the input voltage increases, the d.c. output voltage of the rectifier also increases. Similarly, if the load current increases, the output voltage falls due to the voltage drop in the rectifying element, filter chokes, transformer winding *etc.* In many electronic applications, it is desired that the output voltage should remain constant regardless of the variations in the input voltage or load. In order to ensure this, a voltage stabilising device, called voltage stabiliser is used. Several stabilising circuits have been designed but only *zener diode* as a voltage stabiliser will be discussed.

4.27 Zener Diode

It has already been discussed that when the reverse bias on a crystal diode is increased, a critical voltage, called *breakdown voltage* is reached where the reverse current increases sharply to a high value. The breakdown region is the knee of the reverse characteristic as shown in Fig. 4.61. The satisfactory explanation of this breakdown of the junction was first given by the American scientist C. Zener. Therefore, the breakdown voltage is sometimes called *zener voltage* and the sudden increase in current is known as *zener current*.

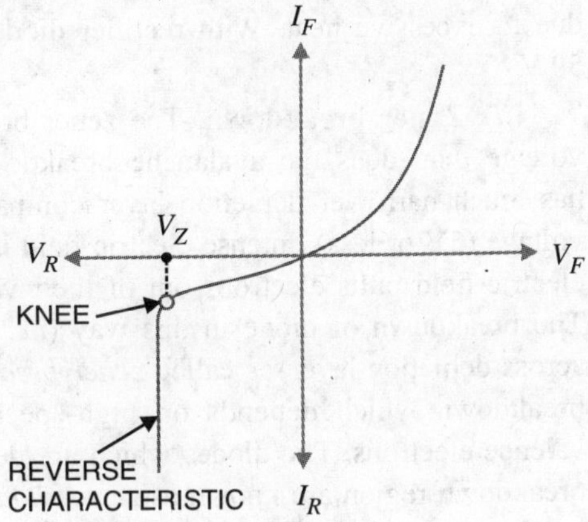

Fig. 4.61

The breakdown or zener voltage depends upon the amount of doping. If the diode is heavily doped, depletion layer will be thin and consequently the breakdown of the junction will occur at a lower reverse voltage. On the other hand, a lightly doped diode has a higher breakdown voltage. When an ordinary crystal diode is properly doped so that it has a sharp breakdown voltage, it is called a zener diode.

A properly doped crystal diode which has a sharp breakdown voltage is known as a **zener diode.**

Fig. 4.62 shows the symbol of a zener diode. It may be seen that it is just like an ordinary diode except that the bar is turned into z-shape. The following points may be noted about the zener diode :

Fig. 4.62

(*i*) A zener diode is like an ordinary diode except that it is properly doped so as to have a sharp breakdown voltage.

(*ii*) A zener diode is always reverse connected *i.e.* it is always reverse biased.

(*iii*) A zener diode has sharp breakdown voltage, called zener voltage V_Z.

(*iv*) When forward biased, its characteristics are just those of ordinary diode.

(*v*) The zener diode is not immediately burnt just because it has entered the *breakdown region. As long as the external circuit connected to the diode limits the diode current to less than *burn out* value, the diode will not burn out.

4.28 Reverse Breakdown of Diodes

There is a limit to how much reverse voltage a diode can withstand before it is destroyed due to excessive heat. There are two types of diode breakdown due to reverse bias *viz.* (*i*) Avalanche breakdown (*ii*) Zener breakdown.

(*i*) **Avalanche breakdown.** The avalanche breakdown occurs at high reverse voltages.

* The current is limited only by both external resistance and the power dissipation of zener diode.

If reverse voltage across the diode is continuously increased, eventually breakdown voltage is reached. Once the breakdown voltage is reached, the thermally generated free electrons acquire enough energy to knock out valence electrons from the covalent bonds. These newly created free electrons further dislodge valence electrons from covalent bonds. The process is geometric and we get an *avalanche* of free electrons. As a result, the reverse current suddenly increases to a large value (called *avalanche current*) and the diode is destroyed due to excessive heat. With rectifier diodes, the breakdown voltage is usually greater than 50 V.

(*ii*) **Zener breakdown.** The zener breakdown occurs at much lower values of reverse voltage than does the avalanche breakdown. The zener diode is heavily doped so that it has much narrower depletion layer compared to a rectifier diode. Even with small reverse voltage (5V or less), intense electric field is produced across the depletion layer. This intense electric field pulls electrons out of their valence orbits and the diode goes into breakdown. The breakdown of diode in this way (*i.e.* producing free electrons by intense electric field across depletion layer) is called *zener breakdown*. This is distinctly different from avalanche breakdown which depends on high-speed thermally generated free electrons dislodging valence electrons. The diodes which are designed to be continuously operated in the reverse breakdown region are known as *zener diodes* regardless of whether the breakdown is caused by the avalanche or zener effect. Note that the zener diodes with low V_Z (zener voltage) ratings experience zener breakdown while the ones with higher V_Z ratings usually experience avalanche breakdown.

4.29 Equivalent Circuit of Zener Diode

The analysis of circuits using zener diodes can be made quite easily by replacing the zener diode by its equivalent circuit.

(*i*) **"ON" state.** When reverse voltage across a zener diode is equal to or more than breakdown voltage V_Z, the current increases very sharply. In this region, the curve is almost vertical. It means that voltage across zener diode is constant at V_Z even though the current through it changes. Therefore, in the breakdown region, an *ideal zener diode can be represented by a battery of voltage V_Z as shown in Fig. 4.63 (*ii*). Under such conditions, the zener diode is said to be in the "ON" state.

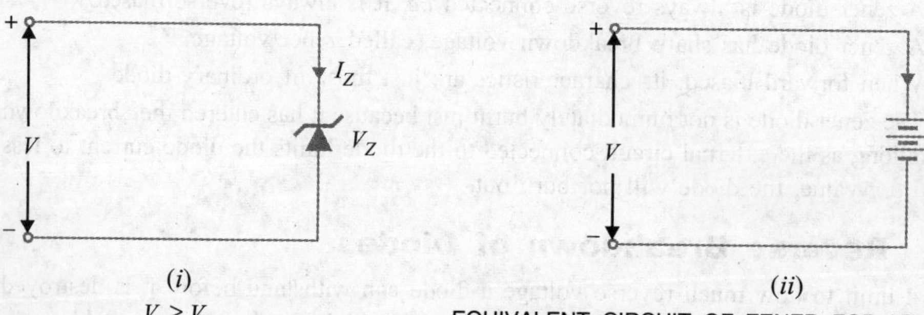

(*i*)
$V \geq V_Z$

(*ii*)
EQUIVALENT CIRCUIT OF ZENER FOR "ON" STATE

Fig. 4.63

* This assumption is fairly reasonable as the impedance of zener diode is quite small in the breakdown region.

(*ii*) **"OFF" state.** When the reverse voltage across the zener diode is less than V_Z but greater than 0 V, the zener diode is in the "OFF" state. Under such conditions, the zener diode can be represented by an open-circuit as shown in Fig. 4.64 (*ii*).

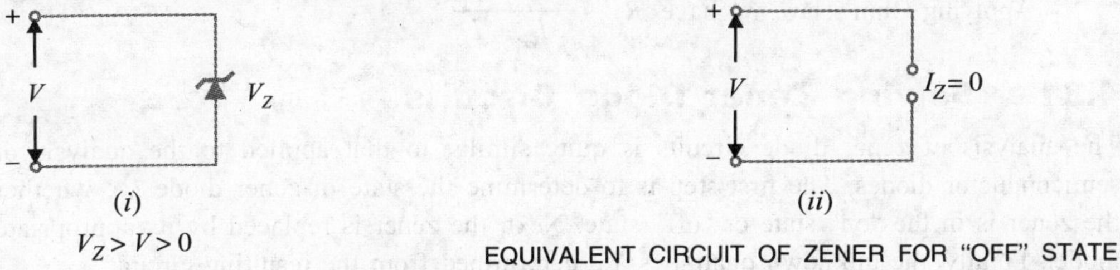

EQUIVALENT CIRCUIT OF ZENER FOR "OFF" STATE

Fig. 4.64

4.30 Zener Diode as Voltage Stabiliser

A zener diode can be used as a voltage regulator to provide a constant voltage from a source whose voltage may vary over sufficient range. The circuit arrangement is shown in Fig. 4.65 (*i*). The zener diode of zener voltage V_Z is reverse connected across the load R_L across which constant output is desired. The series resistance R absorbs the output voltage fluctuations so as to maintain constant voltage across the load. It may be noted that the zener will maintain a constant voltage V_Z (= E_0) across the load so long as the input voltage does not fall below V_Z.

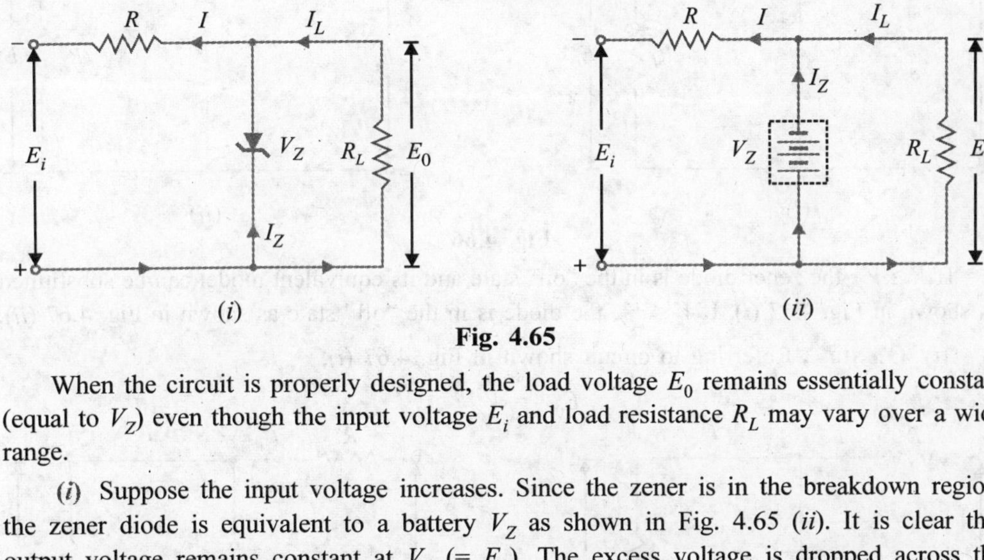

Fig. 4.65

When the circuit is properly designed, the load voltage E_0 remains essentially constant (equal to V_Z) even though the input voltage E_i and load resistance R_L may vary over a wide range.

(*i*) Suppose the input voltage increases. Since the zener is in the breakdown region, the zener diode is equivalent to a battery V_Z as shown in Fig. 4.65 (*ii*). It is clear that output voltage remains constant at V_Z (= E_0). The excess voltage is dropped across the series resistance R. This will cause an increase in the value of total current I. The zener will conduct the increase of current in I while the load current remains constant. Hence, output voltage E_0 remains constant irrespective of the changes in the input voltage E_i.

(*ii*) Now suppose that input voltage is constant but the load resistance R_L decreases. This will cause an increase in load current. The extra current cannot come from the source because drop in R (and hence source current I) will not change as the zener is within its regulating range. The additional load current will come from a decrease in zener current I_Z. Consequently, the output voltage stays at constant value.

$$\text{Voltage drop across } R = E_i - E_0$$

$$\text{Current through } R,\ I = I_Z + I_L$$

$$\text{Applying Ohm's law, we have, } R = \frac{E_i - E_0}{I_Z + I_L}$$

4.31 Solving Zener Diode Circuits

The analysis of zener diode circuits is quite similar to that applied to the analysis of semiconductor diodes. The first step is to determine the state of zener diode *i.e.* whether the zener is in the "on" state or "off" state. Next, the zener is replaced by its appropriate model. Finally, the unknown quantities are determined from the resulting circuit.

1. E_i and R_L fixed. This is the simplest case and is shown in Fig. 4.66 (*i*). Here the applied voltage E_i as well as load R_L is fixed. The first step is to find the state of zener diode. This can be determined by removing the zener from the circuit and calculating the voltage V across the resulting open-circuit as shown in Fig. 4.66 (*ii*).

$$V = E_0 = \frac{R_L E_i}{R + R_L}$$

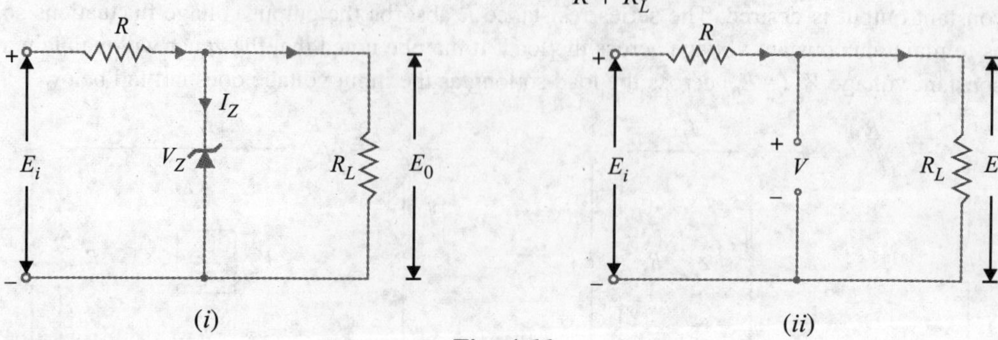

Fig. 4.66

If $V \geq V_Z$, the zener diode is in the "on" state and its equivalent model can be substituted as shown in Fig. 4.67 (*i*). If $V < V_Z$, the diode is in the "off" state as shown in Fig. 4.67 (*ii*).

(*i*) **On state.** Referring to circuit shown in Fig. 4.67 (*i*),

$$E_0 = V_Z$$

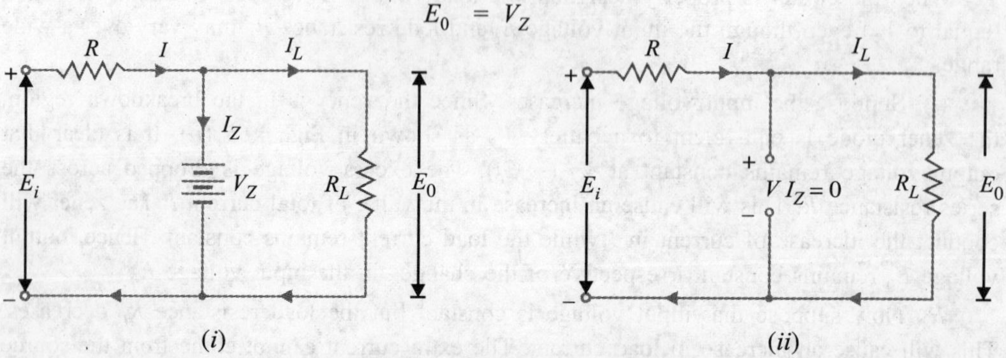

Fig. 4.67

$$I_Z = I - I_L \text{ where } I_L = \frac{E_0}{R_L} \text{ and } I = \frac{E_i - E_0}{R}$$

Power dissipated in zener, $P_Z = V_Z I_Z$

(*ii*) **Off state.** Referring to the circuit shown in Fig. 4.67 (*ii*),

$$I = I_L \quad \text{and} \quad I_Z = 0$$

$$V_R = E_i - E_0 \quad \text{and} \quad V = E_0 \quad (V < V_Z)$$

$$\therefore \quad P_Z = V I_Z = V(0) = 0$$

2. Fixed E_i and variable R_L. This case is shown in Fig. 4.68. Here the applied voltage (E_i) is fixed while load resistance R_L (and hence load current I_L) changes. Note that there is a definite range of R_L values (and hence I_L values) which will ensure the zener diode to be in "on" state. Let us calculate that range of values.

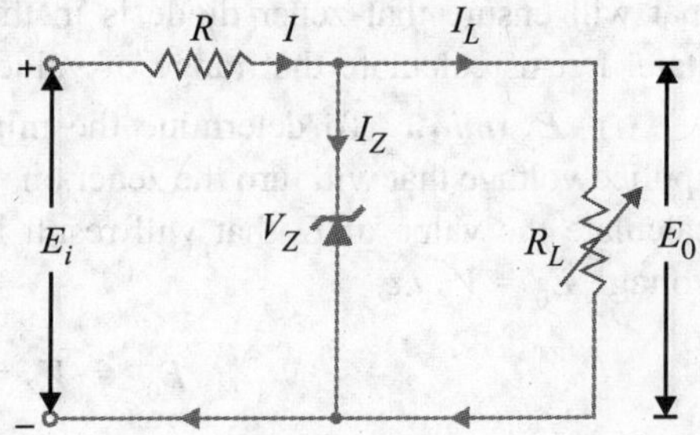

Fig. 4.68

(*i*) **R_{Lmin} and I_{Lmax}.** Once the zener is in the "on" state, load voltage E_0 (= V_Z) is constant. As a result, when load resistance is minimum (*i.e.* R_{Lmin}), load current will be maximum ($I_L = E_0/R_L$). In order to find the minimum load resistance that will turn the zener on, we simply calculate the value of R_L that will result in $E_0 = V_Z$ *i.e.*

$$E_0 = V_Z = {}^{*}\frac{R_L E_i}{R + R_L}$$

$$\therefore \quad R_{Lmin} = \frac{R V_Z}{E_i - V_Z} \qquad ...(i)$$

This is the minimum value of load resistance that will ensure that zener is in the "on" state. Any value of load resistance less than this value will result in a voltage E_0 across the load less than V_Z and the zener will be in the "off" state.

Clearly ; $$I_{Lmax} = \frac{E_0}{R_{Lmin}} = \frac{V_Z}{R_{Lmin}}$$

(*ii*) **I_{Lmin} and R_{Lmax}.** It is easy to see that when load resistance is maximum, load current is minimum.

Now, zener current, $I_Z = I - I_L$

When the zener is in the "on" state, I remains **fixed. This means that when I_L is maximum, I_Z will be minimum. On the other hand, when I_L is minimum, I_Z is maximum. If the maximum current that a zener can carry safely is ***I_{ZM}, then,

$$I_{Lmin} = I - I_{ZM}$$

and $$R_{Lmax} = \frac{E_0}{I_{Lmin}} = \frac{V_Z}{I_{Lmin}}$$

---

* If you remove the zener in the circuit shown in Fig. 4.68, then voltage V across the open-circuit is

$$V = \frac{R_L E_i}{R + R_L}$$

The zener will be turned on when $V = V_Z$.

** Voltage across R, $V_R = E_i - E_0$ and $I = V_R/R$. As E_i and E_0 are fixed, I remains the same.

*** Max. power dissipation in zener, $P_{ZM} = V_Z I_{ZM}$

If the load resistance exceeds this limiting value, the current through zener will exceed I_{ZM} and the device may burn out.

3. Fixed R_L and variable E_i. This case is shown in Fig. 4.69. Here the load resistance R_L is fixed while the applied voltage (E_i) changes. Note that there is a definite range of E_i values that will ensure that zener diode is in the "on" state. Let us calculate that range of values.

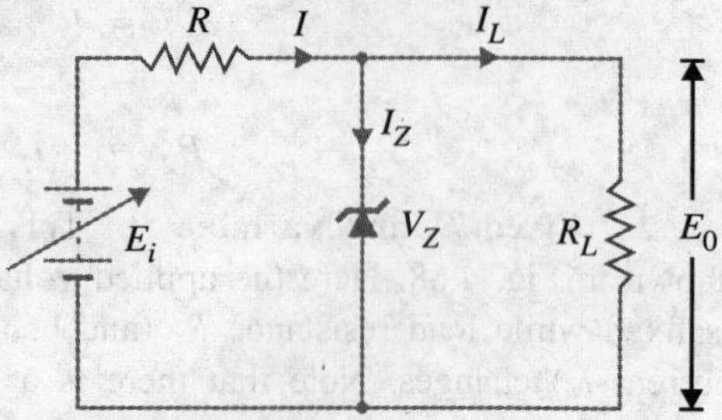

Fig. 4.69

(i) E_i (min). To determine the minimum applied voltage that will turn the zener on, simply calculate the value of E_i that will result in load voltage $E_0 = V_Z$ *i.e.*

$$E_0 = V_Z = \frac{R_L E_i}{R + R_L}$$

$$\therefore \quad E_{i\,(min)} = \frac{(R + R_L)\, V_Z}{R_L}$$

(ii) E_i (max)

Now, current through R, $I = I_Z + I_L$

Since I_L ($= E_0/R_L = V_Z/R_L$) is fixed, the value of I will be maximum when zener current is maximum *i.e.*

$$I_{max} = I_{ZM} + I_L$$

Now, $$E_i = I\,R + E_0$$

Since E_0 ($= V_Z$) is constant, the input voltage will be maximum when I is maximum.

$$\therefore \quad E_{i\,(max)} = I_{max}\, R + V_Z$$

Example 4.40. *For the circuit shown in Fig. 4.70 (i), find :*

(i) the output voltage

(ii) the voltage drop across series resistance

(iii) the current through zener diode.

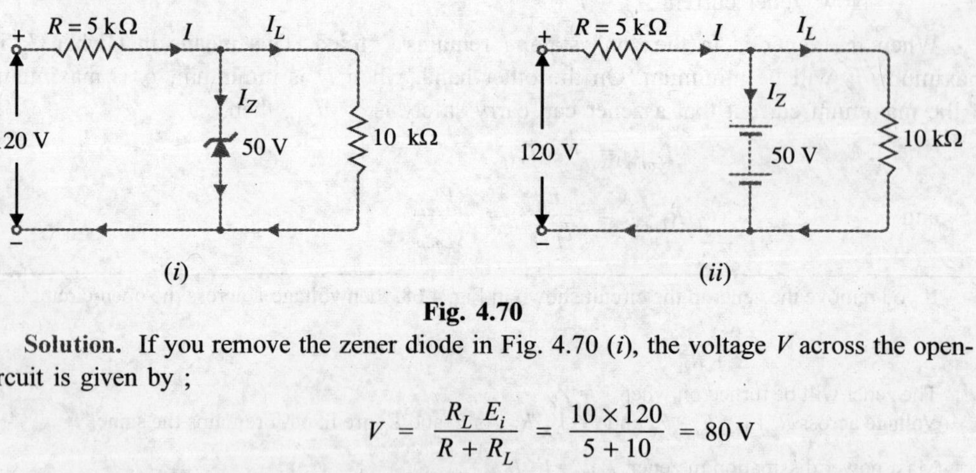

Fig. 4.70

Solution. If you remove the zener diode in Fig. 4.70 (*i*), the voltage V across the open-circuit is given by ;

$$V = \frac{R_L E_i}{R + R_L} = \frac{10 \times 120}{5 + 10} = 80 \text{ V}$$

Since voltage across zener diode is greater than V_Z (= 50 V), the zener is in the "on" state. It can, therefore, be represented by a battery of 50 V as shown in Fig. 4.70 (*ii*).

(*i*) Referring to Fig. 4.70 (*ii*),

$$\text{Output voltage} = V_Z = \mathbf{50\ V}$$

(*ii*) $$\text{Voltage drop across } R = \text{Input voltage} - V_Z = 120 - 50 = \mathbf{70\ V}$$

(*iii*) $$\text{Load current, } I_L = V_Z/R_L = 50\text{ V}/10\text{ k}\Omega = 5\text{ mA}$$

$$\text{Current through } R,\ I = \frac{70\text{ V}}{5\text{ k}\Omega} = 14\text{ mA}$$

Applying Kirchhoff's first law, $I = I_L + I_Z$

$$\therefore \quad \text{Zener current, } I_Z = I - I_L = 14 - 5 = \mathbf{9\ mA}$$

Example 4.41. *For the circuit shown in Fig. 4.71 (i), find the maximum and minimum values of zener diode current.*

Solution. The first step is to determine the state of the zener diode. It is easy to see that for the given range of voltages (80 – 120 V), the voltage across the zener is greater than V_Z (= 50 V). Hence the zener diode will be in the "on" state for this range of applied voltages. Consequently, it can be replaced by a battery of 50 V as shown in Fig. 4.71 (*ii*).

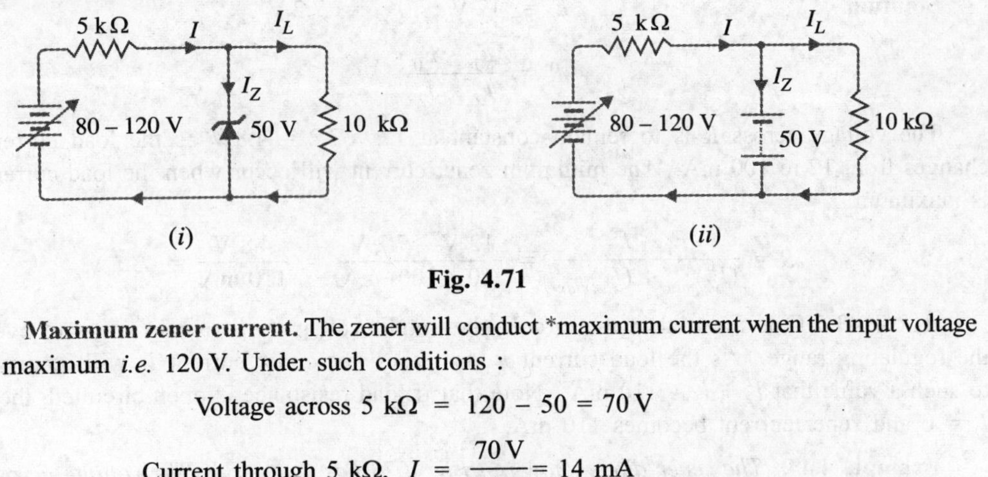

Fig. 4.71

Maximum zener current. The zener will conduct *maximum current when the input voltage is maximum *i.e.* 120 V. Under such conditions :

$$\text{Voltage across 5 k}\Omega = 120 - 50 = 70\text{ V}$$

$$\text{Current through 5 k}\Omega,\ I = \frac{70\text{ V}}{5\text{ k}\Omega} = 14\text{ mA}$$

$$\text{Load current, } I_L = \frac{50\text{ V}}{10\text{ k}\Omega} = 5\text{ mA}$$

Applying Kirchhoff's first law, $I = I_L + I_Z$

$$\therefore \quad \text{Zener current, } I_Z = I - I_L = 14 - 5 = \mathbf{9\ mA}$$

---

* $I_Z = I - I_L$. Since I_L (= V_Z/R_L) is fixed, I_Z will be maximum when I is maximum.

Now, $I = \dfrac{E_i - E_0}{R} = \dfrac{E_i - V_Z}{R}$. Since V_Z (= E_0) and R are fixed, I will be maximum when E_i is maximum and *vice-versa*.

Minimum zener current. The zener will conduct minimum current when the input voltage is minimum *i.e.* 80 V. Under such conditions, we have,

$$\text{Voltage across } 5\text{ k}\Omega = 80 - 50 = 30\text{ V}$$

$$\text{Current through } 5\text{ k}\Omega,\ I = \frac{30\text{ V}}{5\text{ k}\Omega} = 6\text{ mA}$$

$$\text{Load current, } I_L = 5\text{ mA}$$

$$\therefore \quad \text{Zener current, } I_Z = I - I_L = 6 - 5 = \mathbf{1\ mA}$$

Example 4.42. *A 7.2 V zener is used in the circuit shown in Fig. 4.72 and the load current is to vary from 12 to 100 mA. Find the value of series resistance R to maintain a voltage of 7.2 V across the load. The input voltage is constant at 12 V and the minimum zener current is 10 mA.*

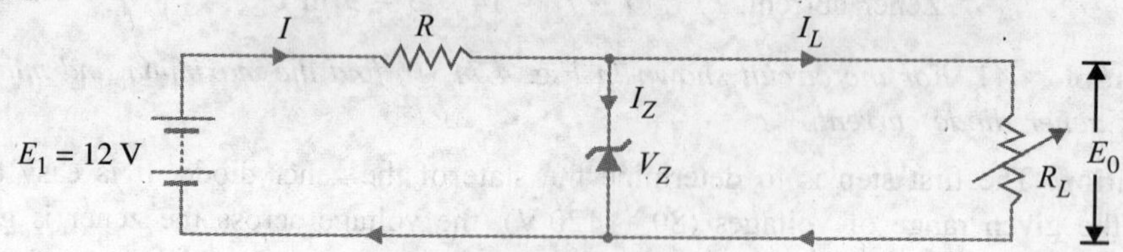

Fig. 4.72

Solution. $E_i = 12\text{ V}\ ;\quad V_Z = 7.2\text{ V}$

$$R = \frac{E_i - E_0}{I_Z + I_L}$$

The voltage across R is to remain constant at 12 – 7.2 = 4.8 V as the load current changes from 12 to 100 mA. The minimum zener current will occur when the load current is maximum.

$$\therefore \quad R = \frac{E_i - E_0}{(I_Z)_{min} + (I_L)_{max}} = \frac{12\text{ V} - 7.2\text{ V}}{(10 + 100)\text{ mA}} = \frac{4.8\text{ V}}{110\text{ mA}} = \mathbf{43.5\ \Omega}$$

If R = 43.5 Ω is inserted in the circuit, the output voltage will remain constant over the regulating range. As the load current I_L decreases, the zener current I_Z will increase to such a value that $I_Z + I_L$ = 110 mA. Note that if load resistance is open-circuited, then I_L = 0 and zener current becomes 110 mA.

Example 4.43. *The zener diode shown in Fig. 4.73 has V_Z = 18 V. The voltage across the load stays at 18 V as long as I_Z is maintained between 200 mA and 2 A. Find the value of series resistance R so that E_0 remains 18 V while input voltage E_i is free to vary between 22 V to 28 V.*

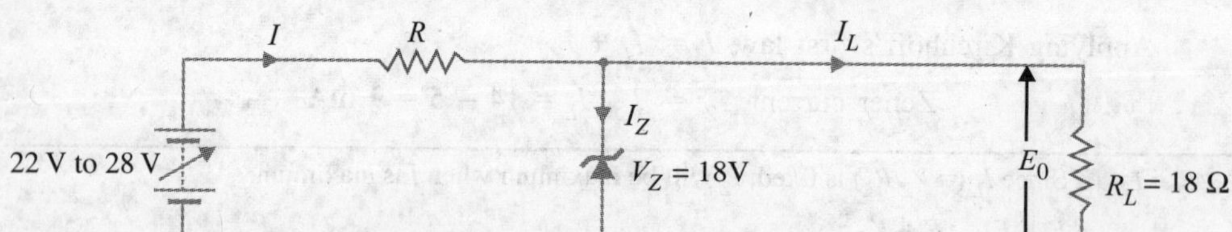

Fig. 4.73

Solution. The zener current will be minimum (*i.e.* 200 mA) when the input voltage is minimum (*i.e.* 22 V). The load current stays at constant value $I_L = V_Z / R_L = 18\ \text{V}/18\ \Omega = 1\ \text{A} = 1000\ \text{mA}$.

$$\therefore \quad R = \frac{E_i - E_0}{(I_Z)_{min} + (I_L)_{max}} = \frac{(22-18)\ \text{V}}{(200+1000)\ \text{mA}} = \frac{4\ \text{V}}{1200\ \text{mA}} = 3.33\ \Omega$$

Example 4.44. *A 10-V zener diode is used to regulate the voltage across a variable load resistor [See Fig. 4.74]. The input voltage varies between 13 V and 16 V and the load current varies between 10 mA and 85 mA. The minimum zener current is 15 mA. Calculate the value of series resistance R.*

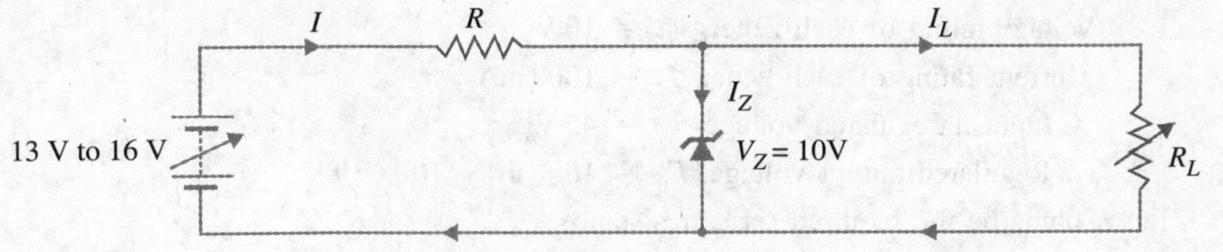

Fig. 4.74

Solution. The zener will conduct minimum current (*i.e.* 15 mA) when input voltage is minimum (*i.e.* 13 V).

$$\therefore \quad R = \frac{E_i - E_0}{(I_Z)_{min} + (I_L)_{max}} = \frac{(13-10)\ \text{V}}{(15+85)\ \text{mA}} = \frac{3\ \text{V}}{100\ \text{mA}} = \mathbf{30\ \Omega}$$

Example 4.45. *The circuit of Fig. 4.75 uses two zener diodes, each rated at 15 V, 200 mA. If the circuit is connected to a 45-volt unregulated supply, determine :*

(i) The regulated output voltage *(ii) The value of series resistance R*

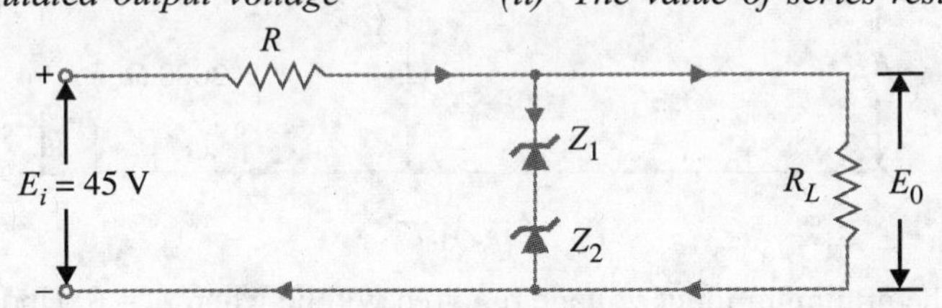

Fig. 4.75

Solution. When the desired regulated output voltage is higher than the rated voltage of the zener, two or more zeners are connected in series as shown in Fig. 4.75. However, in such circuits, care must be taken to select those zeners that have the same current rating.

Current rating of each zener, I_Z = 200 mA

Voltage rating of each zener, V_Z = 15 V

Input voltage, E_i = 45 V

(*i*) Regulated output voltage, E_0 = 15 + 15 = **30 V**

(*ii*) Series resistance, $R = \dfrac{E_i - E_0}{I_Z} = \dfrac{45-30}{200\ \text{mA}} = \dfrac{15\text{V}}{200\ \text{mA}} = 75\ \Omega$

Example 4.46. *What value of series resistance is required when three 10-watt, 10-volt, 1000 mA zener diodes are connected in series to obtain a 30-volt regulated output from a 45 volt d.c. power source?*

Solution. Fig. 4.76 shows the desired circuit. The worst case is at no-load because then zeners carry the maximum current.

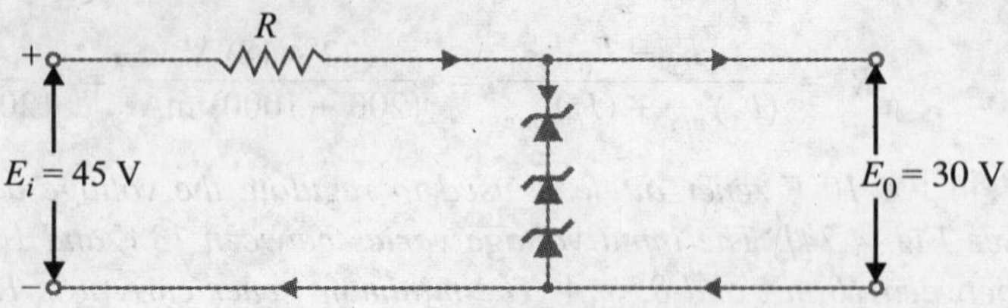

Fig. 4.76

$$\text{Voltage rating of each zener, } V_Z = 10 \text{ V}$$
$$\text{Current rating of each zener, } I_Z = 1000 \text{ mA}$$
$$\text{Input unregulated voltage, } E_i = 45 \text{ V}$$
$$\text{Regulated output voltage, } E_0 = 10 + 10 + 10 = 30 \text{ V}$$

Let R ohms be the required series resistance.

$$\text{Voltage across } R = E_i - E_0 = 45 - 30 = 15 \text{ V}$$

$$\therefore \quad R = \frac{E_i - E_0}{I_Z} = \frac{15 \text{ V}}{1000 \text{ mA}} = \mathbf{15\ \Omega}$$

Example 4.47. *Over what range of input voltages will the zener circuit shown in Fig. 4.77 maintain 30 V across 2000 Ω load, assuming that series resistance R = 200 Ω and zener current rating is 25 mA?*

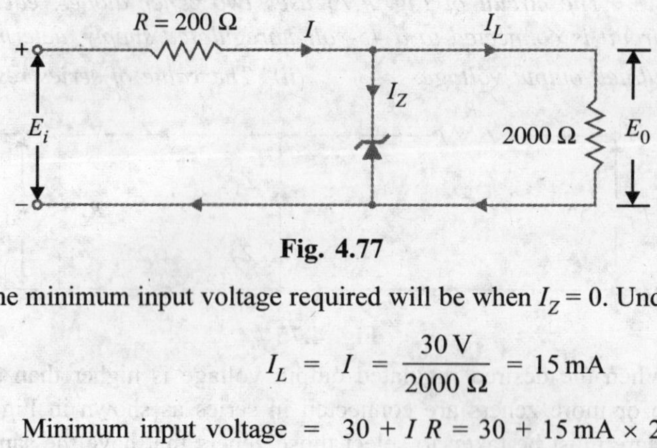

Fig. 4.77

Solution. The minimum input voltage required will be when $I_Z = 0$. Under this condition,

$$I_L = I = \frac{30 \text{ V}}{2000\ \Omega} = 15 \text{ mA}$$

$$\therefore \quad \text{Minimum input voltage} = 30 + I R = 30 + 15 \text{ mA} \times 200\ \Omega$$
$$= 30 + 3 = 33 \text{ V}$$

The maximum input voltage required will be when $I_Z = 25$ mA. Under this condition,

$$I = I_L + I_Z = 15 + 25 = 40 \text{ mA}$$

$$\therefore \quad \text{Max. input voltage} = 30 + I R$$
$$= 30 + 40 \text{ mA} \times 200\ \Omega$$
$$= 30 + 8 = 38 \text{ V}$$

Therefore, the input voltage range over which the circuit will maintain 30 V across the load is **33 V to 38 V.**

Example 4.48. *In the circuit shown in Fig. 4.78, the voltage across the load is to be maintained at 12 V as load current varies from 0 to 200 mA. Design the regulator. Also find the maximum wattage rating of zener diode.*

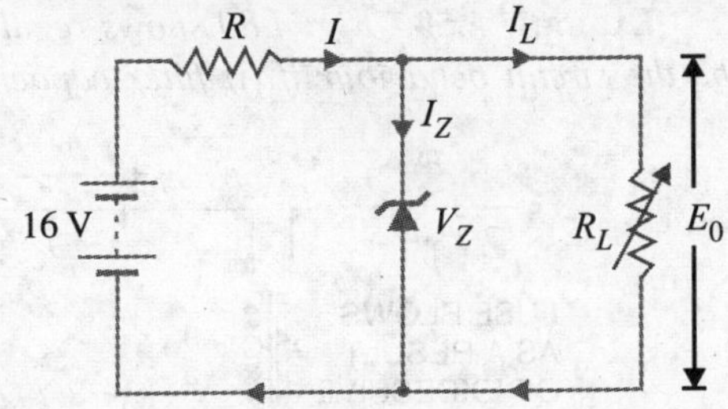

Fig. 4.78

Solution. By designing the regulator here means to find the values of V_Z and R. Since the load voltage is to be maintained at 12 V, we will use a zener diode of zener voltage 12 V *i.e.*

$$V_Z = \mathbf{12\ V}$$

The voltage across R is to remain constant at 16 – 12 = 4 V as the load current changes from 0 to 200 mA. The minimum zener current will occur when the load current is maximum.

$$\therefore \quad R = \frac{E_i - E_0}{(I_Z)_{min} + (I_L)_{max}} = \frac{(16-12)\ V}{(0+200)\ mA} = \frac{4\ V}{200\ mA} = \mathbf{20\ \Omega}$$

Maximum power rating of zener is

$$P_{ZM} = V_Z I_{ZM} = (12\ V)\ (200\ mA) = \mathbf{2.4\ W}$$

Example 4.49. *Fig. 4.79 shows the basic zener diode circuits. What will be the circuit behaviour if the zener is (i) working properly (ii) shorted (iii) open-circuited?*

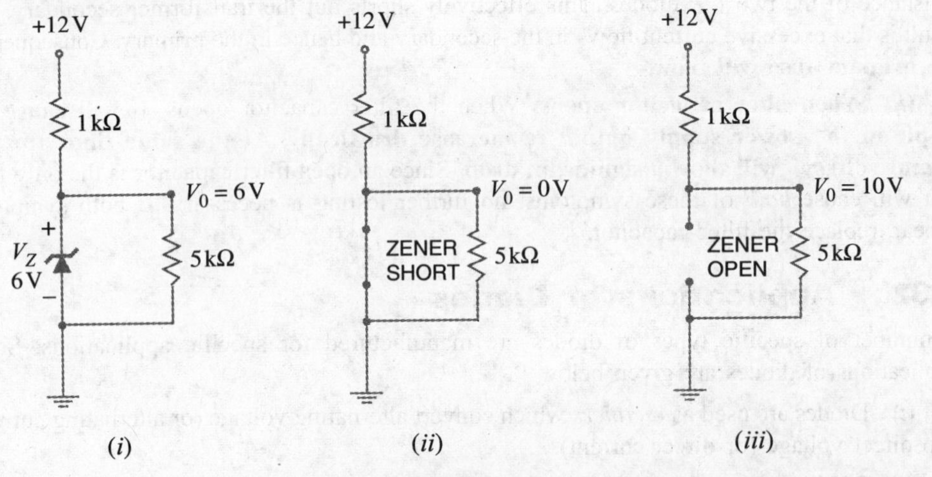

Fig. 4.79

Solution. Zener diodes cannot be tested individually with a multimeter. It is because multimeters usually do not have enough input voltage to put the zener into breakdown region.

(*i*) If the zener diode is working properly, the voltage V_0 **across the load** (= 5 kΩ) **will be nearly 6 V** [See Fig. 4.79 (*i*)].

(*ii*) If the zener diode is short [See Fig. 4.79 (*ii*)], **you will measure V_0 as 0 V.** The same problem could also be caused by a shorted load resistor (= 5 kΩ) or an opened source resistor (= 1 kΩ). The only way to tell which device has failed is to remove the resistors and check them with an ohmmeter. If the resistors are good, then zener diode is bad.

(*iii*) If the zener diode is open-circuited, the voltage V_0 **across the load** (= 5 kΩ) **will be 10 V.**

Example 4.50. *Fig. 4.80 shows regulated power supply using a zener diode. What will be the circuit behaviour if (i) filter capacitor shorts (ii) filter capacitor opens?*

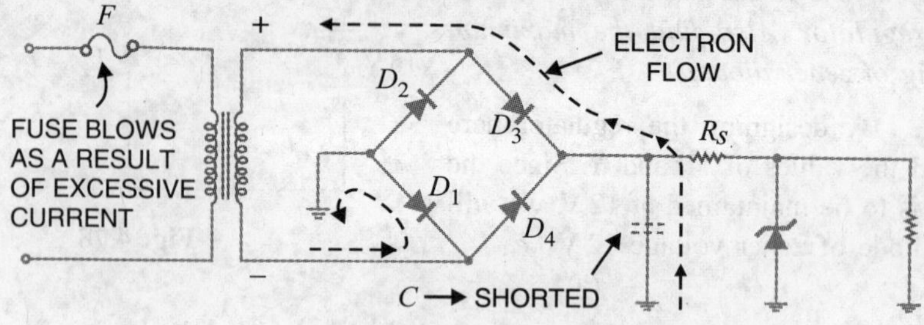

Fig. 4.80

Solution. The common faults in a zener voltage regulator are shorted filter capacitor or opened filter capacitor.

(*i*) **When filter capacitor shorts.** When the filter capacitor shorts, the primary fuse will blow. The reason for this is illustrated in Fig. 4.80. When the filter capacitor shorts, it shorts out the load resistance R_L. This has the same effect as wiring the two sides of the bridge together (See Fig. 4.80). If you trace from the high side of the bridge to the low side, you will see that the only resistance across the secondary of the transformer is the forward resistance of the two *ON* diodes. This effectively shorts out the transformer secondary. The result is that excessive current flows in the secondary and hence in the primary. Consequently, **the primary fuse will blow.**

(*ii*) **When filter capacitor opens.** When the filter capacitor opens, **it will cause the ripple in the power supply output to increase drastically. At the same time, the d.c. output voltage will show a significant drop.** Since an open filter capacitor is the only fault that will cause *both* of these symptoms, no further testing is necessary. If both symptoms appear, replace the filter capacitor.

4.32 Applications of Diodes

A number of specific types of diodes are manufactured for specific applications. Some applications of diodes are given below :

(*i*) Diodes are used as *rectifiers* which convert alternating voltage (or alternating current) into direct voltage (or direct current).

(*ii*) Diodes are used to *regulate d.c. voltages* in a circuit. Such diodes are known as zener diodes.

(*iii*) Diodes are used as *clippers (or limiters)*. A clipper is used to clip off or eliminate a portion of an a.c. signal.

(*iv*) Diodes are used as *clampers*. A clamper is used to change the d.c. level of an a.c. signal.

(*v*) Diodes are used as *voltage multipliers*. A voltage multiplier produces a d.c. voltage that is some multiple of an a.c. peak voltage.

(*vi*) A *LED* (light-emitting diode) is used to display numbers, letters *etc.*

Short Answer Questions

Q.1. *Why do we use transformers in rectifier service?*

Ans. There are two reasons for it. First, a transformer allows us to step-up or step-down the a.c. voltage for rectification. Secondly, it isolates the circuit from power line and thus reduces the risk of electric shock.

Q.2. *Why is the output of a bridge rectifier circuit twice that of the centre-tap circuit for the same secondary voltage?*

Ans. Because in a centre-tap circuit, each diode utilises a.c. voltage appearing across half of the secondary winding. However, a bridge circuit utilises full secondary voltage for rectification.

Q.3. *Why is full-wave rectifier preferred to half-wave rectifier?*

Ans. There are two main reasons for it. First, the output of a full-wave rectifier is twice as great as that of half-wave rectifier. Secondly, the ripple frequency is 100 Hz (twice the a.c. supply frequency of 50 Hz) compared to 50 Hz in a half-wave rectifier. This means a full-wave rectifier can use smaller values of L and C in filter circuits.

Q.4. *Why do we use combinations of L and C for making filter circuits?*

Ans. Because choke (L) blocks a.c. but readily allows d.c. while the capacitor (C) blocks d.c. and allows a.c. to pass through it. By arranging L and C suitably, the a.c. component of rectifier output can be removed, allowing the d.c. component to reach the load.

Q.5. *Why do we prefer centre-tap circuit to bridge circuit in low voltage applications?*

Ans. In a bridge type rectifier, two diodes in series conduct during each half-cycle of input a.c. Consequently, the output voltage is reduced due to two diode drops. However, in the centre-tap circuit, there is only one diode drop.

Q.6. *Why is the average value of current of full-wave rectified wave twice that of half-wave rectified wave?*

Ans. The number of positive alternations that make up the full-wave rectified wave is twice that of half-wave rectified wave for the same time interval. Therefore, the average value of current of full-wave rectified wave is twice that of half-wave rectified wave.

Q.7. *Why is bridge full-wave rectifier the most popular?*

Ans. The advantages of the bridge rectifier include a full-wave output, an ideal peak voltage equal to the peak secondary voltage and no centre-tap on the secondary winding. These advantages have made the bridge rectifier the most popular full-wave rectifier.

Q.8. *In a half-wave rectifier, what will happen if the diode is short-circuited ?*

Ans. If the diode in the half-wave rectifier is shorted, the output from the rectifier will be a sine wave that is identical to the wave of the input primary voltage. Since the diode is shorted, it acts as a piece of wire. Therefore, no half-cycle of input voltage is eliminated and the output will be replica of the input. The remedy is to replace the diode.

Q.9. *Why is PIV the most important consideration in rectifier diodes?*

Ans. Because a diode is used as a rectifier and the reverse voltage which comes across it during negative half-cycle of a.c. input must be less than *PIV* otherwise the diode will be destroyed.

Q.10. *How is that zener does not burn out inspite of the fact that it operates in the breakdown region?*

Ans. Because external resistance is inserted in the circuit to limit the zener diode current to less than the burn out value.

Q.11. *A zener diode is operating in the breakdown region. What will happen if the reverse voltage is reduced below the zener voltage (V_Z)?*

Ans. When the reverse voltage across the zener is reduced below V_Z, it comes out of its breakdown level. As a result, it is restored to pre-breakdown current level.

Q.12. *What is the difference between ordinary diode and zener diode?*

Ans. A zener diode is just like an ordinary diode except that it is properly doped to have a sharp breakdown voltage.

Q.13. *How does a zener diode maintain constant voltage across the load in face of input voltage variations?*

Ans. In the breakdown region, the zener can conduct any amount of current depending upon the external circuit resistance. It conducts such an amount of current so that current through the load remains unchanged. Hence, the voltage across the load remains the same irrespective of the input voltage variations.

Q.14. *Can zener be used as a rectifier?*

Ans. Yes but not as effectively as a rectifier diode.

Multiple-Choice Questions

1. A semiconductor diode has ....
   (*i*) one *pn* junction
   (*ii*) two *pn* junctions
   (*iii*) three *pn* junctions
   (*iv*) none of the above
2. A semiconductor diode has forward resistance of the order of ....
   (*i*) kΩ (*ii*) Ω
   (*iii*) MΩ (*iv*) none of the above
3. If the arrow of semiconductor diode symbol is positive *w.r.t.* bar, then diode is .... biased.
   (*i*) forward
   (*ii*) reverse
   (*iii*) either forward or reverse
   (*iv*) none of the above
4. The reverse current in a diode is of the order of ....
   (*i*) kA (*ii*) mA
   (*iii*) μA (*iv*) A
5. The forward voltage drop across a silicon diode is about ....
   (*i*) 2.5 V (*ii*) 3 V
   (*iii*) 10 V (*iv*) 0.7 V
6. A semiconductor diode is used as ....
   (*i*) an amplifier
   (*ii*) a rectifier
   (*iii*) an oscillator
   (*iv*) a voltage regulator
7. The d.c. resistance of a semiconductor diode is .... its a.c. resistance.
   (*i*) the same as (*ii*) more than
   (*iii*) less than (*iv*) none of the above
8. An ideal crystal diode is one which behaves as a perfect .... when forward biased.

(i) conductor
(ii) insulator
(iii) resistance material
(iv) none of the above

9. The ratio of reverse resistance and forward resistance of a germanium crystal diode is about ....
(i) 1 : 1 (ii) 100 : 1
(iii) 1000 : 1 (iv) 40000 : 1

10. The leakage current in a crystal diode is due to ....
(i) minority carriers
(ii) majority carriers
(iii) junction capacitance
(iv) none of the above

11. If the temperature of a crystal diode increases, then leakage current ....
(i) remains the same
(ii) decreases
(iii) increases
(iv) becomes zero

12. The *PIV* rating of a crystal diode is .... that of equivalent vacuum diode.
(i) the same as (ii) lower than
(iii) more than (iv) none of the above

13. If the doping level of a crystal diode is increased, the breakdown voltage ....
(i) remains the same
(ii) is increased
(iii) is decreased
(iv) none of the above

14. The knee voltage of a crystal diode is approximately equal to ....
(i) applied voltage
(ii) breakdown voltage
(iii) forward voltage
(iv) barrier potential

15. When the graph between current through and voltage across a device is a straight line, the device is referred to as ....
(i) linear (ii) active
(iii) non-linear (iv) passive

16. When the crystal diode current is large, the bias is ....
(i) forward (ii) inverse
(iii) poor (iv) reverse

17. A crystal diode is a .... device.
(i) non-linear (ii) bilateral
(iii) linear (iv) none of the above

18. A crystal diode utilises .... characteristic for rectification.
(i) reverse (ii) forward
(iii) forward or reverse
(iv) none of the above

19. When a crystal diode is used as a rectifier, the most important consideration is ....
(i) forward characteristic
(ii) doping level
(iii) reverse characteristic
(iv) *PIV* rating

20. If the doping level in a crystal diode is increased, the width of depletion layer ....
(i) remains the same
(ii) is decreased
(iii) is increased
(iv) none of the above

21. A zener diode has ....
(i) one *pn* junction
(ii) two *pn* junctions
(iii) three *pn* junctions
(iv) none of the above

22. A zener diode is used as ....
(i) an amplifier
(ii) a voltage regulator
(iii) a rectifier
(iv) a multivibrator

23. The doping level in a zener diode is .... that of a crystal diode.
(i) the same as (ii) less than
(iii) more than (iv) none of the above

24. A zener diode is always .... connected.
(i) reverse

(*ii*) forward
(*iii*) either reverse or forward
(*iv*) none of the above

25. A zener diode utilises .... characteristic for its operation.
(*i*) forward
(*ii*) reverse
(*iii*) both forward and reverse
(*iv*) none of the above

26. In the breakdown region, a zener diode behaves like a .... source.
(*i*) constant voltage
(*ii*) constant current
(*iii*) constant resistance
(*iv*) none of the above

27. A zener diode is destroyed if it ....
(*i*) is forward biased
(*ii*) is reverse biased
(*iii*) carries more than rated current
(*iv*) none of the above

28. A series resistance is connected in the zener circuit to ....
(*i*) properly reverse bias the zener
(*ii*) protect the zener
(*iii*) properly forward bias the zener
(*iv*) none of the above

29. A zener diode is .... device.
(*i*) a non-linear
(*ii*) a linear
(*iii*) an amplifying
(*iv*) none of the above

30. A zener diode has .... breakdown voltage.
(*i*) undefined
(*ii*) sharp
(*iii*) zero
(*iv*) none of the above

31. .... rectifier has the lowest forward resistance.
(*i*) Solid state
(*ii*) Vacuum tube
(*iii*) Gas tube
(*iv*) None of the above

32. Mains a.c. power is converted into d.c. power for ....
(*i*) lighting purposes
(*ii*) heaters
(*iii*) using in electronic equipment
(*iv*) none of the above

33. The disadvantage of a half-wave rectifier is that the ....
(*i*) components are expensive
(*ii*) diodes must have a higher power rating
(*iii*) output is difficult to filter
(*iv*) none of the above

34. If the a.c. input to a half-wave rectifier has an r.m.s. value of $400/\sqrt{2}$ volts, then diode *PIV* rating is ....
(*i*) $400/\sqrt{2}$ V (*ii*) 400 V
(*iii*) $400 \times \sqrt{2}$ V (*iv*) none of the above

35. The ripple factor of a half-wave rectifier is ....
(*i*) 2 (*ii*) 1.21
(*iii*) 2.5 (*iv*) 0.48

36. There is a need of transformer for ....
(*i*) half-wave rectifier
(*ii*) centre-tap full-wave rectifier
(*iii*) bridge full-wave rectifier
(*iv*) none of the above

37. The *PIV* rating of each diode in a bridge rectifier is .... that of the equivalent centre-tap rectifier.
(*i*) one-half (*ii*) the same as
(*iii*) twice (*iv*) four times

38. For the same secondary voltage, the output voltage from a centre-tap rectifier is .... than that of bridge rectifier.
(*i*) twice (*ii*) thrice
(*iii*) four times (*iv*) one-half

39. If the *PIV* rating of a diode is exceeded, ...
(*i*) the diode conducts poorly
(*ii*) the diode is destroyed
(*iii*) the diode behaves as zener diode
(*iv*) none of the above

40. A 10 V power supply would use .... as filter capacitor.
   (*i*) paper capacitor (*ii*) mica capacitor
   (*iii*) electrolytic capacitor
   (*iv*) air capacitor
41. A 1000 V power supply would use .... as a filter capacitor.
   (*i*) paper capacitor
   (*ii*) air capacitor
   (*iii*) mica capacitor
   (*iv*) electrolytic capacitor
42. The ...... filter circuit results in the best voltage regulation.
   (*i*) choke input
   (*ii*) capacitor input
   (*iii*) resistance input
   (*iv*) none of the above
43. A half-wave rectifier has an input voltage of 240 V r.m.s. If the step-down transformer has a turn ratio of 8 : 1, what is the peak load voltage? Ignore diode drop.
   (*i*) 27.5 V (*ii*) 86.5 V
   (*iii*) 30 V (*iv*) 42.5 V
44. The maximum efficiency of a half-wave rectifier is ....
   (*i*) 40.6% (*ii*) 81.2%
   (*iii*) 50% (*iv*) 25%
45. The most widely used rectifier is ....
   (*i*) half-wave rectifier
   (*ii*) centre-tap full-wave rectifier
   (*iii*) bridge full-wave rectifier
   (*iv*) none of the above

Answers to Multiple-Choice Questions

1. (*i*)	2. (*ii*)	3. (*i*)	4. (*iii*)	5. (*iv*)
6. (*ii*)	7. (*iii*)	8. (*i*)	9. (*iv*)	10. (*i*)
11. (*iii*)	12. (*ii*)	13. (*iii*)	14. (*iv*)	15. (*i*)
16. (*i*)	17. (*i*)	18. (*ii*)	19. (*iv*)	20. (*iii*)
21. (*i*)	22. (*ii*)	23. (*iii*)	24. (*i*)	25. (*ii*)
26. (*i*)	27. (*iii*)	28. (*ii*)	29. (*i*)	30. (*ii*)
31. (*i*)	32. (*iii*)	33. (*iii*)	34. (*ii*)	35. (*iv*)
36. (*ii*)	37. (*i*)	38. (*iv*)	39. (*ii*)	40. (*iii*)
41. (*i*)	42. (*i*)	43. (*iv*)	44. (*i*)	45. (*iii*)

Chapter Review Topics

1. What is a semiconductor diode? Explain its rectifying action.
2. Draw the graphic symbol of semiconductor diode and explain its significance. How the polarities of semiconductor diode are identified?
3. What do you understand by the d.c. and a.c. resistance of a semiconductor diode? How will you determine them from the *V-I* characteristic of a semiconductor diode?
4. Draw the equivalent circuit of a semiconductor diode.
5. Discuss the importance of peak inverse voltage in rectifier service.
6. Describe a half-wave rectifier using a crystal diode.
7. Derive an expression for the efficiency of a half-wave rectifier.
8. With a neat sketch, explain the working of (*i*) Centre-tap full-wave rectifier (*ii*) Full-wave bridge rectifier.
9. Derive an expression for the efficiency for a full-wave rectifier.
10. Write a short note about the nature of rectifier output.
11. What is ripple factor? What is its value for a half-wave and full-wave rectifier?

12. Describe the action of the following filter circuits: (*i*) capacitor filter (*ii*) choke input filter (*iii*) capacitor input filter.
13. What is a zener diode? Draw the equivalent circuit of an ideal zener in the breakdown region.
14. Explain how zener diode maintains constant voltage across the load.

Problems

1. Determine the voltage across each diode in Fig. 4.81, assuming that they are germanium diodes with a reverse resistance of 50 MΩ. **[(*i*) – 3 V (*ii*) 0.3 V (*iii*) 0.3 V (*iv*) 0.3 V]**

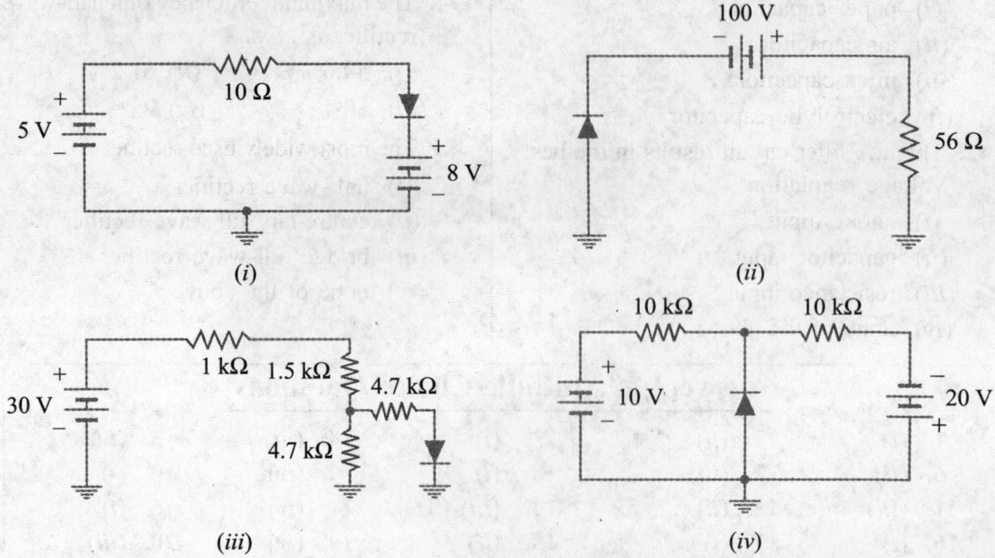

Fig. 4.81

2. Determine the voltage with respect to ground at each point in Fig. 4.82. Assume the diodes are of silicon. **[V_A = 25 V; V_B = 24.3 V; V_C = 8.7 V; V_D = 8 V]**

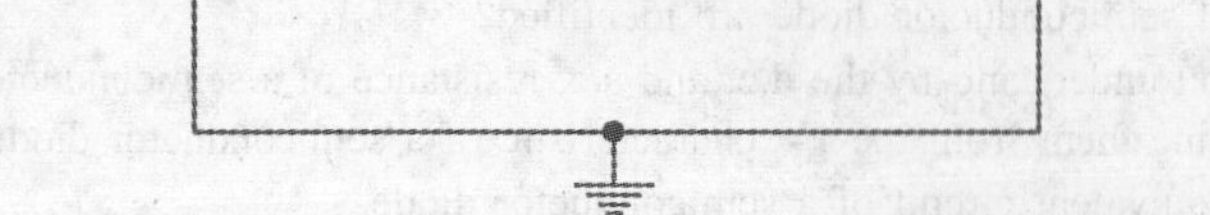

Fig. 4.82

3. What is the current in the circuit in Fig. 4.83? Assume the diode to be ideal. **[10 mA]**

Fig. 4.83

Fig. 4.84

4. Using equivalent circuit, determine the current in the circuit shown in Fig. 4.84. Assume the forward resistance of the diode to be 2 Ω. **[358 mA]**

5. Find the voltage V_A and current I in the circuit shown in Fig. 4.85. Use simplified model. **[14 V; 2 mA]**

6. Determine the magnitude of V_A in the circuit shown in Fig. 4.86. **[9.5 V]**

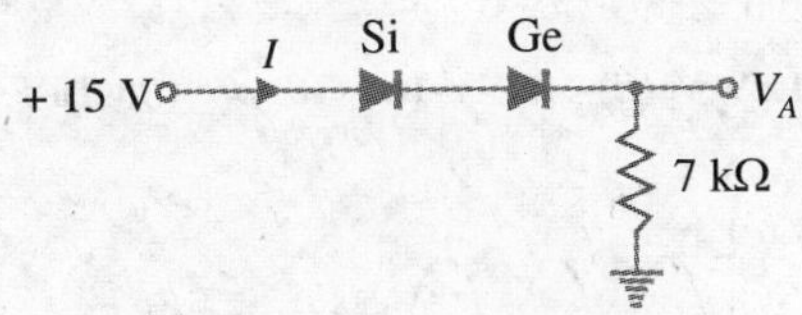

Fig. 4.85

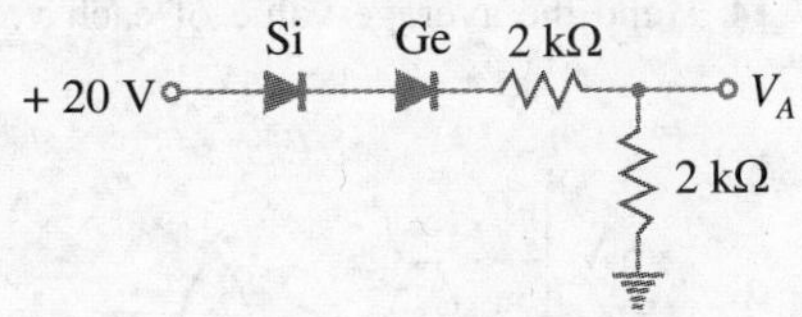

Fig. 4.86

7. A half-wave rectifier uses a transformer of turn ratio 4 : 1. If the primary voltage is 240 V (r.m.s.), find (*i*) d.c. output voltage (*ii*) peak inverse voltage. Assume the diode to be ideal. **[(*i*) 27 V (*ii*) 85 V]**

8. Determine V_{01} and V_{02} for the circuits shown in Fig. 4.87.

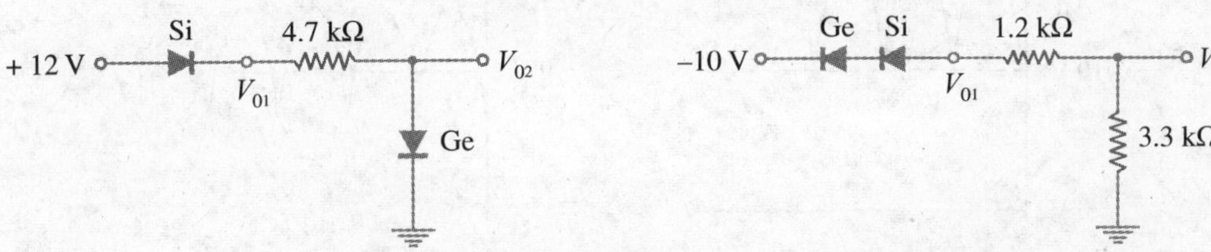

(*i*) (*ii*)

Fig. 4.87

[(*i*) V_{01} = 11.3 V; V_{02} = 0.3 V (*ii*) V_{01} = – 9V; V_{02} = – 6.6 V]

9. A half-wave rectifier uses a transformer of turn ratio 2 : 1. The load resistance is 500 Ω. If the primary voltage (r.m.s.) is 240 V, find (*i*) d.c. output voltage (*ii*) peak inverse voltage. **[(*i*) 54 V (*ii*) 170 V]**

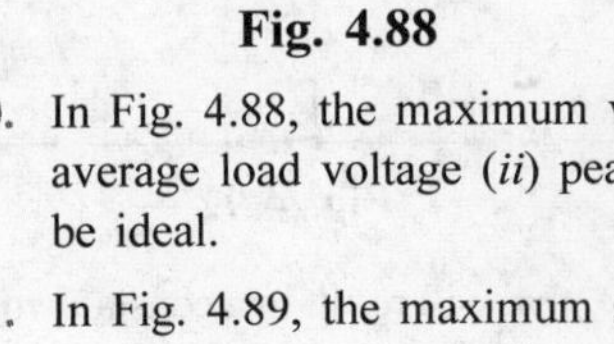

Fig. 4.88

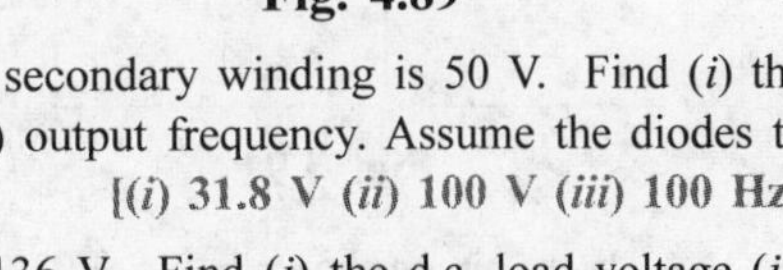

Fig. 4.89

10. In Fig. 4.88, the maximum voltage across half of secondary winding is 50 V. Find (*i*) the average load voltage (*ii*) peak inverse voltage (*iii*) output frequency. Assume the diodes to be ideal. **[(*i*) 31.8 V (*ii*) 100 V (*iii*) 100 Hz]**

11. In Fig. 4.89, the maximum secondary voltage is 136 V. Find (*i*) the d.c. load voltage (*ii*) peak inverse voltage (*iii*) output frequency. **[(*i*) 86.6 V (*ii*) 136 V (*iii*) 100 Hz]**

12. A semiconductor diode having ideal forward and reverse characteristics is used in a half-wave rectifier circuit supplying a resistive load of 1000 Ω. If the r.m.s. value of the sinusoidal supply voltage is 250 V, determine (*i*) the r.m.s. diode current and (*ii*) power dissipated in the load. **[(*i*) 177 mA (*ii*) 31.3 W]**

13. The four semiconductor diodes used in a bridge rectifier circuit have forward resistance which can be considered constant at 0.1Ω and infinite reverse resistance. They supply a mean current of 10 A to a resistive load from a sinusoidally varying alternating supply of 20V r.m.s. Determine the resistance of the load and the efficiency of the circuit.

[1.6Ω; 72%]

14. Find the average value of each voltage in Fig. 4.90.

[(*i*) 1.59 V (*ii*) 63.7 V (*iii*) 16.4 V (*iv*) 10.5 V]

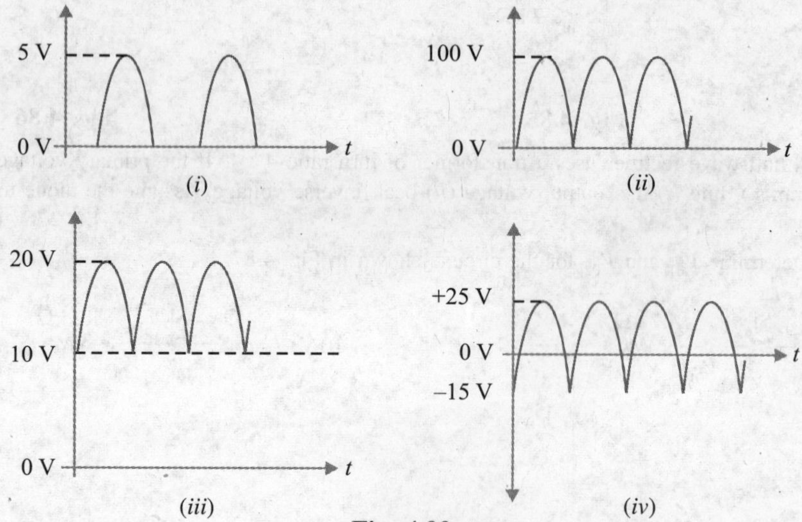

Fig. 4.90

15. Calculate the peak voltage across each half of a centre-tapped transformer used in a full-wave rectifier that has an average output voltage of 110 V. **[173 V]**

16. What *PIV* rating is required for the diodes in a bridge rectifier that produces an average output voltage of 50 V? **[78.5 V]**

17. In the circuit shown in Fig. 4.91, is zener diode in the on or off state? **[Off]**

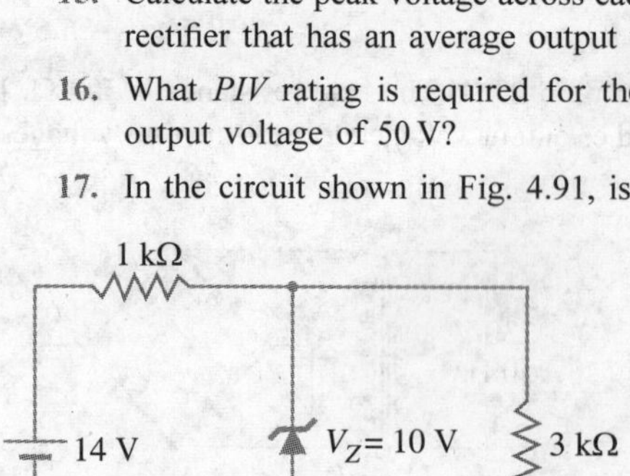

Fig. 4.91

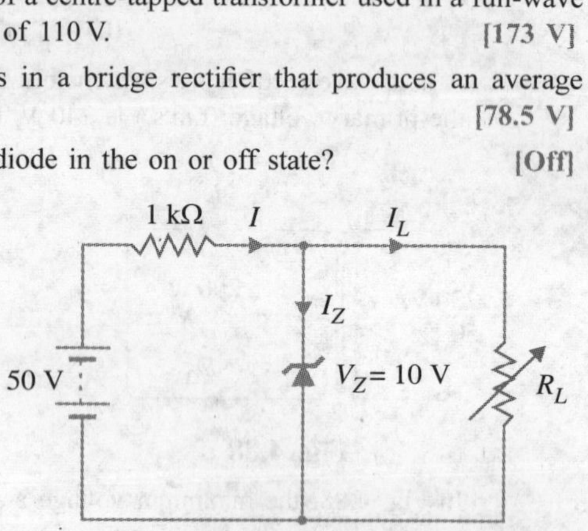

Fig. 4.92

18. In the circuit shown in Fig. 4.92, determine the range of R_L that will result in a constant voltage of 10 V across R_L. **[250 Ω to 1.25 kΩ]**

19. For the circuit shown in Fig. 4.93, determine (*i*) the waveform (*ii*) v_o (*iii*) *PIV* rating of each diode.

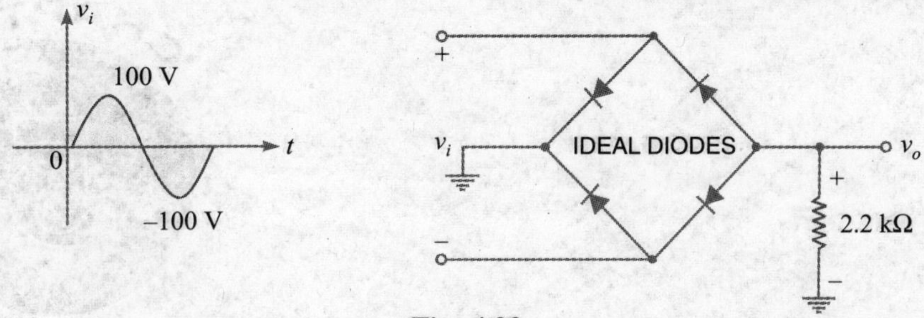

Fig. 4.93

[(*i*) Full-wave rectification, 100 V peak (*ii*) 100 V peak (*iii*) 100 V]

20. For the circuit shown in Fig. 4.94, determine (*i*) the waveform (*ii*) the d.c. voltage available.

[(*i*) Full-wave rectification, 56.67 V peak (*ii*) 36.04 V]

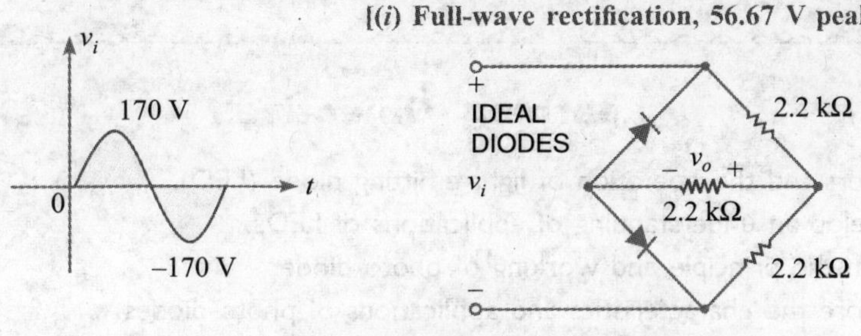

Fig. 4.94

CHAPTER 5

Special-Purpose Diodes

Learning Objectives

- Understand the operation of light-emitting diode (LED).
- Develop an understanding of applications of LEDs.
- Learn the principle and working of photo-diode.
- Explore the characteristics and applications of photo-diodes.
- Develop a strong understanding of theory of tunnel diode.
- Understand *V-I* characteristics of tunnel diode.
- Learn the underlying principle on which varactor diode is based.
- Develop an understanding of the construction and working of Shockley diode.

Introduction

The most common application of diodes is rectification. The rectifier diodes are used in power supplies to convert a.c. voltage into d.c. voltage. But rectification is not all that a diode can do. A number of specific types of diodes are manufactured for specific applications in this fast developing world. Some of the more common special-purpose diodes are (*i*) Zener diode (*ii*) Light-emitting diode (LED) (*iii*) Photo-diode (*iv*) Tunnel diode (*v*) Varactor diode and (*vi*) Shockley diode.

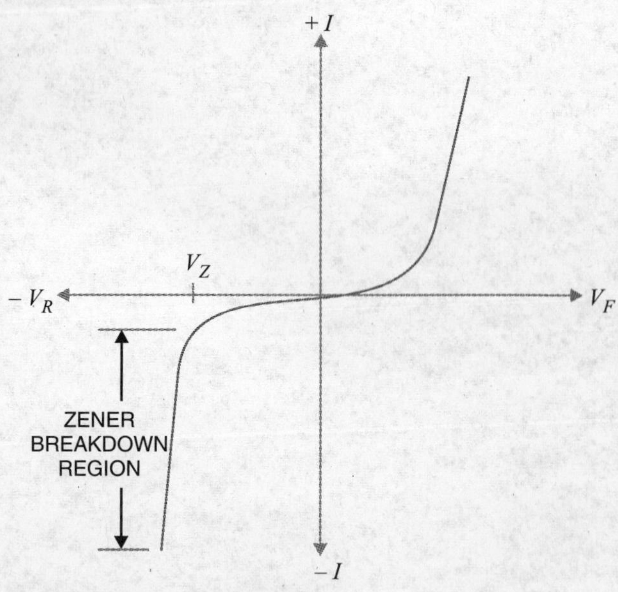

Fig. 5.1

5.1 Zener Diode

A zener diode is a special type of diode that is designed to operate in the reverse breakdown region. An ordinary diode operated in this region will usually be destroyed due to excessive current. This is not the case for the zener diode.

A zener diode is heavily doped to reduce the reverse breakdown voltage. This causes a very thin depletion layer. As a result, a zener diode has a sharp reverse breakdown voltage V_Z. This is clear from the reverse characteristic of zener diode shown in Fig. 5.1. Note that the reverse characteristic drops in an almost vertical manner at reverse voltage V_Z. As the curve reveals, two things happen when V_Z is reached :

(*i*) The diode current increases rapidly.

(*ii*) The reverse voltage V_Z across the diode remains almost constant.

In other words, *the zener diode operated in this region will have a relatively constant voltage across it, regardless of the value of current through the device.* This permits the zener diode to be used as a *voltage regulator.* For detailed discussion on zener diode, the reader may refer to chapter 4 of this book.

5.2 Light-Emitting Diode (LED)

A **light-emitting diode (LED)** *is a diode that gives off visible light when forward biased.*

Light-emitting diodes are not made from silicon or germanium but are made by using elements like gallium, phosphorus and arsenic. By varying the quantities of these elements, it is possible to produce light of different wavelengths with colours that include red, green, yellow and blue. For example, when a LED is manufactured using gallium arsenide, it will produce a red light. If the LED is made with gallium phosphide, it will produce a green light.

Theory. When light-emitting diode (LED) is forward biased as shown in Fig. 5.2 (*i*), the electrons from the *n*-type material cross the *pn* junction and recombine with holes in the *p*-type material. Recall that these free electrons are in the conduction band and at a higher energy level than the holes in the valence band. When recombination takes place, the recombining electrons release energy in the form of heat and light. In germanium and silicon diodes, almost the entire energy is given up in the form of heat and emitted visible light is small. Since germanium and silicon are opaque materials, the emitted light is blocked *i.e.* it is not transmitted to the surroundings. However, in materials like **gallium arsenide** (GaAs) and **gallium phosphide** (GaP), *the number of photons of light energy is sufficient to produce quite intense visible light.* Since these materials are semi-transparent, some of the visible light escapes to the surroundings. LEDs have replaced incandescent lamps in many applications because of their low voltage, long life and fast on-off switching.

Fig. 5.2 (*ii*) shows the schematic symbol for a LED. The arrows are shown as pointing away from the diode, indicating that light is being emitted by the device when forward biased. Although LEDs are available in several colours (red, green, yellow and orange are the most common), the schematic symbol is the same for all LEDs. There is nothing in the symbol to indicate the colour of a particular LED. Fig. 5.3 shows the graph between radiated light and the forward current of the LED. It is clear from the graph that the intensity of radiated light is directly proportional to the forward current of LED.

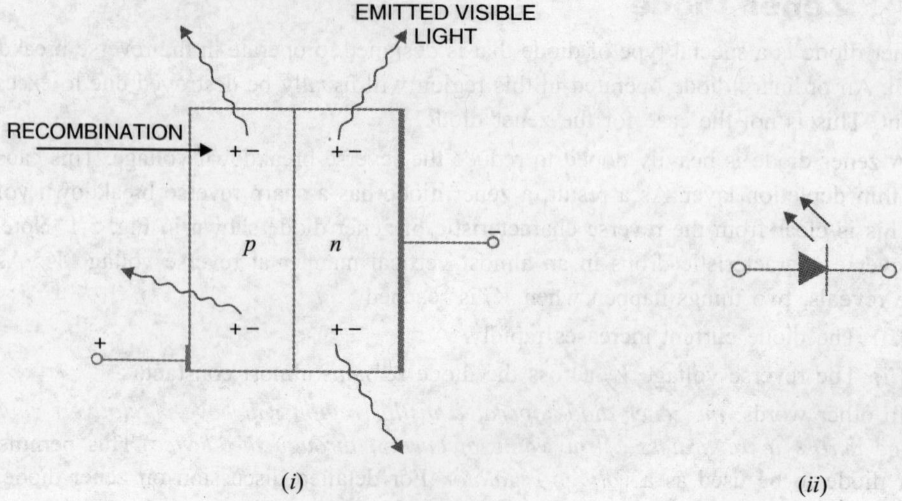

Fig. 5.2

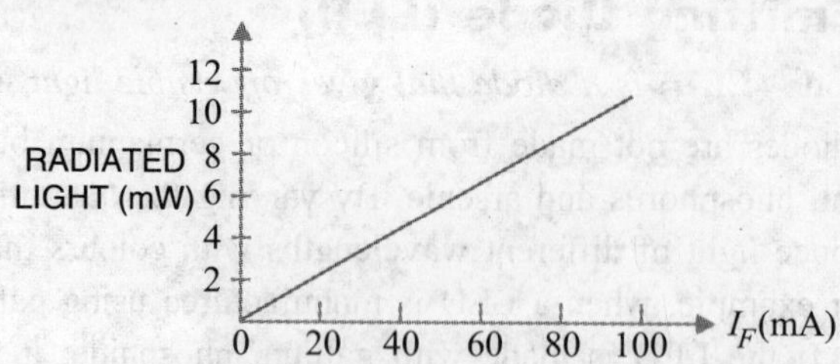

Fig. 5.3

5.3 LED Voltage and Current

The forward voltage ratings of most LEDs is from 1 V to 3 V and forward current ratings range from 20 mA to 100 mA. In order that current through the LED does not exceed the safe value, a resistor R_S is connected in series with it as shown in Fig. 5.4. The input voltage is V_S and the voltage across LED is V_D.

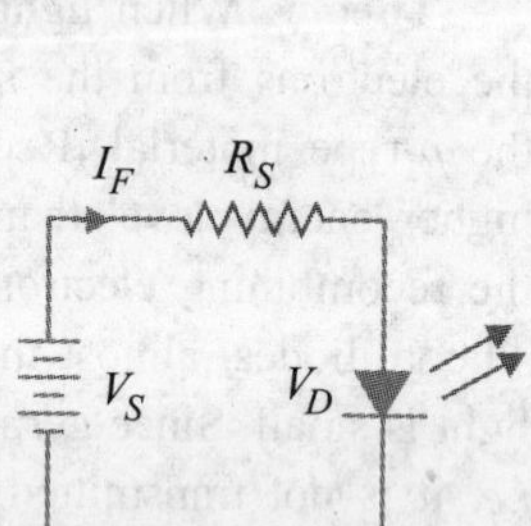

Fig. 5.4

$$\therefore \quad \text{Voltage across } R_S = V_S - V_D$$

$$\therefore \quad \text{Circuit current, } I_F = \frac{V_S - V_D}{R_S}$$

The brightness of a LED is determined by the value of forward current (I_F) of the LED. The more the forward current, the brighter the light is. Therefore, we always keep the value of series **current-limiting resistor** R_S as small as possible to ensure that the LED is as bright as possible.

Note. Note that the forward voltage drop across a LED (2 V to 3 V) is considerably greater than that across a silicon or germanium diode (0.7 V for Si diode and 0.3 V for Ge diode) depending upon the forward current. However, the maximum permissible reverse voltage across a LED (5 V to 10 V) is relatively small.

Example 5.1. *What value of series resistor is required to limit the current through a LED to 20 mA with a forward voltage drop of 1.6 V when connected to a 10 V supply?*

Solution.

$$\text{Series resistor, } R_S = \frac{V_S - V_D}{I_F}$$

Here, $V_S = 10\text{ V}$; $V_D = 1.6\text{ V}$; $I_F = 20\text{ mA} = 20 \times 10^{-3}\text{ A}$

$$\therefore \quad R_S = \frac{10 - 1.6}{20 \times 10^{-3}} = \mathbf{420\ \Omega}$$

Note that resistor R_S is also called *current-limiting resistor.*

Example 5.2. *What is current through the LED in the circuit shown in Fig. 5.5? Assume that voltage drop across the LED is 2 V.*

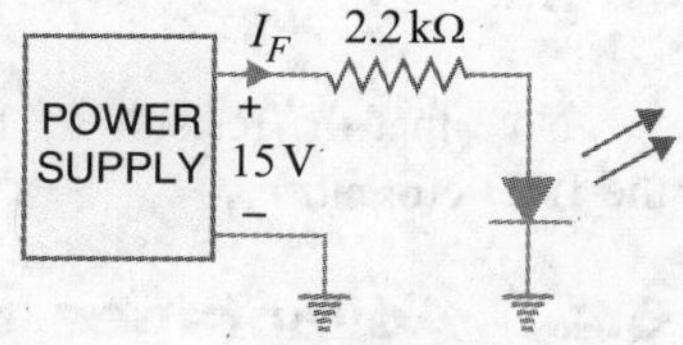

Fig. 5.5

Solution.

$$\text{Current through LED, } I_F = \frac{V_S - V_D}{R_S}$$

Here, $V_S = 15\text{ V}$; $V_D = 2\text{ V}$; $R_S = 2.2\text{ k}\Omega = 2.2 \times 10^3\ \Omega$

$$\therefore \quad I_F = \frac{15 - 2}{2.2 \times 10^3} = 5.91 \times 10^{-3}\text{ A} = \mathbf{5.91\ mA}$$

Example 5.3. *The driving circuit shown in Fig. 5.6 has a peak output voltage of 8 V. The LED has ratings of V_D = 1.8 V to 2.0 V and I_F(max.) = 16 mA. What value of current-limiting resistor R_S is needed in the circuit ?*

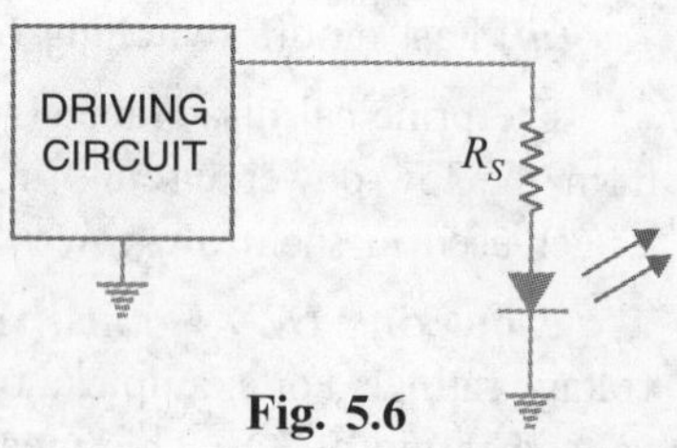

Fig. 5.6

Solution.

$$\text{Forward current, } I_F = \frac{V_S - V_D}{R_S}$$

For maximum brightness of the LED, the forward current I_F should be as large as possible. It is given that $I_F(max)$ = 16 mA and $V_{S(peak)}$ = 8 V. Therefore, we should use the minimum value of forward voltage drop of LED *i.e.* 1.8 V.

$$\therefore \quad 16\text{ mA} = \frac{8\text{ V} - 1.8\text{ V}}{R_S}$$

$$\text{or} \quad R_S = \frac{8\text{ V} - 1.8\text{ V}}{16\text{ mA}} = \mathbf{387.5\ \Omega}$$

Example 5.4. *Bipolar junction transistor (BJT) is often used to switch on the LED. Fig. 5.7 shows a LED driving circuit designed to illuminate the LED when the transistor is turned on (saturated) by a 5V battery at the base. Find the values of R_B and R_C that should be used if the LED is to be illuminated by 20 mA of forward current. Assume that the forward voltage drop across the LED is 2.5 V and that the silicon transistor has β = 50.*

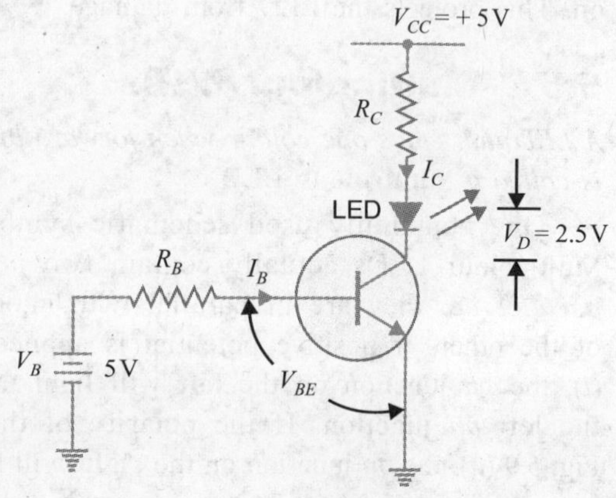

Fig. 5.7

Solution.

When the transistor is saturated, the value of I_C is given by ;

$$I_C = \frac{V_{CC} - V_D}{R_C} = \frac{5\,\text{V} - 2.5\,\text{V}}{R_C} = \frac{2.5\,\text{V}}{R_C}$$

$$\therefore \quad R_C = \frac{2.5\,\text{V}}{I_C} = \frac{2.5\,\text{V}}{20\text{ mA}} = \mathbf{125\ \Omega}$$

Now, Base current, $I_B = I_C/\beta = 20\text{ mA}/50 = 0.4\text{ mA}$

Also, $$R_B = \frac{5 - V_{BE}}{I_B} = \frac{(5-0.7)\,\text{V}}{0.4\text{ mA}} = \mathbf{10.75\ k\Omega}$$

Note that the resistor R_B limits the transistor base current while the resistor R_C limits the LED current.

5.4 Advantages of LED

The light-emitting diode (LED) is a solid-state light source. LEDs have replaced incandescent lamps in many applications because they have the following advantages :

(*i*) Low voltage

(*ii*) Longer life (more than 20 years)

(*iii*) Fast on-off switching (of the order of nanoseconds)

The principal disadvantage of the LED is that it draws considerable amount of current from the low-power circuits in which it is normally used. For this reason, LEDs are no longer used in such low-power devices as calculators and watches.

Protecting LED against reverse bias. The LEDs have low reverse voltage ratings. For example, a typical LED may have a maximum reverse voltage rating of 3 V. This means that if a reverse voltage greater than 3 V is applied to the LED, the LED may be destroyed. Therefore, one must be careful not to use LEDs with a high level of reverse bias. One way to protect a LED is to connect a rectifier diode in parallel with LED as shown in Fig. 5.8. If reverse voltage greater than the reverse voltage rating of LED is accidentally applied, the rectifier diode will be turned on. This protects the LED from damage.

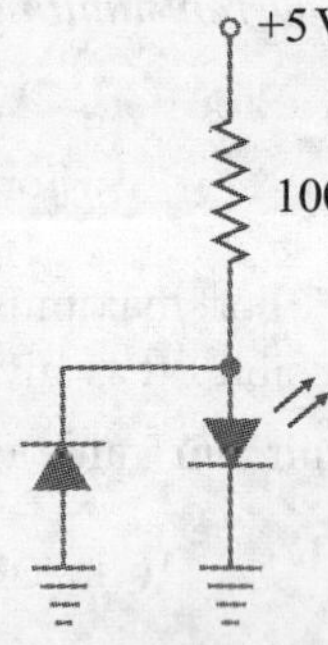

Fig. 5.8

5.5 Multicolour LEDs

A LED that emits one colour when forward biased and another colour when reverse biased is called a **multicolour LED.**

One commonly used schematic symbol for these LEDs is shown in Fig. 5.9. Multicolour LEDs actually contain two *pn* junctions that are connected in *reverse-parallel i.e.* they are in parallel with anode of one being connected to the cathode of the other. If positive potential is applied to the top terminal as shown in Fig. 5.9 (*i*), the *pn* junction on the **left** will light. Note that the device current passes through the left *pn* junction. If the polarity of the voltage source is reversed as shown in Fig. 5.9 (*ii*), the *pn* junction on the **right** will light. Note that the direction of device current has reversed and is now passing through the right *pn* junction.

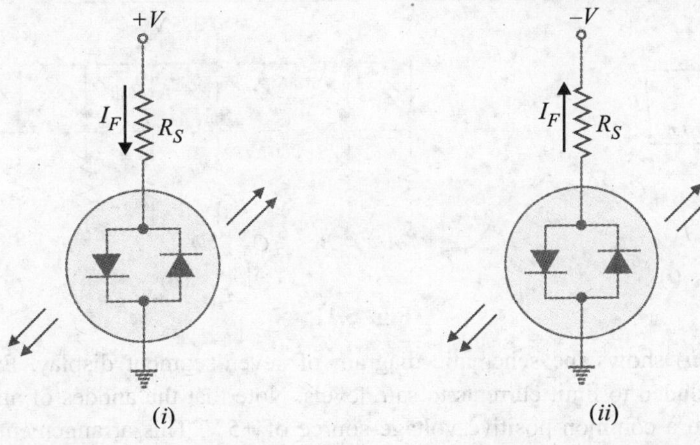

Fig. 5.9

Multicolour LEDs are typically *red* when biased in one direction and *green* when biased in the other. If a multicolour LED is switched fast enough between two polarities, the LED will produce a *third* colour. A red/green LED will produce a *yellow* light when rapidly switched back and forth between biasing polarities.

5.6 Applications of LEDs

The LED is a low-power device. The power rating of a LED is of the order of milliwatts. This means that it is useful as an indicator but not good for illumination. Probably the two most common applications for visible LEDs are (*i*) as a power indicator (*ii*) seven-segment display.

(*i*) **As a power indicator.** A LED can be used to indicate whether the power is on or not. Fig. 5.10 shows the simple use of the LED as a power indicator. When the switch *S* is closed, power is applied to the load. At the same time current also flows through the LED which lights, indicating power is on. The resistor R_S in series with the LED ensures that current rating of the LED is not exceeded.

Fig. 5.10

(*ii*) **Seven-segment display.** LEDs are often grouped to form seven-segment display. Fig. 5.11 (*i*) shows the front of a seven segment display. It contains seven LEDs (*A, B, C, D, E, F* and *G*) shaped in a figure of *8. Each LED is called a **segment. If a particular LED is forward biased, that LED or segment will light and produce a bar of light. By forward biasing various combinations of seven LEDs, it is possible to display any number from 0 to 9. For example, if LEDs *A, B, C, D* and *G* are lit (by forward biasing them), the display will show the number 3. Similarly, if LEDs *C, D, E, F, A* and *G* are lit, the display will show the number 6. To get the number 0, all segments except *G* are lit.

* Note that LEDs *A, B, C, D, E* and *F* are arranged clockwise from the top with LED *G* in the middle.

** Each LED is called a segment because it forms part of the character being displayed.

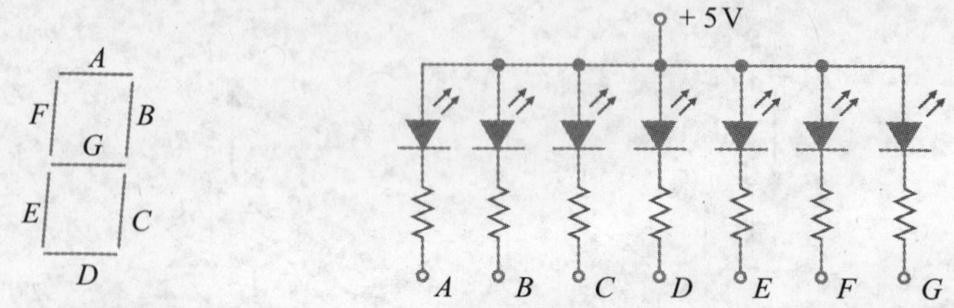

(i) (ii)

Fig. 5.11

Fig. 5.11 (*ii*) shows the schematic diagram of seven-segment display. External series resistors are included to limit currents to safe levels. Note that the anodes of all seven LEDs are connected to a common positive voltage source of +5 V. This arrangement is known as **common-anode type.* In order to light a particular LED, say *A*, we ground the point *A* in Fig. 5.11 (*ii*). It forward biases the LED *A* which will be lit.

5.7 Photo-diode

***A photo-diode** is a reverse-biased silicon or germanium pn junction in which reverse current increases when the junction is exposed to light.*

The reverse current in a photo-diode is directly proportional to the intensity of light falling on its *pn* junction. This means that greater the intensity of light falling on the *pn* junction of photo-diode, the greater will be the reverse current.

Principle. When a rectifier diode is reverse biased, it has a very small reverse leakage current. The same is true for a photo-diode. The reverse current is produced by thermally generated electron-hole pairs which are swept across the junction by the electric field created by the reverse voltage. In a rectifier diode, the reverse current increases with temperature due to an increase in the number of electron-hole pairs. *A photo-diode differs from a rectifier diode in that when its pn junction is exposed to light, the reverse current increases with the increase in light intensity and vice-versa.* This is explained as follows. When light (photons) falls on the ***pn* junction, the energy is imparted by the photons to the atoms in the junction. This will create more free electrons (and more holes). These additional free electrons will increase the reverse current. As the intensity of light incident on the *pn* junction increases, the reverse current also increases. In other words, as the incident light intensity increases, the resistance of the device (photo-diode) ***decreases.

Photo-diode package. Fig. 5.12 (*i*) shows a typical photo-diode package. It consists of a *pn* junction mounted on an insulated substrate and sealed inside a metal case. A glass window is mounted on top of the case to allow light to enter and strike the *pn* junction. The two leads extending from the case are labelled anode and cathode. The cathode is typically identified by a tab extending from the side of the case.

---

* Also available is the *common-cathode type* where all cathodes are connected together.

** This is true only if the light energy is applied at the junction. If it is applied to the crystal at some distance from the junction, the free electrons and holes will recombine (thus neutralising each other) before they can join the flow of reverse current.

*** It is for this reason that semiconductor devices such as diodes and transistors are usually enclosed in opaque case to protect them from light. Those diodes or transistors which are used for light-detecting, on the other hand, must be encased in transparent plastic or glass so that light may fall on them.

Fig. 5.12

Fig. 5.12 (*ii*) shows the schematic symbol of a photo-diode. The inward arrows represent the incoming light.

5.8 Photo-diode Operation

Fig. 5.13 shows the basic photo-diode circuit. The circuit has reverse-biased photo-diode, resistor *R* and d.c. supply. The operation of the photo-diode is as under :

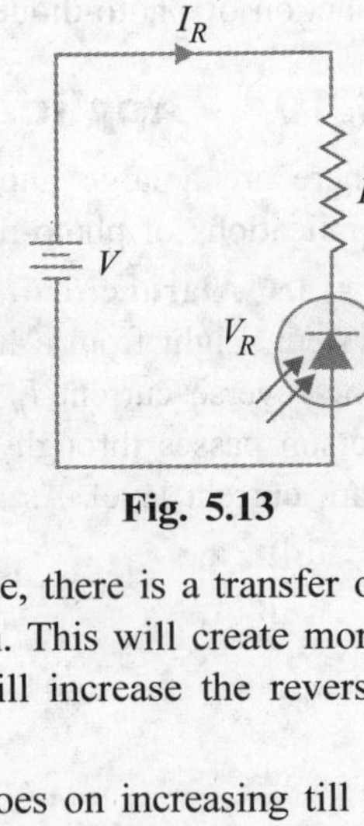

Fig. 5.13

(*i*) When no light is incident on the *pn* junction of photo-diode, the reverse current I_R is extremely small. This is called **dark current.**

The resistance of photo-diode with no incident light is called **dark resistance** (R_R).

$$\text{Dark resistance of photo-diode, } R_R = \frac{V_R}{\text{Dark current}}$$

(*ii*) When light is incident on the *pn* junction of the photo-diode, there is a transfer of energy from the incident light (photons) to the atoms in the junction. This will create more free electrons (and more holes). These additional free electrons will increase the reverse current.

(*iii*) As the intensity of light increases, the reverse current I_R goes on increasing till it becomes maximum. This is called **saturation current.**

5.9 Characteristics of Photo-diode

There are two important characteristics of photo-diode.

(*i*) **Reverse current-Illumination curve.** Fig. 5.14 shows the graph between reverse current (I_R) and illumination (E) of a photo-diode. The reverse current is shown on the vertical axis and is measured in μA. The illumination is indicated on the horizontal axis and is measured in mW/cm^2. Note that graph is a straight line passing through the origin.

$$\therefore \quad I_R = mE$$

where $\quad m$ = Slope of the straight line

The quantity *m* is called the *sensitivity* of the photo-diode.

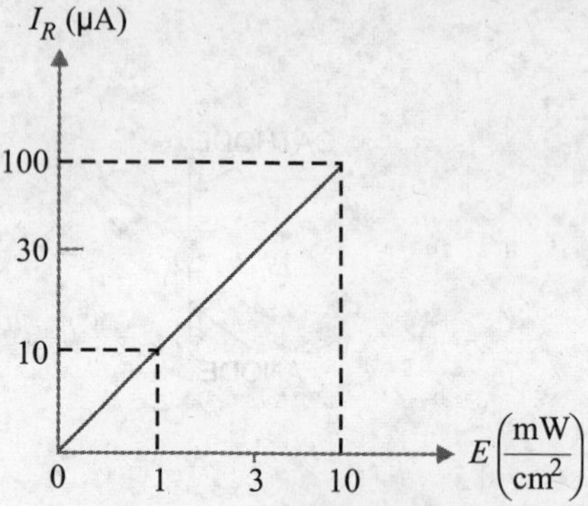

Fig. 5.14

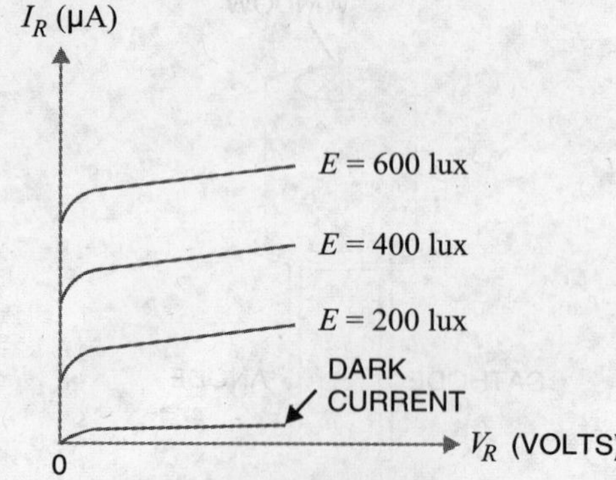

Fig. 5.15

(*ii*) **Reverse voltage-Reverse current curve.** Fig. 5.15 shows the graph between reverse current (I_R) and reverse voltage (V_R) for various illumination levels. It is clear that for a given reverse-biased voltage V_R, the reverse current I_R increases as the illumination (E) on the *pn* junction of photo-diode is increased.

5.10 Applications of Photo-diodes

There are a large number of applications of photo-diodes. However, we shall give two applications of photo-diodes by way of illustration.

(*i*) **Alarm circuit using photo-diode.** Fig. 5.16 shows the use of photo-diode in an alarm system. Light from a light source is allowed to fall on a photo-diode fitted in the doorway. The reverse current I_R will continue to flow so long as the light beam is not broken. If a person passes through the door, light beam is broken and the reverse current drops to the dark current level. As a result, an alarm is sounded.

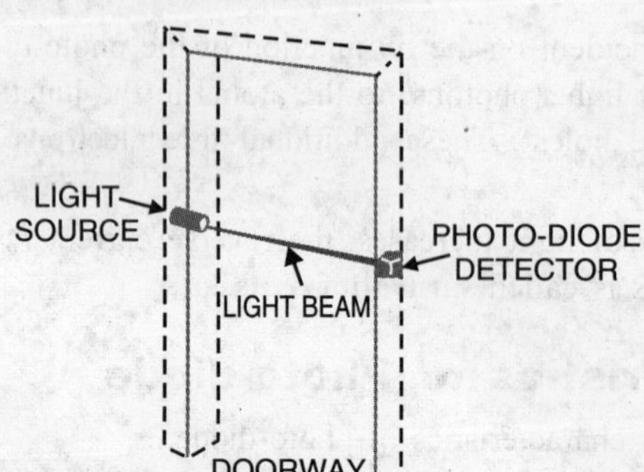

Fig. 5.16

(*ii*) **Counter circuit using photo-diode.** A photo-diode may be used to count items on a conveyor belt. Fig. 5.17 shows a photo-diode circuit used in a system that counts objects as they pass by on a conveyor. In this circuit, a source of light sends a concentrated beam of light across a conveyor to a photo-diode. As the object passes, the light beam is broken, I_R drops to the dark current level and the count is increased by one.

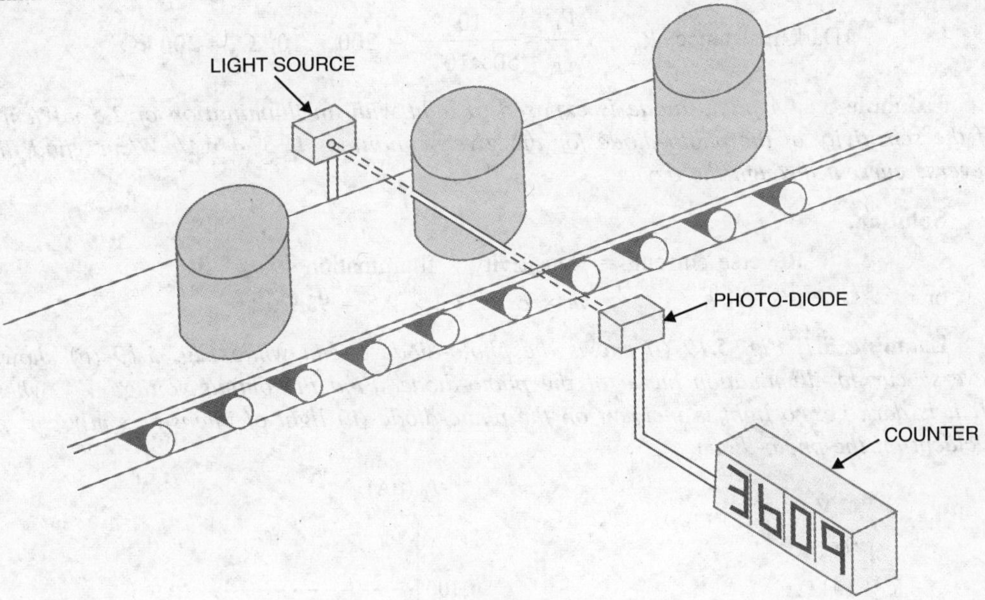

Fig. 5.17

Example 5.5. *From the reverse current-illumination curve for a photo-diode shown in Fig. 5.18, determine the dark resistance. Assume a reverse-biased voltage of 10 V.*

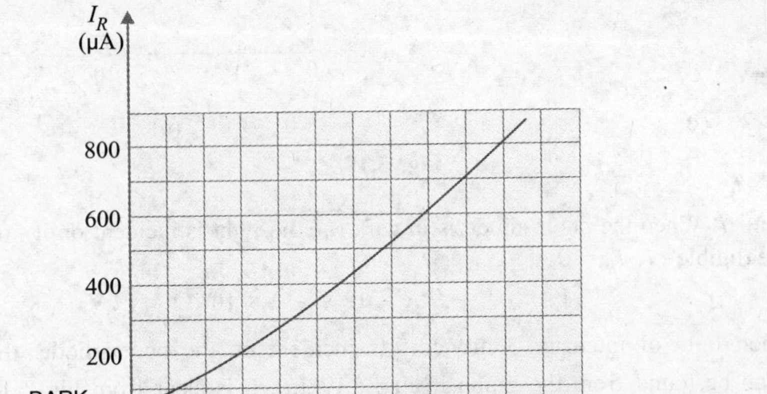

Fig. 5.18

Solution. The current that flows when the incident light is removed from the photo-diode is called *dark current*. The resistance of a photo-diode corresponding to dark current is called *dark resistance*. From the curve shown in Fig. 5.18, it is clear that for zero illumination, the reverse current is 50 μA.

$$\therefore \quad \text{Dark current, } I_R = 50\ \mu\text{A} = 50 \times 10^{-6}\ \text{A}$$

$$\text{Reverse voltage, } V_R = 10\ \text{V}$$

$\therefore$ Dark resistance, $R_R = \frac{V_R}{I_R} = \frac{10}{50 \times 10^{-6}} = 200 \times 10^3\ \Omega = \mathbf{200\ k\Omega}$

Example 5.6. *A photo-diode is exposed to light with an illumination of 2.5 mW/cm². If the sensitivity of the photo-diode for the given conditions is 37.4 μA/mW/cm², find the reverse current through the device.*

Solution.

$$\text{Reverse current} = \text{Sensitivity} \times \text{Illumination}$$

or $$I_R = m \times E = 37.4 \times 2.5 = \mathbf{93.5\ \mu A}$$

Example 5.7. *Fig. 5.19 (i) shows the photo-diode circuit while Fig. 5.19 (ii) shows reverse current-illumination curve of the photo-diode. Find the output voltage V_{out} when (i) it is dark i.e. no light is incident on the photo-diode (ii) light of intensity 3 mW/cm² is incident on the photo-diode.*

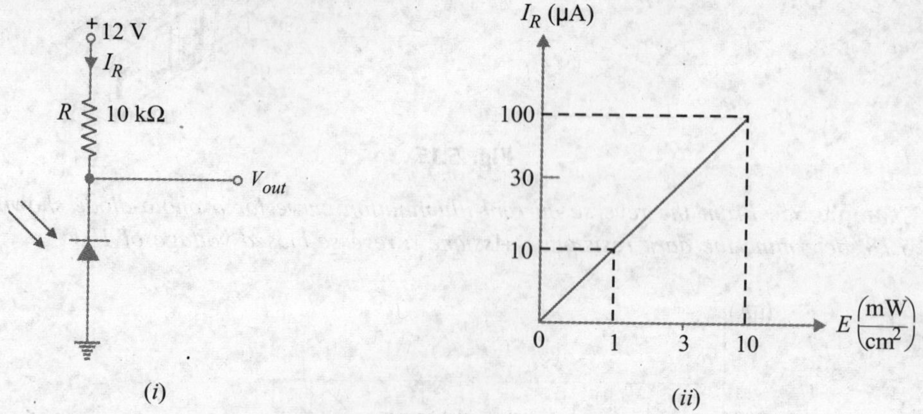

Fig. 5.19

Solution. (*i*) When the photo-diode is in dark (*i.e.* no light is incident on it), the reverse current is negligible *i.e.* $I_R = 0$ A.

$$\therefore \quad V_{out} = V - I_R R = 12\text{ V} - 0 \times 10\text{ k}\Omega = \mathbf{12\ V}$$

(*ii*) When light of intensity 3 mW/cm² is incident on the photo-diode, the reverse current I_R can be found from the graph in Fig. 5.19. (*ii*). It is clear from Fig. 5.19 (*ii*) that corresponding to $E = 3$ mW/cm², $I_R = 30$ μA.

$$\therefore \quad V_{out} = V - I_R R = 12\text{ V} - 30\ \mu\text{A} \times 10\text{ k}\Omega = \mathbf{11.7\ V}$$

Note that V_{out} varies only 0.3 V(= 300 mV) between the light and dark conditions. This is due to the fact that a photo-diode produces small currents ; 30 μA in this case.

5.11 Optoisolator

***An optoisolator** (also called optocoupler) is a device that uses light to couple a signal from its input (a photoemitter e.g., a LED) to its output (a photodetector e.g., a photo-diode).*

Fig. 5.20 shows a LED-photo diode optoisolator. The LED is on the left and the photo-diode is on the right. The arrangement shown in Fig. 5.20 is referred to as *optocoupling* because the output from the LED circuit is coupled *via* light to the photo-diode circuit. When the LED is energised, current flows through the LED. The light from the LED hits the photo-diode and sets up a reverse current through resistor R_2. The voltage across the photo-diode is given by ;

$$V_{out} = V_{SS} - IR_2$$

The output voltage depends on how large the reverse current is. If we vary the LED supply, the amount of light changes and this causes the photo-diode current to change. As a result, V_{out} changes. The key advantage of an optoisolator is the electrical isolation between the input and output circuits ; the only contact between the input and output circuits is the stream of light.

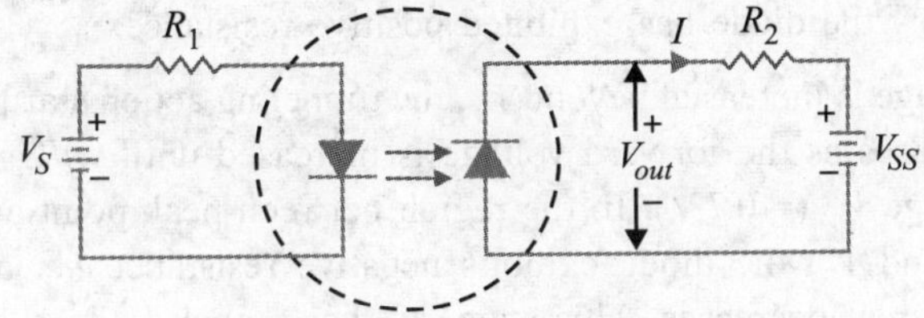

Fig. 5.20

5.12 Tunnel Diode

A **tunnel diode** *is a pn junction that exhibits negative resistance between two values of forward voltage (i.e. between peak-point voltage and valley-point voltage).*

A conventional diode exhibits *positive resistance when it is forward biased or reverse biased. However, if a semiconductor junction diode is heavily doped with impurities, it exhibits negative resistance (*i.e.* current decreases as the voltage is increased) in certain regions in the forward direction. Such a diode is called *tunnel diode.*

Theory. The tunnel diode is basically a *pn* junction with heavy doping of *p*-type and *n*-type semiconductor materials. In fact, a tunnel diode is doped approximately 1000 times as heavily as a conventional diode. This heavy doping results in a large number of majority carriers. Because of the large number of carriers, most are not used during the initial recombination that produces the depletion layer. As a result, the *depletion layer is very narrow.* In comparison with conventional diode, the depletion layer of a tunnel diode is 100 times narrower. The operation of a tunnel diode depends upon the *tunneling effect* and hence the name.

Tunneling effect. The heavy doping provides a large number of majority carriers. Because of the large number of carriers, there is much drift activity in *p* and *n* sections. This causes many valence electrons to have their energy levels raised closer to the conduction region. Therefore, it takes only a very small applied forward voltage to cause conduction.

The movement of valence electrons from the valence energy band to the conduction band with little or no applied forward voltage is called **tunneling.** *Valence electrons seem to tunnel through the forbidden energy band.*

As the forward voltage is first *increased,* the diode current rises rapidly due to tunneling

* If current flowing through a circuit or device increases as the applied voltage is increased, we say that the circuit or device has positive resistance.

effect. Soon the tunneling effect is reduced and current flow starts to *decrease* as the forward voltage across the diode is increased. The tunnel diode is said to have entered the negative resistance region. As the voltage is further increased, the tunneling effect plays less and less part until a valley-point is reached. From now onwards, the tunnel diode behaves as ordinary diode *i.e.* diode current increases with the increase in forward voltage.

V-I Characteristic. Fig. 5.21 (*i*) shows the *V-I* characteristic of a typical tunnel diode.

(*i*) As the forward voltage across the tunnel diode is increased from zero, electrons from the *n*-region "tunnel" through the potential barrier to the *p*-region. As the forward voltage increases, the diode current also increases until the *peak-point P* is reached. The diode current has now reached peak current I_P (= 2.2 mA) at about peak-point voltage V_P (= 0.07 V). Until now the diode has exhibited positive resistance.

(*ii*) As the voltage is increased beyond V_P, the tunneling action starts decreasing and the diode current decreases as the forward voltage is increased until *valley-point V* is reached at valley-point voltage V_V (= 0.7 V). In the region between peak-point and valley-point (*i.e.* between points *P* and *V*), the diode exhibits negative resistance *i.e.* as the forward bias is increased, the current decreases. This suggests that tunnel diode, when operated in the negative resistance region, can be used as an oscillator or a switch.

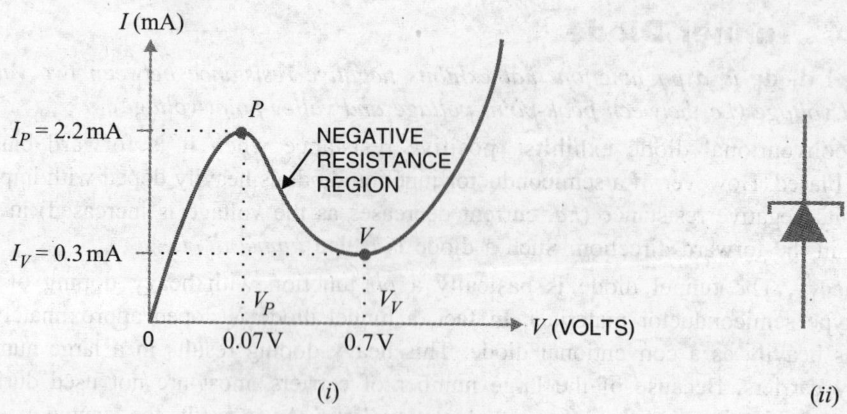

Fig. 5.21

(*iii*) When forward bias is increased beyond valley-point voltage V_V (= 0.7 V), the tunnel diode behaves as a normal diode. In other words, from point *V* onwards, the diode current increases with the increase in forward voltage *i.e.* the diode exhibits positive resistance once again. Fig. 5.21. (*ii*) shows the symbol of tunnel diode. It may be noted that a tunnel diode has a high reverse current but operation under this condition is not generally used.

5.13 Tunnel Diode Oscillator

A tunnel diode is always operated in the negative resistance region. When operated in this region, it works very well in an oscillator. Fig. 5.22 (*i*) shows a parallel resonant circuit. Note that R_P is the parallel equivalent of the series winding resistance of the coil. When the tank circuit is set into oscillations by applying voltage as shown in Fig. 5.22. (*ii*), damped oscillations are produced. It is because energy is lost in the resistance R_P of the tank circuit.

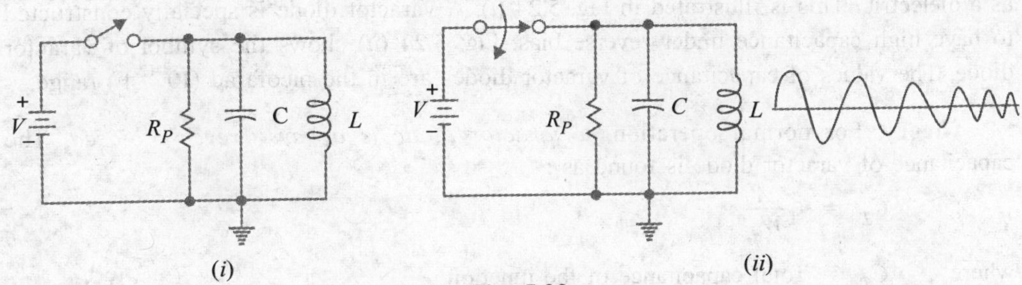

Fig. 5.22

If a tunnel diode is placed in series with the tank circuit and biased at the centre of the negative-resistance portion of its characteristic as shown in Fig. 5.23, undamped oscillations are produced at the output. It is because the negative-resistance characteristic of the tunnel diode counteracts the positive-resistance characteristic of the tank circuit.

The circuit shown in Fig. 5.23 is called *tunnel diode oscillator* or *negative resistance oscillator.* The negative resistance oscillator has one major drawback. While the circuit works very well at extreme high frequencies (upper mega hertz range), it cannot be used efficiently at low frequencies. Low-frequency oscillators generally use transistors.

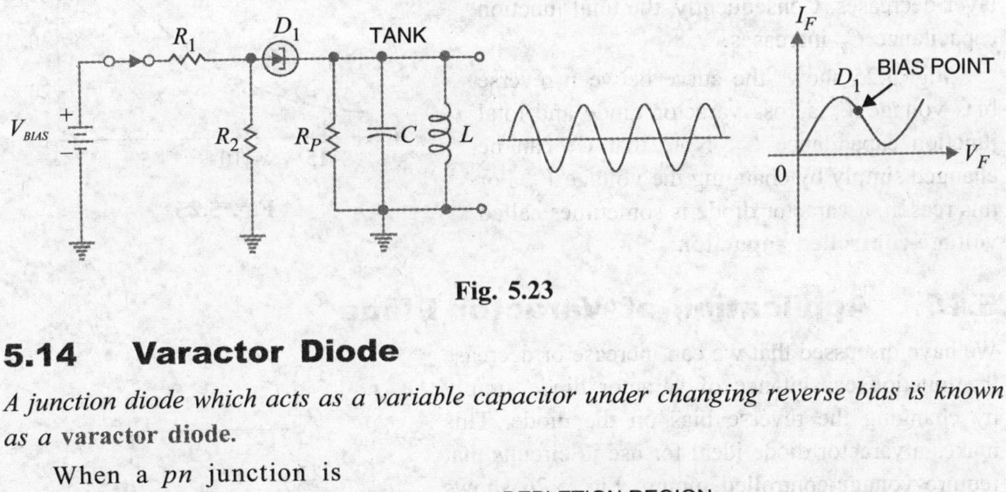

Fig. 5.23

5.14 Varactor Diode

A junction diode which acts as a variable capacitor under changing reverse bias is known as a **varactor diode.**

When a *pn* junction is formed, depletion layer is created in the junction area. Since there are no charge carriers within the depletion zone, the zone acts as an insulator. The *p*-type material with holes (considered positive) as majority carriers and *n*-type material with electrons (–ve charge) as majority carriers act as charged plates. Thus the diode may be considered as a capacitor with *n*-region and *p*-region forming oppositely charged plates and with depletion zone between them acting

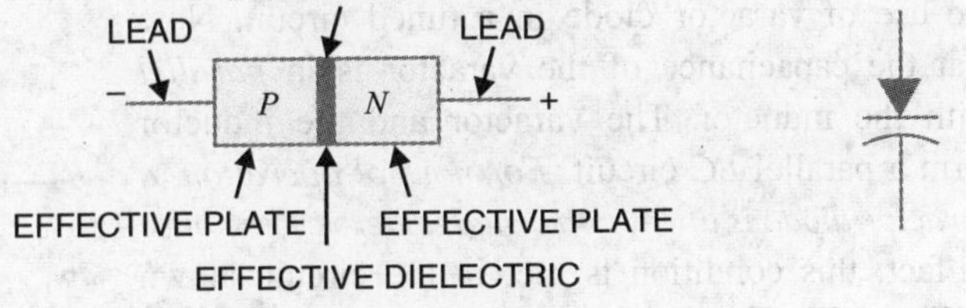

Fig. 5.24

as a dielectric. This is illustrated in Fig. 5.24 (*i*). A varactor diode is specially constructed to have high capacitance under reverse bias. Fig. 5.24 (*ii*) shows the symbol of varactor diode. The values of capacitance of varactor diodes are in the picofarad (10^{-12} F) range.

Theory. For normal operation, a *varactor diode is always *reverse biased.* The capacitance of varactor diode is found as :

$$C_T = \varepsilon \frac{A}{W_d}$$

where C_T = Total capacitance of the junction

ε = Permittivity of the semiconductor material

A = Cross-sectional area of the junction

W_d = Width of the depletion layer

When reverse voltage across a varactor diode is increased, the width W_d of the depletion layer increases. Therefore, the total junction capacitance C_T of the junction decreases. On the other hand, if the reverse voltage across the diode is lowered, the width W_d of the depletion layer decreases. Consequently, the total junction capacitance C_T increases.

Fig. 5.25 shows the curve between reverse bias voltage V_R across varactor diode and total junction capacitance C_T. Note that C_T can be changed simply by changing the voltage V_R. For this reason, a varactor diode is sometimes called **voltage-controlled capacitor.**

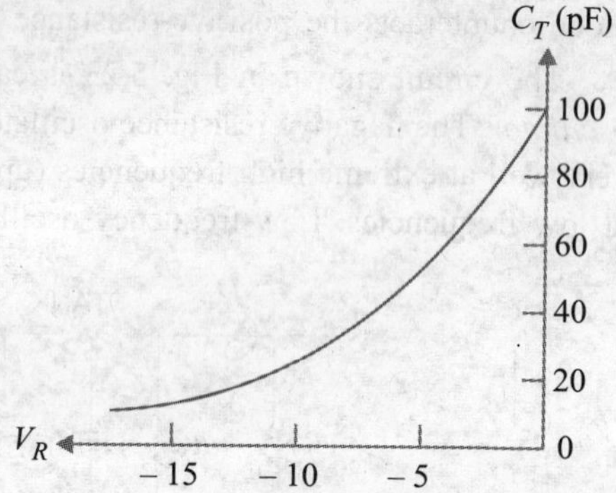

Fig. 5.25

5.15 Application of Varactor Diode

We have discussed that we can increase or decrease the junction capacitance of varactor diode simply by changing the reverse bias on the diode. This makes a varactor diode ideal for use in circuits that require voltage-controlled tuning. Fig. 5.26 shows the use of varactor diode in a tuned circuit. Note that the capacitance of the varactor is in *parallel* with the inductor. The varactor and the inductor form a parallel *LC* circuit. *For normal operation, a varactor diode is always operated under reverse bias.* In fact, this condition is met in the circuit shown in Fig. 5.26. The resistance R_W in the circuit is the *winding resistance* of the inductor. This winding resistance is in series with the potentiometer R_1. Thus R_1 and R_W form a voltage divider that is used to determine the amount of reverse bias across the

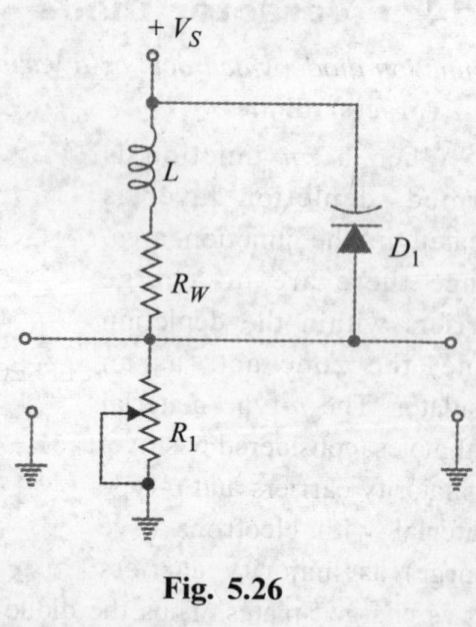

Fig. 5.26

* A forward biased varactor diode would serve no useful purpose.

varactor diode D_1 and therefore its capacitance. By adjusting the setting of R_1, we can vary the diode capacitance. This, in turn, varies the resonant frequency of the *LC* circuit. The resonant frequency f_r of the *LC* circuit is given by ;

$$f_r = \frac{1}{2\pi\sqrt{LC}}$$

If the amount of varactor reverse bias is *decreased,* the value of *C* of the varactor *increases.* The increase in *C* will cause the resonant frequency of the circuit to *decrease.* Thus, *a decrease in reverse bias causes a decrease in resonant frequency and vice-versa.*

Example 5.8. *The LC tank circuit shown in Fig. 5.26 has a 1mH inductor. The varactor has capacitance of 100 pF when reverse bias is 5 V d.c. Determine the resonant frequency of the circuit for this reverse bias.*

Solution.

$$\text{Resonant frequency, } f_r = \frac{1}{2\pi\sqrt{LC}}$$

Here, L = 1mH = 1×10^{-3} H ; C = 100 pF = 100×10^{-12} F

$$\therefore \quad f_r = \frac{1}{2\pi\sqrt{1\times10^{-3}\times100\times10^{-12}}} = 503.3 \times 10^3 \text{ Hz} = \mathbf{503.3\ kHz}$$

5.16 Shockley Diode

Named after its inventor, a Shockley diode is a *PNPN* device having two terminals as shown in Fig. 5.27 (*i*). This *device acts as a switch and consists of four alternate *P*-type and *N*-type layers in a single crystal. The various layers are labelled as P_1, N_1, P_2 and N_2 for identification. Since a *P*-region adjacent to an *N*-region may be considered a junction diode, the Shockley diode is equivalent to three junction diodes connected in series as shown in Fig. 5.27 (*ii*). The symbol of Shockley diode is shown in Fig. 5.27 (*iii*).

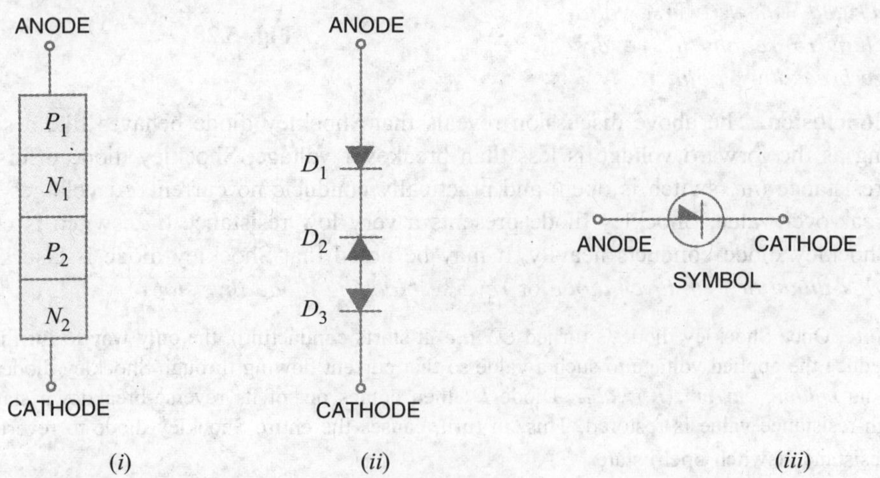

Fig. 5.27

* Note that if we remove the gate terminal of an *SCR*, the resulting device is Shockley diode.

Working

(*i*) When Shockley diode is forward biased (*i.e.* anode is positive *w.r.t.* cathode), diodes D_1 and D_3 would be forward-biased while diode D_2 would be reverse-biased. Since diode D_2 offers very high resistance (being reverse biased) and the three diodes are in series, the Shockley diode presents a very high resistance. As the *forward voltage increases, the reverse bias across D_2 is also increased. At some forward voltage (called *breakover voltage* V_{BO}), reverse breakdown of D_2 occurs. Since this breakdown results in reduced resistance, the Shockley diode presents a very low resistance. From now onwards, the Shockley diode behaves as a conventional forward-biased diode ; the forward current being determined by the applied voltage and external load resistance. This behaviour of Shockley diode is indicated on its *V-I* characteristic in Fig. 5.28.

(*ii*) When Shockley diode is reverse biased (*i.e.* anode is negative *w.r.t.* cathode), diodes D_1 and D_3 would be reverse-biased while diode D_2 would be forward-biased. If reverse voltage is increased sufficiently, the reverse voltage breakdown (point *A* in Fig. 5.28) of Shockley diode is reached. At this point, diodes D_1 and D_3 would go into reverse-voltage breakdown, the reverse current flowing through them would rise rapidly and the heat produced by this current flow could ruin the entire device. For this reason, *Shockley diode should never be operated with a reverse voltage sufficient to reach the reverse-voltage breakdown point.*

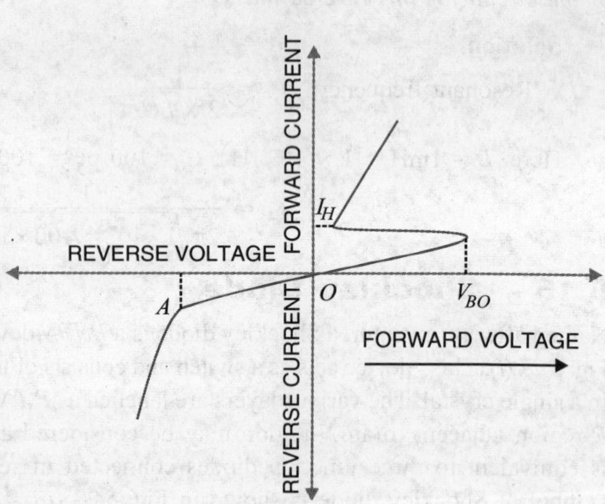

Fig. 5.28

Conclusion. The above discussion reveals that Shockley diode behaves like a switch. So long as the forward voltage is less than breakover voltage, Shockley diode offers very high resistance (*i.e.* switch is open) and practically conducts no current. At voltages above the break-over value, Shockley diode presents a very low resistance (*i.e.* switch is closed) and Shockley diode conducts heavily. It may be noted that Shockley diode is also known as *PNPN diode* or *four layer diode* or *reverse-blocking diode thyristor.*

Note. Once Shockley diode is turned *ON* (*i.e.* it starts conducting), the only way to turn it *OFF* is to reduce the applied voltage to such a value so that current flowing through Shockley diode drops below its *holding current* (I_H) *value.* Diode D_2 then comes out of its reverse-breakdown state and its high-resistance value is restored. This, in turn, causes the entire Shockley diode to revert to its high-resistance (switch open) state.

* Since D_1 and D_3 offer very low resistance (being forward biased), the entire applied voltage appears as reverse voltage across D_2.

Short Answer Questions

Q.1. *An ordinary forward biased diode (Si diode or Ge diode) does not emit light. Why?*

Ans. In an ordinary forward biased diode, free electrons and holes combine at the junction. As a result, energy is *released in the form of heat and light. Since silicon and germanium are opaque materials, none of the light escapes to the surroundings.

Q.2. *Why does a forward biased LED emit light?*

Ans. A light-emitting diode (LED) is made from semi-transparent materials like gallium arsenide. In a forward biased LED, heat and light are again released when free electrons and holes combine at the junction. Since the diode material is semitransparent, some of the light escapes to the surroundings.

Q.3. *When free electrons and holes combine at the junction of a forward biased silicon diode, most of the energy released is in the form of heat. Why?*

Ans. The difference in the energy between the conduction band and the valence band is called *energy gap* (E.G.). For silicon, E.G. = 1.1 eV. Therefore, for silicon, the wavelength of emitted light is

$$\text{Wavelength, } \lambda = \frac{^{**}12400}{\text{E.G. (eV)}}\,\text{Å} = \frac{12400}{1.1}\,\text{Å} \simeq 11000\,\text{Å}$$

This wavelength is in the infrared (heat radiations) region of the light spectrum. For this reason, silicon is not used in the fabrication of LEDs. Also germanium (E.G. = 0.7 eV) is not used for the same reason.

Q.4. *How is the colour of emitted light in LEDs controlled?*

Ans. In the fabrication of LEDs, we use semitransparent materials like *gallium arsenide* (GaAs), *gallium phosphide* (GaP) *etc*. The colour of light emitted by these materials is controlled by the type and degree of doping. Red, green and yellow LEDs are commonly available.

Q.5. *Why is a photo-diode called a photo detector?*

Ans. A photo-diode converts light energy into electrical energy. Devices that convert one form of energy into another form are known as *transducers, sensors* or *detectors*. For this reason, a photo-diode is called a photo detector.

Q.6. *What is an optoelectronic device?*

Ans. A device whose operation depends on light is called an optoelectronic device. For example, solar cells and photo-diodes are optoelectronic devices.

Q.7. *What is dark current in a photo-diode?*

Ans. When no light is incident on a reverse biased photo-diode, the reverse current in the diode is called *dark current*. Its value is negligibly small and may be considered zero. It may be noted that when no light is incident on a reverse-biased photo-diode, it behaves just like an ordinary reverse-biased diode.

---

* Free electrons have more energy (conduction band) than the holes (valence band). As an electron falls into valence band, energy is released.

** This relation can be derived from the following formula :

Energy of photon, $E = hf = \dfrac{hc}{\lambda}$

where h is Planck's constant, c is velocity of light and λ wavelength of light.

Q.8. *Why is an optocoupler called an optoisolator?*

Ans. An optocoupler combines a LED and a photo-diode in a single package. In a LED-photodiode coupler, the light from the LED controls the reverse current in the photo-diode. Therefore, there is electrical isolation between the LED circuit and the photo-diode circuit. For this reason, an optocoupler is also known as an optoisolator.

Q.9. *What is the key characteristic of a tunnel diode?*

Ans. The key characteristic of a tunnel diode is that it has negative resistance region. In this region, an increase in the voltage across the diode results in the reduction of diode current. This unusual behaviour of tunnel diode finds many applications including oscillators, amplifiers and electronic switches.

Q.10. *Give one practical application of varactor diode.*

Ans. A varactor diode can be used for fine tuning of parallel *LC* circuits. We know that resonant frequency of a parallel tuned *LC* circuit is

$$f_r = \frac{1}{2\pi\sqrt{LC}}$$

The capacitance of a varactor diode is small ; of the order of 100 pF. If a varactor diode is connected in parallel with *C*, the fine adjustment of f_r can be made by changing the reverse voltage across the varactor diode.

Multiple-Choice Questions

1. Zener diodes are used primarily as ....
   (*i*) amplifiers (*ii*) voltage regulators
   (*iii*) rectifiers (*iv*) oscillators
2. A *pn* junction that radiates energy as light instead of as heat is called a ....
   (*i*) LED (*ii*) photo-diode
   (*iii*) photocell (*iv*) zener diode
3. The capacitance of a varactor diode increases when reverse voltage across it ....
   (*i*) decreases (*ii*) increases
   (*iii*) breaks down (*iv*) stores charge
4. To display the digit 8 in a seven-segment indicator, ...
   (*i*) *C* must be lighted
   (*ii*) *G* must be off
   (*iii*) *F* must be on
   (*iv*) all segments must be lighted
5. A photo-diode is normally ....
   (*i*) forward-biased
   (*ii*) reverse-biased
   (*iii*) neither forward nor reverse biased
   (*iv*) emitting light
6. When the reverse voltage increases, the junction capacitance ....
   (*i*) decreases
   (*ii*) stays the same
   (*iii*) increases
   (*iv*) has more bandwidth
7. The device associated with voltage-controlled capacitance is a ....
   (*i*) LED (*ii*) photo-diode
   (*iii*) varactor diode (*iv*) zener diode
8. The varactor is usually ....
   (*i*) forward-biased
   (*ii*) reverse-biased
   (*iii*) unbiased
   (*iv*) in the breakdown region

9. When the light increases, the reverse current in a photo-diode ....
   (*i*) increases (*ii*) decreases
   (*iii*) is unaffected
   (*iv*) none of the above
10. To display the digit 0 in a seven segment display, ...
   (*i*) *A* must be lighted
   (*ii*) *F* must be off
   (*iii*) *G* must be on
   (*iv*) all segments except *G* should be lighted

Answers to Multiple-Choice Questions

1. (*ii*)	2. (*i*)	3. (*i*)	4. (*iv*)	5. (*ii*)
6. (*i*)	7. (*iii*)	8. (*ii*)	9. (*i*)	10. (*iv*)

Chapter Review Topics

1. What is a LED?
2. Explain the working of a LED.
3. Give two applications of LEDs.
4. Why do LEDs need series current-limiting resistors?
5. How does LED differ from an ordinary diode?
6. What is a photo-diode?
7. How does photo-diode work?
8. Give two applications of photo-diodes.
9. What is an optoisolator?
10. What is a tunnel diode?
11. Explain the *V-I* characteristics of a tunnel diode.
12. Explain the working of tunnel diode oscillator.
13. What is a varactor diode?
14. Explain the working of varactor diode.
15. Give one application of varactor diode.
16. Explain the working of Shockley diode.

CHAPTER 6

Transistors

Learning Objectives

- Develop an understanding of bipolar junction transistor (BJT) and its operation.
- Learn transistor as an amplifier.
- Develop a strong understanding of common base (CB) transistor connection, common emitter (CE) transistor connection and common collector (CC) transistor connection.
- Understand the input and output characteristics of common base (CB) transistor connection and common emitter (CE) transistor connection.
- Explore the reasons for the widespread use of common emitter (CE) transistor connection.
- Develop an understanding of transistor load line analysis.
- Understand the use of transistor as constant current source.
- Learn the importance of power rating of transistor.
- Explore the applications of transistors.

Introduction

When a third doped element is added to a semiconductor diode in such a way that two *pn* junctions are formed, the resulting device is known as a *bipolar junction transistor (BJT)* or simply *transistor*. Thus a semiconductor diode has one *pn* junction while a transistor has two *pn* junctions. The primary function of a transistor is to raise the strength of a weak signal *i.e.* it acts as an amplifier. Transistors find many other applications in electronic circuits. For example, a transistor is used as a high-speed electronic switch in digital circuits. In this chapter, we shall focus our attention on various aspects of transistors and their increasing applications in electronics industry.

6.1 Transistor

A **transistor** *consists of two pn junctions formed by *sandwiching either p-type or n-type semiconductor between a pair of opposite types.* Accordingly; there are two types of transistors, namely ;

(*i*) *n-p-n* transistor (*ii*) *p-n-p* transistor

* In practice, these three blocks *p, n, p* are grown out of the same crystal by adding corresponding impurities in turn.

An *n-p-n* transistor is composed of two *n*-type semiconductors separated by a thin section of *p*-type as shown in Fig. 6.1 (*i*). However, a *p-n-p* transistor is formed by two *p*-sections separated by a thin section of *n*-type as shown in Fig. 6.1 (*ii*).

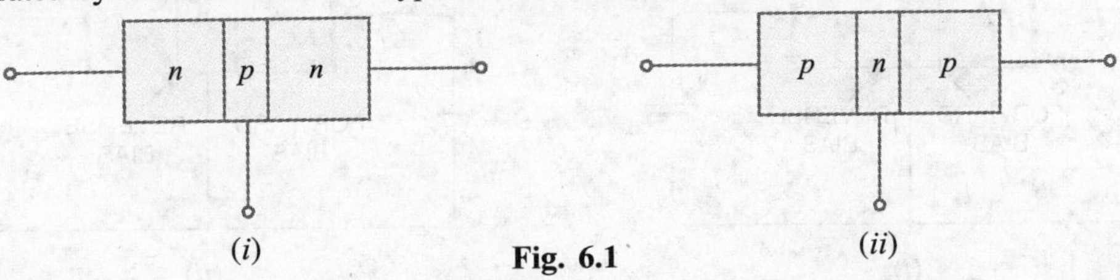

Fig. 6.1

In each type of transistor, the following points may be noted :

(*i*) There are two *pn* junctions. Therefore, a transistor may be regarded as a combination of two diodes connected back to back.

(*ii*) There are three terminals, one taken from each type of semiconductor.

(*iii*) The middle section is a very thin layer. This is the most important factor in the function of a transistor.

Origin of the name "Transistor". When new devices are invented, scientists often try to devise a name that will appropriately describe the device. A transistor has two *pn* junctions. As discussed later, one junction is forward biased and the other is reverse biased. The forward biased junction has a low resistance path whereas a reverse biased junction has a high resistance path. The weak signal is introduced in the low resistance circuit and output is taken from the high resistance circuit. Therefore, a transistor *transfers* a signal from a low resistance to high resistance. The prefix 'trans' means the signal transfer property of the device while 'istor' classifies it as a solid element in the same general family with resistors.

Note. The two junction transistor shown in Fig. 6.1 is called bipolar junction transistor (*BJT*) because two (Bi means two) polarities charge carriers (*viz.* positively charged holes and negatively charged free electrons) are involved in its operation. *Note that in this chapter when we use the terms transistor, it means bipolar junction transistor (BJT) unless stated otherwise.*

6.2 Naming the Transistor Terminals

A transistor (*pnp* or *npn*) has three sections of doped semiconductors. The section on one side is the *emitter* and the section on the opposite side is the *collector*. The middle section is called the *base* and forms two junctions between the emitter and collector.

(*i*) **Emitter.** The section on one side that supplies charge carriers (electrons or holes) is called the *emitter*. *The emitter is always forward biased w.r.t. base* so that it can supply a large number of *majority carriers. In Fig. 6.2 (*i*), the emitter (*p*-type) of *pnp* transistor is forward biased and supplies hole charges to its junction with the base. Similarly, in Fig. 6.2 (*ii*), the emitter (*n*-type) of *npn* transistor has a forward bias and supplies free electrons to its junction with the base.

(*ii*) **Collector.** The section on the other side that collects the charges is called the *collector*. *The collector is always reverse biased.* Its function is to remove charges from its junction with the base. In Fig. 6.2 (*i*), the collector (*p*-type) of *pnp* transistor has a reverse bias and receives hole charges that flow in the output circuit. Similarly, in Fig. 6.2 (*ii*), the collector (*n*-type) of *npn* transistor has reverse bias and receives electrons.

* Holes if emitter is *p*-type and electrons if the emitter is *n*-type.

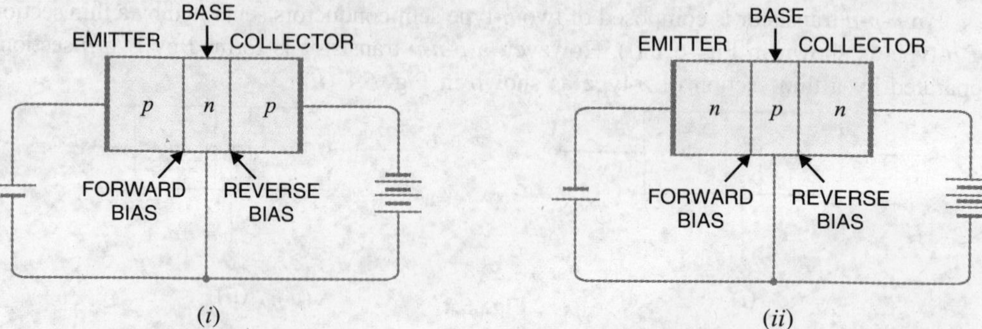

Fig. 6.2

(*iii*) **Base.** The middle section which forms two *pn*-junctions between the emitter and collector is called the *base*. The base-emitter junction is forward biased, allowing low resistance for the emitter circuit. The base-collector junction is reverse biased and provides high resistance in the collector circuit.

6.3 Some Facts about the Transistor

Before discussing transistor action, it is important that the reader may keep in mind the following facts about the transistor :

(*i*) The transistor has three regions, namely; *emitter*, *base* and *collector*. The base is much thinner than the emitter while *collector is wider than both as shown in Fig. 6.3. However, for the sake of convenience, it is customary to show emitter and collector to be of equal size.

(*ii*) The emitter is heavily doped so that it can inject a large number of charge carriers (electrons or holes) into the base. The base is lightly doped and very thin ; it passes most of the emitter injected charge carriers to the collector. The collector is moderately doped.

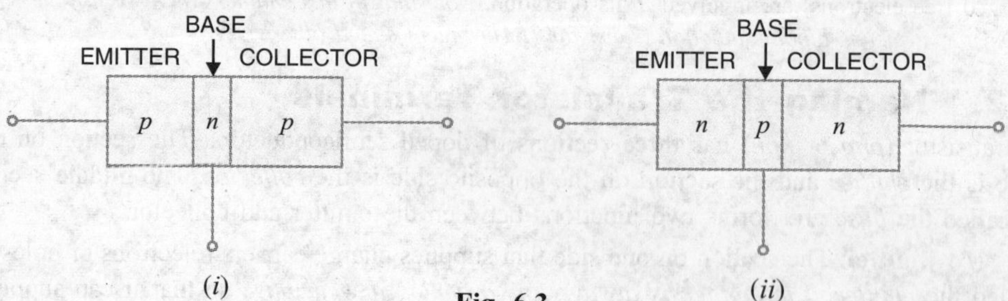

Fig. 6.3

(*iii*) The transistor has two *pn* junctions *i.e.* it is like two diodes. The junction between emitter and base may be called *emitter-base diode* or simply the *emitter diode*. The junction between the base and collector may be called *collector-base diode* or simply *collector diode*.

(*iv*) The emitter diode is always forward biased whereas collector diode is always reverse biased.

(*v*) The resistance of emitter diode (forward biased) is very small as compared to collector diode (reverse biased). Therefore, forward bias applied to the emitter diode is generally very small whereas reverse bias on the collector diode is much higher.

* During transistor operation, much heat is produced at the collector junction. The collector is made larger to dissipate the heat

6.4 Transistor Action

The emitter-base junction of a transistor is forward biased whereas collector-base junction is reverse biased. If for a moment, we ignore the presence of emitter-base junction, then *practically** no current would flow in the collector circuit because of the reverse bias. However, if the emitter-base junction is also present, then forward bias on it causes the emitter current to flow. It is seen that this emitter current almost entirely flows in the collector circuit. Therefore, the current in the collector circuit depends upon the emitter current. If the emitter current is zero, then collector current is nearly zero. However, if the emitter current is 1mA, then collector current is also about 1mA. This is precisely what happens in a transistor. We shall now discuss this transistor action for *npn* and *pnp* transistors.

(*i*) **Working of npn transistor.** Fig. 6.4 shows the *npn* transistor with forward bias to emitter-base junction and reverse bias to collector-base junction. The forward bias causes the electrons in the *n*-type emitter to flow towards the base. This constitutes the emitter current I_E. As these electrons flow through the *p*-type base, they tend to combine with holes. As the base is lightly doped and very thin, therefore, only a few electrons (less than 5%) combine with holes to constitute base** current I_B. The remainder (***more than 95%) cross-over into the collector region to constitute collector current I_C. In this way, almost the entire emitter current flows in the collector circuit. It is clear that emitter current is the sum of collector and base currents *i.e.*

$$I_E = I_B + I_C$$

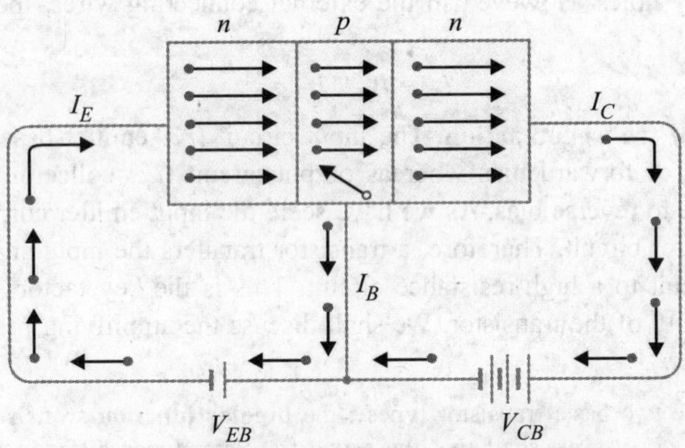

BASIC CONNECTION OF *npn* TRANSISTOR

Fig. 6.4

* In actual practice, a very little current (a few μA) would flow in the collector circuit. This is called collector cut-off current and is due to minority carriers.

** The electrons which combine with holes become valence electrons. Then as valence electrons, they flow down through holes and into the external base lead. This constitutes base current I_B.

*** The reasons that most of the electrons from emitter continue their journey through the base to collector to form collector current are: (*i*) The base is lightly doped and very thin. Therefore, there are a few holes which find enough time to combine with electrons. (*ii*) The reverse bias on collector is quite high and exerts attractive forces on these electrons.

(ii) **Working of pnp transistor.** Fig. 6.5 shows the basic connection of a *pnp* transistor. The forward bias causes the holes in the *p*-type emitter to flow towards the base. This constitutes the emitter current I_E. As these holes cross into *n*-type base, they tend to combine with the electrons. As the base is lightly doped and very thin, therefore, only a few holes (less than 5%) combine with the electrons. The remainder (more than 95%) cross

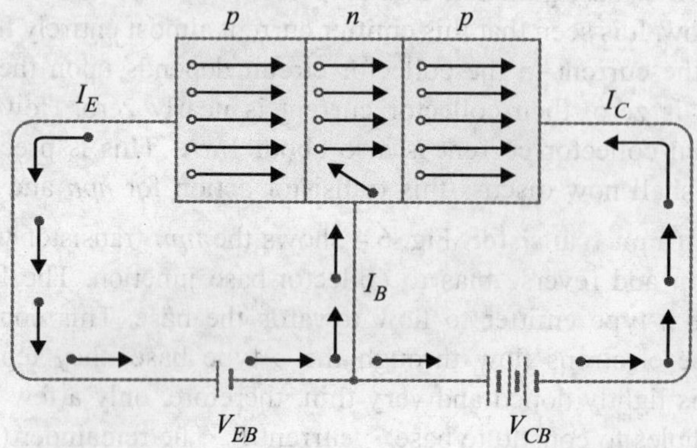

BASIC CONNECTION OF *pnp* TRANSISTOR

Fig. 6.5

into the collector region to constitute collector current I_C. In this way, almost the entire emitter current flows in the collector circuit. It may be noted that current conduction within *pnp* transistor is by holes. However, in the external connecting wires, the current is still by electrons.

Again, $$I_E = I_B + I_C$$

Importance of transistor action. The input circuit (*i.e.* emitter-base junction) has low resistance because of forward bias whereas output circuit (*i.e.* collector-base junction) has high resistance due to reverse bias. As we have seen, the input emitter current almost entirely flows in the collector circuit. Therefore, a transistor transfers the input signal current from a low-resistance circuit to a high-resistance circuit. This is the key factor responsible for the amplifying capability of the transistor. We shall discuss the amplifying property of transistor later in this chapter.

Note. There are two basic transistor types : the **bipolar junction transistor** (*BJT*) and **field-effect transistor** (*FET*). As we shall see, these two transistor types differ in both their operating characteristics and their internal construction. **Note that when we use the term transistor, it means bipolar junction transistor** (*BJT*). The term comes from the fact that in a bipolar transistor, there are *two* types of charge carriers (*viz.* electrons and holes) that play part in conduction. Note that bi means two and polar refers to polarities. The field-effect transistor is simply referred to as *FET.*

6.5 Transistor Biasing

A transistor has two *pn* junctions *viz. emitter-base* junction and *collector-base* junction. *Applying proper d.c. voltages to the two junctions of a transistor is known as* **transistor biasing**. *When a transistor is used as an amplifier, emitter-base junction is always forward biased and collector-base junction is always reverse biased* as shown in Fig. 6.6. When a transistor is operated under these bias conditions, it is said to be operating in the **active region**. *A transistor is normally operated in the active region.*

When both junctions of a transistor are reverse biased, the transistor is said to be operating in the **cut-off region**. Under these bias conditions, the collector current is practically zero (*i.e.* transistor is off) and so are the emitter and base currents. In actual practice, a very small collector current (of the order of μA) flows from emitter to collector. It is called *collector leakage current* and is due to the minority carriers. When both the junctions of a transistor are forward biased, the transistor is said to be operating in the **saturation region.** When the transistor is operated in the saturation region, the collector current becomes maximum. Clearly, saturation is opposite of cut-off. Therefore, a transistor can be operated in one of the following three bias conditions.

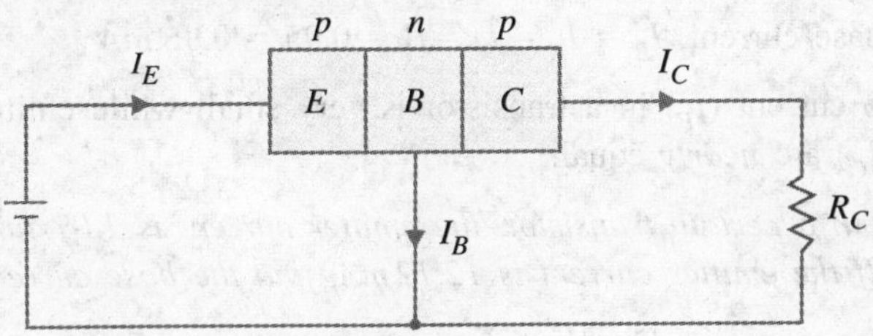

Fig. 6.6

Emitter-Base Junction	Collector-Base Junction	Operating Region
Forward biased	Reverse biased	Active
Reverse biased	Reverse biased	Cut-off
Forward biased	Forward biased	Saturation

Note. *Note that for normal operation of the transistor, the emitter-base junction must be forward biased and the collector-base junction must be reverse biased.* This is an important universal rule for transistor biasing.

6.6 Transistor Symbols

In the earlier diagrams, the transistors have been shown in diagrammatic form. However, for the sake of convenience, the transistors are represented by schematic diagrams. The symbols used for *npn* and *pnp* transistors are shown in Fig. 6.7.

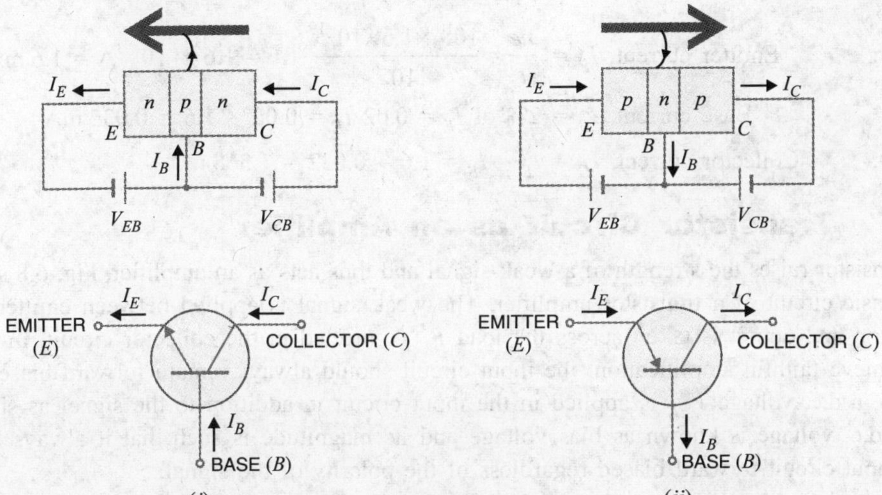

Fig. 6.7

Note that emitter is shown by an arrow which indicates the direction of conventional current flow with forward bias. For *npn* connection, it is clear that conventional current flows out of the emitter as indicated by the outgoing arrow in Fig. 6.7 (*i*). Similarly, for *pnp* connection, the conventional current flows into the emitter as indicated by inward arrow in Fig. 6.7 (*ii*).

Example 6.1. *In a transistor, emitter current is 1 mA and collector current is 0.95 mA. Calculate the value of base current.*

Solution. $I_E = I_B + I_C$

or Base current, $I_B = I_E - I_C = 1 - 0.95 = \mathbf{0.05\ mA}$

Note that base current (I_B) in a transistor is very small while emitter current (I_E) and collector current (I_C) are *nearly* equal.

Example 6.2. *In a certain transistor, the emitter current is 1.01 times as large as the collector current. If the emitter current is 12.12 mA, find the base current.*

Solution. $I_E = I_B + I_C$

or Base current, $I_B = I_E - I_C$

Here, $I_E = 12.12$ mA ; $I_E = 1.01\ I_C$ or $I_C = I_E/1.01 = 12.12/1.01 = 12$ mA

$\therefore$ $I_B = 12.12 - 12 = \mathbf{0.12\ mA}$

Example 6.3. *The emitter current in a certain npn transistor is 10 mA. If 5% of the free electrons injected into the base recombine with holes, find the base current and collector current.*

Solution. $I_B = 5\%$ of $I_E = 0.05\ I_E = 0.05 \times 10 = \mathbf{0.5\ mA}$

Collector current, $I_C = I_E - I_B = 10 - 0.5 = \mathbf{9.5\ mA}$

Example 6.4. *In an npn transistor, 10^{10} electrons enter the emitter in 10^{-6} s. If 2% electrons are lost in the base, find the collector current.*

Solution. Current = Charge/Time

$$\therefore \quad \text{Emitter current, } I_E = \frac{ne}{t} = \frac{10^{10} \times 1.6 \times 10^{-19}}{10^{-6}} \text{ A} = 1.6 \times 10^{-3}\text{ A} = 1.6 \text{ mA}$$

Base current, $I_B = 2\%$ of $I_E = 0.02\ I_E = 0.02 \times 1.6 = 0.032$ mA

$\therefore$ Collector current, $I_C = I_E - I_B = 1.6 - 0.032 = \mathbf{1.568\ mA}$

6.7 Transistor Circuit as an Amplifier

A transistor raises the strength of a weak signal and thus acts as an amplifier. Fig. 6.8 shows the basic circuit of a transistor amplifier. The weak signal is applied between emitter-base junction and output is taken across the load R_C connected in the collector circuit. In order to achieve faithful amplification, the input circuit should always remain forward biased. To do so, a d.c. voltage V_{EE} is applied in the input circuit in addition to the signal as shown. This d.c. voltage is known as bias voltage and its magnitude is such that it always keeps the input circuit forward biased regardless of the polarity of the signal.

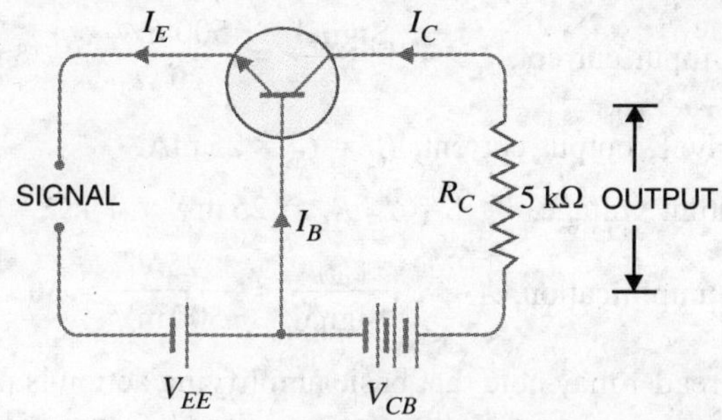

Fig. 6.8

As the input circuit has low resistance, therefore, a small change in signal voltage causes an appreciable change in emitter current. This causes almost the *same change in collector current due to transistor action. The collector current flowing through a high load resistance R_C produces a large voltage across it. Thus, a weak signal applied in the input circuit appears in the amplified form in the collector circuit. It is in this way that a transistor acts as an amplifier.

Illustration. The action of a transistor as an amplifier can be made more illustrative if we consider typical circuit values. Suppose collector load resistance R_C = 5 kΩ. Let us further assume that a change of 0.1 V in signal voltage produces a change of 1 mA in emitter current. Obviously, the change in collector current would also be approximately 1mA. This collector current flowing through collector load R_C would produce a voltage = 5 kΩ × 1 mA = 5 V. Thus, a change of 0.1 V in the signal has caused a change of 5 V in the output circuit. In other words, the transistor has been able to raise the voltage level of the signal from 0.1 V to 5 V *i.e.* voltage amplification is 50.

Example 6.5. *A common base transistor amplifier has an input resistance of 20 Ω and output resistance of 100 kΩ. The collector load is 1 kΩ. If a signal of 500 mV is applied between emitter and base, find the voltage amplification. Assume α_{ac} to be nearly one.*

Solution. **Fig. 6.9 shows the conditions of the problem. Note that output resistance is very high as compared to input resistance. This is not surprising because input junction (base to emitter) of the transistor is forward biased while the output junction (base to collector) is reverse biased.

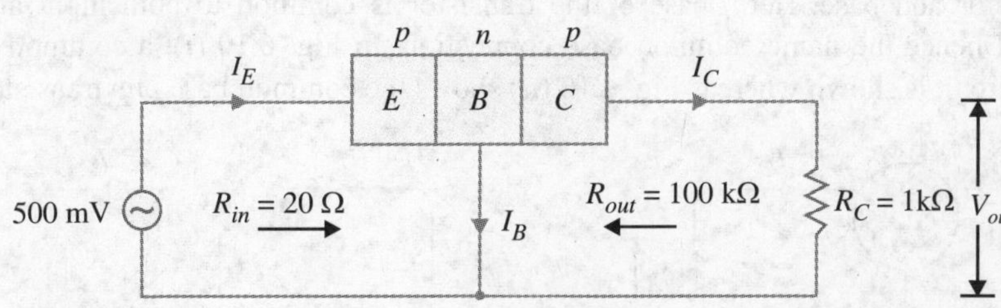

Fig. 6.9

* The reason is as follows. The collector-base junction is reverse biased and has a very high resistance of the order of mega ohms. Thus collector-base voltage has little effect on the collector current. This means that a large resistance R_C can be inserted in series with collector without disturbing the collector current relation to the emitter current *viz.* $I_C = \alpha I_E + I_{CBO}$. Therefore, collector current variations caused by a small base-emitter voltage fluctuations result in voltage changes in R_C that are quite high—often hundreds of times larger than the emitter-base voltage.

** The d.c. biasing is omitted in the figure because our interest is limited to amplification.

$$\text{Input current, } I_E = \frac{\text{Signal}}{R_{in}} = \frac{500 \text{ mV}}{20\ \Omega} = 25 \text{ mA}$$

Since α_{ac} is nearly 1, output current, $I_C = I_E = 25$ mA.

$$\text{Output voltage, } V_{out} = I_C\, R_C = 25 \text{ mA} \times 1 \text{ k}\Omega = 25 \text{ V}$$

$$\therefore \quad \text{Voltage amplification, } A_v = \frac{V_{out}}{\text{Signal}} = \frac{25 \text{ V}}{500 \text{ mV}} = \mathbf{50}$$

Comments. The reader may note that basic amplifying action is produced by transferring a current from a *low-resistance* to a *high-resistance* circuit. Consequently, the name transistor is given to the device by combining the two terms given in magenta letters below :

Transfer + Resistor ⟶ Transistor

6.8 Transistor Connections

There are three leads in a transistor *viz.*, emitter, base and collector terminals. However, when a transistor is to be connected in a circuit, we require four terminals ; two for the input and two for the output. This difficulty is overcome by making one terminal of the transistor common to both input and output terminals. The input is fed between this common terminal and one of the other two terminals. The output is obtained between the common terminal and the remaining terminal. Accordingly ; a transistor can be connected in a circuit in the following three ways :

(*i*) Common base connection
(*ii*) Common emitter connection
(*iii*) Common collector connection

Each circuit connection has specific advantages and disadvantages. It may be noted here that regardless of circuit connection, the emitter is always biased in the forward direction, while the collector always has a reverse bias.

6.9 Common Base Connection

In this circuit arrangement, input is applied between emitter and base and output is taken from collector and base. Here, base of the transistor is common to both input and output circuits and hence the name common base connection. In Fig. 6.10 (*i*), a common base *npn* transistor circuit is shown whereas Fig. 6.10 (*ii*) shows the common base *pnp* transistor circuit.

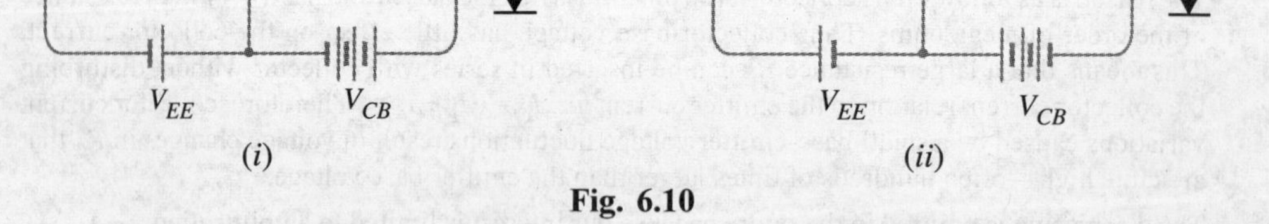

Fig. 6.10

1. Current amplification factor (α). It is the ratio of output current to input current. In a common base connection, the input current is the emitter current I_E and output current is the collector current I_C.

The ratio of change in collector current to the change in emitter current at constant collector-base voltage V_{CB} is known as **current amplification factor** *i.e.*

$$*\alpha = \frac{\Delta I_C}{\Delta I_E} \text{ at constant } V_{CB}$$

It is clear that current amplification factor is less than **unity. This value can be increased (but not more than unity) by decreasing the base current. This is achieved by making the base thin and doping it lightly. Practical values of α in commercial transistors range from 0.9 to 0.99.

2. Expression for collector current. The whole of emitter current does not reach the collector. It is because a small percentage of it, as a result of electron-hole combinations occurring in base area, gives rise to base current. Moreover, as the collector-base junction is reverse biased, therefore, some leakage current flows due to minority carriers. It follows, therefore, that total collector current consists of :

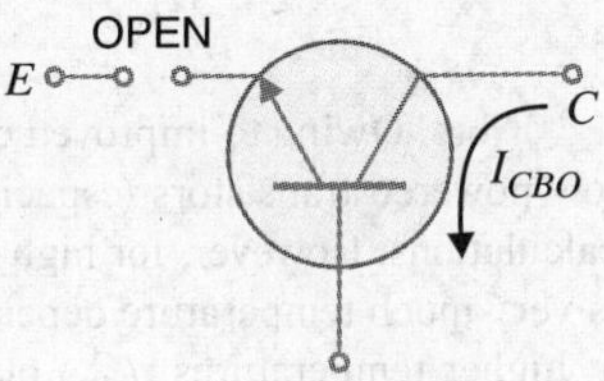

Fig. 6.11

(*i*) That part of emitter current which reaches the collector terminal *i.e.* ***$\alpha\ I_E$.

(*ii*) The leakage current $I_{leakage}$. This current is due to the movement of minority carriers across base-collector junction on account of it being reverse biased. This is generally much smaller than $\alpha\ I_E$.

∴ Total collector current, $I_C = \alpha I_E + I_{leakage}$

It is clear that if $I_E = 0$ (*i.e.* emitter circuit is open), a small leakage current still flows in the collector circuit. This $I_{leakage}$ is abbreviated as I_{CBO}, meaning collector-base current with emitter open. The I_{CBO} is indicated in Fig. 6.11.

∴ $$I_C = \alpha\ I_E + I_{CBO} \quad \text{...(i)}$$

Now, $$I_E = I_C + I_B$$

∴ $$I_C = \alpha\ (I_C + I_B) + I_{CBO}$$

or $$I_C (1 - \alpha) = \alpha\ I_B + I_{CBO}$$

or $$I_C = \frac{\alpha}{1-\alpha} I_B + \frac{I_{CBO}}{1-\alpha} \quad \text{...(ii)}$$

Relation (*i*) or (*ii*) can be used to find I_C. It is further clear from these relations that the collector current of a transistor can be controlled by either the emitter or base current.

---

* If only d.c. values are considered, then, $\alpha = I_C/I_E$.

** At first sight, it might seem that since there is no current gain, no voltage or power amplification could be possible with this arrangement. However, it may be recalled that output circuit resistance is much higher than the input circuit resistance. Therefore, it does give rise to voltage and power gain.

*** $\alpha = \frac{I_C}{I_E}$ $\quad \therefore I_C = \alpha\ I_E$

In other words, $\alpha\ I_E$ part of emitter current reaches the collector terminal.

Fig. 6.12 shows the concept of I_{CBO}. In *CB* configuration, a small collector current flows even when the emitter current is zero. This is the leakage collector current (*i.e.* the collector current when emitter is open) and is denoted by I_{CBO}. When the emitter voltage V_{EE} is also applied, the various currents are as shown in Fig. 6.12 (*ii*).

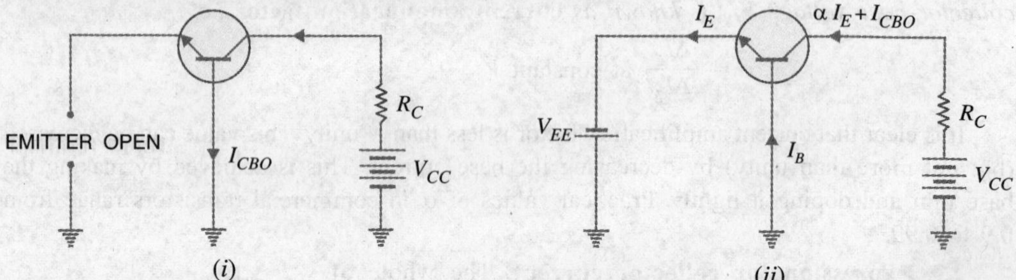

Fig. 6.12

Note. Owing to improved construction techniques, the magnitude of I_{CBO} for general-purpose and low-powered transistors (especially silicon transistors) is usually very small and may be neglected in calculations. However, for high power applications, it will appear in microampere range. Further, I_{CBO} is very much temperature dependent ; it increases rapidly with the increase in temperature. Therefore, at higher temperatures, I_{CBO} plays an important role and must be taken care of in calculations.

Example 6.6. *In a common base connection, current amplification factor is 0.9. If the emitter current is 1mA, determine the value of base current.*

Solution. Here, $\alpha = 0.9$; $I_E = 1$ mA

Now, $$\alpha = \frac{I_C}{I_E}$$

or $$I_C = \alpha I_E = 0.9 \times 1 = 0.9 \text{ mA}$$

Also, $$I_E = I_B + I_C$$

∴ Base current, $$I_B = I_E - I_C = 1 - 0.9 = \mathbf{0.1 \text{ mA}}$$

Example 6.7. *In common base connection, $\alpha = 0.98$ and the base current is 100 μA. Find the collector current and emitter current.*

Solution. $$I_E = I_B + I_C$$

Now, $$I_C = \alpha I_E = \alpha(I_B + I_C)$$

or $$I_C(1 - \alpha) = \alpha I_B$$

∴ $$I_C = \frac{\alpha I_B}{1-\alpha} = \frac{0.98 \times 100 \,\mu\text{A}}{1 - 0.98} = 4900 \,\mu\text{A} = \mathbf{4.9 \text{ mA}}$$

Now, $$I_E = \frac{I_C}{\alpha} = \frac{4.9 \text{ mA}}{0.98} = \mathbf{5 \text{ mA}}$$

Alternatively. $$I_E = I_B + I_C = 100 \,\mu\text{A} + 4.9 \text{ mA} = 5 \text{ mA}$$

Example 6.8. *The emitter current in a common base npn transistor circuit is 8.4 mA. If 0.8% of free electrons injected into the base recombine with holes and the leakage current is 0.1 μA, find (i) base current (ii) collector current (iii) exact value of α (iv) approximate value of α, neglecting I_{CBO}.*

Solution. It is given : $I_E = 8.4\text{ mA} ; I_{CBO} = 0.1\ \mu\text{A} = 10^{-7}\text{ mA}$

(*i*) $I_B = 0.8\%\text{ of } I_E = 0.008 \times 8.4\text{ mA} = \mathbf{0.0672\ mA}$

(*ii*) $I_C = I_E - I_B = 8.4 - 0.0672 = \mathbf{8.3328\ mA}$

(*iii*) Now, $\alpha I_E = I_C - I_{CBO} = 8.3328 - 10^{-7} = 8.3327\text{ mA}$

$\therefore$ Exact value of $\alpha = \frac{8.3327}{8.4} = \mathbf{0.9919}$

(*iv*) Approximate $\alpha = \frac{I_C}{I_E} = \frac{8.3328}{8.4} = \mathbf{0.992}$

Note that the difference between the exact and approximate values of α is so small that it can be neglected without any appreciable error.

Example 6.9. *In a common base connection, the emitter current is 1 mA. If the emitter circuit is open, the collector current is 50 μA. Find the total collector current. Given that α = 0.92.*

Solution. Here, $I_E = 1\text{ mA} ; \alpha = 0.92 ; \ I_{CBO} = 50\ \mu\text{A}$

$\therefore$ Total collector current, $I_C = \alpha I_E + I_{CBO} = 0.92 \times 1 + 50 \times 10^{-3}$

$= 0.92 + 0.05 = \mathbf{0.97\ mA}$

Example 6.10. *In a common base connection, α = 0.95. The voltage drop across 2 kΩ resistance which is connected in the collector is 2 V. Find the base current.*

Solution. Fig. 6.13 shows the required common base connection. The voltage drop across R_C (= 2 kΩ) is 2 V.

$\therefore$ $I_C = 2\text{V}/2\text{k}\Omega = 1\text{ mA}$

Now, $\alpha = I_C/I_E$

$\therefore$ $I_E = \frac{I_C}{\alpha} = \frac{1}{0.95} = 1.05\text{ mA}$

Using the relation, $I_E = I_B + I_C$

$\therefore$ $I_B = I_E - I_C = 1.05 - 1$

$= \mathbf{0.05\ mA}$

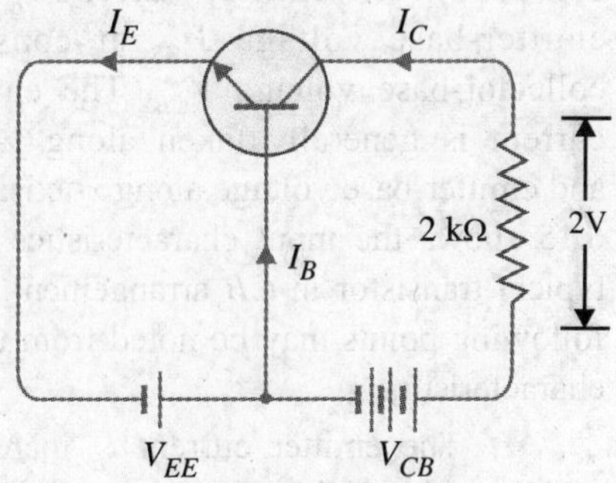

Fig. 6.13

Example 6.11. *For the common base circuit shown in Fig. 6.14, determine* I_C *and* V_{CB}. *Assume the transistor to be of silicon.*

Solution. Since the transistor is of silicon, $V_{BE} = 0.7$ V. Applying Kirchhoff's voltage law to the emitter-side loop, we get,

$$V_{EE} = I_E R_E + V_{BE}$$

or $$I_E = \frac{V_{EE} - V_{BE}}{R_E} = \frac{8\text{ V} - 0.7\text{ V}}{1.5\text{ k}\Omega} = 4.87\text{ mA}$$

$\therefore$ $I_C \simeq I_E = \mathbf{4.87\ mA}$

Applying Kirchhoff's voltage law to the collector-side loop, we have,

$$V_{CC} = I_C R_C + V_{CB}$$

$\therefore \quad V_{CB} = V_{CC} - I_C R_C$

$= 18 \text{ V} - 4.87 \text{ mA} \times 1.2 \text{ k}\Omega = 12.16 \text{ V}$

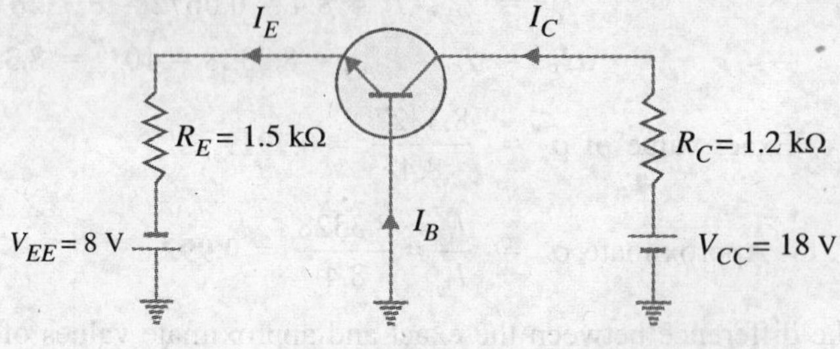

Fig. 6.14

6.10 Characteristics of Common Base Connection

The complete electrical behaviour of a transistor can be described by stating the interrelation of the various currents and voltages. These relationships can be conveniently displayed graphically and the curves thus obtained are known as the characteristics of transistor. The most important characteristics of common base connection are *input characteristics* and *output characteristics.*

1. Input characteristic. It is the curve between emitter current I_E and emitter-base voltage V_{EB} at constant collector-base voltage V_{CB}. The emitter current is generally taken along *y*-axis and emitter-base voltage along *x*-axis. Fig. 6.15 shows the input characteristics of a typical transistor in *CB* arrangement. The following points may be noted from these characteristics:

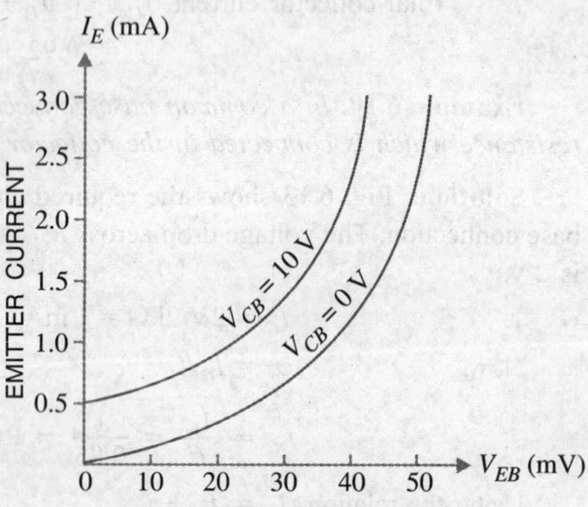

Fig. 6.15

(*i*) The emitter current I_E increases rapidly with small increase in emitter-base voltage V_{EB}. It means that input resistance is very small.

(*ii*) The emitter current is almost independent of collector-base voltage V_{CB}. This leads to the conclusion that emitter current (and hence collector current) is almost independent of collector voltage.

Input resistance. It is the ratio of change in emitter-base voltage (ΔV_{EB}) to the resulting change in emitter current (ΔI_E) at constant collector-base voltage (V_{CB}) *i.e.*

$$\text{Input resistance, } r_i = \frac{\Delta V_{BE}}{\Delta I_E} \text{ at constant } V_{CB}$$

In fact, input resistance is the opposition offered to the signal current. As a very small V_{EB} is sufficient to produce a large flow of emitter current I_E, therefore, input resistance is quite small, of the order of a few ohms.

2. Output characteristic. It is the curve between collector current I_C and collector-base voltage V_{CB} at *constant emitter current I_E. Generally, collector current is taken along *y*-axis and collector-base voltage along *x*-axis. Fig. 6.16 shows the output characteristics of a typical transistor in *CB* arrangement.

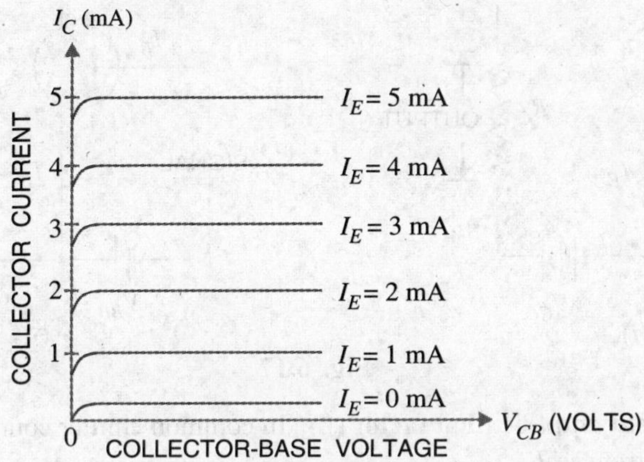

Fig. 6.16

The following points may be noted from the characteristics :

(*i*) The collector current I_C varies with V_{CB} only at very low voltages (< 1 V). The transistor is *never* operated in this region.

(*ii*) When the value of V_{CB} is raised above 1 – 2 V, the collector current becomes constant as indicated by straight horizontal curves. It means that now I_C is independent of V_{CB} and depends upon I_E only. This is consistent with the theory that the emitter current flows *almost* entirely to the collector terminal. The transistor is *always* operated in this region.

(*iii*) A very large change in collector-base voltage produces only a tiny change in collector current. This means that output resistance is very high.

Output resistance. It is the ratio of change in collector-base voltage (ΔV_{CB}) to the resulting change in collector current (ΔI_C) at constant emitter current *i.e.*

$$\text{Output resistance, } r_o = \frac{\Delta V_{CB}}{\Delta I_C} \text{ at constant } I_E$$

The output resistance of *CB* circuit is very high, of the order of several tens of kilo-ohms. This is not surprising because the collector current changes very slightly with the change in V_{CB}.

6.11 Common Emitter Connection

In this circuit arrangement, input is applied between base and emitter and output is taken from the collector and emitter. Here, emitter of the transistor is common to both input and output

* I_E has to be kept constant because any change in I_E will produce corresponding change in I_C. Here, we are interested to see how V_{CB} influences I_C.

circuits and hence the name common emitter connection. Fig. 6.17 (*i*) shows common emitter *npn* transistor circuit whereas Fig. 6.17 (*ii*) shows common emitter *pnp* transistor circuit.

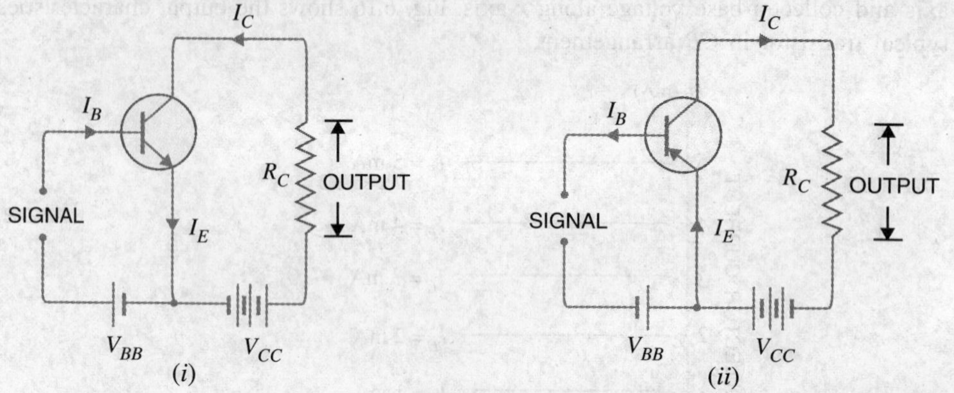

Fig. 6.17

1. Base current amplification factor (β). In common emitter connection, input current is I_B and output current is I_C.

The ratio of change in collector current (ΔI_C) *to the change in base current* (ΔI_B) *is known as* **base current amplification factor** *i.e.*

$$\beta^* = \frac{\Delta I_C}{\Delta I_B}$$

In almost any transistor, less than 5% of emitter current flows as the base current. Therefore, the value of β is generally greater than 20. Usually, its value ranges from 20 to 500. This type of connection is frequently used as it gives appreciable current gain as well as voltage gain.

Relation between β and α. A simple relation exists between β and α. This can be derived as follows :

$$\beta = \frac{\Delta I_C}{\Delta I_B} \quad \text{...(i)}$$

$$\alpha = \frac{\Delta I_C}{\Delta I_E} \quad \text{...(ii)}$$

Now, $I_E = I_B + I_C$

or $\Delta I_E = \Delta I_B + \Delta I_C$

or $\Delta I_B = \Delta I_E - \Delta I_C$

Substituting the value of ΔI_B in exp. (*i*), we get,

$$\beta = \frac{\Delta I_C}{\Delta I_E - \Delta I_C} \quad \text{...(iii)}$$

Dividing the numerator and denominator of R.H.S. of exp. (*iii*) by ΔI_E, we get,

$$\beta = \frac{\Delta I_C / \Delta I_E}{\frac{\Delta I_E}{\Delta I_E} - \frac{\Delta I_C}{\Delta I_E}} = \frac{\alpha}{1 - \alpha} \qquad \left[\because \ \alpha = \frac{\Delta I_C}{\Delta I_E}\right]$$

* If d.c. values are considered, $\beta = I_C/I_B$.

$$\therefore \qquad \beta = \frac{\alpha}{1-\alpha}$$

It is clear that as α approaches unity, β approaches infinity. In other words, the current gain in common emitter connection is very high. It is due to this reason that this circuit arrangement is used in about 90 to 95 percent of all transistor applications.

2. **Expression for collector current.** In common emitter circuit, I_B is the input current and I_C is the output current.

We know $I_E = I_B + I_C$ ...(*i*)

and $I_C = \alpha\, I_E + I_{CBO}$ ...(*ii*)

From exp. (*ii*), we get, $I_C = \alpha\, I_E + I_{CBO} = \alpha\,(I_B + I_C) + I_{CBO}$

or $I_C\,(1-\alpha) = \alpha\, I_B + I_{CBO}$

or $$I_C = \frac{\alpha}{1-\alpha} I_B + \frac{1}{1-\alpha} I_{CBO} \qquad \text{...(iii)}$$

From exp. (*iii*), it is apparent that if $I_B = 0$ (*i.e.* base circuit is open), the collector current will be the current to the emitter. This is abbreviated as I_{CEO}, meaning collector-emitter current with base open.

$$\therefore \qquad I_{CEO} = \frac{1}{1-\alpha} I_{CBO}$$

Substituting the value of $\frac{1}{1-\alpha} I_{CBO} = I_{CEO}$ in exp. (*iii*), we get,

$$I_C = \frac{\alpha}{1-\alpha} I_B + I_{CEO}$$

or $$I_C = \beta\, I_B + I_{CEO} \qquad \left(\because \beta = \frac{\alpha}{1-\alpha}\right)$$

Concept of I_{CEO}. In *CE* configuration, a small collector current flows even when the base current is zero [See Fig. 6.18 (*i*)]. This is the collector cut-off current (*i.e.* the collector current that flows when base is open) and is denoted by I_{CEO}. The value of I_{CEO} is much larger than I_{CBO}.

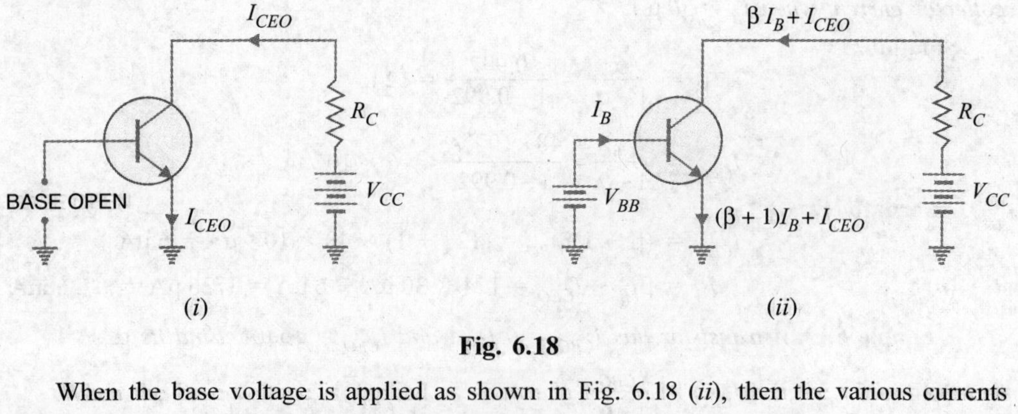

Fig. 6.18

When the base voltage is applied as shown in Fig. 6.18 (*ii*), then the various currents are :

Base current $= I_B$

Collector current $= \beta\, I_B + I_{CEO}$

$$\text{Emitter current} = \text{Collector current} + \text{Base current}$$

$$= (\beta I_B + I_{CEO}) + I_B = (\beta + 1)\, I_B + I_{CEO}$$

It may be noted here that : $I_{CEO} = \frac{1}{1-\alpha} I_{CBO} = (\beta + 1) I_{CBO}$ $\left[\because \frac{1}{1-\alpha} = \beta + 1\right]$

6.12 Measurement of Leakage Current

A very small leakage current flows in all transistor circuits. However, in most cases, it is quite small and can be neglected.

(i) **Circuit for I_{CEO} test.** Fig. 6.19 shows the circuit for measuring I_{CEO}. Since base is open ($I_B = 0$), the transistor is in cut-off. Ideally, $I_C = 0$ but actually there is a small current from collector to emitter due to minority carriers. It is called I_{CEO} (collector-to-emitter current with base open). This current is usually in the nA range for silicon. A faulty transistor will often have excessive leakage current.

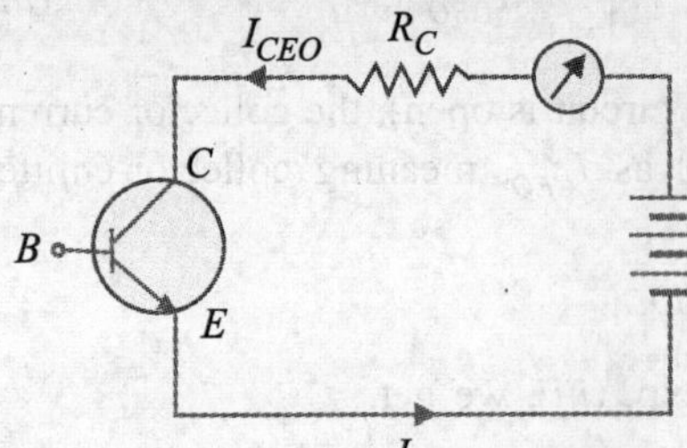

Fig. 6.19

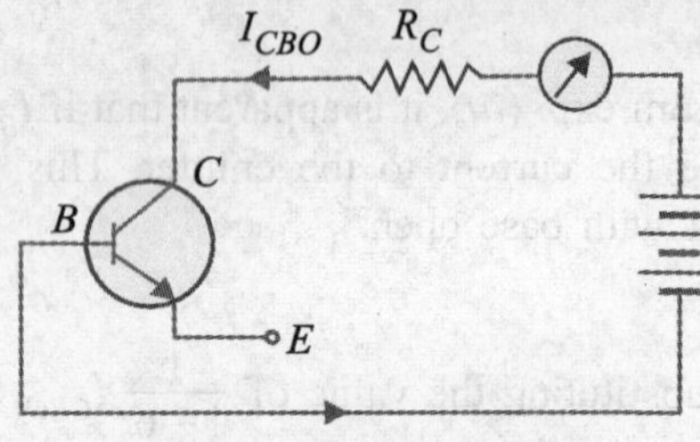

Fig. 6.20

(ii) **Circuit for I_{CBO} test.** Fig. 6.20 shows the circuit for measuring I_{CBO}. Since the emitter is open ($I_E = 0$), there is a small current from collector to base. This is called I_{CBO} (collector-to-base current with emitter open). This current is due to the movement of minority carriers across base-collector junction. The value of I_{CBO} is also small. If in measurement, I_{CBO} is excessive, then there is a possibility that collector-base is shorted.

Example 6.12. *A transistor has I_{CBO} = 48 nA and α = 0.992. Find (i) β and I_{CEO} (ii) collector current when I_B = 30 μA.*

Solution.

(i)
$$\beta = \frac{\alpha}{1-\alpha} = \frac{0.992}{1-0.992} = \mathbf{124}$$

$$I_{CEO} = \frac{I_{CBO}}{1-\alpha} = \frac{48 \times 10^{-9}\text{ A}}{1-0.992} = 6 \times 10^{-6}\text{ A} = \mathbf{6\ \mu A}$$

Alternatively.

$$I_{CEO} = (\beta + 1) I_{CBO} = (124 + 1) \times 48 \times 10^{-3}\ \mu\text{A} = 6\ \mu\text{A}$$

(ii)
$$I_C = \beta I_B + I_{CEO} = 124 \times 30\ \mu\text{A} + 6\ \mu\text{A} = 3726\ \mu\text{A} = \mathbf{3.726\ mA}$$

Example 6.13. *A transistor has I_{CBO} = 0.1 μA and I_{CEO} = 16 μA. Find its α.*

Solution.
$$I_{CEO} = \frac{I_{CBO}}{1-\alpha}$$

or
$$1 - \alpha = \frac{I_{CBO}}{I_{CEO}} = \frac{0.1\ \mu\text{A}}{16\ \mu\text{A}} = 0.00625$$

∴
$$\alpha = 1 - 0.00625 = \mathbf{0.99375}$$

Example 6.14. *Find the α rating of the transistor shown in Fig. 6.21. Hence determine the value of I_C using both α and β ratings of the transistor.*

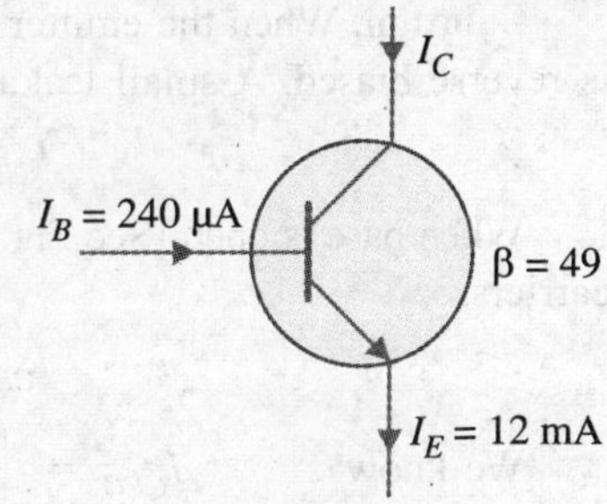

Fig. 6.21

Solution. Fig. 6.21 shows the conditions of the problem.

$$\alpha = \frac{\beta}{1+\beta} = \frac{49}{1+49} = \mathbf{0.98}$$

The value of I_C can be found by using either α or β rating as under :

$$I_C = \alpha\, I_E = 0.98\ (12\ \text{mA}) = \mathbf{11.76\ mA}$$

Also, $$I_C = \beta\, I_B = 49\ (240\ \mu\text{A}) = \mathbf{11.76\ mA}$$

Example 6.15. *A transistor is connected in common emitter (CE) configuration in which collector supply is 8V and the voltage drop across resistance R_C connected in the collector circuit is 0.5 V. The value of R_C = 800 Ω. If α = 0.96, determine :*

(i) collector-emitter voltage

(ii) base current

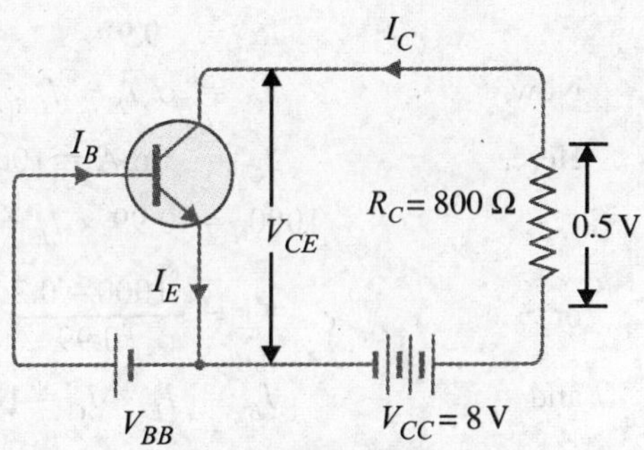

Fig. 6.22

Solution. Fig. 6.22 shows the required common emitter connection with various values.

(i) Collector-emitter voltage is

$$V_{CE} = V_{CC} - 0.5 = 8 - 0.5 = \mathbf{7.5\ V}$$

(ii) The voltage drop across R_C (= 800 Ω) is 0.5 V.

$$\therefore \quad I_C = \frac{0.5\ \text{V}}{800\ \Omega} = \frac{5}{8}\ \text{mA} = 0.625\ \text{mA}$$

Now, $$\beta = \frac{\alpha}{1-\alpha} = \frac{0.96}{1-0.96} = 24$$

$$\therefore \quad \text{Base current, } I_B = \frac{I_C}{\beta} = \frac{0.625}{24} = \mathbf{0.026\ mA}$$

Example 6.16. *An n-p-n transistor at room temperature has its emitter disconnected. A voltage of 5 V is applied between collector and base. With collector positive, a current of 0.2 μA flows. When the base is disconnected and the same voltage is applied between collector and emitter, the current is found to be 20 μA. Find α, I_E and I_B when collector current is 1mA.*

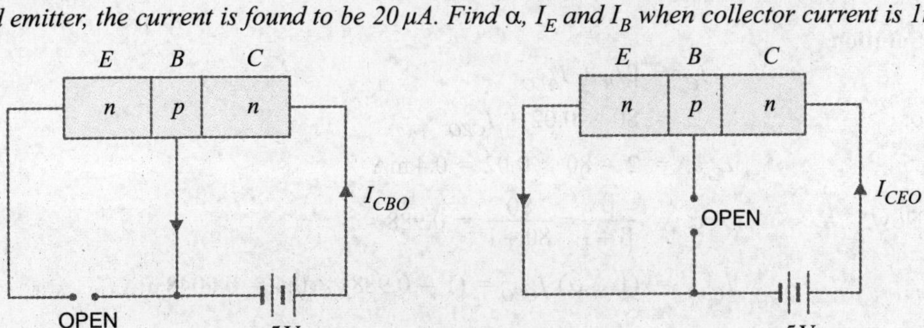

Fig. 6.23

Solution. When the emitter circuit is open [See Fig. 6.23 (*i*)], the collector-base junction is reverse biased. A small leakage current I_{CBO} flows due to minority carriers.

$\therefore \quad I_{CBO} = 0.2\ \mu A$ ...*given*

When base is open [See Fig. 6.23 (*ii*)], a small leakage current I_{CEO} flows due to minority carriers.

$\therefore \quad I_{CEO} = 20\ \mu A$ ...*given*

We know : $$I_{CEO} = \frac{I_{CBO}}{1-\alpha}$$

or $$20 = \frac{0.2}{1-\alpha}$$

$\therefore \quad \alpha = \mathbf{0.99}$

Now, $$I_C = \alpha I_E + I_{CBO}$$

Here, $$I_C = 1\text{ mA} = 1000\ \mu A \;;\; \alpha = 0.99 \;;\; I_{CBO} = 0.2\ \mu A$$

$$\therefore \quad 1000 = 0.99 \times I_E + 0.2$$

or $$I_E = \frac{1000 - 0.2}{0.99} = \mathbf{1010\ \mu A}$$

and $$I_B = I_E - I_C = 1010 - 1000 = \mathbf{10\ \mu A}$$

Example 6.17. *The collector leakage current in a transistor is 300 μA in CE arrangement. If now the transistor is connected in CB arrangement, what will be the leakage current? Given that β = 120.*

Solution. $$I_{CEO} = 300\ \mu A$$

$$\beta = 120 \;;\; \alpha = \frac{\beta}{\beta+1} = \frac{120}{120+1} = 0.992$$

Now, $$I_{CEO} = \frac{I_{CBO}}{1-\alpha}$$

$$\therefore \quad I_{CBO} = (1-\alpha)\, I_{CEO} = (1 - 0.992) \times 300 = \mathbf{2.4\ \mu A}$$

Note that leakage current in *CE* arrangement (*i.e.* I_{CEO}) is much more than in *CB* arrangement (*i.e.* I_{CBO}).

Example 6.18. *For a certain transistor,* $I_B = 20\ \mu A$; $I_C = 2\ mA$ *and* $\beta = 80$. *Calculate* I_{CBO}.

Solution.

$$I_C = \beta I_B + I_{CEO}$$

or $$2 = 80 \times 0.02 + I_{CEO}$$

$$\therefore \quad I_{CEO} = 2 - 80 \times 0.02 = 0.4\text{ mA}$$

Now, $$\alpha = \frac{\beta}{\beta+1} = \frac{80}{80+1} = 0.988$$

$$\therefore \quad I_{CBO} = (1-\alpha)\, I_{CEO} = (1 - 0.988) \times 0.4 = \mathbf{0.0048\ mA}$$

Example 6.19. *Using diagrams, explain the correctness of the relation $I_{CEO} = (\beta + 1)\ I_{CBO}$.*

Solution. The leakage current I_{CBO} is the current that flows through the base-collector junction when emitter is open as shown is Fig. 6.24. When the transistor is in *CE* arrangement, the *base current (*i.e.* I_{CBO}) is multiplied by β in the collector as shown in Fig. 6.25.

$$\therefore \quad I_{CEO} = I_{CBO} + \beta I_{CBO} = (\beta + 1)\ I_{CBO}$$

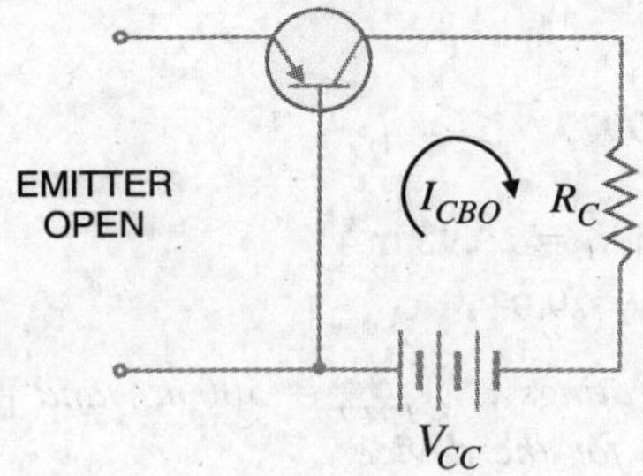

Fig. 6.24

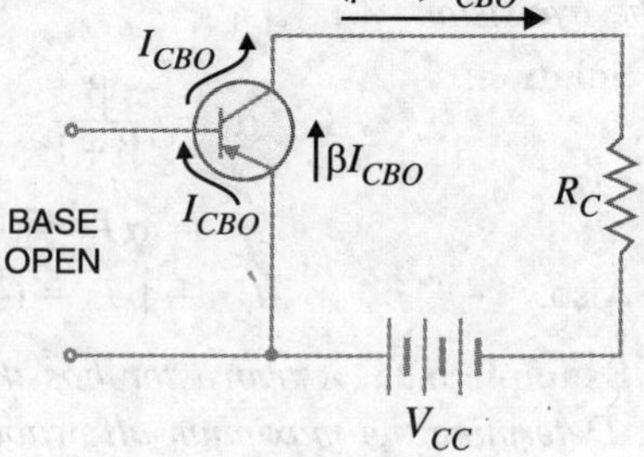

Fig. 6.25

Example 6.20. *Determine V_{CB} in the transistor **circuit shown in Fig. 6.26 (i). The transistor is of silicon and has β = 150.*

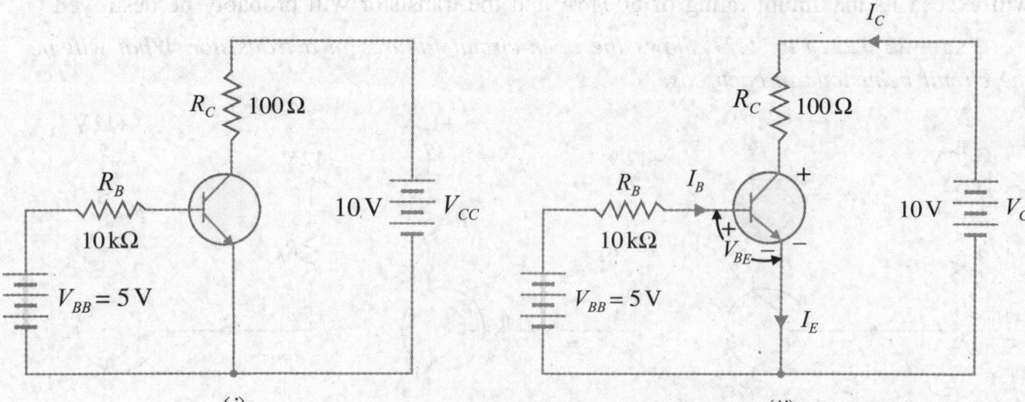

Fig. 6.26

Solution. Fig. 6.26 (*i*) shows the transistor circuit while Fig. 6.26 (*ii*) shows the various currents and voltages alongwith polarities.

Applying Kirchhoff's voltage law to base-emitter loop, we have,

$$V_{BB} - I_B R_B - V_{BE} = 0$$

$$\text{or} \quad I_B = \frac{V_{BB} - V_{BE}}{R_B} = \frac{5\text{ V} - 0.7\text{ V}}{10\text{ k}\Omega} = 430\ \mu\text{A}$$

$$\therefore \quad I_C = \beta I_B = (150)\ (430\ \mu\text{A}) = 64.5\text{ mA}$$

---

* The current I_{CBO} is amplified because it is forced to flow across the base-emitter junction.

** The resistor R_B controls the base current I_B and hence collector current I_C ($= \beta I_B$). If R_B is increased, the base current (I_B) decreases and hence collector current (I_C) will decrease and vice-versa.

Now, $V_{CE} = V_{CC} - I_C R_C$

$= 10 \text{ V} - (64.5 \text{ mA}) (100 \, \Omega) = 10 \text{ V} - 6.45 \text{ V} = 3.55 \text{ V}$

We know that : $V_{CE} = V_{CB} + V_{BE}$

$\therefore \quad V_{CB} = V_{CE} - V_{BE} = 3.55 - 0.7 = \mathbf{2.85 \ V}$

Example 6.21. *In a transistor, I_B = 68 μA, I_E = 30 mA and β = 440. Determine the α rating of the transistor. Then determine the value of I_C using both the α rating and β rating of the transistor.*

Solution.

$$\alpha = \frac{\beta}{\beta + 1} = \frac{440}{440 + 1} = \mathbf{0.9977}$$

$$I_C = \alpha I_E = (0.9977)(30 \text{ mA}) = \mathbf{29.93 \ mA}$$

Also, $I_C = \beta I_B = (440)(68 \, \mu\text{A}) = \mathbf{29.93 \ mA}$

Example 6.22. *A transistor has the following ratings : $I_{C\,(max)}$ = 500 mA and $\beta_{(max)}$ = 300. Determine the maximum allowable value of I_B for the device.*

Solution.

$$I_{B\,(max)} = \frac{I_{C\,(max)}}{\beta_{(max)}} = \frac{500 \text{ mA}}{300} = \mathbf{1.67 \ mA}$$

For this transistor, if the base current is allowed to exceed 1.67 mA, the collector current will exceed its maximum rating of 500 mA and the transistor will probably be destroyed.

Example 6.23. *Fig. 6.27 shows the open-circuit failures in a transistor. What will be the circuit behaviour in each case?*

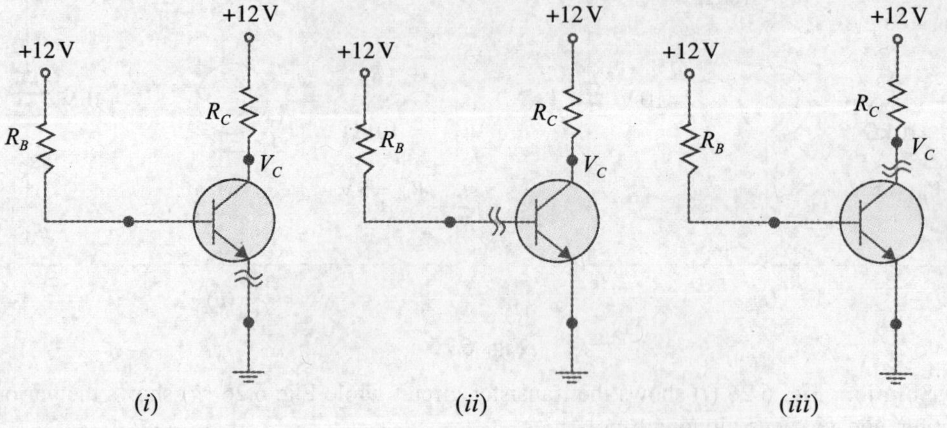

Fig. 6.27

Solution. *Fig. 6.27 shows the open-circuit failures in a transistor. We shall discuss the circuit behaviour in each case.

(*i*) **Open emitter.** Fig. 6.27 (*i*) shows an open emitter failure in a transistor. Since the collector diode is not forward biased, it is *OFF* and there can be neither collector current nor base current. **Therefore, there will be no voltage drops across either resistor and the voltage at the base and at the collector leads of the transistor will be 12 V.**

* The collector resistor R_C controls the collector voltage V_C (= $V_{CC} - I_C R_C$). When R_C increases, V_C decreases and *vice-versa*.

(ii) **Open base.** Fig. 6.27 (*ii*) shows an open base failure in a transistor. Since the base is open, there can be no base current so that the transistor is in *cut-off.* Therefore, all the transistor currents are 0A. **In this case, the base and collector voltages will both be at 12 V.**

Note. It may be noted that an open failure at either the base or emitter will produce similar results.

(iii) **Open collector.** Fig. 6.27 (*iii*) shows an open collector failure in a transistor. In this case, the emitter diode is still *ON*, so we expect to see 0.7 V at the base. **However, we will see 12 V at the collector because there is no collector current.**

Example 6.24. *Fig. 6.28 shows the short-circuit failures in a transistor. What will be the circuit behaviour in each case?*

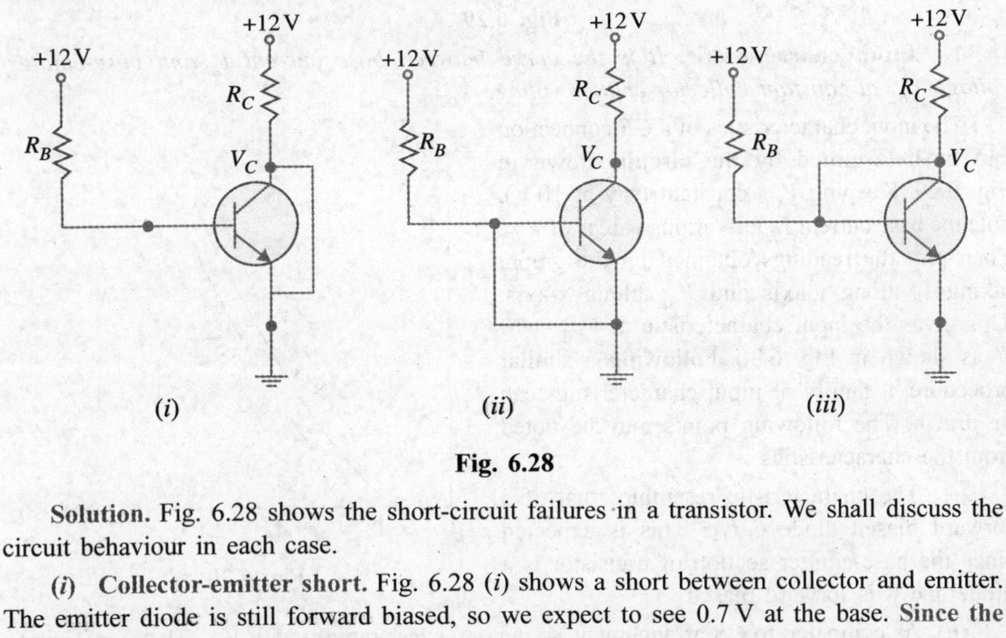

Fig. 6.28

Solution. Fig. 6.28 shows the short-circuit failures in a transistor. We shall discuss the circuit behaviour in each case.

(i) **Collector-emitter short.** Fig. 6.28 (*i*) shows a short between collector and emitter. The emitter diode is still forward biased, so we expect to see 0.7 V at the base. **Since the collector is shorted to the emitter, $V_C = V_E = 0$ V.**

(ii) **Base -emitter short.** Fig 6.28 (*ii*) shows a short between base and emitter. Since the base is now directly connected to ground, $V_B = 0$. Therefore, the current through R_B will be diverted to ground and there is no current to forward bias the emitter diode. As a result, the transistor will be *cut-off* and there is no collector current. **So we will expect the collector voltage to be 12 V.**

(iii) **Collector-base short.** Fig. 6.28 (*iii*) shows a short between the collector and the base. In this case, the emitter diode is still forward biased so $V_B = 0.7$ V. **Now, however, because the collector is shorted to the base, $V_C = V_B = 0.7$ V.**

Note. The collector-emitter short is probably the most common type of fault in a transistor. It is because the collector current (I_C) and collector-emitter voltage (V_{CE}) are responsible for the major part of the power dissipation in the transistor. As we shall see (See Art. 6.25), the power dissipation in a transistor is mainly due to I_C and V_{CE} (*i.e.* $P_D = V_{CE} I_C$). Therefore, the transistor chip between the collector and the emitter is most likely to melt first.

6.13 Characteristics of Common Emitter Connection

The important characteristics of this circuit arrangement are the *input characteristics* and *output characteristics.*

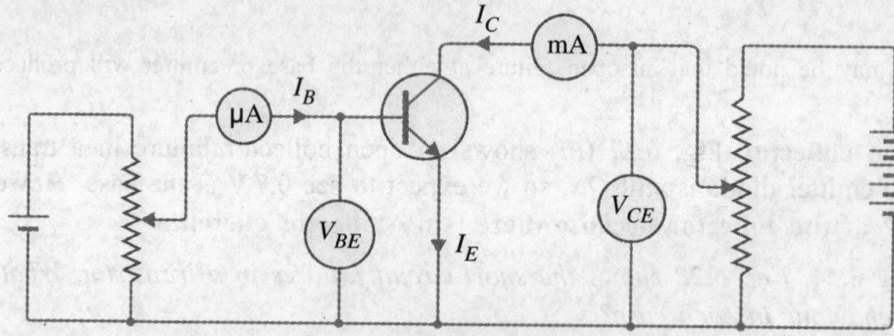

Fig. 6.29

1. Input characteristic. *It is the curve between base current I_B and base-emitter voltage V_{BE} at constant collector-emitter voltage V_{CE}.*

The input characteristics of a *CE* connection can be determined by the circuit shown in Fig. 6.29. Keeping V_{CE} constant (say at 10 V), note the base current I_B for various values of V_{BE}. Then plot the readings obtained on the graph, taking I_B along *y*-axis and V_{BE} along *x*-axis. This gives the input characteristic at V_{CE} = 10 V as shown in Fig. 6.30. Following a similar procedure, a family of input characteristics can be drawn. The following points may be noted from the characteristics :

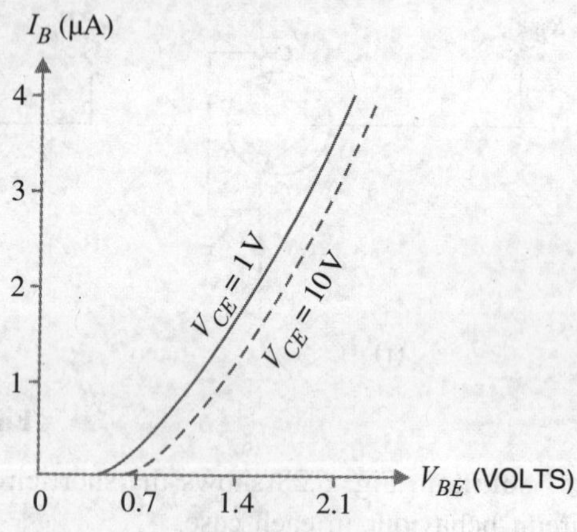

Fig. 6.30

(*i*) The characteristic resembles that of a forward biased diode curve. This is expected since the base-emitter section of transistor is a diode and it is forward biased.

(*ii*) As compared to *CB* arrangement, I_B increases less rapidly with V_{BE}. Therefore, input resistance of a *CE* circuit is higher than that of *CB* circuit.

Input resistance. It is the ratio of change in base-emitter voltage (ΔV_{BE}) to the change in base current (ΔI_B) at constant V_{CE} *i.e.*

$$\text{Input resistance, } r_i = \frac{\Delta V_{BE}}{\Delta I_B} \text{ at constant } V_{CE}$$

The value of input resistance for a *CE* circuit is of the order of a few hundred ohms.

2. Output characteristic. *It is the curve between collector current I_C and collector-emitter voltage V_{CE} at constant base current I_B.*

The output characteristics of a *CE* circuit can be drawn with the help of the circuit shown in Fig. 6.29. Keeping the base current I_B fixed at some value say, 5 μA, note the collector current I_C for various values of V_{CE}. Then plot the readings on a graph, taking I_C along *y*-axis and V_{CE} along *x*-axis. This gives the output characteristic at I_B = 5 μA as shown in Fig. 6.31 (*i*). The test can be repeated for I_B = 10 μA to obtain the new output characteristic as shown in Fig. 6.31 (*ii*). Following similar procedure, a family of output characteristics can be drawn as shown in Fig. 6.31 (*iii*).

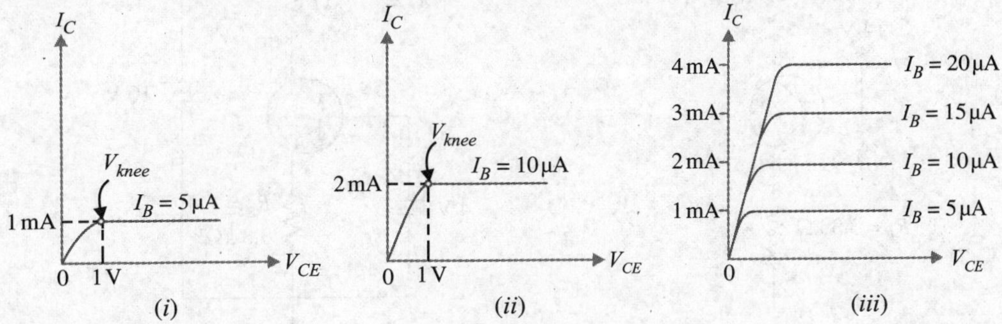

Fig. 6.31

The following points may be noted from the characteristics :

(*i*) The collector current I_C varies with V_{CE} for V_{CE} between 0 and 1 V only. After this, collector current becomes *almost* constant and independent of V_{CE}. This value of V_{CE} upto which collector current I_C changes with V_{CE} is called the *knee voltage* (V_{knee}). *The transistors are always operated in the region above knee voltage.*

(*ii*) Above knee voltage, I_C is almost constant. However, a small increase in I_C with increasing V_{CE} is caused by the collector depletion layer getting wider and capturing a few more majority carriers before electron-hole combinations occur in the base area.

(*iii*) For any value of V_{CE} above knee voltage, the collector current I_C is approximately equal to $\beta \times I_B$.

Output resistance. It is the ratio of change in collector-emitter voltage (ΔV_{CE}) to the change in collector current (ΔI_C) at constant I_B *i.e.*

$$\text{Output resistance, } r_o = \frac{\Delta V_{CE}}{\Delta I_C} \text{ at constant } I_B$$

It may be noted that whereas the output characteristics of *CB* circuit are horizontal, they have noticeable slope for the *CE* circuit. Therefore, the output resistance of a *CE* circuit is less than that of *CB* circuit. Its value is of the order of 50 kΩ.

Example 6.25. *For the transistor circuit shown in Fig. 6.32 (i), determine all node voltages and branch currents. Assume that V_{BE} = 0.7 V and β = 100. Verify that the transistor is operating under normal conditions (i.e. in the active region).*

Solution. The circuit shown in Fig. 6.32 (*i*) can be redrawn as shown in Fig. 6.32 (*ii*).

$$\text{Emitter voltage, } V_E = 4 - 0.7 = 3.3 \text{ V}$$

$$\text{Emitter current, } I_E = \frac{V_E}{R_E} = \frac{3.3 \text{ V}}{3.3 \text{ k}\Omega} = \mathbf{1\ mA}$$

$$\text{Collector current, } I_C = \alpha I_E$$

Now, $$\alpha = \frac{\beta}{\beta+1} = \frac{100}{100+1} = 0.99$$

∴ $$I_C = 0.99 \times 1 \text{ mA} = \mathbf{0.99\ mA}$$

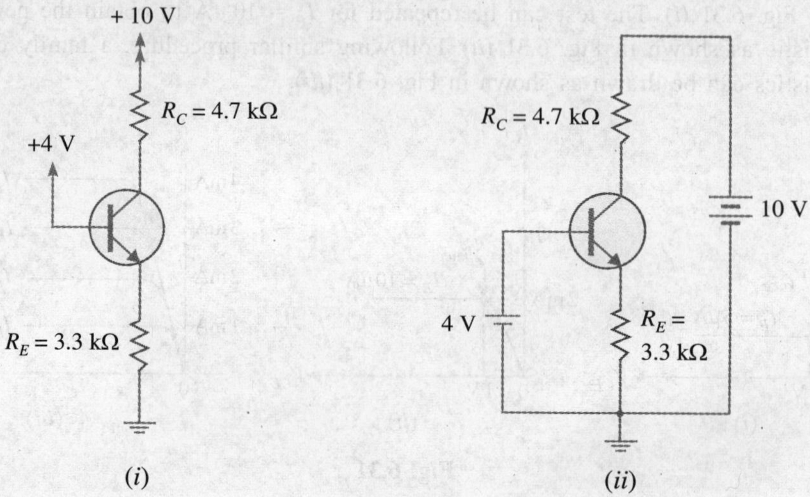

Fig. 6.32

Collector voltage, $V_C = V_{CC} - I_C R_C$

$= 10\text{ V} - 0.99\text{ mA} \times 4.7\text{ k}\Omega$

$= \mathbf{+ 5.3\text{ V}}$

Base current, $I_B = I_E - I_C$

$= 1 - 0.99 = \mathbf{0.01\text{ mA}}$

Base voltage, $V_B = \mathbf{4\text{ V}}$

Since the base is at + 4 V, the collector - base junction is reverse biased by 1.3 V (= 5.3 V – 4 V). Further, the base-emitter junction is forward biased because the base is connected to + 4 V and the emitter is connected to ground through the resistance R_E. Therefore, the transistor is operating under normal conditions *i.e.* the transistor is operating in the active region.

Fig. 6.33 shows the various node voltages and branch currents. This example will enable the reader to *analyse any transistor circuit with a fair degree of ease.

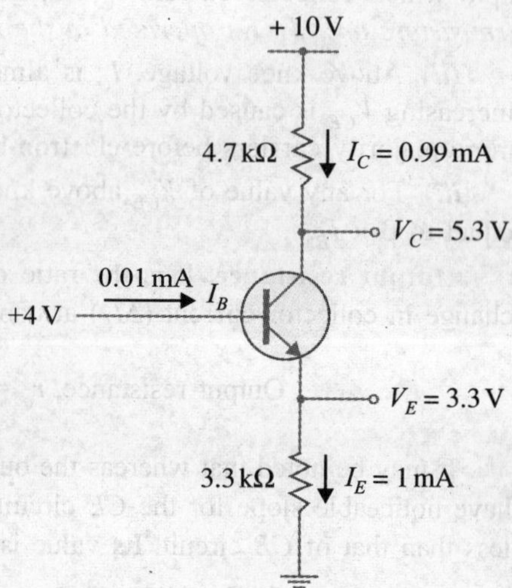

Fig. 6.33

6.14 Common Collector Connection

In this circuit arrangement, input is applied between base and collector while output is taken between the emitter and collector. Here, collector of the transistor is common to both input and output circuits and hence the name common collector connection. Fig. 6.34 (*i*) shows common collector *npn* transistor circuit whereas Fig. 6.34 (*ii*) shows common collector *pnp* circuit.

* Analysing a circuit means finding the various voltages and currents in the circuit.

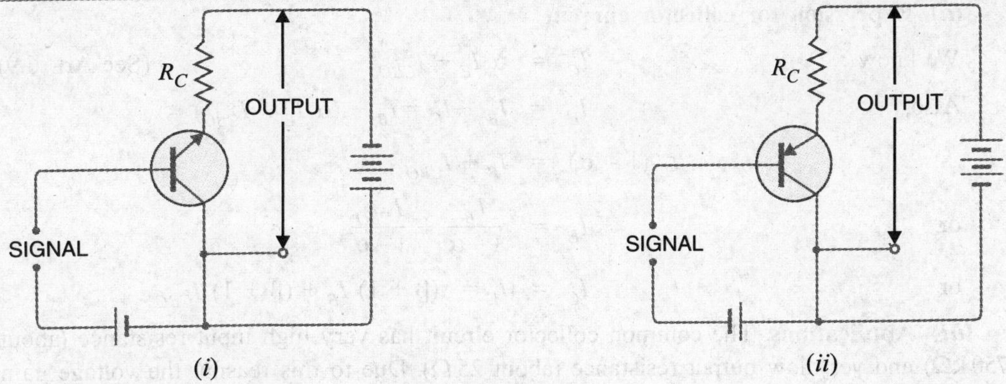

Fig. 6.34

(i) **Current amplification factor γ.** In common collector circuit, input current is the base current I_B and output current is the emitter current I_E. Therefore, current amplification in this circuit arrangement can be defined as under :

The ratio of change in emitter current (ΔI_E) to the change in base current (ΔI_B) is known as **current amplification factor** *in common collector (CC) arrangement i.e.*

$$\gamma = \frac{\Delta I_E}{\Delta I_B}$$

This circuit provides about the same current gain as the common emitter circuit as $\Delta I_E \simeq \Delta I_C$. However, its voltage gain is always less than 1.

Relation between γ and α.

$$\gamma = \frac{\Delta I_E}{\Delta I_B} \quad ...(i)$$

$$\alpha = \frac{\Delta I_C}{\Delta I_E} \quad ...(ii)$$

Now, $I_E = I_B + I_C$

or $\Delta I_E = \Delta I_B + \Delta I_C$

or $\Delta I_B = \Delta I_E - \Delta I_C$

Substituting the value of ΔI_B in exp. *(i)*, we get,

$$\gamma = \frac{\Delta I_E}{\Delta I_E - \Delta I_C}$$

Dividing the numerator and denominator of R.H.S. by ΔI_E, we get,

$$\gamma = \frac{\frac{\Delta I_E}{\Delta I_E}}{\frac{\Delta I_E}{\Delta I_E} - \frac{\Delta I_C}{\Delta I_E}} = \frac{1}{1-\alpha} \qquad \left(\because \alpha = \frac{\Delta I_C}{\Delta I_E}\right)$$

$$\therefore \quad \gamma = \frac{1}{1-\alpha}$$

(ii) Expression for collector current.

We know : $I_C = \alpha I_E + I_{CBO}$ (See Art. 6.9)

Also, $I_E = I_B + I_C = I_B + (\alpha I_E + I_{CBO})$

$\therefore$ $I_E (1 - \alpha) = I_B + I_{CBO}$

or $$I_E = \frac{I_B}{1-\alpha} + \frac{I_{CBO}}{1-\alpha}$$

or $$I_C \simeq I_E = {}^*(\beta + 1) I_B + (\beta + 1) I_{CBO}$$

(iii) Applications. The common collector circuit has very high input resistance (about 750 kΩ) and very low output resistance (about 25 Ω). Due to this reason, the voltage gain provided by this circuit is always less than 1. Therefore, this circuit arrangement is seldom used for amplification. However, due to relatively high input resistance and low output resistance, this circuit is primarily used for impedance matching *i.e.* for driving a low impedance load from a high impedance source.

6.15 Comparison of Transistor Connections

The comparison of various characteristics of the three connections is given below in the tabular form.

S. No.	Characteristic	Common base	Common emitter	Common collector
1.	Input resistance	Low (about 100 Ω)	Low (about 750 Ω)	Very high (about 750 kΩ)
2.	Output resistance	Very high (about 450 kΩ)	High (about 45 kΩ)	Low (about 50 Ω)
3.	Voltage gain	About 150	About 500	Less than 1
4.	Applications	For high frequency applications	For audio frequency applications	For impedance matching
5.	Current gain	No (less than 1)	High (β)	Appreciable

The following points are worth noting about transistor arrangements :

(i) ***CB* circuit.** The input resistance (r_i) of *CB* circuit is low because I_E is high. The output resistance (r_o) is high because of reverse voltage at the collector. It has no current gain ($\alpha < 1$) but voltage gain can be high. The *CB* circuit is seldom used. The only advantage of *CB* circuit is that it provides good stability against increase in temperature.

(ii) ***CE* circuit.** The input resistance (r_i) of a *CE* circuit is high because of small I_B. Therefore, r_i for a *CE* circuit is much higher than that of *CB* circuit. The output resistance (r_o) of *CE* circuit is smaller than that of *CB* circuit. The current gain of *CE* circuit is large because I_C is much larger than I_B. The voltage gain of *CE* circuit is larger than that of *CB* circuit. The *CE* circuit is generally used because it has the best combination of voltage gain

* $\beta = \frac{\alpha}{1-\alpha}$ $\quad\therefore\quad \beta + 1 = \frac{\alpha}{1-\alpha} + 1 = \frac{1}{1-\alpha}$

and current gain. The disadvantage of *CE* circuit is that the leakage current is amplified in the circuit, but bias stabilisation methods can be used.

(iii) **CC circuit.** The input resistance (r_i) and output resistance (r_o) of *CC* circuit are respectively high and low as compared to other circuits. There is no voltage gain ($A_v < 1$) in a *CC* circuit. This circuit is often used for impedance matching.

6.16 Commonly Used Transistor Connection

Out of the three transistor connections, the common emitter circuit is the most efficient. It is used in about 90 to 95 percent of all transistor applications. The main reasons for the widespread use of this circuit arrangement are :

(i) **High current gain.** In a common emitter connection, I_C is the output current and I_B is the input current. In this circuit arrangement, collector current is given by ;

$$I_C = \beta\ I_B + I_{CEO}$$

As the value of β is very large, therefore, the output current I_C is much more than the input current I_B. Hence, the current gain in *CE* arrangement is very high. It may range from 20 to 500.

(ii) **High voltage and power gain.** Due to high current gain, the common emitter circuit has the highest voltage and power gain of three transistor connections. This is the major reason for using the transistor in this circuit arrangement.

(iii) **Moderate output to input impedance ratio.** In a common emitter circuit, the ratio of output impedance to input impedance is small (about 50). This makes this circuit arrangement an ideal one for coupling between various transistor stages. However, in other connections, the ratio of output impedance to input impedance is very large and hence coupling becomes highly inefficient due to gross mismatching.

6.17 Transistor as an Amplifier in CE Arrangement

Fig. 6.35 shows the common emitter *npn* amplifier circuit. Note that a battery V_{BB} is connected in the input circuit in addition to the signal voltage. This d.c. voltage is known as *bias voltage* and its magnitude is such that it always keeps the emitter-base junction forward *biased regardless of the polarity of the signal source.

Operation. During the positive half-cycle of the **signal, the forward bias across the emitter-base junction is increased. Therefore, more electrons flow from the emitter to the collector *via* the base. This causes an increase in collector current. The increased collector current produces a greater voltage drop across the collector load resistance R_C. However, during the negative half-cycle of the signal, the forward bias across emitter-base junction is decreased. Therefore, collector current decreases. This results in the decreased output voltage (in the opposite direction). Hence, an amplified output is obtained across the load.

---

* If d.c. bias voltage is not provided, then during negative half-cycle of the signal, the emitter-base junction will be reverse biased. This will upset the transistor action.

** Throughout the book, we shall use sine wave signals because these are convenient for testing amplifiers. But it must be realised that signals (*e.g.* speech, music *etc.*) with which we work are generally complex having little resemblance to a sine wave. However, Fourier series analysis tells us that such complex signals may be expressed as a sum of sine waves of various frequencies.

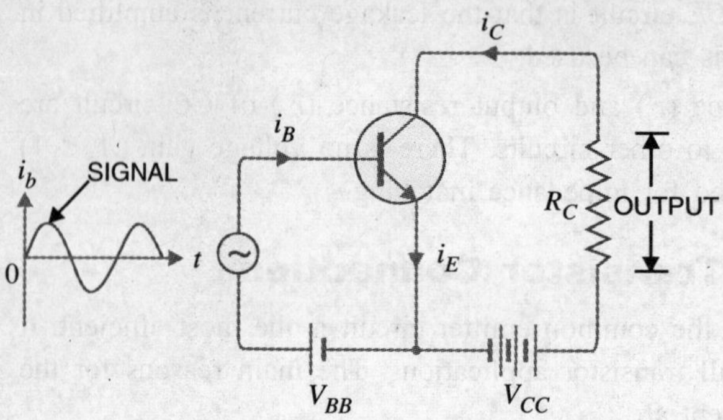

Fig. 6.35

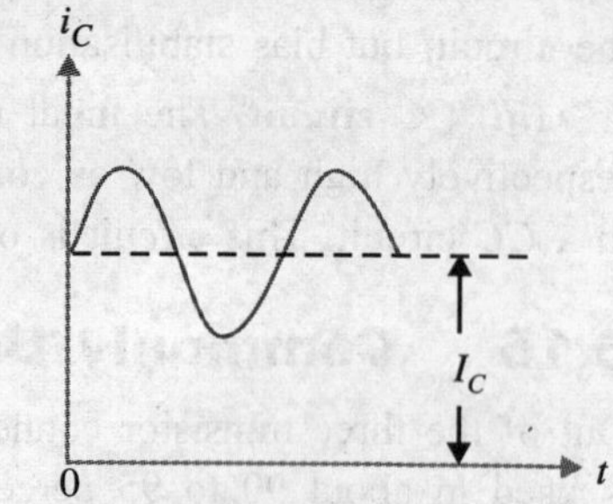

Fig. 6.36

Analysis of collector currents. When no signal is applied, the input circuit is forward biased by the battery V_{BB}. Therefore, a d.c. collector current I_C flows in the collector circuit. This is called *zero signal collector current*. When the signal voltage is applied, the forward bias on the emitter-base junction increases or decreases depending upon whether the signal is positive or negative. During the positive half-cycle of the signal, the forward bias on emitter-base junction is increased, causing total collector current i_C to increase. Reverse will happen for the negative half-cycle of the signal.

Fig. 6.36 shows the graph of total collector current i_C versus time. From the graph, it is clear that total collector current consists of two components, namely ;

(*i*) The d.c. collector current I_C (zero signal collector current) due to bias battery V_{BB}. This is the current that flows in the collector in the absence of signal.

(*ii*) The a.c. collector current i_c due to signal.

$\therefore$ Total collector current, $i_C = i_c + I_C$

The useful output is the voltage drop across collector load R_C due to the a.c. component i_c. The purpose of zero signal collector current is to ensure that the emitter-base junction is forward biased at all times. The table below gives the symbols usually employed for currents and voltages in transistor applications.

S. No.	Particular	Instantaneous a.c.	d.c.	Total
1.	Emitter current	i_e	I_E	i_E
2.	Collector current	i_c	I_C	i_C
3.	Base current	i_b	I_B	i_B
4.	Collector-emitter voltage	v_{ce}	V_{CE}	v_{CE}
5.	Emitter-base voltage	v_{eb}	V_{EB}	v_{EB}

6.18 Transistor Load Line Analysis

In the transistor circuit analysis, it is generally required to determine the collector current for various collector-emitter voltages. One of the methods can be used to plot the output characteristics and determine the collector current at any desired collector-emitter voltage. However, a more convenient method, known as *load line method* can be used to solve such problems. As explained later in this section, this method is quite easy and is frequently used in the analysis of transistor applications.

d.c. load line. Consider a common emitter *npn* transistor circuit shown in Fig. 6.37 (*i*) where no signal is applied. Therefore, d.c. conditions prevail in the circuit. The output characteristics of this circuit are shown in Fig. 6.37 (*ii*).

The value of collector-emitter voltage V_{CE} at any time is given by ;

$$V_{CE} = V_{CC} - I_C R_C \quad ...(i)$$

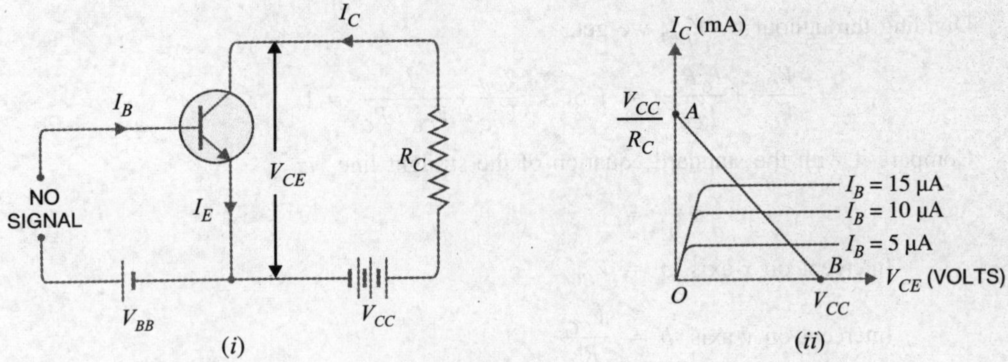

Fig. 6.37

As V_{CC} and R_C are fixed values, therefore, it is a first degree equation and can be represented by a straight line on the output characteristics. This is known as *d.c. load line* and determines the locus of $V_{CE} - I_C$ points for any given value of R_C. To add load line, we need two end points of the straight line. These two points can be located as under :

(*i*) When the collector current $I_C = 0$, then collector-emitter voltage is maximum and is equal to V_{CC} *i.e.*

$$\text{Max. } V_{CE} = V_{CC} - I_C R_C$$
$$= V_{CC} \quad (\because I_C = 0)$$

This gives the first point B ($OB = V_{CC}$) on the collector-emitter voltage axis as shown in Fig. 6.37 (*ii*).

(*ii*) When collector-emitter voltage $V_{CE} = 0$, the collector current is maximum and is equal to V_{CC}/R_C *i.e.*

$$V_{CE} = V_{CC} - I_C R_C$$

or
$$0 = V_{CC} - I_C R_C$$

∴
$$\text{Max. } I_C = V_{CC}/R_C$$

This gives the second point A ($OA = V_{CC}/R_C$) on the collector current axis as shown in Fig. 6.37 (*ii*). By joining these two points, d.c. *load line AB is constructed.

Slope of d.c. load line. We have seen above that for common emitter *npn* transistor circuit:

$$V_{CC} = V_{CE} + I_C R_C$$

or
$$I_C = -\frac{1}{R_C} V_{CE} + \frac{V_{CC}}{R_C}$$

* Why load line? The resistance R_C connected to the device is called load or load resistance for the circuit and, therefore, the line we have just constructed is called the load line.

This is the equation of the d.c. load line of the circuit.

It is clear from the above equation that slope of the d.c. load line is – $1/R_C$. It is measured in **siemens* (S).

Alternate treatement. We have seen above that :

$$V_{CE} + I_C R_C = V_{CC}$$

Dividing throughout by V_{CC}, we get,

$$\frac{V_{CE}}{V_{CC}} + \frac{I_C R_C}{V_{CC}} = 1 \text{ or } \frac{V_{CE}}{V_{CC}} + \frac{I_C}{V_{CC}/R_C} = 1$$

Compare it with the standard equation of the straight line *viz.*

$$\frac{x}{a} + \frac{y}{b} = 1$$

$$\therefore \quad \text{Intercept on } x\text{-axis, } a = V_{CC}$$

$$\text{Intercept on } y\text{-axis, } b = \frac{V_{CC}}{R_C}$$

$$\text{Slope of the line, } m = -\frac{b}{a} = -\frac{1}{R_C}$$

Importance of d.c. load line. The current (I_C) and voltage (V_{CE}) conditions in the transistor circuit are represented by some point on the output characteristics. The same information can be obtained from the load line. Thus when I_C is maximum ($= V_{CC}/R_C$), then $V_{CE} = 0$ as shown in Fig. 6.38. If $I_C = 0$, then V_{CE} is maximum and is equal to V_{CC}. For any other value of collector current say OC, the collector-emitter voltage $V_{CE} = OD$. It follows, therefore, that load line gives a far more convenient and direct solution to the problem.

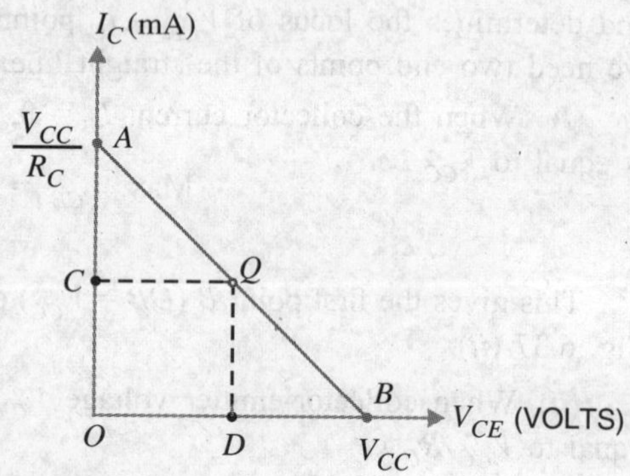

Fig. 6.38

Note. If we plot the load line on the output characteristic of the transistor, we can investigate the behaviour of the transistor amplifier. It is because we have the transistor output current and voltage specified in the form of load line equation and the transistor behaviour itself specified implicitly by the output characteristics.

6.19 Operating Point

The zero signal values of I_C and V_{CE} are known as the **operating point.**

It is called operating point because the variations of I_C and V_{CE} take place about this point when signal is applied. It is also called quiescent (silent) point or *Q-point* because it

* Slope $= -\frac{1}{R_C} = -\frac{1}{\text{ohms}}$

Clearly, the unit of slope is siemens.

is the point on $I_C - V_{CE}$ characteristic when the transistor is silent *i.e.* in the absence of the signal.

Suppose in the absence of signal, the base current is 5 μA. Then I_C and V_{CE} conditions in the circuit must be represented by some point on I_B = 5 μA characteristic. But I_C and V_{CE} conditions in the circuit should also be represented by some point on the d.c. load line *AB*. The point *Q* where the load line and the characteristic intersect is the only point which satisfies both these conditions. Therefore, the point *Q* describes the actual state of affairs in the circuit in the zero signal conditions and is called the operating point. Referring to Fig. 6.39, for I_B = 5 μA, the zero signal values are :

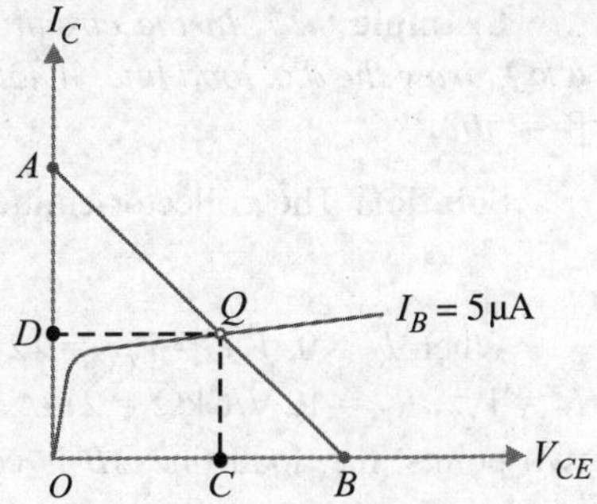

Fig. 6.39

$$V_{CE} = OC \text{ volts}$$
$$I_C = OD \text{ mA}$$

It follows, therefore, that the zero signal values of I_C and V_{CE} (*i.e.* operating point) are determined by the point where d.c. load line intersects the proper base current curve.

Example 6.26. *For the circuit shown in Fig. 6.40 (i), draw the d.c. load line and find its slope.*

Solution. The collector-emitter voltage V_{CE} is given by ;

$$V_{CE} = V_{CC} - I_C R_C \quad ...(i)$$

When I_C = 0, then,

$$V_{CE} = V_{CC} = 12.5 \text{ V}$$

This locates the point *B* of the load line on the collector-emitter voltage axis.

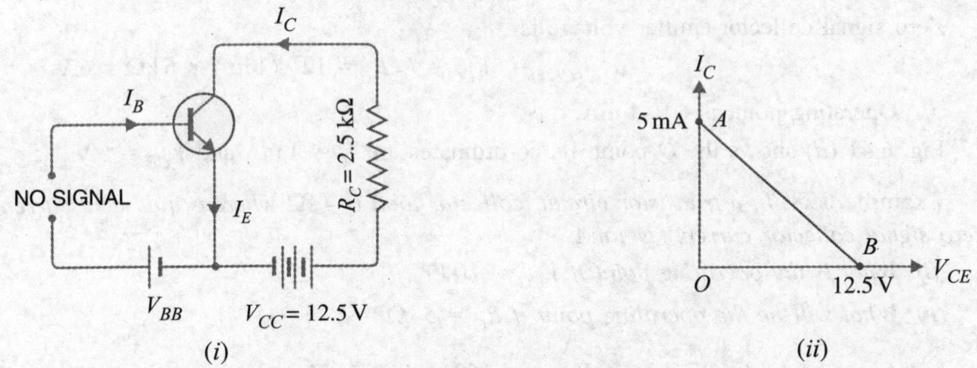

Fig. 6.40

When V_{CE} = 0, then,

$$I_C = V_{CC}/R_C = 12.5 \text{ V}/2.5 \text{ k}\Omega = 5 \text{ mA}$$

This locates the point *A* of the load line on the collector current axis. By joining these two points, we get the d.c. load line *AB* as shown in Fig. 6.40 (*ii*).

$$\text{Slope of d.c. load line} = -\frac{1}{R_C} = -\frac{1}{2.5 \times 10^3 \Omega} = -4 \times 10^{-4} \text{ S}$$

Example 6.27. *In the circuit diagram shown in Fig. 6.41 (i), if V_{CC} = 12 V and R_C = 6 kΩ, draw the d.c. load line. What will be the Q point if zero signal base current is 20 μA and β = 50?*

Solution. The collector-emitter voltage V_{CE} is given by ;

$$V_{CE} = V_{CC} - I_C R_C$$

When $I_C = 0$, $V_{CE} = V_{CC} = 12$ V. This locates the point *B* of the load line. When $V_{CE} = 0$, $I_C = V_{CC}/R_C = 12\text{ V}/6\text{ k}\Omega = 2$ mA. This locates the point *A* of the load line. By joining these two points, d.c. load line *AB* is constructed as shown in Fig. 6.41 (*ii*).

$$\text{Zero signal base current, } I_B = 20\ \mu\text{A} = 0.02\text{ mA}$$

$$\text{Current amplification factor, } \beta = 50$$

$$\therefore \text{ Zero signal collector current, } I_C = \beta I_B = 50 \times 0.02 = 1\text{ mA}$$

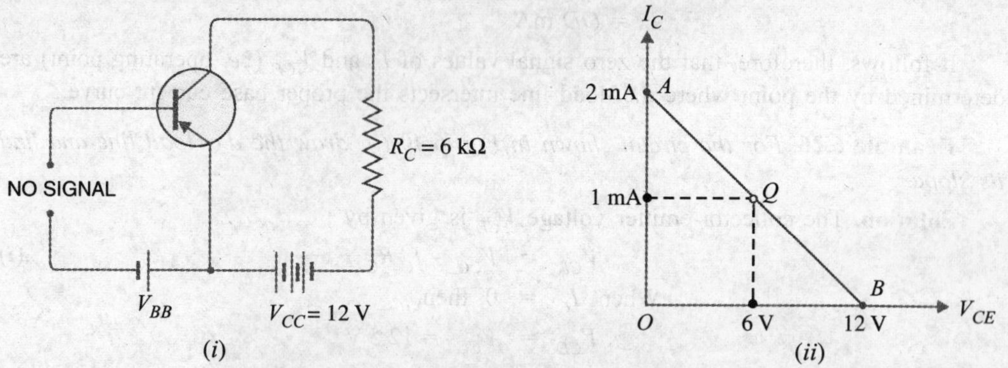

Fig. 6.41

Zero signal collector-emitter voltage is

$$V_{CE} = V_{CC} - I_C R_C = 12 - 1\text{ mA} \times 6\text{ k}\Omega = 6\text{ V}$$

∴ Operating point is **6 V, 1 mA.**

Fig. 6.41 (*ii*) shows the *Q* point. Its co-ordinates are I_C = 1 mA and V_{CE} = 6 V.

Example 6.28. *In a transistor circuit, collector load is 4 kΩ whereas quiescent current (zero signal collector current) is 1 mA.*

(i) What is the operating point if V_{CC} = 10 V?

(ii) What will be the operating point if R_C = 5 kΩ?

Solution. V_{CC} = 10 V ; I_C = 1 mA

(*i*) When collector load R_C = 4 kΩ , then,

$$V_{CE} = V_{CC} - I_C R_C = 10 - 1\text{ mA} \times 4\text{ k}\Omega = 10 - 4 = 6\text{ V}$$

∴ Operating point is **6 V, 1 mA.**

(*ii*) When collector load R_C = 5 kΩ , then,

$$V_{CE} = V_{CC} - I_C R_C = 10 - 1\text{ mA} \times 5\text{ k}\Omega = 10 - 5 = 5\text{ V}$$

∴ Operating point is **5 V, 1 mA.**

Example 6.29. *Determine the Q point of the transistor circuit shown in Fig. 6.42. Also draw the d.c. load line. Given β = 200 and V_{BE} = 0.7 V.*

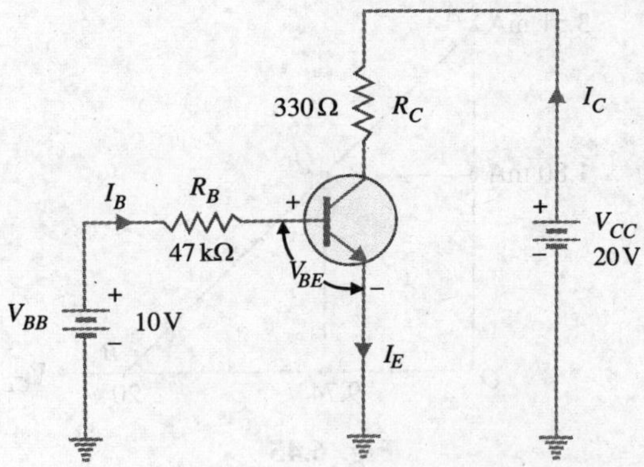

Fig. 6.42

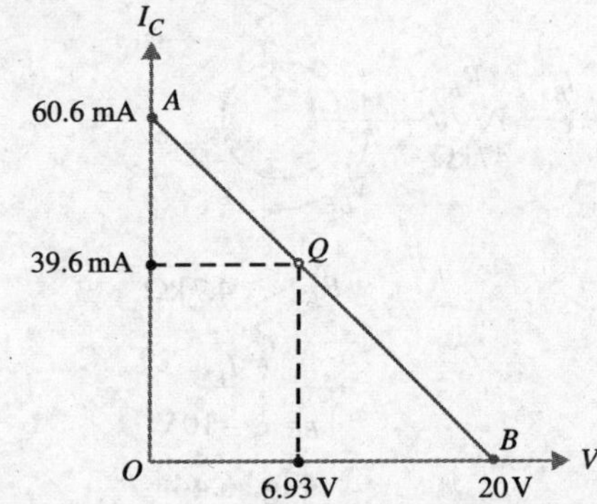

Fig. 6.43

Solution. The presence of resistor R_B in the base circuit should not disturb you because we can apply Kirchhoff's voltage law to find the value of I_B and hence I_C (= βI_B). Referring to Fig. 6.42 and applying Kirchhoff's voltage law to base-emitter loop, we have,

$$V_{BB} - I_B R_B - V_{BE} = 0$$

$$\therefore \quad I_B = \frac{V_{BB} - V_{BE}}{R_B} = \frac{10\text{ V} - 0.7\text{ V}}{47\text{ k}\Omega} = 198\ \mu\text{A}$$

Now, $$I_C = \beta I_B = (200)(198\ \mu\text{A}) = 39.6\text{ mA}$$

Also, $$V_{CE} = V_{CC} - I_C R_C = 20\text{ V} - (39.6\text{ mA})(330\ \Omega)$$

$$= 20\text{ V} - 13.07\text{ V} = 6.93\text{ V}$$

Therefore, the *Q*-point is I_C = **39.6 mA** and V_{CE} = **6.93 V.**

D.C. load line. In order to draw the d.c. load line, we need two end points.

Now, $$V_{CE} = V_{CC} - I_C R_C$$

When I_C = 0, $V_{CE} = V_{CC}$ = 20 V. This locates the point *B* of the d.c. load line on collector-emitter voltage axis as shown in Fig. 6.43. When V_{CE} = 0, $I_C = V_{CC}/R_C$ = 20 V/330 Ω = 60.6 mA. This locates the point *A* of the d.c. load line on collector current axis. By joining these two points, d.c. load line *AB* is constructed as shown in Fig. 6.43.

Example 6.30. *Determine the Q point of the transistor circuit shown in* **Fig. 6.44. Also draw the d.c. load line. Given β = 100 and V_{BE} = 0.7 V.*

* The presence of two power supplies has an effect on the basic equations for I_C and V_{CE} used for single power supply (*i.e.* V_{CC}). Normally, the two supply voltages will be equal. For example, if V_{CC} = + 10 V (d.c.), then V_{EE} = – 10 V (d.c.).

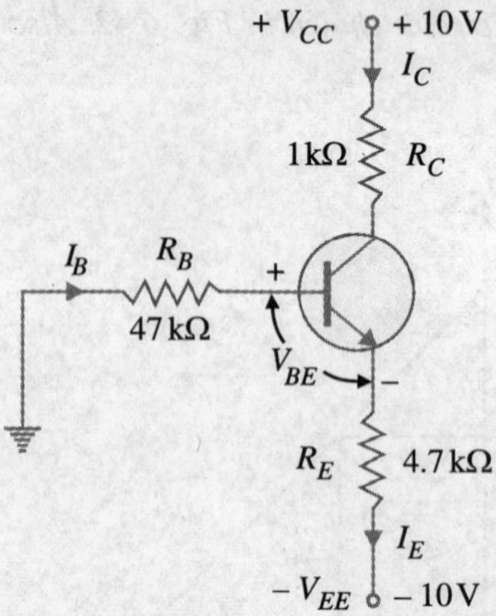

Fig. 6.44

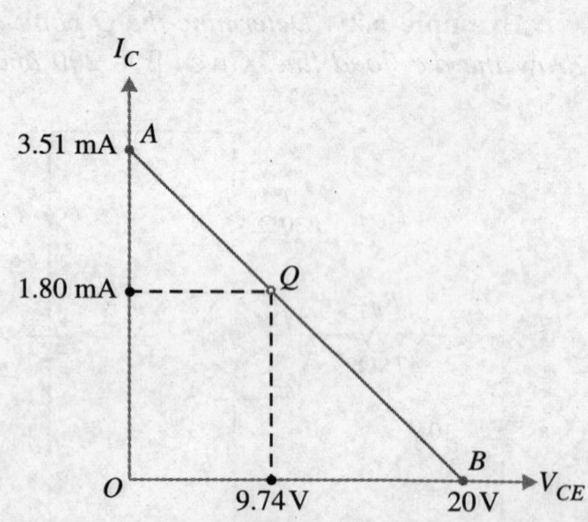

Fig. 6.45

Solution. The transistor circuit shown in Fig. 6.44 may look complex but we can easily apply Kirchhoff's voltage law to find the various voltages and currents in the *circuit.

Applying Kirchhoff's voltage law to the base-emitter loop, we have,

$$-I_B R_B - V_{BE} - I_E R_E + V_{EE} = 0 \quad \text{or} \quad V_{EE} = I_B R_B + I_E R_E + V_{BE}$$

Now, $I_C = \beta I_B$ and $I_C \simeq I_E$. $\therefore$ $I_B = I_E/\beta$. Putting $I_B = I_E/\beta$ in the above equation, we have,

$$V_{EE} = \left(\frac{I_E}{\beta}\right) R_B + I_E R_E + V_{BE}$$

or
$$I_E \left(\frac{R_B}{\beta} + R_E\right) = V_{EE} - V_{BE} \quad \text{or} \quad I_E = \frac{V_{EE} - V_{BE}}{R_E + R_B/\beta}$$

Since $I_C \simeq I_E$,
$$I_C = \frac{V_{EE} - V_{BE}}{R_E + R_B/\beta} = \frac{10\text{ V} - 0.7\text{ V}}{4.7\text{ k}\Omega + 47\text{ k}\Omega/100} = \frac{9.3\text{ V}}{5.17\text{ k}\Omega} = 1.8\text{ mA}$$

Applying Kirchhoff's voltage law to the collector side, we have,

$$V_{CC} - I_C R_C - V_{CE} - I_E R_E + V_{EE} = 0$$

or
$$V_{CE} = V_{CC} + V_{EE} - I_C (R_C + R_E) \qquad (\because I_E \simeq I_C)$$

$$= 10\text{ V} + 10\text{ V} - 1.8\text{ mA}\ (1\text{ k}\Omega + 4.7\text{ k}\Omega) = 9.74\text{ V}$$

Therefore, the operating point of the circuit is I_C = **1.8 mA** and V_{CE} = **9.74 V.**

D.C. load line. The d.c. load line can be constructed as under :

$$V_{CE} = V_{CC} + V_{EE} - I_C (R_C + R_E)$$

When $I_C = 0$; $V_{CE} = V_{CC} + V_{EE}$ = 10 V + 10 V = 20 V. This locates the first point *B* (*OB* = 20 V) of the d.c. load line on the collector-emitter voltage axis. When $V_{CE} = 0$,

$$I_C = \frac{V_{CC} + V_{EE}}{R_C + R_E} = \frac{10\text{ V} + 10\text{ V}}{1\text{ k}\Omega + 4.7\text{ k}\Omega} = \frac{20\text{ V}}{5.7\text{ k}\Omega} = 3.51\text{ mA}$$

This locates the second point *A* (*OA* = 3.51 mA) of the d.c. load line on the collector current axis. By joining points *A* and *B*, d.c. load line *AB* is constructed as shown in Fig. 6.45.

* The emitter resistor R_E provides stabilisation of *Q*-point (See Art. 7.13).

Example 6.31. *In the above example, find (i) emitter voltage w.r.t. ground (ii) base voltage w.r.t. ground (iii) collector voltage w.r.t. ground.*

Solution. Refer to Fig. 6.46.

(*i*) The emitter voltage *w.r.t.* ground is

$$V_E = -V_{EE} + I_E R_E$$
$$= -10\text{ V} + 1.8\text{ mA} \times 4.7\text{ k}\Omega$$
$$= \mathbf{-1.54\ V}$$

(*ii*) The base voltage *w.r.t.* ground is

$$V_B = V_E + V_{BE} = 10\text{ V} + 0.7\text{ V}$$
$$= \mathbf{10.7\ V}$$

(*iii*) The collector voltage *w.r.t.* ground is

$$V_C = V_{CC} - I_C R_C$$
$$= 10\text{ V} - 1.8\text{ mA} \times 1\text{ k}\Omega = \mathbf{8.2\ V}$$

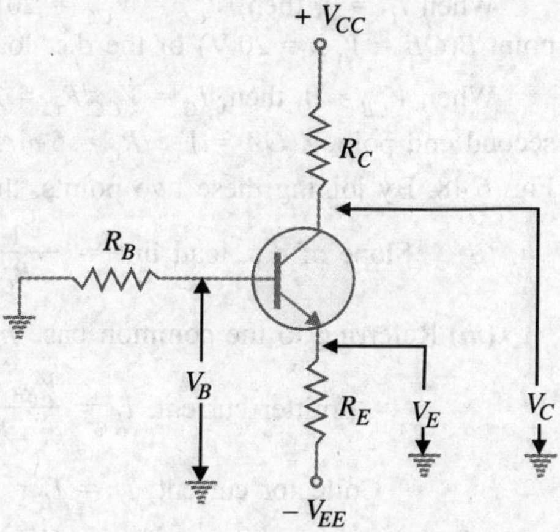

Fig. 6.46

Example 6.32. *For the common base npn transistor circuit shown in Fig. 6.47, determine (i) the equation of the d.c. load line (ii) slope of the d.c. load line and (iii) operating point.*

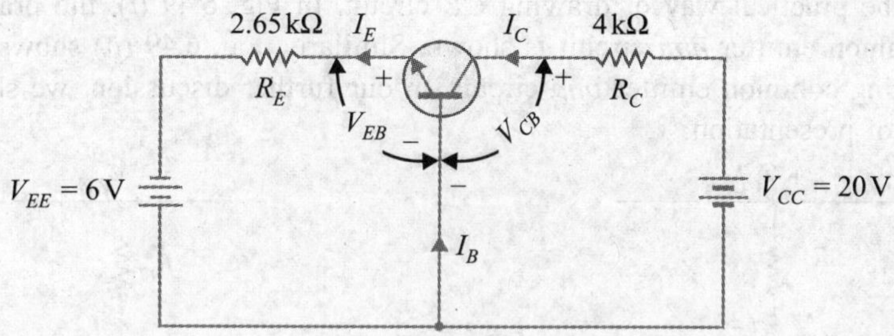

Fig. 6.47

Solution. The operating point (*Q*) of common base transistor circuit means the zero signal values of I_C and V_{CB}.

(*i*) **Equation of d.c. load line.** Referring to Fig. 6.47 and applying Kirchhoff's voltage law to the collector-base loop, we have,

$$V_{CC} = V_{CB} + I_C R_C$$

or
$$I_C = -\frac{1}{R_C}V_{CB} + \frac{V_{CC}}{R_C} \quad \dots (i)$$

As V_{CC} and R_C are constants, therefore, eq. (*i*) is the equation of a straight line. This is called d.c. load line of the common base transistor circuit and can be represented on the $I_C - V_{CB}$ axes as shown in Fig. 6.48. Note that the slope of the d.c. load line is $-1/R_C$ and is measured in siemens (S). To draw the d.c. load line on $I_C - V_{CB}$ axes, we need two end points. These can be located as under :

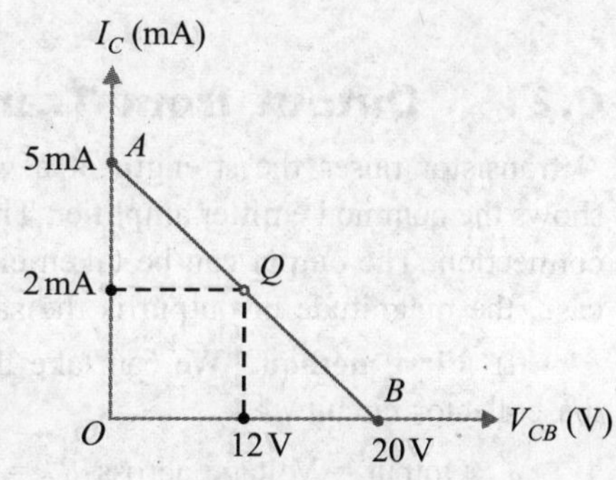

Fig. 6.48

When $I_C = 0$, then, $V_{CB} = V_{CC} = 20\text{V}$ ... from eq. (*i*) above. This locates the first end point $B(OB = V_{CC} = 20\text{ V})$ of the d.c. load line on the V_{CB} axis as shown in Fig. 6.48.

When $V_{CB} = 0$, then, $I_C = V_{CC}/R_C = 20\text{ V}/4\text{ k}\Omega = 5\text{ mA}$ ... from eq. (*i*). This locates the second end point $A(OA = V_{CC}/R_C = 5\text{ mA})$ of the d.c. load line on the I_C axis as shown in Fig. 6.48. By joining these two points, the d.c. load line *AB* of the circuit is constructed.

(*ii*) Slope of d.c. load line $= -\dfrac{1}{R_C} = -\dfrac{1}{4\times10^3\Omega} = -\mathbf{2.5 \times 10^{-4}\ S}$

(*iii*) Referring to the common base *npn* transistor circuit in Fig. 6.47, we have,

$$\text{Emitter current, } I_E = \frac{V_{EE} - V_{EB}}{R_E} = \frac{6\text{ V} - 0.7\text{ V}}{2.65\text{ k}\Omega} = 2\text{ mA}$$

$$\text{Collector current, } I_C \simeq I_E = 2\text{mA}$$

$$\text{Collector-base voltage, } V_{CB} = V_{CC} - I_C R_C = 20\text{ V} - 2\text{ mA} \times 4\text{ k}\Omega = 12\text{ V}$$

Therefore, the operating point (*Q*) of the circuit is **12 V, 2 mA** as shown in Fig. 6.48.

6.20 Practical Way of Drawing CE Circuit

The common emitter circuits drawn so far can be shown in another convenient way. Fig. 6.49 shows the practical way of drawing *CE* circuit. In Fig. 6.49 (*i*), the practical way of drawing common emitter *npn* circuit is shown. Similarly, Fig. 6.49 (*ii*) shows the practical way of drawing common emitter *pnp* circuit. In our further discussion, we shall often use this scheme of presentation.

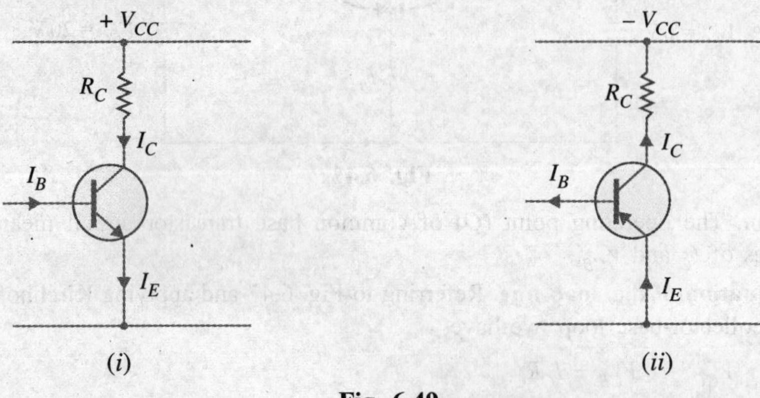

Fig. 6.49

6.21 Output from Transistor Amplifier

A transistor raises the strength of a weak signal and thus acts as an amplifier. Fig. 6.50 shows the common emitter amplifier. There are two ways of taking output from this transistor connection. The output can be taken either across R_C or across terminals 1 and 2. In either case, the magnitude of output is the same. This is clear from the following discussion :

(*i*) **First method.** We can take the output directly by putting a load resistance R_C in the collector circuit *i.e.*

$$\text{Output} = \text{Voltage across } R_C = i_c\, R_C \qquad ...(i)$$

This method of taking output from collector load is used only in single stage of amplification.

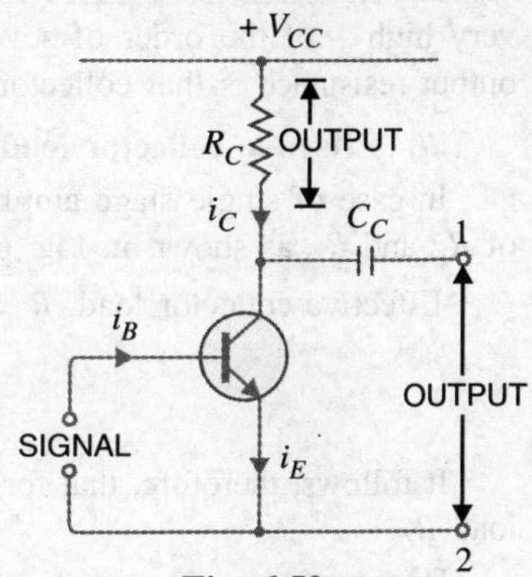

Fig. 6.50

(ii) **Second method.** The output can also be taken across terminals 1 and 2 *i.e.* from collector and emitter end of supply.

$$\text{Output} = \text{Voltage across terminals 1 and 2}$$
$$= V_{CC} - i_c R_C$$

As V_{CC} is a direct voltage and cannot pass through capacitor C_C, therefore, only varying voltage $i_c R_C$ will appear across terminals 1 and 2.

$$\therefore \quad \text{Output} = -i_c R_C \qquad ...(ii)$$

From exps. *(i)* and *(ii)*, it is clear that magnitude of output is the same whether we take output across collector load or terminals 1 and 2. The minus sign in exp. *(ii)* simply indicates the phase reversal. The second method of taking output is used in multistages of amplification.

6.22 Performance of Transistor Amplifier

The performance of a transistor amplifier depends upon input resistance, output resistance, effective collector load, current gain, voltage gain and power gain. As *common emitter connection* is universally adopted, therefore, we shall explain these terms with reference to this mode of connection.

(i) **Input resistance.** *It is the ratio of small change in base-emitter voltage (ΔV_{BE}) to the resulting change in base current (ΔI_B) at constant collector-emitter voltage i.e.*

$$\text{Input resistance, } R_i = \frac{\Delta V_{BE}}{\Delta I_B}$$

The value of input resistance is quite small because the input circuit is always forward biased. It ranges from 500 Ω for small low powered transistors to as low as 5 Ω for high powered transistors. In fact, input resistance is the opposition offered by the base-emitter junction to the signal flow. Fig. 6.51 shows the block diagram of an amplifier. The input voltage V_{BE} causes an input current I_B.

$$\therefore \quad \text{Input resistance, } R_i = \frac{\Delta V_{BE}}{\Delta I_B} = \frac{V_{BE}}{I_B}$$

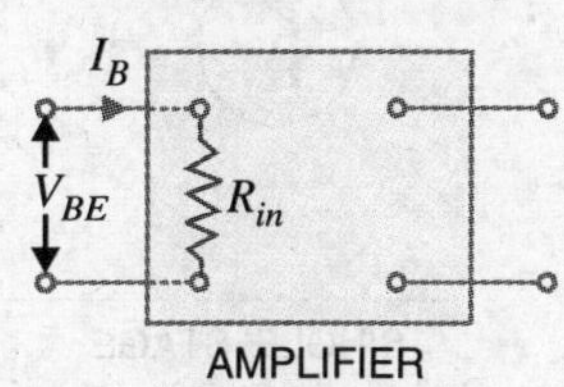

Fig. 6.51

Thus if the input resistance of an amplifier is 500 Ω and the signal voltage at any instant is 1 V, then,

$$\text{Base current, } i_b = \frac{1\text{ V}}{500\ \Omega} = 2\text{ mA}$$

(ii) **Output resistance.** *It is the ratio of change in collector-emitter voltage (ΔV_{CE}) to the resulting change in collector current (ΔI_C) at constant base current i.e.*

$$\text{Output resistance, } R_O = \frac{\Delta V_{CE}}{\Delta I_C}$$

The output characteristics reveal that collector current changes very slightly with the change in collector-emitter voltage. Therefore, output resistance of a transistor amplifier is very high – of the order of several hundred kilo-ohms. The physical explanation of high output resistance is that collector-base junction is reverse biased.

(*iii*) Effective collector load. *It is the total load as seen by the a.c. collector current.*

In case of single stage amplifiers, the effective collector load is a parallel combination of R_C and R_O as shown in Fig. 6.52 (*i*).

$$\text{Effective collector load, } R_{AC} = R_C \parallel R_O$$

$$= \frac{R_C \times R_O}{R_C + R_O} = {}^*R_C$$

It follows, therefore, that for a single stage amplifier, effective load is equal to collector load R_C.

However, in a multistage amplifier (*i.e.* having more than one amplification stage), the input resistance R_i of the next stage also comes into picture as shown in Fig. 6.52 (*ii*). Therefore, effective collector load becomes parallel combination of R_C, R_O and R_i *i.e.*

$$\text{Effective collector load, } R_{AC} = R_C \parallel R_O \parallel R_i$$

$$= {}^{**}R_C \parallel R_i = \frac{R_C\, R_i}{R_C + R_i}$$

As input resistance R_i is quite small (25 Ω to 500 Ω), therefore, effective load is reduced.

(*iv*) Current gain. *It is the ratio of change in collector current* (ΔI_C) *to the change in base current* (ΔI_B) *i.e.*

$$\text{Current gain, } \beta = \frac{\Delta I_C}{\Delta I_B}$$

The value of β ranges from 20 to 500. The current gain indicates that input current becomes β times in the collector circuit.

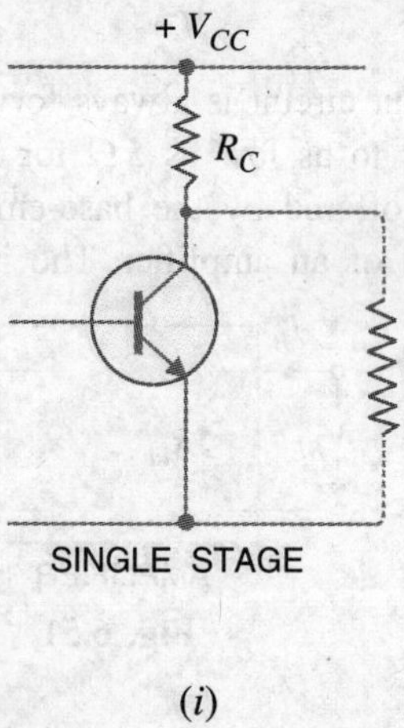

(*i*)

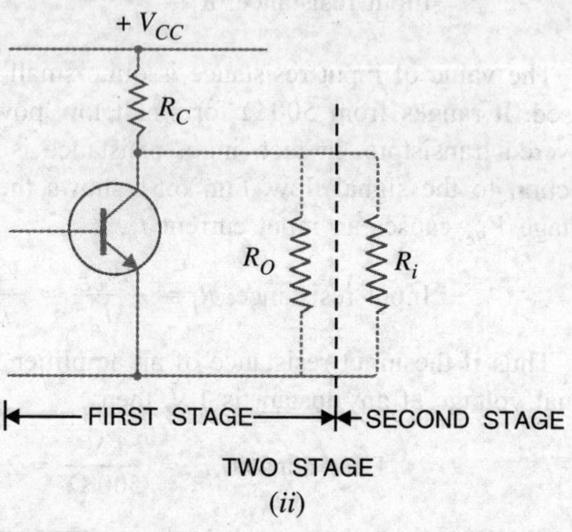

(*ii*)

Fig. 6.52

* As output resistance R_O is several times R_C, therefore, R_C can be neglected as compared to R_O.

$$R_{AC} = \frac{R_C \times R_O}{R_O} = R_C$$

** $R_C \parallel R_O = R_C$ as already explained.

(v) Voltage gain. *It is the ratio of change in output voltage (ΔV_{CE}) to the change in input voltage (ΔV_{BE}) i.e.*

$$\text{Voltage gain, } A_v = \frac{\Delta V_{CE}}{\Delta V_{BE}}$$

$$= \frac{\text{Change in output current} \times \text{Effective load}}{\text{Change in input current} \times \text{Input resistance}}$$

$$= \frac{\Delta I_C \times R_{AC}}{\Delta I_B \times R_i} = \frac{\Delta I_C}{\Delta I_B} \times \frac{R_{AC}}{R_i} = \beta \times \frac{R_{AC}}{R_i}$$

For single stage, $R_{AC} = R_C$. However, for multistage, $R_{AC} = \dfrac{R_C \times R_i}{R_C + R_i}$ where R_i is the input resistance of the next stage.

(vi) Power gain. *It is the ratio of output signal power to the input signal power i.e.*

$$\text{Power gain, } A_p = \frac{(\Delta I_C)^2 \times R_{AC}}{(\Delta I_B)^2 \times R_i} = \left(\frac{\Delta I_C}{\Delta I_B}\right) \times \frac{\Delta I_C \times R_{AC}}{\Delta I_B \times R_i}$$

$$= \text{Current gain} \times \text{Voltage gain}$$

Example 6.33. *A change of 200 mV in base-emitter voltage causes a change of 100 μA in the base current. Find the input resistance of the transistor.*

Solution. Change in base-emitter voltage is

$$\Delta V_{BE} = 200 \text{ mV}$$

$$\text{Change in base current, } \Delta I_B = 100 \text{ μA}$$

$$\therefore \quad \text{Input resistance, } R_i = \frac{\Delta V_{BE}}{\Delta I_B} = \frac{200 \text{ mV}}{100 \text{ μA}} = \mathbf{2\ k\Omega}$$

Example 6.34. *If the collector current changes from 2 mA to 3 mA in a transistor when collector-emitter voltage is increased from 2 V to 10 V, what is the output resistance?*

Solution. Change in collector-emitter voltage is

$$\Delta V_{CE} = 10 - 2 = 8 \text{ V}$$

$$\text{Change in collector current is } \Delta I_C = 3 - 2 = 1 \text{ mA}$$

$$\therefore \quad \text{Output resistance, } R_O = \frac{\Delta V_{CE}}{\Delta I_C} = \frac{8 \text{ V}}{1 \text{ mA}} = \mathbf{8\ k\Omega}$$

Example 6.35. *For a single stage transistor amplifier, the collector load is $R_C = 2\ k\Omega$ and the input resistance $R_i = 1\ k\Omega$. If the current gain is 50, calculate the voltage gain of the amplifier.*

Solution.

$$\text{Collector load, } R_C = 2 \text{ k}\Omega$$

$$\text{Input resistance, } R_i = 1 \text{ k}\Omega$$

$$\text{Current gain, } \beta = 50$$

$$\therefore \quad \text{Voltage gain, } A_v = \beta \times \frac{R_{AC}}{R_i} = \beta \times \frac{R_C}{R_i}$$

[∵ For single stage, $R_{AC} = R_C$]

$$= 50 \times (2/1) = \mathbf{100}$$

Example 6.36. *An amplifier has a voltage gain of 55 and a power gain of 456.5. The a.c. output current is 24.9 mA (r.m.s.) and the a.c. input resistance is 200 Ω. Find (i) current gain (ii) r.m.s. value of a.c. input current (iii) r.m.s. value of a.c. input voltage (iv) r.m.s. value of a.c. output voltage (v) a.c. output resistance and (vi) output power.*

Solution. $A_v = 55$; $A_p = 456.5$; I_{out}(r.m.s.) = 24.9 mA ; $R_{in} = 200\,\Omega$

(*i*) $$A_p = A_v A_i$$

∴ $$\text{Current gain, } A_i = \frac{A_p}{A_v} = \frac{456.5}{55} = \mathbf{8.3}$$

(*ii*) $$\text{AC input current, } I_{in}\text{ (r.m.s.)} = \frac{I_{out}\text{(r.m.s.)}}{A_i} = \frac{24.9\text{ mA}}{8.3} = \mathbf{3\ mA}$$

(*iii*) AC input voltage, V_{in} (r.m.s.) = I_{in} (r.m.s.) R_{in} = 3 mA × 200 Ω = **0.6 V**

(*iv*) AC output voltage, V_{out} (r.m.s.) = $A_v V_{in}$ (r.m.s.) = 55 × 0.6 V = **33 V**

(*v*) $$\text{AC output resistance, } R_{out} = \frac{V_{out}}{I_{out}} = \frac{33\text{ V}}{24.9\text{ mA}} = \mathbf{1.325\ k\Omega}$$

(*vi*) $$\text{AC input power, } P_{in} = V_{in} I_{in} = 0.6\text{ V} \times 3\text{ mA} = 1.8\text{ mW}$$

$$\text{AC output power, } P_{out} = A_p P_{in} = 456.5 \times 1.8\text{ mW} = \mathbf{0.8217\ W}$$

Example 6.37. *The input voltage to an amplifier is 0.008 sin 10^3t volt and the resulting input current is 4 × 10^{-6} sin 10^3t ampere. The a.c. output voltage (r.m.s.) is 0.4 V and current gain of the amplifier is 80. Find (i) voltage gain (ii) input resistance (iii) r.m.s. value of output current (iv) output resistance and (v) power gain.*

Solution. We first find the r.m.s. values of input voltage and input current.

$$\text{Input voltage, } V_{in}\text{ (r.m.s.)} = 0.707 \times \text{Peak value}$$
$$= 0.707 \times 0.008 = 5.66 \times 10^{-3}\text{ V}$$
$$\text{Input current, } I_{in}\text{ (r.m.s.)} = 0.707 \times \text{Peak value}$$
$$= 0.707 \times 4 \times 10^{-6} = 2.828 \times 10^{-6}\text{ A}$$
$$\text{Output voltage, } V_{out}\text{ (r.m.s.)} = 0.4\text{ V (given)}$$

(*i*) $$\text{Voltage gain, } A_v = \frac{V_{out}}{V_{in}} = \frac{0.4\text{ V}}{5.66 \times 10^{-3}\text{ V}} = \mathbf{70.7}$$

(*ii*) $$\text{Input resistance, } R_i = \frac{V_{in}}{I_{in}} = \frac{5.66 \times 10^{-3}\text{ V}}{2.828 \times 10^{-6}\text{ A}} \simeq 2000\ \Omega = \mathbf{2\ k\Omega}$$

(*iii*) $$\text{Output current, } I_{out}\text{(r.m.s.)} = A_i I_{in}\text{(r.m.s.)} = 80 \times 2.828 \times 10^{-6}$$
$$= 226 \times 10^{-6}\text{ A} = \mathbf{0.226\ mA}$$

(*iv*) $$\text{Output resistance, } R_{out} = \frac{V_{out}}{I_{out}} = \frac{0.4\text{ V}}{0.226\text{ mA}} = \mathbf{1770\ \Omega}$$

(*v*) $$\text{Input power, } P_{in} = V_{in} I_{in} = 5.66 \times 10^{-3}\text{ V} \times 2.828 \times 10^{-6}\text{ A}$$
$$= 1.6 \times 10^{-8}\text{ W}$$

$$\text{Output power, } P_{out} = V_{out} I_{out} = 0.4\text{ V} \times 0.226 \times 10^{-3}\text{A} = 9.04 \times 10^{-5}\text{ W}$$

$\therefore$ Power gain, $A_p = \dfrac{P_{out}}{P_{in}} = \dfrac{9.04\times10^{-5}\text{ W}}{1.6\times10^{-8}\text{ W}} = \mathbf{5650}$

Alternatively. Power gain, $A_p = A_v A_i = 70.7 \times 80 = 5650$

6.23 Cut-off and Saturation Points

Fig. 6.53 (*i*) shows *CE* transistor circuit while Fig. 6.53 (*ii*) shows the output characteristics along with the d.c. load line.

(*i*) **Cut-off.** The point where the load line intersects the $I_B = 0$ curve is known as *cut-off*. At this point, $I_B = 0$ and only small collector current (*i.e.* collector leakage current I_{CEO}) exists. At cut-off, the base-emitter junction no longer remains forward biased and normal transistor action is lost. The collector-emitter voltage is nearly equal to V_{CC} *i.e.*

$$V_{CE\ (cut\text{-}off)} = V_{CC}$$

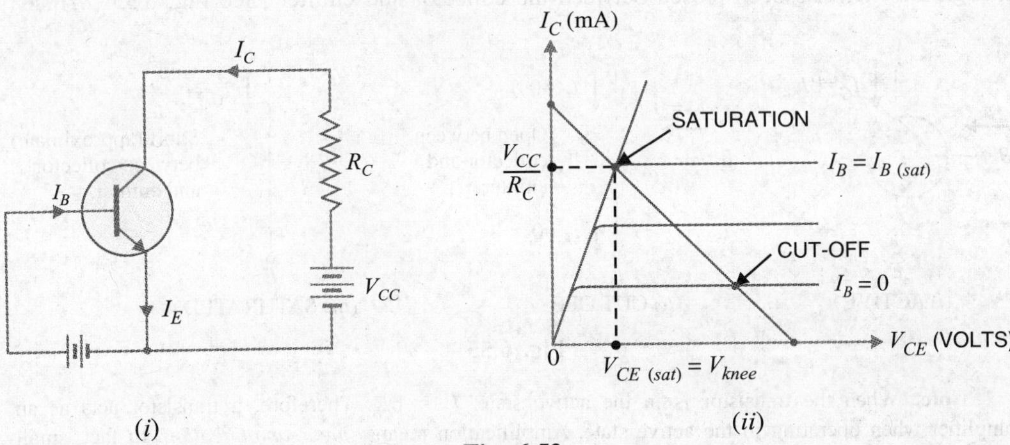

Fig. 6.53

(*ii*) **Saturation.** The point where the load line intersects the $I_B = I_{B(sat)}$ curve is called *saturation*. At this point, the base current is maximum and so is the collector current. At saturation, collector-base junction no longer remains reverse biased and normal transistor action is lost.

$$I_{C\ (sat)} \simeq \frac{V_{CC}}{R_C}\ ;\quad V_{CE} = V_{CE(sat)} = V_{knee}$$

If base current is greater than $I_{B(sat)}$, then collector current cannot increase because collector-base junction is no longer reverse-biased.

(*iii*) **Active region.** The region between cut-off and saturation is known as *active region*. In the active region, collector-base junction remains reverse biased while base-emitter junction remains forward biased. Consequently, the transistor will function normally in this region.

Note. We provide biasing to the transistor to ensure that it operates in the active region. The reader may find the detailed discussion on transistor biasing in the next chapter.

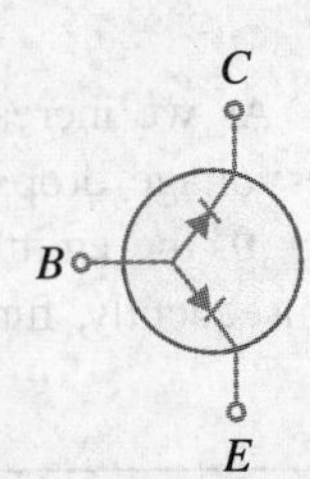

Fig. 6.54

Summary. A transistor has two *pn* junctions *i.e.* it is like two diodes. The junction between base and emitter may be called *emitter diode.* The junction between base and collector may be called *collector diode.* We have seen above that transistor can act in one of the three states : **cut-off, saturated** and **active.** The state of a transistor is entirely determined by the states of the emitter diode and collector diode [See Fig. 6.54]. The relations between the diode states and the transistor states are :

CUT-OFF : Emitter diode and collector diode are **OFF.**

ACTIVE : Emitter diode is **ON** and collector diode is **OFF.**

SATURATED : Emitter diode and collector diode are **ON.**

In the **active state,** collector current [See Fig 6.55 (*i*)] is β times the base current (*i.e.* $I_C = \beta I_B$). If the transistor is **cut-off,** there is no base current, so there is no collector or emitter current. That is collector emitter pathway is open [See Fig. 6.55 (*ii*)]. In **saturation,** the collector and emitter are, in effect, shorted together. That is the transistor behaves as though a switch has been closed between the collector and emitter [See Fig. 6.55 (*iii*)].

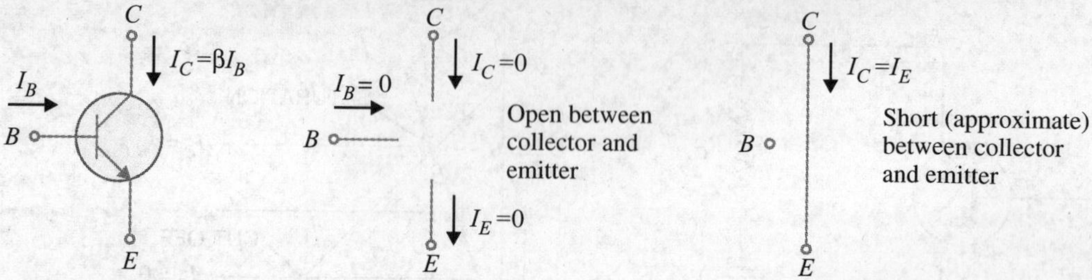

(*i*) ACTIVE (*ii*) CUT-OFF (*iii*) SATURATED

Fig. 6.55

Note. When the transistor is in the active state, $I_C = \beta I_B$. Therefore, a transistor acts as an amplifier when operating in the active state. Amplification means *linear amplification.* In fact, small signal amplifiers are the most common *linear devices.*

Example 6.38. *Find* $I_{C(sat)}$ *and* $V_{CE(cut\text{-}off)}$ *for the circuit shown in Fig. 6.56 (i).*

Solution. As we decrease R_B, base current and hence collector current increases. The increased collector current causes a greater voltage drop across R_C ; this decreases the collector-emitter voltage. Eventually at some value of R_B, V_{CE} decreases to V_{knee}. At this point, collector-base junction is no longer reverse biased and transistor action is lost. Consequently, further increase in collector current is not possible. The transistor conducts maximum collector current ; we say the transistor is saturated.

$$I_{C(sat)} = \frac{V_{CC} - {}^{*}V_{knee}}{R_C} = \frac{V_{CC}}{R_C} = \frac{20\text{ V}}{1\text{ k}\Omega} = \textbf{20 mA}$$

As we increase R_B, base current and hence collector current decreases. This decreases the voltage drop across R_C. This increases the collector-emitter voltage. Eventually, when $I_B = 0$, the emitter-base junction is no longer forward biased and transistor action is lost. Consequently, further increase in V_{CE} is not possible. In fact, V_{CE} now equals to V_{CC}.

$$V_{CE(cut\text{-}off)} = V_{CC} = \textbf{20 V}$$

* V_{knee} is about 0.5 V for Ge transistor and about 1 V for Si transistor. Consequently, V_{knee} can be neglected as compared to V_{CC} (= 20 V in this case).

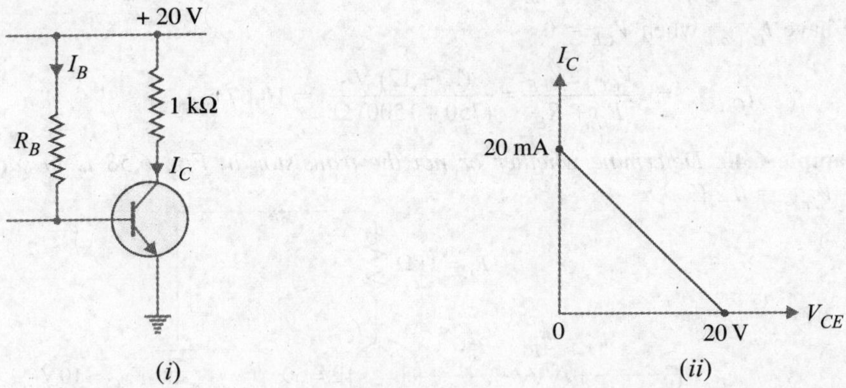

Fig. 6.56

Figure 6.56 (*ii*) shows the saturation and cut-off points. Incidentally, they are end points of the d.c. load line.

Note. The exact value of $V_{CE(cut\text{-}off)} = V_{CC} - I_{CEO}\,R_C$. Since the collector leakage current I_{CEO} is very small, we can neglect $I_{CEO}\,R_C$ as compared to V_{CC}.

Example 6.39. *Determine the values of* $V_{CE\,(off)}$ *and* $I_{C\,(sat)}$ *for the circuit shown in Fig. 6.57.*

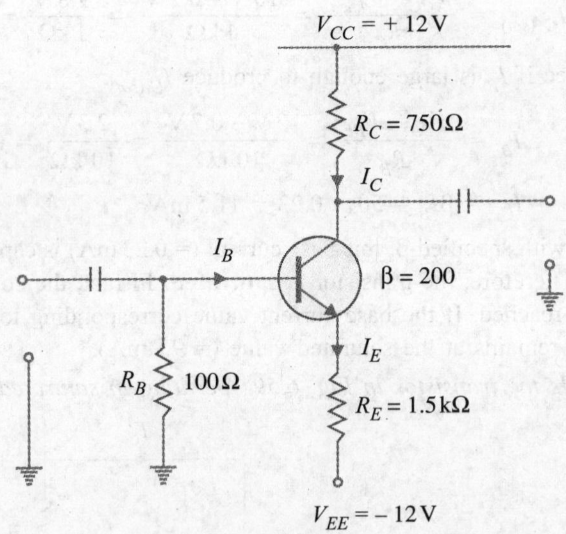

Fig. 6.57

Solution. Applying Kirchhoff's voltage law to the collector side of the circuit in Fig. 6.57, we have,

$$V_{CC} - I_C\,R_C - V_{CE} - {}^*I_C\,R_E + V_{EE} = 0$$

or $$V_{CE} = V_{CC} + V_{EE} - I_C\,(R_C + R_E) \quad \text{... }(i)$$

We have $V_{CE\,(off)}$ when $I_C = 0$. Therefore, putting $I_C = 0$ in eq. (*i*), we have,

* Voltage across $R_E = I_E\,R_E$. Since $I_E \simeq I_C$, voltage across $R_E = I_C\,R_E$.

$$V_{CE\,(off)} = V_{CC} + V_{EE} = 12 + 12 = \mathbf{24\ V}$$

We have $I_{C\,(sat)}$ when $V_{CE} = 0$.

$$\therefore \quad I_{C\,(sat)} = \frac{V_{CC} + V_{EE}}{R_C + R_E} = \frac{(12+12)\ \text{V}}{(750+1500)\ \Omega} = \mathbf{10.67\ mA}$$

Example 6.40. *Determine whether or not the transistor in Fig. 6.58 is in saturation. Assume $V_{knee} = 0.2\ V$.*

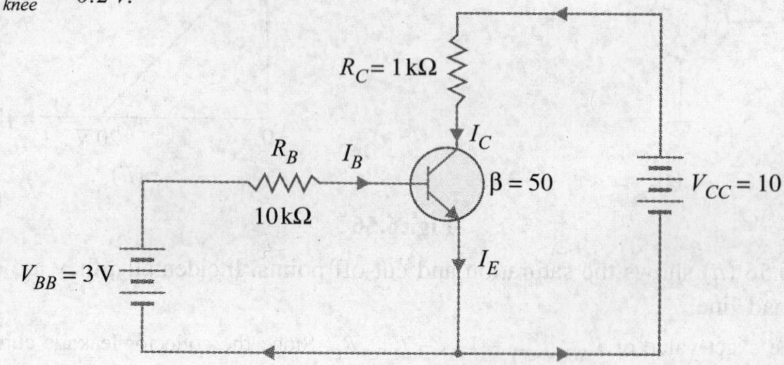

Fig. 6.58

Solution.

$$I_{C\,(sat)} = \frac{V_{CC} - V_{knee}}{R_C} = \frac{10\ \text{V} - 0.2\ \text{V}}{1\ \text{k}\Omega} = \frac{9.8\ \text{V}}{1\ \text{k}\Omega} = 9.8\ \text{mA}$$

Now we shall see if I_B is large enough to produce $I_{C\,(sat)}$.

Now,

$$I_B = \frac{V_{BB} - V_{BE}}{R_B} = \frac{3\ \text{V} - 0.7\ \text{V}}{10\ \text{k}\Omega} = \frac{2.3\ \text{V}}{10\ \text{k}\Omega} = 0.23\ \text{mA}$$

$$\therefore \quad I_C = \beta I_B = 50 \times 0.23 = 11.5\ \text{mA}$$

This shows that with specified β, this base current (= 0.23 mA) is capable of producing I_C greater than $I_{C\,(sat)}$. Therefore, the transistor is **saturated.** In fact, the collector current value of 11.5 mA is never reached. If the base current value corresponding to $I_{C\,(sat)}$ is increased, the collector current remains at the saturated value (= 9.8 mA).

Example 6.41. *Is the transistor in Fig. 6.59 operating in saturated state?*

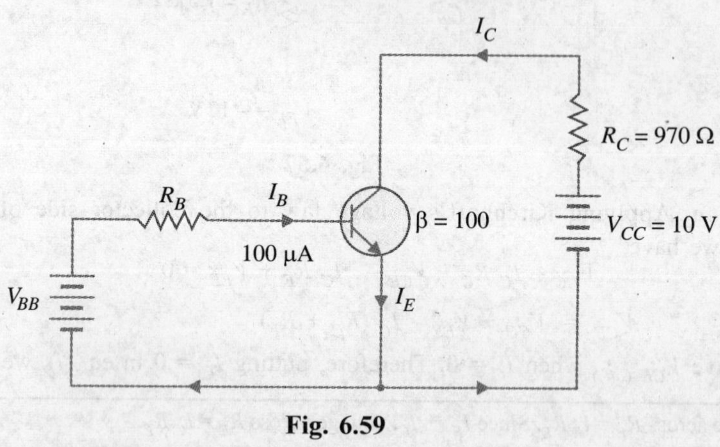

Fig. 6.59

Solution.

$$I_C = \beta I_B = (100)(100\ \mu A) = 10\ mA$$

$$V_{CE} = V_{CC} - I_C R_C$$

$$= 10\ V - (10\ mA)(970\ \Omega) = 0.3\ V$$

Let us relate the values found to the transistor shown in Fig. 6.60. As you can see, the value of V_{BE} is 0.95 V and the value of V_{CE} = 0.3 V. This leaves V_{CB} of 0.65 V (Note that $V_{CE} = V_{CB} + V_{BE}$). In this case, collector - base junction (*i.e.* collector diode) is forward biased as is the emitter-base junction (*i.e.* emitter diode). Therefore, the transistor is operating in the **saturation region.**

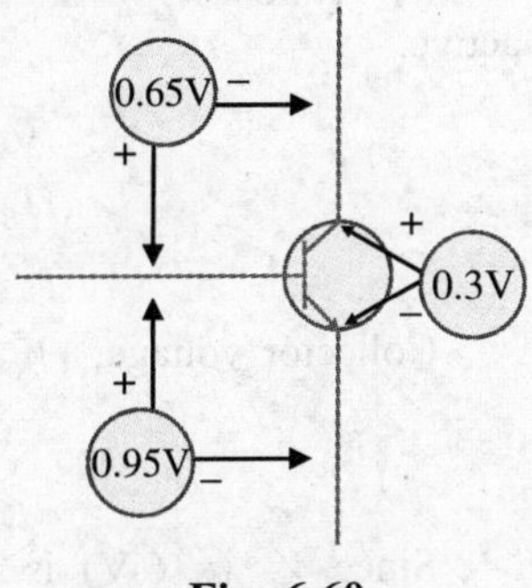

Fig. 6.60

Note. When the transistor is in the saturated state, the base current and collector current are independent of each other. The base current is still (and always is) found only from the base circuit. The collector current is found apporximately by closing the imaginary switch between the collector and the emitter in the collector circuit.

Example 6.42. *For the circuit in Fig. 6.61, find the base supply voltage (V_{BB}) that just puts the transistor into saturation. Assume β = 200.*

Solution. When transistor first goes into saturation, we can assume that the collector shorts to the emitter (*i.e.* V_{CE} = 0) but the collector current is still β times the base current.

$$I_{C(sat)} = \frac{V_{CC} - V_{CE}}{R_C} = \frac{V_{CC} - 0}{R_C}$$

$$= \frac{10\ V - 0}{2\ k\Omega} = 5\ mA$$

The base current I_B corresponding to $I_{C\,(sat)}$ (= 5 mA) is

$$I_B = \frac{I_{C\,(sat)}}{\beta} = \frac{5\ mA}{200} = 0.025\ mA$$

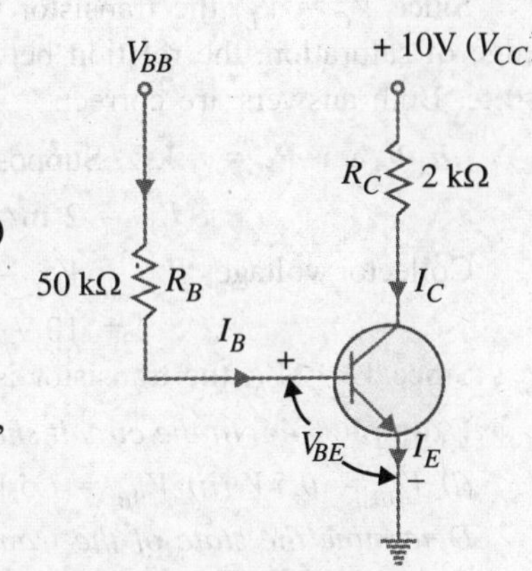

Fig. 6.61

Applying Kirchhoff's voltage law to the base circuit, we have,

$$V_{BB} - I_B R_B - V_{BE} = 0$$

or

$$V_{BB} = V_{BE} + I_B R_B$$

$$= 0.7V + 0.025\ mA \times 50\ k\Omega$$

$$= 0.7 + 1.25 = 1.95\ V$$

Therefore, for $V_{BB} \geq$ **1.95 V**, the transistor will be in *saturation.*

Example. 6.43. *Determine the state of the transistor in Fig. 6.62 for the following values of collector resistor :*

(i) $R_C = 2\ k\Omega$ *(ii)* $R_C = 4\ k\Omega$ *(iii)* $R_C = 8\ k\Omega$

Solution. Since I_E does not depend on the value of the collector resistor R_C, the emitter current (I_E) is the same for all three parts.

Emitter voltage, $V_E = V_B - V_{BE} = V_{BB} - V_{BE}$

$= 2.7\text{ V} - 0.7\text{ V} = 2\text{ V}$

Also, $I_E = \frac{V_E}{R_E} = \frac{2\text{V}}{1\text{ k}\Omega} = 2\text{ mA}$

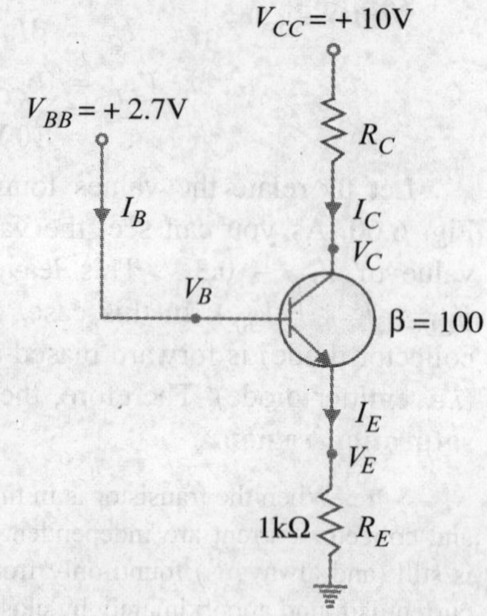

Fig. 6.62

(*i*) **When R_C = 2 kΩ.** Suppose the transistor is active.

$\therefore \quad I_C = I_E = 2\text{ mA}$

$\therefore \quad I_B = I_C/\beta = 2\text{ mA}/100$

$= 0.02\text{ mA}$

Collector voltage, $V_C = V_{CC} - I_C R_C$

$= 10\text{ V} - 2\text{ mA} \times 2\text{ k}\Omega$

$= 10\text{ V} - 4\text{ V} = 6\text{ V}$

Since V_C (= 6 V) is greater than V_E (= 2 V), the transistor is **active.** Therefore, our assumption that transistor is active is correct.

(*ii*) **When R_C = 4 kΩ.** Suppose the transistor is active.

$\therefore \quad I_C = 2\text{ mA}$ and $I_B = 0.02\text{ mA}$ ... as found above

Collector voltage, $V_C = V_{CC} - I_C R_C$

$= 10\text{ V} - 2\text{ mA} \times 4\text{ k}\Omega = 10\text{ V} - 8\text{ V} = 2\text{ V}$

Since $V_C = V_E$, the transistor is just at the edge of **saturation.** We know that at the edge of saturation, the relation between the transistor currents is the same as in the **active state.** Both answers are correct.

(*iii*) **When R_C = 8 kΩ.** Suppose the transistor is active.

$\therefore \quad I_C = 2\text{ mA}$; $I_B = 0.02\text{ mA}$ ... as found earlier

Collector voltage, $V_C = V_{CC} - I_C R_C$

$= 10\text{ V} - 2\text{ mA} \times 8\text{ k}\Omega = 10\text{ V} - 16\text{ V} = -6\text{ V}$

Since $V_C < V_E$, the transistor is **saturated** and our assumption is not correct.

Example 6.44. *In the circuit shown in Fig. 6.63, V_{BB} is set equal to the following values: (i) V_{BB} = 0.5 V (ii) V_{BB} = 1.5 V (iii) V_{BB} = 3 V.*

Determine the state of the transistor for each value of the base supply voltage V_{BB}.

Solution. The state of the transistor also depends on the base supply voltage V_{BB}.

(*i*) **For V_{BB} = 0.5 V.**

Because the base voltage V_B (= V_{BB} = 0.5 V) is less than 0.7 V, the transistor is **cut-off.**

(*ii*) **For V_{BB} = 1.5 V.**

The base voltage V_B controls the emitter voltage V_E which controls the emitter current I_E.

Now, $V_E = V_B - 0.7\text{ V} = 1.5\text{ V} - 0.7\text{ V} = 0.8\text{ V}$

$\therefore \quad I_E = \frac{V_E}{R_E} = \frac{0.8\text{ V}}{1\text{ k}\Omega} = 0.8\text{ mA}$

If the transistor is active, we have,

$I_C = I_E = 0.8$ mA and $I_B = I_C/\beta = 0.8/100$
$= 0.008$ mA

∴ Collector voltage, $V_C = V_{CC} - I_C R_C$
$= 15\text{ V} - 0.8\text{ mA} \times 10\text{ k}\Omega = 15\text{ V} - 8\text{ V} = 7\text{ V}$

Since $V_C > V_E$, the transistor is **active** and our assumption is correct.

(iii) **For V_{BB} = 3 V.**

$$V_E = V_B - 0.7\text{ V} = 3\text{ V} - 0.7\text{ V} = 2.3\text{ V}$$

$$\therefore \quad I_E = \frac{V_E}{R_E} = \frac{2.3\text{ V}}{1\text{ k}\Omega} = 2.3\text{ mA}$$

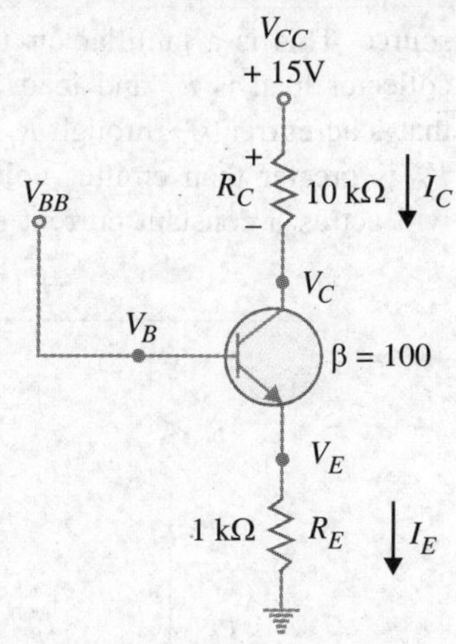

Fig. 6.63

Assuming the transistor is active, we have,

$I_C = I_E = 2.3$ mA ; $I_B = I_C/\beta = 2.3/100 = 0.023$ mA

Collector voltage, $V_C = V_{CC} - I_C R_C$
$= 15\text{ V} - 2.3\text{ mA} \times 10\text{ k}\Omega = 15\text{ V} - 23\text{ V} = -8\text{ V}$

Since $V_C < V_E$, the transistor is **saturated** and our assumption is not correct.

6.24 Use of Transistor as Constant Current Source

Recall that an **ideal* constant current source always supplies the same amount of current to any value of load R_L connected to it. Fig. 6.64 shows an ideal constant current source of 5 mA. This constant current source will always supply 5 mA to the load R_L whatever may be the value of R_L. Note that voltage across R_L is $V_L = I_o R_L = 5\text{ mA} \times R_L$. As the value of R_L is increased, the current through R_L remains 5 mA but voltage V_L across the load R_L increases. However, there is a limit to the increase in the value of R_L because no constant current source is ideal. If the value of R_L exceeds this **limiting value, the current source will no longer behave as a constant current source. For this reason, there is a limit to the maximum load voltage V_L that the current source can provide. Two points are worth noting about a constant current source. First, it will supply constant current to the load R_L so long as the value of R_L does not exceed a certain maximum value. Secondly, there is a limit to the maximum load voltage (V_L) that a constant current source can provide.

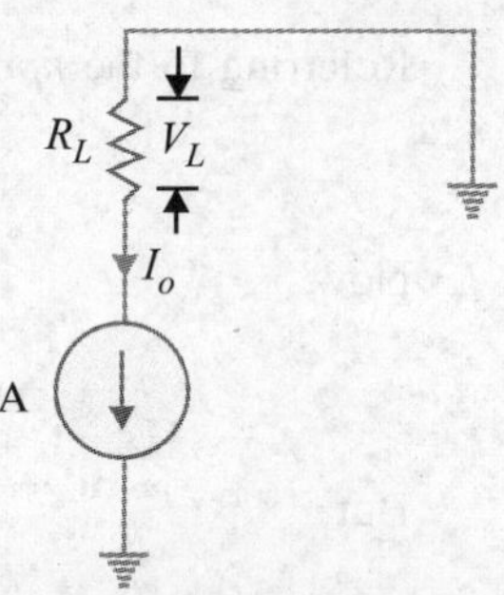

CONSTANT CURRENT SOURCE

Fig. 6.64

Fig. 6.65 (*i*) shows how *npn* transistor can be ***used to act as a constant current (d.c.)

* An ideal constant current source is one that has infinite internal resistance R_i. This internal resistance R_i appears in parallel with the current source (R_i not shown).

** R_L is in parallel with internal resistance R_i of the current source. When R_L becomes sufficiently large, some current flows through R_i and load current (current through R_L) no longer remains constant.

*** Current source is an energy source but transistor is not a source of energy. Therefore, we need d.c. supply V_{CC} and associated circuitry to enable the transistor to act as a constant current source.

source. This is a familiar circuit of transistor biasing by voltage-divider method. Note that collector load is R_L and load current I_o is the collector current *i.e.* $I_o = I_C$. We now show that load current I_o through R_L is independent of the value of R_L, provided collector voltage V_C is greater than emitter voltage V_E. In other words, so long as $V_C > V_E$, this transistor will act as a constant current source.

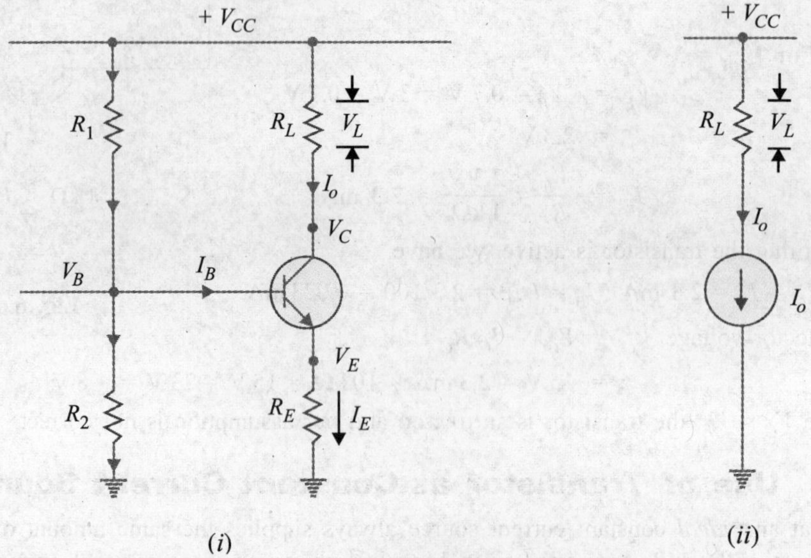

Fig. 6.65

Referring to the *npn* transistor circuit shown in Fig. 6.65 (*i*), we have,

$$V_B = \frac{V_{CC}}{R_1 + R_2} \times R_2$$

Now, $$V_E = V_B - V_{BE} = V_B - 0.7\text{ V} \qquad (\because V_{BE} = 0.7\text{ V})$$

$\therefore$ $$I_E = \frac{V_E}{R_E} = \frac{V_B - V_{BE}}{R_E}$$

But $$I_C \simeq I_E = I_o \qquad (\because I_C = I_o)$$

$\therefore$ $$I_o = \frac{V_B - V_{BE}}{R_E}$$

Since V_B, V_{BE} and R_E are fixed, I_o is constant. Further, the value of load current I_o is independent of the value of R_L. Therefore, the transistor behaves as a constant current source [See Fig. 6.65 (*ii*)]. *It is very important to note that this npn transistor will act as a constant current source so long as* *$V_C > V_E$ *i.e. the transistor operates in the active region.*

Now, Collector voltage, $V_C = V_{CC} - I_o R_L$

Emitter voltage, $V_E = I_o R_E$

The value of V_C depends upon the value of R_L ($\because$ V_{CC} and I_o are constants). As R_L increases, V_C decreases. Therefore, permissible value of R_L is upto which $V_C > V_E$. If the

* For *pnp* transistor, the condition is $V_E > V_C$.

value of R_L exceeds this limiting value, the transistor will not act as a constant current source.

Example 6.45. *For the npn transistor circuit shown in Fig. 6.66, show that the transistor acts as a constant current source upto* $R_L \simeq 40\ k\Omega$.

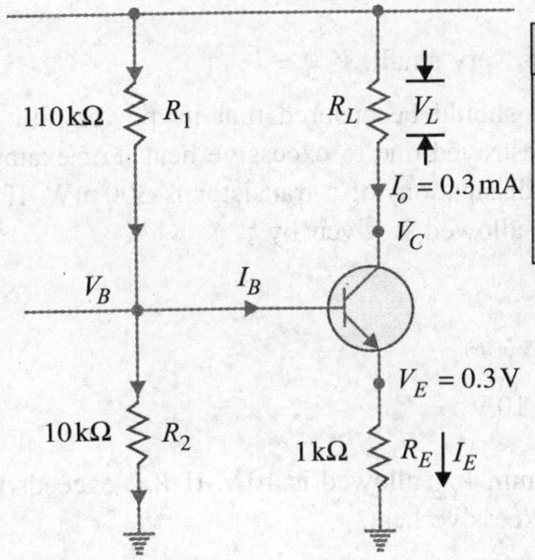

R_L	V_L	V_C	*State*
10 kΩ	3 V	9 V	Active
20 kΩ	6 V	6 V	Active
30 kΩ	9 V	3 V	Active
40 kΩ	12 V	0 V	Saturated

Fig. 6.66

Solution. Referring to the *npn* transistor circuit shown in Fig. 6.66, we have,

$$V_B = \frac{V_{CC}}{R_1 + R_2} \times R_2 = \frac{12V}{(110+10)\ k\Omega} \times 10\ k\Omega = 1\ V$$

Now, $$V_E = V_B - V_{BE} = 1\ V - 0.7\ V = 0.3\ V$$

$\therefore$ $$I_E = \frac{V_E}{R_E} = \frac{0.3\ V}{1k\Omega} = 0.3\ mA$$

$\therefore$ Load current, $I_o = I_C \simeq I_E = 0.3\ mA$

Therefore, the *npn* transistor acts as a constant current source of 0.3 mA so long as $V_C > V_E$ (= 0.3 V). The load voltages (V_L) and collector voltages (V_C) for several values of R_L are tabulated above. The table also shows the state of the transistor for each load. Note that for $R_L = 40\ k\Omega$, $V_L = I_o R_L = 0.3\ mA \times 40\ k\Omega = 12\ V$ and $V_C = V_{CC} - I_o R_L = 12\ V - 0.3\ mA \times 40\ k\Omega = 12\ V - 12\ V = 0$. We note that for $R_L = 40\ k\Omega$, $V_C < V_E$ (= 0.3V) so that the transistor will not act as a constant current source for $R_L = 40\ k\Omega$. However, if $R_L < 40\ k\Omega$ so that $V_C > V_E$, the transistor will behave as a constant current source. *Therefore, it is reasonable to say that the transistor will act as a constant current source for load values upto about 40 kΩ.* This is a pretty good range of load (R_L) values.

6.25 Power Rating of Transistor

The maximum power that a transistor can handle without destruction is known as **power rating** *of the transistor.*

When a transistor is in operation, almost all the power is dissipated at the reverse biased *collector-base junction. The power rating (or maximum power dissipation) is given by ;

$$P_{D\,(max)} = \text{Collector current} \times \text{Collector-base voltage}$$

$$= I_C \times V_{CB}$$

$$\therefore \quad P_{D\,(max)} = I_C \times V_{CE}$$

$$[\because V_{CE} = V_{CB} + V_{BE}. \text{ Since } V_{BE} \text{ is very small, } V_{CB} \simeq V_{CE}]$$

While connecting transistor in a circuit, it should be ensured that its power rating is not exceeded otherwise the transistor may be destroyed due to excessive heat. For example, suppose the power rating (or maximum power dissipation) of a transistor is 300 mW. If the collector current is 30 mA, then maximum V_{CE} allowed is given by ;

$$P_{D\,(max)} = I_C \times V_{CE\,(max)}$$

or $$300 \text{ mW} = 30 \text{ mA} \times V_{CE\,(max)}$$

or $$V_{CE\,(max)} = \frac{300 \text{ mW}}{30 \text{ mA}} = 10 \text{ V}$$

This means that for $I_C = 30$ mA, the maximum V_{CE} allowed is 10 V. If V_{CE} exceeds this value, the transistor will be destroyed due to excessive heat.

Maximum power dissipation curve. For **power transistors, it is sometimes necessary to draw maximum power dissipation curve on the output characteristics. To draw this curve, we should know the power rating (*i.e.* maximum power dissipation) of the transistor. Suppose the power rating of a transistor is 30 mW *i.e.*

$$P_{D\,(max)} = V_{CE} \times I_C$$

or $$30 \text{ mW} = V_{CE} \times I_C$$

Using convenient V_{CE} values, the corresponding collector currents are calculated for the maximum power dissipation. For example, for $V_{CE} = 10$ V,

$$I_{C\,(max)} = \frac{P_{D\,(max)}}{V_{CE}} = \frac{30 \text{ mW}}{10 \text{ V}} = 3 \text{ mA}$$

This locates the point *A* (10 V, 3 mA) on the output characteristics. Similarly, many points such as *B, C, D etc.* can be located on the output characteristics. Now draw a curve through the above points to obtain the maximum power dissipation curve as shown in Fig. 6.67.

In order that transistor may not be destroyed, the transistor voltage and current (*i.e.* V_{CE} and I_C) conditions must at all times be maintained in the portion of the characteristics *below* the maximum power dissipation curve.

---

* The base-emitter junction conducts about the same current as the collector-base junction (*i.e.* $I_E \simeq I_C$). However, V_{BE} is very small (0.3 V for Ge transistor and 0.7 V for Si transistor). For this reason, power dissipated at the base-emitter junction is negligible.

** A transistor that is suitable for large power amplification is called a *power transistor*. It differs from other transistors mostly in size; it is considerably larger to provide for handling the great amount of power.

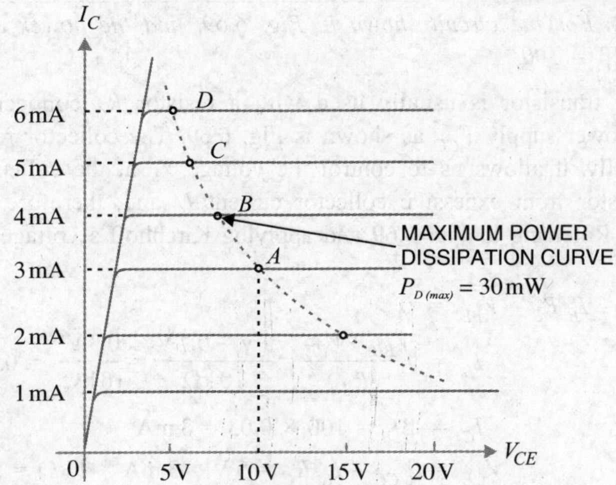

Fig. 6.67

Example 6.46. *The maximum power dissipation of a transistor is 100 mW. If V_{CE} = 20 V, what is the maximum collector current that can be allowed without destruction of the transistor?*

Solution.

$$P_{D\,(max)} = V_{CE} \times I_{C\,(max)}$$

or

$$100\text{ mW} = 20\text{ V} \times I_{C\,(max)}$$

$$\therefore \quad I_{C\,(max)} = \frac{100\text{ mW}}{20\text{ V}} = \mathbf{5mA}$$

Thus for V_{CE} = 20V, the maximum collector current allowed is 5 mA. If collector current exceeds this value, the transistor may be burnt due to excessive heat.

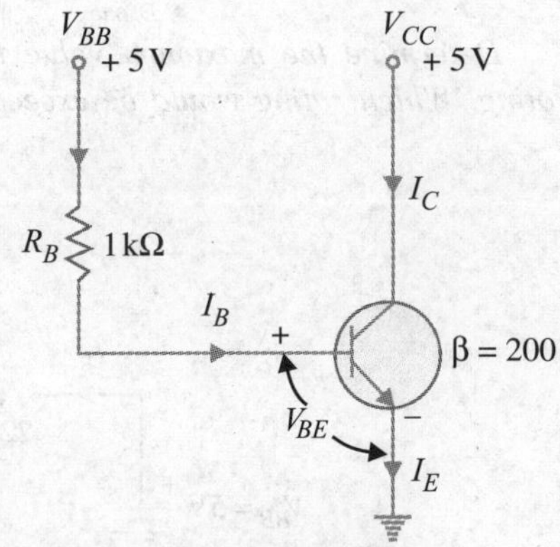

Fig. 6.68

Note. Suppose the collector current becomes 7 mA. The power produced will be 20 V × 7 mA = 140 mW. The transistor can only dissipate 100 mW. The remaining 40 mW will raise the temperature of the transistor and eventually it will be burnt due to excessive heat.

Example 6.47. *For the circuit shown in Fig. 6.68, find the transistor power dissipation. Assume that β = 200.*

Solution.

$$I_B = \frac{V_{BB} - V_{BE}}{R_B} = \frac{(5-0.7)\text{ V}}{1\text{ k}\Omega} = 4.3\text{ mA}$$

$$\therefore \quad I_C = \beta I_B = 200 \times 4.3 = 860\text{ mA}$$

Now, $V_{CE} = V_{CC} - I_C R_C = 5 - I_C \times 0 = 5\text{ V}$

$\therefore$ Power dissipation, $P_D = V_{CE} \times I_C$

$= 5\text{ V} \times 860\text{ mA} = 4300\text{ mW} = \mathbf{4.3\text{ W}}$

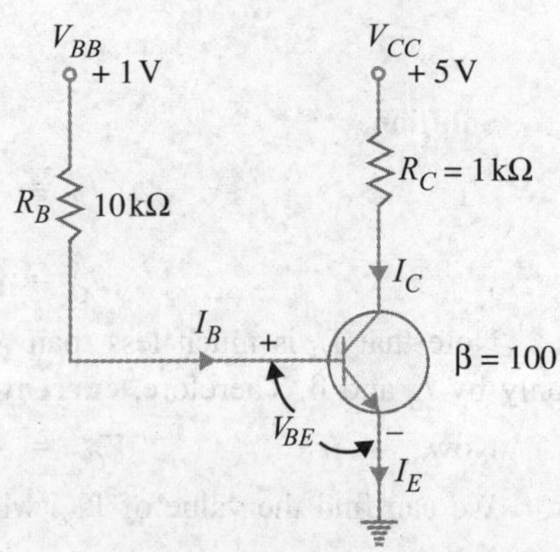

Fig. 6.69

Example 6.48. *For the circuit shown in Fig. 6.69, find the power dissipated in the transistor. Assume β = 100.*

Solution. The transistor is usually used with a resistor R_C connected between the collector and its power supply V_{CC} as shown is Fig. 6.69. The collector resistor R_C serves two purposes. Firstly, it allows us to control the voltage V_C at the collector. Secondly, it protects the transistor from excessive collector current I_C and, therefore, from excessive power dissipation. Referring to Fig. 6.69 and applying Kirchhoff's voltage law to the base side, we have,

$$V_{BB} - I_B R_B - V_{BE} = 0$$

$$\therefore \quad I_B = \frac{V_{BB} - V_{BE}}{R_B} = \frac{1\text{ V} - 0.7\text{ V}}{10\text{ k}\Omega} = \frac{0.3\text{ V}}{10\text{ k}\Omega} = 0.03\text{ mA}$$

Now,
$$I_C = \beta I_B = 100 \times 0.03 = 3\text{ mA}$$

$$\therefore \quad V_{CE} = V_{CC} - I_C R_C = 5\text{ V} - 3\text{ mA} \times 1\text{ k}\Omega = 5\text{ V} - 3\text{ V} = 2\text{ V}$$

∴ Power dissipated in the transistor is

$$P_D = V_{CE} \times I_C = 2\text{ V} \times 3\text{ mA} = \mathbf{6\text{ mW}}$$

Example 6.49. *The transistor in Fig. 6.70 has the following maximum ratings :*

$$P_{D\,(max)} = 800\text{ mW}\ ;\ V_{CE\,(max)} = 15\text{ V}\ ;\ I_{C\,(max)} = 100\text{ mA}$$

Determine the maximum value to which V_{CC} can be adjusted without exceeding any rating. Which rating would be exceeded first?

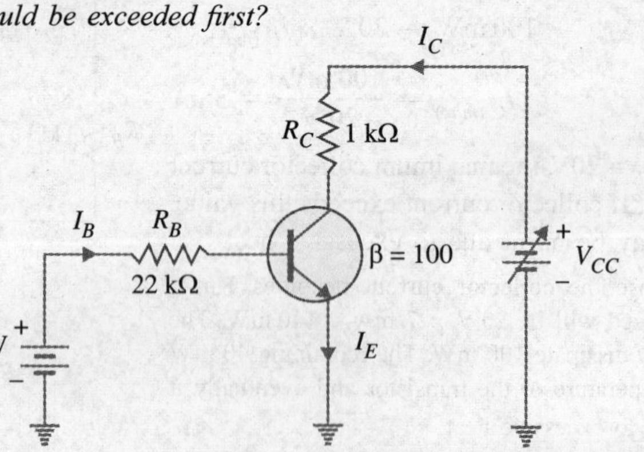

Fig. 6.70

Solution.

$$I_B = \frac{V_{BB} - V_{BE}}{R_B} = \frac{5\text{ V} - 0.7\text{ V}}{22\text{ k}\Omega} = \frac{4.3\text{ V}}{22\text{ k}\Omega} = 195\ \mu\text{A}$$

$$\therefore \quad I_C = \beta I_B = 100 \times 195\ \mu\text{A} = 19.5\text{ mA}$$

Note that I_C is much less than $I_{C(max)}$ and will not change with V_{CC}. It is determined only by I_B and β. Therefore, **current rating is not exceeded.**

Now,
$$V_{CC} = V_{CE} + I_C R_C$$

We can find the value of V_{CC} when $V_{CE\,(max)} = 15$ V.

$$\therefore \quad V_{CC\,(max)} = V_{CE\,(max)} + I_C R_C$$

$$= 15\text{ V} + 19.5\text{ mA} \times 1\text{ k}\Omega = 15\text{ V} + 19.5\text{ V} = 34.5\text{ V}$$

Therefore, we can increase V_{CC} to **34.5 V** before $V_{CE\,(max)}$ is reached.

$$P_D = V_{CE\,(max)}\, I_C = (15\text{ V})\,(19.5\text{ mA}) = 293\text{ mW}$$

Since $P_{D\,(max)} = 800$ mW, **it is not exceeded** when $V_{CC} = 34.5$ V.

If base current is removed causing the transistor to turn off, $V_{CE\,(max)}$ **will be exceeded** because the entire supply voltage V_{CC} will be dropped across the transistor.

6.26 Determination of Transistor Configuration

In practical circuits, you must be able to tell whether a given transistor is connected as a common emitter, common base or common collector. There is an easy way to ascertain it. Just locate the terminals where the input a.c. singal is applied to the transistor and where the a.c. output is taken from the transistor. The remaining third terminal is the common terminal. For instance, if the a.c. input is applied to the base and the a.c. output is taken from the collector, then common terminal is the emitter. Hence the transistor is connected in **common emitter configuration.** If the a.c. input is applied to the base and a.c. output is taken from the emitter, then common terminal is the collector. Therefore, the transistor is connected in **common collector configuration.**

6.27 Semiconductor Devices Numbering System

From the time semiconductor engineering came to existence, several numbering systems were adopted by different countries. However, the accepted numbering system is that announced by Proelectron Standardisation Authority in Belgium. According to this system of numbering semiconductor devices :

(*i*) Every semiconductor device is numbered by five alpha-numeric symbols, comprising either two letters and three numbers (*e.g.* *BF*194) or three letters and two numbers (*e.g.* *BFX*63). When two numbers are included in the symbol (*e.g.* *BFX*63), the device is intended for industrial and professional equipment. When the symbol contains three numbers (*e.g.* *BF*194), the device is intended for entertainment or consumer equipment.

(*ii*) The first letter indicates the nature of semiconductor material. For example :

A = germanium, *B* = silicon, *C* = gallium arsenide, *R* = compound material (*e.g.* cadmium sulphide)

Thus *AC*125 is a germanium transistor whereas *BC*149 is a silicon transistor.

(*iii*) The second letter indicates the device and circuit function.

A = Diode	*B* = Variable capacitance diode
C = A.F. low powered transistor	*D* = A.F. power transistor
E = Tunnel diode	*F* = H.F. low power transistor
G = Multiple device	*H* = Magnetic sensitive diode
K = Hall-effect device	*L* = H.F. power transistor
M = Hall-effect modulator	*P* = Radiation sensitive diode
Q = Radiation generating diode	*R* = Thyristor (*SCR* or Triac)
S = Low power switching transistor	*T* = Thyristor (power)
U = Power switching transistor	*X* = Diode, Multiplier
Y = Power device	*Z* = Zener diode

6.28 Transistor Lead Identification

There are three leads in a transistor *viz. collector, emitter* and *base*. When a transistor is to be connected in a circuit, it is necessary to know which terminal is which. The identification of the leads of transistor varies with manufacturer. However, there are three systems in general use as shown in Fig. 6.71.

(*i*) When the leads of a transistor are in the same plane and unevenly spaced [See Fig. 6.71 (*i*)], they are identified by the positions and spacings of leads. The central lead is the base lead. The collector lead is identified by the larger spacing existing between it and the base lead. The remaining lead is the emitter.

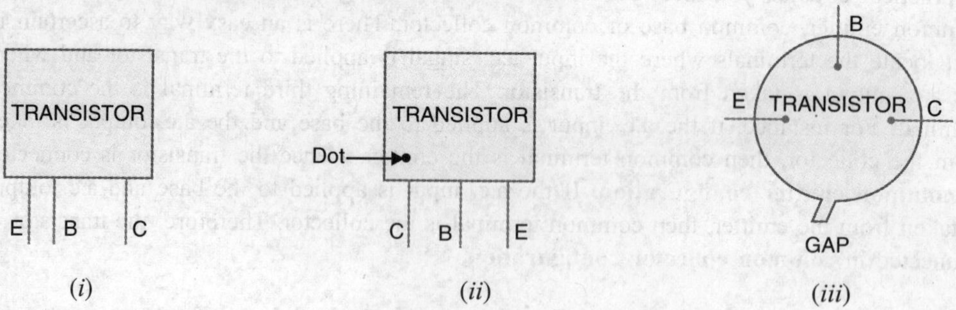

Fig. 6.71

(*ii*) When the leads of a transistor are in the same plane but evenly spaced [See Fig. 6.71 (*ii*)], the central lead is the base, the lead identified by dot is the collector and the remaining lead is the emitter.

(*iii*) When the leads of a transistor are spaced around the circumference of a circle [See Fig. 6.71 (*iii*)], the three leads are generally in *E-B-C* order clockwise from a gap.

6.29 Transistor Testing

An ohmmeter can be used to check the state of a transistor *i.e.* whether the transistor is good or not. We know that base-emitter junction of a transistor is forward biased while collector-base junction is reverse biased. Therefore, forward biased base-emitter junction should have low resistance and reverse biased collector-base junction should register a much higher resistance. Fig. 6.72 shows the process of testing an *npn* transistor with an ohmmeter.

(*i*) The forward biased base-emitter junction (biased by internal supply) should read a low resistance, typically 100 Ω to 1 kΩ as shown in Fig. 6.72 (*i*). If that is so, the transistor is good. However, if it fails this check, the transistor is faulty and it must be replaced.

(*ii*) The reverse biased collector-base junction (again reverse biased by internal supply) should be checked as shown in Fig. 6.72 (*ii*). If the reading of the ohmmeter is 100 kΩ or higher, the transistor is good. If the ohmmeter registers a small resistance, the transistor is faulty and requires replacement.

Note. When testing a *pnp* transistor, the ohmmeter leads must be reversed. The results of the tests, however, will be the same.

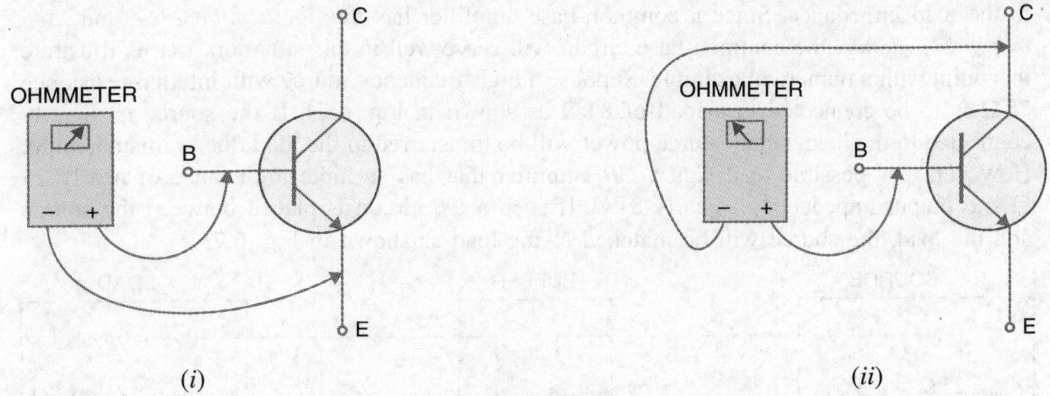

Fig. 6.72

6.30 Applications of Common Base Amplifiers

Common base amplifiers are not used as frequently as the *CE* amplifiers. The two important applications of *CB* amplifiers are : (*i*) to provide voltage gain without current gain and (*ii*) for impedance matching in high frequency applications. Out of the two, the high frequency applications are far more common.

(*i*) To provide voltage gain without current gain. We know that a *CB* amplifier has a high voltage gain while the current gain is nearly 1 (*i.e.* $A_i \simeq 1$). Therefore, this circuit can be used to provide high voltage gain without increasing the value of circuit current. For instance, consider the case where the output current from an amplifier has sufficient value for the required application but the voltage gain needs to be increased. In that case, *CB* amplifier will serve the purpose because it would increase the voltage without increasing the current. This is illustrated in Fig. 6.73. The *CB* amplifier will provide voltage gain without any current gain.

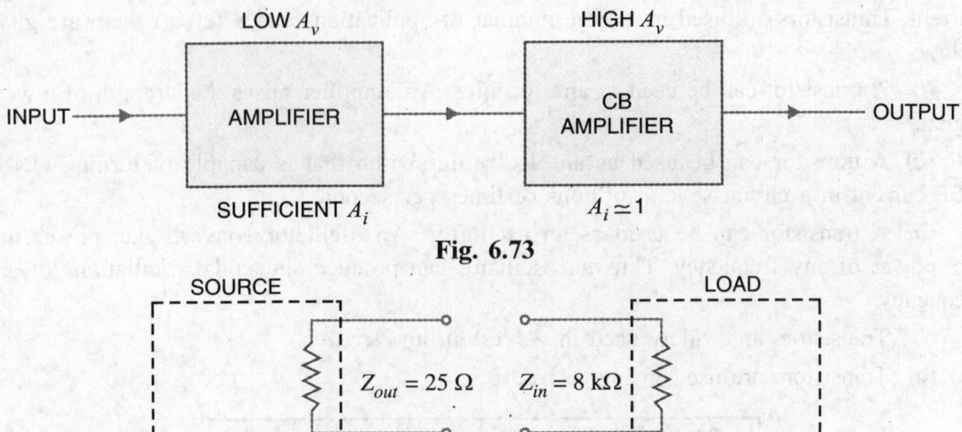

Fig. 6.73

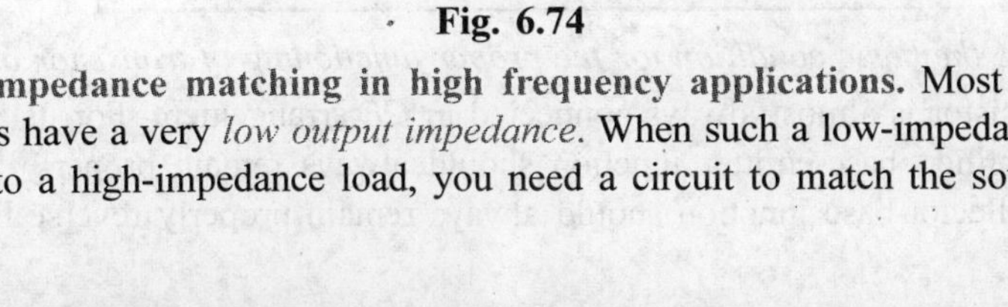

Fig. 6.74

(*ii*) For impedance matching in high frequency applications. Most high-frequency voltage sources have a very *low output impedance*. When such a low-impedance source is to be connected to a high-impedance load, you need a circuit to match the source impedance to the load impedance. Since a common-base amplifier has *low input impedance* and *high output impedance*, the common-base circuit will serve well in this situation. Let us illustrate this point with a numerical example. Suppose a high-frequency source with internal resistance 25 Ω is to be connected to a load of 8 kΩ as shown in Fig. 6.74. If the source is directly connected to the load, small source power will be transferred to the load due to mismatching. However, it is possible to design a *CB* amplifier that has an input impedance of nearly 25 Ω and output impedance of nearly 8 kΩ. If such a *CB* circuit is placed between the source and the load, the source will be matched to the load as shown in Fig. 6.75.

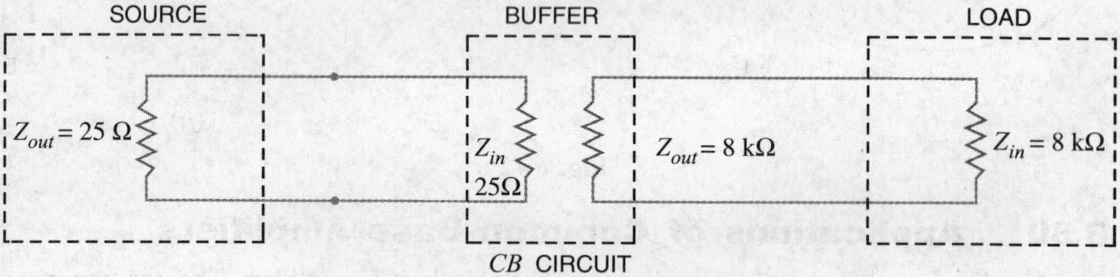

Fig. 6.75

Note that source impedance very closely matches the input impedance of *CB* amplifier. Therefore, there is a maximum power transfer from the source to input of *CB* amplifier. The high output impedance of the amplifier very nearly matches the load resistance. As a result, there is a maximum power transfer from the amplifier to the load. The net result is that maximum power has been transferred from the original source to the original load. A common-base amplifier that is used for this purpose is called a *buffer amplifier.*

6.31 Applications of Transistors

The main building block of modern electronic systems is the transistor. The transistor is a three terminal device whose output current, voltage and/or power are controlled by its input current. Transistors are used in a large number of applications and a few of them are given below.

(*i*) A transistor can be used as an **amplifier**. An amplifier raises the strength of a weak signal.

(*ii*) A transitor can be used as an **electronic switch** that is capable of turning ON or OFF current in a circuit several billions of times per second.

(*iii*) A transistor can be used as an **oscillator**. An oscillator converts d.c. power into a.c. power of any frequency. Thus an oscillator can produce sinusoidal oscillations of any frequency.

(*iv*) Transistors are widely used in **waveshaping circuits.**

(*v*) Transistors are used in **logic circuits.**

Short Answer Questions

Q.1. *What is the basic condition for the proper functioning of transistor as an amplifier?*

Ans. A transistor is almost always connected in *CE* arrangement. For its proper function as amplifier, base-emitter junction should always remain properly forward biased and collector-base junction should always remain properly reverse biased.

Q.2. *Why is the base of a transistor made thin and lightly doped?*

Ans. The base is made thin and lightly doped so that the majority carriers supplied by the emitter do not combine in the base region and most of them pass on to the collector.

Q.3. *What is the key factor responsible for amplifying capability of a transistor?*

Ans. In a transistor, the weak input signal current is introduced in the low resistance input circuit and output is taken from the high resistance output circuit. This property of the transistor to transfer the input signal current from a low resistance circuit to a high resistance circuit is the key factor responsible for amplifying capability of a transistor.

Q.4. *Why is collector of a transistor made wider than emitter and base?*

Ans. During transistor operation, most of heat is produced at the collector junction. The collector is made larger to dissipate heat.

Q.5. *If the emitter of a transistor connected in CB mode is open, will there be any collector current?*

Ans. During normal operation of transistor, the total collector current is given by ;

$$I_C = \alpha I_E + I_{leakage} \quad \text{... } (i)$$

Here, αI_E is the part of the emitter current which reaches the collector terminal and the current $I_{leakage}$ is due to the movement of minority carriers across base-collector junction on account of it being reverse biased. It is clear from eq. (*i*) that when $I_E = 0$ (*i.e.* emitter circuit is open), a small leakage current still flows in the collector circuit. This leakage current is abbreviated as I_{CBO}, meaning collector-base current with emitter open. Thus, when $I_E = 0$, $I_C = I_{CBO}$.

Q.6. *Why does a transistor possess low input resistance and high output resistance?*

Ans. It is because the input circuit is forward biased and the output circuit is reverse biased.

Q.7. *Why is β more than α?*

Ans. Considering the d.c. values, the relations for β and α are given by ;

$$\beta = \frac{I_C}{I_B} \text{ and } \alpha = \frac{I_C}{I_E}$$

As I_B is much smaller than I_C, β has a very large value.

On the other hand I_E is comparable to I_C and as such α has a small value. Therefore, β is much greater than α.

Q.8. *What would be the effect on the value of base current if a transistor was operated with emitter and collector interchanged?*

Ans. In a transistor, the emitter is heavily doped to inject a large number of a charge carriers (holes or electrons) into the base and the collector is moderately doped. If the two are interchanged, the emitter current will decrease and so will be the base current.

Q.9. *What will happen to base current if base of a transistor is made thick?*

Ans. If the base is made thick, a large number of combinations of electrons and holes will take place in the base region. As a result, base current will increase.

Q.10. *Why are most of the transistors npn type and not pnp type?*

Ans. In an *npn* transistor, the current conduction is mainly by free electrons whereas in *pnp* transistor, it is mainly by holes. Since electrons are more mobile than holes, we can have high conduction is an *npn* transistor than in a *pnp* type.

Q.11. *What causes a transistor to be damaged?*

Ans. Excessive heat can damage a transistor. The heat may result from excessive current, ambient temperature or soldering the transistor leads.

Q.12. *Why is power switch turned off when a transistor or supply battery is replaced?*

Ans. Transistor can be damaged by even small transient voltage surges which can produce excessive forward bias. For this reason, the power switch should be turned off when a transistor or the supply battery is replaced.

Q.13. *What are the three main classes of transistors?*

Ans. The three main classes of transistors are :

(*i*) Small signal/general purpose transistors

(*ii*) Power transistors

(*iii*) RF (radio frequency) transistors

Q.14. *A properly connected transistor can do amplification. Is transistor a source of energy?*

Ans. No, transistor is not a source of energy but is merely an energy converter. It receives d.c. power from the supply and converts it into a.c. power which is added to the signal.

Q.15. *What are the three possible states of a transistor?*

Ans. The three possible states of a transistor are :

(*i*) *Cut-off* state in which the transistor is nonconducting *i.e.* collector current is approximately zero.

(*ii*) *Active state* in which the collector current is controlled by the base current and base voltage. A transistor acts as an amplifier in the active state.

(*iii*) *Saturation state* in which the collector current has reached a maximum value and is independent of base current.

Q.16. *What is the relation between collector current I_C and emitter current I_E in terms of β?*

Ans.

$$I_E = I_B + I_C = I_B + \beta I_B = I_B\,(1 + \beta)$$

$$\therefore \quad I_B = \frac{1}{1+\beta} I_E$$

$$\therefore \quad I_C = \beta I_B = \frac{\beta}{1+\beta} I_E$$

Q.17. *What is the effect of temperature on the values of (i) β (ii) V_{BE} (iii) I_{CBO}?*

Ans.

(*i*) The value of β increases with the increase in temperature.

(*ii*) The value of V_{BE} decreases with the increases in temperature. The value of V_{BE} decreases about 7.5 mV/°C increase in temperature.

(*iii*) The value of I_{CBO} approximately doubles for every 10°C rise in temperature.

Q.18. *Why is it desirable for an amplifier to have high input impedance and low output impedance?*

Ans. It is desirable that an *amplifier should have high input impedance (R_{in}) and low output impedance (R_{out}). This is logical, since most of signal sources are not capable of delivering very much current and a high input impedance will not draw a high value of current.

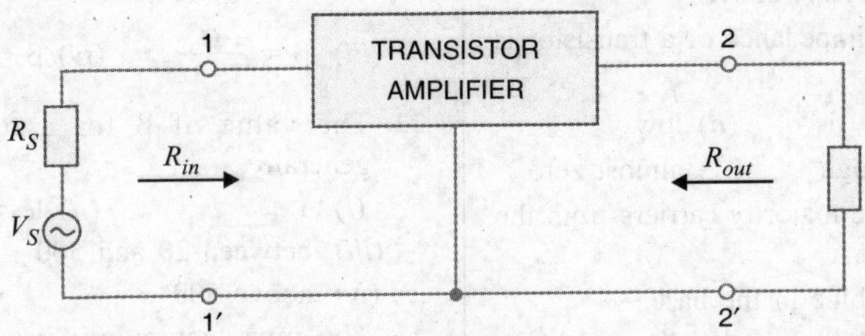

Fig. 6.76

The output impedance is a measure of how much current can be "extracted" from the amplifier to the load (R_L). The lower the value of R_{out}, the greater the value of current supplied by the amplifier to the load. As such, a low value of R_{out} is generally desirable.

Multiple-Choice Questions

1. A transistor has ....
   (*i*) one *pn* junction
   (*ii*) two *pn* junctions
   (*iii*) three *pn* junctions
   (*iv*) four *pn* junctions
2. The number of depletion layers in a transistor is ....
   (*i*) four (*ii*) three
   (*iii*) one (*iv*) two
3. The base of a transistor is .... doped.
   (*i*) heavily (*ii*) moderately
   (*iii*) lightly
   (*iv*) none of the above
4. The element that has the biggest size in a transistor is ....
   (*i*) collector (*ii*) base
   (*iii*) emitter
   (*iv*) collector-base junction
5. In a *pnp* transistor, the current carriers are ....
   (*i*) acceptor ions (*ii*) donor ions
   (*iii*) free electrons (*iv*) holes
6. The collector of a transistor is........ doped.
   (*i*) heavily (*ii*) moderately
   (*iii*) lightly
   (*iv*) none of the above
7. A transistor is a .... operated device.
   (*i*) current
   (*ii*) voltage
   (*iii*) both voltage and current
   (*iv*) none of the above
8. In an *npn* transistor, ... are the minority carriers.
   (*i*) free electrons (*ii*) holes
   (*iii*) donor ions (*iv*) acceptor ions
9. The emitter of a transistor is .... doped.
   (*i*) lightly (*ii*) heavily
   (*iii*) moderately
   (*iv*) none of the above
10. In a transistor, the base current is about .... of emitter current.
    (*i*) 25% (*ii*) 20%
    (*iii*) 35% (*iv*) 5%

* Here V_S is the signal source and R_S its internal resistance.

11. At the base-emitter junction of a transistor, one finds ....
(i) reverse bias
(ii) a wide depletion layer
(iii) low resistance
(iv) none of the above

12. The input impedance of a transistor is ......
(i) high (ii) low
(iii) very high (iv) almost zero

13. Most of the majority carriers from the emitter ....
(i) recombine in the base
(ii) recombine in the emitter
(iii) pass through the base region to the collector
(iv) none of the above

14. The current I_B is ....
(i) electron current
(ii) hole current
(iii) donor ion current
(iv) acceptor ion current

15. In a transistor, ...
(i) $I_C = I_E + I_B$ (ii) $I_B = I_C + I_E$
(iii) $I_E = I_C - I_B$ (iv) $I_E = I_C + I_B$

16. The value of α of a transistor is ....
(i) more than 1 (ii) less than 1
(iii) 1
(iv) none of the above

17. $I_C = \alpha\, I_E +$ ....
(i) I_B (ii) I_{CEO}
(iii) I_{CBO} (iv) $\beta\, I_B$

18. The output impedance of a transistor is ....
(i) high (ii) zero
(iii) low (iv) very low

19. In a transistor, I_C = 100 mA and I_E = 100.5 mA. The value of β is ....
(i) 100 (ii) 50
(iii) about 1 (iv) 200

20. In a transistor if β = 100 and collector current is 10 mA, then I_E is ....
(i) 100 mA (ii) 100.1 mA
(iii) 110 mA
(iv) none of the above

21. The relation between β and α is ....
(i) $\beta = \frac{1}{1-\alpha}$ (ii) $\beta = \frac{1-\alpha}{\alpha}$
(iii) $\beta = \frac{\alpha}{1-\alpha}$ (iv) $\beta = \frac{\alpha}{1+\alpha}$

22. The value of β for a transistor is generally ....
(i) 1 (ii) less than 1
(iii) between 20 and 500
(iv) above 500

23. The most commonly used transistor arrangement is .... arrangement.
(i) common emitter
(ii) common base
(iii) common collector
(iv) none of the above

24. The input impedance of a transistor connected in .... arrangement is the highest.
(i) common emitter
(ii) common collector
(iii) common base
(iv) none of the above

25. The output impedance of a transistor connected in .... arrangement is the highest.
(i) common emitter
(ii) common collector
(iii) common base
(iv) none of the above

26. The phase difference between the input and output voltages in a common base arrangement is ....
(i) 180º (ii) 90º
(iii) 270º (iv) 0º

27. The power gain of a transistor connected in .... arrangement is the highest.
(i) common emitter
(ii) common base
(iii) common collector
(iv) none of the above

28. The phase difference between the input and output voltages of a transistor connected in common emitter arrangement is ....
(*i*) 0° (*ii*) 180°
(*iii*) 90° (*iv*) 270°

29. The voltage gain of a transistor connected in .... arrangement is the highest.
(*i*) common base
(*ii*) common collector
(*iii*) common emitter
(*iv*) none of the above

30. As the temperature of a transistor goes up, the base-emitter resistance ....
(*i*) decreases
(*ii*) increases
(*iii*) remains the same
(*iv*) none of the above

31. The voltage gain of a transistor connected in common collector arrangement is ....
(*i*) equal to 1 (*ii*) more than 10
(*iii*) more than 100 (*iv*) less than 1

32. The phase difference between the input and output voltages of a transistor connected in common collector arrangement is ....
(*i*) 180° (*ii*) 0°
(*iii*) 90° (*iv*) 270°

33. $I_C = \beta I_B + $ ....
(*i*) I_{CBO} (*ii*) I_C
(*iii*) I_{CEO} (*iv*) αI_E

34. $I_C = \frac{\alpha}{1-\alpha} I_B + $ ....
(*i*) I_{CEO} (*ii*) I_{CBO}
(*iii*) I_C (*iv*) $(1-\alpha) I_B$

35. $I_C = \frac{\alpha}{1-\alpha} I_B + \frac{.......}{1-\alpha}$
(*i*) I_{CBO} (*ii*) I_{CEO}
(*iii*) I_C (*iv*) I_E

36. *BC* 147 transistor indicates that it is made of ....
(*i*) germanium (*ii*) silicon
(*iii*) carbon (*iv*) none of the above

37. $I_{CEO} = (........) I_{CBO}$
(*i*) β (*ii*) $1 + \alpha$
(*iii*) $1 + \beta$ (*iv*) none of the above

38. A transistor is connected in *CB* mode. If it is now connected in *CE* mode with same bias voltages, the values of I_E, I_B and I_C will ....
(*i*) remain the same (*ii*) increase
(*iii*) decrease
(*iv*) none of the above

39. If the value of α is 0.9, then value of β is ....
(*i*) 9 (*ii*) 0.9
(*iii*) 900 (*iv*) 90

40. In a transistor, signal is transferred from a .... circuit.
(*i*) high resistance to low resistance
(*ii*) low resistance to high resistance
(*iii*) high resistance to high resistance
(*iv*) low resistance to low resistance

41. The arrow in the symbol of a transistor indicates the direction of ....
(*i*) electron current in the emitter
(*ii*) electron current in the collector
(*iii*) hole current in the emitter
(*iv*) donor ion current

42. The leakage current in *CE* arrangement is .... that in *CB* arrangement.
(*i*) more than (*ii*) less than
(*iii*) the same as (*iv*) none of the above

43. A heat sink is generally used with a transistor to ....
(*i*) increase the forward current
(*ii*) decrease the forward current
(*iii*) compensate for excessive doping
(*iv*) prevent excessive temperature rise

44. The most commonly used semiconductor in the manufacture of a transistor is ....
(*i*) germanium (*ii*) silicon
(*iii*) carbon (*iv*) none of the above

45. The collector-base junction in a transistor has ....
(*i*) forward bias at all times
(*ii*) reverse bias at all times
(*iii*) low resistance
(*iv*) none of the above

Answers to Multiple-Choice Questions

1. (*ii*)	2. (*iv*)	3. (*iii*)	4. (*i*)	5. (*iv*)
6. (*ii*)	7. (*i*)	8. (*ii*)	9. (*ii*)	10. (*iv*)
11. (*iii*)	12. (*ii*)	13. (*iii*)	14. (*i*)	15. (*iv*)
16. (*ii*)	17. (*iii*)	18. (*i*)	19. (*iv*)	20. (*ii*)
21. (*iii*)	22. (*iii*)	23. (*i*)	24. (*ii*)	25. (*iii*)
26. (*iv*)	27. (*i*)	28. (*ii*)	29. (*iii*)	30. (*i*)
31. (*iv*)	32. (*ii*)	33. (*iii*)	34. (*i*)	35. (*i*)
36. (*ii*)	37. (*iii*)	38. (*i*)	39. (*iv*)	40. (*ii*)
41. (*iii*)	42. (*i*)	43. (*iv*)	44. (*ii*)	45. (*ii*)

Chapter Review Topics

1. What is a transistor? Why is it so called?
2. Draw the symbol of *npn* and *pnp* transistor and specify the leads.
3. Show by means of a diagram how you normally connect external batteries in (*i*) *pnp* transistor (*ii*) *npn* transistor.
4. Describe the transistor action in detail.
5. Explain the operation of transistor as an amplifier.
6. Name the three possible transistor connections.
7. Define α. Show that it is always less than unity.
8. Draw the input and output characteristics of *CB* connection. What do you infer from these characteristics?
9. Define β. Show that : $\beta = \dfrac{\alpha}{1-\alpha}$.
10. How will you determine the input and output characteristics of *CE* connection experimentally?
11. Establish the following relations :

    (*i*) $I_C = \alpha\, I_E + I_{CBO}$ (*ii*) $I_C = \dfrac{\alpha}{1-\alpha} I_B + \dfrac{1}{1-\alpha} I_{CBO}$

    (*iii*) $I_C = \beta\, I_B + I_{CEO}$ (*iv*) $\gamma = \dfrac{1}{1-\alpha}$

    (*v*) $I_E = (\beta + 1)\, I_B + (\beta + 1)\, I_{CBO}$
12. How will you draw d.c. load line on the output characteristics of a transistor? What is its importance?
13. Explain the following terms : (*i*) voltage gain (*ii*) power gain (*iii*) effective collector load.
14. Write short notes on the following : (*i*) advantages of transistors (*ii*) operating point (*iii*) d.c. load line.

Problems

1. In a transistor if I_C = 4.9 mA and I_E = 5 mA, what is the value of α? **[0.98]**
2. In a transistor circuit, I_E = 1 mA and I_C = 0.9 mA. What is the value of I_B? **[0.1 mA]**
3. Find the value of β if α = 0.99. **[100]**

4. In a transistor, $\beta = 45$, the voltage across 5 kΩ resistance which is connected in the collector circuit is 5 volts. Find the base current. **[0.022 mA]**

5. In a transistor, $I_B = 68\ \mu A$, $I_E = 30$ mA and $\beta = 440$. Find the value of α. Hence determine the value of I_C. **[0.99; 29.92 mA]**

6. The maximum collector current that a transistor can carry is 500 mA. If $\beta = 300$, what is the maximum allowable base current for the device? **[1.67 mA]**

7. In a certain transistor, 99.5% of the charge carriers injected into the base cross the collector-base junction. If the leakage current is 5 μA and the collector current is 22 mA, find (*i*) the exact α (*ii*) the emitter current (*iii*) the approximate α when I_{CBO} is neglected.

[(*i*) $\alpha = 0.995$ (*ii*) $I_E = 22.1055$ mA (*iii*) $\alpha = 0.9952$]

8. A transistor has $I_{CBO} = 0.1\ \mu A$ and $I_{CEO} = 16\ \mu A$. Find its α. **[0.99375]**

9. Derive the relations : (*i*) $\alpha = \beta/(\beta + 1)$ (*ii*) $I_E = I_B/(1 - \alpha)$.

10. For the transistor circuit in Fig. 6.77, find (*i*) I_C when $V_{CB} = 10$ V (*ii*) V_{CB} when $I_C = 1$ mA.

[(*i*) $I_C = 1.515$ mA (*ii*) $V_{CB} = 11.7$ V]

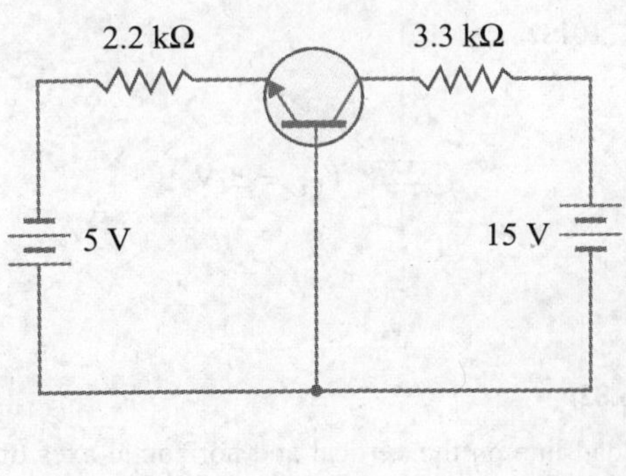

Fig. 6.77

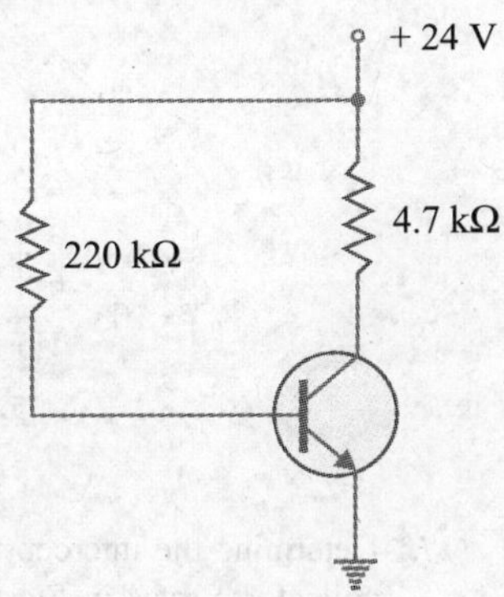

Fig. 6.78

11. In the transistor circuit shown in Fig. 6.78, find (*i*) V_{CE} when $I_C = 1.5$ mA (*ii*) I_C when $V_{CE} =$ 12 V (*iii*) V_{CE} when $I_C = 0$. **[(*i*) $V_{CE} = 16.95$ V (*ii*) $I_C = 2.55$ mA (*iii*) $V_{CE} = 24$ V]**

12. For the circuit shown in Fig. 6.79, draw the d.c. load line.

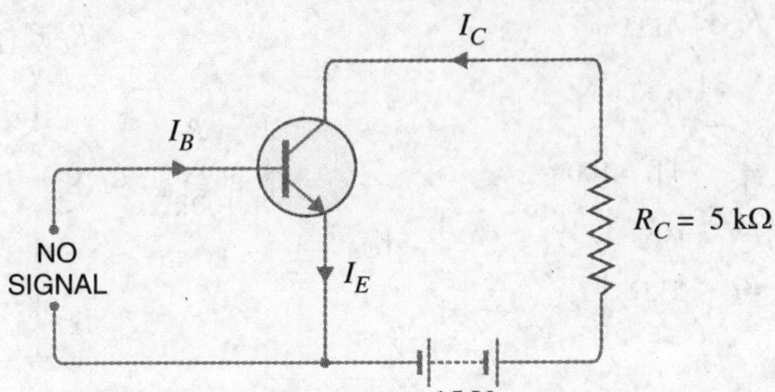

Fig. 6.79

13. Draw the d.c. load line for Fig. 6.80. **[The end points of load line are 6.06 mA and 20 V]**

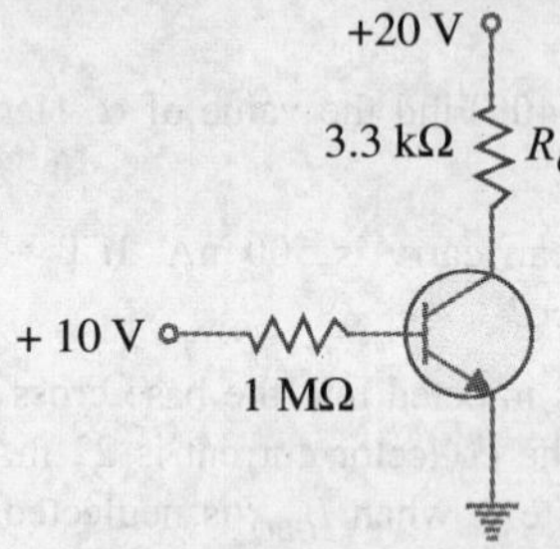

Fig. 6.80

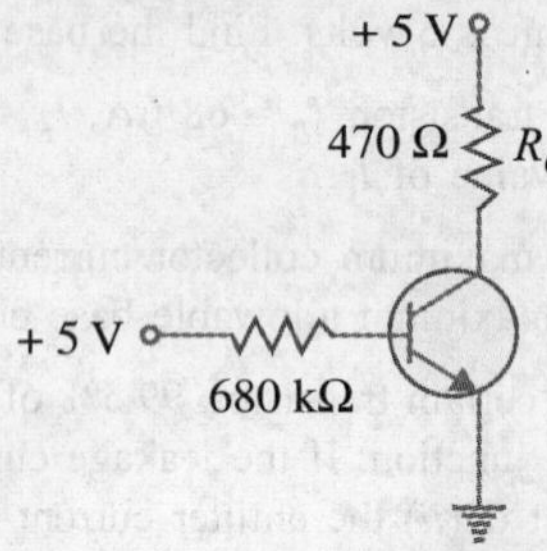

Fig. 6.81

14. If the collector resistance R_C in Fig. 6.80 is reduced to 1 kΩ, what happens to the d.c. load line? **[The end points of d.c. load line are now 20 mA and 20 V]**
15. Draw the d.c. load line for Fig. 6.81. **[The end points of d.c. load line are 10.6 mA and 5 V]**
16. If the collector resistance R_C in Fig. 6.81 is increased to 1 kΩ, what happens to the d.c. load line? **[The end points of d.c. load line are now 5 mA and 5 V]**

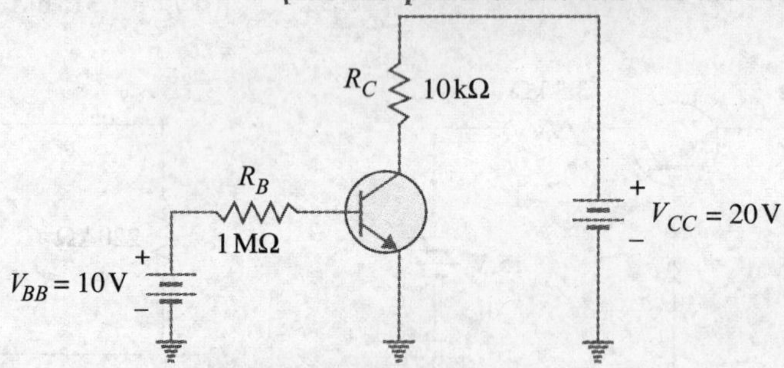

Fig. 6.82

17. Determine the intercept points of the d.c. load line on the vertical and horizontal axes of the collector curves in Fig. 6.82. **[2 mA; 20 V]**
18. For the circuit shown in Fig. 6.83, find (*i*) the state of the transistor and (*ii*) transistor power. **[(*i*) Active (*ii*) 4.52 mW]**

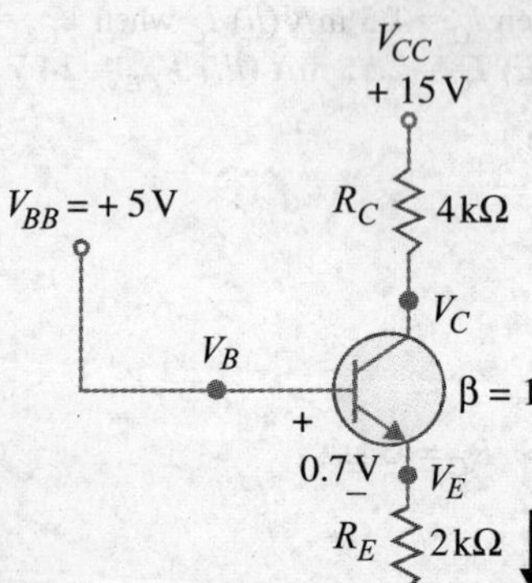

Fig. 6.83

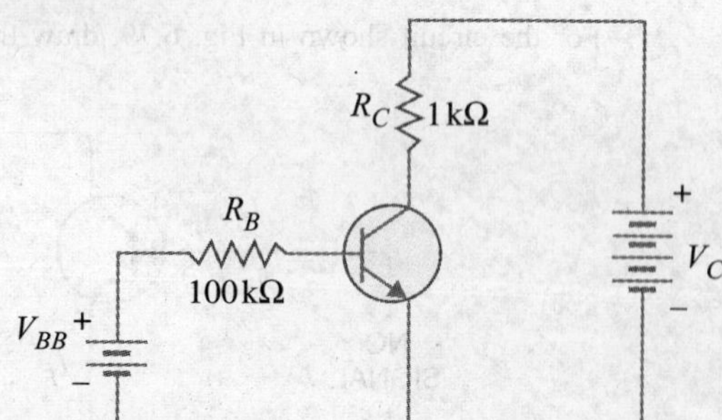

Fig. 6.84

19. A base current of 50 μA is applied to the transistor in Fig. 6.84 and a voltage of 5 V is dropped across R_C. Calculate α for the transistor. **[0.99]**
20. A certain transistor is to be operated at a collector current of 50 mA. How high can V_{CE} go without exceeding $P_{D\,(max)}$ of 1.2 W? **[24 V]**

CHAPTER 7

Transistor Biasing

Learning Objectives

- Develop an understanding of faithful amplification and the conditions to be satisfied by transistor amplifier for this purpose.
- Learn transistor biasing and its need.
- Recognise the need of stabilisation of operating point.
- Understand various methods of transistor biasing.
- Develop a strong understanding of potential-divider bias method.
- Derive an expression of stability factor for potential-divider bias.
- Explore design steps for biasing circuits for low powered transistors.
- Recognise the importance of mid-point biasing of transistor amplifier.
- Learn the comparison of silicon and germanium.

Introduction

The basic function of transistor is amplification. The weak signal is given to the base of the transistor and amplified output is obtained in the collector circuit. One important requirement during amplification is that only the magnitude of the signal should increase and there should be no change in signal shape. This increase in magnitude of the signal without any change in shape is known as *faithful amplification*. In order to achieve this, means are provided to ensure that input circuit (*i.e.* base-emitter junction) of the transistor remains forward biased and output circuit (*i.e.* collector-base junction) always remains reverse biased during all parts of the signal. This is known as transistor biasing. In this chapter, we shall discuss how transistor biasing helps in achieving faithful amplification.

7.1 Faithful Amplification

The process of raising the strength of a weak signal without any change in its general shape is known as **faithful amplification.**

The theory of transistor reveals that it will function properly if its input circuit (*i.e.* base-emitter junction) remains forward biased and output circuit (*i.e.* collector-base junction) remains reverse biased at all times. This is then the key factor for achieving faithful amplification. To ensure this, the following basic conditions must be satisfied :

(*i*) Proper zero signal collector current

(*ii*) Minimum proper base-emitter voltage (V_{BE}) at any instant

(*iii*) Minimum proper collector-emitter voltage (V_{CE}) at any instant

The conditions (*i*) and (*ii*) ensure that base-emitter junction shall remain properly forward biased during all parts of the signal. On the other hand, condition (*iii*) ensures that base-collector junction shall remain properly reverse biased at all times. In other words, the fulfilment of these conditions will ensure that transistor works over the active region of the output characteristics *i.e.* between saturation to cut-off.

(*i*) Proper zero signal collector current. Consider an *npn* transistor circuit shown in Fig. 7.1 (*i*). During the positive half-cycle of the signal, base is positive *w.r.t.* emitter and hence base-emitter junction is forward biased. This will cause a base current and much larger collector current to flow in the circuit. The result is that positive half-cycle of the signal is amplified in the collector as shown. However, during the negative half-cycle of the signal, base-emitter junction is reverse biased and hence no current flows in the circuit. The result is that there is no output due to the negative half-cycle of the signal. Thus we shall get an amplified output of the signal with its negative half-cycles completely cut-off which is unfaithful amplification.

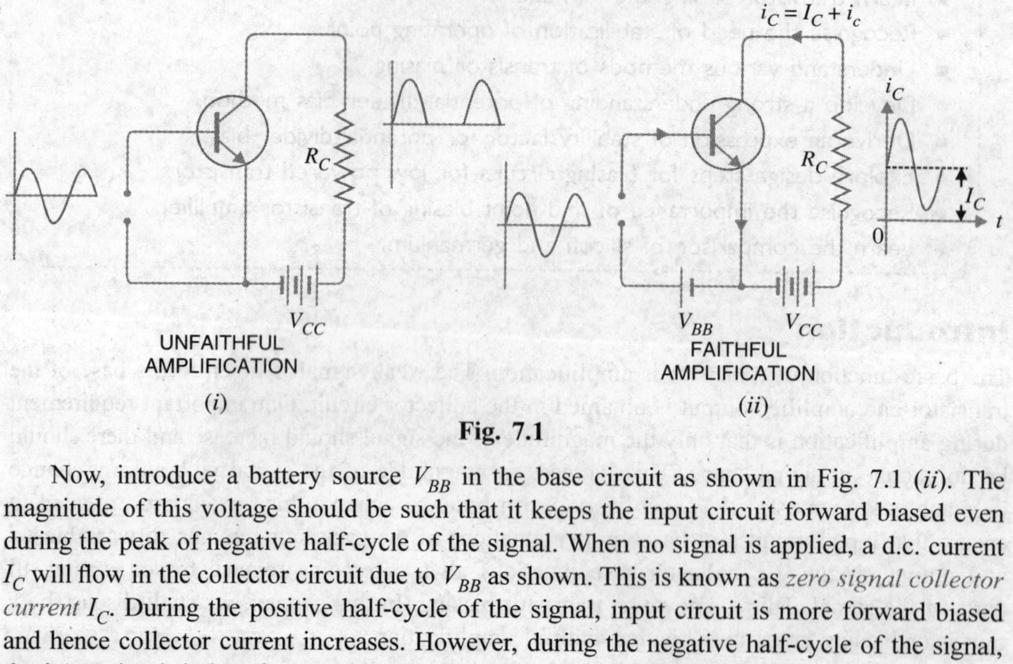

Fig. 7.1

Now, introduce a battery source V_{BB} in the base circuit as shown in Fig. 7.1 (*ii*). The magnitude of this voltage should be such that it keeps the input circuit forward biased even during the peak of negative half-cycle of the signal. When no signal is applied, a d.c. current I_C will flow in the collector circuit due to V_{BB} as shown. This is known as *zero signal collector current* I_C. During the positive half-cycle of the signal, input circuit is more forward biased and hence collector current increases. However, during the negative half-cycle of the signal, the input circuit is less forward biased and collector current decreases. In this way, negative half-cycle of the signal also appears in the output and hence faithful amplification results. It follows, therefore, that for faithful amplification, proper zero signal collector current must flow. *The value of zero signal collector current should be atleast equal to the maximum collector current due to signal alone i.e.*

Zero signal collector current ≥ Max. collector current due to signal alone

Illustration. Suppose a signal applied to the base of a transistor gives a peak collector current of 1mA. Then zero signal collector current must be atleast equal to 1mA so that even during the peak of negative half-cycle of the signal, there is no cut-off as shown in Fig. 7.2 (*i*).

If zero signal collector current is less, say 0.5 mA as shown in Fig. 7.2 (*ii*), then some part (shaded portion) of the negative half-cycle of signal will be cut-off in the output.

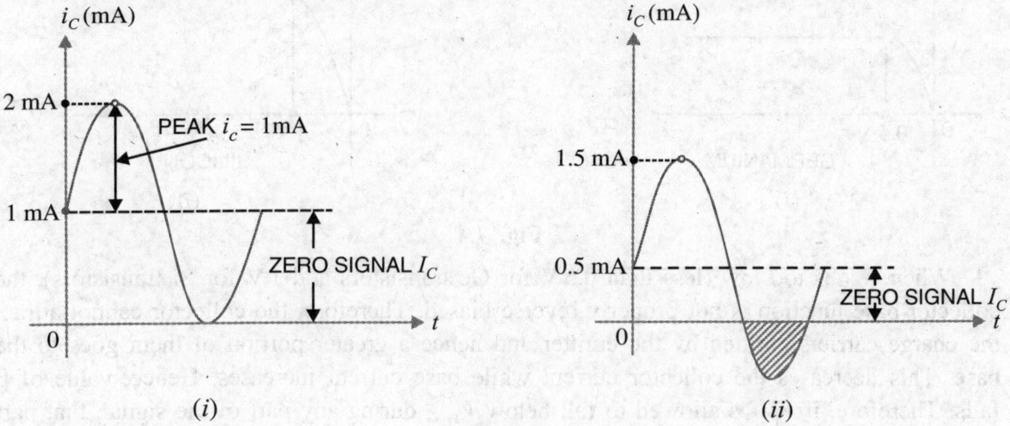

Fig. 7.2

(*ii*) Proper minimum base-emitter voltage. In order to achieve faithful amplification, the base-emitter voltage (V_{BE}) should not fall below 0.5 V for germanium transistors and 0.7 V for Si transistors at any instant.

Fig. 7.3

The base current is very small until the *input voltage overcomes the potential barrier at the base-emitter junction. The value of this potential barrier is 0.5 V for Ge transistors and 0.7 V for Si transistors as shown in Fig. 7.3. Once the potential barrier is overcome, the base current and hence collector current increases sharply. Therefore, if base-emitter voltage V_{BE} falls below these values during any part of the signal, that part will be amplified to lesser extent due to small collector current. This will result in unfaithful amplification.

(*iii*) Proper minimum V_{CE} at any instant. For faithful amplification, the collector-emitter voltage V_{CE} should not fall below 0.5 V for Ge transistors and 1 V for silicon transistors. This is called *knee voltage* (See Fig. 7.4).

* In practice, a.c. signals have small voltage level (< 0.1 V) and if applied directly will not give any collector current.

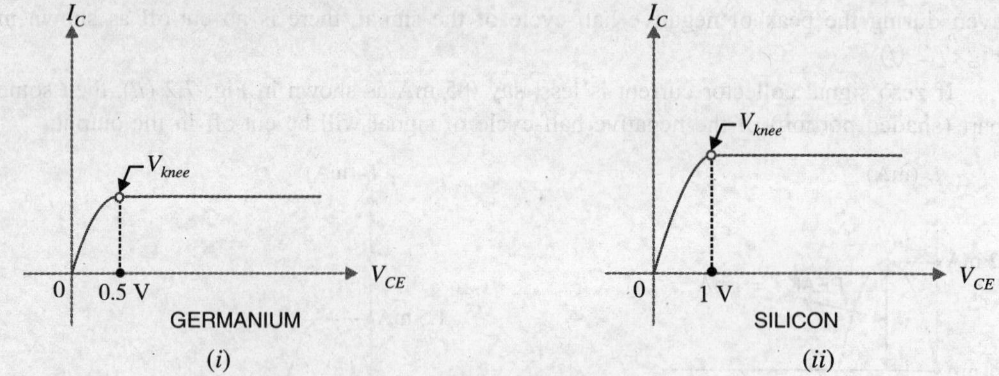

Fig. 7.4

When V_{CE} is too low (less than 0.5 V for Ge transistors and 1 V for Si transistors), the collector-base junction is not properly reverse biased. Therefore, the collector cannot attract the charge carriers emitted by the emitter and hence a greater portion of them goes to the base. This decreases the collector current while base current increases. Hence, value of β falls. Therefore, if V_{CE} is allowed to fall below V_{knee} during any part of the signal, that part will be less amplified due to reduced β. This will result in unfaithful amplification. However, when V_{CE} is greater than V_{knee}, the collector-base junction is properly reverse biased and the value of β remains constant, resulting in faithful amplification.

7.2 Transistor Biasing

It has already been discussed that for faithful amplification, a transistor amplifier must satisfy three basic conditions, namely : (*i*) proper zero signal collector current, (*ii*) proper base-emitter voltage at any instant and (*iii*) proper collector-emitter voltage at any instant. It is the fulfilment of these conditions which is known as transistor biasing.

The proper flow of zero signal collector current and the maintenance of proper collector-emitter voltage during the passage of signal is known as **transistor biasing.**

The basic purpose of transistor biasing is to keep the base-emitter junction properly forward biased and collector-base junction properly reverse biased during the application of signal. This can be achieved with a bias battery or associating a circuit with a transistor. The latter method is more efficient and is frequently employed. The circuit which provides transistor biasing is known as *biasing circuit*. It may be noted that transistor biasing is very essential for the proper operation of transistor in any circuit.

Example 7.1. *An npn silicon transistor has V_{CC} = 6 V and the collector load R_C = 2.5 kΩ. Find:*

(i) The maximum collector current that can be allowed during the application of signal for faithful amplification.

(ii) The minimum zero signal collector current required.

Solution. Collector supply voltage, V_{CC} = 6 V

Collector load, R_C = 2.5 kΩ

(*i*) We know that for faithful amplification, V_{CE} should not be less than 1 V for silicon transistor.

$\therefore$ Max. voltage allowed across $R_C = 6 - 1 = 5$ V

$\therefore$ Max. allowed collector current $= 5\text{ V}/R_C = 5\text{ V}/2.5\text{ k}\Omega = 2$ mA

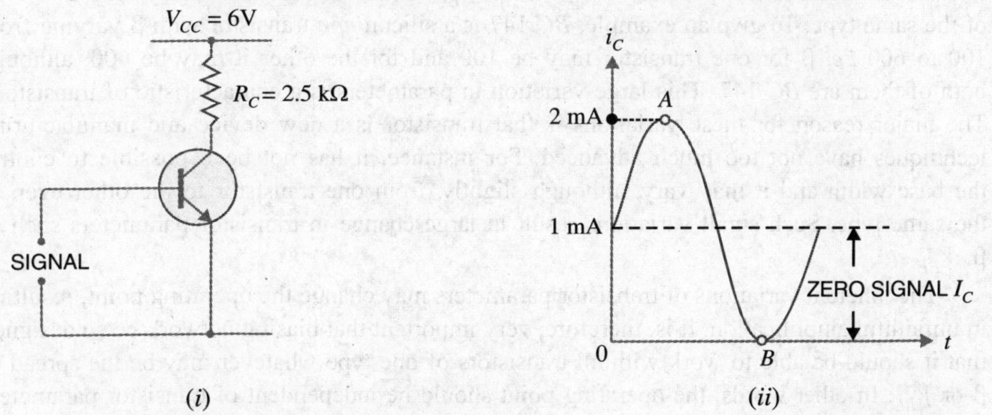

Fig. 7.5

Thus, the maximum collector current allowed during any part of the signal is 2 mA. If the collector current is allowed to rise above this value, V_{CE} will fall below 1 V. Consequently, value of β will fall, resulting in unfaithful amplification.

(*ii*) During the negative peak of the signal, collector current can at the most be allowed to become zero. As the negative and positive half cycles of the signal are equal, therefore, the change in collector current due to these will also be equal but in opposite direction.

$\therefore$ Minimum zero signal collector current required = 2 mA/2 = 1 mA

During the positive peak of the signal [point *A* in Fig. 7.5 (*ii*)], $i_C = 1 + 1 = 2$ mA and during the negative peak (point *B*),

$$i_C = 1 - 1 = 0 \text{ mA}$$

Example 7.2. *A transistor employs a 4 kΩ load and $V_{CC} = 13$ V. What is the maximum input signal if β = 100? Given $V_{knee} = 1$ V and a change of 1 V in V_{BE} causes a change of 5 mA in collector current.*

Solution.

Collector supply voltage, $V_{CC} = 13$ V

Knee voltage, $V_{knee} = 1$ V

Collector load, $R_C = 4$ kΩ

$\therefore$ Max. allowed voltage across $R_C = 13 - 1 = 12$ V

$$\therefore \text{ Max. allowed collector current, } i_C = \frac{12\text{ V}}{R_C} = \frac{12\text{ V}}{4\text{ k}\Omega} = 3\text{ mA}$$

$$\text{Maximum base current, } i_B = \frac{i_C}{\beta} = \frac{3\text{ mA}}{100} = 30\ \mu\text{A}$$

$$\text{Now, } \frac{\text{Collector current}}{\text{Base voltage (Signal voltage)}} = 5\text{ mA/V}$$

$$\therefore \text{ Base voltage (Signal voltage)} = \frac{\text{Collector current}}{5\text{ mA/V}} = \frac{3\text{ mA}}{5\text{ mA/V}} = \mathbf{600\ mV}$$

7.3 Inherent Variations of Transistor Parameters

In practice, the transistor parameters such as β, V_{BE} are not the same for every transistor even of the same type. To give an example, *BC* 147 is a silicon *npn* transistor with β varying from 100 to 600 *i.e.* β for one transistor may be 100 and for the other it may be 600, although both of them are *BC* 147. This large variation in parameters is a characteristic of transistors. The major reason for these variations is that transistor is a new device and manufacturing techniques have not too much advanced. For instance, it has not been possible to control the base width and it may vary, although slightly, from one transistor to the other even of the same type. Such small variations result in large change in transistor parameters such as β, V_{BE} *etc.*

The inherent variations of transistor parameters may change the operating point, resulting in unfaithful amplification. It is, therefore, very important that biasing network be so designed that it should be able to work with all transistors of one type whatever may be the spread in β or V_{BE}. In other words, the operating point should be independent of transistor parameters variations.

7.4 Stabilisation

The collector current in a transistor changes rapidly when

(*i*) the temperature changes,

(*ii*) the transistor is replaced by another of the same type. This is due to the inherent variations of transistor parameters.

When the temperature changes or the transistor is replaced, the operating point (*i.e.* zero signal I_C and V_{CE}) also changes. However, for faithful amplification, it is essential that operating point remains fixed. This necessitates to make the operating point independent of these variations. This is known as stabilisation.

The process of making operating point independent of temperature changes or variations in transistor parameters is known as **stabilisation.**

Once stabilisation is done, the zero signal I_C and V_{CE} become independent of temperature variations or replacement of transistor *i.e.* the operating point is fixed. A good biasing circuit always ensures the stabilisation of operating point.

Need for stabilisation. Stabilisation of the operating point is necessary due to the following reasons :

(*i*) Temperature dependence of I_C (*ii*) Individual variations (*iii*) Thermal runaway

(*i*) **Temperature dependence of I_C.** The collector current I_C for *CE* circuit is given by ;

$$I_C = \beta\, I_B + I_{CEO} \quad = \quad \beta\, I_B + (\beta + 1)\, I_{CBO}$$

The collector leakage current I_{CBO} is greatly influenced (especially in germanium transistors) by temperature changes. *A rise of 10°C doubles the collector leakage current which may be as high as 0.2 mA for low powered germanium transistors.* As biasing conditions in such transistors are generally so set that zero signal I_C = 1 mA, therefore, the change in I_C due to temperature variations cannot be tolerated. This necessitates to stabilise the operating point *i.e.* to hold I_C constant inspite of temperature variations.

(*ii*) **Individual variations.** The value of β and V_{BE} are not exactly the same for any two transistors even of the same type. Further, V_{BE} itself decreases when temperature increases.

When a transistor is replaced by another of the same type, these variations change the operating point. This necessitates to stabilise the operating point *i.e.* to hold I_C constant irrespective of individual variations in transistor parameters.

(*iii*) Thermal runaway. The collector current for a *CE* configuration is given by ;

$$I_C = \beta I_B + (\beta + 1) I_{CBO} \quad ...(i)$$

The collector leakage current I_{CBO} is strongly dependent on temperature. The flow of collector current produces heat within the transistor. This raises the transistor temperature and if no stabilisation is done, the collector leakage current I_{CBO} also increases. It is clear from exp. (*i*) that if I_{CBO} increases, the collector current I_C increases by $(\beta + 1) I_{CBO}$. The increased I_C will raise the temperature of the transistor, which in turn will cause I_{CBO} to increase. This effect is cumulative and in a matter of seconds, the collector current may become very large, causing the transistor to burn out.

The self-destruction of an unstabilised transistor is known as **thermal runaway.**

In order to avoid thermal runaway and consequent destruction of transistor, it is very essential that operating point is stabilised *i.e.* I_C is kept constant. In practice, this is done by causing I_B to decrease automatically with temperature increase by circuit modification. Then decrease in βI_B will compensate for the increase in $(\beta + 1) I_{CBO}$, keeping I_C nearly constant. In fact, this is what is always aimed at while building and designing a biasing circuit.

7.5 Essentials of a Transistor Biasing Circuit

It has already been discussed that transistor biasing is required for faithful amplification. The biasing network associated with the transistor should meet the following requirements :

(*i*) It should ensure proper zero signal collector current.

(*ii*) It should ensure that V_{CE} does not fall below 0.5 V for Ge transistors and 1 V for silicon transistors at any instant.

(*iii*) It should ensure the stabilisation of operating point.

7.6 Stability Factor

It is desirable and necessary to keep I_C constant in the face of variations of I_{CBO} (sometimes represented as I_{CO}). The extent to which a biasing circuit is successful in achieving this goal is measured by stability factor *S*. It is defined as under :

The ratio of change in collector current (ΔI_C) to the change in collector-base leakage current (ΔI_{CBO}) at constant β and I_B is called **stability factor** *i.e.*

$$\text{Stability factor, } S = \frac{\Delta I_C}{\Delta I_{CBO}} \text{ at constant } I_B \text{ and } \beta$$

$$\text{In differential form, } S = \frac{dI_C}{dI_{CBO}} \text{ at constant } I_B \text{ and } \beta$$

The stability factor of a biasing circuit indicates the change in collector current I_C due to change in collector-base leakage current I_{CBO}. Suppose the stability factor (*S*) of a biasing circuit is 50. Then,

$$S = \frac{\Delta I_C}{\Delta I_{CBO}}$$

or Change in collector current, $\Delta I_C = S \times \Delta I_{CBO} = 50 \times \Delta I_{CBO}$

Thus, if I_{CBO} changes by 1 μA (*i.e.*, ΔI_{CBO} = 1 μA), then collector current will change by 50 μA (ΔI_C = 50 × 1 μA = 50 μA). Similarly, if I_{CBO} changes by 4 μA (*i.e.*, ΔI_{CBO} = 4 μA), then collector current will change by 200 μA (ΔI_C = 50 × 4 μA = 200 μA). The following points may be noted :

(*i*) The value of stability factor (*S*) depends on the type of biasing circuit and on the resistor values.

(*ii*) The smaller the value of *S* of a biasing circuit, the *greater is the thermal stability of the circuit. Therefore, it is desirable to have as low value of *S* as possible.

(*iii*) The minimum value of *S* = 1. This is the ideal value of *S* but it is never possible to achieve it in practice. Experience shows that values of *S* exceeding 25 result in unsatisfactory performance of the biasing circuit.

General expression for *S*. The general expression of stability factor for a *CE* configuration can be obtained as under :

$$I_C = \beta I_B + (\beta + 1)\, I_{CBO}$$

**Differentiating above expression *w.r.t.* I_C, we get,

$$1 = \beta \frac{dI_B}{dI_C} + (\beta + 1) \frac{dI_{CBO}}{dI_C}$$

or
$$1 = \beta \frac{dI_B}{dI_C} + \frac{(\beta + 1)}{S} \qquad \left[\because \frac{dI_{CBO}}{dI_C} = \frac{1}{S}\right]$$

or
$$S = \frac{\beta + 1}{1 - \beta\left(\dfrac{dI_B}{dI_C}\right)}$$

We shall use this general expression to determine the stability factor (*S*) of various biasing circuits.

Note. We know that collector current I_C in common base (*CB*) configuration is given by ;

$$I_C = \alpha I_E + I_{CBO}$$

Differentiating the above expression *w.r.t.* I_{CBO}, we have,

$$\frac{dI_C}{dI_{CBO}} = 0 + 1 = 1$$

i.e., Stability factor, $S = 1$

Note that stability factor for *CB* configuration has the ideal value of 1. This means that common base circuits have very high thermal stability. However, we use *CE* configuration in practical circuits due to its many advantages over *CB* configuration.

---

* Smaller *S* means less change in I_C due to change in I_{CBO}. In other words, operating point will *almost* remain fixed.

** Assuming β to be independent of I_C. Note that when I_{CBO} changes due to temperature change, both I_B and I_C change.

7.7 Effect of Temperature on I_{CBO}

The collector-base leakage current I_{CBO} is greatly changed by the change in temperature. If temperature increases, I_{CBO} increases appreciably and *vice-versa.* The change in I_{CBO} over a given temperature range can be calculated by recalling that I_{CBO} *doubles for every 10°C increase in temperature.* The temperature change (ΔT) is divided by 10 to give the number of 10°C changes (n). If the starting level of collector-base leakage current is represented as I_{CBO} (*old*), then the new level of leakage current is

$$I_{CBO}(new) = I_{CBO}(old) \times 2^n$$

If the temperature changes by 30°C (*i.e.*, ΔT = 30°C), then the value of $n = \Delta T/10°C = 30°C/10°C = 3$. Similarly, if temperature changes by 80°C, then $n = \Delta T/10°C = 80°C/10°C = 8$. If we know the stability factor (S) of the biasing circuit, then change in collector current is $\Delta I_C = S \times {}^*\Delta I_{CBO}$.

Example 7.3. *A transistor is biased by a biasing circuit whose stability factor is 56. Find the change in collector current when the circuit temperature increases from 25°C to 105°C. Given that I_{CBO} = 15 nA at 25°C.*

Solution. Temperature change, $\Delta T = 105°C - 25°C = 80°C$

ΔT in 10°C steps, $n = \dfrac{\Delta T}{10°C} = \dfrac{80°C}{10°C} = 8$

$\therefore \quad I_{CBO}(new) = I_{CBO}(old) \times 2^n = {}^{**}15 \text{ nA} \times 2^8 = 3.84\ \mu\text{A}$

Now, $\quad \Delta I_{CBO} = 3.84\ \mu\text{A} - 15 \text{ nA} = 3.83\ \mu\text{A}$

$\therefore \quad$ Change in I_C, $\Delta I_C = S \times \Delta I_{CBO} = 56 \times 3.83 = \mathbf{214\ \mu A}$

7.8 Methods of Transistor Biasing

In the transistor amplifier circuits drawn so far biasing was done with the aid of a battery V_{BB} which was separate from the battery V_{CC} used in the output circuit. However, in the interest of simplicity and economy, it is desirable that transistor circuit should have a single source of supply—the one in the output circuit (*i.e.* V_{CC}). The following are the most commonly used methods of obtaining transistor biasing from one source of supply (*i.e.* V_{CC}) :

(*i*) Base resistor method

(*ii*) Emitter bias method

(*iii*) Biasing with collector-feedback resistor

(*iv*) Voltage-divider bias

In all these methods, the same basic principle is employed *i.e.* required value of base current (and hence I_C) is obtained from V_{CC} in the zero signal conditions. The value of collector load R_C is selected keeping in view that V_{CE} should not fall below 0.5 V for germanium transistors and 1 V for silicon transistors.

For example, if β = 100 and the zero signal collector current I_C is to be set at 1 mA, then I_B is made equal to I_C/β = 1/100 = 10 μA. Thus, the biasing network should be so designed that a base current of 10 μA flows in the zero signal conditions.

* $\Delta I_{CBO} = I_{CBO}(new) - I_{CBO}(old)$

** $1\text{nA} = 10^{-9}\text{A} = 10^{-3}\mu\text{A}$

7.9 Base Resistor Method

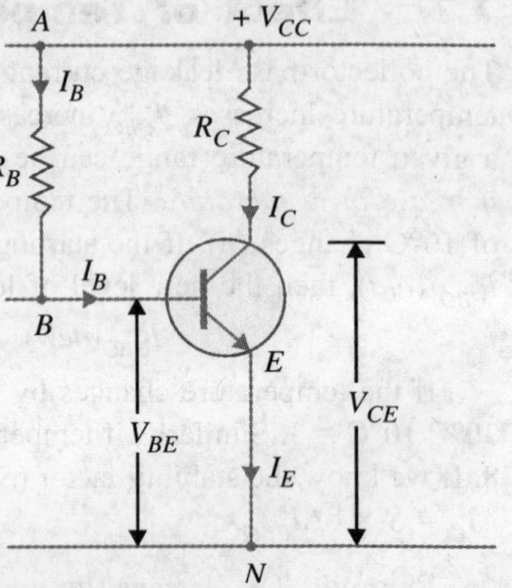

Fig. 7.6

In this method, a high resistance R_B (several hundred kΩ) is connected between the base and +ve end of supply for *npn* transistor (See Fig. 7.6) and between base and negative end of supply for *pnp* transistor. Here, the required zero signal base current is provided by V_{CC} and it flows through R_B. It is because now base is positive *w.r.t.* emitter *i.e.* base-emitter junction is forward biased. The required value of zero signal base current I_B (and hence $I_C = \beta I_B$) can be made to flow by selecting the proper value of base resistor R_B.

Circuit analysis. It is required to find the value of R_B so that required collector current flows in the zero signal conditions. Let I_C be the required zero signal collector current.

$$\therefore \quad I_B = \frac{I_C}{\beta}$$

Considering the closed circuit *ABENA* and applying Kirchhoff's voltage law, we get,

$$V_{CC} = I_B R_B + V_{BE}$$

$$\text{or} \quad I_B R_B = V_{CC} - V_{BE}$$

$$\therefore \quad R_B = \frac{V_{CC} - V_{BE}}{I_B} \quad \text{... } (i)$$

As V_{CC} and I_B are known and V_{BE} can be seen from the transistor manual, therefore, value of R_B can be readily found from exp. (*i*).

Since V_{BE} is generally quite small as compared to V_{CC}, the former can be neglected with little error. It then follows from exp. (*i*) that :

$$R_B = \frac{V_{CC}}{I_B}$$

It may be noted that V_{CC} is a fixed known quantity and I_B is chosen at some suitable value. Hence, R_B can always be found directly, and for this reason, this method is sometimes called *fixed-bias method*.

Stability factor. As shown in Art. 7.6,

$$\text{Stability factor, } S = \frac{\beta + 1}{1 - \beta\left(\frac{dI_B}{dI_C}\right)}$$

In fixed-bias method of biasing, I_B is independent of I_C so that $dI_B/dI_C = 0$. Putting the value of $dI_B/dI_C = 0$ in the above expression, we have,

$$\text{Stability factor, } S = \beta + 1$$

Thus the stability factor in a fixed bias is (β + 1). This means that I_C changes (β + 1) times as much as any change in I_{CO}. For instance, if β = 100, then $S = 101$ which means that I_C increases 101 times faster than I_{CO}. Due to the large value of S in a fixed bias, it has poor thermal stability.

Advantages

(*i*) This biasing circuit is very simple as only one resistance R_B is required.

(*ii*) Biasing conditions can easily be set and the calculations are simple.

(*iii*) There is no loading of the source by the biasing circuit since no resistor is employed across base-emitter junction.

Disadvantages

(*i*) This method provides poor stabilisation. It is because there is no means to stop a self-increase in collector current due to temperature rise and individual variations. For example, if β increases due to transistor replacement, then I_C also increases by the same factor as I_B is constant.

(*ii*) The stability factor is very high. Therefore, there are strong chances of thermal runaway.

Due to these disadvantages, this method of biasing is rarely employed.

Note. We have used *npn* transistor to analyse the base resistor method of transistor biasing. The treatment is similar if we use *pnp* transistor. The only difference is that now voltage polarities and current directions are reversed compared to the *npn* transistor (See Example 7.5).

Example 7.4. *Fig. 7.7 (i) shows biasing with base resistor method. (i) Determine the collector current I_C and collector-emitter voltage V_{CE}. Neglect small base-emitter voltage. Given that β = 50.*

(ii) If R_B in this circuit is changed to 50 kΩ, find the new operating point.

(iii) What is the value of stability factor S in each of the two cases?

Solution.

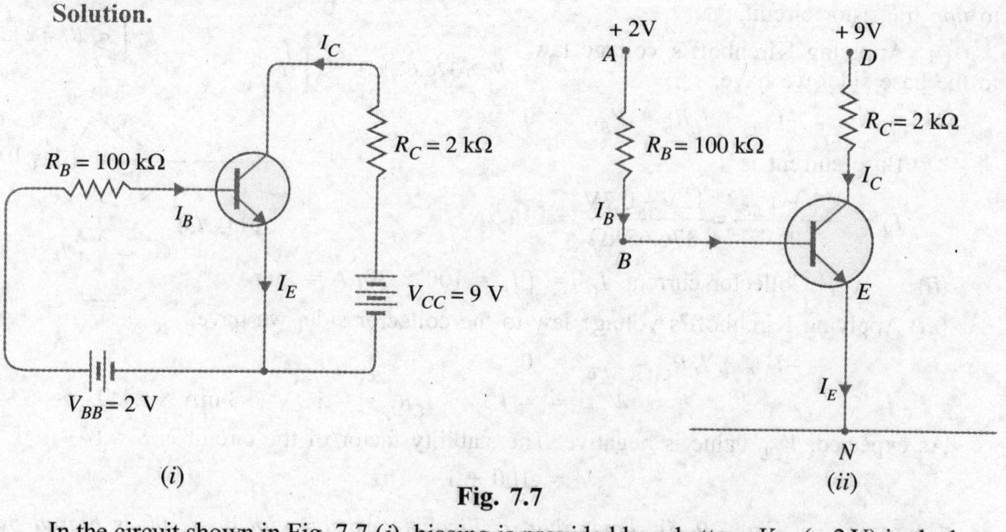

Fig. 7.7

In the circuit shown in Fig. 7.7 (*i*), biasing is provided by a battery V_{BB} (= 2 V) in the base circuit which is separate from the battery V_{CC} (= 9 V) used in the output circuit. The same circuit is shown in a simplified way in Fig. 7.7 (*ii*). Here, we need show only the supply voltages, + 2 V and + 9 V. It may be noted that negative terminals of the power supplies are grounded to get a complete path of current.

(*i*) Referring to Fig.7.7 (*ii*) and applying Kirchhoff's voltage law to the circuit *ABEN*, we get,

$$I_B R_B + V_{BE} = 2\text{ V}$$

As V_{BE} is negligible, $\therefore$ $I_B = \dfrac{2\text{V}}{R_B} = \dfrac{2\text{V}}{100\text{ k}\Omega} = 20\ \mu\text{A}$

Collector current, $I_C = \beta\, I_B = 50 \times 20\ \mu\text{A} = 1000\ \mu\text{A} = \mathbf{1\ mA}$

Applying Kirchhoff's voltage law to the circuit *DEN*, we get,

$$I_C R_C + V_{CE} = 9$$

or $\quad 1\text{ mA} \times 2\text{ k}\Omega + V_{CE} = 9$

or $\quad V_{CE} = 9 - 2 = 7\text{ V}$

(ii) When R_B is made equal to 50 kΩ, then it is easy to see that base current is doubled *i.e.* $I_B = 40\ \mu\text{A}$.

$\therefore$ Collector current, $I_C = \beta\, I_B = 50 \times 40 = 2000\ \mu\text{A} = 2\text{ mA}$

Collector-emitter voltage, $V_{CE} = V_{CC} - I_C R_C = 9 - 2\text{ mA} \times 2\text{ k}\Omega = 5\text{ V}$

$\therefore$ New operating point is **5 V, 2 mA.**

(iii) The value of stability factor *S* in each of the two cases is

$$S = \beta + 1 = 50 + 1 = \mathbf{51}$$

Example 7.5. *For the pnp silicon transistor circuit shown in Fig. 7.8, find (i) base current I_B (ii) collector current I_C (iii) V_{CE} and stability factor S. Assume β = 100.*

Solution. Note that it is a base resistor method of transistor biasing using *pnp* transistor. You can see that voltage polarities and current directions are reversed compared to *npn* transistor circuit.

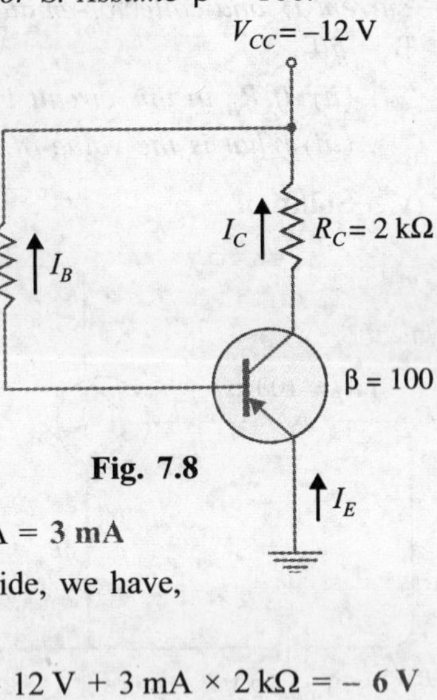

Fig. 7.8

(i) Applying Kirchhoff's voltage law to the base side, we have,

$$-V_{CC} + I_B R_B + V_{BE} = 0$$

$\therefore$ Base current is

$$I_B = \frac{V_{CC} - V_{BE}}{R_B} = \frac{12\text{V} - 0.7\text{V}}{376.67\text{k}\Omega} = \mathbf{30\ \mu A}$$

(ii) Collector current, $I_C = \beta I_B = 100 \times 30\mu\text{A} = \mathbf{3\ mA}$

(iii) Applying Kirchhoff's voltage law to the collector side, we have,

$$-V_{CC} + I_C R_C - V_{CE} = 0$$

$$\therefore \quad V_{CE} = -V_{CC} + I_C R_C = -12\text{ V} + 3\text{ mA} \times 2\text{ k}\Omega = \mathbf{-6\ V}$$

As expected, V_{CE} value is negative. The stability factor of the circuit is $S = \beta + 1$

$$= 100 + 1 = \mathbf{101}$$

Example 7.6. *Fig. 7.9 (i) shows that a silicon transistor with β = 100 is biased by base resistor method. Draw the d.c. load line and determine the operating point. What is the stability factor?*

Solution. $V_{CC} = 6\text{ V}$; $R_B = 530\text{ k}\Omega$; $R_C = 2\text{ k}\Omega$

D.C. load line. Referring to Fig. 7.9 (*i*), $V_{CE} = V_{CC} - I_C R_C$

When $I_C = 0$, $V_{CE} = V_{CC} = 6$ V. This locates the first point *B* (*OB* = 6 V) of the load line on collector-emitter voltage axis as shown in Fig. 7.9 (*ii*).

When $V_{CE} = 0$, $I_C = V_{CC}/R_C = 6\text{V}/2\text{ k}\Omega = 3$ mA. This locates the second point A (OA = 3 mA) of the load line on the collector current axis. By joining points A and B, d.c. load line AB is constructed [See Fig. 7.9 (*ii*)].

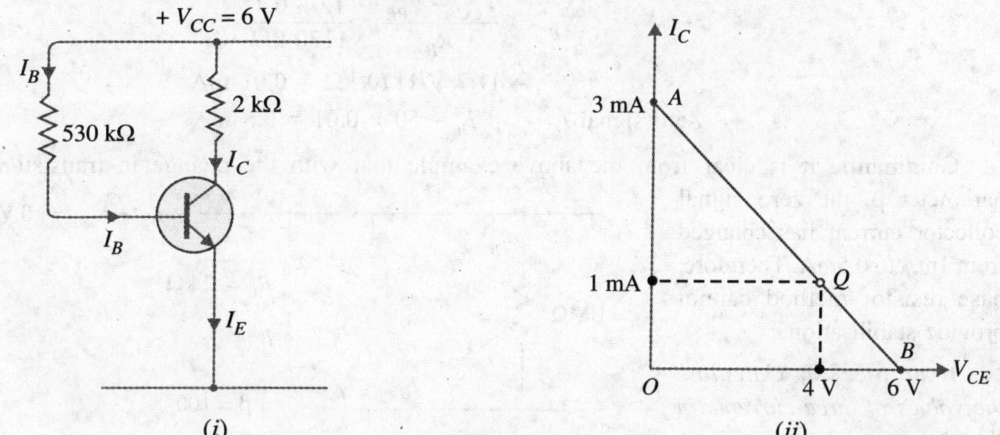

Fig. 7.9

Operating point Q. As it is a silicon transistor, therefore, V_{BE} = 0.7 V. Referring to Fig. 7.9 (*i*), it is clear that :

$$I_B R_B + V_{BE} = V_{CC}$$

or
$$I_B = \frac{V_{CC} - V_{BE}}{R_B} = \frac{(6 - 0.7)\text{ V}}{530\text{ k}\Omega} = 10\ \mu\text{A}$$

$\therefore$ Collector current, $I_C = \beta I_B = 100 \times 10 = 1000\ \mu\text{A} = 1$ mA

Collector-emitter voltage, $V_{CE} = V_{CC} - I_C R_C$

$$= 6 - 1\text{ mA} \times 2\text{ k}\Omega = 6 - 2 = 4\text{ V}$$

$\therefore$ Operating point is **4 V, 1 mA.**

Fig. 7.9 (*ii*) shows the operating point Q on the d.c. load line. Its co-ordinates are I_C = 1 mA and V_{CE} = 4 V.

Stability factor $= \beta + 1 = 100 + 1 =$ **101**

Example 7.7. *(i) A germanium transistor is to be operated at zero signal I_C = 1 mA. If the collector supply V_{CC} = 12 V, what is the value of R_B in the base resistor method? Take β = 100.*

(ii) If another transistor of the same batch with β = 50 is used, what will be the new value of zero signal I_C for the same R_B?

Solution. V_{CC} = 12 V ; β = 100

As it is a Ge transistor, therefore, V_{BE} = 0.3 V.

(*i*) Zero signal I_C = 1 mA

$\therefore$ Zero signal $I_B = I_C/\beta = 1\text{ mA}/100 = 0.01$ mA

Using the relation, $V_{CC} = I_B R_B + V_{BE}$

$$\therefore \quad R_B = \frac{V_{CC} - V_{BE}}{I_B} = \frac{12 - 0.3}{0.01\text{ mA}}$$

$$= 11.7\text{ V}/0.01\text{ mA} = \mathbf{1170\ k\Omega}$$

(*ii*) Now, $\beta = 50$

Again using the relation, $V_{CC} = I_B R_B + V_{BE}$

$$\therefore \quad I_B = \frac{V_{CC} - V_{BE}}{R_B} = \frac{12 - 0.3}{1170 \text{ k}\Omega}$$

$$= 11.7 \text{ V}/1170 \text{ k}\Omega = 0.01 \text{ mA}$$

$$\therefore \quad \text{Zero signal } I_C = \beta I_B = 50 \times 0.01 = \mathbf{0.5 \text{ mA}}$$

Comments. It is clear from the above example that with the change in transistor parameter β, the zero signal collector current has changed from 1mA to 0.5mA. Therefore, base resistor method cannot provide stabilisation.

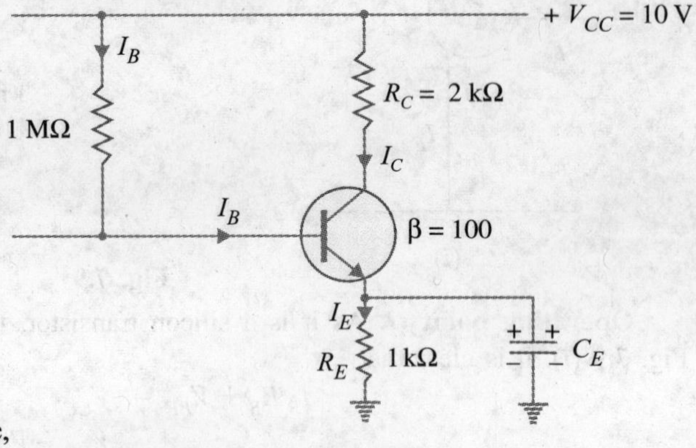

Fig. 7.10

Example 7.8. *Calculate the values of three currents in the circuit shown in Fig. 7.10.*

Solution. Applying Kirchhoff's voltage law to the base side and taking resistances in kΩ and currents in mA, we have,

$$V_{CC} = I_B R_B + V_{BE} + I_E \times 1$$

or $\quad 10 = 1000\, I_B + {}^*0 + (I_C + I_B)$

or $\quad 10 = 1000\, I_B + (\beta\, I_B + I_B)$

or $\quad 10 = 1000\, I_B + (100\, I_B + I_B)$

or $\quad 10 = 1101\, I_B$

$$\therefore \quad I_B = 10/1101 = \mathbf{0.0091 \text{ mA}}$$

$$I_C = \beta I_B = 100 \times 0.0091 = \mathbf{0.91 \text{ mA}}$$

$$I_E = I_C + I_B = 0.91 + 0.0091 = \mathbf{0.919 \text{ mA}}$$

Example 7.9. *Design base resistor bias circuit for a CE amplifier such that operating point is V_{CE} = 8 V and I_C = 2 mA. You are supplied with a fixed 15 V d.c. supply and a silicon transistor with β = 100. Take base-emitter voltage V_{BE} = 0.6 V. Calculate also the value of load resistance that would be employed.*

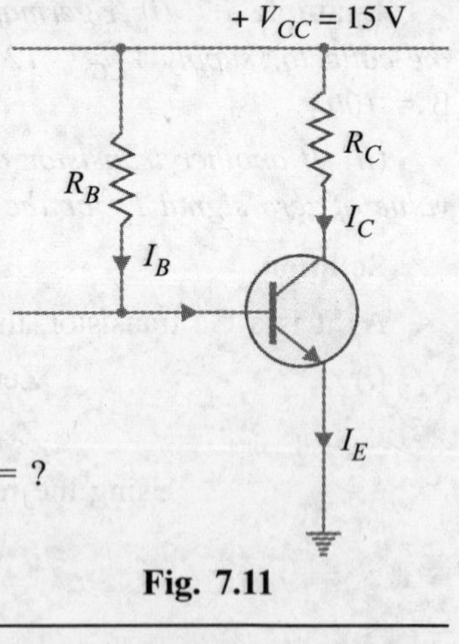

Fig. 7.11

Solution. Fig. 7.11 shows *CE* amplifier using base resistor method of biasing.

$$V_{CC} = 15 \text{ V} \; ; \; \beta = 100 \; ; \; V_{BE} = 0.6 \text{ V}$$

$$V_{CE} = 8 \text{ V} \; ; \; I_C = 2 \text{ mA} \; ; \; R_C = \, ? \; ; \; R_B = \, ?$$

Now, $\quad V_{CC} = V_{CE} + I_C R_C$

or $\quad 15 \text{ V} = 8 \text{ V} + 2 \text{ mA} \times R_C$

* Neglecting V_{BE} as it is generally very small.

$$\therefore \quad R_C = \frac{(15-8)\text{ V}}{2\text{mA}} = 3.5 \text{ k}\Omega$$

$$I_B = I_C/\beta = 2/100 = 0.02 \text{ mA}$$

Also, $$V_{CC} = I_B R_B + V_{BE}$$

$$\therefore \quad R_B = \frac{V_{CC}-V_{BE}}{I_B} = \frac{(15-0.6)\text{ V}}{0.02\text{ mA}} = \mathbf{720\ k\Omega}$$

Example 7.10. *A *base bias circuit in Fig. 7.12 is subjected to an increase in temperature from 25°C to 75°C. If β = 100 at 25°C and 150 at 75°C, determine the percentage change in Q-point values (V_{CE} and I_C) over this temperature range. Neglect any change in V_{BE} and the effects of any leakage current.*

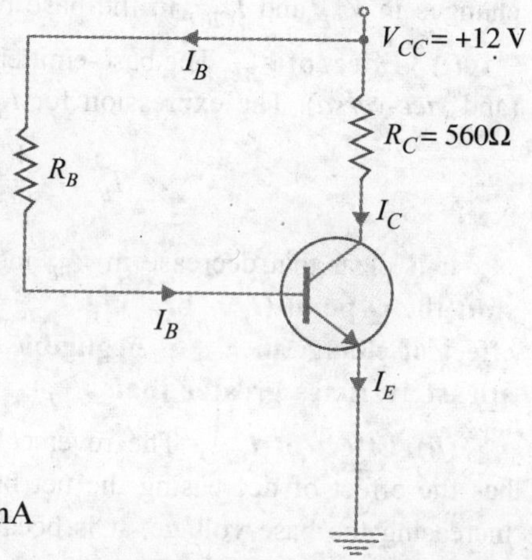

Fig. 7.12

Solution. At 25°C

$$I_B = \frac{V_{CC}-V_{BE}}{R_B} = \frac{12\text{ V}-0.7\text{ V}}{100\text{ k}\Omega} = 0.113 \text{ mA}$$

$$\therefore \quad I_C = \beta I_B = 100 \times 0.113 \text{ mA} = 11.3 \text{ mA}$$

and $$V_{CE} = V_{CC} - I_C R_C = 12\text{ V} - (11.3\text{ mA})(560\ \Omega) = 5.67\text{ V}$$

At 75 °C

$$I_B = \frac{V_{CC}-V_{BE}}{R_B} = \frac{12\text{ V}-0.7\text{ V}}{100\text{ k}\Omega} = 0.113 \text{ mA}$$

$$\therefore \quad I_C = \beta I_B = 150 \times 0.113 \text{ mA} = 17 \text{ mA}$$

and $$V_{CE} = V_{CC} - I_C R_C = 12\text{ V} - (17\text{ mA})(560\ \Omega) = 2.48\text{ V}$$

$$\%\text{ age change in } I_C = \frac{I_{C\,(75°C)} - I_{C\,(25°C)}}{I_{C\,(25°C)}} \times 100 = \frac{17\text{ mA} - 11.3\text{ mA}}{11.3\text{ mA}} \times 100 = \mathbf{50\%}\ \textit{(increase)}$$

Note that I_C changes by the same percentage as β.

$$\%\text{ age change in } V_{CE} = \frac{V_{CE\,(75°C)} - V_{CE\,(25°C)}}{V_{CE\,(25°C)}} \times 100 = \frac{2.48\text{ V} - 5.67\text{ V}}{5.67\text{ V}} \times 100 = -\ \mathbf{56.3\%}\ \textit{(decrease)}$$

* Note that base resistor method is also called *base bias method.*

Comments. It is clear from the above example that *Q*-point is extremely dependent on β in a base bias circuit. Therefore, base bias circuit is very unstable. Consequently, this method is normally not used if linear operation is required. However, it can be used for switching operation.

Example 7.11. *In base bias method, how Q-point is affected by changes in V_{BE} and I_{CBO}?*

Solution. In addition to being affected by change in β, the *Q*-point is also affected by changes in V_{BE} and I_{CBO} in the base bias method.

(*i*) **Effect of V_{BE}.** The base-emitter-voltage V_{BE} decreases with the increase in temperature (and *vice-versa*). The expression for I_B in base bias method is given by ;

$$I_B = \frac{V_{CC} - V_{BE}}{R_B}$$

It is clear that decrease in V_{BE} increases I_B. **This will shift the *Q*-point ($I_C = \beta I_B$ and $V_{CE} = V_{CC} - I_C R_C$). The effect of change in V_{BE} is negligible if $V_{CC} >> V_{BE}$ (V_{CC} atleast 10 times greater than V_{BE}).**

(*ii*) **Effect of I_{CBO}.** The reverse leakage current I_{CBO} has the effect of decreasing the net base current and thus increasing the base voltage. It is because the flow of I_{CBO} creates a voltage drop across R_B that adds to the base voltage as shown in Fig. 7.13. **Therefore, change in I_{CBO} shifts the *Q*-point of the base bias circuit. However, in modern transistors, I_{CBO} is usually less than 100 nA and its effect on the bias is negligible if $V_{BB} >> I_{CBO} R_B$.**

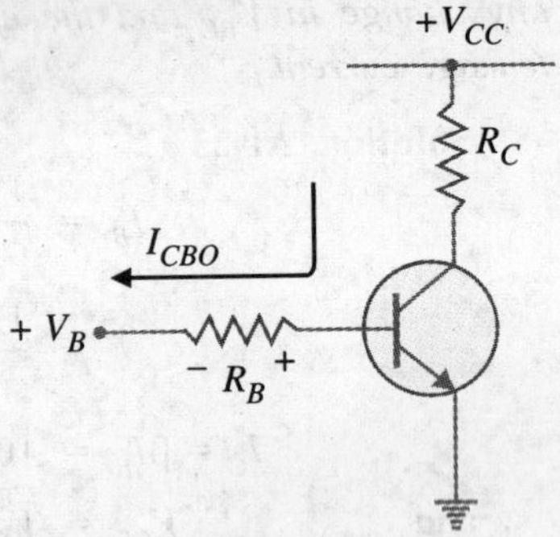

Fig. 7.13

Example 7.12. *Fig. 7.14 (i) shows the base resistor transistor circuit. The device (i.e. transistor) has the characteristics shown in Fig. 7.14 (ii). Determine V_{CC}, R_C and R_B.*

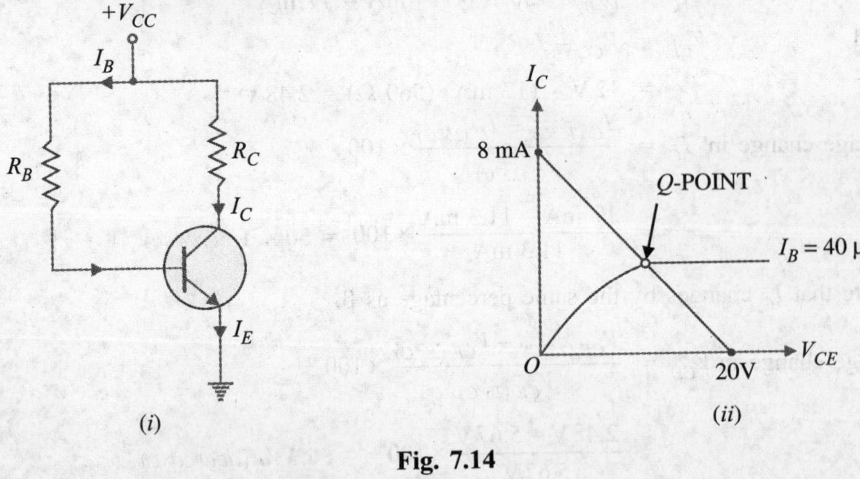

Fig. 7.14

Solution. From the d.c load line, V_{CC} = **20 V.**

$$\text{Max. } I_C = \frac{V_{CC}}{R_C} \quad (\text{when } V_{CE} = 0 \text{ V})$$

$$\therefore \quad R_C = \frac{V_{CC}}{\text{Max. } I_C} = \frac{20\text{V}}{8\text{mA}} = 2.5 \text{ k}\Omega$$

Now,
$$I_B = \frac{V_{CC} - V_{BE}}{R_B}$$

$$\therefore \quad R_B = \frac{V_{CC} - V_{BE}}{I_B} = \frac{20\text{V} - 0.7\text{V}}{40\ \mu\text{A}} = \frac{19.3\text{V}}{40\ \mu\text{A}} = 482.5 \text{ k}\Omega$$

Example 7.13. *What fault is indicated in (i) Fig. 7.15 (i) and (ii) Fig. 7.15 (ii)?*

Solution.

(*i*) The obvious fault in Fig. 7.15 (*i*) is that the **base is internally open.** It is because 3 V at the base and 9 V at the collector mean that transistor is in cut-off state.

(*ii*) The obvious fault in Fig. 7.15 (*ii*) is that **collector is internally open.** The voltage at the base is correct. The voltage of 9 V appears at the collector because the 'open' prevents collector current.

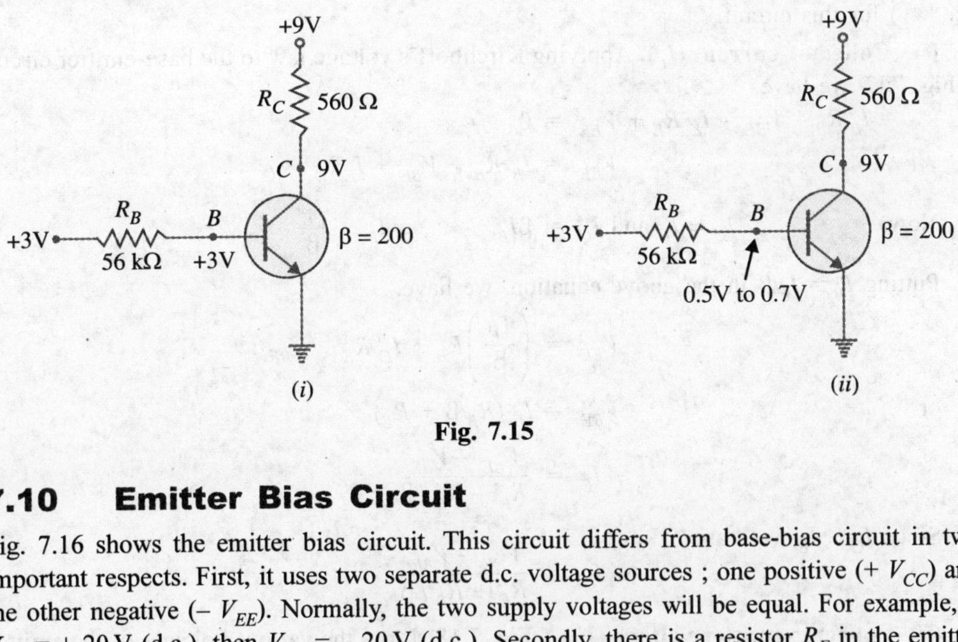

Fig. 7.15

7.10 Emitter Bias Circuit

Fig. 7.16 shows the emitter bias circuit. This circuit differs from base-bias circuit in two important respects. First, it uses two separate d.c. voltage sources ; one positive (+ V_{CC}) and the other negative (– V_{EE}). Normally, the two supply voltages will be equal. For example, if V_{CC} = + 20 V (d.c.), then V_{EE} = – 20 V (d.c.). Secondly, there is a resistor R_E in the emitter circuit.

We shall first redraw the circuit in Fig. 7.16 as it usually appears on schematic diagrams. This means deleting the battery symbols as shown in Fig. 7.17. All the information is still (See Fig. 7.17) on the diagram except that it is in condensed form. That is a negative supply voltage – V_{EE} is applied to the bottom of R_E and a positive voltage of + V_{CC} to the top of R_C.

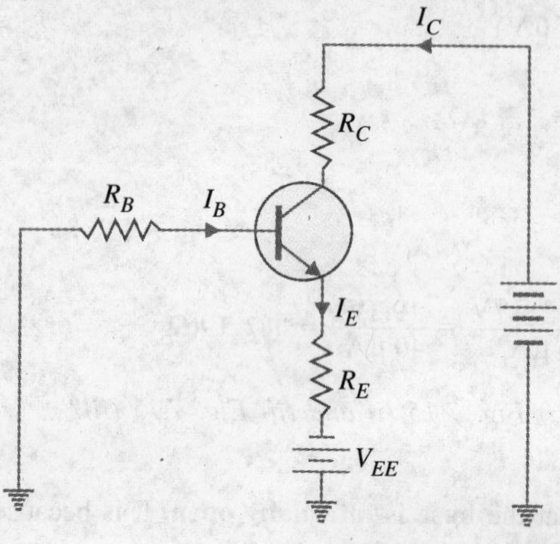

Fig. 7.16

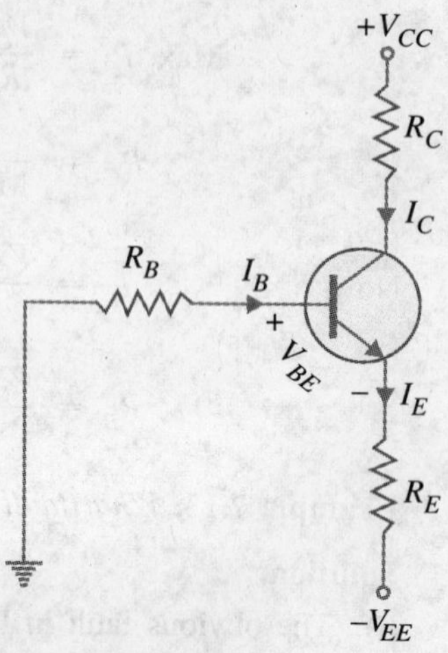

Fig. 7.17

7.11 Circuit Analysis of Emitter Bias

Fig. 7.17 shows the emitter bias circuit. We shall find the *Q*-point values (*i.e.* d.c. I_C and d.c. V_{CE}) for this circuit.

(*i*) **Collector current (I_C).** Applying Kirchhoff's voltage law to the base-emitter circuit in Fig. 7.17, we have,

$$-I_B R_B - V_{BE} - I_E R_E + V_{EE} = 0$$

$$\therefore \quad V_{EE} = I_B R_B + V_{BE} + I_E R_E$$

Now, $\quad I_C \simeq I_E$ and $I_C = \beta I_B \quad \therefore \quad I_B \simeq \dfrac{I_E}{\beta}$

Putting $I_B = I_E/\beta$ in the above equation, we have,

$$V_{EE} = \left(\frac{I_E}{\beta}\right) R_B + I_E R_E + V_{BE}$$

or $$V_{EE} - V_{BE} = I_E (R_B/\beta + R_E)$$

$$\therefore \quad I_E = \frac{V_{EE} - V_{BE}}{R_E + R_B/\beta}$$

Since $I_C \simeq I_E$, we have, $$I_C = \frac{V_{EE} - V_{BE}}{R_E + R_B/\beta}$$

(*ii*) **Collector-emitter voltage (V_{CE}).** Fig. 7.18 shows the various voltages of the emitter bias circuit *w.r.t.* ground.

Emitter voltage *w.r.t.* ground is

$$V_E = -V_{EE} + I_E R_E$$

Base voltage *w.r.t.* ground is

$$V_B = V_E + V_{BE}$$

Collector voltage *w.r.t.* ground is

$$V_C = V_{CC} - I_C R_C$$

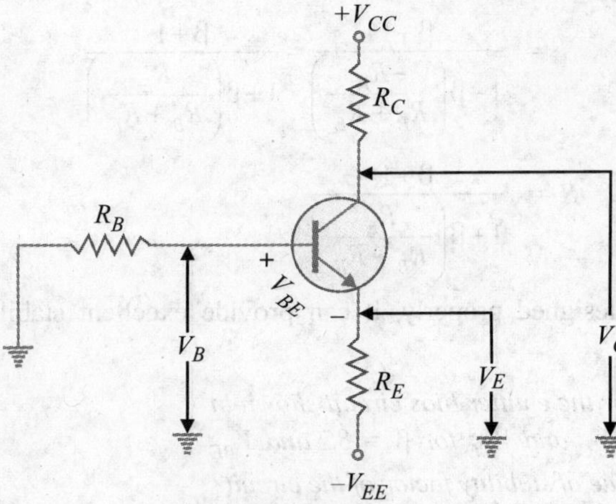

Fig. 7.18

Subtracting V_E from V_C and using the approximation $I_C \simeq I_E$, we have,

$$V_C - V_E = (V_{CC} - I_C R_C) - (-V_{EE} + I_C R_E) \quad (\because I_E \simeq I_C)$$

or $$V_{CE} = V_{CC} + V_{EE} - I_C (R_C + R_E)$$

Alternatively. Applying Kirchhoff's voltage law to the collector side of the emitter bias circuit in Fig. 7.17 (Refer back), we have,

$$V_{CC} - I_C R_C - V_{CE} - I_C{}^* R_E + V_{EE} = 0$$

or $$V_{CE} = V_{CC} + V_{EE} - I_C (R_C + R_E)$$

Stability factor *S*. We have seen above that by applying Kirchhoff's voltage law to the base-emitter circuit in Fig. 7.17, we have,

$$V_{EE} = I_B R_B + V_{BE} + I_E R_E$$

or $$V_{EE} = I_B R_B + V_{BE} + (I_B + I_C) R_E \quad (\because I_E = I_B + I_C)$$

or $$V_{EE} = I_B (R_B + R_E) + I_C R_E + V_{BE} \quad \text{... (i)}$$

**Differentiating exp. (*i*) *w.r.t.* I_C, we have,

$$0 = \frac{dI_B}{dI_C}(R_B + R_E) + 1 \times R_E + 0$$

or $$\frac{dI_B}{dI_C} = -\frac{R_E}{R_B + R_E} \quad \text{... (ii)}$$

The general expression for stability factor S is

$$S = \frac{\beta + 1}{1 - \beta\left(\frac{dI_B}{dI_C}\right)}$$

Putting the value of dI_B/dI_C from exp. (*ii*) into the above expression for S, we have,

---

* $I_C \simeq I_E$.

** Note that V_{EE}, V_{BE}, R_B and R_E are constant. However, I_B and I_C are variable.

$$S = \frac{\beta+1}{1-\beta\left(\frac{-R_E}{R_B+R_E}\right)} = \frac{\beta+1}{1+\beta\left(\frac{R_E}{R_B+R_E}\right)}$$

$$\therefore \quad S = \frac{\beta+1}{1+\beta\left(\frac{R_E}{R_B+R_E}\right)}$$

If the circuit is designed properly, it can provide excellent stability to the operating point.

Example 7.14. *For the emitter bias circuit shown in Fig. 7.19, find I_E, I_C, V_C and V_{CE} for $\beta = 85$ and $V_{BE} = 0.7$ V. What is the value of stability factor of the circuit?*

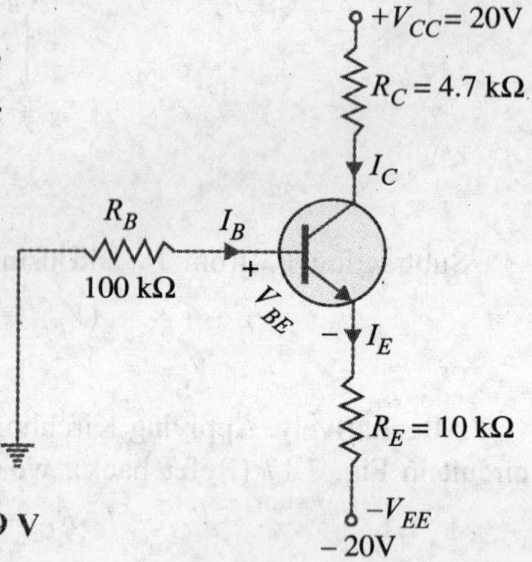

Fig. 7.19

Solution.

$$I_C \simeq I_E = \frac{V_{EE}-V_{BE}}{R_E+R_B/\beta}$$

$$= \frac{20\text{V}-0.7\text{V}}{10\text{ k}\Omega+100\text{ k}\Omega/85} = \mathbf{1.73\text{ mA}}$$

$$V_C = V_{CC} - I_C R_C$$

$$= 20\text{ V} - (1.73\text{ mA})(4.7\text{ k}\Omega) = \mathbf{11.9\text{ V}}$$

$$V_E = -V_{EE} + I_E R_E$$

$$= -20\text{ V} + (1.73\text{ mA})(10\text{ k}\Omega) = -2.7\text{ V}$$

$$\therefore \quad V_{CE} = V_C - V_E = 11.9 - (-2.7\text{ V}) = \mathbf{14.6\text{ V}}$$

Note that operating point (or Q – point) of the circuit is 14.6 V, 1.73 mA.

The stability factor S of the circuit is

$$S = \frac{\beta+1}{1+\beta\left(\frac{R_E}{R_B+R_E}\right)} = \frac{85+1}{1+85\left(\frac{10\text{k}\Omega}{100\text{ k}\Omega+10\text{ k}\Omega}\right)} = \mathbf{9.85}$$

Note that it is a very good value of stability factor. Therefore, change in I_C will be quite small due to temperature change.

Example 7.15. *Determine how much the Q-point in Fig. 7.19 (above) will change over a temperature range where β increases from 85 to 100 and V_{BE} decreases from 0.7 V to 0.6 V.*

Solution.

For $\beta = 85$ and $V_{BE} = 0.7$ V

As calculated in the above example, I_C = 1.73 mA and V_{CE} = 14.6 V.

For $\beta = 100$ and $V_{BE} = 0.6$ V

$$I_C \simeq I_E = \frac{V_{EE}-V_{BE}}{R_E+R_B/\beta} = \frac{20\text{V}-0.6\text{V}}{10\text{ k}\Omega+100\text{ k}\Omega/100} = \frac{19.4\text{V}}{11\text{ k}\Omega} = 1.76\text{ mA}$$

$$V_C = V_{CC} - I_C R_C = 20\text{ V} - (1.76\text{ mA})(4.7\text{ k}\Omega) = 11.7\text{ V}$$

$$V_E = -V_{EE} + I_E R_E = -20\text{ V} + (1.76\text{ mA})\,(10\text{ k}\Omega) = -2.4\text{ V}$$

$$\therefore \quad V_{CE} = V_C - V_E = 11.7 - (-2.4) = 14.1\text{ V}$$

$$\%\text{ age change in } I_C = \frac{1.76\text{ mA} - 1.73\text{ mA}}{1.73\text{ mA}} \times 100 = \mathbf{1.7\%}\ (\textit{increase})$$

$$\%\text{ age change in } V_{CE} = \frac{14.1\text{V} - 14.6\text{V}}{14.1\text{V}} \times 100 = \mathbf{-3.5\%}\ (\textit{decrease})$$

Example 7.16. *For the emitter bias circuit show in Fig. 7.20 (i), β = 100 and V_{BE} = 0.7 V. Find (i) the operating point Q (ii) stability factor of the circuit. Also draw the d.c. load line of the circuit.*

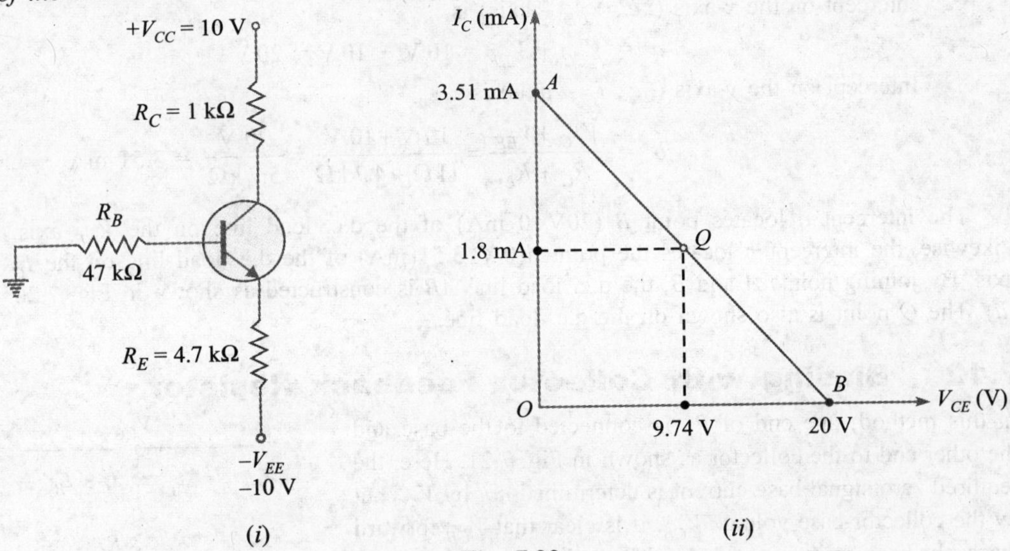

Fig. 7.20

Solution.

(*i*) The operating point *Q* of the circuit is found as under :

$$\text{Collector current, } I_C = \frac{V_{EE} - V_{BE}}{R_E + R_B/\beta} = \frac{10\text{V} - 0.7\text{V}}{4.7\text{ k}\Omega + 47\text{ k}\Omega/100}$$

$$= \frac{9.3\text{V}}{5.17\text{ k}\Omega} = 1.8\text{ mA}$$

$$\text{Collector-emitter voltage, } V_{CE} = V_{CC} + V_{EE} - I_C(R_C + R_E)$$

$$= 10\text{ V} + 10\text{ V} - 1.8\text{ mA} \times (1\text{ k}\Omega + 4.7\text{ k}\Omega) = 9.74\text{ V}$$

Therefore, *Q* point of the circuit is **9.74 V, 1.8 mA.**

(*ii*) The stability factor *S* of the circuit is

$$S = \frac{\beta + 1}{1 + \beta\left(\dfrac{R_E}{R_B + R_E}\right)} = \frac{100 + 1}{1 + 100\left(\dfrac{4.7\text{k}\Omega}{47\text{ k}\Omega + 4.7\text{k}\Omega}\right)} = \mathbf{10}$$

Note that the value of *S* is different in examples 7.14 and 7.16. This is not surprising because the value of *S* also depends on the resistor values (values of R_B and R_E in this case).

D.C. load line. Referring to the circuit shown in Fig. 7.20 (*i*), we have,

$$V_{CE} = V_{CC} + V_{EE} - I_C(R_C + R_E)$$

or $$V_{CE} + I_C(R_C + R_E) = V_{CC} + V_{EE}$$

Dividing throughout by ($V_{CC} + V_{EE}$), we get,

$$\frac{V_{CE}}{V_{CC}+V_{EE}} + \frac{I_C}{(V_{CC}+V_{EE})/(R_C+R_E)} = 1$$

Compare this equation with the standard equation of the straight line *viz.*

$$\frac{x}{a} + \frac{y}{b} = 1$$

∴ Intercept on the *x*-axis (*i.e.*, V_{CE} – axis) is

$$a = V_{CC} + V_{EE} = 10 \text{ V} + 10 \text{ V} = 20 \text{ V}$$

Intercept on the *y*-axis (*i.e.*, I_C – axis) is

$$b = \frac{V_{CC}+V_{EE}}{R_C+R_E} = \frac{10 \text{ V}+10 \text{ V}}{1 \text{ k}\Omega + 4.7 \text{ k}\Omega} = \frac{20 \text{ V}}{5.7 \text{ k}\Omega} = 3.51 \text{ mA}$$

The intercept *a* locates point *B* (20V, 0 mA) of the d.c. load line on the V_{CE} axis. Likewise, the intercept *b* locates the point *A*(0 V, 3.51 mA) of the d.c. load line on the I_C axis. By joining points *A* and *B*, the d.c. load line *AB* is constructed as shown in Fig. 7.20 (*ii*). The *Q* point is also shown on the d.c. load line.

7.12 Biasing with Collector Feedback Resistor

In this method, one end of R_B is connected to the base and the other end to the collector as shown in Fig. 7.21. Here, the required zero signal base current is determined *not* by V_{CC} but by the collector-base voltage V_{CB}. It is clear that V_{CB} forward biases the base-emitter junction and hence base current I_B flows through R_B. This causes the zero signal collector current to flow in the circuit.

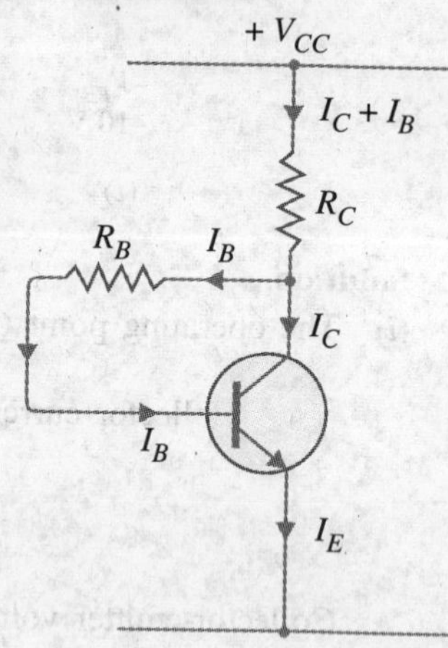

Fig. 7.21

Circuit analysis. The required value of R_B needed to give the zero signal current I_C can be determined as follows. Referring to Fig. 7.21,

$$V_{CC} = {}^*I_C R_C + I_B R_B + V_{BE}$$

or $$R_B = \frac{V_{CC} - V_{BE} - I_C R_C}{I_B}$$

$$= \frac{V_{CC} - V_{BE} - \beta I_B R_C}{I_B} \quad (\because I_C = \beta I_B)$$

Alternatively. $V_{CE} = V_{BE} + V_{CB}$

or $$V_{CB} = V_{CE} - V_{BE}$$

∴ $$R_B = \frac{V_{CB}}{I_B} = \frac{V_{CE} - V_{BE}}{I_B}; \text{ where } I_B = \frac{I_C}{\beta}$$

---

* Actually voltage drop across $R_C = (I_B + I_C) R_C$.

However, $I_B << I_C$. Therefore, as a reasonable approximation, we can say that drop across $R_C = I_C R_C$.

It can be shown mathematically that stability factor S for this method of biasing is less than $(\beta + 1)$ *i.e.*

Stability factor, $S < (\beta + 1)$

Therefore, this method provides better thermal stability than the fixed bias.

Note. It can be easily proved (See *example 7.20) that Q-point values (I_C and V_{CE}) for the circuit shown in Fig. 7.21 are given by ;

$$I_C = \frac{V_{CC} - V_{BE}}{R_B / \beta + R_C}$$

and $$V_{CE} = V_{CC} - I_C R_C$$

Stability factor (S). Referring to the circuit shown in Fig. 7.21 above, we have,

$$V_{CC} = (I_B + I_C)R_C + I_B R_B + V_{BE}$$

or $$V_{CC} = I_C R_C + I_B(R_C + R_B) + V_{BE} \quad \text{... (i)}$$

Differentiating exp. (*i*) *w.r.t.* I_C, we have,

$$0 = 1 \times R_C + \frac{dI_B}{dI_C}(R_C + R_B) + 0$$

or $$\frac{dI_B}{dI_C} = -\frac{R_C}{R_C + R_B} \quad \text{... (ii)}$$

The general expression for the stability factor S is

$$S = \frac{\beta + 1}{1 - \beta\left(\frac{dI_B}{dI_C}\right)}$$

Putting the value of dI_B/dI_C from exp. (*ii*) into the above expression for S, we have,

$$S = \frac{\beta + 1}{1 - \beta\left(\frac{-R_C}{R_C + R_B}\right)} = \frac{\beta + 1}{1 + \beta\left(\frac{R_C}{R_C + R_B}\right)}$$

$$\therefore \quad S = \frac{\beta + 1}{1 + \beta\left(\frac{R_C}{R_C + R_B}\right)}$$

It may be seen that the value of S for this circuit is smaller than $\beta + 1$ (S for the fixed bias) *i.e.*, $S < \beta + 1$. Hence, collector feedback biasing leads to better thermal stability than the fixed bias.

Advantages

(*i*) It is a simple method as it requires only one resistance R_B.

(*ii*) This circuit provides some stabilisation of the operating point as discussed below :

$$V_{CE} = V_{BE} + V_{CB}$$

Suppose the temperature increases. This will increase collector leakage current and hence the total collector current. But as soon as collector current increases, V_{CE} decreases due to

* Put $R_E = 0$ for the expression of I_C in example 7.20. It is because in the present circuit (Fig. 7.21), there is no R_E.

greater drop across R_C. The result is that V_{CB} decreases *i.e.* lesser voltage is available across R_B. Hence the base current I_B decreases. The smaller I_B tends to decrease the collector current to original value.

Disadvantages

(*i*) The circuit does not provide good stabilisation because stability factor is fairly high, though it is lesser than that of fixed bias. Therefore, the operating point does change, although to lesser extent, due to temperature variations and other effects.

(*ii*) This circuit provides a negative feedback which reduces the gain of the amplifier as explained hereafter. During the positive half-cycle of the signal, the collector current increases. The increased collector current would result in greater voltage drop across R_C. This will reduce the base current and hence collector current.

Example 7.17. *Fig. 7.22 shows a silicon transistor biased by collector feedback resistor method. Determine the operating point. Given that β = 100. What is the value of stability factor of the circuit?*

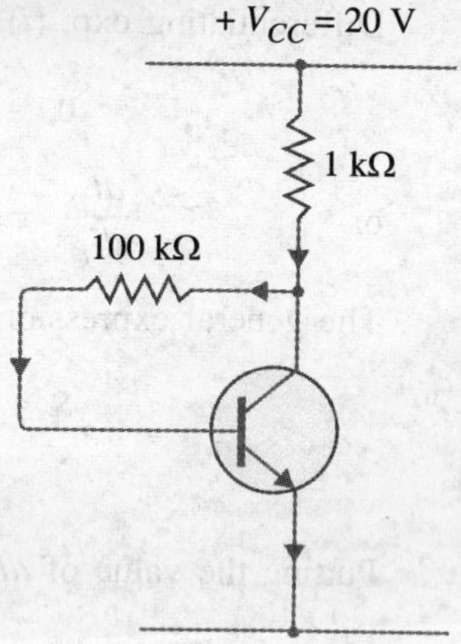

Fig. 7.22

Solution. $V_{CC} = 20$ V ; $R_B = 100$ kΩ ; $R_C = 1$ kΩ

Since it is a silicon transistor, $V_{BE} = 0.7$ V.

Assuming I_B to be in mA and using the relation,

$$R_B = \frac{V_{CC} - V_{BE} - \beta\, I_B R_C}{I_B}$$

or $\quad 100 \times I_B = 20 - 0.7 - 100 \times I_B \times 1$

or $\quad 200\, I_B = 19.3$

or $\quad I_B = \frac{19.3}{200} = 0.096$ mA

∴ Collector current, $I_C = \beta\, I_B = 100 \times 0.096 = 9.6$ mA

Collector-emitter voltage is

$$V_{CE} = V_{CC} - I_C R_C = 20 - 9.6 \text{ mA} \times 1 \text{ k}\Omega = 10.4 \text{ V}$$

∴ Operating point is **10.4 V, 9.6 mA.**

Alternatively.

$$I_C = \frac{V_{CC} - V_{BE}}{R_B/\beta + R_C} = \frac{20\text{V} - 0.7\text{V}}{100\text{ k}\Omega/100 + 1\text{ k}\Omega} = \frac{19.3\text{V}}{2\text{ k}\Omega} = 9.65 \text{ mA}$$

$$V_{CE} = V_{CC} - I_C\, R_C = 20\text{V} - 9.65 \text{ mA} \times 1 \text{ k}\Omega = 10.35 \text{ V}$$

A very slight difference in the values is due to manipulation of calculations.

The stability factor S of the circuit is

$$S = \frac{\beta + 1}{1 + \beta\left(\frac{R_C}{R_C + R_B}\right)} = \frac{100 + 1}{1 + 100\left(\frac{1\text{ k}\Omega}{1\text{ k}\Omega + 100\text{ k}\Omega}\right)} = \mathbf{50.75}$$

Note that this circuit does not provide good thermal stability because the value of S is quite large.

Example 7.18. *(i) It is required to set the operating point by biasing with collector feedback resistor at I_C = 1 mA, V_{CE} = 8 V. If β = 100, V_{CC} = 12 V, V_{BE} = 0.3 V, how will you do it?*

(ii) What will be the new operating point if β = 50, all other circuit values remaining the same?

Solution. $V_{CC} = 12\text{ V}$; $V_{CE} = 8\text{ V}$; $I_C = 1\text{ mA}$; $\beta = 100$; $V_{BE} = 0.3\text{ V}$

(*i*) To obtain the required operating point, we should find the value of R_B.

Now, collector load is given by ;

$$R_C = \frac{V_{CC} - V_{CE}}{I_C} = \frac{(12-8)\text{ V}}{1\text{ mA}} = 4\text{ k}\Omega$$

Also,
$$I_B = \frac{I_C}{\beta} = \frac{1\text{ mA}}{100} = 0.01\text{ mA}$$

Using the relation,
$$R_B = \frac{V_{CC} - V_{BE} - \beta I_B R_C}{I_B}$$
$$= \frac{12 - 0.3 - 100 \times 0.01 \times 4}{0.01} = 770\text{ k}\Omega$$

(*ii*) Now β = 50, and other circuit values remain the same.

$$\therefore \quad V_{CC} = V_{BE} + I_B R_B + \beta I_B R_C$$

or
$$12 = 0.3 + I_B (R_B + \beta R_C)$$

or
$$11.7 = I_B (770 + 50 \times 4)$$

or
$$I_B = \frac{11.7\text{ V}}{970\text{ k}\Omega} = 0.012\text{ mA}$$

∴ Collector current, $I_C = \beta I_B = 50 \times 0.012 = 0.6\text{ mA}$

∴ Collector-emitter voltage, $V_{CE} = V_{CC} - I_C R_C = 12 - 0.6\text{ mA} \times 4\text{ k}\Omega = 9.6\text{ V}$

∴ New operating point is **9.6 V, 0.6 mA.**

Comments. It may be seen that operating point is changed when a new transistor with lesser β is used. Therefore, biasing with collector feedback resistor does not provide very good stabilisation. It may be noted, however, that change in operating point is less than that of base resistor method.

Example 7.19. *It is desired to set the operating point at 2 V, 1 mA by biasing a silicon transistor with collector feedback resistor R_B. If β = 100, find the value of R_B.*

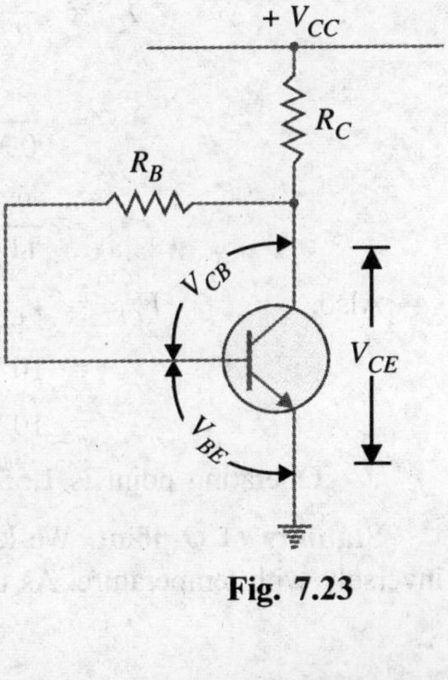

Fig. 7.23

Solution. For a silicon transistor, V_{BE} = 0.7 V.

$$I_B = \frac{I_C}{\beta} = 1/100 = 0.01\text{ mA}$$

Now,
$$V_{CE} = V_{BE} + V_{CB}$$

or
$$2 = 0.7 + V_{CB}$$

$$\therefore \quad V_{CB} = 2 - 0.7 = 1.3\text{ V}$$

$$\therefore \quad R_B = \frac{V_{CB}}{I_B} = \frac{1.3\text{ V}}{0.01\text{ mA}} = 130\text{ k}\Omega$$

Example 7.20. *Find the Q-point values (I_C and V_{CE}) for the collector feedback bias circuit shown in Fig. 7.24.*

Solution. Fig. 7.24 shows the currents in the three resistors (R_C, R_B and R_E) in the circuit. By following the path through V_{CC}, R_C, R_B, V_{BE} and R_E and applying Kirchhoff's voltage law, we have,

$$V_{CC} - (I_C + I_B)\, R_C - I_B\, R_B - V_{BE} - I_E\, R_E = 0$$

Now, $\quad I_B + I_C \simeq I_C\,;\; I_E \simeq I_C$ and $I_B = \dfrac{I_C}{\beta}$

$$\therefore \quad V_{CC} - I_C\, R_C - \frac{I_C}{\beta}\, R_B - V_{BE} - I_C\, R_E = 0$$

or $$I_C \left(R_E + \frac{R_B}{\beta} + R_C\right) = V_{CC} - V_{BE}$$

$$\therefore \quad I_C = \frac{V_{CC} - V_{BE}}{R_E + R_B/\beta + R_C}$$

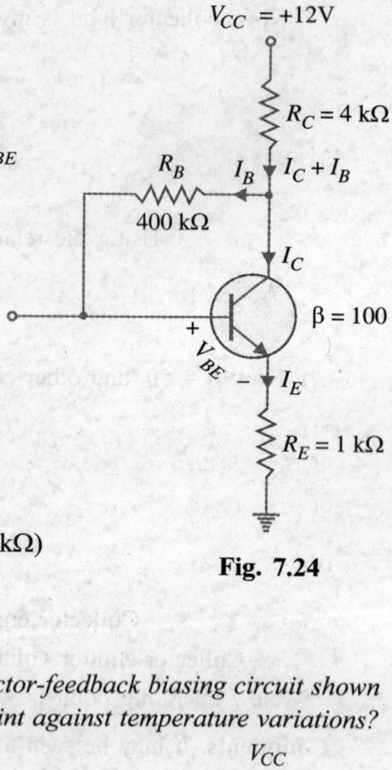

Fig. 7.24

Putting the given circuit values, we have,

$$I_C = \frac{12\text{ V} - 0.7\text{ V}}{1\text{ k}\Omega + 400\text{ k}\Omega/100 + 4\text{ k}\Omega}$$

$$= \frac{11.3\text{ V}}{9\text{ k}\Omega} = 1.26\text{ mA}$$

$$V_{CE} = V_{CC} - I_C\,(R_C + R_E)$$

$$= 12\text{ V} - 1.26\text{ mA}\,(4\text{ k}\Omega + 1\text{ k}\Omega)$$

$$= 12\text{ V} - 6.3\text{ V} = 5.7\text{ V}$$

∴ The operating point is **5.7 V, 1.26 mA.**

Example 7.21. *Find the d.c. bias values for the collector-feedback biasing circuit shown in Fig. 7.25. How does the circuit maintain a stable Q point against temperature variations?*

Solution. The collector current is

$$I_C = \frac{V_{CC} - V_{BE}}{R_E + R_B/\beta + R_C}$$

$$= \frac{10\text{ V} - 0.7\text{ V}}{0 + 100\text{ k}\Omega/100 + 10\text{ k}\Omega}$$

$$= \frac{9.3\text{ V}}{11\text{ k}\Omega} = 0.845\text{ mA}$$

Also, $$V_{CE} = V_{CC} - I_C\, R_C$$

$$= 10\text{ V} - 0.845\text{ mA} \times 10\text{ k}\Omega$$

$$= 10\text{ V} - 8.45\text{ V} = 1.55\text{ V}$$

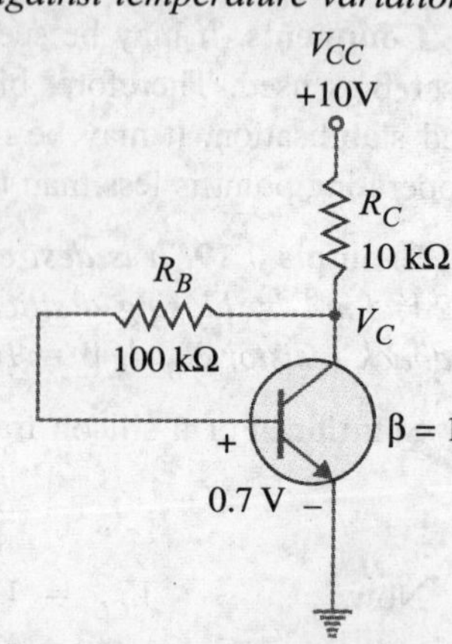

Fig. 7.25

∴ Operating point is **1.55 V, 0.845 mA.**

Stability of *Q*-point. We know that β varies directly with temperature and V_{BE} varies inversely with temperature. As the temperature goes up, β goes up and V_{BE} goes down. The increase in β increases I_C (= βI_B). The decrease in V_{BE} increases I_B which in turn increases I_C. As I_C tries to increase, the voltage drop across R_C (= $I_C R_C$) also tries to increase. This tends to reduce collector voltage V_C (See Fig. 7.25) and, therefore, the voltage across R_B. The reduced voltage across R_B reduces I_B and offsets the attempted increase in I_C and attempted decrease in V_C. The result is that the collector-feedback circuit maintains a stable *Q*-point. The reverse action occurs when the temperature decreases.

Example 7.22. *The collector feedback bias circuit shown in Fig. 7.26 uses a transistor with β = 100. If the circuit temperature increases from 27°C to 107°C, find the change in collector current. Given that I_{CBO} = 15 nA at 27°C.*

Solution. The change in collector current (ΔI_C) due to temperature change is

$$\Delta I_C = S \times \Delta I_{CBO}$$

Here, S = Stability factor of the bias circuit

ΔI_{CBO} = Change in I_{CBO} due to temperature change

The stability factor of this bias circuit is

$$S = \frac{\beta+1}{1+\beta\left(\frac{R_C}{R_C+R_B}\right)} = \frac{100+1}{1+100\left(\frac{2.2\text{ k}\Omega}{2.2\text{ k}\Omega+270\text{ k}\Omega}\right)} \simeq 56$$

Increase in temp., ΔT = 107°C – 27°C = 80°C

ΔT in 10°C steps, $n = \Delta T/10°C$ = 80°C/10°C = 8

$$\therefore \quad I_{CBO}(new) = I_{CBO}(old) \times 2^n$$
$$= 15 \text{ nA} \times 2^8 = 3.84\ \mu\text{A}$$

Now, $$\Delta I_{CBO} = I_{CBO}(new) - I_{CBO}(old)$$
$$= 3.84\ \mu\text{A} - 15 \text{ nA} = 3.83\ \mu\text{A}$$

$$\therefore \quad \Delta I_C = S \times \Delta I_{CBO} = 56 \times 3.83\ \mu\text{A} = 214\ \mu\text{A}$$

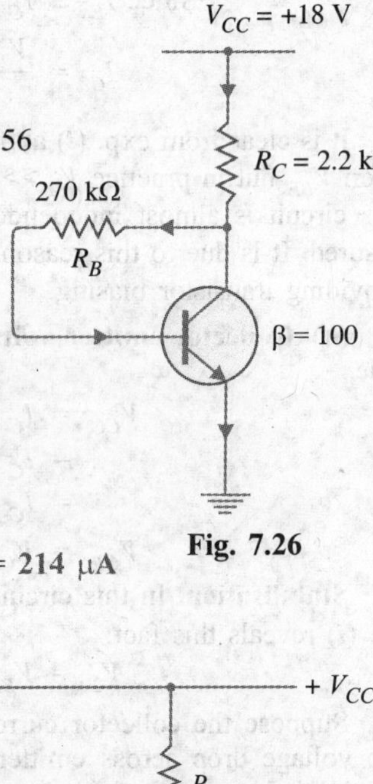

Fig. 7.26

7.13 Voltage Divider Bias Method

This is the most widely used method of providing biasing and stabilisation to a transistor. In this method, two resistances R_1 and R_2 are connected across the supply voltage V_{CC} (See Fig. 7.27) and provide biasing. The emitter resistance R_E provides stabilisation. The name "voltage divider" comes from the voltage divider formed by R_1 and R_2. The voltage drop across R_2 forward biases the base-emitter junction. This causes the base current and hence collector current flow in the zero signal conditions.

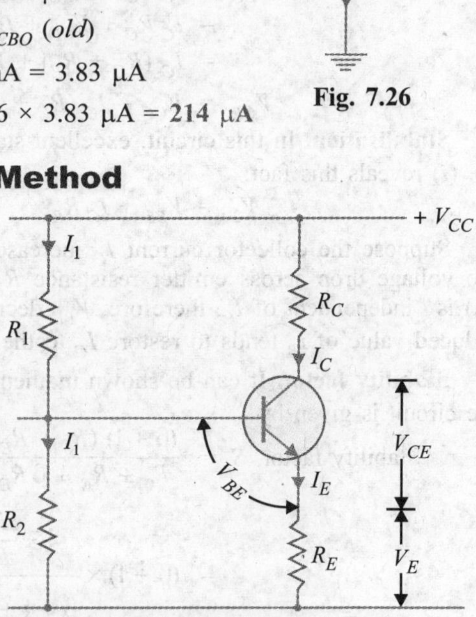

Fig. 7.27

Circuit analysis. Suppose that the current flowing through resistance R_1 is I_1. As base current I_B is very small, therefore, it can be assumed with reasonable accuracy that current flowing through R_2 is also I_1.

(*i*) **Collector current I_C .**

$$I_1 = \frac{V_{CC}}{R_1 + R_2}$$

∴ Voltage across resistance R_2 is

$$V_2 = \left(\frac{V_{CC}}{R_1 + R_2}\right) R_2$$

Applying Kirchhoff's voltage law to the base circuit of Fig. 7.27,

$$V_2 = V_{BE} + V_E$$

or $$V_2 = V_{BE} + I_E R_E$$

or $$I_E = \frac{V_2 - V_{BE}}{R_E}$$

Since $I_E \simeq I_C$

∴ $$I_C = \frac{V_2 - V_{BE}}{R_E} \quad ...(i)$$

It is clear from exp. (*i*) above that I_C does not at all depend upon β. Though I_C depends upon V_{BE} but in practice $V_2 >> V_{BE}$ so that I_C is practically independent of V_{BE}. Thus I_C in this circuit is almost independent of transistor parameters and hence good stabilisation is ensured. It is due to this reason that potential divider bias has become universal method for providing transistor biasing.

(*ii*) **Collector-emitter voltage V_{CE}.** Applying Kirchhoff's voltage law to the collector side,

$$V_{CC} = I_C R_C + V_{CE} + I_E R_E$$

$$= I_C R_C + V_{CE} + I_C R_E \quad (\because I_E \simeq I_C)$$

$$= I_C (R_C + R_E) + V_{CE}$$

∴ $$V_{CE} = V_{CC} - I_C (R_C + R_E)$$

Stabilisation. In this circuit, excellent stabilisation is provided by R_E. Consideration of eq. (*i*) reveals this fact.

$$V_2 = V_{BE} + I_C R_E$$

Suppose the collector current I_C increases due to rise in temperature. This will cause the voltage drop across emitter resistance R_E to increase. As voltage drop across R_2 (*i.e.* V_2) is *independent of I_C, therefore, V_{BE} decreases. This in turn causes I_B to decrease. The reduced value of I_B tends to restore I_C to the original value.

Stability factor. It can be shown mathematically (See Art. 7.14) that stability factor of the circuit is given by ;

$$\text{Stability factor, } S = \frac{(\beta + 1)(R_0 + R_E)}{R_0 + R_E + \beta R_E}$$

$$= (\beta + 1) \times \frac{1 + \dfrac{R_0}{R_E}}{\beta + 1 + \dfrac{R_0}{R_E}} \quad \text{where } R_0 = \frac{R_1 R_2}{R_1 + R_2}$$

---

* Voltage drop across $R_2 = \left(\dfrac{V_{CC}}{R_1 + R_2}\right) R_2$

If the ratio R_0 / R_E is very small, then R_0 / R_E can be neglected as compared to 1 and the stability factor becomes :

$$\text{Stability factor} = (\beta + 1) \times \frac{1}{\beta + 1} = 1$$

This is the smallest possible value of S and leads to the maximum possible thermal stability. Due to design *considerations, R_0 / R_E has a value that cannot be neglected as compared to 1. In actual practice, the circuit may have stability factor around 10.

Example 7.23. *Fig. 7.28 (i) shows the voltage divider bias method. Draw the d.c. load line and determine the operating point. Assume the transistor to be of silicon. What is the value of stability factor S for this bias circuit?*

Solution.

D.C. load line. The collector-emitter voltage V_{CE} is given by ;

$$V_{CE} = V_{CC} - I_C (R_C + R_E)$$

When $I_C = 0$, $V_{CE} = V_{CC} = 15\text{V}$. This locates the first point B ($OB = 15$ V) of the load line on the collector-emitter voltage axis.

$$\text{When } V_{CE} = 0,\ I_C = \frac{V_{CC}}{R_C + R_E} = \frac{15\text{ V}}{(1+2)\text{ k}\Omega} = 5\text{ mA}$$

This locates the second point A ($OA = 5$ mA) of the load line on the collector current axis. By joining points A and B, the d.c. load line AB is constructed as shown in Fig. 7.28 (*ii*).

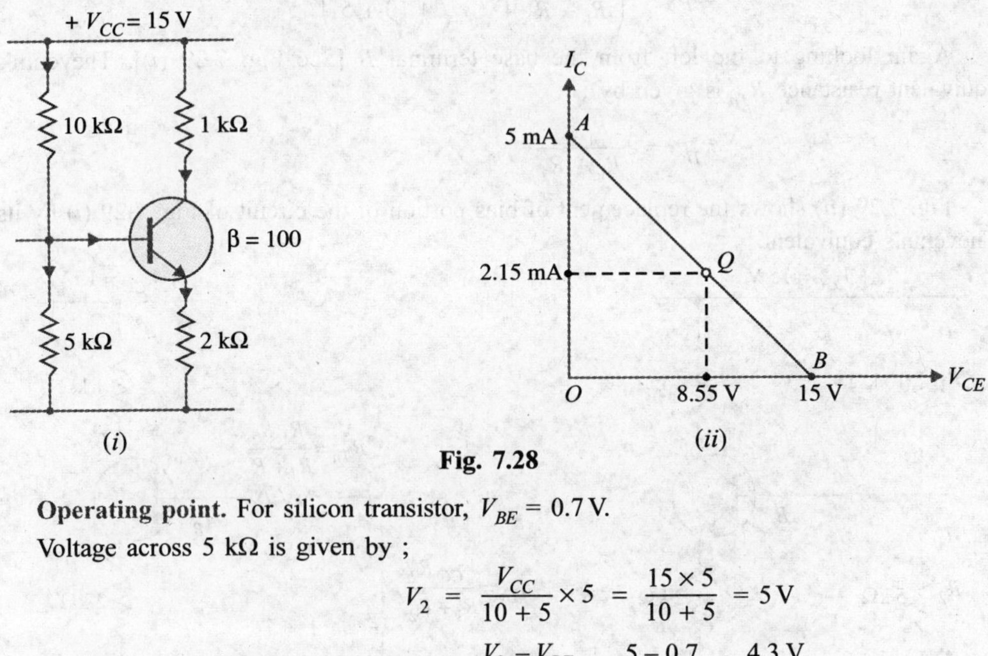

Fig. 7.28

Operating point. For silicon transistor, $V_{BE} = 0.7$ V.

Voltage across 5 kΩ is given by ;

$$V_2 = \frac{V_{CC}}{10+5} \times 5 = \frac{15 \times 5}{10+5} = 5\text{ V}$$

$$\therefore \quad \text{Emitter current, } I_E = \frac{V_2 - V_{BE}}{R_E} = \frac{5 - 0.7}{2\text{ k}\Omega} = \frac{4.3\text{ V}}{2\text{ k}\Omega} = 2.15\text{ mA}$$

* Low value of R_0 can be obtained by making R_2 very small. But with low value of R_2, current drawn from V_{CC} will be large. This puts restrictions on the choice of R_0. Increasing the value of R_E requires greater V_{CC} in order to maintain the same value of zero signal collector current. Therefore, the ratio R_0/R_E cannot be made very small from design point of view.

∴ Collector current is $I_C \simeq I_E = 2.15$ mA.

Collector-emitter voltage, $V_{CE} = V_{CC} - I_C (R_C + R_E)$

$= 15 - 2.15 \text{ mA} \times 3 \text{ k}\Omega = 15 - 6.45 = 8.55$ V

∴ Operating point is **8.55 V, 2.15 mA.**

Fig. 7.28 (*ii*) shows the operating point Q on the load line. Its co-ordinates are I_C = 2.15 mA, V_{CE} = 8.55 V.

$$\text{Stability factor, } S = (\beta + 1) \times \frac{1 + R_0 / R_E}{\beta + 1 + R_0 / R_E}$$

Here, $\beta = 100$; $R_0 = 10 \text{ k}\Omega \parallel 5 \text{ k}\Omega = 3.33 \text{ k}\Omega$; $R_0 / R_E = 3.3 \text{ k}\Omega / 2 \text{ k}\Omega = 1.66$

$$\therefore \quad S = (100 + 1) \times \frac{1 + 1.66}{100 + 1 + 1.66} = \mathbf{2.62}$$

Example 7.24. *Determine the operating point of the circuit shown in the previous problem by using Thevenin's theorem.*

Solution. The circuit is redrawn and shown in Fig. 7.29 (*i*) for facility of reference. The d.c. circuit to the left of base terminal B can be replaced by Thevenin's equivalent circuit shown in Fig. 7.29 (*ii*). Looking to the left from the base terminal B [See Fig. 7.29 (*i*)], Thevenin's equivalent voltage E_{Th} is given by ;

$$E_{Th} = \left(\frac{V_{CC}}{R_1 + R_2}\right) R_2 = \left(\frac{15}{10 + 5}\right) \times 5 = 5 \text{ V}$$

Again looking to the left from the base terminal B [See Fig. 7.29 (*i*)], Thevenin's equivalent resistance R_{Th} is given by ;

$$R_{Th} = \frac{R_1 R_2}{R_1 + R_2}$$

Fig. 7.29 (*ii*) shows the replacement of bias portion of the circuit of Fig. 7.29 (*i*) by its Thevenin's equivalent.

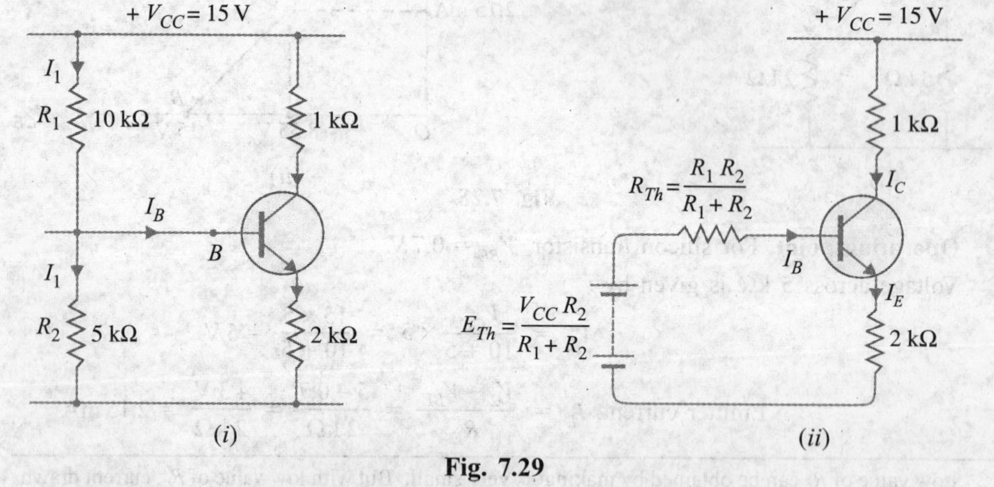

Fig. 7.29

Referring to Fig. 7.29 (*ii*), we have,

$$E_{Th} = I_B R_{Th} + V_{BE} + I_E R_E = I_B R_{Th} + V_{BE} + I_C R_E \quad (\because I_E \simeq I_C)$$

$$= I_B R_{Th} + V_{BE} + \beta I_B R_E = I_B (R_{Th} + \beta R_E) + V_{BE}$$

or $$I_B = \frac{E_{Th} - V_{BE}}{R_{Th} + \beta R_E}$$

$\therefore$ Collector current, $$I_C = \beta I_B = \frac{\beta(E_{Th} - V_{BE})}{R_{Th} + \beta R_E}$$

Dividing the numerator and denominator of R.H.S. by β, we get,

$$I_C = \frac{E_{Th} - V_{BE}}{\frac{R_{Th}}{\beta} + R_E}$$

As $^*R_{Th}/\beta << R_E$, therefore, R_{Th}/β may be neglected as compared to R_E.

$\therefore$ $$I_C = \frac{E_{Th} - V_{BE}}{R_E} = \frac{5 - 0.7}{2\ \text{k}\Omega} = 2.15 \text{ mA}$$

Also, $$V_{CE} = V_{CC} - I_C (R_C + R_E) = 15 - 2.15 \text{ mA} \times 3\ \text{k}\Omega$$
$$= 15 - 6.45 = 8.55 \text{ V}$$

$\therefore$ Operating point is **8.55 V, 2.15 mA.**

Example 7.25. *A transistor uses potential divider method of biasing. $R_1 = 50$ kΩ, $R_2 = 10$ kΩ and $R_E = 1$ kΩ. If $V_{CC} = 12$ V, find :*

(i) the value of I_C ; given $V_{BE} = 0.1$ V

(ii) the value of I_C ; given $V_{BE} = 0.3$ V. Comment on the result.

Solution. $R_1 = 50$ kΩ ; $R_2 = 10$ kΩ ; $R_E = 1$ kΩ ; $V_{CC} = 12$ V

(*i*) When $V_{BE} = 0.1$ V,

Voltage across R_2, $$V_2 = \frac{R_2}{R_1 + R_2} V_{CC} = \frac{10}{50 + 10} \times 12 = 2 \text{ V}$$

$\therefore$ Collector current, $$I_C = \frac{V_2 - V_{BE}}{R_E} = \frac{2 - 0.1}{1\ \text{k}\Omega} = \mathbf{1.9\ mA}$$

(*ii*) When $V_{BE} = 0.3$ V,

Collector current, $$I_C = \frac{V_2 - V_{BE}}{R_E} = \frac{2 - 0.3}{1\ \text{k}\Omega} = \mathbf{1.7\ mA}$$

Comments. From the above example, it is clear that although V_{BE} varies by 300%, the value of I_C changes only by nearly 10%. This explains that in this method, I_C is almost independent of transistor parameter variations.

Example 7.26. *Calculate the emitter current in the voltage divider circuit shown in Fig. 7.30. Also find the value of V_{CE} and collector potential V_C.*

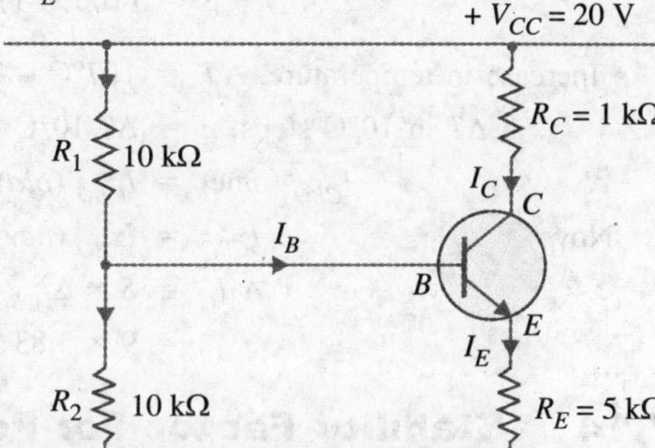

Fig. 7.30

* In fact, this condition means that I_B is very small as compared to I_1, the current flowing through R_1 and R_2.

Solution.

$$\text{Voltage across } R_2,\ V_2 = \left(\frac{V_{CC}}{R_1 + R_2}\right) R_2 = \left(\frac{20}{10+10}\right) 10 = 10 \text{ V}$$

Now, $$V_2 = V_{BE} + I_E R_E$$

As V_{BE} is generally small, therefore, it can be neglected.

$$\therefore \quad I_E = \frac{V_2}{R_E} = \frac{10 \text{ V}}{5 \text{ k}\Omega} = 2 \text{ mA}$$

Now, $$I_C \simeq I_E = 2 \text{ mA}$$

$$\therefore \quad V_{CE} = V_{CC} - I_C (R_C + R_E) = 20 - 2 \text{ mA } (6 \text{ k}\Omega)$$

$$= 20 - 12 = \mathbf{8\ V}$$

$$\text{Collector potential, } V_C = V_{CC} - I_C R_C = 20 - 2 \text{ mA} \times 1 \text{ k}\Omega$$

$$= 20 - 2 = \mathbf{18\ V}$$

Example 7.27. *The voltage divider bias circuit shown in Fig. 7.31 uses a transistor with β = 100. If the circuit temperature increases from 27°C to 107°C, find the change in collector current. Given that I_{CBO} = 15 nA at 27°C.*

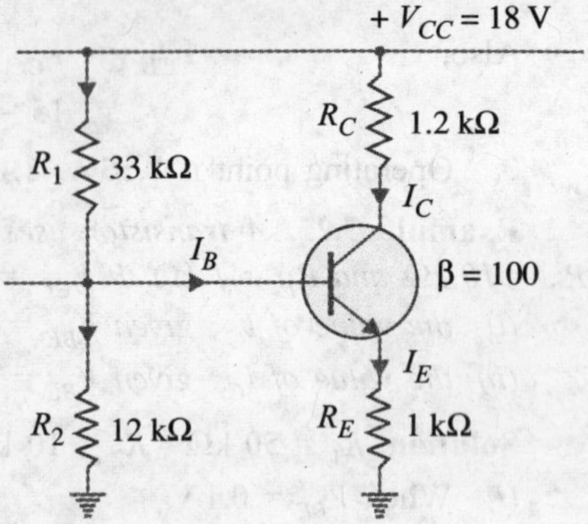

Fig. 7.31

Solution. The change in collector current (ΔI_C) due to temperature change is

$$\Delta I_C = S \times \Delta I_{CBO}$$

Here, S is the stability factor of the bias circuit and ΔI_{CBO} is the change in I_{CBO} due to temperature change. The stability factor S of this bias circuit is given by ;

$$S = (\beta + 1) \times \frac{1 + R_0 / R_E}{\beta + 1 + R_0 / R_E}$$

Here, β = 100 ; $R_0 = R_1 \parallel R_2$ = 33 kΩ||12 kΩ = 8.8 kΩ ; R_0/R_E = 8.8 kΩ/1 kΩ = 8.8

$$\therefore \quad S = (100 + 1) \times \frac{1 + 8.8}{100 + 1 + 8.8} = 9$$

$$\text{Increase in temperature, } \Delta T = 107°\text{C} - 27°\text{C} = 80°\text{C}$$

$$\Delta T \text{ in } 10°\text{C steps, } n = \Delta T/10°\text{C} = 80°\text{C}/10°\text{C} = 8$$

$$\therefore \quad I_{CBO}\ (new) = I_{CBO}\ (old) \times 2^n = 15 \text{ nA} \times 2^8 = 3.84\ \mu\text{A}$$

Now, $$\Delta I_{CBO} = I_{CBO}\ (new) - I_{CBO}\ (old) = 3.84\ \mu\text{A} - 15 \text{ nA} = 3.83\ \mu\text{A}$$

$$\therefore \quad \Delta I_C = S \times \Delta I_{CBO}$$

$$= 9 \times 3.83\ \mu\text{A} = \mathbf{34.47\ \mu A}$$

7.14 Stability Factor For Potential Divider Bias

We have already seen (See example 7.24) how to replace the potential divider circuit of potential divider bias by Thevenin's equivalent circuit. The resulting potential divider bias circuit is redrawn in Fig. 7.32 in order to find the stability factor S for this biasing circuit.

Referring to Fig. 7.32 and applying Kirchhoff's voltage law to the base circuit, we have,

$$E_{Th} - I_B R_{Th} - V_{BE} - I_E R_E = 0$$

or $$E_{Th} = I_B R_{Th} + V_{BE} + (I_B + I_C) R_E$$

Considering V_{BE} to be constant and differentiating the above equation *w.r.t.* I_C, we have,

$$0 = R_{Th} \frac{dI_B}{dI_C} + 0 + R_E \frac{dI_B}{dI_C} + R_E$$

or $$0 = \frac{dI_B}{dI_C}(R_{Th} + R_E) + R_E$$

$$\therefore \quad \frac{dI_B}{dI_C} = \frac{-R_E}{R_{Th} + R_E} \quad ...(i)$$

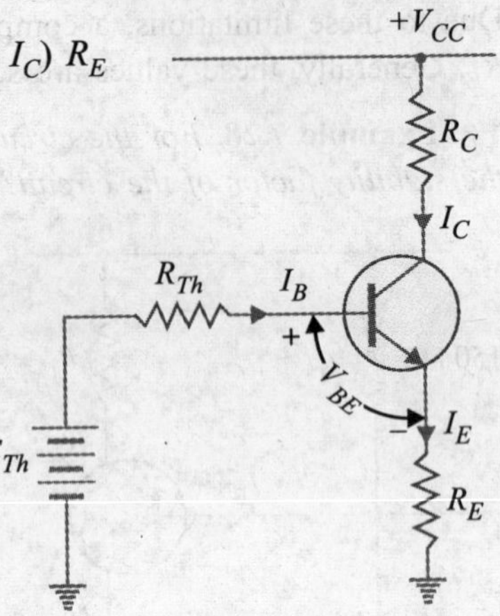

Fig. 7.32

The general expression for stability factor is

$$\text{Stability factor, } S = \frac{\beta + 1}{1 - \beta \frac{dI_B}{dI_C}}$$

Putting the value of dI_B/dI_C from eq. (*i*) into the expression for S, we have,

$$S = \frac{\beta + 1}{1 - \beta \frac{-R_E}{R_{Th} + R_E}} = \frac{\beta + 1}{1 + \left(\frac{\beta R_E}{R_{Th} + R_E}\right)}$$

$$= \frac{(\beta + 1)(R_{Th} + R_E)}{R_{Th} + R_E + \beta R_E} = \frac{(\beta + 1)(R_{Th} + R_E)}{R_{Th} + R_E(\beta + 1)}$$

$$\therefore \quad S = (\beta + 1) \times \frac{R_{Th} + R_E}{R_E(\beta + 1) + R_{Th}}$$

Dividing the numerator and denominator of R.H.S. of the above equation by R_E, we have,

$$S = (\beta + 1) \times \frac{1 + R_{Th}/R_E}{\beta + 1 + R_{Th}/R_E} \quad ...(ii)$$

Eq. (*ii*) gives the formula for the stability factor S for the potential divider bias circuit. The following points may be noted carefully :

(*i*) For greater thermal stability, the value of S should be small. This can be achieved by making R_{Th}/R_E small. If R_{Th}/R_E is made very small, then it can be neglected as compared to 1.

$$\therefore \quad S = (\beta + 1) \times \frac{1}{\beta + 1} = 1$$

This is the ideal value of S and leads to the maximum thermal stability.

(*ii*) The ratio *R_{Th}/R_E can be made very small by decreasing R_{Th} and increasing R_E. Low value of R_{Th} can be obtained by making R_2 very small. But with low value of R_2, current

* Remember, R_{Th} = Thevenin's equivalent resistance = $\frac{R_1 R_2}{R_1 + R_2}$

drawn from V_{CC} will be large. This puts restriction on the choice of R_{Th}. Increasing the value of R_E requires greater V_{CC} in order to maintain the same zero signal collector current. Due to these limitations, a compromise is made in the selection of the values of R_{Th} and R_E. Generally, these values are so selected that $S \simeq 10$.

Example 7.28. *For the circuit shown in Fig. 7.33 (i), find the operating point. What is the stability factor of the circuit? Given that* $\beta = 50$ *and* $V_{BE} = 0.7$ *V.*

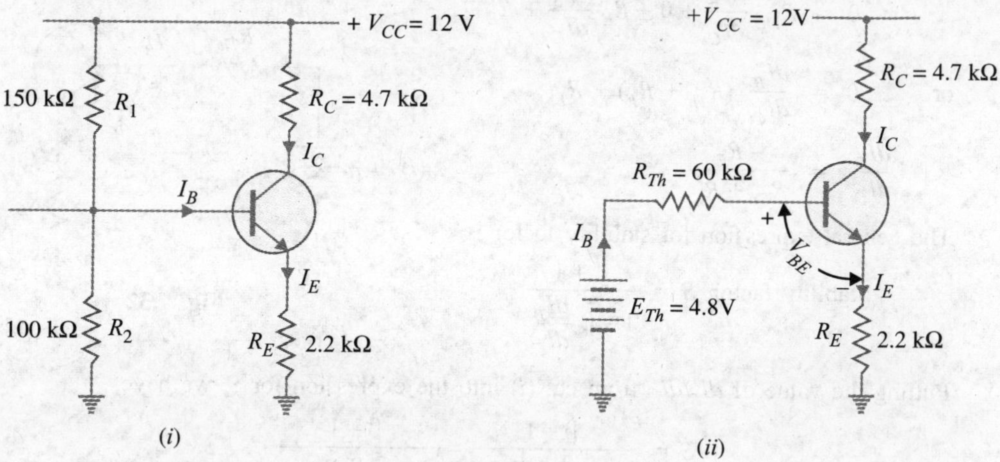

Fig. 7.33

Solution. Fig. 7.33 (*i*) shows the circuit of potential divider bias whereas Fig. 7.33 (*ii*) shows it with potential divider circuit replaced by Thevenin's equivalent circuit.

$$E_{Th} = \frac{V_{CC}}{R_1 + R_2} \times R_2 = \frac{12\text{ V}}{150\text{ k}\Omega + 100\text{ k}\Omega} \times 100\text{ k}\Omega = 4.8\text{ V}$$

$$R_{Th} = \frac{R_1 R_2}{R_1 + R_2} = \frac{150\text{ k}\Omega \times 100\text{ k}\Omega}{150\text{ k}\Omega + 100\text{ k}\Omega} = 60\text{ k}\Omega$$

$$\therefore \quad I_B = \frac{E_{Th} - V_{BE}}{R_{Th} + \beta R_E} \quad \text{... (See Ex. 7.24)}$$

$$= \frac{4.8\text{V} - 0.7\text{V}}{60\text{ k}\Omega + 50 \times 2.2\text{ k}\Omega} = \frac{4.1\text{V}}{170\text{ k}\Omega} = 0.024\text{ mA}$$

Now, $I_C = \beta I_B = 50 \times 0.024 = 1.2$ mA

$\therefore$ $V_{CE} = V_{CC} - I_C (R_C + R_E)$

$= 12\text{ V} - 1.2\text{ mA} (4.7\text{ k}\Omega + 2.2\text{ k}\Omega) = 3.72\text{ V}$

$\therefore$ Operating point is **3.72 V, 1.2 mA.**

Now, $$\frac{R_{Th}}{R_E} = \frac{60\text{ k}\Omega}{2.2\text{ k}\Omega} = 27.3$$

$\therefore$ Stability factor, $$S = (\beta + 1) \times \frac{1 + R_{Th}/R_E}{\beta + 1 + R_{Th}/R_E}$$

$$= (50 + 1) \times \frac{1 + 27.3}{50 + 1 + 27.3} = \mathbf{18.4}$$

Note. We can also find the values of I_C and V_{CE} (See Art. 7.13) as under :

$$I_C = \frac{V_2 - V_{BE}}{R_E} \quad \text{where } V_2 = \frac{V_{CC}}{R_1 + R_2} \times R_2$$

and
$$V_{CE} = V_{CC} - I_C\,(R_C + R_E)$$

However, by replacing the potential divider circuit by Thevenin's equivalent circuit, the expression for I_C can be found more accurately. If not mentioned in the problem, any one of the two methods can be used to obtain the solution.

Example 7.29. *The circuit shown in Fig. 7.34 (i) uses silicon transistor having β = 100. Find the operating point and stability factor.*

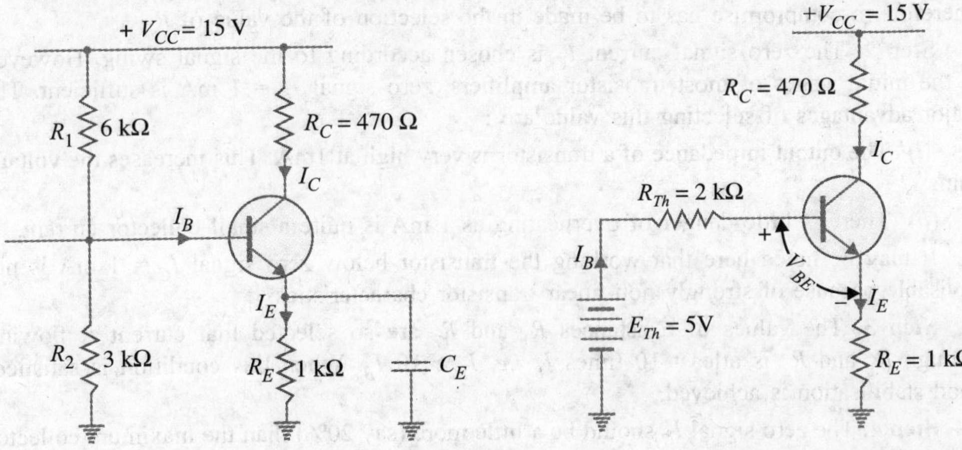

Fig. 7.34

Solution. Fig. 7.34 (*i*) shows the circuit of potential divider bias whereas Fig. 7.34 (*ii*) shows it with potential divider circuit replaced by Thevenin's equivalent circuit.

$$E_{Th} = \frac{V_{CC}}{R_1 + R_2} \times R_2 = \frac{15\text{V}}{6\text{ k}\Omega + 3\text{ k}\Omega} \times 3\text{ k}\Omega = \frac{15\text{V}}{9\text{ k}\Omega} \times 3\text{ k}\Omega = 5\text{ V}$$

$$R_{Th} = \frac{R_1 R_2}{R_1 + R_2} = \frac{6\text{ k}\Omega \times 3\text{ k}\Omega}{6\text{ k}\Omega + 3\text{ k}\Omega} = 2\text{ k}\Omega$$

Now,
$$I_B = \frac{E_{Th} - V_{BE}}{R_{Th} + \beta R_E}$$

$$= \frac{5\text{V} - 0.7\text{V}}{2\text{ k}\Omega + 100 \times 1\text{ k}\Omega} = \frac{4.3\text{V}}{102\text{ k}\Omega} = 0.042\text{ mA}$$

∴
$$I_C = \beta I_B = 100 \times 0.042 = 4.2\text{ mA}$$

and
$$V_{CE} = V_{CC} - I_C\,(R_C + R_E)$$
$$= 15\text{ V} - 4.2\text{ mA}\,(470\,\Omega + 1\text{ k}\Omega) = 8.83\text{ V}$$

∴ Operating point is **8.83V, 4.2 mA.**

Now,
$$R_{Th}/R_E = 2\text{ k}\Omega / 1\text{ k}\Omega = 2$$

∴ Stability factor,
$$S = (\beta + 1) \times \frac{1 + R_{Th}/R_E}{\beta + 1 + R_{Th}/R_E}$$

$$= (100 + 1) \times \frac{1 + 2}{100 + 1 + 2} = 2.94$$

7.15 Design of Transistor Biasing Circuits

(For low powered transistors)

In practice, the following steps are taken to design transistor biasing and stabilisation circuits :

Step 1. It is a common practice to take $R_E = 500 - 1000\ \Omega$. Greater the value of R_E, better is the stabilisation. However, if R_E is very large, higher voltage drop across it leaves reduced voltage drop across the collector load. Consequently, the output is decreased. Therefore, a compromise has to be made in the selection of the value of R_E.

Step 2. The zero signal current I_C is chosen according to the signal swing. However, in the initial stages of most transistor amplifiers, zero signal I_C = 1 mA is sufficient. The major advantages of selecting this value are :

(*i*) The output impedance of a transistor is very high at 1mA. This increases the voltage gain.

(*ii*) There is little danger of overheating as 1 mA is quite a small collector current.

It may be noted here that working the transistor below zero signal I_C = 1 mA is not advisable because of strongly non-linear transistor characteristics.

Step 3. The values of resistances R_1 and R_2 are so selected that current I_1 flowing through R_1 and R_2 is atleast 10 times I_B *i.e.* $I_1 \geq 10\ I_B$. When this condition is satisfied, good stabilisation is achieved.

Step 4. The zero signal I_C should be a little more (say 20%) than the maximum collector current swing due to signal. For example, if collector current change is expected to be 3mA due to signal, then select zero signal $I_C \simeq 3.5$ mA. It is important to note this point. Selecting zero signal I_C below this value may cut off a part of negative half-cycle of a signal. On the other hand, selecting a value much above this value (say 15 mA) may unnecessarily overheat the transistor, resulting in wastage of battery power. Moreover, a higher zero signal I_C will reduce the value of R_C (for same V_{CC}), resulting in reduced voltage gain.

Example 7.30. *In the circuit shown in Fig. 7.35, the operating point is chosen such that I_C = 2 mA, V_{CE} = 3 V. If R_C = 2.2 kΩ, V_{CC} = 9 V and β = 50, determine the values of R_1, R_2 and R_E. Take V_{BE} = 0.3 V and $I_1 = 10 I_B$.*

Solution. R_C = 2.2 kΩ ; V_{CC} = 9 V ; β = 50 ; V_{BE} = 0.3 V ; $I_1 = 10\ I_B$

As I_B is very small as compared to I_1, therefore, we can assume with reasonable accuracy that I_1 flowing through R_1 also flows through R_2.

$$\text{Base current, } I_B = \frac{I_C}{\beta} = \frac{2 \text{ mA}}{50} = 0.04 \text{ mA}$$

Current through R_1 & R_2 is

$$I_1 = 10\ I_B = 10 \times 0.04 = 0.4 \text{ mA}$$

$$\text{Now,} \quad I_1 = \frac{V_{CC}}{R_1 + R_2}$$

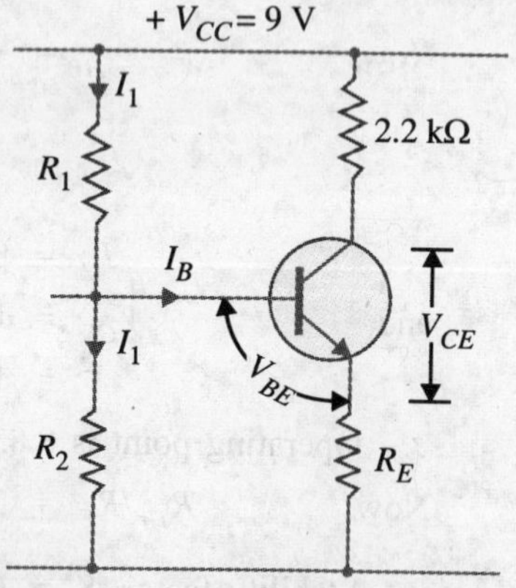

Fig 7.35

$$\therefore \quad R_1 + R_2 = \frac{V_{CC}}{I_1} = \frac{9 \text{ V}}{0.4 \text{ mA}} = 22.5 \text{ k}\Omega$$

Applying Kirchhoff's voltage law to the collector side of the circuit, we get,

$$V_{CC} = I_C R_C + V_{CE} + I_E R_E$$

or $$V_{CC} = I_C R_C + V_{CE} + I_C R_E \qquad (\because I_C \simeq I_E)$$

or $$9 = 2 \text{ mA} \times 2.2 \text{ k}\Omega + 3 + 2 \text{ mA} \times R_E$$

$$\therefore \quad R_E = \frac{9 - 4.4 - 3}{2} = 0.8 \text{ k}\Omega = \mathbf{800\ \Omega}$$

Voltage across R_2, $V_2 = V_{BE} + V_E = 0.3 + 2 \text{ mA} \times 0.8 \text{ k}\Omega$

$$= 0.3 + 1.6 = 1.9 \text{ V}$$

$\therefore$ Resistance $R_2 = V_2 / I_1 = 1.9 \text{ V}/0.4 \text{ mA} = \mathbf{4.75\ k\Omega}$

and $R_1 = 22.5 - 4.75 = \mathbf{17.75\ k\Omega}$

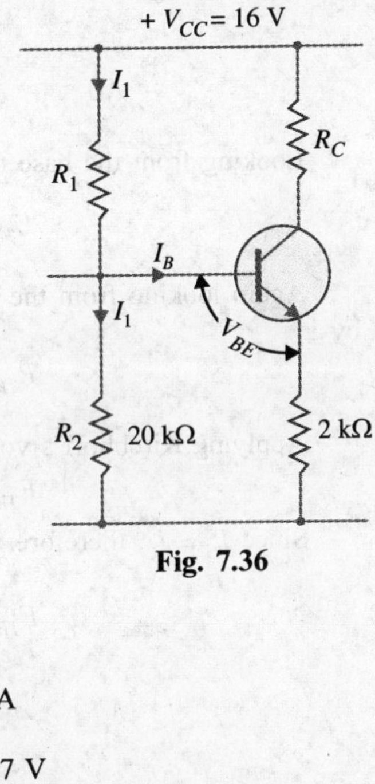

Fig. 7.36

Example 7.31. *An npn transistor circuit (See Fig. 7.36) has $\alpha = 0.985$ and $V_{BE} = 0.3$ V. If $V_{CC} = 16$ V, calculate R_1 and R_C to place Q point at $I_C = 2$ mA, $V_{CE} = 6$ volts.*

Solution. $\alpha = 0.985$; $V_{BE} = 0.3$ V ; $V_{CC} = 16$ V

$$\beta = \frac{\alpha}{1 - \alpha} = \frac{0.985}{1 - 0.985} = 66$$

$$\text{Base current, } I_B = \frac{I_C}{\beta} = \frac{2 \text{ mA}}{66} = 0.03 \text{ mA}$$

Voltage across R_2, $V_2 = V_{BE} + V_E$

$$= 0.3 + 2 \text{ mA} \times 2 \text{ k}\Omega$$

$$= 4.3 \text{ V}$$

$\therefore$ Voltage across $R_1 = V_{CC} - V_2$

$$= 16 - 4.3 = 11.7 \text{ V}$$

$\therefore$ Current through R_1 & R_2 is

$$I_1 = \frac{V_2}{R_2} = \frac{4.3 \text{ V}}{20 \text{ k}\Omega} = 0.215 \text{ mA}$$

$$\therefore \quad \text{Resistance } R_1 = \frac{\text{Voltage across } R_1}{I_1} = \frac{11.7 \text{ V}}{0.215 \text{ mA}}$$

$$= \mathbf{54.4\ k\Omega}$$

Voltage across $R_C = V_{CC} - V_{CE} - V_E$

$$= 16 - 6 - 2 \times 2 = 6 \text{ V}$$

$$\therefore \text{ Collector resistance, } R_C = \frac{\text{Voltage across } R_C}{I_C} = \frac{6 \text{ V}}{2 \text{ mA}} = \mathbf{3\ k\Omega}$$

Example 7.32. *Calculate the exact value of emitter current in the circuit shown in Fig. 7.37 (i). Assume the transistor to be of silicon and $\beta = 100$.*

Solution. In order to obtain accurate value of emitter current I_E, we shall replace the bias portion of the circuit shown in Fig. 7.37 (*i*) by its Thevenin's equivalent. Fig. 7.37 (*ii*) shows the desired circuit.

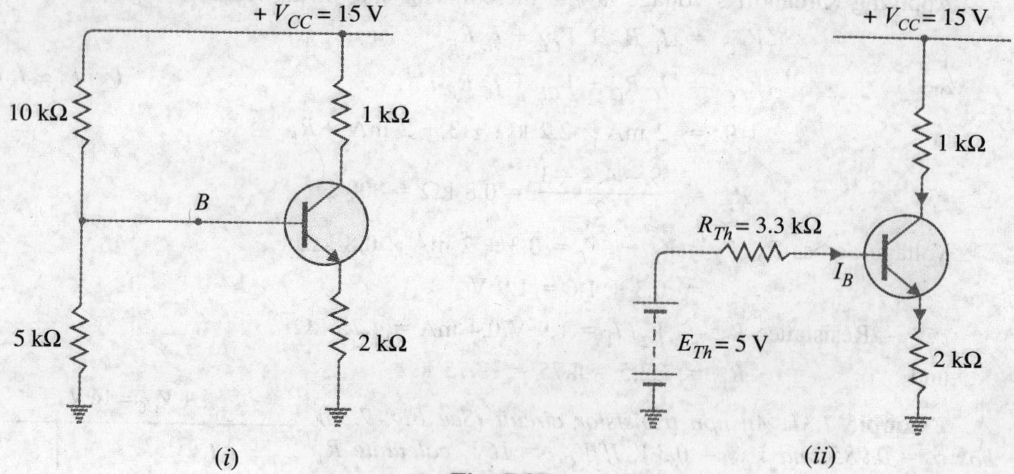

Fig. 7.37

Looking from the base terminal *B* to the left, Thevenin's voltage E_{Th} is given by ;

$$E_{Th} = \frac{R_2}{R_1 + R_2} V_{CC} = \frac{5}{10 + 5} \times 15 = 5 \text{ V}$$

Again looking from the base terminal *B* to the left, Thevenin's resistance R_{Th} is given by ;

$$R_{Th} = \frac{R_1 R_2}{R_1 + R_2} = \frac{10 \times 5}{10 + 5} = \frac{50}{15} = 3.3 \text{ k}\Omega$$

Applying Kirchhoff's voltage law to the base-emitter loop [See Fig. 7.37 (*ii*)],

$$E_{Th} = I_B R_{Th} + V_{BE} + I_E R_E$$

Since $I_E \simeq I_C$, therefore, $I_B = I_E / \beta$.

$$\therefore \quad E_{Th} = \frac{I_E}{\beta} R_{Th} + V_{BE} + I_E R_E$$

$$= I_E \left(\frac{R_{Th}}{\beta} + R_E \right) + V_{BE}$$

$$\therefore \quad I_E = \frac{E_{Th} - V_{BE}}{\frac{R_{Th}}{\beta} + R_E} = \frac{5 - 0.7}{\frac{3.3}{100} + 2} \quad \text{(For Si transistor, } V_{BE} = 0.7 \text{ V)}$$

$$= \frac{4.3 \text{ V}}{2.033 \text{ k}\Omega} = \mathbf{2.11 \text{ mA}}$$

Example 7.33. *The potential divider circuit shown in Fig. 7.38 has the values as follows: I_E = 2 mA, I_B = 50 μA, V_{BE} = 0.2 V, R_E = 1 kΩ, R_2 = 10 kΩ and V_{CC} = 10 V. Find the value of R_1.*

Solution. In this problem, we shall consider that currents through R_1 and R_2 are different, although in practice this difference is very small.

Voltage across R_2, $V_2 = V_{BE} + I_E R_E = 0.2 + 2 \text{ mA} \times 1 \text{ k}\Omega$

$= 0.2 + 2 = 2.2 \text{ V}$

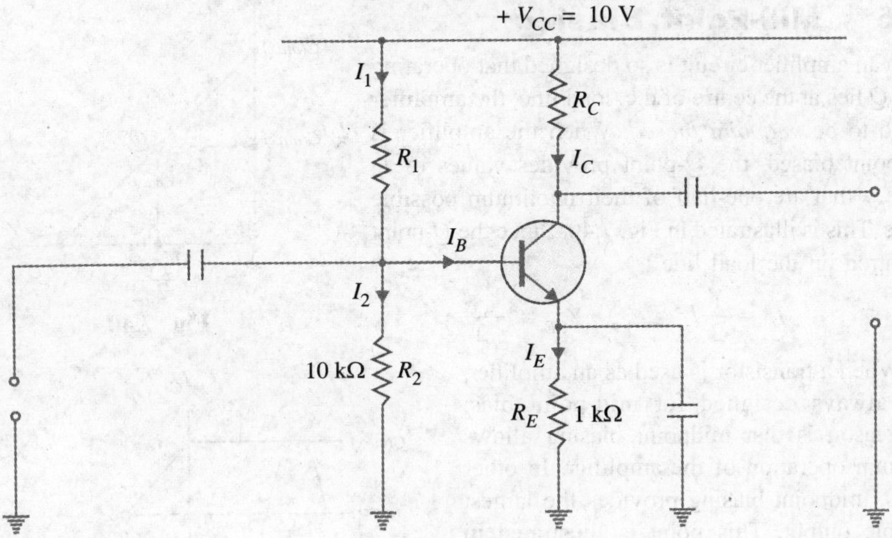

Fig. 7.38

Current through R_2, $I_2 = \dfrac{V_2}{R_2} = \dfrac{2.2 \text{ V}}{10 \text{ k}\Omega} = 0.22 \text{ mA}$

Current through R_1, $I_1 = I_2 + I_B = 0.22 + 0.05 = 0.27 \text{ mA}$

Voltage across R_1, $V_1 = V_{CC} - V_2 = 10 - 2.2 = 7.8 \text{ V}$

$\therefore \quad R_1 = \dfrac{V_1}{I_1} = \dfrac{7.8 \text{ V}}{0.27 \text{ mA}} = \mathbf{28.89 \text{ k}\Omega}$

Example 7.34. *Fig. 7.39 shows the potential divider method of biasing. What will happen if*

(i) resistance R_2 is shorted

(ii) resistance R_2 is open-circuited

(iii) resistance R_1 is shorted

(iv) resistance R_1 is open?

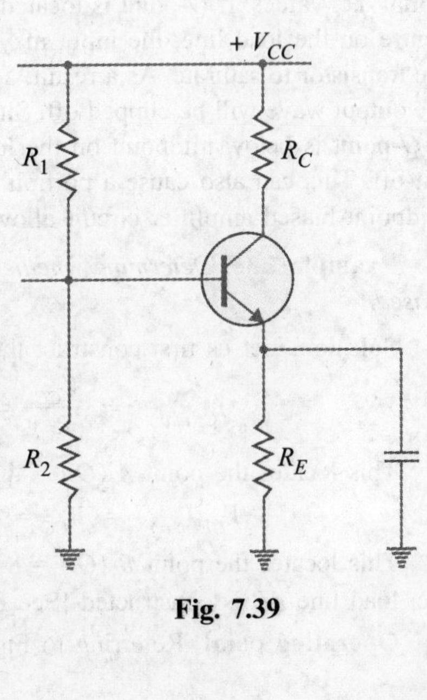

Fig. 7.39

Solution. (*i*) If resistance R_2 is shorted, the base will be grounded. **It will be left without forward bias and the transistor will be cut-off *i.e.*, output will be zero.**

(*ii*) If resistance R_2 is open, the forward bias will be very high. **The collector current will be very high while collector-emitter voltage will be very low.**

(*iii*) If resistance R_1 is shorted, the transistor will be in saturation due to excessive forward bias. **The base will be at V_{CC} and emitter will be only slightly below V_{CC}.**

(*iv*) If R_1 is open, the transistor will be without forward bias. **Hence the transistor will be cut-off *i.e.* output will be zero.**

7.16 Mid-Point Biasing

When an amplifier circuit is so designed that operating point Q lies at the centre of d.c. load line, the amplifier is said to be *midpoint biased.* When the amplifier is mid-point biased, the Q-point provides values of I_C and V_{CE} that are one-half of their maximum possible values. This is illustrated in Fig. 7.40. Since the Q-point is centred on the load line ;

$$I_C = \frac{1}{2} I_{C\,(max)}\,; \quad V_{CE} = \frac{V_{CC}}{2}$$

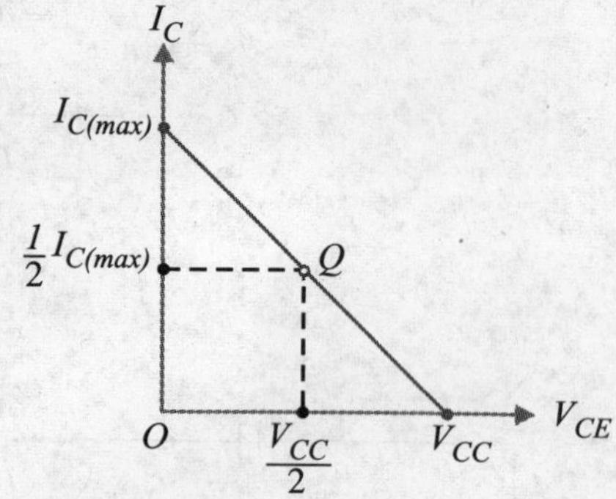

Fig. 7.40

When a transistor is used as an amplifier, it is always designed for mid-point bias. The reason is that midpoint biasing allows optimum operation of the amplifier. In other words, midpoint biasing provides the largest possible output. This point is illustrated in Fig. 7.41 where Q-point is centred on the load line.

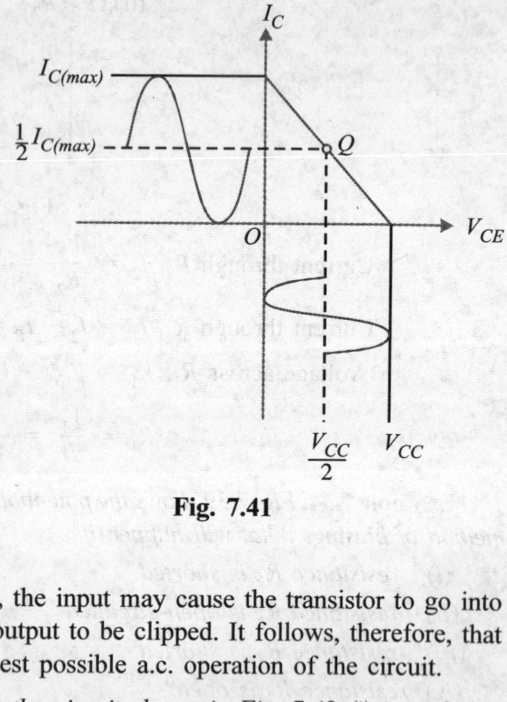

Fig. 7.41

When an a.c. signal is applied to the base of the transistor, collector current and collector- emitter voltage will both vary around their Q-point values. Since Q-point is centred, I_C and V_{CE} can both make the maximum possible transitions above and below their initial d.c. values. If Q-point is located above centre on the load line, the input may cause the transistor to saturate. As a result, a part of the output wave will be clipped off. Similarly, if Q-point is below midpoint on the load line, the input may cause the transistor to go into cut-off. This can also cause a portion of the output to be clipped. It follows, therefore, that midpoint biased amplifier circuit allows the best possible a.c. operation of the circuit.

Example 7.35. *Determine whether or not the circuit shown in Fig. 7.42 (i) is midpoint biased.*

Solution. Let us first construct the d.c. load line.

$$I_{C\,(max)} = \frac{V_{CC}}{R_C} = \frac{8\text{ V}}{2\text{ k}\Omega} = 4\text{ mA}$$

This locates the point A (OA = 4 mA) of the d.c. load line.

$$V_{CE(max)} = V_{CC} = 8\text{ V}$$

This locates the point B (OB = 8 V) of the d.c. load line. By joining these two points, d.c. load line AB is constructed [See Fig. 7.42 (*ii*)].

Operating point. Referring to Fig. 7.42 (*i*), we have,

$$I_B = \frac{V_{CC} - V_{BE}}{R_B} = \frac{8\text{ V} - 0.7\text{ V}}{360\text{ k}\Omega} = 20.28\ \mu\text{A}$$

$$\therefore \quad I_C = \beta I_B = 100\ (20.28\ \mu\text{A}) = 2.028\text{ mA}$$

Also, $$V_{CE} = V_{CC} - I_C R_C = 8\text{ V} - (2.028\text{ mA})\ (2\text{ k}\Omega) = 3.94\text{ V}$$

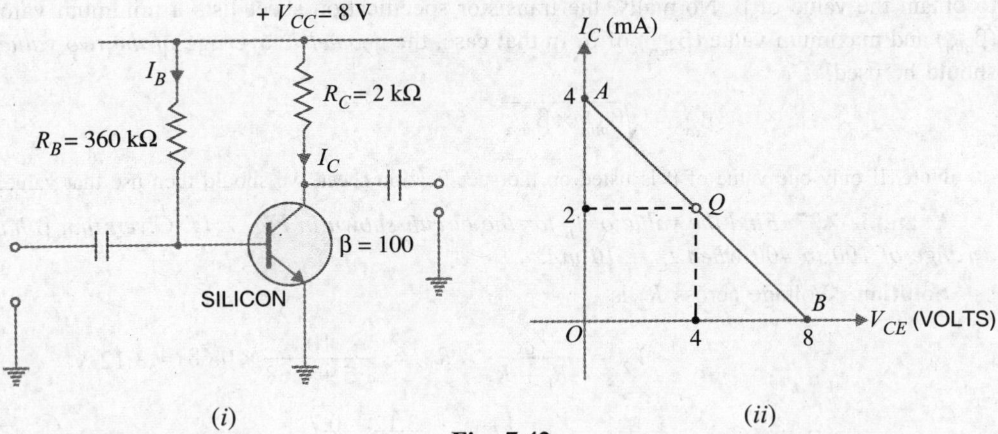

Fig. 7.42

Since V_{CE} is nearly one-half of V_{CC}, *the amplifier circuit is midpoint biased.*

Note. We can determine whether or not the circuit is midpoint biased without drawing the d.c. load line. By definition, a circuit is midpoint biased when the Q-point value of V_{CE} is one-half of V_{CC}. Therefore, all that you have to do is to find the operating point Q of the circuit. If the Q-point value of V_{CE} is one-half of V_{CC}, the circuit is midpoint biased.

Example 7.36. *Determine whether or not the circuit shown in Fig. 7.43 is midpoint biased.*

Solution. In order to determine whether the circuit is midpoint biased or not, we shall first find the operating point of the circuit.

Voltage across R_2 is

$$V_2 = \frac{V_{CC}}{R_1 + R_2} \times R_2$$

$$= \frac{10}{12 + 2.7} \times 2.7 = 1.84\text{ V}$$

∴ Emitter current is

$$I_E = \frac{V_2 - V_{BE}}{R_E}$$

$$= \frac{1.84 - 0.7}{180} = 6.33\text{ mA}$$

∴ Collector current is

$$I_C \simeq I_E = 6.33\text{ mA}$$

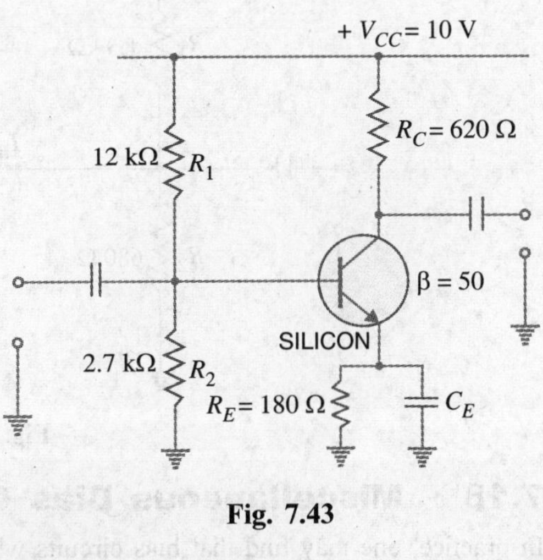

Fig. 7.43

Collector-emitter voltage is

$$V_{CE} = V_{CC} - I_C\ (R_C + R_E)$$

$$= 10 - 6.33\ (0.62 + 0.18) = 4.94\text{ V}$$

Since Q-point value of V_{CE} is approximately one-half of V_{CC} (= 10 V), *the circuit is midpoint biased.* Note that answer has been obtained without the use of a d.c. load line.

7.17 Which Value of β to be used ?

While analysing a biasing circuit, we have to refer to the specification sheet for the transistor to obtain the value of β. Normally, the transistor specification sheet lists a minimum value (β_{min}) and maximum value (β_{max}) of β. In that case, the *geometric average of the two values* should be used.

$$\beta_{av} = \sqrt{\beta_{min} \times \beta_{max}}$$

Note. If only one value of β is listed on the specification sheet, we should then use that value.

Example 7.37. *Find the value of I_B for the circuit shown in Fig. 7.44. Given that β has a range of 100 to 400 when I_C = 10 mA.*

Solution. Voltage across R_2 is

$$V_2 = \frac{V_{CC}}{R_1 + R_2} \times R_2 = \frac{10}{1.5 + 0.68} \times 0.68 = 3.12 \text{ V}$$

$$\therefore \quad \text{Emitter current, } I_E = \frac{V_2 - V_{BE}}{R_E} = \frac{3.12 - 0.7}{0.24} = 10 \text{ mA}$$

$$\therefore \quad \text{Collector current, } I_C \simeq I_E = 10 \text{ mA}$$

It is given that β has a range of 100 to 400 when Q-point value of I_C is 10 mA.

$$\therefore \quad \beta_{av} = \sqrt{\beta_{min} \times \beta_{max}} = \sqrt{100 \times 400} = 200$$

$$\therefore \quad \text{Base current, } I_B = \frac{I_E}{\beta_{av} + 1} = \frac{10 \text{ mA}}{200 + 1} = \mathbf{49.75\ \mu A}$$

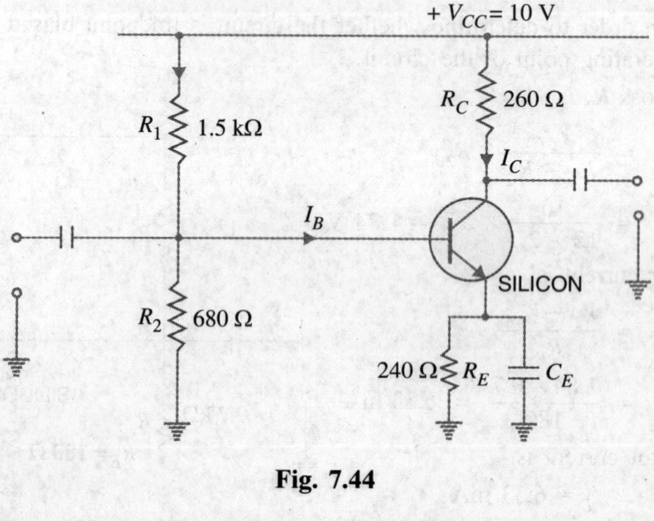

Fig. 7.44

7.18 Miscellaneous Bias Circuits

In practice, one may find that bias circuits which do not always confirm to the basic forms considered in this chapter. There may be slight circuit modifications. However, such bias circuits should not pose any problem if the basic approach to transistor biasing is understood. We shall solve a few examples to show how the basic concepts of biasing can be applied to any biasing circuit.

Example 7.38. *Calculate the operating point of the circuit shown in Fig. 7.45. Given* $\beta = 60$ *and* $V_{BE} = 0.7\ V$.

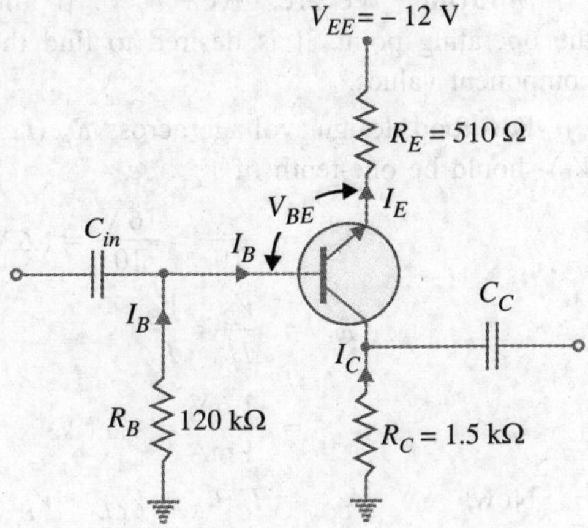

Fig. 7.45

Solution. Such a problem should not pose any difficulty. We are to simply find the d.c. values. Note that capacitors behave as an open to d.c. Applying Kirchhoff's voltage law to the path passing through R_B, V_{BE}, R_E and V_{EE}, we have,

$$-I_B R_B - V_{BE} - I_E R_E + V_{EE} = 0$$

or $$V_{EE} = I_B R_B + V_{BE} + I_C R_E \quad (\because I_E \simeq I_C)$$

or $$V_{EE} - V_{BE} = I_B R_B + \beta I_B R_E \quad (\because I_C = \beta I_B)$$

$$\therefore \quad I_B = \frac{V_{EE} - V_{BE}}{R_B + \beta R_E}$$

$$= \frac{12\text{ V} - 0.7\text{ V}}{120\text{ k}\Omega + 60 \times 0.510\text{ k}\Omega} = \frac{11.3\text{ V}}{150.6\text{ k}\Omega} = 0.075\text{ mA}$$

Now, $$I_C = \beta I_B = 60 \times 0.075\text{ mA} = 4.5\text{ mA}$$

and $$V_{CE} = V_{EE} - I_C (R_C + R_E)$$

$$= 12\text{ V} - 4.5\text{ mA}\ (1.5\text{ k}\Omega + 0.510\text{ k}\Omega) = 2.96\text{ V}$$

$\therefore$ Operating point is **2.96 V, 4.5 mA.**

Example 7.39. *Find the operating point for the circuit shown in Fig. 7.46. Assume* $\beta = 45$ *and* $V_{BE} = 0.7\ V$.

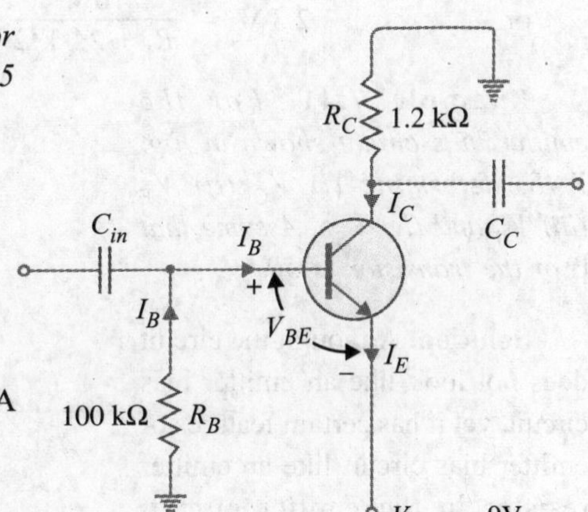

Fig. 7.46

Solution.

$$V_{EE} = I_B R_B + V_{BE}$$

or $$I_B = \frac{V_{EE} - V_{BE}}{R_B}$$

$$= \frac{(9 - 0.7)\text{ V}}{100\text{ k}\Omega} = 0.083\text{ mA}$$

$$\therefore \quad I_C = \beta I_B = 45 \times 0.083\text{ mA}$$

$$= 3.73\text{ mA}$$

Also, $$V_{EE} = I_C R_C + V_{CE}$$

$$\therefore \quad V_{CE} = V_{EE} - I_C R_C$$

$$= 9\text{ V} - (3.73\text{ mA})\ (1.2\text{ k}\Omega) = 4.52\text{ V}$$

$\therefore$ Operating point is **4.52 V, 3.73 mA.**

Example 7.40. *It is desired to design the biasing circuit of an amplifier in Fig. 7.47 in such a way to have an operating point of 6 V, 1 mA. If transistor has β = 150, find R_E, R_C, R_1 and R_2. Assume V_{BE} = 0.7 V.*

Solution. We are given V_{CC}, β and the operating point. It is desired to find the component values.

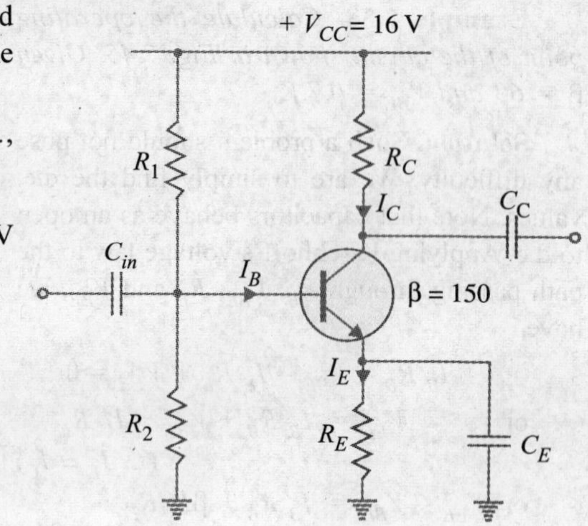

Fig. 7.47

For good design, voltage across R_E (*i.e.*, V_E) should be one-tenth of V_{CC} *i.e.*

$$V_E = \frac{V_{CC}}{10} = \frac{16\text{ V}}{10} = 1.6\text{ V}$$

$$\therefore \quad R_E = \frac{V_E}{I_E} = \frac{V_E}{I_C}$$

$$= \frac{1.6\text{ V}}{1\text{ mA}} = \mathbf{1.6\ k\Omega}$$

Now, $$V_{CC} = I_C R_C + V_{CE} + V_E$$

$$\therefore \quad R_C = \frac{V_{CC} - V_{CE} - V_E}{I_C}$$

$$= \frac{16\text{ V} - 6\text{ V} - 1.6\text{ V}}{1\text{ mA}} = \mathbf{8.4\ k\Omega}$$

$$V_2 = V_E + V_{BE} = 1.6 + 0.7 = 2.3\text{ V}$$

Now, $$R_2 = \frac{1}{10}\ {}^*(\beta R_E) = \frac{1}{10}\ (150 \times 1.6\text{ k}\Omega) = \mathbf{24\ k\Omega}$$

Also, $$V_2 = \frac{V_{CC}}{R_1 + R_2} \times R_2$$

or $$2.3\text{ V} = \frac{16\text{ V}}{R_1 + 24\text{ k}\Omega} \times 24\text{ k}\Omega \qquad \therefore R_1 = \mathbf{143\ k\Omega}$$

Example 7.41. *For the emitter bias circuit shown in Fig. 7.48, determine (i) V_C (ii) V_E (iii) V_B and (iv) V_{BC}. Assume that β of the transistor is 50.*

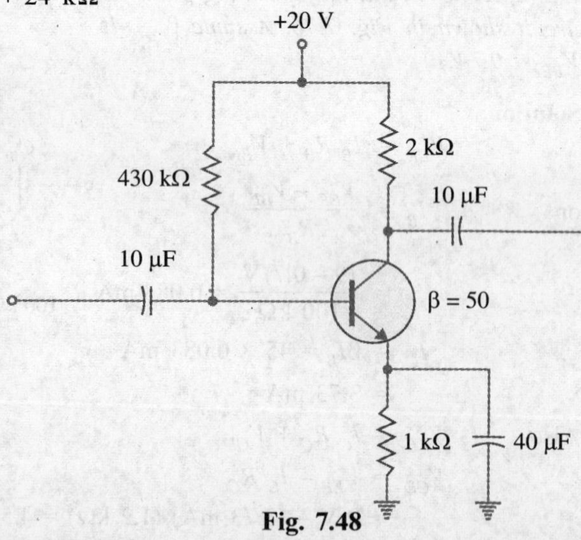

Fig. 7.48

Solution. Although the circuit does not look like an emitter bias circuit, yet it has certain features of emitter bias circuit like an emitter resistor in the emitter circuit. Applying Kirchhoff's voltage law to the base-emitter loop in Fig. 7.48, we have,

* This relation stems from $I_1 = 10\ I_B$.

$$V_{CC} - I_B R_B - V_{BE} - I_E R_E = 0$$

or $$V_{CC} - I_B R_B - V_{BE} - {}^*(\beta + 1)\, I_B R_E = 0$$

$$\therefore \quad I_B = \frac{V_{CC} - V_{BE}}{R_B + (\beta+1)R_E} = \frac{20\text{ V} - 0.7\text{ V}}{430\text{ k}\Omega + (50+1)(1\text{ k}\Omega)} = 40.1\ \mu\text{A}$$

$$I_C = \beta I_B = 50 \times 40.1\ \mu\text{A} \simeq 2.01\text{ mA}$$

$$V_{CE} = V_{CC} - I_C(R_C + R_E) = 20\text{ V} - 2.01\text{ mA }(2\text{ k}\Omega + 1\text{ k}\Omega) = 13.97\text{ V}$$

(*i*) $$V_C = V_{CC} - I_C R_C = 20\text{ V} - 2.01\text{ mA }(2\text{ k}\Omega) = 20\text{ V} - 4.02\text{ V} = \mathbf{15.98\ V}$$

(*ii*) $$V_E = V_C - V_{CE} = 15.98\text{ V} - 13.97\text{ V} = \mathbf{2.01\ V}$$

(*iii*) $$V_B = V_{BE} + V_E = 0.7\text{ V} + 2.01\text{ V} = \mathbf{2.71\ V}$$

(*iv*) $$V_{BC} = V_B - V_C = 2.71\text{ V} - 15.98\text{ V} = \mathbf{-13.27\ V}$$

Note that the negative value of V_{BC} means that the collector-base junction is reverse biased. This is required when a transistor is connected in a circuit for amplification.

Example 7.42. *Use Thevenin theorem to determine V_C and V_B for the circuit shown in Fig. 7.49. Assume that β for the transistor is 120.*

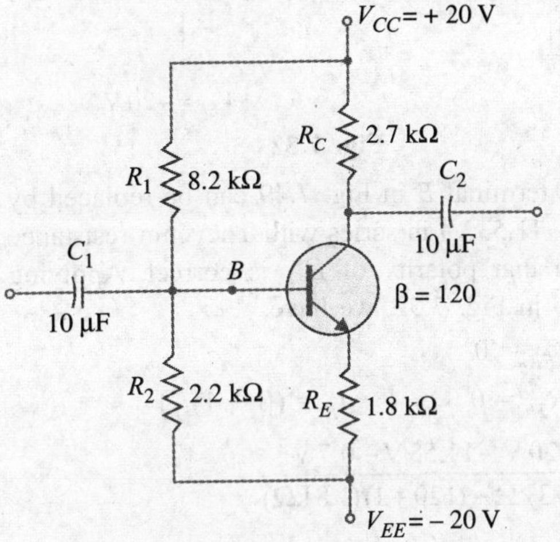

Fig. 7.49

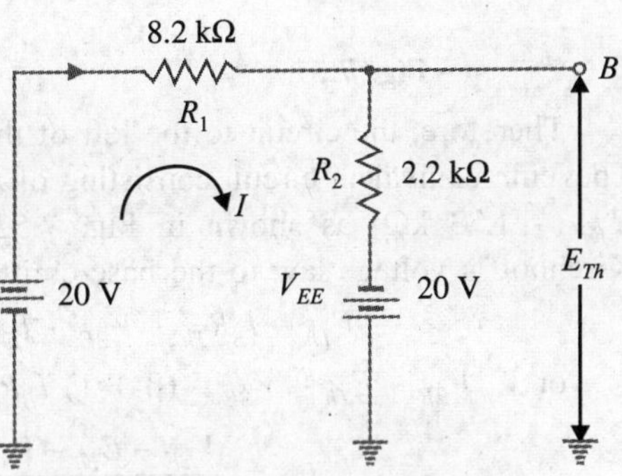

Fig. 7.50

Solution. We shall apply Thevenin theorem to the left of the base terminal *B* in Fig. 7.49 and find Thevenin voltage E_{Th} and Thevenin resistance R_{Th}.

Finding E_{Th}. In order to find E_{Th} to the left of base terminal *B* in Fig. 7.49, we use the circuit shown in Fig. 7.50.

The circuit current *I* in Fig. 7.50 is

$$I = \frac{V_{CC} + V_{EE}}{R_1 + R_2} = \frac{20\text{ V} + 20\text{ V}}{8.2\text{ k}\Omega + 2.2\text{ k}\Omega} = \frac{40\text{ V}}{10.4\text{ k}\Omega} = 3.85\text{ mA}$$

$$\therefore \quad E_{Th} = IR_2 - V_{EE} = 3.85\text{ mA} \times 2.2\text{ k}\Omega - 20\text{ V} = -11.53\text{ V}$$

* $I_E = I_B + I_C = I_B + \beta I_B = (\beta + 1) I_B$

Finding R_{Th}. In order to find R_{Th} to the left of the base terminal B in Fig. 7.49, we use the circuit shown in Fig. 7.51. Note that voltage sources V_{CC} and V_{EE} are replaced by shorts.

$\therefore$ R_{Th} = Resistance between base terminal B and ground in Fig. 7.51

$$= R_1 \parallel R_2 = 8.2 \text{ k}\Omega \parallel 2.2 \text{ k}\Omega = \frac{8.2\text{ k}\Omega \times 2.2\text{ k}\Omega}{8.2\text{ k}\Omega + 2.2\text{ k}\Omega} = 1.73 \text{ k}\Omega$$

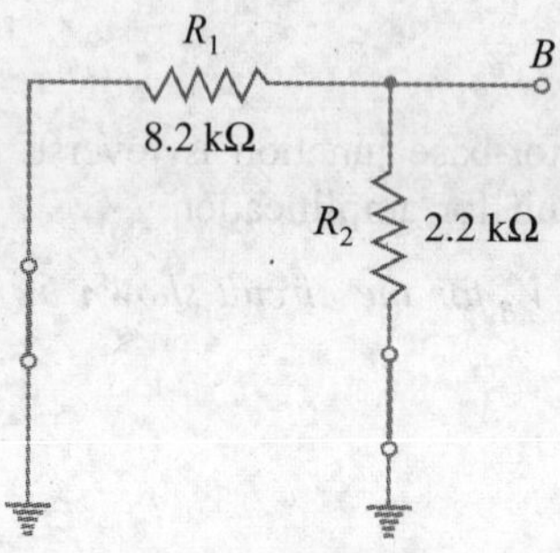

Fig. 7.51

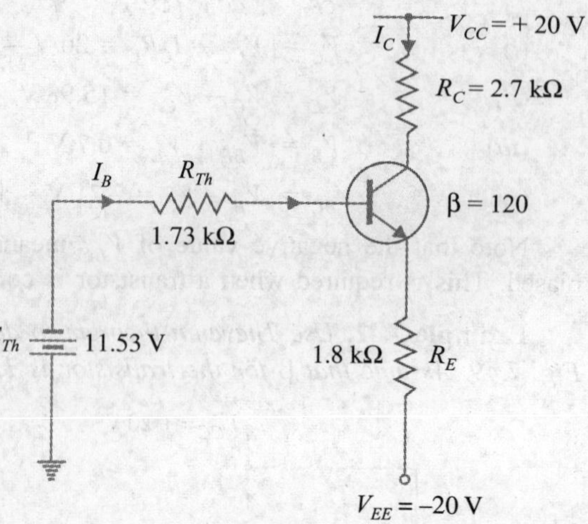

Fig. 7.52

Therefore, the circuit to the left of the base terminal B in Fig. 7.49 can be replaced by Thevenin equivalent circuit consisting of E_{Th}(= – 11.53 V) in series with Thevenin resistance R_{Th} (= 1.73 kΩ) as shown in Fig. 7.52. Note that polarity of E_{Th} is correct. Applying Kirchhoff's voltage law to the base-emitter loop in Fig. 7.52, we have,

$$-E_{Th} - I_B R_{Th} - V_{BE} - I_E R_E + V_{EE} = 0$$

or $$V_{EE} - E_{Th} - V_{BE} - (\beta + 1)\, I_B R_E - I_B R_{Th} = 0 \qquad [\because I_E = (\beta + 1) I_B]$$

$\therefore$ $$I_B = \frac{V_{EE} - E_{Th} - V_{BE}}{R_{Th} + (\beta + 1) R_E} = \frac{20\text{ V} - 11.53\text{ V} - 0.7\text{ V}}{1.73\text{ k}\Omega + (120 + 1)(1.8\text{ k}\Omega)}$$

$$= \frac{7.77\text{ V}}{219.53\text{ k}\Omega} = 35.39\ \mu\text{A}$$

Now, $I_C = \beta I_B = 120 \times 35.39\ \mu\text{A} = 4.25$ mA

(i) $\therefore$ $V_C = V_{CC} - I_C R_C = 20\text{ V} - 4.25\text{ mA} \times 2.7\text{ k}\Omega =$ **8.53 V**

(ii) $V_B = -E_{Th} - I_B R_{Th} = -11.53\text{ V} - 35.39\ \mu\text{A} \times 1.73\text{ k}\Omega =$ **– 11.59 V**

7.19 Silicon Versus Germanium

Although both silicon and germanium are used in semiconductor devices, the present day trend is to use silicon. The main reasons for this are :

(i) **Smaller I_{CBO}.** At room temperature, a silicon crystal has fewer free electrons than a germanium crystal. This implies that silicon will have much smaller collector cut-off current (I_{CBO}) than that of germanium. In general, with germanium, I_{CBO} is 10 to 100 times greater

than with silicon. The typical values of I_{CBO} at 25°C (the figures most often used for normal temperature) for small signal transistors are :

Silicon : 0.01 μA to 1 μA Germanium : 2 to 15 μA

***(ii)* Smaller variation of I_{CBO} with temperature.** The variation of I_{CBO} with temperature is less in silicon as compared to germanium. A rough rule of thumb for germanium is that I_{CBO} approximately doubles with each 8 to 10°C rise while in case of silicon, it approximately doubles with each 12°C rise.

***(iii)* Greater working temperature.** The structure of germanium will be destroyed at a temperature of approximately 100°C. The maximum normal working temperature of germanium is 70°C but silicon can be operated upto 150°C. Therefore, silicon devices are not easily damaged by excess heat.

***(iv)* Higher *PIV* rating.** The *PIV* ratings of silicon diodes are greater than those of germanium diodes. For example, the *PIV* ratings of silicon diodes are in the neighbourhood of 1000 V whereas the *PIV* ratings of germanium diodes are close to 400 V.

The disadvantage of silicon as compared to germanium is that potential barrier of silicon diode (0.7 V) is more than that of germanium diode (0.5 V). This means that higher bias voltage is required to cause current flow in a silicon diode circuit. This drawback of silicon goes to the background in view of the other advantages of silicon mentioned above. Consequently, the modern trend is to use silicon in semiconductor devices.

Example 7.43. *A small signal germanium transistor operating at 25°C has I_{CBO} = 5 μA, β = 40 and zero signal collector current = 2mA.*

(i) Find the collector cut- off current i.e. I_{CEO}.

(ii) Find the percentage change in zero signal collector current if the temperature rises to 55°C. Assume I_{CBO} doubles with every 10°C rise.

(iii) What will be the percentage change in silicon transistor under the same conditions? Given that I_{CBO} for silicon is 0.1μA at 25°C and I_{CBO} doubles for every 10°C rise.

Solution.

(i) $I_{CEO} = (\beta + 1)\, I_{CBO} = (40 + 1)\,(5\ \mu A)$

$= 205\ \mu A = \mathbf{0.205\ mA}$

(ii) Rise in temperature $= 55 - 25 = 30°C$

Since I_{CBO} doubles for every 10°C rise, the new I_{CBO} in Ge transistor at 55°C will be 8 times that at 25°C *i.e.*

New $I_{CBO} = 8 \times 5 = 40\ \mu A$

$\therefore \quad I_{CEO} = (\beta + 1)\, I_{CBO} = (40 + 1)\,(40\ \mu A)$

$= 1640\ \mu A = 1.64\ mA$

∴ Zero signal collector current at 55°C

$= 2 + 1.64 = 3.64\ mA$

Percentage change in zero signal collector current

$$= \frac{3.64 - 2}{2} \times 100 = \mathbf{82\ \%}$$

i.e., zero signal collector current rises 82% above its original value due to 30°C rise in temperature.

(*iii*) With silicon transistor,

$$I_{CBO} = 0.1\ \mu A \text{ at } 25°C \text{ and } \beta = 40$$

$$\therefore \quad I_{CEO} = (\beta + 1)\ I_{CBO} = (40 + 1)\ (0.1\ \mu A)$$

$$= 4.1\ \mu A = 0.0041 \text{ mA}$$

A 30°C rise in temperature would cause I_{CEO} in silicon to increase 8 times.

$$\text{New } I_{CEO} = 8 \times 0.0041 = 0.0328 \text{ mA}$$

∴ Zero signal collector current at 55°C

$$= 2 + 0.0328 = 2.0328 \text{ mA}$$

Percentage change in zero signal collector current

$$= \frac{2.0328 - 2}{2} \times 100 = \mathbf{1.6\ \%}$$

i.e., increase in zero signal collector current is 1.6%.

Comments. The above example shows that change in zero signal collector current with rise in temperature is very small in silicon as compared to germanium. In other words, temperature effects very slightly change the operating point of silicon transistors while they may cause a drastic change in germanium transistors. This is one of the reasons that silicon has become the main semiconductor material in use today.

Example 7.44. *A silicon transistor has I_{CBO} = 0.02 μA at 27°C. The leakage current doubles for every 6°C rise in temperature. Calculate the base current at 57°C when the emitter current is 1 mA. Given that α = 0.99.*

Solution. A 30°C (57 – 27 = 30) rise in temperature would cause I_{CBO} to increase 32 times.

$$\therefore \quad \text{At } 57°C,\ I_{CBO} = 32 \times 0.02 = 0.64\ \mu A = 0.00064 \text{ mA}$$

$$\text{Now,} \quad I_C = \alpha\, I_E + I_{CBO}$$

$$= 0.99 \times 1 + 0.00064 = 0.9906 \text{ mA}$$

$$\therefore \quad I_B = I_E - I_C = 1 - 0.9906 = 0.0094 \text{ mA} = \mathbf{9.4\ \mu A}$$

7.20 Instantaneous Current and Voltage Waveforms

It is worthwhile to show instantaneous current and voltage waveforms in an amplifier. Consider a *CE* amplifier biased by base resistor method as shown in Fig. 7.53. Typical circuit values have been assumed to make the treatment more illustrative. Neglecting V_{BE}, it is clear that zero signal base current $I_B = V_{CC}/R_B$ = 20V/1MΩ = 20 μA. The zero signal collector current $I_C = \beta I_B$ = 100 × 20 μA = 2mA. When signal of peak current 10 μA is applied, alternating current is superimposed on the d.c. base current. The collector current and collector-emitter voltage also vary as the signal changes. The instantaneous waveforms

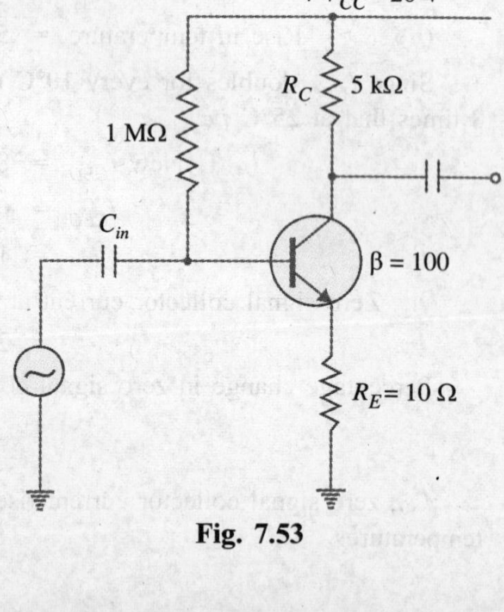

Fig. 7.53

of currents and voltages are shown in Fig. 7.54. Note that base current, collector current and collector-emitter voltage waveforms are composed of (*i*) the d.c. component and (*ii*) the a.c. wave riding on the d.c.

(*i*) At π/2 radians, the base current is composed of 20 μA d.c. component plus 10 μA peak a.c. component, adding to 30 μA *i.e.* $i_B = 20 + 10 = 30$ μA. The corresponding collector current $i_C = 100 \times 30\ \mu A = 3$ mA. The corresponding collector-emitter voltage is

$$v_{CE} = V_{CC} - i_C R_C = 20\text{ V} - 3\text{ mA} \times 5\text{ k}\Omega$$
$$= 20\text{ V} - 15\text{ V} = 5\text{ V}$$

Note that as the input signal goes positive, the collector current increases and collector-emitter voltage decreases. Moreover, during the positive half-cycle of the signal (*i.e.* from 0 to π rad.), the operating point moves from 20 μA to 20 + 10 = 30 μA and then back again *i.e.* operating point follows the path *Q* to *C* and back to *Q* on the load line.

(*ii*) During the negative half-cycle of the signal (from π to 2π rad.), the operating point goes from 20 μA to 20 − 10 = 10 μA and then back again *i.e.* the operating point follows the path *Q* to *D* and back to *Q* on the load line.

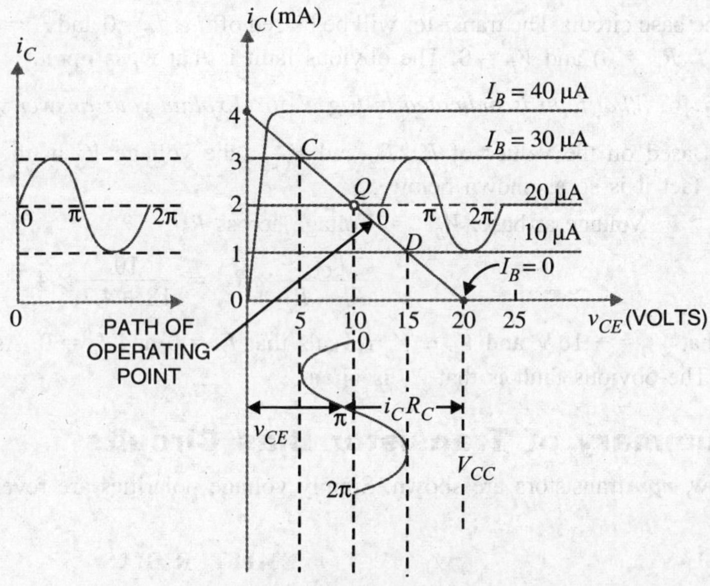

Fig. 7.54

(*iii*) As the operating point moves along the path *CD* or *DC* due to the signal, the base current varies continuously. These variations in the base current cause both collector current and collector-emitter voltage to vary.

(*iv*) Note that when the input signal is maximum positive, the collector-emitter voltage is maximum negative. In other words, input signal voltage and output voltage have a phase difference of 180°. This is an important characteristic of *CE* arrangement.

Example 7.45. *What fault is indicated in Fig. 7.55? Explain your answer with reasons.*

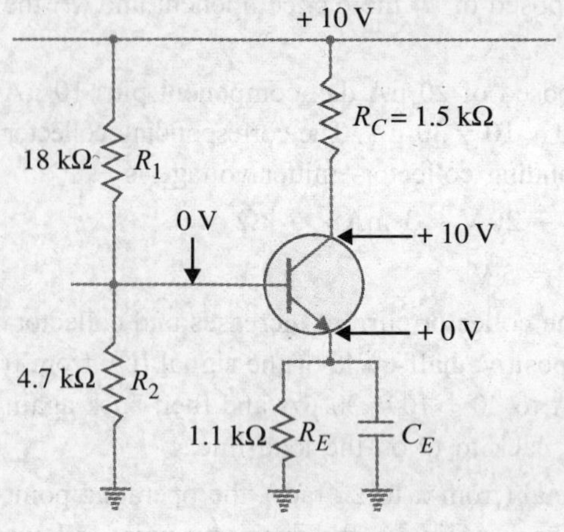

Fig. 7.55

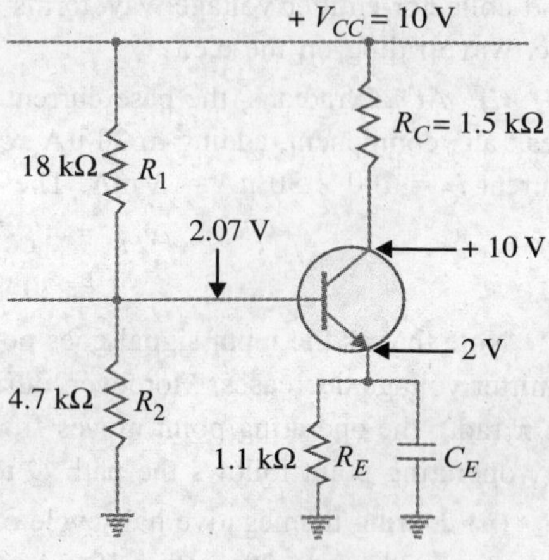

Fig. 7.56

Solution. Since V_B (*i.e.* base voltage *w.r.t.* ground) is zero, it means that there is no path for current in the base circuit. The transistor will be biased off *i.e.* $I_C = 0$ and $I_E = 0$. Therefore, $V_C = 10$ V ($\because I_C R_C = 0$) and $V_E = 0$. The obvious fault is that R_1 **is open.**

Example 7.46. *What fault is indicated in Fig. 7.56? Explain your answer with reasons.*

Solution. Based on the values of R_1, R_2 and V_{CC} , the voltage V_B at the base seems appropriate. In fact it is so as shown below :

$$\text{Voltage at base, } V_B = \text{Voltage across } R_2$$

$$= \frac{V_{CC}}{R_1 + R_2} \times R_2 = \frac{10}{18 + 4.7} \times 4.7 = 2.07 \text{ V}$$

The fact that $V_C = +\ 10$ V and $V_E \simeq V_B$ reveals that $I_C = 0$ and $I_E = 0$. As a result, I_B drops to zero. The obvious fault is that R_E **is open.**

7.21 Summary of Transistor Bias Circuits

In figures below, *npn* transistors are shown. Supply voltage polarities are reversed for *pnp* transistors.

BASE BIAS		EMITTER BIAS	
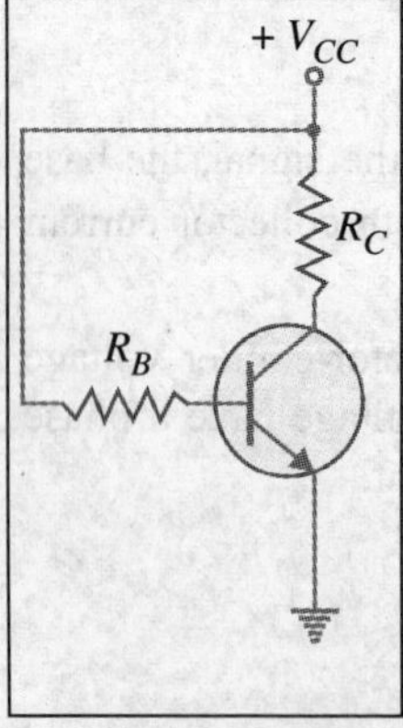	• *Q*-point values ($I_C \simeq I_E$) • Collector current : $I_C = \beta\left(\frac{V_{CC} - V_{BE}}{R_B}\right)$ • Collector-to-emitter voltage : $V_{CE} = V_{CC} - I_C R_C$	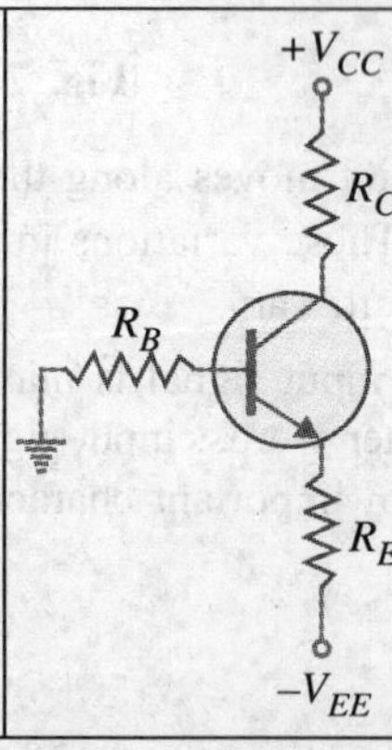	• *Q*-point values ($I_C \simeq I_E$) • Collector current : $I_C \simeq \left(\frac{V_{EE} - V_{BE}}{R_E + R_B / \beta}\right)$ • Collector-to-emitter voltage : $V_{CE} \simeq V_{CC} + V_{EE} - I_C(R_C + R_E)$

VOLTAGE-DIVIDER BIAS		COLLECTOR-FEEDBACK BIAS	
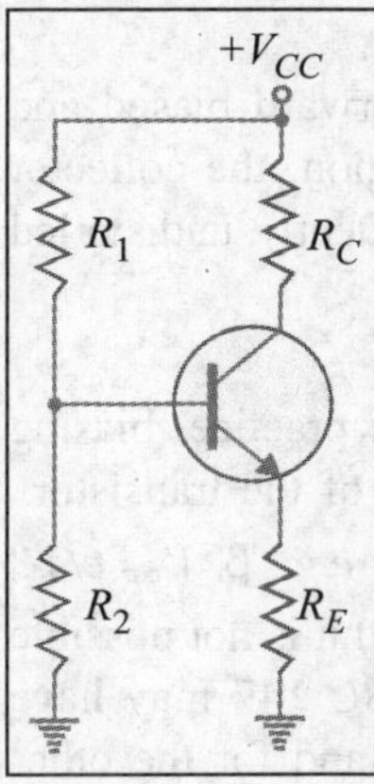	• Q-point values ($I_C \simeq I_E$) • Collector current : $I_C \simeq \dfrac{\left(\dfrac{R_2}{R_1+R_2}\right)V_{CC} - V_{BE}}{R_E}$ • Collector-to-emitter voltage : $V_{CE} \simeq V_{CC} - I_C(R_C + R_E)$	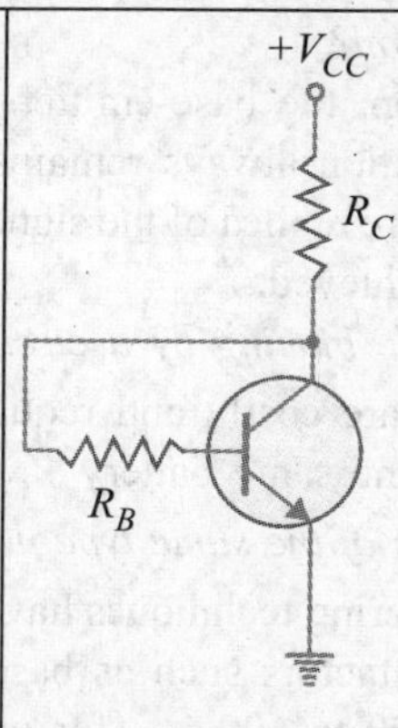	• Q-point values ($I_C \simeq I_E$) • Collector current : $I_C \simeq \left(\dfrac{V_{CC} - V_{BE}}{R_C + R_B/\beta}\right)$ • Collector-to-emitter voltage : $V_{CE} \simeq V_{CC} - I_C R_C$

The following points may be noted :

(*i*) Out of the four bias circuits above, the voltage-divider bias circuit has the lowest value of stability factor $S(= 10)$. Therefore, voltage-divider bias circuit provides excellent *thermal stability and is always preferred. The next bias circuit is the emitter bias circuit. A properly designed emitter bias circuit can provide S value comparable to that of voltage-divider bias circuit. However, fixed bias circuit has very poor thermal stability because its stability factor $S = (\beta + 1)$.

(*ii*) The stability factor (S) of a bias circuit not only depends upon the type of bias circuit but it also depends upon the resistor values.

(*iii*) A low value of stability factor means greater thermal stability. When the stability factor of a bias circuit is high, it not only changes the operating point significantly but it may also lead to *thermal runaway.*

Short Answer Questions

Q.1. *What is the universal rule for biasing a transistor for normal operation?*

Ans. The rule is : The emitter-base junction must be forward biased and the collector-base junction must be reverse biased.

Q.2. *Why should V_{CE} not fall below 0.5 V for Ge transistors and 1 V for Si transistors for faithful amplification?*

Ans. These are the minimum values required to keep the collector-base junction properly reverse biased. Allowing V_{CE} to fall below these values would result in collector not gathering all the charge carriers that reach its depletion layer. This will decrease I_C and increase I_B, making β to fall. Hence that part of signal for which V_{CE} falls below these values (0.5 V for Ge transistors and 1 V for Si transistors) will be amplified less due to reduced β, resulting in unfaithful amplification.

Q.3. *Why should V_{BE} not fall below 0.5 V for Ge transistors and 0.7 V for Si transistors for faithful amplification?*

Ans. The potential barrier has a value of 0.5 V for Ge transistors and 0.7 V for Si transistors. The base-emitter voltage must be more than these values in order to keep the input circuit always properly forward biased.

---

* *i.e.* operating point *almost* remains fixed when temperature changes.

Q.4. *Why should transistor operate in the active region of output characteristics for faithful amplification?*

Ans. In the active region, the base-emitter junction always remains forward biased and collector-base junction always remains reverse biased. In this region, the collector current variations are replica of the signal current variations. Consequently, undistorted amplification is achieved.

Q.5. *Why do we provide biasing by a circuit and not by a battery?*

Ans. Because batteries are costly and require frequent replacement. In practice, biasing is provided from the same battery V_{CC} used in the output circuit of the transistor.

Q.6. *Why do transistors of the same type not have the same parameters e.g., β, V_{BE} etc.?*

Ans. Because manufacturing techniques have not advanced too much and it is not possible to control several factors such as base width *etc.* For example, *BC* 147 may have β ranging from 100 to 600 *i.e.*, for one transistor it may be 100 and for the other it may be 600.

Q.7. *The stability factor of a biasing circuit is 10. What does it mean?*

Ans.
$$S = \frac{dI_C}{dI_{CBO}}$$

or
$$dI_C = S \times dI_{CBO} = 10 \times dI_{CBO}$$

Thus $S = 10$ means that I_C changes 10 times as much as any change in I_{CBO}. For example, if I_{CBO} changes by 2 μA due to temperature variations, then I_C will change by 20 μA.

Q.8. *What is the effect of temperature on transistors?*

Ans. The two most temperature sensitive quantities are : (*i*) base-emitter voltage V_{BE} and (*ii*) collector leakage current I_{CBO}. The temperature co-efficient of V_{BE} is approximately – 2.2 mV/°C for a silicon transistor and – 1.8 mV/°C for a germanium transistor. The leakage current I_{CBO} approximately doubles for each 10°C temperature rise.

Q.9. *Why is voltage divider bias the most widely used method?*

Ans. It is because it provides very good stabilisation of operating point in the face of variations of temperature. The stability factor *S* of this circuit is given by ;

$$S = (\beta + 1) \times \frac{1 + R_0 / R_E}{\beta + 1 + R_0 / R_E} \quad \text{where } R_0 = \frac{R_1 R_2}{R_1 + R_2}$$

By making R_0/R_E very small, it can be neglected as compared to 1.

$$\therefore \quad S = (\beta + 1) \frac{1}{\beta + 1} = 1$$

This means that if I_{CBO} changes by 2 μA, then I_C will also change by 2 μA. Clearly, it is a very good thermal stability.

Q.10. *Why does the point of intersection of d.c. load line and base current curve under study give the operating point?*

Ans. The operating point lies on the d.c. load line. It also lies somewhere on the base current curve under study. Therefore, operating point will be at the point where d.c. load line intersects the base current curve under study.

Q.11. *What should be the minimum values of V_{BE} and V_{CE} for faithful amplification?*

Ans. For faithful amplification, the value of V_{BE} should not fall below 0.5 V for Ge transistors and 0.7 V for silicon transistors. Similarly, V_{CE} should not fall below 1 V for silicon transistors and 0.5 V for Ge transistors. This is shown diagrammatically in Fig. 7.57.

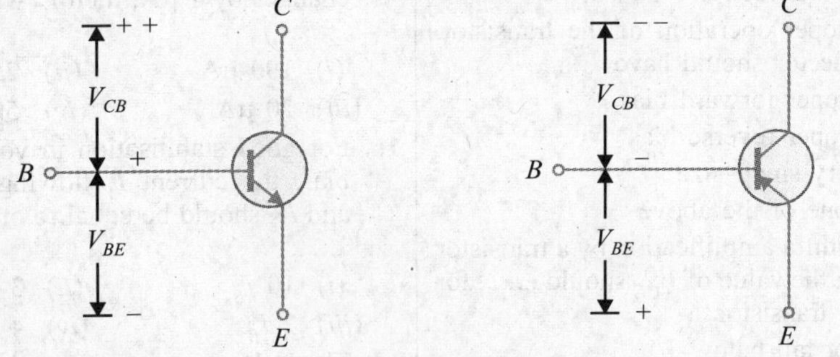

Fig. 7.57

If the voltages do not fall below these limiting values, base-emitter junction will remain forward biased and collector-base junction will effectively remain reverse biased. This will ensure proper functioning of transistor and as a result faithful amplification takes place.

Q.12. *What is thermal runaway?*

Ans. When temperature increases, I_{CBO} increases significantly (I_{CBO} doubles for every 10°C rise in temperature). An increase in I_{CBO} increases the collector current I_C. The I_C increase raises the temperature of collector-base junction. This, in turn, results in a further increase in I_{CBO}. The effect is cummulative so that the end result is a substantial increase in I_C. In the worst case, I_C *might go on increasing until the collector-base junction of the transistor overheats and burns out.* This effect is known as *thermal runaway.*

Multiple-Choice Questions

1. Transistor biasing represents .... conditions.
   (*i*) a.c.
   (*ii*) d.c.
   (*iii*) both a.c. and d.c.
   (*iv*) none of the above
2. Transistor biasing is done to keep .... in the circuit.
   (*i*) proper direct current
   (*ii*) proper alternating current
   (*iii*) the base current small
   (*iv*) collector current small
3. Operating point represents ....
   (*i*) values of I_C and V_{CE} when signal is applied
   (*ii*) the magnitude of signal
   (*iii*) zero signal values of I_C and V_{CE}
   (*iv*) none of the above
4. If biasing is not done in an amplifier circuit, it results in ....
   (*i*) decrease in base current
   (*ii*) unfaithful amplification
   (*iii*) excessive collector bias
   (*iv*) none of the above
5. Transistor biasing is generally provided by a ....
   (*i*) biasing circuit (*ii*) bias battery
   (*iii*) diode (*iv*) none of the above
6. For faithful amplification by a transistor

circuit, the value of V_{BE} should .... for a silicon transistor.

(*i*) be zero
(*ii*) be 0.01 V
(*iii*) not fall below 0.7 V
(*iv*) be between 0 V and 0.1 V

7. For proper operation of the transistor, its collector should have ....

(*i*) proper forward bias
(*ii*) proper reverse bias
(*iii*) very small size
(*iv*) none of the above

8. For faithful amplification by a transistor circuit, the value of V_{CE} should .... for silicon transistor.

(*i*) not fall below 1 V
(*ii*) be zero
(*iii*) be 0.2 V
(*iv*) none of the above

9. The circuit that provides the best stabilisation of operating point is ....

(*i*) base resistor bias
(*ii*) collector-feedback bias
(*iii*) potential-divider bias
(*iv*) none of the above

10. The point of intersection of d.c. and a.c. load lines represents ....

(*i*) operating point (*ii*) current gain
(*iii*) voltage gain (*iv*) none of the above

11. An ideal value of stability factor is ....

(*i*) 100 (*ii*) 200
(*iii*) more than 200 (*iv*) 1

12. The zero signal I_C is generally ...... mA in the initial stages of a transistor amplifier.

(*i*) 4 (*ii*) 1
(*iii*) 3 (*iv*) more than 10

13. If the maximum collector current due to signal alone is 3 mA, then zero signal collector current should be atleast equal to ....

(*i*) 6 mA (*ii*) 1.5 mA
(*iii*) 3 mA (*iv*) 1 mA

14. The disadvantage of base resistor method of transistor biasing is that it ....

(*i*) is complicated
(*ii*) is sensitive to changes in β
(*iii*) provides high stability
(*iv*) is none of the above

15. The biasing circuit has a stability factor of 50. If due to temperature change, I_{CBO} changes by 1 μA, then I_C will change by ....

(*i*) 100 μA (*ii*) 25 μA
(*iii*) 20 μA (*iv*) 50 μA

16. For good stabilisation in voltage-divider bias, the current I_1 flowing through R_1 and R_2 should be equal to or greater than ....

(*i*) 10 I_B (*ii*) 3 I_B
(*iii*) 2 I_B (*iv*) 4 I_B

17. The leakage current in a silicon transistor is about .... the leakage current in a germanium transistor.

(*i*) one hundredth (*ii*) one tenth
(*iii*) one thousandth (*iv*) one millionth

18. The operating point is also called the ....

(*i*) cut-off point
(*ii*) quiescent point
(*iii*) saturation point
(*iv*) none of the above

19. For proper amplification by a transistor circuit, the operating point should be located at .... of the d.c. load line.

(*i*) the end point
(*ii*) the middle
(*iii*) the maximum current point
(*iv*) none of the above

20. The operating point .... on the a.c. load line.

(*i*) also lies (*ii*) does not lie
(*iii*) may or may not lie
(*iv*) data insufficient

21. The disadvantage of voltage-divider bias is that it has ....

(*i*) high stability factor
(*ii*) low base current
(*iii*) many resistors
(*iv*) none of the above

22. Thermal runaway occurs when ....

(*i*) collector is reverse biased
(*ii*) transistor is not biased

(iii) emitter is forward biased
(iv) junction capacitance is high

23. The purpose of resistance in the emitter circuit of a transistor amplifier is to ....
(i) limit the maximum emitter current
(ii) provide base-emitter bias
(iii) limit the change in emitter current
(iv) none of the above

24. In a transistor amplifier circuit, $V_{CE} = V_{CB} +$ ....
(i) V_{BE} (ii) 2 V_{BE}
(iii) 1.5 V_{BE} (iv) none of the above

25. The base resistor method is generally used in ....
(i) amplifier circuits
(ii) switching circuits
(iii) rectifier circuits
(iv) none of the above

26. For germanium transistor amplifier, V_{CE} should .... for faithful amplification.
(i) be zero
(ii) be 0.2 V
(iii) not fall below 0.7 V
(iv) be none of the above

27. In a base resistor method, if the value of β changes by 50, then collector current will change by a factor of ....
(i) 25 (ii) 50
(iii) 100 (iv) 200

28. The stability factor of a collector feedback bias circuit is .... that of base resistor bias.
(i) the same as (ii) more than
(iii) less than (iv) none of the above

29. In the design of a biasing circuit, the value of collector load R_C is determined by ....
(i) V_{CE} consideration
(ii) V_{BE} consideration
(iii) I_B consideration
(iv) none of the above

30. If the value of collector current I_C increases, then value of V_{CE} ....
(i) remains the same
(ii) decreases
(iii) increases
(iv) none of the above

31. If the temperature increases, the value of V_{BE} ....
(i) remains the same
(ii) is increased
(iii) is decreased
(iv) none of the above

32. The stabilisation of operating point in potential-divider method is provided by ....
(i) R_E consideration
(ii) R_C consideration
(iii) V_{CC} consideration
(iv) none of the above

33. The value of V_{BE} ....
(i) depends upon I_C to moderate extent
(ii) is almost independent of I_C
(iii) is strongly dependent on I_C
(iv) is none of the above

34. When the temperature changes, the operating point is shifted due to ....
(i) change in I_{CBO}
(ii) change in V_{CC}
(iii) change in the values of circuit resistances
(iv) none of the above

35. The value of stability factor for a base-resistor bias is ....
(i) R_B (β + 1)
(ii) (β + 1) R_C
(iii) (β + 1)
(iv) 1 − β

36. In a practical biasing circuit, the value of R_E is about ....
(i) 10 kΩ (ii) 1 MΩ
(iii) 100 kΩ (iv) 800 Ω

37. A silicon transistor is biased with base resistor method. If $\beta = 100$, $V_{BE} = 0.7$ V, zero signal collector current $I_C = 1$ mA and $V_{CC} = 6$ V, what is the value of base resistor R_B?
   (*i*) 105 kΩ
   (*ii*) 530 kΩ
   (*iii*) 315 kΩ
   (*iv*) none of the above

38. In voltage-divider bias, $V_{CC} = 25$ V; $R_1 = 10$ kΩ; $R_2 = 2.2$ kΩ; $R_C = 3.6$ kΩ and $R_E = 1$ kΩ. What is the emitter voltage?
   (*i*) 6.7 V (*ii*) 5.3 V
   (*iii*) 4.9 V (*iv*) 3.8 V

39. In the above question, what is the collector voltage?
   (*i*) 12.3 V (*ii*) 14.8 V
   (*iii*) 7.6 V (*iv*) 9.7 V

40. In voltage-divider bias, operating point is 3 V, 2 mA. If $V_{CC} = 9$ V, $R_C = 2.2$ kΩ, what is the value of R_E?
   (*i*) 2000 Ω (*ii*) 1400 Ω
   (*iii*) 800 Ω (*iv*) 1600 Ω

Answers to Multiple-Choice Questions

1. (*ii*)	2. (*i*)	3. (*iii*)	4. (*ii*)	5. (*i*)
6. (*iii*)	7. (*ii*)	8. (*i*)	9. (*iii*)	10. (*i*)
11. (*iv*)	12. (*ii*)	13. (*iii*)	14. (*ii*)	15. (*iv*)
16. (*i*)	17. (*iii*)	18. (*ii*)	19. (*ii*)	20. (*i*)
21. (*iii*)	22. (*ii*)	23. (*iii*)	24. (*i*)	25. (*ii*)
26. (*iii*)	27. (*ii*)	28. (*iii*)	29. (*i*)	30. (*ii*)
31. (*iii*)	32. (*i*)	33. (*ii*)	34. (*i*)	35. (*iii*)
36. (*iv*)	37. (*ii*)	38. (*iv*)	39. (*i*)	40. (*iii*)

Chapter Review Topics

1. What is faithful amplification? Explain the conditions to be fulfilled to achieve faithful amplification in a transistor amplifier.
2. What do you understand by transistor biasing? What is its need?
3. What do you understand by stabilisation of operating point?
4. Mention the essentials of a biasing circuit.
5. Describe the various methods used for transistor biasing. State their advantages and disadvantages.
6. Describe the potential-divider method in detail. How stabilisation of operating point is achieved by this method?
7. Mention the steps that are taken to design the transistor biasing and stabilisation circuits.
8. Write short notes on the following :
   (*i*) Operating point (*ii*) Stabilisation of operating point.

Problems

1. An *npn* silicon transistor has $V_{CC} = 5$ V and the collector load $R_C = 2$ kΩ. Find :
   (*i*) the maximum collector current that can be allowed during the application of signal for faithful amplification
   (*ii*) the minimum zero signal collector current required. **[(*i*) 2 mA (*ii*) 1 mA]**

2. Fig. 7.58 shows biasing with base resistor method. Determine the operating point. Assume the transistor to be of silicon and take $\beta = 100$. [$I_C = 0.93$ mA, $V_{CE} = 17.3$ V]

3. Fig. 7.59 shows biasing by base resistor method. If it is required to set the operating point at 1 mA, 6 V, find the values of R_C and R_B. Given $\beta = 150$, $V_{BE} = 0.3$ V.[$R_C = 3$ kΩ, $R_B = 0.3$ MΩ]

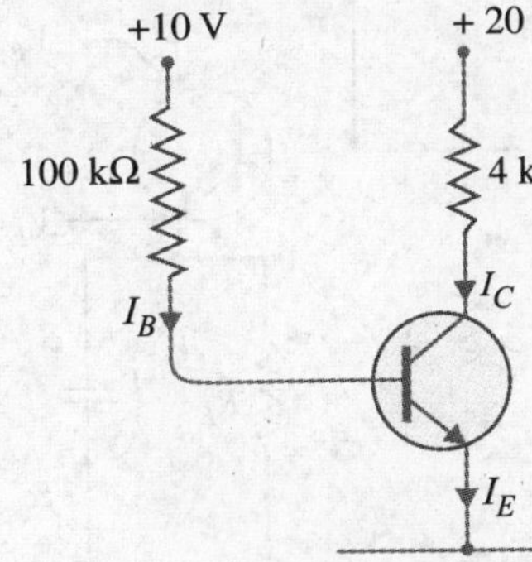

Fig. 7.58

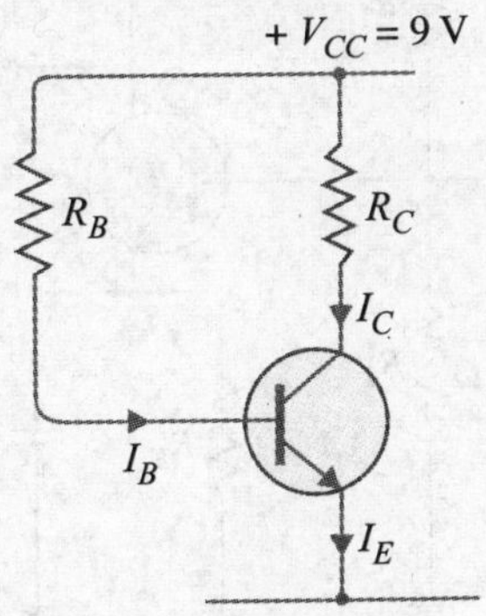

Fig. 7.59

4. A transistor amplifier is biased with feedback resistor R_B of 100 kΩ. If $V_{CC} = 25$ V, $R_C = 1$ kΩ and $\beta = 200$, find the values of zero signal I_C and V_{CE}. [$I_C = 16.2$mA, $V_{CE} = 8.8$V]

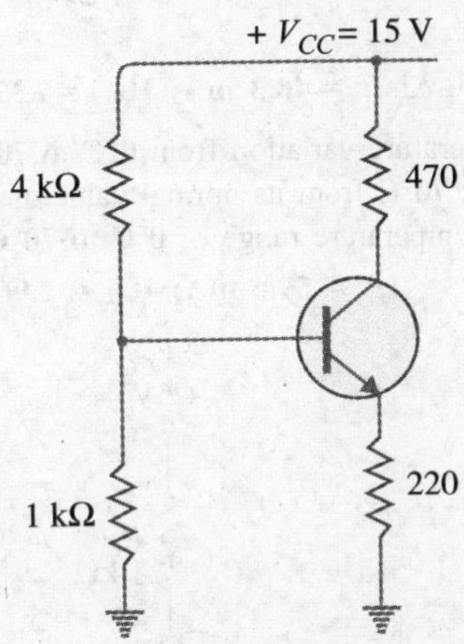

Fig. 7.60

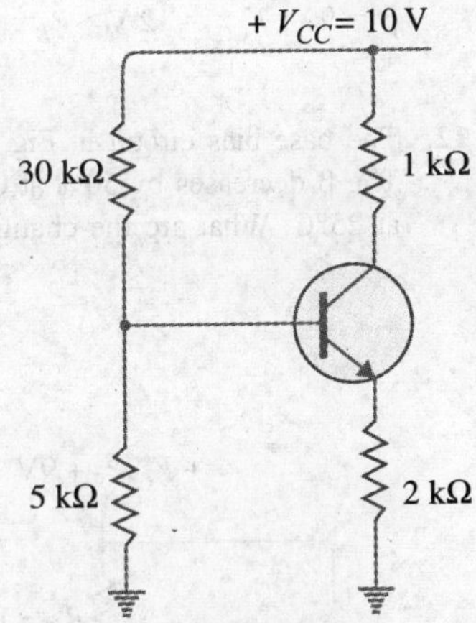

Fig. 7.61

5. Find the value of I_C for potential-divider method if $V_{CC} = 9$ V, $R_E = 1$ kΩ, $R_1 = 39$ kΩ, $R_2 = 10$ kΩ, $R_C = 2.7$ kΩ, $V_{BE} = 0.15$ V and $\beta = 90$. [1.5mA]

6. In an *RC* coupled amplifier, the battery voltage is 16V and collector load $R_C = 4$ kΩ. It is required to set the operating point at $I_C = 1$mA, $V_{CE} = 10$V by potential-divider method. If $V_{BE} = 0.2$ V and $I_1 = 10\ I_B$, $\beta = 100$, find the various circuit values.

7. In the transistor circuit shown in Fig. 7.60, find the operating point. Assume the transistor to be of silicon. [$I_C = 10.5$ mA, $V_{CE} = 7.75$ V]

8. In a transistor circuit shown in Fig. 7.61, find the operating point. Assume silicon transistor is used. [$I_C = 0.365$ mA, $V_{CE} = 8.9$ V]

9. Determine whether or not the circuit shown in Fig. 7.62 is midpoint biased. [Yes]

10. What fault is indicated in Fig. 7.63? Give reasons for your answer. [R_C is open]

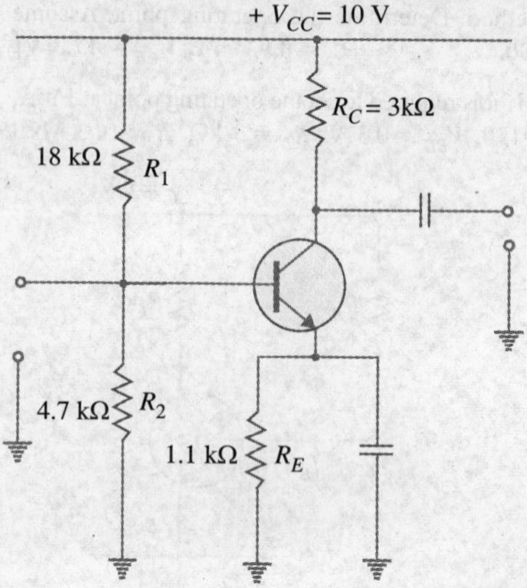

Fig. 7.62

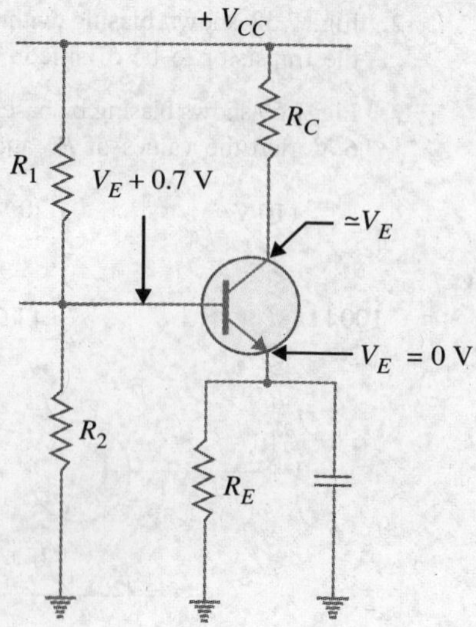

Fig. 7.63

11. Determine I_B, I_C and V_{CE} for a base-biased transistor circuit with the following values : β = 90 ; V_{CC} = 12 V ; R_B = 22 kΩ and R_C = 100Ω.

[I_B = 514 μA; I_C = 46.3 mA; V_{CE} = 7.37 V]

12. The base bias circuit in Fig. 7.64 is subjected to a temperature variation from 0°C to 70°C. The β decreases by 50% at 0°C and increases by 75% at 70°C from its normal value of 110 at 25°C. What are the changes in I_C and V_{CE} over the temperature range of 0°C to 70°C?

[I_C = 59.6 mA; V_{CE} = 5.96 V]

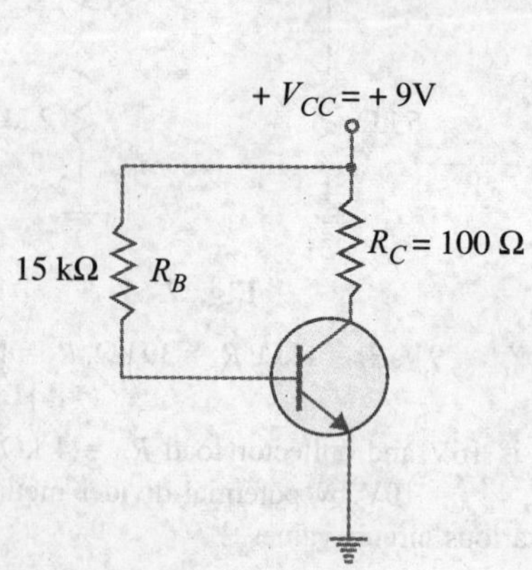

Fig. 7.64

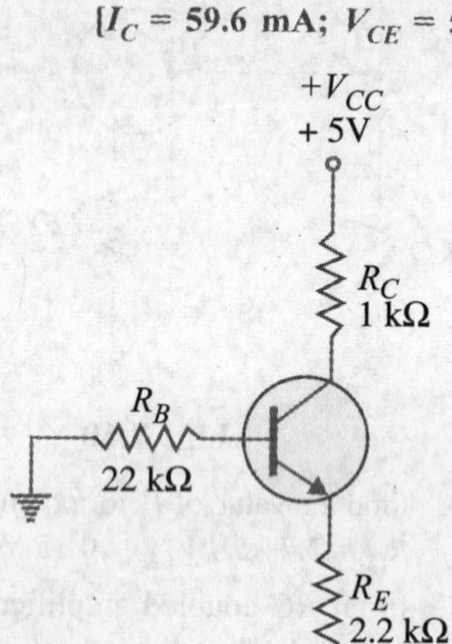

Fig. 7.65

13. To what value can R_E in Fig. 7.65 be reduced without transistor going into saturation?

[639 Ω]

14. When can the effect of β be neglected in the emitter bias circuit? **[When R_E >> R_B/β]**

15. What is the minimum value of β in Fig. 7.66 that makes $R_{in\ (base)} \geq 10\ R_2$? **[69.1]**

16. (*i*) Determine the base voltage V_B in Fig. 7.67.

(*ii*) If R_E is doubled, what will be the value of V_B? **[(*i*) 1.74 V (*ii*) 1.74 V]**

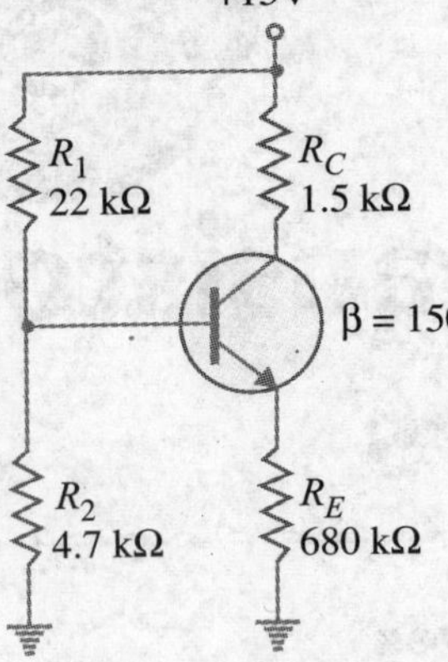

Fig. 7.66

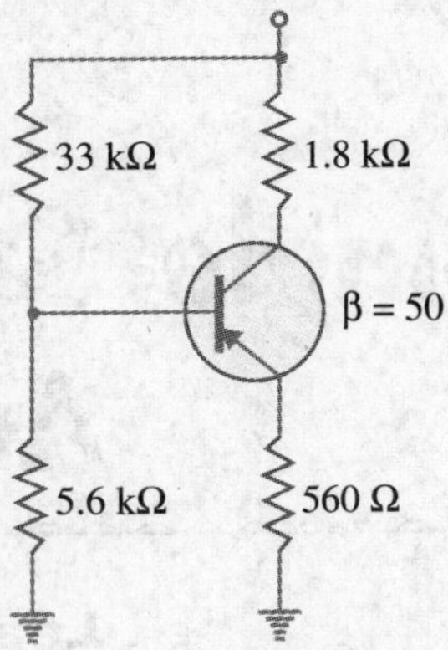

Fig. 7.67

17. (*i*) Find the *Q*-point values for Fig. 7.67.

(*ii*) Find the minimum power rating of transistor in Fig. 7.67.

[(*i*) 1.41 mA; – 8.67 V (*ii*) 12.2 mW]

18. A collector-feedback circuit uses an *npn* transistor with V_{CC} = 12 V, R_C = 1.2 kΩ, R_B = 47 kΩ. Determine the collector voltage and the collector current if β = 200.

[2.56 V; 7.87 mA]

CHAPTER 8

Single Stage Transistor Amplifiers

Learning Objectives

- Develop an understanding of single stage transistor amplifier.
- Recognise the practical circuit of transistor amplifier and the functions of various circuit elements.
- Learn phase reversal in common emitter (*CE*) transistor amplifer.
- Explore the d.c. and a.c. equivalent circuits of transistor amplifier.
- Develop a strong understanding of d.c. and a.c. load lines of transistor amplifier.
- Derive an expression for the voltage gain (A_v) of transistor *CE* amplifier in the form : $A_v = \beta \times R_{AC}/R_{in}$.
- Derive an expression for the voltage gain of *CE* amplifier in the form : $A_v = R_{AC}/r'_e$.
- Derive expressions for input impedance and output impedance of *CE* amplifier.
- Learn about the swamped amplifier.
- Develop an understanding of transistor amplifier equivalent circuit using R_{in} (input resistance), R_{out} (output resistance) and no-load voltage gain A_{v0}.

Introduction

In the previous chapter, it was discussed that a properly biased transistor raises the strength of a weak signal and thus acts as an amplifier. Almost all electronic equipments must include means for amplifying electrical signals. For instance, radio receivers amplify very weak signals—sometimes a few millionth of a volt at antenna—until they are strong enough to fill a room with sound. The transducers used in the medical and scientific investigations generate signals in the microvolt (μV) and millivolt (mV) range. These signals must be amplified thousands and millions of times before they will be strong enough to operate indicating instruments. Therefore, electronic amplifiers are a constant and important ingredient of electronic systems.

Our purpose here will be to discuss *single stage transistor amplifier*. By a *stage* we mean a single transistor with its bias and auxiliary equipment. It may be emphasised here that a practical amplifier is always a multistage amplifier *i.e.* it has a number of stages of amplification. However, it is profitable to consider the multistage amplifier in terms of single stages that are connected together. In this chapter, we shall confine our attention to single stage transistor amplifiers.

8.1 Single Stage Transistor Amplifier

When only one transistor with associated circuitry is used for amplifying a weak signal, the circuit is known as **single stage transistor amplifier.**

A single stage transistor amplifier has one transistor, bias circuit and other auxiliary components. Although a practical amplifier consists of a number of stages, yet such a complex circuit can be conveniently split up into separate single stages. By analysing carefully only a single stage and using this single stage analysis repeatedly, we can effectively analyse the complex circuit. It follows, therefore, that single stage amplifier analysis is of great value in understanding the practical amplifier circuits.

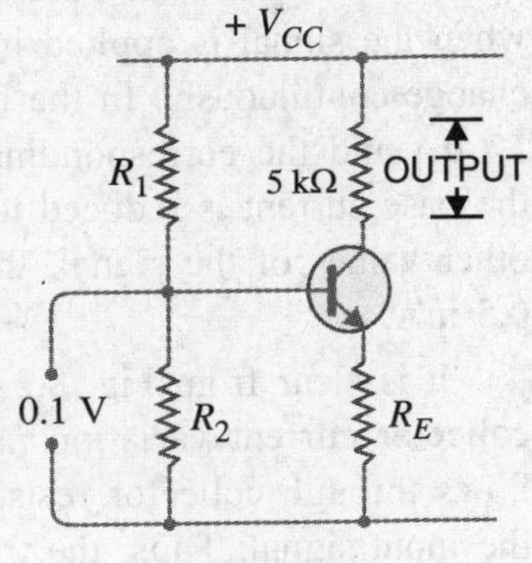

Fig. 8.1

8.2 How Transistor Amplifies ?

Fig. 8.1 shows a single stage transistor amplifier. When a weak a.c. signal is given to the base of transistor, a small base current (which is a.c.) starts flowing. Due to transistor action, a much larger (β times the base current) a.c. current flows through the collector load R_C. As the value of R_C is quite high (usually 4-10 kΩ), therefore, a large voltage appears across R_C. Thus, a weak signal applied in the base circuit appears in amplified form in the collector circuit. It is in this way that a transistor acts as an amplifier.

The action of transistor amplifier can be beautifully explained by referring to Fig. 8.1. Suppose a change of 0.1 V in signal voltage produces a change of 2 mA in the collector current. Obviously, a signal of only 0.1 V applied to the base will give an output voltage = 2 mA × 5 kΩ = 10 V. Thus, the transistor has been able to raise the voltage level of the signal from 0.1 V to 10 V *i.e.* voltage amplification or stage gain is 100.

8.3 Graphical Demonstration of Transistor Amplifier

The function of transistor as an amplifier can also be explained graphically. Fig. 8.2 shows the output characteristics of a transistor in *CE* configuration. Suppose the zero signal base current is 10 μA *i.e.* this is the base current for which the transistor is biased by the biasing network. When an a.c. signal is applied to the base, it makes the base, say positive in the first half-cycle and negative

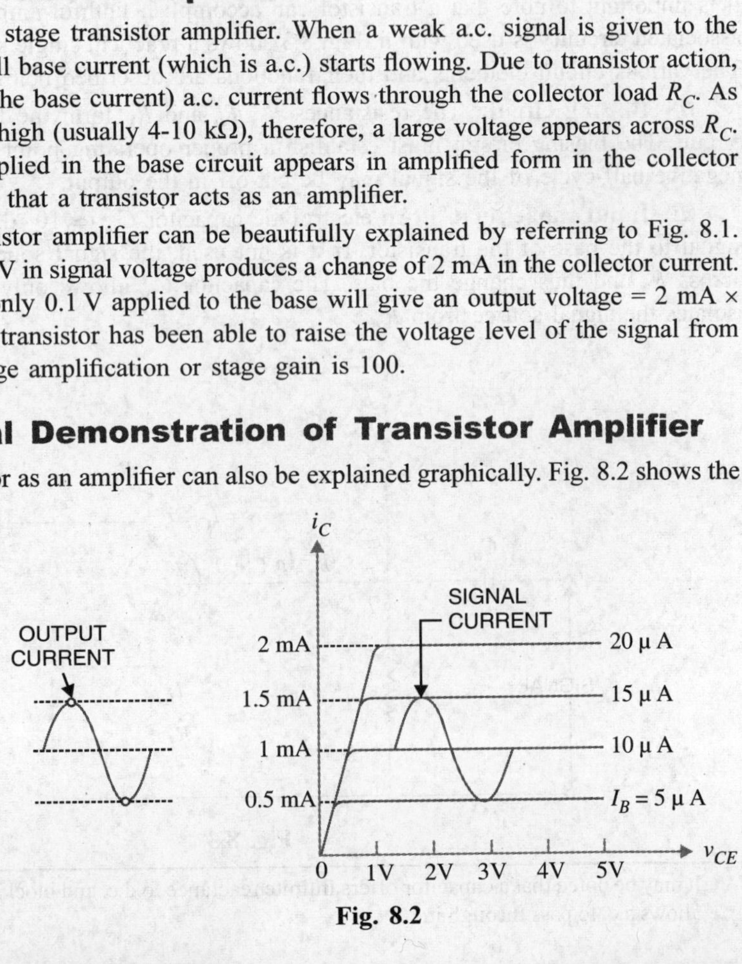

Fig. 8.2

in the second half-cycle. Therefore, the base and collector currents will increase in the first half-cycle when base-emitter junction is more forward-biased. However, they will decrease in the second half-cycle when the base-emitter junction is less forward biased.

For example, consider a sinusoidal signal which increases or decreases the base current by 5 μA in the two half-cycles of the signal. Referring to Fig. 8.2, it is clear that in the absence of signal, the base current is 10μA and the collector current is 1 mA. However, when the signal is applied in the base circuit the base current and hence collector current change continuously. In the first half-cycle peak of the signal, the base current increases to 15 μA and the corresponding collector current is 1.5 mA. In the second half-cycle peak, the base current is reduced to 5 μA and the corresponding collector current is 0.5 mA. For other values of the signal, the collector current is inbetween these values *i.e.* 1.5 mA and 0.5 mA.

It is clear from Fig. 8.2 that 10 μA base current variation results in 1mA (1,000 μA) collector current variation *i.e.* by a factor of 100. This large change in collector current flows through collector resistance R_C. The result is that output signal is much larger than the input signal. Thus, the transistor has done amplification.

8.4 Practical Circuit of Transistor Amplifier

It is important to note that a transistor can accomplish faithful amplification only if proper associated circuitry is used with it. Fig. 8.3 shows a practical single stage transistor amplifier. The various circuit elements and their functions are described below :

(*i*) **Biasing circuit.** The resistances R_1, R_2 and R_E form the biasing and stabilisation circuit. The biasing circuit must establish a proper operating point otherwise a part of the negative half-cycle of the signal may be cut-off in the output.

(*ii*) **Input capacitor C_{in}.** An electrolytic capacitor C_{in} (≃ 10 μF) is used to couple the signal to the base of the transistor. If it is not used, the signal source resistance will come across R_2 and thus change the bias. The capacitor C_{in} allows only a.c. signal to flow but isolates the signal source from R_2.*

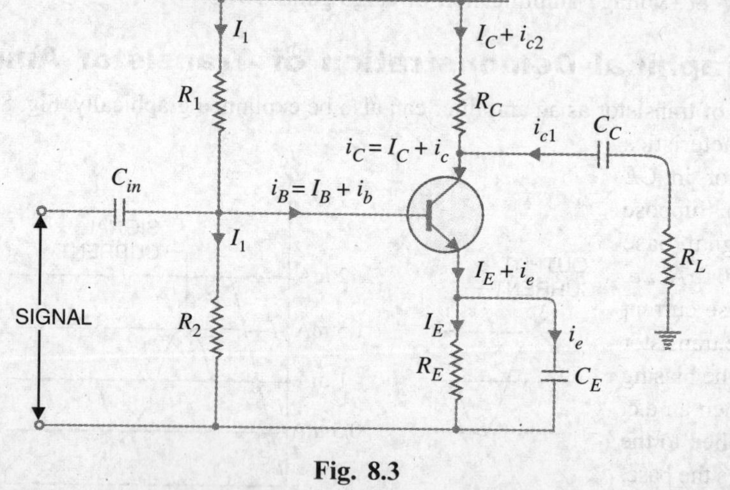

Fig. 8.3

* It may be noted that a capacitor offers infinite reactance to d.c. and blocks it completely whereas it allows a.c. to pass through it.

***(iii)* Emitter bypass capacitor C_E.** An emitter bypass capacitor C_E ($\simeq$ 100 μF) is used in parallel with R_E to provide a low reactance path to the amplified a.c. signal. If it is not used, then amplified a.c. signal flowing through R_E will cause a voltage drop across it, thereby reducing the output voltage.

***(iv)* Coupling capacitor C_C.** The coupling capacitor C_C ($\simeq$ 10 μF) couples one stage of amplification to the next stage. If it is not used, the bias conditions of the next stage will be drastically changed due to the shunting effect of R_C. This is because R_C will come in parallel with the upper resistance R_1 of the biasing network of the next stage, thereby altering the biasing conditions of the latter. In short, the coupling capacitor C_C isolates the d.c. of one stage from the next stage, but allows the passage of a.c. signal.

Various circuit currents. It is useful to mention the various currents in the complete amplifier circuit. These are shown in the circuit of Fig. 8.3.

***(i)* Base current.** When no signal is applied in the base circuit, d.c. base current I_B flows due to biasing circuit. When a.c. signal is applied, a.c. base current i_b also flows. Therefore, with the application of signal, total base current i_B is given by ;

$$i_B = I_B + i_b$$

***(ii)* Collector current.** When no signal is applied, a d.c. collector current I_C flows due to biasing circuit. When a.c. signal is applied, a.c. collector current i_c also flows. Therefore, the total collector current i_C is given by ;

$$i_C = I_C + i_c$$

where $I_C = \beta\, I_B$ = Zero signal collector current

$i_c = \beta\, i_b$ = Collector current due to signal

***(iii)* Emitter current.** When no signal is applied, a d.c. emitter current I_E flows. With the application of signal, total emitter current i_E is given by ;

$$i_E = I_E + i_e$$

It is useful to keep in mind that :

$$I_E = I_B + I_C$$

$$i_e = i_b + i_c$$

Now base current is usually very small, therefore, as a reasonable approximation,

$$I_E \simeq I_C \quad \text{and} \quad i_e \simeq i_c$$

Example 8.1. *What is the role of emitter bypass capacitor C_E in CE amplifier circuit shown in Fig. 8.3? Illustrate with a numerical example.*

Solution. The emitter bypass capacitor C_E (See Fig. 8.3) connected in parallel with R_E plays an important role in the circuit. **If it is not used, the amplified a.c. signal flowing through R_E will cause a voltage drop across it, thereby reducing the a.c. output voltage and hence the voltage gain of the amplifier.**

Let us illustrate the effect of C_E with a numerical example. Suppose R_E = 1000 Ω and capacitive reactance of C_E at the signal frequency is 100 Ω (*i.e.* X_{C_E} = 100 Ω). Then 10/11 of a.c emitter current will flow through C_E and only 1/11 through R_E. The signal voltage developed across R_E is, therefore, only 1/11 of the voltage which would have been developed if C_E were not present. In practical circuits, the value of C_E is so selected that it almost entirely bypasses the a.c. signal (the name for C_E is obvious). *For all practical purposes, we consider C_E to be a short for a.c. signals.*

Example 8.2. *Select a suitable value for the emitter bypass capacitor in Fig. 8.4 if the amplifier is to operate over a frequency range from 2 kHz to 10 kHz.*

Solution. An amplifier usually handles more than one frequency. Therefore, the value of C_E is so selected that it provides adequate bypassing for the *lowest* of all the frequencies. Then it will also be a good bypass ($X_C \propto 1/f$) for all the higher frequencies. Suppose the minimum frequency to be handled by C_E is f_{min}. Then C_E is considered a good bypass if at f_{min},

$$X_{C_E} = \frac{R_E}{10}$$

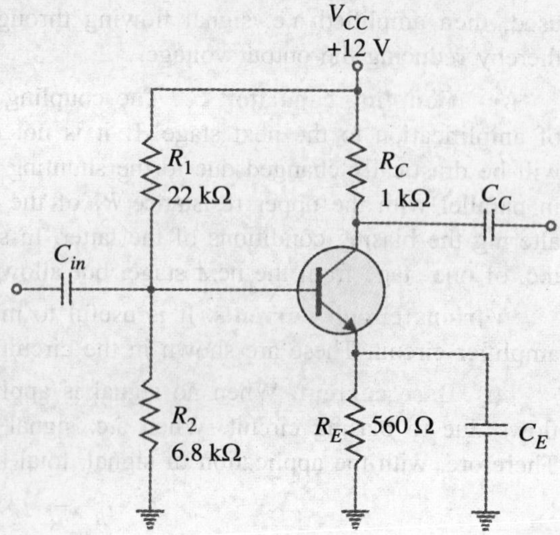

Fig. 8.4

In the given problem, f_{min} = 2kHz ; R_E = 560Ω.

$\therefore \quad 10\, X_{C_E} = 560$

or $\quad X_{C_E} = 560/10 = 56\Omega$

or $$\frac{1}{2\pi\, f_{min}\, C_E} = 56$$

$$\therefore \quad C_E = \frac{1}{2\pi\, f_{min}\, 56} = \frac{1}{2\pi \times (2 \times 10^3) \times 56} = 1.42 \times 10^{-6}\ \text{F} = \mathbf{1.42\ \mu F}$$

Note. While discussing *CE* amplifier, the reader should be very particular about the role of C_E.

8.5 Phase Reversal

In common emitter connection, when the input signal voltage increases in the positive sense, the output voltage increases in the negative direction and *vice-versa*. In other words, there is a phase difference of 180° between the input and output voltage in *CE* connection. This is called phase reversal.*

The phase difference of 180° between the signal voltage and output voltage in a common emitter amplifier is known as **phase reversal.**

Consider a common emitter amplifier circuit shown in Fig. 8.5. The signal is fed at the input terminals (*i.e.* between base and emitter) and output is taken from collector and emitter end of supply. The total instantaneous output voltage v_{CE} is given by;

$$^{**}v_{CE} = V_{CC} - i_C\, R_C \qquad ...(i)$$

* This is so if output is taken from collector and emitter end of supply as is always done. However, if the output is taken across R_C, it will be in phase with the input.

** Reactance of C_C (= 10μF) is negligible at ordinary signal frequencies. Therefore, it can be considered a short for the signal.

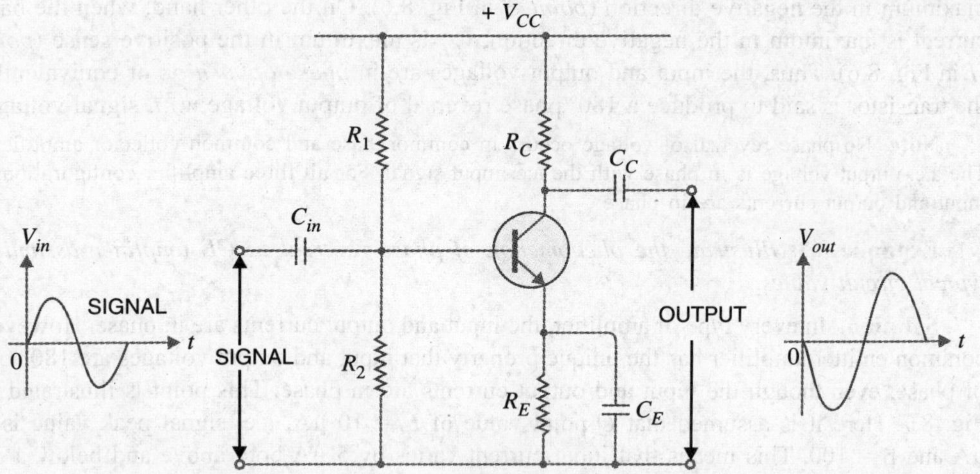

Fig. 8.5

When the signal voltage increases in the positive half-cycle, the base current also increases. The result is that collector current and hence voltage drop $i_C R_C$ increases. As V_{CC} is constant, therefore, output voltage v_{CE} decreases. In other words, as the signal voltage is increasing in the positive half-cycle, the output voltage is increasing in the negative sense *i.e.* output is 180° out of phase with the input. It follows, therefore, that in a common emitter amplifier, the positive half-cycle of the signal appears as amplified negative half-cycle in the output and *vice-versa*. It may be noted that amplification is not affected by this phase reversal.

The fact of phase reversal can be readily proved mathematically. Thus differentiating exp. (*i*), we get,

$$dv_{CE} = 0 - di_c R_C$$

or $$dv_{CE} = - di_c R_C$$

The negative sign shows that output voltage is 180° out of phase with the input signal voltage.

Graphical demonstration. The fact of phase reversal in *CE* connection can be shown graphically with the help of output characteristics and load line (See Fig. 8.6).

In Fig. 8.6, *AB* is the load line. The base current fluctuates between, say ± 5μA with 10μA as the zero signal base current. From the figure, it is clear that when the base current is maximum in the positive direction, v_{CE} becomes

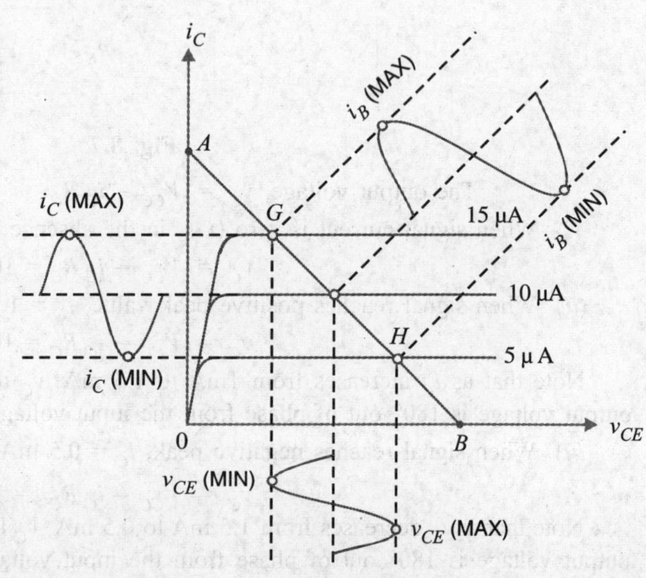

Fig. 8.6

maximum in the negative direction (*point G* in Fig. 8.6). On the other hand, when the base current is maximum in the negative direction, v_{CE} is maximum in the positive sense (*point H* in Fig. 8.6). Thus, the input and output voltages are in *phase opposition* or equivalently, the transistor is said to produce a 180° phase reversal of output voltage *w.r.t.* signal voltage.

Note. No phase reversal of voltage occurs in common base and common collector amplifiers. The a.c. output voltage is in phase with the a.c. input signal. For all three amplifier configurations ; input and output currents are in phase.

Example 8.3. *Illustrate the phenomenon of phase reversal in CE amplifier assuming typical circuit values.*

Solution. In every type of amplifier, the input and output currents are in phase. However, common emitter amplifier has the unique property that input and output voltages are 180° out of phase, even though the input and output currents are in phase. This point is illustrated in Fig. 8.7. Here it is assumed that *Q*-point value of I_B = 10 μA, a.c. signal peak value is 5 μA and β = 100. This means that input current varies by 5 μA both above and below a 10 μA d.c. level. At any instant, the output current will be 100 times the input current at that instant. Thus when the input current is 10 μA, output current is i_C = 100 × 10 μA = 1 mA. However, when the input current is 15 μA, then output current is i_C = 100 × 15 μA = 1.5 mA and so on. Note that input and output currents are in phase.

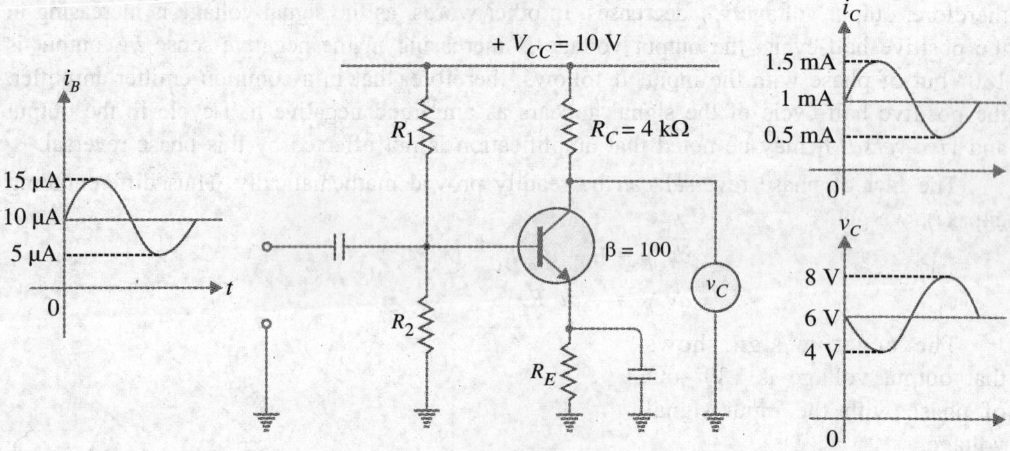

Fig. 8.7

The output voltage, $v_C = V_{CC} - i_C R_C$

(*i*) When signal current is zero (*i.e.*, in the absence of signal), i_C = 1 mA.

$$\therefore \quad v_C = V_{CC} - i_C R_C = 10 \text{ V} - 1 \text{ mA} \times 4 \text{ k}\Omega = 6\text{V}$$

(*ii*) When signal reaches positive peak value, i_C = 1.5 mA.

$$\therefore \quad v_C = V_{CC} - i_C R_C = 10 \text{ V} - 1.5 \text{ mA} \times 4 \text{ k}\Omega = 4 \text{ V}$$

Note that as i_C increases from 1mA to 1.5 mA, v_C decreases from 6V to 4V. Clearly, output voltage is 180° out of phase from the input voltage as shown in Fig. 8.7.

(*iii*) When signal reaches negative peak, i_C = 0.5 mA.

$$\therefore \quad v_C = V_{CC} - i_C R_C = 10 \text{ V} - 0.5 \text{ mA} \times 4 \text{ k}\Omega = 8 \text{ V}$$

Note that as i_C decreases from 1.5 mA to 0.5 mA, v_C increases from 4 V to 8 V. Clearly, output voltage is 180° out of phase from the input voltage. The following points may be noted carefully about *CE* amplifier :

(*a*) The input voltage and input current are in phase.

(*b*) Since the input current and output current are in phase, input voltage and output current are in phase.

(*c*) Output current is 180° out of phase with the output voltage (v_C). *Therefore, input voltage and output voltage are 180° out of phase.*

8.6 Input/Output Phase Relationships

The following points regarding the input / output phase relationships between currents and voltages for the various transistor configurations may be noted :

(*i*) *For every amplifier type (CE, CB and CC), the input and output currents are in phase.* When the input current decreases, the output current also decreases and *vice-versa.*

(*ii*) *Remember that common emitter (CE) circuit is the only configuration that has input and output voltages 180° out of phase.*

(*iii*) *For both common base (CB) and common collector (CC) circuits, the input and output voltages are in phase.* If the input voltage decreases, the output voltage also decreases and vice-versa.

8.7 Important Parameters of an Amplifier

There are many types of amplifiers *e.g.,* voltage amplifiers, power amplifiers, current amplifiers *etc.* A voltage amplifier is designed to raise the voltage level of the signal. Likewise, a power amplifier is designed to raise the power level of the signal. Regardless of their type, all amplifiers have the following five important circuit parameters :

(*i*) Input resistance (R_{in})
(*ii*) Output resistance (R_{out})
(*iii*) Current gain (A_i)
(*iv*) Voltage gain (A_v)
(*v*) Power gain (A_p)

The above circuit parameters play an important role in the design and analysis of an amplifier. Fig. 8.8 shows the block diagram of a transistor amplifier. There are two input terminals and two output terminals. The input side (the side to which the signal V_S is normally applied) is to the left and the output side (where the load R_L is connected) is to the right. It may be added here that we *generally use r.m.s. values of voltages and currents to determine the above parameters of the amplifier.

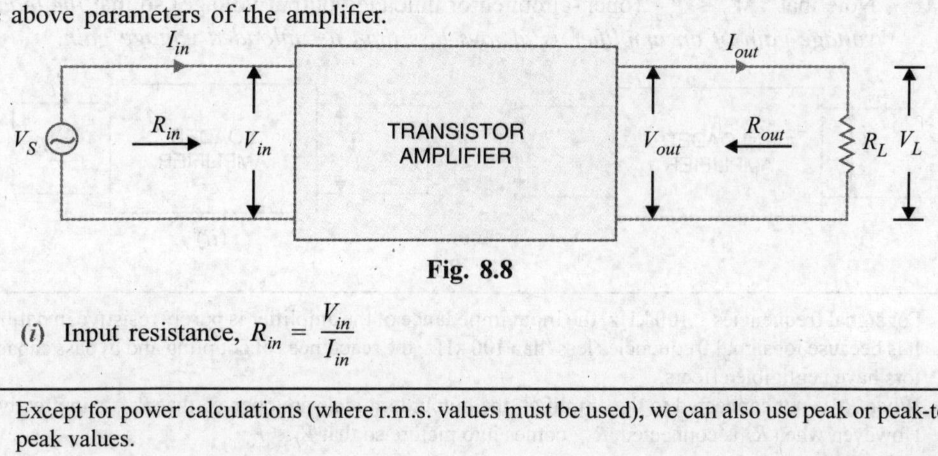

Fig. 8.8

(*i*) Input resistance, $R_{in} = \dfrac{V_{in}}{I_{in}}$

---

* Except for power calculations (where r.m.s. values must be used), we can also use peak or peak-to-peak values.

Note that input *resistance is the resistance "seen" by the input signal source V_S. It is a measure of current (I_{in}) drawn by the amplifier. The input resistance depends upon whether the transistor is in the common base (*CB*), common emitter (*CE*) or common collector (*CC*) arrangement.

(*ii*) Output resistance, $R_{out} = \frac{V_{out}}{I_{out}}$

Note that output resistance (R_{out}) is the resistance "seen" by the load resistance (R_L) when "looking back" into the output terminals of the amplifier. In fact, R_{out} is the internal resistance seen from the output terminals of the amplifier. *It may be thought as Thevenin resistance at the output terminals of the amplifier.*

Note. If reactances of coupling capacitors and bypass capacitors of an amplifier are large and cannot be neglected, then, the terms *input impedance* (Z_{in}) instead of input resistance (R_{in}) and *output impedance* (Z_{out}) instead of output resistance (R_{out}) should be used.

(*iii*) Current gain, $A_i = \frac{I_{out}}{I_{in}}$

Although this parameter receives less attention than voltage gain, it is an important quantity that can have significant impact on the overall performance of an amplifier. The current gain of a transistor amplifier may range from less than 1 to a value that may exceed 100.

(*iv*) Voltage gain, $A_v = \frac{V_{out}}{V_{in}}$

When load resistance R_L is not connected to the amplifier (unloaded condition) as shown in Fig. 8.9 (*i*), then V_{out}/V_{in} is called *no-load voltage gain* A_{v0} of the amplifier *i.e.*

No-load voltage gain, $A_{v0} = \frac{V_{out}}{V_{in}}$ ... in Fig. 8.9 (*i*)

When load resistance R_L is connected to the amplifier (loaded condition) as shown in Fig. 8.9 (*ii*), then V_L/V_{in} is called *loaded voltage gain* A_v of the amplifier *i.e.*

Loaded voltage gain, $A_v = \frac{V_L}{V_{in}}$ ... in Fig. 8.9 (*ii*)

Note that **$V_L < V_{out}$ (open-circuited or unloaded output voltage) so that *the loaded voltage gain of an amplifier is always less than its unloaded voltage gain.*

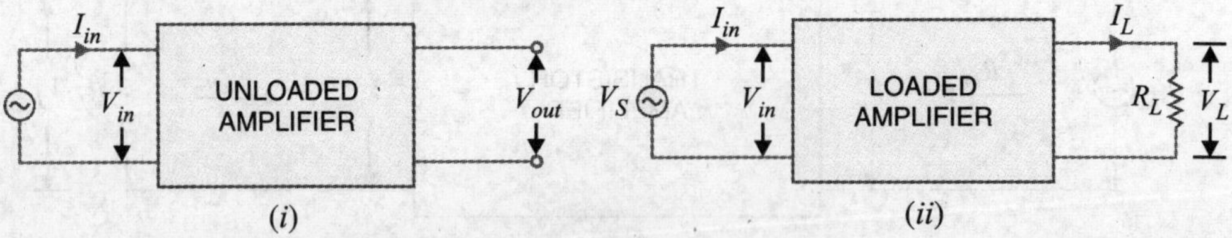

Fig. 8.9

* For signal frequencies < 100 kHz, the input impedance of the amplifier is purely resistive in nature. It is because for signal frequencies less than 100 kHz, the reactances of coupling and bypass capacitors have negligible effects.

** When R_L is not connected to the amplifier, the output terminals are open so that R_{out} is ineffective. However, when R_L is connected, R_{out} comes into picture so that $V_L < V_{out}$.

Voltage gain formula. We can easily derive the loaded voltage gain formula in terms of current gain (A_i).

Input voltage, $V_{in} = I_{in} R_{in}$

Output voltage, $V_L = I_L R_L = I_{out} R_L$ $\quad (\because I_L = I_{out})$

$$\therefore \text{Loaded voltage gain, } A_v = \frac{V_L}{V_{in}} = \frac{I_{out} R_L}{I_{in} R_{in}} = \frac{I_{out}}{I_{in}} \times \frac{R_L}{R_{in}}$$

$$\therefore \quad A_v = A_i \frac{R_L}{R_{in}} \quad \text{... (i)} \quad \left(\because \frac{I_{out}}{I_{in}} = A_i\right)$$

Eq. (*i*) is a general expression for the loaded voltage gain of a transistor amplifier and will always hold true regardless of the type of the amplifier.

(*v*) Power gain, $A_p = \dfrac{P_{out}}{P_{in}}$

Since $P_{in} = V_{in} I_{in}$ and $P_{out} = V_{out} I_{out}$

$$\therefore \quad A_p = \frac{V_{out} I_{out}}{V_{in} I_{in}} = \frac{I_{out}}{I_{in}} \times \frac{V_{out}}{V_{in}} = A_i \times A_v$$

$$\therefore \quad A_p = A_i A_v = \text{Current gain} \times \text{Voltage gain}$$

It may be noted that while carrying out power calculations, the r.m.s. values of currents and voltages must be used.

Example 8.4. *Fig. 8.10 shows the block diagram of a transistor amplifier. The signal source V_S = 12 mV (r.m.s.) and signal source resistance R_S = 1 kΩ. Determine (i) input current (ii) input resistance (iii) output voltage (iv) output current and (v) current gain.*

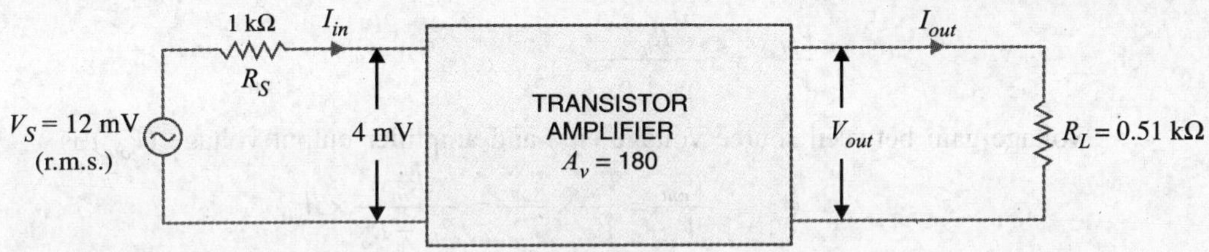

Fig. 8.10

Solution. V_S = 12 mV (r.m.s.) ; R_S = 1 kΩ ; V_{in} = 4 mV (r.m.s.) ; R_L = 0.51 kΩ ; A_v = 180

(*i*) Input current, $I_{in} = \dfrac{V_S - V_{in}}{R_S} = \dfrac{12\text{mV} - 4\text{mV}}{1\text{k}\Omega} = \dfrac{8\text{mV}}{1\text{k}\Omega} = \mathbf{8\ \mu A}$

(*ii*) Input resistance, $R_{in} = \dfrac{V_{in}}{I_{in}} = \dfrac{4\text{ mV}}{8\,\mu\text{A}} = \mathbf{500\ \Omega}$

(*iii*) Output voltage, $V_{out} = A_v V_{in} = 180 \times 4$ mV = **720 mV**

(*iv*) Output current, $I_{out} = \dfrac{V_{out}}{R_L} = \dfrac{720\text{ mV}}{0.51\text{ k}\Omega} = \mathbf{1.41\ mA}$

(*v*) Current gain, $A_i = \dfrac{I_{out}}{I_{in}} = \dfrac{1.41\text{ mA}}{8\,\mu\text{A}} = \mathbf{176.25}$

Example 8.5. *Fig. 8.11 shows the block diagram of a transistor amplifier. Determine (i) input voltage (ii) input current (iii) input resistance and (iv) voltage gain between source voltage and amplifier output.*

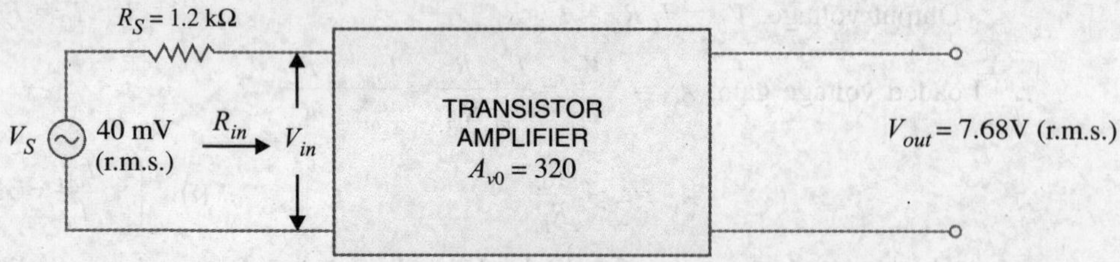

Fig. 8.11

Solution. V_S = 40 mV (r.m.s.) ; R_S = 1.2 kΩ ; A_{v0} = 320 ; V_{out} = 7.68V (r.m.s.)

(*i*) Input voltage, $V_{in} = \dfrac{V_{out}}{A_{v0}} = \dfrac{7.68\text{ V}}{320} = \textbf{24 mV}$

(*ii*) Input current, $I_{in} = \dfrac{V_S - V_{in}}{R_S} = \dfrac{40\text{ mV} - 24\text{ mV}}{1.2\text{ k}\Omega} = \textbf{13.33 μA}$

(*iii*) Input resistance, $R_{in} = \dfrac{V_{in}}{I_{in}} = \dfrac{24\text{ mV}}{13.33\,\mu\text{A}} = \textbf{1.8 kΩ}$

(*iv*) The source resistance R_S is in series with the input resistance R_{in} of the amplifier. Therefore, by voltage-divider rule,

$$V_{in} = \frac{V_S}{R_{in} + R_S} \times R_{in}$$

or

$$\frac{V_{in}}{V_S} = \frac{R_{in}}{R_{in} + R_S}$$

∴ Voltage gain between source voltage (V_S) and amplifier output voltage (V_{out}) is

$$A_{vS} = \frac{V_{out}}{V_S} = \frac{V_{in}}{V_S} \times \frac{V_{out}}{V_{in}} = \frac{R_{in}}{R_{in} + R_S} \times A_{v0}$$

where $A_{v0} = V_{out}/V_{in}$ = No-load voltage gain = 320

∴ $$A_{vS} = \frac{1.8\text{ k}\Omega}{1.8\text{ k}\Omega + 1.2\text{ k}\Omega} \times 320 = \textbf{192}$$

Example 8.6. *The signal source connected to the input of an amplifier has an internal resistance of 1.2 kΩ. The voltage gain of the amplifier from its input to its output is 140. What minimum value of input resistance should the amplifier have if the voltage gain from signal source to amplifier output is to be atleast 100 ?*

Solution. As shown in the above example, the voltage gain A_{vS} between source voltage (V_S) and amplifier output voltage (V_{out}) is

$$A_{vS} = \frac{R_{in}}{R_{in} + R_S} \times A_{v0}$$

Here, A_{vS} = 100 ; A_{v0} = No-load voltage gain = 140 ; R_S = 1.2 kΩ

$$\therefore \qquad 100 = \frac{R_{in}}{R_{in} + 1.2} \times 140$$

$$\text{or} \qquad 100\, R_{in} + 120 = 140\, R_{in} \qquad \therefore\; R_{in} = 3\ \text{k}\Omega$$

Therefore, the amplifier should have minimum input resistance of 3 kΩ.

8.8 Transistor is an Energy Converter

We have seen that a transistor amplifier provides a power gain *i.e.* output a.c. power is greater than the input a.c. signal power. It does not mean that a transistor is a power generating device because it is simply a two *pn* junction device and cannot be a source of power. It appears that law of conservation of energy is violated but that is not so. A transistor receives d.c. power from the d.c. supply (V_{CC}), converts it into a.c. power and adds it to the signal. Therefore, law of conservation of energy holds good. In fact, a transistor is an energy converter *i.e.* it converts d.c. power into a.c. power and the addition of this a.c. power to the signal results in the greater a.c. output power.

8.9 D.C. and A.C. Equivalent Circuits of Transistor Amplifier

In a transistor amplifier, both d.c. and a.c. conditions prevail. The d.c. sources set up d.c. currents and voltages whereas the a.c. source (*i.e.* signal) produces fluctuations in the transistor currents and voltages. Therefore, a simple way to analyse the action of a transistor is to split the analysis into two parts *viz.* a d.c. analysis and an a.c. analysis. In the d.c. analysis, we consider all the d.c. sources at the same time and work out the d.c. currents and voltages in the circuit. On the other hand, for a.c. analysis, we consider all the a.c. sources at the same time and work out the a.c. currents and voltages. By adding the d.c. and a.c. currents and voltages, we get the total currents and voltages in the circuit. For example, consider the amplifier circuit shown in Fig. 8.12. This circuit can be easily analysed by splitting it into *d.c. equivalent circuit* and *a.c. equivalent circuit.*

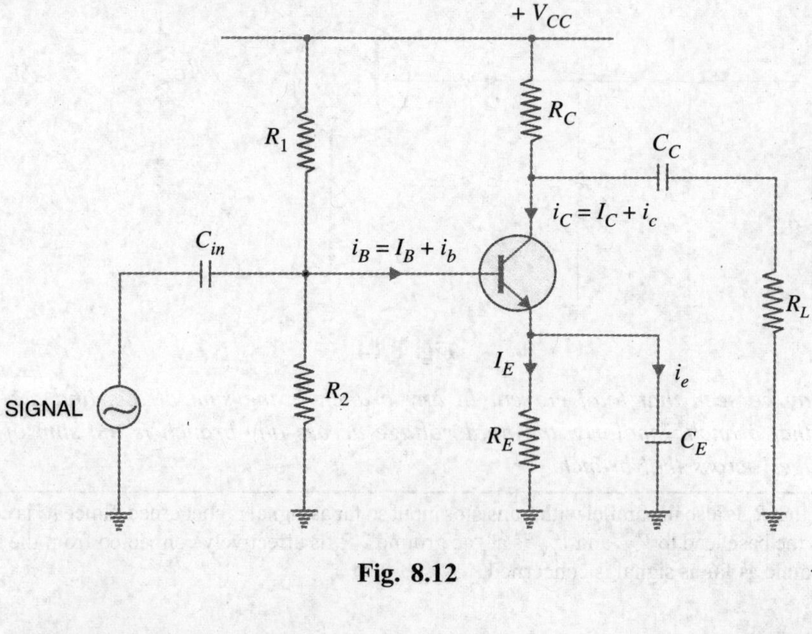

Fig. 8.12

(*i*) **D.C. equivalent circuit.** In the d.c. equivalent circuit of a transistor amplifier, only d.c. conditions are to be considered *i.e.* it is presumed that no signal is applied. As direct current cannot flow through a capacitor, therefore, *all the capacitors look like open circuits in the d.c. equivalent circuit.* It follows, therefore, that in order to draw the equivalent d.c. circuit, the following two steps are applied to the transistor circuit :

(*a*) Reduce all a.c. sources to zero.

(*b*) Open all the capacitors.

Applying these two steps to the circuit shown in Fig. 8.12, we get the d.c. equivalent circuit shown in Fig. 8.13. We can easily calculate the d.c. currents and voltages from this circuit.

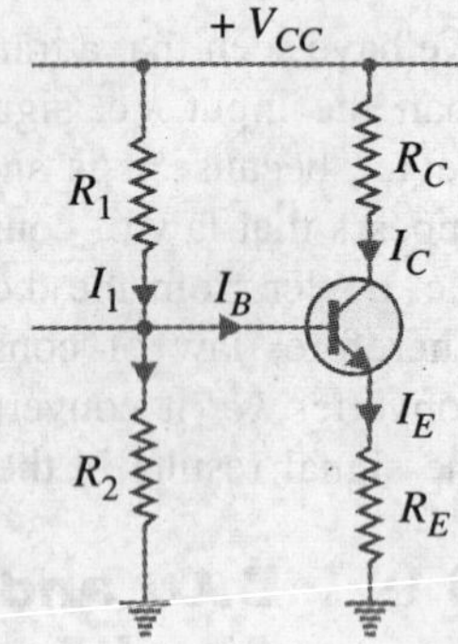

Fig. 8.13

(*ii*) **A.C. equivalent circuit.** In the a.c. equivalent circuit of a transistor amplifier, only a.c. conditions are to be considered. Obviously, the d.c. voltage is not important for such a circuit and may be considered zero. The capacitors are generally used to couple or bypass the a.c. signal. The designer intentionally selects capacitors that are large enough to appear as *short circuits* to the a.c. signal. It follows, therefore, that in order to draw the a.c. equivalent circuit, the following two steps are applied to the transistor circuit :

(*a*) Reduce all d.c. sources to zero (*i.e.* $V_{CC} = 0$).

(*b*) Short all the capacitors.

Applying these two steps to the circuit shown in Fig. 8.12, we get the a.c. *equivalent circuit shown in Fig. 8.14. We can easily calculate the a.c. currents and voltages from this circuit.

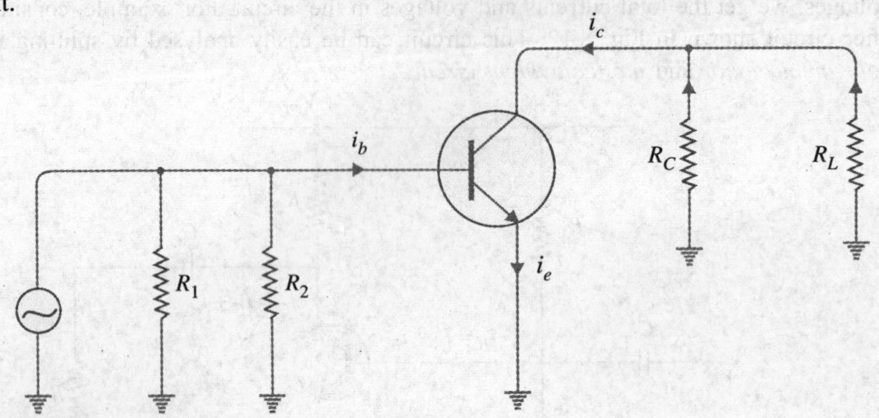

Fig. 8.14

It may be seen that total current in any branch is the sum of d.c. and a.c. currents through that branch. Similarly, the total voltage across any branch is the sum of d.c. and a.c. voltages across that branch.

---

* Note that R_1 is also in parallel with transistor input so far as signal is concerned. Since R_1 is connected from the base lead to V_{CC} and V_{CC} is at "ac ground", R_1 is effectively connected from the base lead to ground as far as signal is concerned.

Example 8.7. *For the transistor amplifier circuit shown in Fig. 8.12, determine :*

(i) d.c. load and a.c. load

(ii) maximum collector-emitter voltage and collector current under d.c. conditions

(iii) maximum collector-emitter voltage and collector current when a.c. signal is applied.

Solution. Refer back to the transistor amplifier circuit shown in Fig. 8.12.

(*i*) The d.c. load for the transistor is Thevenin's equivalent resistance as seen by the collector and emitter terminals. Thus referring to the d.c. equivalent circuit shown in Fig. 8.13, Thevenin's equivalent resistance can be found by shorting the voltage source (*i.e.* V_{CC}) as shown in Fig. 8.15. Because a voltage source looks like a short, it will bypass all other resistances except R_C and R_E which will appear in series.

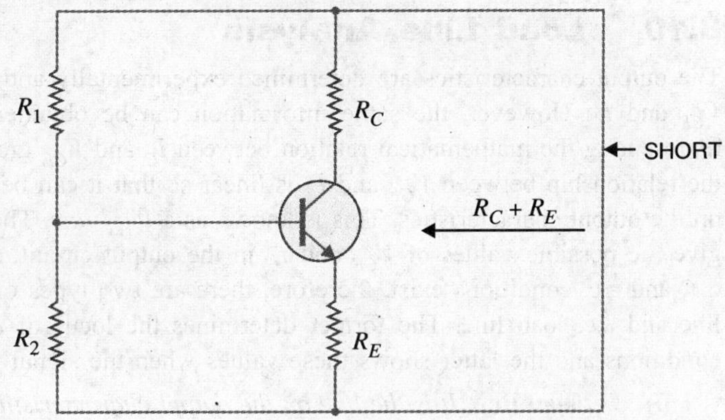

Fig. 8.15

Consequently, transistor amplifier will see a d.c. load of $R_C + R_E$ *i.e.*

$$\text{d.c. load} = R_C + R_E$$

Referring to the a.c. equivalent circuit shown in Fig. 8.14, it is clear that as far as a.c. signal is concerned, resistance R_C appears in parallel with R_L. In other words, transistor amplifier sees an a.c. load equal to $R_C \parallel R_L$ *i.e.*

$$\text{a.c. load, } R_{AC} = R_C \parallel R_L = \frac{R_C\, R_L}{R_C + R_L}$$

(*ii*) Referring to d.c. equivalent circuit of Fig. 8.13,

$$V_{CC} = V_{CE} + I_C\,(R_C + R_E)$$

The maximum value of V_{CE} will occur when there is no collector current *i.e.* $I_C = 0$.

$$\therefore \quad \text{Maximum } V_{CE} = V_{CC}$$

The maximum collector current will flow when $V_{CE} = 0$.

$$\therefore \quad \text{Maximum } I_C = \frac{V_{CC}}{R_C + R_E}$$

(*iii*) When no signal is applied, V_{CE} and I_C are the collector-emitter voltage and collector current respectively. When a.c. signal is applied, it causes changes to take place above and below the operating point Q (*i.e.* V_{CE} and I_C).

Maximum collector current due to a.c. signal = $*I_C$

∴ Maximum positive swing of a.c. collector-emitter voltage

$$= I_C \times R_{AC}$$

* For faithful amplification.

Total maximum collector-emitter voltage

$$= V_{CE} + I_C R_{AC}$$

Maximum positive swing of a.c. collector current

$$= V_{CE}/R_{AC}$$

∴ Total maximum collector current

$$= I_C + V_{CE}/R_{AC}$$

8.10 Load Line Analysis

The output characteristics are determined experimentally and indicate the relation between V_{CE} and I_C. However, the same information can be obtained in a much simpler way by representing the mathematical relation between I_C and V_{CE} graphically. As discussed before, the relationship between V_{CE} and I_C is linear so that it can be represented by a straight line on the output characteristics. This is known as a *load line*. The points lying on the load line give the possible values of V_{CE} and I_C in the output circuit. As in a transistor circuit both d.c. and a.c. conditions exist, therefore, there are two types of load lines, namely; d.c. load line and a.c. load line. The former determines the locus of I_C and V_{CE} in the zero signal conditions and the latter shows these values when the signal is applied.

(*i*) d.c. load line. *It is the line on the output characteristics of a transistor circuit which gives the values of I_C and V_{CE} corresponding to zero signal or d.c. conditions.*

Consider the transistor amplifier shown in Fig. 8.16. In the absence of signal, d.c. conditions prevail in the circuit as shown in Fig. 8.17 (*i*). Referring to this circuit and applying Kirchhoff's voltage law,

$$V_{CE} = V_{CC} - I_C R_C - I_E R_E$$

or

$$V_{CE} = V_{CC} - I_C (R_C + R_E) \qquad ...(i)$$

$$(\because I_E \simeq I_C)$$

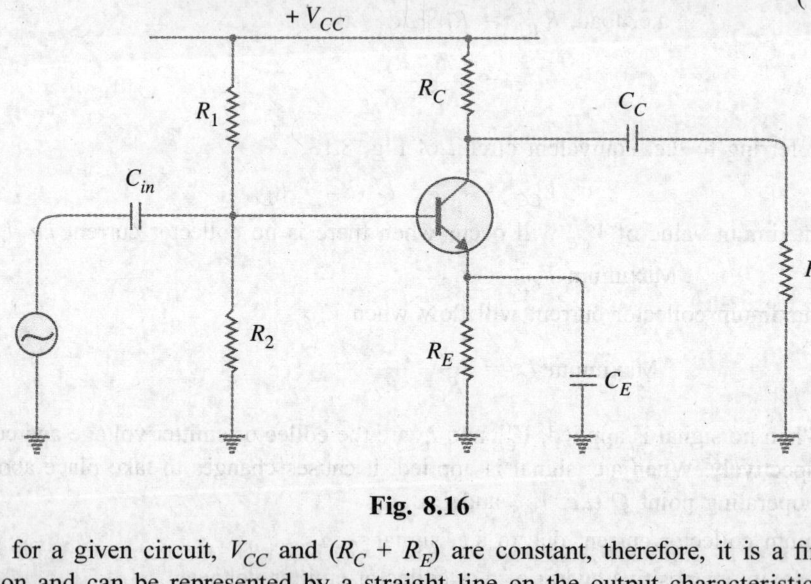

Fig. 8.16

As for a given circuit, V_{CC} and (R_C + R_E) are constant, therefore, it is a first degree *equation and can be represented by a straight line on the output characteristics. This is

* This equation is known as **load line equation** since it relates the collector-emitter voltage (V_{CE}) to the collector current (I_C) flowing through the load.

known as *d.c. load line* and determines the loci of V_{CE} and I_C points in the zero signal conditions. The d.c. load line can be readily plotted by locating two *end points* of the straight line.

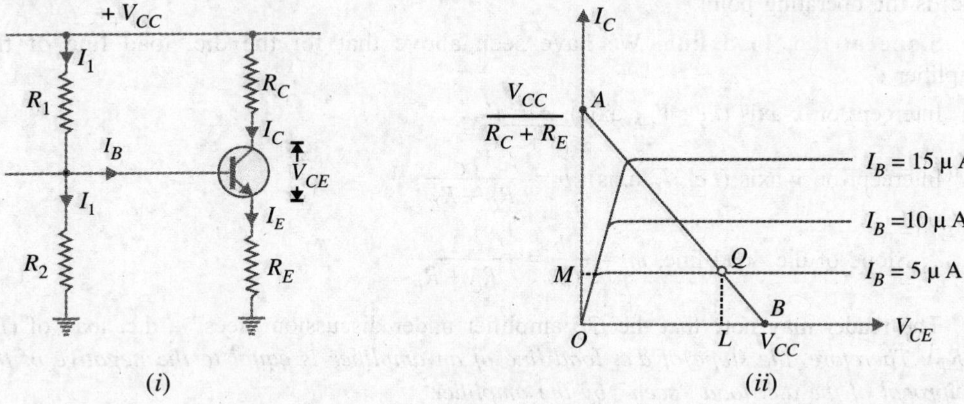

Fig. 8.17

The value of V_{CE} will be maximum when $I_C = 0$. Therefore, by putting $I_C = 0$ in exp. (*i*), we get,

$$\text{Max. } V_{CE} = V_{CC}$$

This locates the first point B ($OB = V_{CC}$) of the d.c. load line.

The value of I_C will be maximum when $V_{CE} = 0$.

$$\therefore \quad \text{Max. } I_C = \frac{V_{CC}}{R_C + R_E}$$

This locates the second point A ($OA = V_{CC}/R_C + R_E$) of the d.c. load line. By joining points A and B, d.c. load line AB is constructed [See Fig. 8.17 (*ii*)].

Alternatively. The two end points of the d.c. load line can also be determined in another way.

$$V_{CE} + I_C (R_C + R_E) = V_{CC}$$

Dividing throughout by V_{CC}, we have,

$$\frac{V_{CE}}{V_{CC}} + \frac{I_C}{(V_{CC} / R_C + R_E)} = 1 \qquad \text{...(i)}$$

The equation of a line having intercepts a and b on x-axis and y-axis respectively is :

$$\frac{x}{a} + \frac{y}{b} = 1 \qquad \text{...(ii)}$$

Comparing eqs. (*i*) and (*ii*), we have,

$$\text{Intercept on } x\text{-axis, } a = V_{CC}$$

$$\text{Intercept on } y\text{-axis, } b = \frac{V_{CC}}{R_C + R_E}$$

With the construction of d.c. load line on the output characteristics, we get the complete information about the output circuit of transistor amplifier in the zero signal conditions. All the points showing zero signal I_C and V_{CE} will obviously lie on the d.c. load line. At the same time I_C and V_{CE} conditions in the circuit are also represented by the output characteristics.

Therefore, actual operating conditions in the circuit will be represented by the point where d.c. load line intersects the base current curve under study. Thus, referring to Fig. 8.17 (*ii*), if I_B = 5 μA is set by the biasing circuit, then Q (*i.e.* intersection of 5 μA curve and load line) is the operating point.

Slope of d.c. load line. We have seen above that for the d.c. load line of the amplifier :

Intercept on *x*-axis (*i.e.*, V_{CE} axis), $a = V_{CC}$

Intercept on *y*-axis (*i.e.*, I_C axis), $b = \dfrac{V_{CC}}{R_C + R_E}$

$$\therefore \text{ Slope of d.c. load line, } m = -\frac{b}{a} = -\frac{1}{R_C + R_E}$$

The reader may note that the *CE* amplifier under discussion "sees" a d.c. load of (R_C + R_E). *Therefore, the slope of d.c. load line of an amplifier is equal to the negative of the reciprocal of the d.c. load "seen" by the amplifier.*

(*ii*) **a.c. load line.** *This is the line on the output characteristics of a transistor circuit which gives the values of i_C and v_{CE} when signal is applied.*

Recall that d.c. load of the *CE* amplifier in Fig. 8.15 is (R_C + R_E). With signal applied to the amplifier, the a.c. load on the amplifier is $R_{AC} = R_C \parallel R_L$. Since the two loads are different, the a.c. load line of the amplifier is *different from its d.c. load line. Referring back to the transistor amplifier shown in Fig. 8.16, its a.c. equivalent circuit as far as output circuit is concerned is as shown in Fig. 8.18 (*i*). To add a.c. load line to the output characteristics, we again require two end points–one maximum collector-emitter voltage point and the other maximum collector current point. Under the application of a.c. signal, these values are (Refer to example 8.7) :

Max. collector-emitter voltage = $V_{CE} + I_C R_{AC}$. This locates the point *C* of the a.c. load line on the collector-emitter voltage axis.

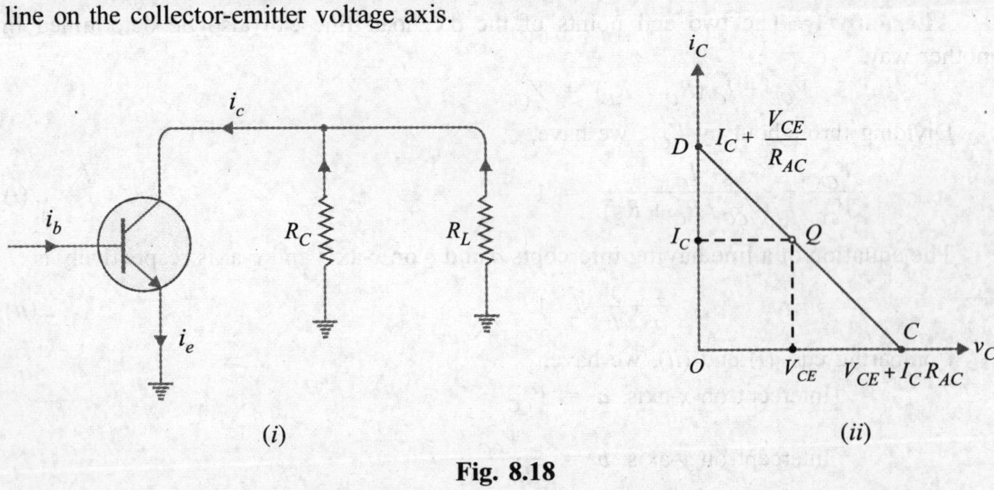

Fig. 8.18

$$\text{Maximum collector current} = I_C + \frac{V_{CE}}{R_{AC}}$$

* Remember, the slope of the load line depends upon load.

$$\text{where} \qquad R_{AC} = R_C \parallel R_L = \frac{R_C R_L}{R_C + R_L}$$

This locates the point D of a.c. load line on the collector-current axis. By joining points C and D, the a.c. load line CD is constructed [See Fig. 8.18 (*ii*)]. Note that Q point values of the amplifier are V_{CE}, I_C.

Slope of a.c. load line. The amplifier shown in Fig. 8.16 "sees" an a.c. load $R_{AC} = R_C \parallel R_L$.

$$\therefore \qquad \text{Slope of the a.c. load line} = -\frac{1}{R_{AC}}$$

Since $R_{AC} < (R_C + R_E)$, the slope of a.c. load line is greater than the slope of the d.c. load line *i.e.* a.c. load line is steeper than the d.c. load line.

Example 8.8. *For the transistor amplifier shown in Fig. 8.19, R_1 = 10 kΩ, R_2 = 5 kΩ, R_C = 1 kΩ, R_E = 2 kΩ and R_L = 1 kΩ.*

(i) Draw d.c. load line (ii) Determine the operating point (iii) Draw a.c. load line (iv) Determine the slope of the two load lines.

Assume V_{BE} = 0.7 V.

Solution. (*i*) d.c. load line.

To draw d.c. load line, we require two end points *viz.* maximum V_{CE} point and maximum I_C point.

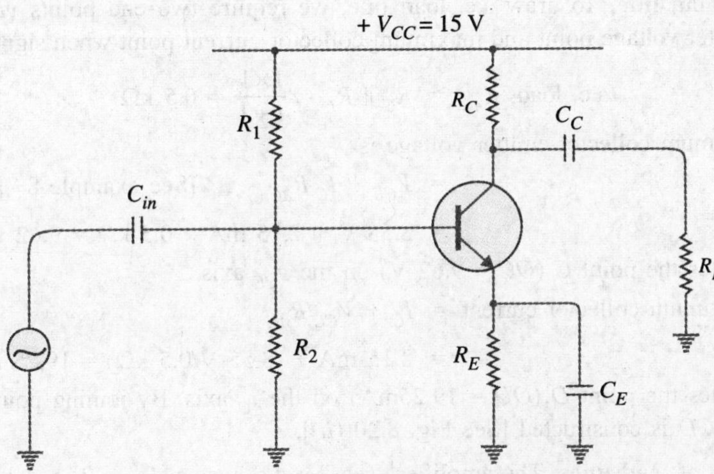

Fig. 8.19

$$\text{Maximum } V_{CE} = V_{CC} = 15 \text{ V [See Art. 8.10]}$$

This locates the point B (OB = 15 V) of the d.c. load line.

$$\text{Maximum } I_C = \frac{V_{CC}}{R_C + R_E} = \frac{15 \text{ V}}{(1 + 2) \text{ k}\Omega} = 5 \text{ mA} \qquad \text{[See Art. 8.10]}$$

This locates the point A (OA = 5 mA) of the d.c. load line. Fig. 8.20 (*i*) shows the d.c. load line AB.

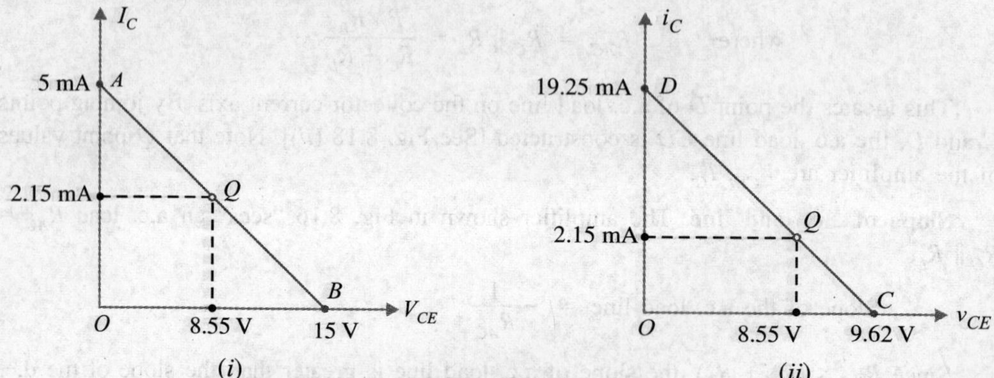

Fig. 8.20

***(ii)* Operating point *Q*.** The voltage across R_2 (= 5 kΩ) is *5 V *i.e.* V_2 = 5 V.

Now, $V_2 = V_{BE} + I_E R_E$

$$\therefore \quad I_E = \frac{V_2 - V_{BE}}{R_E} = \frac{(5-0.7)\text{ V}}{2\text{ k}\Omega} = 2.15\text{ mA}$$

$$\therefore \quad I_C = I_E = 2.15\text{ mA}$$

Now, $V_{CE} = V_{CC} - I_C(R_C + R_E) = 15\text{ V} - 2.15\text{ mA} \times 3\text{ k}\Omega$

$= 8.55$ V

∴ Operating point *Q* is **8.55 V, 2.15 mA.** This is shown on the d.c. load line.

***(iii)* a.c. load line.** To draw a.c. load line, we require two end points *viz.* maximum collector-emitter voltage point and maximum collector current point when signal is applied.

$$\text{a.c. load, } R_{AC} = R_C \parallel R_L = \frac{1\times 1}{1+1} = 0.5\text{ k}\Omega$$

∴ Maximum collector-emitter voltage is

$= V_{CE} + I_C R_{AC}$ [See example 8.7]

$= 8.55\text{ V} + 2.15\text{ mA} \times 0.5\text{ k}\Omega = 9.62$ volts

This locates the point *C* (*OC* = 9.62 V) on the v_{CE} axis.

Maximum collector current $= I_C + V_{CE}/R_{AC}$

$= 2.15\text{ mA} + (8.55\text{ V}/0.5\text{ k}\Omega) = 19.25$ mA

This locates the point *D* (*OD* = 19.25mA) on the i_C axis. By joining points *C* and *D*, a.c. load line *CD* is constructed [See Fig. 8.20 (*ii*)].

***(iv)* Slope of load lines.** The amplifier "sees" a d.c. load $= R_C + R_E = 1\text{ k}\Omega + 2\text{ k}\Omega = 3\text{ k}\Omega = 3\times 10^3\Omega$. Also, the amplifier "sees" an a.c. load $R_{AC} = R_C \parallel R_L = 1\text{ k}\Omega \parallel 1\text{k}\Omega = 0.5\text{ k}\Omega = 0.5\times 10^3\Omega$.

$$\therefore \quad \text{Slope of d.c. load line} = -\frac{1}{R_C + R_E} = -\frac{1}{3\times 10^3\Omega} = -\mathbf{0.33\times 10^{-3}\text{ S}}$$

$$\text{Slope of a.c. load line} = -\frac{1}{R_{AC}} = -\frac{1}{0.5\times 10^3\Omega} = -\mathbf{2\times 10^{-3}\text{ S}}$$

---

* Voltage across series combination of R_1 and R_2 is 15 V. Applying voltage divider rule, voltage across R_2 = 5 V.

Example 8.9. *In the transistor amplifier shown in Fig. 8.19, R_C = 10 kΩ, R_L = 30 kΩ and V_{CC} = 20V. The values R_1 and R_2 are such so as to fix the operating point at 10V, 1mA. Draw the d.c. and a.c. load lines and determine the slope of the two load lines. Assume R_E is negligible.*

Solution. d.c. load line. For drawing d.c. load line, two end points *viz.* maximum V_{CE} point and maximum I_C point are needed. Maximum V_{CE} = 20 V. This locates the point *B* (*OB* = 20V) of the d.c. load line on the V_{CE} axis.

$$\text{Maximum } I_C = \frac{V_{CC}}{R_C + R_E} = \frac{20 \text{ V}}{10 \text{ k}\Omega} = 2 \text{ mA}$$

This locates the point *A* (*OA* = 2 mA) on the I_C axis. By joining points *A* and *B*, the d.c. load line *AB* is constructed (See Fig. 8.21).

a.c. load line. To draw a.c. load line, we require two end points *viz.* maximum collector-emitter voltage point and maximum collector current point when signal is applied.

$$\text{a.c. load, } R_{AC} = R_C \parallel R_L = \frac{10 \times 30}{10 + 30} = 7.5 \text{ k}\Omega$$

Maximum collector-emitter voltage is

$$= V_{CE} + I_C R_{AC}$$

$$= 10 \text{ V} + 1 \text{ mA} \times 7.5 \text{ k}\Omega = 10 + 7.5 = 17.5 \text{ V}$$

This locates the point *D* (*OD* = 17.5 V) on the v_{CE} axis.

$$\text{Maximum collector current} = I_C + V_{CE}/R_{AC}$$

$$= 1 \text{ mA} + 10 \text{ V}/7.5 \text{ k}\Omega = 1 \text{ mA} + 1.33 \text{ mA} = 2.33 \text{ mA}$$

This locates the point *C* (*OC* = 2.33 mA) on the i_C axis. By joining points *C* and *D*, a.c. load line *CD* is constructed (See Fig. 8.21).

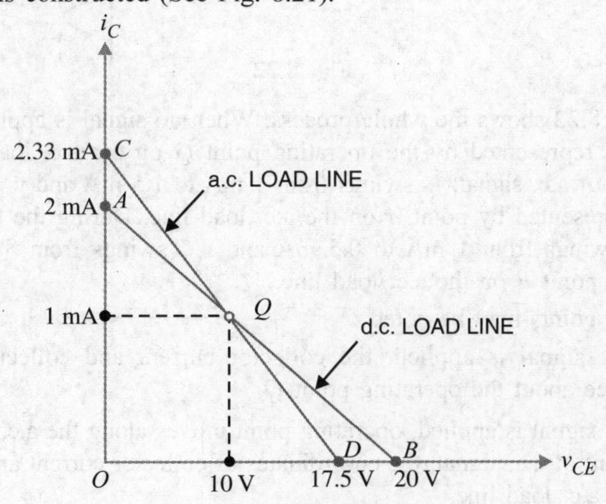

Fig. 8.21

Slope of load lines. The d.c. load for the amplifier = $R_C + R_E$ = 10 kΩ + 0 = 10 kΩ = 10 × 10^3Ω. The a.c. load for the amplifier, $R_{AC} = R_C \parallel R_L$ = 10 kΩ ∥ 30 kΩ = 7.5 kΩ = 7.5 × 10^3 Ω.

$\therefore$ Slope of d.c. load line $= -\dfrac{1}{R_C + R_E} = -\dfrac{1}{10\times10^3\Omega} = -10^{-4}$ **S**

Slope of a.c. load line $= -\dfrac{1}{R_{AC}} = -\dfrac{1}{7.5\times10^3\Omega} = -$ **0.133 × 10^{-3} S**

Comments. The reader may see that the operating point lies on both a.c. and d.c. load lines. It is not surprising because signal is a.c. and it becomes zero after every half-cycle. When the signal is zero, we have the exact d.c. conditions. Therefore, key point to keep in mind is that the point of intersection of d.c. and a.c. load lines is the operating point *Q*.

Example 8.10. *In a transistor amplifier, the operating point Q is fixed at 8V, 1mA. When a.c. signal is applied, the collector current and collector-emitter voltage change about this point. During the positive peak of signal,* i_C *= 1.5 mA and* v_{CE} *= 7V and during negative peak,* i_C *= 0.5mA and* v_{CE} *= 9V. Show this phenomenon with the help of a.c. load line.*

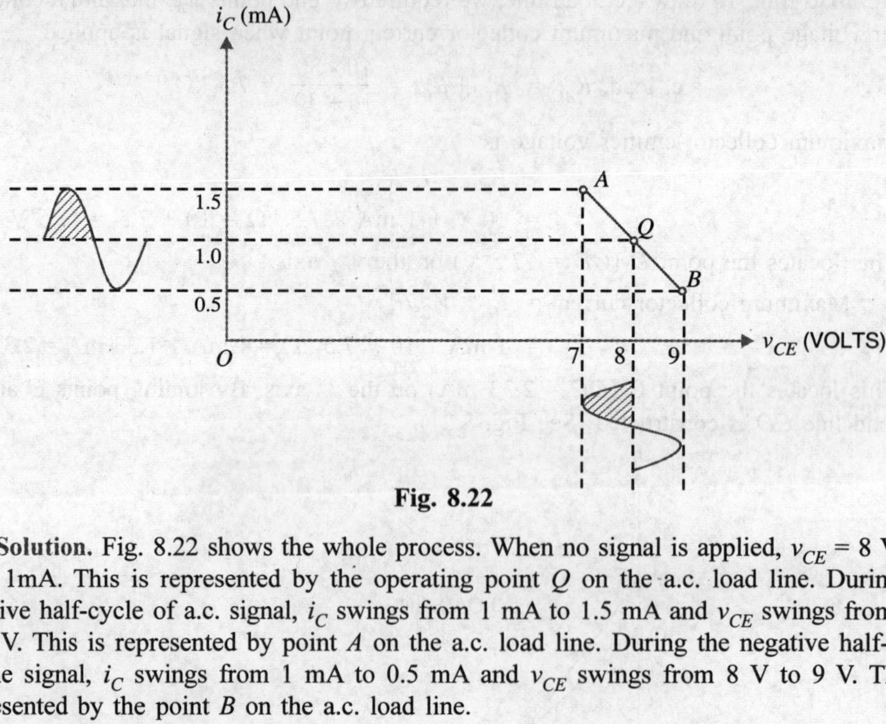

Fig. 8.22

Solution. Fig. 8.22 shows the whole process. When no signal is applied, v_{CE} = 8 V and i_C = 1mA. This is represented by the operating point *Q* on the a.c. load line. During the positive half-cycle of a.c. signal, i_C swings from 1 mA to 1.5 mA and v_{CE} swings from 8 V to 7 V. This is represented by point *A* on the a.c. load line. During the negative half-cycle of the signal, i_C swings from 1 mA to 0.5 mA and v_{CE} swings from 8 V to 9 V. This is represented by the point *B* on the a.c. load line.

The following points may be noted :

(*i*) When a.c. signal is applied, the collector current and collector-emitter voltage variations take place about the operating point *Q*.

(*ii*) When a.c. signal is applied, operating point moves along the a.c. load line. In other words, at any instant of a.c. signal, the co-ordinates of collector current and collector-emitter voltage are on the a.c. load line.

8.11 Important Facts about D.C. and A.C. Load Lines

It is desirable to discuss the important facts about the d.c. and a.c. load lines of the amplifier. For this purpose, consider the *CE* amplifier shown in Fig. 8.16 (Refer back). For this circuit :

d.c. load $= (R_C + R_E)$

$$\text{a.c. load} = R_{AC} = R_C \parallel R_L = \frac{R_C R_L}{R_C + R_L}$$

(*i*) Fig. 8.23 shows the d.c. and a.c. load lines of the *CE* amplifier under consideration. It is clear from the figure that :

For d.c. load line, the two end points are : V_{CC} ; $\dfrac{V_{CC}}{R_C + R_E}$

For a.c. load line, the two end points are : $V_{CE} + I_C R_{AC}$; $I_C + \dfrac{V_{CE}}{R_{AC}}$

Note that operating point Q of the circuit is V_{CE} ; I_C.

(*ii*) As shown in Fig. 8.23, the *Q*-point lies at the point of intersection of the d.c. and a.c. load lines.

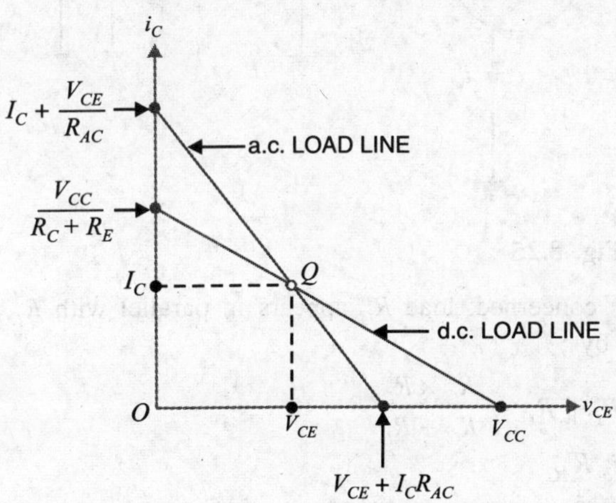

Fig. 8.23

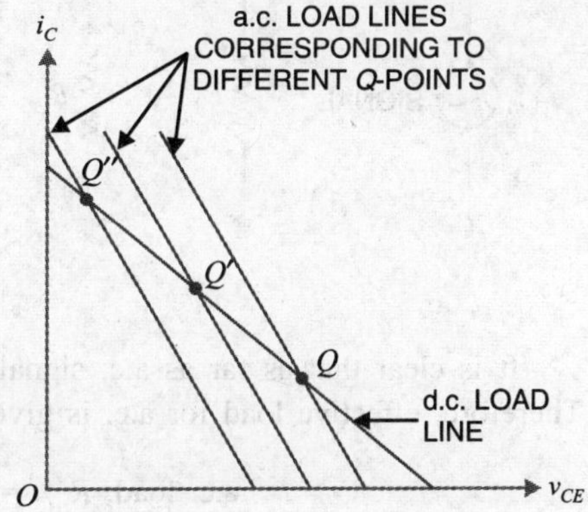

Fig. 8.24

(*iii*) As shown in Fig. 8.23, the a.c. load line is steeper than the d.c. load line. It is because the slope of a.c. load line is greater than that of the d.c. load line.

$$\text{Slope of d.c. load line} = -\frac{1}{R_C + R_E}$$

$$\text{Slope of a.c. load line} = -\frac{1}{R_{AC}} \qquad \text{where } R_{AC} = R_C \parallel R_L$$

Since $R_{AC} < (R_C + R_E)$, the slope of a.c. load line is more than that of the d.c. load line.

(*iv*) If the values of biasing resistors (R_1 and R_2) are changed, the *Q*-point will shift to a new location on the d.c. load line. Since the a.c. load line passes through the *Q*-point, the a.c. load line will also shift. As illustrated in Fig. 8.24, a.c. load lines corresponding to different *Q*-points are parallel to each other because all have the same slope *viz.* $-1/R_{AC}$.

(*v*) As shown in Fig. 8.24, the a.c. load line shifts with the change in *Q*-point. This shift is such that slope of a.c. load line remains the same ($\because R_{AC}$ is same). It is clear that shifting of the a.c. load line causes it to have different intercepts on the v_{CE} and i_C axes.

8.12 Voltage Gain of Transistor Amplifier

The basic function of an amplifier is to raise the strength of an a.c. input signal. The voltage gain of the amplifier is the ratio of a.c. output voltage to the a.c. input signal voltage. Therefore, in order to find the voltage gain, we should consider only the a.c. currents and voltages in the circuit. For this purpose, we should look at the a.c. equivalent circuit of transistor amplifier. For facility of reference, the a.c. equivalent circuit of transistor amplifier is redrawn in Fig. 8.25.

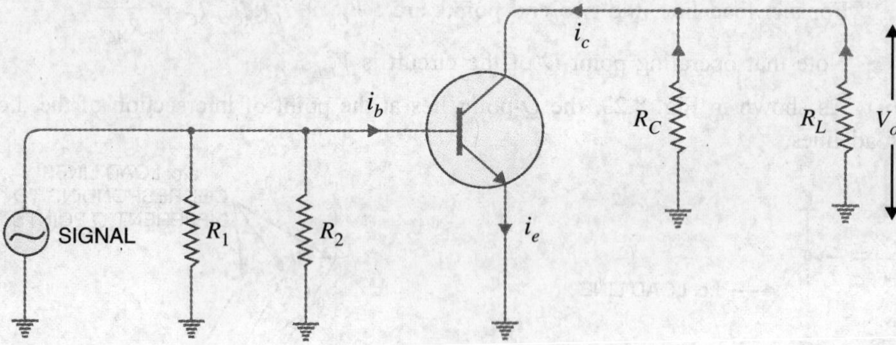

Fig. 8.25

It is clear that as far as a.c. signal is concerned, load R_C appears in parallel with R_L. Therefore, effective load for a.c. is given by ;

$$\text{a.c. load, } R_{AC} = R_C \parallel R_L = \frac{R_C \times R_L}{R_C + R_L}$$

$$\text{Output voltage, } V_{out} = i_c\, R_{AC}$$

$$\text{Input voltage, } V_{in} = i_b\, R_{in}$$

$$\therefore \quad \text{Voltage gain, } A_v = V_{out}/V_{in}$$

$$= \frac{i_c\, R_{AC}}{i_b\, R_{in}} = \beta \times \frac{R_{AC}}{R_{in}} \qquad \left(\because \frac{i_c}{i_b} = \beta\right)$$

Incidentally, power gain is given by ;

$$A_p = \frac{i_c^2\, R_{AC}}{i_b^2\, R_{in}} = \beta^2 \times \frac{R_{AC}}{R_{in}}$$

Example 8.11. *In the circuit shown in Fig. 8.26, find the voltage gain. Given that* $\beta = 60$ *and input resistance* $R_{in} = 1\ k\Omega$.

Solution. So far as voltage gain of the circuit is concerned, we need only R_{AC}, β and R_{in}.

$$\text{Effective a.c. load, } R_{AC} = R_C \parallel R_L$$

$$= \frac{R_C \times R_L}{R_C + R_L} = \frac{2 \times 0.5}{2 + 0.5} = 0.4\ \text{k}\Omega$$

$$\therefore \quad \text{Voltage gain} = \beta \times \frac{R_{AC}}{R_{in}} = \frac{60 \times 0.4\ \text{k}\Omega}{1\ \text{k}\Omega} = 24$$

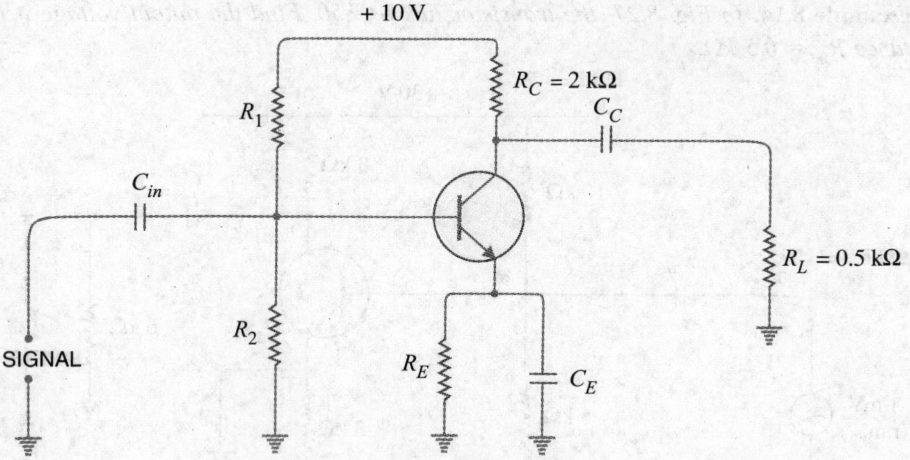

Fig. 8.26

Example 8.12. *In the circuit shown in Fig. 8.26, if R_C = 10 kΩ, R_L = 10 kΩ, R_{in} = 2.5 kΩ, β = 100, find the output voltage for an input voltage of 1mV r.m.s.*

Solution. Effective a.c. load, $R_{AC} = \dfrac{R_C \times R_L}{R_C + R_L} = \dfrac{10 \times 10}{10 + 10} = 5\text{ k}\Omega$

$$\text{Voltage gain} = \beta \times \frac{R_{AC}}{R_{in}} = 100 \times \frac{5\text{ k}\Omega}{2.5\text{ k}\Omega} = 200$$

or $$\frac{V_{out}}{V_{in}} = 200$$

∴ $$V_{out} = 200 \times V_{in} = 200 \times 1\text{ mV} = \mathbf{200\text{ mV}}$$

Example 8.13. *In a transistor amplifier, when the signal changes by 0.02 V, the base current changes by 10 μA and collector current by 1mA. If collector load R_C = 5 kΩ and R_L = 10 kΩ, find : (i) current gain (ii) input impedance (iii) a.c. load (iv) voltage gain (v) power gain.*

Solution. ΔI_B = 10 μA ; ΔI_C = 1 mA ; ΔV_{BE} = 0.02 V ; R_C = 5 kΩ ; R_L = 10 kΩ

(*i*) Current gain, $\beta = \dfrac{\Delta I_C}{\Delta I_B} = \dfrac{1\text{ mA}}{10\text{ μA}} = \mathbf{100}$

(*ii*) Input impedance, $R_{in} = \dfrac{\Delta V_{BE}}{\Delta I_B} = \dfrac{0.02\text{ V}}{10\text{ μA}} = \mathbf{2\text{ k}\Omega}$

(*iii*) a.c. load, $R_{AC} = \dfrac{R_C \times R_L}{R_C + R_L} = \dfrac{5 \times 10}{5 + 10} = \mathbf{3.3\text{ k}\Omega}$

(*iv*) Voltage gain, $A_v = \beta \times \dfrac{R_{AC}}{R_{in}} = 100 \times \dfrac{3.3}{2} = \mathbf{165}$

(*v*) Power gain, A_p = Current gain × Voltage gain

= 100 × 165 = **16500**

Example 8.14. *In Fig. 8.27, the transistor has β = 50. Find the output voltage if input resistance R_{in} = 0.5 kΩ.*

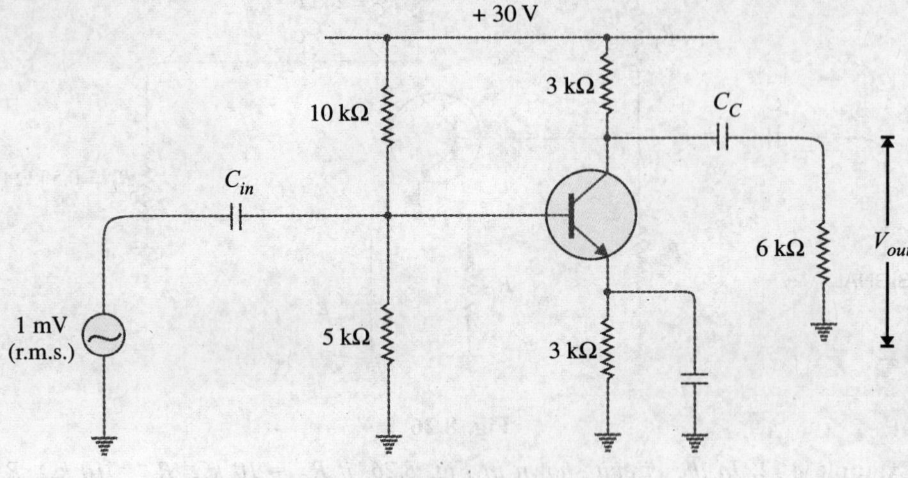

Fig. 8.27

Solution. $\beta = 50 \; ; \; R_{in} = 0.5 \text{ k}\Omega$

$$\text{a.c. load, } R_{AC} = R_C \parallel R_L = \frac{R_C \times R_L}{R_C + R_L} = \frac{3 \times 6}{3 + 6} = 2 \text{ k}\Omega$$

∴ $$\text{Voltage gain} = \beta \times R_{AC}/R_{in} = 50 \times 2/0.5 = 200$$

or $$\frac{V_{out}}{V_{in}} = 200$$

∴ $$\text{Output voltage, } V_{out} = 200 \times V_{in} = 200 \times (1 \text{ mV}) = \mathbf{200 \text{ mV}}$$

Example 8.15. *Fig. 8.28 shows a transistor circuit. The manufacturer of the circuit shows that collector potential is to be + 6 V. The voltage measured at point B by a technician is found to be + 4 V. Is the circuit operating properly?*

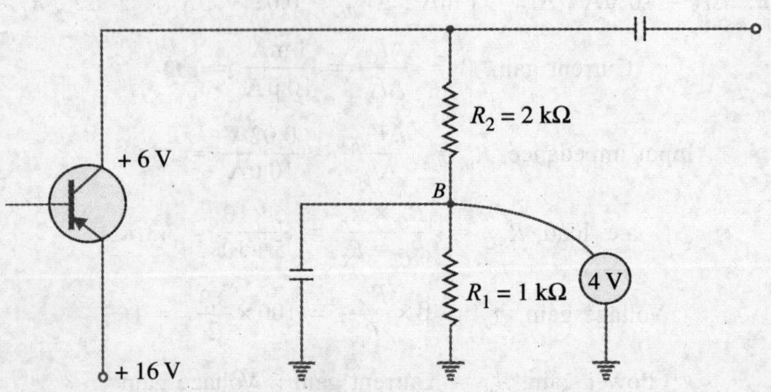

Fig. 8.28

Solution. The voltage at point B is equal to the voltage across R_1. Now total voltage V_T across the series combination of R_1 and R_2 is 6 V. Therefore, using voltage divider method, we have,

$$V_B = \text{Voltage across } R_1$$

$$= \frac{R_1}{R_1 + R_2} \times V_T = \frac{1}{1+2} \times 6 = 2\text{ V}$$

The circuit is **not operating properly**. It is because the voltage at point B should be 2V instead of 4V.

8.13 AC Emitter Resistance

The ac or dynamic resistance of emitter-base junction diode of a transistor is called ac emitter resistance. It is defined as the change in base-emitter voltage divided by change in corresponding emitter current [See Fig. 8.29] *i.e.*

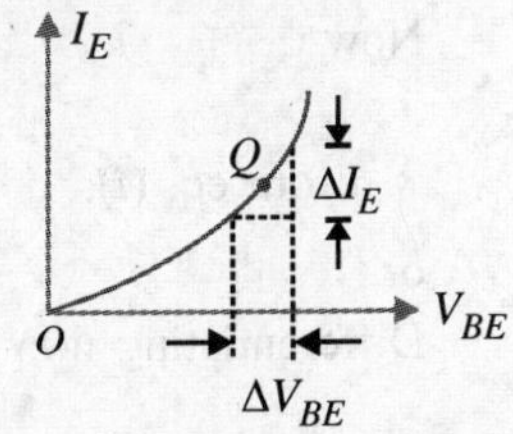

Fig. 8.29

$$R_{ac} = \frac{\Delta V_{BE}}{\Delta I_E}$$

For instance, suppose an a.c. base voltage change of 1 mV produces an a.c. emitter current change of 50 μA. Then emitter diode has an ac resistance of

$$R_{ac} = \frac{1\text{ mV}}{50\text{ μA}} = 20\ \Omega$$

8.14 Formula For AC Emitter Resistance

It can be shown mathematically that the ac resistance of emitter diode is given by ;

$$R_{ac} = \frac{25\text{ mV}}{I_E}$$

where I_E = d.c. emitter current (= V_E/R_E) at Q-point

Note the significance of this formula. It implies that ac emitter resistance can be found simply by substituting the quiescent value of emitter current into the equation. There is no need to have the characteristics available. It is important to keep in mind that this formula is accurate only for small signal operation. It is a usual practice to represent ac emitter resistance by r_e'.

$$\therefore \quad r_e' = \frac{25\text{ mV}}{I_E}$$

The subscript e indicates emitter. The lower case r is used to indicate an ac resistance. The prime shows that it is an internal resistance.

Example 8.16. *Derive the formula for ac resistance of the base-emitter juncton of transistor* $r_e' = 25\ mV/I_E$.

Solution. According to Shockley equation, the forward emitter current I_E across the base-emitter junction of transistor is given by ;

$$*I_E = I_R\,(e^{V_D q/kT} - 1) \quad \text{... (i)}$$

* This is the same diode equation discussed in Art. 4.3 (Chapter 4).

where I_E = Forward current in amperes (A)

I_R = Reverse saturation current in amperes (A)

V_D = Forward voltage across base-emitter junction (V)

q = Charge on electron = 1.6×10^{-19} C

k = Boltzmann's constant = 1.38×10^{-23} J/K

T = Absolute temperature (K)

At room temperature (= 27°C), $T = 273 + 27 = 300$ K

Now, $$\frac{q}{kT} = \frac{1.6\times10^{-19}}{1.38\times10^{-23}\times300} \simeq 40$$

∴ From eq. (*i*), $$I_E = I_R(e^{40V_D} - 1)$$

or $$I_E = I_R e^{40V_D} - I_R \quad \text{... (ii)}$$

Differentiating *w.r.t.* V_D, we get,

$$\frac{dI_E}{dV_D} = 40\, I_R\, e^{40\,V_D}$$

or $$\frac{dI_E}{dV_D} = 40(I_E + I_R) \quad \text{[From eq. (ii)]}$$

Now $I_E >> I_R$ so that I_R can be neglected compared to I_E.

∴ $$\frac{dI_E}{dV_D} = 40 I_E$$

The ac resistance r_e' of the base-emitter junction is

$$r_e' = \frac{dV_D}{dI_E} = \frac{1}{40 I_E} = \frac{1000\,\text{mV}}{40\, I_E} = \frac{25\,\text{mV}}{I_E}$$

∴ $$r_e' = \frac{25\,\text{mV}}{I_E}$$

Example 8.17. *Determine the ac emitter resistance for the transistor circuit shown in Fig. 8.30.*

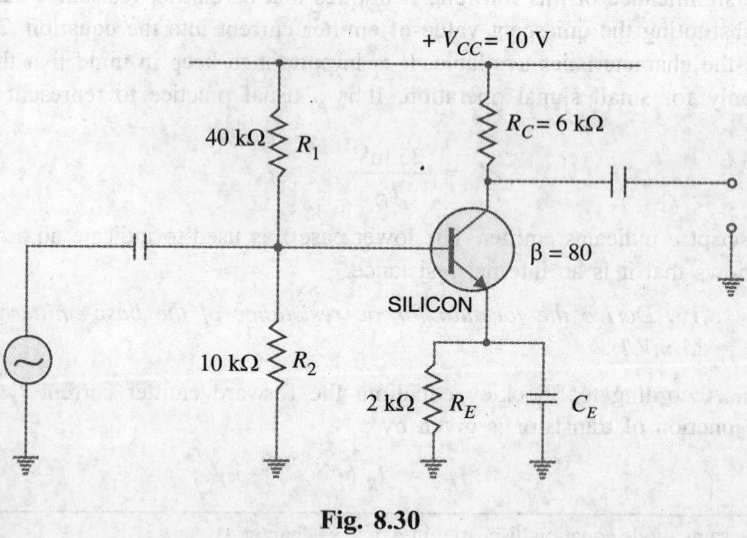

Fig. 8.30

Solution. Voltage across R_2, $V_2 = \dfrac{V_{CC}}{R_1 + R_2} \times R_2 = \dfrac{10}{40+10} \times 10 = 2 \text{ V}$

Voltage across R_E, $V_E = V_2 - V_{BE} = 2 - 0.7 = 1.3 \text{ V}$

Emitter current, $I_E = \dfrac{V_E}{R_E} = \dfrac{1.3 \text{ V}}{2 \text{ k}\Omega} = 0.65 \text{ mA}$

$\therefore$ ac emitter resistance, $r_e' = \dfrac{25 \text{ mV}}{I_E} = \dfrac{25 \text{ mV}}{0.65 \text{ mA}} = \mathbf{38.46\ \Omega}$

8.15 Voltage Gain of CE Amplifier

The voltage gain (A_v) of an amplifier is equal to a.c. output voltage (V_{out}) divided by a.c. input voltage (V_{in}) *i.e.* $A_v = V_{out}/V_{in}$. We have already seen that voltage gain of a *CE* amplifier is given by ;

$$\text{Voltage gain, } A_{v0} = \beta \times \frac{R_C}{R_{in}} \quad \text{... for unloaded amplifier}$$

$$A_v = \beta \times \frac{R_{AC}}{R_{in}} \quad \text{... for loaded amplifier}$$

$$\text{Remember that } R_{AC} = R_C \parallel R_L$$

The above formula for A_v can be used if we know the values of R_C (or R_{AC}), β and R_{in}. Generally, all these values are not known. In that case, we can find the value of A_v in terms of *total a.c. collector resistance* and *total ac emitter resistance.* For the circuit shown in Fig. 8.31 (*with C_E connected across R_E*), it can be proved that the voltage gain is given by ;

$$\text{Voltage gain, } A_{v0} = \frac{R_C}{r_e'} \quad \text{... for unloaded amplifier}$$

$$A_v = \frac{R_{AC}}{r_e'} \quad \text{... for loaded amplifier}$$

8.16 Voltage Gain of Unloaded CE Amplifier

Fig. 8.31 shows the circuit of unloaded *CE* amplifier (*i.e.* no load R_L is connected to the circuit). Note that emitter bypass capacitor C_E is connected in parallel with emitter resistance R_E. The capacitor C_E acts as a *short to the a.c. signal so that it bypasses the a.c. signal to the ground. Therefore, the emitter is effectively at a.c. ground.

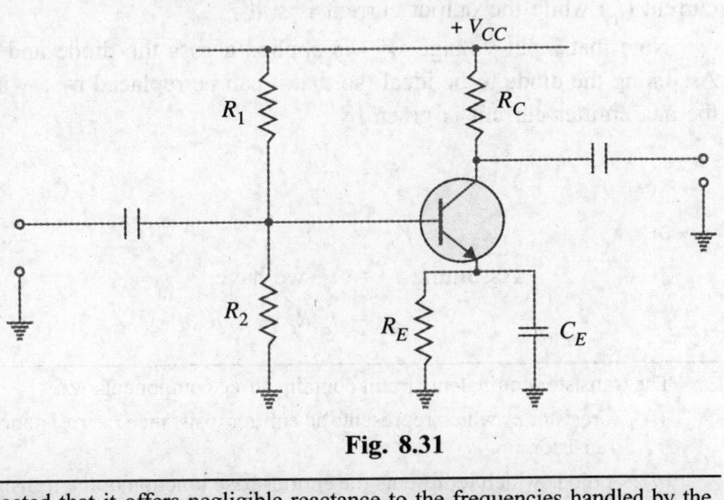

Fig. 8.31

* The size of C_E is so selected that it offers negligible reactance to the frequencies handled by the amplifier (See Example 8.2).

It is important to note that C_E plays an important role in determining the voltage gain (A_{v0}) of the *CE* amplifier. If it is removed, the voltage gain of the amplifier is greatly reduced (soon you will see the reason for it).

$$\text{Voltage gain, } A_{v0} = \frac{R_C}{r_e'}$$

where R_C = a.c. collector resistance

r_e' = ac emitter resistance = 25 mV/I_E

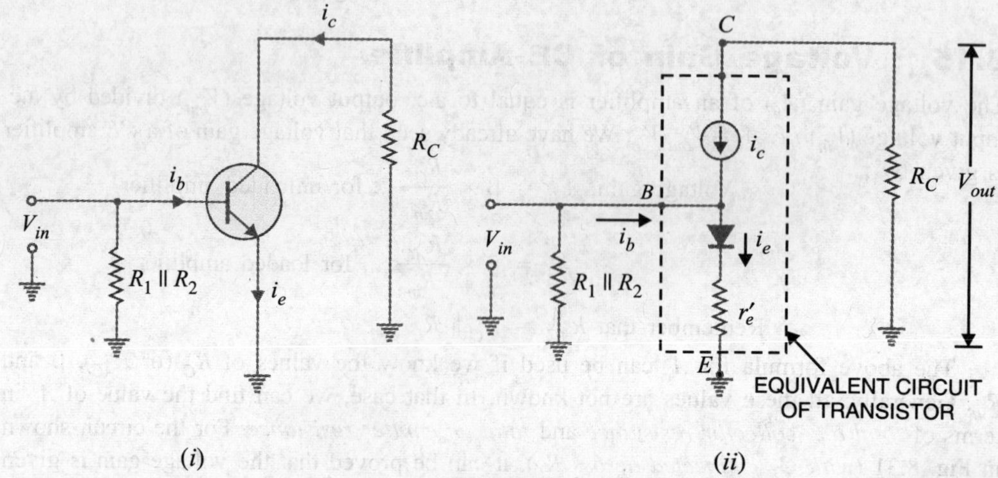

Fig. 8.32

Derivation. Fig. 8.31 shows the common emitter amplifier. The a.c. equivalent circuit of the amplifier is shown in Fig. 8.32. (*i*). Replacing the transistor by its *equivalent circuit, we get the circuit shown in Fig. 8.32 (*ii*). Note that current source is still connected between the collector and base terminals while the diode between the base and emitter terminals. Further, the input current is the base current (i_b) while the output current is still i_c.

Note that input voltage (V_{in}) is applied across the diode and r_e'. Assuming the diode to be ideal (so that it can be replaced by a wire), the a.c. emitter current is given by ;

$$i_e = \frac{V_{in}}{r_e'}$$

or $$V_{in} = i_e\, r_e' \qquad ...(i)$$

Assuming $i_c = i_e$, we have,

$$V_{out} = i_c\, R_C = i_e\, R_C$$

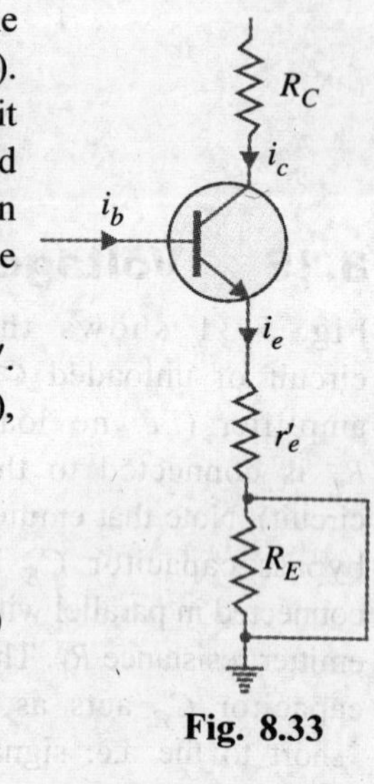

Fig. 8.33

* The transistor equivalent circuit contains three components *viz.*,

(*i*) A resistor r_e' which represents ac emitter resistance *i.e.* resistance of emitter-base diode of the transistor.

(*ii*) A diode which represents the emitter-base junction of the transistor.

(*iii*) A current source which represents the current being supplied to R_C from the collector of the transistor.

$$\therefore \quad \text{Voltage gain, } A_{v0} = \frac{V_{out}}{V_{in}} = \frac{i_e R_C}{i_e r_e'} = \frac{R_C}{r_e'}$$

or

$$A_{v0} = \frac{R_C}{r_e'}$$

where R_C = Total a.c. collector resistance

r_e' = Total ac emitter resistance

Fig. 8.33 shows the simple a.c. model of *CE* amplifier with C_E connected across R_E. Note that C_E behaves as a short so that R_E is cut out from the emitter circuit far a.c. signal. Therefore, as far as a.c. signal is concerned, the total ac emitter resistance is r_e'.

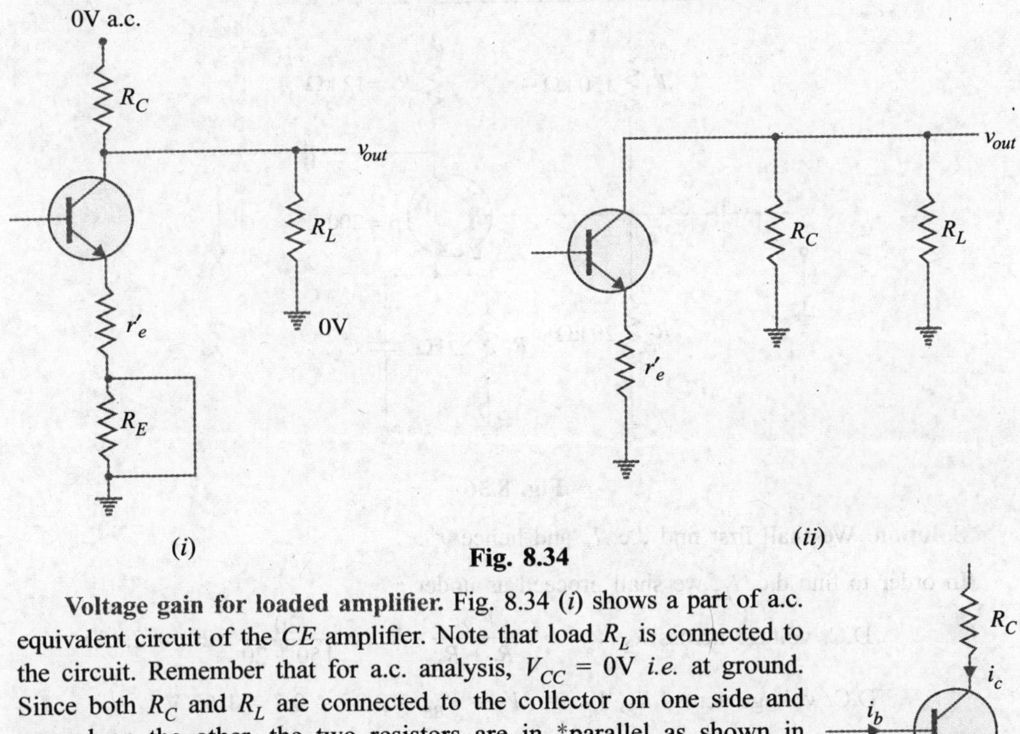

(*i*) (*ii*)

Fig. 8.34

Voltage gain for loaded amplifier. Fig. 8.34 (*i*) shows a part of a.c. equivalent circuit of the *CE* amplifier. Note that load R_L is connected to the circuit. Remember that for a.c. analysis, V_{CC} = 0V *i.e.* at ground. Since both R_C and R_L are connected to the collector on one side and ground on the other, the two resistors are in *parallel as shown in Fig. 8.34 (*ii*).

$$\text{Total a.c. collector resistance, } R_{AC} = R_C \parallel R_L = \frac{R_C R_L}{R_C + R_L}$$

$$\text{Total ac emitter resistance} = r_e'$$

$$\therefore \quad \text{Voltage gain, } A_v = \frac{R_{AC}}{r_e'}$$

8.17 Voltage Gain of CE Amplifier Without C_E

When we remove the emitter bypass capacitor from the *CE* amplifier shown in Fig. 8.31, the voltage gain of the circuit is greatly reduced. The reason is simple. Without the emitter bypass capacitor C_E, the emitter is no longer at the ac ground as shown in Fig. 8.35. Therefore,

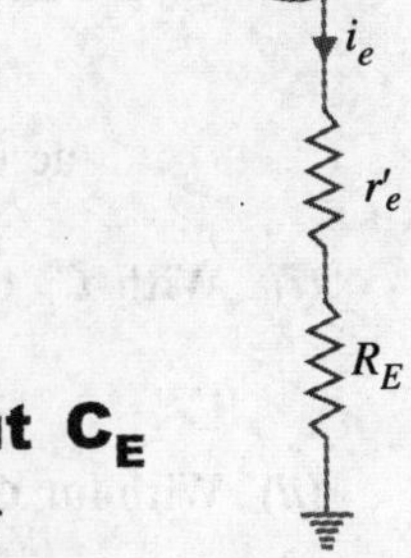

Fig. 8.35

* Note that C_C behaves as a short for a.c. and is replaced by a wire in the two a.c. circuits.

for the a.c. signal, both r_e' and R_E are in series. As a result, the voltage gain of the amplifier becomes:

$$\text{Voltage gain, } A_{v0} = \frac{R_C}{r_e' + R_E} \quad \text{... for unloaded amplifier}$$

$$A_v = \frac{R_{AC}}{r_e' + R_E} \quad \text{... for loaded amplifier}$$

Example 8.18. *For the amplifier circuit shown in Fig. 8.36, find the voltage gain of the amplifier with (i) C_E connected in the circuit (ii) C_E removed from the circuit.*

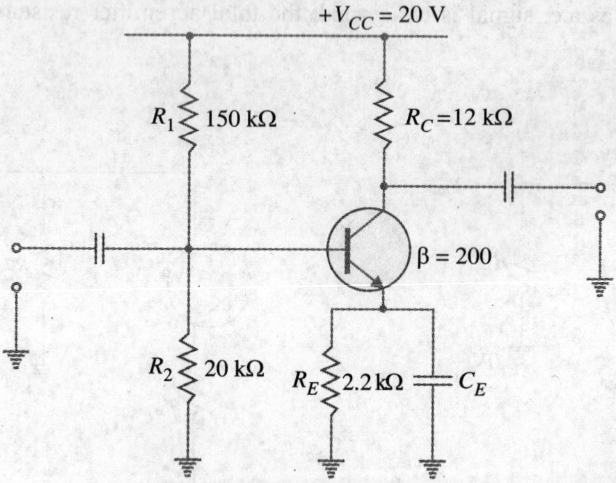

Fig. 8.36

Solution. We shall first find d.c. I_E and hence r_e'.

In order to find d.c. I_E, we shall proceed as under :

$$\text{D.C. voltage across } R_2, V_2 = \frac{V_{CC}}{R_1 + R_2} \times R_2 = \frac{20}{150 + 20} \times 20 = 2.35 \text{ V}$$

$$\text{D.C. voltage across } R_E, V_E = V_2 - V_{BE} = 2.35 - 0.7 = 1.65 \text{ V}$$

$$\therefore \quad \text{D.C. emitter current, } I_E = \frac{V_E}{R_E} = \frac{1.65 \text{ V}}{2.2 \text{ k}\Omega} = 0.75 \text{ mA}$$

$$\therefore \quad \text{ac emitter resistance, } r_e' = \frac{25 \text{ mV}}{I_E} = \frac{25 \text{ mV}}{0.75 \text{ mA}} = 33.3 \ \Omega$$

(*i*) With C_E connected

$$\text{Voltage gain, } A_{v0} = \frac{R_C}{r_e'} = \frac{12 \text{ k}\Omega}{33.3 \ \Omega} = 360$$

(*ii*) Without C_E

$$\text{Voltage gain, } A_{v0} = \frac{R_C}{r_e' + R_E} = \frac{12 \text{ k}\Omega}{33.3 \ \Omega + 2.2 \text{ k}\Omega} = 5.38$$

What a difference the emitter bypass capacitor C_E makes ! With C_E connected, A_{v0} = 360 and when C_E is removed, the voltage gain goes down to 5.38.

Example 8.19. *If in the above example, a load of 6 kΩ is connected (with C_E connected) to the collector terminal through a capacitor, what will be the voltage gain of the amplifier ?*

Solution. Amplifiers are used to provide a.c. power to the load. When load R_L is connected to the collector terminal through a capacitor, the total a.c. resistance of collector changes to :

$$R_{AC} = R_C \parallel R_L = 12\ \text{k}\Omega \parallel 6\ \text{k}\Omega = \frac{12 \times 6}{12 + 6} = 4\ \text{k}\Omega$$

The value of ac emitter resistance remains the same.

$$\therefore \quad \text{Voltage gain, } A_v = \frac{R_{AC}}{r_e'} = \frac{4\ \text{k}\Omega}{33.3\ \Omega} = \mathbf{120}$$

Thus voltage gain of the amplifier is reduced from 360 to 120 when load is connected to the circuit.

Comments. This example shows the fact that voltage gain of the amplifier is *reduced* when load is connected to it. Conversely, if the load is removed from an amplifier, the voltage gain will *increase.* If a load goes open-circuit, the effect will be the same as removing the load entirely. Thus the primary symptom of an open load in an amplifier is an *increase* in the voltage gain of the circuit.

Example 8.20. *For the circuit shown in Fig. 8.37, find (i) ac emitter resistance (ii) voltage gain (iii) d.c. voltage across both capacitors.*

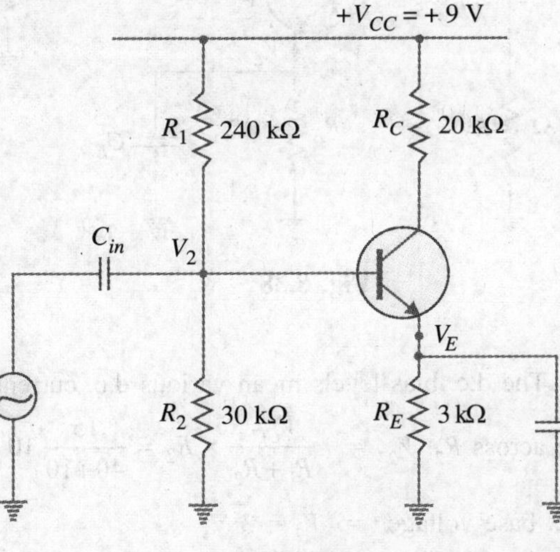

Fig. 8.37

Solution.

(*i*) In order to find ac emitter resistance r_e', we shall first find D.C. emitter current I_E.

To find I_E, we proceed as under :

$$\text{D.C. voltage across } R_2,\ V_2 = \frac{V_{CC}}{R_1 + R_2} \times R_2 = \frac{9}{240 + 30} \times 30 = 1\text{V}$$

$$\text{D.C. voltage across } R_E,\ V_E = V_2 - V_{BE} = 1\text{V} - 0.7\text{V} = 0.3\text{V}$$

$\therefore$ D.C. emitter current, $I_E = \dfrac{V_E}{R_E} = \dfrac{0.3\text{V}}{3\text{ k}\Omega} = 0.1\text{ mA}$

Now, ac emitter resistance, $r_e' = \dfrac{25\text{ mV}}{I_E} = \dfrac{25\text{mV}}{0.1\text{ mA}} = \mathbf{250\ \Omega}$

(*ii*) Voltage gain, $A_{v0} = \dfrac{R_C}{r_e'} = \dfrac{20\text{ k}\Omega}{250\ \Omega} = \mathbf{80}$

(*iii*) The d.c. voltage across input capacitor is equal to the d.c. voltage at the base of the transistor which is V_2 = 1V. Therefore, d.c. voltage across C_{in} is **1 V.**

Similarly, d.c. voltage across C_E = d.c. voltage at the emitter = V_E = **0.3 V.**

Example 8.21. *For the circuit shown in Fig. 8.38, find (i) the d.c. bias levels (ii) d.c. voltages across the capacitors (iii) ac emitter resistance (iv) voltage gain and (v) state of the transistor.*

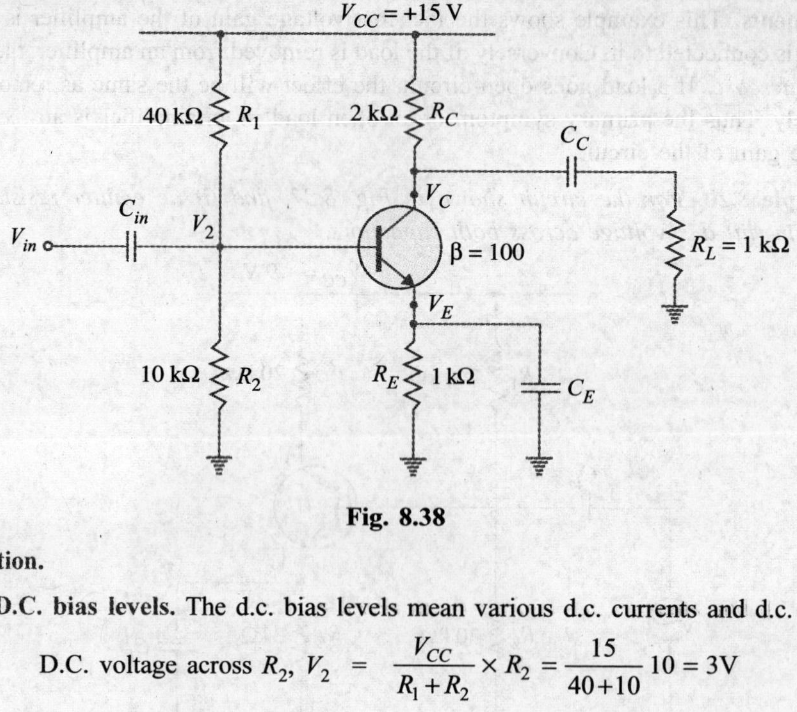

Fig. 8.38

Solution.

(*i*) **D.C. bias levels.** The d.c. bias levels mean various d.c. currents and d.c. voltages.

D.C. voltage across R_2, $V_2 = \dfrac{V_{CC}}{R_1 + R_2} \times R_2 = \dfrac{15}{40+10}\,10 = 3\text{V}$

$\therefore$ D.C. base voltage $= V_2 = \mathbf{3\ V}$

D.C. voltage across R_E, $V_E = V_2 - V_{BE} = 3\text{V} - 0.7\text{V} = \mathbf{2.3\ V}$

D.C. emitter current, $I_E = \dfrac{V_E}{R_E} = \dfrac{2.3\text{V}}{1\text{ k}\Omega} = \mathbf{2.3\ mA}$

D.C. collector current, $I_C = I_E = \mathbf{2.3\ mA}$

D.C. base current, $I_B = I_C/\beta = 2.3\text{ mA}/100 = \mathbf{0.023\ mA}$

D.C. collector voltage, $V_C = V_{CC} - I_C R_C$

$= 15\text{V} - 2.3\text{ mA} \times 2\text{ k}\Omega = \mathbf{10.4\ V}$

Therefore, all d.c. bias levels stand calculated.

(ii) D.C. voltage across C_{in} = $V_2 = 3$ V

D.C. voltage across C_E = $V_E = 2.3$ V

D.C. voltage across C_C = $V_C = 10.4$ V

(iii) ac emitter resistance, $r_e' = \dfrac{25 \text{ mV}}{I_E} = \dfrac{25 \text{ mV}}{2.3 \text{ mA}} = \mathbf{10.9\ \Omega}$

(iv) Total a.c. collector resistance is given by ;

$$R_{AC} = R_C \parallel R_L = \frac{R_C R_L}{R_C + R_L} = \frac{2 \times 1}{2 + 1} = 0.667 \text{ k}\Omega$$

$$\therefore \quad \text{Voltage gain, } A_v = \frac{R_{AC}}{r_e'} = \frac{0.667 \text{ k}\Omega}{10.9\ \Omega} = \mathbf{61.2}$$

(v) As calculated above, $V_C = 10.4$ V and $V_E = 2.3$ V. Since $V_C > V_E$, the transistor is **in active state.**

Example 8.22. *An amplifier has a voltage gain of 132 and β = 200. Determine the power gain and output power of the amplifier if the input power is 60 μW.*

Solution.

Power gain, A_p = Current gain × Voltage gain

$= \beta \times A_v = 200 \times 132 = \mathbf{26400}$

Output power, $P_{out} = A_p \times P_{in} = (26400)(60\ \mu\text{W}) = \mathbf{1.584\ W}$

Example 8.23. *For the circuit shown in Fig. 8.39, determine (i) the current gain (ii) the voltage gain and (iii) the power gain. Neglect the ac emitter resistance for the transistor.*

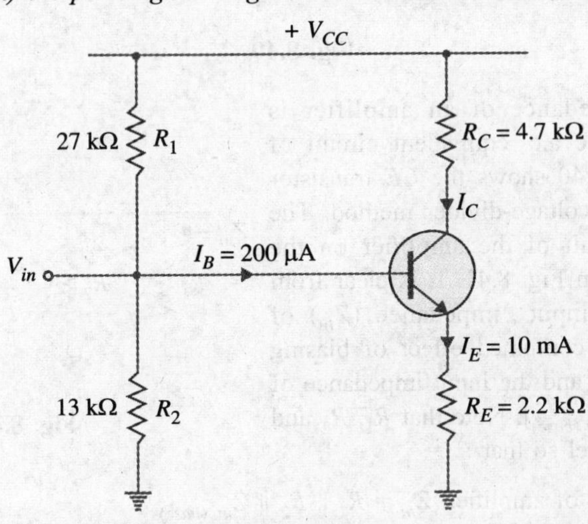

Fig. 8.39

Solution. In most practical circuits, the value of ac emitter resistance r_e' for the transistor is generally quite small as compared to R_E and can be neglected in circuit calculations with reasonable accuracy.

(i) $$I_C = I_E - I_B = 10 \text{ mA} - 200 \text{ μA} = 9.8 \text{ mA}$$

$$\text{Current gain, } A_i = \beta = \frac{I_{out}}{I_{in}} = \frac{I_C}{I_B} = \frac{9.8 \text{ mA}}{200 \text{ μA}} = \mathbf{49}$$

(ii) $$\text{Voltage gain, } A_{v0} = \frac{R_C}{R_E} = \frac{4.7 \text{ k}\Omega}{2.2 \text{ k}\Omega} = \mathbf{2.14}$$

(iii) $$\text{Power gain, } A_p = A_i \times A_{v0} = 49 \times 2.14 = \mathbf{105}$$

8.18 Input Impedance of CE Amplifier

When one *CE* amplifier is being used to drive another, the input impedance of the second amplifier will serve as the load resistance of the first. Therefore, in order to calculate the voltage gain (A_v) of the first amplifier stage correctly, we must calculate the input impedance of the second stage.

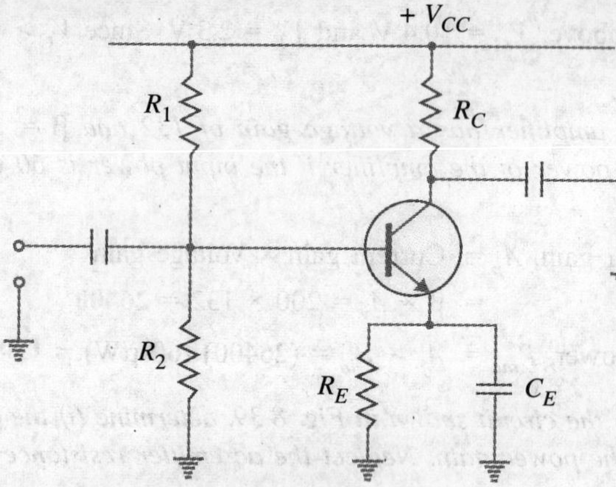

Fig. 8.40

The input impedance of an amplifier is determined using the a.c. equivalent circuit of the amplifier. Fig. 8.40 shows the *CE* transistor amplifier biased by voltage-divider method. The a.c. equivalent circuit of the amplifier on the base side is shown in Fig. 8.41. It is clear from Fig. 8.41 that the input *impedance (Z_{in}) of the amplifier is the combined effect of biasing resistors (R_1 and R_2) and the input impedance of transistor base **[$Z_{in\,(base)}$]. Note that R_1, R_2 and $Z_{in\,(base)}$ are in parallel so that :

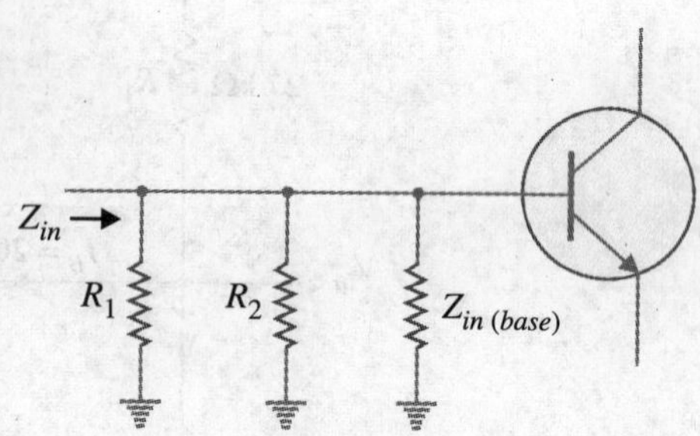

Fig. 8.41

Input impedance of amplifier, $Z_{in} = R_1 \parallel R_2 \parallel Z_{in\,(base)}$

where $Z_{in\,(base)}$ = Input impedance of transistor base

* For signal frequencies < 100 kHz, the input impedance is purely resistive in nature.

** The a.c. resistance looking into the base of the transistor is known as input impedance of transistor base.

It is clear that $Z_{in} < Z_{in\ (base)}$. Once we find $Z_{in\ (base)}$, the input impedance of the amplifier can be easily determined. As we shall see, the value of $Z_{in\ (base)}$ is given by ;

$$Z_{in(base)} = \beta r_e' \quad \text{... with emitter bypass capacitor } C_E$$

$$Z_{in(base)} = \beta(r_e' + R_E) \quad \text{... without emitter bypass capacitor } C_E$$

It may be seen that the presence or absence of the emitter bypass capacitor C_E significantly affects $Z_{in\ (base)}$ and hence the input impedance (Z_{in}) of the transistor amplifier. Previously, we noted that presence or absence of C_E has dramatic effect on the voltage gain (A_v) of the amplifier (See Art. 8.17). *Therefore, emitter bypass capacitor C_E plays an important role in the performance of the transistor amplifier.*

Derivation of formula for $Z_{in(base)}$. We now prove that when emitter bypass capacitor C_E is connected across R_E, then, $Z_{in(base)} = \beta r_e'$. *Fig. 8.42 shows the a.c. equivalent circuit of *CE* amplifier with transistor replaced by its r_e' model.

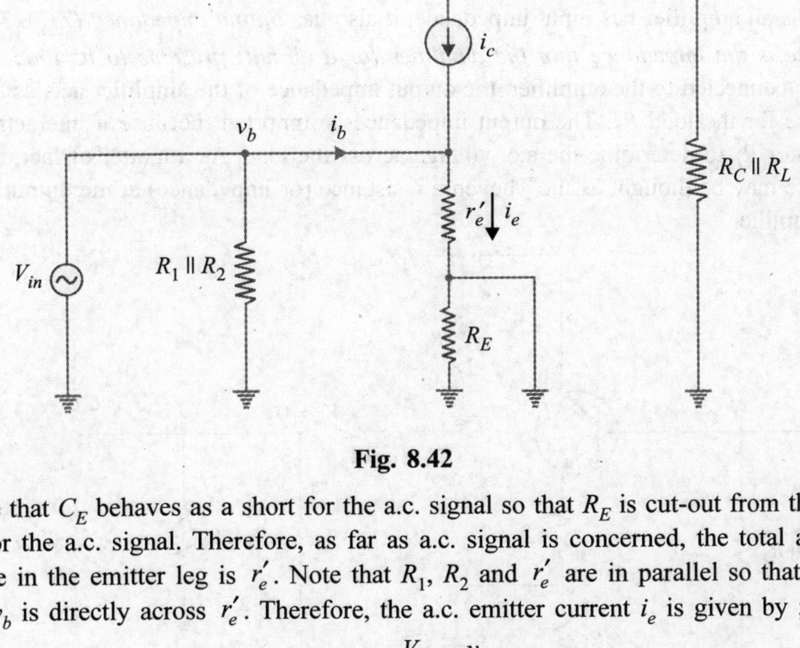

Fig. 8.42

Note that C_E behaves as a short for the a.c. signal so that R_E is cut-out from the emitter circuit for the a.c. signal. Therefore, as far as a.c. signal is concerned, the total ac emitter resistance in the emitter leg is r_e'. Note that R_1, R_2 and r_e' are in parallel so that a.c. base voltage v_b is directly across r_e'. Therefore, the a.c. emitter current i_e is given by ;

$$i_e = \frac{V_{in}}{r_e'} = \frac{v_b}{r_e'} \qquad (\because V_{in} = v_b)$$

or Base voltage, $v_b = i_e r_e'$

The input impedance $Z_{in(base)}$ at the base of the transistor is

$$Z_{in(base)} = \frac{v_b}{i_b} = \frac{i_e r_e'}{i_b} = \frac{i_e}{i_b}(r_e')$$

But i_e/i_b is approximately equal to β.

* Note that this is the same figure as Fig. 8.32 (*ii*) except that the diode, being ideal, is replaced by a wire.

$$\therefore \quad Z_{in(base)} = \beta r'_e \quad \text{... (i)}$$

Note eq. (*i*) very carefully. This equation describes one of the most important properties of the transistor *viz.* a.c. resistance at the base of the transistor is β times the resistance in the emitter. This is an important advantage of the transistor : a small resistance in the emitter (say r'_e) looks very large ($\beta r'_e$) from the base.

$Z_{in(base)}$ without C_E. When emitter bypass capacitor C_E is not connected across R_E, the a.c. signal sees an ac emitter resistance of ($r'_e + R_E$) instead of r'_e.

$$\therefore \quad Z_{in(base)} = \beta(r'_e + R_E)$$

8.19 Output Impedance of CE Amplifier

We have seen above that the input impedance of a *CE* amplifier is $Z_{in} = R_1 \parallel R_2 \parallel Z_{in(base)}$ where R_1 and R_2 are the biasing resistors and $Z_{in(base)}$ is the input impedance of the transistor base. The input impedance is important because it interacts with the source resistance R_S to determine the amount of a.c. voltage available at the input terminals of the amplifier.

Just as an amplifier has input impedance, it also has *output impedance* (Z_{out}). *The output impedance is the impedance that the amplifier (or a circuit) presents to its load* R_L. When load R_L is connected to the amplifier, the output impedance of the amplifier acts as the source impedance for the load R_L. The output impedance is important because it interacts with the load resistor R_L to determine the a.c. voltage across the load. As a matter of fact, the output impedance may be thought as the Thevenin resistance (or impedance) at the output terminals of the amplifier.

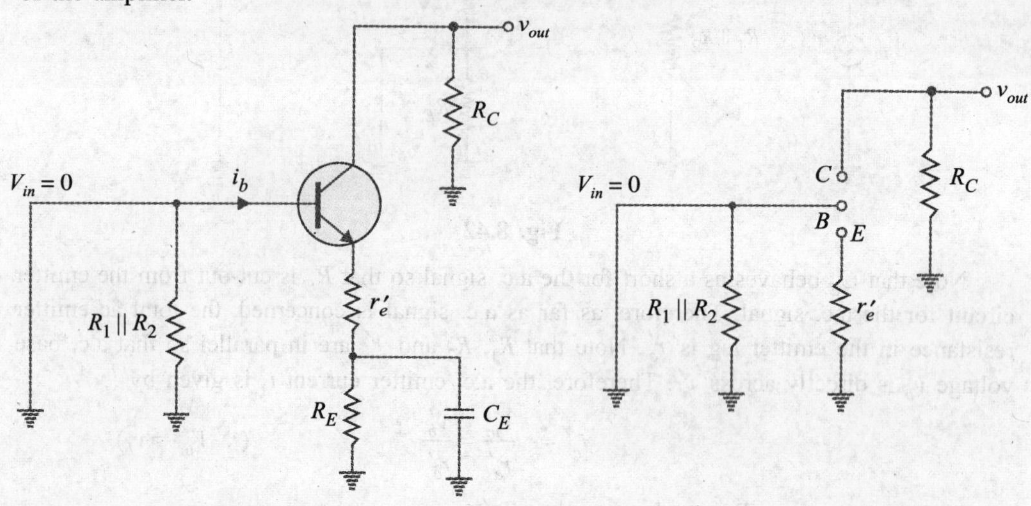

Fig. 8.43

In order to find the output impedance (Z_{out}) of the *CE* amplifier, we should find Thevenin impedance at the output terminals. For this purpose, we replace the voltage source (V_{in}) by a short as shown in Fig. 8.43 (*i*). Since these is no voltage source at the base, the a.c. base current (i_b) will be zero. This means that a.c. collector current as well as a.c. emitter current will be zero. As there are no transistor a.c. currents, therefore, we can replace the transistor with an open-circuit as shown in Fig. 8.43 (*ii*). It is clear from Fig. 8.43(*ii*) that impedance at the output terminals is just the collector resistor R_C.

∴ Output impedance, $Z_{out} = R_C$

It may be noted that the output impedance of the amplifier does not depend on anything in the base or emitter of the transistor.

Example 8.24. *Determine the input impedance of the amplifier circuit shown in Fig. 8.44.*

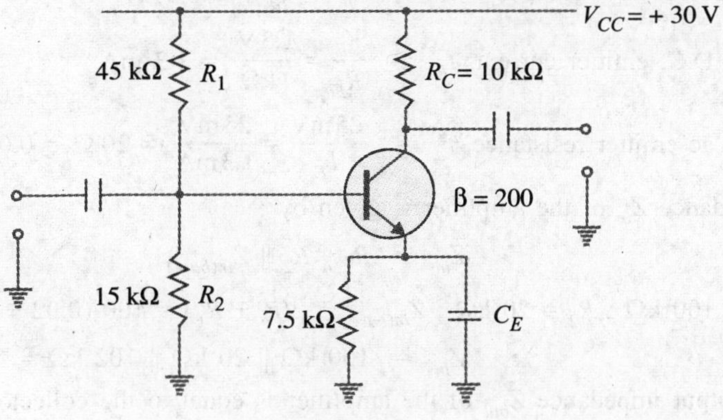

Fig. 8.44

Solution. Voltage across R_2, $V_2 = \frac{V_{CC}}{R_1 + R_2} \times R_2 = \frac{30}{45 + 15} \times 15 = 7.5\text{ V}$

Voltage across R_E, $V_E = V_2 - V_{BE} = 7.5 - 0.7 \simeq 7.5\text{ V}$

Emitter current, $I_E = \frac{V_E}{R_E} = \frac{7.5\text{ V}}{7.5\text{ k}\Omega} = 1\text{ mA}$

ac emitter resistance, $r'_e = 25\text{ mV}/I_E = 25\text{ mV}/1\text{ mA} = 25\ \Omega$

$Z_{in\ (base)} = \beta r'_e = 200 \times 25 = 5 \times 10^3\ \Omega = 5\text{ k}\Omega$

∴ $Z_{in} = R_1 \parallel R_2 \parallel Z_{in\ (base)}$

$= 45\text{ k}\Omega \parallel 15\text{ k}\Omega \parallel 5\text{ k}\Omega = \mathbf{3.45\ k\Omega}$

Example 8.25. *For the CE transistor amplifier shown in Fig. 8.45, find (i) input impedance of the amplifier (ii) output impedance of the amplifier (iii) voltage gain of the amplifier. The transistor has β = 100.*

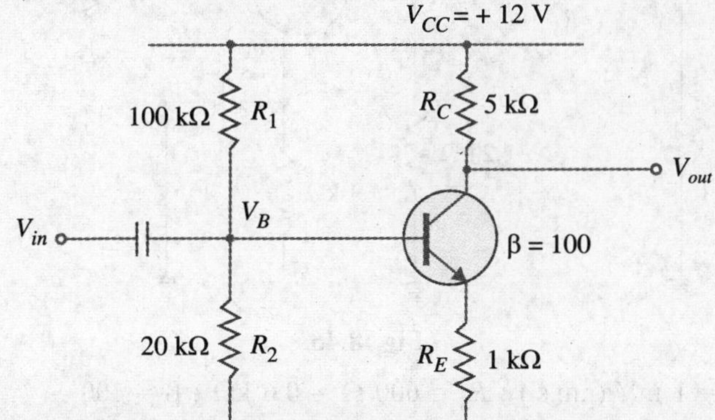

Fig. 8.45

Solution. (*i*) In order to find input impedance Z_{in} of the amplifier, we first find the ac emitter resistance r'_e. Referring to Fig. 8.45, we have,

$$\text{D.C. Voltage } V_B = \frac{V_{CC}}{R_1 + R_2} \times R_2 = \frac{12\text{V}}{100\,\text{k}\Omega + 20\,\text{k}\Omega} \times 20\,\text{k}\Omega = 2\text{ V}$$

$$\text{D.C. voltage across } R_E,\ V_E = V_B - V_{BE} = 2\text{ V} - 0.7\text{ V} = 1.3\text{ V}$$

$$\text{D.C. emitter current, } I_E = \frac{V_E}{R_E} = \frac{1.3\text{V}}{1\,\text{k}\Omega} = 1.3\text{ mA}$$

$$\therefore \quad \text{ac emitter resistance, } r'_e = \frac{25\,\text{mV}}{I_E} = \frac{25\,\text{mV}}{1.3\,\text{mA}} \simeq 20\ \Omega = 0.02\text{ k}\Omega$$

Input impedance Z_{in} of the amplifier is given by ;

$$Z_{in} = R_1 \parallel R_2 \parallel Z_{in(base)}$$

Here, $R_1 = 100$ kΩ ; $R_2 = 20$ kΩ ; $Z_{in(base)} = \beta(r'_e + R_E) = 100\ (0.02 + 1) = 102$ kΩ

$$\therefore \quad Z_{in} = 100\text{ k}\Omega \parallel 20\text{ k}\Omega \parallel 102\text{ k}\Omega = \mathbf{14.3\ k\Omega}$$

(*ii*) The output impedance Z_{out} of the amplifier is equal to the collector resistance R_C *i.e.*,

$$Z_{out} = R_C = \mathbf{5\ k\Omega}$$

(*iii*) The voltage gain (no-load voltage gain) of the amplifier is

$$A_{v0} = \frac{R_C}{r'_e + R_E} = \frac{5\text{k}\Omega}{0.02\,\text{k}\Omega + 1\,\text{k}\Omega} = \mathbf{4.9}$$

Example 8.26. *For the CE transistor amplifier shown in Fig. 8.46, V_S = 1 mV (r.m.s.) and source resistance R_S = 600 Ω. The transistor used has β = 100. What is the output voltage of the amplifier?*

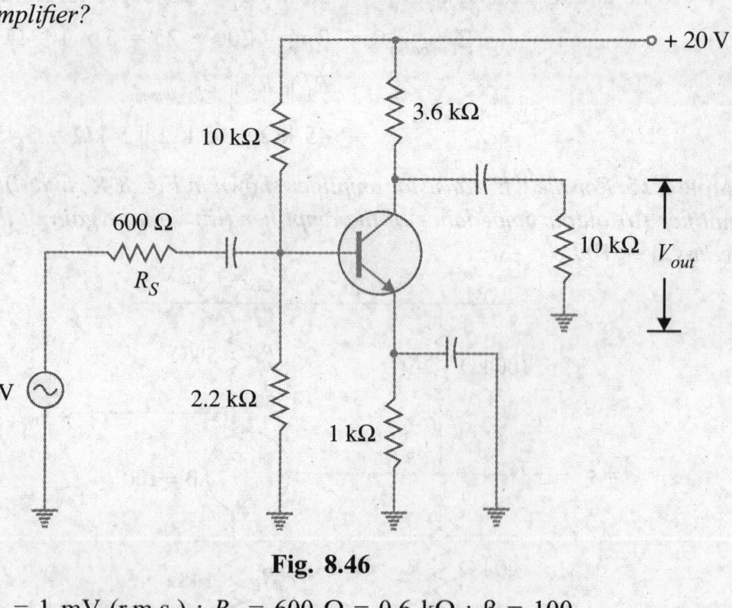

Fig. 8.46

Solution. V_S = 1 mV (r.m.s.) ; R_S = 600 Ω = 0.6 kΩ ; β = 100

$$\text{D.C. base voltage, } V_B = \frac{V_{CC}}{R_1 + R_2} \times R_2 = \frac{20}{10\,\text{k}\Omega + 2.2\,\text{k}\Omega} \times 2.2\text{ k}\Omega = 3.61\text{ V}$$

D.C. voltage across R_E, $V_E = V_B - V_{BE} = 3.61 - 0.7 = 2.91$ V

$$\text{D.C. emitter current, } I_E = \frac{V_E}{R_E} = \frac{2.91\text{V}}{1\text{k}\Omega} = 2.91 \text{ mA}$$

$$\text{ac emitter resistance, } r_e' = \frac{25\text{ mV}}{I_E} = \frac{25\text{ mV}}{2.91\text{mA}} = 8.59\ \Omega$$

$$\text{A.C. load, } R_{AC} = R_C \parallel R_L = 3.6\text{ k}\Omega \parallel 10\text{ k}\Omega = 2.65\text{ k}\Omega$$

$$\text{Voltage gain of amplifier, } A_v = \frac{R_{AC}}{r_e'} = \frac{2.65 \times 10^3\ \Omega}{8.59\ \Omega} = 308$$

Input impedance Z_{in} of the amplifier is given by ;

$$Z_{in} = R_1 \parallel R_2 \parallel Z_{in(base)}$$

Here, $R_1 = 10$ kΩ ; $R_2 = 2.2$ kΩ ; $Z_{in(base)} = \beta r_e' = 100 \times 8.59\ \Omega = 0.859$ kΩ

$$\therefore \quad Z_{in} = 10\text{ k}\Omega \parallel 2.2\text{ k}\Omega \parallel 0.859\text{ k}\Omega = 0.582\text{ k}\Omega$$

Now, the source resistance R_S (= 0.6 kΩ) and input impedance Z_{in} (= 0.582 kΩ) of the amplifier are in series. Therefore, voltage V_1 at the input terminals of the amplifier is

$$V_1 = I_1 Z_{in} = \frac{V_S}{R_S + Z_{in}} \times Z_{in}$$

$$= \frac{1\text{ mV}}{0.6\text{ k}\Omega + 0.582\text{ k}\Omega} \times 0.582\text{ k}\Omega = 0.492\text{ mV}$$

$$\therefore \quad \text{Output voltage, } V_{out} = A_v V_1 = 308 \times 0.492\text{ mV} = \mathbf{152\text{ mV}}$$

8.20 Voltage Gain Stability

One important consideration for an amplifier is the stability of its voltage gain. An amplifier should have voltage gain values that are stable so that the output of the circuit is predictable under all normal conditions. In a standard *CE* amplifier, the entire d.c. emitter resistance R_E is bypassed by the bypass emitter capacitor C_E. Therefore, the total a.c. emitter resistance is r_e'. The voltage gain of such an amplifier at no-load is given by ;

$$\text{Voltage gain, } A_v = \frac{R_C}{r_e'} \qquad \text{where } r_e' = \frac{25\text{ mV}}{I_E}$$

The voltage gain of a standard *CE* amplifier is quite large. However, the drawback of the circuit is that its voltage gain changes with emitter current I_E, temperature variations and transistor replacement. For example, if emitter current I_E increases, the a.c. emitter resistance r_e' decreases. This changes the voltage gain of the amplifier. Similarly, when the temperature varies or when a transistor is replaced, the a.c. current gain β changes. This will also result in the change in voltage gain. In order to stabilise the voltage gain, the emitter resistance R_E is partially bypassed by C_E. Such an amplifier is called a *swamped amplifier.*

Fig. 8.47 (*i*) shows the emitter leg of a standard *CE* amplifier while Fig. 8.47 (*ii*) shows the emitter leg of swamped amplifier. In swamped amplifier, the resistance R_E is split into two parts *viz.* R_{E1} and R_{E2}. Only R_{E2} is bypassed by C_E while R_{E1} is not.

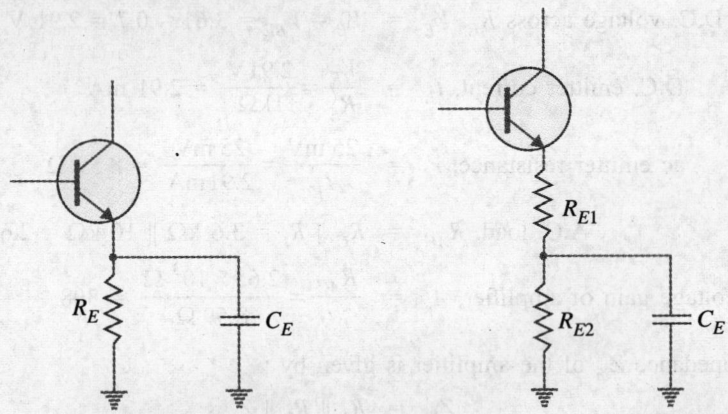

STANDARD *CE* AMPLIFIER (*i*)   SWAMPED AMPLIFIER (*ii*)

Fig. 8.47

8.21 Swamped Amplifier

Fig. 8.48 shows the circuit of a swamped amplifier. Note that d.c. emitter resistance R_E is divided into two parts *viz.* R_{E1} and R_{E2}. Only resistance R_{E2} is bypassed by the capacitor C_E while resistance R_{E1} is not. This method swamps or minimises the effect of r'_e on the voltage gain without reducing the voltage gain too much. Now the total ac emitter resistance is $(r'_e + R_{E1})$ instead of r'_e as in a standard *CE* amplifier. Therefore, the voltage gain of a swamped amplifier at no-load becomes :

$$\text{Voltage gain, } A_{v0} = \frac{R_C}{r'_e + R_{E1}}$$

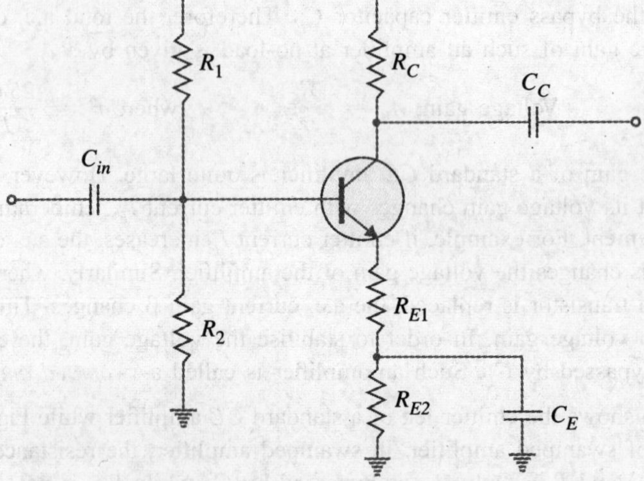

Fig. 8.48

If $R_{E1} \geq 10\ r'_e$, then the effect of r'_e is almost negligible and the voltage gain is given by ;

$$A_{v0} \simeq \frac{R_C}{R_{E1}}$$

Therefore, the voltage gain is essentially independent of r'_e or it is reasonably stabilised.

Effect of swamping on $Z_{in\ (base)}$. The $Z_{in\ (base)}$ with R_E completely bypassed is $Z_{in\ (base)} = \beta r'_e$. When the emitter resistance is partially bypassed, the portion of the resistance that is unbypassed (*i.e.* R_{E1}) is seen by the a.c. signal and appears in series with r'_e. Therefore, for swamped amplifier,

$$Z_{in\ (base)} = \beta\ (r'_e + R_{E1})$$

Example 8.27. *Determine the value of voltage gain (A_{v0}) for the swamped amplifier shown in Fig. 8.49. What will be $Z_{in\ (base)}$ for this circuit ?*

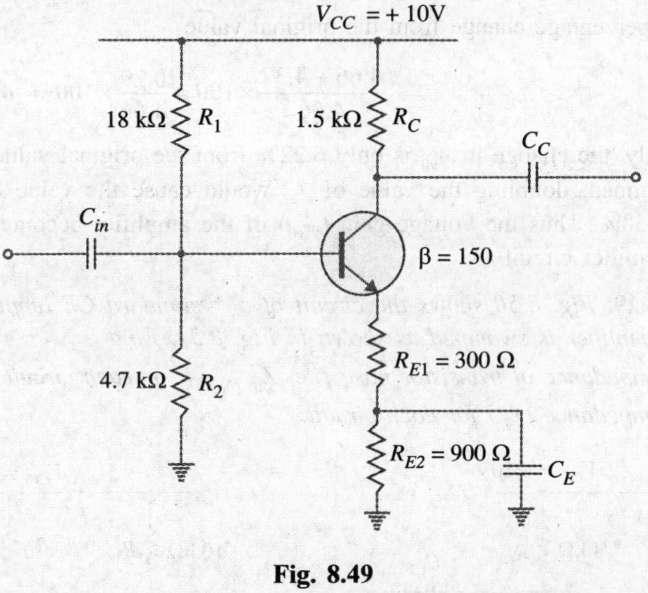

Fig. 8.49

Solution. In order to find voltage gain (A_{v0}), we first determine D.C. emitter current I_E and then ac emitter resistance r'_e. The value of I_E can be determined as under :

D.C. voltage across R_2, $V_2 = \dfrac{V_{CC}}{R_1 + R_2} \times R_2 = \dfrac{10}{18 + 4.7} \times 4.7 = 2.1\text{V}$

D.C. voltage across R_E, $V_E = V_2 - V_{BE} = 2.1\text{V} - 0.7\text{V} = 1.4\text{V}$

D.C. emitter current, $I_E = \dfrac{V_E}{R_{E1} + R_{E2}} = \dfrac{1.4\,\text{V}}{300\ \Omega + 900\ \Omega} = 1.16\text{ mA}$

$\therefore$ ac emitter resistance, $r'_e = \dfrac{25\text{ mV}}{I_E} = \dfrac{25\text{ mV}}{1.16\text{ mA}} = 21.5\ \Omega$

Voltage gain, $A_{v0} = \dfrac{R_C}{r'_e + R_{E1}} = \dfrac{1.5\text{ k}\Omega}{21.5\ \Omega + 300\ \Omega} = \mathbf{4.66}$

Input impedance of transistor base is given by ;

$$Z_{in\ (base)} = \beta\,(r'_e + R_{E1}) = 150\ (21.5\ \Omega + 300\ \Omega) = \mathbf{48.22\ k\Omega}$$

Example 8.28. *Determine the change in voltage gain for the amplifier in example 8.27 when r'_e doubles in value.*

Solution.

$$\text{Voltage gain, } A_{v0} = \frac{R_C}{r'_e + R_{E1}}$$

When r'_e doubles, the value of A_{v0} becomes:

$$A_{v0} = \frac{R_C}{2r'_e + R_{E1}} = \frac{1.5\ \text{k}\Omega}{2 \times 21.5\ \Omega + 300\ \Omega} = 4.37$$

$\therefore$ Change in gain $= 4.66 - 4.37 = 0.29$

Therefore, percentage change from the original value

$$= \frac{4.66 - 4.37}{4.66} \times 100 = \frac{0.29}{4.66} \times 100 = \mathbf{6.22\%}\ \textit{(decrease)}$$

Consequently, the change in A_{v0} is only 6.22% from the original value. In an amplifier that is not swamped, doubling the value of r'_e would cause the value of A_{v0} to change (decrease) by *50%. Thus the voltage gain (A_{v0}) of the amplifier becomes more stable by swamping the emitter circuit.

Example 8.29. *Fig. 8.50 shows the circuit of a **standard CE amplifier. The emitter circuit of this amplifier is swamped as shown in Fig. 8.51. Find :*

(*i*) *input impedance of transistor base [i.e. $Z_{in\ (base)}$] for each circuit.*

(*ii*) *input impedance (Z_{in}) for each circuit.*

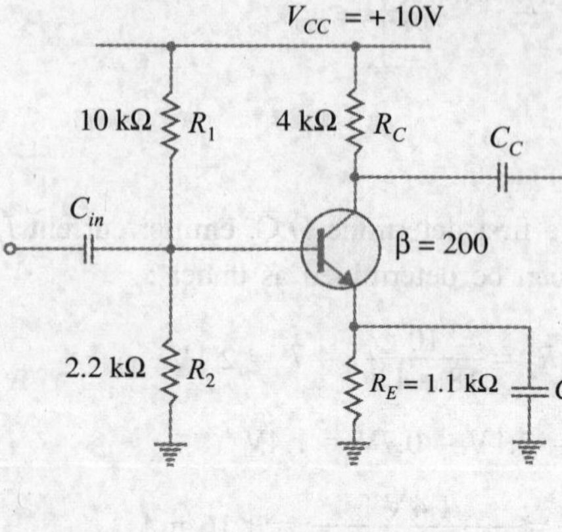

Fig. 8.50

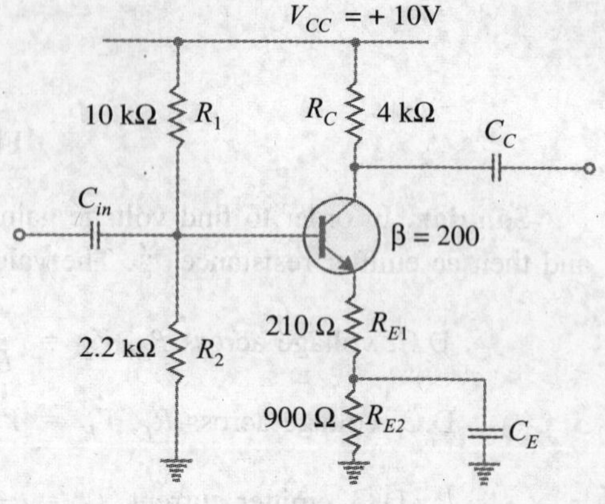

Fig. 8.51

* Original $A_{v0} = \frac{R_C}{r'_e}$; Final $A_{v0} = \frac{R_C}{2\,r'_e}$. Obviously, a change of 50% from the original value.

** Remember that in a standard *CE* amplifier, the emitter resistance R_E is completely bypassed by the capacitor C_E.

Solution. Both the circuits have the same value of ac emitter resistance r'_e. Therefore, following the standard procedure for finding r'_e gives us a value of *25 Ω for both circuits.

(*i*) $Z_{in\ (base)}$

For the standard *CE* amplifier shown in Fig. 8.50, we have,

$$Z_{in\ (base)} = \beta r'_e = 200 \times 25\ \Omega = \mathbf{5\ k\Omega}$$

For the swamped amplifier shown in Fig. 8.51, we have,

$$Z_{in\ (base)} = \beta\ (r'_e + R_{E1})$$

$$= 200\ (25\ \Omega + 210\ \Omega) = 47000\ \Omega = \mathbf{47\ k\Omega}$$

(*ii*) Z_{in}

For the standard *CE* amplifier shown in Fig. 8.50, we have,

$$Z_{in} = R_1 \parallel R_2 \parallel Z_{in\ (base)}$$

$$= 10\ k\Omega \parallel 2.2\ k\Omega \parallel 5\ k\Omega = \mathbf{1.33\ k\Omega}$$

For the swamped amplifier circuit shown in Fig. 8.51, we have,

$$Z_{in} = R_1 \parallel R_2 \parallel Z_{in\ (base)}$$

$$= 10\ k\Omega \parallel 2.2.\ k\Omega \parallel 47\ k\Omega = \mathbf{1.74\ k\Omega}$$

Note that swamping increases the input impedance (Z_{in}) of the amplifier. This reduces the amplifier's loading effects on a previous stage.

Example 8.30. *Find the voltage gain for both circuits of example 8.29.*

Solution.

For the standard *CE* amplifier shown in Fig. 8.50, the voltage gain (A_{v0}) is given by ;

$$A_{v0} = \frac{R_C}{r'_e} = \frac{4\ k\Omega}{25\ \Omega} = \mathbf{160}$$

For the swamped amplifier shown in Fig. 8.51, the voltage gain (A_{v0}) is given by ;

$$A_{v0} = \frac{R_C}{r'_e + R_{E1}} = \frac{4\ k\Omega}{25\ \Omega + 210\ \Omega} = \frac{4\ k\Omega}{235\ \Omega} = \mathbf{17}$$

The following points may be noted :

(*i*) The two circuits are identical for d.c. analysis purposes. Both have a total of 1.1 kΩ d.c. resistance in their emitter circuits.

(*ii*) For a standard *CE* amplifier, the total a.c. emitter resistance is r'_e. When this amplifier is swamped, the total a.c. emitter resistance is increased to ($r'_e + R_{E1}$).

(*iii*) Swamping reduces the voltage gain of the amplifier. However, the gain of a swamped amplifier is more stable than that of a comparable standard *CE* amplifier.

---

* $V_2 = \frac{V_{CC}}{R_1 + R_2} \times R_2 = \frac{10}{10 + 2.2} \times 2.2 = 1.8\ \text{V}$; $V_E = V_2 - V_{BE} = 1.8\text{V} - 0.7\text{V} = 1.1\text{V}$

$I_E = \frac{V_E}{R_E} = \frac{1.1\text{V}}{1.1\ k\Omega} = 1\ \text{mA}$ $\therefore\ r'_e = \frac{25\ \text{mV}}{I_E} = \frac{25\ \text{mV}}{1\ \text{mA}} = 25\ \Omega$

Example 8.31. *For the swamped amplifier shown in Fig. 8.52, the signal source V_S = 50 mV (r.m.s.) and resistance R_S of signal source is 600 Ω. The transistor has β = 100. Find the output voltage. Ignore r'_e.*

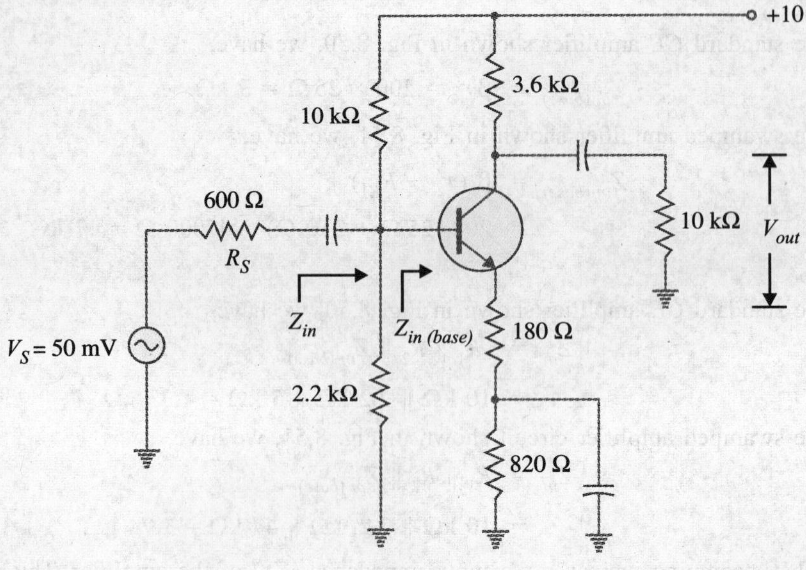

Fig. 8.52

Solution. V_S = 50 mV (r.m.s.) ; R_S = 600 Ω = 0.6 kΩ ; Unbypassed emitter resistance, R_{E1} = 180 Ω ; β = 100 ; R_C = 3.6 kΩ ; R_L = 10 kΩ

For the swamped amplifier in Fig. 8.52, $Z_{in\ (base)}$ is

$$Z_{in\ (base)} = \beta R_{E1} \quad \text{... } r'_e \text{ ignored}$$

$$= 100 \times 180\ \Omega = 18\ \text{k}\Omega$$

∴ Input impedance Z_{in} of the swamped amplifier is

$$Z_{in} = R_1 \parallel R_2 \parallel Z_{in\ (base)} = 10\ \text{k}\Omega \parallel 2.2\ \text{k}\Omega \parallel 18\ \text{k}\Omega = 1.64\ \text{k}\Omega$$

A.C. load, $R_{AC} = R_C \parallel R_L = 3.6\ \text{k}\Omega \parallel 10\ \text{k}\Omega = 2.65\ \text{k}\Omega$

∴ Voltage gain A_v of the swamped amplifier is

$$A_v = \frac{R_{AC}}{R_{E1}} = \frac{2.65 \times 10^3\ \Omega}{180\ \Omega} = 14.7 \quad \text{... } r'_e \text{ ignored}$$

Now, the source resistance R_S (= 0.6 kΩ) and input impedance Z_{in} (= 1.64 kΩ) of the amplifier are in series. Therefore, voltage V_1 at the input terminals of the amplifier is

$$V_1 = I_1 Z_{in} = \frac{V_S}{R_S + Z_{in}} \times Z_{in}$$

$$= \frac{50\ \text{mV}}{0.6\ \text{k}\Omega + 1.64\ \text{k}\Omega} \times 1.64\ \text{k}\Omega = 36.6\ \text{mV}$$

∴ Output voltage, $V_{out} = A_v V_1 = 14.7 \times 36.6\ \text{mV} = \mathbf{0.538V}$

8.22 Classification of Amplifiers

The transistor amplifiers may be classified as to their *usage, frequency capabilities, coupling methods* and *mode of operation.*

(*i*) According to use. The classifications of amplifiers as to usage are basically *voltage amplifiers* and *power amplifiers*. The former primarily increases the voltage level of the signal whereas the latter mainly increases the power level of the signal.

(*ii*) According to frequency capabilities. According to frequency capabilities, amplifiers are classified as *audio amplifiers, radio frequency amplifiers etc.* The former are used to amplify the signals lying in the audio range *i.e.* 20 Hz to 20 kHz whereas the latter are used to amplify signals having very high frequency.

(*iii*) According to coupling methods. The output from a single stage amplifier is usually insufficient to meet the practical requirements. Additional amplification is often necessary. To do this, the output of one stage is coupled to the next stage. Depending upon the coupling device used, the amplifiers are classified as *R-C coupled amplifiers, transformer coupled amplifiers etc.*

(*iv*) According to mode of operation. The amplifiers are frequently classified according to their mode of operation as *class A, class B* and *class C amplifiers*. This classification depends on the portion of the input signal cycle during which collector current is expected to flow. Thus, class *A* amplifier is one in which collector current flows for the entire a.c. signal. Class *B* amplifier is one in which collector current flows for half-cycle of input a.c. signal. Finally, class *C* amplifier is one in which collector current flows for less than half-cycle of a.c. signal.

Example 8.32. *What do you understand by following amplifiers :*

(i) Class A voltage amplifier *(ii) Audio voltage amplifier*

(iii) Class B power amplifier *(iv) Class A transformer coupled power amplifier?*

Solution. (*i*) Class *A* voltage amplifier means that it raises the voltage level of the signal and its mode of operation is such that collector current flows for the whole input signal.

(*ii*) Audio voltage amplifier means that it raises the voltage level of audio signal (*i.e.* one having frequency range 20 Hz to 20 kHz) and its mode of operation is class *A*.

(*iii*) It means that this amplifier raises the power level of the signal and its mode of operation is such that collector current flows for half-cycle of the signal only.

(*iv*) It means that power amplification is being done, coupling is by transformer and mode of operation is class *A*.

8.23 Transistor Amplifier Equivalent Circuit

Fig. 8.53 (*i*) shows the block diagram of a transistor amplifier. In order to find the equivalent circuit of the amplifier, we shall replace the input side and output side of the amplifier by their Thevenin equivalent circuits. Therefore, the input side of the amplifier can be replaced by a voltage source of voltage V_1 in series with input resistance (or *impedance) R_{in} of the amplifier. Similarly, the output side of the amplifier can be replaced by a ***controlled voltage*

* Recall that for signal frequencies < 100 kHz, input and output impedances of the amplifier are resistive in nature.

** Because the magnitude of voltage $A_{v0}V_1$ depends on V_1. If V_1 changes, then $A_{v0}V_1$ also changes. This means that the value of $A_{v0}V_1$ is controlled by V_1.

source of voltage $A_{v0}V_1$ in series with the output resistance R_{out} of the amplifier. Here, A_{v0} is the no-load voltage gain of the amplifier and V_1 is the voltage at the input terminals of the amplifier. Therefore, the equivalent circuit of the transistor amplifier becomes as shown in Fig. 8.53 (*ii*). Note that we have developed the equivalent circuit of the transistor amplifier using the following three parameters of the amplifier :

$$R_{in} = \text{Input resistance of the amplifier}$$
$$R_{out} = \text{Output resistance of the amplifier}$$
$$A_{v0} = \text{No-load voltage gain of the amplifier}$$

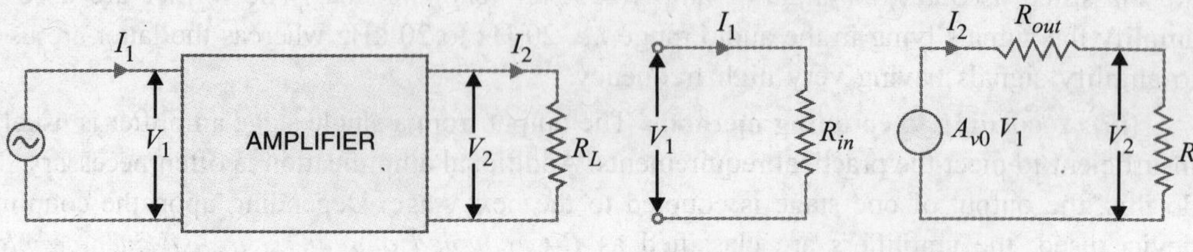

(*i*) **Fig. 8.53** (*ii*)

Note that capability of the amplifier to produce voltage gain is represented by the voltage generator $A_{v0}V_1$. The voltage gain of the loaded amplifier is A_v. Clearly, A_v will be less than A_{v0} due to voltage drop in R_{out}.

8.24 Equivalent Circuit with Resistance R_S of Signal Source

In the equivalent circuit of the transistor amplifier in Fig. 8.53, the resistance R_S of the signal source was assumed zero. Fig. 8.54 shows the equivalent circuit of the transistor amplifier with signal source having internal resistance R_S. Referring to Fig. 8.54, we have,

$$I_1 = \frac{E_S}{R_S + R_{in}}$$

$$\therefore \quad V_1 = I_1 R_{in} = \frac{E_S R_{in}}{R_S + R_{in}}$$

$$I_2 = \frac{A_{v0} V_1}{R_{out} + R_L} \quad \text{...(i)}$$

$$= \frac{A_{v0} I_1 R_{in}}{R_{out} + R_L} \quad \text{...(ii)}$$

$$\therefore \quad V_2 = I_2 R_L = \frac{A_{v0} V_1 R_L}{R_{out} + R_L} \quad \text{...(iii)}$$

(*i*) **Voltage gain.** With load R_L connected to the output terminals of the amplifier, the ratio V_2/V_1 is called the **loaded voltage gain* A_v of the amplifier *i.e.*

$$\text{Loaded voltage gain, } A_v = \frac{V_2}{V_1} = \frac{A_{v0} R_L}{R_{out} + R_L} \quad \text{... From eq. (iii).}$$

It is clear that *loaded voltage gain* (A_v) *of the amplifier is less than its no-load voltage gain* (A_{v0}).

* With R_L disconnected, the ratio V_2/V_1 is called no-load voltage gain A_{v0} of the amplifier.

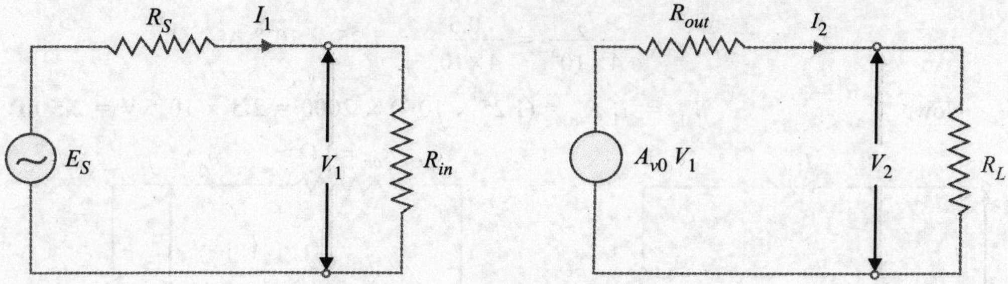

Fig. 8.54

The *overall voltage gain* (A_{vs}) of the amplifier is the voltage gain between source voltage (E_S) and output voltage (V_2) of the amplifier *i.e.*,

$$\text{Overall voltage gain, } A_{vs} = \frac{V_2}{E_S} = \frac{V_2}{V_1} \times \frac{V_1}{E_S}$$

Now,
$$\frac{V_2}{V_1} = \frac{A_{v0} R_L}{R_{out} + R_L} \; ; \; \frac{V_1}{E_S} = \left(\frac{E_S R_{in}}{R_S + R_{in}}\right) \times \frac{1}{E_S} = \frac{R_{in}}{R_S + R_{in}}$$

$$\therefore \quad \text{Overall voltage gain, } A_{vs} = \frac{A_{v0} R_{in} R_L}{(R_{out} + R_L)(R_S + R_{in})}$$

(ii) **Current gain.** The current gain of the amplifier is given by ;

$$\text{Current gain, } A_i = \frac{I_2}{I_1} = \frac{A_{v0}\, R_{in}}{R_{out} + R_L}$$

(iii) **Power gain.** The power gain of the amplifier is given by ;

$$\text{Power gain, } A_p = \frac{I_2^2\, R_L}{I_1^2\, R_{in}} = \frac{(I_2\, R_L)\, I_2}{(I_1\, R_{in})\, I_1}$$

$$= \frac{V_2\, I_2}{V_1\, I_1} = \left(\frac{V_2}{V_1}\right) \times \left(\frac{I_2}{I_1}\right)$$

$$\therefore \quad A_p = A_v \times A_i$$

As expected, the power gain of the amplifier is equal to the product of its voltage gain and current gain.

Note. The use of such an equivalent circuit is restricted to the signal quantities only. Further, in drawing the equivalent circuit, it is assumed that exact linear relationship exists between input and output signals *i.e.* the amplifier produces no waveform distortion.

Example 8.33. *An amplifier has an open-circuit voltage gain of 1000, an input resistance of 2 kΩ and an output resistance of 1 Ω. Determine the input signal voltage required to produce an output signal current of 0.5 A in 4 Ω resistor connected across the output terminals.*

Solution. Fig. 8.55 shows the equivalent circuit of the amplifier. Here A_{v0} = 1000.

$$\frac{I_2}{I_1} = \frac{A_{v0}\, R_{in}}{R_{out} + R_L} \qquad \text{[See Art. 8.24]}$$

$$= \frac{1000 \times 2000}{1 + 4} = 4 \times 10^5$$

$$\therefore \quad I_1 = \frac{I_2}{4\times10^5} = \frac{0.5}{4\times10^5} = 1.25\times10^{-6}\text{ A}$$

Now, $$V_1 = I_1 R_{in} = (1.25\times10^{-6})\times2000 = 2.5\times10^{-3}\text{ V} = \mathbf{2.5\ mV}$$

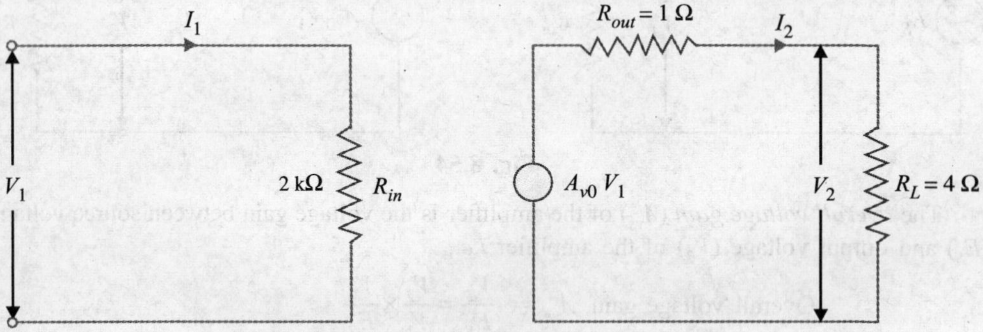

Fig. 8.55

Example 8.34. *An amplifier has an open-circuit voltage gain of 1000, an output resistance of 15 Ω and an input resistance of 7 kΩ. It is supplied from a signal source of e.m.f. 10 mV and internal resistance 3 kΩ. The amplifier feeds a load of 35 Ω. Determine (i) the magnitude of output voltage and (ii) power gain.*

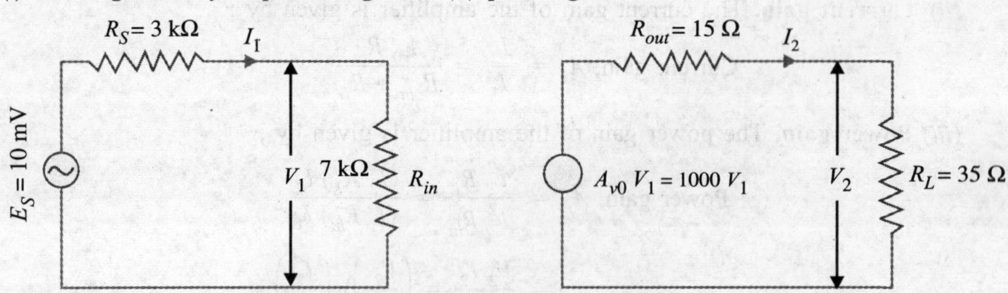

Fig. 8.56

Solution. (*i*) $$I_1 = \frac{E_S}{R_S + R_{in}} = \frac{10\times10^{-3}}{3000+7000} = 10^{-6}\text{ A}$$

$$V_1 = I_1 R_{in} = 10^{-6}\times7000 = 7\times10^{-3}\text{ V}$$

$$A_v = \frac{V_2}{V_1} = \frac{A_{v0} R_L}{R_{out}+R_L} = \frac{1000\times35}{15+35} = 700$$

$$\therefore \quad V_2 = 700\,V_1 = 700\times7\times10^{-3} = \mathbf{4.9\ V}$$

(*ii*) Output power, $$P_2 = \frac{V_2^2}{R_L} = \frac{(4.9)^2}{35} = 0.686\text{ W}$$

Input power, $$P_1 = \frac{V_1^2}{R_{in}} = \frac{(7\times10^{-3})^2}{7000} = 7\times10^{-9}\text{ W}$$

$$\therefore \quad \text{Power gain, } A_p = \frac{P_2}{P_1} = \frac{0.686}{7\times10^{-9}} = \mathbf{98\times10^6}$$

Example 8.35. *An amplifier, when loaded by 2 kΩ resistor, has a voltage gain of 80 and a current gain of 120. Determine the necessary signal voltage and current to give an output voltage of 1 V. What is the power gain of the amplifier?*

Solution. $A_v = \dfrac{V_2}{V_1} = 80$

∴ $V_1 = V_2/80 = 1/80 = 0.0125 \text{ V} = \mathbf{12.5 \text{ mV}}$

$A_v = \dfrac{A_{v0}\, R_L}{R_{out} + R_L}$ ...[See Art. 8.24]

$A_i = \dfrac{A_{v0}\, R_{in}}{R_{out} + R_L}$ ...[See Art. 8.24]

∴ $\dfrac{A_v}{A_i} = \dfrac{R_L}{R_{in}}$

or $\dfrac{80}{120} = \dfrac{2}{R_{in}}$

∴ $R_{in} = 120 \times 2/80 = 3 \text{ k}\Omega$

$I_1 = V_1/R_{in} = 12.5 \text{ mV}/3 \text{ k}\Omega = \mathbf{4.17\ \mu A}$

Power gain $= A_v \times A_i = 80 \times 120 = \mathbf{9600}$

Example 8.36. *Fig. 8.57 shows the conventional symbol for an amplifier : a triangular block with output at the vertex. As shown in the figure, the amplifier has an input resistance of 1500 Ω and drives a 100 Ω load. If the input current is 0.632 mA (r.m.s.) and the load voltage is 30 V (r.m.s.), find (i) current gain (ii) voltage gain and (iii) power gain.*

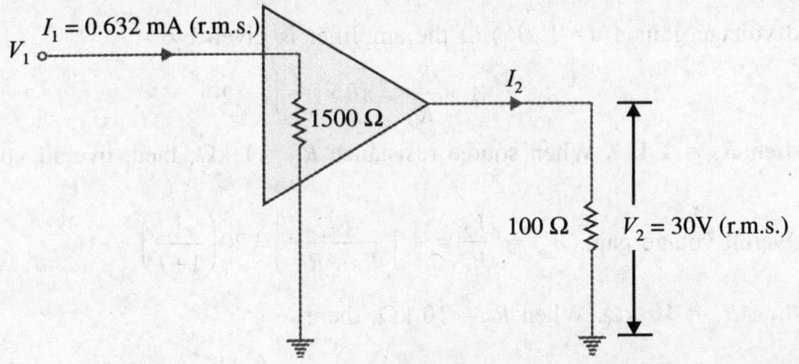

Fig. 8.57

Solution. $I_1 = 0.632$ mA (r.m.s.) ; $R_{in} = 1500\ \Omega$; $R_L = 100\ \Omega$; $V_2 = 30$ V (r.m.s.)

(*i*) Output current, $I_2 = \dfrac{V_2}{R_L} = \dfrac{30 \text{ V}}{100\ \Omega} = 0.3\text{A}$

∴ Current gain, $A_i = \dfrac{I_2}{I_1} = \dfrac{0.3 \text{ A}}{0.632 \text{ mA}} = \mathbf{474.7}$

(*ii*) Input voltage, $V_1 = I_1 R_{in} = 0.632 \text{ mA} \times 1500\ \Omega = 0.948 \text{ V}$

Voltage gain, $A_v = \dfrac{V_2}{V_1} = \dfrac{30 \text{ V}}{0.948 \text{ V}} = \mathbf{31.6}$

(iii) Input power, $P_{in} = I_1^2 R_{in} = (0.632 \times 10^{-3})^2 \times 1500 = 0.6 \times 10^{-3}$ W

$$\text{Output power, } P_{out} = \frac{V_2^2}{R_L} = \frac{(30)^2}{100} = 9\text{W}$$

$$\text{Power gain, } A_p = \frac{P_{out}}{P_{in}} = \frac{9\text{W}}{0.6 \times 10^{-3}\text{W}} = \mathbf{15,000}$$

Alternatively. $A_p = A_i A_v = 474.7 \times 31.6 = 15{,}000$

Example 8.37. *The amplifier shown in Fig. 8.58 has current gain A_i = 10 and input resistance R_{in} = 1 kΩ. It is driven by a signal source that has source resistance R_S. Find the overall voltage gain from source to output when (i) R_S = 1 kΩ (ii) R_S = 10 kΩ.*

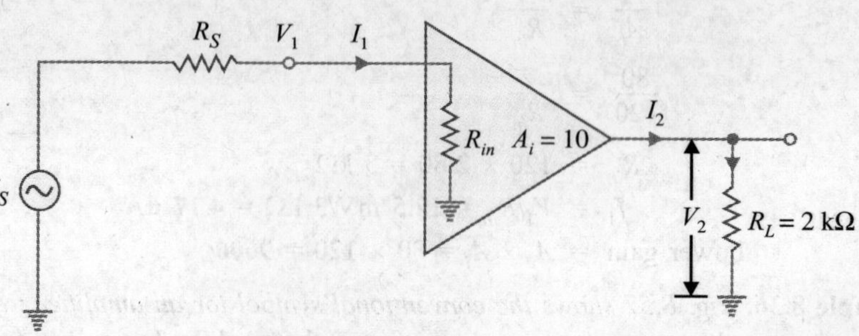

Fig. 8.58

Solution. $A_i = 10 \; ; R_{in} = 1 \text{ k}\Omega \; ; R_L = 2 \text{ k}\Omega$

Loaded voltage gain A_v (= V_2/V_1) of the amplifier is given by ;

$$A_v = A_i \times \frac{R_L}{R_{in}} = 10 \times \left(\frac{2}{1}\right) = 20$$

(*i*) **When R_S = 1 kΩ.** When source resistance R_S = 1 kΩ, then, overall voltage gain is

$$\text{Overall votage gain, } A_{vs} = \frac{V_2}{V_S} = A_v\left(\frac{R_{in}}{R_S + R_{in}}\right) = 20\left(\frac{1}{1+1}\right) = \mathbf{10}$$

(*ii*) **When R_S = 10 kΩ.** When R_S = 10 kΩ, then,

$$\text{Overall voltage gain, } A_{vs} = A_v\left(\frac{R_{in}}{R_S + R_{in}}\right) = 20\left(\frac{1}{10+1}\right) = \mathbf{1.82}$$

Note that when R_S = 1 kΩ, the overall voltage gain is reduced by about 50% and when R_S = 10 kΩ, the overall voltage gain is reduced by 91%. Therefore, it is desirable that source resistance R_S should be as small as possible.

8.25 Gain and Transistor Configurations

We know that the process of raising the strength of an a.c. signal is called amplification and the circuit used to preform this function is called an amplifier. There are three types of gains: *current gain, voltage gain* and *power gain.*

(*i*) The *common emitter (CE) amplifier* exhibits all three types of gains. From input to output, current will increase, voltage will increase and power will increase.

(*ii*) The *common base (CB) amplifier* has voltage gain and power gain but no current gain. Note that the current gain of a *CB* circuit is less than 1.

(*iii*) The *common collector (CC) amplifier* has current gain and power gain but no voltage gain.

It is important to note that the type of gain an amplifier has depends upon the transistor configuration. Consequently, the choice of an amplifier for a given application often depends on the type of gain that is desired. Since *CE* arrangement is widely used (in about 90% applications), we shall be mainly concentrating on this type of circuit.

Short Answer Questions

Q.1. *The output power of a transistor amplifier is more than the input signal power. Does the law of conservation of energy hold good?*

Ans. Yes, the law of conservation of energy is obeyed in this case. The additional energy comes from collector d.c. supply V_{CC}. In fact, a transistor is an energy converter. It receives d.c. power from V_{CC} and converts it into a.c. power for supplying to the a.c. signal.

Q.2. *Why is CB amplifier never used for low frequency applications?*

Ans. The input impedance of a *CB* amplifier is so low ($\simeq 25\ \Omega$) that it overloads most signal sources. In other words, most of the a.c. signal is lost across the source impedance. For this reason, *CB* amplifiers are never used for low frequency applications. However, they are used, though occasionally, in high frequency applications (above 10 MHz) where low source impedances are common.

Q.3. *A CE transistor amplifier causes phase reversal. Why is the message contained in the signal not reversed?*

Ans. Because the speaker is to respond to the amplitude variations in the output and this response will obviously remain unaffected by the phase reversal.

Q.4. *Why do CB and CE amplifiers act almost like constant current sources for load resistances normally used with them?*

Ans. Any device which has low input impedance and very high output impedance acts like a current source for the normal load resistance connected to it. The *CB* and *CE* amplifiers exhibit this characteristic.

Q.5. *What is the advantage of moderate output to input impedance ratio of a CE amplifier?*

Ans. In a *CE* amplifier, $Z_{out} / Z_{in} \simeq 50$. This makes the circuit an ideal one for coupling between various transistor stages. In other arrangements (*CB* or *CC*), the ratio Z_{out} / Z_{in} is very large and hence coupling becomes highly ineffective due to mismatching.

Q.6. *A trouble develops in the amplifier circuit shown in Fig. 8.59. On measurements, it is found that d.c. voltage across terminals 1 and 3 or across 2 and 3 is 20 V. What is the most probable trouble in the circuit?*

Ans. The measurements show that R_C is short-circuited. Therefore, the whole of V_{CC} (= 20 V) appears across terminals 2 and 3.

Q.7. *While troubleshooting amplifier circuit shown in Fig. 8.59, the following d.c. measurements were made :*

Voltage across 1 and 2 = 10 V

Voltage across 2 and 4 = 9.3 V

Voltage across 4 and 5 = 0.7 V

What was the trouble in the circuit?

Ans. This means that emitter resistance R_E is short-circuited because :

Voltage across $R_E = V_{CC} - I_C R_C -$ Voltage between 2 and 5

$= 20 - 10 - (9.3 + 0.7) = 0$ volt

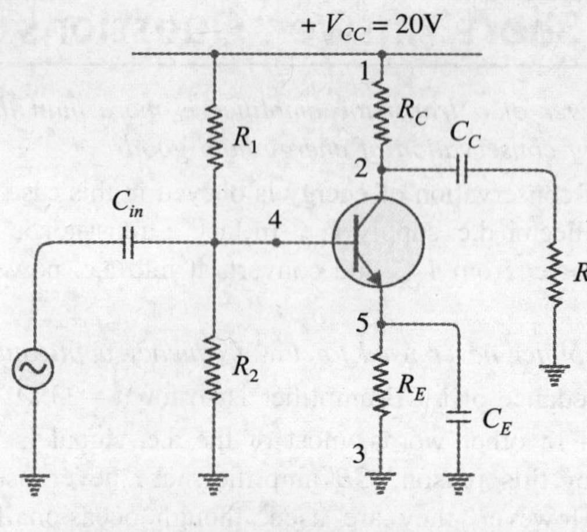

Fig. 8.59

Q.8. *What faults in the amplifier circuit shown in Fig. 8.59 would affect a.c. conditions without disturbing the d.c. conditions?*

Ans. Open-circuits in coupling or bypass capacitors as at C_{in}, C_C or C_E. It is because these capacitors already (in normal operation) act as open-circuits for d.c. and hence this kind of fault does not change d.c. conditions.

Q.9. *In the amplifier circuit shown in Fig. 8.59, what will happen if (i) C_{in} and C_E are open-circuited (ii) C_E is open-circuited?*

Ans. (*i*) If C_{in} and C_E are open-circuited, the signal will be blocked.

(*ii*) An open C_E leaves the emitter resistor R_E unbypassed and may, therefore, reduce the voltage gain drastically.

Q.10. *In the amplifier circuit shown in Fig. 8.59, the d.c. load is $R_C + R_E$ but a.c. load is $R_C \parallel R_L$. Why?*

Ans. The same direct current (d.c.) passes through R_C as well as R_E so that d.c. load is $R_C + R_E$. However, alternating current (a.c.) does not pass through R_E ; it is bypassed by C_E. For a.c., both R_C and R_L are in parallel. Hence a.c. load is $R_C \parallel R_L$.

Multiple-Choice Questions

1. A single stage transistor amplifier contains .... and associated circuitry.
   (i) two transistors
   (ii) one transistor
   (iii) three transistors
   (iv) none of the above
2. The phase difference between the output and input voltages of a *CE* amplifier is ....
   (i) 180° (ii) 0°
   (iii) 90° (iv) 270°
3. It is generally desired that a transistor should have .... input impedance.
   (i) low (ii) very low
   (iii) high (iv) very high
4. When an a.c. signal is applied to an amplifier, the operating point moves along ....
   (i) d.c. load line
   (ii) a.c. load line
   (iii) both d.c. and a.c. load lines
   (iv) none of the above
5. If the collector supply is 10 V, then collector cut-off voltage under d.c. conditions is ....
   (i) 20 V (ii) 5 V
   (iii) 2 V (iv) 10 V
6. In the zero signal conditions, a transistor sees .... load.
   (i) d.c. (ii) a.c.
   (iii) both d.c. and a.c
   (iv) none of the above
7. The input capacitor in an amplifier is the .... capacitor.
   (i) coupling (ii) bypass
   (iii) leakage
   (iv) none of the above
8. The point of intersection of d.c. and a.c. load lines is called ......
   (i) saturation point (ii) cut-off point
   (iii) operating point (iv) none of the above
9. The slope of a.c. load line is .... that of d.c. load line.
   (i) the same as (ii) more than
   (iii) less than
   (iv) none of the above
10. If a transistor amplifier draws 2 mA when input voltage is 10 V, then its input impedance is ....
    (i) 20 kΩ (ii) 0.2 kΩ
    (iii) 10 kΩ (iv) 5 kΩ
11. When a transistor amplifier is operating, the current in any branch is ....
    (i) sum of a.c. and d.c.
    (ii) a.c. only (iii) d.c. only
    (iv) difference of a.c. and d.c.
12. The purpose of capacitors in a transistor amplifier is to ....
    (i) protect the transistor
    (ii) cool the transistor
    (iii) couple or bypass a.c. component
    (iv) provide biasing
13. In the d.c. equivalent circuit of a transistor amplifier, the capacitors are considered ....
    (i) short (ii) open
    (iii) partially short
    (iv) none of the above
14. In a *CE* amplifier, voltage gain $= \ldots\ldots \times \dfrac{R_{AC}}{R_{in}}$.
    (i) α (ii) (1 + α)
    (iii) (1 + β) (iv) β
15. In practice, the voltage gain of an amplifier is expressed ....
    (i) as volts (ii) as a number
    (iii) in dB
    (iv) none of the above
16. If the power and current gains of a transistor amplifier are 16500 and 100 respectively, then voltage gain is ....
    (i) 165 (ii) 165×10^4

(*iii*) 100
(*iv*) none of the above

17. If R_C and R_L represent the collector resistance and load resistance respectively in a single stage transistor amplifier, then a.c. load is ....
(*i*) $R_L + R_C$ (*ii*) $R_C \parallel R_L$
(*iii*) $R_L - R_C$ (*iv*) R_C

18. In a *CE* amplifier, the phase difference between voltage across collector load R_C and signal voltage is ....
(*i*) 180° (*ii*) 270°
(*iii*) 90° (*iv*) 0°

19. In the a.c. equivalent circuit of a transistor amplifier, the capacitors are considered ....
(*i*) short (*ii*) open
(*iii*) partially open
(*iv*) none of the above

20. In a single stage transistor amplifier, R_C and R_L represent collector resistance and load resistance respectively. The transistor sees a d.c. load of ....
(*i*) $R_C + R_L$ (*ii*) $R_C \parallel R_L$
(*iii*) R_L (*iv*) R_C

21. The purpose of d.c. conditions in a transistor is to ....
(*i*) reverse bias the emitter
(*ii*) forward bias the collector
(*iii*) set up operating point
(*iv*) none of the above

22. An amplifier has a power gain of 100. Its dB gain is........
(*i*) 10 dB (*ii*) 20 dB
(*iii*) 40 dB
(*iv*) none of the above

23. In order to get more voltage gain from a transistor amplifier, the transistor used should have ....
(*i*) thin base (*ii*) thin collector
(*iii*) wide emitter
(*iv*) none of the above

24. The purpose of a coupling capacitor in a transistor amplifier is to ....
(*i*) increase the output impedance of transistor
(*ii*) protect the transistor
(*iii*) pass a.c. and block d.c.
(*iv*) provide biasing

25. The purpose of emitter capacitor (*i.e.* capacitor across R_E) is to ....
(*i*) avoid voltage gain drop
(*ii*) forward bias the emitter
(*iii*) reduce noise in the amplifier
(*iv*) none of the above

26. The ratio of output to input impedance of a *CE* amplifier is ....
(*i*) about 1 (*ii*) low
(*iii*) high (*iv*) moderate

27. If a transistor amplifier feeds a load of low resistance (*e.g.* speaker), then voltage gain will be ....
(*i*) high (*ii*) very high
(*iii*) moderate (*iv*) low

28. If the input capacitor of a transistor amplifier is short-circuited, then ....
(*i*) transistor will be destroyed
(*ii*) biasing conditions will change
(*iii*) signal will not reach the base
(*iv*) none of the above

29. The radio wave picked up by the receiving antenna is amplified about .... times to have reasonable sound output.
(*i*) 1000 (*ii*) a million
(*iii*) 100 (*iv*) 10000

30. A *CE* amplifier is also called .... circuit.
(*i*) grounded emitter
(*ii*) grounded base
(*iii*) grounded collector
(*iv*) none of the above

31. The d.c. load of a transistor amplifier is generally ...... that of a.c. load.
(*i*) the same as (*ii*) less than
(*iii*) more than
(*iv*) none of the above

32. The value of collector load R_C in a transistor amplifier is .... the output impedance of the transistor.
    (*i*) the same as (*ii*) less than
    (*iii*) more than
    (*iv*) none of the above

33. A single stage transistor amplifier with collector load R_C and emitter resistance R_E has a d.c. load of ....
    (*i*) R_C (*ii*) $R_C \parallel R_E$
    (*iii*) $R_C - R_E$ (*iv*) $R_C + R_E$

34. In transistor amplifiers, we generally use .... capacitors.
    (*i*) electrolytic (*ii*) mica
    (*iii*) paper (*iv*) air

35. A single stage transistor amplifier with no load sees an a.c. load of ....
    (*i*) $R_C + R_E$ (*ii*) R_C
    (*iii*) $R_C \parallel R_E$ (*iv*) R_C / R_E

36. The output power of a transistor amplifier is more than the input power because the additional power is supplied by ....
    (*i*) transistor (*ii*) biasing circuit
    (*iii*) collector supply V_{CC}
    (*iv*) none of the above

37. A transistor converts ....
    (*i*) d.c. power into a.c. power
    (*ii*) a.c. power into d.c. power
    (*iii*) high resistance into low resistance
    (*iv*) none of the above

38. A transistor amplifier has high output impedance because ....
    (*i*) emitter is heavily doped
    (*ii*) collector has reverse bias
    (*iii*) collector is wider than emitter or base
    (*iv*) none of the above

39. For highest power gain, one would use .... configuration.
    (*i*) CC (*ii*) CB
    (*iii*) CE
    (*iv*) none of the above

40. *CC* configuration is used for impedance matching because its ....
    (*i*) input impedance is very high
    (*ii*) input impedance is low
    (*iii*) output impedance is very low
    (*iv*) none of the above

Answers to Multiple-Choice Questions

1. (*ii*)	2. (*i*)	3. (*iii*)	4. (*ii*)	5. (*iv*)
6. (*i*)	7. (*i*)	8. (*iii*)	9. (*ii*)	10. (*iv*)
11. (*i*)	12. (*iii*)	13. (*ii*)	14. (*iv*)	15. (*iii*)
16. (*i*)	17. (*ii*)	18. (*iv*)	19. (*i*)	20. (*iv*)
21. (*iii*)	22. (*ii*)	23. (*i*)	24. (*iii*)	25. (*i*)
26. (*iv*)	27. (*iv*)	28. (*ii*)	29. (*ii*)	30. (*i*)
31. (*iii*)	32. (*ii*)	33. (*iv*)	34. (*i*)	35. (*ii*)
36. (*iii*)	37. (*i*)	38. (*ii*)	39. (*iii*)	40. (*i*)

Chapter Review Topics

1. What do you understand by single stage transistor amplifiers?
2. Explain with the help of output characteristics how the variations in base current affect collector current variations? Assume the base current varies sinusoidally.
3. Draw the circuit of a practical single stage transistor amplifier. Explain the function of each component.
4. Show the various currents and voltages in a single stage transistor amplifier.

5. Show that the output voltage of a single stage common emitter transistor amplifier is 180° out of phase with the input voltage.
6. What do you understand by d.c. and a.c. load lines? How will you construct them on the output characteristics?
7. Draw the d.c. and a.c. equivalent circuits of a transistor amplifier.
8. Derive an expression for the voltage gain of a transistor amplifier from its a.c. equivalent circuit.
9. Write short notes on the following :
   (*i*) phase reversal (*ii*) d.c. and a.c. load lines
   (*iii*) operating point (*iv*) classification of amplifiers.

Problems

1. In transistor amplifier, the collector current swings from 2 mA to 5 mA as the base current is changed from 5 μA to 15 μA. Find the current gain. **[300]**
2. A transistor amplifier employs a 4 kΩ as collector load. If the input resistance is 1 kΩ, determine the voltage gain. Given β = 100, g_m =10 mA/volt and signal voltage = 50 mV. **[1.04]**
3. Fig. 8.60 shows the transistor amplifier. If R_C = 4 kΩ, R_E = 5 kΩ and V_{CC} = 30 V, draw the d.c. load line.

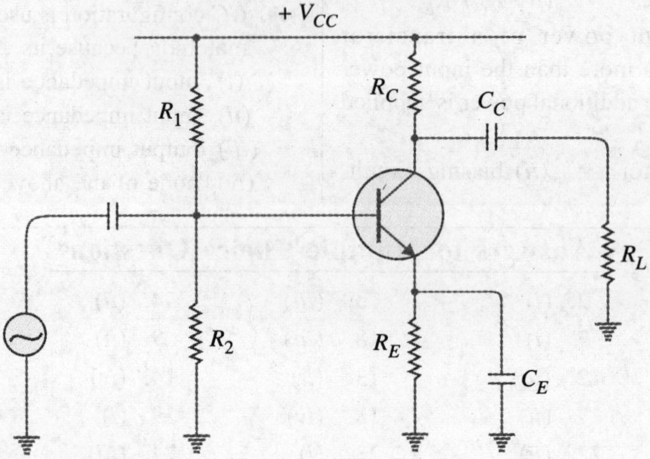

Fig. 8.60

4. Find the operating point for Fig. 8.60, V_{CC} = 30 V, R_1 = 20 kΩ, R_2 = 20 kΩ, R_C = 4 kΩ, R_E =5 kΩ. **[13.2 V, 1.85 mA]**
5. For the circuit shown in Fig. 8.60, find the voltage gain if β = 100, R_C = 3 kΩ, R_L = 6 kΩ and R_{in} = 2 kΩ. **[100]**
6. In the circuit shown in Fig. 8.60, V_{CC} = 30 V, R_1 = 2 kΩ, R_2 = 1 kΩ, R_C = 2 kΩ, R_L = 2 kΩ, R_E = 1 kΩ. Draw the d.c. and a.c. load lines.
7. A voltage-divider biased circuit has an emitter voltage of 2 V and an emitter resistor of 4.7 kΩ. What is the ac resistance of emitter diode? **[58.7 Ω]**
8. A transistor amplifier has a d.c. collector current of 5 mA. What is the a.c. resistance of the base if β = 200? **[1000 Ω]**

9. Determine the voltage gain for the amplifier circuit shown in Fig. 8.61.

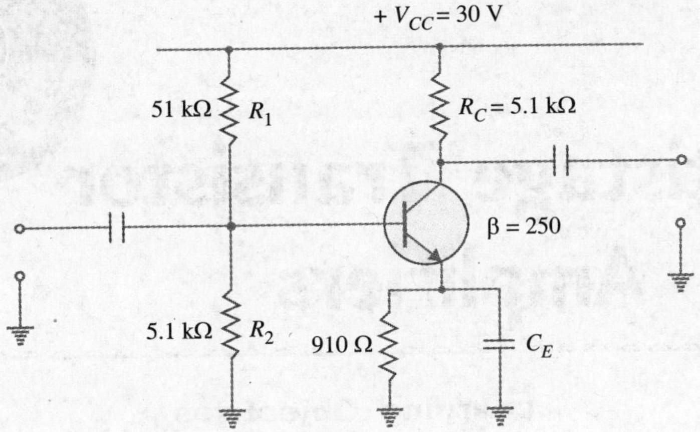

Fig. 8.61

10. What is the input impedance of the amplifier circuit shown in Fig. 8.61? [1.75 kΩ]
11. A voltage-divider biased amplifier has the values of R_1 = 40 kΩ, R_2 = 10 kΩ, R_C = 6 kΩ, R_E = 2 kΩ, V_{CC} = +10 V and β = 80. Determine the ac emitter resistance of the transistor. [38.46Ω]
12. A standard *CE* amplifier has the following values: V_{CC} = 30V, R_1 = 51 kΩ, R_2 = 5.1 kΩ, R_C = 5.1 kΩ, R_E = 910 Ω and β = 250. Determine the voltage gain of the amplifier. [455.4]
13. A *CE* amplifier has a voltage gain A_v = 59.1 and β = 200. Determine the power gain and output power of the amplifier when input power is 80 μW. [11820; 945.6 mW]
14. Determine the voltage gain for the first stage in Fig. 8.62. [53.03]
15. If the value of β for the second stage in Fig. 8.62 is increased to 280, determine the voltage gain of the first amplifier stage. [58.08]

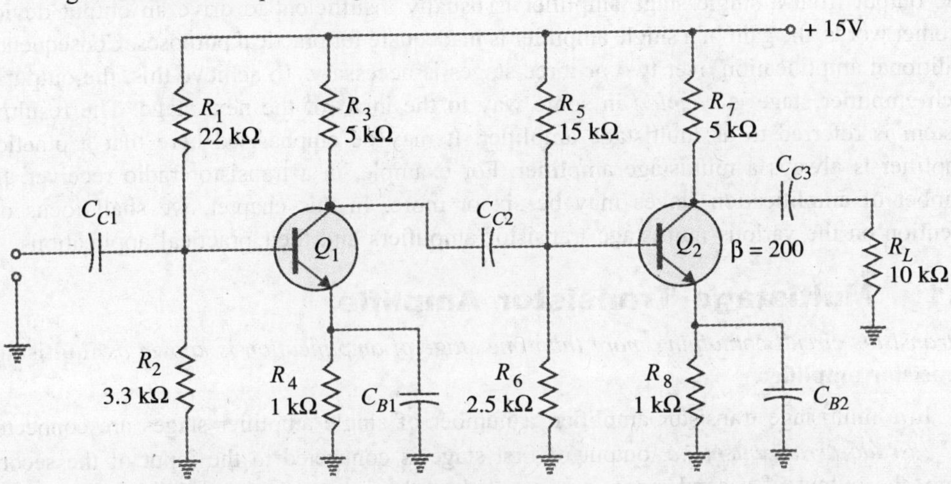

Fig. 8.62

CHAPTER 9

Multistage Transistor Amplifiers

Learning Objectives

- Understand a multistage transistor amplifier and its need.
- Learn the effect of source resistance, load resistance and interstage loading on the voltage gain of multistage transistor amplifier.
- Explore the characteristics of *RC* coupling, transformer coupling and direct coupling in a multistage amplifier.
- Develop an understanding of expressing amplifier gain in decibels (dB).
- Recognise the importance of frequency response and bandwidth of an amplifier.
- Learn the characteristics of *RC* coupled transistor amplifier.
- Understand the important aspects of transformer coupled transistor amplifier.
- Recognise the need of direct coupled transistor amplifier.

Introduction

The output from a single stage amplifier is usually insufficient to drive an output device. In other words, the gain of a single amplifier is inadequate for practical purposes. Consequently, additional amplification over two or three stages is necessary. To achieve this, the output of each amplifier stage is *coupled* in some way to the input of the next stage. The resulting system is referred to as multistage amplifier. It may be emphasised here that a practical amplifier is always a multistage amplifier. For example, in a transistor radio receiver, the number of amplification stages may be six or more. In this chapter, we shall focus our attention on the various multistage transistor amplifiers and their practical applications.

9.1 Multistage Transistor Amplifier

A transistor circuit containing more than one stage of amplification is known as **multistage transistor amplifier.**

In a multistage transistor amplifier, a number of single amplifier stages are connected in **cascade arrangement i.e.* output of first stage is connected to the input of the second stage, the output of second stage is connected to the input of the third stage and so on.

* The term *cascade* means *connected in series.*

Generally, a single amplifier stage cannot provide the required gain (voltage gain or power gain or current gain). The primary need of a multistage amplifier is to provide the additional gain. For example, in an audio amplifier system, we require a large amount of power (of the order of watts) to operate the loudspeaker. However, the signal originating at the microphone has very small power (of the order of mW or less). This requires the use of a multistage amplifier to "boost" the signal to provide the necessary power.

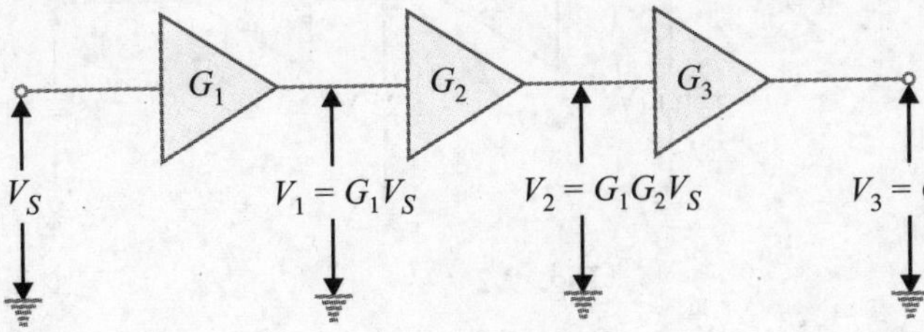

Fig. 9.1

Gain. *The total *gain (G) of a multistage amplifier is equal to the product of the gains of the individual stages (not the sum!).* Consider a 3-stage amplifier shown in Fig. 9.1. If G_1, G_2 and G_3 are the voltage gains of the first stage, second stage and third stage respectively, then total voltage gain G is given by ;

$$G = G_1 \times G_2 \times G_3 \quad \text{... }(i)$$

The above relation can be easily proved. Suppose the input voltage to the first stage is V_S. Then,

$$\text{Output voltage of first stage} = G_1V_S$$

$$\text{Output voltage of second stage} = (G_1V_S)G_2 = G_1G_2V_S$$

$$\text{Output voltage of third stage} = (G_1G_2V_S)G_3 = G_1G_2G_3V_S$$

$$\therefore \quad \text{Total voltage gain, } G = \frac{\text{Output voltage of third stage}}{V_S}$$

or
$$G = \frac{G_1G_2G_3V_S}{V_S} = G_1G_2G_3$$

Thus the total voltage gain is equal to the *product* of individual stage voltage gains. It may be noted that actually the total voltage gain $G < G_1G_2G_3$. It is because while deriving eq. (*i*) above, we have made the following simplyfying assumptions :

(*i*) The resistance R_S of the voltage source connected to the first stage is zero (*i.e.* $R_S = 0$).

(*ii*) The load R_L is not connected to the last stage.

(*iii*) The *interstage loading* effect is neglected. It means that we have ignored the loading effect of the input impedance of the next stage.

In a practical multistage ampifier, all the above factors are present so that $G < G_1G_2G_3$. Nevertheless, the total voltage gain (G) of the multistage amplifier is very large compared to that of the single amplifier stage. This will become quite apparent from the following examples.

* It may be voltage gain (A_v) or power gain (A_p) or current gain (A_i).

Example 9.1. *Fig. 9.2 shows a 3-stage transistor amplifier and the a.c. r.m.s. voltages at several points in the amplifier. Note that V_1 is the input voltage delivered by a signal source having zero resistance (i.e. $R_S = 0$) and that V_3 is the output voltage with no load connected. Find (i) voltage gain of each stage (ii) overall voltage gain V_3/V_1.*

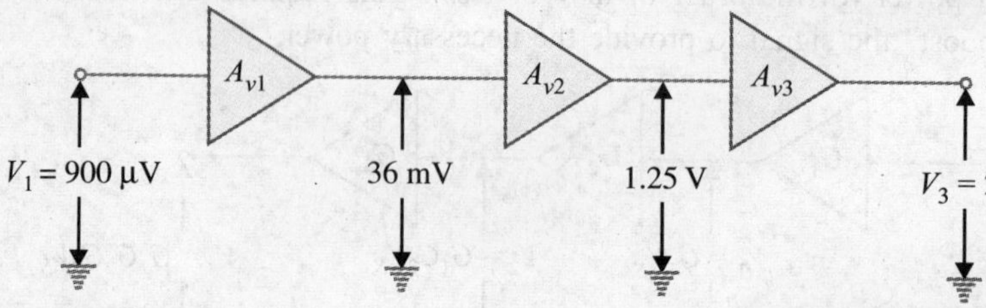

Fig. 9.2

Solution. We neglect the loading effect of the input impedance of next stage in this multistage amplifier.

(*i*) Voltage gain of first stage, $A_{v1} = \dfrac{36\,\text{mV}}{900\,\mu\text{V}} = \mathbf{40}$

Voltage gain of second stage, $A_{v2} = \dfrac{1.25\,\text{V}}{36\,\text{mV}} = \mathbf{34.72}$

Voltage gain of third stage, $A_{v3} = \dfrac{21\,\text{V}}{1.25\,\text{V}} = \mathbf{16.8}$

(*ii*) Overall voltage gain, $A_v = A_{v1} \times A_{v2} \times A_{v3}$

$$= 40 \times 34.72 \times 16.8 = \mathbf{23{,}333}$$

Alternatively. $A_v = \dfrac{V_3}{V_1} = \dfrac{21\,\text{V}}{900\,\mu\text{V}} = 23{,}333$

Example 9.2. *The voltage gains of the stages in a 4-stage transistor amplifier are shown in Fig. 9.3. Find (i) the overall voltage gain, V_{out}/V_{in} (ii) the voltage gain that would be necessary in a fifth stage, which, if added to the cascade, would make the overall voltage gain 10^5.*

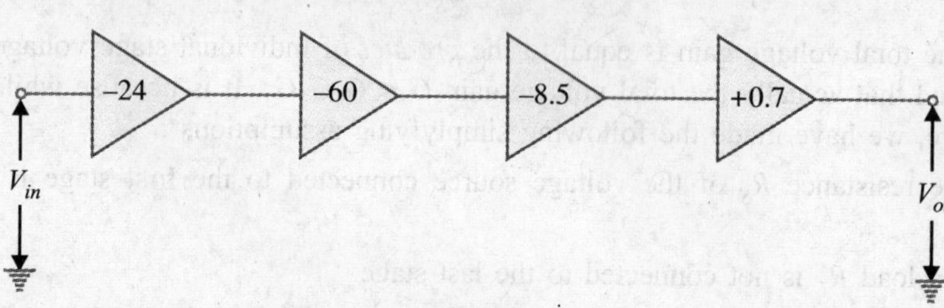

Fig. 9.3

Solution. $A_{v1} = -24$; $A_{v2} = -60$; $A_{v3} = -8.5$; $A_{v4} = +0.7$

(*i*) The overall voltage gain A_v (= V_{out}/V_{in}) of the 4-stage amplifier is equal to the product of voltage gains of the individual stages *i.e.*

$$A_v = A_{v1} \times A_{v2} \times A_{v3} \times A_{v4}$$
$$= (-24) \times (-60) \times (-8.5) \times (+0.7) = \mathbf{-8568}$$

Recall that a *CE* amplifier causes a phase inversion of 180° of the input signal. A multistage amplifier may contain one or more *CE* amplifier stages. If the number of *CE* amplifier stages is *odd*, the final output voltage (V_{out}) will be out of phase with the input voltage (V_{in}). On the other hand, if the number of inverting stages is *even* (or zero) V_{out} will be in phase with V_{in}.

(ii) Suppose the voltage gain of the fifth stage is A'_v. Then as per the conditions of the problem,

$$(-8568) \times A'_v = 10^5$$

$$\therefore \quad A'_v = \frac{10^5}{-8568} = \mathbf{-11.67}$$

Example 9.3. *It is desired to construct a 3-stage amplifier whose overall voltage gain is 500. The *in-circuit voltage gains of the first two stages are to be equal and the voltage gain of the third stage is to be one-half that of each of the first two. What should be the voltage gain of each stage ?*

Solution. Suppose the voltage gain of the first stage is A_{v1}. As per the conditions of the problem,

$$\text{Voltage gain of second stage, } A_{v2} = A_{v1}$$

$$\text{Voltage gain of third stage, } A_{v3} = \frac{A_{v1}}{2}$$

It is given that overall voltage gain of 3-stage amplifier is $A_v = 500$.

$$\text{Now,} \quad A_v = A_{v1} \times A_{v2} \times A_{v3}$$

$$\text{or} \quad 500 = A_{v1} \times A_{v1} \times \frac{A_{v1}}{2}$$

$$\therefore \quad A_{v1} = \sqrt[3]{1000} = 10$$

$$\therefore \quad \text{Voltage gain of first stage} = A_{v1} = \mathbf{10}$$

$$\text{Voltage gain of second stage} = A_{v1} = \mathbf{10}$$

$$\text{Voltage gain of third stage} = A_{v1}/2 = 10/2 = \mathbf{5}$$

9.2 Effect of R_S, R_L and Interstage Loading

Fig. 9.4 shows a 3-stage amplifier. The signal source V_S having internal resistance R_S (also called *source resistance*) is connected to the input of the first stage. The load R_L is connected to the output terminals of the last stage. The load may be a loudspeaker, antenna, an indicating instrument or any other useful device. Further, there is interstage loading *i.e.* output impedance of one stage interacts with the input impedance of the next stage. The presence of R_S, R_L and interstage loading reduces the overall voltage gain of the multistage amplifier.

***(i)* Effect of source resistance R_S.** As shown in Fig. 9.4, the source resistance R_S is in series with the input impedance Z_{in} of the first stage. Therefore, the a.c. voltage V_1 at the input terminals of the first stage is

$$V_1 = I_1 Z_{in} = \frac{V_S}{R_S + Z_{in}} \times Z_{in}$$

* The in-circuit gains mean the gains when all the stages are connected *i.e.* we take into account interstage loading.

or $$V_1 = \left(\frac{Z_{in}}{R_S + Z_{in}}\right) V_S \quad \text{... (i)}$$

It may be seen from eq. (*i*) above that $V_1 < V_S$. Therefore, source resistance R_S reduces the overall voltage gain of the multistage amplifier.

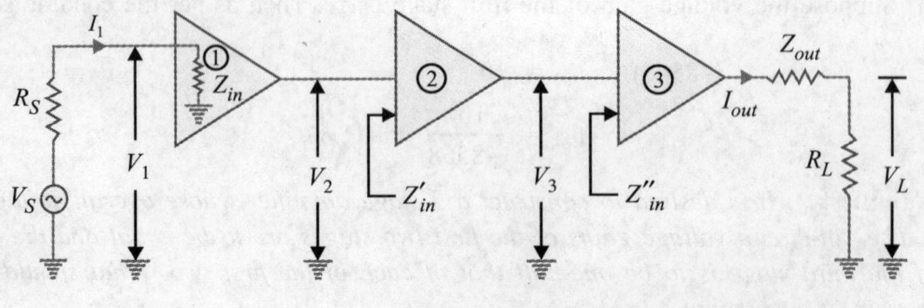

Fig. 9.4

(*ii*) **Effect of load R_L.** As shown in Fig. 9.4, the output impedance Z_{out} of the last stage is in series with the load R_L. Therefore, a.c. voltage V_L across the load R_L is

$$V_L = I_{out} R_L = \frac{A_{v0} V_1}{Z_{out} + R_L} \times R_L$$

or $$V_L = A_{v0} V_1 \left(\frac{R_L}{Z_{out} + R_L}\right) \quad \text{... (ii)}$$

where A_{v0} = No-load voltage gain of multistage amplifier

It may be seen from eq. (*ii*) that $V_L < A_{v0} V_1$ (= unloaded output voltage). Therefore, application of R_L to the last stage reduces the overall voltage gain of the multistage amplifier.

(*iii*) **Effect of interstage loading.** As shown in Fig. 9.4, the output impedance of the first stage comes in parallel with the input impedance Z'_{in} of the second stage. This is called interstage loading and *reduces the voltage gain of the first stage. Likewise, the output impedance of the second stage comes in parallel with the input impedance Z''_{in} of the third stage and as a result, the voltage gain of the second stage is reduced. Therefore, the effect of interstage loading is to reduce the overall voltage gain of the multistage amplifier.

It may be noted that inspite of the fact that the presence of R_S, R_L and interstage loading reduces the voltage gain of the individual stages significantly, the overall voltage gain of a multistage amplifier is still quite large compared to that of a single amplifier stage.

Example 9.4. *Fig. 9.5 shows a 3-stage amplifier where source resistance R_S = 0 and load R_L is not connected to the last stage. Now, suppose that source resistance R_S = 2000 Ω and the load R_L = 25 Ω. Further, stage 1 has input resistance of 1 kΩ and stage 3 has output resistance of 50 Ω. Find :*

(*i*) *the overall voltage gain from signal source to load (i.e. V_L/V_S).*

(*ii*) *power gain between input to the first stage and the load.*

(*iii*) *current gain between input to the first stage and the load.*

---

* Because a.c. load of first stage is reduced.

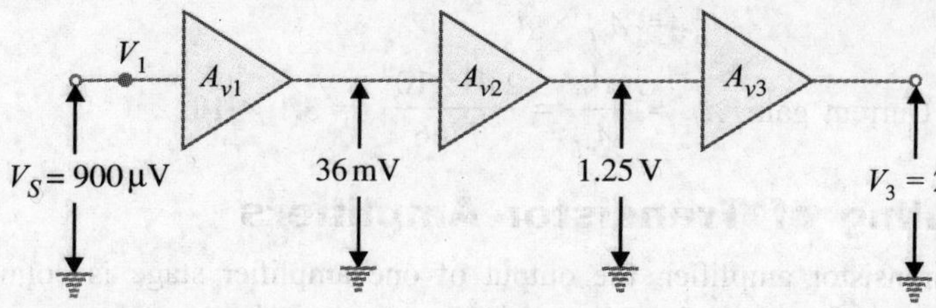

Fig. 9.5

Solution.

(*i*) The no-load voltage gain A_v of the multistage amplifier is given by ;

$$A_{v0} = \frac{V_3}{V_S} = \frac{21\text{ V}}{900\,\mu\text{V}} = 23{,}333$$

The overall voltage gain of the 3-stage amplifier from source to load is (See Art. 8.24)

$$A_{vs} = \frac{V_L}{V_S} = A_{v0}\left(\frac{R_{in}}{R_S + R_{in}}\right)\left(\frac{R_L}{R_{out} + R_L}\right)$$

Here, $A_{v0} = 23{,}333$; $R_{in} = 1\text{ k}\Omega = 1000\,\Omega$; $R_S = 2000\,\Omega$; $R_L = 25\,\Omega$; $R_{out} = 50\,\Omega$

$$\therefore \quad A_{vs} = 23{,}333\left(\frac{1000}{2000+1000}\right)\left(\frac{25}{50+25}\right) = \mathbf{2592.5}$$

(*ii*) The source resistance R_S(= 2000 Ω) is in series with the input resistance R_{in} (= 1000 Ω) of the first stage. Therefore, a.c. voltage V_1 at the input terminals of the first stage is

$$V_1 = \frac{R_{in}}{R_S + R_{in}} \times V_S = \frac{1000}{2000+1000} \times 900\,\mu\text{V} = 300\,\mu\text{V}$$

∴ Input power to the first stage is

$$P_{in} = \frac{V_1^2}{R_{in}} = \frac{(300\times10^{-6})^2}{1000} = 90 \times 10^{-12}\text{ W}$$

The output resistance R_{out} (= 50 Ω) is in series with the load R_L (= 25 Ω). Therefore, a.c. voltage V_L across the load is

$$V_L = A_{v0}V_1\left(\frac{R_L}{R_{out} + R_L}\right)$$

$$= 23{,}333 \times 300\,\mu\text{V}\left(\frac{25}{50+25}\right) = 2.33\text{ V}$$

∴ Output power developed across the load is

$$P_{out} = P_L = \frac{V_L^2}{R_L} = \frac{(2.33)^2}{25} = 0.217\text{ W}$$

$$\therefore \quad \text{Power gain, } A_p = \frac{P_{out}}{P_{in}} = \frac{0.217\text{ W}}{90\times10^{-12}\text{ W}} = \mathbf{2.41 \times 10^9}$$

(*iii*) The voltage gain between the input to the first stage and the load is

$$A_{vL} = \frac{V_L}{V_1} = \frac{2.33\text{ V}}{300\,\mu\text{V}} = 7766$$

Now, $A_p = A_{vL} \times A_i$

$$\therefore \quad \text{Current gain, } A_i = \frac{A_p}{A_{vL}} = \frac{2.41 \times 10^9}{7766} = \mathbf{3.1 \times 10^5}$$

9.3 Coupling of Transistor Amplifiers

In a multistage transistor amplifier, the output of one amplifier stage is connected to the input of the next amplifier stage through a suitable *coupling device* and so on. The purpose of coupling device (*e.g.* a capacitor, transformer *etc.*) is (*i*) to transfer a.c. output of one amplifier stage to the input of the next amplifier stage and (*ii*) to isolate the d.c. conditions of one stage from the next stage. Fig. 9.6 shows the block diagram of a 3-stage amplifier. Each stage consists of one transistor and associated circuitry and is coupled to the next stage through a coupling device. The name of the amplifier is usually given after the type of coupling used *e.g.*

Name of coupling	*Name of multistage amplifier*
Capacitor coupling	Capacitively coupled amplifier
Transformer coupling	Transformer coupled amplifier
Direct coupling	Direct coupled amplifier

Fig. 9.6

Let us briefly examine the characteristics of each of these three types of coupling.

(*i*) **Capacitor coupling.** In this type of coupling, the output of one amplifier stage is connected to the input of the next amplifier stage through a capacitor C (called *coupling capacitor*) as shown in Fig. 9.7. We know that a capacitor blocks d.c. but allows the a.c. to pass through it. Therefore, the capacitor C allows the a.c. signal to pass from the output of one amplifier stage to the input of the next amplifier stage but prevents direct current (d.c.) from flowing between the output of one stage and the input of the next stage. In this way, d.c. bias conditions of one stage are isolated from the d.c. bias conditions of the next stage.

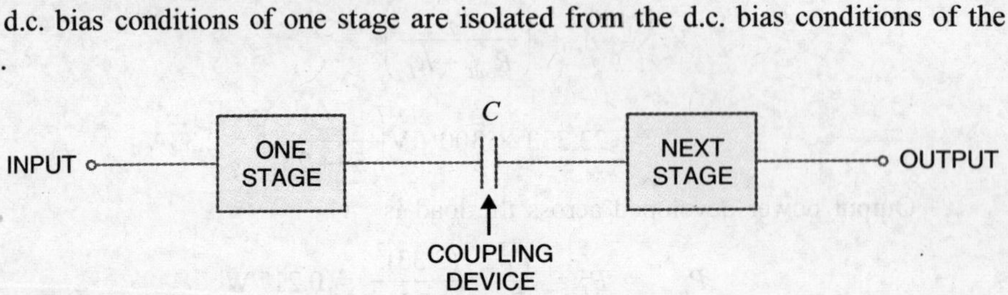

Fig. 9.7

The advantage of capacitor coupling is that it is simple and inexpensive. The disadvantage is that at low signal frequencies, the capacitor poses problems. At *low signal frequencies, the reactance X_C (= $1/2\pi fC$) of the coupling capacitor is quite large and considerable amount

* At high signal frequencies, the capacitor behaves as a short.

of a.c. signal is dropped across it. This reduces the a.c. signal at the input of the next stage and hence output voltage from this stage.

(*ii*) Transformer coupling. In this type of coupling, the output of one amplifier stage is connected to the input of the next amplifier stage through a transformer as shown in Fig. 9.8. We know that a transformer blocks d.c. while it transfers a.c. from primary to the secondary. Therefore, the transformer allows the a.c. signal to pass from the output of one amplifier stage to the input of the next amplifier stage but prevents direct current (d.c.) from flowing between the output of one stage and the input of the next stage. In this way, the transformer provides d.c. isolation between the two stages.

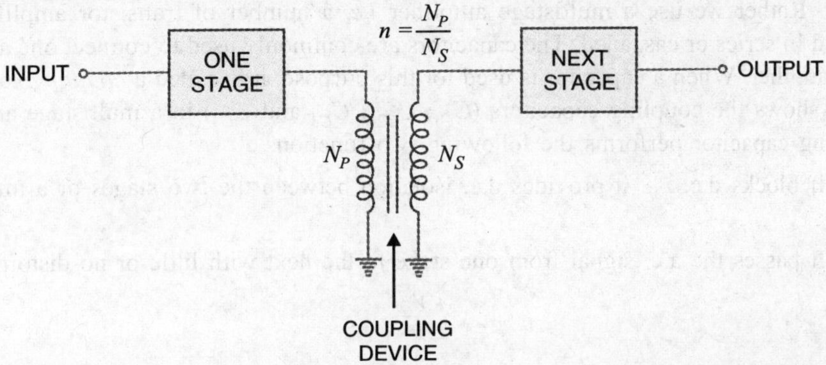

Fig. 9.8

An important advantage of transformer coupling is that it permits impedance matching for maximum power transfer. If a step-down transformer of turn ratio n (= N_P/N_S) is used, then value of primary impedance is *$Z_P = n^2 Z_L$ where Z_L (load) is the input impedance of the second stage. If turn ratio n is so selected to make Z_P equal to the output impedance of the first stage, then maximum power will be transferred to the second stage. The disadvantage of transformer coupling is that there are losses in a transformer.

(*iii*) Direct coupling. In this type of coupling, the output of one amplifier stage is connected *directly* to the input of the next amplifier stage as shown in Fig. 9.9. The major advantage of direct coupling is that there are no losses in the coupling, either at low or high frequencies.

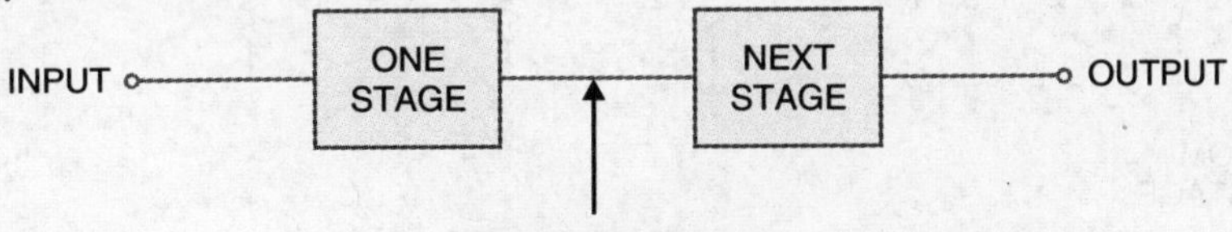

Fig. 9.9

The disadvantage of direct coupling is that the input impedance of the second stage affects the voltage gain as well as Q-point of the first stage.

9.4 Role of Capacitors in Transistor Amplifiers

Regardless of the manner in which a capacitor is connected in a transistor amplifier, its behaviour towards d.c. and a.c. is as follows. *A capacitor blocks d.c. i.e. a capacitor behaves*

* The d.c. resistance of primary is very small and is neglected.

*as an "open**" *to d.c.* Therefore, for d.c. analysis, we can remove the capacitors from the transistor amplifier circuit. A capacitor offers reactance (= $1/2\pi fC$) to a.c. depending upon the values of f and C. In practical transistor circuits, the size of capacitors is so selected that they offer negligible (ideally zero) reactance to the range of frequencies handled by the circuits. Therefore, *for a.c. analysis, we can replace the capacitors by a short i.e. by a wire.* The capacitors serve the following two roles in transistor amplifiers :

1. As coupling capacitors
2. As bypass capacitors

1. As coupling capacitors. In most applications, you will not see a single transistor amplifier. Rather we use a multistage amplifier *i.e.* a number of transistor amplifiers are connected in series or cascaded. The capacitors are commonly used to connect one amplifier stage to another. When a capacitor is used for this purpose, it is called a *coupling capacitor.* Fig. 9.10 shows the coupling capacitors (C_{C1}, C_{C2}, C_{C3} and C_{C4}) in a multistage amplifier. A coupling capacitor performs the following two functions :

(*i*) It blocks d.c. *i.e.* it provides d.c. isolation between the two stages of a multistage amplifier.

(*ii*) It passes the a.c. signal from one stage to the next with little or no distortion.

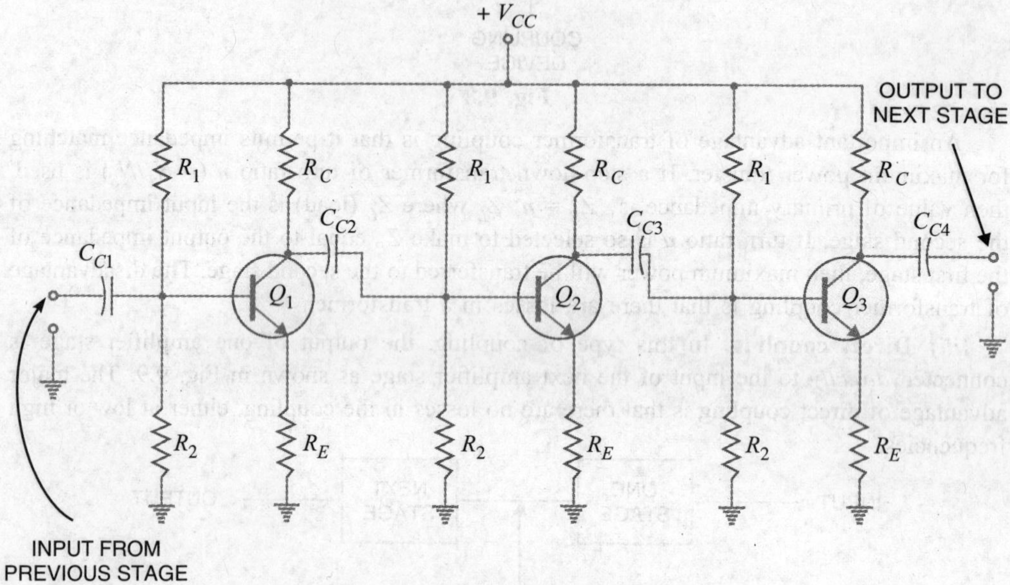

Fig. 9.10

2. As bypass capacitors. Like a coupling capacitor, a bypass capacitor also blocks d.c. and behaves as a short or wire (due to proper selection of capacitor size) to an a.c. signal. But it is used for a different purpose. A bypass capacitor is connected in parallel with a circuit component (*e.g.* resistor) to bypass the a.c. signal and hence the name. Fig. 9.11 shows a bypass capacitor C_E connected across the emitter resistance R_E. Since C_E behaves as a short to the a.c. signal, the whole of a.c. signal (i_e) passes through it. Note that C_E keeps

* $X_C = \dfrac{1}{2\pi fC}$. For d.c., $f = 0$ so that $X_C \to \infty$. Therefore, a capacitor behaves as an open to d.c.

the emitter at a.c. ground. Thus for a.c. purposes, R_E does not exist. We have already seen in the previous chapter that C_E plays an important role in determining the voltage gain of the amplifier circuit. If C_E is removed, the voltage gain of the amplifier is greatly reduced. Note that C_{in} is the coupling capacitor in this circuit.

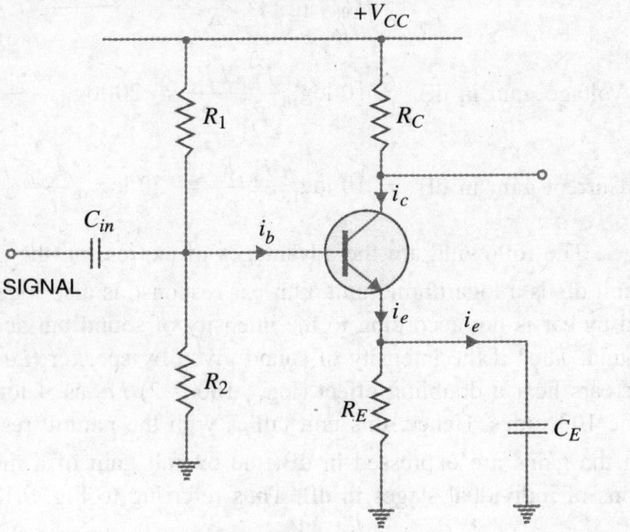

Fig. 9.11

9.5 Amplifier Gain in Decibels (dB)

While designing and analysing a multistage amplifier, the most important consideration is the gain of the individual amplifier stages. Although the gain of an amplifier (voltage gain or power gain or current gain) can be expressed as a number, yet it is of great practical importance to assign it a unit. The unit assigned is *bel or decibel (dB)*.

The common logarithm (log to the base 10) of power gain is known as **bel power gain** *i.e.*

$$\text{Power gain} = \log_{10} \frac{P_{out}}{P_{in}} \text{ bel}$$

$$1 \text{ bel} = 10 \text{ dB}$$

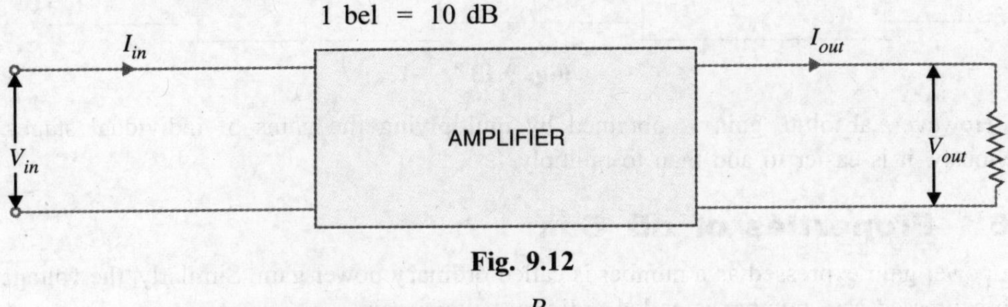

Fig. 9.12

$$\therefore \quad \text{Power gain} = 10 \log_{10} \frac{P_{out}}{P_{in}} \text{ dB}$$

If $P_{out} > P_{in}$, then power gain in dB is *positive*, indicating increase in power. If $P_{out} < P_{in}$, then power given in dB is *negative*, indicating reduction in power. However, if $P_{out} = P_{in}$, then power given is 0 dB since $\log_{10} 1 = 0$.

If the two powers are developed in the same resistance or equal resistances, then,

$$P_1 = \frac{V_{in}^2}{R} = I_{in}^2 R$$

$$P_2 = \frac{V_{out}^2}{R} = I_{out}^2 R$$

$$\therefore \quad \text{Voltage gain in dB} = 10 \log_{10} \frac{V_{out}^2 / R}{V_{in}^2 / R} = 20 \log_{10} \frac{V_{out}}{V_{in}}$$

$$\text{Current gain in dB} = 10 \log_{10} \frac{I_{out}^2 R}{I_{in}^2 R} = 20 \log_{10} \frac{I_{out}}{I_{in}}$$

Advantages. The following are the advantages of expressing the gain in dB :

(*i*) The unit dB is a logarithmic unit. Our ear response is also logarithmic *i.e.* loudness of sound heard by ear is not according to the intensity of sound but according to the log of intensity of sound. Thus if the intensity of sound given by speaker (*i.e.* power) is increased 100 times, our ears hear a doubling effect ($\log_{10} 100 = 2$) *i.e.* as if loudness were doubled instead of made 100 times. Hence, this unit tallies with the natural response of our ears.

(*ii*) When the gains are expressed in dB, the overall gain of a multistage amplifier is the sum of gains of individual stages in dB. Thus referring to Fig. 9.13,

$$\text{Gain as number} = \frac{V_2}{V_1} \times \frac{V_3}{V_2}$$

$$\text{Gain in dB} = 20 \log_{10} \frac{V_2}{V_1} \times \frac{V_3}{V_2}$$

$$= 20 \log_{10} \frac{V_2}{V_1} + 20 \log_{10} \frac{V_3}{V_2}$$

$$= \text{1st stage gain in dB} + \text{2nd stage gain in dB}$$

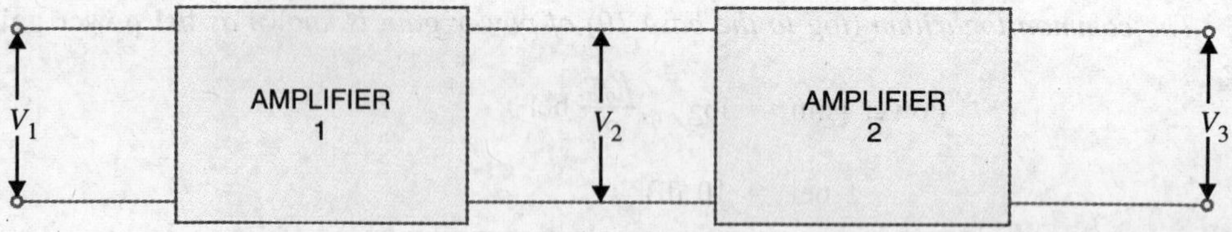

Fig. 9.13

However, absolute gain is obtained by multiplying the gains of individual stages. Obviously, it is easier to add than to multiply.

9.6 Properties of dB Gain

The power gain expressed as a number is called ordinary power gain. Similarly, the voltage gain expressed as a number is called ordinary voltage gain.

1. **Properties of dB power gain.** The following are the useful rules for dB power gain:

(*i*) *Each time the ordinary power gain increases (decreases) by a factor of 10, the dB power gain increases (decreases) by 10 dB.*

For example, suppose the ordinary power gain increases from 100 to 1000 (*i.e.* by a factor of 10).

$\therefore$ Increase in dB power gain $= 10 \log_{10} 1000 - 10 \log_{10} 100$

$= 30 - 20 = 10$ dB

This property also applies for the decrease in power gain.

(ii) *Each time the ordinary power gain increases (decreases) by a factor of 2, the dB power gain increases (decreases) by 3 dB.*

For example, suppose the power gain increases from 100 to 200 (*i.e.* by a factor of 2).

$\therefore$ Increase in dB power gain $= 10 \log_{10} 200 - 10 \log_{10} 100$

$= 23 - 20 = 3$ dB

2. Properties of dB voltage gain. The following are the useful rules for dB voltage gain:

(i) Each time the ordinary voltage gain increases (decreases) by a factor of 10, the dB voltage gain increases (decreases) by 20 dB.

For example, suppose the voltage gain increases from 100 to 1000 (*i.e.* by a factor of 10).

$\therefore$ Increase in dB voltage gain $= 20 \log_{10} 1000 - 20 \log_{10} 100$

$= 60 - 40 = 20$ dB

(ii) *Each time the ordinary voltage gain increases (decreases) by a factor of 2, the dB voltage gain increases (decreases) by 6 dB.*

For example, suppose the voltage gain increases from 100 to 200 (*i.e.* by a factor of 2).

$\therefore$ Increase in dB voltage gain $= 20 \log_{10} 200 - 20 \log_{10} 100$

$= 46 - 40 = 6$ dB

Example 9.5. *Find the gain in dB in the following cases :*

(i) Voltage gain of 30 *(ii) Power gain of 100*

Solution. (*i*) Voltage gain $= 20 \log_{10} 30$ dB $=$ **29.54 dB**

(*ii*) Power gain $= 10 \log_{10} 100$ dB $=$ **20 dB**

Example 9.6. *Express the following gains as a number :*

(i) Power gain of 40 dB *(ii) Power gain of 43 dB*

Solution. (*i*) Power gain $= 40$ dB $= 4$ bel

If we want to find the gain as a number, we should work from logarithm back to the original number.

$\therefore$ Gain $=$ Antilog $4 = 10^4 =$ **10,000**

(*ii*) Power gain $= 43$ dB $= 4.3$ bel

$\therefore$ Power gain $=$ Antilog $4.3 = 2 \times 10^4 =$ **20,000**

Alternatively. $10 \log_{10} \dfrac{P_2}{P_1} = 43$ dB

or $\log_{10} \dfrac{P_2}{P_1} = 43/10 = 4.3$

$\therefore$ $\dfrac{P_2}{P_1} = (10)^{4.3} = 20{,}000$

In general, we have,

$$\frac{V_2}{V_1} = (10)^{gain\ in\ dB/20}$$

$$\frac{P_2}{P_1} = (10)^{gain\ in\ dB/10}$$

Example 9.7. *A three-stage amplifier has a first stage voltage gain of 100, second stage voltage gain of 200 and third stage voltage gain of 400. Find the total voltage gain in dB.*

Solution.

First-stage voltage gain in dB $= 20 \log_{10} 100 = 20 \times 2 = 40$

Second-stage voltage gain in dB $= 20 \log_{10} 200 = 20 \times 2.3 = 46$

Third-stage voltage gain in dB $= 20 \log_{10} 400 = 20 \times 2.6 = 52$

$\therefore$ Total voltage gain $= 40 + 46 + 52 =$ **138 dB**

Example 9.8. (*i*) *A multistage amplifier employs five stages each of which has a power gain of 30. What is the total gain of the amplifier in dB?*

(*ii*) *If a negative feedback of 10 dB is employed, find the resultant gain.*

Solution. Absolute gain of each stage = 30

No. of stages = 5

(*i*) Power gain of one stage in dB $= 10 \log_{10} 30 = 14.77$

$\therefore$ Total power gain $= 5 \times 14.77 =$ **73.85 dB**

(*ii*) Resultant power gain with negative feedback

$= 73.85 - 10 =$ **63.85 dB**

It is clear from the above example that by expressing the gain in dB, calculations have become very simple.

Example 9.9. *In an amplifier, the output power is 1.5 watts at 2 kHz and 0.3 watt at 20 Hz, while the input power is constant at 10 mW. Calculate by how many decibels gain at 20 Hz is below that at 2 kHz?*

Solution.

dB power gain at 2 kHz. At 2 kHz, the output power is 1.5 W and input power is 10 mW.

$\therefore$ Power gain in dB $= 10 \log_{10} \dfrac{1.5\text{ W}}{10\text{ mW}} = 21.76$

dB power gain at 20 Hz. At 20 Hz, the output power is 0.3 W and input power is 10 mW.

$\therefore$ Power gain in dB $= 10 \log_{10} \dfrac{0.3\text{ W}}{10\text{ mW}} = 14.77$

Fall in gain from 2 kHz to 20 Hz $= 21.76 - 14.77 =$ **6.99 dB**

Example 9.10. *A certain amplifier has voltage gain of 15 dB. If the input signal voltage is 0.8 V, what is the output voltage?*

Solution. dB voltage gain $= 20 \log_{10} V_2/V_1$

or $15 = 20 \log_{10} V_2/V_1$

$$\text{or} \qquad 15/20 = \log_{10} V_2/V_1$$

$$\text{or} \qquad 0.75 = \log_{10} V_2/0.8$$

Taking antilogs, we get,

$$\text{Antilog } 0.75 = \text{Antilog } (\log_{10} V_2/0.8)$$

$$\text{or} \qquad 10^{0.75} = V_2/0.8$$

$$\therefore \qquad V_2 = 10^{0.75} \times 0.8 = \mathbf{4.5\ V}$$

Example 9.11. *An amplifier has an open-circuit voltage gain of 70 dB and an output resistance of 1.5 kΩ. Determine the minimum value of load resistance so that voltage gain is not more than 67 dB.*

Solution.

$$A_{v0} = 70\text{ dB}\ ; \quad A_v = 67\text{ dB}$$

$$A_{v0} \text{ in dB} - A_v \text{ in dB} = 70 - 67 = 3\text{ dB}$$

$$\text{or} \qquad 20 \log_{10} A_{v0} - 20 \log_{10} A_v = 3$$

$$\text{or} \qquad 20 \log_{10} \frac{A_{v0}}{A_v} = 3$$

$$\text{or} \qquad \frac{A_{v0}}{A_v} = (10)^{3/20} = 1.41$$

$$\text{But} \qquad \frac{A_v}{A_{v0}} = \frac{R_L}{R_{out} + R_L} \qquad \text{[See Art. 8.24]}$$

$$\therefore \qquad \frac{1}{1.41} = \frac{R_L}{1.5 + R_L}$$

$$\text{or} \qquad R_L = \mathbf{3.65\ k\Omega}$$

Example 9.12. *An amplifier feeding a resistive load of 1 kΩ has a voltage gain of 40 dB. If the input signal is 10 mV, find (i) output voltage (ii) load power.*

Solution.

$$(i) \qquad \frac{V_{out}}{V_{in}} = (10)^{dB\ gain/20} = (10)^{40/20} = 100$$

$$\therefore \qquad V_{out} = 100 \times V_{in} = 100 \times 10\text{ mV} = 1000\text{ mV} = \mathbf{1\ V}$$

$$(ii) \qquad \text{Load power} = \frac{V_{out}^2}{R_L} = \frac{(1)^2}{1000} = 10^{-3}\text{ W} = \mathbf{1\ mW}$$

Example 9.13. *An amplifier rated at 40 W output is connected to a 10 Ω speaker.*

(i) Calculate the input power required for full power output if the power gain is 25 dB.

(ii) Calculate the input voltage for rated output if the amplifier voltage gain is 40 dB.

Solution.

$$(i) \qquad \text{Power gain in dB} = 10 \log_{10} \frac{P_2}{P_1} \qquad \text{or} \qquad 25 = 10 \log_{10} \frac{40\text{ W}}{P_1}$$

$$\therefore \qquad P_1 = \frac{40\text{ W}}{\text{Antilog } 2.5} = \frac{40\text{ W}}{3.16 \times 10^2} = \frac{40\text{ W}}{316} = \mathbf{126.5\ mW}$$

(*ii*) Voltage gain in dB $= 20 \log_{10} \frac{V_2}{V_1}$ or $40 = 20 \log_{10} \frac{V_2}{V_1}$

$$\therefore \quad \frac{V_2}{V_1} = \text{Antilog } 2 = 100$$

Now, $$V_2 = \sqrt{P_2 R} = \sqrt{40\text{W} \times 10\,\Omega} = 20\text{ V}$$

$$\therefore \quad V_1 = \frac{V_2}{100} = \frac{20\text{V}}{100} = \mathbf{200\ mV}$$

Example 9.14. *The input voltage to an amplifier is 4 mV (r.m.s.). At point 1 in the amplifier, the voltage gain w.r.t. the input is – 4.2 dB and at point 2, the voltage gain w.r.t. point 1 is 18.5 dB. Find the (i) voltage at point 1 (ii) voltage at point 2 (iii) voltage gain in dB at point 2 w.r.t. the input.*

Solution. Input voltage, $V_i = 4\text{ mV (r.m.s.)} = 4 \times 10^{-3}\text{ V (r.m.s.)}$

(*i*) Suppose at point 1, the voltage is V_1 (r.m.s.). Then,

$$20 \log_{10} \frac{V_1}{V_i} = -4.2$$

or $$20 \log_{10} \frac{V_1}{4 \times 10^{-3}} = -4.2$$

or $$\log_{10} \frac{V_1}{4 \times 10^{-3}} = -4.2/20 = -0.21$$

Taking antilog of both sides, we get,

$$\frac{V_1}{4 \times 10^{-3}} = 0.617$$

$$\therefore \quad V_1 = 2.47 \times 10^{-3}\text{ V} = \mathbf{2.47\ mV}$$

(*ii*) Suppose the voltage at point 2 is V_2 (r.m.s.). Then,

$$20 \log_{10} \frac{V_2}{V_1} = 18.5$$

or $$20 \log_{10} \frac{V_2}{2.47 \times 10^{-3}} = 18.5$$

or $$\log_{10} \frac{V_2}{2.47 \times 10^{-3}} = \frac{18.5}{20} = 0.925$$

$$\therefore \quad V_2 = 2.47 \times 10^{-3} \text{ antilog } (0.925)$$
$$= 20.75 \times 10^{-3}\text{ V} = \mathbf{20.75\ mV}$$

(*iii*) Voltage gain in dB at point 2 *w.r.t.* the input is

$$\text{Voltage gain} = 20 \log_{10} \frac{V_2}{V_i} = 20 \log_{10} \frac{20.75}{4} = \mathbf{14.3\ dB}$$

Alternatively. Voltage gain = – 4.2 dB + 18.5 dB = 14.3 dB

Note that overall voltage gain (*i.e.* from the input to point 2) in dB is the sum of intermediate gains in dB.

9.7 Frequency Response and Bandwidth of an Amplifier

In the design of an amplifier, the *frequency response* and the *bandwidth* of the amplifier are very important considerations. It is because the performance of an amplifier in a circuit largely depends upon the shape of the frequency response and value of the bandwidth.

1. **Frequency response.** *The curve between voltage gain (or power gain) and the signal frequency of an amplifier is called* **frequency response** *of the amplifier.*

The frequency response of an amplifier indicates how voltage gain (or power gain) of the amplifier *responds* to the changes in the signal frequency. The voltage gain of an amplifier varies with signal frequency. It is because reactance of the capacitors in the circuit changes with signal frequency and hence affects the output voltage. Fig. 9.14 shows the frequency response of a typical amplifier. The gain of the amplifier increases as the frequency increases from zero till it becomes maximum at f_r, called *resonant frequency*. If the frequency of signal increases beyond f_r, the gain decreases. Note that voltage gain of the amplifier is small at both low and high frequencies.

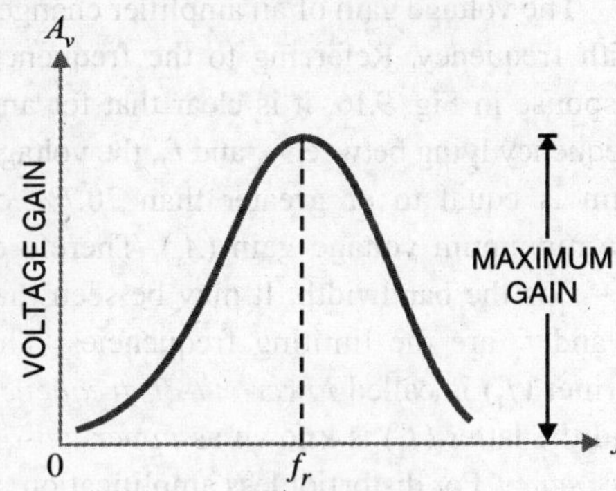

Fig. 9.14

The performance of an amplifier depends to a considerable extent upon its frequency response. While designing an amplifier, appropriate steps must be taken to ensure that gain is essentially uniform over some specified frequency range. For instance, in case of an audio amplifier, which is used to amplify speech or music, it is necessary that all the frequencies in the sound spectrum (*i.e.* 20 Hz to 20 kHz) should be uniformly amplified otherwise speaker will give a distorted sound output.

The following points may be noted :

(*i*) The shape of the frequency response depends upon the type of the amplifier. For example, Fig. 9.15 shows the frequency response of an audio voltage amplifier. Note that the voltage gain of the amplifier is *nearly* constant over the audio frequency range (20 Hz to 20 kHz) but falls off at low and high frequencies. The frequency range over which the gain is more or less constant (*flat*) is called *midband range* (or *mid-frequency region*). You may see that when the signal frequency starts to go outside the mid-frequency region (or midband range), the voltage gain of the amplifier begins to decrease rapidly.

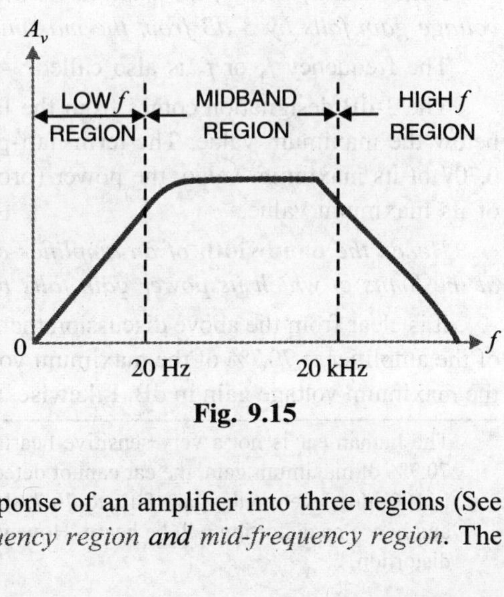

Fig. 9.15

(*ii*) We generally divide the frequency response of an amplifier into three regions (See Fig. 9.15) *viz. low frequency region, high-frequency region and mid-frequency region.* The behaviour of coupling capacitors, bypass capacitors and other reactive elements present in the amplifier circuit is different for these three frequency regions.

(iii) The shape of the frequency response also depends upon the number of amplifier stages in a multistage amplifier.

2. Bandwidth. *The range of frequencies over which the voltage gain of an amplifier is equal to or greater than *70.7% of the maximum voltage gain is known as* **bandwidth** *of the amplifier.*

The voltage gain of an amplifier changes with frequency. Referring to the frequency response in Fig. 9.16, it is clear that for any frequency lying between f_1 and f_2, the voltage gain is equal to or greater than 70.7% of the maximum voltage gain (A_v). Therefore, $f_2 - f_1$ is the bandwidth. It may be seen that f_1 and f_2 are the limiting frequencies. The former (f_1) is called *lower cut-off frequency* and the latter (f_2) is known as *upper cut-off frequency*. For distortionless amplification, it is important that signal frequency range must be within the bandwidth of the amplifier.

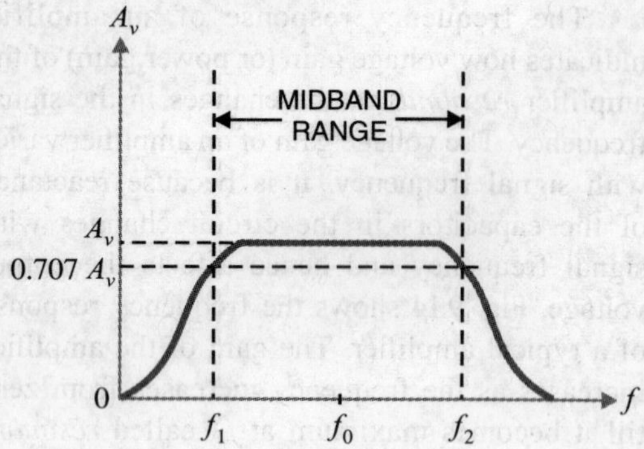

Fig. 9.16

The bandwidth of an amplifier can also be defined in terms of dB. Suppose the maximum voltage gain of an amplifier is 100. Then 70.7% of it is 70.7.

∴ Fall in voltage gain from maximum gain

$$= 20 \log_{10} 100 - 20 \log_{10} 70.7$$

$$= 20 \log_{10} \frac{100}{70.7} \text{ dB}$$

$$= 20 \log_{10} 1.4142 \text{ dB} = 3 \text{ dB}$$

Hence **bandwidth** *of an amplifier is the range of frequencies at the limits of which its voltage gain falls by 3 dB from the maximum voltage gain in dB.*

The frequency f_1 or f_2 is also called *3-dB frequency* or *half-power frequency.*

The 3-dB designation comes from the fact that voltage gain at these frequencies is 3dB below the maximum value. The term half-power is used because when voltage is down to 0.707 of its maximum value, the power (proportional to V^2) is down to $(0.707)^2$ or one-half of its maximum value.

Hence the **bandwidth** *of an amplifier can also be defined as the range of frequencies at the limits of which its power gain falls by 3 dB from the maximum power gain in dB.*

It is clear from the above discussion that at the cut-off frequency f_1 or f_2, the voltage gain of the amplifier is 70.7% of the maximum voltage gain or the voltage gain is 3dB down from the maximum voltage gain in dB. Likewise, at f_1 or f_2, the power gain of the amplifier is 50%

* The human ear is not a very sensitive hearing device. It has been found that if the gain falls to 70.7% of maximum gain, the ear cannot detect the change. For instance, if the gain of an amplifier is 100, then even if the gain falls to 70.7, the ear cannot detect the change in intensity of sound and hence no distortion will be heard. However, if the gain falls below 70.7, the ear will hear clear distortion.

of the maximum power gain or the power gain is 3 dB down from the maximum power gain in dB.

Centre frequency f_0. ***The centre frequency f_0 (also known as resonant frequency) is the frequency in the centre of the bandwidth as shown in Fig. 9.16 above.*** Note that voltage gain (or power gain) of an amplifier is maximum when the signal frequency is equal to f_0. As the signal frequency decreases from f_0 towards f_1 or f_2, the voltage gain (or power gain) begins to decrease slowly. By the time the signal frequency reaches either f_1 or f_2, the voltage gain will drop to 70.7% of maximum voltage gain (or 50% of maximum power gain).

Important relations. We now give important relations among f_1, f_2, BW and f_0.

(*i*) The bandwidth (BW) of an amplifier is given by ;

$$\text{Bandwidth, } BW = f_2 - f_1$$

where f_2 = Upper cut-off frequency

f_1 = Lower cut-off frequency

Thus, if f_1 = 10 kHz and f_2 = 100 kHz, then,

$$\text{Bandwidth, } BW = f_2 - f_1 = 100 \text{ kHz} - 10 \text{ kHz} = 90 \text{ kHz}$$

(*ii*) It can be proved that f_1, f_2 and f_0 are related as :

$$f_0 = \sqrt{f_1 f_2}$$

Thus, the centre frequency f_0 is equal to the geometric mean of the cut-off frequencies (f_1, f_2).

Alternatively. $$\frac{f_0}{f_1} = \frac{f_2}{f_0}$$

(*iii*) The cut-off frequencies (f_1 and f_2) are related to f_0 and BW as :

$$f_1 = f_0 - \frac{BW}{2}\,;\quad f_2 = f_0 + \frac{BW}{2}$$

The above relations are approximately true and lead to resonable accuracy.

Example 9.15. *In an amplifier, the maximum voltage gain is 2000 and occurs at 2 kHz. It falls to 1414 at 10 kHz and 50 Hz. Find :*

(i) Bandwidth (ii) Lower cut-off frequency

(iii) Upper cut-off frequency.

Solution.

(*i*) Referring to the frequency response in Fig. 9.17, the maximum gain is 2000. Then 70.7% of this gain is 0.707 × 2000 = 1414. It is given that gain is 1414 at 50 Hz and 10 kHz. As bandwidth is the range of frequencies over which gain is equal to or greater than 70.7% of maximum gain,

∴ Bandwidth = **50 Hz to 10 kHz**

(*ii*) The frequency (on lower side) at which the voltage gain of the amplifier is exactly 70.7%

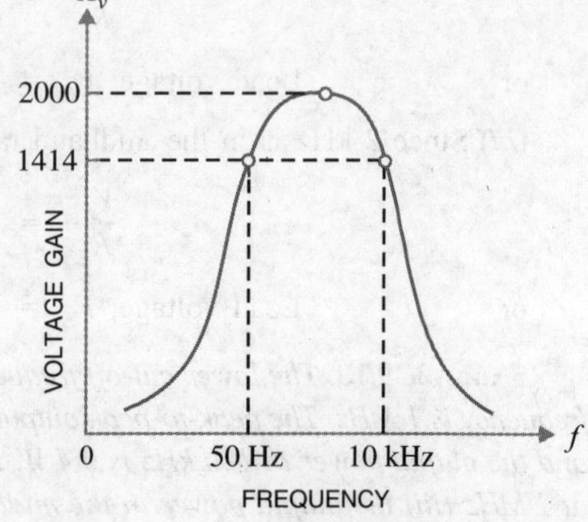

Fig. 9.17

of the maximum gain is known as *lower cut-off frequency.* Referring to Fig. 9.17, it is clear that :

Lower cut-off frequency = **50 Hz**

(*iii*) The frequency (on the higher side) at which the voltage gain of the amplifier is exactly 70.7% of the maximum gain is known as *upper cut-off frequency.* Referring to Fig. 9.17, it is clear that :

Upper cut-off frequency = **10 kHz**

Comments. As bandwidth of the amplifier is 50 Hz to 10 kHz, therefore, it will amplify the signal frequencies lying in this range without any distortion. However, if the signal frequency is not in this range, then there will be distortion in the output.

Note. The dB power rating of communication equipment is normally less than 50 dB.

Example 9.16. *The values of f_0 and f_2 for an amplifier are 60 kHz and 300 kHz respectively. Find the values of f_1 and BW.*

Solution. $f_0 = 60$ kHz ; $f_2 = 300$ kHz ; $f_1 = ?$; $BW = ?$

We know that : $$f_0 = \sqrt{f_1 f_2}$$

or $$f_1 = \frac{f_0^2}{f_2} = \frac{(60)^2}{300} = \mathbf{12\ kHz}$$

$$\text{Bandwidth, } BW = f_2 - f_1 = 300 \text{ kHz} - 12 \text{ kHz} = \mathbf{288\ kHz}$$

Example 9.17. *An audio amplifier has a lower cut-off frequency of 20 Hz and an upper cut-off frequency of 20 kHz. The amplifier delivers 20W to a load of 12 Ω at 1 kHz. Find (i) the bandwidth of the amplifier (ii) the r.m.s. load voltage at 20 Hz (iii) the r.m.s. load voltage at 2 kHz.*

Solution. $$f_1 = 20 \text{ Hz} ;\ f_2 = 20 \text{ kHz}$$

(*i*) $$\text{Bandwidth, } BW = f_2 - f_1 = 20 - 20 \times 10^{-3} = \mathbf{19.98\ kHz}$$

(*ii*) Since 1 kHz is in the midband range, the midband power (*i.e.* maximum power) is 20 W. At 20 Hz (= lower cut-off frequency) the power = 50% of 20 W = 10 W.

∴ $$\frac{V_L^2}{R_L} = 10$$

or $$\text{Load voltage, } V_L = \sqrt{10 \times R_L} = \sqrt{10 \times 12} = \mathbf{10.95\ V\ (r.m.s.)}$$

(*iii*) Since 2 kHz is in the midband range, at 2 kHz the power is 20 W.

∴ $$\frac{V_L^2}{R_L} = 20$$

or $$\text{Load voltage, } V_L = \sqrt{20 \times R_L} = \sqrt{20 \times 12} = \mathbf{15.49\ V\ (r.m.s.)}$$

Example 9.18. *The lower cut-off frequency of an amplifier is 120 kHz and its upper cut-off frequency is 1 MHz. The peak-to-peak output voltage in the midband frequency range is 2.4 V p-p and the output power at 120 kHz is 0.4 W. Find (i) the bandwidth (ii) the r.m.s. output voltage at 1 MHz (iii) the output power in the midband frequency range and (iv) the output power at 1 MHz.*

Solution. f_1 = 120 kHz = 0.12 MHz ; f_2 = 1 MHz

(*i*) Bandwidth, $BW = f_2 - f_1 = 1 - 0.12 =$ **0.88 MHz**

(*ii*) The peak output voltage in the midband frequency range is $V_p = V_{pp}/2$ = 2.4 V/2 = 1.2 V. The r.m.s. output voltage in the midband frequency range is $V_{r.m.s.}$ = 0.707 × 1.2 V = 0.848 V. The r.m.s. output voltage at 1 MHz (= upper cut-off frequency f_2) is 70.7% of 0.848 V *i.e.*

R.M.S. output voltage at 1 MHz = 70.7% of 0.848V = 0.707 × 0.848 = **0.6 V**

(*iii*) The output power at 120 kHz (= lower cut-off frequency f_1) is 0.4 W (given). Therefore, output power in the midband frequency range = 2 × 0.4 = **0.8 W**.

(*iv*) The output power at 1MHz (= f_2) is the same as the output power at 120 kHz (= f_1). Therefore, output power at 1 MHz = **0.4 W**.

9.8 Capacitively Coupled Transistor Amplifier

This is the most popular type of coupling because it is cheap and provides excellent audio fidelity over a wide range of frequencies. It is usually employed for voltage amplification. Fig. 9.18 shows two stages of an *RC* coupled amplifier. A coupling capacitor C_C is used to connect the output of first stage to the base (*i.e. input*) of the second stage and so on. As the coupling from one stage to next is achieved by a coupling capacitor followed by a connection to a shunt resistor, therefore, such amplifiers are called *resistance - capacitance coupled amplifiers.*

The resistances R_1, R_2 and R_E form the biasing and stabilisation network. The emitter bypass capacitor offers low reactance path to the signal. Without it, the voltage gain of each stage would be lost. The coupling capacitor C_C transmits a.c. signal but blocks d.c. This prevents d.c. interference between various stages and the shifting of operating point.

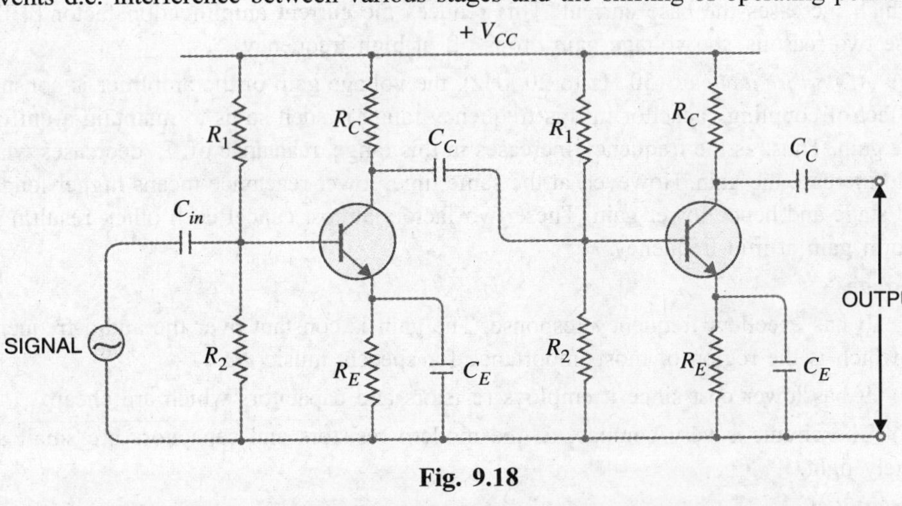

Fig. 9.18

Operation. When a.c. signal is applied to the base of the first transistor, it appears in the amplified form across its collector load R_C. The amplified signal developed across R_C is given to base of next stage through coupling capacitor C_C. The second stage does further amplification of the signal. In this way, the *cascaded* (one after another) stages amplify the signal and the overall gain is considerably increased.

It may be mentioned here that total gain is less than the product of the gains of individual stages. It is because when a second stage is made to follow the first stage, the *effective load resistance* of first stage is reduced due to the shunting effect of the input resistance of second stage. This reduces the gain of the stage which is loaded by the next stage. For instance, in a 3-stage amplifier, the gain of first and second stages will be reduced due to loading effect of next stage. However, the gain of the third stage which has no loading effect of subsequent stage, remains unchanged. The overall gain shall be equal to the product of the gains of three stages.

Frequency response. Fig. 9.19 shows the frequency response of a typical *RC* coupled amplifier. It is clear that voltage gain drops off at low (< 50 Hz) and high (> 20 kHz) frequencies whereas it is uniform over *mid-frequency range* (50 Hz to 20 kHz). This behaviour of the amplifier is briefly explained below :

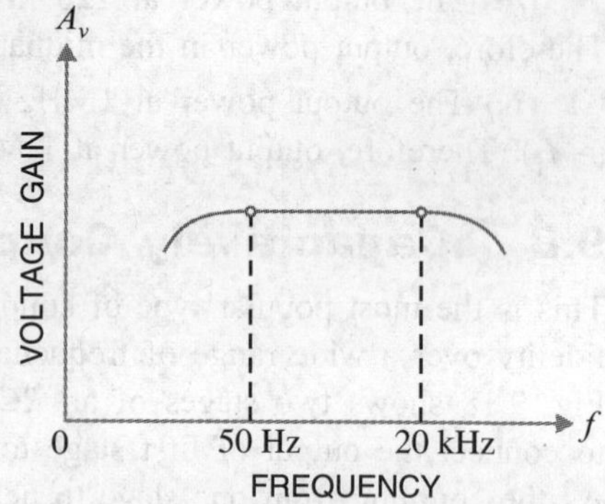

Fig. 9.19

(*i*) *At low frequencies* (< 50 Hz), the reactance of coupling capacitor C_C is quite high and hence very small part of signal will pass from one stage to the next stage. Moreover, C_E cannot shunt the emitter resistance R_E effectively because of its large reactance at low frequencies. These two factors cause a falling of voltage gain at low frequencies.

(*ii*) *At high frequencies* (> 20 kHz), the reactance of C_C is very small and it behaves as a short-circuit. This increases the loading effect of next stage and serves to reduce the voltage gain. Moreover, at high frequency, capacitive reactance of base-emitter junction is low which increases the base current. This reduces the current amplification factor β. Due to these two reasons, the voltage gain drops off at high frequency.

(*iii*) *At mid-frequencies* (50 Hz to 20 kHz), the voltage gain of the amplifier is constant. The effect of coupling capacitor in this frequency range is such so as to maintain a uniform voltage gain. Thus, as the frequency increases in this range, reactance of C_C decreases which tends to increase the gain. However, at the same time, lower reactance means higher loading of first stage and hence lower gain. These two factors almost cancel each other, resulting in a uniform gain at mid-frequency.

Advantages

(*i*) It has excellent frequency response. The gain is constant over the audio frequency range which is the region of most importance for speech, music *etc.*

(*ii*) It has lower cost since it employs resistors and capacitors which are cheap.

(*iii*) The circuit is very compact as the modern resistors and capacitors are small and extremely light.

Disadvantages

(*i*) The *RC* coupled amplifiers have low voltage and power gain. It is because the low resistance presented by the input of each stage to the preceding stage decreases the effective load resistance (R_{AC}) and hence the gain.

(*ii*) They have the tendency to become noisy with age, particularly in moist climates.

(*iii*) Impedance matching is poor. It is because the output impedance of *RC* coupled amplifier is several hundred ohms whereas the input impedance of a speaker is only a few ohms. Hence, little power will be transferred to the speaker.

Applications.

The *RC* coupled amplifiers have excellent audio fidelity over a wide range of frequency. Therefore, they are widely used as voltage amplifiers *e.g.* in the initial stages of public address system. If other type of coupling (*e.g.* transformer coupling) is employed in the initial stages, this results in frequency distortion which may be amplified in next stages. However, because of poor impedance matching, *RC* coupling is rarely used in the final stages.

Note. When there is an even number of cascaded stages (2, 4, 6 *etc.*), the output signal is not inverted from the input. When the number of stages is odd (1, 3, 5 *etc.*), the output signal is inverted from the input.

Example 9.19. *A single stage amplifier has a voltage gain of 60. The collector load R_C = 500 Ω and the input impedance is 1 kΩ. Calculate the overall gain when two such stages are cascaded through R-C coupling. Comment on the result.*

Solution. The gain of second stage remains 60 because it has no loading effect of any stage. However, the gain of first stage is less than 60 due to the loading effect of the input impedance of second stage.

$$\therefore \quad \text{Gain of second stage} = 60$$

$$\text{Effective load of first stage} = R_C \parallel R_{in} = \frac{500 \times 1000}{500 + 1000} = 333\ \Omega$$

$$\text{Gain of first stage} = 60 \times 333/500 = 39.96$$

$$\text{Total gain} = 60 \times 39.96 = \mathbf{2397}$$

Comments. The gain of individual stage is 60. But when two stages are coupled, the gain is *not* 60 × 60 = 3600 as might be expected rather it is less and is equal to 2397 in this case. It is because the first stage has a loading effect of the input impedance of second stage and consequently its gain is reduced. However, the second stage has no loading effect of any subsequent stage. Hence, the gain of second stage remains 60.

Example 9.20. *Fig. 9.20 shows two-stage RC coupled amplifier. If the input resistance R_{in} of each stage is 1 kΩ, find : (i) voltage gain of first stage (ii) voltage gain of second stage (iii) total voltage gain.*

Solution. $R_{in} = 1\ \text{k}\Omega\ ;\quad \beta = 100\ ;\quad R_C = 2\ \text{k}\Omega$

(*i*) The first stage has a loading of input resistance of second stage.

$$\therefore \quad \text{Effective load of first stage, } R_{AC} = R_C \parallel R_{in} = \frac{2 \times 1}{2 + 1} = 0.66\ \text{k}\Omega$$

$$\therefore \quad \text{Voltage gain of first stage} = \beta \times R_{AC} / R_{in} = 100 \times 0.66 /1 = \mathbf{66}$$

(*ii*) The collector of the second stage sees a load of only R_C (= 2 kΩ) as there is no loading effect of any subsequent stage.

$$\therefore \quad \text{Voltage gain of second stage} = \beta \times R_C / R_{in} = 100 \times 2/1 = \mathbf{200}$$

$$(iii) \quad \text{Total voltage gain} = 66 \times 200 = \mathbf{13200}$$

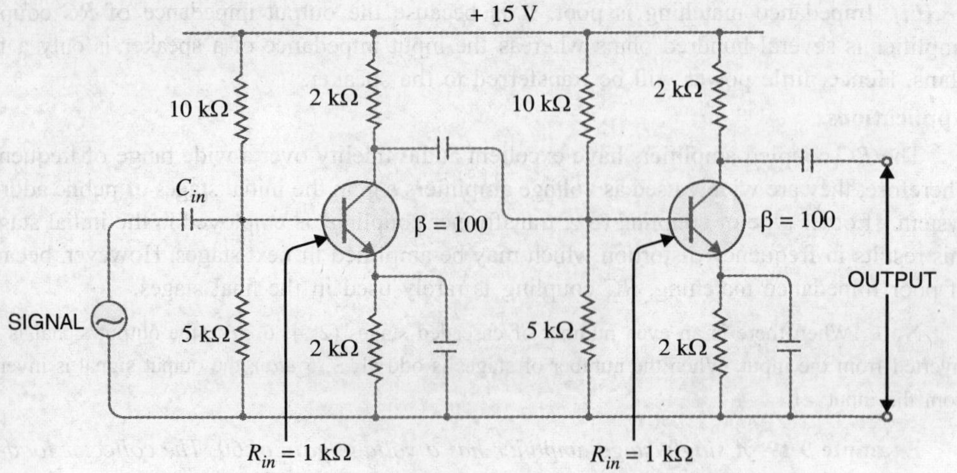

Fig. 9.20

Example 9.21. *A single stage amplifier has collector load R_C = 10 kΩ ; input resistance R_{in} = 1 kΩ and β = 100. If load R_L = 100 Ω, find the voltage gain. Comment on the result.*

Solution. Effective collector load, $R_{AC} = R_C \parallel R_L = 10 \text{ k}\Omega \parallel 100 \ \Omega = {}^*100 \ \Omega$

$$\therefore \quad \text{Voltage gain} = \beta \times \frac{R_{AC}}{R_{in}} = 100 \times \frac{100}{1000} = \mathbf{10}$$

Comments. As the load (*e.g.* speaker) is only of 100 ohms, therefore, effective load of the amplifier is too much reduced. Consequently, voltage gain is quite small. Under such situations, we can use a *transformer* to improve the voltage gain and signal handling capability. For example, if the output to 100 Ω load is delivered through a step-down transformer, the effective collector load and hence voltage gain can be increased.

Example 9.22. *Fig. 9.21 shows a 2-stage RC coupled amplifier. What is the biasing potential for the second stage? If the coupling capacitor C_C is replaced by a wire, what would happen to the circuit?*

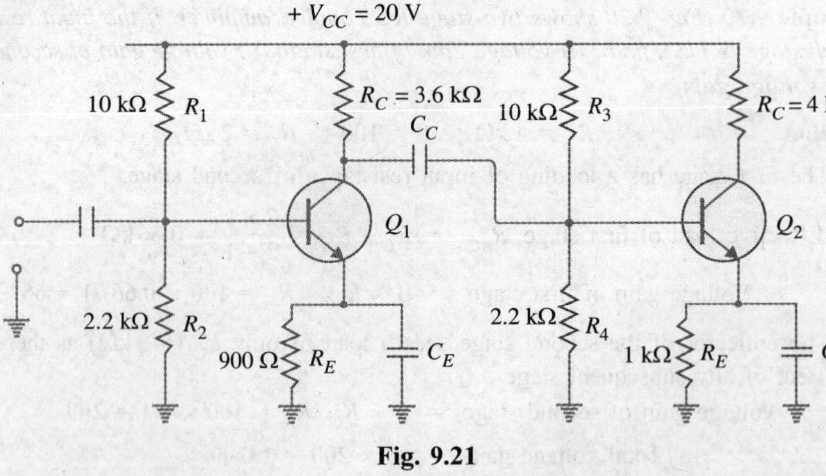

Fig. 9.21

* 10 kΩ || 100 Ω is essentially 100 Ω.

Solution. Referring to Fig. 9.21, we have,

$$\text{Voltage across } R_4,\ V_B = \frac{V_{CC}}{R_3 + R_4} \times R_4 = \frac{20}{10 + 2.2} \times 2.2 = \mathbf{3.6\ V}$$

Thus biasing potential for the second stage is 3.6 V.

When the coupling capacitor C_C is replaced by a wire, this changes the entire picture. It is because now R_C of the first stage is in parallel with R_3 of the second stage as shown in Fig. 9.22 (*i*). The total resistance of R_C (= 3.6 kΩ) and R_3 (= 10 kΩ) is given by ;

$$R_{eq} = \frac{R_3\, R_C}{R_3 + R_C} = \frac{10 \times 3.6}{10 + 3.6} = 2.65\ \text{k}\Omega$$

The circuit shown in Fig. 9.22 (*i*) then reduces to the one shown in Fig. 9.22 (*ii*). Referring to Fig. 9.22 (*ii*), we have,

$$\text{Voltage across } R_4,\ V_B = \frac{V_{CC}}{R_{eq} + R_4} \times R_4 = \frac{20}{2.65 + 2.2} \times 2.2 = 9.07\ \text{V}$$

Thus the biasing potential of second stage is drastically changed. The 9.07 V at the base of Q_2 would undoubtedly cause the transistor to saturate and the device would be rendered useless as an amplifier. This example explains the importance of d.c. isolation in a multistage amplifier. The use of coupling capacitor allows each amplifier stage to maintain its independent biasing potential while allowing the a.c. output from one stage to pass on to the next stage.

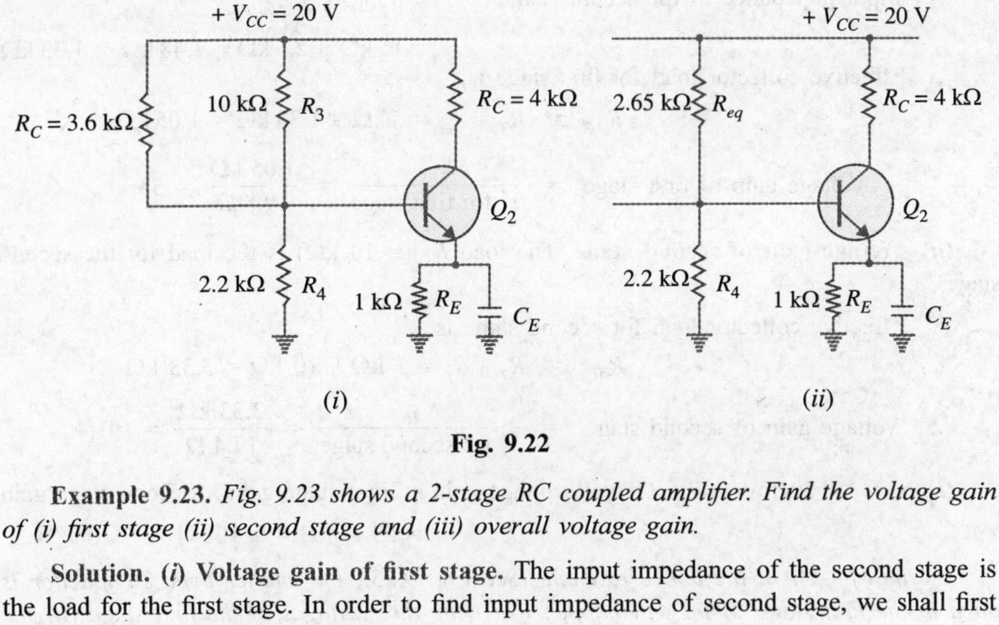

Fig. 9.22

Example 9.23. *Fig. 9.23 shows a 2-stage RC coupled amplifier. Find the voltage gain of (i) first stage (ii) second stage and (iii) overall voltage gain.*

Solution. (*i*) Voltage gain of first stage. The input impedance of the second stage is the load for the first stage. In order to find input impedance of second stage, we shall first find r'_e (ac emitter resistance) for the second stage.

$$\text{Voltage across } R_6 = \frac{V_{CC}}{R_5 + R_6} \times R_6 = \frac{15}{15 + 2.5} \times 2.5 = 2.14\ \text{V}$$

$$\text{Voltage across } R_8 = 2.14 - 0.7 = 1.44\ \text{V}$$

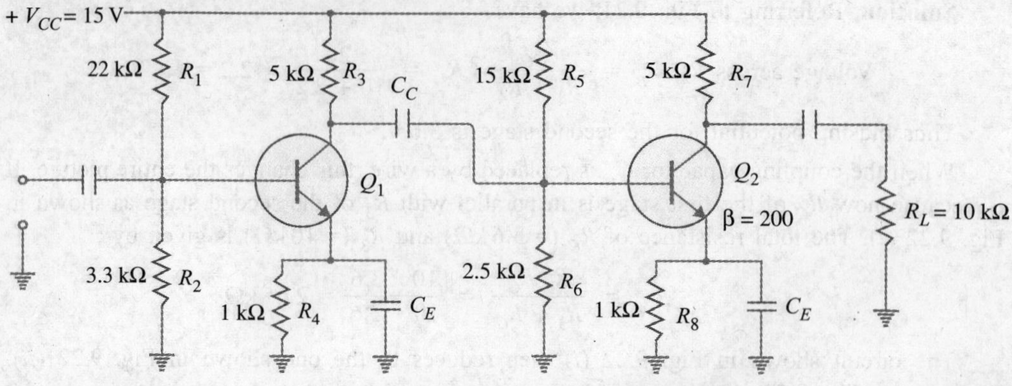

Fig. 9.23

$$\text{Emitter current in } R_8,\ I_E = \frac{1.44\text{ V}}{R_8} = \frac{1.44\text{ V}}{1\text{ k}\Omega} = 1.44\text{ mA}$$

$$r'_e \text{ for second stage} = \frac{25\text{ mV}}{I_E} = \frac{25\text{ mV}}{1.44\text{ mA}} = 17.4\ \Omega$$

Similarly, it can be shown that r'_e for the first stage is 19.8 Ω.

$$Z_{in(base)} \text{ for second stage} = \beta \times r'_e \text{ for second stage} = 200 \times (17.4\ \Omega) = 3.48\text{ k}\Omega$$

$$\text{Input impedance of the second stage, } Z_{in} = R_5 \parallel R_6 \parallel Z_{in(base)}$$

$$= 15\text{ k}\Omega \parallel 2.5\text{ k}\Omega \parallel 3.48\text{ k}\Omega = 1.33\text{ k}\Omega$$

∴ Effective collector load for first stage is

$$R_{AC} = R_3 \parallel Z_{in} = 5\text{ k}\Omega \parallel 1.33\text{ k}\Omega = 1.05\text{ k}\Omega$$

$$\text{Voltage gain of first stage} = \frac{R_{AC}}{r'_e \text{ for first stage}} = \frac{1.05\text{ k}\Omega}{19.8\ \Omega} = 53$$

(*ii*) **Voltage gain of second stage.** The load R_L (= 10 kΩ) is the load for the second stage.

∴ Effective collector load for second stage is

$$R_{AC} = R_7 \parallel R_L = 5\text{ k}\Omega \parallel 10\text{ k}\Omega = 3.33\text{ k}\Omega$$

$$\therefore \text{ Voltage gain of second stage} = \frac{R_{AC}}{r'_e \text{ for second stage}} = \frac{3.33\text{ k}\Omega}{17.4\ \Omega} = \mathbf{191.4}$$

(*iii*) **Overall voltage gain.** Overall voltage gain = First stage gain × Second stage gain

$$= 53 \times 191.4 = \mathbf{10144}$$

Example 9.24. *In the above example (See Fig. 9.23), the emitter bypass capacitor is open in the first stage. What will happen to (i) the d.c. voltages of the first stage (ii) the a.c. input voltage of the second stage (iii) the final output voltage ?*

Solution.

(*i*) We know that a capacitor behaves as an "open" to d.c. Therefore, if the emitter bypass capacitor is open in the first stage, there will be **no effect on the d.c. voltages of the first stage.**

(*ii*) As the emitter bypass capacitor is open in the first stage, the emitter resistor R_4(= 1 kΩ) is no longer a.c. short. Therefore, total ac emitter resistance = $r'_e + R_4$ so that voltage gain of the first stage will be reduced. This means that a.c. output voltage of the first stage and, hence, **a.c. input voltage to the second stage will decrease.**

(*iii*) Since the a.c. input voltage to the second stage is reduced **the final a.c. output voltage will decrease.**

9.9 Transformer-Coupled Amplifier

The main reason for low voltage and power gain of *RC* coupled amplifier is that the effective load (R_{AC}) of each stage is *decreased due to the low resistance presented by the input of each stage to the preceding stage. If the effective load resistance of each stage could be increased, the voltage and power gain could be increased. This can be achieved by transformer coupling. By the use of **impedance-changing properties of transformer, the low resistance of a stage (or load) can be reflected as a high load resistance to the previous stage.

Transformer coupling is generally employed when the load is small. It is mostly used for power amplification. Fig. 9.24 shows two stages of transformer coupled amplifier. A coupling transformer is used to feed the output of one stage to the input of the next stage. The primary *P* of this transformer is made the collector load and its secondary *S* gives input to the next stage.

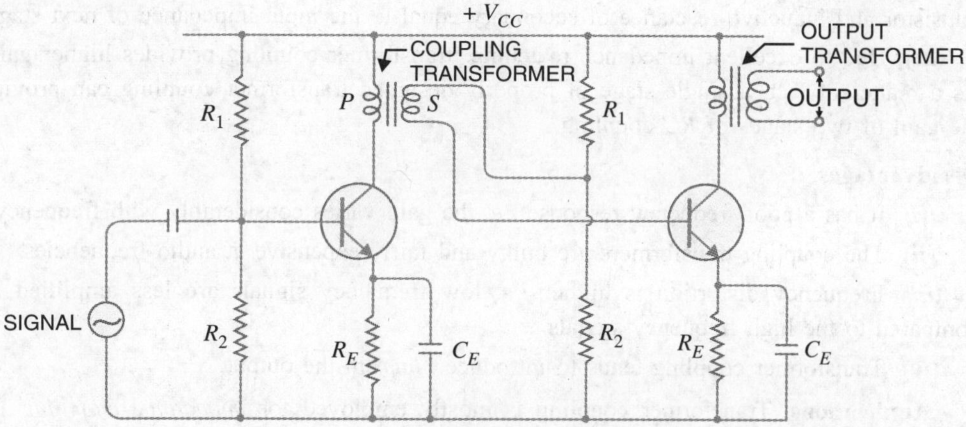

Fig. 9.24

Operation. When an a.c. signal is applied to the base of first transistor, it appears in the amplified form across primary *P* of the coupling transformer. The voltage developed across primary is transferred to the input of the next stage by the transformer secondary as shown in Fig. 9.24. The second stage renders amplification in an exactly similar manner.

Frequency response. The frequency response of a transformer coupled amplifier is shown in Fig. 9.25. It is clear that frequency response is rather poor *i.e.* gain is constant only over

* The input impedance of an amplifier is low while its output impedance is very high. When they are coupled to make a multistage amplifier, the high output impedance of one stage comes in parallel with the low input impedance of next state. Hence effective load (R_{AC}) is decreased.

** The resistance on the secondary side of a transformer reflected on the primary depends upon the turn ratio of the transformer.

a small range of frequencies. The output voltage is equal to the collector current multiplied by reactance of primary. At low frequencies, the reactance of primary begins to fall, resulting in decreased gain. At high frequencies, the capacitance between turns of windings acts as a bypass condenser to reduce the output voltage and hence gain. It follows, therefore, that there will be disproportionate amplification of frequencies in a complete signal such as music, speech *etc.* Hence, transformer-coupled amplifier introduces *frequency distortion.*

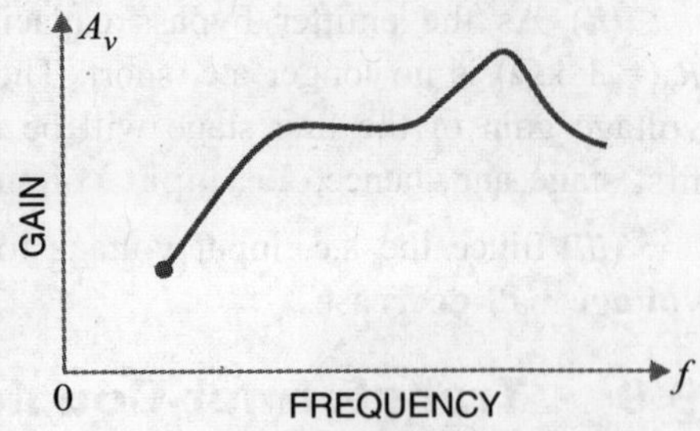

Fig. 9.25

It may be added here that in a properly designed transformer, it is possible to achieve a fairly constant gain over the audio frequency range. But a transformer that achieves a frequency response comparable to *RC* coupling may cost 10 to 20 times as much as the inexpensive *RC* coupled amplifier.

Advantages

(*i*) No signal power is lost in the collector or base resistors.

(*ii*) An excellent impedance matching can be achieved in a transformer coupled amplifier. It is easy to make the inductive reactance of primary equal to the output impedance of the transistor and inductive reactance of secondary equal to the input impedance of next stage.

(*iii*) Due to excellent impedance matching, transformer coupling provides higher gain. As a matter of fact, a single stage of properly designed transformer coupling can provide the gain of two stages of *RC* coupling.

Disadvantages

(*i*) It has a poor frequency response *i.e.* the gain varies considerably with frequency.

(*ii*) The coupling transformers are bulky and fairly expensive at audio frequencies.

(*iii*) Frequency distortion is higher *i.e.* low frequency signals are less amplified as compared to the high frequency signals.

(*iv*) Transformer coupling tends to introduce **hum* in the output.

Applications. Transformer coupling is mostly employed for *impedance matching*. In general, the last stage of a multistage amplifier is the *power stage*. Here, a concentrated effort is made to transfer maximum power to the output device *e.g.* a loudspeaker. For maximum power transfer, the impedance of power source should be equal to that of load. Usually, the impedance of an output device is a few ohms whereas the output impedance of transistor is several hundred times this value. In order to match the impedance, a step-down transformer of proper turn ratio is used. The impedance of secondary of the transformer is made equal to the load impedance and primary impedance equal to the output impedance of transistor. Fig. 9.26 illustrates the impedance matching by a step-down transformer. The output device (*e.g.* speaker) connected to the secondary has a small resistance R_L. The load R'_L appearing on the primary side will be :

* There are hundreds of turns of primary and secondary. These turns will multiply an induced e.m.f. from nearby power wiring. As the transformer is connected in the base circuit, therefore, the induced hum voltage will appear in amplified form in the output.

$$*R'_L = \left(\frac{N_P}{N_S}\right)^2 R_L$$

For instance, suppose the transformer has turn ratio $N_P : N_S :: 10 : 1$. If $R_L = 100\ \Omega$, then load appearing on the primary is :

$$R'_L = \left(\frac{10}{1}\right)^2 \times 100\ \Omega = 10\ \text{k}\Omega$$

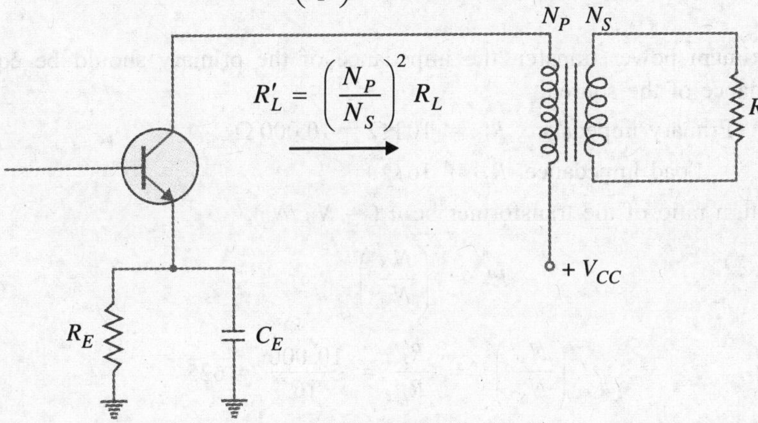

Fig. 9.26

Thus the load on the primary side is comparable to the output impedance of the transistor. This results in maximum power transfer from transistor to the primary of transformer. This shows that low value of load resistance (*e.g.* speaker) can be "stepped-up" to a more favourable value at the collector of transistor by using appropriate turn ratio.

Example 9.25. *A transformer coupling is used in the final stage of a multistage amplifier. If the output impedance of amplifier is 1 kΩ and the speaker has a resistance of 10 Ω, find the turn ratio of the transformer so that maximum power is transferred to the load.*

Solution.

For maximum power transfer, the impedance of the primary should be equal to the output impedance of transistor and impedance of secondary should be equal to load impedance *i.e.*

$$\text{Primary impedance} = 1\ \text{k}\Omega = 1000\ \Omega$$

Let the turn ratio of the transformer be $n\ (= N_P/N_S)$.

$$\text{Primary impedance} = \left(\frac{N_P}{N_S}\right)^2 \times \text{Load impedance}$$

$$\therefore \quad \left(\frac{N_P}{N_S}\right)^2 = \frac{\text{Primary impedance}}{\text{Load impedance}}$$

---

* Suppose primary and secondary of transformer carry currents I_P and I_S respectively. The secondary load R_L can be transferred to primary as R'_L provided the power loss remains the same *i.e.*

$$I_P^2 R'_L = I_S^2 R_L$$

or $$R'_L = \left(\frac{I_S}{I_P}\right)^2 \times R_L = \left(\frac{N_P}{N_S}\right)^2 \times R_L \qquad \left(\because \frac{I_S}{I_P} = \frac{N_P}{N_S}\right)$$

or $$n^2 = 1000/10 = 100$$

∴ $$n = \sqrt{100} = \mathbf{10}$$

A step-down transformer with turn ratio 10 : 1 is required.

Example 9.26. *Determine the necessary transformer turn ratio for transferring maximum power to a 16 Ω load from a source that has an output impedance of 10 kΩ. Also calculate the voltage across the external load if the terminal voltage of the source is 10 V r.m.s.*

Solution.

For maximum power transfer, the impedance of the primary should be equal to the output impedance of the source.

$$\text{Primary impedance, } R'_L = 10\text{ k}\Omega = 10{,}000\ \Omega$$

$$\text{Load impedance, } R_L = 16\ \Omega$$

Let the turn ratio of the transformer be $n\ (= N_P/N_S)$.

∴ $$R'_L = \left(\frac{N_P}{N_S}\right)^2 R_L$$

or $$\left(\frac{N_P}{N_S}\right)^2 = \frac{R'_L}{R_L} = \frac{10{,}000}{16} = 625$$

or $$n^2 = 625$$

or $$n = \sqrt{625} = \mathbf{25}$$

Now, $$\frac{V_S}{V_P} = \frac{N_S}{N_P}$$

∴ $$V_S = \left(\frac{N_S}{N_P}\right) \times V_P = \frac{1}{25} \times 10 = \mathbf{0.4\ V}$$

Example 9.27. *A transistor uses transformer coupling for amplification. The output impedance of amplifier is 10 kΩ while the input impedance of next stage is 2.5 kΩ. Determine the inductance of primary and secondary of the transformer for perfect impedance matching at a frequency of 200 Hz.*

Solution. $$\text{Frequency, } f = 200\text{Hz}$$

$$\text{Output impedance of transistor} = 10\text{ k}\Omega = 10^4\ \Omega$$

$$\text{Input impedance of next stage} = 2.5\text{ k}\Omega = 2.5 \times 10^3\ \Omega$$

Primary inductance. Consider the primary side of the transformer. For perfect impedance matching,

$$\text{Output impedance of transistor} = \text{Primary impedance}$$

or $$10^4 = 2\,\pi\, f\, L_P$$

∴ $$\text{Primary inductance, } L_P = \frac{10^4}{2\pi \times 200} = \mathbf{8\ H}$$

Secondary inductance. Consider the secondary side of transformer. For impedance matching,

$$\text{Input impedance of next stage} = \text{Impedance of secondary}$$

or $$2.5 \times 10^3 = 2\,\pi\, f\, L_S$$

$\therefore$ Secondary inductance, $L_S = \dfrac{2.5 \times 10^3}{2\pi \times 200} = \mathbf{2\ H}$

Example 9.28. *In the above example, find the number of primary and secondary turns. Given that core section of the transformer is such that 1 turn gives an inductance of 10 µH.*

Solution.

We know that inductance of a coil is directly proportional to the square of number of turns of the coil *i.e.*

$$L \propto N^2$$

or $$L = K N^2$$

Now, $$L = 10\ \mu\text{H} = 10^{-5}\ \text{H}\ ; \quad N = 1 \text{ turn}$$

$\therefore$ $$10^{-5} = K(1)^2$$

or $$K = 10^{-5}$$

$$\text{Primary inductance} = K N_P^2$$

or $$8 = 10^{-5}\ N_P^2$$

$\therefore$ Primary turns, $N_P = \sqrt{8 \times 10^5} = \mathbf{894}$

Similarly, Secondary turns, $N_S = \sqrt{2 \times 10^5} = \mathbf{447}$

Example 9.29. *The transformer coupling is used in the final stage of a multistage amplifier. A 3 Ω load is coupled to the amplifier through a step-down transformer. If the d.c. resistance of the primary is 300 Ω, what should be the turn ratio of the transformer so that maximum power is transferred to the load ? The output impedance of the amplifier is 3 kΩ.*

Solution. When no signal is applied, the primary has a resistance of 300 Ω as shown in Fig. 9.27 (*i*). When signal is applied, the load R_L(= 3 Ω) in the secondary appears as

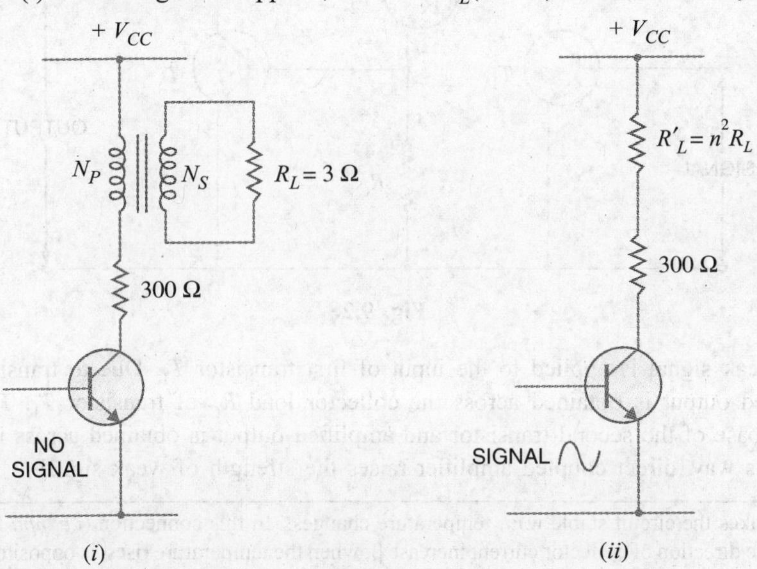

Fig. 9.27

$R'_L = n^2 R_L$ in the primary as shown in Fig. 9.27 (*ii*). Here $n = N_P/N_S$ is the turn ratio of the step-down transformer.

$$\therefore \quad \text{Total primary resistance} = 300 + n^2 R_L$$

In order to transfer maximum power to the load R_L, $300 + n^2 R_L$ should be equal to 3 kΩ (= 3 × 10^3 Ω).

$$\therefore \quad 300 + n^2 R_L = 3000$$

$$\text{or} \quad \text{Turn ratio, } n = \sqrt{\frac{3000 - 300}{R_L}} = \sqrt{\frac{2700}{3}} = \mathbf{30}$$

9.10 Direct-Coupled Amplifier

There are many applications in which extremely low frequency (< 10 Hz) signals are to be amplified *e.g.* amplifying photo-electric current, thermo-couple current *etc.* The coupling devices such as capacitors and transformers cannot be used because the electrical sizes of these components become very large at extremely low frequencies. Under such situations, one stage is *directly* connected to the next stage without any intervening coupling device. This type of coupling is known as *direct coupling*.

Circuit details. Fig. 9.28 shows the circuit of a three-stage direct-coupled amplifier. It uses *complementary transistors. Thus, the first stage uses *npn* transistor, the second stage uses *pnp* transistor and so on. This arrangement makes the design very simple. The output from the collector of first transistor T_1 is fed to the input of the second transistor T_2 and so on.

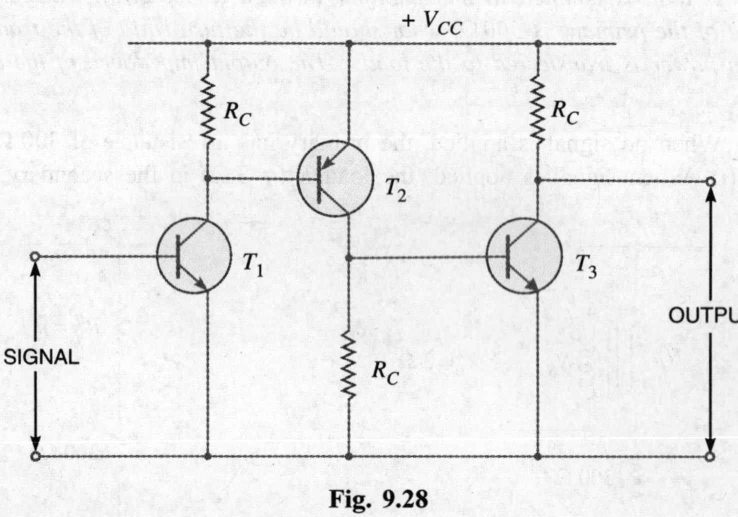

Fig. 9.28

The weak signal is applied to the input of first transistor T_1. Due to transistor action, an amplified output is obtained across the collector load R_C of transistor T_1. This voltage drives the base of the second transistor and amplified output is obtained across its collector load. In this way, direct coupled amplifier raises the strength of weak signal.

* This makes the circuit stable *w.r.t.* temperature changes. In this connection (*i.e. npn* followed by *pnp*), the direction of collector current increase β, when the temperature rises, is opposite for the two transistors. Thus the variation in one transistor tends to cancel that in the other.

Advantages

(*i*) The circuit arrangement is simple because of minimum use of resistors.

(*ii*) The circuit has low cost because of the absence of expensive coupling devices.

Disadvantages

(*i*) It cannot be used for amplifying high frequencies.

(*ii*) The operating point is shifted due to temperature variations.

Example 9.30. *Fig. 9.29 shows a direct coupled two-stage amplifier. Determine (i) d.c. voltages for both stages (ii) voltage gain of each stage and overall voltage gain.*

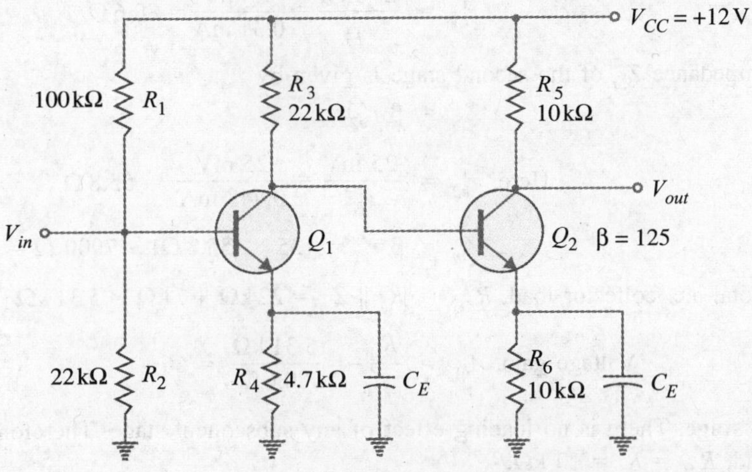

Fig. 9.29

Solution. Note that direct-coupled amplifier has no coupling capacitors between the stages.

(*i*) **D.C. voltages.** We shall now determine the d.c. voltages for both the stages following the established procedure.

First stage

$$\text{D.C. current thro' } R_1 \text{ and } R_2 = \frac{V_{CC}}{R_1 + R_2} = \frac{12 \text{ V}}{100 \text{ k}\Omega + 22 \text{ k}\Omega} = 0.098 \text{ mA}$$

$$\text{D.C. voltage across } R_2 = 0.098 \text{ mA} \times R_2 = 0.098 \text{ mA} \times 22 \text{ k}\Omega = \mathbf{2.16 \text{ V}}$$

This is the d.c. voltage at the base of transistor Q_1.

$$\text{D.C. voltage at the emitter, } V_{E1} = 2.16 - V_{BE} = 2.16 \text{ V} - 0.7 \text{ V} = \mathbf{1.46 \text{ V}}$$

$$\text{D.C. emitter current, } I_{E1} = \frac{V_{E1}}{R_4} = \frac{1.46 \text{ V}}{4.7 \text{ k}\Omega} = 0.31 \text{ mA}$$

$$\text{D.C. collector current, } I_{C1} = 0.31 \text{ mA} \qquad (\because I_{C1} \simeq I_{E1})$$

$$\text{D.C. voltage at collector, } V_{C1} = V_{CC} - I_{C1} R_3$$

$$= 12 \text{ V} - 0.31 \text{ mA} \times 22 \text{ k}\Omega = \mathbf{5.18 \text{ V}}$$

Second stage

$$\text{D.C. base voltage} = V_{C1} = \mathbf{5.18 \text{ V}}$$

D.C. emitter voltage, $V_{E2} = V_{C1} - V_{BE} = 5.18\text{ V} - 0.7\text{ V} = \mathbf{4.48\text{ V}}$

D.C. emitter current, $I_{E2} = \frac{V_{E2}}{R_6} = \frac{4.48\text{ V}}{10\text{ k}\Omega} = 0.448\text{ mA}$

D.C. voltage at collector, $V_{C2} = V_{CC} - I_{C2} R_5$ $\quad (\because I_{E2} \simeq I_{C2})$

$= 12\text{ V} - 0.448\text{ mA} \times 10\text{ k}\Omega = \mathbf{7.52\text{ V}}$

(ii) **Voltage gain.** To find voltage gain, we shall use the standard formula : total a.c. collector load divided by total ac emitter resistance.

First stage

$$r'_{e1} = \frac{25\text{ mV}}{I_{E1}} = \frac{25\text{ mV}}{0.31\text{ mA}} = 80.6\ \Omega$$

Input impedance Z_{in} of the second stage is given by ;

$$Z_{in} = \beta\, r'_{e2}$$

$$\text{Here, } r'_{e2} = \frac{25\text{ mV}}{I_{E2}} = \frac{25\text{ mV}}{0.448\text{ mA}} = 55.8\ \Omega$$

$$\therefore \quad Z_{in} = \beta\, r'_{e2} = 125 \times (55.8\ \Omega) \simeq 7000\ \Omega = 7\text{ k}\Omega$$

Total a.c. collector load, $R_{AC} = R_3 \parallel Z_{in} = 22\text{ k}\Omega \parallel 7\text{ k}\Omega = 5.31\text{ k}\Omega$

$$\therefore \quad \text{Voltage gain, } A_{v1} = \frac{R_{AC}}{r'_{e1}} = \frac{5.31\text{ k}\Omega}{80.6\ \Omega} = \mathbf{66}$$

Second stage. There is no loading effect of any subsequent stage. Therefore, total a.c. collector load, $R_{AC} = R_5 = 10\text{ k}\Omega$.

$$\text{Voltage gain, } A_{v2} = \frac{R_5}{r'_{e2}} = \frac{10\text{ k}\Omega}{55.8\ \Omega} = \mathbf{179}$$

$$\text{Overall voltage gain} = A_{v1} \times A_{v2} = 66 \times 179 = \mathbf{11{,}814}$$

Example 9.31. *For the two-stage direct coupled amplifier shown in Fig. 9.30, find (i) the approximate I_C of the transistor Q_1 (ii) the potential of collector of Q_1 w.r.t. ground (iii) the approximate I_C of transistor Q_2 and (iv) the approximate overall voltage gain of the circuit. The transistors are of silicon.*

Solution. For a silicon transistor, $V_{BE} = 0.7$ V.

(i) Input voltage $= V_{BE}$ + Voltage across 680 Ω

or $\quad 1.4 = 0.7 + I_E \times 680$

or $\quad I_E = \frac{(1.4 - 0.7)\text{ V}}{680\ \Omega} \simeq 1\text{ mA}$

$\therefore \quad I_C \simeq I_E = \mathbf{1\text{ mA}}$

(ii) Drop across 27 kΩ = 1 mA × 27 kΩ = 27 V

∴ Potential of collector of Q_1 *w.r.t.* ground = 30 − 27 = **3 V**

(iii) Now, the input to the transistor Q_2 is 3 V.

Now, $\quad 3 = V_{BE} + I_E \times 2400$

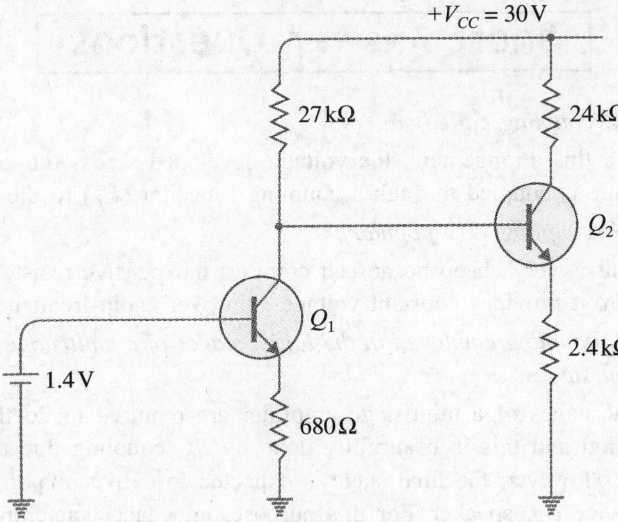

Fig. 9.30

or $$I_E = \frac{(3-0.7)\text{ V}}{2400\ \Omega} = 1\text{ mA}$$

∴ $$I_C \simeq I_E = \mathbf{1\ mA}$$

(iv) The approximate voltage gain of the two stages is given by ;

$$\text{First stage gain} = \frac{R_C}{R_E} = \frac{27\text{ k}\Omega}{680\ \Omega} = 40$$

$$\text{Second stage gain} = \frac{R_C}{R_E} = \frac{24\text{ k}\Omega}{2.4\text{ k}\Omega} = 10$$

Approximate overall voltage gain = 40 × 10 = **400**

9.11 Comparison of Different Types of Coupling

S. No.	Particular	*RC* coupling	Transformer coupling	Direct coupling
1.	*Frequency response*	Excellent in the audio frequency range	Poor	Best
2.	*Cost*	Less	More	Least
3.	*Space and weight*	Less	More	Least
4.	*Impedance matching*	Not good	Excellent	Good
5.	*Use*	For voltage amplification	For power amplification	For amplifiying extremely low frequencies

Short Answer Questions

Q.1. *Why is RC coupling called so?*

Ans. Because in this arrangement, the voltage developed across the collector load (R_C) in one stage is coupled through a coupling capacitor (C_C) to the next stage.

Q.2. *Why is RC coupling very popular?*

Ans. This circuit is very cheap because it employs inexpensive resistors and capacitors. In addition, it provides constant voltage gain over audio-frequency range.

Q.3. *Why is RC coupling employed in the initial stages of a multistage amplifier and not in the final stages?*

Ans. The initial stages of a multistage amplifier are required to do the job of voltage amplification and this is beautifully done by *RC* coupling due to least frequency distortion. However, the final stage is expected to deliver maximum power to the output device *e.g.* speaker. For this purpose, impedance matching is required – a quality which *RC* coupled amplifier does not possess because its output impedance is several hundred ohms compared to a few ohms impedance of the speaker.

Q.4. *Why is transformer coupling not used in the initial stages of a multi-stage amplifier?*

Ans. A transformer coupling has a high frequency distortion *i.e.* output varies considerably with signal frequency. If it is employed in the initial stages, the output may have intolerable frequency distortion due to amplification of distortion by various stages.

Q.5. *Why is transformer coupling invariably employed in the final stage of a multistage amplifier?*

Ans. Transformer coupling provides excellent impedance matching. Usually, the impedance of output device (*e.g.* speaker) is a few ohms whereas the output impedance of a transistor (*CE* arrangement) is several hundred times this value. In order to match impedance, a step-down transformer of proper turn ratio (N_P/N_S) is used. Impedance matching permits the transfer of maximum power to the output device which is really the goal in the final stage.

Q.6. *Why is the total gain of a multistage amplifier less than the product of individual stages?*

Ans. Because of loading effect of next stages. For example, consider two amplifiers each having a stage gain of 80.When these are coupled, the total gain will be less than $80 \times 80 = 6400$. It is because the gain of first stage will be less than 80 due to the loading effect of input impedance of second stage. However, the gain of second stage will remain 80 because it has no loading effect of any subsequent stage.

Q.7. *Why should the coupling capacitors in transistor amplifiers possess high values?*

Ans. A transistor amplifier (*CE* arrangement) has low input impedance of the order of 500 Ω or so. It is desired that reactance offered by the coupling capacitor should not be more than 20 Ω. To achieve this, large values of coupling capacitances must be used. Generally, the capacitance of coupling capacitors ranges from 1 μF to 10 μF.

Q.8. *Why can more gain be obtained by using transformer coupling instead of RC coupling between the stages?*

Ans. A *CE* amplifier has low input impedance but high output impedance (10 kΩ to 20 kΩ). Clearly, a direct connection (or through a coupling capacitor) between two stages will result in a significant loss in gain due to this mismatch. If we accept this reduced gain, it becomes necessary to use more stages in order to obtain the desired amplification. In order to match the higher output impedance of one stage to the lower input impedance of the following stage, a step-down transformer can be used. A 2-stage transformer coupled amplifier provides the same gain as a 3-stage *RC* coupled amplifier. It is because in *RC* coupling, there is mismatch between output impedance of one stage and input impedance of the following stage.

Q.9. *Which factor determines the low frequency response of RC or transformer coupling?*

Ans. The low frequency response is determined by the coupling device. In the low-frequency range, as the frequency decreases, the voltage gain also decreases. This is logical since the coupling elements (C_C or transformer) are designed to block d.c.; hence the closer the signal frequency approaches direct current, the more the gain will be decreased by the coupling element.

Q.10. *Which factor determines the high frequency response of RC or transformer coupling?*

Ans. The high frequency response is determined by the transistor itself. The transistor exhibits low inter-electrode capacitances which tend to decreases the gain.

Q.11. *Why is the frequency response of a multi-stage amplifier usually poorer than that of a single stage?*

Ans. Consider two stages, each having a mid-range gain of 100 and each having an upper cut-off frequency of 20 kHz when connected in cascade.

Mid-range gain of two stages = 100 × 100 = 10,000

Gain of two stages at 20 kHz

$$= (0.707 \times 100) \times (0.707 \times 100)$$

$$= (0.707)^2 \times 10^4 = \frac{1}{2} \times 10^4$$

Gain of single stage at 20 kHz = (0.707) (100)

It is clear that gain at 20 kHz in two-stage amplifier is one-half the mid-range gain. On the other hand, the gain of a single stage is down only to 0.707 of mid-range gain.

Multiple-Choice Questions

1. A radio receiver has .... of amplification.
   (*i*) one stage (*ii*) two stages
   (*iii*) three stages
   (*iv*) more than three stages
2. *RC* coupling is used for .... amplification.
   (*i*) voltage (*ii*) current
   (*iii*) power
   (*iv*) none of the above
3. In an *RC* coupled amplifier, the voltage gain over mid-frequency range ....
   (*i*) changes abruptly with frequency
   (*ii*) is constant
   (*iii*) changes uniformly with frequency
   (*iv*) is none of the above
4. In obtaining the frequency response curve of an amplifier, the ....
   (*i*) amplifier level output is kept constant

(*ii*) amplifier frequency is held constant
(*iii*) generator frequency is held constant
(*iv*) generator output level is held constant

5. An advantage of *RC* coupling scheme is the ....
(*i*) good impedance matching
(*ii*) economy
(*iii*) high efficiency
(*iv*) none of the above

6. The best frequency response is of .... coupling.
(*i*) *RC* (*ii*) transformer
(*iii*) direct
(*iv*) none of the above

7. Transformer coupling is used for .... amplification.
(*i*) power (*ii*) voltage
(*iii*) current
(*iv*) none of the above

8. In an *RC* coupling scheme, the coupling capacitor C_C must be large enough ....
(*i*) to pass d.c. between the stages
(*ii*) not to attenuate the low frequencies
(*iii*) to dissipate high power
(*iv*) none of the above

9. In *RC* coupling, the value of coupling capacitor is about ....
(*i*) 100 pF (*ii*) 0.1 μF
(*iii*) 0.01 μF (*iv*) 10 μF

10. The noise factor of an ideal amplifier expressed in dB is ....
(*i*) 0 (*ii*) 1
(*iii*) 0.1 (*iv*) 10

11. When a multistage amplifier is to amplify d.c. signal, then one must use .... coupling.
(*i*) *RC* (*ii*) transformer
(*iii*) direct
(*iv*) none of the above

12. .... coupling provides the maximum voltage gain.
(*i*) *RC* (*ii*) Transformer
(*iii*) Direct (*iv*) Impedance

13. In practice, voltage gain is expressed ....
(*i*) in dB (*ii*) in volts
(*iii*) as a number
(*iv*) none of the above

14. Transformer coupling provides high efficiency because ....
(*i*) collector voltage is stepped up
(*ii*) d.c. resistance is low
(*iii*) collector voltage is stepped down
(*iv*) none of the above

15. Transformer coupling is generally employed when load resistance is ....
(*i*) large (*ii*) very large
(*iii*) small
(*iv*) none of the above

16. If a three-stage amplifier has individual stage gains of 10 dB, 5 dB and 12 dB, then total gain in dB is ....
(*i*) 600 dB (*ii*) 24 dB
(*iii*) 14 dB (*iv*) 27 dB

17. The final stage of a multistage amplifier uses ....
(*i*) *RC* coupling
(*ii*) transformer coupling
(*iii*) direct coupling
(*iv*) impedance coupling

18. The ear is not sensitive to........
(*i*) frequency distortion
(*ii*) amplitude distortion
(*iii*) frequency as well as amplitude distortion
(*iv*) none of the above

19. *RC* coupling is not used to amplify extremely low frequencies because ....
(*i*) there is considerable power loss
(*ii*) there is hum in the output
(*iii*) electrical size of coupling capacitor becomes very large
(*iv*) none of the above

20. In transistor amplifiers, we use .... transformer for impedance matching.
    (*i*) step-up (*ii*) step-down
    (*iii*) same turn ratio
    (*iv*) none of the above
21. The lower and upper cut-off frequencies are also called .... frequencies.
    (*i*) sideband (*ii*) resonant
    (*iii*) half-resonant
    (*iv*) half-power
22. A gain of 1,000,000 times in power is expressed by ....
    (*i*) 30 dB (*ii*) 60 dB
    (*iii*) 120 dB (*iv*) 600 dB
23. A gain of 1000 times in voltage is expressed by ....
    (*i*) 60 dB (*ii*) 30 dB
    (*iii*) 120 dB (*iv*) 600 dB
24. 1 dB corresponds to .... change in power level.
    (*i*) 50% (*ii*) 35%
    (*iii*) 26% (*iv*) 22%
25. 1 dB corresponds to .... change in voltage or current level.
    (*i*) 40% (*ii*) 80%
    (*iii*) 20% (*iv*) 25%
26. The frequency response of transformer coupling is ....
    (*i*) good (*ii*) very good
    (*iii*) excellent (*iv*) poor
27. In the initial stages of a multistage amplifier, we use ....
    (*i*) *RC* coupling
    (*ii*) transformer coupling
    (*iii*) direct coupling
    (*iv*) none of the above
28. The total gain of a multistage amplifier is less than the product of the gains of individual stages due to ......
    (*i*) power loss in the coupling device
    (*ii*) loading effect of next stage
    (*iii*) the use of many transistors
    (*iv*) the use of many capacitors
29. The gain of an amplifier is expressed in dB because ....
    (*i*) it is a simple unit
    (*ii*) calculations become easy
    (*iii*) human ear response is logarithmic
    (*iv*) none of the above
30. If the power level of an amplifier reduces to half, the dB gain will fall by ....
    (*i*) 0.5 dB (*ii*) 2 dB
    (*iii*) 10 dB (*iv*) 3 dB
31. A current amplification of 2000 is a gain of ....
    (*i*) 3 dB (*ii*) 66 dB
    (*iii*) 20 dB (*iv*) 200 dB
32. An amplifier receives 0.1W of input signal and delivers 15W of signal power. What is the power gain in dB?
    (*i*) 21.8 dB (*ii*) 14.6 dB
    (*iii*) 9.5 dB (*iv*) 17.4 dB
33. The power output of an audio system is 18 W. For a person to notice an increase in the output (loudness or sound intensity) of the system, what must the output power be increased to?
    (*i*) 14.2 W (*ii*) 11.6 W
    (*iii*) 22.68 W
    (*iv*) none of the above
34. The output of a microphone is rated at − 52 dB. The reference level is 1 V under specified sound conditions. What is the output voltage of this microphone under the same sound conditions?
    (*i*) 1.5 mV (*ii*) 6.2 mV
    (*iii*) 3.8 mV (*iv*) 2.5 mV
35. *RC* coupling is generally confined to low power applications because of ....
    (*i*) large value of coupling capacitor
    (*ii*) low efficiency
    (*iii*) large number of components
    (*iv*) none of the above
36. The number of stages that can be directly coupled is limited because ....

(*i*) changes in temperature cause thermal instability
(*ii*) circuit becomes heavy and costly
(*iii*) it becomes difficult to bias the circuit
(*iv*) none of the above

37. The purpose of *RC* or transformer coupling is to ....
(*i*) block a.c.
(*ii*) separate bias of one stage from another
(*iii*) increase thermal stability
(*iv*) none of the above

38. The upper or lower cut-off frequency is also called .... frequency.
(*i*) resonant (*ii*) sideband
(*iii*) 3 dB
(*iv*) none of the above

39. The bandwidth of a single stage amplifier is .... that of a multistage amplifier.
(*i*) more than (*ii*) the same as
(*iii*) less than (*iv*) data insufficient

40. The value of emitter capacitor C_E in a multistage amplifier is about ....
(*i*) 0.1 μF (*ii*) 100 pF
(*iii*) 0.01 μF (*iv*) 50 μF

Answers to Multiple-Choice Questions

1. (*iv*)	2. (*i*)	3. (*ii*)	4. (*iv*)	5. (*ii*)
6. (*iii*)	7. (*i*)	8. (*ii*)	9. (*iv*)	10. (*i*)
11. (*iii*)	12. (*ii*)	13. (*i*)	14. (*ii*)	15. (*iii*)
16. (*iv*)	17. (*ii*)	18. (*i*)	19. (*iii*)	20. (*ii*)
21. (*iv*)	22. (*ii*)	23. (*i*)	24. (*iii*)	25. (*i*)
26. (*iv*)	27. (*i*)	28. (*ii*)	29. (*iii*)	30. (*iv*)
31. (*ii*)	32. (*i*)	33. (*iii*)	34. (*iv*)	35. (*ii*)
36. (*i*)	37. (*ii*)	38. (*iii*)	39. (*i*)	40. (*iv*)

Chapter Review Topics

1. What do you understand by multistage transistor amplifier? Mention its need.
2. Explain the following terms : (*i*) Frequency response (*ii*) Decibel gain (*iii*) Bandwidth.
3. Explain transistor *RC* coupled amplifier with special reference to frequency response, advantages, disadvantages and applications.
4. With a neat circuit diagram, explain the working of transformer-coupled transistor amplifier.
5. How will you achieve impedance matching with transformer coupling?
6. Explain direct coupled transistor amplifier.

Problems

1. The absolute voltage gain of an amplifier is 73. Find its decibel gain. **[37 dB]**
2. The input power to an amplifier is 15 mW while output power is 2 W. Find the decibel gain of the amplifier. **[21.25 dB]**
3. What is the dB gain for an increase of power level from 12 W to 24 W? **[3 dB]**
4. What is the dB gain for an increase of voltage from 4 mV to 8mV? **[6 dB]**
5. A two-stage amplifier has first-stage voltage gain of 20 and second stage voltage gain of 400. Find the total decibel gain. **[78 dB]**
6. A multistage amplifier consists of three stages: the voltage gain of stages are 60, 100 and 160. Calculate the overall gain in dB. **[119.64 dB]**

7. A multistage amplifier consists of three stages: the voltage gains of the stages are 30, 50 and 60. Calculate the overall gain in dB. **[99.1 dB]**
8. In an RC coupled amplifier, the mid-frequency gain is 2000. What will be its value at upper and lower cut-off frequencies? **[1414]**
9. A three-stage amplifier employs RC coupling. The voltage gain of each stage is 50 and R_C = 5 kΩ for each stage. If input impedance of each stage is 2 kΩ, find the overall decibel voltage gain. **[80 dB]**
10. We are to match a 16 Ω speaker load to an amplifier so that the effective load resistance is 10 kΩ. What should be the transformer turn ratio? **[25]**
11. Determine the necessary transformer turn ratio for transferring maximum power to a 50 ohm load from a source that has an output impedance of 5 kΩ. Also find the voltage across the external load if the terminal voltage of the source is 10 V r.m.s. **[10, 1 V]**
12. We are to match an 8 Ω speaker load to an amplifier so that the effective load resistance is 8 kΩ. What should be the transformer turn ratio? **[10]**

CHAPTER 10

Transistor Audio Power Amplifiers

Learning Objectives

- Understand the need of transistor audio power amplifier.
- Learn the difference between voltage and power amplifiers.
- Recognise the use of class *A*, class *B* and class *C* power amplifiers.
- Develop a strong understanding of a.c. operation of transistor amplifier.
- Derive an expression for maximum power efficiency of class *A* (series-fed) power amplifier.
- Derive an expression for maximum power efficiency of transformer coupled class *A* power amplifier.
- Explore the importance of heat sink for power transistors.
- Understand the importance and operation of push-pull amplifier.
- Derive an expression for maximum power efficiency of class *B* power amplifier.
- Understand complementary–symmetry amplifier.

Introduction

A practical amplifier always consists of a number of stages that amplify a weak signal until sufficient power is available to operate a loudspeaker or other output device. The first few stages in this multistage amplifier have the function of only voltage amplification. However, the last stage is designed to provide maximum power. This final stage is known as *power stage*.

The term audio means the range of frequencies which our ears can hear. The range of human hearing extends from 20 Hz to 20 kHz. Therefore, audio amplifiers amplify electrical signals that have a frequency range corresponding to the range of human hearing *i.e.* 20 Hz to 20 kHz. Fig. 10.1 shows the block diagram of an audio amplifier. The early stages build up the voltage level of the signal while the last stage builds up power to a level sufficient to operate the loudspeaker. In this chapter, we shall talk about the final stage in a multistage amplifier—the power amplifier.

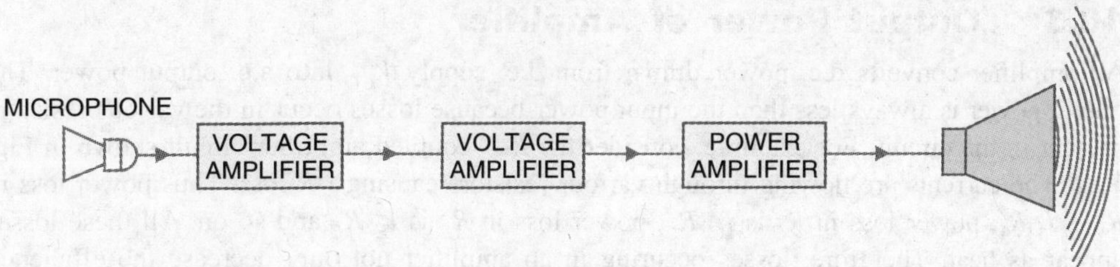

Fig. 10.1

10.1 Transistor Audio Power Amplifier

A transistor amplifier which raises the power level of the signals that have audio frequency range is known as **transistor audio power amplifier.**

In general, the last stage of a multistage amplifier is the *power stage*. The power amplifier differs from all the previous stages in that here a concentrated effort is made to obtain maximum output power. A transistor that is suitable for power amplification is generally called a *power transistor*. It differs from other transistors mostly in size ; it is considerably larger to provide for handling the great amount of power. Audio power amplifiers are used to deliver a large amount of power to a low resistance load. Typical load values range from 300 Ω (for transmission antennas) to 8 Ω (for loudspeakers). Although these load values do not cover every possibility, they do illustrate the fact that audio power amplifiers usually drive low-resistance loads. The typical power output rating of a power amplifier is 1W or more.

10.2 Small-Signal and Large-Signal Amplifiers

The input signal to a multistage amplifier is generally small (a few mV from a cassette or CD or a few μV from an antenna). Therefore, the first few stages of a multistage amplifier handle small signals and have the function of only voltage amplification. However, the last stage handles a large signal and its job is to produce a large amount of power in order to operate the output device (*e.g.* speaker).

(i) **Small-signal amplifiers.** Those amplifiers which handle small input a.c. signals (a few μV or a few mV) are called *small-signal amplifiers.* Voltage amplifiers generally fall in this class. The small-signal amplifiers are designed to operate over the linear portion of the output characteristics. Therefore, the transistor parameters such as current gain, input impedance, output impedance *etc.* do not change as the amplitude of the signal changes. Such amplifiers amplify the signal with little or no distortion.

(ii) **Large-signal amplifiers.** Those amplifiers which handle large input a.c. signals (a few volts) are called *large-signal amplifiers.* Power amplifiers fall in this class. The large-signal amplifiers are designed to provide a large amount of a.c. power output so that they can operate the output device *e.g.* a speaker. The main features of a large-signal amplifier or power amplifier are the circuit's power efficiency, the maximum amount of power that the circuit is capable of handling and the impedance matching to the output device. It may be noted that all large-signal amplifiers are not necessarily power amplifiers but it is safe to say that most are. In general, where amount of power involved is 1W or more, the amplifier is termed as *power amplifier.*

10.3 Output Power of Amplifier

An amplifier converts d.c. power drawn from d.c. supply V_{CC} into a.c. output power. The output power is always less than the input power because losses occur in the various resistors present in the circuit. For example, consider the *R-C* coupled amplifier circuit shown in Fig. 10.2. The currents are flowing through various resistors causing I^2R loss. Thus power loss in R_1 is $I_1^2 R_1$, power loss in R_C is $I_C^2 R_C$, power loss in R_E is $I_E^2 R_E$ and so on. All these losses appear as heat. Therefore, losses occuring in an amplifier not only decrease the efficiency but they also increase the temperature of the circuit.

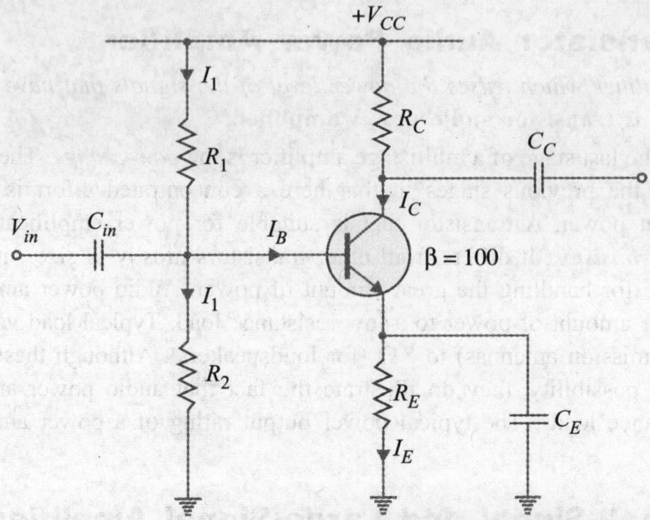

Fig. 10.2

When load R_L is connected to the amplifier, A.C. output power, $P_O = \dfrac{V_L^2}{R_L}$.

where V_L = R.M.S. value of load voltage

Example 10.1. *If in Fig. 10.2 ; R_1 = 10 kΩ ; R_2 = 2.2 kΩ ; R_C = 3.6 kΩ ; R_E = 1.1 kΩ and V_{CC} = + 10 V, find the d.c. power drawn from the supply by the amplifier.*

Solution. The current I_1 flowing through R_1 also flows through R_2 (a reasonable assumption because I_B is small).

$$\therefore \quad I_1 = \frac{V_{CC}}{R_1 + R_2} = \frac{10\text{ V}}{10\text{ k}\Omega + 2.2\text{ k}\Omega} = \frac{10\text{ V}}{12.2\text{ k}\Omega} = 0.82\text{ mA}$$

D.C. voltage across R_2, $V_2 = I_1 R_2 = 0.82\text{ mA} \times 2.2\text{ k}\Omega = 1.8\text{ V}$

D.C. voltage across R_E, $V_E = V_2 - V_{BE} = 1.8\text{ V} - 0.7\text{ V} = 1.1\text{ V}$

D.C. emitter current, $I_E = V_E/R_E = 1.1\text{ V}/1.1\text{ k}\Omega = 1\text{ mA}$

$\therefore \quad I_C \simeq I_E = 1\text{ mA}$

Total d.c. current I_T drawn from the supply is

$$I_T = I_C + I_1 = 1\text{ mA} + 0.82\text{ mA} = 1.82\text{ mA}$$

∴ D.C. power drawn from the supply is

$$P_{dc} = V_{CC}\, I_T = 10\text{ V} \times 1.82\text{ mA} = \mathbf{18.2\text{ mW}}$$

Example 10.2. *Determine the a.c. load power for the circuit shown in Fig. 10.3.*

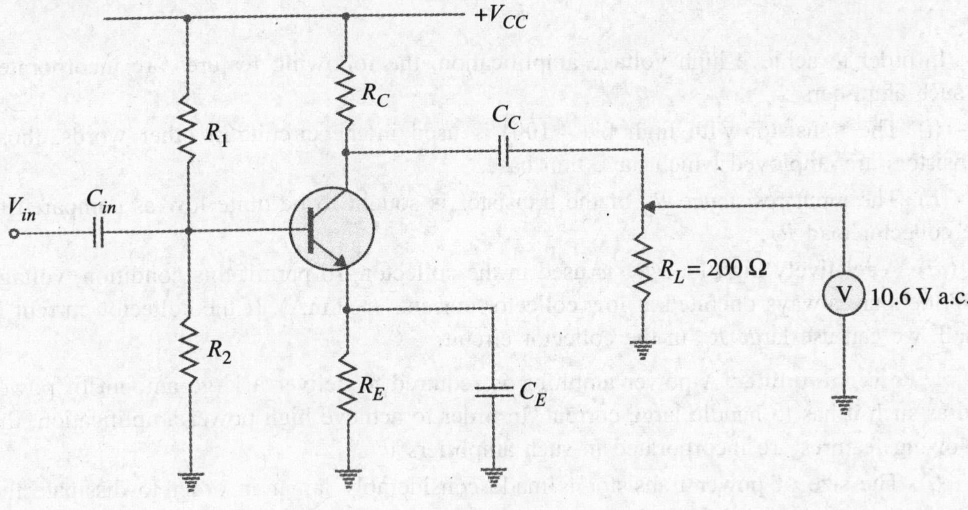

Fig. 10.3

Solution. The reading of a.c. voltmeter is 10.6 V. Since a.c. voltmeters read r.m.s. voltage, we have,

$$\text{A.C. output power, } P_o = \frac{V_L^2}{R_L} = \frac{(10.6)^2}{200\ \Omega} = \mathbf{561.8\ mW}$$

Example 10.3. *In an RC coupled power amplifier, the a.c. voltage across load R_L (= $100\,\Omega$) has a peak- to-peak value of 18 V. Find the maximum possible a.c. load power.*

Solution. The peak-to-peak voltage, V_{pp} = 18 V. Therefore, peak voltage (or maximum voltage) = $V_{pp}/2$ and the r.m.s. value, $V_L = V_{pp}/2\sqrt{2}$.

$$\therefore \quad P_{o\,(max)} = \frac{V_L^2}{R_L} = \frac{(V_{pp}/2\sqrt{2})^2}{R_L} = \frac{V_{pp}^2}{8\,R_L}$$

Here, V_{pp} = 18 V and $R_L = 100\ \Omega$

$$\therefore \quad P_{o\,(max)} = \frac{(18\text{ V})^2}{(8 \times 100)\ \Omega} = 405 \times 10^{-3}\text{ W} = \mathbf{405\ mW}$$

10.4 Difference Between Voltage and Power Amplifiers

The distinction between voltage and power amplifiers is somewhat artificial since useful power (*i.e.* product of voltage and current) is always developed in the load resistance through which current flows. The difference between the two types is really one of degree ; it is a question of how much voltage and how much power. A voltage amplifier is designed to achieve maximum voltage amplification. It is, however, not important to raise the power level. On the other hand, a power amplifier is designed to obtain maximum output power.

1. Voltage amplifier. The voltage gain of an amplifier is given by ;

$$A_v = \beta \times \frac{R_C}{R_{in}}$$

In order to achieve high voltage amplification, the following features are incorporated in such amplifiers :

(*i*) The transistor with high β (>100) is used in the circuit. In other words, those transistors are employed which have thin base.

(*ii*) The input resistance R_{in} of the transistor is sought to be quite low as compared to the collector load R_C.

(*iii*) A relatively high load R_C is used in the collector. To permit this condition, voltage amplifiers are always operated at low collector currents (≃ 1 mA). If the collector current is small, we can use large R_C in the collector circuit.

2. Power amplifier. A power amplifier is required to deliver a large amount of power and as such it has to handle large current. In order to achieve high power amplification, the following features are incorporated in such amplifiers :

(*i*) The size of power transistor is made considerably larger in order to dissipate the heat produced in the transistor during operation.

(*ii*) The base is made thicker to handle large currents. In other words, transistors with comparatively smaller β are used.

(*iii*) Transformer coupling is used for impedance matching.

The comparison between voltage and power amplifiers is given below in the tabular form :

S. No.	Particular	Voltage amplifier	Power amplifier
1.	β	High (> 100)	Low (5 to 20)
2.	R_C	High (4 – 10 kΩ)	Low (5 to 20 Ω)
3.	*Coupling*	Usually *R* – *C* coupling	Invariably transformer coupling
4.	*Input voltage*	Low (a few mV)	High (2 – 4 V)
5.	*Collector current*	Low (≃ 1 mA)	High (> 100 mA)
6.	*Power output*	Low	High
7.	*Output impedance*	High (≃ 12 kΩ)	Low (200 Ω)

Example 10.4. *A power amplifier operated from 12 V battery gives an output of 2 W. Find the maximum collector current in the circuit.*

Solution.

Let I_C be the maximum collector current.

Now, Power = Battery voltage × Collector current

or $2 = 12 \times I_C$

$\therefore \quad I_C = \frac{2}{12} = \frac{1}{6}\text{A} = \mathbf{166.7\ mA}$

This example shows that a power amplifier handles large power as well as large current.

Example 10.5. *A voltage amplifier operated from a 12 V battery has a collector load of 4 kΩ. Find the maximum collector current in the circuit.*

Solution.

The maximum collector current will flow when the whole battery voltage is dropped across R_C.

$$\therefore \quad \text{Max. collector current} = \frac{\text{Battery voltage}}{\text{Collector load}} = \frac{12\text{ V}}{4\text{ k}\Omega} = \mathbf{3\text{ mA}}$$

This example shows that a voltage amplifier handles small current.

Example 10.6. *A power amplifier supplies 50 W to an 8-ohm speaker. Find (i) a.c. output voltage (ii) a.c. output current.*

Solution.

(*i*) $$P = V^2/R$$

$$\therefore \quad \text{A.C. output voltage, } V = \sqrt{PR} = \sqrt{50 \times 8} = \mathbf{20\text{ V}}$$

(*ii*) $$\text{A.C. output current, } I = V/R = 20/8 = \mathbf{2.5\text{ A}}$$

10.5 Classification of Power Amplifiers

Transistor power amplifiers handle large signals. Many of them are driven so hard by the input large signal that collector current is either cut-off or is in the saturation region during a large portion of the input cycle. Therefore, such amplifiers are generally classified according to their mode of operation *i.e.* the portion of the input cycle during which the collector current is expected to flow. On this basis, they are classified as :

1. Class *A* power amplifier 2. Class *B* power amplifier 3. Class *C* power amplifier

1. **Class A power amplifier.** *If the collector current flows at all times during the full cycle of the signal, the power amplifier is known as* **class A power amplifier.**

Obviously, for this to happen, the power amplifier must be biased in such a way that no part of the signal is cut-off. Fig. 10.4 (*i*) shows circuit of class *A* power amplifier. Note that collector has a transformer as the load which is most common for all classes of power amplifiers. The use of transformer permits impedance matching, resulting in the transference of maximum power to the load *e.g.* loudspeaker.

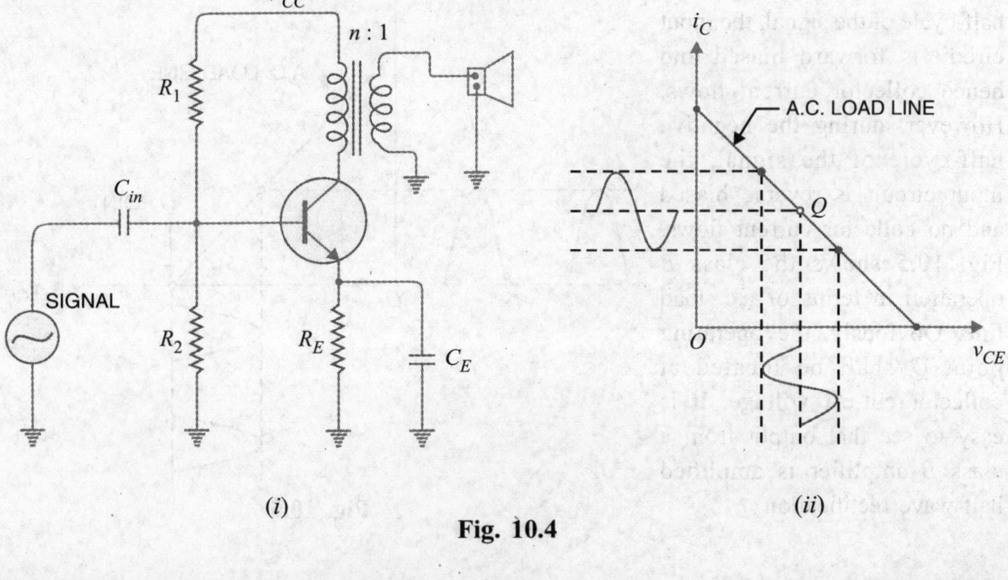

Fig. 10.4

Fig. 10.4 (*ii*) shows the class *A* operation in terms of a.c. load line. The operating point *Q* is so selected that collector current flows at all times throughout the full-cycle of the applied signal. As the output wave shape is exactly similar to the input wave shape, therefore, such amplifiers have least distortion. However, they have the disadvantage of low power output and low power efficiency (about 35%).

The following points may be noted about class *A* amplifiers :

(*i*) Any amplifier operated in class *A* operation conducts during the entire 360° of the input cycle. In other words, during the entire a.c. input cycle of the signal, the output of class *A* amplifier remains in the *active region.*

(*ii*) In a class *A* amplifier, the *Q* point is at or near the centre of the a.c. load line so that neither saturation nor cut-off is reached.

(*iii*) The primary advantage of class *A* amplifier is that there is practically no distortion of the input signal. It is because the output signal is an amplified replica of the input signal.

(*iv*) In a class *A* amplifier, we have to supply d.c. biasing collector current I_C during 360° of the input a.c. signal. This means that proper d.c. level has to be maintained during the entire time the transistor is conducting. Recall that an amplifier converts d.c. power from the d.c. supply (V_{CC}) into a.c. power for supplying to the load. Since considerable amount of d.c. power is used to maintain d.c. level, a class *A* amplifier has low a.c. output and poor efficiency.

(*v*) The poor or low efficiency of class *A* amplifier has another harmful effect. It increases the power dissipation by the transistor, thereby increasing the transistor temperature and reducing its life.

2. Class B power amplifier. *If the collector current flows only during the positive half-cycle of the input signal, it is called a* **class B power amplifier.**

In class *B* operation, the transistor bias is so adjusted that zero signal collector current is zero *i.e.* no biasing circuit is needed at all. During the positive half-cycle of the signal, the input circuit is forward biased and hence collector current flows. However, during the negative half-cycle of the signal, the input circuit is reverse biased and no collector current flows. Fig. 10.5 shows the class *B* operation in terms of a.c. load line. Obviously, the operating point *Q* shall be located at collector cut-off voltage. It is easy to see that output from a class *B* amplifier is amplified half-wave rectification.

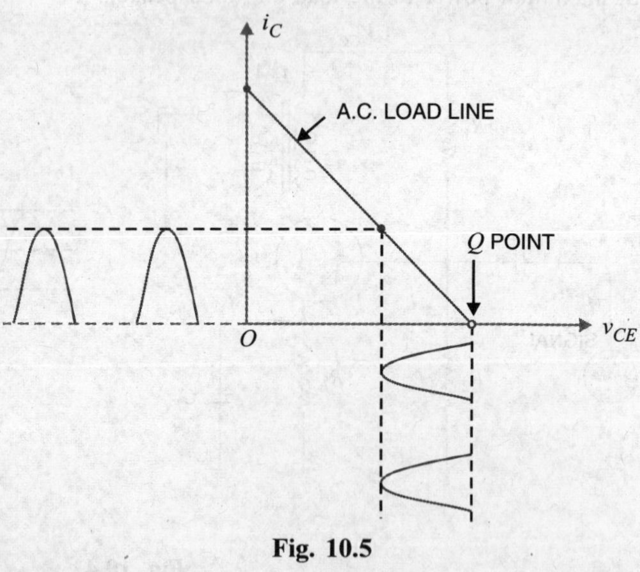

Fig. 10.5

In a class B amplifier, the negative half-cycle of the signal is cut-off and hence a severe distortion occurs. However, class B amplifiers provide higher power output and power efficiency (50 – 60%). Such amplifiers are mostly used for power amplification in push-pull arrangement. In such an arrangement, 2 transistors are used in class B operation. One transistor amplifies the positive half-cycle of the signal while the other amplifies the negative half-cycle.

The following points may be noted about class B amplifiers :

(*i*) A class B amplifier conducts only during 180° of the input cycle of the signal. In other words, transistor conducts current for only one-half of the input signal cycle.

(*ii*) In a class B amplifier, the Q point is at cut-off of the a.c. load line.

(*iii*) The main disadvantage of class B amplifier is that it produces severe distortion of the signal because half of the output signal is missing. To obtain output for the full-cycle of signal, it is necessary to use two transistors. Each transistor conducts on opposite half-cycles ; the combined operation provides a full output signal.

(*iv*) The principal advantage of class B amplifier is that the transistor dissipates power only when it is conducting, which is for 180° of the input signal. During the other half, the transistor is cut-off and there is no output. Therefore, a class B amplifier has higher a.c. output and higher efficiency.

3. **Class C power amplifier.** *If the collector current flows for less than half-cycle of the input signal, it is called* **class C power amplifier.**

In class C amplifier, the base is given some negative bias so that collector current does not flow just when the positive half-cycle of the signal starts. Such amplifiers are never used for power amplification. However, they are used as tuned amplifiers *i.e.* to amplify a narrow band of frequencies near the resonant frequency.

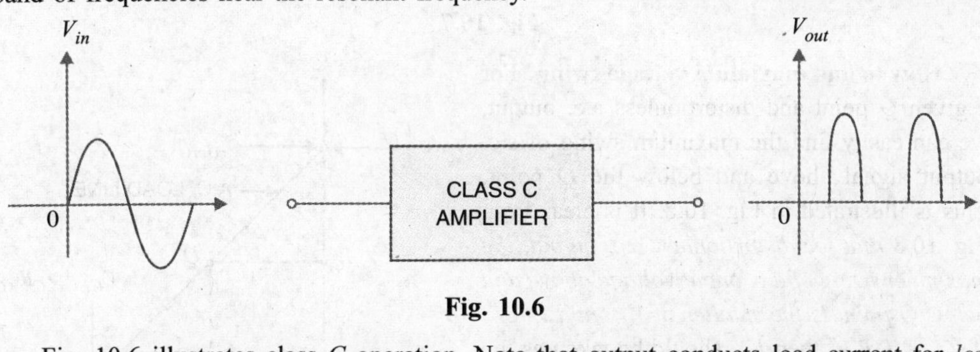

Fig. 10.6

Fig. 10.6 illustrates class C operation. Note that output conducts load current for *less* than half-cycle of the input cycle.

(*i*) The class C amplifier is biased to amplify less than 180° of the input signal.

(*ii*) The output of class C amplifier is highly distorted version of the input signal. Therefore, a class C amplifier cannot be used in applications requiring high fidelity such as an audio power amplifier.

(*iii*) Since the transistor is biased off during most of the input cycle, very little power is dissipated by the transistor. For this reason, a class C amplifier has very high efficiency (nearly 100%).

10.6 A.C. Operation of Transistor Amplifier

The a.c. operation of transistor amplifier can be beautifully explained in terms of a.c. load line. When a.c. signal is applied to the input of the transistor amplifier, the a.c. collector current (i_c) and a.c. collector-emitter voltage (v_{ce}) vary about the operating point Q as shown in Fig. 10.7. For *distortionless a.c. output, i_c can be allowed to vary about Q point between O and $i_{c(sat)}$ ($= I_C + V_{CE}/R_{AC}$) and v_{ce} can be allowed to vary about Q point between O and $v_{ce(off)}$ ($= V_{CE} + I_C R_{AC}$). In order to develop large amount of a.c. power in the load, it is desirable that peak-to-peak values of v_{ce} and i_c should be as large as possible and this depends upon the location of Q point on the a.c. load line [See Fig. 10.7].

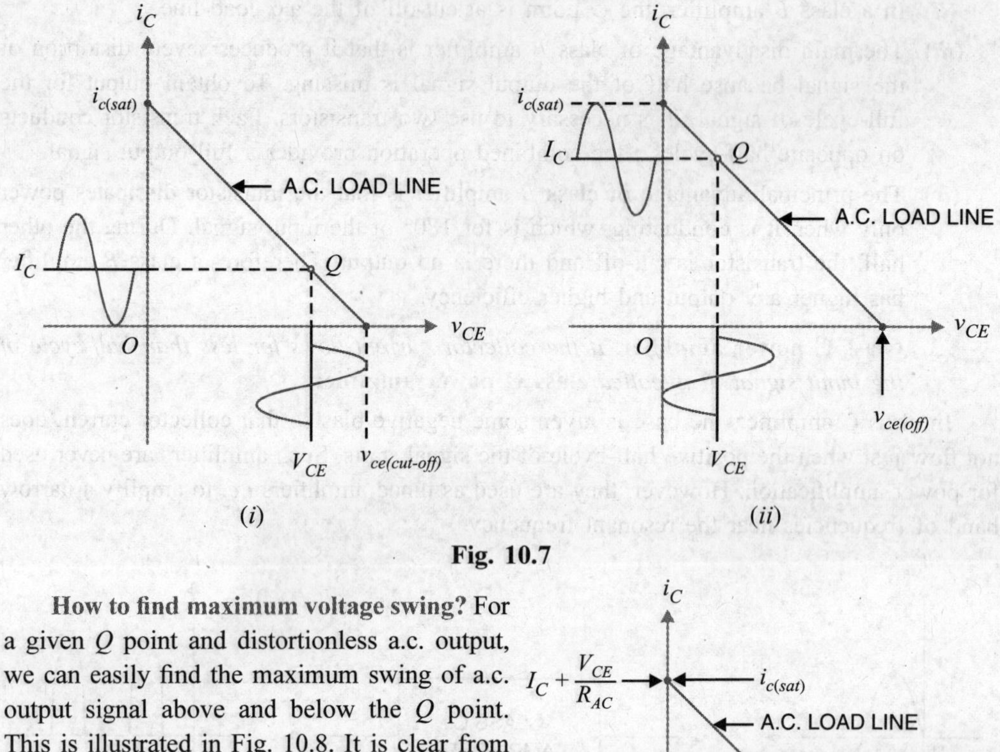

Fig. 10.7

How to find maximum voltage swing? For a given Q point and distortionless a.c. output, we can easily find the maximum swing of a.c. output signal above and below the Q point. This is illustrated in Fig. 10.8. It is clear from Fig. 10.8 *that for distortionless a.c. output, the maximum swing of a.c. ouput voltage above and below Q point is the smaller of* V_{CE} *and* $I_C R_{AC}$. If $V_{CE} < I_C R_{AC}$, then V_{CE} should be taken as the maximum swing from the Q point. If $I_C R_{AC} < V_{CE}$, then $I_C R_{AC}$ should be taken as the maximum swing from the Q point.

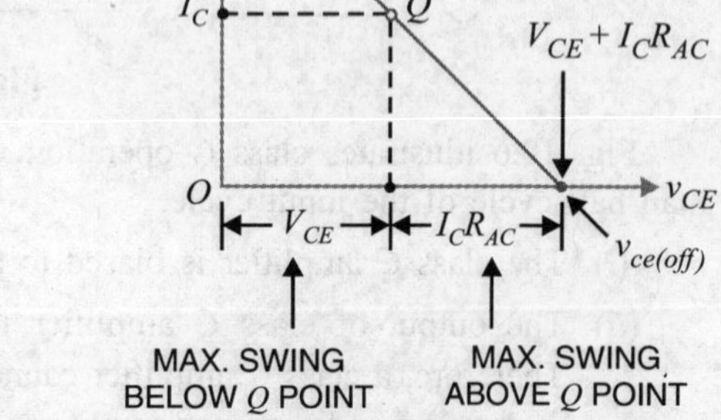

Fig. 10.8

Centred Q point. If Q point is at the centre of the a.c. load line (midway between saturation and cut-off), then we can obtain the *maximum*

* If i_c exceeds $i_{c(sat)}$ (saturation) or v_{ce} exceeds $v_{ce(off)}$ (cut-off), that part of a.c. output signal will be cut-off or clipped.

possible peak-to-peak a.c. output voltage swing. When this condition is met (*i.e.* Q point is at the centre of the a.c. load line), then,

$$V_{CE} = I_C R_{AC}$$

Now, $$V_{CE} = V_{CC} - I_C R_C$$

or $$I_C R_{AC} = V_{CC} - I_C R_C$$

$\therefore$ $$I_C = \frac{V_{CC}}{R_C + R_{AC}} \qquad \text{... (i)}$$

and $$V_{CE} = V_{CC} - I_C R_C \qquad \text{... (ii)}$$

When the Q point is centred on the a.c. load line, the Q point values of I_C and V_{CE} are given by eqs. (*i*) and (*ii*). If the Q point is not at the centre, it can be brought to the centre of the a.c. load line by changing I_C, until $V_{CE} = I_C R_{AC}$.

Example 10.7. *Find the maximum peak-to-peak a.c. output voltage that the amplifier in Fig. 10.9 can produce without causing clipping at the output.*

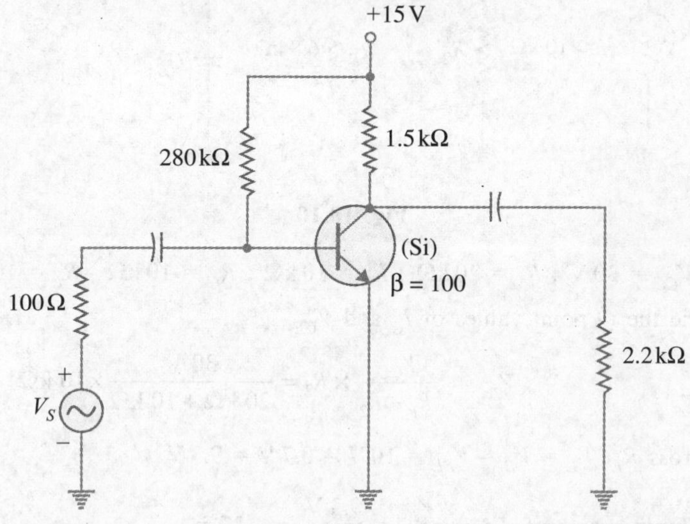

Fig. 10.9

Solution. We first find the Q point values of I_C and V_{CE}.

Now, Base current, $$I_B = \frac{V_{CC} - V_{BE}}{R_B} = \frac{(15 - 0.7)\text{ V}}{280\text{ k}\Omega} = 51.1\ \mu\text{A}$$

Collector current, $$I_C = \beta I_B = 100 \times 51.1\ \mu\text{A} = 5.11\text{ mA}$$

Collector-emitter voltage, $$V_{CE} = V_{CC} - I_C R_C$$

$$= 15\text{ V} - 5.11\text{ mA} \times 1.5\text{ k}\Omega = 7.34\text{ V}$$

A.C. load, $$R_{AC} = R_C \parallel R_L = 1.5\text{ k}\Omega \parallel 2.2\text{ k}\Omega = 892\ \Omega$$

$\therefore$ $$I_C R_{AC} = 5.11\text{ mA} \times 892\ \Omega = 4.56\text{ V}$$

The maximum value of peak a.c. output voltage is the *smaller* of V_{CE}(= 7.34 V) and $I_C R_{AC}$ (= 4.56 V). It is clear that $I_C R_{AC}$ is smaller than V_{CE}.

∴ Maximum peak-to-peak a.c. output voltage

$$= 2 \times I_C R_{AC} = 2 \times 4.56 \text{ V} = \mathbf{9.12 \text{ V (p-p)}}$$

Example 10.8. *For the transistor amplifier circuit shown in Fig. 10.10, find the maximum peak-to-peak a.c. output voltage that the amplifier can produce without distortion at the output.*

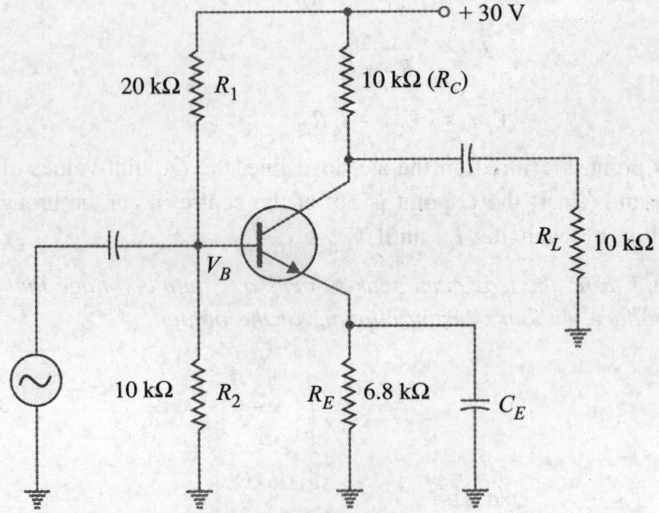

Fig. 10.10

Solution. V_{CC} = 30 V ; R_1 = 20 kΩ ; R_2 = 10 kΩ ; R_C = 10 kΩ ; R_L = 10 kΩ

We first find the Q point values of I_C and V_{CE}.

$$V_B = \frac{V_{CC}}{R_1 + R_2} \times R_2 = \frac{30 \text{ V}}{20 \text{ k}\Omega + 10 \text{ k}\Omega} \times 10 \text{ k}\Omega = 10 \text{ V}$$

Voltage across R_E, $V_E = V_B - V_{BE}$ = 10 V – 0.7 V = 9.3 V

$$\text{Emitter current, } I_E = \frac{V_E}{R_E} = \frac{9.3 \text{ V}}{6.8 \text{ k}\Omega} = 1.37 \text{ mA}$$

$$\text{Collector current, } I_C \simeq I_E = 1.37 \text{ mA}$$

$$\text{Collector-emitter voltage, } V_{CE} = V_{CC} - I_C(R_C + R_E)$$

$$= 30 \text{ V} - 1.37 \text{ mA} \times (10 \text{ k}\Omega + 6.8 \text{ k}\Omega) = 6.98 \text{ V}$$

$$\text{A.C. load, } R_{AC} = R_C \parallel R_L = 10 \text{ k}\Omega \parallel 10 \text{ k}\Omega = 5 \text{ k}\Omega$$

$$\therefore \quad I_C R_{AC} = 1.37 \text{ mA} \times 5 \text{ k}\Omega = 6.85 \text{ V}$$

The maximum value of peak a.c. output voltage is the *smaller* of V_{CE}(= 6.98 V) and $I_C R_{AC}$ (= 6.85 V). It is clear that $I_C R_{AC}$ is smaller than V_{CE}.

∴ Maximum peak-to-peak a.c. output voltage

$$= 2 \times I_C R_{AC} = 2 \times 6.85 \text{ V} = \mathbf{13.7 \text{ V(p – p)}}$$

Example 10.9. *For the transistor amplifier circuit shown in Fig. 10.11, what is the a.c. collector resistance? If the Q point is approximately at the centre of the a.c. load line, what is the a.c. saturation current?*

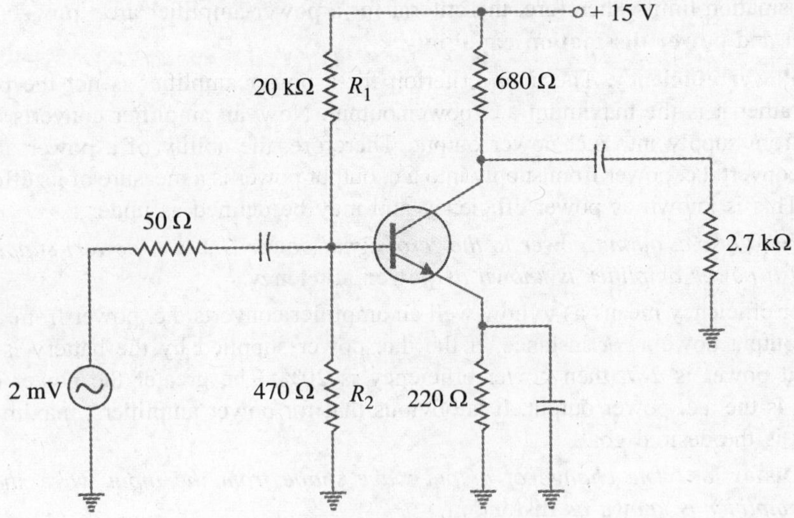

Fig. 10.11

Solution. The a.c. collector resistance R_{AC} is given by ;

$$R_{AC} = R_C \parallel R_L = 680\ \Omega \parallel 2.7\ \text{k}\Omega = \mathbf{543\ \Omega}$$

In order to find a.c. saturation current, we first find the Q point value of I_C. Since the Q point is located at the centre of the a.c. load line, a.c. saturation current is just double of this value (See Fig. 10.12) *i.e.*

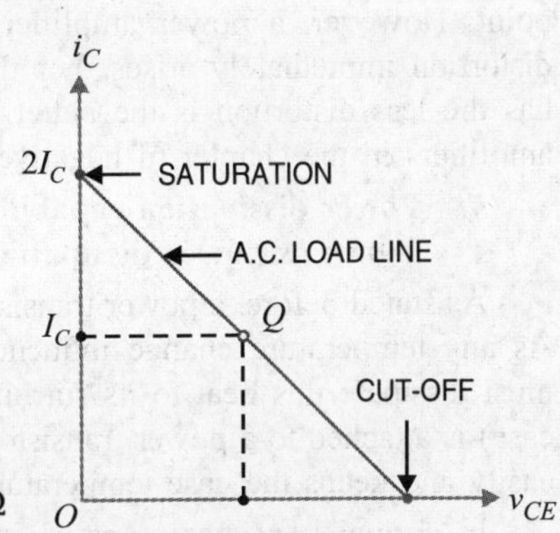

Fig. 10.12

A.C. saturation current $= 2I_C$

For finding I_C, we proceed as under :

$$\text{D.C. base voltage, } V_B = \frac{V_{CC}}{R_1 + R_2} \times R_2 = \frac{15\ \text{V}}{2\ \text{k}\Omega + 470\ \Omega} \times 470\ \Omega = 2.85\ \text{V}$$

$$\text{D.C. emitter voltage, } V_E = V_B - V_{BE} = 2.85\ \text{V} - 0.7\ \text{V} = 2.15\ \text{V}$$

$$\text{D.C. emitter current, } I_E = \frac{V_E}{R_E} = \frac{2.15\ \text{V}}{220\ \Omega} = 9.77\ \text{mA}$$

$$\text{Collector current, } I_C = I_E = 9.77\ \text{mA}$$

$$\therefore \text{ A.C. saturation current} = 2I_C = 2 \times 9.77\ \text{mA} = \mathbf{19.54\ mA}$$

10.7 Performance Quantities of Power Amplifiers

As mentioned previously, the prime objective for a power amplifier is to obtain maximum output power. Since a transistor, like any other electronic device has voltage, current and power dissipation limits, therefore, the criteria for a power amplifier are : **power efficiency, distortion** and **power dissipation capability.**

(*i*) **Power efficiency.** The main criterion for a power amplifier is not the power gain rather it is the maximum a.c. power output. Now, an amplifier converts d.c. power from supply into a.c. power output. Therefore, the ability of a power amplifier to convert d.c. power from supply into a.c. output power is a measure of its effectiveness. This is known as power efficiency and may be defined as under :

The ratio of a.c. output power to the zero signal power (i.e. d.c. power) supplied by the battery of a power amplifier is known as **power efficiency.**

Power efficiency means as to how well an amplifier converts d.c. power from the battery into a.c. output power. For instance, if the d.c. power supplied by the battery is 10W and a.c. output power is 2W, then power efficiency is 20%. The greater the power efficiency, the larger is the a.c. power output. It is obvious that for power amplifiers, maximum power efficiency is the desired goal.

(*ii*) **Distortion.** *The change of output wave shape from the input wave shape of an amplifier is known as* **distortion.**

A transistor like other electronic devices, is essentially a non-linear device. Therefore, whenever a signal is applied to the input of the transistor, the output signal is not exactly like the input signal *i.e.* distortion occurs. Distortion is not a problem for small signals (*i.e.* voltage amplifiers) since transistor is a linear device for small variations about the operating point. However, a power amplifier handles large signals and, therefore, the problem of distortion immediately arises. For the comparison of two power amplifiers, the one which has the less distortion is the better. We shall discuss the method of reducing distortion in amplifiers in the chapter of negative feedback in amplifiers.

(*iii*) **Power dissipation capability.** *The ability of a power transistor to dissipate heat is known as* **power dissipation capability.**

As stated before, a power transistor handles large currents and heats up during operation. As any temperature change influences the operation of transistor, therefore, the transistor must dissipate this heat to its surroundings. To achieve this, generally a *heat sink* (a metal case) is attached to a power transistor case. The increased surface area allows heat to escape easily and keeps the case temperature of the transistor within permissible limits.

In normal transistor operation, the collector-base junction is reverse biased and has a large voltage across it while the base-emitter junction has a small forward-biasing voltage. Consequently, *most of the heat generated in a transistor is produced at the collector-base junction. The total power P_D dissipated at the junctions of a transistor is

$$P_D = V_{CB}I_C + V_{BE}I_E = (V_{CB} + V_{BE})I_C \qquad (\because I_E \simeq I_C)$$

$$\therefore \quad P_D = V_{CE}I_C \qquad \text{... } (i)$$

Here, V_{CE} and I_C are the Q point values of the amplifier. Eq. (*i*) gives the total power dissipated by the transistor. If the power dissipated (thrown off) by the transistor exceeds the power rating of the transistor, the temperature of the transistor will increase. This may damage/destroy the transistor due to overheating.

* Since $V_{CB} >> V_{BE}$.

10.8 Expression for Power Efficiency

An amplifier receives d.c. power from the d.c. supply (V_{CC}) and converts it into a.c. power delivered to the load. Some of the input d.c. power is dissipated in the transistor and in other components. This is wasted power. The power efficiency (η) of a power amplifier indicates how good the amplifier is at converting the d.c. input power (P_{dc}) into the a.c. output power (P_o) delivered to the load *i.e.*

$$\text{Power efficiency, } \eta = \frac{\text{A.C. output power}}{\text{D.C. input power}} = \frac{P_o}{P_{dc}}$$

where $\quad {}^*P_{dc}$ = D.C. input power = $V_{CC} I_C$

P_o = A.C. output power = $V_{ce} I_c$

Here, V_{ce} is the r.m.s. value of signal output voltage and I_c is the r.m.s. value of output signal current. In terms of peak-to-peak values (which are often convenient values in load-line work), the a.c. power output can be expressed as :

$$^{**}P_o = [(0.5 \times 0.707) v_{ce(p-p)}]\,[(0.5 \times 0.707) i_{c(p-p)}]$$

$$= \frac{v_{ce(p-p)} \times i_{c(p-p)}}{8}$$

$$\therefore \quad \text{Power } \eta = \frac{v_{ce(p-p)} \times i_{c(p-p)}}{8 V_{CC}\, I_C}$$

It may be pointed here that while calculating the power efficiency of a power amplifier, we always ignore the small power consumed in the biasing resistors.

Alternate expressions for P_o. We can also express a.c. power delivered to load R_L (*i.e.* a.c. output power P_o) in the following alternate forms :

$$P_o = I_c^2 R_L = \frac{I_p^2 R_L}{2} = \frac{I_{p-p}^2 R_L}{8}$$

$$P_o = \frac{V_{ce}^2}{R_L} = \frac{V_p^2}{2R_L} = \frac{V_{p-p}^2}{8 R_L}$$

Here, the subscripts p and $p-p$ refer to peak and peak-to-peak respectively. Note that $I_c = I_p/\sqrt{2}$ and $V_{ce} = V_p/\sqrt{2}$.

Note. Another type of efficiency that is specified for power amplifiers is the collector efficiency (η_c). It is defined as under :

$$\text{Collector efficiency, } \eta_c = \frac{\text{A.C. output power}}{\text{Power dissipated at collector}} = \frac{P_o}{V_{CE} I_C}$$

10.9 Maximum Power Efficiency of Series-Fed Class A Amplifier

Fig. 10.13 (*i*) shows a ***series – fed class *A* amplifier. This circuit is seldom used for

---

* Note that d.c. input power to the collector circuit of a power amplifier is the product of collector supply V_{CC} (and not the collector-emitter voltage) and the average (*i.e.* d.c.) collector current I_C drawn from d.c. supply V_{CC}.

** $$\text{R.M.S. value} = \frac{1}{2}\left[\frac{\text{Peak-to-peak value}}{\sqrt{2}}\right]$$

$$= 0.5 \times 0.707 \times \text{Peak-to-peak value}$$

*** Note that the input to this circuit is a large signal and that transistor used is a power transistor.

power amplification due to its poor power efficiency. Nevertheless, it will help the reader to understand the class *A* operation. The d.c. load line of the circuit is shown in Fig. 10.13 (*ii*). When an a.c. signal is applied to the amplifier, the output current and voltage will vary about the operating point *Q*. In order to achieve the maximum symmetrical swing of current and voltage (to achieve maximum output power), the *Q* point should be located at the centre of the d.c. load line. In that case, operating point is $I_C = V_{CC}/2R_C$ and $V_{CE} = V_{CC}/2$.

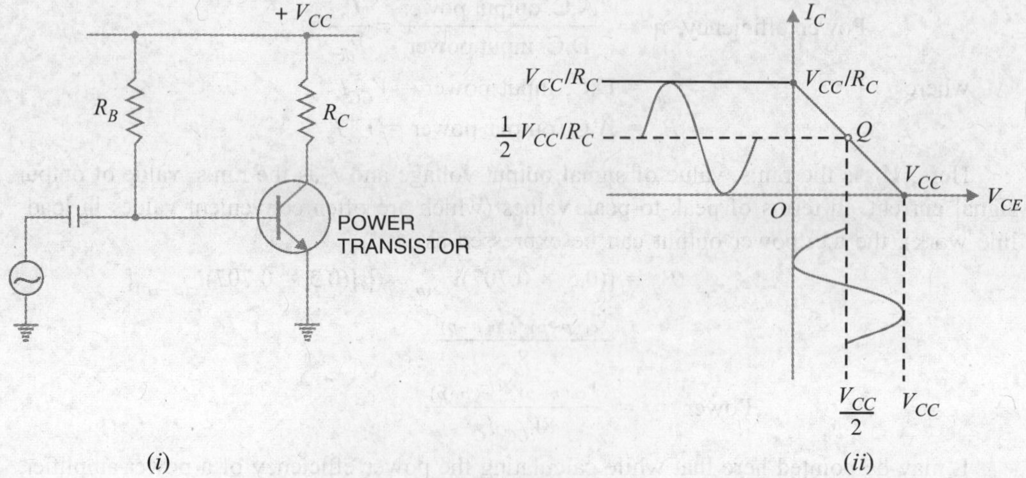

Fig. 10.13

$$\text{Maximum } v_{ce\,(p-p)} = V_{CC}$$

$$\text{Maximum } i_{c\,(p-p)} = V_{CC}/R_C$$

$$\text{Max. a.c. output power, } P_{o\,(max)} = \frac{v_{ce\,(p-p)} \times i_{c\,(p-p)}}{8} = \frac{V_{CC} \times V_{CC}/R_C}{8} = \frac{V_{CC}^2}{8R_C}$$

$$\text{D.C. power supplied, } P_{dc} = V_{CC}\,I_C = V_{CC}\left(\frac{V_{CC}}{2R_C}\right) = \frac{V_{CC}^2}{2R_C}$$

$$\therefore \quad \text{Maximum power } \eta = \frac{P_{o(max)}}{P_{dc}} \times 100 = \frac{V_{CC}^2/8R_C}{V_{CC}^2/2R_C} \times 100 = 25\%$$

Thus the maximum power efficiency of a class *A* series-fed amplifier is 25%. This means that at the most 25% of d.c. power drawn from d.c. supply (V_{CC}) can be converted into a.c. power and delivered to the load. In actual practice, the power efficiency of this type of class *A* amplifier is far less than 25%. For this reason, it is rarely used for large power applications.

Example 10.10. *Calculate the* (*i*) *output power* (*ii*) *input power and* (*iii*) *power efficiency of the amplifier circuit shown in Fig. 10.14* (*i*). *It is given that input voltage results in a base current of 10 mA peak.*

Solution. First draw the d.c. load line by locating the two end points *viz.*, $I_{C\,(sat)} = V_{CC}/R_C = 20\text{ V}/20\,\Omega = 1\text{ A} = 1000\text{ mA}$ and $V_{CE} = V_{CC} = 20\text{ V}$ as shown in Fig. 10.14 (*ii*). The operating point *Q* of the circuit can be located as under :

$$I_B = \frac{V_{CC} - V_{BE}}{R_B} = \frac{(20 - 0.7)\text{ V}}{1\text{ k}\Omega} = 19.3\text{ mA}$$

$\therefore \qquad I_C = \beta I_B = 25\ (19.3\,\text{mA}) = 482\,\text{mA}$

Also, $\qquad V_{CE} = V_{CC} - I_C R_C = 20\,\text{V} - (482\,\text{mA})\,(20\,\Omega) = 10.4\,\text{V}$

The operating point Q (10.4 V, 482 mA) is shown on the d.c. load line.

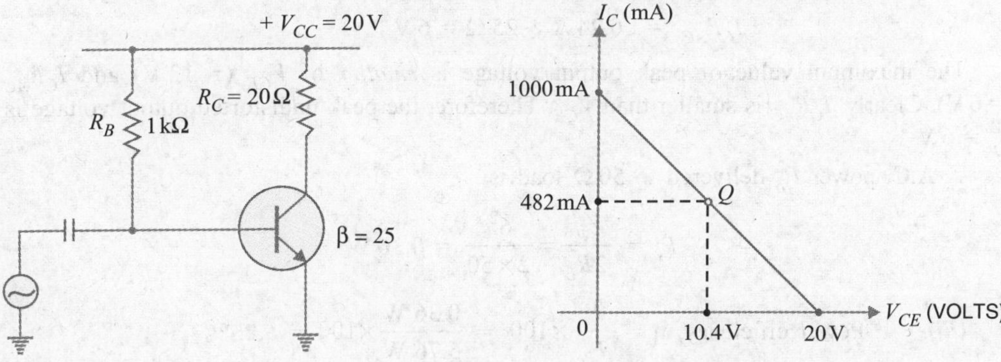

(i) (ii)

Fig. 10.14

(i) $\qquad i_c(peak) = \beta\, i_b\,(peak) = 25 \times (10\,\text{mA}) = 250\,\text{mA}$

$$\therefore \qquad P_o(a.c.) = \frac{i_c^2(peak)}{2} R_C = \frac{(250\times10^{-3})^2}{2}\times 20 = \mathbf{0.625\ W}$$

(ii) $\qquad P_{dc} = V_{CC}\, I_C = (20\,\text{V})\,(482 \times 10^{-3}) = \mathbf{9.6\ W}$

$$\text{(iii)} \qquad \text{Power } \eta = \frac{P_o(a.c.)}{P_{dc}}\times 100 = \frac{0.625}{9.6}\times 100 = \mathbf{6.5\ \%}$$

Example 10.11. *The class A transistor amplifier shown in Fig. 10.15 is biased at V_{CE} = 12 V. The output voltage is maximum possible without distortion. Find (i) average power drawn from the d.c. supply (ii) a.c. power delivered to the load (iii) power efficiency and (iv) collector efficiency.*

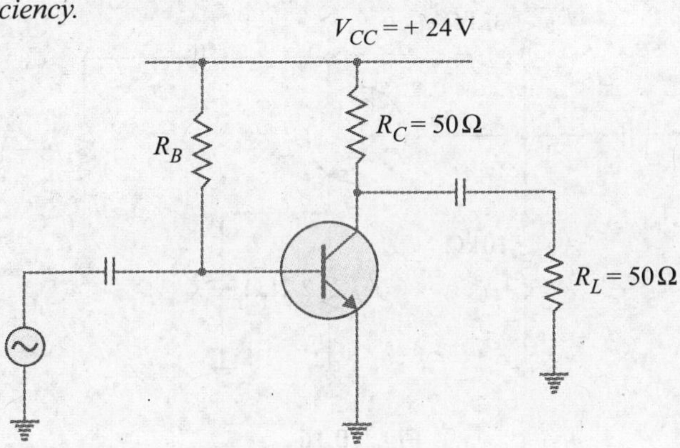

Fig. 10.15

Solution. (*i*) The Q point value of collector current I_C is

$$I_C = \frac{V_{CC} - V_{CE}}{R_C} = \frac{24\,\text{V} - 12\,\text{V}}{50\,\Omega} = 0.24\,\text{A}$$

∴ The d.c. power P_{dc} drawn from the d.c. supply (V_{CC}) is

$$P_{dc} = V_{CC}I_C = 24\text{ V} \times 0.24\text{ A} = \mathbf{5.76\ W}$$

(*ii*) A.C. load, $R_{AC} = R_C \parallel R_L = 50\,\Omega \parallel 50\,\Omega = 25\,\Omega$

$$\therefore \quad I_C R_{AC} = 0.24\text{ A} \times 25\,\Omega = 6\text{ V}$$

The maximum value of peak output voltage is *smaller* of V_{CE} (= 12 V) and $I_C R_{AC}$ (= 6 V). Clearly, $I_C R_{AC}$ is smaller than V_{CE}. Therefore, the peak undistorted output voltage is V_p = 6 V.

∴ A.C. power P_o delivered to 50 Ω load is

$$P_o = \frac{V_p^2}{2R_L} = \frac{6^2}{2 \times 50} = \mathbf{0.36\ W}$$

(*iii*) Power efficiency, $\eta = \dfrac{P_o}{P_{dc}} \times 100 = \dfrac{0.36\text{ W}}{5.76\text{ W}} \times 100 = \mathbf{6.25\%}$

Note that the efficiency is less than the theoretical maximum value (25%) because the bias point permits only a ± 6 V swing (*Q* point is not at the centre of the a.c. load line).

(*iv*) Power dissipated (P_D) at the collector is given by ;

$$P_D = V_{CE}I_C = 12\text{ V} \times 0.24\text{ A} = 2.88\text{ W}$$

∴ Collector efficiency, $\eta_c = \dfrac{P_o}{P_D} \times 100 = \dfrac{0.36\text{ W}}{2.88\text{ W}} \times 100 = \mathbf{12.5\%}$

Example 10.12. *Fig. 10.16 shows CE transistor amplifier. Find the maximum power efficiency of the amplifier without distortion.*

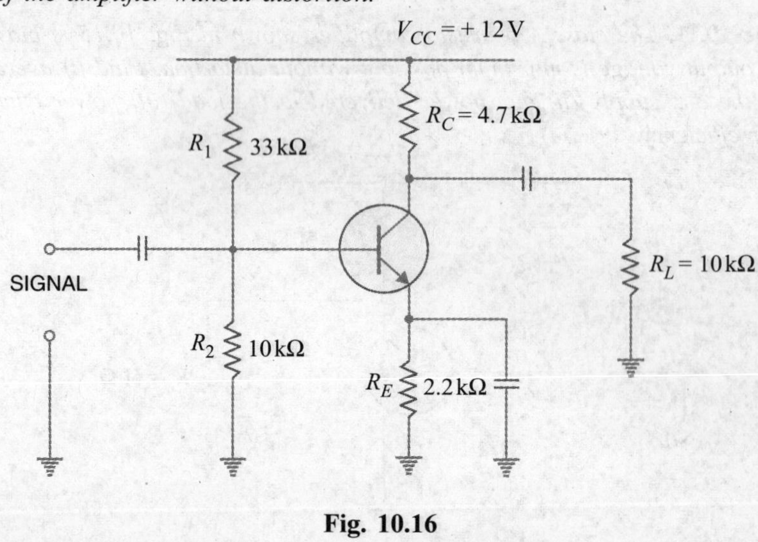

Fig. 10.16

Solution. We first find the *Q* point values of I_C and V_{CE}.

Base voltage, $V_B = \dfrac{V_{CC}}{R_1 + R_2} \times R_2 = \dfrac{12\text{ V}}{33\text{ k}\Omega + 10\text{ k}\Omega} \times 10\text{ k}\Omega = 2.79\text{ V}$

Emitter voltage, $V_E = V_B - V_{BE} = 2.79\text{ V} - 0.7\text{ V} = 2.09\text{ V}$

$$\text{Emitter current, } I_E = \frac{V_E}{R_E} = \frac{2.09\text{ V}}{2.2\text{ k}\Omega} = 0.95\text{ mA}$$

$$\therefore \quad \text{Collector current, } I_C \simeq I_E = 0.95\text{ mA}$$

$$\text{Collector-emitter voltage, } V_{CE} = V_{CC} - I_C(R_C + R_E)$$

$$= 12\text{ V} - 0.95\text{ mA} \times (4.7\text{ k}\Omega + 2.2\text{ k}\Omega) = 5.45\text{ V}$$

$$\text{A.C. load, } R_{AC} = R_C \parallel R_L = 4.7\text{ k}\Omega \parallel 10\text{ k}\Omega = 3.2\text{ k}\Omega$$

$$\therefore \quad I_C R_{AC} = 0.95\text{ mA} \times 3.2\text{ k}\Omega = 3.04\text{ V}$$

The maximum value of peak output voltage is *smaller* of V_{CE} (= 5.45 V) and $I_C R_{AC}$ (=3.04 V). Clearly, $I_C R_{AC}$ is smaller than V_{CE}. Therefore, peak undistorted output voltage is V_p = 3.04 V.

A.C. power P_o delivered to 10 kΩ load is

$$P_o = \frac{V_p^2}{2R_L} = \frac{(3.04)^2}{2 \times 10\text{ k}\Omega} = 462\ \mu\text{W}$$

The d.c. power (P_{dc}) drawn from d.c. supply (V_{CC}) is

$$P_{dc} = V_{CC} I_C = 12\text{ V} \times 0.95\text{ mA} = 11.4\text{ mW}$$

$$\therefore \quad \text{Power efficiency, } \eta = \frac{P_o}{P_{dc}} \times 100 = \frac{462\ \mu\text{W}}{11.4\text{ mW}} \times 100 = 4\%$$

10.10 Maximum Power Efficiency of Transformer Coupled Class A Power Amplifier

In class *A* power amplifier, the load can be either connected directly in the collector or it can be transformer coupled. The latter method is often preferred for two main reasons. First, transformer coupling permits impedance matching and secondly it keeps the d.c. power loss small because of the small resistance of the transformer primary winding.

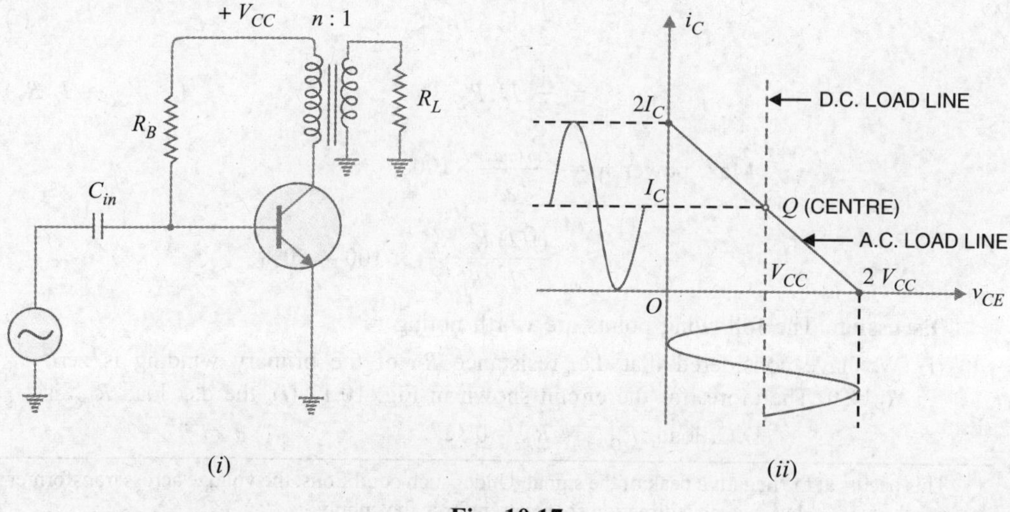

Fig. 10.17

Fig. 10.17 (*i*) shows the transformer coupled class *A* power amplifier. In order to determine maximum power efficiency, refer to the output characteristics shown in Fig. 10.17 (*ii*). Under zero signal conditions, the effective resistance in the collector circuit is that of the primary winding of the transformer. The primary resistance has a very small value and is assumed zero. Therefore, d.c. load line is a vertical line rising from V_{CC} as shown in Fig. 10.17 (*ii*). When signal is applied, the collector current will vary about the operating point *Q*.

In order to get maximum a.c. power output (and hence maximum power η), the peak value of collector current due to signal alone should be equal to the zero signal collector current I_C. In terms of a.c. load line, the operating point *Q* should be located at the centre of a.c. load line.

During the peak of the positive half-cycle of the signal, the total collector current is $2\ I_C$ and $v_{ce} = 0$. During the negative peak of the signal, the collector current is zero and *$v_{ce} = 2V_{CC}$.

∴ Peak-to-peak collector-emitter voltage is

$$v_{ce\ (p-p)} = 2V_{CC}$$

Peak-to-peak collector current, $i_{c\ (p-p)} = 2\ I_C$

$$= \frac{v_{ce(p-p)}}{R'_L} = \frac{2V_{CC}}{R'_L}$$

where R'_L is the reflected value of load R_L and appears in the primary of the transformer. If $n\ (= N_P/N_S)$ is the turn ratio of the transformer, then, $R'_L = n^2 R_L$.

$$\text{D.C. power input, } P_{dc} = V_{CC}\ I_C$$

$$= I_C^2\ R'_L \qquad (\because V_{CC} = I_C R'_L)$$

$$\text{Max. a.c. output power, } P_{o\ (max)} = \frac{v_{ce\ (p-p)} \times i_{c\ (p-p)}}{8}$$

$$= \frac{2V_{CC} \times 2I_C}{8}$$

$$= \frac{1}{2}\ V_{CC}\ I_C \qquad ...(i)$$

$$= \frac{1}{2}\ I_C^2\ R'_L \qquad (\because V_{CC} = I_C R'_L)$$

$$\therefore \quad \text{Max. power } \eta = \frac{P_{o\ (max)}}{P_{dc}} \times 100$$

$$= \frac{(1/2)\ I_C^2\ R'_L}{I_C^2\ R'_L} \times 100 = 50\%$$

Discussion. The following points are worth noting :

(*i*) We have considered that d.c. resistance R_P of the primary winding is zero *i.e.* $R_P = 0$. Therefore, for the circuit shown in Fig. 10.17 (*i*), the d.c. load R_{DC} is

$$\text{D.C. load, } R_{DC} = R_P = 0\ \Omega$$

---

* This occurs at the negative peak of the signal. Under such conditions, the voltage across transformer primary is V_{CC} but in such a direction so as to reinforce the supply.

$$\therefore \quad v_{ce} = 2V_{CC}$$

$$\text{Slope of d.c. load line} = -\frac{1}{R_{DC}} = -\frac{1}{R_P} = -\frac{1}{0} = -\infty$$

Note that slope of the d.c. load line is $-\infty$. For this reason, the d.c. load line is straight vertical line from the voltage point $v_{CE} = V_{CC}$ in Fig. 10.17 (*ii*). In practice, the primary winding has some resistance, so that the d.c. load line is not vertical.

(*ii*) The load R_L across the secondary is reflected to the primary as R'_L and is given by ;

$$R'_L = \left(\frac{N_P}{N_S}\right)^2 R_L$$

Here, $\frac{N_P}{N_S} = n$ = Turn ratio of the transformer

Therefore, total a.c. load for the circuit shown in Fig. 10.17(*i*) is

$$\text{Total a.c. load, } R_{AC} = R'_L + R_P = R'_L + 0 = R'_L$$

$$\text{Slope of a.c. load line} = -\frac{1}{R_{AC}} = -\frac{1}{R'_L}$$

Thus, if $R_L = 8\,\Omega$ and $N_P/N_S = 3$, then, $R'_L = (N_P/N_S)^2\, R_L = (3)^2 \times 8 = 72\,\Omega$ and the slope of the a.c. load line = –1/72S.

(*iii*) We have assumed that the efficiency of the transformer is 100%. In practice, this is not possible.

10.11 Important Points About Class A Power Amplifier

(*i*) A *transformer coupled class *A* power amplifier has a maximum power efficiency of 50% *i.e.* maximum of 50% d.c. supply power is converted into a.c. power output. In practice, the efficiency of such an amplifier is less than 50% (about 35%) due to power losses in the output transformer, power dissipation in the transistor *etc.*

(*ii*) The power dissipated by a transistor is given by ;

$$P_{dis} = P_{dc} - P_{ac}$$

where P_{dc} = Available d.c. power

P_{ac} = Available a.c. power

Clearly, in class *A* operation, the transistor must dissipate less heat when signal is applied and therefore runs cooler.

(*iii*) When no signal is applied to a class *A* power amplifier, $P_{ac} = 0$.

$$\therefore \quad P_{dis} = P_{dc}$$

Thus in class *A* operation, maximum power dissipation in the transistor occurs under zero signal conditions. Therefore, the power dissipation capability of a power transistor (for class *A* operation) must be atleast equal to the zero signal rating. For example, if the zero signal power dissipation of a transistor is 1 W, then transistor needs a rating of atleast 1 W. If the power rating of the transistor is less than 1 W, it is likely to be damaged.

(*iv*) When a class *A* power amplifier is used in the final stage, it is called **single ended class A power amplifier.**

* However, resistance coupled class *A* power amplifier has a maximum power efficiency of 25%.

Example 10.13. *A power transistor working in class A operation has zero signal power dissipation of 10 watts. If the a.c. output power is 4 watts, find :*

(i) power efficiency (ii) power rating of transistor

Solution.

Zero signal power dissipation, P_{dc} = 10 W

A.C. power output, P_o = 4 W

(*i*) $$\text{Power efficiency} = \frac{P_o}{P_{dc}} \times 100 = \frac{4}{10} \times 100 = \mathbf{40\%}$$

(*ii*) The zero signal power represents the worst case *i.e.* maximum power dissipation in a transistor occurs under zero signal conditions.

∴ Power rating of transistor = **10 W**

It means to avoid damage, the transistor must have a power rating of atleast 10 W.

Example 10.14. *A class A power amplifier has a transformer as the load. If the transformer has a turn ratio of 10 and the secondary load is 100 Ω, find the maximum a.c. power output. Given that zero signal collector current is 100 mA.*

Solution.

Secondary load, R_L = 100 Ω

Transformer turn ratio, n = 10

Zero signal collector current, I_C = 100 mA

Load as seen by the primary of the transformer is

$$R'_L = n^2 R_L = (10)^2 \times 100 = 10{,}000\ \Omega$$

$$\therefore \quad \text{Max. a.c. power output} = \frac{1}{2} I_C^2 R'_L = \frac{1}{2}\left(\frac{100}{1000}\right)^2 \times 10{,}000 = \mathbf{50\ W}$$

Example 10.15. *A class A transformer coupled power amplifier has zero signal collector current of 50 mA. If the collector supply voltage is 5 V, find (i) the maximum a.c. power output (ii) the power rating of transistor (iii) the maximum power efficiency.*

Solution.

(*i*) $$\text{Max. a.c. power output, } P_{o\,(max)} = \frac{V_{CC}\, I_C}{2}$$ ...See Art. 10.10

$$= \frac{(5\text{ V}) \times (50\text{ mA})}{2} = \mathbf{125\ mW}$$

(*ii*) D.C input power, $P_{dc} = V_{CC}\, I_C = (5\text{ V}) \times (50\text{ mA}) = 250\text{ mW}$

Since the maximum power is dissipated in the zero signal conditions,

∴ Power rating of transistor = **250 mW**

The reader may note that in class *A* operation :

$$P_{o(max)} = \frac{P_{dis}}{2}$$

or $$P_{dis} = 2\, P_{o\,(max)}$$

It means that power rating of the transistor is twice as great as the maximum a.c. output power. For example, if a transistor dissipates 3 W under no signal conditions, then maximum a.c. output power it can deliver is 1.5 W.

(iii) Max. power $\eta = \frac{P_{o(max)}}{P_{dc}} \times 100 = \frac{125 \text{ mW}}{250 \text{ mW}} \times 100 = 50\%$

Example 10.16. *In a certain transistor amplifier, $i_{c\,(max)} = 160$ mA, $i_{c\,(min)} = 10$ mA, $v_{ce\,(max)} = 12$ V and $v_{ce\,(min)} = 2$ V. Calculate the a.c. output power.*

Solution. A.C. output power, $P_o = \frac{v_{ce\,(p-p)} \times i_{c\,(p-p)}}{8}$

Here, $v_{ce\,(p-p)} = 12\text{ V} - 2\text{ V} = 10\text{ V}$; $i_{c\,(p-p)} = 160\text{ mA} - 10\text{ mA} = 150\text{ mA}$

$$\therefore \quad P_o = \frac{10\text{ V} \times 150\text{ mA}}{8} = \mathbf{187.5\text{ mW}}$$

Example 10.17. *A power transistor working in class A operation is supplied from a 12-volt battery. If the maximum collector current change is 100 mA, find the power transferred to a 5 Ω loudspeaker if it is :*

(i) directly connected in the collector

(ii) transformer-coupled for maximum power transference

Find the turn ratio of the transformer in the second case.

Solution.

Max. collector current change, $\Delta I_C = 100$ mA

Max. collector-emitter voltage change is

$$\Delta V_{CE} = 12\text{ V}$$

Loudspeaker resistance, $R_L = 5\ \Omega$

(i) **Loudspeaker directly connected.** Fig. 10.18 (i) shows the circuit of class A power amplifier with loudspeaker directly connected in the collector.

Max. voltage across loudspeaker $= \Delta I_C \times R_L = 100\text{ mA} \times 5\ \Omega = 0.5\text{ V}$

Power developed in the loudspeaker $= 0.5\text{ V} \times 100\text{ mA}$

$= 0.05\text{ W} = \mathbf{50\text{ mW}}$

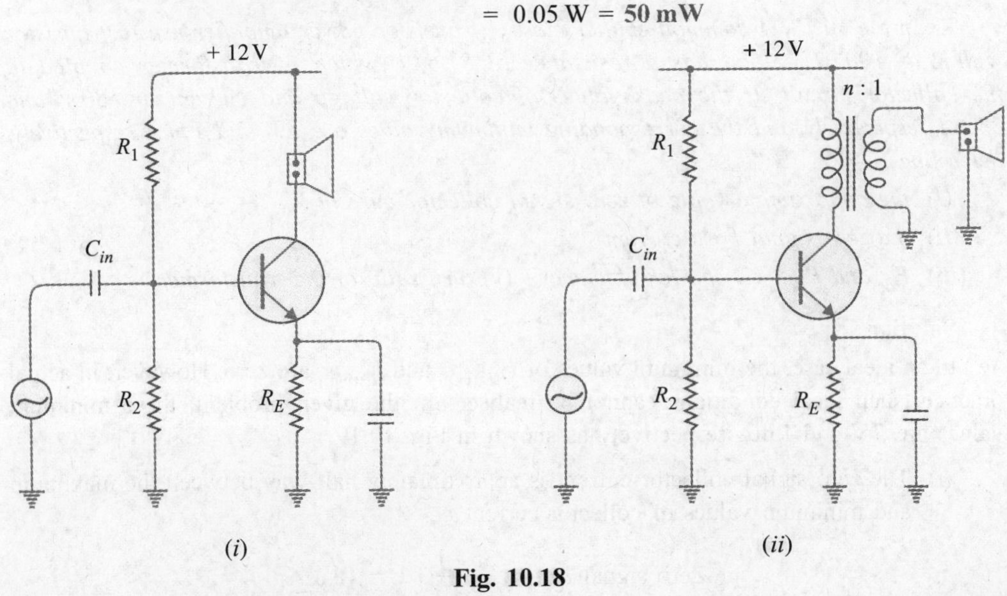

Fig. 10.18

Therefore, when loudspeaker is directly connected in the collector, only 50 mW of power is transferred to the loudspeaker.

(*ii*) **Loudspeaker transformer coupled.** Fig. 10.18 (*ii*) shows the class *A* power amplifier with speaker transformer coupled. As stated before, for impedance matching, step-down transformer is used.

$$\text{Output impedance of transistor} = \frac{\Delta V_{CE}}{\Delta I_C} = 12\text{ V}/100\text{ mA} = 120\ \Omega$$

In order to transfer maximum power, the primary resistance should be 120 Ω.

Now, load R'_L as seen by the primary is

$$R'_L = n^2 R_L$$

or $$120 = n^2 R_L$$

or $$n^2 = \frac{120}{5}$$

$$\therefore \quad \text{Turn ratio, } n = \sqrt{\frac{120}{5}} = \mathbf{4.9}$$

Transformer secondary voltage

$$= \frac{\text{Primary voltage}}{n} = 12/4.9 = 2.47\text{ V}$$

$$\text{Load current, } I_L = \frac{2.47\text{ V}}{5\ \Omega} = 0.49\text{ A}$$

Power transferred to the loudspeaker

$$= I_L^2 R_L$$

$$= (0.49)^2 \times 5 = 1.2\text{ W} = \mathbf{1200\ mW}$$

It is clear that by employing transformer coupling, we have been able to transfer a large amount of power (1200 mW) to the speaker. The main consideration in power amplifiers is the maximum power output and, therefore, transformer coupling is invariably used.

Example 10.18. *A common emitter class A transistor power amplifier uses a transistor with β = 100. The load has a resistance of 81.6 Ω, which is transformer coupled to the collector circuit. If the peak values of collector voltage and current are 30 V and 35 mA respectively and the corresponding minimum values are 5 V and 1 mA respectively, determine :*

(*i*) *the approximate value of zero signal collector current*

(*ii*) *the zero signal base current*

(*iii*) P_{dc} *and* P_{ac} (*iv*) *power efficiency* (*v*) *turn ratio of the transformer.*

Solution.

In an ideal case, the minimum values of $v_{CE\,(min)}$ and $i_{C\,(min)}$ are zero. However, in actual practice, such ideal conditions cannot be realised. In the given problem, these minimum values are 5 V and 1 mA respectively as shown in Fig. 10.19.

(*i*) The zero signal collector current is approximately half-way between the maximum and minimum values of collector current *i.e.*

$$\text{Zero signal } I_C = \frac{35-1}{2} + 1 = \mathbf{18\ mA}$$

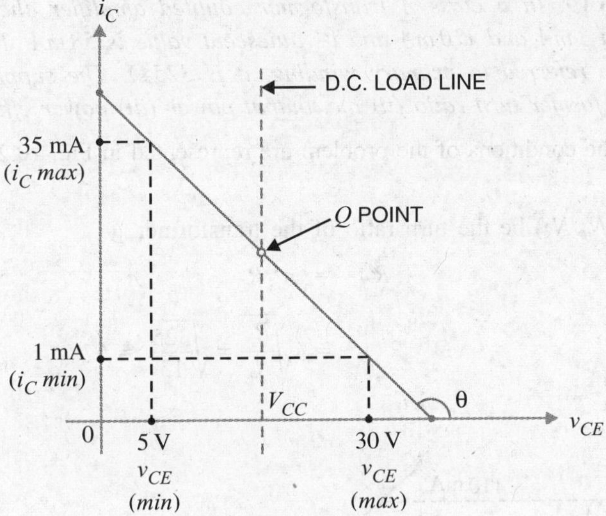

Fig. 10.19

(*ii*) Zero signal $I_B = I_C / \beta = 18/100 = $ **0.18 mA**

(*iii*) Zero signal $V_{CE} = \dfrac{30-5}{2} + 5 = 17.5\text{ V}$

Since the load is transformer coupled, $V_{CC} \simeq 17.5$ V.

$$\text{D.C. input power, } P_{dc} = V_{CC}\, I_C = 17.5\text{ V} \times 18\text{ mA}$$

$$= \mathbf{315\ mW}$$

$$\text{A.C. output voltage, } V_{ce} = \frac{30-5}{2\sqrt{2}} = 8.84\text{ V}$$

$$\text{A.C. output current, } I_c = \frac{35-1}{2\sqrt{2}} = 12\text{ mA}$$

$$\therefore \quad \text{A.C. output power, } P_{ac} = V_{ce} \times I_c$$

$$= 8.84\text{ V} \times 12\text{ mA} = \mathbf{106\ mW}$$

(*iv*) $$\text{Power } \eta = \frac{P_{ac}}{P_{dc}} \times 100 = \frac{106}{315} \times 100 = \mathbf{33.7\%}$$

(*v*) The a.c. resistance R'_L in the collector is determined from the slope of the line.

$$\text{Slope} = -\frac{1}{R'_L} = \frac{35-1}{5-30} = \frac{-34}{25}\text{ kS}$$

$$\therefore \quad R'_L = \frac{25}{34}\text{ k}\Omega = \frac{25}{34} \times 1000\,\Omega = 735\,\Omega$$

$$\therefore \quad \text{Turn ratio, } n = \sqrt{\frac{R'_L}{R_L}} = \sqrt{\frac{735}{81.6}} = 3$$

Example 10.19. *In a class A transformer coupled amplifier, the collector current alternates between 3 mA and 110 mA and its quiescent value is 58 mA. The load resistance is 13 Ω and when referred to primary winding, it is 325 Ω. The supply voltage is 20 V. Calculate (i) transformer turn ratio (ii) a.c. output power (iii) power efficiency.*

Solution. The conditions of the problem are represented in Fig. 10.20. The zero signal I_C = 58 mA.

(*i*) Let n (= N_P/N_S) be the turn ratio of the transformer.

$$\therefore \quad R'_L = n^2 R_L$$

or
$$n = \sqrt{\frac{R'_L}{R_L}} = \sqrt{\frac{325}{13}} = 5$$

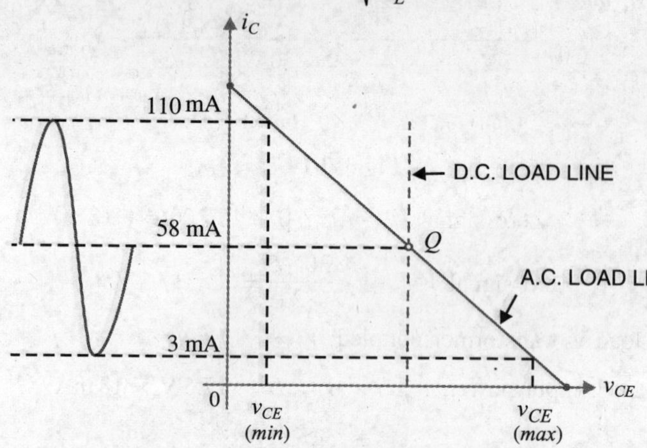

Fig. 10.20

(*ii*) A.C. output power, $P_{ac} = \frac{1}{2} I_C^2 R'_L$

$$= \frac{1}{2} (58 \text{ mA})^2 \times 325\Omega = \textbf{546 mW}$$

(*iii*) D.C. input power, $P_{dc} = V_{CC} I_C = 20 \text{ V} \times 58 \text{ mA} = 1160 \text{ mW}$

$$\therefore \quad \text{Power } \eta = \frac{546}{1160} \times 100 = \textbf{47\%}$$

Example 10.20. *Fig. 10.21 shows transformer-coupled class A amplifier. The d.c. resistance of the primary is R_P = 40 Ω, the number of primary turns is 74 and the number of secondary turns is 14. Find (i) the operating point (ii) d.c. load for the transistor (iii) a.c. load for the transistor and (iv) slope of d.c. and a.c. load lines.*

Solution.

(*i*) The Q point values of I_C and V_{CE} can be found as under :

$$\text{Base voltage, } V_B = \frac{V_{CC}}{R_1 + R_2} \times R_2 = \frac{13 \text{ V}}{4.7 \text{ k}\Omega + 3.7 \text{ k}\Omega} \times 3.7 \text{ k}\Omega = 5.7 \text{ V}$$

$$\text{Emitter voltage, } V_E = V_B - V_{BE} = 5.7 \text{ V} - 0.7 \text{ V} = 5 \text{ V}$$

$$\text{Emitter current, } I_E = \frac{V_E}{R_E} = \frac{5\text{ V}}{1\text{ k}\Omega} = 5\text{ mA}$$

$$\therefore \quad \text{Collector current, } I_C \simeq I_E = 5\text{ mA}$$

$$\text{Collector-emitter voltage, } V_{CE} = V_{CC} - I_C(R_P + R_E)$$

$$= 13\text{ V} - 5\text{ mA}(40\,\Omega + 1000\,\Omega) \simeq 8\text{ V}$$

∴ Operating point of the circuit is **8 V, 5 mA.**

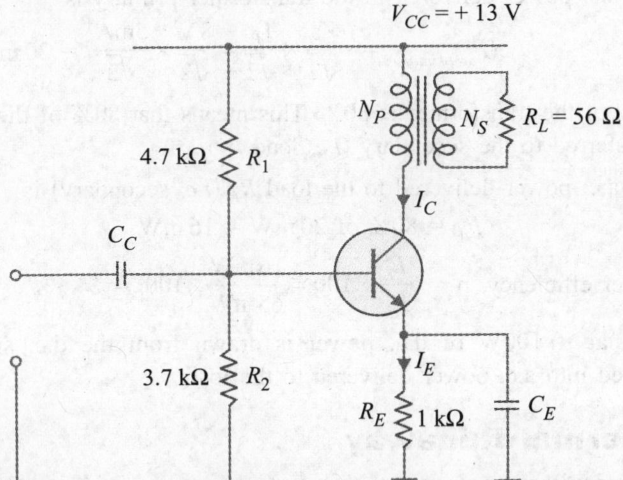

Fig. 10.21

(*ii*) The total d.c. load (R_{DC}) for the transistor is the sum of R_P and R_E *i.e.*

$$\text{Total d.c. load, } R_{DC} = R_P + R_E = 40\,\Omega + 1000\,\Omega = \mathbf{1040\,\Omega}$$

(*iii*) The total a.c. load for the transistor is the sum of R'_L (load R_L referred to the primary) and R_P *i.e.*

$$\text{Total a.c. load, } R_{AC} = R'_L + R_P$$

$$\text{Here, } R_P = 40\Omega\ ;\ R'_L = \left(\frac{N_P}{N_S}\right)^2 R_L = \left(\frac{74}{14}\right)^2 \times 56\Omega = 1565\,\Omega$$

$$\therefore \quad R_{AC} = 1565\,\Omega + 40\Omega = \mathbf{1605\,\Omega}$$

(*iv*)

$$\text{Slope of d.c. load line} = -\frac{1}{R_{DC}} = -\frac{\mathbf{1}}{\mathbf{1040}}\,\mathbf{S}$$

$$\text{Slope of a.c. load line} = -\frac{1}{R_{AC}} = -\frac{\mathbf{1}}{\mathbf{1605}}\,\mathbf{S}$$

Example 10.21. *In the above example, find the maximum power efficiency of the amplifier, assuming that the transformer has an efficiency of 80%.*

Solution. We have found the Q point values in the above example (Example 10.20) as : $I_C = 5$ mA ; $V_{CE} = 8$ V.

∴ D.C. input power (P_{dc}) drawn from the d.c. supply is

$$P_{dc} = V_{CC} I_C = 13\text{ V} \times 5\text{ mA} = 65\text{ mW}$$

Peak value of a.c. voltage (V_P) in the primary is

$$V_P = V_{CE} = 8\text{ V}$$

Peak value of a.c. current (I_P) in the primary is

$$I_P = I_C = 5\text{ mA}$$

∴ Maximum a.c. power delivered to the transformer primary is

$$(P_o)_{primary} = \frac{V_P}{\sqrt{2}} \times \frac{I_P}{\sqrt{2}} = \frac{8\text{ V}}{\sqrt{2}} \times \frac{5\text{mA}}{\sqrt{2}} = 20\text{ mW}$$

The efficiency of the transformer is 80%. This means that 80% of the power delivered to primary is transferred to the secondary (*i.e.* load R_L).

∴ Maximum a.c. power delivered to the load R_L (*i.e.* secondary) is

$$P_L = 80\% \text{ of } 20\text{ mW} = 16\text{ mW}$$

∴ Power efficiency, $\eta = \frac{P_L}{P_{dc}} \times 100 = \frac{16\text{mW}}{65\text{mW}} \times 100 = \mathbf{24.6\%}$

This means that if 100W of d.c. power is drawn from the d.c. supply, then only 24.6 W is converted into a.c. power delivered to the load.

10.12 Thermal Runaway

All semiconductor devices are very sensitive to temperature variations. If the temperature of a transistor exceeds the permissible limit, the transistor may be *permanently damaged. Silicon transistors can withstand temperatures upto 150°C while the germanium transistors can withstand temperatures upto 90°C.

There are two factors which determine the operating temperature of a transistor *viz.* (*i*) surrounding temperature and (*ii*) power dissipated by the transistor.

When the transistor is in operation, almost the entire heat is produced at the collector-base junction. This power dissipation causes the junction temperature to rise. This in turn increases the collector current since more electron-hole pairs are generated due to the rise in temperature. This produces an increased power dissipation in the transistor and consequently a further rise in temperature. Unless adequate cooling is provided or the transistor has built-in temperature compensation circuits to prevent excessive collector current rise, the junction temperature will continue to increase until the maximum permissible temperature is exceeded. If this situation occurs, the transistor will be permanently damaged.

The unstable condition where, owing to rise in temperature, the collector current rises and continues to increase is known as **thermal runaway.**

Thermal runaway must always be avoided. If it occurs, permanent damage is caused and the transistor must be replaced.

10.13 Heat Sink

As power transistors handle large currents, they always heat up during operation. Since transistor is a temperature dependent device, the heat generated must be dissipated to the surroundings

* Almost the entire heat in a transistor is produced at the collector-base junction. If the temperature exceeds the permissible limit, this junction is destroyed and the transistor is rendered useless.

in order to keep the temperature within permissible limits. Generally, the transistor is fixed on a metal sheet (usually aluminium) so that additional heat is transferred to the Al sheet.

The metal sheet that serves to dissipate the additional heat from the power transistor is known as **heat sink.**

Most of the heat within the transistor is produced at the *collector junction. The heat sink increases the surface area and allows heat to escape from the collector junction easily. The result is that temperature of the transistor is sufficiently lowered. Thus heat sink is a direct practical means of combating the undesirable thermal effects *e.g.* thermal runaway. It may be noted that the ability of any heat sink to transfer heat to the surroundings depends upon its material, volume, area, shape, contact between case and sink and movement of air around the sink. Finned aluminium heat sinks yield the best heat transfer per unit cost.

It should be realised that the use of heat sink alone may not be sufficient to prevent thermal runaway under all conditions. In designing a transistor circuit, consideration should also be given to the choice of (*i*) operating point (*ii*) ambient temperatures which are likely to be encountered and (*iii*) the type of transistor *e.g.* metal case transistors are more readily cooled by conduction than plastic ones. Circuits may also be designed to compensate automatically for temperature changes and thus stabilise the operation of the transistor components.

10.14 Mathematical Analysis

When power is dissipated in a transistor, the heat generated at the collector-base junction flows from the junction to the transistor case and then to the atmosphere (ambient). The medium between the junction and the atmosphere offers resistance to the flow of heat and is called *thermal resistance* of the medium. The relation among power dissipation (P_D), junction temperature (T_J) and ambient temperature (T_A) can be expressed as :

$$^{**}P_D = \frac{T_J - T_A}{\theta} \text{ watts}$$

where P_D = Power dissipated in the transistor in watts

T_J = Junction temperature in °C

T_A = Ambient temperature (*i.e.* temperature of surrounding air) in °C

θ = Thermal resistance *i.e.* resistance to heat flow from the junction to the surrounding air

The unit of θ is °C/watt and its value is always given in the transistor manual. A low thermal resistance means that it is easy for heat to flow from the junction to the surrounding air. The larger the transistor case, the lower is the thermal resistance and *vice-versa*. It is then clear that by using heat sink, the value of θ can be decreased considerably, resulting in decreased junction temperature.

---

* Most of power is dissipated at the collector-base junction. This is because collector-base voltage is much greater than the base-emitter voltage, although currents through the two junctions are almost the same.

** Electric current, $I = \frac{V_2 - V_1}{R}$ = Rate of flow of charge

Similarly, the rate of flow of heat energy (*i.e.* power) is equal to *temperature difference* ($T_J - T_A$) divided by thermal resistance θ.

Thermal resistance with heat sink. We have seen above that the use of heat sink reduces the thermal resistance between the junction and the atmosphere. This helps to keep the junction temperature with in the permissible limits. When heat sink is used, the heat generated at the collector-base junction flows from the junction to the transistor case, then from the case to the heat sink and finally from the heat sink to the surrounding air. The heat flow meets the following thermal resistances :

θ_{JC} = *Junction to case* thermal resistance

θ_{CS} = *Case to sink* thermal resistance

θ_{SA} = *Sink to air* thermal resistance

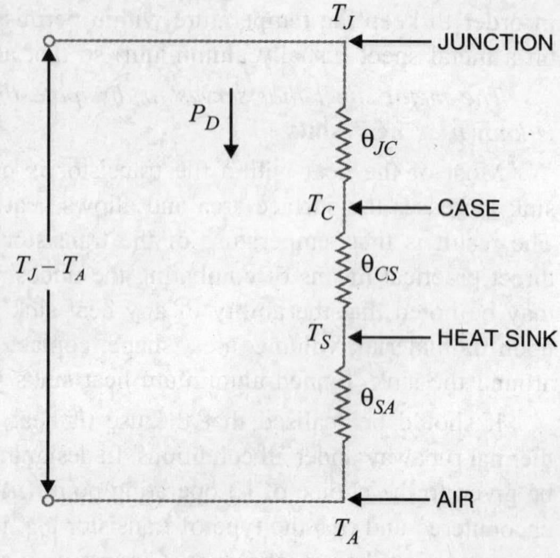

Fig. 10.22

These three thermal resistances are in series as in the case of a series electrical circuit. Fig. 10.22 shows the thermal equivalent circuit for the transistor and heat sink. The total thermal resistance θ_{JA} between the junction and air is

Total thermal resistance, $\theta_{JA} = \theta_{JC} + \theta_{CS} + \theta_{SA}$

Note that power flow in the thermal circuit is similar to the current flow in an electrical series resistive circuit.

Example 10.22. *A power transistor dissipates 4 W. If $T_{J\,max}$ = 90ºC, find the maximum ambient temperature at which it can be operated. Given θ = 10ºC/W.*

Solution. Here, $T_{J\,max}$ is the maximum permissible junction temperature.

$$P_D = 4\text{ W} ; T_{J\,max} = 90°\text{C} ; \theta = 10°\text{C/W}$$

Now,
$$P_D = \frac{T_{J\,max} - T_A}{\theta}$$

or
$$4 = \frac{90 - T_A}{10}$$

∴ Ambient temperature, $T_A = 90 - 40 = $ **50ºC**

The above example shows the effect of ambient temperature on the permissible power dissipation in a transistor. The lower the ambient temperature, the greater is the permissible power dissipation. Thus, a transistor can pass a higher collector current in winter than in summer.

Example 10.23. (*i*) *A power transistor has thermal resistance θ = 300°C/W. If the maximum junction temperature is 90°C and the ambient temperature is 30ºC, find the maximum permissible power dissipation.*

(*ii*) *If a heat sink is used with the above transistor, the value of θ is reduced to 60°C/W. Find the maximum permissible power dissipation.*

Solution.

(*i*) *Without heat sink*

$$T_{J\,max} = 90°C \ ; \ T_A = 30°C \ ; \ \theta = 300°C/W$$

$$\therefore \quad P_D = \frac{T_{J\,max} - T_A}{\theta} = \frac{90-30}{300} = 0.2 \text{ W} = \mathbf{200\ mW}$$

(*ii*) *With heat sink*

$$T_{J\,max} = 90°C \ ; \ T_A = 30°C \ ; \ \theta = 60°C/W$$

$$\therefore \quad P_D = \frac{T_{J\,max} - T_A}{\theta} = \frac{90-30}{60} = 1 \text{ W} = \mathbf{1000\ mW}$$

It is clear from the above example that permissible power dissipation with heat sink is 5 times as compared to the case when no heat sink is used.

Example 10.24. *The total thermal resistance of a power transistor and heat sink is 20°C/W. The ambient temperature is 25°C and $T_{J\,max}$ = 200°C. If V_{CE} = 4 V, find the maximum collector current that the transistor can carry without destruction. What will be the allowed value of collector current if ambient temperature rises to 75°C?*

Solution. $$P_D = \frac{T_{J\,max} - T_A}{\theta} = \frac{200-25}{20} = 8.75 \text{ W}$$

This means that maximum permissible power dissipation of the transistor at ambient temperature of 25°C is 8.75 W *i.e.*

$$V_{CE}\, I_C = 8.75 \qquad \therefore \quad I_C = 8.75/4 = \mathbf{2.19\ A}$$

Again, $$P_D = \frac{T_{J\,max} - T_A}{\theta} = \frac{200-75}{20} = 6.25 \text{ W}$$

$$\therefore \quad I_C = 6.25/4 = \mathbf{1.56\ A}$$

This example clearly shows the effect of ambient temperature.

Example 10.25. *The collector-base junction of a certain transistor dissipates 2 W. The thermal resistance from junction to case is 8°C/W and the thermal resistance from case to air is 20°C/W. The ambient temperature is 25°C. Find the (i) junction temperature (ii) case temperature.*

Solution. Here, θ_{JC} = 8°C/W ; θ_{CA} = 20°C/W ; T_A = 25°C

(*i*) The total thermal resistance θ_{JA} from junction to ambient (atmosphere) is

$$\theta_{JA} = \theta_{JC} + \theta_{CA} = 8°C/W + 20°C/W = 28°C/W$$

Now, $$P_D = \frac{T_J - T_A}{\theta_{JA}}$$

or $$2 = \frac{T_J - 25}{28} \qquad \therefore \quad T_J = \mathbf{81°C}$$

(*ii*) The rate of heat flow (*i.e.* P_D) is constant throughout the series combination of thermal resistances.

Now, $$P_D = \frac{T_J - T_C}{\theta_{JC}}$$

or $$2 = \frac{81 - T_C}{8} \qquad \therefore \quad T_C = \mathbf{65°C}$$

Example 10.26. *The maximum permissible junction temperature of a certain power transistor is 150°C. It is desired to operate the transistor with a power dissipation of 15 W in an ambient temperature of 40°C. The thermal resistances are as under :*

$$\theta_{JC} = 0.5°C/W \text{ (Junction to case)}$$
$$\theta_{CA} = 10°C/W \text{ (Case to ambient)}$$

(*i*) *Determine whether a heat sink is required for this application.*

(*ii*) *If a heat sink is required, determine the maximum thermal resistance it can have. Assume that a mica washer having thermal resistance θ_W = 0.5°C/W must be used between the case and heat sink.*

Solution. Here, θ_{JC} = 0.5°C/W ; θ_{CA} = 10°C/W ; θ_W = 0.5°C/W ; T_A = 40°C ; P_D = 15W ; $T_{J\,max}$ = 150°C.

(*i*) $$P_D = \frac{T_J - T_A}{\theta_{JA}}$$

Here, $\theta_{JA} = \theta_{JC} + \theta_{CA}$ = 0.5°C/W + 10°C/W = 10.5°C/W

$$\therefore \qquad 15 = \frac{T_J - 40}{10.5} \quad \text{or} \quad T_J = 197.5°C$$

Now, the junction temperature is 197.5°C while the maximum permissible junction temperature is 150°C. Therefore, **heat sink must be used** to reduce the total thermal resistance.

(*ii*) Suppose for maximum junction temperature of 150°C (*i.e.* $T_{J\,max}$ = 150°C), the maximum total thermal resistance is θ_T. Then,

$$P_D = \frac{T_{Jmax} - T_A}{\theta_T}$$

or $$15 = \frac{150 - 40}{\theta_T} \qquad \therefore \quad \theta_T = 7.33°C/W$$

Suppose the thermal resistance of heat sink is θ_{SA} (sink to ambient).

$$\therefore \qquad \theta_T = \theta_{JC} + \theta_W + \theta_{SA}$$

or $$7.33 = 0.5 + 0.5 + \theta_{SA}$$

$$\therefore \qquad \theta_{SA} = 7.33 - 0.5 - 0.5 = \mathbf{6.33°C/W}$$

10.15 Stages of a Practical Power Amplifier

The function of a practical power amplifier is to amplify a weak signal until sufficient power is available to operate a loudspeaker or other output device. To achieve this goal, a power amplifier has generally three stages *viz. voltage amplification stage, driver stage* and *output stage.* Fig. 10.23 shows the block diagram of a practical power amplifier.

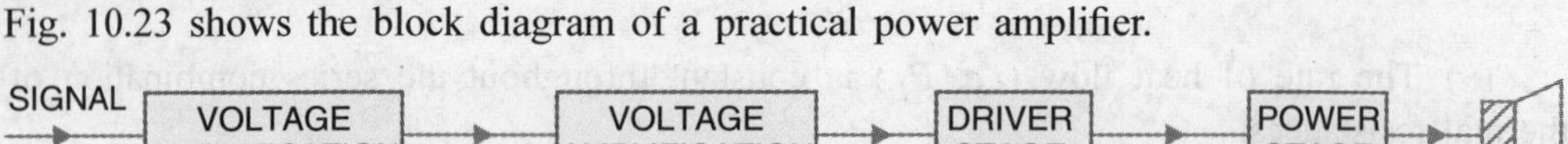

Fig. 10.23

(i) **Voltage amplification stage.** The signals found in practice have extremely low voltage level (< 10 mV). Therefore, the voltage level of the weak signal is raised by two or more voltage amplifiers. Generally, *RC* coupling is employed for this purpose.

(ii) **Driver stage.** The output from the last voltage amplification stage is fed to the driver stage. It supplies the necessary power to the output stage. The driver stage generally employs class *A* transformer coupled power amplifier. Here, concentrated effort is made to obtain *maximum power gain.*

(iii) **Output stage.** The output power from the driver stage is fed to the output stage. It is the final stage and feeds power directly to the speaker or other output device. The output stage is invariably transformer coupled and employs class *B* amplifiers in push-pull arrangement. Here, concentrated effort is made to obtain *maximum power output.*

10.16 Driver Stage

The stage that immediately precedes the output stage is called the *driver stage*. It operates as a class *A* power amplifier and supplies the drive for the output stage. Fig. 10.24 shows the driver stage. Note that transformer coupling is employed. The primary of this transformer is the collector load. The secondary is almost always centre-tapped so as to provide equal and opposite voltages to the input of push-pull amplifier (*i.e.* output stage). The driver transformer is usually a step-down transformer and facilitates impedance matching.

The output from the last voltage amplification stage forms the input to the driver stage. The driver stage renders power amplification in the usual way. It may be added that main consideration here is the maximum power gain. The output of the driver stage is taken from the centre-tapped secondary and is fed to the output stage.

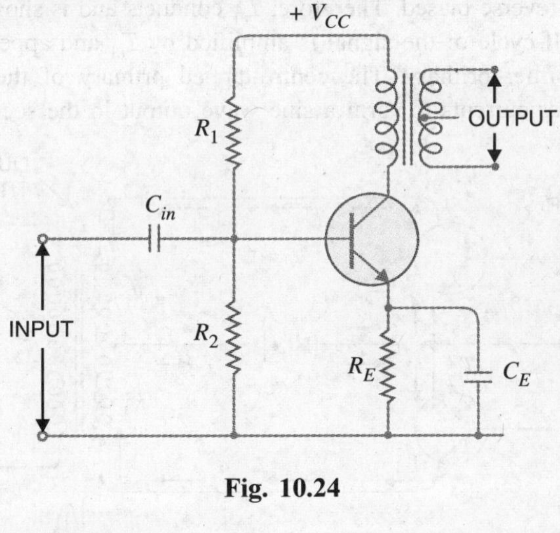

Fig. 10.24

10.17 Output Stage

The output stage essentially consists of a power amplifier and its purpose is to transfer maximum power to the output device. If a single transistor is used in the output stage, it can only be employed as class *A* amplifier for faithful amplification. Unfortunately, the power efficiency of a class *A* amplifier is very low (≃ 35%). As transistor amplifiers are operated from batteries, which is a costly source of power, therefore, such a low efficiency cannot be tolerated.

In order to obtain high output power at high efficiency, push-pull arrangement is used in the output stage. In this arrangement, we employ two transistors in class *B* operation. One transistor amplifies the positive half-cycle of the signal while the other transistor amplifies the negative half-cycle of the signal. In this way, output voltage is a complete sine wave. At the same time, the circuit delivers high output power to the load due to class *B* operation.

10.18 Push-Pull Amplifier

The push-pull amplifier is a power amplifier and is frequently employed in the output stages of electronic circuits. It is used whenever high output power at high efficiency is required. Fig. 10.25 shows the circuit of a push-pull amplifier. Two transistors T_{r1} and T_{r2} placed back to back are employed. Both transistors are operated in class *B* operation *i.e.* collector current is nearly zero in the absence of the signal. The centre-tapped secondary of driver transformer T_1 supplies equal and opposite voltages to the base circuits of two transistors.

The output transformer T_2 has the centre-tapped primary winding. The supply voltage V_{CC} is connected between the bases and this centre tap. The loudspeaker is connected across the secondary of this transformer.

Circuit operation. The input signal appears across the secondary *AB* of driver transformer. Suppose during the first half-cycle (marked 1) of the signal, end *A* becomes positive and end *B* negative. This will make the base-emitter junction of T_{r1} reverse biased and that of T_{r2} forward biased. The circuit will conduct current due to T_{r2} only and is shown by solid arrows. Therefore, this half-cycle of the signal is amplified by T_{r2} and appears in the lower half of the primary of output transformer. In the next half-cycle of the signal, T_{r1} is forward biased whereas T_{r2} is reverse biased. Therefore, T_{r1} conducts and is shown by dotted arrows. Consequently, this half-cycle of the signal is amplified by T_{r1} and appears in the upper half of the output transformer primary. The centre-tapped primary of the output transformer combines two collector currents to form a sine wave output in the secondary.

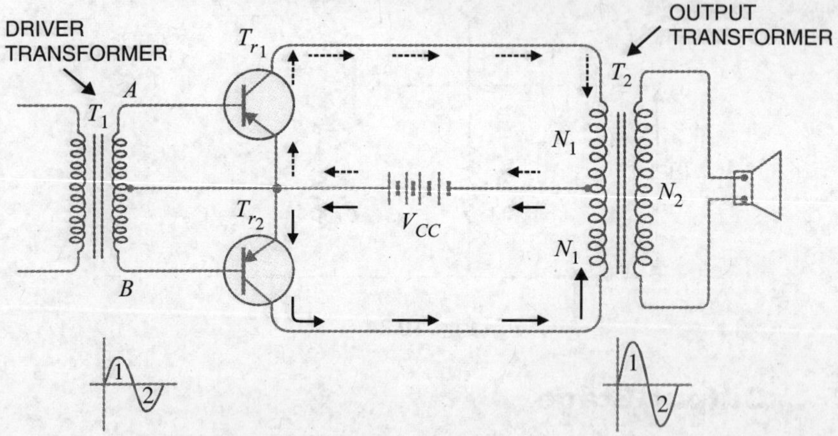

Fig. 10.25

It may be noted here that push-pull arrangement also permits a maximum transfer of power to the load through impedance matching. If R_L is the resistance appearing across secondary of output transformer, then resistance R'_L of primary shall become :

$$R'_L = \left(\frac{2N_1}{N_2}\right)^2 R_L$$

where N_1 = Number of turns between either end of primary winding and centre-tap

N_2 = Number of secondary turns

The name push-pull circuit is derived from the fact that one transistor drives current through the load in one direction (*pushing*) while the other transistor drives current through the load in the opposite direction (*pulling*).

Crossover distortion. One disadvantage of class *B* push-pull amplifier is that the waveform of the output voltage (V_L) across the load is not perfectly sinusoidal as shown in Fig. 10.26. This is called **crossover distortion* and refers to the fact that during the signal crossover from positive to negative (or *vice-versa*), there is some non-linearity in the output signal. The reason for this distortion is that the transistors do not begin to conduct until the input base-emitter voltage is about 0.7 V for a silicon transistor or 0.5 V for a germanium transistor. The crossover distortion can be reduced or eliminated in a class *B* push-pull amplifier by biasing each transistor at a level slightly *above* cut-off instead of being biased at cut-off. By biasing both transistors slightly above cut-off allows the amplifier to provide a linear output that contains no distortion.

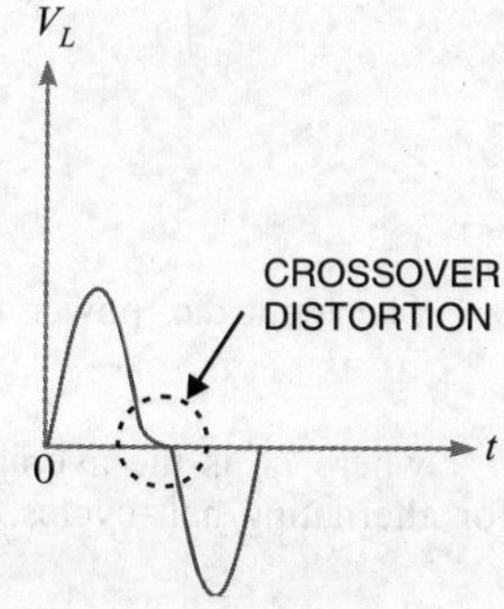

Fig. 10.26

Advantages

(*i*) The power efficiency of the circuit is quite high (≃ 75%) due to class *B* operation.

(*ii*) A high a.c. output power is obtained.

Disadvantages

(*i*) Two transistors have to be used.

(*ii*) It requires two equal and opposite voltages at the input. Therefore, push-pull circuit requires the use of driver stage to furnish these signals.

(*iii*) If the parameters of the two transistors are not the same, there will be unequal amplification of the two halves of the signal.

(*iv*) The circuit gives more distortion.

(*v*) Transformers used are bulky and expensive.

10.19 Maximum Efficiency for Class B Power Amplifier

We have already seen that a push-pull circuit uses two transistors working in class *B* operation. For class *B* operation, the *Q*-point is located at cut-off on both d.c. and a.c. load

* This is called crossover distortion because it occurs where the composite output signal crosses the zero voltage axis.

lines. For maximum signal operation, the two transistors in class B amplifier are alternately driven from cut-off to saturation. This is shown in Fig. 10.27 (*i*). It is clear that a.c. output voltage has a peak value of V_{CE} and a.c. output current has a peak value of $I_{C\,(sat)}$. The same information is also conveyed through the a.c. load line for the circuit [See Fig. 10.27 (*ii*)].

$\therefore$ Peak a.c. output voltage $= V_{CE}$

Peak a.c. output current $= I_{C\,(sat)} = {}^{*}\dfrac{V_{CE}}{R_L} = \dfrac{V_{CC}}{2\,R_L}$ $\qquad \left(\because V_{CE} = \dfrac{V_{CC}}{2}\right)$

Maximum average a.c. output power $P_{o\,(max)}$ is

$P_{o\,(max)}$ = Product of r.m.s. values of a.c. output voltage and a.c. output current

$$= \frac{V_{CE}}{\sqrt{2}} \times \frac{I_{C\,(sat)}}{\sqrt{2}} = \frac{V_{CE}\,I_{C\,(sat)}}{2}$$

$$= \frac{V_{CC}}{2} \times \frac{I_{C\,(sat)}}{2} = \frac{V_{CC}\,I_{C\,(sat)}}{4} \qquad \left(\because V_{CE} = \frac{V_{CC}}{2}\right)$$

$$\therefore \quad P_{o\,(max)} = 0.25\;V_{CC}\,I_{C\,(sat)}$$

The input d.c. power from the supply V_{CC} is

$$P_{dc} = V_{CC}\,I_{dc}$$

where I_{dc} is the average current drawn from the supply V_{CC}. Since the transistor is on for alternating half-cycles, it effectively acts as a half-wave rectifier.

$$\therefore \quad I_{dc} = \frac{I_{C\,(sat)}}{\pi}$$

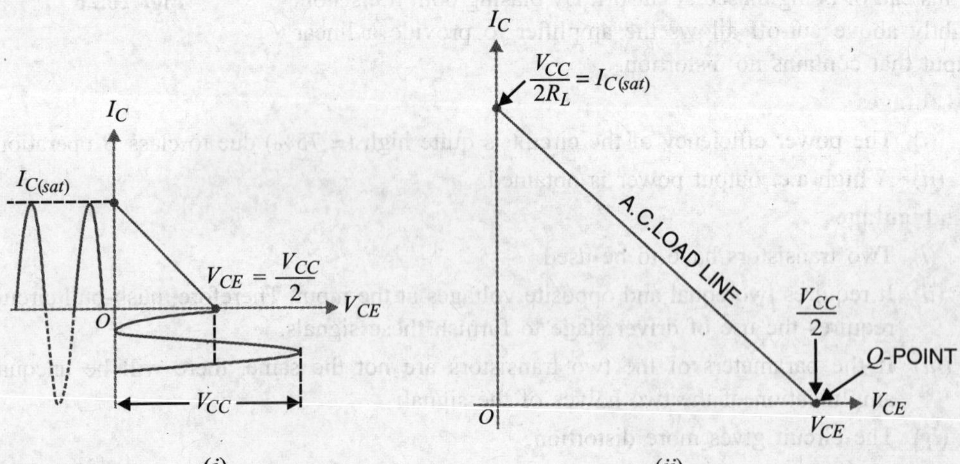

Fig. 10.27

* Since the two transistors are identical, half the supply voltage is dropped across each transistor's collector-emitter terminals *i.e.* $V_{CE} = \dfrac{V_{CC}}{2}$.

Also, peak voltage across each transistor is V_{CE} and it appears across R_L.

$$\therefore \quad I_{C\,(sat)} = \frac{V_{CE}}{R_L} = \frac{V_{CC}}{2} \times \frac{1}{R_L} = \frac{V_{CC}}{2\,R_L}$$

$$\therefore \quad P_{dc} = \frac{V_{CC}\, I_{C\,(sat)}}{\pi}$$

$$\therefore \quad \text{Max. power } \eta = \frac{P_{o\,(max)}}{P_{dc}} = \frac{0.25\, V_{CC}\; I_{C\,(sat)}}{\left(V_{CC}\, I_{C\,(sat)}\right)/\pi} \times 100$$

$$= 0.25\pi \times 100 = 78.5\%$$

Thus the maximum power efficiency of class *B* power amplifier is 78.5%. Recall that maximum power efficiency for class *A* transformer coupled amplifier is 50%.

Power dissipated by transistors. The power dissipated (as heat) by the transistors in class *B* amplifier is the difference between the input power delivered by V_{CC} and the output power delivered to the load *i.e.*

$$P_{2T} = P_{dc} - P_{ac}$$

where P_{2T} = Power dissipated by the two transistors

∴ Power dissipated by each transistor is

$$P_T = \frac{P_{2T}}{2} = \frac{P_{dc} - P_{ac}}{2}$$

Note. For power efficiency of class *C* amplifiers, the reader may refer to Chapter 13 (Transistor tuned amplifiers).

Example 10.27. *For a class B amplifier using a supply of V_{CC} = 12 V and driving a load of 8 Ω, determine (i) maximum load power (ii) d.c. input power (iii) power efficiency.*

Solution. V_{CC} = 12 V ; R_L = 8 Ω

(*i*) Maximum load power, $P_{o\,(max)} = 0.25\, V_{CC}\, I_{C\,(sat)}$

$$= 0.25\, V_{CC} \times \frac{V_{CC}}{2\,R_L} \qquad \left(\because\; I_{C\,(sat)} = \frac{V_{CC}}{2\,R_L}\right)$$

$$= 0.25 \times 12 \times \frac{12}{2 \times 8} = \mathbf{2.25\ W}$$

(*ii*) D.C. input power, $P_{dc} = \dfrac{V_{CC}\, I_{C\,(sat)}}{\pi} = \dfrac{V_{CC}}{\pi} \times \dfrac{V_{CC}}{2\,R_L}$

$$= \frac{12}{\pi} \times \frac{12}{2 \times 8} = \mathbf{2.87\ W}$$

(*iii*) Power $\eta = \dfrac{P_{o\,(max)}}{P_{dc}} \times 100 = \dfrac{2.25}{2.87} \times 100 = \mathbf{78.4\%}$

Example 10.28. *A class B push-pull amplifier with transformer coupled load uses two transistors rated 10 W each. What is the maximum power output one can obtain at the load from the circuit ?*

Solution. The power dissipation by each transistor is P_T = 10 W. Therefore, power dissipated by two transistors is P_{2T} = 2 × 10 = 20W.

Now, $P_{dc} = P_{o\,(max)} + P_{2T}$; Max. Power η = 0.785

$$\therefore \quad \text{Max power } \eta = \frac{P_{o\,(max)}}{P_{dc}} = \frac{P_{o\,(max)}}{P_{o\,(max)} + P_{2T}} = \frac{P_{o\,(max)}}{P_{o\,(max)} + 20}$$

$$\text{or} \quad 0.785 = \frac{P_{o\,(max)}}{P_{o\,(max)} + 20}$$

$$\text{or} \quad 0.785\, P_{o\,(max)} + 15.7 = P_{o\,(max)}$$

$$\text{or} \quad P_{o\,(max)}\,(1 - 0.785) = 15.7$$

$$\therefore \quad P_{o\,(max)} = \frac{15.7}{1 - 0.785} = \frac{15.7}{0.215} = \mathbf{73.02\ W}$$

Example 10.29. *A class B push-pull amplifier has a power efficiency of 60% and each transistor has a rating of 2.5W. Find the a.c. output power and d.c. input power.*

Solution. The power dissipated by each transistor is $P_T = 2.5$ W.

Therefore, power dissipated by the two transistors is $P_{2T} = 2 \times 2.5 = 5$ W.

$$\text{Now,} \quad P_{dc} = P_{ac} + P_{2T}\ ; \ \eta = 0.6$$

$$\therefore \quad \eta = \frac{P_{ac}}{P_{dc}} = \frac{P_{ac}}{P_{ac} + P_{2T}}$$

$$\text{or} \quad 0.6 = \frac{P_{ac}}{P_{ac} + 5} \quad \text{or} \quad 0.6\, P_{ac} + 3 = P_{ac}$$

$$\therefore \quad P_{ac} = \frac{3}{1 - 0.6} = \frac{3}{0.4} = \mathbf{7.5\ W}$$

$$\text{and} \quad P_{dc} = P_{ac} + P_{2T} = 7.5 + 5 = \mathbf{12.5\ W}$$

Example 10.30. *A class B amplifier uses V_{CC} = 10 V and drives a load of 10Ω. Determine the end point values of the a.c. load line.*

Solution.

$$I_{C\,(sat)} = \frac{V_{CC}}{2\,R_L} = \frac{10\text{ V}}{2(10\ \Omega)} = \mathbf{500\ mA}$$

This locates one end-point of the a.c. load line on the collector current axis.

$$V_{CE\,(off)} = \frac{V_{CC}}{2} = \frac{10\text{ V}}{2} = \mathbf{5\ V}$$

This locates the second end-point of the a.c load line on the collector-emitter voltage axis. By joining these two points, the a.c. load line of the amplifier is constructed.

10.20 Complementary-Symmetry Amplifier

By complementary symmetry is meant a principle of assembling push-pull class *B* amplifier without requiring centre-tapped transformers at the input and output stages. Fig. 10.28 shows the transistor push-pull amplifier using complementary symmetry. It employs one *npn* and one *pnp* transistor and requires no centre-tapped transformers. The circuit action is as follows. During the positive-half of the input signal, transistor T_1 (the *npn* transistor) conducts current while T_2 (the *pnp* transistor) is cut-off. During the negative half-cycle of the signal, T_2 conducts while T_1 is cut-off. In this way, *npn* transistor amplifies the positive half-cycles of the signal while the *pnp* transistor amplifies the negative half-cycles of the signal. Note that we generally use an output transformer (not centre-tapped) for impedance matching.

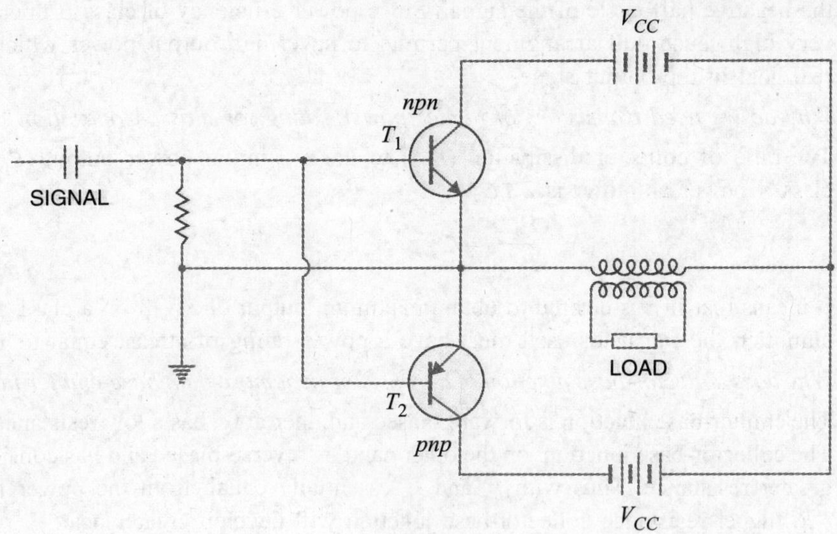

Fig. 10.28

Advantages

(*i*) This circuit does not require transformer. This saves on weight and cost.

(*ii*) Equal and opposite input signal voltages are not required.

Disadvantages

(*i*) It is difficult to get a pair of transistors (*npn* and *pnp*) that have similar characteristics.

(*ii*) We require both positive and negative supply voltages.

Short Answer Questions

Q.1. *Why do power amplifiers handle large signals?*

Ans. A practical amplifier is always multistage ; the first few stages having the job of voltage amplification and the last stage doing power amplification. Due to voltage amplification in the initial stages, the input a.c. signals to power amplifiers have large magnitude.

Q.2. *Why does power efficiency play important part in power amplifiers?*

Ans. Because it tells how well a power amplifier converts d.c. power from supply into a.c. output power. Since the goal in a power amplifier is the maximum a.c. power output, power efficiency is an important consideration in such amplifiers.

Q.3. *Why does the problem of distortion arise in power amplifiers?*

Ans. Because power amplifiers handle large signals. Many of them are driven so hard by the input signal that collector current is either cut-off or is in the saturation region. Since the transistor nonlinearities are greater near the limits of its operation, distortion is a notable feature of power amplifiers.

Q.4. *Why does the output stage employ push-pull arrangement?*

Ans. In push-pull arrangement, two transistors are used in class *B* operation. One transistor amplifies the positive half-cycle of the signal and other transistor amplifies the negative half-cycle of the signal. Since power efficiency of class *B* operation is very high, push-pull arrangment permits to have high output power which is the real goal in the output stage.

Q.5. *Why do we need transistors of higher power rating for class A operation?*

Ans. The ratio of collector dissipation (P_{dis}) to the maximum power output (P_{out}) in a class *A* power amplifier is 2 *i.e.*

$$\frac{P_{dis}}{P_{out}} = 2$$

It means that if it is desired to obtain maximum output of 5W from a class *A* power amplifier, the transistor used must have a power rating of atleast equal to 10W.

Q.6. *Why does collector-base junction of a transistor heat more than base-emitter junction?*

Ans. The emitter-base junction is forward biased and, therefore, has a low resistance value. The collector-base junction, on the other hand, is reverse-biased and has considerably greater resistance. Thus with I_E and I_C essentially equal, from the power formula I^2R, higher-resistance collector-base junction will develop greater heat.

Q.7. *What are the advantages of push-pull circuit over the single ended class A amplifier?*

Ans. The following are the advantages of push-pull circuit over single-ended circuit :

(*i*) The push-pull circuit tends to cancel the even-order harmonics (*i.e.* 2nd, 4th *etc.*) that may be developed in the stage.

(*ii*) The d.c. collector current of a single ended amplifier tends to magnetise the core of the output transformer, resulting in distortion. In push-pull circuit, the collector currents flow in opposite directions, thereby preventing core saturation. Thus push-pull circuit produces less distortion.

(*iii*) Push-pull circuit operates in class *B* operation and, therefore, has much higher power efficiency than the single-ended circuit which operates in class *A* mode. This is the reason why push-pull circuit is preferable to single-ended circuit when large value of audio output power is required.

(*iv*) The push-pull circuit tends to cancel *hum* due to any power supply ripple.

Q.8. *Why is class C operation so efficient?*

Ans. In class *C* operation, the collector current flows for less than half-cycle (*i.e.* < 180°) for the full input cycle of the signal. In other words, collector current flows only while the collector voltage is relatively low. This keeps the power loss to a low value and accounts for high collector efficiency of class *C* amplifiers. Well designed class *C* amplifiers are capable of efficiencies as high as 85%.

Q.9. *Why is a push-pull circuit called so?*

Ans. A push-pull circuit normally employs two transistors in class *B* operation, one transistor amplifies the positive half-cycle of the signal and the other amplifies negative half-cycle. The circuit operation is such that when one transistor is conducting (*pushing*), the other is stopping conduction (*pulling*).

Q.10. *Can impedance matching be always obtained in practice?*

Ans. Generally not. It is because usually transformer of desired turn ratio is not available. However, the problem can be solved by connecting, in addition to the available transformer, a suitable resistance across the secondary.

Multiple-Choice Questions

1. The output stage of a multistage amplifier is also called ....
   (*i*) mixer stage (*ii*) power stage
   (*iii*) detector stage (*iv*) *R.F.* stage
2. .... coupling is generally employed in power amplifiers.
   (*i*) Transformer (*ii*) *RC*
   (*iii*) Direct (*iv*) Impedance
3. A class *A* power amplifier uses ....
   (*i*) two transistors (*ii*) three transistors
   (*iii*) one transistor
   (*iv*) none of the above
4. The maximum efficiency of resistance loaded class *A* power amplifier is ....
   (*i*) 78.5% (*ii*) 50%
   (*iii*) 30% (*iv*) 25%
5. The maximum efficiency of transformer coupled class *A* power amplifier is ....
   (*i*) 30% (*ii*) 50%
   (*iii*) 80% (*iv*) 45%
6. Class .... power amplifier has the highest power efficiency.
   (*i*) *C* (*ii*) *A*
   (*iii*) *B* (*iv*) *AB*
7. Power amplifiers handle .... signals compared to voltage amplifiers.
   (*i*) small (*ii*) very small
   (*iii*) large
   (*iv*) none of the above
8. In class *A* operation, the operating point is generally located .... of the d.c. load line.
   (*i*) at cut-off point
   (*ii*) at the middle
   (*iii*) at saturation point
   (*iv*) none of the above
9. Class *C* amplifiers are used as ....
   (*i*) *A.F.* amplifiers (*ii*) detectors
   (*iii*) *R.F.* amplifiers
   (*iv*) none of the above
10. A power amplifier has comparatively ....β.
   (*i*) small (*ii*) large
   (*iii*) very large
   (*iv*) none of the above
11. The maximum power efficiency of class *B* operation is ....
   (*i*) 50% (*ii*) 90%
   (*iii*) 60.5% (*iv*) 78.5%
12. A 2-transistor class *B* power amplifier is commonly called .... amplifier.
   (*i*) dual (*ii*) push-pull
   (*iii*) symmetrical (*iv*) differential
13. If a transistor is operated in such a way that output current flows for 60° of the input signal, then it is ...... operation.
   (*i*) class *A* (*ii*) class *B*
   (*iii*) class *C*
   (*iv*) none of the above
14. If the zero signal power dissipation of a transistor is 1 W, then power rating of the transistor should be atleast ....
   (*i*) 0.5 W (*ii*) 0.33 W
   (*iii*) 0.75 W (*iv*) 1 W
15. When a transistor is cut-off, ...
   (*i*) maximum voltage appears across transistor
   (*ii*) maximum current flows
   (*iii*) maximum voltage appears across load
   (*iv*) none of the above
16. A class *A* power amplifier is sometimes called .... amplifier.
   (*i*) symmetrical (*ii*) single-ended
   (*iii*) reciprocating (*iv*) differential
17. Class .... operation gives the maximum distortion.
   (*i*) *A* (*ii*) *B*
   (*iii*) *C* (*iv*) *AB*
18. The output stage of a multistage amplifier usually employs ....
   (*i*) push-pull amplifier
   (*ii*) preamplifier

(*iii*) class *A* power amplifier
(*iv*) none of the above

19. The size of a power transistor is made considerably large to ....
(*i*) provide easy handling
(*ii*) dissipate heat
(*iii*) facilitate connections
(*iv*) none of the above

20. Low efficiency of a power amplifier results in ....
(*i*) low forward bias
(*ii*) less battery consumption
(*iii*) more battery consumption
(*iv*) none of the above

21. The driver stage usually employs ....
(*i*) class *A* power amplifier
(*ii*) push-pull amplifier
(*iii*) class *C* amplifier
(*iv*) none of the above

22. If the power rating of a transistor is 1 W and collector current is 100 mA, then maximum allowable collector voltage is ....
(*i*) 1 V (*ii*) 100 V
(*iii*) 20 V (*iv*) 10 V

23. When no signal is applied, the approximate power efficiency of class *A* power amplifier is ....
(*i*) 10% (*ii*) 0%
(*iii*) 25% (*iv*) 50%

24. What will be the power efficiency of a power amplifier having zero signal power dissipation of 5 watts and a.c. power output of 2 watts?
(*i*) 20% (*ii*) 80%
(*iii*) 40% (*iv*) 50%

25. The output signal voltage and current of a power amplifier are 5 V and 200 mA; the values being r.m.s. What is the power output?
(*i*) 1 W (*ii*) 2 W
(*iii*) 4 W
(*iv*) none of the above

26. The maximum a.c. power output from a class *A* power amplifier is 10 W. What should be the minimum power rating of the transistor used?
(*i*) 10 W (*ii*) 15 W
(*iii*) 5 W (*iv*) 20 W

27. For the same a.c. power output as above, what should be the minimum power rating of transistor for class *B* operation?
(*i*) 10 W (*ii*) 4 W
(*iii*) 8 W
(*iv*) none of the above

28. The push-pull circuit must use ..... operation.
(*i*) class *A* (*ii*) class *C*
(*iii*) class *B* (*iv*) class *AB*

29. The class *B* push-pull circuit can deliver 100 W of a.c. output power. What should be the minimum power rating of each transistor?
(*i*) 20 W (*ii*) 40 W
(*iii*) 10 W (*iv*) 80 W

30. What turn ratio (N_P/N_S) of transformer is required to match 4 Ω speaker to a transistor having an output impedance of 8000 Ω?
(*i*) 35.2 (*ii*) 44.7
(*iii*) 54.3
(*iv*) none of the above

31. A transformer coupled class *A* power amplifier has a load of 100 Ω on the secondary. If the turn ratio is 10: 1, what is the value of load appearing on the primary?
(*i*) 5 kΩ (*ii*) 20 kΩ
(*iii*) 100 kΩ (*iv*) 10 kΩ

32. Power amplifiers generally use transformer coupling because transformer permits ....
(*i*) cooling of the circuit
(*ii*) impedance matching
(*iii*) distortionless output
(*iv*) good frequency response

33. Transformer coupling can be used in .... amplifiers.

(*i*) either power or voltage
(*ii*) only power
(*iii*) only voltage
(*iv*) none of the above

34. The output transformer used in a power amplifier is a ...... transformer.

(*i*) 1: 1 ratio (*ii*) step-up
(*iii*) step-down
(*iv*) none of the above

35. The most important consideration in power amplifiers is......

(*i*) biasing the circuit
(*ii*) power efficiency
(*iii*) to keep the transformer cool
(*iv*) none of the above

36. An *A.F.* amplifier is shielded to ....

(*i*) keep the amplifier cool
(*ii*) protect from rusting
(*iii*) prevent induction due to stray magnetic fields
(*iv*) none of the above

37. The pulsating d.c. applied to power amplifier causes ....

(*i*) burning of transistor
(*ii*) hum in the circuit
(*iii*) excessive forward voltage
(*iv*) none of the above

38. The disadvantage of impedance matching is that it ....

(*i*) gives distorted output
(*ii*) gives low power output
(*iii*) requires a transformer
(*iv*) none of the above

39. If the gain versus frequency curve of a transistor amplifier is not flat, then there is .... distortion.

(*i*) amplitude
(*ii*) intermodulation
(*iii*) frequency
(*iv*) none of the above

40. The most costly coupling is .... coupling.

(*i*) *RC* (*ii*) direct
(*iii*) impedance (*iv*) transformer

Answers to Multiple-Choice Questions

1. (*ii*)	**2.** (*i*)	**3.** (*iii*)	**4.** (*iv*)	**5.** (*ii*)
6. (*i*)	**7.** (*iii*)	**8.** (*ii*)	**9.** (*iii*)	**10.** (*i*)
11. (*iv*)	**12.** (*ii*)	**13.** (*iii*)	**14.** (*iv*)	**15.** (*i*)
16. (*ii*)	**17.** (*iii*)	**18.** (*i*)	**19.** (*ii*)	**20.** (*iii*)
21. (*i*)	**22.** (*iv*)	**23.** (*ii*)	**24.** (*iii*)	**25.** (*i*)
26. (*iv*)	**27.** (*ii*)	**28.** (*iii*)	**29.** (*i*)	**30.** (*ii*)
31. (*iv*)	**32.** (*ii*)	**33.** (*i*)	**34.** (*iii*)	**35.** (*ii*)
36. (*iii*)	**37.** (*ii*)	**38.** (*i*)	**39.** (*iii*)	**40.** (*iv*)

Chapter Review Topics

1. What is an audio power amplifier? What is its need?
2. Explain the difference between a voltage and a power amplifier.
3. What do you understand by class *A*, class *B* and class *C* power amplifiers?
4. Define and explain the following terms as applied to power amplifiers :
   (*i*) power efficiency (*ii*) distortion (*iii*) power dissipation capability
5. Show that maximum power efficiency of class *A* transformer coupled power amplifier is 50%.

6. Draw the block diagram of a practical power amplifier.
7. Explain the push-pull circuit with a neat diagram.
8. Write short notes on the following :

   (*i*) Heat sink (*ii*) Driver stage

   (*iii*) Output stage (*iv*) Complementary-symmetry amplifier

Problems

1. The resistance of the secondary of an output transformer is 100 Ω. If the output impedance is 10 kΩ, find the turn ratio of the transformer for maximum power transference. **[*n* = 10]**
2. A power transistor working in class *A* operation has zero signal power dissipation of 5 watts. If a.c. output power is 2 watts, find (*i*) power efficiency (*ii*) power rating of transistor. **[(*i*) 40% (*ii*) 5 watts]**
3. A class *A* power amplifier has a maximum a.c. power output of 30 W. Find the power rating of the transistor. **[60 W]**
4. The a.c. power output of a class *A* power amplifier is 2 W. If the power efficiency is 40%, find the power rating of the transistor. **[5 W]**
5. In a class *A* transformer coupled amplifier, collector current alternates between 3 mA and 110 mA and its quiescent value is 58 mA. The load resistance is 15 Ω and when referred to primary winding is 325 Ω. The supply voltage is 20V. Find (*i*) transformer turn ratio (*ii*) a.c. power output (*iii*) power rating of transistor.
6. A transistor has thermal resistance θ = 80°C/W. If the maximum junction temperature is 90°C and the ambient temperature is 30°C, find the maximum permissible power dissipation. **[750 mW]**
7. A power transistor dissipates 4 W. If $T_{J\,max}$ = 90°C, find the maximum ambient temperature at which it can be operated. Given thermal resistance θ = 8°C/W. **[58°C]**
8. A class *A* transformer-coupled amplifier uses a 25 : 1 transformer to drive a 4 Ω load. Calculate the effective a.c. load (seen by the transistor connected to the larger turns side of the transformer). **[2.5 kΩ]**
9. Calculate the transformer turn ratio required to connect 4 parallel 16 Ω speakers so that they appear as an 8 kΩ effective load. **[44.7]**
10. For a class *B* amplifier with V_{CC} = 25 V driving an 8 Ω load, determine :

    (*i*) maximum input power

    (*ii*) maximum output power

    (*iii*) maximum power efficiency **[(*i*) 49.7 W (*ii*) 39.06 W (*iii*) 78.5 %]**

CHAPTER 11

Amplifiers with Negative Feedback

Learning Objectives

- Understand the importance of negative feedback in amplifiers.
- Derive an expression for voltage gain of negative voltage feedback amplifier.
- Explore the advantages of negative voltage feedback in amplifiers.
- Develop an understanding of negative current feedback in amplifiers.
- Understand the characteristics of emitter follower e.g. voltage gain, d.c. analysis, input impedance and output impedance.
- Explore the importance of Darlington Amplifier.

Introduction

A practical amplifier has a gain of nearly one million *i.e.* its output is one million times the input. Consequently, even a casual disturbance at the input will appear in the amplified form in the output. There is a strong tendency in amplifiers to introduce *hum* due to sudden temperature changes or stray electric and magnetic fields. Therefore, every high gain amplifier tends to give noise along with signal in its output. The noise in the output of an amplifier is undesirable and must be kept to as small a level as possible.

The noise level in amplifiers can be reduced considerably by the use of *negative feedback* *i.e.* by injecting a fraction of output in phase opposition to the input signal. The object of this chapter is to consider the effects and methods of providing negative feedback in transistor amplifiers.

11.1 Feedback

The process of injecting a fraction of output energy of some device back to the input is known as **feedback**.

The principle of feedback is probably as old as the invention of first machine but it is only some 75 years ago that feedback has come into use in connection with electronic circuits. It has been found very useful in reducing noise in amplifiers and making amplifier operation stable. Depending upon whether the feedback energy aids or opposes the input signal, there are two basic types of feedback in amplifiers *viz. positive feedback* and *negative feedback*.

(i) **Positive feedback.** When the feedback energy (voltage or current) is in phase with the input signal and thus aids it, it is called *positive feedback*. This is illustrated in Fig. 11.1. Both amplifier and feedback network introduce a phase shift of 180°. The result is a 360° phase shift around the loop, causing the *feedback voltage* V_f to be in phase with the input signal V_{in}.

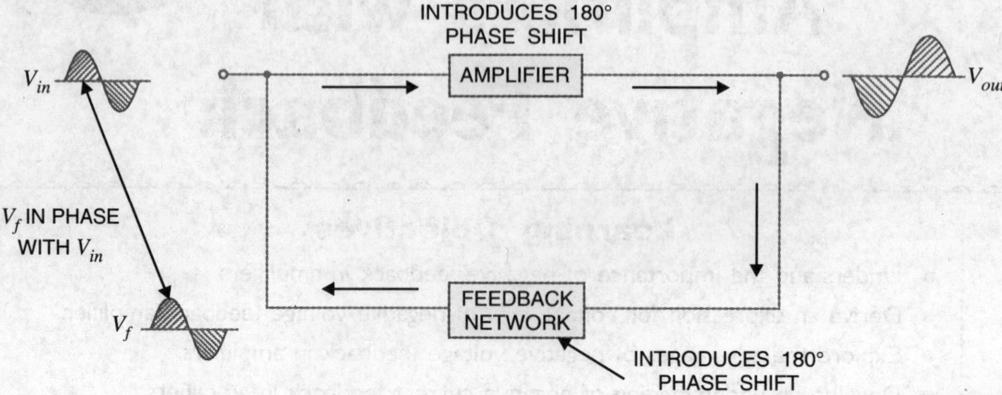

Fig. 11.1

The positive feedback increases the gain of the amplifier. However, it has the disadvantages of increased distortion and instability. Therefore, positive feedback is seldom employed in amplifiers. One important use of positive feedback is in oscillators. As we shall see in the next chapter, if positive feedback is sufficiently large, it leads to oscillations. As a matter of fact, an oscillator is a device that converts d.c. power into a.c. power of any desired frequency.

(ii) **Negative feedback.** When the feedback energy (voltage or current) is out of phase with the input signal and thus opposes it, it is called *negative feedback*. This is illustrated in Fig. 11.2. As you can see, the amplifier introduces a phase shift of 180° into the circuit while the feedback network is so designed that it introduces no phase shift (*i.e.* 0° phase shift). The result is that the *feedback voltage* V_f is 180° out of phase with the input signal V_{in}.

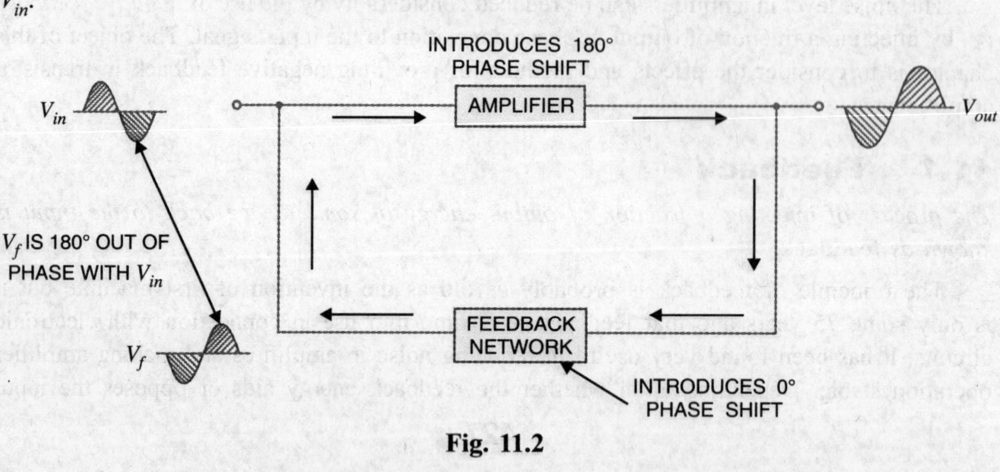

Fig. 11.2

Negative feedback reduces the gain of the amplifier. However, the advantages of negative feedback are : reduction in distortion, stability in gain, increased bandwidth and improved input and output impedances. It is due to these advantages that negative feedback is frequently employed in amplifiers.

11.2 Types of Negative Feedback in Amplifiers

There are two basic types of negative feedback that are used in amplifiers *viz.* (*i*) negative voltage feedback (*ii*) negative current feedback. In negative feedback (voltage or current), a fraction of the output signal is continuously fed back (called *feedback signal*) to the input side and is subtracted from the input signal to create a signal (= input signal–feedback signal) which is then applied to the input of the amplifier. It is clear that negative voltage feedback would decrease input voltage to the amplifier and hence its voltage gain will be reduced. Likewise, negative current feedback reduces the current gain of the amplifier because input current to the amplifier is decreased. However, at the cost of reduced gain (voltage or current), negative feedback in amplifiers offers such advantages as gain stability, modification (increase or decrease) of input and output impedances, reduced distortion, increased bandwidth *etc*. Thus negative feedback improves the performance of the amplifier. It may be mentioned here that loss of gain due to negative feedback can be compensated by adding extra amplifier stage/stages.

(*i*) **Negative voltage feedback.** In negative voltage feedback, a fraction of the output voltage (called *feedback voltage* v_f) of the amplifier is fed back to the signal voltage (v_s) in such a way that the difference of v_s and v_f (*i.e.* $v_s - v_f$) is applied to the input of the amplifier as shown in Fig. 11.3. The output voltage v_0 of the amplifier is applied to the feedback circuit which reduces v_0 to produce the feedback voltage v_f. The feedback circuit is simply a resistive voltage divider. The feedback voltage v_f is applied in series with the signal v_s. As v_f is *180° out of phase with v_s, the input voltage v_i to the amplifier is equal to the *difference* of the two *i.e.*

$$\text{Input voltage, } v_i = v_s - v_f$$

Thus negative voltage feedback has reduced the input voltage to the amplifier. Therefore, the output voltage is lower than that produced when negative feedback is not used. This means that voltage gain of the amplifier is reduced with negative voltage feedback.

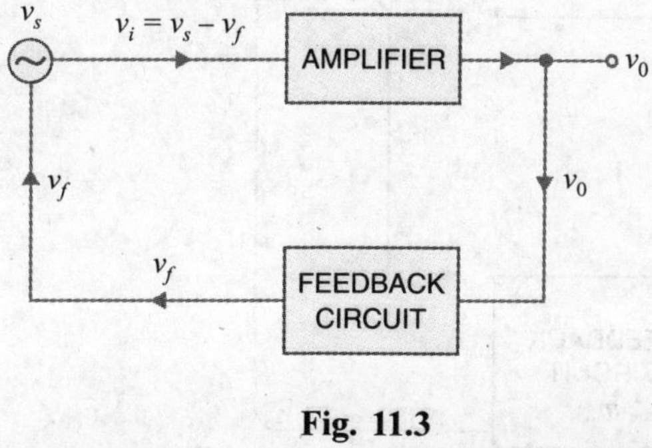

Fig. 11.3

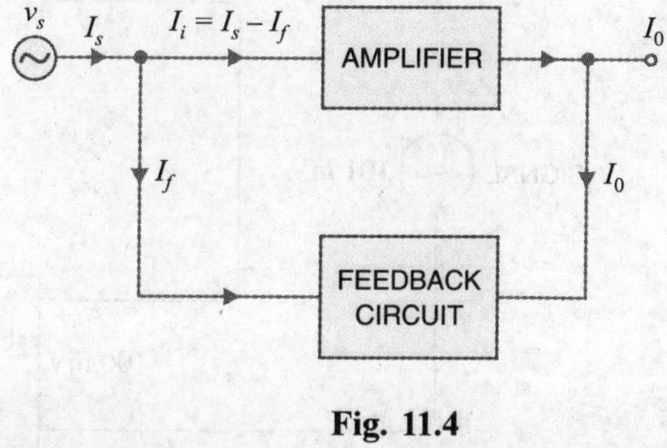

Fig. 11.4

* v_0 is 180° out of phase with v_s. Since feedback circuit is resistive, it produces zero degree phase shift. Therefore, v_f is 180° out of phase with v_s.

(ii) **Negative current feedback.** In negative current feedback, a fraction of the output current (called *feedback current* I_f) of the amplifier is fed back to the signal current (I_s) in such a manner that the difference of I_s and I_f (*i.e.* $I_s - I_f$) is fed to the input of the amplifier as shown in Fig. 11.4. The fact that the arrowhead of the feedback current (I_f) being fed forward is because it is *negative feedback*. It is clear that input current I_i to the amplifier is equal to the *difference* of the two *i.e.*

$$\text{Input current, } I_i = I_s - I_f$$

Again, the negative current feedback has reduced the input current to the amplifier. Therefore, the output current is lower than that produced when negative feedback is not used. This means that current gain of the amplifier is reduced with negative current feedback.

Discussion. The following points may be noted about negative feedback :

(i) The negative voltage feedback decreases the voltage gain (A_v) and output impedance (Z_{out}) of the amplifier. It does not affect the current gain of the amplifier. However, negative voltage feedback will either increase or decrease the input impedance (Z_{in}) of the amplifier, depending on the circuit arrangement.

(ii) The negative current feedback decreases current gain (A_i) and increases output impedance (Z_{out}) of the amplifier. It does not affect the voltage gain of the amplifier. However, negative current feedback will either increase or decrease the input impedance (Z_{in}) of the amplifier, depending on the circuit arrangement.

(iii) Since power gain $A_p = A_v A_i$, both types of negative feedback will decrease the power gain of the amplifier. However, *the drawback of reduced power gain is offset by the advantage of increased bandwidth.*

11.3 Principles of Negative Voltage Feedback in Amplifiers

A feedback amplifier has two parts *viz.* an amplifier and a feedback circuit. The feedback circuit usually consists of resistors and returns a fraction of output energy back to the input. Fig. 11.5 *shows the principles of negative voltage feedback in an amplifier. Typical values

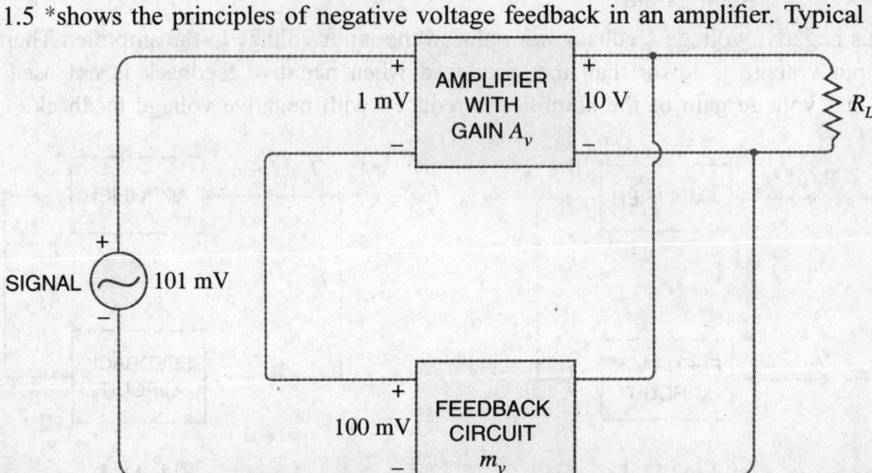

Fig. 11.5

* Note that amplifier and feedback circuits are connected in *series-parallel.* The inputs of amplifier and feedback circuits are in *series* but the outputs are in *parallel.* In practice, this circuit is widely used.

have been assumed to make the treatment more illustrative. The output of the amplifier is 10 V. The fraction m_v of this output *i.e.* 100 mV is fed back to the input where it is applied in series with the input signal of 101 mV. As the feedback is negative, therefore, only 1 mV appears at the input terminals of the amplifier.

Referring to Fig. 11.5, we have,

$$\text{Gain of amplifier without feedback, } A_v = \frac{10\text{ V}}{1\text{ mV}} = 10{,}000$$

$$\text{Fraction of output voltage fed back, } m_v = \frac{100\text{ mV}}{10\text{ V}} = 0.01$$

$$\text{Gain of amplifier with negative feedback, } A_{vf} = \frac{10\text{ V}}{101\text{ mV}} = 100$$

The following points are worth noting :

(*i*) When negative voltage feedback is applied, the gain of the amplifier is *reduced. Thus, the gain of above amplifier without feedback is 10,000 whereas with negative feedback, it is only 100.

(*ii*) When negative voltage feedback is employed, the voltage *actually* applied to the amplifier is extremely small. In this case, the signal voltage is 101 mV and the negative feedback is 100 mV so that voltage applied at the input of the amplifier is only 1 mV.

(*iii*) In a negative voltage feedback circuit, the feedback fraction m_v is always between 0 and 1.

(*iv*) The gain with feedback is sometimes called *closed-loop gain* while the gain without feedback is called *open-loop gain*. These terms come from the fact that amplifier and feedback circuits form a "loop". When the loop is "opened" by disconnecting the feedback circuit from the input, the amplifier's gain is A_v, the "open-loop" gain. When the loop is "closed" by connecting the feedback circuit, the gain decreases to A_{vf}, the "closed-loop" gain.

11.4 Gain of Negative Voltage Feedback Amplifier

Consider the negative voltage feedback amplifier shown in Fig. 11.6. The gain of the amplifier without feedback is A_v. Negative feedback is then applied by feeding a fraction m_v of the output voltage v_0 back to amplifier input. Therefore, the actual input to the amplifier is the signal voltage v_s *minus* feedback voltage $m_v\, v_0$ *i.e.*

$$\text{Actual input to amplifier, } v_i = v_s - m_v\, v_0$$

The output voltage v_0 must be equal to the input voltage $v_s - m_v\, v_0$ multiplied by gain A_v of the amplifier *i.e.*

$$(v_s - m_v\, v_0)\, A_v = v_0$$

or $$A_v\, v_s - A_v\, m_v\, v_0 = v_0$$

or $$v_0\,(1 + A_v\, m_v) = A_v\, v_s$$

or $$\frac{v_0}{v_s} = \frac{A_v}{1 + A_v\, m_v}$$

But v_0/v_s is the voltage gain of the amplifier with negative voltage feedback.

∴ Voltage gain with negative voltage feedback is

* Since with negative voltage feedback the voltage gain is decreased and current gain remains unaffected, the power gain A_p (= $A_v \times A_i$) will decrease. However, the drawback of reduced power gain is offset by the advantage of increased bandwidth.

$$A_{vf} = \frac{A_v}{1 + A_v\ m_v}$$

It may be seen that the gain of the amplifier without feedback is A_v. However, when negative voltage feedback is applied, the gain is reduced by a factor $1 + A_v\ m_v$. It may be noted that negative voltage feedback does not affect the current gain of the circuit.

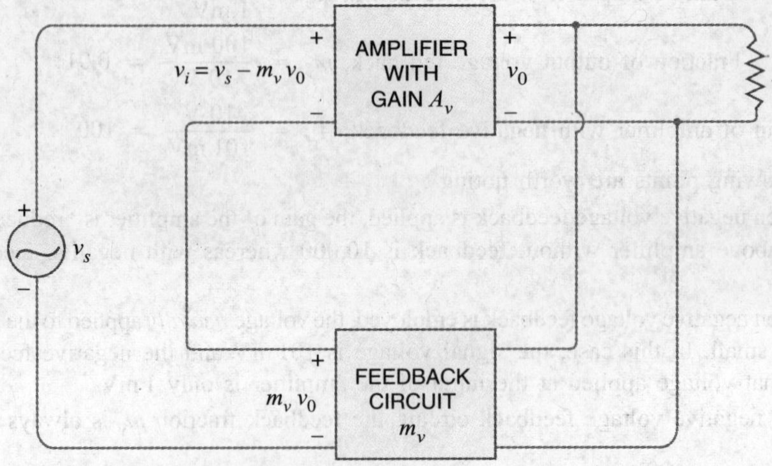

Fig. 11.6

Discussion. The following points are worth noting :

(*i*) The various relations in negative voltage feedback amplifier are :

$$\text{Open-loop voltage gain, } A_v = \frac{v_0}{v_i}$$

$$\text{Feedback fraction, } m_v = \frac{v_f}{v_0}$$

$$\text{Closed-loop voltage gain, } A_{vf} = \frac{A_v}{1+A_v m_v}$$

(*ii*) The quantity $(1 + A_v m_v)$ is called **feedback factor** of the negative voltage feedback amplifier. When negative voltage feedback is applied to an amplifier the voltage gain, input/output impedance and frequency response are controlled by this factor.

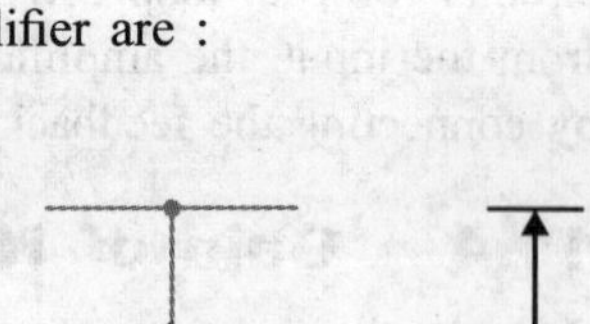

Fig. 11.7

(*iii*) The feedback circuit provides a path from the amplifier output back to the input. The purpose of this path is to "feed" a portion of the output signal back to the amplifier input. The feedback circuit may be as simple as the resistive voltage divider as shown in Fig. 11.7. The output voltage v_0 of the amplifier is applied to the input of the feedback circuit which reduces v_0 by a fraction m_v to produce a feedback voltage $v_f\ (= m_v v_0)$.

$$\text{Clearly,} \quad m_v = \frac{v_f}{v_0}$$

Since feedback circuit is resistive, it produces no phase shift in the output voltage.

Example 11.1. *The voltage gain of an amplifier without feedback is 3000. Calculate the voltage gain of the amplifier if negative voltage feedback is introduced in the circuit. Given that feedback fraction m_v = 0.01.*

Solution. $A_v = 3000 \; ; \quad m_v = 0.01$

∴ Voltage gain with negative voltage feedback is

$$A_{vf} = \frac{A_v}{1 + A_v\, m_v} = \frac{3000}{1 + 3000 \times 0.01} = \frac{3000}{31} = 97$$

Example 11.2. *The overall gain of a multistage amplifier is 140. When negative voltage feedback is applied, the gain is reduced to 17.5. Find the fraction of the output that is fed back to the input.*

Solution. $A_v = 140 \; ; \quad A_{vf} = 17.5$

Let m_v be the feedback fraction. Voltage gain with negative voltage feedback is

$$A_{vf} = \frac{A_v}{1 + A_v\, m_v}$$

or $$17.5 = \frac{140}{1 + 140\, m_v}$$

or $$17.5 + 2450\, m_v = 140$$

∴ $$m_v = \frac{140 - 17.5}{2450} = \frac{1}{20}$$

Example 11.3. *When negative voltage feedback is applied to an amplifier of gain 100, the overall gain falls to 50.*

(i) Calculate the fraction of the output voltage fed back.

(ii) If this fraction is maintained, calculate the value of the amplifier gain required if the overall stage gain is to be 75.

Solution.

(i) Gain without feedback, $A_v = 100$

Gain with feedback, $A_{vf} = 50$

Let m_v be the fraction of the output voltage fed back.

Now, $$A_{vf} = \frac{A_v}{1 + A_v\, m_v}$$

or $$50 = \frac{100}{1 + 100\, m_v}$$

or $$50 + 5000\, m_v = 100$$

or $$m_v = \frac{100 - 50}{5000} = 0.01$$

(ii) $A_{vf} = 75 \; ; \quad m_v = 0.01 \; ; \quad A_v = ?$

$$A_{vf} = \frac{A_v}{1 + A_v\, m_v}$$

or $$75 = \frac{A_v}{1 + 0.01\, A_v}$$

or $$75 + 0.75\,A_v = A_v$$

$$\therefore \quad A_v = \frac{75}{1-0.75} = 300$$

Example 11.4. *With a negative voltage feedback, an amplifier gives an output of 10 V with an input of 0.5 V. When feedback is removed, it requires 0.25 V input for the same output. Calculate (i) gain without feedback (ii) feedback fraction m_v.*

Solution.

(*i*) Gain without feedback, $A_v = 10/0.25 = $ **40**

(*ii*) Gain with feedback, $A_{vf} = 10/0.5 = 20$

Now, $$A_{vf} = \frac{A_v}{1 + A_v\, m_v}$$

or $$20 = \frac{40}{1 + 40\, m_v}$$

or $$20 + 800\, m_v = 40$$

or $$m_v = \frac{40-20}{800} = \mathbf{\frac{1}{40}}$$

Example 11.5. *The gain of an amplifier without feedback is 50 whereas with negative voltage feedback, it falls to 25. If due to ageing, the amplifier gain falls to 40, find the percentage reduction in stage gain (i) without feedback and (ii) with negative feedback.*

Solution. $$A_{vf} = \frac{A_v}{1 + A_v m_v}$$

or $$25 = \frac{50}{1 + 50\, m_v}$$

or $$m_v = 1/50$$

(*i*) **Without feedback.** The gain of the amplifier without feedback is 50. However, due to ageing, it falls to 40.

$$\therefore \quad \%\text{ age reduction in stage gain} = \frac{50-40}{50} \times 100 = \mathbf{20\%}$$

(*ii*) **With negative feedback.** When the gain without feedback was 50, the gain with negative feedback was 25. Now the gain without feedback falls to 40.

$$\therefore \quad \text{New gain with negative feedback} = \frac{A_v}{1 + A_v\, m_v} = \frac{40}{1 + (40 \times 1/50)} = 22.2$$

$$\therefore \quad \%\text{ age reduction in stage gain} = \frac{25-22.2}{25} \times 100 = \mathbf{11.2\%}$$

Example 11.6. *An amplifier has a voltage amplification A_v and a fraction m_v of its output is fed back in opposition to the input. If $m_v = 0.1$ and $A_v = 100$, calculate the percentage change in the gain of the system if A_v falls 6 dB due to ageing.*

Solution. $$A_v = 100\,;\quad m_v = 0.1\,;\quad A_{vf} = ?$$

$$A_{vf} = \frac{A_v}{1 + A_v\, m_v} = \frac{100}{1 + 100 \times 0.1} = 9.09$$

$$\text{Fall in gain} = 6 \text{ dB}$$

Let A_{v1} be the new absolute voltage gain without feedback.

Then, $$20 \log_{10} A_v/A_{v1} = 6$$

or $$\log_{10} A_v/A_{v1} = 6/20 = 0.3$$

or $$\frac{A_v}{A_{v1}} = \text{Antilog } 0.3 = 2$$

or $$A_{v1} = A_v/2 = 100/2 = 50$$

∴ $$\text{New } A_{vf} = \frac{A_{v1}}{1 + A_{v1} m_v} = \frac{50}{1 + 50 \times 0.1} = 8.33$$

$$\%\text{ age change in system gain} = \frac{9.09 - 8.33}{9.09} \times 100 = \mathbf{8.36\%}$$

Example 11.7. *An amplifier has a voltage gain of 500 without feedback. If a negative feedback is applied, the gain is reduced to 100. Calculate the fraction of the output fed back. If, due to ageing of components, the gain without feedback falls by 20%, calculate the percentage fall in gain with feedback.*

Solution. $$A_v = 500 \; ; A_{vf} = 100 \; ; m_v = ?$$

$$A_{vf} = \frac{A_v}{1 + A_v m_v}$$

or $$100 = \frac{500}{1 + 500 m_v}$$

∴ $$m_v = \mathbf{0.008}$$

Now, $$A_v = \frac{80}{100} \times 500 = 400 \; ; m_v = 0.008 \; ; A_{vf} = ?$$

$$A_{vf} = \frac{A_v}{1 + A_v m_v} = \frac{400}{1 + 400 \times 0.008} = \frac{400}{4.2} = 95.3$$

∴ $$\%\text{ age fall in } A_{vf} = \frac{100 - 95.3}{100} \times 100 = \mathbf{4.7\%}$$

Note that without negative feedback, the change in gain is 20%. However, when negative feedback is applied, the change in gain (4.7%) is much less. This shows that negative feedback provides voltage gain stability.

Example 11.8. *An amplifier has an open-loop gain A_v = 100,000. A negative feedback of 10 dB is applied. Find (i) voltage gain with feedback (ii) value of feedback fraction m_v*

Solution.

(*i*) dB voltage gain without feedback

$$= 20 \log_{10} 100{,}000 = 20 \log_{10} 10^5 = 100 \text{ dB}$$

$$\text{Voltage gain with feedback} = 100 - 10 = 90 \text{ dB}$$

Now, $$20 \log_{10} (A_{vf}) = 90$$

or $$\log_{10} (A_{vf}) = 90/20 = 4.5$$

$$\therefore \qquad A_{vf} = \text{Antilog } 4.5 = \mathbf{31622}$$

(ii) $$A_{vf} = \frac{A_v}{1 + A_v\, m_v}$$

or $$31622 = \frac{100{,}000}{1 + 100{,}000 \times m_v}$$

$$\therefore \qquad m_v = \mathbf{2.17 \times 10^{-5}}$$

Example 11.9. *An amplifier with an open-circuit voltage gain of 1000 has an output resistance of 100 Ω and feeds a resistive load of 900 Ω. Negative voltage feedback is provided by connecting a resistive voltage divider across the output and one-fiftieth of the output voltage is fed back in series with the input signal. Determine the voltage gain with negative feedback.*

Solution. Fig. 11.8 shows the equivalent circuit of an amplifier along with the feedback circuit.

Voltage gain of the amplifier without feedback is

$$A_v = \frac{A_{v0}\, R_L}{R_{out} + R_L} \qquad \text{...See Art. 8.24}$$

$$= \frac{1000 \times 900}{100 + 900} = 900$$

$$\therefore \qquad A_{vf} = \frac{A_v}{1 + A_v\, m_v} = \frac{900}{1 + 900 \times (1/50)} = \mathbf{47.4}$$

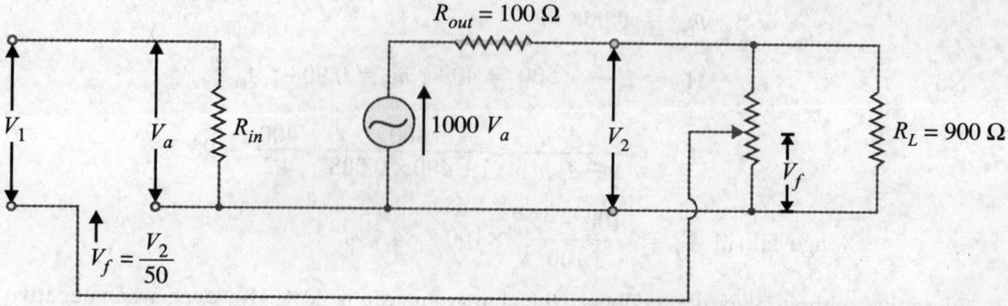

Fig. 11.8

Example 11.10. *An amplifier is required with a voltage gain of 100 which does not vary by more than 1%. If it is to use negative feedback with a basic amplifier the voltage gain of which can vary by 20%, determine the minimum voltage gain required and the feedback factor.*

Solution.

$$100 = \frac{A_v}{1 + A_v\, m_v}$$

or $$100 + 100\, A_v\, m_v = A_v \qquad \text{... (i)}$$

Also, $$99 = \frac{0.8\, A_v}{1 + 0.8\, A_v\, m_v}$$

or $$99 + 79.2\, A_v\, m_v = 0.8\, A_v \qquad \text{... (ii)}$$

Multiplying eq. (*i*) by 0.792, we have,

$$79.2 + 79.2\, A_v\, m_v = 0.792\, A_v \qquad \text{... } (iii)$$

Subtracting [(*ii*) – (*iii*)], we have,

$$19.8 = 0.008\, A_v \quad \therefore \quad A_v = \frac{19.8}{0.008} = \mathbf{2475}$$

Putting the value of A_v (= 2475) in eq. (*i*), we have,

$$100 + 100 \times 2475 \times m_v = 2475$$

$$\therefore \quad m_v = \frac{2475 - 100}{100 \times 2475} = \mathbf{0.0096}$$

11.5 Feedback Circuit

The function of the feedback circuit is to return a fraction of the output voltage to the input of the amplifier. Fig. 11.9 shows the feedback circuit of negative voltage feedback amplifier. It is essentially a potential divider consisting of resistances R_1 and R_2. The output voltage of the amplifier is fed to this potential divider which gives the feedback voltage to the input of the amplifier.

Referring to Fig. 11.9, it is clear that :

$$\text{Voltage across } R_1,\ v_f = \left(\frac{R_1}{R_1 + R_2}\right) v_0$$

$$\text{Feedback fraction, } m_v = \frac{v_f}{v_0} = \frac{R_1}{R_1 + R_2}$$

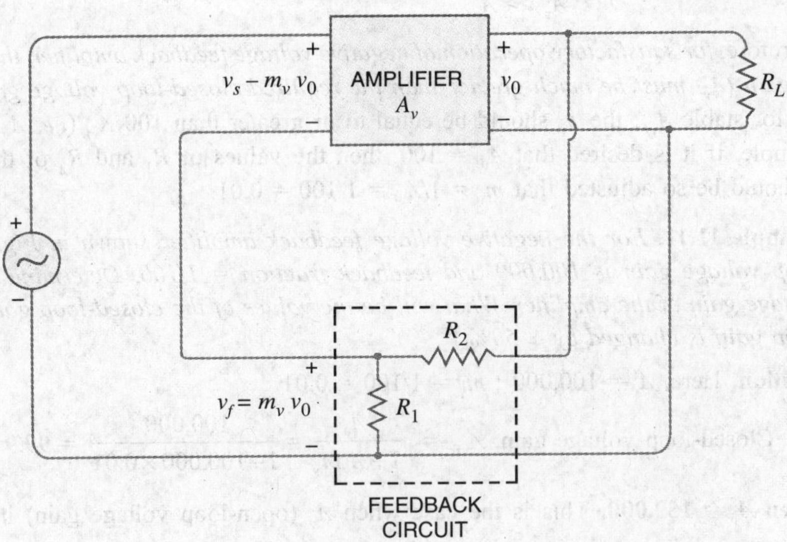

Fig. 11.9

Approximate voltage gain. It has been shown in Art. 11.4 that the voltage gain (A_{vf}) of the negative voltage feedback amplifier is given by ;

$$A_{vf} = \frac{A_v}{1 + A_v\, m_v}$$

If $A_v m_v >> 1$ (as is generally the case), then 1 can be neglected as compared to $A_v m_v$.

$$\therefore \quad A_{vf} = \frac{A_v}{A_v\, m_v} = \frac{1}{m_v} \quad \ldots \quad A_v m_v >> 1$$

Now,
$$m_v = \frac{R_1}{R_1 + R_2}$$

$$\therefore \quad A_{vf} = \frac{1}{m_v} = \frac{R_1 + R_2}{R_1} \quad \ldots (i)$$

Eq. (*i*) gives the expression for the voltage gain of negative voltage feedback amplifier provided $A_v m_v >> 1$.

(*i*) While designing a negative voltage feedback amplifier, we ensure that $A_v m_v >> 1$ so that :

$$A_{vf} = \frac{1}{m_v} = \frac{R_1 + R_2}{R_1}$$

This is a very interesting result. It shows that *the voltage gain of a negative voltage feedback amplifier is almost entirely determined by the feedback circuit*. Since feedback circuit is a resistive network, the voltage gain with negative voltage feedback (A_{vf}) is unaffected by changes in temperature, variations in transistor parameters and signal frequency. Hence, voltage gain with negative feedback is extremely stable.

(*ii*) The condition $A_v m_v >> 1$ means that $A_v >> 1/m_v$ *i.e.*

$$A_v >> A_{vf} \qquad (\because 1/m_v = A_{vf})$$

Therefore, *for satisfactory operation of negative voltage feedback amplifier, the open-loop voltage gain (A_v) must be much greater than the required closed-loop voltage gain (A_{vf})*. In general, for stable A_{vf}, the A_v should be equal to or greater than 100 A_{vf} (*i.e.* $A_v \geq 100\ A_{vf}$). For example, if it is desired that $A_{vf} = 100$, then the values of R_1 and R_2 of the feedback circuit should be so adjusted that $m_v = 1/A_{vf} = 1/100 = 0.01$.

Example 11.11. *For the negative voltage feedback amplifier shown in Fig. 11.10, the open-loop voltage gain is 100,000 and feedback fraction = 1/100. Determine the closed-loop voltage gain of the amplifier. What will be the values of the closed-loop gain when the open-loop gain is changed by ± 50% ?*

Solution. Here, $A_v = 100{,}000$; $m_v = 1/100 = 0.01$

$$\therefore \quad \text{Closed-loop voltage gain, } A_{vf} = \frac{A_v}{1 + A_v m_v} = \frac{100{,}000}{1 + 100{,}000 \times 0.01} = \mathbf{99.9}$$

When A_v = 150,000. This is the case when A_v (open-loop voltage gain) increases by 50%.

$$\therefore \quad A_{vf} = \frac{A_v}{1 + A_v m_v} = \frac{150{,}000}{1 + 150{,}000 \times 0.01} = \mathbf{99.93}$$

When A_v = 50,000. This is the case when A_v decreases by 50%.

$$\therefore \quad A_{vf} = \frac{A_v}{1 + A_v m_v} = \frac{50{,}000}{1 + 50{,}000 \times 0.01} = \mathbf{99.8}$$

This example shows that when the open-loop voltage gain (A_v) changes by $\pm 50\%$, the closed-loop voltage gain (A_{vf}) of the negative voltage feedback amplifier changes within $\pm 0.1\%$. Therefore, the voltage gain of the amplifier with negative voltage feedback is extremely stable.

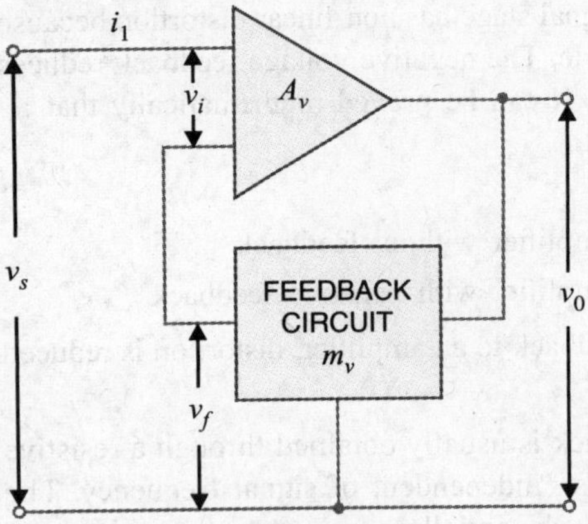

Fig. 11.10

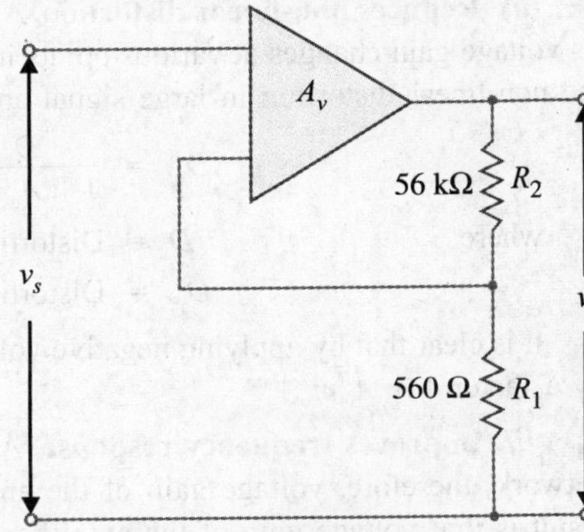

Fig. 11.11

Example 11.12. *For the negative voltage feedback amplifier shown in Fig. 11.11, the open-loop voltage gain is 100,000. Calculate (i) feedback fraction and (ii) closed-loop voltage gain.*

Solution. Here, $A_v = 100{,}000$; $R_1 = 560\,\Omega$; $R_2 = 56\,\text{k}\Omega$

(i) Feedback fraction, $m_v = \dfrac{R_1}{R_1 + R_2} = \dfrac{560\,\Omega}{560\,\Omega + 56\,\text{k}\Omega} = \dfrac{\mathbf{1}}{\mathbf{101}}$

(ii) Closed-loop voltage gain of the amplifier with negative feedback is

$$A_{vf} = \frac{A_v}{1 + A_v m_v} = \frac{100{,}000}{1 + 100{,}000 \times (1/101)} = \mathbf{100.9}$$

Also,

$$A_{vf} \simeq \frac{1}{m_v} = \frac{1}{1/101} = 101$$

Note that error is about 0.1%.

11.6 Advantages of Negative Voltage Feedback

The following are the advantages of negative voltage feedback in amplifiers :

(i) **Gain stability.** An important advantage of negative voltage feedback is that the resultant gain of the amplifier can be made independent of transistor parameters or the supply voltage variations.

$$A_{vf} = \frac{A_v}{1 + A_v\, m_v}$$

For negative voltage feedback in an amplifier to be effective, the designer deliberately makes the product $A_v\, m_v$ much greater than unity. Therefore, in the above relation, 1 can be neglected as compared to $A_v\, m_v$ and the expression becomes :

$$A_{vf} = \frac{A_v}{A_v\, m_v} = \frac{1}{m_v}$$

It may be seen that the gain now depends only upon feedback fraction m_v *i.e.* on the characteristics of feedback circuit. As feedback circuit is usually a voltage divider (a resistive network), therefore, it is unaffected by changes in temperature, variations in transistor parameters and frequency. Hence, the gain of the amplifier is extremely stable.

***(ii)* Reduces non-linear distortion.** A large signal stage has non-linear distortion because its voltage gain changes at various points in the cycle. The negative voltage feedback reduces the non-linear distortion in large signal amplifiers. It can be proved mathematically that :

$$D_{vf} = \frac{D}{1 + A_v\ m_v}$$

where D = Distortion in amplifier without feedback

D_{vf} = Distortion in amplifier with negative feedback

It is clear that by applying negative voltage feedback to an amplifier, distortion is reduced by a factor $1 + A_v\ m_v$.

***(iii)* Improves frequency response.** As feedback is usually obtained through a resistive network, therefore, voltage gain of the amplifier is *independent of signal frequency. The result is that voltage gain of the amplifier will be substantially constant over a wide range of signal frequencies. The negative voltage feedback, therefore, improves the frequency response of the amplifier.

Fig. 11.12 shows the effect of negative voltage feedback on the bandwidth of the amplifier. Here, BW is the bandwidth of the amplifier without feedback and $BW_{(f)}$ is the amplifier bandwidth with negative voltage feedback. Note that with negative voltage feedback, the bandwidth ($BW_{(f)}$) is increased, the lower cut-off frequency [$f_{1(f)}$] is decreased and upper cut-off frequency [$f_{2(f)}$] is increased.

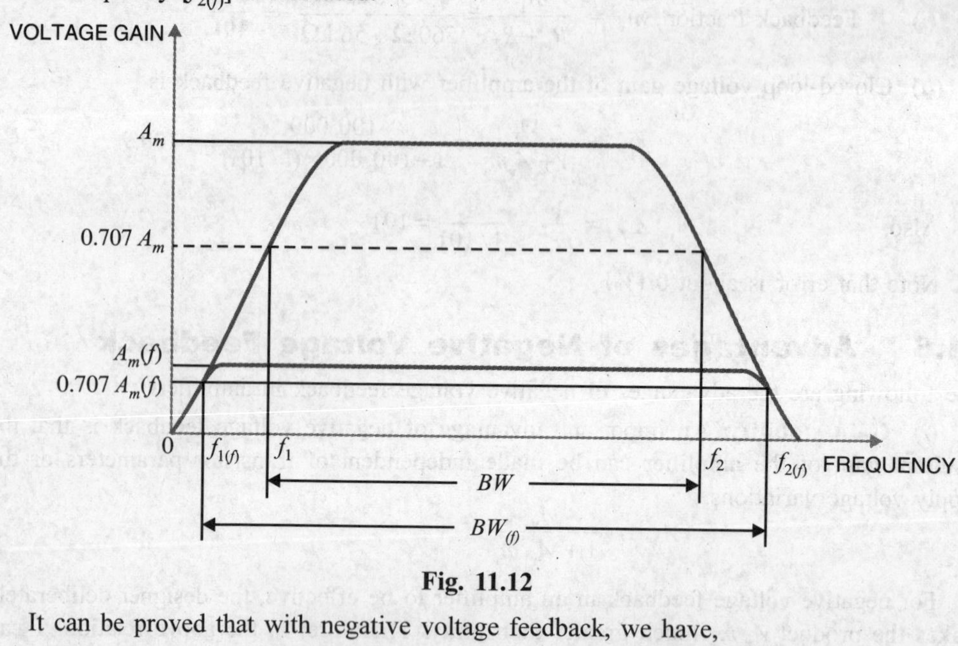

Fig. 11.12

It can be proved that with negative voltage feedback, we have,

---

* $A_{vf} = 1/m_v$. Now m_v depends upon feedback circuit. As feedback circuit consists of resistive network, therefore, value of m_v is unaffected by change in signal frequency.

$$\text{Lower cut-off frequency, } f_{1(f)} = \frac{f_1}{1 + A_v m_v}$$

$$\text{Upper cut-off frequency, } f_{2(f)} = f_2(1 + A_v m_v)$$

$$\text{Bandwidth, } BW_{(f)} = BW\,(1 + A_v m_v)$$

Here, f_1 and f_2 are the lower and upper cut-off frequencies respectively without feedback. Note that with negative voltage feedback, the lower cut-off frequency is *decreased* by $(1 + A_v m_v)$ while upper cut-off frequency is *increased* by the same factor. As a result, the bandwidth of the amplifier is *increased* by $(1 + A_v m_v)$.

***(iv)* Increases circuit stability.** The output of an ordinary amplifier is easily changed due to variations in ambient temperature, frequency and signal amplitude. This changes the gain of the amplifier, resulting in distortion. However, by applying negative voltage feedback, voltage gain of the amplifier is stabilised or accurately fixed in value. This can be easily explained. Suppose the output of a negative voltage feedback amplifier has increased because of temperature change or due to some other reason. This means more negative feedback since feedback is being given from the output. This tends to oppose the increase in amplification and maintains it stable. The same is true should the output voltage decrease. Consequently, the circuit stability is considerably increased.

***(v)* Increases input impedance and decreases output impedance.** The negative voltage feedback increases the input impedance and decreases the output impedance of amplifier. Such a change is profitable in practice as the amplifier can then serve the purpose of impedance matching.

***(a)* Input impedance.** The increase in input impedance with negative voltage feedback can be explained by referring to Fig. 11.13. Suppose the input impedance of the amplifier is Z_{in} without feedback and $Z_{in(f)}$ with negative feedback. Let us further assume that input current is i_1.

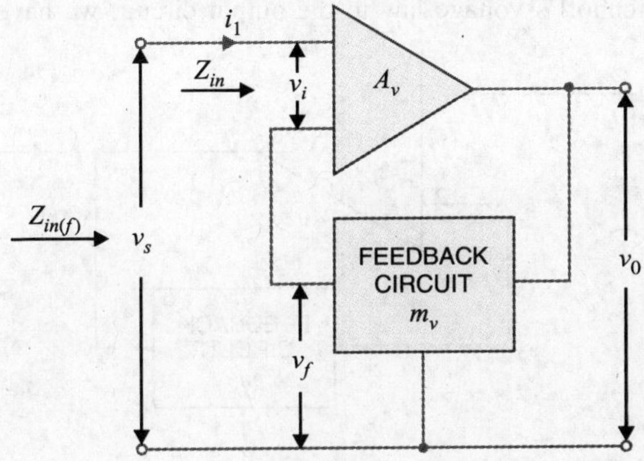

Fig. 11.13

The input impedance of a transistor amplifier without negative feedback is the impedance *looking into* the base of the transistor. If no negative feedback is present in the amplifier shown in Fig. 11.13, the input impedance Z_{in} is

$$Z_{in} = \frac{v_i}{i_1}$$

$\therefore \quad i_1 = \dfrac{v_i}{Z_{in}}$

With negative voltage feedback, the input impedance $Z_{in(f)}$ of the amplifier is

$$Z_{in(f)} = \frac{v_s}{i_1} = \frac{v_s}{v_i} \times Z_{in} \quad \text{... (i)} \quad \left(\because i_1 = \frac{v_i}{Z_{in}}\right)$$

Now, $\quad v_i = v_s - v_f = v_s - m_v\, v_0 \quad (\because v_f = m_v\, v_0)$

or $\quad v_i = v_s - m_v\, A_v\, v_i \quad (\because v_0 = A_v\, v_i)$

or $\quad v_i(1 + A_v\, m_v) = v_s$

or $\quad \dfrac{v_s}{v_i} = (1 + A_v\, m_v)$

From eq. (*i*), $\quad Z_{in(f)} = Z_{in}\,(1 + A_v\, m_v)$

It is clear that by applying negative voltage feedback, the input impedance of the amplifier is increased by a factor $(1 + A_v m_v)$. As $A_v m_v$ is much greater than unity, therefore, input impedance is increased considerably. This is an advantage, since the amplifier will now present less of a load to its source circuit.

(*b*) **Output impedance.** Recall that output impedance of an amplifier is the Thevenin impedance at its output terminals. Therefore, in order to find the output impedance $[Z_{out(f)}]$ of a negative voltage feedback amplifier, short-circuit the input signal source and apply voltage v_0 at the output terminals as shown in Fig. 11.14. Since the input terminals are shorted, the feedback voltage v_f is the only voltage at the amplifier input. Referring to Fig. 11.14, the voltage v_0 produces the feedback voltage v_f and this generates a voltage $A_v v_f$ in series with Z_{out}. Here, Z_{out} is the output impedance of the amplifier without feedback. Assuming an instantaneous positive polarity for v_0, the output current i_0 flows in the direction shown. Applying Kirchhoff's voltage law to the output circuit, we have,

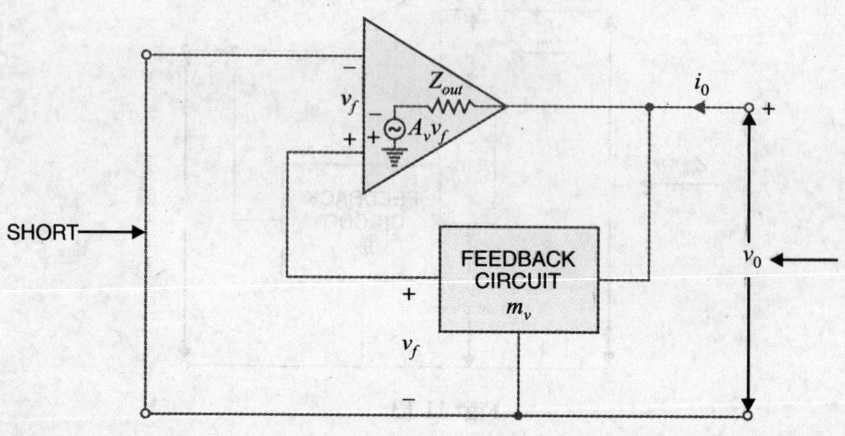

Fig. 11.14

$$v_0 = i_0 Z_{out} - A_v v_f$$

or $\quad i_0 Z_{out} = v_0 + A_v v_f$

or $\quad i_0 Z_{out} = v_0 + A_v\, m_v\, v_0 \quad (\because v_f = m_v v_0)$

or $$i_0 Z_{out} = v_0(1 + A_v\, m_v) \qquad \text{... (i)}$$

With negative voltage feedback, the output impedance of the amplifier is

$$Z_{out(f)} = \frac{v_0}{i_0} = \frac{Z_{out}}{1 + A_v m_v} \qquad \text{... From eq. (i)}$$

$$\therefore \quad Z_{out(f)} = \frac{Z_{out}}{1 + A_v m_v}$$

It is clear that by applying negative feedback, the output impedance of the amplifier is decreased by a factor $(1 + A_v\, m_v)$. This is an added benefit of using negative voltage feedback. With lower value of output impedance, the amplifier is much better suited to drive low impedance loads.

Example 11.13. *Fig. 11.15 shows the negative voltage feedback amplifier. If the gain of the amplifier without feedback is 10,000, find :*

(i) feedback fraction (ii) overall voltage gain (iii) output voltage if input voltage is 1 mV.

Solution. $A_v = 10{,}000\ ;\ R_1 = 2\text{ k}\Omega\ ;\ R_2 = 18\text{ k}\Omega$

(i) Feedback fraction, $m_v = \dfrac{R_1}{R_1 + R_2} = \dfrac{2}{2 + 18} = \mathbf{0.1}$

(ii) Voltage gain with negative feedback is

$$A_{vf} = \frac{A_v}{1 + A_v\, m_v} = \frac{10{,}000}{1 + 10{,}000 \times 0.1} = \mathbf{10}$$

(iii) Output voltage $= A_{vf} \times$ Input voltage

$= 10 \times 1\text{ mV} = \mathbf{10\ mV}$

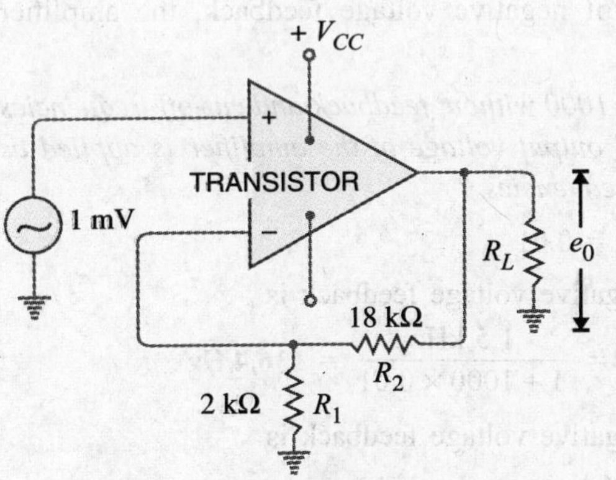

Fig. 11.15

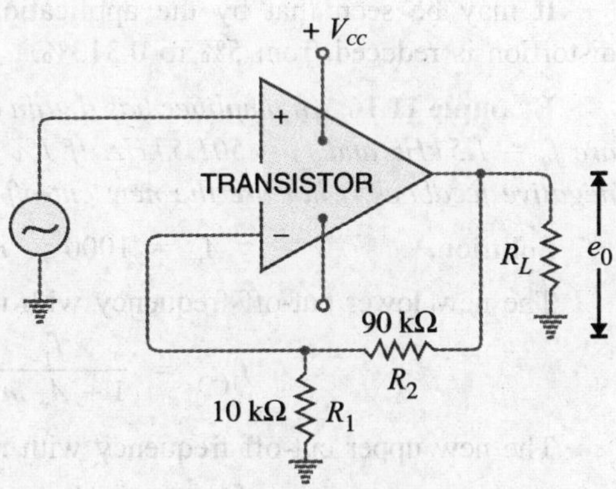

Fig. 11.16

Example 11.14. *Fig. 11.16 shows the circuit of a negative voltage feedback amplifier. If without feedback, A_v = 10,000, Z_{in} = 10 kΩ, Z_{out} = 100 Ω, find :*

(i) feedback fraction *(ii) gain with feedback*

(iii) input impedance with feedback *(iv) output impedance with feedback.*

Solution.

(*i*) Feedback fraction, $m_v = \dfrac{R_1}{R_1 + R_2} = \dfrac{10}{10 + 90} = \mathbf{0.1}$

(*ii*) Gain with negative feedback is

$$A_{vf} = \frac{A_v}{1 + A_v\, m_v} = \frac{10{,}000}{1 + 10{,}000 \times 0.1} = \mathbf{10}$$

(*iii*) With negative voltage feedback, input impedance is increased and is given by ;

$$\begin{aligned} Z_{in(f)} &= (1 + A_v\, m_v)\, Z_{in} \\ &= (1 + 10{,}000 \times 0.1)\; 10\,\text{k}\Omega \\ &= 1001 \times 10\,\text{k}\Omega \\ &= \mathbf{10\,M\Omega} \end{aligned}$$

(*iv*) With negative voltage feedback, output impedance is decreased and is given by ;

$$Z_{out(f)} = \frac{Z_{out}}{1 + A_v\, m_v} = \frac{100\,\Omega}{1 + 10{,}000 \times 0.1} = \frac{100\,\Omega}{1001} = \mathbf{0.1\,\Omega}$$

Example 11.15. *The gain and distortion of an amplifier are 150 and 5% respectively without feedback. If the stage has 10% of its output voltage applied as negative feedback, find the distortion of the amplifier with feedback.*

Solution.

$$\begin{aligned} \text{Gain without feedback, } A_v &= 150 \\ \text{Distortion without feedback, } D &= 5\% = 0.05 \\ \text{Feedback fraction, } m_v &= 10\% = 0.1 \end{aligned}$$

If D_{vf} is the distortion with negative voltage feedback, then,

$$D_{vf} = \frac{D}{1 + A_v\, m_v} = \frac{0.05}{1 + 150 \times 0.1} = 0.00313 = \mathbf{0.313\%}$$

It may be seen that by the application of negative voltage feedback, the amplifier distortion is reduced from 5% to 0.313%.

Example 11.16. *An amplifier has a gain of 1000 without feedback and cut-off frequencies are f_1 = 1.5 kHz and f_2 = 501.5 kHz. If 1% of output voltage of the amplifier is applied as negative feedback, what are the new cut-off frequencies ?*

Solution. $A_v = 1000$; $m_v = 0.01$

The new lower cut-off frequency with negative voltage feedback is

$$f_{1(f)} = \frac{f_1}{1 + A_v\, m_v} = \frac{1.5\text{ kHz}}{1 + 1000 \times 0.01} = \mathbf{136.4\ Hz}$$

The new upper cut-off frequency with negative voltage feedback is

$$\begin{aligned} f_{2(f)} &= f_2\,(1 + m_v\, A_v) \\ &= (501.5\text{ kHz})\,(1 + 1000 \times 0.01) = \mathbf{5.52\ MHz} \end{aligned}$$

Note the effect of negative voltage feedback on the bandwidth of the amplifier. The lower cut-off frequency is decreased by a factor $(1 + m_v\, A_v)$ while upper cut-off frequency is increased by a factor $(1 + m_v\, A_v)$. In other words, the bandwidth of the amplifier is increased approximately by a factor $(1 + m_v\, A_v)$.

$$BW_{(f)} \simeq BW\,(1 + m_v A_v)$$

where BW = Bandwidth of the amplifier without feedback

$BW_{(f)}$ = Bandwidth of the amplifier with negative voltage feedback

Example 11.17. *Fig. 11.17 shows negative voltage feedback amplifier having open-loop voltage gain of 100,000. When negative feedback is not used, the input impedance of the amplifier is 1 kΩ. Find the input impedance of the circuit with negative voltage feedback.*

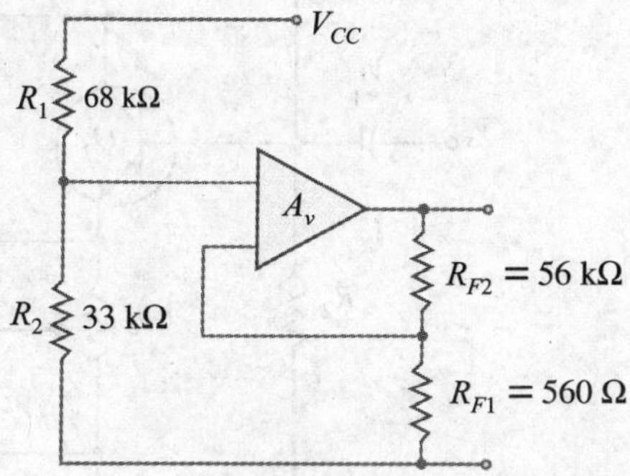

Fig. 11.17

Solution. The input impedance of the circuit means input impedance with biasing resistors (R_1 and R_2). The biasing resistors are not affected by the negative feedback because these are outside the feedback loop.

∴ Input impedance of the circuit is

$$Z_{in\,(circuit)} = R_1 \parallel R_2 \parallel Z_{in(f)}$$

$$\text{Feedback fraction, } m_v = \frac{R_{F1}}{R_{F1}+R_{F2}} = \frac{560\ \Omega}{560\ \Omega + 56\ \text{k}\Omega} = \frac{1}{101}$$

Input impedance of the amplifier without negative feedback is Z_{in} = 1 kΩ. Therefore, with negative voltage feedback, the input impedance of the amplifier is

$$Z_{in(f)} = Z_{in}\,(1 + A_v m_v) = 1\ \text{k}\Omega\ (1 + 100{,}000 \times 1/101) = 991\ \text{k}\Omega$$

$$\therefore \quad Z_{in\,(circuit)} = 68\ \text{k}\Omega \parallel 33\ \text{k}\Omega \parallel 991\ \text{k}\Omega = 21.7\ \text{k}\Omega$$

11.7 Two-stage Amplifier with Negative Voltage Feedback

Fig. 11.18 shows two-stage *CE* amplifier with negative voltage feedback. The feedback path is from the collector of transistor Q_2 to the emitter circuit of transistor Q_1 through the components C_{F1}, R_{F1} and R_{F2}. Here, resistors R_{F1} and R_{F2} form the *feedback circuit* because C_{F1} behaves as a short for a.c. The capacitor C_{F1} is a d.c. blocking capacitor and prevents the d.c. voltage at the collector of Q_2 from affecting the bias conditions of Q_1.

Fig. 11.19 shows the *a.c. equivalent circuit of the feedback circuit.* The capacitor C_2 behaves as a short for a.c. so that C_2 and R_4 (shorted by C_2) are absent from the a.c. equivalent circuit. Referring to Fig. 11.19, the signal voltage (v_s) is applied between the base of Q_1 and ground and the output voltage (v_0) is developed between collector of Q_2 and ground. The feedback voltage (v_f) is developed across resistor R_{F2}. The input voltage (v_i) across the base-emitter of Q_1 is $v_i = v_s - v_f$. This can be easily proved. The output voltage (v_0) of two-stage amplifier is in phase with the input signal (v_s). When v_s is positive going, v_0 is also positive going and consequently v_f is in phase *opposition with v_s.

$$\therefore \quad v_i = v_s - v_f$$

* Since v_s and v_f work in opposition in forcing current through the base emitter junction of Q_1.

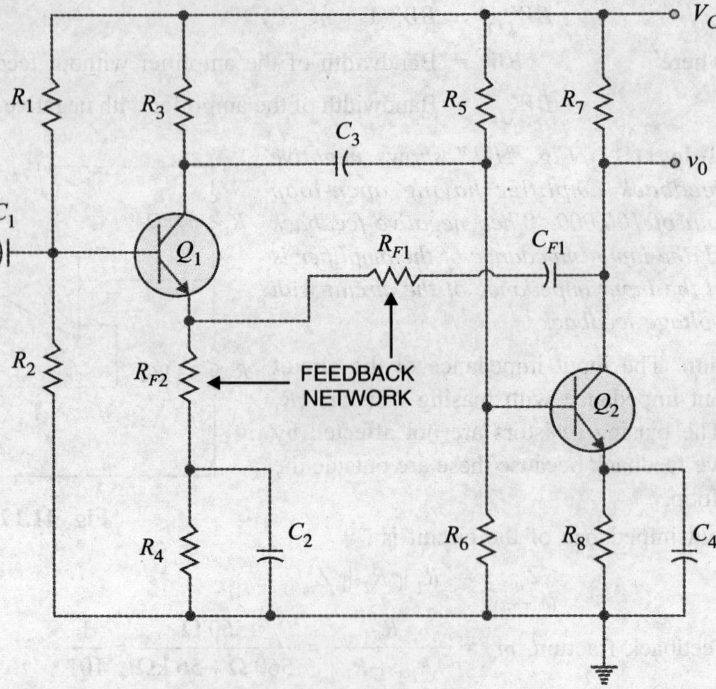

Fig. 11.18

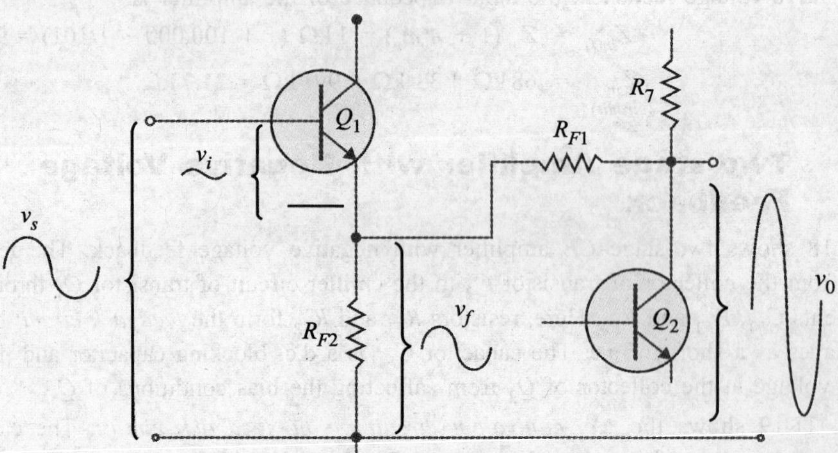

Fig. 11.19

(*i*) The output voltage (v_0) is divided between R_{F2} and R_{F1} to produce feedback voltage v_f (= voltage across R_{F2}).

$$\therefore \quad \text{Feedback fraction, } m_v = \frac{v_f}{v_0} = \frac{R_{F2}}{R_{F1} + R_{F2}}$$

$$\therefore \quad \text{Closed-loop voltage gain, } A_{vf} = \frac{1}{m_v} = \frac{R_{F1} + R_{F2}}{R_{F2}}$$

(*ii*) The biasing resistors at the input of the amplifier circuit are not affected by negative feedback because these are outside the feedback loop. Likewise, the collector resistor in the second transistor is not affected by the negative feedback.

Example 11.18. *For the two-stage negative voltage feedback amplifier circuit shown in Fig. 11.18 above, R_{F1} = 16.2 kΩ, R_{F2} = 220 Ω, R_1 = 120 kΩ, R_2 = 39 kΩ. Without the negative feedback circuit, the circuit has open-loop voltage gain of 58,000 and input impedance at the base of Q_1 is Z_{in} = 1 kΩ. Find (i) feedback fraction (ii) voltage gain and (iii) input impedance of the circuit.*

Solution. R_{F1} = 16.2 kΩ ; R_{F2} = 220 Ω ; R_1 = 120 kΩ ; R_2 = 39 kΩ ; A_v = 58,000 ; Z_{in} = 1 kΩ

(*i*) Feedback fraction,
$$m_v = \frac{R_{F2}}{R_{F1} + R_{F2}} = \frac{220\,\Omega}{16.2\,\text{k}\Omega + 220\,\Omega} = \frac{1}{74.6}$$

(*ii*) Closed-loop voltage gain,
$$A_{vf} = \frac{A_v}{1 + A_v m_v} = \frac{58{,}000}{1 + 58{,}000/74.6} = 74.5$$

Also,
$$A_{vf} \simeq \frac{1}{m_v} = 74.6$$

(*iii*) Input impedance at the base of Q_1 with negative voltage feedback is
$$Z_{in(f)} = Z_{in}(1 + A_v m_v)$$
$$= 1\,\text{k}\Omega\,(1 + 58{,}000/74.6) = 778\,\text{k}\Omega$$

∴ Input impedance of the amplifier circuit is
$$\underset{(circuit)}{Z_{in}} = R_1 \,||\, R_2 \,||\, Z_{in(f)}$$
$$= 120\,\text{k}\Omega \,||\, 39\,\text{k}\Omega \,||\, 778\,\text{k}\Omega = \mathbf{28.4\,k\Omega}$$

Example 11.19. *For the two-stage negative voltage feedback amplifier shown in Fig. 11.20, r'_e = 75 Ω and r_f = 5 kΩ. Find the voltage gain of the amplifier circuit.*

Solution. The feedback path is from the collector of the second stage to the emitter circuit of the first stage. Note that r_f and r'_e form the feedback circuit and the feedback voltage (v_f) is developed across r'_e.

Feedback fraction,
$$m_v = \frac{r'_e}{r'_e + r_f} = \frac{75\,\Omega}{75\,\Omega + 5\,\text{k}\Omega} = \frac{1}{67.7}$$

∴ Voltage gain,
$$A_{vf} = \frac{1}{m_v} = \frac{1}{1/67.7} = 67.7$$

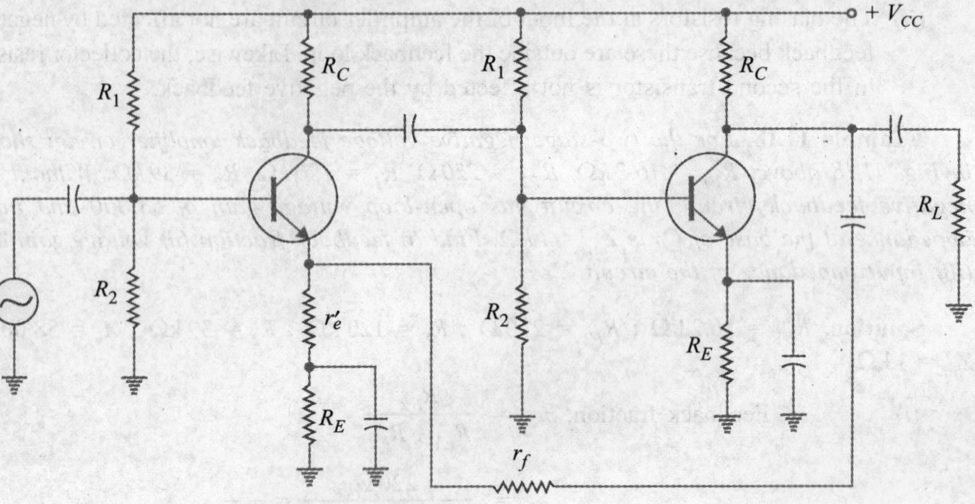

Fig. 11.20

Example 11.20. *Fig. 11.21 shows two-stage negative voltage feedback amplifier. If we want a voltage gain of 70, what should be the value of feedback resistor* r_f?

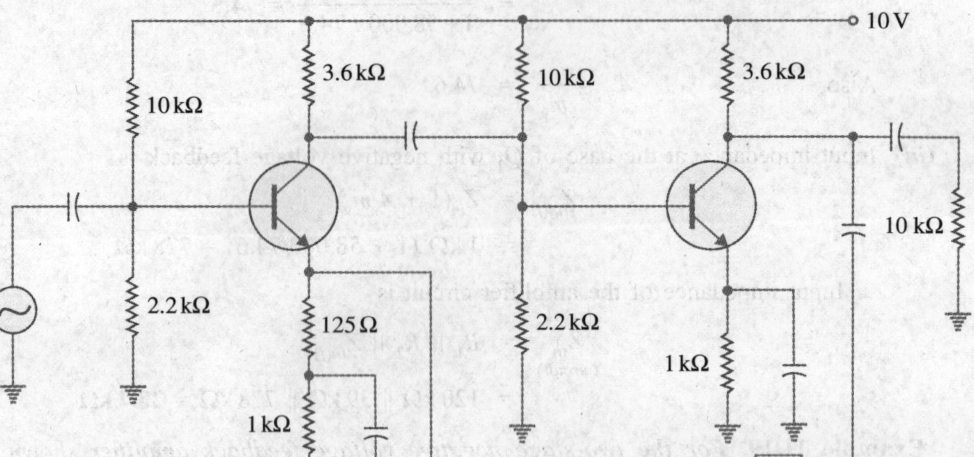

Fig. 11.21

Solution. The feedback resistance in the feedback loop can be changed by adjusting 10 kΩ rheostat. Suppose the feedback resistor is adjusted to a value r_f to obtain voltage gain of 70.

$$\text{Feedback fraction, } m_v = \frac{r'_e}{r'_e + r_f}$$

$$\therefore \quad \text{Voltage gain, } A_{vf} = \frac{1}{m_v} = \frac{r'_e + r_f}{r'_e}$$

or $$A_{vf} = 1+\frac{r_f}{r'_e}$$

or $$70 = 1+\frac{r_f}{r'_e}$$

$\therefore$ $$r_f = (70 - 1)r'_e = 69 \times 125\,\Omega = \mathbf{8.625\,k\Omega}$$

11.8 Principles of Negative Current Feedback

In this method, a fraction of output current is fed back to the input of the amplifier. In other words, the feedback current (I_f) is proportional to the output current (I_{out}) of the amplifier. Fig. 11.22 shows the principles of negative current feedback. This circuit is called *current-shunt feedback circuit*. A feedback resistor R_f is connected between input and output of the amplifier. This amplifier has a current gain of A_i without feedback. It means that a current I_1 at the input terminals of the amplifier will appear as $A_i I_1$ in the output circuit *i.e.* $I_{out} = A_i I_1$. Now a fraction m_i of this output current is fed back to the input through R_f. The fact that arrowhead shows the feed current being fed forward is because it is *negative feedback.*

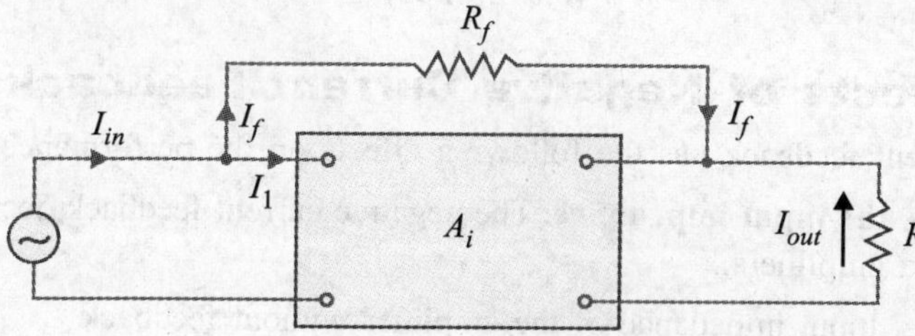

Fig. 11.22

$$\text{Feedback current, } I_f = m_i\, I_{out}$$

$$\therefore \quad \text{Feedback fraction, } m_i = \frac{I_f}{I_{out}} = \frac{\text{Feedback current}}{\text{Output current}}$$

Note that negative current feedback reduces the input current to the amplifier and hence its current gain.

11.9 Current Gain with Negative Current Feedback

Referring to Fig. 11.22, we have,

$$I_{in} = I_1 + I_f = I_1 + m_i\, I_{out}$$

But $I_{out} = A_i I_1$, where A_i is the current gain of the amplifier without feedback.

$\therefore$ $$I_{in} = I_1 + m_i A_i I_1 \qquad (\because I_{out} = A_i I_1)$$

$\therefore$ Current gain with negative current feedback is

$$A_{if} = \frac{I_{out}}{I_{in}} = \frac{A_i I_1}{I_1 + m_i\, A_i\, I_1}$$

or $$A_{if} = \frac{A_i}{1 + m_i\, A_i}$$

This equation looks very much like that for the voltage gain of negative voltage feedback amplifier. The only difference is that we are dealing with current gain rather than the voltage gain. The following points may be noted carefully :

(*i*) The current gain of the amplifier without feedback is A_i. However, when negative current feedback is applied, the current gain is reduced by a factor $(1 + m_i A_i)$.

(*ii*) The feedback fraction (or current attenuation) m_i has a value between 0 and 1.

(*iii*) The negative current feedback does not affect the voltage gain of the amplifier.

Example 11.21. *The current gain of an amplifier is 200 without feedback. When negative current feedback is applied, determine the effective current gain of the amplifier. Given that current attenuation* $m_i = 0.012$.

Solution. $$A_{if} = \frac{A_i}{1 + m_i A_i}$$

Here, $$A_i = 200 \; ; \quad m_i = 0.012$$

$\therefore$ $$A_{if} = \frac{200}{1 + (0.012)(200)} = \mathbf{58.82}$$

11.10 Effects of Negative Current Feedback

The negative current feedback has the following effects on the performance of amplifiers :

(*i*) **Decreases the input impedance.** The negative current feedback decreases the input impedance of most amplifiers.

Let Z_{in} = Input impedance of the amplifier without feedback

Z'_{in} = Input impedance of the amplifier with negative current feedback

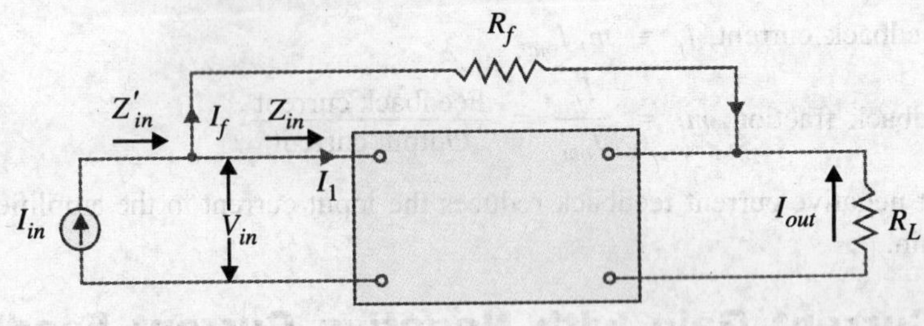

Fig. 11.23

Referring to Fig. 11.23, we have,

$$Z_{in} = \frac{V_{in}}{I_1}$$

and $$Z'_{in} = \frac{V_{in}}{I_{in}}$$

But $$V_{in} = I_1 Z_{in} \text{ and } I_{in} = I_1 + I_f = I_1 + m_i I_{out} = I_1 + m_i A_i I_1$$

$\therefore$ $$Z'_{in} = \frac{I_1 Z_{in}}{I_1 + m_i A_i I_1} = \frac{Z_{in}}{1 + m_i A_i}$$

or $$Z'_{in} = \frac{Z_{in}}{1 + m_i\, A_i}$$

Thus the input impedance of the amplifier is decreased by the factor $(1 + m_i A_i)$. Note the primary difference between negative current feedback and negative voltage feedback. Negative current feedback *decreases* the input impedance of the amplifier while negative voltage feedback *increases* the input impedance of the amplifier.

***(ii)* Increases the output impedance.** It can be proved that with negative current feedback, the output impedance of the amplifier is increased by a factor $(1 + m_i A_i)$.

$$Z'_{out} = Z_{out}\,(1 + m_i A_i)$$

where Z_{out} = Output impedance of the amplifier without feedback

Z'_{out} = Output impedance of the amplifier with negative current feedback

The reader may recall that with negative voltage feedback, the output impedance of the amplifier is decreased.

***(iii)* Increases bandwidth.** It can be shown that with negative current feedback, the bandwidth of the amplifier is increased by the factor $(1 + m_i A_i)$.

$$BW' = BW\,(1 + m_i A_i)$$

where BW = Bandwidth of the amplifier without feedback

BW' = Bandwidth of the amplifier with negative current feedback

Example 11.22. *An amplifier has a current gain of 240 and input impedance of 15 kΩ without feedback. If negative current feedback (m_i = 0.015) is applied, what will be the input impedance of the amplifier ?*

Solution. $$Z'_{in} = \frac{Z_{in}}{1 + m_i A_i}$$

Here, Z_{in} = 15 kΩ ; A_i = 240 ; m_i = 0.015

∴ $$Z'_{in} = \frac{15}{1 + (0.015)\,(240)} = \mathbf{3.26\ k\Omega}$$

Example 11.23. *An amplifier has a current gain of 200 and output impedance of 3 kΩ without feedback. If negative current feedback (m_i = 0.01) is applied ; what is the output impedance of the amplifier ?*

Solution. $Z'_{out} = Z_{out}\,(1 + m_i A_i)$

Here, Z_{out} = 3 kΩ ; A_i = 200 ; m_i = 0.01

∴ Z'_{out} = 3[1 + (0.01) (200)] = **9 kΩ**

Example 11.24. *An amplifier has a current gain of 250 and a bandwidth of 400 kHz without feedback. If negative current feedback (m_i = 0.01) is applied, what is the bandwidth of the amplifier ?*

Solution. $BW' = BW\,(1 + m_i A_i)$

Here, BW = 400 kHz ; m_i = 0.01 ; A_i = 250

∴ BW' = 400 [1 + (0.01) 250] = **1400 kHz**

11.11 Emitter Follower

It is a negative current feedback circuit. The emitter follower is a current amplifier that has no voltage gain. Its most important characteristic is that it has high input impedance and low output impedance. This makes it an ideal circuit for impedance matching.

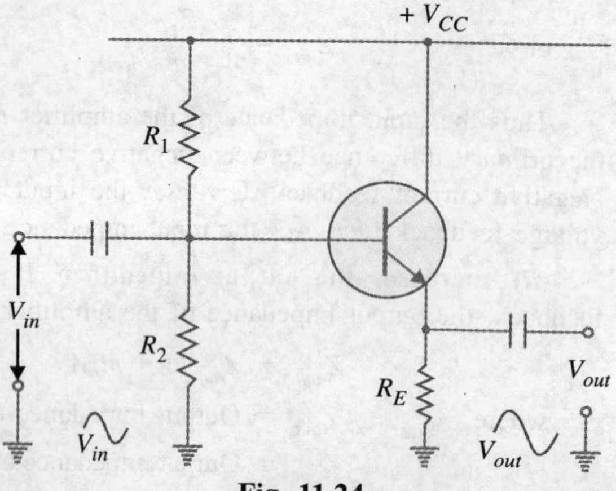

Fig. 11.24

Circuit details. Fig. 11.24 shows the circuit of an emitter follower. As you can see, it differs from the circuitry of a conventional *CE* amplifier by the absence of collector load and emitter bypass capacitor. The emitter resistance R_E itself acts as the load and a.c. output voltage (V_{out}) is taken across R_E. The biasing is generally provided by voltage-divider method or by base resistor method. The following points are worth noting about the emitter follower :

(*i*) There is neither collector resistor in the circuit nor there is emitter bypass capacitor. These are the two circuit recognition features of the emitter follower.

(*ii*) Since the collector is at *ac* ground, this circuit is also known as *common collector (CC) amplifier*.

Operation. The input voltage is applied between base and emitter and the resulting a.c. emitter current produces an output voltage $i_e R_E$ across the emitter resistance. This voltage opposes the input voltage, thus providing negative feedback. Clearly, it is a negative current feedback circuit since the voltage fed back is proportional to the emitter current *i.e.* output current. It is called emitter follower because the output voltage follows the input voltage.

Characteristics. The major characteristics of the emitter follower are :

(*i*) No voltage gain. In fact, the voltage gain of an emitter follower is close to 1.

(*ii*) Relatively high current gain and power gain.

(*iii*) High input impedance and low output impedance.

(*iv*) Input and output a.c. voltages are in phase.

11.12 D.C. Analysis of Emitter Follower

The d.c. analysis of an emitter follower is made in the same way as the voltage divider bias circuit of a *CE* amplifier. Thus referring to Fig. 11.24 above, we have,

$$\text{Voltage across } R_2,\ V_2 = \frac{V_{CC}}{R_1 + R_2} \times R_2$$

$$\text{Emitter current, } I_E = \frac{V_E}{R_E} = \frac{V_2 - V_{BE}}{R_E}$$

$$\text{Collector-emitter voltage, } V_{CE} = V_{CC} - V_E$$

D.C. load line. The d.c. load line of emitter follower can be constructed by locating the two end points *viz.*, $I_{C(sat)}$ and $V_{CE(off)}$.

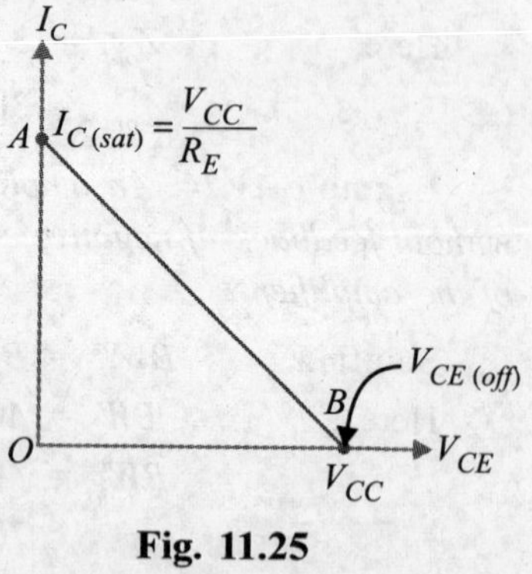

Fig. 11.25

(*i*) When the transistor is saturated, $V_{CE} = 0$.

$$\therefore \quad I_{C(sat)} = \frac{V_{CC}}{R_E}$$

This locates the point A ($OA = V_{CC}/R_E$) of the d.c. load line as shown in Fig. 11.25.

(*ii*) When the transistor is cut-off, $I_C = 0$. Therefore, $V_{CE(off)} = V_{CC}$. This locates the point B ($OB = V_{CC}$) of the d.c. load line.

By joining points A and B, d.c. load line AB is constructed.

Example 11.25. *For the emitter follower circuit shown in Fig. 11.26 (i), find V_E and I_E. Also draw the d.c. load line for this circuit.*

Solution.

$$\text{Voltage across } R_2,\ V_2 = \frac{V_{CC}}{R_1 + R_2} \times R_2 = \frac{18}{16 + 22} \times 22 = 10.42 \text{ V}$$

$$\text{Voltage across } R_E,\ V_E = V_2 - V_{BE} = 10.42 - 0.7 = \mathbf{9.72\ V}$$

$$\text{Emitter current, } I_E = \frac{V_E}{R_E} = \frac{9.72 \text{ V}}{910\ \Omega} = \mathbf{10.68\ mA}$$

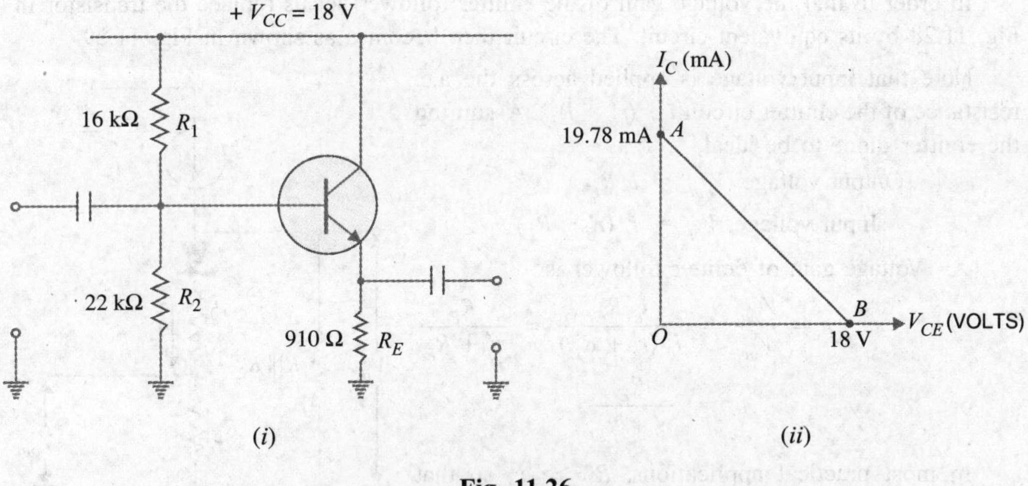

Fig. 11.26

D.C. load line. $$I_{C\,(sat)} = \frac{V_{CC}}{R_E} = \frac{18 \text{ V}}{910\ \Omega} = 19.78 \text{ mA}$$

This locates the point A (OA = 19.78 mA) of the d.c. load line.

$$V_{CE(off)} = V_{CC} = 18 \text{ V}$$

This locates point B (OB = 18 V) of the d.c. load line.

By joining points A and B, d.c. load line AB is constructed [See Fig. 11.26 (*ii*)].

11.13 Voltage Gain of Emitter Follower

Fig. 11.27 shows the emitter follower circuit. Since the emitter resistor is not bypassed by a capacitor, the a.c. equivalent circuit of emitter follower will be as shown in Fig. 11.28. The a.c. resistance r_E of the emitter circuit is given by ;

$$r_E = r'_e + R_E \qquad \text{where } r'_e = \frac{25 \text{ mV}}{I_E}$$

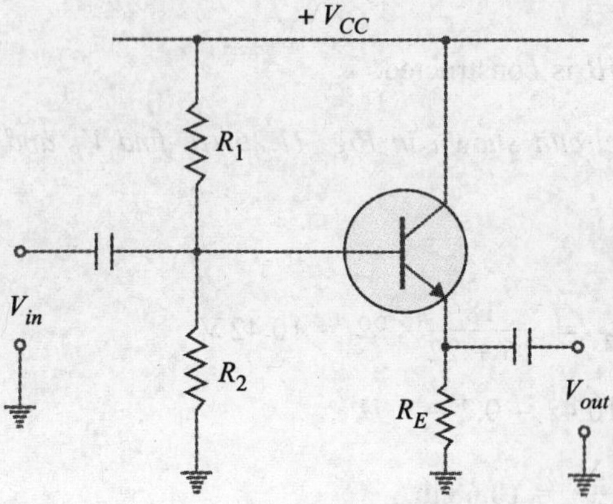

Fig. 11.27

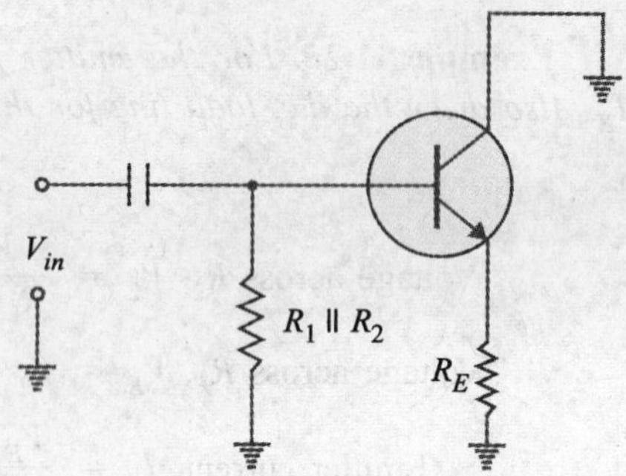

Fig. 11.28

In order to find the voltage gain of the emitter follower, let us replace the transistor in Fig. 11.28 by its equivalent circuit. The circuit then becomes as shown in Fig. 11.29.

Note that input voltage is applied across the a.c. resistance of the emitter circuit *i.e.* $(r'_e + R_E)$. Assuming the emitter diode to be ideal,

$$\text{Output voltage, } V_{out} = i_e R_E$$

$$\text{Input voltage, } V_{in} = i_e (r'_e + R_E)$$

∴ Voltage gain of emitter follower is

$$A_v = \frac{V_{out}}{V_{in}} = \frac{i_e R_E}{i_e (r'_e + R_E)} = \frac{R_E}{r'_e + R_E}$$

or
$$A_v = \frac{R_E}{r'_e + R_E}$$

In most practical applications, $R_E >> r'_e$ so that $A_v \simeq 1$.

In practice, the voltage gain of an emitter follower is between 0.8 and 0.999.

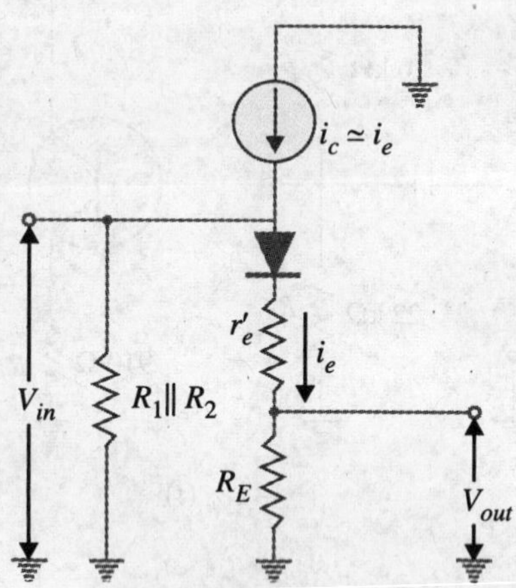

Fig. 11.29

Example 11.26. *Determine the voltage gain of the emitter follower circuit shown in Fig. 11.30.*

Solution.

$$\text{Voltage gain, } A_v = \frac{R_E}{r'_e + R_E}$$

Now,
$$r'_e = \frac{25 \text{ mV}}{I_E}$$

Voltage across R_2, $V_2 = \dfrac{V_{CC}}{R_1 + R_2} \times R_2 = \dfrac{10}{10 + 10} \times 10 = 5\text{ V}$

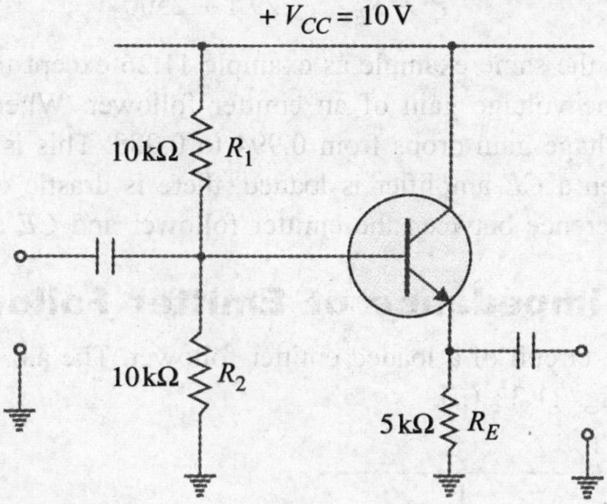

Fig. 11.30

Voltage across R_E, $V_E = V_2 - V_{BE} = 5 - 0.7 = 4.3\text{ V}$

∴ Emitter current, $I_E = \dfrac{V_E}{R_E} = \dfrac{4.3\text{ V}}{5\text{ k}\Omega} = 0.86\text{ mA}$

∴ $r'_e = \dfrac{25\text{ mV}}{I_E} = \dfrac{25\text{ mV}}{0.86\text{ mA}} = 29.1\ \Omega$

∴ Voltage gain, $A_v = \dfrac{R_E}{r'_e + R_E} = \dfrac{5000}{29.1 + 5000} = \mathbf{0.994}$

Example 11.27. *If in the above example, a load of 5 kΩ is added to the emitter follower, what will be the voltage gain of the circuit ?*

Solution. When a load of 5 kΩ is added to the emitter follower, the circuit becomes as shown in Fig. 11.31. The coupling capacitor acts as a short for a.c. signal so that R_E and R_L are in parallel. Therefore, the external emitter resistance R_E changes to R'_E where

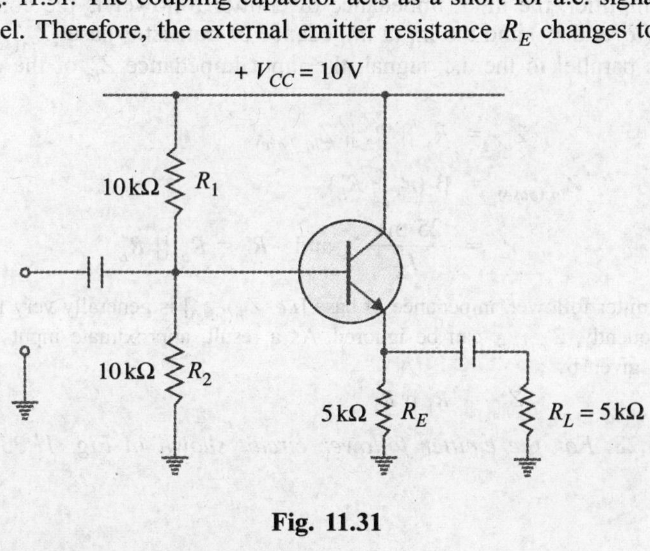

Fig. 11.31

$$R'_E = R_E \,||\, R_L = 5\text{ k}\Omega \,||\, 5\text{ k}\Omega = 2.5\text{ k}\Omega$$

$$\text{Voltage gain, } A_v = \frac{R'_E}{r'_e + R'_E} = \frac{2500}{29.1 + 2500} = \mathbf{0.988}$$

Comments. This is the same example as example 11.26 except that load is added. Note the loading effect on the voltage gain of an emitter follower. When load is added to the emitter follower, the voltage gain drops from 0.994 to 0.988. This is really a small change. On the other hand, when a *CE* amplifier is loaded, there is drastic change in voltage gain. This is yet another difference between the emitter follower and *CE* amplifier.

11.14 Input Impedance of Emitter Follower

Fig. 11.32 (*i*) shows the circuit of a loaded emitter follower. The a.c. equivalent circuit with T model is shown in Fig. 11.32 (*ii*).

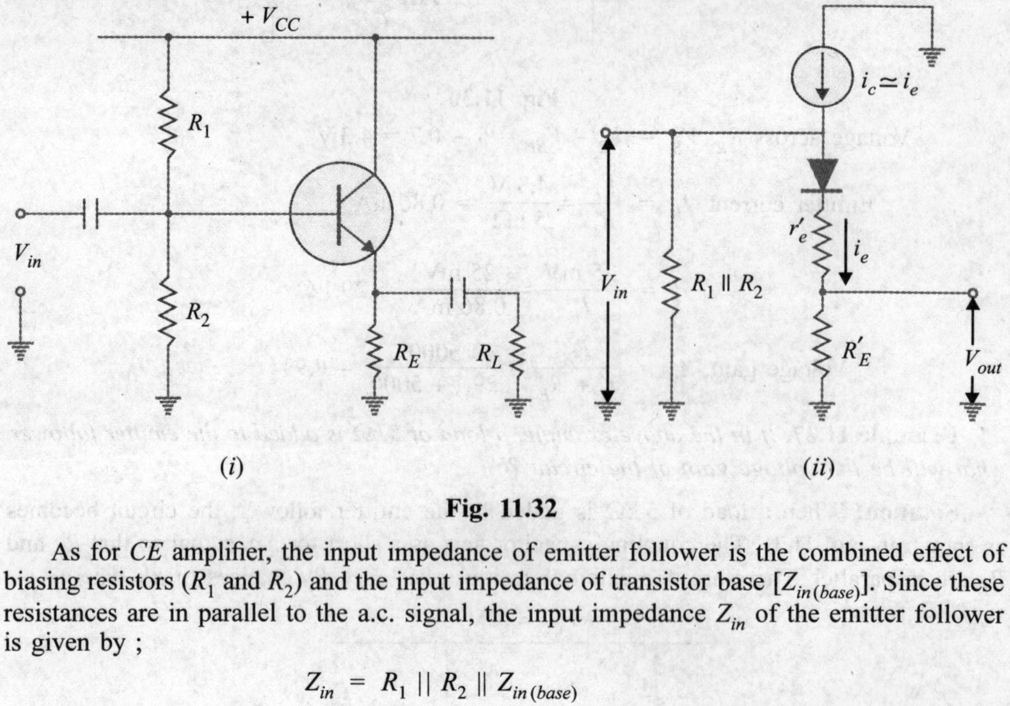

Fig. 11.32

As for *CE* amplifier, the input impedance of emitter follower is the combined effect of biasing resistors (R_1 and R_2) and the input impedance of transistor base [$Z_{in\,(base)}$]. Since these resistances are in parallel to the a.c. signal, the input impedance Z_{in} of the emitter follower is given by ;

$$Z_{in} = R_1 \,||\, R_2 \,||\, Z_{in\,(base)}$$

where $$Z_{in\,(base)} = \beta\,(r'_e + R'_E)$$

Now $$r'_e = \frac{25\text{ mV}}{I_E} \quad \text{and} \quad R'_E = R_E \,||\, R_L$$

Note. In an emitter follower, impedance of base [*i.e.* $Z_{in\,(base)}$] is generally very large as compared to $R_1 \,||\, R_2$. Consequently, $Z_{in\,(base)}$ can be ignored. As a result, approximate input impedance of the emitter follower is given by ;

$$Z_{in} = R_1 \,||\, R_2$$

Example 11.28. *For the emitter follower circuit shown in Fig. 11.33, find the input impedance.*

Solution.

$$\text{Voltage across } R_2,\ V_2 = \frac{V_{CC}}{R_1 + R_2} \times R_2 = \frac{10}{10 + 10} \times 10 = 5\text{ V}$$

$$\text{Voltage across } R_E,\ V_E = V_2 - V_{BE} = 5 - 0.7 = 4.3\text{ V}$$

$$\therefore \quad \text{Emitter current, } I_E = \frac{V_E}{R_E} = \frac{4.3\text{ V}}{4.3\text{ k}\Omega} = 1\text{ mA}$$

$$\therefore \quad \text{ac emitter resistance, } r'_e = \frac{25\text{ mV}}{I_E} = \frac{25\text{ mV}}{1\text{ mA}} = 25\ \Omega$$

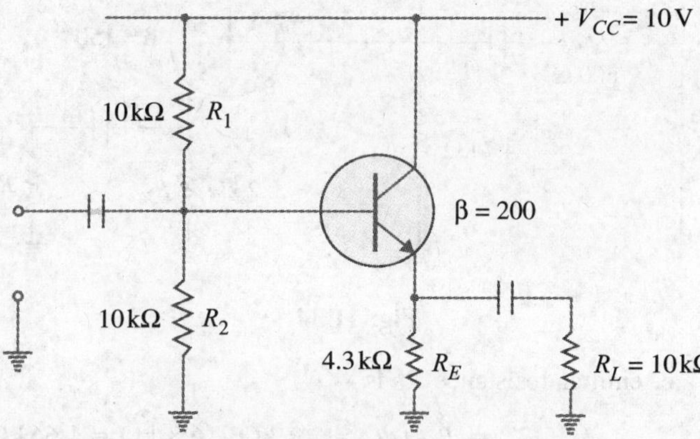

Fig. 11.33

Effective external emitter resistance is

$$R'_E = R_E \parallel R_L = 4.3\text{ k}\Omega \parallel 10\text{ k}\Omega = 3\text{ k}\Omega$$

$$\therefore \quad Z_{in\ (base)} = \beta\,(r'_e + R'_E) = 200\,(0.025 + 3) = 605\text{ k}\Omega$$

$\therefore$ Input impedance of the emitter follower is

$$\begin{aligned} Z_{in} &= R_1 \parallel R_2 \parallel Z_{in\ (base)} \\ &= 10\text{ k}\Omega \parallel 10\text{ k}\Omega \parallel 605\text{ k}\Omega \\ &= 5\text{ k}\Omega \parallel 605\text{ k}\Omega = \mathbf{4.96\text{ k}\Omega} \end{aligned}$$

Note. Since 605 kΩ is much larger than $R_1 \parallel R_2$ (= 5 kΩ), the former can be ignored. Therefore, approximate input impedance of emitter follower is given by ;

$$Z_{in} = R_1 \parallel R_2 = 10\text{ k}\Omega \parallel 10\text{ k}\Omega = 5\text{ k}\Omega$$

Example 11.29. *For the emitter follower circuit shown in Fig. 11.34, find (i) voltage gain (ii) input impedance and (iii) a.c. load voltage. The transistor used has β = 150.*

Solution.

We first find ac emitter resistance r'_e.

$$\text{D.C. voltage across } R_2,\ V_2 = \frac{V_{CC}}{R_1 + R_2} \times R_2 = \frac{15}{4.7 + 4.7} \times 4.7 = 7.5\text{ V}$$

$$\text{D.C. voltage across } R_E,\ V_E = V_2 - V_{BE} = 7.5 - 0.7 = 6.8\text{ V}$$

$$\therefore \quad \text{D.C. emitter current, } I_E = \frac{V_E}{R_E} = \frac{6.8\text{ V}}{2.2\text{ k}\Omega} = 3.09\text{ mA}$$

$$\therefore \quad \text{ac emitter resistance, } r'_e = \frac{25\text{ mV}}{I_E} = \frac{25\text{ mV}}{3.09\text{ mA}} = 8.09\ \Omega$$

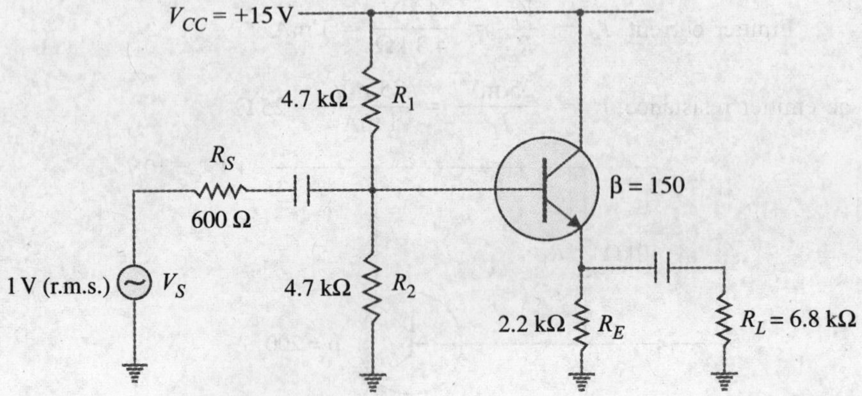

Fig. 11.34

(*i*) External a.c. emitter resistance R'_E is

$$R'_E = R_E \parallel R_L = 2.2\text{ k}\Omega \parallel 6.8\text{ k}\Omega = 1.66\text{ k}\Omega$$

$$\text{Voltage gain, } A_v = \frac{R'_E}{r'_e + R'_E} = \frac{1.66\text{ k}\Omega}{8.09\ \Omega + 1.66\text{ k}\Omega} = \mathbf{0.995}$$

(*ii*) Input impedance of the base is given by ;

$$Z_{in(base)} = \beta\,(r'_e + R'_E)$$

$$= 150(8.09\ \Omega + 1.66\text{ k}\Omega) = 250\text{ k}\Omega$$

Input impedance of the emitter follower is

$$Z_{in} = R_1 \parallel R_2 \parallel Z_{in(base)}$$

$$= 4.7\text{ k}\Omega \parallel 4.7\text{ k}\Omega \parallel 250\text{ k}\Omega$$

Now, $Z_{in(base)}$ (= 250 kΩ) is very large as compared to R_1 (= 4.7 kΩ) or R_2 (= 4.7 kΩ). Therefore, $Z_{in(base)}$ can be neglected with reasonable accuracy.

$$\therefore \quad Z_{in} = 4.7\text{ k}\Omega \parallel 4.7\text{ k}\Omega = \mathbf{2.35\text{ k}\Omega}$$

(*iii*) The signal source resistance R_S (= 600 Ω) is in series with the input impedance Z_{in}(= 2.35 kΩ) of the emitter follower. Therefore, a.c. voltage at the input terminals of the emitter follower is

$$V_{in} = \frac{V_S}{R_S + Z_{in}} \times Z_{in} = \frac{1\text{ V}}{600\ \Omega + 2.35\text{ k}\Omega} \times 2.35\text{ k}\Omega$$

$$= 0.797\text{ V}$$

$$\therefore \quad \text{A.C. output voltage, } V_{out} = A_v V_{in} = 0.995 \times 0.797 = \mathbf{0.793\text{ V}}$$

11.15 Output Impedance of Emitter Follower

The output impedance of a circuit is the impedance that the circuit offers to the load. When load is connected to the circuit, the output impedance acts as the source impedance for the load. Fig. 11.35 shows the circuit of emitter follower. Here R_S is the output resistance of amplifier voltage source.

It can be proved that the output impedance Z_{out} of the emitter follower is given by ;

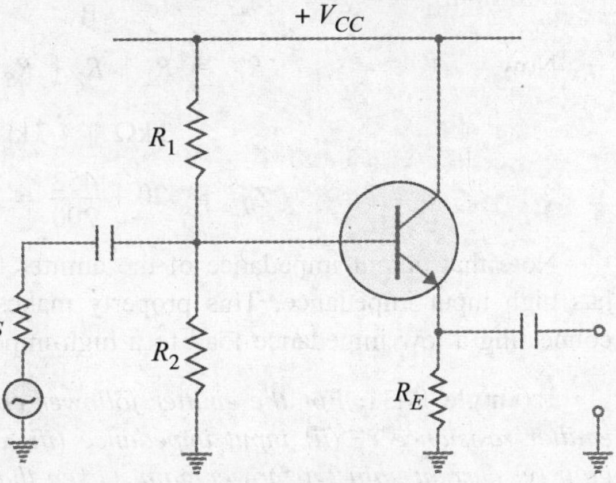

Fig. 11.35

$$Z_{out} = R_E \parallel \left(r'_e + \frac{R'_{in}}{\beta} \right)$$

where $R'_{in} = R_1 \parallel R_2 \parallel R_S$

In practical circuits, the value of R_E is large enough to be ignored. For this reason, the output impedance of emitter follower is approximately given by ;

$$Z_{out} = r'_e + \frac{R'_{in}}{\beta}$$

Note that the a.c. resistance of the base (R'_{in}) is decreased by a factor of β when seen from the emitter. This is due to the property of the transistor. When we move from the emitter to the base, the impedance is stepped up by a factor of β. When we move from the base to the emitter, the impedance is stepped down by a factor of β.

Note. The emitter follower is frequently used for impedance matching purposes. It presents a high impedance at the input and a low impedance at the output, which is the direct opposite of the standard *CE* amplifier. The resulting effect is much the same as that obtained with a transformer where load is matched to the source impedance for maximum power transfer.

Example 11.30. *Determine the output impedance of the emitter follower shown in Fig. 11.36. Given that* $r'_e = 20\ \Omega$.

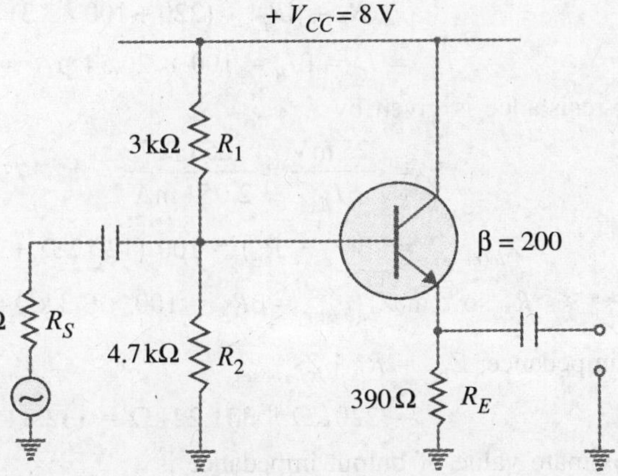

Fig. 11.36

Solution. $Z_{out} = r'_e + \frac{R'_{in}}{\beta}$

Now, $R'_{in} = R_1 \parallel R_2 \parallel R_S$

$= 3\text{ k}\Omega \parallel 4.7\text{ k}\Omega \parallel 600\ \Omega = 452\ \Omega$

$\therefore \quad Z_{out} = 20 + \frac{452}{200} = 20 + 2.3 = \mathbf{22.3\ \Omega}$

Note that output impedance of the emitter follower is very low. On the other hand, it has high input impedance. This property makes the emitter follower a perfect circuit for connecting a low impedance load to a high-impedance source.

Example 11.31. *For the emitter follower circuit shown in Fig. 11.37, determine (i) ac emitter resistance r'_e (ii) input impedance (iii) approximate output impedance (iv) voltage gain (v) current gain (vi) power gain. Given that $\beta = 100$.*

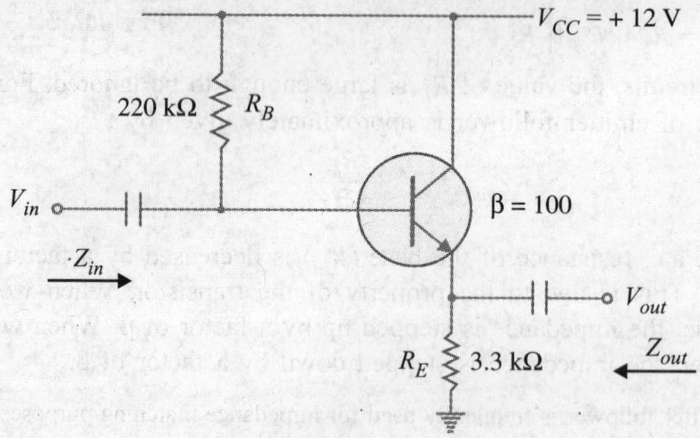

Fig. 11.37

Solution.

We first find d.c. emitter current as under :

$$I_B = \frac{V_{CC} - V_{BE}}{R_B + \beta R_E} = \frac{12\text{ V} - 0.7\text{ V}}{(220 + 100 \times 3.3)\text{ k}\Omega} = 20.54\ \mu\text{A}$$

$$I_E \simeq I_C = \beta I_B = 100 \times 20.54\ \mu\text{A} = 2.054\text{ mA}$$

(*i*) ac emitter resistance is given by ;

$$r'_e = \frac{25\text{ mV}}{I_E} = \frac{25\text{ mV}}{2.054\text{ mA}} = \mathbf{12.17\ \Omega}$$

(*ii*) $Z_{in(base)} = \beta(r'_e + R_E) = 100\ (12.17\ \Omega + 3.3\text{ k}\Omega) = 331.22\text{ k}\Omega$

Note that $r'_e \ll R_E$ so that $Z_{in(base)} \simeq \beta R_E = 100 \times 3.3\text{ k}\Omega = 330\text{ k}\Omega$.

Input impedance, $Z_{in} = R_B \parallel Z_{in(base)}$

$= 220\text{ k}\Omega \parallel 331.22\text{ k}\Omega = \mathbf{132.2\text{ k}\Omega}$

(*iii*) The approximate value of output impedance is

$$Z_{out} \simeq r'_e = \mathbf{12.17\ \Omega}$$

(iv) Voltage gain, $A_v = \dfrac{R_E}{r_e' + R_E} = \dfrac{3.3\ \text{k}\Omega}{12.17\ \Omega + 3.3\ \text{k}\Omega} = \mathbf{0.996}$

(v) It can be shown that current gain (A_i) of the circuit is

$$A_i = \frac{\beta R_B}{R_B + Z_{in(base)}} = \frac{100 \times 220\ \text{k}\Omega}{220\ \text{k}\Omega + 331.22\ \text{k}\Omega} = \mathbf{39.9}$$

(vi) Power gain, $A_p = A_v A_i = 0.996 \times 39.9 = \mathbf{39.74}$

Note that power gain (A_p) is slightly less than current gain (A_i). This is expected because $A_v < 1$.

11.16 Applications of Emitter Follower

The emitter follower has the following principal applications :

(i) To provide current amplification with no voltage gain.

(ii) Impedance matching.

(i) **Current amplification without voltage gain.** We know that an emitter follower is a current amplifier that has no voltage gain ($A_v \simeq 1$). There are many instances (especially in digital electronics) where an increase in current is required but no increase in voltage is needed. In such a situation, an emitter follower can be used. For example, consider the two stage amplifier circuit as shown in Fig. 11.38. Suppose this 2-stage amplifier has the desired voltage gain but current gain of this multistage amplifier is insufficient. In that case, we can use an emitter follower to increase the current gain without increasing the voltage gain.

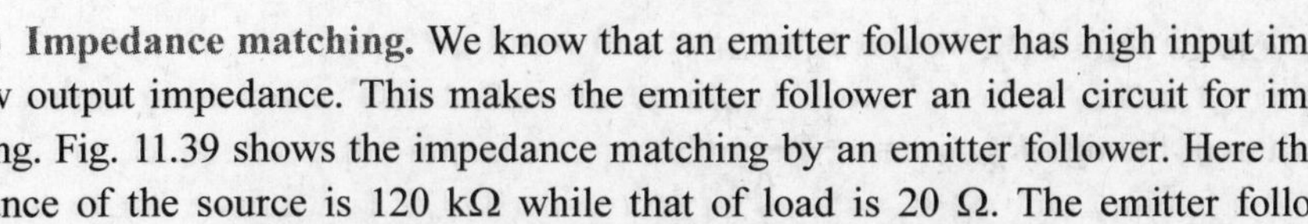

Fig. 11.38

(ii) **Impedance matching.** We know that an emitter follower has high input impedance and low output impedance. This makes the emitter follower an ideal circuit for impedance matching. Fig. 11.39 shows the impedance matching by an emitter follower. Here the output impedance of the source is 120 kΩ while that of load is 20 Ω. The emitter follower has an input impedance of 120 kΩ and output impedance of 22 Ω. It is connected between high-impedance source and low impedance load. The net result of this arrangement is that maximum power is transferred from the original source to the original load. When an emitter follower is used for this purpose, it is called a *buffer amplifier.*

It may be noted that the job of impedance matching can also be accomplished by a transformer. However, emitter follower is preferred for this purpose. It is because emitter follower is not only more convenient than a transformer but it also has much better frequency response *i.e.* it works well over a large frequency range.

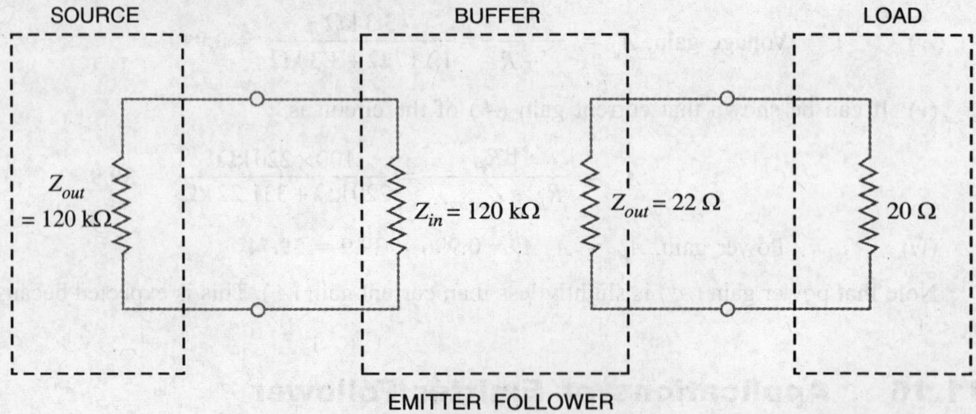

Fig. 11.39

11.17 Darlington Amplifier

Sometimes, the current gain and input impedance of an emitter follower are insufficient to meet the requirement. In order to increase the overall values of circuit current gain (A_i) and input impedance, two transistors are connected in series in emitter follower configuration as shown in Fig. 11.40. Such a circuit is called *Darlington amplifier*. Note that emitter of first transistor is connected to the base of the second transistor and the collector terminals of the two transistors are connected together. The result is that emitter current of the first transistor is the base current of the second transistor. Therefore, the current gain of the pair is equal to product of individual current gains *i.e.*

$$*\beta = \beta_1 \beta_2$$

Note that high current gain is achieved with a minimum use of components.

The biasing analysis is similar to that for one transistor except that two V_{BE} drops are to be considered. Thus referring to Fig. 11.40,

$$\text{Voltage across } R_2,\ V_2 = \frac{V_{CC}}{R_1 + R_2} \times R_2$$

$$\text{Voltage across } R_E,\ V_E = V_2 - 2V_{BE}$$

$$\text{Current through } R_E,\ I_{E2} = \frac{V_2 - 2V_{BE}}{R_E}$$

Since the transistors are directly coupled, $I_{E1} = I_{B2}$. Now $I_{B2} = I_{E2}/\beta_2$.

$$\therefore \quad I_{E1} = \frac{I_{E2}}{\beta_2}$$

---

* $I_{E1} = \beta_1 I_{B1} \quad (\because I_{E1} \simeq I_{C1})$

Now I_{E1} is the base current of Q_2 *i.e.* $I_{E1} = I_{B2}$.

Now, $I_{E2} = \beta_2 I_{B2} = \beta_2 I_{E1} = \beta_2 \beta_1 I_{B1}$

$\therefore$ Overall current gain, $\beta = \dfrac{I_{E2}}{I_{B1}} = \dfrac{\beta_1 \beta_2 I_{B1}}{I_{B1}} = \beta_1 \beta_2$

Input impedance of the darlington amplifier is

$$Z_{in} = \beta_1 \beta_2 R_E \text{ ... neglecting } r'_e$$

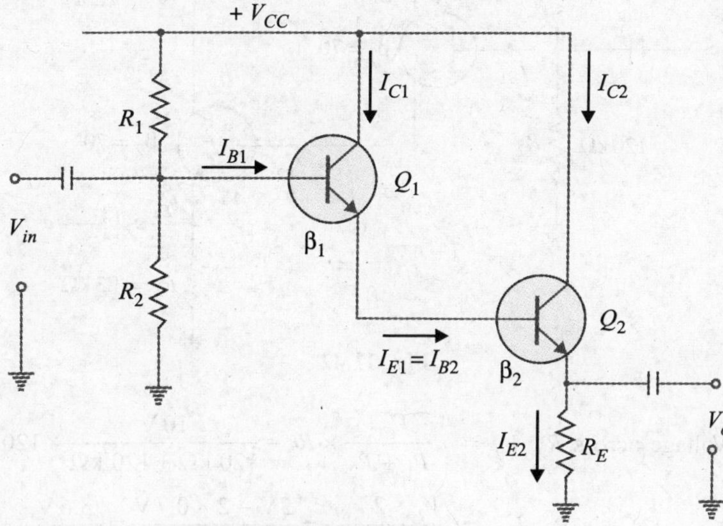

Fig. 11.40

In practice, the two transistors are put inside a single transistor housing and three terminals *E*, *B* and *C* are brought out as shown in Fig. 11.41. This three terminal device is known as a Darlington transistor. The Darlington transistor acts like a single transistor that has high current gain and high input impedance.

Characteristics. The following are the important characteristics of Darlington amplifier :

(*i*) Extremely high input impedance (MΩ).

(*ii*) Extremely high current gain (several thousands).

(*iii*) Extremely low output impedance (a few Ω).

Since the characteristics of the Darlington amplifier are basically the same as those of the emitter follower, the two circuits are used for similar applications. When you need higher input impedance and current gain and/or lower output impedance than the standard emitter follower can provide, you use a Darlington amplifier. Darlington transistors are commonly available. Like standard transistors, they have only three terminals but they have much higher values of current gain and input impedance.

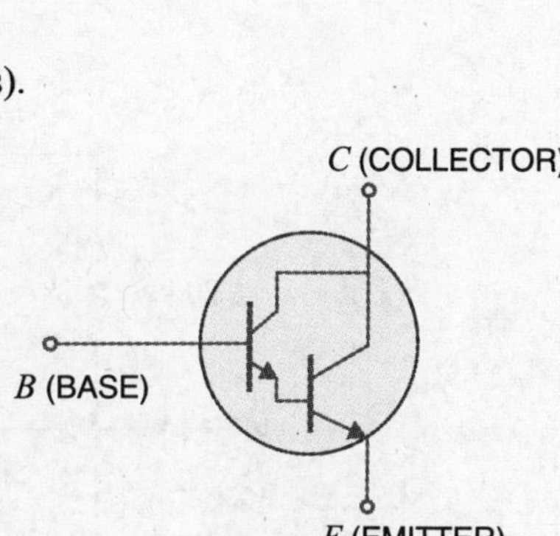

Fig. 11.41

Example 11.32. *Determine (i) d.c. value of current in R_E (ii) input impedance of the Darlington amplifier shown in Fig. 11.42.*

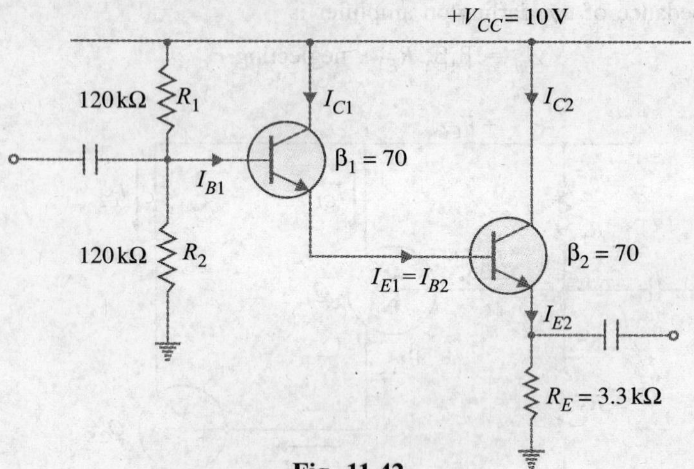

Fig. 11.42

Solution.

(*i*) Voltage across R_2, $V_2 = \frac{V_{CC}}{R_1 + R_2} \times R_2 = \frac{10\text{ V}}{120\text{ k}\Omega + 120\text{ k}\Omega} \times 120\text{ k}\Omega = 5\text{ V}$

D.C. in R_E, $I_{E2} = \frac{V_2 - 2\,V_{BE}}{R_E} = \frac{5\text{ V} - 2 \times 0.7\text{ V}}{3.3\text{ k}\Omega} = \frac{3.6\text{ V}}{3.3\text{ k}\Omega}$ = **1.09 mA**

(*ii*) Input impedance, $Z_{in} = \beta_1\,\beta_2\,R_E$

$= (70)\,(70)\,(3.3\text{ k}\Omega)$ = **16.17 MΩ**

This example illustrates that the input impedance of Darlington amplifier is much higher than that of an ordinary transistor.

Example 11.33. *For the Darlington amplifier in Fig. 11.43, find (i) the d.c. levels of both the transistors and (ii) ac emitter resistances of both transistors.*

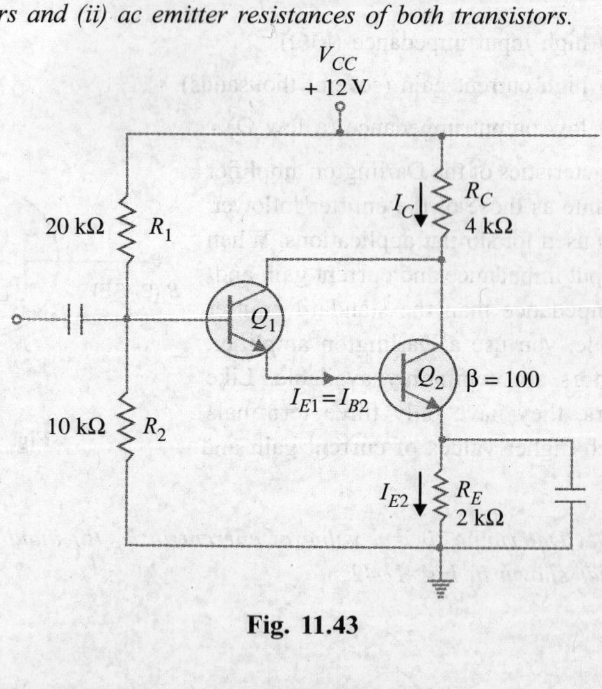

Fig. 11.43

Solution.

(*i*) **D.C. Bias Levels**

$$\text{Base voltage of } Q_1,\ V_{B1} = \frac{V_{CC}}{R_1 + R_2} \times R_2 = \frac{12\text{ V}}{20\text{ k}\Omega + 10\text{ k}\Omega} \times 10\text{ k}\Omega = \mathbf{4\text{ V}}$$

$$\text{Emitter voltage of } Q_1,\ V_{E1} = V_{B1} - V_{BE} = 4\text{ V} - 0.7\text{ V} = \mathbf{3.3\text{ V}}$$

$$\text{Base voltage of } Q_2,\ V_{B2} = V_{E1} = \mathbf{3.3\text{ V}}$$

$$\text{Emitter voltage of } Q_2,\ V_{E2} = V_{B2} - V_{BE} = 3.3\text{ V} - 0.7\text{ V} = \mathbf{2.6\text{ V}}$$

$$\text{Emitter current of } Q_2,\ I_{E2} = \frac{V_{E2}}{R_E} = \frac{2.6\text{ V}}{2\text{ k}\Omega} = \mathbf{1.3\text{ mA}}$$

$$\text{Emitter current of } Q_1,\ I_{E1} = \frac{I_{E2}}{\beta} = \frac{1.3\text{ mA}}{100} = \mathbf{0.013\text{ mA}}$$

(*ii*) **A.C. Analysis**

$$\text{ac emitter resistance of } Q_1,\ r'_{e1} = \frac{25\text{ mV}}{I_{E1}} = \frac{25\text{ mV}}{0.013\text{ mA}} = \mathbf{1923\ \Omega}$$

$$\text{ac emitter resistance of } Q_2,\ r'_{e2} = \frac{25\text{ mV}}{I_{E2}} = \frac{25\text{ mV}}{1.3\text{ mA}} = \mathbf{19.23\ \Omega}$$

Short Answer Questions

Q.1. *What is the effect of negative feedback (voltage or current) on the performance of an amplifier?*

Ans. With negative feedback, the gain of the amplifier is reduced. In exchange for this loss of gain, we have the following advantages :

(*i*) Bandwidth of the amplifier is increased.

(*ii*) Increases gain stability.

(*iii*) Changes amplifier input and output impedances.

Q.2. *What is the effect of negative voltage feedback on the circuit performance of an amplifier?*

Ans. The negative voltage feedback :

(*i*) reduces voltage gain of amplifier

(*ii*) does not affect current gain

(*iii*) increases bandwidth

(*iv*) increases gain stability

(*v*) reduces power gain

(*vi*) decreases output impedance

Q.3. *What is the effect of negative current feedback on the circuit performance of an amplifier?*

Ans. The negative current feedback :

(*i*) reduces current gain of amplifier

(*ii*) does not affect voltage gain

(*iii*) increases bandwidth

(*iv*) increases gain stability

(*v*) reduces power gain

(*vi*) increases output impedance

Q.4. *With negative feedback to an amplifier, the gain is reduced by a factor of 100. By what amount cut-off frequencies will change?*

Ans. The lower cut-off frequency will be decreased by a factor of 100 while upper cut-off frequency will be increased by a factor of 100.

Q.5. *How does negative voltage feedback increase the bandwidth of an amplifier?*

Ans. The voltage gain of an amplifier with negative voltage feedback is given by ;

$$A_{vf} = \frac{A_v}{1 + A_v m_v}$$

For negative feedback to be effective, the designer deliberately makes the product $A_v m_v$ much greater than 1. Therefore, in the above relation, 1 can be neglected as compared to $A_v m_v$ and the relation becomes :

$$A_{vf} = \frac{A_v}{A_v m_v} = \frac{1}{m_v}$$

It may be seen that gain now depends upon feedback fraction m_v *i.e.* characteristics of feedback circuit. As feedback circuit is usually a potential divider (a resistive network), therefore, gain of the amplifier will be independent of signal frequency. In other words, gain will remain substantially constant over a wide range of signal frequencies. Therefore, negative feedback increases the bandwidth of the amplifier.

Q.6. *Negative feedback results in reduced power gain. Do we employ negative feedback in power amplifiers where it may result in loss of power?*

Ans. Negative feedback can be employed to power amplifiers to reduce distortion to a minimum value. For instance, negative feedback is used in the power stage of a radio receiver. As the power output of most radio receivers exceeds the amount usually required for the speaker to operate, the loss of power because of negative feedback is of little consequence.

Q.7. *Does a certain percentage of negative feedback produce a fixed percentage of reduction in gain?*

Ans. No. It depends upon the initial value of gain *i.e.* gain without feedback. For example, suppose an amplifier has $A_v = 40$, $m_v = 0.1$. Then gain with negative feedback is :

$$A_{vf} = \frac{A_v}{1 + A_v m_v} = \frac{40}{1 + 40 \times 0.1} = 8$$

$$\%\text{ age reduction in gain} = \frac{32}{40} \times 100 = 80\%$$

Now suppose the value of A_v is increased to 400 at the same value of m_v (= 0.1). Then,

$$A_{vf} = \frac{A_v}{1 + A_v m_v} = \frac{400}{1 + 400 \times 0.1} = 10$$

$$\%\text{ age reduction in gain} = \frac{390}{400} \times 100 = 97.5\%$$

Q.8. *Do we provide negative feedback in biasing circuits to stabilise the operating point of the amplifier circuit?*

Ans. Yes. In a biasing circuit, d.c. negative feedback is employed to keep the operating point fixed in the face of a change in temperature or a change in β of the transistor.

Q.9. *What is the basic difference between feedback in biasing circuits and amplifier circuits?*

Ans. In biasing circuits, we provide d.c. negative feedback whereas in amplifier circuits, we employ a.c. negative feedback.

Q.10. *Why is emitter follower preferred to transformer for impedance matching?*

Ans. An emitter follower has high input impedance and low output impedance. This permits it to be used for impedance matching. Emitter follower is not only more convenient than a transformer but it has much better frequency response *i.e.* it has a large range of frequencies over which the gain is constant.

Multiple-Choice Questions

1. When negative voltage feedback is applied to an amplifier, its voltage gain ....

(*i*) is increased (*ii*) is reduced
(*iii*) remains the same
(*iv*) none of the above

2. The value of negative feedback fraction is always ....

(*i*) less than 1 (*ii*) more than 1
(*iii*) equal to 1
(*iv*) none of the above

3. If the output of an amplifier is 10 V and 100 mV from the output is fed back to the input, then feedback fraction is ....

(*i*) 10 (*ii*) 0.1
(*iii*) 0.01 (*iv*) 0.15

4. The gain of an amplifier without feedback is 100 dB. If a negative feedback of 3 dB is applied, the gain of the amplifier will become ....

(*i*) 101.5 dB (*ii*) 300 dB
(*iii*) 103 dB (*iv*) 97 dB

5. If the feedback fraction of an amplifier is 0.01, then voltage gain with negative voltage feedback is approximately ......

(*i*) 500 (*ii*) 100
(*iii*) 1000 (*iv*) 5000

6. A feedback circuit usually employs .... network.

(*i*) resistive (*ii*) capacitive
(*iii*) inductive
(*iv*) none of the above

7. The gain of an amplifier with feedback is known as .... gain.

(*i*) resonant (*ii*) open loop
(*iii*) closed loop (*iv*) none of the above

8. When voltage feedback (negative) is applied to an amplifier, its input impedance ......

(*i*) is decreased
(*ii*) is increased
(*iii*) remains the same
(*iv*) none of the above

9. When current feedback (negative) is applied to an amplifier, its input impedance ....

(*i*) is decreased (*ii*) is increased
(*iii*) remains the same
(*iv*) none of the above

10. Negative feedback is employed in ....

(*i*) oscillators (*ii*) rectifiers
(*iii*) amplifiers
(*iv*) none of the above

11. Emitter follower is used for ....

(*i*) current gain
(*ii*) impedance matching
(*iii*) voltage gain
(*iv*) none of the above

12. The voltage gain of an emitter follower is ...
    (*i*) much less than 1
    (*ii*) approximately equal to 1
    (*iii*) greater than 1
    (*iv*) none of the above
13. When current feedback (negative) is applied to an amplifier, its output impedance ......
    (*i*) is increased
    (*ii*) is decreased
    (*iii*) remains the same
    (*iv*) none of the above
14. Emitter follower is a .... circuit.
    (*i*) voltage feedback
    (*ii*) current feedback
    (*iii*) both voltage and current feedback
    (*iv*) none of the above
15. If voltage feedback (negative) is applied to an amplifier, its output impedance ....
    (*i*) remains the same
    (*ii*) is increased (*iii*) is decreased
    (*iv*) none of the above
16. When negative voltage feedback is applied to an amplifier, its bandwidth ....
    (*i*) is increased (*ii*) is decreased
    (*iii*) remains the same
    (*iv*) insufficient data
17. An emitter follower has ..... input impedance.
    (*i*) zero (*ii*) low
    (*iii*) high
    (*iv*) none of the above
18. If voltage gain without feedback and feedback fraction are A_v and m_v respectively, then gain with negative voltage feedback is ....
    (*i*) $\dfrac{A_v}{1 - A_v m_v}$ (*ii*) $\dfrac{A_v}{1 + A_v m_v}$
    (*iii*) $\dfrac{1 + A_v m_v}{A_v}$ (*iv*) $(1 + A_v m_v) A_v$
19. The output impedance of an emitter follower is ....
    (*i*) high (*ii*) very high
    (*iii*) almost zero (*iv*) low
20. The approximate voltage gain of an amplifier with negative voltage feedback (feedback fraction being m_v) is ....
    (*i*) $1/m_v$ (*ii*) m_v
    (*iii*) $\dfrac{1}{1 + m_v}$ (*iv*) $1 - m_v$
21. If A_v and A_{vf} are the voltage gains of an amplifier without feedback and with negative feedback respectively, then feedback fraction is ....
    (*i*) $\dfrac{1}{A_v} - \dfrac{1}{A_{vf}}$ (*ii*) $\dfrac{1}{A_v} + \dfrac{1}{A_{vf}}$
    (*iii*) $\dfrac{A_v}{A_{vf}} + \dfrac{1}{A_v}$ (*iv*) $\dfrac{1}{A_{vf}} - \dfrac{1}{A_v}$
22. In the expression for voltage gain with negative voltage feedback, the term $1 + A_v m_v$ is known as ....
    (*i*) gain factor (*ii*) feedback factor
    (*iii*) sacrifice factor
    (*iv*) none of the above
23. If the output impedance of an amplifier is Z_{out} without feedback, then with negative voltage feedback, its value will be ....
    (*i*) $\dfrac{Z_{out}}{1 + A_v m_v}$ (*ii*) $Z_{out} (1 + A_v m_v)$
    (*iii*) $\dfrac{1 + A_v m_v}{Z_{out}}$ (*iv*) $Z_{out} (1 - A_v m_v)$
24. If the input impedance of an amplifier is Z_{in} without feedback, then with negative voltage feedback, its value will be ....
    (*i*) $\dfrac{Z_{in}}{1 + A_v m_v}$ (*ii*) $Z_{in} (1 + A_v m_v)$
    (*iii*) $\dfrac{1 + A_v m_v}{Z_{in}}$ (*iv*) $Z_{in} (1 - A_v m_v)$
25. Feedback circuit .... frequency.
    (*i*) is independent of
    (*ii*) is strongly dependent on
    (*iii*) is moderately dependent on
    (*iv*) none of the above

26. The basic purpose of applying negative voltage feedback is to ....
(*i*) increase voltage gain
(*ii*) reduce distortion
(*iii*) keep the temperature within limits
(*iv*) none of the above

27. If the voltage gain of an amplifier without feedback is 20 and with negative voltage feedback it is 12, then feedback fraction is ....
(*i*) 5/3 (*ii*) 3/5
(*iii*) 1/5 (*iv*) 0.033

28. In an emitter follower, we employ ...... negative current feedback.
(*i*) 50% (*ii*) 25%
(*iii*) 100% (*iv*) 75%

29. An amplifier has an open loop voltage gain of 1,00,000. With negative voltage feedback, the voltage gain is reduced to 100. What is the sacrifice factor?
(*i*) 1000 (*ii*) 100
(*iii*) 5000
(*iv*) None of the above

30. In the above question, what will happen to circuit performance?
(*i*) Distortion is increased 1000 times
(*ii*) Input impedance is increased 1000 times
(*iii*) Output impedance is increased 1000 times
(*iv*) None of the above

31. The non-linear distortion of an amplifier is D without feedback. The amplifier has an open-loop voltage gain of A_v and feedback fraction is m_v. With negative voltage feedback, the non-linear distortion will be ....
(*i*) $D(1 + A_v m_v)$ (*ii*) $D(1 - A_v m_v)$
(*iii*) $\frac{1 + A_v m_v}{D}$ (*iv*) $\frac{D}{1 + A_v m_v}$

32. The output and input voltages of an emitter follower have a phase difference of ....
(*i*) 180° (*ii*) 90°
(*iii*) 0° (*iv*) 270°

33. It is most necessary to control signal-to-noise ratio at ....
(*i*) initial stage (*ii*) driver stage
(*iii*) output stage (*iv*) detector stage

34. In order to obtain good gain stability in a negative voltage feedback amplifier (A_v = voltage gain without feedback ; m_v = feedback fraction), ...
(*i*) $A_v m_v = 1$ (*ii*) $A_v m_v >> 1$
(*iii*) $A_v m_v < 1$
(*iv*) none of the above

35. Emitter follower is also known as ....
(*i*) grounded emitter circuit
(*ii*) grounded base circuit
(*iii*) grounded collector circuit
(*iv*) none of the above

Answers to Multiple-Choice Questions

1. (*ii*)	2. (*i*)	3. (*iii*)	4. (*iv*)	5. (*ii*)
6. (*i*)	7. (*iii*)	8. (*ii*)	9. (*i*)	10. (*iii*)
11. (*ii*)	12. (*ii*)	13. (*i*)	14. (*ii*)	15. (*iii*)
16. (*i*)	17. (*iii*)	18. (*ii*)	19. (*iv*)	20. (*i*)
21. (*iv*)	22. (*iii*)	23. (*i*)	24. (*ii*)	25. (*i*)
26. (*ii*)	27. (*iv*)	28. (*iii*)	29. (*i*)	30. (*ii*)
31. (*iv*)	32. (*iii*)	33. (*i*)	34. (*ii*)	35. (*iii*)

Chapter Review Topics

1. What do you understand by feedback? Why is negative feedback applied in high gain amplifiers?
2. Discuss the principles of negative voltage feedback in amplifiers with a neat diagram.

3. Derive an expression for the gain of negative voltage feedback amplifier.
4. What is a feedback circuit? Explain how it provides feedback in amplifiers.
5. Describe the action of emitter follower with a neat diagram.
6. Derive the expressions for (*i*) voltage gain (*ii*) input impedance and (*iii*) output impedance of an emitter follower.

Problems

1. An amplifier has a gain of 2×10^5 without feedback. Determine the gain if negative voltage feedback is applied. Take feedback fraction $m_v = 0.02$. **[50]**
2. An amplifier has a gain of 10,000 without feedback. With negative voltage feedback, the gain is reduced to 50. Find the feedback fraction. **[m_v = 0.02]**
3. A feedback amplifier has an internal gain A_v = 40 dB and feedback fraction m_v = 0.05. If the input impedance of this circuit is 12 kΩ, what would have been the input impedance if feedback were not present? **[2 kΩ]**
4. Calculate the gain of a negative voltage feedback amplifier with an internal gain A_v = 75 and feedback fraction m_v = 1/15. What will be the gain if A_v doubles? **[12.5 ; 13.64]**
5. An amplifier with negative feedback has a voltage gain of 100. It is found that without feedback, an input signal of 50 mV is required to produce a given output, whereas with feedback, the input signal must be 0.6 V for the same output. Calculate (*i*) gain without feedback (*ii*) feedback fraction. **[(*i*) 1200 (*ii*) 0.009]**

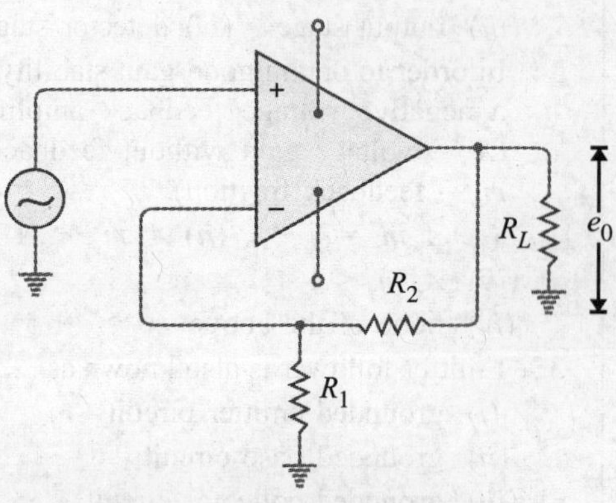

Fig. 11.44

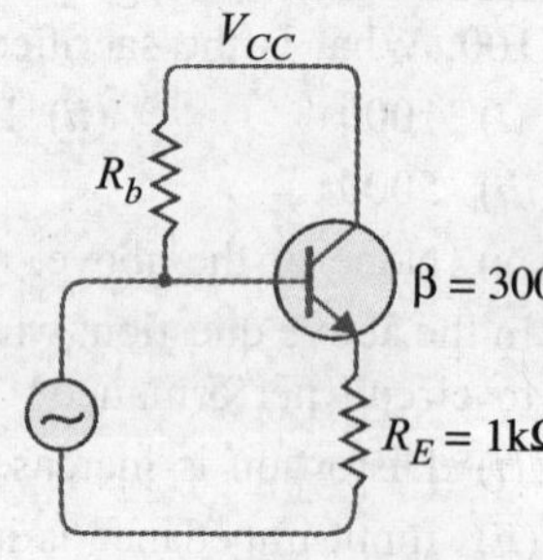

Fig. 11.45

6. Fig. 11.44 shows the negative feedback amplifier. If the gain of the amplifier without feedback is 10^5 and R_1 = 100 Ω, R_2 = 100 kΩ, find (*i*) feedback fraction (*ii*) gain with feedback. **[(*i*) 0.001 (*ii*) 1000]**
7. In Fig. 11.45, if input and output impedances without feedback are 2 MΩ and 500 Ω respectively, find their values after negative voltage feedback. **[302 MΩ ; 1.6 Ω]**
8. An amplifier has a current gain of 240 without feedback. When negative current feedback is applied, determine the effective current gain of the amplifier. Given that current attenuation m_i = 0.015. **[52.7]**
9. An amplifier has an open-loop gain and input impedance of 200 and 15 kΩ respectively. If negative current feedback is applied, what is the effective input impedance of the amplifier? Given that current attenuation m_i = 0.012. **[4.41 kΩ]**
10. An amplifier has A_i = 200 and m_i = 0.012. The open-loop output impedance of the amplifier is 2 kΩ. If negative current feedback is applied, what is the effective output impedance of the amplifier? **[6.8 kΩ]**

CHAPTER 12

Sinusoidal Oscillators

Learning Objectives

- Understand the underlying principle for producing sustained oscillations.
- Develop mathematical condition for proper operation of an oscillator.
- Explore the operation of *LC* oscillators and *RC* oscillators.
- Understand the operation of various types of *LC* oscillators.
- Develop an understanding of *RC* Phase shift oscillator and Wien Bridge oscillator.
- Derive the expression for operating frequency (f_o) for the *RC* Phase Shift oscillator.
- Derive the expression for the balance frequency (f_o) of Wien Bridge oscillator.
- Learn the limitations of *LC* and *RC* oscillators.
- Understand the working and characteristics of Quartz crystal.
- Understand the operation of Transistor Crystal oscillator.

Introduction

In the previous chapter, we have seen that negative feedback in an amplifier improves the circuit performance of the amplifier. However, a proper amount of positive feedback in an amplifier enables it to work as an oscillator *i.e.* it produces continuous oscillations of any desired frequency without any external signal source. The only input power to an oscillator is the d.c. power supply. Therefore, an oscillator is simply a positive feedback amplifier. The frequency of oscillations is determined by the components of the circuit.

Many electronic devices require a source of energy at a specific frequency which may range from a few Hz to several MHz. This is achieved by an electronic device called an *oscillator*. Oscillators are extensively used in electronic equipment. For example, in radio and television receivers, oscillators are used to generate high frequency wave (called *carrier wave*) in the tuning stages. Audio frequency and radio-frequency signals are required for the repair of radio, television and other electronic equipment. Oscillators are also widely used in radar, electronic computers and other electronic devices. Oscillators can produce sinusoidal or non-sinusoidal (*e.g.* square wave) waves. In this chapter, we shall confine our attention to sinusoidal oscillators *i.e.* those which produce sine-wave signals.

12.1 Sinusoidal Oscillator

An electronic device that generates sinusoidal oscillations of desired frequency without any external signal source is known as a **sinusoidal oscillator.**

Note that an oscillator generates oscillations of any desired frequency without any external signal source. The only input power to an oscillator is the d.c. power supply. As we shall see, an oscillator is a *positive feedback amplifier* which provides its own input signal *via* the feedback circuit. Therefore, the oscillator generates or produces oscillations without any externally applied input signal. The frequency of oscillations is determined by the components of the oscillator circuit. Although we speak of an oscillator as "generating" a frequency, it should be noted that it does not create energy, but merely acts as an energy converter. It receives d.c. energy and changes it into a.c. energy of desired frequency. The frequency of oscillations depends upon the constants of the device.

It may be mentioned here that although an alternator produces sinusoidal oscillations of 50Hz, it cannot be called an oscillator. Firstly, an alternator is a mechanical device having rotating parts whereas an oscillator is a non-rotating electronic device. Secondly, an alternator converts mechanical energy into a.c. energy while an oscillator converts d.c. energy into a.c. energy. Thirdly, an alternator cannot produce high frequency oscillations whereas an oscillator can produce oscillations ranging from a few Hz to several MHz.

Advantages

Although oscillations can be produced by mechanical devices (*e.g.* alternators), but electronic oscillators have the following advantages :

(*i*) An oscillator is a non-rotating device. Consequently, there is little wear and tear and hence longer life.

(*ii*) Due to the absence of moving parts, the operation of an oscillator is quite silent.

(*iii*) An oscillator can produce waves from small (20Hz) to extremely high frequencies (> 100 MHz).

(*iv*) The frequency of oscillations can be easily changed when desired.

(*v*) It has good frequency stability *i.e.* frequency once set remains constant for a considerable period of time.

(*vi*) It has very high efficiency.

12.2 Types of Sinusoidal Oscillations

Sinusoidal electrical oscillations can be of two types *viz. damped oscillations* and *undamped oscillations.*

(*i*) Damped oscillations. The electrical oscillations whose amplitude goes on decreasing with time are called *damped oscillations.* Fig. 12.1 (*i*) shows waveform of damped electrical oscillations. Obviously, the electrical system in which these oscillations are generated has losses and some energy is lost during each oscillation. Further, no means are provided to compensate for the losses and consequently the amplitude of the generated wave decreases gradually. It may be noted that frequency of oscillations remains unchanged since it depends upon the constants of the electrical system.

(*ii*) Undamped oscillations. The electrical oscillations whose amplitude remains constant with time are called *undamped oscillations.* Fig. 12.1 (*ii*) shows waveform of undamped electrical oscillations. Although the electrical system in which these oscillations are being generated has also losses, but now right amount of energy is being supplied to overcome the losses. Consequently, the amplitude of the generated wave remains constant. It should be emphasised that an oscillator is required to produce undamped electrical oscillations for utilising in various electronics equipment.

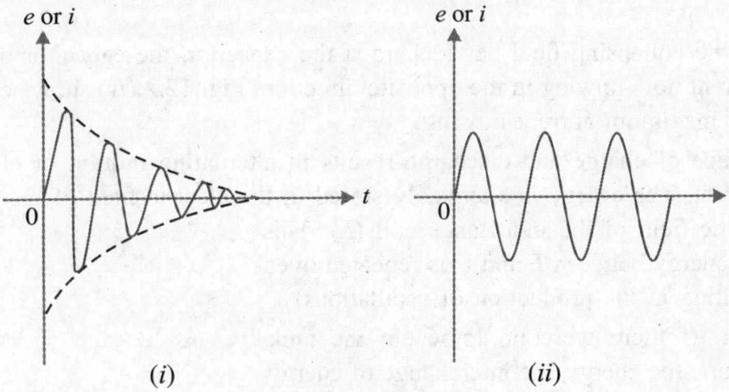

Fig. 12.1

12.3 Oscillatory Circuit

A circuit which produces electrical oscillations of any desired frequency is known as an **oscillatory circuit** *or* **tank circuit.**

A simple oscillatory circuit consists of a capacitor (C) and inductance coil (L) in parallel as shown in Fig. 12.2. This electrical system can produce electrical oscillations of frequency determined by the values of L and C. To understand how this comes about, suppose the capacitor is charged from a d.c. source with a polarity as shown in Fig. 12.2 (*i*).

(*i*) In the position shown in Fig. 12.2 (*i*), the upper plate of capacitor has deficit of electrons and the lower plate has excess of electrons. Therefore, there is a voltage across the capacitor and the capacitor has electrostatic energy.

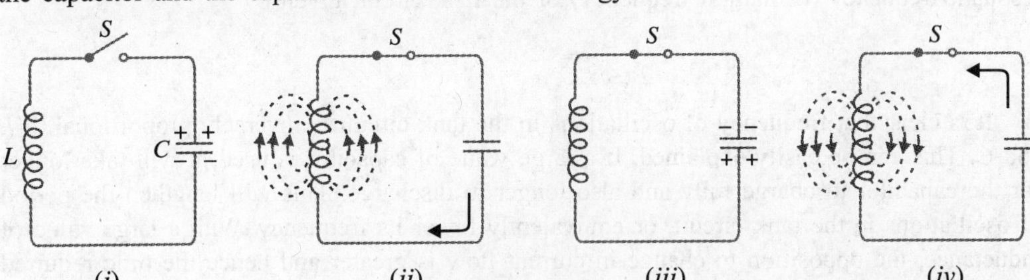

Fig. 12.2

(*ii*) When switch S is closed as shown in Fig. 12.2 (*ii*), the capacitor will discharge through inductance and the electron flow will be in the direction indicated by the arrow. This current flow sets up magnetic field around the coil. Due to the inductive effect, the current builds up slowly towards a maximum value. The circuit current will be maximum when the capacitor is fully discharged. At this instant, electrostatic energy is zero but because electron motion is greatest (*i.e.* maximum current), the magnetic field energy around the coil is maximum. This is shown in Fig. 12.2 (*ii*). Obviously, the electrostatic energy across the capacitor is completely converted into magnetic field energy around the coil.

(*iii*) Once the capacitor is discharged, the magnetic field will begin to collapse and produce a counter e.m.f. According to Lenz's law, the counter e.m.f. will keep the current flowing in the same direction. The result is that the capacitor is now charged with opposite polarity, making upper plate of capacitor negative and lower plate positive as shown in Fig. 12.2 (*iii*).

(*iv*) After the collapsing field has recharged the capacitor, the capacitor now begins to discharge ; current now flowing in the opposite direction. Fig. 12.2 (*iv*) shows capacitor fully discharged and maximum current flowing.

The sequence of charge and discharge results in alternating motion of electrons or an oscillating current. The energy is alternately stored in the electric field of the capacitor (*C*) and the magnetic field of the inductance coil (*L*). This interchange of energy between *L* and *C* is repeated over and again resulting in the production of oscillations.

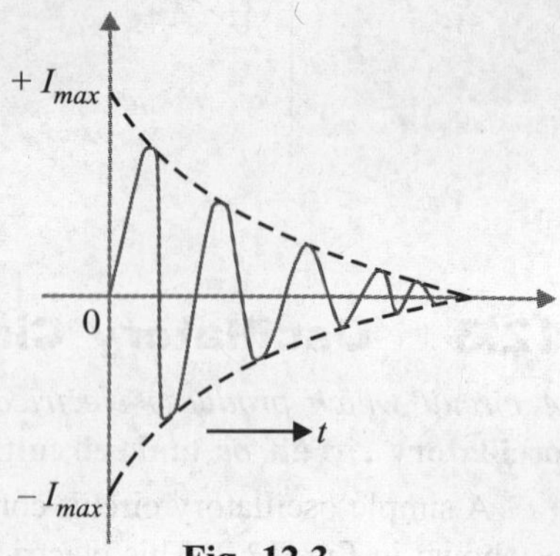

Fig. 12.3

Waveform. If there were no losses in the tank circuit to consume the energy, the interchange of energy between *L* and *C* would continue indefinitely. In a practical tank circuit, there are resistive and radiation losses in the coil and dielectric losses in the capacitor. During each cycle, a small part of the originally imparted energy is used up to overcome these losses. The result is that the amplitude of oscillating current decreases gradually and eventually it becomes zero when all the energy is consumed as losses. Therefore, the tank circuit by itself will produce *damped oscillations* as shown in Fig. 12.3.

Frequency of oscillations. The frequency of oscillations in the tank circuit is determined by the constants of the circuit *viz.* *L* and *C*. The actual frequency of oscillations is the resonant frequency (or natural frequency) of the tank circuit given by ;

$$f_r = \frac{1}{2\pi \sqrt{LC}}$$

It is clear that frequency of oscillations in the tank circuit is inversely proportional to *L* and *C*. This can be easily explained. If a large value of capacitor is used, it will take longer for the capacitor to charge fully and also longer to discharge. This will lengthen the period of oscillations in the tank circuit, or equivalently lower its frequency. With a large value of inductance, the opposition to change in current flow is greater and hence the time required to complete each cycle will be longer. Therefore, the greater the value of inductance, the longer is the period or the lower is the frequency of oscillations in the tank circuit.

12.4 Undamped Oscillations from Tank Circuit

As discussed before, a tank circuit produces damped oscillations. However, in practice, we need continuous undamped oscillations for the successful operation of electronics equipment. In order to make the oscillations in the tank circuit undamped, it is necessary to supply correct amount of energy to the tank circuit at the proper time intervals to meet the losses. Thus referring back to Fig. 12.2, any energy which would be applied to the circuit must have a polarity conforming to the existing polarity at the instant of application of energy. If the applied energy is of opposite polarity, it would oppose the energy in the tank circuit, causing stoppage of oscillations. Therefore, in order to make the oscillations in the tank circuit undamped, the following conditions must be fulfilled :

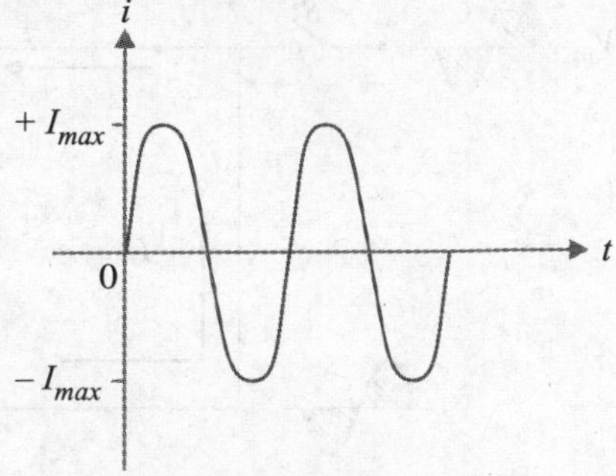

Fig. 12.4

(*i*) The amount of energy supplied should be such so as to meet the losses in the tank circuit and the a.c. energy removed from the circuit by the load. For instance, if losses in *LC* circuit amount to 5 mW and a.c. output being taken is 100 mW, then power of 105 mW should be continuously supplied to the circuit.

(*ii*) The applied energy should have the same frequency as that of the oscillations in the tank circuit.

(*iii*) The applied energy should be in phase with the oscillations set up in the tank circuit *i.e.* it should aid the tank circuit oscillations.

If these conditions are fulfilled, the circuit will produce continuous undamped output as shown in Fig. 12.4.

12.5 Positive Feedback Amplifier — Oscillator

A transistor amplifier with *proper* positive feedback can act as an oscillator *i.e.* it can generate oscillations without any external signal source. Fig. 12.5 shows a transistor amplifier with positive feedback. Remember that a positive feedback amplifier is one that produces a feedback voltage (V_f) that *is in phase* with the original input signal. As you can see, this condition is met in the circuit shown in Fig. 12.5. A phase shift of 180° is produced by the amplifier and a further phase shift of 180° is introduced by feedback network. Consequently, the signal is shifted by 360° and fed to the input *i.e.* feedback voltage is in phase with the input signal.

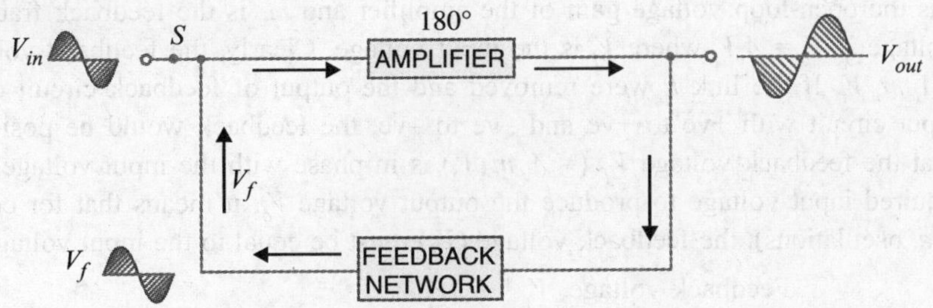

Fig. 12.5

(*i*) We note that the circuit shown in Fig. 12.5 is producing oscillations in the output. However, this circuit has an input signal. This is inconsistent with our definition of an oscillator *i.e.* an oscillator is a circuit that produces oscillations without any external signal source.

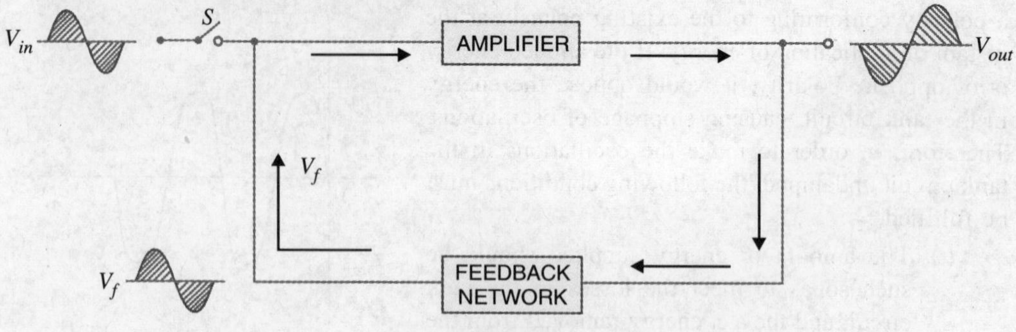

Fig. 12.6

(*ii*) When we open the switch *S* of Fig. 12.5, we get the circuit shown in Fig. 12.6. This means the input signal (V_{in}) is removed. However, V_f (which is in phase with the original signal) is still applied to the input signal. The amplifier will respond to this signal in the same way that it did to V_{in} *i.e.* V_f will be amplified and sent to the output. The feedback network sends a portion of the output back to the input. Therefore, the amplifier receives another input cycle and another output cycle is produced. This process will continue so long as the amplifier is turned on. Therefore, the amplifier will produce sinusoidal output without any external signal source. The only condition is that the amount of positive feedback should be such that the feedback voltage $V_f = V_{in}$. In that case, the circuit will provide its own input signal (V_{in}) *via* the feedback circuit and no externally applied input signal is required. The only input power to the oscillator is the d.c. power supply.

12.6 Condition for Proper Oscillator Operation

An oscillator is an amplifier that uses *positive feedback* for its operation. Therefore, it is a positive feedback amplifier. Proper operation of oscillator means that it should produce continuous oscillations of the desired frequency without any external signal source except the d.c. supply. This is possible only if the positive feedback amplifier provides its own input signal *via* the feedback circuit. Let us find the mathematical condition under which it can happen. Consider the block diagram of positive feedback amplifier shown in Fig. 12.7. Here A_v is the open-loop voltage gain of the amplifier and m_v is the feedback fraction. The output voltage is $V_o = A_v V_i$ where V_i is the input voltage. Clearly, the feedback voltage $V_f = m_v\, V_o = A_v\, m_v\, V_i$. If the link *L* were removed and the output of feedback circuit connected to the input circuit with +ve to +ve and –ve to –ve, the feedback would be positive. This means that the feedback voltage V_f (= $A_v\, m_v\, V_i$) is in phase with the input voltage. Since V_i is the required input voltage to produce the output voltage V_o, it means that for continuous output (*i.e.* oscillations), the feedback voltage (V_f) must be equal to the input voltage (V_i) *i.e.*

Feedback voltage, $V_f = V_i$

or $A_v\, m_v\, V_i = V_i$

∴ $A_v\, m_v = 1$

Under this condition (*i.e.* $A_v\, m_v = 1$), even if these were no input generator (*i.e.* V_i), a continuous output will be obtained across R_L immediately after connecting the necessary d.c. power supply. Fig. 12.8 shows the general layout of an oscillator.

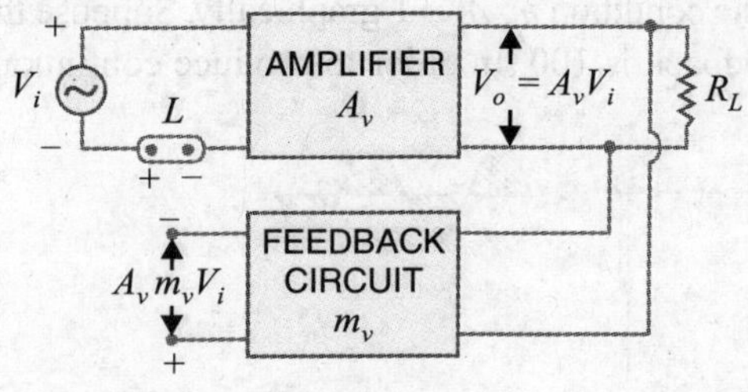

Fig. 12.7

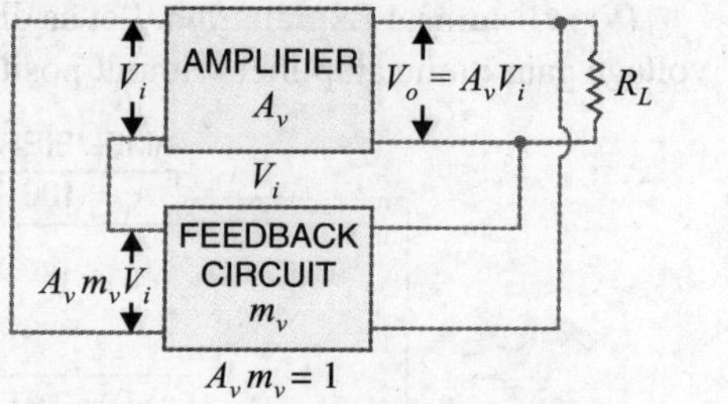

Fig. 12.8

It is clear from Fig. 12.8 that continuous output voltage (V_o) will be obtained with no externally applied input signal. The circuit provides its own input signal *via* the feedback circuit.Thus output voltage $V_o = A_v V_i$ and the feedback voltage, $V_f = A_v m_v V_i = V_i$ ($\because A_v m_v = 1$).

Therefore, for proper operation of oscillator (positive feedback amplifier), the product $A_v m_v$ should be equal to 1 i.e.

$$A_v m_v = 1$$

This relation is called **Barkhausen criterion** and is discussed in detail in the next article.

Note. Let us see what happens if the condition $A_v m_v = 1$ is not met in an oscillator.

(*i*) If $A_v m_v < 1$, the oscillations will die out within a few cycles. It is because the oscillations will lose amplitude on each progressive cycle and will eventually fade out. This loss of signal is called *damping*.

(*ii*) If $A_v m_v > 1$, the output will be distorted. It is because the output will increase from each cycle to the next. Eventually, the output will start to experience saturation and cut-off clipping.

12.7 Explanation of Barkhausen Criterion

Barkhausen criterion is that in order to produce continuous undamped oscillations at the output of an amplifier, the positive feedback should be such that :

$$m_v A_v = 1$$

Once this condition is set in the positive feedback amplifier, continuous undamped oscillations can be obtained at the output immediately after connecting the necessary power supplies.

(*i*) **Mathematical explanation.** The voltage gain of a positive feedback amplifier is given by ;

$$A_{vf} = \frac{A_v}{1 - m_v A_v}$$

$$\text{If } m_v A_v = 1, \text{ then, } A_{vf} \rightarrow \infty.$$

We know that we cannot achieve infinite gain in an amplifier. So what does this result infer in physical terms? It means that a vanishing small input voltage would give rise to finite (*i.e.* a definite amount of) output voltage even when the input signal is zero. Thus once the circuit receives the input trigger, it would become an oscillator, generating oscillations with no external signal source.

(ii) **Graphical Explanation.** Let us discuss the condition $m_v A_v = 1$ graphically. Suppose the voltage gain of the amplifier without positive feedback is 100. In order to produce continuous

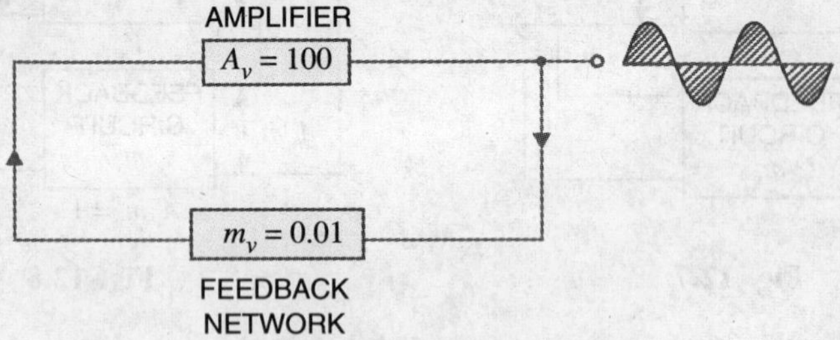

Fig. 12.9

undamped oscillations, $m_v A_v = 1$ or $m_v \times 100 = 1$ or $m_v = 0.01$. This is illustrated in Fig. 12.9. Since the condition $m_v A_v = 1$ is met in the circuit shown in Fig. 12.9, it will produce sustained oscillations.

Suppose the initial triggering voltage is 0.1V peak. Starting with this value, circuit ($A_v = 100$; $m_v = 0.01$) will progress as follows.

Cycle	V_{in}	V_{out}	V_f
1.	0.1Vpk	10Vpk	0.1Vpk
2.	0.1Vpk	10Vpk	0.1Vpk

The same thing will repeat for 3rd, 4th cycles and so on. Note that during each cycle, $V_f = 0.1$Vpk and $V_{out} = 10$ Vpk. Clearly, the oscillator is producing continuous undamped oscillations.

Note. The relation $m_v A_v = 1$ holds good for true ideal circuits. However, practical circuits need an $m_v A_v$ product that is slightly greater than 1. This is to compensate for power loss (*e.g.* in resistors) in the circuit.

12.8 Essential Parts of an Oscillator

An oscillator is a positive feedback amplifier. The amplifier may employ a transistor or any other electronic device capable of doing the job of amplification. Fig. 12.10 shows the block diagram of an oscillator. Its essential components are :

(i) **Frequency-determining circuit.** It consists of either parallel tuned circuit (*LC*) or *RC* circuit. The radio-frequency oscillators generally employ the former while the audio frequency oscillators use the latter. The frequency of oscillations in the circuit depends upon these circuit elements.

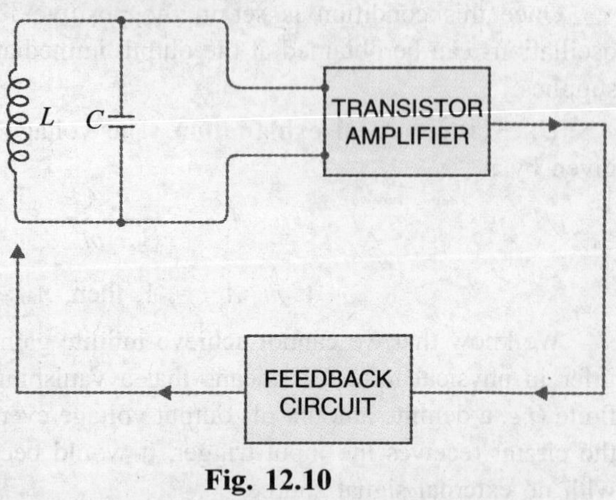

Fig. 12.10

***(ii)* Transistor Amplifier.** The transistor amplifier receives d.c. power from the battery and changes it into a.c. power of desired frequency for supplying to the frequency-determining circuit. The oscillations occurring in the frequency-determining circuit are applied to the input of the transistor amplifier. Because of amplifying properties of the transistor, we get increased output of these oscillations. The output of the transistor is supplied to the frequency-determining circuit to meet the losses.

***(iii)* Feedback Circuit.** The feedback circuit supplies a part of output energy to the frequency determining circuit satisfying the relation $A_v m_v = 1$. The entire circuit arrangement is so designed that energy fed back is in phase with oscillations in the frequency-determining circuit *i.e.* positive feedback is provided.

Discussion. The following points are worth noting about an oscillator :

(i) A transistor amplifier with *proper positive feedback* will work as an oscillator.

(ii) The circuit needs only a quick trigger to start the oscillations. This trigger may be provided when d.c. supply is applied to the oscillator. Once oscillations have started, no external signal source is needed.

(iii) The amount of positive feedback should be such that $A_v\ m_v = 1$.

12.9 Different Types of Oscillators

All oscillators are positive feedback amplifiers in which the amount of positive feedback is such that the condition $A_v\ m_v = 1$ is met. They differ only in the manner in which this objective is achieved. There are two basic types of oscillators *viz.*

(i) LC oscillators *(ii) RC* oscillators

(i) In ***LC* oscillators**, the *LC* circuit determines the frequency of oscillations as well as produces 180° phase shift. A further phase shift of 180° is introduced by the amplifier. In this way, positive feedback is ensured. The important types of *LC* oscillators are :

(a) Tuned collector oscillator
(b) Colpitt's oscillator
(c) Hartley oscillator
(d) Crystal oscillator

(ii) In ***RC* oscillators**, the *RC* network determines the operating frequency f_0. Of course, the *RC* circuit cannot produce oscillations as in *LC* circuit. The *RC* network may produce 180° phase shift or 0° phase shift, depending upon the design of *RC* network. The two important types of *RC* oscillators are :

(a) RC phase shift oscillator
(b) Wien bridge oscillator

12.10 Tuned Collector Oscillator

Fig. 12.11 shows the circuit of tuned collector oscillator. It contains tuned circuit L_1-C_1 in the collector and hence the name. The frequency of oscillations depends upon the values of L_1 and C_1 and is given by ;

$$f = \frac{1}{2\pi\sqrt{L_1 C_1}} \qquad ...(i)$$

The feedback coil L_2 in the base circuit is magnetically coupled to the tank circuit coil L_1. In practice, L_1 and L_2 form the primary and secondary of the transformer respectively. The biasing is provided by potential divider arrangement. The capacitor C connected in the base circuit provides low reactance path to the oscillations.

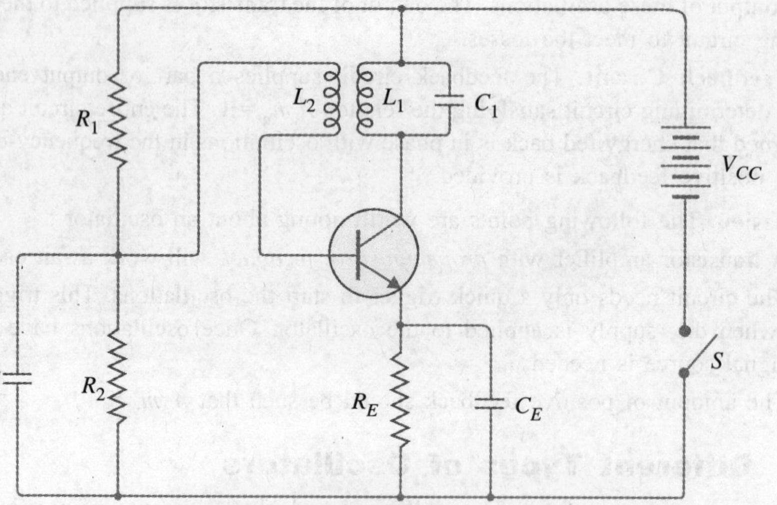

Fig. 12.11

Circuit operation. When switch S is closed, collector current starts increasing and charges the capacitor C_1. When this capacitor is fully charged, it discharges through coil L_1, setting up oscillations of frequency determined by exp. (*i*). These oscillations induce some voltage in coil L_2 by mutual induction. The frequency of voltage in coil L_2 is the same as that of tank circuit but its magnitude depends upon the number of turns of L_2 and coupling between L_1 and L_2. The voltage across L_2 is applied between base and emitter and appears in the amplified form in the collector circuit, thus overcoming the losses occurring in the tank circuit. The number of turns of L_2 and coupling between L_1 and L_2 are so adjusted that oscillations across L_2 are amplified to a level just sufficient to supply losses to the tank circuit.

It may be noted that the phase of feedback is correct *i.e.* energy supplied to the tank circuit is in phase with the generated oscillations. A phase shift of 180° is created between the voltages of L_1 and L_2 due to transformer *action. A further phase shift of 180° takes place between base-emitter and collector circuit due to transistor properties. As a result, the energy fed back to the tank circuit is in phase with the generated oscillations.

Example 12.1. *The tuned collector oscillator circuit used in the local oscillator of a radio receiver makes use of an LC tuned circuit with L_1 = 58.6 μH and C_1 = 300 pF. Calculate the frequency of oscillations.*

Solution.

$$L_1 = 58.6\ \mu\text{H} = 58.6 \times 10^{-6}\ \text{H}$$

$$C_1 = 300\ \text{pF} = 300 \times 10^{-12}\ \text{F}$$

Frequency of oscillations, $f = \dfrac{1}{2\pi\sqrt{L_1 C_1}}$

* All transformers introduce a phase shift of 180° between primary and secondary.

$$= \frac{1}{2\pi\sqrt{58.6\times10^{-6}\times300\times10^{-12}}}\text{ Hz}$$

$$= 1199 \times 10^3 \text{ Hz} = \mathbf{1199\ kHz}$$

Example 12.2. *Find the capacitance of the capacitor required to build an LC oscillator that uses an inductance of L_1 = 1 mH to produce a sine wave of frequency 1 GHz (1 GHz = 1 × 10^{12} Hz).*

Solution.

Frequency of oscillations is given by ;

$$f = \frac{1}{2\pi\sqrt{L_1 C_1}}$$

or
$$C_1 = \frac{1}{L_1(2\pi f)^2} = \frac{1}{(1\times10^{-3})(2\pi\times1\times10^{12})^2}$$

$$= 2{\cdot}53 \times 10^{-23}\text{ F} = 2{\cdot}53 \times 10^{-11}\text{ pF}$$

The *LC* circuit is often called *tuned circuit* or *tank circuit.*

12.11 Colpitt's Oscillator

Fig. 12.12 shows a Colpitt's oscillator. It uses two capacitors and placed across a common inductor *L* and the centre of the two capacitors is tapped. The tank circuit is made up of C_1, C_2 and *L*. The frequency of oscillations is determined by the values of C_1, C_2 and *L* and is given by ;

$$f = \frac{1}{2\pi\sqrt{LC_T}} \quad \text{...}(i)$$

where
$$C_T = \frac{C_1 C_2}{C_1 + C_2}$$

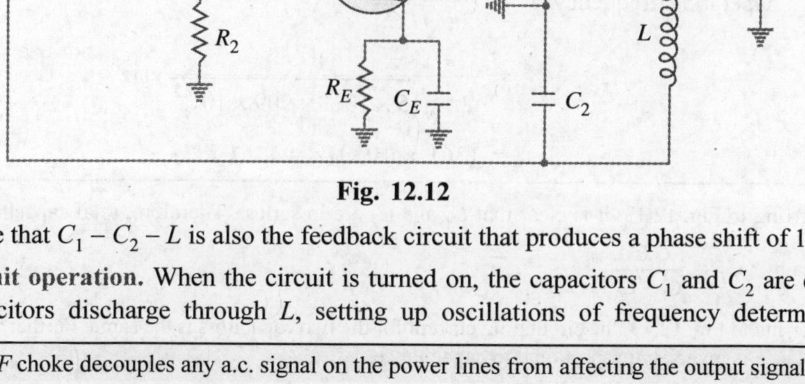

Fig. 12.12

*Note that $C_1 - C_2 - L$ is also the feedback circuit that produces a phase shift of 180°.

Circuit operation. When the circuit is turned on, the capacitors C_1 and C_2 are charged. The capacitors discharge through *L*, setting up oscillations of frequency determined by

* The *RF* choke decouples any a.c. signal on the power lines from affecting the output signal.

exp. *(*i*). The output voltage of the amplifier appears across C_1 and feedback voltage is developed across C_2. The voltage across it is 180° out of phase with the voltage developed across C_1 (V_{out}) as shown in Fig. 12.13. It is easy to see that voltage fed back (voltage across C_2) to the transistor provides positive feedback. A phase shift of 180° is produced by the transistor and a further phase shift of 180° is produced by $C_1 - C_2$ voltage divider. In this way, feedback is properly phased to produce continuous undamped oscillations.

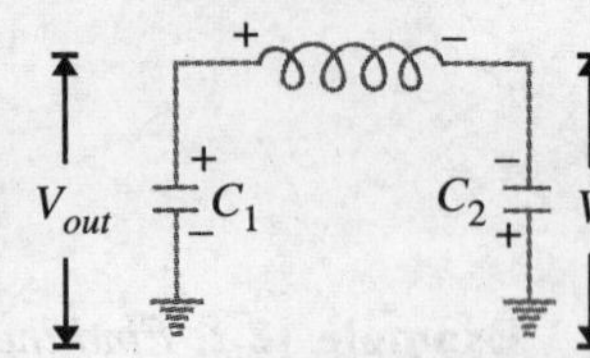

Fig. 12.13

Feedback fraction m_v. The amount of feedback voltage in Colpitt's oscillator depends upon feedback fraction m_v of the circuit. For this circuit,

$$\text{Feedback fraction, } m_v = \frac{V_f}{V_{out}} = \frac{X_{C2}}{X_{C1}} = \frac{C_1}{C_2}^{**}$$

or

$$m_v = \frac{C_1}{C_2}$$

We know that the condition for continuous oscillations in an oscillator is $A_v\, m_v = 1$.

$$\therefore \quad \text{Amplifier voltage gain, } A_v = \frac{1}{m_v} = \frac{C_2}{C_1}$$

Actually, the product $A_v\, m_v$ should be slightly greater than 1 to prevent decrease of oscillations due to power loss within the amplifier.

Example 12.3. *Determine the (i) operating frequency and (ii) feedback fraction for Colpitt's oscillator shown in Fig. 12.14.*

Solution.

(*i*) **Operating Frequency.** The operating frequency of the circuit is always equal to the resonant frequency of the feedback network. As noted previously, the capacitors C_1 and C_2 are in series.

$$\therefore \quad C_T = \frac{C_1 C_2}{C_1 + C_2} = \frac{0.001 \times 0.01}{0.001 + 0.01} = 9.09 \times 10^{-4}\ \mu\text{F}$$

$$= 909 \times 10^{-12}\ \text{F}$$

$$L = 15\ \mu\text{H} = 15 \times 10^{-6}\ \text{H}$$

$$\therefore \quad \text{Operating frequency, } f = \frac{1}{2\pi\sqrt{LC_T}}$$

$$= \frac{1}{2\pi\sqrt{15 \times 10^{-6} \times 909 \times 10^{-12}}}\ \text{Hz}$$

$$= 1361 \times 10^3\ \text{Hz} = \mathbf{1361\ kHz}$$

* Referring to Fig. 12.13, it is clear that C_1 and C_2 are in series. Therefore, total capacitance C_T is given by ; $C_T = \dfrac{C_1 C_2}{C_1 + C_2}$.

** Referring to Fig. 12.13, the circulating current for the two capacitors is the same. Further, capacitive reactance is inversely proportional to capacitance.

(ii) **Feedback fraction**

$$m_v = \frac{C_1}{C_2} = \frac{0.001}{0.01} = \mathbf{0.1}$$

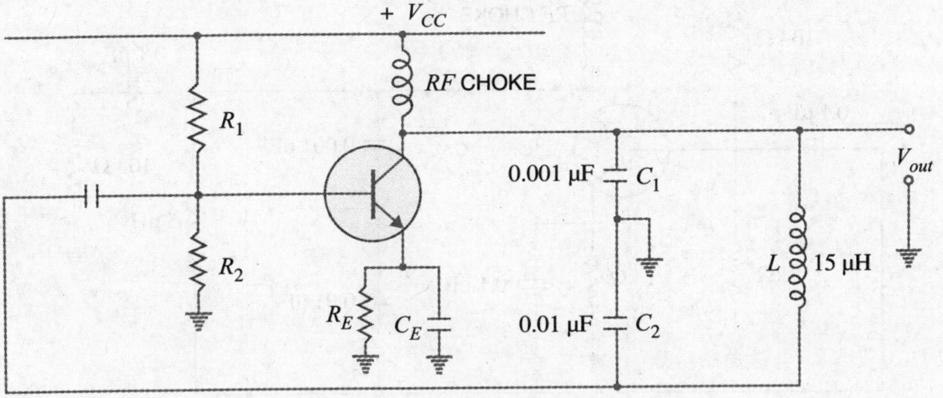

Fig. 12.14

Example 12.4. *A 1 mH inductor is available. Choose the capacitor values in a Colpitt's oscillator so that f = 1 MHz and m_v = 0.25.*

Solution.

$$\text{Feedback fraction, } m_v = \frac{C_1}{C_2}$$

or $$0.25 = \frac{C_1}{C_2} \qquad \therefore\ C_2 = 4C_1$$

Now, $$f = \frac{1}{2\pi\sqrt{LC_T}}$$

or $$C_T = \frac{1}{L(2\pi f)^2} = \frac{1}{(1\times10^{-3})\,(2\pi\times1\times10^6)^2}$$

$$= 25.3 \times 10^{-12}\text{ F}$$

$$= 25.3\text{ pF}$$

or $$\frac{C_1 C_2}{C_1 + C_2} = 25.3\text{ pF} \qquad \left[\because C_T = \frac{C_1 C_2}{C_1 + C_2}\right]$$

or $$\frac{C_2}{1 + \frac{C_2}{C_1}} = 25.3$$

or $$\frac{C_2}{1 + 4} = 25.3 \qquad \therefore\ C_2 = 25.3 \times 5 = \mathbf{126.5\ pF}$$

and $$C_1 = C_2/4 = 126.5/4 = \mathbf{31.6\ pF}$$

Example 12.5. *For the Colpitt's oscillator shown in Fig. 12.15 determine the (i) d.c. emitter current (ii) d.c. collector-emitter voltage (iii) frequency of oscillations (iv) feedback fraction and (v) minimum value of voltage gain to start oscillations.*

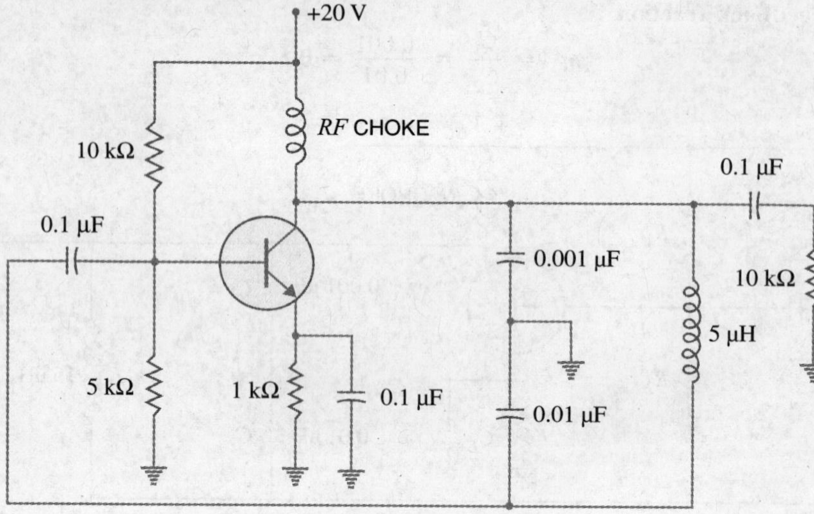

Fig. 12.15

Solution. $R_1 = 10\ k\Omega$; $R_2 = 5\ k\Omega$; $C_1 = 0.001\ \mu F$; $C_2 = 0.01\ \mu F$; $R_E = 1\ k\Omega$

(*i*) The d.c. emitter current I_E is found as under :

$$\text{D.C. base voltage, } V_B = \frac{V_{CC}}{R_1 + R_2} \times R_2 = \frac{20\ V}{(10+5)\ k\Omega} \times 5\ k\Omega = 6.67\ V$$

$$\text{D.C. emitter voltage, } V_E = V_B - V_{BE} = 6.67 - 0.7 = 5.97\ V$$

$$\therefore \quad \text{D.C. emitter current, } I_E = \frac{V_E}{R_E} = \frac{5.97\ V}{1\ k\Omega} = \mathbf{5.97\ mA}$$

(*ii*) Since *RF* choke is short to d.c. ($f = 0$; $X_L = 0$), the collector voltage is $V_C = 20$ V.

$$\therefore \quad V_{CE} = V_C - V_E = 20 - 5.97 = \mathbf{14.03\ V}$$

(*iii*)
$$C_T = \frac{C_1 C_2}{C_1 + C_2} = \frac{(0.001 \times 0.01)}{(0.001 + 0.01)} = 909 \times 10^{-12}\ F$$

∴ Frequency of oscillations of the oscillator is

$$f = \frac{1}{2\pi\sqrt{LC_T}} = \frac{1}{2\pi\sqrt{(5 \times 10^{-6}) \times 909 \times 10^{-12}}}$$

$$= 2.36 \times 10^6\ Hz = \mathbf{2.36\ MHz}$$

(*iv*) Feedback fraction of this type of oscillator is

$$m_v = \frac{C_1}{C_2} = \frac{0.001\ \mu F}{0.01\ \mu F} = \mathbf{0.1}$$

(*v*) For the oscillations to start, the minimum value of voltage gain (A_v) is

$$A_v = \frac{1}{m_v} = \frac{C_2}{C_1} = \frac{1}{0.1} = \mathbf{10}$$

12.12 Hartley Oscillator

The Hartley oscillator is similar to Colpitt's oscillator with minor modifications. Instead of using tapped capacitors, two inductors L_1 and L_2 are placed across a common capacitor C and the centre of the inductors is tapped as shown in Fig. 12.16. The tank circuit is made up of L_1, L_2 and C. The frequency of oscillations is determined by the values of L_1, L_2 and C and is given by ;

$$f = \frac{1}{2\pi\sqrt{CL_T}} \quad \text{...}(i)$$

where $\quad L_T = L_1 + L_2 + 2M$

Here $\quad M$ = Mutual inductance between L_1 and L_2

Note that $L_1 - L_2 - C$ is also the feedback network that produces a phase shift of 180°.

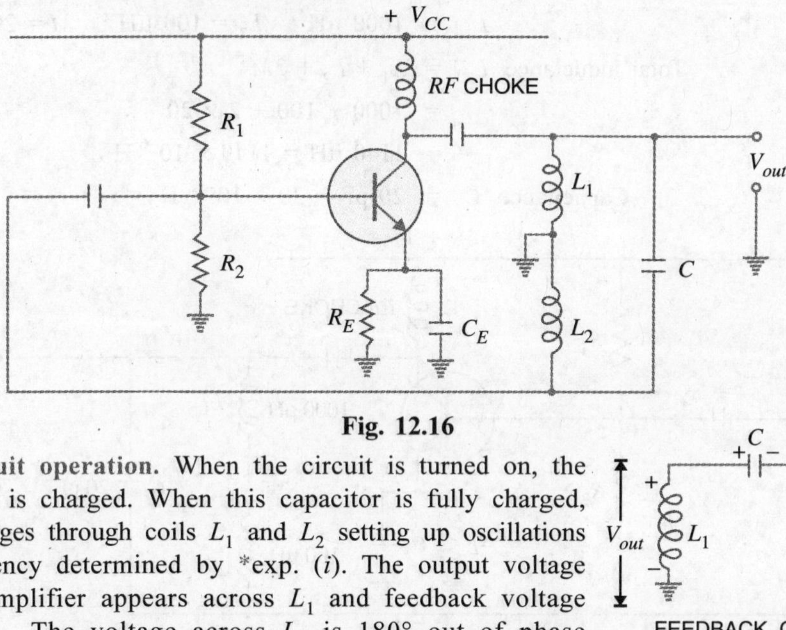

Fig. 12.16

Circuit operation. When the circuit is turned on, the capacitor is charged. When this capacitor is fully charged, it discharges through coils L_1 and L_2 setting up oscillations of frequency determined by *exp. (*i*). The output voltage of the amplifier appears across L_1 and feedback voltage across L_2. The voltage across L_2 is 180° out of phase with the voltage developed across L_1 (V_{out}) as shown in Fig. 12.17. It is easy to see that voltage fed back (*i.e.* voltage across L_2) to the transistor provides positive feedback. A phase shift of 180° is produced by the transistor and a further phase shift of 180° is produced by $L_1 - L_2$ voltage divider. In this way, feedback is properly phased to produce continuous undamped oscillations.

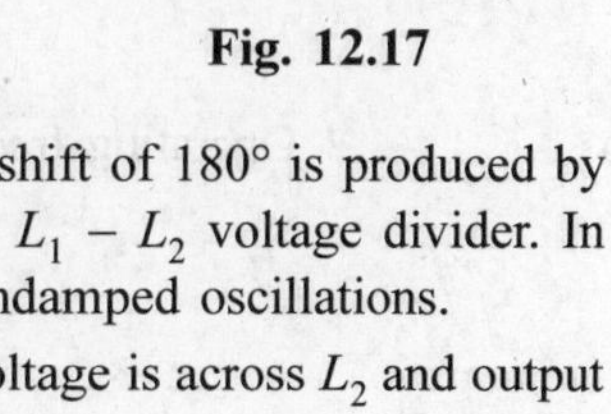

Fig. 12.17

Feedback fraction m_v. In Hartley oscillator, the feedback voltage is across L_2 and output voltage is across L_1.

$$\therefore \quad \text{Feedback fraction, } m_v = \frac{V_f}{V_{out}} = \frac{X_{L_2}}{X_{L_1}} = {}^{**}\frac{L_2}{L_1}$$

* Referring to Fig. 12.17, it is clear that L_1 and L_2 are in series. Therefore, total inductance L_T is given by ; $L_T = L_1 + L_2 + 2M$.

** Referring to Fig. 12.17, the circulating current for the two inductors is the same. Further, inductive reactance is directly proportional to inductance.

or $$m_v = \frac{L_2}{L_1}$$

We know that condition for continuous oscillations in an oscillator is $A_v\, m_v = 1$.

$\therefore$ Amplifier voltage gain, $A_v = \frac{1}{m_v} = \frac{L_1}{L_2}$

Actually, the product $A_v\, m_v$ should be slightly greater than 1 to prevent the loss of oscillations due to power loss within the amplifier.

Example 12.6. *Calculate the (i) operating frequency and (ii) feedback fraction for Hartley oscillator shown in Fig. 12.18. The mutual inductance between the coils, M = 20 μH.*

Solution.

(*i*) $L_1 = 1000\ \mu\text{H}\ ;\ L_2 = 100\ \mu\text{H}\ ;\ M = 20\ \mu\text{H}$

$\therefore$ Total inductance, $L_T = L_1 + L_2 + 2M$

$= 1000 + 100 + 2 \times 20$

$= 1140\ \mu\text{H} = 1140 \times 10^{-6}\ \text{H}$

Capacitance, $C = 20\ \text{pF} = 20 \times 10^{-12}\ \text{F}$

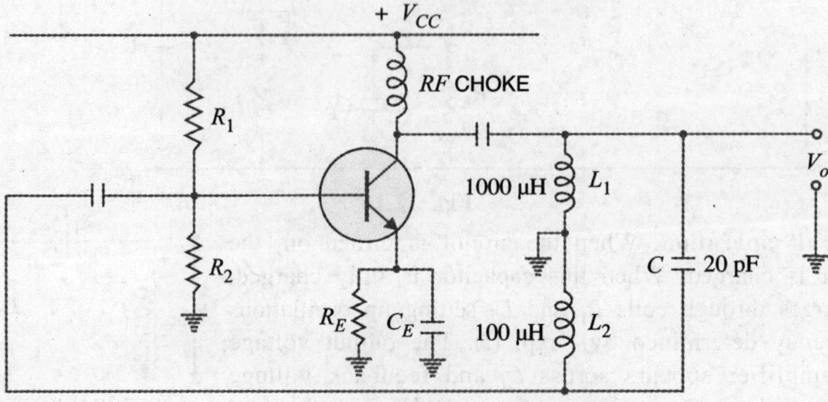

Fig. 12.18

$\therefore$ Operating frequency, $$f = \frac{1}{2\pi\sqrt{L_T C}} = \frac{1}{2\pi\sqrt{1140 \times 10^{-6} \times 20 \times 10^{-12}}}\ \text{Hz}$$

$= 1052 \times 10^3\ \text{Hz} = \mathbf{1052\ kHz}$

(*ii*) Feedback fraction, $$m_v = \frac{L_2}{L_1} = \frac{100\ \mu\text{H}}{1000\ \mu\text{H}} = \mathbf{0.1}$$

Example 12.7. *A 1 pF capacitor is available. Choose the inductor values in a Hartley oscillator so that f = 1 MHz and m_v = 0.2.*

Solution.

Feedback fraction, $$m_v = \frac{L_2}{L_1}$$

or $\qquad 0.2 = \dfrac{L_2}{L_1} \qquad \therefore\ L_1 = 5L_2$

Now, $\qquad f = \dfrac{1}{2\pi\sqrt{L_T C}}$

or $\qquad L_T = \dfrac{1}{C\,(2\pi f)^2} = \dfrac{1}{(1\times 10^{-12})\,(2\pi\times 1\times 10^{6})^2}$

$\qquad = 25.3 \times 10^{-3}\ \text{H} = 25.3\ \text{mH}$

or $\qquad L_1 + L_2 = 25.3\ \text{mH} \qquad (\because\ L_T = L_1 + L_2)$

or $\qquad 5L_2 + L_2 = 25.3 \qquad \therefore\ L_2 = 25.3/6 = \mathbf{4.22\ mH}$

and $\qquad L_1 = 5L_2 = 5 \times 4.22 = \mathbf{21.1\ mH}$

Example 12.8. *The Hartley oscillator shown in Fig. 12.19 has the values as indicated. Find the feedback fraction and the minimum voltage gain needed to start the oscillations. Also find the frequency of oscillations.*

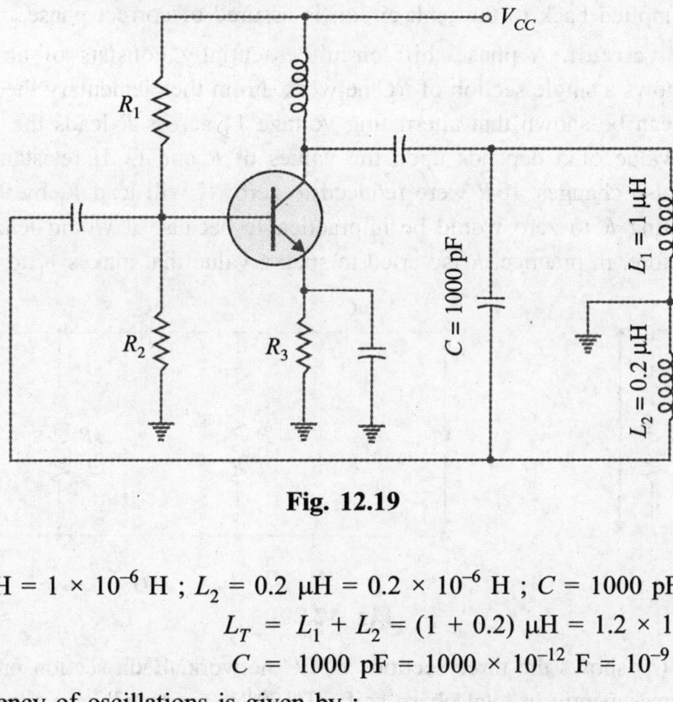

Fig. 12.19

Solution.

$L_1 = 1\ \mu\text{H} = 1 \times 10^{-6}\ \text{H}\ ;\ L_2 = 0.2\ \mu\text{H} = 0.2 \times 10^{-6}\ \text{H}\ ;\ C = 1000\ \text{pF}$

$$L_T = L_1 + L_2 = (1 + 0.2)\ \mu\text{H} = 1.2 \times 10^{-6}\ \text{H}$$

$$C = 1000\ \text{pF} = 1000 \times 10^{-12}\ \text{F} = 10^{-9}\ \text{F}$$

The frequency of oscillations is given by ;

$$f = \frac{1}{2\pi\sqrt{L_T C}} = \frac{1}{2\pi\sqrt{(1.2\times10^{-6})\times10^{-9}}}$$

$$= 4.59 \times 10^{6}\ \text{Hz} = \mathbf{4.59\ MHz}$$

$$\text{Feedback fraction, } m_v = \frac{L_2}{L_1} = \frac{0.2\ \mu\text{H}}{1\ \mu\text{H}} = \mathbf{0.2}$$

$$\text{Minimum voltage gain, } A_v = \frac{1}{m_v} = \frac{L_1}{L_2} = \frac{1}{0.2} = \mathbf{5}$$

12.13 Principle of RC Phase Shift Oscillators

One desirable feature of an oscillator is that it should feed back energy of correct phase to the tank circuit to overcome the losses occurring in it. In the oscillator circuits discussed so far, the tank circuit employed inductive (*L*) and capacitive (*C*) elements. In such circuits, a phase shift of 180° was obtained due to inductive or capacitive coupling and a further phase shift of 180° was obtained due to transistor properties. In this way, energy supplied to the tank circuit was in phase with the generated oscillations. The oscillator circuits employing *L-C* elements have two general drawbacks. Firstly, they suffer from frequency instability and poor waveform. Secondly, they cannot be used for very low frequencies because they become too much bulky and expensive.

Good frequency stability and waveform can be obtained from oscillators employing resistive and capacitive elements. Such amplifiers are called *R-C phase shift oscillators* and have the additional advantage that they can be used for very low frequencies. In a phase shift oscillator, a phase shift of 180° is obtained with a phase shift circuit instead of inductive or capacitive coupling. A further phase shift of 180° is introduced due to the transistor properties. Thus, energy supplied back to the tank circuit is assured of correct phase.

Phase shift circuit. A phase-shift circuit essentially consists of an *R-C* network. Fig. 12.20 (*i*) shows a single section of *RC* network. From the elementary theory of electrical engineering, it can be shown that alternating voltage V'_1 across *R* leads the applied voltage V_1 by ϕ°. The value of ϕ depends upon the values of *R* and *C*. If resistance *R* is varied, the value of ϕ also changes. If *R* were reduced to zero, V'_1 will lead V_1 by 90° *i.e.* $\phi = 90°$. However, adjusting *R* to zero would be impracticable because it would lead to no voltage across *R*. Therefore, in practice, *R* is varied to such a value that makes V'_1 to lead V_1 by 60°.

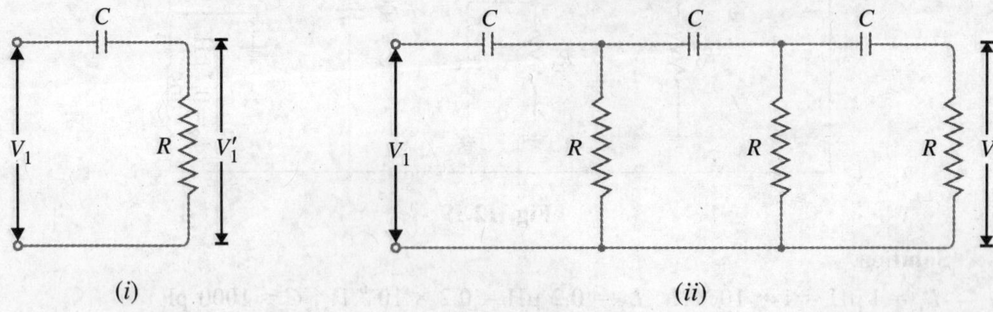

Fig. 12.20

Fig. 12.20 (*ii*) shows the three sections of *RC* network. Each section produces a phase shift of 60°. Consequently, a total phase shift of 180° is produced *i.e.* voltage V_2 leads the voltage V_1 by 180°.

12.14 RC Phase Shift Oscillator

In *RC* phase shift oscillator, the *RC* network (consisting of 3 identical *RC* sections) determines the frequency of oscillations as well as produces 180° phase shift. There is one frequency f_0 at which the *RC* network produces a phase shift of 180° and at the same time the condition $A_v \, m_v = 1$ is met by the oscillator circuit. The value of f_0 depends upon the values of *R* and *C*. By properly changing the values of *R* and *C*, the desired value of f_0 can be obtained.

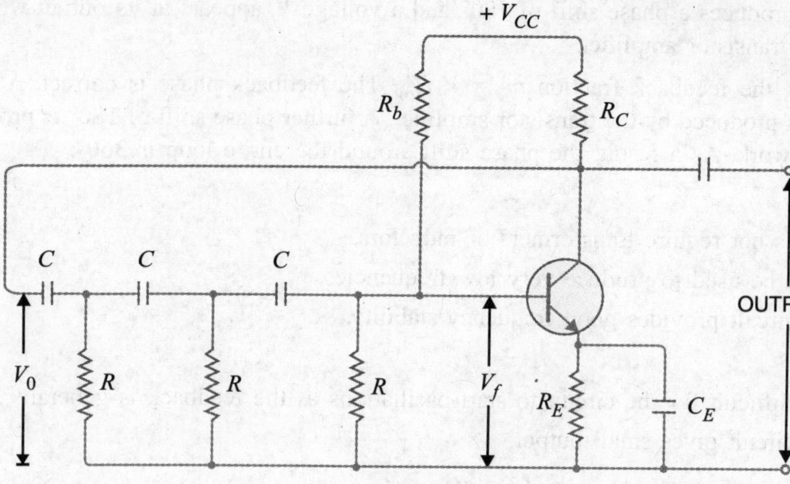

Fig. 12.21

Fig. 12.21 shows the circuit of *RC* phase shift oscillator. It consists of conventional single transistor amplifier and a *RC* phase shift network. The phase shift network consists of three identical *RC* sections (*i.e.* three equal-value resistors *R* and three equal-value capacitors *C*). At some frequency f_0, the phase shift in each *RC* section is 60° so that the total phase shift produced by the *RC* network is 180°. It can be proved mathematically that the frequency of oscillations f_0 is given by ;

$$f_0 = \frac{1}{2\pi RC\sqrt{6}} \quad \text{... (i) [See Example 12.11]}$$

By changing the values of *R* or *C* or both, the desired frequency f_0 can be obtained.

Feedback fraction (m_v). The *RC* network produces phase shift as well as reduces the output voltage (V_0). It can be shown mathematically that at frequency f_0, the *RC* network produces a phase shift of 180° as well as decreases the output voltage by 29 *i.e.* the feedback voltage $V_f = V_0/29$. This feedback voltage V_f is applied to the input of the amplifier as shown in Fig. 12.21.

$$\therefore \quad \text{Feedback fraction, } m_v = \frac{V_f}{V_0} = \frac{V_0/29}{V_0} = \frac{1}{29}$$

We know that the condition for continuous oscillations in an oscillator is $A_v\, m_v = 1$.

$$\therefore \quad \text{Amplifier voltage gain, } A_v = \frac{1}{m_v} = 29$$

Therefore, the amplifier must provide a minimum voltage gain (A_v) of 29 in order to maintain oscillations. If A_v is less than 29, the circuit will not oscillate. However, if A_v is much greater than 29, the output waveform will be distorted. In practice, the voltage gain of the amplifier is made slightly more than 29 to ensure that the circuit will oscillate.

Circuit operation. When the circuit is switched on, it produces oscillations of frequency determined by exp. (*i*). The output V_0 of the amplifier is fed back to *RC* feedback network.

This network produces a phase shift of 180° and a voltage V_f appears at its output which is applied to the transistor amplifier.

Obviously, the feedback fraction $m_v = V_f/V_0$. The feedback phase is correct. A phase shift of 180° is produced by the transistor amplifier. A further phase shift of 180° is produced by the *RC* network. As a result, the phase shift around the entire loop is 360°.

Advantages

(*i*) It does not require transformers or inductors.

(*ii*) It can be used to produce very low frequencies.

(*iii*) The circuit provides good frequency stability.

Disadvantages

(*i*) It is difficult for the circuit to start oscillations as the feedback is generally small.

(*ii*) The circuit gives small output.

Example 12.9. *In the phase shift oscillator shown in Fig. 12.21, R = 1 MΩ and C = 68 pF. At what frequency does the circuit oscillate?*

Solution. $R = 1\ \text{M}\Omega = 10^6\ \Omega$; $C = 68\ \text{pF} = 68 \times 10^{-12}$ F

Frequency of oscillations is given by ;

$$f_0 = \frac{1}{2\pi RC\sqrt{6}}$$

$$= \frac{1}{2\pi \times 10^6 \times 68 \times 10^{-12}\sqrt{6}}\ \text{Hz}$$

$$= \mathbf{954\ Hz}$$

Example 12.10. *A phase shift oscillator uses 5 pF capacitors. Find the value of R to produce a frequency of 800 kHz.*

Solution. Here, $C = 5\ \text{pF} = 5 \times 10^{-12}$ F ; $R = ?$

$$f_0 = \frac{1}{2\pi RC\sqrt{6}}$$

or

$$R = \frac{1}{2\pi f_0 C\sqrt{6}} = \frac{1}{2\pi \times 800 \times 10^3 \times 5 \times 10^{-12} \times \sqrt{6}}$$

$$= 16.2 \times 10^3\ \Omega = \mathbf{16.2\ k\Omega}$$

Example 12.11. *Draw the frequency determining circuit of RC phase shift oscillator and find the expression for the frequency (called operating frequency) at which the phase shift produced by this circuit is exactly 180°. Also find the feedback fraction at this frequency.*

Solution. Fig. 12.22 shows the frequency-determining circuit of *RC* phase shift oscillator. It consists of three identical *RC* sections. The output voltage V_0 of the amplifier is applied to the input of *RC* phase shift circuit and reduced feedback voltage V_f appears across the output terminals of the *RC* network. The frequency f_0 at which the phase shift produced by the *RC* network is exactly 180° and the feedback fraction $m_v (= V_f/V_0)$ at this frequency can be found as follows.

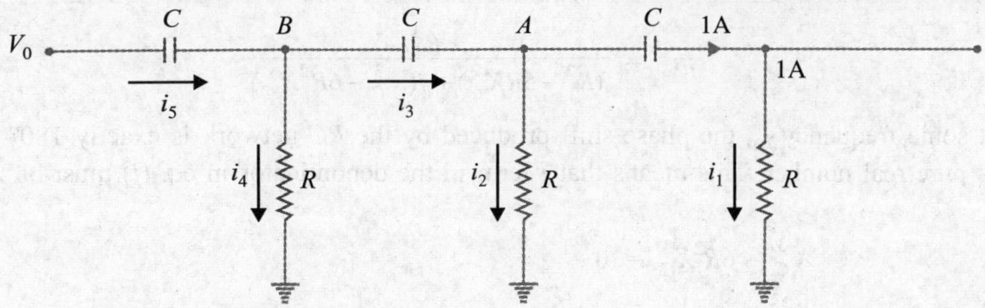

Fig. 12.22

Suppose current i_1 through R in Fig. 12.22 is 1 A *i.e.* i_1 = 1 A. Then,

$$\text{Feedback voltage, } V_f = i_1 R = 1\text{ A} \times R = R \text{ volt}$$

$$\text{Voltage at point } A,\ \boldsymbol{V_A} = V_f + i_1(-jX_C) = V_f + 1(-jX_C) \qquad (\because i_1 = 1\text{ A})$$

or $$\boldsymbol{V_A} = R - jX_C \qquad (\because V_f = R \text{ volt})$$

$$\text{Current } \boldsymbol{i_2} = \frac{\boldsymbol{V_A}}{R} = \frac{R - jX_C}{R} = 1 - \frac{jX_C}{R}$$

$$\text{Current } \boldsymbol{i_3} = i_1 + \boldsymbol{i_2} = 1 + \left(1 - \frac{jX_C}{R}\right) = 2 - \frac{jX_C}{R}$$

$$\text{Voltage at point } B,\ \boldsymbol{V_B} = \boldsymbol{V_A} + \boldsymbol{i_3}(-jX_C)$$

$$= (R - jX_C) + \left(2 - \frac{jX_C}{R}\right)(-jX_C)$$

∴ $$\boldsymbol{V_B} = R - \frac{X_C^2}{R} - j(3X_C)$$

$$\text{Current } \boldsymbol{i_4} = \frac{\boldsymbol{V_B}}{R} = \frac{1}{R}\left[R - \frac{X_C^2}{R} - j(3X_C)\right]$$

or $$\boldsymbol{i_4} = 1 - \frac{X_C^2}{R^2} - j\left(\frac{3X_C}{R}\right)$$

$$\text{Current } \boldsymbol{i_5} = \boldsymbol{i_3} + \boldsymbol{i_4} = 3 - \frac{X_C^2}{R^2} - j\left(\frac{4X_C}{R}\right)$$

∴ $$\text{Voltage } \boldsymbol{V_0} = \boldsymbol{V_B} + \boldsymbol{i_5}(-jX_C)$$

Putting the values of $\boldsymbol{V_B}$ and $\boldsymbol{i_5}$ in the above equation, we have,

$$\boldsymbol{V_0} = R - \frac{5X_C^2}{R} - j\left(6X_C - \frac{X_C^3}{R^2}\right)$$

or $$\boldsymbol{V_0} = \frac{(R^3 - 5RX_C^2) + j(X_C^3 - 6R^2X_C)}{R^2}$$

$$\text{Feedback fraction, } \boldsymbol{m_v} = \frac{V_f}{\boldsymbol{V_0}} = \frac{R}{\boldsymbol{V_0}} \qquad (\because V_f = R \text{ volt})$$

Putting the value of $\boldsymbol{V_0}$, we get,

$$m_v = \frac{R^3}{(R^3 - 5RX_C^2) + j(X_C^3 - 6R^2X_C)} \quad \text{... } (i)$$

At some frequency f_0, the phase shift produced by the *RC* network is exactly 180° and m_v is a pure real number. This means that *j* term in the denominator in eq. (*i*) must be zero *i.e.*

$$X_C^3 - 6R^2X_C = 0$$

or $$X_C = R\sqrt{6}$$

or $$\frac{1}{2\pi f_0 C} = R\sqrt{6}$$

∴ $$\text{Frequency, } f_0 = \frac{1}{2\pi RC\sqrt{6}}$$

This is the expression for the frequency of oscillations for the *RC* phase shift oscillator.

At frequency f_0, the *j* term in the denominator in eq. (*i*) is zero so that m_v is given by ;

$$m_v = \frac{R^3}{R^3 - 5RX_C^2}$$

Putting $X_C = R\sqrt{6}$ in the above expression, we have,

$$m_v = \frac{R^3}{R^3 - 5R(R\sqrt{6})^2} = \frac{R^3}{R^3 - 30R^3} = -\frac{1}{29}$$

The negative sign shows that *RC* network produces a phase shift of 180°.

12.15 Wien Bridge Oscillator

It is an *RC* oscillator that uses a 2-stage amplifier and the Wien bridge. The 2-stage amplifier produces 360° phase shift required for positive feedback. The Wien bridge has *RC* circuit in the two adjacent branches (the other two adjacent branches have pure resistors) and determines the frequency of oscillations as well as acts as a feedback circuit. At one particular frequency f_0, the Wien bridge is balanced and produces 0° phase shift. Therefore, at the balance frequency f_0, the feedback is positive. At all other frequencies, the bridge is not balanced and does not produce 0° phase shift. Therefore, positive feedback is not ensured.

The Wien-bridge oscillator is the standard oscillator circuit for all frequencies in the range of 10 Hz to about 1 MHz. It is the most frequently used type of audio oscillator as the output is free from circuit fluctuations and ambient temperature. Fig. 12.23 shows the circuit of Wien bridge oscillator. It is essentially a two-stage amplifier with *R-C* bridge circuit. The bridge circuit has the arms R_1C_1, R_3, R_2C_2 and tungsten lamp L_P. Resistances R_3 and L_P are used to stabilise the amplitude of the output. The transistor T_1 serves as an oscillator and amplifier while the other transistor T_2 serves as an inverter (*i.e.* to produce a phase shift of 180°). The circuit uses positive and negative feedbacks. The positive feedback is through R_1C_1, C_2R_2 to the transistor T_1. The negative feedback is through the voltage divider to the input of transistor T_2. The frequency of oscillations is determined by the series element R_1C_1 and parallel element R_2C_2 of the bridge. At one particular frequency f_0, the Wien bridge is balanced and produces 0° phase shift. It can be proved mathematically that the balance frequency f_0 is given by ;

$$f_0 = \frac{1}{2\pi\sqrt{R_1C_1R_2C_2}}$$ [See Example 12.13]

If $R_1 = R_2 = R$

and $C_1 = C_2 = C$, then,

$$f_0 = \frac{1}{2\pi RC} \quad \text{... }(i)$$

It can be proved mathematically that at the balance frequency f_0, the feedback fraction m_v is given by ;

$$\text{Feedback fraction, } m_v = \frac{1}{3}$$

We know that the condition for continuous oscillations in an oscillator is $A_v m_v = 1$.

$$\therefore \quad \text{Amplifier voltage gain, } A_v = \frac{1}{m_v} = 3$$

Therefore, the amplifier must provide a minimum voltage gain (A_v) of 3 in order to maintain oscillations.

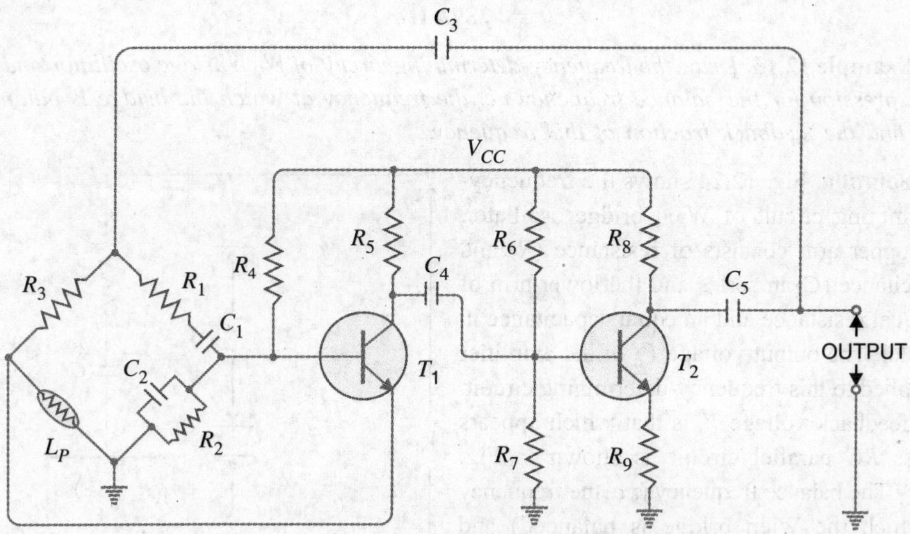

Fig. 12.23

Circuit operation. When the circuit is started, bridge circuit produces oscillations of frequency determined by exp. (*i*). The two transistors produce a total phase shift of 360° so that proper positive feedback is ensured. The negative feedback in the circuit ensures constant output. This is achieved by the temperature sensitive tungsten lamp L_P. Its resistance increases with current. Should the amplitude of output tend to increase, more current would provide more negative feedback. The result is that the output would return to original value. A reverse action would take place if the output tends to decrease.

Advantages

(*i*) It gives constant output.

(*ii*) The circuit works quite easily.

(*iii*) The overall gain is high because of two transistors.

(*iv*) The frequency of oscillations can be easily changed by using a potentiometer.

Disadvantages

(*i*) The circuit requires two transistors and a large number of components.

(*ii*) It cannot generate very high frequencies.

Example 12.12. *In the Wien bridge oscillator shown in Fig. 12.23, $R_1 = R_2 = 220 k\Omega$ and $C_1 = C_2 = 250$ pF. Determine the frequency of oscillations.*

Solution.

$$R_1 = R_2 = R = 220 \text{ k}\Omega = 220 \times 10^3 \ \Omega$$

$$C_1 = C_2 = C = 250 \text{ pF} = 250 \times 10^{-12} \text{ F}$$

$$\text{Frequency of oscillations, } f_0 = \frac{1}{2\pi RC}$$

$$= \frac{1}{2\pi \times 220 \times 10^3 \times 250 \times 10^{-12}} \text{ Hz}$$

$$= \mathbf{2892 \ Hz}$$

Example 12.13. *Draw the frequency-determining circuit of Wien bridge oscillator and find the expression for the balance frequency i.e. the frequency at which the bridge is balanced. Also find the feedback fraction at this frequency.*

Solution. Fig. 12.24 shows the frequency-determining circuit of Wien bridge oscillator. The upper arm consists of resistance (R) and capacitance (C) in series and the lower arm of an equal resistance and an equal capacitance in parallel. The output voltage V_0 of the amplifier is applied to this frequency-determining circuit. The feedback voltage V_f is that which appears across RC parallel circuit as shown in Fig. 12.24. The balance frequency (*i.e.* the frequency at which the Wien bridge is balanced) and feedback fraction at this frequency can be calculated as under.

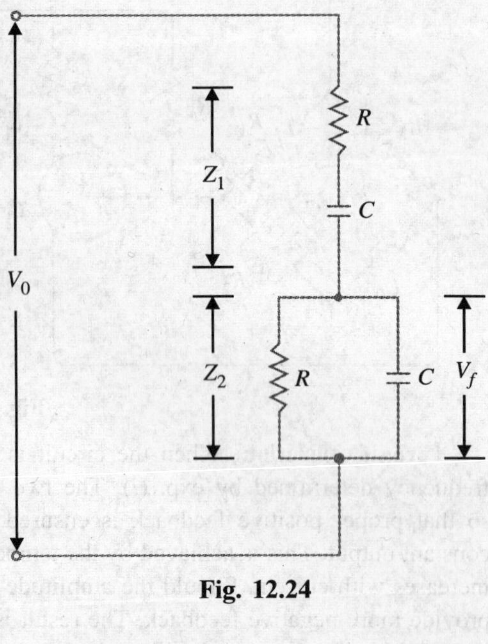

Fig. 12.24

The impedance Z_1 of the series RC circuit is

$$Z_1 = R - jX_C$$

The impedance Z_2 of the parallel RC circuit is

$$Z_2 = R \parallel -jX_C = \frac{-jRX_C}{R - jX_C}$$

$$\text{Feedback fraction, } m_v = \frac{V_f}{V_0} = \frac{Z_2}{Z_1 + Z_2}$$

or $$m_v = \frac{-jRX_C/(R - jX_C)}{R - jX_C - jRX_C/(R - jX_C)}$$

On simplifying the above expression, we have,

$$m_v = \frac{RX_C}{3RX_C + j(R^2 - X_C^2)} \quad ...(i)$$

At the balance frequency f_0, the bridge is balanced and the frequency-determining circuit produces 0° phase shift. Therefore, V_f and V_0 are in phase so that m_v is a pure real number. This means that imaginary term in eq. (*i*) must be zero *i.e.*

$$R^2 - X_C^2 = 0$$

or $$R^2 = X_C^2$$

or $$R = X_C = \frac{1}{2\pi f_0 C}$$

$\therefore$ $$\text{Balance frequency, } f_0 = \frac{1}{2\pi RC}$$

Putting $R = X_C$ in eq. (*i*) above, we have,

$$m_v = \frac{R^2}{3R^2 + j0} = \frac{1}{3}$$

12.16 Limitations of LC and RC Oscillators

The *LC* and *RC* oscillators discussed so far have their own limitations. The major problem in such circuits is that their operating frequency does not remain strictly constant. There are two principal reasons for it *viz.*,

(*i*) As the circuit operates, it will warm up. Consequently, the values of resistors and inductors, which are the frequency determining factors in these circuits, will change with temperature. This causes the change in frequency of the oscillator.

(*ii*) If any component in the feedback network is changed, it will shift the operating frequency of the oscillator.

However, in many applications, it is desirable and necessary to maintain the frequency constant with extreme low tolerances. For example, the frequency tolerance for a broadcasting station should not exceed 0.002% *i.e.* change in frequency due to any reason should not be more than 0.002% of the specified frequency. The broadcasting stations have frequencies which are quite close to each other. In fact, the frequency difference between two broadcasting stations is less than 1%. It is apparent that if we employ *LC* or *RC* circuits, a change of temperature may cause the frequencies of adjacent broadcasting stations to overlap.

In order to maintain constant frequency, *piezoelectric crystals* are used in place of *LC* or *RC* circuits. Oscillators of this type are called *crystal oscillators*. The frequency of a crystal oscillator changes by less than 0.1% due to temperature and other changes. Therefore, such oscillators offer the most satisfactory method of stabilising the frequency and are used in great majority of electronic applications.

12.17 Piezoelectric Crystals

Certain crystalline materials, namely, Rochelle salt, quartz and tourmaline exhibit the *piezoelectric effect i.e.* when we apply an a.c. voltage across them, they vibrate at the frequency of the applied voltage. Conversely, when they are compressed or placed under mechanical strain to vibrate, they produce an a.c. voltage. Such crystals which exhibit piezoelectric effect are called *piezoelectric crystals*. Of the various piezoelectric crystals, quartz is most commonly used because it is inexpensive and readily available in nature.

Quartz crystal. Quartz crystals are generally used in crystal oscillators because of their great mechanical strength and simplicity of manufacture. The natural shape of quartz crystal is hexagonal as shown in Fig. 12.25. The three axes are shown : the *z-axis* is called the *optical axis*, the *x-axis* is called the *electrical axis* and *y-axis* is called the *mechanical axis*. Quartz crystal can be cut in different ways. Crystal cut perpendicular to the *x-axis* is called *x-cut crystal* whereas that cut perpendicular to *y-axis* is called *y-cut crystal*. The piezoelectric properties of a crystal depend upon its cut.

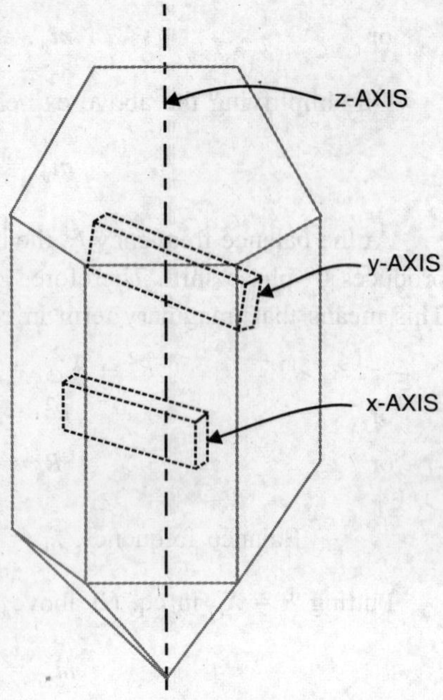

Fig. 12.25

Frequency of crystal. Each crystal has a natural frequency like a pendulum. The natural frequency f of a crystal is given by ;

$$f = \frac{K}{t}$$

where K is a constant that depends upon the cut and t is the thickness of the crystal. It is clear that frequency is inversely proportional to crystal thickness. The thinner the crystal, the greater is its natural frequency and *vice-versa*. However, extremely thin crystal may break because of vibrations. This puts a limit to the frequency obtainable. In practice, frequencies between 25 kHz to 5 MHz have been obtained with crystals.

12.18 Working of Quartz Crystal

In order to use crystal in an electronic circuit, it is placed between two metal plates. The arrangement then forms a capacitor with crystal as the dielectric as shown in Fig. 12.26. If an a.c. voltage is applied across the plates, the crystal will start vibrating at the frequency of applied voltage. However, if the frequency of the applied voltage is made equal to the natural frequency of the crystal, resonance takes place and crystal vibrations reach a maximum value. This natural frequency is almost constant. Effects of temperature change can be eliminated by mounting the crystal in a temperature-controlled oven as in radio and television transmitters.

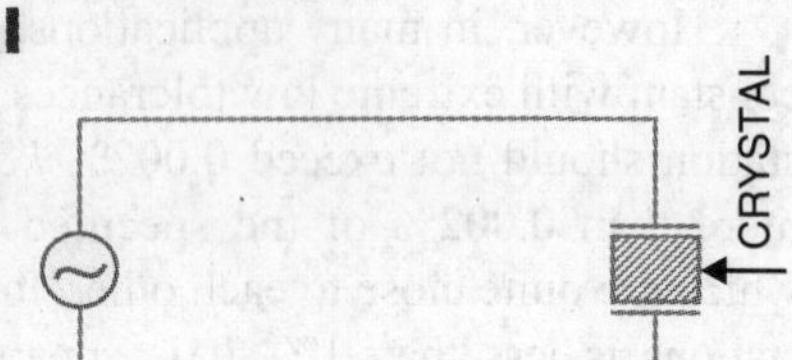

Fig. 12.26

12.19 Equivalent Circuit of Crystal

Although the crystal has electromechanical resonance, we can represent the crystal action by an equivalent electrical circuit.

(*i*) When the crystal is not vibrating, it is equivalent to capacitance C_m because it has two metal plates separated by a dielectric [See Fig. 12.27 (*i*)]. This capacitance is known as *mounting capacitance*.

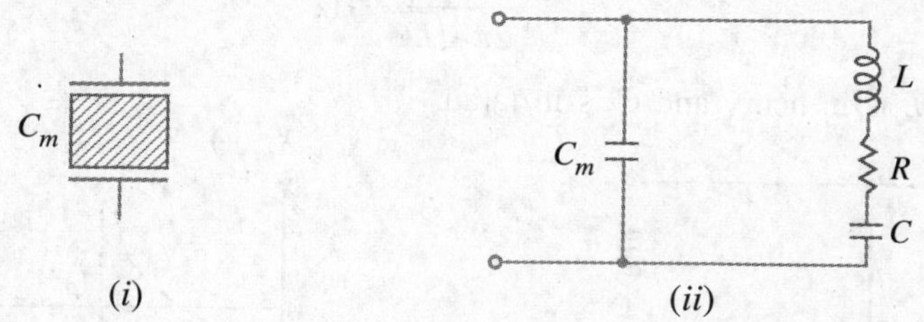

Fig. 12.27

(*ii*) When a crystal vibrates, *it is equivalent to $R - L - C$ series circuit. Therefore, the equivalent circuit of a vibrating crystal is $R - L - C$ series circuit shunted by the mounting capacitance C_m as shown in Fig. 12.27 (*ii*).

C_m = Mounting capacitance

$R - L - C$ = Electrical equivalent of vibrational characteristic of the crystal

Typical values for a 4 MHz crystal are :

$L = 100$ mH ; $R = 100\ \Omega$

$C = 0.015$ pF ; $C_m = 5$ pF

$$\therefore \quad Q\text{-factor of crystal} = \frac{1}{R}\sqrt{\frac{L}{C}}$$

$$= \frac{1}{100}\sqrt{\frac{100 \times 10^{-3}}{0.015 \times 10^{-12}}} = 26{,}000$$

Note that Q of crystal is very high. The extremely high Q of a crystal leads to frequency **stability.

12.20 Frequency Response of Crystal

When the crystal is vibrating, its equivalent electrical circuit is as shown in Fig. 12.28 (*i*). The capacitance values of C and C_m are relatively low (less than 1 pF for C and 4 – 40 pF for C_m). Note that the value of C is much lower than that of C_m.

(*i*) At low frequencies, the impedance of the crystal is controlled by extremely high values of X_{C_m} and X_C. In other words, at low frequencies, the impedance of the network is high and capacitive as shown in Fig. 12.28 (*ii*).

(*ii*) As the frequency is increased, $R - L - C$ branch approaches its resonant frequency. At some definite frequency, the reactance X_L will be equal to X_C. The crystal now acts as a series-resonant circuit. For this condition, the impedance of the crystal is

---

* When the crystal is vibrating, L is the electrical equivalent of crystal mass, C is the electrical equivalent of elasticity and R is electrical equivalent of mechanical friction.

** When Q is high, frequency is primarily determined by L and C of the crystal. Since these values remain fixed for a crystal, the frequency is stable. However, in ordinary LC tank circuit, the values of L and C have large tolerances.

very low ; being equal to R. The frequency at which the vibrating crystal behaves as a series-resonant circuit is called *series-resonant frequency* f_s. Its value is given by :

$$f_s = \frac{1}{2\pi\sqrt{LC}} \text{ Hz}$$

where L is in henry and C is in farad.

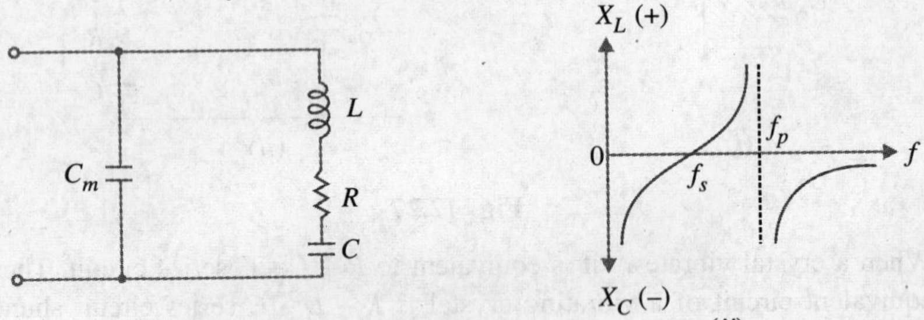

(i) (ii)

Fig. 12.28

(iii) At a slightly higher frequency, the net reactance of branch $R - L - C$ becomes inductive and equal to X_{C_m}. The crystal now acts as a parallel-resonant circuit. For this condition, the crystal offers a very high impedance. The frequency at which the vibrating crystal behaves as a parallel-resonant circuit is called *parallel-resonant frequency* f_p.

$$f_p = \frac{1}{2\pi\sqrt{LC_T}}$$

where $$C_T = \frac{C \times C_m}{C + C_m}$$

Since C_T is less than C, f_p is always greater than f_s. Note that frequencies f_s and f_p are very close to each other.

(iv) At frequencies greater than f_p, the value of X_{C_m} drops and eventually the crystal acts as a short circuit.

Conclusion. The above discussion leads to the following conclusions :

(i) At f_s, the crystal will act as a series-resonant circuit.

(ii) At f_p, the crystal will act as a parallel-resonant circuit.

Therefore, we can use a crystal in place of a series LC circuit or in place of parallel LC circuit. If we use it in place of series LC circuit, the oscillator will operate at f_s. However if we use the crystal in place of parallel LC circuit, the oscillator will operate at f_p. In order to use the crystal properly, it must be connected in a circuit so that its low impedance in the series resonant operating mode or high impedance in the parallel resonant operating mode is selected.

12.21 Transistor Crystal Oscillator

Fig. 12.29 shows the transistor crystal oscillator. Note that it is a Collpit's oscillator modified to act as a crystal oscillator. The only change is the addition of the crystal (Y) in the feedback network. The crystal will act as a parallel-tuned circuit. As you can see in this circuit that instead of resonance caused by L and $(C_1 + C_2)$, we have the parallel resonance of the crystal. At parallel resonance, the impedance of the crystal is maximum. This means that there is a maximum voltage drop across C_1. This in turn will allow the maximum energy transfer through the feedback network at f_p.

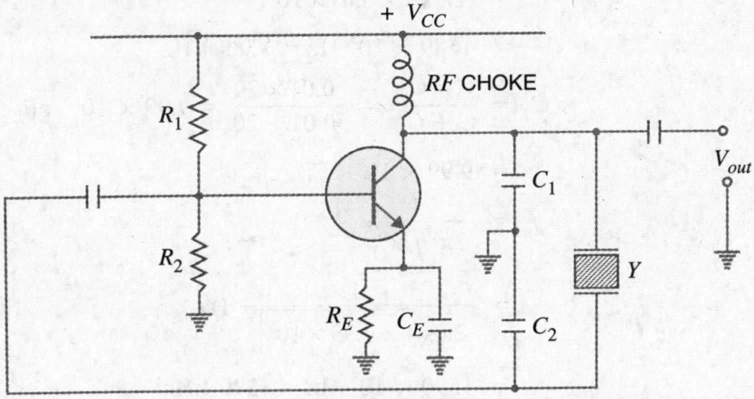

Fig. 12.29

Note that feedback is positive. A phase shift of 180° is produced by the transistor. A further phase shift of 180° is produced by the capacitor voltage divider. This oscillator will oscillate only at f_p. Even the smallest deviation from f_p will cause the oscillator to act as an effective short. Consequently, we have an extremely stable oscillator.

Advantages

(*i*) They have a high order of frequency stability.

(*ii*) The quality factor (Q) of the crystal is very high. The Q factor of the crystal may be as high as 10,000 compared to about 100 of L-C tank.

Disadvantages

(*i*) They are fragile and consequently can only be used in low power circuits.

(*ii*) The frequency of oscillations cannot be changed appreciably.

Example 12.14. *A crystal has a thickness of t mm. If the thickness is reduced by 1%, what happens to frequency of oscillations?*

Solution. Frequency, $f = \frac{K}{t}$

or $f \propto \frac{1}{t}$

If the thickness of the crystal is reduced by 1%, the frequency of oscillations will **increase by 1%.**

Example 12.15. *The a.c. equivalent circuit of a crystal has these values : $L = 1$ H, $C = 0.01$ pF, $R = 1000\ \Omega$ and $C_m = 20$ pF. Calculate f_s and f_p of the crystal.*

Solution.

$$L = 1 \text{ H}$$

$$C = 0.01 \text{ pF} = 0.01 \times 10^{-12} \text{ F}$$

$$C_m = 20 \text{ pF} = 20 \times 10^{-12} \text{ F}$$

$$\therefore \quad f_s = \frac{1}{2\pi\sqrt{LC}}$$

$$= \frac{1}{2\pi\sqrt{1 \times 0.01 \times 10^{-12}}} \text{ Hz}$$

$$= 1589 \times 10^3 \text{ Hz} = \mathbf{1589\ kHz}$$

Now,
$$C_T = \frac{C \times C_m}{C + C_m} = \frac{0.01 \times 20}{0.01 + 20} = 9.99 \times 10^{-3} \text{ pF}$$

$$= 9.99 \times 10^{-15} \text{ F}$$

$$\therefore \quad f_p = \frac{1}{2\pi\sqrt{LC_T}}$$

$$= \frac{1}{2\pi\sqrt{1 \times 9.99 \times 10^{-15}}} \text{ Hz}$$

$$= 1590 \times 10^3 \text{ Hz} = \mathbf{1590\ kHz}$$

If this crystal is used in an oscillator, the frequency of oscillations will lie between 1589 kHz and 1590 kHz.

Short Answer Questions

Q.1. *What is the difference between an alternator and an oscillator?*

Ans. Although an alternator produces sinusoidal oscillations of 50 Hz, it is fundamentally different from an oscillator. First, an alternator is a mechanical device having rotating parts whereas an oscillator is a non-rotating electronic device. Secondly, an alternator converts mechanical energy into electrical energy while an oscillator converts d.c. energy into a.c. energy. Thirdly, an alternator cannot produce high frequency oscillations whereas an oscillator can produce oscillations ranging from a few Hz to several MHz.

Q.2. *What is the difference between an amplifier and an oscillator?*

Ans. An oscillator is an amplifier with positive feedback. The circuit is so designed that :

$$A_v m_v = 1$$

where A_v = Open-loop voltage gain

m_v = Feedback fraction

Q.3. *What are the basic mathematical conditions for sustaining oscillations in an oscillator?*

Ans. An oscillator employs positive feedback. The feedback must be of correct magnitude and phase. To achieve this objective, the following two conditions must be fulfilled :

(*i*) The voltage gain around the loop containing the amplifier and the feedback network should have unity value *i.e.*

$$A_v m_v = 1$$

where A_v = Voltage gain without feedback

m_v = Feedback fraction

(ii) The phase of $A_v\ m_v$ should be zero or any integral multiple of 2π. This is necessary since both A_v and m_v may be complex quantities.

Q.4. *Why do we employ RC oscillators and not LC oscillators for low frequency (1 Hz to 100 kHz) applications?*

Ans. High frequency oscillations (> 100 kHz) generally use *LC* circuits because the physical size of these elements is small at these frequencies. However, at low frequencies (1 Hz to 100 kHz), the physical size of the elements in *LC* resonant circuit becomes much larger and manufacturing difficulties present problems. Therefore, *RC* phase-shift oscillators are commonly used to produce this band of frequencies.

Q.5. *Why do we use three RC sections in RC phase-shift oscillators?*

Ans. In such oscillators, *RC* network is employed to produce a phase-shift of 180°. A single *RC* section can produce a maximum phase-shift of 90° and that too when $R = 0$. This is impracticable because it would lead to zero voltage across *R*. In practice, the value of *R* is so adjusted that a phase-shift of 60° is obtained. Three such sections are used to produce a total phase-shift of 180°.

Q.6. *What is the purpose of providing both positive and negative feedback in Wien bridge oscillator?*

Ans. The positive feedback ensures sustained oscillation—a condition necessary in all oscillators. However, the purpose of providing negative feedback in this type of oscillator is to ensure constant output. Any increase or decrease in the oscillator output is taken care of by the negative feedback.

Q.7. *Why is quartz crystal commonly used in crystal oscillators?*

Ans. Because it is inexpensive and readily available in nature.

Q.8. *Why does crystal oscillator give constant frequency?*

Ans. A crystal can be shown to be equivalent to high-*Q* resonant circuit. Due to the high *Q* of the crystal, the crystal oscillator frequency is very stable or accurately fixed in value.

Q.9. *Why is crystal oscillator used in radio transmitter?*

Ans. A crystal oscillator produces a fixed frequency ; infact change in crystal frequency due to any reason (*e.g.*, temperature change, transistor replacement) is less than 0.1%. On the other hand, in *LC* or *RC* oscillators, the frequency changes considerably (more than 1%) by such variations. Now broadcasting stations have frequencies which are quite close to each other ; the frequency difference between two adjacent broadcasting stations is generally less than 1%. It is apparent that if we employ *LC* or *RC* oscillators, a change in temperature may cause the frequencies of adjacent broadcasting stations to overlap.

Multiple-Choice Questions

1. An oscillator converts ....
   (i) a.c. power into d.c. power
   (ii) d.c. power into a.c. power
   (iii) mechanical power into a.c. power
   (iv) none of the above

2. In an *LC* transistor oscillator, the active device is ....
   (i) *LC* tank circuit (ii) biasing circuit
   (iii) transistor
   (iv) none of the above

3. In an LC circuit, when the capacitor energy is maximum, the inductor energy is ....
   (*i*) minimum (*ii*) maximum
   (*iii*) half-way between maximum and minimum
   (*iv*) none of the above
4. In an LC oscillator, the frequency of oscillator is .... L or C.
   (*i*) proportional to square of
   (*ii*) directly proportional to
   (*iii*) independent of the values of
   (*iv*) inversely proportional to square root of
5. An oscillator produces ...... oscillations.
   (*i*) damped (*ii*) undamped
   (*iii*) modulated
   (*iv*) none of the above
6. An oscillator employs .... feedback.
   (*i*) positive (*ii*) negative
   (*iii*) neither positive nor negative
   (*iv*) data insufficient
7. An LC oscillator cannot be used to produce .... frequencies.
   (*i*) high (*ii*) audio
   (*iii*) very low (*iv*) very high
8. Hartley oscillator is commonly used in ....
   (*i*) radio receivers
   (*ii*) radio transmitters
   (*iii*) TV receivers
   (*iv*) none of the above
9. In a phase shift oscillator, we use .... RC sections.
   (*i*) two (*ii*) three
   (*iii*) four
   (*iv*) none of the above
10. In a phase shift oscillator, the frequency determining elements are ....
    (*i*) L and C (*ii*) R, L and C
    (*iii*) R and C
    (*iv*) none of the above
11. A Wien bridge oscillator uses .... feedback.
    (*i*) only positive (*ii*) only negative
    (*iii*) both positive and negative
    (*iv*) none of the above
12. The piezoelectric effect in a crystal is ....
    (*i*) a voltage developed because of mechanical stress
    (*ii*) a change in resistance because of temperature
    (*iii*) a change of frequency because of temperature
    (*iv*) none of the above
13. If the crystal frequency changes with temperature, we say that crystal has .... temperature coefficient.
    (*i*) positive
    (*ii*) zero
    (*iii*) negative
    (*iv*) none of the above
14. The crystal oscillator frequency is very stable due to .... of the crystal.
    (*i*) rigidity (*ii*) vibrations
    (*iii*) low Q (*iv*) high Q
15. The application where one would most likely find a crystal oscillator is ....
    (*i*) radio receiver (*ii*) radio transmitter
    (*iii*) AF sweep generator
    (*iv*) none of the above
16. An oscillator differs from an amplifier because it ....
    (*i*) has more gain
    (*ii*) requires no input signal
    (*iii*) requires no d.c. supply
    (*iv*) always has the same input
17. One condition for oscillations is ....
    (*i*) a phase shift around the feedback loop of 180º
    (*ii*) a gain around the feedback loop of one-third
    (*iii*) a phase shift around the feedback loop of 0º
    (*iv*) a gain around the feedback loop of less than 1

18. A second condition for oscillations is ....
   (i) a gain of 1 around the feedback loop
   (ii) no gain around the feedback loop
   (iii) the attenuation of the feedback circuit must be one-third
   (iv) the feedback circuit must be capacitive

19. In a certain oscillator, $A_v = 50$. The attenuation of the feedback circuit must be ....
   (i) 1 (ii) 0.01
   (iii) 10 (iv) 0.02

20. For an oscillator to properly start, the gain around the feedback loop must initially be ....
   (i) 1 (ii) greater than 1
   (iii) less than 1
   (iv) equal to attenuation of feedback circuit

21. In a Wien-bridge oscillator, if the resistances in the positive feedback circuit are decreased, the frequency ....
   (i) remains the same
   (ii) decreases
   (iii) increases
   (iv) insufficient data

22. In a Colpitt's oscillator, feedback is obtained ....
   (i) by magnetic induction
   (ii) by a tickler coil
   (iii) from the centre of split capacitors
   (iv) none of the above

23. The Q of a crystal is of the order of ....
   (i) 100 (ii) 1000
   (iii) 50
   (iv) more than 10,000

24. Quartz crystal is most commonly used in crystal oscillators because ....
   (i) it has superior electrical properties
   (ii) it is easily available
   (iii) it is quite inexpensive
   (iv) none of the above

25. In LC oscillators, the frequency of oscillations is given by ....
   (i) $\dfrac{2\pi}{\sqrt{LC}}$ (ii) $\dfrac{1}{2\pi\sqrt{LC}}$
   (iii) $\dfrac{\sqrt{LC}}{2\pi}$ (iv) $\dfrac{2\pi L}{\sqrt{LC}}$

26. The operating frequency of a Wien-bridge oscillator is given by ....
   (i) $\dfrac{1}{2\pi\sqrt{LC}}$ (ii) $\dfrac{1}{4\pi\sqrt{LC}}$
   (iii) $\dfrac{1}{2\pi RC}$ (iv) $\dfrac{1}{29RC}$

27. .... is a fixed frequency oscillator.
   (i) Phase-shift oscillator
   (ii) Hartley oscillator
   (iii) Colpitt's oscillator
   (iv) Crystal oscillator

28. In an LC oscillator, if the value of L is increased four times, the frequency of oscillations is ....
   (i) increased 2 times
   (ii) decreased 4 times
   (iii) increased 4 times
   (iv) decreased 2 times

29. An important limitation of a crystal oscillator is ....
   (i) its low output (ii) its high Q
   (iii) less availability of quartz crystal
   (iv) its high output

30. The signal generator generally used in the laboratories is .... oscillator.
   (i) Wien-bridge (ii) Hartley
   (iii) Crystal (iv) Phase shift

Answers to Multiple-Choice Questions

1. (*ii*)	2. (*iii*)	3. (*i*)	4. (*iv*)	5. (*ii*)
6. (*i*)	7. (*iii*)	8. (*i*)	9. (*ii*)	10. (*iii*)
11. (*iii*)	12. (*i*)	13. (*i*)	14. (*iv*)	15. (*ii*)
16. (*ii*)	17. (*iii*)	18. (*i*)	19. (*iv*)	20. (*ii*)
21. (*iii*)	22. (*iii*)	23. (*iv*)	24. (*i*)	25. (*ii*)
26. (*iii*)	27. (*iv*)	28. (*iv*)	29. (*i*)	30. (*i*)

Chapter Review Topics

1. What is an oscillator? What is its need? Discuss the advantages of oscillators.
2. What do you understand by damped and undamped electrical oscillations? Illustrate your answer with examples.
3. Explain the operation of a tank circuit with neat diagrams.
4. What is the nature of oscillations produced by a tank circuit?
5. How will you get undamped oscillations from a tank circuit?
6. Discuss the essentials of an oscillator.
7. Discuss the circuit operation of tuned collector oscillator.
8. With a neat diagram, explain the action of Hartley and Colpitt's oscillators.
9. What are the drawbacks of *LC* oscillators?
10. Write short notes on the following:

    (*i*) *RC* oscillators (*ii*) Wien bridge oscillator (*iii*) Crystal oscillator

Problems

1. Figure 12.30 shows the Colpitt's oscillator. Determine the (*i*) operating frequency and (*ii*) feedback fraction. **[(*i*) 24.5 kHz (*ii*) 0.1]**

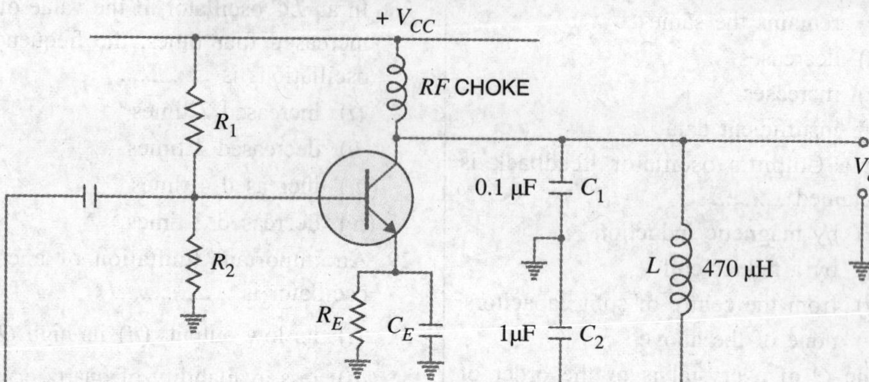

Fig. 12.30

2. Figure 12.31 shows the Hartley oscillator. If L_1 = 1000 μH, L_2 = 100 μH and C = 20 pF, find the (*i*) operating frequency and (*ii*) feedback fraction. **[(*i*) 1052 kHz (*ii*) 0.1]**
3. For the Colpitt's oscillator shown in Fig. 12.30, C_1 = 750 pF, C_2 = 2500 pF and L = 40 μH. Determine (*i*) the operating frequency and (*ii*) feedback fraction. **[(*i*) 1050 kHz (*ii*) 0.3]**
4. For the Hartley oscillator shown in Fig. 12.31, C = 250 pF, L_1 = 1.5 mH, L_2 = 1.5 mH and M = 0.58 mH. Determine the operating frequency. **[159.2 kHz]**

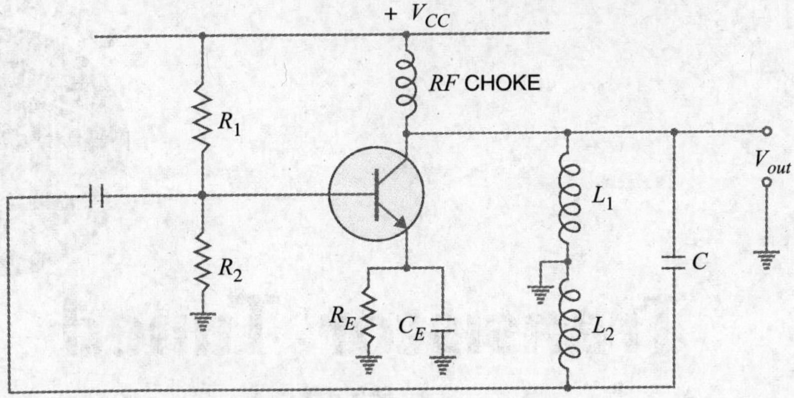

Fig. 12.31

5. A crystal has L = 3H, C = 0.05 pF, R = 2 kΩ and C_m = 10 pF. Calculate the series-resonant and parallel resonant frequencies of the crystal. **[411 kHz; 412 kHz]**

CHAPTER 13

Transistor Tuned Amplifiers

Learning Objectives

- Understand the need of Tuned Amplifiers.
- Learn the difference between tuned amplifiers and other amplifiers.
- Derive the expression for resonant frequency f_r of parallel tuned circuit.
- Develop an understanding of characteristics of parallel resonant circuit.
- Explore the key points about parallel resonance.
- Develop an understanding of the advantages of tuned amplifiers.
- Understand the characteristics of single tuned amplifiers.
- Recognise the operation and advantages of double tuned amplifier.
- Learn the operation of tuned class *C* amplifier.
- Develop a strong understanding of the characteristics of tuned class *C* amplifier.

Introduction

Most of the audio amplifiers we have discussed in the earlier chapters will also work at radio frequencies *i.e.* above 50 kHz. However, they suffer from two major drawbacks. First, they become less efficient at radio frequency. Secondly, such amplifiers have mostly resistive loads and consequently their gain is independent of signal frequency over a large bandwidth. In other words, an audio amplifier amplifies a wide band of frequencies equally well and does not permit the selection of a particular desired frequency while rejecting all other frequencies.

However, sometimes it is desired that an amplifier should be selective *i.e.* it should select a desired frequency or narrow band of frequencies for amplification. For instance, radio and television transmissions are carried on a specific radio frequency assigned to the broadcasting station. The radio receiver is required to pick up and amplify the radio frequency desired while discriminating all others. To achieve this, the simple resistive load is replaced by a parallel tuned circuit whose impedance strongly depends upon frequency. Such a tuned circuit becomes very selective and amplifies very strongly signals of resonant frequency and narrow band on either side. Therefore, the use of tuned circuits in conjunction with a transistor makes possible the selection and efficient amplification of a particular desired radio frequency. Such an amplifier is called a *tuned amplifier*. In this chapter, we shall focus our attention on transistor tuned amplifiers and their increasing applications in high frequency electronic circuits.

13.1 Tuned Amplifiers

Amplifiers which amplify a specific frequency or narrow band of frequencies are called **tuned amplifiers.**

Tuned amplifiers are mostly used for the amplification of high or radio frequencies. It is because radio frequencies are generally single and the tuned circuit permits their selection and efficient amplification. However, such amplifiers are not suitable for the amplification of audio frequencies as they are mixture of frequencies from 20 Hz to 20 kHz and not single. Tuned amplifiers are widely used in radio and television circuits where they are called upon to handle radio frequencies.

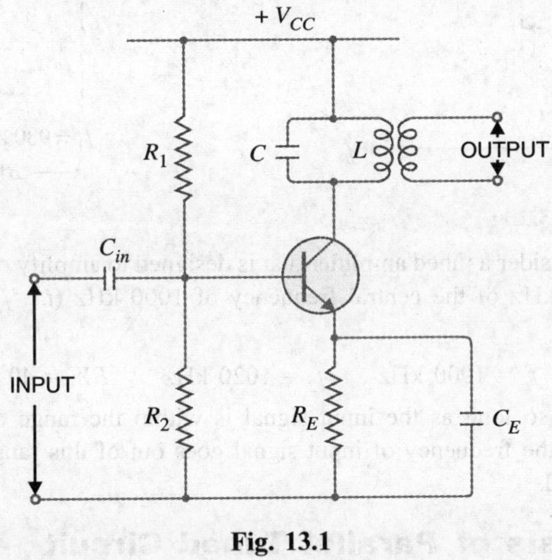

Fig. 13.1

Fig. 13.1 shows the circuit of a simple transistor tuned amplifier. Here, instead of load resistor, we have a parallel tuned circuit in the collector. The impedance of this tuned circuit strongly depends upon frequency. It offers a very high impedance at *resonant frequency* and very small impedance at all other frequencies. If the signal has the same frequency as the resonant frequency of *LC* circuit, large amplification will result due to high impedance of *LC* circuit at this frequency. When signals of many frequencies are present at the input of tuned amplifier, it will select and strongly amplify the signals of resonant frequency while *rejecting all others. Therefore, such amplifiers are very useful in radio receivers to select the signal from one particular broadcasting station when signals of many other frequencies are present at the receiving aerial.

* For all other frequencies, the impedance of *LC* circuit will be very small. Consequently, little amplification will occur for these frequencies.

13.2 Distinction between Tuned Amplifiers and other Amplifiers

We have seen that amplifiers (*e.g.*, voltage amplifier, power amplifier *etc.*) provide the constant gain over a limited band of frequencies *i.e.* from lower cut-off frequency f_1 to upper cut-off frequency f_2. Now bandwidth of the amplifier, $BW = f_2 - f_1$. The reader may wonder, then, what distinguishes a tuned amplifier from other amplifiers? The difference is that tuned amplifiers are designed to have specific, usually narrow bandwidth. This point is illustrated in Fig. 13.2. Note that BW_S is the bandwidth of standard frequency response while BW_T is the bandwidth of the tuned amplifier. In many applications, the narrower the bandwidth of a tuned amplifier, the better it is.

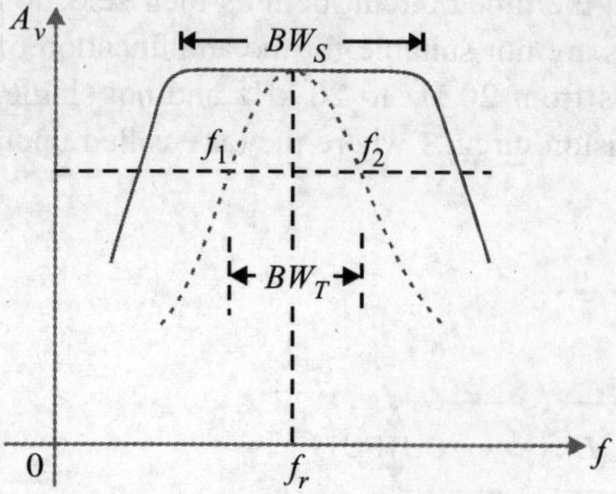

Fig. 13.2

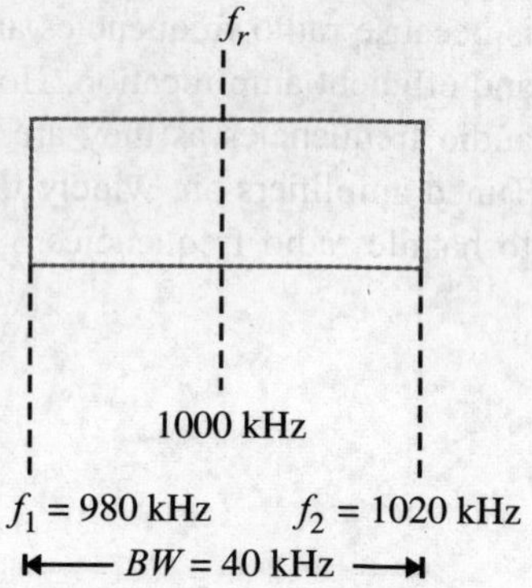

Fig. 13.3

Illustration. Consider a tuned amplifier that is designed to amplify only those frequencies that are within ±20 kHz of the central frequency of 1000 kHz (*i.e.* f_r = 1000 kHz). Here [See Fig. 13.3],

$$f_1 = 980 \text{ kHz}, \quad f_r = 1000 \text{ kHz}, \quad f_2 = 1020 \text{ kHz}, \quad BW = 40 \text{ kHz}$$

This means that so long as the input signal is within the range of 980 – 1020 kHz, it will be amplified. If the frequency of input signal goes out of this range, amplification will be drastically reduced.

13.3 Analysis of Parallel Tuned Circuit

A parallel tuned circuit consists of a capacitor C and inductor L in parallel as shown in Fig. 13.4 (*i*). In practice, some resistance R is always present with the coil. If an alternating voltage is applied across this parallel circuit, the frequency of oscillations will be that of the applied voltage. However, if the frequency of applied voltage is equal to the natural or resonant frequency of LC circuit, then *electrical resonance* will occur. Under such conditions, the impedance of the tuned circuit becomes maximum and the line current is minimum. The circuit then draws just enough energy from a.c. supply necessary to overcome the losses in the resistance R.

Parallel resonance. *A parallel circuit containing reactive elements (L and C) is *resonant when the circuit power factor is unity i.e. applied voltage and the supply current are in*

* Resonance means to be in step with. In an a.c. circuit if applied voltage and supply current are in phase (*i.e.* in step with), resonance is said to occur. If this happens in a parallel a.c. circuit, it is called parallel resonance.

phase. The phasor diagram of the parallel circuit is shown in Fig. 13.4 (*ii*). The coil current I_L has two rectangular components *viz.* active component $I_L \cos \phi_L$ and reactive component $I_L \sin \phi_L$. This parallel circuit will resonate when the circuit power factor is unity. This is possible only when the net reactive component of the circuit current is zero *i.e.*

$$I_C - I_L \sin \phi_L = 0$$

or $$I_C = I_L \sin \phi_L$$

Resonance in parallel circuit can be obtained by changing the supply frequency. At some frequency f_r (called resonant frequency), $I_C = I_L \sin \phi_L$ and resonance occurs.

Resonant frequency. *The frequency at which parallel resonance occurs (i.e. reactive component of circuit current becomes zero) is called the* **resonant frequency** f_r.

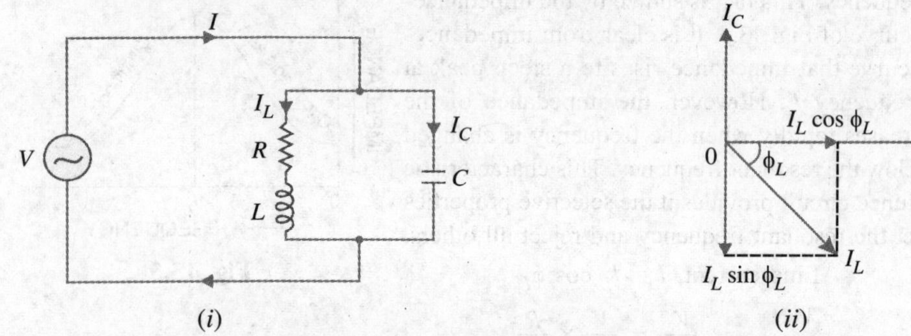

Fig. 13.4

At parallel resonance, we have, $I_C = I_L \sin \phi_L$

Now, $I_L = V/Z_L$; $\sin \phi_L = X_L/Z_L$ and $I_C = V/X_C$

$$\therefore \quad \frac{V}{X_C} = \frac{V}{Z_L} \times \frac{X_L}{Z_L}$$

or $$X_L X_C = Z_L^2$$

or $$\frac{\omega L}{\omega C} = Z_L^2 = R^2 + X_L^2 \qquad ...(i)$$

or $$\frac{L}{C} = R^2 + (2\pi f_r L)^2$$

or $$(2\pi f_r L)^2 = \frac{L}{C} - R^2$$

or $$2\pi f_r L = \sqrt{\frac{L}{C} - R^2}$$

or $$f_r = \frac{1}{2\pi L}\sqrt{\frac{L}{C} - R^2}$$

$$\therefore \quad \text{Resonant frequency, } f_r = \frac{1}{2\pi}\sqrt{\frac{1}{LC} - \frac{R^2}{L^2}} \qquad ...(ii)$$

If coil resistance R is small (as is generally the case), then,

$$f_r = \frac{1}{2\pi\sqrt{LC}} \qquad ...(iii)$$

The resonant frequency will be in Hz if R, L and C are in ohms, henry and farad respectively.

Note. If in the problem, the value of R is given, then eq. (*ii*) should be used to find f_r. However, if R is not given, then eq. (*iii*) may be used to find f_r.

13.4 Characteristics of Parallel Resonant Circuit

It is now desirable to discuss some important characteristics of parallel resonant circuit.

(*i*) **Impedance of resonant circuit.** The impedance offered by the parallel LC circuit is given by the supply voltage divided by the line current *i.e.* V/I. Since at resonance, line current is minimum, therefore, impedance is maximum at resonant frequency. This fact is shown by the impedance-frequency curve of Fig. 13.5. It is clear from impedance-frequency curve that impedance rises to a steep peak at resonant frequency f_r. However, the impedance of the circuit decreases rapidly when the frequency is changed above or below the resonant frequency. This characteristic of parallel tuned circuit provides it the selective properties *i.e.* to select the resonant frequency and reject all others.

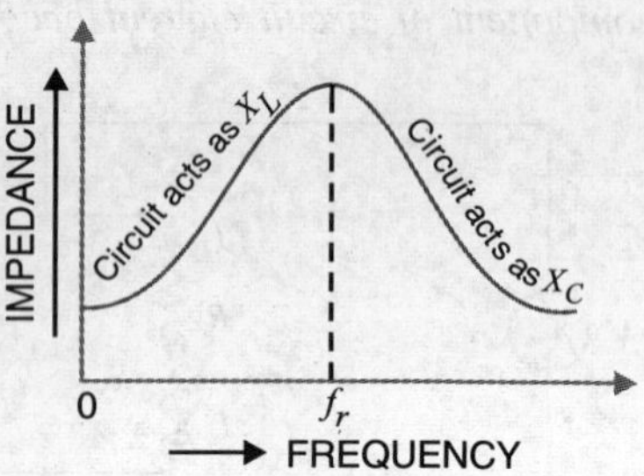

Fig. 13.5

$$\text{Line current, } I = I_L \cos \phi_L$$

or $$\frac{V}{Z_r} = \frac{V}{Z_L} \times \frac{R}{Z_L}$$

or $$\frac{1}{Z_r} = \frac{R}{Z_L^2}$$

or $$\frac{1}{Z_r} = \frac{R}{L/C} = \frac{CR}{L} \qquad \left[\because\ Z_L^2 = \frac{L}{C} \text{ from eq. } (i)\right]$$

$$\therefore \quad \text{Circuit impedance, } Z_r = \frac{L}{CR}$$

Thus at parallel resonance, the circuit impedance is equal to *L/CR. It may be noted that Z_r will be in ohms if R, L and C are measured in ohms, henry and farad respectively.

(*ii*) **Circuit Current.** At parallel resonance, the circuit or line current I is given by the applied voltage divided by the circuit impedance Z_r *i.e.*

$$\text{Line current, } I = \frac{V}{Z_r} \qquad \text{where } Z_r = \frac{L}{CR}$$

Because Z_r is very high, the line current I will be very small.

(*iii*) **Quality Factor Q.** It is desired that resonance curve of a parallel tuned circuit should be as sharp as possible in order to provide selectivity. The sharp resonance curve means that impedance falls rapidly as the frequency is varied from the resonant frequency. The smaller the resistance of coil, the more sharp is the resonance curve. This is due to the fact that a small resistance consumes less power and draws a relatively small line current. The ratio of inductive reactance and resistance of the coil at resonance, therefore, becomes

* Two things are worth noting. First, Z_r (= L/CR) is a pure resistance because there is no frequency term present. Secondly, the value of Z_r is very high because the ratio L/C is very large at parallel resonance.

a measure of the quality of the tuned circuit. This is called *quality factor* of parallel resonant circuit and may be defined as under :

The ratio of inductive reactance of the coil at resonance to its resistance is known as ***quality factor** *Q i.e.*

$$Q = \frac{X_L}{R} = \frac{2\pi f_r L}{R}$$

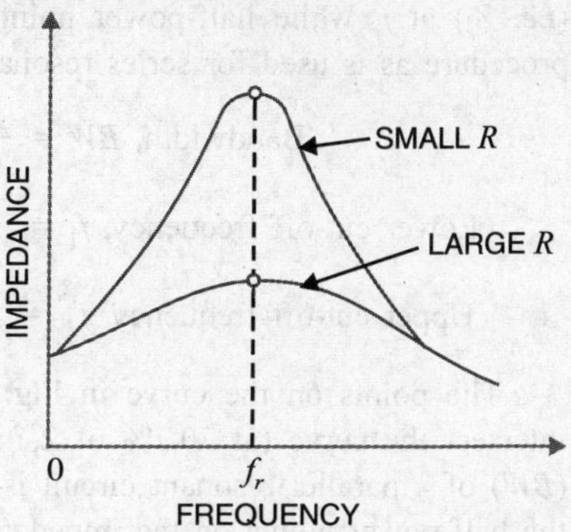

Fig. 13.6

The quality factor Q of a parallel tuned circuit is very important because the sharpness of resonance curve and hence selectivity of the circuit depends upon it. The higher the value of Q, the more selective is the tuned circuit. Fig. 13.6 shows the effect of resistance R of the coil on the sharpness of the resonance curve. It is clear that when the resistance is small, the resonance curve is very sharp. However, if the coil has large resistance, the resonance curve is less sharp. It may be emphasised that where high selectivity is desired, the value of Q should be very large. We can also express the Q factor of a parallel resonant circuit in another useful form. Neglecting resistance R of the coil, the resonant frequency of the circuit is given by ;

$$f_r = \frac{1}{2\pi\sqrt{LC}}$$

$$\therefore \quad Q\text{-factor} = \frac{2\pi f_r L}{R} = 2\pi \times \frac{1}{2\pi\sqrt{LC}} \times \frac{L}{R}$$

or

$$Q\text{-factor} = \frac{1}{R}\sqrt{\frac{L}{C}}$$

This is the expression for the Q-factor in terms of R, L and C.

13.5 Bandwidth of Parallel Resonant Circuit

Just as the bandwidth of a series resonant circuit is defined from current-frequency curve, similarly the bandwidth of a parallel resonant circuit is defined from impedance-frequency curve as shown in Fig. 13.7. The bandwidth of a parallel resonant circuit is defined *as the range of frequencies over which the circuit impedance is equal to or greater than 70.7% of maximum circuit impedance (i.e. Z_r, the impedance at resonance).* The circuit impedance is maximum

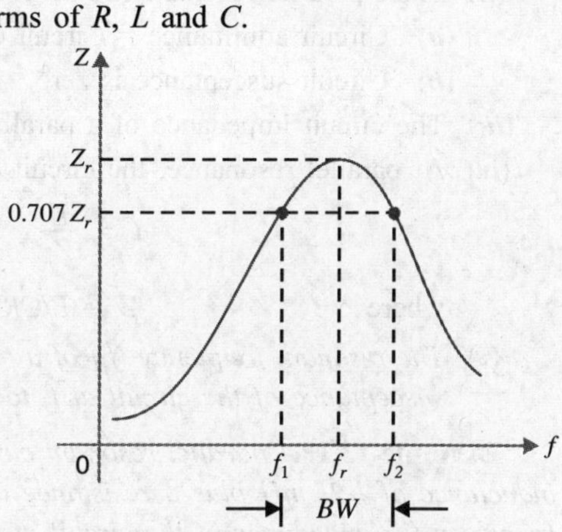

Fig. 13.7

* Strictly speaking, the Q of a tank circuit is defined as the ratio of the energy stored in the circuit to the energy lost in the circuit *i.e.*

$$Q = \frac{\text{Energy stored}}{\text{Energy lost}} = \frac{\text{Reactive Power}}{\text{Resistive Power}} = \frac{I_L^2 X_L}{I_L^2 R} \quad \text{or} \quad Q = \frac{X_L}{R}$$

(*i.e.* Z_r) at f_r while half-power points (*i.e.* f_1 and f_2) occur at 0.707 Z_r. Using the same procedure as is used for series resonant circuit, it can be shown that :

$$\text{Bandwidth, } BW = \frac{f_r}{Q} = f_2 - f_1$$

$$\text{Lower cut-off frequency, } f_1 = f_r - \frac{BW}{2}$$

$$\text{Upper cut-off frequency, } f_2 = f_r + \frac{BW}{2}$$

The points on the curve in Fig. 13.7 where the upper and lower cut-off frequencies intersect the curve (at 70.7% of Z_r) are called the *half-power points*. Thus the bandwidth (*BW*) of a parallel resonant circuit is often referred to as the band of frequencies between the half-power points on the impedance-frequency curve.

At half-power frequencies (*i.e.* f_1 and f_2), the power dissipated in the circuit is one-half of that dissipated at resonant frequency (f_r). A signal at a frequency outside the bandwidth is transmitted with less than 1/2 the output power of a signal of similar strength at the resonant frequency.

Note. The power at cut-off frequencies is half that at the resonant frequency and are known as half-power frequencies. They are also called – 3 dB frequencies because power decreases by 1/2 and in decibels it is

$$10 \log_{10} \frac{1}{2} = -3 \text{ dB}$$

13.6 Key Points About Parallel Resonance

The key points concerning parallel resonant circuits are :

(*i*) The circuit power factor is unity (*i.e.* 1) so that parallel resonant circuit behaves as a resistance.

(*ii*) Since parallel resonant circuit is resistive, it means that :

(*a*) Circuit admittance = Circuit conductance

(*b*) Circuit susceptance is zero.

(*iii*) The circuit impedance of a parallel resonant circuit is maximum.

(*iv*) At parallel resonance, the circuit current is minimum and is given by ;

$$I_r = \frac{V}{Z_r}$$

where $Z_r = L/CR$ = Circuit impedance at parallel resonance

(*v*) *The resonant frequency (f_r) of a parallel resonant circuit can be found by equating susceptance of the circuit at f_r to zero.*

Example 13.1. *A parallel resonant circuit has a capacitor of 250 pF in one branch and inductance of 1.25 mH plus a resistance of 10 Ω in the parallel branch. Find (i) resonant frequency (ii) impedance of the circuit at resonance (iii) Q-factor of the circuit.*

Solution.

$$R = 10\ \Omega\ ;\ L = 1.25 \times 10^{-3}\ \text{H}\ ;\ C = 250 \times 10^{-12}\ \text{F}$$

(*i*) Resonant frequency of the circuit is

$$f_r = \frac{1}{2\pi}\sqrt{\frac{1}{LC} - \frac{R^2}{L^2}}$$

$$= \frac{1}{2\pi}\sqrt{\frac{10^{12}}{1.25\times10^{-3}\times250} - \frac{10^2}{(1.25\times10^{-3})^2}}\ \text{Hz}$$

$$= 284.7 \times 10^3 \text{ Hz} = \mathbf{284.7\ kHz}$$

(*ii*) Impedance of the circuit at resonance is

$$Z_r = \frac{L}{CR} = \frac{1.25\times10^{-3}}{250\times10^{-12}\times10} = 500 \times 10^3\ \Omega = \mathbf{500\ k\Omega}$$

(*iii*) Quality factor of the circuit is

$$Q = \frac{2\pi f_r L}{R} = \frac{2\pi\,(284.7\times10^3)\times1.25\times10^{-3}}{10} = \mathbf{223.6}$$

Example 13.2. *A parallel resonant circuit has a capacitor of 100 pF in one branch and inductance of 100 μH plus a resistance of 10 Ω in parallel branch. If the supply voltage is 10 V, calculate (i) resonant frequency (ii) impedance of the circuit and line current at resonance.*

Solution.

$R = 10\ \Omega\ ;\ L = 100 \times 10^{-6}\ \text{H}\ ;\ C = 100 \times 10^{-12}\ \text{F}$

(*i*) Resonant frequency of the circuit is

$$f_r = \frac{1}{2\pi}\sqrt{\frac{1}{LC} - \frac{R^2}{L^2}}$$

$$= \frac{1}{2\pi}\sqrt{\frac{10^{12}}{100\times10^{-6}\times100} - \frac{10^2}{(100\times10^{-6})^2}}\ \text{Hz}$$

$$= 1592.28 \times 10^3 \text{ Hz} = \mathbf{1592.28\ kHz}$$

(*ii*) Impedance of the circuit at resonance is

$$Z_r = \frac{L}{C\,R} = \frac{L}{C}\times\frac{1}{R} = \frac{100\times10^{-6}}{100\times10^{-12}}\times\frac{1}{R}$$

$$= 10^6 \times \frac{1}{R} = 10^6 \times \frac{1}{10} = 10^5\ \Omega = \mathbf{0.1\ M\Omega}$$

Note that the circuit impedance Z_r is very high at resonance. It is because the ratio L/C is very large at resonance.

Line current at resonance, $I = \dfrac{V}{Z_r} = \dfrac{10\text{ V}}{10^5\ \Omega} = \mathbf{100\ \mu A}$

Example 13.3. *The dynamic impedance of a parallel resonant circuit is 500 kΩ. The circuit consists of a 250 pF capacitor in parallel with a coil of resistance 10 Ω. Calculate (i) the coil inductance (ii) the resonant frequency and (iii) Q-factor of the circuit.*

Solution. Impedance of parallel resonant circuit at resonance is called *dynamic impedance.*

(*i*) Dynamic impedance, $Z_r = \dfrac{L}{CR}$

∴ Inductance of coil, $L = Z_r\,CR = (500 \times 10^3) \times (250 \times 10^{-12}) \times 10$

$$= 1.25 \times 10^{-3}\ \text{H} = \mathbf{1.25\ mH}$$

(*ii*) Resonant frequency, $f_r = \dfrac{1}{2\pi}\sqrt{\dfrac{1}{LC} - \dfrac{R^2}{L^2}}$

$$= \frac{1}{2\pi}\sqrt{\frac{10^{12}}{1.25\times10^{-3}\times 250} - \frac{10^2}{(1.25\times10^{-3})^2}}$$

$$= 284.7 \times 10^3 \text{ Hz} = \mathbf{284.7 \text{ kHz}}$$

(*iii*) *Q*-factor of the circuit $= \dfrac{2\pi f_r L}{R} = \dfrac{2\pi\times(284.7\times10^3)\times(1.25\times10^{-3})}{10} = \mathbf{223.6}$

Example 13.4. *Calculate the impedance of parallel-tuned circuit shown in Fig. 13.8 at a frequency of 500 kHz and for bandwidth of operation equal to 20 kHz.*

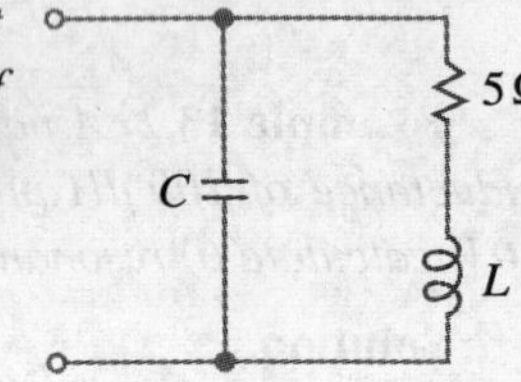

Fig. 13.8

Solution. f_r = 500 kHz = 500 × 10^3 Hz ; *BW* = 20 kHz = 20 × 10^3 Hz

Now, $*BW = \dfrac{R}{2\pi L}$ $\quad\therefore\quad L = \dfrac{R}{2\pi\times BW} = \dfrac{5}{2\pi\times20\times10^3}$

$$= 39 \times 10^{-6} \text{ H}$$

$$f_r = \frac{1}{2\pi}\left(\frac{1}{LC} - \frac{R^2}{L^2}\right)^{1/2} \quad \text{or} \quad \frac{1}{LC} = 4\pi^2 f_r^2 + \frac{R^2}{L^2}$$

$$\therefore \quad \frac{1}{LC} = 4\pi^2\times(500\times10^3)^2 + \frac{(5)^2}{(39\times10^{-6})^2} \simeq 9.87\times10^{12}$$

or $$C = \frac{1}{9.87\times10^{12}\times39\times10^{-6}} = 2\times10^{-9}\text{ F}$$

Circuit impedance, $Z_r = \dfrac{L}{CR} = \dfrac{39\times10^{-6}}{2\times10^{-9}\times5} = 3900\ \Omega = \mathbf{3.9\ k\Omega}$

Example 13.5. *An inductor of resistance 10 Ω and inductance 100 mH is in parallel with a 10 nF capacitor. Find (i) the resonant angular frequency (ii) the Q-factor and (iii) the bandwidth.*

Solution. R = 10 Ω ; L = 100 mH = 100 × 10^{-3} H = 10^{-1} H ; C = 10 nF = 10 × 10^{-9} F = 10^{-8} F

(*i*) The resonant angular frequency ω_r (= $2\pi f_r$) is given by ;

$$\omega_r = \sqrt{\left(\frac{1}{LC} - \frac{R^2}{L^2}\right)} = \sqrt{\left(\frac{1}{10^{-1}\times10^{-8}} - \frac{(10)^2}{(10^{-1})^2}\right)} = \mathbf{3.16\times10^4\ rad/s}$$

---

* $$BW = \frac{f_r}{Q} = f_r\times\frac{1}{Q} = f_r\times\frac{R}{X_L} \qquad \left(\because Q = \frac{X_L}{R}\right)$$

$$\therefore \quad BW = f_r\times\frac{R}{2\pi f_r L} = \frac{R}{2\pi L}$$

(*ii*) $Q\text{-factor} = \frac{\omega_r L}{R} = \frac{3.16\times10^4 \times 10^{-1}}{10} = \mathbf{316}$

(*iii*) $\text{Bandwidth} = \frac{\omega_r}{Q} = \frac{3.16\times10^4}{316} = \mathbf{100\ rad/s}$

Example 13.6. *An inductor of resistance R and inductance L is in parallel with a capacitor C. Show that resonant angular frequency* ω_r *is given by ;*

$$\omega_r = \frac{1}{\sqrt{LC}}\sqrt{\left(1-\frac{1}{Q^2}\right)}$$

Solution. $$\omega_r = \sqrt{\left(\frac{1}{LC}-\frac{R^2}{L^2}\right)} \quad \text{... (i)}$$

This equation can be rearranged as :

$$\omega_r = \frac{1}{\sqrt{LC}}\sqrt{\left(1-\frac{R^2C}{L}\right)}$$

Now, $$Q = \frac{1}{R}\sqrt{\frac{L}{C}} \quad \text{or} \quad Q^2 = \frac{L}{CR^2}$$

$$\therefore \quad \omega_r = \frac{1}{\sqrt{LC}}\sqrt{\left(1-\frac{1}{Q^2}\right)} \quad \text{Q.E.D.}$$

When Q is large, this equation approximates to $\omega_r = \frac{1}{\sqrt{LC}}$.

13.7 Advantages of Tuned Amplifiers

In high frequency applications, it is generally required to amplify a single frequency, rejecting all other frequencies present. For such purposes, tuned amplifiers are used. These amplifiers use tuned parallel circuit as the collector load and offer the following advantages :

(*i*) **Small power loss.** A tuned parallel circuit employs reactive components L and C. Consequently, the power loss in such a circuit is quite low. On the other hand, if a resistive load is used in the collector circuit, there will be considerable loss of power. Therefore, tuned amplifiers are highly efficient.

(*ii*) **High selectivity.** A tuned circuit has the property of selectivity *i.e.* it can select the desired frequency for amplification out of a large number of frequencies simultaneously impressed upon it. For instance, if a mixture of frequencies including f_r is fed to the input of a tuned amplifier, then maximum amplification occurs for f_r. For all other frequencies, the tuned circuit offers very low impedance and hence these are amplified to a little extent and may be thought as rejected by the circuit. On the other hand, if we use resistive load in the collector, all the frequencies will be amplified equally well *i.e.* the circuit will not have the ability to select the desired frequency.

(*iii*) **Smaller collector supply voltage.** Because of little resistance in the parallel tuned circuit, it requires small collector supply voltage V_{CC}. On the other hand, if a high load resistance is used in the collector for amplifying even one frequency, it would mean large voltage drop across it due to zero signal collector current. Consequently, a higher collector supply will be needed.

13.8 Why not Tuned Circuits for Low Frequency Amplification?

The tuned amplifiers are used to select and amplify a specific high frequency or narrow band of frequencies. The reader may be inclined to think as to why tuned circuits are not used to amplify low frequencies. This is due to the following reasons :

(*i*) **Low frequencies are never single.** A tuned amplifier selects and amplifies a single frequency. However, the low frequencies found in practice are the audio frequencies which are a mixture of frequencies from 20 Hz to 20 kHz and are not single. It is desired that all these frequencies should be equally amplified for proper reproduction of the signal. Consequently, tuned amplifiers cannot be used for the purpose.

(*ii*) **High values of *L* and *C*.** The resonant frequency of a parallel tuned circuit is given by ;

$$f_r = \frac{1}{2\pi\sqrt{LC}}$$

For low frequency amplification, we require large values of *L* and *C*. This will make the tuned circuit bulky and expensive. It is worthwhile to mention here that *R-C* and transformer coupled amplifiers, which are comparatively cheap, can be conveniently used for low frequency applications.

13.9 Frequency Response of Tuned Amplifier

The voltage gain of an amplifier depends upon β, input impedance and effective collector load. In a tuned amplifier, tuned circuit is used in the collector. Therefore, voltage gain of such an amplifier is given by ;

$$\text{Voltage gain} = \frac{\beta Z_C}{Z_{in}}$$

where Z_C = Effective collector load

Z_{in} = Input impedance of the amplifier

The value of Z_C and hence gain strongly depends upon frequency in the tuned amplifier. As Z_C is maximum at resonant frequency, therefore, voltage gain will be maximum at this frequency. The value of Z_C and gain decrease as the frequency is varied above and below the resonant frequency. Fig. 13.9 shows the frequency response of a tuned amplifier. It is clear that voltage gain is maximum at resonant frequency and falls off as the frequency is varied in either direction from resonance.

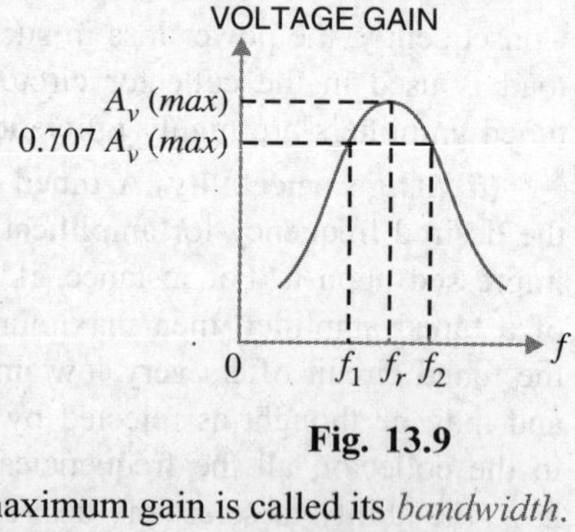

Fig. 13.9

Bandwidth. The range of frequencies at which the voltage gain of the tuned amplifier falls to 70.7% of the maximum gain is called its *bandwidth*. Referring to Fig. 13.9, the bandwidth of tuned amplifier is $f_2 - f_1$. The amplifier will amplify nicely any signal in this frequency range. The bandwidth of tuned amplifier depends upon the value of Q of LC circuit *i.e.* upon the sharpness of the frequency response. The greater the value of Q of tuned circuit, the lesser is the bandwidth of the amplifier and *vice-versa.* In practice, the value of Q of LC circuit is made such so as to permit the amplification of desired narrow band of high frequencies.

The practical importance of bandwidth of tuned amplifiers is found in communication system. In radio and TV transmission, a very high frequency wave, called *carrier wave* is used to carry the audio or picture signal. In radio transmission, the audio signal has a frequency range of 10 kHz. If the carrier wave frequency is 710 kHz, then the resultant radio wave has a frequency range *between (710 –5) kHz and (710 +5) kHz. Consequently, the tuned amplifier must have a bandwidth of 705 kHz to 715 kHz (*i.e.* 10 kHz). The Q of the tuned circuit should be such that bandwidth of the amplifier lies in this range.

13.10 Relation between Q and Bandwidth

The quality factor Q of a tuned amplifier is equal to the ratio of resonant frequency (f_r) to bandwidth (BW) *i.e.*

$$Q = \frac{f_r}{BW}$$

The Q of an amplifier is determined by the circuit component values. It may be noted here that Q of a tuned amplifier is generally greater than 10. When this condition is met, the resonant frequency at parallel resonance is approximately given by ;

$$f_r = \frac{1}{2\pi\sqrt{LC}}$$

Example 13.7. *The Q of a tuned amplifier is 60. If the resonant frequency for the amplifier is 1200 kHz, find (i) bandwidth and (ii) cut-off frequencies.*

Solution.

(*i*) $$BW = \frac{f_r}{Q} = \frac{1200 \text{ kHz}}{60} = \mathbf{20\ kHz}$$

(*ii*) Lower cut-off frequency, f_1 = 1200 – 10 = **1190 kHz**

Upper cut-off frequency, f_2 = 1200 + 10 = **1210 kHz**

Example 13.8. *A tuned amplifier has maximum voltage gain at a frequency of 2 MHz and the bandwidth is 50 kHz. Find the Q factor.*

Solution. The maximum voltage gain occurs at the resonant frequency. Therefore, $f_r = 2$ MHz $= 2 \times 10^6$ Hz and $BW = 50$ kHz $= 50 \times 10^3$ Hz.

Now, $$BW = \frac{f_r}{Q}$$

∴ $$Q = \frac{f_r}{BW} = \frac{2 \times 10^6}{50 \times 10^3} = \mathbf{40}$$

Example 13.9. *Draw the frequency response of an ideal tuned amplifier and discuss its characteristics.*

Solution. Fig. 13.10 shows the frequency response of an ideal tuned amplifier. The ideal tuned amplifier has zero gain for all frequencies from 0 Hz upto the lower cut-off frequency

* See chapter on modulation and demodulation.

f_1. At this point, the gain *instantly* jumps to the maximum value [$A_{v\,(max)}$]. The gain stays at the maximum value until f_2 is reached. At this time, the gain *instantly* drops back to zero. Thus all the frequencies within the bandwidth (f_1 to f_2) of the amplifier would be *passed* by the circuit while all others would be effectively stopped. This is where the terms *pass band* and *stop band* come from. The pass band is the range of frequencies that is passed (amplified) by a tuned amplifier. On the other hand, the stop band is the range of frequencies that is outside the amplifier's pass band.

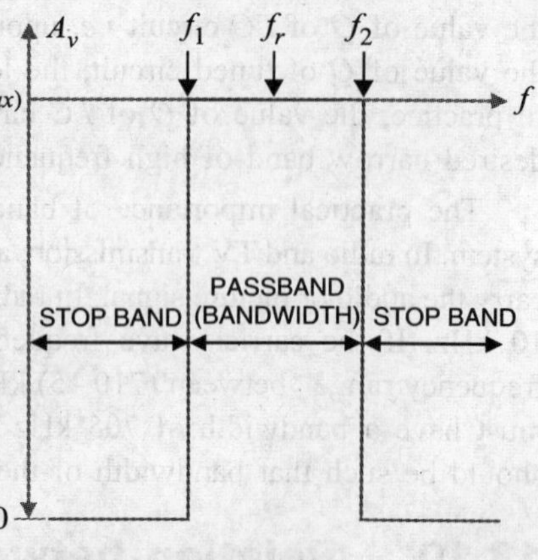

Fig. 13.10

In practice, the ideal characteristics of the tuned amplifier cannot be achieved. In a practical frequency response (Refer back to Fig. 13.9), the gain falls gradually from maximum value as the frequency goes outside the f_1 or f_2 limits. However, the closer the frequency response of a tuned amplifier to that of the ideal, the better.

Example 13.10. *The cut-off frequencies of a tuned amplifier are : lower cut-off frequency f_1 = 1160 kHz and upper cut-off frequency f_2 = 1240 kHz. Determine (i) the resonant frequency (ii) the average of f_1 and f_2 (iii) Q factor of the circuit.*

Solution.

(*i*) It can be proved that resonant frequency f_r (also called *centre frequency*) is equal to the geometric mean of f_1 and f_2 *i.e.*

$$\text{Resonant frequency, } f_r = \sqrt{f_1 f_2} = \sqrt{1160 \times 1240} = \mathbf{1199\ kHz}$$

(*ii*) The average of f_1 and f_2 is given by ;

$$f_{av} = \frac{f_1 + f_2}{2} = \frac{1160 + 1240}{2} = \mathbf{1200\ kHz}$$

Note that the value of f_r is very close to f_{av}. This will always be the case when circuit $Q \geq 2$ (as it is in this case).

(*iii*)
$$\text{Bandwidth, } BW = f_2 - f_1 = 1240 - 1160 = 80 \text{ kHz}$$

$$\therefore \quad Q \text{ factor} = \frac{f_r}{BW} = \frac{1199}{80} = \mathbf{14.99}$$

As you can see that $Q > 2$. Therefore, $f_r \simeq f_{av}$ *i.e.* f_r approaches a value that is exactly halfway between f_1 and f_2.

Note. There is another way to confirm that $f_r \simeq f_{av}$. If the value of f_r for the circuit is equal to or greater than 2 times the bandwidth of the circuit (*i.e.* $f_r \geq 2BW$), then $f_r \simeq f_{av}$.

Example 13.11. *A tuned amplifier has cut-off frequencies as : f_1 = 980 kHz and f_2 = 1080 kHz. Show that resonant frequency $f_r \simeq f_{av}$ (average of f_1 and f_2) and the Q value of the circuit is greater than 2.*

Solution. The resonant frequency f_r of the circuit is given by ;

$$f_r = \sqrt{f_1 f_2} = \sqrt{980 \times 1080} = 1029 \text{ kHz}$$

The average value of f_1 and f_2 is given by ;

$$f_{av} = \frac{f_1 + f_2}{2} = \frac{980 + 1080}{2} = 1030 \text{ kHz}$$

Therefore, $f_r \simeq f_{av}$. We now show that $Q > 2$.

$$\text{Bandwidth, } BW = f_2 - f_1 = 1080 - 980 = 100 \text{ kHz}$$

$$\therefore \quad Q = \frac{f_r}{BW} = \frac{1029}{100} = \mathbf{10.29}$$

13.11 Single Tuned Amplifier

A single tuned amplifier consists of a transistor amplifier containing a parallel tuned circuit as the collector load. The values of capacitance and inductance of the tuned circuit are so selected that its resonant frequency is equal to the frequency to be amplified. The output from a single tuned amplifier can be obtained either (*a*) by a coupling capacitor C_C as shown in Fig. 13.11 (*i*) or (*b*) by a secondary coil as shown in Fig. 13.11 (*ii*).

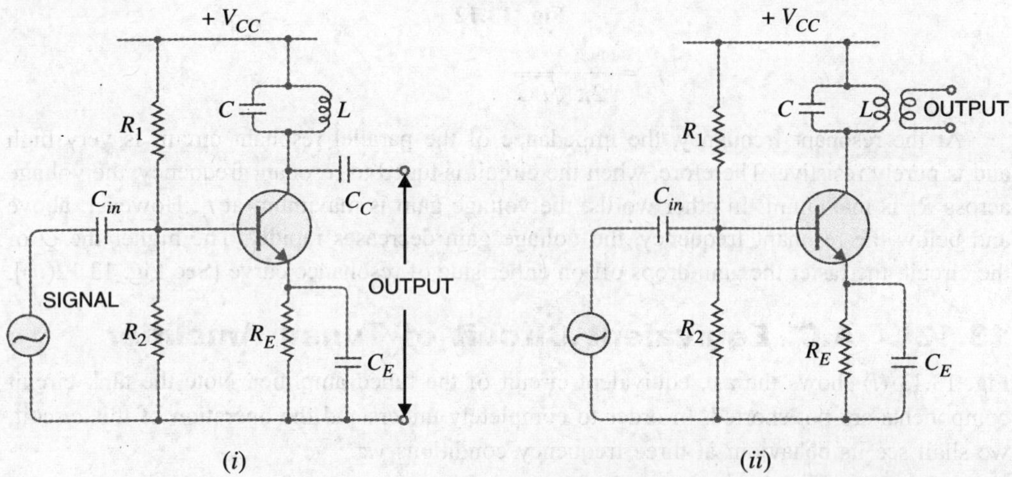

Fig. 13.11

Operation. The high frequency signal to be amplified is given to the input of the amplifier. The resonant frequency of parallel tuned circuit is made equal to the frequency of the signal by changing the value of *C*. Under such conditions, the tuned circuit will offer very high impedance to the signal frequency. Hence a large output appears across the tuned circuit. In case the input signal is complex containing many frequencies, only that frequency which corresponds to the resonant frequency of the tuned circuit will be amplified. All other frequencies will be rejected by the tuned circuit. In this way, a tuned amplifier selects and amplifies the desired frequency.

Note. The fundamental difference between *AF* and tuned (*RF*) amplifiers is the bandwidth they are expected to amplify. The *AF* amplifiers amplify a major portion of *AF* specturm (20 Hz to 20 kHz) equally well throughout. The tuned amplifiers amplify a relatively narrow portion of *RF* spectrum, rejecting all other frequencies.

13.12 Analysis of Tuned Amplifier

Fig. 13.12 (*i*) shows a single tuned amplifier. Note the presence of the parallel *LC* circuit in the collector circuit of the transistor. When the circuit has a high *Q*, the parallel resonance occurs at a frequency f_r given by ;

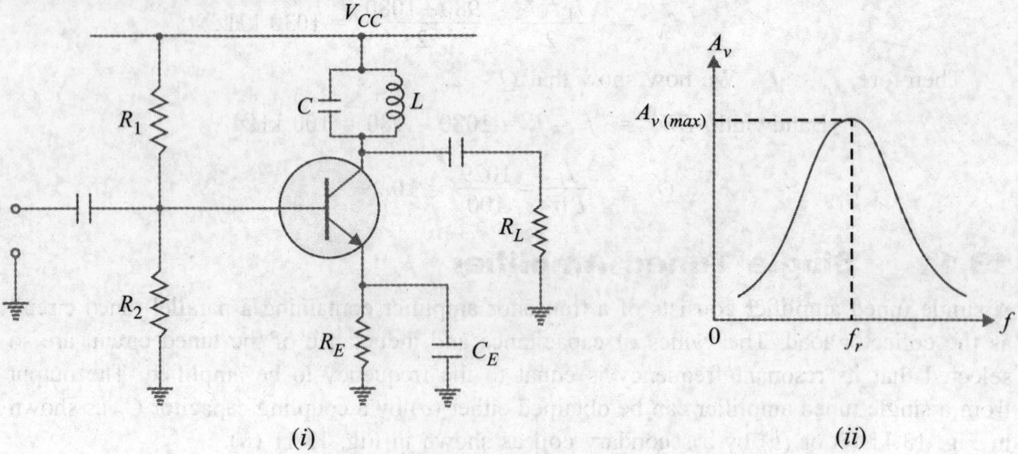

Fig. 13.12

$$f_r = \frac{1}{2\pi\sqrt{LC}}$$

At the resonant frequency, the impedance of the parallel resonant circuit is very high and is purely resistive. Therefore, when the circuit is tuned to resonant frequency, the voltage across R_L is maximum. In other words, the voltage gain is maximum at f_r. However, above and below the resonant frequency, the voltage gain decreases rapidly. The higher the *Q* of the circuit, the faster the gain drops off on either side of resonance curve [See Fig. 13.12(*ii*)].

13.13 A.C. Equivalent Circuit of Tuned Amplifier

Fig. 13.13 (*i*) shows the a.c. equivalent circuit of the tuned amplifier. Note the tank circuit components are not shorted. In order to completely understand the operation of this circuit, we shall see its behaviour at three frequency conditions *viz.*,

(*i*) $f_{in} = f_r$ (*ii*) $f_{in} < f_r$ (*iii*) $f_{in} > f_r$

(*i*) **When input frequency equals f_r** (*i.e.* $f_{in} = f_r$). When the frequency of the input signal is equal to f_r, the parallel *LC* circuit offers a very high impedance *i.e.* it acts as an open. Since R_L represents the only path to ground in the collector circuit, all the a.c. collector current flows through R_L. Therefore, voltage across R_L is maximum *i.e.* the voltage gain is maximum as shown in Fig. 13.13 (*ii*).

(*ii*) **When input frequency is less than f_r** (*i.e.* $f_{in} < f_r$). When the input signal frequency is less than f_r, the circuit is effectively* inductive. As the frequency decreases from f_r, a point is reached when $X_C - X_L = R_L$. When this happens, the voltage gain of the amplifier falls by 3 dB. In other words, the lower cut-off frequency f_1 for the circuit occurs when $X_C - X_L = R_L$.

* At frequencies below f_r, $X_C > X_L$ or $I_L > I_C$. Therefore, the circuit will be inductive.

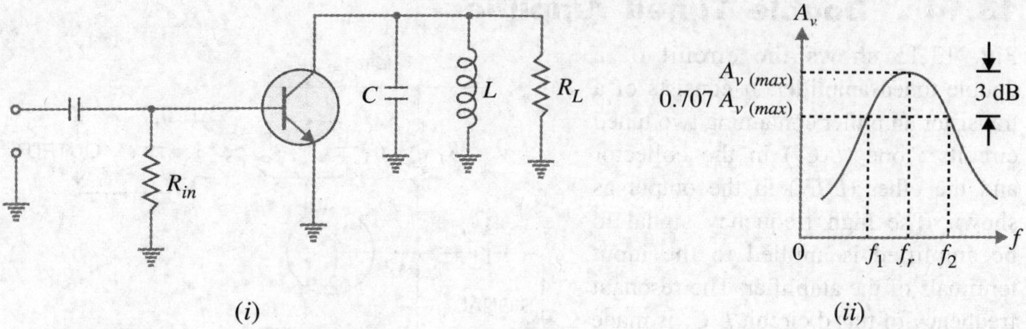

Fig. 13.13

***(iii)* When input frequency is greater than f_r (*i.e.* $f_{in} > f_r$).** When the input signal frequency is greater than f_r, the circuit is effectively capacitive. As f_{in} is increased beyond f_r, a point is reached when $X_L - X_C = R_L$. When this happens, the voltage gain of the amplifier will again fall by 3dB. In other words, the upper cut-off frequency for the circuit will occur when $X_L - X_C = R_L$.

Example 13.12. *For the tuned amplifier shown in Fig. 13.14, determine (i) the resonant frequency (ii) the Q of tank circuit and (iii) bandwidth of the amplifier.*

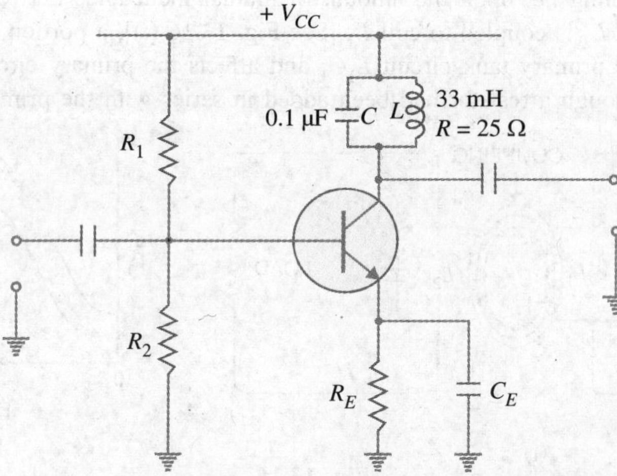

Fig. 13.14

Solution.

(i) Resonant frequency, $f_r = \dfrac{1}{2\pi\sqrt{LC}} = \dfrac{1}{2\pi\sqrt{33 \times 10^{-3} \times 0.1 \times 10^{-6}}}$

$$= 2.77 \times 10^3 \text{ Hz} = \mathbf{2.77 \text{ kHz}}$$

(ii) $X_L = 2\pi f_r L = 2\pi \times (2.77 \times 10^3) \times 33 \times 10^{-3} = 574\ \Omega$

$\therefore$ $Q = \dfrac{X_L}{R} = \dfrac{574}{25} = \mathbf{23}$

(iii) $BW = \dfrac{f_r}{Q} = \dfrac{2.77 \text{ kHz}}{23} = \mathbf{120 \text{ Hz}}$

13.14 Double Tuned Amplifier

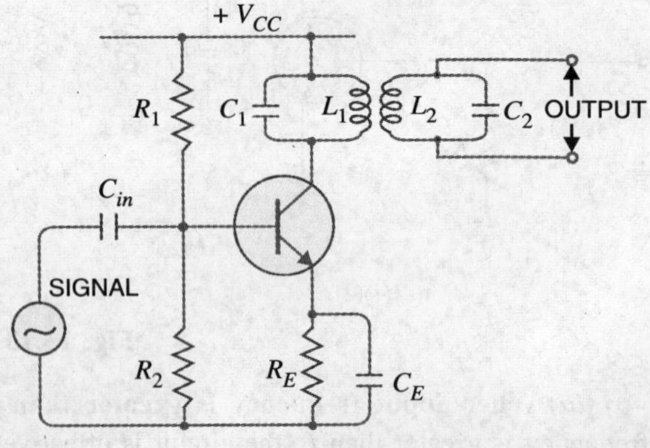

Fig. 13.15

Fig. 13.15 shows the circuit of a double tuned amplifier. It consists of a transistor amplifier containing two tuned circuits ; one (L_1C_1) in the collector and the other (L_2C_2) in the output as shown. The high frequency signal to be amplified is applied to the input terminals of the amplifier. The resonant frequency of tuned circuit L_1C_1 is made equal to the signal frequency. Under such conditions, the tuned circuit offers very high impedance to the signal frequency. Consequently, large output appears across the tuned circuit L_1C_1. The output from this tuned circuit is transferred to the second tuned circuit L_2C_2 through mutual induction. Double tuned circuits are extensively used for coupling the various circuits of radio and television receivers.

Frequency response. The frequency response of a double tuned circuit depends upon the degree of coupling *i.e.* upon the amount of mutual inductance between the two tuned circuits. When coil L_2 is coupled to coil L_1 [See Fig. 13.16 (*i*)], a portion of load resistance is coupled into the primary tank circuit L_1C_1 and affects the primary circuit in exactly the same manner as though a resistor had been added in series with the primary coil L_1.

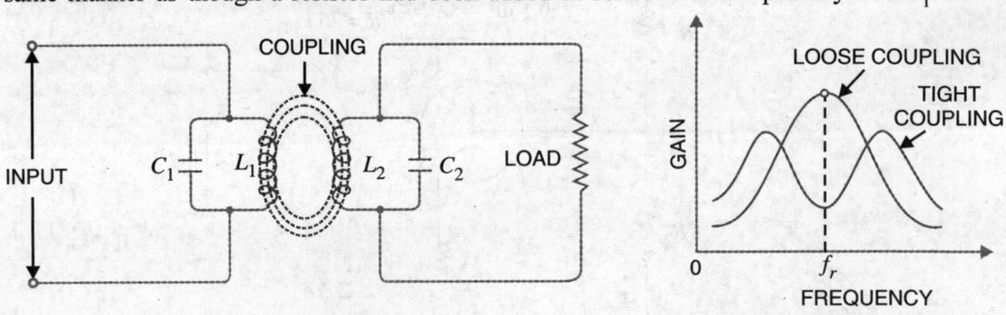

(*i*) (*ii*)

Fig. 13.16

When the coils are spaced apart, all the primary coil L_1 flux will not link the secondary coil L_2. The coils are said to have *loose coupling*. Under such conditions, the resistance reflected from the load (*i.e.* secondary circuit) is small. The resonance curve will be sharp and the circuit Q is high as shown in Fig. 13.16 (*ii*). When the primary and secondary coils are very close together, they are said to have *tight coupling*. Under such conditions, the reflected resistance will be large and the circuit Q is lower. Two positions of gain maxima, one above and the other below the resonant frequency, are obtained.

13.15 Bandwidth of Double-Tuned Circuit

If you refer to the frequency response of double-tuned circuit shown in Fig. 13.16 (*ii*), it is clear that bandwidth increases with the degree of coupling. Obviously, the determining factor in a double-tuned circuit is not Q but the coupling. For a given frequency, the tighter the coupling, the greater is the bandwidth.

$$BW_{dt} = k\, f_r$$

The subscript dt is used to indicate double-tuned circuit. Here k is coefficient of coupling.

Example 13.13. *It is desired to obtain a bandwidth of 200 kHz at an operating frequency of 10 MHz using a double-tuned circuit. What value of co-efficient of coupling should be used ?*

Solution.

$$BW_{dt} = k\, f_r$$

$$\therefore \quad \text{Co-efficient of coupling, } k = \frac{BW_{dt}}{f_r} = \frac{200 \text{ kHz}}{10 \times 10^3 \text{ kHz}} = \mathbf{0.02}$$

13.16 Practical Application of Double Tuned Amplifier

Double tuned amplifiers are used for amplifying radio-frequency (*RF*) signals. One such application is in the radio receiver as shown in Fig. 13.17. This is the *IF* stage using double tuned resonant circuits. Each resonant circuit is tuned to *455 kHz. The critical coupling occurs when the coefficient of coupling is

$$k_{critical} = \frac{1}{\sqrt{Q_1\, Q_2}}$$

where Q_1 = Quality factor of primary resonant circuit (L_1 C_1)

Q_2 = Quality factor of secondary resonant circuit (L_2 C_2)

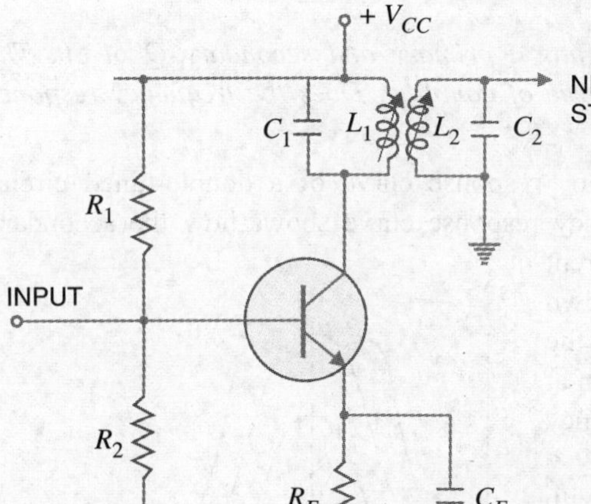

Fig. 13.17

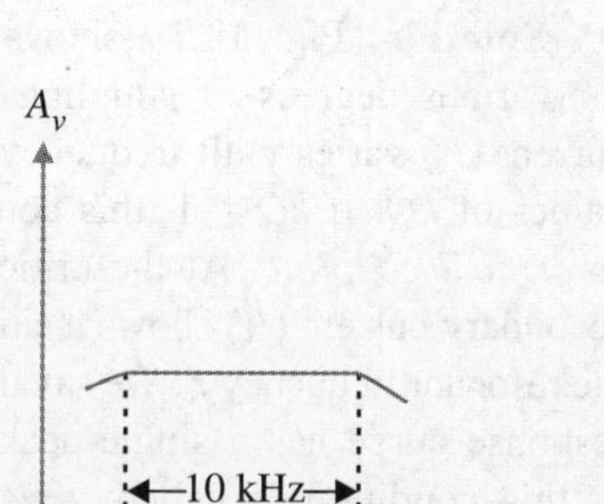

Fig. 13.18

When two resonant circuits are critically coupled, the frequency response becomes flat over a considerable range of frequencies as shown in Fig. 13.18. In other words, the double tuned circuit has better frequency response as compared to that of a single tuned circuit. The use of double tuned circuit offers the following advantages :

* In a radio receiver, the *IF* (intermediate frequency) of 455 kHz is obtained from the mixer circuit regardless of radio station to which the receiver is tuned (See Chapter 14).

(i) Bandwidth is increased.

(ii) Sensitivity (*i.e.* ability to receive weak signals) is increased.

(iii) Selectivity (*i.e.* ability to discriminate against signals in adjacent bands) is increased.

Example 13.14. *The primary and secondary circuits of a double-tuned circuit each (separately) have a Q of 100. The coupling coefficient is 0.02 and the operating frequency is 10 MHz. Find the value of critical coupling coefficient and degree of coupling. What value of secondary Q would produce critical coupling ?*

Solution. $Q_1 = Q_2 = 100$; $f_r = 10$ MHz ; $k = 0.02$

The coefficient of coupling for critical coupling is

$$k_{critical} = \frac{1}{\sqrt{Q_1 Q_2}} = \frac{1}{\sqrt{100 \times 100}} = \mathbf{0.01}$$

Since the given k (= 0.02) is greater than $k_{critical}$ (= 0.01), the circuit is **overcoupled.**

In order that the given k (= 0.02) corresponds to critical coupling, we are to change Q_2 (quality factor of secondary tuned circuit) while maintaining the Q_1 at 100.

$$\therefore \quad 0.02 = \frac{1}{\sqrt{Q_1 Q_2}} = \frac{1}{\sqrt{100 \times Q_2}}$$

$$\therefore \quad \sqrt{100 \times Q_2} = \frac{1}{0.02} = 50$$

or
$$Q_2 = \frac{(50)^2}{100} = 25$$

Example 13.15. *A double-tuned circuit has a primary and secondary Q of 80. The coefficient of coupling is 0.015. Find the degree of coupling. Does the frequency response have a single-peak or double-peak shape ?*

Solution. Fig. 13.19 shows the frequency response curve of a double-tuned circuit for various degrees of coupling. The frequency response curve shows how the secondary current (I_S) varies with frequency (f) for different values of kQ. If $kQ = 1$, this condition is known as *critical coupling*. At the critical coupling, the secondary current (I_S) shows a single maximum at the resonant frequency f_r. We say that the frequency response curve has a single-peak shape. If $kQ < 1$, this condition is called *undercoupling*. With undercoupling, there is also only one maximum (single-peak curve) at the resonant frequency f_r. If $kQ > 1$, this condition is called *overcoupling*. With overcoupling, there are two resonant current peaks. In other words, the frequency response curve for overcoupling has double-peak shape as shown in Fig. 13.19.

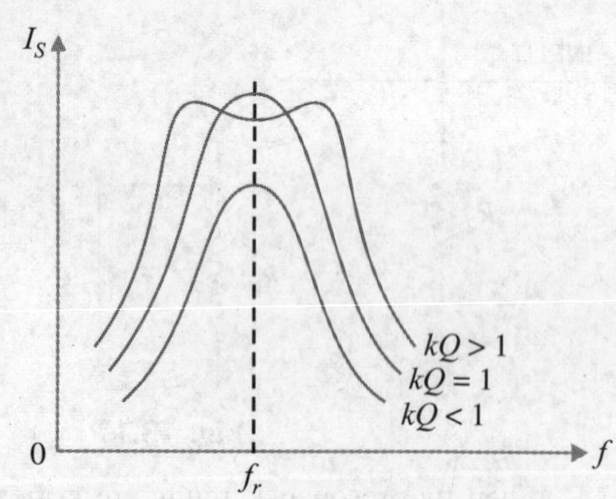

Fig. 13.19

In the given problem, $k = 0.015$ and $Q = 80$.

$$\therefore \quad kQ = 0.015 \times 80 = 1.2$$

Since $kQ > 1$, the circuit is **overcoupled**. As the circuit is overcoupled, therefore, the frequency response curve has **double-peak shape.**

13.17 Tuned Class C Amplifier

So far we have confined our attention to tuned class *A* amplifiers. Such amplifiers are used where *RF* signal has low power level *e.g.* in radio receivers, small signal applications in transmitters. However, owing to low efficiency of class *A* operation, these amplifiers are not employed where large *RF* (radio frequency) power is involved *e.g.* to excite transmitting antenna. In such situations, tuned class *C* power amplifiers are used. Since a class *C* amplifier has a very high efficiency, it can deliver more load power than a class *A* amplifier.

Class *C* operation means that collector current flows for less than 180°. In a practical tuned class *C* amplifier, the collector current flows for much less than 180°; the current looks like narrow pulses as shown in Fig. 13.20. As we shall see later, when narrow current pulses like these drive a high-*Q* resonant (*i.e. LC*) circuit, the voltage across the circuit is almost a perfect sine wave. One very important advantage of class *C* operation is its *high efficiency. Thus 10 W supplied to a class *A* amplifier may produce only about 3.5 W of a.c. output (35% efficiency). The same transistor biased to class *C* may be able to produce 7 W output (70% efficiency). Class *C* power amplifiers normally use *RF* power transistors. The power ratings of such transistors range from 1 W to over 100 W.

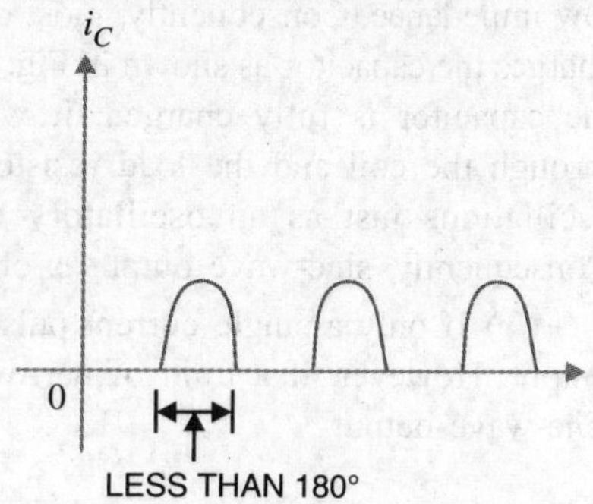

Fig. 13.20

13.18 Class C Operation

Fig. 13.21 (*i*) shows the circuit of tuned class *C* amplifier. The circuit action is as under :

(*i*) When no a.c. input signal is applied, no collector current flows because the emitter diode (*i.e.* base-emitter junction) is unbiased.

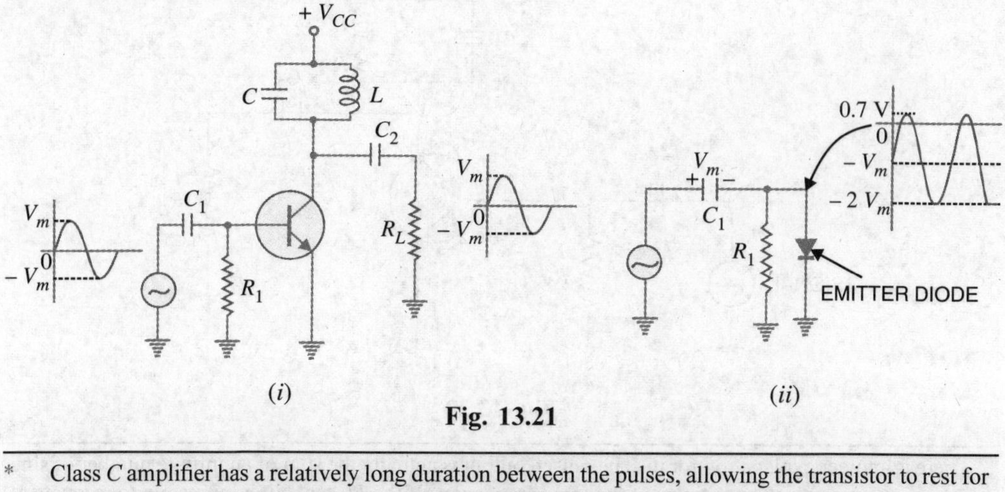

Fig. 13.21

* Class *C* amplifier has a relatively long duration between the pulses, allowing the transistor to rest for a major portion of each input cycle. In other words, very little power is dissipated by the transistor. For this reason, class *C* amplifier has high efficiency.

(*ii*) When an a.c. signal is applied, *clamping action* takes place as shown in Fig. 13.21 (*ii*). The voltage across the emitter diode varies between +0.7 V (during positive peaks of input signal) to about $-2V_m$ (during negative peaks of input signal). This means that conduction of the transistor occurs only for a short period during positive peaks of the signal. This results in the pulsed output *i.e.* collector current waveform is a train of narrow pulses (Refer back to Fig. 13.20).

(*iii*) When this pulsed output is fed to the *LC* circuit, *sine-wave output is obtained. This can be easily explained. Since the pulse is narrow, inductor looks like high impedance and the capacitor like a low impedance. Consequently, most of the current charges the capacitor as shown in Fig. 13.22. When the capacitor is fully charged, it will discharge through the coil and the load resistor, setting up oscillations just as an oscillatory circuit does. Consequently, sine-wave output is obtained.

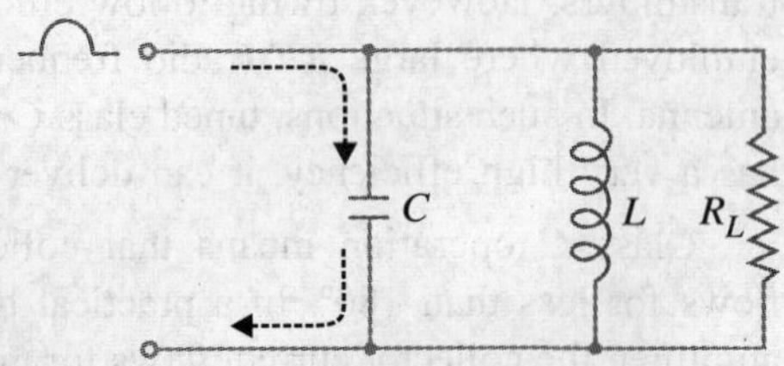

Fig. 13.22

(*iv*) If only a single current pulse drives the *LC* circuit, we will get damped sine-wave output. However, if a train of narrow pulses drive the *LC* circuit, we shall get undamped sine-wave output.

13.19 D.C. and A.C. Loads

Fig. 13.23 shows the circuit of tuned class *C* amplifier. We shall determine the d.c. and a.c. loads of the circuit.

(*i*) The d.c. load of the circuit is just the d.c. resistance *R* of the inductor because the capacitor looks like an open to d.c.

$\therefore$ D.C. load, R_{dc} = D.C. resistance of the inductor = R

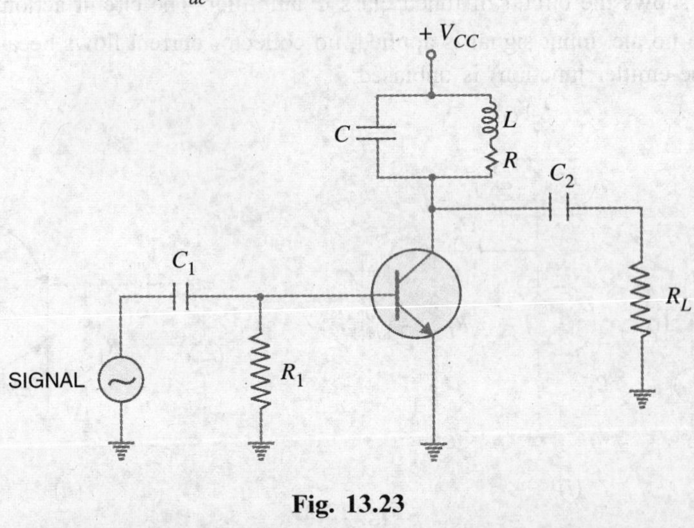

Fig. 13.23

* There is another explanation for it. The pulsed output is actually the sum of an infinite number of sine waves at frequencies in multiples of the input frequency. If the *LC* tank circuit is set up to resonate at the input frequency, it will result in sine-wave output of just the input frequency.

(ii) The a.c. load is a parallel combination of capacitor, coil and load resistance R_L as shown in Fig. 13.24 (*i*). The series resistance R of the inductor can be replaced by its equivalent parallel resistance R_P as shown in Fig. 13.24 (*ii*) where

$$R_P = Q_{coil} \times X_L$$

The a.c. load resistance R_{AC} is the equivalent resistance of the parallel combination of R_P and R_L *i.e.*

$$R_{AC} = R_P \parallel R_L = \frac{R_P \times R_L}{R_P + R_L}$$

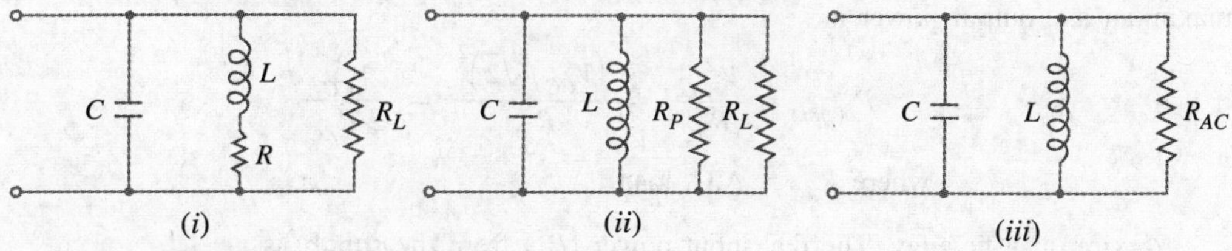

Fig. 13.24

Example 13.16. *In the circuit shown in Fig.13.23, C = 500 pF and the coil has L = 50.7 μH and R = 10 Ω and R_L = 1 MΩ. Find (i) the resonant frequency (ii) d.c. load and a.c. load.*

Solution.

(i) Resonant frequency, $f_r \simeq \dfrac{1}{2\pi\sqrt{LC}} = \dfrac{10^9}{2\pi\sqrt{50.7 \times 500}} = \mathbf{10^6\ Hz}$

(ii) D.C. load, $R_{dc} = R = \mathbf{10\ \Omega}$

$$X_L = 2\pi f_r L = 2\pi \times (10^6) \times (50.7 \times 10^{-6}) = 318\ \Omega$$

$$Q_{coil} = \frac{X_L}{R} = \frac{318}{10} = 31.8$$

The series resistance R (= 10 Ω) of the inductor can be replaced by its equivalent parallel resistance R_P where,

$$R_P = Q_{coil} \times X_L = 31.8 \times 318 = 10^4\ \Omega = 10\ k\Omega$$

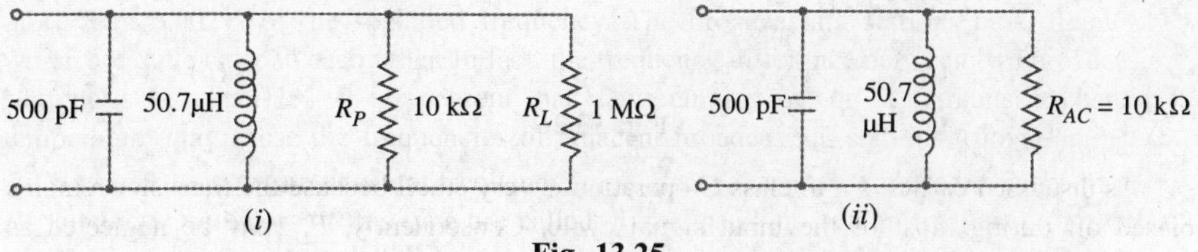

Fig. 13.25

The equivalent circuit is shown in Fig. 13.25 (*i*). This further reduces to the circuit shown in Fig. 13.25 (*ii*).

$$\therefore \quad R_{AC} = R_P \parallel R_L = 10\ k\Omega \parallel 1\ M\Omega \simeq \mathbf{10\ k\Omega}$$

13.20 Maximum A.C. Output Power

Fig. 13.26 (*i*) shows tuned class *C* amplifier. When no signal is applied, the collector-emitter voltage is *V_{CC} *i.e.*

$$v_{CE} = V_{CC}$$

When signal is applied, it causes the total collector-emitter voltage to swing above and below this voltage. The collector-emitter voltage can have a maximum value of $2V_{CC}$ and minimum value 0 (ideally) as shown in Fig. 13.26 (*ii*).

Referring to Fig. 13.26 (*ii*), output voltage has a peak value of V_{CC}. Therefore, the maximum a.c. output power is :

$$P_{o(max)} = \frac{V_{r.m.s.}^2}{R_{AC}} = \frac{\left(V_{CC}/\sqrt{2}\right)^2}{R_{AC}} = \frac{V_{CC}^2}{2\,R_{AC}}$$

where R_{AC} = A.C. load

Maximum efficiency. The d.c. input power (P_{dc}) from the supply is :

$$P_{dc} = P_{o(max)} + P_D$$

where P_D = Power dissipation of the transistor

$$\therefore \quad \text{Max. collector } \eta = \frac{P_{o(max)}}{P_{o(max)} + P_D}$$

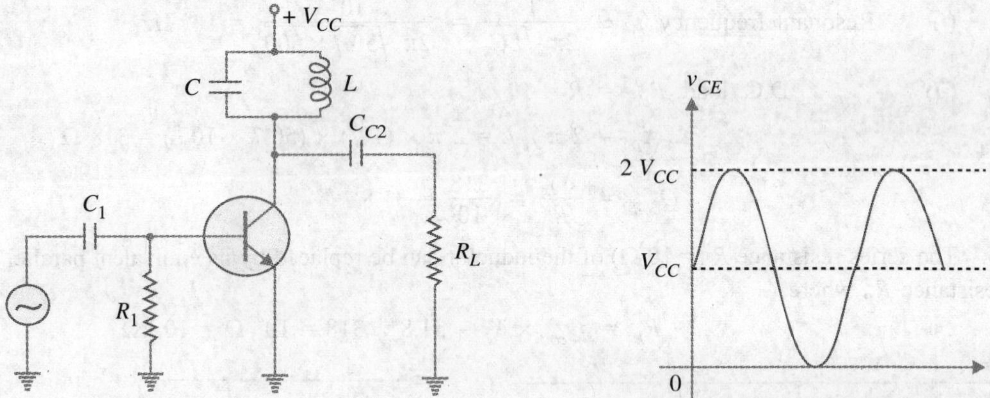

Fig. 13.26

As discussed earlier, P_D in class *C* operation is very small because the transistor remains biased off during most of the input signal cycle. Consequently, P_D may be neglected as compared to $P_{o(max)}$.

$$\therefore \quad \text{Maximum } \eta \simeq \frac{P_{o(max)}}{P_{o(max)}} \simeq 100\%$$

It is worthwhile to give a passing reference about the maximum efficiencies of class *A*, class *B* and class *C* amplifiers. A class *A* amplifier (transformer-coupled) has a maximum

* Because the drop in *L* due to d.c. component is negligible.

efficiency of 50%, class B of 78.5% and class C nearly 100%. It is emphasised here that class C operation is suitable only for *resonant RF applications.

Example 13.17. *Calculate (i) a.c. load (ii) maximum load power in the circuit shown in Fig. 13.27.*

Solution.

(*i*) A.C. load, R_{AC} = Reflected load resistance seen by the collector

$$= (N_P/N_S)^2 \times 50 = (5/1)^2 \times 50 = \mathbf{1250\ \Omega}$$

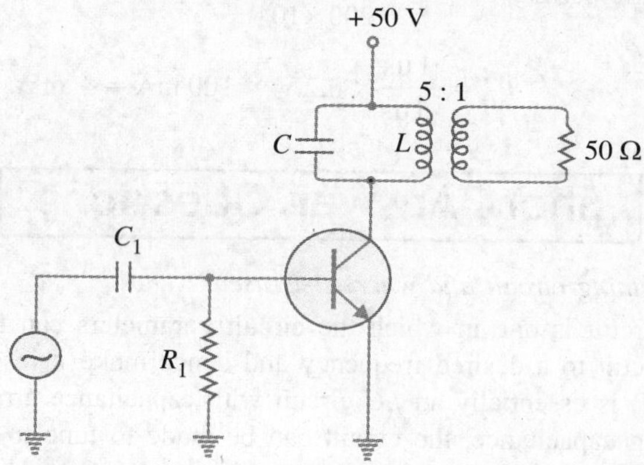

Fig. 13.27

(*ii*) Max. load power, $P_{o(max)} = \dfrac{V_{CC}^2}{2R_{AC}} = \dfrac{(50)^2}{2\times 1250} = \mathbf{1\ W}$

Example 13.18. *In the above example, if power dissipation of the transistor is 4 mW, find the maximum collector efficiency.*

Solution. $P_{o\,(max)} = 1\text{ W}\ ;\ P_D = 4\text{ mW} = 4 \times 10^{-3}\text{ W}$

$\therefore$ Max. collector $\eta = \dfrac{P_{o\,(max)}}{P_{o(max)} + P_D} \times 100$

$$= \frac{1}{1 + 4\times 10^{-3}} \times 100 = \mathbf{99.6\%}$$

Note that maximum collector efficiency is very close to the ideal case of 100% efficiency. Therefore, we can neglect P_D in circuit calculations with reasonable accuracy.

Example 13.19. *A class C amplifier is driven by a 200 kHz signal. The transistor is on for 1 μs and the amplifier is operating over the full range of its load line. If $I_{C(sat)}$ = 100 mA and $V_{CE(sat)}$ = 0.2V, what is the average power dissipation ?*

Solution. In class C amplifier, collector current flows in the form of pulses while the transistor is on. Since the output swings over the entire load line, the collector current and collector-emitter voltage during the time the transistor is on are $I_{C(sat)}$ and $V_{CE(sat)}$ respectively. Therefore, power dissipation during *on time* is

$$P_{D(on)} = V_{CE(sat)}\ I_{C(sat)}$$

* Because power losses are very small (less than 1%) in high - Q resonant circuits. An extremely narrow pulse will compensate the losses.

The transistor is *on* for a short time t_{on} and *off* for the rest of the input cycle. Therefore, average power dissipated during the entire cycle is

$$P_D = \frac{t_{on}}{T} \times V_{CE(sat)} I_{C(sat)}$$

where T = Time period of input signal

In the given problem, f = 200 kHz ; t_{on} = 1 μs.

$$\text{Time period, } T = \frac{1}{f} = \frac{1}{200 \times 10^3} \text{s} = 5 \text{ μs}$$

$$\therefore \quad P_D = \frac{1 \text{ μs}}{5 \text{ μs}} \times 0.2 \text{ V} \times 100 \text{ mA} = \mathbf{4 \text{ mW}}$$

Short Answer Questions

Q.1. *What is a tuning circuit and where it is used?*

Ans. A tuning circuit is one in which the circuit parameters can be varied in order to tune the circuit to a desired frequency and hence make it resonant to that specific frequency. It is essentially an *LC* circuit with capacitance arranged to be variable. By changing capacitance, the circuit can be made to tune to a specific frequency. Tuned circuits are used in communication systems. For example, radio and television receivers use tuned circuits to select one radio signal from the many which are being broadcast. In other fields of electronics, tuned circuits have many applications, some of which are quality control, thickness control, capacitor relay alarms and data processing.

Q.2. *What are the three functions performed by a tuning circuit?*

Ans. A tuning circuit performs the following three functions :

(*i*) Selecting the desired signal.

(*ii*) Reproducing with satisfactory volume the weak signals received by the antenna.

(*iii*) Reproducing faithfully all the frequencies present in the original signal.

Q.3. *What are the characteristics of a parallel tuned circuit?*

Ans. A parallel tuned circuit essentially consists of L and C in parallel. The resonant frequency of the circuit is approximately given by ;

$$f_r = \frac{1}{2\pi\sqrt{LC}} \quad \text{... neglecting resistance}$$

(*i*) At the resonant frequency, the impedance of the circuit is maximum and its value is equal to the product of either reactance and circuit Q.

(*ii*) At the resonant frequency, the line current is minimum and is equal to the applied voltage divided by the impedance of the circuit.

(*iii*) The currents in the inductor and capacitor are approximately 180° out of phase with each other at resonance.

Q.4. *What do you understand by Q of a tuned circuit?*

Ans. The ratio of inductive reactance (X_L) of the tuned circuit to the resistance (R) of the circuit is called Q factor (Q stands for quality) or figure of merit *i.e.*

$$Q = \frac{X_L}{R}$$

The quality factor Q of a tuned circuit is very important because the sharpness of resonance curve and hence selectivity of the circuit depends upon it. The greater the value of Q of tuned circuit, the greater is the selectivity of the circuit. Obviously, Q of the tuned circuit will be large if R is small.

Q.5. *How does a resonant (or tuned) circuit select a particular radio wave?*

Ans. Fig. 13.28 shows how a resonant *LC* circuit selects the desired radio wave in a radio receiver. The receiving antenna intercepts many radio waves of different frequencies. The resonant frequency of the parallel *LC* circuit depends upon the values of *L* and *C*. Suppose $L = 127\ \mu H$ and variable capacitor C is set at 200 pF. For these particular values of *L* and *C*, there is only one frequency (called resonant frequency f_r) for which this *LC* circuit will be in resonance.

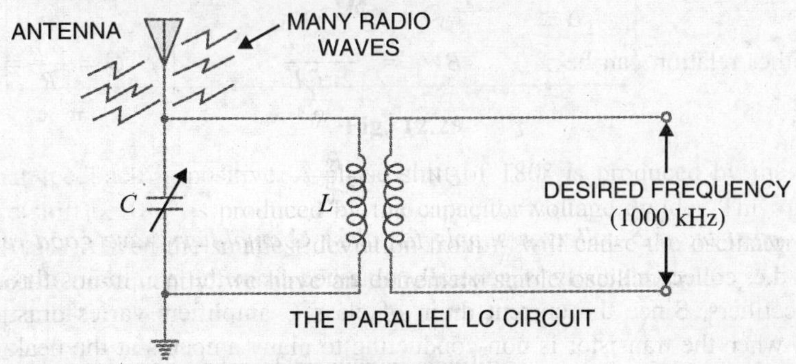

Fig. 13.28

Resonant frequency is

$$f_r = \frac{1}{2\pi\sqrt{LC}} = \frac{1}{2\pi\sqrt{(127\times10^{-6})\times(200\times10^{-12})}}\ \text{Hz}$$

$$= 1000 \times 10^3\ \text{Hz} = 1000\ \text{kHz}$$

The parallel *LC* circuit offers maximum impedance at resonant frequency. Consequently, this circuit will offer maximum impedance to a radio wave of frequency 1000 kHz. For all other frequencies intercepted by the antenna, the impedance of the *LC* circuit is substantially lower. Therefore, the resonant circuit will retain radio wave of 1000 kHz ; while other frequencies will be rejected.

Q.6. *Why is a resonant circuit called a tuned circuit?*

Ans. In a resonant circuit, variation of *C* is commonly used to "tune" the circuit to select the desired frequency when many frequencies are present. For this reason, a resonant circuit is called a tuned circuit.

Q.7. *Why is Q of a coil small at high frequencies?*

Ans. The Q of a coil is given by ;

$$Q = \frac{X_L}{R} = \frac{2\pi f_r L}{R}$$

At higher frequencies, the electrons flowing in the coil travel near the surface of the coil. Thus, only a small portion of the conductor is carrying current. The lessening of the usable cross-sectional area results in an effective higher resistance of the same coil to higher frequency a.c. This increased resistance, known as *skin effect*, lowers the Q of the coil.

Q.8. *Why are tuned amplifiers used for high frequency (RF) amplification?*

Ans. In high frequency applications, it is generally required to amplify a single frequency, rejecting all other frequencies. For this purpose, tuned amplifiers are used. These amplifiers use tuned circuits as the collector load and provide the desired selective property.

Q.9. *What is the relationship among bandwidth BW, resonant frequency f_r and Q of a tuned circuit?*

Ans. The relationship among bandwidth (BW), resonant frequency (f_r) and Q of a tuned circuit is :

$$BW = \frac{f_r}{Q} \quad \text{... (i)}$$

Another relation can be :

$$BW = \frac{f_r}{\frac{2\pi f_r L}{R}} \quad \left(\because \; Q = \frac{X_L}{R} = \frac{2\pi f_r L}{R}\right)$$

or

$$BW = \frac{R}{2\pi L} \quad \text{... (ii)}$$

Q.10. *Why must the d.c. collector supply for class C amplifiers have good regulation?*

Ans. The d.c. collector supply is generally obtained from the a.c. mains through the use of rectifiers. Since the current drain of class C amplifiers varies drastically (from zero when the transistor is non-conducting to many amperes at the peak of collector current pulses), the power supply must be capable of high peak currents and must have good regulation. The use of silicon rectifiers and choke-input filters generally satisfies this condition.

Q.11. *Why are RF (radio frequency) amplifiers operated in class C mode?*

Ans. Most commercial *RF* applications require tremendous power. For example, radio and television transmitters may have output level as high as 50 kW. When dealing with such high power levels, efficiency of operation is of prime importance. Since class C operation has high efficiency, it is always used in *RF* applications. Although this type of operation causes severe distortion, the use of resonant circuit loads results in excellent output waveform.

Multiple-Choice Questions

1. A tuned amplifier uses .... load.
   (*i*) resistive (*ii*) capacitive
   (*iii*) *LC* tank (*iv*) inductive
2. A tuned amplifier is generally operated in .... operation.
   (*i*) class *A*
   (*ii*) class *C*
   (*iii*) class *B*
   (*iv*) none of the above
3. A tuned amplifier is used in .... applications.
   (*i*) radio frequency
   (*ii*) low frequency
   (*iii*) audio frequency
   (*iv*) none of the above
4. Frequencies above .... kHz are called radio frequencies.
   (*i*) 2 (*ii*) 10
   (*iii*) 50 (*iv*) 200

5. At series or parallel resonance, the circuit power factor is ....
   (i) 0 (ii) 0.5
   (iii) 1 (iv) 0.8
6. The voltage gain of a tuned amplifier is .... at resonant frequency.
   (i) minimum (ii) maximum
   (iii) half-way between maximum and minimum
   (iv) zero
7. At parallel resonance, the line current is ....
   (i) minimum (ii) maximum
   (iii) quite large
   (iv) none of the above
8. At series resonance, the circuit offers .... impedance.
   (i) zero (ii) maximum
   (iii) minimum
   (iv) none of the above
9. A resonant circuit contains .... elements.
   (i) *R* and *L* only (ii) *R* and *C* only
   (iii) only *R* (iv) *L* and *C*
10. At series or parallel resonance, the circuit behaves as a .... load.
   (i) capacitive (ii) resistive
   (iii) inductive
   (iv) none of the above
11. At series resonance, voltage across *L* is .... voltage across *C*.
   (i) equal to but opposite in phase to
   (ii) equal to but in phase with
   (iii) greater than but in phase with
   (iv) less than but in phase with
12. When either *L* or *C* is increased, the resonant frequency of *LC* circuit ....
   (i) remains the same
   (ii) increases
   (iii) decreases
   (iv) insufficient data
13. At parallel resonance, the net reactive component of circuit current is ....
   (i) capacitive (ii) zero
   (iii) inductive
   (iv) none of the above
14. At parallel resonance, the circuit impedance is ....
   (i) $\frac{C}{LR}$ (ii) $\frac{R}{LC}$
   (iii) $\frac{CR}{L}$ (iv) $\frac{L}{CR}$
15. In a parallel *LC* circuit, if the input signal frequency is increased above resonant frequency, then ....
   (i) X_L increases and X_C decreases
   (ii) X_L decreases and X_C increases
   (iii) both X_L and X_C increase
   (iv) both X_L and X_C decrease
16. The *Q* of an *LC* circuit is given by ....
   (i) $2\pi f_r \times R$ (ii) $\frac{R}{2\pi f_r L}$
   (iii) $\frac{2\pi f_r L}{R}$ (iv) $\frac{R^2}{2\pi f_r L}$
17. If *Q* of an *LC* circuit increases, then bandwidth ....
   (i) increases (ii) decreases
   (iii) remains the same
   (iv) insufficient data
18. At series resonance, the net reactive component of circuit current is ....
   (i) zero (ii) inductive
   (iii) capacitive
   (iv) none of the above
19. The dimensions of *L/CR* are that of ....
   (i) farad (ii) henry
   (iii) ohm
   (iv) none of the above
20. If *L/C* ratio of a parallel *LC* circuit is increased, the *Q* of the circuit ....
   (i) is decreased (ii) is increased
   (iii) remains the same
   (iv) none of the above
21. At series resonance, the phase angle between applied voltage and circuit current is ....
   (i) 90° (ii) 180°
   (iii) 0°
   (iv) none of the above

22. At parallel resonance, the ratio L/C is ....
   (*i*) very large (*ii*) zero
   (*iii*) small
   (*iv*) none of the above
23. If the resistance of a tuned circuit is increased, the Q of the circuit ....
   (*i*) is increased (*ii*) is decreased
   (*iii*) remains the same
   (*iv*) none of the above
24. The Q of a tuned circuit refers to the property of ....
   (*i*) sensitivity (*ii*) fidelity
   (*iii*) selectivity
   (*iv*) none of the above
25. At parallel resonance, the phase angle between the applied voltage and circuit current is ....
   (*i*) 90° (*ii*) 180°
   (*iii*) 0°
   (*iv*) none of the above
26. In a parallel LC circuit, if the signal frequency is decreased below the resonant frequency, then ....
   (*i*) X_L decreases and X_C increases
   (*ii*) X_L increases and X_C decreases
   (*iii*) line current becomes minimum
   (*iv*) none of the above
27. In series resonance, there is ....
   (*i*) voltage amplification
   (*ii*) current amplification
   (*iii*) both voltage and current amplification
   (*iv*) none of the above
28. The Q of a tuned amplifier is generally ....
   (*i*) less than 5 (*ii*) less than 10
   (*iii*) more than 10
   (*iv*) none of the above
29. The Q of a tuned amplifier is 50. If the resonant frequency for the amplifier is 1000 kHz, then bandwidth is ....
   (*i*) 10 kHz (*ii*) 40 kHz
   (*iii*) 30 kHz (*iv*) 20 kHz
30. In the above question, what are the values of cut-off frequencies?
   (*i*) 140 kHz, 60 kHz
   (*ii*) 1020 kHz, 980 kHz
   (*iii*) 1030 kHz, 970 kHz
   (*iv*) none of the above
31. For frequencies above the resonant frequency, a parallel LC circuit behaves as a .... load.
   (*i*) capacitive (*ii*) resistive
   (*iii*) inductive
   (*iv*) none of the above
32. In parallel resonance, there is ....
   (*i*) both voltage and current amplification
   (*ii*) voltage amplification
   (*iii*) current amplification
   (*iv*) none of the above
33. For frequencies below resonant frequency, a series LC circuit behaves as a .... load.
   (*i*) resistive (*ii*) capacitive
   (*iii*) inductive
   (*iv*) none of the above
34. If a high degree of selectivity is desired, then double-tuned circuit should have .... coupling.
   (*i*) loose (*ii*) tight
   (*iii*) critical
   (*iv*) none of the above
35. In the double tuned circuit, if the mutual inductance between the two tuned circuits is decreased, the level of resonance curve ....
   (*i*) remains the same
   (*ii*) is lowered (*iii*) is raised
   (*iv*) none of the above
36. For frequencies above the resonant frequency, a series LC circuit behaves as a .... load.
   (*i*) resistive (*ii*) inductive
   (*iii*) capacitive
   (*iv*) none of the above
37. Double tuned circuits are used in .... stages of a radio receiver
   (*i*) *IF* (*ii*) audio
   (*iii*) output
   (*iv*) none of the above

38. A class C amplifier always drives .... load.
   (*i*) a pure resistive (*ii*) a pure inductive
   (*iii*) a pure capacitive
   (*iv*) a resonant tank

39. Tuned class C amplifiers are used for *RF* signals of ....
   (*i*) low power
   (*ii*) high power
   (*iii*) very low power
   (*iv*) none of the above

40. For frequencies below the resonant frequency, a parallel *LC* circuit behaves as a .... load.
   (*i*) inductive
   (*ii*) resistive
   (*iii*) capacitive
   (*iv*) none of the above

Answers to Multiple-Choice Questions

1. (*iii*)	2. (*ii*)	3. (*i*)	4. (*iv*)	5. (*iii*)
6. (*ii*)	7. (*i*)	8. (*iii*)	9. (*iv*)	10. (*ii*)
11. (*i*)	12. (*iii*)	13. (*ii*)	14. (*iv*)	15. (*i*)
16. (*iii*)	17. (*ii*)	18. (*i*)	19. (*iii*)	20. (*ii*)
21. (*iii*)	22. (*i*)	23. (*ii*)	24. (*iii*)	25. (*iii*)
26. (*i*)	27. (*i*)	28. (*iii*)	29. (*iv*)	30. (*ii*)
31. (*i*)	32. (*iii*)	33. (*ii*)	34. (*i*)	35. (*iii*)
36. (*ii*)	37. (*i*)	38. (*iv*)	39. (*iv*)	40. (*i*)

Chapter Review Topics

1. What are tuned amplifiers and where are they used?
2. Discuss parallel tuned circuit with special reference to resonant frequency, circuit impedance and frequency response.
3. What do you understand by quality factor Q of parallel tuned circuit?
4. Discuss the advantages of tuned amplifiers.
5. Discuss the circuit operation of a single tuned amplifier.
6. Write short notes on the following:
   (*i*) Double tuned amplifier (*ii*) Bandwidth of tuned amplifier

Problems

1. A parallel circuit has a capacitor of 100 pF in one branch and an inductance of 100 μH plus a resistance of 10 Ω in the second branch. The line voltage is 100 V. Find (*i*) resonant frequency (*ii*) circuit impedance at resonance and (*iii*) line current at resonance. **[(*i*) 1590 kHz (*ii*) 100 kΩ (*iii*) 100 mA]**
2. A tuned amplifier is designed to have a resonant frequency of 1000 kHz and a bandwidth of 40 kHz. What is the Q of this amplifier? **[25]**
3. The Q of a tuned amplifier is 25. If the resonant frequency of the circuit is 1400 kHz, what is its bandwidth? **[56 kHz]**
4. A tuned amplifier has parallel *LC* circuit. One branch of this parallel circuit has a capacitor of 100 pF and the other branch has an inductance of 1 mH plus a resistance of 25 Ω. Determine (*i*) the resonant frequency and (*ii*) Q of the tank circuit. **[(*i*) 503.3 kHz (*ii*) 126.5]**
5. It is desired to obtain a bandwidth of 12 kHz at an operating frequency of 800 kHz, using a double-tuned circuit. What value of co-efficient of coupling should be used? **[0.015]**

CHAPTER 14

Modulation and Demodulation

Learning Objectives

- Understand the underlying principle of radio broadcasting, transmission and reception.
- Learn the need of modulation for radio and TV transmission.
- Develop an understanding of the characteristics of amplitude modulation.
- Explore the importance of modulation factor in amplitude modulation.
- Mathematical analysis of amplitude modulated wave.
- Understand the limitations of amplitude modulation.
- Develop a strong understanding of the various types of radio wave propagation.
- Understand the various aspects of frequency modulation.
- Develop an understanding of demodulation.
- Understand the operation of the various stages of superheterodyne radio receiver.

Introduction

In radio transmission, it is necessary to send audio signal (*e.g.* music, speech *etc.*) from a broadcasting station over great distances to a receiver. This communication of audio signal does not employ any wire and is sometimes called *wireless*. The audio signal cannot be sent directly over the air for appreciable distance. Even if the audio signal is converted into electrical signal, the latter cannot be sent very far without employing large amount of power. The energy of a wave is directly proportional to its frequency. At audio frequencies (20 Hz to 20 kHz), the signal power is quite small and radiation is not practicable.

The radiation of electrical energy is practicable only at high frequencies *e.g.* above 20 kHz. The high frequency signals can be sent thousands of miles even with comparatively small power. Therefore, if audio signal is to be transmitted properly, some means must be devised which will permit transmission to occur at high frequencies while it simultaneously allows the carrying of audio signal. This is achieved by superimposing electrical audio signal on high frequency carrier. The resultant waves are known as *modulated waves* or *radio waves* and the process is called *modulation*. At the radio receiver, the audio signal is extracted from the modulated wave by the process called *demodulation*. The signal is then amplified and reproduced into sound by the loudspeaker. In this chapter, we shall focus our attention on the various aspects of modulation and demodulation.

14.1 Radio Broadcasting, Transmission and Reception

Radio communication means the radiation of radio waves by the transmitting station, the propagation of these waves through space and their reception by the radio receiver. Fig. 14.1 shows the general principles of radio broadcasting, transmission and reception. As a matter of convenience, the entire arrangement can be divided into three parts *viz. transmitter, transmission of radio waves* and *radio receiver*.

1. Transmitter. Transmitter is an extremely important equipment and is housed in the broadcasting station. Its purpose is to produce radio waves for transmission into space. The important components of a transmitter are microphone, audio amplifiers, oscillator and modulator (See Fig. 14.1).

(*i*) **Microphone.** A microphone is a device which converts sound waves into electrical waves. When the speaker speaks or a musical instrument is played, the varying air pressure on the microphone generates an audio electrical signal which corresponds in frequency to the original signal. The output of microphone is fed to a multistage audio amplifier for raising the strength of weak signal.

(*ii*) **Audio amplifier.** The audio signal from the microphone is quite weak and requires amplification. This job is accomplished by cascaded audio amplifiers. The amplified output from the last audio amplifier is fed to the modulator for rendering the process of modulation.

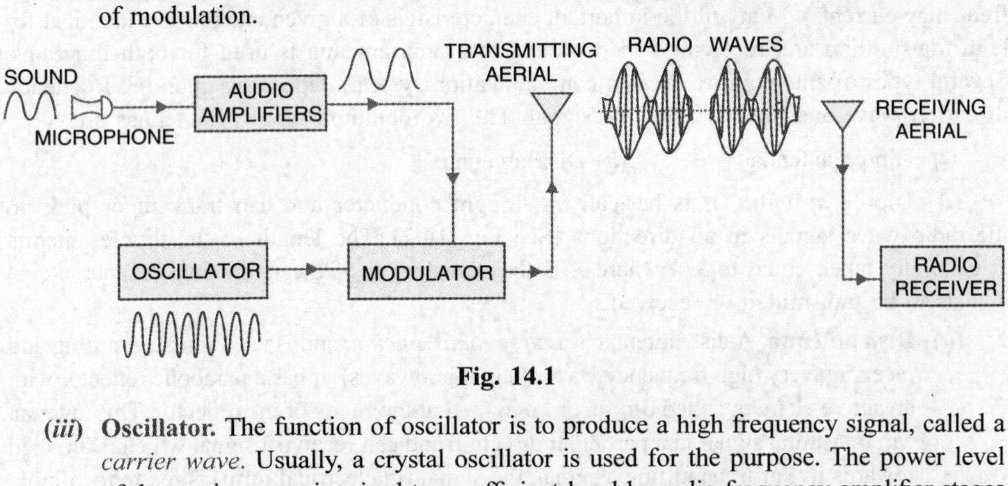

Fig. 14.1

(*iii*) **Oscillator.** The function of oscillator is to produce a high frequency signal, called a *carrier wave*. Usually, a crystal oscillator is used for the purpose. The power level of the carrier wave is raised to a sufficient level by radio frequency amplifier stages (not shown in Fig. 14.1). Most of the broadcasting stations have carrier wave power of several kilowatts. Such high power is necessary for transmitting the signal to the required distances.

(*iv*) **Modulator.** The amplified audio signal and carrier wave are fed to the modulator. Here, the audio signal is superimposed on the carrier wave in a suitable manner. The resultant waves are called *modulated waves or radio waves* and the process is called *modulation*. The process of modulation permits the transmission of audio signal at the carrier frequency. As the carrier frequency is very high, therefore, the audio signal can be transmitted to large distances. The radio waves from the transmitter are fed to the transmitting antenna or aerial from where these are radiated into space.

2. Transmission of radio waves. The transmitting antenna radiates the radio waves in space in all directions. These radio waves travel with the velocity of light *i.e.* 3×10^8 m/sec. The radio waves are electromagnetic waves and possess the same general properties. These are similar to light and heat waves except that they have longer wavelengths. It may be emphasised here that radio waves are sent without employing any wire. It can be easily shown that at high frequency, electrical energy can be radiated into space.

3. Radio receiver. On reaching the receiving antenna, the radio waves induce tiny e.m.f. in it. This small voltage is fed to the radio receiver. Here, the radio waves are first amplified and then signal is extracted from them by the process of *demodulation*. The signal is amplified by audio amplifiers and then fed to the speaker for reproduction into sound waves.

14.2 Antenna or Aerial

***An antenna or aerial** is a metallic structure (e.g. a wire, group of wires etc.) that is used to radiate or receive radio waves (i.e. high frequency electromagnetic waves).*

In communication system, an antenna (or aerial) plays an important role and is used for both transmitting and receiving the radio waves or electromagnetic waves. Its main function is to convert high frequency currents into electromagnetic waves and *vice-versa*. Thus a transmitting antenna converts high frequency currents (electrical energy) into electromagnetic waves while a receiving antenna converts electromagnetic waves into electrical energy (high frequency currents). Many of the important characteristics of a given antenna are identical for both transmitting and receiving functions and the same antenna is used for both functions. Several types of antennas are used in communication system, depending upon the frequency of carrier wave and directivity of the signal. The two common types of antennas are :

(*i*) Dipole antenna (*ii*) Dish antenna

(*i*) **Dipole antenna.** It is basically a straight conductor and can transmit or pick up the radio wave signals in all directions (See Fig. 14.2). The length of the dipole antenna is generally taken equal to $\lambda/4$ where λ is the wavelength of the radiowave (carrier wave) signals to be transmitted or received.

(*ii*) **Dish antenna.** A dish antenna is a *directional antenna* and is used for transmitting and receiving very high frequency waves (*e.g.* microwaves). It has a parabolic reflector with an active element called **dipole** or **horn feed** at the **focus** of the reflector. This antenna can transmit a signal in a particular direction and can receive a signal which is directed towards it. For transmitting a signal, the signal (in electrical form) is fed to the dipole which converts the signal into electromagnetic waves and directs them to the reflector. The reflector then transmits the signal in the form of a parallel beam as shown in Fig. 14.3.

For reception, the electromagnetic waves directed towards the dish are focussed on the dipole which converts them into electrical signals. The electrical signal is fed to the receiver. The dish antennas are used in radars and satellite communication.

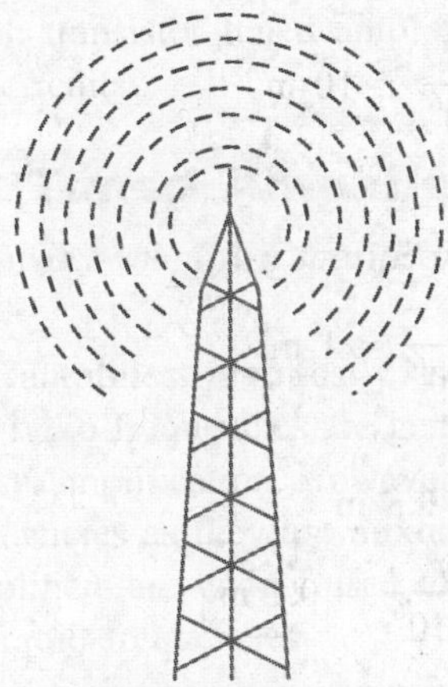

Fig. 14.2

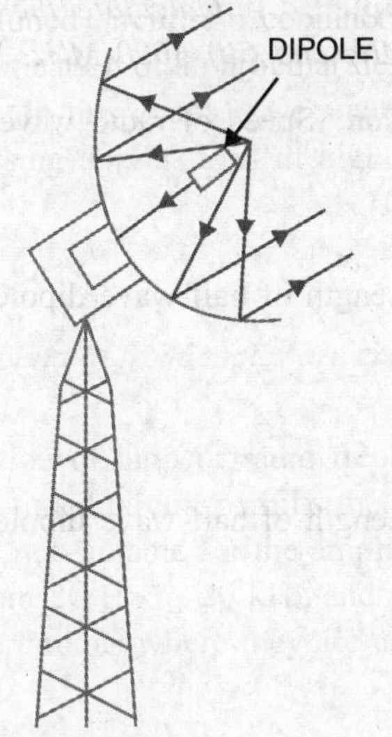

Fig. 14.3

Discussion. The following points may be noted about the antennas :

1. The two fundamental types of antennas are : **Hertz antennas** and **Marconi antennas.** All other types may be considered as derivatives of these two fundamental types :

(*i*) A Hertz antenna (also known as *half-wave isolated antenna*) is a straight conductor of length equal to half the wavelength of radio signals to be transmitted or received *i.e.* antenna length, $l = \lambda/2$. This antenna is grounded.

(*ii*) A Marconi antenna (also known as *ground quarter-wave antenna*) is a straight conductor of length equal to a quarter of wavelength of radio signals to be transmitted or received *i.e.* antenna length, $l = \lambda/4$. This is placed vertically with its lower end touching the ground. The advantage of Marconi antenna over Hertz antenna is that only half length of antenna is required.

2. The same antenna can be used as a transmitting antenna or receiving antenna. The only difference is that the transmitting antenna is excited by high frequency (*i.e.* radio frequency) currents to produce electromagnetic waves while the receiving antenna is excited by electromagnetic waves to produce high frequency currents (Recall that an electromagnetic wave induces an e.m.f. in a conductor).

3. An antenna or aerial in its simplest form is a vertical conductor of height *h*. We know that a straight conductor has inductance distributed along its length and self-capacitance. Therefore, an antenna has inductance L_A and capacitance C_A. The equivalent circuit of an antenna or aerial is a parallel tuned *LC* circuit as shown in Fig. 14.4. Clearly, the resonant frequency of this parallel tuned circuit depends upon the height *h* and is called *natural frequency* of the antenna. In practice, the value of L_A is very small. In order to obtain resonance at the desired frequency, additional inductance is connected in series with the antenna. The additional inductance does not change the physical length of the antenna but increases the inductance of the antenna significantly.

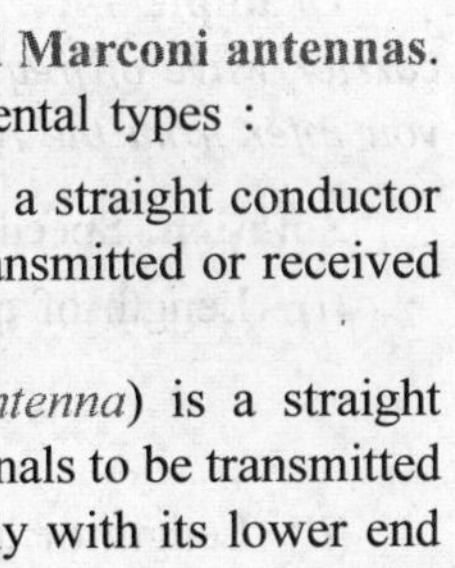

Fig. 14.4

Example 14.1. *Calculate the length of a half-wave dipole antenna at (i) 30 MHz (ii) 300 MHz and (iii) 3000 MHz. What inference do you draw from the results?*

Solution. Speed of radio waves, $c = 3 \times 10^8 \text{ ms}^{-1}$

(*i*) $$\lambda = \frac{c}{f} = \frac{3\times10^8}{30\times10^6} = 10 \text{ m}$$

∴ Length of half-wave dipole antenna is

$$l = \lambda/2 = 10/2 = \mathbf{5\ m}$$

(*ii*) $$\lambda = \frac{c}{f} = \frac{3\times10^8}{300\times10^6} = 1 \text{ m}$$

∴ Length of half-wave dipole antenna is

$$l = \lambda/2 = 1/2 = \mathbf{0.5\ m}$$

(*iii*) $$\lambda = \frac{c}{f} = \frac{3\times10^8}{3000\times10^6} = 0.1 \text{ m}$$

∴ $$l = \lambda/2 = 0.1/2 = \mathbf{0.05\ m}$$

It is clear that length of the dipole antenna decreases as the frequency of carrier wave increases.

Example 14.2. *What would be the length of the quarter-wave dipole antenna for the carrier wave of frequency of (i) 3×10^4 Hz (ii) 6×10^6 Hz and (iii) 5×10^7 Hz? What do you infer from the result?*

Solution. Speed of radio waves, $c = 3 \times 10^8 \text{ ms}^{-1}$

(*i*) Length of quarter-wave dipole antenna is

$$l = \frac{\lambda}{4} = \frac{1}{4}\times\frac{c}{f} = \frac{1}{4}\times\frac{3\times10^8}{3\times10^4} = \mathbf{2.5 \times 10^3\ m}$$

(*ii*) Length of quarter-wave dipole antenna is

$$l = \frac{\lambda}{4} = \frac{1}{4}\times\frac{c}{f} = \frac{1}{4}\times\frac{3\times10^8}{6\times10^6} = \mathbf{12.5\ m}$$

(*iii*) Length of quarter-wave dipole antenna is

$$l = \frac{\lambda}{4} = \frac{1}{4}\times\frac{c}{f} = \frac{1}{4}\times\frac{3\times10^8}{5\times10^7} = \mathbf{1.5\ m}$$

It follows that length of the antenna decreases as the carrier frequency increases.

14.3 Modulation

As discussed earlier, a high frequency carrier wave is used to carry the audio signal. The question arises how the audio signal should be "added" to the carrier wave. The solution lies in changing some characteristic of carrier wave in accordance with the signal. Under such conditions, the audio signal will be contained in the resultant wave. This process is called modulation and may be defined as under :

The process of changing some characteristic (e.g. amplitude, frequency or phase) of a carrier wave in accordance with the intensity of the signal is known as **modulation.**

Modulation means to "change". In modulation, some characteristic of carrier wave is changed in accordance with the intensity (*i.e.* amplitude) of the signal. The resultant wave is called modulated wave or radio wave and contains the audio signal. Therefore, modulation permits the transmission to occur at high frequency while it simultaneously allows the carrying of the audio signal.

Need for modulation. Audio signals have a frequency range from 20 Hz to 20 kHz. These low frequency signals cannot be transmitted directly (*i.e.* without modulation) into space for the following reasons :

(*i*) **Practical antenna length.** For efficient transmission and reception, the transmitting and receiving antennas must have a length atleast equal to $\lambda/4$ where λ is the wavelength of the signal. For an audio signal of 15 kHz, the length of the antenna required will be :

$$\text{Wavelength, } \lambda = \frac{c}{f} = \frac{3\times 10^8}{15\times 10^3} = 20000 \text{ m}$$

$$\therefore \quad \text{Length of antenna, } l = \frac{\lambda}{4} = \frac{20000}{4} = 5000 \text{ m}$$

To set up a vertical antenna of this length is practically impossible.

However, if a carrier wave of 1 MHz is used to carry the signal (*i.e.* modulation is done), the length of the antenna comes out to be 75 m only and this size can be easily constructed.

(*ii*) **Effective power radiated by antenna.** For a linear antenna of length l, it has been found that :

$$\text{Power radiated, } P \propto \left(\frac{l}{\lambda}\right)^2$$

Therefore, for the same antenna length l, the power radiated will be large for signals of shorter wavelength or higher frequency. For good transmission, we need high powers. This requires the transmission to be carried out at high frequencies. For this reason, we use high frequency carrier wave to carry the audio signal.

(*iii*) **Mixing up of signals from different transmitters.** All the audio signals from different transmitters have the same frequency range *i.e.* 20 Hz to 20 kHz. Therefore, if the audio signals from various transmitters are transmitted directly, they will get mixed up and there is no way to distinguish between them. This difficulty is solved by allotting different carrier frequencies to different transmitting stations.

(*iv*) **Operating range.** The energy of a wave depends upon its frequency. The greater the frequency of the wave, the greater the energy possessed by it. As the audio signal frequencies are small, therefore, these cannot be transmitted over large distances if radiated directly into space. The only practical solution is to modulate a high frequency carrier wave with audio signal and permit the transmission to occur at this high frequency (*i.e.* carrier frequency).

(*v*) **Wireless communication.** One desirable feature of radio transmission is that it should be carried without wires *i.e.* radiated into space. At audio frequencies, radiation is not practicable because the efficiency of radiation is poor. However, efficient radiation of electrical energy is possible at high frequencies (> 20 kHz). For this reason, modulation is always done in communication systems.

The above discussion shows the need for high frequency transmission. For this purpose, the audio signal is superimposed on the high frequency carrier wave. This process is called modulation and the resultant waves are called modulated waves. *Hence modulation permits the transmission to occur at high frequency while it simultaneously allows the carrying of the audio signal.*

14.4 Types of Modulation

As you will recall, modulation is the process of changing amplitude or frequency or phase of a carrier wave in accordance with the intensity of the signal. Accordingly, there are three basic types of modulation, namely ;

(*i*) Amplitude modulation (*ii*) Frequency modulation (*iii*) Phase modulation

In India, amplitude modulation is used in radio broadcasting. However, in television transmission, frequency modulation is used for sound signal and amplitude modulation for picture signal. Therefore, our attention in this chapter shall be confined to the first two most important types of modulation.

14.5 Amplitude Modulation

When the amplitude of high frequency carrier wave is changed in accordance with the intensity of the signal, it is called **amplitude modulation.**

In amplitude modulation, only the amplitude of the carrier wave is changed in accordance with the intensity of the signal. However, the frequency of the modulated wave remains the same *i.e.* carrier frequency. Fig. 14.5 shows the principle of amplitude modulation. Fig. 14.5 (*i*) shows the audio electrical signal whereas Fig. 14.5 (*ii*) shows a carrier wave of constant amplitude. Fig. 14.5 (*iii*) shows the amplitude modulated (AM) wave. Note that the amplitudes of both positive and negative half-cycles of carrier wave are changed in accordance with the signal. For instance, when the signal is increasing in the positive sense, the amplitude of carrier wave also increases. On the other hand, during negative half-cycle of the signal, the amplitude of carrier wave decreases. Amplitude modulation is done by an electronic circuit called *modulator*.

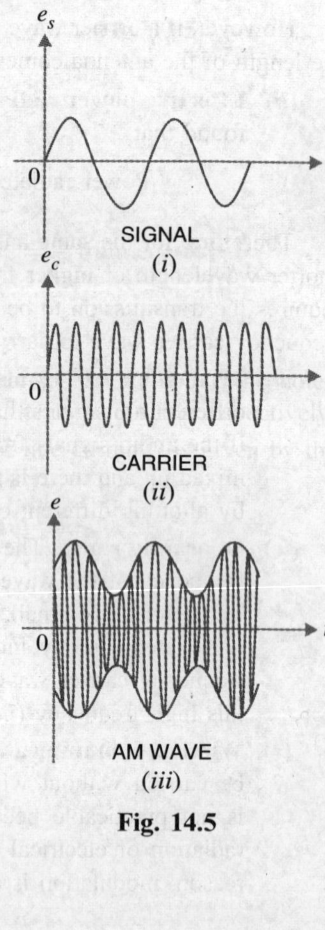

Fig. 14.5

The following points are worth noting in amplitude modulation :

(*i*) The amplitude of the carrier wave changes according to the intensity of the signal.

(*ii*) The amplitude variations of the carrier wave is at the signal frequency f_s.

(*iii*) The frequency of the amplitude modulated wave remains the same *i.e.* carrier frequency f_c.

14.6 Modulation Factor

An important consideration in amplitude modulation is to describe the depth of modulation *i.e.* the extent to which the amplitude of carrier wave is changed by the signal. This is described by a factor called modulation factor which may be defined as under :

The ratio of change of amplitude of carrier wave to the amplitude of normal carrier wave is called the **modulation factor** *m i.e.*

$$\text{Modulation factor, } m = \frac{\text{Amplitude change of carrier wave}}{\text{Normal carrier amplitude (unmodulated)}}$$

The value of modulation factor depends upon the amplitudes of carrier and signal. Fig. 14.6 shows amplitude modulation for different values of modulation factor *m*.

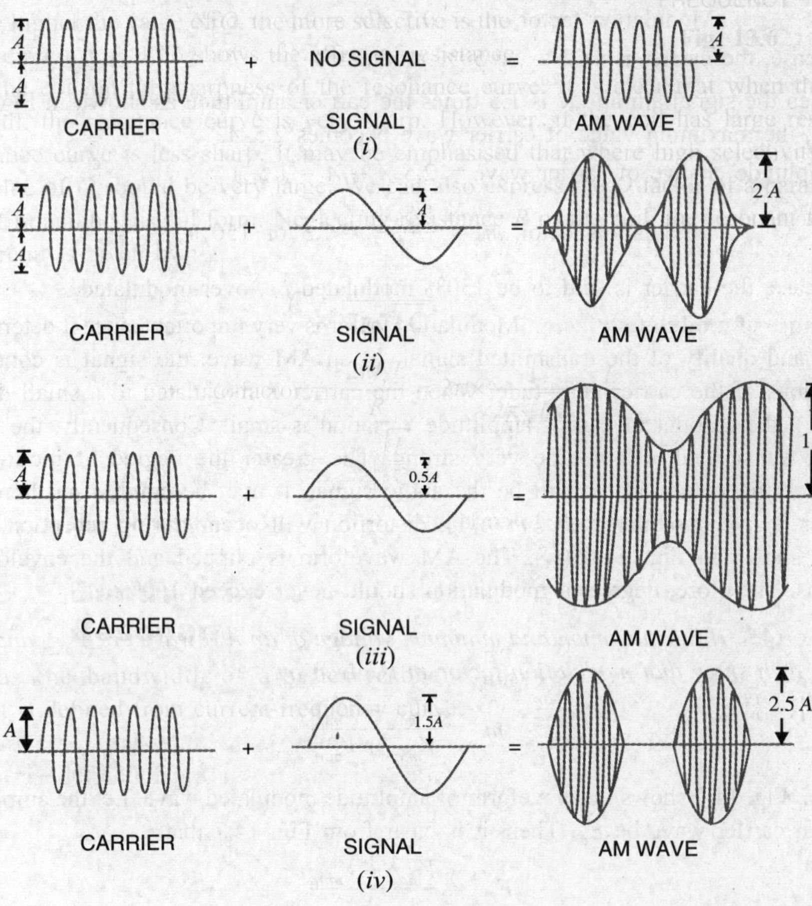

Fig. 14.6

(*i*) When signal amplitude is zero, the carrier wave is not modulated as shown in Fig. 14.6 (*i*). The amplitude of carrier wave remains unchanged.

Amplitude change of carrier $= 0$

Amplitude of normal carrier $= A$

$\therefore$ Modulation factor, $m = 0/A = 0$ or 0%

(*ii*) When signal amplitude is equal to the carrier amplitude as shown in Fig. 14.6 (*ii*), the amplitude of carrier varies between 2*A* and zero.

$$\text{Amplitude change of carrier} = 2A - A = A$$

$$\therefore \quad \text{Modulation factor, } m = \frac{\text{Amplitude change of carrier}}{\text{Amplitude of normal carrier}}$$

$$= A/A = 1 \text{ or } 100\%$$

In this case, the carrier is said to be 100% modulated.

(*iii*) When the signal amplitude is one-half the carrier amplitude as shown in Fig. 14.6 (*iii*), the amplitude of carrier wave varies between 1.5 *A* and 0.5 *A*.

$$\text{Amplitude change of carrier} = 1.5\,A - A = 0.5\,A$$

$$\therefore \quad \text{Modulation factor, } m = 0.5\,A/A = 0.5 \text{ or } 50\%$$

In this case, the carrier is said to be 50% modulated.

(*iv*) When the signal amplitude is 1.5 times the carrier amplitude as shown in Fig. 14.6 (*iv*), the maximum value of carrier wave becomes 2.5 *A*.

$$\text{Amplitude change of carrier wave} = 2.5\,A - A = 1.5\,A$$

$$\therefore \quad \text{Modulation factor, } m = \frac{1.5\,A}{A} = 1.5 \text{ or } 150\%$$

In this case, the carrier is said to be 150% modulated *i.e.* over-modulated.

Importance of modulation factor. Modulation factor is very important since it determines the strength and quality of the transmitted signal. In an AM wave, the signal is contained in the variations of the carrier amplitude. When the carrier is modulated to a small degree (*i.e.* small *m*), the amount of carrier amplitude variation is small. Consequently, the audio signal being transmitted will not be very strong. The greater the degree of modulation (*i.e. m*), the stronger and clearer will be the audio signal. It may be emphasised here that if the carrier is over-modulated (*i.e.* $m > 1$), distortion will occur during reception. This condition is shown in Fig. 14.6 (*iv*). The AM waveform is clipped and the envelope is discontinuous. Therefore, degree of modulation should never exceed 100%.

Example 14.3. *If the maximum and minimum voltage of an AM wave are V_{max} and V_{min} respectively, then show that modulation factor m is given by :*

$$m = \frac{V_{max} - V_{min}}{V_{max} + V_{min}}$$

Solution. Fig. 14.7 shows the waveform of amplitude modulated wave. Let the amplitude of the normal carrier wave be E_C. Then, it is clear from Fig. 14.7 that :

$$E_C = \frac{V_{max} + V_{min}}{2}$$

If E_S is the signal amplitude, then it is clear from Fig. 14.7 that :

$$E_S = \frac{V_{max} - V_{min}}{2}$$

But $$E_S = m\,E_C$$

or $$\frac{V_{max} - V_{min}}{2} = m\,\frac{V_{max} + V_{min}}{2} \quad \text{or} \quad m = \frac{V_{max} - V_{min}}{V_{max} + V_{min}}$$

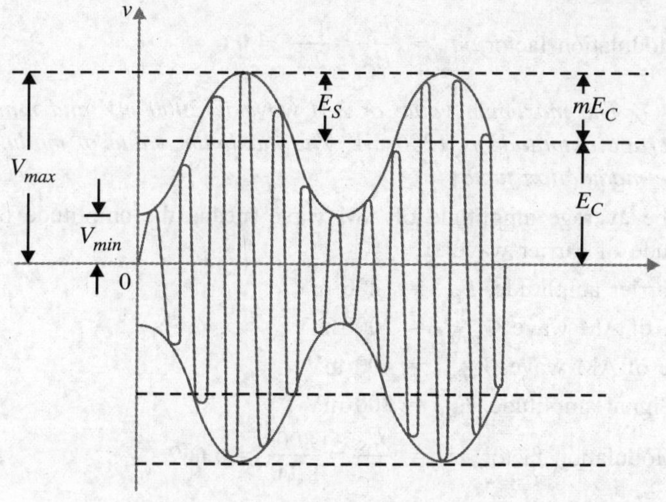

Fig. 14.7

Example 14.4. *The maximum peak-to-peak voltage of an AM wave is 16 mV and the minimum peak-to-peak voltage is 4 mV. Calculate the modulation factor.*

Solution. Fig. 14.8 shows the conditions of the problem.

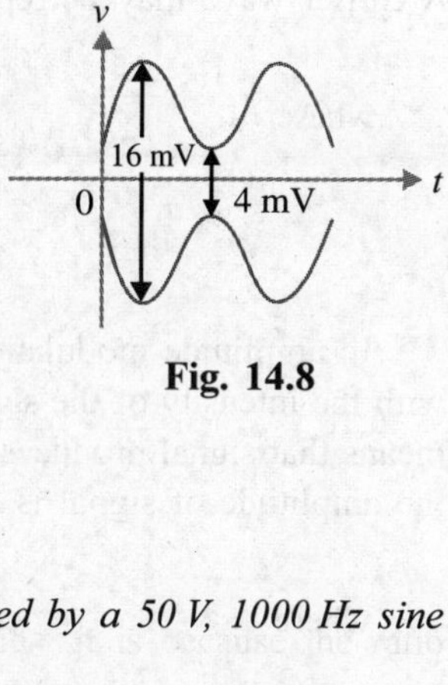

Fig. 14.8

Maximum voltage of AM wave is

$$V_{max} = \frac{16}{2} = 8 \text{ mV}$$

Minimum voltage of AM wave is

$$V_{min} = \frac{4}{2} = 2 \text{ mV}$$

$$\therefore \quad \text{Modulation factor, } m = \frac{V_{max} - V_{min}}{V_{max} + V_{min}}$$

$$= \frac{8-2}{8+2} = \frac{6}{10} = \mathbf{0.6}$$

Example 14.5. *A carrier of 100 V and 1200 kHz is modulated by a 50 V, 1000 Hz sine wave signal. Find the modulation factor.*

Solution.

$$\text{Modulation factor, } m = \frac{E_S}{E_C} = \frac{50 \text{ V}}{100 \text{ V}} = \mathbf{0.5}$$

Example 14.6. *The envelope of a sinusoidally amplitude modulated carrier wave has a maximum value of 4 V and a minimum value of 1 V. Determine the (i) amplitude of carrier (ii) amplitude of modulating signal and (iii) modulation factor.*

Solution.

$$V_{max} = 4 \text{ V}; \; V_{min} = 1 \text{ V}$$

(*i*) Carrier amplitude, $E_C = \frac{V_{max} + V_{min}}{2} = \frac{4+1}{2} = \mathbf{2.5\ V}$

(*ii*) Signal amplitude, $E_S = \frac{V_{max} - V_{min}}{2} = \frac{4-1}{2} = \mathbf{1.5\ V}$

(iii) Modulation factor, $m = \frac{E_S}{E_C} = \frac{1.5}{2.5} = \mathbf{0.6}$

Example 14.7. *The maximum value of AM wave is 2000 mV and minimum value is 400 mV. The average amplitude is 1200 mV. The maximum value of modulating signal is 800 mV. Find the modulation factor.*

Solution. The average amplitude of AM wave means the amplitude of unmodulated wave *i.e.* amplitude of carrier wave.

Carrier amplitude, E_C = 1200 mV

Max. value of AM wave, V_{max} = 2000 mV

Min. value of AM wave, V_{min} = 400 mV

Signal amplitude, E_S = 800 mV

$\therefore$ Modulation factor, $m = \frac{E_S}{E_C} = \frac{800}{1200} = \mathbf{0.667}$

Alternatively. $m = \frac{V_{max} - V_{min}}{V_{max} + V_{min}} = \frac{2000 - 400}{2000 + 400} = 0.667$

14.7 Analysis of Amplitude Modulated Wave

A carrier wave may be represented by ;

$$e_c = E_C \cos \omega_c t$$

where e_c = Instantaneous voltage of carrier

E_C = Amplitude of carrier

$\omega_c = 2 \pi f_c$

= Angular velocity at carrier frequency f_c

In amplitude modulation, the amplitude E_C of the carrier wave is varied in accordance with the intensity of the signal as shown in Fig. 14.9. Suppose the modulation factor is *m*. It means that signal produces a maximum change of $m\ E_C$ in the carrier amplitude. Obviously, the amplitude of signal is mE_C. Therefore, the signal can be represented by ;

$$e_s = m\ E_C \cos \omega_s t$$

where e_s = Instantaneous voltage of signal

mE_C = Amplitude of signal

$\omega_s = 2\pi f_s$ = Angular velocity at signal frequency f_s

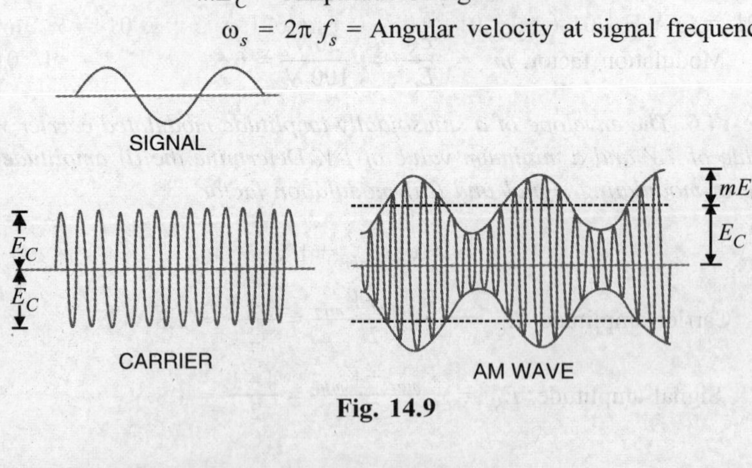

Fig. 14.9

The amplitude of the carrier wave varies at signal frequency f_s. Therefore, the amplitude of AM wave is given by ;

Amplitude of AM wave $= E_C + mE_C \cos \omega_s t = E_C (1 + m \cos \omega_s t)$

The instantaneous voltage of AM wave is :

$$
\begin{aligned}
e &= \text{Amplitude} \times \cos \omega_c t \\
&= E_C (1 + m \cos \omega_s t) \cos \omega_c t \\
&= E_C \cos \omega_c t + m E_C \cos \omega_s t \cos \omega_c t \\
&= E_C \cos \omega_c t + \frac{mE_C}{2} (2 \cos \omega_s t \cos \omega_c t) \\
&= E_C \cos \omega_c t + \frac{mE_C}{2} [\cos (\omega_c + \omega_s) t + \cos (\omega_c - \omega_s) t]^* \\
&= E_C \cos \omega_c t + \frac{mE_C}{2} \cos (\omega_c + \omega_s) t + \frac{mE_C}{2} \cos (\omega_c - \omega_s) t
\end{aligned}
$$

The following points may be noted from the above equation of amplitude modulated wave :

(*i*) The AM wave is equivalent to the summation of three sinusoidal waves ; one having amplitude E_C and frequency **f_c, the second having amplitude $mE_C/2$ and frequency $(f_c + f_s)$ and the third having amplitude $mE_C/2$ and frequency $f_c - f_s$.

(*ii*) The AM wave contains three frequencies *viz* $f_c, f_c + f_s$ and $f_c - f_s$. The first frequency is the carrier frequency. Thus, the process of modulation does not change the original carrier frequency but produces two new frequencies $(f_c + f_s)$ and $(f_c - f_s)$ which are called sideband frequencies.

(*iii*) The sum of carrier frequency and signal frequency *i.e.* $(f_c + f_s)$ is called *upper sideband frequency*. The *lower sideband frequency* is $f_c - f_s$ *i.e.* the difference between carrier and signal frequencies.

14.8 Frequency Spectrum of AM Wave

The equation of amplitude modulated (AM) wave is given by ;

$$e = E_C \cos \omega_c t + \frac{mE_C}{2} \cos(\omega_c + \omega_s)t + \frac{mE_C}{2} \cos(\omega_c - \omega_s) t \quad \text{... } (i)$$

It is clear from eq. (*i*) that an AM wave consists of three separate sinusoidal components which may be considered to be three sinusoidal generators in series as shown in Fig. 14.10 (*i*). The first term on the R.H.S. is the carrier wave of amplitude E_C and frequency f_c. The second term is a cosine wave of amplitude $\frac{mE_C}{2}$ and frequency $(f_c + f_s)$. This component is the *upper sideband frequency component*. The third term is also a cosine wave of amplitude $\frac{mE_C}{2}$ and frequency $(f_c - f_s)$. This component is the *lower sideband frequency component*. Fig. 14.10 (*ii*) shows the frequency spectrum of an amplitude modulated wave.

* From trigonometry, we have the expansion formula:

$$2 \cos A \cos B = \cos (A + B) + \cos (A - B)$$

** $f_c = \frac{\omega_c}{2\pi}, \; f_c + f_s = \frac{\omega_c + \omega_s}{2\pi}, \; f_c - f_s = \frac{\omega_c - \omega_s}{2\pi}$

The frequency components in the AM wave are shown by vertical lines. The height of each vertical line is equal to the amplitude of the components present. It may be added here that in practical radio transmission, carrier frequency f_c is many times greater than signal frequency f_s. Hence, the sideband frequencies are generally close to the carrier frequency. It may be seen that a carrier modulated by a single frequency is equivalent to three simultaneous signals ; the carrier itself and two other sideband frequencies *i.e.* $(f_c + f_s)$ and $(f_c - f_s)$.

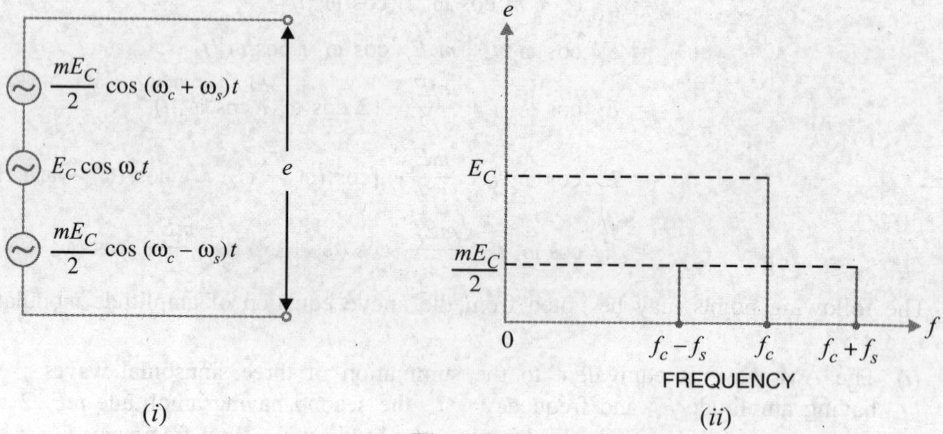

Fig. 14.10

Let us illustrate sideband frequencies with an example. Suppose the carrier frequency is 400 kHz and the signal frequency is 1 kHz. The AM wave will contain three frequencies *viz* 400 kHz, 401 kHz and 399 kHz. It is clear that upper sideband frequency (401 kHz) and lower sideband frequency (399 kHz) are very close to the carrier frequency (400 kHz). Note that in an AM wave, the sideband frequencies are of our interest. It is because the signal frequency f_s is contained in the sideband frequencies.

Bandwidth. In an AM wave, the bandwidth is from $(f_c - f_s)$ to $(f_c + f_s)$ *i.e.* $2 f_s$. Thus in the above example, bandwidth is from 399 to 401 kHz or 2 kHz which is twice the signal frequency. Therefore, we arrive at a very important conclusion that *in amplitude modulation, bandwidth is twice the signal frequency*. The tuned amplifier which is called upon to amplify the modulated wave must have the required bandwidth to include the sideband frequencies. If the tuned amplifier has insufficient bandwidth, the upper sideband frequencies may not be reproduced by the radio receiver.

Example 14.8. *A 2500 kHz carrier is modulated by audio signal with frequency span of 50 – 15000 Hz. What are the frequencies of lower and upper sidebands? What bandwidth of RF amplifier is required to handle the output?*

Solution. The modulating signal (*e.g.* music) has a range of 0.05 to 15 kHz. The sideband frequencies produced range from $f_c \pm 0.05$ kHz to $f_c \pm 15$ kHz. Therefore, upper sideband ranges from **2500.05 to 2515 kHz** and lower sideband ranges from **2499.95 to 2485 kHz.**

The sideband frequencies produced can be approximately expressed as 2500 ± 15 kHz. Therefore, bandwidth requirement = 2515 – 2485 = **30 kHz.** Note that bandwidth of *RF* amplifier required is twice the frequency of highest modulating signal frequency.

14.9 Transistor AM *Modulator

Fig. 14.11 shows the circuit of a simple AM modulator. It is essentially a *CE* amplifier having a voltage gain of *A*. The carrier signal is the input to the amplifier. The modulating signal is applied in the emitter resistance circuit.

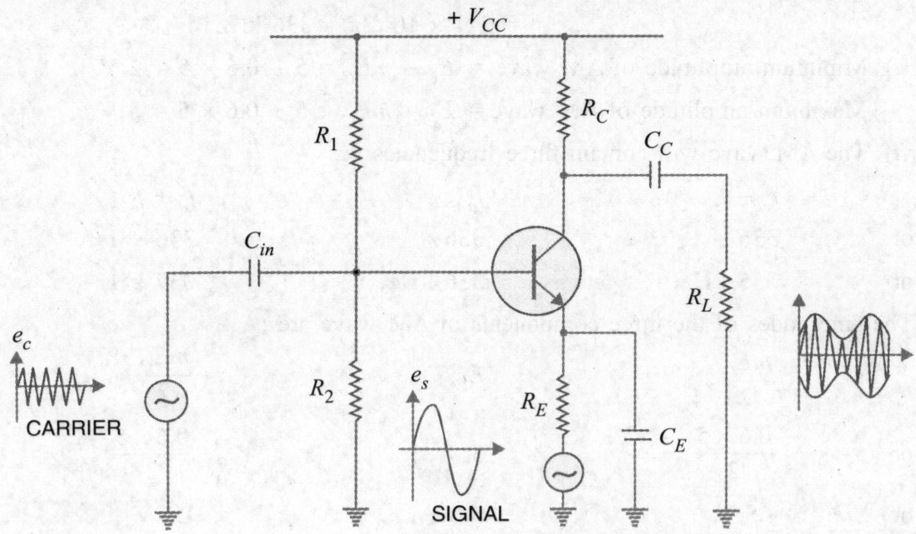

Fig. 14.11

Working. The carrier e_c is applied at the input of the amplifier and the modulating signal e_s is applied in the emitter resistance circuit. The amplifier circuit amplifies the carrier by a factor "*A*" so that the output is Ae_c. Since the modulating signal is a part of the biasing circuit, it produces low-frequency variations in the emitter circuit. This in turn causes **variations in "*A*". The result is that amplitude of the carrier varies in accordance with the strength of the signal. Consequently, amplitude modulated output is obtained across R_L. It may be noted that carrier should not influence the voltage gain *A* ; only the modulating signal should do this. To achieve this objective, carrier should have a small magnitude and signal should have a large magnitude.

Example 14.9. *An AM wave is represented by the expression :*

$$v = 5\,(1 + 0.6 \cos 6280\, t) \sin 211 \times 10^4\, t \text{ volts}$$

(*i*) *What are the minimum and maximum amplitudes of the AM wave?*

(*ii*) *What frequency components are contained in the modulated wave and what is the amplitude of each component?*

Solution.

The AM wave equation is given by :

$$v = 5\,(1 + 0.6 \cos 6280\, t) \sin 211 \times 10^4\, t \text{ volts} \quad ...(i)$$

Compare it with standard AM wave eq.,

$$v = E_C(1 + m \cos \omega_s\, t) \sin \omega_c\, t \quad ...(ii)$$

* A circuit which does amplitude modulation is called AM modulator.

** The principle of this circuit is to change the gain *A* (and hence the amplitude of carrier) by the modulating signal.

From eqs. (*i*) and (*ii*), we get, E_C = Carrier amplitude = 5 V

m = Modulation factor = 0.6

f_s = Signal frequency = $\omega_s/2\pi$ = $6280/2\pi$ = 1 kHz

f_c = Carrier frequency = $\omega_c/2\pi$

= $211 \times 10^4/2\pi$ = 336 kHz

(*i*) Minimum amplitude of AM wave = $E_C - mE_C = 5 - 0.6 \times 5 =$ **2 V**

Maximum amplitude of AM wave = $E_C + mE_C = 5 + 0.6 \times 5 =$ **8 V**

(*ii*) The AM wave will contain three frequencies *viz.*

$f_c - f_s$, f_c, $f_c + f_s$

or 336 − 1, 336, 336 + 1

or **335 kHz, 336 kHz, 337 kHz**

The amplitudes of the three components of AM wave are :

$\frac{mE_C}{2}$, E_C, $\frac{mE_C}{2}$

or $\frac{0.6 \times 5}{2}$, 5, $\frac{0.6 \times 5}{2}$

or **1.5 V, 5 V, 1.5 V**

Example 14.10. *A sinusoidal carrier voltage of frequency 1 MHz and amplitude 100 volts is amplitude modulated by sinusoidal voltage of frequency 5 kHz producing 50% modulation. Calculate the frequency and amplitude of lower and upper sideband terms.*

Solution.

Frequency of carrier, f_c = 1 MHz = 1000 kHz

Frequency of signal, f_s = 5 kHz

Modulation factor, m = 50% = 0.5

Amplitude of carrier, E_C = 100 V

The lower and upper sideband frequencies are :

$f_c - f_s$ and $f_c + f_s$

or (1000 − 5) kHz and (1000 + 5) kHz

or **995 kHz** and **1005 kHz**

Amplitude of each sideband term = $\frac{mE_C}{2} = \frac{0.5 \times 100}{2} =$ **25 V**

Example 14.11. *A carrier wave of frequency 10 MHz and peak value 10 V is amplitude modulated by a 5- kHz sine wave of amplitude 6 V. Determine (i) modulation factor (ii) sideband frequencies and (iii) amplitude of sideband components. Draw the frequency spectrum.*

Solution.

Carrier amplitude, E_C = 10 V

Signal amplitude, E_S = 6 V

Carrier frequency, f_c = 10 MHz

Signal frequency, f_s = 5 kHz = 0.005 MHz

(*i*) Modulation factor, $m = \dfrac{E_S}{E_C} = \dfrac{6}{10} = \mathbf{0.6}$

(*ii*) Sideband frequencies are :

$$f_c - f_s \;;\; f_c + f_s$$

$$10 - 0.005 \;;\; 10 + 0.005$$

$$\mathbf{9.995\ MHz \;;\; 10.005\ MHz}$$

(*iii*) Amplitude of each sideband $= \dfrac{m\,E_C}{2} = \dfrac{0.6 \times 10}{2} = \mathbf{3\ V}$

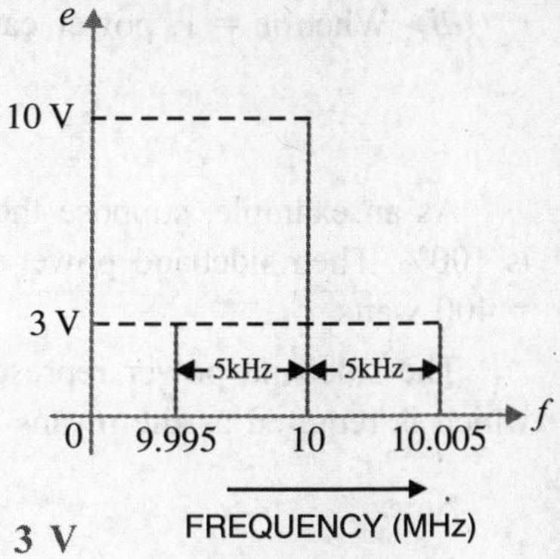

Fig. 14.12

Fig. 14.12 shows the frequency spectrum of the A.M. wave.

14.10 Power in AM Wave

The power dissipated in any circuit is a function of the square of voltage across the circuit and the effective resistance of the circuit. Equation of AM wave reveals that it has three components of amplitude E_C, $mE_C/2$ and $mE_C/2$. Clearly, power output must be distributed among these components.

$$\text{Carrier power, } P_C = \frac{{}^*\left(E_C/\sqrt{2}\right)^2}{R} = \frac{E_C^2}{2R} \qquad \text{...(}i\text{)}$$

$$\text{Total power of sidebands, } P_S = \frac{\left(m\,E_C/2\sqrt{2}\right)^2}{R} + \frac{\left(m\,E_C/2\sqrt{2}\right)^2}{R}$$

$$= \frac{m^2 E_C^2}{8R} + \frac{m^2 E_C^2}{8R} = \frac{m^2 E_C^2}{4R} \qquad \text{...(}ii\text{)}$$

$$\text{Total power of AM wave, } P_T = P_C + P_S$$

$$= \frac{E_C^2}{2R} + \frac{m^2 E_C^2}{4R} = \frac{E_C^2}{2R}\left[1 + \frac{m^2}{2}\right]$$

$$\text{or} \qquad P_T = \frac{E_C^2}{2R}\left[\frac{2 + m^2}{2}\right] \qquad \text{...(}iii\text{)}$$

Fraction of total power carried by sidebands is

$$\frac{P_S}{P_T} = \frac{\text{Exp. (}ii\text{)}}{\text{Exp. (}iii\text{)}} = \frac{m^2}{2 + m^2} \qquad \text{...(}iv\text{)}$$

As the signal is contained in the sideband frequencies, therefore, useful power is in the sidebands. Inspection of exp. (*iv*) reveals that sideband power depends upon the modulation factor *m*. The greater the value of *m*, the greater is the useful power carried by the sidebands. This emphasises the importance of modulation factor.

(*i*) When $m = 0$, power carried by sidebands $= 0^2/2 + 0^2 = 0$

(*ii*) When $m = 0.5$, power carried by sidebands

$$= \frac{(0.5)^2}{2 + (0.5)^2} = 11.1\% \text{ of total power of AM wave}$$

* R.M.S. values are considered.

(*iii*) When $m = 1$, power carried by sidebands

$$= \frac{(1)^2}{2 + (1)^2} = 33.3\% \text{ of total power of AM wave}$$

As an example, suppose the total power of an AM wave is 600 watts and modulation is 100%. Then sideband power is 600/3 = 200 watts and carrier power will be 600 – 200 = 400 watts.

The sideband power represents the signal content and the carrier power is that power which is required as the means of transmission.

Note. $$P_C = \frac{E_C^2}{2R} \quad \text{and } P_S = \frac{m^2 E_C^2}{4R}$$

$$\therefore \quad \frac{P_S}{P_C} = \frac{1}{2} m^2$$

or $$P_S = \frac{1}{2} m^2 P_C \quad \quad ...(v)$$

Expression (*v*) gives the relation between total sideband power (P_S) and carrier power (P_C).

14.11 Limitations of Amplitude Modulation

Although theoretically highly effective, amplitude modulation suffers from the following drawbacks:

(*i*) **Noisy reception.** In an AM wave, the signal is in the amplitude variations of the carrier. Practically all the natural and man made noises consist of electrical amplitude disturbances. As a radio receiver cannot distinguish between amplitude variations that represent noise and those that contain the desired signal, therefore, reception is generally noisy.

(*ii*) **Low efficiency.** In amplitude modulation, useful power is in the sidebands as they contain the signal. As discussed before, an AM wave has low sideband power. For example, if modulation is 100%, the sideband power is only one-third of the total power of AM wave. Hence the efficiency of this type of modulation is low.

(*iii*) **Small operating range.** Due to low efficiency of amplitude modulation, transmitters employing this method have a small operating range *i.e.* messages cannot be transmitted over larger distances.

(*iv*) **Lack of audio quality.** This is a distinct disadvantage of amplitude modulation. In order to attain high-fidelity reception, all audio frequencies upto 15 kHz must be reproduced. This necessitates bandwidth of 30 kHz since both sidebands must be reproduced. But AM broadcasting stations are assigned bandwidth of only 10 kHz to minimise the interference from adjacent broadcasting stations. This means that the highest modulating frequency can be 5 kHz which is hardly sufficient to reproduce the music properly.

Example 14.12. *A carrier wave of 500 watts is subjected to 100% amplitude modulation. Determine:*

(*i*) *power in sidebands* (*ii*) *power of modulated wave.*

Solution. (*i*) Sideband power, $P_S = \frac{1}{2} m^2 P_C = \frac{1}{2} \times 500 = \mathbf{250\ W}$

Thus there are 125 W in upper sideband and 125 W in lower sideband.

(*ii*) Power of AM wave, $P_T = P_C + P_S = 500 + 250 = \mathbf{750\ W}$

Example 14.13. *A 50 kW carrier is to be modulated to a level of (i) 80% (ii) 10%. What is the total sideband power in each case?*

Solution. (*i*)
$$P_S = \frac{1}{2} m^2 P_C = \frac{1}{2} (0.8)^2 \times 50 = \mathbf{16\ kW}$$

(*ii*)
$$P_S = \frac{1}{2} m^2 P_C = \frac{1}{2} (0.1)^2 \times 50 = \mathbf{0.25\ kW}$$

Note the effect of modulation factor on the magnitude of sideband power. In the first case ($m = 80\%$), we generated and transmitted 50 kW carrier in order to send 16 kW of intelligence. In the second case ($m = 10\%$), the same carrier level — 50 kW — is used to send merely 250 W of intelligence. Clearly, the efficiency of operation decreases rapidly as modulation factor decreases. For this reason, in amplitude modulation, the value of m is kept as close to unity as possible.

Example 14.14. *A 40 kW carrier is to be modulated to a level of 100%.*

(*i*) *What is the carrier power after modulation?*

(*ii*) *How much audio power is required if the efficiency of the modulated RF amplifier is 72%?*

Solution. Fig. 14.13 shows the block diagram indicating the power relations.

(*i*) Since the carrier itself is unaffected by the modulating signal, there is no change in the carrier power level.

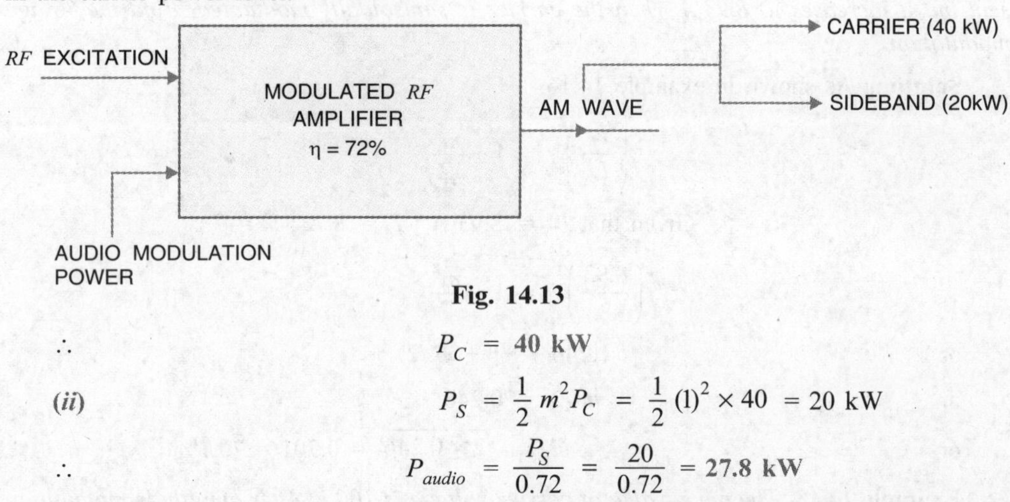

Fig. 14.13

$$\therefore \quad P_C = \mathbf{40\ kW}$$

(*ii*)
$$P_S = \frac{1}{2} m^2 P_C = \frac{1}{2} (1)^2 \times 40 = 20\ \text{kW}$$

$$\therefore \quad P_{audio} = \frac{P_S}{0.72} = \frac{20}{0.72} = \mathbf{27.8\ kW}$$

Example 14.15. *An audio signal of 1 kHz is used to modulate a carrier of 500 kHz. Determine (i) sideband frequencies (ii) bandwidth required.*

Solution. Carrier frequency, f_c = 500 kHz

Signal frequency, f_s = 1 kHz

(*i*) As discussed in Art. 14.7, the AM wave has sideband frequencies of $(f_c + f_s)$ and $(f_c - f_s)$.

∴ Sideband frequencies = (500 + 1) kHz and (500 − 1) kHz

= **501 kHz** and **499 kHz**

(*ii*) Bandwidth required = 499 kHz to 501 kHz = **2 kHz**

Example 14.16. *The load current in the transmitting antenna of an unmodulated AM transmitter is 8A. What will be the antenna current when modulation is 40% ?*

Solution.
$$P_S = \frac{1}{2}m^2 P_C$$

$$P_T = P_C + P_S = P_C\left(1+\frac{m^2}{2}\right)$$

$$\therefore \quad \frac{P_T}{P_C} = 1+\frac{m^2}{2}$$

or
$$\left(\frac{I_T}{I_C}\right)^2 = 1+\frac{m^2}{2}$$

Given that $I_C = 8\text{A}$; $m = 0.4$

$$\therefore \quad \left(\frac{I_T}{8}\right)^2 = 1+\frac{(0.4)^2}{2}$$

or
$$(I_T/8)^2 = 1.08$$

or
$$I_T = 8\sqrt{1.08} = \mathbf{8.31\ A}$$

Example 14.17. *The antenna current of an AM transmitter is 8 A when only carrier is sent but it increases to 8.93 A when the carrier is sinusoidally modulated. Find the % age modulation.*

Solution. As shown in example 14.16,
$$\left(\frac{I_T}{I_C}\right)^2 = 1+\frac{m^2}{2}$$

Given that $I_T = 8.93$ A ; $I_C = 8$ A ; $m = ?$

$$\therefore \quad \left(\frac{8.93}{8}\right)^2 = 1+\frac{m^2}{2}$$

or
$$1.246 = 1 + m^2/2$$

or
$$m^2/2 = 0.246$$

or
$$m = \sqrt{2\times 0.246} = 0.701 = \mathbf{70.1\%}$$

Example 14.18. *The r.m.s. value of carrier voltage is 100 V. After amplitude modulation by a sinusoidal a.f. voltage, the r.m.s. value becomes 110 V. Calculate the modulation index.*

Solution.
$$\frac{P_T}{P_C} = 1+\frac{m^2}{2}$$

or
$$\left(\frac{V_T}{V_C}\right)^2 = 1+\frac{m^2}{2}$$

Given that : $V_T = 110$ V ; $V_C = 100$ V ; $m = ?$

$$\therefore \quad \left(\frac{110}{100}\right)^2 = 1+\frac{m^2}{2}$$

$$\text{or} \qquad 1.21 = 1 + \frac{m^2}{2}$$

$$\text{or} \qquad m^2/2 = 0.21$$

$$\text{or} \qquad m = \sqrt{0.21 \times 2} = \mathbf{0.648}$$

Example 14.19. *An AM wave consists of the following components :*

Carrier component = 5 V peak value

Lower sideband component = 2.5 V peak value

Upper sideband component = 2.5 V peak value

If the AM wave drives a 2 kΩ resistor, find the power delivered to the resistor by (i) carrier (ii) lower sideband component and (iii) upper sideband component. What is the total power delivered?

Solution. Fig. 14.14 (*i*) shows the frequency spectrum of AM wave whereas Fig. 14.14 (*ii*) shows the equivalent circuit.

$$\text{Power} = \frac{(\text{R.M.S. voltage})^2}{R} = \frac{(0.707 \times \text{Peak value})^2}{R}$$

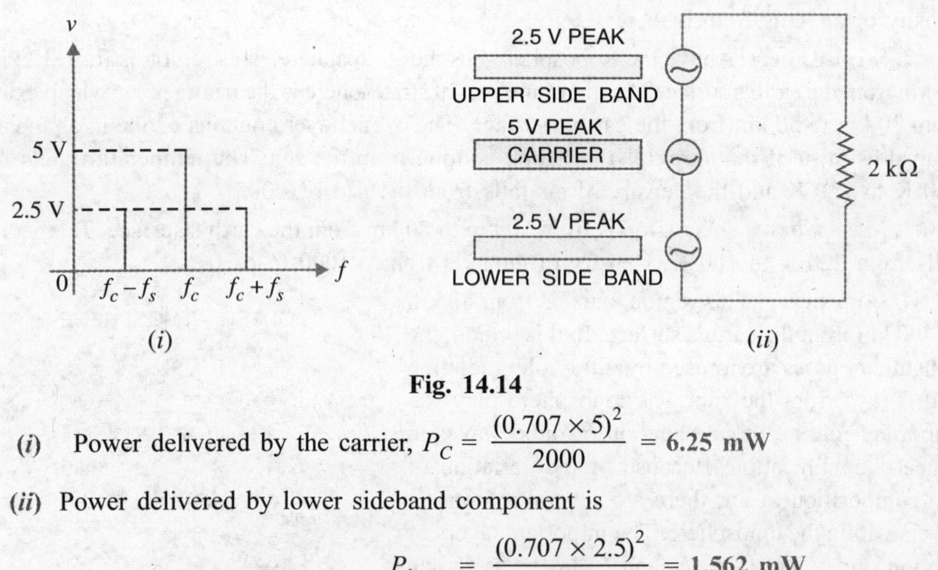

Fig. 14.14

(*i*) Power delivered by the carrier, $P_C = \dfrac{(0.707 \times 5)^2}{2000} = \mathbf{6.25\ mW}$

(*ii*) Power delivered by lower sideband component is

$$P_{lower} = \frac{(0.707 \times 2.5)^2}{2000} = \mathbf{1.562\ mW}$$

(*iii*) Power delivered by upper sideband component is

$$P_{upper} = \frac{(0.707 \times 2.5)^2}{2000} = \mathbf{1.562\ mW}$$

Total power delivered by the AM wave = 6.25 + 1.562 + 1.562 = **9.374 mW**

14.12 Propagation of Radio Waves

We have seen that radio waves or modulated waves are high frequency electromagnetic waves. In radio communication, an antenna at the transmitter radiates the radio waves which travel through the space and reach the receiving antenna at the other end. Several factors influence the propagation of radio waves and the path they follow.

(*i*) As the radio waves travel away from the transmitter, their power goes on decreasing.

(*ii*) The radio waves are little affected by the surrounding atmosphere, rain, snow *etc* and are able to penetrate non-metallic objects easily.

(*iii*) The radio waves are stopped dead by metals or fine mesh screens.

The earth's atmosphere plays an important role in the propagation of radio waves. Therefore, it is desirable to discuss the composition of earth's atmosphere.

14.13 Earth's Atmosphere

The gaseous envelope around the earth is called **atmosphere**. It extends upto a height of about 400 km above the surface of the earth. The earth's atmosphere is mainly composed of nitrogen (78%) and oxygen (21%). It also contains minute quantities of carbon dioxide, neon, water vapours, dust particles, *etc*. The density of air (ρ) goes on decreasing as we go up. The earth's atmosphere has been divided into the following regions :

1. Troposphere. This region is upto a height of 12 km from the earth's surface. It is in this region that clouds are formed. The temperature in troposphere decreases with height at the rate of about 6.5°C per kilometre to a value of about –50°C at its upper boundary. The density of air falls from ρ to $\rho/10$.

2. Stratosphere. Above the troposphere lies the stratosphere. This region is from 12 km to 50 km from the earth's surface. An important part of stratosphere is the **ozone layer** which extends from 30 km to 50 km from the earth's surface. The ozone layer contains ozone in abundance. It absorbs most of the ultraviolet radiation coming from the sun. The temperature falls from 280 K to 220 K and the density of air falls from $\rho/10$ to $\rho/1000$.

3. Mesosphere. This region is from 50 km to 80 km from the earth's surface. Temperature falls from 220 K to 180 K. Density of air falls from $\rho/1000$ to $\rho/10^5$.

4. Ionosphere. This region extends from 80 km to 400 km from the earth's surface. In this region, the constituent gases are ionised by ultraviolet radiation and *X*-rays from the sun. The ionosphere plays an important role in the propagation of radio waves in space communication. Because of the variation of the composition of air, there are several regions or layers within the ionosphere. The important layers of the ionosphere are *D*, *E*, F_1 and F_2 layers as shown in Fig. 14.15. The ionisation density (*i.e.* the number of ions or electrons per unit volume) of each layer varies with time of the day, season, altitude *etc*.

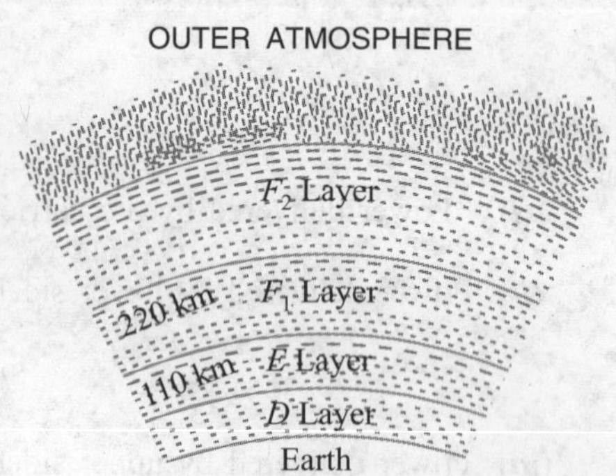

Fig. 14.15

(*i*) The lowest layer, called *D layer*, exists only in the day time at an altitude of 50 to 90 km above the earth's surface. Ionisation in this region is relatively weak and does not affect the direction of travel of radio waves.

(*ii*) The next layer, the *E layer*, is in a region of about 90-140 km above the surface of earth. It has a maximum density at noon but is only weakly ionised at night.

(*iii*) The last layer, the *F layer*, is quite variable. At night, it exists as a single layer in a region of about 140-400 km above the earth's surface. However in day time, it splits into two layers—F_1 and F_2 as shown in Fig. 14.15.

When radio waves from a transmitter are directed towards the ionosphere, the radio waves in the frequency range 2 MHz to 40 MHz are reflected back to earth by the ionosphere. However, radio waves of frequencies greater than 40 MHz are not reflected back to earth by the ionosphere ; they penetrate into ionosphere and escape. Thus ionosphere acts as a reflector for certain types of radio waves and helps in the long distance transmission of radio waves.

14.14 Types of Radio Wave Propagation

The phenomenon of sending, receiving and processing information through space is called **space communication.**

In space communication, the signal is carried by high frequency electromagnetic wave (called *carrier wave*) from the transmitter to the receiver in free space. Since no wires are used, it is also called *wireless communication*. The frequencies used in space communication lie in the range of 10^4 Hz to 10^{11} Hz. Radio, television and satellite communication fall under this category. Depending upon the *frequency of radio waves,* the *distance between the transmitter and receiving antennas* and *the path* (or *paths*) *by which radio waves reach the receiving antenna,* the radio wave propagation can be carried out in the following three ways :

1. Ground or surface wave propagation
2. Sky wave or ionospheric wave propagation
3. Space wave or direct wave or tropospheric propagation

1. Ground wave propagation. *When the radio waves from the transmitting antenna propagate along the surface of earth so as to reach the receiving antenna, it is called ground wave propagation or surface wave propagation.*

In ground wave propagation, the radio waves travel along the surface of earth as shown in Fig. 14.16. This mode of propagation is possible only when the transmitting and receiving antennas are close to the surface of the earth. As the ground wave glides over the surface of earth, it induces current in the earth. This gives rise to resistance losses and dielectric losses in the ground. The energy required to supply these losses must come from the ground wave. Therefore, the energy of a ground wave decreases as it passes over the earth's surface. The ground losses increase rapidly with the increase in frequency. Moreover, ground wave is also attenuated due to *diffraction effect*. As the wave propagates over the earth, it gradually tilts over. At some appreciable distance from the transmitting antenna, the wave lies down and dies.

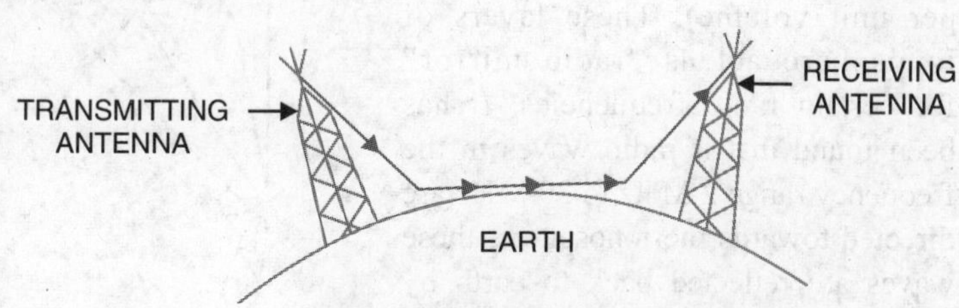

Fig. 14.16

We have seen that loss of power of a ground wave increases rapidly with the increase in frequency as well as with the increase in distance from the transmitting antenna. *Therefore, ground wave propagation is limited to low frequency signals (500 kHz to 1500 kHz) and for short range communication.* It cannot be used for high frequency and long range transmission.

(*i*) Below 500 kHz, reliable communication can be obtained over distance upto 1500 km by ground waves alone.

(*ii*) Amplitude modulated radio waves in the medium frequency band are transmitted primarily *via* ground waves.

Advantages.

Ground wave propagation has the following advantages :

(*a*) Ground wave propagation has excellent reliability.

(*b*) Reception is not affected by changing atmospheric conditions.

(*c*) Given enough transmitting power, ground wave communication can be maintained with any place in the world.

Disadvantages.

Ground wave propagation has the following disadvantages :

(*a*) Ground wave propagation requires high transmitting power.

(*b*) Ground losses increase very rapidly with the increase in frequency of the signal.

(*c*) Ground losses vary with surface material and composition.

2. Sky wave propagation. *When the radio waves from the transmitting antenna reach the receiving antenna after reflection from the ionosphere, it is called sky wave propagation* or *ionospheric propagation.*

We have already discussed that ionosphere has four layers *viz. D*, *E*, F_1 and F_2 layers of varying ionisation density (*i.e.* number of ions or electrons per unit volume). These layers of ionosphere act as "radio mirror" for certain radio frequencies. It has been found that if radio waves in the frequency range 2 MHz to 30 MHz are directed towards the ionosphere, these waves are reflected back to earth by the various layers of the ionosphere as shown in Fig. 14.17. If the frequency of the radio waves is more than 30 MHz, these are not reflected back to earth from the ionosphere and penetrate into the ionosphere and escape. For this reason we use sky wave propagation for radio waves lying in the frequency range 2 MHz to 30 MHz. The sky wave propagation is also known as *ionospheric propagation* because sky waves reach the receiver after reflection from the ionosphere.

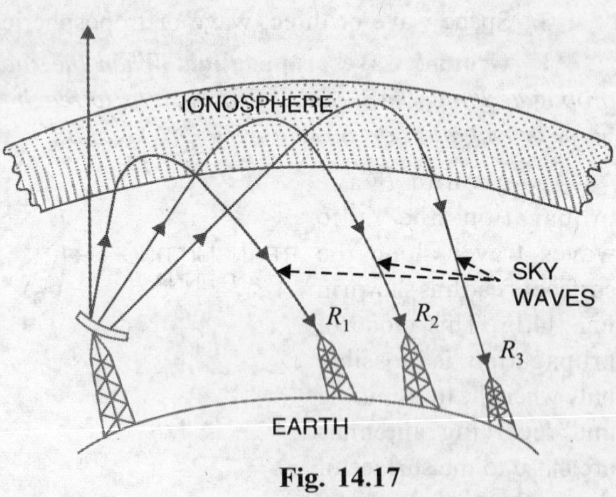

Fig. 14.17

The sky wave propagation is quite unreliable. It is because whether or not a radio wave is reflected back to earth by the ionosphere depends upon several factors including (*i*) frequency of radio wave (*ii*) ionisation density of ionosphere and (*iii*) the angle of incidence at which the radio wave enters the ionosphere. Yet it is the primary means of around-the-world short-wave communication.

(*i*) **Critical frequency.** As the frequency of sky waves increases, the ionosphere becomes progressively less effective in reflecting the waves back to earth. At a certain maximum frequency of radio wave, the wave is not at all reflected back to earth. This is called *critical frequency*.

The highest frequency above which the ionosphere no longer returns the sky wave back to earth when transmitted in vertical direction is called **critical frequency.**

Since the critical frequency depends on the density of ionisation, it will clearly vary with the time of day and season of the year. Furthermore, it is possible for a particular frequency to pierce the E layer but still be returned from the F layer because it has higher density of ionisation. Of course, a still higher frequency will pierce both layers and be lost. The value of critical frequency (f_c) is given by ;

$$f_c = 9\,(N_{,max})^{1/2}$$

where N_{max} = Maximum electron density of ionosphere

(*ii*) **Critical angle.** As the vertical angle of the sky wave *w.r.t.* earth is increased, the ionosphere layers are no longer capable of reflecting sky waves back to earth.

For a given frequency, the vertical angle above which the sky wave no longer returns to earth but travels outward into space is called **critical angle.**

Sky waves at or above critical angle may be refracted (bent) by ionosphere but they are not reflected back to earth. The critical angle primarily depends on density of ionisation and on the frequency of the wave.

3. Space wave propagation (Line of sight propagation). At frequencies above 30 MHz, radio transmission cannot be carried out by ground waves or sky waves. It is because ground waves are quickly attenuated at such high frequencies and at frequencies above 30 MHz, the ionosphere is unable to reflect the sky waves back to earth. Therefore, we use **direct waves** (also called **space waves**) which travel directly in straight lines from the transmitting antenna to the receiving antenna as shown in Fig. 14.18. This mode of propagation is called *space wave propagation* or *line of sight (LOS) propagation* or *tropospheric propagation*.

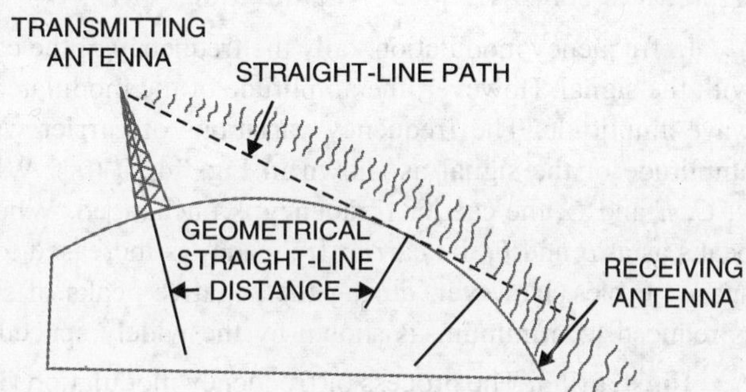

Fig. 14.18

When the radio waves from the transmitting antenna travelling in straight line directly reach the receiving antenna, it is called space wave propagation or line of sight propagation.

As the frequency increases, the radio waves tend to travel more and more in straight lines. Therefore, the receiving antenna must receive the signal directly from the transmitting antenna. The television frequencies lie in the range 100 MHz to 200 MHz. Therefore, reception of TV signals is possible only if the receiving antenna directly intercepts the signal. Similarly, radar, microwave relays and many other services solely depend on space wave propagation. The **range of space wave propagation** is limited by

(*i*) The *line of sight distance i.e.* the distance at which the transmitting and receiving antennas can see each other.

(*ii*) the *curvature of earth.*

The space waves or direct waves travel essentially in a line of sight path. Due to curvature of earth, line of sight distance will depend upon the antenna height. Therefore, the maximum range of communication through space waves is determined by the heights of the transmitting and receiving antennas. This is illustrated in Fig. 14.19.

At receiving location 1, the height of the antenna is within the line of sight distance. For the same antenna height, receiving location 2 would be the maximum range of communication for direct waves. However, if we increase the height of the receiving antenna, the communication range can be increased to reach the location 4.

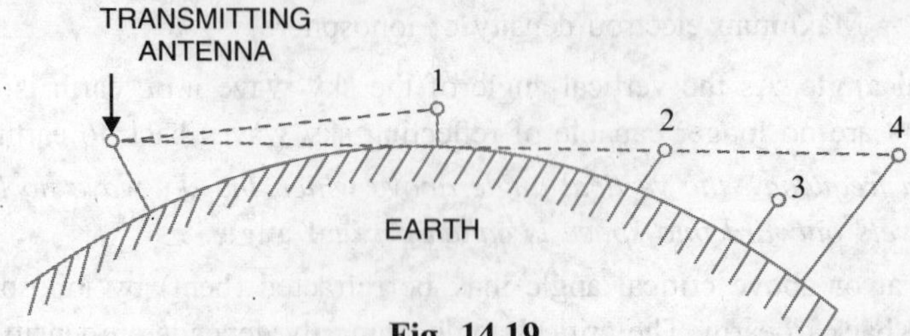

Fig. 14.19

14.15 Frequency Modulation (FM)

When the frequency of carrier wave is changed in accordance with the intensity of the signal, it is called **frequency modulation** (FM).

In frequency modulation, only the frequency of the carrier wave is changed in accordance with the signal. However, the amplitude of the modulated wave remains the same *i.e.* carrier wave amplitude. The frequency variations of carrier wave depend upon the instantaneous amplitude of the signal as shown in Fig. 14.20 (*iii*). When the signal voltage is zero as at *A*, *C*, *E* and *G*, the carrier frequency is unchanged. When the signal approaches its positive peaks as at *B* and *F*, the carrier frequency is increased to maximum as shown by the closely spaced cycles. However, during the negative peaks of signal as at *D*, the carrier frequency is reduced to minimum as shown by the widely spaced cycles.

Illustration. The process of frequency modulation (FM) can be made more illustrative if we consider numerical values. Fig. 14.21 shows the FM signal having carrier frequency f_c = 100 kHz. Note that FM signal has constant amplitude but varying frequencies above and below the carrier frequency of 100 kHz (= f_c). For this reason, f_c (= 100 kHz) is called *centre frequency*. The changes in the carrier frequency are produced by the audio-modulating signal. The amount of change in frequency from f_c (= 100 kHz) or *frequency deviation* depends upon the amplitude of the audio-modulating signal. The frequency deviation increases with the increase in the modulating signal and *vice-versa.* Thus the peak audio voltage will produce maximum frequency deviation. Referring to Fig. 14.21, the centre frequency is 100 kHz and the maximum frequency deviation is 30 kHz. The following points about frequency modulation (FM) may be noted carefully :

(*a*) The frequency deviation of FM signal depends on the amplitude of the modulating signal.

(*b*) The centre frequency is the frequency without modulation or when the modulating voltage is zero.

(*c*) The audio frequency (*i.e.* frequency of modulating signal) does not determine frequency deviation.

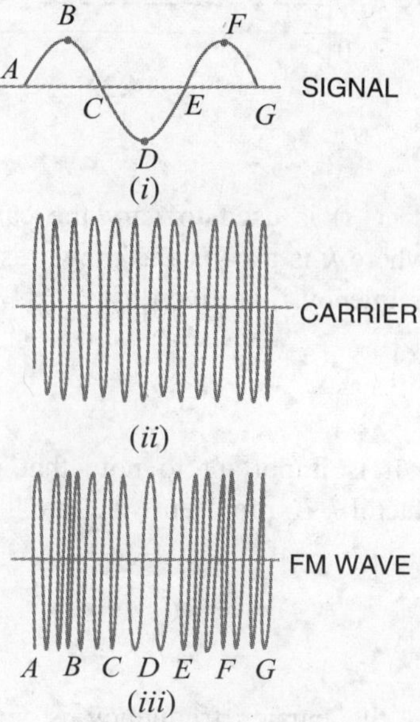

Fig. 14.20

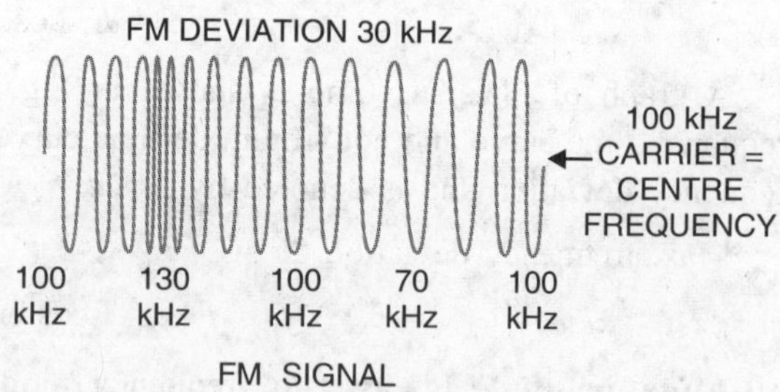

Fig. 14.21

Advantages

The following are the advantages of FM over AM :

(*i*) It gives noiseless reception. As discussed before, noise is a form of amplitude variations and a FM receiver will reject such signals.

(*ii*) The operating range is quite large.

(*iii*) It gives high-fidelity reception.

(*iv*) The efficiency of transmission is very high.

14.16 Theory of Frequency Modulation (FM)

In frequency modulation (FM), the amplitude of the carrier is kept constant but the frequency f_c of the carrier is varied by the modulating signal. The carrier frequency f_c varies at the rate of the *signal frequency f_s ; the frequency deviation being proportional to the instantaneous

* Note this point. It means that modulating frequency is the rate of frequency of deviations in the *RF* carrier. For example, all signals having the same amplitude will deviate the carrier frequency by the same amount, say 50 kHz, no matter what their frequencies. On similar lines, all signals of the same frequency, say, 3 kHz, will deviate the carrier at the same rate of 3000 times per second, no matter what their individual amplitudes.

amplitude of the modulating signal. Note that maximum frequency deviation is ($f_{c\,(max)} - f_c$) and occurs at the peak voltage of the modulating signal. Suppose we modulate a 100 MHz carrier by 1V, 1 kHz modulating signal and the maximum frequency deviation is 25 kHz. This means that the carrier frequency will vary sinusoidally between (100 + 0.025) MHz and (100 – 0.025) MHz at the rate of 1000 times per second. If the amplitude of the modulating signal is increased to 2V, then the maximum frequency deviation will be 50 kHz and the carrier frequency will vary between (100 + 0.05) MHz and (100 – 0.05) MHz at the rate of 1000 times per second.

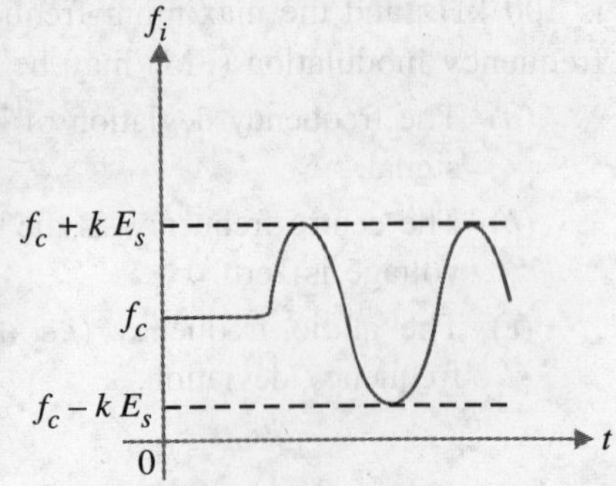

Fig. 14.22

Suppose a modulating sine-wave signal e_s (= $E_s \cos \omega_s t$) is used to vary the carrier frequency f_c. Let the change in carrier frequency be ke_s where k is a constant known as the *frequency deviation constant*. The instantaneous carrier frequency f_i is given by ;

$$f_i = f_c + k\,e_s$$

$$= f_c + k\,E_s \cos \omega_s\,t$$

A graph of f_i versus time is shown in Fig. 14.22. It is important to note that it is frequency-time curve and not amplitude-time curve. The factor $k\,E_s$ represents the maximum frequency deviation and is denoted by Δf *i.e.*

Max. frequency deviation, $\Delta f = {}^*kE_s$

$\therefore$ $$f_i = f_c + \Delta f \cos \omega_s\,t$$

Equation of FM wave. In frequency modulation, the carrier frequency is varied sinusoidally at signal frequency. The instantaneous deviation in frequency from the carrier is proportional to the instantaneous amplitude of the modulating signal. Thus the instantaneous angular frequency of FM is given by ;

$$\omega_i = \omega_c + \Delta\omega_c \cos \omega_s t$$

Total phase angle $\theta = \omega t$ so that if ω is variable, then,

$$\theta = \int_0^t \omega_i\,dt$$

$$= \int_0^t (\omega_c + \Delta\omega_c \cos \omega_s\,t)\,dt$$

$\therefore$ $$\theta = \omega_c t + \frac{\Delta\omega_c}{\omega_s} \sin \omega_s\,t$$

The term $\frac{\Delta\omega_c}{\omega_s}$ is called *modulation index* m_f.

$\therefore$ $$\theta = \omega_c t + m_f \sin \omega_s t$$

* Note that k is in kHz or MHz per volt.

The instantaneous value of FM voltage wave is given by ;

$$e = E_c \cos \theta$$

or
$$e = E_c \cos (\omega_c t + m_f \sin \omega_s t) \qquad ...(i)$$

Exp. (*i*) is the general voltage equation of a FM wave. The following points may be noted carefully :

(*i*) The modulation index m_f is the ratio of maximum frequency deviation (Δf) to the frequency ($= f_s$) of the modulating signal *i.e.*

$$\text{Modulation index, } m_f = \frac{\Delta\omega_c}{\omega_s} = \frac{f_{c\,(max)} - f_c}{f_s} = \frac{\Delta f}{f_s}$$

(*ii*) Unlike amplitude modulation, the modulation index (m_f) for frequency modulation can be greater than unity.

Frequency Spectrum. It requires advanced mathematics to derive the spectrum of FM wave. We will give only the results without derivation. If f_c and f_s are the carrier and signal frequencies respectively, then FM spectrum will have the following frequencies :

f_c ; $f_c \pm f_s$; $f_c \pm 2f_s$; $f_c \pm 3f_s$ and so on.

Note that $f_c + f_s$, $f_c + 2f_s$, $f_c + 3f_s$ ...... are the upper sideband frequencies while $f_c - f_s$, $f_c - 2f_s$, $f_c - 3f_s$ ...... are the lower sideband frequencies.

Example 14.20. *A frequency modulated voltage wave is given by the equation :*

$$e = 12 \cos (6 \times 10^8 t + 5 \sin 1250\, t)$$

Find (i) carrier frequency (ii) signal frequency (iii) modulation index (iv) maximum frequency deviation (v) power dissipated by the FM wave in 10-ohm resistor.

Solution. The given FM voltage wave is

$$e = 12 \cos (6 \times 10^8 t + 5 \sin 1250\, t) \qquad ...(i)$$

The equation of standard FM voltage wave is

$$e = E_c \cos (\omega_c t + m_f \sin \omega_s t) \qquad ...(ii)$$

Comparing eqs. (*i*) and (*ii*), we have,

(*i*) Carrier frequency, $f_c = \dfrac{\omega_c}{2\pi} = \dfrac{6 \times 10^8}{2\pi} = \mathbf{95.5 \times 10^6\ Hz}$

(*ii*) Signal frequency, $f_s = \dfrac{\omega_s}{2\pi} = \dfrac{1250}{2\pi} = \mathbf{199\ Hz}$

(*iii*) Modulation index, $m_f = \mathbf{5}$

(*iv*) Max. frequency deviation, $\Delta f = m_f \times f_s = 5 \times 199 = \mathbf{995\ Hz}$

(*v*) Power dissipated, $P = \dfrac{E_{r.m.s.}^2}{R} = \dfrac{(12/\sqrt{2})^2}{10} = \mathbf{7.2W}$

Example 14.21. *A 25 MHz carrier is modulated by a 400 Hz audio sine wave. If the carrier voltage is 4 V and the maximum frequency deviation is 10 kHz, write down the voltage equation of the FM wave.*

Solution. The voltage equation of the FM wave is

$$e = E_c \cos (\omega_c t + m_f \sin \omega_s t)$$

Here,

$$\omega_c = 2\pi f_c = 2\pi \times 25 \times 10^6 = 1.57 \times 10^8 \text{ rad/s}$$

$$\omega_s = 2\pi f_s = 2\pi \times 400 = 2513 \text{ rad/s}$$

$$m_f = \frac{\Delta f}{f_s} = \frac{10 \text{ kHz}}{400 \text{ Hz}} = \frac{10 \times 10^3 \text{ Hz}}{400 \text{ Hz}} = 25$$

$$\therefore \quad e = 4 \cos (1.57 \times 10^8 t + 25 \sin 2513t) \textbf{ Ans.}$$

Example 14.22. *Calculate the modulation index for an FM wave where the maximum frequency deviation is 50 kHz and the modulating frequency is 5 kHz.*

Solution.

Max. frequency deviation, $\Delta f = 50 \text{ kHz}$

Modulating frequency, $f_s = 5 \text{ kHz}$

$$\therefore \quad \text{Modulation index, } m_f = \frac{\Delta f}{f_s} = \frac{50 \text{ kHz}}{5 \text{ kHz}} = \mathbf{10}$$

Example 14.23. *The carrier frequency in an FM modulator is 1000 kHz. If the modulating frequency is 15 kHz, what are the first three upper sideband and lower sideband frequencies?*

Solution.

Carrier frequency, $f_c = 1000 \text{ kHz}$

Modulating frequency, $f_s = 15 \text{ kHz}$

Upper sideband frequencies

$f_c + f_s$; $f_c + 2 f_s$; $f_c + 3 f_s$

1000 + 15 ; 1000 + 2 × 15 ; 1000 + 3 × 15

1015 kHz ; **1030 kHz** ; **1045 kHz**

Lower sideband frequencies

$f_c - f_s$; $f_c - 2 f_s$; $f_c - 3 f_s$

1000 – 15 ; 1000 – 2 × 15 ; 1000 – 3 × 15

985 kHz ; **970 kHz** ; **955 kHz**

Example 14.24. *The carrier and modulating frequencies of an FM transmitter are 100 MHz and 15 kHz respectively. If the maximum frequency deviation is 75 kHz, find the bandwidth of FM signal.*

Solution. To calculate the exact bandwidth of an FM signal, it requires the use of advanced mathematics (Bessel functions) which is beyond the level of this book. However, the bandwidth of an FM signal is approximately given by ;

$$\text{Bandwidth, } BW = 2 [\Delta f + f_s] = 2 [75 + 15] = \mathbf{180 \text{ kHz}}$$

Example 14.25. *In a frequency modulated wave, frequency deviation constant is 75 kHz/volt and the signal amplitude is 2 V. Find the maximum frequency deviation.*

Solution.

Frequency deviation constant, $k = 75 \text{ kHz/V}$

Amplitude of signal, $E_s = 2 \text{ V}$

$$\therefore \quad \text{Max. frequency deviation, } \Delta f = k E_s = 75 \times 2 = \mathbf{150 \text{ kHz}}$$

Example 14.26. *In an FM system, when the audio frequency (AF) is 500 Hz and the AF voltage is 2.4 V, the frequency deviation is 4.8 kHz. If the AF voltage is now increased to 7.2 V, what is the new frequency deviation? If the AF voltage is raised to 10 V while the AF is dropped to 200 Hz, what is the deviation? Find the modulation index in each case.*

Solution. We know that :

$$\text{Frequency deviation, } \Delta f_1 = k\,E_s$$

$$\therefore \quad \text{Frequency deviation constant, } k = \frac{\Delta f_1}{E_s} = \frac{4.8}{2.4} = 2 \text{ kHz/V}$$

$$\text{For } E_s = 7.2\text{ V}, \quad \Delta f_2 = 2 \times 7.2 = \mathbf{14.4\ kHz}$$

$$\text{For } E_s = 10\text{ V}, \quad \Delta f_3 = 2 \times 10 = \mathbf{20\ kHz}$$

The answer of 20 kHz shows that deviation is independent of modulating frequency. The modulation indices in the three cases are :

$$m_{f1} = \frac{\Delta f_1}{f_{s1}} = \frac{4.8}{0.5} = \mathbf{9.6}$$

$$m_{f2} = \frac{\Delta f_2}{f_{s1}} = \frac{14.4}{0.5} = \mathbf{28.8}$$

$$m_{f3} = \frac{\Delta f_3}{f_{s2}} = \frac{20}{0.2} = \mathbf{100}$$

It is important to note that for calculating modulation index, the modulating frequency change had to be taken into account in the third case.

14.17 Comparison of FM and AM

The comparison of FM and AM is given in the table below.

S. No.	FM	AM
1.	The amplitude of carrier remains constant with modulation.	The amplitude of carrier changes with modulation.
2.	The carrier frequency changes with modulation.	The carrier frequency remains constant with modulation.
3.	The carrier frequency changes according to the strength of the modulating signal.	The carrier amplitude changes according to the strength of the modulating signal.
4.	The value of modulation index (m_f) can be more than 1.	The value of modulation factor (m) cannot be more than 1 for distortionless AM signal.

14.18 Demodulation

The process of recovering the audio signal from the modulated wave is known as **demodulation or detection.**

At the broadcasting station, modulation is done to transmit the audio signal over larger distances to a receiver. When the modulated wave is picked up by the radio receiver, it is necessary to recover the audio signal from it. This process is accomplished in the radio receiver and is called demodulation.

Necessity of demodulation. It was noted previously that amplitude modulated wave consists of carrier and sideband frequencies. The audio signal is contained in the sideband frequencies which are radio frequencies. If the modulated wave after amplification is directly fed to the speaker as shown in Fig. 14.23, no sound will be heard. It is because diaphragm of the speaker is not at all able to respond to such high frequencies. Before the diaphragm is able to move in one direction, the rapid reversal of current tends to move it in the opposite direction *i.e.* diaphragm will not move at all. Consequently, no sound will be heard.

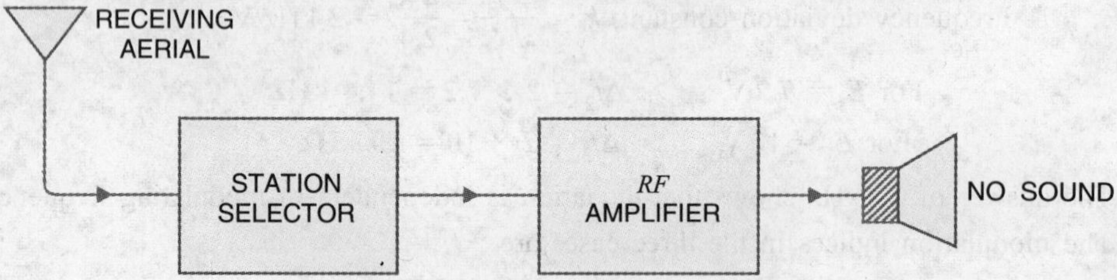

Fig. 14.23

From the above discussion, it follows that audio signal must be separated from the carrier at a suitable stage in the receiver. The recovered audio signal is then amplified and fed to the speaker for conversion into sound.

14.19 Essentials in Demodulation

In order that a modulated wave is audible, it is necessary to change the nature of modulated wave. This is accomplished by a circuit called *detector*. A detector circuit performs the following two functions :

(*i*) **It rectifies the modulated wave** *i.e.* negative half of the modulated wave is eliminated. As shown in Fig. 14.24 (*i*), a modulated wave has positive and negative halves exactly equal. Therefore, average current is zero and speaker cannot respond. If the negative half of this modulated wave is eliminated as shown in Fig. 14.24 (*ii*), the average value of this wave will not be zero since the resultant pulses are now all in one direction. The average value is shown by the dotted line in Fig. 14.24 (*ii*). Therefore, the diaphragm will have definite displacement corresponding to the average value of the wave. It may be seen that shape of the average wave is similar to that of the modulation envelope. As the signal is of the same shape as the envelope, therefore, average wave shape is of the same form as the signal.

(*ii*) **It separates the audio signal from the carrier.** The rectified modulated wave contains the audio signal and the carrier. It is desired to recover the audio signal. This is achieved by a filter circuit which removes the carrier frequency and allows the audio signal to reach the load *i.e.* speaker.

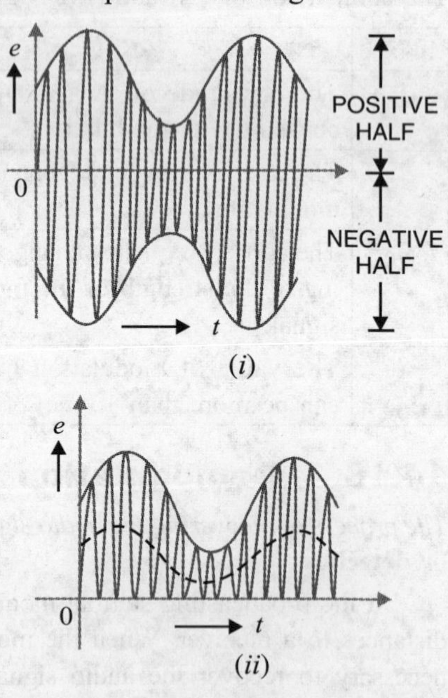

Fig. 14.24

14.20 A.M. Diode Detector

Fig. 14.25 shows a simple detector circuit employing vacuum diode and filter circuit. The modulated wave of desired frequency is selected by the parallel tuned circuit L_1C_1 and is applied to the vacuum diode. During the positive half-cycles of modulated wave, the diode conducts while during negative half-cycles, it does not. The result of this rectifying action is that output of the diode consists of positive half-cycles of modulated wave as shown.

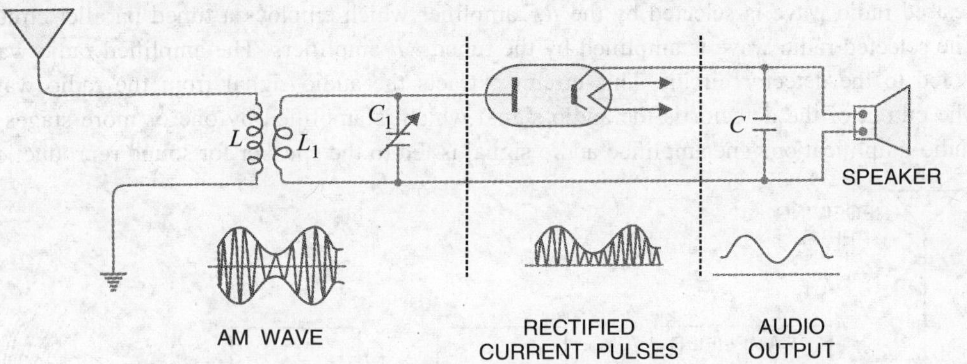

Fig. 14.25

The rectified modulated wave contains radio frequency and the signal and cannot be fed to the speaker for sound reproduction. If done so, no sound will be heard due to the inertia of speaker diaphragm. The *RF* component is filtered by the capacitor *C* shunted across the speaker. The value of this capacitor is sufficiently large to present low reactance to the *RF* component while presenting a relatively high reactance to the audio signal. The result is that the *RF* component is bypassed by the capacitor *C* and the signal is passed on to the speaker for sound reproduction.

Note. If vacuum diode is replaced by a crystal diode, the circuit becomes crystal diode detector.

14.21 A.M. Radio Receivers

A radio receiver is a device which reproduces the modulated or radio waves into sound waves. In India, only amplitude modulation is used for radio transmission and reception. Therefore, such radio receivers are called A.M. radio receivers. In order to reproduce the A.M. wave into sound waves, every radio receiver must perform the following functions :

(*i*) The receiving aerial must intercept a portion of the passing radio waves.

(*ii*) The radio receiver must select the desired radio wave from a number of radio waves intercepted by the receiving aerial. For this purpose, tuned parallel *LC* circuits must be used. These circuits will select only that radio frequency which is in resonance with them.

(*iii*) The selected radio wave must be amplified by the tuned frequency amplifiers.

(*iv*) The audio signal must be recovered from the amplified radio wave.

(*v*) The audio signal must be amplified by suitable number of audio-amplifiers.

(*vi*) The amplified audio signal should be fed to the speaker for sound reproduction.

14.22 Types of A.M. Radio Receivers

A.M. radio receivers can be broadly classified into two types *viz.*, *straight radio receiver* and *superheterodyne radio receiver.* The former was used in the early days of radio communication. However at present, all radio receivers are of superheterodyne type.

1. Straight radio receiver. Fig. 14.26 shows the block diagram of a straight radio receiver. The aerial is receiving radio waves from different broadcasting stations. The desired radio wave is selected by the *RF* amplifier which employs a tuned parallel circuit. The selected radio wave is amplified by the tuned *RF* amplifiers. The amplified radio wave is fed to the detector circuit. This circuit extracts the audio signal from the radio wave. The output of the detector is the audio signal which is amplified by one or more stages of audio-amplification. The amplified audio signal is fed to the speaker for sound reproduction.

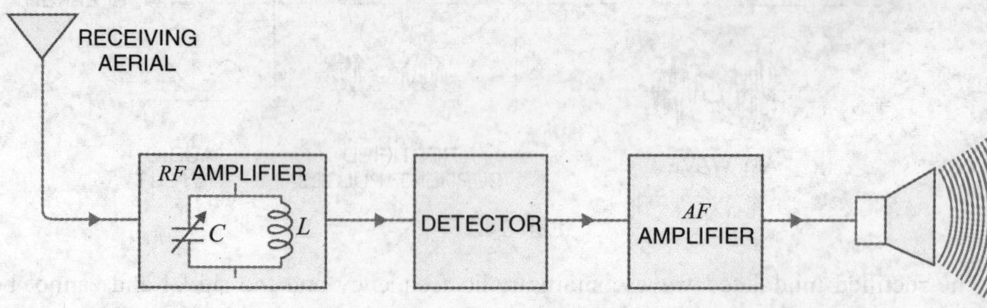

Fig. 14.26

Limitations

(*i*) In straight radio receivers, tuned circuits are used. As it is necessary to change the value of variable capacitors (gang capacitors) for tuning to the desired station, therefore, there is a considerable variation of Q between the closed and open positions of the variable capacitors. This changes the sensitivity and selectivity of the radio receivers.

(*ii*) There is too much interference of adjacent stations.

2. Superheterodyne receiver. The shortcomings of straight radio receiver were overcome by the invention of superheterodyne receiver by Major Edwin H. Armstrong during the First World War. At present, all modern receivers utilise the superheterodyne circuit. In this type of radio receiver, the selected radio frequency is converted to a fixed lower value, called *intermediate frequency* (*IF*). This is achieved by a special electronic circuit called *mixer circuit.* There is a local oscillator in the radio receiver itself. This oscillator produces high frequency waves. The selected radio frequency is mixed with the high frequency wave by the mixer circuit. In this process, beats are produced and the *mixer produces a frequency equal to the difference between local oscillator and radio wave frequency.* As explained later, the circuit is so designed that oscillator always produces a frequency 455 kHz above the selected radio frequency. Therefore, the mixer will always produce an intermediate frequency of 455 kHz regardless of the station to which the receiver is tuned. For instance, if 600 kHz station is tuned, then local oscillator will produce a frequency of 1055 kHz. Consequently, the output from the mixer will have a frequency of 455 kHz. Fig. 14.27 shows the superheterodyne principle with a block diagram. The selected radio frequency f_1 is mixed with a frequency f_2 from a local oscillator. The output from the mixer is a difference (*i.e.* $f_2 - f_1$) and is always 455 kHz regardless of the station to which the receiver is tuned.

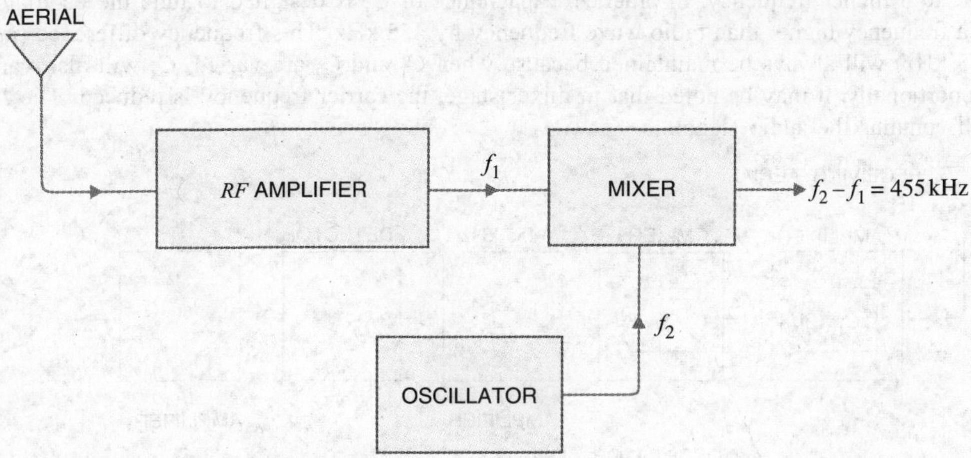

Fig. 14.27

The production of fixed intermediate frequency (455 kHz) is the salient feature of superheterodyne circuit. At this fixed intermediate frequency, the amplifier circuits operate with maximum stability, selectivity and sensitivity. As the conversion of incoming radio frequency to the intermediate frequency is achieved by *heterodyning* or beating the local oscillator against radio frequency, therefore, this circuit is called *superheterodyne circuit.

14.23 Stages of Superheterodyne Radio Receiver

Fig. 14.28 shows the block diagram of a superheterodyne receiver. It may be seen that *RF* amplifier stage, mixer stage and oscillator stage use tuned parallel circuits with variable capacitors. These capacitors are ganged together as shown by the dotted interconnecting lines. The rotation of the common shaft simultaneously changes the capacitance of these tuned circuits.

(*i*) RF amplifier stage. The *RF* amplifier stage uses a tuned parallel circuit L_1C_1 with a variable capacitor C_1. The radio waves from various broadcasting stations are intercepted by the receiving aerial and are coupled to this stage. This stage selects the desired radio wave and raises the strength of the wave to the desired level.

(*ii*) Mixer stage. The amplified output of *RF* amplifier is fed to the mixer stage where it is combined with the output of a local oscillator. The two frequencies beat together and produce an intermediate frequency (*IF*). The intermediate frequency is the difference between oscillator frequency and radio frequency *i.e.*

$$I.F. = \text{Oscillator frequency} - \text{Radio frequency}$$

The *IF* is always 455 kHz regardless of the frequency to which the receiver is tuned. The reason why the mixer will always produce 455 kHz frequency above the radio frequency

* In a super-heterodyne receiver, the heterodyne principle is used to produce an intermediate frequency which is higher than that can be heard *i.e.* supersonic. Superheterodyne is short for supersonic heterodyne.

is that oscillator always produces a frequency 455 kHz *above the selected radio frequency. This is achieved by making C_3 smaller than C_1 and C_2. By making C_3 smaller, oscillator will tune to a higher frequency. In practice, capacitance of C_3 is designed to tune the oscillator to a frequency higher than radio wave frequency by 455 kHz. This frequency difference (*i.e.* 455 kHz) will always be maintained because when C_1 and C_2 are varied, C_3 will also vary proportionally. It may be noted that in mixer stage, the carrier frequency is reduced. The *IF* still contains the audio signal.

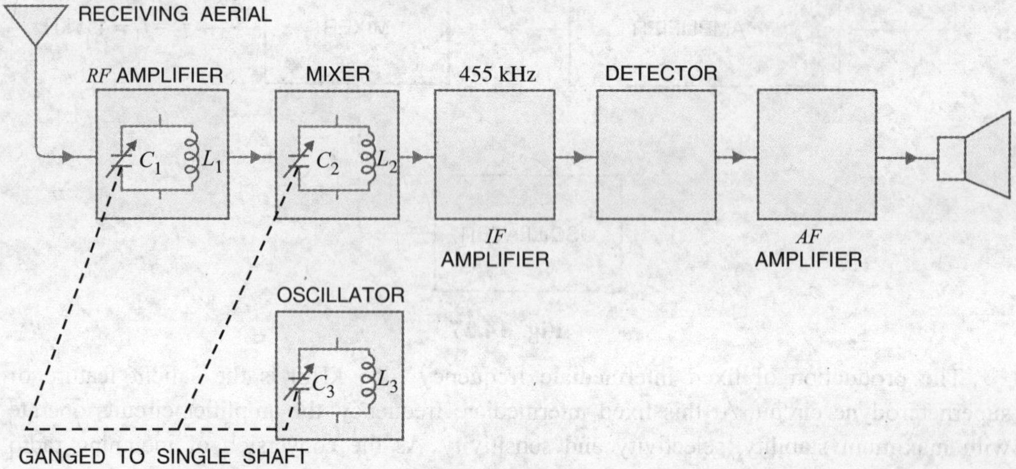

Fig. 14.28

(*iii*) IF amplifier stage. The output of mixer is always 455 kHz and is fed to fixed tuned *IF* amplifiers. These amplifiers are tuned to one frequency (*i.e.* 455 kHz) and render nice amplification.

(*iv*) Detector stage. The output from the last *IF* amplifier stage is coupled to the input of the detector stage. Here, the audio signal is extracted from the *IF* output. Usually, diode detector circuit is used because of its low distortion and excellent audio fidelity.

(*v*) A.F. amplifier stage. The audio signal output of detector stage is fed to a multistage audio amplifier. Here, the signal is amplified until it is sufficiently strong to drive the speaker. The speaker converts the audio signal into sound waves corresponding to the original sound at the broadcasting station.

14.24 Advantages of Superheterodyne Circuit

The basic principle involved in superheterodyne circuit is to obtain a fixed intermediate frequency with the help of a mixer circuit and local oscillator. The superheterodyne principle has the following advantages :

(*i*) High RF amplification. The superheterodyne principle makes it possible to produce an intermediate frequency (*i.e.* 455 kHz) which is much less than the radio frequency. *RF*

* The reason that the oscillator is designed to produce a frequency 455 kHz above and not below the selected frequency is as follows. A radio receiver is required to tune over 550 to 1600 kHz frequency. To provide *IF* of 455 kHz, the oscillator frequency must vary from 1005 to 2055 kHz. If the oscillator is designed to produce a frequency 455 kHz below the selected frequency (of course *IF* will be still 455 kHz), then the frequency range of the oscillator will have to be 95 to 1145 kHz. This frequency ratio is too high to be covered in a single band.

amplification at low frequencies is more stable since feedback through stray and interelectrode capacitance is reduced.

(*ii*) **Improved selectivity.** Losses in the tuned circuits are lower at intermediate frequency. Therefore, the quality factor Q of the tuned circuits is increased. This makes the amplifier circuits to operate with maximum selectivity.

(*iii*) **Lower cost.** In a superheterodyne circuit, a fixed intermediate frequency is obtained regardless of the radio wave selected. This permits the use of fixed *RF* amplifiers. The superheterodyne receiver is thus cheaper than other radio receivers.

14.25 FM Receiver

The FM receiver is more complicated and, therefore, more expensive than the normal AM receiver. As we shall see, an FM receiver also uses superheterodyne principle. The FM broadcast signals lie in the frequency range between 88 MHz and 108 MHz. The *IF* (intermediate frequency) of an FM receiver is 10.7 MHz—much *higher than the *IF* value of 455 kHz in AM receivers. Fig. 14.29 shows the block diagram of an FM receiver. In the interest of understanding, we shall discuss the various sections of the FM receiver.

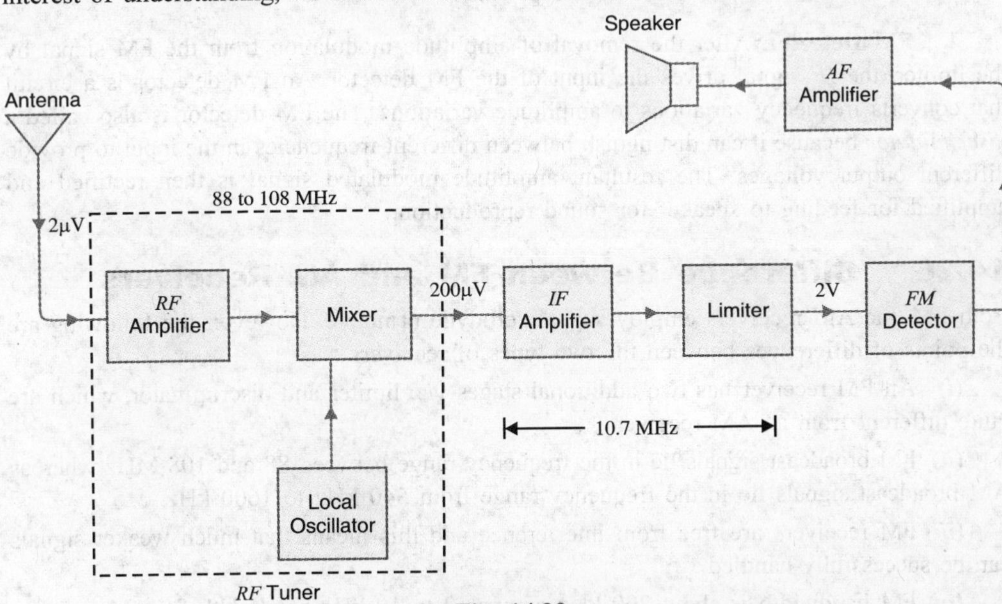

Fig. 14.29

1. **RF Tuner.** The FM signals are in the frequency range of 88 to 108 MHz. The weak FM signal (say 2 μV) is picked up by the antenna and is fed to the *RF* tuner. The *RF* tuner consists of (*i*) *RF* amplifier (*ii*) Mixer and (*iii*) local oscillator. The *RF* amplifier amplifies the selected FM signal (to 200 μV in the present case). The output from the *RF* amplifier is fed to the mixer stage where it is combined with the output signal from a local oscillator. The two frequencies beat together and produce an intermediate frequency (*IF*). The intermediate frequency (*IF*) is equal to the difference between oscillator frequency and the *RF* frequency. The *IF* is always 10.7 MHz (Recall *IF* in AM receiver is 455 kHz) regardless of the frequency to which the FM receiver is tuned.

* It is because the *RF* carrier frequencies for FM radio broadcasting are in the 88 to 108 MHz band.

2. IF Amplifier Stage. The output signal from the mixer always has a frequency of 10.7 MHz and is fed to the *IF* amplifiers. Since *IF* amplifiers are tuned to *IF* (= 10.7 MHz), they render nice amplification. Note that bandwidth of *IF* amplifiers is about 200 kHz or 0.2 MHz. The *IF* gain is very large (assumed 10,000 in this case) so that output is 2 V.

3. Limiter Stage. The output from *IF* stage is fed to the limiter. This circuit is an *IF* amplifier tuned to 10.7 MHz but its main function is to remove AM interference from the FM signal. Fig. 14.30 shows how the limiter removes AM interference from the FM signal. The input is an FM signal, but it has different amplitude levels because of AM interference has been added. However, the limiter circuit keeps the output level constant for different input levels.

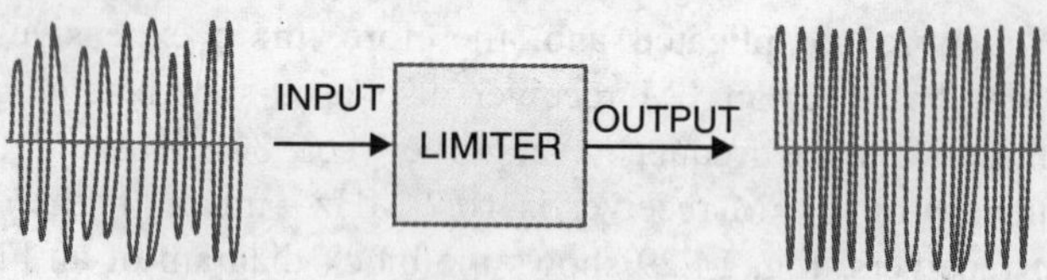

Fig. 14.30

4. FM Detector. After the removal of amplitude modulation from the FM signal by the limiter, the *IF* signal drives the input of the FM detector. An FM detector is a circuit that converts frequency variations to amplitude variations. The FM detector is also called a *discriminator* because it can distinguish between different frequencies in the input to provide different output voltages. The resultant amplitude modulated signal is then rectified and amplified for feeding to speaker for sound reproduction.

14.26 Difference Between FM and AM Receivers

Both FM and AM receivers employ superheterodyne principle. However, the following are the points of differences between the two types of receivers :

(*i*) An FM receiver has two additional stages *viz.* limiter and discriminator, which are quite different from an AM receiver.

(*ii*) FM broadcast signals lie in the frequency range between 88 and 108 MHz whereas AM broadcast signals lie in the frequency range from 540 kHz to 1600 kHz.

(*iii*) FM receivers are free from interference and this means that much weaker signals can be successfully handled.

(*iv*) FM bandwidth is about 200 kHz compared to 10 kHz bandwidth for AM.

(*v*) The *IF* for FM receivers is 10.7 MHz whereas *IF* for AM receivers is 455 kHz.

Short Answer Questions

Q.1. *What is carrier communication?*

Ans. Carrier communication involves the use of a high frequency alternating current (called carrier) as the basic communication agent and the modification of this high frequency current by the audio signal (comparatively low frequency currents) which contains the message to be transmitted.

Q.2. *Why is carrier communication necessary for radio engineering?*

Ans. The audio signal is transmitted by a carrier because of two principal reasons. First, any 2-wire audio frequency communication circuit is limited to one message at a time. If two or more messages are sent over the circuit at a time, their mutual interference render them unintelligible. Secondly, audio frequency signals cannot be sent directly into space, in the form of radio waves, since the radiation of energy from an antenna takes place economically only at frequencies higher than 30 kHz, which is well out of audio-frequency range.

Q.3. *If a broadcasting station is transmitting but not modulating, what is heard in a radio receiver tuned to its frequency?*

Ans. If a broadcasting station is transmitting but not modulating, it means that the station is sending the carrier wave only. The speaker in the radio receiver will produce no sound from this carrier. It is because the carrier frequency is generally very high and the speaker cannot respond to such rapid variations due to mechanical inertia.

Q.4. *When a carrier wave is amplitude modulated, what happens to the power level of the carrier wave itself?*

Ans. The equation of AM wave is given by ;

$$e = E_C \cos \omega_c t + \frac{mE_C}{2} \cos(\omega_c + \omega_s)t + \frac{mE_C}{2} \cos(\omega_c - \omega_s)t$$

If we examine this equation, it is seen that voltage output at the carrier frequency has a peak value of E_C. But if the modulating signal were removed, the output voltage at this frequency would still have a peak value of E_C. Since the carrier itself is unaffected by the modulating signal, *there is no change in the carrier power level.*

Q.5. *Why does the conductivity of earth's atmosphere increase with altitude?*

Ans. Due to low pressure at high altitude, the high energy ultraviolet radiation and *X*-rays from the sun cause ionisation of atoms of the gases present there. As a result, the concentration of electrons and positive ions is very large. This region is called ionosphere. The ionising power of these radiations decreases rapidly as they approach earth. For this reason, the conductivity of earth's atmosphere increases with altitude.

Q.6. *The E-layer of the ionosphere disappears at night. Why?*

Ans. The ionisation in the *E*-layer (90 km to 140 km from the surface of earth) of the ionosphere is caused by ultraviolet radiation and *X*-rays from the sun. At night, sun's radiation is absent and the recombination of ions and electrons takes place. Therefore, the *E*-layer of the ionosphere disappears at night.

Q.7. *The F-layer of ionosphere persists at night. Why?*

Ans. The *F*-layer (140 km to 400 km from the surface of earth) of the ionosphere persists at night due to the following reasons :

(*i*) In the *F*-layer, the density of air is low. Therefore, the ions are widely separated and their recombination is partial.

(*ii*) The *F*-layer of ionosphere is highly ionised so that inspite of recombination of ions, the ionisation persists to some degree.

Q.8. *What is superheterodyne principle?*

Ans. The superheterodyne receiver is based on the principle of converting the selected radio frequency to a lower fixed value, called intermediate frequency (*IF*). The receiver contains an oscillator which is known as local oscillator. The output from this oscillator is mixed with the incoming radio wave in a special electronic circuit called mixer circuit. The output of the mixer contains (*i*) original frequencies (*ii*) sum of two frequencies and (*iii*) difference of two frequencies. By tuning the output of the mixer to difference frequency (*i.e. IF*), *IF* is retained while the others are rejected.

Q.9. *What is the origin of the name superheterodyne?*

Ans. The process of producing difference of two frequencies (*i.e. IF*) by mixing the two is known as heterodyning. The prefix 'super' is a contraction of the word "supersonic" which means frequencies above the range of hearing. Therefore superheterodyne means production of beat frequencies above the range of hearing.

Q.10. *How do we get fixed IF (i.e. 455 kHz) in a superheterodyne receiver?*

Ans. The capacitors of tuned circuit of *RF* amplifier and local oscillator are ganged together. The rotation of the common shaft simultaneously changes the capacitance of these tuned circuits. The values of these capacitors are so selected that the oscillator always produces a frequency 455 kHz above the selected radio frequency. Therefore, the mixer will always produce an intermediate frequency of 455 kHz regardless of the station to which the receiver is tuned.

Q.11. *What are the advantages of obtaining a fixed intermediate frequency (i.e. 455 kHz) in a superheterodyne receiver?*

Ans. There are two principal advantages of obtaining a fixed intermediate frequency. First, since *IF* (*i.e.* 455 kHz) is much lower than radio frequency, higher amplification and better gain stability can be obtained. This is expected because now *IF* amplifiers will work on a fixed lower frequency *i.e. IF*. Secondly, the design of *IF* amplifiers will be simple because they are to handle only one frequency *i.e. IF*.

Q.12. *Why do we always design the oscillator circuit to produce a frequency 455 kHz above and not below the incoming radio wave?*

Ans. A standard radio receiver is capable of tuning to a frequency range of 550 kHz to 1600 kHz. If the oscillator circuit is designed to produce a frequency 455 kHz below the incoming radio wave, then frequency range may become too large for the receiver to handle. Thus to provide an *IF* of 455 kHz, the oscillator must vary from 1005 kHz to 2055 kHz and must always remain 455 kHz above the selected radio frequency. If the local oscillator produces a frequency 455 kHz below the incoming radio wave, then frequency range would become 95 kHz to 1145 kHz. No receiver has yet been designed that will cover this entire range successfully.

Q.13. *Why is the output frequency from the mixer called intermediate frequency?*

Ans. *IF* = Oscillator frequency – *RF* signal frequency

Note that *IF* is still a radio frequency. However, to distinguish it from the incoming signal frequency–and also since it lies between the original carrier and signal frequencies–it is called the intermediate frequency (*IF*).

Q.14. *Why have we selected IF as 455 kHz for radio communication?*

Ans. We know that bandwidth is a function of operating frequency and circuit Q *i.e.*

$$BW = \frac{f}{Q}$$

In communication, we need a bandwidth of 10 kHz in order to have reasonable good reproduction of speech *etc.* The value of Q should be reasonably high–a figure of 50 is good enough. Based on Q of 50, *IF* should be

$$IF = BW \times Q$$

$$= 10 \text{ kHz} \times 50 = 500 \text{ kHz}$$

It has been decided to standardise *IF* at 455 kHz. An advantage of standardising on one fixed value is that this frequency can be cleared (*i.e.* unassigned), thereby preventing direct pickup of signals into a sensitive *IF* system.

Q.15. *A 40 kW carrier is modulated to a level of (i) 100% (ii) 10%. What is the sideband power in each case?*

Ans. (*i*) Sideband power, $P_s = \frac{1}{2} m^2 P_c = \frac{1}{2}(1)^2 \times 40 = 20$ kW

(*ii*) Sideband power, $P_s = \frac{1}{2} m^2 P_c = \frac{1}{2}(0.1)^2 \times 40 = 0.2$ kW = 200W

Note that in case (*i*), 40 kW carrier is sending 20 kW of intelligence (*i.e.* signal). However in case (*ii*), same carrier level – 40 kW– is used to send a mere 200 W of intelligence. Clearly, efficiency of operation decreases rapidly for low degree of modulation.

Multiple-Choice Questions

1. Modulation is done in ....
(*i*) transmitter
(*ii*) radio receiver
(*iii*) between transmitter and radio receiver
(*iv*) none of the above

2. In a transmitter, ... oscillator is used.
(*i*) Hartley (*ii*) *RC* phase-shift
(*iii*) Wien-bridge (*iv*) crystal

3. In India, ... modulation is used for radio transmission.
(*i*) frequency (*ii*) amplitude
(*iii*) phase
(*iv*) none of the above

4. In an AM wave, useful power is carried by ....
(*i*) carrier
(*ii*) sidebands
(*iii*) both sidebands and carrier
(*iv*) none of the above

5. In amplitude modulation, bandwidth is .... the audio signal frequency.
(*i*) thrice (*ii*) four times
(*iii*) twice
(*iv*) none of the above

6. In amplitude modulation, the .... of carrier is varied according to the strength of the signal.
(*i*) amplitude (*ii*) frequency
(*iii*) phase
(*iv*) none of the above

7. Over modulation (amplitude) occurs when signal amplitude is .... carrier amplitude.
(*i*) equal to (*ii*) greater than
(*iii*) less than
(*iv*) none of the above

8. In an AM wave, the majority of the power is in ....
(*i*) lower sideband
(*ii*) upper sideband
(*iii*) carrier
(*iv*) none of the above

9. At 100% modulation, the power in each sideband is .... of that of carrier.
(*i*) 50% (*ii*) 40%
(*iii*) 60% (*iv*) 25%

10. Over modulation results in ....
(*i*) weakening of the signal

(ii) excessive carrier power
(iii) distortion
(iv) none of the above

11. If modulation is 100%, then signal amplitude is .... carrier amplitude.
(i) equal to (ii) greater than
(iii) less than
(iv) none of the above

12. As the modulation level is increased, the carrier power ....
(i) is increased
(ii) remains the same
(iii) is decreased
(iv) none of the above

13. Demodulation is done in ....
(i) receiving antenna
(ii) transmitter
(iii) radio receiver
(iv) transmitting antenna

14. A high Q tuned circuit will permit an amplifier to have high ....
(i) fidelity (ii) frequency range
(iii) sensitivity (iv) selectivity

15. In radio transmission, the medium of transmission is ....
(i) space (ii) an antenna
(iii) a cable
(iv) none of the above

16. If level of modulation is increased, ... power is increased.
(i) carrier
(ii) sideband
(iii) carrier as well as sideband
(iv) none of the above

17. In TV transmission, picture signal is .... modulated.
(i) frequency (ii) phase
(iii) amplitude
(iv) none of the above

18. In a radio receiver, noise is generally developed at ....
(i) IF stage
(ii) receiving antenna
(iii) audio stage (iv) RF stage

19. Man made noises are .... variations.
(i) amplitude (ii) frequency
(iii) phase
(iv) both phase and frequency

20. The signal voltage induced in the aerial of a radio receiver is of the order of ....
(i) mV (ii) μV
(iii) V
(iv) none of the above

21. Superheterodyne principle refers to ....
(i) using a large number of amplifier stages
(ii) using a push-pull circuit
(iii) obtaining lower fixed intermediate frequency
(iv) none of the above

22. If a radio receiver amplifies all the signal frequencies equally well, it is said to have high ....
(i) sensitivity (ii) selectivity
(iii) distortion (iv) fidelity

23. Most of the amplification in a superheterodyne receiver occurs at .... stage.
(i) IF (ii) RF amplifier
(iii) audio amplifier (iv) detector

24. The letters AVC stand for ....
(i) audio voltage control
(ii) abrupt voltage control
(iii) automatic volume control
(iv) automatic voltage control

25. The superheterodyne principle provides selectivity at .... stage.
(i) RF (ii) IF
(iii) audio (iv) before RF

26. In superheterodyne receiver, the input at the mixer stage is ....

(*i*) *IF* and *RF*
(*ii*) *RF* and *AF*
(*iii*) *IF* and *AF*
(*iv*) *RF* and local oscillator signal

27. The major advantage of FM over AM is ....
(*i*) reception is less noisy
(*ii*) higher carrier frequency
(*iii*) smaller bandwidth
(*iv*) small frequency deviation

28. When the modulating signal controls the frequency of the carrier, we get, ...
(*i*) phase modulation
(*ii*) amplitude modulation
(*iii*) frequency modulation
(*iv*) may be any one of the above

29. Modulation refers to a low-frequency signal controlling the ....
(*i*) amplitude of the carrier
(*ii*) frequency of the carrier
(*iii*) phase of the carrier
(*iv*) may be any one of the above

30. The *IF* is 455 kHz. If the radio receiver is tuned to 855 kHz, the local oscillator frequency is ....
(*i*) 455 kHz (*ii*) 1310 kHz
(*iii*) 1500 kHz (*iv*) 1520 kHz

31. If A_{min} = 40 and A_{max} = 60, what is the percentage modulation?
(*i*) 20% (*ii*) 40%
(*iii*) 50% (*iv*) 10%

32. The function of ferrite antenna is to ....
(*i*) reduce stray capacitance
(*ii*) stabilise d.c. bias
(*iii*) increase the *Q* of tuned circuit
(*iv*) reduce noise

33. In a radio receiver, we generally use .... oscillator as a local oscillator.
(*i*) crystal (*ii*) Wien-bridge
(*iii*) phase-shift (*iv*) Hartley

34. A 100 V carrier is made to vary between 160 V and 40 V by the signal. What is the modulation factor?
(*i*) 0.3 (*ii*) 0.6
(*iii*) 0.5
(*iv*) none of the above

35. A 50 kW carrier is to be amplitude modulated to a level of 85%. What is the carrier power after modulation?
(*i*) 50 kW (*ii*) 42.5 kW
(*iii*) 58.8 kW (*iv*) 25 kW

36. In the above question, what is the power in sidebands?
(*i*) 7.8 kW (*ii*) 11.6 kW
(*iii*) 19.06 kW (*iv*) 15.9 kW

37. In a superheterodyne receiver, the difference frequency is chosen as the *IF* rather than the sum frequency because ....
(*i*) the difference frequency is closer to oscillator frequency
(*ii*) lower frequencies are easier to amplify
(*iii*) only the difference frequency can be modulated
(*iv*) none of the above

38. The diode detector in an AM radio receiver is usually found ....
(*i*) before the first *RF* stage
(*ii*) after the first *RF* stage
(*iii*) after several stages of amplification
(*iv*) none of the above

39. In a *TRF* radio receiver, the *RF* and detection stages are tuned to ....
(*i*) radio frequency
(*ii*) *IF*
(*iii*) audio frequency
(*iv*) none of the above

40. In TV transmission, sound signal is .... modulated.
(*i*) amplitude (*ii*) frequency
(*iii*) phase
(*iv*) none of the above

Answers to Multiple-Choice Questions

1. (*i*)	2. (*iv*)	3. (*ii*)	4. (*ii*)	5. (*iii*)
6. (*i*)	7. (*ii*)	8. (*iii*)	9. (*iv*)	10. (*iii*)
11. (*i*)	12. (*ii*)	13. (*iii*)	14. (*iv*)	15. (*i*)
16. (*ii*)	17. (*iii*)	18. (*iv*)	19. (*i*)	20. (*ii*)
21. (*iii*)	22. (*iv*)	23. (*i*)	24. (*iii*)	25. (*ii*)
26. (*iv*)	27. (*i*)	28. (*iii*)	29. (*iv*)	30. (*ii*)
31. (*i*)	32. (*iii*)	33. (*iv*)	34. (*ii*)	35. (*i*)
36. (*iii*)	37. (*ii*)	38. (*iii*)	39. (*i*)	40. (*ii*)

Chapter Review Topics

1. Explain the general principles of radio broadcasting, transmission and reception.
2. What is modulation? Why is modulation necessary in communication system?
3. Explain amplitude modulation. Derive the voltage equation of an AM wave.
4. What do you understand by modulation factor? What is its significance?
5. Draw the waveform of AM wave for the following values of modulation factor :
   (*i*) 0 (*ii*) 0.5 (*iii*) 1 (*iv*) 1.5
6. What do you understand by sideband frequencies in an AM wave?
7. Derive an expression for the fraction of total power carried by the sidebands in amplitude modulation.
8. What are the limitations of amplitude modulation?
9. What do you understand by frequency modulation? Explain its advantages over amplitude modulation.
10. What is demodulation? What are essentials in demodulation?
11. Draw the diode detector circuit and explain its action.
12. What is superheterodyne principle? Explain the function of each stage of superheterodyne receiver with the help of a block diagram.

Problems

1. The maximum peak-to-peak voltage of an AM wave is 16 mV while the minimum peak-to-peak voltage is 8 mV. Find the percentage modulation. **[60%]**
2. A carrier of peak voltage 0.05 V and frequency 200 kHz is amplitude modulated by a signal of peak voltage 10 V and frequency 1 kHz. Find (*i*) frequencies in the output spectrum and (*ii*) the peak values of output components if m = 0.5 and voltage gain A = 100.
   [(*i*) 199 kHz, 200 kHz, 201 kHz (*ii*) 1.25 V, 5 V, 1.25 V]
3. An AM transmitter supplies 10 kW to the antenna when unmodulated. Determine the total power radiated by the transmitter when modulated to 30%. **[10.45 kW]**
4. A certain AM transmitter radiates 8 kW with carrier unmodulated and 9 kW when the carrier is modulated. Find the percentage modulation. **[50%]**
5. A transmitter radiates a total power of 10 kW. The carrier is modulated to a depth of 60%. Calculate (*i*) the power in the carrier and (*ii*) power in each sideband.
   [(*i*) 8.47 kW (*ii*) 0.765 kW]
6. A carrier of 100 V and 1500 kHz is modulated by 60 V, 1200 Hz sinusoidal signal. Calculate modulation factor and express this as percentage. **[0.6; 60%]**

7. A carrier with an amplitude of 140 V is modulated by a signal with an amplitude of 80 V. What is the percentage modulation? What is the amplitude of lower sideband frequency? **[57%; 40 V]**

8. A 50 kW carrier is to be modulated to a level of 85%. What is the carrier power after modulation? What is sideband power? **[50 kW; 19.06 kW]**

9. The r.m.s. antenna current of a radio transmitter is 10 A when unmodulated, rising to 12 A when the carrier is sinusoidally modulated. What is the modulation index? **[0.94]**

10. The r.m.s. antenna current of an AM transmitter increases by 15% over the unmodulated value when sinusoidal modulation is applied. Determine the modulation index. **[0.8]**

11. A 500 Hz modulating voltage produces a frequency deviation of 2.25 kHz. What is the modulation index? If the amplitude of the modulation voltage is kept constant, but its frequency is raised to 6 kHz, what is the new deviation? **[4.5; 54 kHz]**

12. When the modulating frequency in an FM system is 400 Hz and the modulation voltage is 2.4 V, the modulation index is 60. Calculate the maximum deviation. What is the modulation index when the modulating frequency is reduced to 250 Hz and the modulating voltage is simultaneously raised to 3.2 V? **[24 kHz; 128]**

CHAPTER 15

Regulated D.C. Power Supply

Learning Objectives

- Understand the ordinary d.c. power supply and its limitations.
- Learn the important characteristics of d.c. power supply.
- Develop a strong understanding of regulated d.c. power supply and its need.
- Explore the operation of zener diode voltage regulator.
- Develop an understanding of transistor series voltage regulator.
- Understand transistor shunt voltage regulator.
- Learn transistor shunt feedback voltage regulator.
- Understand the various types of IC voltage regulators.

Introduction

In general, electronic circuits using tubes or transistors require a source of d.c. power. For example, in tube amplifiers, d.c. voltage is needed for plate, screen grid and control grid. Similarly, the emitter and collector bias in a transistor must also be a direct current. Batteries are rarely used for this purpose as they are costly and require frequent replacement. In practice, d.c. power for electronic circuits is most conveniently obtained from commercial a.c. lines by using rectifier-filter system, called a *d.c. power supply*.

The rectifier-filter combination constitutes an ordinary d.c. power supply. The d.c. voltage from an ordinary power supply remains constant so long as a.c. mains voltage or load is unaltered. However, in many electronic applications, it is desired that d.c. voltage should remain constant irrespective of changes in a.c. mains or load. Under such situations, *voltage regulating devices* are used with ordinary power supply. This constitutes *regulated d.c. power supply* and keeps the d.c. voltage at fairly constant value. In this chapter, we shall focus our attention on the various voltage regulating circuits used to obtain regulated power supply.

15.1 Ordinary D.C. Power Supply

An ordinary or unregulated d.c. power supply contains a rectifier and a filter circuit as shown in Fig. 15.1. The output from the rectifier is pulsating d.c. These pulsations are due to the presence of a.c. component in the rectifier output. The filter circuit removes the a.c. component so that steady d.c. voltage is obtained across the load.

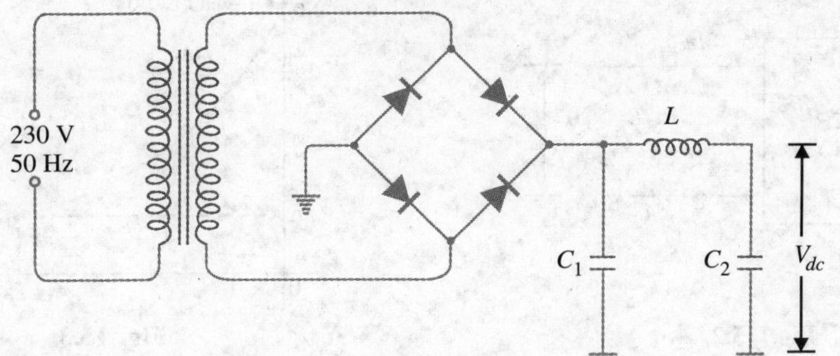

Fig. 15.1

Limitations. An ordinary d.c. power supply has the following drawbacks :

(i) The d.c. output voltage changes directly with input a.c. voltage. For instance, a 5% increase in input a.c. voltage results in approximately 5% increase in d.c. output voltage.

(ii) The d.c. output voltage decreases as the load current increases. This is due to voltage drop in *(a)* transformer windings *(b)* rectifier and *(c)* filter circuit.

These variations in d.c. output voltage may cause inaccurate or erratic operation or even malfunctioning of many electronic circuits. For example, in an oscillator, the frequency will shift and in transmitters, distorted output will result. Therefore, ordinary d.c. power supply is unsuited for many electronic applications and is being replaced by regulated d.c. power supply.

15.2 Characteristics of D.C. Power Supply

The quality of a d.c. power supply depends on its load voltage, load current, voltage regulation, output impedance and other factors. We discuss a few factors that determine the quality of a d.c. power supply.

(i) **Voltage regulation.** The d.c. voltage available across the output terminals of a given power supply depends upon load current. If the load current I_{dc} is increased by decreasing R_L (See Fig. 15.2), there is greater voltage drop in the power supply and hence smaller d.c. output voltage will be available. Reverse will happen if the load current decreases. The variation of output voltage *w.r.t.* the amount of load current drawn from the power supply is known as *voltage regulation* and is expressed by the following relation :

$$\% \text{ Voltage regulation} = \frac{V_{NL} - V_{FL}}{V_{FL}} \times 100$$

where V_{NL} = D.C. output voltage at no-load

V_{FL} = D.C. output voltage at full-load

If the value of full-load voltage (V_{FL}) is the same as the no-load voltage (V_{NL}), the voltage regulation is 0% which is the best expected. This means the d.c. supply is a perfect voltage source for which the output voltage is independent of the load current drawn from the supply.

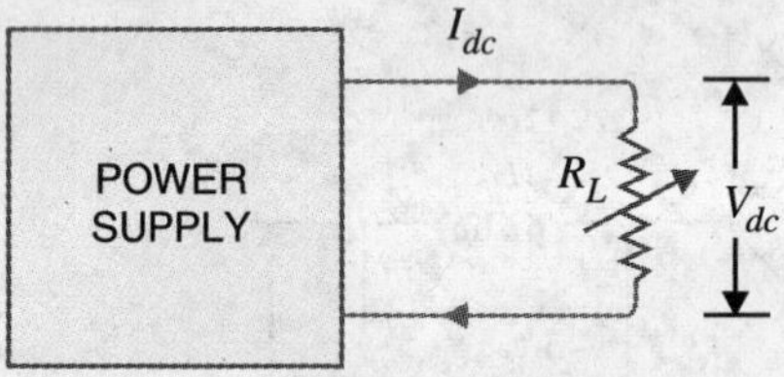

Fig. 15.2

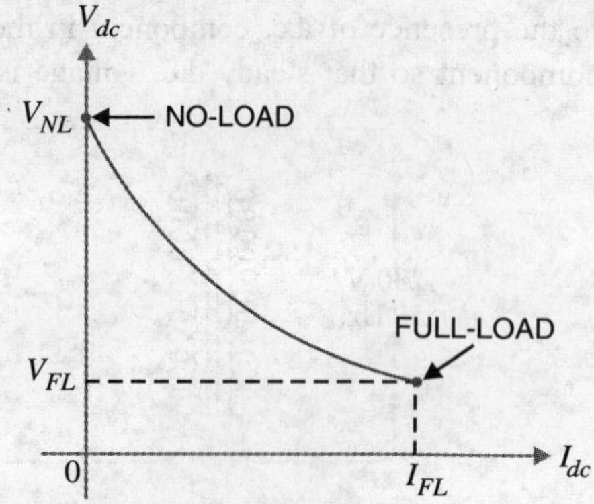

Fig. 15.3

In a well designed d.c. power supply, the full-load voltage is only slightly less than no-load voltage *i.e.* voltage regulation approaches zero. *Therefore, lower the voltage regulation, the lesser the difference between full-load and no-load voltages and better is the d.c. power supply*. Power supplies used in practice have a voltage regulation of 1% *i.e.* full-load voltage is within 1% of the no-load voltage. Fig. 15.3 shows the change of d.c. output voltage with load current. This is known as *voltage regulation curve*.

Note. The above voltage regulation is called *load regulation* because it indicates the change in output voltage due to the change in load current. There is another type of voltage regulation, called *line regulation* and indicates the change in output voltage due to the change in input voltage.

(*ii*) Ripple voltage. The output from rectifier-filter combination has a d.c. value and some a.c. variation (ripple). *The smaller the a.c. variation (ripple) w.r.t. the d.c. value, the better is the d.c. power supply.* In order to find the ripple, the output voltage of the filter circuit is measured using a d.c. voltmeter and an a.c. (r.m.s.) voltmeter. The d.c. voltmeter will read only the d.c. level of the output voltage. The a.c. (r.m.s.) meter will read only the r.m.s. value of the a.c. component of output voltage.

$$\therefore \qquad \% \text{ Ripple, } r = \frac{\text{Ripple voltage(r.m.s.)}}{\text{D.C. voltage}} \times 100 = \frac{V_r(r.m.s.)}{V_{dc}} \times 100$$

(*iii*) Output impedance. Another factor of importance in a d.c. power supply is its output impedance. *The smaller the output impedance of a d.c. power supply, the better is the power supply and vice-versa.* The smaller output impedance of a d.c. power supply means that if we change the load resistance (R_L), there will be practically no change in load voltage. The regulated d.c. power supplies have output impedances in milliohms (mΩ). An ideal d.c. power supply has an output impedance of zero.

(*iv*) Minimum load resistance. The change of load connected to a power supply varies the load current and hence the d.c. output voltage. In order that a power supply gives the rated output voltage and current, there is minimum load resistance allowed. For instance, if a power supply is required to deliver a full-load current I_{FL} at full-load voltage V_{FL}, then,

$$R_{L(min)} = \frac{V_{FL}}{I_{FL}}$$

Thus, if a data sheet specifies that a power supply will give an output voltage of 100V at a maximum rated current of 0.4A, then minimum load resistance you can connect across supply is

R_{min} = 100/0.4 = 250 Ω. If any attempt is made to decrease the value of R_L below this value, the rated d.c. output voltage will not be available.

Example 15.1. *Using a d.c. and a.c. voltmeter to measure the output signal from a filter circuit, we obtain readings of 25 V d.c. and 1.5 V r.m.s. Calculate the ripple of the filter output voltage.*

Solution.

Ripple voltage, V_r = 1.5 V (r.m.s.) ; D.C. voltage, V_{dc} = 25 V

$$\therefore \quad \% \text{ Ripple, } r = \frac{V_r (r.m.s.)}{V_{dc}} \times 100 = \frac{1.5}{25} \times 100 = \mathbf{6\%}$$

Example 15.2. *If the d.c. output voltage is 400 V with no-load attached to power supply but decreases to 300 V at full-load, find the percentage voltage regulation.*

Solution.

$$V_{NL} = 400 \text{ V} ; \quad V_{FL} = 300 \text{ V}$$

$$\therefore \quad \% \text{ Voltage regulation} = \frac{V_{NL} - V_{FL}}{V_{FL}} \times 100 = \frac{400 - 300}{300} \times 100 = \mathbf{33.33\ \%}$$

Example 15.3. *A power supply has a voltage regulation of 1%. If the no-load voltage is 30 V, what is the full-load voltage?*

Solution. Let V_{FL} be the full-load voltage.

$$\% \text{ Voltage regulation} = \frac{V_{NL} - V_{FL}}{V_{FL}} \times 100$$

or

$$1 = \frac{30 - V_{FL}}{V_{FL}} \times 100$$

$$\therefore \quad V_{FL} = \mathbf{29.7\ V}$$

Example 15.4. *Two power supplies A and B are available in the market. Power supply A has no-load and full-load voltages of 30 V and 25 V respectively whereas these values are 30 V and 29 V for power supply B. Which is better power supply?*

Solution. That power supply is better which has lower voltage regulation.

Power supply A

$$V_{NL} = 30 \text{ V} ; \quad V_{FL} = 25 \text{ V}$$

$$\therefore \quad \% \text{ Voltage regulation} = \frac{V_{NL} - V_{FL}}{V_{FL}} \times 100 = \frac{30 - 25}{25} \times 100 = 20\%$$

Power supply B $\quad V_{NL} = 30 \text{ V} ; \quad V_{FL} = 29 \text{ V}$

$$\therefore \quad \% \text{ Voltage regulation} = \frac{V_{NL} - V_{FL}}{V_{FL}} \times 100 = \frac{30 - 29}{29} \times 100 = 3.45\ \%$$

Therefore, power supply B is better than power supply A.

Example 15.5. *Fig. 15.4 shows the regulation curve of a power supply. Find (i) voltage regulation and (ii) minimum load resistance.*

Solution. Referring to the regulation curve shown in Fig. 15.4, it is clear that :

$$V_{NL} = 500 \text{ V} \; ; \quad V_{FL} = 300 \text{ V}$$

$$I_{FL} = 120 \text{ mA} \; ; \quad R_{L\,(min)} = ?$$

(*i*) Regulation $= \dfrac{V_{NL} - V_{FL}}{V_{FL}} \times 100$

$$= \frac{500 - 300}{300} \times 100 = \mathbf{66.7\ \%}$$

(*ii*) $R_{L\,(min)} = \dfrac{V_{FL}}{I_{FL}} = \dfrac{300 \text{ V}}{120 \text{ mA}} = \mathbf{2.5\ k\Omega}$

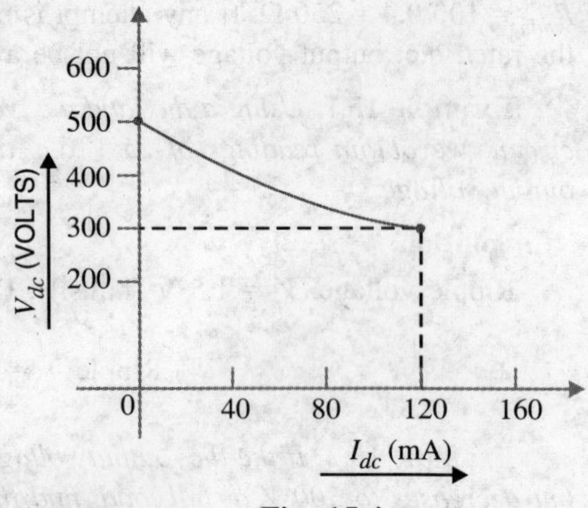

Fig. 15.4

Example 15.6. *In adding 1 A load to an existing 1 A load, the output voltage of a power supply drops from 10.5 V to 10 V. Calculate (i) output impedance of power supply and (ii) no-load voltage of power supply.*

Solution. All practical power supplies always have some internal impedance (often called output impedance) which is denoted by Z_o as shown in Fig. 15.5. It is given by the ratio of change in load voltage to the corresponding change in load current *i.e.*

$$Z_o = \frac{\Delta V_L}{\Delta I_L}$$

(*i*) Output impedance of power supply is

$$Z_o = \frac{\Delta V_L}{\Delta I_L} = \frac{10.5 \text{ V} - 10 \text{ V}}{1 \text{ A}} = \mathbf{0.5\ \Omega}$$

(*ii*) Now, $Z_o = \dfrac{\Delta V_L}{\Delta I_L}$

or $0.5 = \dfrac{V_{NL} - 10.5}{1\text{A}}$

$\therefore \quad V_{NL} = 0.5 \times 1 + 10.5 = \mathbf{11\ V}$

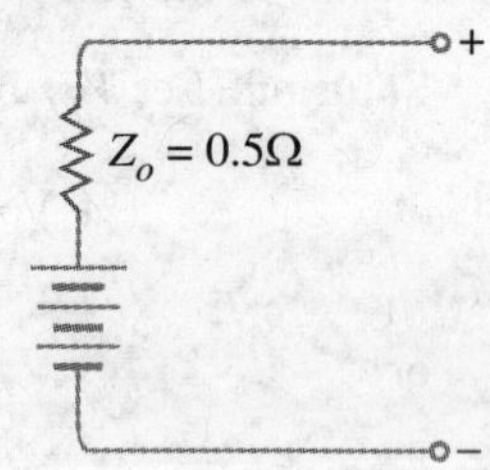

Fig. 15.5

Example 15.7. *A d.c. power supply is delivering 10 V (normally) to a load that is varying sinusoidally between 0.5 A and 1 A at a rate of 10 kHz. If the output impedance of the power supply is 0.01 Ω at 10 kHz, determine the fluctuations in the output voltage caused by this periodic load change.*

Solution. For rapidly changing load levels—the normal situation in electronic systems—the output impedance of power supply varies with the frequency of the load change. This change with frequency occurs because of the impedance of the power supply.

Output impedance of power supply is

$$Z_o = \frac{\Delta V_L}{\Delta I_L}$$

or $\quad \Delta V_L = Z_o \times \Delta I_L = 0.01 \times (1 - 0.5) = 0.005 \text{ V} = 5 \text{ mV}$

Therefore, the *output voltage will have 5 mV p–p fluctuations at a rate of 10 kHz.*

Example 15.8. *A voltage regulator experiences a 10 μV change in its output voltage when its input voltage changes by 5 V. Determine the value of line regulation for the circuit.*

Solution. In practice, a change in input voltage to a voltage regulator will cause a change in its output voltage. The *line regulation* of a voltage regulator indicates the change in output voltage that will occur per unit change in the input voltage *i.e.*

$$\text{Line regulation} = \frac{\Delta V_{out}}{\Delta V_{in}} = \frac{10\ \mu\text{V}}{5\ \text{V}} = 2\ \mu\text{V/V}$$

The 2 μV/V rating of the voltage regulator means that the output voltage will change by 2 μV for every 1 V change in the regulator's input voltage.

15.3 Regulated D.C. Power Supply

A d.c. power supply which maintains the output voltage constant irrespective of a.c. mains fluctuations or load variations is known as **regulated d.c. power supply.**

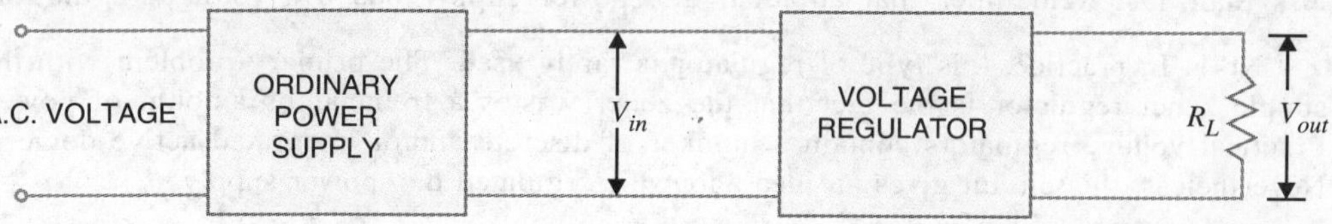

Fig. 15.6

A regulated d.c. power supply consists of an ordinary d.c. power supply and voltage regulating device. Fig. 15.6 shows the block diagram of a regulated d.c. power supply. The output of ordinary d.c. power supply is fed to the voltage regulator which produces the final output. The output voltage remains constant whether the load current changes or there are fluctuations in the input a.c. voltage.

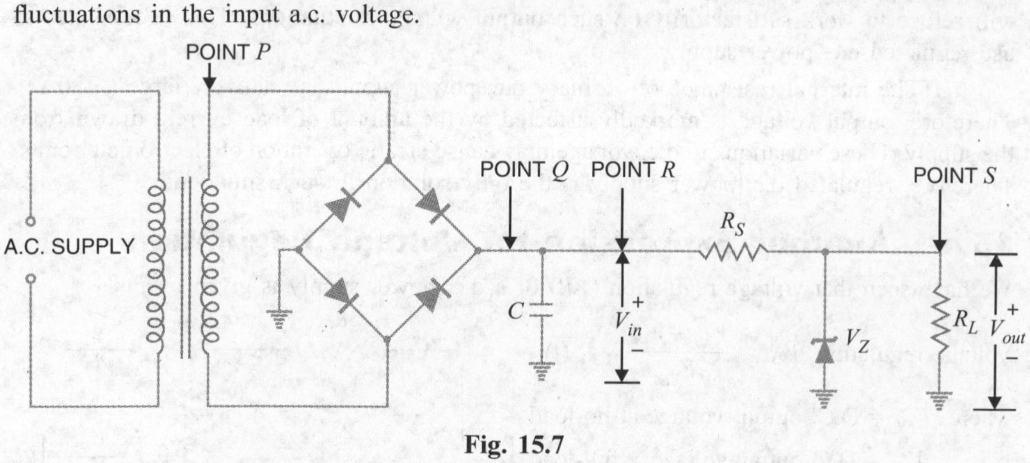

Fig. 15.7

Fig. 15.7 shows the complete circuit of a regulated d.c. power supply using zener diode as a voltage regulating device. As you can see, the regulated d.c. power supply is a combination of three circuits *viz.,* (*i*) bridge rectifier (*ii*) a capacitor filter *C* and (*iii*) zener voltage regulator. The bridge rectifier converts the transformer secondary a.c. voltage (point *P*) into pulsating voltage (point *Q*). The pulsating d.c. voltage is applied to the capacitor filter. This filter reduces the pulsations in the rectifier d.c. output voltage (point *R*). Finally, the zener voltage regulator performs two functions. Firstly, it reduces the variations in the filtered output voltage. Secondly, it keeps the output voltage (V_{out}) nearly constant whether the load current changes or there is change in input a.c. voltage. Fig. 15.8 shows the waveforms at various stages of regulated d.c. power supply. Note that bridge rectifier and capacitor filter constitute an ordinary d.c. power supply. However, when voltage regulating device is added to this ordinary d.c. power supply, it turns into a regulated d.c. power supply. The voltage regulator (Zener diode in this case) maintains a substantially constant d.c. output voltage under variations in the load current drawn from the supply and under variations in line voltage. It may be mentioned here that a well designed regulated d.c. power supply has voltage regulation less than 1%. Remember that an ideal d.c. power supply has 0% voltage regulation.

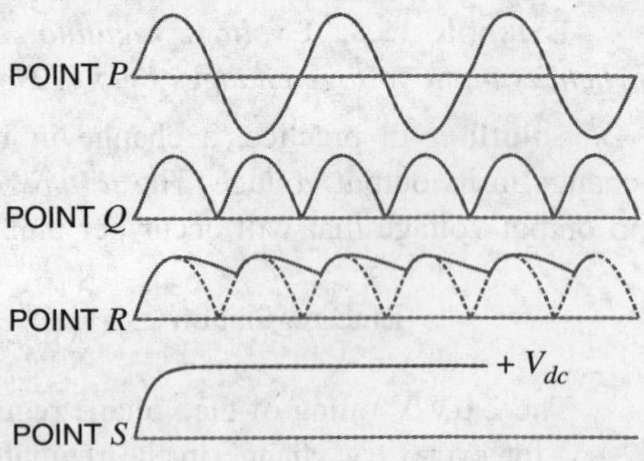

Fig. 15.8

Note. In practice, this type of regulator is rarely used. The primary problem with the simple zener regulator is the fact that the zener wastes a tremendous amount of power. Practical voltage regulators contain a number of discrete and/or integrated active devices. Nevertheless, this circuit gives an idea about the regulated d.c. power supply.

Need of Regulated D.C. Power Supply

In an ordinary d.c. power supply, the voltage regulation is poor *i.e.* d.c. output voltage changes appreciably with load current. Moreover, output voltage also changes due to variations in the input a.c. voltage. This is due to the following reasons :

(*i*) In practice, there are considerable variations in a.c. line voltage caused by outside factors beyond our control. This changes the d.c. output voltage. Most of the electronic circuits will refuse to work satisfactorily on such output voltage fluctuations. This necessitates to use regulated d.c. power supply.

(*ii*) The internal resistance of ordinary d.c. power supply is relatively large (> 30 Ω). Therefore, output voltage is markedly affected by the amount of load current drawn from the supply. These variations in d.c. voltage may cause erratic operation of electronic circuits. Therefore, regulated d.c. power supply is the only solution in such situations.

15.4 Another Expression for Voltage Regulation

We have seen that voltage regulation (*VR*) of a d.c. power supply is given by ;

$$\text{Voltage regulation, } VR = \frac{V_{NL} - V_{FL}}{V_{FL}} \quad \text{... (i)}$$

where, V_{NL} = D.C. output voltage at no-load

V_{FL} = D.C. output voltage at full-load

We can also express voltage regulation of a d.c. power supply in terms of its output resistance (or impedance). Fig. 15.9 shows the Thevenin equivalent circuit of a d.c. power supply. The Thevenin voltage is the no-load

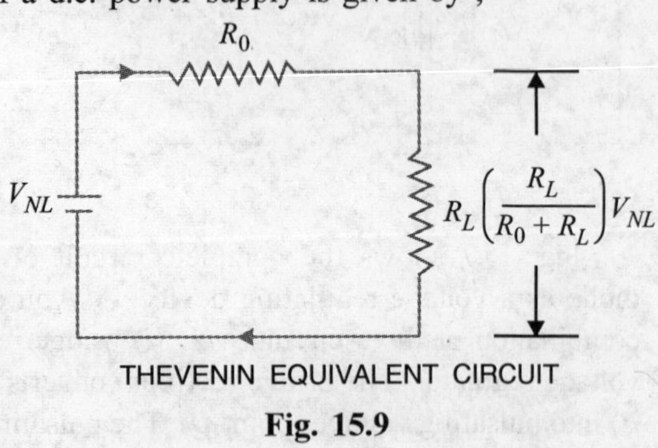

Fig. 15.9

voltage V_{NL} (*i.e.* $V_{Th} = V_{NL}$). The Thevenin resistance is the output resistance R_0 of the power supply (*i.e.* $R_{Th} = R_0$). Referring to Fig. 15.9 and applying voltage-divider rule, we have,

$$\text{Voltage across load } R_L = \frac{V_{NL}}{R_0 + R_L} \times R_L$$

Suppose the full-load voltage is V_{FL} and full-load current is I_{FL}. Then, full-load resistance (R_{FL}) is given by ;

$$R_{FL} = \frac{V_{FL}}{I_{FL}}$$

If $R_L = R_{FL}$, then voltage across load = V_{FL}. Therefore, putting $R_L = R_{FL}$, we have,

$$\text{Voltage across load, } V_{FL} = \frac{V_{NL}}{R_0 + R_{FL}} \times R_{FL} = \left(\frac{R_{FL}}{R_0 + R_{FL}}\right) V_{NL}$$

Putting the value of V_{FL} in eq. (*i*), we have,

$$\text{Voltage regulation, } VR = \frac{V_{NL} - \left(\frac{R_{FL}}{R_0 + R_{FL}}\right) V_{NL}}{\left(\frac{R_{FL}}{R_0 + R_{FL}}\right) V_{NL}} = \frac{R_0}{R_{FL}}$$

But $R_{FL} = V_{FL}/I_{FL}$ so that $1/R_{FL} = I_{FL}/V_{FL}$.

$$\therefore \quad \text{Voltage regulation, } VR = R_0 \left(\frac{I_{FL}}{V_{FL}}\right) \quad \text{... (ii)}$$

Eq. (*ii*) gives the expression for voltage regulation of a d.c. power supply in terms of its output resistance (R_0), full-load voltage (V_{FL}) and full-load current (I_{FL}). This is an important result because very often d.c. power supply manufacturers specify the output resistance of a d.c. power supply rather than its voltage regulation.

Note. We know that an ideal d.c. power supply has zero voltage regulation (VR). If we put VR = 0 in eq. (*ii*) above, then $R_0 = 0$. It is clear that an ideal d.c. power supply has zero output resistance.

Example 15.9. *A d.c. power supply has an output resistance of 1.5 Ω and supplies a full-load current of 500 mA to a load of 50 Ω. Determine (i) the % age voltage regulation of the supply (ii) no-load output voltage of the supply.*

Solution. $R_0 = 1.5\ \Omega$; $I_{FL} = 500$ mA = 0.5 A ; $R_{FL} = 50\ \Omega$

(*i*) Full-load voltage, $V_{FL} = I_{FL} R_{FL} = 0.5 \times 50 = 25$ V

$$\%\ \text{Voltage regulation, } VR(\%) = R_0 \left(\frac{I_{FL}}{V_{FL}}\right) \times 100 = 1.5 \times \left(\frac{0.5}{25}\right) \times 100 = 3\%$$

Alternatively.

$$VR(\%) = \frac{R_0}{R_{FL}} \times 100 = \frac{1.5}{50} \times 100 = 3\%$$

(*ii*)

$$V_{FL} = \left(\frac{R_{FL}}{R_0 + R_{FL}}\right) V_{NL}$$

or

$$V_{NL} = \frac{V_{FL}(R_0 + R_{FL})}{R_{FL}} = \frac{25\,(1.5 + 50)}{50} = \mathbf{25.75\ V}$$

Example 15.10. *A d.c. power supply has 4% voltage regulation and an open-circuit output voltage of 48 V d.c.*

(*i*) *What is the full-load voltage of the supply?*

(*ii*) *If a 120 Ω resistor draws full-load current from the supply, what is its output resistance?*

Solution. $VR = 4\% = 0.04$; $V_{NL} = 48$ V ; $R_{FL} = 120\ \Omega$

(*i*) Voltage regulation, $VR = \dfrac{V_{NL} - V_{FL}}{V_{FL}}$

or $0.04 = \dfrac{48 - V_{FL}}{V_{FL}}$

∴ Full-load voltage, $V_{FL} = \dfrac{48}{1 + 0.04} = \mathbf{46.15\ V}$

(*ii*) Full-load current, $I_{FL} = \dfrac{V_{FL}}{R_{FL}} = \dfrac{46.15}{120} = 0.384$ A

Now, Voltage regulation, $VR = R_0\left(\dfrac{I_{FL}}{V_{FL}}\right)$

or $0.04 = R_0\left(\dfrac{0.384}{46.15}\right)$

∴ Output resistance, $R_0 = \dfrac{46.15 \times 0.04}{0.384} = \mathbf{4.8\ \Omega}$

Example 15.11. *Fig. 15.10 shows the circuit of an unregulated d.c. power supply. The full-load current is 2 A at a full-load voltage of 15 V. Assuming that the transformer and forward biased diode resistances are negligible, find (i) % age voltage regulation of the power supply (ii) no-load voltage of the power supply.*

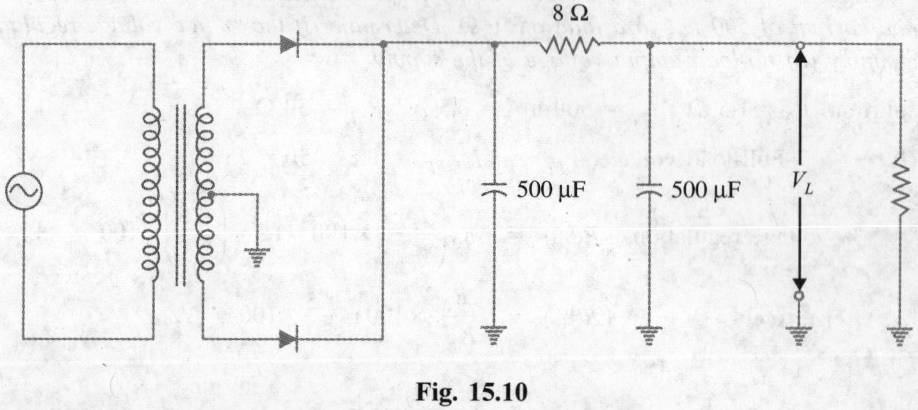

Fig. 15.10

Solution.

(*i*) It is given that the transformer and diode resistances can be neglected. Therefore, d.c. output resistance looking back into the load terminals is $R_0 = 8\ \Omega$. As per the conditions of the problem, $I_{FL} = 2$ A and $V_{FL} = 15$ V.

$$\therefore \quad \% \text{ Voltage regulation} = R_0\left(\frac{I_{FL}}{V_{FL}}\right)\times 100$$

$$= 8\left(\frac{2}{15}\right)\times 100 = \mathbf{106.6\%}$$

This is rather a very high value of voltage regulation. Therefore, such a d.c. power supply would be unacceptable for applications in which load current could very over a wide range.

(*ii*) $$\text{Voltage regulation, } VR = \frac{V_{NL} - V_{FL}}{V_{FL}}$$

$$\therefore \quad \text{No-load voltage, } V_{NL} = VR(V_{FL}) + V_{FL}$$

Here, $VR = 106.6\% = 1.066$; $V_{FL} = 15$ V

$$\therefore \quad V_{NL} = 1.066\ (15) + 15 = \mathbf{31\ V}$$

Note that there is a 16 V change in output voltage between no-load and full-load. This is not surprising because the circuit does not have any voltage regulating device.

15.5 Types of Voltage Regulators

A device which maintains the output voltage of an ordinary d.c. power supply constant irrespective of load variations or changes in input a.c. voltage is known as a *voltage regulator*. A voltage regulator generally employs electronic devices to achieve this objective. There are basic two types of voltage regulators *viz.*, (*i*) series voltage regulator (*ii*) shunt voltage regulator.

The series regulator is placed in series with the load as shown in Fig. 15.11 (*i*). On the other hand, the shunt regulator is placed in parallel with the load as shown in Fig. 15.11 (*ii*). Each type of regulator provides an output voltage that remains constant even if the input voltage varies or the load current changes.

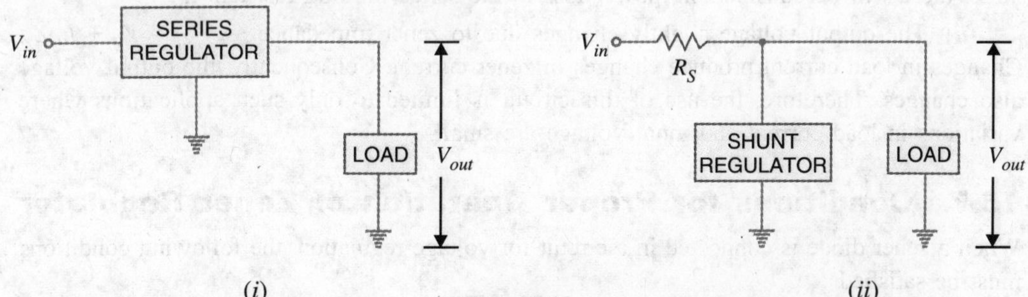

Fig. 15.11

1. For low voltages. For low d.c. output voltages (upto 50 V), either zener diode alone or zener in conjunction with transistor is used. Such supplies are called transistorised power supplies. A transistor power supply can give only low stabilised voltages because the safe value of V_{CE} is about 50 V and if it is increased above this value, the breakdown of the junction may occur.

2. For high voltages. For voltages greater than 50 V, glow tubes are used in conjunction with vacuum tube amplifiers. Such supplies are generally called tube power supplies and are extensively used for the proper operation of vacuum valves.

15.6 Zener Diode Voltage Regulator

As discussed in chapter 4, when the zener diode is operated in the breakdown or zener region, the voltage across it is substantially constant for a large change of current through it. This characteristic permits it to be used as a voltage regulator. Fig. 15.12 shows the circuit of a zener diode regulator. As long as input voltage V_{in} is greater than zener voltage V_Z, the zener operates in the breakdown region and maintains constant voltage across the load. The series limiting resistance R_S limits the input current.

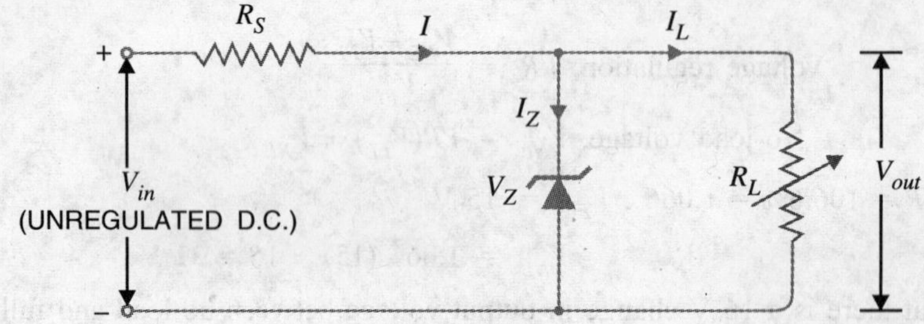

Fig. 15.12

Operation. The zener will maintain constant voltage across the load inspite of changes in load current or input voltage. As the load current increases, the zener current decreases so that current through resistance R_S is constant. As output voltage $= V_{in} - IR_S$, and I is constant, therefore, output voltage remains unchanged. The reverse would be true should the load current decrease. The circuit will also correct for the changes in input voltages. Should the input voltage V_{in} increase, more current will flow through the zener, the voltage drop across R_S will increase but load voltage would remain constant. The reverse would be true should the input voltage decrease.

Limitations. A zener diode regulator has the following drawbacks :

(*i*) It has low efficiency for heavy load currents. It is because if the load current is large, there will be considerable power loss in the series limiting resistance.

(*ii*) The output voltage slightly changes due to zener impedance as $V_{out} = V_Z + I_Z Z_Z$. Changes in load current produce changes in zener current. Consequently, the output voltage also changes. Therefore, the use of this circuit is limited to only such applications where variations in load current and input voltage are small.

15.7 Conditions for Proper Operation of Zener Regulator

When a zener diode is connected in a circuit for voltage regulation, the following conditions must be satisfied :

(*i*) The zener must operate in the breakdown region or regulating region *i.e.* between $I_{Z\,(max)}$ and $I_{Z\,(min)}$. The current $I_{Z\,(min)}$ (generally 10 mA) is the minimum zener current to put the zener diode in the *ON state i.e.* regulating region. The current $I_{Z\,(max)}$ is the maximum zener current that zener diode can conduct without getting destroyed due to excessive heat.

(*ii*) The zener should not be allowed to exceed maximum dissipation power otherwise it will be destroyed due to excessive heat. If maximum power dissipation of a zener is $P_{Z\,(max)}$ and zener voltage is V_Z, then,

$$P_{Z\,(max)} = V_Z I_{Z\,(max)}$$

$$\therefore \quad I_{Z\,(max)} = \frac{P_{Z\,(max)}}{V_Z}$$

(*iii*) There is a minimum value of R_L to ensure that zener diode will remain in the regulating region *i.e.* breakdown region. If the value of R_L falls below this minimum value, the proper voltage will not be available across the zener to drive it into the breakdown region.

Example 15.12. *Fig. 15.13 shows the zener regulator. Calculate (i) current through the series resistance (ii) minimum and maximum load currents and (iii) minimum and maximum zener currents. Comment on the results.*

Solution.

(*i*) $$I_S = \frac{V_{in} - V_{out}}{R_S} = \frac{24-12}{160} = \frac{12\text{ V}}{160\ \Omega} = \mathbf{75\ mA}$$

(*ii*) The minimum load current occurs when $R_L \rightarrow \infty$.

$$\therefore \quad I_{L\,(min)} = \mathbf{0\ A}$$

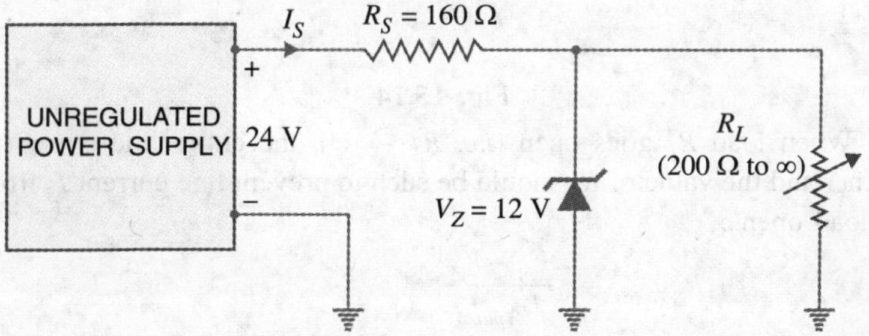

Fig. 15.13

The maximum load current occurs when $R_L = 200\ \Omega$.

$$\therefore \quad I_{L\,(max)} = \frac{V_{out}}{R_{L(min)}} = \frac{12\text{ V}}{200\ \Omega} = \mathbf{60\ mA}$$

(*iii*) $$I_{Z\,(min)} = I_S - I_{L\,(max)} = 75 - 60 = \mathbf{15\ mA}$$

$$I_{Z\,(max)} = I_S - I_{L\,(min)} = 75 - 0 = \mathbf{75\ mA}$$

Comments. The current I_S through the series resistance R_S is constant. When load current increases from 0 to 60 mA, the zener current decreases from 75 mA to 15 mA, maintaining I_S constant in value. This is the normal operation of zener regulator *i.e.* I_S and V_{out} remain constant inspite of changes in load current or source voltage.

Example 15.13. *A zener regulator has $V_Z = 15$ V. The input voltage may vary from 22 V to 40 V and load current from 20 mA to 100 mA. To hold load voltage constant under all conditions, what should be the value of series resistance?*

Solution. In order that zener regulator may hold output voltage constant under all operating conditions, it must operate in the breakdown region. In other words, there must be zener current for all input voltages and load currents. The worst case occurs when the input voltage is minimum and load current is maximum because then zener current drops to a minimum.

$$\therefore \quad R_{S\,(max)} = \frac{V_{in(min)} - V_{out}}{I_{L(max)}}$$

$$= \frac{22 - 15}{0.1} = \frac{7\text{ V}}{0.1\text{ A}} = \mathbf{70\ \Omega}$$

Example 15.14. *Determine the minimum acceptable value of R_S for the zener voltage regulator circuit shown in Fig. 15.14. The zener specifications are :*

$V_Z = 3.3\ V$; $I_{Z\,(min)} = 3\ mA$; $I_{Z\,(max)} = 100\ mA$

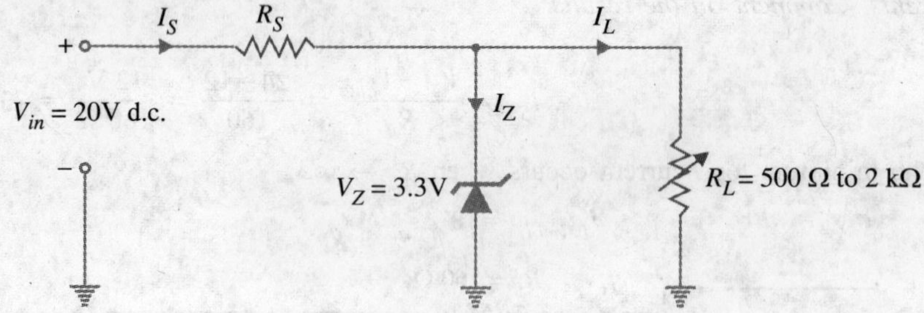

Fig. 15.14

Solution. When load R_L goes open (*i.e.* $R_L \to \infty$), the entire line current I_S will flow through the zener and the value of R_S should be such to prevent line current I_S from exceeding $I_{Z\,(max)}$ if the load opens.

$$\therefore \quad R_{S\,(min)} = \frac{V_{in} - V_Z}{I_{Z(max)}}$$

$$= \frac{20\text{ V} - 3.3\text{ V}}{100\text{ mA}} = \mathbf{167\ \Omega}$$

Example 15.15. *Determine the maximum allowable value of R_S for the zener voltage regulator circuit shown in Fig. 15.14.*

Solution. The maximum value of R_S is limited by the total current requirements in the circuit. The value of R_S must be such so as to allow $I_{Z\,(min)}$ to flow through the zener diode when the load is drawing *maximum* current.

$$R_{S\,(max)} = \frac{V_{in} - V_Z}{I_{L\,(max)} + I_{Z\,(min)}}$$

Now, $$I_{L\,(max)} = \frac{V_Z}{R_{L\,(min)}} = \frac{3.3\text{ V}}{500\ \Omega} = 6.6\text{ mA}$$

$$\therefore \quad R_{S\,(max)} = \frac{20\text{ V} - 3.3\text{ V}}{6.6\text{ mA} + 3\text{ mA}} = \frac{16.7\text{ V}}{9.6\text{ mA}} = \mathbf{1739\ \Omega}$$

15.8 Transistor Series Voltage Regulator

Figure 15.15 shows a simple series voltage regulator using a transistor and zener diode. The circuit is called a series voltage regulator because the load current passes through the series transistor Q_1 as shown in Fig. 15.15. The unregulated d.c. supply is fed to the input terminals and the regulated d.c. output is obtained across the load. The zener diode provides the reference voltage.

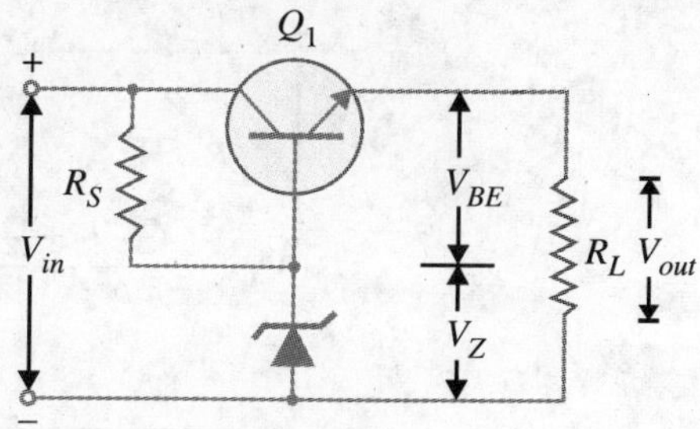

Fig. 15.15

Operation. The base voltage of transistor Q_1 is held to a relatively constant voltage across the zener diode. For example, if 8 V zener (*i.e.* V_Z = 8 V) is used, the base voltage of Q_1 will remain approximately 8 V. Referring to Fig. 15.15,

$$V_{out} = V_Z - V_{BE}$$

(*i*) If the output voltage decreases, the increased base-emitter voltage causes transistor Q_1 to conduct more, thereby raising the output voltage. As a result, the output voltage is maintained at a constant level.

(*ii*) If the output voltage increases, the decreased base-emitter voltage causes transistor Q_1 to conduct less, thereby reducing the output voltage. Consequently, the output voltage is maintained at a constant level.

The advantage of this circuit is that the changes in zener current are reduced by a factor β. Therefore, the effect of zener impedance is greatly reduced and much more stabilised output is obtained.

Limitations

(*i*) Although the changes in zener current are much reduced, yet the output is not absolutely constant. It is because both V_{BE} and V_Z decrease with the increase in room temperature.

(*ii*) The output voltage cannot be changed easily as no such means is provided.

Example 15.16. *For the circuit shown in Fig. 15.15, if V_Z = 10 V, β = 100 and R_L = 1000 Ω, find the load voltage and load current. Assume V_{BE} = 0.5 V and the zener operates in the breakdown region.*

Solution.

$$\text{Output voltage, } V_{out} = V_Z - V_{BE}$$

$$= 10 - 0.5 = \mathbf{9.5\ V}$$

$$\text{Load current, } I_L = \frac{V_{out}}{R_L} = \frac{9.5\text{ V}}{1000\,\Omega} = \mathbf{9.5\ mA}$$

Example 15.17. *A series voltage regulator is required to supply a current of 1 A at a constant voltage of 6 V. If the supply voltage is 10 V and the zener operates in the breakdown region, design the circuit. Assume β = 50, V_{BE} = 0.5 V and minimum zener current = 10 mA.*

Solution. The design steps require the determination of zener breakdown voltage and current limiting resistance R_S. Fig. 15.16 shows the desired circuit of series voltage regulator.

(*i*) **Zener breakdown voltage.** The collector-emitter terminals are in series with the load. Therefore, the load current must pass through the transistor *i.e.*

$$\text{Collector current, } I_C = 1\text{ A}$$

$$\text{Base current, } I_B = I_C/\beta = 1\text{ A}/50 = 20\text{ mA}$$

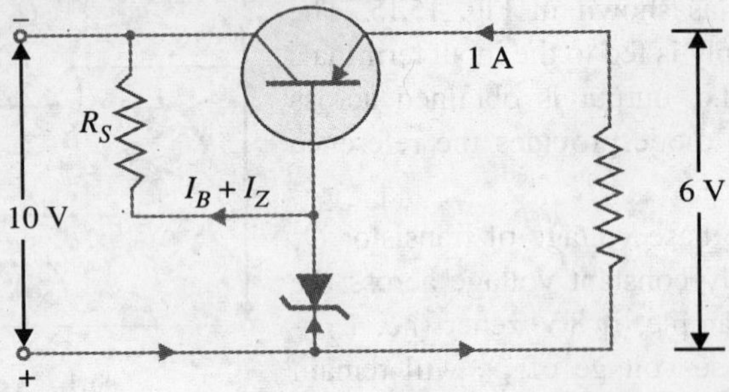

Fig. 15.16

$$\text{Output voltage, } V_{out} = V_Z - V_{BE}$$

or $$6 = V_Z - 0.5$$

∴ $$V_Z = 6 + 0.5 = \mathbf{6.5\ V}$$

Hence zener diode of breakdown voltage 6.5 V is required.

(*ii*) **Value of R_S.**

$$\text{Voltage across } R_S = V_{in} - V_Z = 10 - 6.5 = 3.5\text{ V}$$

∴ $$R_S = \frac{\text{Voltage across } R_S}{I_B + I_Z} = \frac{3.5\text{ V}}{(20+10)\text{ mA}} = \mathbf{117\ \Omega}$$

Example 15.18. *For the series voltage regulator shown in Fig. 15.17, calculate (i) output voltage and (ii) zener current.*

Solution.

(*i*) $$V_{out} = V_Z - V_{BE} = 12 - 0.7 = \mathbf{11.3\ V}$$

(*ii*) $$\text{Voltage across } R_S = V_{in} - V_Z = 20 - 12 = 8\text{ V}$$

$$\text{Current through } R_S,\ I_R = \frac{8\text{ V}}{220\ \Omega} = 36.4\text{ mA}$$

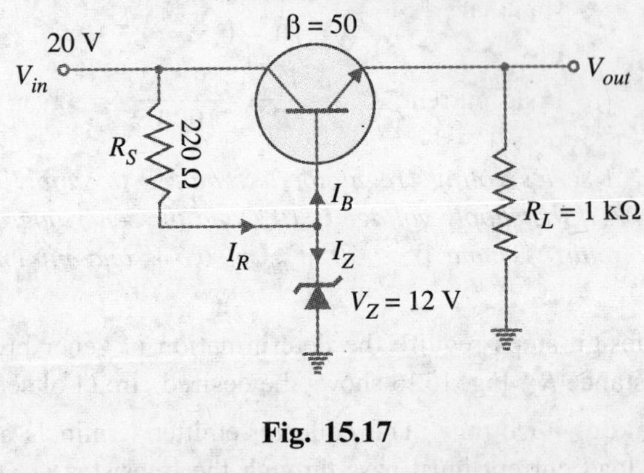

Fig. 15.17

$$\text{Load current, } I_L = \frac{V_{out}}{R_L} = \frac{11.3\text{ V}}{1\text{ k}\Omega} = 11.3\text{ mA}$$

$$\text{Base current, } I_B = \frac{I_C}{\beta} = \frac{11.3}{50} = 0.226 \text{ mA}$$

$$\therefore \quad \text{Zener current, } I_Z = I_R - I_B = 36.4 - 0.226 \simeq \mathbf{36\ mA}$$

Example 15.19. *In a series transistor voltage regulator (See Fig. 15.18), the load current varies from 0 – 1 A and the unregulated d.c. input varies from 12 – 18 V. The 8.5 V zener diode requires atleast 1 mA of current to stay in its regulating region (i.e. $I_{Z\,(min)}$ = 1 mA).*

(i) Determine the value of R_S to ensure proper circuit operation.

(ii) Determine maximum power dissipation in R_S.

(iii) Determine maximum power dissipation in zener doide.

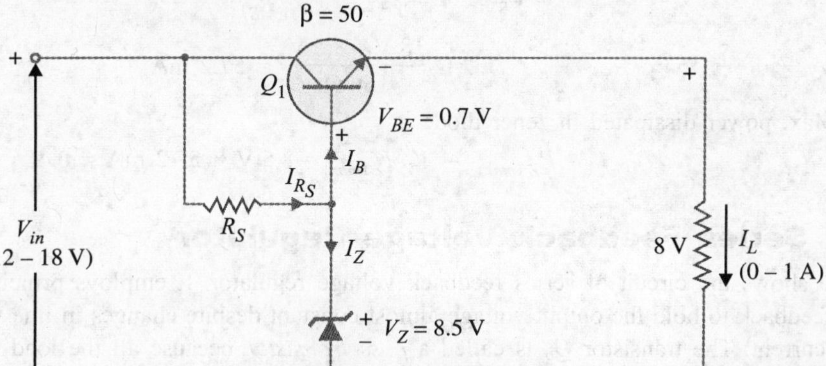

Fig. 15.18

Solution.

(i) The value of R_S should be such that it supplies current for the base of transistor Q_1 and for the zener diode to keep it in the regulating region. The worst condition occurs at the minimum input voltage and maximum load current. This means that under worst condition, the current through R_S must be atleast $I_{Z\,(min)}$ = 1 mA plus maximum base current.

$$I_{B\,(max)} = \frac{I_{L\,(\max)}}{\beta} = \frac{1 \text{ A}}{50} = 20 \text{ mA}$$

$$I_{R_S} = I_{Z\,(min)} + I_{B\,(max)}$$

$$= 1 + 20 = 21 \text{ mA}$$

Now 21 mA must be drawn by R_S under all conditions of input voltage variations—even when the input voltage falls to 12 V which causes the minimum voltage across R_S and hence the lowest value of current it will be able to supply.

$$\therefore \quad R_S = \frac{V_{in\,(min)} - V_Z}{I_{R_S}}$$

$$= \frac{(12 - 8.5) \text{ V}}{21 \text{ mA}} = \frac{3.5 \text{ V}}{21 \text{ mA}} = \mathbf{166\ \Omega}$$

(*ii*) The maximum power dissipation in R_S occurs when the voltage across it is maximum.

$$\text{Max. voltage across } R_S,\ V_{R_{S\,(max)}} = V_{in\,(max)} - V_Z = 18 - 8.5 = 9.5 \text{ V}$$

$$\therefore \quad \text{Max. power dissipated in } R_S = \frac{(V_{R_{S\,(max)}})^2}{R_S} = \frac{(9.5)^2}{166} = \mathbf{0.542\ W}$$

(*iii*) Maximum power dissipation in zener occurs when current through it is maximum. The zener current will be maximum when V_{in} is maximum and load current is minimum (*i.e.* $I_L = 0$). Now $I_L = 0$ means $I_E = 0$ and hence $I_B = 0$. This, in turn, means that all the current passing through R_S will pass through the zener diode.

$$\therefore \quad I_{Z\,(max)} = I_{R_{S(max)}} = \frac{V_{in\,(max)} - V_Z}{R_S}$$

$$= \frac{18\text{ V} - 8.5\text{ V}}{166\ \Omega} = 57.2 \text{ mA}$$

∴ Max. power dissipated in zener diode is

$$P_{Z\,(max)} = V_Z\, I_{Z\,(max)} = 8.5 \text{ V} \times 57.2 \text{ mA} = \mathbf{0.486\ W}$$

15.9 Series Feedback Voltage Regulator

Fig. 15.19 shows the circuit of series feedback voltage regulator. It employs principles of negative feedback to hold the output voltage almost constant despite changes in line voltage and load current. The transistor Q_2 is called a *pass transistor* because all the load current passes through it. The sample and adjust circuit is the voltage divider that consists of R_1 and R_2. The voltage divider samples the output voltage and delivers a negative feedback voltage to the base of Q_1. The feedback voltage V_F controls the collector current of Q_1.

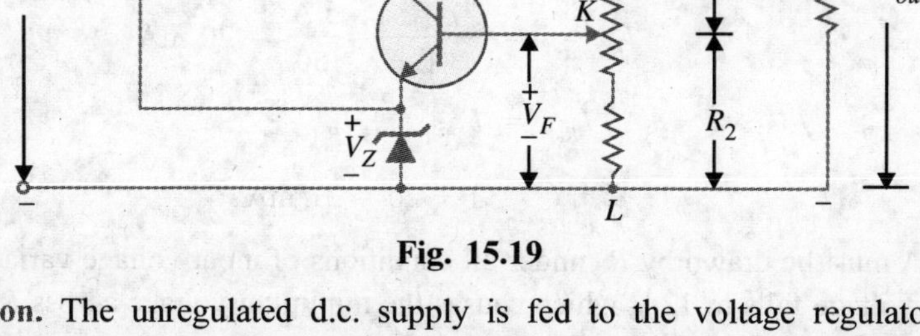

Fig. 15.19

Operation. The unregulated d.c. supply is fed to the voltage regulator. The circuit maintains constant output voltage irrespective of the variations in load or input voltage. Here is how the circuit operates.

(*i*) Suppose the output voltage increases due to any reason. This causes an increase in voltage across *KL* (*i.e.* R_2) as it is a part of the output circuit. This in turn means that more V_F is fed back to the base of transistor Q_1; producing a large collector current of Q_1. Most of this collector current flows through R_3 and causes the base voltage of Q_2 to decrease. This results in less output voltage *i.e.* increase in voltage is offset. Thus output voltage remains constant.

(*ii*) Similarly, if output voltage tries to decrease, the feedback voltage V_F also decreases. This reduces the current through Q_1 and R_3. This means more base voltage at Q_2 and more output voltage. Consequently, the output voltage remains at the original level.

Output Voltage. The voltage divider $R_1 - R_2$ provides the feedback voltage.

$$\text{Feedback fraction, } m = \frac{V_F}{V_{out}} = \frac{R_2}{R_1 + R_2}$$

$$\text{Closed loop voltage gain, } A_{CL} = \frac{1}{m} = \frac{R_1 + R_2}{R_2} = 1 + \frac{R_1}{R_2}$$

Now,
$$V_F = V_Z + V_{BE}$$

or
$$m\,V_{out} = V_Z + V_{BE} \qquad (\because V_F = mV_{out})$$

or
$$V_{out} = \frac{V_Z + V_{BE}}{m}$$

or
$$V_{out} = A_{CL}\,(V_Z + V_{BE}) \qquad (\because 1/m = A_{CL})$$

Therefore, the regulated output voltage is equal to closed-loop voltage gain times the sum of zener voltage and base-emitter voltage.

15.10 Short-Circuit Protection

The main drawback of any series regulator is that the pass transistor can be destroyed by excessive load current if the load is accidentally shorted. To avoid such an eventuality, a current limiting circuit is added to a series regulator as shown in Fig. 15.20. A current limiting circuit consists of a transistor (Q_3) and a series resistor (R_4) that is connected between base and emitter terminals.

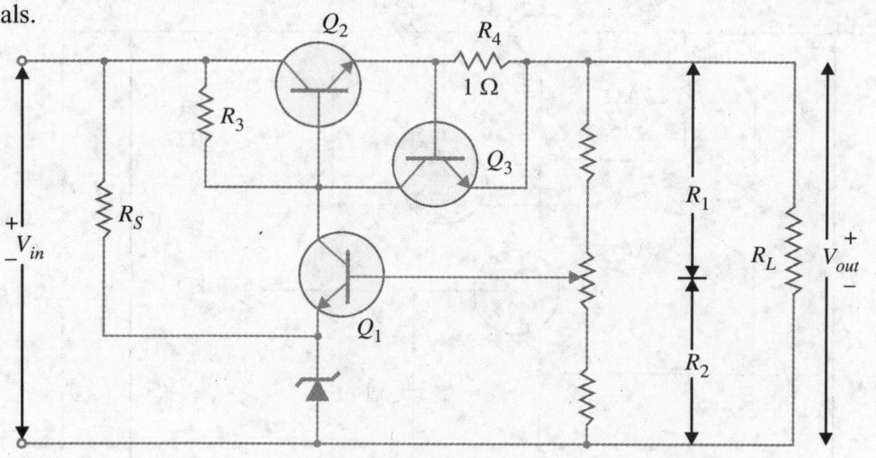

Fig. 15.20

(*i*) When the load current is normal, the voltage across R_4 (= voltage across base-emitter of Q_3) is small and Q_3 is *off. Under this condition, the circuit works as described earlier.

* In order that Q_3 is ON, voltage across R_4 must be about 0.7 V. This means that load current then is I_L = 0.7 V/ 1 Ω = 700 mA. Therefore, if load current is less than 700 mA, Q_3 is off. If load current is more than 700 mA, Q_3 will be turned on.

(*ii*) If load current becomes excessive, the voltage across R_4 becomes large enough to turn on Q_3. The collector current of Q_3 flows through R_3, thereby decreasing the base voltage of Q_2. The decrease in base voltage of Q_2 reduces the conduction of pass transistor (*i.e.* Q_2), preventing any further increase in load current. Thus, the load current for the circuit is limited to about 700 mA.

Example 15.20. *In the series feedback voltage regulator shown in Fig. 15.20,* $R_1 = 2\ k\Omega$, $R_2 = 1\ k\Omega$, $V_Z = 6\ V$ *and* $V_{BE} = 0.7\ V$. *What is the regulated output voltage?*

Solution.

$$\text{Feedback fraction, } m = \frac{R_2}{R_1 + R_2} = \frac{1}{2+1} = \frac{1}{3}$$

$$\therefore \quad \text{Closed-loop voltage gain, } A_{CL} = \frac{1}{m} = 3$$

$$\therefore \quad \text{Regulated output voltage, } V_{out} = A_{CL}\,(V_Z + V_{BE})$$

$$= 3\,(6 + 0.7) = \mathbf{20.1\ V}$$

Example 15.21. *In the series feedback regulator circuit shown in Fig. 15.20,* $R_1 = 30\ k\Omega$ *and* $R_2 = 10\ k\Omega$. *What is the closed loop voltage gain?*

Solution.

$$\text{Feedback fraction, } m = \frac{R_2}{R_1 + R_2} = \frac{10}{30+10} = \frac{1}{4}$$

$$\therefore \quad \text{Closed-loop voltage gain, } A_{CL} = \frac{1}{m} = \mathbf{4}$$

Example 15.22. *The wiper of Fig. 15.21 is at the middle of its travel. If the input voltage is 15 V, what is the output voltage of the regulator? Given that* $V_Z = 5.6\ V$ *and* $V_{BE} = 0.7\ V$.

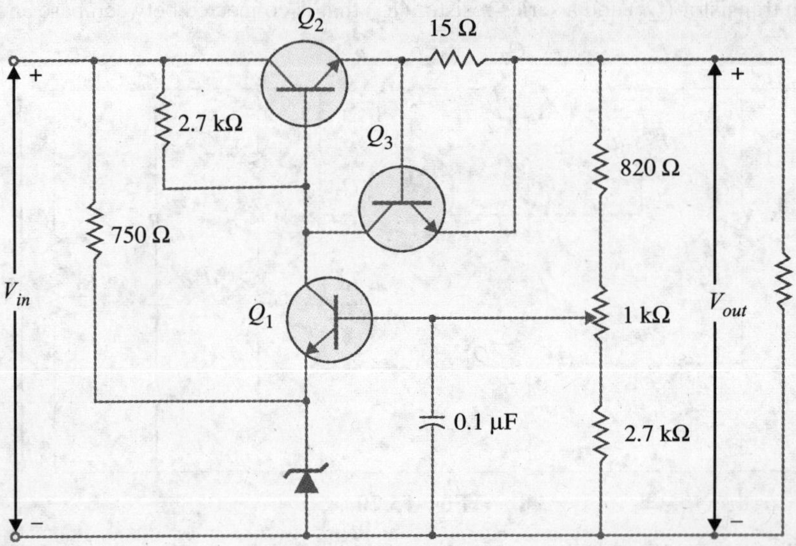

Fig. 15.21

Solution.

The resistance below the wiper is

$$R_2 = 2.7\ k\Omega + \frac{1 k\Omega}{2} = 3.2\ k\Omega$$

$$R_1 + R_2 = 820\ \Omega + 1\ \text{k}\Omega + 2.7\ \text{k}\Omega = 4.52\ \text{k}\Omega$$

$$\text{Feedback fraction, } m = \frac{R_2}{R_1 + R_2} = \frac{3.2}{4.52} = 0.708$$

$$\text{Closed-loop voltage gain, } A_{CL} = \frac{1}{m} = \frac{1}{0.708} = 1.41$$

$$\therefore \quad \text{Regulated output voltage, } V_{out} = A_{CL}(V_Z + V_{BE})$$

$$= 1.41\ (5.6 + 0.7) = \mathbf{8.88\ V}$$

15.11 Transistor Shunt Voltage Regulator

A shunt voltage regulator provides regulation by shunting current away from the load to regulate the output voltage. Fig. 15.22 shows the circuit of shunt voltage regulator. The voltage drop across series resistance depends upon the current supplied to the load R_L. The output voltage is equal to the sum of zener voltage (V_Z) and transistor base-emitter voltage (V_{BE}) *i.e.*

$$V_{out} = V_Z + V_{BE}$$

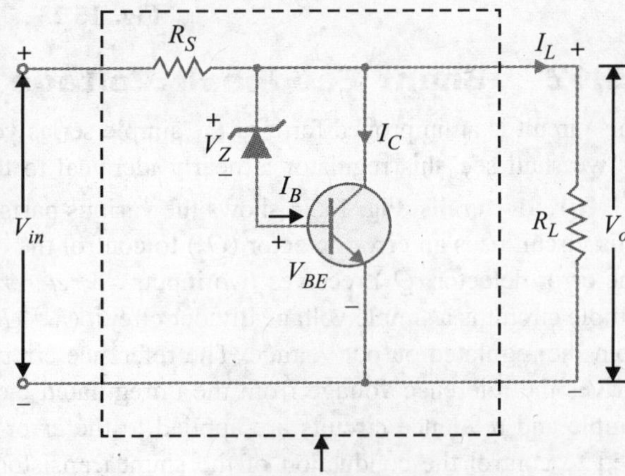

Fig. 15.22

If the load resistance decreases, the current through base of transistor decreases. As a result, less collector current is shunted. Therefore, the load current becomes larger, thereby maintaining the regulated voltage across the load. Reverse happens should the load resistance increase.

Drawbacks. A shunt voltage regulator has the following drawbacks :

(*i*) A large portion of the total current through R_S flows through transistor rather than to the load.

(*ii*) There is considerable power loss in R_S.

(*iii*) There are problems of overvoltage protection in this circuit.

For these reasons, a series voltage regulator is preferred over the shunt voltage regulator.

Example 15.23. *Determine the (i) regulated voltage and (ii) various currents for the shunt regulator shown in Fig. 15.23.*

Solution. (*i*) Output voltage, $V_{out} = V_Z + V_{BE} = 8.3 + 0.7 = \mathbf{9\ V}$

(*ii*) Load current, $I_L = \dfrac{V_{out}}{R_L} = \dfrac{9\ \text{V}}{100\ \Omega} = \mathbf{90\ mA}$

Current through R_S, $I_S = \dfrac{V_{in} - V_{out}}{R_S} = \dfrac{22 - 9}{130} = \dfrac{13\ \text{V}}{130\ \Omega} = \mathbf{100\ mA}$

$\therefore$ Collector current, $I_C = I_S - I_L = 100 - 90 = \mathbf{10\ mA}$

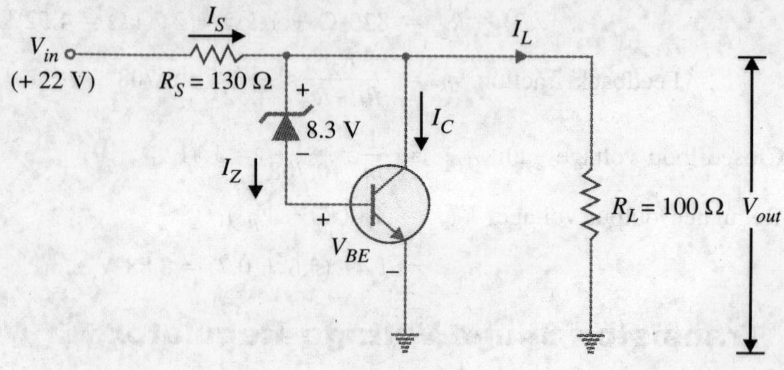

Fig. 15.23

15.12 Shunt Feedback Voltage Regulator

This circuit is an improved form of the simple series voltage regulator discussed in Art. 15.9. As we shall see, this regulator is nearly identical to the series feedback regulator.

Circuit details. Fig. 15.24 shows the various parts of a shunt feedback voltage regulator. This circuit uses an error detector (Q_2) to control the conduction of a **shunt transistor* (Q_1). The error detector (Q_2) receives two inputs : a *reference voltage* and a *sample voltage.* The sample circuit is a simple voltage divider circuit ($R_2 - R_3 - R_4$) and derives the sample voltage from the regulated output voltage. The reference circuit is made up of zener D_1 and R_1 and derives the reference voltage from the unregulated d.c. input voltage. The outputs from the sample and reference circuits are applied to the error detector (Q_2). The output from Q_2 is used to control the conduction of the shunt transistor Q_1. Since Q_1 (shunt transistor) is in parallel with load R_L, the change in the current conduction through Q_1 can control the load voltage.

Circuit operation. In a shunt feedback voltage regulator, the outputs from the sample and reference circuits are applied to the *error detector/amplifier* Q_2. The output from Q_2 controls the conduction current through the shunt transistor Q_1 to maintain the constant load voltage V_L.

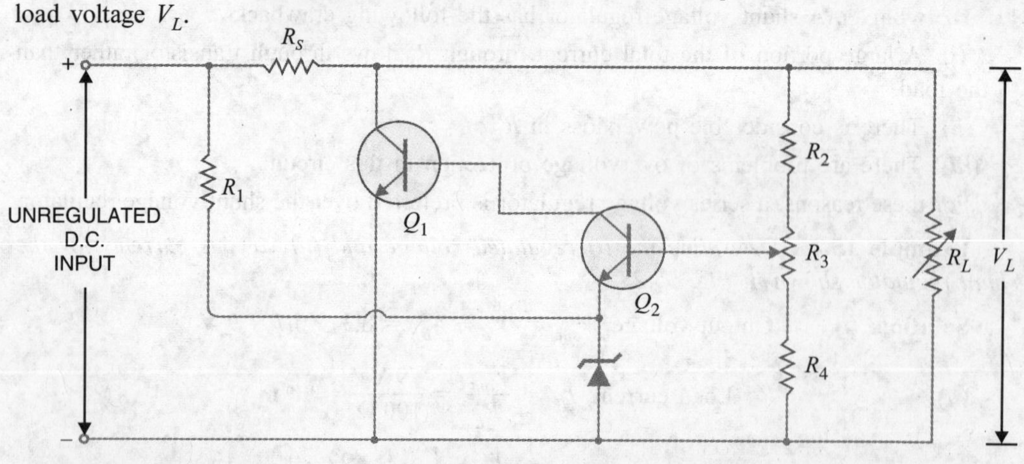

Fig. 15.24

* Note that transistor Q_1 is in parallel with the load, it is called shunt transistor and hence the name of the regulator.

(*i*) Suppose the load voltage V_L decreases due to any reason. This decrease causes the base voltage of Q_2 [$V_{B\,(Q2)}$] to decrease. Since emitter voltage of Q_2 [$V_{E(Q2)}$] is set to a fixed value (V_Z) by the zener diode, $V_{BE\,(Q2)}$ decreases when $V_{B\,(Q2)}$ decreases. This decreases base current $I_{B\,(Q1)}$ through the base of transistor Q_1. This in turn causes I_{C1} to decrease ; thus increasing $V_{C\,(Q1)}$. Since $V_L = V_{C\,(Q1)}$, the increase in $V_{C\,(Q1)}$ offsets the initial decrease in V_L.

(*ii*) Suppose the load voltage increases due to any reason. This increases the conduction through Q_2, causing an increase in $I_{B\,(Q1)}$. The increase in $I_{B\,(Q1)}$ causes the shunt transistor's conduction to increase, decreasing $V_{C\,(Q1)}$. The decrease in $V_{C\,(Q1)}$ offsets the initial increase in V_L.

15.13 IC Voltage Regulators

We can also use integrated circuits (*ICs*) to produce voltage regulators. One advantage of *IC* voltage regulators is that properties like thermal compensation, short circuit protection and surge protection can be built into the device. Most of the commonly used *IC* voltage regulators are three-terminal devices. Fig. 15.25 shows the schematic symbol for a three-terminal *IC* voltage regulator.

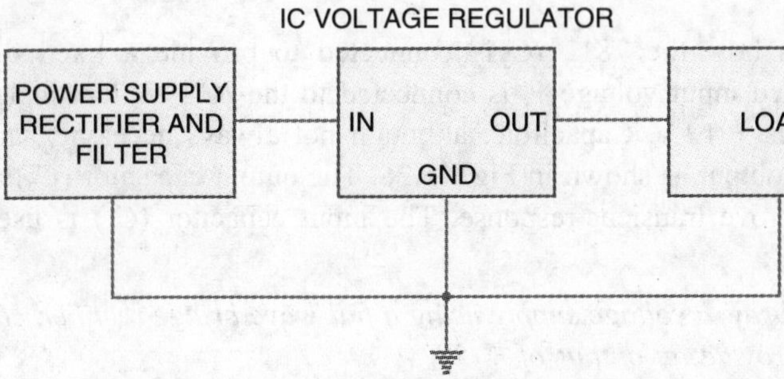

Fig. 15.25

There are basically four types of *IC* voltage regulators *viz.*

(*i*) Fixed positive voltage regulators

(*ii*) Fixed negative voltage regulators

(*iii*) Adjustable voltage regulators

(*iv*) Dual-tracking voltage regulators

The *fixed positive* and *fixed negative IC voltage regulators* are designed to provide specific output voltages. For example, *LM* 309 (fixed positive) provides a + 5 V d.c. output (as long as the regulator input voltages are within the specified ranges). The *adjustable voltage regulator* can be adjusted to provide any d.c. output voltage that is within its two specified limits. For example, the *LM* 317 output can be adjusted to any value between its limits of 1.2 V and 32 V d.c. Both positive and negative adjustable regulators are available. The *dual-tracking regulator* provides equal positive and negative output voltages. For example, the *RC* 4195 provides outputs of + 15 V and – 15 V d.c. Adjustable dual-tracking regulators are also available.

15.14 Fixed Positive Voltage Regulators

This *IC* regulator provides a fixed positive output voltage. Although many types of *IC* regulators are available, the 7800 series of *IC* regulators is the most popular. The last two digits in the part number indicate the d.c. output voltage. For example [See Table below], the 7812 is a + 12 V regulator whereas the 7805 is a + 5 V regulator. Note that this series (7800 series) provides fixed regulated voltages from + 5 V to + 24 V.

Type number	Output voltage
7805	+5.0 V
7806	+6.0 V
7808	+8.0 V
7809	+9.0 V
7812	+12.0 V
7815	+15.0 V
7818	+18.0 V
7824	+24.0 V

The 7800 series

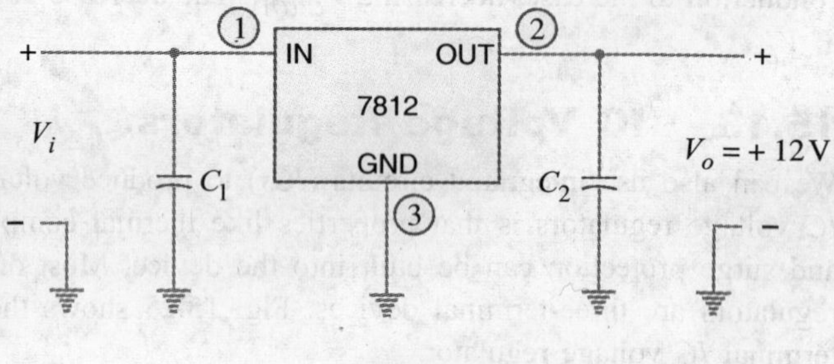

Fig. 15.26

Fig. 15.26 shows how the 7812 *IC* is connected to provide a fixed d.c. output of + 12 V. The unregulated input voltage V_i is connected to the *IC*'s *IN* terminal and the *IC*'s *OUT* terminal provides + 12 V. Capacitors, although not always necessary, are sometimes used on the input and output as shown in Fig. 15.26. The output capacitor (C_2) acts basically as a line filter to improve transient response. The input capacitor (C_1) is used to prevent unwanted oscillations.

Example 15.24. *Draw a voltage supply using a full-wave bridge rectifier, capacitor filter and IC regulator to provide an output of + 5 V.*

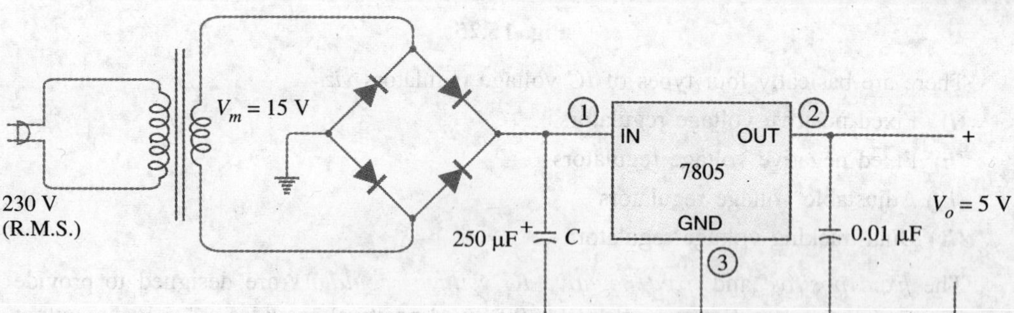

Fig. 15.27

Solution. The resulting circuit is shown in Fig. 15.27. Here we use 7805 *IC* with unregulated power supply.

15.15 Fixed Negative Voltage Regulators

This *IC* regulator provides a fixed negative ouput voltage. The 7900 series of *IC* regulators is commonly used for this purpose. This series (7900) is the negative-voltage counterpart of the 7800 series [See Table below]. Note that 7900 series provides fixed regulated voltages from – 5 V to – 24 V.

Type number	Output voltage
7905	–5.0 V
7905.2	–5.2 V
7906	–6.0 V
7908	–8.0 V
7912	–12.0 V
7915	–15.0 V
7918	–18.0 V
7924	–24.0 V

The 7900 series

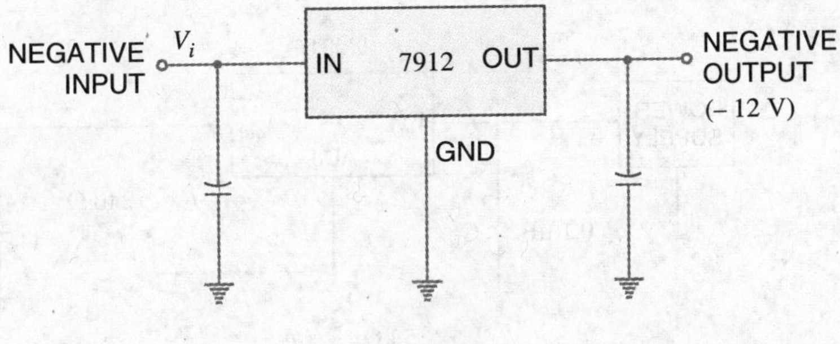

Fig. 15.28

Fig. 15.28 shows how 7912 *IC* is connected to provide a fixed d.c. output of – 12 V. The unregulated negative input voltage V_i is connected to *IC*'s *IN* terminal and the *IC*'s *OUT* terminal provides – 12 V. Capacitors used in the circuit perform the same function as in a fixed positive regulator.

15.16 Adjustable Voltage Regulators

The adjustable voltage regulator can be adjusted to provide any d.c. output voltage that is within its two specified limits. The most popular three-terminal *IC* adjustable voltage regulator is the *LM* 317. It has an input terminal, output terminal and an adjustment terminal. An external voltage divider is used to change the d.c. output voltage of the regulator. By changing R_2, a wide range of output voltages can be obtained.

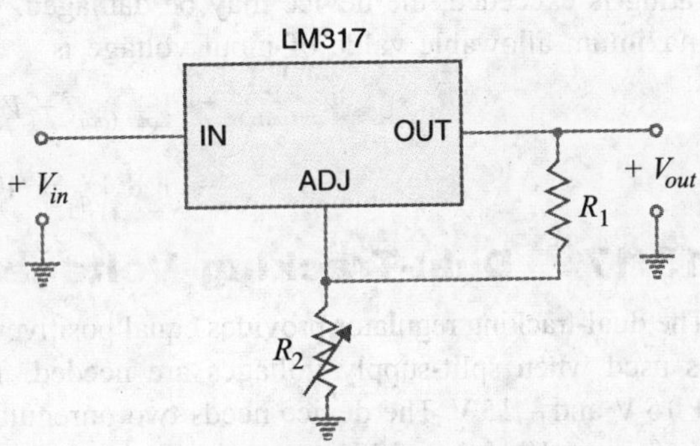

Fig. 15.29

The *LM* 317 is a three-terminal positive adjustable voltage regulator and can supply 1.5 A of load current over an adjustable output range of 1.25 V to 37 V. Fig. 15.29 shows an unregulated power supply driving an *LM* 317 circuit. The data sheet of an *LM* 317 gives the following formula for the output voltage :

$$V_{out} = 1.25\left(\frac{R_2}{R_1}+1\right)$$

This formula is valid from 1.25 V to 37 V.

Example 15.25. *In Fig. 15.30, R_2 is adjusted to 2.4 kΩ. Determine the regulated d.c. output voltage for the circuit.*

Solution. The regulated d.c. output voltage for the circuit is given by ;

$$V_{out} = 1.25\left(\frac{R_2}{R_1}+1\right)$$

$$= 1.25\left(\frac{2.4\text{ k}\Omega}{240\ \Omega}+1\right) = (1.25)\ (11) = \mathbf{13.75\ V}$$

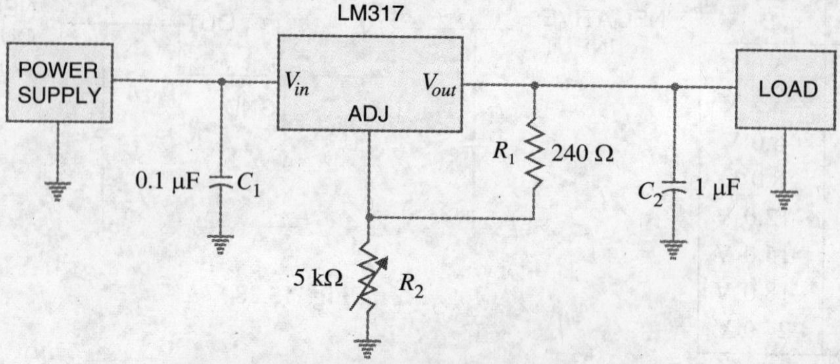

Fig. 15.30

Example 15.26. *The LM 317 is adjusted to provide a + 8 V_{dc} regulated output voltage. Determine the maximum allowable input voltage to the device.*

Solution. The maximum allowable difference between V_{in} and V_{out} of an adjustable voltage regulator is called its input/output differential rating and is denoted by V_d. If this rating is exceeded, the device may be damaged. For the *LM* 317, V_d = 40 V. Therefore, the maximum allowable value of input voltage is

$$V_{in\ (max)} = V_{out\ (adj)} + V_d$$

$$= +\ 8\ V_{dc} + 40 = \mathbf{+\ 48\ V}$$

15.17 Dual-Tracking Voltage Regulators

The dual-tracking regulator provides equal positive and negative output voltages. This regulator is used when split-supply voltages are needed. The *RC* 4195 *IC* provides d.c. outputs of + 15 V and – 15 V. The device needs two unregulated input voltages. The positive input may be from + 18 V to + 30 V and the negative input from – 18 V to –30 V. As shown, the two outputs are ± 15 V. The data sheet of an *RC* 4195 lists a maximum output current of 150 mA for each supply and a load regulation of 3 mV. Adjustable dual-tracking regulators are also available. These regulators have outputs that can be varied between their two rated limits.

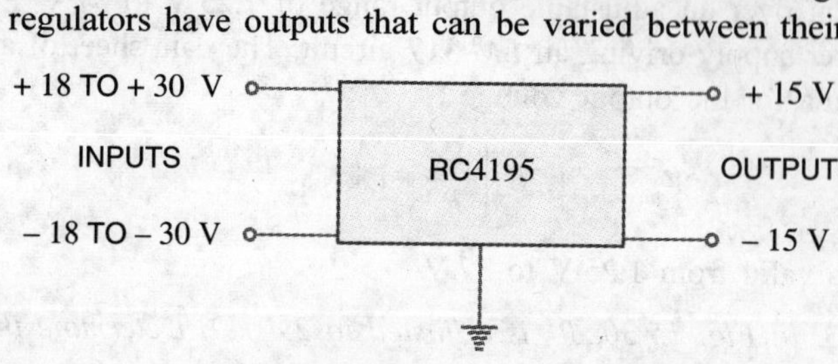

Fig. 15.31

Short Answer Questions

Q.1. *Why do we prefer d.c. power supply to batteries?*

Ans. In practice, d.c. power for electronic circuits is most conveniently obtained from commercial a.c. lines by using rectifier-filter system, called a d.c. power supply. Batteries are rarely used for the purpose as they are costly and require frequent replacement.

Q.2. *What are the causes of poor voltage regulation in a d.c. power supply?*

Ans. The following conditions in a d.c. power supply may cause poor voltage regulation :

(*i*) Resistance in the wires of chokes and transformers. When load current is increased, voltage drop across all resistances increases, decreasing the output voltage of the supply. The reverse would happen should the load current decrease.

(*ii*) Use of capacitive-input filtering instead of inductive-input filtering.

(*iii*) Insufficient capacitance in the filter circuit.

(*iv*) Use of half-wave rectification instead of full-wave rectification.

(*v*) Too high a bleeder resistance value.

Q.3. *What are the limitations of an ordinary d.c. power supply?*

Ans. An ordinary d.c. power supply consists of rectifier-filter combination.The output voltage from such a supply depends upon the input mains voltage and load current. If the input mains voltage or load current changes, the output voltage will also change. These variations in d.c. output voltage may cause inaccurate or erratic operation or even malfunctioning of many electronic circuits. What most of electronic circuits require is the constant d.c. voltage irrespective of mains fluctuations or load current variations. For this reason, electronics equipment require regulated d.c. power supply.

Q.4. *What is a regulated d.c. power supply?*

Ans. A regulated d.c. power supply is the one which maintains a constant d.c. output voltage irrespective of the a.c. mains fluctuations or load variations. It consists of an ordinary d.c. power supply and some voltage regulating device (*e.g.*, zener, zener in conjunction with transistor *etc.*). The output of ordinary d.c. power supply is fed to the voltage regulator which produces the final output. The output voltage remains constant whether the load current changes or there are fluctuations in the input a.c. voltage.

Q.5. *What is zener breakdown?*

Ans. A zener diode that has very narrow depletion layer breaks down by zener mechanism. The zener mechanism can be described as follows. Since the depletion layer is very thin, application of low reverse bias (5 V or less) will develop a high field intensity (of the order of 3×10^5 V/cm). A field of such high magnitude exerts a large force on the valence electrons of semi-conductor atoms, tending to separate them from their respective nuclei. Thus electron-hole pairs are generated in large numbers and a sudden increase in current is observed.

Q.6. *What is avalanche breakdown?*

Ans. A zener diode that has wide depletion layer breaks down by avalanche mechanism. The avalanche mechanism can be described as follows. As the reverse voltage is increased, a point is reached when the minority carriers (electrons) passing across the *pn* junction gain sufficient energy to knock off valence electrons bound to the crystal and raise them to conduction band. One minority electron may thus free several valance electrons through collision.

In turn, these freed valence electrons may gain enough energy to free more valence electrons bound to the crystal. This phenomenon is called *carrier multiplication*. The current carriers are generated so quickly and in such large numbers that there is an apparent *avalanche* or self-sustained multiplication process.

Q.7. *Why does zener voltage decrease with increase in junction temperature?*

Ans. An increase in temperature increases the energies of the valence electrons. This weakens the bonds holding the valence electrons. Consequently, less voltage is required to pull the valence electrons from their position around the nuclei. Thus breakdown voltage decreases as the temperature increases.

Q.8. *What will happen if the collector-base junction is driven into zener effect?*

Ans. The collector of a transistor is reverse biased for proper operation. If excessive reverse bias is applied, the collector junction may be driven into zener effect. Such a situation will have the following effects :

(*i*) This will cause a high current to flow and the transistor action may cease.

(*ii*) The depletion layer will be widened so that emitter may be ohmically connected to the collector.

(*iii*) The reverse current may become high enough to raise the junction temperature to unaccepted level.

Q.9. *How does transistor series voltage regulator remove the objections of zener diode voltage regulator?*

Ans. In this circuit, we use a transistor and zener diode. The circuit is called "series voltage regulator" because collector emitter terminals are in "series with the load". The load voltage equals the zener voltage V_Z (less the V_{BE} drop of the transistor) but the changes in zener current are reduced by a factor β. Therefore, the effect of zener impedance is greatly reduced and much more stabilised output is ensured.

Q.10. *What is the drawback of transistor series voltage regulator?*

Ans. The circuit does not provide protection against overload or short-circuit conditions. If the load terminals accidentally get short-circuited, we get an enormous current through the transistor. Either transistor will be destroyed or a diode in the unregulated d.c. supply will be burnt out or both.

Multiple-Choice Questions

1. In an unregulated d.c. power supply, if load current increases, the output voltage ....

(*i*) remains the same

(*ii*) decreases

(*iii*) increases

(*iv*) none of the above

2. In an unregulated d.c. power supply, if input a.c. voltage increases, the output voltage ....

(*i*) increases

(*ii*) decreases

(*iii*) remains the same

(*iv*) none of the above

3. A d.c. power supply which has a voltage regulation of .... is unregulated d.c. power supply.
   (*i*) 0 % (*ii*) 0.5 %
   (*iii*) 10 % (*iv*) 0.8 %

4. Commercial d.c. power supplies have voltage regulation ....
   (*i*) of 10 % (*ii*) of 15 %
   (*iii*) of 25 % (*iv*) within 1 %

5. An ideal regulated d.c. power supply is one which has voltage regulation of ....
   (*i*) 0 % (*ii*) 5 %
   (*iii*) 10 % (*iv*) 1 %

6. A zener diode utilises .... characteristic for voltage regulation.
   (*i*) forward (*ii*) reverse
   (*iii*) both forward and reverse
   (*iv*) none of the above

7. Zener diode can be used as ....
   (*i*) d.c. voltage regulator only
   (*ii*) a.c. voltage regulator only
   (*iii*) both d.c. and a.c. voltage regulator
   (*iv*) none of the above

8. A zener diode is used as a .... voltage regulating device.
   (*i*) shunt (*ii*) series
   (*iii*) series-shunt
   (*iv*) none of the above

9. As the junction temperature increases, the voltage breakdown point for zener mechanism ....
   (*i*) is increased (*ii*) is decreased
   (*iii*) remains the same
   (*iv*) none of the above

10. The rupture of co-valent bonds will occur when the electric field is ....
   (*i*) 100 V/cm (*ii*) 0.6 V/cm
   (*iii*) 1000 V/cm
   (*iv*) more than 10^5 V/cm

11. In a 15 V zener diode, the breakdown mechanism will occur by ....
   (*i*) avalanche mechanism
   (*ii*) zener mechanism
   (*iii*) both zener and avalanche mechanism
   (*iv*) none of the above

12. A zener diode that has very narrow depletion layer will break down by .... mechanism.
   (*i*) avalanche (*ii*) zener
   (*iii*) both avalanche and zener
   (*iv*) none of the above

13. As the junction temperature increases, the voltage breakdown point for avalanche mechanism ....
   (*i*) remains the same
   (*ii*) decreases (*iii*) increases
   (*iv*) none of the above

14. Another name for zener diode is .... diode.
   (*i*) breakdown (*ii*) voltage
   (*iii*) power (*iv*) current

15. Zener diodes are generally made of ....
   (*i*) germanium (*ii*) silicon
   (*iii*) carbon
   (*iv*) none of the above

16. For increasing the voltage rating, zeners are connected in..........
   (*i*) parallel (*ii*) series-parallel
   (*iii*) series
   (*iv*) none of the above

17. In a zener voltage regulator, the changes in load current produce changes in ....
   (*i*) zener current
   (*ii*) zener voltage
   (*iii*) zener voltage as well as zener current
   (*iv*) none of the above

18. A zener voltage regulator is used for .... load currents.
   (*i*) high (*ii*) very high
   (*iii*) moderate (*iv*) small

19. A zener voltage regulator will cease to act as a voltage regulator if zener current becomes ....
    (*i*) less than load current (*ii*) zero
    (*iii*) more than load current
    (*iv*) none of the above
20. If the doping level is increased, the breakdown voltage of the zener ....
    (*i*) remains the same
    (*ii*) is increased (*iii*) is decreased
    (*iv*) none of the above
21. A 30 V zener will have depletion layer width .... that of 10 V zener.
    (*i*) more than (*ii*) less than
    (*iii*) equal to
    (*iv*) none of the above
22. The current in a zener diode is limited by ....
    (*i*) external resistance
    (*ii*) power dissipation
    (*iii*) both (*i*) and (*ii*)
    (*iv*) none of the above
23. A 5 mA change in zener current produces a 50 mV change in zener voltage. What is the zener impedance?
    (*i*) 1 Ω (*ii*) 0.1 Ω
    (*iii*) 100 Ω (*iv*) 10 Ω
24. A certain regulator has a no-load voltage of 6 V and a full-load output of 5.82 V. What is the load regulation?
    (*i*) 3.09 % (*ii*) 2.87 %
    (*iii*) 5.72 %
    (*iv*) none of the above
25. What is true about the breakdown voltage in a zener diode?
    (*i*) It decreases when load current increases.
    (*ii*) It destroys the diode.
    (*iii*) It equals current times the resistance.
    (*iv*) It is approximately constant.
26. Which of these is the best description for a zener diode?
    (*i*) It is a diode.
    (*ii*) It is a constant-current device.
    (*iii*) It is a constant-voltage device.
    (*iv*) It works in the forward region.
27. A zener diode ....
    (*i*) is a battery
    (*ii*) acts like a battery in the breakdown region
    (*iii*) has a barrier potential of 1 V
    (*iv*) is forward biased
28. The load voltage is approximately constant when a zener diode is ....
    (*i*) forward biased (*ii*) unbiased
    (*iii*) reverse biased
    (*iv*) operating in the breakdown region
29. In a loaded zener regulator, which is the largest current?
    (*i*) Series current (*ii*) Zener current
    (*iii*) Load current
    (*iv*) none of the above
30. If the load resistance decreases in a zener regulator, then zener current ....
    (*i*) decreases (*ii*) stays the same
    (*iii*) increases
    (*iv*) none of the above
31. If the input a.c. voltage to unregulated or ordinary d.c power supply increases by 5 %, what will be the approximate change in d.c. output voltage?
    (*i*) 10 % (*ii*) 20 %
    (*iii*) 15 % (*iv*) 5 %
32. If the load current drawn by unregulated d.c. power supply increases, the d.c. output voltage ....
    (*i*) increases (*ii*) decreases
    (*iii*) stays the same
    (*iv*) none of the above
33. If a d.c. power supply has no-load and full-load voltages of 30 V and 25 V respectively, then percentage voltage regulation is ....
    (*i*) 10 % (*ii*) 20 %

(*iii*) 15 %
(*iv*) none of the above

34. A d.c. power supply has a voltage regulation of 1 %. If the no-load voltage is 20 V, what is the full-load voltage?
(*i*) 20.8 V (*ii*) 15.7 V
(*iii*) 18.6 V (*iv*) 17.2 V

35. Two similar 15 V zeners are connected in series. What is the regulated output voltage?
(*i*) 15 V (*ii*) 7.5 V
(*iii*) 30 V (*iv*) 45 V

36. A d.c. power supply can deliver a maximum rated current of 0.5 A at full-load output voltage of 20 V. What is the minimum load resistance that you can connect across the supply?
(*i*) 10 Ω (*ii*) 20 Ω
(*iii*) 15 Ω (*iv*) 40 Ω

37. In a regulated d.c. power supply, two similar 15 V zeners are connected in series. The input voltage is 45 V d.c. If each zener has a maximum current rating of 300 mA, what should be the value of series resistance?
(*i*) 10 Ω (*ii*) 50 Ω
(*iii*) 25 Ω (*iv*) 40 Ω

38. A zener regulator .... in the power supply.
(*i*) increases the ripple
(*ii*) decreases the ripple
(*iii*) neither increases nor decreases ripple
(*iv*) data insufficient

39. When load current is zero, the zener current will be ....
(*i*) zero (*ii*) minimum
(*iii*) maximum
(*iv*) none of the above

40. The zener current will be minimum when ....
(*i*) load current is maximum
(*ii*) load current is minimum
(*iii*) load current is zero
(*iv*) none of the above

Answers to Multiple-Choice Questions

1. (*ii*)	2. (*i*)	3. (*iii*)	4. (*iv*)	5. (*i*)
6. (*ii*)	7. (*iii*)	8. (*i*)	9. (*ii*)	10. (*iv*)
11. (*i*)	12. (*ii*)	13. (*iii*)	14. (*i*)	15. (*ii*)
16. (*iii*)	17. (*i*)	18. (*iv*)	19. (*ii*)	20. (*iii*)
21. (*i*)	22. (*iii*)	23. (*iv*)	24. (*i*)	25. (*iv*)
26. (*iii*)	27. (*ii*)	28. (*iv*)	29. (*i*)	30. (*i*)
31. (*iv*)	32. (*ii*)	33. (*ii*)	34. (*i*)	35. (*iii*)
36. (*iv*)	37. (*ii*)	38. (*ii*)	39. (*iii*)	40. (*i*)

Chapter Review Topics

1. What do you understand by unregulated d.c. power supply? Draw the circuit of such a supply.
2. What are the limitations of unregulated d.c. power supply?
3. What do you understand by regulated d.c. power supply? Draw the block diagram of such a supply.
4. Write a short note on the need for regulated d.c. power supply.
5. Explain the action of a zener voltage regulator with a neat diagram.

6. Write short notes on the following :
   (*i*) Transistor series voltage regulator
   (*ii*) Negative feedback voltage regulator
7. What are the limitations of transistorised power supplies?

Problems

1. A voltage regulator is rated at an output current of I_L = 0 to 40 mA. Under no-load conditions, the ouput voltage from the circuit is 8 V. Under full-load conditions, the output voltage from the circuit is 7.996 V. Determine the value of load-regulation for the circuit. **[100 μV/mA]**
2. The zener diode in Fig. 15.32 has values of $I_{Z\,(min)}$ = 3 mA and $I_{Z\,(max)}$ = 100 mA. What is the minimum allowable value of R_L? **[241 Ω]**

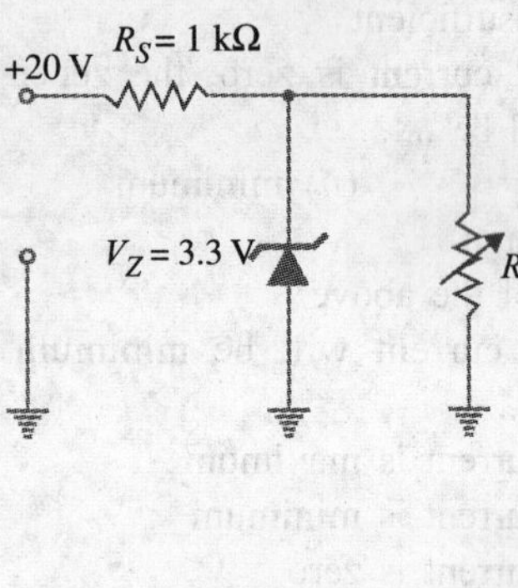

Fig. 15.32

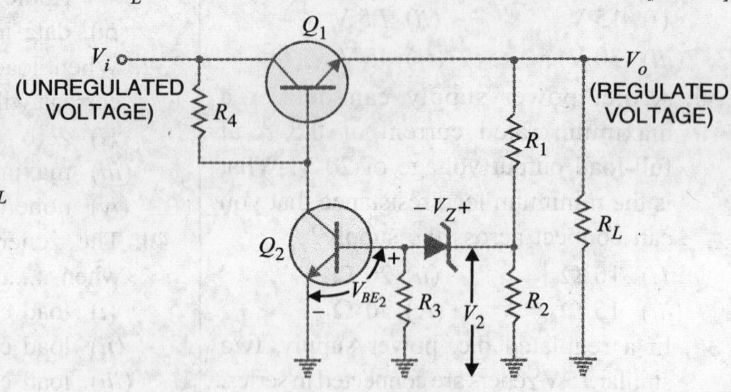

Fig. 15.33

3. What regulated output voltage is provided by the circuit of Fig. 15.33 for circuit elements : R_1 = 20 kΩ, R_2 = 30 kΩ and V_Z = 8.3 V? **[15 V]**
4. Calculate the output voltage and zener diode current in the regulator circuit of Fig. 15.34. **[7.6 V, 3.66 mA]**

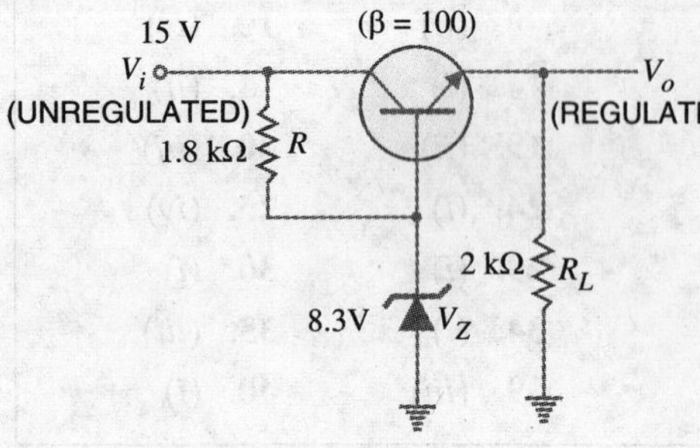

Fig. 15.34

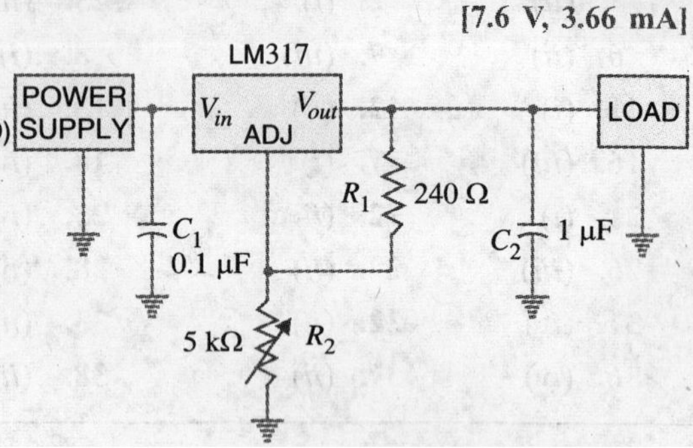

Fig. 15.35

5. If R_2 in Fig. 15.35 is adjusted to 1.68 kΩ, determine the regulated d.c. output voltage for the *LM* 317. **[10 V]**

CHAPTER 16

Solid-State Switching Circuits

Learning Objectives

- Understand the limitations of mechanical switch.
- Explore the advantages of electromechanical switch or relay over mechanical switch.
- Develop an understanding of electronic switch.
- Learn the advantages of electronic switches.
- Understand the operation of the various types of multivibrators.
- Develop an understanding of differentiating and integrating circuits.
- Develop a strong understanding of diode clippers.
- Learn the application of diodes as the clamping circuits.

Introduction

In practice, it is often required to make or break an electrical circuit in many operations. In some applications, it is desirable and necessary that this make and break should be very quick and without sparking. The mechanical switches cannot be used for the purpose for two main reasons. Firstly, a mechanical switch has high inertia which limits its speed of operation. Secondly, there is sparking at the contacts during breaking operation which results in the burning of the contacts.

The researches in the past years have revealed that tubes and transistors can serve as switching devices. They can turn ON or OFF power in an electrical circuit at a very high speed without any sparking. Such switches are known as electronic switches. The electronic switches are being extensively used to produce non-sinusoidal waves *e.g.*, square, rectangular, triangular or saw-tooth waves. Solid-state switching circuits are finding increasing applications. For example, solid-state switching circuits are the fundamental components of modern computer systems. In this chapter, we shall confine our attention to transistor as a switch. Once the reader gets acquainted with the switching action of a transistor, he/she can continue to study digital electronics on his/her own.

16.1 Switching Circuit

A circuit which can turn ON or OFF current in an electrical circuit is known as a **switching circuit.**

A switching circuit essentially consists of two parts *viz.* (*i*) a switch and (*ii*) associated circuitry. The switch is the most important part of the switching circuit. It actually makes or breaks the electrical circuit. The function of associated circuitry is to help the switch in turning ON or OFF current in the circuit. It may be worthwhile to mention here that associated circuitry is particularly used with electronic switches.

16.2 Switch

A switch is a device that can turn ON or OFF current in an electrical circuit. It is the most important part of a switching circuit. The switches can be broadly classified into the following three types :

(*i*) Mechanical switch

(*ii*) Electro-mechanical switch or Relay

(*iii*) Electronic switch

Although the basic purpose of this chapter is to discuss the switching action of a transistor, yet a brief description of mechanical and electromechanical switches is being presented. This will help the reader to understand the importance of transistor as a switch.

16.3 Mechanical Switch

A switch which is operated mechanically to turn ON or OFF current in an electrical circuit is known as a **mechanical switch.**

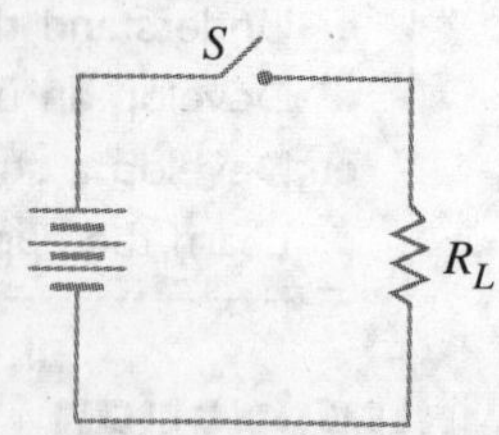

Fig. 16.1

The familiar example of a mechanical switch is the tumbler switch used in homes to turn ON or OFF power supply to various appliances such as fans, heaters, bulbs *etc*. The action of a mechanical switch can be beautifully understood by referring to Fig. 16.1 where a load R_L is connected in series with a battery and a mechanical switch *S*. As long as the switch is open, there is no current in the circuit. When switch is closed, the current flow is established in the circuit. It is easy to see that the whole current flows through the load as well as the switch.

Limitations. A mechanical switch suffers from the following drawbacks :

(*i*) In the closed position, the switch carries the whole of the load current. For a large load current, the switch contacts have to be made heavy to enable them to carry the necessary current without overheating. This increases the size of the switch.

(*ii*) If the load current carried by the circuit is large, there will be sparking at the contacts of the switch during breaking operation. This results in the wear and tear of the contacts.

(*iii*) Due to high inertia of a mechanical switch, the speed of operation is very small.

Due to above limitations, the use of mechanical switches is restricted to situations where switching speed is small and the load current to be handled is not very heavy.

16.4 Electro-mechanical Switch or Relay

It is a mechanical switch which is operated electrically to turn ON or OFF current in an electrical circuit.

The electro-mechanical switch or relay is an improved form of simple mechanical switch. Fig. 16.2 shows the schematic diagram of a typical relay. It consists of lever *L* carrying armature A and a solenoid C. The spring pulls the lever upwards while the solenoid when energised pulls it downwards. The solenoid circuit is so designed that when switch S is closed, the downward pull of the solenoid exceeds the upward pull of the spring.

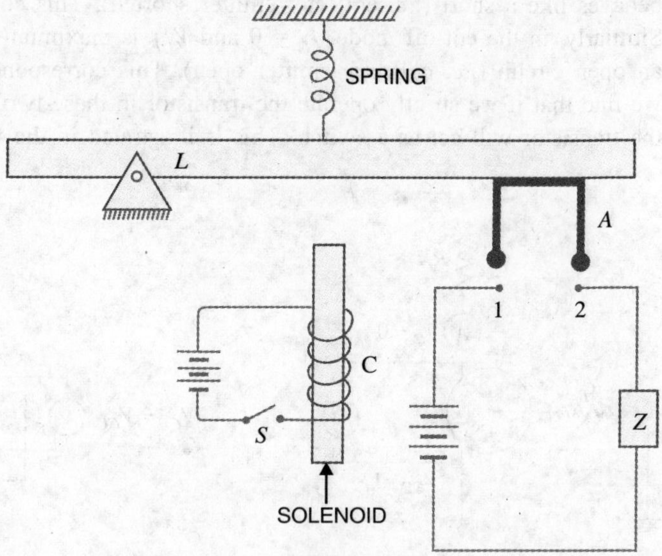

Fig. 16.2

When the switch S is closed, the lever is pulled downward and the armature A closes the relay contacts 1 and 2. This turns ON current in the circuit. However, when switch S is opened, the solenoid is de-energised and the spring pulls the lever and hence the armature A upwards. Consequently, the relay contacts 1 and 2 are opened and current flow in the circuit is interrupted. In this way, a relay acts as a switch.

Advantages. A relay possesses the following advantages over a simple mechanical switch :

(*i*) The relay or electro-mechanical switch requires a small power for its operation. This permits to control a large power in the load by a small power to the relay circuit. Thus a relay acts as a power amplifier *i.e.* it combines control with power amplification.

(*ii*) The switch in the relay coil carries a small current as compared to the load current. This permits the use of a smaller switch in the relay coil circuit.

(*iii*) The operator can turn ON or OFF power to a load even from a distance. This is a very important advantage when high voltages are to be handled.

(*iv*) There is no danger of sparking as the turning ON or OFF is carried by the relay coil switch which carries a small current.

However, a relay has two principal limitations. First, the speed of operation is very small; less than 5 operations per second. Secondly, a relay has moving parts and hence there is considerable wear and tear.

16.5 Electronic Switches

It is a device which can turn ON or OFF current in an electrical circuit with the help of electronic devices e.g., transistors.

So far we have been using the transistor to amplify weak a.c. signals. In the amplification mode, the transistor operates in the active region of its output characteristics. However, a transistor can also be used as a *switch* under proper operating conditions. Thus, when a transistor is operated in *saturation mode* and *cut-off mode,* it will behave as a switch. We know that in the saturation mode or state, I_C is maximum and $V_{CE} = 0$ so that the transistor behaves like a short (*i.e.* collector-emitter shorted). This corresponds to closure of switch. Similarly, in the cut-off mode, $I_C = 0$ and V_{CE} is maximum and the transistor behaves like an open circuit (*i.e.* collector-emitter open). This corresponds to opening of switch. Thus, we find that if we strictly operate the transistor in these two modes (saturation and cut-off), the transistor will act as a switch. This is illustrated in the following discussion.

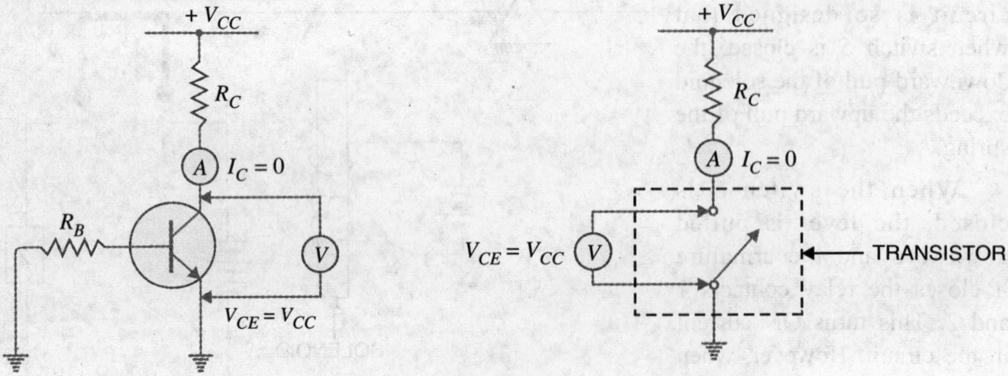

(*i*) (*ii*)

Fig. 16.3

(*i*) When the base input voltage is enough negative, the transistor is cut-off and no current flows in collector load [See Fig. 16.3 (*i*)]. As a result, there is no voltage drop across R_C and the output voltage is *ideally V_{CC} *i.e.*

$$I_C = 0 \quad \text{and} \quad V_{CE} = V_{CC}$$

This condition is similar to that of an open switch (*i.e.* OFF state) as shown in Fig. 16.3 (*ii*).

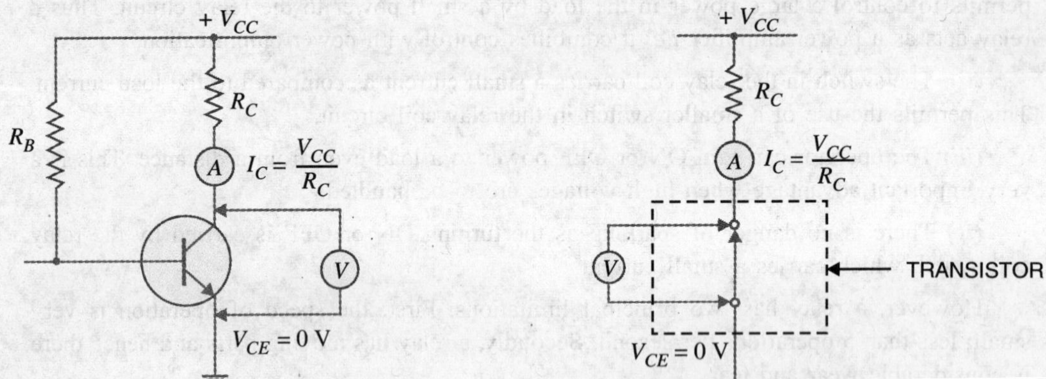

(*i*) (*ii*)

Fig. 16.4

(*ii*) When the input base voltage is positive enough that transistor saturates, then $I_{C(sat)}$ will flow through R_C. Under such conditions, the entire V_{CC} will drop across collector load R_C and output voltage is ideally zero *i.e.*

* The collector current will not be zero since a little leakage current always flows even when the base input voltage is negative or zero.

$\therefore$ Output voltage $= V_{CC} - I_{leakage} R_C$. If $I_{leakage} = 0$, then output voltage $= V_{CC}$.

$$I_C = I_{C(sat)} = \frac{V_{CC}}{R_C} \quad \text{and} \quad V_{CE} = 0$$

This condition is similar to that of a closed switch (*i.e.* ON state) as shown in Fig. 16.4 (*ii*).

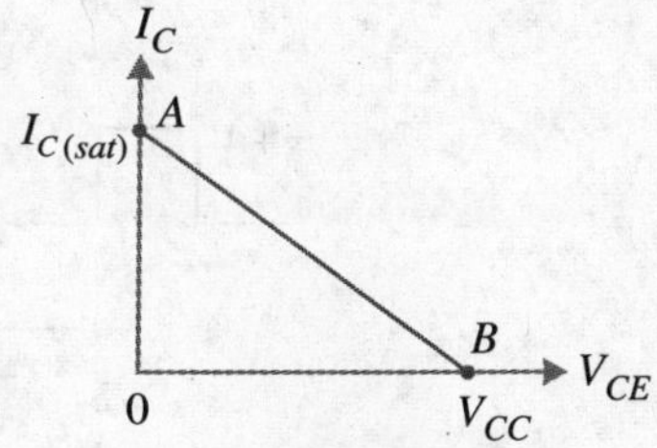

Fig. 16.5

Conclusion. The above discussion leads to the conclusion that a transistor can act as a switch if it is operated in two modes, namely ; *saturation* and *cut-off*. Note that saturation (ON state) and cut-off (OFF state) are stable states. This means that if the transistor is in saturation state, it will remain in this state until it is switched to cut-off state by reducing the base current to zero.

Similarly, if the transistor is in cut-off state, it will remain in this state until it is switched to saturation state by providing sufficient base current. Fig. 16.5 shows the switching action of a transistor in terms of d.c. load line. The point *A* of the load line represents the ON condition (saturation state) and point *B* represents the OFF condition (cut-off state).

16.6 Simple Application of Transistor Switch

There are many applications of transistor as a switch *e.g.*, for computer and control applications. Fig. 16.6 shows a simple application of a transistor as a switch. Here transistor switch is used to turn ON or OFF a LED (Light-Emitting Diode). The input voltage (V_{in}) is a square wave that has a time period of 2 s (say). When input voltage $V_{in} = 0$ V, the transistor is cut-off so that the transistor acts as an open switch. The collector current is zero (*i.e.* $I_C = 0$) and the LED does not emit light. When the square wave goes to its high level (*i.e.* $V_{in} = + 5$ V), the transistor is saturated so that the transistor acts as a closed switch. Therefore, large collector current [$= I_{C(sat)}$] flowing through the LED causes it to emit light. Since the time period of the input square wave is assumed to be 2 s, the LED is ON for 1 s and OFF for 1 s. As a result, we have a blinking LED.

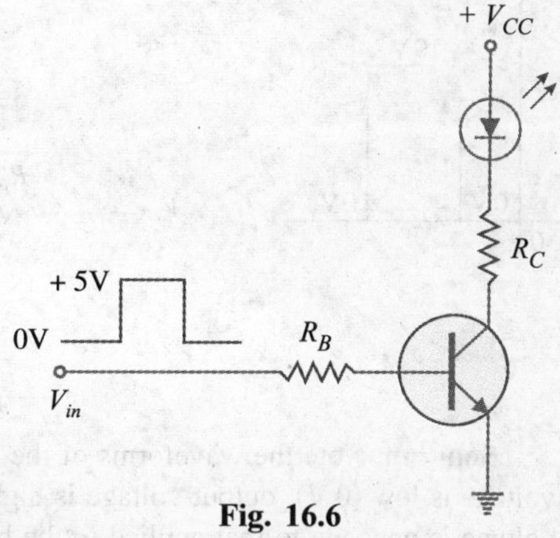

Fig. 16.6

Example 16.1. *Determine the minimum high input voltage (+V) required to saturate the transistor switch shown in Fig. 16.7.*

Solution. Assuming the transistor to be ideal,

$$I_{C(sat)} = V_{CC}/R_C = 10 \text{ V}/1 \text{ k}\Omega = 10 \text{ mA}$$

$$\therefore \quad I_B = \frac{I_{C(sat)}}{\beta} = \frac{10 \text{ mA}}{100} = 0.1 \text{ mA}$$

Now,

$$+V = I_B R_B + V_{BE}$$

$$= (0.1 \text{ mA}) (47 \text{ k}\Omega) + 0.7$$

$$= 4.7 + 0.7 = \mathbf{5.4 \text{ V}}$$

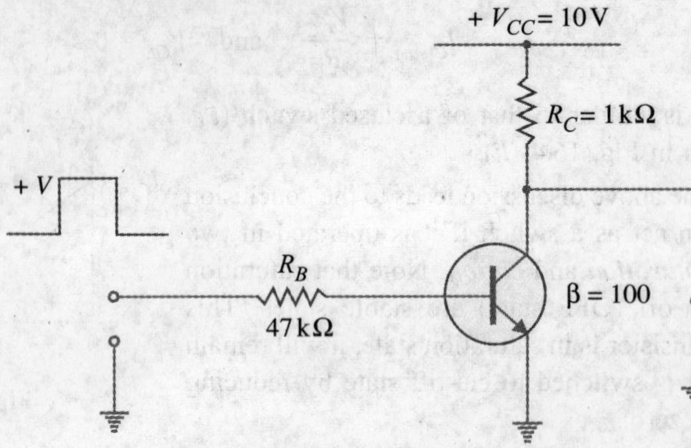

Fig. 16.7

Hence in order to saturate the transistor, we require + 5.4 V.

Example 16.2. *Determine the values of R_C and R_B for the transistor switch of Fig. 16.8 if $I_{C(sat)}$ = 10 mA.*

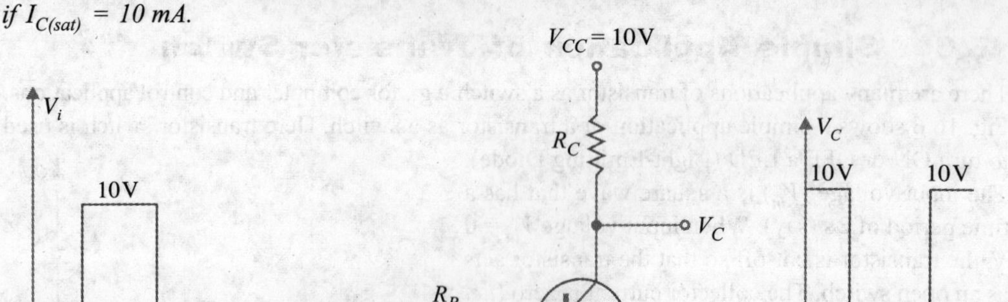

Fig. 16.8

Solution. Note the waveforms of the input and output voltages. We find that when input voltage is low (0 V), output voltage is high (10 V) and *vice-versa*. In other words, the output voltage is opposite to that applied to the base or input terminals. For this reason, a transistor switch is also called *transistor inverter.*

At saturation, the collector current $I_{C(sat)}$ is given by ;

$$I_{C(sat)} = \frac{V_{CC}}{R_C}$$

$$\therefore \quad R_C = \frac{V_{CC}}{I_{C(sat)}} = \frac{10 \text{ V}}{10 \text{ mA}} = \mathbf{1 \text{ k}\Omega}$$

At saturation, the base current I_B is given by ;

$$I_B = \frac{I_{C(sat)}}{\beta} = \frac{10 \text{ mA}}{250} = 40 \ \mu\text{A}$$

Now, $$R_B = \frac{V_i - V_{BE}}{I_B} = \frac{(10 - 0.7)\text{ V}}{40\ \mu\text{A}} \simeq 232\ \text{k}\Omega$$

Example 16.3. *Show that the circuit shown in Fig. 16.9 behaves like an inverter when the input voltage switches between 0 V and + 5 V. Assume that the transistor is of silicon and that β = 100.*

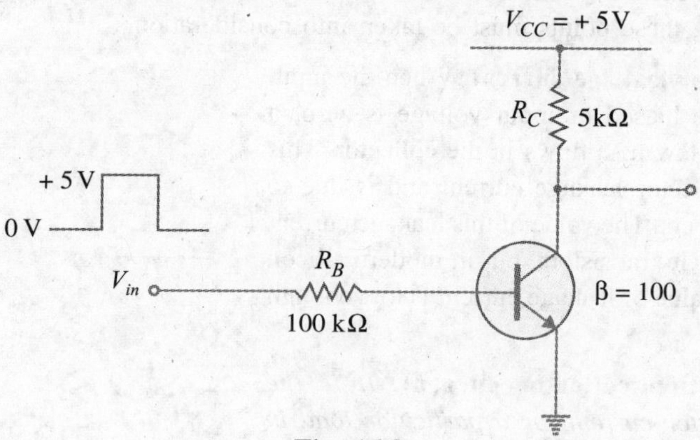

Fig. 16.9

Solution. Transistor switch is also called transistor inverter. Therefore, in order to verify that the circuit in Fig. 16.9 behaves as an inverter, it is only necessary to show that the transistor is saturated when input voltage V_{in} = + 5 V. Referring to Fig. 16.9, we have,

$$\text{Base current, } I_B = \frac{V_{in} - V_{BE}}{R_B}$$

Here, V_{in} = + 5 V ; V_{BE} = 0.7 V ; R_B = 100 kΩ

$$\therefore \quad I_B = \frac{(5 - 0.7)\text{ V}}{100\ \text{k}\Omega} = 43\ \mu\text{A}$$

$$\text{Collector current, } I_C = \beta I_B = 100 \times 43\ \mu\text{A} = 4.3\ \text{mA}$$

Collector saturation current $I_{C(sat)}$ is given by ;

$$I_{C(sat)} = \frac{V_{CC}}{R_C} = \frac{5\text{ V}}{5\ \text{k}\Omega} = 1\ \text{mA}$$

Since $I_C > I_{C(sat)}$, the **transistor is saturated**. Therefore, we conclude that the circuit of Fig. 16.9 behaves as an inverter.

16.7 Advantages of Electronic Switches

The following are the advantages of transistor switch over other types of switches :

(*i*) It has no moving parts and hence there is little wear and tear. Therefore, it gives noiseless operation.

(*ii*) It has smaller size and weight.

(*iii*) It gives troublefree service because of solid state.

(*iv*) It is cheaper than other switches and requires little maintenance.

(*v*) It has a very fast speed of operation say upto 10^9 operations per second. On the other hand, the mechanical switches have a small speed of operation *e.g.*, less than 5 operations in a second.

16.8 Important Terms

So far we have considered the transistor to be an ideal one. An ideal transistor has $V_{CE} = V_{CC}$ (or $I_C = 0$) in the OFF state and $V_{CE} = 0$ (or $I_C = I_{C(sat)}$) in the ON state. However, such ideal conditions are not realised in practice. In a practical transistor, the output voltage is neither V_{CC} in the OFF state nor it is zero in the ON state. While designing a transistor switching circuit, these points must be taken into consideration.

(*i*) **Collector leakage current.** When the input circuit is reverse biased or input voltage is zero, a small current (a few μA) flows in the collector. This is known as collector leakage current and is due to the minority carriers. The value of this leakage current is quite large in Ge transistors, but in modern silicon transistors, the value of leakage current is low enough to be ignored.

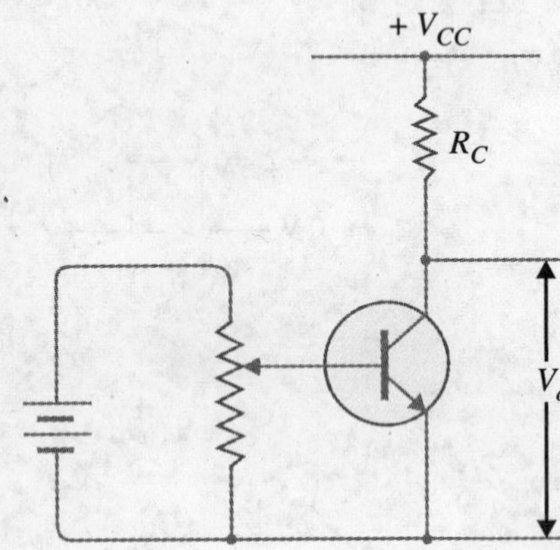

Fig. 16.10

(*ii*) **Saturation collector current.** *It is the maximum collector current for a particular load in a transistor.*

Consider an *npn* transistor having a load R_C in its collector circuit as shown in Fig. 16.10. As the input forward bias is increased, the collector current I_C also increases because $I_C = \beta I_B$. However, with the increase in I_C, the voltage drop across R_C increases. This results in the *decrease of V_{CE}. When V_{CE} drops to knee voltage (V_{knee}), any further increase in collector current is not possible since β decreases sharply when V_{CE} falls below knee voltage. This maximum current is known as saturation collector current.

$$\therefore \quad \text{Saturation collector current, } I_{C\,(sat)} = \frac{V_{CC} - V_{knee}}{R_C} = \frac{V_{CC} - V_{CE(sat)}}{R_C}$$

Note that $V_{knee} = V_{CE(sat)}$.

16.9 Switching Transistors

A transistor which is used as a switch is known as a **switching transistor.**

In general, switching transistor is fabricated by the same process as an ordinary transistor except that it has special design features to reduce switch-off time and saturation voltage. It is so arranged in the circuit that either maximum current (called saturation collector current) flows through the load or minimum current (called collector leakage current) flows through the load. In other words, a switching transistor has two states *viz.* (*i*) ON state or when collector saturation current flows through the load (*ii*) OFF state or when collector leakage current flows through the load. In the discussion that follows transistor means the switching transistor.

16.10 Switching Action of a Transistor

The switching action of a transistor can also be explained with the help of output characteristics. Fig. 16.11 shows the output characteristics of a typical transistor for a *CE* configuration. The

---

* $V_{CE} = V_{CC} - I_C R_C$

load line is drawn for load R_C and collector supply V_{CC}. The characteristics are arranged in three regions : OFF, ON or saturation and active regions.

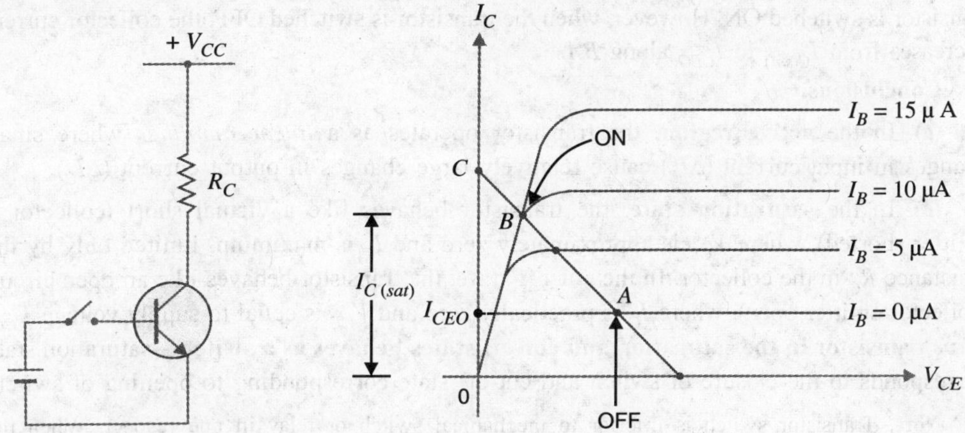

Fig. 16.11

(i) **OFF region.** When the input base voltage is zero or *negative, the transistor is said to be in the OFF condition. In this condition, $I_B = 0$ and the collector current is equal to the collector leakage current I_{CEO}. The value of I_{CEO} can be obtained from the characteristics if we know V_{CE}.

$$\text{Power loss} = \text{Output voltage} \times \text{Output current}$$

As already noted, in the OFF condition, the output voltage = V_{CC} since voltage drop in the load due to I_{CEO} is negligible.

$$\therefore \quad \text{Power loss} = V_{CC} \times I_{CEO}$$

Since I_{CEO} is very small as compared to full-load current that flows in the ON condition, power loss in the transistor is quite small in the OFF condition. It means that the transistor has a high efficiency as a switch in the OFF condition.

(ii) **ON or Saturation region.** When the input voltage is made so much positive that saturation collector current flows, the transistor is said to be in the ON condition. In this condition, the saturation collector current is given by ;

$$I_{C(sat)} = \frac{V_{CC} - V_{knee}}{R_C}$$

$$\text{Power loss} = \text{Output voltage} \times \text{Output current}$$

The output voltage in the ON condition is equal to V_{knee} and output current is $I_{C(sat)}$.

$$\therefore \quad \text{Power loss} = V_{knee} \times I_{C(sat)}$$

Again the efficiency of transistor as a switch in the ON condition is high. It is because the power loss in this condition is quite low due to small value of V_{knee}.

(iii) **Active region.** *It is the region that lies between OFF and ON conditions.*

The OFF and ON regions are the stable regions of operation. The active region is the unstable (or transient) region through which the operation of the transistor passes while

---

* If a small negative voltage is given to the input, the base-emitter junction is reverse biased and load current becomes I_{CBO} instead of I_{CEO}.

changing from OFF state to the ON state. Thus referring to Fig. 16.11, the path *AB* is the active region. The collector current increases from I_{CEO} to $I_{C(sat)}$ along the path *AB* as the transistor is switched ON. However, when the transistor is switched OFF, the collector current decreases from $I_{C(sat)}$ to I_{CEO} along *BA*.

Conclusions.

(*a*) In the **active region**, the transistor operates as a *linear amplifier* where small changes in input current (ΔI_B) cause relatively large changes in output current (ΔI_C).

(*b*) In the **saturation state**, the transistor behaves like a virtual short (collector – emitter shorted), where V_{CE} is approximately zero and I_C is maximum, limited only by the resistance R_C in the collector. In the **cut-off state**, the transistor behaves like an open circuit (collector-emitter open), where I_C is practically zero and V_{CE} is equal to supply voltage V_{CC}. **Thus transistor in the saturation and cut-off states behaves as a switch**—saturation state corresponds to the closure of switch and cut-off state corresponding to opening of switch.

Note. Transistor switch is inferior to mechanical switch or relay in one respect. When the mechanical switch or relay is in the OFF condition, the load current is exactly zero. But when transistor switch is in the OFF condition, the collector current is not zero but is equal to small collector leakage current. However, for all practical purposes, this small collector leakage current may be neglected.

Example 16.4. *A transistor is used as a switch. If V_{CC} = 10 V, R_C = 1 kΩ and I_{CBO} = 10 μA, determine the value of V_{CE} when the transistor is (i) cut-off and (ii) saturated.*

Solution.

(*i*) *At cut-off*

$$I_C = I_{CBO} = 10 \ \mu A$$

$$\therefore \quad V_{CE} = V_{CC} - I_{CBO} R_C$$

$$= 10 \text{ V} - 10 \ \mu\text{A} \times 1 \text{ k}\Omega = 10 \text{ V} - 10 \text{ mV} = \mathbf{9.99 \text{ V}}$$

(*ii*) *At saturation*

$$I_{C\,(sat)} = \frac{V_{CC} - V_{knee}}{R_C}$$

$$\therefore \quad V_{CE} = V_{knee} = \mathbf{0.7 \text{ V}}$$

Example 16.5. *Fig. 16.12 shows the transistor switching circuit. Given that R_B = 2.7 kΩ, V_{BB} = 2 V, V_{BE} = 0.7 V and V_{knee} = 0.7 V.*

(*i*) *Calculate the minimum value of β for saturation.*

(*ii*) *If V_{BB} is changed to 1 V and transistor has minimum β = 50, will the transistor be saturated?*

Solution. (*i*)

$$I_B = \frac{V_{BB} - V_{BE}}{R_B}$$

$$= \frac{2 \text{ V} - 0.7 \text{ V}}{2.7 \text{ k}\Omega} = \frac{1.3 \text{ V}}{2.7 \text{ k}\Omega} = 0.48 \text{ mA}$$

Now,

$$I_{C\,(sat)} = \frac{V_{CC} - V_{knee}}{R_C}$$

$$= \frac{10\text{ V} - 0.7\text{ V}}{1\text{ k}\Omega} = \frac{9.3\text{ V}}{1\text{ k}\Omega} = 9.3\text{ mA}$$

$$\therefore \quad \text{Minimum } \beta = \frac{I_{C(sat)}}{I_B} = \frac{9.3\text{ mA}}{0.48\text{ mA}} = \mathbf{19.4}$$

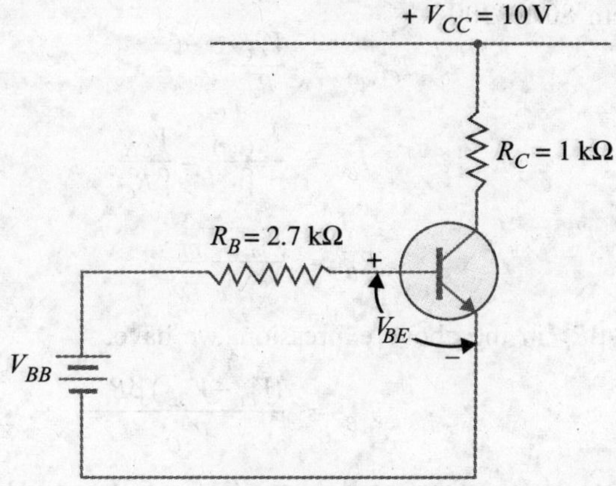

Fig. 16.12

(*ii*)
$$I_B = \frac{V_{BB} - V_{BE}}{R_B}$$

$$= \frac{1\text{ V} - 0.7\text{ V}}{2.7\text{ k}\Omega} = \frac{0.3\text{ V}}{2.7\text{ k}\Omega} = 0.111\text{ mA}$$

$$\therefore \quad I_C = \beta I_B = 50 \times 0.111 = 5.55\text{ mA}$$

Since the collector current is less than saturation current (= 9.3 mA), **the transistor will not be saturated.**

Example 16.6. *Design the transistor inverter shown in Fig. 16.13 so that it will operate satisfactorily with silicon transistors whose β-values range from 80 to 200. Assume $V_{CE(sat)}$ = 0 V.*

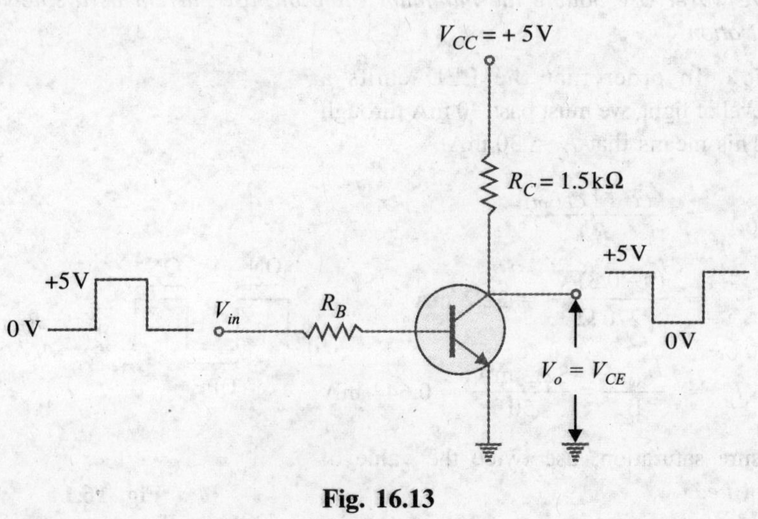

Fig. 16.13

Solution. When the input to the inverter is low (0 V), the transistor is cut-off and its output is high (+5 V). However, when the input to the inverter is high (+5 V), the transistor is saturated and its output is low (0 V). Note that output voltage is opposite to that applied at the input or base terminals. Hence the name inverter. Here, designing the inverter means to find the value of R_B so that the transistor remains saturated when V_{in} = + 5 V and β values may range from 80 to 200.

$$I_{C(sat)} = \frac{V_{CC}}{R_C} \qquad (\because V_{CE(sat)} = 0 \text{ V})$$

$$\therefore \quad I_B = \frac{I_{C(sat)}}{\beta} = \frac{V_{CC}}{\beta R_C}$$

Now,
$$R_B = \frac{V_{in} - V_{BE}}{I_B}$$

Putting $I_B = V_{CC}/\beta R_C$ in the above expression, we have,

$$R_B = \frac{(V_{in} - V_{BE})\,\beta R_C}{V_{CC}} \qquad \text{... } (i)$$

In order that the transistor is saturated when V_{in} = + 5 V and values of β range from 80 to 200, the value of R_B should be equal to or less than that given by eq. (*i*). This is possible if we use minimum value of β, *i.e.* $\beta = \beta_{min}$.

$$\therefore \quad R_B = \frac{(V_{in} - V_{BE})\,\beta_{min} R_C}{V_{CC}}$$

Here, V_{in} = + 5 V ; V_{BE} = 0.7 V ; V_{CC} = + 5 V ; β_{min} = 80 ; R_C = 1.5 kΩ

$$\therefore \quad R_B = \frac{(5 - 0.7)\text{ V} \times 80 \times 1.5 \text{ k}\Omega}{5 \text{ V}} = \mathbf{103.2\ k\Omega}$$

Example 16.7. *Fig. 16.14 shows the use of a transistor to switch ON or OFF the LED. The LED requires 30 mA to emit sufficient level of light. The circuit values are : V_{CC} = 9 V ; $V_{CE(sat)}$ = 0.3 V ; R_C = 270 Ω ; R_B = 3.3 kΩ and β = 50. Find the amplitude of the input square wave. Use double the minimum value of base current as a safety measure to ensure saturation.*

Solution. In order that the LED emits a sufficient level of light, we must pass 30 mA through the LED. This means that I_C = 30 mA.

$$I_{C(sat)} = \frac{V_{CC} - V_{CE(sat)}}{R_C} = \frac{(9 - 0.3)\text{ V}}{270\ \Omega} = 32.2 \text{ mA}$$

$$\therefore \quad I_B = \frac{I_{C(sat)}}{\beta} = \frac{32.2 \text{ mA}}{50} = 0.644 \text{ mA}$$

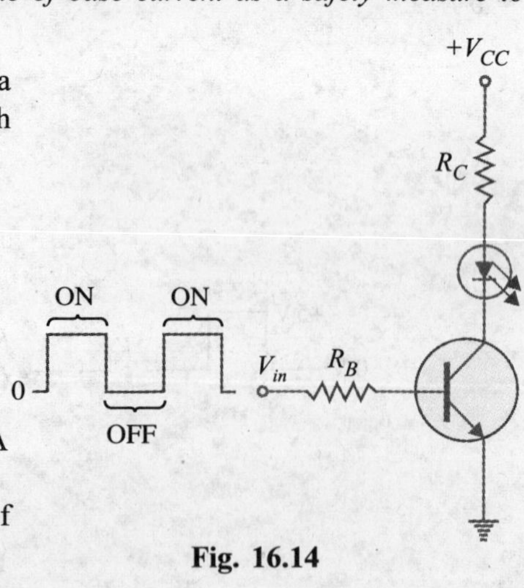

Fig. 16.14

To ensure saturation, use twice the value of base current *i.e.*

$$2I_B = 2 \times 0.644 = 1.29 \text{ mA}$$

Amplitude V_{in} of input square wave is given by ;

$$V_{in} = V_{BE} + 2I_B R_B$$
$$= 0.7 \text{ V} + 1.29 \text{ mA} \times 3.3 \text{ k}\Omega = \mathbf{4.96 \text{ V}}$$
$$(\because 2I_B = 1.29 \text{ mA})$$

16.11 Multivibrators

An electronic circuit that generates square waves (or other non-sinusoidals such as rectangular, saw-tooth waves) is known as a ***multivibrator.**

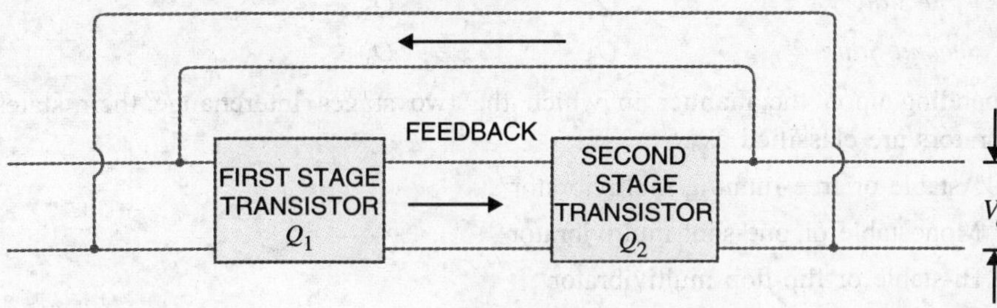

Fig. 16.15

A multivibrator is a switching circuit which depends for operation on positive feedback. It is basically a two-stage amplifier with output of one fed back to the input of the other as shown in Fig. 16.15.

The circuit operates in two states (*viz.* ON and OFF) controlled by circuit conditions. Each amplifier stage supplies feedback to the other in such a manner that will drive the transistor of one stage to saturation (ON state) and the other to cut-off (OFF state).

After a certain time controlled by circuit conditions, the action is reversed *i.e.* saturated stage is driven to cut-off and the cut-off stage is driven to saturation. The output can be taken across either stage and may be rectangular or square wave depending upon the circuit conditions.

Fig. 16.15 shows the block diagram of a multivibrator. It is a two-stage amplifier with 100% positive feedback. Suppose output is taken across the transistor Q_2. At any particular instant, one transistor is ON and conducts $I_{C(sat)}$ while the other is OFF. Suppose Q_2 is ON and Q_1 is OFF. The collector current in Q_2 will be $I_{C(sat)}$ as shown in Fig. 16.16. This condition will prevail for a time (*bc* in this case) determined by circuit conditions. After this time, transistor Q_2 is cut-off and Q_1 is turned ON. The collector current in Q_2 is now I_{CEO} as shown. The circuit

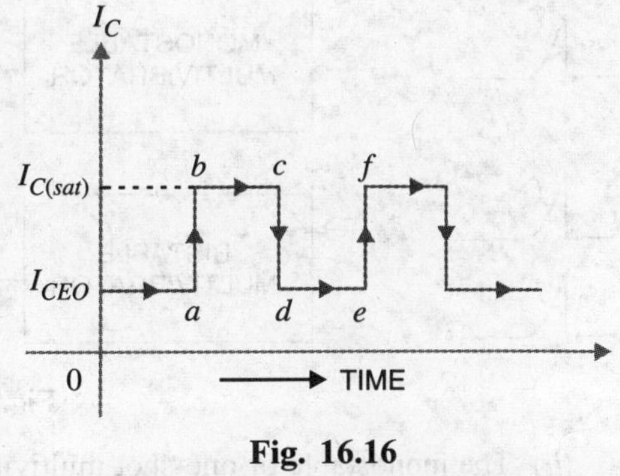

Fig. 16.16

* The name multivibrator is derived from the fact that a square wave actually consists of a large number of (Fourier series analysis) sinusoidals of different frequencies.

will stay in this condition for a time *de*. Again Q_2 is turned ON and Q_1 is driven to cut-off. In this way, the output will be a square wave.

16.12 Types of Multivibrators

A multivibrator is basically a two-stage amplifier with output of one fed back to the input of the other. At any particular instant, one transistor is ON and the other is OFF. After a certain time depending upon the circuit components, the stages reverse their conditions – the conducting stage suddenly cuts off and the non-conducting stage suddenly starts to conduct. The two possible states of a multivibrator are :

	ON	OFF
First State	Q_1	Q_2
Second State	Q_2	Q_1

Depending upon the manner in which the two stages interchange their states, the multivibrators are classified as :

(*i*) Astable or free running multivibrator

(*ii*) Monostable or one-shot multivibrator

(*iii*) Bi-stable or flip-flop multivibrator

Fig. 16.17 shows the input/output relations for the three types of multivibrators.

(*i*) The astable or free running multivibrator alternates automatically between the two states and remains in each for a time dependent upon the circuit constants. Thus it is just an oscillator since it requires no external pulse for its operation. Of course, it does require a source of d.c. power. Because it continuously produces the square-wave output, it is often referred to as a *free running multivibrator.*

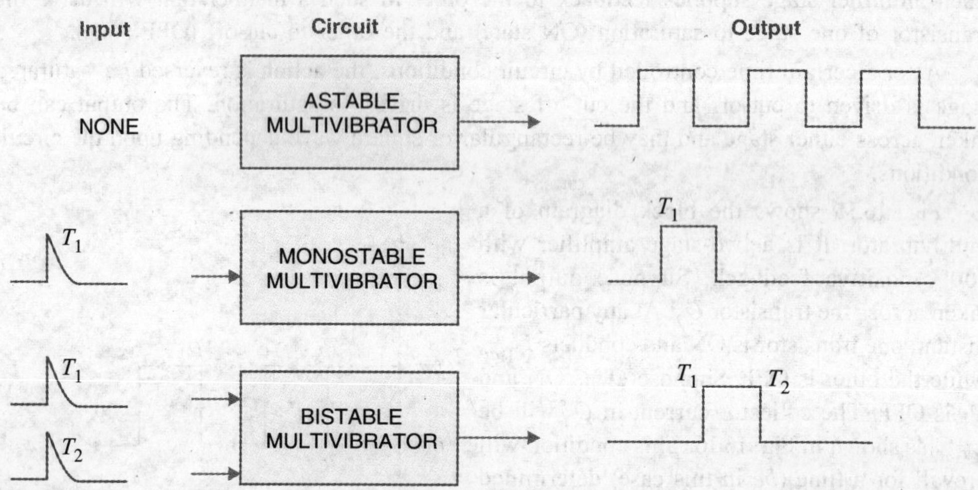

Fig. 16.17

(*ii*) The monostable or one-shot multivibrator has one state stable and one quasi-stable (*i.e.* half-stable) state. The application of input pulse triggers the circuit into its quasi-stable state, in which it remains for a period determined by circuit constants. After this period of time, the circuit returns to its initial stable state, the process is repeated upon the application of each trigger pulse. Since the monostable multivibrator produces a single output pulse for each input trigger pulse, it is generally called *one-shot multivibrator.*

(*iii*) The bistable multivibrator has both the two states stable. It requires the application of an external triggering pulse to change the operation from either one state to the other. Thus one pulse is used to generate half-cycle of square wave and another pulse to generate the next half-cycle of square wave. It is also known as a *flip-flop multivibrator* because of the two possible states it can assume.

16.13 Transistor Astable Multivibrator

A multivibrator which generates square waves of its own (i.e. without any external triggering pulse) is known as an **astable** *or* **free running multivibrator.**

The *astable multivibrator has no stable state. It switches back and forth from one state to the other, remaining in each state for a time determined by circuit constants. In other words, at first one transistor conducts (*i.e.* ON state) and the other stays in the OFF state for some time. After this period of time, the second transistor is automatically turned ON and the first transistor is turned OFF. Thus the multivibrator will generate a square wave output of its own. The width of the square wave and its frequency will depend upon the circuit constants.

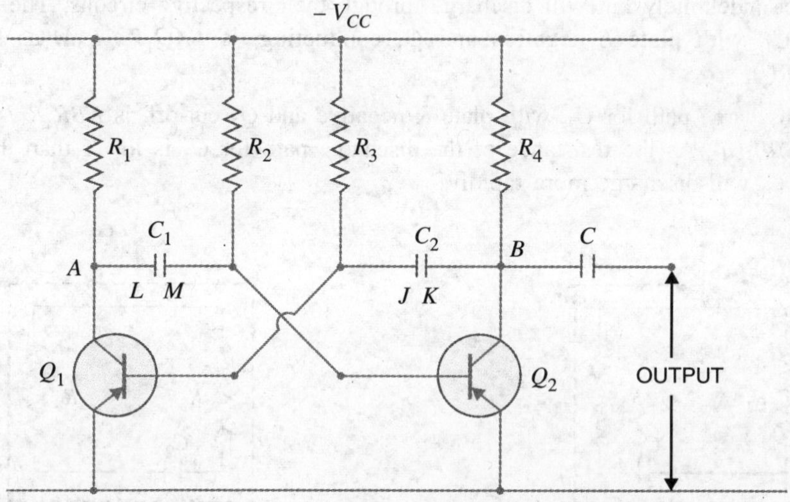

Fig. 16.18

Circuit details. Fig. 16.18 shows the circuit of a typical transistor astable multivibrator using two identical transistors Q_1 and Q_2. The circuit essentially consists of two symmetrical *CE* amplifier stages, each providing a feedback to the other. Thus collector loads of the two stages are equal *i.e.* $R_1 = R_4$ and the biasing resistors are also equal *i.e.* $R_2 = R_3$. The output of transistor Q_1 is coupled to the input of Q_2 through C_1 while the output of Q_2 is fed to the input of Q_1 through C_2. The square wave output can be taken from Q_1 or Q_2.

Operation. When V_{CC} is applied, collector currents start flowing in Q_1 and Q_2. In addition, the coupling capacitors C_1 and C_2 also start charging up. As the characteristics of no two transistors

* A means not. Hence astable means that it has no stable state.

(*i.e.* β, V_{BE}) are *exactly* alike, therefore, one transistor, say Q_1, will conduct more rapidly than the other. The rising collector current in Q_1 drives its collector more and more positive. The increasing positive output at point *A* is applied to the base of transistor Q_2 through C_1. This establishes a reverse bias on Q_2 and its collector current starts decreasing. As the collector of Q_2 is connected to the base of Q_1 through C_2, therefore, base of Q_1 becomes more negative *i.e.* Q_1 is more forward biased. This further increases the collector current in Q_1 and causes a further decrease of collector current in Q_2. This series of actions is repeated until the circuit drives Q_1 to saturation and Q_2 to cut-off. These actions occur very rapidly and may be considered practically instantaneous. The output of Q_1 (ON state) is approximately zero and that of Q_2 (OFF state) is approximately V_{CC}. This is shown by *ab* in Fig. 16.19.

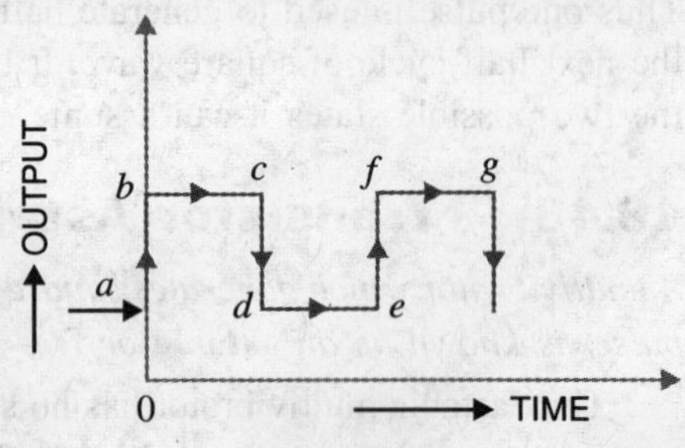

Fig. 16.19

When Q_1 is at saturation and Q_2 is cut-off, the full voltage V_{CC} appears across R_1 and voltage across R_4 will be zero. The charges developed across C_1 and C_2 are sufficient to maintain the saturation and cut-off conditions at Q_1 and Q_2 respectively. This condition is represented by time interval *bc* in Fig. 16.19. However, the capacitors will not retain the charges indefinitely but will discharge through their respective circuits. The discharge path for C_1, with plate *L* negative and Q_1 conducting, is $LAQ_1V_{CC}R_2M$ as shown in Fig. 16.20 (*i*).

The discharge path for C_2, with plate *K* negative and Q_2 cut-off, is KBR_4R_3J as shown in Fig. 16.20 (*ii*). As the resistance of the discharge path for C_1 is lower than that of C_2, therefore, C_1 will discharge more rapidly.

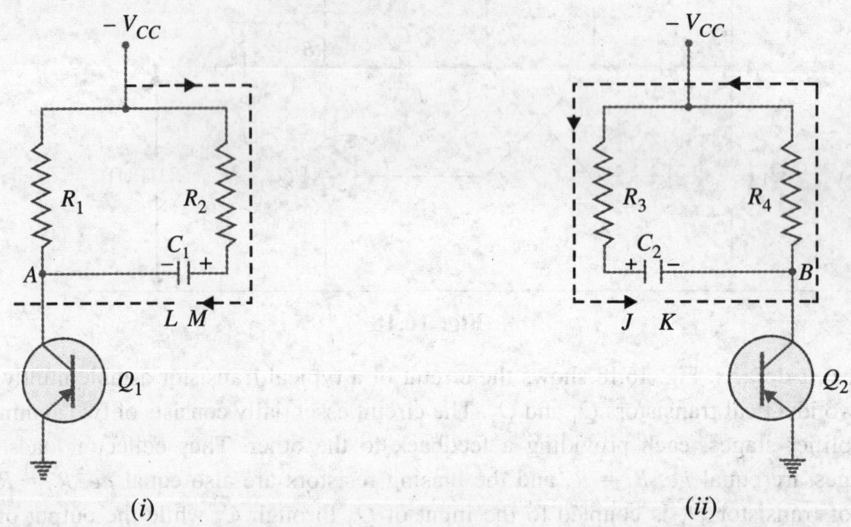

Fig. 16.20

As C_1 discharges, the base bias at Q_2 becomes less positive and at a time determined by R_2 and C_1, forward bias is re-established at Q_2. This causes the collector current to start in Q_2. The increasing positive potential at collector of Q_2 is applied to the base of Q_1 through the capacitor

C_2. Hence the base of Q_1 will become more positive *i.e.* Q_1 is reverse biased. The decrease in collector current in Q_1 sends a negative voltage to the base of Q_2 through C_1, thereby causing further increase in the collector current of Q_2. With this set of actions taking place, Q_2 is quickly driven to saturation and Q_1 to cut-off. This condition is represented by *cd* in Fig. 16.19. The period of time during which Q_2 remains at saturation and Q_1 at cut-off is determined by C_2 and R_3.

ON or OFF time. The time for which either transistor remains ON or OFF is given by ;

ON time for Q_1 (or OFF time for Q_2) is

$$T_1 = 0.694\, R_2 C_1$$

OFF time for Q_1 (or ON time for Q_2) is

$$T_2 = 0.694\, R_3 C_2$$

Total time period of the square wave is

$$T = T_1 + T_2 = 0.694\,(R_2 C_1 + R_3 C_2)$$

As $R_2 = R_3 = R$ and $C_1 = C_2 = C$,

$$\therefore \quad T = 0.694\,(RC + RC) \simeq 1.4\, RC \text{ seconds}$$

Frequency of the square wave is

$$f = \frac{1}{T} \simeq \frac{0.7}{RC} \text{ Hz}$$

It may be noted that in these expressions, R is in ohms and C in farad.

Applications. Astable multivibrators find wide applications in electronic systems. Some are :

(*i*) *Timing oscillator.* The symmetrical astable multivibrator can be used as a clock generator producing *X* and *Y* pulses, each transistor providing one output.

(*ii*) *Variable frequency oscillator.* The frequency of oscillations can be changed by varying the circuit components. Thus square waves of any frequency can be obtained.

(*iii*) *Frequency divider.* An astable multivibrator can be readily synchronised to an input signal and used for counting down.

(*iv*) *Harmonic generation.* Because of the rapid transitions of the waveform, high order harmonics are generated.

Example 16.8. *In the astable multivibrator shown in Fig. 16.18, $R_2 = R_3 = 10\text{ k}\Omega$ and $C_1 = C_2 = 0.01\ \mu F$. Determine the time period and frequency of the square wave.*

Solution.

Here, $R = 10\text{ k}\Omega = 10^4\ \Omega$; $C = 0.01\ \mu\text{F} = 10^{-8}$ F

Time period of the square wave is

$$T = 1.4\, RC = 1.4 \times 10^4 \times 10^{-8} \text{ second}$$

$$= 1.4 \times 10^{-4} \text{ second} = 1.4 \times 10^{-4} \times 10^3 \text{ m sec}$$

$$= \mathbf{0.14\ m\ sec}$$

Frequency of the square wave is given by ;

$$f = \frac{1}{T \text{ in second}} \text{ Hz} = \frac{1}{1.4 \times 10^{-4}} \text{ Hz}$$

$$= 7 \times 10^3 \text{ Hz} = \mathbf{7\ kHz}$$

16.14 Transistor Monostable Multivibrator

A multivibrator in which one transistor is always conducting (i.e. in the ON state) and the other is non-conducting (i.e. in the OFF state) is called a **monostable multivibrator.**

A *monostable multivibrator has only one state stable. In other words, if one transistor is conducting and the other is non-conducting, the circuit will remain in this position. It is only with the application of external pulse that the circuit will interchange the states. However, after a certain time, the circuit will automatically switch back to the original stable state and remains there until another pulse is applied. Thus a monostable multivibrator cannot generate square waves of its own like an astable multivibrator. Only external pulse will cause it to generate the square wave.

Circuit details. Fig. 16.21 shows the circuit of a transistor monostable multivibrator. It consists of two similar transistors Q_1 and Q_2 with equal collector loads *i.e.* $R_1 = R_4$. The values of V_{BB} and R_5 are such as to reverse bias Q_1 and keep it at cut-off. The collector supply V_{CC} and R_2 forward bias Q_2 and keep it at saturation. The input pulse is given through C_2 to obtain the square wave. Again output can be taken from Q_1 or Q_2.

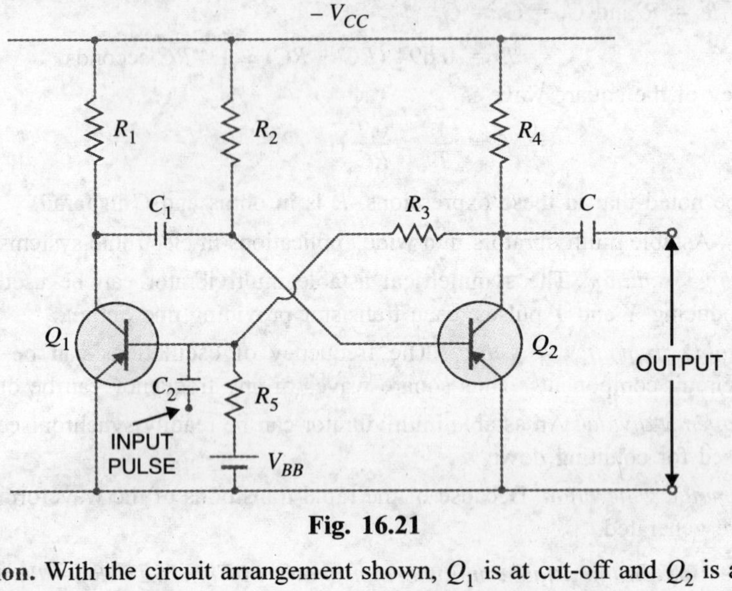

Fig. 16.21

Operation. With the circuit arrangement shown, Q_1 is at cut-off and Q_2 is at saturation. This is the stable state for the circuit and it will continue to stay in this state until a triggering pulse is applied at C_2. When a negative pulse of short duration and sufficient magnitude is applied to the base of Q_1 through C_2, the transistor Q_1 starts conducting and positive potential is established at its collector. The positive potential at the collector of Q_1 is coupled to the base of Q_2 through capacitor C_1. This decreases the forward bias on Q_2 and its collector current decreases. The increasing negative potential on the collector of Q_2 is applied to the base of Q_1 through R_3. This further increases the forward bias on Q_1 and hence its collector current. With this set of actions taking place, Q_1 is quickly driven to saturation and Q_2 to cut-off.

With Q_1 at saturation and Q_2 at cut-off, the circuit will come back to the original stage (*i.e.* Q_2 at saturation and Q_1 at cut-off) after some time as explained in the following discussion.

* Mono means single.

The capacitor C_1 (charged to approximately V_{CC}) discharges through the path $R_2 V_{CC}$ Q_1. As C_1 discharges, it sends a voltage to the base of Q_2 to make it less positive. This goes on until a point is reached when forward bias is re-established on Q_2 and collector current starts to flow in Q_2. The step by step events already explained occur and Q_2 is quickly driven to saturation and Q_1 to cut-off. This is the stable state for the circuit and it remains in this condition until another pulse causes the circuit to switch over the states.

Applications. Monostable multivibrators are commonly used for the following applications :

(i) Pulse forming. In a monostable multivibrator, an input pulse can be transformed into a pulse of controlled duration and amplitude. This property makes the circuit useful for pulse forming.

(ii) Counting. The monostable multivibrator having once been triggered is insensitive to further pulses until it reverts to its former state. This property makes the device useful as a counter.

(iii) Delay. The trailing edge of the output waveform may be used to provide a pulse delayed from the input pulse. Since the duration of the output waveform can be controlled, as in astable multivibrator, a variable delay is obtained.

16.15 Transistor Bistable Multivibrator

A multivibrator which has both the states stable is called a **bistable multivibrator.**

The bistable multivibrator has both the states stable. It will remain in whichever state it happens to be until a trigger pulse causes it to switch to the other state. For instance, suppose at any particular instant, transistor Q_1 is conducting and transistor Q_2 is at cut-off. If left to itself, the bistable multivibrator will stay in this position forever. However, if an external pulse is applied to the circuit in such a way that Q_1 is cut-off and Q_2 is turned on, the circuit will stay in the new position. Another trigger pulse is then required to switch the circuit back to its original state.

Circuit details. Fig. 16.22 shows the circuit of a typical transistor bistable multivibrator. It consists of two identical *CE* amplifier stages with output of one fed to the input of the other. The feedback is coupled through resistors (R_2, R_3) shunted by capacitors C_1 and C_2. The main purpose of capacitors C_1 and C_2 is to improve the switching characteristics of the circuit by passing the high frequency components of the square wave. This allows fast rise and fall times and hence distortionless square wave output. The output can be taken across either transistor.

Operation. When V_{CC} is applied, one transistor will start conducting slightly ahead of the other due to some differences in the characteristics of the transistors. This will drive one transistor to saturation and the other to cut-off in a manner described for the astable multivibrator. Assume that Q_1 is turned ON and Q_2 is cut-off. If left to itself, the circuit will stay in this condition. In order to switch the multivibrator to its other state, a trigger pulse must be applied. A negative pulse applied to the base of Q_1 through C_3 will cut it off or a positive pulse applied to the base of Q_2 through C_4 will cause it to conduct.

Suppose a negative pulse of sufficient magnitude is applied to the base of Q_1 through C_3. This will reduce the forward bias on Q_1 and cause a decrease in its collector current and an increase in collector voltage. The rising collector voltage is coupled to the base of Q_2 where it forward biases the base-emitter junction of Q_2. This will cause an increase in its collector current and decrease in collector voltage. The decreasing collector voltage is applied to the base of Q_1 where it further reverse biases the base-emitter junction of Q_1 to decrease its collector current. With this set of actions taking place, Q_2 is quickly driven to saturation and Q_1 to cut-off. The circuit will now remain stable in this state until a negative trigger pulse at Q_2 (or a positive trigger pulse at Q_1) changes this state.

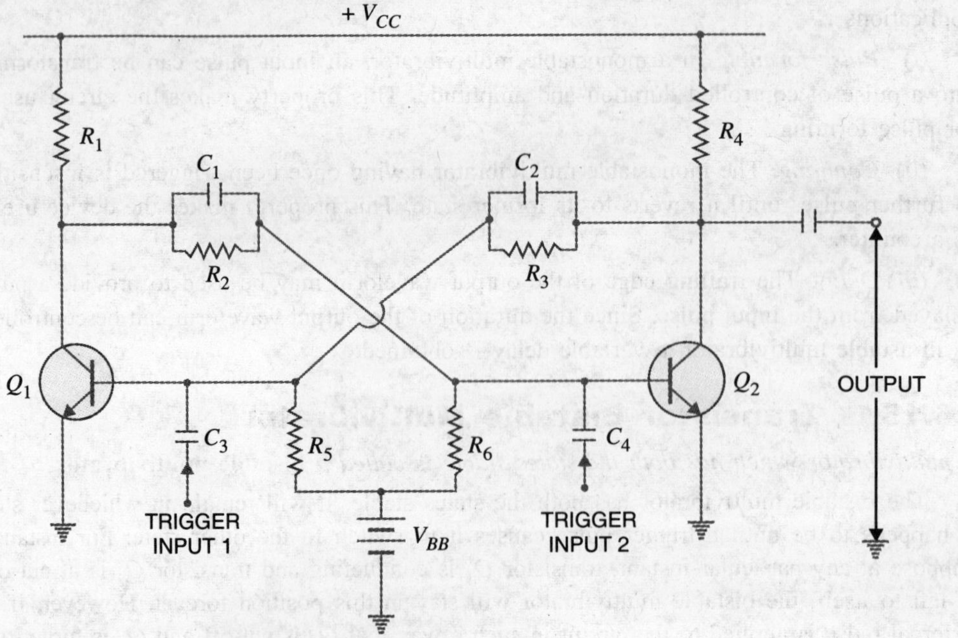

Fig. 16.22

Applications. Two important applications of bistable multivibrator are :

(*i*) *Counting.* As two successive input signals are necessary to restore the circuit to its original state, it may be used for counting by two.

(*ii*) *Memory element.* The two alternative states of bistable multivibrator may be designed "no" (or "0") and yes (or "1") and this facility is used in shift register circuits.

16.16 Differentiating Circuit

A circuit in which output voltage is directly proportional to the derivative of the input is known as a **differentiating circuit.**

$$\text{Output} \propto \frac{d}{dt}(\text{Input})$$

A differentiating circuit is a simple *RC* series circuit with output taken across the resistor *R*. The circuit is suitably designed so that output is proportional to the derivative of the input. Thus if a d.c. or constant input is applied to such a circuit, the output will be zero. It is because the derivative of a constant is zero.

Fig. 16.23 shows a typical differentiating circuit. The output across R will be the derivative of the input. It is important to note that merely using voltage across R does not make the circuit a differentiator ; it is also necessary to set the proper circuit values. In order to achieve good differentiation, the following two conditions should be satisfied :

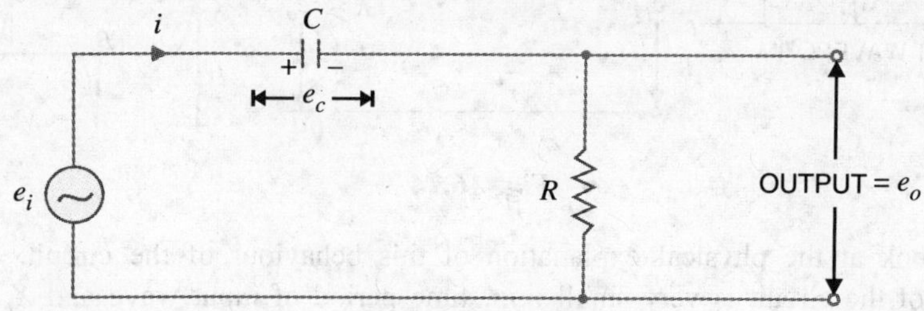

Fig. 16.23

(*a*) The time constant RC of the circuit should be much smaller than the time period of the input wave.

(*b*) The value of X_C should be 10 or more times larger than R at the operating frequency.

Fulfilled these conditions, the output across R in Fig. 16.23 will be the derivative of the input.

Let e_i be the input alternating voltage and let i be the resulting alternating current. The charge q on the capacitor at any instant is

$$q = C\, e_c$$

Now, $$i = \frac{dq}{dt} = \frac{d}{dt}(q) = \frac{d}{dt}(C\, e_c)$$

or $$i = C\frac{d}{dt}(e_c)$$

Since the capacitive reactance is very much larger than R, the input voltage can be considered equal to the capacitor voltage with negligible error *i.e.* $e_c = e_i$.

$\therefore$ $$i = C\frac{d}{dt}(e_i)$$

Output voltage, $$e_o = i\, R$$

$$= RC\frac{d}{dt}(e_i)$$

$$\propto \frac{d}{dt}(e_i) \qquad (\because RC \text{ is constant})$$

$\therefore$ $$\text{Output voltage} \propto \frac{d}{dt}(\text{Input})$$

Output waveforms. The output waveform from a differentiating circuit depends upon the time constant and shape of the input wave. Three important cases will be considered.

(*i*) **When input is a square wave.** When the input fed to a differentiating circuit is a square wave, output will consist of sharp narrow pulses as shown in Fig. 16.24. During the OC part of input wave, its amplitude changes abruptly and hence the differentiated wave will be a sharp narrow pulse as shown in Fig. 16.24. However, during the constant part CB of the input, the output will be zero because the derivative of a constant is zero.

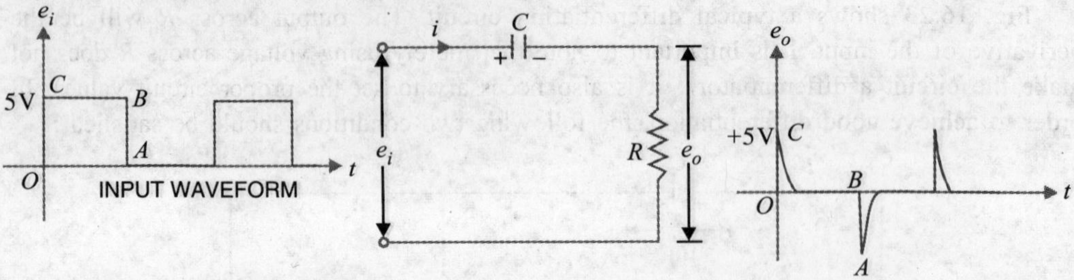

Fig. 16.24

Let us look at the physical explanation of this behaviour of the circuit. Since time constant *RC* of the circuit is very small *w.r.t.* time period of input wave and $X_C >> R$, the capacitor will become fully charged during the early part of each half-cycle of the input wave. During the remainder part of the half-cycle, the output of the circuit will be zero because the capacitor voltage (e_c) neutralises the input voltage and there can be no current flow through *R*. Thus we shall get sharp pulse at the output during the start of each half-cycle of input wave while for the remainder part of the half-cycle of input wave, the output will be zero. In this way, a symmetrical output wave with sharp positive and negative peaks is produced. Such pulses are used in many ways in electronic circuits *e.g.* in television transmitters and receivers, in multivibrators to initiate action *etc.*

***(ii)* When input is a triangular wave.** When the input fed to a differentiating circuit is a triangular wave, the output will be a rectangular wave as shown in Fig. 16.25. During the period *OA* of the input wave, its amplitude changes at a constant rate and, therefore, the differentiated wave has a constant value for each constant rate of change. During the period *AB* of the input wave, the change is less abrupt so that the output will be a very narrow pulse of rectangular form. Thus when a triangular wave is fed to a differentiating circuit, the output consists of a succession of rectangular waves of equal or unequal duration depending upon the shape of the input wave.

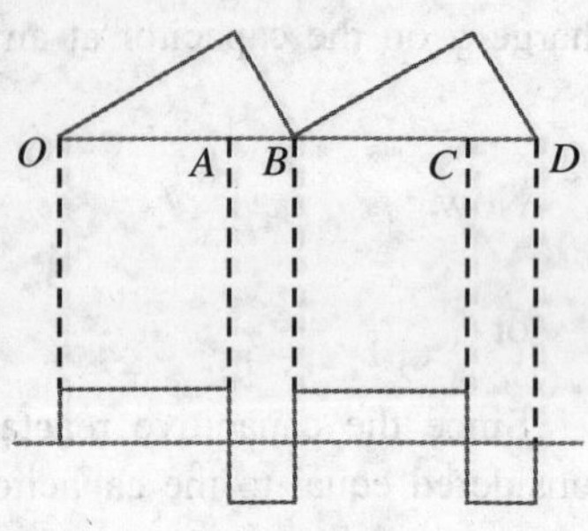

Fig. 16.25

***(iii)* When input is a sine wave.** A sine wave input becomes a cosine wave and a cosine wave input becomes an inverted sine wave at the output.

***(iv)* When input is ramp voltage.** When voltage increases or decreases at a constant rate (*i.e.* voltage changes linearly with time), it is called a *ramp voltage.* In Fig. 16.26 (*i*), the input is the ramp voltage $e_i = Et$. The slope of this ramp is constant *i.e. E* volts/second. Since the rate of change (= slope of ramp) of the input is constant, the output of the differentiator is constant = *E* as shown in Fig. 16.26 (*ii*).

$$e_o = \frac{de_i}{dt} = \frac{d}{dt}(e_i) = \frac{d}{dt}(Et) = E$$

***(v)* When input is constant direct voltage.** When input to a differentiator is constant direct voltage *E*, then $e_i = E$. Therefore, the output voltage (e_o) will be zero because $e_o = dE/dt = 0$. The derivative of a constant is zero.

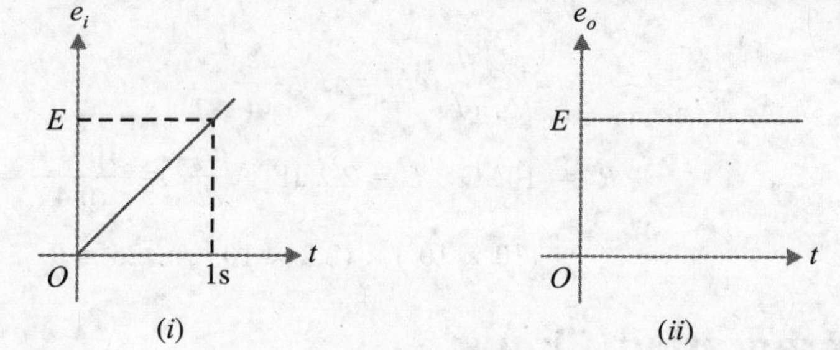

Fig. 16.26

Example 16.9. (*i*) *What is the effect of time constant of an RC circuit on the differentiated wave?*

(*ii*) *Sketch the output waveform from the differentiating circuit when input is square wave for T = 100 RC, T = 10 RC, T = RC.*

Solution.

(*i*) In an *RC* differentiating circuit, the output voltage is taken across *R* and the waveform of the output depends upon the time constant of the circuit. The circuit will function as a differentiator if the product *RC* is many times smaller than the time period of the input wave.

(*ii*) **Square wave input.** Fig. 16.27 shows the input square wave fed to a differentiating circuit. Fig. 16.28 shows the output waveforms for different values of time period of the input wave.

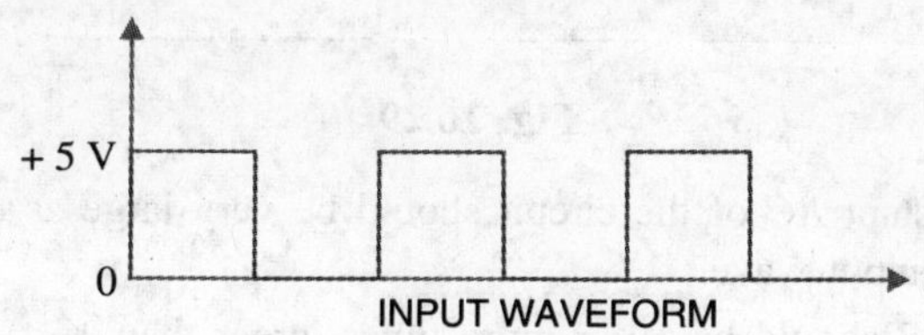

Fig. 16.27

It may be noted that *RC* coupling circuit is the same as a differentiating circuit except that it has a long time constant—in excess of 5 *RC*. Therefore, a *RC* coupling circuit does not noticeably differentiate the input wave.

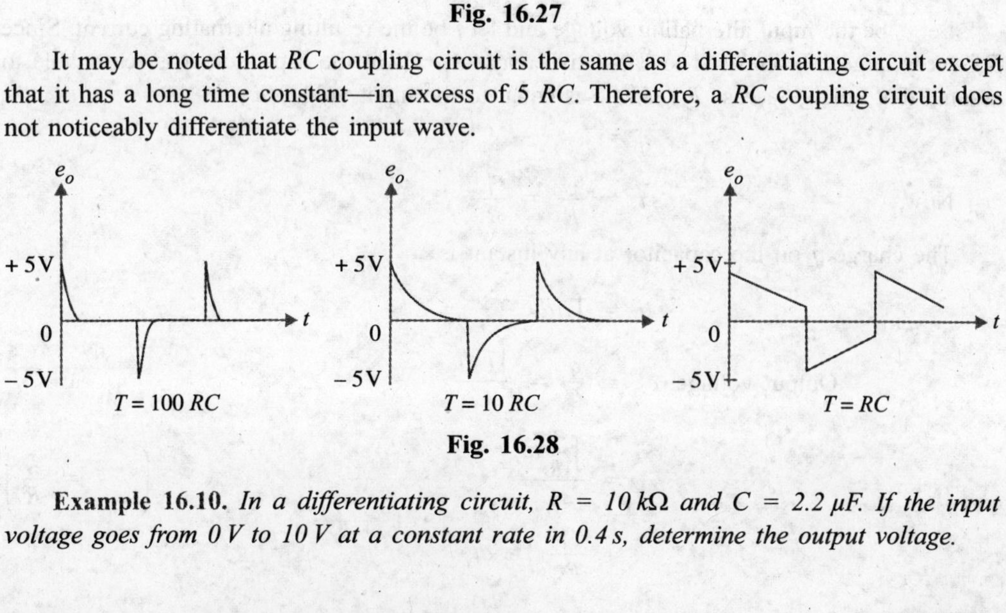

Fig. 16.28

Example 16.10. *In a differentiating circuit, R = 10 kΩ and C = 2.2 μF. If the input voltage goes from 0 V to 10 V at a constant rate in 0.4 s, determine the output voltage.*

Solution.

$$e_o = RC\frac{d}{dt}(e_i) = RC\frac{de_i}{dt}$$ ... See Art. 16.16

Here, $R = 10\,k\Omega\ ;\ C = 2.2\,\mu F\ ;\ \frac{de_i}{dt} = \frac{10-0}{0.4} = 25\ V/s$

$\therefore$ $e_o = (10 \times 10^3) \times (2.2 \times 10^{-6}) \times 25 = \mathbf{0.55\ V}$

16.17 Integrating Circuit

A circuit in which output voltage is directly proportional to the integral of the input is known as an **integrating circuit** *i.e.*

$$\text{Output} \propto \int \text{Input}$$

An integrating circuit is a simple *RC* series circuit with output taken across the capacitor *C* as shown in Fig. 16.29. It may be seen that *R* and *C* of the differentiating circuit have changed places. In order that the circuit renders good integration, the following conditions should be fulfilled :

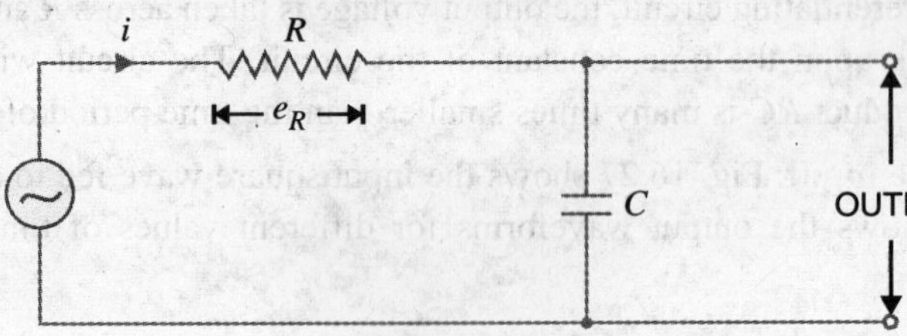

Fig. 16.29

(*a*) The time constant *RC* of the circuit should be very large as compared to the time period of the input wave.

(*b*) The value of *R* should be 10 or more times larger than X_C.

Let e_i be the input alternating voltage and let *i* be the resulting alternating current. Since *R* is very large as compared to capacitive reactance X_C of the capacitor, it is reasonable to assume that voltage across *R* (*i.e.* e_R) is equal to the input voltage *i.e.*

$$e_i = e_R$$

Now, $i = \frac{e_R}{R} = \frac{e_i}{R}$

The charge *q* on the capacitor at any instant is

$$q = \int i\,dt$$

Output voltage, $e_o = \frac{q}{C} = \frac{\int i\,dt}{C}$

$$= \frac{\int \frac{e_i}{R}dt}{C} \qquad \left(\because i = \frac{e_i}{R}\right)$$

$$= \frac{1}{RC}\int e_i\,dt$$

$$\propto \int e_i \, dt \qquad (\because RC \text{ is constant})$$

$$\therefore \quad \text{Output voltage} \propto \int \text{Input}$$

Output waveforms. The output waveform from an integrating circuit depends upon time constant and shape of the input wave. Two important cases will be discussed :

(*i*) **When input is a square wave.** When the input fed to an integrating circuit is a square wave, the output will be a triangular wave as shown in Fig. 16.30 (*i*). As integration means summation, therefore, output from an integrating circuit will be the sum of all the input waves at any instant. This sum is zero at *A* and goes on increasing till it becomes maximum at *C*. After this, the summation goes on decreasing to the onset of negative movement *CD* of the input.

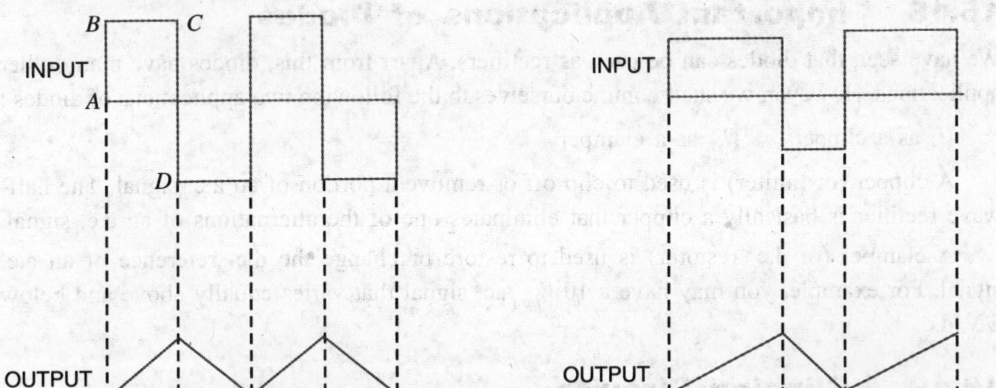

Fig. 16.30

(*ii*) **When input is rectangular wave.** When the input fed to an integrating circuit is a rectangular wave, the output will be a triangular wave as shown in Fig. 16.30 (*ii*).

Example 16.11. *The input to a properly designed integrator is a constant d.c. voltage of E volts. Show that the output is a ramp voltage.*

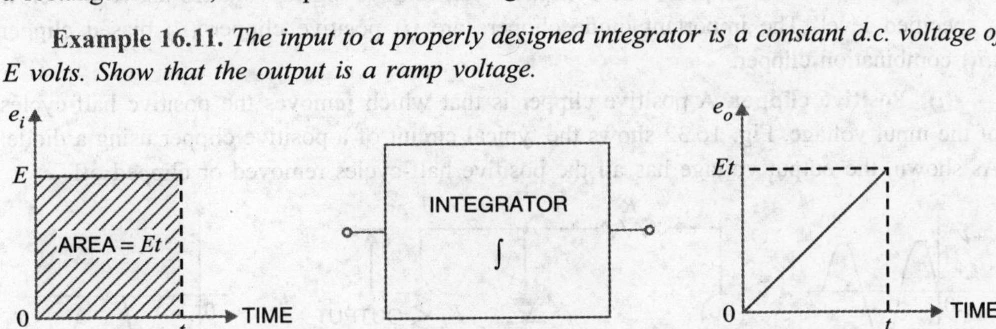

Fig. 16.31

Solution. We know that a properly designed integrator produces an output waveform whose value at any instant of time is equal to the *total area under the input waveform* upto that point in time. In Fig. 16.31, input to the integrator is steady d.c. voltage of *E* volts. The graph of the input d.c. voltage versus time is simply a horizontal line at the level of *E* volts because the d.c. voltage is constant. As the time passes, the area that accumulates under the d.c. waveform increases. At any time t, the total area under the input waveform between time 0 and t is Et as shown. Therefore, at any time t, the output voltage $e_o = Et$. For example, suppose $E = 5$ volts. Then at $t = 1$ s, the output voltage $e_o = Et = 5 \times 1 = 5$ V. Similarily, at $t = 2$ s, $e_o = 10$ V ; at $t = 3$ s, $e_o = 15$ V and so on. Thus the output is the **ramp voltage** $e_o = Et$. Recall that a voltage that changes (increases or decreases) linearly with time is called a ramp voltage.

Now, Output voltage, $e_o = \int_0^t e_i dt = \int_0^t E dt = Et$

Note. If the input voltage goes negative for a certain interval of time, the total area under that interval is *negative* and is subtracted from whatever positive area had previously accumulated, thus reducing the output voltage.

16.18 Important Applications of Diodes

We have seen that diodes can be used as rectifiers. Apart from this, diodes have many other applications. However, we shall confine ourselves to the following two applications of diodes :

(*i*) as a clipper (*ii*) as a clamper

A clipper (or limiter) is used to clip off or remove a portion of an a.c. signal. The half-wave rectifier is basically a clipper that eliminates one of the alternations of an a.c. signal.

A clamper (or d.c. restorer) is used to restore or change the d.c. reference of an a.c. signal. For example, you may have a 10 V_{pp} a.c. signal that varies equally above and below 2 V d.c.

16.19 Clipping Circuits

The circuit with which the waveform is shaped by removing (or clipping) a portion of the applied wave is known as a ***clipping circuit.***

Clippers find extensive use in radar, digital and other electronic systems. Although several clipping circuits have been developed to change the wave shape, we shall confine our attention to diode clippers. These clippers can remove signal voltages above or below a specified level. The important diode clippers are (*i*) positive clipper (*ii*) biased clipper (*iii*) combination clipper.

(*i*) **Positive clipper.** A positive clipper is that which removes the positive half-cycles of the input voltage. Fig. 16.32 shows the typical circuit of a positive clipper using a diode. As shown, the output voltage has all the positive half-cycles removed or clipped off.

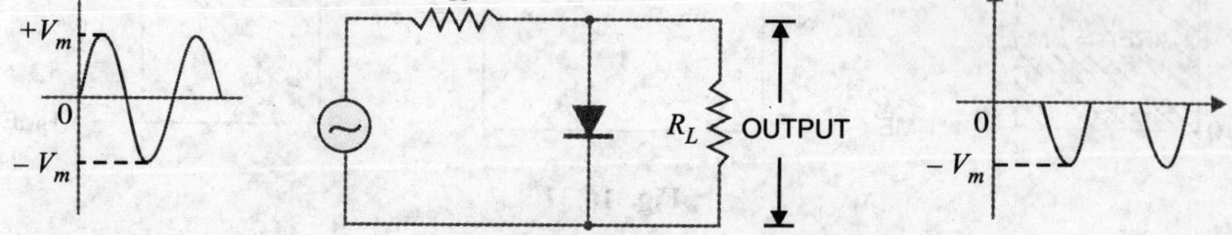

Fig. 16.32

The circuit action is as follows. During the positive half-cycle of the input voltage, the diode is forward biased and conducts heavily. Therefore, the voltage across the diode (which behaves as a short) and hence across the load R_L is zero. Hence *output voltage during positive half-cycles is zero.

During the negative half-cycle of the input voltage, the diode is reverse biased and behaves as an open. In this condition, the circuit behaves as a voltage divider with an output given by ;

$$\text{Output voltage} = -\frac{R_L}{R + R_L} V_m$$

Generally, R_L is much greater than R.

$$\therefore \quad \text{Output voltage} = -V_m$$

It may be noted that if it is desired to remove the negative half-cycle of the input, the only thing to be done is to reverse the polarities of the diode in the circuit shown in Fig. 16.32. Such a clipper is then called a *negative clipper*.

***(ii)* Biased clipper.** Sometimes it is desired to remove a small portion of positive or negative half-cycle of the signal voltage. For this purpose, biased clipper is used. Fig. 16.33 shows the circuit of a biased clipper using a diode with a battery of V volts. With the polarities of battery shown, a portion of each positive half-cycle will be clipped. However, the negative half-cycles will appear as such across the load. Such a clipper is called *biased positive clipper*.

The circuit action is as follows. The diode will conduct heavily so long as input voltage is greater than $+V$. When input voltage is greater than $+V$, the diode behaves as a short and the output equals $+V$. The output will stay at $+V$ so long as the input voltage is greater than $+V$. During the period the input voltage is less than $+V$, the diode is reverse biased and behaves as an open. Therefore, most of the input voltage appears across the output. In this way, the biased positive clipper removes input voltage above $+V$.

During the negative half-cycle of the input voltage, the diode remains reverse biased. Therefore, almost entire negative half-cycle appears across the load.

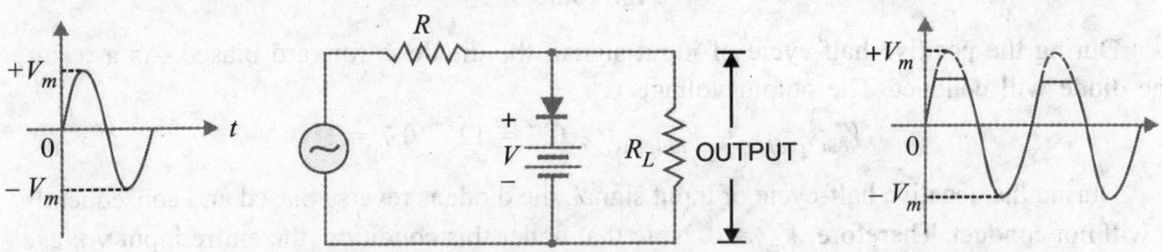

Fig. 16.33

If it is desired to clip a portion of negative half-cycles of input voltage, the only thing to be done is to reverse the polarities of diode or battery. Such a circuit is then called a *biased negative clipper*.

***(iii)* Combination clipper.** It is a combination of biased positive and negative clippers. With a combination clipper, a portion of both positive and negative half-cycles of input voltage can be removed or clipped as shown in Fig. 16.34.

The circuit action is as follows. When positive input voltage is greater than $+V_1$, diode D_1 conducts heavily while diode D_2 remains reverse biased. Therefore, a voltage $+V_1$ appears across

* It may be noted that all the input voltage during this half-cycle is dropped across R.

the load. This output stays at $+V_1$ so long as the input voltage exceeds $+V_1$. On the other hand, during the negative half-cycle, the diode D_2 will conduct heavily and the output stays at $-V_2$ so long as the input voltage is greater than $-V_2$. Note that $+V_1$ and $-V_2$ are less than $+V_m$ and $-V_m$ respectively.

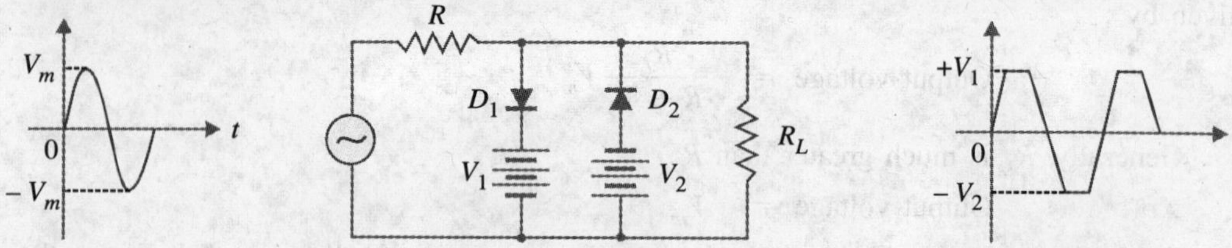

Fig. 16.34

Between $+V_1$ and $-V_2$ neither diode is on. Therefore, in this condition, most of the input voltage appears across the load. It is interesting to note that this clipping circuit can give square wave output if V_m is much greater than the clipping levels.

Example 16.12. *For the negative series clipper shown in Fig.16.35, what is the peak output voltage from the circuit?*

Solution. When the diode is connected in series with the load, it is called a **series clipper**. Since it is a negative clipper, it will remove negative portion of input a.c. signal.

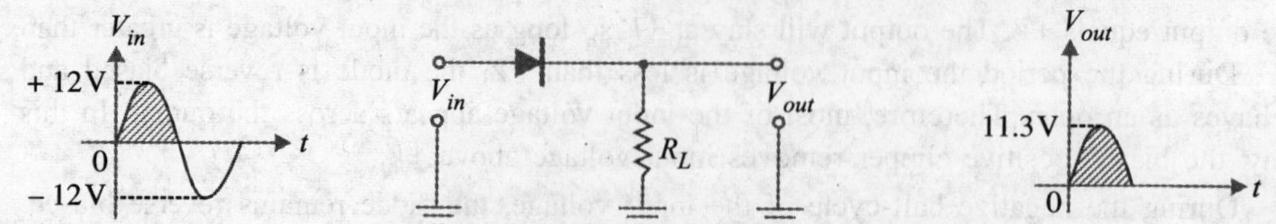

Fig. 16.35

During the positive half-cycle of input signal, the dioide is forward biased. As a result, the diode will conduct. The output voltage is

$$V_{out\ (peak)} = V_{in\ (peak)} - 0.7 = 12 - 0.7 = \mathbf{11.3\ V}$$

During the negative half-cycle of input signal, the diode is reverse biased and consequently it will not conduct. Therefore, $V_{out} = 0$. Note that under this condition, the entire input voltage will appear across the diode.

Example 16.13. *The negative shunt clipper shown in Fig. 16.36 (i) has a peak input voltage of + 10 V. What is the peak output voltage from this circuit?*

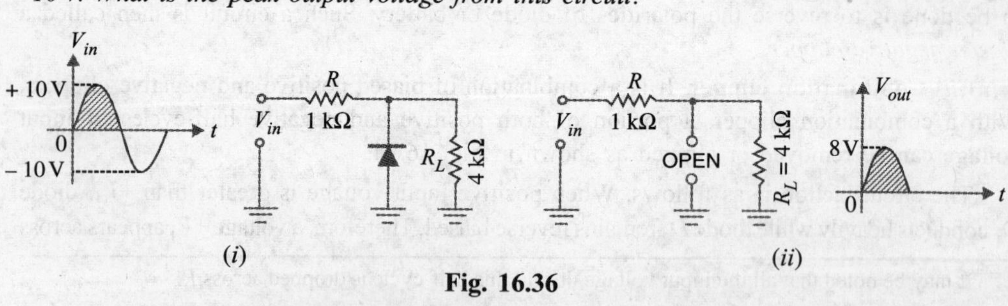

Fig. 16.36

Solution. When the diode is connected in parallel with the load, it is called a **shunt clipper**. During the positive half-cycle of input a.c. signal, the diode is reverse biased and it will behave as an open. This is shown in Fig. 16.36 (*ii*). With diode as an open,

$$V_{out\ (peak)} = \text{Peak voltage across } R_L$$

$$= \frac{R_L}{R + R_L} V_{in\,(peak)} = \frac{4}{1+4} \times 10 = \mathbf{8\ V}$$

Note that peak output voltage is somewhat less than the peak input voltage.

Example 16.14. *In example 16.13, what will be the output voltage and voltage across R when the input voltage is –10 V?*

Solution. During the negative half-cycle of input signal, the diode is forward biased. Therefore, diode can be replaced by its simplified equivalent circuit as shown in Fig. 16.37. Since load is connected in parallel with the diode,

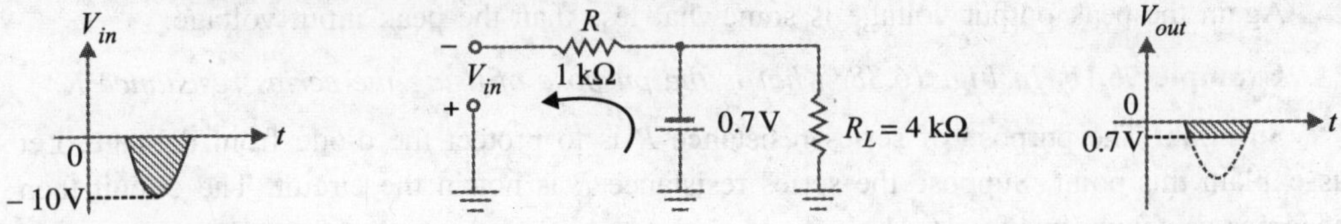

Fig. 16.37

$$\therefore \quad V_{out} = \mathbf{-\ 0.7\ V}$$

$$\text{Voltage across } R,\ V_R = (-\ 10) - (-\ 0.7) = -\ 10 + 0.7 = \mathbf{-\ 9.3\ V}$$

Example 16.15. *The positive shunt clipper shown in Fig. 16.38 has the input waveform as indicated. Determine the value of V_{out} for each of the input alternations.*

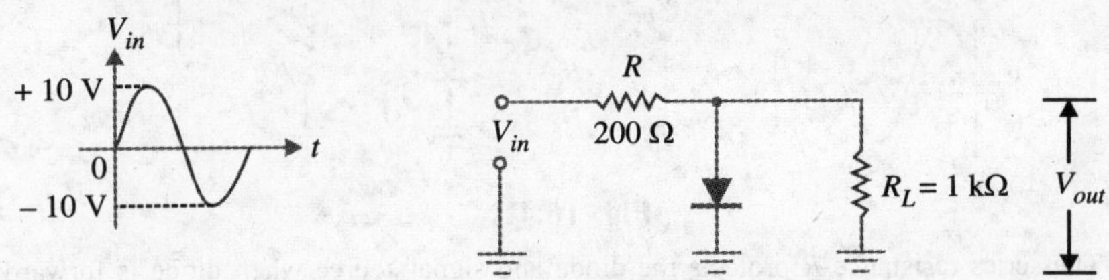

Fig. 16.38

Solution.

Positive half-cycle. During the positive half-cycle of the input a.c. signal, the diode is forward biased. Therefore, diode can be replaced by its simplified equivalent circuit as shown in Fig. 16.39. Since the load is connected in parallel with the diode,

$$\therefore \quad V_{out} = \mathbf{0.7\ V}$$

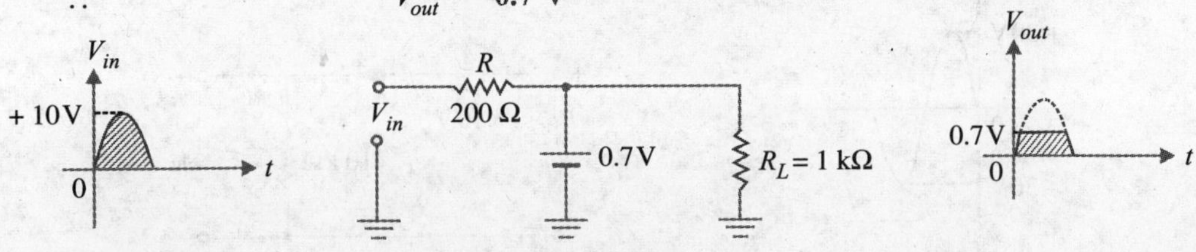

Fig. 16.39

Negative half-cycle. During the negative half-cycle of the input a.c. signal, the diode is reverse biased and it conducts no current. Therefore, the diode will behave as an open as shown in Fig. 16.40.

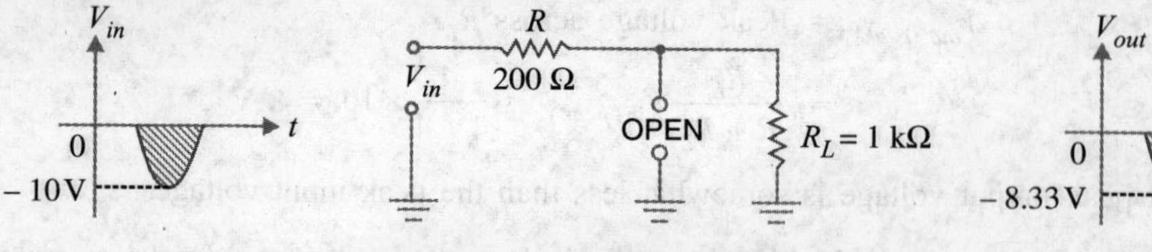

Fig. 16.40

$$\therefore \quad V_{out\,(peak)} = \frac{R_L}{R + R_L} V_{in\,(peak)} = \left(\frac{1000}{200 + 1000}\right)(-10\text{V}) = -\,8.33\text{ V}$$

Again the peak output voltage is somewhat less than the peak input voltage.

Example 16.16. *In Fig. 16.38, what is the purpose of using the series resistance R?*

Solution. The purpose of series resistance R is to protect the diode from damage. Let us explain this point. Suppose the series resistance R is not in the circuit. The circuit then becomes as shown in Fig. 16.41.

During the positive half-cycle of the input signal, the diode is forward biased. Since series resistance R is not present, it is easy to see that the diode will short the signal source to the ground. As a result, excessive current will flow through the diode as well as through the signal source. This large current may damage/destroy either the diode or the signal source.

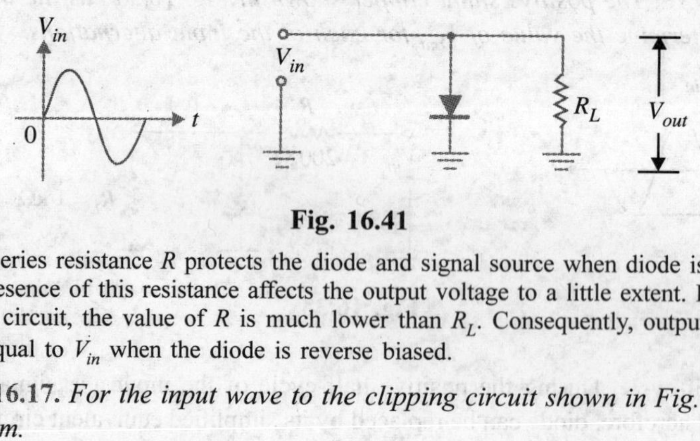

Fig. 16.41

Note. The series resistance R protects the diode and signal source when diode is forward biased. However, the presence of this resistance affects the output voltage to a little extent. It is because in a practical clipper circuit, the value of R is much lower than R_L. Consequently, output voltage will be approximately equal to V_{in} when the diode is reverse biased.

Example 16.17. *For the input wave to the clipping circuit shown in Fig. 16.42, find the output waveform.*

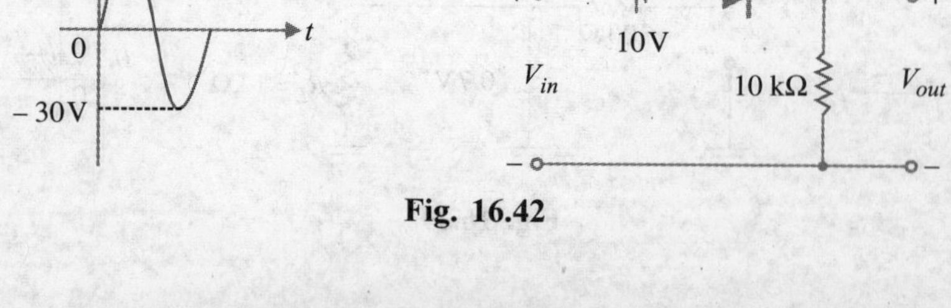

Fig. 16.42

Solution. For any value of $V_{in} > 10$ V, the ideal diode is forward biased and $V_{out} = V_{in} - 10$. For example, at $V_{in} = 15$ V [See Fig. 16.43 (*i*)], $V_{out} = 15 - 10 = 5$ V.

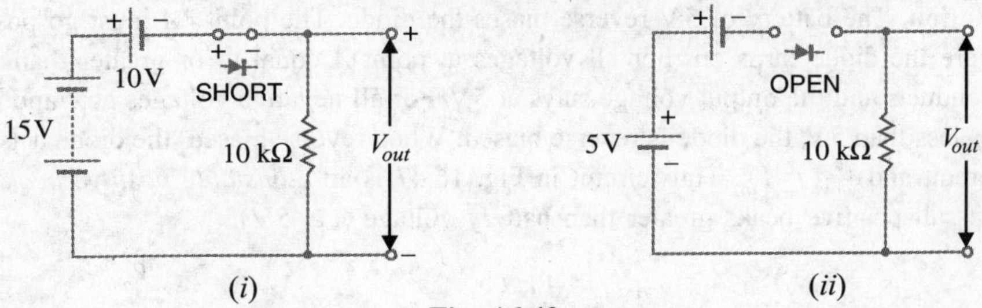

Fig. 16.43

For any value of $V_{in} < 10$ V, the ideal diode is reverse biased. Therefore, circuit current is zero and hence $V_{out} = 0$. For example, with $V_{in} = 5$ V [See Fig. 16.43 (*ii*)], $V_{out} = 0$ and V_d (drop across diode) = 5 V.

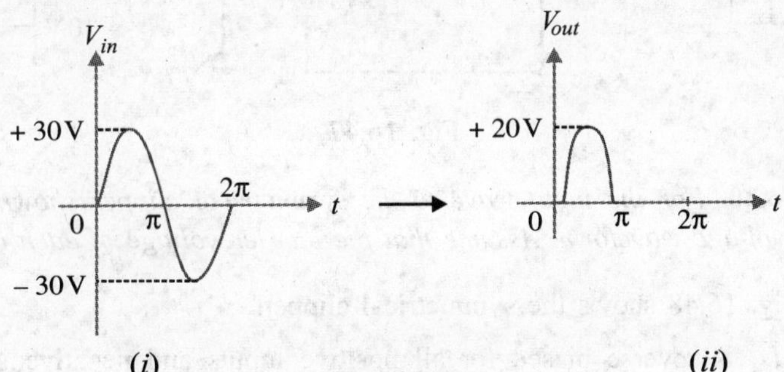

Fig. 16.44

The output waveform will be as shown in Fig. 16.44 (*ii*).

Example 16.18. *For the input wave to the clipping circuit in Fig. 16.45, find the output waveform.*

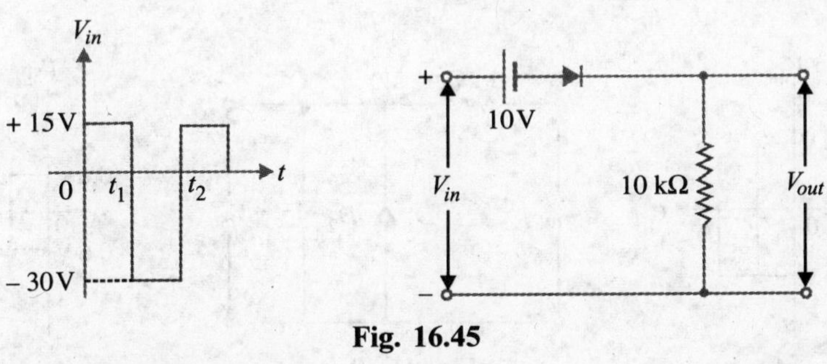

Fig. 16.45

Solution. For any value of $V_{in} > 10$ V, the ideal diode is forward biased and $V_{out} = V_{in} - 10$. For any value of $V_{in} < 10$ V, the ideal diode is reverse biased and $V_{out} = 0$.

The output waveform will be as shown in Fig. 16.46 (*ii*).

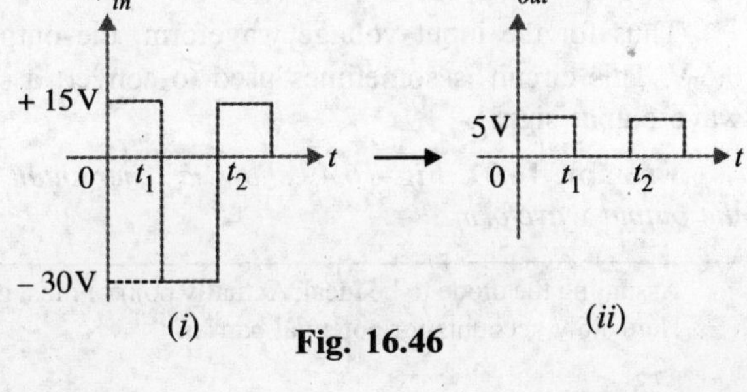

Fig. 16.46

Example 16.19. *For the input waveform to the clipping circuit in Fig. 16.47, find the output voltage waveform.*

Solution. The battery of 5 V reverse biases the diode. The point *A must go positive to 5 V before the diode turns on. For all voltages at point A equal to or greater than 5 V, the diode conducts and the output voltage stays at 5 V. For all negative voltages at A and positive voltages less than 5 V, the diode is reverse biased. When reverse biased, the diode acts like an open circuit and $V_{out} = V_{in}$. Thus circuit in Fig. 16.47 is an *adjustable positive peak clipper* that clips all positive peaks greater than battery voltage (*i.e.* 5 V).

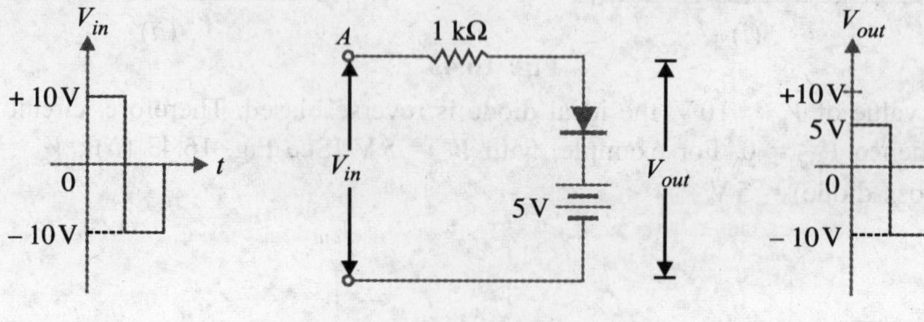

Fig. 16.47

Example 16.20. *For the input wave to the symmetrical clipper shown in Fig. 16.48, find the output voltage waveform. Assume that the barrier voltage of each diode is 0.6 V.*

Solution. Fig. 16.48 shows the symmetrical clipper.

(*i*) Diode D_1 is reverse biased for all positive inputs and negative inputs less than 0.6 V. Diode D_2 is reverse biased for all negative inputs and positive inputs less than 0.6 V.

(*ii*) For all positive inputs greater than 0.6V, D_2 is forward biased and output voltage stays at 0.6 V. For all negative inputs greater than –0.6 V, D_1 is forward biased and output stays at – 0.6 V.

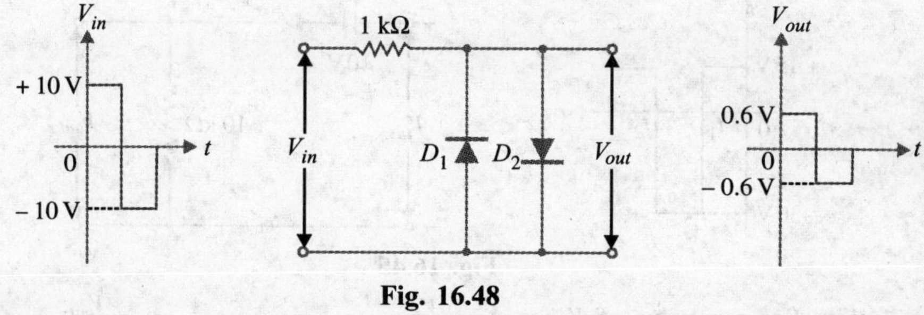

Fig. 16.48

Thus for the input voltage waveform, the output voltage varies between – 0.6 V and 0.6 V. This circuit is sometimes used to convert a sine-wave input to a reasonably square-wave output signal.

Example 16.21. *Fig. 16.49 shows a zener shunt clipper with sine wave input. Determine the output waveform.*

* Assuming the diode to be ideal. Actually point A must go positive to 5.6 V before the diode turns on. Here 0.6 V accounts for potential barrier.

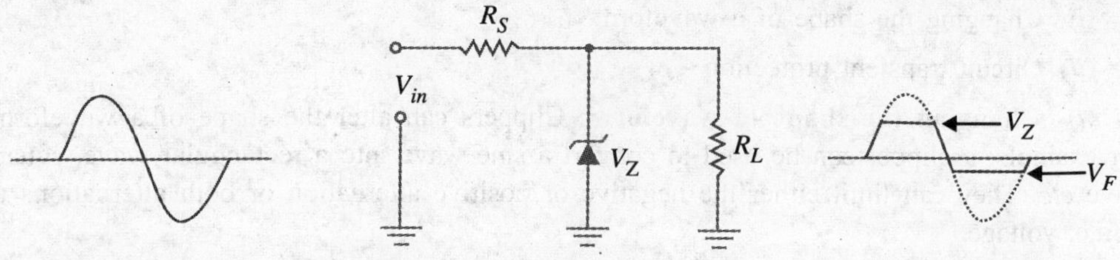

Fig. 16.49

Solution. The *zener shunt clipper* uses a zener diode in place of the ordinary diode. The zener shunt clipper uses both the forward and reverse characteristics of the zener diode. Thus when the zener diode is forward biased, this clipper acts just like the standard shunt clipper.

When the input signal in Fig. 16.49 goes positive, the zener is reverse biased. Therefore, the zener diode will remain OFF until the value of V_{in} reaches the value of V_Z. At that time, the zener diode will turn ON, clipping the input signal. As long as V_{in} is greater than V_Z, the zener will remain ON and the output voltage is

$$V_{out} = V_Z$$

When the input signal goes negative, the zener is forward biased. Therefore, the zener is clipped off at 0.7 V.

Example 16.22. *Fig. 16.50 shows a symmetrical zener shunt clipper with sine wave input. Determine the output waveform.*

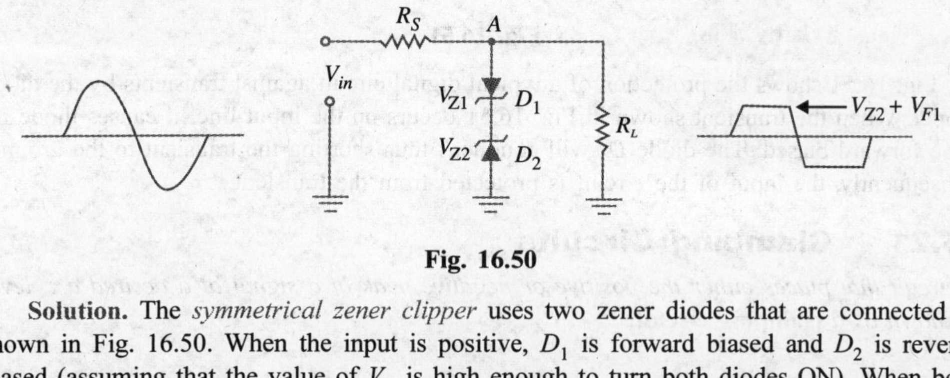

Fig. 16.50

Solution. The *symmetrical zener clipper* uses two zener diodes that are connected as shown in Fig. 16.50. When the input is positive, D_1 is forward biased and D_2 is reverse biased (assuming that the value of V_{in} is high enough to turn both diodes ON). When both diodes are conducting, the voltage from point A to ground will be equal to the sum of V_{Z2} and the forward voltage drop of D_1 *i.e.* V_{F1} = 0.7 V.

$$\therefore \quad V_{out} = V_{Z2} + V_{F1} = V_{Z2} + 0.7 \text{ V}$$

When the input is negative, D_1 is reverse biased and D_2 is forward biased. The output voltage in this case is

$$V_{out} = -V_{Z1} - V_{F2} = -(V_{Z1} + 0.7 \text{ V})$$

In practice, the two zeners have the same voltage rating *i.e.* $V_{Z1} = V_{Z2} = V_Z$.

$$\therefore \quad V_{out} = \pm(V_Z + 0.7 \text{ V})$$

16.20 Applications of Clippers

There are numerous clipper applications and it is not possible to discuss all of them. However, in general, clippers are used to perform one of the following two functions :

(*i*) Changing the shape of a waveform

(*ii*) Circuit transient protection

(*i*) **Changing the shape of waveform.** Clippers can alter the shape of a waveform. For example, a clipper can be used to convert a sine wave into a rectangular wave, square wave *etc.* They can limit either the negative or positive alternation or both alternations of an a.c. voltage.

(*ii*) **Circuit transient protection.** *Transients can cause considerable damage to many types of circuits *e.g.*, a digital circuit. In that case, a clipper diode can be used to prevent the transient from reaching that circuit.

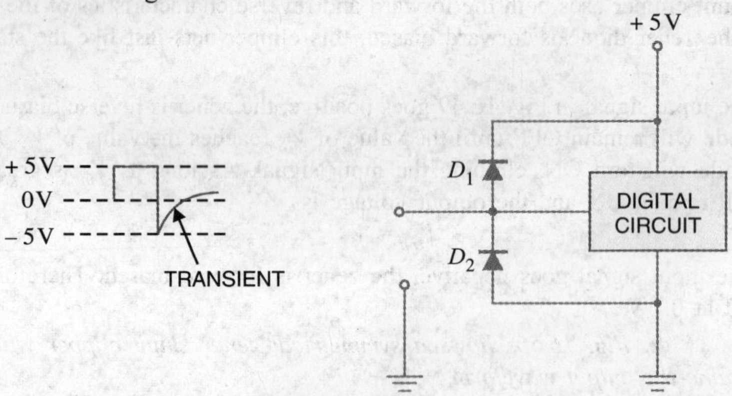

Fig. 16.51

Fig. 16.51 shows the protection of a typical digital circuit against transients by the diode clipper. When the transient shown in Fig. 16.51 occurs on the input line, it causes diode D_2 to be forward biased. The diode D_2 will conduct ; thus shorting the transient to the ground. Consequently, the input of the circuit is protected from the transient.

16.21 Clamping Circuits

A circuit that places either the positive or negative peak of a signal at a desired d.c. level is known as a **clamping circuit.**

Fig. 16.52

A clamping circuit (or a clamper) essentially adds a d.c. component to the signal. Fig. 16.52 shows the key idea behind clamping. The input signal is a sine wave having a peak-to-peak value of 10 V. The clamper adds the d.c. component and pushes the signal upwards so that the negative peaks fall on the zero level. As you can see, the waveform now has peak values of +10 V and 0 V.

---

* A transient is a sudden current or voltage rise that has an extremely short duration.

It may be seen that the shape of the original signal has not changed ; only there is vertical shift in the signal. Such a clamper is called a *positive clamper*. The *negative clamper* does the reverse *i.e.* it pushes the signal downwards so that the positive peaks fall on the zero level.

The following points may be noted carefully :

(*i*) The clamping circuit does not change the peak-to-peak or r.m.s. value of the waveform. Thus referring to Fig. 16.52 above, the input waveform and clamped output have the same peak-to-peak value *i.e.* 10 V in this case. If you measure the input voltage and clamped output with an a.c. voltmeter, the readings will be the same.

(*ii*) A clamping circuit changes the peak and average values of a waveform. This point needs explanation. Thus in the above circuit, it is easy to see that input waveform has a peak value of 5 V and average value over a cycle is zero. The clamped output varies between 10 V and 0 V. Therefore, the peak value of clamped output is 10 V and *average value is 5 V. Hence we arrive at a very important conclusion that *a clamper changes the peak value as well as the average value of a wave-form.*

16.22 Basic Idea of a Clamper

A clamping circuit should not change peak-to-peak value of the signal ; it should only change the d.c. level. To do so, a clamping circuit uses a capacitor, together with a diode and a load resistor R_L. Fig. 16.53 shows the circuit of a positive clamper. *The operation of a clamper is based on the principle that charging time of a capacitor is made very small as compared to its discharging time.* Thus referring to Fig. 16.53,

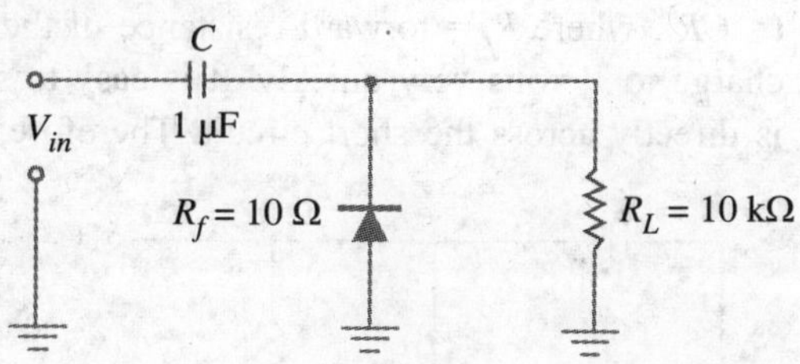

Fig. 16.53

**Charging time constant, $\tau = R_f C = (10\ \Omega) \times (10^{-6}\ F) = 10\ \mu s$

Total charging time, $\tau_C =$ ***$5R_f C = 5 \times 10 = 50\ \mu s$

****Discharging time constant, $\tau = R_L C = (10 \times 10^3) \times (1 \times 10^{-6}) = 10$ ms

Total discharging time, $\tau_D = 5\ R_L C = 5 \times 10 = 50$ ms

It may be noted that charging time (*i.e.* 50 μs) is very small as compared to the discharging time (*i.e.* 50 ms). This is the basis of clamper circuit operation. In a practical clamping circuit, the values of C and R_L are so chosen that discharging time is very large.

---

* Average value (or d.c. value) $= \dfrac{10+0}{2} = 5$ V

** When diode is forward biased.

*** From the knowledge of electrical engineering, we know that charging time of a capacitor is $\simeq 5\ RC$.

**** When diode is reverse biased.

16.23 Positive Clamper

Fig. 16.54 shows the circuit of a *positive clamper. The input signal is assumed to be a square wave with time period T. The clamped output is obtained across R_L. The circuit design incorporates two main features. Firstly, the values of C and R_L are so selected that time constant $\tau = CR_L$ is very large. This means that voltage across the capacitor will not discharge significantly during the interval the diode is non-conducting. Secondly, R_LC time constant is deliberately made much greater than the time period T of the incoming signal.

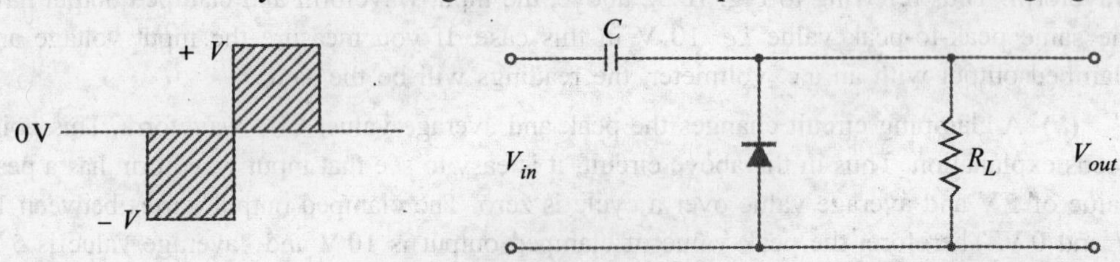

Fig. 16.54

Operation

(*i*) During the negative half-cycle of the input signal, the diode is forward biased. Therefore, the diode behaves as a short as shown in Fig. 16.55. The charging time constant (= CR_f, where R_f = forward resistance of the diode) is very small so that the capacitor will charge to V volts very quickly. It is easy to see that during this interval, the output voltage is directly across the short circuit. Therefore, $V_{out} = 0$.

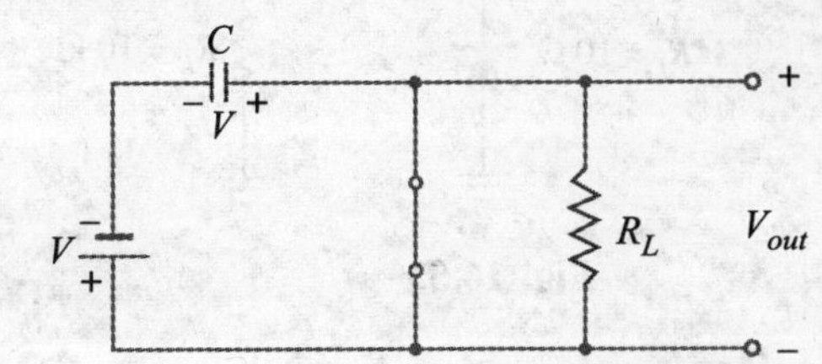

Fig. 16.55

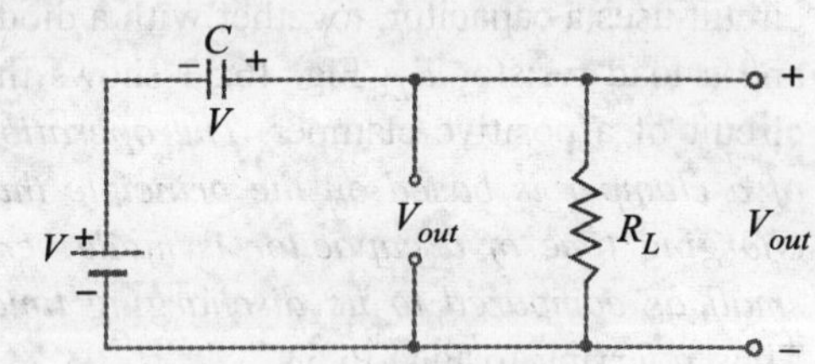

Fig. 16.56

(*ii*) When the input switches to $+V$ state (*i.e.* positive half-cycle), the diode is reverse biased and behaves as an open as shown in Fig. 16.56. Since the discharging time constant (= CR_L) is much greater than the time period of the input signal, the capacitor remains almost fully charged to V volts during the off time of the diode. Referring to Fig. 16.56 and applying Kirchhoff's voltage law to the input loop, we have,

$$V + V - V_{out} = 0$$

or $$V_{out} = 2\,V$$

The resulting waveform is shown in Fig. 16.57. It is clear that it is a positively clamped output. That is to say the input signal has been pushed upward by V volts so that negative peaks fall on the zero level.

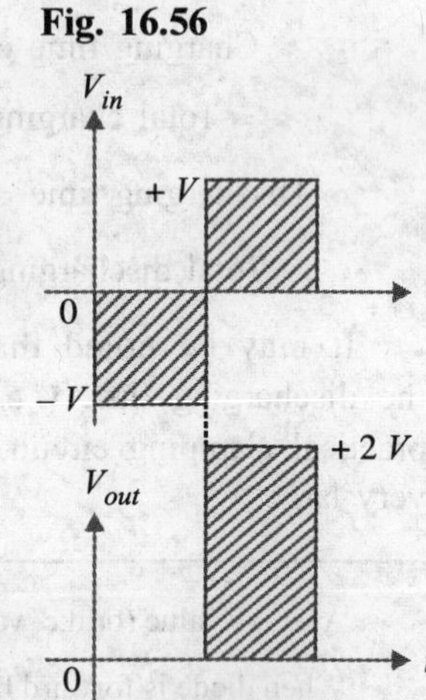

Fig. 16.57

* If you want to determine what type of clamper you are dealing with, here is an easy memory trick. If the diode is pointing up (away from ground), the circuit is a positive clamper. On the other hand, if diode is pointing down (towards ground), the circuit is a negative clamper.

16.24 Negative Clamper

Fig. 16.58 shows the circuit of a negative clamper. The clamped output is taken across R_L. Note that only change from the positive clamper is that the connections of diode are reversed.

(*i*) During the positive half-cycle of the input signal, the diode is forward biased. Therefore, the diode behaves as a short as shown in Fig. 16.59. The charging time constant (= CR_f) is very small so that the capacitor will charge to V volts very quickly. It is easy to see that during this interval, the output voltage is directly across the short circuit. Therefore, V_{out} = 0.

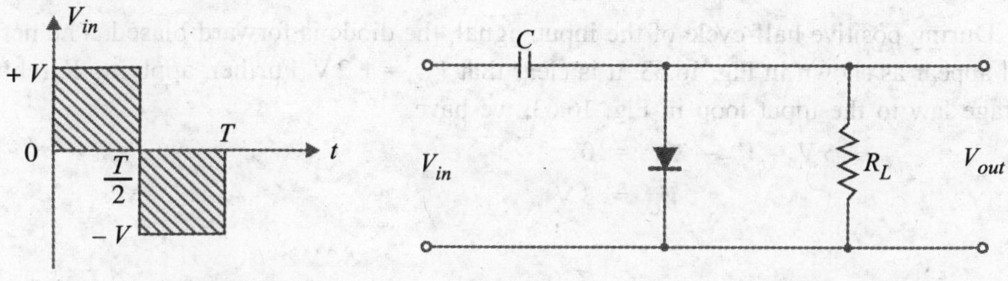

Fig. 16.58

(*ii*) When the input switches to $-V$ state (*i.e.* negative half-cycle), the diode is reverse biased

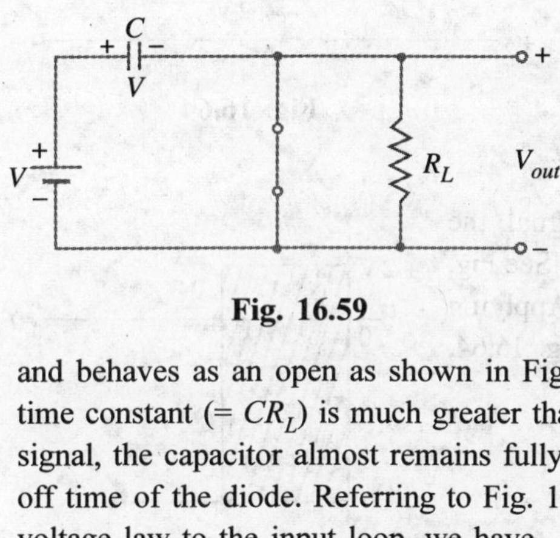

Fig. 16.59

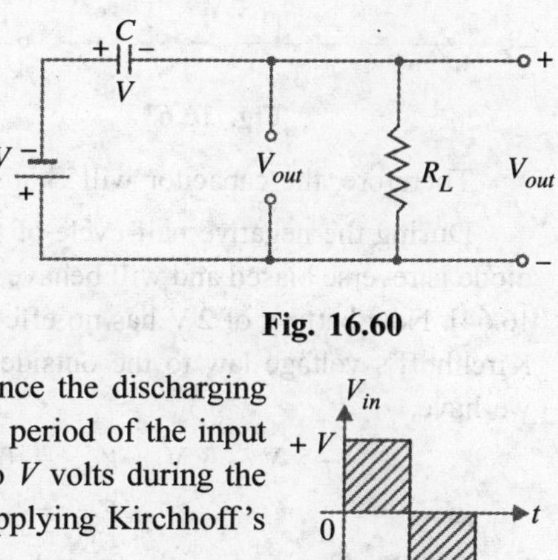

Fig. 16.60

and behaves as an open as shown in Fig. 16.60. Since the discharging time constant (= CR_L) is much greater than the time period of the input signal, the capacitor almost remains fully charged to V volts during the off time of the diode. Referring to Fig. 16.60 and applying Kirchhoff's voltage law to the input loop, we have,

$$-V - V - V_{out} = 0$$

or

$$V_{out} = -2V$$

The resulting waveform is shown in Fig. 16.61. Note that total swing of the output signal is equal to the total swing of the input signal.

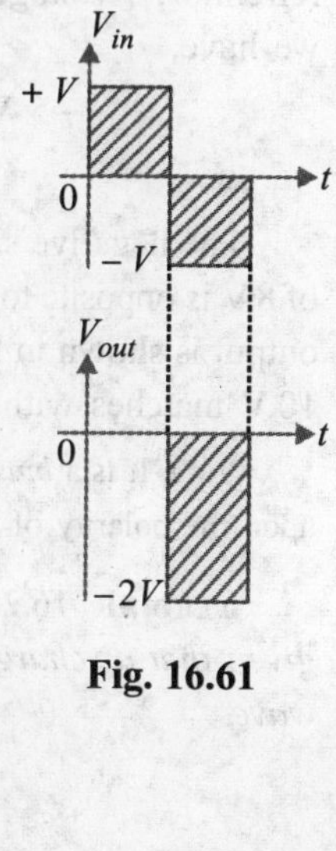

Fig. 16.61

Example 16.23. *Sketch the output waveform for the circuit shown in Fig. 16.62. It is given that discharging time constant (CR_L) is much greater than the time period of input wave.*

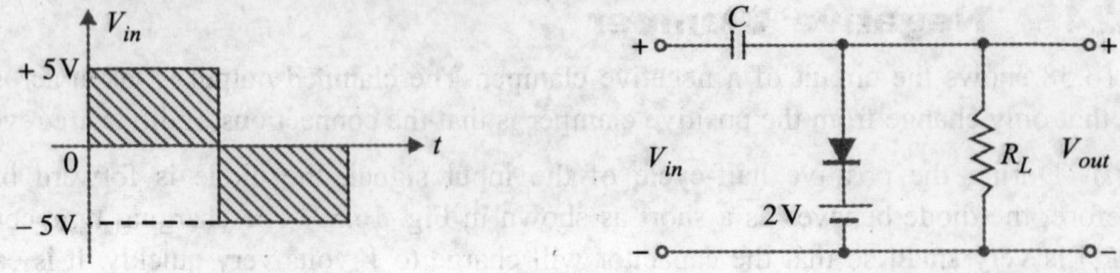

Fig. 16.62

Solution.

During positive half-cycle of the input signal, the diode is forward biased. The network will appear as shown in Fig. 16.63. It is clear that $V_{out} = +2$ V. Further, applying Kirchhoff's voltage law to the input loop in Fig. 16.63, we have,

$$5\text{ V} - V_C - 2\text{ V} = 0$$

$$\therefore \quad V_C = 3\text{ V}$$

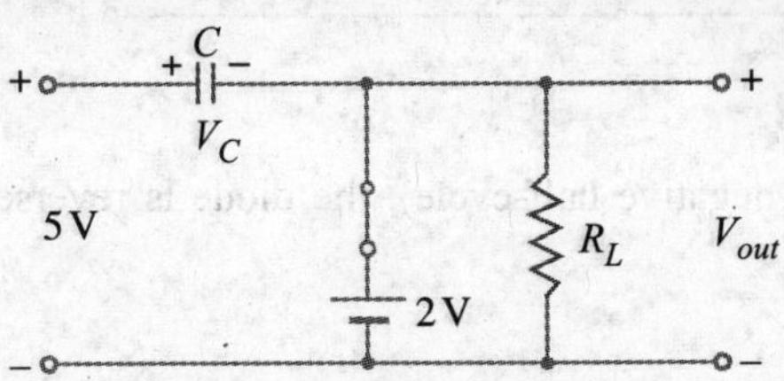

Fig. 16.63

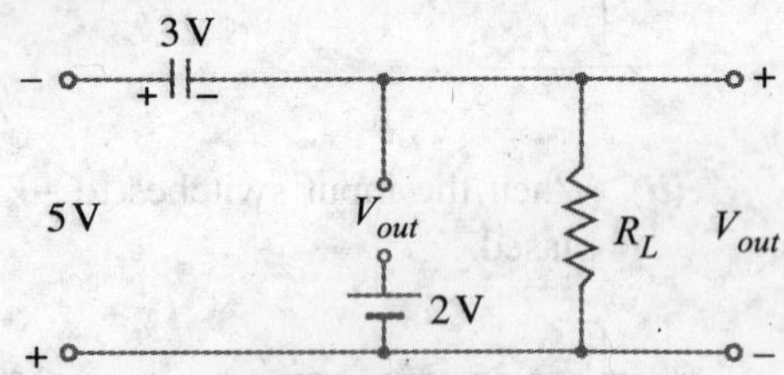

Fig. 16.64

Therefore, the capacitor will charge upto 3 V.

During the negative half-cycle of the input signal, the diode is reverse biased and will behave as an open [See Fig. 16.64]. Now battery of 2 V has no effect on V_{out}. Applying Kirchhoff's voltage law to the outside loop of Fig. 16.64, we have,

$$-5\text{ V} - 3\text{ V} - V_{out} = 0$$

or $$V_{out} = -8\text{ V}$$

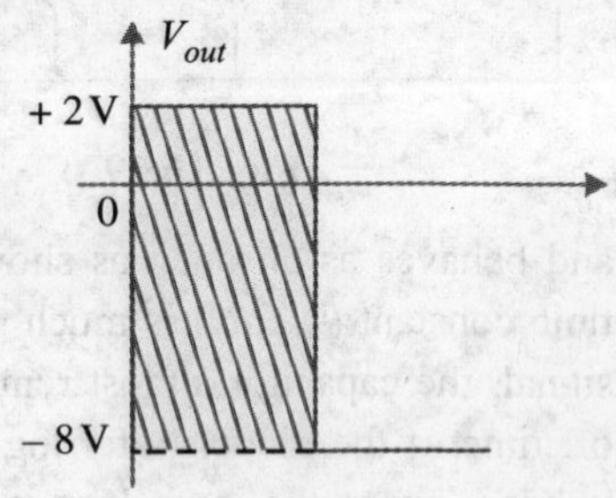

Fig. 16.65

The negative sign results from the fact that the polarity of 8 V is opposite to the polarity defined for V_{out}. The clamped output is shown in Fig. 16.65. Note that the output swing of 10 V matches with the input swing.

Note. It is a *biased clamper circuit*. It allows a waveform to be shifted above or below (depending upon the polarity of 2 V battery) a d.c. reference other than 0 V.

Example 16.24. *Sketch the output waveform for the circuit shown in Fig. 16.66. It is given that discharging time constant (= CR_L) is much greater than the time period of input wave.*

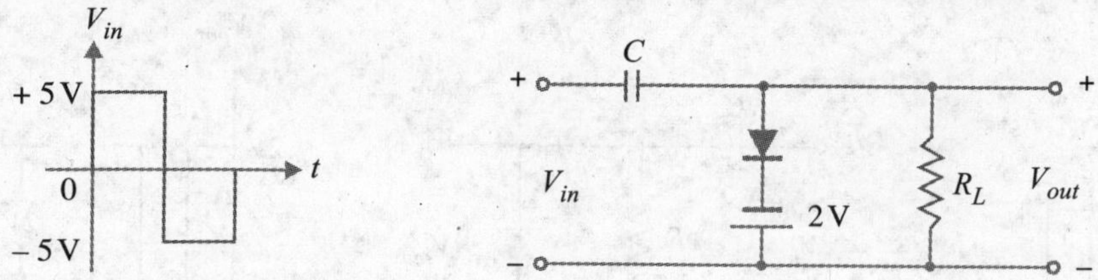

Fig. 16.66

Solution.

During the positive half-cycle of input signal, the diode is forward biased. Therefore, the diode behaves as a short [See Fig. 16.67]. It is easy to see that V_{out} = –2 V. Further, applying Kirchhoff's voltage law to the input loop [See Fig. 16.67], we have,

$$5\text{ V} - V_C + 2\text{ V} = 0$$

or $$V_C = 7\text{ V}$$

Therefore, the capacitor will charge upto 7 V.

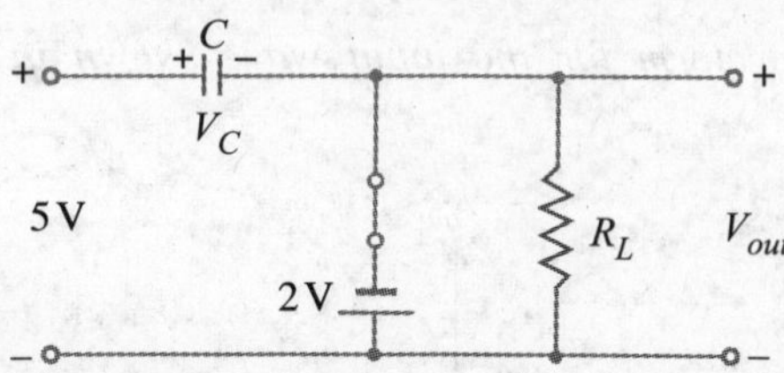

Fig. 16.67

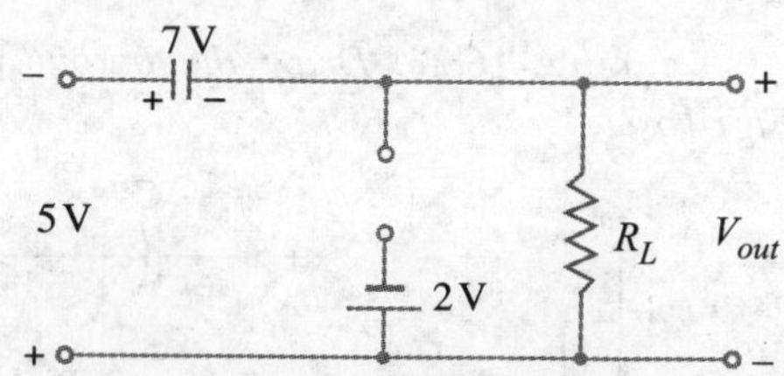

Fig. 16.68

During the negative half-cycle of the input signal, the diode is reverse biased and behaves as an open as shown in Fig. 16.68. Now battery of 2 V has no effect on V_{out}. Applying Kirchhoff's voltage law to the outside loop of Fig. 16.68, we have,

$$-5\text{ V} - 7\text{ V} - V_{out} = 0$$

or $$V_{out} = -12\text{ V}$$

The negative sign results from the fact that the polarity of 12 V is opposite to the polarity defined for V_{out}. The clamped output is shown in Fig. 16.69. Note that output and input swings are the same.

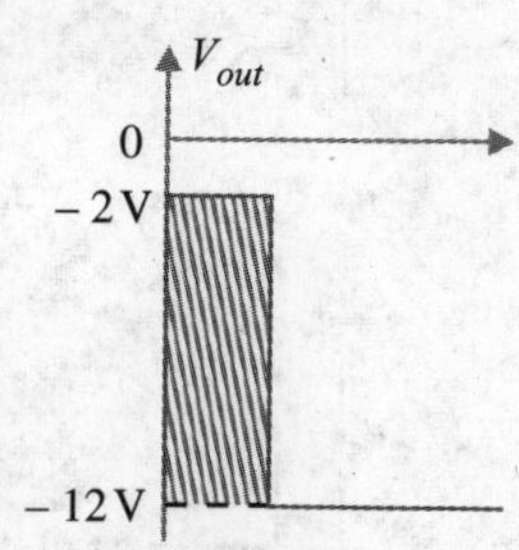

Fig. 16.69

Example 16.25. *Draw the output voltage waveform for the input shown in Fig. 16.70.*

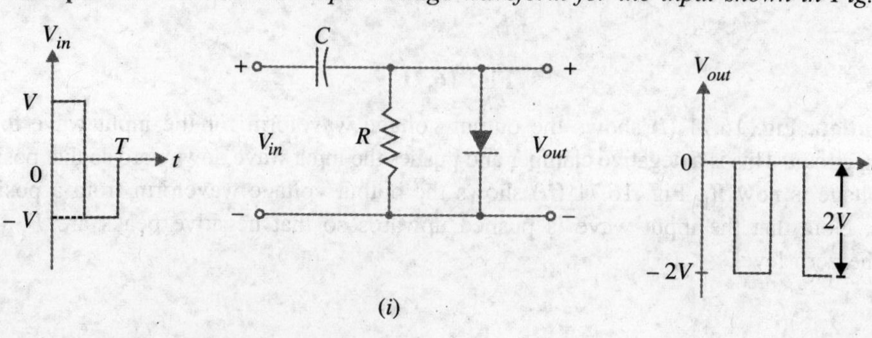

(i)

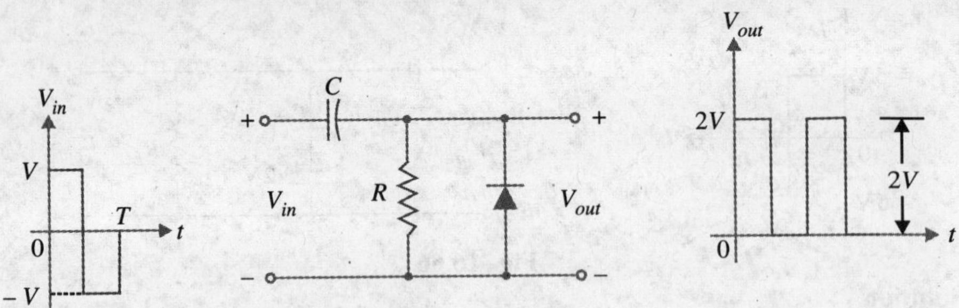

(ii)

Fig. 16.70

Solution. Fig. 16.70 (*i*) shows the output voltage waveform for the input wave to the clamping circuit. Note that it is a *negative clamper* and pushes the input wave downwards so that positive peaks fall on the zero level. Fig. 16.70 (*ii*) shows the output voltage waveform for the input wave to a *positive clamper*. The clamper pushes the input wave upwards so that the negative peaks fall on the zero level.

Example 16.26. *Draw the output voltage waveform for the input wave shown in Fig. 16.71.*

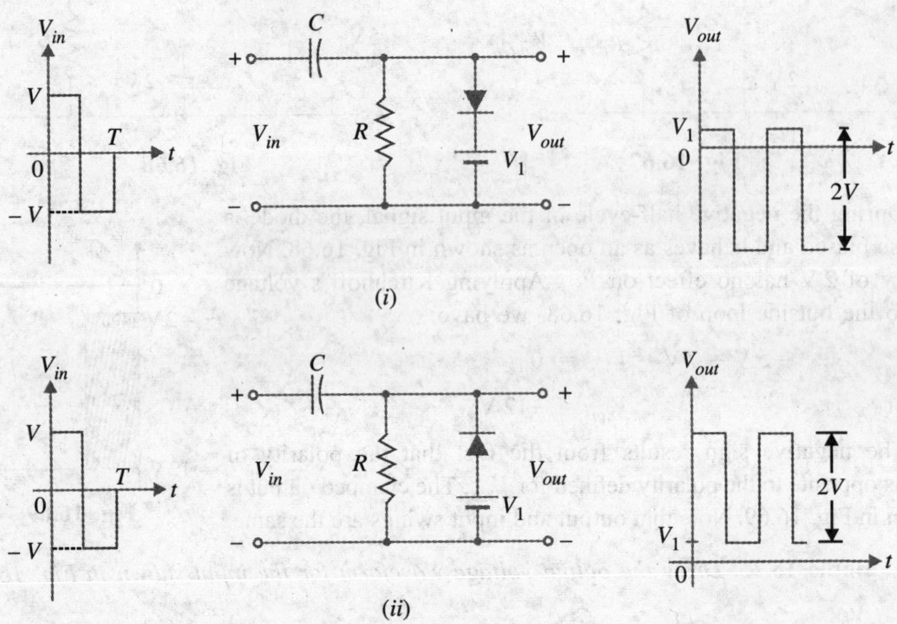

Fig. 16.71

Solution. Fig. 16.71 (*i*) shows the output voltage waveform for the input wave to the clamping circuit. This is a negative clamper and pushes the input wave downward so that positive peak voltage is now V_1. Fig. 16.71 (*ii*) shows the output voltage waveform from a positive clamper. Note that the input wave is pushed upwards so that negative peaks are V_1 volts above the zero level.

Example 16.27. *Draw the output voltage waveform for the input wave shown in Fig. 16.72.*

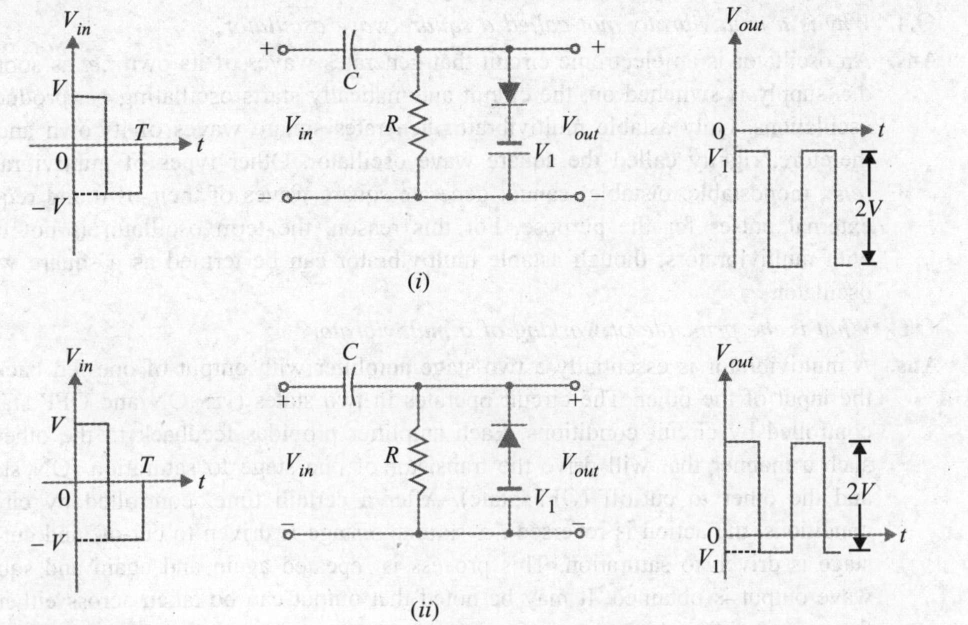

Fig. 16.72

Solution. Fig. 16.72 (*i*) shows the output voltage waveform for the input wave to the clamping circuit. This is a negative clamper and pushes the input wave downward so that positive peaks lie V_1 volts below the zero level. Fig. 16.72 (*ii*) shows the output voltage waveform from a positive clamper. Note that the input waveform is pushed upwards so that negative peaks lie V_1 volts below zero level.

Short Answer Questions

Q.1. *What is the difference between an ordinary transistor and a transistor used as a switch?*

Ans. In general, a switching transistor is fabricated by the same process as an ordinary transistor except that it has special design features to reduce switch-off time and saturation voltage.

Q.2. *How does a transistor act as a switch?*

Ans. When a transistor is to be used as a switch, it is so arranged in the circuit that either maximum current (called saturation collector current $I_{C(sat)}$) flows through the load or minimum collector current (called collector leakage current) flows through the load. Consequently, the transistor will have two states *viz.* (*i*) ON state or when saturation collector current flows (*ii*) OFF state or when collector leakage current flows through the load. Obviously, the transistor will behave as a switch.

Q.3. *What is the origin of the name multivibrator?*

Ans. A multivibrator is an electronic circuit that generates square waves. It can be shown by Fourier series analysis that a square wave (infact, there may be any waveform) can be resolved into a large number of sinusoidals of different frequencies. The prefix "multi" means many and "vibrator" signifies sinusoidals of different frequencies.

Q.4. *Why is a multivibrator not called a square wave oscillator?*

Ans. An oscillator is an electronic circuit that generates waves of its own *i.e.* as soon as d.c. supply is switched on, the circuit automatically starts oscillating *i.e.* producing oscillations. Only astable multivibrator generates square waves of its own and is, therefore, rightly called the square wave oscillator. Other types of multivibrators (*e.g.*, monostable, bistable) cannot generate square waves of their own and require external pulses for the purpose. For this reason, the term oscillator is not used with multivibrators, though astable multivibrator can be termed as a square wave oscillator.

Q.5. *What is the principle of working of a multivibrator?*

Ans. A multivibrator is essentially a two-stage amplifier with output of one fed back to the input of the other. The circuit operates in two states (*viz.* ON and OFF states) controlled by circuit conditions. Each amplifier provides feedback to the other in such a manner that will drive the transistor of one stage to saturation (ON state) and the other to cut-off (OFF state). After a certain time, controlled by circuit conditions, the action is reversed *i.e.* saturated stage is driven to cut-off and cut-off stage is driven to saturation. This process is repeated again and again and square wave output is obtained. It may be noted that output can be taken across either of the two amplifier stages.

Q.6. *Why is monostable multivibrator called one-shot multivibrator?*

Ans. Because this multivibrator produces one output pulse for every input trigger pulse it receives, it is often called one-shot multivibrator.

Q.7. *Why is bistable multivibrator often called flip-flop circuit?*

Ans. In a bistable multivibrator, both the states are stable. The circuit will maintain one of its states indefinitely (as long as supply voltage is applied), changing to the other state only when it receives a triggering pulse from some external source. When it receives another triggering pulse, the circuit changes back to the original state. Because one trigger pulse causes the multivibrator to "flip" from one state to the other and the next pulse causes it to "flop" back to its original state, the bistable multivibrator is popularly known as flip-flop circuit.

Q.8. *What are the applications of rectangular pulses and spiked pulses?*

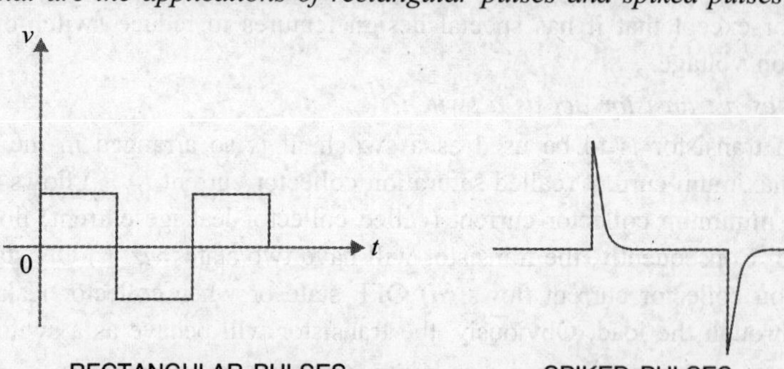

Fig. 16.73

Ans. Pulses are brief voltage or current changes which occur once and may not occur again for a comparatively long time. The most common waveforms of pulses are *rectangular* (as produced by astable multivibrator) and a sharp pulse of very short duration known as **spiked pulse*.

The sinusoidal wave is used extensively in radio communication. For other purposes, such as television, radar and industrial applications, although sinusoidal and other types of periodic waves are employed, the major emphasis is on pulses. For example, rectangular pulses are used in computers to convey information and instructions for processing it. Spiked pulses are used extensively for triggering purposes *e.g.,* to trigger the various oscillators.

Q.9. *Give one practical application or a limiter or clipper.*

Ans. In an FM receiver, we use a limiter stage before demodulator to remove any amplitude modulation which may have crept into the signal. Such amplitude modulation, generally due to external sources, consists of sharp, narrow, high-level pulses appearing in the signal. The limiter (or clipping circuit) clips off these pulses, thus restoring the signal to its normal, constant level.

Q.10. *What is the effect of time constant of an RC circuit on the differentiated wave?*

Ans. In an *RC* differentiating circuit, the output voltage is taken across *R* and the waveform of the output depends upon the time constant of the circuit. The circuit will function as a differentiator if the product *RC* (*i.e.* time constant) is many times (atleast 10 times) smaller than the time period of the input wave.

Q.11. *A differentiating circuit is essentially an RC series circuit. Why is output from RC coupling not a differentiated wave?*

Ans. *RC* coupling circuit is the same as a differentiating circuit except that it has a long time constant *i.e.* time constant is more than 5 *RC*. Therefore, *RC* coupling circuit does not noticeably differentiate the input wave.

* If the input to a differentiating circuit is a square wave, then output will be spiked pulses.

Multiple-Choice Questions

1. A switch has ....
   (*i*) one state (*ii*) two states
   (*iii*) three states
   (*iv*) none of the above
2. A relay is .... switch.
   (*i*) a mechanical (*ii*) an electronic
   (*iii*) an electromechanical
   (*iv*) none of the above
3. The switch that has the fastest speed of operation is .... switch.
   (*i*) electronic
   (*ii*) mechanical
   (*iii*) electromechanical
   (*iv*) none of the above
4. The most inexpensive switch is .... switch.
   (*i*) electronic
   (*ii*) mechanical
   (*iii*) electromechanical
   (*iv*) none of the above
5. The main disadvantage of a mechanical switch is that it ....
   (*i*) is operated mechanically
   (*ii*) is costly
   (*iii*) has high inertia
   (*iv*) none of the above
6. When a transistor is driven to saturation, ideally the output is ....

(i) V_{CC} (ii) 0
(iii) $V_{CC}/2$ (iv) $2V_{CC}$

7. The maximum speed of electronic switch can be .... operations per second.
(i) 10^4 (ii) 10
(iii) 1000 (iv) 10^9

8. A relay is superior to a mechanical switch because it ....
(i) is relatively inexpensive
(ii) does not require moving contacts
(iii) combines control with power amplification
(iv) none of the above

9. When a transistor is driven to cut-off, ideally the output is ....
(i) V_{CC} (ii) 0
(iii) $V_{CC}/2$ (iv) $V_{CC}/3$

10. .... multivibrator is a square wave oscillator.
(i) Monostable (ii) Astable
(iii) Bistable
(iv) None of the above

11. An astable multivibrator has ....
(i) one stable state (ii) two stable states
(iii) no stable state (iv) none of the above

12. If d.c. supply of 10 V is fed to a differentiating circuit, then output will be ....
(i) 20 V (ii) 10 V
(iii) 0 V
(iv) none of the above

13. If the input to a differentiating circuit is a saw-tooth wave, then output will be .... wave.
(i) square (ii) triangular
(iii) sine (iv) rectangular

14. A bistable multivibrator has ....
(i) two stable states
(ii) one stable state
(iii) no stable state
(iv) none of the above

15. If a square wave is fed to a differentiating circuit, the output will be ....
(i) sine wave
(ii) sharp narrow pulses
(iii) rectangular wave
(iv) triangular wave

16. An integrating circuit is a simple *RC* series circuit with output taken across ....
(i) both *R* and *C* (ii) *R*
(iii) *C*
(iv) none of the above

17. For an integrating circuit to be effective, the *RC* product should be .... the time period of the input wave.
(i) 5 times greater than
(ii) 5 times smaller than
(iii) equal to
(iv) atleast 10 times greater than

18. A differentiating circuit is a simple *RC* circuit with output taken across ....
(i) *R* (ii) *C*
(iii) both *R* and *C*
(iv) none of the above

19. A monostable multivibrator has ....
(i) no stable state
(ii) one stable state
(iii) two stable states
(iv) none of the above

20. The multivibrator which generates square wave of its own is the .... multivibrator.
(i) monostable (ii) bistable
(iii) astable
(iv) none of the above

21. For a differentiating circuit to be effective, the *RC* product should be .... the time period of the input wave.
(i) equal to
(ii) 5 times greater than
(iii) 5 times smaller than
(iv) atleast 10 times smaller than

22. When a rectangular voltage waveform is applied to a capacitor, then the current waveform is ....

(*i*) rectangular (*ii*) sinusoidal
(*iii*) sawtooth (*iv*) square

23. The positive clipper is that which removes the .... half-cycles of the input voltage.
(*i*) negative
(*ii*) positive
(*iii*) both positive and negative
(*iv*) none of the above

24. A clamping circuit adds .... component to the signal.
(*i*) d.c.
(*ii*) a.c.
(*iii*) both d.c. and a.c.
(*iv*) none of the above

25. One would find a clamping circuit in ....
(*i*) receiving antenna
(*ii*) radio transmitter
(*iii*) radio receiver
(*iv*) television receiver

26. When transistor is used as an amplifier, it is operated in the .... region.
(*i*) off (*ii*) saturation
(*iii*) active
(*iv*) none of the above

27. When the transistor (*CE* arrangement) is in the cut-off region, the collector current is ....
(*i*) I_{CBO} (*ii*) I_{CEO}
(*iii*) $(\beta + 1)\ I_{CEO}$ (*iv*) $I_{C\ (sat)}$

28. A negative clipper removes the .... half-cycles of the input voltage.
(*i*) negative
(*ii*) positive
(*iii*) both positive and negative
(*iv*) none of the above

29. If the input to an integrating circuit is a succession of alternating positive and negative pulses of very short duration, the output will be .... wave.
(*i*) rectangular (*ii*) triangular
(*iii*) sine (*iv*) square

30. In a multivibrator, we have .... feedback.
(*i*) negative
(*ii*) 100 % positive
(*iii*) both positive and negative
(*iv*) none of the above

Answers to Multiple-Choice Questions

1. (*ii*)	2. (*iii*)	3. (*i*)	4. (*i*)	5. (*iii*)
6. (*ii*)	7. (*iv*)	8. (*iii*)	9. (*i*)	10. (*ii*)
11. (*iii*)	12. (*iii*)	13. (*iv*)	14. (*i*)	15. (*ii*)
16. (*iii*)	17. (*iv*)	18. (*i*)	19. (*ii*)	20. (*iii*)
21. (*iv*)	22. (*i*)	23. (*ii*)	24. (*i*)	25. (*iv*)
26. (*iii*)	27. (*ii*)	28. (*i*)	29. (*iv*)	30. (*ii*)

Chapter Review Topics

1. What is a switching circuit?
2. Discuss the advantages of an electronic switch over a mechanical or electro-mechanical switch.
3. Explain the terms collector leakage current and saturation collector current.
4. Explain the switching action of a transistor with the help of output characteristics.
5. What is a multivibrator? Explain the principle on which it works.
6. With a neat sketch, explain the working of (*i*) astable multivibrator (*ii*) monostable multivibrator (*iii*) bistable multivibrator.
7. What is the basic difference among the three types of multivibrators?

8. Show that the output from a differentiating circuit is derivative of the input. What assumptions are made in the derivation?
9. Sketch the output waveforms from a differentiating circuit when input is (*i*) a square wave (*ii*) a saw-tooth wave.
10. Show that the output from an integrating circuit is the integral of the input.
11. What is a clipper? Describe (*i*) positive clipper (*ii*) biased clipper and (*iii*) combination clipper.
12. What do you understand by a clamping circuit? With neat diagrams explain the action of a (*i*) positive clamper (*ii*) negative clamper.

Problems

1. The negative shunt clipper shown in Fig. 16.74 has peak input voltage of +15 V. What is the output peak voltage? **[12.54 V]**

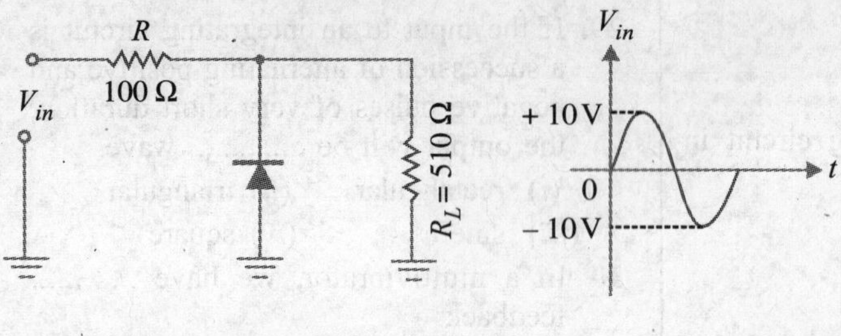

Fig. 16.74

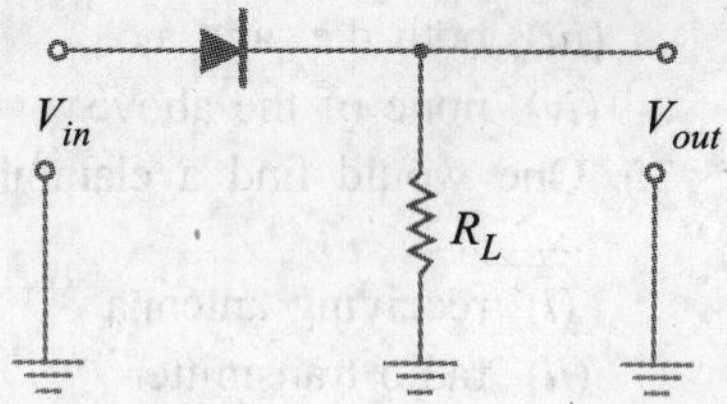

Fig. 16.75

2. In the negative series clipper shown in Fig. 16.75, what is the peak output voltage? **[9.3 V]**
3. In the circuit shown in Fig. 16.76, what are the minimum and peak values of the clamped output? **[0 V; –20 V]**

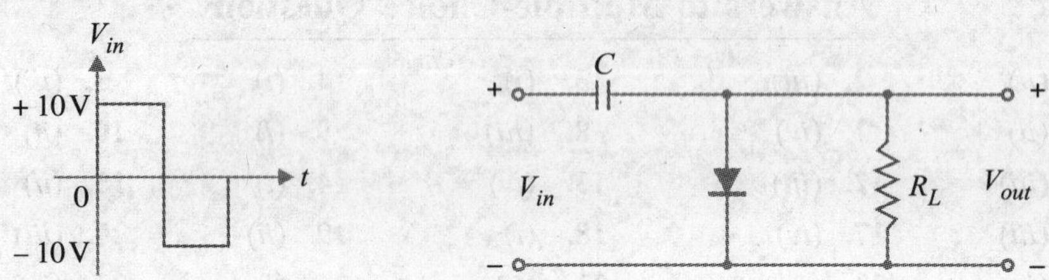

Fig. 16.76

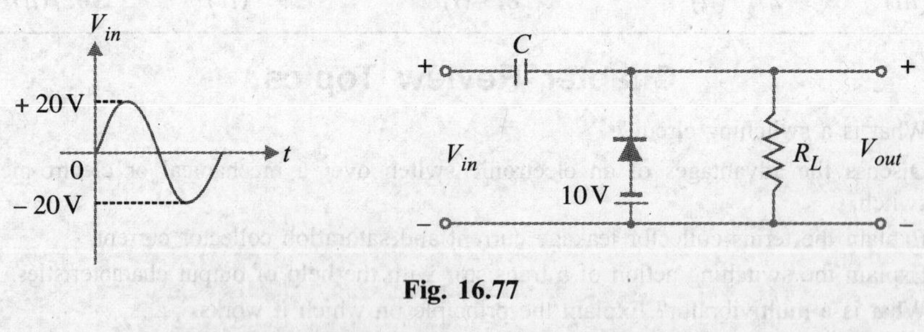

Fig. 16.77

4. Sketch the wave shape of clamped output in Fig. 16.77.
5. Determine the output waveform for the clipper network shown in Fig. 16.78.

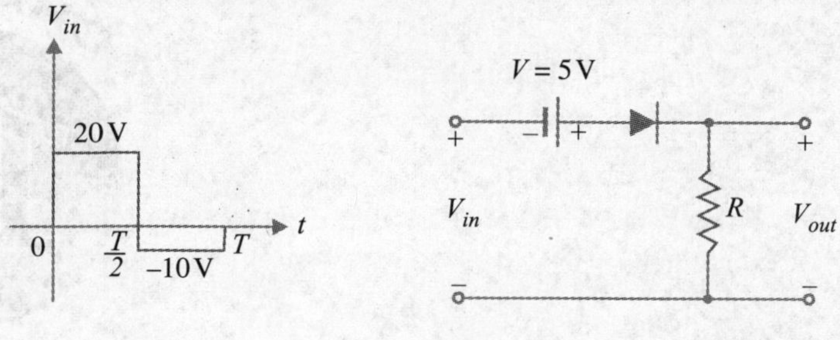

Fig. 16.78

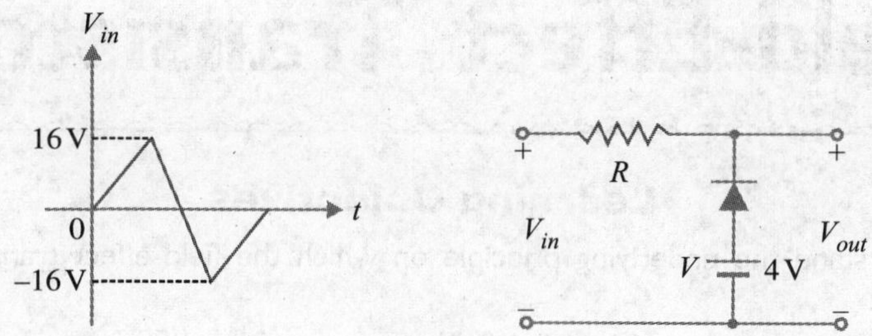

Fig. 16.79

6. Determine the output waveform for the clipper network shown in Fig. 16.79.

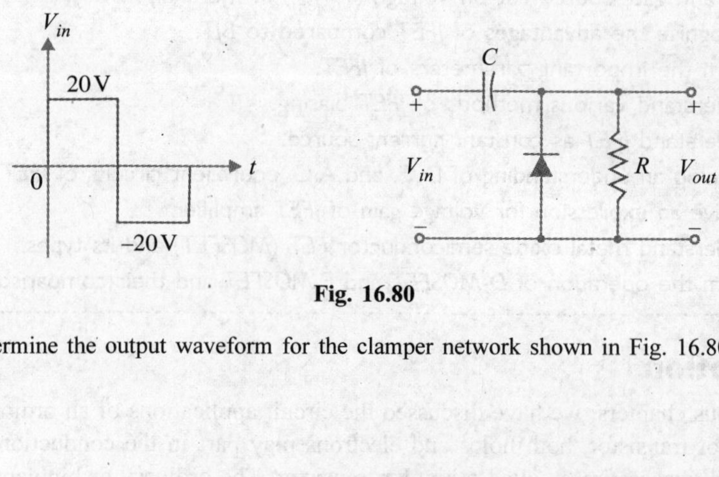

Fig. 16.80

7. Determine the output waveform for the clamper network shown in Fig. 16.80.

CHAPTER 17

Field-Effect Transistors

Learning Objectives

- Understand the underlying principle on which the field-effect transistor (*FET*) is based.
- Learn the operation of junction field-effect transistor (*JFET*).
- Explore the difference between *JFET* and *BJT*.
- Develop an understanding of shorted-gate drain current (I_{DSS}), pinch off voltage (V_P) and gate-source cut-off voltage ($V_{GS(off)}$) in the analysis of *JFET*.
- Recognise the advantages of *JFET* compared to *BJT*.
- Learn the important parameters of *JFET*.
- Understand various methods of *JFET* biasing.
- Understand *JFET* as constant current source.
- Develop an understanding of D.C. and A.C. equivalent circuits of *JFET* amplifier.
- Derive an expression for voltage gain of *JFET* amplifier.
- Understand metal oxide semiconductor *FET* (*MOSFET*) and its types.
- Learn the operation of *D-MOSFET* and *E-MOSFET* and their comparison.

Introduction

In the previous chapters, we have discussed the circuit applications of an ordinary transistor. In this type of transistor, both holes and electrons play part in the conduction process. For this reason, it is sometimes called a bipolar transistor. The ordinary or bipolar transistor has two principal disadvantages. First, it has a low input impedance because of forward biased emitter-base junction. Secondly, it has considerable noise level. Although low input impedance problem may be improved by careful design and use of more than one transistor, yet it is difficult to achieve input impedance more than a few megaohms. The field-effect transistor (*FET*) has, by virtue of its construction and biasing, large input impedance which may be more than 100 megaohms. The *FET* is generally much less noisy than the ordinary or bipolar transistor. The rapidly expanding *FET* market has led many semiconductor marketing managers to believe that this device will soon become the most important electronic device, primarily because of its integrated-circuit applications. In this chapter, we shall focus our attention on the construction, working and circuit applications of field-effect transistors.

17.1 Field-Effect Transistor (FET)

***A field-effect transistor (FET)** is a three-terminal, voltage-controlled semiconductor device in which current conduction is due to the flow of only one of the two kinds of charge carriers viz. electrons or holes.*

A field-effect transistor (*FET*), like the bipolar junction transistor (*BJT*), is a three-terminal semiconductor device. The three terminals of a *FET* are named as *drain (D), source (S)* and *gate (G)* corresponding to *collector (C), emitter (E)* and *base (B)* of a *BJT*. A *FET* can perform the duties of amplification and switching operations far better than the *BJT*. Although both *FET* and *BJT* are three-terminal devices, they have quite different construction, operating principle and characteristics. Thus current conduction in a *FET* is due to the flow of only one of the two types of charge carriers : electrons or holes. For this reason, a *FET* is also called *unipolar transistor* (the prefix *uni* means single). On the other hand, current conduction in a *BJT* is due to the flow of both kinds of charges (electrons and holes) and hence the name bipolar junction transistor. Another important difference between *BJT* and *FET* is that *BJT* is a *current-controlled device* whereas *FET* is a *voltage-controlled device*. This is illustrated in Fig. 17.1. In a *BJT*, the base current (I_B) controls the output current I_C (collector current) while in a *FET*, the input gate-source voltage (V_{GS}) controls the output current I_D (drain current). In each case, the output current is being controlled by a parameter of the input circuit–base current (I_B) in *BJT* and gate-source voltage (V_{GS}) in a *FET*.

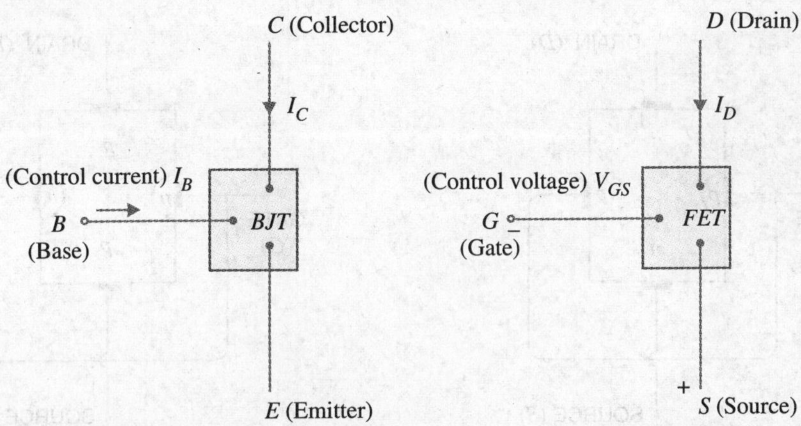

Fig. 17.1

The name *field-effect* is derived for *FET* from the fact that the current flow in a *FET* is controlled by the externally applied input voltage V_{GS} (Recall that voltage sets up electric field).

17.2 Types of Field-Effect Transistors

A bipolar junction transistor (*BJT*) is a current-controlled device *i.e.*, output characteristics of the device are controlled by base current and not by base voltage. However, in a field-effect transistor (*FET*), the output characteristics are controlled by input voltage (*i.e.*, electric field) and not by input current. This is probably the biggest difference between *BJT* and *FET*. There are two basic types of field-effect transistors :

(*i*) Junction field-effect transistor (*JFET*)

(*ii*) Metal oxide semiconductor field-effect transistor (*MOSFET*)

To begin with, we shall study about *JFET* and then improved form of *JFET*, namely ; *MOSFET*.

17.3 Junction Field-Effect Transistor (JFET)

***A junction field-effect transistor** is a three-terminal semiconductor device in which current conduction is by one type of charge carriers i.e., electrons or holes.*

The *JFET* was developed about the same time as the transistor but it came into general use only in the late 1960s. In a *JFET*, the current conduction is either by electrons or holes and is controlled by means of an electric field between the gate electrode and the conducting channel of the device. The *JFET* has high input impedance and low noise level.

Constructional details. A *JFET* consists of a *p*-type or *n*-type silicon bar containing two *pn* junctions at the sides as shown in Fig. 17.2. The bar forms the conducting channel for the charge carriers. If the bar is of *n*-type, it is called an *n-channel JFET* as shown in Fig. 17.2 (*i*) and if the bar is of *p*-type, it is called a *p-channel JFET* as shown in Fig. 17.2 (*ii*). The two *pn* junctions forming diodes are connected *internally and a common terminal called *gate* is taken out. Other terminals are *source* and *drain* taken out from the bar as shown. Thus a *JFET* has essentially three terminals *viz*., *gate* (*G*), *source* (*S*) and *drain* (*D*).

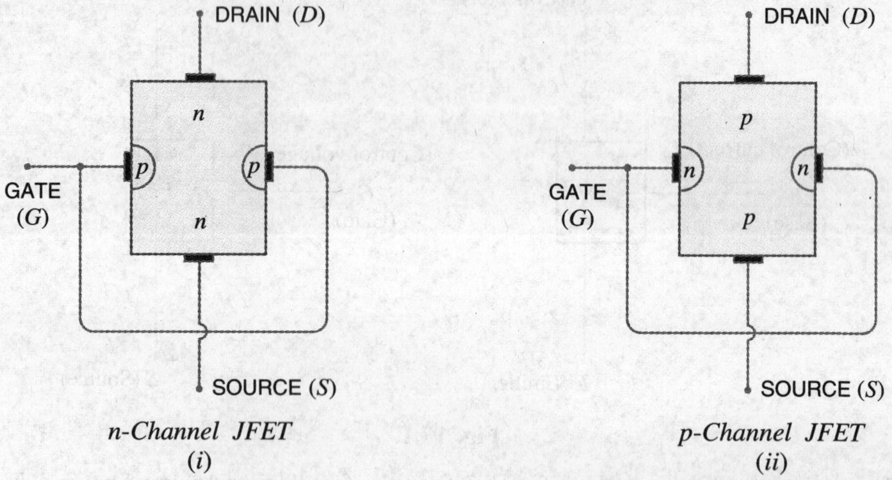

Fig. 17.2

JFET polarities. Fig. 17.3 (*i*) shows an *n*-channel *JFET* polarities whereas Fig. 17.3 (*ii*) shows the *p*-channel *JFET* polarities. Note that in each case, the voltage between the gate and source is such that the gate is reverse biased. This is the normal way of *JFET* connection. The drain and source terminals are interchangeable *i.e.*, either end can be used as source and the other end as drain.

* It would seem from Fig. 17.2 that there are three doped material regions. However, this is not the case. The gate material *surrounds* the channel in the same manner as a belt surrounding your waist.

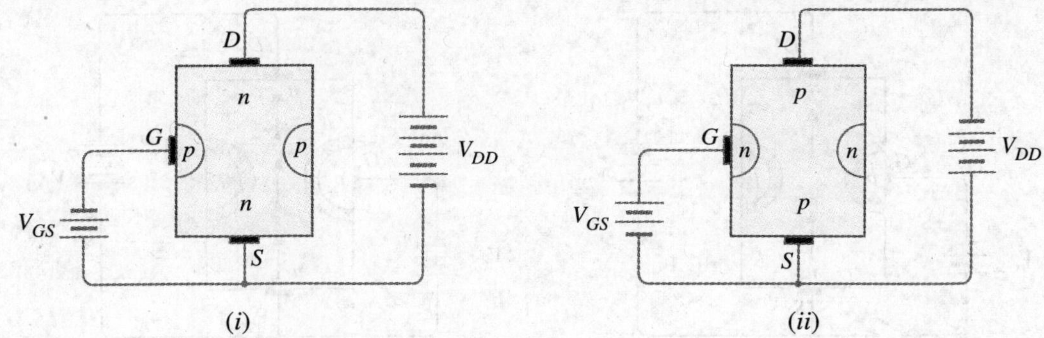

Fig. 17.3

The following points may be noted :

(i) *The input circuit (i.e. gate to source) of a JFET is always reverse biased. This means that the device has high input impedance.*

(ii) *The drain is so biased w.r.t. source that drain current I_D flows from the source to drain.*

(iii) *In all JFETs, source current I_S is equal to the drain current i.e. $I_S = I_D$.*

17.4 Principle and Working of JFET

Fig. 17.4 shows the circuit of an *n*-channel *JFET* with normal polarities. Note that the gate is reverse biased.

Principle. The two *pn* junctions at the sides form two depletion layers. The current conduction by charge carriers (*i.e.* free electrons in this case) is through the channel between the two depletion layers and out of the drain. The width and hence *resistance of this channel can be controlled by changing the input voltage V_{GS}. The greater the reverse voltage V_{GS}, the wider will be the depletion layers and narrower will be the conducting channel. The narrower channel means greater resistance and hence source to drain current decreases. Reverse will happen should V_{GS} decrease. *Thus JFET operates on the principle that width and hence resistance of the conducting channel can be varied by changing the reverse voltage V_{GS}.* In other words, the magnitude of drain current (I_D) can be changed by altering V_{GS}.

Working. The working of *JFET* will be discussed considering an *n*-channel *JFET*.

(i) When voltage V_{DS} (= V_{DD} in this case) is applied between drain and source terminals (drain being positive *w.r.t.* source) and voltage on the gate is zero (*i.e.* V_{GS} = 0V) as shown in Fig. 17.4 (*i*), the two *pn* junctions at the sides of the *n*-type bar establish depletion layers. The electrons will flow from source to drain through a channel between the depletion layers (The source is so named because it is regarded as the origin of charge carriers *i.e.* electrons in this case). Thus, *conventional current* flows from drain to source and is limited by the resistance of the *n*-type bar. The size of the depletion layers determines the width of the channel and hence current conduction through the bar. If the channel is narrow, its resistance increases and current through the bar is decreased and *vice-versa*.

* The resistance of the channel depends upon its area of *X*-section. The greater the *X*-sectional area of this channel, the lower will be its resistance and the greater will be the current flow through it.

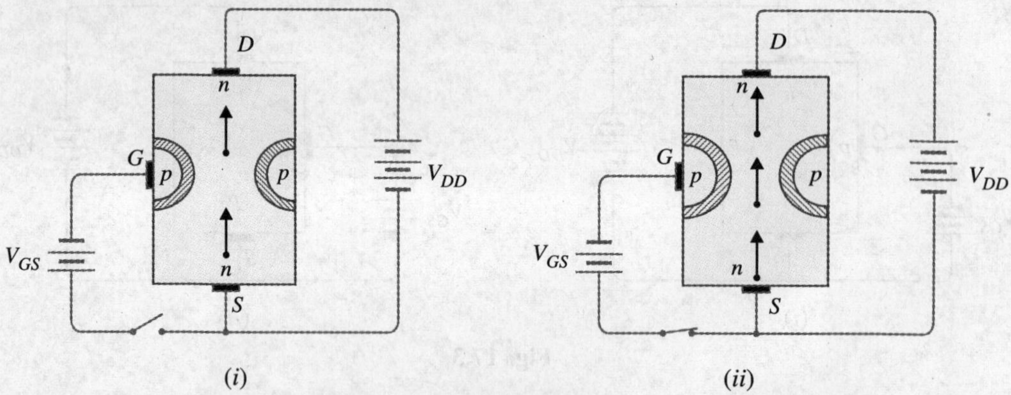

Fig. 17.4

(*ii*) When switch is closed as in Fig. 17.4 (*ii*), a reverse voltage V_{GS} is applied between gate and source. *This is the normal way of operating a JFET*. Under the reverse voltage V_{GS}, the width of the depletion layers is increased. This reduces the width of conducting channel, thereby increasing the resistance of *n*-type bar. Consequently, the current from source to drain is decreased. On the other hand, if the reverse voltage on the gate is decreased, the width of the depletion layers also decreases. This increases the width of the conducting channel and hence source to drain current.

It is clear from the above discussion that current from source to drain can be controlled by the application of potential (*i.e.* electric field) on the gate. For this reason, the device is called *field-effect transistor*. It may be noted that a *p*-channel *JFET* operates in the same manner as an *n*-channel *JFET* except that channel current carriers will be the holes instead of electrons and the polarities of V_{GS} and V_{DS} are reversed.

Note. If the reverse voltage V_{GS} on the gate is continuously increased, a state is reached when the two depletion layers touch each other and the channel is cut-off. Under such conditions, the channel becomes a non-conductor.

Water Analogy for JFET Operation. Analogies are seldom perfect and at times can be misleading. Nevertheless, we do make use of analogies to provide simple explanation for the device/ process. Fig. 17.5 shows the water analogy for the operation of *JFET*. The source of water pressure can be regarded as the applied voltage between the drain the source that will establish a flow of water (electrons for *n*-type bar) from the spigot (source). The "gate" through an applied signal (potential) controls the flow of water (charge) to the "drain". The drain and source terminals are at the opposite ends of the channel.

Fig. 17.5

17.5 Schematic Symbol of JFET

Fig. 17.6 shows the schematic symbol of *JFET*. The vertical line in the symbol may be thought as channel and source (*S*) and drain (*D*) connected to this line. As in the case of every semiconductor device symbol, the arrowhead points from *p*-type to *n*-type material (This is the direction of conventional current flow if the *pn* junction were forward biased). For an *n*-channel *JFET*, the arrowhead points from the *p*-type gate to the *n*-type channel as shown in Fig. 17.6 (*i*). For the *p*-channel *JFET*, the only difference in the symbol is the direction of the arrowhead as shown in Fig. 17.6 (*ii*) *i.e.* now the arrowhead points from channel to gate.

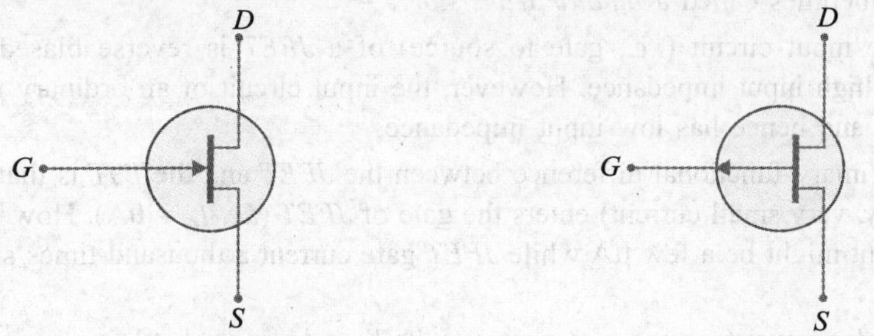

Fig. 17.6

17.6 JFET Biasing

The gate-source junction of a JFET is always reverse biased. Thus for an *n*-channel *JFET*, the gate should be negative *w.r.t.* source and for *p*-channel *JFET*, the gate should be positive *w.r.t.* source. This is an important point to keep in mind. Even when the gate and source are shorted, the drain current (I_D) will cause the gate-source junction to be reverse biased. *The gate-source junction of a JFET should never be allowed to become forward biased.* If gate-source junction is allowed to become forward biased, current will start to flow in the gate circuit. Under most circumstances, this current will destroy the *JFET*. There are several methods of *JFET* biasing *i.e.* keeping gate-source junction reverse biased. We shall discuss these methods in Art. 17.21.

17.7 Importance of JFET

A *JFET* acts like a voltage-controlled device *i.e.* input voltage (V_{GS}) controls the output current. This is different from ordinary transistor (or bipolar transistor) where input current controls the output current. Thus *JFET* is a semiconductor device acting *like a vacuum tube. The need for *JFET* arose because as modern electronic equipment became increasingly transistorised, it became apparent that there were many functions in which bipolar transistors were unable to replace vacuum tubes. Owing to their extremely high input impedance, *JFET* devices are more like vacuum tubes than are the bipolar transistors and hence are able to take over many vacuum-tube functions. Thus, because of *JFET*, electronic equipment is closer today to being completely solid state.

The *JFET* devices have not only taken over the functions of vacuum tubes but they now also threaten to depose the bipolar transistors as the most widely used semiconductor devices. As an amplifier, the *JFET* has higher input impedance than that of a conventional transistor, generates less noise and has greater resistance to nuclear radiations.

17.8 Difference Between JFET and Bipolar Transistor

The *JFET* differs from an ordinary or bipolar transistor in the following ways :

(*i*) In a *JFET*, there is only one type of carrier, holes in *p*-type channel and electrons

---

* The gate, source and drain of a *JFET* correspond to grid, cathode and anode of a vacuum tube.

in *n*-type channel. For this reason, it is also called a *unipolar transistor*. However, in an ordinary transistor, both holes and electrons play part in conduction. Therefore, an ordinary transistor is sometimes called a *bipolar transistor*.

(*ii*) As the input circuit (*i.e.*, gate to source) of a *JFET* is reverse biased, therefore, the device has high input impedance. However, the input circuit of an ordinary transistor is forward biased and hence has low input impedance.

(*iii*) The primary functional difference between the *JFET* and the *BJT* is that no current (actually, a very, very small current) enters the gate of *JFET* (*i.e.* I_G = 0A). However, typical *BJT* base current might be a few μA while *JFET* gate current a thousand times smaller [See Fig. 17.7].

(*iv*) A bipolar transistor uses a current into its base to control a large current between collector and emitter whereas a *JFET* uses voltage on the 'gate' (= base) terminal to control the current between drain (= collector) and source (= emitter). Thus a bipolar transistor gain is characterised by current gain whereas the *JFET* gain is characterised as a transconductance *i.e.*, the ratio of change in output current (drain current) to the input (gate) voltage.

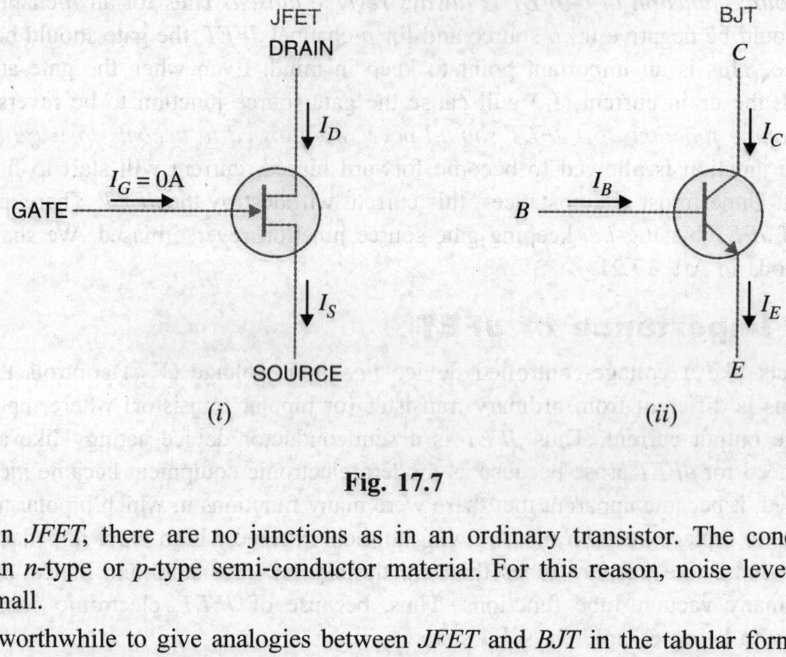

Fig. 17.7

(*v*) In *JFET*, there are no junctions as in an ordinary transistor. The conduction is through an *n*-type or *p*-type semi-conductor material. For this reason, noise level in *JFET* is very small.

It is worthwhile to give analogies between *JFET* and *BJT* in the tabular form.

S.No.	JFET	BJT
1.	Source *S*	Emitter *E*
2.	Drain *D*	Collector *C*
3.	Gate *G*	Base *B*
4.	Drain supply V_{DD}	Collector supply V_{CC}
5.	Gate supply V_{GG}	Base supply V_{BB}
6.	Drain current I_D	Collector current I_C

17.9 Output or Drain Characteristics of JFET

The curve between drain current (I_D) and drain-source voltage (V_{DS}) of a JFET at constant gate-source voltage (V_{GS}) is known as **output or drain characteristics** *of JFET.* Fig. 17.8 shows the circuit for determining the output characteristics of an *n*-channel *JFET*. Keeping V_{GS} fixed at some value, say –1V*, the drian-source voltage is changed in steps. Corresponding to each value of V_{DS}, the drain current I_D is noted. A plot of these values gives the output characteristic of *JFET* at V_{GS} = –1V. Repeating similar procedure, output characteristics at other gate-source voltages can be drawn. Fig. 17.9 shows a family of output characteristics.

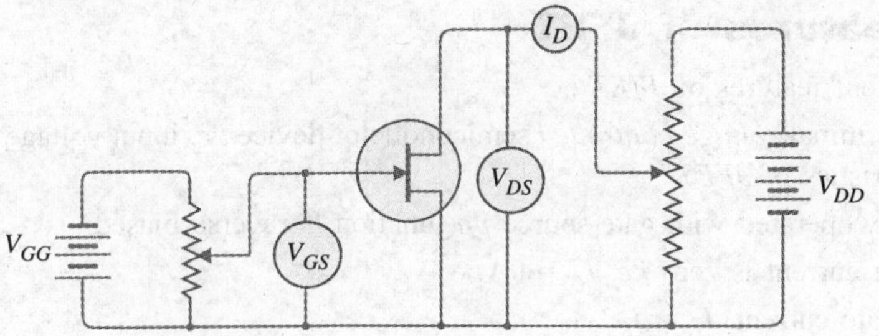

Fig. 17.8

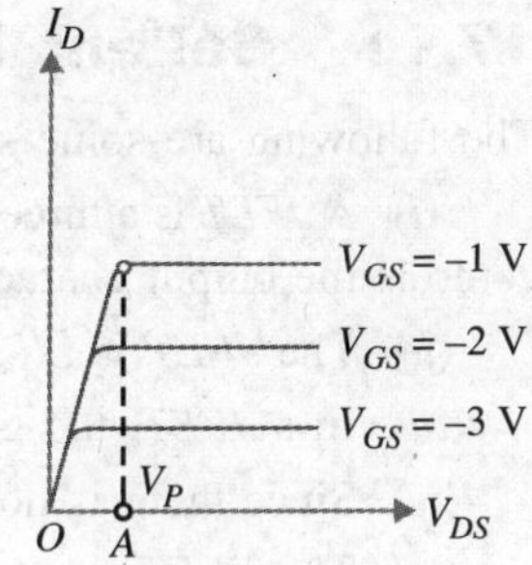

Fig. 17.9

The following points may be noted from the characteristics :

(*i*) At first, the drain current I_D rises rapidly with drain-source voltage V_{DS} but then becomes constant. The drain-source voltage above which drain current becomes constant is known as *pinch off voltage*. Thus in Fig. 17.9, *OA* is the *pinch off voltage* V_P.

(*ii*) After pinch off voltage, the channel width becomes so narrow that depletion layers almost touch each other. The drain current passes through the small passage between these layers. Therefore, increase in drain current is very small with V_{DS} above pinch off voltage. Consequently, drain current remains constant.

(*iii*) The characteristics resemble that of a pentode valve.

17.10 Transconductance Curve of JFET

The curve between drain current (I_D) and gate-source voltage (V_{GS}) at constant drain-source voltage (V_{DS}) is called **transconductance curve** *of JFET.*

Another way to graphically display the characteristics of an *n*-channel *JFET* is with transconductance curve shown in Fig 17.10. The circuit shown in Fig. 17.8 can be used to draw the transconductance curve. Keeping V_{DS} fixed at some value, say 8V, the gate-source voltage (V_{GS}) is changed in steps. Corresponding to each value of V_{GS}, the drain current I_D is noted. A plot of these values gives the transconductance curve of *JFET* at V_{DS} = 8V.

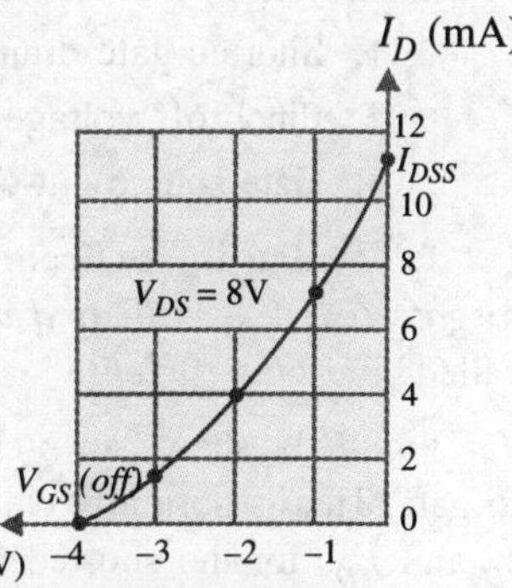

Fig. 17.10

* Note that V_{GS} is negative *i.e.* gate-source junction is reverse biased. This is the normal way of operating *JFET*.

It is clear that *transconductance curve of a *JFET* is not constant. Transconductance of a *JFET* is an important parameter for calculating the gain of a *JFET*. The following points may be noted from the transconductance curve :

(*i*) The drain current I_D is *maximum* when V_{GS} = 0. This maximum value is called I_{DSS} (*shorted-gate drain current*).

(*ii*) The drain current I_D decreases when $-V_{GS}$ is increased.

(*iii*) The drain current I_D becomes *zero* when $-V_{GS}$ is increased to cut-off value. This value of $-V_{GS}$ at which I_D is zero is called *gate-source cut-off voltage* $V_{GS(off)}$.

17.11 Salient Features of JFET

The following are some salient features of *JFET* :

(*i*) A *JFET* is a three-terminal *voltage-controlled* semiconductor device *i.e.* input voltage controls the output characteristics of *JFET*.

(*ii*) The *JFET* is *always* operated with gate-source *pn* junction **reverse biased.

(*iii*) In a *JFET*, the gate current is zero *i.e.* I_G = 0A.

(*iv*) Since there is no gate current, $I_D = I_S$.

(*v*) The *JFET* must be operated between V_{GS} = 0V and $V_{GS\ (off)}$. For this range of gate-to-source voltages, I_D will vary from a maximum of I_{DSS} to a minimum of almost zero [See Fig. 17.10].

(*vi*) Because the two gates are at the same potential, both depletion layers widen or narrow down by an equal amount.

(*vii*) The *JFET* is not subjected to thermal runaway when the temperature of the device increases.

(*viii*) The drain current I_D is controlled by changing the channel width.

(*ix*) Since *JFET* has no gate current, there is no β rating of the device. We can find drain current I_D by using the eq. mentioned in Art. 17.13.

17.12 Important Terms about JFET

In the analysis of a *JFET* circuit, the following important terms are often used :

1. Shorted-gate drain current (I_{DSS})
2. Pinch off voltage (V_P)
3. Gate-source cut-off voltage [$V_{GS\ (off)}$]

1. Shorted-gate drain current (I_{DSS}). *It is the drain current with source short-circuited to gate (i.e. V_{GS} = 0) and drain voltage (V_{DS}) equal to pinch off voltage.* It is sometimes called *zero-bias current*.

Fig 17.11 shows an *n*-channel *JFET* circuit with V_{GS} = 0 *i.e.*, source shorted-circuited to gate. This is normally called shorted-gate condition. Fig. 17.12 shows the graph between I_D and V_{DS} for the shorted gate condition. The drain current rises rapidly at first and then levels off at pinch off voltage V_P. The drain current has now reached the maximum value

* The control that the gate voltage has over the drain current is measured by transconductance of *JFET*.

** Forward biasing gate-source *pn* junction may destroy the device.

I_{DSS}. When V_{DS} is increased beyond V_P, the depletion layers expand at the top of the channel. The channel now acts as a current limiter and *holds drain current constant at I_{DSS}.

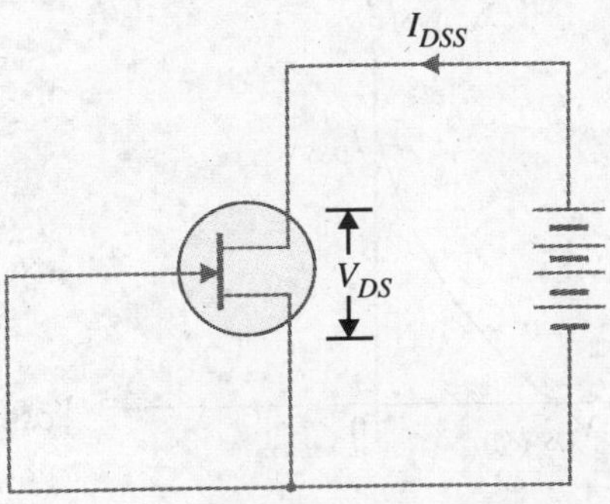

Fig. 17.11

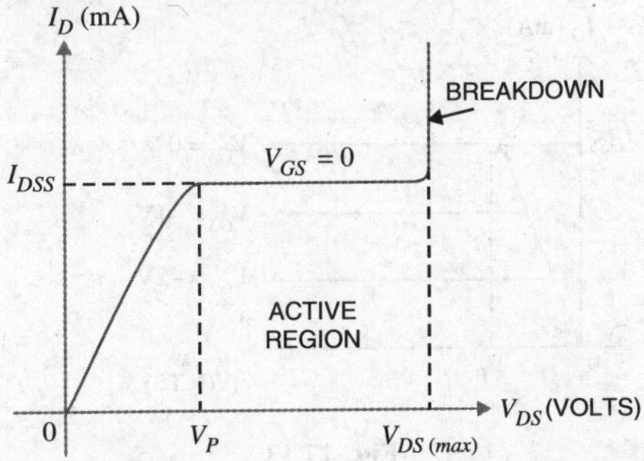

Fig. 17.12

The following points may be noted carefully :

(*i*) Since I_{DSS} is measured under shorted gate conditions, it is the maximum drain current that you can get with normal operation of *JFET*.

(*ii*) There is a maximum drain voltage [$V_{DS\ (max)}$] that can be applied to a *JFET*. If the drain voltage exceeds $V_{DS\ (max)}$, *JFET* would breakdown as shown in Fig. 17.12.

(*iii*) The region between V_P and $V_{DS\ (max)}$ (breakdown voltage) is called *constant-current region* or *active region*. As long as V_{DS} is kept within this range, I_D will remain constant for a constant value of V_{GS}. In other words, in the active region, *JFET* behaves as a constant–current device. For proper working of *JFET*, it must be operated in the active region.

2. Pinch off Voltage (V_P). *It is the minimum drain-source voltage at which the drain current essentially becomes constant.*

Figure 17.13 shows the drain curves of an *n*-channel *JFET*. Note that pinch off voltage is V_P. The highest curve is for V_{GS} = 0V, the shorted-gate condition. For values of V_{DS} greater than V_P, the drain current is almost constant. It is because when V_{DS} equals V_P, the channel is effectively closed and does not allow further increase in drain current. It may be noted that for proper function of *JFET*, it is always operated for $V_{DS} > V_P$. However, V_{DS} should not exceed $V_{DS\ (max)}$ otherwise *JFET* may breakdown.

3. Gate-source cut-off voltage [$V_{GS\ (off)}$]. *It is the gate-source voltage where the channel is completely cut-off and the drain current becomes zero.*

The idea of gate-source cut-off voltage can be easily understood if we refer to the transfer characteristic of an *n*-channel *JFET* shown in Fig. 17.14. As the reverse gate-source voltage is increased, the cross-sectional area of the channel decreases. This in turn decreases the drain current. At some reverse gate-source voltage, the depletion layers extend completely

* When drain voltage equals V_P, the channel becomes narrow and the depletion layers almost touch each other. The channel now acts as a current limiter and holds drain current at a constant value of I_{DSS}.

across the channel. In this condition, the channel is cut-off and the drain current reduces to zero. The gate voltage at which the channel is cut-off (*i.e.* channel becomes non-conducting) is called gate-source cut-off voltage $V_{GS\,(off)}$.

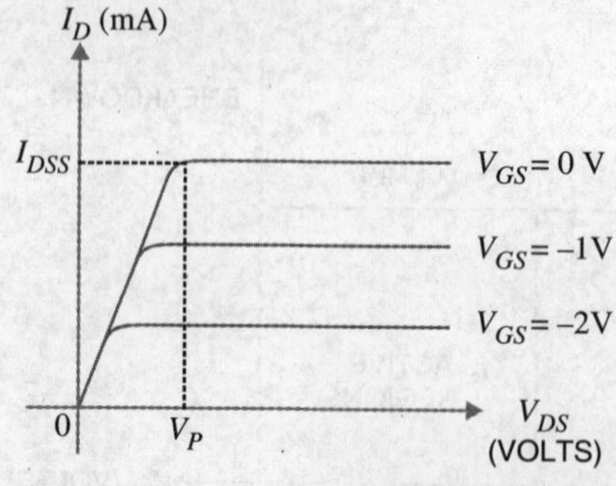

Fig. 17.13

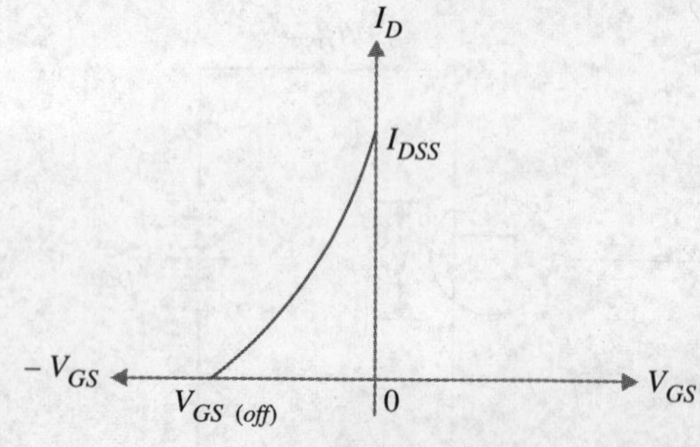

Fig. 17.14

Notes. (*i*) It is interesting to note that $V_{GS\,(off)}$ will always have the same magnitude value as V_P. For example if V_P = 6 V, then $V_{GS\,(off)}$ = –6 V. Since these two values are always equal and opposite, only one is listed on the specification sheet for a given *JFET*.

$$|V_P| = |V_{GS(off)}|$$

For an *n*-channel *JFET*, V_{GS} is negative and V_P is positive. However, for *p*-channel *JFET*, V_{GS} is positive and V_P is negative.

(*ii*) There is a distinct difference between V_P and $V_{GS\,(off)}$. Note that V_P is the value of V_{DS} that causes the *JFET* to become a constant-current device. It is measured at V_{GS} = 0V and will have a constant drain current = I_{DSS}. However, $V_{GS\,(off)}$ is the value of V_{GS} that causes I_D to drop to nearly zero.

17.13 Expression for Drain Current (I_D)

In a *JFET*, the output current I_D (drain current) is controlled by the input voltage V_{GS} (gate-source voltage). Unfortunately, no linear relation exists between I_D and V_{GS}. It can be proved mathematically that relation between I_D and V_{GS} is given by ;

$$I_D = I_{DSS}\left[1 - \frac{V_{GS}}{V_{GS\,(off)}}\right]^2 \quad ...(i)$$

where I_D = Drain current at given V_{GS}

I_{DSS} = Shorted–gate drain current

V_{GS} = Gate–source voltage

$V_{GS\,(off)}$ = Gate–source cut-off voltage

Eq. (*i*) is known as *Shockley's equation* and shows that curve between I_D and V_{GS} is nearly parabolic as shown in Fig. 17.15. We know that this is the *transconductance curve* or *transfer characteristic* of the *JFET*. Note that the bottom end of the curve is at a point on the V_{GS} axis equal to $V_{GS(off)}$ and the top end of the curve is at a point on the I_D axis equal to I_{DSS}. This curve shows that the operating limits of a *JFET* are :

$$I_D = 0 \quad \text{when } V_{GS} = V_{GS(off)}$$

and $$I_D = I_{DSS} \text{ when } V_{GS} = 0$$

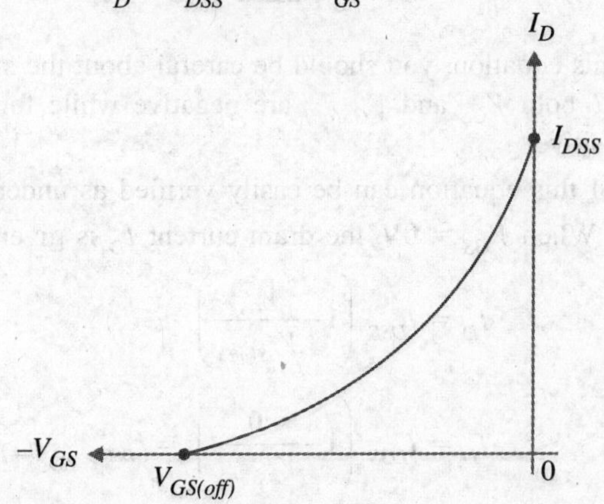

Fig. 17.15

Fig. 17.16 shows the transconductance curve and drain characteristic of a *JFET*. Note that the magnitude of gate-source cut-off voltage [$V_{GS(off)}$] on the transfer characteristic is equal to the magnitude of pinch off voltage V_P on the drain characteristic *i.e.*

$$|V_{GS(off)}| = |V_P|$$

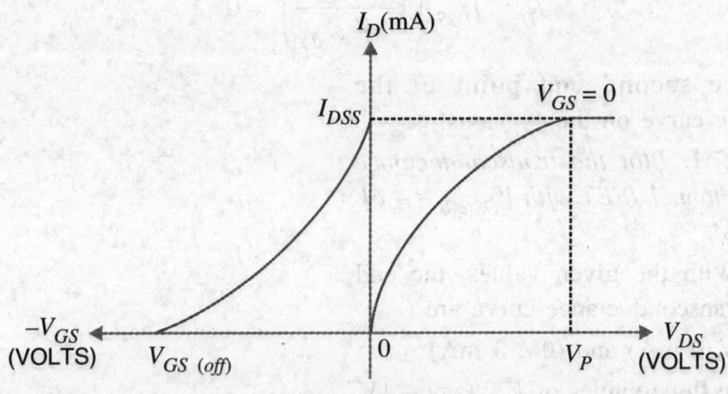

Fig. 17.16

17.14 Explanation of Shockley's Equation

The Shockley's equation for *JFET* gives the relation between I_D and V_{GS} and is given by ;

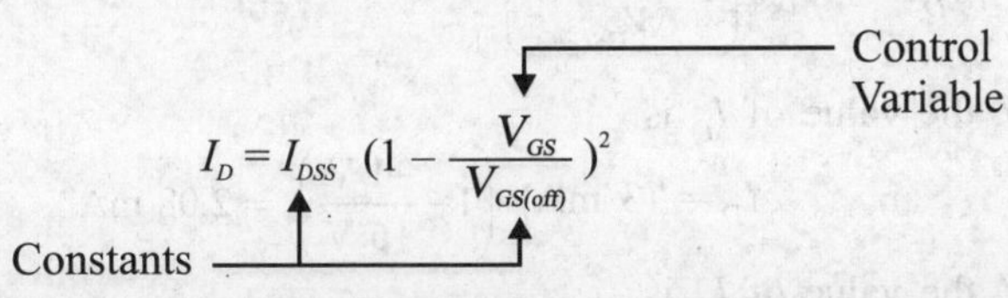

(*i*) The terms I_{DSS} and $V_{GS(off)}$ are constants for a given *JFET* and are given in the manual for the *JFET*. Therefore, the value of I_D depends upon V_{GS}. As V_{GS} changes, I_D also changes.

(*ii*) Because of the squared term in the equation, the curve between I_D and V_{GS} (*i.e.* transconductance curve) is nearly parabolic. For this reason, a *JFET* is offen referred to as a *square-law device.*

(*iii*) While using this equation, you should be careful about the signs of V_{GS} and $V_{GS(off)}$. For an *n*-channel *JFET* both V_{GS} and $V_{GS(off)}$ are negative while for *p*-channel *JFET*, both these quantities are positive.

(*iv*) The validity of this equation can be easily verified as under :

When V_{GS} = 0V, the drain current I_D is given by ;

$$I_D = I_{DSS}\left(1-\frac{V_{GS}}{V_{GS(off)}}\right)^2$$

or
$$I_D = I_{DSS}\left(1-\frac{0}{V_{GS(off)}}\right)^2 = I_{DSS}$$

This is one end point of the transconductance curve on the I_D axis.

When $V_{GS} = V_{GS(off)}$, the drain current I_D is given by ;

$$I_D = I_{DSS}\left(1-\frac{V_{GS}}{V_{GS(off)}}\right)^2$$

or
$$I_D = I_{DSS}\left(1-\frac{V_{GS(off)}}{V_{GS(off)}}\right)^2 = 0$$

This is the second end point of the transconductance curve on the V_{GS} axis.

Example 17.1. *Plot the transconductance curve for an n-channel JFET with* $V_{GS(off)} = -6V$ *and* $I_{DSS} = 3$ *mA.*

Solution. With the given values, the end points for the transconductance curve are :

(– 6V, 0 mA) and (0V, 3 mA)

We now use three values of V_{GS} (say – 1V, – 3V and – 5V) and calculate the corresponding value of drain current I_D for each. We know that for a *JFET*, the drain current I_D is given by ;

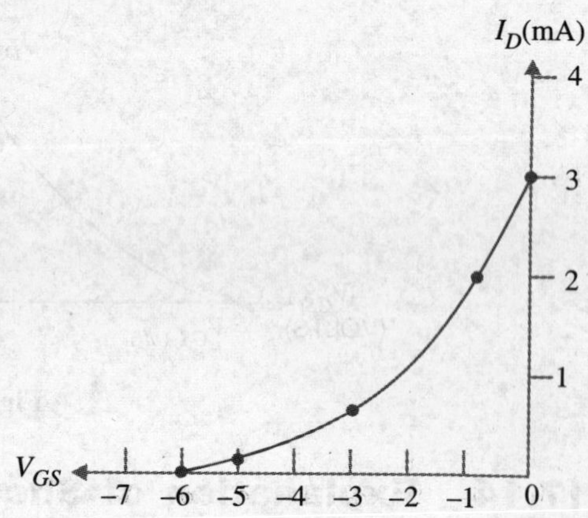

Fig. 17.17

$$I_D = I_{DSS}\left(I-\frac{V_{GS}}{V_{GS(off)}}\right)^2$$

At V_{GS} = –1V, the value of I_D is

$$I_D = (3\text{ mA})\left(1-\frac{-1\text{ V}}{-6\text{ V}}\right)^2 = 2.08\text{ mA}$$

At V_{GS} = –3V, the value of I_D is

$$I_D = (3\text{ mA})\left(1-\frac{-3\text{ V}}{-6\text{ V}}\right)^2 = 0.75\text{ mA}$$

At V_{GS} = –5V, the value of I_D is

$$I_D = (3\text{ mA})\left(1-\frac{-5\text{ V}}{-6\text{ V}}\right)^2 = 0.083\text{ mA}$$

The three points are now plotted on $I_D - V_{GS}$ graph and connected to form the transconductance curve as shown in Fig. 17.17.

Example 17.2. *Fig. 17.18 shows the transfer characteristic curve of an n-channel JFET. Write the equation for drain current.*

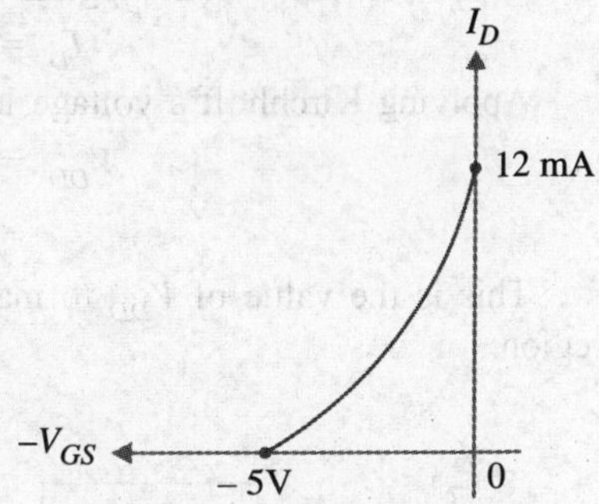

Fig. 17.18

Solution. Referring to the transfer characteristic curve in Fig. 17.18, we have,

$$I_{DSS} = 12\text{ mA}$$

$$V_{GS(off)} = -5\text{V}$$

$$\therefore \quad I_D = I_{DSS}\left[1-\frac{V_{GS}}{V_{GS(off)}}\right]^2$$

$$\text{or} \quad I_D = 12\left[1+\frac{V_{GS}}{5}\right]^2 \text{ mA}$$

Since it is an *n*-channel *JFET*, V_{GS} is negative in this equation.

Example 17.3. *An n-channel JFET has the following parameters : I_{DSS} = 32mA ; $V_{GS(off)}$ = – 8V ; V_{GS} = – 4.5 V. Find the value of drain current.*

Solution.

$$I_D = I_{DSS}\left[1-\frac{V_{GS}}{V_{GS(off)}}\right]^2$$

$$= 32\left[1-\frac{(-4.5)}{-8}\right]^2 \text{ mA}$$

$$= \mathbf{6.12\text{ mA}}$$

Example 17.4. *An n-channel JFET has a drain current of 5 mA. If I_{DSS} = 10 mA and $V_{GS\ (off)}$ = – 6 V, find the values of (i) V_{GS} and (ii) V_P.*

Solution.

$$I_D = I_{DSS}\left[1-\frac{V_{GS}}{V_{GS\ (off)}}\right]^2$$

$$\text{or} \quad 5 = 10\left[1+\frac{V_{GS}}{6}\right]^2$$

$$\text{or} \quad 1+\frac{V_{GS}}{6} = \sqrt{5/10} = 0.707$$

(*i*) ∴ $V_{GS} = \mathbf{-\ 1.76\ V}$

(*ii*) and $V_P = -\ V_{GS\ (off)} = \mathbf{6\ V}$

Example 17.5. *For an n-channel JFET in Fig. 17.19, $V_{GS\ (off)}$ = – 4V and I_{DSS} = 12 mA. Determine the minimum value of V_{DD} required to put the device in the constant-current region of operation.*

Solution. Since $V_{GS\,(off)} = -4V$, $V_P = 4V$. The minimum value of V_{DS} for the *n*-channel *JFET* to be in constant-current region is

$$V_{DS} = V_P = 4V$$

In the constant-current region with $V_{GS} = 0V$,

$$I_D = I_{DSS} = 12 \text{ mA}$$

Applying Kirchhoff's voltage law around the drain circuit, we have,

$$V_{DD} = V_{DS} + V_{R_D} = V_{DS} + I_D R_D$$

$$= 4V + (12 \text{ mA})(560\Omega) = 4V + 6.72V = \mathbf{10.72V}$$

This is the value of V_{DD} to make $V_{DS} = V_P$ and put the device in the constant-current region.

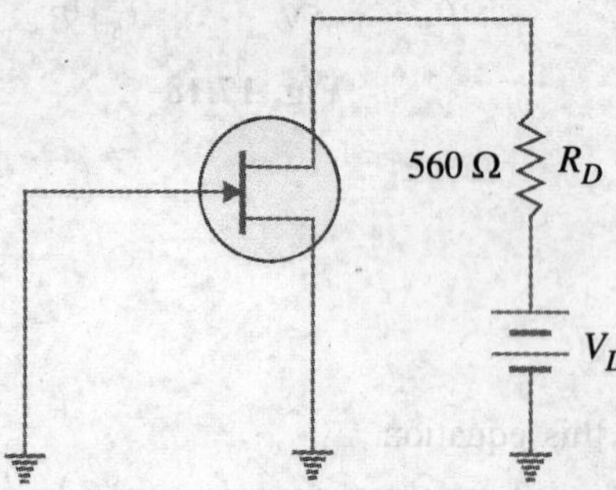

Fig. 17.19

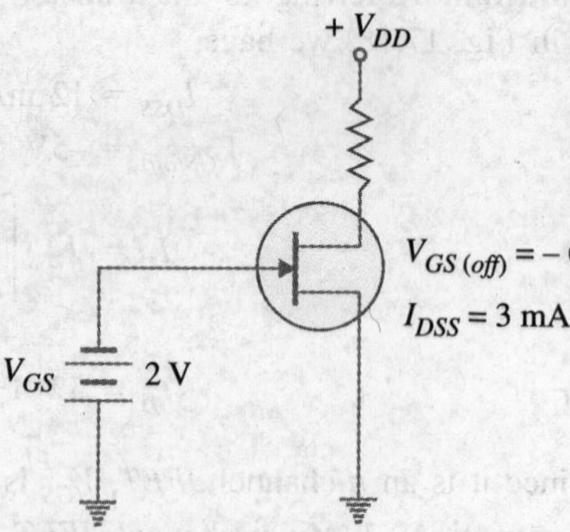

Fig. 17.20

Example 17.6. *Determine the value of drain current for the circuit shown in Fig. 17.20.*

Solution. It is clear from Fig. 17.20 that $V_{GS} = -2V$. The drain current for the circuit is given by ;

$$I_D = I_{DSS}\left(1 - \frac{V_{GS}}{V_{GS\,(off)}}\right)^2$$

$$= 3 \text{ mA}\left(1 - \frac{-2V}{-6V}\right)^2$$

$$= (3 \text{ mA})(0.444) = \mathbf{1.33 \text{ mA}}$$

Example 17.7. *A particular p-channel JFET has a $V_{GS\,(off)} = +4V$. What is I_D when $V_{GS} = +6V$?*

Solution. The *p*-channel *JFET* requires a positive gate-to-source voltage to pass drain current I_D. The more the positive voltage, the less the drain current. When $V_{GS} = 4V$, $I_D = 0$ and *JFET* is cut-off. Any further increase in V_{GS} keeps the *JFET* cut-off. Therefore, at $V_{GS} = +6V$, $I_D = \mathbf{0A}$.

17.15 JFET Connections

There are three leads in a *JFET viz.*, *source, gate* and *drain* terminals. However, when *JFET* is to be connected in a circuit, we require four terminals ; two for the input and two for the output. This difficulty is overcome by making one terminal of the *JFET* common to both input and output terminals. Accordingly, a *JFET* can be connected in a circuit in the following three ways :

(*i*) Common source (*CS*) connection

(*ii*) Common gate (*CG*) connection

(*iii*) Common drain (*CD*) connection

These connections may be viewed as equivalent to *CE* (common emitter), *CB* (common base) and *CC* (common collector) in transistor connections. *The common source connection is the most widely used arrangement.* It is because this connection provides high input impedance, good voltage gain and a moderate output impedance. However, the circuit produces a phase reversal *i.e.*, output signal is 180° out of phase with the input signal. Fig. 17.21 shows a common source *n*-channel *JFET* amplifier. Note that source terminal is common to both input and output.

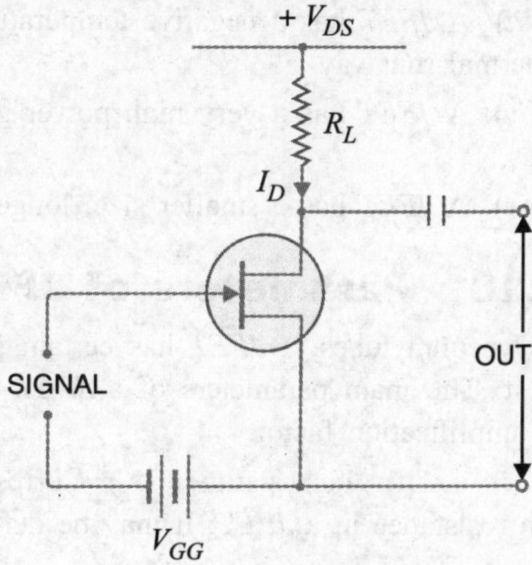

Fig. 17.21

Note. A common source *JFET* amplifier is the *JFET* equivalent of common emitter amplifier. Both amplifiers have a 180° phase shift from input to output. Although the two amplifiers serve the same basic purpose, the means by which they operate are quite different.

17.16 JFET as an Amplifier

Fig. 17.21 shows common source *n*-channel *JFET* amplifier circuit. The weak signal is applied between gate and source and amplified output is obtained in the drain-source circuit. For the proper operation of *JFET*, the gate must be negative *w.r.t.* source *i.e.*, input circuit should always be reverse biased. This is achieved either by inserting a battery V_{GG} in the gate circuit or by a circuit known as biasing circuit. In the present case, we are providing biasing by the battery V_{GG}.

A small change in the reverse bias on the gate produces a large change in drain current. This fact makes *JFET* capable of raising the strength of a weak signal. During the positive half of signal, the reverse bias on the gate decreases. This increases the channel width and hence the drain current. During the negative half-cycle of the signal, the reverse voltage on the gate increases. Consequently, the drain current decreases. The result is that a small change in voltage at the gate produces a large change in drain current. These large variations in drain current produce large output across the load R_L. In this way, *JFET* acts as an amplifier.

17.17 Advantages of JFET

A *JFET* is a voltage-controlled, constant-current device (similar to a vacuum pentode) in which variations in input voltage control the output current. It combines the many advantages of both bipolar transistor and vacuum pentode. Some of the advantages of a *JFET* are :

(*i*) It has a very high input impedance (of the order of 100 MΩ). This permits high degree of isolation between the input and output circuits.

(*ii*) The operation of a *JFET* depends upon the bulk material current carriers that do not cross junctions. Therefore, the inherent noise of tubes (due to high-temperature operation) and those of transistors (due to junction transitions) are not present in a *JFET*.

(*iii*) A *JFET* has a negative temperature co-efficient of resistance. This avoids the risk of thermal runaway.

(*iv*) A *JFET* has a very high power gain. This eliminates the necessity of using driver stages.

(*v*) A *JFET* has a smaller size, longer life and high efficiency.

17.18 Parameters of JFET

Like vacuum tubes, a *JFET* has certain parameters which determine its performance in a circuit. The main parameters of a *JFET* are (*i*) a.c. drain resistance (*ii*) transconductance (*iii*) amplification factor.

(*i*) **a.c. drain resistance (r_d).** Corresponding to the a.c. plate resistance, we have a.c. drain resistance in a *JFET*. It may be defined as follows :

It is the ratio of change in drain-source voltage (ΔV_{DS}) to the change in drain current (ΔI_D) at constant gate-source voltage i.e.

$$\text{a.c. drain resistance, } r_d = \frac{\Delta V_{DS}}{\Delta I_D} \text{ at constant } V_{GS}$$

For instance, if a change in drain voltage of 2 V produces a change in drain current of 0.02 mA, then,

$$\text{a.c. drain resistance, } r_d = \frac{2\text{ V}}{0.02\text{ mA}} = 100\text{ k}\Omega$$

Referring to the output characteristics of an *n*-channel *JFET* in Fig. 17.9, it is clear that above the pinch-off voltage, the change in I_D is small for a change in V_{DS} because the curve is almost flat. Therefore, drain resistance of a *JFET* has a large value, ranging from 10 kΩ to 1 MΩ.

(*ii*) **Transconductance (g_{fs}).** The control that the gate voltage has over the drain current is measured by transconductance g_{fs} and is similar to the transconductance g_m of the tube. It may be defined as follows :

It is the ratio of change in drain current (ΔI_D) to the change in gate-source voltage (ΔV_{GS}) at constant drain-source voltage i.e.

$$\text{Transconductance, } g_{fs} = \frac{\Delta I_D}{\Delta V_{GS}} \text{ at constant } V_{DS}$$

The transconductance of a *JFET* is usually expressed either in mA/volt or micromho. As an example, if a change in gate voltage of 0.1 V causes a change in drain current of 0.3 mA, then,

$$\text{Transconductance, } g_{fs} = \frac{0.3\text{ mA}}{0.1\text{ V}} = 3\text{ mA/V} = 3 \times 10^{-3}\text{ A/V or mho or S (siemens)}$$

$$= 3 \times 10^{-3} \times 10^{6}\ \mu\text{ mho} = 3000\ \mu\text{ mho (or } \mu\text{S)}$$

(*iii*) **Amplification factor (μ).** *It is the ratio of change in drain-source voltage (ΔV_{DS}) to the change in gate-source voltage (ΔV_{GS}) at constant drain current i.e.*

$$\text{Amplification factor, } \mu = \frac{\Delta V_{DS}}{\Delta V_{GS}} \text{ at constant } I_D$$

Amplification factor of a *JFET* indicates how much more control the gate voltage has over drain current than has the drain voltage. For instance, if the amplification factor of a *JFET* is 50, it means that gate voltage is 50 times as effective as the drain voltage in controlling the drain current.

17.19 Relation Among JFET Parameters

The relationship among *JFET* parameters can be established as under :

We know $$\mu = \frac{\Delta V_{DS}}{\Delta V_{GS}}$$

Multiplying the numerator and denominator on R.H.S. by ΔI_D, we get,

$$\mu = \frac{\Delta V_{DS}}{\Delta V_{GS}} \times \frac{\Delta I_D}{\Delta I_D} = \frac{\Delta V_{DS}}{\Delta I_D} \times \frac{\Delta I_D}{\Delta V_{GS}}$$

$$\therefore \quad \mu = r_d \times g_{fs}$$

i.e. amplification factor = a.c. drain resistance × transconductance

Example 17.8. *When a reverse gate voltage of 15 V is applied to a JFET, the gate current is* 10^{-3} *μA. Find the resistance between gate and source.*

Solution. $V_{GS} = 15\text{ V}\ ;\ I_G = 10^{-3}\ \mu\text{A} = 10^{-9}\text{ A}$

$$\therefore \quad \text{Gate to source resistance} = \frac{V_{GS}}{I_G} = \frac{15\text{ V}}{10^{-9}\text{ A}} = 15 \times 10^9\ \Omega = \mathbf{15{,}000\ M\Omega}$$

This example shows the major difference between a *JFET* and a bipolar transistor. Whereas the input impedance of a *JFET* is several hundred MΩ, the input impedance of a bipolar transistor is only hundreds or thousands of ohms. The large input impedance of a *JFET* permits high degree of isolation between the input and output.

Example 17.9. *When* V_{GS} *of an n-channel JFET changes from –3.1 V to –3 V, the drain current changes from 1 mA to 1.3 mA. What is the value of transconductance ?*

Solution. $\Delta V_{GS} = 3.1 - 3 = 0.1\text{ V}$ ... magnitude

$\Delta I_D = 1.3 - 1 = 0.3\text{ mA}$

$$\therefore \quad \text{Transconductance, } g_{fs} = \frac{\Delta I_D}{\Delta V_{GS}} = \frac{0.3\text{ mA}}{0.1\text{ V}} = 3\text{ mA/V} = \mathbf{3000\ \mu\ mho}$$

Example 17.10. *The following readings were obtained experimentally from a JFET :*

V_{GS}	*0 V*	*0 V*	*– 0.2 V*
V_{DS}	*7 V*	*15 V*	*15 V*
I_D	*10 mA*	*10.25 mA*	*9.65 mA*

Determine (i) a.c. drain resistance (ii) transconductance and (iii) amplification factor.

Solution. (*i*) With V_{GS} constant at 0V, the increase in V_{DS} from 7 V to 15 V increases the drain current from 10 mA to 10.25 mA *i.e.*

Change in drain-source voltage, $\Delta V_{DS} = 15 - 7 = 8\text{ V}$

Change in drain current, $\Delta I_D = 10.25 - 10 = 0.25\text{ mA}$

$$\therefore \quad \text{a.c. drain resistance, } r_d = \frac{\Delta V_{DS}}{\Delta I_D} = \frac{8\text{ V}}{0.25\text{ mA}} = \mathbf{32\ k\Omega}$$

(*ii*) With V_{DS} constant at 15 V, drain current changes from 10.25 mA to 9.65 mA as V_{GS} is changed from 0 V to – 0.2 V.

$$\Delta V_{GS} = 0.2 - 0 = 0.2 \text{ V}$$

$$\Delta I_D = 10.25 - 9.65 = 0.6 \text{ mA}$$

$$\therefore \quad \text{Transconductance, } g_{fs} = \frac{\Delta I_D}{\Delta V_{GS}} = \frac{0.6 \text{ mA}}{0.2 \text{ V}} = 3 \text{ mA/V} = \mathbf{3000\ \mu\ mho}$$

(*iii*) Amplification factor, $\mu = r_d \times g_{fs} = (32 \times 10^3) \times (3000 \times 10^{-6}) = \mathbf{96}$

17.20 Variation of Transconductance (g_m or g_{fs}) of JFET

We have seen that transconductance g_m of a *JFET* is the ratio of a change in drain current (ΔI_D) to a change in gate-source voltage (ΔV_{GS}) at constant V_{DS} *i.e.*

$$g_m = \frac{\Delta I_D}{\Delta V_{GS}}$$

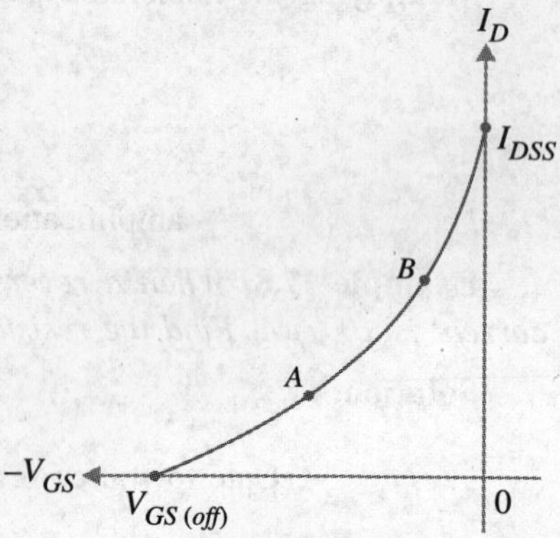

Fig. 17.22

The transconductance g_m of a *JFET* is an important parameter because it is a major factor in determining the voltage gain of *JFET* amplifiers. However, the transfer characteristic curve for a *JFET* is nonlinear so that the value of g_m depends upon the location on the curve. Thus the value of g_m at point *A* in Fig. 17.22 will be different from that at point *B*. Luckily, there is following equation to determine the value of g_m at a specified value of V_{GS} :

$$g_m = g_{mo}\left(1 - \frac{V_{GS}}{V_{GS\,(off)}}\right)$$

where g_m = Value of transconductance at any point on the transfer characteristic curve

g_{mo} = Value of transconductance (maximum) at $V_{GS} = 0$

Normally, the data sheet provides the value of g_{mo}. When the value of g_{mo} is not available, you can approximately calculate g_{mo} using the following relation :

$$g_{mo} = \frac{2I_{DSS}}{|V_{GS\,(off)}|}$$

Example 17.11. *An n-channel JFET has a value of g_{mo} = 4000 μS. Determine the value of g_m at V_{GS} = – 3V. Given that $V_{GS\,(off)}$ = – 8V.*

Solution.

$$g_m = g_{mo}\left(1 - \frac{V_{GS}}{V_{GS\,(off)}}\right)$$

$$= 4000\ \mu\text{S}\left(1 - \frac{-3\text{V}}{-8\text{V}}\right)$$

$$= 4000\ \mu\text{S}\ (0.625) = \mathbf{2500\ \mu S}$$

Example 17.12. *The data sheet of an n-channel JFET gives the following information : I_{DSS} = 3 mA, $V_{GS\,(off)}$ = – 6V and $g_{m\,(max)}$ = 5000 μS. Determine the transconductance for V_{GS} = – 4V and find drain current I_D at this point.*

Solution. At $V_{GS} = 0$, the value of g_m is maximum *i.e.* g_{mo}.

$$\therefore \quad g_{mo} = 5000 \ \mu S$$

Now,
$$g_m = g_{mo}\left(1 - \frac{V_{GS}}{V_{GS\,(off)}}\right)$$

$$= 5000 \ \mu S \left(1 - \frac{-4V}{-6V}\right)$$

$$= 5000 \ \mu S \ (1/3) = \mathbf{1667 \ \mu S}$$

Also,
$$I_D = I_{DSS}\left(1 - \frac{V_{GS}}{V_{GS\,(off)}}\right)^2 = 3 \text{ mA}\left(1 - \frac{-4}{-6}\right)^2$$

$$= \mathbf{333 \ \mu A}$$

17.21 Methods of JFET Biasing

For proper operation of *JFET* amplifier, the gate-source junction should be reverse biased. Therefore, for *n*-channel *JFET*, the gate must be negative *w.r.t.* source. This can be achieved either by inserting a battery in the gate circuit or by a circuit known as biasing circuit. The latter method is preferred because batteries are costly and require frequent replacement.

1. Bias battery. In this method, *JFET* is biased by a bias battery V_{GG}. This battery ensures that gate is always negative *w.r.t.* source during all parts of the signal.

2. Biasing circuit. The biasing circuit uses supply voltage V_{DD} to provide the necessary bias. Two most commonly used methods are (*i*) self-bias (*ii*) potential-divider method. We shall discuss each method in turn.

17.22 JFET Biasing by Bias Battery

Fig. 17.23 shows the biasing of an *n*-channel *JFET* by a bias battery – V_{GG}. This method is also called *gate bias*. The battery voltage – V_{GG} ensures that gate – source junction remains reverse biased.

Since there is no gate current, there will be no voltage drop across R_G.

$$\therefore \quad V_{GS} = V_{GG}$$

We can find the value of drain current I_D from the following relation :

$$I_D = I_{DSS}\left(1 - \frac{V_{GS}}{V_{GS\,(off)}}\right)^2$$

The value of V_{DS} is given by ;

$$V_{DS} = V_{DD} - I_D R_D$$

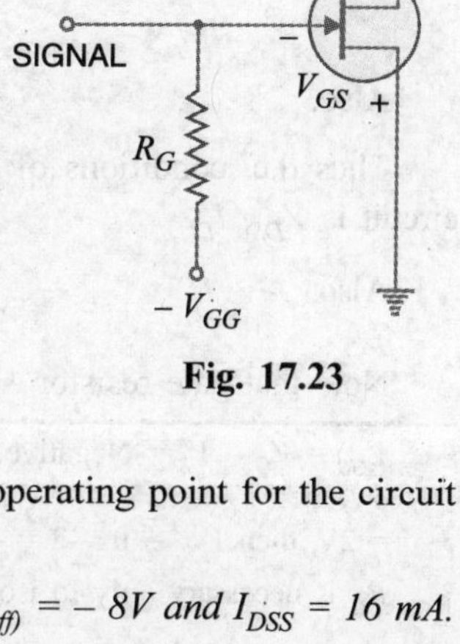

Fig. 17.23

Thus the d.c. values of I_D and V_{DS} stand determined. The operating point for the circuit is V_{DS}, I_D.

Example 17.13. *A JFET in Fig. 17.24 has values of* $V_{GS\,(off)} = -8V$ *and* $I_{DSS} = 16$ *mA. Determine the values of* V_{GS}, I_D *and* V_{DS} *for the circuit.*

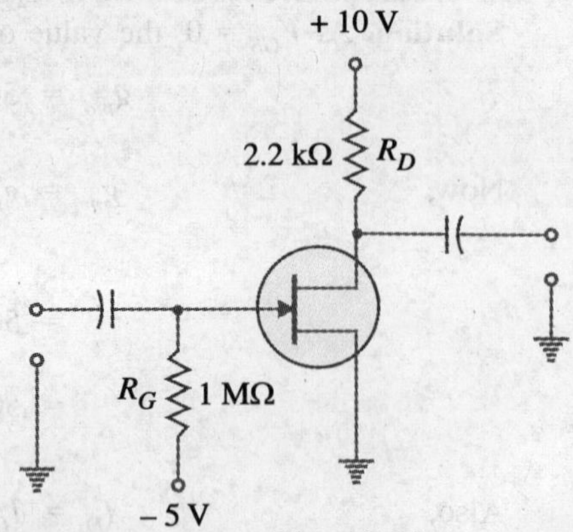

Fig. 17.24

Solution. Since there is no gate current, there will be no voltage drop across R_G.

$\therefore \quad V_{GS} = V_{GG} = -$ **5V**

Now, $\quad I_D = I_{DSS}\left(1 - \frac{V_{GS}}{V_{GS\,(off)}}\right)^2$

$= 16 \text{ mA}\left(1 - \frac{-5}{-8}\right)^2$

$= 16 \text{ mA } (0.1406) =$ **2.25 mA**

Also, $\quad V_{DS} = V_{DD} - I_D\, R_D$

$= 10 \text{ V} - 2.25 \text{ mA} \times 2.2 \text{ k}\Omega =$ **5.05 V**

Note that operating point for the circuit is 5.05V, 2.25 mA.

17.23 Self-Bias for JFET

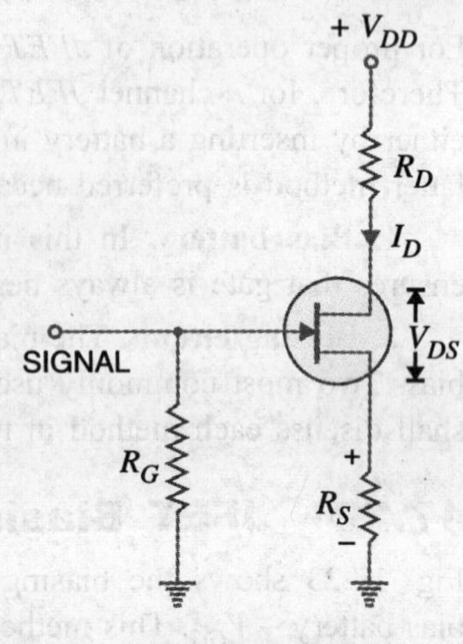

Fig. 17.25

Fig. 17.25 shows the self-bias method for an *n*-channel *JFET*. The resistor R_S is the bias resistor. The d.c. component of drain current flowing through R_S produces the desired bias voltage.

Voltage across R_S, $V_S = I_D R_S$

Since gate current is negligibly small, the gate terminal is at d.c. ground *i.e.*, $V_G = 0$.

$\therefore \quad V_{GS} = V_G - V_S = 0 - I_D R_S$

or $\quad V_{GS} = -\,{}^{*}I_D R_S$

Thus bias voltage V_{GS} keeps gate negative *w.r.t.* source.

Operating point. The operating point (*i.e.*, zero signal I_D and V_{DS}) can be easily determined. Since the parameters of the *JFET* are usually known, zero signal I_D can be calculated from the following relation :

$$I_D = I_{DSS}\left(1 - \frac{V_{GS}}{V_{GS\,(off)}}\right)^2$$

Also, $\quad V_{DS} = V_{DD} - I_D\,(R_D + R_S)$

Thus d.c. conditions of *JFET* amplifier are fully specified *i.e.* operating point for the circuit is V_{DS}, I_D.

Also, $\quad R_S = \frac{|V_{GS}|}{|I_D|}$

Note that gate resistor **R_G does not affect bias because voltage across it is zero.

* $V_{GS} = V_G - V_S$ = Negative. This means that V_G is negative *w.r.t.* V_S. Thus if V_G = 2V and V_S = 4V, then V_{GS} = 2 – 4 = – 2V *i.e.* gate is less positive than the source. Again if V_G = 0V and V_S = 2V, then V_{GS} = 0 – 2 = – 2V. Note that V_G is less positive than V_S.

** R_G is necessary only to isolate an a.c. signal from ground in amplifier applications.

Midpoint Bias. It is often desirable to bias a *JFET* near the midpoint of its transfer characteristic curve where $I_D = I_{DSS}/2$. When signal is applied, the midpoint bias allows a maximum amount of drain current swing between I_{DSS} and 0. It can be proved that when $V_{GS} = V_{GS\,(off)}$ / 3.4, midpoint bias conditions are obtained for I_D.

$$I_D = I_{DSS}\left(1-\frac{V_{GS}}{V_{GS\,(off)}}\right)^2 = I_{DSS}\left(1-\frac{V_{GS\,(off)}/3.4}{V_{GS\,(off)}}\right)^2 = 0.5\,I_{DSS}$$

To set the drain voltage at midpoint ($V_D = V_{DD}/2$), select a value of R_D to produce the desired voltage drop.

Example 17.14. *Find V_{DS} and V_{GS} in Fig. 17.26, given that I_D = 5 mA.*

Solution.

$$V_S = I_D R_S = (5\text{ mA})\,(470\ \Omega) = 2.35\text{ V}$$

and
$$V_D = V_{DD} - I_D R_D$$
$$= 15\text{V} - (5\text{ mA}) \times (1\text{ k}\Omega) = 10\text{V}$$

$\therefore$
$$V_{DS} = V_D - V_S = 10\text{V} - 2.35\text{ V} = \mathbf{7.65V}$$

Since there is no gate current, there will be no voltage drop across R_G and V_G = 0.

Now,
$$V_{GS} = V_G - V_S = 0 - 2.35\text{V} = \mathbf{-\,2.35\text{ V}}$$

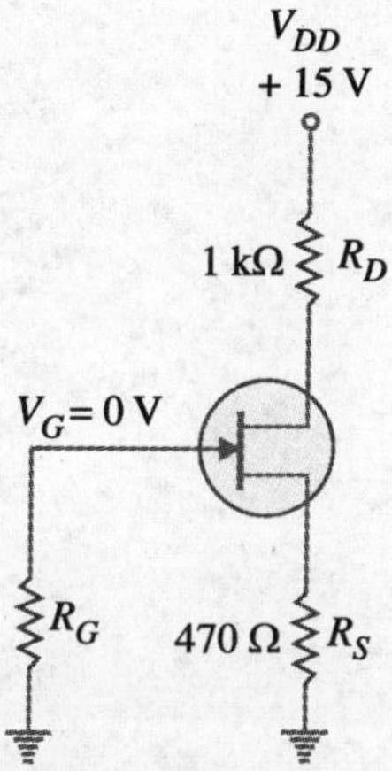

Fig. 17.26

Example 17.15. *The transfer characteristic of an n-channel JFET reveals that when V_{GS} = – 5V, I_D = 6.25 mA. Determine the value of R_S required.*

Solution.

$$R_S = \frac{|V_{GS}|}{|I_D|} = \frac{5\text{V}}{6.25\text{ mA}} = \mathbf{800\ \Omega}$$

Example 17.16. *Determine the value of R_S required to self-bias a p-channel JFET with I_{DSS} = 25 mA, $V_{GS\,(off)}$ = 15 V and V_{GS} = 5V.*

Solution.

$$I_D = I_{DSS}\left(1-\frac{V_{GS}}{V_{GS\,(off)}}\right)^2 = 25\text{ mA}\left(1-\frac{5\text{ V}}{15\text{ V}}\right)^2 = 25\text{mA}\,(1 - 0.333)^2 = 11.1\text{ mA}$$

$\therefore$
$$R_S = \frac{|V_{GS}|}{|I_D|} = \frac{5\text{V}}{11.1\text{ mA}} = \mathbf{450\ \Omega}$$

Example 17.17. *Select resistor values in Fig. 17.27 to set up an approximate midpoint bias. The JFET parameters are : I_{DSS} = 15 mA and $V_{GS\,(off)}$ = – 8V. The voltage V_D should be 6V (one-half of V_{DD}).*

Solution. For midpoint bias, we have,

$$I_D \simeq \frac{I_{DSS}}{2} = \frac{15\text{ mA}}{2} = 7.5\text{ mA}$$

and
$$V_{GS} = \frac{V_{GS\,(off)}}{3.4} = \frac{-8}{3.4} = -\,2.35\text{ V}$$

$\therefore$
$$R_S = \frac{|V_{GS}|}{|I_D|} = \frac{2.35\text{V}}{7.5\text{ mA}} = \mathbf{313\Omega}$$

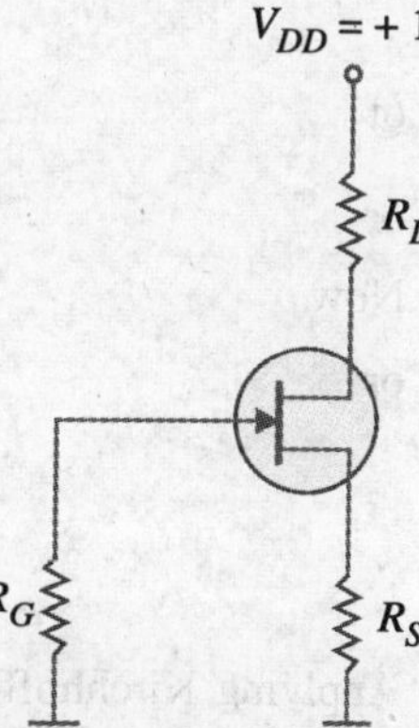

Fig. 17.27

Now, $$V_D = V_{DD} - I_D R_D$$

$\therefore$ $$R_D = \frac{V_{DD} - V_D}{I_D} = \frac{12\text{V} - 6\text{V}}{7.5\text{ mA}} = \mathbf{800\Omega}$$

Example 17.18. *In a self-bias n-channel JFET, the operating point is to be set at I_D = 1.5 mA and V_{DS} =10 V. The JFET parameters are I_{DSS} = 5 mA and $V_{GS\,(off)}$ = – 2 V. Find the values of R_S and R_D. Given that V_{DD} = 20 V.*

Solution. Fig. 17.28 shows the circuit arrangement.

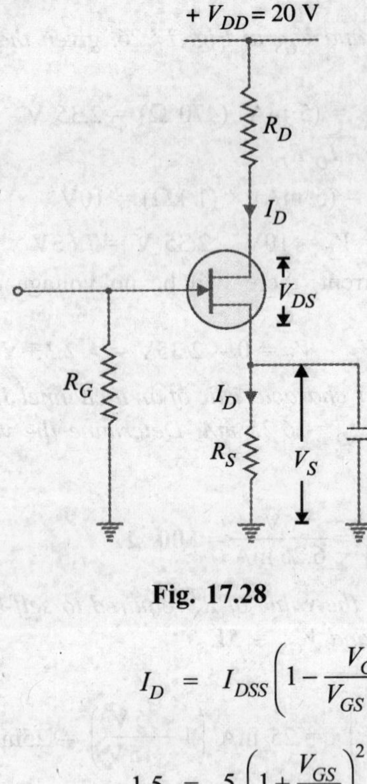

Fig. 17.28

$$I_D = I_{DSS}\left(1 - \frac{V_{GS}}{V_{GS\,(off)}}\right)^2$$

or $$1.5 = 5\left(1 + \frac{V_{GS}}{2}\right)^2$$

or $$1 + \frac{V_{GS}}{2} = \sqrt{1.5/5} = 0.55$$

or $$V_{GS} = -\ 0.9\text{ V}$$

Now, $$V_{GS} = V_G - V_S$$

or $$V_S = V_G - V_{GS}$$

$$= 0 - (-\ 0.9) = 0.9\text{ V}$$

$\therefore$ $$R_S = \frac{V_S}{I_D} = \frac{0.9\text{ V}}{1.5\text{ mA}} = \mathbf{0.6\ k\Omega}$$

Applying Kirchhoff's voltage law to the drain circuit, we have,

$$V_{DD} = I_D R_D + V_{DS} + I_D R_S$$

or $\quad 20 = 1.5 \text{ mA} \times R_D + 10 + 0.9$

$\therefore \quad R_D = \frac{(20 - 10 - 0.9) \text{ V}}{1.5 \text{ mA}} = 6 \text{ k}\Omega$

Example 17.19. *In an n-channel JFET circuit shown in Fig. 17.29, find (i) V_{DS} and (ii) V_{GS}.*

Solution.

(*i*) $\quad V_{DS} = V_{DD} - I_D (R_D + R_S) = 30 - 2.5 \text{ mA} (5 + 0.2) = 30 - 13 = \mathbf{17 \text{ V}}$

(*ii*) $\quad V_{GS} = - I_D R_S = - (2.5 \times 10^{-3}) \times 200 = \mathbf{- 0.5 \text{ V}}$

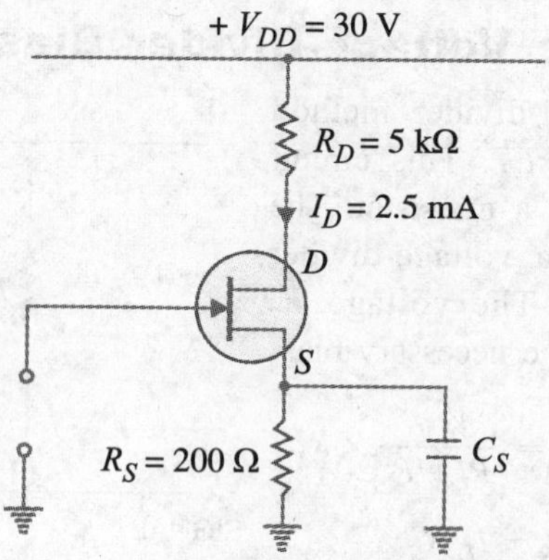

Fig. 17.29

Example 17.20. *Figure 17.30 shows two stages of an n-channel JFET amplifier. The first stage has I_D = 2.15 mA and the second stage has I_D = 9.15 mA. Find the d.c. voltage of drain and source of each stage w.r.t. ground.*

Solution. Voltage drop in 8.2 kΩ = 2.15 mA × 8.2 kΩ = 17.63 V

D.C. potential of drain of first stage *w.r.t.* ground is

$$V_D = V_{DD} - 17.63 = 30 - 17.63 = \mathbf{12.37 \text{ V}}$$

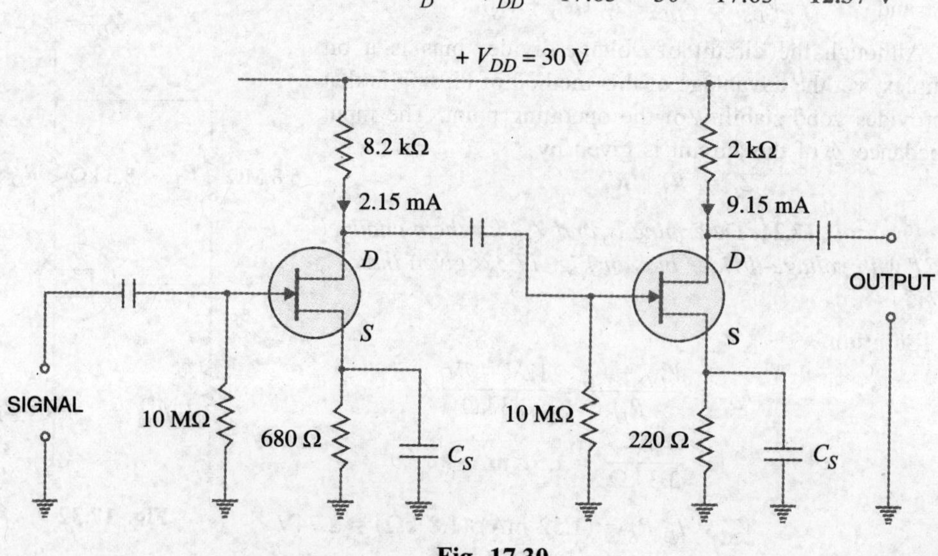

Fig. 17.30

D.C. potential of source of first stage to ground is

$$V_S = I_D R_S = 2.15 \text{ mA} \times 0.68 \text{ k}\Omega = \mathbf{1.46 \text{ V}}$$

$$\text{Voltage drop in 2 k}\Omega = 9.15 \text{ mA} \times 2 \text{ k}\Omega = 18.3 \text{ V}$$

D.C. potential of drain of second stage to ground is

$$V_D = V_{DD} - 18.3 = 30 - 18.3 = \mathbf{11.7 \text{ V}}$$

D.C. potential of source of second stage to ground is

$$V_S = I_D R_S = 9.15 \text{ mA} \times 0.22 \text{ k}\Omega = \mathbf{2.01 \text{ V}}$$

17.24 JFET with Voltage-Divider Bias

Fig. 17.31 shows potential-divider method of biasing an *n*-channel *JFET*. This circuit is identical to that used for a transistor. The resistors R_1 and R_2 form a voltage-divider across drain supply V_{DD}. The voltage V_2 (= V_G) across R_2 provides the necessary bias.

$$V_2 = V_G = \frac{V_{DD}}{R_1 + R_2} \times R_2$$

Now, $$V_2 = V_{GS} + I_D R_S$$

or $$V_{GS} = V_2 - I_D R_S$$

The circuit is so designed that $I_D R_S$ is larger than V_2 so that V_{GS} is negative. This provides correct bias voltage. We can find the operating point as under :

$$I_D = \frac{V_2 - V_{GS}}{R_S}$$

and $$V_{DS} = V_{DD} - I_D (R_D + R_S)$$

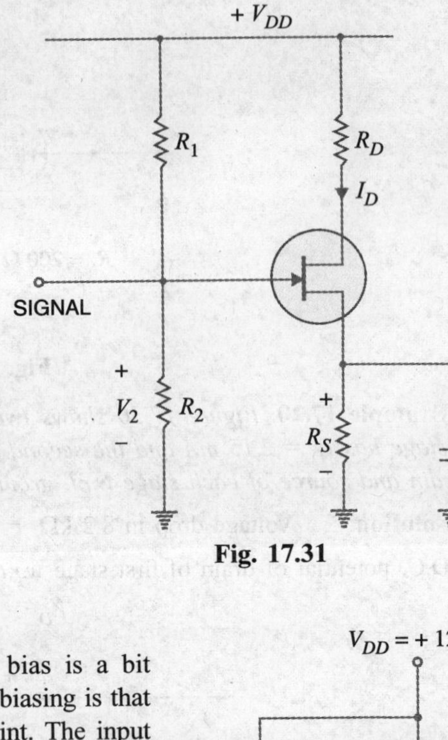

Fig. 17.31

Although the circuit of voltage-divider bias is a bit complex, yet the advantage of this method of biasing is that it provides good stability of the operating point. The input impedance Z_i of this circuit is given by ;

$$Z_i = R_1 \parallel R_2$$

Example 17.21. *Determine I_D and V_{GS} for the n-channel JFET with voltage-divider bias in Fig. 17.32, given that V_D = 7V.*

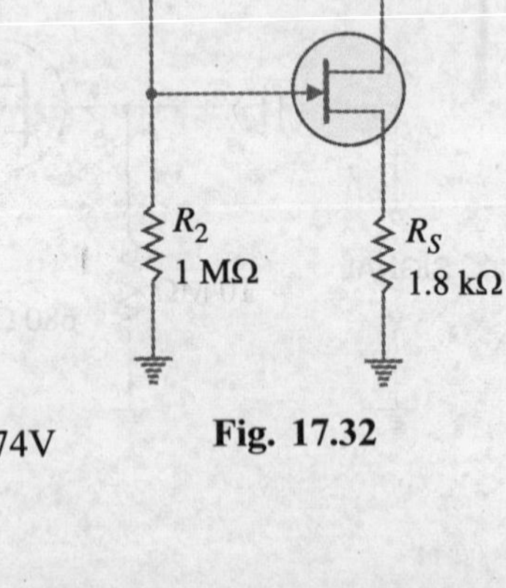

Fig. 17.32

Solution.

$$I_D = \frac{V_{DD} - V_D}{R_D} = \frac{12\text{V} - 7\text{V}}{3.3 \text{ k}\Omega}$$

$$= \frac{5\text{V}}{3.3 \text{ k}\Omega} = \mathbf{1.52 \text{ mA}}$$

$$V_S = I_D R_S = (1.52 \text{ mA}) (1.8 \text{ k}\Omega) = 2.74\text{V}$$

$$V_G = \frac{V_{DD}}{R_1 + R_2} \times R_2 = \frac{12\text{V}}{7.8\text{ M}\Omega} \times 1\text{ M}\Omega = 1.54\text{V}$$

$$\therefore \quad V_{GS} = V_G - V_S = 1.54\text{ V} - 2.74\text{ V} = \mathbf{-1.2V}$$

Example 17.22. *In an n-channel JFET biased by potential-divider method, it is desired to set the operating point at I_D = 2.5 mA and V_{DS} = 8V. If V_{DD} = 30 V, R_1 = 1 MΩ and R_2 = 500 kΩ, find the value of R_S. The parameters of JFET are I_{DSS} = 10 mA and $V_{GS\,(off)}$ = – 5 V.*

Solution. Fig. 17.33 shows the conditions of the problem.

$$I_D = I_{DSS}\left(1 - \frac{V_{GS}}{V_{GS\,(off)}}\right)^2$$

or $$2.5 = 10\left(1 + \frac{V_{GS}}{5}\right)^2$$

or $$1 + \frac{V_{GS}}{5} = \sqrt{2.5/10} = 0.5$$

or $$V_{GS} = -2.5\text{ V}$$

Now, $$V_2 = \frac{V_{DD}}{R_1 + R_2} \times R_2 = \frac{30}{1000 + 500} \times 500 = 10\text{ V}$$

Now, $$V_2 = V_{GS} + I_D R_S$$

or $$10\text{ V} = -2.5\text{ V} + 2.5\text{ mA} \times R_S$$

$$\therefore \quad R_S = \frac{10\text{ V} + 2.5\text{ V}}{2.5\text{ mA}} = \frac{12.5\text{ V}}{2.5\text{ mA}} = \mathbf{5\ k\Omega}$$

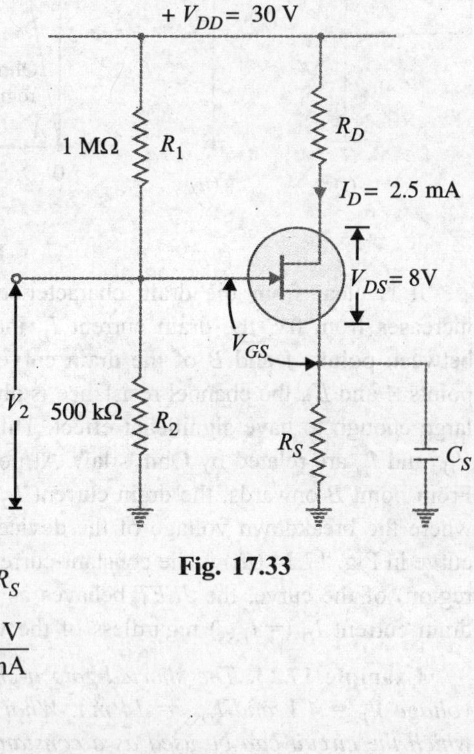

Fig. 17.33

17.25 JFET as Constant-Current Source

A *shorted-gate *JFET* acts as a constant-current source of magnitude I_{DSS} (= shorted-gate drain current) when it operates in the pinch-off region (*i.e.* between pinch off voltage V_P and breakdown voltage $V_{DS(max.)}$). Consider a shorted-gate *n*-channel *JFET* operating in the pinch-off region as shown in Fig. 17.34 (*i*). In this condition, the *JFET* will supply maximum and constant drain current equal to I_{DSS} to load R_D whatever may be the value of R_D (See Fig. 17.34 (*ii*)). Note that condition for shorted-gate *JFET* to act as a constant-current source is $V_{DS} > V_P$ *i.e.* it should operate in the pinch-off region. This can be easily explained by referring to the drain characteristic of *JFET* for V_{GS} = 0V (shorted-gate condition) in Fig. 17.34(*iii*).

* *i.e.* when gate of *JFET* is directly connected to its source.

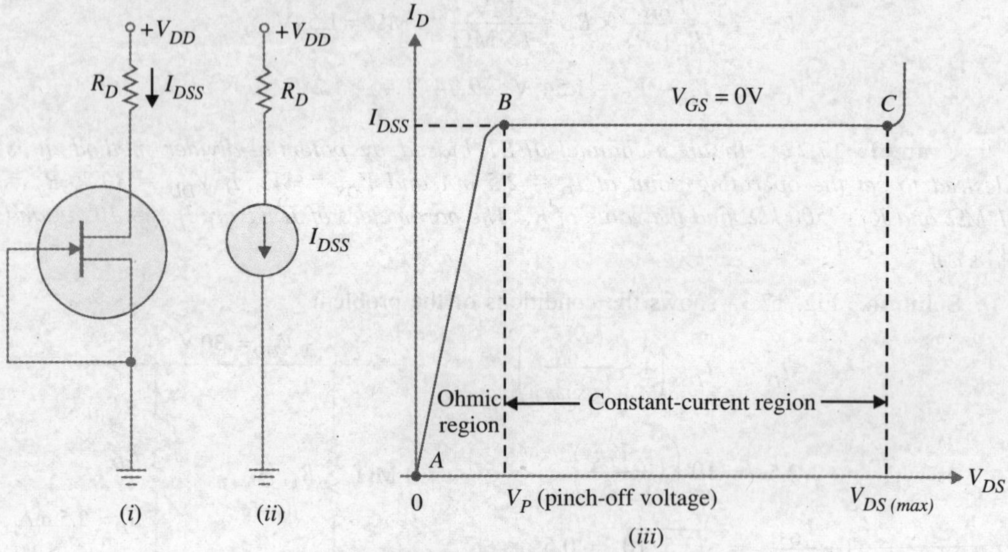

Fig. 17.34

It is clear from the drain characteristic for V_{GS} = 0V that as V_{DD} (and hence V_{DS}) increases from 0V, the drain current I_D increases linearly with V_{DS} through the *n*-channel between points *A* and *B* of the drain curve in Fig. 17.34 (*iii*). In this region (*i.e.* between points *A* and *B*), the channel resistance is almost constant because the depletion region is not large enough to have significant effect. This is called Ohmic region because in this region, V_{DS} and I_D are related by Ohm's law. At point *B* in Fig. 17.34 (*iii*), $V_{DS} = V_P$ and $I_D = I_{DSS}$. From point *B* onwards, the drain current I_D (= I_{DSS}) remains constant till point *C* is reached where the breakdown voltage of the device occurs. The horizontal portion *BC* of the drain curve in Fig. 17.34 (*iii*) is the constant-current region. Therefore, in the portion *BC* (pinch-off region) of the curve, the *JFET* behaves as a constant-current source and supplies the same drain current I_D (= I_{DSS}) regardless of the value of load resistance R_D.

Example 17.23. *The shorted-gate n-channel JFET shown in Fig. 17.35 has pinch-off voltage V_P = 4 V and I_{DSS} = 14 mA. What is the maximum value of load resistance R_D for which the circuit can be used as a constant-current source ?*

Solution. Here, V_{DD} = 18 V ; $I_D = I_{DSS}$ = 14 mA = 14 × 10^{-3}A ; V_P = 4 V. In order that the *JFET* behaves as a constant-current source, it must operate in the pinch-off region

i.e. $$V_{DS} > V_P$$

or $$V_{DD} - I_{DSS}\,R_D > V_P$$

or $$18 - (14 \times 10^{-3})\,R_D > 4$$

or $$-14 \times 10^{-3}\,R_D > -14$$

∴ $$R_D < \frac{14}{14 \times 10^{-3}}\,\Omega = 1000\ \Omega = \mathbf{1\ k\Omega}$$

Therefore, the value of load resistance R_D must be less than 1 kΩ. This will ensure that $V_{DS} > V_P$ and put the device in the constant-current region.

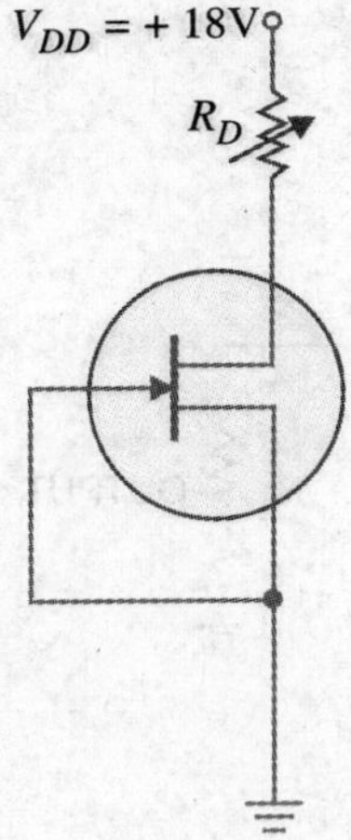

Fig. 17.35

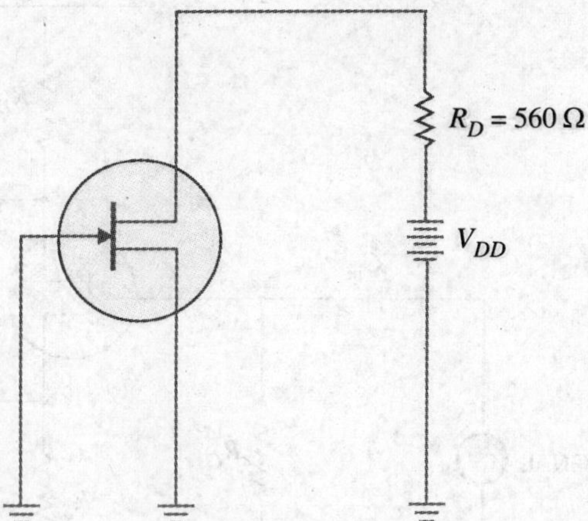

Fig. 17.36

Example 17.24. *For the n-channel JFET shown in Fig. 17.36, $V_{GS(off)}$ = – 4 V and I_{DSS} = 12 mA. Determine the minimum value of V_{DD} required to put the device in the constant-current region of operation.*

Solution. As $V_{GS(off)} = -4$ V, therefore, $V_P = +4$ V. The minimum value of V_{DS} for the *JFET* to be in its constant-current region is

$$V_{DS} = V_P = 4 \text{ V}$$

In the constant-current region with $V_{GS} = 0$ V (shorted-gate condition), the drain current I_D is

$$I_D = I_{DSS} = 12 \text{ mA}$$

Applying Kirchhoff's voltage law to the drain circuit, we have,

$$V_{DD} = I_D R_D + V_{DS}$$

or $$V_{DD} = 12 \text{ mA} \times 560 \ \Omega + 4 \text{ V} = 6.72 \text{ V} + 4 \text{ V}$$

$$= \mathbf{10.72 \ V}$$

This is the minimum value of V_{DD} to make $V_{DS} = V_P$ and put the device in the constant-current region.

17.26 Practical JFET Amplifier

It is important to note that a *JFET* can accomplish faithful amplification only if proper associated circuitry is used. Fig. 17.37 shows the practical circuit of an *n*-channel *JFET*. The gate resistor R_G serves two purposes. It keeps the gate at approximately 0V d.c. (∵ gate current is nearly zero) and its large value (usually several megaohms) prevents loading of the a.c. signal source. The bias voltage is created by the drop across R_S. The bypass capacitor C_S bypasses the a.c. signal and thus keeps the source of the *JFET* effectively at a.c. ground. The coupling capacitor C_{in} couples the signal to the input of *JFET* amplifier.

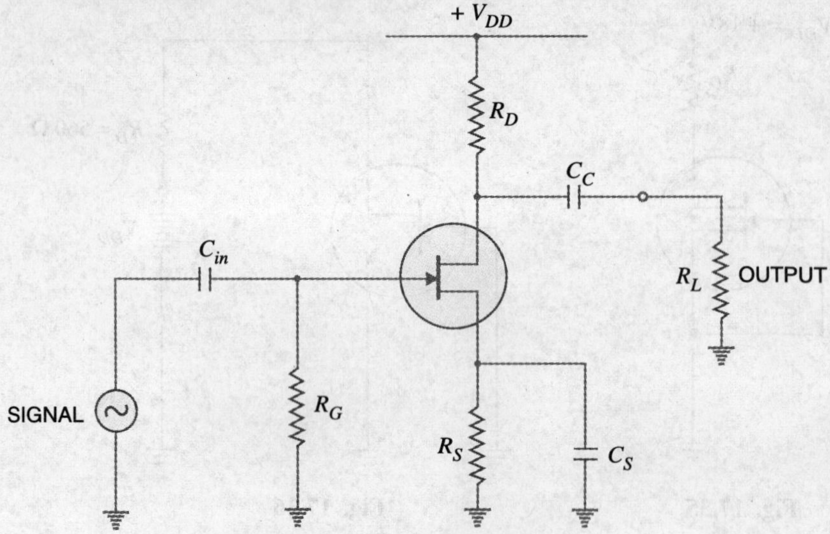

COMMON SOURCE CONNECTION

Fig. 17.37

17.27 D.C. and A.C. Equivalent Circuits of JFET Amplifier

Like in a transistor amplifier, both d.c. and a.c. conditions prevail in a *JFET* amplifier. The d.c. sources set up d.c. currents and voltages whereas the a.c. source (*i.e.* signal) produces fluctuations in the *JFET* currents and voltages. Therefore, a simple way to analyse the action of a *JFET* amplifier is to split the circuit into two parts *viz. d.c. equivalent circuit* and *a.c. equivalent circuit.* The d.c. equivalent circuit will determine the operating point (d.c. bias levels) for the circuit while a.c. equivalent circuit determines the output voltage and hence voltage gain of the circuit.

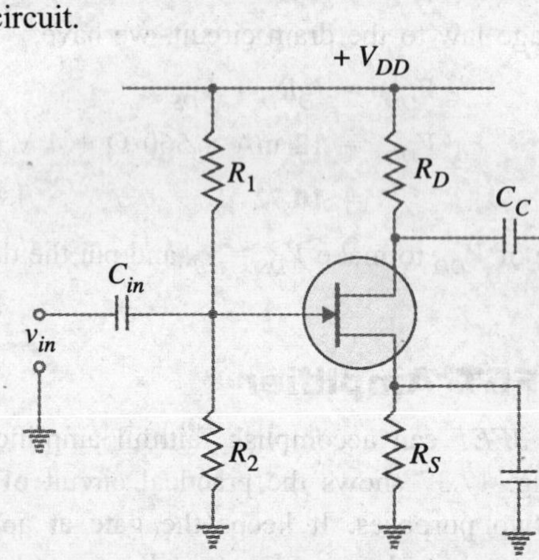

Fig. 17.38

We shall split the *n*-channel *JFET* amplifier shown in Fig. 17.38 into d.c. and a.c. equivalent circuits. Note that biasing is provided by voltage-divider circuit.

1. **D.C. equivalent circuit.** In the d.c. equivalent circuit of a *JFET* amplifier, only d.c. conditions are considered *i.e.* it is presumed that no signal is applied. As direct

current cannot flow through a capacitor, *all the capacitors look like open circuits in the d.c. equivalent circuit.* It follows, therefore, that in order to draw the d.c. equivalent circuit, the following two steps are applied to the *JFET* amplifier circuit :

(*i*) Reduce all a.c. sources to zero.

(*ii*) Open all the capacitors.

Applying these two steps to the *JFET* amplifier circuit shown in Fig. 17.38, we get the d.c. equivalent circuit shown in Fig. 17.39. We can easily calculate the d.c. currents and voltages from this circuit.

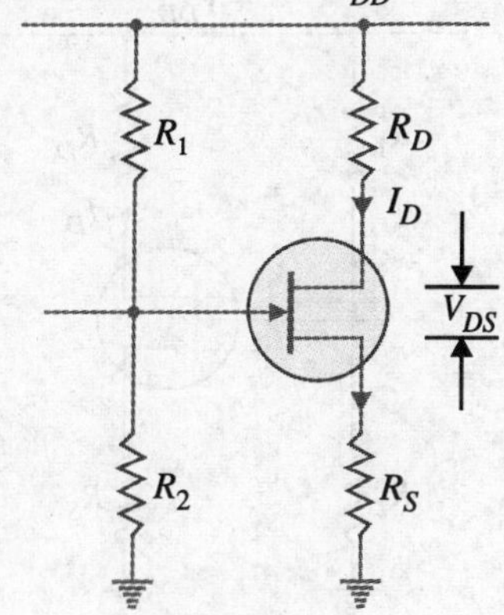

Fig. 17.39

2. **A.C. equivalent circuit.** In the a.c. equivalent circuit of a *JFET* amplifier, only a.c. conditions are to be considered. Obviously, the d.c. voltage is not important for such a circuit and may be considered zero. The capacitors are generally used to couple or bypass the a.c. signal. The designer intentionally selects capacitors that are large enough to appear as *short circuits* to the a.c. signal. It follows, therefore, that in order to draw the a.c. equivalent circuit, the following two steps are applied to the *JFET* amplifier circuit :

(*i*) Reduce all d.c. sources to zero (*i.e.* $V_{DD} = 0$).

(*ii*) Short all the capacitors.

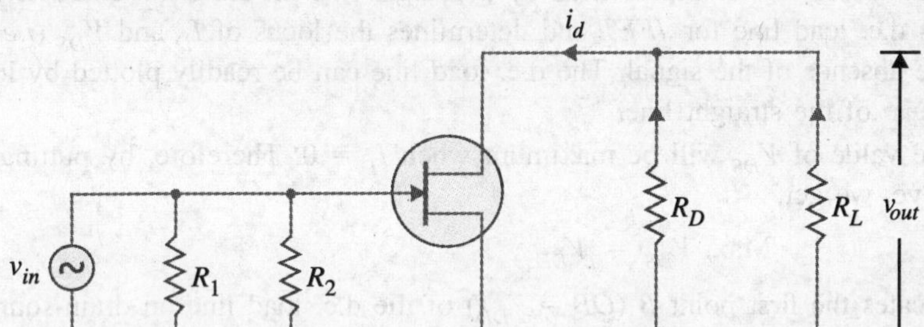

Fig. 17.40

Applying these two steps to the circuit shown in Fig. 17.38, we get the a.c. *equivalent circuit shown in Fig. 17.40. Note that here *JFET* is considered ideal. We can easily calculate the a.c. currents and voltages from this circuit.

17.28 D.C. Load Line Analysis of JFET

The operating point of a *JFET* amplifier can be determined graphically by drawing d.c. load line on the drain characteristics ($V_{DS} - I_D$ curves). This method is identical to that used for transistors.

* Note that one end of R_1 and R_2 is connected to one point (See Fig. 17.38) and the other end of R_1 and R_2 is connected to ground. Therefore, $R_1 \parallel R_2$. Similar is the case with R_D and R_L so that $R_D \parallel R_L$.

The d.c. equivalent circuit of an *n*-channel *JFET* amplifier using voltage-divider bias is shown in Fig. 17.41 (*i*). It is clear that :

$$V_{DD} = V_{DS} + I_D (R_D + R_S)$$

or

$$V_{DS} = V_{DD} - I_D (R_D + R_S) \quad \text{... } (i)$$

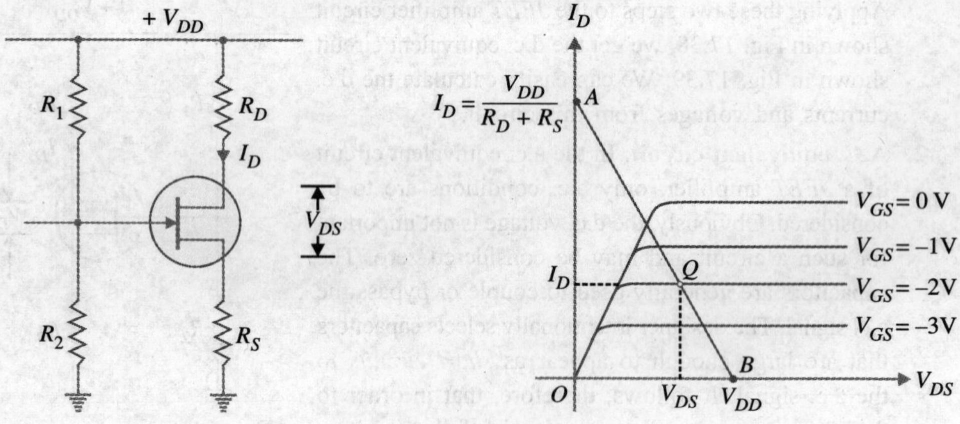

(*i*) (*ii*)

Fig. 17.41

As for a given circuit, V_{DD} and $(R_D + R_S)$ are constant, therefore, exp. (*i*) is a first degree equation and can be represented by a straight line on the drain characteristics. This is known as d.c. load line for *JFET* and determines the locus of I_D and V_{DS} (*i.e.* operating point) in the absence of the signal. The d.c. load line can be readily plotted by locating the *two end points* of the straight line.

(*i*) The value of V_{DS} will be maximum when $I_D = 0$. Therefore, by putting $I_D = 0$ in exp. (*i*) above, we get,

$$\text{Max. } V_{DS} = V_{DD}$$

This locates the first point *B* ($OB = V_{DD}$) of the d.c. load line on drain-source voltage axis.

(*ii*) The value of I_D will be maximum when $V_{DS} = 0$.

$$\therefore \quad \text{Max. } I_D = \frac{V_{DD}}{R_D + R_S}$$

This locates the second point *A* ($OA = V_{DD} / R_D + R_S$) of the d.c. load line on drain current axis.

By joining points *A* and *B*, d.c. load line *AB* is constructed [See Fig. 17.41 (*ii*)].

The operating point *Q* is located at the intersection of the d.c. load line and the drain curve which corresponds to V_{GS} provided by biasing. If we assume in Fig. 17.41 (*i*) that $V_{GS} = -2\text{V}$, then point *Q* is located at the intersection of the d.c. load line and the $V_{GS} = -2\text{V}$ curve as shown in Fig. 17.41 (*ii*). The I_D and V_{DS} of *Q* point are marked on the graph.

Example 17.25. *Draw the d.c. load line for the JFET amplifier shown in Fig. 17.42 (i).*

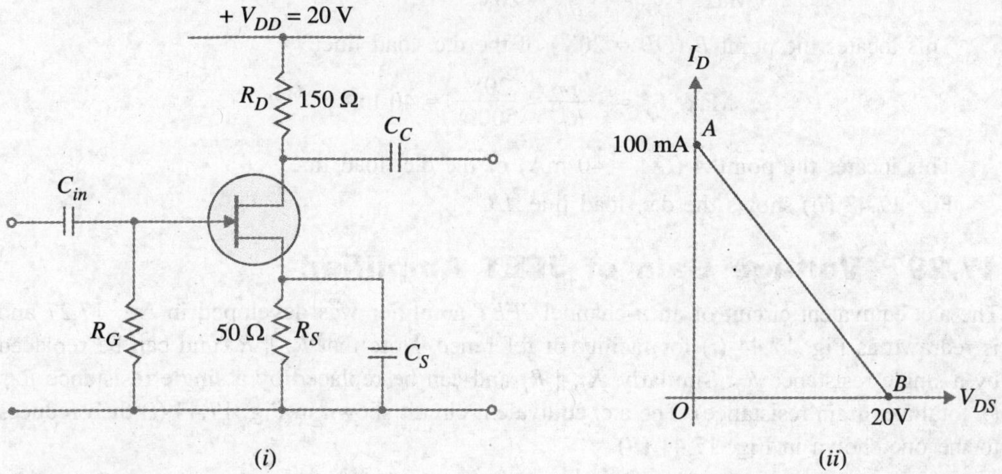

Fig. 17.42

Solution. To draw d.c. load line, we require two end points *viz.*, max. V_{DS} and max. I_D points.

$$\text{Max. } V_{DS} = V_{DD} = 20\text{V}$$

This locates point *B* (*OB* = 20V) of the d.c. load line.

$$\text{Max. } I_D = \frac{V_{DD}}{R_D + R_S} = \frac{20\text{V}}{(150 + 50)\ \Omega}$$

$$= \frac{20\text{V}}{200\Omega} = 100 \text{ mA}$$

This locates point *A* (*OA* = 100 mA) of the d.c. load line. Joining *A* and *B*, d.c. load line *AB* is constructed as shown in Fig. 17.42 (*ii*).

Example 17.26. *Draw the d.c. load line for the JFET amplifier shown in Fig. 17.43 (i).*

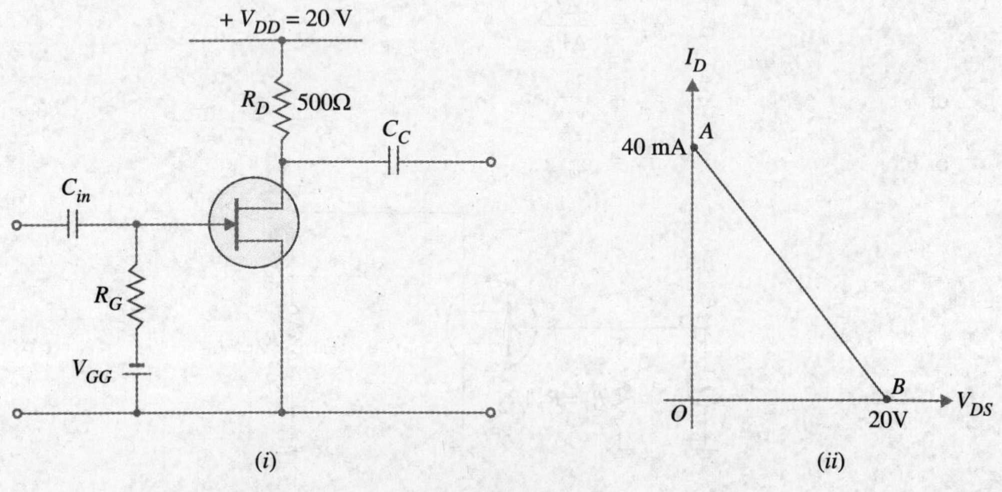

Fig. 17.43

Solution.

$$\text{Max. } V_{DS} = V_{DD} = 20\text{V}$$

This locates the point *B* (*OB* = 20V) of the d.c. load line.

$$\text{Max. } I_D = \frac{V_{DD}}{R_D} = \frac{20\text{V}}{500\Omega} = 40 \text{ mA}$$

This locates the point *A* (*OA* = 40 mA) of the d.c. load line.

Fig. 17.43 (*ii*) shows the d.c. load line *AB*.

17.29 Voltage Gain of JFET Amplifier

The a.c. equivalent circuit of an *n*-channel *JFET* amplifier was developed in Art. 17.27 and is redrawn as Fig. 17.44 (*i*) for facility of reference. Note that $R_1 \parallel R_2$ and can be replaced by a single resistance R_T. Similarly, $R_D \parallel R_L$ and can be replaced by a single resistance R_{AC} (= total a.c. drain resistance). The a.c. equivalent circuit shown in Fig. 17.44 (*i*) then reduces to the one shown in Fig. 17.44 (*ii*).

We now find the expression for voltage gain of this amplifier. Referring to Fig. 17.44 (*ii*), output voltage (v_{out}) is given by ;

$$v_{out} = i_d R_{AC} \qquad \text{... (i)}$$

Remember that we define g_m as :

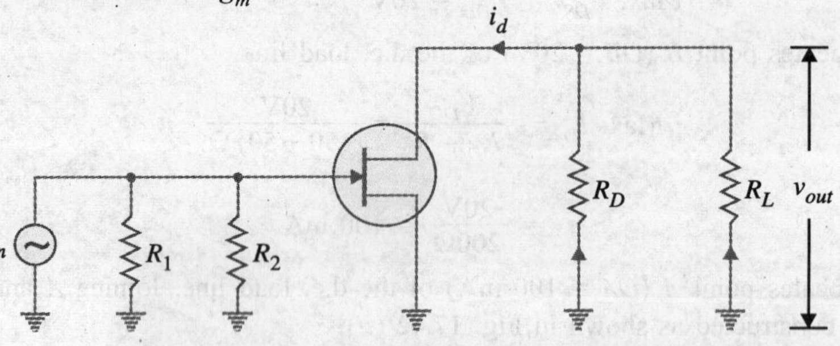

Fig. 17.44 (*i*)

$$g_m = \frac{\Delta I_D}{\Delta V_{GS}}$$

or $$g_m = \frac{i_d}{v_{gs}}$$

or $$i_d = g_m v_{gs}$$

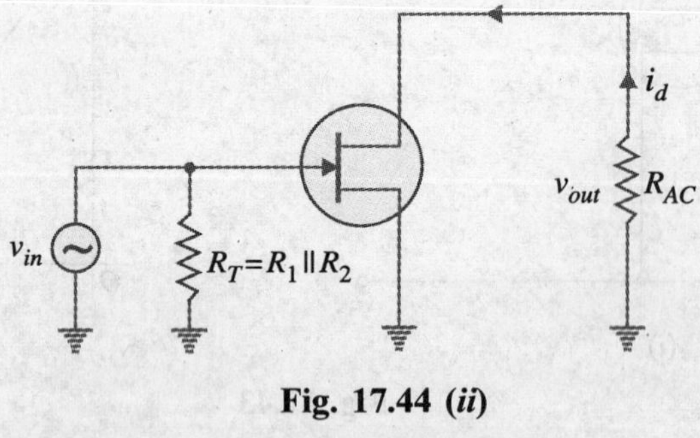

Fig. 17.44 (*ii*)

Putting the value of i_d (= g_m v_{gs}) in eq. (*i*), we have,

$$v_{out} = g_m \, v_{gs} \, R_{AC}$$

Now, $v_{in} = v_{gs}$ so that a.c. output voltage is

$$v_{out} = g_m \, v_{in} \, R_{AC}$$

or
$$v_{out}/v_{in} = g_m \, R_{AC}$$

But v_{out}/v_{in} is the voltage gain (A_v) of the amplifier.

$\therefore$ Voltage gain, $A_v = g_m \, R_{AC}$ ... for loaded amplifier

$= g_m \, R_D$ ... for unloaded amplifier

Example 17.27. *The n-channel JFET amplifier of Fig. 17.45 has a transconductance g_m = 1 mA/V. If the source resistance R_S is very small as compared to R_G, find the voltage gain of the amplifier.*

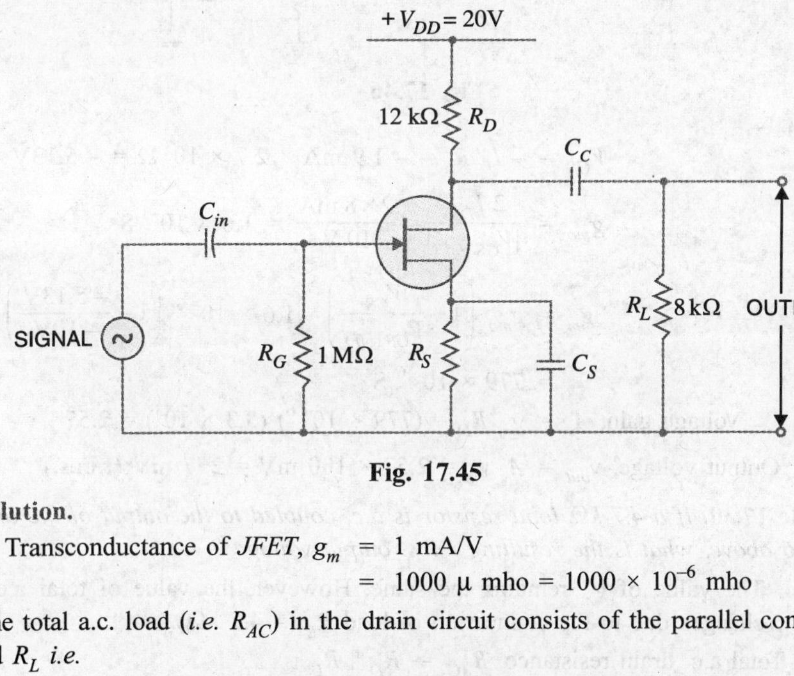

Fig. 17.45

Solution.

Transconductance of *JFET*, g_m = 1 mA/V

= 1000 μ mho = 1000 × 10^{-6} mho

The total a.c. load (*i.e.* R_{AC}) in the drain circuit consists of the parallel combination of R_D and R_L *i.e.*

Total a.c. load, $R_{AC} = R_D \parallel R_L$

$= 12 \text{ k}\Omega \parallel 8 \text{ k}\Omega = \dfrac{12 \times 8}{12+8} = 4.8 \text{ k}\Omega$

$\therefore$ Voltage gain, $A_v = g_m \times R_{AC}$

$= (1000 \times 10^{-6}) \times (4.8 \times 10^3) = \mathbf{4.8}$

Example 17.28. *The transconductance of a JFET used as a voltage amplifier is 3000 μmho and drain resistance is 10 kΩ. Calculate the voltage gain of the amplifier.*

Solution.

Transconductance of *JFET*, g_m = 3000 μmho = 3000 × 10^{-6} mho

Drain resistance, R_D = 10 kΩ = 10 × 10^3Ω

$\therefore$ Voltage gain, $A_v = g_m \, R_D = (3000 \times 10^{-6}) \, (10 \times 10^3) = \mathbf{30}$

Example 17.29. *What is the r.m.s. output voltage of the unloaded amplifier in Fig. 17.46 ? The* I_{DSS} *= 8 mA,* $V_{GS\,(off)}$ *= – 10V and* I_D *= 1.9 mA.*

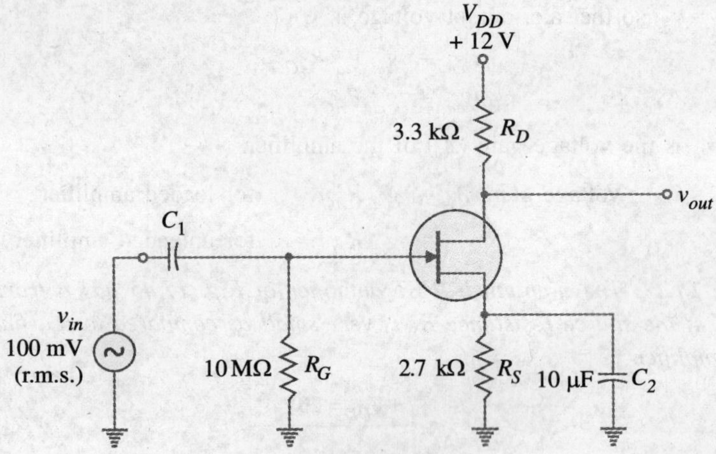

Fig. 17.46

Solution.

$$V_{GS} = -I_D R_S = -1.9 \text{ mA} \times 2.7 \times 10^3 \ \Omega = -5.13\text{V}$$

$$g_{mo} = \frac{2\,I_{DSS}}{|V_{GS\,(off)}|} = \frac{2 \times 8 \text{ mA}}{10 \text{ V}} = 1.6 \times 10^{-3} \text{ S}$$

$$\therefore \quad g_m = g_{mo}\left(1 - \frac{V_{GS}}{V_{GS\,(off)}}\right) = 1.6 \times 10^{-3}\left(1 - \frac{-5.13\text{V}}{-10\text{V}}\right)$$

$$= 779 \times 10^{-6} \text{ S}$$

$$\text{Voltage gain, } A_v = g_m R_D = (779 \times 10^{-6})\,(3.3 \times 10^3) = 2.57$$

$$\therefore \quad \text{Output voltage, } v_{out} = A_v\, v_{in} = 2.57 \times 100 \text{ mV} = \mathbf{257 \text{ mV (r.m.s.)}}$$

Example 17.30. *If a 4.7 kΩ load resistor is a.c. coupled to the output of the amplifier in Fig. 17.46 above, what is the resulting r.m.s. output voltage ?*

Solution. The value of g_m remains the same. However, the value of total a.c. drain resistance R_{AC} changes due to the connection of load R_L (= 4.7 kΩ).

$$\therefore \quad \text{Total a.c. drain resistance, } R_{AC} = R_D \,||\, R_L$$

$$= \frac{R_D\, R_L}{R_D + R_L} = \frac{(3.3 \text{ k}\Omega)\,(4.7 \text{ k}\Omega)}{3.3 \text{ k}\Omega + 4.7 \text{ k}\Omega} = 1.94 \text{ k}\Omega$$

$$\therefore \quad \text{Voltage gain, } A_v = g_m R_{AC} = (779 \times 10^{-6})\,(1.94 \times 10^3) = 1.51$$

$$\text{Output voltage, } v_{out} = A_v\, v_{in} = 1.51 \times 100 \text{ mV} = \mathbf{151 \text{ mV (r.m.s.)}}$$

17.30 Voltage Gain of JFET Amplifier (*With Source Resistance* R_S)

Fig. 17.47 (*i*) shows an *n*-channel *JFET* amplifier with source resistor R_S unbypassed. This means that a.c. signal will not be bypassed by the capacitor C_S.

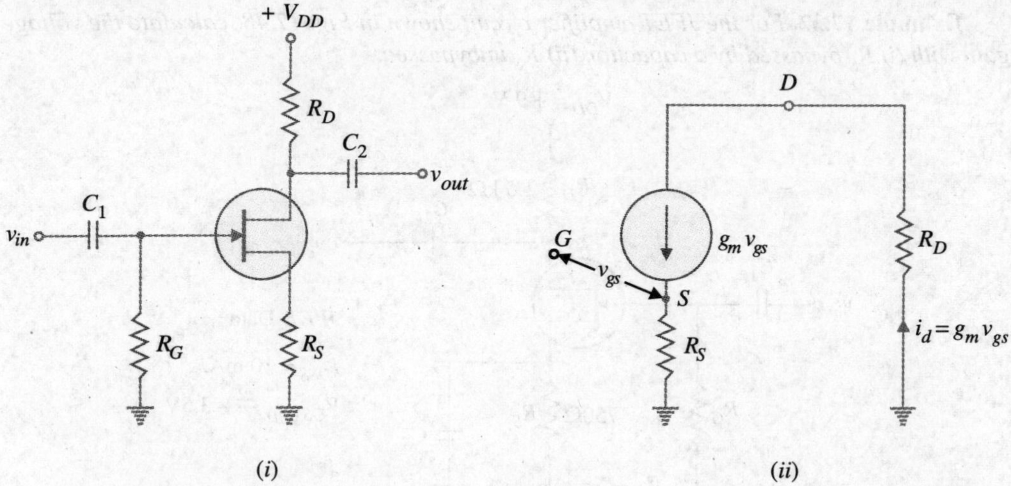

Fig. 17.47

Fig. 17.47 (*ii*) shows the simplified a.c. equivalent circuit of the above *JFET* amplifier. Since $g_m = i_d / v_{gs}$, a current source $i_d = g_m v_{gs}$ appears between drain and source. Referring to Fig. 17.47 (*ii*),

$$v_{in} = v_{gs} + i_d R_S$$

$$v_{out} = i_d R_D$$

$$\therefore \quad \text{Voltage gain, } A_v = \frac{v_{out}}{v_{in}} = \frac{i_d R_D}{v_{gs} + i_d R_S}$$

$$= \frac{g_m v_{gs} R_D}{v_{gs} + g_m v_{gs} R_S} = \frac{g_m v_{gs} R_D}{v_{gs}(1 + g_m R_S)} \qquad (\because i_d = g_m v_{gs})$$

$$\therefore \quad A_v = \frac{g_m R_D}{1 + g_m R_S} \quad \text{... for unloaded amplifier}$$

$$= \frac{g_m R_{AC}}{1 + g_m R_S} \quad \text{... for loaded amplifier}$$

Note that R_{AC} ($= R_D \parallel R_L$) is the total a.c. drain resistance.

Example 17.31. *In a JFET amplifier, the source resistance R_S is unbypassed. Find the voltage gain of the amplifier. Given g_m = 4 mS ; R_D = 1.5 kΩ and R_S = 560Ω.*

Solution. Voltage gain, $A_v = \dfrac{g_m R_D}{1 + g_m R_S}$

Here, $g_m = 4 \text{ mS} = 4 \times 10^{-3} \text{ S}$; $R_D = 1.5 \text{ k}\Omega = 1.5 \times 10^3 \Omega$; $R_S = 560\Omega$

$$\therefore \quad A_v = \frac{(4 \times 10^{-3})(1.5 \times 10^3)}{1 + (4 \times 10^{-3})(560)} = \frac{6}{1 + 2.24} = \mathbf{1.85}$$

If R_S is bypassed by a capacitor, then,

$$A_v = g_m R_D = (4 \times 10^{-3})(1.5 \times 10^3) = 6$$

Thus with unbypassed R_S, the gain = 1.85 whereas with R_S bypassed by a capacitor, the gain is 6. Therefore, voltage gain is reduced when R_S is unbypassed.

Example 17.32. *For the JFET amplifier circuit shown in Fig. 17.48, calculate the voltage gain with (i) R_S bypassed by a capacitor (ii) R_S unbypassed.*

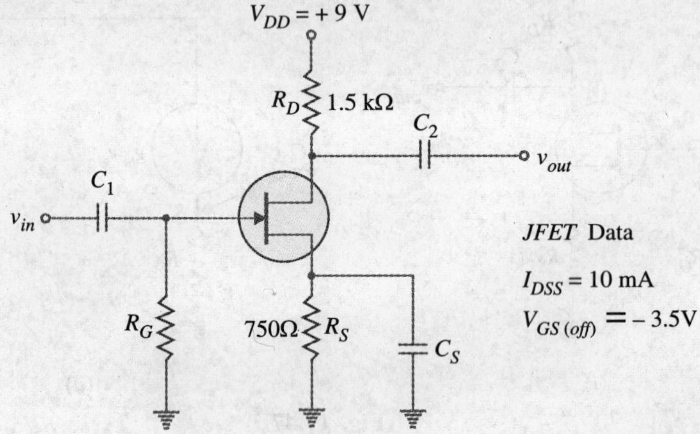

Fig. 17.48

Solution. From the d.c. bias analysis, we get, *$I_D = 2.3$ mA and $V_{GS} = -1.8$V.

The value of g_m is given by ;

$$g_m = \frac{2\,I_{DSS}}{|V_{GS\,(off)}|}\left(1 - \frac{V_{GS}}{V_{GS\,(off)}}\right)$$

$$= \frac{2 \times 10}{3.5}\left(1 - \frac{-1.8}{-3.5}\right) = (5.7 \text{ mS})\ (0.486) = 2.77 \text{ mS}$$

(*i*) The voltage gain with R_S bypassed is

$$A_v = g_m\, R_D = (2.77 \text{ mS})\ (1.5 \text{ k}\Omega) = \mathbf{4.155}$$

(*ii*) The voltage gain with R_S unbypassed is

$$A_v = \frac{g_m\, R_D}{1 + g_m\, R_S} = \frac{4.155}{1 + (2.77 \text{ mS})\ (0.75 \text{ k}\Omega)} = \mathbf{1.35}$$

17.31 JFET Applications

The high input impedance and low output impedance and low noise level make *JFET* far superior to the bipolar transistor. Some of the circuit applications of *JFET* are :

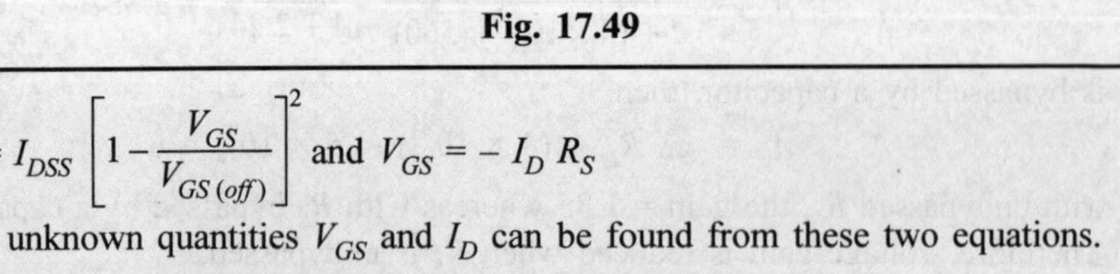

Fig. 17.49

---

* $I_D = I_{DSS}\left[1 - \dfrac{V_{GS}}{V_{GS\,(off)}}\right]^2$ and $V_{GS} = -\,I_D\, R_S$

The unknown quantities V_{GS} and I_D can be found from these two equations.

(*i*) **As a buffer amplifier.** A buffer amplifier is a stage of amplification that isolates the preceding stage from the following stage. Because of the high input impedance and low output impedance, a *JFET* can act as an excellent buffer amplifier (See Fig. 17.49). The high input impedance of *JFET* means light loading of the preceding stage. This permits almost the entire output from first stage to appear at the buffer input. The low output impedance of *JFET* can drive heavy loads (or small load resistances). This ensures that all the output from the buffer reaches the input of the second stage.

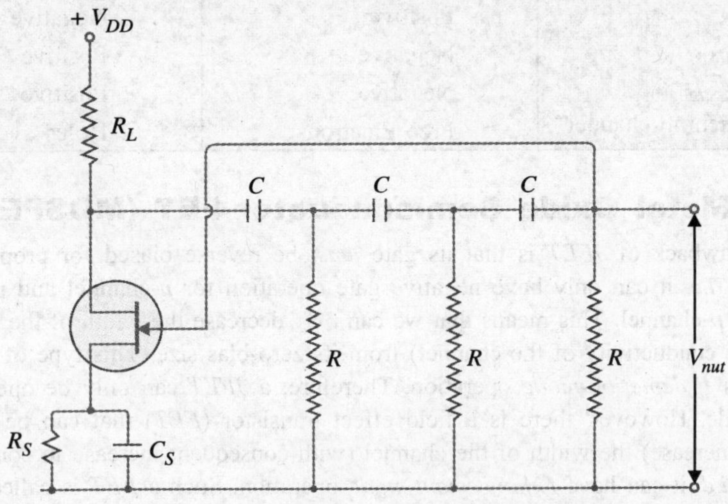

Fig. 17.50

(*ii*) **Phase-shift oscillators.** The oscillators discussed in chapter 12 will also work with *JFETs*. However, the high input impedance of *JFET* is especially valuable in phase-shift oscillators to minimise the loading effect. Fig. 17.50 shows the phase-shift oscillator using *n*-channel *JFET*.

(*iii*) **As RF amplifier.** In communication electronics, we have to use *JFET RF* amplifier in a receiver instead of *BJT* amplifier for the following reasons :

(*a*) The noise level of *JFET* is very low. The *JFET* will not generate significant amount of noise and is thus useful as an *RF* amplifier.

(*b*) The antenna of the receiver receives a very weak signal that has an extremely low amount of current. Since *JFET* is a voltage-controlled device, it will well respond to low current signal provided by the antenna.

17.32 Comparison between *n*-channel JFET and *p*-channel JFET

While discussing *JFET*, we have been using *n*-channel *JFET*. This is due to the fact that *n*-channel *JFETs* are used more commonly than *p*-channel *JFETs*. However, it must be remembered that all of the *n*-channel *JFET* principles apply equally to *p*-channel *JFET*. The only differences are the voltage polarities and the direction of the drain and source currents. The *n*-channel and *p*-channel *JFET* are compared in the table given below :

JFET type	*n*-channel	*p*-channel
Schematic symbol	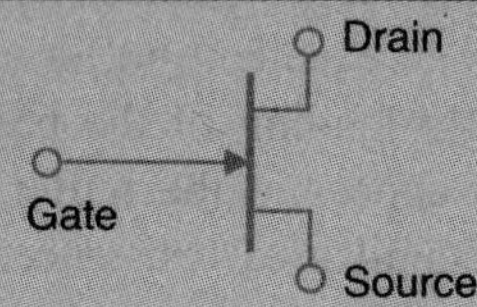	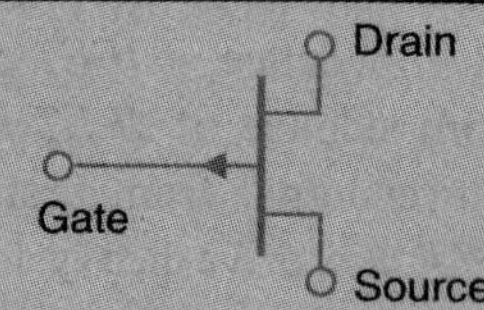
V_{DS} :	Positive	Negative
V_P :	Positive	Negative
V_{GS} :	Negative	Positive
$V_{GS(off)}$:	Negative	Positive
Current carriers in channel	Free Electrons	Holes

17.33 Metal Oxide Semiconductor FET (MOSFET)

The main drawback of *JFET* is that its gate *must* be reverse biased for proper operation of the device *i.e.* it can only have negative gate operation for *n*-channel and positive gate operation for *p*-channel. This means that we can *only* decrease the width of the channel (*i.e.* decrease the *conductivity of the channel) from its zero-bias size. This type of operation is referred to as ***depletion-mode* operation. Therefore, a *JFET* can only be operated in the depletion-mode. However, there is a field effect transistor (*FET*) that can be operated to enhance (or increase) the width of the channel (with consequent increase in conductivity of the channel) *i.e.* it can have *enhancement-mode* operation. Such a *FET* is called *MOSFET*.

A field effect transistor (FET) that can be operated in the enhancement-mode is called a MOSFET.

A *MOSFET* is an important semiconductor device and can be used in any of the circuits covered for *JFET*. However, a *MOSFET* has several advantages over *JFET* including high input impedance and low cost of production.

17.34 Types of MOSFETs

There are two basic types of *MOSFETs viz.*

1. **Depletion-type MOSFET or D-MOSFET.** The *D-MOSFET* can be operated in both the depletion-mode and the enhancement-mode. For this reason, a *D-MOSFET* is sometimes called *depletion/enhancement MOSFET*.
2. **Enhancement-type MOSFET or E-MOSFET.** The *E-MOSFET* can be operated *only* in enhancement-mode.

The manner in which a *MOSFET* is constructed determines whether it is *D-MOSFET* or *E-MOSFET*.

* With the decrease in channel width, the *X*-sectional area of the channel decreases and hence its resistance increases. This means that conductivity of the channel will decrease. Reverse happens if channel width increases.

** With gate reverse biased, the channel is depleted (*i.e.* emptied) of charge carriers (free electrons for *n*-channel and holes for *p*-channel) and hence the name depletion-mode. Note that depletion means decrease. In this mode of operation, conductivity decreases from the zero-bias level.

1. D-MOSFET. Fig. 17.51 shows the constructional details of *n*-channel *D-MOSFET*. It is similar to *n*-channel *JFET* except with the following modifications/remarks :

(*i*) The *n*-channel *D-MOSFET* is a piece of *n*-type material with a *p*-type region (called *substrate*) on the right and an *insulated gate* on the left as shown in Fig. 17.51. The free electrons (∵ it is *n*-channel) flowing from source to drain must pass through the narrow channel between the gate and the *p*-type region (*i.e.* substrate).

(*ii*) Note carefully the gate construction of *D-MOSFET*. A thin layer of metal oxide (usually silicon dioxide, SiO_2) is deposited over a small portion of the channel. A metallic gate is deposited over the oxide layer. As SiO_2 is an insulator, therefore, gate is insulated from the channel. Note that the arrangement forms a capacitor. One plate of this capacitor is the gate and the other plate is the channel with SiO_2 as the dielectric. Recall that we have a gate diode in a *JFET*.

(*iii*) It is a usual practice to connect the substrate to the source (*S*) internally so that a *MOSFET* has three terminals *viz source* (*S*), *gate* (*G*) and *drain* (*D*).

(*iv*) Since the gate is insulated from the channel, we can apply either negative or positive voltage to the gate. Therefore, *D-MOSFET* can be operated in both depletion-mode and enhancement-mode. However, *JFET* can be operated only in depletion-mode.

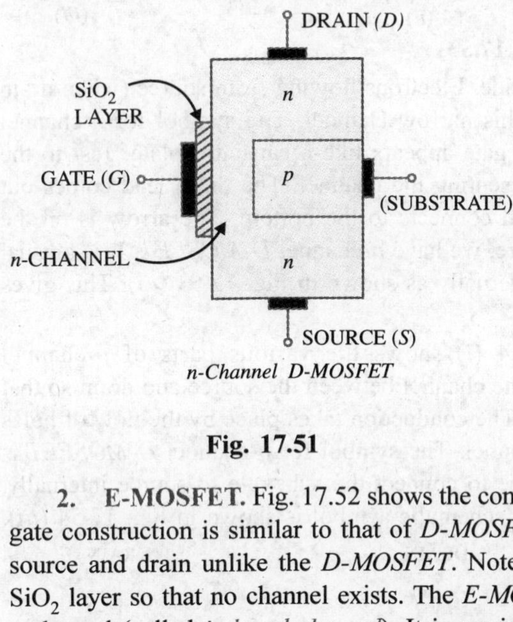

Fig. 17.51

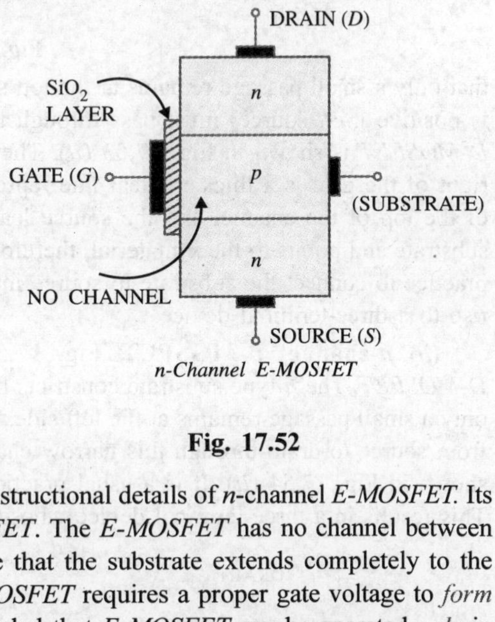

Fig. 17.52

2. E-MOSFET. Fig. 17.52 shows the constructional details of *n*-channel *E-MOSFET*. Its gate construction is similar to that of *D-MOSFET*. The *E-MOSFET* has no channel between source and drain unlike the *D-MOSFET*. Note that the substrate extends completely to the SiO_2 layer so that no channel exists. The *E-MOSFET* requires a proper gate voltage to *form* a channel (called *induced channel*). It is reminded that *E-MOSFET* can be operated *only* in enhancement mode. In short, the construction of *E-MOSFET* is quite similar to that of the *D-MOSFET* except for the absence of a channel between the drain and source terminals.

Why the name MOSFET ? The reader may wonder why is the device called *MOSFET?* The answer is simple. The SiO_2 layer is an insulator. The gate terminal is made of a metal conductor. Thus, going from gate to substrate, you have a *metal oxide semiconductor* and hence the name *MOSFET*. Since the gate is insulated from the channel, the *MOSFET* is sometimes called *insulated-gate FET (IGFET)*. However, this term is rarely used in place of the term *MOSFET*.

17.35 Symbols for D-MOSFET

There are two types of *D-MOSFETs viz* (*i*) *n*-channel *D-MOSFET* and (*ii*) *p*-channel *D-MOSFET*.

(*i*) ***n*-channel D-MOSFET.** Fig. 17.53 (*i*) shows the various parts of *n*-channel *D-MOSFET*. The *p*-type substrate constricts the channel between the source and drain so

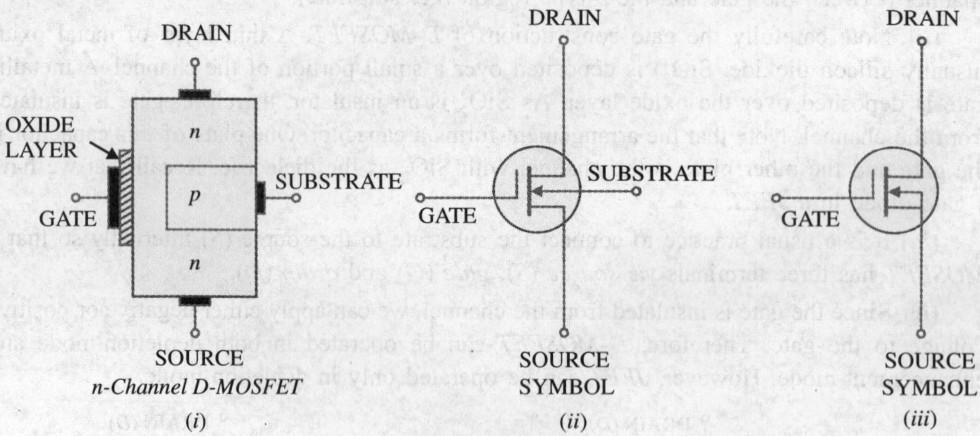

Fig. 17.53

that only a small passage remains at the left side. Electrons flowing from source (when drain is positive *w.r.t.* source) must pass through this narrow channel. The symbol for *n*-channel *D-MOSFET* is shown in Fig. 17.53 (*ii*). The gate appears like a capacitor plate. Just to the right of the gate is a thick vertical line representing the channel. The drain lead comes out of the top of the channel and the source lead connects to the bottom. The arrow is on the substrate and points to the *n*-material, therefore, we have *n*-channel *D-MOSFET*. It is a usual practice to connect the substrate to source internally as shown in Fig. 17.53 (*iii*). This gives rise to a three-terminal device.

(*ii*) ***p*-channel D-MOSFET.** Fig. 17.54 (*i*) shows the various parts of *p*-channel *D-MOSFET*. The *n*-type substrate constricts the channel between the source and drain so that only a small passage remains at the left side. The conduction takes place by the flow of holes from source to drain through this narrow channel. The symbol for *p*-channel *D-MOSFET* is shown in Fig. 17.54 (*ii*). It is a usual practice to connect the substrate to source internally. This results in a three-terminal device whose schematic symbol is shown in Fig. 17.54 (*iii*).

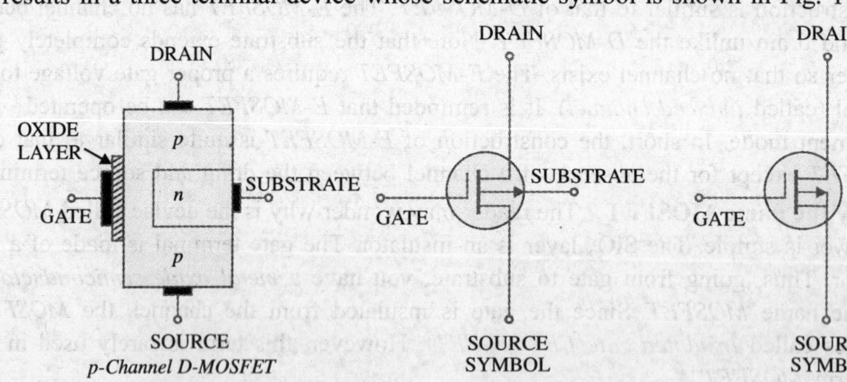

Fig. 17.54

17.36 Circuit Operation of D-MOSFET

Fig. 17.55 (*i*) shows the circuit of *n*-channel *D-MOSFET*. The gate forms a small capacitor. One plate of this capacitor is the gate and the other plate is the channel with metal oxide layer as the dielectric. When gate voltage is changed, the electric field of the capacitor changes which in turn changes the resistance of the *n*-channel. Since the gate is insulated from the channel, we can apply either negative or positive voltage to the gate. The negative-gate operation is called *depletion mode* whereas positive-gate operation is known as *enhancement mode.*

(*i*) **Depletion mode.** Fig. 17.55 (*i*) shows depletion-mode operation of *n*-channel *D-MOSFET*. Since gate is negative, it means electrons are on the gate as shown is Fig. 17.55 (*ii*). These electrons *repel the free electrons in the *n*-channel, leaving a layer of positive ions in a part of the channel as shown in Fig. 17.55 (*ii*). In other words, we have depleted (*i.e.* emptied) the *n*-channel of some of its free electrons. Therefore, lesser number of free electrons are made available for current conduction through the *n*-channel. This is the same thing as if the resistance of the channel is increased. The greater the negative voltage on the gate, the lesser is the current from source to drain.

Thus by changing the negative voltage on the gate, we can vary the resistance of the *n*-channel and hence the current from source to drain. Note that with negative voltage to the gate, the action of *D-MOSFET* is similar to *JFET*. Because the action with negative gate depends upon depleting (*i.e.* emptying) the channel of free electrons, the negative-gate operation is called *depletion mode.*

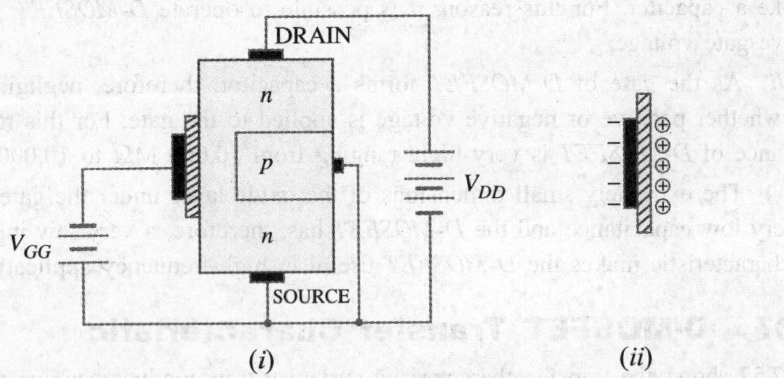

Fig. 17.55

(*ii*) **Enhancement mode.** Fig. 17.56 (*i*) shows enhancement-mode operation of *n*-channel *D-MOSFET*. Again, the gate acts like a capacitor. Since the gate is positive, it induces negative charges in the *n*-channel as shown in Fig. 17.56 (*ii*). These negative charges are the free electrons drawn into the channel. Because these free electrons are added to those already in the channel, the total number of free electrons in the channel is increased. Thus a positive gate voltage *enhances* or *increases* the conductivity of the channel. The greater the positive voltage on the gate, greater the conduction from source to drain.

Thus by changing the positive voltage on the gate, we can change the conductivity of the channel. The main difference between *D-MOSFET* and *JFET* is that we can apply positive

* If one plate of the capacitor is negatively charged, it induces positive charge on the other plate.

gate voltage to *D-MOSFET* and still have essentially *zero current. Because the action with a positive gate depends upon *enhancing* the conductivity of the channel, the positive gate operation is called *enhancement mode.*

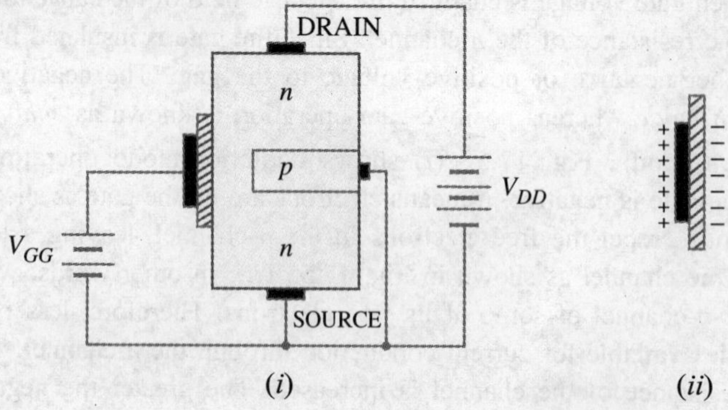

Fig. 17.56

The following points may be noted about *D-MOSFET* operation :

(*i*) In a *D-MOSFET*, the source to drain current is controlled by the electric field of capacitor formed at the gate.

(*ii*) The gate of *JFET* behaves as a reverse-biased diode whereas the gate of a *D-MOSFET* acts like a capacitor. For this reason, it is possible to operate *D-MOSFET* with positive or negative gate voltage.

(*iii*) As the gate of *D-MOSFET* forms a capacitor, therefore, negligible gate current flows whether positive or negative voltage is applied to the gate. For this reason, the input impedance of *D-MOSFET* is very high, ranging from 10,000 MΩ to 10,000,00 MΩ.

(*iv*) The extremely small dimensions of the oxide layer under the gate terminal result in a very low capacitance and the *D-MOSFET* has, therefore, a very low input capacitance. This characteristic makes the *D-MOSFET* useful in high-frequency applications.

17.37 D-MOSFET Transfer Characteristic

Fig. 17.57 shows the transfer characteristic curve (or transconductance curve) for *n*-channel *D-MOSFET*. The behaviour of this device can be beautifully explained with the help of this curve as under :

(*i*) The point on the curve where $V_{GS} = 0$, $I_D = I_{DSS}$. It is expected because I_{DSS} is the value of I_D when gate and source terminals are shorted *i.e.* $V_{GS} = 0$.

(*ii*) As V_{GS} goes *negative*, I_D decreases below the value of I_{DSS} till I_D reaches zero when $V_{GS} = V_{GS\,(off)}$ just as with *JFET*.

(*iii*) When V_{GS} is *positive*, I_D increases above the value of I_{DSS}. The maximum allowable value of I_D is given on the data sheet of *D-MOSFET*.

* Note that gate of *JFET* is always reverse biased for proper operation. However, in a *MOSFET*, because of the insulating layer, a negligible gate current flows whether we apply negative or positive voltage to gate.

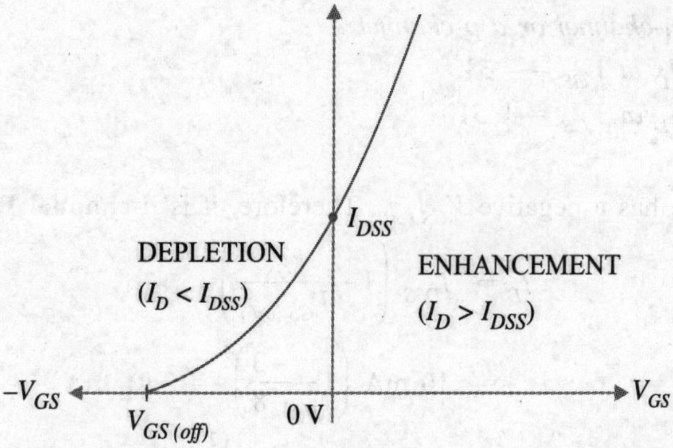

Fig. 17.57

Note that the transconductance curve for the *n*-channel *D-MOSFET* is very similar to the curve for an *n*-channel *JFET*. Because of this similarity, the *JFET* and the *D-MOSFET* have the same transconductance equation *viz.*

$$I_D = I_{DSS}\left(1 - \frac{V_{GS}}{V_{GS\,(off)}}\right)^2$$

17.38 D-MOSFET Drain Characteristics

The drain characteristics of a D-MOSFET show the graphs between drain currents (I_D) and drain-source voltages (V_{DS}) for various positive and negative gate-source voltages (V_{GS}). Fig. 17.58 shows the drain characteristics of a typical *n*-channel *D-MOSFET*. For positive values of V_{GS}, the drain current I_D increases and *D-MOSFET* operates in the *enhancement mode*. It is because for *n*-channel *D-MOSFET*, a positive gate voltage pulls more charge carriers (*i.e.* free electrons in this case) into the channel. This increases the drain current and puts the device in the enhancement mode. For negative values of V_{GS}, the drain current I_D decreases and *D-MOSFET* operates in the *depletion mode*. This is due to the fact that negative gate voltage pushes the charge carriers (*i.e.* free electrons) out of the *n*-channel and drain current I_D decreases. As a result, *D-MOSFET* is forced into depletion mode operation. Note that a significant part of drain curves is horizontal, showing constant drain current. It means that during this part of the curve, a *D-MOSFET behaves as a constant-current source*. A *D-MOSFET* is normally operated in this region.

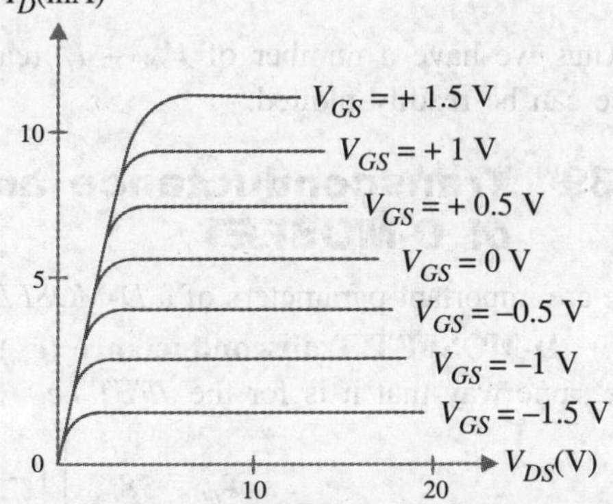

Fig. 17.58

Example 17.33. *For a certain D-MOSFET, I_{DSS} = 10 mA and $V_{GS\,(off)}$ = – 8V.*

(i) Is this an n-channel or a p-channel ?

(ii) Calculate I_D at V_{GS} = – 3V.

(iii) Calculate I_D at V_{GS} = + 3V.

Solution.

(*i*) The device has a negative $V_{GS\,(off)}$. Therefore, it is ***n*-channel D-MOSFET.**

(*ii*)

$$I_D = I_{DSS}\left(1-\frac{V_{GS}}{V_{GS\,(off)}}\right)^2$$

$$= 10\text{ mA}\left(1-\frac{-3}{-8}\right)^2 = \mathbf{3.91\ mA}$$

(*iii*)

$$I_D = I_{DSS}\left(1-\frac{V_{GS}}{V_{GS\,(off)}}\right)^2$$

$$= 10\text{ mA}\left(1-\frac{+3V}{-8V}\right)^2 = \mathbf{18.9\ mA}$$

Example 17.34. *A D-MOSFET has parameters of $V_{GS\,(off)}$ = – 6V and I_{DSS} = 1 mA. How will you plot the transconductance curve for the device ?*

Solution. When V_{GS} = 0 V, $I_D = I_{DSS}$ = 1 mA and when $V_{GS} = V_{GS\,(off)}$, I_D = 0A. This locates two points *viz* I_{DSS} and $V_{GS\,(off)}$ on the transconductance curve. We can locate more points of the curve by *changing V_{GS} values.

When V_{GS} = – 3V ; $I_D = 1\text{ mA}\left(1-\frac{-3V}{-6V}\right)^2$ = 0.25 mA

When V_{GS} = – 1V ; $I_D = 1\text{ mA}\left(1-\frac{-1V}{-6V}\right)^2$ = 0.694 mA

When V_{GS} = + 1V ; $I_D = 1\text{ mA}\left(1-\frac{+1V}{-6V}\right)^2$ = 1.36 mA

When V_{GS} = + 3V ; $I_D = 1\text{ mA}\left(1-\frac{+3V}{-6V}\right)^2$ = 2.25 mA

Thus we have a number of $V_{GS} - I_D$ readings so that transconductance curve for the device can be readily plotted.

17.39 Transconductance and Input Impedance of D-MOSFET

These are important parameters of a *D-MOSFET* and a brief discussion on them is desirable.

(*i*) **D-MOSFET Transconductance (g_m).** The value of g_m is found for a *D-MOSFET* in the same way that it is for the *JFET i.e.*

$$g_m = g_{mo}\left(1-\frac{V_{GS}}{V_{GS\,(off)}}\right)$$

* We can only change V_{GS} because the values of I_{DSS} and $V_{GS\,(off)}$ are *constant* for a given *D-MOSFET.*

(ii) **D-MOSFET Input Impedance.** The gate impedance of a *D-MOSFET* is extremely high. For example, a typical *D-MOSFET* may have a maximum gate current of 10 pA when V_{GS} = 35V.

$$\therefore \quad \text{Input impedance} = \frac{35\,\text{V}}{10\,\text{pA}} = \frac{35\,\text{V}}{10 \times 10^{-12}\,\text{A}} = 3.5 \times 10^{12}\,\Omega$$

With an input impedance in this range, *D-MOSFET* would present virtually no load to a source circuit.

17.40 D-MOSFET Biasing

The following methods may be used for *D-MOSFET* biasing :

(i) Gate bias *(ii)* Self-bias

(iii) Voltage-divider bias *(iv)* Zero-bias

The first three methods are exactly the same as those used for *JFETs* and are not discussed here. However, the last method of zero-bias is widely used in *D-MOSFET* circuits.

Zero-bias. Since a *D-MOSFET* can be operated with either positive or negative values of V_{GS}, we can set its *Q*-point at V_{GS} = 0V as shown in Fig. 17.59. Then an input a.c. signal to the gate can produce variations above and below the *Q*-point.

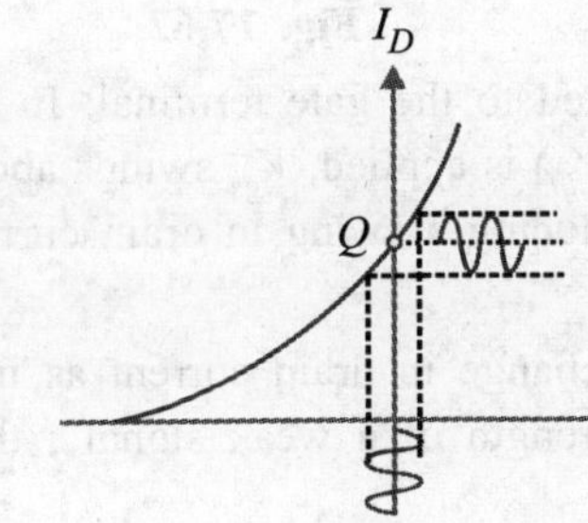

Fig. 17.59

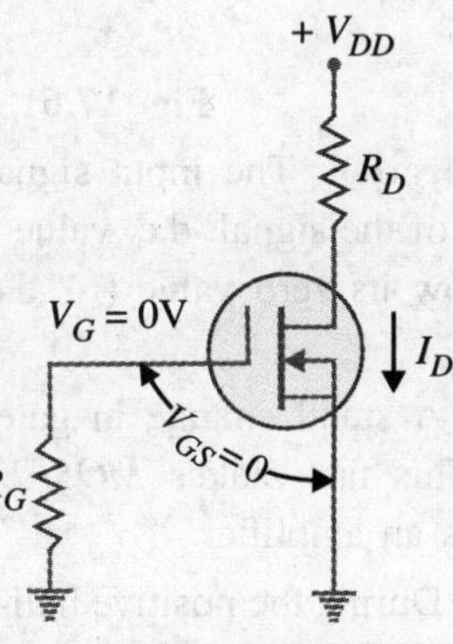

Fig. 17.60

We can use the simple circuit of Fig. 17.60 to provide zero-bias. This circuit has V_{GS} = 0V and $I_D = I_{DSS}$. We can find V_{DS} as under :

$$V_{DS} = V_{DD} - I_{DSS}\, R_D$$

Note that for the *D-MOSFET* zero-bias circuit, the source resistor (R_S) is not necessary. With no source resistor, the value of V_S is 0V. This gives us a value of V_{GS} = 0V. This biases the circuit at $I_D = I_{DSS}$ and V_{GS} = 0V. For mid-point biasing, the value of R_D is so selected that $V_{DS} = V_{DD}/2$.

Example 17.35. *Determine the drain-to-source voltage (V_{DS}) in the circuit shown in Fig. 17.60 above if V_{DD} = +18V and R_D = 620Ω. The MOSFET data sheet gives $V_{GS\,(off)}$ = – 8V and I_{DSS} = 12 mA.*

Solution. Since $I_D = I_{DSS}$ = 12 mA, the V_{DS} is given by ;

$$V_{DS} = V_{DD} - I_{DSS}\, R_D$$

$$= 18\text{V} - (12\text{ mA})(0.62\text{ k}\Omega) = \mathbf{10.6V}$$

17.41 Common-Source D-MOSFET Amplifier

Fig. 17.61 shows a common-source amplifier using *n*-channel *D-MOSFET.* Since the source terminal is common to the input and output terminals, the circuit is called *common-source amplifier. The circuit is zero-biased with an a.c. source coupled to the gate through the coupling capacitor C_1. The gate is at approximately 0V d.c. and the source terminal is grounded, thus making V_{GS} = 0V.

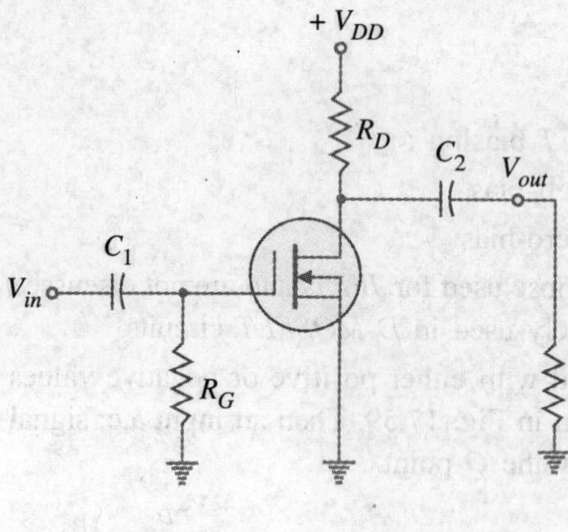

Fig. 17.61

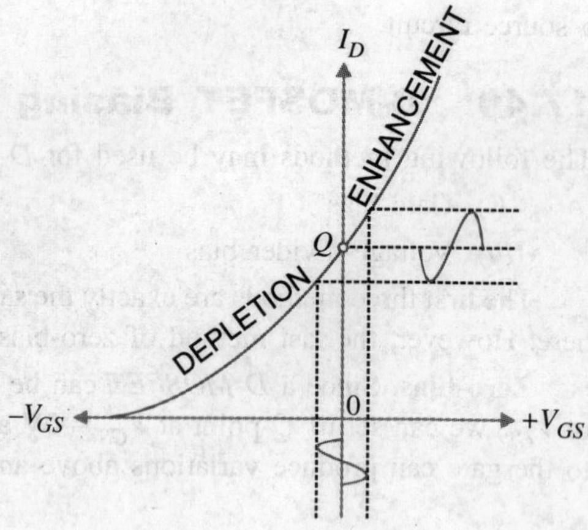

Fig. 17.62

Operation. The input signal (V_{in}) is capacitively coupled to the gate terminal. In the absence of the signal, d.c. value of V_{GS} = 0V. When signal (V_{in}) is applied, V_{gs} swings above and below its zero value ($\because$ d.c. value of V_{GS} = 0V), producing a swing in drain current I_d.

(*i*) A small change in gate voltage produces a large change in drain current as in a *JFET*. This fact makes *MOSFET* capable of raising the strength of a weak signal ; thus acting as an amplifier.

(*ii*) During the positive half-cycle of the signal, the positive voltage on the gate increases and produces the enhancement-mode. This increases the channel conductivity and hence the drain current.

(*iii*) During the negative half-cycle of the signal, the positive voltage on the gate decreases and produces depletion-mode. This decreases the conductivity and hence the drain current.

The result of above action is that a small change in gate voltage produces a large change in the drain current. This large variation in drain current produces a large a.c. output voltage across drain resistance R_D. In this way, *D-MOSFET* acts as an amplifier. Fig. 17.62 shows the amplifying action of *D-MOSFET* on transconductance curve.

Voltage gain. The a.c. analysis of *D-MOSFET* is similar to that of the *JFET*. Therefore, voltage gain expressions derived for *JFET* are also applicable to *D-MOSFET*.

Voltage gain, $A_v = g_m R_D$ ... for unloaded *D-MOSFET* amplifier

$= g_m R_{AC}$ ... for loaded *D-MOSFET* amplifier

Note the total a.c. drain resistance $R_{AC} = R_D \parallel R_L$.

* It is comparable to common-emitter transistor amplifier.

Example 17.36. *The D-MOSFET used in the amplifier of Fig. 17.63 has an I_{DSS} = 12 mA and g_m = 3.2 mS. Determine (i) d.c. drain-to-source voltage V_{DS} and (ii) a.c. output voltage. Given v_{in} = 500 mV.*

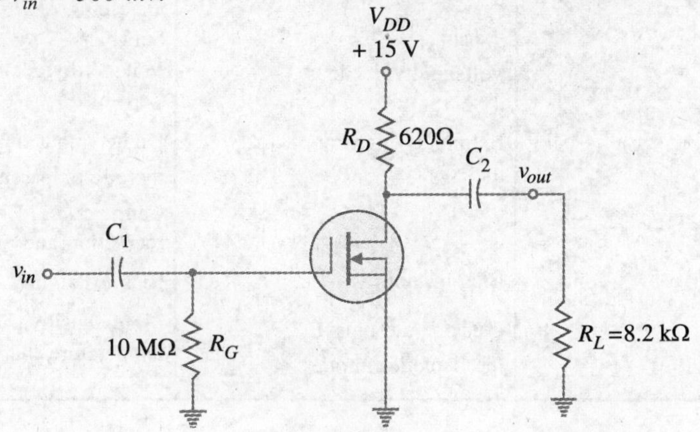

Fig. 17.63

Solution.

(*i*) Since the amplifier is zero-biased, $I_D = I_{DSS}$ = 12 mA.

$$\therefore \quad V_{DS} = V_{DD} - I_{DSS}\, R_D$$

$$= 15\text{V} - (12 \text{ mA})\ (0.62 \text{ k}\Omega) = \mathbf{7.56V}$$

(*ii*) Total a.c. drain resistance R_{AC} of the circuit is

$$R_{AC} = R_D \parallel R_L = 620\Omega \parallel 8.2 \text{ k}\Omega = 576\Omega$$

$$\therefore \quad v_{out} = A_v \times v_{in} = (g_m\, R_{AC})\ (v_{in})$$

$$= (3.2 \times 10^{-3} \text{ S} \times 576\ \Omega)\ (500 \text{ mV}) = \mathbf{922 \text{ mV}}$$

17.42 D-MOSFETs Versus JFETs

Table below summarises many of the characteristics of *JFET*s and *D-MOSFET*s.

Devices :	JFETs	D-MOSFETs
Schematic symbol :	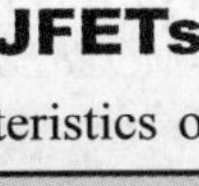	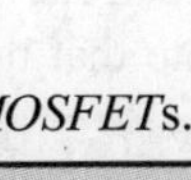
Transconductance curve :	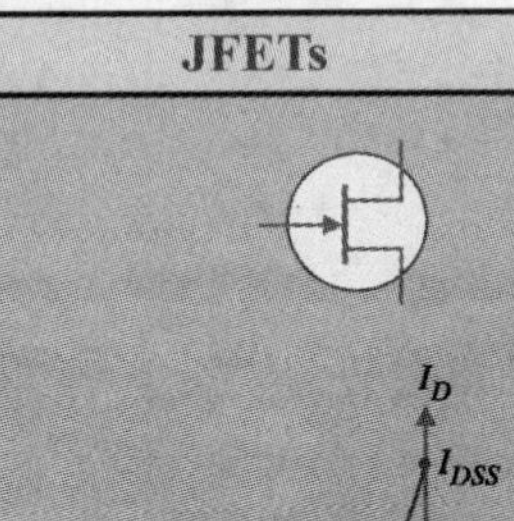	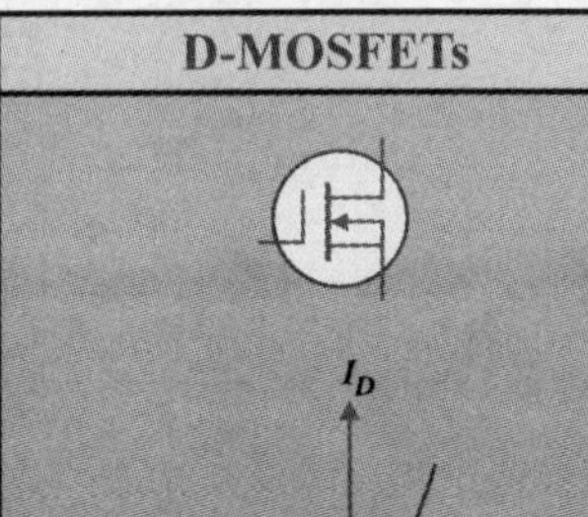

Modes of operation :	Depletion only	Depletion and enhancement
Commonly used bias circuits :	Gate bias Self-bias Voltage-divider bias	Gate bias Self-bias Voltage-divider bias Zero-bias
Advantages :	Extremely high input impedance.	Higher input impedance than a comparable *JFET*. Can operate in both modes (depletion and enhancement).
Disadvantages :	Bias instability. Can operate only in the depletion mode.	Bias instability. More sensitive to changes in temperature than the *JFET*.

17.43 E-MOSFET

Two things are worth noting about *E-MOSFET*. First, *E-MOSFET* operates *only* in the enhancement mode and has no depletion mode. Secondly, the *E-MOSFET* has no physical channel from source to drain because the substrate extends completely to the SiO_2 layer [See Fig. 17.64 (*i*)]. It is only by the application of V_{GS} (gate-to-source voltage) of proper magnitude and polarity that the device starts conducting. The minimum value of V_{GS} of proper polarity that turns on the *E-MOSFET* is called *Threshold voltage* [$V_{GS\ (th)}$]. The *n*-channel device requires positive V_{GS} ($\geqslant V_{GS\ (th)}$) and the *p*-channel device requires negative V_{GS} ($\geqslant V_{GS\ (th)}$).

Operation. Fig. 17.64 (*i*) shows the circuit of *n*-channel *E-MOSFET*. The circuit action is as under :

(*i*) When V_{GS} = 0V [See Fig. 17.64(*i*)], there is no channel connecting the source and drain. The *p* substrate has only a few thermally produced free electrons (minority carriers) so that drain current is essentially zero. For this reason, *E-MOSFET* is normally OFF when V_{GS} = 0 V. Note that this behaviour of *E-MOSFET* is quite different from *JFET* or *D-MOSFET*.

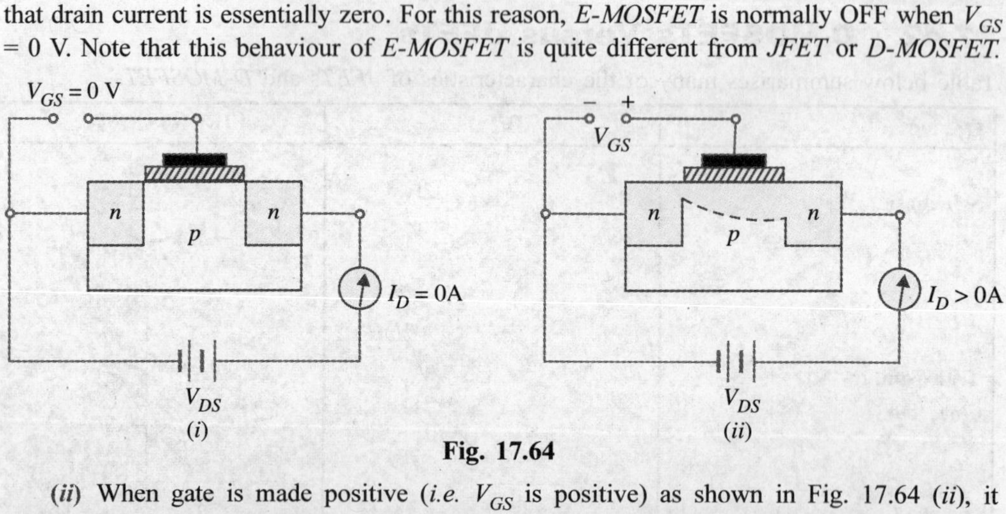

Fig. 17.64

(*ii*) When gate is made positive (*i.e.* V_{GS} is positive) as shown in Fig. 17.64 (*ii*), it attracts free electrons into the *p* region. The free electrons combine with the holes next to the SiO_2 layer. If V_{GS} is positive enough, all the holes touching the SiO_2 layer are filled and free electrons begin to flow from the source to drain. The effect is the same as creating a thin layer of *n*-type material (*i.e.* inducing a thin *n*-channel) adjacent to the SiO_2 layer. Thus the *E-MOSFET* is turned ON and drain current I_D starts flowing form the source to the drain.

The minimum value of V_{GS} *that turns the E-MOSFET ON is called* **threshold voltage** $[V_{GS\,(th)}]$.

(*iii*) When V_{GS} is less than $V_{GS\,(th)}$, there is no induced channel and the drain current I_D is zero. When V_{GS} is equal to $V_{GS\,(th)}$, the *E-MOSFET* is turned ON and the induced channel conducts drain current from the source to the drain. Beyond $V_{GS\,(th)}$, if the value of V_{GS} is *increased*, the newly formed channel becomes wider, causing I_D to increase. If the value of V_{GS} decreases [not less than $V_{GS\,(th)}$], the channel becomes narrower and I_D will decrease. This fact is revealed by the transconductance curve of *n*-channel *E-MOSFET* shown in Fig. 17.65. As you can see, $I_D = 0$ when $V_{GS} = 0$. Therefore, the value of I_{DSS} for the *E-MOSFET* is zero. Note also that there is no drain current until V_{GS} reaches $V_{GS\,(th)}$.

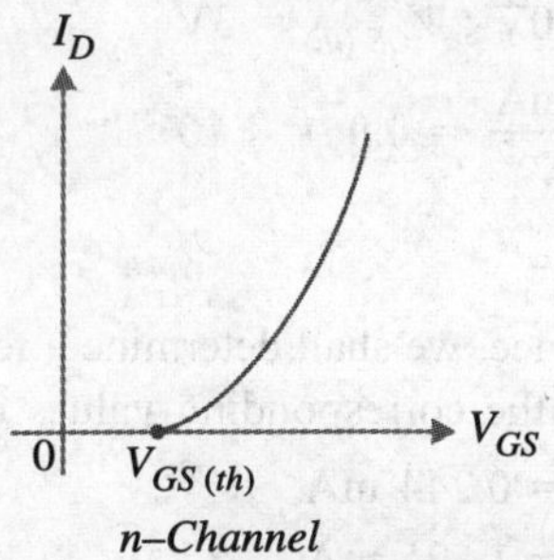

Fig. 17.65

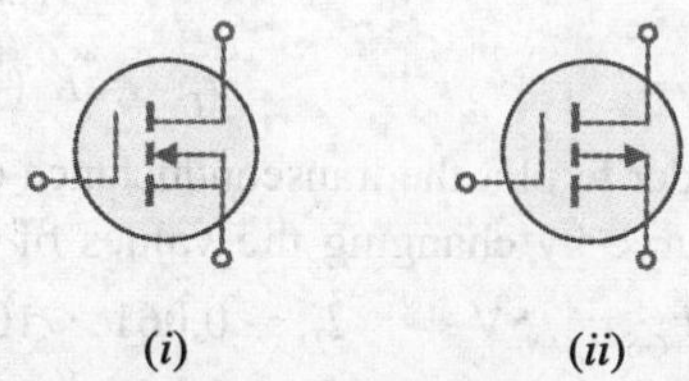

Fig. 17.66

Schematic Symbols. Fig. 17.66 (*i*) shows the schematic symbols for *n*-channel *E-MOSFET* whereas Fig. 17.66 (*ii*) shows the schematic symbol for *p*-channel *E-MOSFET.* When $V_{GS} = 0$, the *E-MOSFET* is OFF because there is no conducting channel between source and drain. The broken channel line in the symbols indicates the normally OFF condition.

Equation for Transconductance Curve. Fig. 17.67 shows the transconductance curve for *n*-channel *E-MOSFET*. Note that this curve is different from the transconductance curve for *n*-channel *JFET* or *n*-channel *D-MOSFET*. It is because it starts at $V_{GS\,(th)}$ rather than $V_{GS\,(off)}$ on the horizontal axis and never intersects the vertical axis. The equation for the *E-MOSFET* transconductance curve (for $V_{GS} > V_{GS\,(th)}$) is

$$I_D = K\,(V_{GS} - V_{GS\,(th)})^2$$

The constant *K* depends on the particular *E-MOSFET* and its value is determined from the following equation :

$$K = \frac{I_{D\,(on)}}{(V_{GS\,(on)} - V_{GS\,(th)})^2}$$

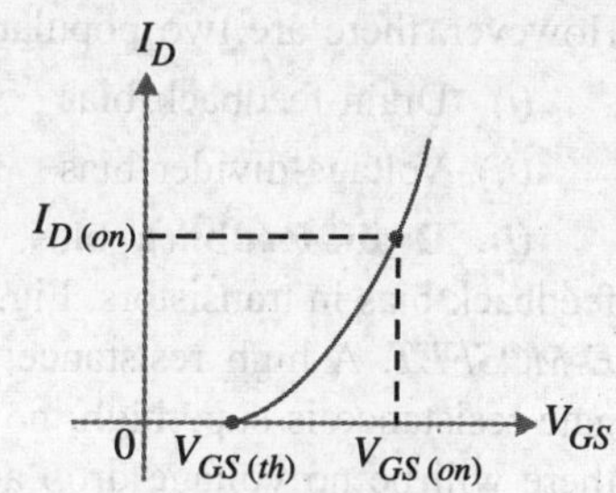

Fig. 17.67

Any data sheet for an *E-MOSFET* will include the current $I_{D(on)}$ and the voltage $V_{GS\,(on)}$ for one point well above the threshold voltage as shown in Fig. 17.67.

Example 17.37. *The data sheet for an E-MOSFET gives* $I_{D(on)}$ = *500 mA at* V_{GS} = *10V and* $V_{GS\,(th)}$ = *1V. Determine the drain current for* V_{GS} = *5V.*

Solution. Here, $V_{GS\,(on)} = 10$ V.

$$I_D = K\,(V_{GS} - V_{GS\,(th)})^2 \qquad ...\,(i)$$

Here,
$$K = \frac{I_{D\,(on)}}{(V_{GS\,(on)} - V_{GS\,(th)})^2} = \frac{500\text{ mA}}{(10\text{V} - 1\text{V})^2} = 6.17\text{ mA/V}^2$$

Putting the various values in eq. (*i*), we have,

$$I_D = 6.17\,(5\text{V} - 1\text{V})^2 = \mathbf{98.7\ mA}$$

Example 17.38. *The data sheet for an E-MOSFET gives $I_{D\,(on)} = 3$ mA at $V_{GS} = 10V$ and $V_{GS\,(th)} = 3V$. Determine the resulting value of K for the device. How will you plot the transconductance curve for this MOSFET ?*

Solution. The value of K can be determined from the following equation :

$$K = \frac{I_{D\,(on)}}{(V_{GS\,(on)} - V_{GS\,(th)})^2}$$

Here, $\quad I_{D\,(on)} = 3\text{ mA}\ ;\ V_{GS\,(on)} = 10\text{V}\ ;\ V_{GS\,(th)} = 3\text{V}$

$$\therefore \quad K = \frac{3\text{ mA}}{(10\text{V} - 3\text{V})^2} = \frac{3\text{ mA}}{(7\text{V})^2} = \mathbf{0.061 \times 10^{-3}\ A/V^2}$$

Now, $\quad I_D = K\,(V_{GS} - V_{GS\,(th)})^2$

In order to plot the transconductance curve for the device, we shall determine a few points for the curve by changing the values of V_{GS} and noting the corresponding values of I_D.

For $V_{GS} = 5\text{V}$; $I_D = 0.061 \times 10^{-3}\,(5\text{V} - 3\text{V})^2 = 0.244$ mA

For $V_{GS} = 8\text{V}$; $I_D = 0.061 \times 10^{-3}\,(8\text{V} - 3\text{V})^2 = 1.525$ mA

For $V_{GS} = 10\text{V}$; $I_D = 0.061 \times 10^{-3}\,(10\text{V} - 3\text{V})^2 = 3$ mA

For $V_{GS} = 12\text{V}$; $I_D = 0.061 \times 10^{-3}\,(12\text{V} - 3\text{V})^2 = 4.94$ mA

Thus we can plot the transconductance curve for the *E-MOSFET* from these V_{GS}/I_D points.

17.44 E-MOSFET Biasing Circuits

One of the problems with *E-MOSFET* is the fact that many of the biasing circuits used for *JFETs* and *D-MOSFETs* cannot be used with this device. For example, *E-MOSFETs* must have V_{GS} greater than the threshold value ($V_{GS\,(th)}$) so that zero bias cannot be used. However, there are two popular methods for *E-MOSFET* biasing *viz.*

(*i*) Drain-feedback bias

(*ii*) Voltage-divider bias

(*i*) Drain-feedback bias. This method of *E-MOSFET* bias is equivalent to collector-feedback bias in transistors. Fig. 17.68 (*i*) shows the drain-feedback bias circuit for *n*-channel *E-MOSFET*. A high resistance R_G is connected between the drain and the gate. Since the gate resistance is superhigh, no current will flow in the gate circuit (*i.e.* $I_G = 0$). Therefore, there will be no voltage drop across R_G. Since there is no voltage drop across R_G, the gate will be at the same potential as the drain. This fact is illustrated in the d.c. equivalent circuit of drain-feedback bias as in Fig. 17.68 (*ii*).

$$\therefore \quad V_D = V_G \quad \text{and} \quad V_{DS} = V_{GS}$$

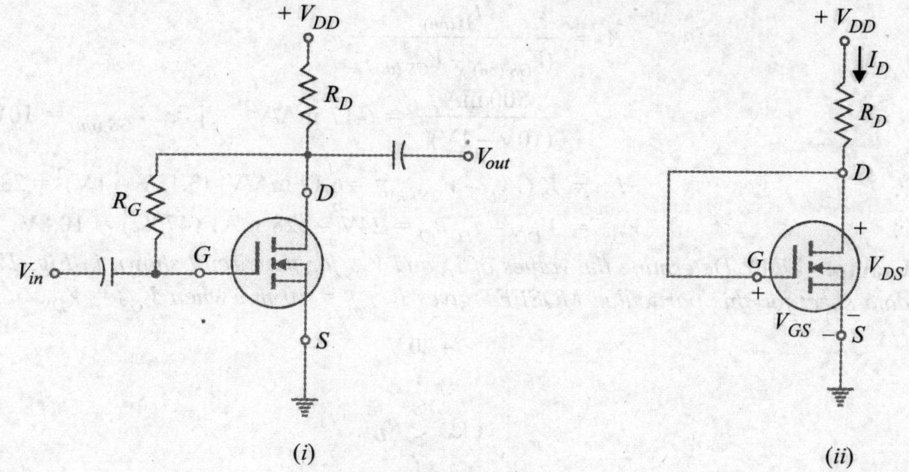

Fig. 17.68

The value of drain-source voltage V_{DS} for the drain-feedback circuit is

$$V_{DS} = V_{DD} - I_D R_D$$

$$\text{Since } V_{DS} = V_{GS},\ V_{GS} = V_{DD} - I_D R_D$$

Since in this circuit $V_{DS} = V_{GS}$; $I_D = I_{D\,(on)}$.

Therefore, the Q-point of the circuit stands determined.

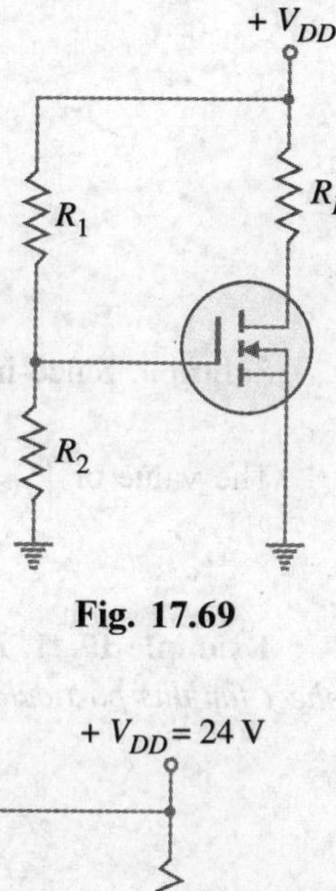

Fig. 17.69

(*ii*) Voltage-divider Bias. Fig. 17.69 shows voltage-divider biasing arrangement for *n*-channel *E-MOSFET*. Since $I_G = 0$, the analysis of the method is as follows :

$$V_{GS} = \frac{V_{DD}}{R_1 + R_2} \times R_2$$

and
$$V_{DS} = V_{DD} - I_D R_D$$

where
$$I_D = K (V_{GS} - V_{GS\,(th)})^2$$

Once I_D and V_{DS} are known, all the remaining quantities of the circuit such as V_D *etc.* can be determined.

Example 17.39. *Determine V_{GS} and V_{DS} for the E-MOSFET circuit in Fig. 17.70. The data sheet for this particular MOSFET gives $I_{D\,(on)}$ = 500 mA at V_{GS} = 10V and $V_{GS\,(th)}$ = 1V.*

Solution. Referring to the circuit shown in Fig. 17.70, we have,

$$V_{GS} = \frac{V_{DD}}{R_1 + R_2} \times R_2$$

$$= \frac{24\text{V}}{(100 + 15)\ \text{k}\Omega} \times 15\ \text{k}\Omega = 3.13\text{V}$$

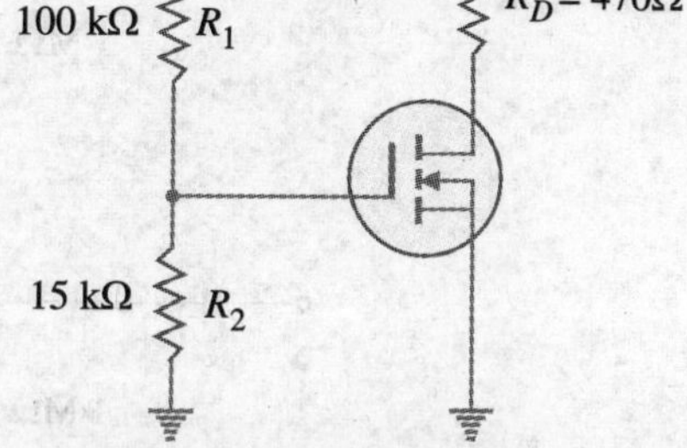

Fig. 17.70

The value of K can be determined from the following equation :

$$K = \frac{I_{D(on)}}{(V_{GS(on)} - V_{GS(th)})^2}$$

$$= \frac{500 \text{ mA}}{(10\text{V} - 1\text{V})^2} = 6.17 \text{ mA/V}^2 \quad [\because V_{GS(on)} = 10\text{V}]$$

$$\therefore \quad I_D = K(V_{GS} - V_{GS(th)})^2 = 6.17 \text{ mA/V}^2 \, (3.13\text{V} - 1\text{V})^2 = 28 \text{ mA}$$

$$\therefore \quad V_{DS} = V_{DD} - I_D R_D = 24\text{V} - (28 \text{ mA})(470\Omega) = \mathbf{10.8V}$$

Example 17.40. *Determine the values of I_D and V_{DS} for the circuit shown in Fig. 17.71. The data sheet for this particular MOSFET gives $I_{D(on)}$ = 10 mA when $V_{GS} = V_{DS}$.*

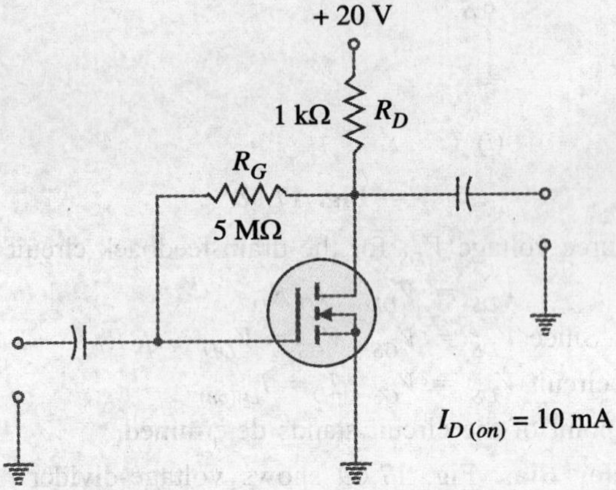

Fig. 17.71

Solution. Since in the drain-feedback circuit $V_{GS} = V_{DS}$,

$$\therefore \quad I_D = I_{D(on)} = \mathbf{10 \text{ mA}}$$

The value of V_{DS} (and thus V_{GS}) is given by ;

$$V_{DS} = V_{DD} - I_D R_D$$

$$= 20\text{V} - (10 \text{ mA})(1 \text{ k}\Omega) = 20\text{V} - 10\text{V} = \mathbf{10V}$$

Example 17.41. *Determine the value of I_D for the circuit shown in Fig. 17.72. The data sheet for this particular MOSFET gives $I_{D(on)}$ = 10 mA at V_{GS} = 10 V and $V_{GS(th)}$ = 1.5 V.*

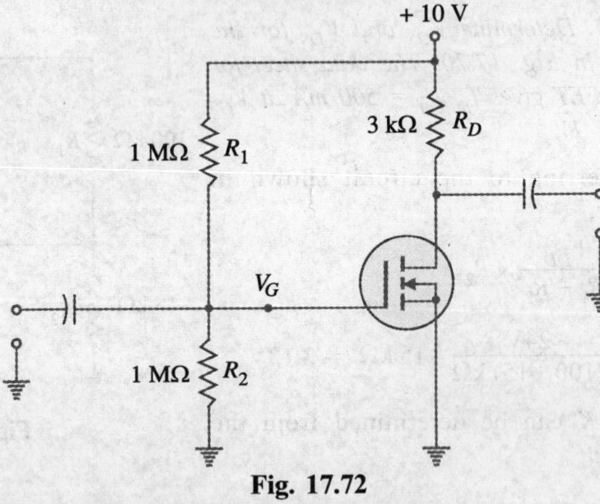

Fig. 17.72

Solution. The value of K can be determined from the following equation :

$$K = \frac{I_{D(on)}}{(V_{GS(on)} - V_{GS(th)})^2}$$

$$= \frac{10 \text{ mA}}{(10\text{ V} - 1.5\text{V})^2} = 1.38 \times 10^{-1} \text{ mA/V}^2$$

$$[\because V_{GS(on)} = 10\text{V}]$$

From the circuit, the source voltage is seen to be 0V. Therefore, $V_{GS} = V_G - V_S = V_G - 0 = V_G$. The value of V_G (= V_{GS}) is given by ;

$$V_G \text{ (or } V_{GS}) = \frac{V_{DD}}{R_1 + R_2} \times R_2 = \frac{10\text{V}}{(1+1)\text{ M}\Omega} \times 1\text{M}\Omega = 5\text{V}$$

$$\therefore \quad I_D = K\,(V_{GS} - V_{GS(th)})^2$$

$$= (1.38 \times 10^{-1} \text{ mA/V}^2)\,(5\text{V} - 1.5\text{V})^2 = \mathbf{1.69\ mA}$$

17.45 D-MOSFETs Versus E-MOSFETs

Table below summarises many of the characteristics of *D-MOSFETs* and *E-MOSFETs*.

Devices:	D-MOSFETs	E-MOSFETs
Schematic symbol :	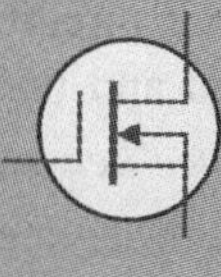	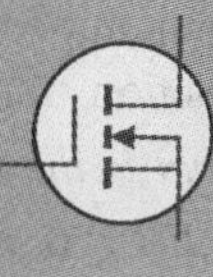
Transconductance curve :	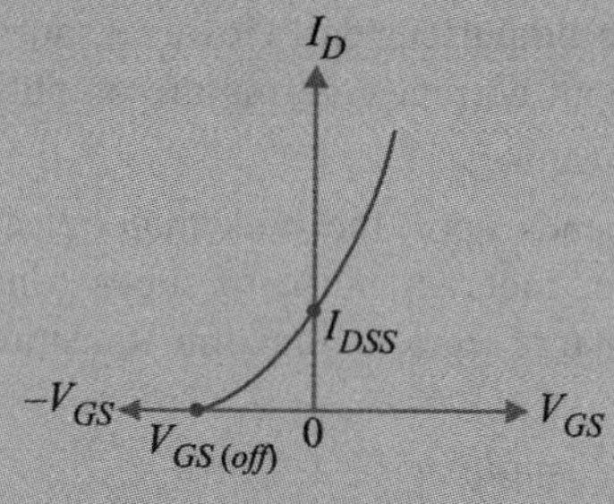	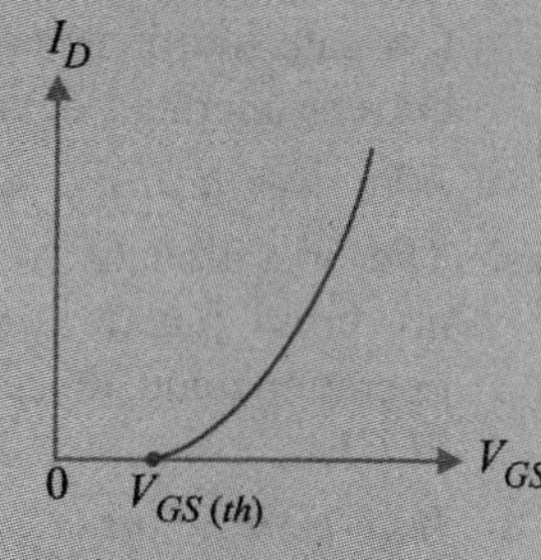
Modes of operation :	Depletion and enhancement.	Enhancement only.
Commonly used bias circuits :	Gate bias Self-bias Voltage-divider bias Zero-bias	Gate bias Voltage-divider bias Drain-feedback bias

Short Answer Questions

Q. 1. *What is the working principle of JFET ?*

Ans. The conductivity of the channel between source and drain is controlled by the width of depletion layers at either side of the channel. If the reverse voltage on the gate increases, the channel width decreases. This increases the resistance of the channel and hence source to drain current decreases. Reverse happens should the reverse gate voltage decrease.

Q. 2. *What are the JFET polarities in normal operation ?*

Ans. In normal operation of *JFET*, the voltage between gate and source is such that gate is reverse biased. The drain and source terminals are interchangeable *i.e.* either end of silicon bar can be used as the source and the other as the drain.

Q. 3. *Why is JFET called a unipolar transistor ?*

Ans. In a *JFET,* current conduction is by one type of carrier *viz.* holes in *p*-channel *JFET* and electrons in *n*-channel *JFET*. '*Uni*' means single and polar indicates charge carrier. Hence *JFET* is sometimes called a unipolar transistor. However, in an ordinary transistor, both holes and electrons play part in conduction. Therefore, an ordinary transistor is sometimes called a bipolar transistor.

Q. 4. *What is the origin of the name JFET ?*

Ans. In a *JFET,* current from source to drain can be controlled by the application of potential (*i.e.*, electric field) on the gate. For this reason, the device is called junction field effect transistor (*JFET*).

Q. 5. *Why is the input impedance of a JFET higher than that of ordinary transistor ?*

Ans. The input circuit of the *JFET* is reverse-biased. Consequently, it offers a very high input impedance. However, the input circuit of an ordinary transistor (or bipolar junction transistor) is forward biased and hence has low input impedance.

Q. 6. *What is the basic operational difference between a JFET and an ordinary transistor ?*

Ans. A *JFET* is essentially a voltage-driven device like a vacuum tube *i.e.* input voltage (on gate terminal) controls the output current. However, an ordinary transistor is a current-operated device *i.e.* input current controls the output current.

Q. 7. *Why does a JFET has low noise level ?*

Ans. The operation of a *JFET* depends upon the bulk material current carriers that do not cross junctions. Therefore, inherent noise of tubes (due to high temperature operation) and those of transistors (due to junction transitions) are not present in *JFET*.

Q. 8. *Why is JFET used as a buffer amplifier ?*

Ans. A buffer amplifier is a stage of amplification that isolates the preceding stage from the following stage. Because of the high input impedance and low output impedance, a *JFET* can act as an excellent buffer amplifier.

Q. 9. *What are the advantages of JFET over ordinary transistors ?*

Ans. The advantages of *JFET* over ordinary transistors (or bipolar junction transistors) are :

(*i*) High input impedance and low output impedance. This permits high degree of isolation between input and output circuits.

(*ii*) A low noise level.

(*iii*) Very high power gain. This eliminates the necessity of using driver stages.

(*iv*) Smaller size, longer life and higher efficiency.

Q. 10. *What is the working principle of MOSFET ?*

Ans. The *MOSFET* operates on the principle of controlling the conductivity of the channel between source and drain by the electric field of a capacitor formed at the gate.

Q. 11. *Why is the input impedance of a MOSFET higher than that of a JFET ?*

Ans. The gate of a *JFET* behaves like a reverse biased diode while the gate of a *MOSFET* acts like a capacitor. For this reason, the gate current in a *MOSFET* is essentially zero whether we apply positive or negative voltage to the gate, Hence, the input impedance of a *MOSFET* is very large, larger than that of *JFET*.

Q. 12. *Why does JFET has less voltage gain than that of an ordinary transistor ?*

Ans. Due to large input impedance of *JFET*, it has smaller control over output current. In other words, a *JFET* is less sensitive to changes in input voltage than a bipolar transistor. For example, an input change of 0.2 V in almost any *JFET* will hardly produce a change of 5 mA in the output current. But in a bipolar transistor, 0.1 V change in input voltage can easily produce more than 10 mA change in the output current. This means that a *JFET* will have less voltage gain than a bipolar transistor.

Multiple-Choice Questions

1. A *JFET* has three terminals, namely ....
(*i*) cathode, anode, grid
(*ii*) emitter, base, collector
(*iii*) source, gate, drain
(*iv*) none of the above

2. A *JFET* is similar in operation to .... valve.
(*i*) diode (*ii*) pentode
(*iii*) triode (*iv*) tetrode

3. A *JFET* is also called .... transistor.
(*i*) unipolar (*ii*) bipolar
(*iii*) unijunction (*iv*) none of the above

4. A *JFET* is a .... driven device.
(*i*) current
(*ii*) voltage
(*iii*) both current and voltage
(*iv*) none of the above

5. The gate of a *JFET* is .... biased.
(*i*) reverse
(*ii*) forward
(*iii*) reverse as well as forward
(*iv*) none of the above

6. The input impedance of a *JFET* is .... that of an ordinary transistor.
(*i*) equal to (*ii*) less than
(*iii*) more than (*iv*) none of the above

7. In a *p*-channel *JFET*, the charge carriers are ....
(*i*) electrons (*ii*) holes
(*iii*) both electrons and holes
(*iv*) none of the above

8. When drain voltage equals the pinch-off voltage, then drain current .... with the increase in drain voltage.
(*i*) decreases
(*ii*) increases
(*iii*) remains constant
(*iv*) none of the above

9. If the reverse bias on the gate of a *JFET* is increased, then width of the conducting channel ....
(*i*) is decreased
(*ii*) is increased
(*iii*) remains the same
(*iv*) none of the above

10. A *MOSFET* has .... terminals.
(*i*) two (*ii*) five
(*iii*) four (*iv*) three

11. A *MOSFET* can be operated with ....
(*i*) negative gate voltage only
(*ii*) positive gate voltage only

(*iii*) positive as well as negative gate voltages

(*iv*) none of the above

12. A *JFET* has .... power gain.

(*i*) small (*ii*) very high

(*iii*) very small (*iv*) none of the above

13. The input control parameter of a *JFET* is ....

(*i*) gate voltage (*ii*) source voltage

(*iii*) drain voltage (*iv*) gate current

14. A common base configuration of a *pnp* transistor is analogous to .... of a *JFET*.

(*i*) common source configuration

(*ii*) common drain configuration

(*iii*) common gate configuration

(*iv*) none of the above

15. A *JFET* has high input impedance because ....

(*i*) it is made of semiconductor material

(*ii*) input is reverse biased

(*iii*) of impurity atoms

(*iv*) none of the above

16. In a *JFET*, when drain voltage is equal to pinch-off voltage, the depletion layers ....

(*i*) almost touch each other

(*ii*) have large gap

(*iii*) have moderate gap

(*iv*) none of the above

17. In a *JFET*, I_{DSS} is known as ....

(*i*) drain to source current

(*ii*) drain to source current with gate shorted

(*iii*) drain to source current with gate open

(*iv*) none of the above

18. The two important advantages of a *JFET* are ....

(*i*) high input impedance and square-law property

(*ii*) inexpensive and high output impedance

(*iii*) low input impedance and high output impedance

(*iv*) none of the above

19. .... has the lowest noise-level.

(*i*) Triode

(*ii*) Ordinary transistor

(*iii*) Tetrode

(*iv*) *JFET*

20. A *MOSFET* is sometimes called .... *JFET*.

(*i*) many gate (*ii*) open gate

(*iii*) insulated gate (*iv*) shorted gate

21. Which of the following devices has the highest input impedance ?

(*i*) *JFET*

(*ii*) *MOSFET*

(*iii*) Crystal diode

(*iv*) Ordinary transistor

22. A *MOSFET* uses the electric field of a .... to control the channel current.

(*i*) capacitor

(*ii*) battery

(*iii*) generator

(*iv*) none of the above

23. The pinch-off voltage in a *JFET* is analogous to .... voltage in a vacuum tube.

(*i*) anode (*ii*) cathode

(*iii*) grid cut-off

(*iv*) none of the above

24. The formula for a.c. drain resistance of a *JFET* is ....

(*i*) $\dfrac{\Delta V_{DS}}{\Delta I_D}$ at constant V_{GS}

(*ii*) $\dfrac{\Delta V_{GS}}{\Delta I_D}$ at constant V_{DS}

(*iii*) $\dfrac{\Delta I_D}{\Delta V_{GS}}$ at constant V_{DS}

(*iv*) $\dfrac{\Delta I_D}{\Delta V_{DS}}$ at constant V_{GS}

25. In class *A* operation, the input circuit of a *JFET* is .... biased.

(*i*) forward (*ii*) reverse
(*iii*) not
(*iv*) none of the above

26. If the gate of a *JFET* is made less negative, the width of the conducting channel ....

(*i*) remains the same
(*ii*) is decreased
(*iii*) is increased
(*iv*) none of the above

27. The pinch-off voltage of a *JFET* is about ....

(*i*) 5 V (*ii*) 0.6 V
(*iii*) 15 V (*iv*) 25 V

28. The input impedance of a *MOSFET* is of the order of ....

(*i*) Ω
(*ii*) a few hundred Ω
(*iii*) kΩ
(*iv*) several MΩ

29. The gate voltage in a *JFET* at which drain current becomes zero is called .... voltage.

(*i*) saturation (*ii*) pinch-off
(*iii*) active (*iv*) cut-off

30. The drain current I_D in a *JFET* is given by ; ....

(*i*) $I_D = I_{DSS}\left(1 - \frac{V_{GS}}{V_P}\right)^2$

(*ii*) $I_D = I_{DSS}\left(1 + \frac{V_{GS}}{V_P}\right)^2$

(*iii*) $I_D = I_{DSS}\left(1 - \frac{V_P}{V_{GS}}\right)^2$

(*iv*) $I_D = I_{DSS}\left(1 + \frac{V_P}{V_{GS}}\right)^{1/2}$

31. In a *JFET*, there are .... *pn* junctions at the sides.

(*i*) three (*ii*) four
(*iii*) five (*iv*) two

32. The transconductance of a *JFET* ranges from ....

(*i*) 100 to 500 mA/V
(*ii*) 500 to 1000 mA/V
(*iii*) 0.5 to 30 mA/V
(*iv*) above 1000 mA/V

33. The source terminal of a *JFET* corresponds to.......... of a vacuum tube.

(*i*) plate
(*ii*) cathode
(*iii*) grid
(*iv*) none of the above

34. The output characteristics of a *JFET* closely resemble the output characteristics of a .... valve.

(*i*) pentode (*ii*) tetrode
(*iii*) triode (*iv*) diode

35. If the cross-sectional area of the channel in *n*-channel *JFET* increases, the drain current ....

(*i*) is increased
(*ii*) is decreased
(*iii*) remains the same
(*iv*) none of the above

36. The channel of a *JFET* is between the ....

(*i*) gate and drain
(*ii*) drain and source
(*iii*) gate and source
(*iv*) input and output

37. For $V_{GS} = 0$ V, the drain current becomes constant when V_{DS} exceeds ....

(*i*) cut-off (*ii*) V_{DD}
(*iii*) V_P (*iv*) 0 V

38. A certain *JFET* data sheet gives $V_{GS(off)} = -4$ V. The pinch-off voltage V_P is ......

(*i*) + 4 V (*ii*) − 4 V
(*iii*) dependent on V_{GS}
(*iv*) data insufficient

39. The constant-current region of a *JFET* lies between ....

(*i*) cut-off and saturation
(*ii*) cut-off and pinch-off
(*iii*) 0 and I_{DSS}
(*iv*) pinch-off and breakdown

40. At cut-off, the *JFET* channel is ....
   (*i*) at its widest point
   (*ii*) completely closed by the depletion region
   (*iii*) extremely narrow
   (*iv*) reverse biased

41. A *MOSFET* differs from a *JFET* mainly because ....
   (*i*) of power rating
   (*ii*) the *MOSFET* has two gates
   (*iii*) the *JFET* has a *pn* junction
   (*iv*) none of above

42. A certain *D-MOSFET* is biased at V_{GS} = 0V. Its data sheet specifies I_{DSS} = 20 mA and $V_{GS\ (off)}$ = – 5V. The value of the drain current is ....
   (*i*) 20 mA (*ii*) 0 mA
   (*iii*) 40 mA (*iv*) 10 mA

43. An *n*-channel *D-MOSFET* with a positive V_{GS} is operating in ....
   (*i*) the depletion-mode
   (*ii*) the enhancement-mode
   (*iii*) cut-off (*iv*) saturation

44. A certain *p*-channel *E-MOSFET* has a $V_{GS\,(th)}$ = – 2V. If V_{GS} = 0V, the drain current is ....
   (*i*) 0 mA (*ii*) $I_{D\ (on)}$
   (*iii*) maximum (*iv*) I_{DSS}

45. In a common-source *JFET* amplifier, the output voltage is ....
   (*i*) 180° out of phase with the input
   (*ii*) in phase with the input
   (*iii*) 90° out of phase with the input
   (*iv*) taken at the source

46. In a certain common-source *D-MOSFET* amplifier, V_{ds} = 3.2 V r.m.s. and V_{gs} = 280 mV r.m.s. The voltage gain is ....
   (*i*) 1 (*ii*) 11.4
   (*iii*) 8.75 (*iv*) 3.2

47. In a certain *CS JFET* amplifier, R_D = 1 kΩ, R_S = 560Ω, V_{DD} = 10 V and g_m = 4500 μS. If the source resistor is completely bypassed, the voltage gain is ....
   (*i*) 450 (*ii*) 45
   (*iii*) 2.52 (*iv*) 4.5

48. A certain common-source *JFET* has a voltage gain of 10. If the source bypass capacitor is removed, ...
   (*i*) the voltage gain will increase
   (*ii*) the transconductance will increase
   (*iii*) the voltage gain will decrease
   (*iv*) the *Q*-point will shift

49. A *CS JFET* amplifier has a load resistance of 10 kΩ and R_D = 820Ω. If g_m = 5 mS and V_{in} = 500 mV, the output signal voltage is ...
   (*i*) 2.05 V (*ii*) 25 V
   (*iii*) 0.5 V (*iv*) 1.89 V

50. If load resistance in *Q*. 49 is removed, the output voltage will ....
   (*i*) increase (*ii*) decrease
   (*iii*) stay the same (*iv*) be zero

Answers to Multiple-Choice Questions

1. (*iii*)	2. (*ii*)	3. (*i*)	4. (*ii*)	5. (*i*)
6. (*iii*)	7. (*ii*)	8. (*iii*)	9. (*i*)	10. (*iv*)
11. (*iii*)	12. (*ii*)	13. (*i*)	14. (*iii*)	15. (*ii*)
16. (*i*)	17. (*ii*)	18. (*i*)	19. (*iv*)	20. (*iii*)
21. (*ii*)	22. (*i*)	23. (*iii*)	24. (*i*)	25. (*ii*)
26. (*iii*)	27. (*i*)	28. (*iv*)	29. (*ii*)	30. (*i*)
31. (*iv*)	32. (*iii*)	33. (*ii*)	34. (*i*)	35. (*i*)
36. (*ii*)	37. (*iii*)	38. (*i*)	39. (*iv*)	40. (*ii*)
41. (*iii*)	42. (*i*)	43. (*ii*)	44. (*i*)	45. (*i*)
46. (*ii*)	47. (*iv*)	48. (*iii*)	49. (*iv*)	50. (*i*)

Chapter Review Topics

1. Explain the construction and working of a *JFET*.
2. What is the difference between a *JFET* and a bipolar transistor ?
3. How will you determine the drain characteristics of *JFET* ? What do they indicate?
4. Define the *JFET* parameters and establish the relationship between them.
5. Briefly describe some practical applications of *JFET*.
6. Explain the construction and working of *MOSFET*.
7. Write short notes on the following :
   (*i*) Advantages of *JFET* (*ii*) Difference between *MOSFET* and *JFET*

Problems

1. A *JFET* has a drain current of 5 mA. If I_{DSS} = 10 mA and $V_{GS(off)}$ is – 6 V, find the value of (*i*) V_{GS} and (*ii*) V_P. **[(*i*) – 1.5 V (*ii*) 6 V]**
2. A *JFET* has an I_{DSS} of 9 mA and a $V_{GS\ (off)}$ of – 3V. Find the value of drain current when V_{GS} = –1.5V. **[2.25mA]**
3. In the *JFET* circuit shown in Fig. 17.73 if I_D = 1.9 mA, find V_{GS} and V_{DS}. **[– 1.56V; 13.5V]**

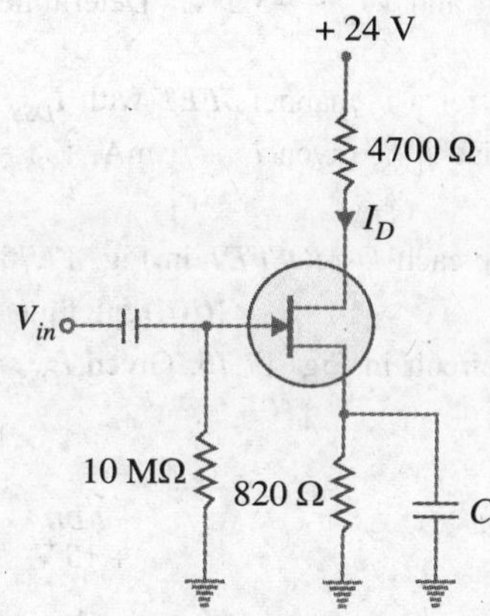

Fig. 17.73

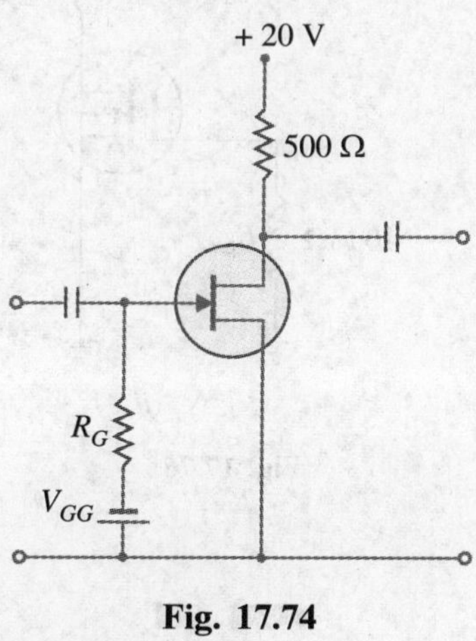

Fig. 17.74

4. For the *JFET* amplifier shown in Fig. 17.74, draw the d.c. load line.

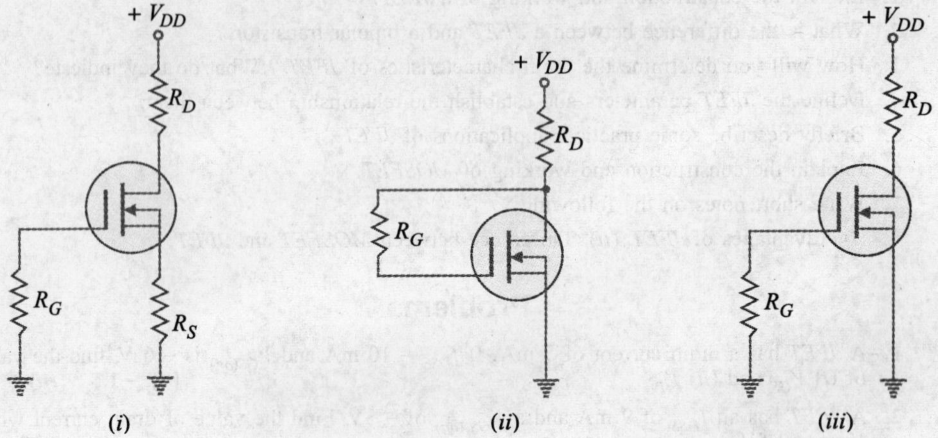

Fig. 17.75

5. For a *JFET*, I_{DSS} = 9 mA and V_{GS} = −3.5 V. Determine I_D when (*i*) V_{GS} = 0 V (*ii*) V_{GS} = − 2V. **[(*i*) 9mA (*ii*) 1.65 mA]**
6. Sketch the transfer curve for a *p*-channel *JFET* with I_{DSS} = 4 mA and V_P = 3 V.
7. In a *D-MOSFET*, determine I_{DSS}, given I_D = 3 mA, V_{GS} = − 2V and $V_{GS\,(off)}$ = − 10V. **[4.69 mA]**
8. Determine in which mode each *D-MOSFET* in Fig. 17.75 is biased. **[(*i*) Depletion (*ii*) Enhancement (*iii*) Zero bias]**
9. Determine V_{DS} for each circuit in Fig. 17.76. Given I_{DSS} = 8 mA. **[(*i*) 4V (*ii*) 5.4V (*iii*) − 4.52V]**

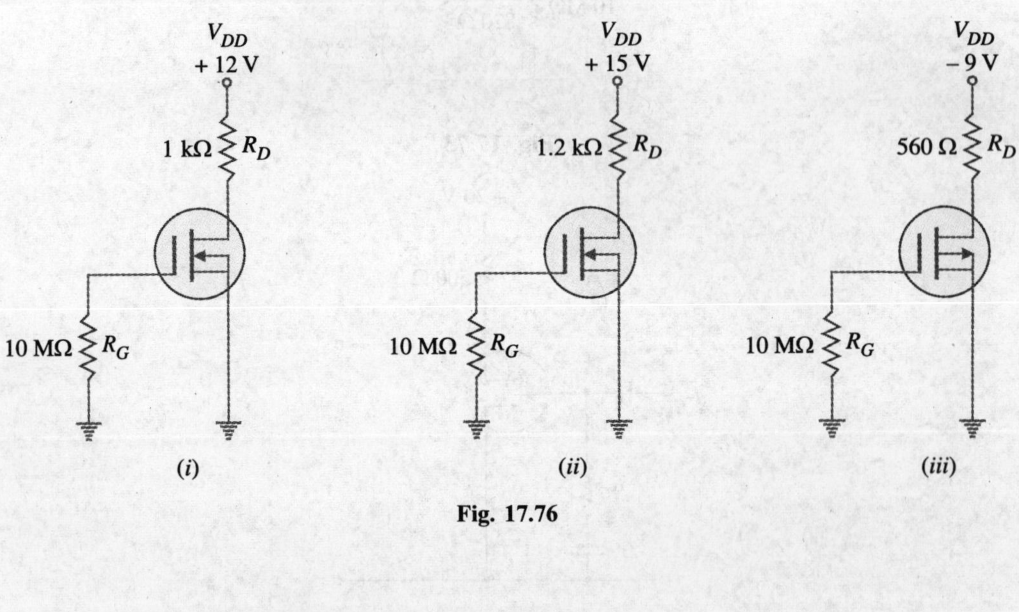

Fig. 17.76

10. If a 50 mV r.m.s. input signal is applied to the amplifier in Fig. 17.77, what is the peak-to-peak output voltage? Given that g_m = 5000 μS. [920 mV]

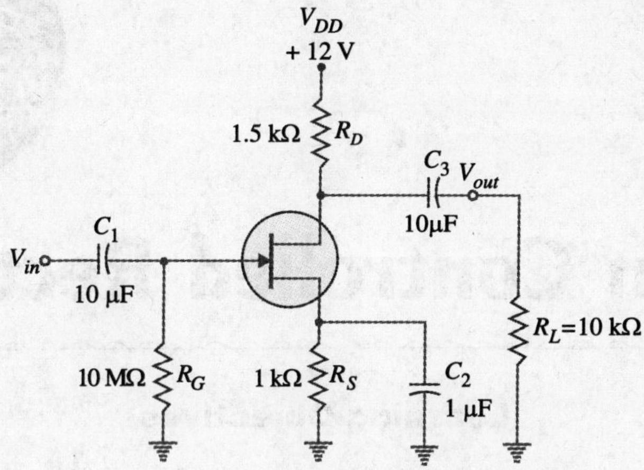

Fig. 17.77

CHAPTER 18

Silicon Controlled Rectifiers

Learning Objectives

- Understand the construction, working and applications of silicon controlled rectifier (*SCR*).
- Develop an understanding of two-transistor equivalent circuit of *SCR*.
- Learn the important terms used in the study of *SCR e.g.* forward breakover voltage, peak reverse voltage, forward current rating, gate trigger current, holding current *etc*.
- Explore the characteristics of *SCR*.
- Recognise the advantages and limitations of *SCR* as a switch.
- Study the various aspects of *SCR* half-wave rectifier and *SCR* full-wave rectifier.
- Understand the working of single-phase *SCR* inverter circuit.
- Learn the important applications of *SCR*.

Introduction

We know that a semiconductor diode is a two-terminal, single *pn* junction device while a bipolar junction transistor or *BJT* (*npn* or *pnp*) is a three-terminal, two *pn* junction device. However, the silicon controlled rectifier (abbreviated as *SCR*) is a three-terminal three *pn* junction (*pnpn*) device that is used as an electronic switch. Invented in 1957, an *SCR* can be used as a controlled switch to perform various functions such as rectification, inversion and regulation of power flow. The *SCR* has assumed paramount importance in electronics because it can be produced in versions to handle currents upto several thousand amperes and voltages upto more than 1 kV.

The *SCR* has appeared in the market under different names such as thyristor, thyrode transistor. It is a unidirectional power switch and is being extensively used in switching d.c. and a.c., rectifying a.c. to give controlled d.c. output, converting d.c. into a.c. *etc*. In this chapter, we shall examine the various characteristics of silicon controlled rectifiers and their increasing applications in power electronics.

18.1 Silicon Controlled Rectifier (SCR)

A silicon ****controlled rectifier (SCR)*** *is a three-terminal, three pn junction semiconductor **device that acts as a true electronic switch.*

The *SCR* is a four-layer semiconductor device made of alternate layers of *p*-type and *n*-type silicon to form *pnpn* device. Silicon (and not germanium) was chosen because of its high temperature and power capabilities. As the name indicates, the *SCR* can change alternating current into direct current and at the same time can control the amount of power fed to the load. Thus *SCR* combines the features of a rectifier and a transistor.

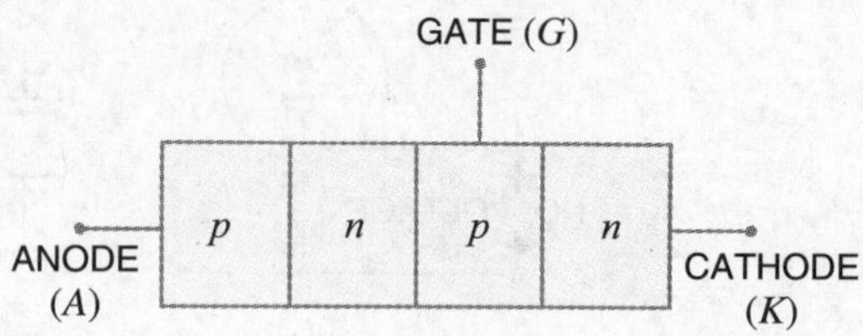

(*i*)

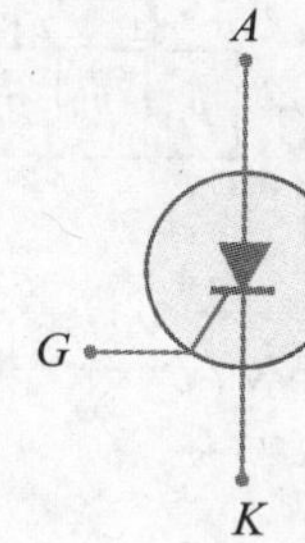

(*ii*)

Fig. 18.1

Constructional details. When a *pn* junction is added to a junction transistor, the resulting three *pn* junction device is called a silicon controlled rectifier. Fig. 18.1 (*i*) shows its construction. It is clear that it is essentially an ordinary rectifier (*pn*) and a junction transistor (*npn*) combined in one unit to form *pnpn device*. Three terminals are taken; one from the outer *p-type* material called *anode A*, second from the outer *n*-type material called *cathode K* and the third from the base of transistor section and is called *gate G*. In the normal operating conditions of *SCR*, anode is held at high positive potential *w.r.t.* cathode and gate at small positive potential *w.r.t.* cathode. Fig. 18.1 (*ii*) shows the symbol of *SCR*.

The silicon controlled rectifier is a solid state equivalent of thyratron. The gate, anode and cathode of *SCR* correspond to the grid, plate and cathode of thyratron. For this reason, *SCR* is sometimes called *thyristor*.

18.2 Working of SCR

In a silicon controlled rectifier, load is connected in series with anode. *The anode is always kept at positive potential w.r.t. cathode*. The working of *SCR* can be studied under the following two heads :

(*i*) **When gate is open.** Fig. 18.2 shows the *SCR* circuit with gate open *i.e.* no voltage applied to the gate. Under this condition, junction J_2 is reverse biased while junctions J_1 and J_3 are forward biased. Hence, the situation in the junctions J_1 and J_3 is just as in a *npn* transistor with base open. Consequently, no current flows through the load R_L and the *SCR*

* **Why not germanium controlled rectifier ?** The device is made of silicon because leakage current in silicon is very small as compared to germanium. Since the device is used as a switch, it will carry leakage current in the off condition which should be as small as possible.

** It got this name because it is a silicon device and is used as a rectifier and that rectification can be controlled.

is *cut-off*. However, if the applied voltage is gradually increased, a stage is reached when * reverse biased junction J_2 breaks down. The *SCR* now conducts ** heavily and is said to be in the *ON* state. The applied voltage at which *SCR* conducts heavily without gate voltage is called *forward breakover voltage*.

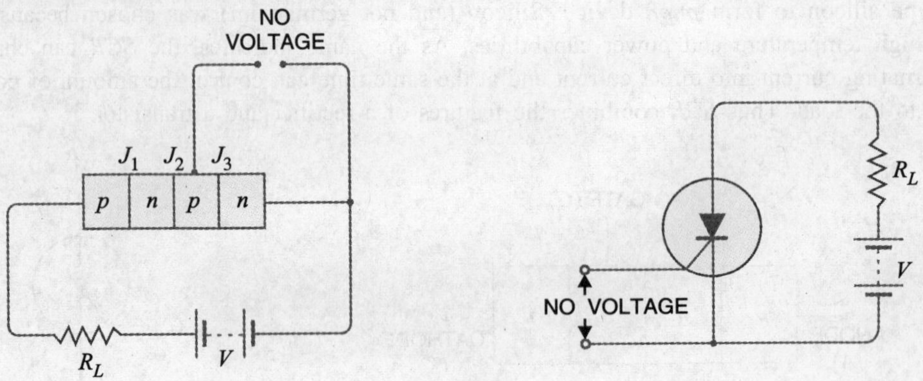

Fig. 18.2

(ii) **When gate is positive w.r.t. cathode.** The *SCR* can be made to conduct heavily at smaller applied voltage by applying a small positive potential to the gate as shown in Fig. 18.3. Now junction J_3 is forward biased and junction J_2 is reverse biased. The electrons from *n*-type material start moving across junction J_3 towards left whereas holes from *p*-type towards the right. Consequently, the electrons from junction J_3 are attracted across junction J_2 and gate current starts flowing. As soon as the gate current flows, anode current increases. The increased anode current in turn makes more electrons available at junction J_2. This process continues and in an extremely small time, junction J_2 breaks down and the *SCR* starts conducting heavily. *Once SCR starts conducting, the gate* (the reason for this name is obvious) *loses all control*. Even if gate voltage is removed, the anode current does not decrease at all. The only way to stop conduction (*i.e.* bring *SCR* in off condition) is to reduce the applied voltage to zero.

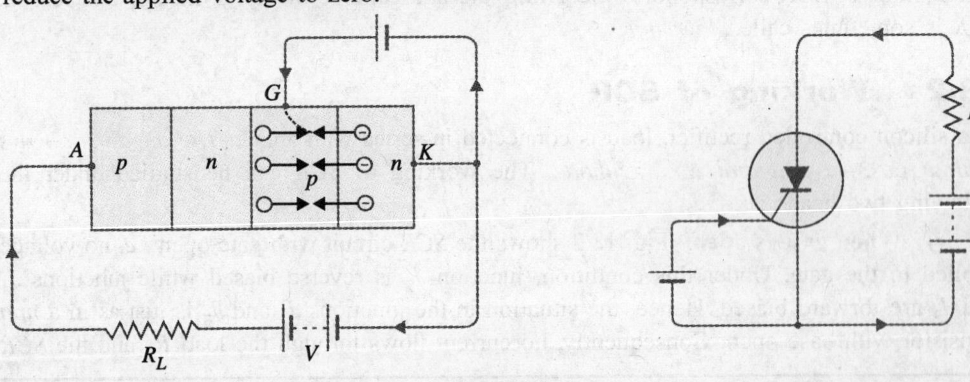

Fig. 18.3

* The whole applied voltage V appears as reverse bias across junction J_2 as junctions J_1 and J_3 are forward biased.

** Because J_1 and J_3 are forward biased and J_2 has broken down.

Conclusion. The following conclusions are drawn from the working of *SCR* :

(*i*) An *SCR* has two states *i.e.* either it does not conduct or it conducts heavily. There is no state inbetween. Therefore, *SCR* behaves like a switch.

(*ii*) There are two ways to turn on the *SCR*. The first method is to keep the gate open and make the supply voltage equal to the forward breakover voltage. The second method is to operate *SCR* with supply voltage less than forward breakover voltage and then turn it on by means of a small voltage (typically 1.5 V, 30 mA) applied to the gate.

(*iii*) Applying small positive voltage to the gate is the normal way to close an *SCR* because the forward breakover voltage is usually much greater than supply voltage.

(*iv*) To open the *SCR* (*i.e.* to make it non-conducting), reduce the supply voltage to zero.

18.3 Two-Transistor Equivalent Circuit of SCR

The *SCR* shown in Fig. 18.4 (*i*) can be visualised as separated into two transistors as shown in Fig. 18.4 (*ii*). Thus, the equivalent circuit of *SCR* is composed of *pnp* transistor and *npn* transistor connected as shown in Fig. 18.4. (*iii*). It is clear that collector of each transistor is coupled to the base of the other, thereby making a positive feedback loop.

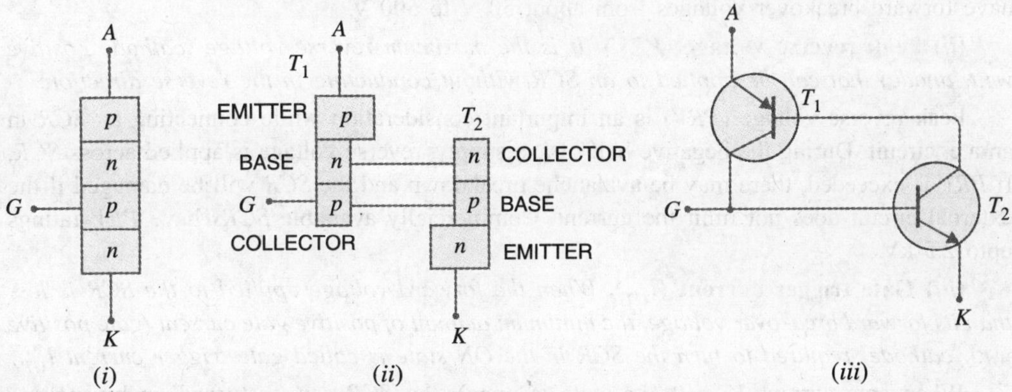

Fig. 18.4

Working. The working of *SCR* can be easily explained from its equivalent circuit. Fig. 18.5 shows the equivalent circuit of *SCR* with supply voltage *V* and load resistance R_L. Assume the supply voltage *V* is less than forward breakover voltage as is usually the case. With gate open (*i.e.* switch *S* open), there is no base current in transistor T_2. Therefore, no current flows in the collector of T_2 and hence that of T_1. Under such conditions, the *SCR* is open. However, if switch *S* is closed, a small gate current will flow through the base of T_2 which means its collector current will increase. The collector current of T_2 is the base current of T_1. Therefore, collector current of T_1 increases. But collector current of T_1 is the base current

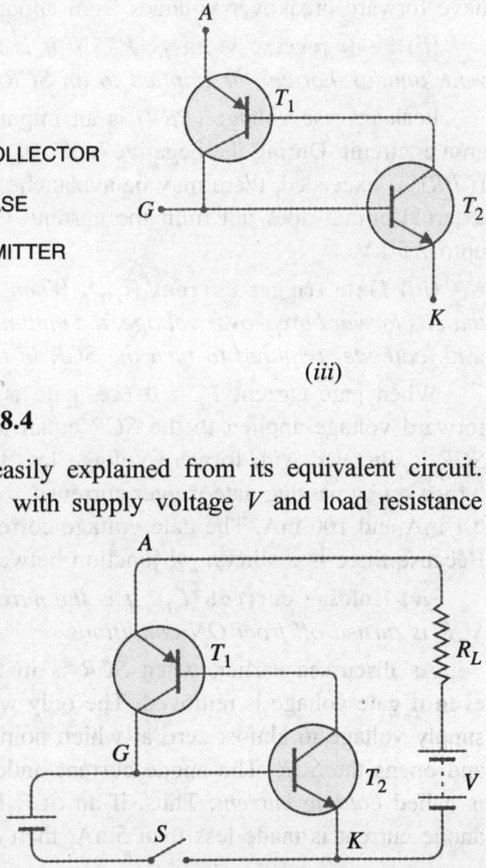

Fig. 18.5

of T_2. This action is accumulative since an increase of current in one transistor causes an increase of current in the other transistor. As a result of this action, both transistors are driven to saturation, and heavy current flows through the load R_L. Under such conditions, the *SCR* closes.

18.4 Important Terms

The following terms are much used in the study of *SCR* :

(*i*) Forward Breakover voltage	(*ii*) Peak reverse voltage
(*iii*) Gate trigger current	(*iv*) Holding current
(*v*) Forward current rating	(*vi*) Circuit fusing rating

(*i*) Forward Breakover voltage. *It is the minimum forward voltage, gate being open, at which SCR starts conducting heavily i.e. turned on.*

Thus, if the forward breakover voltage of an *SCR* is 200 V, it means that it can block a forward voltage (*i.e. SCR* remains open) as long as the supply voltage is less than 200 V. If the supply voltage is more than this value, then *SCR* will be turned on. In practice, the *SCR* is operated with supply voltage less than forward breakover voltage and it is then turned on by means of a small voltage applied to the gate. Commercially available *SCRs* have forward breakover voltages from about 50 V to 500 V.

(*ii*) Peak reverse voltage (*PRV*). *It is the maximum reverse voltage (cathode positive w.r.t. anode) that can be applied to an SCR without conducting in the reverse direction.*

Peak reverse voltage (*PRV*) is an important consideration while connecting an *SCR* in an a.c. circuit. During the negative half of a.c. supply, reverse voltage is applied across *SCR*. If *PRV* is exceeded, there may be avalanche breakdown and the *SCR* will be damaged if the external circuit does not limit the current. Commercially available *SCRs* have *PRV* ratings upto 2.5 kV.

(*iii*) Gate trigger current (I_{GT}). *When the forward voltage applied to the SCR is less than its forward breakover voltage, the minimum amount of positive gate current (gate positive w.r.t. cathode) required to turn the SCR in the ON state is called gate trigger current* I_{GT}.

When gate current $I_G = 0$ (*i.e.* gate is open), the *SCR* can be turned on by making forward voltage applied to the *SCR* equal to its forward breakover voltage. Normally, the *SCR* is operated with forward voltage less than forward breakover voltage and the *SCR* is turned on by passing gate trigger current I_{GT}. The value of gate trigger current I_{GT} is between 0.1 mA and 100 mA. The gate voltage corresponding to I_{GT} is the gate trigger voltage V_{GT}. Because there is a silicon *pn* junction between gate and cathode, $V_{GT} \simeq 0.7$ V.

(*iv*) Holding current (I_H). *It is the maximum anode current, gate being open, at which SCR is turned off from ON conditions.*

As discussed earlier, when *SCR* is in the conducting state, it cannot be turned *OFF* even if gate voltage is removed. The only way to turn off or open the *SCR* is to reduce the supply voltage to almost zero at which point the internal transistor comes out of saturation and opens the *SCR*. The anode current under this condition is very small (a few mA) and is called *holding current*. Thus, if an *SCR* has a holding current of 5mA, it means that if anode current is made less than 5mA, then *SCR* will be turned off.

(*v*) Forward current rating. *It is the maximum anode current that an SCR is capable of passing without destruction.*

Every *SCR* has a safe value of forward current which it can conduct. If the value of current exceeds this value, the *SCR* may be destroyed due to intensive heating at the junctions. For example, if an *SCR* has a forward current rating of 40A, it means that the *SCR* can safely carry only 40 A. Any attempt to exceed this value will result in the destruction of the *SCR*. Commercially available *SCRs* have forward current ratings from about 30A to 100A.

(*vi*) Circuit fusing (I^2t) rating. *It is the product of square of forward surge current and the time of duration of the surge i.e.,*

$$\text{Circuit fusing rating} = I^2t$$

The circuit fusing rating indicates the maximum forward surge current capability of *SCR*. For example, consider an *SCR* having circuit fusing rating of 90 A^2s. If this rating is exceeded in the *SCR* circuit, the device will be destroyed by excessive power dissipation.

Example 18.1. *An SCR has a forward breakover voltage of 400 V, a gate trigger current of 10 mA and holding current of 10 mA. What do you infer from it ? What will happen if gate current is made 15 mA ?*

Solution. (*i*) *Forward breakover voltage of 400 V.* It means that if gate is open and the supply voltage is 400 V, then *SCR* will start conducting heavily. However, as long as the supply voltage is less than 400 V, the *SCR* stays open *i.e.* it does not conduct.

(*ii*) *Gate trigger current of 10 mA.* It means that if the supply voltage is less than forward breakover voltage (*i.e.* 400 V) and a minimum gate current of 10 mA is passed, the *SCR* will close *i.e.* starts conducting heavily. The *SCR* will not conduct if the gate current is less than 10 mA. It may be emphasised that triggering is the normal way to close an *SCR* as the supply voltage is normally much less than the forward breakover voltage.

(*iii*) *Holding current of 10 mA.* When the *SCR* is conducting, it will not open (*i.e.* stop conducting) even if triggering current is removed. However, if supply voltage is reduced, the anode current also decreases. When the anode current drops to 10 mA, the holding current, the *SCR* is turned off.

(*iv*) If gate current is increased to 15 mA, the *SCR* will be turned on lower supply voltage.

Example 18.2. *An SCR in a circuit is subjected to a 50 A surge that lasts for 12 ms. Determine whether or not this surge will destroy the device. Given that circuit fusing rating is 90 A^2s.*

Solution. Circuit fusing rating = $I^2t = (50)^2 \times (12 \times 10^{-3}) = 30\ A^2s$

Since this value is well below the maximum rating of 90 A^2s, the device **will not be destroyed.**

Example 18.3. *An SCR has a circuit fusing rating of 50 A^2s. The device is being used in a circuit where it could be subjected to a 100 A surge. Determine the maximum allowable duration of such a surge.*

Solution. $$t_{max} = \frac{I^2t\ (\text{rating})}{I_s^2}$$ where I_s = Known value of surge current

$$\therefore \quad t_{max} = \frac{50}{(100)^2} = 5 \times 10^{-3}\text{s} = \mathbf{5\ ms}$$

Example 18.4. *A 220Ω resistor is connected in series with the gate of an SCR as shown in Fig. 18.6. The gate current required to fire the SCR is 7mA. What is the input voltage (V_{in}) required to fire the SCR ?*

Solution. The input voltage must overcome the junction voltage between the gate and cathode (0.7 V) and also cause 7 mA to flow through the 220 Ω resistor. According to Kirchhoff's voltage law, V_{in} is given by ;

$$V_{in} = V_{GT} + I_{GT} R_G$$
$$= 0.7 \text{ V} + (7 \text{ mA}) (220 \text{ } \Omega) = \mathbf{2.24 \text{ V}}$$

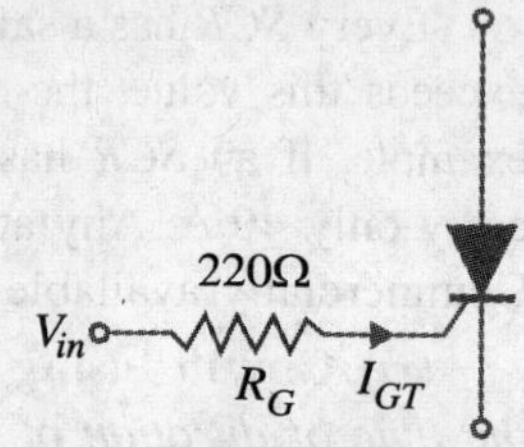

Fig. 18.6

Therefore, the input voltage required to trigger the *SCR* is 2.24 V.

Example 18.5. *The SCR of Fig. 18.7 has V_{GT} = 0.7 V, I_{GT} = 2 mA and I_H = 2 mA. What is the output voltage when the SCR is off ? What is the input voltage that triggers the SCR ? If V_{CC} is decreased until the SCR opens, what is the value of V_{CC} ?*

Solution. Here, V_{GT} = 0.7 V ; I_{GT} = 2 mA ; I_H = 2 mA ; V_{CC} = 12 V ; R_L = 47 Ω.

When the *SCR* is off, no current flows in the circuit, *i.e.* current through load R_L is zero. Therefore, output voltage, V_{out} = **12 V**.

The gate triggering voltage V_{GT} = 0.7 V and gate triggering current I_{GT} = 2 mA. Therefore, when the *SCR* is triggered, the required input voltage V_{in} is

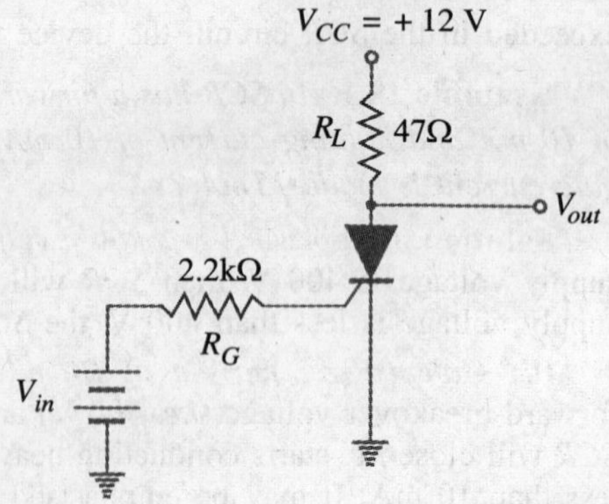

Fig. 18.7

$$V_{in} = V_{GT} + I_{GT} R_G$$
$$= 0.7 \text{ V} + (2 \text{ mA}) (2.2 \text{ k}\Omega) = \mathbf{5.1 \text{ V}}$$

Once the *SCR* is turned on, it can be turned off by reducing the supply voltage V_{CC} to a value V given by ;

$$V = I_H R_L + V_{GT}$$
$$= 2 \text{ mA} \times 47 \text{ } \Omega + 0.7 \text{ V} = \mathbf{0.79 \text{ V}}$$

Example 18.6. *The value of R in Fig. 18.8 is adjusted to 500 Ω. What is the peak output voltage ?*

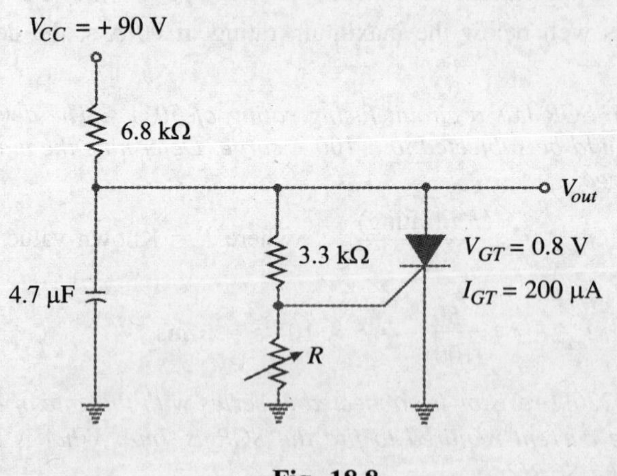

Fig. 18.8

Solution. Here, $V_{GT} = 0.8$ V ; $I_{GT} = 200$ μA ; $V_{CC} = 90$ V ; $R = 500\ \Omega$

Current through R due to V_{GT} is

$$I_R = \frac{V_{GT}}{R} = \frac{0.8\text{ V}}{500\ \Omega} = 1.6\text{ mA}$$

This 1.6 mA must also flow through 3.3 kΩ. The gate trigger current of 200 μA also flows through 3.3 kΩ resistor. Therefore, output voltage V_{out} is

$$V_{out} = V_{GT} + (1.6\text{ mA} + 200\ \mu\text{A})\ (3.3\text{ k}\Omega)$$
$$= 0.8\text{ V} + 5.94\text{ V} = \mathbf{6.74\ V}$$

Example 18.7. *The SCR in Fig. 18.9 has a holding current I_H = 0.5 mA. If it is triggered when the input voltage is 30 V, find (i) the maximum current in the circuit (ii) the input voltage when the SCR turns off. Assume that voltage across the SCR is 1 V when it is conducting.*

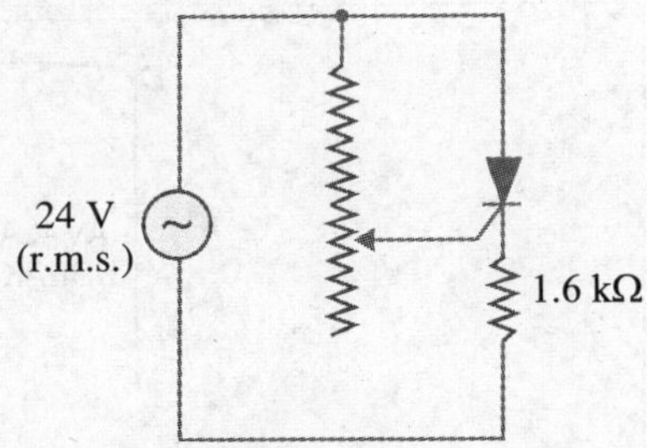

Fig. 18.9

Solution. V_{in} = 24 V (r.m.s.) ; I_H = 0.5 mA ; R_L = 1.6 kΩ.

(*i*) The input voltage is 24 V (r.m.s.). Therefore, peak or maximum input voltage is

$$V_{in(max.)} = \sqrt{2}\ V_{in}(\text{r.m.s.}) = \sqrt{2} \times 24 = 34\text{ V}$$

Since the *SCR* is triggered (*i.e.* turned on) when the input voltage is 30 V, the *SCR* is *ON* when the input voltage reaches its maximum value *i.e.* 34 V. The voltage drop across *SCR* is 1 V and the remaining 33 V appears across the load R_L (= 1.6 kΩ). Therefore, maximum current $I_{max.}$ in the circuit is

$$I_{max.} = \frac{33\text{ V}}{1.6\text{ k}\Omega} = \mathbf{20.6\ mA}$$

(*ii*) To determine the input voltage just after the *SCR* turns *OFF*, we find the voltage drop across the load R_L at the holding current. This is = $I_H R_L$ = 0.5 mA × 1.6 kΩ = 0.8 V. Assuming 1 V across *SCR* when it is *OFF*, the input voltage when the *SCR* is off = 0.8 V + 1 V = **1.8 V.**

18.5 V-I Characteristics of SCR

It is the curve between anode-cathode voltage (*V*) and anode current (*I*) of an *SCR* at constant gate current. Fig. 18.10 shows the *V-I* characteristics of a typical *SCR*.

(*i*) **Forward characteristics.** When anode is positive *w.r.t.* cathode, the curve between *V* and *I* is called the *forward characteristic*. In Fig. 18.10, *OABC* is the forward characteristic of *SCR* at I_G=0. If the supply voltage is increased from zero, a point is reached (point *A*) when the *SCR* starts conducting. Under this condition, the voltage across *SCR* suddenly drops as shown by dotted curve *AB* and most of supply voltage appears across the load resistance R_L. If proper gate current is made to flow, *SCR* can close at much smaller supply voltage.

(*ii*) **Reverse characteristics.** When anode is negative *w.r.t.* cathode, the curve between *V* and *I* is known as *reverse characteristic.* The reverse voltage does come across *SCR* when it is operated with a.c. supply. If the reverse voltage is gradually increased, at first the anode current remains small (*i.e.* leakage current) and at some reverse voltage, avalanche breakdown occurs and the *SCR* starts conducting heavily in the reverse direction as shown by the curve *DE*. This maximum reverse voltage at which *SCR* starts conducting heavily is known as *reverse breakdown voltage*.

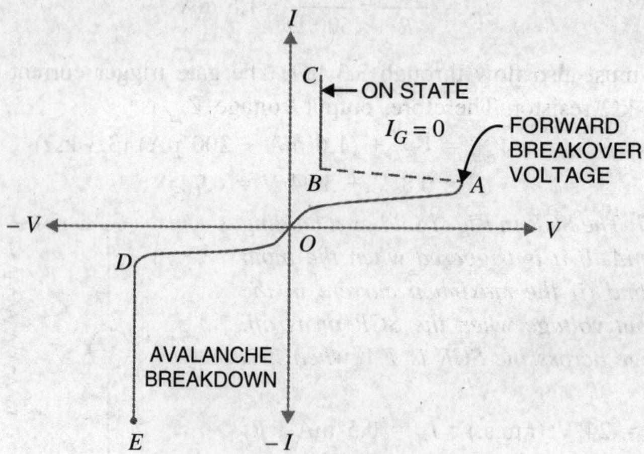

Fig. 18.10

18.6 SCR in Normal Operation

In order to operate the *SCR* in normal operation, the following points are kept in view :

(*i*) The supply voltage is generally much less than forward breakover voltage.

(*ii*) The *SCR* is turned on by passing an appropriate amount of gate current (a few mA) and not by forward breakover voltage.

(*iii*) When *SCR* is operated from a.c. supply, the peak reverse voltage which comes during negative half-cycle should not exceed the reverse breakdown voltage.

(*iv*) When *SCR* is to be turned *OFF* from the *ON* state, anode current should be reduced to holding current.

(*v*) If gate current is increased above the required value, the *SCR* will close at much reduced supply voltage.

18.7 SCR as a Switch

The *SCR* has only two states, namely; *ON* state and *OFF* state and no state inbetween. When appropriate gate current is passed, the *SCR* starts conducting heavily and remains in this position indefinitely even if gate voltage is removed. This corresponds to the *ON* condition. However, when the anode current is reduced to the holding current, the *SCR* is turned *OFF*. It is clear that behaviour of *SCR* is similar to a mechanical switch. As *SCR* is an electronic device, therefore, it is more appropriate to call it an *electronic switch*.

Advantages of SCR as a switch. An *SCR* has the following advantages over a mechanical or electromechanical switch (relay) :

(*i*) It has no moving parts. Consequently, it gives noiseless operation at high efficiency.

(*ii*) The switching speed is very high upto 10^9 operations per second.

(*iii*) It permits control over large current (30–100 A) in the load by means of a small gate current (a few mA).

(*iv*) It has small size and gives trouble free service.

Limitations. The following are the difficulties in the use of *SCRs* as switch :

(*i*) One difficulty in the use of *SCR* is the presence of *transient pulses*. These transients are accidental and unwanted pulses of high amplitude and short duration which may appear in the anode voltage. If the pulse is in the forward direction, it may cause the *SCR* to fire at an undesired time. If it is in the reverse direction, it may cause a reverse-voltage breakdown which may ruin the *SCR*. In circuits employing ordinary *SCR*, especially where high voltages and currents are involved, special precautions must be taken to guard against these transient pulses.

(*ii*) Another limitation of an ordinary *SCR* is that it is triggered by a positive pulse and may be turned off by reducing the anode voltage to about zero. A negative pulse applied to the gate will not turn off the *SCR* because the relatively heavy feedback current keeps the rectifier in its conducting state.

18.8 SCR Switching

We have seen that *SCR* behaves as a switch *i.e.* it has only two states *viz. ON* state and *OFF* state. It is profitable to discuss the methods employed to turn-on or turn-off an *SCR*.

1. SCR turn-on methods. In order to turn on the *SCR*, the gate voltage V_G is increased upto a minimum value to initiate triggering. This minimum value of gate voltage at which *SCR* is turned *ON* is called *gate triggering voltage* V_{GT}. The resulting gate current is called *gate triggering current* I_{GT}. Thus to turn on an *SCR* all that we have to do is to apply positive gate voltage equal to V_{GT} or pass a gate current equal to I_{GT}. For most of the *SCRs*, V_{GT} = 2 to 10 V and I_{GT} = 100 μA to 1500 mA. We shall discuss two methods to turn on an *SCR*.

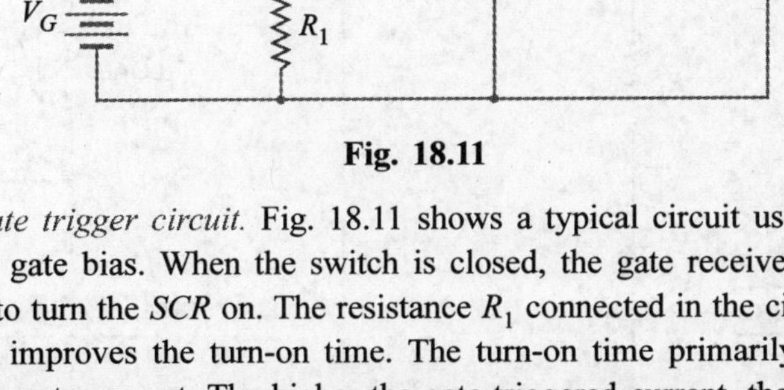

Fig. 18.11

(*i*) *D.C. gate trigger circuit.* Fig. 18.11 shows a typical circuit used for triggering an *SCR* with a d.c. gate bias. When the switch is closed, the gate receives sufficient positive voltage (= V_{GT}) to turn the *SCR* on. The resistance R_1 connected in the circuit provides noise suppression and improves the turn-on time. The turn-on time primarily depends upon the magnitude of the gate current. The higher the gate-triggered current, the shorter the turn-on time.

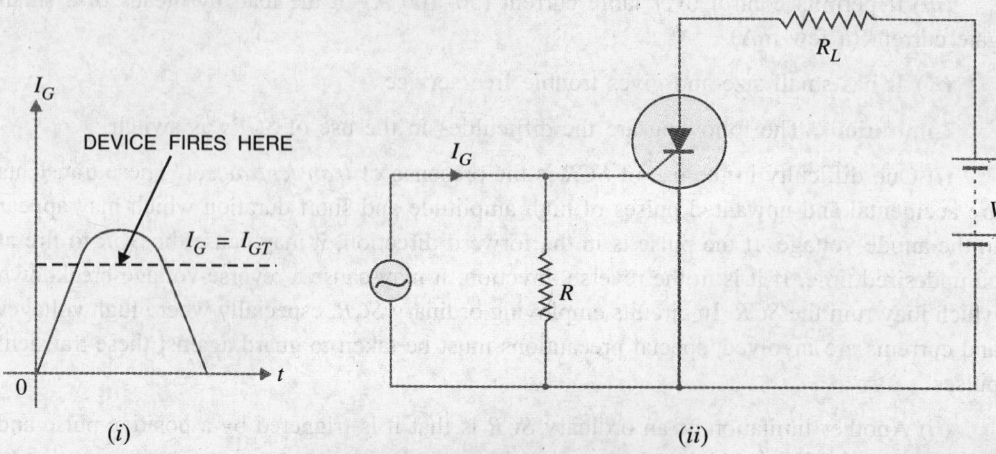

Fig. 18.12

(*ii*) *A.C. trigger circuit.* An *SCR* can also be turned on with positive cycle of a.c. gate current. Fig. 18.12 (*ii*) shows such a circuit. During the positive half-cycle of the gate current, at some point $I_G = I_{GT}$, the device is turned on as shown in Fig. 18.12 (*i*).

2. **SCR turn-off methods.** The *SCR* turn-off poses more problems than *SCR* turn-on. It is because once the device is *ON*, the gate loses all control. There are many methods of *SCR* turn-off but only two will be discussed.

(*i*) *Anode current interruption.* When the anode current is reduced below a minimum value called holding current, the *SCR* turns off. The simple way to turn off the *SCR* is to open the line switch *S* as shown in Fig. 18.13.

(*ii*) *Forced commutation.* The method of discharging a capacitor in parallel with an *SCR* to turn off the *SCR* is called forced commutation. Fig. 18.14 shows the forced commutation of *SCR* where capacitor *C* performs the commutation. Assuming the *SCRs* are switches with *SCR*1 *ON* and *SCR*2 *OFF*, current flows through the load and *C* as shown in Fig. 18.14. When *SCR*2 is triggered on, *C* is effectively paralleled across *SCR*1. The charge on *C* is then opposite to *SCR*1's forward voltage, *SCR*1 is thus turned off and the current is transferred to *R*–*SCR*2 path.

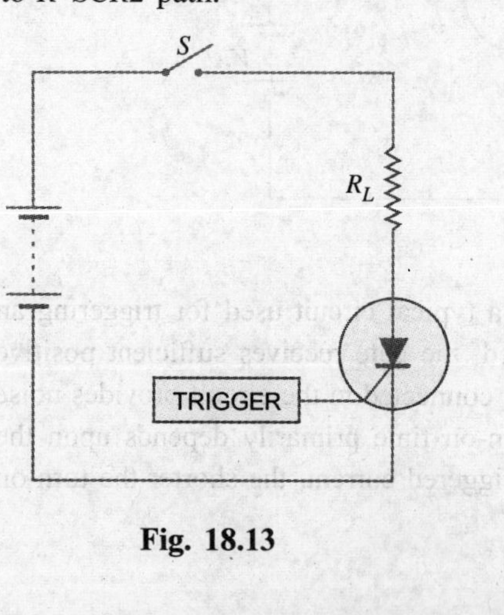

Fig. 18.13

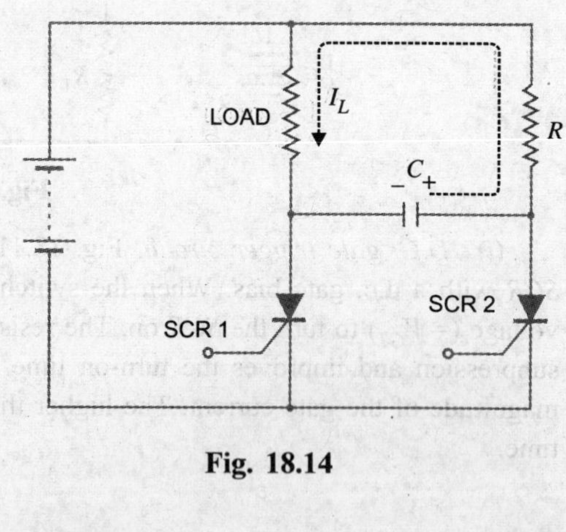

Fig. 18.14

18.9 SCR Half-Wave Rectifier

One important application of an *SCR* is the controlled half-wave rectification. Fig. 18.15 (*i*) shows the circuit of an *SCR* half-wave rectifier. The a.c. supply to be rectified is supplied through the transformer. The load resistance R_L is connected in series with the anode. A variable resistance *r* is inserted in the gate circuit to control the gate current.

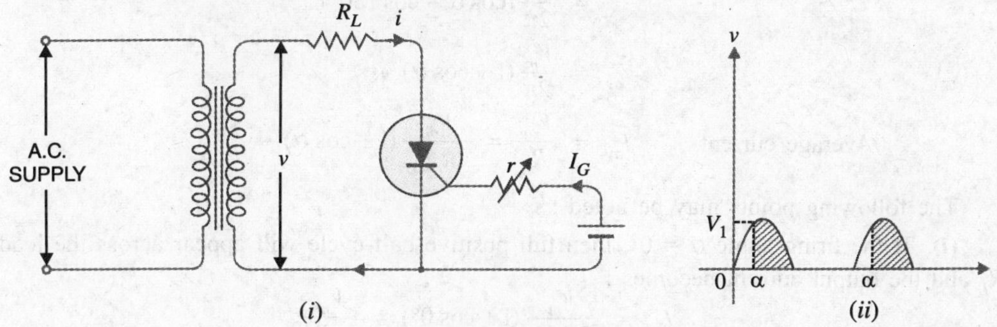

Fig. 18.15

Operation. The a.c. supply to be converted into d.c. supply is applied to the primary of the transformer. Suppose the peak reverse voltage appearing across secondary is less than the reverse breakdown voltage of the *SCR*. This condition ensures that *SCR* will not break down during negative half-cycles of a.c. supply. The circuit action is as follows :

(*i*) During the negative half-cycles of a.c. voltage appearing across secondary, the *SCR* does not conduct regardless of the gate voltage. It is because in this condition, anode is negative *w.r.t.* cathode and also *PRV* is less than the reverse breakdown voltage.

(*ii*) The *SCR* will conduct during the positive half-cycles provided proper gate current is made to flow. The greater the gate current, the lesser the supply voltage at which *SCR* is turned *ON*. The gate current can be changed by the variable resistance *r* as shown in Fig. 18.15 (*i*).

(*iii*) Suppose that gate current is adjusted to such a value that *SCR* closes at a positive voltage V_1 which is less than the peak voltage V_m. Referring to Fig. 18.15 (*ii*), it is clear that *SCR* will start conducting when secondary a.c. voltage becomes V_1 in the positive half-cycle. Beyond this, the *SCR* will continue to conduct till voltage becomes zero at which point it is turned *OFF*. Again at the start of the next positive half-cycle, *SCR* will start conducting when secondary voltage becomes V_1.

(*iv*) Referring to Fig. 18.15 (*ii*), it is clear that firing angle is α *i.e.* at this angle in the positive half-cycle, *SCR* starts conduction. The conduction angle is ϕ (= 180° − α).

It is worthwhile to distinguish between an ordinary half-wave rectifier and *SCR* half-wave rectifier. Whereas an ordinary half-wave rectifier will conduct full positive half-cycle, an *SCR* half-wave rectifier can be made to conduct full or part of a positive half-cycle by proper adjustment of gate current. Therefore, an *SCR* can control power fed to the load and hence the name *controlled rectifier*.

Mathematical treatment. Referring to Fig. 18.15 (*i*), let $v = V_m \sin\theta$ be the alternating voltage that appears across the secondary. Let α be the firing angle. It means that rectifier will conduct from α to 180° during the positive half-cycles.

$$\therefore \quad \text{Average output voltage, } V_{av} = \frac{1}{2\pi}\int_{\alpha}^{180°} V_m \sin\theta\, d\theta = \frac{V_m}{2\pi}\int_{\alpha}^{180°} \sin\theta\, d\theta$$

$$= \frac{V_m}{2\pi}\left[-\cos\theta\right]_{\alpha}^{180°}$$

$$= \frac{V_m}{2\pi}(\cos\alpha - \cos 180°)$$

$$\therefore \quad V_{av} = \frac{V_m}{2\pi}(1+\cos\alpha)$$

$$\text{Average current, } \quad I_{av} = \frac{V_{av}}{R_L} = \frac{V_m}{2\pi R_L}(1+\cos\alpha)$$

The following points may be noted :

(*i*) If the firing angle $\alpha = 0°$, then full positive half-cycle will appear across the load R_L and the output current becomes :

$$I_{av} = \frac{V_m}{2\pi R_L}(1+\cos 0°) = \frac{V_m}{\pi R_L}$$

This is the value of average current for ordinary half-wave rectifier. This is expected since the full positive half-cycle is being conducted.

(*ii*) If $\alpha = 90°$, then average current is given by ;

$$I_{av} = \frac{V_m}{2\pi R_L}(1+\cos 90°) = \frac{V_m}{2\pi R_L}$$

This shows that greater the firing angle α, the smaller is the average current and *vice-versa.*

Example 18.8. *A half-wave rectifier circuit employing an SCR is adjusted to have a gate current of 1mA. The forward breakdown voltage of SCR is 100 V for I_g = 1mA. If a sinusoidal voltage of 200 V peak is applied, find :*

(*i*) *firing angle* (*ii*) *conduction angle* (*iii*) *average current.*

Assume load resistance = 100Ω and the holding current to be zero.

Solution. $v = V_m \sin\theta$

Here, $v = 100$ V ; $V_m = 200$ V

(*i*) $\therefore \quad 100 = 200 \sin\theta$

or $\sin\theta = \frac{100}{200} = 0.5$

$\therefore \quad \theta = \sin^{-1}(0.5) = 30°$ *i.e.* Firing angle, $\alpha = \theta =$ **30°**

(*ii*) Conduction angle, $\phi = 180° - \alpha = 180° - 30° =$ **150°**

(*iii*) Average voltage $= \frac{V_m}{2\pi}(1+\cos\alpha) = \frac{200}{2\pi}(1+\cos 30°) \doteq 59.25$ V

$\therefore$ Average current $= \frac{\text{Average voltage}}{R_L} = \frac{59.25}{100} =$ **0.5925 A**

Example 18.9. *An SCR half-wave rectifier has a forward breakdown voltage of 150 V when a gate current of 1 mA flows in the gate circuit. If a sinusoidal voltage of 400 V peak is applied, find :*

(i) firing angle *(ii) average output voltage*

(iii) average current for a load resistance of 200Ω *(iv) power output*

Assume that the gate current is 1mA throughout and the forward breakdown voltage is more than 400 V when I_g = 1 mA.

Solution. Here, $V_m = 400 \text{ V}; \quad v = 150 \text{ V}; \quad R_L = 200\ \Omega$

(i) Now, $v = V_m \sin\theta$

or $$\sin\theta = \frac{v}{V_m} = \frac{150}{400} = 0.375$$

i.e. Firing angle, $\alpha\ (=\theta) = \sin^{-1} 0.375 = 22°$

(ii) Average output voltage is

$$V_{av} = \frac{V_m}{2\pi}(1+\cos 22°) = \frac{400}{2\pi}(1+\cos 22°) = \mathbf{122.6\ V}$$

(iii) Average current, $$I_{av} = \frac{\text{Average output voltage}}{R_L} = \frac{122.6}{200} = \mathbf{0.613\ A}$$

(iv) Output power $= V_{av} \times I_{av} = 122.6 \times 0.613 = \mathbf{75.15\ W}$

Example 18.10. *An a.c. voltage v = 240 sin314t is applied to an SCR half-wave rectifier. If the SCR has a forward breakdown voltage of 180 V, find the time during which SCR remains off.*

Solution. The *SCR* will remain off till the voltage across it reaches 180 V. This is shown in Fig. 18.16. Clearly, *SCR* will remain off for *t* second.

Now, $v = V_m \sin 314\,t$

Here, $v = 180 \text{ V}; \quad V_m = 240 \text{ V}$

$\therefore \quad 180 = 240 \sin(314\,t)$

or $$\sin 314\,t = \frac{180}{240} = 0.75$$

or $$314\,t = \sin^{-1}(0.75)$$
$$= 48.6° = 0.848 \text{ radian}$$

$$\therefore \quad t = \frac{0.848}{314} = 0.0027 \text{ sec}$$
$$= \mathbf{2.7\ millisecond}$$

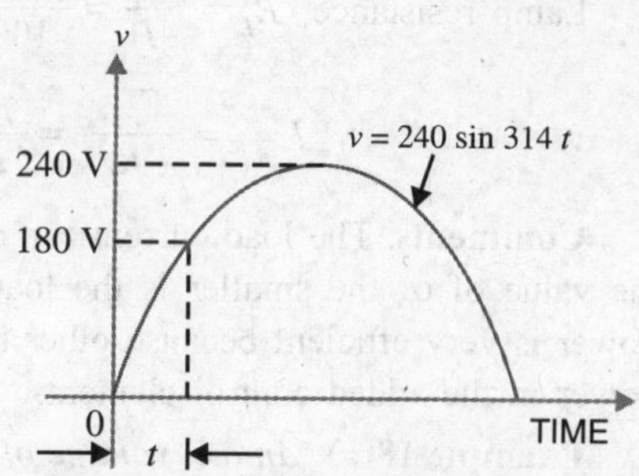

Fig. 18.16

Example 18.11. *In an SCR half-wave rectifier circuit, what peak-load current will occur if we measure an average (d.c.) load current of 1 A at a firing angle of 30° ?*

Solution. Let I_m be the peak load current.

Now, $$I_{av} = \frac{V_m}{2\pi R_L}(1+\cos\alpha)$$
$$= \frac{I_m}{2\pi}(1+\cos\alpha) \qquad \left(\because I_m = \frac{V_m}{R_L}\right)$$

$$\therefore \quad I_m = \frac{2\pi I_{av}}{1+\cos\alpha}$$

Here, $I_{av} = I_{dc} = 1\text{A}; \ \alpha = 30°$

$$\therefore \quad I_m = \frac{2\pi \times 1}{1 + \cos 30°} = \mathbf{3.36 A}$$

Example 18.12. *Power (brightness) of a 100W, 110 V tungsten lamp is to be varied by controlling the firing angle of an SCR in a half-wave rectifier circuit supplied with 110 V a.c. What r.m.s. voltage and current are developed in the lamp at firing angle α = 60°?*

Solution. The a.c. voltage is given by ;

$$v = V_m \sin \theta$$

Let α be the firing angle as shown in Fig. 18.17. This means that the *SCR* will fire (*i.e.* start conducting) at θ = α. Clearly, *SCR* will conduct from α to 180°.

$$E^2_{r.m.s.} = \frac{1}{2\pi}\int_{\alpha}^{\pi} V_m^2 \sin^2\theta \, d\theta$$

$$= V_m^2 \; {}^{*}\frac{2(\pi - \alpha) + \sin 2\alpha}{8\pi}$$

$$\therefore \quad E_{r.m.s.} = V_m \sqrt{\frac{2(\pi - \alpha) + \sin 2\alpha}{8\pi}}$$

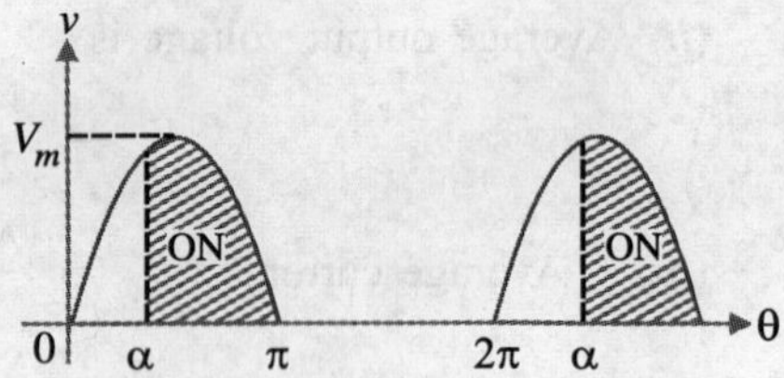

Fig. 18.17

Here, $V_m = \sqrt{2} \times 110 = 156\text{V}$; $\alpha = 60° = \pi/3$

$$\therefore \quad E_{r.m.s.} = 156\sqrt{\frac{2(\pi - \pi/3) + \sin 120°}{8\pi}} = \mathbf{70 V}$$

Lamp resistance, $R_L = \dfrac{V^2}{P} = \dfrac{(110)^2}{100} = 121\Omega$

$$\therefore \quad I_{r.m.s.} = \frac{E_{r.m.s.}}{R_L} = \frac{70}{121} = \mathbf{0.58 A}$$

Comments. The load current can be decreased by increasing the firing angle. The larger the value of α, the smaller is the load current and *vice-versa.* This method of controlling power is very efficient because other methods, such as added series resistance, waste much power in the added control element.

Example 18.13. *An a.c. voltage of 120 V (r.m.s.) is applied to an SCR half-wave rectifier and load R_L = 40 Ω. Assume that voltage drop across SCR is 1 V when it is conducting.*

(i) What should be the firing angle if it is desired to deliver an average current of 1 A to the load ?

(ii) What is the r.m.s. value of current in the load ?

(iii) What is the average power delivered to the load ?

Solution. (*i*) When the *SCR* conducts, the maximum value of load current I_m is

$$I_m = \frac{V_m - V_{AK}}{R_L}$$

where V_m = Maximum a.c. voltage

$= 120 \times \sqrt{2} = 169.7$ V

* On carrying out the integration.

$$V_{AK} = \text{Voltage drop across } SCR = 1 \text{ V (given)}$$

$$R_L = \text{Load resistance} = 40\ \Omega$$

$$\therefore \quad I_m = \frac{(169.7 - 1)\text{ V}}{40\ \Omega} = 4.2175 \text{ A}$$

The average load current I_{av} in an SCR half-wave rectifier is

$$I_{av} = \frac{V_m}{2\pi R_L}(1 + \cos\alpha) \text{ .... } \alpha = \text{firing angle}$$

or
$$I_{av} = \frac{I_m}{2\pi}(1 + \cos\alpha) \qquad \left[\because I_m = \frac{V_m}{R_L}\right]$$

Here, $I_{av} = 1$ A ; $I_m = 4.2175$ A

$$\therefore \quad 1 = \frac{4.2175}{2\pi}(1 + \cos\alpha)$$

or $\cos\alpha = 0.4898 \quad \therefore$ Firing angle, $\alpha =$ **60.67°**

(*ii*) It can be shown mathematically that r.m.s. value of load current is given by ;

$$I_{r.m.s.} = \frac{I_m}{2}\sqrt{\left(1 - \frac{\alpha}{180^\circ}\right) + \frac{\sin 2\alpha}{2\pi}} \text{ .... } \alpha \text{ in degrees}$$

$$= \frac{4.2175}{2}\sqrt{\left(1 - \frac{60.67^\circ}{180^\circ}\right) + \frac{\sin 2(60.67^\circ)}{2\pi}} = \mathbf{1.885\ A}$$

(*iii*) Average power P delivered to the load is

$$P = I^2_{r.m.s.} R_L = (1.885)^2 \times 40 = \mathbf{142.1\ W}$$

18.10 SCR Full-Wave Rectifier

Fig. 18.10 (*i*) shows the circuit of *SCR* full-wave rectifier. It is exactly like an ordinary centre-tap circuit except that the two diodes have been replaced by two *SCRs*. The gates of both *SCR*s get their supply from two gate controls. One *SCR* conducts during the positive

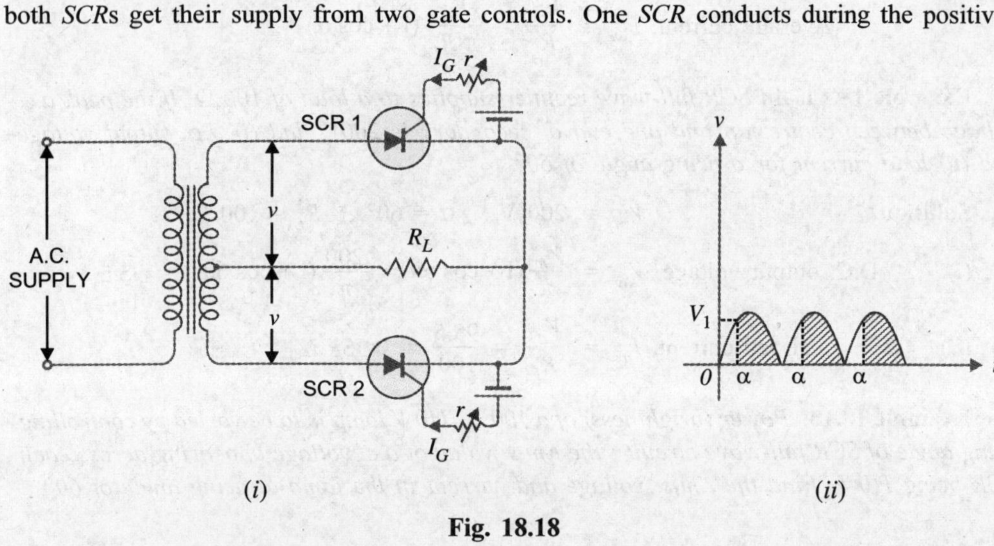

Fig. 18.18

half-cycle and the other during the negative half-cycle. Consequently, full-wave rectified output is obtained across the load.

Operation. The angle of conduction can be changed by adjusting the gate currents. Suppose the gate currents are so adjusted that *SCRs* conduct as the secondary voltage (across half winding) becomes V_1. During the positive half-cycle of a.c. across secondary, the upper end of secondary is positive and the lower end negative. This will cause *SCR*1 to conduct. However, the conduction will start only when the voltage across the upper half of secondary becomes V_1 as shown in Fig. 18.18 (*ii*). In this way, only shaded portion of positive half-cycle will pass through the load.

During the negative half-cycle of a.c. input, the upper end of secondary becomes negative and the lower end positive. This will cause *SCR*2 to conduct when the voltage across the lower half of secondary becomes V_1. It may be seen that current through the load is in the same direction (d.c.) on both half-cycles of input a.c. The obvious advantage of this circuit over ordinary full-wave rectifier circuit is that by adjusting the gate currents, we can change the conduction angle and hence the output voltage.

Mathematical treatment. Referring to Fig. 18.18 (*i*), let $v = V_m \sin\theta$ be the alternating voltage that appears between centre tap and either end of secondary. Let α be the firing angle.

$$\text{Average output voltage, } V_{av} = \frac{1}{\pi}\int_{\alpha}^{180°} V_m \sin\theta\, d\theta = \frac{V_m}{\pi}\int_{\alpha}^{180°} \sin\theta\, d\theta$$

$$= \frac{V_m}{\pi}[-\cos\theta]_{\alpha}^{180°} = \frac{V_m}{\pi}(\cos\alpha - \cos 180°)$$

$$\therefore \quad V_{av} = \frac{V_m}{\pi}(1 + \cos\alpha)$$

This value is double that of a half-wave rectifier. It is expected since now negative half-cycle is also rectified.

$$\text{Average current, } I_{av} = \frac{V_{av}}{R_L} = \frac{V_m}{\pi R_L}(1 + \cos\alpha)$$

Example 18.14. *An SCR full-wave rectifier supplies to a load of 100 Ω. If the peak a.c. voltage between centre tap and one end of secondary is 200V, find (i) d.c. output voltage and (ii) load current for a firing angle of 60°.*

Solution. $V_m = 200 \text{ V}\ ;\quad \alpha = 60°\ ;\quad R_L = 100\ \Omega$

(*i*) D.C. output voltage, $V_{av} = \frac{V_m}{\pi}(1 + \cos\alpha) = \frac{200}{\pi}(1 + \cos 60°) = $ **95.5 V**

(*ii*) Load current, $I_{av} = \frac{V_{av}}{R_L} = \frac{95.5}{100} = $ **0.955 A**

Example 18.15. *Power (brightness) of a 100 W, 110 V lamp is to be varied by controlling firing angle of SCR full-wave circuit ; the r.m.s. value of a.c. voltage appearing across each SCR being 110 V. Find the r.m.s. voltage and current in the lamp at firing angle of 60°.*

Solution. Let $v = V_m \sin\theta$ be the alternating voltage that appears between centre tap and either end of the secondary. Let α be the firing angle as shown in Fig. 18.19. This means that *SCR* will conduct at θ = α. Clearly, *SCR* circuit will conduct from α to 180°.

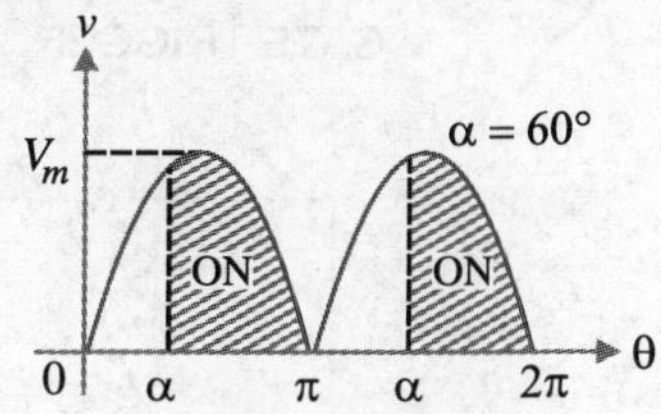

Fig. 18.19

$$E^2_{r.m.s.} = \frac{1}{\pi}\int_{\alpha}^{\pi} V_m^2 \sin^2\theta\, d\theta$$

$$= V_m^2\,\frac{2(\pi-\alpha)+\sin 2\alpha}{4\pi}\;^{*}$$

$$\therefore \quad E_{r.m.s.} = V_m\sqrt{\frac{2(\pi-\alpha)+\sin 2\alpha}{4\pi}}$$

Here, $V_m = 110 \times \sqrt{2} = 156\text{V}\;;\quad \alpha = 60°$

$$\therefore \quad E_{r.m.s.} = 156\sqrt{\frac{2(\pi-\pi/3)+\sin 120°}{4\pi}}$$

$$= \mathbf{98.9V}$$

$$\text{Lamp resistance}, R_L = \frac{V^2}{P} = \frac{(110)^2}{100} = 121\Omega$$

$$\therefore \quad I_{r.m.s.} = \frac{E_{r.m.s.}}{R_L} = \frac{98.9}{121}$$

$$= \mathbf{0.82\ A}$$

18.11 Single-Phase SCR Inverter Circuit

SCR inverter provides an efficient and economical way of converting direct current or voltage into alternating current or voltage. In this application, *SCR* acts as a controlled switch, alternately opening and closing a d.c. circuit. Fig. 18.20 shows the basic inverter circuit. Here, a.c. voltage is generated by alternately closing and opening switches S_1 and S_2. Replacing, the mechanical switches with *SCR*s, whose gates are triggered by an external pulse generator, we get the practical *SCR* inverter as shown in Fig. 18.21.

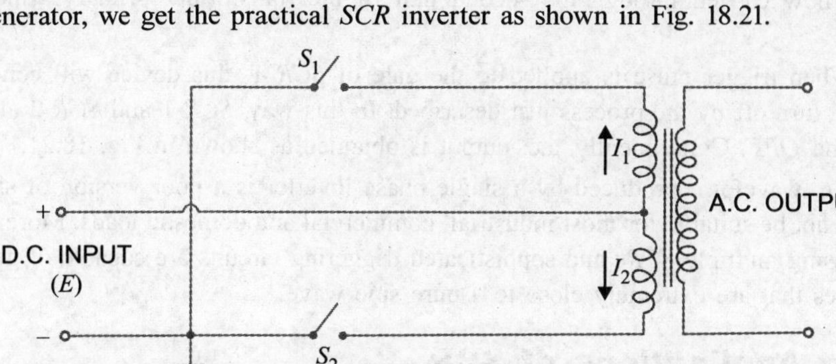

Fig. 18.20

* On carrying out the integration.

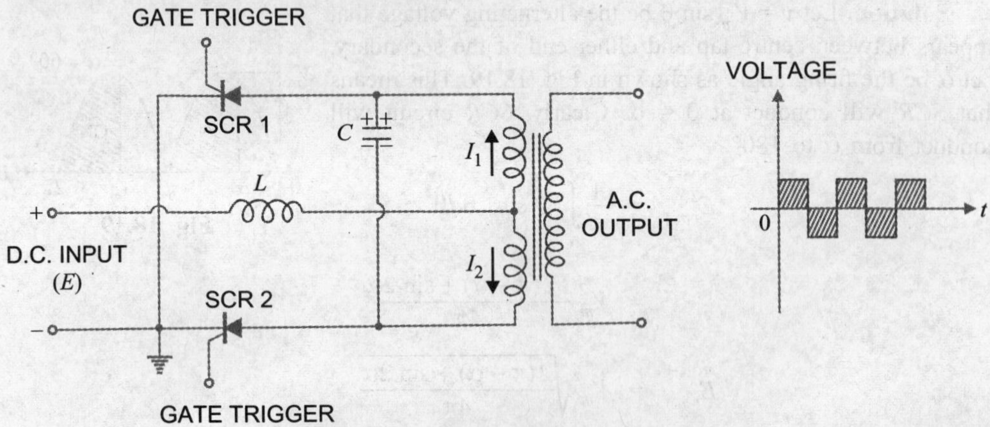

Fig. 18.21

Circuit Action

The circuit action is as under :

(*i*) When conduction is initiated by applying a positive trigger pulse to *SCR* 1 (*SCR* 2 is assumed *OFF*), the voltage across *SCR* decreases rapidly as the current through it increases. At the same time, the capacitor *C* charges through *SCR* 1 in the polarity shown. The load current flows through inductor *L*, upper half of the transformer primary winding and *SCR*1.

(*ii*) When a firing pulse is applied to the gate of *SCR* 2, this *SCR* turns on and conducts current. At this instant, capacitor *C* begins to discharge through *SCR* 1 and *SCR* 2. This discharge current flows through *SCR* 1 in a *reverse* direction. This reverse current turns off *SCR*1. At this time, with *SCR* 1 turned off, the capacitor voltage (approximately–*2*E*) appears across *SCR* 1 as a reverse voltage, long enough for this *SCR* to recover for forward blocking.

Simultaneously, during this interval, conducting *SCR* 2 allows the capacitor to discharge through the transformer primary winding and inductor *L*. The function of *L* is to control the discharge rate of *C* to allow sufficient time for *SCR* 1 to turn *OFF*. Capacitor *C* discharges rapidly from – 2*E* to zero and then charges up in the opposite direction to + 2*E*. The load current is now carried through the second half of the transformer primary winding and *SCR* 2.

(*iii*) When trigger pulse is applied to the gate of *SCR* 1, this device will conduct and *SCR* 2 will turn off by the process just described. In this way, *SCR* 1 and *SCR* 2 alternately turn *ON* and *OFF*. Consequently, a.c. output is obtained as shown in Fig. 18.21.

The a.c. waveform produced by a single-phase inverter is a poor version of sine wave and would not be suitable for most industrial, commercial and domestic loads. More complex inverters using multiple *SCR*s and sophisticated triggering circuits are capable of generating a.c. voltages that are extremely close to a pure sine wave.

18.12 Applications of SCR

The ability of an *SCR* to control large currents in a load by means of small gate current makes

* The capacitor voltage will charge to double the suppply voltage (*E*) as a result of transformer action between the two primary windings.

this device useful in switching and control applications. Some of the important applications of *SCR* are discussed below :

(i) SCR as static contactor. An important application of *SCR* is for switching operations. As *SCR* has no moving parts, therefore, when it is used as a switch, it is often called a *static contactor*.

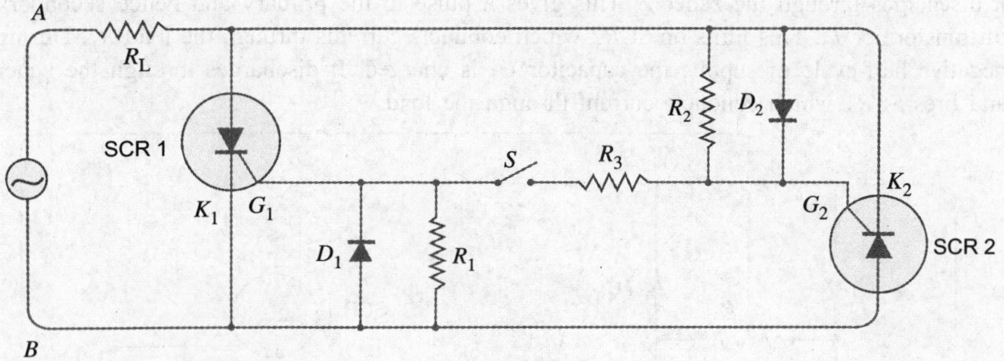

Fig. 18.22

Fig. 18.22 shows the use of *SCR* to switch *ON* or *OFF* a.c. power to a load R_L. Resistances R_1 and R_2 are for the protection of diodes D_1 and D_2 respectively. Resistance R_3 is the gate current limiting resistor. To start the circuit, switch is closed. During the positive half-cycle of a.c. supply, end *A* is positive and end *B* is negative. Then diode D_2 sends gate current through *SCR*1. Therefore, *SCR*1 is turned *ON* while *SCR*2 remains *OFF* as its anode is negative *w.r.t.* cathode. The current conduction by *SCR*1 follows the path AR_LK_1BA. Similarly, in the next half-cycle, *SCR*2 is turned *ON* and conducts current through the load. It may be seen that switch *S* handles only a few mA of gate current to switch *ON* several hundred amperes in the load R_L. This is a distinct advantage over a mechanical switch.

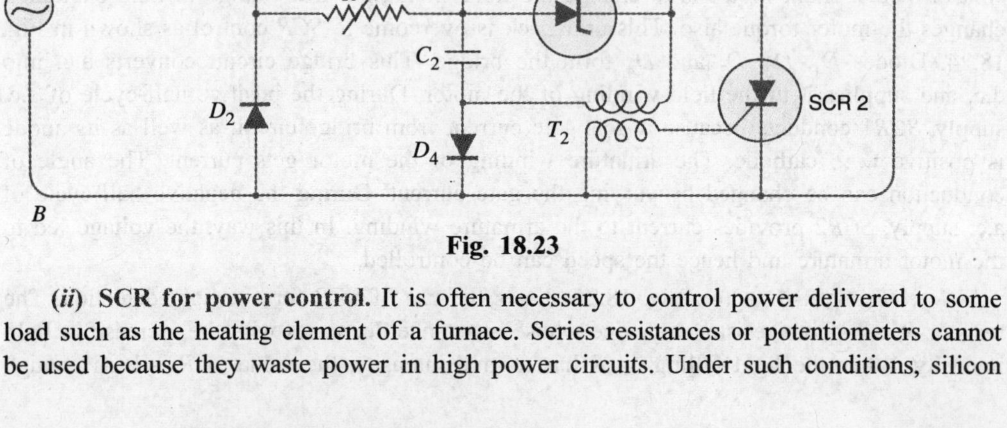

Fig. 18.23

(ii) SCR for power control. It is often necessary to control power delivered to some load such as the heating element of a furnace. Series resistances or potentiometers cannot be used because they waste power in high power circuits. Under such conditions, silicon controlled rectifiers are used which are capable of adjusting the transmitted power with little waste. Fig. 18.23 shows a common circuit for controlling power in the load R_L. During the positive half-cycle of a.c. supply, end A is positive and end B is negative. Therefore, capacitor C_2 is charged through $AD_1\ RC_2\ D_4B$. The charge on the capacitor C_2 depends upon the value of potentiometer R. When the capacitor C_2 is charged through a sufficient voltage, it discharges through the zener Z. This gives a pulse to the primary and hence secondary of transformer T_2. This turns on $SCR2$ which conducts currents through the load R_L. During negative half-cycle of supply, the capacitor C_1 is charged. It discharges through the zener and fires $SCR1$ which conducts current through the load.

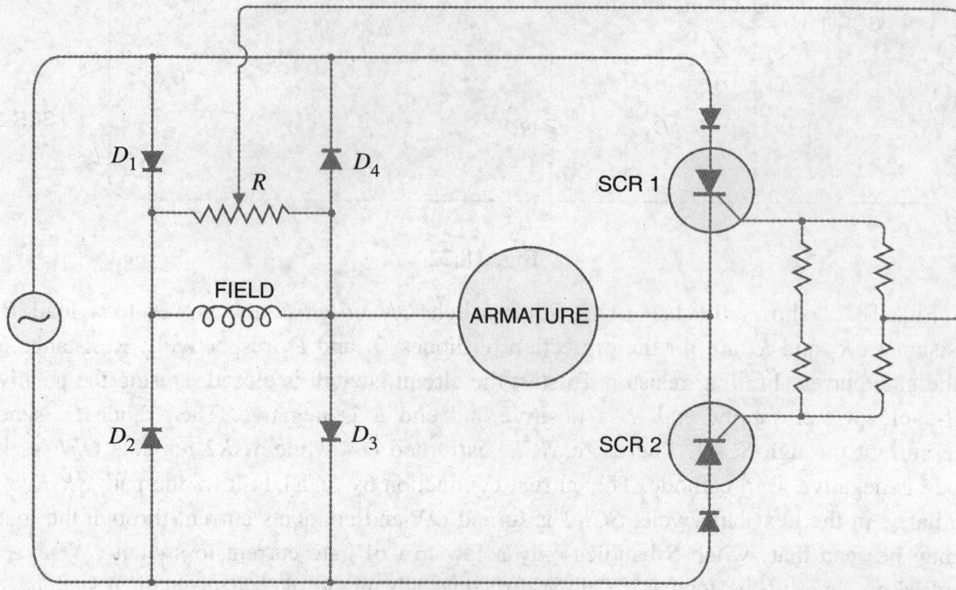

Fig. 18.24

The angle of conduction can be controlled by the potentiometer R. The greater the resistance of R, lesser is the voltage across C_1 or C_2 and hence smaller will be the time during which $SCR1$ and $SCR2$ will conduct in a full cycle. In this way, we can control a large power of several kW in the load R_L with the help of a small potentiometer R.

(*iii*) SCRs for speed control of d.c. shunt motor. The conventional method of speed control of d.c. shunt motor is to change the field excitation. But change in field excitation changes the motor torque also. This drawback is overcome in *SCR* control as shown in Fig. 18.24. Diodes D_1, D_2, D_3 and D_4 form the bridge. This bridge circuit converts a.c. into d.c. and supplies it to the field winding of the motor. During the positive half-cycle of a.c. supply, $SCR1$ conducts because it gets gate current from bridge circuit as well as its anode is positive *w.r.t.* cathode. The armature winding of the motor gets current. The angle of conduction can be changed by varying the gate current. During the negative half-cycle of a.c. supply, $SCR2$ provides current to the armature winding. In this way, the voltage fed to the motor armature and hence the speed can be controlled.

(*iv*) Overlight detector. Fig. 18.25 shows the use of *SCR* for overlight detection. The resistor R is a photo-resistor, a device whose resistance decreases with the increase in light intensity. When the light falling on R has normal intensity, the value of R is high enough and the voltage across R_1 is insufficient to trigger the *SCR*. However, when *R* is in strong light, its resistance decreases and the voltage drop across R_1 becomes high enough to trigger the *SCR*. Consequently, the buzzer sounds the alarm. It may be noted that even if the strong light disappears, the buzzer continues to sound the alarm. It is because once the *SCR* is fired, the gate loses all control.

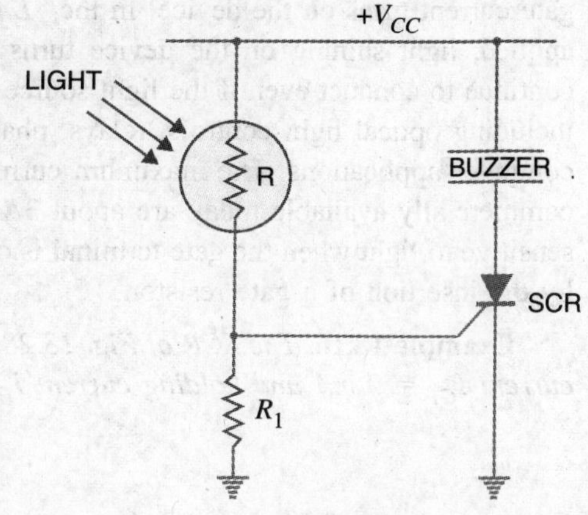

Fig. 18.25

(*v*) **SCR Crowbar.** A crowbar is a circuit that is used to protect a voltage-sensitive load from excessive d.c. power supply output voltages. Fig. 18.26 shows the *SCR* crowbar circuit. It consists of a zener diode, a gate resistor R_G and an *SCR*. It also contains a *snubber to prevent false triggering.

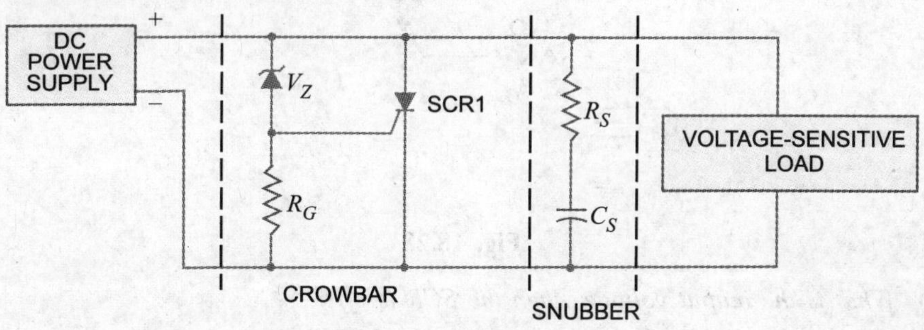

Fig. 18.26

Operation. The circuit action is as under :

(*a*) Under normal conditions, the zener diode and the *SCR* are *OFF*. With zener diode being in cut-off, there is no current through R_G and no voltage drop occurs across this resistor. This means that the gate of *SCR* is at 0V so that the *SCR* is in the off state. Therefore, as long as zener diode is off, the *SCR* behaves as an open and will not affect either the d.c. power supply or the load.

(*b*) Suppose the output voltage from the d.c. power supply suddenly *increases*. This causes the zener diode to break down and conduct current. As the current flows through the zener diode, voltage is developed across resistor R_G which causes the *SCR* to conduct current. When the *SCR* conducts, the voltage source is shorted by the *SCR*. The supply voltage fuse blows out and the load is protected from overvoltage.

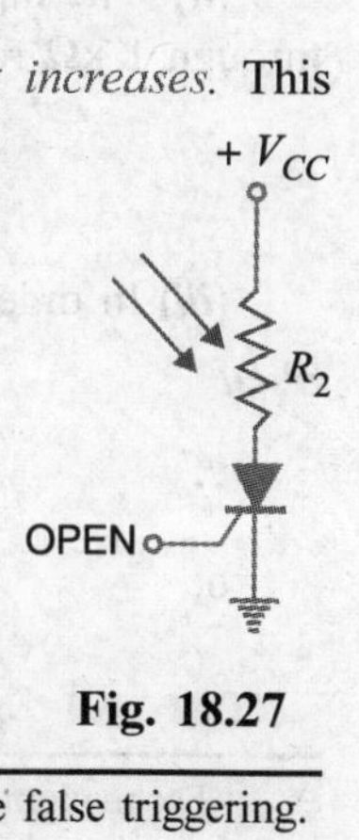

Fig. 18.27

18.13 Light-Activated SCR

The light-activated *SCR* (*LASCR*) is the light sensitive equivalent of the normal *SCR* and is shown in Fig. 18.27. As the name suggests, its state is controlled by the light falling on depletion layers. In a normal *SCR*,

* It is an *RC* circuit connected between the *SCR* anode and cathode to eliminate false triggering.

gate current turns on the device. In the **LASCR*, instead of having the external gate current applied, light shining on the device turns it *ON*. Just as a normal *SCR*, the *LASCR* will continue to conduct even if the light source is removed. The *LASCRs* find many applications including optical light controls, relays, phase control, motor control and a large number of computer applications. The maximum current (r.m.s.) and power (gate) ratings for *LASCRs* commercially available today are about 3A and 0.1W. It may be noted that *LASCR* is most sensitive to light when the gate terminal is open. Its sensitivity can be reduced and controlled by the insertion of a gate resistor.

Example 18.16. *The SCR of Fig. 18.28 has gate trigger voltage V_{GT} = 0.7V, gate trigger current I_{GT} = 7 mA and holding current I_H = 6 mA.*

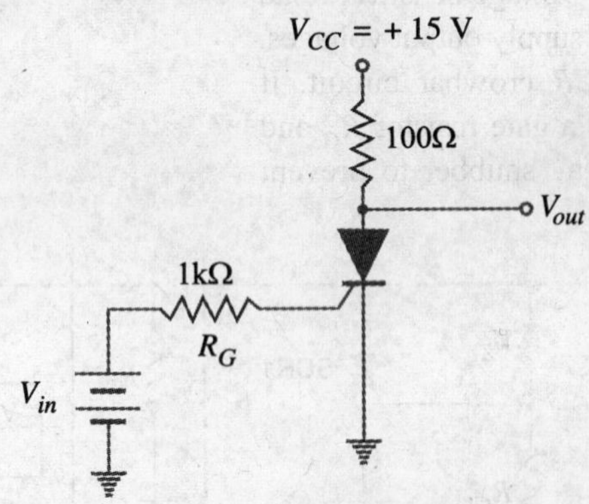

Fig. 18.28

(*i*) *What is the output voltage when the SCR is off ?*

(*ii*) *What is the input voltage that triggers the SCR ?*

(*iii*) *If V_{CC} is decreased until the SCR opens, what is the value of V_{CC} ?*

Solution.

(*i*) When the *SCR* is off (*i.e.* it is not conducting), there is no current through the 100 Ω resistor.

$$\therefore \quad V_{out} = \text{Supply voltage } V_{CC} = 15 \text{ V}$$

(*ii*) The input voltage V_{in} must overcome V_{GT} (= 0.7 V) and also cause 7 mA to flow through 1 kΩ resistor.

$$\therefore \quad V_{in} = V_{GT} + I_{GT} R_G = 0.7 + (7 \text{ mA}) (1 \text{ k}\Omega)$$

$$= 7.7\text{V}$$

(*iii*) In order to open the *SCR*, the V_{CC} must be reduced so that anode current is equal to I_H.

$$\therefore \quad I_H = \frac{V_{CC} - V_{GT}}{100\ \Omega}$$

$$\text{or} \quad V_{CC} = (100\Omega)\,(I_H) + V_{GT}$$

$$= (100\Omega)\,(6 \text{ mA}) + 0.7 = 1.3 \text{ V}$$

* For maximum sensitivity to light, the gate is left open.

Example 18.17. *In Fig. 18.29, the SCR has a trigger voltage of 0.7 V. Calculate the supply voltage that turns on the crowbar. Ignore zener diode resistance.*

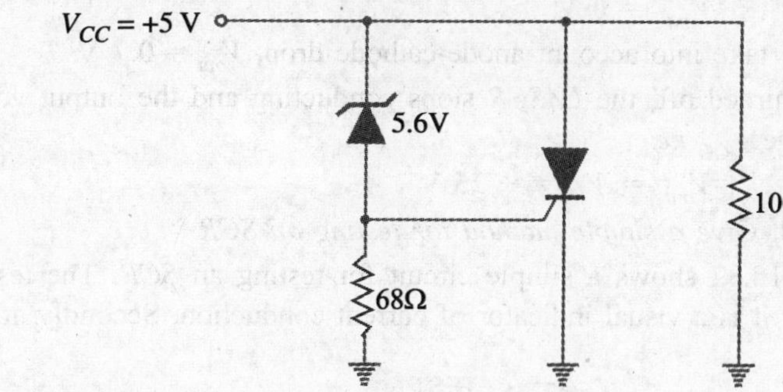

Fig. 18.29

Solution. The breakdown voltage of the zener is 5.6 V. To turn on the *SCR*, the voltage across 68Ω has to be equal to V_{GT} (= 0.7 V).

$$\therefore \quad V_{CC} = V_Z + V_{GT} = 5.6 + 0.7 = \mathbf{6.3\ V}$$

When the supply voltage becomes 6.3 V, the zener breaks down and starts conducting. The voltage V_{GT} (= 0.7 V) across 68 Ω forces the *SCR* into conduction. When the *SCR* conducts, the supply voltage is shorted by the *SCR* and the fuse in the supply voltage burns out. Thus the load (100Ω) is protected from overvoltage.

Example. 18.18. *The zener diode of Fig. 18.30 has a tolerance of ± 10% and the trigger voltage can be as high as 1.5 V. What is the maximum supply voltage where crowbarring takes place?*

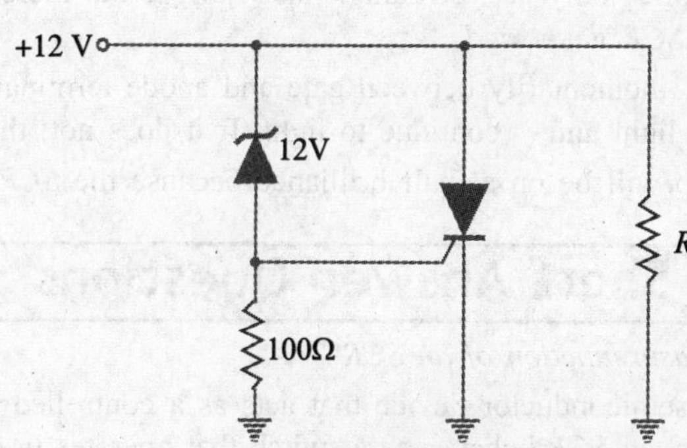

Fig. 18.30

Solution. The breakdown voltage of the zener diode is 12 V and it has a tolerance of ± 10%. It means that breakdown voltage of zener can vary from 10.8V to 13.2 V. Since the trigger voltage of *SCR* has a maximum value of 1.5 V,

∴ Maximum value of supply voltage for crowbarring = 13.2 V + 1.5 V = **14.7 V.**

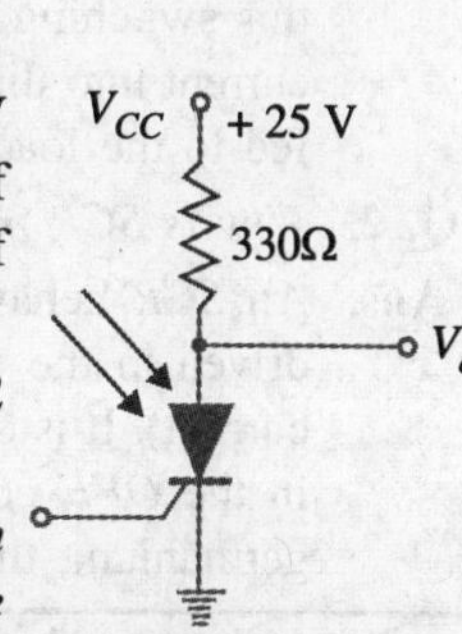

Fig. 18.31

Example. 18.19. *The circuit of Fig. 18.31 is in a dark room. When a bright light is turned on, the LASCR fires. What is the approximate output voltage? If the bright light is turned off, what is the output voltage ?*

Solution. Fig. 18.31 shows a light-activated *SCR*, also known as a *photo-SCR*. When light falls on the device, it starts conducting and the output voltage is ideally,

$$V_{out} = 0 \text{ V}$$

However, if we take into account anode-cathode drop, $V_{out} = 0.7$ V.

When light is turned off, the *LASCR* stops conducting and the output voltage is equal to the supply voltage V_{CC} *i.e.*

$$V_{out} = V_{CC} = + 25 \text{ V}$$

Example 18.20. *Give a simple method for testing an SCR.*

Solution. Fig. 18.32 shows a simple circuit for testing an *SCR*. The test lamp serves two purposes. First, it is a visual indicator of current conduction. Secondly, it limits current through the *SCR*.

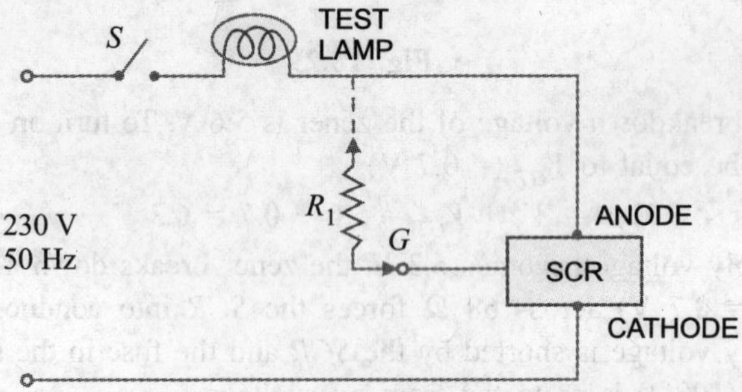

Fig. 18.32

(*i*) When switch *S* is closed, the lamp should not light for the *SCR* to be good. It is because voltage is *applied only between anode and cathode but there is no trigger voltage. If the lamp lights, the *SCR* is *shorted.*

(*ii*) Now touch R_1 momentarily between gate and anode terminals. For the *SCR* to be good, the lamp should light and **continue to light. If it does not, the *SCR* is *open.*

Note that the lamp will be on at half brilliance because the *SCR* conducts only every other half-cycle.

Short Answer Questions

Q. 1. *What is the basic function of an SCR ?*

Ans. An *SCR* is a semiconductor device that acts as a controlled unidirectional switch. In other words, an *SCR* behaves as a switch that operates in one direction and that this switching action can be controlled. This property permits to change alternating current into direct current and at the same time we can control the amount of power fed to the load. Thus, an *SCR* combines the features of a rectifier and a transistor.

Q. 2. *Why is SCR made of silicon and not germanium ?*

Ans. An *SCR* behaves as a switch, *i.e.*, either it is *ON* or it if *OFF*. When an *SCR* is driven in the *OFF* condition, it conducts an extremely small current (*i.e.*, leakage current). It is desirable that leakage current in an *SCR* should be as small as possible in the *OFF* condition. As leakage current in silicon is very small as compared to germanium, therefore, *SCRs* are invariably made of silicon.

* It is understood that the applied voltage is less than the forward breakover voltage of the *SCR*.

** Recall that once the *SCR* is fired by the gate voltage, it continues to conduct current even if the gate voltage is removed.

Q. 3. *Why is SCR sometimes called thyristor ?*

Ans. An *SCR* is sometimes called a thyristor. 'Thy' is a Greek word which means switch while 'ristor' means an association with transistor family.

Q. 4. *How does SCR differ from an ordinary rectifier ?*

Ans. An ordinary rectifier acts as a unidirectional switch whereas an *SCR* acts as controlled unidirectional switch. Let us take the example of an ordinary half-wave rectifier and *SCR* half-wave rectifier. Whereas the ordinary rectifier will conduct full positive half-cycle of a.c. supply, an *SCR* half-wave rectifier can be made to conduct full or part of a positive half-cycle by proper adjustment of gate current. Thus an *SCR* can control the power fed to the load and hence the name controlled rectifier.

Q. 5. *Why is SCR always turned ON by gate current ?*

Ans. Although *SCR* can be turned *ON* by breakover voltage (gate being open), this method reduces the range of control of the device. By proper adjustment of gate current, the *SCR* can be turned *ON* for any supply voltage and thus this method permits wide range of control for power fed to the load than the breakover voltage method.

Q. 6. *Why SCR cannot be used as a bidirectional switch ?*

Ans. An *SCR* can conduct (or turn on) only when anode is positive *w.r.t.* cathode and that too when proper gate current is made to flow. Therefore, it can operate only in one direction *i.e.*, it can act as a unidirectional switch.

Q. 7. *What is the normal way of operating an SCR ?*

Ans. The normal way of operating an *SCR* is to keep the supply voltage less than breakover voltage (of course, anode being positive *w.r.t.* cathode) and then turning *SCR ON* by passing proper gate current.

Q. 8. *What is the origin of the name SCR ?*

Ans. The abbreviation *SCR* stands for silicon controlled rectifier. The device is made of silicon and it can act as a controlled rectifier *i.e.*, it can change a.c. into d.c. and at the same time can control the amount of power fed to the load and hence the name silicon controlled rectifier.

Q. 9. *Why is the gate terminal of an SCR called so ?*

Ans. An *SCR* is turned *ON* by passing appropriate gate current. Once *SCR* is turned *ON* (*i.e.* starts conducting), the gate loses all control. The reason for naming this terminal as gate is obvious from this action of *SCR*.

Q. 10. *An SCR is controlled rectifier. Now, a semiconductor device that provides rectification is a diode and that provides control is a transistor. Does the construction of SCR indicate the presence of these devices ?*

Ans. An *SCR* is a three junction *pnpn* device. This construction makes it clear that the device is essentially an ordinary rectifier (*pn*) and a junction transistor (*npn*). The rectifier unit permits to change a.c. into d.c. and the transistor unit exercises control over the power fed to the load. Thus *SCR* combines the features of a rectifier and a transistor, hence the name controlled rectifier.

Q. 11. *What is the basic operational difference between a thyratron and an SCR ?*

Ans. An *SCR* is a solid-state equivalent of thyratron (*i.e.* gas-filled triode). The basic operational difference between the two is that whereas an *SCR* is a current-triggered device, a thyratron is a voltage-triggered device.

Multiple-Choice Questions

1. An *SCR* has .... *pn* junctions.
   (*i*) two (*ii*) three
   (*iii*) four
   (*iv*) none of the above
2. An *SCR* is a solid state equivalent of....
   (*i*) triode (*ii*) pentode
   (*iii*) gas-filled triode
   (*iv*) tetrode
3. An *SCR* has .... semiconductor layers.
   (*i*) two (*ii*) three
   (*iii*) four
   (*iv*) none of the above
4. An *SCR* has three terminals *viz.* ....
   (*i*) cathode, anode, gate
   (*ii*) anode, cathode, grid
   (*iii*) anode, cathode, drain
   (*iv*) none of the above
5. An *SCR* behaves as a .... switch.
   (*i*) unidirectional
   (*ii*) bidirectional
   (*iii*) mechanical
   (*iv*) none of the above
6. An *SCR* is sometimes called a ....
   (*i*) triac (*ii*) diac
   (*iii*) unijunction transistor
   (*iv*) thyristor
7. An *SCR* is made of ....
   (*i*) germanium (*ii*) silicon
   (*iii*) carbon
   (*iv*) none of the above
8. In the normal operation of an *SCR*, anode is .... *w.r.t.* cathode.
   (*i*) at zero potential
   (*ii*) negative
   (*iii*) positive
   (*iv*) none of the above
9. In normal operation of an *SCR*, gate is .... *w.r.t.* cathode.
   (*i*) positive (*ii*) negative
   (*iii*) at zero potential
   (*iv*) none of the above
10. An *SCR* combines the features of ....
    (*i*) a rectifier and resistance
    (*ii*) a rectifier and transistor
    (*iii*) a rectifier and capacitor
    (*iv*) none of the above
11. The control element in an *SCR* is ....
    (*i*) cathode (*ii*) anode
    (*iii*) anode supply (*iv*) gate
12. The normal way to turn on an *SCR* is by ....
    (*i*) forward breakover voltage
    (*ii*) appropriate anode current
    (*iii*) appropriate gate current
    (*iv*) none of the above
13. An *SCR* is turned off by ....
    (*i*) reducing anode voltage to zero
    (*ii*) reducing gate voltage to zero
    (*iii*) reverse biasing the gate
    (*iv*) none of the above
14. An *SCR* is a .... triggered device.
    (*i*) voltage (*ii*) current
    (*iii*) voltage as well as current
    (*iv*) none of the above
15. In an *SCR* circuit, the supply voltage is generally .... that of forward breakover voltage.
    (*i*) equal to (*ii*) less than
    (*iii*) greater than
    (*iv*) none of the above
16. When an *SCR* is turned on, the voltage across it is about ....
    (*i*) zero (*ii*) 10 V
    (*iii*) 0.1 V (*iv*) 1V
17. An *SCR* is made of silicon and not germanium because silicon ....
    (*i*) is inexpensive
    (*ii*) is mechanically strong
    (*iii*) has small leakage current
    (*iv*) is tetravalent
18. An *SCR* is turned off when ....
    (*i*) anode current is reduced to zero
    (*ii*) gate voltage is reduced to zero
    (*iii*) gate is reverse biased
    (*iv*) none of the above

19. In an *SCR* circuit, the angle of conduction can be changed by ....
   (*i*) changing anode voltage
   (*ii*) changing gate voltage
   (*iii*) reverse biasing the gate
   (*iv*) none of the above
20. If firing angle in an *SCR* circuit is increased, the output ....
   (*i*) remains the same
   (*ii*) is increased
   (*iii*) is decreased
   (*iv*) none of the above
21. If gate current is increased, then anode-cathode voltage at which *SCR* closes ....
   (*i*) is decreased (*ii*) is increased
   (*iii*) remains the same
   (*iv*) none of the above
22. When *SCR* is *OFF*, the current in the circuit is ....
   (*i*) exactly zero
   (*ii*) small leakage current
   (*iii*) large leakage current
   (*iv*) none of the above
23. An *SCR* can exercise control over .... of a.c. supply.
   (*i*) positive half-cycles only
   (*ii*) negative half-cycles only
   (*iii*) both positive and negative half-cycles
   (*iv*) positive or negative half-cycles
24. We can control a.c. power in a load by connecting ....
   (*i*) two *SCRs* in series
   (*ii*) two *SCRs* in parallel
   (*iii*) two *SCRs* in parallel opposition
   (*iv*) none of the above
25. When *SCR* starts conducting, then .... loses all control.
   (*i*) gate (*ii*) cathode
   (*iii*) anode
   (*iv*) none of the above

Answers to Multiple-Choice Questions

1. (*ii*)	2. (*iii*)	3. (*iii*)	4. (*i*)	5. (*i*)
6. (*iv*)	7. (*ii*)	8. (*iii*)	9. (*i*)	10. (*ii*)
11. (*iv*)	12. (*iii*)	13. (*i*)	14. (*ii*)	15. (*ii*)
16. (*iv*)	17. (*iii*)	18. (*i*)	19. (*ii*)	20. (*iii*)
21. (*i*)	22. (*ii*)	23. (*iv*)	24. (*iii*)	25. (*i*)

Chapter Review Topics

1. Explain the construction and working of an *SCR*.
2. Draw the equivalent circuit of an *SCR* and explain its working from this equivalent circuit.
3. Explain the terms forward breakover voltage, holding current and forward current rating as used in connection with *SCR* analysis.
4. Draw the *V-I* characteristics of an *SCR*. What do you infer from them ?
5. Explain the action of an *SCR* as a switch. What are the advantages of *SCR* switch over a mechanical or electro-mechanical switch ?
6. Discuss some important applications of *SCR*.

Problems

1. An *SCR* has a forward breakover voltage of 450 V, a gate trigger current of 15 mA and holding current of 10 mA. What do you infer from it?
2. An *SCR* in a circuit is subjected to a 50 A current surge that lasts for 10 ms. Determine whether or not this surge will destroy the device. Given that circuit fusing rating of *SCR* is 90 A^2s. **[will not be destroyed]**

3. An *SCR* has a circuit fusing rating of 70 A^2s. The device is being used in a circuit where it could be subjected to a 100 A surge. Determine the limit on the duration of such a surge. **[7ms]**
4. An *SCR* has a circuit fusing rating of 60 A^2s. Determine the highest surge current value that *SCR* can withstand for a period of 20 ms. **[54.77 A]**
5. In Fig. 18.33, what value of input voltage would be required to cause the *SCR* to break down if the gate current required for firing is 10 mA ? **[3.7 V]**

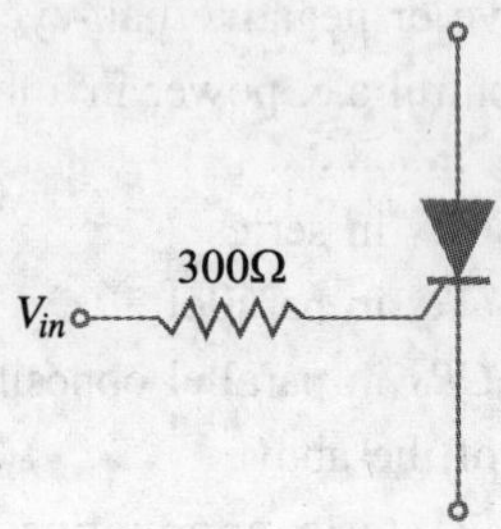

Fig. 18.33

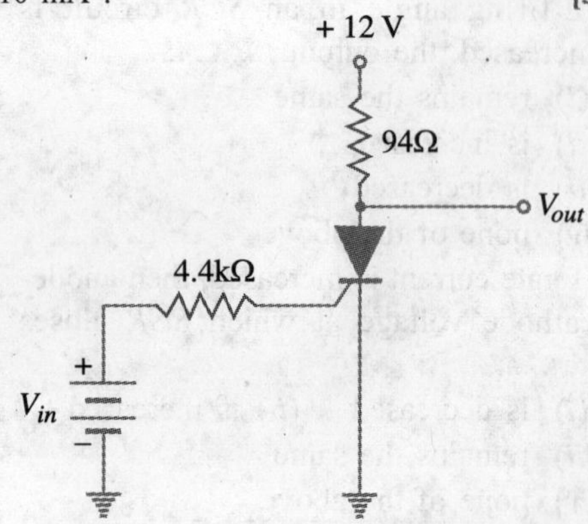

Fig. 18.34

6. In Fig. 18.34, if the gate trigger current of the *SCR* is 1.5 mA, what is the input voltage that triggers the *SCR*? Given V_{GT} = 0.7 V. **[7.3 V]**
7. A 24 V r.m.s. supply is connected to a half-wave *SCR* circuit that is triggered at 50°. What is the d.c. voltage delivered to the load ? **[8.88 V]**
8. An a.c. voltage of 125 V (r.m.s.) is applied to an *SCR* half-wave rectifier. The load resistance R_L = 50 Ω. Assuming that forward voltage drop across the *SCR* is 0.9 V when it is conducting, find the average power dissipated by the *SCR* when the firing angle is 30°. **[0.375 W]**
9. The *SCR* in Fig. 18.35 has V_{GT} = 0.75 V, I_{GT} = 7 mA and I_H = 6 mA. What is the output voltage when the *SCR* is off ? What is the input voltage that triggers the *SCR* ? If V_{CC} is decreased until the *SCR* opens, what is the value of V_{CC} ? **[15 V ; 7.75 V ; 1.3 V]**

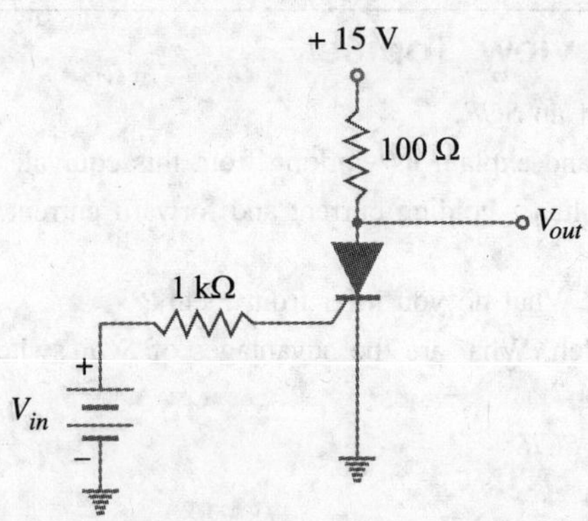

Fig. 18.35

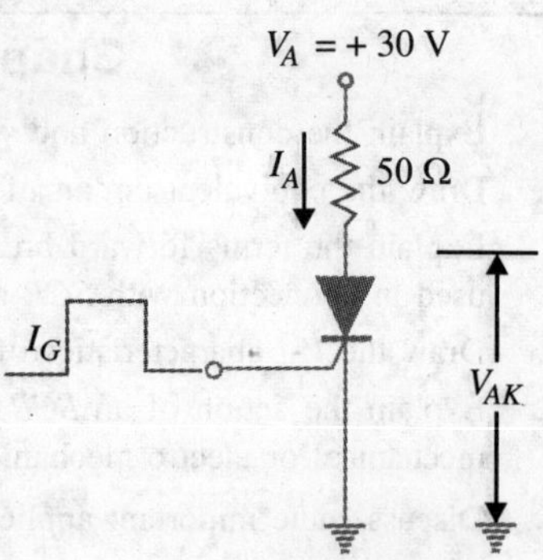

Fig. 18.36

10. The voltage drop across the *SCR* in Fig. 18.36 is 1 V when it is conducting. It has a holding current of 2 mA when I_G = 0. If the *SCR* is triggered on by a momentary pulse of gate current, to what value must V_A be reduced to turn the *SCR* off ? **[1.1 V]**

CHAPTER 19

Power Electronics

Learning Objectives

- Understand the importance of power electronics in the modern world.
- Develop a strong understanding of construction and working of triac.
- Learn the *SCR* equivalent circuit of triac.
- Explore the transistor equivalent circuit of triac.
- Study the important applications of triac.
- Learn the construction, working and applications of diac.
- Understand the construction and operation of unijunction transistor (*UJT*).
- Develop an understanding of the equivalent circuit of *UJT*.
- Explore the characteristics of *UJT*.
- Study the applications of *UJT*.

Introduction

The electric power system is a.c. and operates at 50 Hz. It is often desirable to control or regulate the flow of a.c. power in a load such as an electric motor, lighting system *etc*. This can be achieved by using electronic switches. We know that semiconductor diodes, bipolar junction transistors (*BJTs*) and field-effect transistors (*FETs*) act as electronic switches under proper operating conditions. However, the major drawback of these electronic switches is that their current and power handling capabilities are small. In order to control or regulate large a.c. power, we have developed high-power semiconductor electronic switches. These high-power electronic switches are called **thyristors* and the branch of electronics which deals with the use and application of these high-power electronic switches is called *power electronics*. The thyristors (*i.e.* high-power electronic switches) can be used to handle hundreds of amperes and thousands of watts. The purpose of this chapter is to discuss some important high-power switching devices much used in power electronics.

---

* The name thyristor is derived from the Greek word ; the first part (*i.e.* thy) means switch and the second part (*i.e.* ristor) means an association with transistor family.

19.1 Power Electronics

The branch of electronics which deals with the control of power at 50 Hz (i.e. supply frequency) is known as **power electronics.**

There are many applications where it is desired to control (or regulate) the power fed to a load *e.g.* to change the speed of a fan or motor. So far we have been using electrical methods to exercise such a control. However, electrical methods do not permit a *fine control over the flow of power in a system. Moreover, there is a considerable wastage of power. In the recent years, such semiconductor devices have been developed which can exercise fine control over the flow of large blocks of power in a system. Such devices act as controlled switches and can perform the duties of controlled rectification, inversion and regulation of power in a load. The important semiconductor switching devices are:

(*i*) Silicon controlled rectifier (*SCR*) (*ii*) Triac

(*iii*) Diac (*iv*) Unijunction transistor (*UJT*)

The silicon controlled rectifier (*SCR*) has already been discussed in the previous chapter. Therefore, we shall deal with the other three switching devices in the following discussion.

19.2 The Triac

The major drawback of an *SCR* is that it can conduct current in one direction only. Therefore, an *SCR* can only control d.c. power or forward biased half-cycles of a.c. in a load. However, in an a.c. system, it is often desirable and necessary to exercise control over both positive and negative half-cycles. For this purpose, a semiconductor device called *triac* is used.

A **triac** *is a three-terminal semiconductor switching device which can control alternating current in a load.*

Triac is an abbreviation for *triode a.c. switch.* 'Tri'– indicates that the device has three terminals and 'ac' means that the device controls alternating current or can conduct current in either direction.

The key function of a triac may be understood by referring to the simplified Fig. 19.1. The **control circuit of triac can be adjusted to pass the desired portions of positive and negative half-cycle of a.c. supply through the load R_L. Thus referring to Fig. 19.1 (*ii*), the

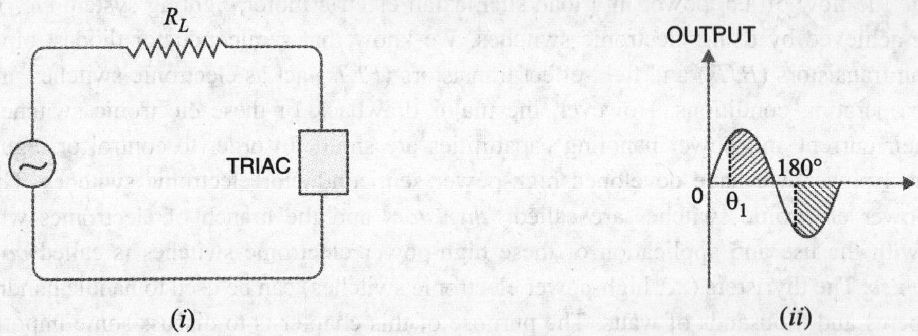

Fig. 19.1

* For example, the speed of a ceiling fan can be changed in four to five steps by electrical method.

** Although it appears that 'triac' has two terminals, there is also a third terminal connected to the control circuit.

triac passes the positive half-cycle of the supply from θ_1 to 180° *i.e.* the shaded portion of positive half-cycle. Similarly, the shaded portion of negative half-cycle will pass through the load. In this way, the alternating current and hence a.c. power flowing through the load can be controlled.

Since a triac can control conduction of both positive and negative half-cycles of a.c. supply, it is sometimes called a bidirectional semi-conductor triode switch. The above action of a triac is certainly not a rectifying action (as in an **SCR*) so that the triac makes no mention of rectification in its name.

19.3 Triac Construction

A **triac** *is a three-terminal, five-layer semiconductor switching device whose forward and reverse characteristics are indentical to the forward characteristics of the SCR.* The three terminals are designated as main terminal *MT*1, main terminal *MT*2 and gate *G*.

Fig. 19.2 (*i*) shows the basic structure of a triac. As we shall see, a triac is equivalent to two separate *SCR*s connected in inverse parallel (*i.e.* anode of each connected to the cathode of the other) with gates commoned as shown in Fig. 19.2 (*ii*). Therefore, a triac acts like a bidirectional switch *i.e.* it can conduct current in either direction. This is unlike an *SCR* which can conduct current only in one direction. Fig. 19.2 (*iii*) shows the schematic symbol of a triac. The symbol consists of two parallel diodes connected in opposite directions with a single gate lead. It can be seen that even the symbol of triac indicates that it can conduct current for either polarity of the main terminals (*MT*1 and *MT*2) *i.e.* it can act as a bidirectional switch. The gate provides control over conduction in either direction.

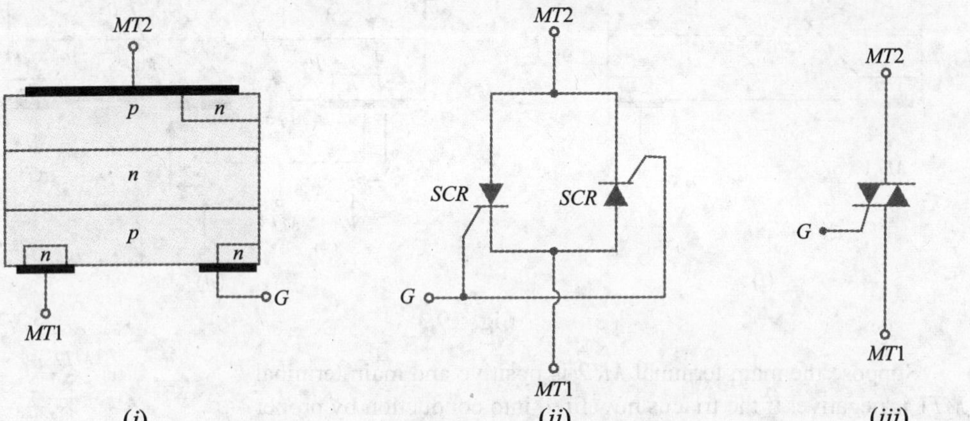

Fig. 19.2

The following points many be noted about the triac :

(*i*) The triac can conduct current (of course with proper gate current) regardless of the polarities of the main terminals *MT*1 and *MT*2. Since there is no longer a specific anode or cathode, the main leads are referred to as *MT*1 and *MT*2.

(*ii*) A triac can be turned on either with a positive or negative voltage at the gate of the device.

* *SCR* is a controlled rectifier. It is a unidirectional switch and can conduct only in one direction. Therefore, it can control only one half-cycle (positive or negative) of a.c. supply.

(*iii*) Like the *SCR*, once the triac is fired into conduction, the gate loses all control. The triac can be turned off by reducing the circuit current to the value of holding current.

(*iv*) The main disadvantage of triacs over *SCR*s is that triacs have considerably lower current-handling capabilities. Most triacs are available in ratings of less than 40A at voltages up to 600V.

19.4 SCR Equivalent Circuit of Triac

We shall now see that a triac is equivalent to two separate *SCRs* connected in inverse parallel (*i.e.* anode of each connected to the cathode of the other) with gates commoned. Fig. 19.3 (*i*) shows the basic structure of a triac. If we split the basic structure of a triac into two halves as shown in Fig. 19.3 (*ii*), it is easy to see that we have two *SCR*s connected in inverse parallel. The left half in Fig. 19.3 (*ii*) consists of a *pnpn* device ($p_1n_2\ p_2n_4$) having three *pn* junctions and constitutes *SCR*1. Similarly, the right half in Fig. 19.3 (*ii*) consists of *pnpn* device ($p_2n_3p_1n_1$) having three *pn* junctions and constitutes *SCR*2. The *SCR* equivalent circuit of the triac is shown in Fig. 19.4.

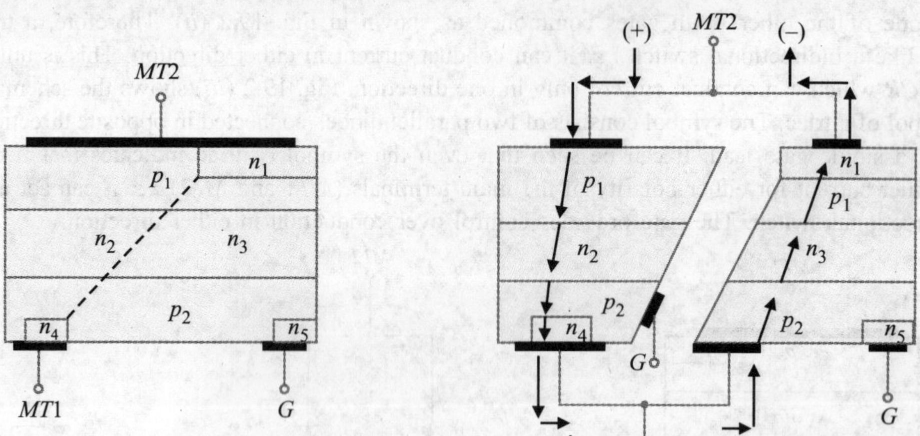

Fig. 19.3

Suppose the main terminal *MT*2 is positive and main terminal *MT*1 is negative. If the triac is now fired into conduction by proper gate current, the triac will conduct current following the path (left half) shown in Fig. 19.3 (*ii*). In relation to Fig. 19.4, the *SCR*1 is *ON* and the *SCR*2 is *OFF*. Now suppose that *MT*2 is negative and *MT*1 is positive. With proper gate current, the triac will be fired into conduction. The current through the devices follows the path (right half) as shown in Fig. 19.3 (*ii*). In relation to Fig. 19.4, the *SCR*2 is *ON* and the *SCR*1 is *OFF*. Note that the triac will conduct current in the appropriate direction as long as the current through the device is greater than its holding current.

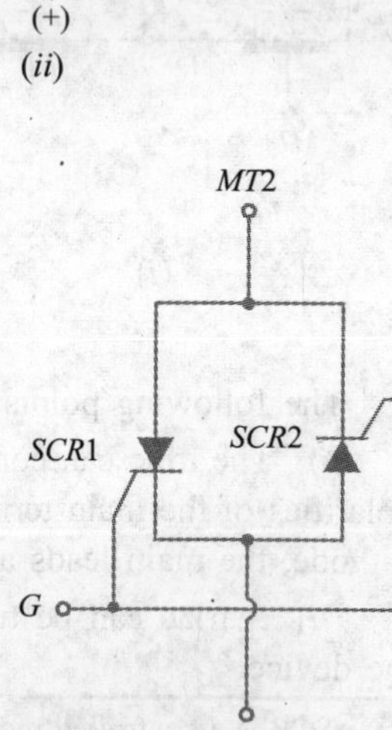

Fig. 19.4

19.5 Triac Operation

Fig. 19.5 shows the simple triac circuit. The a.c. supply to be controlled is connected across the main terminals of triac through a load resistance R_L. The gate circuit consists of battery, a current limiting resistor R and a switch S. The circuit action is as follows :

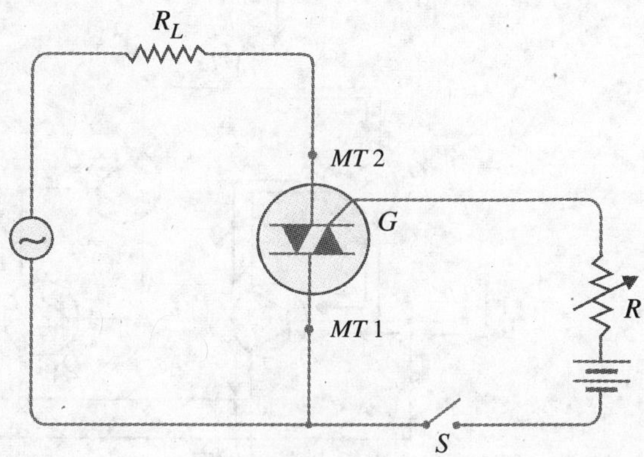

Fig. 19.5

(*i*) With switch *S* open, there will be no gate current and the triac is cut-off. Even with no gate current, the triac can be turned on provided the supply voltage becomes equal to the breakover voltage of triac. However, the normal way to turn on a triac is by introducing a proper gate current.

(*ii*) When switch *S* is closed, the gate current starts flowing in the gate circuit. In a similar manner to *SCR*, the breakover voltage of the triac can be varied by making proper gate current to flow. With a few milliamperes introduced at the gate, the triac will start conducting whether terminal *MT*2 is positive or negative *w.r.t. MT*1.

(*iii*) If terminal *MT*2 is positive *w.r.t. MT*1, the triac turns on and the conventional current will flow from *MT*2 to *MT*1. If the terminal *MT*2 is negative *w.r.t. MT*1, the triac is again turned on but this time the conventional current flows from *MT*1 to *MT*2.

The above action of triac reveals that it can act as an a.c. contactor to switch on or off alternating current to a load. The additional advantage of triac is that by adjusting the gate current to a proper value, any portion of both positive and negative half-cycles of a.c. supply can be made to flow through the load. This permits to adjust the transfer of a.c. power from the source to the load.

Example 19.1. *Draw the transistor equivalent circuit of a triac and explain its operation from this equivalent circuit.*

Solution. We have seen that transistor equivalent circuit of an *SCR* is composed of *pnp* transistor and *npn* transistor with collector of each transistor coupled to the base of the other. Since a triac is equivalent to two *SCR*s connected in inverse parallel (Refer back to Fig. 19.4), the transistor equivalent circuit of triac will be composed of four transistors arranged as shown in Fig. 19.6 (*i*). The transistors Q_1 and Q_2 constitute the equivalent circuit of *SCR*1 while the transistors Q_3 and Q_4 constitute the equivalent circuit of *SCR*2. We can explain the action of triac from its transistor equivalent circuit as under :

(*i*) When *MT*2 is positive *w.r.t. MT*1 and appropriate gate current is allowed in the gate circuit, *SCR*1 is turned *ON* while *SCR*2 remains *OFF*. In terms of transistor equivalent circuit, Q_1 and Q_2 are forward biased while Q_3 and Q_4 are reverse biased. Therefore, transistors Q_1 and Q_2 conduct current as shown in Fig. 19.6 (*i*). Since Q_1 and Q_2 form a positive feedback, both transistors are quickly driven to stauration and a large current flows through the load R_L. This is as if switch between *MT*2 and *MT*1 were closed.

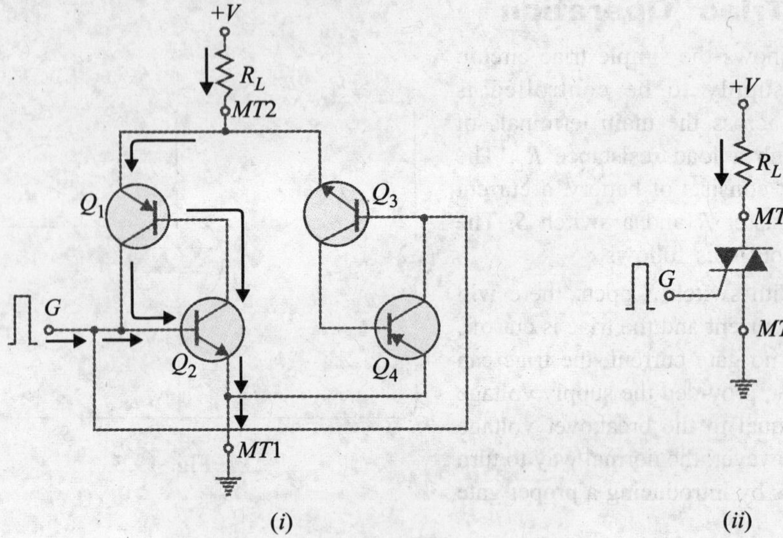

Fig. 19.6

Fig. 19.6 (*ii*) shows the action of triac (*MT*2 positive *w.r.t. MT*1) by replacing the triac with its symbol.

(*ii*) When *MT*2 is negative *w.r.t. MT*1 and appropriate gate current is allowed in the gate circuit, *SCR*2 is turned *ON* and *SCR*1 is *OFF*. In terms of transistor equivalent circuit, Q_3 an Q_4 are forward biased while Q_1 and Q_2 are reverse biased. Therefore, transistors Q_3 and Q_4 will conduct as shown in Fig. 19.7 (*i*). As explained above, the current in load R_L will quickly attain a large value. The circuit will behave as if a switch is closed between *MT*2 and *MT*1.

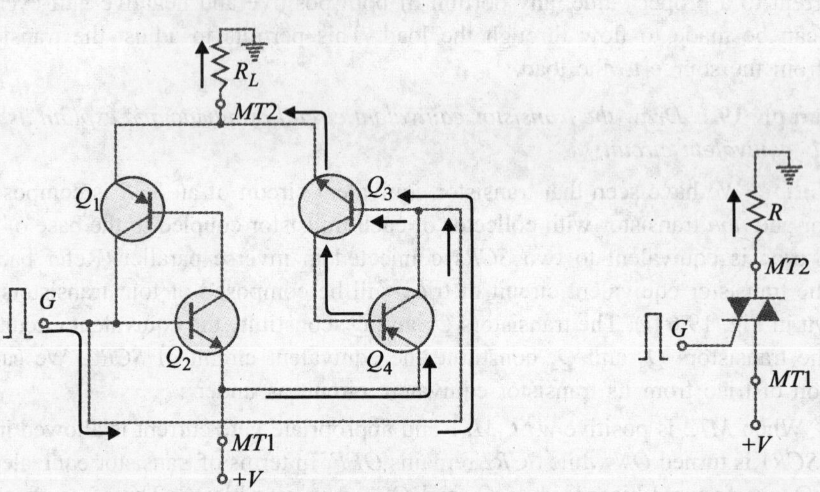

Fig. 19.7

Fig. 19.7 (*ii*) shows the action of triac (*MT*2 negative *w.r.t. MT*1) by replacing the triac with its symbol.

19.6 Traic Characteristics

Fig. 19.8 shows the *V-I* characteristics of a triac. Because the triac essentially consists of two *SCRs* of opposite orientation fabricated in the same crystal, its operating characteristics in the first and third quadrants are the same except for the direction of applied voltage and current flow. The following points may be noted from the triac characteristics :

(*i*) The *V-I* characteristics for triac in the Ist and IIIrd quadrants are essentially identical to those of an *SCR* in the Ist quadrant.

(*ii*) The triac can be operated with either positive or negative gate control voltage but in *normal operation usually the gate voltage is positive in quadrant I and negative in quadrant III.

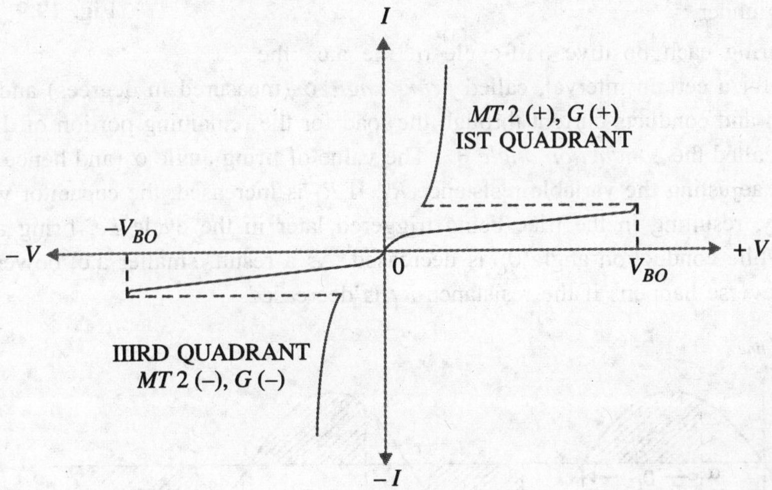

Fig. 19.8

(*iii*) Like an *SCR*, the triac can be switched from the *OFF* state to *ON* state by the breakover voltage (V_{BO}). The breakover point can be reached at lower supply voltage by applying a positive or a negative trigger pulse to the gate.

(*iv*) The supply voltage at which the triac is turned *ON* depends upon the gate current. The greater the gate current, the smaller the supply voltage at which the triac is turned on. This permits to use a triac to control a.c. power in a load from zero to full power in a smooth and continuous manner with no loss in the controlling device.

(*v*) Like an *SCR*, the magnitudes of the breakover voltage and holding current become smaller as the value of gate current increases. Whatever may be the direction of current through the triac, its value must be reduced below the holding current to turn the triac *OFF*.

19.7 Triac Phase Control Circuit

A **triac can be used to control the average a.c. power to a load by passing a portion of positive and negative half-cycles of input a.c. This is achieved by changing the conduction

* With this arrangement, less charge is required to turn on the triac.

** An *SCR* can only control the positive half-cycle or negative half-cycle of a.c.

angle through the load. Fig. 19.9 shows the basic triac phase control circuit. This circuit uses a capacitor C and variable resistance R_1 to shift the phase angle of the gate signal. Because of this phase shift, the gate voltage lags the line voltage by an angle between 0° and 90°. By adjusting the variable resistance R_1, the conduction angle through the load can be changed. Thus any portion of positive and negative half-cycles of the a.c. can be passed through the load. This action of triac permits it to be used as a *controlled* * *bidirectional switch.*

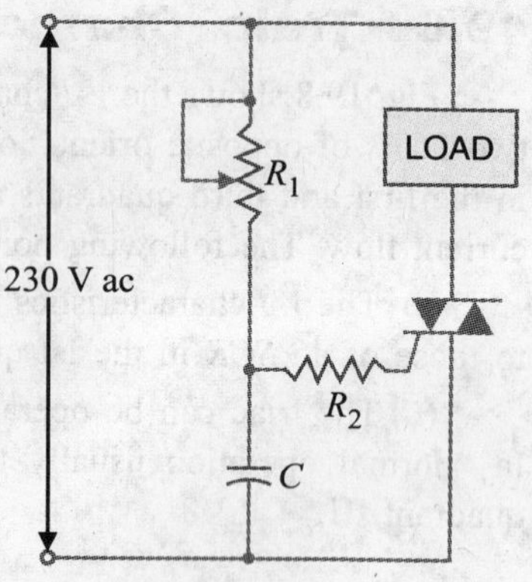

Fig. 19.9

Circuit action. The operation of triac phase control circuit is as under :

(*i*) During each positive half-cycle of the a.c., the triac is off for a certain interval, called *firing angle* α (measured in degrees) and then it is triggered on and conducts current through the load for the remaining portion of the positive half-cycle, called the *conduction angle* θ_C. The value of firing angle α (and hence θ_C)can be changed by adjusting the variable resistance R_1. If R_1 is increased, the capacitor will charge more slowly, resulting in the triac being triggered later in the cycle *i.e.* firing angle α is increased while conduction angle θ_C is decreased. As a result, smaller a.c. power is fed to the load. Reverse happens if the resistance R_1 is decreased.

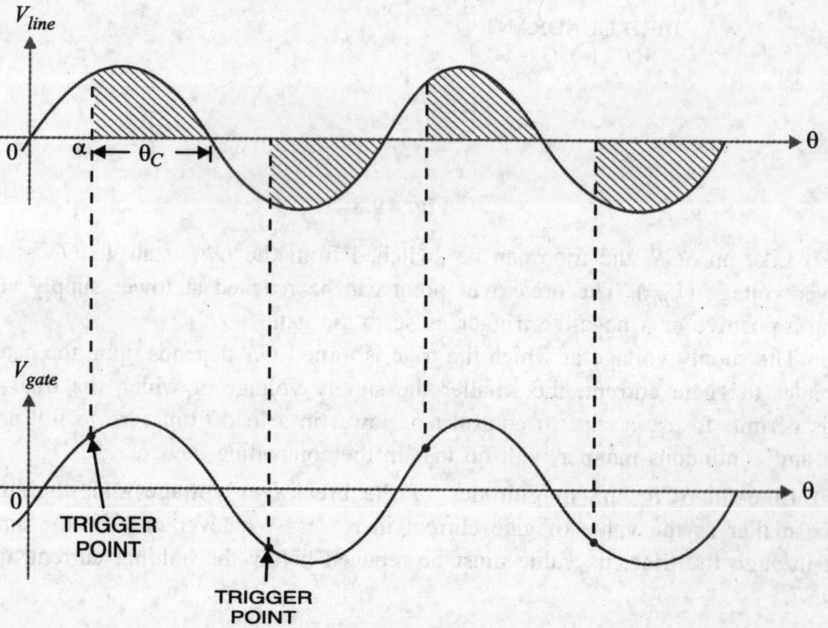

Fig. 19.10

(*ii*) During each negative half-cycle of the a.c., a similar action occurs except that now current in the load is in the opposite direction.

* An *SCR* is a controlled unidirectional switch because it can conduct only in one direction.

Fig. 19.10 shows the waveforms of the line voltage and gate voltage. Only the shaded portion of the positive and negative half-cycles pass through the load. We can change the phase angle of gate voltage by adjusting the variable resistance R_1. Thus a triac can control the a.c. power fed to a load. This control of a.c. power is useful in many applications such as industrial heating, lighting *etc.*

19.8 Applications of Triac

As low gate currents and voltages can be used to control large load currents and voltages, therefore, triac is often used as an electronic on/off switch controlled by a low-current mechanical switch.

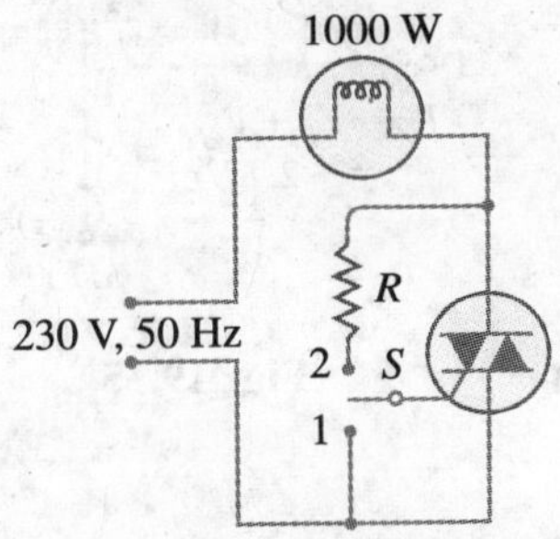

Fig. 19.11

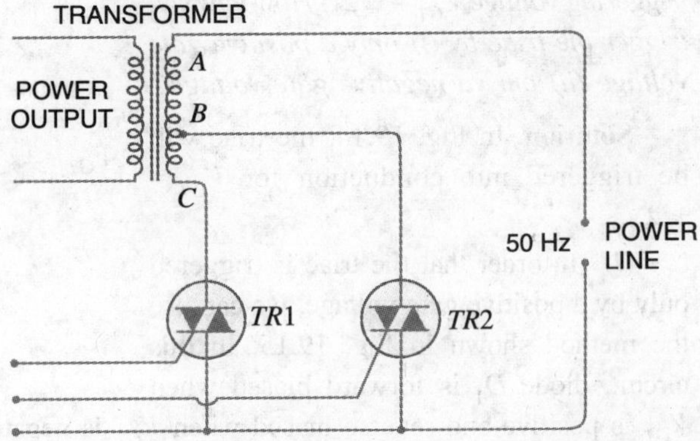

Fig. 19.12

(*i*) **As a high-power lamp switch.** Fig. 19.11 shows the use of a triac as an a.c. on/off switch. When switch *S* is thrown to position 1, the triac is cut-off and the output power of lamp is zero. But as the switch is thrown to position 2, a small gate current (a few mA) flowing through the gate turns the triac on. Consequently, the lamp is switched on to give full output of 1000 watts.

(*ii*) **Electronic change-over of transformer taps.** Fig. 19.12 shows the circuit of electronic change-over of power transformer input taps. Two triacs *TR*1 and *TR*2 are used for the purpose. When triac *TR*1 is turned on and *TR*2 is turned off, the line input is connected across the full transformer primary *AC*. However, if it is desired to change the tapping so that input appears across part *AB* of the primary, then *TR*2 is turned on and *TR*1 is turned off. The gate control signals are so controlled that both triacs are never switched on together. This avoids a dangerous short-circuit on the section *BC* of the primary.

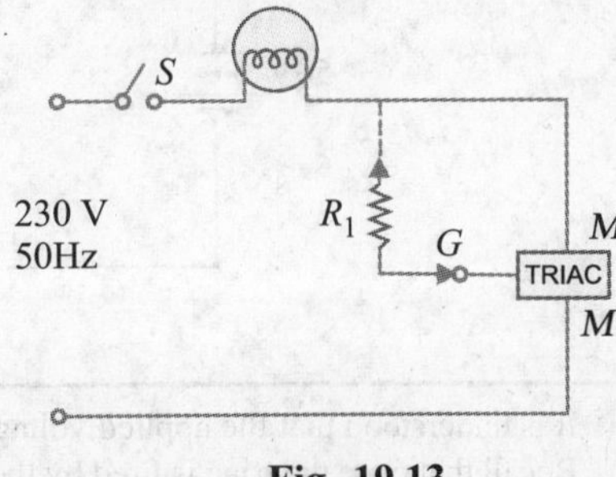

Fig. 19.13

Example 19.2. *Give a simple method for testing a triac.*

Solution. Fig. 19.13 shows a simple circuit for testing a triac. The test lamp serves two purposes. First, it is a visual indicator of current conduction. Secondly, it limits current through the triac.

(*i*) When switch *S* is closed, the lamp should not light for the triac to be good. It is because voltage is *applied only between $MT2$ and $MT1$ but there is no trigger voltage. If the lamp lights, the triac is *shorted.*

(ii) Now touch R_1 momentarily between gate and $MT2$ terminal. For the triac to be good, the lamp should light and **continue to light. If it does not, the triac is *open.*

Note that the lamp is on at full brilliance because the triac conducts both half-cycles (positive and negative) of a.c. This method of triac testing has two main advantages. First, it is a very simple method. Secondly, it does not require expensive apparatus.

Example 19.3. *The triac shown in Fig. 19.14 can be triggered by the gate triggering voltage $V_{GT} = \pm 2V$. How will you trigger the triac by (i) only a positive gate voltage (ii) only a negative gate voltage?*

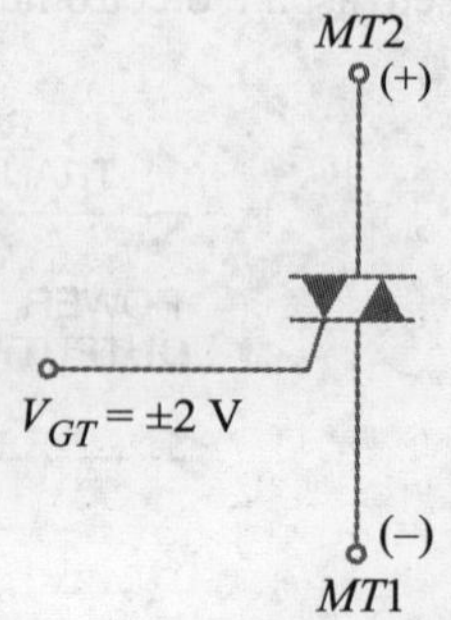

Fig. 19.14

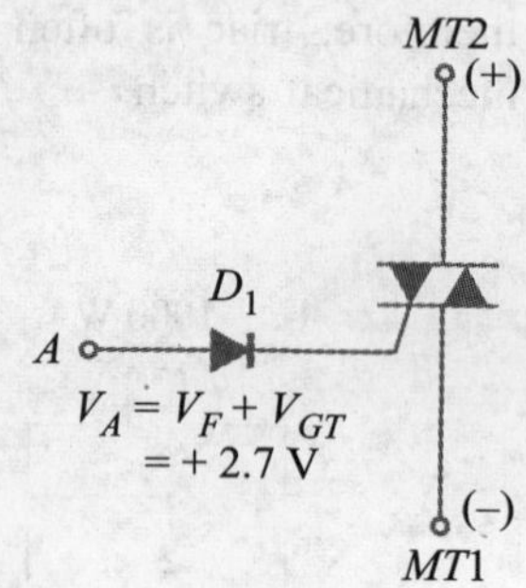

Fig. 19.15

Solution. In Fig. 19.14, the triac will be triggered into conduction for $V_{GT} = \pm 2V$.

(i) In order that the triac is triggered only by a positive gate voltage, we can use the method shown in Fig. 19.15. In this circuit, diode D_1 is forward biased when V_{GT} is positive and reverse biased when V_{GT} is negative. Since D_1 will conduct only when V_{GT} is positive, the triac can only be triggered by a positive gate signal. The voltage V_A required to trigger the device is equal to the sum of V_F for diode D_1 and the required gate triggering voltage *i.e.*

$$V_A = V_F + V_{GT} = 0.7V + 2V = 2.7V$$

(ii) In order that the triac is triggered only by the negative voltage, **reverse the direction of diode D_1.**

Example 19.4. *In Fig. 19.16, the switch is closed. If the triac has fired, what is the current through 50Ω resistor when (i) triac is ideal (ii) triac has a drop of 1V ?*

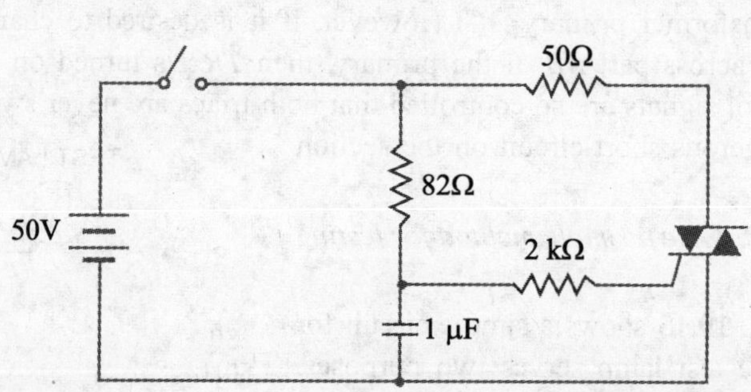

Fig. 19.16

* It is understood that the applied voltage is less than the breakover voltage of the triac.

** Recall that once the triac is fired by the gate voltage, it continues to conduct even if the gate voltage is removed.

Solution.

(*i*) Since the triac is ideal and it is fired into conduction, the voltage across triac is 0V. Therefore, the entire supply voltage of 50V appears across 50Ω resistor.

$$\therefore \quad \text{Current in } 50\Omega = \frac{50\text{V}}{50\Omega} = \textbf{1 A}$$

(*ii*) When triac is fired into conduction, voltage across 50Ω resistor = 50V – 1V = 49V.

$$\therefore \quad \text{Current in } 50\Omega = \frac{49\text{V}}{50\Omega} = \textbf{0.98 A}$$

Example 19.5. *The power controller circuit utilising a triac shown in Fig. 19.17 (i) has input voltage V_{in} = 60 V (r.m.s.) and load resistance R_L = 15 Ω. Neglecting the voltage drop across the triac when it is conducting, find (i) the maximum possible power that can be delivered to the load (ii) the % age of maximum load power delivered when firing angle α = 45°.*

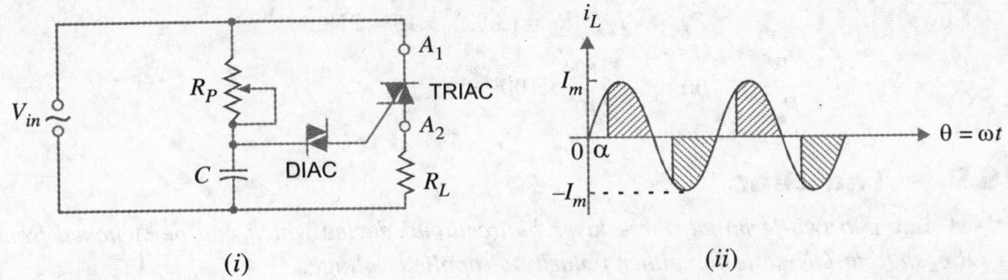

Fig. 19.17

Solution. The triac in Fig. 19.17(*i*) is triggered by a positive gate current when terminal A_1 is positive *w.r.t.* terminal A_2 and by a negative gate current when terminal A_1 is negative *w.r.t.* terminal A_2. When the capacitor voltage rises to a value to overcome the sum of the breakover voltage of the diac and the forward voltage drop of the gate, the diac fires. As a result, the capacitor discharges rapidly into the gate and the triac is turned on *i.e.* it starts conducting. Since R_P controls the time constant at which the capacitor charges, it is used to control the firing angle α [See Fig. 19.17(*ii*)]. A large value of R_P delays the charging and hence increases the firing angle. Therefore, firing angle can be adjusted from 0° to 180°. When firing angle α = 0°, the load current is a full sine wave. The average load current is zero because the positive and negative areas of load current waveform are equal. It can be proved mathematically that the r.m.s. value of load current ($I_{r.m.s.}$) is given by ;

$$I_{r.m.s.} = \frac{I_m}{\sqrt{2}}\sqrt{\left(1-\frac{\alpha}{180°}\right)+\left(\frac{\sin 2\alpha}{2\pi}\right)}$$

where I_m = Maximum value of load current

α = Firing angle in degrees

(*i*) Maximum power is delivered to the load when the firing angle is α = 0°. Under this condition, the load current is a full sine wave and the load voltage, V_L = 60 V (r.m.s.).

∴ Maximum power (P_{max}) delivered to the load is

$$P_{max} = \frac{V_L^2}{R_L} = \frac{(60)^2}{15} = \textbf{240 W}$$

(*ii*) Maximum value of load voltage (V_m) is

$$V_m = \sqrt{2}V_L = \sqrt{2}\times 60 = 84.85 \text{ V}$$

∴ Maximum value of load current is

$$I_m = \frac{V_m}{R_L} = \frac{84.85}{15} = 5.66 \text{ A}$$

The r.m.s. value of load current ($I_{r.m.s.}$) is

$$I_{r.m.s.} = \frac{I_m}{\sqrt{2}}\sqrt{\left(1-\frac{\alpha}{180°}\right)+\left(\frac{\sin 2\alpha}{2\pi}\right)}$$

$$= \frac{5.66}{\sqrt{2}}\sqrt{\left(1-\frac{45°}{180°}\right)+\frac{\sin(2\times 45°)}{2\pi}} = 3.82 \text{ A (r.m.s.)}$$

Now, $$P_L = I_{r.m.s.}^2 R_L = (3.82)^2 \times 15 = 218.8 \text{ W}$$

∴ $$\frac{P_L}{P_{max}}\times 100 = \frac{218.8}{240}\times 100 = \mathbf{91.2\%}$$

19.9 The Diac

***A diac** is a two-terminal, three layer bidirectional device which can be switched from its OFF state to ON state for either polarity of applied voltage.*

Fig. 19.18

The diac can be constructed in either *npn* or *pnp* form. Fig. 19.18 (*i*) shows the basic structure of a diac in *pnp* form. The two leads are connected to *p*-regions of silicon separated by an *n*-region. The structure of diac is very much similar to that of a transistor. However, there are several imporant differences :

(*i*) There is no terminal attached to the base layer.

(*ii*) The three regions are nearly identical in size.

(*iii*) The doping concentrations are identical (unlike a bipolar transistor) to give the device symmetrical properties.

Fig. 19.18 (*ii*) shows the symbol of a diac.

Operation. When a positive or negative voltage is applied across the terminals of a diac, only a small leakage current I_{BO} will flow through the device. As the applied voltage is increased, the leakage current will continue to flow until the voltage reaches the breakover voltage V_{BO}. At this point, avalanche breakdown of the reverse-biased junction occurs and the device exhibits negative resistance *i.e.* current through the device increases with the decreasing values of applied voltage. The voltage across the device then drops to 'breakback' voltage V_W.

Fig. 19.19 shows the *V-I* characteristics of a diac. For applied positive voltage less than $+ V_{BO}$ and negative voltage less than $- V_{BO}$, a small leakage current ($\pm I_{BO}$) flows through the device. Under such conditions, the diac blocks the flow of current and effectively behaves as an open-circuit. The voltages $+ V_{BO}$ and $- V_{BO}$ are the breakdown voltages and usually have a range of 30 to 50 volts.

When the positive or negative applied voltage is equal to or greater than the breakdown voltage, diac begins to conduct and the voltage drop across it becomes a few volts. Conduction then continues until the device current drops below its holding current. Note that the breakover voltage and holding current values are identical for the forward and reverse regions of operation.

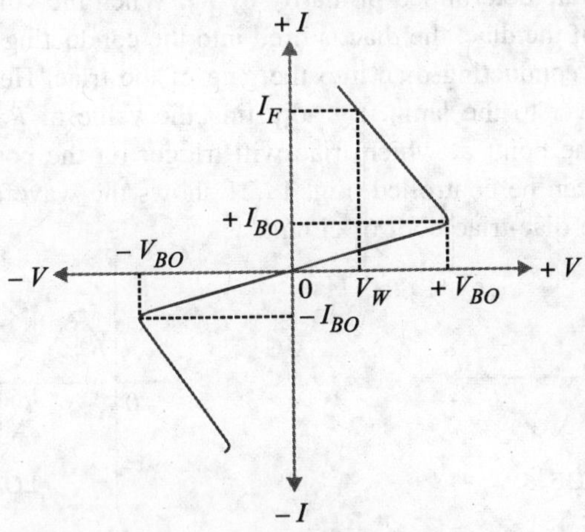

Fig. 19.19

Diacs are used primarily for triggering of triacs in adjustable phase control of a.c. mains power. Some of the circuit applications of diac are (*i*) light dimming (*ii*) heat control and (*iii*) universal motor speed control.

19.10 Applications of Diac

Although a triac may be fired into the conducting state by a simple resistive triggering circuit, more reliable and faster turn-on may be had if a switching device is used in series with the gate. One of the switching devices that can trigger a triac is the diac. This is illustrated in the following applications.

(*i*) Lamp dimmer. Fig. 19.20 shows a typical circuit that may be used for smooth control of a.c. power fed to a lamp. This permits to control the light output from the lamp. The basic control is by an *RC* variable gate voltage arrangement. The series $R_4 - C_1$ circuit across the triac is designed to limit the rate of voltage rise across the device during switch off.

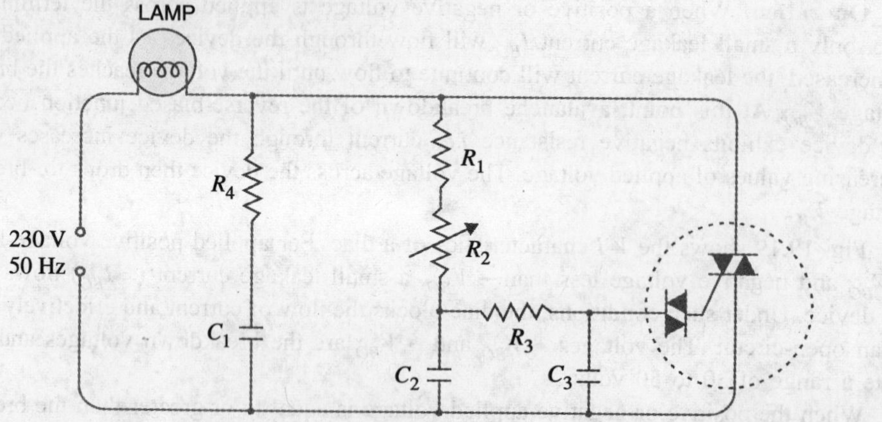

Fig. 19.20

The circuit action is as follows: As the input voltage increases positively or negatively, C_1 and C_2 charge at a rate determined primarily by R_2. When the voltage across C_3 exceeds the breakover voltage of the diac, the diac is fired into the conducting state. The capacitor C_3 discharges through the conducting diac into the gate of the triac. Hence, the triac is turned on to pass the a.c. power to the lamp. By adjusting the value of R_2, the rate of charge of capacitors and hence the point at which triac will trigger on the positive or negative half-cycle of input voltage can be controlled. Fig. 19.21 shows the waveforms of supply voltage and load voltage in the diac-triac control circuit.

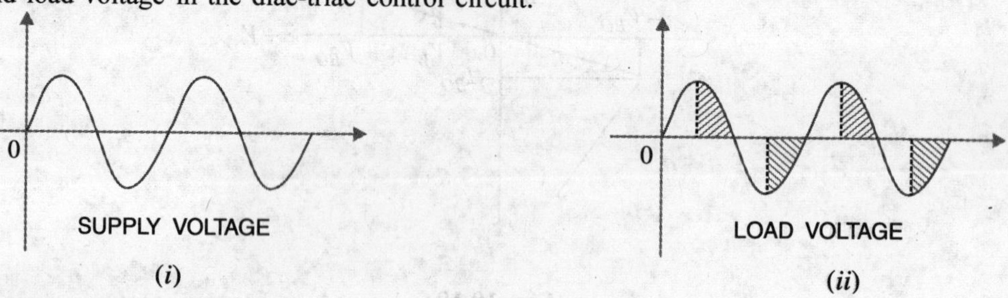

Fig. 19.21

The firing of triac can be controlled upto a maximum of 180°. In this way, we can provide a continuous control of load voltage from practically zero to full r.m.s. value.

***(ii)* Heat control.** Fig. 19.22 shows a typical diac-triac circuit that may be used for the smooth control of a.c. power in a heater. This is similar to the circuit shown in Fig. 19.20. The capacitor C_1 in series with choke L across the triac helps to slow-up the voltage rise across the device during switch-off. The resistor R_4 in parallel with the diac ensures smooth control at all positions of variable resistance R_2.

The circuit action is as follows: As the input voltage increases positively or negatively, C_1 and C_2 charge at a rate determined primarily by R_2. When the voltage across C_3 exceeds the breakover voltage of the diac, the diac conducts. The capacitor C_3 discharges through the conducting diac into the gate of the triac. This turns on the triac and hence a.c. power to the heater. By adjusting the value of R_2, any portion of positive and negative half-cycles of the supply voltage can be passed through the heater. This permits a smooth control of the heat output from the heater.

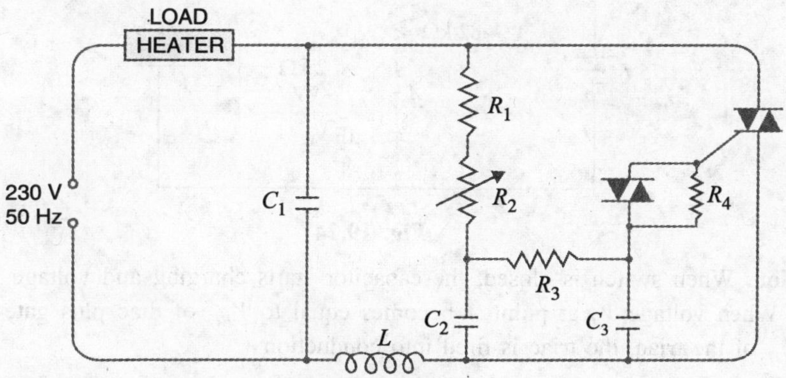

Fig. 19.22

Example 19.6. *We require a small gate triggering voltage V_{GT} (say ± 2V) to fire a triac into conduction. How will you raise the trigger level of the triac ?*

Solution. The rated values of V_{GT} for triacs are generally low. For example, the traic shown in Fig. 19.23 (*i*) has V_{GT} = ± 2V. However, we can raise the triggering level of the triac by using a *diac* in the gate circuit as shown in Fig. 19.23 (*ii*). With the diac present, the triac can still be triggered by both positive and negative gate voltages. However, to fire the triac into conduction, the potential V_A at point *A* in Fig. 19.23 (*ii*) must overcome the V_{BO} (breakover voltage) for the diac plus V_{GT} *i.e*

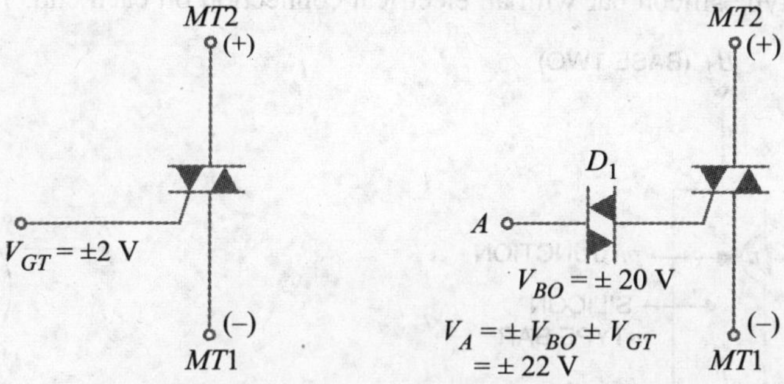

Fig. 19.23

$$V_A = \pm V_{BO} \pm V_{GT}$$
$$= \pm 20\text{V} \pm 2\text{V} = \pm 22\text{V}$$

Therefore, in order to turn on the triac, the gate trigger signal V_A must be ± 22V. Note that triac triggering control is the primary function of a diac.

Example 19.7. *In Fig. 19.24, the switch is closed. A diac with breakover voltage V_{BO} = 30V is connected in the circuit. If the triac has a trigger voltage of 1V and a trigger current of 10 mA, what is the capacitor voltage that triggers the triac?*

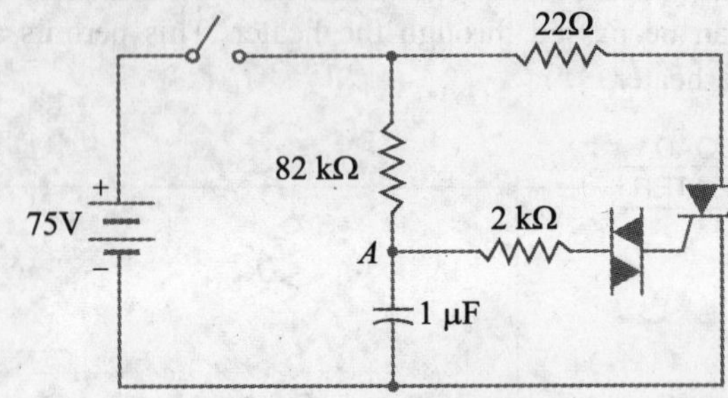

Fig. 19.24

Solution. When switch is closed, the capacitor starts charging and voltage at point *A* increases. When voltage V_A at point *A* becomes equal to V_{BO} of diac plus gate triggering voltage V_{GT} of the triac, the triac is fired into conduction.

$$\therefore \quad V_A = V_{BO} + V_{GT} = 30V + 1V = \mathbf{31V}$$

This is the minimum capacitor voltage that will trigger the triac.

19.11 Unijunction Transistor (UJT)

A unijunction transistor (abbreviated as *UJT*) is a three-terminal semiconductor switching device. This device has a unique characteristic that when it is triggered, the emitter current increases regeneratively until it is limited by emitter power supply. Due to this characteristic, the unijunction transistor can be employed in a variety of applications *e.g.*, switching, pulse generator, saw-tooth generator *etc.*

Construction. Fig. 19.25 (*i*) shows the basic *structure of a unijunction transistor. It consists of an *n*-type silicon bar with an electrical connection on each end. The leads to these

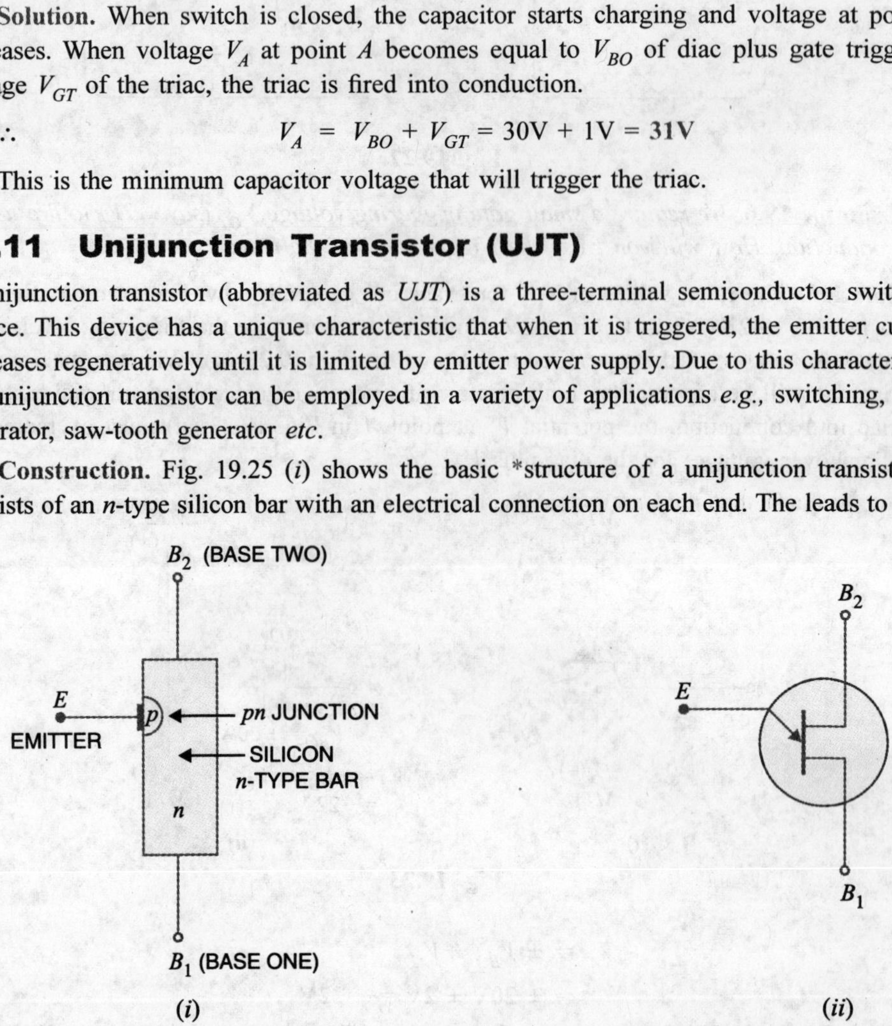

Fig. 19.25

* Note that structure of *UJT* is very much similar to that of the *n*-channel *JFET*. The only difference in the two components is that *p*-type (gate) material of the *JFET surrounds* the *n*-type (channel) material.

connections are called base leads *base-one* B_1 and *base two* B_2. Part way along the bar between the two bases, nearer to B_2 than B_1, a *pn* junction is formed between a *p*-type emitter and the bar. The lead to this junction is called the *emitter lead E*. Fig. 19.25 (*ii*) shows the symbol of unijunction transistor. Note that emitter is shown closer to B_2 than B_1. The following points are worth noting :

(*i*) Since the device has one *pn* junction and three leads, it is *commonly called a unijunction transistor (*uni* means single).

(*ii*) With only one *pn*-junction, the device is really a form of diode. Because the two base terminals are taken from one section of the diode, this device is also called *double-based diode.*

(*iii*) The emitter is heavily doped having many holes. The *n* region, however, is lightly doped. For this reason, the resistance between the base terminals is very high (5 to 10 kΩ) when emitter lead is open.

Operation. Fig. 19.26 shows the basic circuit operation of a unijunction transistor. The device has normally B_2 positive *w.r.t.* B_1.

(*i*) If voltage V_{BB} is applied between B_2 and B_1 with emitter open [See Fig. 19.26 (*i*)], a voltage gradient is established along the *n*-type bar. Since the emitter is located nearer to B_2, more than **half of V_{BB} appears between the emitter and B_1. The voltage V_1 between emitter and B_1 establishes a reverse bias on the *pn* junction and the emitter current is cut-off. Of course, a small leakage current flows from B_2 to emitter due to minority carriers.

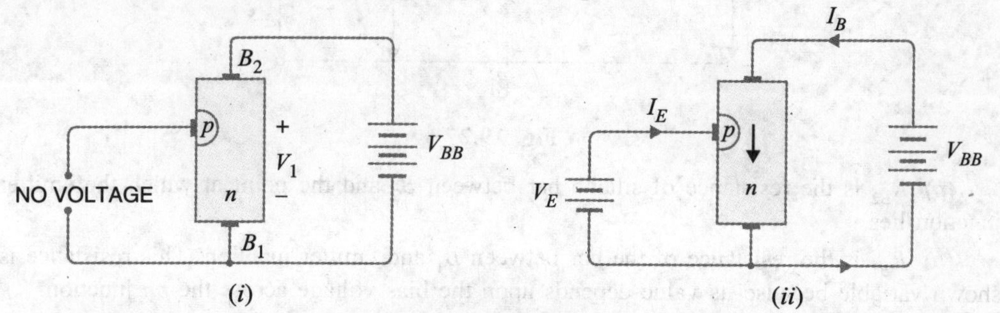

Fig. 19.26

(*ii*) If a positive voltage is applied at the emitter [See Fig. 19.26 (*ii*)], the *pn* junction will remain reverse biased so long as the input voltage is less than V_1. If the input voltage to the emitter exceeds V_1, the *pn* junction becomes ***forward biased. Under these conditions, holes are injected from *p*-type material into the *n*-type bar. These holes are repelled by positive B_2 terminal and they are attracted towards B_1 terminal of the bar. This accumulation of holes in the emitter to B_1 region results in the decrease of resistance in this section of the bar. The result is that internal voltage drop from emitter to B_1 is decreased and hence

* In packaged form, a *UJT* looks very much like a small signal transistor. As a *UJT* has only one *pn* junction, therefore, naming it a 'transistor' is really a misnomer.

** The *n*-type silicon bar has a high resistance. The resistance between emitter and B_1 is greater than between B_2 and emitter. It is because emitter is nearer to B_2 than B_1.

*** The main operational difference between the *JFET* and the *UJT* is that the *JFET* is normally operated with the gate junction reverse biased whereas the useful behaviour of the *UJT* occurs when the emitter is forward biased.

the emitter current I_E increases. As more holes are injected, a condition of saturation will eventually be reached. At this point, the emitter current is limited by emitter power supply only. The device is now in the *ON* state.

(*iii*) If a negative pulse is applied to the emitter, the *pn* junction is reverse biased and the emitter current is cut-off. The device is then said to be in the *OFF* state.

19.12 Equivalent Circuit of a UJT

Fig. 19.27 shows the equivalent circuit of a *UJT*. The resistance of the silicon bar is called the inter-base resistance R_{BB}. The inter-base resistance is represented by two resistors in series *viz*

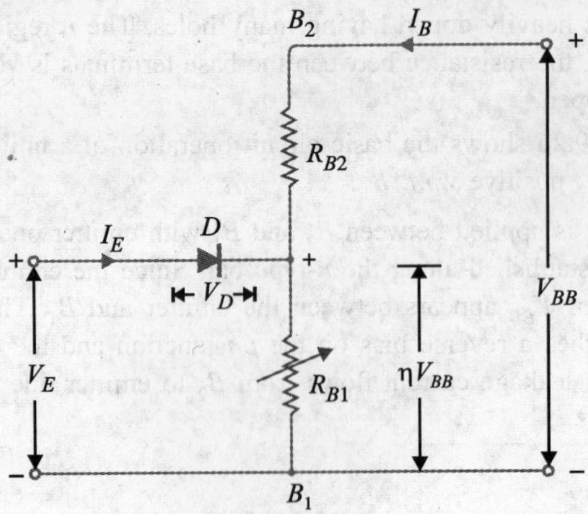

Fig. 19.27

(*a*) R_{B2} is the resistance of silicon bar between B_2 and the point at which the emitter junction lies.

(*b*) R_{B1} is the resistance of the bar between B_1 and emitter junction. This resistance is shown variable because its value depends upon the bias voltage across the *pn* junction.

The *pn* junction is represented in the emitter by a diode *D*.

The circuit action of a *UJT* can be explained more clearly from its equivalent circuit.

(*i*) With no voltage applied to the *UJT*, the inter-base resistance is given by ;

$$R_{BB} = R_{B1} + R_{B2}$$

The value of R_{BB} generally lies between 4 kΩ and 10 kΩ.

(*ii*) If a voltage V_{BB} is applied between the bases with emitter open, the voltage will divide up across R_{B1} and R_{B2}.

$$\text{Voltage across } R_{B1}, \quad V_1 = \frac{R_{B1}}{R_{B1} + R_{B2}} V_{BB}$$

or

$$V_1/V_{BB} = \frac{R_{B1}}{R_{B1} + R_{B2}}$$

The ratio V_1/V_{BB} is called *intrinsic stand-off ratio* and is represented by Greek letter η.

$$\text{Obviously,} \quad \eta = \frac{R_{B1}}{R_{B1} + R_{B2}}$$

The value of η usually lies between 0.51 and 0.82.

∴ Voltage across $R_{B1} = \eta V_{BB}$

The voltage ηV_{BB} appearing across R_{B1} reverse biases the diode. Therefore, the emitter current is zero.

(iii) If now a progressively rising positive voltage is applied to the emitter, the diode will become forward biased when input voltage exceeds ηV_{BB} by V_D, the forward voltage drop across the silicon diode *i.e.*

$$V_P = \eta V_{BB} + V_D$$

where V_P = 'Peak-Point voltage'

V_D = Forward voltage drop across silicon diode (≃ 0.7 V)

When the diode *D* starts conducting, holes are injected from *p*-type material to the *n*-type bar. These holes are swept down towards the terminal B_1. This decreases the resistance between emitter and B_1 (indicated by variable resistance symbol for R_{B1}) and hence the internal drop from emitter to B_1. The emitter current now increases regeneratively until it is limited by the emitter power supply.

Conclusion. The above discussion leads to the conclusion that when input positive voltage to the emitter is less than peak-point voltage V_P, the *pn*-junction remains reverse biased and the emitter current is practically zero. However, when the input voltage exceeds V_P, R_{B1} falls from several thousand ohms to a small value. The diode is now forward biased and the emitter current quickly reaches to a saturation value limited by R_{B1} (about 20 Ω) and forward resistance of *pn*-junction (about 200 Ω).

19.13 Characteristics of UJT

Fig. 19.28 shows the curve between emitter voltage (V_E) and emitter current (I_E) of a *UJT* at a given voltage V_{BB} between the bases. This is known as the emitter characteristic of *UJT*. The following points may be noted from the characteristics :

(i) Initially, in the cut-off region, as V_E increases from zero, slight leakage current flows from terminal B_2 to the emitter. This current is due to the minority carriers in the reverse biased diode.

(ii) Above a certain value of V_E, forward I_E begins to flow, increasing until the peak voltage V_P and current I_P are reached at point *P*.

(iii) After the peak point *P*, an attempt to increase V_E is followed by a sudden increase in emitter current I_E with a corresponding decrease in V_E. This is a *negative resistance* portion of the curve because with increase in I_E, V_E decreases. The device, therefore, has a negative resistance region which is stable enough to be used

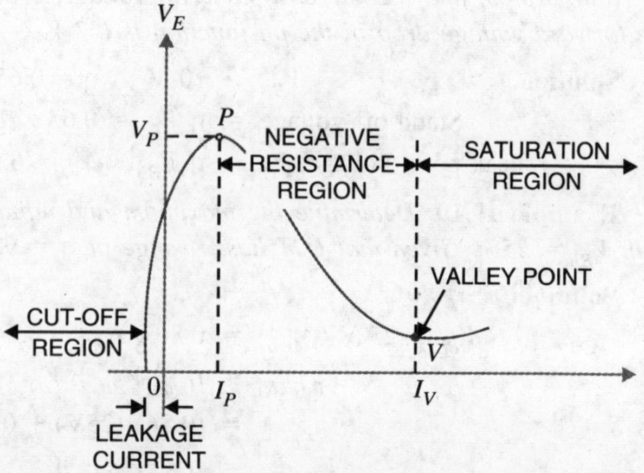

Fig. 19.28

with a great deal of reliability in many areas *e.g.*, trigger circuits, saw-tooth generators, timing circuits.

(*iv*) The negative portion of the curve lasts until the valley point V is reached with valley-point voltage V_V and valley-point current I_V. After the valley point, the device is driven to saturation.

Fig. 19.29 shows the typical family of V_E/I_E characteristics of a *UJT* at different voltages between the bases. It is clear that peak-point voltage ($= \eta\ V_{BB} + V_D$) falls steadily with reducing V_{BB} and so does the valley point voltage V_V. The difference $V_P - V_V$ is a measure of the switching efficiency of *UJT* and can be seen to fall-off as V_{BB} decreases. For a general purpose *UJT*, the peak - point current is of the order of 1 μA at V_{BB} = 20 V with a valley-point voltage of about 2.5 V at 6 mA.

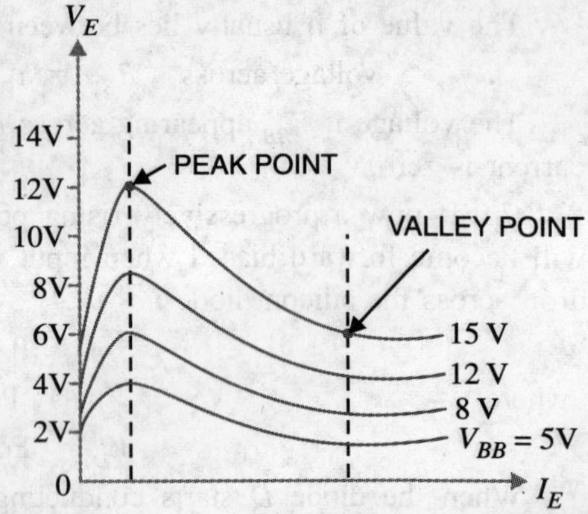

Fig. 19.29

Example 19.8. *The intrinsic stand-off ratio for a UJT is determined to be 0.6. If the inter-base resistance is 10 kΩ, what are the values of R_{B1} and R_{B2} ?*

Solution. $R_{BB} = 10\ k\Omega\ ;\quad \eta = 0.6$

Now, $R_{BB} = R_{B1} + R_{B2}$

or $10 = R_{B1} + R_{B2}$

Also, $$\eta = \frac{R_{B1}}{R_{B1} + R_{B2}}$$

or $$0.6 = \frac{R_{B1}}{10} \qquad (\because R_{B1} + R_{B2} = 10\ k\Omega)$$

∴ $R_{B1} = 10 \times 0.6 = \mathbf{6\ k\Omega}$

and $R_{B2} = 10 - 6 = \mathbf{4\ k\Omega}$

Example 19.9. *A unijunction transistor has 10 V between the bases. If the intrinsic stand off ratio is 0.65, find the value of stand-off voltage. What will be the peak-point voltage if the forward voltage drop in the pn junction is 0.7 V ?*

Solution. $V_{BB} = 10\ V\ ;\quad \eta = 0.65\ ;\quad V_D = 0.7\ V$

Stand-off voltage $= \eta\ V_{BB} = 0.65 \times 10 = \mathbf{6.5\ V}$

Peak-point voltage, $V_P = \eta\ V_{BB} + V_D = 6.5 + 0.7 = \mathbf{7.2\ V}$

Example 19.10. *Determine the maximum and minimum peak-point voltage for a UJT with V_{BB} = 25 V. Given that UJT has a range of η = 0.74 to 0.86.*

Solution.

$$V_{BB} = 25\ V\ ;\ \eta_{max} = 0.86\ ;\ \eta_{min} = 0.74$$

∴ $$V_{P\ (max)} = \eta_{max}\ V_{BB} + V_D$$
$$= (0.86)\ (25V) + 0.7\ V = \mathbf{22.2\ V}$$
$$V_{P\ (min)} = \eta_{min}\ V_{BB} + V_D$$
$$= (0.74)\ (25\ V) + 0.7\ V = \mathbf{19.2\ V}$$

Example 19.11. *Fig. 19.30 (i) shows the UJT circuit. The parameters of UJT are* $\eta = 0.65$ *and* $R_{BB} = 7\ k\Omega$*. Find (i)* R_{B1} *and* R_{B2} *(ii)* V_{B2B1} *and* V_P.

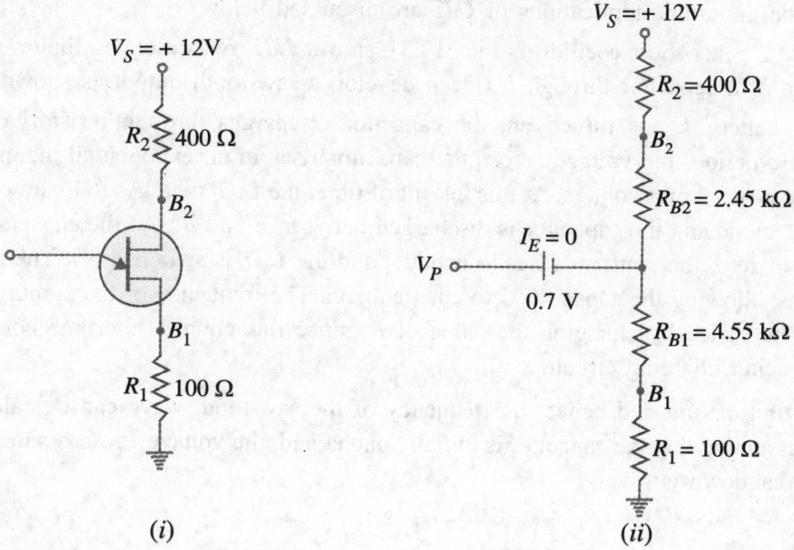

Fig. 19.30

Solution.

$$\eta = 0.65\ ;\ V_S = 12\ V\ ;\ R_{BB} = 7\ k\Omega$$

(i) $$\eta = \frac{R_{B1}}{R_{BB}} \quad \therefore \quad R_{B1} = \eta\ R_{BB} = 0.65 \times 7\ k\Omega = \mathbf{4.55\ k\Omega}$$

Also, $$R_{B2} = R_{BB} - R_{B1} = 7\ k\Omega - 4.55\ k\Omega = \mathbf{2.45\ k\Omega}$$

(ii) Fig. 19.30 *(ii)* shows the *UJT* circuit model. Because *UJT* is *OFF*, $I_E = 0$. We can find V_{B2B1} by the voltage-divider rule.

$$V_{B2B1} = \frac{V_S}{R_1 + R_2 + R_{BB}} \times R_{BB}$$

$$= \frac{12V}{(100 + 400 + 7000)\Omega} \times 7000\Omega = \mathbf{11.2\ V}$$

The voltage V_P required to turn on the *UJT* is

$$V_P = \eta V_{B2B1} + V_D = 0.65 \times 11.2\ V + 0.7\ V = \mathbf{7.98\ V}$$

19.14 Advantages of UJT

The *UJT* was introduced in 1948 but did not become commercially available until 1952. Since then, the device has achieved great popularity due to the following reasons :

(i) It is a low cost device.

(ii) It has excellent characteristics.

(iii) It is a low-power absorbing device under normal operating conditions.

Due to above reasons, this device is being used in a variety of applications. A few include oscillators, trigger circuits, saw-tooth generators, bistable network *etc.*

19.15 Applications of UJT

Unijunction transistors are used extensively in oscillator, pulse and voltage sensing circuits. Some of the important applications of *UJT* are discussed below :

(*i*) **UJT relaxation oscillator.** Fig. 19.31 shows *UJT* relaxation oscillator where the discharging of a capacitor through *UJT* can develop a saw-tooth output as shown.

When battery V_{BB} is turned on, the capacitor C charges through resistor R_1. During the charging period, the voltage across the capacitor rises in an exponential manner until it reaches the peak - point voltage. At this instant of time, the *UJT* switches to its low resistance conducting mode and the capacitor is discharged between E and B_1. As the capacitor voltage flies back to zero, the emitter ceases to conduct and the *UJT* is switched off. The next cycle then begins, allowing the capacitor C to charge again. The frequency of the output saw-tooth wave can be varied by changing the value of R_1 since this controls the time constant R_1C of the capacitor charging circuit.

The time period and hence the frequency of the saw-tooth wave can be calculated as follows. Assuming that the capacitor is initially uncharged, the voltage V_C across the capacitor prior to breakdown is given ;

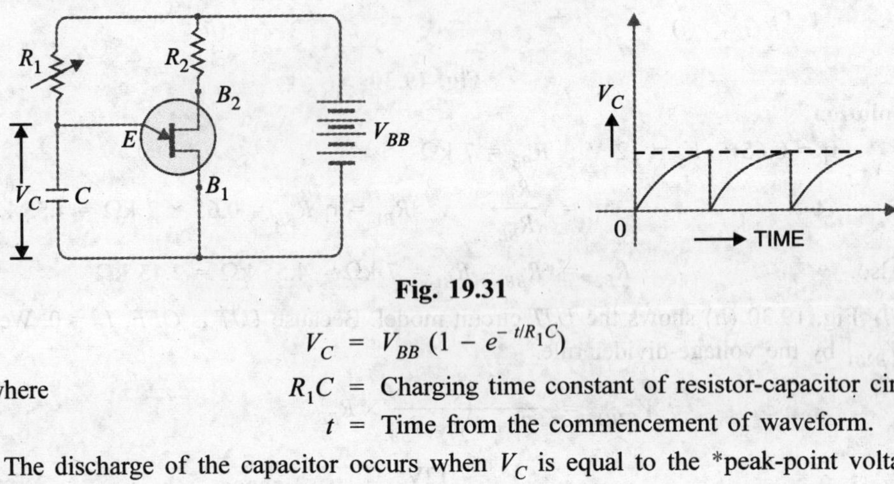

Fig. 19.31

$$V_C = V_{BB}\,(1 - e^{-t/R_1C})$$

where R_1C = Charging time constant of resistor-capacitor circuit

t = Time from the commencement of waveform.

The discharge of the capacitor occurs when V_C is equal to the *peak-point voltage $\eta\, V_{BB}$ *i.e.*

$$\eta\, V_{BB} = V_{BB}\,(1 - e^{-t/R_1C})$$

or $$\eta = 1 - e^{-t/R_1C}$$

or $$e^{-t/R_1C} = 1 - \eta$$

or $$t = R_1\, C \log_e \frac{1}{1-\eta}$$

∴ $$\text{Time period, } t = 2.3\, R_1\, C \log_{10} \frac{1}{1-\eta}$$

$$\text{Frequency of saw-tooth wave, } f = \frac{1}{t \text{ in seconds}}\ \text{Hz}$$

* Actually, peak-point voltage, $V_P = \eta V_{BB} + V_D$. As V_D, the forward voltage drop across emitter diode is generally small, it can be neglected with reasonable accuracy.

∴ $$V_P = \eta\, V_{BB}$$

(ii) **Overvoltage detector.** Fig. 19.32 shows a simple d.c. over-voltage indicator. A warning pilot-lamp L is connected between the emitter and B_1 circuit. So long as the input voltage is less than the peak-point voltage (V_P) of the *UJT*, the device remains switched off. However, when the input voltage exceeds V_P, the *UJT* is switched on and the capacitor discharges through the low resistance path between terminals E and B_1. The current flowing in the pilot lamp L lights it, thereby indicating the overvoltage in the circuit.

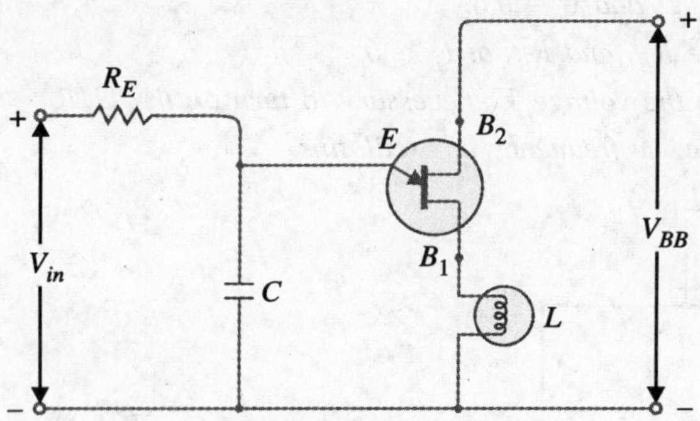

Fig. 19.32

Example 19.12. *The circuit shown in Fig. 19.33 uses variable resistor R_E to change the frequency of pulses delivered at V_{out}. The variable resistor is initially set at 5 kΩ and is then adjusted to 10 kΩ. Determine the frequency of the voltage spikes produced for (i) 5 kΩ setting and (ii) 10 kΩ setting.*

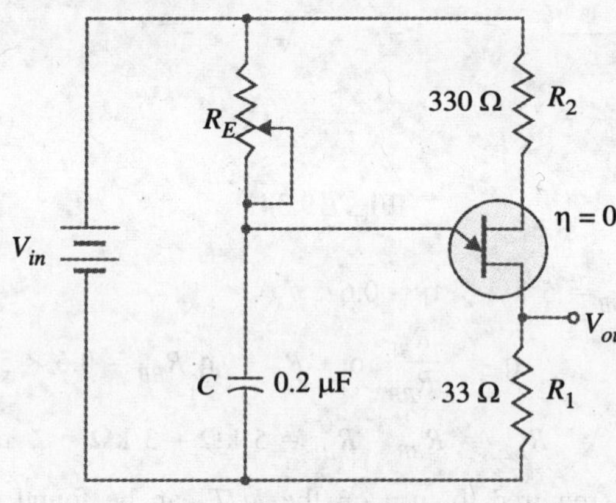

Fig. 19.33

Solution.

(i) Time period, $t = R_E\, C \log_e \dfrac{1}{1-\eta}$

Here, $R_E = 5\ \text{k}\Omega = 5 \times 10^3 \Omega$; $C = 0.2\ \mu\text{F} = 0.2 \times 10^{-6}$ F ; $\eta = 0.54$

$$\therefore \quad t = (5 \times 10^3)\,(0.2 \times 10^{-6}) \log_e \frac{1}{1-0.54}$$

$$= 0.78 \times 10^{-3}\text{s} = 0.78 \text{ ms}$$

$\therefore$ Frequency, $f = 1/t = 1/0.78 \times 10^{-3}\text{s} =$ **1282 Hz** (*ii*) $$t = (10 \times 10^3)(0.2 \times 10^{-6}) \log_e \frac{1}{1-0.54}$$

$$= 1.55 \times 10^{-3}\text{s} = 1.55 \text{ ms}$$

$\therefore$ Frequency, $f = 1/t = 1/1.55 \times 10^{-3}\text{s} = \mathbf{645\ Hz}$

Example 19.13. *Fig. 19.34 (i) shows the relaxation oscillator. The parameters of the UJT are $R_{BB} = 5\ k\Omega$ and $\eta = 0.6$.*

(i) Determine R_{B1} and R_{B2} at $I_E = 0$.

(ii) Calculate the voltage V_P necessary to turn on the UJT.

(iii) Determine the frequency of oscillatins.

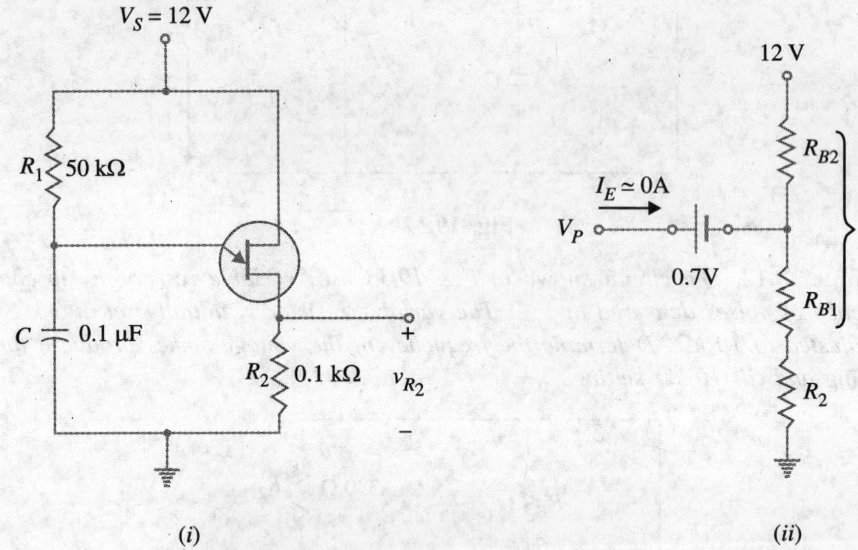

Fig. 19.34

Solution.

$$V_S = 12\text{V}\ ;\ R_{BB} = 5\ \text{k}\Omega\ ;\ \eta = 0.6$$

(*i*) $$\eta = \frac{R_{B1}}{R_{BB}} \text{ or } R_{B1} = \eta\, R_{BB} = 0.6 \times 5\ \text{k}\Omega = \mathbf{3\ k\Omega}$$

Also, $$R_{B2} = R_{BB} - R_{B1} = 5\ \text{k}\Omega - 3\ \text{k}\Omega = \mathbf{2\ k\Omega}$$

(*ii*) The voltage V_P required to turn on the *UJT* can be found from the *UJT* circuit model shown is Fig. 19.34 (*ii*). Referring to Fig. 19.34 (*ii*), we have,

$$V_P = V_D + \text{Voltage drop across } (R_{B1} + R_2)$$

$$= V_D + \frac{V_S}{R_{BB} + R_2} \times (R_{B1} + R_2)$$

$$= 0.7\text{V} + \frac{12\text{V}}{5\ \text{k}\Omega + 0.1\ \text{k}\Omega} \times (3 + 0.1)\ \text{k}\Omega$$

$$= 0.7\text{V} + 7.294\ \text{V} \simeq \mathbf{8V}$$

(*iii*) Time period, $$t = R_1 C \log_e \frac{1}{1-\eta}$$

Here, $R_1 = 50\ \text{k}\Omega = 50 \times 10^3\ \Omega$; $C = 0.1\ \mu\text{F} = 0.1 \times 10^{-6}\text{F}$; $\eta = 0.6$

$$\therefore \quad t = (50 \times 10^3)(0.1 \times 10^{-6}) \log_e \frac{1}{1-0.6}$$

$$= 4.58 \times 10^{-3}\text{s}$$

$$= 4.58 \text{ ms}$$

$$\therefore \quad \text{Frequency, } f = \frac{1}{t} = \frac{1}{4.58 \times 10^{-3}\text{s}}$$

$$= \mathbf{218\ Hz}$$

Example 19.14. *Determine the value of R_1 in the relaxation oscillator of Fig. 19.35 that will ensure proper turn-on and turn-off of the UJT. The characteristic of the UJT shows the following values :*

$\eta = 0.5$; $V_V = 1\ V$; $I_V = 10\ mA$; $I_P = 20\ \mu A$ and $V_P = 14 V$.

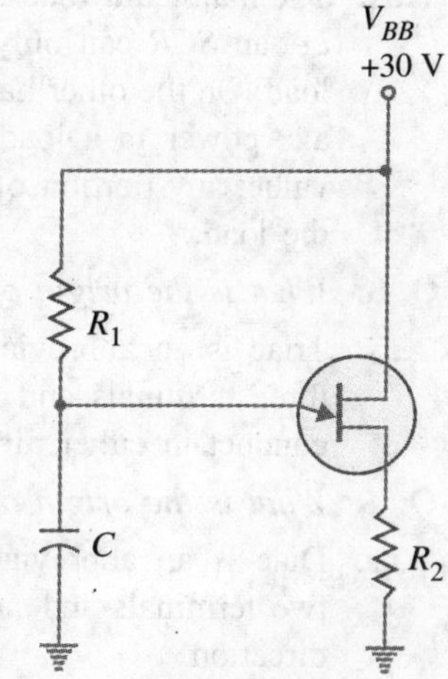

Fig. 19.35

Solution. In the relaxation oscillator of Fig. 19.35, certain conditions must be met for the *UJT* to reliably turn-on and turn-off. To ensure turn-on, R_1 must not limit I_E at the peak point to less than I_P. This will be so if the voltage drop across R_1 at the peak point is greater than $I_P R_1$. Therefore, the condition for turn-on is

$$V_{BB} - V_P > I_P R_1$$

or
$$R_1 < \frac{V_{BB} - V_P}{I_P}$$

Here, $V_{BB} = +30\ \text{V}$; $I_P = 20\ \mu\text{A}$; $V_P = 14\ \text{V}$

$$R_1 < \frac{30\ \text{V} - 14\ \text{V}}{20\ \mu\text{A}} \quad \text{or } R_1 < \mathbf{800\ k\Omega}.$$

To ensure turn-off the *UJT* at the valley point, R_1 must be large enough that I_E (at the valley point) can decrease below the specified value of I_V. This means that the voltage drop across R_1 at the valley point must be less than $I_V R_1$. Therefore, the condition for turn-off is

$$V_{BB} - V_V < I_V R_1$$

or
$$R_1 > \frac{V_{BB} - V_V}{I_V}$$

Here, $V_{BB} = +30\ \text{V}$; $V_V = 1\ \text{V}$; $I_V = 10\ \text{mA}$

$$\therefore \quad R_1 > \frac{30\ \text{V} - 1\ \text{V}}{10\ \text{mA}} \quad \text{or } R_1 > \mathbf{2.9\ k\Omega}.$$

Therefore, the value of R_1 should be between 2.9 kΩ and 800 kΩ. Note that R_1 has a wide range of possible values.

Short Answer Questions

Q. 1. *What is power electronics ?*

Ans. The modern power system is a.c. and operates at 50 Hz. Power electronics essentially deals with the control of 50 Hz power. The various electronic devices used for this purpose are triac, diac, *UJT etc.*

Q. 2. *What is the advantage of a triac over an SCR ?*

Ans. The major drawback of an *SCR* is that it can conduct current in one direction only *i.e.* an *SCR* can only control d.c. power or forward-biased half-cycles of a.c. in a load. On the other hand, a triac can conduct in both directions and hence can control a.c. power in a load. In fact, by adjusting the gate current in a triac to a proper value, any portion of both positive and negative half-cycles can be passed through the load.

Q. 3. *What is the origin of the name triac ?*

Ans. Triac is an abbreviation for triode a.c. switch. 'Tri' indicates that the device has three terminals and 'ac' means that the device controls alternating current or can conduct in either direction.

Q. 4. *What is the origin of the name diac ?*

Ans. Diac is an abbreviation for diode a.c. switch. 'Di' indicates that the device has two terminals and 'ac' means that the device controls a.c. or can conduct in either direction.

Q. 5. *Why is a diac preferred to trigger a triac ?*

Ans. Although a triac may be fired into the conducting state by a simple resistive triggering circuit, yet a diac is preferred for this purpose. It is because a diac is more reliable and provides faster turn-on than the resistive triggering circuit.

Q. 6. *What is the origin of the name UJT ?*

Ans. *UJT* stands for unijunction transistor. The prefix "Uni" means that the device has one junction (uni-means single) and transistor signifies that the device has three terminals.

Q. 7. *Is it appropriate to mention transistor in the name of UJT ?*

Ans. No. Although the device has three terminals and in package form it looks very much like a small transistor, yet naming it a 'transistor' is really a misnomer. It is because a transistor has two *pn* junctions whereas a *UJT* has only one *pn* junction. Because of one *pn* junction, the device is really a form of diode. As the two base terminals are taken from one section of the diode, the device is rightly called *double-based diode.*

Q. 8. *Why is the resistance between the base terminals of a UJT high when the emitter lead is open ?*

Ans. A *UJT* consists of an *n*-type silicon bar with an electrical connection on each end. Part way along the bar, between the bases, a *pn* junction is formed between a *p*-type emitter and the bar. The emitter is heavily doped having many holes. The *n*-type bar, however, is lightly doped. Due to light doping of *n*-type bar, the resistance between its two ends is very high when the emitter lead is open.

Multiple-Choice Questions

1. A triac has three terminals *viz.* ....
   (*i*) drain, source, gate
   (*ii*) two main terminal and a gate terminal
   (*iii*) cathode, anode, gate
   (*iv*) none of the above
2. A triac is equivalent to two *SCRs* ....
   (*i*) in parallel (*ii*) in series
   (*iii*) in inverse-parallel
   (*iv*) none of the above
3. A triac is a .... switch.
   (*i*) bidirectional (*ii*) unidirectional
   (*iii*) mechanical
   (*iv*) none of the above
4. The *V-I* characteristics for a triac in the first and third quadrants are essentially identical to those of......in the first quadrant.
   (*i*) transistor (*ii*) *SCR*
   (*iii*) *UJT*
   (*iv*) none of the above
5. A triac can pass a portion of .... half-cycle through the load.
   (*i*) only positive (*ii*) only negative
   (*iii*) both positive and negative
   (*iv*) none of the above
6. A diac has .... terminals.
   (*i*) two (*ii*) three
   (*iii*) four
   (*iv*) none of the above
7. A triac has .... semiconductor layers.
   (*i*) two (*ii*) three
   (*iii*) four (*iv*) five
8. A diac has .... *pn* junctions.
   (*i*) four (*ii*) two
   (*iii*) three
   (*iv*) none of the above
9. The device that does not have the gate terminal is ....
   (*i*) triac (*ii*) *FET*
   (*iii*) *SCR* (*iv*) diac
10. A diac has .... semiconductor layers.
    (*i*) three (*ii*) two
    (*iii*) four
    (*iv*) none of the above
11. A *UJT* has ....
    (*i*) two *pn* junctions
    (*ii*) one *pn* junction
    (*iii*) three *pn* junctions
    (*iv*) none of the above
12. The normal way to turn-on a diac is by....
    (*i*) gate current (*ii*) gate voltage
    (*iii*) breakover voltage
    (*iv*) none of the above
13. A diac is .... switch.
    (*i*) an a.c. (*ii*) a d.c.
    (*iii*) a mechanical
    (*iv*) none of the above
14. In a *UJT,* the *p*-type emitter is .... doped.
    (*i*) lightly (*ii*) heavily
    (*iii*) moderately
    (*iv*) none of the above
15. Power electronics essentially deals with control of a.c. power at ....
    (*i*) frequencies above 20 kHz
    (*ii*) frequencies above 1000 kHz
    (*iii*) frequencies less than 10 Hz
    (*iv*) 50 Hz frequency
16. When the emitter terminal of a *UJT* is open, the resistance between the base terminals is generally ....
    (*i*) high (*ii*) low
    (*iii*) extremely low
    (*iv*) none of the above
17. When a *UJT* is turned *ON*, the resistance between emitter terminal and lower base terminal ....
    (*i*) remains the same
    (*ii*) is decreased
    (*iii*) is increased
    (*iv*) none of the above
18. To turn on *UJT*, the forward bias on the emitter diode should be .... the peak-point voltage.

(*i*) less than (*ii*) equal to
(*iii*) more than
(*iv*) none of the above

19. A *UJT* is sometimes called .... diode.
(*i*) low resistance (*ii*) high resistance
(*iii*) single-base (*iv*) double-based

20. When the temperature increases, the inter-base resistance (R_{BB}) of a *UJT* ....
(*i*) increases
(*ii*) decreases
(*iii*) remains the same
(*iv*) none of the above

21. The intrinsic stand-off ratio (η) of a *UJT* is given by ....
(*i*) $R_{B1} + R_{B2}$ (*ii*) $\frac{R_{B1} + R_{B2}}{R_{B1}}$
(*ii*) $\frac{R_{B1}}{R_{B1} + R_{B2}}$ (*iv*) $\frac{R_{B1} + R_{B2}}{R_{B2}}$

22. When the temperature increases, the intrinsic stand-off ratio ....
(*i*) increases
(*ii*) decreases
(*iii*) essentially remains the same
(*iv*) none of the above

23. Between the peak-point and the valley-point of *UJT* emitter characteristics we have .... region.
(*i*) saturation
(*ii*) negative resistance
(*iii*) cut-off
(*iv*) none of the above

24. A diac is turned on by ....
(*i*) breakover voltage
(*ii*) gate voltage
(*iii*) gate current
(*iv*) none of the above

25. The device that exhibits negative resistance region is ....
(*i*) diac (*ii*) triac
(*iii*) transistor (*iv*) *UJT*

26. The *UJT* may be used as ....
(*i*) an amplifier
(*ii*) a sawtooth generator
(*iii*) a rectifier
(*iv*) none of the above

27. A diac is simply ....
(*i*) a single junction device
(*ii*) a three junction device
(*iii*) a triac without gate terminal
(*iv*) none of the above

28. After peak-point, the *UJT* operates in the .... region.
(*i*) cut-off
(*ii*) saturation
(*iii*) negative resistance
(*iv*) none of the above

29. Which of the following is not a characteristic of *UJT* ?
(*i*) Intrinsic stand-off ratio
(*ii*) Negative resistance
(*iii*) Peak-point voltage
(*iv*) Bilateral conduction

30. The triac is ....
(*i*) like a bidirectional *SCR*
(*ii*) a four-terminal device
(*iii*) not a thyristor
(*iv*) answers (*i*) and (*ii*)

Answers to Multiple-Choice Questions

1. (*ii*)	2. (*iii*)	3. (*i*)	4. (*ii*)	5. (*iii*)
6. (*i*)	7. (*iii*)	8. (*ii*)	9. (*iv*)	10. (*i*)
11. (*ii*)	12. (*iii*)	13. (*i*)	14. (*ii*)	15. (*iv*)
16. (*i*)	17. (*ii*)	18. (*iii*)	19. (*iv*)	20. (*i*)
21. (*iii*)	22. (*iii*)	23. (*ii*)	24. (*i*)	25. (*iv*)
26. (*ii*)	27. (*iii*)	28. (*iii*)	29. (*iv*)	30. (*i*)

Chapter Review Topics

1. Discuss the importance of power electronics.
2. Explain the construction and working of a triac.
3. Sketch the *V-I* characteristics of a triac. What do you infer from them ?
4. Describe some important applications of a triac.
5. Explain the construction and working of a diac.
6. Discuss the applications of a diac.
7. Explain the construction and working of a *UJT*.
8. Draw the equivalent circuit of a *UJT* and discuss its working from the circuit.
9. Describe some important applications of a *UJT*.
10. Write short notes on the following :

    (*i*) *UJT* relaxation oscillator (*ii*) Triac as an a.c. switch (*iii*) Diac as a triggering device

Problems

1. The intrinsic stand-off ratio for a *UJT* is determined to be 0.6. If the inter-base resistance is 5 kΩ, what are the values of R_{B1} and R_{B2} ? **[R_{B1} = 3 kΩ ; R_{B2} = 2 kΩ]**
2. A unijunction transistor has 18 V between the bases. If the intrinsic stand-off ratio is 0.8, find the value of stand-off voltage. What will be the peak-point voltage if the forward voltage drop in the *pn* junction is 0.7 V ? **[14.4 V ; 15.1 V]**
3. In a unijunction transistor, η = 0.8, V_P = 10.3 V and R_{B2} = 5 kΩ. Determine R_{B1} and V_{BB}. **[20 kΩ ; 12 V]**
4. The instrinsic stand-off ratio for a *UJT* is 0.75 and V_{BB} = 12 V. If the forward drop in the *pn*-junction is 0.7 V, find the peak-point voltage. **[9.7 V]**
5. A unijunction transistor has 12 V between the bases. If the intrinsic stand-off ratio is 2/3, find the value of stand-off voltage. What will be the peak-point voltage if the forward drop in the *pn* junction is 0.7 V ? **[8 V ; 8.7 V]**
6. In Fig. 19.36, the switch is closed. If the triac has fired, what is the current through the 22Ω ? **[3.41 A]**

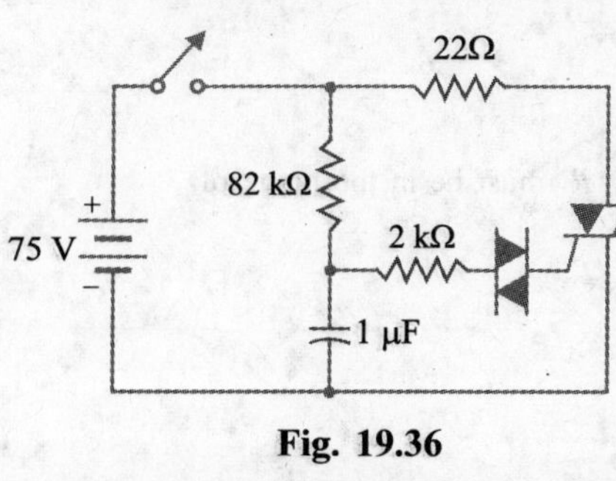

Fig. 19.36

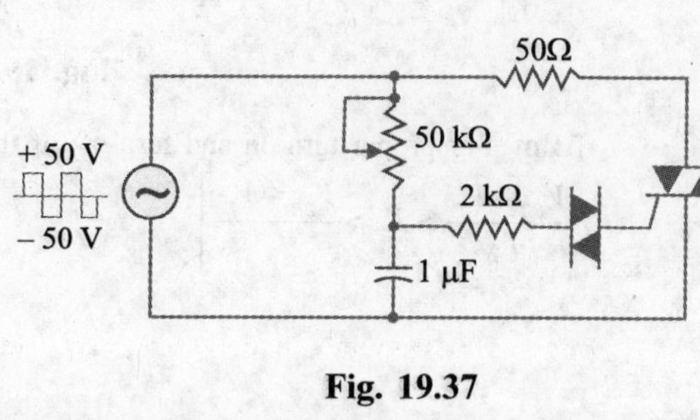

Fig. 19.37

7. If the triac of Fig. 19.37 has1V across it when it is conducting, what is the maximum current through the 50Ω ? **[0.98 A]**

8. The *UJT* of Fig. 19.38 has $\eta = 0.85$. What is the ideal emitter current ? **[50 mA]**

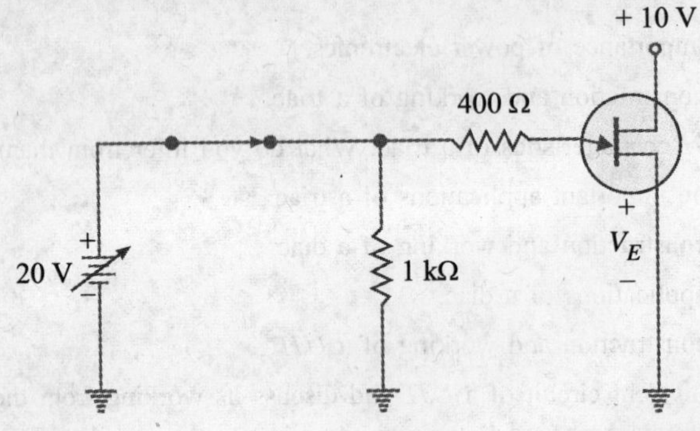

Fig. 19.38

9. The valley current of *UJT* in Fig. 19.38 is 7 mA and the emitter voltage is 1 V at this point. At what value of the emitter supply voltage does the *UJT* open ? **[3.8 V]**

10. In the relaxation oscillator of Fig. 19.39, determine the value of R_1 that will ensure proper turn-on and turn-off of the *UJT*. The characteristic of the *UJT* shows the following values : $\eta = 0.33$; $V_V = 0.8$ V ; $I_V = 15$ mA ; $I_P = 35$ μA and $V_P = 18$ V.

[343 kΩ > R_1 > 1.95 kΩ]

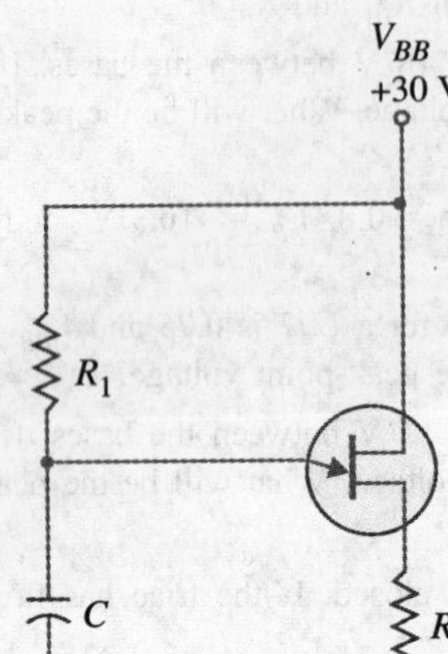

Fig. 19.39

[**Hint.** For proper turn-on and turn-off of the *UJT*, R_1 must be in the range of

$$\frac{V_{BB} - V_P}{I_P} > R_1 > \frac{V_{BB} - V_V}{I_V}\Bigg]$$

CHAPTER 20

Electronic Instruments

Learning Objectives

- Develop an understanding of importance of electronic instruments.
- Understand the construction and functions of multimeter.
- Learn the applications of multimeter.
- Explore the advantages and disadvantages of multimeter.
- Develop a strong understanding of various types of electronic voltmeters.
- Study the construction and working of Cathode Ray Oscilloscope (*CRO*).
- Learn the various controls of *CRO*.
- Develop an understanding of applications of *CRO*.

Introduction

In recent years, the rapid strides and remarkable advances in the field of electronics is partly due to modern electronic instruments. By using these instruments, we can gather much information regarding the performance of specific electronic circuit. Electronic instruments are also used for troubleshooting since they permit readings to be taken so that circuit faults can be located by ascertaining which component values do not coincide with the pre-established values indicated by the manufacturer. In fact, electronic instruments are playing a vital role in the fast developing field of electronics. It is with this view that they have been treated in a separate chapter.

20.1 Electronic Instruments

Those instruments which employ electronic devices for measuring various electrical quantities (*e.g. voltage, current, resistance etc.*) *are known as* **electronic instruments**.

There are a large number of electronic instruments available for completion of various tests and measurements. However, in this chapter, we shall confine our attention to the following electronic instruments :

(*i*) Multimeter

(*ii*) Electronic Voltmeters

(*iii*) Cathode ray oscilloscope

The knowledge of the manner in which each instrument is used plus an understanding of the applications and limitations of each instrument will enable the reader to utilise such instruments successfully.

20.2 Multimeter

A *multimeter* is an electronic instrument which can measure resistances, currents and voltages. It is an indispensable instrument and can be used for measuring d.c. as well as a.c. voltages and currents. Multimeter is the most inexpensive equipment and can make various electrical measurements with reasonable accuracy.

Construction. A multimeter consists of an ordinary pivoted type of permanent magnet moving coil (*PMMC*) galvanometer. This galvanometer consists of a coil pivoted on jeweled bearings between the poles of a permanent magnet. The indicating needle is fastened to the coil. When electric current is passed through the coil, mechanical force acts and the pointer moves over the scale.

Functions. A multimeter can measure voltages, currents and resistances. To achieve this objective, proper circuits are incorporated with the galvanometer. The galvanometer in a multimeter is always of *left zero type i.e.* normally its needle rests in extreme left position as compared to centre zero position of ordinary galvanometers.

(*i*) **Multimeter as voltmeter.** When a high resistance is connected in series with a galvanometer, it becomes a voltmeter. Fig. 20.1 (*i*) shows a high resistance R connected in series with the galvanometer of resistance G. If I_g is the *full-scale deflection current*, then the galvanometer becomes a voltmeter of range $0 - V$ volts. The required value of series resistance R is given by;

$$V = I_g R + I_g G$$

or $$V/I_g = R + G$$

or $$R = V/I_g - G$$

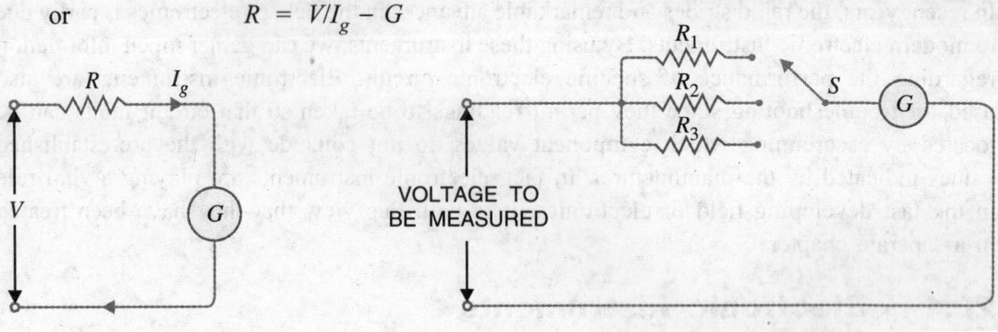

(*i*) (*ii*)

Fig. 20.1

For maximum accuracy, a multimeter is always provided with a number of voltage ranges. This is achieved by providing a number of high resistances in the multimeter as shown in Fig. 20.1 (*ii*). Each resistance corresponds to one voltage range. With the help of selector switch S, we can put any resistance (R_1, R_2 and R_3) in series with the galvanometer. When d.c. voltages are to be measured, the multimeter switch is turned on to d.c. position. This puts the circuit shown in Fig. 20.1 (*ii*) in action. By throwing the range selector switch S to a suitable position, the given d.c. voltage can be measured.

The multimeter can also measure a.c. voltages. To permit it to perform this function, a full-wave rectifier is used as shown in Fig. 20.2. The rectifier converts a.c. into d.c. for application to the galvanometer. The desired a.c. voltage range can be selected by the switch S. When a.c. voltage is to be measured, the multimeter switch is thrown to a.c. position. This puts the circuit shown in Fig. 20.2 in action. By throwing the range selector switch S to a suitable position, the given a.c. voltage can be measured. It may be mentioned here that a.c. voltage scale is calibrated in r.m.s. values. Therefore, the meter will give the r.m.s. value of the a.c. voltage under measurement.

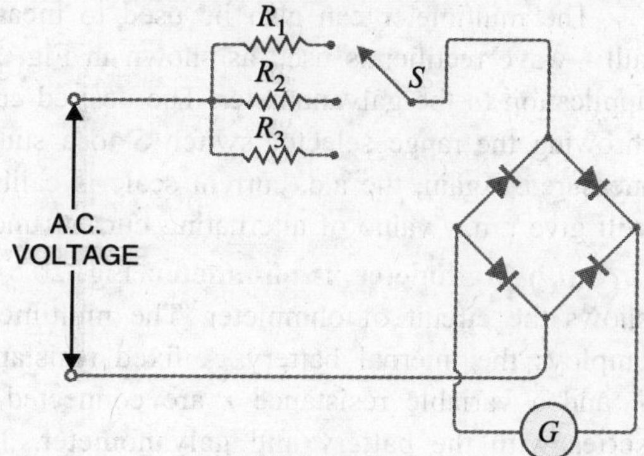

Fig. 20.2

(*ii*) **Multimeter as ammeter.** When low resistance is connected in parallel with a galvanometer, it becomes an ammeter. Fig. 20.3 (*i*) shows a low resistance S (generally called *shunt*) connected in parallel with the galvanometer of resistance G. If I_g is the full- scale deflection current, then the galvanometer becomes an ammeter of range $0 - I$ amperes. The required value of shunt resistance S is given by ;

$$I_s S = I_g G$$

or $$I_s/I_g = G/S \quad \text{or} \quad \frac{I_s}{I_g} + 1 = \frac{G}{S} + 1$$

or $$\frac{I_s + I_g}{I_g} = \frac{G+S}{S} \quad \text{or} \quad \frac{I}{I_g} = \frac{G+S}{S}$$

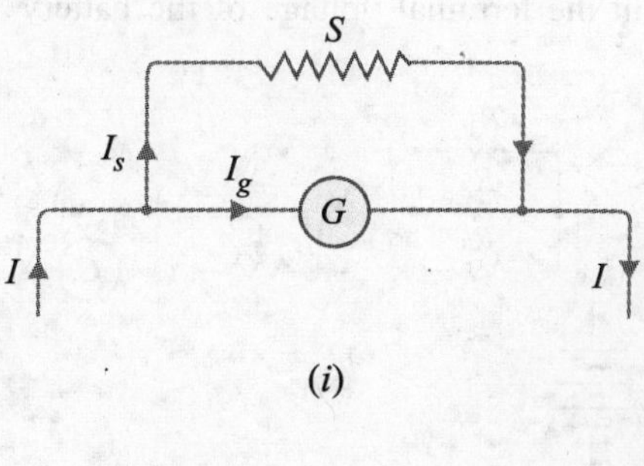

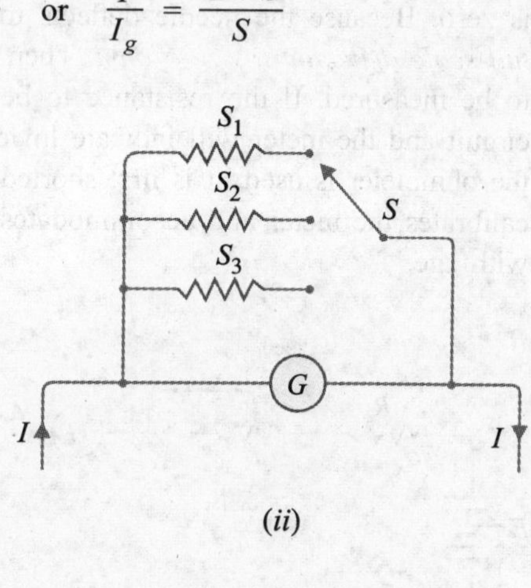

Fig. 20.3

In practice, a number of low resistances are connected in parallel with the galvanometer to provide a number of current ranges as shown in Fig. 20.3 (*ii*). With the help of range selector switch S, any shunt can be put in parallel with the galvanometer. When d.c. current is to be measured, the multimeter switch is turned on to d.c. position. This puts the circuit shown in Fig. 20.3 (*ii*) in action. By throwing the range selector switch S to a suitable position, the desired d.c. current can be measured.

The multimeter can also be used to measure alternating current. For this purpose, a full - wave rectifier is used as shown in Fig. 20.4. The rectifier converts a.c. into d.c. for application to the galvanometer. The desired current range can be selected by switch *S*. By throwing the range selector switch *S* to a suitable position, the given a.c. current can be measured. Again, the a.c. current scale is calibrated in r.m.s. values so that the instrument will give r.m.s. value of alternating current under measurement.

(*iii*) **Multimeter as ohmmeter.** Fig. 20.5 (*i*) shows the circuit of ohmmeter. The multimeter employs the internal battery. A fixed resistance *R* and a variable resistance *r* are connected in series with the battery and galvanometer. The fixed resistance *R* limits the current within the range desired and variable resistance *r* is for zero-adjustment reading. The resistance to be measured is connected between terminals *A* and *B*. The current flowing through the circuit will depend upon the value of resistor connected across the terminals. The ohmmeter scale is calibrated in terms of ohms. The ohmmeter is generally made multirange instrument by using different values of *R* as shown in Fig. 20.5 (*ii*).

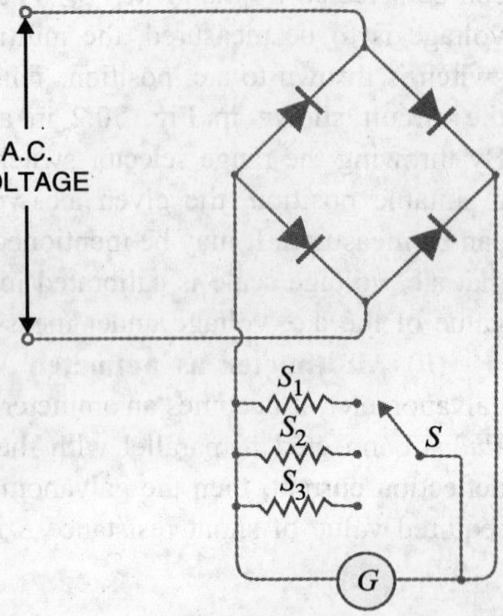

Fig. 20.4

To use ohmmeter, terminals *A* and *B* are shorted and resistance *r* is adjusted to give full-scale deflection of the galvanometer. Under this condition, the resistance under measurement is zero. Because the needle deflects to full scale, the ohmmeter scale must then indicate *full-scale deflection as zero ohm*. Then probes *A* and *B* are connected across the resistance to be measured. If the resistance to be measured is high, lower current flows through the circuit and the meter will indicate lower reading. It may be mentioned here that each time the ohmmeter is used, it is first shorted across *AB* and *r* is adjusted to zero the meter. This calibrates the meter and accommodates any decrease in the terminal voltage of the battery with age.

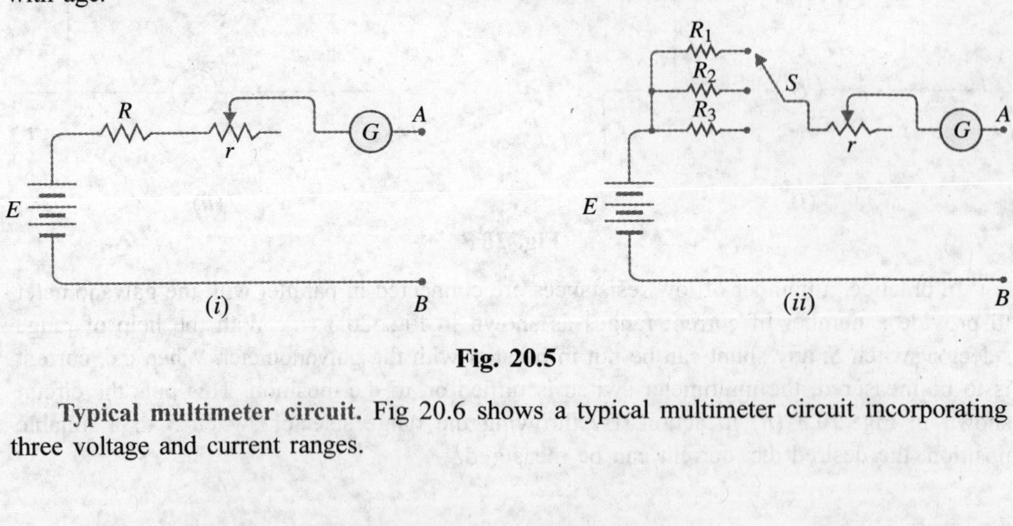

Fig. 20.5

Typical multimeter circuit. Fig 20.6 shows a typical multimeter circuit incorporating three voltage and current ranges.

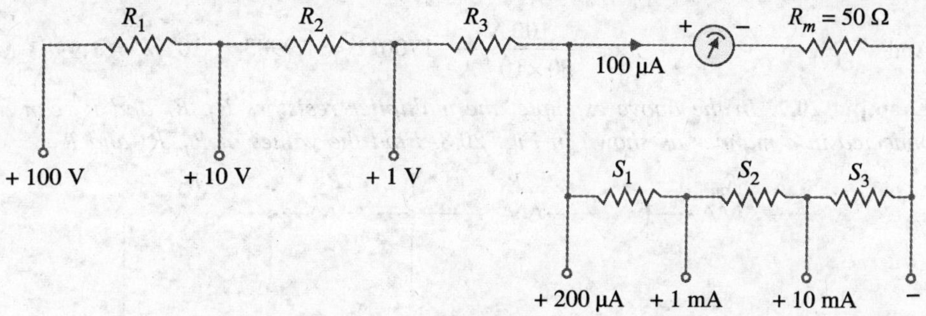

Fig. 20.6

Here the full-scale deflection (f.s.d.) current of the meter is 100 μA and meter resistance is 50 Ω. The design of this multimeter means finding the values of various resistances.

Example 20.1. *A PMMC (permanent-magnet moving coil) instrument with full-scale deflection (f.s.d.) current of 50 μA and meter resistance R_m = 1700 Ω is to be employed as a voltmeter with ranges of 10 V, 50 V and 100 V. Calculate the required values of multiplier resistors for the circuit shown in Fig. 20.7.*

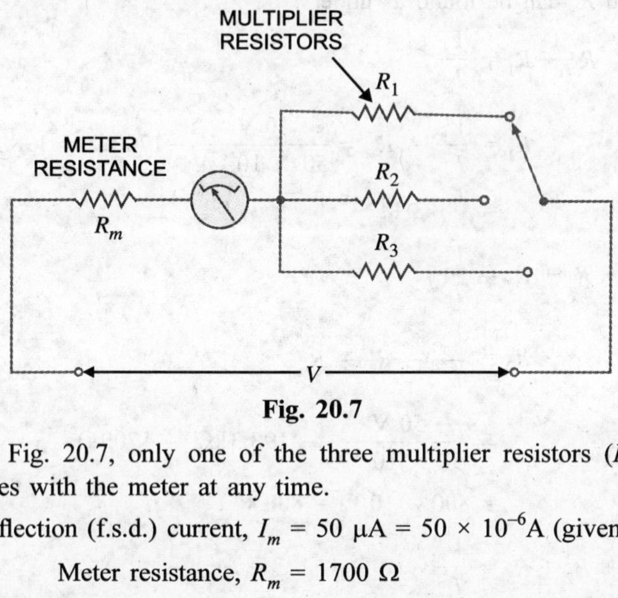

Fig. 20.7

Solution. In Fig. 20.7, only one of the three multiplier resistors (R_1, R_2 and R_3) is connected in series with the meter at any time.

Full-scale deflection (f.s.d.) current, I_m = 50 μA = 50 × 10^{-6}A (given)

Meter resistance, R_m = 1700 Ω

Referring to Fig. 20.7, the values of multiplier resistors R_1, R_2 and R_3 can be found as under :

$$R_m + R_1 = \frac{V_1}{I_m}$$

or
$$R_1 = \frac{V_1}{I_m} - R_m$$

∴
$$R_1 = \frac{10\ \text{V}}{50 \times 10^{-6}\ \text{A}} - 1700\,\Omega = 198.3 \times 10^3\,\Omega = \mathbf{198.3\ k\Omega}$$

Similarly,
$$R_2 = \frac{50\ \text{V}}{50 \times 10^{-6}\ \text{A}} - 1700\,\Omega = 998.3 \times 10^3\,\Omega = \mathbf{998.3\ k\Omega}$$

And $$R_3 = \frac{100\text{ V}}{50\times10^{-6}\text{ A}} - 1700\,\Omega = 1.9983\times10^6\,\Omega = \mathbf{1.9983\ M\Omega}$$

Example 20.2. *In the above example, the multiplier resistors R_1, R_2 and R_3 can also be connected in a manner as shown in Fig. 20.8. Find the values of R_1, R_2 and R_3.*

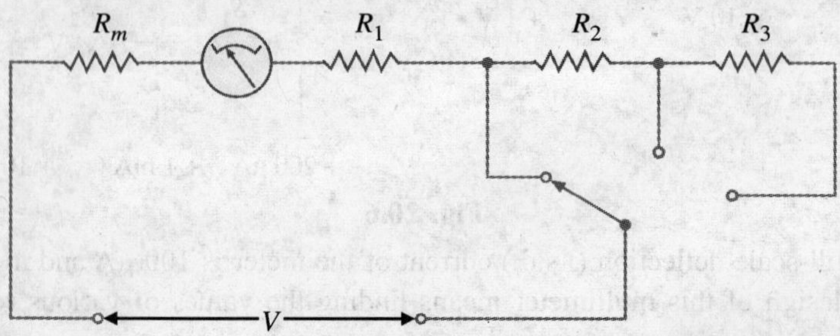

Fig. 20.8

Solution. In Fig. 20.8, the multiplier resistors (R_1, R_2 and R_3) are connected in series and each junction is connected to one of the switch terminals. Referring to Fig. 20.8, the values of R_1, R_2 and R_3 can be found as under :

$$R_m + R_1 = \frac{V_1}{I_m}$$

$$\therefore \quad R_1 = \frac{V_1}{I_m} - R_m = \frac{10\text{ V}}{50\times10^{-6}\text{ A}} - 1700\,\Omega = 198.3\times10^3\,\Omega$$

$$= \mathbf{198.3\ k\Omega}$$

Also, $$R_m + R_1 + R_2 = \frac{V_2}{I_m}$$

$$\therefore \quad R_2 = \frac{V_2}{I_m} - R_1 - R_m$$

$$= \frac{50\text{ V}}{50\times10^{-6}\text{ A}} - 198.3\text{k}\Omega - 1700\,\Omega$$

$$= 800\times10^3\,\Omega = \mathbf{800\ k\Omega}$$

And $$R_m + R_1 + R_2 + R_3 = \frac{V_3}{I_m}$$

$$\therefore \quad R_3 = \frac{V_3}{I_m} - R_2 - R_1 - R_m$$

$$= \frac{100\text{ V}}{50\times10^{-6}\text{ A}} - 800\text{k}\Omega - 198.3\text{k}\Omega - 1700\,\Omega$$

$$= 1\times10^6\,\Omega = \mathbf{1\ M\Omega}$$

Note. We have seen that a multirange voltmeter can be constructed by using multiplier resistors R_1, R_2 and R_3 in two possible ways [See Figs. 20.7 and 20.8]. However, the circuit shown in Fig. 20.8 is least expensive to construct. It is because all the multiplier resistors in Fig. 20.7 must be special (non-standard) values while in Fig. 20.8, only R_1 is a special (non-standard) resistor and all other multipliers are standard-value resistors.

20.3 Applications of Multimeter

A multimeter is an extremely important electronic instrument and is extensively used for carrying out various tests and measurements in electronic circuits. It is used :

(*i*) For checking the circuit continuity. When the multimeter is employed as continuity-checking device, the ohmmeter scale is utilised and the equipment to be checked is shut off or disconnected from the power mains.

(*ii*) For measuring d.c. current flowing through the cathode, plate, screen and other vacuum tube circuits.

(*iii*) For measuring d.c. voltages across various resistors in electronic circuits.

(*iv*) For measuring a.c. voltages across power supply transformers.

(*v*) For ascertaining whether or not open or short circuit exists in the circuit under study.

20.4 Sensitivity of Multimeter

The resistance offered per volt of full-scale deflection by the multimeter is known as **multimeter sensitivity.**

Multimeter sensitivity indicates the internal resistance of the multimeter. For example, if the total resistance of the meter is 5000 ohms and the meter is to read 5 volts full-scale, then internal resistance of the meter is 1000 Ω per volt *i.e.* meter sensitivity is 1000 Ω per volt. Conversely, if the meter sensitivity is 400 Ω per volt which reads from 0 to 100 V, then meter resistance is 40,000 ohms. If the meter is to read V volts and I_g is the full-scale deflection current, then,

$$\text{Meter resistance} = \frac{V}{I_g}$$

$$\text{Meter sensitivity} = \text{Resistance per volt full-scale deflection}$$

$$= \frac{V}{I_g}\Big/V = \frac{1}{I_g}$$

Sensitivity is the most important characteristic of a multimeter. If the sensitivity of a multimeter is high, it means that it has high internal resistance. When such a meter is connected in the circuit to read voltage, it will draw a very small current. Consequently, there will be no change in the circuit current due to the introduction of the meter. Hence, it will measure the voltage correctly. On the other hand, if the sensitivity of multimeter is low, it would cause serious error in voltage measurement. The sensitivity of multimeters available in the market range from 5 kΩ per volt to 20 kΩ per volt.

20.5 Merits and Demerits of Multimeter

Although multimeter is widely used for manufacturing and servicing of electronics equipment, it has its own merits and demerits.

Merits

(*i*) It is a single meter that performs several measuring functions.

(*ii*) It has a small size and is easily portable.

(*iii*) It can make measurements with reasonable accuracy.

Demerits

(*i*) It is a costly instrument. The cost of a multimeter having sensitivity of 20 kΩ per volt is about Rs. 1000.

(*ii*) It cannot make precise and accurate measurements due to the loading effect.

(*iii*) Technical skill is required to handle it.

20.6 Meter Protection

It is important to provide protection for the meter in the event of an accidental overload. This is achieved by connecting a diode in parallel with the voltmeter as shown in Fig. 20.9.

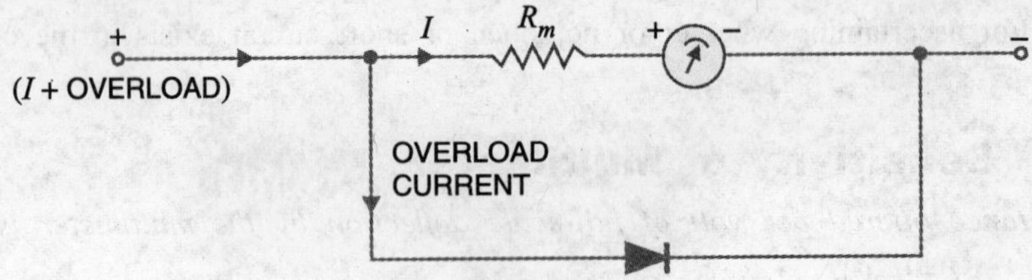

Fig. 20.9

Let us see how diode across the meter enables it to withstand overload without destroying the expensive movement. If I is the normal f.s.d. current, a potential difference of IR_m is developed across the diode. The circuit is so designed that IR_m does not turn on the diode. In the event of an accidental overload (say 5 I), the voltage across diode becomes 5 times greater and it is immediately turned on. Consequently, diode diverts most of the overload current in the same manner as a shunt. Thus protection of the meter against overload is ensured. Silicon diodes are perhaps the best to use in such circuits.

Example 20.3. *A multimeter has full-scale deflection current of 1 mA. Determine its sensitivity.*

Solution. Full-scale deflection current, I_g = 1 mA = 10^{-3} A

∴ Multimeter sensitivity = $1/I_g$ = $1/10^{-3}$ = **1000 Ω per volt**

Example 20.4. *A multimeter has a sensitivity of 1000 Ω per volt and reads 50 V full-scale. If the meter is to be used to measure the voltage across 50000 Ω resistor, will it read correctly ?*

Solution. Meter sensitivity = 1000 Ω per volt

Full-scale volts = 50 V

∴ Meter resistance = 50 × 1000 = 50,000 Ω

When the meter is used to measure the voltage across the resistance as shown in Fig. 20.10, the total resistance of the circuit is a parallel combination of two 50,000 Ω resistors. Therefore, the circuit resistance would be reduced to 25000 Ω and double the amount of current would be drawn than would otherwise be the case.

∴ **Meter will give highly incorrect reading.**

Comments. This example shows the limitation of multimeter. The multimeter will read correctly only if its resistance is very high as compared to the resistance across which voltage is to be measured.

As a rule, the resistance of the multimeter should be atleast 100 times the resistance across which voltage is to be measured.

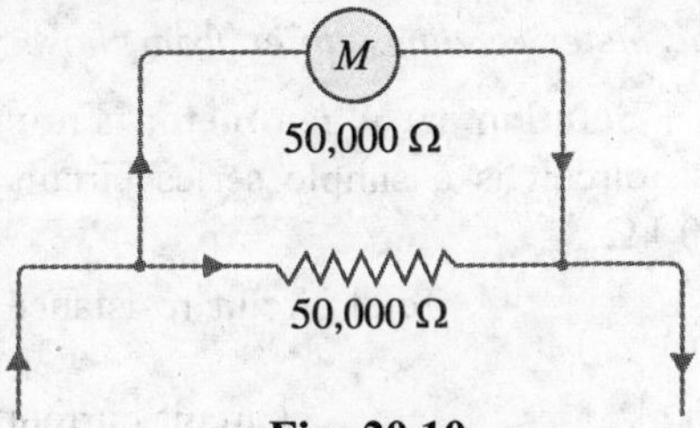

Fig. 20.10

Example 20.5. *In the circuit shown in Fig. 20.11 (i), it is desired to measure the voltage across 10 kΩ resistance. If a multimeter of sensitivity 4 kΩ/volt and range 0-10 V is used for the purpose, what will be the reading ?*

Solution. In the circuit shown in Fig. 20.11 (*i*), the circuit current by Ohm's law is 1 mA. Therefore, voltage across 10 kΩ resistance is 10 V. Let us see whether the given multimeter reads this value. Fig. 20.11 (*ii*) shows the multimeter connected across 10 kΩ resistance. The introduction of multimeter will change the circuit resistance and hence circuit current.

$$\text{Resistance of meter} = 4\ \text{k}\Omega \times 10 = 40\ \text{k}\Omega$$

$$\text{Total circuit resistance} = 40\ \text{k}\Omega \parallel 10\ \text{k}\Omega + 10\ \text{k}\Omega$$

$$= \frac{40 \times 10}{40 + 10} + 10 = 8 + 10 = 18\ \text{k}\Omega$$

$$\therefore \quad \text{Circuit current} = \frac{20\ \text{V}}{18\ \text{k}\Omega} = 1.11\ \text{mA}$$

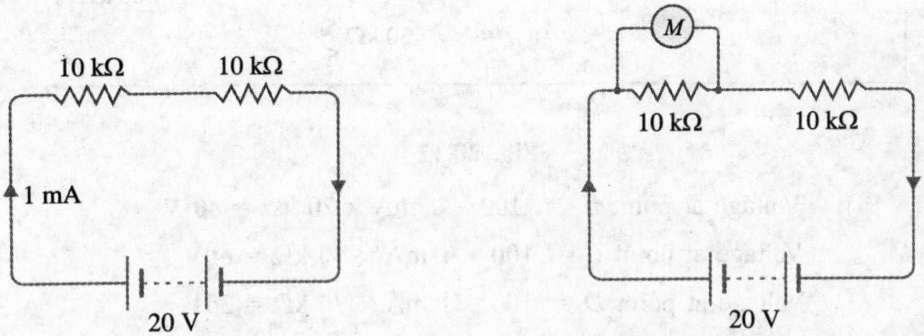

Fig. 20.11

$$\text{Voltage read by multimeter} = 8\ \text{k}\Omega \times 1.11\ \text{mA} = \mathbf{8.88\ V}$$

Example 20.6. *If in the above example, a multimeter of sensitivity 20 kΩ per volt is used, what will be the reading ?*

Solution.

$$\text{Meter resistance} = 20\ \text{k}\Omega \times 10 = 200\ \text{k}\Omega$$

$$\text{Total circuit resistance} = 200\ \text{k}\Omega \parallel 10\ \text{k}\Omega + 10\ \text{k}\Omega$$

$$= \frac{200 \times 10}{200 + 10} + 10 = 9.5 + 10 = 19.5\ \text{k}\Omega$$

$$\text{Circuit current} = \frac{20\ \text{V}}{19.5\ \text{k}\Omega} = 1.04\ \text{mA}$$

$$\therefore \quad \text{Voltage read by multimeter} = 9.5\ \text{k}\Omega \times 1.04\ \text{mA} = \mathbf{9.88\ V}$$

A comparison of examples 20.5 and 20.6 shows that a multimeter with higher sensitivity gives more correct reading.

Example 20.7. *In the circuit shown in Fig. 20.12, find the voltage at points A, B, C and D (i) before the meter is connected and (ii) after the meter is connected. Explain why the meter readings differ from those without the meter connected.*

Solution. (*i*) When meter is not connected. When meter is not connected in the circuit, the circuit is a simple series circuit consisting of resistances 20 kΩ, 20 kΩ, 30 kΩ and 30 kΩ.

$$\text{Total circuit resistance} = 20 + 20 + 30 + 30 = 100 \text{ k}\Omega$$

$$\therefore \quad \text{Circuit current} = \frac{100 \text{ V}}{100 \text{ k}\Omega} = 1 \text{ mA}$$

$$\text{Voltage at point } A = \mathbf{100 \text{ V}}$$

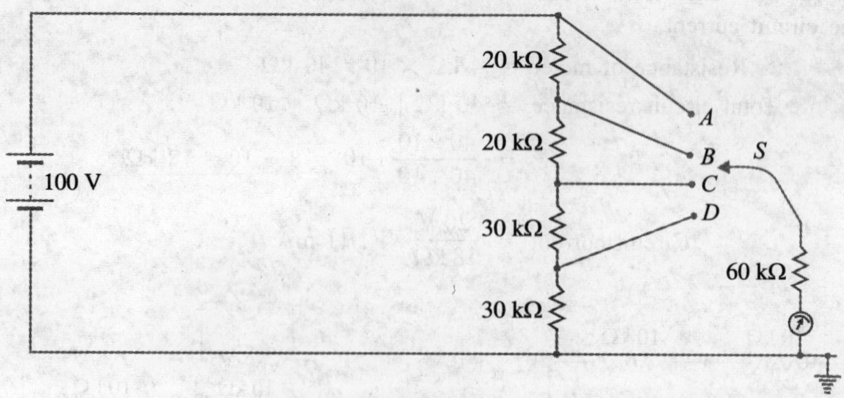

Fig. 20.12

$$\text{Voltage at point } B = 100 - 1 \text{ mA} \times 20 \text{ k}\Omega = \mathbf{80 \text{ V}}$$

$$\text{Voltage at point } C = 100 - 1 \text{ mA} \times 40 \text{ k}\Omega = \mathbf{60 \text{V}}$$

$$\text{Voltage at point } D = 100 - 1 \text{ mA} \times 70 \text{ k}\Omega = \mathbf{30 \text{V}}$$

(*ii*) When meter is connected. When meter is connected in the circuit, the circuit becomes a series-parallel circuit. The total circuit resistance would depend upon the position of switch *S*.

(*a*) *When switch is at position A*

The voltage at point *A* is 100 V because point *A* is directly connected to the voltage source.

$$\therefore \quad \text{Voltage at point } A = \mathbf{100 \text{ V}}$$

(*b*) *When switch is at position B*

$$\text{Total circuit resistance} = 20 + \frac{80 \times 60}{80 + 60} = 20 + 34.28 = 54.28 \text{ k}\Omega$$

$$\therefore \quad \text{Circuit current} = \frac{100 \text{ V}}{54.28 \text{ k}\Omega}$$

$$\therefore \quad \text{Voltage at point } B = \frac{100 \text{ V}}{54.28 \text{ k}\Omega} \times 34.28 \text{ k}\Omega = \mathbf{63 \text{ V}}$$

(*c*) *When switch is at point C*

$$\text{Total circuit resistance} = 40 + \frac{60 \times 60}{60 + 60} = 40 + 30 = 70 \text{ k}\Omega$$

$$\therefore \qquad \text{Circuit current} = \frac{100 \text{ V}}{70 \text{ k}\Omega}$$

$$\therefore \qquad \text{Voltage at point } C = \frac{100 \text{ V}}{70 \text{ k}\Omega} \times 30 \text{ k}\Omega = \mathbf{42.8 \text{ V}}$$

(*d*) *When switch is at point D*

$$\text{Total circuit resistance} = 70 + \frac{30 \times 60}{30 + 60} = 70 + 20 = 90 \text{ k}\Omega$$

$$\therefore \qquad \text{Circuit current} = \frac{100 \text{ V}}{90 \text{ k}\Omega}$$

$$\therefore \qquad \text{Voltage at point } D = \frac{100 \text{ V}}{90 \text{ k}\Omega} \times 20 \text{ k}\Omega = \mathbf{22.2 \text{ V}}$$

Comments. Note that potential measurements are being made in a high-impedance circuit ; the circuit resistance is comparable to meter resistance. As a rule, the resistance of the voltmeter should be 100 times the resistance across which voltage is to be measured. Since such a condition is not realised in this problem, the meter readings differ appreciably from those without the meter connected.

Example 20.8. *It is desired to measure the voltage across a 50 kΩ resistor in the circuit shown in Fig. 20.13. Two multimeters are available for this purpose :*

Multimeter A with a sensitivity of 1000 Ω/V

Multimeter B with a sensitivity of 20,000 Ω/V

Both meters have 0 – 50 V range. Calculate (i) the reading of each voltmeter (ii) the error in each reading expressed as a percentage of true value.

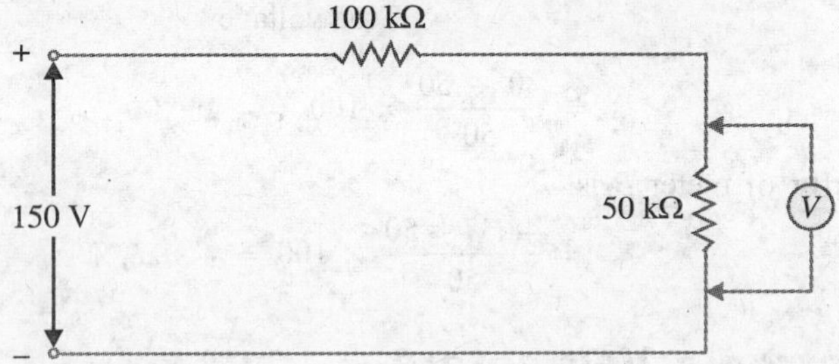

Fig. 20.13

Solution. The true value of voltage across 50 kΩ resistor is

$$\text{True voltage} = \text{Current through } 50 \text{ k}\Omega \times 50 \text{ k}\Omega$$

$$= \frac{150 \text{ V}}{(100 + 50) \text{ k}\Omega} \times 50 \text{ k}\Omega = 50 \text{ V}$$

(*i*) Multimeter *A*. Resistance R_A of the meter is

$$R_A = S_V V = 1000 \times 50 \ \Omega = 50 \text{ k}\Omega$$

When the meter is connected across 50 kΩ resistor, the resistance R_P of parallel combination of meter and 50 kΩ resistor is

$$R_P = R_A \parallel 50\ \text{k}\Omega = \frac{50 \times 50}{50 + 50} = 25\ \text{k}\Omega$$

Total circuit resistance = 100 kΩ + R_P = 125 kΩ

∴ Reading of meter = Voltage across R_P

$$= \frac{150\ \text{V}}{125\ \text{k}\Omega} \times 25\,\text{k}\Omega = \mathbf{30\ V}$$

Therefore, the meter A will indicate a voltage of 30 V.

Multimeter *B*. Resistance R_B of the meter is

$$R_B = S_V\, V = 20{,}000 \times 50\ \Omega = 1000\ \text{k}\Omega$$

When the meter is connected across 50 kΩ resistor, the resistance R'_P of the parallel combination of meter and 50 kΩ resistor is

$$R'_P = R_B \parallel 50\ \text{k}\Omega = \frac{100 \times 50}{1000 + 50} = 47.6\ \text{k}\Omega$$

Total circuit resistance = 100 kΩ + R'_P = 147.6 kΩ

∴ Reading of meter = Voltage across R'_P

$$= \frac{150\ \text{V}}{147.6\ \text{k}\Omega} \times 47.6\,\text{k}\Omega = \mathbf{48.37\ V}$$

Therefore, the meter B will indicate 48.37 V.

(*ii*) Percentage error of meter A is

$$= \frac{\text{Indicated voltage} - \text{True voltage}}{\text{True voltage}} \times 100$$

$$= \frac{30 - 50}{50} \times 100 = \mathbf{-\ 40\ \%}$$

Percentage error of meter B is

$$= \frac{48.37 - 50}{50} \times 100 = \mathbf{-\ 3.26\ \%}$$

20.7 Electronic Voltmeters

The electromagnetic and electrostatic voltmeters have two main drawbacks. First, the input resistance/impedance of these instruments is not very high so that there is a considerable *loading effect of the instrument. Secondly, considerable power is drawn from the circuit under measurement. Both these drawbacks are overcome in electronic voltmeters. The electronic devices (*e.g.* vacuum tubes, transistors *etc.*) have very high input resistance/impedance and possess the property of amplification. The latter property permits the input signal to be amplified so that the power to operate the indicating mechanism comes from a source other than the measured circuit. There are a large number of electronic voltmeters. However, we

* When a voltmeter is connected across a resistance R to measure voltage, the measured voltage will be less than the actual value. It is because the resistance R is shunted by the voltmeter. This is called loading effect of the meter. The greater the input resistance of voltmeter, the smaller will be the loading effect and more accurate is the reading.

shall discuss the following three types of electronic voltmeters :

(*i*) Vacuum Tube Voltmeter (*VTVM*)

(*ii*) Transistor Voltmeter

(*iii*) Bridge Rectifier Voltmeter

20.8 Vacuum Tube Voltmeter (VTVM)

A vacuum tube voltmeter consists of any ordinary voltmeter and electron tubes. It is extensively used for measuring both a.c. and d.c. voltages. The vacuum tube voltmeter has high internal resistance (>10 MΩ) and draws extremely small current from the circuit across which it is connected. In other words, the loading effect of this instrument is very small. Therefore, a *VTVM* measures the exact voltage even across a high resistance. In fact, the ability of *VTVM* to measure the voltages accurately has made this instrument the most popular with technicians for troubleshooting radio and television receivers as well as for laboratory work involving research and design.

(*i*) **Simple VTVM circuit.** Fig. 20.14 shows the simple circuit of a vacuum tube voltmeter. It consists of a triode having meter *M* connected in the plate circuit. The meter is calibrated in volts. R_1 is the grid leak resistor. The voltage to be measured is applied at the grid of triode in such a way that grid is always negative *w.r.t.* cathode. This voltage at the grid is transformed by the triode into corresponding plate current. The meter *M* connected in the plate circuit directly gives the value of the voltage under measurement. It may be seen that as grid draws extremely small current (< 1 μA), therefore, internal resistance of *VTVM* is very large. This circuit has the disadvantage that if the applied voltages change (especially filament voltage), the plate current will also change. Consequently, the meter will give wrong reading.

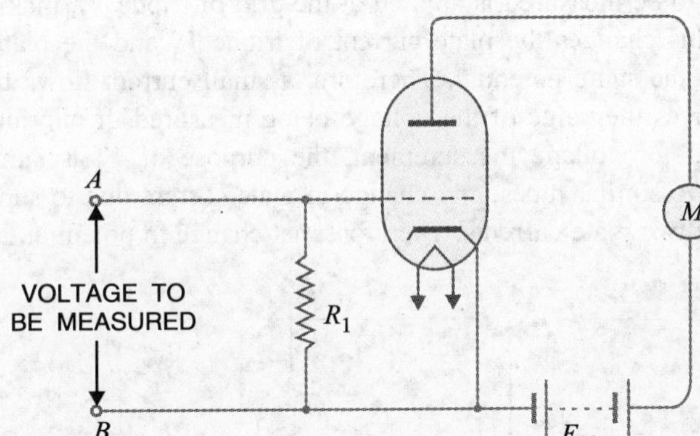

Fig. 20.14

(*ii*) **Balanced bridge Type VTVM.** The disadvantage of above circuit is overcome in the balanced bridge type *VTVM* shown in Fig. 20.15. Here, two similar triodes V_1 and V_2 are used. The meter *M* is connected between the plates of triodes and indicates the voltage to be measured. The variable resistance *r* in the plate circuit of V_2 is for zero adjustment of the meter. The voltage to be measured is applied at the grid of triode V_1 in such a way that grid is always negative with respect to cathode.

Operation. When no voltage is applied at the input terminals *AB*, the plate currents flowing in both valves are equal as the triodes are similar. Therefore, plates of both valves are at the same potential. Consequently, the current through the meter *M* is zero and the meter reads zero volt. However, in actual practice, there are always some constructional differences in plates, grids and cathodes of the two valves. The result is that two plate currents differ slightly and the meter may give some reading. In such a case, the meter needle is brought to zero by changing resistance *r*.

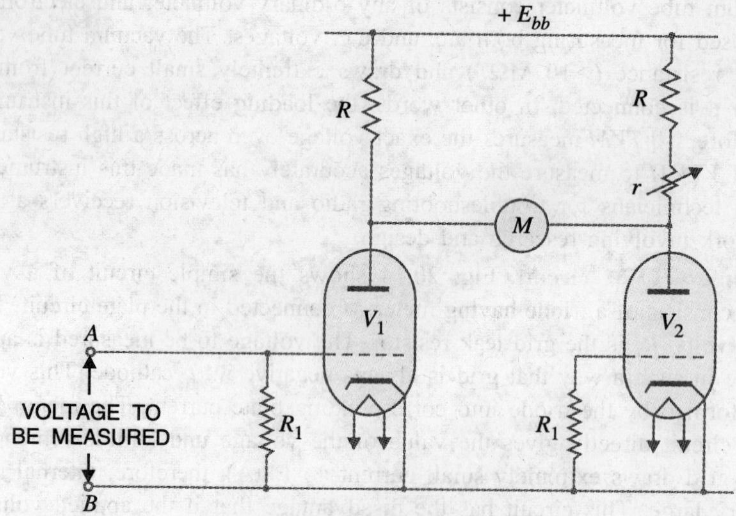

Fig. 20.15

The voltage to be measured is applied at the grid of triode V_1, making the grid negative *w.r.t.* cathode. This changes the plate current of triode V_1 and the plates of two valves no longer remain at the same potential. Therefore, a small current flows through the meter *M* which directly gives the value of the voltage being measured. It may be noted that actually triode V_1 is used for voltage measurement, the purpose of V_2 is simply to prevent zero drift. By using two similar tubes, any change in plate current due to supply fluctuations will equally affect the two plate currents. Therefore, net change in potential drop across voltmeter is zero.

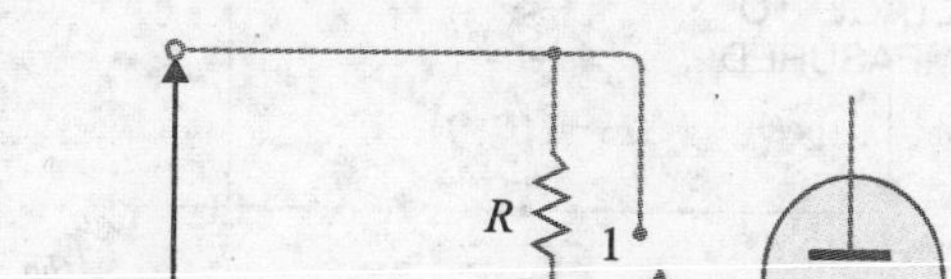

Fig. 20.16

Range selection. In practice, a *VTVM* is made a multirange instrument by employing a potentiometer at the input circuit as shown in Fig. 20.16. By throwing the range selector switch *S* to a suitable position, the desired voltage range can be obtained. Thus when the range selector switch *S* is thrown to position 1, the voltage applied to the grid is three times as compared to position 3. Although only three voltage ranges have been considered, a commercial *VTVM* may have more ranges.

20.9 Applications of VTVM

A *VTVM* is far superior to a multimeter and performs a number of measuring functions. A few important applications of *VTVM* are discussed below :

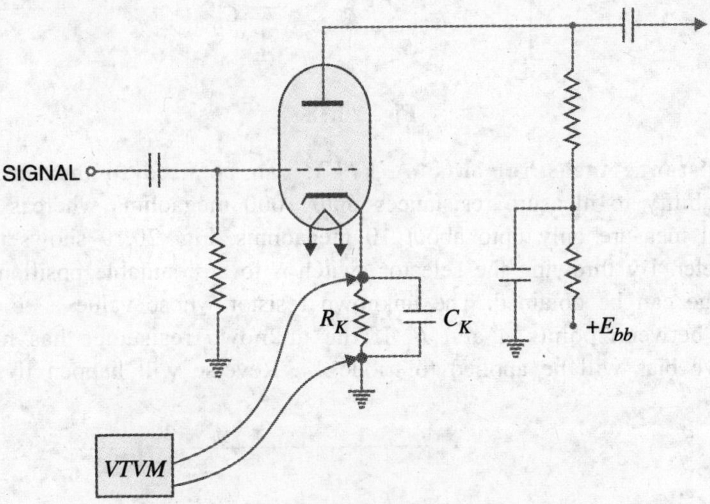

Fig. 20.17

***(i)* D.C. voltage measurements.** A *VTVM* can accurately measure the D.C. voltages in an electronic circuit. The d.c. voltage to be measured is applied at the input (*i.e.* grid of V_1) terminals in such a way that grid of the input valve V_1 is always negative. Fig. 20.17 shows the circuit of an amplifier stage and measurement of D.C. voltage across cathode resistor R_K.

***(ii)* D.C. current measurements.** A conventional *VTVM* does not incorporate a current scale. However, current values can be found indirectly. For instance, in Fig. 20.17, the D.C. current through R_K can be found by noting the voltage across R_K and dividing it by the resistance R_K.

***(iii)* A.C. voltage measurements.** For measuring a.c. voltage, a rectifier is used in conjunction with a *VTVM.* The rectifier converts a.c. into d.c. for application to the grid of valve V_1. In fact, rectifier circuit is a part of *VTVM.* Fig. 20.18 shows the transistor power amplifier stage and measurement of a.c. voltage across the speaker.

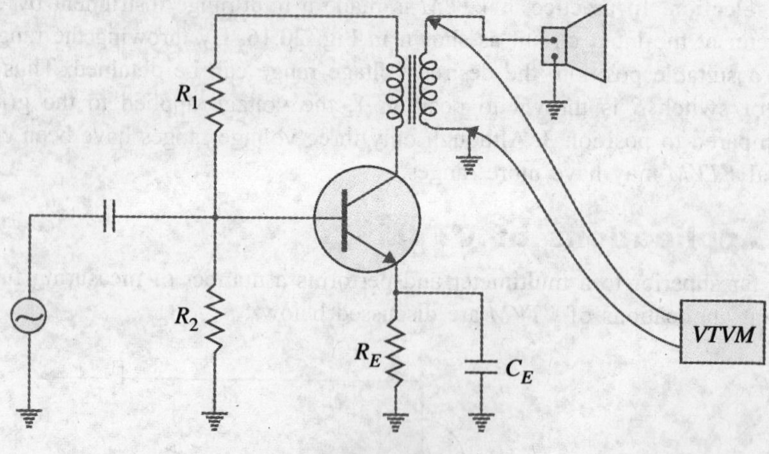

Fig. 20.18

(*iv*) **Resistance measurements.** A *VTVM* can be used to measure resistances and has the ability to measure resistances upto 1000 megaohms whereas the ordinary ohmmeter will measure only upto about 10 megaohms. Fig. 20.19 shows the circuit of *VTVM* ohmmeter. By throwing the selector switch *S* to any suitable position, the desired resistance range can be obtained. The unknown resistor whose value is to be measured is connected between points *A* and *B*. If the unknown resistance has high value, a higher negative bias will be applied to triode V_1. Reverse will happen if the unknown

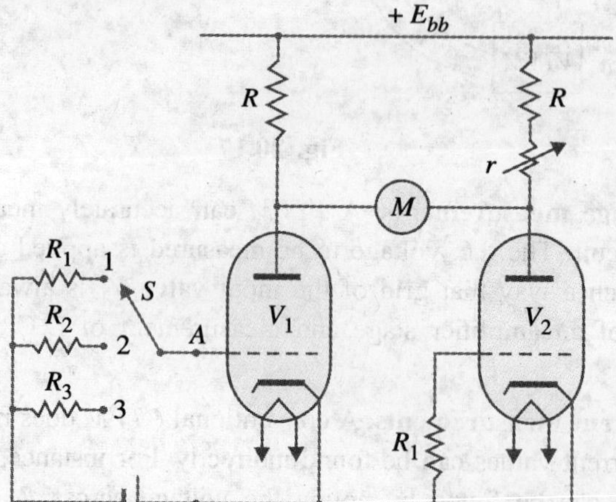

Fig. 20.19

resistance has low value. The imbalance in the plate currents of the two valves will cause a current through the meter *M* which will directly give the value over the resistance scale of the meter.

20.10 Merits and Demerits of VTVM

A *VTVM* is an extremely important electronic equipment and is widely used for making different measurements in electronic circuits.

Merits

(*i*) A *VTVM* draws extremely small current from the measuring circuit. Therefore, it gives accurate readings.

(*ii*) There is little effect of temperature variations.

(*iii*) Because a *VTVM* uses triodes, the voltage to be measured is amplified. This permits the use of less sensitive meter.

(*iv*) It has a wide frequency response.

Demerits

(*i*) It cannot make current measurements directly.

(*ii*) Accurate readings can be obtained only for sine waves.

20.11 Transistor Voltmeter Circuit

Since vacuum tubes have become obsolete, these have been replaced by transistors and other semiconductor devices. Fig. 20.20 shows the circuit of an emitter-follower voltmeter. The voltage E to be measured is applied between base and emitter. A permanent-magnet moving coil (PMMC) instrument and a multiplier resistor R_S are connected in series with the transistor emitter. The circuit measures the voltage quite accurately because the emitter follower offers high input resistance to the voltage being measured and provides a low output resistance to drive current through the coil of PMMC meter.

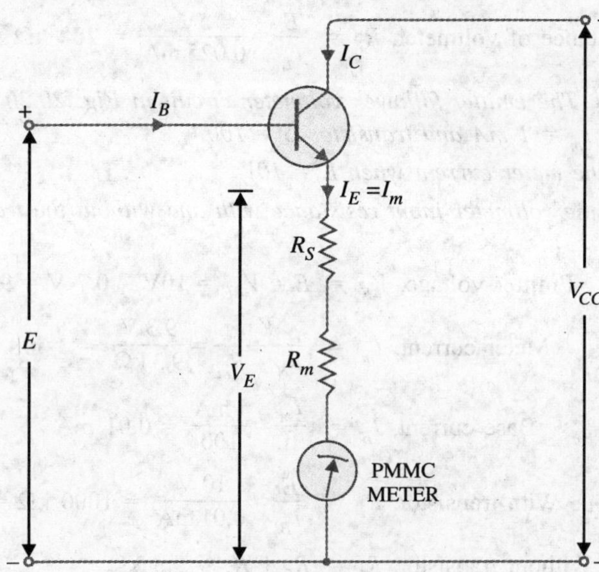

Fig. 20.20

Operation. The voltage E to be measured is applied between base and emitter of the transistor and causes a base current I_B to flow through the base circuit. Therefore, collector current $I_C = \beta I_B$ where β is the current amplification factor of the transistor. Since $I_E \simeq I_C$ and the meter is connected in the emitter, the meter current $I_m = I_E = \beta I_B$. Now the meter current I_m depends upon the input voltage to be measured. Therefore, the PMMC meter can be calibrated to read the input voltage directly.

$$\text{Emitter voltage, } V_E = E - V_{BE}$$

$$\text{Meter current, } I_m = \frac{V_E}{R_S + R_m}$$

Here, R_S = Multiplier resistor ; R_m = Meter resistance

$$\text{Input resistance of voltmeter, } R_i = \frac{E}{I_B}$$

Example 20.9. *The emitter follower circuit shown in Fig. 20.20 has* $V_{CC} = 12\ V$; $R_m = 1k\Omega$ *and a 2 mA meter. If transistor* $\beta = 80$, *calculate (i) the suitable resistance for* R_S *to give full -scale deflection when* $E = 5V$ *(ii) the voltmeter input resistance.*

Solution.

$$\text{Meter resistance, } R_m = 1\ k\Omega$$

$$\text{F.S.D. current of meter, } I_{m\ (f.s.d.)} = 2\ mA = 2 \times 10^{-3} A$$

(*i*) $$\text{Emitter voltage, } V_E = E - V_{BE} = 5\ V - 0.7\ V = 4.3\ V$$

$\therefore$ $$I_{m\ (f.s.d.)} = \frac{V_E}{R_S + R_m}$$

or $$2 \times 10^{-3} = \frac{4.3V}{1000\Omega + R_S} \qquad \therefore\ R_S = \mathbf{1150\ \Omega}$$

(*ii*) $$\text{Base current, } I_B = \frac{I_{m\ (f.s.d.)}}{\beta} = \frac{2\ mA}{80} = 0.025\ mA$$

$\therefore$ $$\text{Input resistance of voltmeter, } R_i = \frac{E}{I_B} = \frac{5V}{0.025\ mA} = \mathbf{200\ k\Omega}$$

Example 20.10. *The emitter-follower voltmeter circuit in Fig. 20.20 has* $V_{CC} = 20\ V$, $R_S + R_m = 9.3\ k\Omega$, $I_m = 1\ mA$ *and transistor* $\beta = 100$.

(*i*) *Calculate the meter current when* $E = 10V$.

(*ii*) *Determine the voltmeter input resistance with and without the transistor.*

Solution.

(*i*) $$\text{Emitter voltage, } V_E = E - V_{BE} = 10V - 0.7\ V = 9.3\ V$$

$\therefore$ $$\text{Meter current, } I_m = \frac{V_E}{R_S + R_m} = \frac{9.3\ V}{9.3\ k\Omega} = \mathbf{1\ mA}$$

(*ii*) $$\text{Base current, } I_B = \frac{I_m}{\beta} = \frac{1\ mA}{100} = 0.01\ mA$$

$$\text{With transistor, } R_i = \frac{E}{I_B} = \frac{10\ V}{0.01\ mA} = 1000\ k\Omega = \mathbf{1\ M\Omega}$$

$$\text{Without transistor, } R_i = R_S + R_m = \mathbf{9.3\ k\Omega}$$

Note that without transistor, the voltmeter input resistance = $R_S + R_m$ = 9.3 kΩ. However, with transistor, the voltmeter input resistance = 1 MΩ = 1000 kΩ. The obvious advantage of the electronic voltmeter is that its loading effect in voltage measurement will be very small.

Example 20.11. *In the above example, if* $E = 5V$, *all other values remaining the same, what will be the value of meter current? Comment on the result.*

Solution.

$$\text{Meter current, } I_m = \frac{E - V_{BE}}{R_S + R_m} = \frac{5V - 0.7V}{9.3\ k\Omega} = \frac{4.3V}{9.3\ k\Omega} = \mathbf{0.46\ mA}$$

With $E = 5$ V, the meter should read half of full-scale reading *i.e.* 0.5 mA. However, the meter current is actually 0.46 mA. This error is due to V_{BE} and can be eliminated by the modification of the circuit.

20.12 Bridge Rectifier Voltmeter

A permanent-magnet moving coil (PMMC) instrument responds to *average* or d.c. value of current through the moving coil. If alternating current is passed through the moving coil, the driving torque would be *zero. It is because the average value of a sine wave over one cycle is zero. Therefore, a PMMC instrument connected directly to measure a.c. indicates zero reading. In order to measure a.c. with a PMMC instrument, the given a.c. is converted into d.c. by using a bridge rectifier. The instrument is then called rectifier type instrument.

Circuit details. Fig. 20.21 shows bridge rectifier voltmeter for the measurement of a.c. voltages. A multiplier resistor R_S is connected in **series with the PMMC instrument having resistance R_m. When a.c. voltage to be measured is applied to the circuit, full-wave rectification will be obtained as shown in Fig. 20.22. The meter deflection will be proportional to the †average current. Since there is a definite relationship between the average value and r.m.s. value of a sine-wave (r.m.s. value = 1.11 × average value), the meter scale can be calibrated to read the r.m.s. value directly.

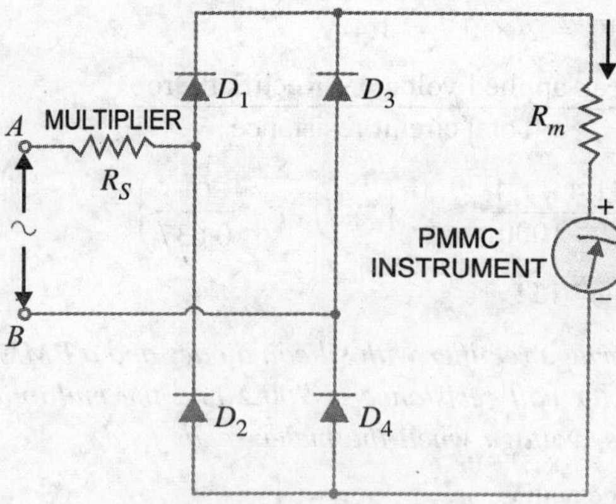

Fig. 20.21

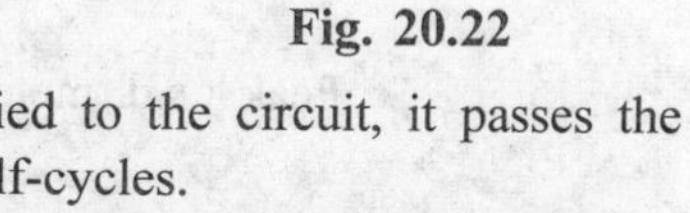

Fig. 20.22

Operation. When a.c. voltage to be measured is applied to the circuit, it passes the positive half-cycles of the input and inverts the negative half-cycles.

(*i*) During the positive half-cycle of the a.c. input voltage, point A is positive *w.r.t.* point B. Therefore, diodes D_1 and D_4 are forward biased while diodes D_2 and D_3 are reverse biased. As a result, diodes D_1 and D_4 conduct and the current follows the path R_S – D_1 – PMMC meter – D_4 – back to point B. Note that multiplier resistor R_S and the meter are in series.

* The driving torque would be in one direction for the positive half-cycle and in the other direction for the negative half-cycle. The inertia of the coil is so great that at supply frequency (50 Hz), the pointer cannot follow the rapid reversals of the driving torque. Therefore, pointer of the meter remains stationary at zero mark.

** If you see carefully, the four diodes (D_1, D_2, D_3 and D_4) form the bridge. Note that the same current flows through the R_S and R_m.

† Note that a voltmeter is a current-operated device.

(*ii*) During the negative half-cycle of the a.c. input voltage, diodes D_2 and D_3 are forward biased while diodes D_1 and D_4 are reverse biased. As a result, diodes D_2 and D_3 conduct and the current follows path D_3 – PMMC meter – D_2 – R_S back to point A. Note that current through the meter is in the same direction as for the positive half-cycle. Consequently, full-wave rectification results.

The scale of the PMMC meter is calibrated to read directly the r.m.s. value of a.c. voltage being measured. It may be noted that rectifier voltmeter can be used only to measure pure sine-wave voltages. When other than pure sine-waves are applied, the meter will *not* indicate the r.m.s. voltage.

Example 20.12. *A PMMC instrument with a full-scale deflection (f.s.d.) current of 100 μA and R_m = 1 kΩ is to be used as a voltmeter of range 0 – 100 V (r.m.s.). The diodes used in the bridge rectifier circuit are of silicon. Calculate the value of multiplier resistor R_S required.*

Solution. Note that 100 μA is the average current.

$$\text{F.S.D. current of meter, } I_{m\,(f.s.d.)} = 100\ \mu\text{A} = 100 \times 10^{-6}\ \text{A} = 10^{-4}\ \text{A}$$

$$\text{Total circuit resistance, } R_T = R_S + R_m = (R_S + 1000)\ \Omega$$

$$\text{Peak value of applied voltage, } V_m = \sqrt{2}\,V_{r.m.s.} = \sqrt{2} \times 100\ \text{V} = 141.4\ \text{V}$$

$$^{*}\ \text{Total rectifier drop} = 2\,V_F = 2 \times 0.7 = 1.4\ \text{V}$$

$$\text{Peak f.s.d. current of meter} = \frac{\text{Peak applied voltage} - \text{Rectifier Drop}}{\text{Total circuit resistance}}$$

$$\text{or} \qquad \frac{10^{-4}}{0.637} = \frac{141.4 - 1.4}{R_S + 1000} \qquad \left(\because\ I_{peak} = \frac{I_{av}}{0.637}\right)$$

$$\therefore \qquad R_S = \mathbf{890.7\ k\Omega}$$

Example 20.13. *An a.c. voltmeter uses a bridge rectifier with silicon diodes and a PMMC instrument with f.s.d. current of 75 μA. If meter coil resistance is 900Ω and the multiplier resistor is 708 kΩ, calculate the applied r.m.s. voltage when the meter reads f.s.d.*

Solution. The PMMC meter reads average value.

$$\therefore \qquad \text{Peak f.s.d. meter current} = \frac{75 \times 10^{-6}}{0.637}\ \text{A} \qquad \left(\because\ I_{peak} = \frac{I_{av}}{0.637}\right)$$

$$\text{Now,} \qquad \text{Peak f.s.d. meter current} = \frac{\text{Peak applied voltage} - \text{Rectifier drop}}{\text{Total circuit resistance}}$$

$$\text{or} \qquad \frac{75 \times 10^{-6}}{0.637} = \frac{\sqrt{2}\,V_{r.m.s.} - 2 \times 0.7}{R_S + R_m}$$

$$\text{or} \qquad 117.74 \times 10^{-6} = \frac{1.414\,V_{r.m.s.} - 1.4}{708 \times 10^3 + 900}$$

$$\therefore \qquad V_{r.m.s.} = \frac{(117.74 \times 10^{-6})\,(708 \times 10^3 + 900) + 1.4}{1.414} = \mathbf{60\ V}$$

---

* During positive or negative half-cycle of input a.c. voltage, two diodes (D_1 and D_4 or D_2 and D_3) are in series.

20.13 Cathode Ray Oscilloscope

The cathode ray oscilloscope (commonly abbreviated as *CRO*) is an electronic device which is capable of giving a visual indication of a signal waveform. No other instrument used in the electronic industry is as versatile as the cathode ray oscilloscope. It is widely used for troubleshooting radio and television receivers as well as for laboratory work involving research and design. With an oscilloscope, the waveshape of a signal can be studied with respect to amplitude distortion and deviation from the normal. In addition, the oscilloscope can also be used for measuring voltage, frequency and phase shift.

In an oscilloscope, the electrons are emitted from a cathode accelerated to a high velocity and brought to focus on a fluorescent screen. The screen produces a visible spot where the electron beam strikes. By deflecting the electron beam over the screen in response to the electrical signal, the electrons can be made to act as an *electrical pencil of light* which produces a spot of light wherever it strikes. An oscilloscope obtains its remarkable properties as a measuring instrument from the fact that it uses as an indicating needle a beam of electrons. As electrons have negligible mass, therefore, they respond almost instantaneously when acted upon by an electrical signal and can trace almost any electrical variation no matter how rapid. A cathode ray oscilloscope contains a *cathode ray tube* and necessary power equipment to make it operate.

20.14 Cathode Ray Tube

A cathode ray tube (commonly abbreviated as *CRT*) is the heart of the oscilloscope. It is a vacuum tube of special geometrical shape and converts an electrical signal into visual one. A cathode ray tube makes available plenty of electrons. These electrons are accelerated to high velocity and are brought to focus on a fluorescent screen. The electron beam produces a spot of light wherever it strikes. The electron beam is deflected on its journey in response to the electrical signal under study. The result is that electrical signal waveform is displayed visually. Fig. 20.23 shows the various parts of cathode ray tube.

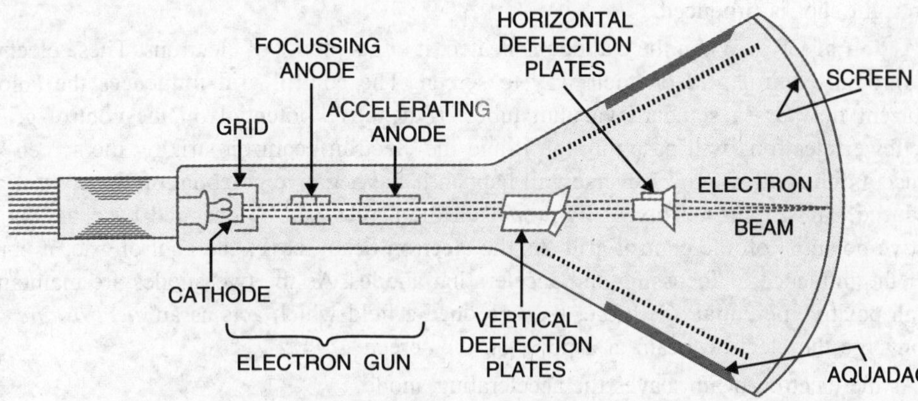

Fig. 20.23

(*i*) **Glass envelope.** It is conical highly evacuated glass housing and maintains vacuum inside and supports the various electrodes. The inner walls of *CRT* between neck and screen are usually coated with a conducting material, called *aquadag.* This coating is electrically connected to the accelerating anode so that electrons which accidentally strike the walls are returned to the anode. This prevents the walls of the tube from charging to a high negative potential.

(ii) **Electron gun assembly.** The arrangement of electrodes which produce a focussed beam of electrons is called the *electron gun.* It essentially consists of an indirectly heated *cathode*, a *control grid*, a *focussing anode* and an *accelerating anode*. The control grid is held at negative potential *w.r.t.* cathode whereas the two anodes are maintained at high positive potential *w.r.t.* cathode.

The cathode consists of a nickel cylinder coated with oxide coating and provides plenty of electrons. The control grid encloses the cathode and consists of a metal cylinder with a tiny circular opening to keep the electron beam small in size. The focussing anode focuses the electron beam into a sharp pin-point by controlling the positive potential on it. The positive potential (about 10,000 V) on the accelerating anode is much higher than on the focussing anode. For this reason, this anode accelerates the narrow beam to a high velocity. Therefore, the electron gun assembly forms a narrow, accelerated beam of electrons which produces a spot of light when it strikes the screen.

(iii) **Deflection plate assembly.** The deflection of the beam is accomplished by two sets of deflecting plates placed within the tube beyond the accelerating anode as shown in Fig. 20.23. One set is the *vertical deflection plates* and the other set is the *horizontal deflection plates*.

The vertical deflection plates are mounted horizontally in the tube. By applying proper potential to these plates, the electron beam can be made to move up and down vertically on the fluorescent screen. The horizontal deflection plates are mounted in the vertical plane. An appropriate potential on these plates can cause the electron beam to move right and left horizontally on the screen.

(iv) **Screen.** The screen is the inside face of the tube and is coated with some fluorescent material such as zinc orthosilicate, zinc oxide *etc.* When high velocity electron beam strikes the screen, a spot of light is produced at the point of impact. The colour of the spot depends upon the nature of fluorescent material. If zinc orthosilicate is used as the fluorescent material, green light spot is produced.

Action of CRT. When the cathode is heated, it emits plenty of electrons. These electrons pass through control grid on their way to screen. The control grid influences the amount of current flow as in standard vacuum tubes. If negative potential on the control grid is high, fewer electrons will pass through it and the electron beam on striking the screen will produce a dim spot of light. Reverse will happen if the negative potential on the control grid is reduced. Thus, the intensity of light spot on the screen can be changed by changing the negative potential on the control grid. As the electron beam leaves the control grid, it comes under the influence of focussing and accelerating anodes. As the two anodes are maintained at high positive potential, therefore, they produce a field which acts as an *electrostatic lens* to converge the electron beam at a point on the screen.

As the electron beam leaves the accelerating anode, it comes under the influence of vertical and horizontal deflection plates. If no voltage is applied to the deflection plates, the electron beam will produce spot of light at the centre (point O in Fig. 20.24) of the screen. If the voltage is applied to vertical plates *only* as shown in Fig. 20.24, the electron beam and hence the spot of light

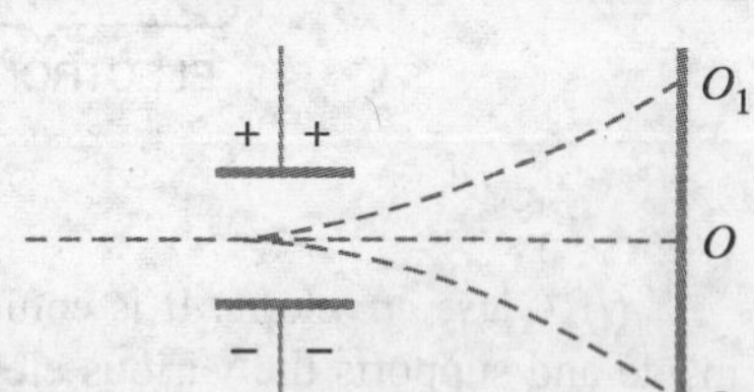

Fig. 20.24

will be deflected upwards (point O_1). The spot of light will be deflected downwards (point O_2) if the potential on the plates is reversed. Similarly, the spot of light can be moved horizontally by applying voltage across the horizontal plates.

20.15 Deflection Sensitivity of CRT

The shift of the spot of light on the screen per unit change in voltage across the deflection plates is known as **deflection sensitivity of CRT**. For instance, if a voltage of 100 V applied to the vertical plates produces a vertical shift of 3 mm in the spot, then deflection sensitivity is 0.03 mm/V. In general,

$$\text{Spot deflection} = \text{Deflection sensitivity} \times \text{Applied voltage}$$

The deflection sensitivity depends not only on the design of the tube but also on the voltage applied to the accelerating anode. The deflection sensitivity is low at high accelerating voltages and vice-versa.

Example 20.14. *The deflection sensitivity of a CRT is 0.01 mm/V. Find the shift produced in the spot when 400 V are applied to the vertical plates.*

Solution. As voltage is applied to the vertical plates only, therefore, the spot will be shifted vertically.

$$\text{Spot shift} = \text{Deflection sensitivity} \times \text{Applied voltage}$$

$$= 0.01 \times 400 = \mathbf{4\ mm}$$

Example 20.15. *The deflection sensitivity of a CRT is 0.03 mm/V. If an unknown voltage is applied to the horizontal plates, the spot shifts 3 mm horizontally. Find the value of unknown voltage.*

Solution. Deflection sensitivity = 0.03 mm/V

Spot shift = 3 mm

Now, Spot shift = Deflection sensitivity × Applied Voltage

$$\therefore \quad \text{Applied voltage} = \frac{\text{Spot shift}}{\text{Deflection sensitivity}} = \frac{3\text{ mm}}{0.03\text{ mm/V}} = \mathbf{100\ V}$$

Example 20.16. *Calculate the maximum velocity of the beam of electrons in a CRT having cathode-anode voltage of 800 V. Assume that the electrons leave the cathode with zero velocity. Charge on electron = 1.6×10^{-19}C and mass of electron = 9.1×10^{-31} kg.*

Solution. Consider an electron of charge e and mass m at rest. If this electron is accelerated through a potential difference of V volts, then potential energy (P.E.) of electron is

$$\text{P.E. of electron} = eV \quad \text{...}(i)$$

The P.E. of electron is converted into kinetic energy (K.E.) of electron and imparts velocity v to the electron.

$$\text{K.E. of electron} = \frac{1}{2}mv^2 \quad \text{...}(ii)$$

From eqs. (*i*) and (*ii*), $\frac{1}{2}mv^2 = eV$

$$\therefore \quad \text{Velocity of electron beam, } v = \sqrt{\frac{2eV}{m}}$$

Here, $e = 1.6 \times 10^{-19}$C ; $m = 9.1 \times 10^{-31}$ kg ; $V = 800$ volts

$$\therefore \quad v = \sqrt{\frac{2 \times 1.6 \times 10^{-19} \times 800}{9.1 \times 10^{-31}}} = \mathbf{16.8 \times 10^6\ m/s}$$

Example 20.17. *An electron moving with a velocity of 10^7 ms^{-1} enters mid-way between two horizontal plates P, Q in a direction parallel to the plates as shown in Fig. 20.25. The length of the plates is 5 cm and their separation is 2 cm. If a p.d. of 90 V is applied between the plates, calculate the transverse deflection produced by the electric field when the electron just passes the field. Assume e/m = 1.8×10^{11} C kg^{-1}.*

Solution. Fig. 20.25 shows the conditions of the problem.

$$\text{Electric field, } E = \frac{V}{d} = \frac{90}{2 \times 10^{-2}} = 45 \times 10^2\ \text{Vm}^{-1}$$

Downward force on the electron = $e\,E$

Downward acceleration of the electron is

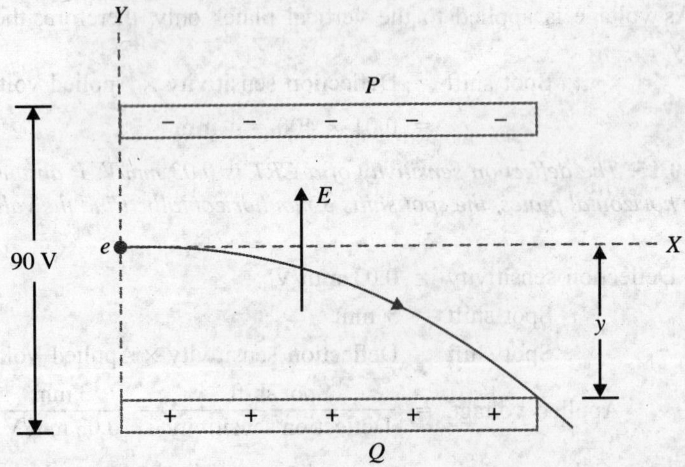

Fig. 20.25

$$a = \frac{eE}{m} = (1.8 \times 10^{11}) \times (45 \times 10^2) = 81 \times 10^{13}\ \text{ms}^{-2}$$

$$\text{Time taken to cross the field, } t = \frac{x}{v} = \frac{5 \times 10^{-2}}{10^7} = 5 \times 10^{-9}\ \text{s}$$

$$\therefore \quad \text{Transverse deflection, } y = \frac{1}{2}at^2 = \frac{1}{2}(81 \times 10^{13}) \times (5 \times 10^{-9})^2 = 0.01\ \text{m}$$

$$= \mathbf{1\ cm}$$

Example 20.18. *An electron with a velocity of 3×10^7 ms^{-1} enters a uniform electric field at right angles to the lines of force. If the field is produced by applying 500 V between two parallel plates 6 cm long and 2 cm apart, at what angle will the electron leave the uniform electric field at the end of the parallel plates ? Assume e/m = 1.8×10^{11} C kg^{-1}.*

Solution. Fig. 20.26 shows the conditions of the problem.

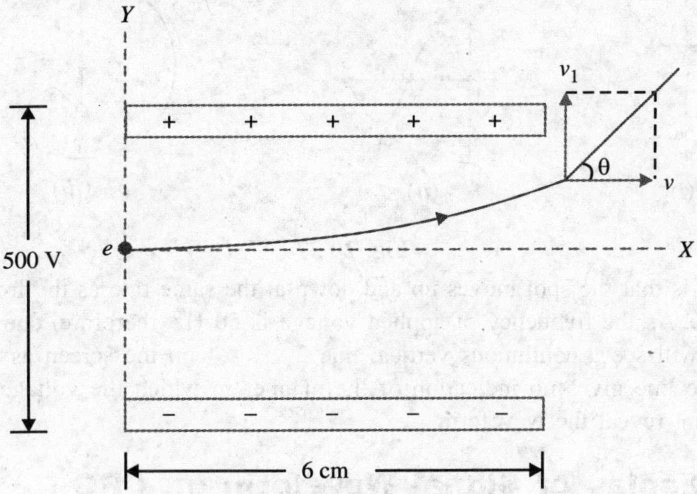

Fig. 20.26

$$\text{Electric field, } E = \frac{V}{d} = \frac{500}{2 \times 10^{-2}} = 2.5 \times 10^{4}\ \text{Vm}^{-1}$$

$$\text{Upward force on the electron} = eE$$

$$\text{Upward acceleration of the electron, } a = \frac{eE}{m} = (1.8 \times 10^{11}) \times 2.5 \times 10^{4}$$

$$= 4.5 \times 10^{15}\ \text{ms}^{-2}$$

$$\text{Time taken to cross the field, } t = \frac{x}{v} = \frac{6 \times 10^{-2}}{3 \times 10^{7}} = 2 \times 10^{-9}\ \text{s}$$

The vertical velocity gained by electron in time t is

$$v_1 = a\,t = (4.5 \times 10^{15}) \times 2 \times 10^{-9}$$

$$= 9 \times 10^{6}\ \text{ms}^{-1}$$

$$\tan\theta = \frac{v_1}{v} = \frac{9 \times 10^{6}}{3 \times 10^{7}} = 0.$$

$$\therefore \quad \theta = \tan^{-1} 0.3 = \mathbf{16.7°}$$

20.16 Applying Signal Across Vertical Plates

If a sinusoidal voltage is applied to the vertical deflection plates, it will make the plates alternately positive and negative. Thus, in the positive half of the signal, upper plate will be positive and lower plate negative while in the negative half-cycle, the plate polarities will be reversed.

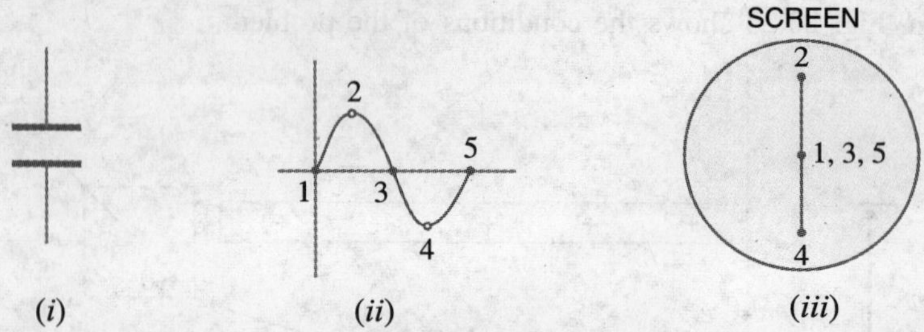

Fig. 20.27

The result is that the spot moves up and down at the same rate as the frequency of the applied voltage. As the frequency of applied voltage is 50 Hz, therefore, due to persistence of vision, we will see a continuous vertical line 2 - 1 - 4 on the screen as shown in Fig. 20.27 (*iii*). The line gives no indication of the manner in which the voltage is alternating since it does not reveal the waveform.

20.17 Display of Signal Waveform on CRO

One interesting application of *CRO* is to present the wave-shape of the signal on the screen. As discussed before, if sinusoidal signal is applied to the vertical deflection plates, we get a vertical line. However, it is desired to see the signal voltage variations with time on the screen. This is possible only if we could also move the beam horizontally from left to right at a uniform speed while it is moving up and down. Further, as soon as a full cycle of the signal is traced, the beam should return quickly to the left hand side of the screen so that it can start tracing the second cycle.

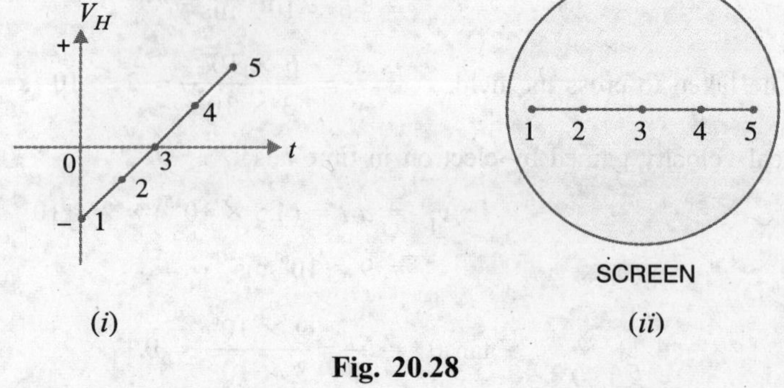

Fig. 20.28

In order that the beam moves from left to right at a uniform rate, a voltage that varies linearly with time should be applied to the horizontal plates. This condition is exactly met in the saw-tooth wave shown in Fig. 20.28 (*i*).

When time $t = 0$, the negative voltage on the horizontal plates keep the beam to the extreme left on the screen as shown in Fig. 20.28 (*ii*). As the time progresses, the negative voltage decreases linearly with time and the beam moves towards right forming a horizontal line. In this way, the saw-tooth wave applied to horizontal plates moves the beam from left to right at a uniform rate.

20.18 Signal Pattern on Screen

If the signal voltage is applied to the vertical plates and saw-tooth wave to the horizontal plates, we get the exact pattern of the signal as shown in Fig. 20.29. When the signal is at the instant 1, its amplitude is zero. But at this instant, maximum negative voltage is applied to horizontal plates. The result is that the beam is at the extreme left on the screen as shown. When the signal is at the instant 2, its amplitude is maximum. However, the negative voltage on the horizontal plates is decreased. Therefore, the beam is deflected upwards by the signal and towards the right by the saw-tooth wave. The result is that the beam now strikes the screen at point 2. On similar reasoning, the beam strikes the screen at points 3, 4 and 5. In this way, we have the exact signal pattern on the screen.

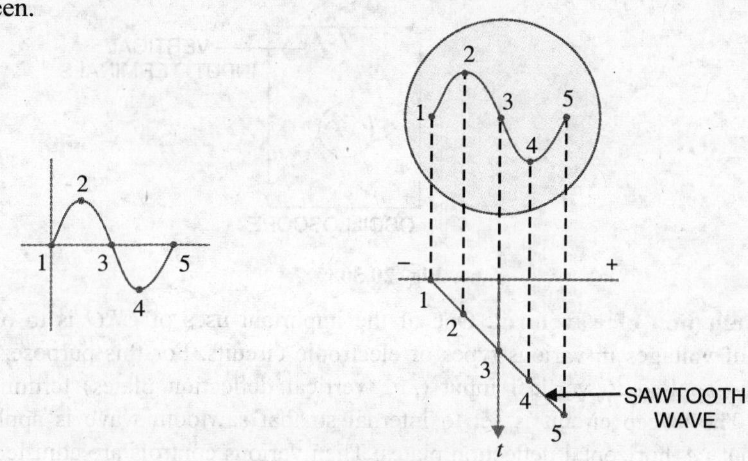

Fig. 20.29

20.19 Various Controls of CRO

In order to facilitate the proper functioning of *CRO*, various controls are provided on the face of *CRO*. A few of them are given below :

(*i*) **Intensity control.** The knob of intensity control regulates the bias on the control grid and affects the electron beam intensity. If the negative bias on the grid is increased, the intensity of electron beam is decreased, thus reducing the brightness of the spot.

(*ii*) **Focus control.** The knob of focus control regulates the positive potential on the focussing anode. If the positive potential on this anode is increased, the electron beam becomes quite narrow and the spot on the screen is a pin-point.

(*iii*) **Horizontal position control.** The knob of horizontal position control regulates the amplitude of d.c. potential which is applied to the horizontal deflection plates, in addition to the usual saw-tooth wave. By adjusting this control, the spot can be moved to right or left as required.

(*iv*) **Vertical position control.** The knob of vertical position control regulates the amplitude of d.c. potential which is applied to the vertical deflection plates in addition to the signal. By adjusting this control, the image can be moved up or down as required.

20.20 Applications of CRO

The modern cathode ray oscilloscope provides a powerful tool for solving problems in electrical measurements. Some important applications of *CRO* are :

1. Examination of waveforms
2. Voltage measurement
3. Frequency measurement

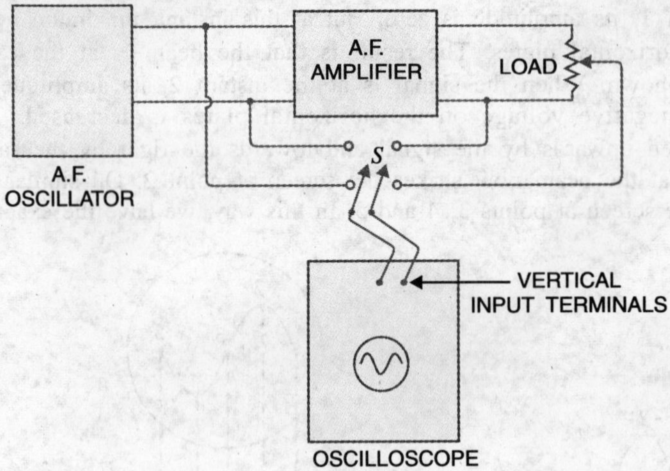

Fig. 20.30

1. **Examination of waveform.** One of the important uses of *CRO* is to observe the wave-shapes of voltages in various types of electronic circuits. For this purpose, the signal under study is applied to vertical input (*i.e.,* vertical deflection plates) terminals of the oscilloscope. The sweep circuit is set to internal so that sawtooth wave is applied to the horizontal input *i.e.* horizontal deflection plates. Then various controls are adjusted to obtain sharp and well defined signal waveform on the screen.

Fig. 20.30 shows the circuit for studying the performance of an audio amplifier. With the help of switch *S*, the output and input of amplifier is applied in turn to the vertical input terminals. If the waveforms are identical in shape, the fidelity of the amplifier is excellent.

2. **Voltage measurement.** As discussed before, if the signal is applied to the vertical deflection plates only, a vertical line appears on the screen. The height of the line is proportional to peak-to-peak voltage of the applied signal. The following procedure is adopted for measuring voltages with *CRO*.

(*i*) Shut off the internal horizontal sweep generator.

(*ii*) Attach a transparent plastic screen to the face of oscilloscope. Mark off the screen with vertical and horizontal lines in the form of graph.

(*iii*) Now, calibrate the oscilloscope against a known voltage. Apply the known voltage, say 10 V, to the vertical input terminals of the oscilloscope. Since the sweep circuit is shut-off, you will get a vertical line. Adjust the vertical gain till a good deflection is obtained. Let the deflection sensitivity be *V* volts/mm.

(*iv*) Keeping the vertical gain unchanged, apply the unknown voltage to be measured to the vertical input terminals of *CRO*.

(*v*) Measure the length of the vertical line obtained. Let it be *l* mm.

Then, Unknown voltage $= l \times V$ volts

3. **Frequency measurement.** The unknown frequency can be accurately determined with the help of a *CRO*. The steps of the procedure are as under :

(*i*) A known frequency is applied to horizontal input and unknown frequency to the vertical input.

(*ii*) The various controls are adjusted.

(*iii*) A pattern with loops is obtained.

(*iv*) The number of loops cut by the horizontal line gives the frequency on the vertical plates (f_v) and the number of loops cut by the vertical line gives the frequency on the horizontal plates (f_H).

$$\therefore \quad \frac{f_v}{f_H} = \frac{\text{No. of loops cut by horizontal line}}{\text{No. of loops cut by vertical line}}$$

For instance, suppose during the frequency measurement test, a pattern shown in Fig. 20.31 is obtained. Let us further assume that frequency applied to horizontal plates is 2000 Hz. If we draw horizontal and vertical lines, we find that one loop is cut by the horizontal line and two loops by the vertical line. Therefore,

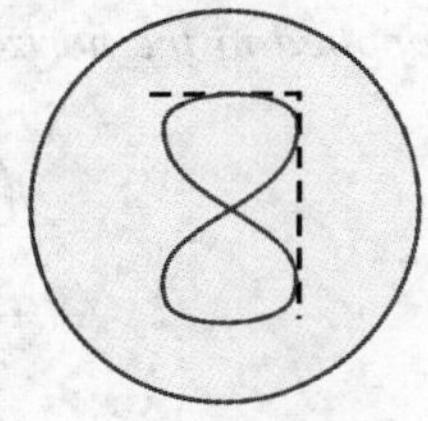

Fig. 20.31

$$\frac{f_v}{f_H} = \frac{\text{No. of loops cut by horizontal line}}{\text{No. of loops cut by vertical line}}$$

or $$\frac{f_v}{2000} = \frac{1}{2}$$

or $$f_v = 2000 \times 1/2 = 1000 \text{ Hz}$$

i.e. Unknown frequency is 1000 Hz.

Example 20.19. *In an oscilloscope, 200 V, 50 Hz signal produces a deflection of 2 cm corresponding to a certain setting of vertical gain control. If another voltage produces 3 cm deflection, what is the value of this voltage ?*

Solution. Deflection sensitivity = 200 V/2 cm = 100 V/cm

Unknown voltage = D.S. × Deflection = 100 × 3 = **300 V**

Example 20.20. *When signals of different frequencies were applied to the vertical input terminals of oscilloscope, the patterns shown in Fig. 20.32 were obtained. If the frequency applied to horizontal plates in each case is 1000Hz, determine the unknown frequency.*

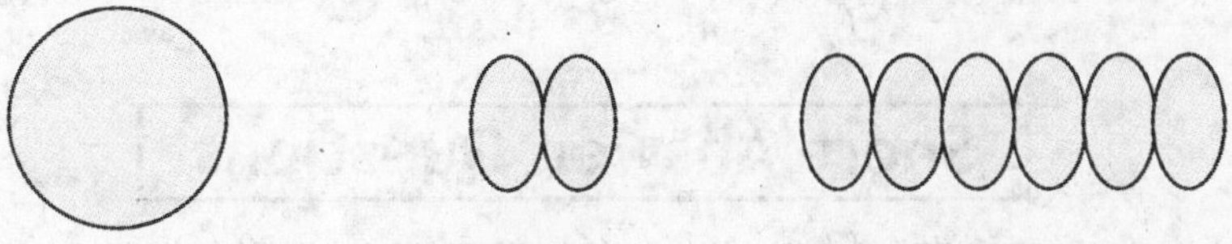

(*i*) (*ii*) (*iii*)

Fig. 20.32

Solution.

(*i*) The number of loops cut by horizontal and vertical lines is one.

$$\therefore \quad \frac{f_v}{f_H} = \frac{1}{1} \quad \text{or} \quad f_v = f_H = \mathbf{1000\ Hz}$$

(*ii*) The number of loops cut by horizontal line is 2 and the number of loops cut by vertical line is 1.

$$\therefore \quad \frac{f_v}{f_H} = \frac{2}{1} \quad \text{or} \quad f_v = 2 \times f_H = 2 \times 1000 = \mathbf{2000\ Hz}$$

(*iii*) The number of loops cut by the horizontal line is 6 and that by vertical line is 1.

$$\therefore \quad \frac{f_v}{f_H} = \frac{6}{1}$$

$$\text{or} \quad f_v = 6\, f_H = 6 \times 1000 = \mathbf{6000\ Hz}$$

Example 20.21. *When signals of different frequencies were applied to the vertical input terminals of oscilloscope, the patterns obtained were as shown in Fig. 20.33. If the frequency applied to the horizontal plates in each case is 1500 Hz, determine the unknown frequency.*

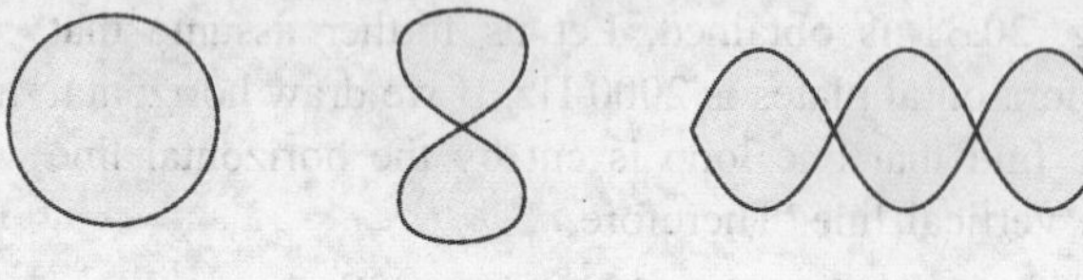

(*i*) (*ii*) (*iii*)

Fig. 20.33

Solution. (*i*) The number of loops cut by the horizontal and vertical lines is one [See Fig. 20.33 (*i*)].

$$\therefore \quad \frac{f_v}{f_H} = \frac{1}{1} \quad \text{or } f_v = f_H = \mathbf{1500\ Hz}$$

(*ii*) The number of loops cut by the horizontal line is one and the number of loops cut by the vertical line is two [See Fig. 20.33 (*ii*)].

$$\therefore \quad \frac{f_v}{f_H} = \frac{1}{2} \quad \text{or} \quad f_v = \frac{f_H}{2} = \frac{1500}{2} = \mathbf{750\ Hz}$$

(*iii*) The number of loops cut by the horizontal line is three and the number of loops cut by the vertical line is one [See Fig. 20.33 (*iii*)].

$$\therefore \quad \frac{f_v}{f_H} = \frac{3}{1} \quad \text{or} \quad f_v = 3f_H = 3 \times 1500 = \mathbf{4500\ Hz}$$

Short Answer Questions

Q. 1. *Why is the sensitivity of best multimeter not more than 20 kΩ per volt ?*

Ans. There are two reasons for it. First, the input resistance/impedance of a multimeter is not very high and there is considerable loading effect of the meter. Secondly, considerable amount of power is drawn from the circuit under measurement.

Q. 2. *Why do we generally prefer VTVM to multimeter for measurements in electronic circuits ?*

Ans. It is because *VTVM* has very high input resistance/impedance and possesses the property of amplification. The latter property permits the input signal to be amplified so that the power to operate the indicating mechanism comes from a source other than the measured circuit.

Q. 3. *Why does oscilloscope give more accurate measurements than a VTVM ?*

Ans. An oscilloscope obtains its remarkable properties as a measuring instrument from the fact that it uses a beam of electrons as an indicating needle. As electrons have negligible mass, therefore, they respond almost instantaneously when acted upon by an electrical signal and can trace almost any electrical variation no matter how rapid.

Q. 4. *What are the three main advantages of VTVM over the conventional type voltmeters ?*

Ans. The following are the three main advantages of *VTVM* over the conventional type voltmeters :

(*i*) A *VTVM* draws an extremely small current from the measuring circuit. Therefore, it gives accurate readings.

(*ii*) There is little effect of temperature variations.

(*iii*) A *VTVM* has a wide frequency response.

Q. 5. *What are the advantages of balanced bridge type VTVM over a single triode voltmeter ?*

Ans. The following are the advantages of balanced bridge type *VTVM* over a single triode voltmeter :

(*i*) Effects of changes of variations of tube characteristics are minimised.

(*ii*) Meter zero has less tendency of shifting during operation.

(*iii*) Power supply fluctuations have little effect on the measuring circuit.

Multiple-Choice Questions

1. An ammeter is connected in .... with the circuit element whose current we wish to measure.

(*i*) series (*ii*) parallel

(*iii*) series or parallel

(*iv*) none of the above

2. A galvanometer in series with a high resistance is called ....

(*i*) an ammeter (*ii*) a voltmeter

(*iii*) a wattmeter

(*iv*) none of the above

3. An ammeter should have .... resistance.

(*i*) infinite (*ii*) very large

(*iii*) very low

(*iv*) none of the above

4. A voltmeter is connected in .... with the circuit component across which potential difference is to be measured.

(*i*) parallel (*ii*) series

(*iii*) series or parallel

(*iv*) none of the above

5. A voltmeter should have .... resistance.

(*i*) zero (*ii*) very high

(*iii*) very low

(*iv*) none of the above

6. The sensitivity of a multimeter is given in ....

(*i*) Ω (*ii*) amperes

(*iii*) kΩ/V

(*iv*) none of the above

7. If the full-scale deflection current of a multimeter is 50 μA, its sensitivity is ....

(*i*) 10 kΩ/V (*ii*) 100 kΩ/V

(*iii*) 50 kΩ/V (*iv*) 20 kΩ/V

8. If a multimeter has a sensitivity of 1000 Ω per volt and reads 50 V full-scale, its internal resistance is..............

(*i*) 20 kΩ (*ii*) 50 kΩ

(*iii*) 10 kΩ

(*iv*) none of the above

9. A *VTVM* has .... input resistance than that of a multimeter.
   (*i*) more (*ii*) less
   (*iii*) same
   (*iv*) none of the above
10. The input resistance of a *VTVM* is about ....
   (*i*) 1000 Ω (*ii*) 10 kΩ
   (*iii*) 20 kΩ (*iv*) 10 MΩ
11. If the negative potential on the control grid of *CRT* is increased, the intensity of spot ....
   (*i*) is increased (*ii*) is decreased
   (*iii*) remains the same
   (*iv*) none of the above
12. For display of signal pattern, ... voltage is applied to the horizontal plates of a *CRO*.
   (*i*) sinusoidal (*ii*) rectangular
   (*iii*) sawtooth
   (*iv*) none of the above
13. Two multimeters *A* and *B* have sensitivities of 10 kΩ/V and 30 kΩ/V respectively. Then, ...
   (*i*) multimeter *A* is more sensitive
   (*ii*) multimeter *B* is more sensitive
   (*iii*) both are equally sensitive
   (*iv*) none of the above
14. A galvanometer of resistance *G* is shunted by a very small resistance *S*. The resistance of the resulting ammeter is ....
   (*i*) $\frac{GS}{G+S}$ (*ii*) $G + S$
   (*iii*) $G - S$
   (*iv*) none of the above
15. A *VTVM* is never used to measure ....
   (*i*) voltage (*ii*) current
   (*iii*) resistance
   (*iv*) none of the above
16. The sensitivity of a voltmeter which uses a 100 μA meter movement is ....
   (*i*) 1 kΩ/V (*ii*) 10 kΩ/V
   (*iii*) 5 kΩ/V (*iv*) data insufficient
17. What is the total resistance of a voltmeter on the 10 V range when the meter movement is rated for 50 μA of full-scale current?
   (*i*) 10 kΩ (*ii*) 20 kΩ
   (*iii*) 200 kΩ
   (*iv*) None of the above
18. The material used to coat inside face of *CRT* is ....
   (*i*) carbon (*ii*) sulphur
   (*iii*) silicon (*iv*) phosphorus
19. When an ammeter is inserted in the circuit, the circuit current will ....
   (*i*) increase (*ii*) decrease
   (*iii*) remain the same
   (*iv*) none of the above
20. A series ohmmeter circuit uses a 3 V battery and a 1 mA meter movement. What is the half-scale resistance for this movement?
   (*i*) 3 kΩ (*ii*) 1.5 kΩ
   (*iii*) 4.5 kΩ (*iv*) 6 kΩ
21. The most accurate device for measuring voltage is ....
   (*i*) voltmeter (*ii*) multimeter
   (*iii*) *CRO* (*iv*) *VTVM*
22. The horizontal plates of a *CRO* are supplied with .... to observe the waveform of a signal.
   (*i*) sinusoidal wave
   (*ii*) cosine wave (*iii*) sawtooth wave
   (*iv*) none of the above
23. A *CRO* is used to measure ....
   (*i*) voltage (*ii*) frequency
   (*iii*) phase (*iv*) all of above
24. If 2 % of the main current is to be passed through a galvanometer of resistance *G*, then resistance of the shunt required is....
   (*i*) *G*/50 (*ii*) *G*/49
   (*iii*) 49 *G* (*iv*) 50 *G*

25. Which of the following is likely to have the largest resistance ?
    (*i*) Voltmeter of range 10 V
    (*ii*) Moving coil galvanometer
    (*iii*) Ammeter of range 1 A
    (*iv*) A copper wire of length 1 m and diameter 3 mm

26. An ideal ammeter has .... resistance.
    (*i*) low (*ii*) infinite
    (*iii*) zero (*iv*) high

27. The resistance of an ideal voltmeter is ....
    (*i*) low (*ii*) infinite
    (*iii*) zero (*iv*) high

28. To send 10% of the main current through a moving coil galvanometer of resistance 99 Ω, the shunt required is ....
    (*i*) 11 Ω (*ii*) 9.9 Ω
    (*iii*) 100 Ω (*iv*) 9 Ω

29. A voltmeter has a resistance of *G* ohms and range *V* volts. The value of resistance required in series to convert it into voltmeter of range *nV* is ....
    (*i*) nG (*ii*) $\frac{G}{n}$
    (*iii*) $\frac{G}{n-1}$ (*iv*) $(n-1)\,G$

30. An ammeter has a resistance of *G* ohms and range of *I* amperes. The value of resistance required in parallel to convert it into an ammeter of range *nI* is ....
    (*i*) nG (*ii*) $(n-1)\,G$
    (*iii*) $\frac{G}{n-1}$ (*iv*) $\frac{G}{n}$

Answers to Multiple-Choice Questions

1. (*i*)	2. (*ii*)	3. (*iii*)	4. (*i*)	5. (*ii*)
6. (*iii*)	7. (*iv*)	8. (*ii*)	9. (*i*)	10. (*iv*)
11. (*ii*)	12. (*iii*)	13. (*ii*)	14. (*i*)	15. (*ii*)
16. (*ii*)	17. (*iii*)	18. (*iv*)	19. (*ii*)	20. (*i*)
21. (*iii*)	22. (*iii*)	23. (*iv*)	24. (*ii*)	25. (*i*)
26. (*iii*)	27. (*ii*)	28. (*i*)	29. (*iv*)	30. (*iii*)

Chapter Review Topics

1. What is a multimeter ? How does it work ?
2. What type of measurements can be made with a multimeter ? Explain with suitable diagrams.
3. Briefly explain the advantages of 20 kΩ/volt multimeter as compared to a 10 kΩ/volt multimeter.
4. What are the applications of a multimeter ?
5. Discuss the advantages and disadvantages of a multimeter.
6. What is a *VTVM* ? Explain balanced bridge Type *VTVM* with a neat circuit diagram.
7. What are the applications of *VTVM* ?
8. Discuss the advantages and disadvantages of *VTVM*.
9. Briefly explain the differences between a *VTVM* and a multimeter.
10. Explain the construction and working of a cathode ray tube.
11. How will you make the following measurements with a *CRO* :
    (*i*) voltage (*ii*) frequency ?
12. Write short notes on the following :
    (*i*) Limitations of multimeter

(*ii*) Advantages of oscilloscope

(*iii*) Vacuum tube voltmeter

(*iv*) Oscilloscope controls

Problems

1. A voltmeter is used to measure voltage across 20 kΩ resistor as shown in Fig. 20.34. What will be the voltage value if (*i*) voltmeter has infinite resistance (*ii*) voltmeter has a sensitivity of 1000 Ω per volt and reads 100 V full-scale ? **[(*i*) 50 V (*ii*) 45 V]**

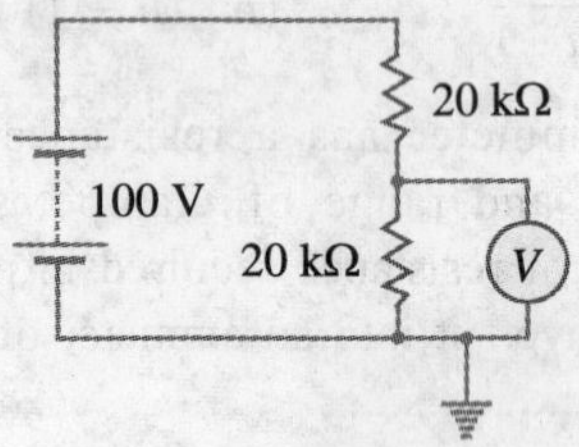

Fig. 20.34

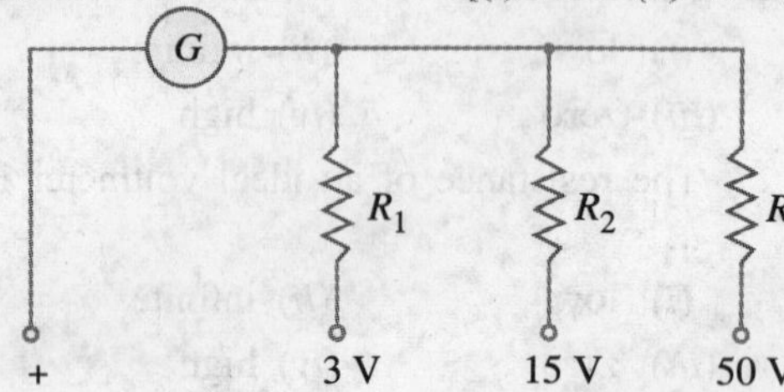

Fig. 20.35

2. The three range voltmeter is arranged as shown in Fig. 20.35. The ranges are 0 to 3 V, 0 to 15 V and 0 to 50 V as marked. If the full-scale deflection current is 10 mA, what should be the values of R_1, R_2 and R_3 ? The resistance of the meter is 5 Ω. **[305 Ω, 1505 Ω, 5005 Ω]**
3. If the sensitivity of voltmeter in Fig. 20.34 is 500 Ω/volt (Full-scale reading being 100 V), what will be the reading of the voltmeter ? **[41.7 V]**
4. What is the lowest full-scale voltage that could be displayed with a 100 μA meter movement with an internal resistance of 150 Ω ? What would be the sensitivity of this meter in ohms per volt ? **[15 mV, 10,000 Ω/V]**
5. If a 20,000 Ω/V meter with 5 kΩ internal resistance is used in an ohmmeter with a 3-V-battery, what internal resistance is required in the meter to produce proper zeroing? **[60 kΩ]**
6. A PMMC instrument with f.s.d. = 100 μA and R_m = 1 kΩ is to be used as an a.c. voltmeter with f.s.d. = 100 V (r.m.s.) as shown in Fig. 20.21. Silicon diodes are used in the bridge rectifier circuit. Calculate the pointer indications for the voltmeter when the r.m.s. input voltage is (*i*) 75 V (*ii*) 50 V. **[0.75 f.s.d. ; 0.5 f.s.d.]**
7. In the above example, calculate the voltmeter sensitivity. **[9 kΩ/V]**
8. Two metal plates are 4 cm long and 3 cm apart in vacuum, one being vertically above the other. The p.d. between the plates is 300 V. An electron having a velocity of 10^7 ms^{-1} is injected horizontally midway between the plates and in a direction parallel to 4 cm edge. Calculate the transverse deflection of the electron as it leaves the plates. Assume *e*/*m* = 1.8×10^{11} C kg^{-1}. **[1.44 cm]**
9. A beam of electrons moving with a velocity 1.0×10^7 ms^{-1} enters midway between two horizontal plates in a direction parallel to the plates. The plates are 5 cm long, 2 cm apart and have a p.d. *V* applied between them. Calculate the value of *V* if the beam is deflected so that it just grazes the edge of lower plate. Assume *e*/*m* = 1.8×10^{11}C kg^{-1}. **[89 V]**

   [Hint : Transverse deflection *y* = 1 cm.]
10. A beam of electrons moving with a speed of 3×10^7 ms^{-1} is deflected by 2 mm in traversing through a distance of 0.1 m in a uniform electric field of 1800 Vm^{-1}. Calculate *e*/*m* for electrons. **[2×10^{11}C kg^{-1}]**

CHAPTER 21

Integrated Circuits

Learning Objectives

- Understand the need of integrated circuits (*ICs*).
- Develop an understanding of monolithic *IC* and its fabrication.
- Learn the different types of *IC* packings.
- Explore the working of some simple circuits using *ICs*.

Introduction

The circuits discussed so far in the text consisted of separately manufactured components (*e.g.* resistors, capacitors, diodes, transistors *etc.*) joined by wires or plated conductors on printed boards. Such circuits are known as *discrete circuits* because each component added to the circuit is discrete (*i.e.* distinct or separate) from the others. Discrete circuits have two main disadvantages. Firstly, in a large circuit (*e.g.* TV circuit, computer circuit) there may be hundreds of components and consequently discrete assembly would occupy a large space. Secondly, there will be hundreds of soldered points posing a considerable problem of reliability. To meet these problems of space conservation and reliability, engineers started a drive for miniaturized circuits. This led to the development of *microelectronics* in the late 1950's.

Microelectronics is the branch of electronics engineering which deals with micro-circuits. A micro-circuit is simply a miniature assembly of electronic components. One type of such circuit is the *integrated circuit*, generally abbreviated as *IC*. An integrated circuit has various components such as resistors, capacitors, diodes, transistors *etc.* fabricated on a small semiconductor chip. How circuits containing hundreds of components are fabricated on a small semiconductor chip to produce an *IC* is a fascinating feat of microelectronics. This has not only fulfilled the everincreasing demand of industries for electronic equipment of smaller size, lighter weight and low power requirements, but it has also resulted in high degree of reliability. In this chapter, we shall focus our attention on the various aspects of integrated circuits.

21.1 Integrated Circuit

An **integrated circuit** *is one in which circuit components such as transistors, diodes, resistors, capacitors etc. are automatically part of a small semiconductor chip.*

An integrated circuit consists of a number of circuit components (*e.g.* transistors, diodes, resistors *etc.*) and their inter connections in a single small package to perform a complete electronic function. These components are formed and connected within a small chip of semi-conductor material. The following points are worth noting about integrated circuits :

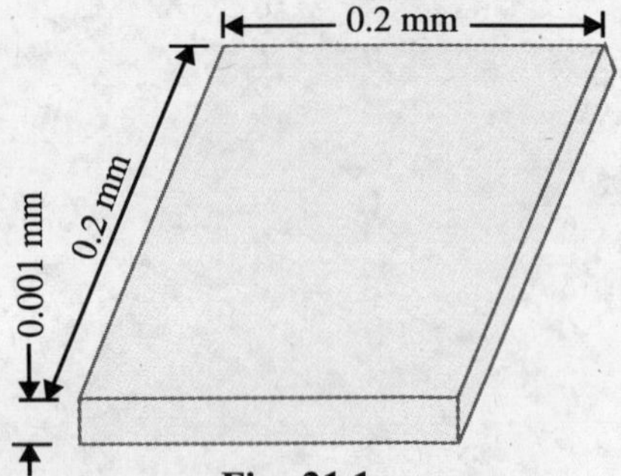

Fig. 21.1

(*i*) In an *IC*, the various components are automatically part of a small semi-conductor chip and the individual components cannot be removed or replaced. This is in contrast to discrete assembly in which individual components can be removed or replaced if necessary.

(*ii*) The size of an **IC* is extremely small. In fact, *ICs* are so small that you normally need a microscope to see the connections between the components. Fig. 21.1 shows a typical semi-conductor chip having dimensions 0.2 mm × 0.2 mm × 0.001 mm. It is possible to produce circuits containing many transistors, diodes, resistors *etc.* on the surface of this small chip.

(*iii*) No components of an *IC* are seen to project above the surface of the chip. This is because all the components are formed within the chip.

21.2 Advantages and Disadvantages of Integrated Circuits

Integrated circuits free the equipment designer from the need to construct circuits with individual discrete components such as transistors, diodes and resistors. With the exception of a few very simple circuits, the availability of a large number of low-cost integrated circuits have largely rendered discrete circuitry obsolete. It is, therefore, desirable to mention the significant advantages of integrated circuits over discrete circuits. However, integrated circuits have some disadvantages and continuous efforts are on to overcome them.

Advantages. Integrated circuits possess the following advantages over discrete circuits :

(*i*) Increased reliability due to lesser number of connections.

(*ii*) Extremely small size due to the fabrication of various circuit elements in a single chip of semi-conductor material.

(*iii*) Lesser weight and **space requirement due to miniaturized circuit.

(*iv*) Low power requirements.

(*v*) Greater ability to operate at extreme values of temperature.

(*vi*) Low cost because of simultaneous production of hundreds of alike circuits on a small semi-conductor wafer.

(*vii*) The circuit layout is greatly simplified because integrated circuits are constrained to use minimum number of external connections.

Disadvantages. The disadvantages of integrated circuits are :

(*i*) If any component in an *IC* goes out of order, the whole *IC* has to be replaced by the new one.

---

* Since it combines both active (*e.g.*, transistors, diodes *etc.*) and passive elements (*e.g.*, resistors, capacitors *etc.*) in a monolithic structure, the complete unit is called an integrated circuit.

** Typically, this is about 10% of the space required by comparable discrete assembly.

(*ii*) In an *IC*, it is neither convenient nor economical to fabricate capacitances exceeding 30 pF. Therefore, for high values of capacitance, discrete components exterior to *IC* chip are connected.

(*iii*) It is not possible to fabricate inductors and transformers on the surface of semi-conductor chip. Therefore, these components are connected exterior to the semi-conductor chip.

(*iv*) It is not possible to produce high power *ICs* (greater than 10 W).

(*v*) There is a lack of flexibility in an *IC i.e.*, it is generally not possible to modify the parameters within which an integrated circuit will operate.

21.3 Inside an IC Package

The *IC* units are fast replacing the discrete components in all electronic equipment. These are similar to the discrete circuits that they replace. However, there are some points to be noted. An integrated circuit (*IC*) usually contains only transistors, diodes and resistors. It is usually very difficult to form inductors in an *IC*. Also, only very small capacitors, in the picofarad range, can be included. When inductors and large values of *C* are needed, they are connected externally to an *IC*. The various components in an *IC* are so small that they cannot be seen with a naked eye. Therefore, individual components cannot be removed or replaced. If a single component within an *IC* fails, the complete *IC* is replaced. When studying circuits using *IC*s, we are more concerned with the external connections to the *IC*s than with what is actually going on inside. We cannot get into an *IC* to repair its internal circuitry.

21.4 IC Classifications

Four basic types of constructions are employed in the manufacture of integrated circuits, namely ;

(*i*) Mono-lithic (*ii*) Thin-film (*iii*) Thick-film (*iv*) Hybrid.

Monolithic *ICs* are by far the most common type used in practice. Therefore, in this chapter we shall confine our attention to the construction of this type of *ICs* only. It may be worthwhile to mention here that regardless of the type of method used to fabricate active and passive components, the basic characteristics and circuit operation of an *IC* are the same as for any of their counterparts in a similar circuit using separate circuit components.

21.5 Making Monolithic IC

A **monolithic** IC *is one in which all circuit components and their inter-connections are formed on a single thin wafer called the substrate.*

The word monolithic is from Greek language and means "one stone." The word is appropriate because all the components are part of one chip. Although we are mainly interested in using *ICs*, yet it is profitable to know something about their fabrication. The basic production processes for the monolithic *ICs* are as follow :

(*i*) **p-Substrate.** This is the first step in the making of an *IC*. A cylindrical *p*-type *silicon crystal is grown having typical dimensions 25 cm long and 2.5 cm diameter [See Fig. 21.2 (*i*)]. The crystal is then cut by a diamond saw into many thin wafers like Fig. 21.2 (*ii*), the typical thickness of the wafer being 200 μm. One side of wafer is polished to get rid of surface imperfections. This wafer is called the substrate. The *ICs* are produced on this wafer.

* Since silicon possesses characteristics which are best suited to *IC* manufacturing processes.

(i) (ii)

Fig. 21.2

***(ii)* Epitaxial n layer.** The next step is to put the wafers in a diffusion furnace. A gas mixture of silicon atoms and pentavalent atoms is passed over the wafers. This forms a thin layer of *n*-type semi-conductor on the heated surface of substrate [See Fig. 21.3 (*i*)]. This thin layer is called the **epitaxial layer* and is about 10 μm thick. It is in this layer that the whole integrated circuit is formed.

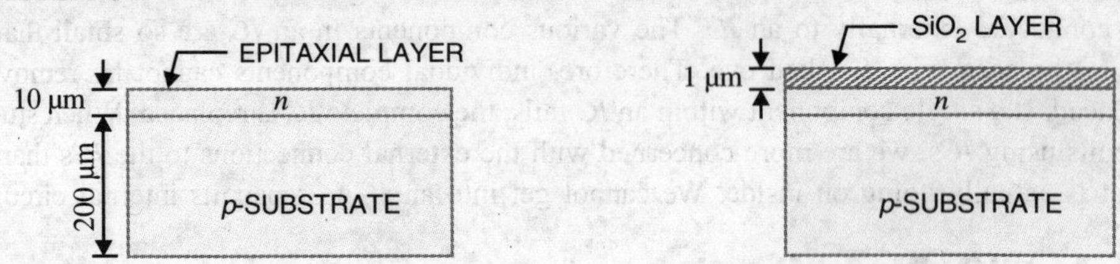

(i) (ii)

Fig. 21.3

***(iii)* Insulating layer.** In order to prevent the contamination of the epitaxial layer, a thin SiO_2 layer about 1μm thick is deposited over the entire surface as shown in Fig. 21.3 (*ii*). This is achieved by passing pure oxygen over the epitaxial layer. The oxygen atoms combine with silicon atoms to form a layer of silicon dioxide (SiO_2).

***(iv)* Producing components.** By the process of **diffusion, appropriate materials are added to the substrate at specific locations to produce diodes, transistors, resistors and capacitors. The production of these components on the wafer is discussed in Art. 21.6.

***(v)* Etching.** Before any impurity is added to the substrate, the oxide layer (*i.e.* SiO_2 layer) is etched. The process of etching exposes the epitaxial layer and permits the production of desired components. The terminals are processed by etching the oxide layer at the desired locations.

***(vi)* Chips.** In practice, the wafer shown in Fig. 21.4 is divided into a large number of areas. Each of these areas will be a separate chip. The manufacturer produces hundreds of alike *ICs* on the wafer over each area. To separate the individual *ICs*,

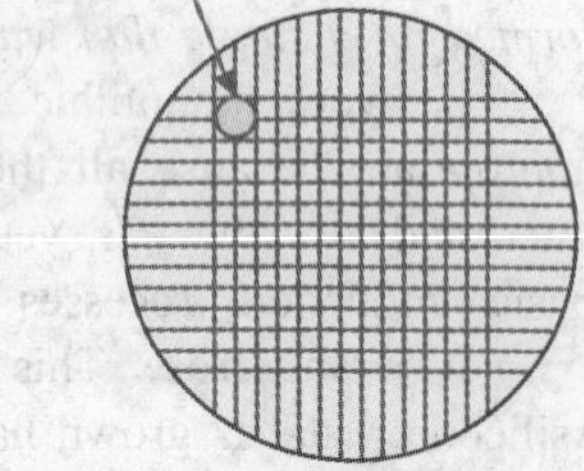

Fig. 21.4

* The word "epitaxial" is derived from the Greek language and means arranged upon.

** In *IC* construction, diffusion is the process of deliberately adding controlled impurities at specific locations of substrate by thermal processes.

the wafer is divided into small chips by a process similar to glass cutting. This is illustrated in Fig. 21.4. It may be seen that hundreds of alike *ICs* can be produced from a small wafer. This simultaneous mass production is the reason for the low cost of integrated circuits.

After the chip is cut, it is bonded to its mounting and connections are made between the *IC* and external leads. The *IC* is then encapsulated to prevent it from becoming contaminated by the surrounding atmosphere.

21.6 Fabrication of Components on Monolithic IC

The notable feature of an *IC* is that it comprises a number of circuit elements inseparably associated in a single small package to perform a complete electronic function. This differs from discrete assembly where separately manufactured components are joined by wires. We shall now see how various circuit elements (*e.g.* diodes, transistors, resistors *etc.*) can be constructed in an *IC* form.

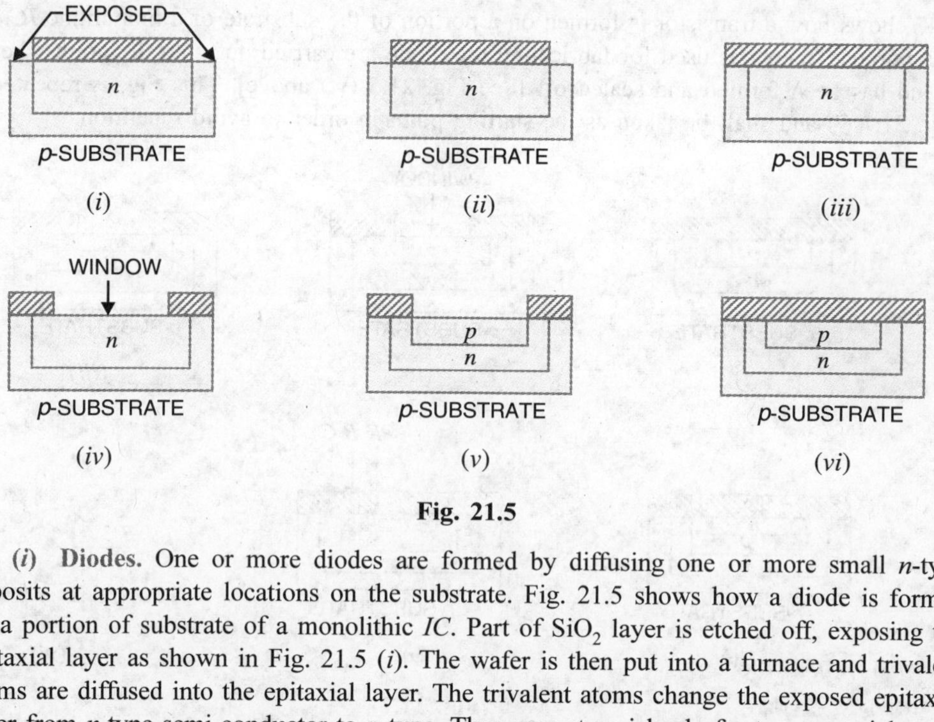

Fig. 21.5

(*i*) **Diodes.** One or more diodes are formed by diffusing one or more small *n*-type deposits at appropriate locations on the substrate. Fig. 21.5 shows how a diode is formed on a portion of substrate of a monolithic *IC*. Part of SiO_2 layer is etched off, exposing the epitaxial layer as shown in Fig. 21.5 (*i*). The wafer is then put into a furnace and trivalent atoms are diffused into the epitaxial layer. The trivalent atoms change the exposed epitaxial layer from *n*-type semi-conductor to *p*-type. Thus we get an island of *n*-type material under the SiO_2 layer as shown in Fig. 21.5 (*ii*).

Next pure oxygen is passed over the wafer to form a complete SiO_2 layer as shown in Fig. 21.5 (*iii*). A hole is then etched at the centre of this layer ; thus exposing the *n*-epitaxial layer [See Fig. 21.5 (*iv*)]. This hole in SiO_2 layer is called a **window**. Now we pass trivalent atoms through the window. The trivalent atoms diffuse into the epitaxial layer to form an island of *p*-type material as shown in Fig. 21.5 (*v*). The SiO_2 layer is again formed on the wafer by blowing pure oxygen over the wafer [See Fig. 21.5 (*vi*)]. Thus a *p-n* junction diode is formed on the substrate.

The last step is to attach the terminals. For this purpose, we etch the SiO_2 layer at the desired locations as shown in Fig 21.6 (*i*). By depositing metal at these locations, we make electrical contact with the anode and cathode of the integrated diode. Fig. 21.6 (*ii*) shows the electrical circuit of the diode.

Fig. 21.6

(*ii*) **Transistors.** Transistors are formed by using the same principle as for diodes. Fig. 21.7 shows how a transistor is formed on a portion of the substrate of a monolithic *IC*. For this purpose, the steps used for fabricating the diode are carried out upto the point where *p* island has been formed and sealed off [See Fig. 21.5 (*vi*) above]. This Fig. is repeated as Fig. 21.7 (*i*) and shall be taken as the starting point in order to avoid repetition.

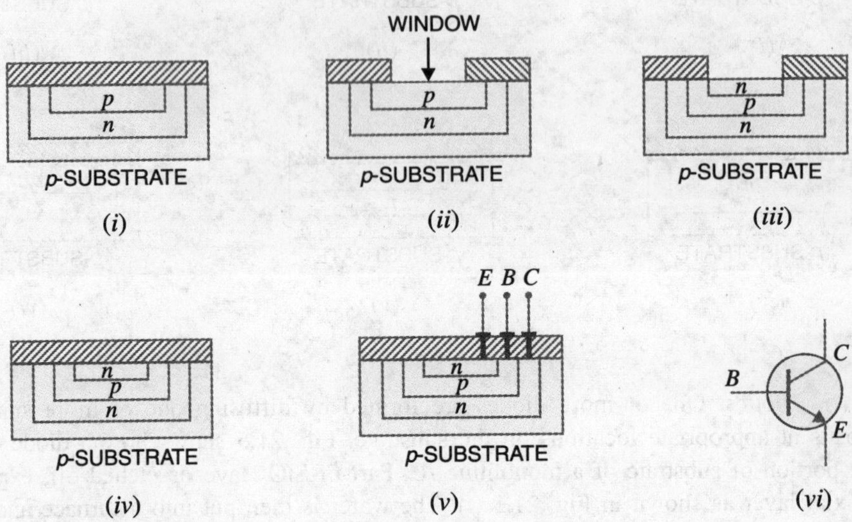

Fig. 21.7

A window is now formed at the centre of SiO_2 layer, thus exposing the *p*-epitaxial layer as shown in Fig. 21.7 (*ii*). Then we pass pentavalent atoms through the window. The pentavalent atoms diffuse into the epitaxial layer to form an island of *n*-type material as shown in Fig. 21.7 (*iii*). The SiO_2 layer is re-formed over the wafer by passing pure oxygen [See Fig. 21.7 (*iv*)]. The terminals are processed by etching the SiO_2 layer at appropriate locations and depositing the metal at these locations as shown in Fig. 21.7 (*v*). In this way, we get the integrated transistor. Fig. 21.7 (*vi*) shows the electrical circuit of a transistor.

(*iii*) **Resistors.** Fig. 21.8 shows how a resistor is formed on a portion of the substrate of a monolithic *IC*. For this purpose, the steps used for fabricating diode are carried out upto the point where n island has been formed and sealed off [Refer back to Fig. 21.5 (*iii*)]. This figure is repeated as Fig. 21.8 (*i*) and shall be taken as the starting point in order to avoid repetition.

A window is now formed at the centre of SiO_2 layer, thus exposing the n-epitaxial layer as shown in Fig. 21.8 (*ii*). Then we diffuse a p-type material into the n-type area as shown in Fig. 21.8 (*iii*). The SiO_2 layer is re-formed over the wafer by passing pure oxygen [See Fig. 21.8 (*iv*)]. The terminals are processed by etching SiO_2 layer at two points above the p island and depositing the metal at these locations [See Fig. 21.8 (*v*)]. In this way, we get an integrated resistor. Fig. 21.8 (*vi*) shows the electrical circuit of a resistor.

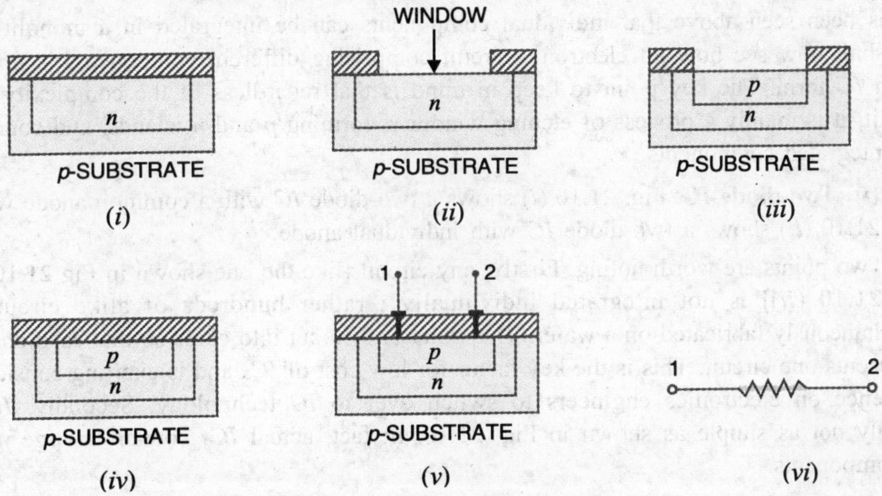

Fig. 21.8

The value of resistor is determined by the material, its length and area of cross-section. The high-resistance resistors are long and narrow while low-resistance resistors are short and of greater cross-section.

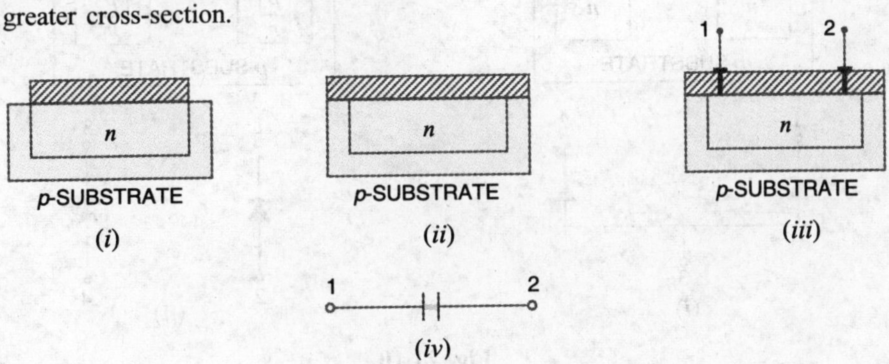

Fig. 21.9

(*iv*) Capacitors. Fig. 21.9 shows the process of fabricating a capacitor in the monolithic *IC*. The first step is to diffuse an n-type material into the substrate which forms one plate of the capacitor as shown in Fig. 21.9 (*i*). Then SiO_2 layer is re-formed over the wafer by passing pure oxygen as shown in Fig. 21.9 (*ii*).

The SiO_2 layer formed acts as the dielectric of the capacitor. The oxide layer is etched and terminal 1 is added as shown in Fig. 21.9 (*iii*). Next a large (compared to the electrode at terminal 1) metallic electrode is deposited on the SiO_2 layer and forms the second plate of the capacitor. The oxide layer is etched and terminal 2 is added. This gives an integrated capacitor. The value of capacitor formed depends upon the dielectric constant of SiO_2 layer, thickness of SiO_2 layer and the area of cross-section of the smaller of the two electrodes.

21.7 Simple Monolithic ICs

It has been seen above that individual components can be integrated in a monolithic *IC*. We shall now see how an electronic circuit comprising different components is produced in an *IC* form. The key point to keep in mind is that regardless of the complexity of the circuit, it is mainly a process of etching windows, forming *p* and *n* islands, and connecting the integrated components.

(*i*) Two-diode IC. Fig. 21.10 (*i*) shows a two-diode *IC* with a common anode whereas Fig. 21.10 (*ii*) shows a two-diode *IC* with individual anode.

Two points are worth noting. Firstly, any circuit [like the one shown in Fig 21.10 (*i*) or Fig 21.10 (*ii*)] is not integrated individually ; rather hundreds of alike circuits are simultaneously fabricated on a wafer. The wafer is then cut into chips so that each chip area represents one circuit. This is the key factor for low cost of *ICs* and is exerting considerable influence on electronics engineers to switch over to *IC* technology. Secondly, *ICs* are usually not as simple as shown in Fig. 21.10. In fact, actual *ICs* contain a large number of components.

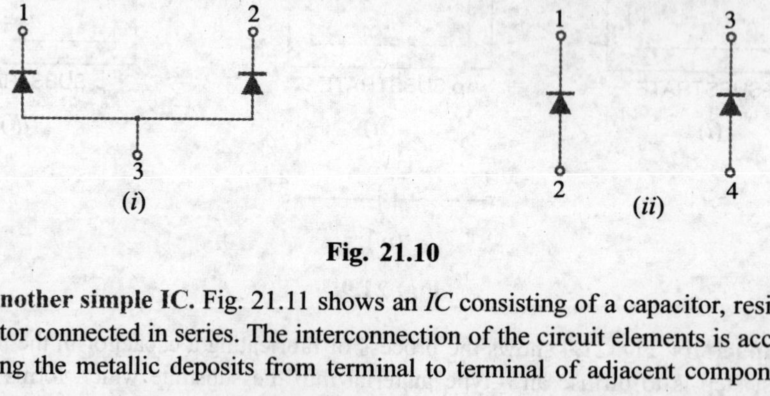

Fig. 21.10

(*ii*) Another simple IC. Fig. 21.11 shows an *IC* consisting of a capacitor, resistor, diode and transistor connected in series. The interconnection of the circuit elements is accomplished by extending the metallic deposits from terminal to terminal of adjacent components.

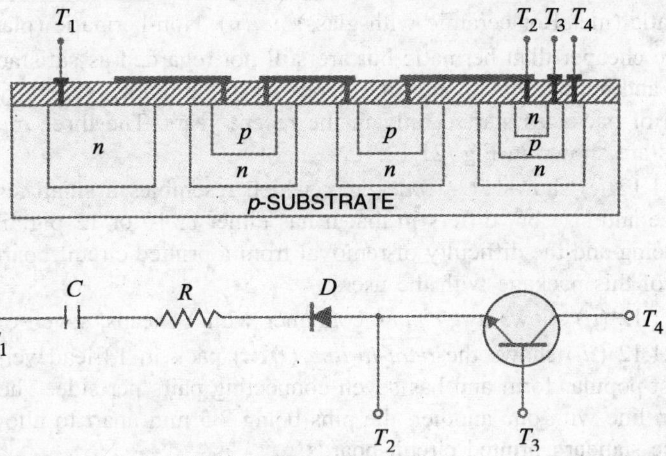

Fig. 21.11

It is interesting to see that p substrate isolates the integrated components from each other. Thus referring to Fig. 21.11, depletion layers exist between p substrate and the four n islands touching it. As the depletion layers have virtually no current carriers, therefore, the integrated components are insulated from each other.

21.8 IC Packings

In order to protect *ICs* from external environment and to provide mechanical protection, various forms of encapsulation are used for integrated circuits. Just as with semi-conductor devices, *IC* packages are of two types *viz.*

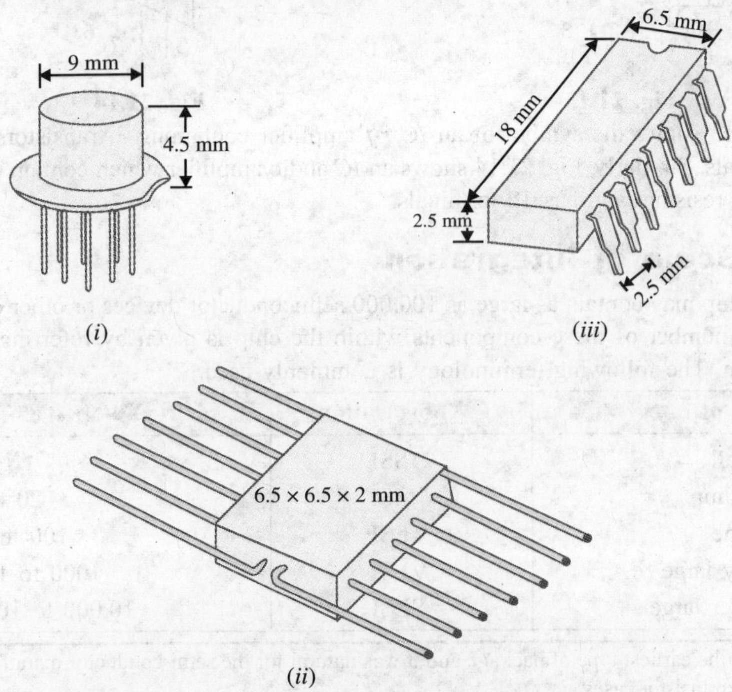

Fig. 21.12

(*a*) Hermatic (metal or ceramic with glass) (*b*) Non-hermatic (plastics)

Plastics are cheaper than hermatic but are still not regarded as satisfactory in extremes of temperature and humidity. Although *ICs* appeared in the market several years ago, yet the standardisation of packages started only in the recent years. The three most popular types of *IC* packages are shown in Fig. 21.12.

(*i*) Fig. 21.12 (*i*) shows *TO-5 package** which resembles a small signal transistor in both appearance and size but differs in that it has either 8, 10 or 12 pigtail-type leads. The close leads spacing and the difficulty of removal from a printed circuit board has diminished the popularity of this package with the users.

(*ii*) Fig. 21.12 (*ii*) shows a *flat pack* container with 14 leads, seven on each side.

(*iii*) Fig. 21.12 (*iii*) shows the *dual-in-line* (**DIL**) pack in 14-lead version. The 14-pin DIL is the most popular form and has seven connecting pairs per side. The pairs of pins of this pack are in line with one another, the pins being 2.5 mm apart to allow *IC* to be fitted directly into the standard printed circuit boards.

21.9 IC Symbols

In general, no standard symbols exist for *ICs*. Often the circuit diagram merely shows a block with numbered terminals. However, sometimes standard symbols are used for operational amplifiers or digital logic gates. Some of the symbols used with *ICs* are shown below.

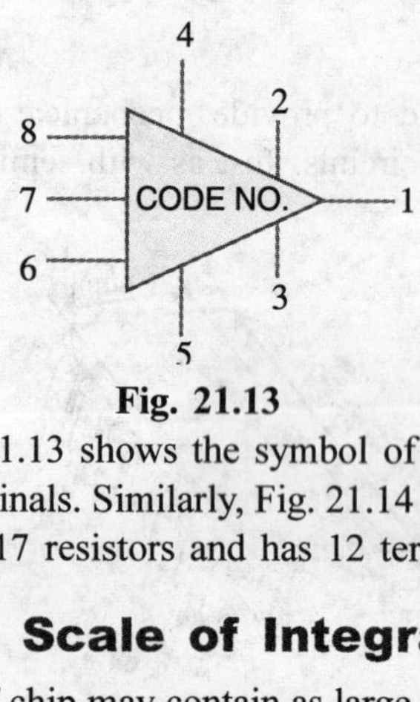

Fig. 21.13

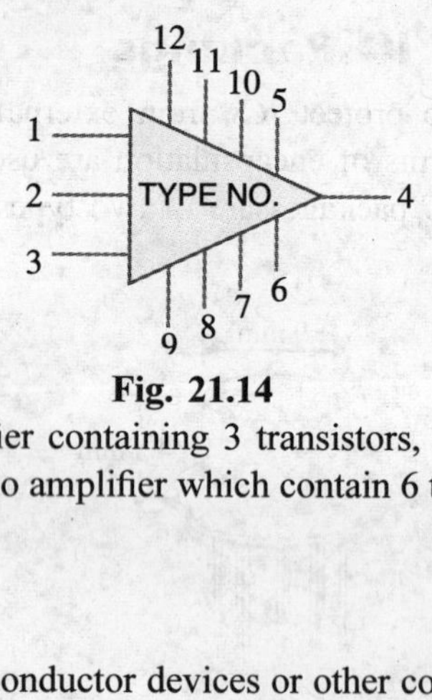

Fig. 21.14

Fig. 21.13 shows the symbol of an *IC r-f* amplifier containing 3 transistors, 3 resistors and 8 terminals. Similarly, Fig. 21.14 shows an *IC* audio amplifier which contain 6 transistors, 2 diodes, 17 resistors and has 12 terminals.

21.10 Scale of Integration

An *IC* chip may contain as large as 100,000 semiconductor devices or other components. The relative number of these components within the chip is given by referring to its scale of integration. The following terminology is commonly used.

Scale of integration	Abbreviation	Number of components
Small	**SSI	1 to 20
Medium	MSI	20 to 100
Large	LSI	100 to 1000
Very large	VLSI	1000 to 10,000
Super large	SLSI	10,000 to 100,000

* This was the earliest type of package and it was natural for the semi-conductor manufactures to use modified transistor cases.

** SSI stands for small scale integration.

21.11 Some Circuits Using ICs

Integrated circuits are fairly complex because they contain a large number of circuit components within a small semiconductor chip. While studying circuits using *ICs*, we are more concerned with the external connections to the *IC* rather than what is actually going on inside.

(i) **IC Fixed 5-volt Voltage Regulator.** The *IC* voltage regulator is a device that is used to hold the output voltage from a d.c. power supply constant as the input voltage or load current changes. For example, LM 309 (fixed positive) provides a + 5 V d.c. output. This regulator is frequently used in digital circuits. Fig. 21.15 shows the circuit of the voltage regulator using LM 309. It is a three terminal device with terminals labelled as input, output and ground terminal. It provides a fixed 5 V between the output and ground terminals.

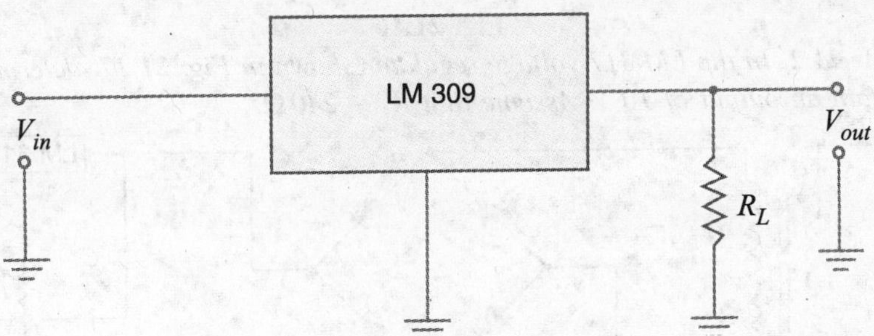

Fig. 21.15

The LM 309 has a number of advantages over the zener diode. First, it is much more accurate than the zener diode. Secondly, there is built-in overload protection. The LM 309 also has overheating protection. If the internal temperature becomes excessive, it shuts off until the temperature is reduced, at which point it will start up again.

(ii) **IC Adjustable Voltage Regulator.** Sometimes, we want a voltage regulator whose voltage we can vary. An example of such a voltage regulator is LM 317 whose schematic diagram is shown in Fig. 21.16. By varying the value of R_2, the output voltage of the regulator can be adjusted. The following equation is used to determine the regulated d.c. output voltage for an LM 317 regulator circuit :

$$V_{out} = 1.25\left(\frac{R_2}{R_1}+1\right)$$

A typical value for R_1 is 240 Ω. Thus, if $R_2 = 0$, then putting $R_1 = 240\ \Omega$ and $R_2 = 0$ in the above equation, we have, $V_{out} = 1.25$ V.

Similarly, if $R_2 = 1.2$ kΩ, then $V_{out} = 7.5$ V.

Example 21.1. *In LM 317 voltage regulator shown in Fig. 21.16, R_2 is adjusted to 2.4 kΩ. If the value of R_1 is 240 Ω, determine the regulated d.c. output voltage for the circuit.*

Solution.

$$V_{out} = 1.25\left(\frac{R_2}{R_1}+1\right)$$

$$= 1.25\left(\frac{2.4\text{ k}\Omega}{240\ \Omega}+1\right) = \mathbf{13.75\ V}$$

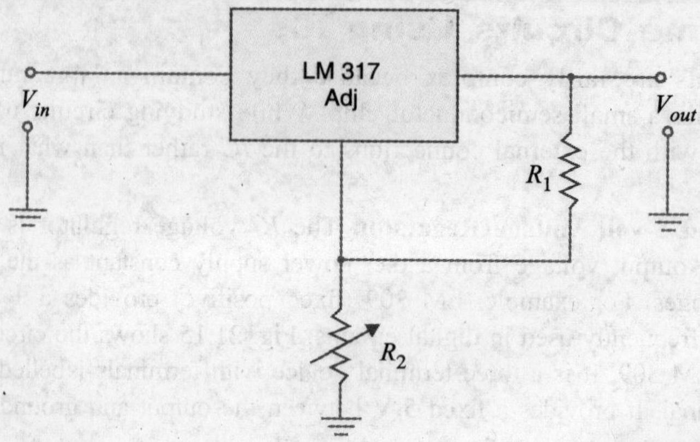

Fig. 21.16

Example 21.2. *In the LM 317 voltage regulator shown in Fig. 21.17, determine the size of R_2 to obtain an output of 10 V. Assume that R_1 = 240 Ω.*

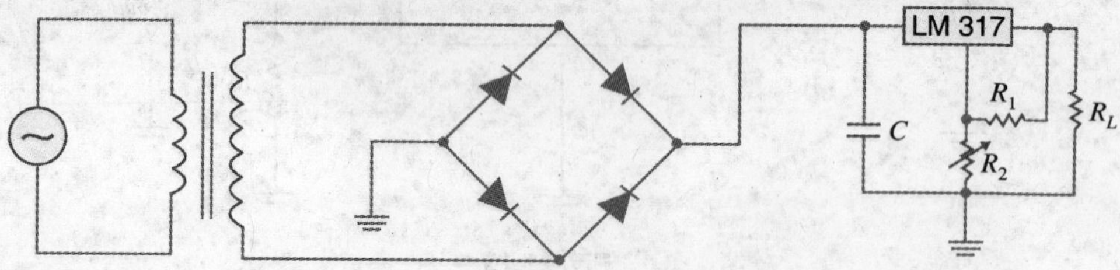

Fig. 21.17

Solution. Here, $R_1 = 240\ \Omega\ ;\ V_{out} = 10\ \text{V}\ ;\ R_2 = ?$

Now, $$V_{out} = 1.25\left(1 + \frac{R_2}{R_1}\right)$$

or $$10 = 1.25\left(1 + \frac{R_2}{R_1}\right)$$

$$\therefore \quad \frac{R_2}{R_1} = 7 \text{ or } R_2 = 7R_1 = 7 \times 240\ \Omega = \mathbf{1680\ \Omega}$$

Note. The LM 317 is a very popular device and is very often used in commercial power supplies where a variable output is required. An example of this would be a universal a.c. adapter which may be used to replace from 1 to eight 1.5 V batteries (1.5 V to 12 V).

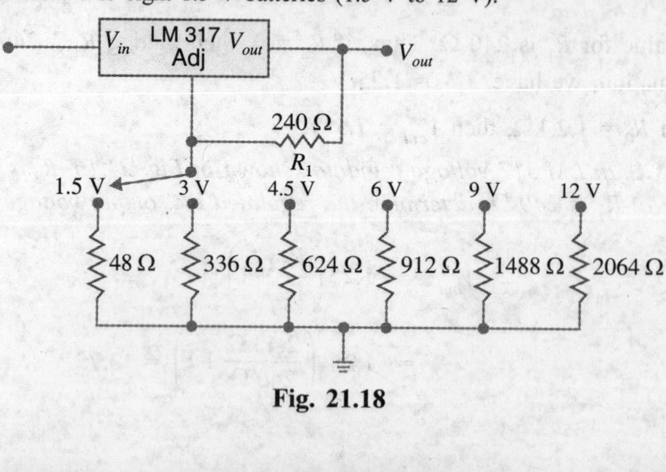

Fig. 21.18

Fig. 21.18 shows LM 317 in power pack for battery replacement. The transformer, filter capacitor and regulator remain the same. All that is necessary is to have available different fixed values for R_2.

(*iii*) The 555 Timer as monostable multivibrator. Fig. 21.19 shows the circuit of the 555 timer as a monostable multivibrator. The R and C are the external components whose values determine the time T (in seconds) for which the circuit is on. This time is given by ;

$$T = 1.1\ RC$$

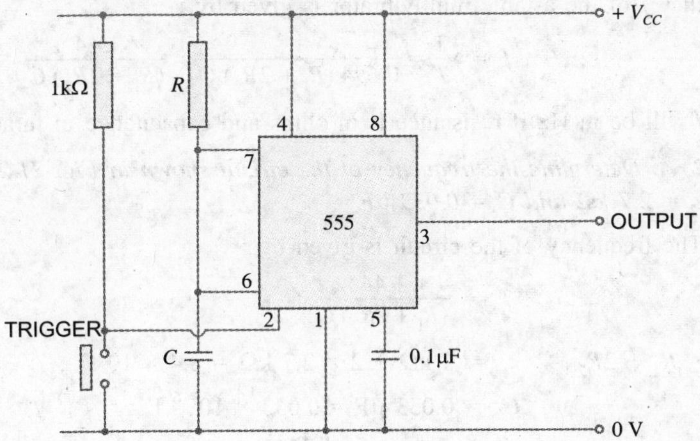

Fig. 21.19

Example 21.3. *The monostable multivibrator like the one in Fig. 21.19 has the values of R = 1.2 kΩ and C = 0.1 μF. Determine the time T for which the circuit is on.*

Solution. The time T for which the circuit is on is given by ;

$$T = 1.1\ RC = 1.1(1.2 \times 10^3)\ (0.1 \times 10^{-6})$$

$$= 132 \times 10^{-6} \text{s} = \mathbf{132\ \mu s}$$

(*iv*) The 555 Timer as astable multivibrator. Fig. 21.20 shows the 555 timer as an astable multivibrator. Note that the circuit contains *two resistors (R_1 and R_2) and one capacitor (C)* and *does not have an input from any other circuit.* The lack of a triggering signal from an external source is the circuit recognition feature of the astable multivibrator.

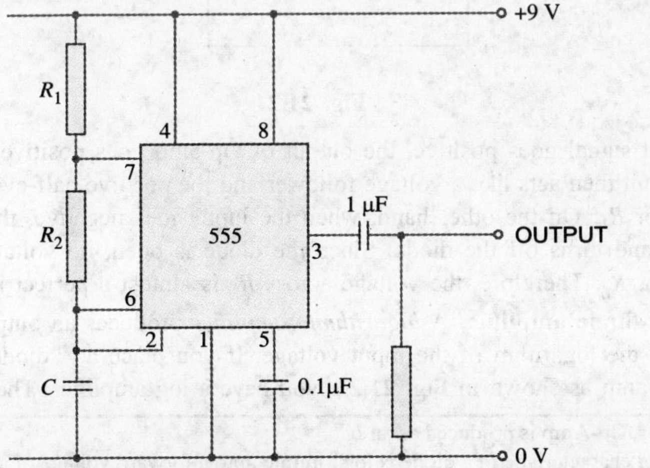

Fig. 21.20

The time T_1 for which the output is 'high' is given by ;

$$T_1 = 0.694\,(R_1 + R_2)\,C$$

The time T_2 for which the output is 'low' is given by ;

$$T_2 = 0.694\,R_2 C$$

∴ Total period T for the oscillations is

$$T = T_1 + T_2 = 0.694\,(R_1 + 2\,R_2)\,C$$

The frequency f of the astable multivibrator is given by ;

$$f = \frac{1}{T} = \frac{1}{0.694\,(R_1 + 2R_2)\,C} = \frac{1.44}{(R_1 + 2R_2)\,C}$$

Note that f will be in Hz if resistance is in ohms and capacitance in farads.

Example 21.4. *Determine the frequency of the circuit shown in Fig. 21.20. Given that $R_1 = 3\ k\Omega$; $R_2 = 2.7\ k\Omega$ and $C = 0.033\ \mu F$.*

Solution. The frequency of the circuit is given by ;

$$f = \frac{1.44}{(R_1 + 2R_2)\,C}$$

Here, $R_1 + 2R_2 = 3\ k\Omega + 2 \times 2.7\ k\Omega = 8.4 \times 10^3\Omega$;

$$C = 0.033\ \mu F = 0.033 \times 10^{-6}\ F$$

$$\therefore \quad f = \frac{1.44}{(8{\cdot}4 \times 10^3)\,(0.033 \times 10^{-6})} = 5.19 \times 10^3\ Hz = \mathbf{5.19\ kHz}$$

(v) Op-Amp Half-wave Rectifier. Fig. 21.21 shows the half-wave rectifier using an *Op-amp. The use of Op-amp greatly reduces the effect of diode offset voltage and allows the circuit to be used in the millivolt region.

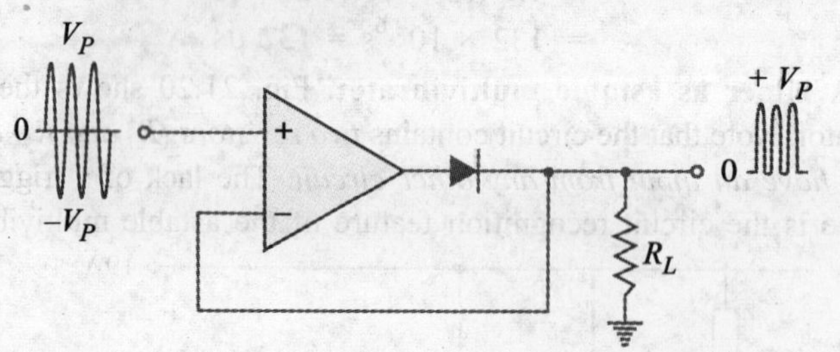

Fig. 21.21

When the input signal goes positive, the output of Op-amp goes positive and turns on the diode. The circuit then acts like a voltage follower and the positive half-cycle appears across the load resistor R_L. On the other hand, when the input goes negative, the Op-amp output goes negative and turns off the diode. Since the diode is open, no voltage appears across the load resistor R_L. Therefore, the voltage across R_L is almost a perfect half-wave signal.

(vi) Logarithmic amplifier. A *logarithmic amplifier* produces an output voltage that is proportional to the logarithm of the input voltage. If you place a **diode in the feedback loop of an Op-amp as shown in Fig. 21.22, you have a log amplifier. The output is limited

* Now-a-days, Op-Amp is produced as an *IC*.

** The forward characteristic of a diode is logarithmic upto a forward voltage of about 0.7 V.

to a maximum value of about 0.7 V because the diodes logarithmic characteristic is limited to voltages below 0.7 V. Also, the input must be positive when the diode is connected in the direction shown in Fig. 21.22. To handle negative inputs, you should reverse the direction of the diode. It can be shown that the output voltage of the circuit shown is given by ;

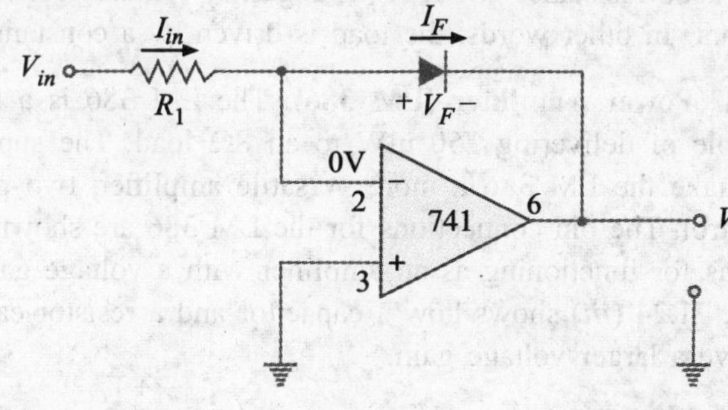

Fig. 21.22

$$V_{out} = -(0.025\text{ V})\log_e\left(\frac{V_{in}}{I_R R_1}\right)$$

where I_R = Reverse leakage current of the diode

For example, if $V_{in} = +2$ V, $R_1 = 100$ kΩ and $I_R = 50$ nA, then,

$$V_{out} = -(0.025)\log_e\left(\frac{2\text{ V}}{(50\times10^{-9})(100\times10^{3})}\right)$$

$$= -(0.025)\log_e(400) = -0.15\text{ V}$$

***(vii)* Constant-current source.** A constant current source delivers a load current I_L that remains constant when the load resistance R_L changes. Fig. 21.23 shows the basic circuit of a constant current source. Since the inverting (–) input of the Op-amp is at virtual ground (0 V), the value of I_i is determined by V_{in} and R_i *i.e.*

$$I_i = \frac{V_{in}}{R_i}$$

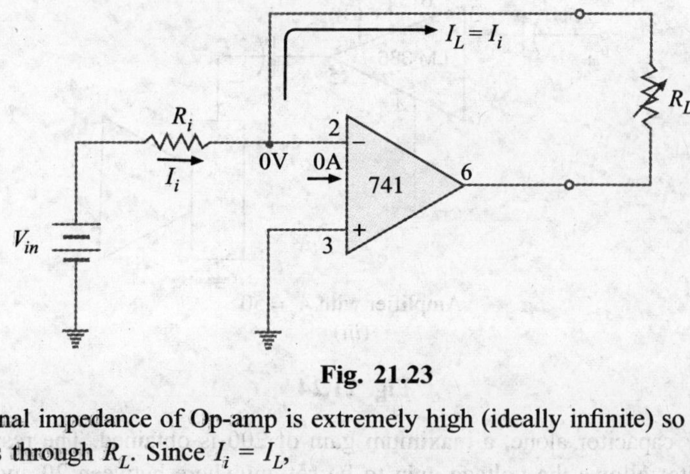

Fig. 21.23

The internal impedance of Op-amp is extremely high (ideally infinite) so that practically all of I_i flows through R_L. Since $I_i = I_L$,

$$\therefore \quad I_L = \frac{V_{in}}{R_i}$$

Note that load resistance R_L does not appear in this equation. Therefore, the load current (I_L) is independent of load resistance R_L. If R_L changes, I_L remains constant as long as V_{in} and R_i are held constant. In other words, the load is driven by a constant current source.

(*viii*) 250 mW IC Power Amplifier (LM 386). The LM 386 is a low voltage audio power amplifier capable of delivering 250 mW to an 8Ω load. The supply voltage range is 5 V to 18 V. To make the LM 386 a more versatile amplifier, two pins (1 and 8) are provided for gain control. The pin connections for the LM 386 are shown in Fig. 21.24 (*i*). The circuit connections for functioning as an amplifier with a voltage gain of 20 is shown in Fig. 21.24 (*ii*). Fig. 21.24 (*iii*) shows how a capacitor and a resistor can be connected at pins 1 and 8 to achieve a larger voltage gain.

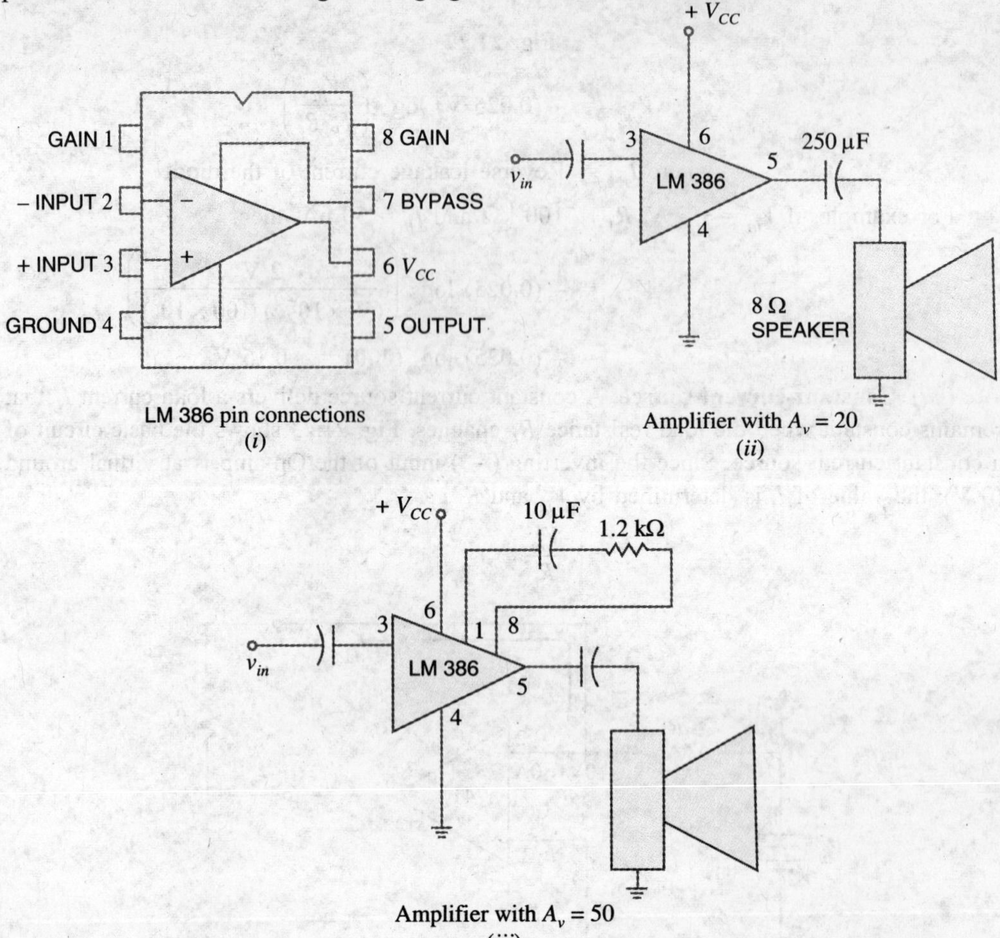

Fig. 21.24

With the 10 μF capacitor alone, a maximum gain of 200 is obtained. The resistor in series with the capacitor allows the voltage gain to be set anywhere between 20 and 200.

Short Answer Questions

Q. 1. *What are discrete circuits ?*

Ans. The circuits consisting of separately manufactured components (*e.g.* resistors, capacitors, diodes, transistors *etc.*) joined by wires or plated conductors on printed boards are called discrete circuits. These are called so because each component added to the circuit is discrete (*i.e.* distinct or separate) from the others.

Q. 2. *What are the two main disadvantages of discrete circuits ?*

Ans. The two main disadvantages of discrete circuits are :

(*i*) In a large circuit (*e.g.*, TV circuit, computer circuit), there may be hundreds of components and consequently discrete assembly would occupy a large space.

(*ii*) There will be hundreds of soldered points posing a considerable problem of reliability.

Q. 3. *Give three important advantages of integrated circuits over discrete circuits.*

Ans. The three important advantages of integrated circuits over the discrete circuits are :

(*i*) Increased reliability due to lesser number of connections.

(*ii*) Low power requirements.

(*iii*) Lesser space and weight requirement.

Q. 4. *What is the size of an IC ?*

Ans. The size of an *IC* is very small ; it is about 10 % of the space required by comparable discrete assembly.

Q. 5. *Give three disadvantages of integrated circuits over the discrete circuits.*

Ans. The three disadvantages of integrated circuits over the discrete circuits are :

(*i*) It is not possible to produce high power *ICs*.

(*ii*) If any component of an *IC* goes out of order, the whole *IC* has to be replaced by the new one.

(*iii*) It is not possible to fabricate inductors and transformers on the surface of semiconductor chip.

Multiple-Choice Questions

1. An *IC* has .... size.

(*i*) very large
(*ii*) large
(*iii*) extremely small
(*iv*) none of the above

2. *ICs* are generally made of ....

(*i*) silicon (*ii*) germanium
(*iii*) copper
(*iv*) none of the above

3. .... *ICs* are the most commonly used.

(*i*) Thin film (*ii*) Monolithic
(*iii*) Hybrid
(*iv*) None of the above

4. The most popular form of *IC* package is ....

(*i*) DIL (*ii*) flatpack
(*iii*) TO-5
(*iv*) none of the above

5. .... cannot be fabricated on an *IC*.

(*i*) Transistors
(*ii*) Diodes
(*iii*) Resistors
(*iv*) Large inductors and transformers

6. An audio amplifier is an example of ....
   (*i*) digital *IC*
   (*ii*) linear *IC*
   (*iii*) both digitial and linear *IC*
   (*iv*) none of the above
7. The active components in an *IC* are ....
   (*i*) resistors
   (*ii*) capacitors
   (*iii*) transistors and diodes
   (*iv*) none of the above
8. We use .... *ICs* in computers.
   (*i*) digital
   (*ii*) linear
   (*iii*) both digital and linear
   (*iv*) none of the above
9. The SiO_2 layer in an *IC* acts as ....
   (*i*) a resistor
   (*ii*) an insulating layer
   (*iii*) mechanical output
   (*iv*) none of the above
10. *ICs* are used in ....
   (*i*) linear devices only
   (*ii*) digital devices only
   (*iii*) both linear and digital devices
   (*iv*) none of the above
11. A transistor takes .... inductor on a silicon *IC* chip.
   (*i*) less space than
   (*ii*) more space than
   (*iii*) same space as
   (*iv*) none of the above
12. The most popular types of *ICs* are ....
   (*i*) thin-film (*ii*) hybrid
   (*iii*) thick-film (*iv*) monolithic
13. Digital *ICs* process ....
   (*i*) linear signals only
   (*ii*) digital signals only
   (*iii*) both digital and linear signals
   (*iv*) none of the above
14. Operational amplifiers use ....
   (*i*) linear *ICs*
   (*ii*) digital *ICs*
   (*iii*) both linear and digital *ICs*
   (*iv*) none of the above
15. Which of the following is most difficult to fabricate in an *IC* ?
   (*i*) Diode (*ii*) Transistor
   (*iii*) *FET* (*iv*) Capacitor

Answers to Multiple-Choice Questions

1. (*iii*)	2. (*i*)	3. (*ii*)	4. (*ii*)	5. (*iv*)
6. (*ii*)	7. (*iii*)	8. (*i*)	9. (*ii*)	10. (*iii*)
11. (*i*)	12. (*iv*)	13. (*iii*)	14. (*ii*)	15. (*iv*)

Chapter Review Topics

1. What is an integrated circuit ? Discuss the relative advantages and disadvantages of *ICs* over discrete assembly.
2. How will you make a monolithic *IC* ?
3. Explain how (*i*) a diode (*ii*) a transistor (*iii*) a resistor and (*iv*) a capacitor can be constructed in a monolithic integrated circuit.
4. Explain how electronic circuit consisting of different components can be constructed in a monolithic *IC*.
5. Write short notes on the following :
   (*i*) Epitaxial layer (*ii*) *IC* packages (*iii*) *IC* symbols

CHAPTER 22

Hybrid Parameters

Learning Objectives

- Understand the need of developing hybrid parameters.
- Develop mathematical expressions for the four *h* parameters.
- Explore the *h* parameters of a transistor.
- Learn the approximate hybrid formulas for a transistor amplifier.
- Develop an understanding of experimental determination of transistor *h* parameters.

Introduction

In order to predict the behaviour of a small-signal transistor amplifier, it is important to know its operating characteristics *e.g.*, input impedance, output impedance, voltage gain *etc*. In the text so far, these characteristics were determined by using *β and circuit resistance values. This method of analysis has two principal advantages. Firstly, the values of circuit components are readily available and secondly the procedure followed is easily understood. However, the major drawback of this method is that accurate results cannot be obtained. It is because the input and output circuits of a transistor amplifier are not completely independent. For example, output current is affected by the value of load resistance rather than being constant at the value βI_b. Similarly, output voltage has an effect on the input circuit so that changes in the output cause changes in the input.

One of the methods that takes into account all the effects in a transistor amplifier is the hybrid parameter approach. In this method, four parameters (one measured in ohm, one in siemens, two dimensionless) of a transistor are measured experimentally. These are called hybrid or *h* parameters of the transistor. Once these parameters for a transistor are known, formulas can be developed for input impedance, voltage gain *etc.* in terms of *h* parameters. There are two main reasons for using *h* parameter method in describing the characteristics of a transistor. Firstly, it yields exact results because the inter-effects of input and output circuits are taken into account. Secondly, these parameters can be measured very easily. To begin with, we shall apply *h* parameter approach to general circuits and then extend it to transistor amplifiers.

* Since transistor is generally connected in *CE* arrangement, current amplification factor β is mentioned here.

22.1 Hybrid Parameters of a Linear Circuit

*Every *linear circuit having input and output terminals can be analysed by four parameters (one measured in ohm, one in siemens and two dimensionless) called* **hybrid** *or* **h Parameters.**

Hybrid means "mixed". Since these parameters have mixed dimensions, they are called hybrid parameters. Consider a linear circuit shown in Fig. 22.1. This circuit has input voltage and current labelled v_1 and i_1. This circuit also has output voltage and current labelled v_2 and i_2. Note that both input and output currents (i_1 and i_2) are assumed to flow *into* the box ; input and output voltages (v_1 and v_2) are assumed *positive* from the upper to the lower terminals. These are standard conventions and do not necessarily correspond to the actual directions and polarities. When we analyse circuits in which the voltages are of opposite polarity or where the currents flow out of the box, we simply treat these voltages and currents as negative quantities.

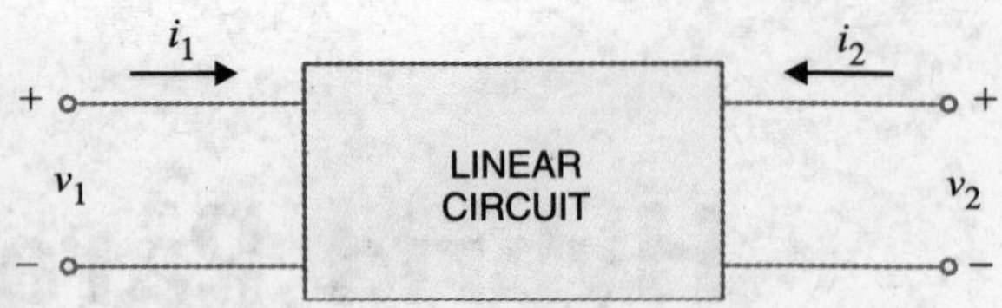

Fig. 22.1

It can be proved by advanced circuit theory that voltages and currents in Fig. 22.1 can be related by the following sets of equations :

$$v_1 = h_{11}\, i_1 + h_{12}\, v_2 \qquad ...(i)$$

$$i_2 = h_{21}\, i_1 + h_{22}\, v_2 \qquad ...(ii)$$

In these equations, the *hs* are fixed constants for a given circuit and are called *h* parameters. Once these parameters are known, we can use equations (*i*) and (*ii*) to find the voltages and currents in the circuit. If we look at eq. (*i*), it is clear that **h_{11} has the dimension of ohm and h_{12} is dimensionless. Similarly, from eq. (*ii*), h_{21} is dimensionless and h_{22} has the dimension of siemens. The following points may be noted about *h* parameters :

(*i*) The *h* parameters of a linear circuit describe the input voltage (v_1) and output current (i_2) in terms of input current (i_1) and output voltage (v_2).

(*ii*) Every linear circuit has four *h* parmeters having different units. Thus h_{11} has units of resistance (Ω), h_{12} unitless, h_{21} is unitless and h_{22} has units of conductance (siemens).

(*iii*) The *h* parameters of a given circuit are constant. If we change the circuit, *h* parameters would also change.

(*iv*) Suppose that in a particular linear circuit, voltages and currents are related as under :

$$v_1 = 10 i_1 + 6 v_2$$

$$i_2 = 4 i_1 + 3 v_2$$

Here we can say that the circuit has *h* parameters given by $h_{11} = 10\ \Omega$; $h_{12} = 6$; $h_{21} = 4$ and $h_{22} = 3\text{S}$.

---

* A linear circuit is one in which resistances, inductances and capacitances remain fixed when voltage across them changes.

** The two parts on the R.H.S. of eq. (*i*) must have the unit of voltage. Since current (amperes) must be multiplied by resistance (ohms) to get voltage (volts), h_{11} should have the dimension of resistance *i.e.* ohms.

22.2 Determination of *h* Parameters

The major reason for the use of *h* parameters is the relative ease with which they can be measured. The *h* parameters of a circuit shown in Fig. 22.1 can be found out as under :

(*i*) If we short-circuit the output terminals (See Fig. 22.2), we can say that output voltage $v_2 = 0$. Putting $v_2 = 0$ in equations (*i*) and (*ii*), we get,

$$v_1 = h_{11}\, i_1 + h_{12} \times 0$$

$$i_2 = h_{21}\, i_1 + h_{22} \times 0$$

$$\therefore \quad h_{11} = \frac{v_1}{i_1} \text{ for } v_2 = 0 \text{ i.e. output shorted}$$

$$\text{and} \quad h_{21} = \frac{i_2}{i_1} \text{ for } v_2 = 0 \text{ i.e. output shorted}$$

Let us now turn to the physical meaning of h_{11} and h_{21} . Since h_{11} is a ratio of voltage and current (*i.e.* v_1/i_1), it is an impedance and is called * "*input impedance with output shorted*". Similarly, h_{21} is the ratio of output and input current (*i.e.*, i_2/i_1), it will be dimensionless and is called "*current gain with output shorted*".

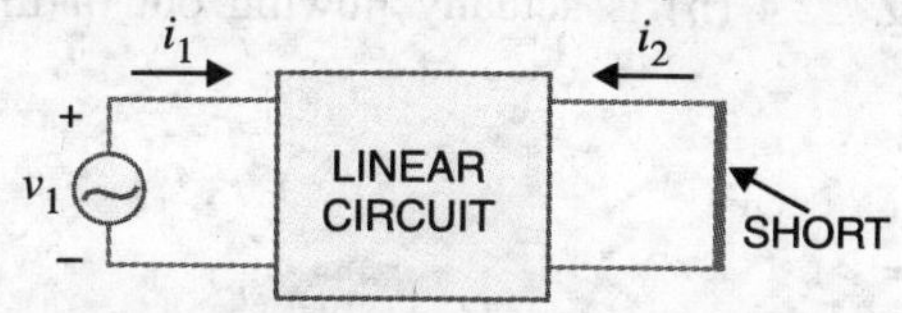

Fig. 22.2

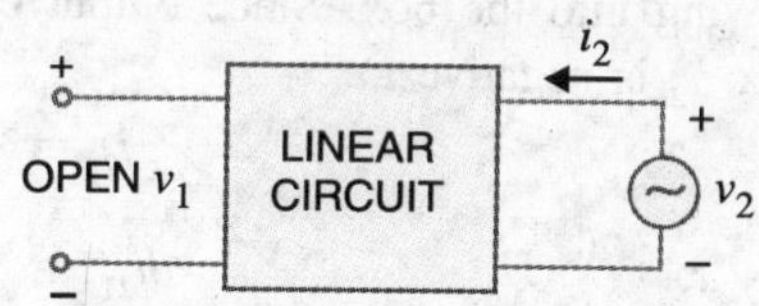

Fig. 22.3

(*ii*) The other two *h* parameters (*viz.* h_{12} and h_{22}) can be found by making $i_1 = 0$. This can be done by the arrangement shown in Fig. 22.3. Here, we drive the output terminals with voltage v_2, keeping the input terminals open. With this set up, $i_1 = 0$ and the equations become :

$$v_1 = h_{11} \times 0 + h_{12}\, v_2$$

$$i_2 = h_{21} \times 0 + h_{22}\, v_2$$

$$\therefore \quad h_{12} = \frac{v_1}{v_2} \text{ for } i_1 = 0 \text{ i.e. input open}$$

$$\text{and} \quad h_{22} = \frac{i_2}{v_2} \text{ for } i_1 = 0 \text{ i.e. input open}$$

Since h_{12} is a ratio of input and output voltages (*i.e.* v_1/v_2), it is dimensionless and is called "*reverse voltage feedback ratio with input terminals open*". Similarly, h_{22} is a ratio of output current and output voltage (*i.e.* i_2/v_2), it will be admittance and is called "*output conductance with input terminals open*".

The table below gives the meaning of each *h* parameter and the required condition for determining it.

S.No.	Parameter	Meaning	Condition
1.	h_{11}	Input resistance	Output shorted
2.	h_{12}	Reverse voltage feedback ratio	Input open
3.	h_{21}	Current gain	Output shorted
4.	h_{22}	Output conductance	Input open

* Note that v_1 is the input voltage and i_1 is the input current. Hence v_1/i_1 is given the name input impedance.

Example 22.1. *Find the h parameters of the circuit shown in Fig. 22.4 (i).*

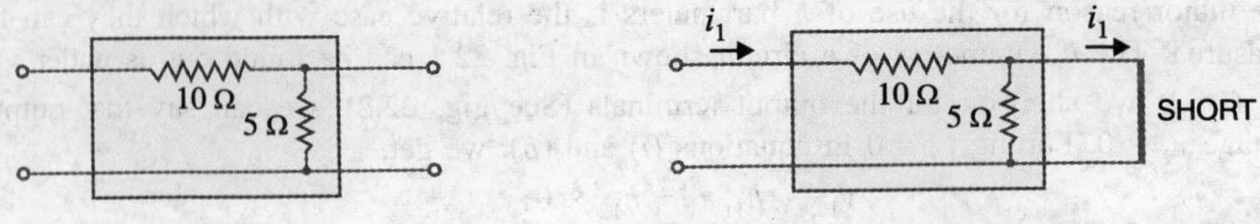

Fig. 22.4

Solution. The h parameters of the circuit shown in Fig. 22.4 (*i*) can be found as under :

To find h_{11} and h_{21}, short - circuit the output terminals as shown in Fig. 22.4 (*ii*). It is clear that input impedance of the circuit is 10 Ω because 5 Ω resistance is shorted out.

$$\therefore \quad h_{11} = 10\ \Omega$$

Now current i_1 flowing into the box will flow through 10 Ω resistor and then through the shorted path as shown. It may be noted that in our discussion, i_2 is the output current flowing into the box. Since output current in Fig. 22.4 (*ii*) is actually flowing out of the box, i_2 is negative *i.e.*,

$$i_2 = -i_1$$

$$\therefore \quad h_{21} = \frac{i_2}{i_1} = \frac{-i_1}{i_1} = -1$$

To find h_{12} and h_{22}, make the arrangement as shown in Fig. 22.4 (*iii*). Here we are driving the output terminals with a voltage v_2. This sets up a current i_2. Note that input terminals are open. Under this condition, there will be no current in 10 Ω resistor and, therefore, there can be no voltage drop across it. Consequently, all the voltage appears across input terminals *i.e.*

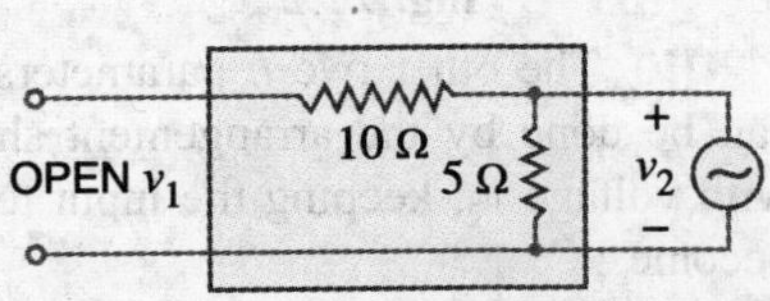

Fig. 22.4 (*iii*)

$$v_1 = v_2$$

$$\therefore \quad h_{12} = \frac{v_1}{v_2} = \frac{v_2}{v_2} = 1$$

Now the output impedance looking into the output terminals with input terminals open is simply 5 Ω. Then h_{22} will be the reciprocal of it because h_{22} is the output conductance with input terminals open.

$$\therefore \quad h_{22} = 1/5 = 0.2\ \text{S}$$

The h parameters of the circuit are :

$$h_{11} = 10\ \Omega\ ;\ h_{21} = -1$$

$$h_{12} = 1\ ;\ h_{22} = 0.2\ \text{S}$$

It may be mentioned here that in practice, dimensions are not written with h parameters. It is because it is understood that h_{11} is always in ohms, h_{12} and h_{21} are dimensionless and h_{22} is in siemens.

Example 22.2. *Find the h parameters of the circuit shown in Fig. 22.5 (i).*

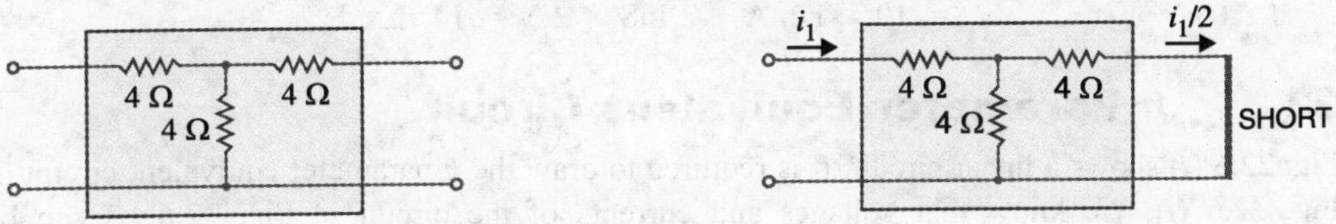

(*i*) (*ii*)

Fig. 22.5

Solution. First of all imagine that output terminals are short-circuited as shown in Fig. 22.5 (*ii*). The input impedance under this condition is the parameter h_{11}.

Obviously, $h_{11} = 4 + 4 \parallel 4$

$$= 4 + \frac{4 \times 4}{4 + 4} = 6\ \Omega$$

Now the input current i_1 in Fig. 22.5 (*ii*) will divide equally at the junction of 4 Ω resistors so that output current is $i_1/2$ *i.e.*

$$i_2 = -i_1/2 = -\,0.5\ i_1$$

$$\therefore \quad h_{21} = \frac{i_2}{i_1} = \frac{-\,0.5\ i_1}{i_1} = \mathbf{-0.5}$$

In order to find h_{12} and h_{22}, imagine the arrangement as shown in Fig. 22.5 (*iii*). Here we are driving the output terminals with voltage v_2, keeping the input terminals open. Under this condition, any voltage v_2 applied to the output will divide by a factor 2 *i.e.*

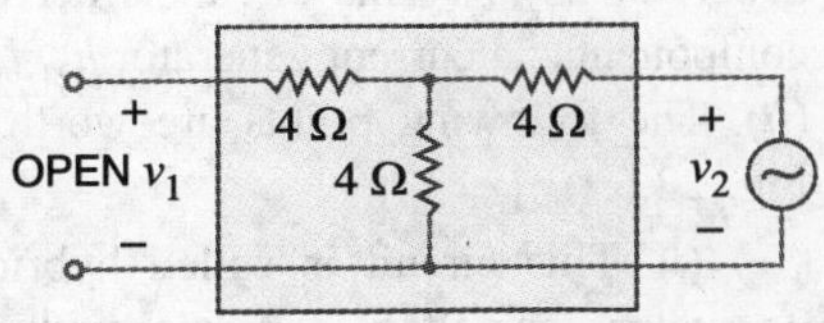

Fig. 22.5 (*iii*)

$$v_1 = \frac{v_2}{2} = 0.5\ v_2$$

$$\therefore \quad h_{12} = \frac{v_1}{v_2} = \frac{0.5\ v_2}{v_2} = \mathbf{0.5}$$

Now the output impedance looking into the output terminals with input terminals open is simply 8 Ω. Then h_{22} will be the reciprocal of this *i.e.*

$$h_{22} = \frac{1}{8} = \mathbf{0.125\ S}$$

Example 22.3. *The h parameters of a linear circuit are measured as under :*

$$h_{11} = 10\ k\Omega\ ;\ h_{12} = 0.5\ ;\ h_{21} = 10\ ;\ h_{22} = 2\ mS$$

Find the input voltage v_1 and output current i_2 if input current i_1 = 1 mA and output voltage v_2 = 2 V.

Solution. The *h* parameters of the circuit describe the relation of v_1 (input voltage) in terms of i_1 and v_2 as under :

$$v_1 = h_{11}\ i_1 + h_{12}\ v_2$$

Putting the various values, we have,

$$v_1 = 10\ k\Omega \times 1\ mA + 0.5 \times 2\ V = \mathbf{11\ V}$$

Also, $i_2 = h_{21}\, i_1 + h_{22}\, v_2$

$= 10 \times 1 \text{ mA} + 2 \text{ mS} \times 2 \text{ V} = \mathbf{14\ mA}$

22.3 *h* Parameter Equivalent Circuit

Fig. 22.6 (*i*) shows a linear circuit. It is required to draw the *h* parameter equivalent circuit of Fig. 22.6 (*i*). We know that voltages and currents of the circuit in Fig. 22.6 (*i*) can be expressed in terms of *h* parameters as under :

$$v_1 = h_{11}\, i_1 + h_{12}\, v_2 \qquad ...(i)$$

$$i_2 = h_{21}\, i_1 + h_{22}\, v_2 \qquad ...(ii)$$

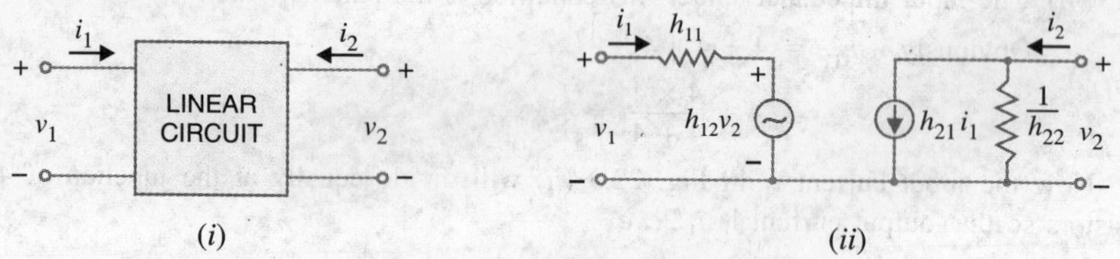

Fig. 22.6

Fig. 22.6 (*ii*) shows *h* parameter equivalent circuit of Fig. 22.6 (*i*) and is derived from equations (*i*) and (*ii*). The *input circuit* appears as a resistance h_{11} in series with a voltage generator $h_{12}\, v_2$. This circuit is derived from equation (*i*). The *output circuit* involves two components ; a current generator $h_{21}\, i_1$ and shunt resistance $1/h_{22}$ and is derived from equation (*ii*). The following points are worth noting about the *h* parameter equivalent circuit [See Fig. 22.6 (*ii*)] :

(*i*) This circuit is called hybrid equivalent because its input portion is a Thevenin equivalent, or voltage generator with series resistance, while output side is Norton equivalent, or current generator with shunt resistance. Thus it is a mixture or a hybrid. The symbol '*h*' is simply the abbreviation of the word hybrid (hybrid means "mixed").

(*ii*) The different hybrid parameters are distinguished by different number subscripts. The notation shown in Fig. 22.6 (*ii*) is used in general circuit analysis. The first number designates the circuit in which the effect takes place and the second number designates the circuit from which the effect comes. For instance, h_{21} is the "short-circuit forward current gain" or the ratio of the current in the output (*circuit* 2) to the current in the input (*circuit* 1).

(*iii*) The equivalent circuit of Fig. 22.6 (*ii*) is extremely useful for two main reasons. First, it isolates the input and output circuits, their interaction being accounted for by the two controlled sources. Thus, the effect of output upon input is represented by the equivalent voltage generator $h_{12}v_2$ and its value depends upon output voltage. Similarly, the effect of input upon output is represented by current generator $h_{21}i_1$ and its value depends upon input current. Secondly, the two parts of the circuit are in a form which makes it simple to take into account source and load circuits.

22.4 Performance of a Linear Circuit in *h* Parameters

We have already seen that any linear circuit with input and output has a set of *h* parameters. We shall now develop formulas for input impedance, current gain, voltage gain *etc.* of a linear circuit in terms of *h* parameters.

(*i*) **Input impedance.** Consider a linear circuit with a load resistance r_L across its terminals as shown in Fig. 22.7. The input impedance Z_{in} of this circuit is the ratio of input voltage to input current *i.e.*

$$Z_{in} = \frac{v_1}{i_1}$$

Now $v_1 = h_{11}\, i_1 + h_{12}\, v_2$ in terms of h parameters. Substituting the value of v_1 in the above expression, we get,

$$Z_{in} = \frac{h_{11}\, i_1 + h_{12}\, v_2}{i_1} = h_{11} + \frac{h_{12}\, v_2}{i_1} \qquad ...(i)$$

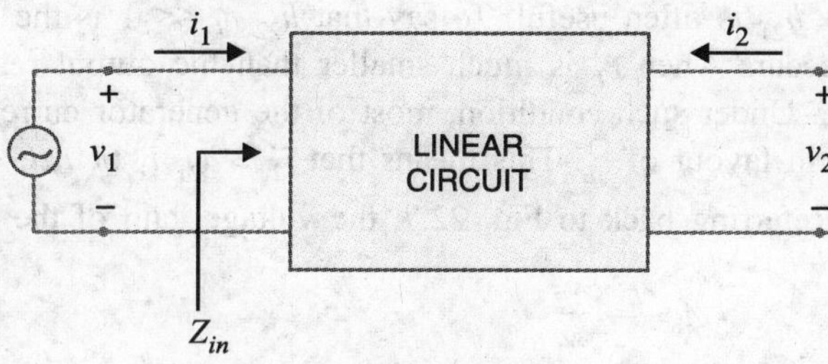

Fig. 22.7

Now, $i_2 = h_{21}\, i_1 + h_{22}\, v_2$ in terms of h parameters. Further from Fig. 22.7, it is clear that $i_2 = -v_2/r_L$. The minus sign is used here because the actual load current is opposite to the direction of i_2.

$$\therefore \quad \frac{-v_2}{r_L} = h_{21}\, i_1 + h_{22}\, v_2 \qquad \left[\because \; i_2 = \frac{-v_2}{r_L}\right]$$

$$\text{or} \quad -h_{21}\, i_1 = h_{22}\, v_2 + \frac{v_2}{r_L} = v_2\left(h_{22} + \frac{1}{r_L}\right)$$

$$\therefore \quad \frac{v_2}{i_1} = \frac{-h_{21}}{h_{22} + \frac{1}{r_L}} \qquad ...(ii)$$

Substituting the value of v_2/i_1 from exp. (*ii*) into exp. (*i*), we get,

$$\mathbf{Z}_{in} = h_{11} - \frac{h_{12}\, h_{21}}{h_{22} + \frac{1}{r_L}} \qquad ...(iii)$$

This is the expression for input impedance of a linear circuit in terms of h parameters and load connected to the output terminals. If either h_{12} or r_L is very small, the second term in exp. (*iii*) can be neglected and input impedance becomes :

$$\mathbf{Z}_{in} \simeq h_{11}$$

(*ii*) **Current Gain.** Referring to Fig. 22.7, the current gain A_i of the circuit is given by ;

$$A_i = \frac{i_2}{i_1}$$

Now, $\quad i_2 = h_{21}\, i_1 + h_{22}\, v_2$

and $\quad v_2 = -\, i_2\, r_L$

$$\therefore \quad i_2 = h_{21}\, i_1 - h_{22}\, i_2\, r_L$$

or $$i_2\,(1 + h_{22}\, r_L) = h_{21}\, i_1$$

or $$\frac{i_2}{i_1} = \frac{h_{21}}{1 + h_{22}\, r_L}$$

But $i_2/i_1 = A_i$, the current gain of the circuit.

$$\therefore \quad A_i = \frac{h_{21}}{1 + h_{22}\, r_L}$$

If $h_{22}\, r_L << 1$, then $A_i \simeq h_{21}$.

The expression $A_i \simeq h_{21}$ is often useful. To say that $h_{22}\, r_L << 1$ is the same as saying that $r_L << 1/h_{22}$. This occurs when r_L is much smaller than the output resistance $(1/h_{22})$, shunting $h_{21}\, i_1$ generator. Under such condition, most of the generator current bypasses the circuit output resistance in favour of r_L. This means that $i_2 \simeq h_{21}\, i_1$ or $i_2/i_1 \simeq h_{21}$.

***(iii)* Voltage gain.** Referring back to Fig. 22.7, the voltage gain of the circuit is given by ;

$$A_v = \frac{v_2}{v_1}$$

$$= \frac{v_2}{i_1\, Z_{in}} \qquad (\because\ v_1 = i_1 Z_{in}) \qquad \text{...(iv)}$$

While developing expression for input impedance, we found that :

$$\frac{v_2}{i_1} = \frac{-h_{21}}{h_{22} + \frac{1}{r_L}}$$

Substituting the value of v_2/i_1 in exp. (*iv*), we get,

$$A_v = \frac{-h_{21}}{Z_{in}\left(h_{22} + \frac{1}{r_L}\right)}$$

***(iv)* Output impedance.** In order to find the output impedance, remove the load r_L, set the signal voltage v_1 to zero and connect a generator of voltage v_2 at the output terminals. Then *h* parameter equivalent circuit becomes as shown in Fig. 22.8. By definition, the output impedance Z_{out} is

$$Z_{out} = \frac{v_2}{i_2}$$

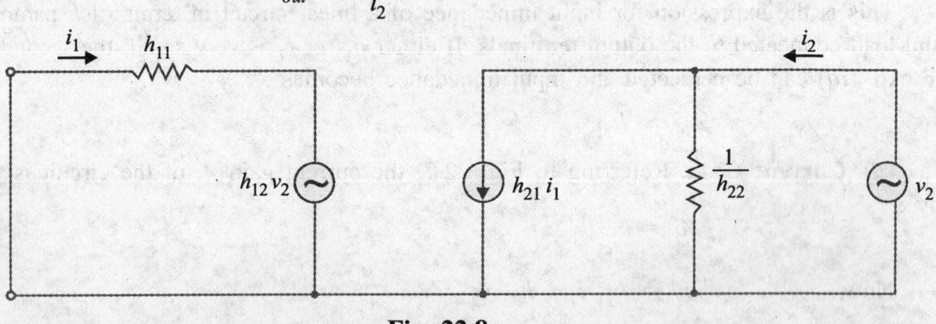

Fig. 22.8

With $v_1 = 0$ and applying Kirchhoff's voltage law to the input circuit, we have,

$$0 = i_1 h_{11} + h_{12} v_2$$

$$\therefore \quad i_1 = -\frac{h_{12} v_2}{h_{11}}$$

Now, $\quad i_2 = h_{21} i_1 + h_{22} v_2$

Putting the value of i_1 $(= -h_{12} v_2/h_{11})$ in this equation, we get,

$$i_2 = h_{21}\left(-\frac{h_{12} v_2}{h_{11}}\right) + h_{22} v_2$$

or $$i_2 = -\frac{h_{21} h_{12} v_2}{h_{11}} + h_{22} v_2$$

Dividing throughout by v_2, we have,

$$\frac{i_2}{v_2} = \frac{-h_{21} h_{12}}{h_{11}} + h_{22}$$

$$\therefore \quad Z_{out} = \frac{v_2}{i_2} = \frac{1}{h_{22} - \frac{h_{21} h_{12}}{h_{11}}}$$

Example 22.4. *Find the (i) input impedance and (ii) voltage gain for the circuit shown in Fig. 22.9.*

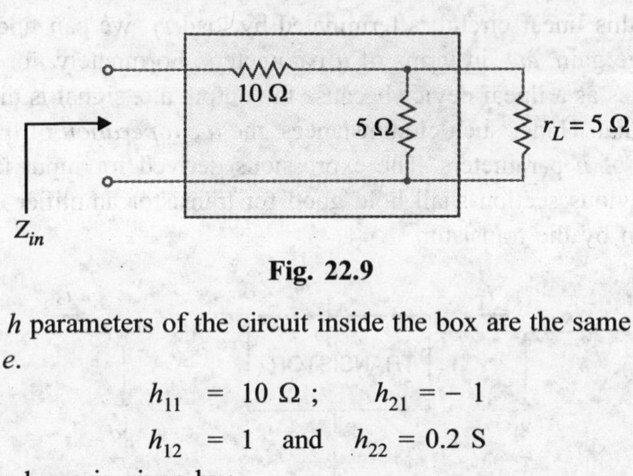

Fig. 22.9

Solution. The h parameters of the circuit inside the box are the same as those calculated in example 22.1 *i.e.*

$$h_{11} = 10\ \Omega\ ; \qquad h_{21} = -1$$

$$h_{12} = 1 \quad \text{and} \quad h_{22} = 0.2\ \text{S}$$

(*i*) Input impedance is given by;

$$Z_{in} = h_{11} - \frac{h_{12} h_{21}}{h_{22} + \frac{1}{r_L}} = 10 - \frac{1 \times -1}{0.2 + \frac{1}{5}}$$

$$= 10 + 2.5 = \mathbf{12.5\ \Omega}$$

By inspection, we can see that input impedance is equal to 10 Ω plus two 5 Ω resistances in parallel *i.e.*

$$Z_{in} = 10 + 5 \parallel 5$$

$$= 10 + \frac{5 \times 5}{5 + 5} = 12.5\ \Omega$$

(*ii*) Voltage gain, $A_v = \dfrac{-h_{21}}{Z_{in}\left(h_{22} + \dfrac{1}{r_L}\right)} = \dfrac{1}{12.5\left(0.2 + \dfrac{1}{5}\right)} = \dfrac{1}{5}$

It means that output voltage is one-fifth of the input voltage. This can be readily established by inspection of Fig. 22.9. The two 5 Ω resistors in parallel give a net resistance of 2.5 Ω. Therefore, we have a voltage divider consisting of 10 Ω resistor in series with 2.5 Ω resistor.

$$\therefore \quad \text{Output voltage} = \frac{2.5}{12.5} \times \text{Input voltage}$$

$$\text{or} \quad \frac{\text{Output voltage}}{\text{Input voltage}} = \frac{2.5}{12.5} = \frac{1}{5}$$

$$\text{or} \quad A_v = \frac{1}{5}$$

Comments. The reader may note that in a simple circuit like that of Fig. 22.9, it is not advisable to use *h* parameters to find the input impedance and voltage gain. It is because answers of such circuits can be found directly by inspection. However, in complicated circuits, inspection method becomes cumbersome and the use of *h* parameters yields quick results.

22.5 The *h* Parameters of a Transistor

It has been seen in the previous sections that every linear circuit is associated with *h* parameters. When this linear circuit is terminated by load r_L, we can find input impedance, current gain, voltage gain, *etc.* in terms of *h* parameters. Fortunately, for *small a.c. signals*, the transistor behaves as a linear device because the output a.c. signal is directly proportional to the input a.c. signal. Under such circumstances, the *a.c. operation* of the transistor can be described in terms of *h* parameters. The expressions derived for input impedance, voltage gain *etc.* in the previous section shall hold good for transistor amplifier except that here r_L is the a.c. load seen by the transistor.

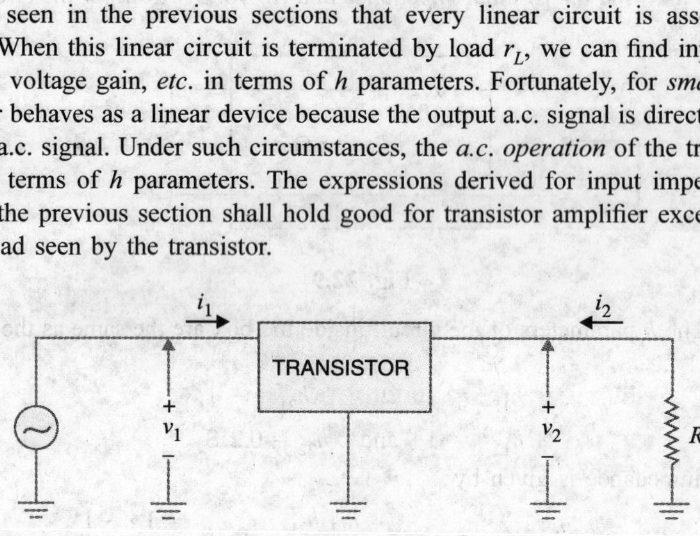

Fig. 22.10

Fig. 22.10 shows the transistor amplifier circuit. There are four quantities required to describe the external behaviour of the transistor amplifier. These are v_1, i_1, v_2 and i_2 shown on the diagram of Fig. 22.10. These voltages and currents can be related by the following sets of equations :

$$v_1 = h_{11}\, i_1 + h_{12}\, v_2$$

$$i_2 = h_{21}\, i_1 + h_{22}\, v_2$$

The following points are worth noting while considering the behaviour of transistor in terms of h parameters :

(*i*) For small a.c. signals, a transistor behaves as a linear circuit. Therefore, its a.c. operation can be described in terms of h parameters.

(*ii*) The value of h parameters of a transistor will depend upon the transistor connection (*i.e. CB, CE* or *CC*) used. For instance, a transistor used in *CB* arrangement may have h_{11} = 20 Ω. If we use the same transistor in *CE* arrangement, h_{11} will have a different value. Same is the case with other h parameters.

(*iii*) The expressions for input impedance, voltage gain *etc.* derived in Art. 22.4 are also applicable to transistor amplifier except that r_L is the a.c. load seen by the transistor *i.e.*

$$r_L = R_C \parallel R_L$$

(*iv*) The values of h parameters depend upon the operating point. If the operating point is changed, parameter values are also changed.

(*v*) The notations v_1, i_1, v_2 and i_2 are used for general circuit analysis. In a transistor amplifier, we use the notation depending upon the configuration in which transistor is used. Thus for *CE* arrangement,

$$v_1 = V_{be} \quad ; \quad i_1 = I_b \quad ; \quad v_2 = V_{ce} \quad ; \quad i_2 = I_c$$

Here V_{be}, I_b, V_{ce} and I_c are the R.M.S. values.

22.6 Nomenclature for Transistor *h* Parameters

The numerical subscript notation for h parameters (*viz.* h_{11}, h_{21}, h_{12} and h_{22}) is used in general circuit analysis. However, this nomenclature has been modified for a transistor to indicate the nature of parameter and the transistor configuration used. The h parameters of a transistor are represented by the following notation :

(*i*) The numerical subscripts are replaced by letter subscripts.

(*ii*) The first letter in the double subscript notation indicates the nature of parameter.

(*iii*) The second letter in the double subscript notation indicates the circuit arrangement (*i.e. CB, CE* or *CC*) used.

Table below shows the h parameter nomenclature of a transistor :

S.No.	*h* parameter	Notation in CB	Notation in CE	Notation in CC
1.	h_{11}	h_{ib}	h_{ie}	h_{ic}
2.	h_{12}	h_{rb}	h_{re}	h_{rc}
3.	h_{21}	h_{fb}	h_{fe}	h_{fc}
4.	h_{22}	h_{ob}	h_{oe}	h_{oc}

Note that first letter *i, r, f* or *o* indicates the nature of parameter. Thus h_{11} indicates input impedance and this parameter is designated by the subscript *i*. Similarly, letters *r, f* and *o* respectively indicate reverse voltage feedback ratio, forward current transfer ratio and output admittance. The second letters *b, e* and *c* respectively indicate *CB, CE* and *CC* arrangement.

22.7 Transistor *h* Parameter Model

We can visualise a transistor as a two-port or 4-terminal device because one of its three terminals is common to both input and output circuits. Fig. 22.11(*i*) shows the transistor in common emitter (*CE*) arrangement. The *h* parameter *a.c.* equations for this circuit are :

$$V_{be} = h_{ie}\, I_b + h_{re}\, V_{ce} \qquad ...(i)$$

$$I_c = h_{fe}\, I_b + h_{oe}\, V_{ce} \qquad ...(ii)$$

Here V_{be}, V_{ce}, I_b and I_c are the r.m.s. values. It should be noted that h_{ie} has the units of resistance (ohms) while h_{oe} has the units of conductance (siemens). The other two h_{re} and h_{fe} are pure numbers. Further, h_{ie} and h_{fe} are measured with $V_{ce} = 0$ (*i.e.* output short-circuited to a.c.) while h_{re} and h_{oe} are measured with $I_b = 0$ (*i.e.* input open-circuited to a.c.).

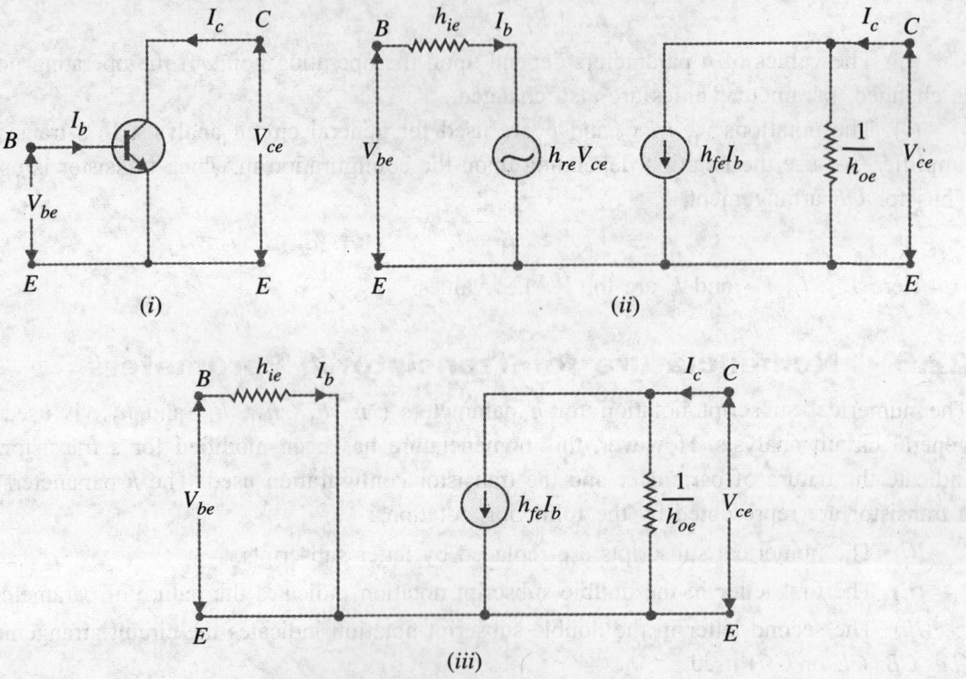

Fig. 22.11

Relations (*i*) and (*ii*) above can be represented by the circuit shown in Fig. 22.11 (*ii*) and is called *transistor h parameter model* or *hybrid equivalent circuit of transistor*. The input resistance h_{ie} appears in series at the input. The reverse voltage feedback ratio h_{re} is multiplied by the output voltage (V_{ce}) to produce an equivalent voltage source ($h_{re}\, V_{ce}$) in series with the input. The current gain h_{fe} is multiplied by the input current (I_b) to produce an equivalent current source ($h_{fe}\, I_b$) in the output. The output conductance h_{oe} appears across the output terminals. Although V_{ce} can be very much greater than V_{be} due to voltage gain obtainable from the transistor, h_{re} is very small (typically 5×10^{-4}) and hence $h_{re}\, V_{ce}$ is small in comparison to V_{be}. Therefore, this voltage source (*i.e.* $h_{re}\, V_{ce}$) can be removed from the input side without any large error. This produces the *simplified hybrid equivalent circuit of transistor* as shown in Fig. 22.11 (*iii*). The following points may be noted about the hybrid equivalent circuit of a transistor :

(*i*) The *h* parameter method is an analytical method. Therefore, it is more accurate than the graphical method for the analysis of a transistor circuit.

(*ii*) The hybrid a.c. equivalent circuit of the input circuit of *CE* circuit is derived using Thevenin's theorem. Thevenin's theorem states that any circuit can be replaced by a single voltage source in series with a single resistance. Here, the input circuit is replaced by a single voltage source $h_{re}\ V_{ce}$ in series with a single resistance h_{ie}.

(*iii*) The hybrid a.c. equivalent circuit of the output circuit of *CE* circuit is derived using Norton's theorem. Norton's theorem states that any circuit can be replaced by a single current source in parallel with a single resistance. Here, the output circuit is *replaced by a single current source $h_{fe}\ I_b$ in parallel with a single resistance $1/h_{oe}$. Note that h_{oe} is output conductance so that $1/h_{oe}$ is the output resistance.

Example 22.5. *The transistor used in the circuit shown in Fig. 22.12 has the following h parameters :*

$$h_{ie} = 2\ k\Omega\ ;\ h_{fe} = 60\ ;\ h_{oe} = 20\ \mu S\ ;\ h_{re} \text{ is negligible}$$

The input signal is supplied from a source is e.m.f. E in series with a resistance of 10 kΩ. Determine the necessary value of E to give a signal power of 10 mW in a load of 10 kΩ (resistive) connected across the output terminals. The reactances of the coupling capacitors may be neglected.

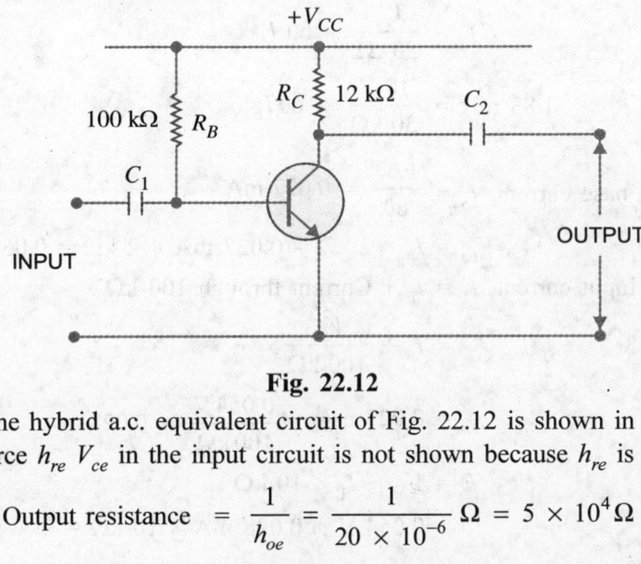

Fig. 22.12

Solution. The hybrid a.c. equivalent circuit of Fig. 22.12 is shown in Fig. 22.13. Note that voltage source $h_{re}\ V_{ce}$ in the input circuit is not shown because h_{re} is negligible.

$$\text{Output resistance} = \frac{1}{h_{oe}} = \frac{1}{20 \times 10^{-6}}\ \Omega = 5 \times 10^4\ \Omega = 50\ \text{k}\Omega$$

$$\text{Load resistance, } R_L = 10\ \text{k}\Omega$$

$$\text{Current source} = h_{fe}\ I_b = 60\ I_b$$

$$\text{Load power, } P = 10\ \text{mW} = 10 \times 10^{-3}\ \text{W}$$

$$\text{Now,} \quad P = \frac{V_{ce}^2}{R_L} \quad \text{or} \quad 10 \times 10^{-3} = \frac{V_{ce}^2}{10 \times 10^3}$$

$$\therefore \quad V_{ce} = \sqrt{100} = 10\ \text{V}$$

* The output circuit is replaced by Norton equivalent circuit because I_c depends upon factors outside of the collector-emitter circuit under normal conditions. Thus changing the value of collector resistance will hardly change I_c significantly. However, changing I_b *will.*

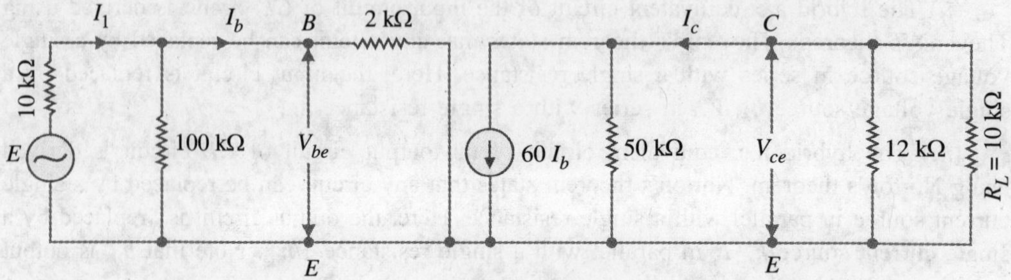

Fig. 22.13

Effective collection load, $r_L = R_C \parallel R_L = \dfrac{R_C\, R_L}{R_C + R_L}$

$$= \frac{12 \times 10}{12 + 10} = \frac{120}{22} \text{ k}\Omega$$

$\therefore$ A.C. collector current, $I_c = \dfrac{V_{ce}}{r_L} = \dfrac{10\text{ V}}{(120/22)\text{k}\Omega} = 1.83 \text{ mA}$

Now, $$I_c = \frac{V_{ce}}{50\text{ k}\Omega} + 60\, I_b$$

or $$1.83 \text{ mA} = \frac{10 \text{ V}}{50 \text{ k}\Omega} + 60 I_b$$

$\therefore$ A.C. base current, $I_b = \dfrac{1.63}{60} = 0.027 \text{ mA}$

$\therefore$ $$V_{be} = I_b \times 2 \text{ k}\Omega = 0.027 \text{ mA} \times 2 \text{ k}\Omega = 0.054 \text{ V}$$

$\therefore$ Input current, $I_1 = I_b$ + Current through 100 kΩ

$$= I_b + \frac{V_{be}}{100 \text{ k}\Omega}$$

$$= 0.027 \text{ mA} + \frac{0.054 \text{ V}}{100 \text{ k}\Omega} = 0.028 \text{ mA}$$

$\therefore$ $$E = V_{be} + I_1 \times 10 \text{ k}\Omega$$

$$= 0.054 \text{ V} + 0.028 \text{ mA} \times 10 \text{ k}\Omega = 0.334 \text{ V}$$

22.8 Transistor Circuit Performance in *h* Parameters

The expressions for input impedance, voltage gain *etc.* in terms of *h* parameters derived in Art. 22.4 for general circuit analysis apply equally for transistor analysis. However, it is profitable to rewrite them in standard transistor *h* parameter nomenclature.

(*i*) **Input impedance.** The general expression for input impedance is

$$Z_{in} = h_{11} - \frac{h_{12}\, h_{21}}{h_{22} + \dfrac{1}{r_L}}$$

Using standard *h* parameter nomenclature for transistor, its value for *CE* arrangement will be :

$$Z_{in} = h_{ie} - \frac{h_{re}\, h_{fe}}{h_{oe} + \frac{1}{r_L}}$$

Similarly, expressions for input impedance in *CB* and *CC* arrangements can be written. It may be noted that r_L is the a.c. load seen by the transistor.

(ii) **Current gain.** The general expression for current gain is

$$A_i = \frac{h_{21}}{1 + h_{22}\, r_L}$$

Using standard transistor *h* parameter nomenclature, its value for *CE* arrangement is

$$A_i = \frac{h_{fe}}{1 + h_{oe}\, r_L}$$

The reader can readily write down the expressions for *CB* and *CC* arrangements.

(iii) **Voltage gain.** The general expression for voltage gain is

$$A_v = \frac{-h_{21}}{Z_{in}\left(h_{22} + \frac{1}{r_L}\right)}$$

Using standard transistor *h* parameter nomenclature, its value for *CE* arrangement is

$$A_v = \frac{-h_{fe}}{Z_{in}\left(h_{oe} + \frac{1}{r_L}\right)}$$

In the same way, expressions for voltage gain in *CB* and *CC* arrangements can be written.

(iv) **Output impedance.** The general expression for output impedance is

$$Z_{out} = \frac{1}{h_{22} - \frac{h_{21}\, h_{12}}{h_{11}}}$$

Using standard transistors *h* parameter nomenclature, its value for *CE* arrangement is

$$Z_{out} = \frac{1}{h_{oe} - \frac{h_{fe}\, h_{re}}{h_{ie}}}$$

In the same way, expression for output impedance in *CB* and *CC* arrangements can be written.

The above expression for Z_{out} is for the transistor. If the transistor is connected in a circuit to form a single stage amplifier, then output impedance of the stage = $Z_{out} \parallel r_L$ where $r_L = R_C \parallel R_L$.

Example 22.6. *A transistor used in CE arrangement has the following set of h parameters when the d.c. operating point is* V_{CE} *= 10 volts and* I_C *= 1 mA :*

$h_{ie} = 2000\ \Omega$; $h_{oe} = 10^{-4}$ S ; $h_{re} = 10^{-3}$; $h_{fe} = 50$

Determine (i) input impedance (ii) current gain and (iii) voltage gain. The a.c. load seen by the transistor is $r_L = 600\ \Omega$. *What will be approximate values using reasonable approximations ?*

Solution. (*i*) Input impedance is given by ;

$$Z_{in} = h_{ie} - \frac{h_{re}\, h_{fe}}{h_{oe} + \frac{1}{r_L}} = 2000 - \frac{10^{-3} \times 50}{10^{-4} + \frac{1}{600}} \quad \text{.... } (i)$$

$$= 2000 - 28 = \mathbf{1972\ \Omega}$$

The second term in eq. (*i*) is quite small as compared to the first.

$$\therefore \quad Z_{in} \simeq h_{ie} = \mathbf{2000\ \Omega}$$

(*ii*) Current gain, $A_i = \dfrac{h_{fe}}{1 + h_{oe} \times r_L} = \dfrac{50}{1 + (600 \times 10^{-4})} = \mathbf{47}$

If $h_{oe}\, r_L << 1$, then $A_i \simeq h_{fe} = \mathbf{50.}$

(*iii*) Voltage gain, $A_v = \dfrac{-h_{fe}}{Z_{in}\left(h_{oe} + \dfrac{1}{r_L}\right)} = \dfrac{-50}{1972\left(10^{-4} + \dfrac{1}{600}\right)} = \mathbf{-14.4}$

The negative sign indicates that there is 180° phase shift between input and output. The magnitude of gain is 14.4. In other words, the output signal is 14.4 times greater than the input and it is 180° out of phase with the input.

Example 22.7. *A transistor used in CE connection has the following set of h parameters when the d.c. operating point is V_{CE} = 5 volts and I_C = 1 mA :*

h_{ie} = 1700 Ω ; h_{re} = 1.3 × 10^{-4} ; h_{fe} = 38 ; h_{oe} = 6 × 10^{-6} S

If the a.c. load r_L seen by the transistor is 2 kΩ, find (i) the input impedance (ii) current gain and (iii) voltage gain.

Solution. (*i*) The input impedance looking into the base of transistor is

$$Z_{in} = h_{ie} - \frac{h_{re}\, h_{fe}}{h_{oe} + \frac{1}{r_L}} = 1700 - \frac{1.3 \times 10^{-4} \times 38}{6 \times 10^{-6} + \frac{1}{2000}} \simeq \mathbf{1690\ \Omega}$$

(*ii*) Current gain, $A_i = \dfrac{h_{fe}}{1 + h_{oe} r_L} = \dfrac{38}{1 + 6 \times 10^{-6} \times 2000} = \dfrac{38}{1.012} \simeq \mathbf{37.6}$

(*iii*) Voltage gain, $A_v = \dfrac{-h_{fe}}{Z_{in}\left(h_{oe} + \dfrac{1}{r_L}\right)} = \dfrac{-38}{1690\left(6 \times 10^{-6} + \dfrac{1}{2000}\right)} = \mathbf{-44.4}$

Example 22.8. *Fig. 22.14 shows the transistor amplifier in CE arrangement. The h parameters of transistor are as under :*

h_{ie} = 1500 Ω ; h_{fe} = 50 ; h_{re} = 4 × 10^{-4} ; h_{oe} = 5 × 10^{-5} S

Find (i) a.c. input impedance of the amplifier (ii) voltage gain and (iii) output impedance.

Solution. The a.c. load r_L seen by the transistor is equivalent of the parallel combination of R_C (= 10 kΩ) and R_L (= 30 kΩ) *i.e.*

$$r_L = \frac{R_C\, R_L}{R_C + R_L} = \frac{10 \times 30}{10 + 30} = 7.5\ \text{k}\Omega = 7500\ \Omega$$

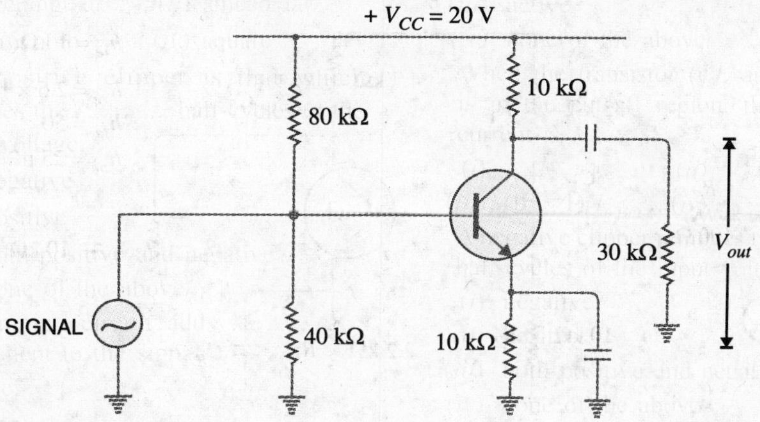

Fig. 22.14

(*i*) The input impedance looking into the base of transistor is given by ;

$$Z_{in} = h_{ie} - \frac{h_{re}\ h_{fe}}{h_{oe} + \dfrac{1}{r_L}} = 1500 - \frac{4 \times 10^{-4} \times 50}{5 \times 10^{-5} + \dfrac{1}{7500}} = 1390\ \Omega$$

This is only the input impedance looking into the base of transistor. The a.c. input impedance of the entire stage will be Z_{in} in parallel with bias resistors *i.e.*

Input impedance of stage = $80 \times 10^3 \parallel 40 \times 10^3 \parallel 1390$ = **1320 Ω**

(*ii*) Voltage gain,
$$A_v = \frac{-h_{fe}}{Z_{in}\left(h_{oe} + \dfrac{1}{r_L}\right)} = \frac{-50}{1390\left(5 \times 10^{-5} + \dfrac{1}{7500}\right)} = \mathbf{-196}$$

The negative sign indicates phase reversal. The magnitude of gain is 196.

(*iii*) Output impedance of transistor is

$$Z_{out} = \frac{1}{h_{oe} - \dfrac{h_{fe}\ h_{re}}{h_{ie}}}$$

$$= \frac{1}{5 \times 10^{-5} - \dfrac{50 \times 4 \times 10^{-4}}{1500}}$$

$$= 27270\ \Omega = 27.27\ \text{k}\Omega$$

∴ Output impedance of the stage

$$= Z_{out} \parallel R_L \parallel R_C$$

$$= 27.27\ \text{k}\Omega \parallel 30\ \text{k}\Omega \parallel 10\ \text{k}\Omega = \mathbf{5.88\ k\Omega}$$

Example 22.9. *Find the value of current gain for the circuit shown in Fig. 22.15. The h parameter values of the transistor are given alongside the figure.*

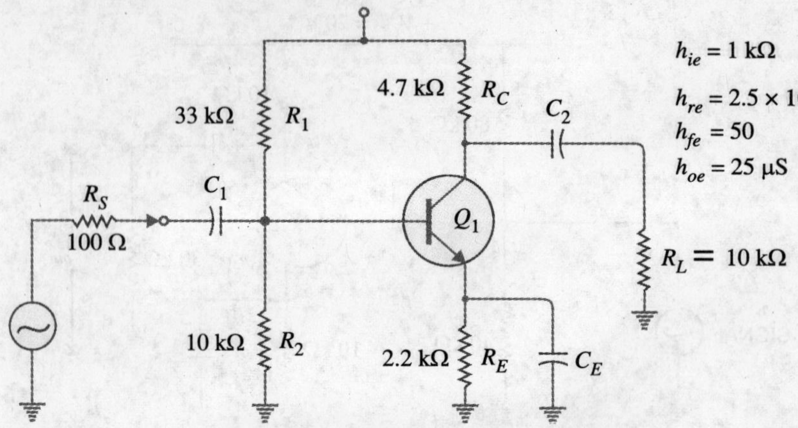

Fig. 22.15

Solution. The current gain A_i for the circuit is given by ;

$$A_i = \frac{h_{fe}}{1 + h_{oe}\, r_L}$$

$$\text{Here, } r_L = R_C \parallel R_L = 4.7 \text{ k}\Omega \parallel 10 \text{ k}\Omega = 3.2 \text{ k}\Omega$$

$$\therefore \quad A_i = \frac{50}{1 + (25 \times 10^{-6})\,(3.2 \times 10^3)} = \mathbf{46.3}$$

Note that current gain of the circuit is very close to the value h_{fe}. The reason for this is that $h_{oe}\, r_L << 1$. Since this is normally the case, $A_i \simeq h_{fe}$.

Example 22.10. *In the above example, determine the output impedance of the transistor.*

Solution. Note that the signal source (See Fig. 22.15) has resistance $R_S = 100\ \Omega$.

∴ Output impedance Z_{out} of the transistor is

$$Z_{out} = \frac{1}{h_{oe} - \left(\dfrac{h_{fe}\, h_{re}}{h_{ie} + R_S}\right)}$$

$$= \frac{1}{(25 \times 10^{-6}) - \dfrac{(50)\,(2.5 \times 10^{-4})}{(1 \times 10^3) + 100}}$$

$$= 73{\cdot}3 \times 10^3\ \Omega = 73.3 \text{ k}\Omega$$

Example 22.11. *A transistor has the following small-signal hybrid parameters :*

$h_{ie} = 1\ k\Omega\ ;\ h_{fe} = 50\ ;\ h_{oe} = 25\ \mu S\ ;\ h_{re} = 0$

The total effective load connected between the collector and emitter can be considered as a 5 kΩ resistor. The signal is supplied from a source which can be considered as an e.m.f. of 100 mV in series with a resistance of 9 kΩ. Determine (i) current gain (ii) voltage gain (iii) power gain of the transistor and (iv) signal power fed to the load. The effect of the bias components and the reactances of the coupling capacitors may be neglected.

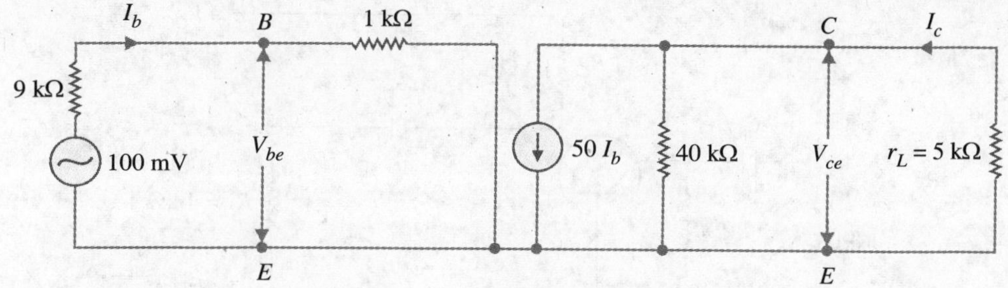

Fig. 22.16

Solution. Fig. 22.16 shows the a.c. equivalent circuit of the transistor amplifier where transistor has been replaced by its *simplified* hybrid equivalent circuit.

$$\text{Output resistance} = \frac{1}{h_{oe}} = \frac{1}{25 \times 10^{-6}} \Omega = 4 \times 10^4 \Omega = 40 \text{ k}\Omega$$

$$\text{Effective a.c. load, } r_L = 5 \text{ k}\Omega$$

$$\text{Current source} = h_{fe} I_b = 50 \, I_b$$

By current-divider rule, a.c. collector current I_c is given by ;

$$I_c = 50 \, I_b \times \frac{40}{40 + 5} \quad \text{...(i)}$$

(*i*) $\quad$ Current gain, $A_i = \dfrac{I_c}{I_b} = \dfrac{40 \times 50}{45} =$ **44.4** [From eq. (*i*)]

(*ii*) $\quad$ Voltage gain, $A_v = \dfrac{V_{ce}}{V_{be}} = \dfrac{I_c \times 5}{I_b \times 1} = \dfrac{I_c}{I_b} \times 5 = 44.4 \times 5 =$ **222**

(*iii*) $\quad$ Power gain, $A_p = A_v \, A_i = 222 \times 44.4 =$ **9860**

(*iv*) $\quad$ A.C. base current, $I_b = \dfrac{E}{(9+1) \text{ k}\Omega} = \dfrac{100 \text{ mV}}{10 \text{ k}\Omega} = 10 \, \mu\text{A}$

$\therefore$ $\quad$ A.C. collector current, $I_c = A_i \, I_b = 44.4 \times 10 \, \mu\text{A} = 444 \, \mu\text{A}$

$\therefore$ $\quad$ Signal power fed to load $= I_c^2 \, r_L = (444 \times 10^{-6})^2 \times 5 \times 10^3$

$$= 9.85 \times 10^{-4} \text{ W} = \mathbf{0.985 \text{ mW}}$$

Example 22.12. *The transistor used in the common base circuit shown in Fig. 22.17 has the following small-signal h parameters :*

$h_{ib} = 50 \ \Omega$; $h_{fb} = -0.96$; $h_{ob} = 1 \ \mu S$; h_{rb} *is negligible*

The signal source can be considered as a constant-current generator of 100 μA and an internal resistance of 450 Ω. Calculate the voltage and power gains of the stage when the load is a 1 kΩ resistor connected across the output terminals. The effect of the bias resistor and the reactances of coupling capacitors may be neglected.

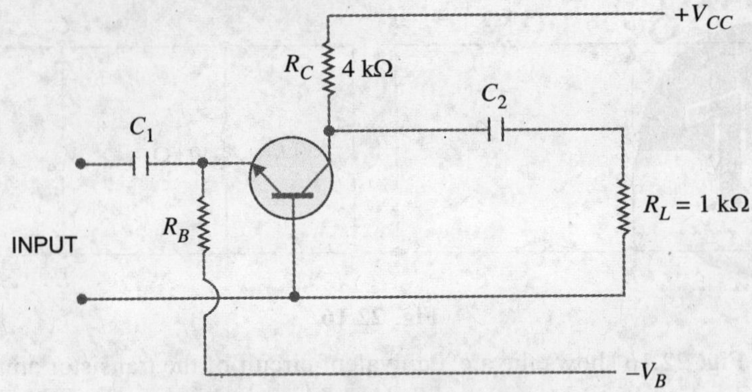

Fig. 22.17

Solution. Fig. 22.18 shows the a.c. equivalent circuit of the transistor amplifier where the transistor has been replaced by its simplified hybrid equivalent circuit.

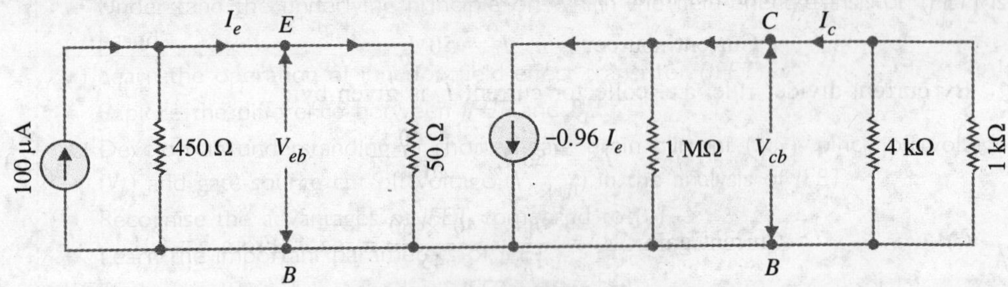

Fig. 22.18

$$\text{Output resistance} = \frac{1}{h_{ob}} = \frac{1}{1 \times 10^{-6}}\ \Omega = 10^6\ \Omega = 1\ \text{M}\Omega$$

$$\text{Load resistance, } R_L = 1\ \text{k}\Omega$$

Referring to the a.c. equivalent circuit in Fig. 22.18, current of 100 μA divides between 450 Ω and 50 Ω resistances. By current-divider rule, the a.c. emitter current I_e is given by ;

$$I_e = 100\ \mu\text{A} \times \frac{450}{450 + 50} = 90\ \mu\text{A}$$

Therefore, a.c. emitter-base voltage V_{eb} is

$$V_{eb} = I_e \times 50\ \Omega = 90\ \mu\text{A} \times 50\ \Omega = 4.5\ \text{mV}$$

$$\text{A.C. collector current, } I_c = -0.96^*\ I_e = -0.96 \times 90\ \mu\text{A} = -86\ \mu\text{A}$$

$$\text{Effective a.c. load, } r_L = R_C \parallel R_L = 4\ \text{k}\Omega \parallel 1\ \text{k}\Omega$$

$$= \frac{4\text{k}\Omega \times 1\text{k}\Omega}{4\text{k}\Omega + 1\text{k}\Omega} = \frac{4}{5}\ \text{k}\Omega$$

* The output resistance (= 1 MΩ) is very large as compared to 4 kΩ || 1 kΩ. Therefore, $I_c \simeq -0.96\ I_e$.

∴ A.C. collector-base voltage is

$$V_{cb} = -I_c r_L = -(-86\ \mu A) \times \frac{4}{5}\ k\Omega = 69\ mV$$

$$\therefore \quad \text{Voltage gain, } A_v = \frac{V_{cb}}{V_{eb}} = \frac{69\,mV}{4.5\,mV} = \mathbf{15}$$

$$\text{Signal power in load} = \frac{V_{cb}^2}{R_L} = \frac{(69 \times 10^{-3})^2}{1 \times 10^3} = 4.8 \times 10^{-6}\ W$$

$$\text{Signal power into stage} = I_e^2 \times 50\ \Omega = (90 \times 10^{-6})^2 \times 50\ \Omega = 0.41 \times 10^{-6}\ W$$

$$\therefore \quad \text{Power gain, } A_p = \frac{4.8 \times 10^{-6}}{0.41 \times 10^{-6}} = \mathbf{12}$$

Note. Power gain of transistor = 15 × 0.96 = 14.4

22.9 Approximate Hybrid Formulas for Transistor Amplifier

The h parameter formulas (CE configuration) covered in Art. 22.8 can be approximated to a form that is easier to handle. While these approximate formulas will not give results that are as accurate as the original formulas, they can be used for many applications.

(*i*) Input impedance

$$\text{Input impedance, } Z_{in} = h_{ie} - \frac{h_{re}\, h_{fe}}{h_{oe} + \frac{1}{r_L}}$$

In actual practice, the second term in this expression is very small as compared to the first term.

$$\therefore \quad Z_{in} = h_{ie} \quad \text{... approximate formula}$$

(*ii*) Current gain

$$\text{Current gain, } A_i = \frac{h_{fe}}{1 + h_{oe}\, r_L}$$

In actual practice, $h_{oe}\, r_L$ is very small as compared to 1.

$$\therefore \quad A_i = h_{fe} \quad \text{... approximate formula}$$

(*iii*) Voltage gain

$$\text{Voltage gain, } A_v = \frac{-h_{fe}}{Z_{in}\left(h_{oe} + \frac{1}{r_L}\right)}$$

$$= \frac{-h_{fe}\, r_L}{Z_{in}\,(h_{oe}\, r_L + 1)}$$

Now approximate formula for Z_{in} is h_{ie}. Also $h_{oe}\, r_L$ is very small as compared to 1.

$$\therefore \quad A_v = \frac{h_{fe}\, r_L}{h_{ie}} \quad \text{... approximate formula}$$

(*iv*) Output impedance

$$\text{Output impedance of transistor, } Z_{out} = \frac{1}{h_{oe} - \frac{h_{fe}\, h_{re}}{h_{ie}}}$$

The second term in the denominator is very small as compared to h_{oe}.

$$\therefore \quad Z_{out} = \frac{1}{h_{oe}} \quad \text{... approximate formula}$$

The output impedance of transistor amplifier

$$= Z_{out} \parallel r_L \quad \text{where } {}^*r_L = R_C \parallel R_L$$

Example 22.13. *For the circuit shown in Fig. 22.19, use approximate hybrid formulas to determine (i) the input impedance (ii) voltage gain. The h parameters of the transistor are h_{ie} = 1.94 kΩ and h_{fe} = 71.*

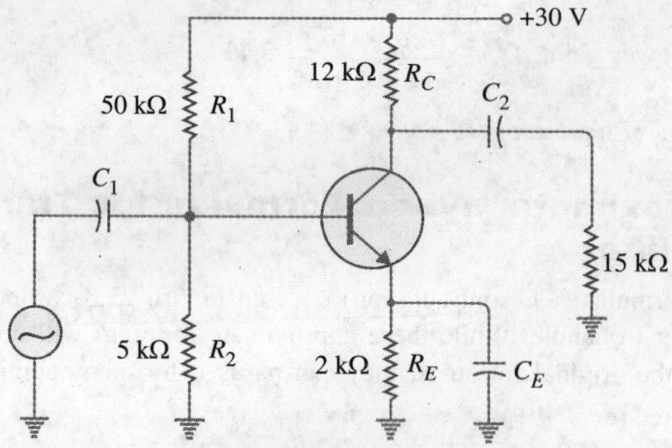

Fig. 22.19

Solution.

A.C. collector load, $r_L = R_C \parallel R_L = 12\ \text{k}\Omega \parallel 15\ \text{k}\Omega = 6.67\ \text{k}\Omega$

(*i*) Transistor input impedance is

$$Z_{in\ (base)} = h_{ie} = 1.94\ \text{k}\Omega$$

$$\therefore \quad \text{Circuit input impedance} = Z_{in\ (base)} \parallel R_1 \parallel R_2$$

$$= 1.94\ \text{k}\Omega \parallel 50\ \text{k}\Omega \parallel 5\ \text{k}\Omega = \mathbf{1.35\ k\Omega}$$

(*ii*) $$\text{Voltage gain, } A_v = \frac{h_{fe}\, r_L}{h_{ie}} = \frac{71 \times 6.67\ \text{k}\Omega}{1.94\ \text{k}\Omega} = \mathbf{244}$$

Example 22.14. *A transistor used in an amplifier has h parameter values of h_{ie} = 600 Ω to 800Ω and h_{fe} = 110 to 140. Using approximate hybrid formula, determine the voltage gain for the circuit. The a.c. collector load, r_L = 460 Ω.*

Solution. When *minimum* and *maximum h* parameter values are given, we should determine the *geometric average* of the two values. Thus the values that we would use in the analysis of circuit are found as under :

$$h_{ie} = \sqrt{h_{ie\ (min)} \times h_{ie\ (max)}}$$

$$= \sqrt{(600\ \Omega)\ (800\ \Omega)} = 693\Omega$$

$$h_{fe} = \sqrt{h_{fe\ (min)} \times h_{fe\ (max)}}$$

* If the amplifier is unloaded (*i.e.* $R_L = \infty$), $r_L = R_C$.

$$= \sqrt{(110)\,(140)} = 124$$

$$\text{Voltage gain, } A_v = \frac{h_{fe}\, r_L}{h_{ie}} = \frac{(124)\,(460)}{693} = 82.3$$

22.10 Experimental Determination of Transistor *h* Parameters

The determination of *h* parameters of a general linear circuit has already been discussed in Art. 22.2. To illustrate how such a procedure is carried out for a *CE* transistor amplifier, consider the circuit of Fig. 22.20. The r.m.s. values will be considered in the discussion. Using standard transistor nomenclature :

$$V_{be} = h_{ie}\, I_b + h_{re}\, V_{ce} \qquad ...(i)$$

$$I_c = h_{fe}\, I_b + h_{oe}\, V_{ce} \qquad ...(ii)$$

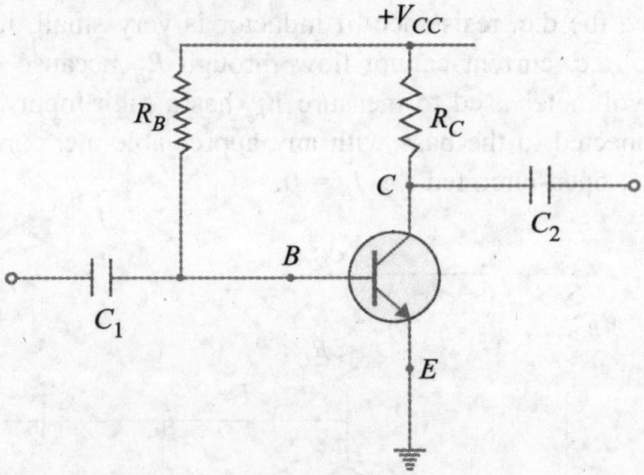

Fig. 22.20

(*i*) **Determination of h_{fe} and h_{ie}.** In order to determine these parameters, the output is a.c. short-circuited as shown in Fig. 22.21 (*i*). This is accomplished by making the capacitance of C_2 deliberately large. The result is that changing component of collector current flows through C_2 instead of R_C and a.c. voltage developed across C_2 is zero *i.e.* *$V_{ce} = 0$.

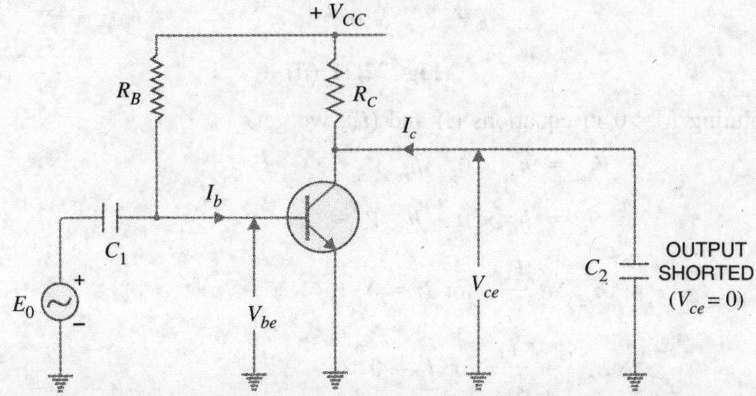

Fig. 22.21 (*i*)

* Note that setting $V_{ce} = 0$ does not mean that V_{CE} (the d.c. collector-emitter voltage) is zero. Only a.c. output is short-circuited.

Substituting $V_{ce} = 0$ in equations (*i*) and (*ii*) above, we get,

$$V_{be} = h_{ie} I_b + h_{re} \times 0$$

$$I_c = h_{fe} I_b + h_{oe} \times 0$$

$$\therefore \quad h_{fe} = \frac{I_c}{I_b} \text{ for } V_{ce} = 0$$

$$\text{and} \quad h_{ie} = \frac{V_{be}}{I_b} \text{ for } V_{ce} = 0$$

Note that I_c and I_b are the a.c. r.m.s. collector and base currents respectively. Also V_{be} is the a.c. r.m.s. base-emitter voltage.

(*ii*) Determination of h_{re} and h_{oe}. In order to determine these two parameters, the input is a.c. open-circuited, a signal generator is applied across the output and resulting V_{be}, V_{ce} and I_c are measured. This is illustrated in Fig. 22.21 (*ii*). A large inductor *L* is connected in series with R_B. Since the d.c. resistance of inductor is very small, it does not disturb the operating point. Again, a.c. current cannot flow through R_B because of large reactance of inductor. Further, the voltmeter used to measure V_{be} has a high input impedance and hence there are no paths connected to the base with any appreciable a.c. current. This means that base is *effectively a.c. open-circuited *i.e.* $I_b = 0$.

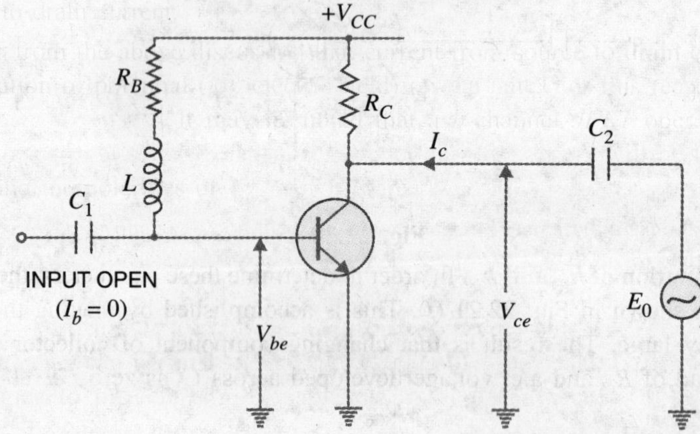

Fig. 22.21 (*ii*)

Substituting $I_b = 0$ in equations (*i*) and (*ii*), we get,

$$V_{be} = h_{ie} \times 0 + h_{re} V_{ce}$$

$$I_c = h_{fe} \times 0 + h_{oe} V_{ce}$$

$$\therefore \quad h_{re} = \frac{V_{be}}{V_{ce}} \text{ for } I_b = 0$$

$$\text{and} \quad h_{oe} = \frac{I_c}{V_{ce}} \text{ for } I_b = 0$$

* How effectively the base is a.c. open-circuited depends upon the reactance *L* and the input impedance of the voltmeter used to measure V_{be}.

Example 22.15. *The following quantities are measured in a CE amplifier circuit :*

(*a*) *With output a.c. short-circuited (i.e.* $V_{ce} = 0$*)*

$$I_b = 10\ \mu A\ ;\ I_c = 1\ mA\ ;\ V_{be} = 10\ mV$$

(*b*) *With input a.c. open-circuited (i.e.* $I_b = 0$*)*

$$V_{be} = 0.65\ mV\ ;\ I_c = 60\ \mu A\ ;\ V_{ce} = 1\ V$$

Determine all the four h parameters.

Solution.

$$h_{ie} = \frac{V_{be}}{I_b} = \frac{10 \times 10^{-3}}{10 \times 10^{-6}} = \mathbf{1000\ \Omega}$$

$$h_{fe} = \frac{I_c}{I_b} = \frac{1 \times 10^{-3}}{10 \times 10^{-6}} = \mathbf{100}$$

$$h_{re} = \frac{V_{be}}{V_{ce}} = \frac{0.65 \times 10^{-3}}{1} = \mathbf{0.65 \times 10^{-3}}$$

$$h_{oe} = \frac{I_c}{V_{ce}} = \frac{60 \times 10^{-6}}{1} = \mathbf{60\ \mu S}$$

22.11 Limitations of *h* Parameters

The *h* parameter approach provides accurate information regarding the current gain, voltage gain, input impedance and output impedance of a transistor amplifier. However, there are two major limitations on the use of these parameters.

(*i*) It is very difficult to get the exact values of *h* parameters for a particular transistor. It is because these parameters are subject to considerable variation—unit to unit variation, variation due to change in temperature and variation due to change in operating point. In predicting an amplifier performance, care must be taken to use *h* parameter values that are correct for the operating point being considered.

(*ii*) The *h* parameter approach gives correct answers for small a.c. signals only. It is because a transistor behaves as a linear device for small signals only.

Short Answer Questions

Q. 1. *What is a one-port network ?*

Ans. A one-port network is a two-terminal device or element (such as resistors, capacitors and inductors) through which a current may enter or leave the network. A one-port circuit is represented in Fig. 22.22.

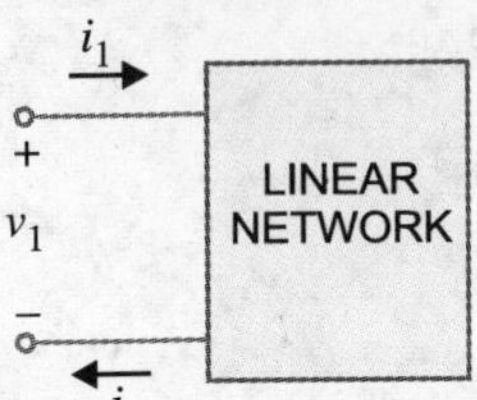

One-port Network

Fig. 22.22

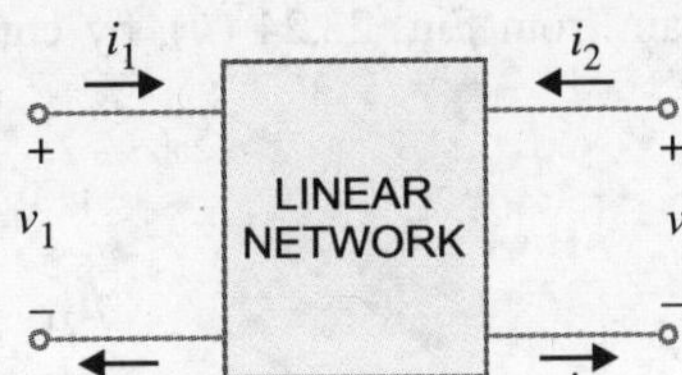

Two-port Network

Fig. 22.23

Q. 2. *What is a two-port network ?*

Ans. A two-port network is an electrical network with two separate ports for input and output. Thus a two-port network has four terminals—two for input and two for output as shown in Fig. 22.23. When voltage v_1 is applied to the input terminals, then input and output currents i_1 and i_2 flow and an output v_2 is produced. Three-terminal devices such as transistors can be arranged as a two-port network by making one of its terminals common to both input and output terminals.

Q. 3. *Why do we study two-port networks ?*

Ans. We study two-port networks for two main reasons. First, such networks are useful in communications, control systems, power systems and electronics. For example, they are used in electronics to model transistors and to facilitate cascaded design. Second, knowing the parameters of a two-port network enables us to treat it as a "black box" when embedded within a larger network.

Q. 4. *Why are h parameters of a two-port linear network called hybrid parameters ?*

Ans. Hybrid means mixed. The four parameters of a two-port linear circuit (*viz* h_{11}, h_{12}, h_{21} and h_{22}) do not have the same dimensions and hence they are called hybrid parameters.

Q. 5. *What are the h parameters h_{11} and h_{21} for the two-port network shown in Fig. 22.24 (i) ?*

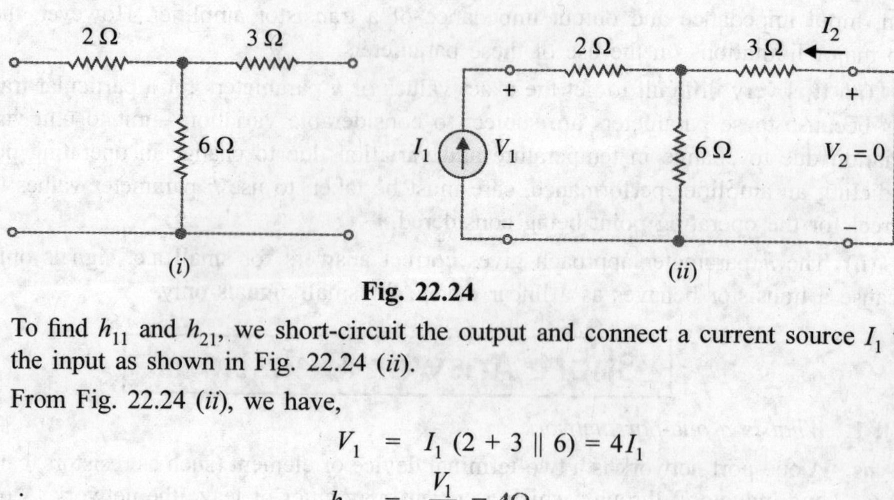

Fig. 22.24

Ans. To find h_{11} and h_{21}, we short-circuit the output and connect a current source I_1 to the input as shown in Fig. 22.24 (*ii*).

From Fig. 22.24 (*ii*), we have,

$$V_1 = I_1 (2 + 3 \parallel 6) = 4I_1$$

$$\therefore \quad h_{11} = \frac{V_1}{I_1} = 4\Omega$$

Again from Fig. 22.24 (*ii*), by current-divider rule

$$-I_2 = \frac{6}{6+3} I_1$$

$$\therefore \quad h_{21} = \frac{I_2}{I_1} = \frac{-2}{3}$$

Q. 6. *Why are hybrid parameters of a transistor named so ?*

Ans. Hybrid means "mixed". A transistor has four *h* parameters. One of them (input resistance) has units of ohms and the other (output conductance) has the units of siemens. The remaining two *h* parameters (reverse voltage feedback ratio and current gain) are pure numbers. Since h parameters have mixed units, they are called hybrid parameters.

Q. 7. *What are the two conditions for the measurement of h parameters ?*

Ans. A transistor has four h parameters. Two of them (*i.e.* input resistance and current gain) are measured with $V_{ce} = 0$ (*i.e.* output short-circuited to a.c.). The remaining two (*i.e.* reverse voltage feedback ratio and output conductance) are measured with $I_b = 0$ (*i.e.* input open-circuited to a.c.).

Q. 8. *While determining the h parameter input resistance (h_{ie}) of a transistor, why is output short-circuited ?*

Ans. You may recall that any resistance in the emitter circuit is reflected back to the base. This condition is described in the following equation :

$$Z_{in(base)} = h_{fe}\,(r_e' + R_E) \text{............See Art. 8.18}$$

By shorting the collector and emitter terminals, the measured value of input resistance does not reflect any external resistance in the circuit.

Q. 9. *While determining the h parameter current gain (h_{fe}) of a transistor, why is output short-circuited ?*

Ans. A shorted output represents full-load so current gain (h_{fe}) represents the gain under full-load conditions. If the output was left open, I_c would be zero. Shorting the output gives us a measurable value of I_c that can be reproduced in a particular test.

Q. 10. *While determining the h parameter reverse voltage feedback ratio (h_{re}), why is input open-circuited ?*

Ans. The reverse voltage feedback ratio (h_{re}) is the amount of output voltage that is reflected back to the input. This value is measured with the input open. The value of h_{re} (= V_{be}/V_{ce}) is always less than 1. By measuring h_{re} with input open, you ensure that voltage fed back to the base will always be at its maximum possible value because maximum voltage is always developed across an open-circuit.

Multiple-Choice Questions

1. Hybrid means ....
   (*i*) mixed (*ii*) single
   (*iii*) unique
   (*iv*) none of the above

2. There are .... h parameters of a transistor.
   (*i*) two (*ii*) four
   (*iii*) three
   (*iv*) none of the above

3. The h parameters approach gives correct results for ....
   (*i*) large signals only
   (*ii*) small signals only
   (*iii*) both small and large signals
   (*iv*) none of the above

4. A transistor behaves as a linear device for ....
   (*i*) small signals only
   (*ii*) large signals only
   (*iii*) both small and large signals
   (*iv*) none of the above

5. The parameter h_{ie} stands for input impedance in ....
   (*i*) *CB* arrangement with output shorted
   (*ii*) *CC* arrangement with output shorted
   (*iii*) *CE* arrangement with output shorted
   (*iv*) none of the above

6. The dimensions of h_{ie} parameter are ....
   (*i*) siemens (*ii*) ohms
   (*iii*) farads
   (*iv*) none of the above

7. The parameter h_{fe} is called .... in *CE* arrangement with output shorted.
   (*i*) voltage gain
   (*ii*) current gain
   (*iii*) input impedance
   (*iv*) none of the above
8. If the operating point changes, the *h* parameters of a transistor ....
   (*i*) also change
   (*ii*) do not change
   (*iii*) may or may not change
   (*iv*) none of the above
9. The values of *h* parameters of a transistor in *CE* arrangement are .... arrangement.
   (*i*) the same as for *CB*
   (*ii*) the same as for *CC*
   (*iii*) different from that in *CB*
   (*iv*) none of the above
10. In order to determine h_{fe} and h_{ie} parameters of a transistor, ... is a.c. short-circuited.
    (*i*) input
    (*ii*) output
    (*iii*) input as well as output
    (*iv*) none of the above
11. If temperature changes, *h* parameters of a transistor ....
    (*i*) may or may not change
    (*ii*) do not change
    (*iii*) also change
    (*iv*) none of the above
12. In *CE* arrangement, the value of input impedance is approximately equal to......
    (*i*) h_{ie} (*ii*) h_{oe}
    (*iii*) h_{re}
    (*iv*) none of the above
13. Using standard transistor *h* parameter nomenclature, the voltage gain in *CE* arrangement is ....
    (*i*) $\dfrac{-h_{fe}}{Z_{in}\left(h_{oe}+\dfrac{1}{r_L}\right)}$
    (*ii*) $\dfrac{-h_{fe}}{Z_{out}\left(h_{oe}+1\right)}$
    (*iii*) $\dfrac{-h_{fe}}{h_{oe}+h_{re}}$
    (*iv*) none of the above
14. $Z_{in} = h_{ie} - \dfrac{\cdots\cdots}{h_{oe}+\dfrac{1}{r_L}}$
    (*i*) $h_{re}\ h_{oe}$
    (*ii*) $h_{re}\ h_{fe}$
    (*iii*) $r_L\ h_{oe}$
    (*iv*) none of the above
15. .... *h* parameters of a transistor are dimensionless.
    (*i*) Four (*ii*) Three
    (*iii*) Two
    (*iv*) None of the above

Answers to Multiple-Choice Questions

1. (*i*)	2. (*ii*)	3. (*ii*)	4. (*i*)	5. (*iii*)
6. (*ii*)	7. (*ii*)	8. (*i*)	9. (*iii*)	10. (*ii*)
11. (*iii*)	12. (*i*)	13. (*i*)	14. (*ii*)	15. (*iii*)

Chapter Review Topics

1. What do you understand by hybrid parameters ? What are their dimensions ?
2. How will you measure *h* parameters of a linear circuit ?
3. Draw the *h* parameter equivalent circuit of a linear circuit.
4. What is the physical meaning of *h* parameters ?
5. Derive the general formula for :
   (*i*) input impedance (*ii*) current gain and (*iii*) voltage gain in terms of *h* parameters and the load.

6. What are the notations for h parameters of a transistor when used in (*i*) *CB* (*ii*) *CE* and (*iii*) *CC* arrangement ?
7. How are h parameters of a transistor measured ?
8. What are the drawbacks of h parameter approach in the design of a transistor amplifier ?

Problems

1. Determine the h-parameter values for the circuit shown in Fig. 22.25.

[h_{11} = 8Ω ; h_{21} = – 0.5 ; h_{12} = 0.5 ; h_{22} = 0.125 S]

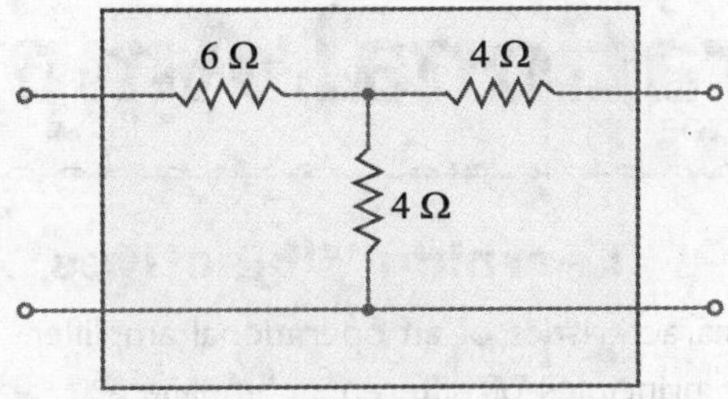

Fig. 22.25

2. Measurement of a circuit gives the following h-parameters :
h_{11} = 10 kΩ ; h_{12} = 0.5 ; h_{21} = 100 ; h_{22} = 2 mS
Find v_1 and i_2 if i_1 = 1 mA and v_2 = 2V.

[v_1 = 11V ; i_2 = 14 mA]

3. A *CE* amplifier has h_{ie} = 1000Ω, $h_{re} = 10^{-4}$, h_{fe} = 100 and $h_{oe} = 12 \times 10^{-6}$S. The load resistance = 2000 Ω. Find (*i*) current gain (*ii*) voltage gain (*iii*) output resistance.

[(*i*) 97.7 (*ii*) – 199.2 (*iii*) 5 × 10^5Ω]

4. A *CE* amplifier has h_{ie} = 1500Ω, h_{fe} = – 60 and $h_{oe} = 12.5 \times 10^{-6}$S. The load resistance varies between 5×10^3Ω and 10×10^3Ω. Find the maximum and minimum values of (*i*) current gain (*ii*) voltage gain.

[(*i*) 36.9, 26.7 (*ii*) 178, 123]

5. An amplifier has values of R_C = 12 kΩ, R_L = 4.7 kΩ, R_1 = 33 kΩ, R_2 = 4.7 kΩ and I_C = 1 mA. At 1 mA, the transistor has h-parameter values of h_{ie} = 1 kΩ to 5 kΩ and h_{fe} = 70 to 350. Determine the values of (*i*) input impedance (*ii*) voltage gain for the circuit. Use approximate hybrid formulas. **[(*i*) 1.45 kΩ (*ii*) 236.9]**

CHAPTER 23

Operational Amplifiers

Learning Objectives

- Establish the characteristics of an operational amplifier.
- Learn the basic principles of differential amplifier.
- Understand the operation of differential amplifier on the application of common-mode and differential mode signals.
- Explore the importance of *CMRR*.
- Recognise that the input offset parameters in an op-amp cause a d.c. offset on the output voltage.
- Explore the various factors on which the output voltage of op-amp depends.
- Explore the difference between practical op-amp and an ideal op-amp.
- Learn how slew rate in an op-amp can distort an amplified waveform.
- Develop an understanding of inverting and noninverting amplifier.
- Learn the principles of op-amp circuits for performing integration and differentiation.

Introduction

Historically, an operational amplifier (*OP*-amp) was designed to perform such mathematical operations as addition, subtraction, integration and differentiation. Hence, the name operational amplifier. An operational amplifier is a multistage amplifier and consists of a differential amplifier stage, a high-gain *CE* amplifier stage and class *B* push-pull emitter follower. An operational amplifier (*OP*-amp) is an *integrated circuit and is widely used in computers, as video and audio amplifiers in communication electronics. Because of their multi-purpose use, *OP*-amps are used in all branches of electronics, both digital and linear circuits. In this chapter, we shall discuss the various aspects of operational amplifiers.

23.1 Operational Amplifier

An **operational amplifier** *(OP-amp) is a circuit that can perform such mathematical operations as addition, subtraction, integration and differentiation.*

* All the components of an *OP*-amp (*e.g.*, transistors, resistors *etc.*) are fabricated on a small chip called integrated circuit.

Fig. 23.1 shows the block diagram of an operational amplifier. Note that *OP*-amp is a multistage amplifier. The three stages are : differential amplifier input stage followed by a high-gain *CE* amplifier and finally the output stage. *The key electronic circuit in an OP-amp is the* **differential amplifier.** A differential amplifier (*DA*) can accept two input signals and amplifies the difference between these two input signals.

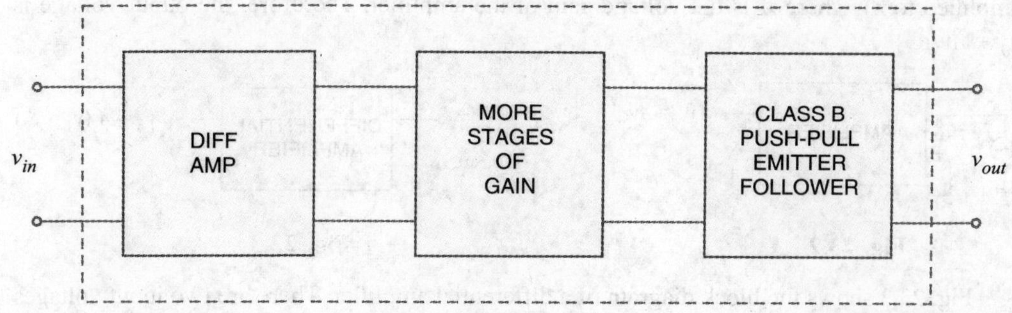

Block diagram of OP-Amp

Fig. 23.1

The following points may be noted about operational amplifiers (*OP*-amps) :

(*i*) The input stage of an *OP*-amp is a *differential amplifier* (*DA*) and the output stage is typically a class *B* push-pull emitter follower.

(*ii*) The internal stages of an *OP*-amp are *direct-coupled i.e.*, no coupling capacitors are used. The direct coupling allows the *OP*-amp to amplify d.c. as well as a.c. signals.

(*iii*) An *OP*-amp has *very high input impedance* (ideally infinite) and *very low output impedance* (ideally zero). The effect of high input impedance is that the amplifier will draw a very small current (ideally zero) from the signal source. The effect of very low output impedance is that the amplifier will provide a constant output voltage independent of current drawn from the source.

(*iv*) An *OP*-amp has *very high *open-loop voltage gain* (ideally infinite); typically more than 200,000.

(*v*) The *OP*-amps are almost always operated with negative feedback. It is because the open-loop voltage gain of these amplifiers is very high and we can sacrifice the gain to achieve the advantages of negative feedback including large bandwidth (*BW*) and gain stability (Refer to chapter 11 of the book).

23.2 Differential Amplifier (DA)

Since differential amplifier (*DA*) is key to the operation of *OP*-amp, we shall discuss this circuit in detail. So far in the book we have considered general-purpose amplifiers. In these conventional amplifiers, the signal (generally single input) is applied at the input terminals and amplified output is obtained at the output terminals. However, we can design an amplifier

* The gain of an *OP*-amp without feedback circuit is called open-loop gain. The gain of an *OP*-amp with feedback circuit is called closed-loop gain. These terms were discussed in Chapter 11.

circuit that accepts two input signals and amplifies the difference between these two signals. Such an amplifier is called a ***differential amplifier* (*DA*).

***A differential amplifier** is a circuit that can accept two input signals and amplify the difference between these two input signals.*

Fig. 23.2 shows the block diagram of an ordinary amplifier. The input voltage v is amplified to Av where A is the voltage gain of the amplifier. Therefore, the ouput voltage is $v_o = Av$.

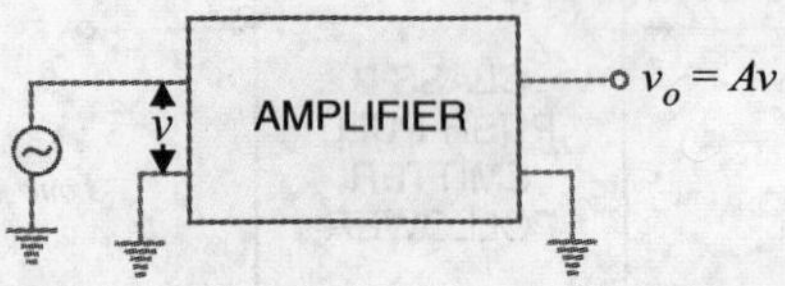

Fig. 23.2

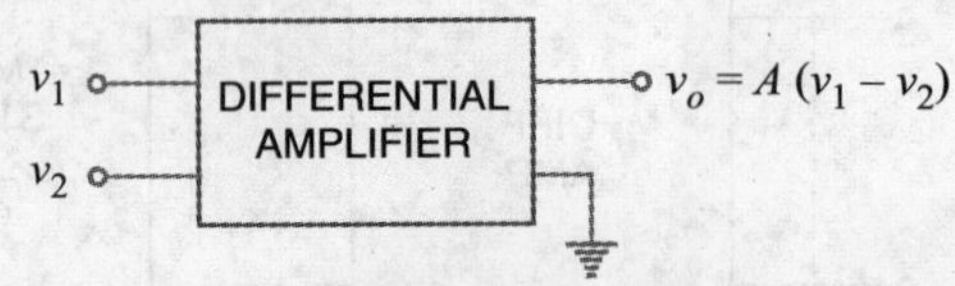

Fig. 23.3

Fig. 23.3 shows the block diagram of a differential amplifier. There are two input voltages v_1 and v_2. This amplifier amplifies the difference between the two input voltages. Therefore, the output voltage is $v_o = A(v_1 - v_2)$ where A is the voltage gain of the amplifier.

Example 23.1. *A differential amplifier has an open-circuit voltage gain of 100. The input signals are 3.25 V and 3.15 V. Determine the output voltage.*

Solution.

Output voltage, $v_o = A(v_1 - v_2)$

Here, $A = 100$; $v_1 = 3.25$ V ; $v_2 = 3.15$ V

$\therefore$ $v_o = 100(3.25 - 3.15) = \mathbf{10\ V}$

23.3 Basic Circuit of Differential Amplifier

Fig. 23.4(*i*) shows the basic circuit of a differential amplifier. It consists of two transistors Q_1 and Q_2 that have identical (ideally) characteristics. They share a common positive supply **V_{CC}, common emitter resistor R_E and common negative supply V_{EE}. Note that the circuit is symmetrical. Fig. 23.4(*ii*) shows the symbol of differential amplifier.

The following points may be noted about the differential amplifier :

(*i*) The differential amplifier (*DA*) is a two-input terminal device using atleast two transistors. There are two output terminals marked 1 ($v_{out\,1}$) and 2($v_{out\,2}$).

(*ii*) The *DA* transistors Q_1 and Q_2 are matched so that their characteristics are the same. The collector resistors (R_{C1} and R_{C2}) are also equal. The equality of the matched circuit components makes the *DA* circuit arrangement completely symmetrical.

(*iii*) We can apply signal to a differential amplifier (*DA*) in the following two ways :

(*a*) The signal is applied to one input of *DA* and the other input is grounded. In that case, it is called *single-ended input* arrangement.

* The name is appropriate because the amplifier is amplifying the difference between the two input signals.

** Note that for this circuit, we need two supply voltages *viz.* $+V_{CC}$ and $-V_{EE}$. The negative terminal of V_{CC} is grounded and positive terminal of V_{EE} is grounded.

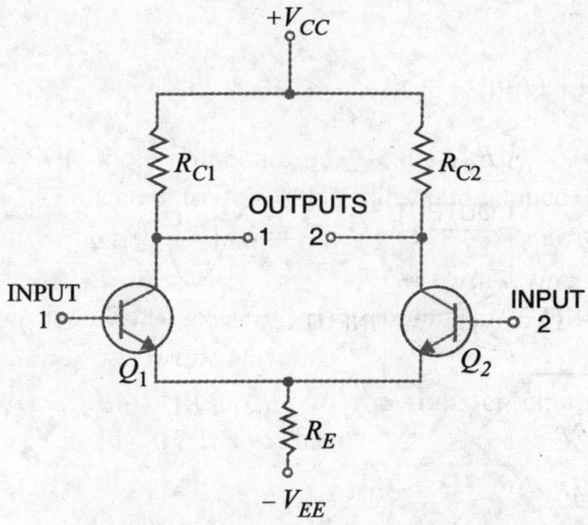

(i)

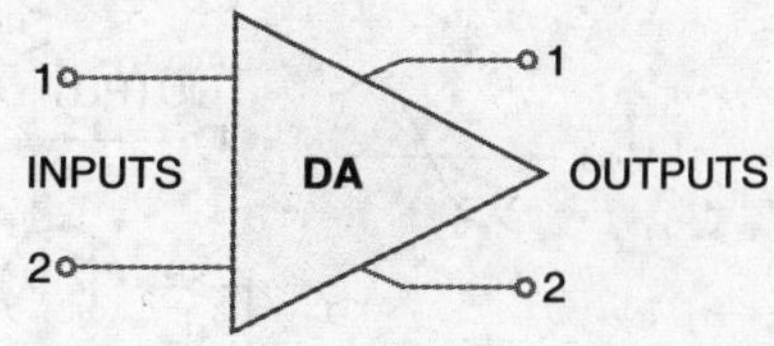

(ii)

Fig. 23.4

(*b*) The signals are applied to both inputs of *DA*. In that case, it is called *dual-ended* or *double-ended input* arrangement.

(*iv*) We can take output from *DA* in the following two ways :

(*a*) The output can be taken from one of the output terminals and the ground. In that case, it is called *single-ended output* arrangement.

(*b*) The output can be taken between the two output terminals (*i.e.*, between the collectors of Q_1 and Q_2). In that case, it is called *double-ended output* arrangement or *differential output*.

(*v*) *Generally, the differential amplifier (DA) is operated for single-ended output.* In other words, we take the output either from output terminal 1 and ground or from output terminal 2 and ground. Any input/output terminal that is grounded is at 0V.

23.4 Operation of Differential Amplifier

For *simplicity, we shall discuss the operation of single-ended input (*i.e.*, signal is applied to one input of *DA* and the other input is grounded) and double-ended output *DA*.

(*i*) Suppose the signal is applied to input 1 (*i.e.*, base of transistor Q_1) and input 2 (*i.e.*, base of transistor Q_2) is grounded as shown in Fig. 23.5. The transistor Q_1 will act in two ways : as a common emitter amplifier and as a common collector amplifier. As a common emitter amplifier, the input signal to Q_1 (input 1) will appear at output 1 (*i.e.*, collector of Q_1) as amplified inverted signal as shown in Fig. 23.5. As a common collector amplifier, the signal appears on the emitter of Q_1 in phase with the input and only slightly smaller. Since the emitters of Q_1 and Q_2 are common, the emitter signal becomes input to Q_2. Therefore, Q_2 functions as a **common base amplifier. As a result, the signal on the emitter of Q_2 will be amplified and appears on output 2 (*i.e.*, collector of Q_2) in phase with the emitter signal and hence in phase with the input signal (signal at input 1). This is illustrated in Fig. 23.5.

* The operation of double-ended input *DA* will then be easier to understand.

** In a common base amplifier, output signal is in phase with the input signal. Recall that only in *CE* amplifier, the output voltage is 180° out of phase with the input voltage.

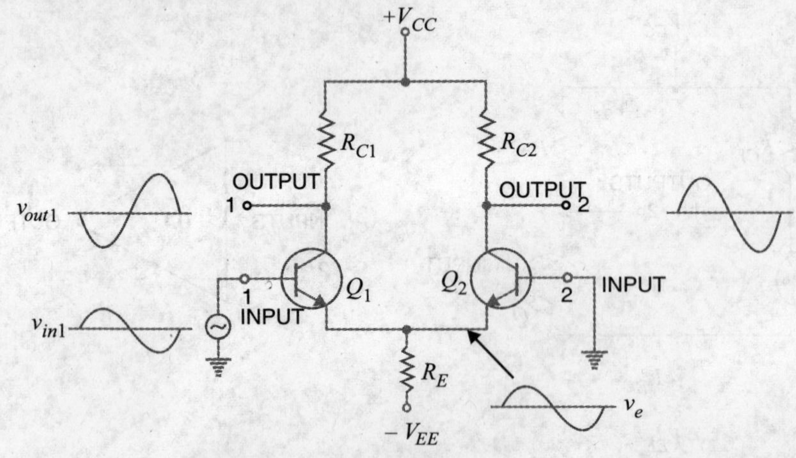

Fig. 23.5

(*ii*) Now suppose the signal is applied to input 2(*i.e.*, base of transistor Q_2) and input 1(base of transistor Q_1) is grounded. As explained above, now Q_2 acts as a common emitter amplifier and common collector amplifier while Q_1 functions as a common base amplifier. Therefore, an inverted and amplified signal appears at output 2 (*i.e.*, at collector of Q_2) and non-inverted, amplified signal appears at output 1(*i.e.*, at collector of Q_1). This is illustrated in Fig. 23.6.

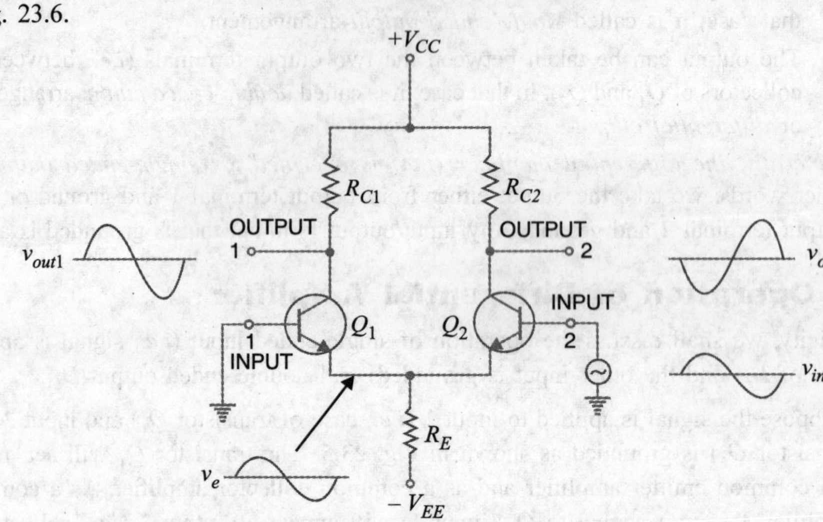

Fig. 23.6

The following points are worth noting about single-ended input *DA* :

(*a*) When signal is applied to input 1 (*i.e.*, base of transistor Q_1 in Fig. 23.5), an inverted, amplified signal appears at output 1 and non-inverted, amplified signal appears at output 2. Reverse happens when signal is applied to input 2 and input 1 is grounded.

(*b*) When only one output terminal is available, the phase of the output of single-ended input *DA* depends on which input receives the input signal. This is illustrated in Fig. 23.7.

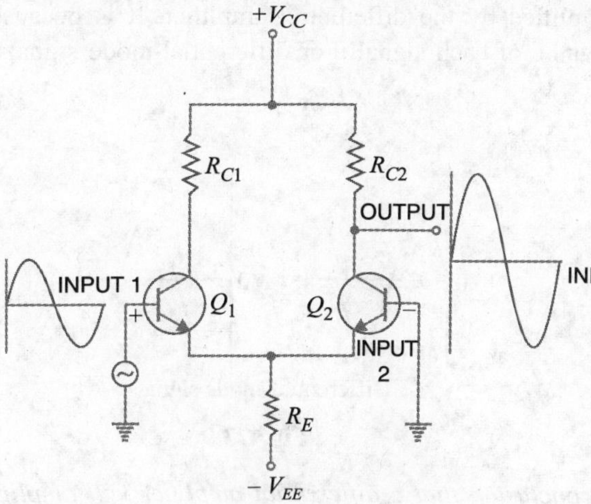

(*i*) Noninverting input

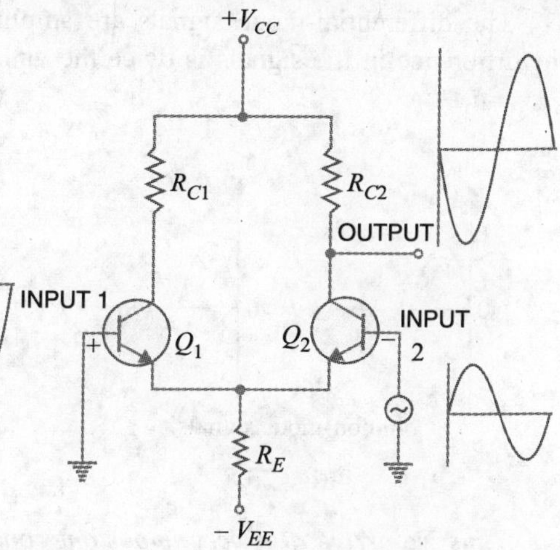

(*ii*) Inverting input

Fig. 23.7

When signal applied to the input of *DA* produces no phase shift in the output, it is called *non-inverting input* [See Fig. 23.7 (*i*)]. In other words, for noninverting input, the output signal is in phase with the input signal. When the signal applied to the input of *DA* produces 180° phase shift, it is called *inverting input* [See Fig. 23.7 (*ii*)]. In other words, for inverting input, the output signal is 180° out of phase with the input signal. Since inverting input provides 180° phase shift, it is often identified with –sign. The noninverting input is then represented by +*sign. It may be noted that terms noninverting input and inverting input are meaningful when only one output terminal of *DA* is available.

23.5 Common-mode and Differential-mode Signals

The importance of a differential amplifier lies in the fact that the outputs are proportional to the *difference* between the two input signals. Thus the circuit can be used to amplify the difference between the two input signals or amplify only one input signal simply by grounding the other input. The input signals to a *DA* are defined as :

(*i*) Common-mode signals (*ii*) Differential-mode signals

(*i*) Common-mode signals. When the input signals to a *DA* are in phase and exactly equal in amplitude, they are called *common-mode signals* as shown in Fig. 23.8. The common-mode signals are rejected (not amplified) by the differential amplifier. It is because a differential amplifier amplifies the difference between the two signals ($v_1 - v_2$) and for common-mode signals, this difference is zero. Note that for common-mode operations, $v_1 = v_2$.

(*ii*) Differential-mode signals. When the input signals to a *DA* are 180° out of phase and exactly equal in amplitude, they are called *differential-mode signals* as shown in Fig. 23.9.

* Note that in Fig. 23.7, the noninverting input terminal is given the +ve sign while the inverting input terminal is given the –ve sign.

The differential-mode signals are amplified by the differential amplifier. It is because the difference in the signals is twice the value of each signal. For differential-mode signals, $v_1 = -v_2$.

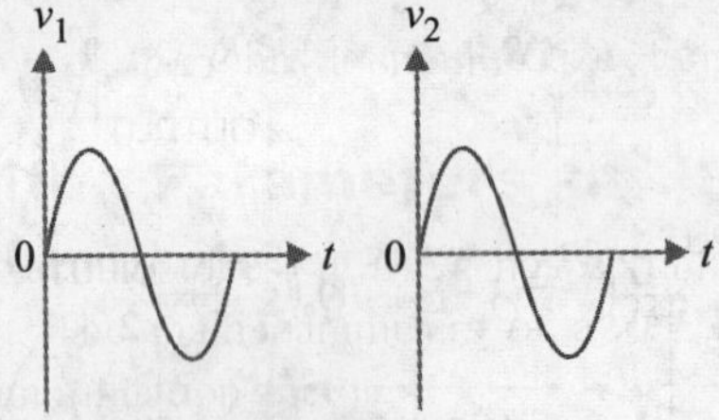

Common-mode signals

Fig. 23.8

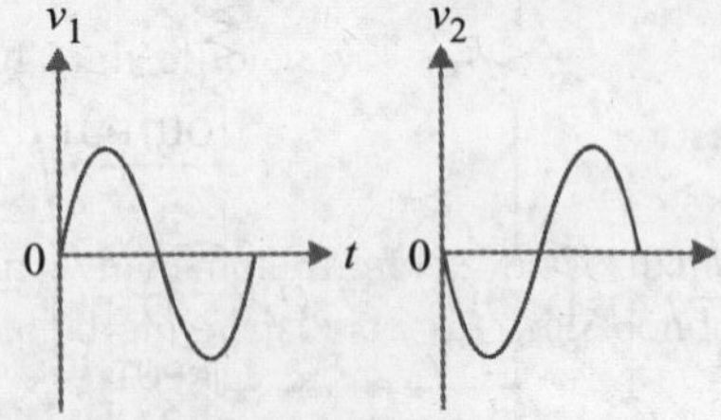

Differential-mode signals

Fig. 23.9

Thus we arrive at a very important conclusion that a differential amplifier will amplify the differential-mode signals while it will reject the common-mode signals.

23.6 Double-ended Input Operation of DA

A differential amplifier (*DA*) has two inputs so that it can simultaneously receive two signals. The input signals to a *DA* are defined as :

(*i*) Differential-mode signals (*ii*) Common-mode signals

The differential-mode signals are equal in amplitude but 180° out of phase. The common-mode signals are equal in magnitude and have the same phase.

(*i*) **Differential input.** In this mode (arrangement), two opposite-polarity (180° out of phase) signals are applied to the inputs of *DA* as shown in Fig. 23.10 (*i*).

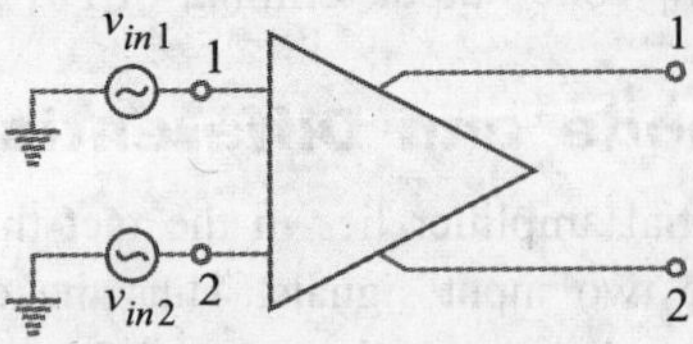

(*i*) Differential inputs

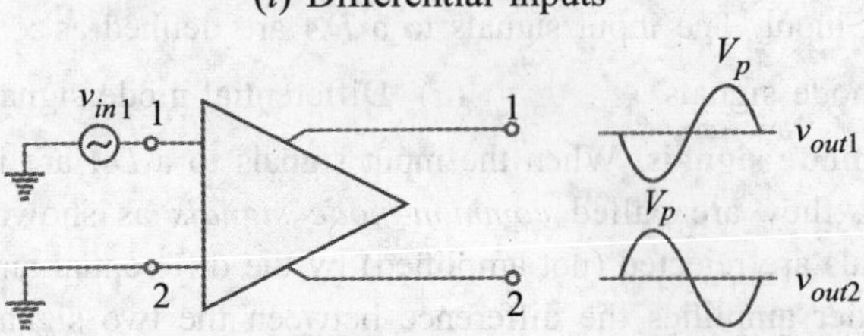

(*ii*) Outputs due to v_{in1}

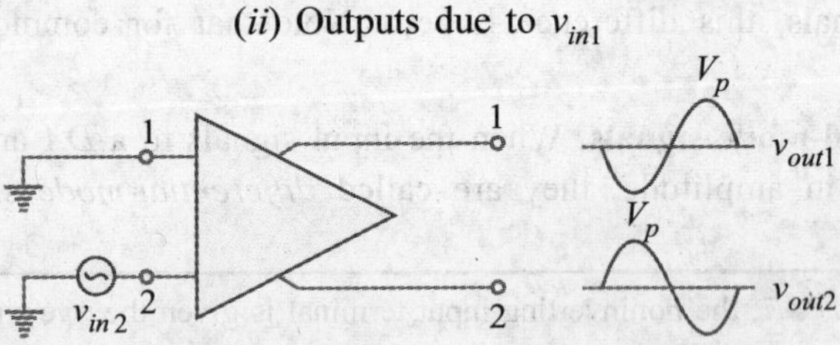

(*iii*) Outputs due to v_{in2}

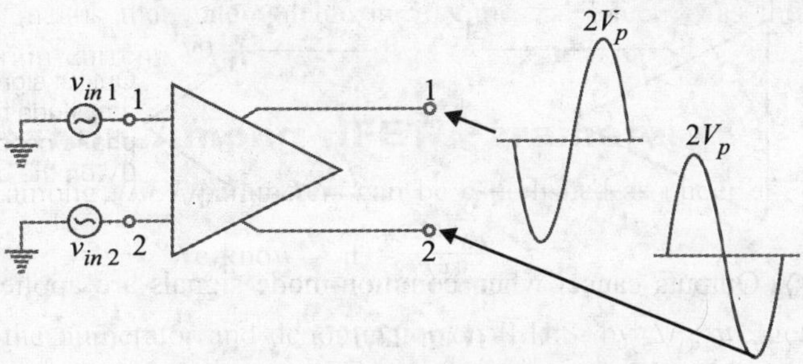

(*iv*) Total outputs due to differential inputs.

Fig. 23.10

As we shall see, each input affects the *outputs. Fig. 23.10 (*ii*) shows the output signals due to the signal on input 1 acting alone as a single-ended input. Fig. 23.10 (*iii*) shows the output signals due to the signal on **input 2 acting alone as a single-ended input. Note that in Figs. 23.10 (*ii*) and (*iii*), the signals on output 1 are of the same polarity. The same is also true for output 2. By superimposing both output 1 signals and both output 2 signals, we get the total outputs due to differential inputs [See Fig. 23.10 (*iv*)].

(*ii*) **Common-mode input.** In this mode, two signals equal in amplitude and having the same phase are applied to the inputs of *DA* as shown in Fig. 23.11 (*i*).

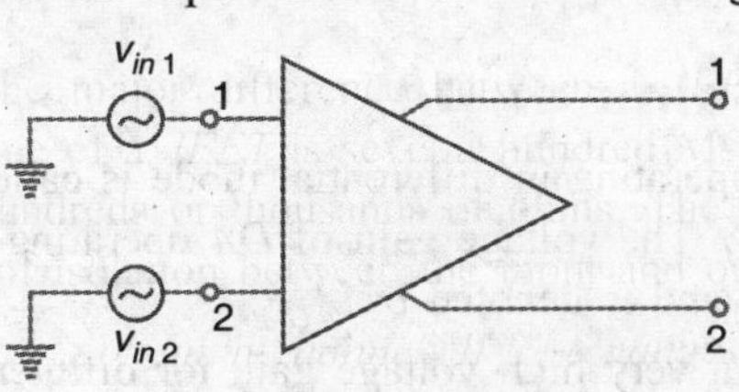

(*i*) Common-mode inputs

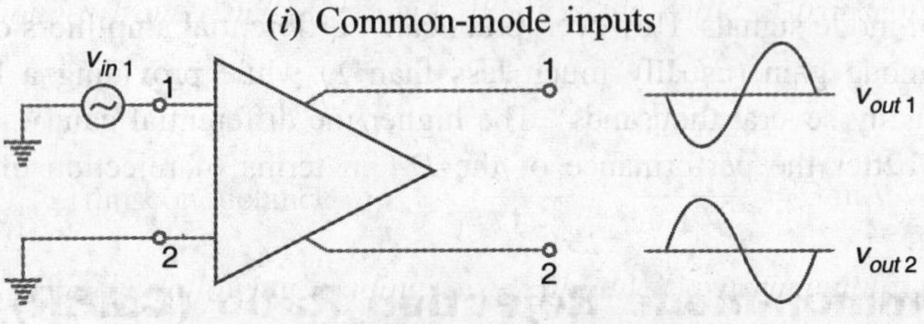

(*ii*) Outputs due to v_{in1}

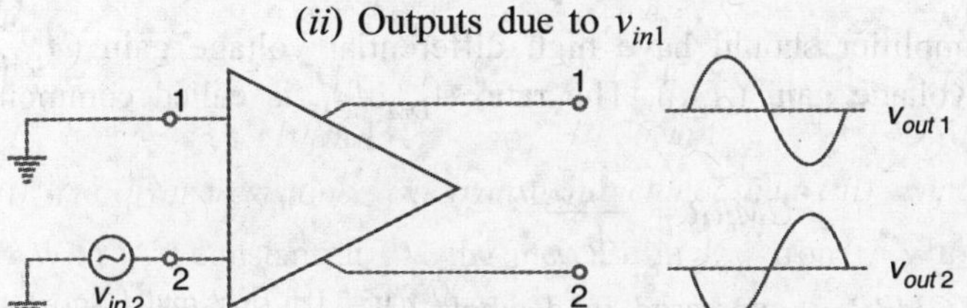

(*iii*) Outputs due to v_{in2}

* Note that in all the figures in Fig. 23.10, the phase of the input signal is given in the circle and the phase shift in the output signal is given at the output terminals. Thus in Fig. 23.10 (*ii*), the output signal at terminal 1 is 180° out of phase with the input signal. On the other hand, the output signal at terminal 2 is in phase with the input signal.

** Note that a *DA* amplifies $(v_1 - v_2)$. For Fig. 23.10 (*ii*), $v_{in1} - v_{in2} = v_{in1} - 0 = v_{in1}$. For Fig. 23.10 (*iii*), $v_{in1} - v_{in2} = 0 - v_{in2} = -v_{in2}$.

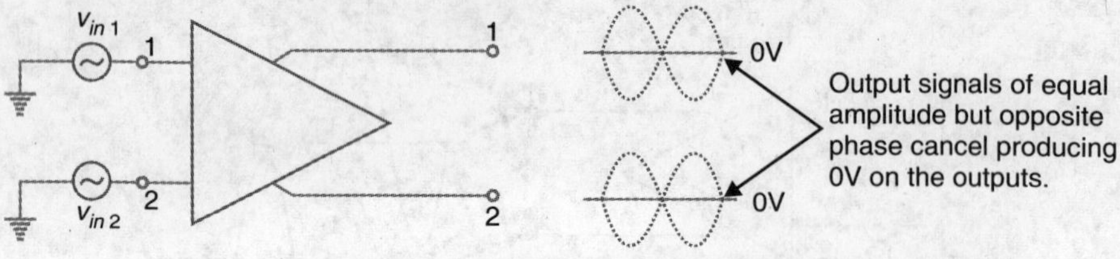

(*iv*) Outputs cancel when common-mode signals are applied.

Fig. 23.11

Again, by considering each input signal acting alone, the basic operation of *DA* in this mode can be explained. Fig. 23.11 (*ii*) shows the output signals due to the signal on only input 1 while Fig. 23.11 (*iii*) shows the output signals due to the signal on only input 2. Note that the corresponding signals on output 1 are of the opposite polarity and so are the ones on output 2. When these are superimposed, they cancel, resulting in zero output voltage as shown in Fig. 23.11 (*iv*).

It is important to note that common-mode signals are rejected by *DA*. This action is called *common-mode rejection.* Most of noises and other unwanted signals are common-mode signals. When these unwanted signals appear on the inputs of a *DA*, they are virtually eliminated on the output.

23.7 Voltage Gains of DA

The voltage gain of a *DA* operating in differential mode is called *differential-mode voltage gain* and is denoted by A_{DM}. The voltage gain of *DA* operating in common-mode is called *common-mode voltage gain* and is denoted by A_{CM}.

Ideally, a *DA* provides a very high voltage gain for differential-mode signals and zero gain for common-mode signals. However, practically, differential amplifiers do exhibit a very small common-mode gain (usually much less than 1) while providing a high differential voltage gain (usually several thousands). The higher the differential gain *w.r.t.* the common-mode gain, the better the performance of the *DA* in terms of rejection of common-mode signals.

23.8 Common-mode Rejection Ratio (CMRR)

A differential amplifier should have high differential voltage gain (A_{DM}) and very low common-mode voltage gain (A_{CM}). The ratio A_{DM}/A_{CM} is called common-mode rejection ratio (*CMRR*) *i.e.*,

$$CMRR = \frac{A_{DM}}{A_{CM}}$$

Very often, the *CMRR* is expressed in decibels (dB). The decibel measure for *CMRR* is given by ;

$$CMRR_{dB} = 20\log_{10}\frac{A_{DM}}{A_{CM}} = 20\log_{10}CMRR$$

The following table shows the relation between the two measurements :

$CMRR$	$CMRR_{dB}$
10	20dB
10^3	60dB
10^5	100dB
10^7	140dB

Importance of CMRR. The *CMRR* is the ability of a *DA* to reject the common-mode signals. The larger the *CMRR*, the better the *DA* is at eliminating common-mode signals. Let us illustrate this point. Suppose the differential amplifier in Fig. 23.12 has a differential voltage gain of 1500 (*i.e.*, $A_{DM} = 1500$) and a common-mode gain of 0.01 (*i.e.*, $A_{CM} = 0.01$).

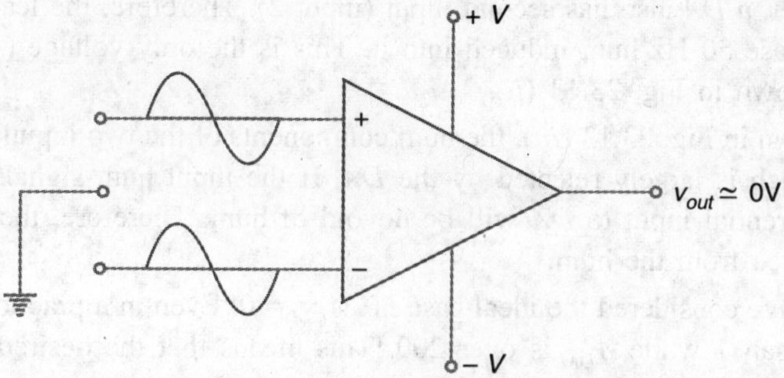

Fig. 23.12

$$\therefore \qquad CMRR = \frac{1500}{0.01} = 150,000$$

This means that the output produced by a difference between the inputs would be 150,000 times as great as an output produced by a common-mode signal.

The ability of the *DA* to reject common-mode signals is one of its main advantages. Common-mode signals are usually *undesired signals* caused by external interference. For example, any *RF* signals picked up by the *DA* inputs would be considered undesirable. The *CMRR* indicates the *DA*'s ability to reject such unwanted signals.

Practical Illustrations. Fig. 23.13 shows how a differential amplifier (*DA*) rejects hum and static voltages induced into its input leads.

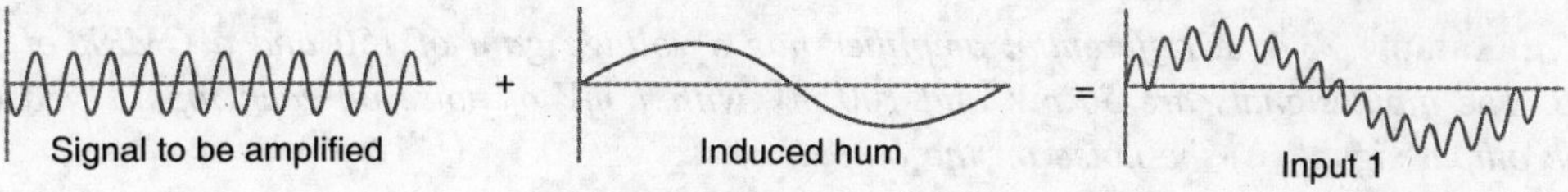

(*i*) Signal input to input 1

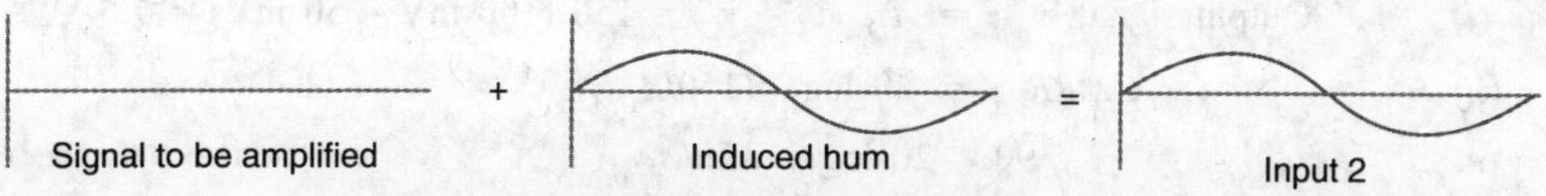

(*ii*) Signal input to input 2

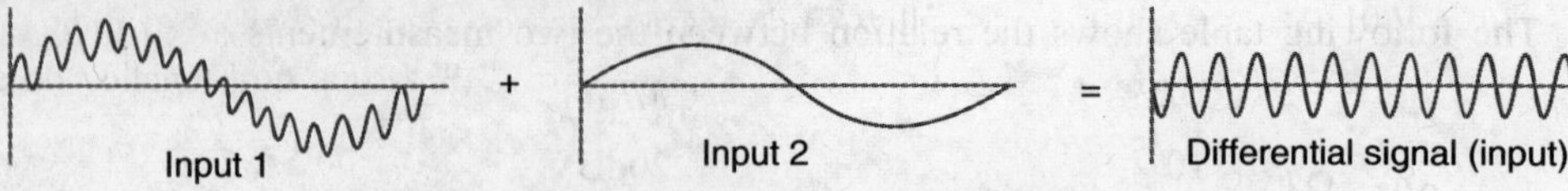

(*iii*) Differential input signal

Fig. 23.13

(*i*) In Fig. 23.13 (*i*), the signal is applied to input 1 of the *DA*. However, a low frequency *hum* voltage is also induced into the lead wire. This hum is produced due to building and collapsing magnetic field generated by adjacent conductor carrying 50Hz current. The resultant waveform is shown in Fig. 23.13 (*i*). If this waveform is amplified by a conventional amplifier, the 50 Hz hum in the output will be stronger than the desired signal.

(*ii*) However, a *DA* also has second input (input 2). Therefore, the lead of second input has the same phase 50 Hz hum induced into it. This is the only voltage (*i.e.*, hum) applied to input 2 as shown in Fig. 23.13 (*ii*).

(*iii*) As shown in Fig. 23.13 (*iii*), the hum components of the two inputs form a common-mode signal which is largely rejected by the *DA*. If the input hum signals are equal at the input, then differential input to *DA* will be devoid of hum. Therefore, the amplified output of *DA* will be free from the hum.

Note. We have considered the ideal case *i.e.* $A_{CM} = 0$. Even in a practical case, the value of A_{CM} is less than 1 while A_{DM} is over 200. This means that the desired signal would be over 200 times larger than the hum at the output terminal.

Example 23.2. *A certain differential amplifier has a differential voltage gain of 2000 and a common mode gain of 0.2. Determine CMRR and express it in dB.*

Solution.
$$CMRR = \frac{A_{DM}}{A_{CM}} = \frac{2000}{0.2} = \mathbf{10,000}$$
$$CMRR_{dB} = 20 \log_{10} 10,000 = \mathbf{80\ dB}$$

Example 23.3. *A differential amplifier has an output of 1V with a differential input of 10 mV and an output of 5 mV with a common-mode input of 10 mV. Find the CMRR in dB.*

Solution. Differential gain is
$$A_{DM} = 1\text{V}/10\text{ mV} = 100$$
Common-mode gain, $A_{CM} = 5\text{ mV}/10\text{ mV} = 0.5$
$$\therefore \quad CMRR_{dB} = 20 \log_{10} (100/0.5) = \mathbf{46\ dB}$$

Example 23.4. *A differential amplifier has a voltage gain of 150 and a CMRR of 90 dB. The input signals are 50 mV and 100 mV with 1 mV of noise on each input. Find* (*i*) *the output signal* (*ii*) *the noise on the output.*

Solution.

(*i*) Output signal, $v_{out} = A_{DM}(v_1 - v_2) = 150\ (100\text{ mV} - 50\text{ mV}) = \mathbf{7.5\ V}$

(*ii*) $CMRR_{dB} = 20 \log_{10} (150/A_{CM})$

or $90 = 20 \log_{10} (150/A_{CM})$

$\therefore \quad A_{CM} = 4.7 \times 10^{-3}$

Noise on output $= A_{CM} \times 1\text{ mV} = 4.7 \times 10^{-3} \times 1\text{mV} = \mathbf{4.7 \times 10^{-6}\ V}$

Example 23.5. *The differential amplifier shown in Fig. 23.14 has a differential voltage gain of 2500 and a CMRR of 30,000. A single-ended input signal of 500 μV r.m.s. is applied. At the same time, 1 V, 50 Hz interference signal appears on both inputs as a result of radiated pick-up from the a.c. power system.*

(i) Determine the common-mode gain.

(ii) Express the CMRR in dB.

(iii) Determine the r.m.s. output signal.

(iv) Determine the r.m.s. interference voltage on the output.

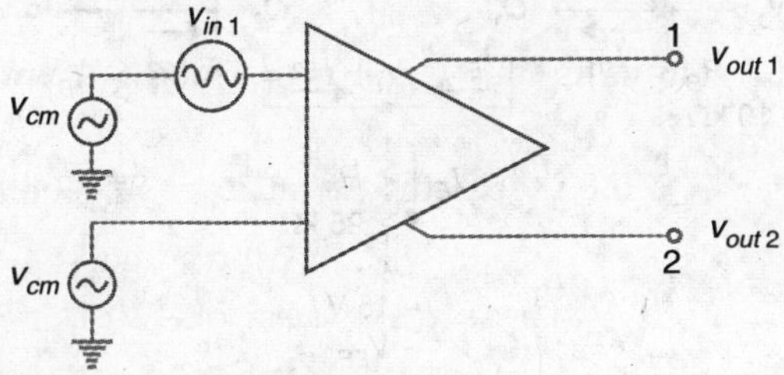

Fig. 23.14

Solution. (*i*)

$$CMRR = \frac{A_{DM}}{A_{CM}}$$

$$\therefore \quad A_{CM} = \frac{A_{DM}}{CMRR} = \frac{2500}{30,000} = \mathbf{0.083}$$

(*ii*)

$$CMRR_{dB} = 20 \log_{10} (30,000) = \mathbf{89.5\ dB}$$

(*iii*) In Fig. 23.14, the differential input voltage is the difference between the voltages on input 1 and that on input 2. Since input 2 is grounded, its voltage is zero.

∴ Differential input voltage = 500 μV – 0 = 500 μV

The output signal in this case is taken at output 1.

$$\therefore \quad v_{out1} = A_{DM} \times (500\ \mu V) = (2500 \times 500)\ \mu V = \mathbf{1.25\ V}$$

(*iv*) The common-mode input is 1V r.m.s. and the common-mode gain is $A_{CM} = 0.083$.

$$\therefore \quad \text{Noise on the output} = A_{CM} \times (1V) = (0.083)\ (1V) = \mathbf{83\ mV}$$

23.9 D.C. Analysis of Differential Amplifier (DA)

When no signal is applied to a *DA,* d.c. or quiescent conditions prevail in the circuit. From the transistor circuit theory, we can find *bias voltages and bias currents in the circuit. Fig. 23.15 shows the basic arrangement for an *npn* differential amplifier. Typical circuit values have been assumed to make the treatment illustrative. The circuit is symmetrical *i.e.*, the transistors Q_1 and Q_2 are identical ; collector loads are equal ($R_{C1} = R_{C2}$) and base resistors R_B are equal. We assume that base current is very small so that we can ignore the voltage drops across base resistors *i.e.*,

$$V_{B1} = V_{B2} \simeq {}^{**}0V$$

* Bias voltages and currents mean d.c. values.

** As we shall see, for the considered circuit values, $I_{B1} = I_{B2} = 2.86\ \mu A$.

∴ Base voltage, $V_B = I_B \times R_B = 2.86\ \mu A \times 10\ k\Omega = 28.6$ mV. Compared to the –15V of V_{EE}, this is a negligible amount of voltage.

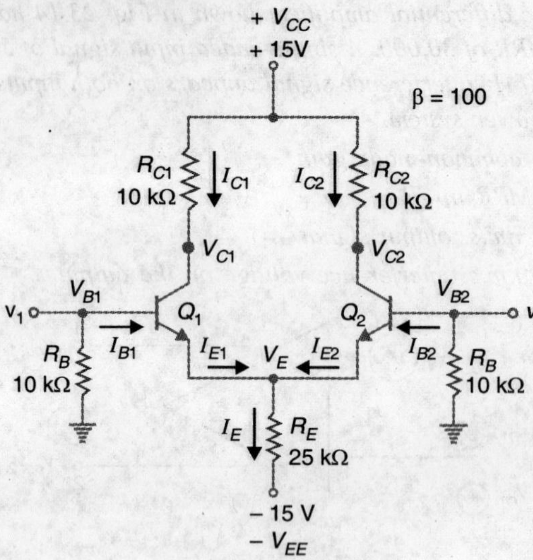

Fig. 23.15

Now, $V_E = V_{B1} - 0.7 = V_{B2} - 0.7 = {}^*0 - 0.7 = -0.7$ V

∴ $V_{E1} = V_{E2} = V_E = \mathbf{-0.7\ V}$

† Voltage across $R_E = V_{EE} - V_{BE}$

Current in R_E, $I_E = \dfrac{V_{EE} - V_{BE}}{R_E}$

The current through the emitter resistor R_E is called ††**tail current.** For the circuit values considered in Fig. 23.15, we have,

$$\text{Tail current, } I_E = \frac{V_{EE} - V_{BE}}{R_E} = \frac{(15 - 0.7)\text{V}}{25\ \text{k}\Omega} = \mathbf{0.572\ mA}$$

Because of the symmetry in the circuit, I_E must split equally between Q_1 and Q_2.

$$\therefore \quad I_{E1} = I_{E2} = \frac{I_E}{2} = \frac{0.572\ \text{mA}}{2} = \mathbf{0.286\ mA}$$

Now, $I_{C1} \simeq I_{E1} = \mathbf{0.286\ mA}$; $I_{C2} \simeq I_{E2} = \mathbf{0.286\ mA}$

Also, $I_{B1} = \dfrac{I_{C1}}{\beta} = \dfrac{0.286\ \text{mA}}{100} = \mathbf{2.86\ \mu A}$; $I_{B2} = \dfrac{I_{C2}}{\beta} = \mathbf{2.86\ \mu A}$

$$V_{C1} = V_{CC} - I_{C1} R_{C1} = 15\ \text{V} - 0.286\ \text{mA} \times 10\ \text{k}\Omega = \mathbf{12.1\ V}$$

$$V_{C2} = V_{CC} - I_{C2} R_{C2} = 15\ \text{V} - 0.286\ \text{mA} \times 10\ \text{k}\Omega = \mathbf{12.1\ V}$$

* $V_{BE} = 0.7$ V. There is a plus-to-minus drop in going from the base to emitter. Since base voltage is 0V, $V_E = -0.7$ V.

† Applying Kirchhoff's voltage law to the loop consisting of V_{EE}, the base-emitter junction of Q_1 and R_B, we have,

$$V_{RE} = V_{EE} - V_{BE} - V_{RB} = V_{EE} - V_{BE} - 0 = V_{EE} - V_{BE}$$

†† If you look at the circuit, you can see that emitter resistor R_E is like a tail.

Note that bold type results are bias voltages or bias currents. An important point to see is that $V_{C1} = V_{C2} = 12.1$ V. It means that there is no potential difference between the collectors. Therefore, the differential d.c. output for a balanced *DA* is zero.

Example 23.6. *Find the bias voltages and currents for the differential amplifier circuit shown in Fig. 23.16.*

Solution. Note that there is no resistor in the collector circuit of Q_1. This makes no difference in the values of the two collector currents because the collector currents are determined by the emitter circuit, not the collector circuit. This arrangement will only affect V_{C1}. Because of symmetry in the base-emitter circuits of Q_1 and Q_2, we have,

$$I_{E1} = I_{E2} = \frac{I_E}{2}$$

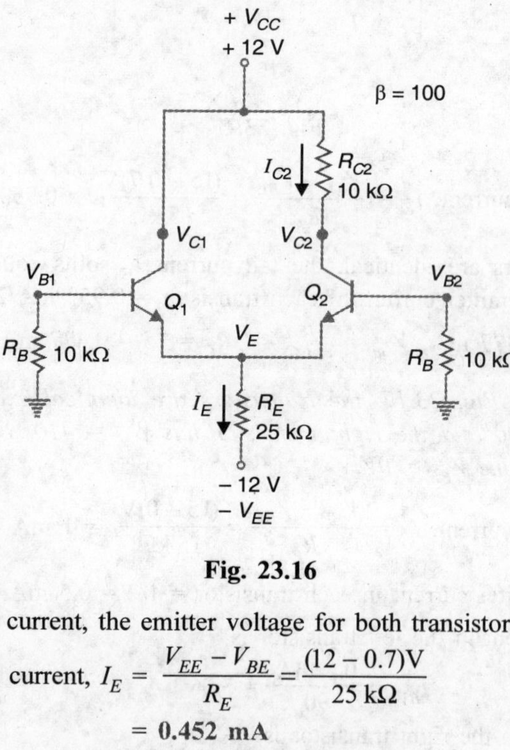

Fig. 23.16

Ignoring the base current, the emitter voltage for both transistors is V_E = **– 0.7 V.**

Now, Tail current, $I_E = \frac{V_{EE} - V_{BE}}{R_E} = \frac{(12 - 0.7)\text{V}}{25\text{ k}\Omega}$

$= \mathbf{0.452\ mA}$

∴ $I_{E1} = I_{E2} = I_E/2 = 0.452\text{ mA}/2$

$= \mathbf{0.226\ mA}$

Now, $I_{C1} = I_{C2} = \mathbf{0.226\ mA}$ $(\because I_{C1} \simeq I_{E1} \text{ and } I_{C2} \simeq I_{E2})$

∴ $I_{B1} = I_{B2} = 0.226\text{ mA}/\beta$

$= 0.226\text{ mA}/100 = \mathbf{2.26\ \mu A}$

$V_{C1} = V_{CC} = \mathbf{12V}$

$V_{C2} = V_{CC} - I_{C2}R_{C2} = 12 - 0.226\text{ mA} \times 10\text{ k}\Omega = \mathbf{9.7\ V}$

Example 23.7. *In Fig. 23.17, the transistors are identical with* β_{dc} = *100. Find the output voltage.*

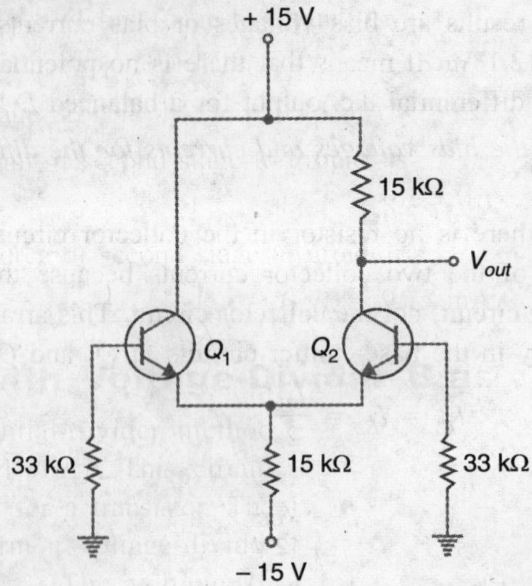

Fig. 23.17

Solution. Tail current, $I_E = \dfrac{V_{EE} - V_{BE}}{R_E} = \dfrac{(15 - 0.7)\text{V}}{15 \text{ k}\Omega} = 0.953 \text{ mA}$

Since the transistors are identical, the tail current I_E splits equally between the two transistors. Therefore, emitter current of each transistor = 0.953 mA/2 = 0.477 mA.

Since $I_C \simeq I_E = 0.477$ mA, $V_{out} = V_{CC} - I_C R_C = 15 - 0.477 \text{ mA} \times 15 \text{ k}\Omega =$ **7.85 V**

Example 23.8. *In Fig. 23.17, the transistors are identical except for β_{dc}. The left transistor has β_{dc} = 90 and the right transistor has β_{dc} = 110. Find (i) base currents (ii) base voltages. Assume V_{BE} = 0V.*

Solution. (*i*) Tail current, $I_E = \dfrac{V_{EE} - V_{BE}}{R_E} = \dfrac{(15 - 0)\text{V}}{15 \text{ k}\Omega} = 1 \text{ mA}$

Therefore, the emitter current in each transistor = 1/2 = 0.5 mA.

The base current in the left transistor is

$$I_{B1} = \frac{0.5 \text{ mA}}{90} = \mathbf{5.56\ \mu A}$$

The base current in the right transistor is

$$I_{B2} = 0.5 \text{ mA}/110 = \mathbf{4.55\ \mu A}$$

(*ii*) Voltage on left base, V_{B1} = –*(5.56 μA) × (33 kΩ) = **–0.183 V**

Voltage on right base, V_{B2} = –(4.55 μA) × (33 kΩ) = **–0.15 V**

Since the transistors are not identical, the base currents and base voltages are different.

Example 23.9. *Most differential amplifiers are biased with a current source as shown in Fig. 23.18. Find the bias currents and voltages.*

* Note that for the ideal case, emitter of each transistor is at 0 V.

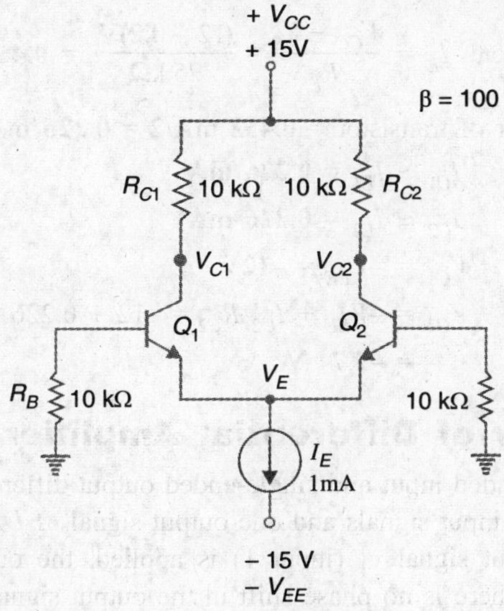

Fig. 23.18

Solution. In this circuit, the current I_E is given by the current source. Ignoring base current, the potential at each emitter is V_E = **–0.7 V.**

Emitter current in each transistor = $I_E/2$ = 1mA/2 = **0.5 mA**

Also, $I_{C1} \simeq I_{E1}$ = **0.5 mA** ; $I_{C2} \simeq I_{E2}$ = **0.5 mA**

∴ $V_{C1} = V_{C2}$ = 15 – 0.5 mA × 10 kΩ = **10 V**

Example 23.10. *Fig. 23.19 shows a differential amplifier constructed with pnp transistors. Find the bias currents and voltages.*

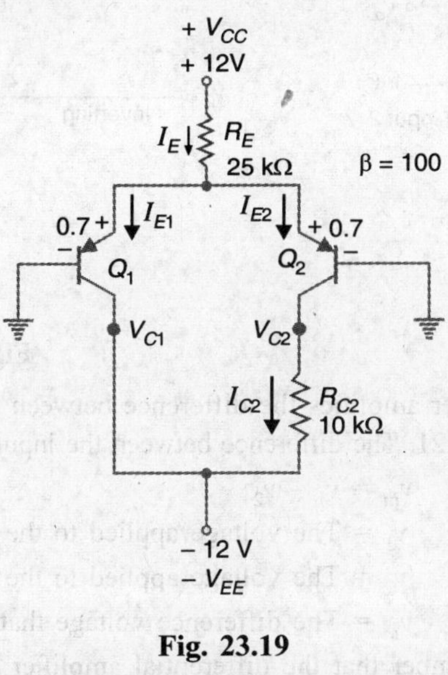

Fig. 23.19

Solution. For the *pnp* transistor, the emitter voltage (V_E) is a diode drop above the base voltage. Therefore, V_E = **+ 0.7 V.**

$$\text{Tail current, } I_E = \frac{V_{CC} - V_E}{R_E} = \frac{(12 - 0.7)\text{V}}{25 \text{ k}\Omega} = \mathbf{0.452 \text{ mA}}$$

Current in each emitter of transistor = 0.452 mA/2 = **0.226 mA**

Now, $I_{C1} \simeq I_{E1} = \mathbf{0.226 \text{ mA}}$

And $I_{C2} \simeq I_{E2} = \mathbf{0.226 \text{ mA}}$

$$V_{C1} = -V_{EE} = \mathbf{-12V}$$

$$V_{C2} = -V_{EE} + I_{C2} R_{C2} = -12 + 0.226 \text{ mA} \times 10 \text{ k}\Omega$$

$$= \mathbf{-9.74 \text{ V}}$$

23.10 Overview of Differential Amplifier

Fig. 23.20 shows double-ended input and single-ended output differential amplifier (*DA*). In other words, there are two input signals and one output signal. *A DA is *normally operated in this fashion.* When input signal v_1 (input 1) is applied, the output signal is in phase with the input signal *i.e.*, there is no phase shift in the output signal. For this reason, input signal v_1 is called *non-inverting input*. When input signal v_2 (input 2) is applied, the output signal is 180° out of phase with the input signal. For this reason, input signal v_2 is called *inverting input*.

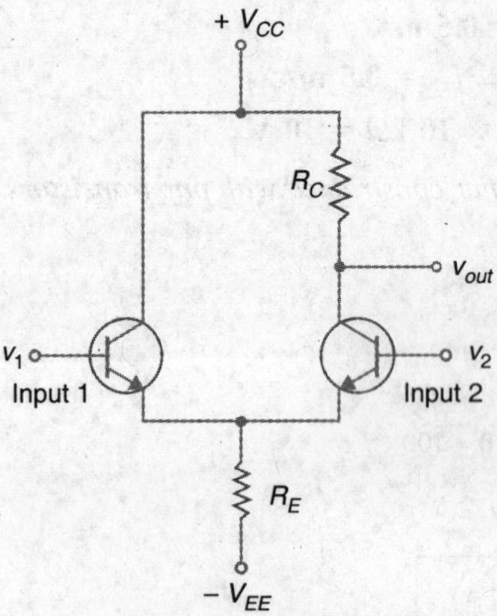

Fig. 23.20

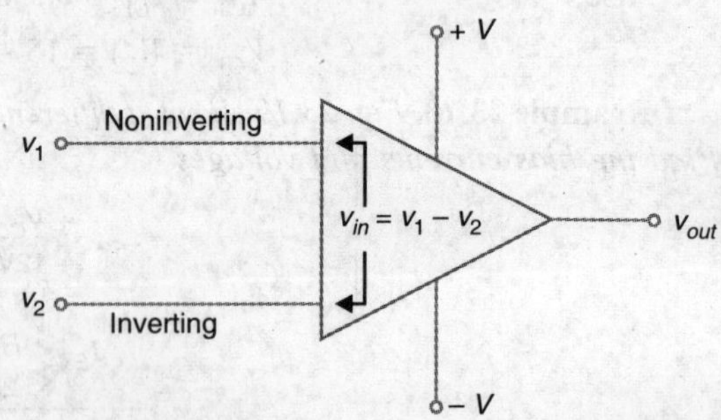

Fig. 23.21

The differential amplifier amplifies the difference between the two input voltages. This point is illustrated in Fig. 23.21. The difference between the input voltages is $v_{in} = v_1 - v_2$ *i.e.*,

$$v_{in} = v_1 - v_2$$

where v_1 = The voltage applied to the noninverting input

v_2 = The voltage applied to the inverting input

v_{in} = The difference voltage that will be amplified

It is important to remember that the differential amplifier is amplifying the difference between the input terminal voltages.

* One input can be grounded (*i.e.*, at 0V).

23.11 Parameters of DA (or OP–amp) due to Mismatch of Transistors

Our discussion on the differential amplifier (*DA*) has been based on the assumption that the transistors are *perfectly matched i.e.*, they have exactly the same electrical characteristics. In practice, this cannot happen. There will always be *some* difference between the characteristics of the two transistors. This leads us to the following two parameters of *DA* (or *OP*– amp) :

1. Output offset voltage   2. Input offset current

1. Output Offset Voltage. Even though the transistors in the differential amplifier are very closely matched, there are some differences in their electrical characteristics. One of these differences is found in the values of V_{BE} for the two transistors. When $V_{BE1} \neq V_{BE2}$, an imbalance is created in the differential amplifier. The *DA* (or *OP*-amp) may show some voltage at the output even when the voltage applied between two input terminals is zero. This is called *output offset voltage*. This point is illustrated in Fig. 23.22. Note that with the inputs of *DA* grounded, the output shows a measurable voltage. This voltage is a result of the imbalance in the differential amplifier, which causes one of the transistors to conduct harder than the other.

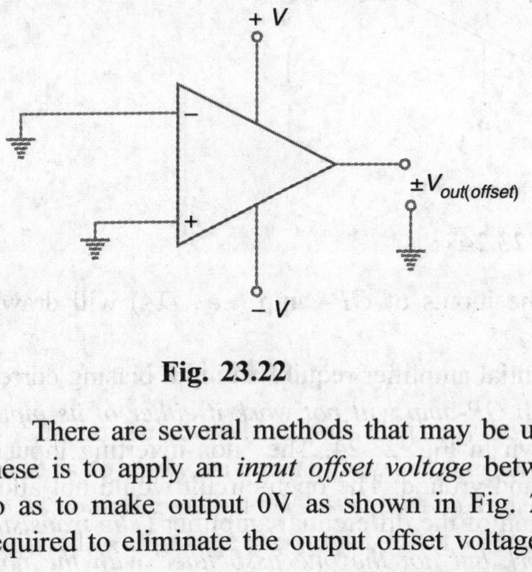

Fig. 23.22

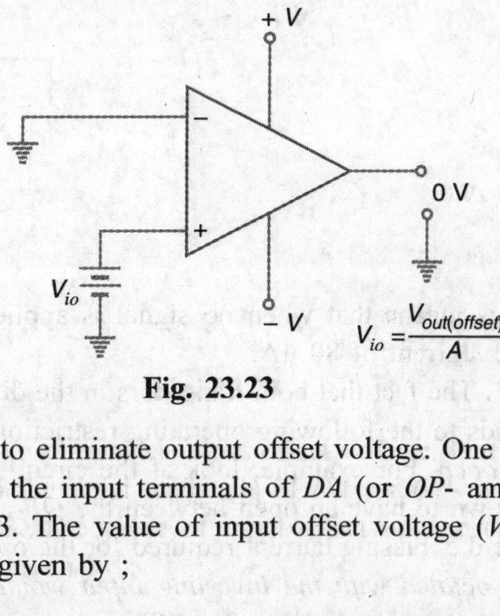

Fig. 23.23

There are several methods that may be used to eliminate output offset voltage. One of these is to apply an *input offset voltage* between the input terminals of *DA* (or *OP*- amp) so as to make output 0V as shown in Fig. 23.23. The value of input offset voltage (V_{io}) required to eliminate the output offset voltage is given by ;

$$V_{io} = \frac{V_{out}(offset)}{A} \qquad \text{...A is voltage gain.}$$

2. Input Offset Current. When the output offset voltage of a *DA* (or *OP*- amp) is eliminated, there will be a slight difference between the input currents to the noninverting and inverting inputs of the device. This slight difference in input currents is called *input offset current* and is caused by a beta (β) mismatch between the transistors in the differential amplifier. As an example, suppose I_{B1} = 75 μA and I_{B2} = 65 μA. Then,

$$I_{in\ (offset)} = 75 - 65 = 10\ \mu A$$

The difference in the base currents indicates how closely matched the transistors are. If the transistors are identical, the input offset current is zero because both base currents will be equal. But in practice, the two transistors are different and the two base currents are not equal.

23.12 Input Bias Current

The inputs to an *OP*-amp require some amount of d.c. biasing current for the transistors in the differential amplifier. The *input bias current* is defined as the average of the two d.c. base currents *i.e.*,

$$I_{in(bias)} = \frac{I_{B1} + I_{B2}}{2}$$

For example, if I_{B1} = 85 μA and I_{B2} = 75 μA, then the input bias current is

$$I_{in(bias)} = \frac{85 + 75}{2} = 80\ \mu\text{A}$$

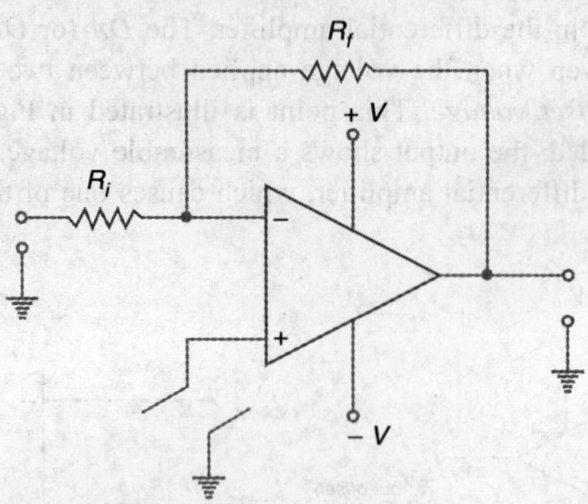

Fig. 23.24

This means that when no signal is applied, the inputs of *OP*- amp (*i.e., DA*) will draw a d.c. current of 80 μA.

The fact that both transistors in the differential amplifier require an input biasing current leads to the following operating restriction : *An OP-amp will not work if either of its inputs is open.* For example, look at the circuit shown in Fig. 23.24. The *non-inverting input is shown to have an open between the *OP*-amp and ground. The open-circuit would not allow the d.c. biasing current required for the operation of the differential amplifier (*The transistor associated with the inverting input would work but not the one associated with the non-inverting input*). Since the differential amplifier would not work, the overall *OP*-amp circuit would not work. Thus an input bias current path must always be provided for *both OP*-amp inputs.

Example 23.11. *In Fig. 23.25, the left transistor has* β_{dc} *= 90 and the right transistor has* β_{dc} *= 110. Find (i) the input offset current (ii) input bias current. Neglect* V_{BE}.

Solution.

(*i*) Tail current, $I_E = \dfrac{V_{EE} - V_{BE}}{R_E} = \dfrac{(15 - 0)\text{ V}}{1\text{ M}\Omega} = 15\ \mu\text{A}$

The emitter current in each transistor is

$$I_{E1} = I_{E2} = I_E/2 = 15\mu\text{A}/2 = 7.5\ \mu\text{A}$$

* Recall that +sign indicates noninverting input and –sign indicates inverting input.

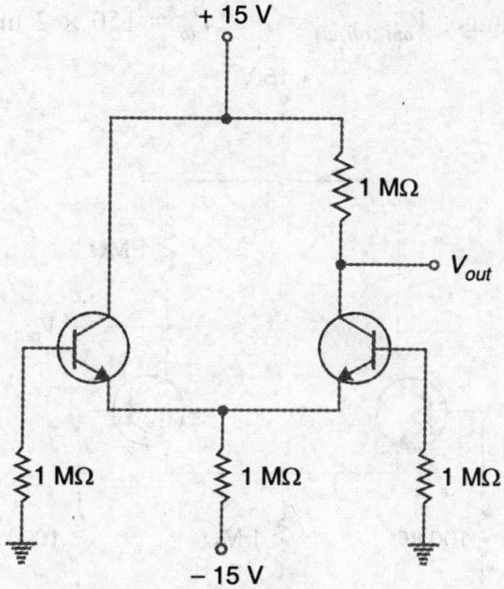

Fig. 23.25

The base current in the left transistor is

$$I_{B1} = \frac{I_{E1}}{\beta} = \frac{7.5\ \mu A}{90} = 83.3\ nA$$

The base current in the right transistor is

$$I_{B2} = \frac{I_{E2}}{\beta} = \frac{7.5\ \mu A}{110} = 68.2\ nA$$

∴ Input offset current is given by ;

$$I_{in(offset)} = I_{B1} - I_{B2} = 83.3 - 68.2 = \mathbf{15.1\ nA}$$

(*ii*) The input bias current is the average of the two base currents.

∴ Input bias current, $I_{in(bias)} = \frac{I_{B1} + I_{B2}}{2} = \frac{83.3 + 68.2}{2} = \mathbf{75.8\ nA}$

Example 23.12. *The data sheet of an IC OP-amp gives these values: $I_{in(offset)}$ = 20 nA and $I_{in(bias)}$ = 80 nA. Find the values of two base currents.*

Solution. An $I_{in(offset)}$ of 20 nA means that one base current is 20 nA greater than the other. There is no way to tell which of the two base currents will be greater. It can go either way in mass production. Assume that I_{B1} is greater than I_{B2}. Then,

$$I_{B1} = I_{in(bias)} + \frac{I_{in(offset)}}{2} = 80\ nA + \frac{20 nA}{2} = \mathbf{90\ nA}$$

$$I_{B2} = I_{in(bias)} - \frac{I_{in(offset)}}{2} = 80\ nA - \frac{20 nA}{2} = \mathbf{70\ nA}$$

If I_{B2} is greater than I_{B1}, then the values are reversed *i.e.*, I_{B1} = 70 nA and I_{B2} = 90 nA.

Example 23.13. *In Fig. 23.26, what is the output offset voltage if $I_{in(bias)}$ = 80 nA and $I_{in(offset)}$ = 20 nA? Assume that voltage gain is A = 150. Assume only β_{dc} differences exist.*

Solution. The two base resistors are equal ; each being 100 kΩ.

∴ Input offset voltage, $V_{io} = I_{in(offset)} \times R_B$ = (20 nA) (100 kΩ) = 2 mV

∴ Output offset voltage, $V_{out\ (offset)} = A \times V_{io} = 150 \times 2 \text{ mV} = 0.3 \text{ V}$

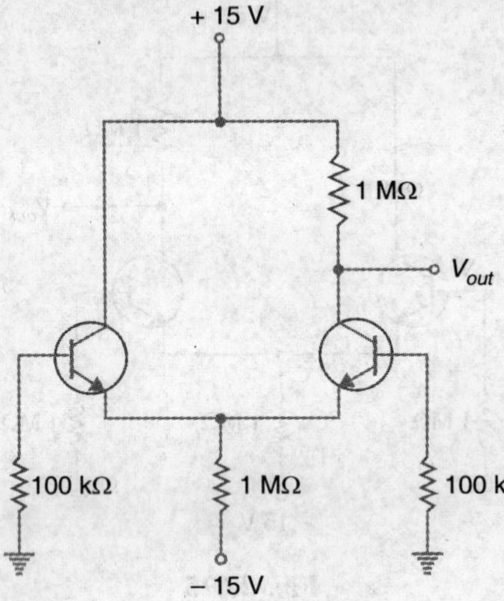

Fig. 23.26

23.13 A.C. Analysis of Differential Amplifier

A differential amplifier (*DA*) has a noninverting input and an inverting input. Fig. 23.27 shows the differential amplifier (*DA*) that is used in *IC OP-* amps. Note the circuit has double-ended input and single-ended output.

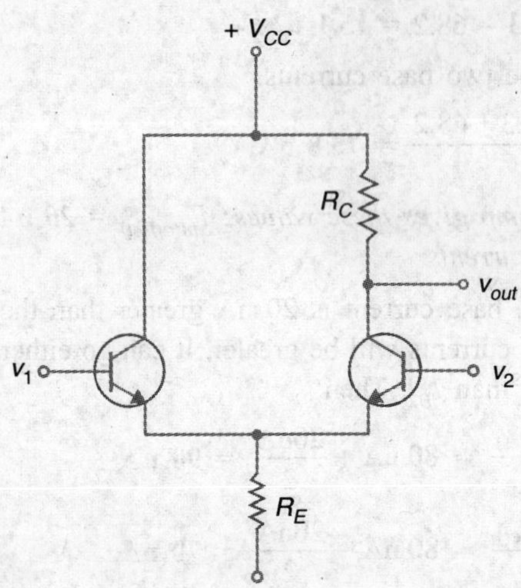

Fig. 23.27

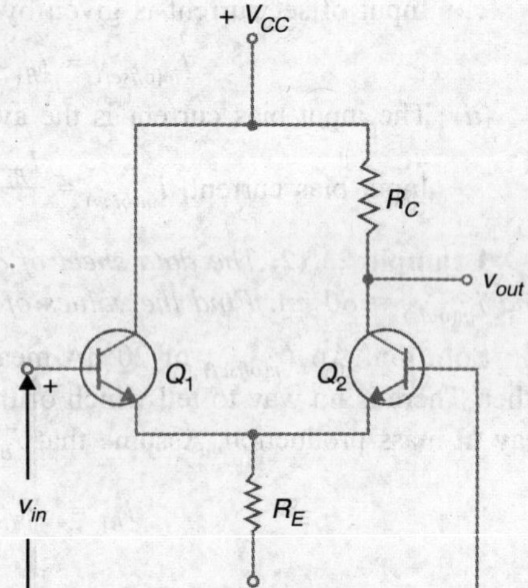

Fig. 23.28

If you look at the differential amplifier circuit in Fig. 23.27, it responds to the difference between the voltages at the two input terminals. In other words, *DA* responds to v_{in} (= $v_1 - v_2$). Therefore, the circuit shown in Fig. 23.27 can be drawn as shown in Fig. 23.28.

Ideally (*i.e.*, V_{BE} is negligible), the tail current I_E in Fig. 23.27 is $I_E = V_{EE}/R_E$. Once the values of V_{EE} and R_E are set, the tail current is constant. Therefore, to simplify a.c. analysis, we can replace the tail current by a current source (I_E) as shown in Fig. 23.29. This will simplify the a.c. analysis of the circuit with almost no loss of accuracy.

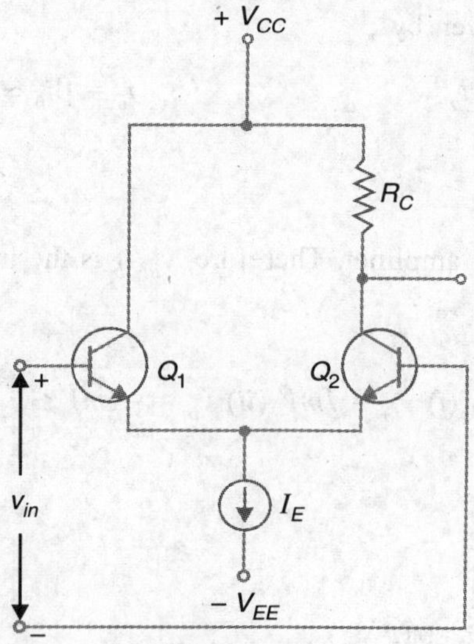

Fig. 23.29

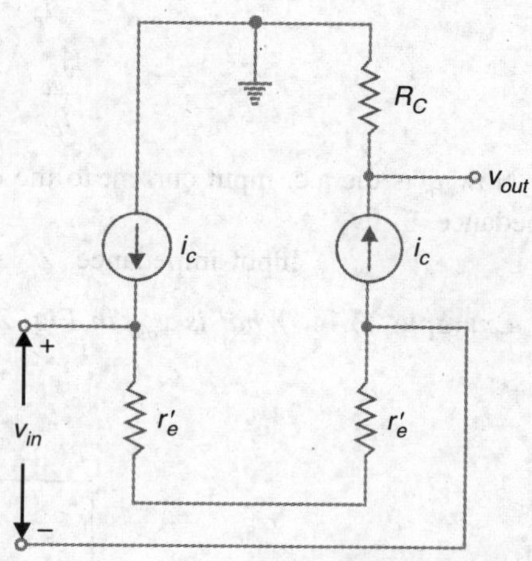

Fig. 23.30

A.C. Equivalent Circuit. We can easily find the a.c. equivalent circuit of differential amplifier by applying the usual rules to Fig. 23.29. The rules are to short all the capacitors, and reduce all d.c. sources to zero. Since a differential amplifier has no capacitors, all that we have to do is to reduce the d.c. sources to zero. Reducing a voltage source to zero is equivalent to replacing it by a short. Reducing a current source to zero is equivalent to opening it. In Fig. 23.29, this means ground the V_{CC} point, ground the V_{EE} point and open the current source. Applying these conditions to Fig. 23.29, we get the a.c. equivalent circuit of differential amplifier shown in Fig. 23.30.

Voltage gain. Fig. 23.30 shows the a.c. equivalent circuit of a differential amplifier. This is how a differential amplifier looks to an a.c. signal. Note that r'_e (= 25 mV/d.c. emitter current) is the a.c. emitter resistance. Since the two r_e's are in series, the same a.c. emitter current exists in both transistors. The a.c. emitter current is given by ;

$$i_e = \frac{v_{in}}{2r'_e}$$

This expression is easy to remember because it is almost identical to a *CE* amplifier where $i_e = v_{in}/r'_e$. The only difference is the factor 2 because a differential amplifier uses two transistors.

The a.c. collector current is approximately equal to the a.c. emitter current *i.e.* $i_c \simeq i_e$.

$$\therefore \quad \text{Output voltage, } v_{out} = i_c R_C = \frac{v_{in}}{2r'_e} R_C \qquad \left[\because\ i_e \simeq i_c = \frac{v_{in}}{2r'_e}\right]$$

$$\therefore \quad \text{Voltage gain, } A = \frac{v_{out}}{v_{in}} = \frac{R_C}{2r'_e}$$

This gain of *DA* is referred to as differential-mode voltage gain and is usually denoted by A_{DM}.

$$\therefore \quad \text{Differential voltage gain, } A_{DM} = \frac{R_C}{2r'_e}$$

Input impedance. The a.c. emitter current is given by ;

$$i_e = \frac{v_{in}}{2r'_e} \simeq \beta\, i_b \qquad (\because\ i_c = \beta i_b \simeq i_e)$$

$$\therefore \quad \frac{v_{in}}{i_b} = 2\beta r'_e$$

Now i_b is the a.c. input current to the differential amplifier. Therefore, v_{in}/i_b is the input impedance.

$$\therefore \quad \text{Input impedance, } Z_i = 2\beta r'_e$$

Example 23.14. *What is v_{out} in Fig. 23.31 when (i) v_{in} = 1mV (ii) v_{in} = –1mV?*

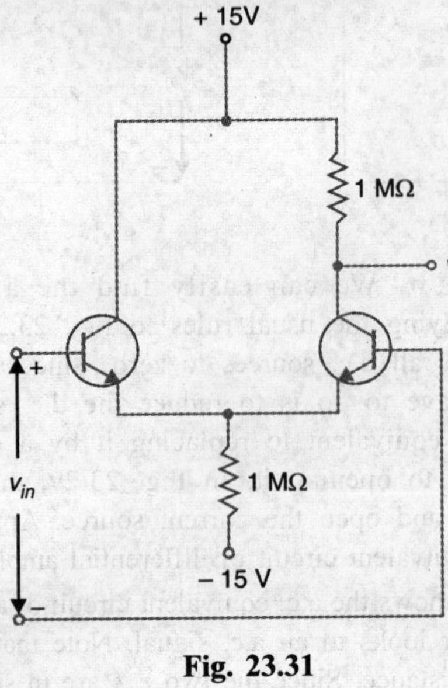

Fig. 23.31

Solution. The ideal tail current, $I_E = V_{EE}/R_E$ = 15V/1MΩ = 15 μA. Therefore, d.c. emitter current in each transistor = $I_{E1} = I_{E2} = I_E/2$ = 15 μA/2 = 7.5 μA.

$$\therefore \quad \text{A.C. emitter resistance, } r'_e = \frac{25\text{ mV}}{I_{E1}\text{ or }I_{E2}} = \frac{25\text{ mV}}{7.5\ \mu\text{A}} = 3.33\ \text{k}\Omega$$

$$\therefore \quad \text{Differential voltage gain, } A_{DM} = \frac{R_C}{2r'_e} = \frac{1\text{ M}\Omega}{2 \times 3.33\text{ k}\Omega} = 150$$

(*i*) Output voltage = $A_{DM} \times v_{in}$ = 150 × (1mV) = 150 mV = **0.15 V**

(*ii*) Output voltage, $v_{out} = A_{DM} \times v_{in}$ = 150 × (–1 mV) = **–0.15 V**

Example 23.15. *For the circuit shown in Fig. 23.32, find (i) input impedance (ii) differential voltage gain.*

Solution.

(*i*) Tail current, $I_E = \dfrac{V_{EE} - V_{BE}}{R_E}$

$$= \frac{(12 - 0.7)\text{ V}}{100\text{ k}\Omega} = 113\ \mu\text{A}$$

The d.c. emitter current in each transistor is

$$I_{E1} = I_{E2} = I_E/2 = 113\ \mu\text{A}/2 = 56.5\ \mu\text{A}$$

∴ A.C. emitter resistance, $r'_e = \dfrac{25\text{ mV}}{I_{E1}\text{ or }I_{E2}} = \dfrac{25\text{ mV}}{56.5\ \mu\text{A}} = 442\ \Omega$

∴ Input impedance, $Z_i = 2\beta r'_e = 2 \times 220 \times 442\ \Omega$

$$= 194 \times 10^3\Omega = \mathbf{194\ k\Omega}$$

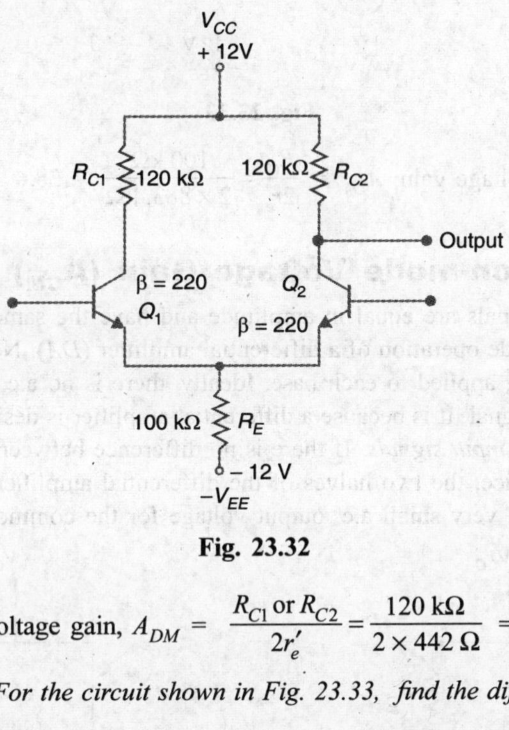

Fig. 23.32

(*ii*) Differential voltage gain, $A_{DM} = \dfrac{R_{C1}\text{ or }R_{C2}}{2r'_e} = \dfrac{120\text{ k}\Omega}{2 \times 442\ \Omega} = \mathbf{136}$

Example 23.16. *For the circuit shown in Fig. 23.33, find the differential-mode voltage gain.*

Solution. Because we are taking the output from Q_2, there is no need for collector resistor on Q_1. To find the relevant bias currents, we set the input sources to 0V, connecting the two bases to ground.

Tail current, $I_E = \dfrac{V_{EE} - V_{BE}}{R_E} = \dfrac{(12 - 0.7)\text{ V}}{200\text{ k}\Omega} = 0.0565\text{ mA}$

Therefore, emitter current in each transistor is

$$I_{E1} = I_{E2} = I_E/2 = 0.0565\text{ mA}/2 = 0.0283\text{ mA}$$

∴ A.C. emitter resistance, $r'_e = \dfrac{25\text{ mV}}{I_{E1}} = \dfrac{25\text{ mV}}{0.0283\text{ mA}} = 883.4\ \Omega$

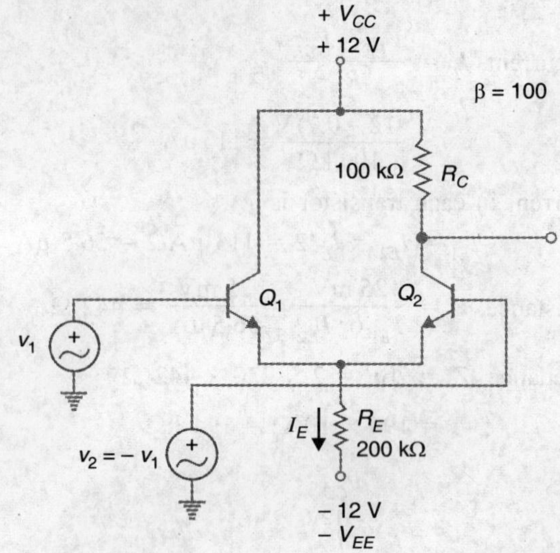

Fig. 23.33

$$\therefore \quad \text{Differential voltage gain, } A_{DM} = \frac{R_C}{2r_e'} = \frac{100 \text{ k}\Omega}{2 \times 883.4\ \Omega} = \mathbf{56.6}$$

23.14 Common-mode Voltage Gain (A_{CM})

The common-mode signals are equal in amplitude and have the same phase. Fig. 23.34 (*i*) shows the common-mode operation of a differential amplifier (*DA*). Note that the same input voltage, $v_{in(CM)}$ is being applied to each base. Ideally, there is no a.c. output voltage with a common-mode input signal. It is because a differential amplifier is designed to respond to *the difference between two input signals.* If there is no difference between the inputs, the output of *DA* is zero. In practice, the two halves of the differential amplifier are never completely balanced and there is a very small a.c. output voltage for the common-mode signal.

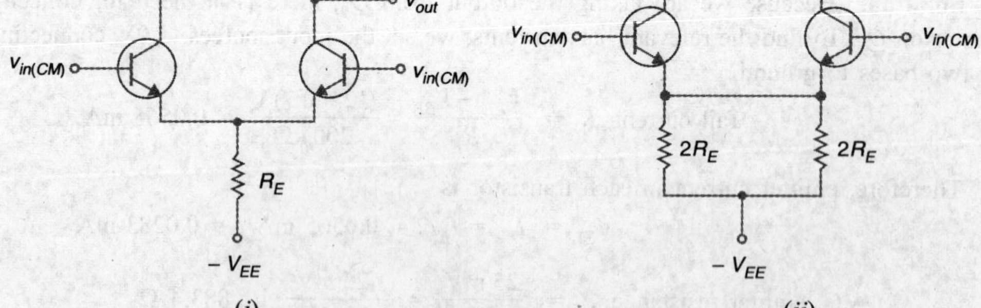

Fig. 23.34

The circuit shown in Fig. 23.34 (*i*) can be redrawn as shown in Fig. 23.34 (*ii*). In this equivalent circuit, the two parallel resistances of 2 R_E produce an equivalent resistance of R_E. Therefore, this equivalent circuit will not affect the output voltage. Assuming identical transistors, the two emitter currents will be equal and produce the same voltage across emitter resistors. Therefore, there is no current through the wire between the emitters. We can remove this wire and the circuit becomes as shown in Fig. 23.35 (*i*).

A.C. equivalent circuit. To get the a.c. equivalent circuit, we can reduce both supply voltages to zero *i.e.*, we ground each supply point. Replacing the transistors by their a.c. equivalent circuits, we get a.c. equivalent circuit of differential amplifier for commom-mode operation as shown in Fig. 23.35 (*ii*). We can derive the voltage gain of an *unbypassed *CE* circuit. Here we will use 2 R_E in place of R_E.

$\therefore$ Common-mode voltage gain, $A_{CM} = \dfrac{v_{out}}{v_{in(CM)}} = \dfrac{R_C}{r'_e + 2R_E}$

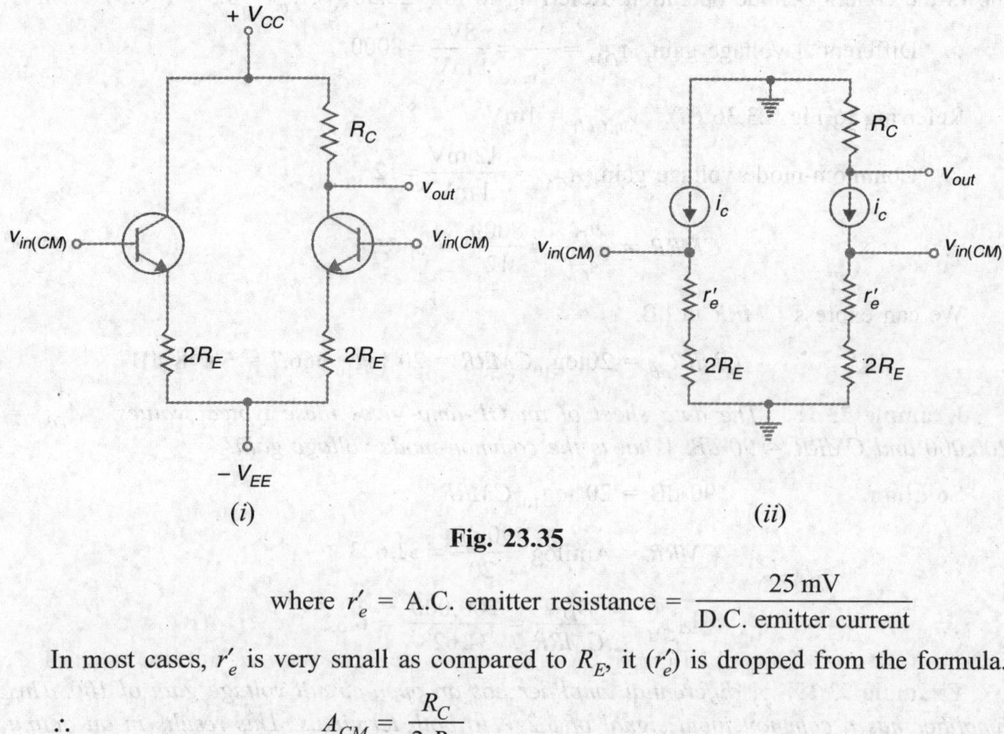

Fig. 23.35

where r'_e = A.C. emitter resistance = $\dfrac{25 \text{ mV}}{\text{D.C. emitter current}}$

In most cases, r'_e is very small as compared to R_E, it (r'_e) is dropped from the formula.

$$\therefore \quad A_{CM} = \frac{R_C}{2\,R_E}$$

The common-mode voltage gain (A_{CM}) is very small. For example, a typical *DA* may have R_C = 150 kΩ and R_E = 143 kΩ.

$$\therefore \quad A_{CM} = \frac{R_C}{2\,R_E} = \frac{150 \text{ k}\Omega}{2 \times 143 \text{ k}\Omega} = 0.52$$

* We have derived the voltage gain of bypassed capacitor *CE* amplifier in Art. 8.16 as : $A_v = R_C / r'_e$. Without bypass capacitor, the emitter is no longer at a.c. ground. Instead, R_E is seen by the a.c. signal between the emitter and ground and effectively adds to r'_e in the above formula. Therefore, voltage gain without the bypass capacitor becomes : $A_v = \dfrac{R_C}{r'_e + R_E}$

Example 23.17. *Calculate the CMRR for the circuit measurements shown in Fig. 23.36.*

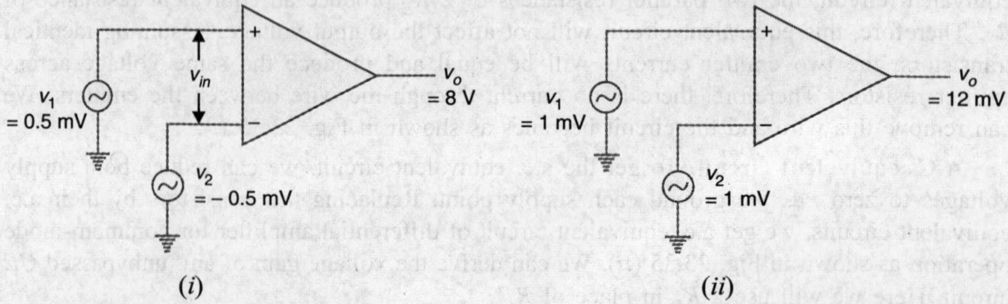

Fig. 23.36

Solution. Fig. 23.36 (*i*) shows the differential-mode operation whereas Fig. 23.36 (*ii*) shows the common-mode operation. Referring to Fig. 23.36 (*i*), v_{in} = 0.5 – (–0.5) = 1mV.

$$\therefore \quad \text{Differential voltage gain, } A_{DM} = \frac{v_o}{v_{in}} = \frac{8\text{V}}{1\text{ mV}} = 8000$$

Referring to Fig. 23.36 (*ii*), $^*v_{in(CM)}$ = 1mV

$$\therefore \quad \text{Commom-mode voltage gain, } A_{CM} = \frac{12\text{ mV}}{1\text{ mV}} = 12$$

$$\therefore \quad CMRR = \frac{A_{DM}}{A_{CM}} = \frac{8000}{12} = \mathbf{666.7}$$

We can express *CMRR* in dB.

$$CMRR_{dB} = 20\log_{10} CMRR = 20 \log_{10} 666.7 = \mathbf{56.48\ dB}$$

Example 23.18. *The data sheet of an OP-amp gives these typical values : A_{DM} = 200,000 and CMRR = 90 dB. What is the common-mode voltage gain?*

Solution. $\quad 90 \text{ dB} = 20 \log_{10} CMRR$

$$\therefore \quad CMRR = \text{Antilog } \frac{90 \text{ dB}}{20} = 31,623$$

$$\therefore \quad A_{CM} = \frac{A_{DM}}{CMRR} = \frac{200,000}{31,623} = \mathbf{6.32}$$

Example 23.19. *A differential amplifier has an open-circuit voltage gain of 100. This amplifier has a common input signal of 3.2 V to both terminals. This results in an output signal of 26 mV. Determine (i) common-mode voltage gain (ii) the CMRR in dB.*

Solution.

(*i*) $\quad v_{in(CM)} = 3.2 \text{ V} ; v_{out} = 26 \text{ mV} = 26 \times 10^{-3} \text{ V}$

$$\therefore \quad \text{Common-mode voltage gain, } A_{CM} = \frac{v_{out}}{v_{in(CM)}} = \frac{26 \times 10^{-3}}{3.2} = \mathbf{0.0081}$$

(*ii*) $$CMRR_{dB} = 20\log_{10}\frac{A_{DM}}{A_{CM}} = 20\log_{10}\frac{100}{0.0081} = \mathbf{81.8\ dB}$$

* Since the output voltage is 12mV, the common-mode signals are not exactly equal in magnitude.

Example 23.20. *For the circuit shown in Fig. 23.37, find (i) the common-mode voltage gain (ii) the CMRR in dB.*

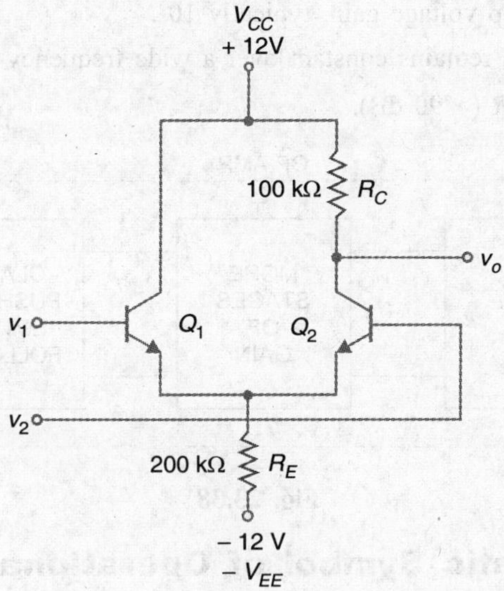

Fig. 23.37

Solution. (*i*) Common-mode voltage gain, $A_{CM} = \dfrac{R_C}{2R_E} = \dfrac{100\text{ k}\Omega}{2 \times 200\text{k}\Omega} = \mathbf{0.25}$

(*ii*) In order to find differential voltage gain, we should first find d.c. emitter current.

$$\text{Tail current, } I_E = \frac{V_{EE} - V_{BE}}{R_E} = \frac{(12 - 0.7)\text{ V}}{200\text{ k}\Omega} = 0.0565\text{ mA}$$

The d.c. emitter current in each transistor is

$$I_{E1} = I_{E2} = I_E/2 = 0.0565\text{ mA}/2 = 0.0283\text{ mA}$$

$$\therefore \text{ A.C. emitter resistance, } r_e' = \frac{25\text{ mV}}{I_{E1}} = \frac{25\text{ mV}}{0.0283} = 883.4\ \Omega$$

$$\therefore \quad \text{Differential voltage gain, } A_{DM} = \frac{R_C}{2r_e'} = \frac{100\text{ k}\Omega}{2 \times 883.4\ \Omega} = 56.6$$

$$\therefore \quad CMRR_{dB} = 20\log_{10}\frac{A_{DM}}{A_{CM}} = 20\log_{10}\frac{56.6}{0.25} = \mathbf{47.09\text{ dB}}$$

23.15 Properties of Operational Amplifier (OP-Amp)

Fig. 23.38 shows the block diagram of an operational amplifier (*OP*-amp). The input stage of an *OP*- amp is a differential stage followed by more stages of gain and a class *B* push-pull emitter follower.

The following are the important properties common to all operational amplifiers (*OP*-amps) :

(*i*) An operational amplifier is a multistage amplifier. The input stage of an *OP*-amp is a differential amplifier stage.

(*ii*) An inverting input and a noninverting input.

(iii) A high input impedance (usually assumed infinite) at both inputs.

(iv) A low output impedance (< 200 Ω).

(v) A large open-loop voltage gain, typically 10^5.

(vi) The voltage gain remains constant over a wide frequency range.

(vii) Very large *CMRR* (> 90 dB).

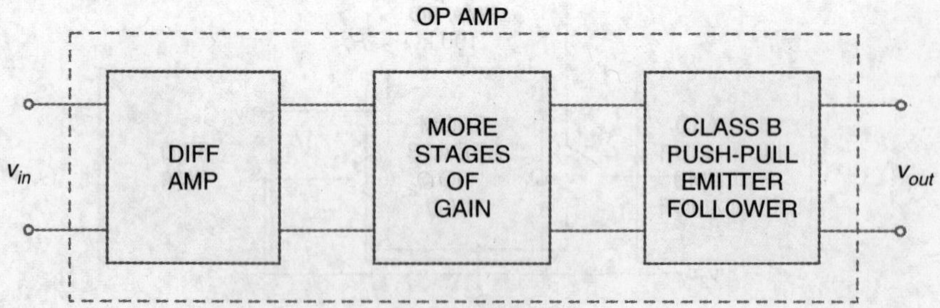

Fig. 23.38

23.16 Schematic Symbol of Operational Amplifier

Fig. 23.39 (*i*) shows the schematic symbol of an operational amplifier. The following points are worth noting :

(*i*) The basic operational amplifier has *five terminals : two terminals for supply voltages +*V* and –*V* ; two input terminals (inverting input and noninverting input) and one output terminal.

(*ii*) Note that the input terminals are marked + and –. These are not polarity signs. The – sign indicates the *inverting input* while the + sign indicates the *noninverting input.* A signal applied to plus terminal will appear in the same phase at the output as at the input. A signal applied to the minus terminal will be shifted in phase 180° at the output.

(*iii*) The voltages v_1, v_2 and v_{out} are node voltages. This means that they are always measured *w.r.t.* ground. The differential input v_{in} is the difference of two node voltages v_1 and v_2. We normally do not show the ground in the symbol.

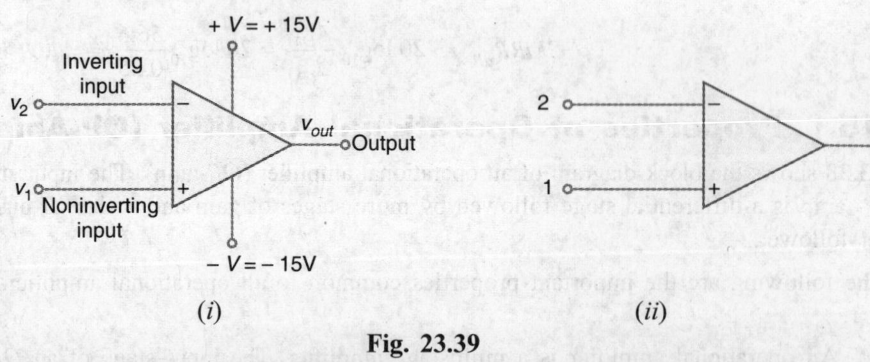

Fig. 23.39

* Two other terminals, the *offset null terminals*, are used to ensure zero output when the two inputs are equal. These are normally used when small d.c. signals are involved.

(*iv*) For the sake of *simplicity, + *V* and – *V* terminals are often omitted from the symbol as shown in Fig. 23.39 (*ii*). The two input leads are always shown on the symbol regardless of whether they are both used.

(*v*) In most cases, if only one input is required for an *OP*-amp circuit, the input not in use will be shown connected to ground. A single-input *OP*-amp is generally classified as either inverting or noninverting.

(*vi*) The *OP*-amp is produced as an integrated circuit (*IC*). Because of the complexity of the internal circuitry of an *OP*-amp, the *OP*-amp symbol is used exclusively in circuit diagrams.

23.17 Output Voltage From OP-Amp

The output voltage from an *OP*-amp for a given pair of input voltages depends mainly on the following factors :

1. The voltage gain of *OP*-amp.
2. The polarity relationship between v_1 and v_2.
3. The values of supply voltages, $+V$ and $-V$.

1. Voltage gain of OP-amp. The *maximum* possible voltage gain from a given *OP*-amp is called *open-loop voltage gain* and is denoted by the symbol A_{OL}. The value of A_{OL} for an *OP*-amp is generally greater than 10,000.

The term open-loop indicates a circuit condition where there is *no feedback path from the output to the input of OP-amp.* The *OP*-amps are almost always operated with negative feedback *i.e.*, a part of the output signal is fed back in phase opposition to the input. Such a condition is illustrated in Fig. 23.40. Here R_i is the input resistance and R_f is the feedback resistor. Consequently, the voltage gain of *OP*-amp is reduced. When a feedback path is present such as R_f connection in Fig. 23.40, the resulting circuit gain is referred to as *closed-loop voltage gain* (A_{CL}). The following points may be noted :

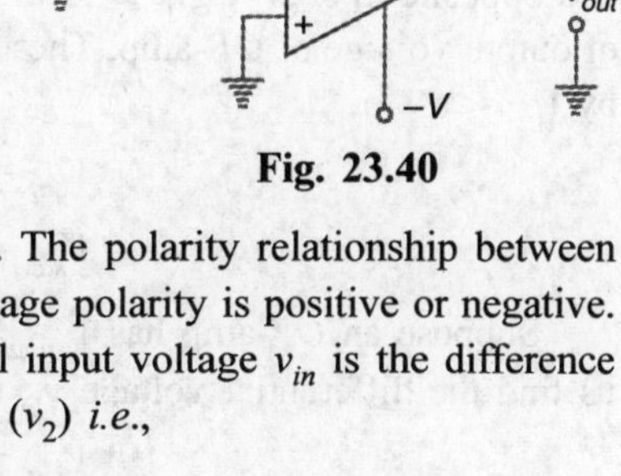

Fig. 23.40

(*i*) The maximum voltage gain of given *OP*-amp is A_{OL}. Its value is generally greater than 10,000.

(*ii*) The actual gain (A_{CL}) of an *OP*-amplifier is reduced when negative feedback path exists between output and input.

2. OP-Amp Input/Output Polarity Relationship. The polarity relationship between v_1 and v_2 will determine whether the *OP*-amp output voltage polarity is positive or negative. There is an easy method for it. We know the differential input voltage v_{in} is the difference between the non-inverting input (v_1) and inverting input (v_2) *i.e.*,

$$v_{in} = v_1 - v_2$$

When the result of this equation is *positive,* the *OP*-amp output voltage will be *positive.* When the result of this equation is *negative,* the output voltage will be *negative.*

* Since two or more *OP*-amps are often contained in a single *IC* package, eliminating these terminals on the symbol eliminates unnecessary duplication.

Illustration. Let us illustrate *OP*-amp input/output polarity relationship with numerical values.

(*i*) In Fig. 23.41 (*i*), v_1 = + 4V and v_2 = + 2V so that $v_{in} = v_1 - v_2$ = (+ 4V) – (+ 2V) = 2V. Since v_{in} is *positive*, the *OP*-amp output voltage will be *positive.*

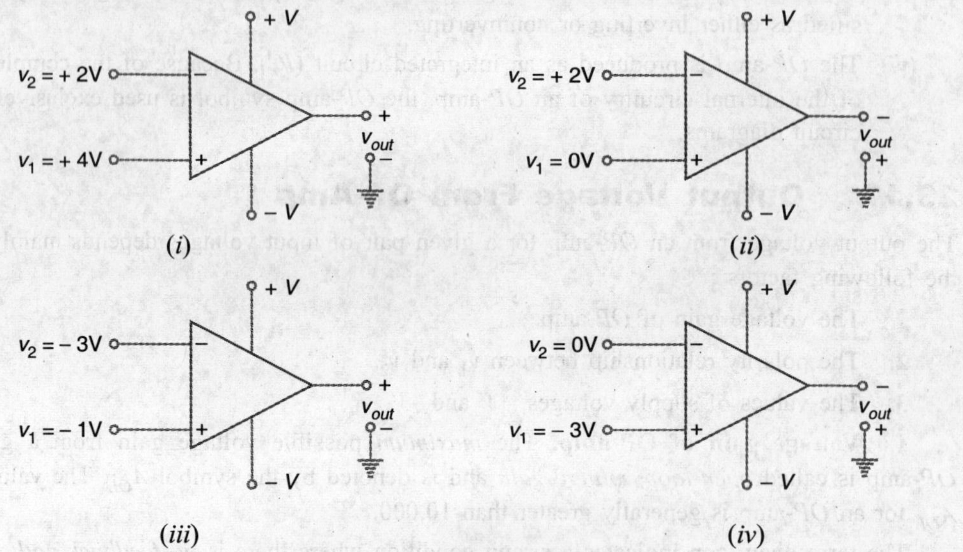

Fig. 23.41

(*ii*) In Fig. 23.41 (*ii*), v_1 = 0V and v_2 = + 2V so that $v_{in} = v_1 - v_2$ = (0V) – (+2V) = –2V. Since v_{in} is *negative*, the *OP*-amp output voltage will be *negative.*

(*iii*) In Fig. 23.41 (*iii*), v_1 = –1V and v_2 = –3V so that $v_{in} = v_1 - v_2$ = (–1V) – (–3V) = 2V. Clearly, the *OP*-amp output voltage will be *positive.*

(*iv*) In Fig. 23.41 (*iv*), v_1 = –3V and v_2 = 0V so that $v_{in} = v_1 - v_2$ = (–3V) – (0V) = –3V. Therefore, the *OP*-amp output voltage will be *negative.*

3. Supply Voltages. The supply voltages for an *OP*-amp are normally equal in magnitude and opposite in sign *e.g.*, ± 15V, ± 12V, ± 18V. These supply voltages determine the limits of output voltage of *OP*-amp. These limits, known as *saturation voltages,* are generally given by ;

$$+V_{sat} = +V_{supply} - 2\text{ V}$$

$$-V_{sat} = -V_{supply} + 2\text{ V}$$

Suppose an *OP*-amp has V_{supply} = ± 15V and open-loop voltage gain A_{OL} = 20,000. Let us find the differential voltage v_{in} to avoid saturation.

$$V_{sat} = V_{supply} - 2 = 15 - 2 = 13\text{ V}$$

$$\therefore \quad v_{in} = \frac{V_{sat}}{A_{OL}} = \frac{13\text{ V}}{20{,}000} = 650\ \mu\text{V}$$

If the differential input voltage v_{in} exceeds this value in an *OP*-amp, it will be driven into saturation and the device will become non-linear.

Note. Although input terminals of an *OP*-amp are labelled as + and –, this does not mean you have to apply positive voltages to the + terminal and negative voltages to the –terminal. Any voltages can be applied to either terminal. The true meaning of the input terminal labels (+ and –) is that a *positive voltage applied to the + terminal drives the output voltage towards $+V$ of d.c. supply; a positive voltage applied to the – terminal drives the output voltage towards $-V$ of d.c. supply.

23.18 A.C. Analysis of OP-Amp

The basic *OP*-amp has two input terminals and one output terminal. The input terminals are labelled as + (noninverting input) and –(inverting input). As discussed earlier, a signal applied to the non-inverting input (+) will produce an output voltage that is in phase with the input voltage. However, a signal applied to the inverting input (–) will produce an output voltage that is 180° out of phase with the input signal.

(*i*) Practical OP–amp. Fig. 23.42 shows the a.c. equivalent circuit of a practical *OP*-amp. The characteristics of a practical *OP*-amp are : *very high voltage gain, very high input impedance* and *very low output impedance.* The input voltage v_{in} appears between the two input terminals and the output voltage is $A_v v_{in}$ taken through the output impedance Z_{out}. The consequences of these properties of a practical *OP*–amp are :

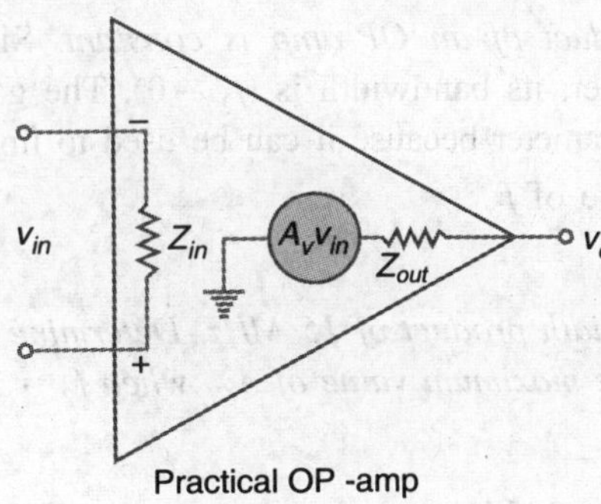

Fig. 23.42

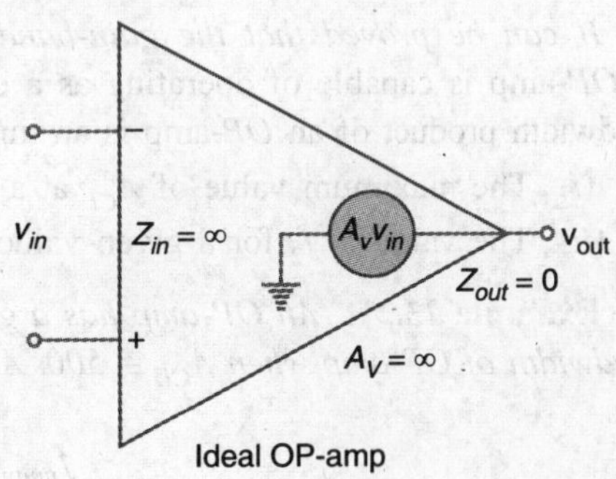

Fig. 23.43

(*a*) Since the voltage gain (A_v) of a practical *OP*-amp is very high, an extremely small input voltage (v_{in}) will produce a large output voltage (v_{out}).

(*b*) Since the input impedance (Z_{in}) is very high, a practical *OP*-amp has very small input current.

(*c*) Since the output impedance (Z_{out}) of a practical *OP*-amp is very low, it means that output voltage is practically independent of the value of load connected to *OP*-amp.

(*ii*) Ideal OP-amp. Fig. 23.43 shows the a.c. equivalent circuit of an ideal *OP*-amp. The characteristics of an ideal *OP*-amp are : *infinite voltage gain, infinite input impedance* and *zero output impedance.* The consequences of these properties of an ideal *OP*-amp are :

(*a*) Since the voltage gain (A_v) of an ideal *OP*-amp is infinite, it means that we can set v_{in} = 0V.

(*b*) Since the input impedance (Z_{in}) is infinite, an ideal *OP*-amp has zero input current.

(*c*) Since the output impedance (Z_{out}) of an ideal *OP*-amp is zero, it means the output voltage does not depend on the value of load connected to *OP*-amp.

---

* Note that positive and negative are relative terms. Thus if +4V is applied at +input terminal and +2V at –input terminal, then + terminal is at more positive potential. Therefore, the output voltage will swing towards $+V$ of d.c. supply.

We can sum up the values of parameters of a practical *OP*-amp and an ideal *OP*-amp as under :

Practical *OP*-amp	**Ideal *OP*-amp**
$Z_{in} = 2\ M\Omega$	$Z_{in} \to \infty$ (Open-circuit)
$A_v = 1 \times 10^5$	$A_v \to \infty$
$Z_{out} = 100\ \Omega$	$Z_{out} = 0\Omega$

23.19 Bandwidth of an OP-Amp

All electronic devices work only over a limited range of frequencies. This range of frequencies is called **bandwidth.** Every *OP*-amp has a bandwidth *i.e.*, the range of frequencies over which it will work properly. The bandwidth of an *OP*-amp depends upon the closed-loop gain of the *OP*-amp circuit. One important parameter is **gain-bandwidth product (GBW).** It is defined as under :

$$A_{CL} \times f_2 = f_{unity} = GBW$$

where A_{CL} = Closed-loop gain at frequency f_2

f_{unity} = Frequency at which the closed-loop gain is unity

It can be proved that the gain-bandwidth product of an OP-amp is constant. Since an *OP*-amp is capable of operating as a d.c. amplifier, its bandwidth is (f_2 – 0). The gain-bandwidth product of an *OP*-amp is an important parameter because it can be used to find :

(*i*) The maximum value of A_{CL} at a given value of f_2.

(*ii*) The value of f_2 for a given value of A_{CL}.

Example 23.21. *An OP-amp has a gain-bandwidth product of 15 MHz. Determine the bandwidth of OP-amp when A_{CL} = 500. Also find the maximum value of A_{CL} when f_2 = 200 kHz.*

Solution.

$$f_2 = \frac{f_{unity}}{A_{CL}} = \frac{15\text{MHz}}{500} = 30\text{ kHz}$$

Since the *OP*-amp is capable of operating as a d.c. amplifier, bandwidth BW = **30 kHz.**

$$A_{CL} = \frac{f_{unity}}{f_2} = \frac{15\text{MHz}}{200\text{ kHz}} = \mathbf{75\ or\ 37.5\ dB}$$

Example 23.22. *An OP-amp has a gain-bandwidth product of 1.5 MHz. Find the operating bandwidth for the following closed-loop gains* (*i*) A_{CL} = 1 (*ii*) A_{CL} = 10 (*iii*) A_{CL} = 100.

Solution. Bandwidth, $BW = \dfrac{GBW}{A_{CL}}$

(*i*) For $A_{CL} = 1$, $BW = \dfrac{1.5\text{ MHz}}{1}$ = **1.5 MHz**

(*ii*) For $A_{CL} = 10$, $BW = \dfrac{1.5\text{ MHz}}{10}$ = **150 kHz**

(*iii*) For $A_{CL} = 100$, $BW = \dfrac{1.5\text{ MHz}}{100}$ = **15 kHz**

From this example, we conclude that :

(*a*) The higher the gain (A_{CL}) of an *OP*-amp, the narrower its bandwidth.

(*b*) The lower the gain of an *OP*-amp, the wider its bandwidth.

23.20 Slew Rate

The slew rate of an *OP*-amp is a measure of *how fast the output voltage can change* and is measured in volts per microsecond (V/µs). If the slew rate of an *OP*-amp is 0.5V/µs, it means that the output from the amplifier can change by 0.5 V every µs. Since frequency is a function of time, the *slew rate can be used to determine the maximum operating frequency of the OP-amp* as follows :

$$\text{Maximum operating frequency, } f_{max} = \frac{\text{Slew rate}}{2\pi\, V_{pk}}$$

Here V_{pk} is the peak output voltage.

Example 23.23. *Determine the maximum operating frequency for the circuit shown in Fig. 23.44. The slew rate is 0.5 V/µs.*

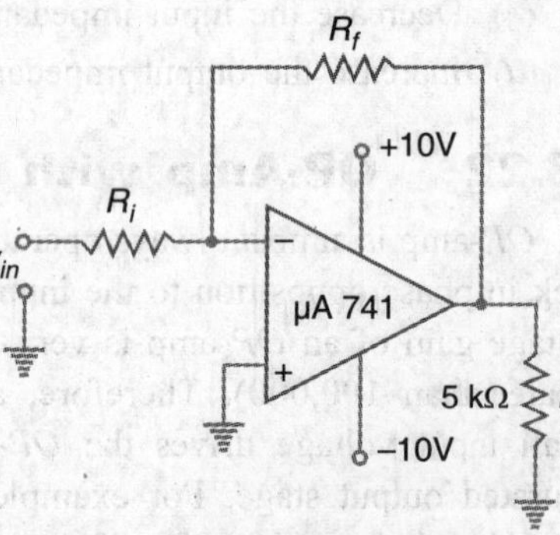

Fig. 23.44

Solution. The maximum peak output voltage (V_{pk}) is approximately *8V. Therefore, maximum operating frequency (f_{max}) is given by ;

$$f_{max} = \frac{\text{Slew rate}}{2\pi V_{pk}} = \frac{0.5\text{ V}/\mu\text{s}}{2\pi \times 8}$$

$$= \frac{500 \text{ kHz}}{2\pi \times 8}$$

$$(\because 0.5 \text{ V}/\mu\text{s} = 500 \text{ kHz})$$

$$= \mathbf{9.95\ kHz}$$

While 9.95 kHz may not seem to be a very high output frequency, you must realise that the amplifier was assumed to be operating at its maximum output voltage. Let us see what happens when peak output voltage is reduced (See example 23.24).

Example 23.24. *The amplifier in Fig. 23.44 is being used to amplify an input signal to a peak output voltage of 100 mV. What is the maximum operating frequency of the amplifier?*

Solution. The maximum operating frequency (f_{max}) of the amplifier is given by ;

$$f_{max} = \frac{\text{Slew rate}}{2\pi V_{pk}} = \frac{0.5\text{ V}/\mu\text{s}}{2\pi \times 0.1} \qquad (\because 100 \text{ mV} = 0.1\text{V})$$

$$= \frac{500 \text{ kHz}}{2\pi \times 0.1} = \mathbf{796\ kHz} \qquad (\because 0.5\text{V}/\mu\text{s} = 500 \text{ kHz})$$

The above examples show that an *OP*-amp can be operated at a much higher frequency when being used as a small-signal amplifier than when being used as a large-signal amplifier.

* $+V_{sat} = +V_{supply} - 2 = 10 - 2 = 8\text{V}$

23.21 Frequency Response of an OP-Amp

The operating frequency has a significant effect on the operation of an *OP*-amp. The following are the important points regarding the frequency response of an *OP*-amp :

(*i*) The maximum operating frequency of an *OP*-amp is given by ;

$$f_{max} = \frac{\text{Slew rate}}{2\pi V_{pk}}$$

Thus, the *peak output voltage limits the maximum operating frequency.*

(*ii*) When the maximum operating frequency of an *OP*-amp is exceeded, the result is a distorted output waveform.

(*iii*) Increasing the operating frequency of an *OP*-amp beyond a certain point will :

(*a*) Decrease the maximum output voltage swing.

(*b*) Decrease the open-loop voltage gain.

(*c*) Decrease the input impedance.

(*d*) Increase the output impedance.

23.22 OP-Amp with Negative Feedback

An *OP*-amp is almost always operated with negative feedback *i.e.*, a part of the output is fed back in phase opposition to the input (See Fig. 23.45). The reason is simple. The open-loop voltage gain of an *OP*-amp is very high (usually greater than 100,000). Therefore, an extremely small input voltage drives the *OP*-amp into its saturated output stage. For example, assume v_{in} = 1mV and A_{OL} = 100,000. Then,

$$v_{out} = A_{OL}\, v_{in} = (100{,}000) \times (1 \text{ mV}) = 100 \text{ V}$$

Since the output level of an *OP*-amp can never reach 100 V, it is driven deep into saturation and the device becomes non-linear.

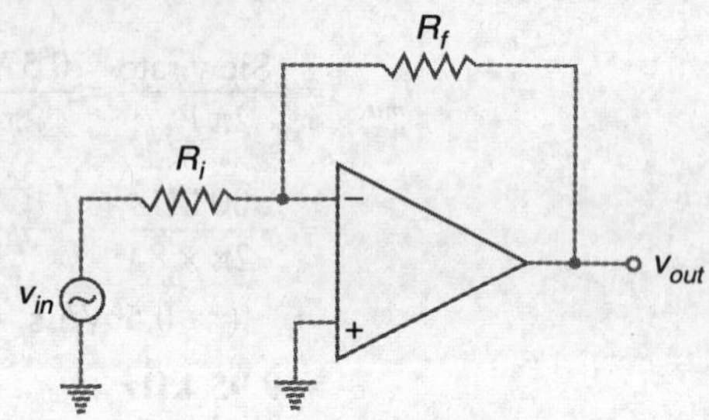

Fig. 23.45

With negative feedback, the voltage gain (A_{CL}) can be reduced and controlled so that *OP*-amp can function as a linear amplifier. In addition to providing a controlled and stable gain, negative feedback also provides for control of the input and output impedances and amplifier bandwidth. The table below shows the general effects of negative feedback on the performance of *OP*-amps.

	Voltage gain	Input Z	Output Z	Bandwidth
Without negative feedback	A_{OL} is too high for linear amplifier applications	Relatively high	Relatively low	Relatively narrow
With negative feedback	A_{CL} is set by the feedback circuit to desired value	Can be increased or reduced to a desired value depending on type of circuit	Can be reduced to a desired value	Significantly wider

23.23 Applications of OP-Amps

The operational amplifiers have many practical applications. The *OP*-amp can be connected in a large number of circuits to provide various operating characteristics. In the sections to follow, we shall discuss important applications of *OP*-amps.

23.24 Inverting Amplifier

An *OP*-amp can be operated as an inverting amplifier as shown in Fig. 23.46. An input signal v_{in} is applied through input resistor R_i to the minus input (inverting input). The output is fed back to the same minus input through feedback resistor R_f. The plus input (noninverting input) is grounded. Note that the resistor R_f provides the **negative feedback*. Since the input signal is applied to the inverting input (–), the output will be inverted (*i.e.* 180° out of phase) as compared to the input. Hence the name inverting amplifier.

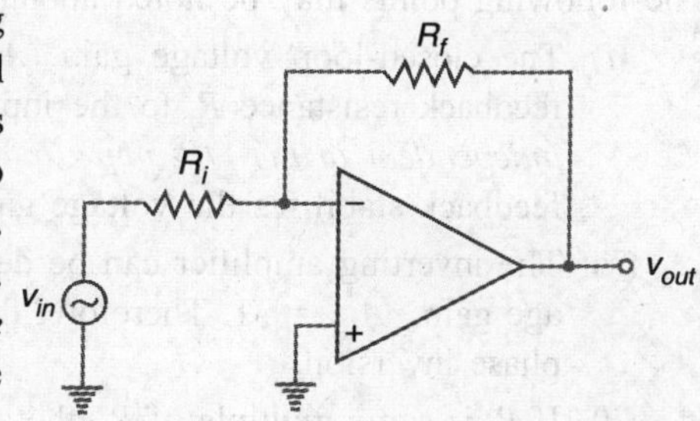

Fig. 23.46

Voltage gain. An *OP*-amp has an infinite input impedance. This means that there is zero current at the inverting input. If there is zero current through the input impedance, then there must be *no* voltage drop between the inverting and non-inverting inputs. This means that voltage at the inverting input (–) is zero (point *A*) because the other input (+) is grounded. The 0V at the inverting input terminal (point *A*) is referred to as **virtual ground.** This condition is illustrated in Fig. 23.47. The point *A* is said to be at virtual ground because it is at 0V but is not physically connected to the ground (*i.e.* V_A = 0V).

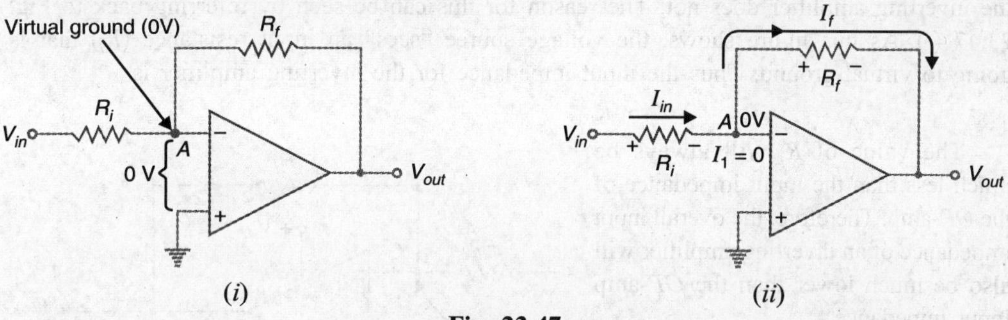

Fig. 23.47

Referring to Fig. 23.47 (*ii*), the current I_1 to the inverting input is zero. Therefore, current I_{in} flowing through R_i entirely flows through feedback resistor R_f. In other words, $I_f = I_{in}$.

Now, $$I_{in} = \frac{\text{Voltage across } R_i}{R_i} = \frac{V_{in} - V_A}{R_i} = \frac{V_{in} - 0}{R_i} = \frac{V_{in}}{R_i}$$

and $$I_f = \frac{\text{Voltage across } R_f}{R_f} = \frac{V_A - V_{out}}{R_f} = \frac{0 - V_{out}}{R_f} = \frac{-V_{out}}{R_f}$$

* The output voltage is 180° out of phase with the input. Since the voltage drop across R_f is of the opposite polarity to the applied voltage, the circuit is providing negative feedback.

Since $I_f = I_{in}$, $\quad -\dfrac{V_{out}}{R_f} = \dfrac{V_{in}}{R_i}$

$$\therefore \quad \text{Voltage gain, } A_{CL} = \frac{V_{out}}{V_{in}} = -\frac{R_f}{R_i}$$

The negative sign indicates that output signal is inverted as compared to the input signal. The following points may be noted about the inverting amplifier :

(*i*) The closed-loop voltage gain (A_{CL}) of an inverting amplifier is the ratio of the feedback resistance R_f to the input resistance R_i. *The closed-loop voltage gain is independent of the OP-amp's internal open-loop voltage gain.* Thus the negative feedback stabilises the voltage gain.

(*ii*) The inverting amplifier can be designed for unity gain. Thus if $R_f = R_i$, then voltage gain, $A_{CL} = -1$. Therefore, the circuit provides a unity voltage gain with 180° phase inversion.

(*iii*) If R_f is some multiple of R_i, the amplifier gain is constant. For example, if $R_f = 10\ R_i$, then $A_{CL} = -10$ and the circuit provides a voltage gain of exactly 10 along with a 180° phase inversion from the input signal. If we select precise resistor values for R_f and R_i, we can obtain a wide range of voltage gains. *Thus the inverting amplifier provides constant voltage gain.*

23.25 Input and Output Impedance of Inverting Amplifier

It is worthwhile to give a brief discussion about the input impedance and output impedance of inverting amplifier.

(*i*) **Input impedance.** While an *OP*-amp has an extremely high input impedance, the inverting amplifier does not. The reason for this can be seen by referring back to Fig. 23.47 (*i*). As this figure shows, the voltage source "sees" an input resistance (R_i) that is going to virtual ground. Thus the input impedance for the inverting amplifier is

$$Z_i \simeq R_i$$

The value of R_i will always be much less than the input impedance of the *OP*-amp. Therefore, the overall input impedance of an inverting amplifier will also be much lower than the *OP*-amp input impedance.

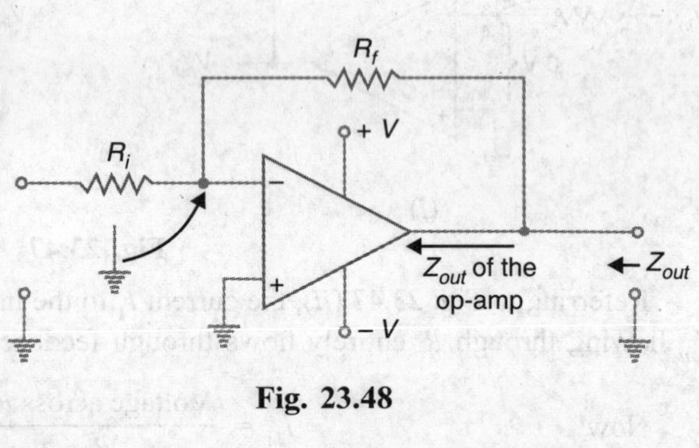

Fig. 23.48

(*ii*) **Output impedance.** Fig. 23.48 shows the inverting amplifier circuit. You can see from this figure that the output impedance of the inverting amplifier is the parallel combination of R_f and the output impedance of *OP*-amp itself.

The presence of the negative feedback circuit reduces the output impedance of the amplifier to a value that is less than the output impedance of *OP*-amp.

Example 23.25. *Given the OP-amp configuration in Fig. 23.49, determine the value of R_f required to produce a closed-loop voltage gain of −100.*

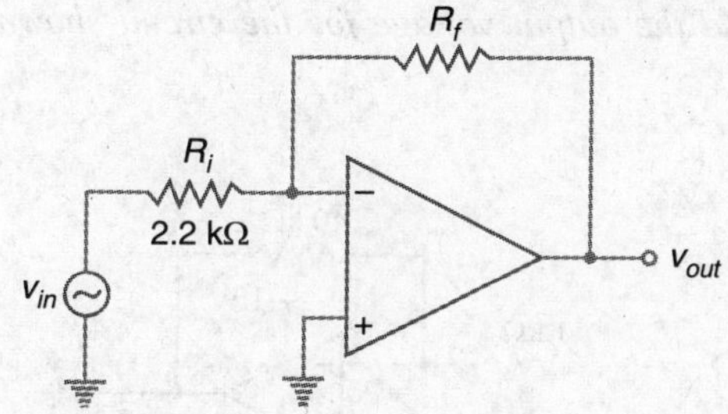

Fig. 23.49

Solution. $$A_{CL} = -\frac{R_f}{R_i} \quad \text{or} \quad -100 = -\frac{R_f}{2.2}$$

$\therefore$ $$R_f = 100 \times 2.2 = \mathbf{220\ k\Omega}$$

Example 23.26. *Determine the output voltage for the circuit of Fig. 23.50.*

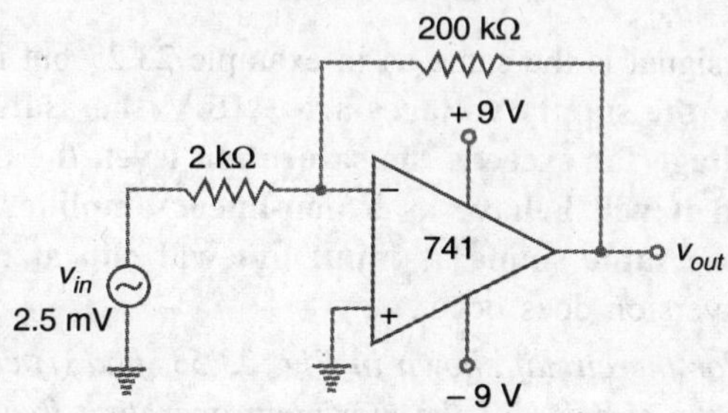

Fig. 23.50

Solution. $$A_{CL} = -\frac{R_f}{R_i} = -\frac{200\ k\Omega}{2\ k\Omega} = -100$$

$\therefore$ Output voltage, $v_{out} = A_{CL} \times v_{in} = (-100) \times (2.5\ mV) = -250\ mV = \mathbf{-0.25\ V}$

Example 23.27. *Find the output voltage for the circuit shown in Fig. 23.51.*

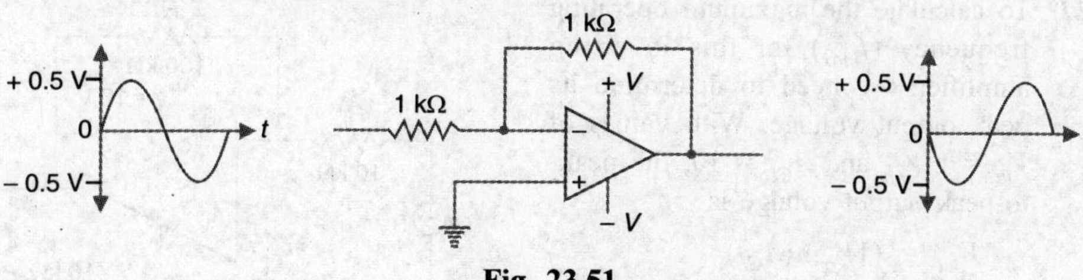

Fig. 23.51

Solution. Voltage gain, $A_{CL} = -\frac{R_f}{R_i} = -\frac{1\ k\Omega}{1\ k\Omega} = -1$

Since the voltage gain of the circuit is –1, **the output will have the same amplitude but with 180° phase inversion.**

Example 23.28. *Find the output voltage for the circuit shown in Fig. 23.52.*

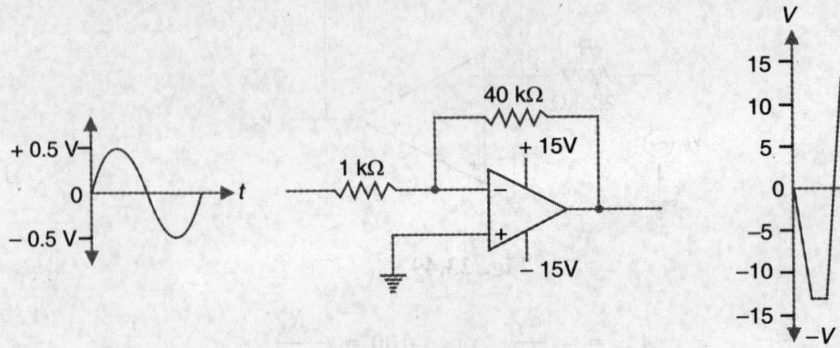

Fig. 23.52

Solution. Voltage gain, $A_{CL} = -\frac{R_f}{R_i} = -\frac{40\ \text{k}\Omega}{1\ \text{k}\Omega} = -40$

Note that the input signal is the same as in example 23.27 but now the voltage gain is –40 instead of –1. **Since the supply voltages are ± 15 V, the *saturation occurs at ± 13 V. Since the output voltage far exceeds the saturation level, the *OP*-amp will be driven to deep saturation and it will behave as a non-linear amplifier. This means that the output will not have the same shape as input but will clip at the saturation voltage.** Note that 180° phase inversion does occur.

Example 23.29. *For the circuit shown in Fig. 23.53, find (i) closed-loop voltage gain (ii) input impedance of the circuit (iii) the maximum operating frequency. The slew rate is 0.5V/μs.*

Solution.

(*i*) Closed-loop voltage gain, $A_{CL} = -\frac{R_f}{R_i} = \frac{100\ \text{k}\Omega}{10\ \text{k}\Omega} = \mathbf{-10}$

(*ii*) The input impedance Z_i of the circuit is

$$Z_i \simeq R_i = \mathbf{10\ k\Omega}$$

(*iii*) To calculate the maximum operating frequency (f_{max}) for this inverting amplifier, we need to determine its peak output voltage. With values of $V_{in} = 1\ V_{pp}$ and $A_{CL} = 10$, the peak-to-peak output voltage is

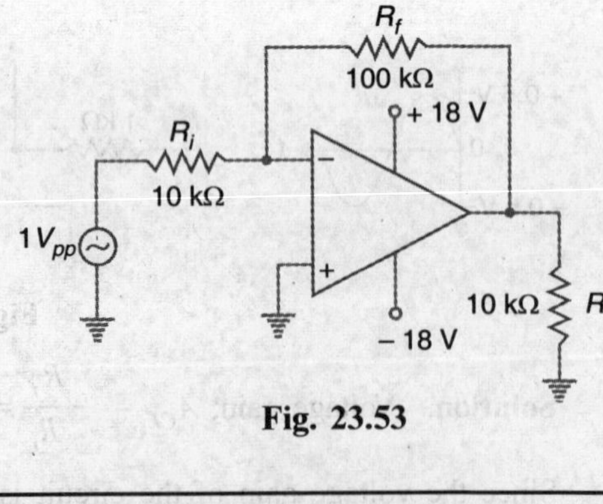

Fig. 23.53

$$V_{out} = (1V_{pp})\,(A_{CL})$$
$$= (1V_{pp}) \times 10 = 10\ V_{pp}$$

Therefore, the peak output voltage is

$$V_{pk} = 10/2 = 5\ \text{V}$$

* $+V_{sat} = +V_{supply} - 2\ \text{V} = +15\ \text{V} - 2\ \text{V} = +13\ \text{V}$

$-V_{sat} = -V_{supply} + 2\ \text{V} = -15\ \text{V} + 2\ \text{V} = -13\ \text{V}$

$$\therefore \quad f_{max} = \frac{\text{Slew rate}}{2\pi V_{pk}} = \frac{0.5\ \text{V}/\mu\text{s}}{2\pi \times 5}$$

$$= \frac{500\ \text{kHz}}{2\pi \times 5} = \mathbf{15.9\ kHz} \quad (\because 0.5\ \text{V}/\mu\text{s} = 500\ \text{kHz})$$

Example 23.30. *You have the following resistor values available :*

1 kΩ ; 5 kΩ ; 10 kΩ and 20 kΩ

Design the OP-amp circuit to have a voltage gain of –4.

Solution. Since the voltage gain is negative, the *OP*-amp is operating as an inverting amplifier.

Now,
$$A_{CL} = -\frac{R_f}{R_i} = -4$$

We need to use resistors that have a ratio of 4 : 1. The two resistors which satisfy this requirement are : R_f = **20 kΩ** and R_i = **5kΩ.**

Example 23.31. *Fig. 23.54 shows an inverting OP-amp. Find the closed-loop gain if (i)* R_{source} = 0Ω *(ii)* R_{source} = 1 kΩ.

Solution. (*i*) When $R_{source} = 0\ \Omega$; $\quad A_{CL} = -\frac{R_f}{R_i} = -\frac{100\ \text{k}\Omega}{1\ \text{k}\Omega} = \mathbf{-100}$

(*ii*) When $R_{source} = 1\ \text{k}\Omega$; $\quad A_{CL} = -\frac{R_f}{R_{source} + R_i} = -\frac{100\ \text{k}\Omega}{1\ \text{k}\Omega + 1\ \text{k}\Omega} = \mathbf{-50}$

Note that we have lost half of the voltage gain.

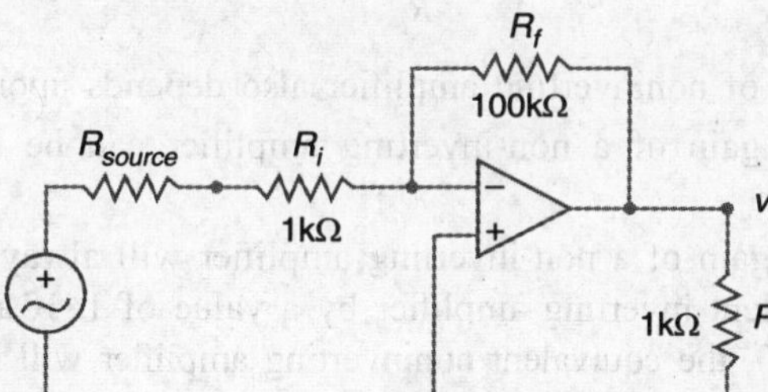

Fig. 23.54

23.26 Noninverting Amplifier

There are times when we wish to have an output signal of the same polarity as the input signal. In this case, the *OP*-amp is connected as noninverting amplifier as shown in Fig. 23.55. The input signal is applied to the noninverting input (+). The output is applied back to the input through the feedback circuit formed by feedback resistor R_f and input resistance R_i. Note that resistors R_f and R_i form a

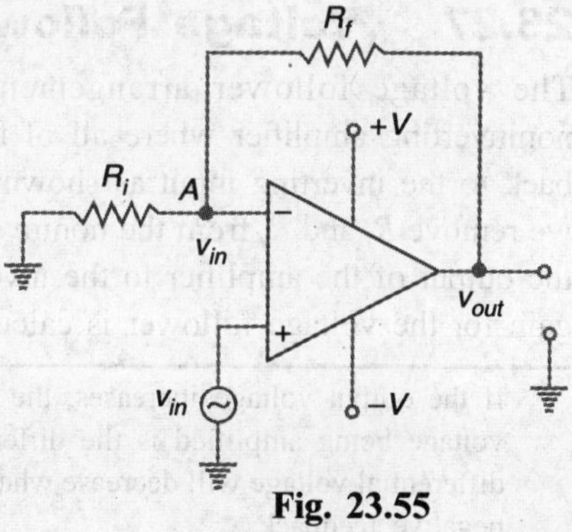

Fig. 23.55

voltage divider at the inverting input (–). This produces *negative feedback* in the circuit. Note that R_i is grounded. Since the input signal is applied to the noninverting input (+), the output signal will be noninverted *i.e.*, the output signal will be in phase with the input signal. Hence, the name non-inverting amplifier.

Voltage gain. If we assume that we are not at saturation, the potential at point A is the same as V_{in}. Since the input impedance of OP-amp is very high, all of the current that flows through R_f also flows through R_i. Keeping these things in mind, we have,

$$\text{Voltage across } R_i = V_{in} - 0 \text{ ; Voltage across } R_f = V_{out} - V_{in}$$

Now, $\quad$ Current through R_i = Current through R_f

or $$\frac{V_{in} - 0}{R_i} = \frac{V_{out} - V_{in}}{R_f}$$

or $$V_{in} R_f = V_{out} R_i - V_{in} R_i$$

or $$V_{in} (R_f + R_i) = V_{out} R_i$$

or $$\frac{V_{out}}{V_{in}} = \frac{R_f + R_i}{R_i} = 1 + \frac{R_f}{R_i}$$

$\therefore$ Closed-loop voltage gain, $$A_{CL} = \frac{V_{out}}{V_{in}} = 1 + \frac{R_f}{R_i}$$

The following points may be noted about the noninverting amplifier :

(i) $$A_{CL} = 1 + \frac{R_f}{R_i}$$

The voltage gain of noninverting amplifier also depends upon the values of R_f and R_i.

(ii) The voltage gain of a non-inverting amplifier can be made equal to or greater than 1.

(iii) The voltage gain of a non-inverting amplifier will always be greater than the gain of an equivalent inverting amplifier by a value of 1. If an inverting amplifier has a gain of 150, the equivalent noninverting amplifier will have a gain of 151.

(iv) The voltage gain is positive. This is not surprising because output signal is in phase with the input signal.

23.27 Voltage Follower

The voltage follower arrangement is a special case of noninverting amplifier where all of the output voltage is fed back to the inverting input as shown in Fig. 23.56. Note that we remove R_i and R_f from the noninverting amplifier and short the output of the amplifier to the inverting input. The voltage gain for the voltage follower is calculated as under :

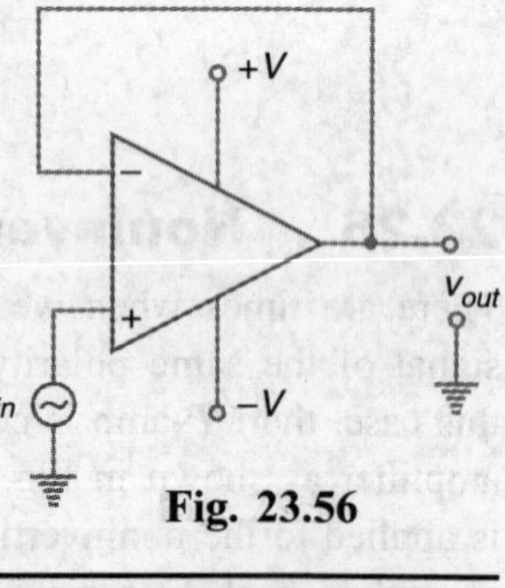

Fig. 23.56

* If the output voltage increases, the voltage at the inverting input will also increase. Since the voltage being amplified is the difference between the voltages at the two input terminals, the differential voltage will decrease when the output voltage increases. Therefore, the circuit provides negative feedback.

$$A_{CL} = 1 + \frac{R_f}{R_i} = 1 + \frac{0}{R_i} = 1 \qquad (\because R_f = 0\Omega)$$

Thus the closed-loop voltage gain of the voltage follower is 1. The most important features of the voltage follower configuration are its *very high input impedance* and its *very low output impedance.* These features make it a nearly ideal buffer amplifier to be connected between high-impedance sources and low-impedance loads.

Example 23.32. *Calculate the output voltage from the noninverting amplifier circuit shown in Fig. 23.57 for an input of 120 μV.*

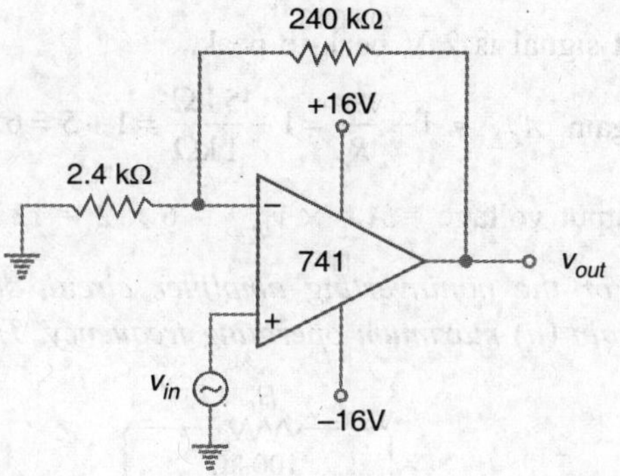

Fig. 23.57

Solution. Voltage gain, $A_{CL} = 1 + \frac{R_f}{R_i} = 1 + \frac{240\text{ k}\Omega}{2.4\text{ k}\Omega} = 1 + 100 = 101$

Output voltage, $v_{out} = A_{CL} \times v_{in} = (101) \times (120\ \mu\text{V}) = \mathbf{12.12\ mV}$

Example 23.33. *For the noninverting amplifier circuit shown in Fig. 23.58, find the output voltage for an input voltage of (i) 1 V (ii) –1 V.*

Solution. Voltage gain, $A_{CL} = 1 + \frac{R_f}{R_i} = 1 + \frac{10\text{ k}\Omega}{1\text{ k}\Omega} = 1 + 10 = 11$

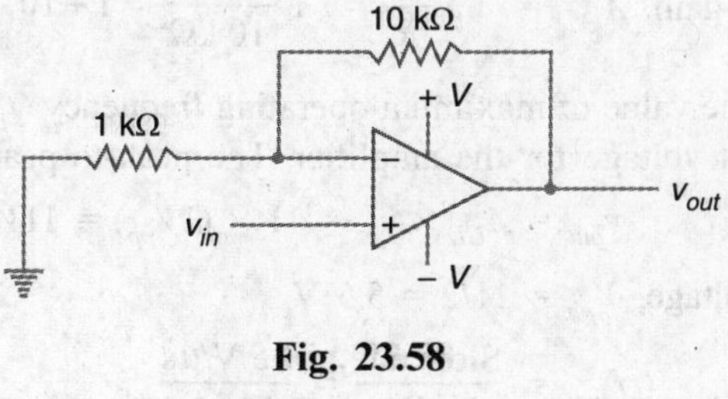

Fig. 23.58

(*i*) For $v_{in} = 1$ V ; $v_{out} = A_{CL} \times v_{in} = 11 \times 1\text{ V} = \mathbf{11\ V}$

(*ii*) For $v_{in} = -1$ V ; $v_{out} = A_{CL} \times v_{in} = 11 \times (-1\text{ V}) = \mathbf{-11\ V}$

Example 23.34. *For the noninverting amplifier circuit shown in Fig. 23.59, find peak-to-peak output voltage.*

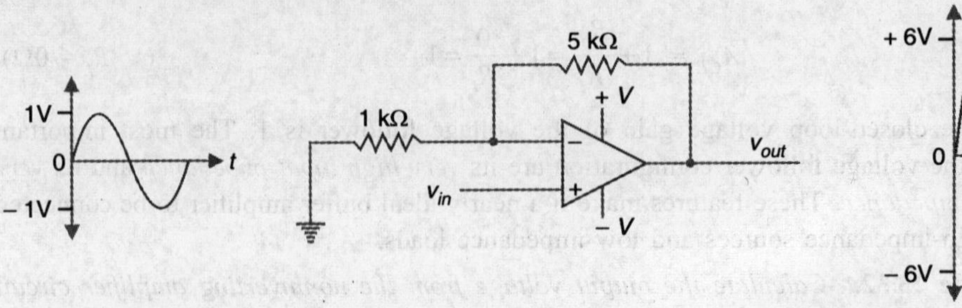

Fig. 23.59

Solution. The input signal is 2 V peak-to-peak.

$$\text{Voltage gain, } A_{CL} = 1 + \frac{R_f}{R_i} = 1 + \frac{5\ \text{k}\Omega}{1\ \text{k}\Omega} = 1 + 5 = 6$$

∴ Peak-to-peak output voltage $= A_{CL} \times v_{inpp} = 6 \times 2 = $ **12 V**

Example 23.35. *For the noninverting amplifier circuit shown in Fig. 23.60, find (i) closed-loop voltage gain (ii) maximum operating frequency. The slew rate is 0.5 V/μs.*

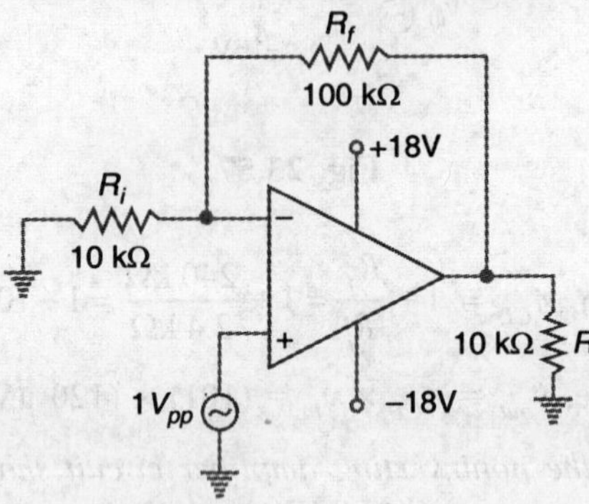

Fig. 23.60

Solution.

(*i*) $$\text{Voltage gain, } A_{CL} = 1 + \frac{R_f}{R_i} = 1 + \frac{100\ \text{k}\Omega}{10\ \text{k}\Omega} = 1 + 10 = \mathbf{11}$$

(*ii*) To determine the value of maximum operating frequency (f_{max}), we need to calculate the peak output voltage for the amplifier. The peak-to-peak output voltage is

$$v_{out} = A_{CL} \times v_{in} = 11 \times (1 V_{pp}) = 11 V_{pp}$$

∴ Peak output voltage, $V_{pk} = 11/2 = 5.5$ V

∴ $$f_{max} = \frac{\text{Slew rate}}{2\pi V_{pk}} = \frac{0.5\ \text{V}/\mu\text{s}}{2\pi \times 5.5}$$

$$= \frac{500\ \text{kHz}}{2\pi \times 5.5} = \mathbf{14.47\ kHz} \qquad (\because 0.5\ \text{V}/\mu\text{s} = 500\ \text{kHz})$$

Example 23.36. *Determine the bandwidth of each of the amplifiers in Fig. 23.61. Both OP-amps have an open-loop voltage gain of 100 dB and a unity-gain bandwidth of 3MHz.*

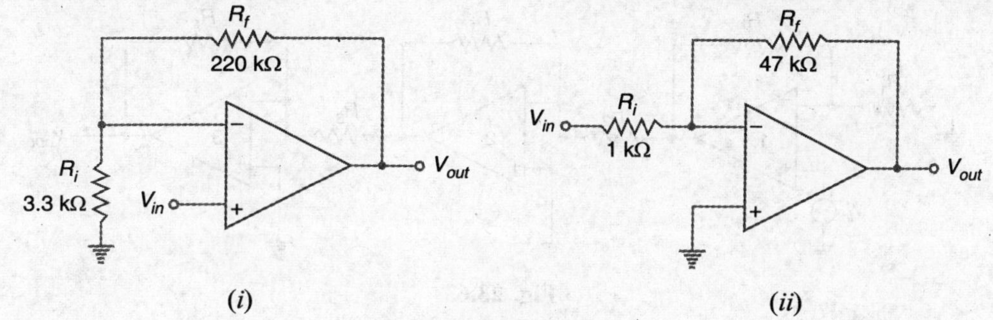

Fig. 23.61

Solution.

(*i*) For the noninverting amplifier shown in Fig. 23.61 (*i*), the closed-loop voltage gain (A_{CL}) is

$$A_{CL} = 1 + \frac{R_f}{R_i} = 1 + \frac{220 \text{ k}\Omega}{3.3 \text{ k}\Omega} = 1 + 66.7 = 67.7$$

$$\therefore \quad \text{Bandwidth, } BW = \frac{\text{Unity-gain } BW}{A_{CL}} = \frac{3 \text{ MHz}}{67.7} = \mathbf{44.3 \text{ kHz}}$$

(*ii*) For the inverting amplifier shown in Fig. 23.61 (*ii*),

$$A_{CL} = -\frac{R_f}{R_i} = -\frac{47 \text{ k}\Omega}{1 \text{ k}\Omega} = -47$$

$$\therefore \quad \text{Bandwidth, } BW = \frac{3 \text{ MHz}}{47} = \mathbf{63.8 \text{ kHz}}$$

Example 23.37. *Fig. 23.62 shows the circuit of voltage follower. Find* (*i*) *the closed-loop voltage gain and* (*ii*) *maximum operating frequency. The slew rate is 0.5 V/μs.*

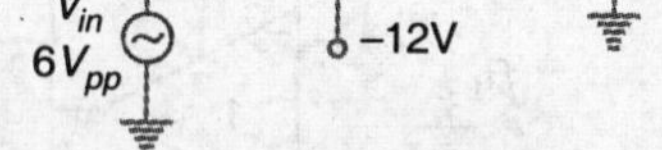

Fig. 23.62

Solution.

(*i*) For the voltage follower, A_{CL} = **1.**

(*ii*) Since $A_{CL} = 1$ for the circuit, $v_{out} = v_{in}$. Therefore, peak output voltage (V_{pk}) is one-half of $6V_{pp}$ *i.e.*, $V_{pk} = 6/2 = 3$ V. The maximum operating frequency (f_{max}) is given by ;

$$f_{max} = \frac{\text{Slew rate}}{2\pi V_{pk}} = \frac{0.5 \text{ V}/\mu\text{s}}{2\pi \times 3}$$

$$= \frac{500 \text{ kHz}}{2\pi \times 3} = \mathbf{26.53 \text{ kHz}} \qquad (\because 0.5 \text{ V}/\mu\text{s} = 500 \text{ kHz})$$

23.28 Multi-stage OP-Amp Circuits

When a number of *OP*-amp stages are connected in series, the overall voltage gain is equal to the product of individual stage gains. Fig. 23.63 shows connection of three stages. The first stage is connected to provide noninverting gain. The next two stages provide inverting gains.

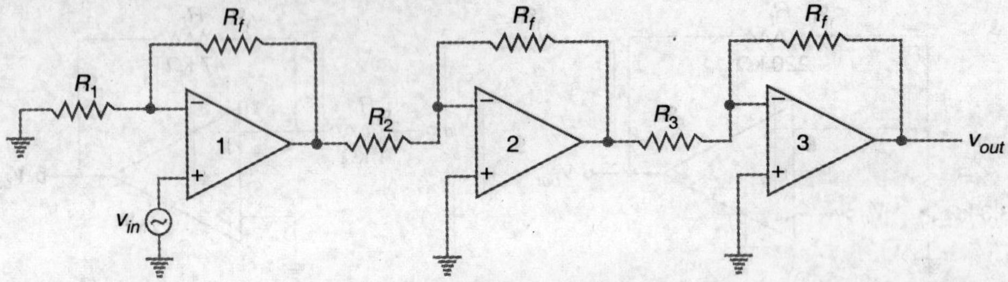

Fig. 23.63

The overall voltage gain A of this circuit is given by ;

$$A = A_1 A_2 A_3$$

where A_1 = Voltage gain of first stage = $1 + (R_f/R_1)$

A_2 = Voltage gain of second stage = $-R_f/R_2$

A_3 = Voltage gain of third stage = $-R_f/R_3$

Since the overall voltage gain is positive, the circuit behaves as a noninverting amplifier.

Example 23.38. *Fig. 23.63 shows the multi-stage OP-amp circuit. The resistor values are : R_f = 470 kΩ ; R_1 = 4.3 kΩ ; R_2 = 33 kΩ and R_3 = 33 kΩ. Find the output voltage for an input of 80 μV.*

Solution. Voltage gain of first stage, $A_1 = 1 + (R_f/R_1) = 1 + (470\ k\Omega/4.3\ k\Omega) = 110.3$

Voltage gain of second stage, $A_2 = -R_f/R_2 = -470\ k\Omega/33\ k\Omega = -14.2$

Voltage gain of third stage, $A_3 = -R_f/R_3 = -470\ k\Omega/33\ k\Omega = -14.2$

∴ Overall voltage gain, $A = A_1A_2A_3 = (110.3) \times (-14.2) \times (-14.2) = 22.2 \times 10^3$

Output voltage, $v_{out} = A \times v_{in} = 22.2 \times 10^3 \times (80\ \mu V) = \mathbf{1.78\ V}$

Example 23.39. *A three-stage OP-amp circuit is required to provide voltage gains of +10, –18 and –27. Design the OP-amp circuit. Use a 270 kΩ feedback resistor for all three circuits. What output voltage will result for an input of 150 μV?*

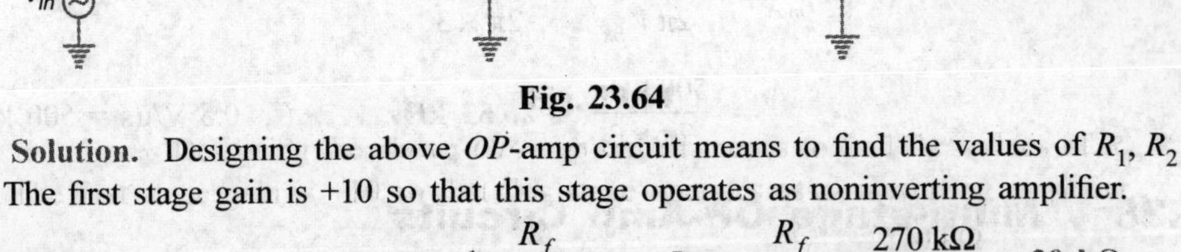

Fig. 23.64

Solution. Designing the above *OP*-amp circuit means to find the values of R_1, R_2 and R_3. The first stage gain is +10 so that this stage operates as noninverting amplifier.

Now,
$$+10 = 1 + \frac{R_f}{R_1} \quad \therefore \quad R_1 = \frac{R_f}{10-1} = \frac{270\ k\Omega}{9} = \mathbf{30\ k\Omega}$$

The second stage gain is –18 so that this stage operates as an inverting amplifier.

$$\therefore \quad -18 = -\frac{R_f}{R_2} \quad \text{or} \quad R_2 = \frac{R_f}{18} = \frac{270\ k\Omega}{18} = \mathbf{15\ k\Omega}$$

The third stage gain is –27 so that this stage operates as an inverting amplifier.

$$\therefore \quad -27 = -\frac{R_f}{R_3} \quad \text{or} \quad R_3 = \frac{R_f}{27} = \frac{270\ \text{k}\Omega}{27} = \mathbf{10\ k\Omega}$$

$$\text{Overall voltage gain, } A = A_1 A_2 A_3 = (10) \times (-18) \times (-27) = 4860$$

$$\text{Output voltage, } v_{out} = A \times v_{in} = (4860) \times (150\ \mu\text{V}) = \mathbf{0.729\ V}$$

Example 23.40. *Show the connection of three OP-amp stages using an LM 348 IC to provide outputs that are 10, 20 and 50 times larger than the input and 180° out of phase w.r.t. input. Use a feedback resistor of R_f = 500 kΩ in all stages.*

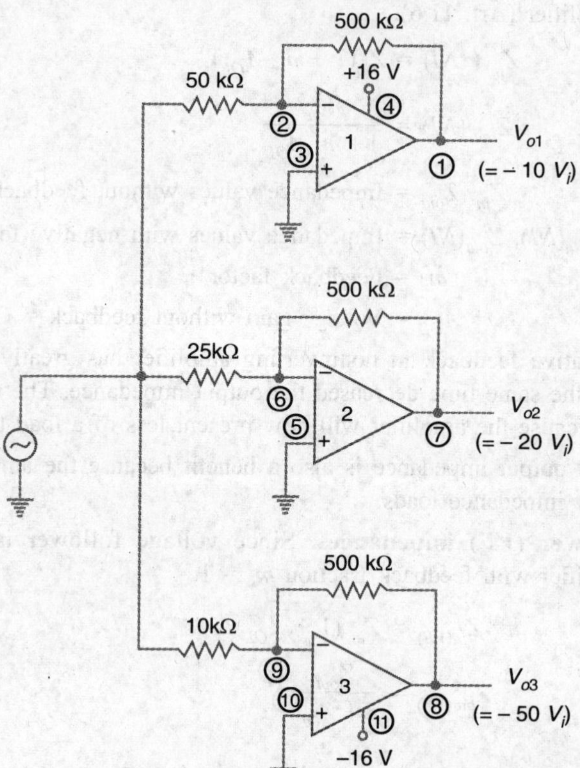

Fig. 23.65

Solution. The resistor component for each stage will be :

$$R_1 = -\frac{R_f}{A_1} = -\frac{500\ \text{k}\Omega}{-10} = \mathbf{50\ k\Omega}$$

$$R_2 = -\frac{R_f}{A_2} = -\frac{500\ \text{k}\Omega}{-20} = \mathbf{25\ k\Omega}$$

$$R_3 = -\frac{R_f}{A_3} = -\frac{500\ \text{k}\Omega}{-50} = \mathbf{10\ k\Omega}$$

The resulting circuit is shown in Fig. 23.65.

23.29 Effect of Negative Feedback on OP-Amp Impedances

In a negative feedback amplifier, a part of the output is fed in phase opposition to the input. The negative feedback produces remarkable changes in the circuit performance. The advantages of negative feedback are : stable gain, less distortion, increased bandwidth and affecting the input impedance and output impedance of the circuit. We now discuss the effect of negative feedback on the impedances of both noninverting and inverting amplifiers.

(i) Noninverting Amplifier (Fig. 23.66). The expressions for the input and output impedances on account of negative voltage feedback in noninverting amplifier are the same as for discrete amplifier (Art. 11.6).

$$Z_{in}\ (^*NI) = Z_{in}(1 + m_v\ A_{OL})$$

$$Z_{out}\ (NI) = \frac{Z_{out}}{1 + m_v\ A_{OL}}$$

where Z_{in}, Z_{out} = Impedance values without feedback

$Z_{in}(NI), Z_{out}(NI)$ = Impedance values with negative feedback

m_v = Feedback factor

A_{OL} = Voltage gain without feedback = Open-loop gain

Note that negative feedback in noninverting amplifier has greatly increased the input impedance and at the same time decreased the output impedance. The increased impedance is an advantage because the amplifier will now present less of a load to its source circuit.

The decreased output impedance is also a benefit because the amplifier will be better suited to drive low impedance loads.

Voltage-follower (*VF*) impedances. Since voltage follower is a special case of noninverting amplifier with feedback fraction $m_v = 1$,

$$\therefore \quad Z_{in(VF)} = Z_{in}\ (1 + A_{OL})$$

$$Z_{out(VF)} = \frac{Z_{out}}{1 + A_{OL}}$$

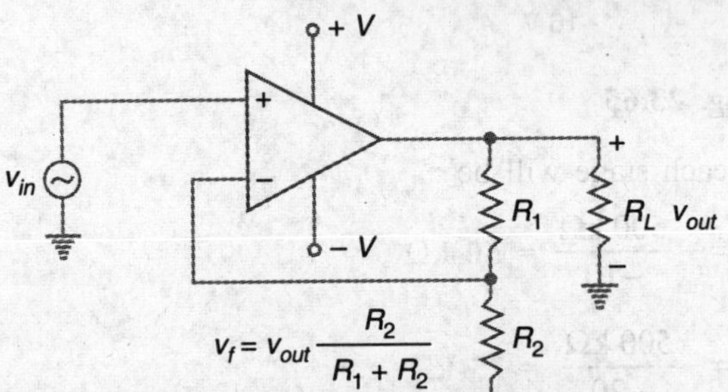

Fig. 23.66
Noninverting feedback amplifier.

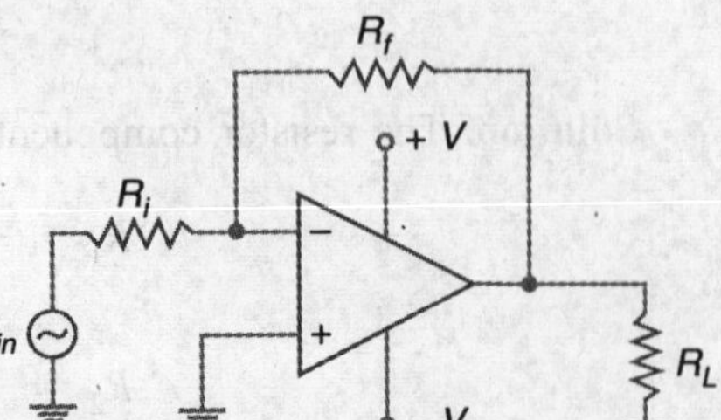

Fig. 23.67
Inverting feedback amplifier.

* Note that '*NI*' means noninverting amplifier.

Note that the voltage-follower input impedance is greater for given Z_{in} and A_{OL} than for the non-inverting configuration with the voltage-divider feedback circuit. Also, its output impedance is much smaller.

(*ii*) **Inverting Amplifier.** Fig. 23.67 shows the inverting amplifier. It can be shown that :

Input impedance, $Z_{in(*I)} \simeq R_i$

Output impedance, $Z_{out(I)} \simeq Z_{out}$ of *OP*-amp

Note that *the addition of negative voltage feedback to the inverting OP-amp reduces the input impedance of the circuit.* The reduction of Z_{in} is the primary difference betwen the inverting and the noninverting negative feedback circuits. Otherwise, the effects of negative voltage feedback are nearly identical for the two circuits.

Example 23.41. (*i*) *Determine the input and output impedances of the amplifier in Fig. 23.68. The OP-amp data sheet gives Z_{in} = 2 MΩ, Z_{out} = 75 Ω and open-loop gain of 200,000.*

(*ii*) *Find the closed-loop voltage gain.*

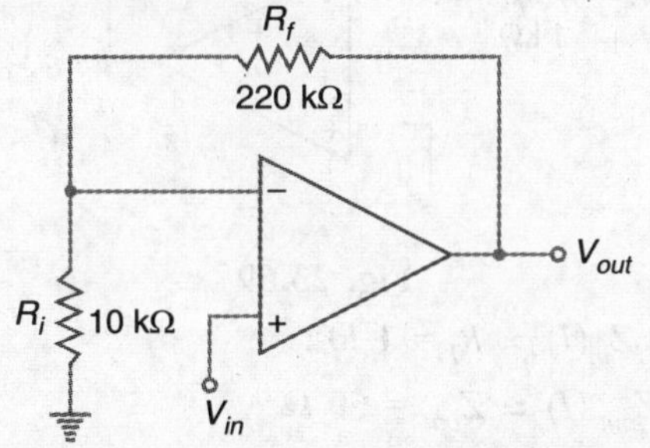

Fig. 23.68

Solution.

(*i*) Feedback fraction, $m_v = \dfrac{R_i}{R_i + R_f} = \dfrac{10\text{ k}\Omega}{10\text{ k}\Omega + 220\text{ k}\Omega} = \dfrac{10\text{ k}\Omega}{230\text{ k}\Omega} = 0.043$

Input impedance, $Z_{in(NI)} = Z_{in}\,(1 + A_{OL} m_v)$

$= (2\text{M}\Omega)\,[(1 + 200,000 \times 0.043)]$

$= (2\text{M}\Omega)\,[1 + 8600] = \mathbf{17{,}202\ M\Omega}$

Output impedance, $Z_{out(NI)} = \dfrac{Z_{out}}{1 + A_{OL} m_v} = \dfrac{75\ \Omega}{1 + 8600} = \mathbf{8.7 \times 10^{-3}\ \Omega}$

(*ii*) Closed-loop voltage gain, $A_{CL} = 1 + \dfrac{R_f}{R_i} = 1 + \dfrac{220\text{ k}\Omega}{10\text{ k}\Omega} = \mathbf{23}$

Comments. Note the effect of negative voltage feedback on noninverting amplifier.

(*a*) Input and output signals are in phase.

(*b*) A virtually infinite input impedance.

(*c*) Virtually zero output impedance.

Example 23.42. *The same OP-amp in example 23.41 is used in a voltage-follower arrangement. Determine the input and output impedances.*

* Note that "*I*" means inverting amplifier.

Solution. For voltage follower, feedback factor $m_v = 1$.

$$\therefore \quad Z_{in(VF)} = Z_{in}(1 + A_{OL}) = 2\text{M}\Omega\ (1 + 200{,}000) = \mathbf{400{,}002\ M\Omega}$$

$$Z_{out(VF)} = \frac{Z_{out}}{1 + A_{OL}} = \frac{75\ \Omega}{1 + 200{,}000} = \mathbf{0.38 \times 10^{-3}\ \Omega}$$

Note that $Z_{in(VF)}$ is much greater than $Z_{in(NI)}$ and $Z_{out(VF)}$ is much less than $Z_{out(NI)}$ from example 23.41.

Example 23.43. *Find the values of the input and output impedances in Fig. 23.69. Also determine the closed-loop voltage gain. The OP-amp has the following parameters : Z_{in} = 4 MΩ ; Z_{out} = 50 Ω and open-loop voltage gain = 50,000.*

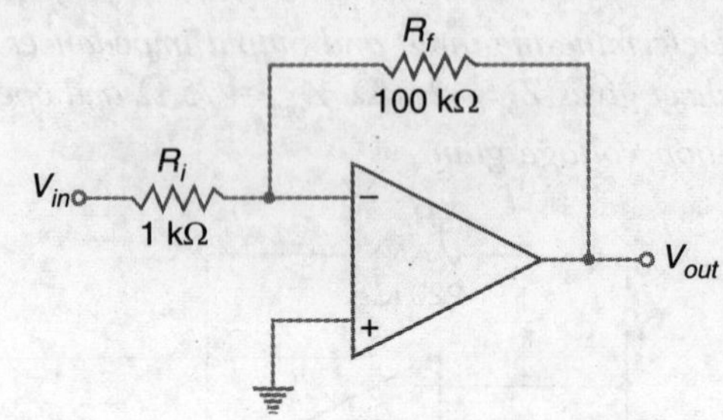

Fig. 23.69

Solution.

$$Z_{in}(I) \simeq R_i = \mathbf{1\ k\Omega}$$

$$Z_{out}(I) \simeq Z_{out} = \mathbf{50\ \Omega}$$

$$A_{CL} = -\frac{R_f}{R_i} = -\frac{100\ \text{k}\Omega}{1\ \text{k}\Omega} = \mathbf{-100}$$

23.30 Faults in Feedback Circuit

A failure of the feedback circuit in an *OP*-amp is one of the easiest problems in the world to locate. The most noticeable effect is that *voltage gain of the amplifier will change drastically.* Sometimes the gain will increase ; sometimes it will decrease. It all depends on which component goes bad. For example, consider the circuit shown in Fig. 23.70.

(*i*) **Under normal conditions.** Under normal conditions, the output from the amplifier is $v_{out} = A_{CL}\ v_{in}$. The waveform would be correct. This is shown in Fig. 23.70 (*i*)

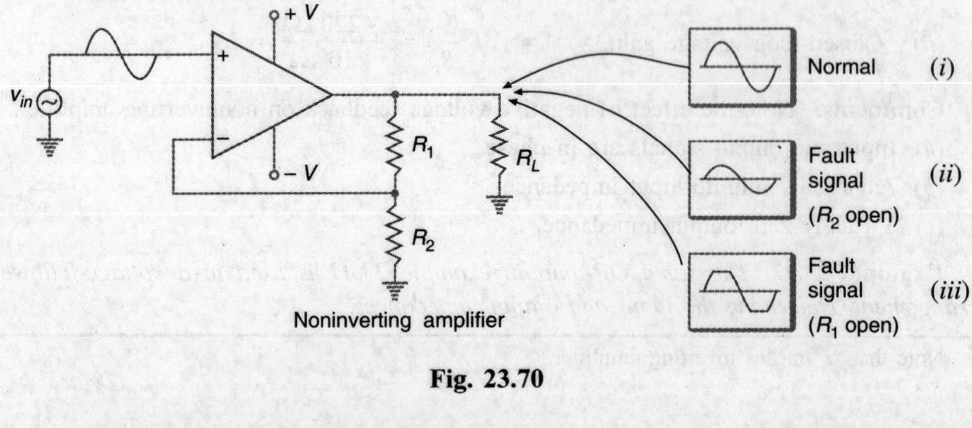

Fig. 23.70

(*ii*) **When R_2 is open.** If R_2 opens, the feedback circuit would consist solely of R_1. In this case, the gain would *drop*. It is because the circuit would now act as a *voltage follower. In other words, the circuit would now be a buffer with an output voltage that is equal to the input voltage. Thus we would have the output signal as shown in Fig. 23.70 (*ii*). The waveform would be correct but we would have unity gain.

(*iii*) **When R_1 is open.** When R_1 opens, the entire feedback circuit would be effectively removed. This would cause the gain of the amplifier to increase to the value of open-loop gain A_{OL}. Clearly, the output voltage will clip at or near the values of $+V$ and $-V$. This results in the distorted output signal as shown in Fig. 23.70 (*iii*).

23.31 Summary of OP-AMP Configurations

Basic OP-Amp

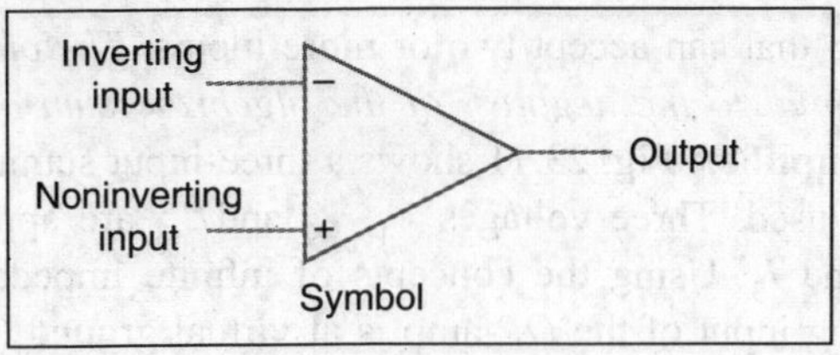

- Very high open-loop voltage gain
- Very high input impedance
- Very low output impedance

Noninverting Amplifier

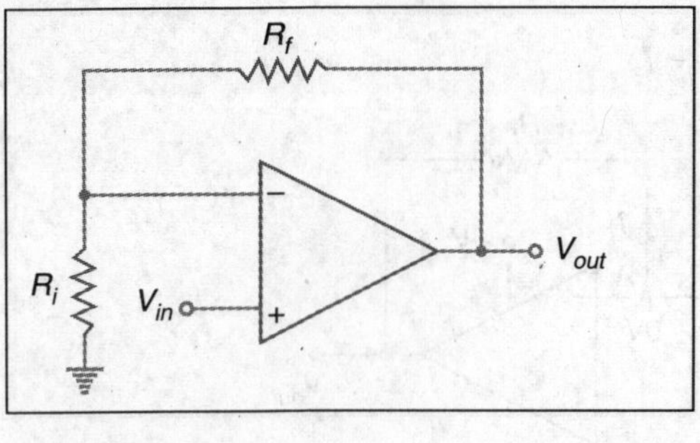

- Voltage gain :

$$A_{CL}(NI) = 1 + \frac{R_f}{R_i}$$

- Input impedance :

$$Z_{in(NI)} = (1 + A_{OL} m_v)\, Z_{in}$$

- Output impedance :

$$Z_{out(NI)} = \frac{Z_{out}}{1 + A_{OL} m_v}$$

Voltage Follower

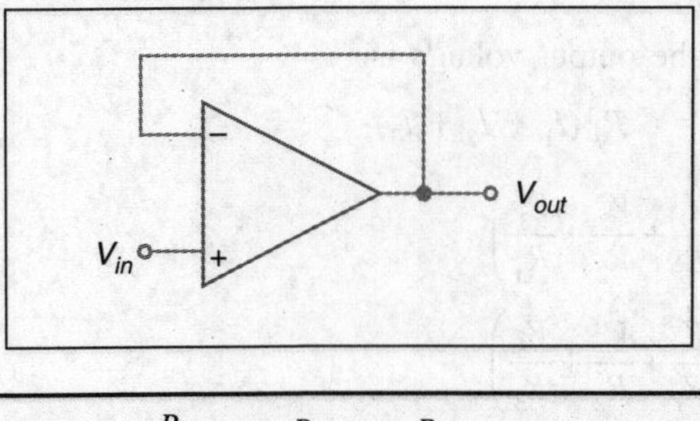

- Voltage gain :

$$A_{CL(VF)} = 1$$

- Input impedance :

$$Z_{in(VF)} = (1 + A_{OL})\, Z_{in}$$

- Output impedance :

$$Z_{out(VF)} = \frac{Z_{out}}{1 + A_{OL}}$$

---

* $A_{CL} = 1 + \frac{R_f}{R_i} = 1 + \frac{R_1}{R_2} = 1 + \frac{R_1}{\infty} = 1 + 0 = 1$

Under these conditions, the closed-loop voltage gain would be unity.

Inverting Amplifier

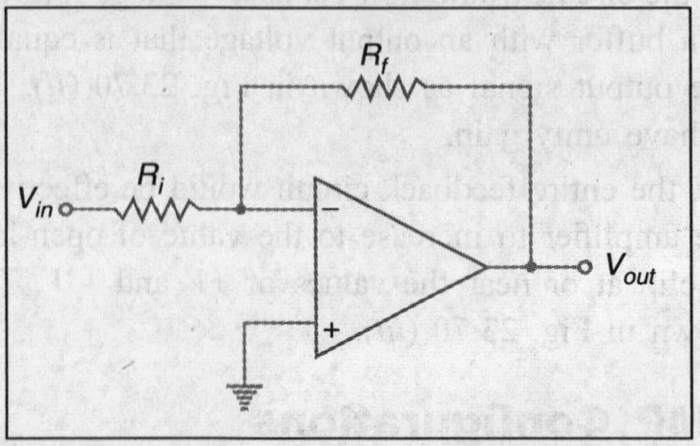

- Voltage gain :

$$A_{CL} = -\frac{R_f}{R_i}$$

- Input impedance :

$$Z_{in(I)} \simeq R_i$$

- Output impedance :

$$Z_{out(I)} \simeq Z_{out}$$

23.32 Summing Amplifiers

A summing amplifier is an inverted *OP*-amp that can accept two or more inputs. *The output voltage of a summing amplifier* is *proportional to the negative of the algebraic sum of its input voltages.* Hence the name **summing amplifier**. Fig. 23.71 shows a three-input summing amplifier but any number of inputs can be used. Three voltages V_1, V_2 and V_3 are applied to the inputs and produce currents I_1, I_2 and I_3. Using the concepts of infinite impedance and virtual ground, you can see that inverting input of the *OP*-amp is at virtual ground (0V) and there is no current to the input. This means that the three input currents I_1, I_2 and I_3 combine at the summing point *A* and form the total current (I_f) which goes through R_f as shown in Fig. 23.71.

$$\therefore \quad I_f = I_1 + I_2 + I_3$$

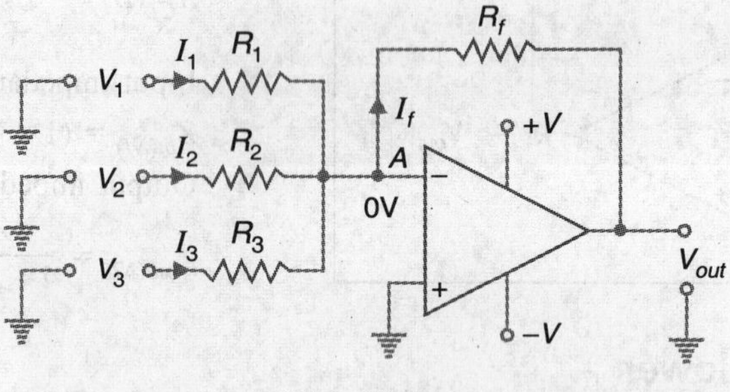

Fig. 23.71

When all the three inputs are applied, the output voltage is :

$$\text{Output voltage, } V_{out} = -I_f R_f = -R_f (I_1 + I_2 + I_3)$$

$$= -R_f\left(\frac{V_1}{R_1}+\frac{V_2}{R_2}+\frac{V_3}{R_3}\right)$$

$$\therefore \quad V_{out} = -R_f\left(\frac{V_1}{R_1}+\frac{V_2}{R_2}+\frac{V_3}{R_3}\right)$$

If $R_1 = R_2 = R_3 = R$, then, we have,

$$V_{out} = -\frac{R_f}{R}(V_1+V_2+V_3)$$

Thus the output voltage is proportional to the algebraic sum of the input voltages (of course neglecting negative sign). An interesting case results when the **gain of the amplifier is unity.** In that case, $R_f = R_1 = R_2 = R_3$ and output voltage is

$$V_{out} = -(V_1 + V_2 + V_3)$$

Thus, when the gain of summing amplifier is unity, the output voltage is the algebraic sum of the input voltages.

Summing amplifier with gain greater than unity. When R_f is larger than the input resistors, the amplifer has a gain of R_f/R where R is the value of each input resistor. The general expression for the output voltage is

$$V_{out} = -\frac{R_f}{R}(V_1 + V_2 + V_3 + \ldots)$$

As you can see, the output voltage is the sum of input voltages multiplied by a constant determined by the ratio R_f/R.

Example 23.44. *Determine the output voltage for the summing amplifier in Fig. 23.72.*

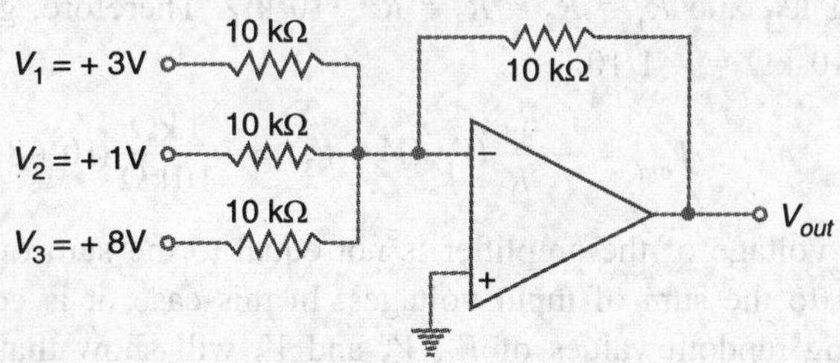

Fig. 23.72

Solution. Referring to Fig. 23.72, all the three input resistor values are equal and each is equal to the value of feedback resistor. Therefore, the gain of the summing amplifier is 1. As a result, the output voltage is the algebraic sum of three input voltages.

$$\therefore \quad V_{out} = -(V_1 + V_2 + V_3) = -(3 + 1 + 8) = -12 \text{ V}$$

Example 23.45. *Determine the output voltage for the summing amplifier shown in Fig. 23.73.*

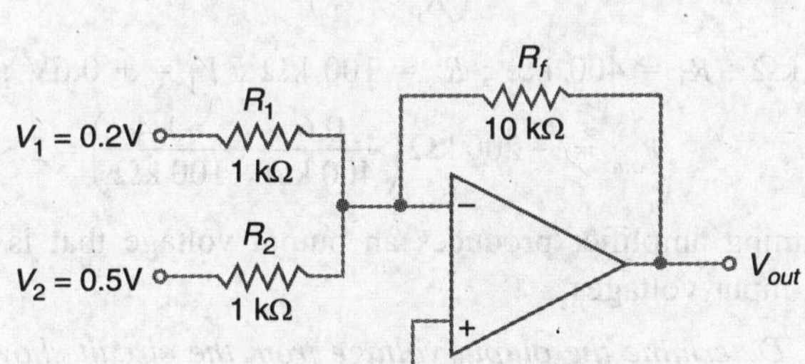

Fig. 23.73

Solution. $R_f = 10$ kΩ and $R_1 = R_2 = R = 1$ kΩ. Therefore, gain of the amplifier $= -R_f/R = -10 \text{ k}\Omega/1\text{k}\Omega = -10$.

Now, $$V_{out} = -\frac{R_f}{R}(V_1 + V_2) = -\frac{10 \text{ k}\Omega}{1\text{k}\Omega}(0.2 + 0.5) = -7\text{V}$$

Note that the output voltage is not equal to the sum of input voltages. Rather it is equal to the sum of input voltages multiplied by the amplifier gain. In other words, the output voltage is proportional to the sum of the input voltages.

Example 23.46. *Determine the output voltage for the summing amplifier shown in Fig. 23.74.*

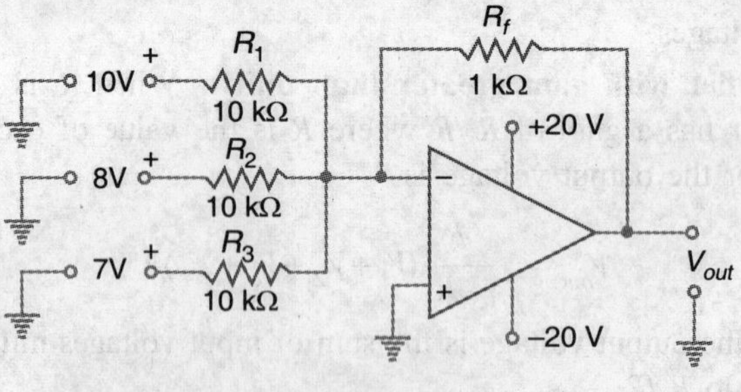

Fig. 23.74

Solution. $R_f = 1$ kΩ and $R_1 = R_2 = R_3 = R = 10$ kΩ. Therefore, gain of the amplifier $= -R_f/R = -1$ kΩ/10 kΩ $= -1/10$.

Now,
$$V_{out} = -\frac{R_f}{R}(V_1 + V_2 + V_3) = -\frac{1\text{ k}\Omega}{10\text{k}\Omega}(10+8+7) = -\mathbf{2.5\ V}$$

Note that output voltage of the amplifier is not equal to the sum of input voltages, but rather is proportional to the sum of input voltages. In this case, it is equal to one-tenth of the input sum. Picking random values of V_1, V_2 and V_3 will show that this circuit always provides an output voltage that is one-tenth of the sum of input voltages.

Example 23.47. *Two voltages of + 0.6V and – 1.4 V are applied to the two input resistors of a summing amplifier. The respective input resistors are 400 kΩ and 100 kΩ and feedback resistor is 200 kΩ. Determine the output voltage.*

Solution. The output voltage of the summing amplifier is given by ;

$$V_{out} = -R_f\left(\frac{V_1}{R_1} + \frac{V_2}{R_2}\right)$$

Here, $R_f = 200$ kΩ ; $R_1 = 400$ kΩ ; $R_2 = 100$ kΩ ; $V_1 = +0.6$V ; $V_2 = -1.4$ V

$$\therefore \quad V_{out} = -200\text{ k}\Omega\left(\frac{0.6}{400\text{ k}\Omega} + \frac{-1.4}{100\text{ k}\Omega}\right) = \mathbf{2.5\ V}$$

Note that a summing amplifier produces an output voltage that is proportional to the *algebraic sum of* the input voltages.

Example 23.48. *Determine the output voltage from the circuit shown in Fig. 23.75 for each of the following input combinations :*

V_1(V)	V_2(V)	V_3(V)
+10	0	+10
0	+10	+10
+10	+10	+10

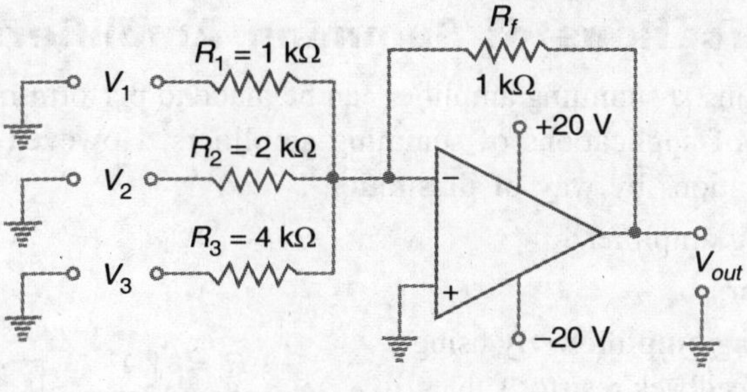

Fig. 23.75

Solution. The output voltage from the circuit is given by ;

$$V_{out} = -\left(\frac{R_f}{R_1}V_1 + \frac{R_f}{R_2}V_2 + \frac{R_f}{R_3}V_3\right)$$

$$= -\left(\frac{1\,\text{k}\Omega}{1\,\text{k}\Omega}V_1 + \frac{1\,\text{k}\Omega}{2\,\text{k}\Omega}V_2 + \frac{1\,\text{k}\Omega}{4\,\text{k}\Omega}V_3\right)$$

$$\therefore \quad V_{out} = -(V_1 + 0.5\ V_2 + 0.25\ V_3)$$

The output voltage for the first set of inputs is

$$V_{out} = -(10 + 0.5 \times 0 + 0.25 \times 10) = \mathbf{-12.5\ V}$$

The output voltage for the second set of inputs is

$$V_{out} = -(0 + 0.5 \times 10 + 0.25 \times 10) = \mathbf{-7.5\ V}$$

The output voltage for the third set of inputs is

$$V_{out} = -(10 + 0.5 \times 10 + 0.25 \times 10) = \mathbf{-17.5\ V}$$

Example 23.49. *Calculate the output voltage for the circuit of Fig. 23.76. The inputs are V_1 = 50 sin (1000 t)mV and V_2 = 10 sin (3000 t) mV.*

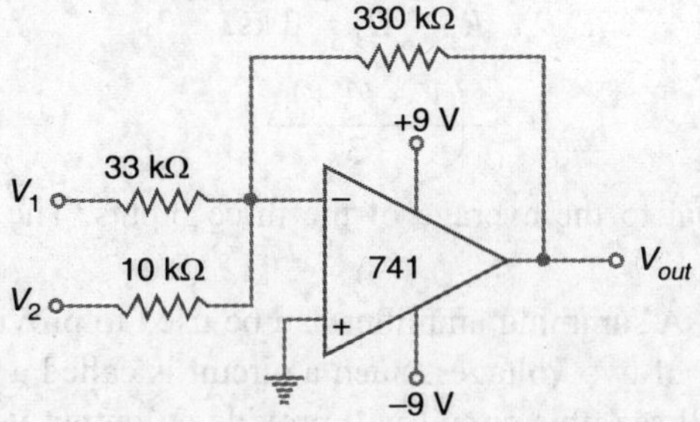

Fig. 23.76

Solution. The output voltage for the circuit is

$$V_{out} = -\left(\frac{R_f}{R_1}V_1 + \frac{R_f}{R_2}V_2\right) = -\left(\frac{330\text{k}\Omega}{33\text{k}\Omega}V_1 + \frac{330\text{k}\Omega}{10\text{k}\Omega}V_2\right)$$

$$= -(10\ V_1 + 33\ V_2) = -[10 \times 50 \sin(1000t) + 33 \times 10 \sin(3000t)]\ \text{mV}$$

$$= \mathbf{-\ [0.5\ sin\ (1000\ \mathit{t}) + 0.33\ sin\ (3000\mathit{t})]\ V}$$

23.33 Applications of Summing Amplifiers

By proper modifications, a summing amplifier can be made to perform many useful functions. There are a number of applications of summing amplifiers. However, we shall discuss the following two applications by way of illustration :

1. As averaging amplifier
2. As subtractor

1. As averaging amplifier. By using the proper input and feedback resistor values, a summing amplifier can be designed to provide an output voltage that is equal to the *average* of input voltages. A summing amplifier will act as an averaging amplifier when *both* of the following conditions are met :

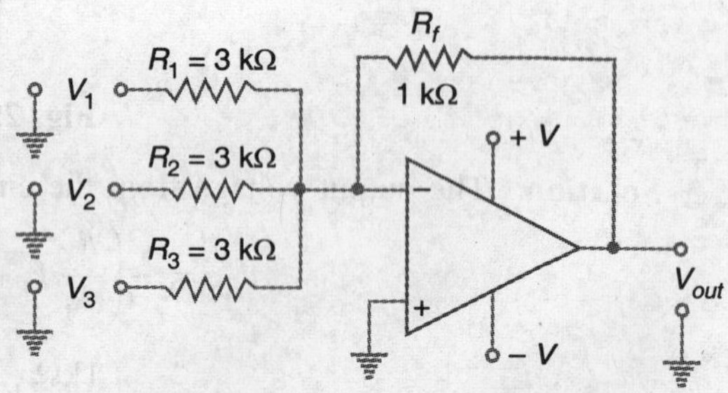

Fig. 23.77

(*i*) All input resistors (R_1, R_2 and so on) are *equal in value.*

(*ii*) The ratio of any input resistor to the feedback resistor is equal to the number of input circuits.

Fig. 23.77 shows the circuit of averaging amplifier. Note that it is a summing amplifier meeting the above two conditions. All input resistors are equal in value (3 kΩ). If we take the ratio of any input resistor to the feedback resistor, we get 3 kΩ/ 1 kΩ = 3. This is equal to the number of inputs to the circuit. Referring to the circuit in Fig. 23.77, the output voltage is given by ;

$$V_{out} = -\left(\frac{R_f}{R_1}V_1 + \frac{R_f}{R_2}V_2 + \frac{R_f}{R_3}V_3\right)$$

Now, $$\frac{R_f}{R_1} = \frac{R_f}{R_2} = \frac{R_f}{R_3} = \frac{1\text{ k}\Omega}{3\text{ k}\Omega} = \frac{1}{3}$$

$\therefore$ $$V_{out} = -\left(\frac{V_1 + V_2 + V_3}{3}\right)$$

Note that V_{out} is equal to the average of the three inputs. The negative sign shows the phase reversal.

2. As subtractor. A summing amplifier can be used to provide an output voltage that is equal to the difference of two voltages. Such a circuit is called a **subtractor** and is shown in Fig. 23.78. As we shall see, this circuit will provide an output voltage that is equal to the difference between V_1 and V_2.

The voltage V_1 is applied to a standard inverting amplifier that has *unity gain.* Because of this, the output from the inverting amplifier will be equal to $-V_1$. This output is then applied to the summing amplifier (also having unity gain) along with V_2. Thus output from second *OP*-amp is given by ;

$$V_{out} = -(V_A + V_B) = -(-V_1 + V_2) = V_1 - V_2$$

It may be noted that the gain of the second stage in the subtractor can be varied to provide an output that is proportional to (rather than equal to) the difference between the input voltages. However, if the circuit is to act as a subtractor, the input inverting amplifier *must* have unity gain. Otherwise, the output will not be proportional to the true difference between V_1 and V_2.

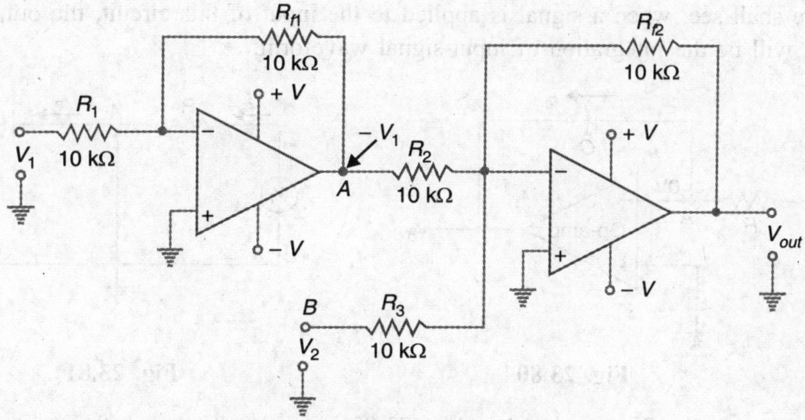

Fig. 23.78

23.34 OP-Amp Integrators and Differentiators

A circuit that performs the mathematical integration of input signal is called an *integrator*. The output of an integrator is proportional to the area of the input waveform over a period of time. A circuit that performs the mathematical differentiation of input signal is called a *differentiator*. The output of a differentiator is proportional to the rate of change of its input signal. Note that the two operations are opposite.

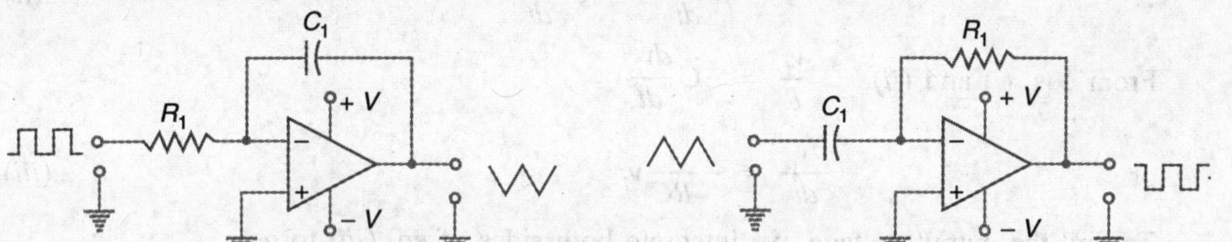

OP-amp Integrator (*i*)

OP-amp differentiator (*ii*)

Fig. 23.79

Fig. 23.79 shows *OP*-amp integrator and differentiator. As you can see, the two circuits are nearly identical in terms of their construction. Each contains a single *OP*-amp and an *RC* circuit. However, the difference in resistor/capacitor placement in the two circuits causes them to have input/output relationships that are exact opposites. For example, if the input to the integrator is a square wave, the output will be a triangular wave as shown in Fig. 23.79(*i*). However, the differentiator will convert a triangular wave into square wave as shown in Fig. 23.79 (*ii*).

23.35 OP-Amp Integrator

As discussed above, an integrator is a circuit that performs integration of the input signal. The most popular application of an integrator is to produce a *ramp* output voltage (*i.e.* a linearly increasing or decreasing voltage). Fig. 23.80 shows the circuit of an *OP*-amp integrator. It consists of an *OP*-amp, input resistor *R* and feedback capacitor *C*. Note that the feedback component is a capacitor instead of a resistor.

As we shall see, when a signal is applied to the input of this circuit, the output-signal waveform will be the integration of input-signal waveform.

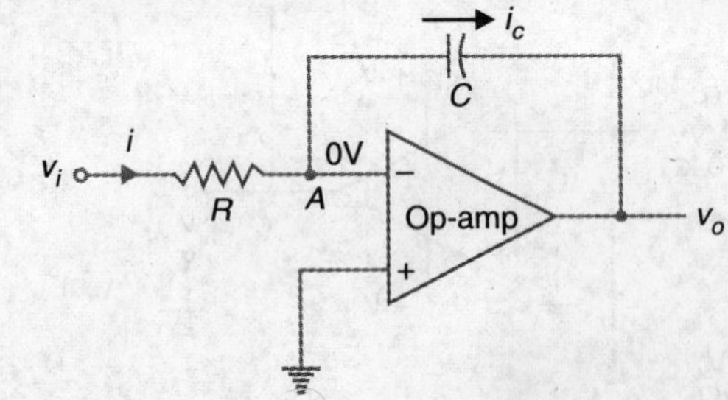

Fig. 23.80

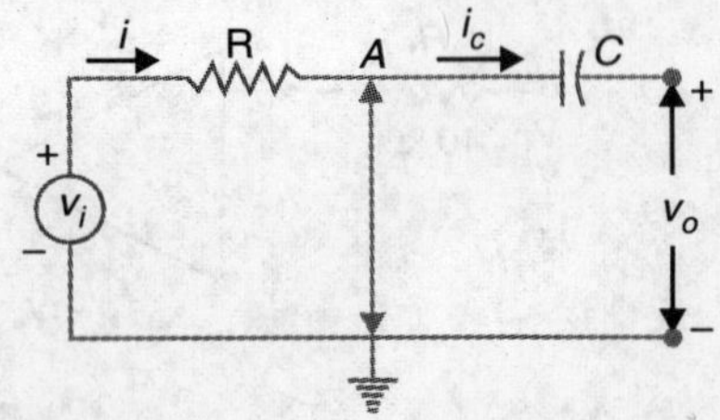

Fig. 23.81

Circuit Analysis. Since point *A* in Fig. 23.80 is at virtual ground, the *virtual-ground equivalent circuit of operational integrator will be as shown in Fig. 23.81. Because of virtual ground and infinite impedance of the *OP*-amp, all of the input current *i* flows through the capacitor *i.e.* $i = i_C$.

Now, $$i = \frac{v_i - 0}{R} = \frac{v_i}{R} \qquad ...(i)$$

Also, voltage across capacitor is $v_c = 0 - v_o = -v_o$

$$\therefore \quad i_c = \frac{C\,dv_c}{dt} = -C\frac{dv_o}{dt} \qquad ...(ii)$$

From eqs. (*i*) and (*ii*), $$\frac{v_i}{R} = -C\frac{dv_o}{dt}$$

or $$\frac{dv_o}{dt} = -\frac{1}{RC}v_i \qquad ...(iii)$$

To find the output voltage, we integrate both sides of eq. (*iii*) to get,

$$v_o = -\frac{1}{RC}\int_0^t v_i\,dt \qquad ...(iv)$$

Eq. (*iv*) shows that the output is the integral of the input with an inversion and scale multiplier of $1/RC$.

Output voltage. If a fixed voltage is applied to the input of an integrator, eq. (*iv*) shows that the output voltage grows over a period of time, providing a ramp voltage. Eq. (*iv*) also shows that the output voltage ramp (for a fixed input voltage) is opposite in polarity to the input voltage and is multiplied by the factor $1/RC$. As an example, consider an input voltage v_1 = 1V to the integrator circuit of Fig. 23.82 (*i*). The scale factor of $1/RC$ is

$$-\frac{1}{RC} = -\frac{1}{(1\ \text{M}\Omega)\,(1\ \mu\text{F})} = -1$$

so that the output is a negative ramp voltage as shown in Fig. 23.82 (*ii*).

* Recall that virtual ground means that point *A* is 0V but it is not mechanically grounded. Therefore, no current flows from point *A* to ground.

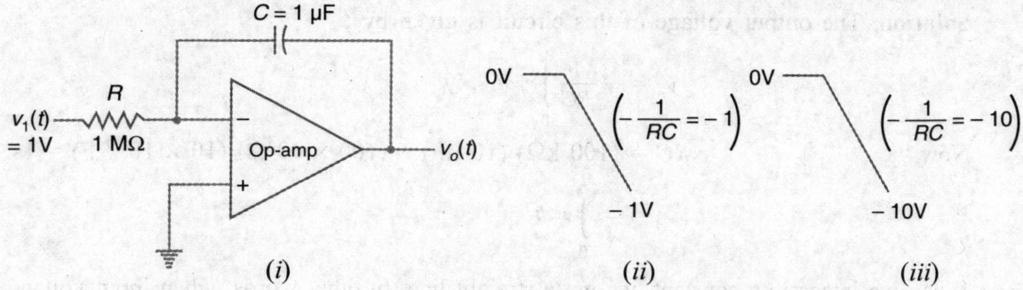

Fig. 23.82

If the scale factor is changed by making R = 100 kΩ, then,

$$-\frac{1}{RC} = -\frac{1}{(100\,\text{k}\Omega)\,(1\,\mu\text{F})} = -10$$

and output is then a steeper ramp voltage as shown in Fig. 23.82 (*iii*).

23.36 Critical Frequency of Integrators

The integrator shown in Fig. 23.80 (Refer back) has no feedback at 0 Hz. This is a serious disadvantage in low-frequency applications. By connecting a feedback resistor R_f in parallel with the capacitor, precise closed-loop voltage gain is possible. The circuit shown in Fig. 23.83 is an integrator with a feedback resistor R_f to provide increased stability.

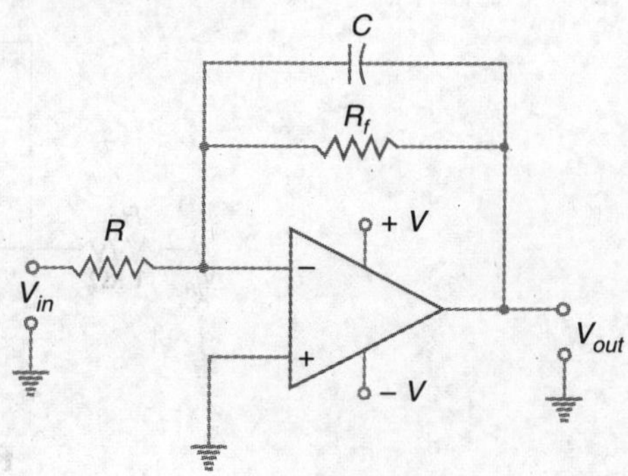

Fig. 23.83

All integrators have a critical frequency f_c below which they do not perform proper integration. If the input frequency is less than f_c, the circuit behaves like a simple inverting amplifier and no integration occurs. The following equation is used to calculate the critical frequency of an integrator :

$$f_c = \frac{1}{2\pi R_f C}$$

Example 23.50. *Fig. 23.84 (i) shows the OP-amp integrator and the square wave input. Find the output voltage.*

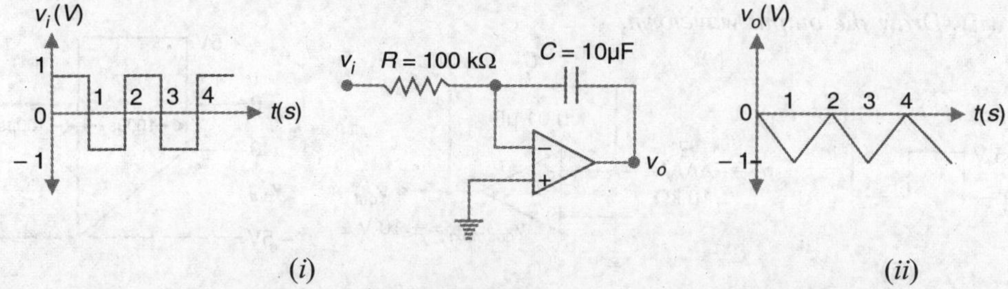

Fig. 23.84

Solution. The output voltage of this circuit is given by ;

$$v_o = -\frac{1}{RC}\int_0^t v_i\,dt$$

Now, $$RC = (100\text{ k}\Omega)\,(10\ \mu\text{F}) = (100 \times 10^3\Omega)\,(10 \times 10^{-6}\text{ F}) = 1\text{s}$$

$\therefore$ $$v_o = -\int_0^t v_i\,dt$$

When we integrate a constant, we get a straight line. In other words, when input voltage to an integrator is constant, the output is a linear ramp. **Therefore, the integration of the square wave results in the triangular wave as shown in Fig. 23.84 (*ii*). Since the input to the integrator is applied to the inverting input, the output of the circuit will be 180° out of phase with the input.** Thus, when the input goes positive, the output will be a negative ramp. When the input is negative, the output will be a positive ramp. Fig. 23.84 (*ii*) shows this relationship.

Example 23.51. *Determine the lower frequency limit (critical frequency) for the integrator circuit shown in Fig. 23.85.*

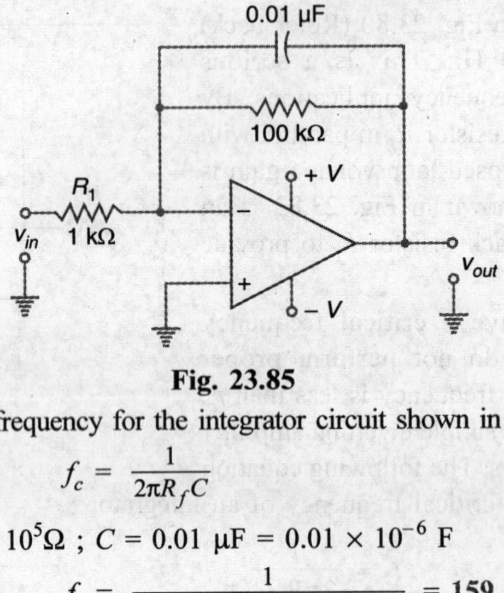

Fig. 23.85

Solution. The critical frequency for the integrator circuit shown in Fig. 23.85 is given by ;

$$f_c = \frac{1}{2\pi R_f C}$$

Here, $R_f = 100\text{ k}\Omega = 10^5\Omega$; $C = 0.01\ \mu\text{F} = 0.01 \times 10^{-6}$ F

$\therefore$ $$f_c = \frac{1}{2\pi \times (10^5) \times (0.01 \times 10^{-6})} = \mathbf{159\ Hz}$$

Example 23.52. (*i*) *Determine the rate of change of the output voltage in response to a single pulse input to the integrator circuit shown in Fig. 23.86 (i).*

(*ii*) *Draw the output waveform.*

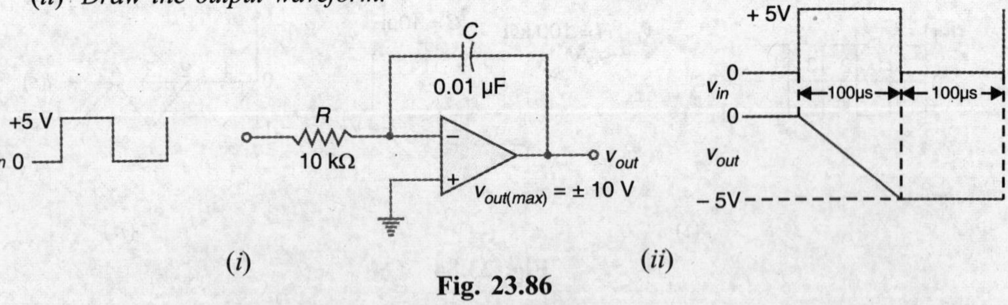

Fig. 23.86

Solution.

(*i*) Output voltage, $v_{out} = -\frac{1}{RC}\int_0^t v_{in}\,dt$

Therefore, the rate of change of output voltage is

$$\frac{\Delta v_{out}}{dt} = -\frac{v_{in}}{RC} = -\frac{5\text{ V}}{(10\text{ k}\Omega)(0.01\,\mu\text{F})} = -50\text{ kV/s} = -50\text{ mV/}\mu\text{s}$$

(*ii*) The rate of change of output voltage is – 50 mV/μs. When the input is at + 5 V, the output is a negative-going ramp. When the input is at 0 V, the output is a constant level. In 100 μs, the output voltage decreases.

$$\therefore \quad \Delta v_{out} = \frac{\Delta v_{out}}{dt}\times dt = -\frac{50\text{ mV}}{\mu\text{s}}\times 100\,\mu\text{s} = -5\text{V}$$

Therefore, the negative-going ramp reaches – 5 V at the end of the pulse (*i.e.* after 100 μs from the initial condition). The output voltage then remains constant at – 5 V for the time the input is zero. **Fig. 23.86 (*ii*) shows the output waveform.**

Example 23.53. *For the integrator circuit shown in Fig. 23.87 (i), how long does it take for the output to reach saturation ?*

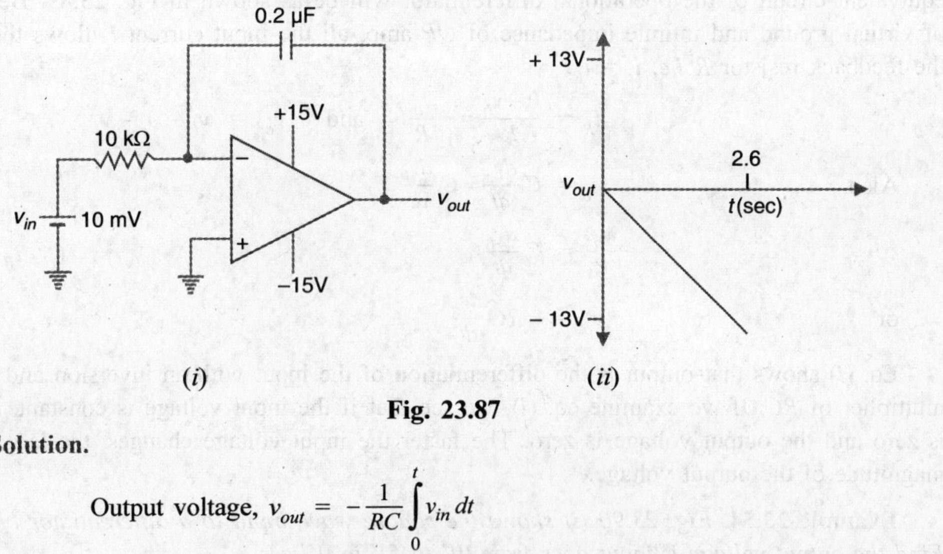

Fig. 23.87

Solution.

Output voltage, $v_{out} = -\frac{1}{RC}\int_0^t v_{in}\,dt$

Since the input voltage v_{in} (= 10 mV) is constant,

$$\therefore \quad v_{out} = -\frac{1}{RC}v_{in}t = -\frac{1}{(10\text{ k}\Omega)(0.2\,\mu\text{F})}\times(10\text{ mV})\times t$$

or $\quad v_{out} = -5t$ volts

Now, Saturation voltage, $V_{sat} = -V_{supply} + 2 = -15 + 2 = -13$ V

$$\therefore \quad \text{Time required, } t = \frac{V_{sat}}{-5} = \frac{-13}{-5} = \textbf{2.6 seconds}$$

Fig. 23.87 (*ii*) shows the output waveform.

23.37 OP-Amp Differentiator

A differentiator is a circuit that performs differentiation of the input signal. In other words, a differentiator produces an output voltage that is proportional to the rate of change of the input voltage. Its important application is to produce a rectangular output from a ramp input. Fig. 23.88 shows the circuit of *OP*-amp differentiator. It consists of an *OP*-amp, an input capacitor *C* and feedback resistor *R*. Note how the placement of the capacitor and resistor differs from the integrator. The capacitor is now the input element.

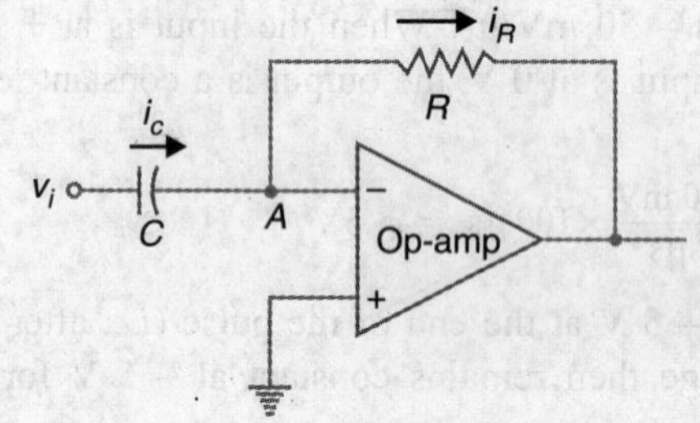

Fig. 23.88

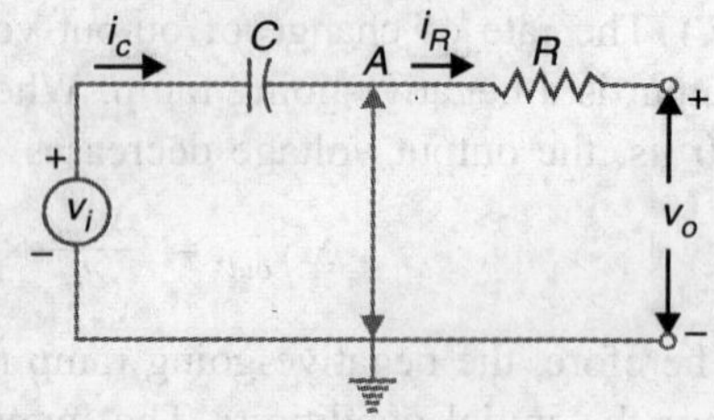

Fig. 23.89

Circuit analysis. Since point *A* in Fig. 23.88 is at virtual ground, the virtual-ground equivalent circuit of the operational differentiator will be as shown in Fig. 23.89. Because of virtual ground and infinite impedance of *OP*-amp, all the input current i_c flows through the feedback resistor *R* *i.e.* $i_c = i_R$.

$$\therefore \quad i_R = \frac{0 - v_o}{R} = -\frac{v_o}{R} \quad \text{and} \quad v_c = v_i - 0 = v_i$$

Also,
$$i_c = C\frac{dv_c}{dt} = C\frac{dv_i}{dt}$$

$$\therefore \quad -\frac{v_o}{R} = C\frac{dv_i}{dt} \qquad (\because i_R = i_c)$$

or
$$v_o = -RC\frac{dv_i}{dt} \qquad ...(i)$$

Eq. (*i*) shows that output is the differentiation of the input with an inversion and scale multiplier of *RC*. If we examine eq. (*i*), we see that if the input voltage is constant, dv_i/dt is zero and the output voltage is zero. The faster the input voltage changes, the larger the magnitude of the output voltage.

Example 23.54. *Fig. 23.90 (i) shows the square wave input to a differentiator circuit. Find the output voltage if input goes from 0V to 5V in 0.1 ms.*

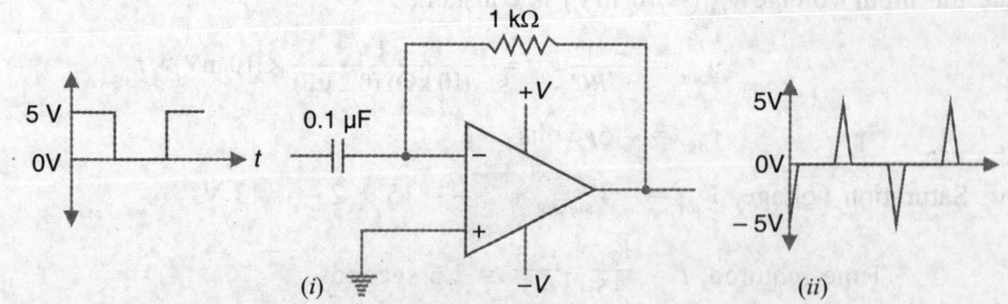

Fig. 23.90

Solution. Output voltage, $v_o = -RC\dfrac{dv_i}{dt}$

Now, $RC = (1\ \text{k}\Omega) \times (0.1\ \mu\text{F}) = (10^3\ \Omega)\ (0.1 \times 10^{-6}\ \text{F}) = 0.1 \times 10^{-3}\ \text{s}$

Also, $\dfrac{dv_i}{dt} = \dfrac{5\text{V}}{0.1\ \text{ms}} = \dfrac{5 \times 10^4\ \text{V}}{1\ \text{s}} = 5 \times 10^4\ \text{V/s}$

$\therefore\ v_o = -(0.1 \times 10^{-3})\ (5 \times 10^4) = -\mathbf{5\ V}$

The signal quickly returns to zero as the input signal becomes constant. The output will be as shown in Fig. 23.90 (*ii*).

Example 23.55. *For the differentiator circuit shown in Fig. 23.91, determine the output voltage if the input goes from 0V to 10V in 0.4s. Assume the input voltage changes at constant rate.*

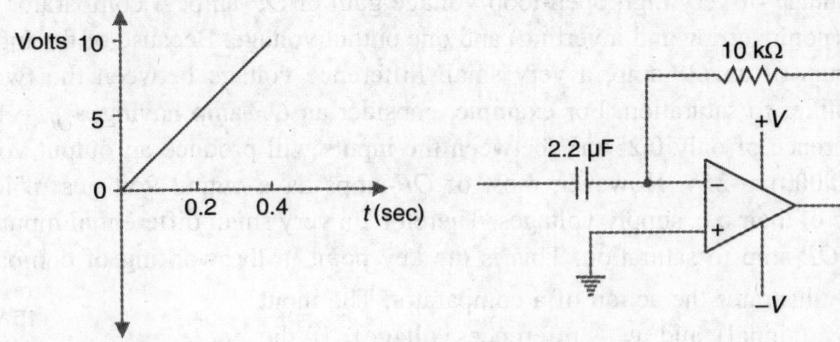

Fig. 23.91

Solution. Output voltage, $v_o = -RC\dfrac{dv_i}{dt}$

Now, $RC = (10\ \text{k}\Omega) \times (2.2\ \mu\text{F}) = (10^4\ \Omega)\ (2.2 \times 10^{-6}\ \text{F}) = 2.2 \times 10^{-2}\ \text{s}$

Also, $\dfrac{dv_i}{dt} = \dfrac{(10-0)\text{V}}{0.4\text{s}} = \dfrac{10\text{V}}{0.4\text{s}} = 25\ \text{V/s}$

$\therefore\ v_o = -(2.2 \times 10^{-2}) \times 25 = -\mathbf{0.55\ V}$

The output voltage stays constant at – 0.55 V.

Example 23.56. *For the differentiator circuit shown in Fig. 23.92(i), determine (i) the expression for the output voltage (ii) the output voltage for the given input.*

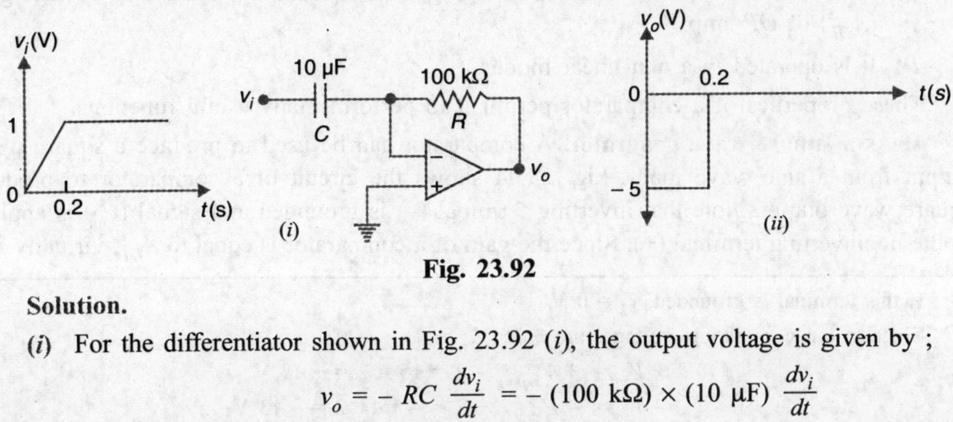

Fig. 23.92

Solution.

(*i*) For the differentiator shown in Fig. 23.92 (*i*), the output voltage is given by ;

$$v_o = -RC\frac{dv_i}{dt} = -(100\ \text{k}\Omega) \times (10\ \mu\text{F})\frac{dv_i}{dt}$$

$$= -(100 \times 10^3\ \Omega) \times (10 \times 10^{-6}\ \text{F})\ \frac{dv_i}{dt} = -\frac{dv_i}{dt}$$

(*ii*) Since the input voltage is a straight line between 0 and 0.2s, the output voltage is

$$v_o = -\frac{dv_i}{dt} = -\frac{(1-0)}{0.2} = -5\ \text{V}$$

Therefore, between 0 to 0.2s, the output voltage is constant at – 5 V. For $t > 0.2$s, the input is constant so that output voltage is zero. Fig. 23.92 (*ii*) shows the output waveform.

23.38 Comparators

Often we want to compare one voltage to another to see which is larger. In this situation, a *comparator* may be used. *A comparator is an OP-amp circuit without negative feedback* and takes advantage of very high open-loop voltage gain of *OP*-amp. A comparator has two input voltages (noninverting and inverting) and one output voltage. Because of the high open-loop voltage gain of an *OP*-amp, a very small difference voltage between the two inputs drives the amplifier to saturation. For example, consider an *OP*-amp having A_{OL} = 100,000. A voltage difference of only 0.25 mV between the inputs will produce an output voltage of (0.25 mV)(100,000) = 25V. However, most of *OP*-amps have output voltages of less than ± 15 V because of their d.c. supply voltages. Therefore, a very small differential input voltage will drive the *OP*-amp to saturation. This is the key point in the working of comparator.

Fig. 23.93 illustrates the action of a comparator. The input voltages are v_1 (signal) and v_2 (*reference voltage). If the differential input is positive, the circuit is driven to saturation and output goes to maximum positive value (**$+V_{sat}$ = + 13 V). Reverse happens when the differential input goes negative *i.e.* now output is maximum negative ($-V_{sat}$ = – 13 V). This circuit is called comparator because it compares v_1 to v_2 to produce a saturated positive or negative output voltage. Note that output voltage rapidly changes from – 13 V to + 13 V and *vice-versa*.

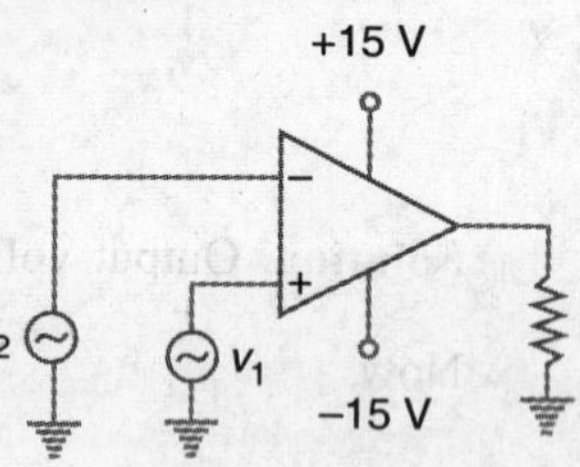

Fig. 23.93

23.39 Comparator Circuits

A comparator circuit has the following two characteristics :

(*i*) It uses no feedback so that the voltage gain is equal to the open-loop voltage gain (A_{OL}) of *OP*-amp.

(*ii*) It is operated in a non-linear mode.

These properties of a comparator permit it to perform many useful functions.

1. As a square wave generator. A comparator can be used to produce a square wave output from a sine wave input. Fig. 23.94 shows the circuit of a comparator to produce square wave output. Note that inverting terminal (–) is grounded and signal (v_{in}) is applied to the noninverting terminal (+). Since the gain of a comparator is equal to A_{OL}, virtually any

* If this terminal is grounded, v_2 = 0 V.

** Since in our case supply voltages are ± 15 V,

$$+V_{sat} = +V_{supply} - 2 = 15 - 2 = +13\ \text{V}$$
$$-V_{sat} = -V_{supply} + 2 = -15 + 2 = -13\ \text{V}$$

difference voltage at the inputs will cause the output to go to one of the voltage extremes ($+V_{sat}$ or $-V_{sat}$) and stay there until the voltage difference is removed. The polarity of the input difference voltage will determine to which extreme ($+V_{sat}$ or $-V_{sat}$) the output of the comparator goes.

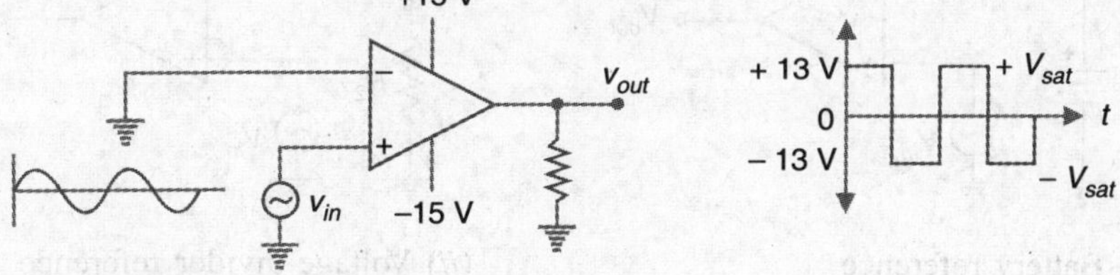

Fig. 23.94

When the input signal goes positive, the output jumps to about + 13 V. When the input goes negative, the output jumps to about − 13 V. The output changes rapidly from − 13 V to + 13 V and *vice-versa*. This change is so rapid that we get a square wave output for a sine wave input.

2. As a zero-crossing detector. When one input of a comparator is connected to ground, it is known as zero-crossing detector because the output changes when the input crosses 0 V. The zero-crossing circuit is shown in Fig. 23.95. The input and output waveforms are also shown. When the input signal is positive-going, the output is driven to positive maximum value (*i.e.* $+V_{sat} = +13$ V). When the input crosses the zero axis and begins to go negative, the output is driven to negative maximum value (*i.e.* $-V_{sat} = -13$ V).

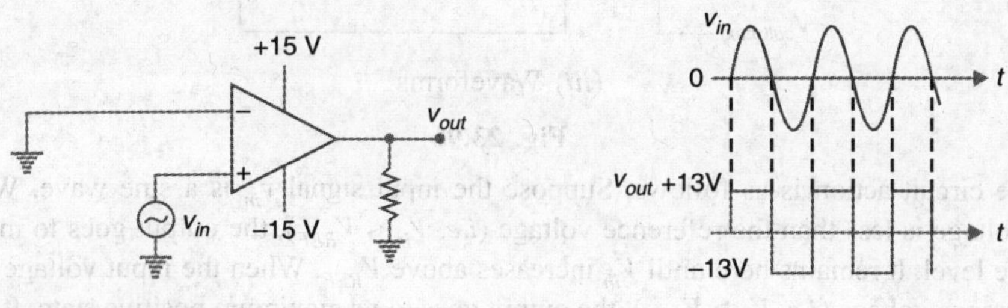

Fig. 23.95

From the input/output waveforms, you can see that every time the input crosses 0 V going positive, the output jumps to + 13 V. Similarly, every time the input crosses 0 V going negative, the output jumps to − 13 V. Since the change (+ 13 V or − 13 V) occurs every time the input crosses 0 V, we can tell when the input signal has crossed 0 V. Hence the name zero-crossing detector.

3. As a level detector. When a comparator is used to compare a signal amplitude to a fixed d.c. level (reference voltage), the circuit is referred to as a level detector. We can modify zero-crossing detector circuit to construct level detector. This can be done by connecting a fixed reference voltage V_{REF} to the inverting input as shown in Fig. 23.96 (*i*). A more practical arrangement is shown in Fig. 23.96 (*ii*) using a voltage divider to set the reference voltage as follows :

$$V_{REF} = \frac{R_2}{R_1 + R_2}(+V)$$

where $+V$ is the positive *OP*-amp d.c. supply voltage.

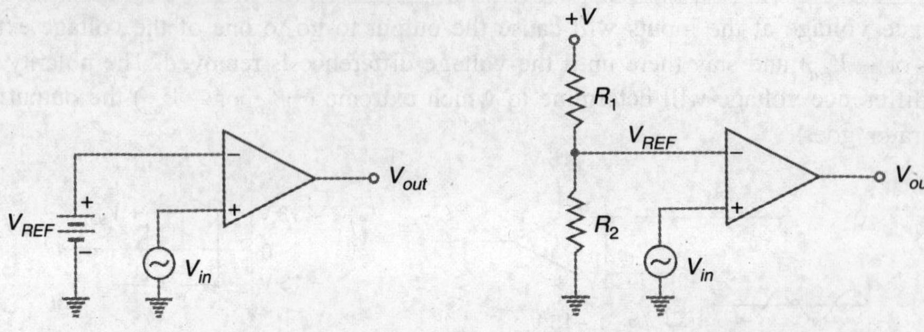

(*i*) Battery reference (*ii*) Voltage-divider reference

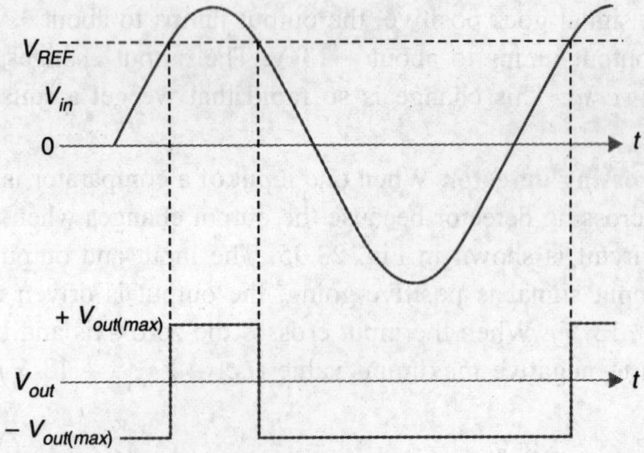

(*iii*) Waveforms

Fig. 23.96

The circuit action is as follows. Suppose the input signal v_{in} is a sine wave. When the input voltage is less than the reference voltage (*i.e.* $V_{in} < V_{REF}$), the output goes to maximum negative level. It remains here until V_{in} increases above V_{REF}. When the input voltage exceeds the reference voltage (*i.e.* $V_{in} > V_{REF}$), the output goes to its maximum positive state. It remains here until V_{in} decreases below V_{REF}. Fig. 23.96 (*iii*) shows the input/output waveforms. Note that this circuit is used for non zero-level detection.

Short Answer Questions

Q. 1. *What is an operational amplifier ?*

Ans. An operational amplifier (*OP*-amp) is a circuit that can perform such mathematical operations as addition, subtraction, integration and differentiation.

Q. 2. *What is the key electronic circuit in an OP-amp ?*

Ans. The key electronic circuit in an *OP*-amp is the differential amplifier. The input stage of an *OP*-amp is a differential amplifier and the output stage is typically a class *B* push-pull emitter follower.

Q. 3. *What is a differential amplifier ?*

Ans. A differential amplifier is a circuit that can accept two input signals and amplify the difference between these two input signals.

Q. 4. *What is the difference between the conventional amplifier and the differential amplifier ?*

Ans. In a conventional amplifier, the signal (generally single input) is applied at the input terminals and amplified output is obtained at the output terminals. But a differential amplifier accepts two input signals and amplifies the difference between these two input signals.

Q. 5. *Give important properties of an OP-amp.*

Ans. The following are the important properties common to all operational amplifiers (*OP*-amp) :

(*i*) An operational amplifier is a multistage amplifier. The input stage of an *OP*-amp is a differential amplifier stage.

(*ii*) An inverting input and a noninverting input.

(*iii*) A high input impedance (usually assumed infinite) at both inputs.

(*iv*) A low output impedance (< 200 Ω).

(*v*) A large open-loop voltage gain, typically 10^5.

(*vi*) The voltage gain remains constant over a wide frequency range.

(*vii*) Very large *CMRR* (> 90 dB).

Q. 6. *What do + and – signs on the symbol of an OP-amp indicate ?*

Ans. The + and – signs on the symbol of an *OP*-amp are not polarity signs. The – sign indicates the *inverting input* while the + sign indicates the *noninverting input*. A signal applied to the plus terminal will appear in the same phase at the output as at the input. A signal applied to the minus terminal will be shifted in phase by 180° at the output.

Q. 7. *What is an ideal OP-amp ?*

Ans. An ideal *OP*-amp has the following properties :

(*i*) Infinite open-loop voltage gain.

(*ii*) Infinite input impedance.

(*iii*) Zero output impedance.

Multiple-Choice Questions

1. A differential amplifier ....
   (*i*) is a part of an *OP*-amp
   (*ii*) has one input and one output
   (*iii*) has two outputs
   (*iv*) answers (*i*) and (*iii*)

2. When a differential amplifier is operated single-ended, ...
   (*i*) the output is grounded
   (*ii*) one input is grounded and signal is applied to the other
   (*iii*) both inputs are connected together
   (*iv*) the output is not inverted

3. In differential-mode, ...
   (*i*) opposite polarity signals are applied to the inputs
   (*ii*) the gain is one
   (*iii*) the outputs are of different amplitudes
   (*iv*) only one supply voltage is used

4. In the common-mode, ...
   (*i*) both inputs are grounded
   (*ii*) the outputs are connected together

(*iii*) an identical signal appears on both inputs
(*iv*) the output signals are in-phase

5. The common-mode gain is ....
(*i*) very high (*ii*) very low
(*iii*) always unity (*iv*) unpredictable

6. The differential gain is ....
(*i*) very high
(*ii*) very low
(*iii*) dependent on input voltage
(*iv*) about 100

7. If A_{DM} = 3500 and A_{CM} = 0.35, the *CMRR* is ....
(*i*) 1225
(*ii*) 10,000
(*iii*) 80 dB
(*iv*) answers (*ii*) and (*iii*)

8. With zero volts on both inputs, an *OP*-amp ideally should have an output ....
(*i*) equal to the positive supply voltage
(*ii*) equal to the negative supply voltage
(*iii*) equal to zero
(*iv*) equal to the *CMRR*

9. Of the values listed, the most realistic value for open-loop voltage gain of an *OP*-amp is ....
(*i*) 1 (*ii*) 2000
(*iii*) 80 dB (*iv*) 100,000

10. A certain *OP*-amp has bias currents of 50 μA and 49.3 μA. The input offset current is ....
(*i*) 700 nA (*ii*) 99.3 μA
(*iii*) 49.7 μA (*iv*) none of these

11. The output of a particular *OP*-amp increases 8 V in 12 μs. The slew rate is ....
(*i*) 90 V/μs (*ii*) 0.67 V/μs
(*iii*) 1.5 V/μs (*iv*) none of these

12. For an *OP*-amp with negative feedback, the output is ....
(*i*) equal to the input
(*ii*) increased
(*iii*) fed back to the inverting input
(*iv*) fed back to the noninverting input

13. The use of negative feedback ....
(*i*) reduces the voltage gain of an *OP*-amp
(*ii*) makes the *OP*-amp oscillate
(*iii*) makes linear operation possible
(*iv*) answers (*i*) and (*iii*)

14. Negative feedback ....
(*i*) increases the input and output impedances
(*ii*) increases the input impedance and bandwidth
(*iii*) decreases the output impedance and bandwidth
(*iv*) does not affect impedance or bandwidth

15. A certain noninverting amplifier has R_i of 1 kΩ and R_f of 100 kΩ. The closed-loop voltage gain is ....
(*i*) 100,000 (*ii*) 1000
(*iii*) 101 (*iv*) 100

16. If feedback resistor in *Q*.15 is open, the voltage gain ....
(*i*) increases
(*ii*) decreases
(*iii*) is not affected
(*iv*) depends on R_i

17. A certain inverting amplifier has a closed-loop voltage gain of 25. The *OP*-amp has an open-loop voltage gain of 100,000. If an *OP*-amp with an open-loop voltage gain of 200,000 is substituted in the arrangement, the closed-loop gain ....
(*i*) doubles
(*ii*) drops to 12.5
(*iii*) remains at 25
(*iv*) increases slightly

18. A voltage follower ....
(*i*) has a voltage gain of 1
(*ii*) is noninverting
(*iii*) has no feedback resistor
(*iv*) has all of these

19. The *OP*-amp can amplify ....
    (*i*) a.c. signals only
    (*ii*) d.c. signals only
    (*iii*) both a.c. and d.c. signals
    (*iv*) neither d.c. nor a.c. signals
20. The input offset current equals the ....
    (*i*) difference between two base currents
    (*ii*) average of two base currents
    (*iii*) collector current divided by current gain
    (*iv*) none of these
21. The tail current of a differential amplifier is....
    (*i*) half of either collector current
    (*ii*) equal to either collector current
    (*iii*) two times either collector current
    (*iv*) equal to the difference in base currents
22. The node voltage at the top of the tail resistor is closest to....
    (*i*) collector supply voltage
    (*ii*) zero
    (*iii*) emitter supply voltage
    (*iv*) tail current times base resistance
23. The tail current in a differential amplifier equals ....
    (*i*) difference between two emitter currents
    (*ii*) sum of two emitter currents
    (*iii*) collector current divided by current gain
    (*iv*) collector voltage divided by collector resistance
24. The differential voltage gain of a differential amplifier is equal to R_C divided by ....
    (*i*) r'_e (*ii*) $r'_e/2$
    (*iii*) $2\ r'_e$ (*iv*) R_E
25. The input impedance of a differential amplifier equals r'_e times ....
    (*i*) β (*ii*) R_E
    (*iii*) R_C (*iv*) 2β
26. A common-mode signal is applied to ....
    (*i*) the noninverting input
    (*ii*) the inverting input
    (*iii*) both inputs
    (*iv*) top of the tail resistor
27. The common-mode voltage gain is ....
    (*i*) smaller than differential voltage gain
    (*ii*) equal to differential voltage gain
    (*iii*) greater than differential voltage gain
    (*iv*) none of the above
28. The input stage of an *OP*-amp is usually a ....
    (*i*) differential amplifier
    (*ii*) class *B* push-pull amplifier
    (*iii*) *CE* amplifier
    (*iv*) swamped amplifier
29. The common-mode voltage gain of a differential amplifier is equal to R_C divided by ....
    (*i*) r'_e (*ii*) $2\ r'_e$
    (*iii*) $r'_e/2$ (*iv*) $2\ R_E$
30. Current cannot flow to ground through ....
    (*i*) a mechanical ground
    (*ii*) an a.c. ground
    (*iii*) a virtual ground
    (*iv*) an ordinary ground

Answers to Multiple-Choice Questions

1. (*iv*)	2. (*ii*)	3. (*i*)	4. (*iii*)	5. (*ii*)
6. (*i*)	7. (*iv*)	8. (*iii*)	9. (*iv*)	10. (*i*)
11. (*ii*)	12. (*iii*)	13. (*iv*)	14. (*ii*)	15. (*iii*)
16. (*i*)	17. (*iii*)	18. (*iv*)	19. (*iii*)	20. (*i*)
21. (*iii*)	22. (*ii*)	23. (*ii*)	24. (*iii*)	25. (*iv*)
26. (*iii*)	27. (*i*)	28. (*i*)	29. (*iv*)	30. (*iii*)

Chapter Review Topics

1. What is an operational amplifier (*OP*-amp)?
2. Give the block diagram of an operational amplifier.
3. What is a differential amplifier?
4. Draw the basic circuit of a differential amplifier.
5. Discuss the operation of a differential amplifier.
6. What do you mean by noninverting and inverting input of a differential amplifier?
7. What are common-mode and differential-mode signals?
8. What do you mean by *CMRR*?
9. What is the importance of *CMRR*?
10. Explain the d.c. analysis of a differential amplifier.
11. What do you mean by (*i*) output offset voltage (*ii*) input offset current?
12. Derive an expression for differential-mode voltage gain of a differential amplifier.
13. Derive an expression for the common-mode voltage gian of a differential amplifier.
14. Draw the schematic symbol of an operational amplifier indicating the various terminals.
15. What do you-mean by (*i*) open-loop voltage gain (*ii*) closed-loop voltage gain of an *OP*-amp?
16. Discuss *OP*-amp input/output polarity relationship.
17. What do you mean by bandwidth of an *OP*-amp?
18. What do you mean by slew rate of an *OP*-amp?
19. Discuss the frequency response of an *OP*-amp.
20. What is the need of negative feedback in an *OP*-amp?
21. Derive an expression for the voltage gain of an inverting amplifier.
22. Derive an expression for the voltage gain of a noninverting amplifier.
23. What is a voltage follower?
24. Draw the circuit of multistage *OP*-amp.
25. What is the effect of negative feedback on (*i*) noninverting amplifier (*ii*) inverting amplifier?
26. Discuss the operation of a summing amplifier.
27. Discuss two applications of summing amplifiers.
28. Discuss the operation of an *OP*-amp integrator.
29. What is the most important application of an *OP*-amp integrator?
30. Discuss the operation of *OP*-amp differentiator.

Problems

1. In Fig. 23.97, the transistors are identical with β_{dc} = 200. What is the output voltage?

[7.5 V]

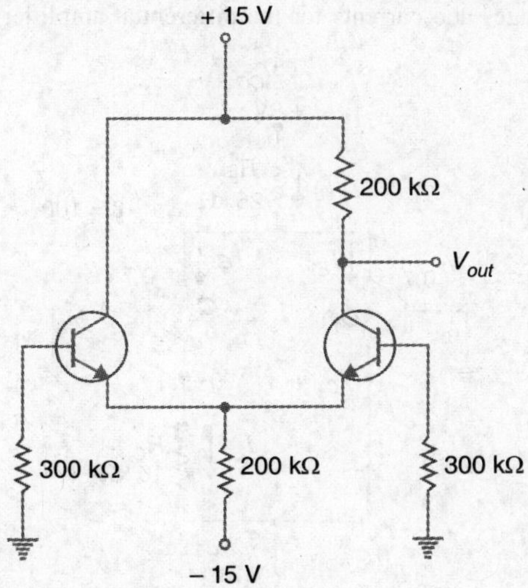

Fig. 23.97

2. In Fig. 23.97, the left transistor has β_{dc} = 225 and the right transistor has β_{dc} = 275. What are the base voltages? **[– 0.05 V ; – 0.0409 V]**
3. A data sheet gives an input bias current of 20 nA and an input offset current of 3 nA. What are the base currents? **[18.5 nA ; 21.5 nA]**
4. Find bias voltages and currents for the differential amplifier circuit in Fig. 23.98.

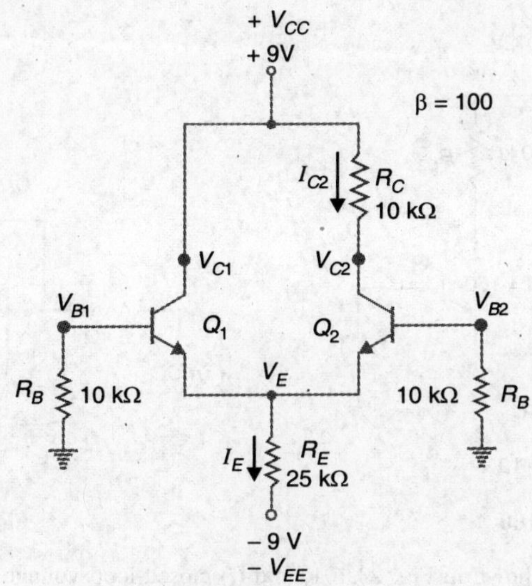

Fig. 23.98

[V_E = – 0.7 V ; I_E = 0.332 mA ; I_{E1} = I_{E2} = 0.166 mA ; I_{C1} = I_{C2} = 0.166 mA ; I_{B1} = I_{B2} = 1.66 μA; V_{C1} = 9 V ; V_{C2} = 7.34 V]

5. Find the bias voltages and currents for the differential amplifier circuit shown in Fig. 23.99.

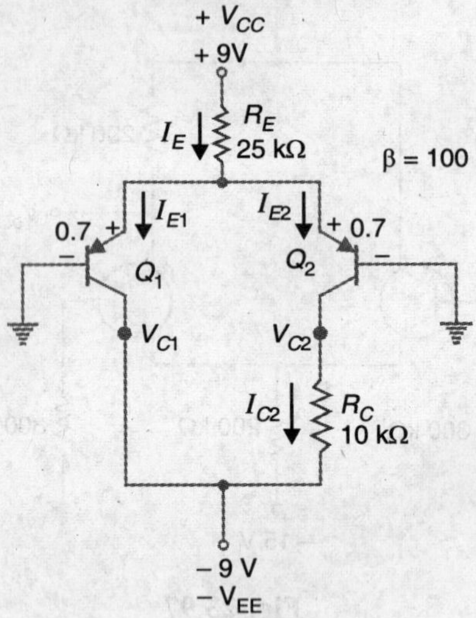

Fig. 23.99

[V_E = 0.7 V ; I_E = 0.332 mA ; I_{E1} = I_{E2} = 0.166 mA ; I_{C1} = I_{C2} = 0.166 mA ; I_{B1} = I_{B2} = 1.66 μA; V_{C1} = – 9 V ; V_{C2} = – 7.34 V]

6. For the circuit shown in Fig. 23.100, determine (*i*) common-mode voltage gain (*ii*) differential-mode voltage gain (*iii*) *CMRR*. **[(*i*) 0.42 (*ii*) 90.6 (*iii*) 216]**

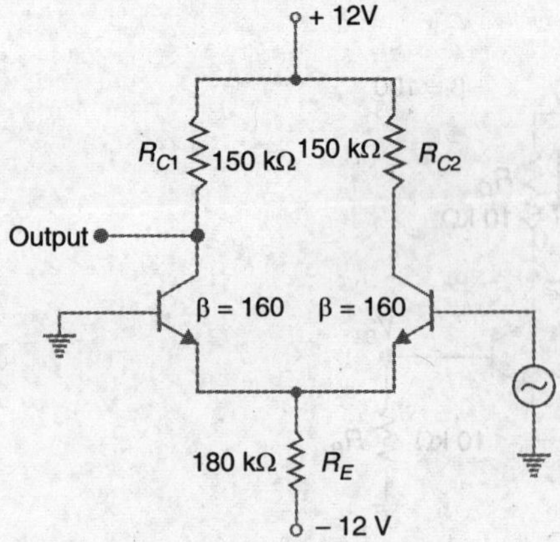

Fig. 23.100

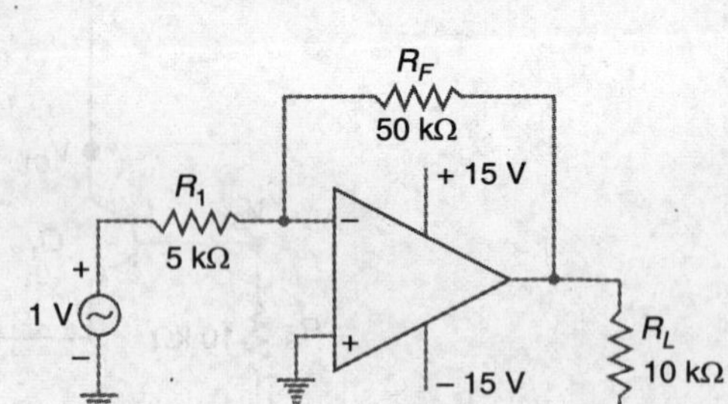

Fig. 23.101

7. For the circuit shown in Fig. 23.101, find (*i*) closed-loop voltage gain (*ii*) the instantaneous voltage across R_F when the signal voltage is + 1V (*iii*) the instantaneous voltage on the – terminal when the signal voltage is + 1V. **[(*i*) 10 (*ii*) 10 V (*iii*) 0 V]**

8. A noninverting amplifier has an R_i of 1 kΩ and an R_f of 100 kΩ. Determine (*i*) V_f (*ii*) feedback factor if V_{out} = 5 V. **[(*i*) 49.5 mV (*ii*) 9.9 × 10^{-3}]**

9. Determine the closed-loop voltage gain for the circuit shown in Fig. 23.102. **[11]**

10. Determine the closed-loop voltage gain for the circuit shown in Fig. 23.103. **[101]**

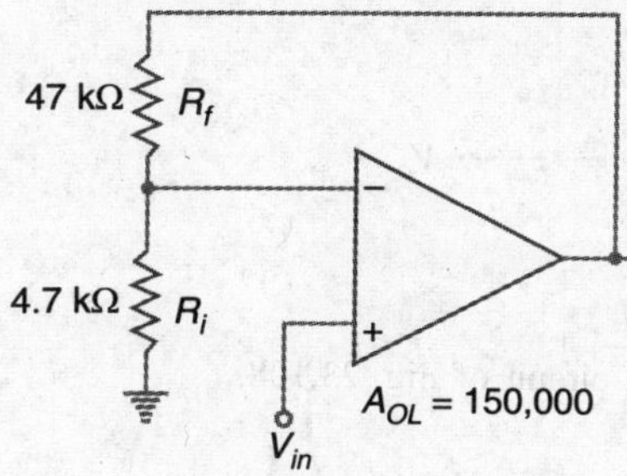

Fig. 23.102

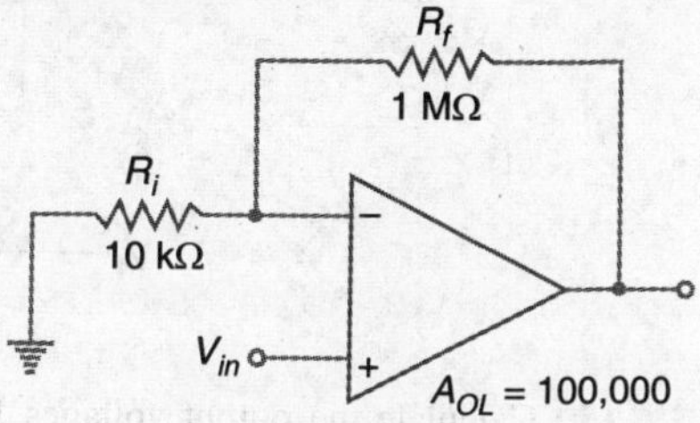

Fig. 23.103

11. Find the closed-loop voltage gain for each of the circuits shown in Fig. 23.104. **[1 ; – 1]**

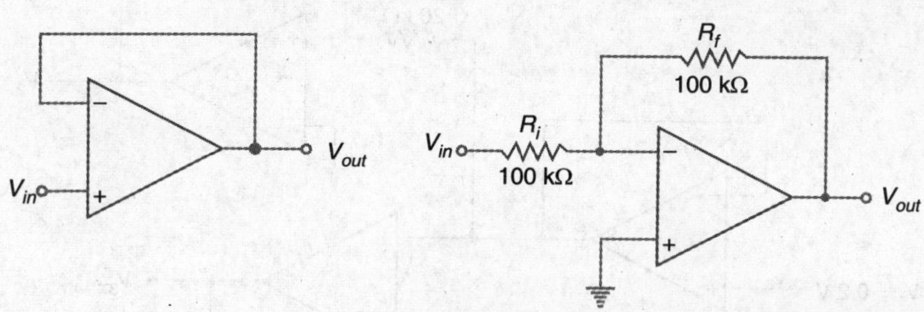

Fig. 23.104

12. Determine the approximate values of (*i*) I_{in} (*ii*) I_f (*iii*) V_{out} (*iv*) closed-loop voltage gain for the circuit in Fig. 23.105. **[(*i*) 455 μA (*ii*) 455 μA (*iii*) – 10V (*iv*) – 10]**

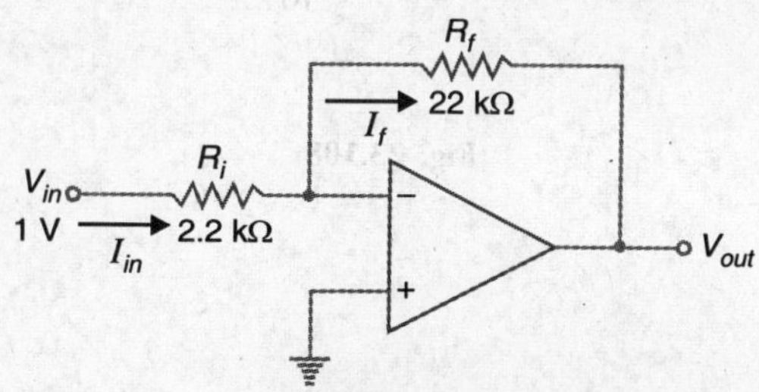

Fig. 23.105

13. Calculate the output voltage of the circuit in Fig. 23.106 for R_f = 68 kΩ. **[V_o = – 3.39 V]**

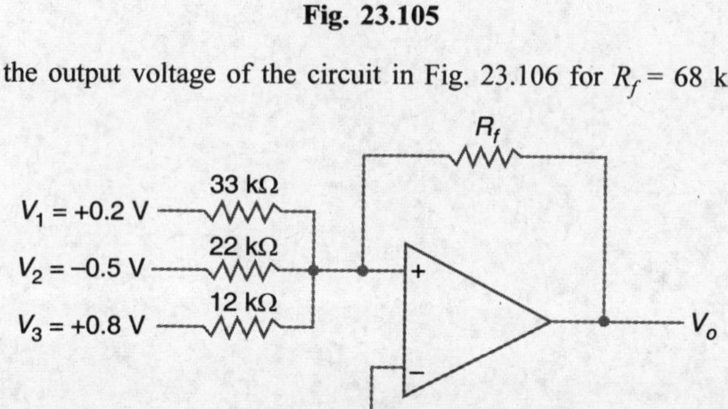

Fig. 23.106

14. What output voltage results in the circuit of Fig. 23.107 for $V_1 = +0.5$ V ? [$V_o = 0.5$ V]

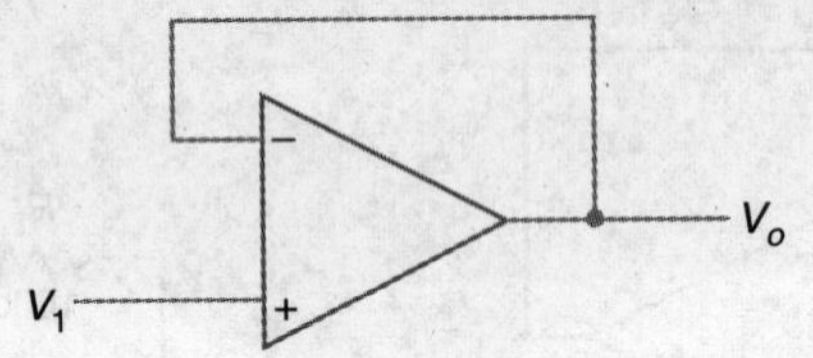

Fig. 23.107

15. Calculate the output voltages V_2 and V_3 in the circuit of Fig. 23.108.

[$V_2 = -2$ V ; $V_3 = 4.2$ V]

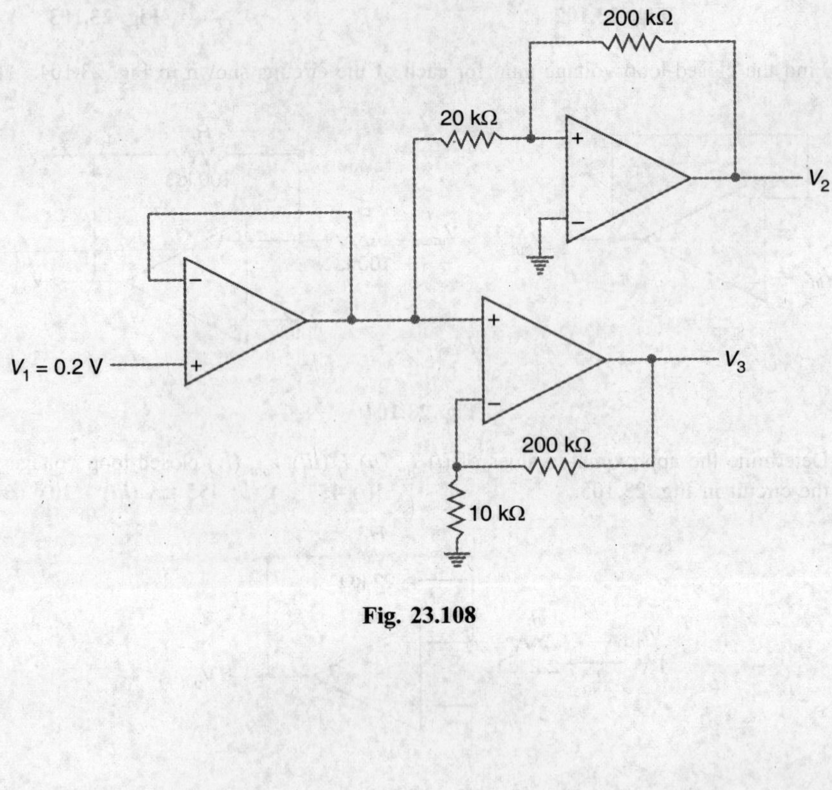

Fig. 23.108

CHAPTER 24

Digital Electronics

Learning Objectives

- Recognise the advantages of digital signal over analog signal.
- Learn the various number systems used by digital circuits.
- Develop an understanding of logic gates.
- Recognise that OR, AND and NOT gates are building blocks of all digital circuits.
- Understand the algebra used to symbolically describe logic functions *i.e.* Boolean algebra.
- Develop an understanding of combinational and sequential logic circuits (*i.e.* the two basic logic circuits).
- Learn to construct truth table from a logic circuit.
- Explore the operation of logic circuits which perform binary addition *viz.* half adders and full adders.
- Learn that combinational logic circuits involve AND, OR and NOT gates as building blocks and sequential logic circuits involve flip-flops (*FFs*) as building blocks.
- Understand the working of *SR FF,* clocked *SR FF, D FF, JK FF* and *T FF*.
- Learn that *JK FF* is the most widely used *FF*.

Introduction

A continuously varying signal (voltage or current) is called an *analog signal*. For example, a sinusoidal voltage is an analog signal. In the previous chapters, we studied the behaviour of diodes and transistors primarily from the analog or continuous-signal point of view. In an analog electronic circuit, the output voltage changes continuously according to the input voltage variations. In other words, the output voltage can have an infinite number of values. A signal (voltage or current) which can have only two discrete values is called a *digital signal*. For example, a square wave is a digital signal. The semiconductors devices (*e.g.* diodes, transistors *etc.*) can be designed for two-state operation *viz.*, saturation and cut-off. In that case, the output voltage can have only two states (*i.e.*, values), either *low or high. An electronic circuit that is designed for two-state operation is called a **digital circuit**.

* The exact value of voltage is unimportant if the voltage is distinguishable as low or high.

The branch of electronics which deals with digital circuits is called **digital electronics.** When most of us hear the term *digital,* we immediately think of "digital calculator" or "digital computer". This is attributed to the dramatic way the low-cost, powerful calculators and computers have become accessible to an average person. Now digital circuits are being used in many electronic products such as video games, microwave ovens and oscilloscopes. Digital techniques have also replaced a lot of the older "analog circuits" used in consumer products such as radios, TV sets and high-fidelity sound recording and playback equipment. In this chapter, we shall discuss the fundamental aspects of digital electronics.

24.1 Analog and Digital Signals

(*i*) Analog signal. *A continuously varying signal (voltage or current) is called an* **analog signal.** For example, an alternating voltage varying sinusoidally is an analog signal [See Fig. 24.1]. If such an analog signal is applied to the input of a transistor amplifier, the output voltage will also vary sinusoidally. This is the analog operation *i.e.,* the output voltage can have an infinite number of values. Due to many-valued output, the analog operation is less reliable.

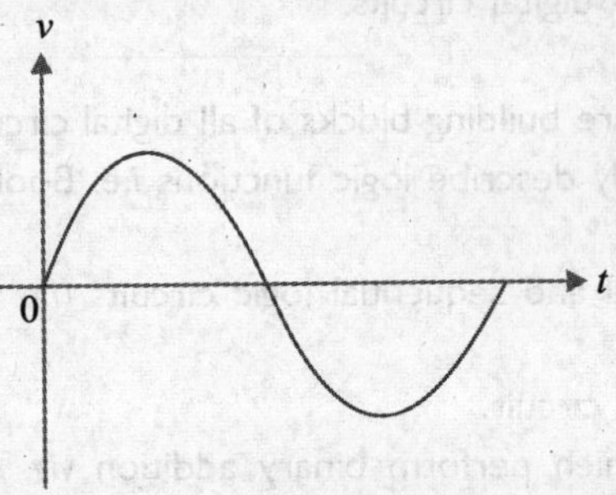

Fig. 24.1

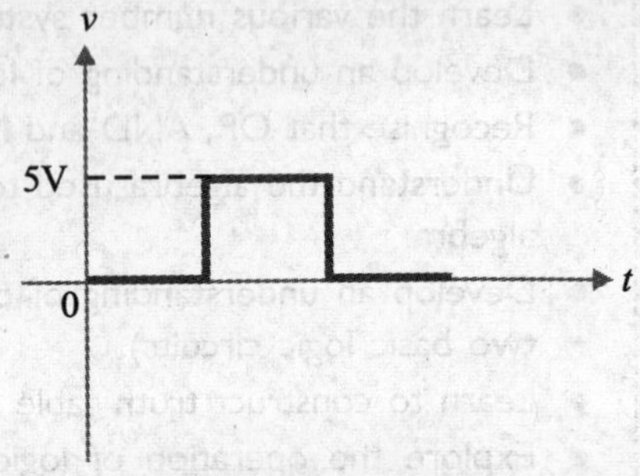

Fig. 24.2

(*ii*) Digital signal. *A signal (voltage or current) that can have only two discrete values is called a* **digital signal.** For example, a square wave is a digital signal [See Fig. 24.2]. It is because this signal has only two values *viz,* +5 V and 0 V and no other value. These values are labelled as *High* and *Low.* The High voltage is + 5 V and the Low voltage is 0 V. If proper digital signal is applied to the input of a transistor, the transistor can be driven between *cut-off* and *saturation.* In other words, the transistor will have two-states operation *i.e.,* output is either low or high. Since digital operation has only two states (*i.e., ON* or *OFF*), it is far more reliable than many-valued analog operation. It is because with two-states operation, all the signals are easily recognised as either low or high.

24.2 Digital Circuit

An electronic circuit that handles only a digital signal is called a **digital circuit.**

The output voltage of a digital circuit is either low or high and no other value. In other words, digital operation is a two-states operation. These states are expressed as (*High* or *Low*) or (*ON* or *OFF*) or (1 or 0). Therefore, a digital circuit is one that expresses the values in digits 1's or 0's. Hence the name digital. The numbering concept that uses only the two digits 1 and 0 is the *binary numbering system.* Therefore, the first step would be to discuss the various number systems.

24.3 Number Systems

A number system is a code that uses symbols to count the number of items. The total number of digits used to count number of items in a number system is called **radix** *or* **base** of the system. Thus decimal number system uses 10 digits (0 to 9) to count number of items. Therefore, the radix or base of decimal system is 10. There are many number systems but the most common are :

(*i*) Decimal number system (*ii*) Binary number system

(*iii*) Octal number system (*iv*) Hexadecimal number system

(*i*) **Decimal number system.** The most common and familiar number system is the decimal number system. The decimal number system uses 10 digits (0, 1, 2, 3, 4, 5, 6, 7, 8 and 9) for counting items. Therefore, the radix or base of decimal system is 10. The decimal system has evolved naturally as a result of the fact that man has 10 fingers. In fact, the word "digit" is derived from the Latin word for "finger".

(*ii*) **Binary number system.** The binary system uses only two digits (0 and 1) for counting items. Therefore, the radix or base of binary system is 2. Although binary system uses only two digits, yet it can be used to represent any quantity that can be represented in decimal or any other number system.

(*iii*) **Octal number system.** The octal number system has a radix of 8 so that it uses eight digits : 0, 1, 2, 3, 4, 5, 6 and 7. Thus each digit of an octal number can have any value from 0 to 7. The ease with which conversions can be made between octal and binary makes the octal system attractive as a "shorthand" means of expressing large binary numbers.

(*iv*) **Hexadecimal number system.** The hexadecimal system has a radix of 16. Therefore, it has 16 possible digit symbols. The hexadecimal system uses the first 10 digits of the decimal system (0, 1, 2, 3, 4, 5, 6, 7, 8 and 9) as well as first 6 letters of the alphabet A, B, C, D, E and F. The hexadecimal system is popular in computer applications because large numbers can be stored and the code easily converts to binary.

24.4 Radix or Base of Number System

The number of digits used in a number system is called **radix or base** *of that system*. Thus the radix or base of decimal system is 10 and that of binary system is 2. Similarly, the radix or base of octal system is 8 while that of hexadecimal system is 16.

All number systems are based on the principle of *position weighting i.e.* the value of a digit depends upon its position in the number. For example, consider the decimal number 642. This can be expressed as :

$$642 = 600 + 40 + 2$$

We know that the digit 6 actually represents *6 hundreds*, the 4 represents *4 tens* and 2 represents *2 units*. This decimal number can be written as :

$$642 = 6 \times 10^2 + 4 \times 10^1 + 2 \times 10^0$$

Thus in a multidigit decimal number, each position has a value that is 10 times (*i.e.* radix or base times) the value of the next position to its immediate right. In the decimal number under consideration, we find that 6 carries the most weight of the three digits and is referred to as the *most significant digit* (*MSD*). The 2 carries the least weight and is called the *least significant digit* (*LSD*).

Radix point. All number systems centre around the *radix point* (or *base point*) as shown in Fig. 24.3. Digits to the left of radix point represent a whole number while digits to the right of the radix point represent a fraction. In the decimal system (radix = 10), the radix point is called the *decimal point*. For the binary system (radix = 2), the radix point is called the *binary point*.

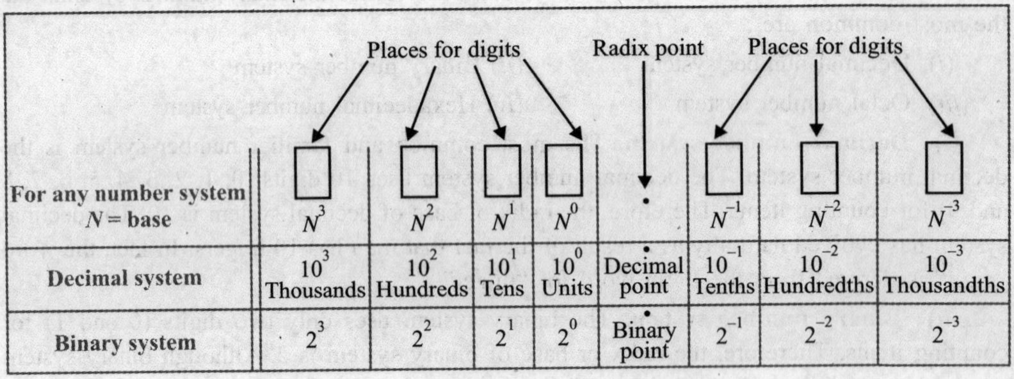

Fig. 24.3

In the decimal system, the places to the left of radix point (*i.e.,* decimal point) are positive powers of 10 and places to the right are negative powers of 10. Similarly, in the binary system, the places to the left of radix point (*i.e.,* binary point) are positive powers of 2 and places to the right of radix point are negative powers of 2.

24.5 Counting in Decimal and Binary Systems

A binary system uses only two digits (0 and 1) for counting items. The reader may wonder how to count the items in a binary system. Let us see how it is done. Fig. 24.4 shows the counting of stones in decimal as well as binary system. As you will see, the counting in the binary number system is performed much the same way as in the decimal number system.

Stones	Decimal	Binary
No stone	0	0
•	1	1
• •	2	10
• • •	3	11
• • • •	4	100
• • • • •	5	101
• • • • • •	6	110
• • • • • • •	7	111
• • • • • • • •	8	1000
• • • • • • • • •	9	1001

Fig. 24.4

(*i*) Let us first see how items are counted in decimal system. In this system, the count starts as 0, 1, ..., 9. After 9, we are to write the next number. To do so, we use the second digit of the decimal system (*i.e.,* 1) followed by the first digit (*i.e.,* 0). So after 9, the next number is 10. The count again continues as 10, 11, 12 ....., 19. After 19, we use the third digit of the system (*i.e.*, 2) followed by the first digit (*i.e.*, 0) and the count continues as 20, 21, ... *etc.* In this way, we get the number upto 99. In order to represent a number next to 99, we use three decimal digits (100). That is to say second digit of the decimal system (*i.e.*, 1) followed by two first digits (*i.e*, two zeros).

(*ii*) Let us now turn to binary system. Note that 0 and 1 count in the binary system is the same as in the decimal counting. To represent 2 stones, we use the second binary digit (*i.e.*, 1) followed by the first (*i.e.*, 0). This gives binary number 10 (read as one-zero and not ten) as an equivalent of 2 in the decimal system. Likewise, 3 in the decimal system can be represented by the binary number 11 (read as one-one and not eleven). After this, the two binary digits are exhausted. We shall use three digits to represent the next binary number. Thus, to represent 4 (four), we use the second binary digit followed by two first binary digits. This gives the binary *100 (read as one-zero-zero) as equivalent to 4 in the decimal system. Here is a simple way to find binary equivalents. *Each time the two digits 1 and 0 in one position are exhausted (counted as high as they will go), a 1 is added at the left, all digits to the right are made 0, and the count continues.* The reader may apply this simple rule to find next binary numbers.

Notes.

(*i*) Each binary digit (0 or 1) is referred to as a ***bit*. A string of four bits is called as a *nibble* and eight bits make a *byte*. Thus, 1001 is a nibble and 10010110 is a binary byte.

(*ii*) The binary number system is the most useful in digital circuits because there are only two digits (0 and 1).

24.6 Place Value

Consider the decimal number 642. This can be expressed as :

$$642 = 600 + 40 + 2$$

Note that in a multidigit decimal number (*i.e.*, 642 in the present case), each position has a value that is 10 times the value of the next position to its immediate right. In other words, every position can be expressed as :

$$642 = 6 \times 10^2 + 4 \times 10^1 + 2 \times 10^0$$

Thus, we find that values of various positions in a decimal number system are powers of 10 *i.e.*, equal to the number of digits used in the system. This number is called *base* or *radix* of the system. Thus, the decimal system has base of 10 (ten).

For the decimals, the digit to the extreme right is referred to as the *least significant digit* (*LSD*) because its positional value or weight is the lowest. For the decimal number 642, 2 is the *LSD*. The left-most digit in the decimal number is the *most significant digit* (*MSD*) because its positional value or weight is the highest. For the decimal number 642, 6 is the *MSD* with a value of 600.

Binary number system. In the binary number system, only two digits (0 and 1) are used. Therefore, the *base* of this system is 2. In a binary number, each position has a value

* Note that the procedure is similar to that which was used to write 100 (hundred) in the decimal system.

** **Binary digit**. Note that bit is a shorthand notation for "binary digit".

that is 2 times the value of the next position to its immediate right. In other words, every position can be expressed by 2 raised to some power. We know that binary number 1001 is equal to the decimal number 9. This can be readily shown as under :

$$1001 \quad = \quad 1 \times 2^3 + 0 \times 2^2 + 0 \times 2^1 + 1 \times 2^0 = 9$$

For binary numbers, the digit at the extreme right is referred to as *least significant bit* (*LSB*). In the binary number 1001, the 1 at the right is the *LSB*. The left-most digit is called the *most significant bit* (*MSB*). In the binary number 1001, the 1 at the left is the *MSB*. The *MSB* and *LSB* are illustrated in Fig. 24.5.

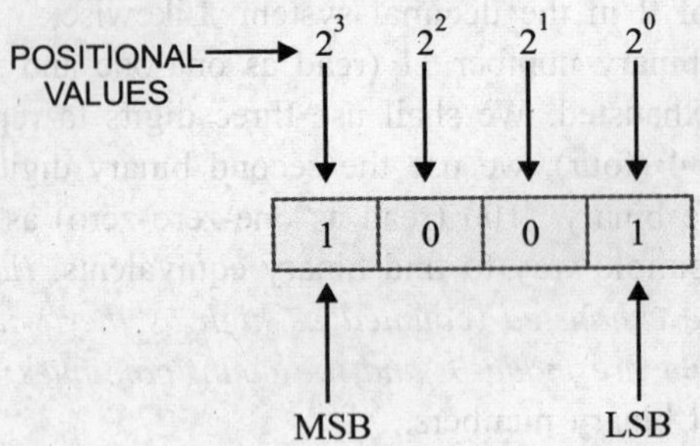

Fig. 24.5

24.7 Decimal to Binary Conversion

Decimal to binary conversion is a more lengthy process. Also there is more than one method that can be used to perform the conversion.

(*i*) **When decimal number is an integer.** When decimal number is an integer (*i.e.* whole number), we generally use *double dabble method.* It is so called because it requires successive divisions by 2. The method can be summarised as under :

Divide progressively the decimal number by 2 and write down the remainder after each division. Continue this process till you get a quotient of 0 and remainder of 1, the conversion is now complete. The remainders, taken in reverse order form the binary number.

Illustration. Suppose we are to convert decimal number 13 (whole number) into equivalent binary number. The method is illustrated in Fig. 24.6. Note that 13 is first divided by 2, giving a quotient of 6 with a remainder of 1. This remainder becomes the 2^0 position in the binary number. The 6 is then divided by 2, giving a quotient of 3 with a remainder of 0. This remainder becomes the 2^1 position in the binary number.

Continuing this procedure, the equivalent binary number is 1101.

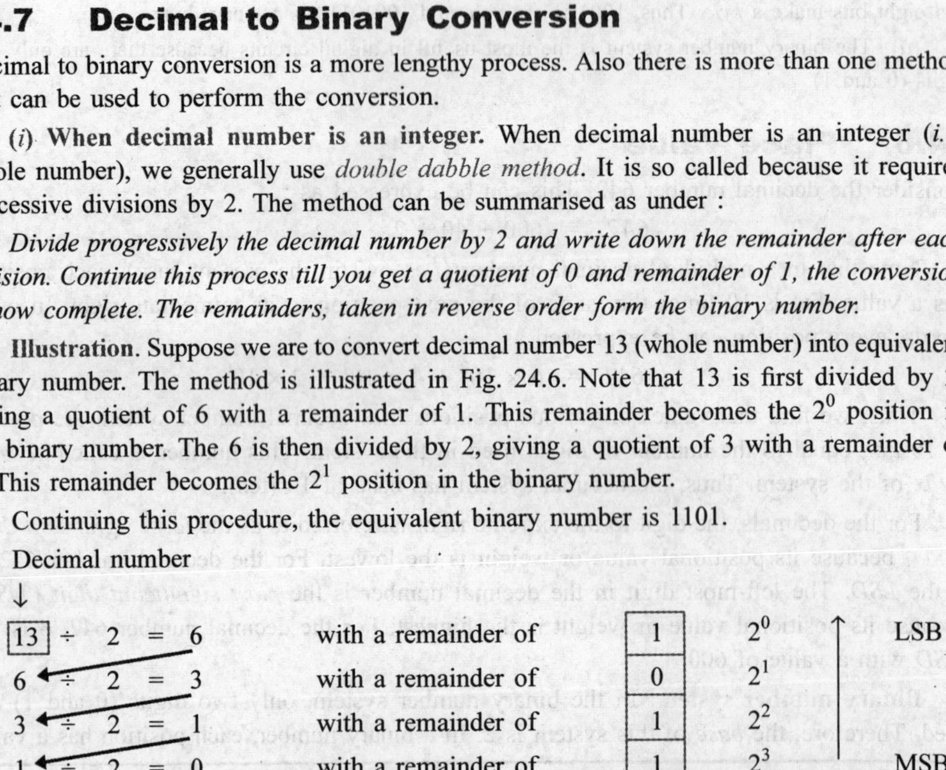

Fig. 24.6

(ii) When decimal number contains fraction. When the decimal number contains a fraction (*e.g.* 1.25), we use two-part method to convert it into equivalent binary number. We first convert the whole part of the decimal number by double dabble method explained above. Next, the fractional part of the decimal number is converted into equivalent binary number by repeatedly multiplying by 2. If the multiplication produces a whole number equal to 1 plus some fractional part, write a 1. The multiplications by 2 are repeated until the fractional part is exactly zero or until a sufficient number of digits have been obtained.

Illustration. Let us convert the decimal number 53.3125 into equivalent binary number. We first convert 53 (whole number part) into equivalent binary number and then convert 0.3125 (fractional part) into equivalent binary number. The two equivalent binary numbers are combined through binary point.

Conversion of 53

2	53	
2	26 – 1	LSB ↑
2	13 – 0	
2	6 – 1	
2	3 – 0	
2	1 – 1	
	0 – 1	MSB

Conversion of 0.3125

0.3125		
× 2		
0.6250	0	MSB ↓
× 2		
1.2500	1	
× 2		
0.5000	0	
× 2		
1.0000	1	LSB

$\therefore\ (53)_{10} = (110101)_2$

$\therefore\ (0.3125)_{10} = (0.0101)_2$

$\therefore\ (53.3125)_{10} = (110101.0101)_2$

Example 24.1. *Convert* $(1245)_{10}$ *into binary number system.*

Solution. For this conversion, we use double dabble method illustrated below.

2	1245	
2	622 – 1	LSB ↑
2	311 – 0	
2	155 – 1	
2	77 – 1	
2	38 – 1	
2	19 – 0	
2	9 – 1	
2	4 – 1	
2	2 – 0	
2	1 – 0	
	0 – 1	MSB

$\therefore\ \mathbf{(1245)_{10} = (10011011101)_2}$

Example 24.2. *Convert the fractional decimal number* $(0.625)_{10}$ *into binary number.*

Solution. The conversion of fractional decimal number is given below :

0.625		
× 2		
1.25	1	MSB
× 2		↓
0.50	0	↓
× 2		↓
1.00	1	LSB

$$\therefore \quad (0.625)_{10} = (0.101)_2$$

Example 24.3. *Convert the decimal number* $(100.625)_{10}$ *into binary number.*

Solution. The conversion is carried out as under :

Conversion of 100

2	100	
2	50 – 0	LSB
2	25 – 0	↑
2	12 – 1	↑
2	6 – 0	↑
2	3 – 0	↑
2	1 – 1	↑
	0 – 1	MSB

Conversion of 0.625

0.625		
× 2		
1.25	1	MSB
× 2		↓
0.50	0	↓
× 2		↓
1.00	1	LSB

$$\therefore \quad (0.625)_{10} = (0.101)_2$$

$$\therefore \quad (100)_{10} = (1100100)_2$$

$$\therefore \quad (100.625)_{10} = (1100100.101)_2$$

24.8 Binary to Decimal Conversion

Binary numbers can be converted to equivalent decimal numbers quite easily.

***(i)* When binary number is an integer.** When binary number is an integer (whole number), its conversion to equivalent decimal number involves the following two steps :

(*a*) Place the decimal value of each position of the binary number.

(*b*) Add all the decimal values to get the equivalent decimal number.

Illustration. Suppose we are to convert binary number 110011 (integer) into equivalent decimal number. The binary number along with its decimal values of various positions is shown.

$$\begin{array}{cccccc} 1 & 1 & 0 & 0 & 1 & 1 \\ 2^5 & 2^4 & 2^3 & 2^2 & 2^1 & 2^0 \end{array}$$

$$\therefore \quad (110011)_2 = 1 \times 2^5 + 1 \times 2^4 + 0 \times 2^3 + 0 \times 2^2 + 1 \times 2^1 + 1 \times 2^0$$
$$= 32 + 16 + 0 + 0 + 2 + 1 = 51$$

or $\quad (110011)_2 = (51)_{10}$

Note. In binary to decimal conversion, all positions containing 0 can be ignored. Only add the decimal values of the positions where 1 appears. Thus, in case of the above binary number,

$$(110011)_2 = 1 \times 2^5 + 1 \times 2^4 + 1 \times 2^1 + 1 \times 2^0$$
$$= 32 + 16 + 2 + 1 = 51$$

***(ii)* When binary number contains a fraction.** When binary number contains a fraction (*e.g.* 110.111), the weights of digit positions to the left as well as right of the binary point are indicated. By adding all the decimal values, we get the equivalent decimal number.

Illustration. Suppose we are to convert binary number 110.111 into equivalent decimal number. The binary number along with its decimal values is shown.

$$\begin{array}{ccccc} 1 & 1 & 0.1 & 1 & 1 \\ 2^2 & 2^1 & 2^0.2^{-1} & 2^{-2} & 2^{-3} \end{array}$$

$$\therefore \quad (110.111)_2 = 1 \times 2^2 + 1 \times 2^1 + 0 \times 2^0 + 1 \times 2^{-1} + 1 \times 2^{-2} + 1 \times 2^{-3}$$
$$= 4 + 2 + 0 + 0.5 + 0.25 + 0.125 = 6.875$$

or $\quad (110.111)_2 = (6.875)_{10}$

Example 24.4. *Convert the binary number $(110001)_2$ to its equivalent decimal number.*

Solution. The binary number along with its decimal values of various positions is shown.

$$\begin{array}{cccccc} 1 & 1 & 0 & 0 & 0 & 1 \\ 2^5 & 2^4 & 2^3 & 2^2 & 2^1 & 2^0 \end{array}$$

$$\therefore \quad (110001)_2 = 1 \times 2^5 + 1 \times 2^4 + 1 \times 2^0$$
$$= 32 + 16 + 1 = 49$$

or $\quad \mathbf{(110001)_2 = (49)_{10}}$

Example 24.5. *Convert the binary number $(1011.101)_2$ to its equivalent decimal number.*

Solution. The binary number along with its decimal values of various positions is shown.

$$\begin{array}{cccccc} 1 & 0 & 1 & 1.1 & 0 & 1 \\ 2^3 & 2^2 & 2^1 & 2^0.2^{-1} & 2^{-2} & 2^{-3} \end{array}$$

$$\therefore \quad (1011.101)_2 = 1 \times 2^3 + 1 \times 2^1 + 1 \times 2^0 + 1 \times 2^{-1} + 1 \times 2^{-3}$$
$$= 8 + 2 + 1 + 0.5 + 0.125 = 11.625$$

$\therefore \quad \mathbf{(1011.101)_2 = (11.625)_{10}}$

24.9 Binary Arithmetic

The reader is familiar with arithmetic operations such as addition, subtraction, multiplication and division in the decimal system. These operations on binary numbers are performed in the same manner as with decimal numbers. As a matter of fact, binary arithmetic is simpler than decimal arithmetic because here we use only two digits (0 and 1).

1. Binary addition. The addition of two binary numbers is performed in exactly the same manner as the addition of decimal numbers. In fact, binary addition is simpler, because we have only two digits (0 and 1). When adding binary numbers, the following rules apply :

$$0 + 0 = 0$$
$$1 + 0 = 1$$
$$1 + 1 = 10 = 0 + \text{carry of 1 into next position}$$
$$1 + 1 + 1 = 11 = 1 + \text{carry of 1 into next position}$$

It is not necessary to consider the addition of more than two binary numbers at a time. It is because the circuitry that actually performs the addition in a digital system can handle only two numbers at a time. When more than two numbers are to be added, the first two are added together and then their sum is added to the third number and so on.

Examples. Let us illustrate binary addition with two examples.

(*i*) Let us add the following two binary numbers :

$$\begin{array}{r} 1\,0\,1 \\ +\,0\,1\,0 \\ \hline 1\,1\,1 \end{array}$$

The addition is done column by column starting at the right as it is in the decimal number system. For the first column, 1 + 0 = 1 and for the second column, 0 + 1 = 1. For the third or last column, 1 + 0 = 1. This produces a sum of 111.

(*ii*) Let us now add the following two binary numbers :

$$\begin{array}{r} 1\,0\,1\,1\,1 \\ +\,1\,0\,0\,1\,0 \\ \hline 1\,0\,1\,0\,0\,1 \end{array}$$

For the first column, 1 + 0 = 1. In the second column, we have, 1 + 1 = 10 = 0 with a carry of 1. For the third column, we have, 1 + 0 + carry of 1 = 1 + 1 = 10 = 0 with a carry of 1. For the fourth column, 1 + 0 + 0 = 1 + 0 = 1. For the fifth column, we have, 1 + 1 = 10 = 0 with a carry of 1. This produces a sum of 101001.

2. Binary subtraction. For subtraction of binary numbers, the following rules apply :

$$0 - 0 = 0$$
$$1 - 0 = 1$$
$$1 - 1 = 0$$
$$10 - 1 = 1$$

Examples. Let us illustrate binary subtraction with two examples.

(*i*) Let us subtract binary number 101 from binary number 111.

$$\begin{array}{r} 1\,1\,1 \\ -\,1\,0\,1 \\ \hline 0\,1\,0 \end{array}$$

The subtraction is done column by column starting at right as in the decimal system. For the first column, 1 – 1 = 0 and for the second column, 1 – 0 = 1. For the third or last column, 1 – 1 = 0. Therefore, the result of binary subtraction is 010.

(*ii*) Let us now subtract binary number 1001 from binary number 1010.

```
  1 0 1 0
– 1 0 0 1
---------
  0 0 0 1
```

We again start from the least significant column (*i.e.,* from right as in the decimal system). For the first column, 0 – 1 = 1. It is because we borrow 1 from the next column of binary number 1010 so that 10 – 1 = 1. For the second column, 0 – 0 = 0 (because 1 has already been borrowed so that we have 0 in second column). For the third column, 0 – 0 = 0. Finally, for the fourth column, 1 – 1 = 0. The result of binary subtraction is 0001.

3. Binary multiplication. The multiplication of binary numbers is done in the same manner as the multiplication of decimal numbers. The process is actually simpler because the multiplier digits are either 0 or 1 and no other digits. For multiplication of binary numbers, the following rules apply :

$$0 \times 0 = 0$$
$$0 \times 1 = 0$$
$$1 \times 0 = 0$$
$$1 \times 1 = 1$$

While carrying out the multiplication of binary numbers, we take each digit in turn and shift the resulting each successive product (called partial product) one place to the left. The partial products are summed to produce the final product. Note that this is the same procedure that we follow for the multiplication of decimal numbers.

Examples. Let us illustrate the multiplication of binary numbers with two examples.

(*i*) The multiplication of binary number 111 by binary number 101 is given below. The top number is called the *multiplicand* and the bottom number the *multiplier*. The multiplicand is first multiplied by the least significant bit (LSB) of the multiplier. In our example, it is 1. This 1 is multiplied with multiplicand (111) to produce 111 which is written down as the first partial product. Next, the second bit of the multiplier (*i.e.,* 0 in this case) is multiplied with multiplicand to produce 000 which is written as the second partial product. Note that the second partial product is *shifted* one place to the left relative to the first one. Similarly, third partial product is obtained and shifted one place to the left relative to the second one. The three partial products are then summed up to produce the final product 100011.

```
     Binary
      1 1 1   Multiplicand
    × 1 0 1   Multiplier
    -------
      1 1 1   First partial product
    0 0 0     Second partial product
  1 1 1       Third partial product
-----------
1 0 0 0 1 1   Final product
```

(*ii*) Let us now multiply binary number 11011 by binary number 1100. Here, 11011 is the multiplicand and 1100 is the multiplier. Note that steps which involve multiplication with 0 (first and second partial products) result in 00000 and have been omitted.

Binary	
1 1 0 1 1	Multiplicand
× 1 1 0 0	Multiplier
1 1 0 1 1 0 0	Third partial product
1 1 0 1 1	Fourth partial product
1 0 1 0 0 0 1 0 0	Final Product

4. Binary division. Binary division is obtained using the same procedure as the decimal division. Let us discuss two examples.

(*i*) Let us divide $(63)_{10}$ by $(9)_{10}$ by means of binary numbers.

Now, $(63)_{10} = (111111)_2$ and $(9)_{10} = (1001)_2$

```
                      1 1 1  ←— Quotient
Divisor ——→ 1 0 0 1 ) 1 1 1 1 1 1  ←— Dividend
                      1 0 0 1↓
                      0 1 1 0 1
                        1 0 0 1↓
                        0 1 0 0 1
                          1 0 0 1
                          0 0 0 0 ←— Remainder
```

Hence, $(111111)_2 \div (1001)_2 = (111)_2$

and $(63)_{10} \div (9)_{10} = (7)_{10}$

There is no remainder.

(*ii*) Let us now divide $(61)_{10}$ by $(9)_{10}$ by means of binary numbers.

Now, $(61)_{10} = (111101)_2$ and $(9)_{10} = (1001)_2$

```
                1 1 0
    1 0 0 1 ) 1 1 1 1 0 1
              1 0 0 1↓
              0 1 1 0 0
                1 0 0 1↓
                0 0 1 1 1
```

Hence, $(111101)_2 \div (1001)_2 = (110)_2$ with a remainder of $(111)_2$

and $(61)_{10} \div (9)_{10} = (6)_{10}$ with a remainder of $(7)_{10}$

24.10 1's and 2's Complement of Binary Numbers

The complement (*opposite*) of a number is a number which is of the same magnitude but with opposite sign. Let us see how we form the 1's complement and 2's complement of a binary number.

(*i*) **1's complement representation.** The 1's complement of a binary number is obtained by changing each 0 to a 1 and each 1 to a 0. In other words, we change each bit in the binary to its complement. The process of obtaining 1's complement of binary number 101101 is shown below.

```
1 0 1 1 0 1    Original binary number
↓ ↓ ↓ ↓ ↓ ↓
0 1 0 0 1 0    1's complement
```

Thus, we say that 1's complement of binary number 101101 is 010010. More examples of converting a binary number to its 1's complement form are :

Binary number	*1's complement*
100	011
1110	0001
1010	0101
10111	01000

(*ii*) 2's complement representation. The 2's complement of a binary number is obtained by first taking the 1's complement of the number and then adding 1. As a formula :

2's complement = 1's complement + 1

Thus to find 2's complement of binary number 101101, the procedure is as under :

```
  1 0 1 1 0 1    Original binary number
  0 1 0 0 1 0    1's complement
+           1    Add 1 to form 2's complement
─────────────
  0 1 0 0 1 1    2's complement of original binary number
```

More examples of converting a binary number to its 2's complement form are :

Binary number	*1's complement*	*2's complement*
0 0 0 1 0 1 1 0	1 1 1 0 1 0 0 1	1 1 1 0 1 0 1 0
1 1 1 1 1 1 0 0	0 0 0 0 0 0 1 1	0 0 0 0 0 1 0 0
1 0 0 1 0 0 0 1	0 1 1 0 1 1 1 0	0 1 1 0 1 1 1 1

24.11 Signed Binary Numbers

In the decimal number system, a plus sign (+) is used to denote a positive number and a minus sign (–) to denote a negative number. This representation of numbers is called *signed numbers*. A natural question arises, how to indicate the sign of positive and negative binary numbers. The obvious choice would be to use the digits 0 and 1 because digital circuits can understand only these symbols. Generally, an additional bit (called *sign bit*) is placed in front of the MSB (most significant bit) of the binary number to indicate the sign of the number. *A 0 in the sign bit represents a positive binary number and a 1 in the sign bit represents a negative binary number.* This notation is called *sign-magnitude notation* because the sign bit is given first and then the magnitude of binary number.

Illustration. Fig. 24.7 shows the representation of signed binary numbers in sign-magnitude form. The leftmost bit is the sign bit and the other six digits are the magnitude of the binary number.

Fig. 24.7 (*i*) shows that the signed binary number is 0110100. The 0 in the leftmost bit (*i.e.,* sign bit) indicates that the binary number is positive. The other six digits (*i.e.,* 110100) indicate the magnitude of the binary number which is equal to the decimal number 52.

$\therefore \quad (0110100)_2 = (+52)_{10}$

Fig. 24.7 (*ii*) shows that the signed binary number is 1110100. The 1 in the leftmost bit (*i.e.*, sign bit) indicates that the binary number is negative. The other six digits (*i.e.* 110100) indicate the magnitude of the binary number which is equal to decimal number 52.

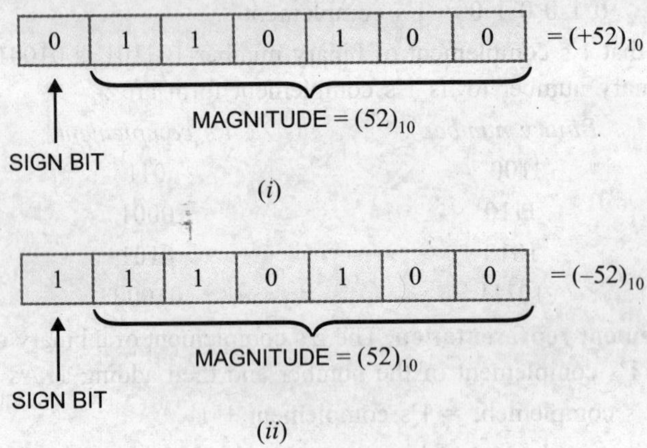

Fig. 24.7

$\therefore \quad (1110100)_2 = (-52)_{10}$

24.12 Representation of Signed Binary Numbers Using 2's Complement

The 2's complement system for representing signed binary numbers is as under :

(*i*) If the binary number is positive, the magnitude is represented in its true binary form and sign bit of 0 is placed in front of the MSB. This is shown in Fig. 24.8 (*i*) for $(+45)_{10}$.

(*ii*) If the binary number is negative, the magnitude is represented in its 2's complement form and a sign bit of 1 is placed in front of the MSB. This is shown in Fig. 24.8 (*ii*) for $(-45)_{10}$.

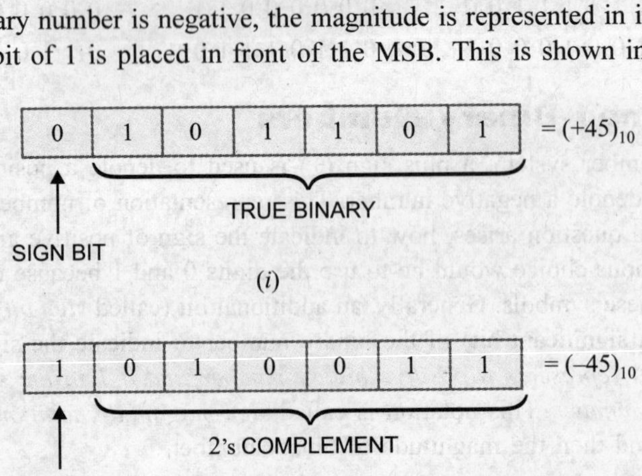

Fig. 24.8

The 2's complement system is used to represent signed binary numbers because it allows us to perform the operation of subtraction by actually performing addition. This is significant because it means that digial computer can use the same circuitry for addition and subtraction.

Example 24.6. *Represent the signed decimal numbers (i) + 13 (ii) – 9 as signed binary numbers in the 2's complement system.*

Solution. (*i*) Since the number is positive, the magnitude (13) will be represented in its true magnitude form *i.e.*, $(13)_{10} = (1101)_2$. Attaching the sign bit of zero, we have,

$$(+13)_{10} = (01101)_2$$

Sign bit (the leading 0)

(*ii*) Since the number is negative, the magnitude (9) has to be represented in 2's complement form :

$$(+9)_{10} = (1001)_2$$

0110	1's complement
+ 1	Add 1 to form 2's complement
0111	2's complement

When we attach the sign bit of 1, the complete signed number becomes :

$$(-9)_{10} = (10111)_2$$

Sign bit (the leading 1)

Example 24.7. *Each of the following numbers is a signed binary number in the 2's complement system. Determine the decimal value in each case : (i) 01100 (ii) 11010.*

Solution. (*i*) The sign bit is zero, so the number is positive. The other 4 bits (*i.e.*, 1100) represent the magnitude of the number *i.e.*, $(1100)_2 = (12)_{10}$. Thus the decimal number is +12.

(*ii*) The sign bit of 11010 is 1. Therefore, the number is negative but we cannot tell what the magnitude is. We can find the magnitude by negating (2's complementing) the number to convert it to its positive equivalence.

1 1 0 1 0	Original negative number
0 0 1 0 1	1's complement
+ 1	Add 1 to form 2's complement
0 0 1 1 0	(+ 6)

Since the result is 00110 = +6, the original number 11010 must be equivalent to –6.

24.13 Binary Subtraction Using 2's Complement

In this method, subtraction is obtained by using 2's complement of the binary number (thereby making it a negative number) to be subtracted and adding it to the other binary number taking care to discard any carry bit. Thus binary subtraction operation using 2's complement system actually involves the operation of addition.

(*i*) **Subtraction of smaller number from larger one.** In order to subtract smaller binary number from larger binary number, the procedure is as under :

(*a*) Find the 2's complement of the smaller binary number, thereby making it a negative number.

(*b*) Add the 2's complement of the smaller binary number to the larger one.

(*c*) Discard the carry bit (there is always a carry bit in this case).

Illustration. Let us subtract $(1010)_2$ from $(1111)_2$ using 2's complement method. We shall also show direct subtraction for comparison.

Direction subtraction	*2's complement method*	
1 1 1 1	1 1 1 1	
– 1 0 1 0	0 1 1 0	2's complement
0 1 0 1	Carry → 1 0 1 0 1	

Discarding carry bit, the answer is 0101.

(*ii*) **Subtraction of larger number from smaller one.** In order to subtract larger binary number from smaller binary number, the procedure is as under :

(*a*) Find the 2's complement of the larger binary number, thereby making it a negative number.

(*b*) Add 2's complement of the larger binary number to the smaller one.

(*c*) There is no carry bit. The result is in 2's complement form and is negative.

(*d*) To get the answer in true form, take 2's complement and change the sign.

Illustration. Let us subtract $(1010)_2$ from $(1000)_2$ using 2's complement method. We shall also show direct subtraction for comparison.

Direction subtraction	*2's complement method*	
1 0 0 0	1 0 0 0	
– 1 0 1 0	0 1 1 0	2's complement
– 0 0 1 0	1 1 1 0	

There is no carry. The 2's complement of $(1110)_2$ is $(0010)_2$. Therefore, the answer is –0010.

24.14 Octal Number System

The octal number system has a *radix* of eight so that it uses eight digits : 0, 1, 2, 3, 4, 5, 6 and 7. Thus, each digit of an octal number can have any value from 0 to 7. The position weights in the octal system are powers of 8 as shown in Fig. 24.9. Note that the places to the left of octal point (radix point) are the positive powers of 8 and places to the right are negative powers of 8.

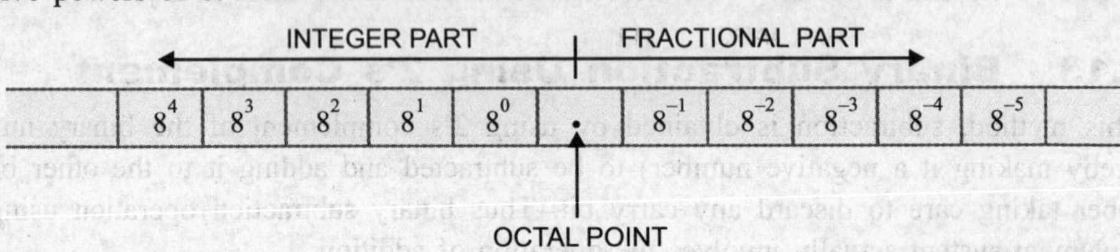

Fig. 24.9

The octal number system is frequently used in digital circuits due to two principal reasons. First, it can be easily converted to binary. Secondly, there are significantly fewer digits in any given octal number than in the corresponding binary number so that it is much easier to work with shorter octal numbers.

1. Decimal-to-octal conversion. To convert a decimal number to octal, we employ the same repeated-division method that we used in decimal-to-binary conversion. However, here the division factor is 8 instead of two. The following examples illustrate decimal-to-octal conversion.

(*i*) To convert decimal number 91 to octal number, the procedure is as under :

Division	*Remainder*
91 ÷ 8 = 11	3 (LSB)
11 ÷ 8 = 1	3
1 ÷ 8 = 0	1 (MSB)

$\therefore \quad (91)_{10} = (133)_8$

(*ii*) Let us now convert decimal fractional number $(0.3125)_{10}$ to octal number.

0.3125 × 8 = 2.5 = 0.5 with a carry of 2

0.5 × 8 = 4.0 = 0 with a carry of 4

$\therefore \quad (0.3125)_{10} = (0.24)_8$

2. Octal-to-decimal conversion. An octal-to-decimal conversion can be done in the same manner as a binary-to-decimal conversion *i.e.,* simply add up the position weights to obtain the decimal number. The following examples illustrate octal-to-decimal conversion.

(*i*) To convert octal number $(133)_8$ to decimal number, the procedure is as under :

Position weights $\quad 8^2 \; 8^1 \; 8^0$

Octal number $\quad 1 \;\; 3 \;\; 3$

$$\therefore \quad (133)_8 = (8^2 \times 1) + (8^1 \times 3) + (8^0 \times 3)$$

$$= 64 + 24 + 3 = 91$$

$$\therefore \quad (133)_8 = (91)_{10}$$

(*ii*) Let us now convert octal fractional number $(0.74)_8$ to decimal number.

$$(0.74)_8 = 7 \times 8^{-1} + 4 \times 8^{-2} = 0.875 + 0.0625 = 0.9375$$

$$\therefore \quad (0.74)_8 = (0.9375)_{10}$$

3. Octal-to-binary conversion. The advantage of octal number system is the ease with which an octal number can be converted to a binary number and vice-versa. It is because eight is the third power of two, providing a direct correlation between three-bit groups in a binary number and the octal digits *i.e.,* each three-bit group of binary bits can be represented by one octal digit. Therefore, conversion from octal to binary is performed by converting *each* octal digit to its 3-bit binary equivalent. The eight possible digits are converted as shown in the adjoining table.

Octal and Binary Equivalents	
Octal Digit	Binary Bits
0	000
1	001
2	010
3	011
4	100
5	101
6	110
7	111

(*i*) The conversion of octal number $(472)_8$ to binary number is done as under :

4	7	2
↓	↓	↓
100	111	010

Therefore, octal 472 is equivalent to binary 100111010 *i.e.,*

$$(472)_8 = (100111010)_2$$

(*ii*) As another example, consider the conversion of octal number $(5431)_8$ to binary number.

5	4	3	1
101	100	011	001

Therefore, octal 5431 is equivalent to binary 101100011001 *i.e.*,

$$(5431)_8 = (101100011001)_2$$

4. Binary-to-octal conversion. The conversion of binary number to octal number is simply the reverse of the above process. The bits of the binary number are grouped into groups of *three* bits starting at the LSB. Then each group is converted to its octal equivalent. To illustrate this method, consider the conversion of binary number $(100111010)_2$ to octal number. The procedure is as under :

$$\underbrace{100}_{\downarrow \atop 4} \quad \underbrace{111}_{\downarrow \atop 7} \quad \underbrace{010}_{\downarrow \atop 2}$$

$$\therefore \quad (100111010)_2 = (472)_8$$

Note that there are fewer digits in the octal number than in the corresponding binary number. Therefore, it is much easier to work with shorter octal numbers.

Sometimes the binary number will not have even groups of 3 bits. In that case, we can add one or two 0s to the left of the MSB of the binary number to fill the last group. This point is illustrated below for the binary number 11010110.

$$\underbrace{011}_{3} \quad \underbrace{010}_{2} \quad \underbrace{110}_{6}$$

Note that a 0 is placed to the left of the MSB to produce even groups of 3 bits.

Example 24.8. *Convert the following decimal numbers to octal equivalent.*

(i) $(76)_{10}$ *(ii)* $(255)_{10}$ *(iii)* $(372)_{10}$

Solution.

(*i*)

Division	Remainder
76 ÷ 8 = 9	4 (LSB) ↑
9 ÷ 8 = 1	1
1 ÷ 8 = 0	1 (MSB)

$$\therefore \quad (76)_{10} = (114)_8$$

(*ii*)

Division	Remainder
255 ÷ 8 = 31	7 (LSB) ↑
31 ÷ 8 = 3	7
3 ÷ 8 = 0	3 (MSB)

$$\therefore \quad (255)_{10} = (377)_8$$

(*iii*)

Division	Remainder
372 ÷ 8 = 46	4 (LSB) ↑
46 ÷ 8 = 5	6
5 ÷ 8 = 0	5 (MSB)

$$\therefore \quad (372)_{10} = (564)_8$$

Example 24.9. *Convert octal number $(24.6)_8$ to the equivalent decimal number.*

Solution.

$$(24.6)_8 = (2 \times 8^1) + (4 \times 8^0) + (6 \times 8^{-1})$$
$$= 16 + 4 + 0.75 = 20.75$$
$$\therefore \quad \mathbf{(24.6)_8 = (20.75)_{10}}$$

Example 24.10. *Convert $(177)_{10}$ to its 8-bit binary equivalent by first converting to octal.*

Solution. We shall first convert $(177)_{10}$ to its equivalent octal number as under :

Division	*Remainder*
$177 \div 8 = 22$	1 (LSB) ↑
$22 \div 8 = 2$	6
$2 \div 8 = 0$	2 (MSB)

$$\therefore \quad \mathbf{(177)_{10} = (261)_8}$$

We now convert the octal number $(261)_8$ to its equivalent binary number as under :

2	6	1
↓	↓	↓
010	110	001

Therefore, the binary equivalent is 010110001. We remove the leading zero to express the result as 8 bits.

$$\therefore \quad \mathbf{(177)_{10} = (10110001)_2}$$

Example 24.11. *Convert the decimal number $(39.75)_{10}$ to octal.*

Solution. Given decimal number is $(39.75)_{10}$.

Integer part

8	39
8	4 – 7 ↑
	0 – 4

$\therefore \quad (39)_{10} = (47)_8$

Fractional part

$0.75 \times 8 = 6.0$ with a carry of 6

$\therefore \quad (0.75)_{10} = (0.6)_8$

$$\therefore \quad \mathbf{(39.75)_{10} = (47.6)_8}$$

Example 24.12. *Convert the octal number $(375.37)_8$ to binary number.*

Solution. The conversion is done as under :

				Binary point ↓		
$(375.37)_8 =$	3	7	5	.	3	7
	↓	↓	↓		↓	↓
	011	111	101		011	111

$$\therefore \quad \mathbf{(375.37)_8 = (011111101.011111)_2}$$

Example 24.13. *Convert the binary number $(0.101010110)_2$ to octal number.*

Solution. The conversion is done as under :

$$(0.101010110)_2 = 0.101 \quad 010 \quad 110$$
$$\downarrow \quad \downarrow \quad \downarrow$$
$$5 \quad 2 \quad 6$$

$$\therefore \quad \mathbf{(0.101010110)_2 = (0.526)_8}$$

Example 24.14. *Convert $(101011)_2$ into decimal system and octal system.*

Solution. The conversion into decimal system is as under :

$$(101011)_2 = 1 \times 2^5 + 0 \times 2^4 + 1 \times 2^3 + 0 \times 2^2 + 1 \times 2^1 + 1 \times 2^0$$
$$= 32 + 0 + 8 + 0 + 2 + 1 = 43$$

$$\therefore \quad \mathbf{(101011)_2 = (43)_{10}}$$

The conversion into octal number is as under :

$$(101011)_2 = 101 \quad 011$$
$$\downarrow \quad \downarrow$$
$$5 \quad 3$$

$$\therefore \quad \mathbf{(101011)_2 = (53)_8}$$

24.15 Hexadecimal Number System

The hexadecimal system uses a *radix* of 16. Therefore, it has 16 possible digit symbols. The first ten digits in the hexadecimal system are represented by the numbers 0 through 9 (0, 1, 2, 3, 4, 5, 6, 7, 8 and 9) and the letters A though F are used to represent the numbers 10, 11, 12, 13, 14 and 15 respectively. The adjoining table shows the relationships among hexadecimal, decimal and binary. Note that each hexadecimal digit represents a group of four binary digits.

Hexadecimal	Decimal	Binary
0	0	0000
1	1	0001
2	2	0010
3	3	0011
4	4	0100
5	5	0101
6	6	0110
7	7	0111
8	8	1000
9	9	1001
A	10	1010
B	11	1011
C	12	1100
D	13	1101
E	14	1110
F	15	1111

As is true for binary and decimal numbers, each digit in the hexadecimal system has a positional value or weight. For the right most digit of a hex (abbreviation for hexadecimal) number, the positional weight is 16^0 (= 1), the next digit to the left has a positional weight of 16^1 (= 16) and so on. The positional weight distribution of a hex number system is given below :

	16^3	16^2	16^1	16^0
etc. ←	4096	256	16	1

1. Decimal-to-hex conversion. To convert a decimal number to hex number, the technique is the same as used for decimal-to-binary conversion or decimal-to-octal conversion. Recall that we did decimal-to-binary conversion using repeated division by 2 and decimal-to-octal conversion using repeated division by 8. Likewise, decimal-to-hex conversion is done using repeated division by 16. Let us illustrate the decimal-to-hex conversion procedure. Suppose we are to convert the decimal number 423 to hex number.

Division		Remainder	
$423 \div 16 = 26$		7	(LSB) ↑
$26 \div 16 = 1$		10	
$1 \div 16 = 0$		1	(MSB)

$$\therefore \quad (423)_{10} = (1A7)_{16}$$

Note that 10 is represented by the letter A.

2. **Hex-to-decimal conversion.** In order to convert a hex number to its decimal equivalent, simply add up the position weight of each digit in the hex number. The following example illustrates this conversion.

$$(356)_{16} = (3 \times 16^2) + (5 \times 16^1) + (6 \times 16^0)$$

$$= 768 + 80 + 6 = 854$$

$$\therefore \quad (356)_{16} = (854)_{10}$$

3. **Hex-to-binary conversion.** The conversion from hex to binary is performed by converting each hex digit to its 4-bit binary equivalent (See above table). The following example illustrates this point. Here, we shall convert hex number $(9F2)_{16}$ to its binary equivalent.

```
  9       F       2
  ↓       ↓       ↓
1001    1111    0010
```

$$\therefore \quad (9F2)_{16} = (100111110010)_2$$

4. **Binary-to-hex conversion.** The conversion from binary to hex is just the reverse of the above process. The binary number is grouped into groups of *four* bits and each group is converted to its equivalent hex digit. The following example illustrates this point. Here, we shall convert binary number $(1110100110)_2$ to its *equivalent hex number.

$$\underbrace{0011}_{3} \quad \underbrace{1010}_{A} \quad \underbrace{0110}_{6}$$

$$\therefore \quad (1110100110)_2 = (3A6)_{16}$$

5. **Octal-to-hexadecimal conversion.** The method generally used is to first convert the given octal number to its binary equivalent and then from binary to hexadecimal. Therefore, this method involves two steps. First, each octal digit is replaced by its 3-bit binary equivalent. This gives us the binary equivalent of the given octal number. Second, we form 4-bit groups of the binary number starting from the right-most position. Then we replace each 4-bit group by its hexadecimal equivalent. This will give us the hexadecimal equivalent of the given octal number.

(*i*) The conversion of the octal number $(321)_8$ to hexadecimal number is done as under :

```
                 3      2      1
      Step 1     ↓      ↓      ↓
                011    010    001

      Step 2      1101   0001
                   ↓      ↓
                   D      1
```

* Zeros are added, as needed, to complete 4-bit group.

Thus the hexadecimal equivalent of the given octal number is $(D1)_{16}$.

$\therefore$ $(321)_8 = (D1)_{16}$

(*ii*) Let us now convert the octal number $(1024)_8$ to its hexadecimal equivalent.

	1	0	2	4
Step 1	↓	↓	↓	↓
	001	000	010	100
Step 2	0010	0001	0100	
	↓	↓	↓	
	2	1	4	

$\therefore$ $(1024)_8 = (214)_{16}$

6. Hexadecimal-to-octal conversion. The method generally used is to first convert the given hexadecimal number to its binary equivalent and then from binary to octal. Therefore, this method involves two steps. First, each hexadecimal digit is replaced by its 4-bit binary equivalent. This gives us the binary equivalent of the given hexadecimal number. Second, we form 3-bit groups of the binary number starting from the right-most position. Then we replace each 3-bit group by its octal equivalent. This will give us the octal equivalent of the given hexadecimal number.

(*i*) The conversion of the hexadecimal number $(5C2)_{16}$ to octal number is done as under :

	5	C	2	
Step 1	↓	↓	↓	
	0101	1100	0010	
Step 2	010	111	000	010
	↓	↓	↓	↓
	2	7	0	2

$\therefore$ $(5C2)_{16} = (2702)_8$

(*ii*) Let us now convert the hexadecimal number $(A72E)_{16}$ to its octal equivalent.

	A	7	2	E		
Step 1	↓	↓	↓	↓		
	1010	0111	0010	1110		
Step 2	001	010	011	100	101	110
	↓	↓	↓	↓	↓	↓
	1	2	3	4	5	6

$\therefore$ $(A72E)_{16} = (123456)_8$

Example 24.15. *Convert decimal number $(541)_{10}$ to hexadecimal.*

Solution.

Division	*Remainder*
541 ÷ 16 = 33	13 (LSB)
33 ÷ 16 = 2	1
2 ÷ 16 = 0	2 (MSB)

$\therefore \quad \mathbf{(541)_{10} = (21D)_{16}}$

Example 24.16. *Convert decimal number $(378)_{10}$ to a 16-bit number by first converting to hexadecimal.*

Solution.

Division	*Remainder*
378 ÷ 16 = 23	10 (LSB)
23 ÷ 16 = 1	7
1 ÷ 16 = 0	1 (MSB)

Thus $\mathbf{(378)_{10} = (17A)_{16}}$. We can easily convert this hex number to binary 000101111010. Therefore, we can express $(378)_{10}$ as a 16-bit binary number by adding four leading 0s.

$\therefore \quad \mathbf{(378)_{10} = (0000000101111010)_2}$

Example 24.17. *Convert $(B2F)_{16}$ to octal.*

Solution. It is easier to first convert hex to binary and then to octal.

$(B2F)_{16}$ = 1011 0010 1111 .... Conversion to binary

= 101 100 101 111 .... 3-bit groupings

= 5 4 5 7

$\therefore \quad \mathbf{(B2F)_{16} = (5457)_8}$

Example 24.18. *Convert the decimal number $(150)_{10}$ into its equivalent binary, hexadecimal and octal number.*

Solution. The various conversions are carried out as under :

Binary

2	150
2	75 – 0
2	37 – 1
2	18 – 1
2	9 – 0
2	4 – 1
2	2 – 0
2	1 – 0
	0 – 1

$\therefore \quad \mathbf{(150)_{10} = (10010110)_2}$

Hexadecimal

16	150
16	9 – 6
	0 – 9

$\therefore \quad \mathbf{(150)_{10} = (96)_{16}}$

Octal

8	150
8	18 – 6
8	2 – 2
	0 – 2

$\therefore \quad \mathbf{(150)_{10} = (226)_8}$

Example 24.19. *Convert the hexadecimal number 8A3D into decimal and binary equivalent. Convert the decimal number 5796.12 into hexadecimal.*

Solution. The conversion of hexadecimal number 8A3D into equivalent decimal number is done as under :

$$(8A3D)_{16} = 8 \times 16^3 + 10 \times 16^2 + 3 \times 16^1 + 13 \times 16^0$$

$$= 32768 + 2560 + 48 + 13 = 35389$$

$$\therefore \quad \mathbf{(8A3D)_{16} = (35389)_{10}}$$

The conversion of hexadecimal number $(8A3D)_{16}$ into equivalent binary number is done as under :

$$(8A3D)_{16} = \begin{matrix} 8 & A & 3 & D \\ \downarrow & \downarrow & \downarrow & \downarrow \\ 1000 & 1010 & 0011 & 1101 \end{matrix}$$

$$\therefore \quad \mathbf{(8A3D)_{16} = (1000\ 1010\ 0011\ 1101)_2}$$

The conversion of decimal number $(5796.12)_{10}$ into equivalent hexadecimal number is done as under :

Integer		*Fraction*
16	5796	0.12 × 16 = 1.92 = 0.92 with a carry of 1
16	362 – 4	0.92 × 16 = 14.72 = 0.72 with a carry of 14
16	22 – 10	0.72 × 16 = 11.52 = 0.52 with a carry of 11
16	1 – 6	0.52 × 16 = 8.32 = 0.32 with a carry of 8
	0 – 1	

$\therefore \quad (5796)_{10} = (16A4)_{16} \qquad \therefore \quad (0.12)_{10} = (0.1EB8)_{16}$

$$\therefore \quad \mathbf{(5796.12)_{10} = (16A4.1EB8)_{16}}$$

Example 24.20. *Convert the hexadecimal number $(3A.2F)_{16}$ into decimal.*

Solution. The conversion is done as under :

$$(3A.2F)_{16} = 3 \times 16^1 + 10 \times 16^0 + 2 \times 16^{-1} + 15 \times 16^{-2}$$

$$= 48 + 10 + 0.125 + 0.05859375 = (58.18359375)_{10}$$

$$\therefore \quad \mathbf{(3A.2F)_{16} = (58.18359375)_{10}}$$

Example 24.21. *Convert the number $(11011.011)_{10}$ into hexadecimal number.*

Solution. The conversion is carried out as under :

Integer part		*Fractional part*
16	11011	0.011 × 16 = 0.176 with a carry of 0
16	688 – 3	0.176 × 16 = 0.816 with a carry of 2
16	43 – 0	0.816 × 16 = 0.056 with a carry of 13
16	2 – 11	0.056 × 16 = 0.896 with a carry of 0
	0 – 2	0.896 × 16 = 0.336 with a carry of 14

$\therefore \quad (11011)_{10} = (2B03)_{16} \qquad \therefore \quad (0.011)_{10} = (0.02D0E)_{16}$

$\therefore \quad \mathbf{(11011.011)_{10} = (2B03.02D0E)_{16}}$

Example 24.22. *Convert the decimal number $(39.75)_{10}$ to hexadecimal.*

Solution. The conversion is carried out as under :

Integer part		*Fractional part*
16	39	
16	2 – 7	0.75 × 16 = 0 with a carry of 12 *i.e.*, C
	0 – 2	

$\therefore \quad (39)_{10} = (27)_{16} \qquad \therefore \quad (0.75)_{10} = (0.C)_{16}$

$\therefore \quad (39.75)_{10} = (27.C)_{16}$

24.16 Binary-Coded Decimal Code (BCD Code)

Circuits and machines can deal readily with binary numbers, but people are used to working with decimal numbers. Moreover, there are considerably fewer decimal digits required to represent a number than there are binary. It is much easier to remember just a few digits than it is to remember many. Thus whenever there is an interface between digital circuits and people, the interface data usually takes the decimal from. As a result, the digital circuits must utilise some binary code to conveniently represent the decimal numbers. The code used for this purpose is called BCD code. *In a BCD code, each decimal number is represented by a 4-bit binary number*. For example, to convert decimal number $(489)_{10}$ to BCD, the procedure is as under :

BCD code	Decimal digit
0000	0
0001	1
0010	2
0011	3
0100	4
0101	5
0110	6
0111	7
1000	8
1001	9

4	8	9
0100	1000	1001

Note that the highest BCD value that a 4-bit binary number could represent is 9 which would be $(1001)_2$ in binary. Clearly, only the 4-bit binary numbers from 0000 through 1001 are used.

The adjoining table on page 957 shows the BCD code. Each of the decimal digits (0 through 9) is represented by its binary equivalent. Since a decimal digit can be as large as 9, four bits are required to code each decimal digit (the binary code for 9 is 1001).

Note that each decimal digit is assigned a 4-bit binary number even though the binary equivalent may require fewer than four binary places. This way, circuits which use BCD code always handle the string of binary bits in four - place groups. When using BCD code, remember that all zeros must be retained, unlike a binary number where leading zeros can be dropped. The BCD code is used when it is necessary to transfer decimal information into and out of a digital machine. Examples of digital machines include the digital clocks, calculators, digital voltmeters and frequency counters.

Example 24.23. *What decimal number is represented by the BCD string given below ?*

0100 0000 0010

Solution. Divide the BCD number into 4-bit groups and convert each to decimal.

$$\underbrace{0100}_{4} \quad \underbrace{0000}_{0} \quad \underbrace{0010}_{2}$$

Therefore, the equivalent decimal number is $(402)_{10}$.

Note. To avoid confusion between BCD and true binary, a BCD string is often separated into groups of 4 binary bits or a subscript BCD is sometimes attached to the string as illustrated under :

0100 0000 0010 or 010000000010_{BCD}

24.17 Logic Gate

A logic gate is a digital circuit that produces one output voltage which is either high or low, depending upon the levels (high or low) of input voltage/voltages.

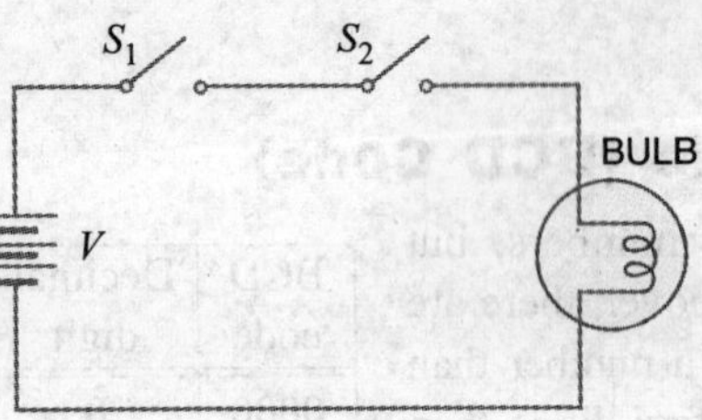

(*i*)

S_1	S_2	BULB
OPEN	OPEN	OFF
OPEN	CLOSED	OFF
CLOSED	OPEN	OFF
CLOSED	CLOSED	ON

(*ii*)

Truth Table

S_1	S_2	OUTPUT
0	0	0
0	1	0
1	0	0
1	1	1

(*iii*)

Fig. 24.10

A logic gate may have one or more inputs but only one output. Since a logic gate is a switching circuit (*i.e.*, a digital circuit), its output can have only one of the two possible states *viz.*, either a high voltage (1) or a low voltage (0) – it is either *ON* or *OFF*. Whether the output voltage of a logic gate is high (1) or low (0) will depend upon the conditions at its input. Fig. 24.10 (*i*) shows the basic idea of a *logic gate using switches.

(*i*) When S_1 and S_2 are open, the bulb is *OFF*.

* In itself, the circuit is not actually a logic gate but the logic is similar. The actual gate circuits are made with diodes and transistors. In other words, switches S_1 and S_2 are replaced by diodes or transistors.

(*ii*) When S_1 is open and S_2 closed, the bulb is *OFF*.

(*iii*) When S_2 is open and S_1 closed, the bulb is *OFF*.

(*iv*) When both S_1 and S_2 are closed, the bulb is *ON*.

Note that output (*OFF* or *ON*) depends upon the conditions at the input.

The four possible combinations of switches S_1 and S_2 are shown in the table in Fig. 24.10 (*ii*). It is clear that when either of the switches (S_1 or S_2) or both are open, the bulb is *OFF*. In binary language, when either of the inputs or both the inputs are low (0), the output is low. When both switches are closed, the bulb is *ON*. In terms of binary language, when both the inputs are high (1), the output is high. It is usual practice to show the conditions at the input and output of a logic gate in the binary form as shown in the table in Fig. 24.10 (*iii*). Such a table is called **truth table**.

The term "logic" is usually used to refer to a decision-making process. A logic gate makes logical decisions regarding the existence of output depending upon the nature of the input. Hence, such circuits are called logic circuits. The following points may be noted about the logic *gates :

(*i*) A binary 0 represents 0 V and binary 1 represents + 5 V**. It is common to refer to binary 0 as LOW input or output and binary 1 as HIGH input or output.

(*ii*) A logic gate has only one output signal. The output will depend upon the input signal/signals and the type of gate.

(*iii*) The operation of a logic gate may be described either by *truth table* or *Boolean algebra*.

24.18 Three Basic Logic Gates

The three basic logic gates that make up all the digital circuits are :

(*i*) OR gate

(*ii*) AND gate

(*iii*) NOT gate

In logic gates, we use junction diodes and transistors to perform switching functions. Each logic gate is represented by a graphic symbol (called logic symbol) and its function is defined by truth table or by a Boolean expression.

24.19 OR Gate

An **OR gate** *is a logic gate that has two or more inputs but only one output. It gives high output (1) if any or all the inputs are high (1) otherwise the output is low (0).*

It is called OR gate because the output is HIGH if any or all the inputs are HIGH. For the same reason, an OR gate is sometimes called "any or all gate". For example, consider a 2-input OR gate. The output *Y* will be HIGH if either or both inputs are HIGH.

---

* A *gate* can be regarded as a *barrier* which when closed prevents the passage of information but if open allows the signal/signals to pass through freely.

** In digital systems, the binary information is represented by two voltage levels, generally + 5 V and 0 V. So 5 V is used to represent binary 1 and 0 V is used to represent binary 0.

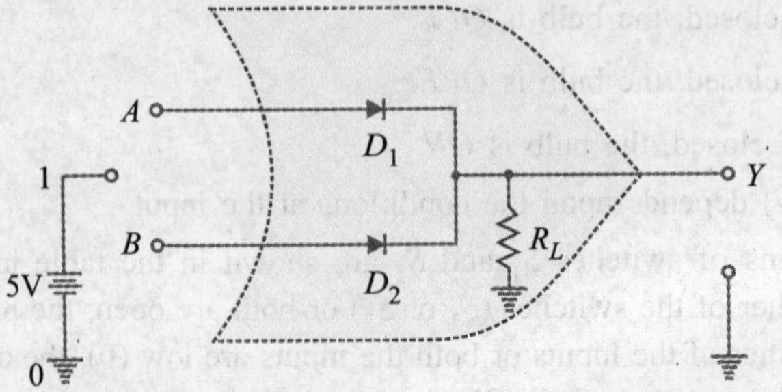

(i)

Truth table

A	B	Y
0	0	0
0	1	1
1	0	1
1	1	1

(ii)

Fig. 24.11

OR gate operation. Fig. 24.11 (*i*) shows two-input OR gate built by using two ideal diodes D_1 and D_2 and a resistor R_L. The input voltages are labelled as *A* and *B* while the output voltage is *Y*. Note that negative terminal of the battery is grounded and corresponds to 0 state (LOW level). The positive terminal of the battery (+ 5 V) corresponds to 1 state (HIGH level). There are only four input-output possibilities.

(*i*) When *A* = 0 and *B* = 0. This means that both diodes D_1 and D_2 are connected to ground (0V). Therefore, both the diodes are reverse biased and non-conducting. As a result, output voltage across R_L is zero *i.e.*, *Y* = 0 as shown in the truth table in Fig. 24.11 (*ii*).

(*ii*) When *A* = 0 and *B* = 1. This means that D_1 is connected to ground (0 V) and D_2 is connected to + 5 V. Therefore, D_1 does not conduct (reverse biased) while D_2 conducts (forward biased). As a result, full voltage of + 5 V appears across the load R_L. In other words, output is high *i.e.*, *Y* = 1 as shown in the truth table in Fig. 24.11 (*ii*).

(*iii*) When *A* = 1 and *B* = 0. This means that diode D_1 is connected to + 5 V and diode D_2 to ground (0 V). Therefore, D_1 is forward biased and conducts while D_2 is reverse biased and does not conduct. As a result, voltage across R_L is again + 5 V. In other words, output is high *i.e.*, *Y* = 1 as shown in the truth table in Fig. 24.11 (*ii*).

(*iv*) When *A* = 1 and *B* = 1. This means that both diodes D_1 and D_2 are connected to + 5 V. Therefore, both diodes are forward biased and conduct. Since the diodes are in parallel, the output voltage is + 5 V. In other words, the output is high *i.e.*, *Y* = 1 as shown in the truth table in Fig. 24.11 (*ii*).

It is clear from the truth table that *for OR gate, the output is high if any or all of the inputs are high. The only way to get a low output is by having all inputs low*. Fig. 24.12 shows the logic symbol of OR gate. Note that the symbol has curved line at the input.

Boolean expression. The algebra used to symbolically describe logic functions is called Boolean algebra. The "+" sign in Boolean algebra refers to the logical OR function. The Boolean expression for OR function is

$$A + B = Y$$

↑

OR symbol

Fig. 24.12

A + B	=	Y
0 + 0	=	0
0 + 1	=	1
1 + 0	=	1
1 + 1	=	1

This expression is read as *Y equals A or B* **not** *Y equals A plus B.*

The adjoining truth table to Fig. 24.12 shows how OR gate functions for the two inputs. It is clear from the truth table that when 0 is ORed with 0, the result equals 0. Also, any variable ORed with 1 equals 1. The OR function can be summed up as under :

0 ORed with 0 equals 0
0 ORed with 1 equals 1
1 ORed with 1 equals 1

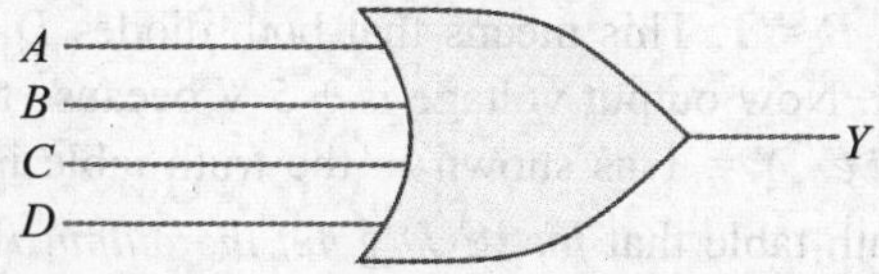

Fig. 24.13

An OR gate may have more than two inputs but there is only one output. Thus Fig. 24.13 shows four-input OR gate. The boolean expression for this four-input OR gate is

$$A + B + C + D = Y$$

Note. The number of possible input combinations (rows) in a truth table is equal to 2^n where n is the number of input variables. For example, an eight-input gate would require $2^8 = 256$ rows for its truth table.

24.20 AND Gate

***An AND gate** is a logic gate that has two or more inputs but only one output, it gives high output (1) when all the inputs are high (1), otherwise the output is low (0).*

It is called AND gate because output is HIGH only when all the inputs are HIGH. For this reason, the AND gate is sometimes called "all or nothing gate". For example, consider a 2-input AND gate. The output will be HIGH when both the inputs are HIGH.

AND gate operation. Fig. 24.14 (*i*) shows two-input AND gate built by using two ideal diodes D_1 and D_2 and a resistor R_L. There are four input-output possibilities.

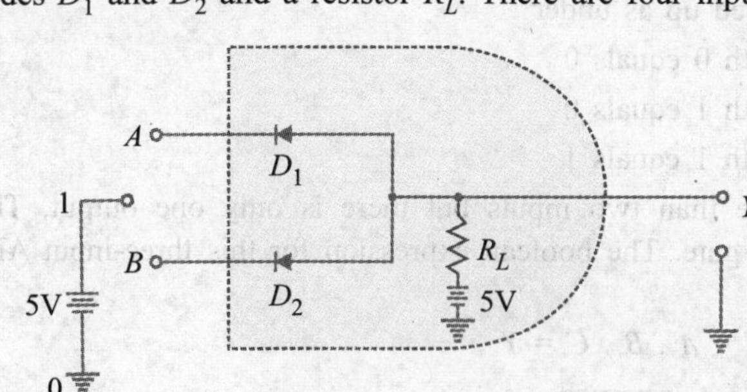

(*i*)

A	*B*	*Y*
0	0	0
0	1	0
1	0	0
1	1	1

(*ii*)

Fig. 24.14

(*i*) When *A* = 0 and *B* = 0. This means that both diodes D_1 and D_2 are connected to the ground (0 V). Both the diodes are forward biased and conduct current. As a result, the two diodes are shorted so that *Y* also gets grounded through these shorted diodes. Hence output is zero *i.e.*, *Y* = 0 as shown in the truth table in Fig. 24.14. (*ii*).

(*ii*) When $A = 0$ and $B = 1$. This means that diode D_1 is connected to ground (0 V) and diode D_2 is connected to + 5 V. Therefore, D_1 conducts (forward biased) and is grounded while D_2 does not conduct. Hence output is zero *i.e.*, $Y = 0$ as shown in the truth table in Fig. 24.14 (*ii*).

(*iii*) When $A = 1$ and $B = 0$. This means that diode D_1 is connected to + 5 V and diode D_2 is connected to ground (0 V). Therefore, the roles of diodes are interchanged as compared to the case (*ii*). Hence output is again zero *i.e.*, $Y = 0$ as shown in the truth table in Fig. 24.14 (*ii*).

(*iv*) When $A = 1$ and $B = 1$. This means that both diodes D_1 and D_2 are at + 5 V and do not conduct any current. Now output voltage is + 5 V because there is no current through R_L. Hence output is high *i.e.*, $Y = 1$ as shown in the truth table in Fig. 24.14 (*ii*).

It is clear from the truth table that *for AND gate, the output is high if all the inputs are high. However, the output is low if any or all the inputs are low.*

Fig. 24.15 shows the logic symbol of AND gate having two inputs A and B and output Y. The Boolean expression for two-input AND gate shown in Fig. 24.15 is

$$A \,.\, B = Y$$

↑

AND Symbol

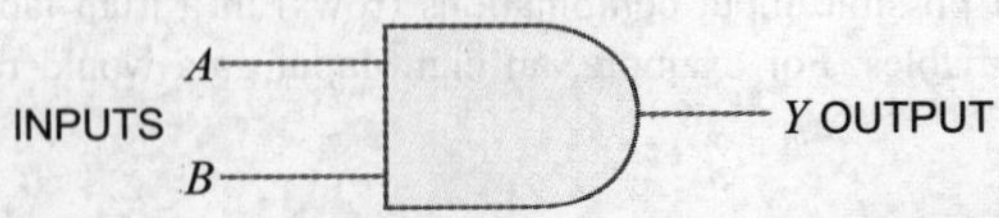

Fig. 24.15

The dot between A and B is the symbol for AND operation. This expression is read as "Y equals A and B". Quite often the dot is omitted and the expression is written as $AB = Y$. However, the expression $AB = Y$ is still read as Y equals A and B.

$A \,.\, B = Y$
$0 \,.\, 0 = 0$
$0 \,.\, 1 = 0$
$1 \,.\, 0 = 0$
$1 \,.\, 1 = 1$

The adjoining truth table shows how AND gate functions for two inputs. It is clear that 0 ANDed with any variable equals 0. Also 1 ANDed with 1 equals 1. The AND function can be summed up as under :

0 ANDed with 0 equals 0

0 ANDed with 1 equals 0

1 ANDed with 1 equals 1

An AND gate may have more than two inputs but there is only one output. Thus Fig. 24.16 shows three-input AND gate. The boolean expression for this three-input AND gate is

$$A \,.\, B \,.\, C = Y$$

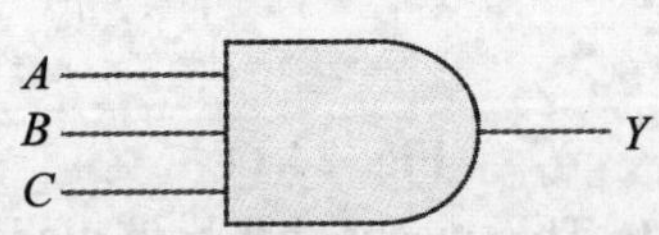

Fig. 24.16

Note that the output Y of AND gate will be high (1) when all of the inputs are high (1) otherwise the output will be low.

24.21 NOT Gate (Or Inverter)

A NOT gate is a logic gate with one input and one output. It gives high output (1) if the input is low (0) and vice-versa.

NOT gate operation. The NOT gate is often called *inverter* because it inverts the input *i.e.,* output is opposite of the input. Fig. 24.17 (*i*) shows the inverter or NOT circuit. Here R_K and R_B form a voltage divider between ground and the negative voltage ($-V$). The operation of NOT gate circuit is as under :

(*i*) When input $A = 0$. When A (the base of the transistor) is connected to ground, the base of the transistor Q_1 will become negative. This makes base-emitter junction reverse biased so that base current and hence collector current is zero. Therefore, output voltage is $+V$ volts. Hence the output is high *i.e.,* $Y = 1$ as shown in the truth table in Fig. 24.17 (*ii*).

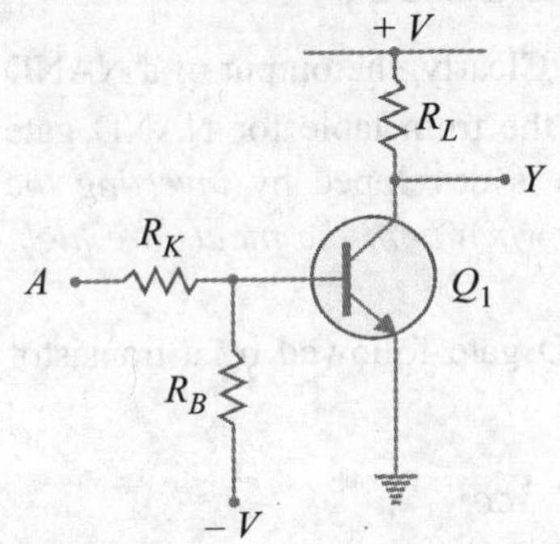

(*i*)

A	Y
0	1
1	0

(*ii*)

(*iii*)

Fig. 24.17

(*ii*) When input $A = 1$. If sufficiently large positive voltage is applied at A, the base-emitter junction and collector-base junction become forward biased and the transistor comes in the *saturated mode*. Under this condition, maximum collector current flows. Therefore, output voltage becomes zero. Hence the output is low *i.e.,* $Y = 0$ as shown in the truth table in Fig. 24.17 (*ii*). It is clear from the truth table that *for a NOT gate, the output is high (1) if input is low (0) and vice-versa.*

Fig. 24.17 (*iii*) shows the logic symbol for NOT gate or inverter. Note that small bubble on the inverter symbol represents inversion. The Boolean expression for NOT function is

$$Y = \overline{A}$$

Note that bar above the input A represents inversion.

If $A = 0$, then $Y = \overline{0}$ or $Y = 1$

If $A = 1$, then $Y = \overline{1}$ or $Y = 0$

24.22 Combination of Basic Logic Gates

The OR, AND and NOT gates are the three basic circuits that make up all digital circuits. We shall discuss a few combinations of these basic circuits.

(*i*) NAND gate. *When the output of an AND gate is connected to the input of a NOT gate, the resulting gate is called* **NAND gate** as shown in Fig. 24.18 (*i*). The Boolean expression for NAND function is

$$Y = \overline{A.B}$$

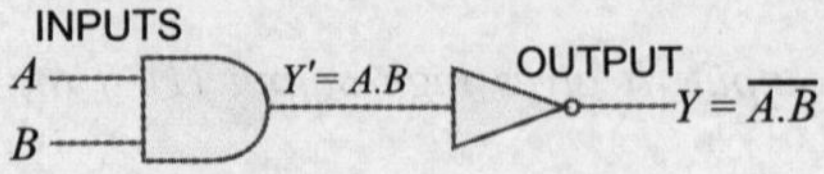

(i)

INPUTS		OUTPUT	
A	B	AND (Y′)	NAND (Y)
0	0	0	1
1	0	0	1
0	1	0	1
1	1	1	0

(ii)

(iii)

Fig. 24.18

To perform the Boolean algebra operation, first the inputs must be ANDed and then the inversion is performed. Fig. 24.18 (*iii*) shows the logic symbol for a NAND gate. The little bubble (small circle) on the right end of the symbol means to invert the AND.

Truth Table for NAND gate. For NAND gate, $Y = \overline{A.B}$. Clearly, the output of a NAND gate is opposite to that of AND gate. This fact is shown in the truth table for NAND gate in Fig. 24.18 (*ii*). Note that truth table for the NAND gate is developed by *inverting the outputs* of the AND gate. *The output of a NAND gate is low when both the inputs are high, otherwise the output is high.*

A NAND gate can be made by connecting a diode AND gate followed by a transistor inverter as shown in Fig. 24.19.

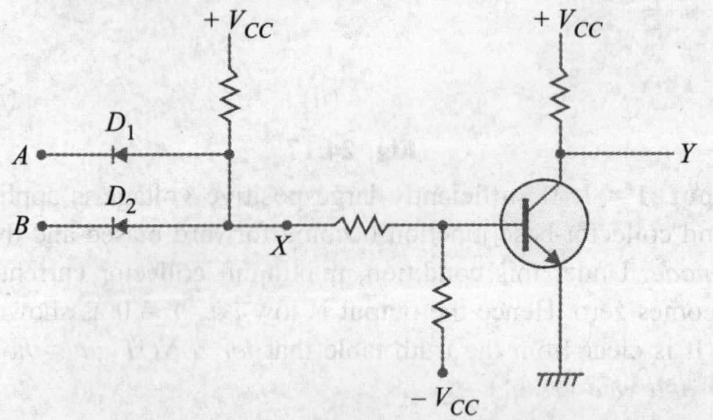

Fig. 24.19

It is clear from the figure that point X would be driven to ground when either of the diodes D_1 or D_2 or both D_1 and D_2 conduct. Under these conditions, the transistor is cut-off and the output Y becomes *high*. The output is *low* only when both input voltages are high (at $+ V_{CC}$) so that X is at $+ V_{CC}$ and transistor operates in saturation.

(*ii*) NOR gate. *When the output of an OR gate is connected to the input of a NOT gate, the resulting gate is called* **NOR** gate as shown in Fig. 24.20 (*i*). The Boolean expression for NOR function is

$$Y = \overline{A + B}$$

To perform the Boolean algebra operation, first the inputs must be ORed and then the inversion is performed. Fig. 24.20 (*iii*) shows the logic symbol for a NOR gate. The bubble (small circle) at the Y output indicates inversion.

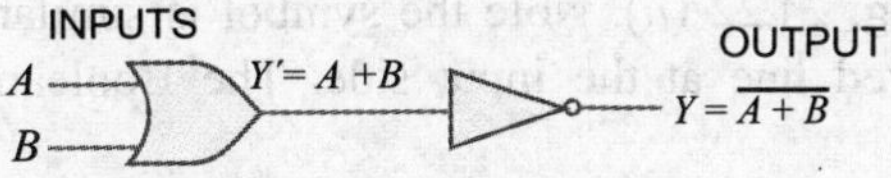

(i)

INPUTS		OUTPUT	
A	B	OR (Y′)	NOR (Y)
0	0	0	1
1	0	1	0
0	1	1	0
1	1	1	0

(ii)

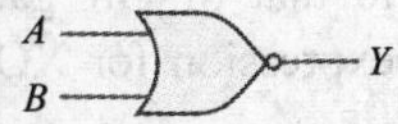

(iii)

Fig. 24.20

Truth table for NOR gate. For NOR gate, $Y = \overline{A+B}$. Clearly, the output of a NOR gate is opposite to that of OR gate. This fact is shown in the truth table for NOR gate in Fig. 24.20 (*ii*). Note that the truth table for NOR gate is developed by inverting the outputs of an OR gate. *The output of a NOR gate is high when both the inputs are low, otherwise the output is low.* A NOR gate can be made by connecting a diode OR gate followed by a transistor inverter as shown in Fig. 24.21.

It is clear from the figure that output *Y* from the NOR gate is *high* only when all the inputs are *low*. If any of the inputs is *high*, the output is *low*.

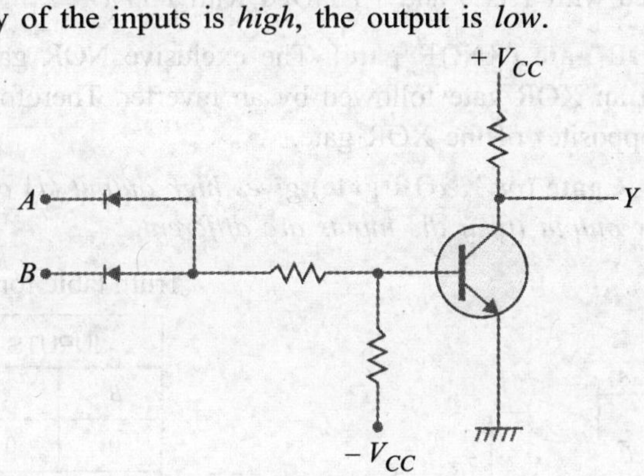

Fig. 24.21

(*iii*) Exclusive OR Gate (XOR Gate). Many digital systems require a circuit which *provides a high output (1) only when one input is high (1) and the other input is low (0).* This job can be done by **exclusive OR** gate usually shortened to XOR gate.

***The exclusive OR gate (or XOR gate)** gives high output (1) if either input A or input B is high but not when both inputs are high.*

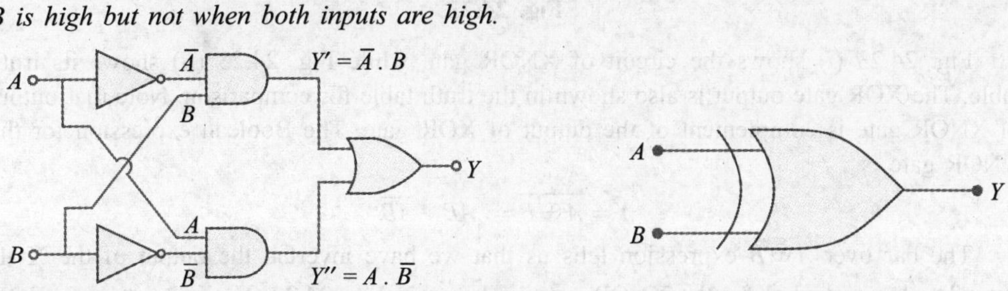

(i) (ii)

Fig. 24.22

The XOR gate is obtained by using NOT, AND and OR gates as shown in Fig. 24.22 (*i*). The logic symbol for XOR gate is shown in Fig. 24.22 (*ii*). Note the symbol is similar to that of OR gate except for the additional curved line at the input side. The Boolean expression for XOR gate is :

$$Y = \bar{A}.B + A.\bar{B}$$

or

$$Y = A \oplus B$$

Note the unique XOR symbol with a circle round the + symbol of OR. Read the expression as *A* exclusively ORed with *B* equals output *Y*.

The logic operations in the circuit are as under :

A	B	$\bar{A}$	$\bar{B}$	$\bar{A}.B = Y'$	$A.\bar{B} = Y''$	$Y = Y' + Y''$
0	0	1	1	0	0	0
1	0	0	1	0	1	1
0	1	1	0	1	0	1
1	1	0	0	0	0	0

Note that 0 ANDed with 1 is 0 and 1 ANDed with 1 is 1.

(*iv*) Exclusive NOR gate (XNOR gate). The exclusive NOR gate (usually shortened to XNOR gate) is just an XOR gate followed by an inverter. Therefore, the job of XNOR gate is complement (opposite) of the XOR gate.

The **exclusive NOR gate** (*or* **XNOR gate**) *gives high output (1) only if both inputs are same (0 or 1) and low output (0) if the inputs are different.*

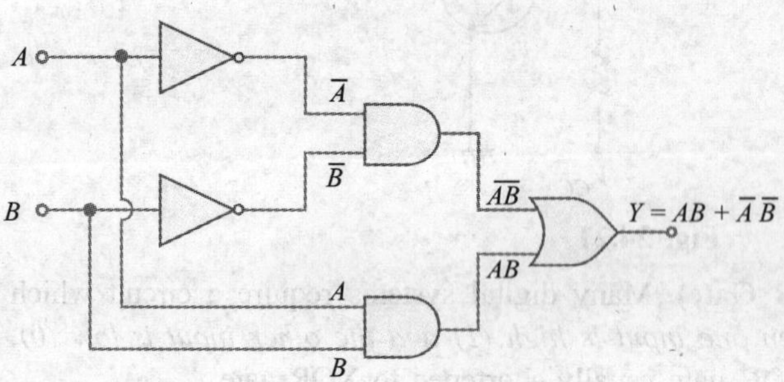

(*i*)

Truth table for XOR and XNOR gate

INPUTS		OUTPUT	
B	*A*	XOR	XNOR
0	0	0	1
0	1	1	0
1	0	1	0
1	1	0	1

(*ii*)

Fig. 24.23

Fig. 24.23 (*i*) shows the circuit of XNOR gate while Fig. 24.23 (*ii*) shows its truth table. The XOR gate output is also shown in the truth table for comparison. Note that output of XNOR gate is complement of the output of XOR gate. The Boolean expression for the XNOR gate is

$$Y = \overline{A \oplus B} = {}^{*}AB + \bar{A}\bar{B}$$

The bar over $A \oplus B$ expression tells us that we have inverted the output of the XOR gate. The logic symbol for the XNOR gate is shown in Fig. 24.24. Note that it is the XOR

* Boolean algebra (Refer to Arts. 24.29 and 24.30).

symbol with bubble at the output side and indicates the inversion of the XOR function. The XNOR gate is also called *coincidence gate* because the output is high (1) when its inputs coincide (either 0, 0 or 1, 1).

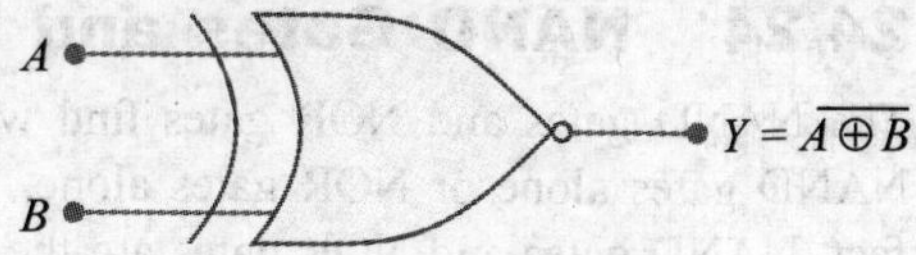

Symbol for XNOR gate

Fig. 24.24

24.23 Summary of Basic Logic Gates

Fig. 24.25 shows the summary of *seven basic* logic gates. This information should be memorised.

LOGIC FUNCTION	LOGIC SYMBOL	BOOLEAN EXPRESSION	TRUTH TABLE INPUTS B	INPUTS A	OUTPUT Y
AND	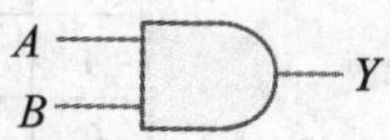	$A . B = Y$	0	0	0
			0	1	0
			1	0	0
			1	1	1
OR		$A + B = Y$	0	0	0
			0	1	1
			1	0	1
			1	1	1
Inverter		$A = \overline{A}$		0	1
				1	0
NAND	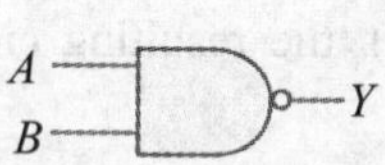	$\overline{A . B} = Y$	0	0	1
			0	1	1
			1	0	1
			1	1	0
NOR	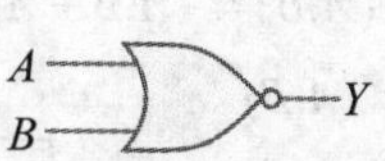	$\overline{A + B} = Y$	0	0	1
			0	1	0
			1	0	0
			1	1	0
XOR	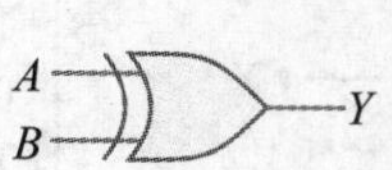	$A \oplus B = Y$	0	0	0
			0	1	1
			1	0	1
			1	1	0
XNOR	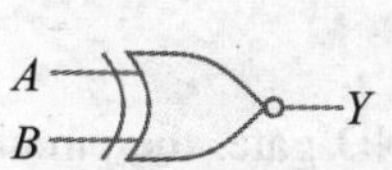	$\overline{A \oplus B} = Y$	0	0	1
			0	1	0
			1	0	0
			1	1	1

Fig. 24.25

24.24 NAND Gates and NOR Gates as Universal Gates

The NAND gates and NOR gates find wide applications in digital circuits. By connecting NAND gates alone or NOR gates alone, we can produce any of other basic logic gates. In fact, NAND gates and NOR gates are the building blocks in digital circuits. For this reason, NAND gates and NOR gates are called *universal gates*.

24.25 Logic Gates Using NAND Gates Only

By using NAND gates only, we can produce any of the other logic gates. By way of illustration, we now show how NAND gates alone can be wired to produce NOT gate, AND gate and OR gate.

(*i*) NOT gate from NAND gate. When the two inputs of a NAND gate are joined together so that it has one input, the resulting circuit is NOT gate [See Fig. 24.26 (*i*)]. It is because $A = B$ so that $Y = \overline{A.A} = {}^{*}\overline{A} + \overline{A} = \overline{A}$.

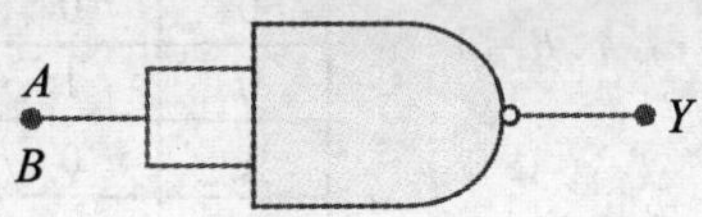

(*i*)

A	B	Y
0	0	1
1	1	0

(*ii*)

Fig. 24.26

In the truth table of NAND gate if we put $B = A$, we get the truth table of NOT gate as shown in Fig. 24.26 (*ii*). Therefore, the circuit shown in Fig. 24.26 (*i*) works as a NOT gate.

(*ii*) AND gate from NAND gates. For this purpose, we use two NAND gates. If the output of first NAND gate (I) is used as the input to the second NAND gate (II) acting as an inverter or NOT gate (*i.e.,* its two inputs joined), the resulting circuit is the AND gate [See Fig. 24.27 (*i*)]. This is proved as under :

$$\text{Output of gate I, } Y' = \overline{A.B}$$

$$\text{Output of gate II, } Y = \overline{Y'.Y'} = \overline{(\overline{A.B}).(\overline{A.B})} = {}^{*}\overline{\overline{A.B}} + \overline{\overline{A.B}}$$

$$= (A.B) + (A.B) = (A.B)$$

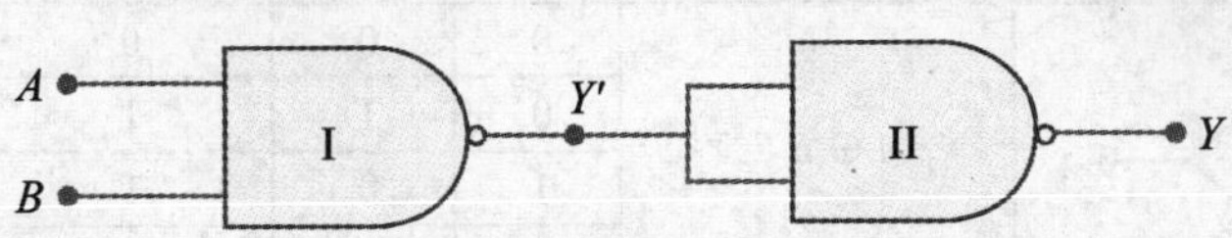

(*i*)

A	B	Y'	Y
0	0	1	0
1	0	1	0
0	1	1	0
1	1	0	1

(*ii*)

Fig. 24.27

Since $Y = A.B$ is the Boolean expression for AND gate, the circuit shown in Fig. 24.27 (*i*) is AND gate. Its truth table is shown in Fig. 24.27 (*ii*).

(*iii*) OR gate from NAND gates. For this purpose, we use three NAND gates ; the

* Boolean algebra (Refer to Arts. 24.29 and 24.30).

first two NAND gates (I and II) are operated as NOT gates and their outputs are fed to the third NAND gate (III). The resulting circuit is OR gate [See Fig. 24.28 (*i*)]. This is proved as under :

$$\text{Output of gate I} = \overline{A.A} = {}^{*}\overline{A}+\overline{A} = \overline{A}$$

$$\text{Output of gate II} = \overline{B.B} = {}^{*}\overline{B}+\overline{B} = \overline{B}$$

$$\text{Output of gate III, } Y = \overline{\overline{A}.\overline{B}} = {}^{*}\overline{\overline{A}}+\overline{\overline{B}} = A+B$$

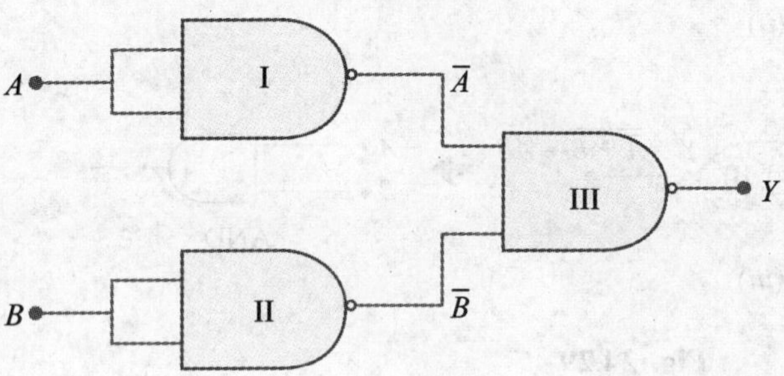

(*i*)

A	B	$\overline{A}$	$\overline{B}$	Y
0	0	1	1	0
0	1	1	0	1
1	0	0	1	1
1	1	0	0	1

(*ii*)

Fig. 24.28

Since $Y = A + B$ is the Boolean expression for OR gate, the circuit shown in Fig. 24.28 (*i*) is OR gate. Its truth table is shown in Fig. 24.28 (*ii*).

24.26 Logic Gates Using NOR Gates only

By using NOR gates only, we can produce any of the other logic gates. By way of illustrating, we now shown how NOR gates alone can be wired to produce NOT gate, OR gate and AND gate.

***(i)* NOT gate from NOR gate.** Fig. 24.29 (*i*) shows that a NOR gate with its inputs connected together behaves as a NOT gate since $Y = \overline{A+A} = {}^{*}\overline{A}$.

***(ii)* OR gate from NOR gates.** Fig. 24.29 (*ii*) shows how two NOR gates are connected to produce an OR gate. The NOR gate II is used as an inverter to change $\overline{A+B}$ to $\overline{\overline{A+B}} = {}^{*}A+B$ which is OR function.

***(iii)* AND gate from NOR gates.** Fig. 24.29 (*iii*) shows how three NOR gates are connected to produce the AND gate. Here, NOR gates I and II are used as inverters. The final output is

* Boolean algebra (Refer to Arts. 24.29 and 24.30).

$$Y = \overline{\overline{A} + \overline{B}} = {}^{*}A.B. \text{ This is AND function.}$$

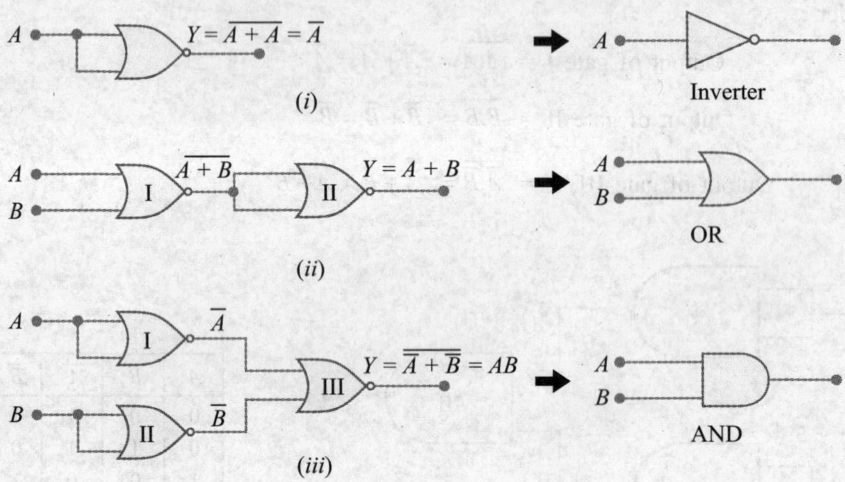

Fig. 24.29

Example 24.24. *Obtain the truth table for the circuit shown in Fig. 24.30 (i).*

Solution. Figure 24.30 (*ii*) shows the truth table for the circuit. The truth table can be obtained very easily if the reader remembers the following simple Boolean operations :

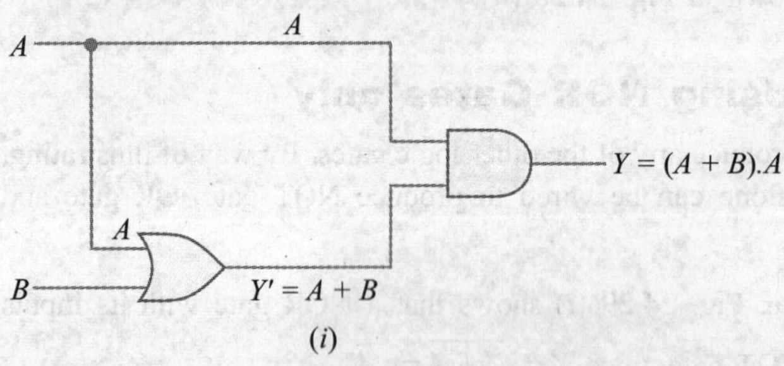

A	B	$Y' = A + B$	$Y = Y'.A$
0	0	0	0
1	0	1	1
0	1	1	0
1	1	1	1

(*ii*)

Fig. 24.30

(*i*) ***ORed with 0 = 0 ; 1 ORed with 1 = 1 ; 1 ORed with 0 = 1.*

(*ii*) *0 !ANDed with 0 = 0 ; 0 ANDed with 1 = 0 ; 1 ANDed with 1 = 1.*

Thus, when $A = 0$ and $B = 0$, then A ORed with $B = 0$ *i.e.,* $Y' = 0$. When $Y'(= 0)$ is ANDed with A (= 0), the result is 0. Again when $A = 1$ and $B = 0$, then A ORed with B is 1 *i.e.,* $Y' = 1$. Now Y' (= 1) ANDed with A (= 1), the result is 1.

* Boolean algebra (Refer to Arts. 24.29 and 24.30).

** Note that $A + B$ means A ORed with B.

! Note that $A.B$ means A ANDed with B.

Example 24.25. *Obtain the truth table for the circuit shown in Fig. 24.31.*

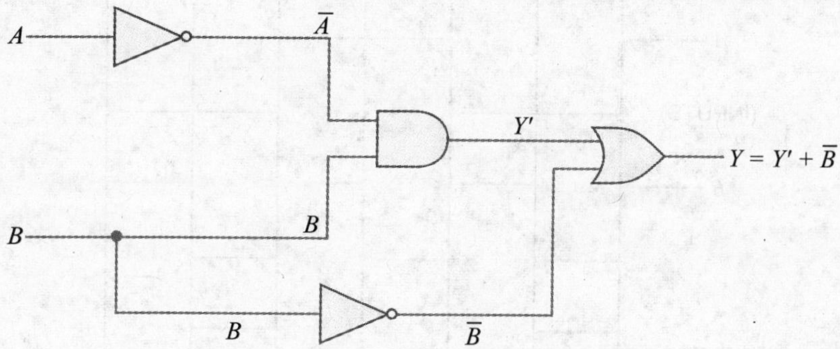

Fig. 24.31

Solution. The truth table for the circuit is shown below :

A	B	$\overline{A}$	$Y' = \overline{A}.B$	$\overline{B}$	$Y = Y' + \overline{B}$
0	0	1	0	1	1
1	0	0	0	1	1
0	1	1	1	0	1
1	1	0	0	0	0

(*i*) When $A = 0$ and $B = 0$, then $\overline{A} = 1$. Now Y' is equal to $\overline{A}$ (= 1) ANDed with B (= 0). The result is 0. Then Y' (= 0) ORed with $\overline{B}$ (= 1) is 1 *i.e.*, $Y = 1$.

(*ii*) When $A = 1$ and $B = 0$, then $\overline{A} = 0$. Now Y' is equal to $\overline{A}$ (= 0) ANDed with B (= 0) and the result is 0 *i.e.*, $Y' = 0$. Then Y' (= 0) ORed with $\overline{B}$ (= 1) is 1 *i.e.*, $Y = 1$. The reader can proceed in a similar way to find the other output values.

Example 24.26. *Identify the logic gates marked X, Y in Fig. 24.32. Write down the output at y, when A = 1, B = 1 and A = 0, B = 0.*

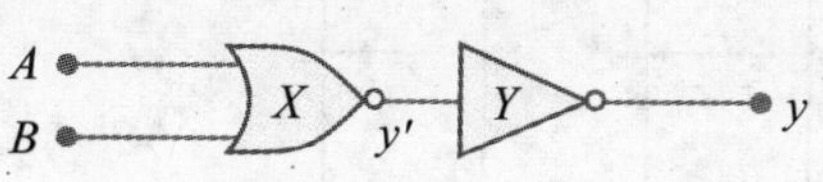

A	B	y'	y
1	1	0	1
0	0	1	0

Fig. 24.32

Solution. The logic gate marked as X is NOR gate while the gate marked as Y is NOT gate. It is clear that :

$$y' = \overline{A + B} \text{ and } y = \overline{y'}$$

Therefore, the output at y will be as shown in the truth table.

Example 24.27. *Sketch the output (Y) waveform from a NOR gate for the input A and B shown in Fig. 24.33.*

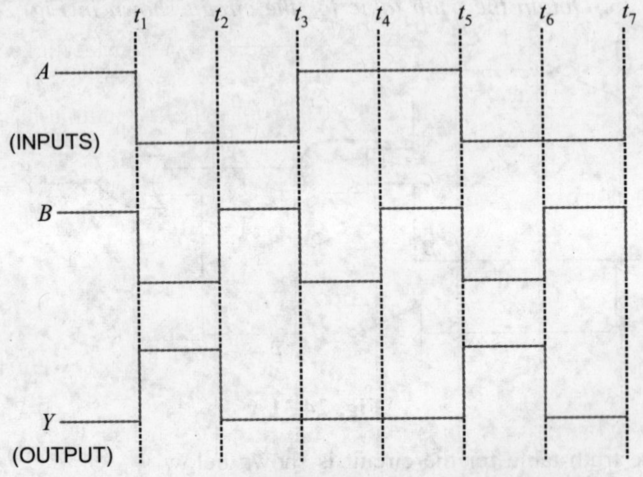

Fig. 24.33

Solution. The output from a NOR gate is high (1) only when all the inputs are low (0). If any of the inputs is high (1), the output is low (0).

For $t < t_1$; $A = 1$, $B = 1$; Hence $Y = 0$.

For t_1 to t_2 ; $A = 0$, $B = 0$; Hence $Y = 1$.

For t_2 to t_3 ; $A = 0$, $B = 1$; Hence $Y = 0$.

For t_3 to t_4 ; $A = 1$, $B = 0$; Hence $Y = 0$.

For t_4 to t_5 ; $A = 1$, $B = 1$; Hence $Y = 0$.

For t_5 to t_6 ; $A = 0$, $B = 0$; Hence $Y = 1$.

For t_6 to t_7 ; $A = 0$, $B = 1$; Hence $Y = 0$.

Example 24.28. *Sketch the output (Y) waveform from a NOT gate having input A as shown in Fig. 24.34 (i).*

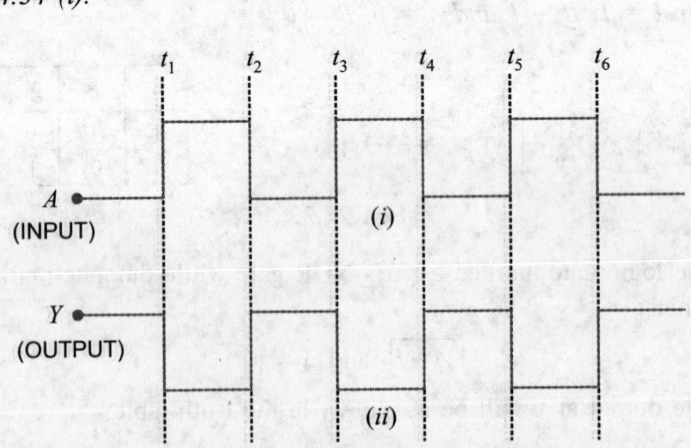

Fig. 24.34

Solution. The output (Y) of a NOT gate is high (1) when input is low (0) and *vice-versa* ($Y = \overline{A}$).

For $t \leq t_1$; $A = 0$, Hence $Y = 1$.

For t_1 to t_2 ; $A = 1$, Hence $Y = 0$.

For t_2 to t_3 ; $A = 0$, Hence $Y = 1$.

For t_3 to t_4 ; $A = 1$, Hence $Y = 0$.

For t_4 to t_5 ; $A = 0$, Hence $Y = 1$.

For t_5 to t_6 ; $A = 1$, Hence $Y = 0$.

For $t > t_6$; $A = 0$, Hence $Y = 1$.

The waveform of the output Y for the given waveform will be as shown in Fig. 24.34 (*ii*).

Example 24.29. *Realise XOR gate using NAND gates.*

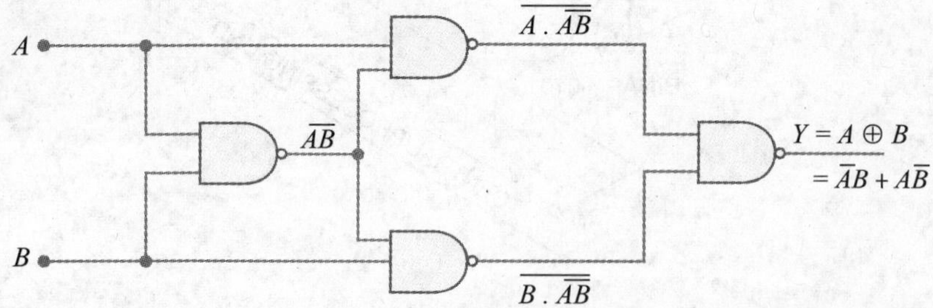

Fig. 24.35

Solution. The XOR gate can be produced by using four NAND gates as shown in Fig. 24.35.

24.27 Encoders and Decoders

A digital circuit can process numbers in binary form. However, most of the information we handle is in decimal form. Therefore, a digital machine must perform the following functions :

(*i*) Convert the information from decimal to digital (binary) form.

(*ii*) Process the digital information.

(*iii*) Convert the digital output back to decimal form.

The circuit that converts decimal form to digital (binary) form is called **encoder** and the circuit that converts digital form to decimal form is called **decoder**. Fig. 24.36 shows encoding and decoding in a digital calculator. Here the input is the decimal number 5 punched in at the keyboard. The encoder changes the decimal number 5 to the digital form as the binary digit 0101. The central processing unit (CPU) contains digital logic circuits for necessary calculations. Here all operations are carried out in binary form. The output of CPU is fed to the decoder which changes the binary signal back to the decimal form. The output display is in the decimal form, showing the original number 5.

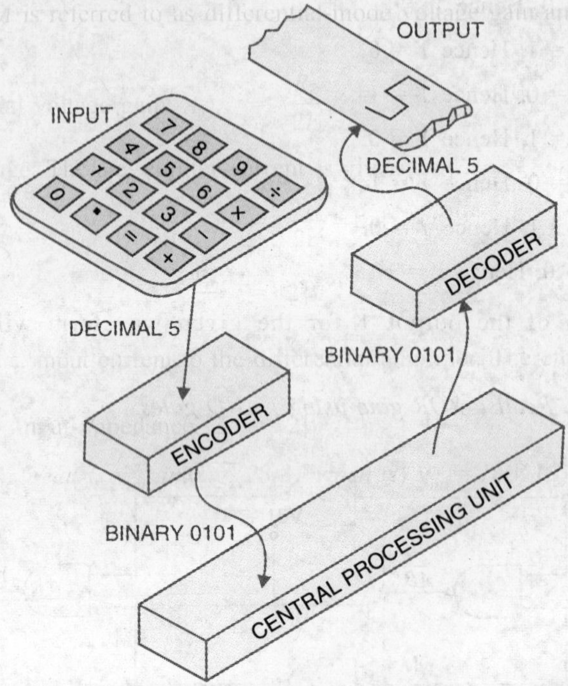

Fig. 24.36

24.28 Advantages and Disadvantages of Digital Electronics

The world of electronics can be classified as either digital or analog circuits. An increasing majority of applications in electronics use digital techniques to perform operations that were once performed using analog methods. It is worthwhile to give advantages and disadvantages of digital electronics.

Advantages. The chief reasons for the shift to digital technology are :

(*i*) *Digital systems are generally easier to design.* It is because the circuits that are used are *switching circuits* where *exact* values of voltages or currents are not important, only the range (HIGH or LOW) in which they fall is important.

(*ii*) *Digital circuits provide greater accuracy and precision.* It is because digital circuits can handle as many digits of precision as you need simply by adding more switching circuits. In analog systems, precision is usually limited to three or four digits because the values of voltage and current are directly dependent on the circuit components.

(*iii*) *Digital circuits are less affected by noise.* Spurious fluctuations in voltage (noise) are not as critical in digital systems as in analog systems. It is because in a digital circuit, the exact value of a voltage is not important as long as the noise is not large enough to prevent us from distinguishing a HIGH from a LOW.

(*iv*) *More digital circuitry can be fabricated on IC chips.* Analog system uses such devices (high-value capacitors, inductors, transformers) that cannot be economically integrated. For this reason, analog systems cannot achieve the same degree of integration as digital circuits.

(*v*) Information storage is easy with digital circuits.

Disadvantages. (*i*) The real world is mainly analog. However, the digital circuits can handle only digital signals. This necessitates encoders and decoders which increase the cost of the equipment.

(*ii*) There are situations where using only analog techniques is simpler and more economical. For example, the process of signal amplification is most easily accomplished using analog circuitry.

However, advantages of digital techniques outweigh the disadvantages. For this reason, we are fast switching to digital techniques.

24.29 Boolean Algebra

Digital circuits perform the binary arithmetic operations with binary digits 1 and 0. These operations are called logic functions or logical operations. *The algebra used to symbolically describe logic functions is called* **Boolean algebra.** Boolean algebra is a set of rules and theorems by which logical operations can be expressed symbolically in equation form and be manipulated mathematically. As with the ordinary algebra, the *letters of alphabet (*e.g.* A, B, C *etc.*) can be used to represent the variables. Boolean algebra differs from ordinary algebra in that Boolean constants and variables can have only two values ; 0 and 1. There are four connecting symbols used in Boolean algebra *viz.*

(*i*) equals sign (=) (*ii*) plus sign (+)
(*iii*) multiply sign (·) (*iv*) bar (–)

(*i*) **Equals sign** (=). The equals sign in Boolean algebra refers to the standard mathematical equality. In other words, the logical value on one side of the sign is identical to the logical value on the other side of the sign. Suppose we are given two logical variables such that $A = B$. Then if $A = 1$, then $B = 1$ and if $A = 0$, then $B = 0$.

(*ii*) **Plus sign** (+). The plus sign in Boolean algebra refers to the logical *OR operation.* Thus, when the statement $A + B = 1$ appears in Boolean algebra, it means A ORed with B equals 1. Consequently, either $A = 1$ or $B = 1$ or both equal 1.

(*iii*) **Multiply sign** (·). The multiply sign in Boolean algebra refers to *AND operation.* Thus, when the statement $A \cdot B = 1$ appears in Boolean algebra, it means A ANDed with B equals 1. Consequently, $A = 1$ and $B = 1$. The function $A \cdot B$ is often written as AB, omitting the dot for convenience.

(*iv*) **Bar sign** (–). The bar sign in Boolean algebra refers to *NOT operation.* The NOT has the effect of inverting (complementing) the logical value. Thus, if $A = 1$, then $\bar{A} = 0$.

24.30 Boolean Theorems

We now discuss the basic Boolean theorems that are useful in manipulating and simplifying Boolean expressions. For convenience, we divide the theorems into two groups :

(*i*) Single variable theorems

(*ii*) Multivariable theorems

* For example, A might represent a certain digital circuit input or output and at any time, we must have either $A = 0$ or $A = 1$.

(*i*) **Single variable theorems.** These theorems refer to the condition when only one input to the logic gate is variable. Table 24.1 gives single variable Boolean theorems.

Table 24.1

Theorem 1 :	$A + 0 = A$
Theorem 2 :	$A \cdot 1 = A$
Theorem 3 :	$A + \overline{A} = 1$
Theorem 4 :	$A \cdot \overline{A} = 0$
Theorem 5 :	$A + A = A$
Theorem 6 :	$A \cdot A = A$
Theorem 7 :	$A + 1 = 1$
Theorem 8 :	$A \cdot 0 = 0$
Theorem 9 :	$\overline{\overline{A}} = A$

Theorem 1. ($A + 0 = A$). This theorem can be verified by ORing a variable A with a 0 and is illustrated in Fig. 24.37. Here one input to OR gate is always 0 and the other input A can be a value 1 or 0. When A is at 1, the output is 1 which is equal to A. When A is at 0, the output is 0 which is also equal to A (= 0). Therefore, a variable ORed with 0 is equal to the value of the variable. This is easy to remember since 0 added to anything does not effect the value of the variable, either in regular addition or OR addition.

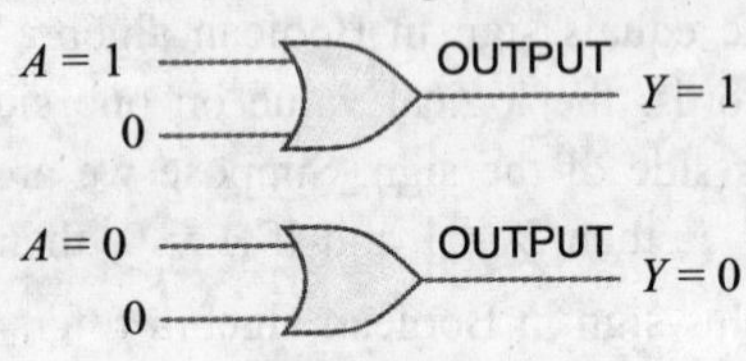

Fig. 24.37

Theorem 2. ($A \cdot 1 = A$). This theorem can be verified by ANDing a variable A with a 1 and is illustrated in Fig. 24.38. Here one input to AND gate is always 1 and the other can be a value 1 or 0. If A is 1, the output of the AND gate is 1 because both the inputs are now 1's. If A is 0, the output of the AND gate is a 0. Therefore, a variable ANDed with a 1 is equal to the value of the variable ($A \cdot 1 = A$). This is easy to remember because AND operation is just like ordinary multiplication.

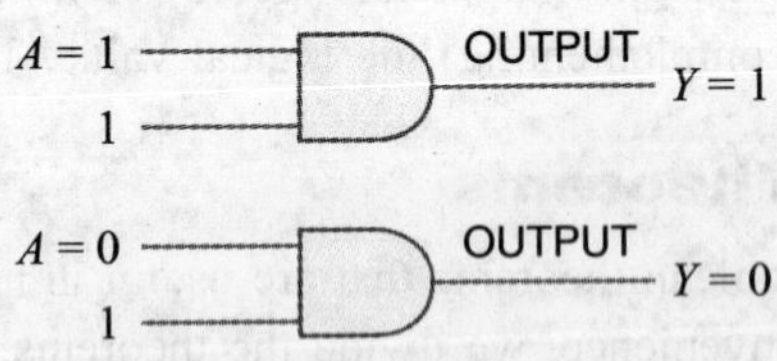

Fig. 24.38

Theorem 3. ($A + \overline{A} = 1$). This theorem can be easily explained. If a variable A and its complement ($\overline{A}$) are ORed, the result is always 1. If A is a 0, then $0 + \overline{0} = 0 + 1 = 1$. If A is a 1, then $1 + \overline{1} = 1 + 0 = 1$. Fig. 24.39 illustrates this theorem.

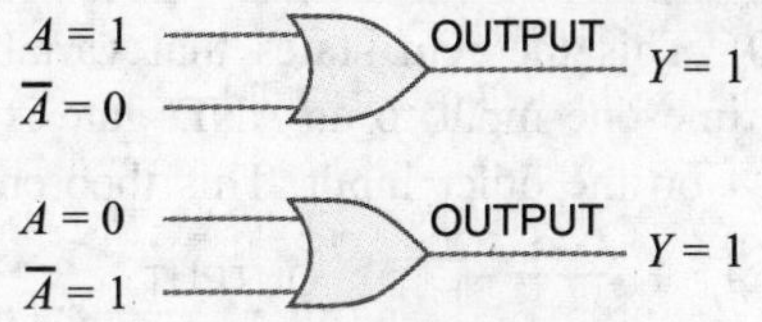

Fig. 24.39

Theorem 4. ($A \cdot \overline{A} = 0$). This theorem states that if a variable A is ANDed with its complement, the result is zero. This is readily apparent because either A or $\overline{A}$ will always be 0. Therefore, when one of the inputs to an AND gate is 0, the output is always 0. Fig. 24.40 illustrates this theorem.

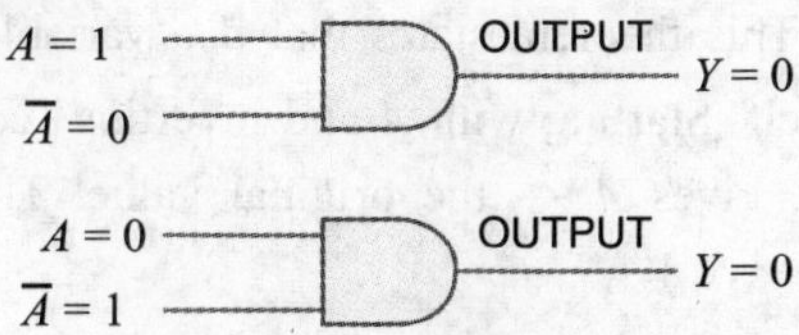

Fig. 24.40

Theorem 5. ($A + A = A$). This theorem states that when a variable A is ORed with itself, the output is equal to the variable. Thus, if A is a 0, then $0 + 0 = 0$ and if A is a 1, then $1 + 1 = 1$. Fig. 24.41 illustrates this theorem.

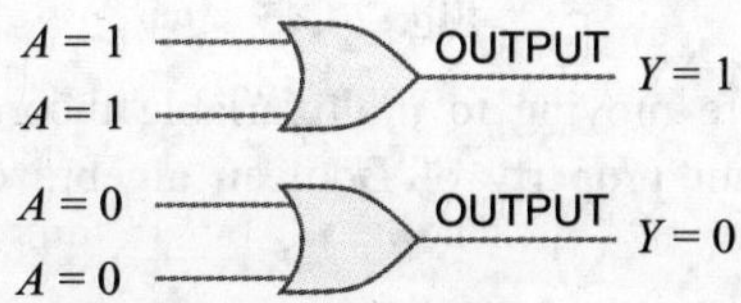

Fig. 24.41

Theorem 6. ($A \cdot A = A$). This theorem states that if a variable A is ANDed with itself, the result is equal to the variable. For example, if $A = 0$, then $0 \cdot 0 = 0$ and if $A = 1$, then $1 \cdot 1 = 1$. For either case, the output of an AND gate is equal to the value of the input variable A. Fig. 24.42 illustrates this theorem.

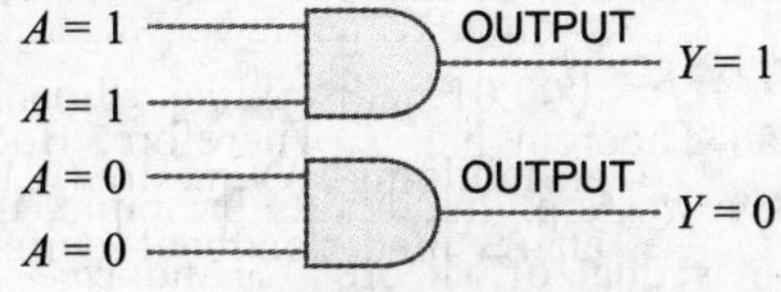

Fig. 24.42

Theorem 7. ($A + 1 = 1$). This theorem states that when a variable A is ORed with 1, the outuput is always equal to 1. Fig. 24.43 illustrates this theorem. One input to an OR gate is always 1 and the other input A can be either 1 or 0. Now 1 on an input to OR gate produces 1 on the output regardless of the value of the variable on the other input.

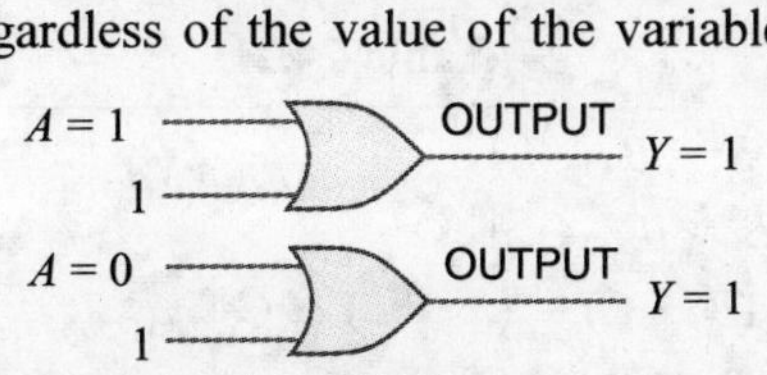

Fig. 24.43

Theorem 8. ($A \cdot 0 = 0$). This theorem states that variable A ANDed with 0 always produces 0. Recall that any time one input to an AND gate is 0, the output is 0 regardless of the value of the variable A on the other input. This theorem is illustrated in Fig. 24.44.

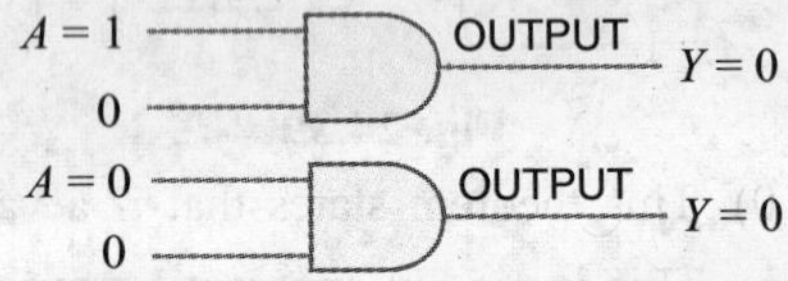

Fig. 24.44

Theorem 9. ($\overline{\overline{A}} = A$). This theorem states that if a variable A is complemented twice, the result is the variable itself. Starting with A and inverting (complementing) it once gives $\overline{A}$. Inverting it once more gives A — the original value. This theorem is illustrated in Fig. 24.45.

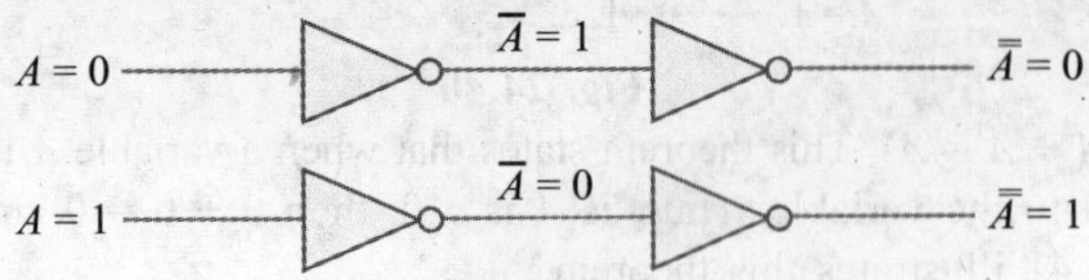

Fig. 24.45

Duality Principle. Before moving to multivariable theorems, this would be the right place to mention an important property of Boolean algebra called *duality principle*. It is stated below :

The duality principle states that a Boolean expression remains valid if operators OR and AND are interchanged and 1's and 0's in the expression are also interchanged.

In order to understand this principle, consider the Boolean Theorem 1 *viz.*

$$A + 0 = A$$

According to duality principle, this Boolean expression remains valid if OR function is replaced by AND function and 0 by 1. In that case, the Boolean expression becomes :

$$A \cdot 1 = A$$

Note that this is Boolean Theorem No. 2. Therefore, Boolean Theorem 2 is dual of Boolean Theorem 1 and *vice-versa*. Applying duality principle, Theorem 4 is dual of Theorem 3 and *vice-versa,* Theorem 6 is dual of Theorem 5 and *vice-versa*, Theorem 8 is dual of Theorem 7 and *vice-versa*. To apply duality principle to a Boolean expression, we simply interchange OR and AND operator and replace 1's by 0's and 0's by 1's.

(*ii*) Multivariable theorems. These theorems refer to the condition when more than one input to the logic gate are variable. Table 24.2 gives multivariable Boolean theorems.

Table 24.2

Theorem 10 :	$A + B = B + A$	*Commutative Law*
Theorem 11 :	$A \cdot B = B \cdot A$	
Theorem 12 :	$A + (B + C) = (A + B) + C$	*Associative Law*
Theorem 13 :	$A \cdot (B \cdot C) = (A \cdot B) \cdot C$	

Theorem 14 :	$A \cdot (B + C) = A \cdot B + A \cdot C$	Distributive Law
Theorem 15 :	$(A + B) \cdot (C + D) = A \cdot C + B \cdot C + A \cdot D + B \cdot D$	
Theorem 16 :	$A + A \cdot B = A$	De Morgan's Theorems
Theorem 17 :	$\overline{(A + B)} = \overline{A} \cdot \overline{B}$	
Theorem 18 :	$\overline{(A \cdot B)} = \overline{A} + \overline{B}$	

The following points may be noted about these theorems :

(*a*) Theorems 10 and 11 obey **commutative law.** *This law states that the order in which the variables are ORed or ANDed makes no difference.*

Fig. 24.46

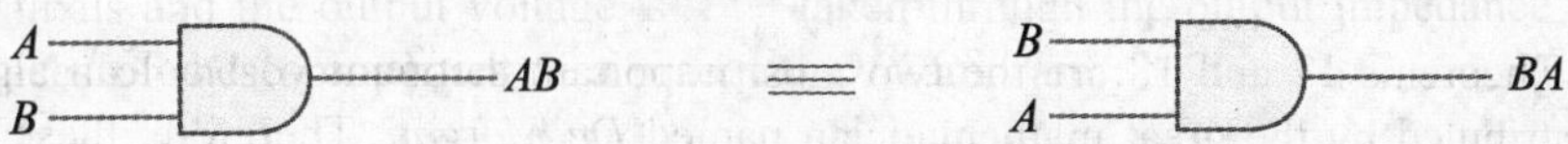

Fig. 24.47

Figure 24.46 illustrates the commutative law as applied to the OR gate while Fig. 24.47 illustrates the commutative law as applied to an AND gate.

(*b*) Theorems 12 and 13 obey **associative law.** *This law states that in the ORing or ANDing of several variables, the result is the same regardless of the grouping of the variables.*

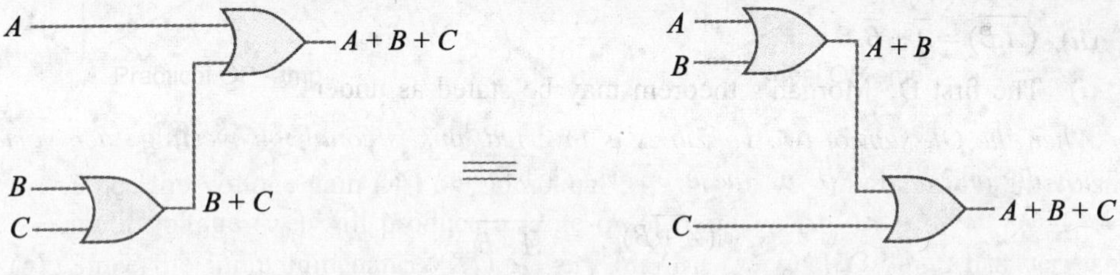

Fig. 24.48

Fig. 24.49

Figure 24.48 illustrates the associative law as applied to the OR gate while Fig. 24.49 illustrates the associative law as applied to an AND gate.

(*c*) Theorems 14 and 15 obey **distributive law.** *This law states that a Boolean expression can be expanded by multiplying term-by-term just the same as in ordinary algebra.*

Fig. 24.50

Fig. 24.50 illustrates the distributive law in terms of gate implementation.

(*d*) We will prove Theorem 16 by factoring and using Theorems 2, 7, 10 and 14.

$$A + A \cdot B = A \cdot 1 + A \cdot B \quad \text{...Theorem 2}$$
$$= A \cdot (1 + B) \quad \text{...Theorem 14}$$
$$= A \cdot (B + 1) \quad \text{...Theorem 10}$$
$$= A \cdot 1 \quad \text{...Theorem 7}$$
$$= A \quad \text{...Theorem 2}$$

(*e*) Theorems 17 and 18 are the two most important theorems of Boolean algebra and were contributed by the great mathematician named *De Morgan*. Therefore, these theorems are called De Morgan's theorems.

24.31 De Morgan's Theorems

De Morgan's theorems are extremely useful in simplifying expressions in which a product or sum of variables is inverted. The two theorems are :

(*i*) $\overline{(A+B)} = \overline{A} \cdot \overline{B}$

(*ii*) $\overline{(A.B)} = \overline{A} + \overline{B}$

(*i*) The first De Morgan's theorem may be stated as under :

When the OR sum of two variables is inverted, this is equal to inverting each variable individually and then ANDing these inverted variables i.e.,

$$\overline{(A+B)} = \overline{A \cdot B}$$

In this expression, *A* and *B* are the two variables. The L.H.S. is the complement of the OR sum of the two variables. The R.H.S. is the AND product of individual inverted variables.

(*ii*) The second De Morgan's theorem may be stated as under :

When the AND product of two variables is inverted, this is equal to inverting each variable individually and then ORing them i.e.,

$$(\overline{A \cdot B}) = \overline{A} + \overline{B}$$

In this expression, *A* and *B* are the two variables. The L.H.S. is the complement of the AND product of the two variables. The R.H.S. is the OR sum of the individual inverted variables.

24.32 Operator Precedence

The operator precedence for evaluating Boolean expression is (*i*) parenthesis (*ii*) NOT (*iii*) AND and (*iv*) OR. In other words, the expression inside the parenthesis must be evaluated before all other operations. The next operation that holds precedence is the complement, then follows the AND and finally the OR. For example, consider the Boolean expression :

$$A+\overline{B}\,.\,(C+D)$$

The sequence of operations will be :

(*i*) The expression inside the parenthesis (*i.e.*, $C + D$) will be evaluated first.

(*ii*) Then $\overline{B}$ will be evaluated.

(*iii*) Then the results of the two (*i.e.*, $\overline{B}$ and $C + D$) will be ANDed.

(*iv*) Finally, the result of the product will be ORed with A.

Example 24.30. *Using Boolean algebraic techniques, simplify the following expression :*

$$Y = A\cdot B\cdot\overline{C}\cdot\overline{D}+\overline{A}\cdot B\cdot\overline{C}\cdot\overline{D}+\overline{A}\cdot B\cdot C\cdot\overline{D}+A\cdot B\cdot C\cdot\overline{D}$$

Solution. $$Y = A\cdot B\cdot\overline{C}\cdot\overline{D}+\overline{A}\cdot B\cdot\overline{C}\cdot\overline{D}+\overline{A}\cdot B\cdot C\cdot\overline{D}+A\cdot B\cdot C\cdot\overline{D} \quad ...(i)$$

Step 1 : Take out the common factors as below :

$$Y = B\,\overline{C}\,\overline{D}\,(A+\overline{A})+B\,C\,\overline{D}\,(A+\overline{A})$$

Step 2 : Apply Theorem 3 $(A+\overline{A}=1)$:

$$Y = B\,\overline{C}\,\overline{D}+B\,C\,\overline{D}$$

Step 3 : Again factorise :

$$Y = B\,\overline{D}\,(C+\overline{C})$$

Step 4 : Apply Theorem 3 $(C+\overline{C}=1)$:

$$Y = B\,\overline{D}\cdot 1 = B\overline{D}$$

This is the simplified form of exp. (*i*).

Example 24.31. *Using Boolean techniques, simplify the following expression* :

$$Y = AB + A\,(B + C) + B\,(B + C)$$

Solution. $$Y = AB + A\,(B + C) + B\,(B + C) \quad ...(i)$$

Step 1 : Apply Theorem 14 (distributive law) to second and third terms :

$$Y = AB + AB + AC + BB + BC$$

Step 2 : Apply Theorem 6 $(B\cdot B = B)$:

$$Y = AB + AB + AC + B + BC$$

Step 3 : Apply Theorem 5 $(AB + AB = AB)$:

$$Y = AB + AC + B + BC$$

Step 4 : Factor B out of last 2 terms :

$$Y = AB + AC + B\,(1 + C)$$

Step 5 : Apply commutative law and Theorem 7 $(1 + C = C + 1 = 1)$:

$$Y = AB + AC + B\cdot 1$$

Step 6 : Apply Theorem 2 $(B\cdot 1 = B)$:

$$Y = AB + AC + B$$

Step 7 : Factor B out of first and third terms :

$$Y = B\,(A + 1) + AC$$

Step 8 : Apply Theorem 7 ($A + 1 = 1$) :

$$Y = B \cdot 1 + AC$$

Step 9 : Apply Theorem 2 ($B \cdot 1 = B$) :

$$Y = B + AC$$

This is the simplified form of exp. (*i*).

Example 24.32. *Simplify the following Boolean expressions to a minimum number of literals :*

(*i*) $Y = A + \overline{A}B$ (*ii*) $Y = AB + \overline{A}C + BC$

Solution. (*i*)

$$Y = A + \overline{A}B$$

$$= A + AB + \overline{A}B \qquad [\because A = A + AB \text{ from Theorem 16}]$$

$$= A + B(A + \overline{A})$$

$$= A + B \qquad [\because A + \overline{A} = 1 \text{ from Theorem 3}]$$

$$\therefore \quad Y = A + B$$

(*ii*)

$$Y = AB + \overline{A}C + BC$$

$$= AB + \overline{A}C + BC \cdot (A + \overline{A})$$

$$= AB + \overline{A}C + ABC + \overline{A}BC$$

$$= AB(1 + C) + \overline{A}C(1 + B)$$

$$= AB + \overline{A}C$$

$$\therefore \quad Y = AB + \overline{A}C$$

Example 24.33. *Determine output expression for the circuit shown below and simplify it using De Morgan's theorem.*

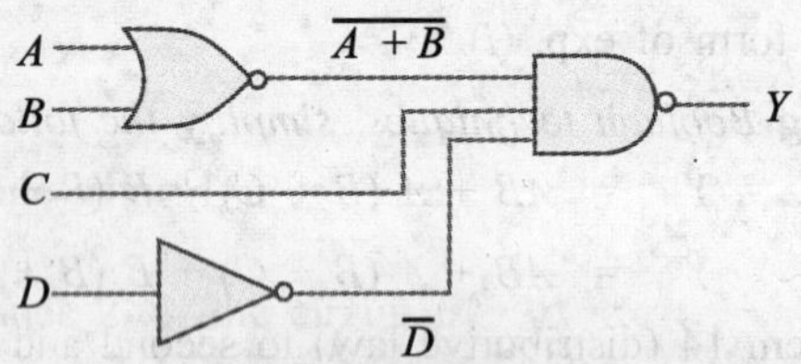

Fig. 24.51

Solution. The output expression for the circuit shown above is :

$$Y = \overline{[\overline{(A + B)} \cdot C \cdot \overline{D}]}$$

Using De Morgan's theorem :

$$Y = (A + B) + \overline{C} + D$$

$$\therefore \quad Y = A + B + \overline{C} + D$$

Example 24.34. *Find the complement of the expressions given below :*

(*i*) $Y = AB\overline{C} + A\,\overline{BC}$

(*ii*) $Y = \overline{A} + (B\overline{C} + \overline{B}C)$

Solution. (*i*)

$$Y = AB\overline{C} + A\,\overline{BC}$$

$$\overline{Y} = \overline{(AB\overline{C} + A\,\overline{BC})}$$

Applying De Morgan's theorem :

$$\overline{Y} = \overline{(AB\overline{C})} \cdot \overline{(A\overline{B}\,\overline{C})}$$

Again applying De Morgan's theorem to the each expression inside the brackets :

$$\overline{Y} = (\overline{A} + \overline{B} + C) \cdot (\overline{A} + B + C)$$

(ii)

$$Y = \overline{A}\,(B\overline{C} + \overline{B}C)$$

$$\overline{Y} = \overline{\overline{A}\,(B\overline{C} + \overline{B}C)}$$

Applying De Morgan's theorem :

$$\overline{Y} = A + \overline{(B\overline{C} + \overline{B}C)}$$

Again applying De Morgan's theorem to the expression inside the bracket :

$$\overline{Y} = A + \overline{(B\overline{C})} \cdot \overline{(\overline{B}C)}$$

Applying De Morgan's theorem for the third time, we get,

$$\overline{Y} = A + (\overline{B} + C) \cdot (B + \overline{C})$$

or

$$\overline{Y} = A + \overline{BC} + BC$$

Example 24.35. *Simplify the following Boolean expressions* :

(*i*) $Y = (A + B + C) \cdot (A + B)$

(*ii*) $Y = AB + ABC + AB\overline{C}$

(*iii*) $Y = 1 + A\,(B \cdot \overline{C} + BC + \overline{B}\,\overline{C}) + A\overline{B}C + AC$

(*iv*) $Y = \overline{(A + \overline{B} + C) + (B + \overline{C})}$

Solution. (*i*)

$$\begin{aligned} Y &= (A + B + C) \cdot (A + B) \\ &= A \cdot A + A \cdot B + B \cdot A + B \cdot B + C \cdot A + C \cdot B \end{aligned}$$

Using $A \cdot A = A$, we get,

$$\begin{aligned} Y &= A + AB + AB + B + AC + BC \\ &= A + AB + B + AC + BC & [\because AB + AB = AB] \\ &= A + B + AC + BC & [\because A + AB = A] \\ &= A\,(1 + C) + B\,(1 + C) \\ &= A \cdot 1 + B \cdot 1 & [\because 1 + C = 1] \end{aligned}$$

$$\therefore \quad Y = A + B$$

(*ii*)

$$\begin{aligned} Y &= AB + ABC + AB\overline{C} \\ &= AB + AB\,(C + \overline{C}) \\ &= AB + AB & [\because C + \overline{C} = 1] \end{aligned}$$

$$\therefore \quad Y = AB$$

(*iii*)

$$Y = 1 + A\,(B \cdot \overline{C} + BC + \overline{B}\,\overline{C}) + A\overline{B}C + AC$$

Using $1 + A = 1$, we get,

$$\begin{aligned} Y &= 1 + A\overline{B}C + AC & [\because 1 + A\,(B\overline{C} + BC + \overline{BC}) = 1] \\ &= 1 + AC \end{aligned}$$

$\therefore \qquad Y = 1$

Thus, because of the first term Y reduces to 1. Therefore, any Boolean expression ORed with 1, results in 1.

(*iv*) $$Y = \overline{(A+\overline{B}+C)+(B+\overline{C})}$$

Applying De Morgan's theorem :

$$Y = \overline{(A+\overline{B}+C)}\cdot\overline{(B+\overline{C})}$$

Again applying De Morgan's theorem :

$$Y = (\overline{A}\cdot B\cdot\overline{C})\cdot(\overline{B}\cdot C) = 0 \quad [\because\ B\cdot\overline{B} = 0,\ C\cdot\overline{C} = 0]$$

Example 24.36. *Simplify the following Boolean expression* :

$$Y = A\,\overline{B}\,D + A\overline{B}\,\overline{D}$$

Solution. $$Y = A\,\overline{B}\,D + A\overline{B}\,\overline{D}$$

Factoring out the common variables $A\,\overline{B}$ (using Theorem 14), we get,

$$Y = A\,\overline{B}\,(D+\overline{D})$$

Using Theorem 3, $D+\overline{D} = 1.$

$$\therefore \qquad Y = A\,\overline{B}\cdot 1$$

Using Theorem 2, we get,

$$Y = A\,\overline{B}$$

Example 24.37. *Simplify the following Boolean expression* :

$$Y = (\overline{A}+B)\,(A+B)$$

Solution. $$Y = (\overline{A}+B)\,(A+B)$$

The expression can be expanded by multiplying out the terms [Theorem 15].

$$Y = \overline{A}\cdot A + \overline{A}\cdot B + B\cdot A + B\cdot B$$

Using Theorem 4, $\overline{A}\cdot A = 0$. Also $B\,.\,B = B$ [Theorem 6].

$$\therefore \qquad Y = 0+\overline{A}\cdot B + B\cdot A + B$$

$$= \overline{A}\cdot B + A\,B + B$$

Factoring out the variable B [Theorem 14], we have,

$$Y = B\,(\overline{A}+A+1)$$

Using Theorem 7, $A + 1 = 1.$

$$\therefore \qquad Y = B\,(\overline{A}+1)$$

Again using Theorem 7, $\overline{A}+1 = 1.$

$$\therefore \qquad Y = B\cdot 1$$

Finally, using Theorem 2, we have,

$$Y = B$$

24.33 Combinational and Sequential Logic Circuits

We classify logic circuits in two groups *viz*.

(*i*) Combinational logic circuits

(*ii*) Sequential logic circuits

(*i*) **Combinational logic circuits.** We connect two or more than two logic gates (*e.g.* OR gate, AND gate, NOT gate *etc.*) to perform such functions as binary addition, subtraction *etc*. When such a logic circuit has no storage capability, it is called a combinational logic circuit. This type of circuit *combines* the input variables in such a way that output is always dependent on the combination of inputs.

(*ii*) **Sequential logic circuits.** In this type of logic circuit, the logic gates are connected in such a way that binary information can be stored. The output of a sequential logic circuit depends on both past and present input conditions. A good example of a sequential system is a telephone. When the last number is dialed, the party reached (output) is dependent on this last number (present value of input) and the past history of the system (previously dialed numbers).

24.34 Boolean Expressions for Combinational Logic Circuits

A combinational logic circuit (or digital circuit) often consists of several different logic gates, interconnected in such a way as to perform a specific logic function. By using the laws, theorems and techniques of Boolean algebra, we can find the Boolean expression for any combinational logic circuit. Let us find Boolean expression for some logic circuits.

(*i*) Fig. 24.52 shows a logic circuit. It consists of two AND gates and one OR gate. Each of the three gates has two input variables. Because gate G_1 is an AND gate and its two inputs are A and B, its output is AB. The gate G_2 is also an AND gate and its two inputs are C and D so that its output is CD. The gate G_3 is an OR gate so that its output is the ORing of AB and CD, producing $AB + CD$. Therefore, the Boolean expression for the output of this logic circuit is

$$Y = AB + CD$$

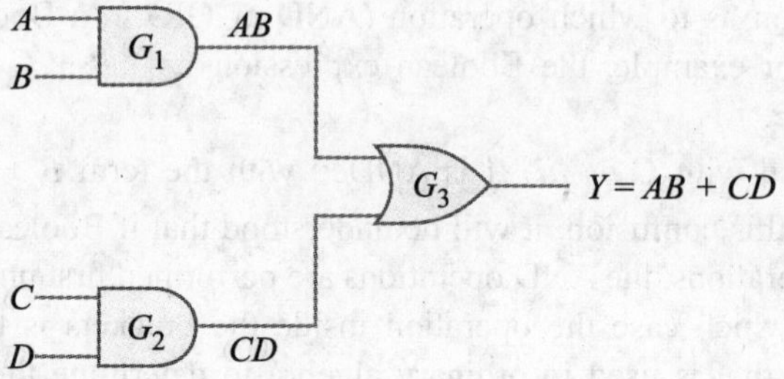

Fig. 24.52

(*ii*) Fig. 24.53 shows a logic circuit consisting of two AND gates and one OR gate. There are four (A, B, C and D) input variables.

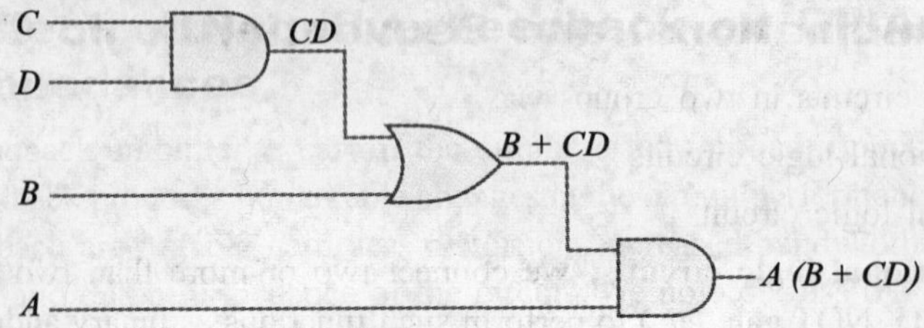

Fig. 24.53

The variable C is ANDed with D and its output is CD. Then CD is ORed with B, giving an output $(B + CD)$. This sum is then ANDed with A, resulting in the following Boolean expression :

$$Y = A\ (B + CD)$$

(*iii*) Fig. 24.54 shows a logic circuit consisting of three AND gates and one OR gate. There are six input variables *viz.* A, B, C, D, E and F.

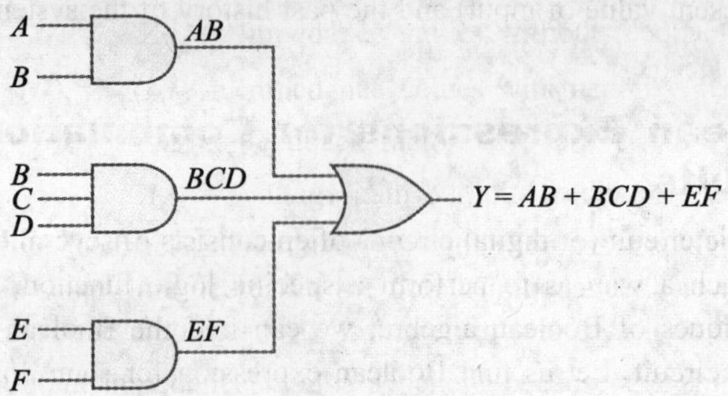

Fig. 24.54

The variables A and B are ANDed to produce AB, B, C and D are ANDed to produce BCD and E and F are ANDed to produce EF. The input to OR gate is AB, BCD and EF so the output of this gate is the ORing of the inputs *i.e.*,

$$Y = AB + BCD + EF$$

24.35 AND and OR Operations in Boolean Expression

It is very important to interpret AND and OR operations in the Boolean expression. Sometimes, there may be confusion as to which operation (AND or OR) in a Boolean expression is to be performed first. For example, the Boolean expression *$A \cdot B + C$ can be interpreted in two different ways *viz.*

(*i*) $A \cdot B$ is ORed with C *or* (*ii*) A is ANDed with the term $B + C$

In order to avoid this confusion, it will be understood that if Boolean expression contains both AND and OR operations, the AND operations are performed first unless there are brackets in the expression in which case the operation inside the brackets is to be performed first. This is the same rule that is used in ordinary algebra to determine the order of operations. Let us illustrate this important point with examples.

(*a*) Consider the Boolean expression : $Y = A \cdot B + C$. Since there are no brackets in this expression, AND operation is to be performed first. In other words, first A and B

* Recall that it indicates A ANDed with B and can be expressed as $A \cdot B$ or AB.

are ANDed to produce $A \cdot B$ and then $A \cdot B$ is ORed with C to give the final function : $Y = A \cdot B + C$. The logic circuit representing this Boolean expression is shown in Fig. 24.55.

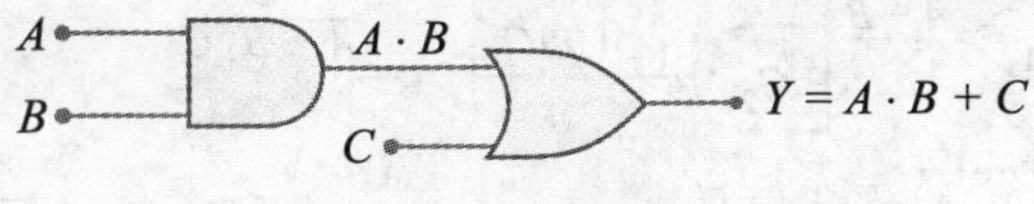

Fig. 24.55

(*b*) Consider the Boolean expression : $Y = (A + B) \cdot C$. Since $A + B$ is in the bracket, the OR operation is to be performed first. In other words, A is ORed with B to give $(A + B)$ and then $(A + B)$ is ANDed with C to produce $Y = (A + B) \cdot C$. The logic circuit representing this Boolean expression is shown in Fig. 24.56.

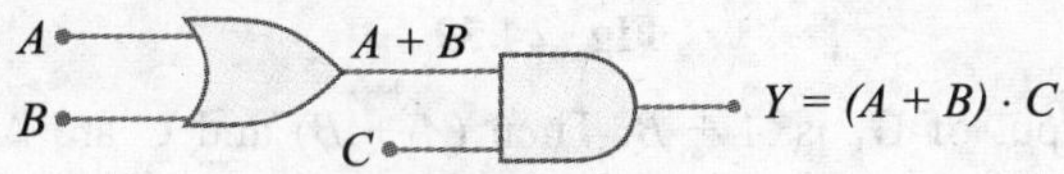

Fig. 24.56

Example 24.38. *Write the Boolean expression for the digital circuit shown in Fig. 24.57.*

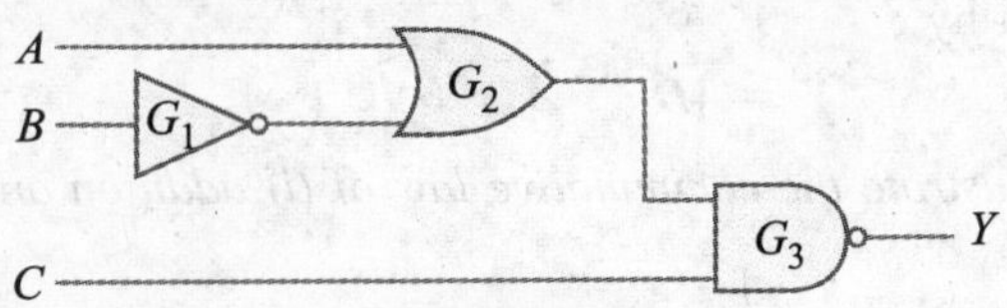

Fig. 24.57

Solution. The output of G_2 is $A + \overline{B}$. This output of G_2 is NANDed with C to yield $Y = \overline{(A + \overline{B})\, C}$. Note that bracket around $(A + \overline{B})$ is required otherwise the expression may be interpreted that A is NORed with $\overline{B}\, C$.

Example 24.39. *Write the Boolean expression for the logic circuit (digital circuit) shown in Fig. 24.58.*

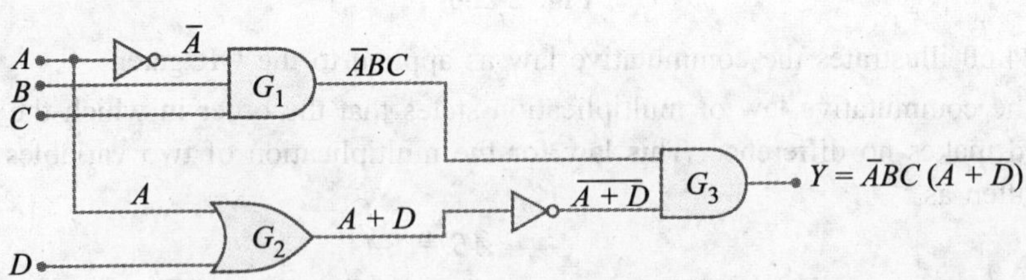

Fig. 24.58

Solution. There are two AND gates (*viz.* G_1 and G_3) and one OR gate (G_2). Also the circuit has two inverters. The input A is fed to G_1 through the inverter. Therefore, $\overline{A}$, B and C are ANDed by G_1 to produce an output of $\overline{A}\,BC$. The input A is ORed with D by G_2 to produce an output of $(A + D)$. The inputs $\overline{A}BC$ and $\overline{(A + D)}$ to G_3 are ANDed to produce the final output :

$$Y = \overline{A}BC\ \overline{(A + D)}$$

Example 24.40. *Write the Boolean expression for the digital circuit shown in Fig. 24.59.*

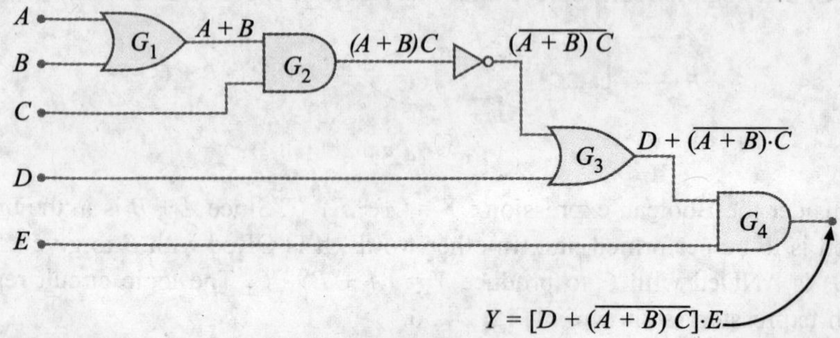

Fig. 24.59

Solution. The output of G_1 is $A + B$. Then $(A + B)$ and C are ANDed to produce an output of $(A + B)\,C$. It is easy to see $\overline{(A + B)\,C}$ and D are ORed by G_3 to produce an output of $D + \overline{(A + B) \cdot C}$. Finally, $D + \overline{(A + B) \cdot C}$ and E are ANDed by G_4 to produce the final output Y given by ;

$$Y = [D + \overline{(A + B)\,C}\,] \cdot E$$

Example 24.41. *Illustrate the commutative law of (i) addition and (ii) multiplication of two variables.*

Solution.

(*i*) The commutative law of addition states that the order in which the variables are ORed makes no difference. This law for the addition of two variables A and B can be written as :

$$A + B = B + A$$

Fig. 24.60

Fig. 24.60 illustrates the commutative law as applied to the OR gate.

(*ii*) The commutative law of multiplication states that the order in which the variables are ANDed makes no difference. This law for the multiplication of two variables A and B can be written as :

$$AB = BA$$

Fig. 24.61

Fig. 24.61 illustrates the commutative law as applied to the AND gate.

Example 24.42. *Illustrate associative law of (i) addition and (ii) multiplication as applied to Boolean algebra.*

Solution.

(*i*) The associative law of addition states that in the ORing of several variables, the result is the same regardless of the grouping of the variables. This law for the addition of the three variables A, B and C can be written as :

$$A + (B + C) = (A + B) + C$$

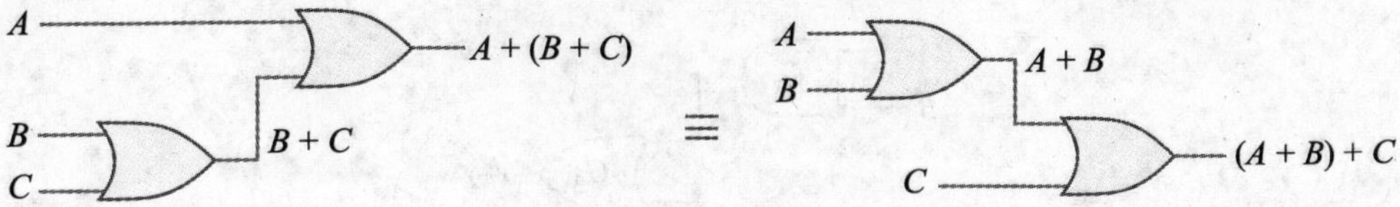

Fig. 24.62

Fig. 24.62. illustrates this law as applied to OR gates.

(*ii*) The associative law of multiplication states that it makes no difference in what order the variables are grouped when ANDing several variables. This law for the multiplication of the three variables A, B and C can be written as :

$$A\ (BC) = (AB)C$$

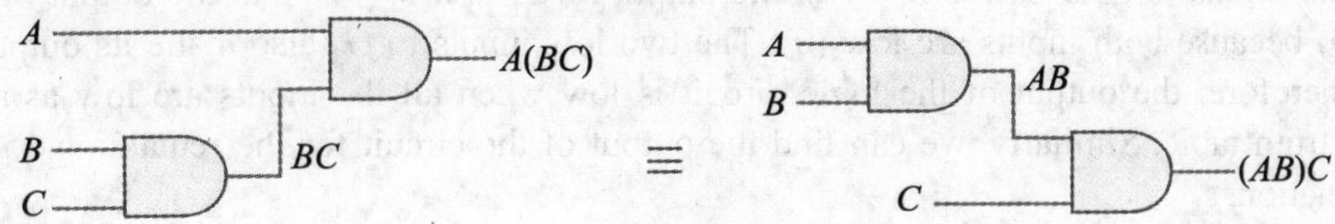

Fig. 24.63

Fig. 24.63 illustrates this law as applied to AND gates.

Example 24.43. *Illustrate distributive law as applied to Boolean algebra.*

Solution. This law states that ORing several variables and ANDing the result with a single variable is equivalent to ANDing the single variable with each of the several variables and ORing the products. The distributive law can be written for three variables (A, B and C) as under :

$$A\ (B + C) = AB + BC$$

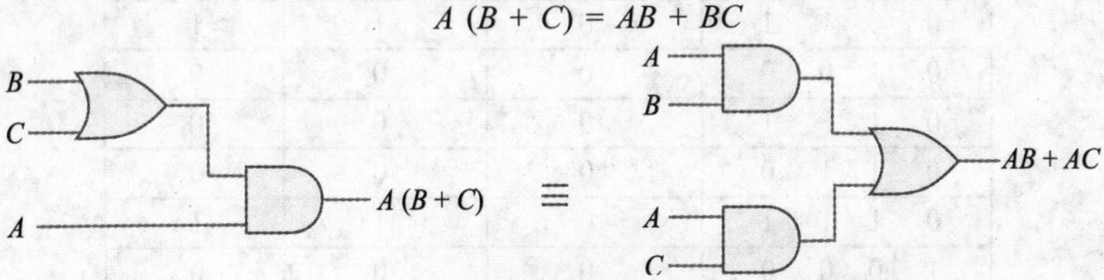

Fig. 24.64

Fig. 24.64 illustrates the distributive law in terms of gate implementation.

Note. The commutative law, the associative law and the distributive law of Boolean algebra are applicable regardless of the number of variables involved.

24.36 Truth Table from Logic Circuit

The output of a logic circuit for the given inputs can be determined directly from the circuit *without* using the Boolean expression. The procedure is to take a given set of inputs and work through each gate of the circuit to determine the final output. Let us illustrate the procedure with an example. Consider the logic circuit shown in Fig. 24.65. It consists of two AND gates and one OR gate. Each of the three gates has two input variables. Each of the input variables can be either a high (1) or a low (0). Since there are four input variables, there are 16 possible combinations of the input variables ($2^4 = 16$). We shall develop the truth table for the circuit without using the Boolean expression for the circuit.

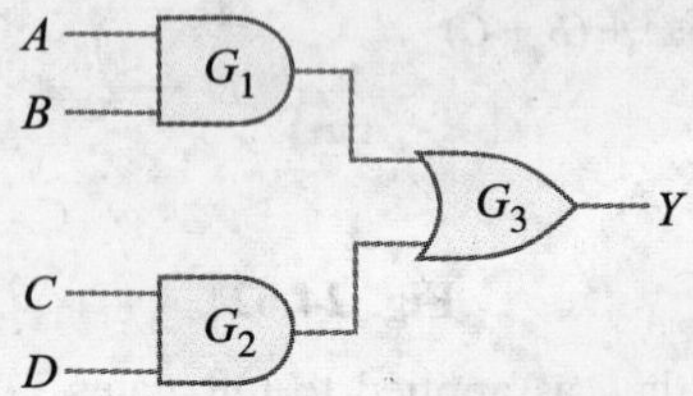

Fig. 24.65

First, suppose each input is low (*i.e.* $A = 0$, $B = 0$, $C = 0$ and $D = 0$). Under this condition, we shall examine the output of each gate in the circuit to arrive at the final output Y. If the inputs to gate G_1 are low (0), the output of G_1 is low (0). Also the output of G_2 is low (0) because both inputs are low (0). The two low inputs to G_3 also make its output low (0). Therefore, the output of the logic circuit is low when all the inputs are low as shown in the truth table. Similarly, we can find the output of the circuit for the remaining 15 input combinations.

Truth Table

$$Y = AB + CD$$

Inputs				G_1 Output	G_2 Output	G_3 Output
A	B	C	D	(AB)	(CD)	Y
0	0	0	0	0	0	0
0	0	0	1	0	0	0
0	0	1	0	0	0	0
0	0	1	1	0	1	1
0	1	0	0	0	0	0
0	1	0	1	0	0	0
0	1	1	0	0	0	0
0	1	1	1	0	1	1
1	0	0	0	0	0	0
1	0	0	1	0	0	0
1	0	1	0	0	0	0
1	0	1	1	0	1	1
1	1	0	0	1	0	1
1	1	0	1	1	0	1
1	1	1	0	1	0	1
1	1	1	1	1	1	1

Note. We can also construct the truth table by developing the Boolean expression from the logic circuit. The logic circuit shown above (See Fig. 24.65) is redrawn as shown in Fig. 24.66. The Boolean expression for this logic circuit is

$$Y = AB + CD$$

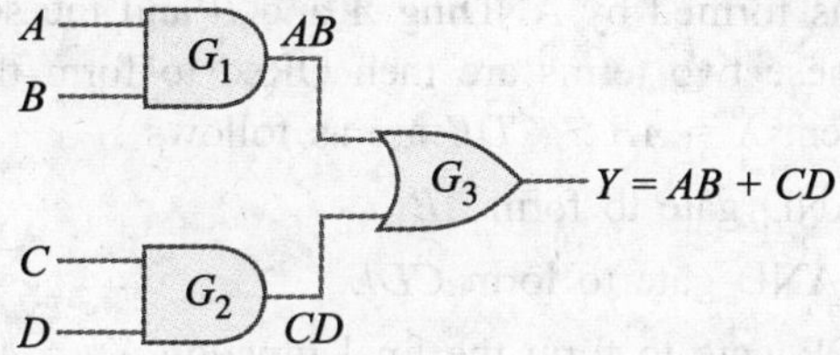

Fig. 24.66

Now simply put in the values for each combination of inputs and use the Boolean rules and laws to determine the final output Y. For example, when $A = 1$, $B = 1$, $C = 1$ and $D = 0$, the final output is

$$\begin{aligned} Y &= AB + CD \\ &= 1.1 + 1.0 \\ &= 1 + 0 \\ &= 1 \end{aligned}$$

Similary, we can find Y for the remaining 15 input combinations to get the truth table shown above.

24.37 Developing Logic Circuit from its Boolean Expression

We have seen how we can derive the Boolean expression for a given logic circuit. We now do the opposite *i.e.* create the logic circuit from its Boolean expression. Consider the following Boolean expression :

$$X = AB + CD$$

The function is composed of two terms, AB and CD, and it contains four variables. The first term is formed by ANDing A and B and the second term is formed by ANDing C and D. These two terms are then ORed to form the function X. The logic gates required to implement $X = AB + CD$ are as follows :

(*i*) Two two-input AND gates to form AB and CD.

(*ii*) One two-input OR gate to form the final function X ($= AB + CD$).

The logic circuit that performs the function $X = AB + CD$ is shown in Fig. 24.67.

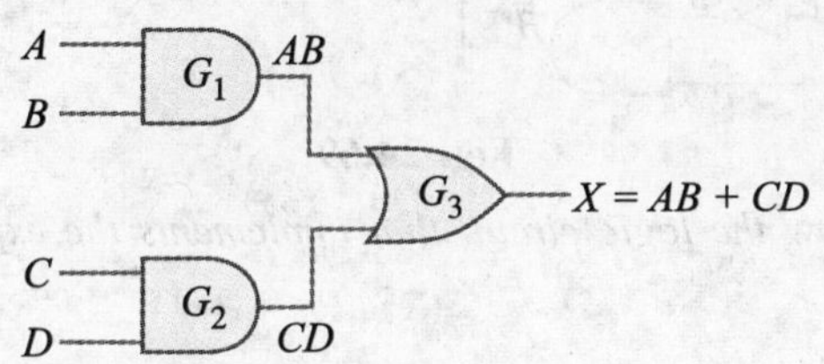

Fig. 24.67

Example 24.44. *Draw the logic circuit that implements the expression $X = AB + CDE$.*

Solution.

$$X = AB + CDE$$

The function is composed of two terms, AB and CDE, and it contains a total of five variables. The first term is formed by ANDing A and B and the second term is formed by ANDing C, D and E. These two terms are then ORed to form the function X. The logic gates required to implement $X = AB + CDE$ are as follows :

(*i*) One two-input AND gate to form AB.

(*ii*) One three-input AND gate to form CDE.

(*iii*) One two-input OR gate to form the final function X (= $AB + CDE$).

The logic circuit that performs the function $X = AB + CDE$ is shown in Fig. 24.68.

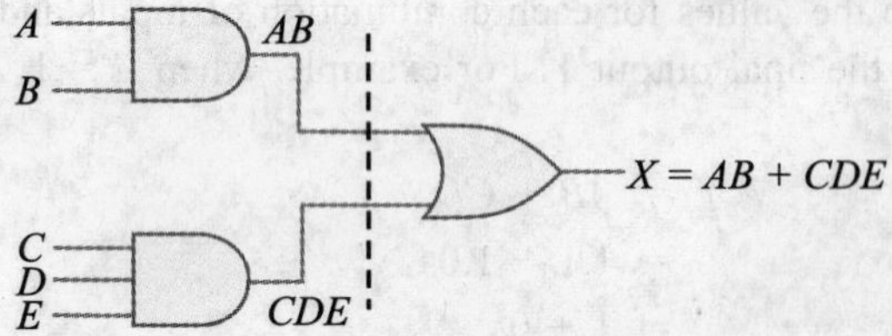

Fig. 24.68

Example 24.45. *Draw the logic circuit that implements the expression* $X = AB + \overline{B}C$.

Solution.

$$X = AB + \overline{B}C$$

The function is composed of two terms, AB and $\overline{B}C$, and it contains three variables. The first term is formed by ANDing A and B and the second term is formed by ANDing $\overline{B}$ and C. These two terms are then ORed to form the final function. The logic gates required to implement $X = AB + \overline{B}C$ are as follows :

(*i*) One inverter to form $\overline{B}$.

(*ii*) Two two-input AND gates to form AB and $\overline{B}C$.

(*iii*) One two-input OR gate to form the final function X (= $AB + \overline{B}C$).

The logic circuit that performs the function $X = AB + \overline{B}C$ is shown in Fig. 24.69.

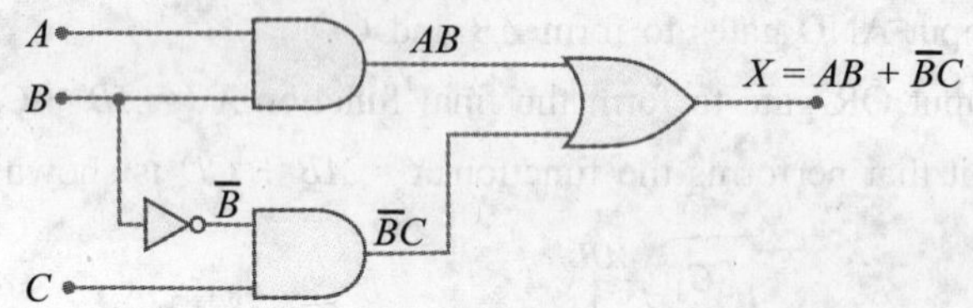

Fig. 24.69

Example 24.46. *Draw the logic circuit that implements the expression* $X = AB\,(C\overline{D} + EF)$.

Solution.

$$X = AB\,(C\overline{D} + EF)$$

A breakdown of this equation shows that the term AB and the term $C\overline{D} + EF$ are ANDed. The term AB is formed by ANDing the variables A and B. The term $C\overline{D} + EF$ is formed first by ANDing C and $\overline{D}$, ANDing E and F and then ORing these two terms. The logic gates required to implement $X = AB\,(C\overline{D} + EF)$ are as follows :

(*i*) One inverter to form $\overline{D}$.

(*ii*) Two two-input AND gates to form $C\overline{D}$ and EF.

(*iii*) One two-input OR gate to form $C\overline{D} + EF$.

(*iv*) One two-input AND gate to form AB.

(*v*) One two-input AND gate to form $X\ [= AB\,(C\overline{D} + EF)]$.

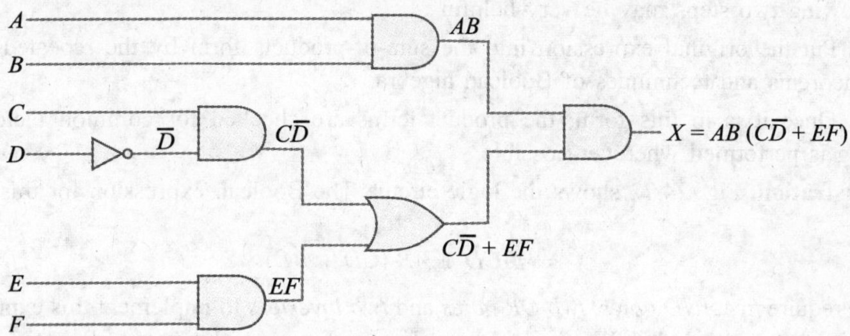

Fig. 24.70

The logic circuit that forms the function $X = AB\,(C\overline{D} + EF)$ is shown in Fig. 24.70.

24.38 Sum-of-Products Form

The sum-of-products form of a Boolean expression consists of two or more AND terms (*i.e.* products) that are ORed together. For example, $AB + CD$ is a sum-of-products expression. Here AND terms AB and CD are ORed (added). Other examples of this form are :

(*i*) $ABC + \overline{A}B\overline{C}$

(*ii*) $A\overline{B}C + DEF + A\,\overline{EF}$

(*iii*) $\overline{A} + BC\overline{D} + EFG$

The sum-of-products is a very useful form of a Boolean expression due to the straightforward manner in which it can be implemented in logic gates. For example, Fig. 24.71 shows the logic circuit that results in a sum-of-products form. It has simply two steps : ANDing and then ORing. Therefore, this form is always only a *two-level* gate network *i.e.,* the maximum number of gates through which a signal must pass in going form input to output is two (excluding inversions, but these can also be worked in).

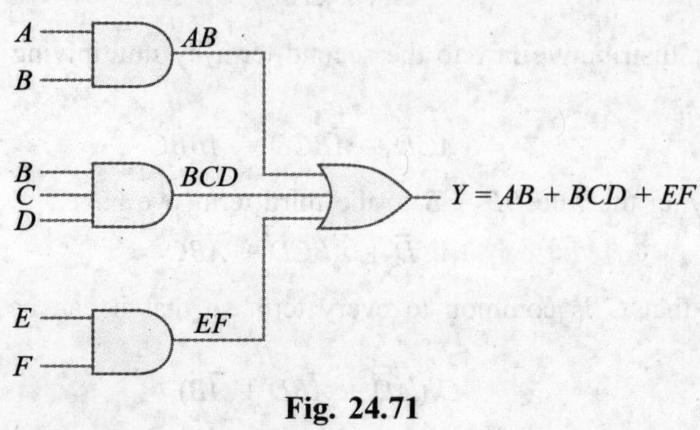

Fig. 24.71

24.39 Simplification of Boolean Expressions

The form of Boolean expression determines how many and which types of logic gates are needed as well as how they are connected together. The more complicated the Boolean expression, the more complex the logic circuit will be. It is, therefore, desirable to simplify an expression as much as possible to get the simplest logic circuit. Note that the new expression can be used to implement a logic circuit that is equivalent to the original logic circuit but contains fewer gates and connections. While simplifying a Boolean expression, the following two steps may be very helpful :

(*i*) Put the original expression into the sum-of-products form by the repeated use of rules, theorems and techniques of Boolean algebra.

(*ii*) Once it is in this form, the product terms are checked for common factors and factoring is performed wherever possible.

Illustration. Fig. 24.72 shows the logic circuit. The Boolean expression for this circuit is

$$X = AC\overline{D} + \overline{A}B\,(CD + BC)$$

We require *five AND gates*, *two OR gates* and *two inverters* to implement this expression. In all, nine gates are needed. We shall now use laws, rules and techniques of Boolean algebra to get the simplest expression for the given function.

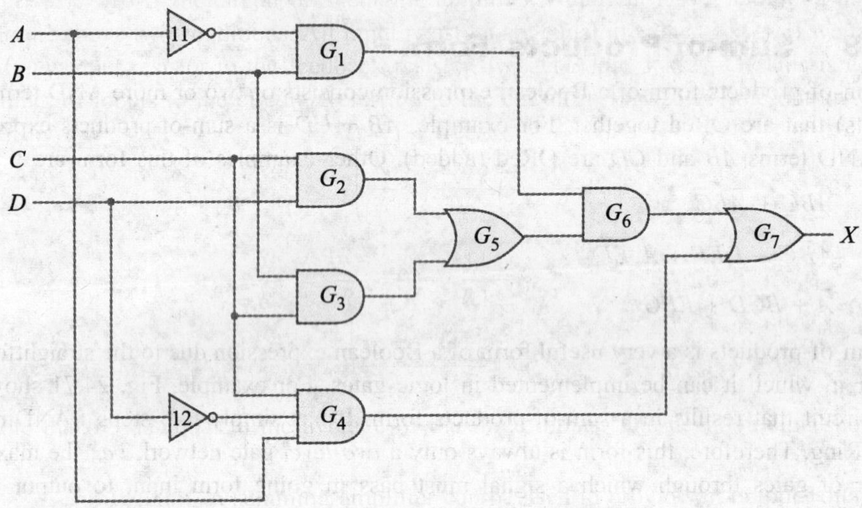

Fig. 24.72

Step 1. Apply distributive law to the second term by multiplying the term $CD + BC$ by $\overline{A}B$. The result is :

$$X = AC\overline{D} + \overline{A}BCD + \overline{A}BBC$$

Step 2. Applying the rule $BB = B$ to the third term, we have,

$$X = AC\overline{D} + \overline{A}BCD + \overline{A}BC$$

Step 3. Note that C is common to every term so that it can be factored out using distributive law.

$$\therefore \quad X = C\,(A\overline{D} + \overline{A}BD + \overline{A}B)$$

Step 4. We see that the term $\overline{A}B$ appears in the last two terms within the bracket and can be factored out of those two terms.

$$\therefore \quad X = C\,[A\overline{D} + \overline{A}B\,(D + 1)]$$

$$\text{Since } D + 1 = 1,\ X = C\,(A\overline{D} + \overline{A}B)$$

It appears that this equation cannot be simplified any further, but it can be written in a slightly different way by applying the distributive law (this results in the sum-of-products form) :

$$X = AC\overline{D} + \overline{A}BC$$

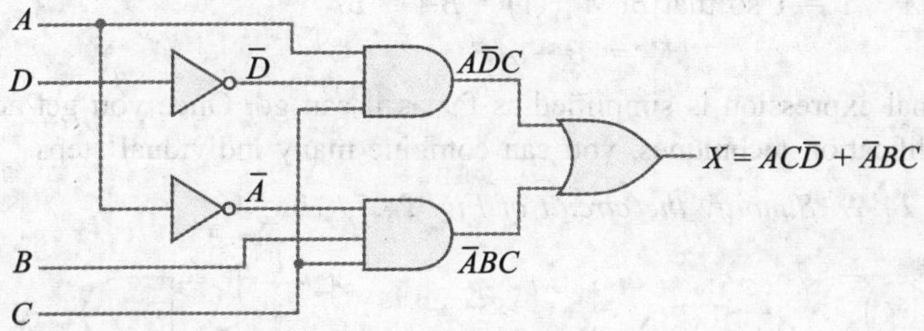

Fig. 24.73

Implementing this equation (*i.e.* $X = AC\overline{D} + \overline{A}BC$) into the logic circuit, it only requires *two three-input AND gates*, *two inverters* and *one two-input OR gate* as shown in Fig. 24.73. Note that this minimised circuit is equivalent to the original circuit of Fig. 24.72 but only requires five gates (instead of nine).

Example 24.47. *Simplify the expression :*

$$X = \overline{A}\overline{B}C + A\overline{B}C + AB\overline{C} + ABC$$

Solution.

$$X = \overline{A}\overline{B}C + A\overline{B}C + AB\overline{C} + ABC$$

Step 1. Note that the first two terms have $\overline{B}C$ as common factors while the last two terms have AB as common factors.

$$\therefore \quad X = \overline{B}C\,(A + \overline{A}) + AB\,(C + \overline{C})$$

Step 2. $A + \overline{A} = 1$ and $C + \overline{C} = 1$ so that :

$$X = \overline{B}C \cdot 1 + AB \cdot 1$$

Step 3. Since $\overline{B}C \cdot 1 = \overline{B}C$ and $AB \cdot 1 = AB$ so that :

$$X = AB + \overline{B}C$$

Note that not only is the Boolean expression simplified, but so is the resultant logic circuit.

Example 24.48. *Simplify the expression :*

$$X = AB + A\,(B + C) + B\,(B + C)$$

Solution.

$$X = AB + A\,(B + C) + B\,(B + C)$$

Step 1. Applying distributive law to the second and third terms, we have,

$$X = AB + AB + AC + BB + BC$$

Step 2. Now $BB = B$ and $AB + AB = AB$ so that :

$$X = AB + AC + B + BC$$

Step 3. $B + BC = B(1 + C) = B \cdot 1 = B$

$\therefore$ $X = AB + AC + B$

Step 4. Factoring B out, we have,

$$X = B(A + 1) + AC$$

Step 5. $A + 1 = 1$ so that $B(A + 1) = B \cdot 1 = B$.

$\therefore$ $X = B + AC$

The original expression is simplified as far as it can go. Once you get acquainted with Boolean simplification techniques, you can combine many individual steps.

Example 24.49. *Simplify the circuit of Fig. 24.74 (i).*

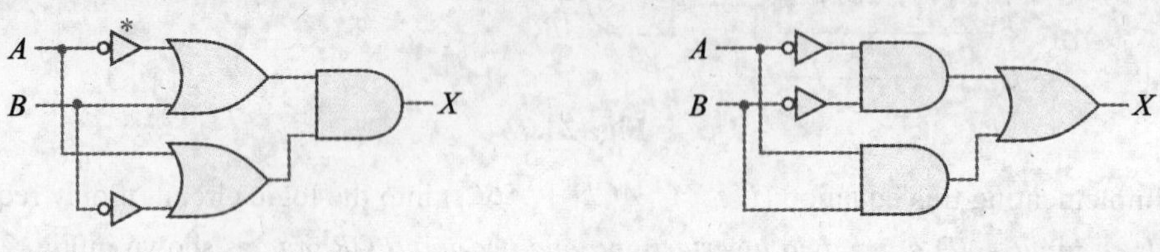

Fig. 24.74

Solution. The Boolean expression for the circuit shown in Fig. 24.74 (*i*) is

$$X = (\overline{A} + B)(A + \overline{B})$$

Step 1. Multiplying out to get the sum-of-products form, we have,

$$X = \overline{A}A + \overline{A}\overline{B} + BA + B\overline{B}$$

Step 2. Now $\overline{A}A = 0$ and $B\overline{B} = 0$ so that :

$$X = \overline{A}\overline{B} + AB$$

This expression is implemented in Fig. 24.74 (*ii*). If you compare the circuit with the original circuit, you see that both circuits contain the same number of gates and connections. **Therefore, the simplification process has not produced a simpler circuit but it has produced an alternative circuit.**

24.40 Electronic Adders

A logic circuit that performs the function of binary addition is called **electronic adder** *or* **adder.** The adder circuit consists of properly connected logic gates. There are usually two forms of the adder in common applications *viz.* (*i*) half adder (*ii*) full adder.

(*i*) **Half Adder (HA).** A logic circuit that can add two binary bits is called a half adder (*HA*). Fig. 24.75 (*i*) shows the block symbol for the half adder. The adder circuit would need two inputs and two outputs. The two inputs are for the two digits to be added,

* This is the alternate symbol for the inverter.

either 0 or 1. One output terminal is for the sum of the two inputs and the other output is for the carry, if necessary. Fig. 24.75 (*ii*) shows the addition table of the adder and can be thought of as a truth table. The numbers being added are on the input side of the table. The truth table has two output columns, one column for the sum and one column for the carry. The sum column is labelled with summation symbol Σ. The carry column is labelled with C_o. The C_o stands for *carry output* or *carry out*. Thus the half adder circuit has two inputs (A, B) and two outputs (Σ, C_o).

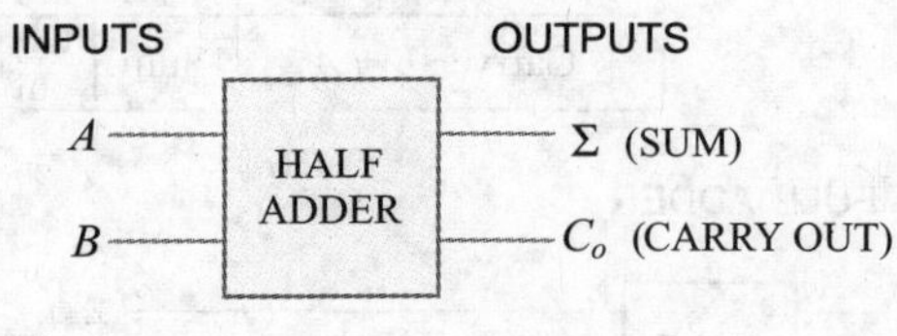

(*i*)

TRUTH TABLE

INPUTS		OUTPUTS	
B	A	Σ	C_o
0	0	0	0
0	1	1	0
1	0	1	0
1	1	0	1
BINARY DIGITS TO BE ADDED		SUM	CARRY OUT
		XOR	AND

(*ii*)

Fig. 24.75

The half adder would behave according to the truth table shown in Fig. 24.75 (*ii*). Take a careful look at the truth table. The output columns (sum and carry) can be produced by using two gates as under :

(*a*) The sum column is the output of XOR gate. Remember that XOR gate has HIGH output when either input is HIGH but not when both inputs are the same.

(*b*) The carry column is the output of the AND gate. Both inputs must be HIGH for there to be a HIGH in the output.

Thus we can produce half adder using a two-input AND gate and a two-input XOR gate as shown in Fig. 24.76. The half adder circuit adds only the LSB column (1s column) in a binary addition problem. The reason is that it has no input for a carry-in.

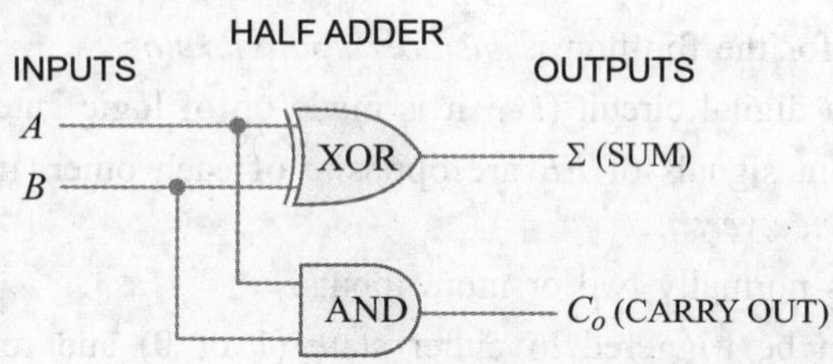

Fig. 24.76

(*ii*) **Full Adder (FA).** A full adder adds two binary bits plus a carry input (C_{in}) to produce the sum (Σ) and carry-out (C_o) outputs. Fig. 24.77 (*i*) shows the block diagram of a full adder. It is formed by using two half adder circuits and an OR gate as shown in Fig. 24.77 (*iii*). Note the carry-in input which requires the extra half adder. The output of the OR gate forms the carry-out (C_o) output.

The full adder has three inputs : C_{in}, A and B. These three inputs must be added to get the Σ and C_o outputs. Fig. 24.77 (*ii*) shows the truth table for a full adder. It shows all the possible combinations of A, B and C_{in}. The full adder is used for binary addition in all places except the 1s place which means 2s, 4s, 8s *etc.*

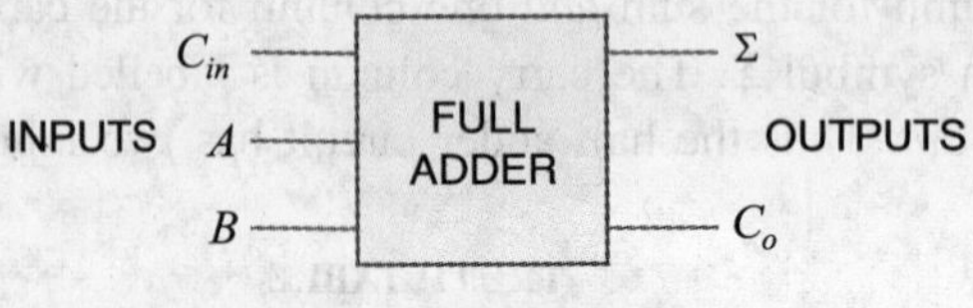

(i)

Truth Table

INPUTS			OUTPUTS	
C_{in}	B	A	Σ	C_o
0	0	0	0	0
0	0	1	1	0
0	1	0	1	0
0	1	1	0	1
1	0	0	1	0
1	0	1	0	1
1	1	0	0	1
1	1	1	1	1
Carry + B + A			Sum	Carry out

(ii)

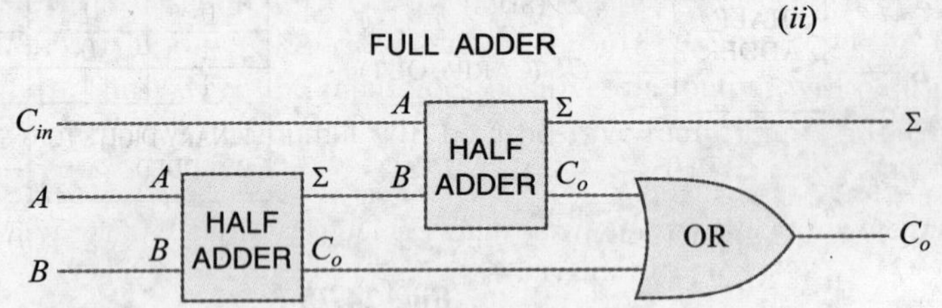

(iii)

Fig. 24.77

24.41 Flip-Flop (FF)

The most important memory or storage element is the *flip-flop* (abbreviated as *FF*) which is made up of an assembly of logic gates. *A* **flip-flop** *is a bistable circuit that can exist in either of two stable states (1 and 0) indefinitely and can be made to change its state by means of some external signal.* Because the flip-flop's output remains at either a 1 or a 0 depending on the last input, the flip-flop can be said to "remember" *i.e.,* it can act as a memory or storage element. The following points may be noted about flip-flops :

(*i*) Another name for the flip-flop is *bistable multivibrator*.

(*ii*) A flip-flop is a digital circuit (*i.e.,* it is made up of logic gates) with 2 outputs.

(*iii*) The two output signals of *FF* are opposite of each other. If one output is 1, the other output is 0 and *vice-versa*.

(*iv*) A flip-flop has normally two or more inputs.

(*v*) A flip-flop can be triggered to either state (1 or 0) and remains there after the triggering signal is removed. This property of *FF* accounts for the storage capability of *FF*.

24.42 Flip-Flop Terminology

There are several unique terms used to describe flip-flop operation. Fig. 24.78 shows the general type of symbol used for a flip-flop.

(*i*) A flip-flop has normally two or more inputs and two outputs. The two outputs, labelled Q and $\overline{Q}$, are opposite of each other. If Q = 1 (HIGH), then $\overline{Q}$ = 0 (LOW). If Q = 0, then $\overline{Q}$ = 1.

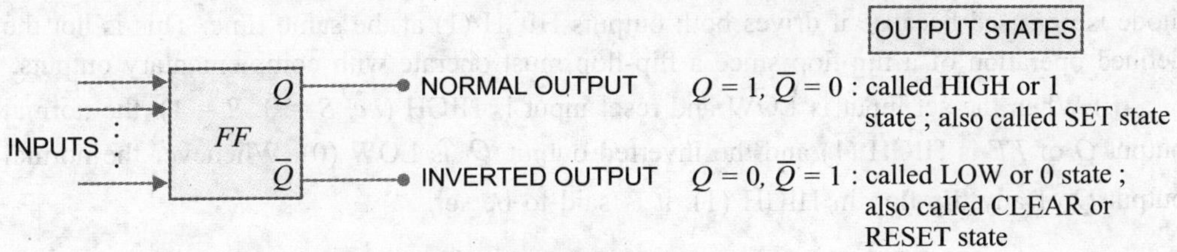

Fig. 24.78

(*ii*) The Q output is called the *normal FF output* and $\overline{Q}$ is the *inverted FF output*. Whenever we refer to the state of a flip-flop, we are referring to the state of its normal (Q) output ; it is understood that its inverted output ($\overline{Q}$) is in the opposite state. Thus when we say that *FF* is in HIGH state (1), it means $Q = 1$. If we say that *FF* is in LOW state (0), it means $Q = 0$. Of course, $\overline{Q}$ state will always be the opposite of Q.

(*iii*) A flip-flop is said to be *set* when $Q = 1$ and $\overline{Q} = 0$. Conversely, the flip-flop is *reset* when $Q = 0$ and $\overline{Q} = 1$. Thus the process of causing the flip-flop to go to the $Q = 1$, $\overline{Q} = 0$ state is called *setting* the flip-flop. On the other hand, causing the flip-flop to go to the $Q = 0$, $\overline{Q} = 1$ state is called *resetting* (this action is also referred to as "*clearing*") the flip-flop.

The circuit is called flip-flop because external triggering signals at the input of *FF* cause the *FF* to switch back and forth ("flip-flop") between its two possible stable output states.

24.43 SR Flip-Flop

The *SR* flip-flop is built using two cross-coupled NAND gates or NOR gates. The NAND gate version is called NAND gate *SR* flip-flop. Since the NAND and NOR *SR* flip-flops perform the same basic function, we shall limit our discussion of *SR* flip-flop to the NAND-type. Fig. 24.79 shows the circuit diagram of *SR* flip-flop using two NAND gates. The two NAND gates 1 and 2 are cross-coupled so that the output of NAND-1 is connected to one of the inputs of NAND-2 and *vice-versa*. The two inputs are labelled *S*(*set*) and *R*(*reset*) and the two outputs are Q and $\overline{Q}$. The logic symbol for *SR* flip-flop is shown in Fig. 24.80. The exact behaviour of a flip-flop is defined by its truth table. The truth table for a NAND *SR* flip-flop is shown in Fig. 24.81.

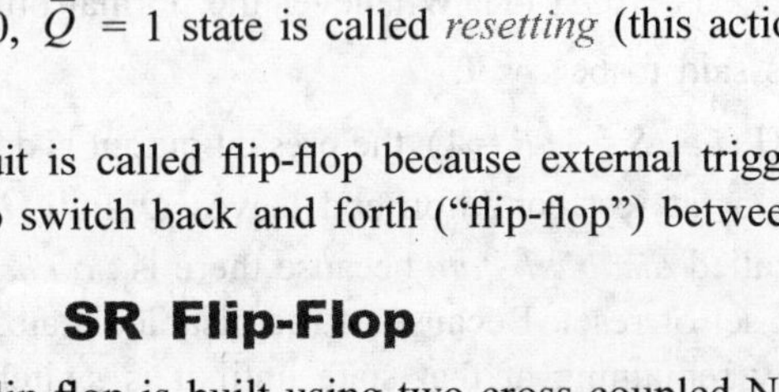

Fig. 24.79

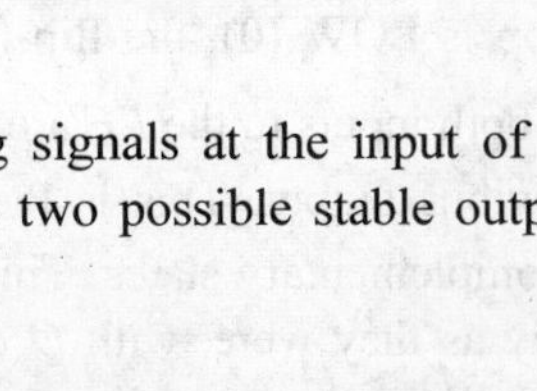

Fig. 24.80

(*i*) In Fig. 24.81, if both inputs to the flip-flop are LOW (*i.e.*, $S = 0$ and $R = 0$), both outputs will be HIGH (1). This is identified in the truth table as being "non allowed". This mode is not used because it drives both outputs HIGH (1) at the same time. This is not the defined operation of a flip-flop since a flip-flop must operate with complementary outputs.

(*ii*) When the set input is LOW and reset input is HIGH (*i.e.* $S = 0$, $R = 1$), the normal output Q of *FF* is HIGH (1) and the inverted output $\overline{Q}$ is LOW (0). Whenever the normal output Q of any flip-flop in HIGH (1), it is said to be **set**.

Truth table for NAND SR flip-flop

S	R	Q	$\overline{Q}$	State
0	0	1	1	Not allowed
0	1	1	0	Set
1	0	0	1	Reset
1	1	No change		Previous

Fig. 24.81

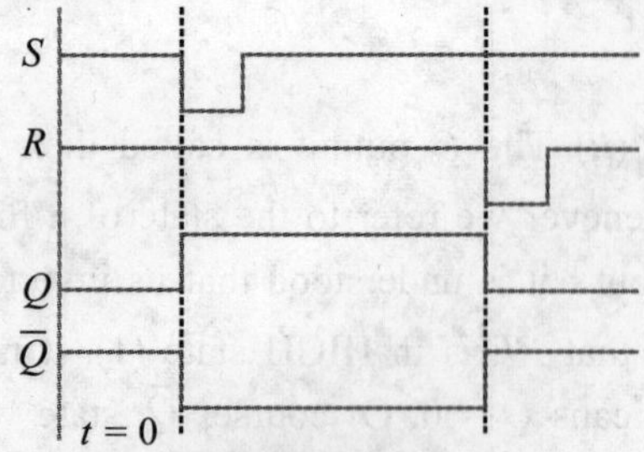

Timing diagram

Fig. 24.82

(*iii*) When the set input is HIGH and reset input is LOW (*i.e.*, $S = 1$, $R = 0$), the normal output Q of *FF* is LOW (0) and output $\overline{Q}$ is HIGH (1). Whenever the normal output Q of any flip-flop is LOW (0), the flip-flop is said to be **reset**.

(*iv*) If both inputs to the *FF* are HIGH (*i.e.*, $S = 1$, $R = 1$), the present output is determined by the previous output. This is the idle or at rest condition and leaves Q and $\overline{Q}$ in their previous complementary states. This is called *disabled state* because there is *no change*. The outputs stay as they were, with Q either set or reset. Because of the disabled state, the flip-flop "remembers" the preceding state by remaining at that state until it is switched. This operation is possible because a flip-flop is a bistable circuit.

Fig. 24.82 shows the *timing diagram* or *waveform diagram* for NAND *SR* flip-flop. This diagram shows the voltage level and timing between inputs and outputs. The horizontal distance is *time* and vertical distance *voltage*.

Note. Fig. 24.83 shows the circuit of *SR* flip-flop using NOR gates. This circuit is implemented by cross-connecting two NOR gates in a manner that is very similar to the NAND gate version described above. The truth table for NOR gate *SR* flip-flop is shown in Fig. 24.84.

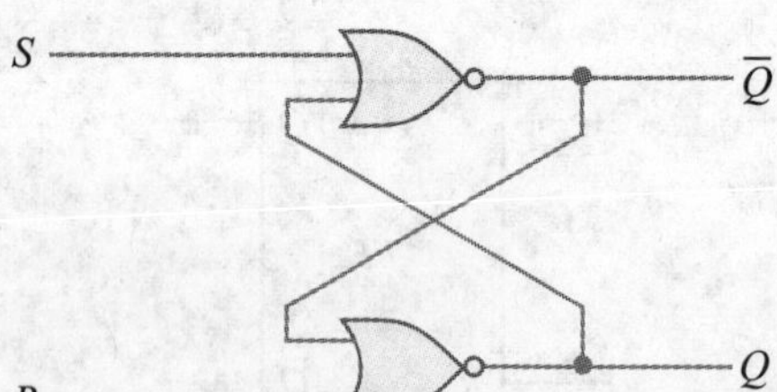

Logic symbol

Fig. 24.83

Truth table for NOR SR flip-flop

S	R	Q	$\overline{Q}$	State
0	0	No change		Previous
0	1	0	1	Reset
1	0	1	0	Set
1	1	1	1	Not allowed

Fig. 24.84

The main difference between the two (NAND type and NOR type) is that as the activity changes at the S and R inputs logic are 1's instead of logic 0's. Therefore, for the NOR gate configuration, the quiescent state is $S = 0$, $R = 0$ such that no change will occur and the *not allowed* state is $S = 1$, $R = 1$. Since this state is not allowed, the output for this condition is often referred to as undefined. This state becomes important when considering logic minimisation techniques. Fig. 24.85 shows the timing diagram of NOR *SR* flip-flop.

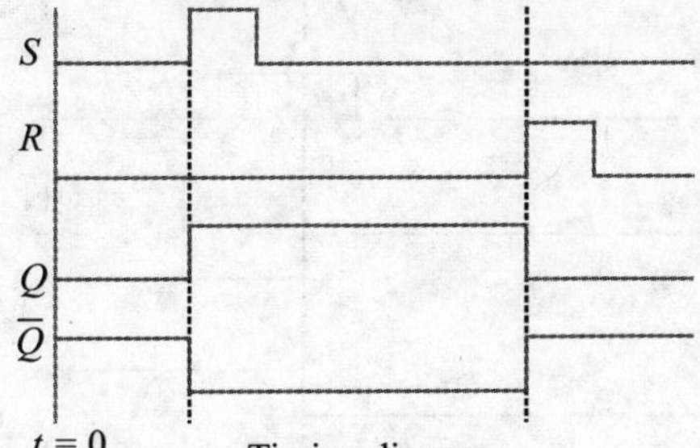

Timing diagram

Fig. 24.85

24.44 Clocked SR Flip-Flop

Basically, a clocked *SR* flip-flop operates in the same manner as the NAND and NOR gate versions with the exception that the clocked pulse (square wave) controls the times at which state changes can happen. The logic symbol for a clocked *SR* flip-flop is shown in Fig. 24.86. Note that it looks almost like *SR* flip-flop except that it has one extra input labelled *CLK* (for clock). With the simple *SR* flip-flop, the outputs change at the same time as the *R* and *S* inputs change. However, with a clocked flip-flop, the outputs do not change until a clock pulse is received on the *CLK* input. These clock pulses occur at some fixed interval (for example, every 10 microseconds) and all logic state changes are synchronised to occur at the time when the clock pulse occurs.

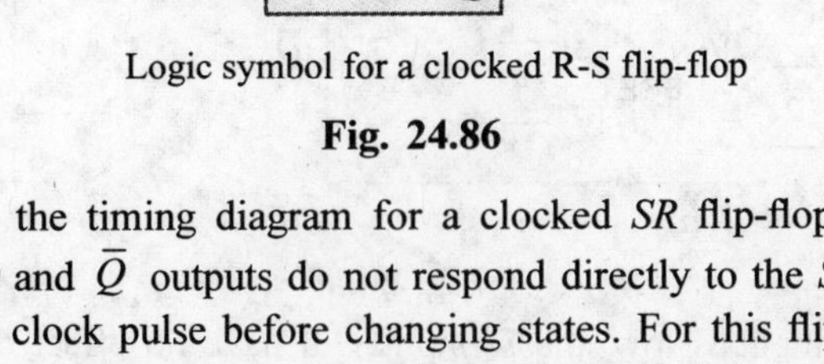

Logic symbol for a clocked R-S flip-flop

Fig. 24.86

Fig. 24.87 shows the timing diagram for a clocked *SR* flip-flop. It is clear from the timing diagram that Q and $\overline{Q}$ outputs do not respond directly to the S and R inputs ; rather they wait for the next clock pulse before changing states. For this flip-flop, if the output is in Q state when clock pulse n occurs, it will remain in that state for clock pulse $n + 1$ provided that inputs S and R are both zero. If either S or R goes to the logic 1 state, then the flip-flop will assume the appropriate state upon the occurrence of the next clock pulse. Finally, as with other *SR* flip-flops, there is a *not allowed* condition, namely $S = 1$, $R = 1$ for which the output is undefined. This condition is represented by a question mark in the truth table in Fig. 24.88.

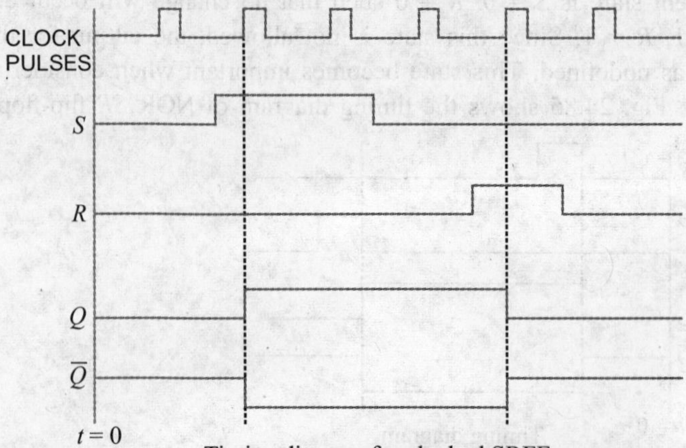

Timing diagram for clocked SRFF

Fig. 24.87

S	R	Q^{n+1}
0	0	Q^n
0	1	0
1	0	1
1	1	?

Truth table

Fig. 24.88

Note that the output of the clocked *SR* flip-flop *changes only on a clock pulse*. We say that this flip-flop operates *synchronously i.e.,* it operates *in step with* the clock. Synchronous operation is very important in calculators and computers where each step must happen in a very exact order.

24.45 D Flip-Flop

The *D* flip-flop is a slight modification of the *SR* flip-flop. If we add two NAND gates and an inverter to a NAND *SR* flip-flop as shown in Fig. 24.89 (*i*), the resulting circuit is called *D* flip-flop. The logic symbol for *D* flip-flop is shown in Fig. 24.89 (*ii*). The inputs for this type of flip-flop are labelled *D* and *T*. The outputs are labelled Q and $\overline{Q}$. The *T* input represents the input for a stable clock pulse. The *D* input represents the data input. As we shall see, when clock pulse is applied, the *Q* output of the flip-flop assumes the same logic level as the *D* input. If $D = 1$, then $Q = 1$ and if $D = 0$, then $Q = 0$.

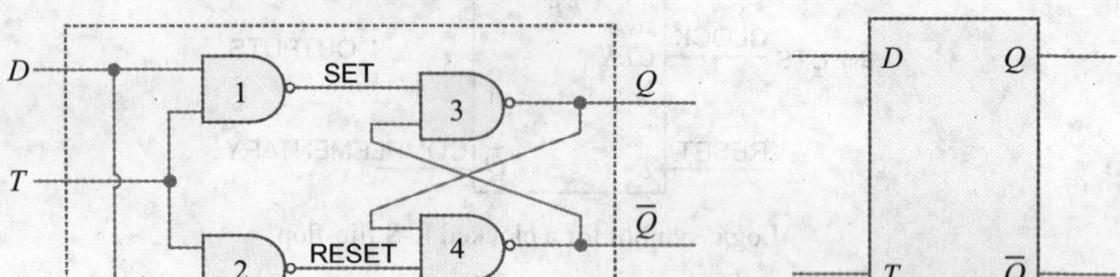

Fig. 24.89

When the clock input is low, the *D* input will have no effect on the output, since the low at the *T* input causes the output of both NAND 1 and NAND 2 to go high. When the clock input is high, the *D* input will produce a low at either the set or reset of gates 3 and 4, depending on whether *D* input is high or low. When clock pulse is applied, the flip-flop follows the state of whatever data is present at the *D* input. In other words, when $T = 1$, Q follows the *D* input. If $D = 1$, the *Q* output of the flip-flop goes to the 1 state but if $D = 0$, the *Q* output of the flip-flop goes to 0 state. This fact is shown in the truth table for the *D* flip-flop in Fig. 24.90.

T	D	Q	$\overline{Q}$	State
0	0	No change		Previous
0	1	No change		Previous
1	0	0	1	Reset
1	1	1	0	Set

Fig. 24.90

The D flip-flop is often called a *delay flip-flop*. The word "delay" describes what happens to the data at input D. The data (a 0 or 1) at input D is *delayed one clock pulse* from getting to the output Q. In other words, the flip-flop can be considered to have stored the data from the time of the first clock pulse until such time as another clock pulse occurs.

D filp-flops are connected in series to form *shift-registers* or in parallel to form *data storage registers*. In the shift-register arrangement, delays of many clock periods can be obtained. When used as storage registers, the flip-flops can store data indefinitely or until required. Shift registers and storage registers are widely used in digital data handling circuits.

24.46 JK Flip-Flop

The JK flip-flop is probably the most widely used and is considered the universal flip-flop because it can be used in many ways. This type of device has a clock input as well as a J and K input. The JK flip-flop is a refinement of SR flip-flop in that JK flip-flop has four different operating states while SR flip-flop (NAND type of NOR type) has only three usable states. Inputs J and K behave like inputs S and R to set and reset the flip-flop respectively. The logic symbol for the JK flip-flop is shown in Fig. 24.91 (*i*). The inputs labelled J and K are the data inputs. The input labelled CLK is the clock input. Outputs Q and $\overline{Q}$ are the usual normal and complementary outputs. Note that J input is equivalent to the S (set) input of SR flip-flop and the K input is equivalent to the R (reset) input.

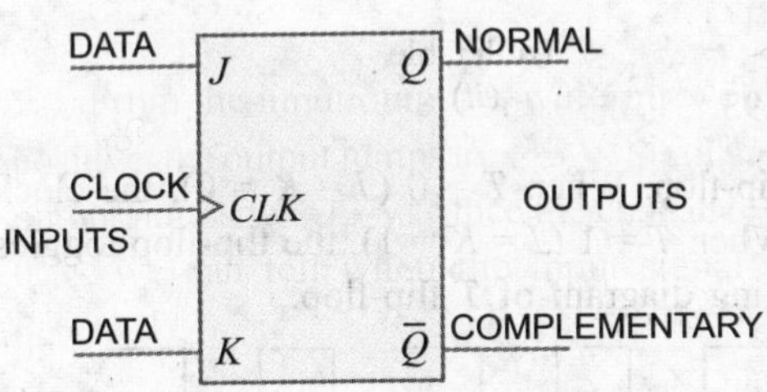

(*i*)

Truth table

Mode of operation	*Inputs*			*Outputs*		
	CLK	*J*	*K*	*Q*	$\overline{Q}$	*Effect on output Q*
Hold	⎍	0	0	No change		No change – disable
Reset	⎍	0	1	0	1	Reset or cleared to 0
Set	⎍	1	0	1	0	Set to 1
Toggle	⎍	1	1	Toggle		Changes to opposite state

(*ii*)

Fig. 24.91

Fig. 24.91 (*ii*) shown the truth table for the J-K flip-flop. This table shows four useful modes of operation. When the J and K inputs are both 0, the flip-flop is in the *hold* (or *disabled*) *mode*. In the hold mode, the data inputs have no effect on the outputs. The outputs "hold" the last data present. Lines 2 and 3 of the truth table show the reset and set conditions for the Q output. Line 4 of the truth table illustrates the useful *toggle* position of the J–K flip-flop. When both data inputs J and K are at 1, repeated *clock pulses cause the output to turn off-on-off-on-off-on and so on. This off-on action is like a toggle switch and is called *toggling*. Each clock pulse toggles the outputs to switch to their opposite states. *This condition is known as race around condition* Fig. 24.92 shows the timing diagram of JK flip-flop.

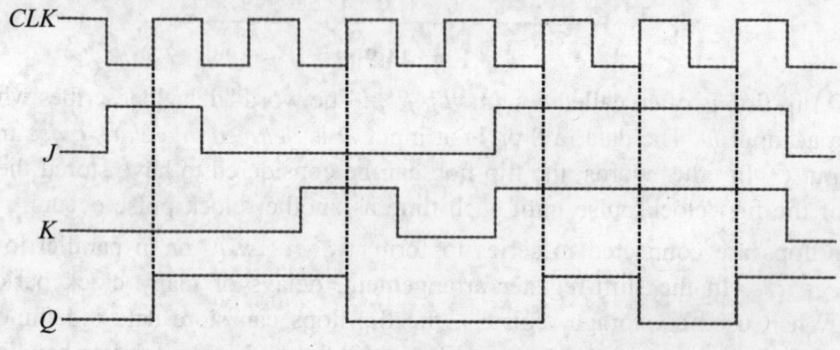

Timing diagram

Fig. 24.92

24.47 T Flip-Flop

Another type of flip-flop found in practice is the *T* (toggle) flip-flop. This flip-flop has the ability to change states (toggle) each time a clock pulse occurs. This flip-flop is obtained from the *JK* type when inputs *J* and *K* are connected to provide a single input designated by *T* as shown in Fig. 24.93 (*i*). The input *T* determines whether or not the toggling action will occur.

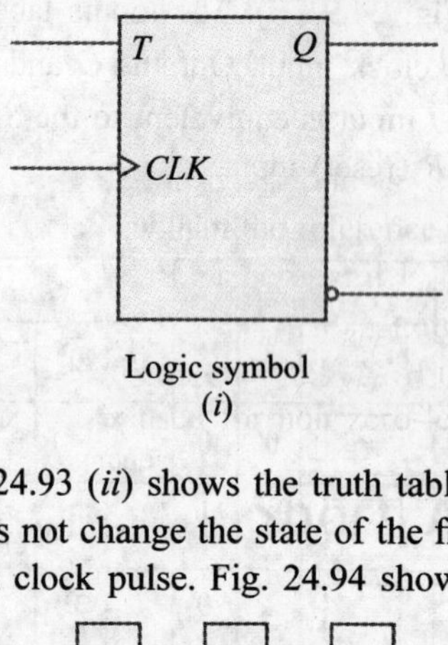

Logic symbol
(*i*)

T	Q (t + 1)	
0	Q(t)	No change
1	Q′(t)	Complement

Truth table
(*ii*)

Fig. 24.93

Fig. 24.93 (*ii*) shows the truth table for *T* flip-flop. When $T = 0$ ($J = K = 0$), the clock pulse does not change the state of the flip-flop. When $T = 1$ ($J = K = 1$), the flip-flop toggles with each clock pulse. Fig. 24.94 shows the timing diagram of *T* flip-flop.

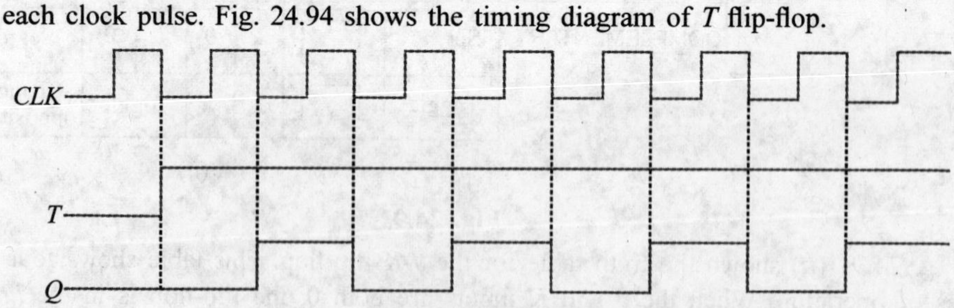

Timing diagram

Fig. 24.94

* The clock pulses occur at some fixed interval, say after every 10 microsecond.

Short Answer Questions

Q. 1. *What do you mean by encoding and decoding ?*

Ans. Digital circuits understand only binary numbers *i.e.,* input to a digital circuit must be in binary form. *Encoding* means changing decimal numbers to a binary form. In *decoding*, the digital information is converted back into decimal numbers.

Q. 2. *Give two advantages of digital circuits over the analog circuits.*

Ans. The two advantages of digital circuits over analog circuits are :

(*i*) Information can be stored for short and indefinite period.

(*ii*) Digital circuits are very accurate.

Q. 3. *What are the limitations of digital circuits ?*

Ans. The following are the limitations of digital circuits as compared to analog circuits:

(*i*) Most of "real world" information dealing with time, speed, pressure and position measurements are analog in nature.

(*ii*) Analog processing is generally simpler and faster.

Q. 4. *What is a truth table ?*

Ans. The action of a logic gate is usually summarised in the form of a truth table. A truth table defines the output of a logic gate for all possible combinations of inputs.

Q. 5. *What is an inverter ?*

Ans. The NOT gate is also known as an inverter. It has only one input and one output. The output of an inverter is always opposite to its input *i.e.,* if input is 1 (HIGH), the output is 0 (LOW) and *vice-versa*.

Q. 6. *What is the importance of digital techniques ?*

Ans. A continuously varying signal (voltage or current) is called an *analog signal*. Due to many-valued output, the analog operation is less reliable. On the other hand, a signal that can have only two discrete values is called a *digital signal*. Since digital operation has only two states, it is far more reliable than many-valued analog operation.

Q. 7. *Give three examples of (i) analog devices (ii) digital devices.*

Ans. (*i*) Amplifiers, oscillators and radio receivers.

(*ii*) Calculators, electronic watches and computers.

Q. 8. *What is a logic gate ?*

Ans. A logic gate is a digital circuit which works according to some logical relationship between input and output voltages. It has one or more input signals but only one output signal.

Q. 9. *State the rule used in the operation of (i) OR gate (ii) AND gate (iii) NOT gate.*

Ans. (*i*) The output of an OR gate is high (1) if one or more inputs are high (1). However, the output is low (0) when all inputs are low (0).

(*ii*) The output of an AND gate is high (1) if all the inputs are high (1). However, the output is low (0) if any or all inputs are low (0).

(*iii*) The output of a NOT gate is high (1) if input is low (0) and *vice-versa*.

Q. 10. *Write the Boolean expression for (i) OR gate (ii) AND gate (iii) NOT gate.*

Ans. (*i*) $A + B = Y$ (*ii*) $A \, . \, B = Y$ (*iii*) $\overline{A} = Y$.

Q. 11. *Write the truth table for (i) OR gate (ii) AND gate and (iii) NOT gate.*

Ans. (*i*) Truth Table for OR gate

Inputs		Output
A	*B*	*Y = A + B*
0	0	0
0	1	1
1	0	1
1	1	1

(*ii*) Truth Table for AND gate

Inputs		Output
A	*B*	*Y = A . B*
0	0	0
0	1	0
1	0	0
1	1	1

(*iii*) Truth Table for NOT gate

Input	Output
A	*Y*
0	1
1	0

Fig. 24.95

The required truth tables for the logic gates are shown in Fig. 24.95.

Q. 12. *When is the output of (i) OR gate high (ii) NAND gate low ?*

Ans. (*i*) The output of OR gate is high (1) when any or all the inputs are high (1).
(*ii*) The output of NAND gate is low (0) when both the inputs are high (1).

Q. 13. *How 2-input NAND gate can be converted into a NOT gate ?*

Ans. When two inputs of NAND gate are connected together so that it has one input, the NAND works as a NOT gate. Symbolically, it is shown in Fig. 24.96.

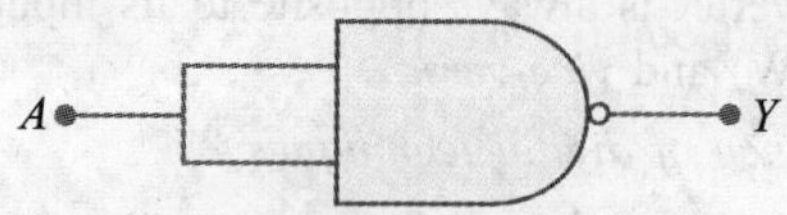

Fig. 24.96

Q. 14. *Why is logic gate called so ?*

Ans. The term "logic" is usually used to refer to a decision-making process. A logic gate makes logical decisions regarding the existence of output depending upon the nature of the input. For this reason, it is called a logic gate.

Q. 15. *Why are the NAND (or NOR) gates called as digital building blocks ?*

Ans. The repeated use of NAND (or NOR) gates can produce all the three basic gates *i.e.,* OR, AND and NOT gates. Hence the NAND and NOR gates are called universal gates.

Q. 16. *Identify the logic gates marked P and Q in Fig. 24.97. Write down the output at X for the inputs (i) A = 0, B = 0 and (ii) A = 1, B = 1.*

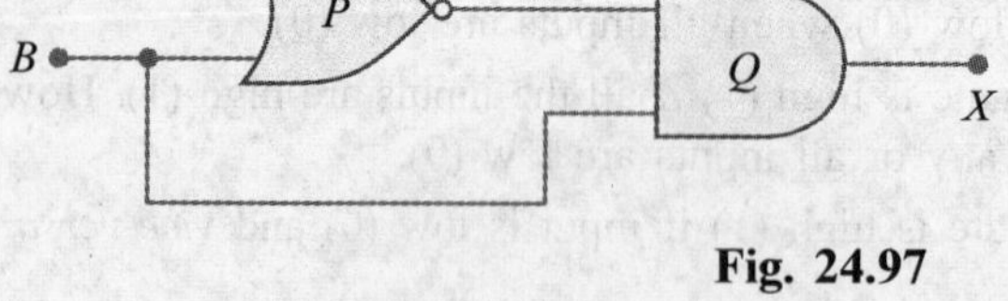

A	*B*	*Y*	*X*
0	0	1	0
1	1	0	0

Fig. 24.97

Ans. The logic gate marked as P is NOR gate while the gate marked as Q is AND gate. It is clear that :

$$Y = \overline{A+B} \text{ and } X = B \cdot Y$$

Therefore, the output at X will be as shown in the truth table.

Q. 17. *Identify the logic gates marked P and Q in Fig. 24.98. Write down the output at X for the inputs (i) A = 0, B = 0 and (ii) A = 1, B = 1.*

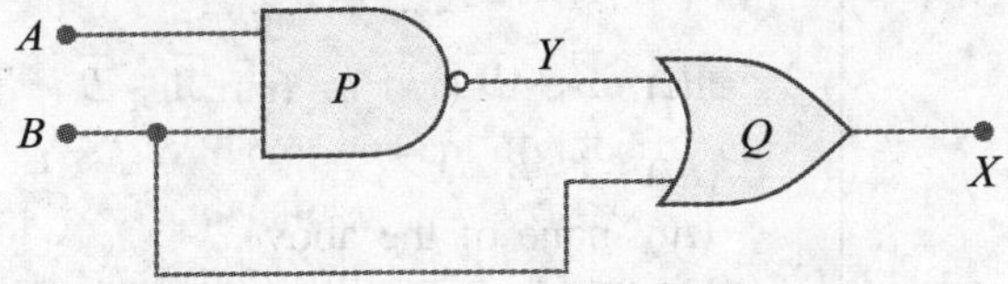

A	B	Y	X
0	0	1	1
1	1	0	1

Fig. 24.98

Ans. The logic gate marked as P is NAND gate while the gate marked as Q is OR gate. It is clear that :

$$Y = \overline{A \cdot B} \text{ and } X = B + Y$$

Therefore, the output at X will be as shown in the truth table.

Q. 18. *How will you obtain T flip-flop from JK flip-flop ?*

Ans. The T flip-flop is obtained from JK flip-flop when inputs J and K are connected to provide a single input designated by T.

Multiple-Choice Questions

1. The binary number 10101 is equivalent to decimal number ....
   (*i*) 19 (*ii*) 12
   (*iii*) 27 (*iv*) 21
2. The universal gate is ....
   (*i*) NAND gate (*ii*) OR gate
   (*iii*) NOT gate
   (*iv*) none of the above
3. The inverter is ....
   (*i*) NOT gate (*ii*) OR gate
   (*iii*) AND gate
   (*iv*) none of the above
4. The inputs of a NAND gate are connected together. The resulting circuit is ....
   (*i*) OR gate (*ii*) AND gate
   (*iii*) NOT gate
   (*iv*) none of the above
5. The NOR gate is OR gate followed by ....
   (*i*) AND gate (*ii*) NAND gate
   (*iii*) NOT gate
   (*iv*) none of the above
6. The NAND gate is AND gate followed by ....
   (*i*) NOT gate (*ii*) OR gate
   (*iii*) AND gate
   (*iv*) none of the above
7. Digital circuit can be made by the repeated use of ....
   (*i*) OR gates (*ii*) NOT gates
   (*iii*) NAND gates
   (*iv*) none of the above
8. The only function of NOT gate is to ....
   (*i*) stop a signal
   (*ii*) invert input signal
   (*iii*) act as a universal gate
   (*iv*) none of the above

9. When an input signal 1 is applied to a NOT gate, the output is ....
   (*i*) 0
   (*ii*) 1
   (*iii*) either 0 or 1
   (*iv*) none of the above

10. In Boolean algebra, the bar sign (–) indicates ....
   (*i*) OR operation
   (*ii*) AND operation
   (*iii*) NOT operation
   (*iv*) none of the above

11. The given Boolean expression is :

$$Y = A\overline{B} + B\overline{A}$$

If $A = 1$ and $B = 1$, then $Y =$ ....
   (*i*) 1 (*ii*) 0
   (*iii*) either 1 or 0
   (*iv*) none of the above

12. In Boolean algebra, the plus sign (+) indicates ....
   (*i*) AND operation
   (*ii*) OR operation
   (*iii*) NOT operation
   (*iv*) none of the above

13. $\overline{(A + B)}$ = ...
   (*i*) $\overline{A} + \overline{B}$ (*ii*) $\overline{A} - \overline{B}$
   (*iii*) $\overline{A} \cdot \overline{B}$
   (*iv*) none of the above

14. $\overline{(A \cdot B)}$ = ...
   (*i*) $\overline{A} + \overline{B}$ (*ii*) $\overline{A} \cdot \overline{B}$
   (*iii*) $\overline{A} - \overline{B}$
   (*iv*) none of the above

15. $A + A \cdot B =$ ....
   (*i*) B (*ii*) A
   (*iii*) $\overline{A} + B$
   (*iv*) none of the above

Answers to Multiple-Choice Questions

1. (*iv*)	2. (*i*)	3. (*i*)	4. (*iii*)	5. (*iii*)
6. (*i*)	7. (*iii*)	8. (*ii*)	9. (*i*)	10. (*iii*)
11. (*ii*)	12. (*ii*)	13. (*iii*)	14. (*i*)	15. (*ii*)

Chapter Review Topics

1. Write a short note on analog and digital signals.
2. What is a digital circuit ?
3. What is binary number system ?
4. How will you make decimal to binary conversion ?
5. How will you make binary to decimal conversion ?
6. What is a logic gate ?
7. What are the three basic logic gates ?
8. Describe OR function with a 2-input OR gate.
9. Explain AND function with a 2-input AND gate.
10. What is a NAND gate ?
11. What is a NOR gate ?
12. How will you obtain NOT gate from NAND gate ?
13. What is indicated by plus (+), dot (.) and bar (—) in a Boolean expression ?
14. State De Morgan's theorems.
15. What are encoders and decoders ?

Problems

1. Convert decimal number 23 into equivalent binary number. $[(10111)_2]$
2. Simplify the expression $Y = ACD + \bar{A}BCD$. $[Y = A\bar{C} + \bar{B}D]$
3. Simplify the expression $Y = \overline{(\bar{A} + C)\cdot(B + \bar{D})}$ to one having only single variables inverted. $[Y = A\bar{C} + \bar{B}D]$
4. Find the complement function of $Y = \bar{A}B\bar{C} + \bar{A}\bar{B}C$. $[(A + \bar{B} + C)(A + B + \bar{C})]$
5. Simplify the expression $Y = A \cdot B + A \cdot \bar{B}$. $[Y = A]$
6. Simplify the expression $Y = A \cdot B \cdot C + B \cdot C$. $[Y = B \cdot C]$
7. Simplify the following Boolean expression to a minimum number of literals :
$$Y = A(\bar{A} + B)$$ $[Y = AB]$
8. Simplify the following Boolean function to a minimum number of literals :
$$Y = \bar{A}\bar{B}C + \bar{A}BC + A\bar{B}$$ $[Y = \bar{A}C + A\bar{B}]$
9. Simplify the expression : $Y = (A + B)(\bar{A} + C)(B + C)$ $[Y = (A + B)(\bar{A} + C)]$
10. Find the complement of the function :
$$Y = A(\bar{B}\bar{C} + BC)H$$ $[\bar{Y} = \bar{A} + (B + C)(\bar{B} + \bar{C})]$
11. Draw the logic circuit for the following Boolean expressions :
(i) $Y = ABC + \bar{D}$ (ii) $Y = (\bar{E}G + B)H$

[Ans.
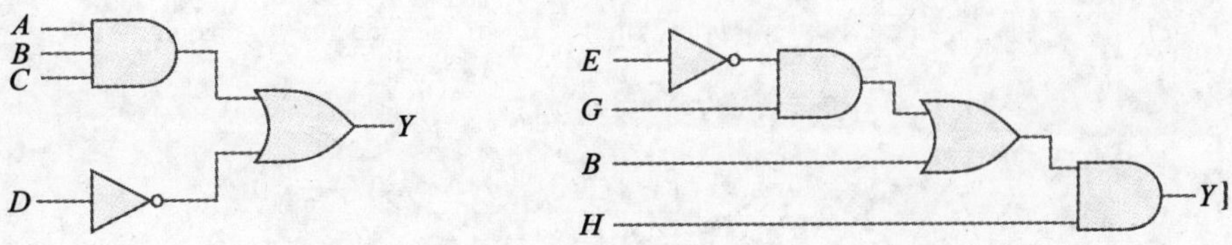

(i) (ii)
]

12. A logic circuit is given by the Boolean escperssion : $Y = \overline{\overline{AB} + C}$. Draw the logic circuit for this expression.

[Ans.
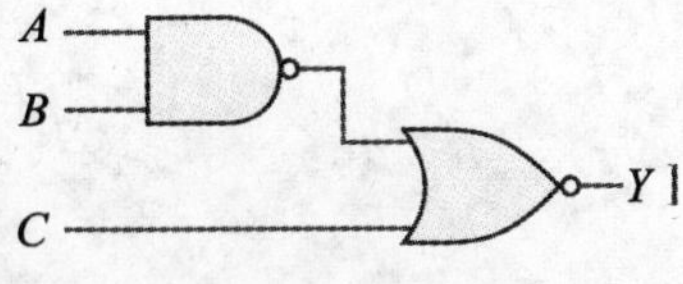

(iii)
]

13. Write the Boolean expressions for the logic circuits shown in Fig. 24.99.

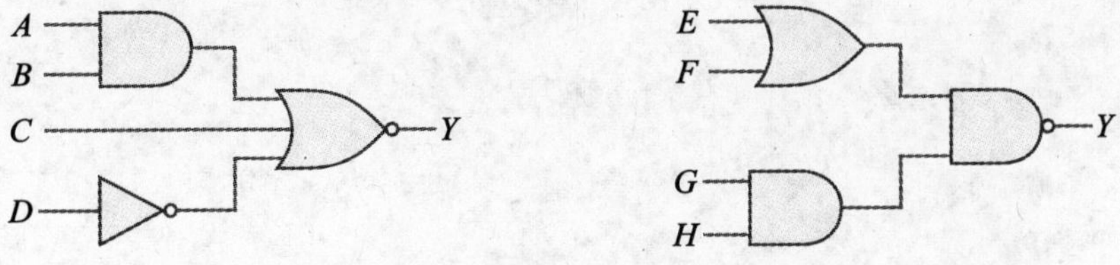

(i) (ii)

Fig. 24.99

$[(i)\ Y = \overline{AB + C + \bar{D}}\quad (ii)\ Y = \overline{(E + F)(GH)}]$

14. Write the Boolean expression for the logic circuit shown in Fig. 24.100.

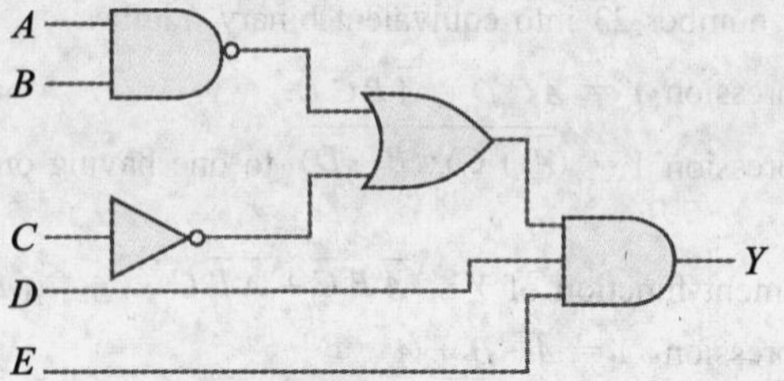

Fig. 24.100

[$Y = (\overline{\overline{AB} + \overline{C}})\,DE$]

15. Write the BCD code for 829. [100000101001_{BCD}]

16. What is the decimal number for 10000111000 BCD ? [$(438)_{10}$]

Index

For Notes and Formulae

For Notes and Formulae

For Notes and Formulae

For Notes and Formulae